Discrete Mathematics and Its Applications

2025 Release

Kenneth H. Rosen
formerly AT&T Laboratories

Mc Graw Hill

DISCRETE MATHEMATICS AND ITS APPLICATIONS

Published by McGraw Hill LLC, 1325 Avenue of the Americas, New York, NY 10019.

This book is printed on acid-free paper.

1 2 3 4 5 6 7 8 9 LWI 30 29 28 27 26 25

ISBN 978-1-266-19154-1
MHID 1-266-19154-2

Cover Image: *Karl Dehnam/Alamy Stock Photo*

mheducation.com/highered

Contents

About the Author vi
Preface vii
Online Resources xvii
To the Student xx

1 The Foundations: Logic and Proofs 1

1.1 Propositional Logic 1
1.2 Applications of Propositional Logic 17
1.3 Propositional Equivalences 26
1.4 Predicates and Quantifiers 40
1.5 Nested Quantifiers 61
1.6 Rules of Inference 73
1.7 Introduction to Proofs 85
1.8 Proof Methods and Strategy 97
End-of-Chapter Material 116

2 Basic Structures: Sets, Functions, Sequences, Sums, and Matrices 123

2.1 Sets 123
2.2 Set Operations 137
2.3 Functions 151
2.4 Sequences and Summations 169
2.5 Cardinality of Sets 183
2.6 Matrices 192
End-of-Chapter Material 200

3 Algorithms 205

3.1 Algorithms 205
3.2 The Growth of Functions 220
3.3 Complexity of Algorithms 235
End-of-Chapter Material 248

4 Number Theory and Cryptography 255

4.1 Divisibility and Modular Arithmetic 255
4.2 Integer Representations and Algorithms 264
4.3 Primes and Greatest Common Divisors 275
4.4 Solving Congruences 294
4.5 Applications of Congruences 307
4.6 Cryptography 314
End-of-Chapter Material 328

5 Induction and Recursion 335

5.1 Mathematical Induction 335
5.2 Strong Induction and Well-Ordering 358
5.3 Recursive Definitions and Structural Induction 370
5.4 Recursive Algorithms 385
5.5 Program Correctness 397
End-of-Chapter Material 403

6 Counting 411

6.1 The Basics of Counting 411
6.2 The Pigeonhole Principle 426
6.3 Permutations and Combinations 434
6.4 Binomial Coefficients and Identities 443
6.5 Generalized Permutations and Combinations 451
6.6 Generating Permutations and Combinations 463
End-of-Chapter Material 468

7 Discrete Probability 475

7.1 An Introduction to Discrete Probability 475
7.2 Probability Theory 483
7.3 Bayes' Theorem 500
7.4 Expected Value and Variance 509
End-of-Chapter Material 526

8 Advanced Counting Techniques 533

8.1 Applications of Recurrence Relations 533
8.2 Solving Linear Recurrence Relations 546
8.3 Divide-and-Conquer Algorithms and Recurrence Relations 559
8.4 Generating Functions 569
8.5 Inclusion–Exclusion 585
8.6 Applications of Inclusion–Exclusion 591
End-of-Chapter Material 598

9 Relations 605

9.1 Relations and Their Properties 605
9.2 n-ary Relations and Their Applications 617
9.3 Representing Relations 627
9.4 Closures of Relations 634
9.5 Equivalence Relations 644
9.6 Partial Orderings 656
End-of-Chapter Material 671

10 Graphs 679

10.1 Graphs and Graph Models 679
10.2 Graph Terminology and Special Types of Graphs 691
10.3 Representing Graphs and Graph Isomorphism 709
10.4 Connectivity 720
10.5 Euler and Hamilton Paths 734
10.6 Shortest-Path Problems 749
10.7 Planar Graphs 759
10.8 Graph Coloring 768
End-of-Chapter Material 777

11 Trees 787

11.1 Introduction to Trees 787
11.2 Applications of Trees 799
11.3 Tree Traversal 815
11.4 Spanning Trees 828
11.5 Minimum Spanning Trees 842
End-of-Chapter Material 848

12 Boolean Algebra 855

12.1 Boolean Functions 855
12.2 Representing Boolean Functions 863
12.3 Logic Gates 866
12.4 Minimization of Circuits 872
End-of-Chapter Material 887

13 Modeling Computation 893

13.1 Languages and Grammars 893
13.2 Finite-State Machines with Output 905
13.3 Finite-State Machines with No Output 912
13.4 Language Recognition 925
13.5 Turing Machines 935
End-of-Chapter Material 946

Appendices A-1

1 Axioms for the Real Numbers and the Positive Integers A-1
2 Exponential and Logarithmic Functions A-7
3 Pseudocode A-11

***Suggested Readings* B-1**
***Answers to Odd-Numbered Exercises* S-1**
***Index of Biographies* I-1**
***Index* I-2**

About the Author

Kenneth H. Rosen received his B.S. in Mathematics from the University of Michigan, Ann Arbor (1972), and his Ph.D. in Mathematics from M.I.T. (1976), where he wrote his thesis in number theory under the direction of Harold Stark. Before joining Bell Laboratories in 1982, he held positions at the University of Colorado, Boulder; The Ohio State University, Columbus; and the University of Maine, Orono, where he was an associate professor of mathematics. He enjoyed a long career as a Distinguished Member of the Technical Staff at AT&T Bell Laboratories (and AT&T Laboratories) in Monmouth County, New Jersey. While working at Bell Labs, he taught at Monmouth University, teaching courses in discrete mathematics, coding theory, and data security. After leaving AT&T Labs, he became a visiting research professor of computer science at Monmouth University, where he has taught courses in algorithm design, computer security and cryptography, and discrete mathematics.

Dr. Rosen has published numerous articles in professional journals on number theory and on mathematical modeling. He is the author of the widely used *Elementary Number Theory and Its Applications*, published by Pearson, currently in its seventh edition, which has been translated into Chinese. He is also the author of *Discrete Mathematics and Its Applications*, published by McGraw Hill, currently in its 2025 Release, which followed the eighth edition. *Discrete Mathematics and Its Applications* has sold more than 450,000 copies in North America during its lifetime, and hundreds of thousands of copies throughout the rest of the world. This book has also been translated into many languages, including Spanish, French, Portuguese, Greek, Chinese, Vietnamese, and Korean. He is also co-author of *UNIX: The Complete Reference*; *UNIX System V Release 4: An Introduction*; and *Best UNIX Tips Ever*, all published by Osborne McGraw Hill. These books have sold more than 150,000 copies, with translations into Chinese, German, Spanish, and Italian. Dr. Rosen is also the editor of both the first and second editions (published in 1999 and 2018, respectively) of the *Handbook of Discrete and Combinatorial Mathematics*, published by CRC Press. He has served as the advisory editor of the CRC series of books in discrete mathematics, sponsoring more than 70 volumes on diverse aspects of discrete mathematics, many of which are introduced in this book. He is an advisory editor for the CRC series of mathematics textbooks, where he has helped more than 35 authors write better texts. Dr. Rosen serves as an Associate Editor for the journal *Discrete Mathematics*, where he has handled more than 1500 papers in many areas, including graph theory, enumeration, number theory, and cryptography.

Dr. Rosen has had a longstanding interest in integrating mathematical software into the educational and professional environments. He has worked on several projects with Waterloo Maple Inc.'s Maple™ software in both these areas. Dr. Rosen has devoted a great deal of energy to ensuring that the online homework for *Discrete Mathematics and its Applications* is a superior teaching tool. Dr. Rosen has also worked with several publishing companies on their homework delivery platforms.

At Bell Laboratories and AT&T Laboratories, Dr. Rosen worked on a wide range of projects, including operations research studies, product line planning for computers and data communications equipment, technology assessment and innovation, and many other efforts. He helped plan AT&T's products and services in the area of multimedia, including video communications, speech recognition, speech synthesis, and image networking. He evaluated new technology for use by AT&T and did standards work in the area of image networking. He also invented many new services, and holds more than 70 patents. One of his more interesting projects involved helping evaluate technology for the AT&T attraction that was part of EPCOT Center. After leaving AT&T, Dr. Rosen has worked as a technology consultant for Google and for AT&T. Dr. Rosen enjoys international travel and has visited approximately 70 countries. His goal is to visit at least 100.

Preface

In writing this book, I was guided by my long-standing experience and interest in teaching discrete mathematics. For the student, my purpose was to present material in a precise, readable manner, with the concepts and techniques of discrete mathematics clearly presented and demonstrated. My goal was to show the relevance and practicality of discrete mathematics to students, who are often skeptical. I wanted to give students studying computer science all of the mathematical foundations they need for their future studies. I wanted to give mathematics students an understanding of important mathematical concepts together with a sense of why these concepts are important for applications. And most importantly, I wanted to accomplish these goals without watering down the material.

For the instructor, my purpose was to design a flexible, comprehensive teaching tool using proven pedagogical techniques in mathematics. I wanted to provide instructors with a package of materials that they could use to teach discrete mathematics effectively and efficiently in the most appropriate manner for their particular set of students. I hope that I have achieved these goals.

I have been extremely gratified by the tremendous success of this text, including its use by more than one million students around the world over the last 30 years and its translation into many different languages. The many improvements in the eighth edition and 2025 release have been made possible by the feedback and suggestions of a large number of instructors and students at many of the more than 600 North American schools, and at many universities in different parts of the world, where this book has been successfully used. I have been able to significantly improve the appeal and effectiveness of this book edition to edition because of the feedback I have received and the significant investments that have been made in the evolution of the book.

This text is designed for a one- or two-term introductory discrete mathematics course taken by students in a wide variety of majors, including mathematics, computer science, and engineering. College algebra is the only explicit prerequisite, although a certain degree of mathematical maturity is needed to study discrete mathematics in a meaningful way. This book has been designed to meet the needs of almost all types of introductory discrete mathematics courses. It is highly flexible and extremely comprehensive. The book is designed not only to be a successful textbook, but also to serve as a valuable resource students can consult throughout their studies and professional life.

Goals of a Discrete Mathematics Course

A discrete mathematics course has more than one purpose. Students should learn a particular set of mathematical facts and how to apply them; more importantly, such a course should teach students how to think logically and mathematically. To achieve these goals, this text stresses mathematical reasoning and the different ways problems are solved. Five important themes are interwoven in this text: mathematical reasoning, combinatorial analysis, discrete structures, algorithmic thinking, and applications and modeling. A successful discrete mathematics course should carefully blend and balance all five themes.

1. *Mathematical Reasoning:* Students must understand mathematical reasoning in order to read, comprehend, and construct mathematical arguments. This text starts with a discussion of mathematical logic, which serves as the foundation for the subsequent discussions of methods of proof. Both the science and the art of constructing proofs are addressed. The technique of

mathematical induction is stressed through many different types of examples of such proofs and a careful explanation of why mathematical induction is a valid proof technique.

2. *Combinatorial Analysis:* An important problem-solving skill is the ability to count or enumerate objects. The discussion of enumeration in this book begins with the basic techniques of counting. The stress is on performing combinatorial analysis to solve counting problems and analyze algorithms, not on applying formulae.
3. *Discrete Structures:* A course in discrete mathematics should teach students how to work with discrete structures, which are the abstract mathematical structures used to represent discrete objects and relationships between these objects. These discrete structures include sets, permutations, relations, graphs, trees, and finite-state machines.
4. *Algorithmic Thinking:* Certain classes of problems are solved by the specification of an algorithm. After an algorithm has been described, a computer program can be constructed implementing it. The mathematical portions of this activity, which include the specification of the algorithm, the verification that it works properly, and the analysis of the computer memory and time required to perform it, are all covered in this text. Algorithms are described using both English and an easily understood form of pseudocode.
5. *Applications and Modeling:* Discrete mathematics has applications to almost every conceivable area of study. There are many applications to computer science and data networking in this text, as well as applications to such diverse areas as chemistry, biology, linguistics, geography, business, and the Internet. These applications are natural and important uses of discrete mathematics and are not contrived. Modeling with discrete mathematics is an extremely important problem-solving skill, which students have the opportunity to develop by constructing their own models in some of the exercises.

Changes in the 2025 Release

This book has been the leading discrete mathematics textbook worldwide for several decades. One of the keys to its success has been its evolution over time to incorporate new ideas and to respond to the needs of instructors and students. This new release provides many enhancements and additional features that address these needs.

Many of the changes in this new release have directly resulted from satisfying the requests of reviewers and users, while other changes have been made because of the evolution of technology. Moreover, after an extensive review, careful alterations have been made to increase support for diversity, equality, and inclusion for all students. Instructors moving to this new release should consult information on the Instructor's website about where new exercises have been added to migrate their homework assignments.

In particular, overall changes in the book include

- Additional coverage of selected topics
- Inclusion of computational discoveries
- More coverage of countability
- Expanded coverage of mathematical induction
- Revised examples and additional examples
- Revised exercises and additional exercises
- New instructive videos embedded in relevant sections of the eBook with expanded discussions of topics and step-by-step solutions
- New biographies
- Numerous updates of various sorts

We will now provide a chapter-by-chapter breakout of noteworthy changes:

Chapter 1

- A biography of Alan Colmerauer, the father of the programming language Prolog, has been added.
- More information about Prolog has been provided.

Chapter 2

- Coverage of set builder notation has been expanded, including an expanded example with new exercises.
- An example is used to show how relations can be defined using set builder notation.
- A short summary of the history of axiomatic set theory has been added.
- The countability of the set of algebraic numbers and the uncountability of transcendental numbers are discussed in the text and in the exercises.

Chapter 3

- The distinction is made between the meaning of a social media algorithm and the meaning of an algorithm.

Chapter 4

- The value of the largest known Mersenne prime has been updated and recently computed values of the prime counting function are given.

Chapter 5

- Remarks have been added to make it clear that the basis case of a proof by mathematical induction can start at any integer.
- Two new exercises ask students to find the flaw in incorrect proofs by strong induction.
- The notion of downward (or backward) induction has been introduced in the exercises.

Chapter 6

- Practical information about relatively safe passwords is provided.
- Set partitions are introduced and Bell numbers are now covered.
- New writing projects added on the subjects: password requirements, lockscreen patterns, and set partitions.

Chapter 7

- The application of probability theory to genetics is now discussed, with an example involving eye color used to illustrate dominant and recessive genes.
- Additional exercises involving probability in genetics are provided.

Chapter 8

- A biography of Michael Shamos, one of the pioneers of computational geometry, has been added to Section 8.3.

Chapter 10

- Modeling social networks that may include both symmetric and asymmetric relationships between people is done using directed graphs. In particular, this is applied to model social networks between users of X (Twitter), LinkedIn, and Instagram.
- The density of graphs is introduced in the exercises in Section 10.2.
- The table of the number of actors with a given Bacon number in Section 10.4 has been updated.

Chapter 11

- The Tree of Fractions, which is used to produce a list of all the positive rational numbers without repetition, is introduced and used (in the exercises) to show that the set of rational numbers is countable.
- A biography of Herbert Wilf has been added.

Changes in the Eighth Edition

The eighth edition was a significant improvement from previous editions with major changes to the narrative and extensive additions to the online exercises and course offerings. The changes made in the eighth edition are repeated here for users unfamiliar with it.

Noteworthy Changes

The exposition in the eighth edition was improved with a focus on adding clarity and comprehension. Many proofs were enhanced by adding details and explanation. Moreover, many new examples and exercises were included.

The topic coverage was extended in various ways, including these important changes:

- **Number theory** Numerical and theoretical advancements were added.
- **Cryptography** Coverage of homomorphic encryption was added.
- **Mathematical induction** The template from proofs by mathematical induction was expanded.
- **Counting methods** The coverage of the division rule for counting was enhanced.
- **Data mining** An important aspect of data mining, association rules, were introduced and studied.
- **Graphs theory applications** The modeling of semantic networks, an important concept in artificial intelligence, was described.
- **Biographies** Seven new biographies were added.

Features of the Book

ACCESSIBILITY This text has proven to be easily read and understood by many beginning students. There are no mathematical prerequisites beyond college algebra for almost all the contents of the text. Students needing extra help will find tools on the companion website for bringing their mathematical maturity up to the level of the text. The few places in the book where calculus is referred to are explicitly noted. Most students should easily understand the pseudocode used in the text to express algorithms, regardless of whether they have formally studied programming languages. There is no formal computer science prerequisite.

Each chapter begins at an easily understood and accessible level. Once basic mathematical concepts have been carefully developed, more difficult material and applications to other areas of study are presented.

FLEXIBILITY This text has been carefully designed for flexible use. The dependence of chapters on previous material has been minimized. Each chapter is divided into sections of approximately the same length, and each section is divided into subsections that form natural blocks of material for teaching. Instructors can easily pace their lectures using these blocks.

WRITING STYLE The writing style in this book is direct and pragmatic. Precise mathematical language is used without excessive formalism and abstraction. Care has been taken to balance the mix of notation and words in mathematical statements.

MATHEMATICAL RIGOR AND PRECISION All definitions and theorems in this text are stated extremely carefully so that students will appreciate the precision of language and rigor needed in mathematics. Proofs are motivated and developed slowly; their steps are all carefully justified. The axioms used in proofs and the basic properties that follow from them are explicitly described in an appendix, giving students a clear idea of what they can assume in a proof. Recursive definitions are explained and used extensively.

WORKED EXAMPLES Over 800 examples are used to illustrate concepts, relate different topics, and introduce applications. In most examples, a question is first posed, then its solution is presented with the appropriate amount of detail.

APPLICATIONS The applications included in this text demonstrate the utility of discrete mathematics in the solution of real-world problems. This text includes applications to a wide variety of areas, including computer science, data networking, psychology, chemistry, engineering, linguistics, biology, business, and the Internet.

ALGORITHMS Results in discrete mathematics are often expressed in terms of algorithms; hence, key algorithms are introduced in most chapters of the book. These algorithms are expressed in words and in an easily understood form of structured pseudocode, which is described and specified in Appendix 3. The computational complexity of the algorithms in the text is also analyzed at an elementary level.

HISTORICAL INFORMATION The background of many topics is succinctly described in the text. Brief biographies of 92 mathematicians and computer scientists are included as footnotes. These biographies include information about the lives, careers, and accomplishments of these important contributors to discrete mathematics, and images, when available, are displayed.

In addition, numerous historical footnotes are included that supplement the historical information in the main body of the text. Efforts have been made to keep the book up-to-date by reflecting the latest discoveries.

KEY TERMS AND RESULTS A list of key terms and results follows each chapter. The key terms include only the most important that students should learn, and not every term defined in the chapter.

EXERCISES There are over 4200 exercises in the text, with many different types of questions posed. There is an ample supply of straightforward exercises that develop basic skills, a large number of intermediate exercises, and many challenging exercises. Exercises are stated clearly and unambiguously, and all are carefully graded for level of difficulty. Exercise sets contain special discussions that develop new concepts not covered in the text, enabling students to discover new ideas through their own work.

Exercises that are somewhat more difficult than average are marked with a single star, *; those that are much more challenging are marked with two stars, **. Exercises whose solutions require calculus are explicitly noted. Exercises that develop results used in the text are clearly identified with the right pointing hand symbol, ☞. Answers or outlined solutions to all odd-numbered exercises are provided at the back of the text. The solutions include proofs in which most of the steps are clearly spelled out.

REVIEW QUESTIONS A set of review questions is provided at the end of each chapter. These questions are designed to help students focus their study on the most important concepts and techniques of that chapter. To answer these questions students need to write long answers, rather than just perform calculations or give short replies.

SUPPLEMENTARY EXERCISE SETS Each chapter is followed by a rich and varied set of supplementary exercises. These exercises are generally more difficult than those in the exercise sets following the sections. The supplementary exercises reinforce the concepts of the chapter and integrate different topics more effectively.

COMPUTER PROJECTS Each chapter is followed by a set of computer projects. The approximately 150 computer projects tie together what students may have learned in computing and in discrete mathematics. Computer projects that are more difficult than average, from both a mathematical and a programming point of view, are marked with a star, and those that are extremely challenging are marked with two stars.

COMPUTATIONS AND EXPLORATIONS A set of computations and explorations is included at the conclusion of each chapter. These exercises (approximately 120 in total) are designed to be completed using existing software tools, such as programs that students or instructors have written or mathematical computation packages such as Maple™ or Mathematica™. Many of these exercises give students the opportunity to uncover new facts and ideas through computation. (Some of these exercises are discussed in the *Exploring Discrete Mathematics* supplements available online.)

WRITING PROJECTS Each chapter is followed by a set of writing projects. To do these projects students need to consult the mathematical literature. Some of these projects are historical in nature and may involve looking up original sources. Others are designed to serve as gateways to new topics and ideas. All are designed to expose students to ideas not covered in depth in the text. These projects tie mathematical concepts together with the writing process and help expose students to possible areas for future study. (Suggested references for these projects can be found online.)

APPENDICES There are three appendices to the text. The first introduces axioms for real numbers and the positive integers, and illustrates how facts are proved directly from these axioms. The second covers exponential and logarithmic functions, reviewing some basic material used heavily in the course. The third specifies the pseudocode used to describe algorithms in this text.

SUGGESTED READINGS A list of suggested readings for the overall book and for each chapter is provided after the appendices. These suggested readings include books at or below the level of this text, more difficult books, expository articles, and articles in which discoveries in discrete mathematics were originally published. Some of these publications are classics, published many years ago, while others have been published in the last few years. These suggested readings are complemented by the many links to valuable resources available on the web that can be found on the website for this book.

How to Use This Book

This text has been carefully written and constructed to support discrete mathematics courses at several levels and with differing foci. The following table identifies the core and optional sections. An introductory one-term course in discrete mathematics at the sophomore level can be based on the core sections of the text, with other sections covered at the discretion of the instructor. A two-term introductory course can include all the optional mathematics sections in addition to the core sections. A course with a strong computer science emphasis can be taught by covering some or all of the optional computer science sections. Instructors can find sample syllabi for a wide range of discrete mathematics courses and teaching suggestions for using each section of the text can be found in the online instructor resources for this book.

Chapter	*Core*	*Optional CS*	*Optional Math*
1	1.1–1.8 (as needed)		
2	2.1–2.4, 2.6 (as needed)		2.5
3		3.1–3.3 (as needed)	
4	4.1–4.4 (as needed)	4.5, 4.6	
5	5.1–5.3	5.4, 5.5	
6	6.1–6.3	6.6	6.4, 6.5
7	7.1	7.4	7.2, 7.3
8	8.1, 8.5	8.3	8.2, 8.4, 8.6
9	9.1, 9.3, 9.5	9.2	9.4, 9.6
10	10.1–10.5		10.6–10.8
11	11.1	11.2, 11.3	11.4, 11.5
12		12.1–12.4	
13		13.1–13.5	

Instructors using this book can adjust the level of difficulty of their course by choosing either to cover or to omit the more challenging examples at the end of sections, as well as the more challenging exercises. The chapter dependency chart shown here displays the strong dependencies. A star indicates that only relevant sections of the chapter are needed for study of a later chapter. Weak dependencies have been ignored. More details can be found in the instructor resources online.

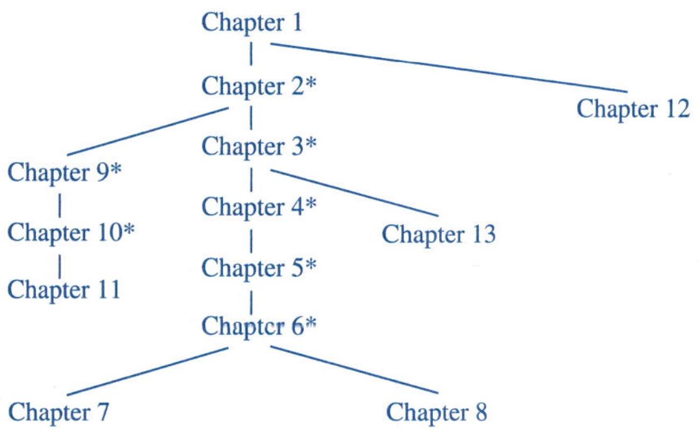

Acknowledgments

I would like to thank the many instructors and students at a variety of schools who have used this book and provided me with their valuable feedback and helpful suggestions. Their input has made this a much better book than it would have been otherwise. I am indebted to Dan Jordan for carefully reviewing this new release and his many useful suggestions. I also thank Jerrold Grossman and Dan Jordan for their technical reviews of the eighth edition and their "eagle eyes," which have helped ensure the accuracy and quality of this book. Both have proofread every part of the book many times as it has gone through the different steps of production and have helped eliminate old errata and prevented the insertion of new errata.

Thanks go to Dan Jordan for his work on the student solutions manual and instructor's resource guide. He has done an admirable job updating these ancillaries. Jerrold Grossman, the author of these ancillaries for the first seven editions of the book, has provided valuable assistance to Dan. I would also like to express my gratitude to the many people who have helped create and maintain the online resources for this book. In particular, special thanks go to Chunlei Liu and Rochus Boerner for their extensive work improving online questions for the Connect Site, described later in this preface.

I thank the reviewers of this 2025 release and all previous editions. These reviewers have provided much helpful criticism and encouragement to me. I hope this release lives up to their high expectations. There have been well in excess of 200 reviewers of this book since its first edition, with many from countries other than the United States. The most recent reviewers are listed here.

Recent Reviewers

Barbara Anthony
Southwestern University

Philip Barry
University of Minnesota, Minneapolis

Benkam Bobga
University of North Georgia

Miklos Bona
University of Florida

Steve Brick
University of South Alabama

Kirby Brown
Queens College

John Carter
University of Toronto

Narendra Chaudhari
Nanyang Technological University

Tim Chappell
Penn Valley Community College

Allan Cochran
University of Arkansas

Daniel Cunningham
Buffalo State College

H.K. Dai
Oklahoma State University

George Davis
Georgia State University

Andrzej Derdzinski
The Ohio State University

Ronald Dotzel
University of Missouri-St. Louis

T.J. Duda
Columbus State Community College

Bruce Elenbogen
University of Michigan, Dearborn

Norma Elias
Purdue University, Calumet-Hammond

Herbert Enderton
University of California, Los Angeles

Anthony Evans
Wright State University

Kim Factor
Marquette University

Margaret Fleck
University of Illinois, Champaign

Melissa Gaddini
Robert Morris University

William Garrison
University of Pittsburgh

Peter Gillespie
Fayetteville State University

Johannes Hattingh
Georgia State University

James Helmreich
Marist College

Kimberly Hess
Polk State College

Ken Holladay
University of New Orleans

Jerry Ianni
LaGuardia Community College

Milagros Izquierdo
Linköping University

Ravi Janardan
University of Minnesota, Minneapolis

Norliza Katuk
University of Utara Malaysia

Monika Kiss
Saint Leo University

William Klostermeyer
University of North Florida

Przemo Kranz
University of Mississippi

Jaromy Kuhl
University of West Florida

Loredana Lanzani
University of Arkansas, Fayetteville

Frederic Latour
Central Connecticut State University

Steven Leonhardi
Winona State University

Chunlei Liu
Valdosta State University

Xu Liutong
Beijing University of Posts and Telecommunications

Vladimir Logvinenko
De Anza Community College

Tamsen McGinley
Santa Clara University

Ramon A. Mata-Toledo
James Madison University

Tamara Melnik
Computer Systems Institute

Osvaldo Mendez
University of Texas at El Paso

Darrell Minor
Columbus State Community College

Kathleen O'Connor
Quinsigamond Community College

Keith Olson
Utah Valley University

Dimitris Papamichail
The College of New Jersey

Yongyuth Permpoontanalarp
King Mongkut's University of Technology, Thonburi

Galin Piatniskaia
University of Missouri, St. Louis

Shawon Rahman
University of Hawaii at Hilo

Eric Rawdon
University of St. Thomas

Stefan Robila
Montclair State University

Chris Rodger
Auburn University

Sukhit Singh
Texas State University, San Marcos

David Snyder
Texas State University, San Marcos

Wasin So
San Jose State University

Bogdan Suceava
California State University, Fullerton

Christopher Swanson
Ashland University

Bon Sy
Queens College

Fereja Tahir
Illinois Central College

K.A. Venkatesh
Presidency University

Matthew Walsh
Indiana-Purdue University, Fort Wayne

Sheri Wang
University of Phoenix

Gideon Weinstein
Western Governors University

David Wilczynski
University of Southern California

James Wooland
Florida State University

I also want to thank the many students who have provided suggestions and reported errata. The students in the discrete mathematics courses I have taught at Monmouth University, both undergraduate and graduate computer science students, have helped me improve the book in many ways.

There are many people to thank at McGraw Hill Higher Education, the publisher of this book, as well as people who did the production work at Aptara. I would also like to thank the original editor at Random House, Wayne Yuhasz, whose insights and skills helped ensure the book's success, as well as all the many other previous editors of this book.

I want to express my deep appreciation to Katie Peterson, the Product Developer who has gone far beyond her assigned duties to support the author. She has displayed many skills and virtues working to solve a wide variety of problems that have arisen in the development of this new release.

I am also grateful to Traci Vaske, the Content Product Manager, who managed the production process. She has kept the production on track and has helped resolve many issues that have arisen during this process. Thanks go to Santosh Kumar Verma, Product Manager at Aptara, and his colleagues at Aptara, who worked diligently to ensure the production quality of this release.

I also want to express my appreciation to the many others in the Science, Engineering, and Mathematics Division of McGraw Hill Higher Education for their valuable support for this new edition and the associated media content, including

- Ian Townsend, Portfolio Director, Math & Chemistry
- Deb Harden, Sr. Portfolio Manager
- Joanna Confalone, Marketing Director
- Courtney Cozzy, Marketing Manager
- Sandy Ludovissey, Buyer
- Egzon Shaqiri, Designer
- Ron Nelms, Assessment Content Project Manager
- Lola Briseno, Editorial Coordinator
- Fran Simon, Program Manager
- Amanda Ramsey, Content Licensing Specialist

Kenneth H. Rosen

Online Resources

Extensive effort has been devoted to producing valuable web resources for this book. Instructors should make a special effort to explore these resources to identify those they feel will help their students learn and explore discrete mathematics. These resources are available in the Online Learning Center, which is available to all students and instructors, and the Connect Site, designed for interactive instruction, which instructors can choose to use. To use Connect, students purchase online access for a specific time period.

0.1 The Online Learning Center

The Online Learning Center (OLC), accessible at *www.mhhe.com/rosen*, comprises the *Student Site* and the *Instructor Site*. Key features of each area are described here.

0.1.1 Student Site

The Student Site contains a wealth of resources available for student use, including the following, tied into the text wherever the special icons displayed below are found in the text:

Extra Examples

- ***Extra Examples*** You can find a large number of additional examples on the site, covering all chapters of the book. These examples are concentrated in areas where students often ask for additional material. Although most of these examples amplify the basic concepts, more-challenging examples can also be found here. Many new extra examples have been recently added for the eighth edition. Each icon in the book corresponds to one or more extra examples on the website.

Demo

- ***Interactive Demonstration Applets*** These applets enable you to interactively explore how important algorithms work, and are tied directly to material in the text with linkages to examples and exercises. Additional resources are provided on how to use and apply these applets.

Assessment

- ***Self Assessments*** These interactive guides help you assess your understanding of 14 key concepts, providing a question bank where each question includes a brief tutorial followed by a multiple-choice question. If you select an incorrect answer, advice is provided to help you understand your error. Using these Self Assessments, you should be able to diagnose your problems and find appropriate help.

Links

- ***Web Resources Guide*** This guide provides annotated links to hundreds of external websites containing relevant material such as historical and biographical information, puzzles and problems, discussions, applets, programs, and more. These links are keyed to the text by page number.

Additional resources in the Student Site include:

- ***Solutions to Odd-Numbered Exercises*** This ancillary provides complete, worked solutions for all odd-numbered exercises in the book. Students can also purchase a print version of the solutions manual through Create.
- ***Exploring Discrete Mathematics*** This ancillary provides help for using a computer algebra system to do a wide range of computations in discrete mathematics. Each chapter provides

a description of relevant functions in the computer algebra system and how they are used, programs to carry out computations in discrete mathematics, examples, and exercises that can be worked using this computer algebra system. Two versions, *Exploring Discrete Mathematics with Maple*™ and *Exploring Discrete Mathematics with Mathematica*™, are available.

- ***Applications of Discrete Mathematics*** This ancillary contains 24 chapters—each with its own set of exercises—presenting a wide variety of interesting and important applications covering three general areas in discrete mathematics: discrete structures, combinatorics, and graph theory. These applications are ideal for supplementing the text or for independent study.
- ***A Guide to Proof-Writing*** This guide provides additional help for writing proofs, a skill that many students find difficult to master. By reading this guide at the beginning of the course and periodically thereafter when proof writing is required, you will be rewarded as your proof-writing ability grows.
- ***Common Mistakes in Discrete Mathematics*** This guide includes a detailed list of common misconceptions that students of discrete mathematics often have and the kinds of errors they tend to make. You are encouraged to review this list from time to time to help avoid these common traps.
- ***Advice on Writing Projects*** This guide offers helpful hints and suggestions for the Writing Projects in the text, including an extensive bibliography of helpful books and articles for research, discussion of various resources available in print and online, tips on doing library research, and suggestions on how to write well.

0.1.2 Instructor Site

This part of the website provides access to all of the resources on the Student Site, as well as these resources for instructors:

- ***Solutions to Even-Numbered Exercises*** This ancillary provides complete, worked solutions for all even-numbered exercises in the book.
- ***Suggested Syllabi*** Detailed course outlines are shown, offering suggestions for courses with different emphases and different student backgrounds and ability levels.
- ***Teaching Suggestions*** This guide contains detailed teaching suggestions for instructors, including chapter overviews for the entire text, detailed remarks on each section, and comments on the exercise sets.
- ***Printable Tests*** Printable tests are offered in TeX and Word format for every chapter, and can be customized by instructors.
- ***PowerPoint Lecture Slides and PowerPoint Figures and Tables*** An extensive collection of PowerPoint lecture notes for all chapters of the text are provided for instructor use. In addition, images of all figures and tables from the text are provided as PowerPoint slides.

0.1.3 Connect

A comprehensive online learning package has been developed in conjunction with the text. A high-level description of this site will be provided here. Interested instructors and students can find out more about Connect from McGraw Hill Higher Education. When instructors choose to use this option, students in their classes must obtain access to Connect for this text, either by purchasing a copy of the textbook that also includes access privileges or by purchasing access only with the option of buying a loose-leaf version of the textbook.

Instructors who adopt Connect have access to a full-featured course management system. Course management capabilities are provided that allow instructors to create assignments, automatically assign and grade homework, quiz, and test questions from a bank of questions tied directly to the text, create and edit their own questions, manage course announcements and due dates, and track student progress.

Instructors can create their own assignments using Connect. They select the particular exercises from each section of the book that they want to assign. They can also assign chapters from the SmartBook version of the text, which provides an adaptive learning environment. Students have access to the interactive version of the textbook, the online homework exercises, and SmartBook.

eBook Students have access to an easy-to-use interactive version of the textbook when they use Connect. The interactive site provides the full content of the text, as well as the many extra resources that enrich the book. The resources include extra examples, interactive demonstrations, and self-assessments.

New to the 2025 release are instructive videos embedded in relevant sections of the eBook. The videos include expanded discussions of topics and step-by-step solutions. The new videos cover a range of topics for beginner and advanced students and will be expanded in future releases.

Homework and Learning Management Solution An extensive learning management solution has been developed that instructors can use to construct homework assignments. Approximately 800 online questions are available, including questions from every section of the text. These questions are tied to the most commonly assigned exercises in the book.

These online questions have been constructed to support the same objectives as the corresponding written homework questions. This challenge has been met by stretching the capabilities of different question types supported by the Connect platform.

Evergreen The original edition of this text was published in 1988; new editions were published approximately every four years. Because of this schedule, desired changes in the book had to wait up to four years to be made. Today, many students use the eBook and wealth of online content that make up the McGraw Hill Connect product for their studies, and important updates can be delivered more quickly to these online products.

To provide more seamless updates, McGraw Hill has introduced the new Evergreen delivery model that allows updates to be made more frequently to eBooks and the accompanying courseware. The 2025 Release of this text and Connect content was prepared using this Evergreen delivery model. This allows the online version of the text to be updated, along with online homework and other associated online materials. Instructors will receive the updates and new content in their discrete mathematics Connect course without having to switch editions or build a new course. Plans are underway to take advantage of this new model.

SmartBook Connect also provides another extended online version of the text in the McGraw Hill SmartBook platform. The SmartBook version of the text includes a set of objectives for each chapter of the text. A collection of questions, called probes, is provided to assess student understanding of each objective. Students are directed to the appropriate part of the text to review the material needed for each of these objectives. SmartBook provides an adaptive learning environment; it selects probes for students based on their performance answering earlier probes. Instructors can assign SmartBook as assignments or can have their students use SmartBook as a learning tool.

To the Student

What is discrete mathematics? Discrete mathematics is the part of mathematics devoted to the study of discrete objects. (Here *discrete* means consisting of distinct or unconnected elements.) The kinds of problems solved using discrete mathematics include:

- How many ways are there to choose a valid password on a computer system?
- What is the probability of winning a lottery?
- Is there a link between two computers in a network?
- How can I identify spam email messages?
- How can I encrypt a message so that no unintended recipient can read it?
- What is the shortest path between two cities using a transportation system?
- How can a list of integers be sorted so that the integers are in increasing order?
- How many steps are required to do such a sorting?
- How can it be proved that a sorting algorithm correctly sorts a list?
- How can a circuit that adds two integers be designed?
- How many valid Internet addresses are there?

You will learn the discrete structures and techniques needed to solve problems such as these.

More generally, discrete mathematics is used whenever objects are counted, when relationships between finite (or countable) sets are studied, and when processes involving a finite number of steps are analyzed. A key reason for the growth in the importance of discrete mathematics is that information is stored and manipulated by computing machines in a discrete fashion.

WHY STUDY DISCRETE MATHEMATICS? There are several important reasons for studying discrete mathematics. First, through this course you can develop your mathematical maturity: that is, your ability to understand and create mathematical arguments. You will not get very far in your studies in the mathematical sciences without these skills.

Second, discrete mathematics is the gateway to more advanced courses in all parts of the mathematical sciences. Discrete mathematics provides the mathematical foundations for many computer science courses, including data structures, algorithms, database theory, artificial intelligence, automata theory, formal languages, compiler theory, computer security, and operating systems. Students find these courses much more difficult when they have not had the appropriate mathematical foundations from discrete mathematics. One student sent me an email message saying that she used the contents of this book in every computer science course she took!

Math courses based on the material studied in discrete mathematics include logic, set theory, number theory, linear algebra, abstract algebra, combinatorics, graph theory, and probability theory (the discrete part of the subject).

Also, discrete mathematics contains the necessary mathematical background for solving problems in operations research (including discrete optimization), chemistry, engineering, biology, and so on. In the text, we will study applications to some of these areas.

Many students find their introductory discrete mathematics course to be significantly more challenging than courses they have previously taken. One reason for this is that one of the primary goals of this course is to teach mathematical reasoning and problem solving, rather than a discrete set of skills. The exercises in this book are designed to reflect this goal. Although there are plenty of exercises in this text similar to those addressed in the examples, a large percentage of the exercises require original thought. This is intentional. The material discussed in the text provides the tools needed to solve these exercises, but your job is to successfully apply these tools using your own creativity. One of the primary goals of this course is to learn how to attack problems that may be somewhat different from any you may have previously seen. Unfortunately, learning how to solve only particular types of exercises is not sufficient for success in developing the problem-solving skills needed in subsequent courses and professional work. This text addresses many different topics, but discrete mathematics is an extremely diverse and large area of study. One of my goals as an author is to help you develop the skills needed to master the additional material you will need in your own future pursuits.

Finally, discrete mathematics is an excellent environment in which to learn how to read and write mathematical proofs. In addition to explicit material on proofs in Chapter 1 and Chapter 5, this textbook contains throughout many proofs of theorems and many exercises asking the student to prove statements. This not only deepens one's understanding of the subject matter but is also valuable preparation for more advanced courses in mathematics and theoretical computer science.

THE EXERCISES I would like to offer some advice about how you can best learn discrete mathematics (and other subjects in the mathematical and computing sciences). You will learn the most by actively working exercises. I suggest that you solve as many as you possibly can. After working the exercises your instructor has assigned, I encourage you to solve additional exercises such as those in the exercise sets following each section of the text and in the supplementary exercises at the end of each chapter. (Note the key explaining the markings preceding exercises.)

Key to the Exercises

no marking	A routine exercise
*	A difficult exercise
**	An extremely challenging exercise
☞	An exercise containing a result used in the book (Table 1 on the following page shows where these exercises are used.)
(*Requires calculus*)	An exercise whose solution requires the use of limits or concepts from differential or integral calculus

The best approach is to try exercises yourself before you consult the answer section at the end of this book. Note that the odd-numbered exercise answers provided in the text are answers only and not full solutions; in particular, the reasoning required to obtain answers is omitted in these answers. The *Student's Solutions Guide*, available online provides complete, worked solutions to all odd-numbered exercises in this text. When you hit an impasse trying to solve an odd-numbered exercise, I suggest you consult the *Student's Solutions Manual* and look for some guidance as to how to solve the problem. The more work you do yourself rather than passively reading or copying solutions, the more you will learn. The answers and solutions to the even-numbered exercises are intentionally not available from the publisher; ask your instructor if you have trouble with these.

WEB RESOURCES All users of the book are able to access the online resources accessible via the Online Learning Center (OLC) for the book. You will find many Extra Examples designed to clarify key concepts, Self Assessments for gauging how well you understand core topics, Interactive Demonstrations that explore key algorithms and other concepts, a Web Resources Guide containing an extensive selection of links to external sites relevant to the world of discrete mathematics, extra explanations and practice to help you master core concepts, added instruction on writing proofs and on avoiding common mistakes in discrete mathematics, in-depth discussions of important applications, and guidance on utilizing Maple™ and

TABLE 1 Hand-Icon Exercises and Where They Are Used

Section	*Exercise*	*Section Where Used*	*Pages Where Used*
1.1	42	1.3	33
1.1	43	1.3	33
1.3	11	1.6	76
1.3	12	1.6	74, 76
1.3	19	1.6	76
1.3	33	1.6	76
1.3	34	1.6	76, 79
1.3	46	12.2	864
1.7	18	1.7	92
2.3	74	2.3	157
2.3	81	2.5	184
2.5	15	2.5	188
2.5	16	2.5	187
3.1	45	3.1	212
3.2	74	11.2	804
4.3	37	4.4	301
4.4	2	4.6	322
4.4	44	7.2	496
6.4	21	7.2	497
6.4	25	7.4	512
7.2	15	7.2	497
9.1	26	9.4	635
10.4	59	11.1	789
11.1	15	11.1	792
11.1	30	11.1	797
11.1	48	11.2	804
12.1	12	12.3	869
A.2	4	8.3	563

Mathematica™ software to explore the computational aspects of discrete mathematics. Places in the text where these additional online resources are available are identified in the margins by special icons. For more details on these and other online resources, see the description of the companion website immediately preceding this "To the Student" message.

THE VALUE OF THIS BOOK My intention is to make your substantial investment in this text an excellent value. The book, the associated ancillaries, and companion website have taken many years of effort to develop and refine. I am confident that most of you will find that the text and associated materials will help you master discrete mathematics, just as so many previous students have. Even though it is likely that you will not cover some chapters in your current course, you should find it helpful—as many other students have—to read the relevant sections of the book as you take additional courses. Most of you will return to this book as a useful tool throughout your future studies, especially for those of you who continue in computer science, mathematics, and engineering. I have designed this book to be a gateway for future studies and explorations, and to be comprehensive reference, and I wish you luck as you begin your journey.

Kenneth H. Rosen

CHAPTER

1 The Foundations: Logic and Proofs

1.1 Propositional Logic

1.2 Applications of Propositional Logic

1.3 Propositional Equivalences

1.4 Predicates and Quantifiers

1.5 Nested Quantifiers

1.6 Rules of Inference

1.7 Introduction to Proofs

1.8 Proof Methods and Strategy

The rules of logic specify the meaning of mathematical statements. For instance, these rules help us understand and reason with statements such as "There exists an integer that is not the sum of two squares" and "For every positive integer n, the sum of the positive integers not exceeding n is $n(n + 1)/2$." Logic is the basis of all mathematical reasoning, and of all automated reasoning. It has practical applications to the design of computing machines, to the specification of systems, to artificial intelligence, to computer programming, to programming languages, and to other areas of computer science, as well as to many other fields of study.

To understand mathematics, we must understand what makes up a correct mathematical argument, that is, a proof. Once we prove a mathematical statement is true, we call it a theorem. A collection of theorems on a topic organizes what we know about this topic. To learn a mathematical topic, a person needs to actively construct mathematical arguments on this topic, and not just read exposition. Moreover, knowing the proof of a theorem often makes it possible to modify the result to fit new situations.

Everyone knows that proofs are important throughout mathematics, but many people find it surprising how important proofs are in computer science. In fact, proofs are used to verify that computer programs produce the correct output for all possible input values, to show that algorithms always produce the correct result, to establish the security of a system, and to create artificial intelligence. Furthermore, automated reasoning systems have been created to allow computers to construct their own proofs.

In this chapter, we will explain what makes up a correct mathematical argument and introduce tools to construct these arguments. We will develop an arsenal of different proof methods that will enable us to prove many different types of results. After introducing many different methods of proof, we will introduce several strategies for constructing proofs. We will introduce the notion of a conjecture and explain the process of developing mathematics by studying conjectures.

1.1 Propositional Logic

1.1.1 Introduction

The rules of logic give precise meaning to mathematical statements. These rules are used to distinguish between valid and invalid mathematical arguments. Because a major goal of this book is to teach the reader how to understand and how to construct correct mathematical arguments, we begin our study of discrete mathematics with an introduction to logic.

Besides the importance of logic in understanding mathematical reasoning, logic has numerous applications to computer science. These rules are used in the design of computer circuits, the construction of computer programs, the verification of the correctness of programs, and in many other ways. Furthermore, software systems have been developed for constructing some, but not all, types of proofs automatically. We will discuss these applications of logic in this and later chapters.

1.1.2 Propositions

Our discussion begins with an introduction to the basic building blocks of logic—propositions. A **proposition** is a declarative sentence (that is, a sentence that declares a fact) that is either true or false, but not both.

EXAMPLE 1 All the following declarative sentences are propositions.

Extra Examples

1. Washington, D.C., is the capital of the United States of America.
2. Toronto is the capital of Canada.
3. $1 + 1 = 2$.
4. $2 + 2 = 3$.

Propositions 1 and 3 are true, whereas 2 and 4 are false. ◀

Some sentences that are not propositions are given in Example 2.

EXAMPLE 2 Consider the following sentences.

1. What time is it?
2. Read this carefully.
3. $x + 1 = 2$.
4. $x + y = z$.

Sentences 1 and 2 are not propositions because they are not declarative sentences. Sentences 3 and 4 are not propositions because they are neither true nor false. Note that each of sentences 3 and 4 can be turned into a proposition if we assign values to the variables. We will also discuss other ways to turn sentences such as these into propositions in Section 1.4. ◀

We use letters to denote **propositional variables** (or **sentential variables**), that is, variables that represent propositions, just as letters are used to denote numerical variables. The conventional letters used for propositional variables are $p, q, r, s, \ldots$. The **truth value** of a proposition

Links

Source: National Library of Medicine

ARISTOTLE (384 B.C.E.–322 B.C.E.) Aristotle was born in Stagirus (Stagira) in northern Greece. His father was the personal physician of the King of Macedonia. Because his father died when Aristotle was young, Aristotle could not follow the custom of following his father's profession. Aristotle became an orphan at a young age when his mother also died. His guardian who raised him taught him poetry, rhetoric, and Greek. At the age of 17, his guardian sent him to Athens to further his education. Aristotle joined Plato's Academy, where for 20 years he attended Plato's lectures, later presenting his own lectures on rhetoric. When Plato died in 347 B.C.E., Aristotle was not chosen to succeed him because his views differed too much from those of Plato. Instead, Aristotle joined the court of King Hermeas where he remained for three years, and married the niece of the King. When the Persians defeated Hermeas, Aristotle moved to Mytilene and, at the invitation of King Philip of Macedonia, he tutored Alexander, Philip's son, who later became Alexander the Great. Aristotle tutored Alexander for five years and after the death of King Philip, he returned to Athens and set up his own school, called the Lyceum.

Aristotle's followers were called the peripatetics, which means "to walk about," because Aristotle often walked around as he discussed philosophical questions. Aristotle taught at the Lyceum for 13 years where he lectured to his advanced students in the morning and gave popular lectures to a broad audience in the evening. When Alexander the Great died in 323 B.C.E., a backlash against anything related to Alexander led to trumped-up charges of impiety against Aristotle. Aristotle fled to Chalcis to avoid prosecution. He only lived one year in Chalcis, dying of a stomach ailment in 322 B.C.E.

Aristotle wrote three types of works: those written for a popular audience, compilations of scientific facts, and systematic treatises. The systematic treatises included works on logic, philosophy, psychology, physics, and natural history. Aristotle's writings were preserved by a student and were hidden in a vault where a wealthy book collector discovered them about 200 years later. They were taken to Rome, where they were studied by scholars and issued in new editions, preserving them for posterity.

is true, denoted by T, if it is a true proposition, and the truth value of a proposition is false, denoted by F, if it is a false proposition. Propositions that cannot be expressed in terms of simpler propositions are called **atomic propositions**.

The area of logic that deals with propositions is called the **propositional calculus** or **propositional logic**. It was first developed systematically by the Greek philosopher Aristotle more than 2300 years ago.

Links

We now turn our attention to methods for producing new propositions from those that we already have. These methods were discussed by the English mathematician George Boole in 1854 in his book *The Laws of Thought*. Many mathematical statements are constructed by combining one or more propositions. New propositions, called **compound propositions**, are formed from existing propositions using **logical operators**.

Definition 1 Let p be a proposition. The *negation of* p, denoted by $\neg p$ (also denoted by $\overline{p}$), is the statement

"It is not the case that p."

The proposition $\neg p$ is read "not p." The truth value of the negation of p, $\neg p$, is the opposite of the truth value of p.

Remark: The notation for the negation operator is not standardized. Although $\neg p$ and $\overline{p}$ are the most common notations used in mathematics to express the negation of p, other notations you might see are $\sim p$, $-p$, p', Np, and $!p$.

EXAMPLE 3 Find the negation of the proposition

"Michael's PC runs Linux"

Extra Examples

and express this in simple English.

Solution: The negation is

"It is not the case that Michael's PC runs Linux."

This negation can be more simply expressed as

"Michael's PC does not run Linux." ◀

EXAMPLE 4 Find the negation of the proposition

"Vandana's smartphone has at least 128 GB of memory"

and express this in simple English.

Solution: The negation is

"It is not the case that Vandana's smartphone has at least 128 GB of memory."

This negation can also be expressed as

"Vandana's smartphone does not have at least 128 GB of memory"

or even more simply as

"Vandana's smartphone has less than 128 GB of memory." ◀

TABLE 1 The Truth Table for the Negation of a Proposition.

p	$\neg p$
T	F
F	T

Table 1 displays the **truth table** for the negation of a proposition p. This table has a row for each of the two possible truth values of p. Each row shows the truth value of $\neg p$ corresponding to the truth value of p for this row.

The negation of a proposition can also be considered the result of the operation of the **negation operator** on a proposition. The negation operator constructs a new proposition from a single existing proposition. We will now introduce the logical operators that are used to form new propositions from two or more existing propositions. These logical operators are also called **connectives**.

Definition 2 Let p and q be propositions. The *conjunction* of p and q, denoted by $p \wedge q$, is the proposition "p and q." The conjunction $p \wedge q$ is true when both p and q are true and is false otherwise.

Table 2 displays the truth table of $p \wedge q$. This table has a row for each of the four possible combinations of truth values of p and q. The four rows correspond to the pairs of truth values TT, TF, FT, and FF, where the first truth value in the pair is the truth value of p and the second truth value is the truth value of q.

Note that in logic the word "but" sometimes is used instead of "and" in a conjunction. For example, the statement "The sun is shining, but it is raining" is another way of saying "The sun is shining and it is raining." (In natural language, there is a subtle difference in meaning between "and" and "but"; we will not be concerned with this nuance here.)

EXAMPLE 5 Find the conjunction of the propositions p and q where p is the proposition "Rebecca's PC has more than 16 GB free hard disk space" and q is the proposition "The processor in Rebecca's PC runs faster than 1 GHz."

Solution: The conjunction of these propositions, $p \wedge q$, is the proposition "Rebecca's PC has more than 16 GB free hard disk space, and the processor in Rebecca's PC runs faster than 1 GHz." This conjunction can be expressed more simply as "Rebecca's PC has more than 16 GB free hard disk space, and its processor runs faster than 1 GHz." For this conjunction to be true, both conditions given must be true. It is false when one or both of these conditions are false. ◀

Definition 3 Let p and q be propositions. The *disjunction* of p and q, denoted by $p \vee q$, is the proposition "p or q." The disjunction $p \vee q$ is false when both p and q are false and is true otherwise.

Table 3 displays the truth table for $p \vee q$.

TABLE 2 The Truth Table for the Conjunction of Two Propositions.

p	q	$p \wedge q$
T	T	T
T	F	F
F	T	F
F	F	F

TABLE 3 The Truth Table for the Disjunction of Two Propositions.

p	q	$p \vee q$
T	T	T
T	F	T
F	T	T
F	F	F

The use of the connective *or* in a disjunction corresponds to one of the two ways the word *or* is used in English, namely, as an **inclusive or**. A disjunction is true when at least one of the two propositions is true. That is, $p \vee q$ is true when both p and q are true or when exactly one of p and q is true.

EXAMPLE 6 Translate the statement "Students who have taken calculus or introductory computer science can take this class" in a statement in propositional logic using the propositions p: "A student who has taken calculus can take this class" and q: "A student who has taken introductory computer science can take this class."

Solution: We assume that this statement means that students who have taken both calculus and introductory computer science can take the class, as well as the students who have taken only one of the two subjects. Hence, this statement can be expressed as $p \vee q$, the inclusive or, or disjunction, of p and q. ◀

EXAMPLE 7 What is the disjunction of the propositions p and q, where p and q are the same propositions as in Example 5?

Extra Examples

Solution: The disjunction of p and q, $p \vee q$, is the proposition

"Rebecca's PC has at least 16 GB free hard disk space, or the processor in Rebecca's PC runs faster than 1 GHz."

This proposition is true when Rebecca's PC has at least 16 GB free hard disk space, when the PC's processor runs faster than 1 GHz, and when both conditions are true. It is false when both of these conditions are false, that is, when Rebecca's PC has less than 16 GB free hard disk space and the processor in her PC runs at 1 GHz or slower. ◀

Besides its use in disjunctions, the connective *or* is also used to express an *exclusive or*. Unlike the disjunction of two propositions p and q, the exclusive or of these two propositions is true when exactly one of p and q is true; it is false when both p and q are true (and when both are false).

Definition 4 Let p and q be propositions. The *exclusive or* of p and q, denoted by $p \oplus q$ (or p XOR q), is the proposition that is true when exactly one of p and q is true and is false otherwise.

Links

Source: Library of Congress Prints and Photographs Division [LC-USZ62-61664]

GEORGE BOOLE (1815–1864) George Boole, the son of a cobbler, was born in Lincoln, England, in November 1815. Because of his family's difficult financial situation, Boole struggled to educate himself while supporting his family. Nevertheless, he became one of the most important mathematicians of the 1800s. Although he considered a career as a clergyman, he decided instead to go into teaching, and soon afterward opened a school of his own. In his preparation for teaching mathematics, Boole—unsatisfied with textbooks of his day—decided to read the works of the great mathematicians. While reading papers of the great French mathematician Lagrange, Boole made discoveries in the calculus of variations, the branch of analysis dealing with finding curves and surfaces by optimizing certain parameters.

In 1848 Boole published *The Mathematical Analysis of Logic*, the first of his contributions to symbolic logic. In 1849 he was appointed professor of mathematics at Queen's College in Cork, Ireland. In 1854 he published *The Laws of Thought*, his most famous work. In this book, Boole introduced what is now called *Boolean algebra* in his honor. Boole wrote textbooks on differential equations and on difference equations that were used in Great Britain until the end of the nineteenth century. Boole married in 1855; his wife was the niece of the professor of Greek at Queen's College. In 1864 Boole died from pneumonia, which he contracted as a result of keeping a lecture engagement even though he was soaking wet from a rainstorm.

The truth table for the exclusive or of two propositions is displayed in Table 4.

EXAMPLE 8 Let p and q be the propositions that state "A student can have a salad with dinner" and "A student can have soup with dinner," respectively. What is $p \oplus q$, the exclusive or of p and q?

Solution: The exclusive or of p and q is the statement that is true when exactly one of p and q is true. That is, $p \oplus q$ is the statement "A student can have soup or salad, but not both, with dinner." Note that this is often stated as "A student can have soup or a salad with dinner," without explicitly stating that taking both is not permitted. ◀

EXAMPLE 9 Express the statement "I will use all my savings to travel to Europe or to buy an electric car" in propositional logic using the statement p: "I will use all my savings to travel to Europe" and the statement q: "I will use all my savings to buy an electric car."

Solution: To translate this statement, we first note that the or in this statement must be an exclusive or because this student can either use all their savings to travel to Europe or use all these savings to buy an electric car, but cannot both go to Europe and buy an electric car. (This is clear because either option requires all their savings.) Hence, this statement can be expressed as $p \oplus q$. ◀

1.1.3 Conditional Statements

We will discuss several other important ways in which propositions can be combined.

Definition 5 Let p and q be propositions. The *conditional statement* $p \rightarrow q$ is the proposition "if p, then q." The conditional statement $p \rightarrow q$ is false when p is true and q is false, and true otherwise. In the conditional statement $p \rightarrow q$, p is called the *hypothesis* (or *antecedent* or *premise*) and q is called the *conclusion* (or *consequence*).

Assessment ›

The statement $p \rightarrow q$ is called a conditional statement because $p \rightarrow q$ asserts that q is true on the condition that p holds. A conditional statement is also called an **implication**.

The truth table for the conditional statement $p \rightarrow q$ is shown in Table 5. Note that the statement $p \rightarrow q$ is true when both p and q are true and when p is false (no matter what truth value q has).

TABLE 4 The Truth Table for the Exclusive Or of Two Propositions.

p	q	$p \oplus q$
T	T	F
T	F	T
F	T	T
F	F	F

TABLE 5 The Truth Table for the Conditional Statement $p \rightarrow q$.

p	q	$p \rightarrow q$
T	T	T
T	F	F
F	T	T
F	F	T

Because conditional statements play such an essential role in mathematical reasoning, a variety of terminology is used to express $p \to q$. You will encounter most if not all of the following ways to express this conditional statement:

"if p, then q"	"p implies q"
"if p, q"	"p only if q"
"p is sufficient for q"	"a sufficient condition for q is p"
"q if p"	"q whenever p"
"q when p"	"q is necessary for p"
"a necessary condition for p is q"	"q follows from p"
"q unless $\neg p$"	"q provided that p"

A useful way to understand the truth value of a conditional statement is to think of an obligation or a contract. For example, the pledge many politicians make when running for office is

"If I am elected, then I will lower taxes."

If the politician is elected, voters would expect this politician to lower taxes. Furthermore, if the politician is not elected, then voters will not have any expectation that this person will lower taxes, although the person may have sufficient influence to cause those in power to lower taxes. It is only when the politician is elected but does not lower taxes that voters can say that the politician has broken the campaign pledge. This last scenario corresponds to the case when p is true but q is false in $p \to q$.

Similarly, consider a statement that a professor might make:

"If you get 100% on the final, then you will get an A."

If you manage to get 100% on the final, then you would expect to receive an A. If you do not get 100%, you may or may not receive an A depending on other factors. However, if you do get 100%, but the professor does not give you an A, you will feel cheated.

Remark: Because some of the different ways to express the implication p implies q can be confusing, we will provide some extra guidance. To remember that "p only if q" expresses the same thing as "if p, then q," note that "p only if q" says that p cannot be true when q is not true. That is, the statement is false if p is true, but q is false. When p is false, q may be either true or false, because the statement says nothing about the truth value of q.

For example, suppose your professor tells you

"You can receive an A in the course only if your score on the final is at least 90%."

Then, if you receive an A in the course, then you know that your score on the final is at least 90%. If you do not receive an A, you may or may not have scored at least 90% on the final. Be careful not to use "q only if p" to express $p \to q$ because this is incorrect. The word "only" plays an essential role here. To see this, note that the truth values of "q only if p" and $p \to q$ are different when p and q have different truth values. To see why "q is necessary for p" is equivalent to "if p, then q," observe that "q is necessary for p" means that p cannot be true unless q is true, or that if q is false, then p is false. This is the same as saying that if p is true, then q is true. To see why "p is sufficient for q" is equivalent to "if p, then q," note that "p is sufficient for q" means if p is true, it must be the case that q is also true. This is the same as saying that if p is true, then q is also true.

You might have trouble understanding how "unless" is used in conditional statements unless you read this paragraph carefully.

To remember that "q unless $\neg p$" expresses the same conditional statement as "if p, then q," note that "q unless $\neg p$" means that if $\neg p$ is false, then q must be true. That is, the statement "q unless $\neg p$" is false when p is true but q is false, but it is true otherwise. Consequently, "q unless $\neg p$" and $p \to q$ always have the same truth value.

We illustrate the translation between conditional statements and English statements in Example 10.

EXAMPLE 10 Let p be the statement "Maria learns discrete mathematics" and q the statement "Maria will find a good job." Express the statement $p \rightarrow q$ as a statement in English.

Extra Examples

Solution: From the definition of conditional statements, we see that when p is the statement "Maria learns discrete mathematics" and q is the statement "Maria will find a good job," $p \rightarrow q$ represents the statement

"If Maria learns discrete mathematics, then she will find a good job."

There are many other ways to express this conditional statement in English. Among the most natural of these are

"Maria will find a good job when she learns discrete mathematics."

"For Maria to get a good job, it is sufficient for her to learn discrete mathematics."

and

"Maria will find a good job unless she does not learn discrete mathematics." ◀

Note that the way we have defined conditional statements is more general than the meaning attached to such statements in the English language. For instance, the conditional statement in Example 10 and the statement

"If it is sunny, then we will go to the beach"

are statements used in normal language where there is a relationship between the hypothesis and the conclusion. Further, the first of these statements is true unless Maria learns discrete mathematics, but she does not get a good job, and the second is true unless it is indeed sunny, but we do not go to the beach. On the other hand, the statement

"If Juan has a smartphone, then $2 + 3 = 5$"

is true from the definition of a conditional statement, because its conclusion is true. (The truth value of the hypothesis does not matter then.) The conditional statement

"If Juan has a smartphone, then $2 + 3 = 6$"

is true if Juan does not have a smartphone, even though $2 + 3 = 6$ is false. We would not use these last two conditional statements in natural language (except perhaps in sarcasm), because there is no relationship between the hypothesis and the conclusion in either statement. In mathematical reasoning, we consider conditional statements of a more general sort than we use in English. The mathematical concept of a conditional statement is independent of a cause-and-effect relationship between hypothesis and conclusion. Our definition of a conditional statement specifies its truth values; it is not based on English usage. Propositional language is an artificial language; we only parallel English usage to make it easy to use and remember.

The if-then construction used in many programming languages is different from that used in logic. Most programming languages contain statements such as **if** p **then** S, where p is a proposition and S is a program segment (one or more statements to be executed). (Although this looks as if it might be a conditional statement, S is not a proposition, but rather is a set of executable instructions.) When execution of a program encounters such a statement, S is executed if p is true, but S is not executed if p is false, as illustrated in Example 11.

EXAMPLE 11 What is the value of the variable x after the statement

if $2 + 2 = 4$ **then** $x := x + 1$

if $x = 0$ before this statement is encountered? (The symbol := stands for assignment. The statement $x := x + 1$ means the assignment of the value of $x + 1$ to x.)

Solution: Because $2 + 2 = 4$ is true, the assignment statement $x := x + 1$ is executed. Hence, x has the value $0 + 1 = 1$ after this statement is encountered. ◀

CONVERSE, CONTRAPOSITIVE, AND INVERSE We can form some new conditional statements starting with a conditional statement $p \to q$. In particular, there are three related conditional statements that occur so often that they have special names. The proposition $q \to p$ is called the **converse** of $p \to q$. The **contrapositive** of $p \to q$ is the proposition $\neg q \to \neg p$. The proposition $\neg p \to \neg q$ is called the **inverse** of $p \to q$. We will see that of these three conditional statements formed from $p \to q$, only the contrapositive always has the same truth value as $p \to q$.

We first show that the contrapositive, $\neg q \to \neg p$, of a conditional statement $p \to q$ always has the same truth value as $p \to q$. To see this, note that the contrapositive is false only when $\neg p$ is false and $\neg q$ is true, that is, only when p is true and q is false. We now show that neither the converse, $q \to p$, nor the inverse, $\neg p \to \neg q$, has the same truth value as $p \to q$ for all possible truth values of p and q. Note that when p is true and q is false, the original conditional statement is false, but the converse and the inverse are both true.

Remember that the contrapositive, but neither the converse or inverse, of a conditional statement is equivalent to it.

When two compound propositions always have the same truth values, regardless of the truth values of its propositional variables, we call them **equivalent**. Hence, a conditional statement and its contrapositive are equivalent. The converse and the inverse of a conditional statement are also equivalent, as the reader can verify, but neither is equivalent to the original conditional statement. (We will study equivalent propositions in Section 1.3.) Take note that one of the most common logical errors is to assume that the converse or the inverse of a conditional statement is equivalent to this conditional statement.

We illustrate the use of conditional statements in Example 12.

EXAMPLE 12 Find the contrapositive, the converse, and the inverse of the conditional statement

Extra Examples ›

"The home team wins whenever it is raining."

Solution: Because "q whenever p" is one of the ways to express the conditional statement $p \to q$, the original statement can be rewritten as

"If it is raining, then the home team wins."

Consequently, the contrapositive of this conditional statement is

"If the home team does not win, then it is not raining."

The converse is

"If the home team wins, then it is raining."

The inverse is

"If it is not raining, then the home team does not win."

Only the contrapositive is equivalent to the original statement. ◀

TABLE 6 The Truth Table for the Biconditional $p \leftrightarrow q$.

p	q	$p \leftrightarrow q$
T	T	T
T	F	F
F	T	F
F	F	T

BICONDITIONALS We now introduce another way to combine propositions that expresses that two propositions have the same truth value.

Definition 6 Let p and q be propositions. The *biconditional statement* $p \leftrightarrow q$ is the proposition "p if and only if q." The biconditional statement $p \leftrightarrow q$ is true when p and q have the same truth values, and is false otherwise. Biconditional statements are also called *bi-implications*.

The truth table for $p \leftrightarrow q$ is shown in Table 6. Note that the statement $p \leftrightarrow q$ is true when both the conditional statements $p \to q$ and $q \to p$ are true and is false otherwise. That is why we use the words "if and only if" to express this logical connective and why it is symbolically written by combining the symbols $\to$ and $\leftarrow$. There are some other common ways to express $p \leftrightarrow q$:

"p is necessary and sufficient for q"
"if p then q, and conversely"
"p iff q"
"p exactly when q."

The last way of expressing the biconditional statement $p \leftrightarrow q$ uses the abbreviation "iff" for "if and only if." Note that $p \leftrightarrow q$ has exactly the same truth value as $(p \to q) \wedge (q \to p)$.

EXAMPLE 13 Let p be the statement "You can take the flight," and let q be the statement "You buy a ticket." Then $p \leftrightarrow q$ is the statement

Extra Examples

"You can take the flight if and only if you buy a ticket."

This statement is true if p and q are either both true or both false, that is, if you buy a ticket and can take the flight or if you do not buy a ticket and you cannot take the flight. It is false when p and q have opposite truth values, that is, when you do not buy a ticket, but you can take the flight (such as when you get a free trip) and when you buy a ticket but you cannot take the flight (such as when the airline bumps you). ◀

IMPLICIT USE OF BICONDITIONALS You should be aware that biconditionals are not always explicit in natural language. In particular, the "if and only if" construction used in biconditionals is rarely used in common language. Instead, biconditionals are often expressed using an "if, then" or an "only if" construction. The other part of the "if and only if" is implicit. That is, the converse is implied, but not stated. For example, consider the statement in English "If you finish your meal, then you can have dessert." What is really meant is "You can have dessert if and only if you finish your meal." This last statement is logically equivalent to the two statements "If you finish your meal, then you can have dessert" and "You can have dessert only if you finish your meal." Because of this imprecision in natural language, we need to make an assumption whether a conditional statement in natural language implicitly includes its converse. Because precision is essential in mathematics and in logic, we will always distinguish between the conditional statement $p \to q$ and the biconditional statement $p \leftrightarrow q$.

1.1.4 Truth Tables of Compound Propositions

Demo

We have now introduced five important logical connectives—conjunction, disjunction, exclusive or, implication, and the biconditional operator—as well as negation. We can use these connectives to build up complicated compound propositions involving any number of propositional variables. We can use truth tables to determine the truth values of these compound propositions, as Example 14 illustrates. We use a separate column to find the truth value of each compound expression that occurs in the compound proposition as it is built up. The truth values of the compound proposition for each combination of truth values of the propositional variables in it is found in the final column of the table.

EXAMPLE 14 Construct the truth table of the compound proposition

$$(p \vee \neg q) \rightarrow (p \wedge q).$$

Solution: Because this truth table involves two propositional variables p and q, there are four rows in this truth table, one for each of the pairs of truth values TT, TF, FT, and FF. The first two columns are used for the truth values of p and q, respectively. In the third column we find the truth value of $\neg q$, needed to find the truth value of $p \vee \neg q$, found in the fourth column. The fifth column gives the truth value of $p \wedge q$. Finally, the truth value of $(p \vee \neg q) \rightarrow (p \wedge q)$ is found in the last column. The resulting truth table is shown in Table 7. ◀

TABLE 7 The Truth Table of $(p \vee \neg q) \rightarrow (p \wedge q)$.

p	q	$\neg q$	$p \vee \neg q$	$p \wedge q$	$(p \vee \neg q) \rightarrow (p \wedge q)$
T	T	F	T	T	T
T	F	T	T	F	F
F	T	F	F	F	T
F	F	T	T	F	F

1.1.5 Precedence of Logical Operators

We can construct compound propositions using the negation operator and the logical operators defined so far. We will generally use parentheses to specify the order in which logical operators in a compound proposition are to be applied. For instance, $(p \vee q) \wedge (\neg r)$ is the conjunction of $p \vee q$ and $\neg r$. However, to reduce the number of parentheses, we specify that the negation operator is applied before all other logical operators. This means that $\neg p \wedge q$ is the conjunction of $\neg p$ and q, namely, $(\neg p) \wedge q$, not the negation of the conjunction of p and q, namely $\neg(p \wedge q)$. (It is generally the case that unary operators that involve only one object precede binary operators.)

Another general rule of precedence is that the conjunction operator takes precedence over the disjunction operator, so that $p \vee q \wedge r$ means $p \vee (q \wedge r)$ rather than $(p \vee q) \wedge r$ and $p \wedge q \vee r$ means $(p \wedge q) \vee r$ rather than $p \wedge (q \vee r)$. Because this rule may be difficult to remember, we will continue to use parentheses so that the order of the disjunction and conjunction operators is clear.

Finally, it is an accepted rule that the conditional and biconditional operators, $\rightarrow$ and $\leftrightarrow$, have lower precedence than the conjunction and disjunction operators, $\wedge$ and $\vee$. Consequently, $p \rightarrow q \vee r$ means $p \rightarrow (q \vee r)$ rather than $(p \rightarrow q) \vee r$ and $p \vee q \rightarrow r$ means $(p \vee q) \rightarrow r$ rather than $p \vee (q \rightarrow r)$. We will use parentheses when the order of the conditional operator and biconditional operator is at issue, although the conditional operator has precedence over the biconditional operator. Table 8 displays the precedence levels of the logical operators, $\neg$, $\wedge$, $\vee$, $\rightarrow$, and $\leftrightarrow$.

TABLE 8 Precedence of Logical Operators.

Operator	*Precedence*
$\neg$	1
$\wedge$	2
$\vee$	3
$\rightarrow$	4
$\leftrightarrow$	5

1.1.6 Logic and Bit Operations

Truth Value	*Bit*
T	1
F	0

Computers represent information using bits. A **bit** is a symbol with two possible values, namely, 0 (zero) and 1 (one). This meaning of the word bit comes from *b*inary dig*it*, because zeros and ones are the digits used in binary representations of numbers. The well-known statistician John Tukey introduced this terminology in 1946. A bit can be used to represent a truth value, because there are two truth values, namely, *true* and *false*. As is customarily done, we will use a 1 bit to represent true and a 0 bit to represent false. That is, 1 represents T (true), 0 represents F (false). A variable is called a **Boolean variable** if its value is either true or false. Consequently, a Boolean variable can be represented using a bit.

Links

Computer **bit operations** correspond to the logical connectives. By replacing true by a one and false by a zero in the truth tables for the operators $\wedge$, $\vee$, and $\oplus$, the columns in Table 9 for the corresponding bit operations are obtained. We will also use the notation *OR*, *AND*, and *XOR* for the operators $\vee$, $\wedge$, and $\oplus$, as is done in various programming languages.

TABLE 9 Table for the Bit Operators *OR*, *AND*, and *XOR*.

x	y	$x \vee y$	$x \wedge y$	$x \oplus y$
0	0	0	0	0
0	1	1	0	1
1	0	1	0	1
1	1	1	1	0

Information is often represented using bit strings, which are lists of zeros and ones. When this is done, operations on the bit strings can be used to manipulate this information.

Definition 7 A *bit string* is a sequence of zero or more bits. The *length* of this string is the number of bits in the string.

EXAMPLE 15 101010011 is a bit string of length nine. ◀

Links

Alfred Eisenstaedt/ The LIFE Picture Collection/Shutterstock

JOHN WILDER TUKEY (1915–2000) Tukey, born in New Bedford, Massachusetts, was an only child. His parents, both teachers, decided home schooling would best develop his potential. His formal education began at Brown University, where he studied mathematics and chemistry. He received a master's degree in chemistry from Brown and continued his studies at Princeton University, changing his field of study from chemistry to mathematics. He received his Ph.D. from Princeton in 1939 for work in topology, when he was appointed an instructor in mathematics at Princeton. With the start of World War II, he joined the Fire Control Research Office, where he began working in statistics. Tukey found statistical research to his liking and impressed several leading statisticians with his skills. In 1945, at the conclusion of the war, Tukey returned to the mathematics department at Princeton as a professor of statistics, and he also took a position at AT&T Bell Laboratories. Tukey founded the Statistics Department at Princeton in 1966 and was its first chairman. Tukey made significant contributions to many areas of statistics, including the analysis of variance, the estimation of spectra of time series, inferences about the values of a set of parameters from a single experiment, and the philosophy of statistics. However, he is best known for his invention, with J. W. Cooley, of the fast Fourier transform. In addition to his contributions to statistics, Tukey was noted as a skilled wordsmith; he is credited with coining the terms *bit* and *software*.

Tukey contributed his insight and expertise by serving on the President's Science Advisory Committee. He chaired several important committees dealing with the environment, education, and chemicals and health. He also served on committees working on nuclear disarmament. Tukey received many awards, including the National Medal of Science.

HISTORICAL NOTE There were several other suggested words for a binary digit, including *binit* and *bigit*, that never were widely accepted. The adoption of the word *bit* may be due to its meaning as a common English word. For an account of Tukey's coining of the word *bit*, see the April 1984 issue of *Annals of the History of Computing*.

We can extend bit operations to bit strings. We define the **bitwise *OR***, **bitwise *AND***, and **bitwise *XOR*** of two strings of the same length to be the strings that have as their bits the *OR*, *AND*, and *XOR* of the corresponding bits in the two strings, respectively. We use the symbols $\vee$, $\wedge$, and $\oplus$ to represent the bitwise *OR*, bitwise *AND*, and bitwise *XOR* operations, respectively. We illustrate bitwise operations on bit strings with Example 16.

EXAMPLE 16 Find the bitwise *OR*, bitwise *AND*, and bitwise *XOR* of the bit strings 01 1011 0110 and 11 0001 1101. (Here, and throughout this book, bit strings will be split into blocks of four bits to make them easier to read.)

Solution: The bitwise *OR*, bitwise *AND*, and bitwise *XOR* of these strings are obtained by taking the *OR*, *AND*, and *XOR* of the corresponding bits, respectively. This gives us

```
01 1011 0110
11 0001 1101
------------
11 1011 1111   bitwise OR
01 0001 0100   bitwise AND
10 1010 1011   bitwise XOR
```

◀

Exercises

1. Which of these sentences are propositions? What are the truth values of those that are propositions?
- **a)** Boston is the capital of Massachusetts.
- **b)** Miami is the capital of Florida.
- **c)** $2 + 3 = 5$.
- **d)** $5 + 7 = 10$.
- **e)** $x + 2 = 11$.
- **f)** Answer this question.

2. Which of these are propositions? What are the truth values of those that are propositions?
- **a)** Do not pass go.
- **b)** What time is it?
- **c)** There are no black flies in Maine.
- **d)** $4 + x = 5$.
- **e)** The moon is made of green cheese.
- **f)** $2^n \geq 100$.

3. What is the negation of each of these propositions?
- **a)** Linda is younger than Sanjay.
- **b)** Mei makes more money than Isabella.
- **c)** Moshe is taller than Monica.
- **d)** Abby is richer than Ricardo.

4. What is the negation of each of these propositions?
- **a)** Janice has more Facebook friends than Juan.
- **b)** Quincy is smarter than Venkat.
- **c)** Zelda drives more miles to school than Paola.
- **d)** Briana sleeps longer than Gloria.

5. What is the negation of each of these propositions?
- **a)** Mei has an MP3 player.
- **b)** There is no pollution in New Jersey.
- **c)** $2 + 1 = 3$.
- **d)** The summer in Maine is hot and sunny.

6. What is the negation of each of these propositions?
- **a)** Jennifer and Teja are friends.
- **b)** There are 13 items in a baker's dozen.
- **c)** Abby sent more than 100 text messages yesterday.
- **d)** 121 is a perfect square.

7. What is the negation of each of these propositions?
- **a)** Steve has more than 100 GB free disk space on his laptop.
- **b)** Zach blocks emails and texts from Jennifer.
- **c)** $7 \cdot 11 \cdot 13 = 999$.
- **d)** Diane rode her bicycle 100 miles on Sunday.

8. Suppose that Smartphone A has 8 GB RAM and 256 GB ROM, and the resolution of its camera is 200 MP; Smartphone B has 12 GB RAM and 512 GB ROM, and the resolution of its camera is 24 MP; and Smartphone C has 4 GB RAM and 256 GB ROM, and the resolution of its camera is 48 MP. Determine the truth value of each of these propositions.
- **a)** Smartphone B has the most RAM of these three smartphones.
- **b)** Smartphone C has more ROM or a higher resolution camera than Smartphone B.
- **c)** Smartphone B has more RAM, more ROM, and a higher resolution camera than Smartphone A.
- **d)** If Smartphone B has more RAM and more ROM than Smartphone C, then it also has a higher resolution camera.
- **e)** Smartphone A has more RAM than Smartphone B if and only if Smartphone B has more RAM than Smartphone A.

9. Suppose that during the most recent fiscal year, the annual revenue of Acme Computer was 138 billion dollars and its net profit was 8 billion dollars, the annual revenue of Nadir Software was 87 billion dollars and its net profit was 5 billion dollars, and the annual revenue of Quixote Media was 111 billion dollars and its net profit was 13 billion dollars. Determine the truth value of each of these propositions for the most recent fiscal year.

a) Quixote Media had the largest annual revenue.
b) Nadir Software had the lowest net profit and Acme Computer had the largest annual revenue.
c) Acme Computer had the largest net profit or Quixote Media had the largest net profit.
d) If Quixote Media had the smallest net profit, then Acme Computer had the largest annual revenue.
e) Nadir Software had the smallest net profit if and only if Acme Computer had the largest annual revenue.

10. Let p and q be the propositions

p: I bought a lottery ticket this week.
q: I won the million dollar jackpot.

Express each of these propositions as an English sentence.

a) $\neg p$ **b)** $p \vee q$ **c)** $p \rightarrow q$
d) $p \wedge q$ **e)** $p \leftrightarrow q$ **f)** $\neg p \rightarrow \neg q$
g) $\neg p \wedge \neg q$ **h)** $\neg p \vee (p \wedge q)$

11. Let p and q be the propositions "Swimming at the New Jersey shore is allowed" and "Sharks have been spotted near the shore," respectively. Express each of these compound propositions as an English sentence.

a) $\neg q$ **b)** $p \wedge q$ **c)** $\neg p \vee q$
d) $p \rightarrow \neg q$ **e)** $\neg q \rightarrow p$ **f)** $\neg p \rightarrow \neg q$
g) $p \leftrightarrow \neg q$ **h)** $\neg p \wedge (p \vee \neg q)$

12. Let p and q be the propositions "The election is decided" and "The votes have been counted," respectively. Express each of these compound propositions as an English sentence.

a) $\neg p$ **b)** $p \vee q$
c) $\neg p \wedge q$ **d)** $q \rightarrow p$
e) $\neg q \rightarrow \neg p$ **f)** $\neg p \rightarrow \neg q$
g) $p \leftrightarrow q$ **h)** $\neg q \vee (\neg p \wedge q)$

13. Let p and q be the propositions

p: It is below freezing.
q: It is snowing.

Write these propositions using p and q and logical connectives (including negations).

a) It is below freezing and snowing.
b) It is below freezing but not snowing.
c) It is not below freezing and it is not snowing.
d) It is either snowing or below freezing (or both).
e) If it is below freezing, it is also snowing.
f) Either it is below freezing or it is snowing, but it is not snowing if it is below freezing.
g) That it is below freezing is necessary and sufficient for it to be snowing.

14. Let p, q, and r be the propositions

p: You have the flu.
q: You miss the final examination.
r: You pass the course.

Express each of these propositions as an English sentence.

a) $p \rightarrow q$ **b)** $\neg q \leftrightarrow r$
c) $q \rightarrow \neg r$ **d)** $p \vee q \vee r$
e) $(p \rightarrow \neg r) \vee (q \rightarrow \neg r)$ **f)** $(p \wedge q) \vee (\neg q \wedge r)$

15. Let p and q be the propositions

p: You drive over 65 miles per hour.
q: You get a speeding ticket.

Write these propositions using p and q and logical connectives (including negations).

a) You do not drive over 65 miles per hour.
b) You drive over 65 miles per hour, but you do not get a speeding ticket.
c) You will get a speeding ticket if you drive over 65 miles per hour.
d) If you do not drive over 65 miles per hour, then you will not get a speeding ticket.
e) Driving over 65 miles per hour is sufficient for getting a speeding ticket.
f) You get a speeding ticket, but you do not drive over 65 miles per hour.
g) Whenever you get a speeding ticket, you are driving over 65 miles per hour.

16. Let p, q, and r be the propositions

p: You get an A on the final exam.
q: You do every exercise in this book.
r: You get an A in this class.

Write these propositions using p, q, and r and logical connectives (including negations).

a) You get an A in this class, but you do not do every exercise in this book.
b) You get an A on the final, you do every exercise in this book, and you get an A in this class.
c) To get an A in this class, it is necessary for you to get an A on the final.
d) You get an A on the final, but you don't do every exercise in this book; nevertheless, you get an A in this class.
e) Getting an A on the final and doing every exercise in this book is sufficient for getting an A in this class.
f) You will get an A in this class if and only if you either do every exercise in this book or you get an A on the final.

17. Let p, q, and r be the propositions

p: Grizzly bears have been seen in the area.
q: Hiking is safe on the trail.
r: Berries are ripe along the trail.

Write these propositions using p, q, and r and logical connectives (including negations).

a) Berries are ripe along the trail, but grizzly bears have not been seen in the area.
b) Grizzly bears have not been seen in the area and hiking on the trail is safe, but berries are ripe along the trail.
c) If berries are ripe along the trail, hiking is safe if and only if grizzly bears have not been seen in the area.
d) It is not safe to hike on the trail, but grizzly bears have not been seen in the area and the berries along the trail are ripe.
e) For hiking on the trail to be safe, it is necessary but not sufficient that berries not be ripe along the trail and for grizzly bears not to have been seen in the area.

f) Hiking is not safe on the trail whenever grizzly bears have been seen in the area and berries are ripe along the trail.

18. Determine whether these biconditionals are true or false.

a) $2 + 2 = 4$ if and only if $1 + 1 = 2$.
b) $1 + 1 = 2$ if and only if $2 + 3 = 4$.
c) $1 + 1 = 3$ if and only if monkeys can fly.
d) $0 > 1$ if and only if $2 > 1$.

19. Determine whether each of these conditional statements is true or false.

a) If $1 + 1 = 2$, then $2 + 2 = 5$.
b) If $1 + 1 = 3$, then $2 + 2 = 4$.
c) If $1 + 1 = 3$, then $2 + 2 = 5$.
d) If monkeys can fly, then $1 + 1 = 3$.

20. Determine whether each of these conditional statements is true or false.

a) If $1 + 1 = 3$, then unicorns exist.
b) If $1 + 1 = 3$, then dogs can fly.
c) If $1 + 1 = 2$, then dogs can fly.
d) If $2 + 2 = 4$, then $1 + 2 = 3$.

21. For each of these sentences, determine whether an inclusive or, or an exclusive or, is intended. Explain your answer.

a) Coffee or tea comes with dinner.
b) A password must have at least three digits or be at least eight characters long.
c) The prerequisite for the course is a course in number theory or a course in cryptography.
d) You can pay using U.S. dollars or euros.

22. For each of these sentences, determine whether an inclusive or, or an exclusive or, is intended. Explain your answer.

a) Experience with C++ or Java is required.
b) Lunch includes soup or salad.
c) To enter the country you need a passport or a voter registration card.
d) Publish or perish.

23. For each of these sentences, state what the sentence means if the logical connective or is an inclusive or (that is, a disjunction) versus an exclusive or. Which of these meanings of or do you think is intended?

a) To take discrete mathematics, you must have taken calculus or a course in computer science.
b) When you buy a new car from Acme Motor Company, you get \$2000 back in cash or a 2% car loan.
c) Dinner for two includes two items from column A or three items from column B.
d) School is closed if more than two feet of snow falls or if the wind chill is below −100 °F.

24. Write each of these statements in the form "if p, then q" in English. [*Hint:* Refer to the list of common ways to express conditional statements provided in this section.]

a) It is necessary to wash the boss's car to get promoted.
b) Winds from the south imply a spring thaw.
c) A sufficient condition for the warranty to be good is that you bought the computer less than a year ago.
d) Willy gets caught whenever he cheats.
e) You can access the website only if you pay a subscription fee.
f) Getting elected follows from knowing the right people.
g) Carol gets seasick whenever she is on a boat.

25. Write each of these statements in the form "if p, then q" in English. [*Hint:* Refer to the list of common ways to express conditional statements.]

a) It snows whenever the wind blows from the northeast.
b) The apple trees will bloom if it stays warm for a week.
c) That the Pistons win the championship implies that they beat the Lakers.
d) It is necessary to walk eight miles to get to the top of Long's Peak.
e) To get tenure as a professor, it is sufficient to be world famous.
f) If you drive more than 400 miles, you will need to buy gasoline.
g) Your guarantee is good only if you bought your CD player less than 90 days ago.
h) Jan will go swimming unless the water is too cold.
i) We will have a future, provided that people believe in science.

26. Write each of these statements in the form "if p, then q" in English. [*Hint:* Refer to the list of common ways to express conditional statements provided in this section.]

a) I will remember to send you the address only if you send me an email message.
b) To be a citizen of this country, it is sufficient that you were born in the United States.
c) If you keep your textbook, it will be a useful reference in your future courses.
d) The Red Wings will win the Stanley Cup if their goalie plays well.
e) That you get the job implies that you had the best credentials.
f) The beach erodes whenever there is a storm.
g) It is necessary to have a valid password to log on to the server.
h) You will reach the summit unless you begin your climb too late.
i) You will get a free ice cream cone, provided that you are among the first 100 customers tomorrow.

27. Write each of these propositions in the form "p if and only if q" in English.

a) If it is hot outside you buy an ice cream cone, and if you buy an ice cream cone it is hot outside.
b) For you to win the contest it is necessary and sufficient that you have the only winning ticket.
c) You get promoted only if you have connections, and you have connections only if you get promoted.
d) If you watch television your mind will decay, and conversely.
e) The trains run late on exactly those days when I take it.

28. Write each of these propositions in the form "p if and only if q" in English.

a) For you to get an A in this course, it is necessary and sufficient that you learn how to solve discrete mathematics problems.

b) If you read the newspaper every day, you will be informed, and conversely.

c) It rains if it is a weekend day, and it is a weekend day if it rains.

d) You can see the wizard only if the wizard is not in, and the wizard is not in only if you can see him.

e) My airplane flight is late exactly when I have to catch a connecting flight.

29. State the converse, contrapositive, and inverse of each of these conditional statements.

a) If it snows today, I will ski tomorrow.

b) I come to class whenever there is going to be a quiz.

c) A positive integer is a prime only if it has no divisors other than 1 and itself.

30. State the converse, contrapositive, and inverse of each of these conditional statements.

a) If it snows tonight, then I will stay at home.

b) I go to the beach whenever it is a sunny summer day.

c) When I stay up late, it is necessary that I sleep until noon.

31. How many rows appear in a truth table for each of these compound propositions?

a) $p \rightarrow \neg p$

b) $(p \vee \neg r) \wedge (q \vee \neg s)$

c) $q \vee p \vee \neg s \vee \neg r \vee \neg t \vee u$

d) $(p \wedge r \wedge t) \leftrightarrow (q \wedge t)$

32. How many rows appear in a truth table for each of these compound propositions?

a) $(q \rightarrow \neg p) \vee (\neg p \rightarrow \neg q)$

b) $(p \vee \neg t) \wedge (p \vee \neg s)$

c) $(p \rightarrow r) \vee (\neg s \rightarrow \neg t) \vee (\neg u \rightarrow v)$

d) $(p \wedge r \wedge s) \vee (q \wedge t) \vee (r \wedge \neg t)$

33. Construct a truth table for each of these compound propositions.

a) $p \wedge \neg p$ **b)** $p \vee \neg p$

c) $(p \vee \neg q) \rightarrow q$ **d)** $(p \vee q) \rightarrow (p \wedge q)$

e) $(p \rightarrow q) \leftrightarrow (\neg q \rightarrow \neg p)$

f) $(p \rightarrow q) \rightarrow (q \rightarrow p)$

34. Construct a truth table for each of these compound propositions.

a) $p \rightarrow \neg p$ **b)** $p \leftrightarrow \neg p$

c) $p \oplus (p \vee q)$ **d)** $(p \wedge q) \rightarrow (p \vee q)$

e) $(q \rightarrow \neg p) \leftrightarrow (p \leftrightarrow q)$

f) $(p \leftrightarrow q) \oplus (p \leftrightarrow \neg q)$

35. Construct a truth table for each of these compound propositions.

a) $(p \vee q) \rightarrow (p \oplus q)$ **b)** $(p \oplus q) \rightarrow (p \wedge q)$

c) $(p \vee q) \oplus (p \wedge q)$ **d)** $(p \leftrightarrow q) \oplus (\neg p \leftrightarrow q)$

e) $(p \leftrightarrow q) \oplus (\neg p \leftrightarrow \neg r)$

f) $(p \oplus q) \rightarrow (p \oplus \neg q)$

36. Construct a truth table for each of these compound propositions.

a) $p \oplus p$ **b)** $p \oplus \neg p$

c) $p \oplus \neg q$ **d)** $\neg p \oplus \neg q$

e) $(p \oplus q) \vee (p \oplus \neg q)$ **f)** $(p \oplus q) \wedge (p \oplus \neg q)$

37. Construct a truth table for each of these compound propositions.

a) $p \rightarrow \neg q$ **b)** $\neg p \leftrightarrow q$

c) $(p \rightarrow q) \vee (\neg p \rightarrow q)$ **d)** $(p \rightarrow q) \wedge (\neg p \rightarrow q)$

e) $(p \leftrightarrow q) \vee (\neg p \leftrightarrow q)$

f) $(\neg p \leftrightarrow \neg q) \leftrightarrow (p \leftrightarrow q)$

38. Construct a truth table for each of these compound propositions.

a) $(p \vee q) \vee r$ **b)** $(p \vee q) \wedge r$

c) $(p \wedge q) \vee r$ **d)** $(p \wedge q) \wedge r$

e) $(p \vee q) \wedge \neg r$ **f)** $(p \wedge q) \vee \neg r$

39. Construct a truth table for each of these compound propositions.

a) $p \rightarrow (\neg q \vee r)$

b) $\neg p \rightarrow (q \rightarrow r)$

c) $(p \rightarrow q) \vee (\neg p \rightarrow r)$

d) $(p \rightarrow q) \wedge (\neg p \rightarrow r)$

e) $(p \leftrightarrow q) \vee (\neg q \leftrightarrow r)$

f) $(\neg p \leftrightarrow \neg q) \leftrightarrow (q \leftrightarrow r)$

40. Construct a truth table for $((p \rightarrow q) \rightarrow r) \rightarrow s$.

41. Construct a truth table for $(p \leftrightarrow q) \leftrightarrow (r \leftrightarrow s)$.

42. Explain, without using a truth table, why $(p \vee \neg q) \wedge (q \vee \neg r) \wedge (r \vee \neg p)$ is true when p, q, and r have the same truth value and it is false otherwise.

43. Explain, without using a truth table, why $(p \vee q \vee r) \wedge (\neg p \vee \neg q \vee \neg r)$ is true when at least one of p, q, and r is true and at least one is false, but is false when all three variables have the same truth value.

44. If $p_1, p_2, \ldots, p_n$ are n propositions, explain why

$$\bigwedge_{i=1}^{n-1} \bigwedge_{j=i+1}^{n} (\neg p_i \vee \neg p_j)$$

is true if and only if at most one of $p_1, p_2, \ldots, p_n$ is true.

45. Use Exercise 44 to construct a compound proposition that is true if and only if exactly one of the propositions $p_1, p_2, \ldots, p_n$ is true. [*Hint*: Combine the compound proposition in Exercise 44 and a compound proposition that is true if and only if at least one of $p_1, p_2, \ldots, p_n$ is true.]

46. What is the value of x after each of these statements is encountered in a computer program, if $x = 1$ before the statement is reached?

a) **if** $x + 2 = 3$ **then** $x := x + 1$

b) **if** $(x + 1 = 3)$ *OR* $(2x + 2 = 3)$ **then** $x := x + 1$

c) **if** $(2x + 3 = 5)$ *AND* $(3x + 4 = 7)$ **then** $x := x + 1$

d) **if** $(x + 1 = 2)$ *XOR* $(x + 2 = 3)$ **then** $x := x + 1$

e) **if** $x < 2$ **then** $x := x + 1$

47. Find the bitwise *OR*, bitwise *AND*, and bitwise *XOR* of each of these pairs of bit strings.

a) 101 1110, 010 0001

b) 1111 0000, 1010 1010

c) 00 0111 0001, 10 0100 1000

d) 11 1111 1111, 00 0000 0000

48. Evaluate each of these expressions.

a) 1 1000 ∧ (0 1011 ∨ 1 1011)
b) (0 1111 ∧ 1 0101) ∨ 0 1000
c) (0 1010 ⊕ 1 1011) ⊕ 0 1000
d) (1 1011 ∨ 0 1010) ∧ (1 0001 ∨ 1 1011)

Fuzzy logic is used in artificial intelligence. In fuzzy logic, a proposition has a truth value that is a number between 0 and 1, inclusive. A proposition with a truth value of 0 is false and one with a truth value of 1 is true. Truth values that are between 0 and 1 indicate varying degrees of truth. For instance, the truth value 0.8 can be assigned to the statement "Fred is happy," because Fred is happy most of the time, and the truth value 0.4 can be assigned to the statement "John is happy," because John is happy slightly less than half the time. Use these truth values to solve Exercises 49–51.

49. The truth value of the negation of a proposition in fuzzy logic is 1 minus the truth value of the proposition. What are the truth values of the statements "Fred is not happy" and "John is not happy"?

50. The truth value of the conjunction of two propositions in fuzzy logic is the minimum of the truth values of the two propositions. What are the truth values of the statements "Fred and John are happy" and "Neither Fred nor John is happy"?

51. The truth value of the disjunction of two propositions in fuzzy logic is the maximum of the truth values of the two propositions. What are the truth values of the statements "Fred is happy, or John is happy" and "Fred is not happy, or John is not happy"?

***52.** Is the assertion "This statement is false" a proposition?

***53.** The *n*th statement in a list of 100 statements is "Exactly *n* of the statements in this list are false."

a) What conclusions can you draw from these statements?
b) Answer part (a) if the *n*th statement is "At least *n* of the statements in this list are false."
c) Answer part (b) assuming that the list contains 99 statements.

54. An ancient Sicilian legend says that the barber in a remote town who can be reached only by traveling a dangerous mountain road shaves those people, and only those people, who do not shave themselves. Can there be such a barber?

1.2 Applications of Propositional Logic

1.2.1 Introduction

Logic has many important applications to mathematics, computer science, and numerous other disciplines. Statements in mathematics and the sciences and in natural language often are imprecise or ambiguous. To make such statements precise, they can be translated into the language of logic. For example, logic is used in the specification of software and hardware, because these specifications need to be precise before development begins. Furthermore, propositional logic and its rules can be used to design computer circuits, to construct computer programs, to verify the correctness of programs, and to build expert systems. Logic can be used to analyze and solve many familiar puzzles. Software systems based on the rules of logic have been developed for constructing some, but not all, types of proofs automatically. We will discuss some of these applications of propositional logic in this section and in later chapters.

1.2.2 Translating English Sentences

There are many reasons to translate English sentences into expressions involving propositional variables and logical connectives. In particular, English (and every other human language) is often ambiguous. Translating sentences into compound statements (and other types of logical expressions, which we will introduce later in this chapter) removes the ambiguity. Note that this may involve making a set of reasonable assumptions based on the intended meaning of the sentence. Moreover, once we have translated sentences from English into logical expressions, we can analyze these logical expressions to determine their truth values, we can manipulate them, and we can use rules of inference (which are discussed in Section 1.6) to reason about them.

To illustrate the process of translating an English sentence into a logical expression, consider Examples 1 and 2.

EXAMPLE 1 How can this English sentence be translated into a logical expression?

Extra Examples

"You can access the Internet from campus only if you are a computer science major or you are not a freshman."

Solution: There are many ways to translate this sentence into a logical expression. Although it is possible to represent the sentence by a single propositional variable, such as p, this would not be useful when analyzing its meaning or reasoning with it. Instead, we will use propositional variables to represent each sentence part and determine the appropriate logical connectives between them. In particular, we let a, c, and f represent "You can access the Internet from campus," "You are a computer science major," and "You are a freshman," respectively. Noting that "only if" is one way a conditional statement can be expressed, this sentence can be represented as

$$a \to (c \vee \neg f).$$ ◀

EXAMPLE 2 How can this English sentence be translated into a logical expression?

"You cannot ride the roller coaster if you are under 4 feet tall unless you are older than 16 years old."

Solution: Let q, r, and s represent "You can ride the roller coaster," "You are under 4 feet tall," and "You are older than 16 years old," respectively. Then the sentence can be translated to

$$(r \wedge \neg s) \to \neg q.$$

There are other ways to represent the original sentence as a logical expression, but the one we have used should meet our needs. ◀

1.2.3 System Specifications

Translating sentences in natural language (such as English) into logical expressions is an essential part of specifying both hardware and software systems. System and software engineers take requirements in natural language and produce precise and unambiguous specifications that can be used as the basis for system development. Example 3 shows how compound propositions can be used in this process.

EXAMPLE 3 Express the specification "The automated reply cannot be sent when the file system is full" using logical connectives.

Extra Examples

Solution: One way to translate this is to let p denote "The automated reply can be sent" and q denote "The file system is full." Then $\neg p$ represents "It is not the case that the automated reply can be sent," which can also be expressed as "The automated reply cannot be sent." Consequently, our specification can be represented by the conditional statement $q \to \neg p$. ◀

System specifications should be **consistent**, that is, they should not contain conflicting requirements that could be used to derive a contradiction. When specifications are not consistent, there would be no way to develop a system that satisfies all specifications.

EXAMPLE 4 Determine whether these system specifications are consistent:

"The diagnostic message is stored in the buffer or it is retransmitted."
"The diagnostic message is not stored in the buffer."
"If the diagnostic message is stored in the buffer, then it is retransmitted."

Solution: To determine whether these specifications are consistent, we first express them using logical expressions. Let p denote "The diagnostic message is stored in the buffer" and let q denote "The diagnostic message is retransmitted." The specifications can then be written as $p \vee q$, $\neg p$, and $p \to q$. An assignment of truth values that makes all three specifications true must have p false to make $\neg p$ true. Because we want $p \vee q$ to be true but p must be false, q must be true. Because $p \to q$ is true when p is false and q is true, we conclude that these specifications are consistent, because they are all true when p is false and q is true. We could come to the same conclusion by use of a truth table to examine the four possible assignments of truth values to p and q. ◀

EXAMPLE 5 Do the system specifications in Example 4 remain consistent if the specification "The diagnostic message is not retransmitted" is added?

Solution: By the reasoning in Example 4, the three specifications from that example are true only in the case when p is false and q is true. However, this new specification is $\neg q$, which is false when q is true. Consequently, these four specifications are inconsistent. ◀

1.2.4 Boolean Searches

Links ›

Logical connectives are used extensively in searches of large collections of information, such as indexes of Web pages. Because these searches employ techniques from propositional logic, they are called **Boolean searches**.

In Boolean searches, the connective *AND* is used to match records that contain both of two search terms, the connective *OR* is used to match one or both of two search terms, and the connective *NOT* (sometimes written as *AND NOT*) is used to exclude a particular search term. Careful planning of how logical connectives are used is often required when Boolean searches are used to locate information of potential interest. Example 6 illustrates how Boolean searches are carried out.

EXAMPLE 6 **Web Page Searching** Most Web search engines support Boolean searching techniques, which is useful for finding Web pages about particular subjects. For instance, using Boolean searching to find Web pages about universities in New Mexico, we can look for pages matching NEW *AND* MEXICO *AND* UNIVERSITIES. The results of this search will include those pages that contain the three words NEW, MEXICO, and UNIVERSITIES. This will include all of the pages of interest, together with others such as a page about new universities in Mexico. (Note that Google, and many other search engines, do not require the use of "AND" because such search engines use all search terms by default.) Most search engines support the use of quotation marks to search for specific phrases. So, it may be more effective to search for pages matching "NEW MEXICO" *AND* UNIVERSITIES.

Extra Examples ›

Next, to find pages that deal with universities in New Mexico or Arizona, we can search for pages matching (NEW *AND* MEXICO *OR* ARIZONA) *AND* UNIVERSITIES. (*Note:* Here the *AND* operator takes precedence over the *OR* operator. Also, in Google, the terms used for this search would be NEW MEXICO *OR* ARIZONA.) The results of this search will include all pages that contain the word UNIVERSITIES and either both the words NEW and MEXICO or the word ARIZONA. Again, pages besides those of interest will be listed. Finally, to find Web pages that deal with universities in Mexico (and not New Mexico), we might first look for pages matching MEXICO *AND* UNIVERSITIES, but because the results of this search will include pages about universities in New Mexico, as well as universities in Mexico, it might be better to search for pages matching (MEXICO *AND* UNIVERSITIES) *NOT* NEW. The results of this search include pages that contain both the words MEXICO and UNIVERSITIES but do not contain the word NEW. (In Google, and many other search engines, the word "NOT" is replaced by the symbol "-". In Google, the terms used for this last search would be MEXICO UNIVERSITIES -NEW.) ◀

1.2.5 Logic Puzzles

Links

Puzzles that can be solved using logical reasoning are known as **logic puzzles**. Solving logic puzzles is an excellent way to practice working with the rules of logic. Also, computer programs designed to carry out logical reasoning often use well-known logic puzzles to illustrate their capabilities. Many people enjoy solving logic puzzles, published in periodicals, books, and on the Web, as a recreational activity.

The next three examples present logic puzzles, in increasing level of difficulty. Many others can be found in the exercises. In Section 1.3 we will discuss the n-queens puzzle and the game of Sudoku.

EXAMPLE 7 As a reward for saving his daughter from pirates, the King has given you the opportunity to win a treasure hidden inside one of three trunks. The two trunks that do not hold the treasure are empty. To win, you must select the correct trunk. Trunks 1 and 2 are each inscribed with the message "This trunk is empty," and Trunk 3 is inscribed with the message "The treasure is in Trunk 2." The Queen, who never lies, tells you that only one of these inscriptions is true, while the other two are wrong. Which trunk should you select to win?

Solution: Let p_i be the proposition that the treasure is in Trunk i, for $i = 1, 2, 3$. To translate into propositional logic the Queen's statement that exactly one of the inscriptions is true, we observe that the inscriptions on Trunk 1, Trunk 2, and Trunk 3, are $\neg p_1$, $\neg p_2$, and p_2, respectively. So, her statement can be translated to

$$(\neg p_1 \wedge \neg(\neg p_2) \wedge \neg p_2) \vee (\neg(\neg p_1) \wedge \neg p_2 \wedge \neg p_2) \vee (\neg(\neg p_1) \wedge \neg(\neg p_2) \wedge p_2)).$$

Using the rules for propositional logic, we see that this is equivalent to $(p_1 \wedge \neg p_2) \vee (p_1 \wedge p_2)$. By the distributive law, $(p_1 \wedge \neg p_2) \vee (p_1 \wedge p_2)$ is equivalent to $p_1 \wedge (\neg p_2 \vee p_2)$. But because $\neg p_2 \vee p_2$ must be true, this is then equivalent to $p_1 \wedge \mathbf{T}$, which is in turn equivalent to p_1. So the treasure is in Trunk 1 (that is, p_1 is true), and p_2 and p_3 are false; and the inscription on Trunk 2 is the only true one. (Here, we have used the concept of propositional equivalence, which is discussed in Section 1.3.) ◀

Next, we introduce a puzzle originally posed by Raymond Smullyan, a master of logic puzzles, who has published more than a dozen books containing challenging puzzles that involve logical reasoning.

EXAMPLE 8 In [Sm78] Smullyan posed many puzzles about an island that has two kinds of inhabitants, knights, who always tell the truth, and their opposites, knaves, who always lie. You encounter two people A and B. What are A and B if A says "B is a knight" and B says "The two of us are opposite types"?

Extra Examples

Solution: Let p and q be the statements that A is a knight and B is a knight, respectively, so that $\neg p$ and $\neg q$ are the statements that A is a knave and B is a knave, respectively.

We first consider the possibility that A is a knight; this is the statement that p is true. If A is a knight, then he is telling the truth when he says that B is a knight, so that q is true, and A and B are the same type. However, if B is a knight, then B's statement that A and B are of opposite types, the statement $(p \wedge \neg q) \vee (\neg p \wedge q)$, would have to be true, which it is not, because A and B are both knights. Consequently, we can conclude that A is not a knight, that is, that p is false.

If A is a knave, then because everything a knave says is false, A's statement that B is a knight, that is, that q is true, is a lie. This means that q is false and B is also a knave. Furthermore, if B is a knave, then B's statement that A and B are opposite types is a lie, which is consistent with both A and B being knaves. We can conclude that both A and B are knaves. ◀

We pose more of Smullyan's puzzles about knights and knaves in Exercises 23–27. In Exercises 28–35 we introduce related puzzles where we have three types of people, knights and knaves as in this puzzle together with spies who can lie.

Next, we pose a puzzle known as the **muddy children puzzle** for the case of two children.

EXAMPLE 9 A father tells his two children, a boy and a girl, to play in their backyard without getting dirty. However, while playing, both children get mud on their foreheads. When the children stop playing, the father says "At least one of you has a muddy forehead," and then asks the children to answer "Yes" or "No" to the question: "Do you know whether you have a muddy forehead?" The father asks this question twice. What will the children answer each time this question is asked, assuming that a child can see whether his or her sibling has a muddy forehead, but cannot see his or her own forehead? Assume that both children are honest and that the children answer each question simultaneously.

Solution: Let s be the statement that the son has a muddy forehead and let d be the statement that the daughter has a muddy forehead. When the father says that at least one of the two children has a muddy forehead, he is stating that the disjunction $s \vee d$ is true. Both children will answer "No" the first time the question is asked because each sees mud on the other child's forehead. That is, the son knows that d is true, but does not know whether s is true, and the daughter knows that s is true, but does not know whether d is true.

After the son has answered "No" to the first question, the daughter can determine that d must be true. This follows because when the first question is asked, the son knows that $s \vee d$ is true, but cannot determine whether s is true. Using this information, the daughter can conclude that d must be true, for if d were false, the son could have reasoned that because $s \vee d$ is true,

Links

Courtesy of Indiana University Archives

RAYMOND SMULLYAN (1919–2017) Raymond Smullyan, the son of a businessman and a homemaker, was born in Far Rockaway, Queens, New York. He dropped out of high school because he wanted to study what he was really interested in and not standard high school material. After attending Pacific College and Reed College in Oregon, he earned an undergraduate degree in mathematics at the University of Chicago in 1955. He paid his college expenses by performing magic tricks at parties and clubs, using the stage name Five-Ace Merrill. He obtained a Ph.D. in logic in 1959 at Princeton, studying under Alonzo Church. After graduating from Princeton, he taught mathematics and logic at Dartmouth College, Princeton University, Yeshiva University, and the City University of New York. He joined the philosophy department at Indiana University in 1981, where be became an emeritus professor.

Smullyan wrote many books on recreational logic and mathematics, including *Satan, Cantor, and Infinity*; *What Is the Name of This Book?*; *The Lady or the Tiger?*; *Alice in Puzzleland*; *To Mock a Mockingbird*; *Forever Undecided*; and *The Riddle of Scheherazade: Amazing Logic Puzzles, Ancient and Modern*. Because his logic puzzles are challenging, entertaining, and thought-provoking, he was considered to be a modern-day Lewis Carroll. Smullyan also wrote several books about the application of deductive logic to chess, three collections of philosophical essays and aphorisms, and several advanced books on mathematical logic and set theory. He was particularly interested in self-reference and worked on extending some of Gödel's results that show that it is impossible to write a computer program that can solve all mathematical problems. He was also particularly interested in explaining ideas from mathematical logic to the public.

Smullyan was a talented musician and often played piano with his second wife, who was a concert-level pianist. Making telescopes was one of his hobbies and he was interested in optics and stereo photography. He said, "I've never had a conflict between teaching and research as some people do because when I'm teaching, I'm doing research." Smullyan is the subject of a documentary short film entitled *This Film Needs No Title*.

then s must be true, and he would have answered "Yes" to the first question. The son can reason in a similar way to determine that s must be true. It follows that both children answer "Yes" the second time the question is asked. ◀

1.2.6 Logic Circuits

Propositional logic can be applied to the design of computer hardware. This was first observed in 1938 by Claude Shannon in his MIT master's thesis. In Chapter 12 we will study this topic in depth. (See that chapter for a biography of Shannon.) We give a brief introduction to this application here.

In Chapter 12 we will design some useful circuits.

A **logic circuit** (or **digital circuit**) receives input signals $p_1, p_2, \ldots, p_n$, each a bit [either 0 (off) or 1 (on)], and produces output signals $s_1, s_2, \ldots, s_n$, each a bit. In this section we will restrict our attention to logic circuits with a single output signal; in general, digital circuits may have multiple outputs.

Complicated digital circuits can be constructed from three basic circuits, called **gates**, shown in Figure 1. The **inverter**, or **NOT gate**, takes an input bit p, and produces as output $\neg p$. The **OR gate** takes two input signals p and q, each a bit, and produces as output the signal $p \vee q$. Finally, the **AND gate** takes two input signals p and q, each a bit, and produces as output the signal $p \wedge q$. We use combinations of these three basic gates to build more complicated circuits, such as that shown in Figure 2.

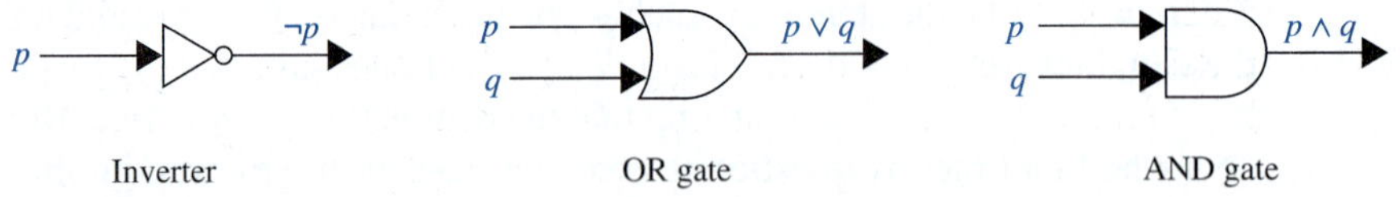

FIGURE 1 Basic logic gates.

Given a circuit built from the basic logic gates and the inputs to the circuit, we determine the output by tracing through the circuit, as Example 10 shows.

EXAMPLE 10 Determine the output for the combinatorial circuit in Figure 2.

Solution: In Figure 2 we display the output of each logic gate in the circuit. We see that the AND gate takes input of p and $\neg q$, the output of the inverter with input q, and produces $p \wedge \neg q$. Next, we note that the OR gate takes input $p \wedge \neg q$ and $\neg r$, the output of the inverter with input r, and produces the final output $(p \wedge \neg q) \vee \neg r$. ◀

Suppose that we have a formula for the output of a digital circuit in terms of negations, disjunctions, and conjunctions. Then, we can systematically build a digital circuit with the desired output, as illustrated in Example 11.

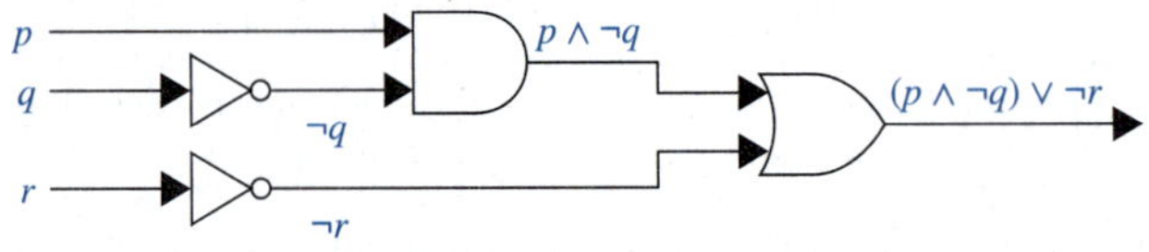

FIGURE 2 A combinatorial circuit.

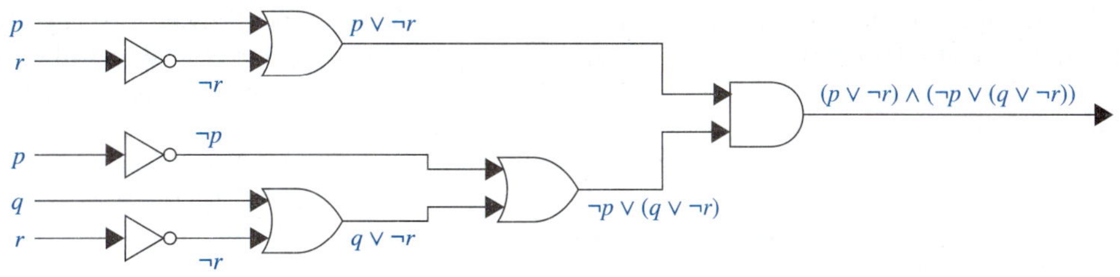

FIGURE 3 **The circuit for $(p \vee \neg r) \wedge (\neg p \vee (q \vee \neg r))$.**

EXAMPLE 11 Build a digital circuit that produces the output $(p \vee \neg r) \wedge (\neg p \vee (q \vee \neg r))$ when given input bits p, q, and r.

Solution: To construct the desired circuit, we build separate circuits for $p \vee \neg r$ and for $\neg p \vee (q \vee \neg r)$ and combine them using an AND gate. To construct a circuit for $p \vee \neg r$, we use an inverter to produce $\neg r$ from the input r. Then, we use an OR gate to combine p and $\neg r$. To build a circuit for $\neg p \vee (q \vee \neg r)$, we first use an inverter to obtain $\neg r$. Then we use an OR gate with inputs q and $\neg r$ to obtain $q \vee \neg r$. Finally, we use another inverter and an OR gate to get $\neg p \vee (q \vee \neg r)$ from the inputs p and $q \vee \neg r$.

To complete the construction, we employ a final AND gate, with inputs $p \vee \neg r$ and $\neg p \vee (q \vee \neg r)$. The resulting circuit is displayed in Figure 3. ◀

We will study logic circuits in great detail in Chapter 12 in the context of Boolean algebra, and with different notation.

Exercises

In Exercises 1–6, translate the given statement into propositional logic using the propositions provided.

1. You cannot edit a protected Wikipedia entry unless you are an administrator. Express your answer in terms of e: "You can edit a protected Wikipedia entry" and a: "You are an administrator."
2. You can see the movie only if you are over 18 years old or you have the permission of a parent. Express your answer in terms of m: "You can see the movie," e: "You are over 18 years old," and p: "You have the permission of a parent."
3. You can graduate only if you have completed the requirements of your major and you do not owe money to the university and you do not have an overdue library book. Express your answer in terms of g: "You can graduate," m: "You owe money to the university," r: "You have completed the requirements of your major," and b: "You have an overdue library book."
4. To use the wireless network in the airport you must pay the daily fee unless you are a subscriber to the service. Express your answer in terms of w: "You can use the wireless network in the airport," d: "You pay the daily fee," and s: "You are a subscriber to the service."
5. You are eligible to be President of the U.S.A. only if you are at least 35 years old, were born in the U.S.A., or at the time of your birth both of your parents were citizens, and you have lived at least 14 years in the country. Express your answer in terms of e: "You are eligible to be President of the U.S.A.," a: "You are at least 35 years old," b: "You were born in the U.S.A.," p: "At the time of your birth, both of your parents were citizens," and r: "You have lived at least 14 years in the U.S.A."
6. You can upgrade your operating system only if you have a 32-bit processor running at 1 GHz or faster, at least 1 GB RAM, and 16 GB free hard disk space, or a 64-bit processor running at 2 GHz or faster, at least 2 GB RAM, and at least 32 GB free hard disk space. Express your answer in terms of u: "You can upgrade your operating system," b_{32}: "You have a 32-bit processor," b_{64}: "You have a 64-bit processor," g_1: "Your processor runs at 1 GHz or faster," g_2: "Your processor runs at 2 GHz or faster," r_1: "Your processor has at least 1 GB RAM," r_2: "Your processor has at least 2 GB RAM," h_{16}: "You have at least 16 GB free hard disk space," and h_{32}: "You have at least 32 GB free hard disk space."
7. Express these system specifications using the propositions p: "The message is scanned for viruses" and q: "The message was sent from an unknown system" together with logical connectives (including negations).
   a) "The message is scanned for viruses whenever the message was sent from an unknown system."

b) "The message was sent from an unknown system but it was not scanned for viruses."

c) "It is necessary to scan the message for viruses whenever it was sent from an unknown system."

d) "When a message is not sent from an unknown system it is not scanned for viruses."

8. Express these system specifications using the propositions p: "The user enters a valid password," q: "Access is granted," and r: "The user has paid the subscription fee" and logical connectives (including negations).

a) "The user has paid the subscription fee, but does not enter a valid password."

b) "Access is granted whenever the user has paid the subscription fee and enters a valid password."

c) "Access is denied if the user has not paid the subscription fee."

d) "If the user has not entered a valid password but has paid the subscription fee, then access is granted."

9. Are these system specifications consistent? "The system is in multiuser state if and only if it is operating normally. If the system is operating normally, the kernel is functioning. The kernel is not functioning or the system is in interrupt mode. If the system is not in multiuser state, then it is in interrupt mode. The system is not in interrupt mode."

10. Are these system specifications consistent? "Whenever the system software is being upgraded, users cannot access the file system. If users can access the file system, then they can save new files. If users cannot save new files, then the system software is not being upgraded."

11. Are these system specifications consistent? "The router can send packets to the edge system only if it supports the new address space. For the router to support the new address space it is necessary that the latest software release be installed. The router can send packets to the edge system if the latest software release is installed. The router does not support the new address space."

12. Are these system specifications consistent? "If the file system is not locked, then new messages will be queued. If the file system is not locked, then the system is functioning normally, and conversely. If new messages are not queued, then they will be sent to the message buffer. If the file system is not locked, then new messages will be sent to the message buffer. New messages will not be sent to the message buffer."

13. What Boolean search would you use to look for Web pages about beaches in New Jersey? What if you wanted to find Web pages about beaches on the isle of Jersey (in the English Channel)?

14. What Boolean search would you use to look for Web pages about hiking in West Virginia? What if you wanted to find Web pages about hiking in Virginia, but not in West Virginia?

15. What Google search would you use to look for Web pages relating to Ethiopian restaurants in New York or New Jersey?

16. What Google search would you use to look for men's shoes or boots not designed for work?

17. Suppose that in Example 7, the inscriptions on Trunks 1, 2, and 3 are "The treasure is in Trunk 3," "The treasure is in Trunk 1," and "This trunk is empty." For each of these statements, determine whether the Queen who never lies could state this, and if so, which trunk the treasure is in.

a) "All the inscriptions are false."

b) "Exactly one of the inscriptions is true."

c) "Exactly two of the inscriptions are true."

d) "All three inscriptions are true."

18. Suppose that in Example 7 there are treasures in two of the three trunks. The inscriptions on Trunks 1, 2, and 3 are "This trunk is empty," "There is a treasure in Trunk 1," and "There is a treasure in Trunk 2." For each of these statements, determine whether the Queen who never lies could state this, and if so, which two trunks the treasures are in.

a) "All the inscriptions are false."

b) "Exactly one of the inscriptions is true."

c) "Exactly two of the inscriptions are true."

d) "All three inscriptions are true."

***19.** Each inhabitant of a remote village always tells the truth or always lies. A villager will give only a "Yes" or a "No" response to a question a tourist asks. Suppose you are a tourist visiting this area and come to a fork in the road. One branch leads to the ruins you want to visit; the other branch leads deep into the jungle. A villager is standing at the fork in the road. What one question can you ask the villager to determine which branch to take?

20. An explorer is captured by a group of cannibals. There are two types of cannibals—those who always tell the truth and those who always lie. The cannibals will barbecue the explorer unless she can determine whether a particular cannibal always lies or always tells the truth. She is allowed to ask the cannibal exactly one question.

a) Explain why the question "Are you a liar?" does not work.

b) Find a question that the explorer can use to determine whether the cannibal always lies or always tells the truth.

21. When three professors are seated in a restaurant, the waiter asks them: "Does everyone want coffee?" The first professor says "I do not know." The second professor then says "I do not know." Finally, the third professor says "No, not everyone wants coffee." The waiter comes back and gives coffee to the professors who want it. How did he figure out who wanted coffee?

22. When planning a party you want to know whom to invite. Among the people you would like to invite are three testy friends. You know that if Jasmine attends, she will become unhappy if Samir is there, Samir will attend only if Kanti will be there, and Kanti will not attend unless Jasmine also does. Which combinations of these three friends can you invite so as not to make someone unhappy?

Exercises 23–27 relate to inhabitants of the island of knights and knaves created by Smullyan, where knights always tell the truth and knaves always lie. You encounter two people, A and B. Determine, if possible, what A and B are if they address you in the ways described. If you cannot determine what these two people are, can you draw any conclusions?

23. A says "At least one of us is a knave" and B says nothing.

24. A says "The two of us are both knights" and B says "A is a knave."

25. A says "I am a knave or B is a knight" and B says nothing.

26. Both A and B say "I am a knight."

27. A says "We are both knaves" and B says nothing.

Exercises 28–35 relate to inhabitants of an island on which there are three kinds of people: knights who always tell the truth, knaves who always lie, and spies (called normals by Smullyan [Sm78]) who can either lie or tell the truth. You encounter three people, A, B, and C. You know one of these people is a knight, one is a knave, and one is a spy. Each of the three people knows the type of person each of other two is. For each of these situations, if possible, determine whether there is a unique solution and determine who the knave, knight, and spy are. When there is no unique solution, list all possible solutions or state that there are no solutions.

28. A says "C is the knave," B says "A is the knight," and C says "I am the spy."

29. A says "I am the knight," B says "I am the knave," and C says "B is the knight."

30. A says "I am the knave," B says "I am the knave," and C says "I am the knave."

31. A says "I am the knight," B says "A is telling the truth," and C says "I am the spy."

32. A says "I am the knight," B says "A is not the knave," and C says "B is not the knave."

33. A says "I am the knight," B says "I am the knight," and C says "I am the knight."

34. A says "I am not the spy," B says "I am not the spy," and C says "A is the spy."

35. A says "I am not the spy," B says "I am not the spy," and C says "I am not the spy."

Exercises 36–42 are puzzles that can be solved by translating statements into logical expressions and reasoning from these expressions using truth tables.

36. The police have three suspects for the murder of Mr. Cooper: Mr. Smith, Mr. Jones, and Mr. Williams. Smith, Jones, and Williams each declare that they did not kill Cooper. Smith also states that Cooper was a friend of Jones and that Williams disliked him. Jones also states that he did not know Cooper and that he was out of town the day Cooper was killed. Williams also states that he saw both Smith and Jones with Cooper the day of the killing and that either Smith or Jones must have killed him. Can you determine who the murderer was if

a) one of the three men is guilty, the two innocent men are telling the truth, but the statements of the guilty man may or may not be true?

b) innocent men do not lie?

37. Steve would like to determine the relative salaries of three coworkers using two facts. First, he knows that if Fred is not the highest paid of the three, then Janice is. Second, he knows that if Janice is not the lowest paid, then Maggie is paid the most. Is it possible to determine the relative salaries of Fred, Maggie, and Janice from what Steve knows? If so, who is paid the most and who the least? Explain your reasoning.

38. Five friends have access to a chat room. Is it possible to determine who is chatting if the following information is known? Either Kevin or Heather, or both, are chatting. Either Randy or Vijay, but not both, are chatting. If Abby is chatting, so is Randy. Vijay and Kevin are either both chatting or neither is. If Heather is chatting, then so are Abby and Kevin. Explain your reasoning.

39. A detective has interviewed four witnesses to a crime. From the stories of the witnesses the detective has concluded that if the butler is telling the truth then so is the cook; the cook and the gardener cannot both be telling the truth; the gardener and the property manager are not both lying; and if the property manager is telling the truth then the cook is lying. For each of the four witnesses, can the detective determine whether that person is telling the truth or lying? Explain your reasoning.

40. Four friends have been identified as suspects for an unauthorized access into a computer system. They have made statements to the investigating authorities. Alice said, "Carlos did it." John said, "I did not do it." Carlos said, "Diana did it." Diana said, "Carlos lied when he said that I did it."

a) If the authorities also know that exactly one of the four suspects is telling the truth, who did it? Explain your reasoning.

b) If the authorities also know that exactly one is lying, who did it? Explain your reasoning.

41. Suppose there are signs on the doors to two rooms. The sign on the first door reads "In this room there is a lady, and in the other one there is a tiger"; and the sign on the second door reads "In one of these rooms, there is a lady, and in one of them there is a tiger." Suppose that you know that one of these signs is true and the other is false. Behind which door is the lady?

***42.** Solve this famous logic puzzle, attributed to Albert Einstein, and known as the **zebra puzzle**. Five men with different nationalities and with different jobs live in consecutive houses on a street. These houses are painted different colors. The men have different pets and have different favorite drinks. Determine who owns a zebra and whose favorite drink is mineral water (which is one of the favorite drinks) given these clues: The Englishman lives in the red house. The Spaniard owns a dog. The Japanese man is a painter. The Italian drinks tea. The Norwegian lives in the first house on the left. The green house is immediately to the right of the white one. The photographer breeds snails. The diplomat lives in the yellow house. Milk is drunk in the middle house. The owner of the green house drinks coffee. The Norwegian's house is next to the

Links

blue one. The violinist drinks orange juice. The fox is in a house next to that of the physician. The horse is in a house next to that of the diplomat. [*Hint:* Make a table where the rows represent the men and columns represent the color of their houses, their jobs, their pets, and their favorite drinks and use logical reasoning to determine the correct entries in the table.]

43. Freedonia has 50 senators. Each senator is either honest or corrupt. Suppose you know that at least one of the Freedonian senators is honest and that, given any two Freedonian senators, at least one is corrupt. Based on these facts, can you determine how many Freedonian senators are honest and how many are corrupt? If so, what is the answer?

44. Find the output of each of these combinatorial circuits.

a) 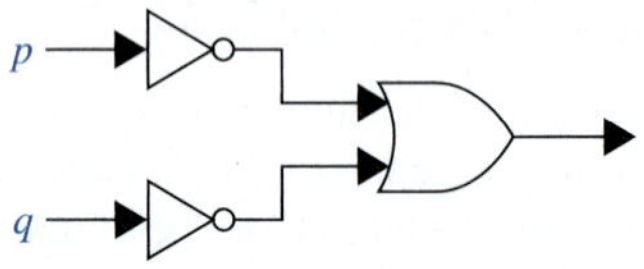

b) 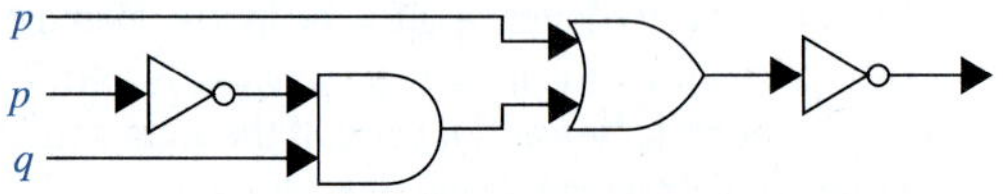

45. Find the output of each of these combinatorial circuits.

a) 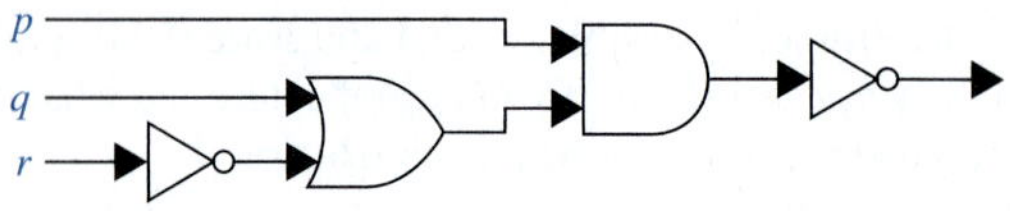

b) 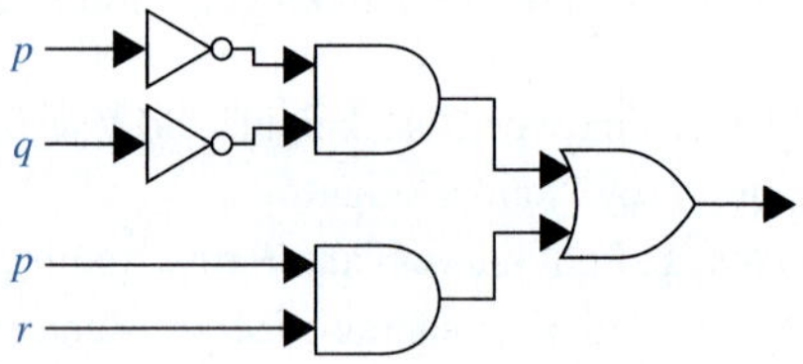

46. Construct a combinatorial circuit using inverters, OR gates, and AND gates that produces the output $(p \wedge \neg r) \vee (\neg q \wedge r)$ from input bits p, q, and r.

47. Construct a combinatorial circuit using inverters, OR gates, and AND gates that produces the output $((\neg p \vee \neg r) \wedge \neg q) \vee (\neg p \wedge (q \vee r))$ from input bits p, q, and r.

1.3 Propositional Equivalences

1.3.1 Introduction

An important type of step used in a mathematical argument is the replacement of a statement with another statement with the same truth value. Because of this, methods that produce propositions with the same truth value as a given compound proposition are used extensively in the construction of mathematical arguments. Note that we will use the term "compound proposition" to refer to an expression formed from propositional variables using logical operators, such as $p \wedge q$.

We begin our discussion with a classification of compound propositions according to their possible truth values.

Definition 1 A compound proposition that is always true, no matter what the truth values of the propositional variables that occur in it, is called a *tautology*. A compound proposition that is always false is called a *contradiction*. A compound proposition that is neither a tautology nor a contradiction is called a *contingency*.

Tautologies and contradictions are often important in mathematical reasoning. Example 1 illustrates these types of compound propositions.

EXAMPLE 1 We can construct examples of tautologies and contradictions using just one propositional variable. Consider the truth tables of $p \vee \neg p$ and $p \wedge \neg p$, shown in Table 1. Because $p \vee \neg p$ is always true, it is a tautology. Because $p \wedge \neg p$ is always false, it is a contradiction. ◀

TABLE 1 Examples of a Tautology and a Contradiction.			
p	$\neg p$	$p \vee \neg p$	$p \wedge \neg p$
T	F	T	F
F	T	T	F

1.3.2 Logical Equivalences

Demo

Compound propositions that have the same truth values in all possible cases are called **logically equivalent**. We can also define this notion as follows.

Definition 2 The compound propositions p and q are called *logically equivalent* if $p \leftrightarrow q$ is a tautology. The notation $p \equiv q$ denotes that p and q are logically equivalent.

Remark: The symbol $\equiv$ is not a logical connective, and $p \equiv q$ is not a compound proposition but rather is the statement that $p \leftrightarrow q$ is a tautology. The symbol $\Leftrightarrow$ is sometimes used instead of $\equiv$ to denote logical equivalence.

Extra Examples

One way to determine whether two compound propositions are equivalent is to use a truth table. In particular, the compound propositions p and q are equivalent if and only if the columns giving their truth values agree. Example 2 illustrates this method to establish an extremely important and useful logical equivalence, namely, that of $\neg(p \vee q)$ with $\neg p \wedge \neg q$. This logical equivalence is one of the two **De Morgan laws**, shown in Table 2, named after the English mathematician Augustus De Morgan, of the mid-nineteenth century.

TABLE 2 De Morgan's Laws.
$\neg(p \wedge q) \equiv \neg p \vee \neg q$
$\neg(p \vee q) \equiv \neg p \wedge \neg q$

EXAMPLE 2 Show that $\neg(p \vee q)$ and $\neg p \wedge \neg q$ are logically equivalent.

Solution: The truth tables for these compound propositions are displayed in Table 3. Because the truth values of the compound propositions $\neg(p \vee q)$ and $\neg p \wedge \neg q$ agree for all possible combinations of the truth values of p and q, it follows that $\neg(p \vee q) \leftrightarrow (\neg p \wedge \neg q)$ is a tautology and that these compound propositions are logically equivalent. ◀

TABLE 3 Truth Tables for $\neg(p \vee q)$ and $\neg p \wedge \neg q$.						
p	q	$p \vee q$	$\neg(p \vee q)$	$\neg p$	$\neg q$	$\neg p \wedge \neg q$
T	T	T	F	F	F	F
T	F	T	F	F	T	F
F	T	T	F	T	F	F
F	F	F	T	T	T	T

The next example establishes an extremely important equivalence. It allows us to replace conditional statements with negations and disjunctions.

EXAMPLE 3 Show that $p \rightarrow q$ and $\neg p \vee q$ are logically equivalent. (This is known as the **conditional-disjunction equivalence**.)

Solution: We construct the truth table for these compound propositions in Table 4. Because the truth values of $\neg p \vee q$ and $p \rightarrow q$ agree, they are logically equivalent. ◀

TABLE 4 **Truth Tables for $\neg p \vee q$ and $p \rightarrow q$.**

p	q	$\neg p$	$\neg p \vee q$	$p \rightarrow q$
T	T	F	T	T
T	F	F	F	F
F	T	T	T	T
F	F	T	T	T

We will now establish a logical equivalence of two compound propositions involving three different propositional variables p, q, and r. To use a truth table to establish such a logical equivalence, we need eight rows, one for each possible combination of truth values of these three variables. We symbolically represent these combinations by listing the truth values of p, q, and r, respectively. These eight combinations of truth values are TTT, TTF, TFT, TFF, FTT, FTF, FFT, and FFF; we use this order when we display the rows of the truth table. Note that we need to double the number of rows in the truth tables we use to show that compound propositions are equivalent for each additional propositional variable, so that 16 rows are needed to establish the logical equivalence of two compound propositions involving four propositional variables, and so on. In general, 2^n rows are required if a compound proposition involves n propositional variables. Because of the rapid growth of 2^n, more efficient ways are needed to establish logical equivalences, such as by using ones we already know. This technique will be discussed later.

EXAMPLE 4 Show that $p \vee (q \wedge r)$ and $(p \vee q) \wedge (p \vee r)$ are logically equivalent. This is the **distributive law** of disjunction over conjunction.

Solution: We construct the truth table for these compound propositions in Table 5. Because the truth values of $p \vee (q \wedge r)$ and $(p \vee q) \wedge (p \vee r)$ agree, these compound propositions are logically equivalent. ◀

TABLE 5 **A Demonstration That $p \vee (q \wedge r)$ and $(p \vee q) \wedge (p \vee r)$ Are Logically Equivalent.**

p	q	r	$q \wedge r$	$p \vee (q \wedge r)$	$p \vee q$	$p \vee r$	$(p \vee q) \wedge (p \vee r)$
T	T	T	T	T	T	T	T
T	T	F	F	T	T	T	T
T	F	T	F	T	T	T	T
T	F	F	F	T	T	T	T
F	T	T	T	T	T	T	T
F	T	F	F	F	T	F	F
F	F	T	F	F	F	T	F
F	F	F	F	F	F	F	F

TABLE 6 Logical Equivalences.

Equivalence	***Name***
$p \wedge \mathbf{T} \equiv p$ $p \vee \mathbf{F} \equiv p$	Identity laws
$p \vee \mathbf{T} \equiv \mathbf{T}$ $p \wedge \mathbf{F} \equiv \mathbf{F}$	Domination laws
$p \vee p \equiv p$ $p \wedge p \equiv p$	Idempotent laws
$\neg(\neg p) \equiv p$	Double negation law
$p \vee q \equiv q \vee p$ $p \wedge q \equiv q \wedge p$	Commutative laws
$(p \vee q) \vee r \equiv p \vee (q \vee r)$ $(p \wedge q) \wedge r \equiv p \wedge (q \wedge r)$	Associative laws
$p \vee (q \wedge r) \equiv (p \vee q) \wedge (p \vee r)$ $p \wedge (q \vee r) \equiv (p \wedge q) \vee (p \wedge r)$	Distributive laws
$\neg(p \wedge q) \equiv \neg p \vee \neg q$ $\neg(p \vee q) \equiv \neg p \wedge \neg q$	De Morgan's laws
$p \vee (p \wedge q) \equiv p$ $p \wedge (p \vee q) \equiv p$	Absorption laws
$p \vee \neg p \equiv \mathbf{T}$ $p \wedge \neg p \equiv \mathbf{F}$	Negation laws

The identities in Table 6 are a special case of Boolean algebra identities found in Table 5 of Section 12.1. See Table 1 in Section 2.2 for analogous set identities.

Table 6 contains some important equivalences. In these equivalences, **T** denotes the compound proposition that is always true and **F** denotes the compound proposition that is always false. We also display some useful equivalences for compound propositions involving conditional statements and biconditional statements in Tables 7 and 8, respectively. The reader is asked to verify the equivalences in Tables 6–8 in the exercises.

TABLE 7 Logical Equivalences Involving Conditional Statements.

$p \to q \equiv \neg p \vee q$
$p \to q \equiv \neg q \to \neg p$
$p \vee q \equiv \neg p \to q$
$p \wedge q \equiv \neg(p \to \neg q)$
$\neg(p \to q) \equiv p \wedge \neg q$
$(p \to q) \wedge (p \to r) \equiv p \to (q \wedge r)$
$(p \to r) \wedge (q \to r) \equiv (p \vee q) \to r$
$(p \to q) \vee (p \to r) \equiv p \to (q \vee r)$
$(p \to r) \vee (q \to r) \equiv (p \wedge q) \to r$

TABLE 8 Logical Equivalences Involving Biconditional Statements.

$p \leftrightarrow q \equiv (p \to q) \wedge (q \to p)$
$p \leftrightarrow q \equiv \neg p \leftrightarrow \neg q$
$p \leftrightarrow q \equiv (p \wedge q) \vee (\neg p \wedge \neg q)$
$\neg(p \leftrightarrow q) \equiv p \leftrightarrow \neg q$

Be careful not to apply logical identities, such as associative laws, distributive laws, or De Morgan's laws, to expressions that have a mix of conjunctions and disjunctions when the identities only apply when all these operators are the same.

The associative law for disjunction shows that the expression $p \vee q \vee r$ is well defined, in the sense that it does not matter whether we first take the disjunction of p with q and then the disjunction of $p \vee q$ with r, or if we first take the disjunction of q and r and then take the disjunction of p with $q \vee r$. Similarly, the expression $p \wedge q \wedge r$ is well defined. By extending this reasoning, it follows that $p_1 \vee p_2 \vee \cdots \vee p_n$ and $p_1 \wedge p_2 \wedge \cdots \wedge p_n$ are well defined whenever $p_1, p_2, \ldots, p_n$ are propositions.

Furthermore, note that De Morgan's laws extend to

$$\neg(p_1 \vee p_2 \vee \cdots \vee p_n) \equiv (\neg p_1 \wedge \neg p_2 \wedge \cdots \wedge \neg p_n)$$

and

$$\neg(p_1 \wedge p_2 \wedge \cdots \wedge p_n) \equiv (\neg p_1 \vee \neg p_2 \vee \cdots \vee \neg p_n).$$

We will sometimes use the notation $\bigvee_{j=1}^{n} p_j$ for $p_1 \vee p_2 \vee \cdots \vee p_n$ and $\bigwedge_{j=1}^{n} p_j$ for $p_1 \wedge p_2 \wedge \cdots \wedge p_n$. Using this notation, the extended version of De Morgan's laws can be written concisely as $\neg\left(\bigvee_{j=1}^{n} p_j\right) \equiv \bigwedge_{j=1}^{n} \neg p_j$ and $\neg\left(\bigwedge_{j=1}^{n} p_j\right) \equiv \bigvee_{j=1}^{n} \neg p_j$. (Methods for proving these identities will be given in Section 5.1.)

A truth table with 2^n rows is needed to prove the equivalence of two compound propositions in n variables. (Note that the number of rows doubles for each additional propositional variable added. See Chapter 6 for details about solving counting problems such as this.) Because 2^n grows extremely rapidly as n increases (see Section 3.2), the use of truth tables to establish equivalences becomes impractical as the number of variables grows. It is quicker to use other methods, such as employing logical equivalences that we already know. How that can be done is discussed later in this section.

1.3.3 Using De Morgan's Laws

When using De Morgan's laws, remember to change the logical connective after you negate.

The two logical equivalences known as De Morgan's laws are particularly important. They tell us how to negate conjunctions and how to negate disjunctions. In particular, the equivalence $\neg(p \vee q) \equiv \neg p \wedge \neg q$ tells us that the negation of a disjunction is formed by taking the conjunction of the negations of the component propositions. Similarly, the equivalence $\neg(p \wedge q) \equiv \neg p \vee \neg q$ tells us that the negation of a conjunction is formed by taking the disjunction of the negations of the component propositions. Example 5 illustrates the use of De Morgan's laws.

EXAMPLE 5 Use De Morgan's laws to express the negations of "Miguel has a cellphone and he has a laptop computer" and "Heather will go to the concert or Steve will go to the concert."

Assessment

Solution: Let p be "Miguel has a cellphone" and q be "Miguel has a laptop computer." Then "Miguel has a cellphone and he has a laptop computer" can be represented by $p \wedge q$. By the first of De Morgan's laws, $\neg(p \wedge q)$ is equivalent to $\neg p \vee \neg q$. Consequently, we can express the negation of our original statement as "Miguel does not have a cellphone or he does not have a laptop computer."

Let r be "Heather will go to the concert" and s be "Steve will go to the concert." Then "Heather will go to the concert or Steve will go to the concert" can be represented by $r \vee s$. By the second of De Morgan's laws, $\neg(r \vee s)$ is equivalent to $\neg r \wedge \neg s$. Consequently, we can express the negation of our original statement as "Heather will not go to the concert and Steve will not go to the concert." ◀

1.3.4 Constructing New Logical Equivalences

The logical equivalences in Table 6, as well as any others that have been established (such as those shown in Tables 7 and 8), can be used to construct additional logical equivalences. The reason for this is that a proposition in a compound proposition can be replaced by a compound proposition that is logically equivalent to it without changing the truth value of the original compound proposition. This technique is illustrated in Examples 6–8, where we also use the fact that if p and q are logically equivalent and q and r are logically equivalent, then p and r are logically equivalent (see Exercise 60).

EXAMPLE 6 Show that $\neg(p \to q)$ and $p \wedge \neg q$ are logically equivalent.

Extra Examples

Solution: We could use a truth table to show that these compound propositions are equivalent (similar to what we did in Example 4). Indeed, it would not be hard to do so. However, we want to illustrate how to use logical identities that we already know to establish new logical identities, something that is of practical importance for establishing equivalences of compound propositions with a large number of variables. So, we will establish this equivalence by developing a series of logical equivalences, using one of the equivalences in Table 6 at a time, starting with $\neg(p \to q)$ and ending with $p \wedge \neg q$. We have the following equivalences.

$$\begin{aligned} \neg(p \to q) &\equiv \neg(\neg p \vee q) && \text{by the conditional-disjunction equivalence (Example 3)} \\ &\equiv \neg(\neg p) \wedge \neg q && \text{by the second De Morgan law} \\ &\equiv p \wedge \neg q && \text{by the double negation law} \end{aligned}$$

◀

EXAMPLE 7 Show that $\neg(p \vee (\neg p \wedge q))$ and $\neg p \wedge \neg q$ are logically equivalent by developing a series of logical equivalences.

Solution: We will use one of the equivalences in Table 6 at a time, starting with $\neg(p \vee (\neg p \wedge q))$ and ending with $\neg p \wedge \neg q$. (*Note:* we could also easily establish this equivalence using a truth table.) We have the following equivalences.

Links

Bettmann/Getty Images

AUGUSTUS DE MORGAN (1806–1871) Augustus De Morgan was born in India, where his father was a colonel in the Indian army. De Morgan's family moved to England when he was 7 months old. He attended private schools, where in his early teens he developed a strong interest in mathematics. De Morgan studied at Trinity College, Cambridge, graduating in 1827. Although he considered medicine or law, he decided on mathematics for his career. He won a position at University College, London, in 1828, but resigned after the college dismissed a fellow professor without giving reasons. However, he resumed this position in 1836 when his successor died, remaining until 1866.

De Morgan was a noted teacher who stressed principles over techniques. His students included many famous mathematicians, including Augusta Ada, Countess of Lovelace, who was Charles Babbage's collaborator in his work on computing machines (see page 32 for biographical notes on Augusta Ada). (De Morgan cautioned the countess against studying too much mathematics, because it might interfere with her childbearing abilities!)

De Morgan was an extremely prolific writer, publishing more than 1000 articles in more than 15 periodicals. De Morgan also wrote textbooks on many subjects, including logic, probability, calculus, and algebra. In 1838 he presented what was perhaps the first clear explanation of an important proof technique known as *mathematical induction* (discussed in Section 5.1 of this text), a term he coined. In the 1840s De Morgan made fundamental contributions to the development of symbolic logic. He invented notations that helped him prove propositional equivalences, such as the laws that are named after him. In 1842 De Morgan presented what is considered to be the first precise definition of a limit and developed new tests for convergence of infinite series. De Morgan was also interested in the history of mathematics and wrote biographies of Newton and Halley.

In 1837 De Morgan married Sophia Frend, who wrote his biography in 1882. De Morgan's research, writing, and teaching left little time for his family or social life. Nevertheless, he was noted for his kindness, humor, and wide range of knowledge.

$$
\begin{aligned}
\neg(p \lor (\neg p \land q)) &\equiv \neg p \land \neg(\neg p \land q) && \text{by the second De Morgan law}\\
&\equiv \neg p \land [\neg(\neg p) \lor \neg q] && \text{by the first De Morgan law}\\
&\equiv \neg p \land (p \lor \neg q) && \text{by the double negation law}\\
&\equiv (\neg p \land p) \lor (\neg p \land \neg q) && \text{by the second distributive law}\\
&\equiv \mathbf{F} \lor (\neg p \land \neg q) && \text{because } \neg p \land p \equiv \mathbf{F}\\
&\equiv (\neg p \land \neg q) \lor \mathbf{F} && \text{by the commutative law for disjunction}\\
&\equiv \neg p \land \neg q && \text{by the identity law for } \mathbf{F}
\end{aligned}
$$

Consequently $\neg(p \lor (\neg p \land q))$ and $\neg p \land \neg q$ are logically equivalent. ◀

EXAMPLE 8 Show that $(p \land q) \to (p \lor q)$ is a tautology.

Solution: To show that this statement is a tautology, we will use logical equivalences to demonstrate that it is logically equivalent to **T**. (*Note:* This could also be done using a truth table.)

$$
\begin{aligned}
(p \land q) \to (p \lor q) &\equiv \neg(p \land q) \lor (p \lor q) && \text{by Example 3}\\
&\equiv (\neg p \lor \neg q) \lor (p \lor q) && \text{by the first De Morgan law}\\
&\equiv (\neg p \lor p) \lor (\neg q \lor q) && \text{by the associative and commutative laws for disjunction}\\
&\equiv \mathbf{T} \lor \mathbf{T} && \text{by Example 1 and the commutative law for disjunction}\\
&\equiv \mathbf{T} && \text{by the domination law}
\end{aligned}
$$
◀

Links

Hulton Archive/Getty Images

AUGUSTA ADA, COUNTESS OF LOVELACE (1815–1852) Augusta Ada was the only child from the marriage of the flamboyant and notorious poet Lord Byron and Lady Byron, Annabella Milbanke, who separated when Ada was 1 month old, because of Lord Byron's scandalous affair with his half sister. The Lord Byron had quite a reputation, being described by one of his lovers as "mad, bad, and dangerous to know." Lady Byron was noted for her intellect and had a passion for mathematics; she was called by Lord Byron "The Princess of Parallelograms." Augusta was raised by her mother, who encouraged her intellectual talents especially in music and mathematics, to counter what Lady Byron considered dangerous poetic tendencies. At this time, women were not allowed to attend universities and could not join learned societies. Nevertheless, Augusta pursued her mathematical studies independently and with mathematicians, including William Frend. She was also encouraged by another female mathematician, Mary Somerville, and in 1834 at a dinner party hosted by Mary Somerville, she learned about Charles Babbage's ideas for a calculating machine, called the Analytic Engine. In 1838 Augusta Ada married Lord King, later elevated to Earl of Lovelace. Together they had three children.

Augusta Ada continued her mathematical studies after her marriage. Charles Babbage had continued work on his Analytic Engine and lectured on this in Europe. In 1842 Babbage asked Augusta Ada to translate an article in French describing Babbage's invention. When Babbage saw her translation, he suggested she add her own notes, and the resulting work was three times the length of the original. The most complete accounts of the Analytic Engine are found in Augusta Ada's notes. In her notes, she compared the working of the Analytic Engine to that of the Jacquard loom, with Babbage's punch cards analogous to the cards used to create patterns on the loom. Furthermore, she recognized the promise of the machine as a general purpose computer much better than Babbage did. She stated that the "engine is the material expression of any indefinite function of any degree of generality and complexity." Her notes on the Analytic Engine anticipate many future developments, including computer-generated music. Augusta Ada published her writings under her initials A.A.L., concealing her identity as a woman as did many women at a time when women were not considered to be the intellectual equals of men. After 1845 she and Babbage worked toward the development of a system to predict horse races. Unfortunately, their system did not work well, leaving Augusta Ada heavily in debt at the time of her death at an unfortunately young age from uterine cancer.

In 1953 Augusta Ada's notes on the Analytic Engine were republished more than 100 years after they were written, and after they had been long forgotten. In his work in the 1950s on the capacity of computers to think (and his influential Turing test for determining whether a machine is intelligent), Alan Turing responded to Augusta Ada's statement that "The Analytic Engine has no pretensions whatever to originate anything. It can do whatever we know how to order it to perform." This "dialogue" between Turing and Augusta Ada is still the subject of controversy. Because of her fundamental contributions to computing, the programming language Ada is named in honor of the Countess of Lovelace.

1.3.5 Satisfiability

A compound proposition is **satisfiable** if there is an assignment of truth values to its variables that makes it true (that is, when it is a tautology or a contingency). When no such assignments exists, that is, when the compound proposition is false for all assignments of truth values to its variables, the compound proposition is **unsatisfiable**. Note that a compound proposition is unsatisfiable if and only if its negation is true for all assignments of truth values to the variables, that is, if and only if its negation is a tautology.

When we find a particular assignment of truth values that makes a compound proposition true, we have shown that it is satisfiable; such an assignment is called a **solution** of this particular satisfiability problem. However, to show that a compound proposition is unsatisfiable, we need to show that *every* assignment of truth values to its variables makes it false. Although we can always use a truth table to determine whether a compound proposition is satisfiable, it is often more efficient not to, as Example 9 demonstrates.

EXAMPLE 9 Determine whether each of the compound propositions $(p \vee \neg q) \wedge (q \vee \neg r) \wedge (r \vee \neg p)$, $(p \vee q \vee r) \wedge (\neg p \vee \neg q \vee \neg r)$, and $(p \vee \neg q) \wedge (q \vee \neg r) \wedge (r \vee \neg p) \wedge (p \vee q \vee r) \wedge (\neg p \vee \neg q \vee \neg r)$ is satisfiable.

Solution: Instead of using a truth table to solve this problem, we will reason about truth values. Note that $(p \vee \neg q) \wedge (q \vee \neg r) \wedge (r \vee \neg p)$ is true when the three variables p, q, and r have the same truth value (see Exercise 42 of Section 1.1). Hence, it is satisfiable as there is at least one assignment of truth values for p, q, and r that makes it true. Similarly, note that $(p \vee q \vee r) \wedge (\neg p \vee \neg q \vee \neg r)$ is true when at least one of p, q, and r is true and at least one is false (see Exercise 43 of Section 1.1). Hence, $(p \vee q \vee r) \wedge (\neg p \vee \neg q \vee \neg r)$ is satisfiable, as there is at least one assignment of truth values for p, q, and r that makes it true.

Finally, note that for $(p \vee \neg q) \wedge (q \vee \neg r) \wedge (r \vee \neg p) \wedge (p \vee q \vee r) \wedge (\neg p \vee \neg q \vee \neg r)$ to be true, $(p \vee \neg q) \wedge (q \vee \neg r) \wedge (r \vee \neg p)$ and $(p \vee q \vee r) \wedge (\neg p \vee \neg q \vee \neg r)$ must both be true. For the first to be true, the three variables must have the same truth values, and for the second to be true, at least one of the three variables must be true and at least one must be false. However, these conditions are contradictory. From these observations we conclude that no assignment of truth values to p, q, and r makes $(p \vee \neg q) \wedge (q \vee \neg r) \wedge (r \vee \neg p) \wedge (p \vee q \vee r) \wedge (\neg p \vee \neg q \vee \neg r)$ true. Hence, it is unsatisfiable. ◀

1.3.6 Applications of Satisfiability

Many problems, in diverse areas such as robotics, software testing, artificial intelligence planning, computer-aided design, machine vision, integrated circuit design, scheduling, computer networking, and genetics, can be modeled in terms of propositional satisfiability. Although most applications are quite complex and beyond the scope of this book, we can illustrate how two puzzles can be modeled as satisfiability problems.

EXAMPLE 10 **The *n*-Queens Problem** The n-queens problem asks for a placement of n queens on an $n \times n$ chessboard so that no queen can attack another queen. This means that no two queens can be placed in the same row, in the same column, or on the same diagonal. We display a solution to the eight-queens problem in Figure 1. (The eight-queens problem dates back to 1848 when it was proposed by Max Bezzel and was completely solved by Franz Nauck in 1850. We will return to the n-queens problem in Section 11.4.)

To model the n-queens problem as a satisfiability problem, we introduce n^2 variables, $p(i, j)$ for $i = 1, 2, \ldots, n$ and $j = 1, 2, \ldots, n$. For a given placement of queens on the chessboard, $p(i, j)$ is true when there is a queen on the square in the ith row and jth column, and is false

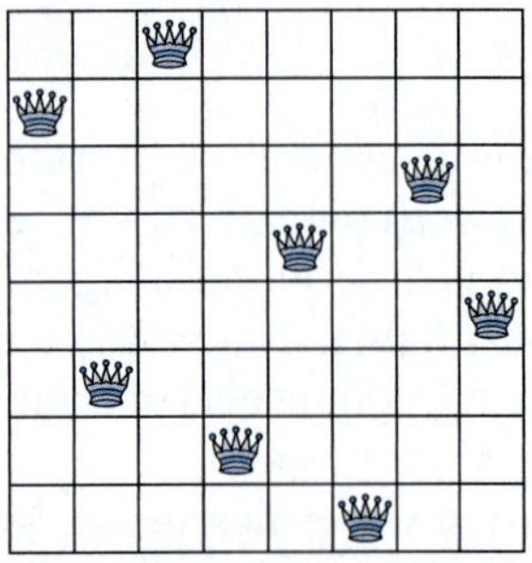

FIGURE 1 A solution to the eight-queens problem.

otherwise. Note that squares (i, j) and (i', j') are on the same diagonal if either $i + i' = j + j'$ or $i - i' = j - j'$. In the chessboard in Figure 1, $p(6, 2)$ and $p(2, 1)$ are true, while $p(3, 4)$ and $p(5, 4)$ are false.

For no two of the n queens to be in the same row, there must be one queen in each row. We can show that there is one queen in each row by verifying that every row contains at least one queen and that every row contains at most one queen. We first note that $\bigvee_{j=1}^{n} p(i, j)$ asserts that row i contains at least one queen, and

$$Q_1 = \bigwedge_{i=1}^{n} \bigvee_{j=1}^{n} p(i, j)$$

asserts that every row contains at least one queen.

For every row to include at most one queen, it must be the case that $p(i, j)$ and $p(k, j)$ are not both true for integers j and k with $1 \le j < k \le n$. Observe that $\neg p(i, j) \vee \neg p(i, k)$ asserts that at least one of $\neg p(i, j)$ and $\neg p(i, k)$ is true, which means that at least one of $p(i, j)$ and $p(i, k)$ is false. So, to check that there is at most one queen in each row, we assert

$$Q_2 = \bigwedge_{i=1}^{n} \bigwedge_{j=1}^{n-1} \bigwedge_{k=j+1}^{n} (\neg p(i, j) \vee \neg p(k, j)).$$

To assert that no column contains more than one queen, we assert that

$$Q_3 = \bigwedge_{j=1}^{n} \bigwedge_{i=1}^{n-1} \bigwedge_{k=i+1}^{n} (\neg p(i, j) \vee \neg p(k, j)).$$

(This assertion, together with the previous assertion that every row contains a queen, implies that every column contains a queen.)

To assert that no diagonal contains two queens, we assert

$$Q_4 = \bigwedge_{i=2}^{n} \bigwedge_{j=1}^{n-1} \bigwedge_{k=1}^{\min(i-1,n-j)} (\neg p(i, j) \vee \neg p(i - k, k + j))$$

and

$$Q_5 = \bigwedge_{i=1}^{n-1} \bigwedge_{j=1}^{n-1} \bigwedge_{k=1}^{\min(n-i,n-j)} (\neg p(i, j) \vee \neg p(i + k, j + k)).$$

The innermost conjunction in Q_4 and in Q_5 for a pair (i, j) runs through the positions on a diagonal that begin at (i, j) and runs rightward along this diagonal. The upper limits on these innermost conjunctions identify the last cell in the board on each diagonal.

Putting all this together, we find that the solutions of the n-queens problem are given by the assignments of truth values to the variables $p(i, j)$, $i = 1, 2, \dots, n$ and $j = 1, 2, \dots, n$ that make

$$Q = Q_1 \wedge Q_2 \wedge Q_3 \wedge Q_4 \wedge Q_5$$

true.

Using this and other approaches, the number of ways n queens can be placed on a chessboard so that no queen can attack another has been computed for $n \leq 27$. When $n = 8$ there are 92 such placements, while for $n = 16$ this number grows to 14,772,512. (See the OEIS discussed in Section 2.4 for details.) ◀

EXAMPLE 11 **Sudoku** Sudoku puzzles are constructed using a 9×9 grid made up of nine 3×3 subgrids, known as **blocks**, as shown in Figure 2. For each puzzle, some of the 81 cells, called **givens**, are assigned one of the numbers $1, 2, \dots, 9$, and the other cells are blank. The puzzle is solved by assigning a number to each blank cell so that every row, every column, and every one of the nine 3×3 blocks contains each of the nine possible numbers. Note that instead of using a 9×9 grid, Sudoku puzzles can be based on $n^2 \times n^2$ grids, for any positive integer n, with the $n^2 \times n^2$ grid made up of n^2 $n \times n$ subgrids.

Links ›

The popularity of Sudoku dates back to the 1980s when it was introduced in Japan. It took 20 years for Sudoku to spread to rest of the world, but by 2005, Sudoku puzzles were a worldwide craze. The name Sudoku is short for the Japanese *suuji wa dokushin ni kagiru*, which means "the digits must remain single." The modern game of Sudoku was apparently designed in the late 1970s by an American puzzle designer. The basic ideas of Sudoku date back even further; puzzles printed in French newspapers in the 1890s were quite similar, but not identical, to modern Sudoku.

Sudoku puzzles designed for entertainment have two additional important properties. First, they have exactly one solution. Second, they can be solved using reasoning alone, that is, without resorting to searching all possible assignments of numbers to the cells. As a Sudoku puzzle is solved, entries in blank cells are successively determined by already known values. For instance, in the grid in Figure 2, the number 4 must appear in exactly one cell in the second row. How can we determine in which of the seven blank cells it must appear? First, we observe that 4 cannot appear in one of the first three cells or in one of the last three cells of this row, because it already appears in another cell in the block each of these cells is in. We can also see that 4

	2	9				4		
			5			1		
	4							
				4	2			
6							7	
5								
7			3					5
	1			9				
							6	

FIGURE 2 A 9 × 9 Sudoku puzzle.

cannot appear in the fifth cell in this row, as it already appears in the fifth column in the fourth row. This means that 4 must appear in the sixth cell of the second row.

Many strategies based on logic and mathematics have been devised for solving Sudoku puzzles (see [Da10], for example). Here, we discuss one of the ways that have been developed for solving Sudoku puzzles with the aid of a computer, which depends on modeling the puzzle as a propositional satisfiability problem. Using the model we describe, particular Sudoku puzzles can be solved using software developed to solve satisfiability problems. Currently, Sudoku puzzles can be solved in less than 10 milliseconds this way. It should be noted that there are many other approaches for solving Sudoku puzzles via computers using other techniques.

To encode a Sudoku puzzle, let $p(i, j, n)$ denote the proposition that is true when the number n is in the cell in the ith row and jth column. There are $9 \times 9 \times 9 = 729$ such propositions, as i, j, and n all range from 1 to 9. For example, for the puzzle in Figure 2, the number 6 is given as the value in the fifth row and first column. Hence, we see that $p(5, 1, 6)$ is true, but $p(5, j, 6)$ is false for $j = 2, 3, \ldots, 9$.

Given a particular Sudoku puzzle, we begin by encoding each of the given values. Then, we construct compound propositions that assert that every row contains every number, every column contains every number, every 3×3 block contains every number, and each cell contains no more than one number. It follows, as the reader should verify, that the Sudoku puzzle is solved by finding an assignment of truth values to the 729 propositions $p(i, j, n)$ with i, j, and n each ranging from 1 to 9 that makes the conjunction of all these compound propositions true. After listing these assertions, we will explain how to construct the assertion that every row contains every integer from 1 to 9. We will leave the construction of the other assertions that every column contains every number and each of the nine 3×3 blocks contains every number to the exercises.

- For each cell with a given value, we assert $p(i, j, n)$ when the cell in row i and column j has the given value n.
- We assert that every row contains every number:

$$\bigwedge_{i=1}^{9} \bigwedge_{n=1}^{9} \bigvee_{j=1}^{9} p(i, j, n)$$

- We assert that every column contains every number:

$$\bigwedge_{j=1}^{9} \bigwedge_{n=1}^{9} \bigvee_{i=1}^{9} p(i, j, n)$$

- We assert that each of the nine 3×3 blocks contains every number:

It is tricky setting up the two inner indices so that all nine cells in each square block are examined.

$$\bigwedge_{r=0}^{2} \bigwedge_{s=0}^{2} \bigwedge_{n=1}^{9} \bigvee_{i=1}^{3} \bigvee_{j=1}^{3} p(3r + i, 3s + j, n)$$

- To assert that no cell contains more than one number, we take the conjunction over all values of n, n', i, and j, where each variable ranges from 1 to 9 and $n \neq n'$ of $p(i, j, n) \rightarrow \neg p(i, j, n')$.

We now explain how to construct the assertion that every row contains every number. First, to assert that row i contains the number n, we form $\bigvee_{j=1}^{9} p(i, j, n)$. To assert that row i contains all n numbers, we form the conjunction of these disjunctions over all nine possible values of n, giving us $\bigwedge_{n=1}^{9} \bigvee_{j=1}^{9} p(i, j, n)$. Finally, to assert that every row contains every number, we take the conjunction of $\bigwedge_{n=1}^{9} \bigvee_{j=1}^{9} p(i, j, n)$ over all nine rows. This gives us $\bigwedge_{i=1}^{9} \bigwedge_{n=1}^{9} \bigvee_{j=1}^{9} p(i, j, n)$. (Exercises 71 and 72 ask for explanations of the assertions that every column contains every number and that each of the nine 3×3 blocks contains every number.)

Given a particular Sudoku puzzle, to solve this puzzle we can find a solution to the satisfiability problems that asks for a set of truth values for the 729 variables $p(i, j, n)$ that makes the conjunction of all the listed assertions true. ◀

1.3.7 Solving Satisfiability Problems

A truth table can be used to determine whether a compound proposition is satisfiable, or equivalently, whether its negation is a tautology (see Exercise 64). This can be done by hand for a compound proposition with a small number of variables, but when the number of variables grows, this becomes impractical. For instance, there are $2^{20} = 1{,}048{,}576$ rows in the truth table for a compound proposition with 20 variables. Thus, you need a computer to help you determine, in this way, whether a compound proposition in 20 variables is satisfiable.

When many applications are modeled, questions concerning the satisfiability of compound propositions with hundreds, thousands, or millions of variables arise. Note, for example, that when there are 1000 variables, checking every one of the 2^{1000} (a number with more than 300 decimal digits) possible combinations of truth values of the variables in a compound proposition cannot be done by a computer in even trillions of years. No procedure is known that a computer can follow to determine in a reasonable amount of time whether an arbitrary compound proposition in such a large number of variables is satisfiable. However, progress has been made developing methods for solving the satisfiability problem for the particular types of compound propositions that arise in practical applications, such as for the solution of Sudoku puzzles. Many computer programs have been developed for solving satisfiability problems which have practical use. In our discussion of the subject of algorithms in Chapter 3, we will discuss this question further. In particular, we will explain the important role the propositional satisfiability problem plays in the study of the complexity of algorithms.

Links

Links

©*Harvard University Portrait Collection, Department of Philosophy*

HENRY MAURICE SHEFFER (1883–1964) Henry Maurice Sheffer, born to Jewish parents in the western Ukraine, emigrated to the United States in 1892 with his parents and six siblings. He studied at the Boston Latin School before entering Harvard, where he completed his undergraduate degree in 1905, his master's in 1907, and his Ph.D. in philosophy in 1908. After holding a postdoctoral position at Harvard, Henry traveled to Europe on a fellowship. Upon returning to the United States, he became an academic nomad, spending one year each at the University of Washington, Cornell, the University of Minnesota, the University of Missouri, and City College in New York. In 1916 he returned to Harvard as a faculty member in the philosophy department. He remained at Harvard until his retirement in 1952.

Sheffer introduced what is now known as the Sheffer stroke in 1913; it became well known only after its use in the 1925 edition of Whitehead and Russell's *Principia Mathematica*. In this same edition Russell wrote that Sheffer had invented a powerful method that could be used to simplify the *Principia*. Because of this comment, Sheffer was something of a mystery man to logicians, especially because Sheffer, who published little in his career, never published the details of this method, only describing it in mimeographed notes and in a brief published abstract.

Sheffer was a dedicated teacher of mathematical logic. He liked his classes to be small and did not like auditors. When strangers appeared in his classroom, Sheffer would order them to leave, even his colleagues or distinguished guests visiting Harvard. Sheffer was barely five feet tall; he was noted for his wit and vigor, as well as for his nervousness and irritability. Although widely liked, he was quite lonely. He is noted for a quip he spoke at his retirement: "Old professors never die, they just become emeriti." Sheffer is also credited with coining the term "Boolean algebra" (the subject of Chapter 12 of this text). Sheffer was briefly married and lived most of his later life in small rooms at a hotel packed with his logic books and vast files of slips of paper he used to jot down his ideas. Unfortunately, Sheffer suffered from severe depression during the last two decades of his life.

Exercises

1. Use truth tables to verify these equivalences.
   a) $p \land \mathbf{T} \equiv p$
   b) $p \lor \mathbf{F} \equiv p$
   c) $p \land \mathbf{F} \equiv \mathbf{F}$
   d) $p \lor \mathbf{T} \equiv \mathbf{T}$
   e) $p \lor p \equiv p$
   f) $p \land p \equiv p$
2. Show that $\neg(\neg p)$ and p are logically equivalent.
3. Use truth tables to verify the commutative laws
   a) $p \lor q \equiv q \lor p$.
   b) $p \land q \equiv q \land p$.
4. Use truth tables to verify the associative laws
   a) $(p \lor q) \lor r \equiv p \lor (q \lor r)$.
   b) $(p \land q) \land r \equiv p \land (q \land r)$.
5. Use a truth table to verify the distributive law $p \land (q \lor r) \equiv (p \land q) \lor (p \land r)$.
6. Use a truth table to verify the first De Morgan law $\neg(p \land q) \equiv \neg p \lor \neg q$.
7. Use De Morgan's laws to find the negation of each of the following statements.
   a) Jan is rich and happy.
   b) Carlos will bicycle or run tomorrow.
   c) Mei walks or takes the bus to class.
   d) Ibrahim is smart and hard working.
8. Use De Morgan's laws to find the negation of each of the following statements.
   a) Kwame will take a job in industry or go to graduate school.
   b) Yoshiko knows Java and calculus.
   c) James is young and strong.
   d) Rita will move to Oregon or Washington.
9. For each of these compound propositions, use the conditional-disjunction equivalence (Example 3) to find an equivalent compound proposition that does not involve conditionals.
   a) $p \to \neg q$
   b) $(p \to q) \to r$
   c) $(\neg q \to p) \to (p \to \neg q)$
10. For each of these compound propositions, use the conditional-disjunction equivalence (Example 3) to find an equivalent compound proposition that does not involve conditionals.
   a) $\neg p \to \neg q$
   b) $(p \lor q) \to \neg p$
   c) $(p \to \neg q) \to (\neg p \to q)$
11. Show that each of these conditional statements is a tautology by using truth tables.
   a) $(p \land q) \to p$
   b) $p \to (p \lor q)$
   c) $\neg p \to (p \to q)$
   d) $(p \land q) \to (p \to q)$
   e) $\neg(p \to q) \to p$
   f) $\neg(p \to q) \to \neg q$
12. Show that each of these conditional statements is a tautology by using truth tables.
   a) $[\neg p \land (p \lor q)] \to q$
   b) $[(p \to q) \land (q \to r)] \to (p \to r)$
   c) $[p \land (p \to q)] \to q$
   d) $[(p \lor q) \land (p \to r) \land (q \to r)] \to r$
13. Show that each conditional statement in Exercise 11 is a tautology using the fact that a conditional statement is false exactly when the hypothesis is true and the conclusion is false. (Do not use truth tables.)
14. Show that each conditional statement in Exercise 12 is a tautology using the fact that a conditional statement is false exactly when the hypothesis is true and the conclusion is false. (Do not use truth tables.)
15. Show that each conditional statement in Exercise 11 is a tautology by applying a chain of logical identities as in Example 8. (Do not use truth tables.)
16. Show that each conditional statement in Exercise 12 is a tautology by applying a chain of logical identities as in Example 8. (Do not use truth tables.)
17. Use truth tables to verify the absorption laws.
   a) $p \lor (p \land q) \equiv p$
   b) $p \land (p \lor q) \equiv p$
18. Determine whether $(\neg p \land (p \to q)) \to \neg q$ is a tautology.
19. Determine whether $(\neg q \land (p \to q)) \to \neg p$ is a tautology.

Each of Exercises 20–32 asks you to show that two compound propositions are logically equivalent. To do this, either show that both sides are true, or that both sides are false, for exactly the same combinations of truth values of the propositional variables in these expressions (whichever is easier).

20. Show that $p \leftrightarrow q$ and $(p \land q) \lor (\neg p \land \neg q)$ are logically equivalent.
21. Show that $\neg(p \leftrightarrow q)$ and $p \leftrightarrow \neg q$ are logically equivalent.
22. Show that $p \to q$ and $\neg q \to \neg p$ are logically equivalent.
23. Show that $\neg p \leftrightarrow q$ and $p \leftrightarrow \neg q$ are logically equivalent.
24. Show that $\neg(p \oplus q)$ and $p \leftrightarrow q$ are logically equivalent.
25. Show that $\neg(p \leftrightarrow q)$ and $\neg p \leftrightarrow q$ are logically equivalent.
26. Show that $(p \to q) \land (p \to r)$ and $p \to (q \land r)$ are logically equivalent.
27. Show that $(p \to r) \land (q \to r)$ and $(p \lor q) \to r$ are logically equivalent.
28. Show that $(p \to q) \lor (p \to r)$ and $p \to (q \lor r)$ are logically equivalent.
29. Show that $(p \to r) \lor (q \to r)$ and $(p \land q) \to r$ are logically equivalent.
30. Show that $\neg p \to (q \to r)$ and $q \to (p \lor r)$ are logically equivalent.
31. Show that $p \leftrightarrow q$ and $(p \to q) \land (q \to p)$ are logically equivalent.
32. Show that $p \leftrightarrow q$ and $\neg p \leftrightarrow \neg q$ are logically equivalent.
33. Show that $(p \to q) \land (q \to r) \to (p \to r)$ is a tautology.
34. Show that $(p \lor q) \land (\neg p \lor r) \to (q \lor r)$ is a tautology.
35. Show that $(p \to q) \to r$ and $p \to (q \to r)$ are not logically equivalent.
36. Show that $(p \land q) \to r$ and $(p \to r) \land (q \to r)$ are not logically equivalent.

37. Show that $(p \to q) \to (r \to s)$ and $(p \to r) \to (q \to s)$ are not logically equivalent.

The **dual** of a compound proposition that contains only the logical operators $\vee$, $\wedge$, and $\neg$ is the compound proposition obtained by replacing each $\vee$ by $\wedge$, each $\wedge$ by $\vee$, each **T** by **F**, and each **F** by **T**. The dual of s is denoted by s^*.

38. Find the dual of each of these compound propositions.

a) $p \vee \neg q$ **b)** $p \wedge (q \vee (r \wedge \mathbf{T}))$
c) $(p \wedge \neg q) \vee (q \wedge \mathbf{F})$

39. Find the dual of each of these compound propositions.

a) $p \wedge \neg q \wedge \neg r$ **b)** $(p \wedge q \wedge r) \vee s$
c) $(p \vee \mathbf{F}) \wedge (q \vee \mathbf{T})$

40. When does $s^* = s$, where s is a compound proposition?

41. Show that $(s^*)^* = s$ when s is a compound proposition.

42. Show that the logical equivalences in Table 6, except for the double negation law, come in pairs, where each pair contains compound propositions that are duals of each other.

****43.** Why are the duals of two equivalent compound propositions also equivalent, where these compound propositions contain only the operators $\wedge$, $\vee$, and $\neg$?

44. Find a compound proposition involving the propositional variables p, q, and r that is true when p and q are true and r is false, but is false otherwise. [*Hint:* Use a conjunction of each propositional variable or its negation.]

45. Find a compound proposition involving the propositional variables p, q, and r that is true when exactly two of p, q, and r are true and is false otherwise. [*Hint:* Form a disjunction of conjunctions. Include a conjunction for each combination of values for which the compound proposition is true. Each conjunction should include each of the three propositional variables or its negations.]

46. Suppose that a truth table in n propositional variables is specified. Show that a compound proposition with this truth table can be formed by taking the disjunction of conjunctions of the variables or their negations, with one conjunction included for each combination of values for which the compound proposition is true. The resulting compound proposition is said to be in **disjunctive normal form**.

A collection of logical operators is called **functionally complete** if every compound proposition is logically equivalent to a compound proposition involving only these logical operators.

47. Show that $\neg$, $\wedge$, and $\vee$ form a functionally complete collection of logical operators. [*Hint:* Use the fact that every compound proposition is logically equivalent to one in disjunctive normal form, as shown in Exercise 46.]

***48.** Show that $\neg$ and $\wedge$ form a functionally complete collection of logical operators. [*Hint:* First use a De Morgan law to show that $p \vee q$ is logically equivalent to $\neg(\neg p \wedge \neg q)$.]

***49.** Show that $\neg$ and $\vee$ form a functionally complete collection of logical operators.

We now present a group of exercises that involve the logical operators *NAND* and *NOR*. The proposition p *NAND* q is true when either p or q, or both, are false; and it is false when both p and q are true. The proposition p *NOR* q is true when both p and q are false, and it is false otherwise. The propositions p *NAND* q and p *NOR* q are denoted by $p \mid q$ and $p \downarrow q$, respectively. (The operators $\mid$ and $\downarrow$ are called the **Sheffer stroke** and the **Peirce arrow** after H. M. Sheffer and C. S. Peirce, respectively.)

50. Construct a truth table for the logical operator *NAND*.

51. Show that $p \mid q$ is logically equivalent to $\neg(p \wedge q)$.

52. Construct a truth table for the logical operator *NOR*.

53. Show that $p \downarrow q$ is logically equivalent to $\neg(p \vee q)$.

54. In this exercise we will show that $\{\downarrow\}$ is a functionally complete collection of logical operators.

a) Show that $p \downarrow p$ is logically equivalent to $\neg p$.
b) Show that $(p \downarrow q) \downarrow (p \downarrow q)$ is logically equivalent to $p \vee q$.
c) Conclude from parts (a) and (b), and Exercise 49, that $\{\downarrow\}$ is a functionally complete collection of logical operators.

***55.** Find a compound proposition logically equivalent to $p \to q$ using only the logical operator $\downarrow$.

56. Show that $\{\mid\}$ is a functionally complete collection of logical operators.

57. Show that $p \mid q$ and $q \mid p$ are equivalent.

58. Show that $p \mid (q \mid r)$ and $(p \mid q) \mid r$ are not equivalent, so that the logical operator $\mid$ is not associative.

***59.** How many different truth tables of compound propositions are there that involve the propositional variables p and q?

60. Show that if p, q, and r are compound propositions such that p and q are logically equivalent and q and r are logically equivalent, then p and r are logically equivalent.

61. The following sentence is taken from the specification of a telephone system: "If the directory database is opened, then the monitor is put in a closed state, if the system is not in its initial state." This specification is hard to understand because it involves two conditional statements. Find an equivalent, easier-to-understand specification that involves disjunctions and negations but not conditional statements.

62. How many of the disjunctions $p \vee \neg q$, $\neg p \vee q$, $q \vee r$, $q \vee \neg r$, and $\neg q \vee \neg r$ can be made simultaneously true by an assignment of truth values to p, q, and r?

63. How many of the disjunctions $p \vee \neg q \vee s$, $\neg p \vee \neg r \vee s$, $\neg p \vee \neg r \vee \neg s$, $\neg p \vee q \vee \neg s$, $q \vee r \vee \neg s$, $q \vee \neg r \vee \neg s$, $\neg p \vee \neg q \vee \neg s$, $p \vee r \vee s$, and $p \vee r \vee \neg s$ can be made simultaneously true by an assignment of truth values to p, q, r, and s?

64. Show that the negation of an unsatisfiable compound proposition is a tautology and the negation of a compound proposition that is a tautology is unsatisfiable.

65. Determine whether each of these compound propositions is satisfiable.

a) $(p \vee \neg q) \wedge (\neg p \vee q) \wedge (\neg p \vee \neg q)$
b) $(p \to q) \wedge (p \to \neg q) \wedge (\neg p \to q) \wedge (\neg p \to \neg q)$
c) $(p \leftrightarrow q) \wedge (\neg p \leftrightarrow q)$

66. Determine whether each of these compound propositions is satisfiable.

a) $(p \vee q \vee \neg r) \wedge (p \vee \neg q \vee \neg s) \wedge (p \vee \neg r \vee \neg s) \wedge (\neg p \vee \neg q \vee \neg s) \wedge (p \vee q \vee \neg s)$

b) $(\neg p \vee \neg q \vee r) \wedge (\neg p \vee q \vee \neg s) \wedge (p \vee \neg q \vee \neg s) \wedge (\neg p \vee \neg r \vee \neg s) \wedge (p \vee q \vee \neg r) \wedge (p \vee \neg r \vee \neg s)$

c) $(p \vee q \vee r) \wedge (p \vee \neg q \vee \neg s) \wedge (q \vee \neg r \vee s) \wedge (\neg p \vee r \vee s) \wedge (\neg p \vee q \vee \neg s) \wedge (p \vee \neg q \vee \neg r) \wedge (\neg p \vee \neg q \vee s) \wedge (\neg p \vee \neg r \vee \neg s)$

67. Find the compound proposition Q constructed in Example 10 for the n-queens problem, and use it to find all the ways that n queens can be placed on an $n \times n$ chessboard, so that no queen can attack another when n is

a) 2. **b)** 3. **c)** 4.

68. Starting with the compound proposition Q found in Example 10, construct a compound proposition that can be used to find all solutions of the n-queens problem where the queen in the first column is in an odd-numbered row.

69. Show how the solution of a given 4×4 Sudoku puzzle can be found by solving a satisfiability problem.

70. Construct a compound proposition that asserts that every cell of a 9×9 Sudoku puzzle contains at least one number.

71. Explain the steps in the construction of the compound proposition given in the text that asserts that every column of a 9×9 Sudoku puzzle contains every number.

***72.** Explain the steps in the construction of the compound proposition given in the text that asserts that each of the nine 3×3 blocks of a 9×9 Sudoku puzzle contains every number.

1.4 Predicates and Quantifiers

1.4.1 Introduction

Propositional logic, studied in Sections 1.1–1.3, cannot adequately express the meaning of all statements in mathematics and in natural language. For example, suppose that we know that

"Every computer connected to the university network is functioning properly."

No rules of propositional logic allow us to conclude the truth of the statement

"MATH3 is functioning properly,"

where MATH3 is one of the computers connected to the university network. Likewise, we cannot use the rules of propositional logic to conclude from the statement

"CS2 is under attack by an intruder,"

where CS2 is a computer on the university network, to conclude the truth of

"There is a computer on the university network that is under attack by an intruder."

In this section we will introduce a more powerful type of logic called **predicate logic**. We will see how predicate logic can be used to express the meaning of a wide range of statements in mathematics and computer science in ways that permit us to reason and explore relationships between objects. To understand predicate logic, we first need to introduce the concept of a predicate. Afterward, we will introduce the notion of quantifiers, which enable us to reason with statements that assert that a certain property holds for all objects of a certain type and with statements that assert the existence of an object with a particular property.

1.4.2 Predicates

Statements involving variables, such as

"$x > 3$," "$x = y + 3$," "$x + y = z$,"

and

"Computer x is under attack by an intruder,"

and

"Computer x is functioning properly,"

are often found in mathematical assertions, in computer programs, and in system specifications. These statements are neither true nor false when the values of the variables are not specified. In this section, we will discuss the ways that propositions can be produced from such statements.

The statement "x is greater than 3" has two parts. The first part, the variable x, is the subject of the statement. The second part—the **predicate**, "is greater than 3"—refers to a property that the subject of the statement can have. We can denote the statement "x is greater than 3" by $P(x)$, where P denotes the predicate "is greater than 3" and x is the variable. The statement $P(x)$ is also said to be the value of the **propositional function** P at x. Once a value has been assigned to the variable x, the statement $P(x)$ becomes a proposition and has a truth value. Consider Examples 1 and 2.

EXAMPLE 1 Let $P(x)$ denote the statement "$x > 3$." What are the truth values of $P(4)$ and $P(2)$?

Solution: We obtain the statement $P(4)$ by setting $x = 4$ in the statement "$x > 3$." Hence, $P(4)$, which is the statement "$4 > 3$," is true. However, $P(2)$, which is the statement "$2 > 3$," is false. ◀

EXAMPLE 2 Let $A(x)$ denote the statement "Computer x is under attack by an intruder." Suppose that of the computers on campus, only CS2 and MATH1 are currently under attack by intruders. What are truth values of $A(\text{CS1})$, $A(\text{CS2})$, and $A(\text{MATH1})$?

Solution: We obtain the statement $A(\text{CS1})$ by setting $x = \text{CS1}$ in the statement "Computer x is under attack by an intruder." Because CS1 is not on the list of computers currently under attack, we conclude that $A(\text{CS1})$ is false. Similarly, because CS2 and MATH1 are on the list of computers under attack, we know that $A(\text{CS2})$ and $A(\text{MATH1})$ are true. ◀

We can also have statements that involve more than one variable. For instance, consider the statement "$x = y + 3$." We can denote this statement by $Q(x, y)$, where x and y are variables and Q is the predicate. When values are assigned to the variables x and y, the statement $Q(x, y)$ has a truth value.

EXAMPLE 3 Let $Q(x, y)$ denote the statement "$x = y + 3$." What are the truth values of the propositions $Q(1, 2)$ and $Q(3, 0)$?

Extra Examples ❯

Solution: To obtain $Q(1, 2)$, set $x = 1$ and $y = 2$ in the statement $Q(x, y)$. Hence, $Q(1, 2)$ is the statement "$1 = 2 + 3$," which is false. The statement $Q(3, 0)$ is the proposition "$3 = 0 + 3$," which is true. ◀

EXAMPLE 4 Let $A(c, n)$ denote the statement "Computer c is connected to network n," where c is a variable representing a computer and n is a variable representing a network. Suppose that the computer MATH1 is connected to network CAMPUS2, but not to network CAMPUS1. What are the values of $A(\text{MATH1, CAMPUS1})$ and $A(\text{MATH1, CAMPUS2})$?

Solution: Because MATH1 is not connected to the CAMPUS1 network, we see that A(MATH1, CAMPUS1) is false. However, because MATH1 is connected to the CAMPUS2 network, we see that A(MATH1, CAMPUS2) is true. ◀

Similarly, we can let $R(x, y, z)$ denote the statement "$x + y = z$." When values are assigned to the variables x, y, and z, this statement has a truth value.

EXAMPLE 5 What are the truth values of the propositions $R(1, 2, 3)$ and $R(0, 0, 1)$?

Solution: The proposition $R(1, 2, 3)$ is obtained by setting $x = 1$, $y = 2$, and $z = 3$ in the statement $R(x, y, z)$. We see that $R(1, 2, 3)$ is the statement "$1 + 2 = 3$," which is true. Also note that $R(0, 0, 1)$, which is the statement "$0 + 0 = 1$," is false. ◀

In general, a statement involving the n variables $x_1, x_2, \ldots, x_n$ can be denoted by

$$P(x_1, x_2, \ldots, x_n).$$

A statement of the form $P(x_1, x_2, \ldots, x_n)$ is the value of the **propositional function** P at the n-tuple $(x_1, x_2, \ldots, x_n)$, and P is also called an **n-place predicate** or an **n-ary predicate**.

Propositional functions occur in computer programs, as Example 6 demonstrates.

EXAMPLE 6 Consider the statement

if $x > 0$ **then** $x := x + 1$.

Links

Bettmann/Getty Images

CHARLES SANDERS PEIRCE (1839–1914) Many consider Charles Peirce, born in Cambridge, Massachusetts, to be the most original and versatile American intellect. He made important contributions to an amazing number of disciplines, including mathematics, astronomy, chemistry, geodesy, metrology, engineering, psychology, philology, the history of science, and economics. Peirce was also an inventor, a lifelong student of medicine, a book reviewer, a dramatist and an actor, a short story writer, a phenomenologist, a logician, and a metaphysician. He is noted as the preeminent system-building philosopher competent and productive in logic, mathematics, and a wide range of sciences. He was encouraged by his father, Benjamin Peirce, a professor of mathematics and natural philosophy at Harvard, to pursue a career in science. Instead, he decided to study logic and scientific methodology. Peirce attended Harvard (1855–1859) and received a Harvard master of arts degree (1862) and an advanced degree in chemistry from the Lawrence Scientific School (1863).

In 1861, Peirce became an aide in the U.S. Coast Survey, with the goal of better understanding scientific methodology. His service for the Survey exempted him from military service during the Civil War. While working for the Survey, Peirce did astronomical and geodesic work. He made fundamental contributions to the design of pendulums and to map projections, applying new mathematical developments in the theory of elliptic functions. He was the first person to use the wavelength of light as a unit of measurement. Peirce rose to the position of Assistant for the Survey, a position he held until forced to resign in 1891 when he disagreed with the direction taken by the Survey's new administration.

While making his living from work in the physical sciences, Peirce developed a hierarchy of sciences, with mathematics at the top rung, in which the methods of one science could be adapted for use by those sciences under it in the hierarchy. During this time, he also founded the American philosophical theory of pragmatism.

The only academic position Peirce ever held was lecturer in logic at Johns Hopkins University in Baltimore (1879–1884). His mathematical work during this time included contributions to logic, set theory, abstract algebra, and the philosophy of mathematics. His work is still relevant today, with recent applications to artificial intelligence. Peirce believed that the study of mathematics could develop the mind's powers of imagination, abstraction, and generalization. His diverse activities after retiring from the Survey included writing for periodicals, contributing to scholarly dictionaries, translating scientific papers, guest lecturing, and textbook writing. Unfortunately, his income from these pursuits was insufficient to protect him and his second wife from abject poverty. He was supported in his later years by a fund created by his many admirers and administered by the philosopher William James, his lifelong friend. Although Peirce wrote and published voluminously in a vast range of subjects, he left more than 100,000 pages of unpublished manuscripts. Because of the difficulty of studying his unpublished writings, scholars have only recently started to understand some of his varied contributions. A group of people is devoted to making his work available over the Internet to bring a better appreciation of Peirce's accomplishments to the world.

When this statement is encountered in a program, the value of the variable x at that point in the execution of the program is inserted into $P(x)$, which is "$x > 0$." If $P(x)$ is true for this value of x, the assignment statement $x := x + 1$ is executed, so the value of x is increased by 1. If $P(x)$ is false for this value of x, the assignment statement is not executed, so the value of x is not changed. ◀

PRECONDITIONS AND POSTCONDITIONS Predicates are also used to establish the correctness of computer programs, that is, to show that computer programs always produce the desired output when given valid input. (Note that unless the correctness of a computer program is established, no amount of testing can show that it produces the desired output for all input values, unless every input value is tested.) The statements that describe valid input are known as **preconditions** and the conditions that the output should satisfy when the program has run are known as **postconditions**. As Example 7 illustrates, we use predicates to describe both preconditions and postconditions. We will study this process in greater detail in Section 5.5.

EXAMPLE 7 Consider the following program, designed to interchange the values of two variables x and y.

```
temp := x
x := y
y := temp
```

Find predicates that we can use as the precondition and the postcondition to verify the correctness of this program. Then explain how to use them to verify that for all valid input the program does what is intended.

Solution: For the precondition, we need to express that x and y have particular values before we run the program. So, for this precondition we can use the predicate $P(x, y)$, where $P(x, y)$ is the statement "$x = a$ and $y = b$," where a and b are the values of x and y before we run the program. Because we want to verify that the program swaps the values of x and y for all input values, for the postcondition we can use $Q(x, y)$, where $Q(x, y)$ is the statement "$x = b$ and $y = a$."

To verify that the program always does what it is supposed to do, suppose that the precondition $P(x, y)$ holds. That is, we suppose that the statement "$x = a$ and $y = b$" is true. This means that $x = a$ and $y = b$. The first step of the program, $temp := x$, assigns the value of x to the variable $temp$, so after this step we know that $x = a$, $temp = a$, and $y = b$. After the second step of the program, $x := y$, we know that $x = b$, $temp = a$, and $y = b$. Finally, after the third step, we know that $x = b$, $temp = a$, and $y = a$. Consequently, after this program is run, the postcondition $Q(x, y)$ holds, that is, the statement "$x = b$ and $y = a$" is true. ◀

1.4.3 Quantifiers

When the variables in a propositional function are assigned values, the resulting statement becomes a proposition with a certain truth value. However, there is another important way, called **quantification**, to create a proposition from a propositional function. Quantification expresses the extent to which a predicate is true over a range of elements. In English, the words *all*, *some*, *many*, *none*, and *few* are used in quantifications. We will focus on two types of quantification here: universal quantification, which tells us that a predicate is true for every element under consideration, and existential quantification, which tells us that there is one or more element under consideration for which the predicate is true. The area of logic that deals with predicates and quantifiers is called the **predicate calculus**.

Assessment ❯

Assessment

THE UNIVERSAL QUANTIFIER Many mathematical statements assert that a property is true for all values of a variable in a particular domain, called the **domain of discourse** (or the **universe of discourse**), often just referred to as the **domain**. Such a statement is expressed using universal quantification. The universal quantification of $P(x)$ for a particular domain is the proposition that asserts that $P(x)$ is true for all values of x in this domain. Note that the domain specifies the possible values of the variable x. The meaning of the universal quantification of $P(x)$ changes when we change the domain. The domain must always be specified when a universal quantifier is used; without it, the universal quantification of a statement is not defined.

Definition 1

The *universal quantification* of $P(x)$ is the statement

"$P(x)$ for all values of x in the domain."

The notation $\forall x P(x)$ denotes the universal quantification of $P(x)$. Here $\forall$ is called the **universal quantifier**. We read $\forall x P(x)$ as "for all $x P(x)$" or "for every $x P(x)$." An element for which $P(x)$ is false is called a **counterexample** to $\forall x P(x)$.

The meaning of the universal quantifier is summarized in the first row of Table 1. We illustrate the use of the universal quantifier in Examples 8–12 and 15.

EXAMPLE 8 Let $P(x)$ be the statement "$x + 1 > x$." What is the truth value of the quantification $\forall x P(x)$, where the domain consists of all real numbers?

Extra Examples

Solution: Because $P(x)$ is true for all real numbers x, the quantification

$$\forall x P(x)$$

is true. ◀

Remark: Generally, an implicit assumption is made that all domains of discourse for quantifiers are nonempty. Note that if the domain is empty, then $\forall x P(x)$ is true for any propositional function $P(x)$ because there are no elements x in the domain for which $P(x)$ is false.

Remember that the truth value of $\forall x P(x)$ depends on the domain!

Besides "for all" and "for every," universal quantification can be expressed in many other ways, including "all of," "for each," "given any," "for arbitrary," "for each," and "for any."

Remark: It is best to avoid using "for any x" because it is often ambiguous as to whether "any" means "every" or "some." In some cases, "any" is unambiguous, such as when it is used in negatives: "There is not any reason to avoid studying."

A statement $\forall x P(x)$ is false, where $P(x)$ is a propositional function, if and only if $P(x)$ is not always true when x is in the domain. One way to show that $P(x)$ is not always true when x is in the domain is to find a counterexample to the statement $\forall x P(x)$. Note that a single counterexample is all we need to establish that $\forall x P(x)$ is false. Example 9 illustrates how counterexamples are used.

TABLE 1 Quantifiers.

Statement	*When True?*	*When False?*
$\forall x P(x)$	$P(x)$ is true for every x.	There is an x for which $P(x)$ is false.
$\exists x P(x)$	There is an x for which $P(x)$ is true.	$P(x)$ is false for every x.

EXAMPLE 9 Let $Q(x)$ be the statement "$x < 2$." What is the truth value of the quantification $\forall x Q(x)$, where the domain consists of all real numbers?

Solution: $Q(x)$ is not true for every real number x, because, for instance, $Q(3)$ is false. That is, $x = 3$ is a counterexample for the statement $\forall x Q(x)$. Thus,

$$\forall x Q(x)$$

is false. ◀

EXAMPLE 10 Suppose that $P(x)$ is "$x^2 > 0$." To show that the statement $\forall x P(x)$ is false where the universe of discourse consists of all integers, we give a counterexample. We see that $x = 0$ is a counterexample because $x^2 = 0$ when $x = 0$, so that x^2 is not greater than 0 when $x = 0$. ◀

Looking for counterexamples to universally quantified statements is an important activity in the study of mathematics, as we will see in subsequent sections of this book.

EXAMPLE 11 What does the statement $\forall x N(x)$ mean if $N(x)$ is "Computer x is connected to the network" and the domain consists of all computers on campus?

Solution: The statement $\forall x N(x)$ means that for every computer x on campus, that computer x is connected to the network. This statement can be expressed in English as "Every computer on campus is connected to the network." ◀

As we have pointed out, specifying the domain is mandatory when quantifiers are used. The truth value of a quantified statement often depends on which elements are in this domain, as Example 12 shows.

EXAMPLE 12 What is the truth value of $\forall x(x^2 \geq x)$ if the domain consists of all real numbers? What is the truth value of this statement if the domain consists of all integers?

Solution: The universal quantification $\forall x(x^2 \geq x)$, where the domain consists of all real numbers, is false. For example, $(\frac{1}{2})^2 \not\geq \frac{1}{2}$. Note that $x^2 \geq x$ if and only if $x^2 - x = x(x-1) \geq 0$. Consequently, $x^2 \geq x$ if and only if $x \leq 0$ or $x \geq 1$. It follows that $\forall x(x^2 \geq x)$ is false if the domain consists of all real numbers (because the inequality is false for all real numbers x with $0 < x < 1$). However, if the domain consists of the integers, $\forall x(x^2 \geq x)$ is true, because there are no integers x with $0 < x < 1$. ◀

THE EXISTENTIAL QUANTIFIER Many mathematical statements assert that there is an element with a certain property. Such statements are expressed using existential quantification. With existential quantification, we form a proposition that is true if and only if $P(x)$ is true for at least one value of x in the domain.

Definition 2 The *existential quantification* of $P(x)$ is the proposition

"There exists an element x in the domain such that $P(x)$."

We use the notation $\exists x P(x)$ for the existential quantification of $P(x)$. Here $\exists$ is called the *existential quantifier*.

A domain must always be specified when a statement $\exists x P(x)$ is used. Furthermore, the meaning of $\exists x P(x)$ changes when the domain changes. Without specifying the domain, the statement $\exists x P(x)$ has no meaning.

Besides the phrase "there exists," we can also express existential quantification in many other ways, such as by using the words "for some," "for at least one," or "there is." The existential quantification $\exists x P(x)$ is read as

"There is an x such that $P(x)$,"
"There is at least one x such that $P(x)$,"

or

"For some $x P(x)$."

The meaning of the existential quantifier is summarized in the second row of Table 1. We illustrate the use of the existential quantifier in Examples 13, 14, and 16.

EXAMPLE 13 Let $P(x)$ denote the statement "$x > 3$." What is the truth value of the quantification $\exists x P(x)$, where the domain consists of all real numbers?

Extra Examples

Solution: Because "$x > 3$" is sometimes true—for instance, when $x = 4$—the existential quantification of $P(x)$, which is $\exists x P(x)$, is true. ◀

Observe that the statement $\exists x P(x)$ is false if and only if there is no element x in the domain for which $P(x)$ is true. That is, $\exists x P(x)$ is false if and only if $P(x)$ is false for every element of the domain. We illustrate this observation in Example 14.

EXAMPLE 14 Let $Q(x)$ denote the statement "$x = x + 1$." What is the truth value of the quantification $\exists x Q(x)$, where the domain consists of all real numbers?

Solution: Because $Q(x)$ is false for every real number x, the existential quantification of $Q(x)$, which is $\exists x Q(x)$, is false. ◀

Remember that the truth value of $\exists x P(x)$ depends on the domain!

Remark: Generally, an implicit assumption is made that all domains of discourse for quantifiers are nonempty. If the domain is empty, then $\exists x Q(x)$ is false whenever $Q(x)$ is a propositional function because when the domain is empty, there can be no element x in the domain for which $Q(x)$ is true.

THE UNIQUENESS QUANTIFIER We have now introduced universal and existential quantifiers. These are the most important quantifiers in mathematics and computer science. However, there is no limitation on the number of different quantifiers we can define, such as "there are exactly two," "there are no more than three," "there are at least 100," and so on. Of these other quantifiers, the one that is most often seen is the **uniqueness quantifier**, denoted by $\exists!$ or $\exists_1$. The notation $\exists! x P(x)$ [or $\exists_1 x P(x)$] states "There exists a unique x such that $P(x)$ is true." (Other phrases for uniqueness quantification include "there is exactly one" and "there is one and only one.") For instance, $\exists! x(x - 1 = 0)$, where the domain is the set of real numbers, states that there is a unique real number x such that $x - 1 = 0$. This is a true statement, as $x = 1$ is the unique real number such that $x - 1 = 0$. Observe that we can use quantifiers and propositional logic to express uniqueness (see Exercise 52 in Section 1.5), so the uniqueness quantifier can be avoided. Generally, it is best to stick with existential and universal quantifiers so that rules of inference for these quantifiers can be used.

1.4.4 QUANTIFIERS OVER FINITE DOMAINS

When the domain of a quantifier is finite, that is, when all its elements can be listed, quantified statements can be expressed using propositional logic. In particular, when the elements of the domain are $x_1, x_2, \ldots, x_n$, where n is a positive integer, the universal quantification $\forall x P(x)$ is the same as the conjunction

$$P(x_1) \wedge P(x_2) \wedge \cdots \wedge P(x_n),$$

because this conjunction is true if and only if $P(x_1), P(x_2), \ldots, P(x_n)$ are all true.

EXAMPLE 15 What is the truth value of $\forall x P(x)$, where $P(x)$ is the statement "$x^2 < 10$" and the domain consists of the positive integers not exceeding 4?

Solution: The statement $\forall x P(x)$ is the same as the conjunction

$$P(1) \wedge P(2) \wedge P(3) \wedge P(4),$$

because the domain consists of the integers 1, 2, 3, and 4. Because $P(4)$, which is the statement "$4^2 < 10$," is false, it follows that $\forall x P(x)$ is false. ◀

Similarly, when the elements of the domain are $x_1, x_2, \ldots, x_n$, where n is a positive integer, the existential quantification $\exists x P(x)$ is the same as the disjunction

$$P(x_1) \vee P(x_2) \vee \cdots \vee P(x_n),$$

because this disjunction is true if and only if at least one of $P(x_1), P(x_2), \ldots, P(x_n)$ is true.

EXAMPLE 16 What is the truth value of $\exists x P(x)$, where $P(x)$ is the statement "$x^2 > 10$" and the universe of discourse consists of the positive integers not exceeding 4?

Solution: Because the domain is $\{1, 2, 3, 4\}$, the proposition $\exists x P(x)$ is the same as the disjunction

$$P(1) \vee P(2) \vee P(3) \vee P(4).$$

Because $P(4)$, which is the statement "$4^2 > 10$," is true, it follows that $\exists x P(x)$ is true. ◀

CONNECTIONS BETWEEN QUANTIFICATION AND LOOPING It is sometimes helpful to think in terms of looping and searching when determining the truth value of a quantification. Suppose that there are n objects in the domain for the variable x. To determine whether $\forall x P(x)$ is true, we can loop through all n values of x to see whether $P(x)$ is always true. If we encounter a value x for which $P(x)$ is false, then we have shown that $\forall x P(x)$ is false. Otherwise, $\forall x P(x)$ is true. To see whether $\exists x P(x)$ is true, we loop through the n values of x searching for a value for which $P(x)$ is true. If we find one, then $\exists x P(x)$ is true. If we never find such an x, then we have determined that $\exists x P(x)$ is false. (Note that this searching procedure does not apply if there are infinitely many values in the domain. However, it is still a useful way of thinking about the truth values of quantifications.)

1.4.5 Quantifiers with Restricted Domains

An abbreviated notation is often used to restrict the domain of a quantifier. In this notation, a condition a variable must satisfy is included after the quantifier. This is illustrated in Example 17. We will also describe other forms of this notation involving set membership in Section 2.1.

EXAMPLE 17 What do the statements $\forall x < 0\,(x^2 > 0)$, $\forall y \neq 0\,(y^3 \neq 0)$, and $\exists z > 0\,(z^2 = 2)$ mean, where the domain in each case consists of the real numbers?

Solution: The statement $\forall x < 0\ (x^2 > 0)$ states that for every real number x with $x < 0$, $x^2 > 0$. That is, it states "The square of a negative real number is positive." This statement is the same as $\forall x(x < 0 \rightarrow x^2 > 0)$.

The statement $\forall y \neq 0\,(y^3 \neq 0)$ states that for every real number y with $y \neq 0$, we have $y^3 \neq 0$. That is, it states "The cube of every nonzero real number is nonzero." This statement is equivalent to $\forall y(y \neq 0 \rightarrow y^3 \neq 0)$.

Finally, the statement $\exists z > 0\,(z^2 = 2)$ states that there exists a real number z with $z > 0$ such that $z^2 = 2$. That is, it states "There is a positive square root of 2." This statement is equivalent to $\exists z(z > 0 \wedge z^2 = 2)$. ◀

Note that the restriction of a universal quantification is the same as the universal quantification of a conditional statement. For instance, $\forall x < 0\ (x^2 > 0)$ is another way of expressing $\forall x(x < 0 \rightarrow x^2 > 0)$. On the other hand, the restriction of an existential quantification is the same as the existential quantification of a conjunction. For instance, $\exists z > 0\ (z^2 = 2)$ is another way of expressing $\exists z(z > 0 \wedge z^2 = 2)$.

1.4.6 Precedence of Quantifiers

The quantifiers $\forall$ and $\exists$ have higher precedence than all logical operators from propositional calculus. For example, $\forall x P(x) \vee Q(x)$ is the disjunction of $\forall x P(x)$ and $Q(x)$. In other words, it means $(\forall x P(x)) \vee Q(x)$ rather than $\forall x(P(x) \vee Q(x))$.

1.4.7 Binding Variables

When a quantifier is used on the variable x, we say that this occurrence of the variable is **bound**. An occurrence of a variable that is not bound by a quantifier or set equal to a particular value is said to be **free**. All the variables that occur in a propositional function must be bound or set equal to a particular value to turn it into a proposition. This can be done using a combination of universal quantifiers, existential quantifiers, and value assignments.

The part of a logical expression to which a quantifier is applied is called the **scope** of this quantifier. Consequently, a variable is free if it is outside the scope of all quantifiers in the formula that specify this variable.

EXAMPLE 18 In the statement $\exists x(x + y = 1)$, the variable x is bound by the existential quantification $\exists x$, but the variable y is free because it is not bound by a quantifier and no value is assigned to this variable. This illustrates that in the statement $\exists x(x + y = 1)$, x is bound, but y is free.

In the statement $\exists x(P(x) \wedge Q(x)) \vee \forall x R(x)$, all variables are bound. The scope of the first quantifier, $\exists x$, is the expression $P(x) \wedge Q(x)$, because $\exists x$ is applied only to $P(x) \wedge Q(x)$ and not to the rest of the statement. Similarly, the scope of the second quantifier, $\forall x$, is the expression $R(x)$. That is, the existential quantifier binds the variable x in $P(x) \wedge Q(x)$ and the universal quantifier $\forall x$ binds the variable x in $R(x)$. Observe that we could have written our statement using two different variables x and y, as $\exists x(P(x) \wedge Q(x)) \vee \forall y R(y)$, because the scopes of the two quantifiers do not overlap. The reader should be aware that in common usage, the same letter is often used to represent variables bound by different quantifiers with scopes that do not overlap. ◀

1.4.8 Logical Equivalences Involving Quantifiers

In Section 1.3 we introduced the notion of logical equivalences of compound propositions. We can extend this notion to expressions involving predicates and quantifiers.

Definition 3 Statements involving predicates and quantifiers are *logically equivalent* if and only if they have the same truth value no matter which predicates are substituted into these statements and which domain of discourse is used for the variables in these propositional functions. We use the notation $S \equiv T$ to indicate that two statements S and T involving predicates and quantifiers are logically equivalent.

Example 19 illustrates how to show that two statements involving predicates and quantifiers are logically equivalent.

EXAMPLE 19 Show that $\forall x(P(x) \wedge Q(x))$ and $\forall xP(x) \wedge \forall xQ(x)$ are logically equivalent (where the same domain is used throughout). This logical equivalence shows that we can distribute a universal quantifier over a conjunction. Furthermore, we can also distribute an existential quantifier over a disjunction. However, we cannot distribute a universal quantifier over a disjunction, nor can we distribute an existential quantifier over a conjunction. (See Exercises 52 and 53.)

Solution: To show that these statements are logically equivalent, we must show that they always take the same truth value, no matter what the predicates P and Q are, and no matter which domain of discourse is used. Suppose that we have particular predicates P and Q, with a common domain. We can show that $\forall x(P(x) \wedge Q(x))$ and $\forall xP(x) \wedge \forall xQ(x)$ are logically equivalent by doing two things. First, we show that if $\forall x(P(x) \wedge Q(x))$ is true, then $\forall xP(x) \wedge \forall xQ(x)$ is true. Second, we show that if $\forall xP(x) \wedge \forall xQ(x)$ is true, then $\forall x(P(x) \wedge Q(x))$ is true.

So, suppose that $\forall x(P(x) \wedge Q(x))$ is true. This means that if a is in the domain, then $P(a) \wedge Q(a)$ is true. Hence, $P(a)$ is true and $Q(a)$ is true. Because $P(a)$ is true and $Q(a)$ is true for every element a in the domain, we can conclude that $\forall xP(x)$ and $\forall xQ(x)$ are both true. This means that $\forall xP(x) \wedge \forall xQ(x)$ is true.

Next, suppose that $\forall xP(x) \wedge \forall xQ(x)$ is true. It follows that $\forall xP(x)$ is true and $\forall xQ(x)$ is true. Hence, if a is in the domain, then $P(a)$ is true and $Q(a)$ is true [because $P(x)$ and $Q(x)$ are both true for all elements in the domain, there is no conflict using the same value of a here]. It follows that for all a, $P(a) \wedge Q(a)$ is true. It follows that $\forall x(P(x) \wedge Q(x))$ is true. We can now conclude that

$$\forall x(P(x) \wedge Q(x)) \equiv \forall xP(x) \wedge \forall xQ(x).$$ ◀

1.4.9 Negating Quantified Expressions

We will often want to consider the negation of a quantified expression. For instance, consider the negation of the statement

"Every student in your class has taken a course in calculus."

This statement is a universal quantification, namely,

$$\forall xP(x),$$

Assessment

where $P(x)$ is the statement "x has taken a course in calculus" and the domain consists of the students in your class. The negation of this statement is "It is not the case that every student in your class has taken a course in calculus." This is equivalent to "There is a student in your class who has not taken a course in calculus." And this is simply the existential quantification of the negation of the original propositional function, namely,

$$\exists x \neg P(x).$$

This example illustrates the following logical equivalence:

$$\neg \forall x P(x) \equiv \exists x \neg P(x).$$

To show that $\neg \forall x P(x)$ and $\exists x \neg P(x)$ are logically equivalent no matter what the propositional function $P(x)$ is and what the domain is, first note that $\neg \forall x P(x)$ is true if and only if $\forall x P(x)$ is false. Next, note that $\forall x P(x)$ is false if and only if there is an element x in the domain for which $P(x)$ is false. This holds if and only if there is an element x in the domain for which $\neg P(x)$ is true. Finally, note that there is an element x in the domain for which $\neg P(x)$ is true if and only if $\exists x \neg P(x)$ is true. Putting these steps together, we can conclude that $\neg \forall x P(x)$ is true if and only if $\exists x \neg P(x)$ is true. It follows that $\neg \forall x P(x)$ and $\exists x \neg P(x)$ are logically equivalent.

Suppose we wish to negate an existential quantification. For instance, consider the proposition "There is a student in this class who has taken a course in calculus." This is the existential quantification

$$\exists x Q(x),$$

where $Q(x)$ is the statement "x has taken a course in calculus." The negation of this statement is the proposition "It is not the case that there is a student in this class who has taken a course in calculus." This is equivalent to "Every student in this class has not taken calculus," which is just the universal quantification of the negation of the original propositional function, or, phrased in the language of quantifiers,

$$\forall x \neg Q(x).$$

This example illustrates the equivalence

$$\neg \exists x Q(x) \equiv \forall x \neg Q(x).$$

To show that $\neg \exists x Q(x)$ and $\forall x \neg Q(x)$ are logically equivalent no matter what $Q(x)$ is and what the domain is, first note that $\neg \exists x Q(x)$ is true if and only if $\exists x Q(x)$ is false. This is true if and only if no x exists in the domain for which $Q(x)$ is true. Next, note that no x exists in the domain for which $Q(x)$ is true if and only if $Q(x)$ is false for every x in the domain. Finally, note that $Q(x)$ is false for every x in the domain if and only if $\neg Q(x)$ is true for all x in the domain, which holds if and only if $\forall x \neg Q(x)$ is true. Putting these steps together, we see that $\neg \exists x Q(x)$ is true if and only if $\forall x \neg Q(x)$ is true. We conclude that $\neg \exists x Q(x)$ and $\forall x \neg Q(x)$ are logically equivalent.

The rules for negations for quantifiers are called **De Morgan's laws for quantifiers**. These rules are summarized in Table 2.

Remark: When the domain of a predicate $P(x)$ consists of n elements, where n is a positive integer greater than one, the rules for negating quantified statements are exactly the same as De Morgan's laws discussed in Section 1.3. This is why these rules are called De Morgan's laws for quantifiers. When the domain has n elements $x_1, x_2, \dots, x_n$, it follows that $\neg \forall x P(x)$ is the same as $\neg(P(x_1) \wedge P(x_2) \wedge \cdots \wedge P(x_n))$, which is equivalent to $\neg P(x_1) \vee \neg P(x_2) \vee \cdots \vee \neg P(x_n)$ by De Morgan's laws, and this is the same as $\exists x \neg P(x)$. Similarly, $\neg \exists x P(x)$ is the same as $\neg(P(x_1) \vee$

TABLE 2 De Morgan's Laws for Quantifiers.

Negation	*Equivalent Statement*	*When Is Negation True?*	*When False?*
$\neg\exists x P(x)$	$\forall x \neg P(x)$	For every x, $P(x)$ is false.	There is an x for which $P(x)$ is true.
$\neg\forall x P(x)$	$\exists x \neg P(x)$	There is an x for which $P(x)$ is false.	$P(x)$ is true for every x.

$P(x_2) \vee \cdots \vee P(x_n))$, which by De Morgan's laws is equivalent to $\neg P(x_1) \wedge \neg P(x_2) \wedge \cdots \wedge \neg P(x_n)$, and this is the same as $\forall x \neg P(x)$.

We illustrate the negation of quantified statements in Examples 20 and 21.

EXAMPLE 20 What are the negations of the statements "There is an honest politician" and "All Americans eat cheeseburgers"?

Solution: Let $H(x)$ denote "x is honest." Then the statement "There is an honest politician" is represented by $\exists x H(x)$, where the domain consists of all politicians. The negation of this statement is $\neg\exists x H(x)$, which is equivalent to $\forall x \neg H(x)$. This negation can be expressed as "Every politician is dishonest." (*Note:* In English, the statement "All politicians are not honest" is ambiguous. In common usage, this statement often means "Not all politicians are honest." Consequently, we do not use this statement to express this negation.)

Extra Examples

Let $C(x)$ denote "x eats cheeseburgers." Then the statement "All Americans eat cheeseburgers" is represented by $\forall x C(x)$, where the domain consists of all Americans. The negation of this statement is $\neg\forall x C(x)$, which is equivalent to $\exists x \neg C(x)$. This negation can be expressed in several different ways, including "Some American does not eat cheeseburgers" and "There is an American who does not eat cheeseburgers." ◀

EXAMPLE 21 What are the negations of the statements $\forall x(x^2 > x)$ and $\exists x(x^2 = 2)$?

Solution: The negation of $\forall x(x^2 > x)$ is the statement $\neg\forall x(x^2 > x)$, which is equivalent to $\exists x \neg(x^2 > x)$. This can be rewritten as $\exists x(x^2 \leq x)$. The negation of $\exists x(x^2 = 2)$ is the statement $\neg\exists x(x^2 = 2)$, which is equivalent to $\forall x \neg(x^2 = 2)$. This can be rewritten as $\forall x(x^2 \neq 2)$. The truth values of these statements depend on the domain. ◀

We use De Morgan's laws for quantifiers in Example 22.

EXAMPLE 22 Show that $\neg\forall x(P(x) \to Q(x))$ and $\exists x(P(x) \wedge \neg Q(x))$ are logically equivalent.

Solution: By De Morgan's law for universal quantifiers, we know that $\neg\forall x(P(x) \to Q(x))$ and $\exists x(\neg(P(x) \to Q(x)))$ are logically equivalent. By the fifth logical equivalence in Table 7 in Section 1.3, we know that $\neg(P(x) \to Q(x))$ and $P(x) \wedge \neg Q(x)$ are logically equivalent for every x. Because we can substitute one logically equivalent expression for another in a logical equivalence, it follows that $\neg\forall x(P(x) \to Q(x))$ and $\exists x(P(x) \wedge \neg Q(x))$ are logically equivalent. ◀

1.4.10 Translating from English into Logical Expressions

Translating sentences in English (or other natural languages) into logical expressions is a crucial task in mathematics, logic programming, artificial intelligence, software engineering, and many other disciplines. We began studying this topic in Section 1.1, where we used propositions to express sentences in logical expressions. In that discussion, we purposely avoided sentences whose translations required predicates and quantifiers. Translating from English to logical expressions becomes even more complex when quantifiers are needed. Furthermore, there can be many ways to translate a particular sentence. (As a consequence, there is no "cookbook" approach that can be followed step by step.) We will use some examples to illustrate how to translate sentences from English into logical expressions. The goal in this translation is to produce simple and useful logical expressions. In this section, we restrict ourselves to sentences that can be translated into logical expressions using a single quantifier; in the next section, we will look at more complicated sentences that require multiple quantifiers.

EXAMPLE 23 Express the statement "Every student in this class has studied calculus" using predicates and quantifiers.

Extra Examples

Solution: First, we rewrite the statement so that we can clearly identify the appropriate quantifiers to use. Doing so, we obtain:

"For every student in this class, that student has studied calculus."

Next, we introduce a variable x so that our statement becomes

"For every student x in this class, x has studied calculus."

Continuing, we introduce $C(x)$, which is the statement "x has studied calculus." Consequently, if the domain for x consists of the students in the class, we can translate our statement as $\forall x C(x)$.

However, there are other correct approaches; different domains of discourse and other predicates can be used. The approach we select depends on the subsequent reasoning we want to carry out. For example, we may be interested in a wider group of people than only those in this class. If we change the domain to consist of all people, we will need to express our statement as

"For every person x, if person x is a student in this class, then x has studied calculus."

If $S(x)$ represents the statement that person x is in this class, we see that our statement can be expressed as $\forall x(S(x) \to C(x))$. [*Caution!* Our statement *cannot* be expressed as $\forall x(S(x) \land C(x))$ because this statement says that all people are students in this class and have studied calculus!]

Finally, when we are interested in the background of people in subjects besides calculus, we may prefer to use the two-variable quantifier $Q(x, y)$ for the statement "Student x has studied subject y." Then we would replace $C(x)$ by $Q(x, \text{calculus})$ in both approaches to obtain $\forall x Q(x, \text{calculus})$ or $\forall x(S(x) \to Q(x, \text{calculus}))$. ◀

In Example 23 we displayed different approaches for expressing the same statement using predicates and quantifiers. However, we should always adopt the simplest approach that is adequate for use in subsequent reasoning.

EXAMPLE 24 Express the statements "Some student in this class has visited Mexico" and "Every student in this class has visited either Canada or Mexico" using predicates and quantifiers.

Solution: The statement "Some student in this class has visited Mexico" means that

"There is a student in this class with the property that the student has visited Mexico."

We can introduce a variable x, so that our statement becomes

"There is a student x in this class having the property that x has visited Mexico."

We introduce $M(x)$, which is the statement "x has visited Mexico." If the domain for x consists of the students in this class, we can translate this first statement as $\exists x M(x)$.

However, if we are interested in people other than those in this class, we look at the statement a little differently. Our statement can be expressed as

> "There is a person x having the properties that x is a student in this class and x has visited Mexico."

In this case, the domain for the variable x consists of all people. We introduce $S(x)$ to represent "x is a student in this class." Our solution becomes $\exists x(S(x) \wedge M(x))$ because the statement is that there is a person x who is a student in this class and who has visited Mexico. [*Caution!* Our statement cannot be expressed as $\exists x(S(x) \rightarrow M(x))$, which is true when there is someone not in the class because, in that case, for such a person x, $S(x) \rightarrow M(x)$ becomes either **F** $\rightarrow$ **T** or **F** $\rightarrow$ **F**, both of which are true.]

Similarly, the second statement can be expressed as

> "For every x in this class, x has the property that x has visited Mexico or x has visited Canada."

(Note that we are assuming the inclusive, rather than the exclusive, or here.) We let $C(x)$ be "x has visited Canada." Following our earlier reasoning, we see that if the domain for x consists of the students in this class, this second statement can be expressed as $\forall x(C(x) \vee M(x))$. However, if the domain for x consists of all people, our statement can be expressed as

> "For every person x, if x is a student in this class, then x has visited Mexico or x has visited Canada."

In this case, the statement can be expressed as $\forall x(S(x) \rightarrow (C(x) \vee M(x)))$.

Instead of using $M(x)$ and $C(x)$ to represent that x has visited Mexico and x has visited Canada, respectively, we could use a two-place predicate $V(x, y)$ to represent "x has visited country y." In this case, $V(x, \text{Mexico})$ and $V(x, \text{Canada})$ would have the same meaning as $M(x)$ and $C(x)$ and could replace them in our answers. If we are working with many statements that involve people visiting different countries, we might prefer to use this two-variable approach. Otherwise, for simplicity, we would stick with the one-variable predicates $M(x)$ and $C(x)$. ◀

1.4.11 Using Quantifiers in System Specifications

In Section 1.2 we used propositions to represent system specifications. However, many system specifications involve predicates and quantifications. This is illustrated in Example 25.

EXAMPLE 25 Use predicates and quantifiers to express the system specifications "Every mail message larger than one megabyte will be compressed" and "If a user is active, at least one network link will be available."

Extra Examples

Solution: Let $S(m, y)$ be "Mail message m is larger than y megabytes," where the variable m has the domain of all mail messages and the variable y is a positive real number, and let $C(m)$ denote "Mail message m will be compressed." Then the specification "Every mail message larger than one megabyte will be compressed" can be represented as $\forall m(S(m, 1) \rightarrow C(m))$.

Remember the rules of precedence for quantifiers and logical connectives!

Let $A(u)$ represent "User u is active," where the variable u has the domain of all users, let $S(n, x)$ denote "Network link n is in state x," where n has the domain of all network links and x has the domain of all possible states for a network link. Then the specification "If a user is active, at least one network link will be available" can be represented by $\exists u A(u) \rightarrow \exists n S(n, \text{available})$. ◀

1.4.12 Examples from Lewis Carroll

Lewis Carroll (really C. L. Dodgson writing under a pseudonym), the author of *Alice in Wonderland*, is also the author of several works on symbolic logic. His books contain many examples of reasoning using quantifiers. Examples 26 and 27 come from his book *Symbolic Logic;* other examples from that book are given in the exercises at the end of this section. These examples illustrate how quantifiers are used to express various types of statements.

EXAMPLE 26 Consider these statements. The first two are called *premises* and the third is called the *conclusion*. The entire set is called an *argument*.

"All lions are fierce."
"Some lions do not drink coffee."
"Some fierce creatures do not drink coffee."

(In Section 1.6 we will discuss the issue of determining whether the conclusion is a valid consequence of the premises. In this example, it is.) Let $P(x)$, $Q(x)$, and $R(x)$ be the statements "x is a lion," "x is fierce," and "x drinks coffee," respectively. Assuming that the domain consists of all creatures, express the statements in the argument using quantifiers and $P(x)$, $Q(x)$, and $R(x)$.

Solution: We can express these statements as

$$\forall x(P(x) \to Q(x)).$$
$$\exists x(P(x) \land \neg R(x)).$$
$$\exists x(Q(x) \land \neg R(x)).$$

Notice that the second statement cannot be written as $\exists x(P(x) \to \neg R(x))$. The reason is that $P(x) \to \neg R(x)$ is true whenever x is not a lion, so that $\exists x(P(x) \to \neg R(x))$ is true as long as there is at least one creature that is not a lion, even if every lion drinks coffee. Similarly, the third statement cannot be written as

$$\exists x(Q(x) \to \neg R(x)).$$ ◀

EXAMPLE 27 Consider these statements, of which the first three are premises and the fourth is a valid conclusion.

"All hummingbirds are richly colored."
"No large birds live on honey."
"Birds that do not live on honey are dull in color."
"Hummingbirds are small."

Links

Oscar Gustav Rejlander/Hulton Archive/Getty Images

CHARLES LUTWIDGE DODGSON (1832–1898) We know Charles Dodgson as Lewis Carroll—the pseudonym he used in his literary works. Dodgson, the son of a clergyman, was the third of 11 children, all of whom stuttered. He was uncomfortable in the company of adults and is said to have spoken without stuttering only to young girls, many of whom he entertained, corresponded with, and photographed (sometimes in poses that today would be considered inappropriate). Although attracted to young girls, he was extremely puritanical and religious. His friendship with the three young daughters of Dean Liddell led to his writing *Alice in Wonderland*, which brought him money and fame.

Dodgson graduated from Oxford in 1854 and obtained his master of arts degree in 1857. He was appointed lecturer in mathematics at Christ Church College, Oxford, in 1855. He was ordained in the Church of England in 1861 but never practiced his ministry. His writings published under this real name include articles and books on geometry, determinants, and the mathematics of tournaments and elections. (He also used the pseudonym Lewis Carroll for his many works on recreational logic.)

Let $P(x)$, $Q(x)$, $R(x)$, and $S(x)$ be the statements "x is a hummingbird," "x is large," "x lives on honey," and "x is richly colored," respectively. Assuming that the domain consists of all birds, express the statements in the argument using quantifiers and $P(x)$, $Q(x)$, $R(x)$, and $S(x)$.

Solution: We can express the statements in the argument as

$$\forall x(P(x) \to S(x)).$$
$$\neg\exists x(Q(x) \land R(x)).$$
$$\forall x(\neg R(x) \to \neg S(x)).$$
$$\forall x(P(x) \to \neg Q(x)).$$

(Note we have assumed that "small" is the same as "not large" and that "dull in color" is the same as "not richly colored." To show that the fourth statement is a valid conclusion of the first three, we need to use rules of inference that will be discussed in Section 1.6.) ◀

1.4.13 Logic Programming

An important type of programming language is designed to reason using the rules of predicate logic. Prolog (from *Pro*gramming in *Log*ic), developed in the 1970s by computer scientists working in the area of artificial intelligence, is an example of such a language. Prolog programs include a set of declarations consisting of two types of statements, **Prolog facts** and **Prolog rules**. Prolog facts define predicates by specifying the elements that satisfy these predicates. Prolog rules are used to define new predicates using those already defined by Prolog facts. Example 28 illustrates these notions.

Links

EXAMPLE 28 Consider a Prolog program given facts telling it the instructor of each class and in which classes students are enrolled. The program uses these facts to answer queries concerning the professors who teach particular students. Such a program could use the predicates *instructor*(p, c) and *enrolled*(s, c) to represent that professor p is the instructor of course c and that

Links

Alain Colmerauer (Colette Colmerauer)

ALAIN COLMERAUER (1941–2017) A French computer scientist and mathematician known as the father of Prolog, a programming language based on predicate logic, and a pioneer of artificial intelligence, was born in Carcassonne, France. He earned a degree in computer science from the Polytechnic Institute of Grenoble in 1963, and a doctorate from the Ensimag in Grenoble in 1967. He served as an assistant professor at the University of Montreal from 1967–1970, and was promoted to associate and full professor. From 1993 to 1995, he headed the Laboratory of Informatics of Marseille. In 2006 he retired as an emeritus professor, but remained a member of the artificial intelligence taskforce in Luminy.

Colmerauer played a key role in the development and applications of Prolog. At the University of Montreal he invented Q-Systems, an early linguistic formalism used in a prototype of machine translation. He was the center of a research group that originally conceived of Prolog. In 1972, Colmerauer and Philippe Roussel developed the first Prolog system; its first implementations were in interpreters. In 1984, Colmerauer created the company Prologia to exploit the development of Prolog III, later purchased by Experian, which made it part of the company's decision support business. In 1987, he founded the field of constraint logic programming, a form of programming in which logic programming is extended to include concepts from constraint satisfaction.

Colmerauer won many awards. We mention a few of these here. In 1985 he received the Michel Monpetit Award from the French Academy of Sciences. In 1986, he was made knight in the legion of honor Chevalier by the French government. He became a fellow of the American Association of Artificial Intelligence in 1991, and in 1997 the Association of Logic Programming bestowed upon him and other researchers the title of Founder of Logic Programming. He was also a Correspondent of the French Academy of Sciences in the area of Mathematics.

Among his colleagues and students, he was noted for his great sense of humor. He supervised the Ph.D. work of 51 students. For enjoyment, he enjoyed sailing the blue waters of the Mediterranean Sea. He and his wife Colette Coursaget had three children.

student *s* is enrolled in course *c*, respectively. For example, the Prolog facts in such a program might include:

```
instructor(chan,math273)
instructor(patel,ee222)
instructor(grossman,cs301)
enrolled(kevin,math273)
enrolled(juana,ee222)
enrolled(juana,cs301)
enrolled(kiko,math273)
enrolled(kiko,cs301)
```

(Lowercase letters have been used for entries because Prolog considers names beginning with an uppercase letter to be variables.)

A new predicate *teaches*(*p, s*), representing that professor *p* teaches student *s*, can be defined using the Prolog rule

```
teaches(P,S) :- instructor(P,C), enrolled(S,C)
```

which means that *teaches*(*p, s*) is true if there exists a class *c* such that professor *p* is the instructor of class *c* and student *s* is enrolled in class *c*. (Note that a comma is used to represent a conjunction of predicates in Prolog. Similarly, a semicolon is used to represent a disjunction of predicates.)

Prolog answers queries using the facts and rules it is given. For example, using the facts and rules listed, the query

```
?enrolled(kevin,math273)
```

produces the response

```
yes
```

because the fact *enrolled*(kevin, math273) was provided as input. The query

```
?enrolled(X,math273)
```

produces the response

```
kevin
kiko
```

To produce this response, Prolog determines all possible values of *X* for which *enrolled*(*X*, math273) has been included as a Prolog fact. Similarly, to find all the professors who are instructors in classes being taken by Juana, we use the query

```
?teaches(X,juana)
```

This query returns

```
patel
grossman
```

Prolog has played an important role in work on natural language understanding, symbolic computing, and probabilistic programming. Moreover, Prolog was used to build the Watson question answering system built by IBM, the winner of the *Jeopardy!* Man vs. Machine Challenge where it beat almost all human competitors (with Congressman Rush Holt a noteworthy exception). ◀

Exercises

1. Let $P(x)$ denote the statement "$x \leq 4$." What are these truth values?

a) $P(0)$ **b)** $P(4)$ **c)** $P(6)$

2. Let $P(x)$ be the statement "The word x contains the letter a." What are these truth values?

a) $P(\text{orange})$ **b)** $P(\text{lemon})$
c) $P(\text{true})$ **d)** $P(\text{false})$

3. Let $Q(x, y)$ denote the statement "x is the capital of y." What are these truth values?

a) $Q(\text{Denver, Colorado})$
b) $Q(\text{Detroit, Michigan})$
c) $Q(\text{Massachusetts, Boston})$
d) $Q(\text{New York, New York})$

4. State the value of x after the statement **if** $P(x)$ **then** $x := 1$ is executed, where $P(x)$ is the statement "$x > 1$," if the value of x when this statement is reached is

a) $x = 0$. **b)** $x = 1$.
c) $x = 2$.

5. Let $P(x)$ be the statement "x spends more than five hours every weekday in class," where the domain for x consists of all students. Express each of these quantifications in English.

a) $\exists x P(x)$ **b)** $\forall x P(x)$
c) $\exists x \neg P(x)$ **d)** $\forall x \neg P(x)$

6. Let $N(x)$ be the statement "x has visited North Dakota," where the domain consists of the students in your school. Express each of these quantifications in English.

a) $\exists x N(x)$ **b)** $\forall x N(x)$ **c)** $\neg \exists x N(x)$
d) $\exists x \neg N(x)$ **e)** $\neg \forall x N(x)$ **f)** $\forall x \neg N(x)$

7. Translate these statements into English, where $C(x)$ is "x is a comedian" and $F(x)$ is "x is funny" and the domain consists of all people.

a) $\forall x(C(x) \rightarrow F(x))$ **b)** $\forall x(C(x) \wedge F(x))$
c) $\exists x(C(x) \rightarrow F(x))$ **d)** $\exists x(C(x) \wedge F(x))$

8. Translate these statements into English, where $R(x)$ is "x is a rabbit" and $H(x)$ is "x hops" and the domain consists of all animals.

a) $\forall x(R(x) \rightarrow H(x))$ **b)** $\forall x(R(x) \wedge H(x))$
c) $\exists x(R(x) \rightarrow H(x))$ **d)** $\exists x(R(x) \wedge H(x))$

9. Let $P(x)$ be the statement "x can speak Russian" and let $Q(x)$ be the statement "x knows the computer language C++." Express each of these sentences in terms of $P(x)$, $Q(x)$, quantifiers, and logical connectives. The domain for quantifiers consists of all students at your school.

a) There is a student at your school who can speak Russian and who knows C++.

b) There is a student at your school who can speak Russian but who doesn't know C++.

c) Every student at your school either can speak Russian or knows C++.

d) No student at your school can speak Russian or knows C++.

10. Let $C(x)$ be the statement "x has a cat," let $D(x)$ be the statement "x has a dog," and let $F(x)$ be the statement "x has a ferret." Express each of these statements in terms of $C(x)$, $D(x)$, $F(x)$, quantifiers, and logical connectives. Let the domain consist of all students in your class.

a) A student in your class has a cat, a dog, and a ferret.

b) All students in your class have a cat, a dog, or a ferret.

c) Some student in your class has a cat and a ferret, but not a dog.

d) No student in your class has a cat, a dog, and a ferret.

e) For each of the three animals, cats, dogs, and ferrets, there is a student in your class who has this animal as a pet.

11. Let $P(x)$ be the statement "$x = x^2$." If the domain consists of the integers, what are these truth values?

a) $P(0)$ **b)** $P(1)$ **c)** $P(2)$
d) $P(-1)$ **e)** $\exists x P(x)$ **f)** $\forall x P(x)$

12. Let $Q(x)$ be the statement "$x + 1 > 2x$." If the domain consists of all integers, what are these truth values?

a) $Q(0)$ **b)** $Q(-1)$ **c)** $Q(1)$
d) $\exists x Q(x)$ **e)** $\forall x Q(x)$ **f)** $\exists x \neg Q(x)$
g) $\forall x \neg Q(x)$

13. Determine the truth value of each of these statements if the domain consists of all integers.

a) $\forall n(n + 1 > n)$ **b)** $\exists n(2n = 3n)$
c) $\exists n(n = -n)$ **d)** $\forall n(3n \leq 4n)$

14. Determine the truth value of each of these statements if the domain consists of all real numbers.

a) $\exists x(x^3 = -1)$ **b)** $\exists x(x^4 < x^2)$
c) $\forall x((-x)^2 = x^2)$ **d)** $\forall x(2x > x)$

15. Determine the truth value of each of these statements if the domain for all variables consists of all integers.

a) $\forall n(n^2 \geq 0)$ **b)** $\exists n(n^2 = 2)$
c) $\forall n(n^2 \geq n)$ **d)** $\exists n(n^2 < 0)$

16. Determine the truth value of each of these statements if the domain of each variable consists of all real numbers.

a) $\exists x(x^2 = 2)$ **b)** $\exists x(x^2 = -1)$
c) $\forall x(x^2 + 2 \geq 1)$ **d)** $\forall x(x^2 \neq x)$

17. Suppose that the domain of the propositional function $P(x)$ consists of the integers 0, 1, 2, 3, and 4. Write out each of these propositions using disjunctions, conjunctions, and negations.

a) $\exists x P(x)$ **b)** $\forall x P(x)$ **c)** $\exists x \neg P(x)$
d) $\forall x \neg P(x)$ **e)** $\neg \exists x P(x)$ **f)** $\neg \forall x P(x)$

18. Suppose that the domain of the propositional function $P(x)$ consists of the integers -2, -1, 0, 1, and 2. Write out each of these propositions using disjunctions, conjunctions, and negations.

a) $\exists x P(x)$ **b)** $\forall x P(x)$ **c)** $\exists x \neg P(x)$
d) $\forall x \neg P(x)$ **e)** $\neg \exists x P(x)$ **f)** $\neg \forall x P(x)$

19. Suppose that the domain of the propositional function $P(x)$ consists of the integers 1, 2, 3, 4, and 5. Express these statements without using quantifiers, instead using only negations, disjunctions, and conjunctions.

a) $\exists x P(x)$ **b)** $\forall x P(x)$
c) $\neg\exists x P(x)$ **d)** $\neg\forall x P(x)$
e) $\forall x((x \neq 3) \rightarrow P(x)) \vee \exists x \neg P(x)$

20. Suppose that the domain of the propositional function $P(x)$ consists of -5, -3, -1, 1, 3, and 5. Express these statements without using quantifiers, instead using only negations, disjunctions, and conjunctions.

a) $\exists x P(x)$ **b)** $\forall x P(x)$
c) $\forall x((x \neq 1) \rightarrow P(x))$
d) $\exists x((x \geq 0) \wedge P(x))$
e) $\exists x(\neg P(x)) \wedge \forall x((x < 0) \rightarrow P(x))$

21. For each of these statements find a domain for which the statement is true and a domain for which the statement is false.

a) Everyone is studying discrete mathematics.
b) Everyone is older than 21 years.
c) Every two people have the same mother.
d) No two different people have the same grandmother.

22. For each of these statements find a domain for which the statement is true and a domain for which the statement is false.

a) Everyone speaks Hindi.
b) There is someone older than 21 years.
c) Every two people have the same first name.
d) Someone knows more than two other people.

23. Translate in two ways each of these statements into logical expressions using predicates, quantifiers, and logical connectives. First, let the domain consist of the students in your class and second, let it consist of all people.

a) Someone in your class can speak Hindi.
b) Everyone in your class is friendly.
c) There is a person in your class who was not born in California.
d) A student in your class has been in a movie.
e) No student in your class has taken a course in logic programming.

24. Translate in two ways each of these statements into logical expressions using predicates, quantifiers, and logical connectives. First, let the domain consist of the students in your class and second, let it consist of all people.

a) Everyone in your class has a cellular phone.
b) Somebody in your class has seen a foreign movie.
c) There is a person in your class who cannot swim.
d) All students in your class can solve quadratic equations.
e) Some student in your class does not want to be rich.

25. Translate each of these statements into logical expressions using predicates, quantifiers, and logical connectives.

a) No one is perfect.
b) Not everyone is perfect.
c) All your friends are perfect.
d) At least one of your friends is perfect.
e) Everyone is your friend and is perfect.
f) Not everybody is your friend or someone is not perfect.

26. Translate each of these statements into logical expressions in three different ways by varying the domain and by using predicates with one and with two variables.

a) Someone in your school has visited Uzbekistan.
b) Everyone in your class has studied calculus and C++.
c) No one in your school owns both a bicycle and a motorcycle.
d) There is a person in your school who is not happy.
e) Everyone in your school was born in the twentieth century.

27. Translate each of these statements into logical expressions in three different ways by varying the domain and by using predicates with one and with two variables.

a) A student in your school has lived in Vietnam.
b) There is a student in your school who cannot speak Hindi.
c) A student in your school knows Java, Prolog, and C++.
d) Everyone in your class enjoys Thai food.
e) Someone in your class does not play hockey.

28. Translate each of these statements into logical expressions using predicates, quantifiers, and logical connectives.

a) Something is not in the correct place.
b) All tools are in the correct place and are in excellent condition.
c) Everything is in the correct place and in excellent condition.
d) Nothing is in the correct place and is in excellent condition.
e) One of your tools is not in the correct place, but it is in excellent condition.

29. Express each of these statements using logical operators, predicates, and quantifiers.

a) Some propositions are tautologies.
b) The negation of a contradiction is a tautology.
c) The disjunction of two contingencies can be a tautology.
d) The conjunction of two tautologies is a tautology.

30. Suppose the domain of the propositional function $P(x, y)$ consists of pairs x and y, where x is 1, 2, or 3 and y is 1, 2, or 3. Write out these propositions using disjunctions and conjunctions.

a) $\exists x\, P(x, 3)$ **b)** $\forall y\, P(1, y)$
c) $\exists y \neg P(2, y)$ **d)** $\forall x \neg P(x, 2)$

31. Suppose that the domain of $Q(x, y, z)$ consists of triples x, y, z, where $x = 0$, 1, or 2, $y = 0$ or 1, and $z = 0$ or 1. Write out these propositions using disjunctions and conjunctions.

a) $\forall y Q(0, y, 0)$ **b)** $\exists x Q(x, 1, 1)$
c) $\exists z \neg Q(0, 0, z)$ **d)** $\exists x \neg Q(x, 0, 1)$

32. Express each of these statements using quantifiers. Then form the negation of the statement so that no negation is to the left of a quantifier. Next, express the negation in simple English. (Do not simply use the phrase "It is not the case that.")

a) All dogs have fleas.
b) There is a horse that can add.
c) Every koala can climb.
d) No monkey can speak French.
e) There exists a pig that can swim and catch fish.

33. Express each of these statements using quantifiers. Then form the negation of the statement, so that no negation is to the left of a quantifier. Next, express the negation in simple English. (Do not simply use the phrase "It is not the case that.")

a) Some old dogs can learn new tricks.
b) No rabbit knows calculus.
c) Every bird can fly.
d) There is no dog that can talk.
e) There is no one in this class who knows French and Russian.

34. Express the negation of these propositions using quantifiers, and then express the negation in English.

a) Some drivers do not obey the speed limit.
b) All Swedish movies are serious.
c) No one can keep a secret.
d) There is someone in this class who does not have a good attitude.

35. Express the negation of each of these statements in terms of quantifiers without using the negation symbol.

a) $\forall x(x > 1)$
b) $\forall x(x \leq 2)$
c) $\exists x(x \geq 4)$
d) $\exists x(x < 0)$
e) $\forall x((x < -1) \vee (x > 2))$
f) $\exists x((x < 4) \vee (x > 7))$

36. Express the negation of each of these statements in terms of quantifiers without using the negation symbol.

a) $\forall x(-2 < x < 3)$
b) $\forall x(0 \leq x < 5)$
c) $\exists x(-4 \leq x \leq 1)$
d) $\exists x(-5 < x < -1)$

37. Find a counterexample, if possible, to these universally quantified statements, where the domain for all variables consists of all integers.

a) $\forall x(x^2 \geq x)$
b) $\forall x(x > 0 \vee x < 0)$
c) $\forall x(x = 1)$

38. Find a counterexample, if possible, to these universally quantified statements, where the domain for all variables consists of all real numbers.

a) $\forall x(x^2 \neq x)$
b) $\forall x(x^2 \neq 2)$
c) $\forall x(|x| > 0)$

39. Express each of these statements using predicates and quantifiers.

a) A passenger on an airline qualifies as an elite flyer if the passenger flies more than 25,000 miles in a year or takes more than 25 flights during that year.
b) A man qualifies for the marathon if his best previous time is less than 3 hours and a woman qualifies for the marathon if her best previous time is less than 3.5 hours.
c) A student must take at least 60 course hours, or at least 45 course hours and write a master's thesis, and receive a grade no lower than a B in all required courses, to receive a master's degree.
d) There is a student who has taken more than 21 credit hours in a semester and received all A's.

Exercises 40–44 deal with the translation between system specification and logical expressions involving quantifiers.

40. Translate these system specifications into English, where the predicate $S(x, y)$ is "x is in state y" and where the domain for x and y consists of all systems and all possible states, respectively.

a) $\exists x S(x, \text{open})$
b) $\forall x(S(x, \text{malfunctioning}) \vee S(x, \text{diagnostic}))$
c) $\exists x S(x, \text{open}) \vee \exists x S(x, \text{diagnostic})$
d) $\exists x \neg S(x, \text{available})$
e) $\forall x \neg S(x, \text{working})$

41. Translate these specifications into English, where $F(p)$ is "Printer p is out of service," $B(p)$ is "Printer p is busy," $L(j)$ is "Print job j is lost," and $Q(j)$ is "Print job j is queued."

a) $\exists p(F(p) \wedge B(p)) \rightarrow \exists j L(j)$
b) $\forall p B(p) \rightarrow \exists j Q(j)$
c) $\exists j(Q(j) \wedge L(j)) \rightarrow \exists p F(p)$
d) $(\forall p B(p) \wedge \forall j Q(j)) \rightarrow \exists j L(j)$

42. Express each of these system specifications using predicates, quantifiers, and logical connectives.

a) When there is less than 30 megabytes free on the hard disk, a warning message is sent to all users.
b) No directories in the file system can be opened and no files can be closed when system errors have been detected.
c) The file system cannot be backed up if there is a user currently logged on.
d) Video on demand can be delivered when there are at least 8 megabytes of memory available and the connection speed is at least 56 kilobits per second.

43. Express each of these system specifications using predicates, quantifiers, and logical connectives.

a) At least one mail message, among the nonempty set of messages, can be saved if there is a disk with more than 10 kilobytes of free space.
b) Whenever there is an active alert, all queued messages are transmitted.

c) The diagnostic monitor tracks the status of all systems except the main console.

d) Each participant on the conference call whom the host of the call did not put on a special list was billed.

44. Express each of these system specifications using predicates, quantifiers, and logical connectives.

a) Every user has access to an electronic mailbox.

b) The system mailbox can be accessed by everyone in the group if the file system is locked.

c) The firewall is in a diagnostic state only if the proxy server is in a diagnostic state.

d) At least one router is functioning normally if the throughput is between 100 kbps and 500 kbps and the proxy server is not in diagnostic mode.

45. Determine whether $\forall x(P(x) \to Q(x))$ and $\forall xP(x) \to \forall xQ(x)$ are logically equivalent. Justify your answer.

46. Determine whether $\forall x(P(x) \leftrightarrow Q(x))$ and $\forall x\ P(x) \leftrightarrow \forall xQ(x)$ are logically equivalent. Justify your answer.

47. Show that $\exists x(P(x) \vee Q(x))$ and $\exists xP(x) \vee \exists xQ(x)$ are logically equivalent.

Exercises 48–51 establish rules for **null quantification** that we can use when a quantified variable does not appear in part of a statement.

48. Establish these logical equivalences, where x does not occur as a free variable in A. Assume that the domain is nonempty.

a) $(\forall xP(x)) \vee A \equiv \forall x(P(x) \vee A)$

b) $(\exists xP(x)) \vee A \equiv \exists x(P(x) \vee A)$

49. Establish these logical equivalences, where x does not occur as a free variable in A. Assume that the domain is nonempty.

a) $(\forall xP(x)) \wedge A \equiv \forall x(P(x) \wedge A)$

b) $(\exists xP(x)) \wedge A \equiv \exists x(P(x) \wedge A)$

50. Establish these logical equivalences, where x does not occur as a free variable in A. Assume that the domain is nonempty.

a) $\forall x(A \to P(x)) \equiv A \to \forall xP(x)$

b) $\exists x(A \to P(x)) \equiv A \to \exists xP(x)$

51. Establish these logical equivalences, where x does not occur as a free variable in A. Assume that the domain is nonempty.

a) $\forall x(P(x) \to A) \equiv \exists xP(x) \to A$

b) $\exists x(P(x) \to A) \equiv \forall xP(x) \to A$

52. Show that $\forall xP(x) \vee \forall xQ(x)$ and $\forall x(P(x) \vee Q(x))$ are not logically equivalent.

53. Show that $\exists xP(x) \wedge \exists xQ(x)$ and $\exists x(P(x) \wedge Q(x))$ are not logically equivalent.

54. As mentioned in the text, the notation $\exists!xP(x)$ denotes

"There exists a unique x such that $P(x)$ is true."

If the domain consists of all integers, what are the truth values of these statements?

a) $\exists!x(x > 1)$

b) $\exists!x(x^2 = 1)$

c) $\exists!x(x + 3 = 2x)$

d) $\exists!x(x = x + 1)$

55. What are the truth values of these statements?

a) $\exists!xP(x) \to \exists xP(x)$

b) $\forall xP(x) \to \exists!xP(x)$

c) $\exists!x\neg P(x) \to \neg\forall xP(x)$

56. Write out $\exists!xP(x)$, where the domain consists of the integers 1, 2, and 3, in terms of negations, conjunctions, and disjunctions.

57. Given the Prolog facts in Example 28, what would Prolog return given these queries?

a) `?instructor(chan,math273)`

b) `?instructor(patel,cs301)`

c) `?enrolled(X,cs301)`

d) `?enrolled(kiko,Y)`

e) `?teaches(grossman,Y)`

58. Given the Prolog facts in Example 28, what would Prolog return when given these queries?

a) `?enrolled(kevin,ee222)`

b) `?enrolled(kiko,math273)`

c) `?instructor(grossman,X)`

d) `?instructor(X,cs301)`

e) `?teaches(X,kevin)`

59. Suppose that Prolog facts are used to define the predicates *mother*(M, Y) and *father*(F, X), which represent that M is the mother of Y and F is the father of X, respectively. Give a Prolog rule to define the predicate *sibling*(X, Y), which represents that X and Y are siblings (that is, have the same mother and the same father).

60. Suppose that Prolog facts are used to define the predicates *mother*(M, Y) and *father*(F, X), which represent that M is the mother of Y and F is the father of X, respectively. Give a Prolog rule to define the predicate *grandfather*(X, Y), which represents that X is the grandfather of Y. [*Hint:* You can write a disjunction in Prolog either by using a semicolon to separate predicates or by putting these predicates on separate lines.]

Exercises 61–64 are based on questions found in the book *Symbolic Logic* by Lewis Carroll.

61. Let $P(x)$, $Q(x)$, and $R(x)$ be the statements "x is a professor," "x is ignorant," and "x is vain," respectively. Express each of these statements using quantifiers; logical connectives; and $P(x)$, $Q(x)$, and $R(x)$, where the domain consists of all people.

a) No professors are ignorant.

b) All ignorant people are vain.

c) No professors are vain.

d) Does (c) follow from (a) and (b)?

62. Let $P(x)$, $Q(x)$, and $R(x)$ be the statements "x is a clear explanation," "x is satisfactory," and "x is an excuse," respectively. Suppose that the domain for x consists of all English text. Express each of these statements using quantifiers, logical connectives, and $P(x)$, $Q(x)$, and $R(x)$.

a) All clear explanations are satisfactory.

b) Some excuses are unsatisfactory.

c) Some excuses are not clear explanations.

***d)** Does (c) follow from (a) and (b)?

63. Let $P(x)$, $Q(x)$, $R(x)$, and $S(x)$ be the statements "x is a baby," "x is logical," "x is able to manage a crocodile," and "x is despised," respectively. Suppose that the domain consists of all people. Express each of these statements using quantifiers; logical connectives; and $P(x)$, $Q(x)$, $R(x)$, and $S(x)$.

a) Babies are illogical.
b) Nobody is despised who can manage a crocodile.
c) Illogical persons are despised.
d) Babies cannot manage crocodiles.
***e)** Does (d) follow from (a), (b), and (c)? If not, is there a correct conclusion?

64. Let $P(x)$, $Q(x)$, $R(x)$, and $S(x)$ be the statements "x is a duck," "x is one of my poultry," "x is an officer," and "x is willing to waltz," respectively. Express each of these statements using quantifiers; logical connectives; and $P(x)$, $Q(x)$, $R(x)$, and $S(x)$.

a) No ducks are willing to waltz.
b) No officers ever decline to waltz.
c) All my poultry are ducks.
d) My poultry are not officers.
***e)** Does (d) follow from (a), (b), and (c)? If not, is there a correct conclusion?

1.5 Nested Quantifiers

1.5.1 Introduction

In Section 1.4 we defined the existential and universal quantifiers and showed how they can be used to represent mathematical statements. We also explained how they can be used to translate English sentences into logical expressions. However, in Section 1.4 we avoided **nested quantifiers**, where one quantifier is within the scope of another, such as

$$\forall x \exists y(x + y = 0).$$

Note that everything within the scope of a quantifier can be thought of as a propositional function. For example,

$$\forall x \exists y(x + y = 0)$$

is the same thing as $\forall x Q(x)$, where $Q(x)$ is $\exists y P(x, y)$, where $P(x, y)$ is $x + y = 0$.

Nested quantifiers commonly occur in mathematics and computer science. Although nested quantifiers can sometimes be difficult to understand, the rules we have already studied in Section 1.4 can help us use them. In this section we will gain experience working with nested quantifiers. We will see how to use nested quantifiers to express mathematical statements such as "The sum of two positive integers is always positive." We will show how nested quantifiers can be used to translate English sentences such as "Everyone has exactly one best friend" into logical statements. Moreover, we will gain experience working with the negations of statements involving nested quantifiers.

1.5.2 Understanding Statements Involving Nested Quantifiers

To understand statements involving nested quantifiers, we need to unravel what the quantifiers and predicates that appear mean. This is illustrated in Examples 1 and 2.

EXAMPLE 1 Assume that the domain for the variables x and y consists of all real numbers. The statement

$$\forall x \forall y(x + y = y + x)$$

Extra Examples

says that $x + y = y + x$ for all real numbers x and y. This is the commutative law for addition of real numbers. Likewise, the statement

$$\forall x \exists y(x + y = 0)$$

says that for every real number x there is a real number y such that $x + y = 0$. This states that every real number has an additive inverse. Similarly, the statement

$$\forall x \forall y \forall z(x + (y + z) = (x + y) + z)$$

is the associative law for addition of real numbers. ◀

EXAMPLE 2 Translate into English the statement

$$\forall x \forall y((x > 0) \wedge (y < 0) \rightarrow (xy < 0)),$$

where the domain for both variables consists of all real numbers.

Solution: This statement says that for every real number x and for every real number y, if $x > 0$ and $y < 0$, then $xy < 0$. That is, this statement says that for real numbers x and y, if x is positive and y is negative, then xy is negative. This can be stated more succinctly as "The product of a positive real number and a negative real number is always a negative real number." ◀

THINKING OF QUANTIFICATION AS LOOPS In working with quantifications of more than one variable, it is sometimes helpful to think in terms of nested loops. (If there are infinitely many elements in the domain of some variable, we cannot actually loop through all values. Nevertheless, this way of thinking is helpful in understanding nested quantifiers.) For example, to see whether $\forall x \forall y P(x, y)$ is true, we loop through the values for x, and for each x we loop through the values for y. If we find that for all values of x that $P(x, y)$ is true for all values of y, we have determined that $\forall x \forall y P(x, y)$ is true. If we ever hit a value x for which we hit a value y for which $P(x, y)$ is false, we have shown that $\forall x \forall y P(x, y)$ is false.

Similarly, to determine whether $\forall x \exists y P(x, y)$ is true, we loop through the values for x. For each x we loop through the values for y until we find a y for which $P(x, y)$ is true. If for every x we hit such a y, then $\forall x \exists y P(x, y)$ is true; if for some x we never hit such a y, then $\forall x \exists y P(x, y)$ is false.

To see whether $\exists x \forall y P(x, y)$ is true, we loop through the values for x until we find an x for which $P(x, y)$ is always true when we loop through all values for y. Once we find such an x, we know that $\exists x \forall y P(x, y)$ is true. If we never hit such an x, then we know that $\exists x \forall y P(x, y)$ is false.

Finally, to see whether $\exists x \exists y P(x, y)$ is true, we loop through the values for x, where for each x we loop through the values for y until we hit an x for which we hit a y for which $P(x, y)$ is true. The statement $\exists x \exists y P(x, y)$ is false only if we never hit an x for which we hit a y such that $P(x, y)$ is true.

1.5.3 The Order of Quantifiers

Many mathematical statements involve multiple quantifications of propositional functions involving more than one variable. It is important to note that the order of the quantifiers is important, unless all the quantifiers are universal quantifiers or all are existential quantifiers.

These remarks are illustrated by Examples 3–5.

EXAMPLE 3 Let $P(x, y)$ be the statement "$x + y = y + x$." What are the truth values of the quantifications $\forall x \forall y P(x, y)$ and $\forall y \forall x P(x, y)$, where the domain for all variables consists of all real numbers?

Extra Examples

Solution: The quantification

$$\forall x \forall y P(x, y)$$

denotes the proposition

"For all real numbers x, for all real numbers y, $x + y = y + x$."

Because $P(x, y)$ is true for all real numbers x and y (it is the commutative law for addition, which is an axiom for the real numbers—see Appendix 1), the proposition $\forall x \forall y P(x, y)$ is true. Note that the statement $\forall y \forall x P(x, y)$ says "For all real numbers y, for all real numbers x, $x + y = y + x$." This has the same meaning as the statement "For all real numbers x, for all real numbers y, $x + y = y + x$." That is, $\forall x \forall y P(x, y)$ and $\forall y \forall x P(x, y)$ have the same meaning, and both are true. This illustrates the principle that the order of nested universal quantifiers in a statement without other quantifiers can be changed without changing the meaning of the quantified statement. ◀

EXAMPLE 4 Let $Q(x, y)$ denote "$x + y = 0$." What are the truth values of the quantifications $\exists y \forall x Q(x, y)$ and $\forall x \exists y Q(x, y)$, where the domain for all variables consists of all real numbers?

Solution: The quantification

$$\exists y \forall x Q(x, y)$$

denotes the proposition

"There is a real number y such that for every real number x, $Q(x, y)$."

No matter what value of y is chosen, there is only one value of x for which $x + y = 0$. Because there is no real number y such that $x + y = 0$ for all real numbers x, the statement $\exists y \forall x Q(x, y)$ is false.

The quantification

$$\forall x \exists y Q(x, y)$$

denotes the proposition

"For every real number x there is a real number y such that $Q(x, y)$."

Given a real number x, there is a real number y such that $x + y = 0$; namely, $y = -x$. Hence, the statement $\forall x \exists y Q(x, y)$ is true. ◀

Be careful with the order of existential and universal quantifiers!

Example 4 illustrates that the order in which quantifiers appear makes a difference. The statements $\exists y \forall x P(x, y)$ and $\forall x \exists y P(x, y)$ are not logically equivalent. The statement $\exists y \forall x P(x, y)$ is true if and only if there is a y that makes $P(x, y)$ true for every x. So, for this statement to be true, there must be a particular value of y for which $P(x, y)$ is true regardless of the choice of x. On the other hand, $\forall x \exists y P(x, y)$ is true if and only if for every value of x there is a value of y for which $P(x, y)$ is true. So, for this statement to be true, no matter which x you choose, there must be a value of y (possibly depending on the x you choose) for which $P(x, y)$ is true. In other words, in the second case, y can depend on x, whereas in the first case, y is a constant independent of x.

From these observations, it follows that if $\exists y \forall x P(x, y)$ is true, then $\forall x \exists y P(x, y)$ must also be true. However, if $\forall x \exists y P(x, y)$ is true, it is not necessary for $\exists y \forall x P(x, y)$ to be true. (See Supplementary Exercises 30 and 31.)

Table 1 summarizes the meanings of the different possible quantifications involving two variables.

Quantifications of more than two variables are also common, as Example 5 illustrates.

TABLE 1 Quantifications of Two Variables.

Statement	*When True?*	*When False?*
$\forall x \forall y P(x, y)$ $\forall y \forall x P(x, y)$	$P(x, y)$ is true for every pair x, y.	There is a pair x, y for which $P(x, y)$ is false.
$\forall x \exists y P(x, y)$	For every x there is a y for which $P(x, y)$ is true.	There is an x such that $P(x, y)$ is false for every y.
$\exists x \forall y P(x, y)$	There is an x for which $P(x, y)$ is true for every y.	For every x there is a y for which $P(x, y)$ is false.
$\exists x \exists y P(x, y)$ $\exists y \exists x P(x, y)$	There is a pair x, y for which $P(x, y)$ is true.	$P(x, y)$ is false for every pair x, y.

EXAMPLE 5 Let $Q(x, y, z)$ be the statement "$x + y = z$." What are the truth values of the statements $\forall x \forall y \exists z Q(x, y, z)$ and $\exists z \forall x \forall y Q(x, y, z)$, where the domain of all variables consists of all real numbers?

Solution: Suppose that x and y are assigned values. Then, there exists a real number z such that $x + y = z$. Consequently, the quantification

$$\forall x \forall y \exists z Q(x, y, z),$$

which is the statement

"For all real numbers x and for all real numbers y there is a real number z such that $x + y = z$,"

is true. The order of the quantification here is important, because the quantification

$$\exists z \forall x \forall y Q(x, y, z),$$

which is the statement

"There is a real number z such that for all real numbers x and for all real numbers y it is true that $x + y = z$,"

is false, because there is no value of z that satisfies the equation $x + y = z$ for all values of x and y. ◀

1.5.4 Translating Mathematical Statements into Statements Involving Nested Quantifiers

Mathematical statements expressed in English can be translated into logical expressions, as Examples 6–8 show.

EXAMPLE 6 Translate the statement "The sum of two positive integers is always positive" into a logical expression.

Extra Examples

Solution: To translate this statement into a logical expression, we first rewrite it so that the implied quantifiers and a domain are shown: "For every two integers, if these integers are both positive, then the sum of these integers is positive." Next, we introduce the variables x and y to obtain "For all positive integers x and y, $x + y$ is positive." Consequently, we can express this statement as

$$\forall x \forall y((x > 0) \land (y > 0) \rightarrow (x + y > 0)),$$

where the domain for both variables consists of all integers. Note that we could also translate this using the positive integers as the domain. Then the statement "The sum of two positive integers is always positive" becomes "For every two positive integers, the sum of these integers is positive." We can express this as

$$\forall x \forall y(x + y > 0),$$

where the domain for both variables consists of all positive integers. ◀

EXAMPLE 7 Translate the statement "Every real number except zero has a multiplicative inverse." (A **multiplicative inverse** of a real number x is a real number y such that $xy = 1$.)

Solution: We first rewrite this as "For every real number x except zero, x has a multiplicative inverse." We can rewrite this as "For every real number x, if $x \neq 0$, then there exists a real number y such that $xy = 1$." This can be rewritten as

$$\forall x((x \neq 0) \rightarrow \exists y(xy = 1)).$$ ◀

One example that you may be familiar with is the concept of limit, which is important in calculus.

EXAMPLE 8 *(Requires calculus)* Use quantifiers to express the definition of the limit of a real-valued function $f(x)$ of a real variable x at a point a in its domain.

Solution: Recall that the definition of the statement

$$\lim_{x \to a} f(x) = L$$

is: For every real number $\epsilon > 0$ there exists a real number $\delta > 0$ such that $|f(x) - L| < \epsilon$ whenever $0 < |x - a| < \delta$. This definition of a limit can be phrased in terms of quantifiers by

$$\forall c \exists \delta \forall x(0 < |x - a| < \delta \rightarrow |f(x) - L| < c),$$

where the domain for the variables δ and ϵ consists of all positive real numbers and for x consists of all real numbers.

This definition can also be expressed as

$$\forall \epsilon > 0\ \exists \delta > 0\ \forall x(0 < |x - a| < \delta \rightarrow |f(x) - L| < \epsilon)$$

when the domain for the variables ϵ and δ consists of all real numbers, rather than just the positive real numbers. [Here, restricted quantifiers have been used. Recall that $\forall x > 0\ P(x)$ means that for all x with $x > 0$, $P(x)$ is true.] ◀

1.5.5 Translating from Nested Quantifiers into English

Expressions with nested quantifiers expressing statements in English can be quite complicated. The first step in translating such an expression is to write out what the quantifiers and predicates in the expression mean. The next step is to express this meaning in a simpler sentence. This process is illustrated in Examples 9 and 10.

EXAMPLE 9 Translate the statement

$$\forall x(C(x) \vee \exists y(C(y) \wedge F(x, y)))$$

into English, where $C(x)$ is "x has a computer," $F(x, y)$ is "x and y are friends," and the domain for both x and y consists of all students in your school.

Solution: The statement says that for every student x in your school, x has a computer or there is a student y such that y has a computer and x and y are friends. In other words, every student in your school has a computer or has a friend who has a computer. ◀

EXAMPLE 10 Translate the statement

$$\exists x \forall y \forall z((F(x, y) \wedge F(x, z) \wedge (y \neq z)) \to \neg F(y, z))$$

into English, where $F(a, b)$ means a and b are friends and the domain for x, y, and z consists of all students in your school.

Solution: We first examine the expression $(F(x, y) \wedge F(x, z) \wedge (y \neq z)) \to \neg F(y, z)$. This expression says that if students x and y are friends, and students x and z are friends, and furthermore, if y and z are not the same student, then y and z are not friends. It follows that the original statement, which is triply quantified, says that there is a student x such that for all students y and all students z other than y, if x and y are friends and x and z are friends, then y and z are not friends. In other words, there is a student none of whose friends are also friends with each other. ◀

1.5.6 Translating English Sentences into Logical Expressions

In Section 1.4 we showed how quantifiers can be used to translate sentences into logical expressions. However, we avoided sentences whose translation into logical expressions required the use of nested quantifiers. We now address the translation of such sentences.

EXAMPLE 11 Express the statement "If a person is female and is a parent, then this person is someone's mother" as a logical expression involving predicates, quantifiers with a domain consisting of all people, and logical connectives.

Solution: The statement "If a person is female and is a parent, then this person is someone's mother" can be expressed as "For every person x, if person x is female and person x is a parent, then there exists a person y such that person x is the mother of person y." We introduce the propositional functions $F(x)$ to represent "x is female," $P(x)$ to represent "x is a parent," and $M(x, y)$ to represent "x is the mother of y." The original statement can be represented as

$$\forall x((F(x) \wedge P(x)) \to \exists y M(x, y)).$$

Using the null quantification rule in part (b) of Exercise 49 in Section 1.4, we can move $\exists y$ to the left so that it appears just after $\forall x$, because y does not appear in $F(x) \wedge P(x)$. We obtain the logically equivalent expression

$$\forall x \exists y((F(x) \wedge P(x)) \rightarrow M(x, y)).$$

◀

EXAMPLE 12 Express the statement "Everyone has exactly one best friend" as a logical expression involving predicates, quantifiers with a domain consisting of all people, and logical connectives.

Solution: The statement "Everyone has exactly one best friend" can be expressed as "For every person x, person x has exactly one best friend." Introducing the universal quantifier, we see that this statement is the same as "$\forall x$(person x has exactly one best friend)," where the domain consists of all people.

To say that x has exactly one best friend means that there is a person y who is the best friend of x, and furthermore, that for every person z, if person z is not person y, then z is not the best friend of x. When we introduce the predicate $B(x, y)$ to be the statement "y is the best friend of x," the statement that x has exactly one best friend can be represented as

$$\exists y(B(x, y) \wedge \forall z((z \neq y) \rightarrow \neg B(x, z))).$$

Consequently, our original statement can be expressed as

$$\forall x \exists y(B(x, y) \wedge \forall z((z \neq y) \rightarrow \neg B(x, z))).$$

[Note that we can write this statement as $\forall x \exists ! y B(x, y)$, where $\exists !$ is the "uniqueness quantifier" defined in Section 1.4.] ◀

EXAMPLE 13 Use quantifiers to express the statement "There is a woman who has taken a flight on every airline in the world."

Solution: Let $P(w, f)$ be "w has taken f" and $Q(f, a)$ be "f is a flight on a." We can express the statement as

$$\exists w \forall a \exists f(P(w, f) \wedge Q(f, a)),$$

where the domains of discourse for w, f, and a consist of all the women in the world, all airplane flights, and all airlines, respectively.

The statement could also be expressed as

$$\exists w \forall a \exists f R(w, f, a),$$

where $R(w, f, a)$ is "w has taken f on a." Although this is more compact, it somewhat obscures the relationships among the variables. Consequently, the first solution is usually preferable. ◀

1.5.7 Negating Nested Quantifiers

Assessment ❯ Statements involving nested quantifiers can be negated by successively applying the rules for negating statements involving a single quantifier. This is illustrated in Examples 14–16.

EXAMPLE 14 Express the negation of the statement $\forall x \exists y(xy = 1)$ so that no negation precedes a quantifier.

Extra Examples

Solution: By successively applying De Morgan's laws for quantifiers in Table 2 of Section 1.4, we can move the negation in $\neg\forall x \exists y(xy = 1)$ inside all the quantifiers. We find that $\neg\forall x \exists y(xy = 1)$ is equivalent to $\exists x \neg\exists y(xy = 1)$, which is equivalent to $\exists x \forall y \neg(xy = 1)$. Because $\neg(xy = 1)$ can be expressed more simply as $xy \neq 1$, we conclude that our negated statement can be expressed as $\exists x \forall y(xy \neq 1)$. ◀

EXAMPLE 15 Use quantifiers to express the statement that "There does not exist a woman who has taken a flight on every airline in the world."

Solution: This statement is the negation of the statement "There is a woman who has taken a flight on every airline in the world" from Example 13. By Example 13, our statement can be expressed as $\neg\exists w \forall a \exists f(P(w, f) \wedge Q(f, a))$, where $P(w, f)$ is "w has taken f" and $Q(f, a)$ is "f is a flight on a." By successively applying De Morgan's laws for quantifiers in Table 2 of Section 1.4 to move the negation inside successive quantifiers and by applying De Morgan's law for negating a conjunction in the last step, we find that our statement is equivalent to each of this sequence of statements:

$$\begin{aligned}\forall w \neg\forall a \exists f(P(w, f) \wedge Q(f, a)) &\equiv \forall w \exists a \neg\exists f(P(w, f) \wedge Q(f, a))\\ &\equiv \forall w \exists a \forall f \neg(P(w, f) \wedge Q(f, a))\\ &\equiv \forall w \exists a \forall f(\neg P(w, f) \vee \neg Q(f, a)).\end{aligned}$$

This last statement states "For every woman there is an airline such that for all flights, this woman has not taken that flight or that flight is not on this airline." ◀

EXAMPLE 16 (*Requires calculus*) Use quantifiers and predicates to express the fact that $\lim_{x\to a} f(x)$ does not exist where $f(x)$ is a real-valued function of a real variable x and a belongs to the domain of f

Solution: To say that $\lim_{x\to a} f(x)$ does not exist means that for all real numbers L, $\lim_{x\to a} f(x) \neq L$. By using Example 8, the statement $\lim_{x\to a} f(x) \neq L$ can be expressed as

$$\neg\forall\epsilon > 0\ \exists\delta > 0\ \forall x(0 < |x - a| < \delta \rightarrow |f(x) - L| < \epsilon).$$

Successively applying the rules for negating quantified expressions, we construct this sequence of equivalent statements:

$$\begin{aligned}&\neg\forall\epsilon > 0\ \exists\delta > 0\ \forall x(0 < |x - a| < \delta \rightarrow |f(x) - L| < \epsilon)\\ &\quad\equiv \exists\epsilon > 0\ \neg\exists\delta > 0\ \forall x(0 < |x - a| < \delta \rightarrow |f(x) - L| < \epsilon)\\ &\quad\equiv \exists\epsilon > 0\ \forall\delta > 0\ \neg\forall x(0 < |x - a| < \delta \rightarrow |f(x) - L| < \epsilon)\\ &\quad\equiv \exists\epsilon > 0\ \forall\delta > 0\ \exists x\ \neg(0 < |x - a| < \delta \rightarrow |f(x) - L| < \epsilon)\\ &\quad\equiv \exists\epsilon > 0\ \forall\delta > 0\ \exists x(0 < |x - a| < \delta \wedge |f(x) - L| \geq \epsilon).\end{aligned}$$

In the last step we used the equivalence $\neg(p \rightarrow q) \equiv p \wedge \neg q$, which follows from the fifth equivalence in Table 7 of Section 1.3.

Because the statement "$\lim_{x\to a} f(x)$ does not exist" means for all real numbers L, $\lim_{x\to a} f(x) \neq L$, this can be expressed as

$$\forall L \exists\epsilon > 0\ \forall\delta > 0\ \exists x(0 < |x - a| < \delta \wedge |f(x) - L| \geq \epsilon).$$

This last statement says that for every real number L there is a real number $\epsilon > 0$ such that for every real number $\delta > 0$, there exists a real number x such that $0 < |x - a| < \delta$ and $|f(x) - L| \geq \epsilon$. ◀

Exercises

1. Translate these statements into English, where the domain for each variable consists of all real numbers.

a) $\forall x \exists y(x < y)$
b) $\forall x \forall y(((x \geq 0) \wedge (y \geq 0)) \rightarrow (xy \geq 0))$
c) $\forall x \forall y \exists z(xy = z)$

2. Translate these statements into English, where the domain for each variable consists of all real numbers.

a) $\exists x \forall y(xy = y)$
b) $\forall x \forall y(((x \geq 0) \wedge (y < 0)) \rightarrow (x - y > 0))$
c) $\forall x \forall y \exists z(x = y + z)$

3. Let $Q(x, y)$ be the statement "x has sent an email message to y," where the domain for both x and y consists of all students in your class. Express each of these quantifications in English.

a) $\exists x \exists y Q(x, y)$
b) $\exists x \forall y Q(x, y)$
c) $\forall x \exists y Q(x, y)$
d) $\exists y \forall x Q(x, y)$
e) $\forall y \exists x Q(x, y)$
f) $\forall x \forall y Q(x, y)$

4. Let $P(x, y)$ be the statement "Student x has taken class y," where the domain for x consists of all students in your class and for y consists of all computer science courses at your school. Express each of these quantifications in English.

a) $\exists x \exists y P(x, y)$
b) $\exists x \forall y P(x, y)$
c) $\forall x \exists y P(x, y)$
d) $\exists y \forall x P(x, y)$
e) $\forall y \exists x P(x, y)$
f) $\forall x \forall y P(x, y)$

5. Let $W(x, y)$ mean that student x has visited website y, where the domain for x consists of all students in your school and the domain for y consists of all websites. Express each of these statements by a simple English sentence.

a) W(Sarah Smith, www.att.com)
b) $\exists x W(x,$ www.imdb.org)
c) $\exists y W$(José Orez, y)
d) $\exists y(W$(Ashok Puri, $y) \wedge W$(Cindy Yoon, y))
e) $\exists y \forall z(y \neq$ (David Belcher) $\wedge$ (W(David Belcher, $z) \rightarrow W(y, z)$))
f) $\exists x \exists y \forall z((x \neq y) \wedge (W(x, z) \leftrightarrow W(y, z)))$

6. Let $C(x, y)$ mean that student x is enrolled in class y, where the domain for x consists of all students in your school and the domain for y consists of all classes being given at your school. Express each of these statements by a simple English sentence.

a) C(Randy Goldberg, CS 252)
b) $\exists x C(x,$ Math 695)
c) $\exists y C$(Carol Sitea, y)
d) $\exists x(C(x,$ Math 222) $\wedge$ $C(x,$ CS 252))
e) $\exists x \exists y \forall z((x \neq y) \wedge (C(x, z) \rightarrow C(y, z)))$
f) $\exists x \exists y \forall z((x \neq y) \wedge (C(x, z) \leftrightarrow C(y, z)))$

7. Let $T(x, y)$ mean that student x likes cuisine y, where the domain for x consists of all students at your school and the domain for y consists of all cuisines. Express each of these statements by a simple English sentence.

a) $\neg T$(Abdallah Hussein, Japanese)
b) $\exists x T(x,$ Korean) $\wedge$ $\forall x T(x,$ Mexican)
c) $\exists y(T$(Monique Arsenault, $y) \vee T$(Jay Johnson, y))
d) $\forall x \forall z \exists y((x \neq z) \rightarrow \neg(T(x, y) \wedge T(z, y)))$
e) $\exists x \exists z \forall y(T(x, y) \leftrightarrow T(z, y))$
f) $\forall x \forall z \exists y(T(x, y) \leftrightarrow T(z, y))$

8. Let $Q(x, y)$ be the statement "Student x has been a contestant on quiz show y." Express each of these sentences in terms of $Q(x, y)$, quantifiers, and logical connectives, where the domain for x consists of all students at your school and for y consists of all quiz shows on television.

a) There is a student at your school who has been a contestant on a television quiz show.
b) No student at your school has ever been a contestant on a television quiz show.
c) There is a student at your school who has been a contestant on *Jeopardy!* and on *Wheel of Fortune*.
d) Every television quiz show has had a student from your school as a contestant.
e) At least two students from your school have been contestants on *Jeopardy!*.

9. Let $L(x, y)$ be the statement "x loves y," where the domain for both x and y consists of all people in the world. Use quantifiers to express each of these statements.

a) Everybody loves Jerry.
b) Everybody loves somebody.
c) There is somebody whom everybody loves.
d) Nobody loves everybody.
e) There is somebody whom Lydia does not love.
f) There is somebody whom no one loves.
g) There is exactly one person whom everybody loves.
h) There are exactly two people whom Lynn loves.
i) Everyone loves himself or herself.
j) There is someone who loves no one besides himself or herself.

10. Let $F(x, y)$ be the statement "x can fool y," where the domain consists of all people in the world. Use quantifiers to express each of these statements.

a) Everybody can fool Fred.
b) Evelyn can fool everybody.
c) Everybody can fool somebody.
d) There is no one who can fool everybody.
e) Everyone can be fooled by somebody.

f) No one can fool both Fred and Jerry.
g) Nancy can fool exactly two people.
h) There is exactly one person whom everybody can fool.
i) No one can fool himself or herself.
j) There is someone who can fool exactly one person besides himself or herself.

11. Let $S(x)$ be the predicate "x is a student," $F(x)$ the predicate "x is a faculty member," and $A(x, y)$ the predicate "x has asked y a question," where the domain consists of all people associated with your school. Use quantifiers to express each of these statements.

a) Lois has asked Professor Michaels a question.
b) Every student has asked Professor Gross a question.
c) Every faculty member has either asked Professor Miller a question or been asked a question by Professor Miller.
d) Some student has not asked any faculty member a question.
e) There is a faculty member who has never been asked a question by a student.
f) Some student has asked every faculty member a question.
g) There is a faculty member who has asked every other faculty member a question.
h) Some student has never been asked a question by a faculty member.

12. Let $I(x)$ be the statement "x has an Internet connection" and $C(x, y)$ be the statement "x and y have chatted over the Internet," where the domain for the variables x and y consists of all students in your class. Use quantifiers to express each of these statements.

a) Jerry does not have an Internet connection.
b) Rachel has not chatted over the Internet with Chelsea.
c) Jan and Sharon have never chatted over the Internet.
d) No one in the class has chatted with Bob.
e) Sanjay has chatted with everyone except Joseph.
f) Someone in your class does not have an Internet connection.
g) Not everyone in your class has an Internet connection.
h) Exactly one student in your class has an Internet connection.
i) Everyone except one student in your class has an Internet connection.
j) Everyone in your class with an Internet connection has chatted over the Internet with at least one other student in your class.
k) Someone in your class has an Internet connection but has not chatted with anyone else in your class.
l) There are two students in your class who have not chatted with each other over the Internet.
m) There is a student in your class who has chatted with everyone in your class over the Internet.
n) There are at least two students in your class who have not chatted with the same person in your class.
o) There are two students in the class who between them have chatted with everyone else in the class.

13. Let $M(x, y)$ be "x has sent y an email message" and $T(x, y)$ be "x has telephoned y," where the domain consists of all students in your class. Use quantifiers to express each of these statements. (Assume that all email messages that were sent are received, which is not the way things often work.)

a) Chou has never sent an email message to Koko.
b) Arlene has never sent an email message to or telephoned Sarah.
c) José has never received an email message from Deborah.
d) Every student in your class has sent an email message to Ken.
e) No one in your class has telephoned Nina.
f) Everyone in your class has either telephoned Avi or sent him an email message.
g) There is a student in your class who has sent everyone else in your class an email message.
h) There is someone in your class who has either sent an email message or telephoned everyone else in your class.
i) There are two different students in your class who have sent each other email messages.
j) There is a student who has sent himself or herself an email message.
k) There is a student in your class who has not received an email message from anyone else in the class and who has not been called by any other student in the class.
l) Every student in the class has either received an email message or received a telephone call from another student in the class.
m) There are at least two students in your class such that one student has sent the other email and the second student has telephoned the first student.
n) There are two different students in your class who between them have sent an email message to or telephoned everyone else in the class.

14. Use quantifiers and predicates with more than one variable to express these statements.

a) There is a student in this class who can speak Hindi.
b) Every student in this class plays some sport.
c) Some student in this class has visited Alaska but has not visited Hawaii.
d) All students in this class have learned at least one programming language.
e) There is a student in this class who has taken every course offered by one of the departments in this school.
f) Some student in this class grew up in the same town as exactly one other student in this class.
g) Every student in this class has chatted with at least one other student in at least one chat group.

15. Use quantifiers and predicates with more than one variable to express these statements.

a) Every computer science student needs a course in discrete mathematics.

b) There is a student in this class who owns a personal computer.
c) Every student in this class has taken at least one computer science course.
d) There is a student in this class who has taken at least one course in computer science.
e) Every student in this class has been in every building on campus.
f) There is a student in this class who has been in every room of at least one building on campus.
g) Every student in this class has been in at least one room of every building on campus.

16. A discrete mathematics class contains 1 mathematics major who is a freshman, 12 mathematics majors who are sophomores, 15 computer science majors who are sophomores, 2 mathematics majors who are juniors, 2 computer science majors who are juniors, and 1 computer science major who is a senior. Express each of these statements in terms of quantifiers and then determine its truth value.

a) There is a student in the class who is a junior.
b) Every student in the class is a computer science major.
c) There is a student in the class who is neither a mathematics major nor a junior.
d) Every student in the class is either a sophomore or a computer science major.
e) There is a major such that there is a student in the class in every year of study with that major.

17. Express each of these system specifications using predicates, quantifiers, and logical connectives, if necessary.

a) Every user has access to exactly one mailbox.
b) There is a process that continues to run during all error conditions only if the kernel is working correctly.
c) All users on the campus network can access all websites whose url has a .edu extension.
***d)** There are exactly two systems that monitor every remote server.

18. Express each of these system specifications using predicates, quantifiers, and logical connectives, if necessary.

a) At least one console must be accessible during every fault condition.
b) The email address of every user can be retrieved whenever the archive contains at least one message sent by every user on the system.
c) For every security breach there is at least one mechanism that can detect that breach if and only if there is a process that has not been compromised.
d) There are at least two paths connecting every two distinct endpoints on the network.
e) No one knows the password of every user on the system except for the system administrator, who knows all passwords.

19. Express each of these statements using mathematical and logical operators, predicates, and quantifiers, where the domain consists of all integers.

a) The sum of two negative integers is negative.
b) The difference of two positive integers is not necessarily positive.
c) The sum of the squares of two integers is greater than or equal to the square of their sum.
d) The absolute value of the product of two integers is the product of their absolute values.

20. Express each of these statements using predicates, quantifiers, logical connectives, and mathematical operators where the domain consists of all integers.

a) The product of two negative integers is positive.
b) The average of two positive integers is positive.
c) The difference of two negative integers is not necessarily negative.
d) The absolute value of the sum of two integers does not exceed the sum of the absolute values of these integers.

21. Use predicates, quantifiers, logical connectives, and mathematical operators to express the statement that every positive integer is the sum of the squares of four integers.

22. Use predicates, quantifiers, logical connectives, and mathematical operators to express the statement that there is a positive integer that is not the sum of three squares.

23. Express each of these mathematical statements using predicates, quantifiers, logical connectives, and mathematical operators.

a) The product of two negative real numbers is positive.
b) The difference of a real number and itself is zero.
c) Every positive real number has exactly two square roots.
d) A negative real number does not have a square root that is a real number.

24. Translate each of these nested quantifications into an English statement that expresses a mathematical fact. The domain in each case consists of all real numbers.

a) $\exists x \forall y(x + y = y)$
b) $\forall x \forall y(((x \geq 0) \wedge (y < 0)) \rightarrow (x - y > 0))$
c) $\exists x \exists y(((x \leq 0) \wedge (y \leq 0)) \wedge (x - y > 0))$
d) $\forall x \forall y((x \neq 0) \wedge (y \neq 0) \leftrightarrow (xy \neq 0))$

25. Translate each of these nested quantifications into an English statement that expresses a mathematical fact. The domain in each case consists of all real numbers.

a) $\exists x \forall y(xy = y)$
b) $\forall x \forall y(((x < 0) \wedge (y < 0)) \rightarrow (xy > 0))$
c) $\exists x \exists y((x^2 > y) \wedge (x < y))$
d) $\forall x \forall y \exists z(x + y = z)$

26. Let $Q(x, y)$ be the statement "$x + y = x - y$." If the domain for both variables consists of all integers, what are the truth values?

a) $Q(1, 1)$
b) $Q(2, 0)$
c) $\forall y Q(1, y)$
d) $\exists x Q(x, 2)$
e) $\exists x \exists y Q(x, y)$
f) $\forall x \exists y Q(x, y)$
g) $\exists y \forall x Q(x, y)$
h) $\forall y \exists x Q(x, y)$
i) $\forall x \forall y Q(x, y)$

27. Determine the truth value of each of these statements if the domain for all variables consists of all integers.

a) $\forall n \exists m(n^2 < m)$
b) $\exists n \forall m(n < m^2)$
c) $\forall n \exists m(n + m = 0)$
d) $\exists n \forall m(nm = m)$
e) $\exists n \exists m(n^2 + m^2 = 5)$
f) $\exists n \exists m(n^2 + m^2 = 6)$
g) $\exists n \exists m(n + m = 4 \wedge n - m = 1)$
h) $\exists n \exists m(n + m = 4 \wedge n - m = 2)$
i) $\forall n \forall m \exists p(p = (m + n)/2)$

28. Determine the truth value of each of these statements if the domain of each variable consists of all real numbers.

a) $\forall x \exists y(x^2 = y)$
b) $\forall x \exists y(x = y^2)$
c) $\exists x \forall y(xy = 0)$
d) $\exists x \exists y(x + y \neq y + x)$
e) $\forall x(x \neq 0 \rightarrow \exists y(xy = 1))$
f) $\exists x \forall y(y \neq 0 \rightarrow xy = 1)$
g) $\forall x \exists y(x + y = 1)$
h) $\exists x \exists y(x + 2y = 2 \wedge 2x + 4y = 5)$
i) $\forall x \exists y(x + y = 2 \wedge 2x - y = 1)$
j) $\forall x \forall y \exists z(z = (x + y)/2)$

29. Suppose the domain of the propositional function $P(x, y)$ consists of pairs x and y, where x is 1, 2, or 3 and y is 1, 2, or 3. Write out these propositions using disjunctions and conjunctions.

a) $\forall x \forall y P(x, y)$
b) $\exists x \exists y P(x, y)$
c) $\exists x \forall y P(x, y)$
d) $\forall y \exists x P(x, y)$

30. Rewrite each of these statements so that negations appear only within predicates (that is, so that no negation is outside a quantifier or an expression involving logical connectives).

a) $\neg \exists y \exists x P(x, y)$
b) $\neg \forall x \exists y P(x, y)$
c) $\neg \exists y(Q(y) \wedge \forall x \neg R(x, y))$
d) $\neg \exists y(\exists x R(x, y) \vee \forall x S(x, y))$
e) $\neg \exists y(\forall x \exists z T(x, y, z) \vee \exists x \forall z U(x, y, z))$

31. Express the negations of each of these statements so that all negation symbols immediately precede predicates.

a) $\forall x \exists y \forall z T(x, y, z)$
b) $\forall x \exists y P(x, y) \vee \forall x \exists y Q(x, y)$
c) $\forall x \exists y(P(x, y) \wedge \exists z R(x, y, z))$
d) $\forall x \exists y(P(x, y) \rightarrow Q(x, y))$

32. Express the negations of each of these statements so that all negation symbols immediately precede predicates.

a) $\exists z \forall y \forall x T(x, y, z)$
b) $\exists x \exists y P(x, y) \wedge \forall x \forall y Q(x, y)$
c) $\exists x \exists y(Q(x, y) \leftrightarrow Q(y, x))$
d) $\forall y \exists x \exists z(T(x, y, z) \vee Q(x, y))$

33. Rewrite each of these statements so that negations appear only within predicates (that is, so that no negation is outside a quantifier or an expression involving logical connectives).

a) $\neg \forall x \forall y P(x, y)$
b) $\neg \forall y \exists x P(x, y)$
c) $\neg \forall y \forall x(P(x, y) \vee Q(x, y))$
d) $\neg(\exists x \exists y \neg P(x, y) \wedge \forall x \forall y Q(x, y))$
e) $\neg \forall x(\exists y \forall z P(x, y, z) \wedge \exists z \forall y P(x, y, z))$

34. Find a common domain for the variables x, y, and z for which the statement $\forall x \forall y((x \neq y) \rightarrow \forall z((z = x) \vee (z = y)))$ is true and another domain for which it is false.

35. Find a common domain for the variables x, y, z, and w for which the statement $\forall x \forall y \forall z \exists w((w \neq x) \wedge (w \neq y) \wedge (w \neq z))$ is true and another common domain for these variables for which it is false.

36. Express each of these statements using quantifiers. Then form the negation of the statement so that no negation is to the left of a quantifier. Next, express the negation in simple English. (Do not simply use the phrase "It is not the case that.")

a) No one has lost more than one thousand dollars playing the lottery.
b) There is a student in this class who has chatted with exactly one other student.
c) No student in this class has sent email to exactly two other students in this class.
d) Some student has solved every exercise in this book.
e) No student has solved at least one exercise in every section of this book.

37. Express each of these statements using quantifiers. Then form the negation of the statement so that no negation is to the left of a quantifier. Next, express the negation in simple English. (Do not simply use the phrase "It is not the case that.")

a) Every student in this class has taken exactly two mathematics classes at this school.
b) Someone has visited every country in the world except Libya.
c) No one has climbed every mountain in the Himalayas.
d) Every movie actor has either been in a movie with Kevin Bacon or has been in a movie with someone who has been in a movie with Kevin Bacon.

38. Express the negations of these propositions using quantifiers, and in English.

a) Every student in this class likes mathematics.
b) There is a student in this class who has never seen a computer.
c) There is a student in this class who has taken every mathematics course offered at this school.
d) There is a student in this class who has been in at least one room of every building on campus.

39. Find a counterexample, if possible, to these universally quantified statements, where the domain for all variables consists of all integers.

a) $\forall x \forall y(x^2 = y^2 \rightarrow x = y)$
b) $\forall x \exists y(y^2 = x)$
c) $\forall x \forall y(xy \geq x)$

40. Find a counterexample, if possible, to these universally quantified statements, where the domain for all variables consists of all integers.

a) $\forall x \exists y(x = 1/y)$
b) $\forall x \exists y(y^2 - x < 100)$
c) $\forall x \forall y(x^2 \neq y^3)$

41. Use quantifiers to express the associative law for multiplication of real numbers.

42. Use quantifiers to express the distributive laws of multiplication over addition for real numbers.

43. Use quantifiers and logical connectives to express the fact that every linear polynomial (that is, polynomial of degree 1) with real coefficients and where the coefficient of x is nonzero, has exactly one real root.

44. Use quantifiers and logical connectives to express the fact that a quadratic polynomial with real number coefficients has at most two real roots.

45. Determine the truth value of the statement $\forall x \exists y(xy = 1)$ if the domain for the variables consists of

a) the nonzero real numbers.
b) the nonzero integers.
c) the positive real numbers.

46. Determine the truth value of the statement $\exists x \forall y(x \leq y^2)$ if the domain for the variables consists of

a) the positive real numbers.
b) the integers.
c) the nonzero real numbers.

47. Show that the two statements $\neg\exists x \forall y P(x, y)$ and $\forall x \exists y \neg P(x, y)$, where both quantifiers over the first variable in $P(x, y)$ have the same domain, and both quantifiers over the second variable in $P(x, y)$ have the same domain, are logically equivalent.

***48.** Show that $\forall x P(x) \vee \forall x Q(x)$ and $\forall x \forall y(P(x) \vee Q(y))$, where all quantifiers have the same nonempty domain, are logically equivalent. (The new variable y is used to combine the quantifications correctly.)

***49. a)** Show that $\forall x P(x) \wedge \exists x Q(x)$ is logically equivalent to $\forall x \exists y\ (P(x) \wedge Q(y))$, where all quantifiers have the same nonempty domain.

b) Show that $\forall x P(x) \vee \exists x Q(x)$ is equivalent to $\forall x \exists y\ (P(x) \vee Q(y))$, where all quantifiers have the same nonempty domain.

A statement is in **prenex normal form (PNF)** if and only if it is of the form

$$Q_1 x_1 Q_2 x_2 \cdots Q_k x_k P(x_1, x_2, \ldots, x_k),$$

where each Q_i, $i = 1, 2, \ldots, k$, is either the existential quantifier or the universal quantifier, and $P(x_1, \ldots, x_k)$ is a predicate involving no quantifiers. For example, $\exists x \forall y(P(x, y) \wedge Q(y))$ is in prenex normal form, whereas $\exists x P(x) \vee \forall x Q(x)$ is not (because the quantifiers do not all occur first).

Every statement formed from propositional variables, predicates, **T**, and **F** using logical connectives and quantifiers is equivalent to a statement in prenex normal form. Exercise 51 asks for a proof of this fact.

***50.** Put these statements in prenex normal form. [*Hint:* Use logical equivalence from Tables 6 and 7 in Section 1.3, Table 2 in Section 1.4, Example 19 in Section 1.4, Exercises 47 and 48 in Section 1.4, and Exercises 48 and 49.]

a) $\exists x P(x) \vee \exists x Q(x) \vee A$, where A is a proposition not involving any quantifiers

b) $\neg(\forall x P(x) \vee \forall x Q(x))$

c) $\exists x P(x) \to \exists x Q(x)$

****51.** Show how to transform an arbitrary statement to a statement in prenex normal form that is equivalent to the given statement. (Note: A formal solution of this exercise requires use of structural induction, covered in Section 5.3.)

***52.** Express the quantification $\exists! x P(x)$, introduced in Section 1.4, using universal quantifications, existential quantifications, and logical operators.

1.6 Rules of Inference

1.6.1 Introduction

Later in this chapter we will study proofs. Proofs in mathematics are valid arguments that establish the truth of mathematical statements. By an **argument**, we mean a sequence of statements that end with a conclusion. By **valid**, we mean that the conclusion, or final statement of the argument, must follow from the truth of the preceding statements, or **premises**, of the argument. That is, an argument is valid if and only if it is impossible for all the premises to be true and the conclusion to be false. To deduce new statements from statements we already have, we use rules of inference which are templates for constructing valid arguments. Rules of inference are our basic tools for establishing the truth of statements.

Before we study mathematical proofs, we will look at arguments that involve only compound propositions. We will define what it means for an argument involving compound propositions to be valid. Then we will introduce a collection of rules of inference in propositional logic. These rules of inference are among the most important ingredients in producing valid arguments. After we illustrate how rules of inference are used to produce valid arguments, we will describe some common forms of incorrect reasoning, called **fallacies**, which lead to invalid arguments.

After studying rules of inference in propositional logic, we will introduce rules of inference for quantified statements. We will describe how these rules of inference can be used to produce valid arguments. These rules of inference for statements involving existential and universal quantifiers play an important role in proofs in computer science and mathematics, although they are often used without being explicitly mentioned.

Finally, we will show how rules of inference for propositions and for quantified statements can be combined. These combinations of rule of inference are often used together in complicated arguments.

1.6.2 Valid Arguments in Propositional Logic

Consider the following argument involving propositions (which, by definition, is a sequence of propositions):

"If you have a current password, then you can log onto the network."

"You have a current password."

Therefore,

"You can log onto the network."

We would like to determine whether this is a valid argument. That is, we would like to determine whether the conclusion "You can log onto the network" must be true when the premises "If you have a current password, then you can log onto the network" and "You have a current password" are both true.

Before we discuss the validity of this particular argument, we will look at its form. Use p to represent "You have a current password" and q to represent "You can log onto the network." Then, the argument has the form

$$\begin{array}{l} \quad p \to q \\ \quad p \\ \hline \therefore q \end{array}$$

where $\therefore$ is the symbol that denotes "therefore."

We know that when p and q are propositional variables, the statement $((p \to q) \land p) \to q$ is a tautology (see Exercise 12(c) in Section 1.3). In particular, when both $p \to q$ and p are true, we know that q must also be true. We say this form of argument is **valid** because whenever all its premises (all statements in the argument other than the final one, the conclusion) are true, the conclusion must also be true. Now suppose that both "If you have a current password, then you can log onto the network" and "You have a current password" are true statements. When we replace p by "You have a current password" and q by "You can log onto the network," it necessarily follows that the conclusion "You can log onto the network" is true. This argument is **valid** because its form is valid. Note that whenever we replace p and q by propositions where $p \to q$ and p are both true, then q must also be true.

What happens when we replace p and q in this argument form by propositions where not both p and $p \to q$ are true? For example, suppose that p represents "You have access to the network" and q represents "You can change your grade" and that p is true, but $p \to q$ is false. The argument we obtain by substituting these values of p and q into the argument form is

"If you have access to the network, then you can change your grade."
"You have access to the network."

$\therefore$ "You can change your grade."

The argument we obtained is a valid argument, but because one of the premises, namely the first premise, is false, we cannot conclude that the conclusion is true. (Most likely, this conclusion is false.)

In our discussion, to analyze an argument, we replaced propositions by propositional variables. This changed an argument to an **argument form**. We saw that the validity of an argument follows from the validity of the form of the argument. We summarize the terminology used to discuss the validity of arguments with our definition of the key notions.

Definition 1 An *argument* in propositional logic is a sequence of propositions. All but the final proposition in the argument are called *premises* and the final proposition is called the *conclusion*. An argument is *valid* if the truth of all its premises implies that the conclusion is true.

An *argument form* in propositional logic is a sequence of compound propositions involving propositional variables. An argument form is *valid* if no matter which particular propositions are substituted for the propositional variables in its premises, the conclusion is true if the premises are all true.

Remark: From the definition of a valid argument form we see that the argument form with premises $p_1, p_2, \ldots, p_n$ and conclusion q is valid exactly when $(p_1 \wedge p_2 \wedge \cdots \wedge p_n) \rightarrow q$ is a tautology.

The key to showing that an argument in propositional logic is valid is to show that its argument form is valid. Consequently, we would like techniques to show that argument forms are valid. We will now develop methods for accomplishing this task.

1.6.3 Rules of Inference for Propositional Logic

We can always use a truth table to show that an argument form is valid. We do this by showing that whenever the premises are true, the conclusion must also be true. However, this can be a tedious approach. For example, when an argument form involves 10 different propositional variables, to use a truth table to show this argument form is valid requires $2^{10} = 1024$ different rows. Fortunately, we do not have to resort to truth tables. Instead, we can first establish the validity of some relatively simple argument forms, called **rules of inference**. These rules of inference can be used as building blocks to construct more complicated valid argument forms. We will now introduce the most important rules of inference in propositional logic.

The tautology $(p \wedge (p \rightarrow q)) \rightarrow q$ is the basis of the rule of inference called **modus ponens**, or the **law of detachment**. (Modus ponens is Latin for *mode that affirms*.) This tautology leads to the following valid argument form, which we have already seen in our initial discussion about arguments (where, as before, the symbol $\therefore$ denotes "therefore"):

$$\begin{array}{rl} & p \\ & p \rightarrow q \\ \hline \therefore & q \end{array}$$

Using this notation, the hypotheses are written in a column, followed by a horizontal bar, followed by a line that begins with the therefore symbol and ends with the conclusion. In particular, modus ponens tells us that if a conditional statement and the hypothesis of this conditional statement are both true, then the conclusion must also be true. Example 1 illustrates the use of modus ponens.

EXAMPLE 1 Suppose that the conditional statement "If it snows today, then we will go skiing" and its hypothesis, "It is snowing today," are true. Then, by modus ponens, it follows that the conclusion of the conditional statement, "We will go skiing," is true. ◀

As we mentioned earlier, a valid argument can lead to an incorrect conclusion if one or more of its premises is false. We illustrate this again in Example 2.

EXAMPLE 2 Determine whether the argument given here is valid and determine whether its conclusion must be true because of the validity of the argument.

"If $\sqrt{2} > \frac{3}{2}$, then $\left(\sqrt{2}\right)^2 > \left(\frac{3}{2}\right)^2$. We know that $\sqrt{2} > \frac{3}{2}$. Consequently, $\left(\sqrt{2}\right)^2 = 2 > \left(\frac{3}{2}\right)^2 = \frac{9}{4}$."

Solution: Let p be the proposition "$\sqrt{2} > \frac{3}{2}$" and q the proposition "$2 > (\frac{3}{2})^2$." The premises of the argument are $p \to q$ and p, and q is its conclusion. This argument is valid because it is constructed by using modus ponens, a valid argument form. However, one of its premises, $\sqrt{2} > \frac{3}{2}$, is false. Consequently, we cannot conclude that the conclusion is true. Furthermore, note that the conclusion of this argument is false, because $2 < \frac{9}{4}$. ◀

There are many useful rules of inference for propositional logic. Perhaps the most widely used of these are listed in Table 1. Exercises 11–16, 19, 33, and 34 in Section 1.3 ask for the verifications that these rules of inference are valid argument forms. We now give examples of

TABLE 1 Rules of Inference.

Rule of Inference	*Tautology*	*Name*
p $p \to q$ $\therefore q$	$(p \land (p \to q)) \to q$	Modus ponens
$\neg q$ $p \to q$ $\therefore \neg p$	$(\neg q \land (p \to q)) \to \neg p$	Modus tollens
$p \to q$ $q \to r$ $\therefore p \to r$	$((p \to q) \land (q \to r)) \to (p \to r)$	Hypothetical syllogism
$p \lor q$ $\neg p$ $\therefore q$	$((p \lor q) \land \neg p) \to q$	Disjunctive syllogism
p $\therefore p \lor q$	$p \to (p \lor q)$	Addition
$p \land q$ $\therefore p$	$(p \land q) \to p$	Simplification
p q $\therefore p \land q$	$((p) \land (q)) \to (p \land q)$	Conjunction
$p \lor q$ $\neg p \lor r$ $\therefore q \lor r$	$((p \lor q) \land (\neg p \lor r)) \to (q \lor r)$	Resolution

arguments that use these rules of inference. In each argument, we first use propositional variables to express the propositions in the argument. We then show that the resulting argument form is a rule of inference from Table 1.

EXAMPLE 3 State which rule of inference is the basis of the following argument: "It is below freezing now. Therefore, it is below freezing or raining now."

Solution: Let p be the proposition "It is below freezing now," and let q be the proposition "It is raining now." Then this argument is of the form

$$\begin{array}{l} \quad p \\ \hline \therefore p \vee q \end{array}$$

This is an argument that uses the addition rule. ◀

EXAMPLE 4 State which rule of inference is the basis of the following argument: "It is below freezing and raining now. Therefore, it is below freezing now."

Solution: Let p be the proposition "It is below freezing now," and let q be the proposition "It is raining now." This argument is of the form

$$\begin{array}{l} \quad p \wedge q \\ \hline \therefore p \end{array}$$

This argument uses the simplification rule. ◀

EXAMPLE 5 State which rule of inference is used in the argument:

> If it rains today, then we will not have a barbecue today. If we do not have a barbecue today, then we will have a barbecue tomorrow. Therefore, if it rains today, then we will have a barbecue tomorrow.

Solution: Let p be the proposition "It is raining today," let q be the proposition "We will not have a barbecue today," and let r be the proposition "We will have a barbecue tomorrow." Then this argument is of the form

$$\begin{array}{l} \quad p \to q \\ \quad q \to r \\ \hline \therefore p \to r \end{array}$$

Hence, this argument is a hypothetical syllogism. ◀

1.6.4 Using Rules of Inference to Build Arguments

When there are many premises, several rules of inference are often needed to show that an argument is valid. This is illustrated by Examples 6 and 7, where the steps of arguments are displayed on separate lines, with the reason for each step explicitly stated. These examples also show how arguments in English can be analyzed using rules of inference.

EXAMPLE 6 Show that the premises "It is not sunny this afternoon and it is colder than yesterday," "We will go swimming only if it is sunny," "If we do not go swimming, then we will take a canoe trip," and "If we take a canoe trip, then we will be home by sunset" lead to the conclusion "We will be home by sunset."

Extra Examples

Solution: Let p be the proposition "It is sunny this afternoon," q the proposition "It is colder than yesterday," r the proposition "We will go swimming," s the proposition "We will take a canoe trip," and t the proposition "We will be home by sunset." Then the premises become $\neg p \wedge q$, $r \to p$, $\neg r \to s$, and $s \to t$. The conclusion is simply t. We need to give a valid argument with premises $\neg p \wedge q$, $r \to p$, $\neg r \to s$, and $s \to t$ and conclusion t.

We construct an argument to show that our premises lead to the desired conclusion as follows.

Step	Reason
1. $\neg p \wedge q$	Premise
2. $\neg p$	Simplification using (1)
3. $r \to p$	Premise
4. $\neg r$	Modus tollens using (2) and (3)
5. $\neg r \to s$	Premise
6. s	Modus ponens using (4) and (5)
7. $s \to t$	Premise
8. t	Modus ponens using (6) and (7)

Note that we could have used a truth table to show that whenever each of the four hypotheses is true, the conclusion is also true. However, because we are working with five propositional variables, p, q, r, s, and t, such a truth table would have 32 rows. ◀

EXAMPLE 7 Show that the premises "If you send me an email message, then I will finish writing the program," "If you do not send me an email message, then I will go to sleep early," and "If I go to sleep early, then I will wake up feeling refreshed" lead to the conclusion "If I do not finish writing the program, then I will wake up feeling refreshed."

Solution: Let p be the proposition "You send me an email message," q the proposition "I will finish writing the program," r the proposition "I will go to sleep early," and s the proposition "I will wake up feeling refreshed." Then the premises are $p \to q$, $\neg p \to r$, and $r \to s$. The desired conclusion is $\neg q \to s$. We need to give a valid argument with premises $p \to q$, $\neg p \to r$, and $r \to s$ and conclusion $\neg q \to s$.

This argument form shows that the premises lead to the desired conclusion.

Step	Reason
1. $p \to q$	Premise
2. $\neg q \to \neg p$	Contrapositive of (1)
3. $\neg p \to r$	Premise
4. $\neg q \to r$	Hypothetical syllogism using (2) and (3)
5. $r \to s$	Premise
6. $\neg q \to s$	Hypothetical syllogism using (4) and (5)

◀

1.6.5 Resolution

Links

Computer programs have been developed to automate the task of reasoning and proving theorems. Many of these programs make use of a rule of inference known as **resolution**. This rule of inference is based on the tautology

$$((p \vee q) \wedge (\neg p \vee r)) \rightarrow (q \vee r).$$

(Exercise 34 in Section 1.3 asks for the verification that this is a tautology.) The final disjunction in the resolution rule, $q \vee r$, is called the **resolvent**. When we let $q = r$ in this tautology, we obtain $(p \vee q) \wedge (\neg p \vee q) \rightarrow q$. Furthermore, when we let $r = \mathbf{F}$, we obtain $(p \vee q) \wedge (\neg p) \rightarrow q$ (because $q \vee \mathbf{F} \equiv q$), which is the tautology on which the rule of disjunctive syllogism is based.

EXAMPLE 8 Use resolution to show that the hypotheses "Jasmine is skiing or it is not snowing" and "It is snowing or Bart is playing hockey" imply that "Jasmine is skiing or Bart is playing hockey."

Extra Examples

Solution: Let p be the proposition "It is snowing," q the proposition "Jasmine is skiing," and r the proposition "Bart is playing hockey." We can represent the hypotheses as $\neg p \vee q$ and $p \vee r$, respectively. Using resolution, the proposition $q \vee r$, "Jasmine is skiing or Bart is playing hockey," follows. ◀

Resolution plays an important role in programming languages based on the rules of logic, such as Prolog (where resolution rules for quantified statements are applied). Furthermore, it can be used to build automatic theorem proving systems. To construct proofs in propositional logic using resolution as the only rule of inference, the hypotheses and the conclusion must be expressed as **clauses**, where a clause is a disjunction of variables or negations of these variables. We can replace a statement in propositional logic that is not a clause by one or more equivalent statements that are clauses. For example, suppose we have a statement of the form $p \vee (q \wedge r)$. Because $p \vee (q \wedge r) \equiv (p \vee q) \wedge (p \vee r)$, we can replace the single statement $p \vee (q \wedge r)$ by two statements $p \vee q$ and $p \vee r$, each of which is a clause. We can replace a statement of the form $\neg(p \vee q)$ by the two statements $\neg p$ and $\neg q$ because De Morgan's law tells us that $\neg(p \vee q) \equiv \neg p \wedge \neg q$. We can also replace a conditional statement $p \rightarrow q$ with the equivalent disjunction $\neg p \vee q$.

EXAMPLE 9 Show that the premises $(p \wedge q) \vee r$ and $r \rightarrow s$ imply the conclusion $p \vee s$.

Solution: We can rewrite the premises $(p \wedge q) \vee r$ as two clauses, $p \vee r$ and $q \vee r$. We can also replace $r \rightarrow s$ by the equivalent clause $\neg r \vee s$. Using the two clauses $p \vee r$ and $\neg r \vee s$, we can use resolution to conclude $p \vee s$. ◀

1.6.6 Fallacies

Several common fallacies arise in incorrect arguments. These fallacies resemble rules of inference, but are based on contingencies rather than tautologies. These are discussed here to show the distinction between correct and incorrect reasoning.

Links

The proposition $((p \rightarrow q) \wedge q) \rightarrow p$ is not a tautology, because it is false when p is false and q is true. However, there are many incorrect arguments that treat this as a tautology. In other words, they treat the argument with premises $p \rightarrow q$ and q and conclusion p as a valid argument form, which it is not. This type of incorrect reasoning is called the **fallacy of affirming the conclusion**.

EXAMPLE 10 Is the following argument valid?

If you do every problem in this book, then you will learn discrete mathematics. You learned discrete mathematics.

Therefore, you did every problem in this book.

Solution: Let p be the proposition "You did every problem in this book." Let q be the proposition "You learned discrete mathematics." Then this argument is of the form: if $p \to q$ and q, then p. This is an example of an incorrect argument using the fallacy of affirming the conclusion. Indeed, it is possible for you to learn discrete mathematics in some way other than by doing every problem in this book. (You may learn discrete mathematics by reading, listening to lectures, doing some, but not all, the problems in this book, and so on.) ◀

The proposition $((p \to q) \wedge \neg p) \to \neg q$ is not a tautology, because it is false when p is false and q is true. Many incorrect arguments use this incorrectly as a rule of inference. This type of incorrect reasoning is called the **fallacy of denying the hypothesis**.

EXAMPLE 11 Let p and q be as in Example 10. If the conditional statement $p \to q$ is true, and $\neg p$ is true, is it correct to conclude that $\neg q$ is true? In other words, is it correct to assume that you did not learn discrete mathematics if you did not do every problem in the book, assuming that if you do every problem in this book, then you will learn discrete mathematics?

Solution: It is possible that you learned discrete mathematics even if you did not do every problem in this book. This incorrect argument is of the form $p \to q$ and $\neg p$ imply $\neg q$, which is an example of the fallacy of denying the hypothesis. ◀

1.6.7 Rules of Inference for Quantified Statements

We have discussed rules of inference for propositions. We will now describe some important rules of inference for statements involving quantifiers. These rules of inference are used extensively in mathematical arguments, often without being explicitly mentioned.

Universal instantiation is the rule of inference used to conclude that $P(c)$ is true, where c is a particular member of the domain, given the premise $\forall x P(x)$. Universal instantiation is used when we conclude from the statement "All women are wise" that "Lisa is wise," where Lisa is a member of the domain of all women.

Universal generalization is the rule of inference that states that $\forall x P(x)$ is true, given the premise that $P(c)$ is true for all elements c in the domain. Universal generalization is used when we show that $\forall x P(x)$ is true by taking an arbitrary element c from the domain and showing that $P(c)$ is true. The element c that we select must be an arbitrary, and not a specific, element of the domain. That is, when we assert from $\forall x P(x)$ the existence of an element c in the domain, we have no control over c and cannot make any other assumptions about c other than it comes from the domain. Universal generalization is used implicitly in many proofs in mathematics and is seldom mentioned explicitly. However, the error of adding unwarranted assumptions about the arbitrary element c when universal generalization is used is all too common in incorrect reasoning.

Existential instantiation is the rule that allows us to conclude that there is an element c in the domain for which $P(c)$ is true if we know that $\exists x P(x)$ is true. We cannot select an arbitrary value of c here, but rather it must be a c for which $P(c)$ is true. Usually we have no knowledge of what c is, only that it exists. Because it exists, we may give it a name (c) and continue our argument.

Existential generalization is the rule of inference that is used to conclude that $\exists x P(x)$ is true when a particular element c with $P(c)$ true is known. That is, if we know one element c in the domain for which $P(c)$ is true, then we know that $\exists x P(x)$ is true.

TABLE 2 Rules of Inference for Quantified Statements.

Rule of Inference	*Name*
$\forall x P(x)$ $\therefore P(c)$	Universal instantiation
$P(c)$ for an arbitrary c $\therefore \forall x P(x)$	Universal generalization
$\exists x P(x)$ $\therefore P(c)$ for some element c	Existential instantiation
$P(c)$ for some element c $\therefore \exists x P(x)$	Existential generalization

We summarize these rules of inference in Table 2. We will illustrate how some of these rules of inference for quantified statements are used in Examples 12 and 13.

EXAMPLE 12 Show that the premises "Everyone in this discrete mathematics class has taken a course in computer science" and "Marla is a student in this class" imply the conclusion "Marla has taken a course in computer science."

Extra Examples

Solution: Let $D(x)$ denote "x is in this discrete mathematics class," and let $C(x)$ denote "x has taken a course in computer science." Then the premises are $\forall x(D(x) \to C(x))$ and $D(\text{Marla})$. The conclusion is $C(\text{Marla})$.

The following steps can be used to establish the conclusion from the premises.

Step	**Reason**
1. $\forall x(D(x) \to C(x))$	Premise
2. $D(\text{Marla}) \to C(\text{Marla})$	Universal instantiation from (1)
3. $D(\text{Marla})$	Premise
4. $C(\text{Marla})$	Modus ponens from (2) and (3)

◀

EXAMPLE 13 Show that the premises "A student in this class has not read the book," and "Everyone in this class passed the first exam" imply the conclusion "Someone who passed the first exam has not read the book."

Solution: Let $C(x)$ be "x is in this class," $B(x)$ be "x has read the book," and $P(x)$ be "x passed the first exam." The premises are $\exists x(C(x) \wedge \neg B(x))$ and $\forall x(C(x) \to P(x))$. The conclusion is $\exists x(P(x) \wedge \neg B(x))$. These steps can be used to establish the conclusion from the premises.

Step	**Reason**
1. $\exists x(C(x) \wedge \neg B(x))$	Premise
2. $C(a) \wedge \neg B(a)$	Existential instantiation from (1)
3. $C(a)$	Simplification from (2)
4. $\forall x(C(x) \to P(x))$	Premise
5. $C(a) \to P(a)$	Universal instantiation from (4)
6. $P(a)$	Modus ponens from (3) and (5)
7. $\neg B(a)$	Simplification from (2)
8. $P(a) \wedge \neg B(a)$	Conjunction from (6) and (7)
9. $\exists x(P(x) \wedge \neg B(x))$	Existential generalization from (8)

◀

1.6.8 Combining Rules of Inference for Propositions and Quantified Statements

We have developed rules of inference both for propositions and for quantified statements. Note that in our arguments in Examples 12 and 13 we used both universal instantiation, a rule of inference for quantified statements, and modus ponens, a rule of inference for propositional logic. We will often need to use this combination of rules of inference. Because universal instantiation and modus ponens are used so often together, this combination of rules is sometimes called **universal modus ponens**. This rule tells us that if $\forall x(P(x) \to Q(x))$ is true, and if $P(a)$ is true for a particular element a in the domain of the universal quantifier, then $Q(a)$ must also be true. To see this, note that by universal instantiation, $P(a) \to Q(a)$ is true. Then, by modus ponens, $Q(a)$ must also be true. We can describe universal modus ponens as follows:

$\forall x(P(x) \to Q(x))$
$P(a)$, where a is a particular element in the domain

$\therefore Q(a)$

Universal modus ponens is commonly used in mathematical arguments. This is illustrated in Example 14.

EXAMPLE 14 Assume that "For all positive integers n, if n is greater than 4, then n^2 is less than 2^n" is true. Use universal modus ponens to show that $100^2 < 2^{100}$.

Solution: Let $P(n)$ denote "$n > 4$" and $Q(n)$ denote "$n^2 < 2^n$." The statement "For all positive integers n, if n is greater than 4, then n^2 is less than 2^n" can be represented by $\forall n(P(n) \to Q(n))$, where the domain consists of all positive integers. We are assuming that $\forall n(P(n) \to Q(n))$ is true. Note that $P(100)$ is true because $100 > 4$. It follows by universal modus ponens that $Q(100)$ is true, namely, that $100^2 < 2^{100}$. ◀

Another useful combination of a rule of inference from propositional logic and a rule of inference for quantified statements is **universal modus tollens**. Universal modus tollens combines universal instantiation and modus tollens and can be expressed in the following way:

$\forall x(P(x) \to Q(x))$
$\neg Q(a)$, where a is a particular element in the domain

$\therefore \neg P(a)$

The verification of universal modus tollens is left as Exercise 25. Exercises 26–29 develop additional combinations of rules of inference in propositional logic and quantified statements.

Exercises

1. Find the argument form for the following argument and determine whether it is valid. Can we conclude that the conclusion is true if the premises are true?

If Socrates is human, then Socrates is mortal.
Socrates is human.

∴ Socrates is mortal.

2. Find the argument form for the following argument and determine whether it is valid. Can we conclude that the conclusion is true if the premises are true?

If George does not have eight legs, then he is not a spider.
George is a spider.

∴ George has eight legs.

3. What rule of inference is used in each of these arguments?

a) Alice is a mathematics major. Therefore, Alice is either a mathematics major or a computer science major.

b) Jerry is a mathematics major and a computer science major. Therefore, Jerry is a mathematics major.

c) If it is rainy, then the pool will be closed. It is rainy. Therefore, the pool is closed.

d) If it snows today, the university will close. The university is not closed today. Therefore, it did not snow today.

e) If I go swimming, then I will stay in the sun too long. If I stay in the sun too long, then I will sunburn. Therefore, if I go swimming, then I will sunburn.

4. What rule of inference is used in each of these arguments?

a) Kangaroos live in Australia and are marsupials. Therefore, kangaroos are marsupials.

b) It is either hotter than 100 degrees today or the pollution is dangerous. It is less than 100 degrees outside today. Therefore, the pollution is dangerous.

c) Linda is an excellent swimmer. If Linda is an excellent swimmer, then she can work as a lifeguard. Therefore, Linda can work as a lifeguard.

d) Steve will work at a computer company this summer. Therefore, this summer Steve will work at a computer company or he will be a beach bum.

e) If I work all night on this homework, then I can answer all the exercises. If I answer all the exercises, I will understand the material. Therefore, if I work all night on this homework, then I will understand the material.

5. Use rules of inference to show that the hypotheses "Randy works hard," "If Randy works hard, then he is a dull boy," and "If Randy is a dull boy, then he will not get the job" imply the conclusion "Randy will not get the job."

6. Use rules of inference to show that the hypotheses "If it does not rain or if it is not foggy, then the sailing race will be held and the lifesaving demonstration will go on," "If the sailing race is held, then the trophy will be awarded," and "The trophy was not awarded" imply the conclusion "It rained."

7. What rules of inference are used in this famous argument? "All men are mortal. Socrates is a man. Therefore, Socrates is mortal."

8. What rules of inference are used in this argument? "No man is an island. Manhattan is an island. Therefore, Manhattan is not a man."

9. For each of these collections of premises, what relevant conclusion or conclusions can be drawn? Explain the rules of inference used to obtain each conclusion from the premises.

a) "If I take the day off, it either rains or snows." "I took Tuesday off or I took Thursday off." "It was sunny on Tuesday." "It did not snow on Thursday."

b) "If I eat spicy foods, then I have strange dreams." "I have strange dreams if there is thunder while I sleep." "I did not have strange dreams."

c) "I am either clever or lucky." "I am not lucky." "If I am lucky, then I will win the lottery."

d) "Every computer science major has a personal computer." "Ralph does not have a personal computer." "Ann has a personal computer."

e) "What is good for corporations is good for the United States." "What is good for the United States is good for you." "What is good for corporations is for you to buy lots of stuff."

f) "All rodents gnaw their food." "Mice are rodents." "Rabbits do not gnaw their food." "Bats are not rodents."

10. For each of these sets of premises, what relevant conclusion or conclusions can be drawn? Explain the rules of inference used to obtain each conclusion from the premises.

a) "If I play hockey, then I am sore the next day." "I use the whirlpool if I am sore." "I did not use the whirlpool."

b) "If I work, it is either sunny or partly sunny." "I worked last Monday or I worked last Friday." "It was not sunny on Tuesday." "It was not partly sunny on Friday."

c) "All insects have six legs." "Dragonflies are insects." "Spiders do not have six legs." "Spiders eat dragonflies."

d) "Every student has an Internet account." "Homer does not have an Internet account." "Maggie has an Internet account."

e) "All foods that are healthy to eat do not taste good." "Tofu is healthy to eat." "You only eat what tastes good." "You do not eat tofu." "Cheeseburgers are not healthy to eat."

f) "I am either dreaming or hallucinating." "I am not dreaming." "If I am hallucinating, I see elephants running down the road."

11. Show that the argument form with premises $p_1, p_2, \ldots, p_n$ and conclusion $q \to r$ is valid if the argument form with premises $p_1, p_2, \ldots, p_n, q$, and conclusion r is valid.

12. Show that the argument form with premises $(p \wedge t) \to (r \vee s)$, $q \to (u \wedge t)$, $u \to p$, and $\neg s$ and conclusion $q \to r$ is valid by first using Exercise 11 and then using rules of inference from Table 1.

13. For each of these arguments, explain which rules of inference are used for each step.

a) "Doug, a student in this class, knows how to write programs in JAVA. Everyone who knows how to write programs in JAVA can get a high-paying job. Therefore, someone in this class can get a high-paying job."

b) "Somebody in this class enjoys whale watching. Every person who enjoys whale watching cares about ocean pollution. Therefore, there is a person in this class who cares about ocean pollution."

c) "Each of the 93 students in this class owns a personal computer. Everyone who owns a personal computer can use a word processing program. Therefore, Zeke, a student in this class, can use a word processing program."

d) "Everyone in New Jersey lives within 50 miles of the ocean. Someone in New Jersey has never seen the ocean. Therefore, someone who lives within 50 miles of the ocean has never seen the ocean."

14. For each of these arguments, explain which rules of inference are used for each step.

a) "Linda, a student in this class, owns a red convertible. Everyone who owns a red convertible has gotten at least one speeding ticket. Therefore, someone in this class has gotten a speeding ticket."

b) "Each of five roommates, Melissa, Aaron, Ralph, Veneesha, and Keeshawn, has taken a course in discrete mathematics. Every student who has taken a course in discrete mathematics can take a course in algorithms. Therefore, all five roommates can take a course in algorithms next year."

c) "All movies produced by John Sayles are wonderful. John Sayles produced a movie about coal miners. Therefore, there is a wonderful movie about coal miners."

d) "There is someone in this class who has been to France. Everyone who goes to France visits the Louvre. Therefore, someone in this class has visited the Louvre."

15. For each of these arguments determine whether the argument is correct or incorrect and explain why.

a) All students in this class understand logic. Xavier is a student in this class. Therefore, Xavier understands logic.

b) Every computer science major takes discrete mathematics. Natasha is taking discrete mathematics. Therefore, Natasha is a computer science major.

c) All parrots like fruit. My pet bird is not a parrot. Therefore, my pet bird does not like fruit.

d) Everyone who eats granola every day is healthy. Linda is not healthy. Therefore, Linda does not eat granola every day.

16. For each of these arguments determine whether the argument is correct or incorrect and explain why.

a) Everyone enrolled in the university has lived in a dormitory. Mia has never lived in a dormitory. Therefore, Mia is not enrolled in the university.

b) A convertible car is fun to drive. Isaac's car is not a convertible. Therefore, Isaac's car is not fun to drive.

c) Quincy likes all action movies. Quincy likes the movie *Eight Men Out*. Therefore, *Eight Men Out* is an action movie.

d) All lobstermen set at least a dozen traps. Hamilton is a lobsterman. Therefore, Hamilton sets at least a dozen traps.

17. What is wrong with this argument? Let $H(x)$ be "x is happy." Given the premise $\exists x H(x)$, we conclude that $H(\text{Lola})$. Therefore, Lola is happy.

18. What is wrong with this argument? Let $S(x, y)$ be "x is shorter than y." Given the premise $\exists s S(s, \text{Max})$, it follows that $S(\text{Max}, \text{Max})$. Then by existential generalization it follows that $\exists x S(x, x)$, so that someone is shorter than himself.

19. Determine whether each of these arguments is valid. If an argument is correct, what rule of inference is being used? If it is not, what logical error occurs?

a) If n is a real number such that $n > 1$, then $n^2 > 1$. Suppose that $n^2 > 1$. Then $n > 1$.

b) If n is a real number with $n > 3$, then $n^2 > 9$. Suppose that $n^2 \leq 9$. Then $n \leq 3$.

c) If n is a real number with $n > 2$, then $n^2 > 4$. Suppose that $n \leq 2$. Then $n^2 \leq 4$.

20. Determine whether these are valid arguments.

a) If x is a positive real number, then x^2 is a positive real number. Therefore, if a^2 is positive, where a is a real number, then a is a positive real number.

b) If $x^2 \neq 0$, where x is a real number, then $x \neq 0$. Let a be a real number with $a^2 \neq 0$; then $a \neq 0$.

21. Which rules of inference are used to establish the conclusion of Lewis Carroll's argument described in Example 26 of Section 1.4?

22. Which rules of inference are used to establish the conclusion of Lewis Carroll's argument described in Example 27 of Section 1.4?

23. Identify the error or errors in this argument that supposedly shows that if $\exists x P(x) \wedge \exists x Q(x)$ is true then $\exists x (P(x) \wedge Q(x))$ is true.

1. $\exists x P(x) \vee \exists x Q(x)$	Premise
2. $\exists x P(x)$	Simplification from (1)
3. $P(c)$	Existential instantiation from (2)
4. $\exists x Q(x)$	Simplification from (1)
5. $Q(c)$	Existential instantiation from (4)
6. $P(c) \wedge Q(c)$	Conjunction from (3) and (5)
7. $\exists x (P(x) \wedge Q(x))$	Existential generalization

24. Identify the error or errors in this argument that supposedly shows that if $\forall x (P(x) \vee Q(x))$ is true then $\forall x P(x) \vee \forall x Q(x)$ is true.

1. $\forall x (P(x) \vee Q(x))$	Premise
2. $P(c) \vee Q(c)$	Universal instantiation from (1)
3. $P(c)$	Simplification from (2)
4. $\forall x P(x)$	Universal generalization from (3)
5. $Q(c)$	Simplification from (2)
6. $\forall x Q(x)$	Universal generalization from (5)
7. $\forall x (P(x) \vee \forall x Q(x))$	Conjunction from (4) and (6)

25. Justify the rule of universal modus tollens by showing that the premises $\forall x (P(x) \rightarrow Q(x))$ and $\neg Q(a)$ for a particular element a in the domain, imply $\neg P(a)$.

26. Justify the rule of **universal transitivity**, which states that if $\forall x (P(x) \rightarrow Q(x))$ and $\forall x (Q(x) \rightarrow R(x))$ are true, then $\forall x (P(x) \rightarrow R(x))$ is true, where the domains of all quantifiers are the same.

27. Use rules of inference to show that if $\forall x (P(x) \rightarrow (Q(x) \wedge S(x)))$ and $\forall x (P(x) \wedge R(x))$ are true, then $\forall x (R(x) \wedge S(x))$ is true.

28. Use rules of inference to show that if $\forall x(P(x) \vee Q(x))$ and $\forall x((\neg P(x) \wedge Q(x)) \rightarrow R(x))$ are true, then $\forall x(\neg R(x) \rightarrow P(x))$ is also true, where the domains of all quantifiers are the same.

29. Use rules of inference to show that if $\forall x(P(x) \vee Q(x))$, $\forall x(\neg Q(x) \vee S(x))$, $\forall x(R(x) \rightarrow \neg S(x))$, and $\exists x \neg P(x)$ are true, then $\exists x \neg R(x)$ is true.

30. Use resolution to show the hypotheses "Allen is a bad boy or Hillary is a good girl" and "Allen is a good boy or David is happy" imply the conclusion "Hillary is a good girl or David is happy."

31. Use resolution to show that the hypotheses "It is not raining or Yvette has her umbrella," "Yvette does not have her umbrella or she does not get wet," and "It is raining or Yvette does not get wet" imply that "Yvette does not get wet."

32. Show that the equivalence $p \wedge \neg p \equiv \mathbf{F}$ can be derived using resolution together with the fact that a conditional statement with a false hypothesis is true. [*Hint:* Let $q = r = \mathbf{F}$ in resolution.]

33. Use resolution to show that the compound proposition $(p \vee q) \wedge (\neg p \vee q) \wedge (p \vee \neg q) \wedge (\neg p \vee \neg q)$ is not satisfiable.

***34.** The Logic Problem, taken from *WFF'N PROOF, The Game of Logic*, has these two assumptions:
1. "Logic is difficult or not many students like logic."
2. "If mathematics is easy, then logic is not difficult."

By translating these assumptions into statements involving propositional variables and logical connectives, determine whether each of the following are valid conclusions of these assumptions:

a) That mathematics is not easy, if many students like logic.
b) That not many students like logic, if mathematics is not easy.
c) That mathematics is not easy or logic is difficult.
d) That logic is not difficult or mathematics is not easy.
e) That if not many students like logic, then either mathematics is not easy or logic is not difficult.

***35.** Determine whether this argument, taken from Kalish and Montague [KaMo64], is valid.

> If Superman were able and willing to prevent evil, he would do so. If Superman were unable to prevent evil, he would be impotent; if he were unwilling to prevent evil, he would be malevolent. Superman does not prevent evil. If Superman exists, he is neither impotent nor malevolent. Therefore, Superman does not exist.

1.7 Introduction to Proofs

1.7.1 Introduction

In this section we introduce the notion of a proof and describe methods for constructing proofs. A proof is a valid argument that establishes the truth of a mathematical statement. A proof can use the hypotheses of the theorem, if any, axioms assumed to be true, and previously proven theorems. Using these ingredients and rules of inference, the final step of the proof establishes the truth of the statement being proved.

In our discussion we move from formal proofs of theorems toward more informal proofs. The arguments we introduced in Section 1.6 to show that statements involving propositions and quantified statements are true were formal proofs, where all steps were supplied, and the rules for each step in the argument were given. However, formal proofs of useful theorems can be extremely long and hard to follow. In practice, the proofs of theorems designed for human consumption are almost always **informal proofs**, where more than one rule of inference may be used in each step, where steps may be skipped, where the axioms being assumed and the rules of inference used are not explicitly stated. Informal proofs can often explain to humans why theorems are true, while computers are perfectly happy producing formal proofs using automated reasoning systems.

The methods of proof discussed in this chapter are important not only because they are used to prove mathematical theorems, but also for their many applications to computer science. These applications include verifying that computer programs are correct, establishing that operating systems are secure, making inferences in artificial intelligence, showing that system specifications are consistent, and so on. Consequently, understanding the techniques used in proofs is essential both in mathematics and in computer science.

1.7.2 Some Terminology

Links

Formally, a **theorem** is a statement that can be shown to be true. In mathematical writing, the term theorem is usually reserved for a statement that is considered at least somewhat important. Less important theorems sometimes are called **propositions**. (Theorems can also be referred to as **facts** or **results**.) A theorem may be the universal quantification of a conditional statement with one or more premises and a conclusion. However, it may be some other type of logical statement, as the examples later in this chapter will show. We demonstrate that a theorem is true with a **proof**. A proof is a valid argument that establishes the truth of a theorem. The statements used in a proof can include **axioms** (or **postulates**), which are statements we assume to be true (for example, the axioms for the real numbers, given in Appendix 1, and the axioms of plane geometry), the premises, if any, of the theorem, and previously proven theorems. Axioms may be stated using primitive terms that do not require definition, but all other terms used in theorems and their proofs must be defined. Rules of inference, together with definitions of terms, are used to draw conclusions from other assertions, tying together the steps of a proof. In practice, the final step of a proof is usually just the conclusion of the theorem. However, for clarity, we will often recap the statement of the theorem as the final step of a proof.

A less important theorem that is helpful in the proof of other results is called a **lemma** (plural *lemmas* or *lemmata*). Complicated proofs are usually easier to understand when they are proved using a series of lemmas, where each lemma is proved individually. A **corollary** is a theorem that can be established directly from a theorem that has been proved. A **conjecture** is a statement that is being proposed to be a true statement, usually on the basis of some partial evidence, a heuristic argument, or the intuition of an expert. When a proof of a conjecture is found, the conjecture becomes a theorem. Many times conjectures are shown to be false, so they are not theorems.

1.7.3 Understanding How Theorems Are Stated

Extra Examples

Before we introduce methods for proving theorems, we need to understand how many mathematical theorems are stated. Many theorems assert that a property holds for all elements in a domain, such as the integers or the real numbers. Although the precise statement of such theorems needs to include a universal quantifier, the standard convention in mathematics is to omit it. For example, the statement

"If $x > y$, where x and y are positive real numbers, then $x^2 > y^2$"

really means

"For all positive real numbers x and y, if $x > y$, then $x^2 > y^2$."

Furthermore, when theorems of this type are proved, the first step of the proof usually involves selecting a general element of the domain. Subsequent steps show that this element has the property in question. Finally, universal generalization implies that the theorem holds for all members of the domain.

1.7.4 Methods of Proving Theorems

Assessment

Proving mathematical theorems can be difficult. To construct proofs we need all available ammunition, including a powerful battery of different proof methods. These methods provide the overall approach and strategy of proofs. Understanding these methods is a key component of learning how to read and construct mathematical proofs. Once we have chosen a proof method, we use axioms, definitions of terms, previously proved results, and rules of inference to complete the proof. Note that in this book we will always assume the axioms for real numbers found in Appendix 1. We will also assume the usual axioms whenever we prove a result about geometry. When you construct your own proofs, be careful not to use anything but these axioms, definitions, and previously proved results as facts!

To prove a theorem of the form $\forall x(P(x) \rightarrow Q(x))$, our goal is to show that $P(c) \rightarrow Q(c)$ is true, where c is an arbitrary element of the domain, and then apply universal generalization. In this proof, we need to show that a conditional statement is true. Because of this, we now focus on methods that show that conditional statements are true. Recall that $p \rightarrow q$ is true unless p is true but q is false. Note that to prove the statement $p \rightarrow q$, we need only show that q is true if p is true. The following discussion will give the most common techniques for proving conditional statements. Later we will discuss methods for proving other types of statements. In this section, and in Section 1.8, we will develop a large arsenal of proof techniques that can be used to prove a wide variety of theorems.

When you read proofs, you will often find the words "obviously" or "clearly." These words indicate that steps have been omitted that the author expects the reader to be able to fill in. Unfortunately, this assumption is often not warranted and readers are not at all sure how to fill in the gaps. We will assiduously try to avoid using these words and try not to omit too many steps. However, if we included all steps in proofs, our proofs would often be excruciatingly long.

1.7.5 Direct Proofs

A **direct proof** of a conditional statement $p \rightarrow q$ is constructed when the first step is the assumption that p is true; subsequent steps are constructed using rules of inference, with the final step showing that q must also be true. A direct proof shows that a conditional statement $p \rightarrow q$ is true by showing that if p is true, then q must also be true, so that the combination p true and q false never occurs. In a direct proof, we assume that p is true and use axioms, definitions, and previously proven theorems, together with rules of inference, to show that q must also be true. You will find that direct proofs of many results are quite straightforward. Starting with the hypothesis and leading to the conclusion, the way forward is essentially dictated by the premises available at that step. However, direct proofs sometimes require particular insights and can be quite tricky. The first direct proofs we present here are quite straightforward; later in the text you will see some that require some insight.

We will provide examples of several different direct proofs. Before we give the first example, we need to define some terminology.

Definition 1 The integer n is *even* if there exists an integer k such that $n = 2k$, and n is *odd* if there exists an integer k such that $n = 2k + 1$. (Note that every integer is either even or odd, and no integer is both even and odd.) Two integers have the *same parity* when both are even or both are odd; they have *opposite parity* when one is even and the other is odd.

EXAMPLE 1 Give a direct proof of the theorem "If n is an odd integer, then n^2 is odd."

Extra Examples

Solution: Note that this theorem states $\forall n(P(n) \rightarrow Q(n))$, where $P(n)$ is "n is an odd integer" and $Q(n)$ is "n^2 is odd." As we have said, we will follow the usual convention in mathematical proofs by showing that $P(n)$ implies $Q(n)$, and not explicitly using universal instantiation. To begin a direct proof of this theorem, we assume that the hypothesis of this conditional statement is true, namely, we assume that n is odd. By the definition of an odd integer, it follows that $n = 2k + 1$, where k is some integer. We want to show that n^2 is also odd. We can square both sides of the equation $n = 2k + 1$ to obtain a new equation that expresses n^2. When we do this, we find that $n^2 = (2k + 1)^2 = 4k^2 + 4k + 1 = 2(2k^2 + 2k) + 1$. By the definition of an odd integer, we can conclude that n^2 is an odd integer (it is one more than twice an integer). Consequently, we have proved that if n is an odd integer, then n^2 is an odd integer. ◄

EXAMPLE 2 Give a direct proof that if m and n are both perfect squares, then nm is also a perfect square. (An integer a is a **perfect square** if there is an integer b such that $a = b^2$.)

Solution: To produce a direct proof of this theorem, we assume that the hypothesis of this conditional statement is true, namely, we assume that m and n are both perfect squares. By the definition of a perfect square, it follows that there are integers s and t such that $m = s^2$ and $n = t^2$. The goal of the proof is to show that mn must also be a perfect square when m and n are; looking ahead we see how we can show this by substituting s^2 for m and t^2 for n into mn. This tells us that $mn = s^2t^2$. Hence, $mn = s^2t^2 = (ss)(tt) = (st)(st) = (st)^2$, using commutativity and associativity of multiplication. By the definition of perfect square, it follows that mn is also a perfect square, because it is the square of st, which is an integer. We have proved that if m and n are both perfect squares, then mn is also a perfect square. ◀

1.7.6 Proof by Contraposition

Direct proofs lead from the premises of a theorem to the conclusion. They begin with the premises, continue with a sequence of deductions, and end with the conclusion. However, we will see that attempts at direct proofs often reach dead ends. We need other methods of proving theorems of the form $\forall x(P(x) \to Q(x))$. Proofs of theorems of this type that are not direct proofs, that is, that do not start with the premises and end with the conclusion, are called **indirect proofs**.

An extremely useful type of indirect proof is known as **proof by contraposition**. Proofs by contraposition make use of the fact that the conditional statement $p \to q$ is equivalent to its contrapositive, $\neg q \to \neg p$. This means that the conditional statement $p \to q$ can be proved by showing that its contrapositive, $\neg q \to \neg p$, is true. In a proof by contraposition of $p \to q$, we take $\neg q$ as a premise, and using axioms, definitions, and previously proven theorems, together with rules of inference, we show that $\neg p$ must follow. We will illustrate proof by contraposition with two examples. These examples show that proof by contraposition can succeed when we cannot easily find a direct proof.

EXAMPLE 3 Prove that if n is an integer and $3n + 2$ is odd, then n is odd.

Extra Examples

Solution: We first attempt a direct proof. To construct a direct proof, we first assume that $3n + 2$ is an odd integer. From the definition of an odd integer, we know that $3n + 2 = 2k + 1$ for some integer k. Can we use this fact to show that n is odd? We see that $3n + 1 = 2k$, but there does not seem to be any direct way to conclude that n is odd. Because our attempt at a direct proof failed, we next try a proof by contraposition.

The first step in a proof by contraposition is to assume that the conclusion of the conditional statement "If $3n + 2$ is odd, then n is odd" is false; namely, assume that n is even. Then, by the definition of an even integer, $n = 2k$ for some integer k. Substituting $2k$ for n, we find that $3n + 2 = 3(2k) + 2 = 6k + 2 = 2(3k + 1)$. This tells us that $3n + 2$ is even (because it is a multiple of 2), and therefore not odd. This is the negation of the premise of the theorem. Because the negation of the conclusion of the conditional statement implies that the hypothesis is false, the original conditional statement is true. Our proof by contraposition succeeded; we have proved the theorem "If $3n + 2$ is odd, then n is odd." ◀

EXAMPLE 4 Prove that if $n = ab$, where a and b are positive integers, then $a \leq \sqrt{n}$ or $b \leq \sqrt{n}$.

Solution: Because there is no obvious way of showing that $a \leq \sqrt{n}$ or $b \leq \sqrt{n}$ directly from the equation $n = ab$, where a and b are positive integers, we attempt a proof by contraposition.

The first step in a proof by contraposition is to assume that the conclusion of the conditional statement "If $n = ab$, where a and b are positive integers, then $a \leq \sqrt{n}$ or $b \leq \sqrt{n}$" is false. That is, we assume that the statement $(a \leq \sqrt{n}) \vee (b \leq \sqrt{n})$ is false. Using the meaning of disjunction together with De Morgan's law, we see that this implies that both $a \leq \sqrt{n}$ and $b \leq \sqrt{n}$ are false. This implies that $a > \sqrt{n}$ and $b > \sqrt{n}$. We can multiply these inequalities together (using the fact that if $0 < s < t$ and $0 < u < v$, then $su < tv$) to obtain $ab > \sqrt{n} \cdot \sqrt{n} = n$. This shows that $ab \neq n$, which contradicts the statement $n = ab$.

Because the negation of the conclusion of the conditional statement implies that the hypothesis is false, the original conditional statement is true. Our proof by contraposition succeeded; we have proved that if $n = ab$, where a and b are positive integers, then $a \leq \sqrt{n}$ or $b \leq \sqrt{n}$. ◀

VACUOUS AND TRIVIAL PROOFS We can quickly prove that a conditional statement $p \rightarrow q$ is true when we know that p is false, because $p \rightarrow q$ must be true when p is false. Consequently, if we can show that p is false, then we have a proof, called a **vacuous proof**, of the conditional statement $p \rightarrow q$. Vacuous proofs are often used to establish special cases of theorems that state that a conditional statement is true for all positive integers [i.e., a theorem of the kind $\forall n P(n)$, where $P(n)$ is a propositional function]. Proof techniques for theorems of this kind will be discussed in Section 5.1.

EXAMPLE 5 Show that the proposition $P(0)$ is true, where $P(n)$ is "If $n > 1$, then $n^2 > n$" and the domain consists of all integers.

Solution: Note that $P(0)$ is "If $0 > 1$, then $0^2 > 0$." We can show $P(0)$ using a vacuous proof. Indeed, the hypothesis $0 > 1$ is false. This tells us that $P(0)$ is automatically true. ◀

Remark: The fact that the conclusion of this conditional statement, $0^2 > 0$, is false is irrelevant to the truth value of the conditional statement, because a conditional statement with a false hypothesis is guaranteed to be true.

EXAMPLE 6 Prove that if n is an integer with $10 \leq n \leq 15$ which is a perfect square, then n is also a perfect cube.

Solution: Note that there are no perfect squares n with $10 \leq n \leq 15$, because $3^2 = 9$ and $4^2 = 16$. Hence, the statement that n is an integer with $10 \leq n \leq 15$ which is a perfect square is false for all integers n. Consequently, the statement to be proved is true for all integers n. ◀

We can also quickly prove a conditional statement $p \rightarrow q$ if we know that the conclusion q is true. By showing that q is true, it follows that $p \rightarrow q$ must also be true. A proof of $p \rightarrow q$ that uses the fact that q is true is called a **trivial proof**. Trivial proofs are often important when special cases of theorems are proved (see the discussion of proof by cases in Section 1.8) and in mathematical induction, which is a proof technique discussed in Section 5.1.

EXAMPLE 7 Let $P(n)$ be "If a and b are positive integers with $a \geq b$, then $a^n \geq b^n$," where the domain consists of all nonnegative integers. Show that $P(0)$ is true.

Solution: The proposition $P(0)$ is "If $a \geq b$, then $a^0 \geq b^0$." Because $a^0 = b^0 = 1$, the conclusion of the conditional statement "If $a \geq b$, then $a^0 \geq b^0$" is true. Hence, this conditional statement, which is $P(0)$, is true. This is an example of a trivial proof. Note that the hypothesis, which is the statement "$a \geq b$," was not needed in this proof. ◀

A LITTLE PROOF STRATEGY We have described two important approaches for proving theorems of the form $\forall x(P(x) \rightarrow Q(x))$: direct proof and proof by contraposition. We have also given examples that show how each is used. However, when you are presented with a theorem of the form $\forall x(P(x) \rightarrow Q(x))$, which method should you use to attempt to prove it? We will provide a few rules of thumb here; in Section 1.8 we will discuss proof strategy at greater length.

When you want to prove a statement of the form $\forall x(P(x) \rightarrow Q(x))$, first evaluate whether a direct proof looks promising. Begin by expanding the definitions in the hypotheses. Start to reason using these hypotheses, together with axioms and available theorems. If a direct proof does not seem to go anywhere, for instance when there is no clear way to use hypotheses as in Examples 3 and 4 to reach the conclusion, try the same thing with a proof by contraposition.

(Hypotheses such as x is irrational or $x \neq 0$ that are difficult to reason from are a clue that an indirect proof might be your best best.)

Recall that in a proof by contraposition you assume that the conclusion of the conditional statement is false and use a direct proof to show this implies that the hypothesis must be false. Often, you will find that a proof by contraposition is easily constructed from the negation of the conclusion. We illustrate this strategy in Examples 7 and 8. In each example, note how straightforward a proof by contraposition is, while there is no clear way to provide a direct proof.

Before we present our next example, we need a definition.

Definition 2 The real number r is *rational* if there exist integers p and q with $q \neq 0$ such that $r = p/q$. A real number that is not rational is called *irrational*.

EXAMPLE 8 Prove that the sum of two rational numbers is rational. (Note that if we include the implicit quantifiers here, the theorem we want to prove is "For every real number r and every real number s, if r and s are rational numbers, then $r + s$ is rational.")

Extra Examples

Solution: We first attempt a direct proof. To begin, suppose that r and s are rational numbers. From the definition of a rational number, it follows that there are integers p and q, with $q \neq 0$, such that $r = p/q$, and integers t and u, with $u \neq 0$, such that $s = t/u$. Can we use this information to show that $r + s$ is rational? That is, can we find integers v and w such that $r + s = v/w$ and $w \neq 0$?

With the goal of finding these integers v and w, we add $r = p/q$ and $s = t/u$, using qu as the common denominator. We find that

$$r + s = \frac{p}{q} + \frac{t}{u} = \frac{pu + qt}{qu}.$$

Because $q \neq 0$ and $u \neq 0$, it follows that $qu \neq 0$. Consequently, we have expressed $r + s$ as the ratio of two integers, $v = pu + qt$ and $w = qu$, where $w \neq 0$. This means that $r + s$ is rational. We have proved that the sum of two rational numbers is rational; our attempt to find a direct proof succeeded. ◀

EXAMPLE 9 Prove that if n is an integer and n^2 is odd, then n is odd.

Solution: We first attempt a direct proof. Suppose that n is an integer and n^2 is odd. From the definition of an odd integer, there exists an integer k such that $n^2 = 2k + 1$. Can we use this information to show that n is odd? There seems to be no obvious approach to show that n is odd because solving for n produces the equation $n = \pm\sqrt{2k + 1}$, which is not terribly useful.

Because this attempt to use a direct proof did not bear fruit, we next attempt a proof by contraposition. We take as our hypothesis the statement that n is not odd. Because every integer is odd or even, this means that n is even. This implies that there exists an integer k such that $n = 2k$. To prove the theorem, we need to show that this hypothesis implies the conclusion that n^2 is not odd, that is, that n^2 is even. Can we use the equation $n = 2k$ to achieve this? By squaring both sides of this equation, we obtain $n^2 = 4k^2 = 2(2k^2)$, which implies that n^2 is also even because $n^2 = 2t$, where $t = 2k^2$. We have proved that if n is an integer and n^2 is odd, then n is odd. Our attempt to find a proof by contraposition succeeded. ◀

1.7.7 Proofs by Contradiction

Suppose we want to prove that a statement p is true. Furthermore, suppose that we can find a contradiction q such that $\neg p \rightarrow q$ is true. Because q is false, but $\neg p \rightarrow q$ is true, we can conclude that $\neg p$ is false, which means that p is true. How can we find a contradiction q that might help us prove that p is true in this way?

Because the statement $r \wedge \neg r$ is a contradiction whenever r is a proposition, we can prove that p is true if we can show that $\neg p \rightarrow (r \wedge \neg r)$ is true for some proposition r. Proofs of this type are called **proofs by contradiction**. Because a proof by contradiction does not prove a result directly, it is another type of indirect proof. We provide three examples of proof by contradiction. The first is an example of an application of the pigeonhole principle, a combinatorial technique that we will cover in depth in Section 6.2.

EXAMPLE 10 Show that at least four of any 22 days must fall on the same day of the week.

Extra Examples

Solution: Let p be the proposition "At least four of 22 chosen days fall on the same day of the week." Suppose that $\neg p$ is true. This means that at most three of the 22 days fall on the same day of the week. Because there are seven days of the week, this implies that at most 21 days could have been chosen, as for each of the days of the week, at most three of the chosen days could fall on that day. This contradicts the premise that we have 22 days under consideration. That is, if r is the statement that 22 days are chosen, then we have shown that $\neg p \rightarrow (r \wedge \neg r)$. Consequently, we know that p is true. We have proved that at least four of 22 chosen days fall on the same day of the week. ◀

EXAMPLE 11 Prove that $\sqrt{2}$ is irrational by giving a proof by contradiction.

Solution: Let p be the proposition "$\sqrt{2}$ is irrational." To start a proof by contradiction, we suppose that $\neg p$ is true. Note that $\neg p$ is the statement "It is not the case that $\sqrt{2}$ is irrational," which says that $\sqrt{2}$ is rational. We will show that assuming that $\neg p$ is true leads to a contradiction.

If $\sqrt{2}$ is rational, there exist integers a and b with $\sqrt{2} - a/b$, where $b \neq 0$ and a and b have no common factors (so that the fraction a/b is in lowest terms). (Here, we are using the fact that every rational number can be written in lowest terms.) Because $\sqrt{2} = a/b$, when both sides of this equation are squared, it follows that

$$2 = \frac{a^2}{b^2}.$$

Hence,

$$2b^2 = a^2.$$

By the definition of an even integer it follows that a^2 is even. We next use the fact that if a^2 is even, a must also be even, which follows by Exercise 18. Furthermore, because a is even, by the definition of an even integer, $a = 2c$ for some integer c. Thus,

$$2b^2 = 4c^2.$$

Dividing both sides of this equation by 2 gives

$$b^2 = 2c^2.$$

By the definition of even, this means that b^2 is even. Again using the fact that if the square of an integer is even, then the integer itself must be even, we conclude that b must be even as well.

We have now shown that the assumption of $\neg p$ leads to the equation $\sqrt{2} = a/b$, where a and b have no common factors, but both a and b are even, that is, 2 divides both a and b. Note that the statement that $\sqrt{2} = a/b$, where a and b have no common factors, means, in particular, that 2 does not divide both a and b. Because our assumption of $\neg p$ leads to the contradiction that 2 divides both a and b and 2 does not divide both a and b, $\neg p$ must be false. That is, the statement p, "$\sqrt{2}$ is irrational," is true. We have proved that $\sqrt{2}$ is irrational. ◀

Proof by contradiction can be used to prove conditional statements. In such proofs, we first assume the negation of the conclusion. We then use the premises of the theorem and the negation of the conclusion to arrive at a contradiction. (The reason that such proofs are valid rests on the logical equivalence of $p \to q$ and $(p \wedge \neg q) \to \mathbf{F}$. To see that these statements are equivalent, simply note that each is false in exactly one case, namely, when p is true and q is false.)

Note that we can rewrite a proof by contraposition of a conditional statement as a proof by contradiction. In a proof of $p \to q$ by contraposition, we assume that $\neg q$ is true. We then show that $\neg p$ must also be true. To rewrite a proof by contraposition of $p \to q$ as a proof by contradiction, we suppose that both p and $\neg q$ are true. Then, we use the steps from the proof of $\neg q \to \neg p$ to show that $\neg p$ is true. This leads to the contradiction $p \wedge \neg p$, completing the proof. Example 12 illustrates how a proof by contraposition of a conditional statement can be rewritten as a proof by contradiction.

EXAMPLE 12 Give a proof by contradiction of the theorem "If $3n + 2$ is odd, then n is odd."

Solution: Let p be "$3n + 2$ is odd" and q be "n is odd." To construct a proof by contradiction, assume that both p and $\neg q$ are true. That is, assume that $3n + 2$ is odd and that n is not odd. Because n is not odd, we know that it is even. Because n is even, there is an integer k such that $n = 2k$. This implies that $3n + 2 = 3(2k) + 2 = 6k + 2 = 2(3k + 1)$. Because $3n + 2$ is $2t$, where $t = 3k + 1$, $3n + 2$ is even. Note that the statement "$3n + 2$ is even" is equivalent to the statement $\neg p$, because an integer is even if and only if it is not odd. Because both p and $\neg p$ are true, we have a contradiction. This completes the proof by contradiction, proving that if $3n + 2$ is odd, then n is odd. ◀

Note that we can also prove by contradiction that $p \to q$ is true by assuming that p and $\neg q$ are true, and showing that q must be also be true. This implies that $\neg q$ and q are both true, a contradiction. This observation tells us that we can turn a direct proof into a proof by contradiction.

PROOFS OF EQUIVALENCE To prove a theorem that is a biconditional statement, that is, a statement of the form $p \leftrightarrow q$, we show that $p \rightarrow q$ and $q \rightarrow p$ are both true. The validity of this approach is based on the tautology

$$(p \leftrightarrow q) \leftrightarrow (p \rightarrow q) \wedge (q \rightarrow p).$$

EXAMPLE 13 Prove the theorem "If n is an integer, then n is odd if and only if n^2 is odd."

Extra Examples

Solution: This theorem has the form "p if and only if q," where p is "n is odd" and q is "n^2 is odd." (As usual, we do not explicitly deal with the universal quantification.) To prove this theorem, we need to show that $p \rightarrow q$ and $q \rightarrow p$ are true.

We have already shown (in Example 1) that $p \rightarrow q$ is true and (in Example 9) that $q \rightarrow p$ is true.

Because we have shown that both $p \rightarrow q$ and $q \rightarrow p$ are true, we have shown that the theorem is true. ◀

Sometimes a theorem states that several propositions are equivalent. Such a theorem states that propositions $p_1, p_2, p_3, \ldots, p_n$ are equivalent. This can be written as

$$p_1 \leftrightarrow p_2 \leftrightarrow \cdots \leftrightarrow p_n,$$

which states that all n propositions have the same truth values, and consequently, that for all i and j with $1 \leq i \leq n$ and $1 \leq j \leq n$, p_i and p_j are equivalent. One way to prove these are mutually equivalent is to use the tautology

$$p_1 \leftrightarrow p_2 \leftrightarrow \cdots \leftrightarrow p_n \leftrightarrow (p_1 \rightarrow p_2) \wedge (p_2 \rightarrow p_3) \wedge \cdots \wedge (p_n \rightarrow p_1).$$

This shows that if the n conditional statements $p_1 \rightarrow p_2, p_2 \rightarrow p_3, \ldots, p_n \rightarrow p_1$ can be shown to be true, then the propositions $p_1, p_2, \ldots, p_n$ are all equivalent.

This is much more efficient than proving that $p_i \rightarrow p_j$ for all $i \neq j$ with $1 \leq i \leq n$ and $1 \leq j \leq n$. (Note that there are $n^2 - n$ such conditional statements.)

When we prove that a group of statements are equivalent, we can establish any chain of conditional statements we choose as long as it is possible to work through the chain to go from any one of these statements to any other statement. For example, we can show that p_1, p_2, and p_3 are equivalent by showing that $p_1 \rightarrow p_3$, $p_3 \rightarrow p_2$, and $p_2 \rightarrow p_1$.

EXAMPLE 14 Show that these statements about the integer n are equivalent:

p_1: n is even.
p_2: $n - 1$ is odd.
p_3: n^2 is even.

Solution: We will show that these three statements are equivalent by showing that the conditional statements $p_1 \rightarrow p_2$, $p_2 \rightarrow p_3$, and $p_3 \rightarrow p_1$ are true.

We use a direct proof to show that $p_1 \rightarrow p_2$. Suppose that n is even. Then $n = 2k$ for some integer k. Consequently, $n - 1 = 2k - 1 = 2(k - 1) + 1$. This means that $n - 1$ is odd because it is of the form $2m + 1$, where m is the integer $k - 1$.

We also use a direct proof to show that $p_2 \rightarrow p_3$. Now suppose $n - 1$ is odd. Then $n - 1 = 2k + 1$ for some integer k. Hence, $n = 2k + 2$ so that $n^2 = (2k + 2)^2 = 4k^2 + 8k + 4 = 2(2k^2 + 4k + 2)$. This means that n^2 is twice the integer $2k^2 + 4k + 2$, and hence is even.

To prove $p_3 \rightarrow p_1$, we use a proof by contraposition. That is, we prove that if n is not even, then n^2 is not even. This is the same as proving that if n is odd, then n^2 is odd, which we have already done in Example 1. This completes the proof. ◀

COUNTEREXAMPLES In Section 1.4 we stated that to show that a statement of the form $\forall x P(x)$ is false, we need only find a **counterexample**, that is, an example x for which $P(x)$ is false. When presented with a statement of the form $\forall x P(x)$, which we believe to be false or which has resisted all proof attempts, we look for a counterexample. We illustrate the use of counterexamples in Example 15.

EXAMPLE 15 Show that the statement "Every positive integer is the sum of the squares of two integers" is false.

Extra Examples

Solution: To show that this statement is false, we look for a counterexample, which is a particular integer that is not the sum of the squares of two integers. It does not take long to find a counterexample, because 3 cannot be written as the sum of the squares of two integers. To show this is the case, note that the only perfect squares not exceeding 3 are $0^2 = 0$ and $1^2 = 1$. Furthermore, there is no way to get 3 as the sum of two terms each of which is 0 or 1. Consequently, we have shown that "Every positive integer is the sum of the squares of two integers" is false. ◀

1.7.8 Mistakes in Proofs

There are many common errors made in constructing mathematical proofs. We will briefly describe some of these here. Among the most common errors are mistakes in arithmetic and basic algebra. Even professional mathematicians make such errors, especially when working with complicated formulae. Whenever you use such computations you should check them as carefully as possible. (You should also review any troublesome aspects of basic algebra, especially before you study Section 5.1.)

Links

Each step of a mathematical proof needs to be correct and the conclusion needs to follow logically from the steps that precede it. Many mistakes result from the introduction of steps that do not logically follow from those that precede it. This is illustrated in Examples 16–18.

EXAMPLE 16 What is wrong with this famous supposed "proof" that 1 = 2?

"Proof": We use these steps, where a and b are two equal positive integers.

Step	**Reason**
1. $a = b$	Given
2. $a^2 = ab$	Multiply both sides of (1) by a
3. $a^2 - b^2 = ab - b^2$	Subtract b^2 from both sides of (2)
4. $(a - b)(a + b) = b(a - b)$	Factor both sides of (3)
5. $a + b = b$	Divide both sides of (4) by $a - b$
6. $2b = b$	Replace a by b in (5) because $a = b$ and simplify
7. $2 = 1$	Divide both sides of (6) by b

Solution: Every step is valid except for step 5, where we divided both sides by $a - b$. The error is that $a - b$ equals zero; division of both sides of an equation by the same quantity is valid as long as this quantity is not zero. ◀

EXAMPLE 17 What is wrong with this "proof"?

"Theorem": If n^2 is positive, then n is positive.

"Proof": Suppose that n^2 is positive. Because the conditional statement "If n is positive, then n^2 is positive" is true, we can conclude that n is positive.

Solution: Let $P(n)$ be "n is positive" and $Q(n)$ be "n^2 is positive." Then our hypothesis is $Q(n)$. The statement "If n is positive, then n^2 is positive" is the statement $\forall n(P(n) \to Q(n))$. From the hypothesis $Q(n)$ and the statement $\forall n(P(n) \to Q(n))$ we cannot conclude $P(n)$, because we are not using a valid rule of inference. Instead, this is an example of the fallacy of affirming the conclusion. A counterexample is supplied by $n = -1$ for which $n^2 = 1$ is positive, but n is negative. ◀

EXAMPLE 18 What is wrong with this "proof"?

"Theorem": If n is not positive, then n^2 is not positive. (This is the contrapositive of the "theorem" in Example 17.)

"Proof": Suppose that n is not positive. Because the conditional statement "If n is positive, then n^2 is positive" is true, we can conclude that n^2 is not positive.

Solution: Let $P(n)$ and $Q(n)$ be as in the solution of Example 17. Then our hypothesis is $\neg P(n)$ and the statement "If n is positive, then n^2 is positive" is the statement $\forall n(P(n) \to Q(n))$. From the hypothesis $\neg P(n)$ and the statement $\forall n(P(n) \to Q(n))$ we cannot conclude $\neg Q(n)$, because we are not using a valid rule of inference. Instead, this is an example of the fallacy of denying the hypothesis. A counterexample is supplied by $n = -1$, as in Example 17. ◀

Finally, we briefly discuss a particularly nasty type of error. Many incorrect arguments are based on a fallacy called **begging the question**. This fallacy occurs when one or more steps of a proof are based on the truth of the statement being proved. In other words, this fallacy arises when a statement is proved using itself, or a statement equivalent to it. That is why this fallacy is also called **circular reasoning**.

EXAMPLE 19 Is the following argument correct? It supposedly shows that n is an even integer whenever n^2 is an even integer.

Suppose that n^2 is even. Then $n^2 = 2k$ for some integer k. Let $n = 2l$ for some integer l. This shows that n is even.

Solution: This argument is incorrect. The statement "let $n = 2l$ for some integer l" occurs in the proof. No argument has been given to show that n can be written as $2l$ for some integer l. This is circular reasoning because this statement is equivalent to the statement being proved, namely, "n is even." The result itself is correct; only the method of proof is wrong. ◀

Making mistakes in proofs is part of the learning process. When you make a mistake that someone else finds, you should carefully analyze where you went wrong and make sure that you do not make the same mistake again. Even professional mathematicians make mistakes in proofs. More than a few incorrect proofs of important results have fooled people for many years before subtle errors in them were found.

1.7.9 Just a Beginning

We have now developed a basic arsenal of proof methods. In the next section we will introduce other important proof methods. We will also introduce several important proof techniques in Chapter 5, including mathematical induction, which can be used to prove results that hold for all positive integers. In Chapter 6 we will introduce the notion of combinatorial proofs.

In this section we introduced several methods for proving theorems of the form $\forall x(P(x) \rightarrow Q(x))$, including direct proofs and proofs by contraposition. There are many theorems of this type whose proofs are easy to construct by directly working through the hypotheses and definitions of the terms of the theorem. However, it is often difficult to prove a theorem without resorting to a clever use of a proof by contraposition or a proof by contradiction, or some other proof technique. In Section 1.8 we will address proof strategy. We will describe various approaches that can be used to find proofs when straightforward approaches do not work. Constructing proofs is an art that can be learned only through experience, including writing proofs, having your proofs critiqued, and reading and analyzing other proofs.

Exercises

1. Use a direct proof to show that the sum of two odd integers is even.

2. Use a direct proof to show that the sum of two even integers is even.

3. Show that the square of an even number is an even number using a direct proof.

4. Show that the additive inverse, or negative, of an even number is an even number using a direct proof.

5. Prove that if $m + n$ and $n + p$ are even integers, where m, n, and p are integers, then $m + p$ is even. What kind of proof did you use?

6. Use a direct proof to show that the product of two odd numbers is odd.

7. Use a direct proof to show that every odd integer is the difference of two squares. [*Hint:* Find the difference of the squares of $k + 1$ and k where k is a positive integer.]

8. Prove that if n is a perfect square, then $n + 2$ is not a perfect square.

9. Use a proof by contradiction to prove that the sum of an irrational number and a rational number is irrational.

10. Use a direct proof to show that the product of two rational numbers is rational.

11. Prove or disprove that the product of two irrational numbers is irrational.

12. Prove or disprove that the product of a nonzero rational number and an irrational number is irrational.

13. Prove that if x is irrational, then $1/x$ is irrational.

14. Prove that if x is rational and $x \neq 0$, then $1/x$ is rational.

15. Prove that if x is an irrational number and $x > 0$, then $\sqrt{x}$ is also irrational.

16. Prove that if x, y, and z are integers and $x + y + z$ is odd, then at least one of x, y, and z is odd.

17. Use a proof by contraposition to show that if $x + y \geq 2$, where x and y are real numbers, then $x \geq 1$ or $y \geq 1$.

18. Prove that if m and n are integers and mn is even, then m is even or n is even.

19. Show that if n is an integer and $n^3 + 5$ is odd, then n is even using

a) a proof by contraposition.
b) a proof by contradiction.

20. Prove that if n is an integer and $3n + 2$ is even, then n is even using

a) a proof by contraposition.
b) a proof by contradiction.

21. Prove the proposition $P(0)$, where $P(n)$ is the proposition "If n is a positive integer greater than 1, then $n^2 > n$." What kind of proof did you use?

22. Prove the proposition $P(1)$, where $P(n)$ is the proposition "If n is a positive integer, then $n^2 \geq n$." What kind of proof did you use?

23. Let $P(n)$ be the proposition "If a and b are positive real numbers, then $(a + b)^n \geq a^n + b^n$." Prove that $P(1)$ is true. What kind of proof did you use?

24. Show that if you pick three socks from a drawer containing just blue socks and black socks, you must get either a pair of blue socks or a pair of black socks.

25. Show that at least ten of any 64 days chosen must fall on the same day of the week.

26. Show that at least three of any 25 days chosen must fall in the same month of the year.

27. Use a proof by contradiction to show that there is no rational number r for which $r^3 + r + 1 = 0$. [*Hint:* Assume that $r = a/b$ is a root, where a and b are integers and a/b is in lowest terms. Obtain an equation involving integers by multiplying by b^3. Then look at whether a and b are each odd or even.]

28. Prove that if n is a positive integer, then n is even if and only if $7n + 4$ is even.

29. Prove that if n is a positive integer, then n is odd if and only if $5n + 6$ is odd.

30. Prove that $m^2 = n^2$ if and only if $m = n$ or $m = -n$.

31. Prove or disprove that if m and n are integers such that $mn = 1$, then either $m = 1$ and $n = 1$, or else $m = -1$ and $n = -1$.

32. Show that these three statements are equivalent, where a and b are real numbers: (*i*) a is less than b, (*ii*) the average of a and b is greater than a, and (*iii*) the average of a and b is less than b.

33. Show that these statements about the integer x are equivalent: (*i*) $3x + 2$ is even, (*ii*) $x + 5$ is odd, (*iii*) x^2 is even.

34. Show that these statements about the real number x are equivalent: (*i*) x is rational, (*ii*) $x/2$ is rational, (*iii*) $3x - 1$ is rational.

35. Show that these statements about the real number x are equivalent: (*i*) x is irrational, (*ii*) $3x + 2$ is irrational, (*iii*) $x/2$ is irrational.

36. Is this reasoning for finding the solutions of the equation $\sqrt{2x^2 - 1} = x$ correct? (1) $\sqrt{2x^2 - 1} = x$ is given; (2) $2x^2 - 1 = x^2$, obtained by squaring both sides of (1); (3) $x^2 - 1 = 0$, obtained by subtracting x^2 from both sides of (2); (4) $(x - 1)(x + 1) = 0$, obtained by factoring the left-hand side of $x^2 - 1$; (5) $x = 1$ or $x = -1$, which follows because $ab = 0$ implies that $a = 0$ or $b = 0$.

37. Are these steps for finding the solutions of $\sqrt{x + 3} = 3 - x$ correct? (1) $\sqrt{x + 3} = 3 - x$ is given; (2) $x + 3 = x^2 - 6x + 9$, obtained by squaring both sides of (1); (3) $0 = x^2 - 7x + 6$, obtained by subtracting $x + 3$ from both sides of (2); (4) $0 = (x - 1)(x - 6)$, obtained by factoring the right-hand side of (3); (5) $x = 1$ or $x = 6$, which follows from (4) because $ab = 0$ implies that $a = 0$ or $b = 0$.

38. Show that the propositions p_1, p_2, p_3, and p_4 can be shown to be equivalent by showing that $p_1 \leftrightarrow p_4$, $p_2 \leftrightarrow p_3$, and $p_1 \leftrightarrow p_3$.

39. Show that the propositions p_1, p_2, p_3, p_4, and p_5 can be shown to be equivalent by proving that the conditional statements $p_1 \rightarrow p_4$, $p_3 \rightarrow p_1$, $p_4 \rightarrow p_2$, $p_2 \rightarrow p_5$, and $p_5 \rightarrow p_3$ are true.

40. Find a counterexample to the statement that every positive integer can be written as the sum of the squares of three integers.

41. Prove that at least one of the real numbers $a_1, a_2, \ldots, a_n$ is greater than or equal to the average of these numbers. What kind of proof did you use?

42. Use Exercise 41 to show that if the first 10 positive integers are placed around a circle, in any order, there exist three integers in consecutive locations around the circle that have a sum greater than or equal to 17.

43. Prove that if n is an integer, these four statements are equivalent: (*i*) n is even, (*ii*) $n + 1$ is odd, (*iii*) $3n + 1$ is odd, (*iv*) $3n$ is even.

44. Prove that these four statements about the integer n are equivalent: (*i*) n^2 is odd, (*ii*) $1 - n$ is even, (*iii*) n^3 is odd, (*iv*) $n^2 + 1$ is even.

1.8 Proof Methods and Strategy

1.8.1 Introduction

Assessment

In Section 1.7 we introduced many methods of proof and illustrated how each method can be used. In this section we continue this effort. We will introduce several other commonly used proof methods, including the method of proving a theorem by considering different cases separately. We will also discuss proofs where we prove the existence of objects with desired properties.

In Section 1.7 we briefly discussed the strategy behind constructing proofs. This strategy includes selecting a proof method and then successfully constructing an argument step by step, based on this method. In this section, after we have developed a versatile arsenal of proof methods, we will study some aspects of the art and science of proofs. We will provide advice on how to find a proof of a theorem. We will describe some tricks of the trade, including how proofs can be found by working backward and by adapting existing proofs.

When mathematicians work, they formulate conjectures and attempt to prove or disprove them. We will briefly describe this process here by proving results about tiling checkerboards with dominoes and other types of pieces. Looking at tilings of this kind, we will be able to quickly formulate conjectures and prove theorems without first developing a theory.

We will conclude the section by discussing the role of open questions. In particular, we will discuss some interesting problems either that have been solved after remaining open for hundreds of years or that still remain open.

1.8.2 Exhaustive Proof and Proof by Cases

Sometimes we cannot prove a theorem using a single argument that holds for all possible cases. We now introduce a method that can be used to prove a theorem by considering different cases separately. This method is based on a rule of inference that we will now introduce. To prove a conditional statement of the form

$$(p_1 \vee p_2 \vee \cdots \vee p_n) \rightarrow q$$

the tautology

$$[(p_1 \vee p_2 \vee \cdots \vee p_n) \rightarrow q] \leftrightarrow [(p_1 \rightarrow q) \wedge (p_2 \rightarrow q) \wedge \cdots \wedge (p_n \rightarrow q)]$$

can be used as a rule of inference. This shows that the original conditional statement with a hypothesis made up of a disjunction of the propositions $p_1, p_2, \ldots, p_n$ can be proved by proving each of the n conditional statements $p_i \rightarrow q$, $i = 1, 2, \ldots, n$, individually. Such an argument is called a **proof by cases**. Sometimes to prove that a conditional statement $p \rightarrow q$ is true, it is convenient to use a disjunction $p_1 \vee p_2 \vee \cdots \vee p_n$ instead of p as the hypothesis of the conditional statement, where p and $p_1 \vee p_2 \vee \cdots \vee p_n$ are equivalent.

EXHAUSTIVE PROOF Some theorems can be proved by examining a relatively small number of examples. Such proofs are called **exhaustive proofs**, or **proofs by exhaustion** because these proofs proceed by exhausting all possibilities. An exhaustive proof is a special type of proof by cases where each case involves checking a single example. We now provide some illustrations of exhaustive proofs.

EXAMPLE 1 Prove that $(n + 1)^3 \geq 3^n$ if n is a positive integer with $n \leq 4$.

Extra Examples

Solution: We use a proof by exhaustion. We only need verify the inequality $(n + 1)^3 \geq 3^n$ when $n = 1, 2, 3$, and 4. For $n = 1$, we have $(n + 1)^3 = 2^3 = 8$ and $3^n = 3^1 = 3$; for $n = 2$, we have $(n + 1)^3 = 3^3 = 27$ and $3^n = 3^2 = 9$; for $n = 3$, we have $(n + 1)^3 = 4^3 = 64$ and $3^n = 3^3 = 27$; and for $n = 4$, we have $(n + 1)^3 = 5^3 = 125$ and $3^n = 3^4 = 81$. In each of these four cases, we see that $(n + 1)^3 \geq 3^n$. We have used the method of exhaustion to prove that $(n + 1)^3 \geq 3^n$ if n is a positive integer with $n \leq 4$. ◀

EXAMPLE 2 Prove that the only consecutive positive integers not exceeding 100 that are perfect powers are 8 and 9. (An integer n is a **perfect power** if it equals m^a, where m is an integer and a is an integer greater than 1.)

Solution: We use a proof by exhaustion. In particular, we can prove this fact by examining positive integers n not exceeding 100, first checking whether n is a perfect power, and if it is, checking whether $n + 1$ is also a perfect power. A quicker way to do this is simply to look at all perfect powers not exceeding 100 and checking whether the next largest integer is also a perfect power. The squares of positive integers not exceeding 100 are 1, 4, 9, 16, 25, 36, 49, 64, 81, and

100. The cubes of positive integers not exceeding 100 are 1, 8, 27, and 64. The fourth powers of positive integers not exceeding 100 are 1, 16, and 81. The fifth powers of positive integers not exceeding 100 are 1 and 32. The sixth powers of positive integers not exceeding 100 are 1 and 64. There are no powers of positive integers higher than the sixth power not exceeding 100, other than 1. Looking at this list of perfect powers not exceeding 100, we see that $n = 8$ is the only perfect power n for which $n + 1$ is also a perfect power. That is, $2^3 = 8$ and $3^2 = 9$ are the only two consecutive perfect powers not exceeding 100. ◀

Proofs by exhaustion can tire out people and computers when the number of cases challenges the available processing power!

People can carry out exhaustive proofs when it is necessary to check only a relatively small number of instances of a statement. Computers do not complain when they are asked to check a much larger number of instances of a statement, but they still have limitations. Note that not even a computer can check all instances when it is impossible to list all instances to check.

PROOF BY CASES A proof by cases must cover all possible cases that arise in a theorem. We illustrate proof by cases with a couple of examples. In each example, you should check that all possible cases are covered.

EXAMPLE 3 Prove that if n is an integer, then $n^2 \geq n$.

Extra Examples

Solution: We can prove that $n^2 \geq n$ for every integer by considering three cases, when $n = 0$, when $n \geq 1$, and when $n \leq -1$. We split the proof into three cases because it is straightforward to prove the result by considering zero, positive integers, and negative integers separately.

Case (i): When $n = 0$, because $0^2 = 0$, we see that $0^2 \geq 0$. It follows that $n^2 \geq n$ is true in this case.

Case (ii): When $n \geq 1$, when we multiply both sides of the inequality $n \geq 1$ by the positive integer n, we obtain $n \cdot n \geq n \cdot 1$. This implies that $n^2 \geq n$ for $n \geq 1$.

Case (iii): In this case $n \leq -1$. However, $n^2 \geq 0$. It follows that $n^2 \geq n$.

Because the inequality $n^2 \geq n$ holds in all three cases, we can conclude that if n is an integer, then $n^2 \geq n$. ◀

EXAMPLE 4 Use a proof by cases to show that $|xy| = |x||y|$, where x and y are real numbers. (Recall that $|a|$, the absolute value of a, equals a when $a \geq 0$ and equals $-a$ when $a \leq 0$.)

Solution: In our proof of this theorem, we remove absolute values using the fact that $|a| = a$ when $a \geq 0$ and $|a| = -a$ when $a \leq 0$. Because both $|x|$ and $|y|$ occur in our formula, we will need four cases: *(i)* x and y both nonnegative, *(ii)* x nonnegative and y negative, *(iii)* x negative and y nonnegative, and *(iv)* x negative and y negative. We denote by p_1, p_2, p_3, and p_4, the proposition stating the assumption for each of these four cases, respectively.

(Note that we can remove the absolute value signs by making the appropriate choice of signs within each case.)

Case (i): We see that $p_1 \to q$ because $xy \geq 0$ when $x \geq 0$ and $y \geq 0$, so that $|xy| = xy = |x||y|$.

Case (ii): To see that $p_2 \to q$, note that if $x \geq 0$ and $y < 0$, then $xy \leq 0$, so that $|xy| = -xy = x(-y) = |x||y|$. (Here, because $y < 0$, we have $|y| = -y$.)

Case (iii): To see that $p_3 \to q$, we follow the same reasoning as the previous case with the roles of x and y reversed.

Case (iv): To see that $p_4 \to q$, note that when $x < 0$ and $y < 0$, it follows that $xy > 0$. Hence, $|xy| = xy = (-x)(-y) = |x||y|$.

Because $|xy| = |x||y|$ holds in each of the four cases and these cases exhaust all possibilities, we can conclude that $|xy| = |x||y|$, whenever x and y are real numbers. ◀

LEVERAGING PROOF BY CASES The examples we have presented illustrating proof by cases provide some insight into when to use this method of proof. In particular, when it is not possible to consider all cases of a proof at the same time, a proof by cases should be considered. When should you use such a proof? Generally, look for a proof by cases when there is no obvious way to begin a proof, but when extra information in each case helps move the proof forward. Example 5 illustrates how the method of proof by cases can be used effectively.

EXAMPLE 5 Formulate a conjecture about the final decimal digit of the square of an integer and prove your result.

Solution: The smallest perfect squares are 1, 4, 9, 16, 25, 36, 49, 64, 81, 100, 121, 144, 169, 196, 225, and so on. We notice that the digits that occur as the final digit of a square are 0, 1, 4, 5, 6, and 9, with 2, 3, 7, and 8 never appearing as the final digit of a square. We conjecture this theorem: The final decimal digit of a perfect square is 0, 1, 4, 5, 6, or 9. How can we prove this theorem?

We first note that we can express an integer n as $10a + b$, where a and b are positive integers and b is 0, 1, 2, 3, 4, 5, 6, 7, 8, or 9. Here a is the integer obtained by subtracting the final decimal digit of n from n and dividing by 10. Next, note that $(10a + b)^2 = 100a^2 + 20ab + b^2 = 10(10a^2 + 2b) + b^2$, so that the final decimal digit of n^2 is the same as the final decimal digit of b^2. Furthermore, note that the final decimal digit of b^2 is the same as the final decimal digit of $(10 - b)^2 = 100 - 20b + b^2$. Consequently, we can reduce our proof to the consideration of six cases.

Case (i): The final digit of n is 1 or 9. Then the final decimal digit of n^2 is the final decimal digit of $1^2 = 1$ or $9^2 = 81$, namely, 1.

Case (ii): The final digit of n is 2 or 8. Then the final decimal digit of n^2 is the final decimal digit of $2^2 = 4$ or $8^2 = 64$, namely, 4.

Case (iii): The final digit of n is 3 or 7. Then the final decimal digit of n^2 is the final decimal digit of $3^2 = 9$ or $7^2 = 49$, namely, 9.

Case (iv): The final digit of n is 4 or 6. Then the final decimal digit of n^2 is the final decimal digit of $4^2 = 16$ or $6^2 = 36$, namely, 6.

Case (v): The final decimal digit of n is 5. Then the final decimal digit of n^2 is the final decimal digit of $5^2 = 25$, namely, 5.

Case (vi): The final decimal digit of n is 0. Then the final decimal digit of n^2 is the final decimal digit of $0^2 = 0$, namely, 0.

Because we have considered all six cases, we can conclude that the final decimal digit of n^2, where n is an integer is either 0, 1, 4, 5, 6, or 9. ◀

Sometimes we can eliminate all but a few examples in a proof by cases, as Example 6 illustrates.

EXAMPLE 6 Show that there are no solutions in integers x and y of $x^2 + 3y^2 = 8$.

Solution: We can quickly reduce a proof to checking just a few simple cases because $x^2 > 8$ when $|x| \geq 3$ and $3y^2 > 8$ when $|y| \geq 2$. This leaves the cases when x equals $-2, -1, 0, 1$, or 2 and y equals $-1, 0$, or 1. We can finish using an exhaustive proof. To dispense with the remaining cases, we note that possible values for x^2 are 0, 1, and 4, and possible values for $3y^2$ are 0 and 3, and the largest sum of possible values for x^2 and $3y^2$ is 7. Consequently, it is impossible for $x^2 + 3y^2 = 8$ to hold when x and y are integers. ◀

WITHOUT LOSS OF GENERALITY In the proof in Example 4, we dismissed case (*iii*), where $x < 0$ and $y \geq 0$, because it is the same as case (*ii*), where $x \geq 0$ and $y < 0$, with the roles of x and y reversed. To shorten the proof, we could have proved cases (*ii*) and (*iii*) together by assuming, **without loss of generality**, that $x \geq 0$ and $y < 0$. Implicit in this statement is that we can complete the case with $x < 0$ and $y \geq 0$ using the same argument as we used for the case with $x \geq 0$ and $y < 0$, but with the obvious changes.

In general, when the phrase "without loss of generality" is used in a proof (often abbreviated as WLOG), we assert that by proving one case of a theorem, no additional argument is required to prove other specified cases. That is, other cases follow by making straightforward changes to the argument, or by filling in some straightforward initial step. Proofs by cases can often be made much more efficient when the notion of without loss of generality is employed. Incorrect use of this principle, however, can lead to unfortunate errors. Sometimes assumptions are made that lead to a loss in generality. Such assumptions can be made that do not take into account that one case may be substantially different from others. This can lead to an incomplete, and possibly unsalvageable, proof. In fact, many incorrect proofs of famous theorems turned out to rely on arguments that used the idea of "without loss of generality" to establish cases that could not be quickly proved from simpler cases.

In a proof by cases be sure not to omit any cases and check that you have proved all cases correctly!

We now illustrate a proof where without loss of generality is used effectively together with other proof techniques.

EXAMPLE 7 Show that if x and y are integers and both xy and $x + y$ are even, then both x and y are even.

Solution: We will use proof by contraposition, the notion of without loss of generality, and proof by cases. First, suppose that x and y are not both even. That is, assume that x is odd or that y is odd (or both). Without loss of generality, we assume that x is odd, so that $x = 2m + 1$ for some integer k.

To complete the proof, we need to show that xy is odd or $x + y$ is odd. Consider two cases: (*i*) y is even, and (*ii*) y is odd. In (*i*), $y = 2n$ for some integer n, so that $x + y = (2m + 1) + 2n = 2(m + n) + 1$ is odd. In (*ii*), $y = 2n + 1$ for some integer n, so that $xy = (2m + 1)(2n + 1) = 4mn + 2m + 2n + 1 = 2(2mn + m + n) + 1$ is odd. This completes the proof by contraposition. (Note that our use of without loss of generality within the proof is justified because the proof when y is odd can be obtained by simply interchanging the roles of x and y in the proof we have given.) ◀

COMMON ERRORS WITH EXHAUSTIVE PROOF AND PROOF BY CASES A common error of reasoning is to draw incorrect conclusions from examples. No matter how many separate examples are considered, a theorem is not proved by considering examples unless every possible case is covered. The problem of proving a theorem is analogous to showing that a computer program always produces the output desired. No matter how many input values are tested, unless all input values are tested, we cannot conclude that the program always produces the correct output.

EXAMPLE 8 Is it true that every positive integer is the sum of 18 fourth powers of integers?

Solution: To determine whether a positive integer n can be written as the sum of 18 fourth powers of integers, we might begin by examining whether n is the sum of 18 fourth powers of integers for the smallest positive integers. Because the fourth powers of integers are 0, 1, 16, 81, …, if we can select 18 terms from these numbers that add up to n, then n is the sum of 18 fourth powers. We can show that all positive integers up to 78 can be written as the sum of 18 fourth powers. (The details are left to the reader.) However, if we decided this was enough checking, we would come to the wrong conclusion. It is not true that every positive integer is the sum of 18 fourth powers because 79 is not the sum of 18 fourth powers (as the reader can verify). ◀

Another common error involves making unwarranted assumptions that lead to incorrect proofs by cases where not all cases are considered. This is illustrated in Example 9.

EXAMPLE 9 What is wrong with this "proof"?

"Theorem": If x is a real number, then x^2 is a positive real number.

"Proof": Let p_1 be "x is positive," let p_2 be "x is negative," and let q be "x^2 is positive." To show that $p_1 \rightarrow q$ is true, note that when x is positive, x^2 is positive because it is the product of two positive numbers, x and x. To show that $p_2 \rightarrow q$, note that when x is negative, x^2 is positive because it is the product of two negative numbers, x and x. This completes the proof.

Solution: The problem with this "proof" is that we missed the case of $x = 0$. When $x = 0$, $x^2 = 0$ is not positive, so the supposed theorem is false. If p is "x is a real number," then we can prove results where p is the hypothesis with three cases, p_1, p_2, and p_3, where p_1 is "x is positive," p_2 is "x is negative," and p_3 is "$x = 0$" because of the equivalence $p \leftrightarrow p_1 \vee p_2 \vee p_3$. ◀

1.8.3 Existence Proofs

Many theorems are assertions that objects of a particular type exist. A theorem of this type is a proposition of the form $\exists x P(x)$, where P is a predicate. A proof of a proposition of the form $\exists x P(x)$ is called an **existence proof**. There are several ways to prove a theorem of this type. Sometimes an existence proof of $\exists x P(x)$ can be given by finding an element a, called a **witness**, such that $P(a)$ is true. This type of existence proof is called **constructive**. It is also possible to give an existence proof that is **nonconstructive**; that is, we do not find an element a such that $P(a)$ is true, but rather prove that $\exists x P(x)$ is true in some other way. One common method of giving a nonconstructive existence proof is to use proof by contradiction and show that the negation of the existential quantification implies a contradiction. The concept of a constructive existence proof is illustrated by Example 10 and the concept of a nonconstructive existence proof is illustrated by Example 11.

EXAMPLE 10 **A Constructive Existence Proof** Show that there is a positive integer that can be written as the sum of cubes of positive integers in two different ways.

Extra Examples ›

Solution: After considerable computation (such as a computer search) we find that

$$1729 = 10^3 + 9^3 = 12^3 + 1^3.$$

Because we have displayed a positive integer that can be written as the sum of cubes in two different ways, we are done.

There is an interesting story pertaining to this example. The English mathematician G. H. Hardy, when visiting the ailing Indian prodigy Ramanujan in the hospital, remarked that 1729, the number of the cab he took, was rather dull. Ramanujan replied "No, it is a very interesting number; it is the smallest number expressible as the sum of cubes in two different ways." ◀

EXAMPLE 11 **A Nonconstructive Existence Proof** Show that there exist irrational numbers x and y such that x^y is rational.

Solution: By Example 11 in Section 1.7 we know that $\sqrt{2}$ is irrational. Consider the number $\sqrt{2}^{\sqrt{2}}$. If it is rational, we have two irrational numbers x and y with x^y rational, namely, $x = \sqrt{2}$ and $y = \sqrt{2}$. On the other hand if $\sqrt{2}^{\sqrt{2}}$ is irrational, then we can let $x = \sqrt{2}^{\sqrt{2}}$ and $y = \sqrt{2}$ so that $x^y = (\sqrt{2}^{\sqrt{2}})^{\sqrt{2}} = \sqrt{2}^{(\sqrt{2}\cdot\sqrt{2})} = \sqrt{2}^2 = 2$.

This proof is an example of a nonconstructive existence proof because we have not found irrational numbers x and y such that x^y is rational. Rather, we have shown that either the pair $x = \sqrt{2}$, $y = \sqrt{2}$ or the pair $x = \sqrt{2}^{\sqrt{2}}$, $y = \sqrt{2}$ have the desired property, but we do not know which of these two pairs works! ◀

Nonconstructive existence proofs often are quite subtle, as Example 12 illustrates.

EXAMPLE 12 **Chomp** is a game played by two players. In this game, cookies are laid out on a rectangular grid. The cookie in the top left position is poisoned, as shown in Figure 1(a). The two players take turns making moves; at each move, a player is required to eat a remaining cookie, together with all cookies to the right and/or below it (see Figure 1(b), for example). The loser is the player who has no choice but to eat the poisoned cookie. We ask whether one of the two players has a winning strategy. That is, can one of the players always make moves that are guaranteed to lead to a win?

Links

Links

AMERICAN PHILOSOPHICAL SOCIETY/Science Source

GODFREY HAROLD HARDY (1877–1947) Hardy, born in Cranleigh, Surrey, England, was the older of two children of Isaac Hardy and Sophia Hall Hardy. His father was the geography and drawing master at the Cranleigh School and also gave singing lessons and played soccer. His mother gave piano lessons and helped run a boardinghouse for young students. Hardy's parents were devoted to their children's education. Hardy demonstrated his numerical ability at the early age of two when he began writing down numbers into the millions. He had a private mathematics tutor rather than attending regular classes at the Cranleigh School. He moved to Winchester College, a private high school, when he was 13 and was awarded a scholarship. He excelled in his studies and demonstrated a strong interest in mathematics. He entered Trinity College, Cambridge, in 1896 on a scholarship and won several prizes during his time there, graduating in 1899.

Hardy held the position of lecturer in mathematics at Trinity College at Cambridge University from 1906 to 1919, when he was appointed to the Sullivan chair of geometry at Oxford. He had become unhappy with Cambridge over the dismissal of the noted philosopher and mathematician Bertrand Russell from Trinity for antiwar activities and did not like a heavy load of administrative duties. In 1931 he returned to Cambridge as the Sadleirian professor of pure mathematics, where he remained until his retirement in 1942. He was a pure mathematician and held an elitist view of mathematics, hoping that his research could never be applied. Ironically, he is perhaps best known as one of the developers of the Hardy–Weinberg law, which predicts patterns of inheritance. His work in this area appeared as a letter to the journal *Science* in which he used simple algebraic ideas to demonstrate errors in an article on genetics. Hardy worked primarily in number theory and function theory, exploring such topics as the Riemann zeta function, Fourier series, and the distribution of primes. He made many important contributions to many important problems, such as Waring's problem about representing positive integers as sums of kth powers and the problem of representing odd integers as sums of three primes. Hardy is also remembered for his collaborations with John E. Littlewood, a colleague at Cambridge, with whom he wrote more than 100 papers, and the celebrated Indian mathematical prodigy Srinivasa Ramanujan. His collaboration with Littlewood led to the joke that there were only three important English mathematicians at that time, Hardy, Littlewood, and Hardy–Littlewood. However, some people believed that Hardy had invented a fictitious person, Littlewood, because Littlewood was seldom seen away from Cambridge. Hardy had the wisdom of recognizing Ramanujan's genius from unconventional but extremely creative writings Ramanujan sent him, while other mathematicians failed to see the genius. Hardy brought Ramanujan to Cambridge and collaborated on important joint papers, establishing new results on the number of partitions of an integer. Hardy was interested in mathematics education, and his book *A Course of Pure Mathematics* had a profound effect on undergraduate instruction in mathematics in the first half of the twentieth century. Hardy also wrote *A Mathematician's Apology*, in which he gives his answer to the question of whether it is worthwhile to devote one's life to the study of mathematics. It presents Hardy's view of what mathematics is and what a mathematician does.

Hardy had a strong interest in sports. He was an avid cricket fan and followed scores closely. One peculiar trait he had was that he did not like his picture taken (only five snapshots are known) and disliked mirrors, covering them with towels immediately upon entering a hotel room.

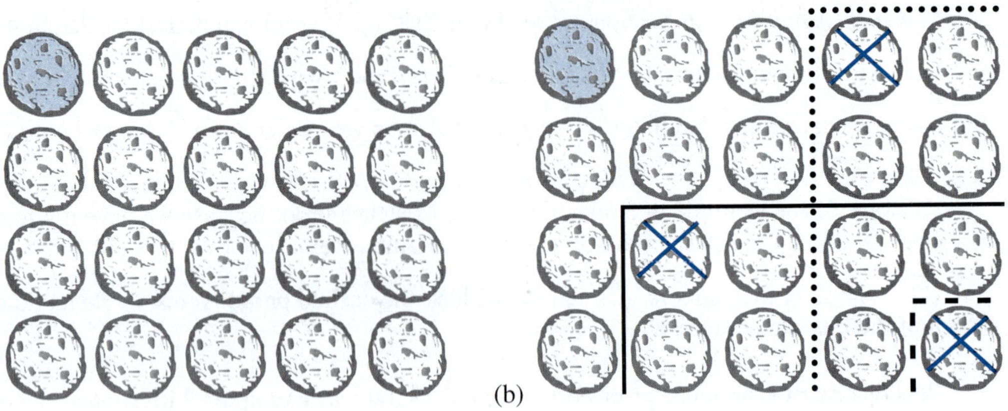

(a) (b)

FIGURE 1 **(a) Chomp (top left cookie poisoned). (b) Three possible moves.**

Solution: We will give a nonconstructive existence proof of a winning strategy for the first player. That is, we will show that the first player always has a winning strategy without explicitly describing the moves this player must follow.

First, note that the game ends and cannot finish in a draw because with each move at least one cookie is eaten, so after no more than $m \times n$ moves the game ends, where the initial grid is $m \times n$. Now, suppose that the first player begins the game by eating just the cookie in the bottom right corner. There are two possibilities, this is the first move of a winning strategy for the first player, or the second player can make a move that is the first move of a winning strategy for the second player. In this second case, instead of eating just the cookie in the bottom right corner, the first player could have made the same move that the second player made as the first move of a winning strategy (and then continued to follow that winning strategy). This would guarantee a win for the first player.

Note that we showed that a winning strategy exists, but we did not specify an actual winning strategy. Consequently, the proof is a nonconstructive existence proof. In fact, no one has been able to describe a winning strategy for Chomp that applies for all rectangular grids by describing the moves that the first player should follow. However, winning strategies can be described for certain special cases, such as when the grid is square and when the grid only has two rows of cookies (see Exercises 17 and 18 in Section 5.2). ◀

1.8.4 Uniqueness Proofs

Some theorems assert the existence of a unique element with a particular property. In other words, these theorems assert that there is exactly one element with this property. To prove a statement of this type we need to show that an element with this property exists and that no other element has this property. The two parts of a **uniqueness proof** are:

Existence: We show that an element x with the desired property exists.
Uniqueness: We show that if x and y both have the desired property, then $x = y$.

Remark: Showing that there is a unique element x such that $P(x)$ is the same as proving the statement $\exists x(P(x) \wedge \forall y(y \neq x \rightarrow \neg P(y)))$.

We illustrate the elements of a uniqueness proof in Example 13.

EXAMPLE 13 Show that if a and b are real numbers and $a \neq 0$, then there is a unique real number r such that $ar + b = 0$.

Solution: First, note that the real number $r = -b/a$ is a solution of $ar + b = 0$ because $a(-b/a) + b = -b + b = 0$. Consequently, a real number r exists for which $ar + b = 0$. This is the existence part of the proof.

Extra Examples

Second, suppose that s is a real number such that $as + b = 0$. Then $ar + b = as + b$, where $r = -b/a$. Subtracting b from both sides, we find that $ar = as$. Dividing both sides of this last equation by a, which is nonzero, we see that $r = s$. This establishes the uniqueness part of the proof. ◀

1.8.5 Proof Strategies

Finding proofs can be a challenging business. When you are confronted with a statement to prove, you should first replace terms by their definitions and then carefully analyze what the hypotheses and the conclusion mean. After doing so, you can attempt to prove the result using one of the available methods of proof. We have already provided some proof strategies in Section 1.7 for theorems of the form $\forall x(P(x) \rightarrow Q(x))$, when we introduced direct proof, proof by contraposition, and proof by contradiction. If the statement is a conditional statement, try a direct proof first as long as the hypotheses provide a good starting point; if this fails, try a proof

Links

Nick Higham/Alamy Stock Photo

SRINIVASA RAMANUJAN (1887–1920) The renowned mathematical prodigy Ramanujan was born and raised in southern India near the city of Madras (now called Chennai). His father was a clerk in a cloth shop. His mother contributed to the family income by singing at a local temple. Ramanujan studied at the local English language school, displaying his talent and interest for mathematics. At the age of 13 he mastered a textbook used by college students. When he was 15, a university student lent him a copy of *Synopsis of Pure Mathematics*. Ramanujan decided to work out the over 6000 results in this book, stated without proof or explanation, writing on sheets later collected to form notebooks. He graduated from high school in 1904, winning a scholarship to the University of Madras. Enrolling in a fine arts curriculum, he neglected his subjects other than mathematics and lost his scholarship. He failed to pass examinations at the university four times from 1904 to 1907, doing well only in mathematics. During this time he filled his notebooks with original writings, sometimes rediscovering already published work and at other times making new discoveries.

Without a university degree, it was difficult for Ramanujan to find a decent job. To survive, he had to depend on the goodwill of his friends. He tutored students in mathematics, but his unconventional ways of thinking and failure to stick to the syllabus caused problems. He was married in 1909 in an arranged marriage to a young woman nine years his junior. Needing to support himself and his wife, he moved to Madras and sought a job. He showed his notebooks of mathematical writings to his potential employers, but the books bewildered them. However, a professor at the Presidency College recognized his genius and supported him, and in 1912 he found work as an accounts clerk, earning a small salary.

Ramanujan continued his mathematical work during this time and published his first paper in 1910 in an Indian journal. He realized that his work was beyond that of the Indian mathematicians of his day and decided to write to leading English mathematicians. The first mathematicians he wrote to turned down his request for help. But in January 1913 he wrote to G. H. Hardy, who was inclined to turn Ramanujan down, but the mathematical statements in the letter, although stated without proof, puzzled Hardy. He decided to examine them closely with the help of his colleague and collaborator J. E. Littlewood. They decided, after careful study, that Ramanujan was probably a genius, because his statements "could only be written down by a mathematician of the highest class; they must be true, because if they were not true, no one would have the imagination to invent them."

Hardy arranged a scholarship for Ramanujan, bringing him to England in 1914. Hardy personally tutored him in mathematical analysis, and they collaborated for five years, proving significant theorems about the number of partitions of integers. During this time, Ramanujan made important contributions to number theory and also worked on continued fractions, infinite series, and elliptic functions. Ramanujan had amazing insight involving certain types of functions and series, but his purported theorems on prime numbers were often wrong, illustrating his vague idea of what constitutes a correct proof. He was one of the youngest members ever appointed a Fellow of the Royal Society. Unfortunately, in 1917 Ramanujan became extremely ill. At the time, it was thought that he had trouble with the English climate and had contracted tuberculosis. It is now thought that he suffered from a vitamin deficiency, brought on by Ramanujan's strict vegetarianism and shortages in wartime England. He returned to India in 1919, continuing to do mathematics even when confined to his bed. He was religious and thought his mathematical talent came from his family deity, Namagiri. He considered mathematics and religion to be linked. He said that "an equation for me has no meaning unless it expresses a thought of God." His short life came to an end in April 1920, when he was 32 years old. Ramanujan left several notebooks of unpublished results. The writings in these notebooks illustrate Ramanujan's insights but are quite sketchy. Many mathematicians have devoted many years of work to explaining and justifying the results in these notebooks. An excellent movie, *The Man Who Knew Infinity*, about the life of Ramanujan was released in 2015.

by contrapostion. If neither of these approaches works, you might try a proof by contradiction. However, we did not provide any further guidance how to create such proofs. We now present some strategies that you can use to develop new proofs.

FORWARD AND BACKWARD REASONING Whichever method you choose, you need a starting point for your proof. To begin a direct proof of a conditional statement, you start with the premises. Using these premises, together with axioms and known theorems, you can construct a proof using a sequence of steps that leads to the conclusion. This type of reasoning, called *forward reasoning*, is the most common type of reasoning used to prove relatively simple results. Similarly, with indirect reasoning you can start with the negation of the conclusion and, using a sequence of steps, obtain the negation of the premises.

Unfortunately, forward reasoning is often difficult to use to prove more complicated results, because the reasoning needed to reach the desired conclusion may be far from obvious. In such cases it may be helpful to use *backward reasoning*. To reason backward to prove a statement q, we find a statement p that we can prove with the property that $p \to q$. (Note that it is not helpful to find a statement r that you can prove such that $q \to r$, because it is the fallacy of begging the question to conclude from $q \to r$ and r that q is true.) Backward reasoning is illustrated in Examples 14 and 15.

EXAMPLE 14

Extra Examples

Given two positive real numbers x and y, their **arithmetic mean** is $(x + y)/2$ and their **geometric mean** is $\sqrt{xy}$. When we compare the arithmetic and geometric means of pairs of distinct positive real numbers, we find that the arithmetic mean is always greater than the geometric mean. [For example, when $x = 4$ and $y = 6$, we have $5 = (4 + 6)/2 > \sqrt{4 \cdot 6} = \sqrt{24}$.] Can we prove that this inequality is always true?

Solution: To prove that $(x + y)/2 > \sqrt{xy}$ when x and y are distinct positive real numbers, we can work backward. We construct a sequence of equivalent inequalities. The equivalent inequalities are

$$\begin{aligned}
(x + y)/2 &> \sqrt{xy},\\
(x + y)^2/4 &> xy,\\
(x + y)^2 &> 4xy,\\
x^2 + 2xy + y^2 &> 4xy,\\
x^2 - 2xy + y^2 &> 0,\\
(x - y)^2 &> 0.
\end{aligned}$$

Because $(x - y)^2 > 0$ when $x \neq y$, it follows that the final inequality is true. Because all these inequalities are equivalent, it follows that $(x + y)/2 > \sqrt{xy}$ when $x \neq y$. Once we have carried out this backward reasoning, we can build a proof based on reversing the steps. This produces a proof using forward reasoning. (Note that the steps of our backward reasoning will not be part of the final proof. These steps serve as our guide for putting this proof together.)

Proof: Suppose that x and y are distinct positive real numbers. Then $(x - y)^2 > 0$ because the square of a nonzero real number is positive (see Appendix 1). Because $(x - y)^2 = x^2 - 2xy + y^2$, this implies that $x^2 - 2xy + y^2 > 0$. Adding $4xy$ to both sides, we obtain $x^2 + 2xy + y^2 > 4xy$. Because $x^2 + 2xy + y^2 = (x + y)^2$, this means that $(x + y)^2 \geq 4xy$. Dividing both sides of this equation by 4, we see that $(x + y)^2/4 > xy$. Finally, taking square roots of both sides (which preserves the inequality because both sides are positive) yields $(x + y)/2 > \sqrt{xy}$. We conclude that if x and y are distinct positive real numbers, then their arithmetic mean $(x + y)/2$ is greater than their geometric mean $\sqrt{xy}$. ◀

EXAMPLE 15 Suppose that two people play a game taking turns removing one, two, or three stones at a time from a pile that begins with 15 stones. The person who removes the last stone wins the game. Show that the first player can win the game no matter what the second player does.

Solution: To prove that the first player can always win the game, we work backward. At the last step, the first player can win if this player is left with a pile containing one, two, or three stones. The second player will be forced to leave one, two, or three stones if this player has to remove stones from a pile containing four stones. Consequently, one way for the first person to win is to leave four stones for the second player on the next-to-last move. The first person can leave four stones when there are five, six, or seven stones left at the beginning of this player's move, which happens when the second player has to remove stones from a pile with eight stones. Consequently, to force the second player to leave five, six, or seven stones, the first player should leave eight stones for the second player at the second-to-last move for the first player. This means that there are nine, ten, or eleven stones when the first player makes this move. Similarly, the first player should leave twelve stones when this player makes the first move. We can reverse this argument to show that the first player can always make moves so that this player wins the game no matter what the second player does. These moves successively leave twelve, eight, and four stones for the second player. ◀

ADAPTING EXISTING PROOFS An excellent way to look for possible approaches that can be used to prove a statement is to take advantage of existing proofs of similar results. Often an existing proof can be adapted to prove other facts. Even when this is not the case, some of the ideas used in existing proofs may be helpful. Because existing proofs provide clues for new proofs, you should read and understand the proofs you encounter in your studies. This process is illustrated in Example 16.

EXAMPLE 16 In Example 11 of Section 1.7 we proved that $\sqrt{2}$ is irrational. We now conjecture that $\sqrt{3}$ is irrational. Can we adapt the proof in Example 11 in Section 1.7 to show that $\sqrt{3}$ is irrational?

Extra Examples

Solution: To adapt the proof in Example 11 in Section 1.7, we begin by mimicking the steps in that proof, but with $\sqrt{2}$ replaced with $\sqrt{3}$. First, we suppose that $\sqrt{3} = c/d$ where the fraction c/d is in lowest terms. Squaring both sides tells us that $3 = c^2/d^2$, so that $3d^2 = c^2$. Can we use this equation to show that 3 must be a factor of both c and d, similar to how we used the equation $2b^2 = a^2$ in Example 11 in Section 1.7 to show that 2 must be a factor of both a and b? (Recall that an integer s is a factor of the integer t if t/s is an integer. An integer n is even if and only if 2 is a factor of n.) In turns out that we can, but we need some ammunition from number theory, which we will develop in Chapter 4. We sketch out the remainder of the proof, but leave the justification of these steps until Chapter 4. Because 3 is a factor of c^2, it must also be a factor of c. Furthermore, because 3 is a factor of c, 9 is a factor of c^2, which means that 9 is a factor of $3d^2$. This implies that 3 is a factor of d^2, which means that 3 is a factor of that d. This makes 3 a factor of both c and d, which contradicts the assumption that c/d is in lowest terms. After we have filled in the justification for these steps, we will have shown that $\sqrt{3}$ is irrational by adapting the proof that $\sqrt{2}$ is irrational. Note that this proof can be extended to show that $\sqrt{n}$ is irrational whenever n is a positive integer that is not a perfect square. We leave the details of this to Chapter 4. ◀

A good tip is to look for existing proofs that you might adapt when you are confronted with proving a new theorem, particularly when the new theorem seems similar to one you have already proved.

1.8.6 Looking for Counterexamples

In Section 1.7 we introduced the use of counterexamples to show that certain statements are false. When confronted with a conjecture, you might first try to prove this conjecture, and if your attempts are unsuccessful, you might try to find a counterexample, first by looking at the simplest, smallest examples. If you cannot find a counterexample, you might again try to prove the statement. In any case, looking for counterexamples is an extremely important pursuit, which often provides insights into problems. We will illustrate the role of counterexamples in Example 17.

EXAMPLE 17 In Example 15 in Section 1.7 we showed that the statement "Every positive integer is the sum of two squares of integers" is false by finding a counterexample. That is, there are positive integers that cannot be written as the sum of the squares of two integers. Although we cannot write every positive integer as the sum of the squares of two integers, maybe we can write every positive integer as the sum of the squares of three integers. That is, is the statement "Every positive integer is the sum of the squares of three integers" true or false?

Extra Examples

Solution: Because we know that not every positive integer can be written as the sum of two squares of integers, we might initially be skeptical that every positive integer can be written as the sum of three squares of integers. So, we first look for a counterexample. That is, we can show that the statement "Every positive integer is the sum of three squares of integers" is false if we can find a particular integer that is not the sum of the squares of three integers. To look for a counterexample, we try to write successive positive integers as a sum of three squares. We find that $1 = 0^2 + 0^2 + 1^2$, $2 = 0^2 + 1^2 + 1^2$, $3 = 1^2 + 1^2 + 1^2$, $4 = 0^2 + 0^2 + 2^2$, $5 = 0^2 + 1^2 + 2^2$, $6 = 1^2 + 1^2 + 2^2$, but we cannot find a way to write 7 as the sum of three squares. To show that there are not three squares that add up to 7, we note that the only possible squares we can use are those not exceeding 7, namely, 0, 1, and 4. Because no three terms where each term is 0, 1, or 4 add up to 7, it follows that 7 is a counterexample. We conclude that the statement "Every positive integer is the sum of the squares of three integers" is false.

We have shown that not every positive integer is the sum of the squares of three integers. The next question to ask is whether every positive integer is the sum of the squares of four positive integers. Some experimentation provides evidence that the answer is yes. For example, $7 = 1^2 + 1^2 + 1^2 + 2^2$, $25 = 4^2 + 2^2 + 2^2 + 1^2$, and $87 = 9^2 + 2^2 + 1^2 + 1^2$. It turns out the conjecture "Every positive integer is the sum of the squares of four integers" is true. For a proof, see [Ro10]. ◀

1.8.7 Proof Strategy in Action

Mathematics is generally taught as if mathematical facts were carved in stone. Mathematics texts (including the bulk of this book) formally present theorems and their proofs. Such presentations do not convey the discovery process in mathematics. This process begins with exploring concepts and examples, asking questions, formulating conjectures, and attempting to settle these conjectures either by proof or by counterexample. These are the day-to-day activities of mathematicians. Believe it or not, the material taught in textbooks was originally developed in this way.

Extra Examples

People formulate conjectures on the basis of many types of possible evidence. The examination of special cases can lead to a conjecture, as can the identification of possible patterns. Altering the hypotheses and conclusions of known theorems also can lead to plausible conjectures. At other times, conjectures are made based on intuition or a belief that a result holds. No matter how a conjecture was made, once it has been formulated, the goal is to prove or disprove it. When mathematicians believe that a conjecture may be true, they try to find a proof. If they cannot find a proof, they may look for a counterexample. When they cannot find a counterexample, they may switch gears and once again try to prove the conjecture. Although many

FIGURE 2 **The standard checkerboard.**

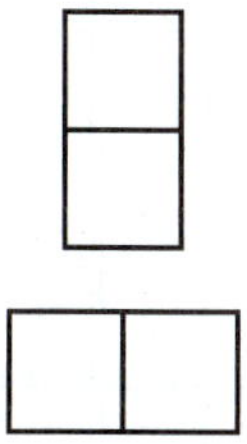

FIGURE 3
Two dominoes.

conjectures are quickly settled, a few conjectures resist attack for hundreds of years and lead to the development of new parts of mathematics. We will mention a few famous conjectures later in this section.

1.8.8 Tilings

Links

We can illustrate aspects of proof strategy through a brief study of tilings of checkerboards. Looking at tilings of checkerboards is a fruitful way to quickly discover many different results and construct their proofs using a variety of proof methods. There are almost an endless number of conjectures that can be made and studied in this area, too. To begin, we need to define some terms. A **checkerboard** is a rectangle divided into squares of the same size by horizontal and vertical lines. The game of checkers is played on a board with 8 rows and 8 columns; this board is called the **standard checkerboard** and is shown in Figure 2. In this section we use the term **board** to refer to a checkerboard of any rectangular size as well as parts of checkerboards obtained by removing one or more squares. A **domino** is a rectangular piece that is one square by two squares, as shown in Figure 3. We say that a board is **tiled** by dominoes when all its squares are covered with no overlapping dominoes and no dominoes overhanging the board. We now develop some results about tiling boards using dominoes.

EXAMPLE 18 Can we tile the standard checkerboard using dominoes?

Solution: We can find many ways to tile the standard checkerboard using dominoes. For example, we can tile it by placing 32 dominoes horizontally, as shown in Figure 4. The existence of one such tiling completes a constructive existence proof. There are a large number of other ways to do this tiling. We can place 32 dominoes vertically on the board or we can place some tiles vertically and some horizontally. But for a constructive existence proof we needed to find just one such tiling. ◀

EXAMPLE 19 Can we tile a board obtained by removing one of the four corner squares of a standard checkerboard?

Extra Examples

Solution: To answer this question, note that a standard checkerboard has 64 squares, so removing a square produces a board with 63 squares. Now suppose that we could tile a board obtained from the standard checkerboard by removing a corner square. The board has an even number of squares because each domino covers two squares and no two dominoes overlap and no dominoes

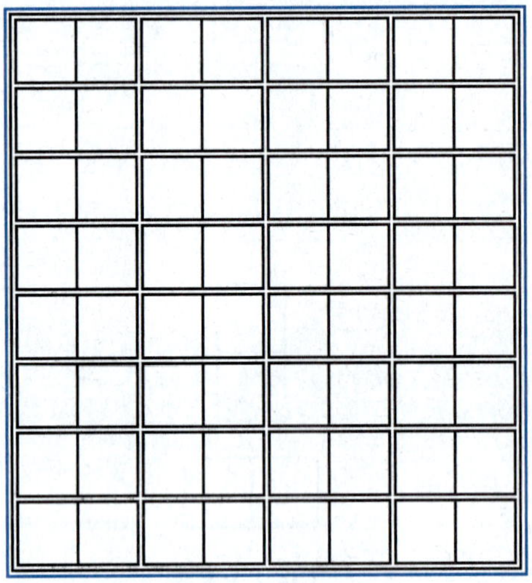

FIGURE 4 **Tiling the standard checkerboard.**

FIGURE 5 **The standard checkerboard with the upper left and lower right squares removed.**

overhang the board. Consequently, we can prove by contradiction that a standard checkerboard with one square removed cannot be tiled using dominoes because such a board has an odd number of squares. ◀

We now consider a trickier situation.

EXAMPLE 20 Can we tile the board obtained by deleting the upper left and lower right corner squares of a standard checkerboard, shown in Figure 5?

Solution: A board obtained by deleting two squares of a standard checkerboard contains $64 - 2 = 62$ squares. Because 62 is even, we cannot quickly rule out the existence of a tiling of the standard checkerboard with its upper left and lower right squares removed, unlike Example 19, where we ruled out the existence of a tiling of the standard checkerboard with one corner square removed. Trying to construct a tiling of this board by successively placing dominoes might be a first approach, as the reader should attempt. However, no matter how much we try, we cannot find such a tiling. Because our efforts do not produce a tiling, we are led to conjecture that no tiling exists.

We might try to prove that no tiling exists by showing that we reach a dead end however we successively place dominoes on the board. To construct such a proof, we would have to consider all possible cases that arise as we run through all possible choices of successively placing dominoes. For example, we have two choices for covering the square in the second column of the first row, next to the removed top left corner. We could cover it with a horizontally placed tile or a vertically placed tile. Each of these two choices leads to further choices, and so on. It does not take long to see that this is not a fruitful plan of attack for a person, although a computer could be used to complete such a proof by exhaustion. (Exercise 47 asks you to supply such a proof to show that a 4×4 checkerboard with opposite corners removed cannot be tiled.)

We need another approach. Perhaps there is an easier way to prove there is no tiling of a standard checkerboard with two opposite corners removed. As with many proofs, a key observation can help. We color the squares of this checkerboard using alternating white and black squares, as in Figure 2. Observe that a domino in a tiling of such a board covers one white square and one black square. Next, note that this board has unequal numbers of white squares and black squares. We can use these observations to prove by contradiction that a standard checkerboard with opposite corners removed cannot be tiled using dominoes. We now present such a proof.

Proof: Suppose we can use dominoes to tile a standard checkerboard with opposite corners removed. Note that the standard checkerboard with opposite corners removed contains $64 - 2 = 62$ squares. The tiling would use $62/2 = 31$ dominoes. Note that each domino in this tiling covers one white and one black square. Consequently, the tiling covers 31 white squares and 31 black squares. However, when we remove two opposite corner squares, either 32 of the remaining squares are white and 30 are black or else 30 are white and 32 are black. This contradicts the assumption that we can use dominoes to cover a standard checkerboard with opposite corners removed, completing the proof. ◀

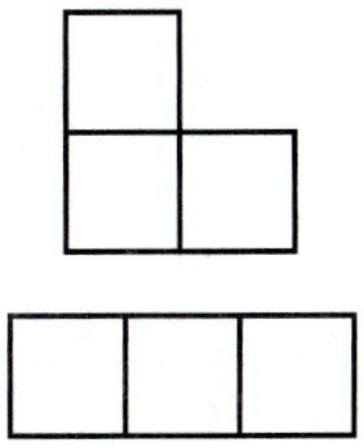

FIGURE 6 A right triomino and a straight triomino.

We can use other types of pieces besides dominoes in tilings. Instead of dominoes we can study tilings that use identically shaped pieces constructed from congruent squares that are connected along their edges. Such pieces are called **polyominoes**, a term coined in 1953 by the mathematician Solomon Golomb, the author of an entertaining book about them [Go94]. We will consider two polyominoes with the same number of squares the same if we can rotate and/or flip one of the polyominoes to get the other one. For example, there are two types of triominoes (see Figure 6), which are polyominoes made up of three squares connected by their sides. One type of triomino, the **straight triomino**, has three horizontally connected squares; the other type, **right triominoes**, resembles the letter L in shape, flipped and/or rotated, if necessary. We will study the tilings of a checkerboard by straight triominoes here; we will study tilings by right triominoes in Section 5.1.

EXAMPLE 21 Can you use straight triominoes to tile a standard checkerboard?

Solution: The standard checkerboard contains 64 squares and each triomino covers three squares. Consequently, if triominoes tile a board, the number of squares of the board must be a multiple of 3. Because 64 is not a multiple of 3, triominoes cannot be used to cover an 8×8 checkerboard. ◀

In Example 22, we consider the problem of using straight triominoes to tile a standard checkerboard with one corner missing.

EXAMPLE 22 Can we use straight triominoes to tile a standard checkerboard with one of its four corners removed? An 8×8 checkerboard with one corner removed contains $64 - 1 = 63$ squares. Any tiling by straight triominoes of one of these four boards uses $63/3 = 21$ triominoes. However, when we experiment, we cannot find a tiling of one of these boards using straight triominoes. A proof by exhaustion does not appear promising. Can we adapt our proof from Example 20 to prove that no such tiling exists?

Solution: We will color the squares of the checkerboard in an attempt to adapt the proof by contradiction we gave in Example 20 of the impossibility of using dominoes to tile a standard checkerboard with opposite corners removed. Because we are using straight triominoes rather than dominoes, we color the squares using three colors rather than two colors, as shown in Figure 7. Note that there are 21 blue squares, 21 black squares, and 22 white squares in this coloring. Next, we make the crucial observation that when a straight triomino covers three squares of the checkerboard, it covers one blue square, one black square, and one white square. Next, note that each of the three colors appears in a corner square. Thus, without loss of generality, we may assume that we have rotated the coloring so that the missing square is colored blue. Therefore, we assume that the remaining board contains 20 blue squares, 21 black squares, and 22 white squares.

FIGURE 7 Coloring the squares of the standard checkerboard with three colors.

If we could tile this board using straight triominoes, then we would use $63/3 = 21$ straight triominoes. These triominoes would cover 21 blue squares, 21 black squares, and 21 white squares. This contradicts the fact that this board contains 20 blue squares, 21 black squares, and 22 white squares. Therefore, we cannot tile this board using straight triominoes. ◀

1.8.9 The Role of Open Problems

Many advances in mathematics have been made by people trying to solve famous unsolved problems. In the past 20 years, many unsolved problems have finally been resolved, such as the proof of a conjecture in number theory made more than 300 years ago. This conjecture asserts the truth of the statement known as **Fermat's last theorem**.

THEOREM 1 **FERMAT'S LAST THEOREM** The equation

$$x^n + y^n = z^n$$

has no solutions in integers x, y, and z with $xyz \neq 0$ whenever n is an integer with $n > 2$.

Links

Remark: The equation $x^2 + y^2 = z^2$ has infinitely many solutions in integers x, y, and z; these solutions are called Pythagorean triples and correspond to the lengths of the sides of right triangles with integer lengths. See Exercise 34.

This problem has a fascinating history. In the seventeenth century, Fermat jotted in the margin of his copy of the works of Diophantus that he had a "wondrous proof" that there are no integer solutions of $x^n + y^n = z^n$ when n is an integer greater than 2 with $xyz \neq 0$. However, he never published a proof (Fermat published almost nothing), and no proof could be found in the papers he left when he died. Mathematicians looked for a proof for three centuries without success, although many people were convinced that a relatively simple proof could be found. (Proofs of special cases were found, such as the proof of the case when $n = 3$ by Euler and the proof of the $n = 4$ case by Fermat himself.) Over the years, several established mathematicians thought that they had proved this theorem. In the nineteenth century, one of these failed attempts led to the development of the part of number theory called algebraic number theory. A correct proof, requiring hundreds of pages of advanced mathematics, was not found until the 1990s, when Andrew Wiles used recently developed ideas from a sophisticated area of number theory called the theory of elliptic curves to prove Fermat's last theorem. Wiles's quest to find a proof of Fermat's last theorem using this powerful theory, described in a program in the *Nova* series on public television, took close to ten years! Moreover, his proof was based on major contributions of many mathematicians. (The interested reader should consult [Ro10] for more information about Fermat's last theorem and for additional references concerning this problem and its resolution.)

We now state an open problem that is simple to describe, but that seems quite difficult to resolve.

EXAMPLE 23 ***The 3x + 1 Conjecture*** Let T be the transformation that sends an even integer x to $x/2$ and an odd integer x to $3x + 1$. A famous conjecture, sometimes known as the **3x + 1 conjecture**, states that for all positive integers x, when we repeatedly apply the transformation T, we will eventually reach the integer 1. For example, starting with $x = 13$, we find $T(13) = 3 \cdot 13 + 1 = 40$, $T(40) = 40/2 = 20$, $T(20) = 20/2 = 10$, $T(10) = 10/2 = 5$, $T(5) = 3 \cdot 5 + 1 = 16$, $T(16) = 8$, $T(8) = 4$, $T(4) = 2$, and $T(2) = 1$. The $3x + 1$ conjecture has been verified using computers for all integers x up to $5.9 \cdot 10^{20}$.

Links

The $3x + 1$ conjecture has an interesting history and has attracted the attention of mathematicians since the 1950s. The conjecture has been raised many times and goes by many other names, including the Collatz problem, Hasse's algorithm, Ulam's problem, the Syracuse problem, and Kakutani's problem. Many mathematicians have been diverted from their work to spend time attacking this conjecture. This led to the joke that this problem was part of a conspiracy to slow down American mathematical research. See the article by Jeffrey Lagarias [La10] for a fascinating discussion of this problem and the results that have been found by mathematicians attacking it. ◀

Watch out! Working on the $3x + 1$ problem can be addictive.

There are a surprising number of important open problems throughout discrete mathematics. For instance, in Chapter 4 you will encounter many open questions about prime numbers. (Students already familiar with the basic notions about primes might want to explore Section

Charles Rex Arbogast/AP Images

ANDREW WILES (born 1953) Andrew Wiles was born in Cambridge, England. His father was a Professor of Divinity. Wiles attended the King's College School and the Leys School in Cambridge. Wiles become interested in Fermat's last theorem when at age ten he read a book stating the problem. He knew then that he would never let this problem go, as it looked simple but none of the great mathematicians could solve it. Wiles entered Merton College, Oxford in 1971. He received his B.A. in 1974, and then entered Clare College, Cambridge, for his graduate studies. He received his Ph.D. in 1980; his graduate research was on the theory of elliptic curves. He was a Benjamin Peirce Assistant Professor at Harvard University from 1977 until 1980. In 1981, he held a post at the Institute for Advanced Study in Princeton, and in 1982 he was appointed to a professorship at Princeton University. He was awarded a Guggenheim Fellowship in 1985 and spent a year at the Institut des Hautes Études Scientifiques and the École Normale Supérieure in Paris. Ironically, he did not realize that during his years working on elliptic curves he was learning techniques that would later help him solve the problem that obsessed him.

In 1986 when Wiles learned of work that showed that Fermat's last theorem follows from a conjecture in the theory of elliptic curves, he realized that this led to a possible strategy for a poof. He abandoned his ongoing research and devoted himself entirely to working on Fermat's last theorem. It took him more than seven years to complete his proof and two more years for some parts of the proof to be corrected. During this time he spent time only on this problem and with his young daughters. In 1988 he took a position as a research professor at Oxford University, returning to Princeton in 1990, where he remained until 2011, when he rejoined Oxford University as the Royal Society Research Professor.

Not only did Wiles become famous when he proved Fermat's last theorem, he also won the Wolfskehl Prize, which was established in 1908 for the first correct proof. This prize included 100,000 German marks (in the currency of the day), which would have been worth over $1,500,000 today. Although it was common wisdom that this prize had become worthless because of the two world wars, currency changes, and hyperinflation, Wiles received approximately $50,000. Wiles has won many of the top awards in mathematics, including the Abel Prize, the Fermat Prize, and the Wolf Prize. In 2000, he was made a Knight Commander of the Order of the British Empire by the Queen of England, making him Sir Andrew Wiles.

4.3, where these open questions are discussed.) You will encounter many other open questions as you read this book. The study of such problems has played and continues to play an important role in the development of many parts of discrete mathematics.

Build up your arsenal of proof methods as you work through this book.

1.8.10 Additional Proof Methods

In this chapter we introduced the basic methods used in proofs. We also described how to leverage these methods to prove a variety of results. We will use these proof methods in all subsequent chapters. In particular, we will use them in Chapters 2, 3, and 4 to prove results about sets, functions, algorithms, and number theory and in Chapters 9, 10, and 11 to prove results in graph theory. Among the theorems we will prove is the famous halting theorem, which states that there is a problem that cannot be solved using any procedure. However, there are many important proof methods besides those we have covered. We will introduce some of these methods later in this book. In particular, in Section 5.1 we will discuss mathematical induction, which is an extremely useful method for proving statements of the form $\forall n P(n)$, where the domain consists of all positive integers. In Section 5.3 we will introduce structural induction, which can be used to prove results about recursively defined sets. We will use the Cantor diagonalization method, which can be used to prove results about the size of infinite sets, in Section 2.5. In Chapter 6 we will introduce the notion of combinatorial proofs, which can be used to prove results by counting arguments. The reader should note that entire books have been devoted to the activities discussed in this section, including many excellent works by George Pólya ([Po61], [Po71], [Po90]).

Finally, note that we have not given a procedure that can be used for proving theorems in mathematics. It is a deep theorem of mathematical logic that there is no such procedure.

Exercises

1. Prove that $n^2 + 1 \geq 2^n$ when n is a positive integer with $1 \leq n \leq 4$.

2. Use a proof by cases to show that 10 is not the square of a positive integer. [*Hint*: Consider two cases: (*i*) $1 \leq x \leq 3$, (*ii*) $x \geq 4$.]

3. Use a proof by cases to show that 100 is not the cube of a positive integer. [*Hint*: Consider two cases: (*i*) $1 \leq x \leq 4$, (*ii*) $x \geq 5$.]

4. Prove that there are no positive perfect cubes less than 1000 that are the sum of the cubes of two positive integers.

5. Prove that if x and y are real numbers, then $\max(x, y) + \min(x, y) = x + y$. [*Hint:* Use a proof by cases, with the two cases corresponding to $x \geq y$ and $x < y$, respectively.]

6. Use a proof by cases to show that $\min(a, \min(b, c)) = \min(\min(a, b), c)$ whenever a, b, and c are real numbers.

7. Prove using the notion of without loss of generality that $\min(x, y) = (x + y - |x - y|)/2$ and $\max(x, y) = (x + y + |x - y|)/2$ whenever x and y are real numbers.

8. Prove using the notion of without loss of generality that $5x + 5y$ is an odd integer when x and y are integers of opposite parity.

9. Prove the **triangle inequality**, which states that if x and y are real numbers, then $|x| + |y| \geq |x + y|$ (where $|x|$ represents the absolute value of x, which equals x if $x \geq 0$ and equals $-x$ if $x < 0$).

10. Prove that there is a positive integer that equals the sum of the positive integers not exceeding it. Is your proof constructive or nonconstructive?

11. Prove that there are 100 consecutive positive integers that are not perfect squares. Is your proof constructive or nonconstructive?

12. Prove that either $2 \cdot 10^{500} + 15$ or $2 \cdot 10^{500} + 16$ is not a perfect square. Is your proof constructive or nonconstructive?

13. Prove that there exists a pair of consecutive integers such that one of these integers is a perfect square and the other is a perfect cube.

14. Show that the product of two of the numbers $65^{1000} - 8^{2001} + 3^{177}$, $79^{1212} - 9^{2399} + 2^{2001}$, and $24^{4493} - 5^{8192} + 7^{1777}$ is nonnegative. Is your proof constructive or nonconstructive? [*Hint:* Do not try to evaluate these numbers!]

15. Prove or disprove that there is a rational number x and an irrational number y such that x^y is irrational.

16. Prove or disprove that if a and b are rational numbers, then a^b is also rational.

17. Show that each of these statements can be used to express the fact that there is a unique element x such that $P(x)$ is true. [Note that we can also write this statement as $\exists!xP(x)$.]

a) $\exists x\forall y(P(y) \leftrightarrow x = y)$
b) $\exists xP(x) \wedge \forall x\forall y(P(x) \wedge P(y) \rightarrow x = y)$
c) $\exists x(P(x) \wedge \forall y(P(y) \rightarrow x = y))$

18. Show that if a, b, and c are real numbers and $a \neq 0$, then there is a unique solution of the equation $ax + b = c$.

19. Suppose that a and b are odd integers with $a \neq b$. Show there is a unique integer c such that $|a - c| = |b - c|$.

20. Show that if r is an irrational number, there is a unique integer n such that the distance between r and n is less than $1/2$.

21. Show that if n is an odd integer, then there is a unique integer k such that n is the sum of $k - 2$ and $k + 3$.

22. Prove that given a real number x there exist unique numbers n and ϵ such that $x = n + \epsilon$, n is an integer, and $0 \leq \epsilon < 1$.

23. Prove that given a real number x there exist unique numbers n and ϵ such that $x = n - \epsilon$, n is an integer, and $0 \leq \epsilon < 1$.

24. Use forward reasoning to show that if x is a nonzero real number, then $x^2 + 1/x^2 \geq 2$. [*Hint:* Start with the inequality $(x - 1/x)^2 \geq 0$, which holds for all nonzero real numbers x.]

25. The **harmonic mean** of two real numbers x and y equals $2xy/(x + y)$. By computing the harmonic and geometric means of different pairs of positive real numbers, formulate a conjecture about their relative sizes and prove your conjecture.

26. The **quadratic mean** of two real numbers x and y equals $\sqrt{(x^2 + y^2)/2}$. By computing the arithmetic and quadratic means of different pairs of positive real numbers, formulate a conjecture about their relative sizes and prove your conjecture.

***27.** Write the numbers $1, 2, \ldots, 2n$ on a blackboard, where n is an odd integer. Pick any two of the numbers, j and k, write $|j - k|$ on the board and erase j and k. Continue this process until only one integer is written on the board. Prove that this integer must be odd.

***28.** Suppose that five ones and four zeros are arranged around a circle. Between any two equal bits you insert a 0 and between any two unequal bits you insert a 1 to produce nine new bits. Then you erase the nine original bits. Show that when you iterate this procedure, you can never get nine zeros. [*Hint:* Work backward, assuming that you did end up with nine zeros.]

29. Formulate a conjecture about the decimal digits that appear as the final decimal digit of the fourth power of an integer. Prove your conjecture using a proof by cases.

30. Formulate a conjecture about the final two decimal digits of the square of an integer. Prove your conjecture using a proof by cases.

31. Prove that there is no positive integer n such that $n^2 + n^3 = 100$.

32. Prove that there are no solutions in integers x and y to the equation $2x^2 + 5y^2 = 14$.

33. Prove that there are no solutions in positive integers x and y to the equation $x^4 + y^4 = 625$.

34. Prove that there are infinitely many solutions in positive integers x, y, and z to the equation $x^2 + y^2 = z^2$. [*Hint:* Let $x = m^2 - n^2$, $y = 2mn$, and $z = m^2 + n^2$, where m and n are integers.]

35. Adapt the proof in Example 4 in Section 1.7 to prove that if $n = abc$, where a, b, and c are positive integers, then $a \leq \sqrt[3]{n}$, $b \leq \sqrt[3]{n}$, or $c \leq \sqrt[3]{n}$.

36. Prove that $\sqrt[3]{2}$ is irrational.

37. Prove that between every two rational numbers there is an irrational number.

38. Prove that between every rational number and every irrational number there is an irrational number.

***39.** Let $S = x_1y_1 + x_2y_2 + \cdots + x_ny_n$, where $x_1, x_2, \ldots, x_n$ and $y_1, y_2, \ldots, y_n$ are orderings of two different sequences of positive real numbers, each containing n elements.

a) Show that S takes its maximum value over all orderings of the two sequences when both sequences are sorted (so that the elements in each sequence are in nondecreasing order).

b) Show that S takes its minimum value over all orderings of the two sequences when one sequence is sorted into nondecreasing order and the other is sorted into nonincreasing order.

40. Prove or disprove that if you have an 8-gallon jug of water and two empty jugs with capacities of 5 gallons and 3 gallons, respectively, then you can measure 4 gallons by successively pouring some of or all of the water in a jug into another jug.

41. Verify the $3x + 1$ conjecture for these integers.

a) 6 **b)** 7 **c)** 17 **d)** 21

42. Verify the $3x + 1$ conjecture for these integers.

a) 16 **b)** 11 **c)** 35 **d)** 113

43. Prove or disprove that you can use dominoes to tile the standard checkerboard with two adjacent corners removed (that is, corners that are not opposite).

44. Prove or disprove that you can use dominoes to tile a standard checkerboard with all four corners removed.

45. Prove that you can use dominoes to tile a rectangular checkerboard with an even number of squares.

46. Prove or disprove that you can use dominoes to tile a 5×5 checkerboard with three corners removed.

47. Use a proof by exhaustion to show that a tiling using dominoes of a 4×4 checkerboard with opposite corners removed does not exist. [*Hint:* First show that you can assume that the squares in the upper left and lower right corners are removed. Number the squares of the original checkerboard from 1 to 16, starting in the first row, moving right in this row, then starting in the leftmost square in the second row and moving right, and so on. Remove squares 1 and 16. To begin the proof, note that square 2 is covered either by a domino laid horizontally, which covers squares 2 and 3, or vertically, which covers squares 2 and 6. Consider each of these cases separately, and work through all the subcases that arise.]

***48.** Prove that when a white square and a black square are removed from an 8×8 checkerboard (colored as in the text) you can tile the remaining squares of the checkerboard using dominoes. [*Hint:* Show that when one black and one white square are removed, each part of the partition of the remaining cells formed by inserting the barriers shown in the figure can be covered by dominoes.]

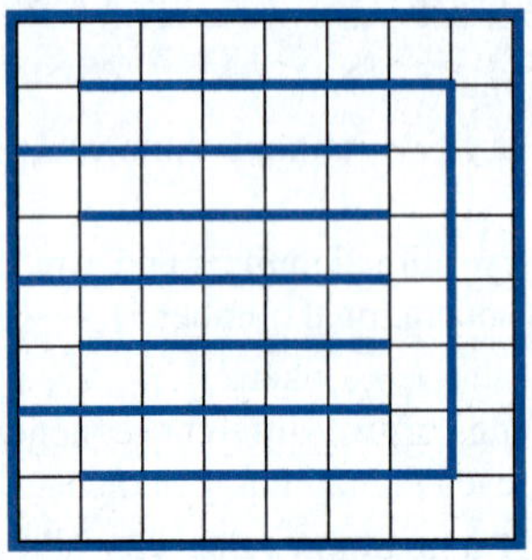

49. Show that by removing two white squares and two black squares from an 8×8 checkerboard (colored as in the text) you can make it impossible to tile the remaining squares using dominoes.

***50.** Find all squares, if they exist, on an 8×8 checkerboard such that the board obtained by removing one of these squares can be tiled using straight triominoes. [*Hint:* First use arguments based on coloring and rotations to eliminate as many squares as possible from consideration.]

***51. a)** Draw each of the five different tetrominoes, where a tetromino is a polyomino consisting of four squares.

b) For each of the five different tetrominoes, prove or disprove that you can tile a standard checkerboard using these tetrominoes.

***52.** Prove or disprove that you can tile a 10×10 checkerboard using straight tetrominoes.

Key Terms and Results

TERMS

proposition: a statement that is true or false

propositional variable: a variable that represents a proposition

truth value: true or false

$\neg p$ (negation of p): the proposition with truth value opposite to the truth value of p

logical operators: operators used to combine propositions

compound proposition: a proposition constructed by combining propositions using logical operators

truth table: a table displaying all possible truth values of propositions

$p \vee q$ (disjunction of p and q): the proposition "p or q," which is true if and only if at least one of p and q is true

$p \wedge q$ (conjunction of p and q): the proposition "p and q," which is true if and only if both p and q are true

$p \oplus q$ (exclusive or of p and q): the proposition "p *XOR* q," which is true when exactly one of p and q is true

$p \rightarrow q$ (p implies q): the proposition "if p, then q," which is false if and only if p is true and q is false

converse of $p \rightarrow q$: the conditional statement $q \rightarrow p$

contrapositive of $p \rightarrow q$: the conditional statement $\neg q \rightarrow \neg p$

inverse of $p \rightarrow q$: the conditional statement $\neg p \rightarrow \neg q$

$p \leftrightarrow q$ (biconditional): the proposition "p if and only if q," which is true if and only if p and q have the same truth value

bit: either a 0 or a 1

Boolean variable: a variable that has a value of 0 or 1

bit operation: an operation on a bit or bits

bit string: a list of bits

bitwise operations: operations on bit strings that operate on each bit in one string and the corresponding bit in the other string

logic gate: a logic element that performs a logical operation on one or more bits to produce an output bit

logic circuit: a switching circuit made up of logic gates that produces one or more output bits

tautology: a compound proposition that is always true

contradiction: a compound proposition that is always false

contingency: a compound proposition that is sometimes true and sometimes false

consistent compound propositions: compound propositions for which there is an assignment of truth values to the variables that makes all these propositions true

satisfiable compound proposition: a compound proposition for which there is an assignment of truth values to its variables that makes it true

logically equivalent compound propositions: compound propositions that always have the same truth values

predicate: part of a sentence that attributes a property to the subject

propositional function: a statement containing one or more variables that becomes a proposition when each of its variables is assigned a value or is bound by a quantifier

domain (or universe) of discourse: the values a variable in a propositional function may take

$\exists x P(x)$ (existential quantification of $P(x)$): the proposition that is true if and only if there exists an x in the domain such that $P(x)$ is true

$\forall x P(x)$ (universal quantification of $P(x)$): the proposition that is true if and only if $P(x)$ is true for every x in the domain

logically equivalent expressions: expressions that have the same truth value no matter which propositional functions and domains are used

free variable: a variable not bound in a propositional function

bound variable: a variable that is quantified

scope of a quantifier: portion of a statement where the quantifier binds its variable

argument: a sequence of statements

argument form: a sequence of compound propositions involving propositional variables

premise: a statement, in an argument, or argument form, other than the final one

conclusion: the final statement in an argument or argument form

valid argument form: a sequence of compound propositions involving propositional variables where the truth of all the premises implies the truth of the conclusion

valid argument: an argument with a valid argument form

rule of inference: a valid argument form that can be used in the demonstration that arguments are valid

fallacy: an invalid argument form often used incorrectly as a rule of inference (or sometimes, more generally, an incorrect argument)

circular reasoning or begging the question: reasoning where one or more steps are based on the truth of the statement being proved

theorem: a mathematical assertion that can be shown to be true

conjecture: a mathematical assertion proposed to be true, but that has not been proved

proof: a demonstration that a theorem is true

axiom: a statement that is assumed to be true and that can be used as a basis for proving theorems

lemma: a theorem used to prove other theorems

corollary: a proposition that can be proved as a consequence of a theorem that has just been proved

vacuous proof: a proof that $p \to q$ is true based on the fact that p is false

trivial proof: a proof that $p \to q$ is true based on the fact that q is true

direct proof: a proof that $p \to q$ is true that proceeds by showing that q must be true when p is true

proof by contraposition: a proof that $p \to q$ is true that proceeds by showing that p must be false when q is false

proof by contradiction: a proof that p is true based on the truth of the conditional statement $\neg p \to q$, where q is a contradiction

exhaustive proof: a proof that establishes a result by checking a list of all possible cases

proof by cases: a proof broken into separate cases, where these cases cover all possibilities

without loss of generality: an assumption in a proof that makes it possible to prove a theorem by reducing the number of cases to consider in the proof

counterexample: an element x such that $P(x)$ is false

constructive existence proof: a proof that an element with a specified property exists that explicitly finds such an element

nonconstructive existence proof: a proof that an element with a specified property exists that does not explicitly find such an element

rational number: a number that can be expressed as the ratio of two integers p and q such that $q \neq 0$

uniqueness proof: a proof that there is exactly one element satisfying a specified property

RESULTS

The logical equivalences given in Tables 6, 7, and 8 in Section 1.3.

De Morgan's laws for quantifiers.

Rules of inference for propositional calculus.

Rules of inference for quantified statements.

Review Questions

1. a) Define the negation of a proposition.
   b) What is the negation of "This is a boring course"?
2. a) Define (using truth tables) the disjunction, conjunction, exclusive or, conditional, and biconditional of the propositions p and q.
   b) What are the disjunction, conjunction, exclusive or, conditional, and biconditional of the propositions "I'll go to the movies tonight" and "I'll finish my discrete mathematics homework"?
3. a) Describe at least five different ways to write the conditional statement $p \to q$ in English.
   b) Define the converse and contrapositive of a conditional statement.
   c) State the converse and the contrapositive of the conditional statement "If it is sunny tomorrow, then I will go for a walk in the woods."
4. a) What does it mean for two propositions to be logically equivalent?
   b) Describe the different ways to show that two compound propositions are logically equivalent.
   c) Show in at least two different ways that the compound propositions $\neg p \vee (r \to \neg q)$ and $\neg p \vee \neg q \vee \neg r$ are equivalent.

5. *(Depends on the Exercise Set in Section 1.3)*
   a) Given a truth table, explain how to use disjunctive normal form to construct a compound proposition with this truth table.
   b) Explain why part (a) shows that the operators $\wedge$, $\vee$, and $\neg$ are functionally complete.
   c) Is there an operator such that the set containing just this operator is functionally complete?
6. What are the universal and existential quantifications of a predicate $P(x)$? What are their negations?
7. a) What is the difference between the quantification $\exists x \forall y P(x, y)$ and $\forall y \exists x P(x, y)$, where $P(x, y)$ is a predicate?
   b) Give an example of a predicate $P(x, y)$ such that $\exists x \forall y P(x, y)$ and $\forall y \exists x P(x, y)$ have different truth values.
8. Describe what is meant by a valid argument in propositional logic and show that the argument "If the earth is flat, then you can sail off the edge of the earth," "You cannot sail off the edge of the earth," therefore, "The earth is not flat" is a valid argument.
9. Use rules of inference to show that if the premises "All zebras have stripes" and "Mark is a zebra" are true, then the conclusion "Mark has stripes" is true.
10. a) Describe what is meant by a direct proof, a proof by contraposition, and a proof by contradiction of a conditional statement $p \to q$.
    b) Give a direct proof, a proof by contraposition, and a proof by contradiction of the statement: "If n is even, then $n + 4$ is even."
11. a) Describe a way to prove the biconditional $p \leftrightarrow q$.
    b) Prove the statement: "The integer $3n + 2$ is odd if and only if the integer $9n + 5$ is even, where n is an integer."
12. To prove that the statements p_1, p_2, p_3, and p_4 are equivalent, is it sufficient to show that the conditional statements $p_4 \to p_2$, $p_3 \to p_1$, and $p_1 \to p_2$ are valid? If not, provide another collection of conditional statements that can be used to show that the four statements are equivalent.
13. a) Suppose that a statement of the form $\forall x P(x)$ is false. How can this be proved?
    b) Show that the statement "For every positive integer n, $n^2 \geq 2n$" is false.
14. What is the difference between a constructive and nonconstructive existence proof? Give an example of each.
15. What are the elements of a proof that there is a unique element x such that $P(x)$, where $P(x)$ is a propositional function?
16. Explain how a proof by cases can be used to prove a result about absolute values, such as the fact that $|xy| = |x||y|$ for all real numbers x and y.

Supplementary Exercises

1. Let p be the proposition "I will do every exercise in this book" and q be the proposition "I will get an A in this course." Express each of these as a combination of p and q.
   a) I will get an A in this course only if I do every exercise in this book.
   b) I will get an A in this course and I will do every exercise in this book.
   c) Either I will not get an A in this course or I will not do every exercise in this book.
   d) For me to get an A in this course it is necessary and sufficient that I do every exercise in this book.
2. Find the truth table of the compound proposition $(p \vee q) \to (p \wedge \neg r)$.
3. Show that these compound propositions are tautologies.
   a) $(\neg q \wedge (p \to q)) \to \neg p$
   b) $((p \vee q) \wedge \neg p) \to q$
4. Give the converse, the contrapositive, and the inverse of these conditional statements.
   a) If it rains today, then I will drive to work.
   b) If $|x| = x$, then $x \geq 0$.
   c) If n is greater than 3, then n^2 is greater than 9.
5. Given a conditional statement $p \to q$, find the converse of its inverse, the converse of its converse, and the converse of its contrapositive.
6. Given a conditional statement $p \to q$, find the inverse of its inverse, the inverse of its converse, and the inverse of its contrapositive.
7. Find a compound proposition involving the propositional variables p, q, r, and s that is true when exactly three of these propositional variables are true and is false otherwise.
8. Show that these statements are inconsistent: "If Sergei takes the job offer, then he will get a signing bonus." "If Sergei takes the job offer, then he will receive a higher salary." "If Sergei gets a signing bonus, then he will not receive a higher salary." "Sergei takes the job offer."
9. Show that these statements are inconsistent: "If Miranda does not take a course in discrete mathematics, then she will not graduate." "If Miranda does not graduate, then she is not qualified for the job." "If Miranda reads this book, then she is qualified for the job." "Miranda does not take a course in discrete mathematics but she reads this book."

Teachers in the Middle Ages supposedly tested the realtime propositional logic ability of a student via a technique known as an **obligato game**. In an obligato game, a number of rounds is set and in each round the teacher gives the student successive assertions that the student must either accept or reject as they are given. When the student accepts an assertion, it is added as a commitment; when the student rejects an assertion its negation is added as a commitment. The student passes the test if the consistency of all commitments is maintained throughout the test.

10. Suppose that in a three-round obligato game, the teacher first gives the student the proposition $p \to q$, then the proposition $\neg(p \vee r) \vee q$, and finally, the proposition q. For which of the eight possible sequences of three answers will the student pass the test?

11. Suppose that in a four-round obligato game, the teacher first gives the student the proposition $\neg(p \to (q \wedge r))$, then the proposition $p \vee \neg q$, then the proposition $\neg r$, and finally, the proposition $(p \wedge r) \vee (q \to p)$. For which of the 16 possible sequences of four answers will the student pass the test?

12. Explain why every obligato game has a winning strategy.

Exercises 13 and 14 are set on the island of knights and knaves described in Example 8 in Section 1.2.

13. Suppose that you meet three people, Aaron, Bohan, and Crystal. Can you determine what Aaron, Bohan, and Crystal are if Aaron says "All of us are knaves" and Bohan says "Exactly one of us is a knave"?

14. Suppose that you meet three people, Anita, Boris, and Carmen. What are Anita, Boris, and Carmen if Anita says "I am a knave and Boris is a knight" and Boris says "Exactly one of the three of us is a knight"?

15. (Adapted from [Sm78]) Suppose that on an island there are three types of people, knights, knaves, and normals (also known as spies). Knights always tell the truth, knaves always lie, and normals sometimes lie and sometimes tell the truth. Detectives questioned three inhabitants of the island—Amy, Brenda, and Claire—as part of the investigation of a crime. The detectives knew that one of the three committed the crime, but not which one. They also knew that the criminal was a knight, and that the other two were not. Additionally, the detectives recorded these statements: Amy: "I am innocent." Brenda: "What Amy says is true." Claire: "Brenda is not a normal." After analyzing their information, the detectives positively identified the guilty party. Who was it?

16. Show that if S is a proposition, where S is the conditional statement "If S is true, then unicorns live," then "Unicorns live" is true. Show that it follows that S cannot be a proposition. (This paradox is known as *Löb's paradox.*)

17. Show that the argument with premises "The tooth fairy is a real person" and "The tooth fairy is not a real person" and conclusion "You can find gold at the end of the rainbow" is a valid argument. Does this show that the conclusion is true?

18. Suppose that the truth value of the proposition p_i is **T** whenever i is an odd positive integer and is **F** whenever i is an even positive integer. Find the truth values of $\bigvee_{i=1}^{100}(p_i \wedge p_{i+1})$ and $\bigwedge_{i=1}^{100}(p_i \vee p_{i+1})$.

***19.** Model 16×16 Sudoku puzzles (with 4×4 blocks) as satisfiability problems.

20. Let $P(x)$ be the statement "Student x knows calculus" and let $Q(y)$ be the statement "Class y contains a student who knows calculus." Express each of these as quantifications of $P(x)$ and $Q(y)$.

a) Some students know calculus.
b) Not every student knows calculus.
c) Every class has a student in it who knows calculus.
d) Every student in every class knows calculus.
e) There is at least one class with no students who know calculus.

21. Let $P(m, n)$ be the statement "m divides n," where the domain for both variables consists of all positive integers. (By "m divides n" we mean that $n = km$ for some integer k.) Determine the truth values of each of these statements.

a) $P(4, 5)$
b) $P(2, 4)$
c) $\forall m \forall n\, P(m, n)$
d) $\exists m \forall n\, P(m, n)$
e) $\exists n \forall m\, P(m, n)$
f) $\forall n\, P(1, n)$

22. Find a domain for the quantifiers in $\exists x \exists y (x \neq y \wedge \forall z((z = x) \vee (z = y)))$ such that this statement is true.

23. Find a domain for the quantifiers in $\exists x \exists y (x \neq y \wedge \forall z((z = x) \vee (z = y)))$ such that this statement is false.

24. Use existential and universal quantifiers to express the statement "No one has more than three grandmothers" using the propositional function $G(x, y)$, which represents "x is the grandmother of y."

25. Use existential and universal quantifiers to express the statement "Everyone has exactly two biological parents" using the propositional function $P(x, y)$, which represents "x is the biological parent of y."

26. The quantifier $\exists_n$ denotes "there exists exactly n," so that $\exists_n x P(x)$ means there exist exactly n values in the domain such that $P(x)$ is true. Determine the true value of these statements where the domain consists of all real numbers.

a) $\exists_0 x(x^2 = -1)$
b) $\exists_1 x(|x| = 0)$
c) $\exists_2 x(x^2 = 2)$
d) $\exists_3 x(x = |x|)$

27. Express each of these statements using existential and universal quantifiers and propositional logic, where $\exists_n$ is defined in Exercise 26.

a) $\exists_0 x P(x)$
b) $\exists_1 x P(x)$
c) $\exists_2 x P(x)$
d) $\exists_3 x P(x)$

28. Let $P(x, y)$ be a propositional function. Show that $\exists x \forall y\, P(x, y) \to \forall y \exists x\, P(x, y)$ is a tautology.

29. Let $P(x)$ and $Q(x)$ be propositional functions. Show that $\exists x\, (P(x) \to Q(x))$ and $\forall x\, P(x) \to \exists x\, Q(x)$ always have the same truth value.

30. If $\forall y \exists x\, P(x, y)$ is true, does it necessarily follow that $\exists x \forall y\, P(x, y)$ is true?

31. If $\forall x \exists y\, P(x, y)$ is true, does it necessarily follow that $\exists x \forall y\, P(x, y)$ is true?

32. Find the negations of these statements.

a) If it snows today, then I will go skiing tomorrow.

b) Every person in this class understands mathematical induction.

c) Some students in this class do not like discrete mathematics.

d) In every mathematics class there is some student who falls asleep during lectures.

33. Express this statement using quantifiers: "Every student in this class has taken some course in every department in the school of mathematical sciences."

34. Express this statement using quantifiers: "There is a building on the campus of some college in the United States in which every room is painted white."

35. Express the statement "There is exactly one student in this class who has taken exactly one mathematics class at this school" using the uniqueness quantifier. Then express this statement using quantifiers, without using the uniqueness quantifier.

36. Describe a rule of inference that can be used to prove that there are exactly two elements x and y in a domain such that $P(x)$ and $P(y)$ are true. Express this rule of inference as a statement in English.

37. Use rules of inference to show that if the premises $\forall x(P(x) \rightarrow Q(x))$, $\forall x(Q(x) \rightarrow R(x))$, and $\neg R(a)$, where a is in the domain, are true, then the conclusion $\neg P(a)$ is true.

38. Prove that if x^3 is irrational, then x is irrational.

39. Prove or disprove that if x^2 is irrational, then x^3 is irrational.

40. Prove that given a nonnegative integer n, there is a unique nonnegative integer m such that $m^2 \leq n < (m + 1)^2$.

41. Prove that there exists an integer m such that $m^2 > 10^{1000}$. Is your proof constructive or nonconstructive?

42. Prove that there is a positive integer that can be written as the sum of squares of positive integers in two different ways. (Use a computer or calculator to speed up your work.)

43. Disprove the statement that every positive integer is the sum of the cubes of eight nonnegative integers.

44. Disprove the statement that every positive integer is the sum of at most two squares and a cube of nonnegative integers.

45. Disprove the statement that every positive integer is the sum of 36 fifth powers of nonnegative integers.

46. Assuming the truth of the theorem that states that $\sqrt{n}$ is irrational whenever n is a positive integer that is not a perfect square, prove that $\sqrt{2} + \sqrt{3}$ is irrational.

Computer Projects

Write programs with the specified input and output.

1. Given the truth values of the propositions p and q, find the truth values of the conjunction, disjunction, exclusive or, conditional statement, and biconditional of these propositions.

2. Given two bit strings of length n, find the bitwise *AND*, bitwise *OR*, and bitwise *XOR* of these strings.

***3.** Give a compound proposition, determine whether it is satisfiable by checking its truth value for all positive assignments of truth values to its propositional variables.

4. Given the truth values of the propositions p and q in fuzzy logic, find the truth value of the disjunction and the conjunction of p and q (see Exercises 50 and 51 of Section 1.1).

***5.** Given positive integers m and n, interactively play the game of Chomp.

***6.** Given a portion of a checkerboard, look for tilings of this checkerboard with various types of polyominoes, including dominoes, the two types of triominoes, and larger polyominoes.

Computations and Explorations

Use a computational program or programs you have written to do these exercises.

1. Look for positive integers that are not the sum of the cubes of nine different positive integers.

2. Look for positive integers greater than 79 that are not the sum of the fourth powers of 18 positive integers.

3. Find as many positive integers as you can that can be written as the sum of cubes of positive integers, in two different ways, sharing this property with 1729.

***4.** Try to find winning strategies for the game of Chomp for different initial configurations of cookies.

5. Construct the 12 different pentominoes, where a pentomino is a polyomino consisting of five squares.

6. Find all the rectangles of 60 squares that can be tiled using every one of the 12 different pentominoes.

Writing Projects

Respond to these with essays using outside sources.

1. Discuss logical paradoxes, including the paradox of Epimenides the Cretan, Jourdain's card paradox, and the barber paradox, and how they are resolved.
2. Describe how fuzzy logic is being applied to practical applications. Consult one or more of the recent books on fuzzy logic written for general audiences.
3. Describe some of the practical problems that can be modeled as satisfiability problems.
4. Explain how satisfiability can be used to model a round robin tournament.
5. Describe some of the techniques that have been devised to help people solve Sudoku puzzles without the use of a computer.
6. Describe the basic rules of *WFF'N PROOF, The Game of Modern Logic*, developed by Layman Allen. Give examples of some of the games included in *WFF'N PROOF*.
7. Read some of the writings of Lewis Carroll on symbolic logic. Describe in detail some of the models he used to represent logical arguments and the rules of inference he used in these arguments.
8. Extend the discussion of Prolog given in Section 1.4, explaining in more depth how Prolog employs resolution.
9. Discuss some of the techniques used in computational logic, including Skolem's rule.
10. "Automated theorem proving" is the task of using computers to mechanically prove theorems. Discuss the goals and applications of automated theorem proving and the progress made in developing automated theorem provers.
11. Describe how DNA computing has been used to solve instances of the satisfiability problem.
12. Look up some of the incorrect proofs of famous open questions and open questions that were solved since 1970 and describe the type of error made in each proof.
13. Discuss what is known about winning strategies in the game of Chomp.
14. Describe various aspects of proof strategy discussed by George Pólya in his writings on reasoning, including [Po62], [Po71], and [Po90].
15. Describe a few problems and results about tilings with polyominoes, as described in [Go94] and [Ma91], for example.

CHAPTER

2 Basic Structures: Sets, Functions, Sequences, Sums, and Matrices

2.1 Sets

2.2 Set Operations

2.3 Functions

2.4 Sequences and Summations

2.5 Cardinality of Sets

2.6 Matrices

Much of discrete mathematics is devoted to the study of discrete structures, used to represent discrete objects. Many important discrete structures are built using sets, which are collections of objects. Among the discrete structures built from sets are combinations, unordered collections of objects used extensively in counting; relations, sets of ordered pairs that represent relationships between objects; graphs, sets of vertices and edges that connect vertices; and finite state machines, used to model computing machines. These are some of the topics we will study in later chapters.

The concept of a function is extremely important in discrete mathematics. A function assigns to each element of a first set exactly one element of a second set, where the two sets are not necessarily distinct. Functions play important roles throughout discrete mathematics. They are used to represent the computational complexity of algorithms, to study the size of sets, to count objects, and in a myriad of other ways. Useful structures such as sequences and strings are special types of functions. In this chapter, we will introduce the notion of sequences, which represent ordered lists of elements. Furthermore, we will introduce some important types of sequences and we will show how to define the terms of a sequence using earlier terms. We will also address the problem of identifying a sequence from its first few terms.

In our study of discrete mathematics, we will often add consecutive terms of a sequence of numbers. Because adding terms from a sequence, as well as other indexed sets of numbers, is such a common occurrence, a special notation has been developed for adding such terms. In this chapter, we will introduce the notation used to express summations. We will develop formulae for certain types of summations that appear throughout the study of discrete mathematics. For instance, we will encounter such summations in the analysis of the number of steps used by an algorithm to sort a list of numbers so that its terms are in increasing order.

The relative sizes of infinite sets can be studied by introducing the notion of the size, or cardinality, of a set. We say that a set is countable when it is finite or has the same size as the set of positive integers. In this chapter we will establish the surprising result that the set of rational numbers is countable, while the set of real numbers is not. We will also show how the concepts we discuss can be used to show that there are functions that cannot be computed using a computer program in any programming language.

Matrices are used in discrete mathematics to represent a variety of discrete structures. We will review the basic material about matrices and matrix arithmetic needed to represent relations and graphs. The matrix arithmetic we study will be used to solve a variety of problems involving these structures.

2.1 Sets

2.1.1 Introduction

In this section, we study the fundamental discrete structure on which all other discrete structures are built, namely, the set. Sets are used to group objects together. Often, but not always, the objects in a set have similar properties. For instance, all the students who are currently enrolled in your school make up a set. Likewise, all the students currently taking a course in discrete mathematics at any school make up a set. In addition, those students enrolled in your school who are taking a course in discrete mathematics form a set that can be obtained by taking the elements common to the first two collections. The language of sets is a means to study such collections in an organized fashion. We now provide a definition of a set. This definition is an intuitive definition, which is not part of a formal theory of sets.

Definition 1 A *set* is an unordered collection of distinct objects, called *elements* or *members* of the set. A set is said to *contain* its elements. We write $a \in A$ to denote that a is an element of the set A. The notation $a \notin A$ denotes that a is not an element of the set A.

It is common for sets to be denoted using uppercase letters. Lowercase letters are usually used to denote elements of sets.

There are several ways to describe a set. One way is to list all the members of a set, when this is possible. We use a notation where all members of the set are listed between braces. For example, the notation $\{a, b, c, d\}$ represents the set with the four elements a, b, c, and d. This way of describing a set is known as the **roster method**.

EXAMPLE 1 The set V of all vowels in the English alphabet can be written as $V = \{a, e, i, o, u\}$. ◀

EXAMPLE 2 The set O of odd positive integers less than 10 can be expressed by $O = \{1, 3, 5, 7, 9\}$. ◀

EXAMPLE 3 Although sets are usually used to group together elements with common properties, there is nothing that prevents a set from having seemingly unrelated elements. For instance, $\{a$, 2, Fred, New Jersey$\}$ is the set containing the four elements a, 2, Fred, and New Jersey. ◀

Sometimes the roster method is used to describe a set without listing all its members. Some members of the set are listed, and then *ellipses* (...) are used when the general pattern of the elements is obvious.

EXAMPLE 4 The set of positive integers less than 100 can be denoted by $\{1, 2, 3, \ldots, 99\}$. ◀

Extra Examples ❯

Another way to describe sets is to use **set builder notation**. We characterize the elements of a set by stating the properties objects must have to be members. The basic form of this notation is $\{x \mid x$ has property $P\}$ and is read "the set of all x such that x has property P." Here the vertical bar | is the symbol for "such that." We can also specify the universe U for a variable x using the notation $\{x \in U \mid x$ has property $P\}$. Furthermore, we can specify that the elements of set have a particular form, such as $f(x)$ where f is a function, giving us the form $\{f(x) \mid x \in U\}$. (Here, "x has property P" is a predicate, as discussed in Section 1.4.) We illustrate the use of set builder notation in Example 5.

Remark: Sometimes a colon is used in set builder notation instead of a vertical bar to represent "such that."

EXAMPLE 5 Use set builder notation to describe the set O of odd positive integers not exceeding 10.

Solution: We can specify this set in three different ways. First, using the basic form we have $O = \{x \mid x$ is an odd positive number not exceeding $10\}$.

Second, we first specify that x is a positive integer and then write $O = \{x \in \mathbf{Z}^+ \mid x \text{ is odd and } x \leq 10\}$.

Third, we can describe the same set by $O = \{2x - 1 \mid x \in \mathbf{Z}^+ \text{ and } x \leq 5\}$. ◀

We often use this type of notation to describe sets when it is impossible to list all the elements of the set. For instance, the set $\mathbf{Q}^+$ of all positive rational numbers can be written as

$$\mathbf{Q}^+ = \{x \in \mathbf{R} \mid x = \tfrac{p}{q}, \text{ for some positive integers } p \text{ and } q\}.$$

Beware that mathematicians disagree whether 0 is a natural number. We consider it quite natural.

These sets, each denoted using a boldface letter, play an important role in discrete mathematics:

$\mathbf{N} = \{0, 1, 2, 3, \ldots\}$, the set of all **natural numbers**
$\mathbf{Z} = \{\ldots, -2, -1, 0, 1, 2, \ldots\}$, the set of all **integers**
$\mathbf{Z}^+ = \{1, 2, 3, \ldots\}$, the set of all **positive integers**
$\mathbf{Q} = \{p/q \mid p \in \mathbf{Z}, q \in \mathbf{Z}, \text{ and } q \neq 0\}$, the set of all **rational numbers**
$\mathbf{R}$, the set of all **real numbers**
$\mathbf{R}^+$, the set of all **positive real numbers**
$\mathbf{C}$, the set of all **complex numbers**.

(Note that some people do not consider 0 a natural number, so be careful to check how the term *natural numbers* is used when you read other books.)

Among the sets studied in calculus and other subjects are **intervals**, sets of all the real numbers between two numbers a and b, with or without a and b. If a and b are real numbers with $a \leq b$, we denote these intervals by

$$[a, b] = \{x \mid a \leq x \leq b\}$$
$$[a, b) = \{x \mid a \leq x < b\}$$
$$(a, b] = \{x \mid a < x \leq b\}$$
$$(a, b) = \{x \mid a < x < b\}.$$

Note that $[a, b]$ is called the **closed interval** from a to b and (a, b) is called the **open interval** from a to b. Each of the intervals $[a, b]$, $[a, b)$, $(a, b]$, and (a, b) contains all the real numbers strictly between a and b. The first two of these contain a and the first and third contain b.

Remark: Some books use the notations $[a, b[$, $]a, b]$, and $]a, b[$ for $[a, b)$, $(a, b]$, and (a, b), respectively.

Sets can have other sets as members, as Example 6 illustrates.

EXAMPLE 6 The set $\{\mathbf{N}, \mathbf{Z}, \mathbf{Q}, \mathbf{R}\}$ is a set containing four elements, each of which is a set. The four elements of this set are $\mathbf{N}$, the set of natural numbers; $\mathbf{Z}$, the set of integers; $\mathbf{Q}$, the set of rational numbers; and $\mathbf{R}$, the set of real numbers. ◀

Remark: Note that the concept of a datatype, or type, in computer science is built upon the concept of a set. In particular, a **datatype** or **type** is the name of a set, together with a set of operations that can be performed on objects from that set. For example, *boolean* is the name of the set $\{0, 1\}$, together with operators on one or more elements of this set, such as AND, OR, and NOT.

Because many mathematical statements assert that two differently specified collections of objects are really the same set, we need to understand what it means for two sets to be equal.

Definition 2 Two sets are *equal* if and only if they have the same elements. Therefore, if A and B are sets, then A and B are equal if and only if $\forall x(x \in A \leftrightarrow x \in B)$. We write $A = B$ if A and B are equal sets.

EXAMPLE 7 The sets $\{1, 3, 5\}$ and $\{3, 5, 1\}$ are equal, because they have the same elements. Note that the order in which the elements of a set are listed does not matter. Note also that it does not matter if an element of a set is listed more than once, so $\{1, 3, 3, 3, 5, 5, 5, 5\}$ is the same as the set $\{1, 3, 5\}$ because they have the same elements. (Note that while it is not technically incorrect to list elements multiple times, it is redundant. Because this may confuse some readers and defies convention, it is best to list extra copies of elements only when there is a particular reason to do so.) ◀

THE EMPTY SET There is a special set that has no elements. This set is called the **empty set**, or **null set**, and is denoted by $\emptyset$. The empty set can also be denoted by $\{\ \}$ (that is, we represent the empty set with a pair of braces that encloses all the elements in this set). Often, a set of elements with certain properties turns out to be the null set. For instance, the set of all positive integers that are greater than their squares is the null set.

$\{\emptyset\}$ has one more element than $\emptyset$.

A set with one element is called a **singleton set**. A common error is to confuse the empty set $\emptyset$ with the set $\{\emptyset\}$, which is a singleton set. The single element of the set $\{\emptyset\}$ is the empty set itself! A useful analogy for remembering this difference is to think of folders in a computer file system. The empty set can be thought of as an empty folder and the set consisting of just the empty set can be thought of as a folder with exactly one folder inside, namely, the empty folder.

Links

NAIVE SET THEORY Note that the term *object* has been used in the definition of a set, Definition 1, without specifying what an object is. This description of a set as a collection of objects, based on the intuitive notion of an object, was first stated in 1895 by the German mathematician Georg Cantor. The theory that results from this intuitive definition of a set, and the use of the intuitive notion that for any property whatever, there is a set consisting of exactly the objects with this property, leads to **paradoxes**, or logical inconsistencies. This was shown by the English philosopher Bertrand Russell in 1902 (see Exercise 54 for a description of one of these paradoxes). These logical inconsistencies can be avoided by building set theory beginning

Links

Source: Library of Congress Prints and Photographs Division [LC-USZ62-74393]

GEORG CANTOR (1845–1918) Georg Cantor was born in St. Petersburg, Russia, where his father was a successful merchant. Cantor developed his interest in mathematics in his teens. He began his university studies in Zurich in 1862, but when his father died he left Zurich. He continued his university studies at the University of Berlin in 1863, where he studied under the eminent mathematicians Weierstrass, Kummer, and Kronecker. He received his doctor's degree in 1867, after having written a dissertation on number theory. Cantor assumed a position at the University of Halle in 1869, where he continued working until his death.

Cantor is considered the founder of set theory. His contributions in this area include the discovery that the set of real numbers is uncountable. He is also noted for his many important contributions to analysis. Cantor also was interested in philosophy and wrote papers relating his theory of sets with metaphysics.

Cantor married in 1874 and had six children. His melancholy temperament was balanced by his wife's happy disposition. Although he received a large inheritance from his father, he was poorly paid as a professor. To mitigate this, he tried to obtain a better-paying position at the University of Berlin. His appointment there was blocked by Kronecker, who did not agree with Cantor's views on set theory. Cantor suffered from mental illness throughout the later years of his life. He died in 1918 from a heart attack.

with axioms. However, we will use Cantor's original version of set theory, known as **naive set theory**, in this book because all sets considered in this book can be treated consistently using Cantor's original theory. Students will find familiarity with naive set theory helpful if they go on to learn about axiomatic set theory. They will also find the development of axiomatic set theory much more abstract than the material in this text. We refer the interested reader to [Su72] to learn more about axiomatic set theory.

Because naive set theory leads to paradoxes, and set theory plays such an important role in mathematics, mathematicians developed axiomatic versions of set theory to avoid these paradoxes. Zermelo–Fraenkel set theory developed early in the 20th century, named after Ernst Zermelo and Abraham Fraenkel, denoted by **ZFC** where the **C** comes from an important axiom, called the *axiom of choice,* is the most widely adopted form of set theory (see [Su99]). However, many other alternate axiomatic versions of set theory have been developed. Another approach, known as *type theory,* was developed by Bertrand Russell, Gottlob Frege, Alonzo Church, and others, to avoid the paradoxes of naive set theory. In such a theory, a hierarchy of types is defined and objects are assigned to be of a particular type.

2.1.2 Venn Diagrams

Sets can be represented graphically using Venn diagrams, named after the English mathematician John Venn, who introduced their use in 1881. In Venn diagrams the **universal set** U, which contains all the objects under consideration, is represented by a rectangle. (Note that the universal set varies depending on which objects are of interest.) Inside this rectangle, circles or other geometrical figures are used to represent sets. Sometimes points are used to represent the particular elements of the set. Venn diagrams are often used to indicate the relationships between sets. We show how a Venn diagram can be used in Example 8.

Assessment

EXAMPLE 8 Draw a Venn diagram that represents V, the set of vowels in the English alphabet.

Solution: We draw a rectangle to indicate the universal set U, which is the set of the 26 letters of the English alphabet. Inside this rectangle we draw a circle to represent V. Inside this circle we indicate the elements of V with points (see Figure 1). ◀

2.1.3 Subsets

It is common to encounter situations where the elements of one set are also the elements of a second set. We now introduce some terminology and notation to express such relationships between sets.

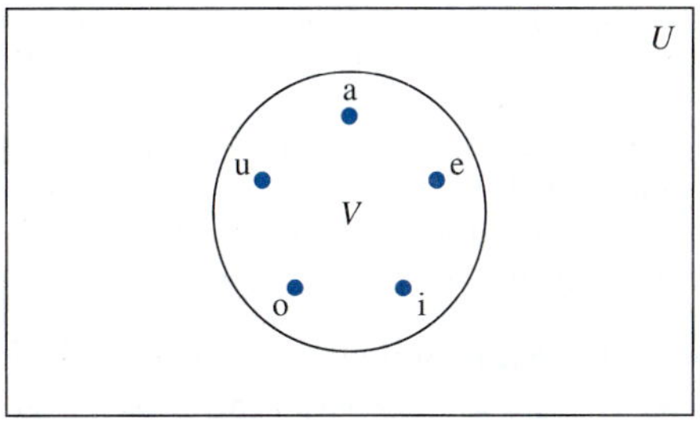

FIGURE 1 Venn diagram for the set of vowels.

Definition 3 The set A is a *subset* of B, and B is a *superset* of A, if and only if every element of A is also an element of B. We use the notation $A \subseteq B$ to indicate that A is a subset of the set B. If, instead, we want to stress that B is a superset of A, we use the equivalent notation $B \supseteq A$. (So, $A \subseteq B$ and $B \supseteq A$ are equivalent statements.)

We see that $A \subseteq B$ if and only if the quantification

$$\forall x(x \in A \rightarrow x \in B)$$

is true. Note that to show that A is not a subset of B we need only find one element $x \in A$ with $x \notin B$. Such an x is a counterexample to the claim that $x \in A$ implies $x \in B$.

We have these useful rules for determining whether one set is a subset of another:

Showing that A is a Subset of B To show that $A \subseteq B$, show that if x belongs to A then x also belongs to B.

Showing that A is Not a Subset of B To show that $A \nsubseteq B$, find a single $x \in A$ such that $x \notin B$.

EXAMPLE 9 The set of all odd positive integers less than 10 is a subset of the set of all positive integers less than 10, the set of rational numbers is a subset of the set of real numbers, the set of all computer science majors at your school is a subset of the set of all students at your school, and the set of all people in China is a subset of the set of all people in China (that is, it is a subset of itself). Each of these facts follows immediately by noting that an element that belongs to the first set in each pair of sets also belongs to the second set in that pair. ◀

EXAMPLE 10 The set of integers with squares less than 100 is not a subset of the set of nonnegative integers because -1 is in the former set [as $(-1)^2 < 100$], but not the latter set. The set of people who have taken discrete mathematics at your school is not a subset of the set of all computer science majors at your school if there is at least one student who has taken discrete mathematics who is not a computer science major. ◀

Links

Source: Library of Congress Prints and Photographs Division [LC-USZ62-49535]

BERTRAND RUSSELL (1872–1970) Bertrand Russell was born into a prominent English family active in the progressive movement and having a strong commitment to liberty. He became an orphan at an early age and was placed in the care of his father's parents, who had him educated at home. He entered Trinity College, Cambridge, in 1890, where he excelled in mathematics and in moral science. He won a fellowship on the basis of his work on the foundations of geometry. In 1910, Trinity College appointed him to a lectureship in logic and the philosophy of mathematics.

Russell fought for progressive causes throughout his life. He held strong pacifist views, and his protests against World War I led to dismissal from his position at Trinity College. He was imprisoned for 6 months in 1918 because of an article he wrote that was branded as seditious. Russell fought for women's suffrage in Great Britain. In 1961, at the age of 89, he was imprisoned for the second time for his protests advocating nuclear disarmament.

Russell's greatest work was in his development of principles that could be used as a foundation for all of mathematics. His most famous work is *Principia Mathematica,* written with Alfred North Whitehead, which attempts to deduce all of mathematics using a set of primitive axioms. He wrote many books on philosophy, physics, and his political ideas. Russell won the Nobel Prize for Literature in 1950.

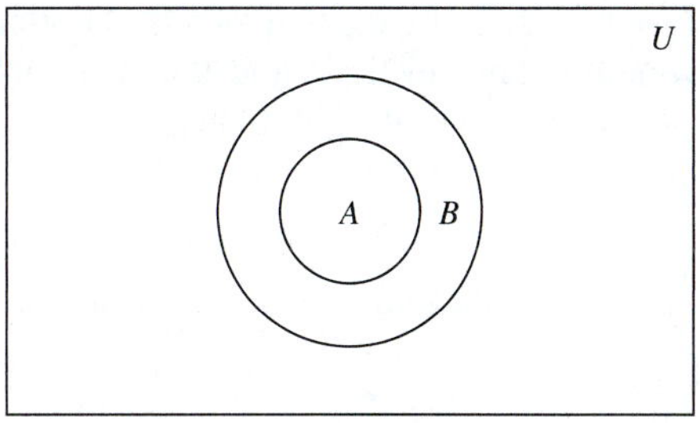

FIGURE 2 **Venn diagram showing that A is a subset of B.**

Theorem 1 shows that every nonempty set S is guaranteed to have at least two subsets, the empty set and the set S itself, that is, $\emptyset \subseteq S$ and $S \subseteq S$.

THEOREM 1 For every set S, $(i)\ \emptyset \subseteq S$ and $(ii)\ S \subseteq S$.

Proof: We will prove (i) and leave the proof of (ii) as an exercise.

Let S be a set. To show that $\emptyset \subseteq S$, we must show that $\forall x(x \in \emptyset \rightarrow x \in S)$ is true. Because the empty set contains no elements, it follows that $x \in \emptyset$ is always false. It follows that the conditional statement $x \in \emptyset \rightarrow x \in S$ is always true, because its hypothesis is always false and a conditional statement with a false hypothesis is true. Therefore, $\forall x(x \in \emptyset \rightarrow x \in S)$ is true. This completes the proof of (i). Note that this is an example of a vacuous proof. ◁

When we wish to emphasize that a set A is a subset of a set B but that $A \neq B$, we write $A \subset B$ and say that A is a **proper subset** of B. For $A \subset B$ to be true, it must be the case that $A \subseteq B$ and there must exist an element x of B that is not an element of A. That is, A is a proper subset of B if and only if

$$\forall x(x \in A \rightarrow x \in B) \wedge \exists x(x \in B \wedge x \notin A)$$

is true.

Venn diagrams can be used to illustrate that a set A is a subset of a set B. We draw the universal set U as a rectangle. Within this rectangle we draw a circle for B. Because A is a subset of B, we draw the circle for A within the circle for B. This relationship is shown in Figure 2.

Recall from Definition 2 that sets are equal if they have the same elements. A useful way to show that two sets have the same elements is to show that each set is a subset of the other. In other words, we can show that if A and B are sets with $A \subseteq B$ and $B \subseteq A$, then $A = B$. That

Links

Alpha Historica/Alamy Stock Photo

JOHN VENN (1834–1923) John Venn was born into a London suburban family noted for its philanthropy. He attended London schools and got his mathematics degree from Caius College, Cambridge, in 1857. He was elected a fellow of this college and held his fellowship there until his death. He took holy orders in 1859 and, after a brief stint of religious work, returned to Cambridge, where he developed programs in the moral sciences. Besides his mathematical work, Venn had an interest in history and wrote extensively about his college and family.

Venn's book *Symbolic Logic* clarifies ideas originally presented by Boole. In this book, Venn presents a systematic development of a method that uses geometric figures, known now as *Venn diagrams*. Today these diagrams are primarily used to analyze logical arguments and to illustrate relationships between sets. In addition to his work on symbolic logic, Venn made contributions to probability theory described in his widely used textbook on that subject.

is, $A = B$ if and only if $\forall x(x \in A \rightarrow x \in B)$ and $\forall x(x \in B \rightarrow x \in A)$ or equivalently if and only if $\forall x(x \in A \leftrightarrow x \in B)$, which is what it means for the A and B to be equal. Because this method of showing two sets are equal is so useful, we highlight it here.

Showing Two Sets are Equal To show that two sets A and B are equal, show that $A \subseteq B$ and $B \subseteq A$.

Sets may have other sets as members. For instance, we have the sets

$$A = \{\emptyset, \{a\}, \{b\}, \{a, b\}\} \quad \text{and} \quad B = \{x \mid x \text{ is a subset of the set } \{a, b\}\}.$$

Note that these two sets are equal, that is, $A = B$. Also note that $\{a\} \in A$, but $a \notin A$.

2.1.4 The Size of a Set

Sets are used extensively in counting problems, and for such applications we need to discuss the sizes of sets.

Definition 4 Let S be a set. If there are exactly n distinct elements in S where n is a nonnegative integer, we say that S is a *finite set* and that n is the *cardinality* of S. The cardinality of S is denoted by $|S|$.

Remark: The term *cardinality* comes from the common usage of the term *cardinal number* as the size of a finite set.

EXAMPLE 11 Let A be the set of odd positive integers less than 10. Then $|A| = 5$. ◀

EXAMPLE 12 Let S be the set of letters in the English alphabet. Then $|S| = 26$. ◀

EXAMPLE 13 Because the null set has no elements, it follows that $|\emptyset| = 0$. ◀

We will also be interested in sets that are not finite.

Definition 5 A set is said to be *infinite* if it is not finite.

EXAMPLE 14 The set of positive integers is infinite. ◀

Extra Examples ❯

We will extend the notion of cardinality to infinite sets in Section 2.5, a challenging topic full of surprising results.

2.1.5 Power Sets

Many problems involve testing all combinations of elements of a set to see if they satisfy some property. To consider all such combinations of elements of a set S, we build a new set that has as its members all the subsets of S.

Definition 6 Given a set S, the *power set* of S is the set of all subsets of the set S. The power set of S is denoted by $\mathcal{P}(S)$.

EXAMPLE 15 What is the power set of the set $\{0, 1, 2\}$?

Extra Examples

Solution: The power set $\mathcal{P}(\{0, 1, 2\})$ is the set of all subsets of $\{0, 1, 2\}$. Hence,

$$\mathcal{P}(\{0, 1, 2\}) = \{\emptyset, \{0\}, \{1\}, \{2\}, \{0, 1\}, \{0, 2\}, \{1, 2\}, \{0, 1, 2\}\}.$$

Note that the empty set and the set itself are members of this set of subsets. ◀

EXAMPLE 16 What is the power set of the empty set? What is the power set of the set $\{\emptyset\}$?

Solution: The empty set has exactly one subset, namely, itself. Consequently,

$$\mathcal{P}(\emptyset) = \{\emptyset\}.$$

The set $\{\emptyset\}$ has exactly two subsets, namely, $\emptyset$ and the set $\{\emptyset\}$ itself. Therefore,

$$\mathcal{P}(\{\emptyset\}) = \{\emptyset, \{\emptyset\}\}.$$

◀

If a set has n elements, then its power set has 2^n elements. We will demonstrate this fact in several ways in subsequent sections of the text.

2.1.6 Cartesian Products

The order of elements in a collection is often important. Because sets are unordered, a different structure is needed to represent ordered collections. This is provided by **ordered n-tuples**.

Definition 7 The *ordered n-tuple* $(a_1, a_2, \ldots, a_n)$ is the ordered collection that has a_1 as its first element, a_2 as its second element, …, and a_n as its nth element.

We say that two ordered n-tuples are equal if and only if each corresponding pair of their elements is equal. In other words, $(a_1, a_2, \ldots, a_n) = (b_1, b_2, \ldots, b_n)$ if and only if $a_i = b_i$, for $i = 1, 2, \ldots, n$. In particular, ordered 2-tuples are called **ordered pairs**. The ordered pairs (a, b) and (c, d) are equal if and only if $a = c$ and $b = d$. Note that (a, b) and (b, a) are not equal unless $a = b$.

Many of the discrete structures we will study in later chapters are based on the notion of the *Cartesian product* of sets (named after René Descartes). We first define the Cartesian product of two sets.

Definition 8 Let A and B be sets. The *Cartesian product* of A and B, denoted by $A \times B$, is the set of all ordered pairs (a, b), where $a \in A$ and $b \in B$. Hence,

$$A \times B = \{(a, b) \mid a \in A \land b \in B\}.$$

EXAMPLE 17 Let A represent the set of all students at a university, and let B represent the set of all courses offered at the university. What is the Cartesian product $A \times B$ and how can it be used?

Extra Examples

Solution: The Cartesian product $A \times B$ consists of all the ordered pairs of the form (a, b), where a is a student at the university and b is a course offered at the university. One way to use the set $A \times B$ is to represent all possible enrollments of students in courses at the university. Furthermore, observe that each subset of $A \times B$ represents one possible total enrollment configuration, and $\mathcal{P}(A \times B)$ represents all possible enrollment configurations. ◀

EXAMPLE 18 What is the Cartesian product of $A = \{1, 2\}$ and $B = \{a, b, c\}$?

Solution: The Cartesian product $A \times B$ is

$$A \times B = \{(1, a), (1, b), (1, c), (2, a), (2, b), (2, c)\}.$$ ◀

Note that the Cartesian products $A \times B$ and $B \times A$ are not equal unless $A = \emptyset$ or $B = \emptyset$ (so that $A \times B = \emptyset$) or $A = B$ (see Exercises 37 and 44). This is illustrated in Example 19.

EXAMPLE 19 Show that the Cartesian product $B \times A$ is not equal to the Cartesian product $A \times B$, where A and B are as in Example 18.

Links

Stock Montage/Archive Photos/Getty Images

RENÉ DESCARTES (1596–1650) René Descartes was born into a noble family near Tours, France, about 130 miles southwest of Paris. He was the third child of his father's first wife; she died several days after his birth. Because of René's poor health, his father, a provincial judge, let his son's formal lessons slide until, at the age of 8, René entered the Jesuit college at La Flèche. The rector of the school took a liking to him and permitted him to stay in bed until late in the morning because of his frail health. From then on, Descartes spent his mornings in bed; he considered these times his most productive hours for thinking.

Descartes left school in 1612, moving to Paris, where he spent 2 years studying mathematics. He earned a law degree in 1616 from the University of Poitiers. At 18 Descartes became disgusted with studying and decided to see the world. He moved to Paris and became a successful gambler. However, he grew tired of bawdy living and moved to the suburb of Saint-Germain, where he devoted himself to mathematical study. When his gambling friends found him, he decided to leave France and undertake a military career. However, he never did any fighting. One day, while escaping the cold in an overheated room at a military encampment, he had several feverish dreams, which revealed his future career as a mathematician and philosopher.

After ending his military career, he traveled throughout Europe. He then spent several years in Paris, where he studied mathematics and philosophy and constructed optical instruments. Descartes decided to move to Holland, where he spent 20 years wandering around the country, accomplishing his most important work. During this time he wrote several books, including the *Discours,* which contains his contributions to analytic geometry, for which he is best known. He also made fundamental contributions to philosophy.

In 1649 Descartes was invited by Queen Christina to visit her court in Sweden to tutor her in philosophy. Although he was reluctant to live in what he called "the land of bears amongst rocks and ice," he finally accepted the invitation and moved to Sweden. Unfortunately, the winter of 1649–1650 was extremely bitter. Descartes caught pneumonia and died in mid-February.

Solution: The Cartesian product $B \times A$ is

$$B \times A = \{(a, 1), (a, 2), (b, 1), (b, 2), (c, 1), (c, 2)\}.$$

This is not equal to $A \times B$, which was found in Example 18. ◀

The Cartesian product of more than two sets can also be defined.

Definition 9 The *Cartesian product* of the sets $A_1, A_2, \ldots, A_n$, denoted by $A_1 \times A_2 \times \cdots \times A_n$, is the set of ordered n-tuples $(a_1, a_2, \ldots, a_n)$, where a_i belongs to A_i for $i = 1, 2, \ldots, n$. In other words,

$$A_1 \times A_2 \times \cdots \times A_n = \{(a_1, a_2, \ldots, a_n) \mid a_i \in A_i \text{ for } i = 1, 2, \ldots, n\}.$$

EXAMPLE 20 What is the Cartesian product $A \times B \times C$, where $A = \{0, 1\}$, $B = \{1, 2\}$, and $C = \{0, 1, 2\}$?

Solution: The Cartesian product $A \times B \times C$ consists of all ordered triples (a, b, c), where $a \in A$, $b \in B$, and $c \in C$. Hence,

$$A \times B \times C = \{(0, 1, 0), (0, 1, 1), (0, 1, 2), (0, 2, 0), (0, 2, 1), (0, 2, 2), (1, 1, 0), (1, 1, 1), (1, 1, 2), (1, 2, 0), (1, 2, 1), (1, 2, 2)\}.$$ ◀

Remark: Note that when A, B, and C are sets, $(A \times B) \times C$ is not the same as $A \times B \times C$ (see Exercise 45).

We use the notation A^2 to denote $A \times A$, the Cartesian product of the set A with itself. Similarly, $A^3 = A \times A \times A$, $A^4 = A \times A \times A \times A$, and so on. More generally,

$$A^n = \{(a_1, a_2, \ldots, a_n) \mid a_i \in A \text{ for } i = 1, 2, \ldots, n\}.$$

EXAMPLE 21 Suppose that $A = \{1, 2\}$. It follows that $A^2 = \{(1, 1), (1, 2), (2, 1), (2, 2)\}$ and $A^3 = \{(1, 1, 1), (1, 1, 2), (1, 2, 1), (1, 2, 2), (2, 1, 1), (2, 1, 2), (2, 2, 1), (2, 2, 2)\}$. ◀

A subset R of the Cartesian product $A \times B$ is called a **relation** from the set A to the set B. The elements of R are ordered pairs, where the first element belongs to A and the second to B. For example, $R = \{(a, 0), (a, 1), (a, 3), (b, 1), (b, 2), (c, 0), (c, 3)\}$ is a relation from the set $\{a, b, c\}$ to the set $\{0, 1, 2, 3\}$, and it is also a relation from the set $\{a, b, c, d, e\}$ to the set $\{0, 1, 3, 4\}$. (This illustrates that a relation need not contain a pair (x, y) for every element x of A.) A relation from a set A to itself is called a relation on A.

EXAMPLE 22 What are the ordered pairs in the less than or equal to relation, which contains (a, b) if $a \le b$, on the set $\{0, 1, 2, 3\}$?

Solution: The ordered pair (a, b) belongs to R if and only if both a and b belong to $\{0, 1, 2, 3\}$ and $a \le b$. Consequently, $R = \{(0, 0), (0, 1), (0, 2), (0, 3), (1, 1), (1, 2), (1, 3), (2, 2), (2, 3), (3, 3)\}$. ◀

We can use set builder notation to describe relations, as Example 23 illustrates.

EXAMPLE 23 Use set builder notation to describe the relation R consisting of ordered pairs (a, b) where a and b are positive integers and where $b = a^2 + 1$ and list the pairs with $b < 100$.

Solution: We can specify R in set builder notation as $R = \{(a, b) \mid a \in \mathbf{Z}^+, b \in \mathbf{Z}^+ \text{ and } b = a^2 + 1\}$. The members of R with $b < 100$ are (1, 2), (2, 5), (3, 10), (4, 17), (5, 26), (6, 37), (7, 50), (8, 65), and (9, 82). ◀

We will study relations and their properties at length in Chapter 9.

2.1.7 Using Set Notation with Quantifiers

Sometimes we restrict the domain of a quantified statement explicitly by making use of a particular notation. For example, $\forall x \in S(P(x))$ denotes the universal quantification of $P(x)$ over all elements in the set S. In other words, $\forall x \in S(P(x))$ is shorthand for $\forall x(x \in S \rightarrow P(x))$. Similarly, $\exists x \in S(P(x))$ denotes the existential quantification of $P(x)$ over all elements in S. That is, $\exists x \in S(P(x))$ is shorthand for $\exists x(x \in S \wedge P(x))$.

EXAMPLE 24 What do the statements $\forall x \in \mathbf{R}\ (x^2 \geq 0)$ and $\exists x \in \mathbf{Z}\ (x^2 = 1)$ mean?

Solution: The statement $\forall x \in \mathbf{R}(x^2 \geq 0)$ states that for every real number x, $x^2 \geq 0$. This statement can be expressed as "The square of every real number is nonnegative." This is a true statement.

The statement $\exists x \in \mathbf{Z}(x^2 = 1)$ states that there exists an integer x such that $x^2 = 1$. This statement can be expressed as "There is an integer whose square is 1." This is also a true statement because $x = 1$ is such an integer (as is -1). ◀

2.1.8 Truth Sets and Quantifiers

We will now tie together concepts from set theory and from predicate logic. Given a predicate P and a domain D, we define the **truth set** of P to be the set of elements x in D for which $P(x)$ is true. The truth set of $P(x)$ is denoted by $\{x \in D \mid P(x)\}$.

EXAMPLE 25 What are the truth sets of the predicates $P(x)$, $Q(x)$, and $R(x)$, where the domain is the set of integers and $P(x)$ is "$|x| = 1$," $Q(x)$ is "$x^2 = 2$," and $R(x)$ is "$|x| = x$."

Solution: The truth set of P, $\{x \in \mathbf{Z} \mid |x| = 1\}$, is the set of integers for which $|x| = 1$. Because $|x| = 1$ when $x = 1$ or $x = -1$, and for no other integers x, we see that the truth set of P is the set $\{-1, 1\}$.

The truth set of Q, $\{x \in \mathbf{Z} \mid x^2 = 2\}$, is the set of integers for which $x^2 = 2$. This is the empty set because there are no integers x for which $x^2 = 2$.

The truth set of R, $\{x \in \mathbf{Z} \mid |x| = x\}$, is the set of integers for which $|x| = x$. Because $|x| = x$ if and only if $x \geq 0$, it follows that the truth set of R is $\mathbf{N}$, the set of nonnegative integers. ◀

Note that $\forall x P(x)$ is true over the domain U if and only if the truth set of P is the set U. Likewise, $\exists x P(x)$ is true over the domain U if and only if the truth set of P is nonempty.

Exercises

1. List the members of these sets.

a) $\{x \mid x \text{ is a real number such that } x^2 = 1\}$

b) $\{x \mid x \text{ is a positive integer less than } 12\}$

c) $\{x \mid x \text{ is the square of an integer and } x < 100\}$

d) $\{x \mid x \text{ is an integer such that } x^2 = 2\}$

2. Use set builder notation to give a description of each of these sets.

a) $\{0, 3, 6, 9, 12\}$

b) $\{-3, -2, -1, 0, 1, 2, 3\}$

c) $\{m, n, o, p\}$

3. List the elements of each of these sets described by set builder notation

a) $\{a \mid a \text{ is the square of a positive integer less than } 12\}$

b) $\{a \in \mathbf{Z}^+ \mid a < 20 \text{ and } a \text{ is not a power of } 2\}$

c) $\{7a + 4 \mid a \in \mathbf{Z}^+ \text{ and } a \text{ is odd with } a < 10\}$

4. List the elements of each of these sets described by set builder notation

a) $\{x \mid x \text{ is an even positive integer less than } 20\}$

b) $\{x \in \mathbf{Z}^+ \mid x \text{ is odd with } 10 < x < 30\}$

c) $\{5x + 3 \mid x \in \mathbf{Z}^+ \text{ and } x \text{ is even with } 40 < x < 50\}$

5. Which of the intervals $(0, 5)$, $(0, 5]$, $[0, 5)$, $[0, 5]$, $(1, 4]$, $[2, 3]$, $(2, 3)$ contains

a) 0? **b)** 1?

c) 2? **d)** 3?

e) 4? **f)** 5?

6. For each of these intervals, list all its elements or explain why it is empty.

a) $[a, a]$ **b)** $[a, a)$

c) $(a, a]$ **d)** (a, a)

e) (a, b), where $a > b$ **f)** $[a, b]$, where $a > b$

7. Use set builder notation to describe the set of ordered pairs (a, b) where a and b are positive integers such that

a) $a \geq b$

b) a and b have different parity

c) b exceeds twice a

8. Use set builder notation to describe the set of ordered pairs (m, n) where m and n are positive integers such that

a) there is an integer q such that $n = qm$

b) there are integers s and t such that the product of m and n is the sum of the squares of s and t

c) the integers m and n differ by more than 2

9. For each of these pairs of sets, determine whether the first is a subset of the second, the second is a subset of the first, or neither is a subset of the other.

a) the set of airline flights from New York to New Delhi, the set of nonstop airline flights from New York to New Delhi

b) the set of people who speak English, the set of people who speak Chinese

c) the set of flying squirrels, the set of living creatures that can fly

10. For each of these pairs of sets, determine whether the first is a subset of the second, the second is a subset of the first, or neither is a subset of the other.

a) the set of people who speak English, the set of people who speak English with an Australian accent

b) the set of fruits, the set of citrus fruits

c) the set of students studying discrete mathematics, the set of students studying data structures

11. Determine whether each of these pairs of sets are equal.

a) $\{1, 3, 3, 3, 5, 5, 5, 5, 5\}$, $\{5, 3, 1\}$

b) $\{\{1\}\}$, $\{1, \{1\}\}$ **c)** $\emptyset$, $\{\emptyset\}$

12. Suppose that $A = \{2, 4, 6\}$, $B = \{2, 6\}$, $C = \{4, 6\}$, and $D = \{4, 6, 8\}$. Determine which of these sets are subsets of which other of these sets.

13. For each of the following sets, determine whether 2 is an element of that set.

a) $\{x \in \mathbf{R} \mid x \text{ is an integer greater than } 1\}$

b) $\{x \in \mathbf{R} \mid x \text{ is the square of an integer}\}$

c) $\{2, \{2\}\}$ **d)** $\{\{2\}, \{\{2\}\}\}$

e) $\{\{2\}, \{2, \{2\}\}\}$ **f)** $\{\{\{2\}\}\}$

14. For each of the sets in Exercise 13, determine whether $\{2\}$ is an element of that set.

15. Determine whether each of these statements is true or false.

a) $0 \in \emptyset$ **b)** $\emptyset \in \{0\}$

c) $\{0\} \subset \emptyset$ **d)** $\emptyset \subset \{0\}$

e) $\{0\} \in \{0\}$ **f)** $\{0\} \subset \{0\}$

g) $\{\emptyset\} \subseteq \{\emptyset\}$

16. Determine whether these statements are true or false.

a) $\emptyset \in \{\emptyset\}$ **b)** $\emptyset \in \{\emptyset, \{\emptyset\}\}$

c) $\{\emptyset\} \in \{\emptyset\}$ **d)** $\{\emptyset\} \in \{\{\emptyset\}\}$

e) $\{\emptyset\} \subset \{\emptyset, \{\emptyset\}\}$ **f)** $\{\{\emptyset\}\} \subset \{\emptyset, \{\emptyset\}\}$

g) $\{\{\emptyset\}\} \subset \{\{\emptyset\}, \{\emptyset\}\}$

17. Determine whether each of these statements is true or false.

a) $x \in \{x\}$ **b)** $\{x\} \subseteq \{x\}$ **c)** $\{x\} \in \{x\}$

d) $\{x\} \in \{\{x\}\}$ **e)** $\emptyset \subseteq \{x\}$ **f)** $\emptyset \in \{x\}$

18. Use a Venn diagram to illustrate the subset of odd integers in the set of all positive integers not exceeding 10.

19. Use a Venn diagram to illustrate the set of all months of the year whose names do not contain the letter R in the set of all months of the year.

20. Use a Venn diagram to illustrate the relationship $A \subseteq B$ and $B \subseteq C$.

21. Use a Venn diagram to illustrate the relationships $A \subset B$ and $B \subset C$.

22. Use a Venn diagram to illustrate the relationships $A \subset B$ and $A \subset C$.

23. Suppose that A, B, and C are sets such that $A \subseteq B$ and $B \subseteq C$. Show that $A \subseteq C$.

24. Find two sets A and B such that $A \in B$ and $A \subseteq B$.

25. What is the cardinality of each of these sets?

a) $\{a\}$ **b)** $\{\{a\}\}$
c) $\{a, \{a\}\}$ **d)** $\{a, \{a\}, \{a, \{a\}\}\}$

26. What is the cardinality of each of these sets?

a) $\emptyset$ **b)** $\{\emptyset\}$
c) $\{\emptyset, \{\emptyset\}\}$ **d)** $\{\emptyset, \{\emptyset\}, \{\emptyset, \{\emptyset\}\}\}$

27. Find the power set of each of these sets, where a and b are distinct elements.

a) $\{a\}$ **b)** $\{a, b\}$ **c)** $\{\emptyset, \{\emptyset\}\}$

28. Can you conclude that $A = B$ if A and B are two sets with the same power set?

29. How many elements does each of these sets have where a and b are distinct elements?

a) $\mathcal{P}(\{a, b, \{a, b\}\})$
b) $\mathcal{P}(\{\emptyset, a, \{a\}, \{\{a\}\}\})$
c) $\mathcal{P}(\mathcal{P}(\emptyset))$

30. Determine whether each of these sets is the power set of a set, where a and b are distinct elements.

a) $\emptyset$ **b)** $\{\emptyset, \{a\}\}$
c) $\{\emptyset, \{a\}, \{\emptyset, a\}\}$ **d)** $\{\emptyset, \{a\}, \{b\}, \{a, b\}\}$

31. Prove that $\mathcal{P}(A) \subseteq \mathcal{P}(B)$ if and only if $A \subseteq B$.

32. Show that if $A \subseteq C$ and $B \subseteq D$, then $A \times B \subseteq C \times D$

33. Let $A = \{a, b, c, d\}$ and $B = \{y, z\}$. Find

a) $A \times B$. **b)** $B \times A$.

34. What is the Cartesian product $A \times B$, where A is the set of courses offered by the mathematics department at a university and B is the set of mathematics professors at this university? Give an example of how this Cartesian product can be used.

35. What is the Cartesian product $A \times B \times C$, where A is the set of all airlines and B and C are both the set of all cities in the United States? Give an example of how this Cartesian product can be used.

36. Suppose that $A \times B = \emptyset$, where A and B are sets. What can you conclude?

37. Let A be a set. Show that $\emptyset \times A = A \times \emptyset = \emptyset$.

38. Let $A = \{a, b, c\}$, $B = \{x, y\}$, and $C = \{0, 1\}$. Find

a) $A \times B \times C$. **b)** $C \times B \times A$.
c) $C \times A \times B$. **d)** $B \times B \times B$.

39. Find A^2 if

a) $A = \{0, 1, 3\}$. **b)** $A = \{1, 2, a, b\}$.

40. Find A^3 if

a) $A = \{a\}$. **b)** $A = \{0, a\}$.

41. How many different elements does $A \times B$ have if A has m elements and B has n elements?

42. How many different elements does $A \times B \times C$ have if A has m elements, B has n elements, and C has p elements?

43. How many different elements does A^n have when A has m elements and n is a positive integer?

44. Show that $A \times B \neq B \times A$, when A and B are nonempty, unless $A = B$.

45. Explain why $A \times B \times C$ and $(A \times B) \times C$ are not the same.

46. Explain why $(A \times B) \times (C \times D)$ and $A \times (B \times C) \times D$ are not the same.

47. Prove or disprove that if A and B are sets, then $\mathcal{P}(A \times B) = \mathcal{P}(A) \times \mathcal{P}(B)$.

48. Prove or disprove that if A, B, and C are nonempty sets and $A \times B = A \times C$, then $B = C$.

49. Translate each of these quantifications into English and determine its truth value.

a) $\forall x \in \mathbf{R}\ (x^2 \neq -1)$ **b)** $\exists x \in \mathbf{Z}\ (x^2 = 2)$
c) $\forall x \in \mathbf{Z}\ (x^2 > 0)$ **d)** $\exists x \in \mathbf{R}\ (x^2 = x)$

50. Translate each of these quantifications into English and determine its truth value.

a) $\exists x \in \mathbf{R}\ (x^3 = -1)$ **b)** $\exists x \in \mathbf{Z}\ (x + 1 > x)$
c) $\forall x \in \mathbf{Z}\ (x - 1 \in \mathbf{Z})$ **d)** $\forall x \in \mathbf{Z}\ (x^2 \in \mathbf{Z})$

51. Find the truth set of each of these predicates where the domain is the set of integers.

a) $P(x)$: $x^2 < 3$ **b)** $Q(x)$: $x^2 > x$
c) $R(x)$: $2x + 1 = 0$

52. Find the truth set of each of these predicates where the domain is the set of integers.

a) $P(x)$: $x^3 \geq 1$ **b)** $Q(x)$: $x^2 = 2$
c) $R(x)$: $x < x^2$

***53.** The defining property of an ordered pair is that two ordered pairs are equal if and only if their first elements are equal and their second elements are equal. Surprisingly, instead of taking the ordered pair as a primitive concept, we can construct ordered pairs using basic notions from set theory. Show that if we define the ordered pair (a, b) to be $\{\{a\}, \{a, b\}\}$, then $(a, b) = (c, d)$ if and only if $a = c$ and $b = d$. [*Hint:* First show that $\{\{a\}, \{a, b\}\} = \{\{c\}, \{c, d\}\}$ if and only if $a = c$ and $b = d$.]

***54.** This exercise presents **Russell's paradox**. Let S be the set that contains a set x if the set x does not belong to itself, so that $S = \{x \mid x \notin x\}$.

Links

a) Show the assumption that S is a member of S leads to a contradiction.

b) Show the assumption that S is not a member of S leads to a contradiction.

By parts (a) and (b) it follows that the set S cannot be defined as it was. This paradox can be avoided by restricting the types of elements that sets can have.

***55.** Describe a procedure for listing all the subsets of a finite set.

2.2 Set Operations

2.2.1 Introduction

Two, or more, sets can be combined in many different ways. For instance, starting with the set of mathematics majors at your school and the set of computer science majors at your school, we can form the set of students who are mathematics majors or computer science majors, the set of students who are joint majors in mathematics and computer science, the set of all students not majoring in mathematics, and so on.

Links

Definition 1 Let A and B be sets. The *union* of the sets A and B, denoted by $A \cup B$, is the set that contains those elements that are either in A or in B, or in both.

An element x belongs to the union of the sets A and B if and only if x belongs to A or x belongs to B. This tells us that

$$A \cup B = \{x \mid x \in A \vee x \in B\}.$$

The Venn diagram shown in Figure 1 represents the union of two sets A and B. The area that represents $A \cup B$ is the shaded area within either the circle representing A or the circle representing B.

We will give some examples of the union of sets.

EXAMPLE 1 The union of the sets $\{1, 3, 5\}$ and $\{1, 2, 3\}$ is the set $\{1, 2, 3, 5\}$; that is, $\{1, 3, 5\} \cup \{1, 2, 3\} = \{1, 2, 3, 5\}$. ◀

EXAMPLE 2 The union of the set of all computer science majors at your school and the set of all mathematics majors at your school is the set of students at your school who are majoring either in mathematics or in computer science (or in both). ◀

Definition 2 Let A and B be sets. The *intersection* of the sets A and B, denoted by $A \cap B$, is the set containing those elements in both A and B.

An element x belongs to the intersection of the sets A and B if and only if x belongs to A and x belongs to B. This tells us that

$$A \cap B = \{x \mid x \in A \wedge x \in B\}.$$

The Venn diagram shown in Figure 2 represents the intersection of two sets A and B. The shaded area that is within both the circles representing the sets A and B is the area that represents the intersection of A and B.

We give some examples of the intersection of sets.

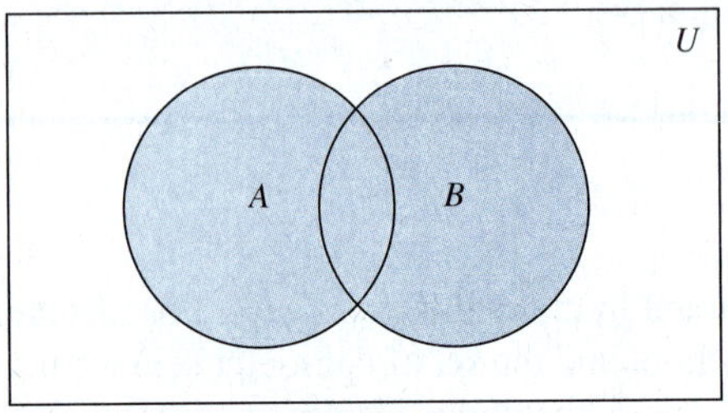

$A \cup B$ is shaded.

FIGURE 1 Venn diagram of the union of A and B.

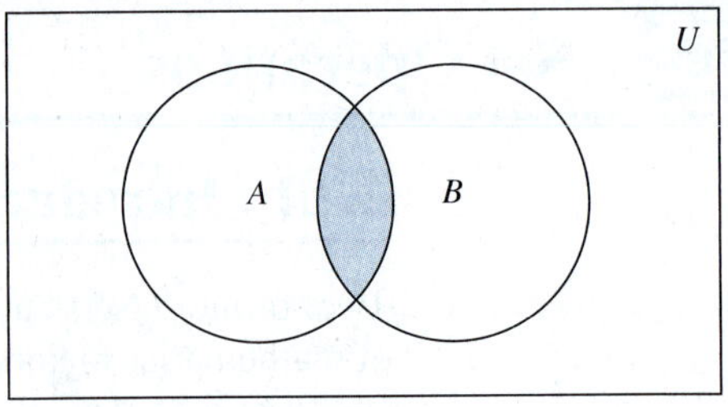

$A \cap B$ is shaded.

FIGURE 2 Venn diagram of the intersection of A and B.

EXAMPLE 3 The intersection of the sets $\{1, 3, 5\}$ and $\{1, 2, 3\}$ is the set $\{1, 3\}$; that is, $\{1, 3, 5\} \cap \{1, 2, 3\} = \{1, 3\}$. ◀

EXAMPLE 4 The intersection of the set of all computer science majors at your school and the set of all mathematics majors is the set of all students who are joint majors in mathematics and computer science. ◀

Definition 3 Two sets are called *disjoint* if their intersection is the empty set.

EXAMPLE 5 Let $A = \{1, 3, 5, 7, 9\}$ and $B = \{2, 4, 6, 8, 10\}$. Because $A \cap B = \emptyset$, A and B are disjoint. ◀

Be careful not to overcount!

We are often interested in finding the cardinality of a union of two finite sets A and B. Note that $|A| + |B|$ counts each element that is in A but not in B or in B but not in A exactly once, and each element that is in both A and B exactly twice. Thus, if the number of elements that are in both A and B is subtracted from $|A| + |B|$, elements in $A \cap B$ will be counted only once. Hence,

$$|A \cup B| = |A| + |B| - |A \cap B|.$$

The generalization of this result to unions of an arbitrary number of sets is called the **principle of inclusion–exclusion**. The principle of inclusion–exclusion is an important technique used in enumeration. We will discuss this principle and other counting techniques in detail in Chapters 6 and 8.

There are other important ways to combine sets.

Definition 4 Let A and B be sets. The *difference* of A and B, denoted by $A - B$, is the set containing those elements that are in A but not in B. The difference of A and B is also called the *complement of B with respect to A*.

Remark: The difference of sets A and B is sometimes denoted by $A \backslash B$.

An element x belongs to the difference of A and B if and only if $x \in A$ and $x \notin B$. This tells us that

$$A - B = \{x \mid x \in A \wedge x \notin B\}.$$

The Venn diagram shown in Figure 3 represents the difference of the sets A and B. The shaded area inside the circle that represents A and outside the circle that represents B is the area that represents $A - B$.

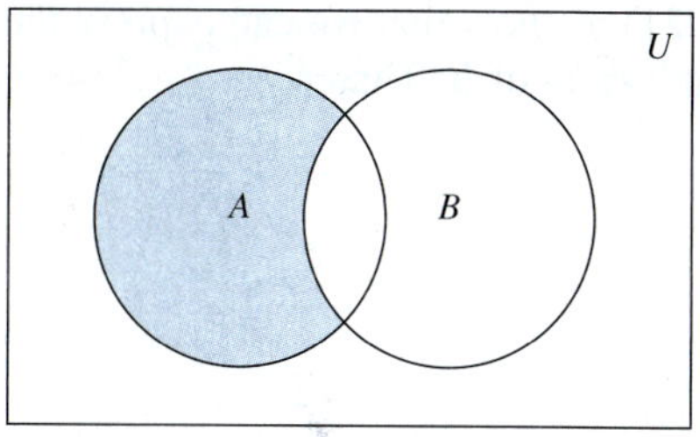

$A - B$ is shaded.

FIGURE 3 Venn diagram for the difference of A and B.

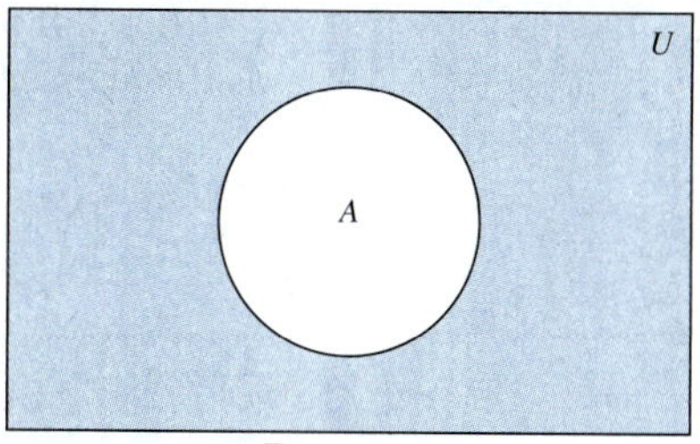

$\overline{A}$ is shaded.

FIGURE 4 Venn diagram for the complement of the set A.

We give some examples of differences of sets.

EXAMPLE 6 The difference of $\{1, 3, 5\}$ and $\{1, 2, 3\}$ is the set $\{5\}$; that is, $\{1, 3, 5\} - \{1, 2, 3\} = \{5\}$. This is different from the difference of $\{1, 2, 3\}$ and $\{1, 3, 5\}$, which is the set $\{2\}$. ◀

EXAMPLE 7 The difference of the set of computer science majors at your school and the set of mathematics majors at your school is the set of all computer science majors at your school who are not also mathematics majors. ◀

Once the universal set U has been specified, the **complement** of a set can be defined.

Definition 5 Let U be the universal set. The *complement* of the set A, denoted by $\overline{A}$, is the complement of A with respect to U. Therefore, the complement of the set A is $U - A$.

Remark: The definition of the complement of A depends on a particular universal set U. This definition makes sense for any superset U of A. If we want to identify the universal set U, we would write "the complement of A with respect to the set U."

An element belongs to $\overline{A}$ if and only if $x \notin A$. This tells us that

$$\overline{A} = \{x \in U \mid x \notin A\}.$$

In Figure 4 the shaded area outside the circle representing A is the area representing $\overline{A}$.
We give some examples of the complement of a set.

EXAMPLE 8 Let $A = \{a, e, i, o, u\}$ (where the universal set is the set of letters of the English alphabet). Then $\overline{A} = \{b, c, d, f, g, h, j, k, l, m, n, p, q, r, s, t, v, w, x, y, z\}$. ◀

EXAMPLE 9 Let A be the set of positive integers greater than 10 (with universal set the set of all positive integers). Then $\overline{A} = \{1, 2, 3, 4, 5, 6, 7, 8, 9, 10\}$. ◀

It is left to the reader (Exercise 21) to show that we can express the difference of A and B as the intersection of A and the complement of B. That is,

$$A - B = A \cap \overline{B}.$$

2.2.2 Set Identities

Set identities and propositional equivalences are just special cases of identities for Boolean algebra.

Table 1 lists the most important identities of unions, intersections, and complements of sets. We will prove several of these identities here, using three different methods. These methods are presented to illustrate that there are often many different approaches to the solution of a problem. The proofs of the remaining identities will be left as exercises. The reader should note the similarity between these set identities and the logical equivalences discussed in Section 1.3. (Compare Table 6 of Section 1.3 and Table 1.) In fact, the set identities given can be proved directly from the corresponding logical equivalences. Furthermore, both are special cases of identities that hold for Boolean algebra (discussed in Chapter 12).

Before we discuss different approaches for proving set identities, we briefly discuss the role of Venn diagrams. Although these diagrams can help us understand sets constructed using two or three **atomic sets** (the sets used to construct more complicated combinations of these sets), they provide far less insight when four or more atomic sets are involved. Venn diagrams for

TABLE 1 Set Identities.

Identity	*Name*
$A \cap U = A$ $A \cup \emptyset = A$	Identity laws
$A \cup U = U$ $A \cap \emptyset = \emptyset$	Domination laws
$A \cup A = A$ $A \cap A = A$	Idempotent laws
$\overline{(\overline{A})} = A$	Complementation law
$A \cup B = B \cup A$ $A \cap B = B \cap A$	Commutative laws
$A \cup (B \cup C) = (A \cup B) \cup C$ $A \cap (B \cap C) = (A \cap B) \cap C$	Associative laws
$A \cup (B \cap C) = (A \cup B) \cap (A \cup C)$ $A \cap (B \cup C) = (A \cap B) \cup (A \cap C)$	Distributive laws
$\overline{A \cap B} = \overline{A} \cup \overline{B}$ $\overline{A \cup B} = \overline{A} \cap \overline{B}$	De Morgan's laws
$A \cup (A \cap B) = A$ $A \cap (A \cup B) = A$	Absorption laws
$A \cup \overline{A} = U$ $A \cap \overline{A} = \emptyset$	Complement laws

four or more sets are quite complex because it is necessary to use ellipses rather than circles to represent the sets. This is necessary to ensure that every possible combination of the sets is represented by a nonempty region. Although Venn diagrams can provide an informal proof for some identities, such proofs should be formalized using one of the three methods we will now describe.

This identity says that the complement of the intersection of two sets is the union of their complements.

One way to show that two sets are equal is to show that each is a subset of the other. Recall that to show that one set is a subset of a second set, we can show that if an element belongs to the first set, then it must also belong to the second set. We generally use a direct proof to do this. We illustrate this type of proof by establishing the first of De Morgan's laws.

EXAMPLE 10 Prove that $\overline{A \cap B} = \overline{A} \cup \overline{B}$.

Extra Examples

Solution: We will prove that the two sets $\overline{A \cap B}$ and $\overline{A} \cup \overline{B}$ are equal by showing that each set is a subset of the other.

First, we will show that $\overline{A \cap B} \subseteq \overline{A} \cup \overline{B}$. We do this by showing that if x is in $\overline{A \cap B}$, then it must also be in $\overline{A} \cup \overline{B}$. Now suppose that $x \in \overline{A \cap B}$. By the definition of complement, $x \notin A \cap B$. Using the definition of intersection, we see that the proposition $\neg((x \in A) \wedge (x \in B))$ is true.

By applying De Morgan's law for propositions, we see that $\neg(x \in A)$ or $\neg(x \in B)$. Using the definition of negation of propositions, we have $x \notin A$ or $x \notin B$. Using the definition of the complement of a set, we see that this implies that $x \in \overline{A}$ or $x \in \overline{B}$. Consequently, by the definition of union, we see that $x \in \overline{A} \cup \overline{B}$. We have now shown that $\overline{A \cap B} \subseteq \overline{A} \cup \overline{B}$.

Next, we will show that $\overline{A} \cup \overline{B} \subseteq \overline{A \cap B}$. We do this by showing that if x is in $\overline{A} \cup \overline{B}$, then it must also be in $\overline{A \cap B}$. Now suppose that $x \in \overline{A} \cup \overline{B}$. By the definition of union, we know that $x \in \overline{A}$ or $x \in \overline{B}$. Using the definition of complement, we see that $x \notin A$ or $x \notin B$. Consequently, the proposition $\neg(x \in A) \vee \neg(x \in B)$ is true.

By De Morgan's law for propositions, we conclude that $\neg((x \in A) \wedge (x \in B))$ is true. By the definition of intersection, it follows that $\neg(x \in A \cap B)$. We now use the definition of complement to conclude that $x \in \overline{A \cap B}$. This shows that $\overline{A} \cup \overline{B} \subseteq \overline{A \cap B}$.

Because we have shown that each set is a subset of the other, the two sets are equal, and the identity is proved. ◀

We can more succinctly express the reasoning used in Example 10 using set builder notation, as Example 11 illustrates.

EXAMPLE 11 Use set builder notation and logical equivalences to establish the first De Morgan law $\overline{A \cap B} = \overline{A} \cup \overline{B}$.

Solution: We can prove this identity with the following steps.

$\overline{A \cap B} = \{x \mid x \notin A \cap B\}$	by definition of complement
$= \{x \mid \neg(x \in (A \cap B))\}$	by definition of does not belong symbol
$= \{x \mid \neg(x \in A \wedge x \in B)\}$	by definition of intersection
$= \{x \mid \neg(x \in A) \vee \neg(x \in B)\}$	by the first De Morgan law for logical equivalences
$= \{x \mid x \notin A \vee x \notin B\}$	by definition of does not belong symbol
$= \{x \mid x \in \overline{A} \vee x \in \overline{B}\}$	by definition of complement
$= \{x \mid x \in \overline{A} \cup \overline{B}\}$	by definition of union
$= \overline{A} \cup \overline{B}$	by meaning of set builder notation

Note that besides the definitions of complement, union, set membership, and set builder notation, this proof uses the second De Morgan law for logical equivalences. ◀

Proving a set identity involving more than two sets by showing each side of the identity is a subset of the other often requires that we keep track of different cases, as illustrated by the proof in Example 12 of one of the distributive laws for sets.

EXAMPLE 12 Prove the second distributive law from Table 1, which states that $A \cap (B \cup C) = (A \cap B) \cup (A \cap C)$ for all sets A, B, and C.

Solution: We will prove this identity by showing that each side is a subset of the other side.

Suppose that $x \in A \cap (B \cup C)$. Then $x \in A$ and $x \in B \cup C$. By the definition of union, it follows that $x \in A$, and $x \in B$ or $x \in C$ (or both). In other words, we know that the compound proposition $(x \in A) \wedge ((x \in B) \vee (x \in C))$ is true. By the distributive law for conjunction over disjunction, it follows that $((x \in A) \wedge (x \in B)) \vee ((x \in A) \wedge (x \in C))$. We conclude that either $x \in A$ and $x \in B$, or $x \in A$ and $x \in C$. By the definition of intersection, it follows that $x \in A \cap B$ or $x \in A \cap C$. Using the definition of union, we conclude that $x \in (A \cap B) \cup (A \cap C)$. We conclude that $A \cap (B \cup C) \subseteq (A \cap B) \cup (A \cap C)$.

Now suppose that $x \in (A \cap B) \cup (A \cap C)$. Then, by the definition of union, $x \in A \cap B$ or $x \in A \cap C$. By the definition of intersection, it follows that $x \in A$ and $x \in B$ or that $x \in A$ and $x \in C$. From this we see that $x \in A$, and $x \in B$ or $x \in C$. Consequently, by the definition of union we see that $x \in A$ and $x \in B \cup C$. Furthermore, by the definition of intersection, it follows that $x \in A \cap (B \cup C)$. We conclude that $(A \cap B) \cup (A \cap C) \subseteq A \cap (B \cup C)$. This completes the proof of the identity. ◀

Set identities can also be proved using **membership tables.** We consider each combination of the atomic sets (that is, the original sets used to produce the sets on each side) that an element can belong to and verify that elements in the same combinations of sets belong to both the sets in the identity. To indicate that an element is in a set, a 1 is used; to indicate that an element is not in a set, a 0 is used. (The reader should note the similarity between membership tables and truth tables.)

EXAMPLE 13 Use a membership table to show that $A \cap (B \cup C) = (A \cap B) \cup (A \cap C)$.

Solution: The membership table for these combinations of sets is shown in Table 2. This table has eight rows. Because the columns for $A \cap (B \cup C)$ and $(A \cap B) \cup (A \cap C)$ are the same, the identity is valid. ◀

TABLE 2 A Membership Table for the Distributive Property.

A	B	C	$B \cup C$	$A \cap (B \cup C)$	$A \cap B$	$A \cap C$	$(A \cap B) \cup (A \cap C)$
1	1	1	1	1	1	1	1
1	1	0	1	1	1	0	1
1	0	1	1	1	0	1	1
1	0	0	0	0	0	0	0
0	1	1	1	0	0	0	0
0	1	0	1	0	0	0	0
0	0	1	1	0	0	0	0
0	0	0	0	0	0	0	0

Once we have proved set identities, we can use them to prove new identities. In particular, we can apply a string of identities, one in each step, to take us from one side of a desired identity to the other. It is helpful to explicitly state the identity that is used in each step, as we do in Example 14.

EXAMPLE 14 Let A, B, and C be sets. Show that

$$\overline{A \cup (B \cap C)} = (\overline{C} \cup \overline{B}) \cap \overline{A}.$$

Solution: We have

$$\begin{aligned}\overline{A \cup (B \cap C)} &= \overline{A} \cap \overline{(B \cap C)} && \text{by the first De Morgan law}\\ &= \overline{A} \cap (\overline{B} \cup \overline{C}) && \text{by the second De Morgan law}\\ &= (\overline{B} \cup \overline{C}) \cap \overline{A} && \text{by the commutative law for intersections}\\ &= (\overline{C} \cup \overline{B}) \cap \overline{A} && \text{by the commutative law for unions}\end{aligned}$$

◀

We summarize the three different ways for proving set identities in Table 3.

TABLE 3 **Methods of Proving Set Identities.**

Description	*Method*
Subset method	Show that each side of the identity is a subset of the other side.
Membership table	For each possible combination of the atomic sets, show that an element in exactly these atomic sets must either belong to both sides or belong to neither side.
Apply existing identities	Start with one side, transform it into the other side using a sequence of steps by applying an established identity.

2.2.3 Generalized Unions and Intersections

Because unions and intersections of sets satisfy associative laws, the sets $A \cup B \cup C$ and $A \cap B \cap C$ are well defined; that is, the meaning of this notation is unambiguous when A, B, and C are sets. That is, we do not have to use parentheses to indicate which operation comes first because $A \cup (B \cup C) = (A \cup B) \cup C$ and $A \cap (B \cap C) = (A \cap B) \cap C$. Note that $A \cup B \cup C$ contains those elements that are in at least one of the sets A, B, and C, and that $A \cap B \cap C$ contains those elements that are in all of A, B, and C. These combinations of the three sets, A, B, and C, are shown in Figure 5.

EXAMPLE 15 Let $A = \{0, 2, 4, 6, 8\}$, $B = \{0, 1, 2, 3, 4\}$, and $C = \{0, 3, 6, 9\}$. What are $A \cup B \cup C$ and $A \cap B \cap C$?

Solution: The set $A \cup B \cup C$ contains those elements in at least one of A, B, and C. Hence,

$$A \cup B \cup C = \{0, 1, 2, 3, 4, 6, 8, 9\}.$$

The set $A \cap B \cap C$ contains those elements in all three of A, B, and C. Thus,

$$A \cap B \cap C = \{0\}.$$

◀

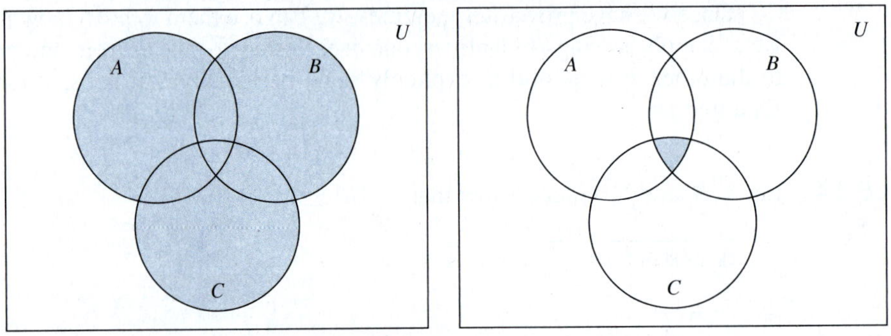

(a) $A \cup B \cup C$ is shaded.

(b) $A \cap B \cap C$ is shaded.

FIGURE 5 The union and intersection of A, B, and C.

We can also consider unions and intersections of an arbitrary number of sets. We introduce these definitions.

Definition 6 The *union* of a collection of sets is the set that contains those elements that are members of at least one set in the collection.

We use the notation

$$A_1 \cup A_2 \cup \cdots \cup A_n = \bigcup_{i=1}^{n} A_i$$

to denote the union of the sets $A_1, A_2, \ldots, A_n$.

Definition 7 The *intersection* of a collection of sets is the set that contains those elements that are members of all the sets in the collection.

We use the notation

$$A_1 \cap A_2 \cap \cdots \cap A_n = \bigcap_{i=1}^{n} A_i$$

to denote the intersection of the sets $A_1, A_2, \ldots, A_n$. We illustrate generalized unions and intersections with Example 16.

EXAMPLE 16 For $i = 1, 2, \ldots$, let $A_i = \{i, i+1, i+2, \ldots\}$. Then,

Extra Examples

$$\bigcup_{i=1}^{n} A_i = \bigcup_{i=1}^{n} \{i, i+1, i+2, \ldots\} = \{1, 2, 3, \ldots\},$$

and

$$\bigcap_{i=1}^{n} A_i = \bigcap_{i=1}^{n} \{i, i+1, i+2, \ldots\} = \{n, n+1, n+2, \ldots\} = A_n.$$

◀

We can extend the notation we have introduced for unions and intersections to other families of sets. In particular, to denote the union of the infinite family of sets $A_1, A_2, \ldots, A_n, \ldots$, we use the notation

$$A_1 \cup A_2 \cup \cdots \cup A_n \cup \cdots = \bigcup_{i=1}^{\infty} A_i.$$

Similarly, the intersection of these sets is denoted by

$$A_1 \cap A_2 \cap \cdots \cap A_n \cap \cdots = \bigcap_{i=1}^{\infty} A_i.$$

More generally, when I is a set, the notations $\bigcap_{i \in I} A_i$ and $\bigcup_{i \in I} A_i$ are used to denote the intersection and union of the sets A_i for $i \in I$, respectively. Note that we have $\bigcap_{i \in I} A_i = \{x \mid \forall i \in I\,(x \in A_i)\}$ and $\bigcup_{i \in I} A_i = \{x \mid \exists i \in I\,(x \in A_i)\}$.

EXAMPLE 17 Suppose that $A_i = \{1, 2, 3, \ldots, i\}$ for $i = 1, 2, 3, \ldots$. Then,

$$\bigcup_{i=1}^{\infty} A_i = \bigcup_{i=1}^{\infty} \{1, 2, 3, \ldots, i\} = \{1, 2, 3, \ldots\} = \mathbf{Z}^+$$

and

$$\bigcap_{i=1}^{\infty} A_i = \bigcap_{i=1}^{\infty} \{1, 2, 3, \ldots, i\} = \{1\}.$$

To see that the union of these sets is the set of positive integers, note that every positive integer n is in at least one of the sets, because it belongs to $A_n = \{1, 2, \ldots, n\}$, and every element of the sets in the union is a positive integer. To see that the intersection of these sets is the set $\{1\}$, note that the only element that belongs to all the sets $A_1, A_2, \ldots$ is 1. To see this note that $A_1 = \{1\}$ and $1 \in A_i$ for $i = 1, 2, \ldots$. ◀

2.2.4 Computer Representation of Sets

There are various ways to represent sets using a computer. One method is to store the elements of the set in an unordered fashion. However, if this is done, the operations of computing the union, intersection, or difference of two sets would be time consuming, because each of these operations would require a large amount of searching for elements. We will present a method for storing elements using an arbitrary ordering of the elements of the universal set. This method of representing sets makes computing combinations of sets easy.

Assume that the universal set U is finite (and of reasonable size so that the number of elements of U is not larger than the memory size of the computer being used). First, specify an arbitrary ordering of the elements of U, for instance $a_1, a_2, \ldots, a_n$. Represent a subset A of U with the bit string of length n, where the ith bit in this string is 1 if a_i belongs to A and is 0 if a_i does not belong to A. Example 18 illustrates this technique.

EXAMPLE 18 Let $U = \{1, 2, 3, 4, 5, 6, 7, 8, 9, 10\}$, and the ordering of elements of U has the elements in increasing order; that is, $a_i = i$. What bit strings represent the subset of all odd integers in U, the subset of all even integers in U, and the subset of integers not exceeding 5 in U?

Solution: The bit string that represents the set of odd integers in U, namely, $\{1, 3, 5, 7, 9\}$, has a one bit in the first, third, fifth, seventh, and ninth positions, and a zero elsewhere. It is

$$10\ 1010\ 1010.$$

(We have split this bit string of length ten into blocks of length four for easy reading.) Similarly, we represent the subset of all even integers in U, namely, $\{2, 4, 6, 8, 10\}$, by the string

$$01\ 0101\ 0101.$$

The set of all integers in U that do not exceed 5, namely, $\{1, 2, 3, 4, 5\}$, is represented by the string

$$11\ 1110\ 0000.$$ ◀

Using bit strings to represent sets, it is easy to find complements of sets and unions, intersections, and differences of sets. To find the bit string for the complement of a set from the bit string for that set, we simply change each 1 to a 0 and each 0 to 1, because $x \in A$ if and only if $x \notin \overline{A}$. Note that this operation corresponds to taking the negation of each bit when we associate a bit with a truth value—with 1 representing true and 0 representing false.

EXAMPLE 19 We have seen that the bit string for the set $\{1, 3, 5, 7, 9\}$ (with universal set $\{1, 2, 3, 4, 5, 6, 7, 8, 9, 10\}$) is

$$10\ 1010\ 1010.$$

What is the bit string for the complement of this set?

Solution: The bit string for the complement of this set is obtained by replacing 0s with 1s and vice versa. This yields the string

$$01\ 0101\ 0101,$$

which corresponds to the set $\{2, 4, 6, 8, 10\}$. ◀

To obtain the bit string for the union and intersection of two sets we perform bitwise Boolean operations on the bit strings representing the two sets. The bit in the ith position of the bit string of the union is 1 if either of the bits in the ith position in the two strings is 1 (or both are 1), and is 0 when both bits are 0. Hence, the bit string for the union is the bitwise *OR* of the bit strings for the two sets. The bit in the ith position of the bit string of the intersection is 1 when the bits in the corresponding position in the two strings are both 1, and is 0 when either of the two bits is 0 (or both are). Hence, the bit string for the intersection is the bitwise *AND* of the bit strings for the two sets.

EXAMPLE 20 The bit strings for the sets $\{1, 2, 3, 4, 5\}$ and $\{1, 3, 5, 7, 9\}$ are 11 1110 0000 and 10 1010 1010, respectively. Use bit strings to find the union and intersection of these sets.

Solution: The bit string for the union of these sets is

$$11\ 1110\ 0000 \vee 10\ 1010\ 1010 = 11\ 1110\ 1010,$$

which corresponds to the set $\{1, 2, 3, 4, 5, 7, 9\}$. The bit string for the intersection of these sets is

$$11\ 1110\ 0000 \wedge 10\ 1010\ 1010 = 10\ 1010\ 0000,$$

which corresponds to the set $\{1, 3, 5\}$. ◀

2.2.5 Multisets

Sometimes the number of times that an element occurs in an unordered collection matters. A **multiset** (short for multiple-membership set) is an unordered collection of elements where an element can occur as a member more than once. We can use the same notation for a multiset as we do for a set, but each element is listed the number of times it occurs. (Recall that in a set, an element either belongs to a set or it does not. Listing it more than once does not affect the membership of this element in the set.) So, the multiset denoted by $\{a, a, a, b, b\}$ is the multiset that contains the element a thrice and the element b twice. When we use this notation, it must be clear that we are working with multisets and not ordinary sets. We can avoid this ambiguity by using an alternate notation for multisets. The notation $\{m_1 \cdot a_1, m_2 \cdot a_2, \ldots, m_r \cdot a_r\}$ denotes the multiset with element a_1 occurring m_1 times, element a_2 occurring m_2 times, and so on. The numbers m_i, $i = 1, 2, \ldots, r$, are called the **multiplicities** of the elements a_i, $i = 1, 2, \ldots, r$. (Elements not in a multiset are assigned 0 as their multiplicity in this set.) The cardinality of a multiset is defined to be the sum of the multiplicities of its elements. The word *multiset* was introduced by Nicolaas Govert de Bruijn in the 1970s, but the concept dates back to the 12th century work of the Indian mathematician Bhaskaracharya.

Let P and Q be multisets. The **union** of the multisets P and Q is the multiset in which the multiplicity of an element is the maximum of its multiplicities in P and Q. The **intersection** of P and Q is the multiset in which the multiplicity of an element is the minimum of its multiplicities in P and Q. The **difference** of P and Q is the multiset in which the multiplicity of an element is the multiplicity of the element in P less its multiplicity in Q unless this difference is negative, in which case the multiplicity is 0. The **sum** of P and Q is the multiset in which the multiplicity of an element is the sum of multiplicities in P and Q. The union, intersection, and difference of P and Q are denoted by $P \cup Q$, $P \cap Q$, and $P - Q$, respectively (where these operations should not be confused with the analogous operations for sets). The sum of P and Q is denoted by $P + Q$.

EXAMPLE 21 Suppose that P and Q are the multisets $\{4 \cdot a, 1 \cdot b, 3 \cdot c\}$ and $\{3 \cdot a, 4 \cdot b, 2 \cdot d\}$, respectively. Find $P \cup Q$, $P \cap Q$, $P - Q$, and $P + Q$.

Solution: We have

$$\begin{aligned} P \cup Q &= \{\max(4, 3) \cdot a, \max(1, 4) \cdot b, \max(3, 0) \cdot c, \max(0, 2) \cdot d\} \\ &= \{4 \cdot a, 4 \cdot b, 3 \cdot c, 2 \cdot d\}, \end{aligned}$$

$$\begin{aligned} P \cap Q &= \{\min(4, 3) \cdot a, \min(1, 4) \cdot b, \min(3, 0) \cdot c, \min(0, 2) \cdot d\} \\ &= \{3 \cdot a, 1 \cdot b, 0 \cdot c, 0 \cdot d\} = \{3 \cdot a, 1 \cdot b\}, \end{aligned}$$

Links

Dinodia Photos/Alamy Stock Photo

BHASKARACHARYA (1114–1185) Bhaskaracharya was born in Bijapur in the Indian state of Karnataka. (Bhaskaracharya's name was actually Bhaskara, but the title Acharya, which means teacher, was added honorifically.) His father was a well-known scholar and a famous astrologer. Bhaskaracharya was head of the astronomical observatory at Ujjain, the leading Indian mathematical center of the day. He is considered to be the greatest mathematician of medieval India. Bhaskaracharya made discoveries in many parts of mathematics, including geometry, plane and spherical trigonometry, algebra, number theory, and combinatorics. Bhaskaracharya described the principles of differential calculus, which he applied to astronomical problems, predating the works of Newton and Leibniz by more than 500 years. In number theory he made many discoveries about Diophantine equations, the study of the solution in integers of equations, which were rediscovered more than 600 years later. His greatest work is the *Crown of Treatises* (*Siddhanta Shiromani*), which includes four main parts, covering arithmetic, algebra, mathematics of the planets, and spheres.

$$P - Q = \{\max(4-3,0)\cdot a, \max(1-4,0)\cdot b, \max(3-0,0)\cdot c, \max(0-2,0)\cdot d\}$$
$$= \{1\cdot a, 0\cdot b, 3\cdot c, 0\cdot d\} = \{1\cdot a, 3\cdot c\}, \text{ and}$$

$$P + Q = \{(4+3)\cdot a, (1+4)\cdot b, (3+0)\cdot c, (0+2)\cdot d\}$$
$$= \{7\cdot a, 5\cdot b, 3\cdot c, 2\cdot d\}.$$

◀

Exercises

1. Let A be the set of students who live within one mile of school and let B be the set of students who walk to classes. Describe the students in each of these sets.
   a) $A \cap B$ b) $A \cup B$
   c) $A - B$ d) $B - A$
2. Suppose that A is the set of sophomores at your school and B is the set of students in discrete mathematics at your school. Express each of these sets in terms of A and B.
   a) the set of sophomores taking discrete mathematics in your school
   b) the set of sophomores at your school who are not taking discrete mathematics
   c) the set of students at your school who either are sophomores or are taking discrete mathematics
   d) the set of students at your school who either are not sophomores or are not taking discrete mathematics
3. Let $A = \{1, 2, 3, 4, 5\}$ and $B = \{0, 3, 6\}$. Find
   a) $A \cup B$. b) $A \cap B$.
   c) $A - B$. d) $B - A$.
4. Let $A = \{a, b, c, d, e\}$ and $B = \{a, b, c, d, e, f, g, h\}$. Find
   a) $A \cup B$. b) $A \cap B$.
   c) $A - B$. d) $B - A$.

In Exercises 5–10 assume that A is a subset of some underlying universal set U.

5. Prove the complementation law in Table 1 by showing that $\overline{\overline{A}} = A$.
6. Prove the identity laws in Table 1 by showing that
   a) $A \cup \emptyset = A$. b) $A \cap U = A$.
7. Prove the domination laws in Table 1 by showing that
   a) $A \cup U = U$. b) $A \cap \emptyset = \emptyset$.
8. Prove the idempotent laws in Table 1 by showing that
   a) $A \cup A = A$. b) $A \cap A = A$.
9. Prove the complement laws in Table 1 by showing that
   a) $A \cup \overline{A} = U$. b) $A \cap \overline{A} = \emptyset$.
10. Show that
    a) $A - \emptyset = A$. b) $\emptyset - A = \emptyset$.
11. Let A and B be sets. Prove the commutative laws from Table 1 by showing that
    a) $A \cup B = B \cup A$.
    b) $A \cap B = B \cap A$.
12. Prove the first absorption law from Table 1 by showing that if A and B are sets, then $A \cup (A \cap B) = A$.
13. Prove the second absorption law from Table 1 by showing that if A and B are sets, then $A \cap (A \cup B) = A$.
14. Find the sets A and B if $A - B = \{1, 5, 7, 8\}$, $B - A = \{2, 10\}$, and $A \cap B = \{3, 6, 9\}$.
15. Prove the second De Morgan law in Table 1 by showing that if A and B are sets, then $\overline{A \cup B} = \overline{A} \cap \overline{B}$
    a) by showing each side is a subset of the other side.
    b) using a membership table.
16. Let A and B be sets. Show that
    a) $(A \cap B) \subseteq A$. b) $A \subseteq (A \cup B)$.
    c) $A - B \subseteq A$. d) $A \cap (B - A) = \emptyset$.
    e) $A \cup (B - A) = A \cup B$.
17. Show that if A and B are sets in a universe U then $A \subseteq B$ if and only if $\overline{A} \cup B = U$.
18. Given sets A and B in a universe U, draw the Venn diagrams of each of these sets.
    a) $A \to B = \{x \in U \mid x \in A \to x \in B\}$
    b) $A \leftrightarrow B = \{x \in U \mid x \in A \leftrightarrow x \in B\}$
19. Show that if A, B, and C are sets, then $\overline{A \cap B \cap C} = \overline{A} \cup \overline{B} \cup \overline{C}$
    a) by showing each side is a subset of the other side.
    b) using a membership table.
20. Let A, B, and C be sets. Show that
    a) $(A \cup B) \subseteq (A \cup B \cup C)$.
    b) $(A \cap B \cap C) \subseteq (A \cap B)$.
    c) $(A - B) - C \subseteq A - C$.
    d) $(A - C) \cap (C - B) = \emptyset$.
    e) $(B - A) \cup (C - A) = (B \cup C) - A$.
21. Show that if A and B are sets, then
    a) $A - B = A \cap \overline{B}$.
    b) $(A \cap B) \cup (A \cap \overline{B}) = A$.
22. Show that if A and B are sets with $A \subseteq B$, then
    a) $A \cup B = B$.
    b) $A \cap B = A$.
23. Prove the first associative law from Table 1 by showing that if A, B, and C are sets, then $A \cup (B \cup C) = (A \cup B) \cup C$.
24. Prove the second associative law from Table 1 by showing that if A, B, and C are sets, then $A \cap (B \cap C) = (A \cap B) \cap C$.
25. Prove the first distributive law from Table 1 by showing that if A, B, and C are sets, then $A \cup (B \cap C) = (A \cup B) \cap (A \cup C)$.

26. Let A, B, and C be sets. Show that $(A - B) - C = (A - C) - (B - C)$.

27. Let $A = \{0, 2, 4, 6, 8, 10\}$, $B = \{0, 1, 2, 3, 4, 5, 6\}$, and $C = \{4, 5, 6, 7, 8, 9, 10\}$. Find

a) $A \cap B \cap C$. **b)** $A \cup B \cup C$.
c) $(A \cup B) \cap C$. **d)** $(A \cap B) \cup C$.

28. Draw the Venn diagrams for each of these combinations of the sets A, B, and C.

a) $A \cap (B \cup C)$ **b)** $\overline{A} \cap \overline{B} \cap \overline{C}$
c) $(A - B) \cup (A - C) \cup (B - C)$

29. Draw the Venn diagrams for each of these combinations of the sets A, B, and C.

a) $A \cap (B - C)$ **b)** $(A \cap B) \cup (A \cap C)$
c) $(A \cap \overline{B}) \cup (A \cap \overline{C})$

30. Draw the Venn diagrams for each of these combinations of the sets A, B, C, and D.

a) $(A \cap B) \cup (C \cap D)$ **b)** $\overline{A} \cup \overline{B} \cup \overline{C} \cup \overline{D}$
c) $A - (B \cap C \cap D)$

31. What can you say about the sets A and B if we know that

a) $A \cup B = A$? **b)** $A \cap B = A$?
c) $A - B = A$? **d)** $A \cap B = B \cap A$?
e) $A - B = B - A$?

32. Can you conclude that $A = B$ if A, B, and C are sets such that

a) $A \cup C = B \cup C$? **b)** $A \cap C = B \cap C$?
c) $A \cup C = B \cup C$ and $A \cap C = B \cap C$?

33. Let A and B be subsets of a universal set U. Show that $A \subseteq B$ if and only if $\overline{B} \subseteq \overline{A}$.

34. Let A, B, and C be sets. Use the the identity $A - B = A \cap \overline{B}$, which holds for any sets A and B, and the identities from Table 1 to show that $(A - B) \cap (B - C) \cap (A - C) = \emptyset$.

35. Let A, B, and C be sets. Use the identities in Table 1 to show that $\overline{(A \cup B) \cap (B \cup C) \cap (A \cup C)} = \overline{A} \cap \overline{B} \cap \overline{C}$.

36. Prove or disprove that for all sets A, B, and C, we have

a) $A \times (B \cup C) = (A \times B) \cup (A \times C)$.
b) $A \times (B \cap C) = (A \times B) \cap (A \times C)$.

37. Prove or disprove that for all sets A, B, and C, we have

a) $A \times (B - C) = (A \times B) - (A \times C)$.
b) $\overline{A} \times \overline{(B \cup C)} = \overline{A \times (B \cup C)}$.

The **symmetric difference** of A and B, denoted by $A \oplus B$, is the set containing those elements in either A or B, but not in both A and B.

38. Find the symmetric difference of $\{1, 3, 5\}$ and $\{1, 2, 3\}$.

39. Find the symmetric difference of the set of computer science majors at a school and the set of mathematics majors at this school.

40. Draw a Venn diagram for the symmetric difference of the sets A and B.

41. Show that $A \oplus B = (A \cup B) - (A \cap B)$.

42. Show that $A \oplus B = (A - B) \cup (B - A)$.

43. Show that if A is a subset of a universal set U, then

a) $A \oplus A = \emptyset$. **b)** $A \oplus \emptyset = A$.
c) $A \oplus U = \overline{A}$. **d)** $A \oplus \overline{A} = U$.

44. Show that if A and B are sets, then

a) $A \oplus B = B \oplus A$. **b)** $(A \oplus B) \oplus B = A$.

45. What can you say about the sets A and B if $A \oplus B = A$?

***46.** Determine whether the symmetric difference is associative; that is, if A, B, and C are sets, does it follow that $A \oplus (B \oplus C) = (A \oplus B) \oplus C$?

***47.** Suppose that A, B, and C are sets such that $A \oplus C = B \oplus C$. Must it be the case that $A = B$?

48. If A, B, C, and D are sets, does it follow that $(A \oplus B) \oplus (C \oplus D) = (A \oplus C) \oplus (B \oplus D)$?

49. If A, B, C, and D are sets, does it follow that $(A \oplus B) \oplus (C \oplus D) = (A \oplus D) \oplus (B \oplus C)$?

50. Show that if A and B are finite sets, then $A \cup B$ is a finite set.

51. Show that if A is an infinite set, then whenever B is a set, $A \cup B$ is also an infinite set.

***52.** Show that if A, B, and C are sets, then

$$|A \cup B \cup C| = |A| + |B| + |C| - |A \cap B| - |A \cap C| - |B \cap C| + |A \cap B \cap C|.$$

(This is a special case of the inclusion–exclusion principle, which will be studied in Chapter 8.)

53. Let $A_i = \{1, 2, 3, \ldots, i\}$ for $i = 1, 2, 3, \ldots$. Find

a) $\bigcup_{i=1}^{n} A_i$. **b)** $\bigcap_{i=1}^{n} A_i$.

54. Let $A_i = \{\ldots, -2, -1, 0, 1, \ldots, i\}$. Find

a) $\bigcup_{i=1}^{n} A_i$. **b)** $\bigcap_{i=1}^{n} A_i$.

55. Let A_i be the set of all nonempty bit strings (that is, bit strings of length at least one) of length not exceeding i. Find

a) $\bigcup_{i=1}^{n} A_i$. **b)** $\bigcap_{i=1}^{n} A_i$.

56. Find $\bigcup_{i=1}^{\infty} A_i$ and $\bigcap_{i=1}^{\infty} A_i$ if for every positive integer i,

a) $A_i = \{i, i + 1, i + 2, \ldots\}$.
b) $A_i = \{0, i\}$.
c) $A_i = (0, i)$, that is, the set of real numbers x with $0 < x < i$.
d) $A_i = (i, \infty)$, that is, the set of real numbers x with $x > i$.

57. Find $\bigcup_{i=1}^{\infty} A_i$ and $\bigcap_{i=1}^{\infty} A_i$ if for every positive integer i,

a) $A_i = \{-i, -i + 1, \ldots, -1, 0, 1, \ldots, i - 1, i\}$.
b) $A_i = \{-i, i\}$.
c) $A_i = [-i, i]$, that is, the set of real numbers x with $-i \leq x \leq i$.
d) $A_i = [i, \infty)$, that is, the set of real numbers x with $x \geq i$.

58. Suppose that the universal set is $U = \{1, 2, 3, 4, 5, 6, 7, 8, 9, 10\}$. Express each of these sets with bit strings where the ith bit in the string is 1 if i is in the set and 0 otherwise.

a) $\{3, 4, 5\}$
b) $\{1, 3, 6, 10\}$
c) $\{2, 3, 4, 7, 8, 9\}$

59. Using the same universal set as in the last exercise, find the set specified by each of these bit strings.

a) 11 1100 1111
b) 01 0111 1000
c) 10 0000 0001

60. What subsets of a finite universal set do these bit strings represent?

a) the string with all zeros
b) the string with all ones

61. What is the bit string corresponding to the difference of two sets?

62. What is the bit string corresponding to the symmetric difference of two sets?

63. Show how bitwise operations on bit strings can be used to find these combinations of $A = \{a, b, c, d, e\}$, $B = \{b, c, d, g, p, t, v\}$, $C = \{c, e, i, o, u, x, y, z\}$, and $D = \{d, e, h, i, n, o, t, u, x, y\}$.

a) $A \cup B$
b) $A \cap B$
c) $(A \cup D) \cap (B \cup C)$
d) $A \cup B \cup C \cup D$

64. How can the union and intersection of n sets that all are subsets of the universal set U be found using bit strings?

The **successor** of the set A is the set $A \cup \{A\}$.

65. Find the successors of the following sets.

a) $\{1, 2, 3\}$
b) $\emptyset$
c) $\{\emptyset\}$
d) $\{\emptyset, \{\emptyset\}\}$

66. How many elements does the successor of a set with n elements have?

67. Let A and B be the multisets $\{3 \cdot a, 2 \cdot b, 1 \cdot c\}$ and $\{2 \cdot a, 3 \cdot b, 4 \cdot d\}$, respectively. Find

a) $A \cup B$.
b) $A \cap B$.
c) $A - B$.
d) $B - A$.
e) $A + B$.

68. Assume that $a \in A$, where A is a set. Which of these statements are true and which are false, where all sets shown are ordinary sets, and not multisets. Explain each answer.

a) $\{a, a\} \cup \{a, a, a\} = \{a, a, a, a, a\}$
b) $\{a, a\} \cup \{a, a, a\} = \{a\}$
c) $\{a, a\} \cap \{a, a, a\} = \{a, a\}$
d) $\{a, a\} \cap \{a, a, a\} = \{a\}$
e) $\{a, a, a\} - \{a, a\} = \{a\}$

69. Answer the same questions as posed in Exercise 68 where all sets are multisets and not ordinary sets.

70. Suppose that A is the multiset that has as its elements the types of computer equipment needed by one department of a university and the multiplicities are the number of pieces of each type needed, and B is the analogous multiset for a second department of the university. For instance, A could be the multiset $\{107 \cdot$ personal computers, $44 \cdot$ routers, $6 \cdot$ servers$\}$ and B could be the multiset $\{14 \cdot$ personal computers, $6 \cdot$ routers, $2 \cdot$ mainframes$\}$.

a) What combination of A and B represents the equipment the university should buy assuming both departments use the same equipment?
b) What combination of A and B represents the equipment that will be used by both departments if both departments use the same equipment?
c) What combination of A and B represents the equipment that the second department uses, but the first department does not, if both departments use the same equipment?
d) What combination of A and B represents the equipment that the university should purchase if the departments do not share equipment?

The **Jaccard similarity** $J(A, B)$ of the finite sets A and B is $J(A, B) = |A \cap B|/|A \cup B|$, with $J(\emptyset, \emptyset) = 1$. The **Jaccard distance** $d_J(A, B)$ between A and B equals $d_J(A, B) = 1 - J(A, B)$.

71. Find $J(A, B)$ and $d_J(A, B)$ for these pairs of sets.

a) $A = \{1, 3, 5\}$, $B = \{2, 4, 6\}$
b) $A = \{1, 2, 3, 4\}$, $B = \{3, 4, 5, 6\}$
c) $A = \{1, 2, 3, 4, 5, 6\}$, $B = \{1, 2, 3, 4, 5, 6\}$
d) $A = \{1\}$, $B = \{1, 2, 3, 4, 5, 6\}$

72. Prove that each of the properties in parts (a)–(d) holds whenever A and B are finite sets.

a) $J(A, A) = 1$ and $d_J(A, A) = 0$
b) $J(A, B) = J(B, A)$ and $d_J(A, B) = d_J(B, A)$
c) $J(A, B) = 1$ and $d_J(A, B) = 0$ if and only if $A = B$
d) $0 \leq J(A, B) \leq 1$ and $0 \leq d_J(A, B) \leq 1$
e) Show that if A, B, and C are sets, then $d_J(A, C) \leq d_J(A, B) + d_J(B, C)$. (This inequality is known as the **triangle inequality and together with parts (a), (b), and (c) implies that d_J is a **metric**.)

Links

Fuzzy sets are used in artificial intelligence. Each element in the universal set U has a **degree of membership**, which is a real number between 0 and 1 (including 0 and 1), in a fuzzy set S. The fuzzy set S is denoted by listing the elements with their degrees of membership (elements with 0 degree of membership are not listed). For instance, we write {0.6 Alice, 0.9 Brian, 0.4 Fred, 0.1 Oscar, 0.5 Rita} for the set F (of famous people) to indicate that Alice has a 0.6 degree of membership in F, Brian has a 0.9 degree of membership in F, Fred has a 0.4 degree of membership in F, Oscar has a 0.1 degree of membership in F, and Rita has a 0.5 degree of membership in F (so that Brian is the most famous and Oscar is the least famous of these people). Also suppose that R is the set of rich people with $R =$ {0.4 Alice, 0.8 Brian, 0.2 Fred, 0.9 Oscar, 0.7 Rita}.

73. The **complement** of a fuzzy set S is the set $\overline{S}$, with the degree of the membership of an element in $\overline{S}$ equal to 1 minus the degree of membership of this element in S. Find $\overline{F}$ (the fuzzy set of people who are not famous) and $\overline{R}$ (the fuzzy set of people who are not rich).

74. The **union** of two fuzzy sets S and T is the fuzzy set $S \cup T$, where the degree of membership of an element in $S \cup T$ is the maximum of the degrees of membership of this element in S and in T. Find the fuzzy set $F \cup R$ of rich or famous people.

75. The **intersection** of two fuzzy sets S and T is the fuzzy set $S \cap T$, where the degree of membership of an element in $S \cap T$ is the minimum of the degrees of membership of this element in S and in T. Find the fuzzy set $F \cap R$ of rich and famous people.

2.3 Functions

2.3.1 Introduction

In many instances we assign to each element of a set a particular element of a second set (which may be the same as the first). For example, suppose that each student in a discrete mathematics class is assigned a letter grade from the set $\{A, B, C, D, F\}$. And suppose that the grades are A for Adams, C for Chou, B for Goodfriend, A for Rodriguez, and F for Stevens. This assignment of grades is illustrated in Figure 1.

This assignment is an example of a function. The concept of a function is extremely important in mathematics and computer science. For example, in discrete mathematics functions are used in the definition of such discrete structures as sequences and strings. Functions are also used to represent how long it takes a computer to solve problems of a given size. Many computer programs and subroutines are designed to calculate values of functions. Recursive functions, which are functions defined in terms of themselves, are used throughout computer science; they will be studied in Chapter 5. This section reviews the basic concepts involving functions needed in discrete mathematics.

Definition 1 Let A and B be nonempty sets. A *function* f from A to B is an assignment of exactly one element of B to each element of A. We write $f(a) = b$ if b is the unique element of B assigned by the function f to the element a of A. If f is a function from A to B, we write $f : A \to B$.

Remark: Functions are sometimes also called **mappings** or **transformations**.

Assessment

Functions are specified in many different ways. Sometimes we explicitly state the assignments, as in Figure 1. Often we give a formula, such as $f(x) = x + 1$, to define a function. Other times we use a computer program to specify a function.

A function $f: A \to B$ can also be defined in terms of a relation from A to B. Recall from Section 2.1 that a relation from A to B is just a subset of $A \times B$. A relation from A to B that contains one, and only one, ordered pair (a, b) for every element $a \in A$, defines a function f from A to B. This function is defined by the assignment $f(a) = b$, where (a, b) is the unique ordered pair in the relation that has a as its first element.

Definition 2 If f is a function from A to B, we say that A is the *domain* of f and B is the *codomain* of f. If $f(a) = b$, we say that b is the *image* of a and a is a *preimage* of b. The *range*, or *image*, of f is the set of all images of elements of A. Also, if f is a function from A to B, we say that f *maps* A to B.

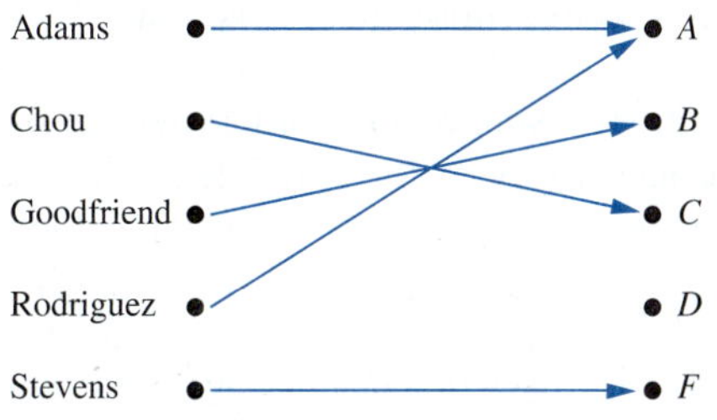

FIGURE 1 Assignment of grades in a discrete mathematics class.

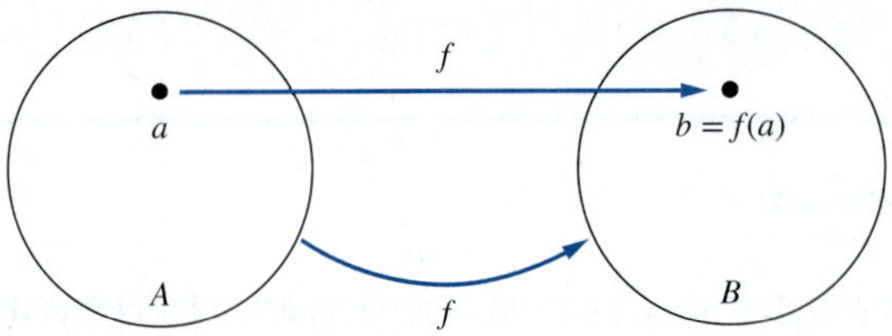

FIGURE 2 The function f maps A to B.

Figure 2 represents a function f from A to B.

Remark: Note that the codomain of a function from A to B is the set of all possible values of such a function (that is, all elements of B), and the range is the set of all values of $f(a)$ for $a \in A$, and is always a subset of the codomain. That is, the codomain is the set of possible values of the function and the range is the set of all elements of the codomain that are achieved as the value of f for at least one element of the domain.

When we define a function we specify its domain, its codomain, and the mapping of elements of the domain to elements in the codomain. Two functions are **equal** when they have the same domain, have the same codomain, and map each element of their common domain to the same element in their common codomain. Note that if we change either the domain or the codomain of a function, then we obtain a different function. If we change the mapping of elements, then we also obtain a different function.

Examples 1–5 provide examples of functions. In each case, we describe the domain, the codomain, the range, and the assignment of values to elements of the domain.

EXAMPLE 1 What are the domain, codomain, and range of the function that assigns grades to students described in the first paragraph of the introduction of this section?

Solution: Let G be the function that assigns a grade to a student in our discrete mathematics class. Note that $G(\text{Adams}) = A$, for instance. The domain of G is the set {Adams, Chou, Goodfriend, Rodriguez, Stevens}, and the codomain is the set $\{A, B, C, D, F\}$. The range of G is the set $\{A, B, C, F\}$, because each grade except D is assigned to some student. ◀

EXAMPLE 2 Let R be the relation with ordered pairs (Abdul, 22), (Brenda, 24), (Carla, 21), (Desire, 22), (Eddie, 24), and (Felicia, 22). Here each pair consists of a graduate student and this student's age. Specify a function determined by this relation.

Solution: If f is a function specified by R, then $f(\text{Abdul}) = 22$, $f(\text{Brenda}) = 24$, $f(\text{Carla}) = 21$, $f(\text{Desire}) = 22$, $f(\text{Eddie}) = 24$, and $f(\text{Felicia}) = 22$. [Here, $f(x)$ is the age of x, where x is a student.] For the domain, we take the set {Abdul, Brenda, Carla, Desire, Eddie, Felicia}. We also need to specify a codomain, which needs to contain all possible ages of students. Because it is highly likely that all students are less than 100 years old, we can take the set of positive integers less than 100 as the codomain. (Note that we could choose a different codomain, such as the set of all positive integers or the set of positive integers between 10 and 90, but that would change the function. Using this codomain will also allow us to extend the function by adding the names and ages of more students later.) The range of the function we have specified is the set of different ages of these students, which is the set $\{21, 22, 24\}$. ◀

EXAMPLE 3 Let f be the function that assigns the last two bits of a bit string of length 2 or greater to that string. For example, $f(11010) = 10$. Then, the domain of f is the set of all bit strings of length 2 or greater, and both the codomain and range are the set $\{00, 01, 10, 11\}$. ◀

Extra Examples

EXAMPLE 4 Let $f\colon \mathbf{Z} \to \mathbf{Z}$ assign the square of an integer to this integer. Then, $f(x) = x^2$, where the domain of f is the set of all integers, the codomain of f is the set of all integers, and the range of f is the set of all integers that are perfect squares, namely, $\{0, 1, 4, 9, \dots\}$. ◀

Remark: The notation for function is often abused in both computer science and mathematics. For instance, the notation $f\colon \mathbf{Z} \to \mathbf{Z}$, $f(x) = \sqrt{x}$ does not define a function. See Section 2.3 for an explanation.

EXAMPLE 5 The domain and codomain of functions are often specified in programming languages. For instance,

int **floor**(float x)

in both Java and C++ tells us that the domain of the floor function is the set of real numbers (represented by floating point numbers) and its codomain is the set of integers. ◀

A function is called **real-valued** if its codomain is the set of real numbers, and it is called **integer-valued** if its codomain is the set of integers. Two real-valued functions or two integer-valued functions with the same domain can be added as well as multiplied.

Definition 3 Let f_1 and f_2 be functions from A to $\mathbf{R}$. Then $f_1 + f_2$ and $f_1 f_2$ are also functions from A to $\mathbf{R}$ defined for all $x \in A$ by

$$(f_1 + f_2)(x) = f_1(x) + f_2(x),$$
$$(f_1 f_2)(x) = f_1(x) f_2(x).$$

Note that the functions $f_1 + f_2$ and $f_1 f_2$ have been defined by specifying their values at x in terms of the values of f_1 and f_2 at x.

EXAMPLE 6 Let f_1 and f_2 be functions from $\mathbf{R}$ to $\mathbf{R}$ such that $f_1(x) = x^2$ and $f_2(x) = x - x^2$. What are the functions $f_1 + f_2$ and $f_1 f_2$?

Solution: From the definition of the sum and product of functions, it follows that

$$(f_1 + f_2)(x) = f_1(x) + f_2(x) = x^2 + (x - x^2) = x$$

and

$$(f_1 f_2)(x) = x^2(x - x^2) = x^3 - x^4.$$ ◀

When f is a function from A to B, the image of a subset of A can also be defined.

Definition 4 Let f be a function from A to B and let S be a subset of A. The *image* of S under the function f is the subset of B that consists of the images of the elements of S. We denote the image of S by $f(S)$, so

$$f(S) = \{t \mid \exists s \in S\ (t = f(s))\}.$$

We also use the shorthand $\{f(s) \mid s \in S\}$ to denote this set.

Remark: The notation $f(S)$ for the image of the set S under the function f is potentially ambiguous. Here, $f(S)$ denotes a set and not the value of the function f for the set S.

EXAMPLE 7 Let $A = \{a, b, c, d, e\}$ and $B = \{1, 2, 3, 4\}$ with $f(a) = 2, f(b) = 1, f(c) = 4, f(d) = 1$, and $f(e) = 1$. The image of the subset $S = \{b, c, d\}$ is the set $f(S) = \{1, 4\}$. ◀

2.3.2 One-to-One and Onto Functions

Some functions never assign the same value to two different domain elements. These functions are said to be **one-to-one**.

Definition 5 A function f is said to be *one-to-one*, or an *injection*, if and only if $f(a) = f(b)$ implies that $a = b$ for all a and b in the domain of f. A function is said to be *injective* if it is one-to-one.

Note that a function f is one-to-one if and only if $f(a) \neq f(b)$ whenever $a \neq b$. This way of expressing that f is one-to-one is obtained by taking the contrapositive of the implication in the definition.

Remark: We can express that f is one-to-one using quantifiers as $\forall a \forall b(f(a) = f(b) \rightarrow a = b)$ or equivalently $\forall a \forall b(a \neq b \rightarrow f(a) \neq f(b))$, where the universe of discourse is the domain of the function.

Assessment ❯

We illustrate this concept by giving examples of functions that are one-to-one and other functions that are not one-to-one.

EXAMPLE 8 Determine whether the function f from $\{a, b, c, d\}$ to $\{1, 2, 3, 4, 5\}$ with $f(a) = 4$, $f(b) = 5$, $f(c) = 1$, and $f(d) = 3$ is one-to-one.

Extra Examples ❯

Solution: The function f is one-to-one because f takes on different values at the four elements of its domain. This is illustrated in Figure 3. ◀

EXAMPLE 9 Determine whether the function $f(x) = x^2$ from the set of integers to the set of integers is one-to-one.

Solution: The function $f(x) = x^2$ is not one-to-one because, for instance, $f(1) = f(-1) = 1$, but $1 \neq -1$. ◀

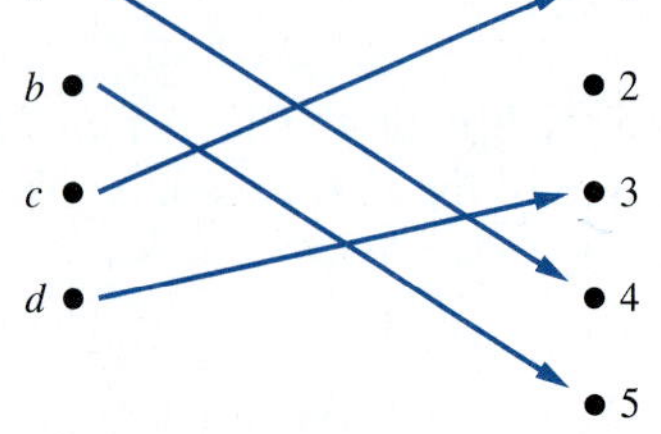

FIGURE 3 A one-to-one function.

Remark: The function $f(x) = x^2$ with domain $\mathbf{Z}^+$ is one-to-one. (See the explanation in Example 12 to see why.) This is a different function from the function in Example 9 because of the difference in their domains.

EXAMPLE 10 Determine whether the function $f(x) = x + 1$ from the set of real numbers to itself is one-to-one.

Solution: Suppose that x and y are real numbers with $f(x) = f(y)$, so that $x + 1 = y + 1$. This means that $x = y$. Hence, $f(x) = x + 1$ is a one-to-one function from $\mathbf{R}$ to $\mathbf{R}$. ◀

EXAMPLE 11 Suppose that each worker in a group of employees is assigned a job from a set of possible jobs, each to be done by a single worker. In this situation, the function f that assigns a job to each worker is one-to-one. To see this, note that if x and y are two different workers, then $f(x) \neq f(y)$ because the two workers x and y must be assigned different jobs. ◀

We now give some conditions that guarantee that a function is one-to-one.

Definition 6 A function f whose domain and codomain are subsets of the set of real numbers is called *increasing* if $f(x) \leq f(y)$, and *strictly increasing* if $f(x) < f(y)$, whenever $x < y$ and x and y are in the domain of f. Similarly, f is called *decreasing* if $f(x) \geq f(y)$, and *strictly decreasing* if $f(x) > f(y)$, whenever $x < y$ and x and y are in the domain of f. (The word *strictly* in this definition indicates a strict inequality.)

Remark: A function f is increasing if $\forall x \forall y(x < y \rightarrow f(x) \leq f(y))$, strictly increasing if $\forall x \forall y(x < y \rightarrow f(x) < f(y))$, decreasing if $\forall x \forall y(x < y \rightarrow f(x) \geq f(y))$, and strictly decreasing if $\forall x \forall y(x < y \rightarrow f(x) > f(y))$, where the universe of discourse is the domain of f.

EXAMPLE 12 The function $f(x) = x^2$ from $\mathbf{R}^+$ to $\mathbf{R}^+$ is strictly increasing. To see this, suppose that x and y are positive real numbers with $x < y$. Multiplying both sides of this inequality by x gives $x^2 < xy$. Similarly, multiplying both sides by y gives $xy < y^2$. Hence, $f(x) = x^2 < xy < y^2 = f(y)$. However, the function $f(x) = x^2$ from $\mathbf{R}$ to the set of nonnegative real numbers is not strictly increasing because $-1 < 0$, but $f(-1) = (-1)^2 = 1$ is not less than $f(0) = 0^2 = 0$. ◀

From these definitions, it can be shown (see Exercises 26 and 27) that a function that is either strictly increasing or strictly decreasing must be one-to-one. However, a function that is increasing, but not strictly increasing, or decreasing, but not strictly decreasing, is not one-to-one.

For some functions the range and the codomain are equal. That is, every member of the codomain is the image of some element of the domain. Functions with this property are called **onto** functions.

Definition 7 A function f from A to B is called *onto,* or a *surjection,* if and only if for every element $b \in B$ there is an element $a \in A$ with $f(a) = b$. A function f is called *surjective* if it is onto.

Remark: A function f is onto if $\forall y \exists x(f(x) = y)$, where the domain for x is the domain of the function and the domain for y is the codomain of the function.

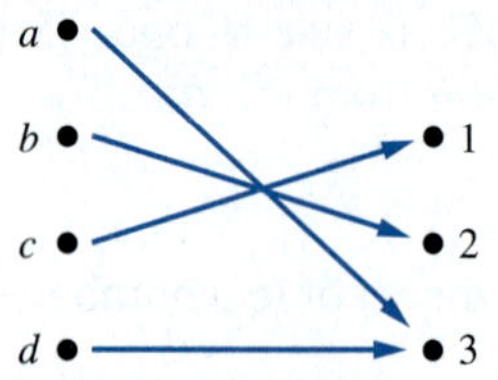

FIGURE 4 **An onto function.**

We now give examples of onto functions and functions that are not onto.

EXAMPLE 13 Let f be the function from $\{a, b, c, d\}$ to $\{1, 2, 3\}$ defined by $f(a) = 3, f(b) = 2, f(c) = 1$, and $f(d) = 3$. Is f an onto function?

Extra Examples

Solution: Because all three elements of the codomain are images of elements in the domain, we see that f is onto. This is illustrated in Figure 4. Note that if the codomain were $\{1, 2, 3, 4\}$, then f would not be onto. ◀

EXAMPLE 14 Is the function $f(x) = x^2$ from the set of integers to the set of integers onto?

Solution: The function f is not onto because there is no integer x with $x^2 = -1$, for instance. ◀

EXAMPLE 15 Is the function $f(x) = x + 1$ from the set of integers to the set of integers onto?

Solution: This function is onto, because for every integer y there is an integer x such that $f(x) = y$. To see this, note that $f(x) = y$ if and only if $x + 1 = y$, which holds if and only if $x = y - 1$. (Note that $y - 1$ is also an integer, and so, is in the domain of f.) ◀

EXAMPLE 16 Consider the function f in Example 11 that assigns jobs to workers. The function f is onto if for every job there is a worker assigned this job. The function f is not onto when there is at least one job that has no worker assigned it. ◀

Definition 8 The function f is a *one-to-one correspondence,* or a *bijection,* if it is both one-to-one and onto. We also say that such a function is *bijective.*

Examples 16 and 17 illustrate the concept of a bijection.

EXAMPLE 17 Let f be the function from $\{a, b, c, d\}$ to $\{1, 2, 3, 4\}$ with $f(a) = 4, f(b) = 2, f(c) = 1$, and $f(d) = 3$. Is f a bijection?

Solution: The function f is one-to-one and onto. It is one-to-one because no two values in the domain are assigned the same function value. It is onto because all four elements of the codomain are images of elements in the domain. Hence, f is a bijection. ◀

Figure 5 displays four functions where the first is one-to-one but not onto, the second is onto but not one-to-one, the third is both one-to-one and onto, and the fourth is neither one-to-one nor onto. The fifth correspondence in Figure 5 is not a function, because it sends an element to two different elements.

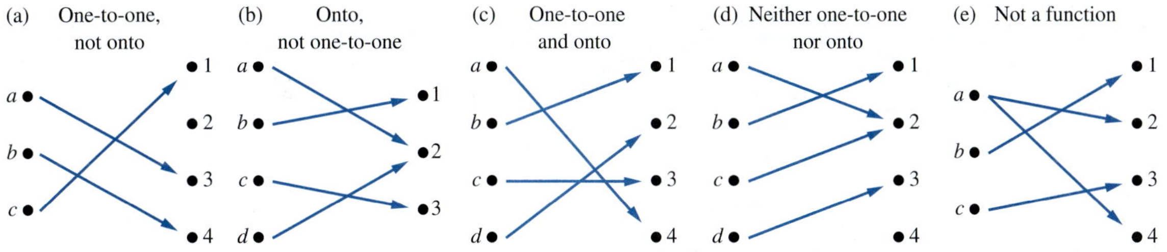

FIGURE 5 **Examples of different types of correspondences.**

Suppose that f is a function from a set A to itself. If A is finite, then f is one-to-one if and only if it is onto. (This follows from the result in Exercise 74.) This is not necessarily the case if A is infinite (as will be shown in Section 2.5).

EXAMPLE 18 Let A be a set. The *identity function* on A is the function $\iota_A : A \to A$, where

$$\iota_A(x) = x$$

for all $x \in A$. In other words, the identity function ι_A is the function that assigns each element to itself. The function ι_A is one-to-one and onto, so it is a bijection. (Note that ι is the Greek letter iota.) ◀

For future reference, we summarize what needs be to shown to establish whether a function is one-to-one and whether it is onto. It is instructive to review Examples 8–17 in light of this summary.

Suppose that $f: A \to B$.

To show that f is injective Show that if $f(x) = f(y)$ for arbitrary $x, y \in A$, then $x = y$.

To show that f is not injective Find particular elements $x, y \in A$ such that $x \neq y$ and $f(x) = f(y)$.

To show that f is surjective Consider an arbitrary element $y \in B$ and find an element $x \in A$ such that $f(x) = y$.

To show that f is not surjective Find a particular $y \in B$ such that $f(x) \neq y$ for all $x \in A$.

2.3.3 Inverse Functions and Compositions of Functions

Now consider a one-to-one correspondence f from the set A to the set B. Because f is an onto function, every element of B is the image of some element in A. Furthermore, because f is also a one-to-one function, every element of B is the image of a *unique* element of A. Consequently, we can define a new function from B to A that reverses the correspondence given by f. This leads to Definition 9.

Definition 9 Let f be a one-to-one correspondence from the set A to the set B. The *inverse function* of f is the function that assigns to an element b belonging to B the unique element a in A such that $f(a) = b$. The inverse function of f is denoted by f^{-1}. Hence, $f^{-1}(b) = a$ when $f(a) = b$.

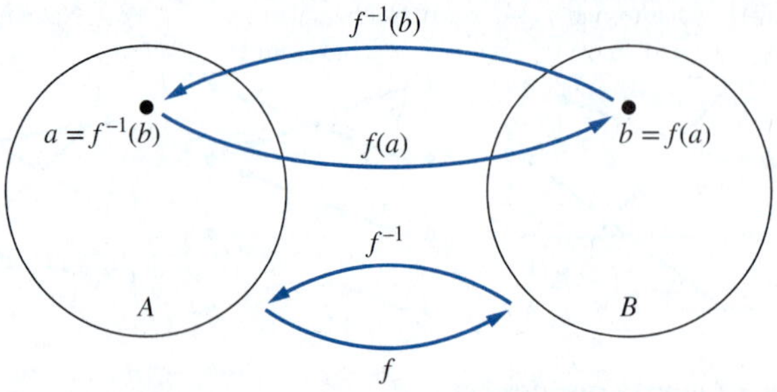

FIGURE 6 The function f^{-1} is the inverse of function f.

Remark: Be sure not to confuse the function f^{-1} with the function $1/f$, which is the function that assigns to each x in the domain the value $1/f(x)$. Notice that the latter makes sense only when $f(x)$ is a nonzero real number.

Figure 6 illustrates the concept of an inverse function.

If a function f is not a one-to-one correspondence, we cannot define an inverse function of f. When f is not a one-to-one correspondence, either it is not one-to-one or it is not onto. If f is not one-to-one, some element b in the codomain is the image of more than one element in the domain. If f is not onto, for some element b in the codomain, no element a in the domain exists for which $f(a) = b$. Consequently, if f is not a one-to-one correspondence, we cannot assign to each element b in the codomain a unique element a in the domain such that $f(a) = b$ (because for some b there is either more than one such a or no such a).

A one-to-one correspondence is called **invertible** because we can define an inverse of this function. A function is **not invertible** if it is not a one-to-one correspondence, because the inverse of such a function does not exist.

EXAMPLE 19 Let f be the function from $\{a, b, c\}$ to $\{1, 2, 3\}$ such that $f(a) = 2$, $f(b) = 3$, and $f(c) = 1$. Is f invertible, and if it is, what is its inverse?

Solution: The function f is invertible because it is a one-to-one correspondence. The inverse function f^{-1} reverses the correspondence given by f, so $f^{-1}(1) = c$, $f^{-1}(2) = a$, and $f^{-1}(3) = b$. ◀

EXAMPLE 20 Let $f: \mathbf{Z} \to \mathbf{Z}$ be such that $f(x) = x + 1$. Is f invertible, and if it is, what is its inverse?

Solution: The function f has an inverse because it is a one-to-one correspondence, as follows from Examples 10 and 15. To reverse the correspondence, suppose that y is the image of x, so that $y = x + 1$. Then $x = y - 1$. This means that $y - 1$ is the unique element of $\mathbf{Z}$ that is sent to y by f. Consequently, $f^{-1}(y) = y - 1$. ◀

EXAMPLE 21 Let f be the function from $\mathbf{R}$ to $\mathbf{R}$ with $f(x) = x^2$. Is f invertible?

Solution: Because $f(-2) = f(2) = 4$, f is not one-to-one. If an inverse function were defined, it would have to assign two elements to 4. Hence, f is not invertible. (Note we can also show that f is not invertible because it is not onto.) ◀

Sometimes we can restrict the domain or the codomain of a function, or both, to obtain an invertible function, as Example 22 illustrates.

EXAMPLE 22 Show that if we restrict the function $f(x) = x^2$ in Example 21 to a function from the set of all nonnegative real numbers to the set of all nonnegative real numbers, then f is invertible.

Solution: The function $f(x) = x^2$ from the set of nonnegative real numbers to the set of nonnegative real numbers is one-to-one. To see this, note that if $f(x) = f(y)$, then $x^2 = y^2$, so $x^2 - y^2 = (x + y)(x - y) = 0$. This means that $x + y = 0$ or $x - y = 0$, so $x = -y$ or $x = y$. Because both x and y are nonnegative, we must have $x = y$. So, this function is one-to-one. Furthermore, $f(x) = x^2$ is onto when the codomain is the set of all nonnegative real numbers, because each nonnegative real number has a square root. That is, if y is a nonnegative real number, there exists a nonnegative real number x such that $x = \sqrt{y}$, which means that $x^2 = y$. Because the function $f(x) = x^2$ from the set of nonnegative real numbers to the set of nonnegative real numbers is one-to-one and onto, it is invertible. Its inverse is given by the rule $f^{-1}(y) = \sqrt{y}$. ◀

Definition 10 Let g be a function from the set A to the set B and let f be a function from the set B to the set C. The *composition* of the functions f and g, denoted for all $a \in A$ by $f \circ g$, is the function from A to C defined by

$$(f \circ g)(a) = f(g(a)).$$

In other words, $f \circ g$ is the function that assigns to the element a of A the element assigned by f to $g(a)$. The domain of $f \circ g$ is the domain of g. The range of $f \circ g$ is the image of the range of g with respect to the function f. That is, to find $(f \circ g)(a)$ we first apply the function g to a to obtain $g(a)$ and then we apply the function f to the result $g(a)$ to obtain $(f \circ g)(a) = f(g(a))$. Note that the composition $f \circ g$ cannot be defined unless the range of g is a subset of the domain of f. In Figure 7 the composition of functions is shown.

EXAMPLE 23 Let g be the function from the set $\{a, b, c\}$ to itself such that $g(a) = b$, $g(b) = c$, and $g(c) = a$. Let f be the function from the set $\{a, b, c\}$ to the set $\{1, 2, 3\}$ such that $f(a) = 3$, $f(b) = 2$, and $f(c) = 1$. What is the composition of f and g, and what is the composition of g and f?

Solution: The composition $f \circ g$ is defined by $(f \circ g)(a) = f(g(a)) = f(b) = 2$, $(f \circ g)(b) = f(g(b)) = f(c) = 1$, and $(f \circ g)(c) = f(g(c)) = f(a) = 3$.

Note that $g \circ f$ is not defined, because the range of f is not a subset of the domain of g. ◀

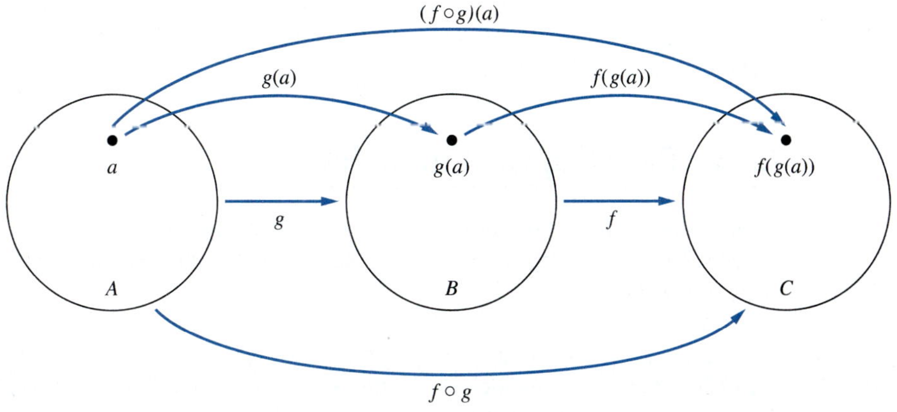

FIGURE 7 **The composition of the functions f and g.**

EXAMPLE 24 Let f and g be the functions from the set of integers to the set of integers defined by $f(x) = 2x + 3$ and $g(x) = 3x + 2$. What is the composition of f and g? What is the composition of g and f?

Solution: Both the compositions $f \circ g$ and $g \circ f$ are defined. Moreover,

$$(f \circ g)(x) = f(g(x)) = f(3x + 2) = 2(3x + 2) + 3 = 6x + 7$$

and

$$(g \circ f)(x) = g(f(x)) = g(2x + 3) = 3(2x + 3) + 2 = 6x + 11.$$ ◀

Remark: Note that even though $f \circ g$ and $g \circ f$ are defined for the functions f and g in Example 24, $f \circ g$ and $g \circ f$ are not equal. In other words, the commutative law does not hold for the composition of functions.

EXAMPLE 25 Let f and g be the functions defined by $f: \mathbf{R} \rightarrow \mathbf{R}^+ \cup \{0\}$ with $f(x) = x^2$ and $g : \mathbf{R}^+ \cup \{0\} \rightarrow \mathbf{R}$ with $g(x) = \sqrt{x}$ (where $\sqrt{x}$ is the nonnegative square root of x). What is the function $(f \circ g)(x)$?

Solution: The domain of $(f \circ g)(x) = f(g(x))$ is the domain of g, which is $\mathbf{R}^+ \cup \{0\}$, the set of nonnegative real numbers. If x is a nonnegative real number, we have $(f \circ g)(x) = f(g(x)) = f(\sqrt{x}) = \left(\sqrt{x}\right)^2 = x$. The range of $f \circ g$ is the image of the range of g with respect to the function f. This is the set $\mathbf{R}^+ \cup \{0\}$, the set of nonnegative real numbers. Summarizing, $f \circ g$: $\mathbf{R}^+ \cup \{0\} \rightarrow \mathbf{R}^+ \cup \{0\}$ and $(f \circ g)(x) = x$ for all x. ◀

When the composition of a function and its inverse is formed, in either order, an identity function is obtained. To see this, suppose that f is a one-to-one correspondence from the set A to the set B. Then the inverse function f^{-1} exists and is a one-to-one correspondence from B to A. The inverse function reverses the correspondence of the original function, so $f^{-1}(b) = a$ when $f(a) = b$, and $f(a) = b$ when $f^{-1}(b) = a$. Hence,

$$(f^{-1} \circ f)(a) = f^{-1}(f(a)) = f^{-1}(b) = a,$$

and

$$(f \circ f^{-1})(b) = f(f^{-1}(b)) = f(a) = b.$$

Consequently $f^{-1} \circ f = \iota_A$ and $f \circ f^{-1} = \iota_B$, where ι_A and ι_B are the identity functions on the sets A and B, respectively. That is, $(f^{-1})^{-1} = f$.

2.3.4 The Graphs of Functions

We can associate a set of pairs in $A \times B$ to each function from A to B. This set of pairs is called the **graph** of the function and is often displayed pictorially to aid in understanding the behavior of the function.

Definition 11 Let f be a function from the set A to the set B. The *graph* of the function f is the set of ordered pairs $\{(a, b) \mid a \in A \text{ and } f(a) = b\}$.

From the definition, the graph of a function f from A to B is the subset of $A \times B$ containing the ordered pairs with the second entry equal to the element of B assigned by f to the first entry. Also, note that the graph of a function f from A to B is the same as the relation from A to B determined by the function f, as described in Section 2.3.

EXAMPLE 26 Display the graph of the function $f(n) = 2n + 1$ from the set of integers to the set of integers.

Solution: The graph of f is the set of ordered pairs of the form $(n, 2n + 1)$, where n is an integer. This graph is displayed in Figure 8. ◀

EXAMPLE 27 Display the graph of the function $f(x) = x^2$ from the set of integers to the set of integers.

Solution: The graph of f is the set of ordered pairs of the form $(x, f(x)) = (x, x^2)$, where x is an integer. This graph is displayed in Figure 9. ◀

2.3.5 Some Important Functions

Next, we introduce two important functions in discrete mathematics, namely, the floor and ceiling functions. Let x be a real number. The floor function rounds x down to the closest integer less than or equal to x, and the ceiling function rounds x up to the closest integer greater than or equal to x. These functions are often used when objects are counted. They play an important role in the analysis of the number of steps used by procedures to solve problems of a particular size.

Definition 12 The *floor function* assigns to the real number x the largest integer that is less than or equal to x. The value of the floor function at x is denoted by $\lfloor x \rfloor$. The *ceiling function* assigns to the real number x the smallest integer that is greater than or equal to x. The value of the ceiling function at x is denoted by $\lceil x \rceil$.

Remark: The floor function is often also called the *greatest integer function*. It is often denoted by $[x]$.

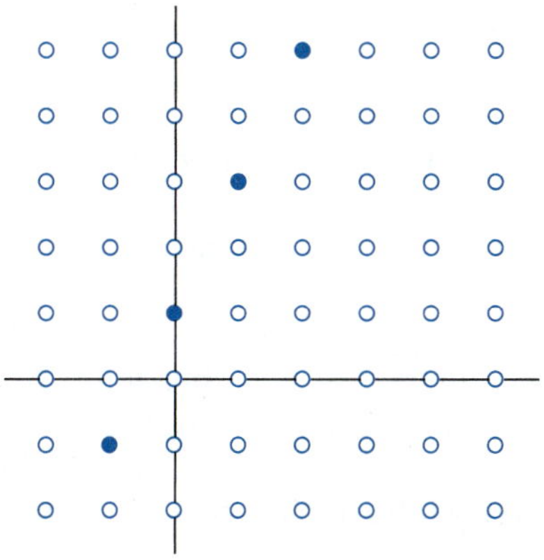

FIGURE 8 The graph of $f(n) = 2n + 1$ from Z to Z.

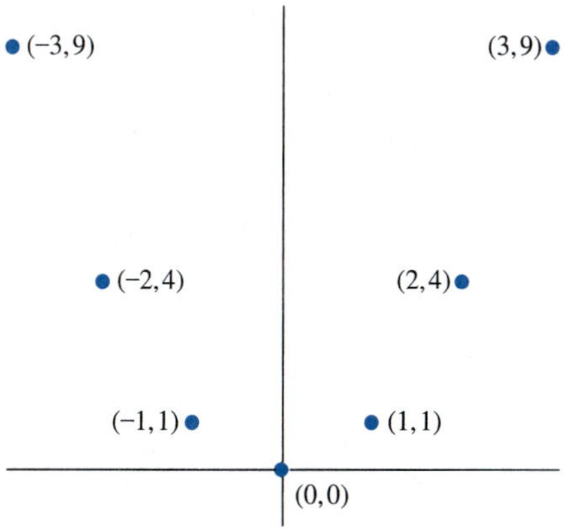

FIGURE 9 The graph of $f(x) = x^2$ from Z to Z.

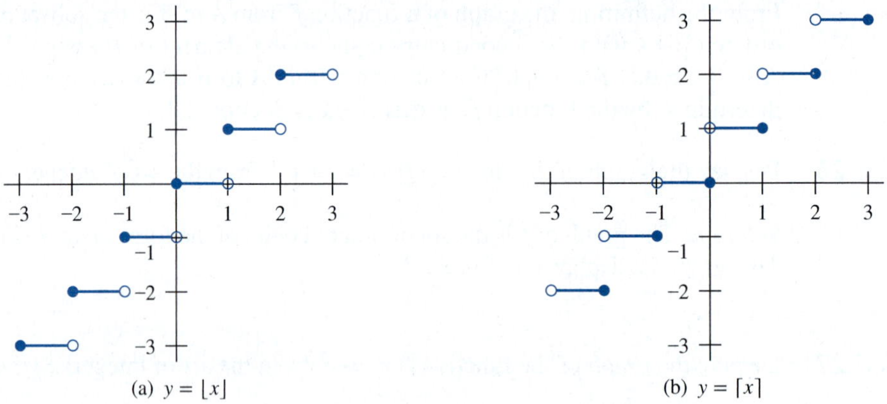

FIGURE 10 Graphs of the (a) floor and (b) ceiling functions.

EXAMPLE 28 These are some values of the floor and ceiling functions:

$$\lfloor \tfrac{1}{2} \rfloor = 0,\ \lceil \tfrac{1}{2} \rceil = 1,\ \lfloor -\tfrac{1}{2} \rfloor = -1,\ \lceil -\tfrac{1}{2} \rceil = 0,\ \lfloor 3.1 \rfloor = 3,\ \lceil 3.1 \rceil = 4,\ \lfloor 7 \rfloor = 7,\ \lceil 7 \rceil = 7.$$ ◀

We display the graphs of the floor and ceiling functions in Figure 10. In Figure 10(a) we display the graph of the floor function $\lfloor x \rfloor$. Note that this function has the same value throughout the interval $[n, n+1)$, namely n, and then it jumps up to $n+1$ when $x = n+1$. In Figure 10(b) we display the graph of the ceiling function $\lceil x \rceil$. Note that this function has the same value throughout the interval $(n, n+1]$, namely $n+1$, and then jumps to $n+2$ when x is a little larger than $n+1$.

Links

The floor and ceiling functions are useful in a wide variety of applications, including those involving data storage and data transmission. Consider Examples 29 and 30, typical of basic calculations done when database and data communications problems are studied.

EXAMPLE 29 Data stored on a computer disk or transmitted over a data network are usually represented as a string of bytes. Each byte is made up of 8 bits. How many bytes are required to encode 100 bits of data?

Solution: To determine the number of bytes needed, we determine the smallest integer that is at least as large as the quotient when 100 is divided by 8, the number of bits in a byte. Consequently, $\lceil 100/8 \rceil = \lceil 12.5 \rceil = 13$ bytes are required. ◀

EXAMPLE 30 In asynchronous transfer mode (ATM) (a communications protocol used for high-speed trading in financial industries), data are organized into cells of 53 bytes. How many ATM cells can be transmitted in 1 minute over a connection that transmits data at the rate of 500 kilobits per second?

Solution: In 1 minute, this connection can transmit $500{,}000 \cdot 60 = 30{,}000{,}000$ bits. Each ATM cell is 53 bytes long, which means that it is $53 \cdot 8 = 424$ bits long. To determine the number of cells that can be transmitted in 1 minute, we determine the largest integer not exceeding the quotient when 30,000,000 is divided by 424. Consequently, $\lfloor 30{,}000{,}000/424 \rfloor = 70{,}754$ ATM cells can be transmitted in 1 minute over a 500 kilobit per second connection. ◀

Table 1, with x denoting a real number, displays some simple but important properties of the floor and ceiling functions. Because these functions appear so frequently in discrete

TABLE 1 Useful Properties of the Floor and Ceiling Functions.
(n is an integer, x is a real number)

(1a)	$\lfloor x \rfloor = n$ if and only if $n \leq x < n+1$
(1b)	$\lceil x \rceil = n$ if and only if $n-1 < x \leq n$
(1c)	$\lfloor x \rfloor = n$ if and only if $x-1 < n \leq x$
(1d)	$\lceil x \rceil = n$ if and only if $x \leq n < x+1$
(2)	$x-1 < \lfloor x \rfloor \leq x \leq \lceil x \rceil < x+1$
(3a)	$\lfloor -x \rfloor = -\lceil x \rceil$
(3b)	$\lceil -x \rceil = -\lfloor x \rfloor$
(4a)	$\lfloor x+n \rfloor = \lfloor x \rfloor + n$
(4b)	$\lceil x+n \rceil = \lceil x \rceil + n$

mathematics, it is useful to look over these identities. Each property in this table can be established using the definitions of the floor and ceiling functions. Properties (1a), (1b), (1c), and (1d) follow directly from these definitions. For example, (1a) states that $\lfloor x \rfloor = n$ if and only if the integer n is less than or equal to x and $n+1$ is larger than x. This is precisely what it means for n to be the greatest integer not exceeding x, which is the definition of $\lfloor x \rfloor = n$. Properties (1b), (1c), and (1d) can be established similarly. We will prove property (4a) using a direct proof.

Proof: Suppose that $\lfloor x \rfloor = m$, where m is a positive integer. By property (1a), it follows that $m \leq x < m+1$. Adding n to all three quantities in this chain of two inequalities shows that $m+n \leq x+n < m+n+1$. Using property (1a) again, we see that $\lfloor x+n \rfloor = m+n = \lfloor x \rfloor + n$. This completes the proof. Proofs of the other properties are left as exercises. ◁

The floor and ceiling functions enjoy many other useful properties besides those displayed in Table 1. There are also many statements about these functions that may appear to be correct, but actually are not. We will consider statements about the floor and ceiling functions in Examples 31 and 32.

A useful approach for considering statements about the floor function is to let $x = n + \epsilon$, where $n = \lfloor x \rfloor$ is an integer, and ϵ, the fractional part of x, satisfies the inequality $0 \leq \epsilon < 1$. Similarly, when considering statements about the ceiling function, it is useful to write $x = n - \epsilon$, where $n = \lceil x \rceil$ is an integer and $0 \leq \epsilon < 1$.

EXAMPLE 31 Prove that if x is a real number, then $\lfloor 2x \rfloor = \lfloor x \rfloor + \lfloor x + \frac{1}{2} \rfloor$.

Extra Examples

Solution: To prove this statement we let $x = n + \epsilon$, where n is an integer and $0 \leq \epsilon < 1$. There are two cases to consider, depending on whether ϵ is less than, or greater than or equal to $\frac{1}{2}$. (The reason we choose these two cases will be made clear in the proof.)

Links

We first consider the case when $0 \leq \epsilon < \frac{1}{2}$. In this case, $2x = 2n + 2\epsilon$ and $\lfloor 2x \rfloor = 2n$ because $0 \leq 2\epsilon < 1$. Similarly, $x + \frac{1}{2} = n + (\frac{1}{2} + \epsilon)$, so $\lfloor x + \frac{1}{2} \rfloor = n$, because $0 < \frac{1}{2} + \epsilon < 1$. Consequently, $\lfloor 2x \rfloor = 2n$ and $\lfloor x \rfloor + \lfloor x + \frac{1}{2} \rfloor = n + n = 2n$.

Next, we consider the case when $\frac{1}{2} \leq \epsilon < 1$. In this case, $2x = 2n + 2\epsilon = (2n+1) + (2\epsilon - 1)$. Because $0 \leq 2\epsilon - 1 < 1$, it follows that $\lfloor 2x \rfloor = 2n + 1$. Because

$\lfloor x + \frac{1}{2} \rfloor = \lfloor n + (\frac{1}{2} + \epsilon) \rfloor = \lfloor n + 1 + (\epsilon - \frac{1}{2}) \rfloor$ and $0 \leq \epsilon - \frac{1}{2} < 1$, it follows that $\lfloor x + \frac{1}{2} \rfloor = n + 1$. Consequently, $\lfloor 2x \rfloor = 2n + 1$ and $\lfloor x \rfloor + \lfloor x + \frac{1}{2} \rfloor = n + (n + 1) = 2n + 1$. This concludes the proof. ◀

EXAMPLE 32 Prove or disprove that $\lceil x + y \rceil = \lceil x \rceil + \lceil y \rceil$ for all real numbers x and y.

Solution: Although this statement may appear reasonable, it is false. A counterexample is supplied by $x = \frac{1}{2}$ and $y = \frac{1}{2}$. With these values we find that $\lceil x + y \rceil = \lceil \frac{1}{2} + \frac{1}{2} \rceil = \lceil 1 \rceil = 1$, but $\lceil x \rceil + \lceil y \rceil = \lceil \frac{1}{2} \rceil + \lceil \frac{1}{2} \rceil = 1 + 1 = 2$. ◀

There are certain types of functions that will be used throughout the text. These include polynomial, logarithmic, and exponential functions. A brief review of the properties of these functions needed in this text is given in Appendix 2. In this book the notation $\log x$ will be used to denote the logarithm to the base 2 of x, because 2 is the base that we will usually use for logarithms. We will denote logarithms to the base b, where b is any real number greater than 1, by $\log_b x$, and the natural logarithm by $\ln x$.

Another function we will use throughout this text is the **factorial function** $f\colon \mathbf{N} \to \mathbf{Z}^+$, denoted by $f(n) = n!$. The value of $f(n) = n!$ is the product of the first n positive integers, so $f(n) = 1 \cdot 2 \cdots (n-1) \cdot n$ [and $f(0) = 0! = 1$].

EXAMPLE 33 We have $f(1) = 1! = 1$, $f(2) = 2! = 1 \cdot 2 = 2$, $f(6) = 6! = 1 \cdot 2 \cdot 3 \cdot 4 \cdot 5 \cdot 6 = 720$, and $f(20) = 1 \cdot 2 \cdot 3 \cdot 4 \cdot 5 \cdot 6 \cdot 7 \cdot 8 \cdot 9 \cdot 10 \cdot 11 \cdot 12 \cdot 13 \cdot 14 \cdot 15 \cdot 16 \cdot 17 \cdot 18 \cdot 19 \cdot 20 =$ 2,432,902,008,176,640,000. ◀

Example 33 illustrates that the factorial function grows extremely rapidly as n grows. The rapid growth of the factorial function is made clearer by Stirling's formula, a result from higher mathematics that tells us that $n! \sim \sqrt{2\pi n}(n/e)^n$. Here, we have used the notation $f(n) \sim g(n)$, which means that the ratio $f(n)/g(n)$ approaches 1 as n grows without bound (that is, $\lim_{n\to\infty} f(n)/g(n) = 1$). The symbol $\sim$ is read "is asymptotic to." Stirling's formula is named after James Stirling, a Scottish mathematician of the eighteenth century.

JAMES STIRLING (1692–1770) James Stirling was born near the town of Stirling, Scotland. His family strongly supported the Jacobite cause of the Stuarts as an alternative to the British crown. The first information known about James is that he entered Balliol College, Oxford, on a scholarship in 1711. However, he later lost his scholarship when he refused to pledge his allegiance to the British crown. The first Jacobean rebellion took place in 1715, and Stirling was accused of communicating with rebels. He was charged with cursing King George, but he was acquitted of these charges. Even though he could not graduate from Oxford because of his politics, he remained there for several years. Stirling published his first work, which extended Newton's work on plane curves, in 1717. He traveled to Venice, where a chair of mathematics had been promised to him, an appointment that unfortunately fell through. Nevertheless, Stirling stayed in Venice, continuing his mathematical work. He attended the University of Padua in 1721, and in 1722 he returned to Glasgow. Stirling apparently fled Italy after learning the secrets of the Italian glass industry, avoiding the efforts of Italian glass makers to assassinate him to protect their secrets.

In late 1724 Stirling moved to London, staying there 10 years teaching mathematics and actively engaging in research. In 1730 he published *Methodus Differentialis*, his most important work, presenting results on infinite series, summations, interpolation, and quadrature. It is in this book that his asymptotic formula for $n!$ appears. Stirling also worked on gravitation and the shape of the earth; he stated, but did not prove, that the earth is an oblate spheroid. Stirling returned to Scotland in 1735, when he was appointed manager of a Scottish mining company. He was very successful in this role and even published a paper on the ventilation of mine shafts. He continued his mathematical research, but at a reduced pace, during his years in the mining industry. Stirling is also noted for surveying the River Clyde with the goal of creating a series of locks to make it navigable. In 1752 the citizens of Glasgow presented him with a silver teakettle as a reward for this work.

2.3.6 Partial Functions

A program designed to evaluate a function may not produce the correct value of the function for all elements in the domain of this function. For example, a program may not produce a correct value because evaluating the function may lead to an infinite loop or an overflow. Similarly, in abstract mathematics, we often want to discuss functions that are defined only for a subset of the real numbers, such as $1/x$, $\sqrt{x}$, and $\arcsin(x)$. Also, we may want to use such notions as the "youngest child" function, which is undefined for someone without children, or the "time of sunrise," which is undefined for some days above the Arctic Circle. To study such situations, we use the concept of a partial function.

Definition 13 A *partial function* f from a set A to a set B is an assignment to each element a in a subset of A, called the *domain of definition* of f, of a unique element b in B. The sets A and B are called the *domain* and *codomain* of f, respectively. We say that f is *undefined* for elements in A that are not in the domain of definition of f. When the domain of definition of f equals A, we say that f is a *total function.*

Remark: We write $f : A \to B$ to denote that f is a partial function from A to B. Note that this is the same notation as is used for functions. The context in which the notation is used determines whether f is a partial function or a total function.

EXAMPLE 34 The function $f : \mathbf{Z} \to \mathbf{R}$ where $f(n) = \sqrt{n}$ is a partial function from $\mathbf{Z}$ to $\mathbf{R}$ where the domain of definition is the set of nonnegative integers. Note that f is undefined for negative integers. ◀

Exercises

1. Why is f not a function from $\mathbf{R}$ to $\mathbf{R}$ if
   a) $f(x) = 1/x$?
   b) $f(x) = \sqrt{x}$?
   c) $f(x) = \pm\sqrt{(x^2 + 1)}$?
2. Determine whether f is a function from $\mathbf{Z}$ to $\mathbf{R}$ if
   a) $f(n) = \pm n$.
   b) $f(n) = \sqrt{n^2 + 1}$.
   c) $f(n) = 1/(n^2 - 4)$.
3. Determine whether f is a function from the set of all bit strings to the set of integers if
   a) $f(S)$ is the position of a 0 bit in S.
   b) $f(S)$ is the number of 1 bits in S.
   c) $f(S)$ is the smallest integer i such that the ith bit of S is 1 and $f(S) = 0$ when S is the empty string, the string with no bits.
4. Find the domain and range of these functions. Note that in each case, to find the domain, determine the set of elements assigned values by the function.
   a) the function that assigns to each nonnegative integer its last digit
   b) the function that assigns the next largest integer to a positive integer
   c) the function that assigns to a bit string the number of one bits in the string
   d) the function that assigns to a bit string the number of bits in the string
5. Find the domain and range of these functions. Note that in each case, to find the domain, determine the set of elements assigned values by the function.
   a) the function that assigns to each bit string the number of ones in the string minus the number of zeros in the string
   b) the function that assigns to each bit string twice the number of zeros in that string
   c) the function that assigns the number of bits left over when a bit string is split into bytes (which are blocks of 8 bits)
   d) the function that assigns to each positive integer the largest perfect square not exceeding this integer
6. Find the domain and range of these functions.
   a) the function that assigns to each pair of positive integers the first integer of the pair
   b) the function that assigns to each positive integer its largest decimal digit
   c) the function that assigns to a bit string the number of ones minus the number of zeros in the string
   d) the function that assigns to each positive integer the largest integer not exceeding the square root of the integer
   e) the function that assigns to a bit string the longest string of ones in the string

7. Find the domain and range of these functions.

a) the function that assigns to each pair of positive integers the maximum of these two integers

b) the function that assigns to each positive integer the number of the digits 0, 1, 2, 3, 4, 5, 6, 7, 8, 9 that do not appear as decimal digits of the integer

c) the function that assigns to a bit string the number of times the block 11 appears

d) the function that assigns to a bit string the numerical position of the first 1 in the string and that assigns the value 0 to a bit string consisting of all 0s

8. Find these values.

a) $\lfloor 1.1 \rfloor$ **b)** $\lceil 1.1 \rceil$

c) $\lfloor -0.1 \rfloor$ **d)** $\lceil -0.1 \rceil$

e) $\lceil 2.99 \rceil$ **f)** $\lceil -2.99 \rceil$

g) $\lfloor \frac{1}{2} + \lceil \frac{1}{2} \rceil \rfloor$ **h)** $\lceil \lfloor \frac{1}{2} \rfloor + \lceil \frac{1}{2} \rceil + \frac{1}{2} \rceil$

9. Find these values.

a) $\lceil \frac{3}{4} \rceil$ **b)** $\lfloor \frac{7}{8} \rfloor$

c) $\lceil -\frac{3}{4} \rceil$ **d)** $\lfloor -\frac{7}{8} \rfloor$

e) $\lceil 3 \rceil$ **f)** $\lfloor -1 \rfloor$

g) $\lfloor \frac{1}{2} + \lceil \frac{3}{2} \rceil \rfloor$ **h)** $\lfloor \frac{1}{2} \cdot \lfloor \frac{5}{2} \rfloor \rfloor$

10. Determine whether each of these functions from $\{a, b, c, d\}$ to itself is one-to-one.

a) $f(a) = b, f(b) = a, f(c) = c, f(d) = d$

b) $f(a) = b, f(b) = b, f(c) = d, f(d) = c$

c) $f(a) = d, f(b) = b, f(c) = c, f(d) = d$

11. Which functions in Exercise 10 are onto?

12. Determine whether each of these functions from **Z** to **Z** is one-to-one.

a) $f(n) = n - 1$ **b)** $f(n) = n^2 + 1$

c) $f(n) = n^3$ **d)** $f(n) = \lceil n/2 \rceil$

13. Which functions in Exercise 12 are onto?

14. Determine whether $f: \mathbf{Z} \times \mathbf{Z} \to \mathbf{Z}$ is onto if

a) $f(m, n) = 2m - n.$

b) $f(m, n) = m^2 - n^2.$

c) $f(m, n) = m + n + 1.$

d) $f(m, n) = |m| - |n|.$

e) $f(m, n) = m^2 - 4.$

15. Determine whether the function $f: \mathbf{Z} \times \mathbf{Z} \to \mathbf{Z}$ is onto if

a) $f(m, n) = m + n.$

b) $f(m, n) = m^2 + n^2.$

c) $f(m, n) = m.$

d) $f(m, n) = |n|.$

e) $f(m, n) = m - n.$

16. Consider these functions from the set of students in a discrete mathematics class. Under what conditions is the function one-to-one if it assigns to a student the student's

a) mobile phone number.

b) student identification number.

c) final grade in the class.

d) home town.

17. Consider these functions from the set of teachers in a school. Under what conditions is the function one-to-one if it assigns to a teacher this teacher's

a) office.

b) assigned bus to chaperone in a group of buses taking students on a field trip.

c) salary.

d) social security number.

18. Specify a codomain for each of the functions in Exercise 16. Under what conditions is each of these functions with the codomain you specified onto?

19. Specify a codomain for each of the functions in Exercise 17. Under what conditions is each of the functions with the codomain you specified onto?

20. Give an example of a function from **N** to **N** that is

a) one-to-one but not onto.

b) onto but not one-to-one.

c) both onto and one-to-one (but different from the identity function).

d) neither one-to-one nor onto.

21. Give an explicit formula for a function from the set of integers to the set of positive integers that is

a) one-to-one, but not onto.

b) onto, but not one-to-one.

c) one-to-one and onto.

d) neither one-to-one nor onto.

22. Determine whether each of these functions is a bijection from **R** to **R**.

a) $f(x) = -3x + 4$

b) $f(x) = -3x^2 + 7$

c) $f(x) = (x + 1)/(x + 2)$

d) $f(x) = x^5 + 1$

23. Determine whether each of these functions is a bijection from **R** to **R**.

a) $f(x) = 2x + 1$

b) $f(x) = x^2 + 1$

c) $f(x) = x^3$

d) $f(x) = (x^2 + 1)/(x^2 + 2)$

24. Let $f: \mathbf{R} \to \mathbf{R}$ and let $f(x) > 0$ for all $x \in \mathbf{R}$. Show that $f(x)$ is strictly increasing if and only if the function $g(x) = 1/f(x)$ is strictly decreasing.

25. Let $f: \mathbf{R} \to \mathbf{R}$ and let $f(x) > 0$ for all $x \in \mathbf{R}$. Show that $f(x)$ is strictly decreasing if and only if the function $g(x) = 1/f(x)$ is strictly increasing.

26. a) Prove that a strictly increasing function from **R** to itself is one-to-one.

b) Give an example of an increasing function from **R** to itself that is not one-to-one.

27. a) Prove that a strictly decreasing function from **R** to itself is one-to-one.

b) Give an example of a decreasing function from **R** to itself that is not one-to-one.

28. Show that the function $f(x) = e^x$ from the set of real numbers to the set of real numbers is not invertible, but if the codomain is restricted to the set of positive real numbers, the resulting function is invertible.

29. Show that the function $f(x) = |x|$ from the set of real numbers to the set of nonnegative real numbers is not invertible, but if the domain is restricted to the set of nonnegative real numbers, the resulting function is invertible.

30. Let $S = \{-1, 0, 2, 4, 7\}$. Find $f(S)$ if

a) $f(x) = 1$. b) $f(x) = 2x + 1$.

c) $f(x) = \lceil x/5 \rceil$. d) $f(x) = \lfloor (x^2 + 1)/3 \rfloor$.

31. Let $f(x) = \lfloor x^2/3 \rfloor$. Find $f(S)$ if

a) $S = \{-2, -1, 0, 1, 2, 3\}$.

b) $S = \{0, 1, 2, 3, 4, 5\}$.

c) $S = \{1, 5, 7, 11\}$.

d) $S = \{2, 6, 10, 14\}$.

32. Let $f(x) = 2x$ where the domain is the set of real numbers. What is

a) $f(\mathbf{Z})$? b) $f(\mathbf{N})$? c) $f(\mathbf{R})$?

33. Suppose that g is a function from A to B and f is a function from B to C.

a) Show that if both f and g are one-to-one functions, then $f \circ g$ is also one-to-one.

b) Show that if both f and g are onto functions, then $f \circ g$ is also onto.

34. Suppose that g is a function from A to B and f is a function from B to C. Prove each of these statements.

a) If $f \circ g$ is onto, then f must also be onto.

b) If $f \circ g$ is one-to-one, then g must also be one-to-one.

c) If $f \circ g$ is a bijection, then g is onto if and only if f is one-to-one.

35. Find an example of functions f and g such that $f \circ g$ is a bijection, but g is not onto and f is not one-to-one.

***36.** If f and $f \circ g$ are one-to-one, does it follow that g is one-to-one? Justify your answer.

***37.** If f and $f \circ g$ are onto, does it follow that g is onto? Justify your answer.

38. Find $f \circ g$ and $g \circ f$, where $f(x) = x^2 + 1$ and $g(x) = x + 2$, are functions from $\mathbf{R}$ to $\mathbf{R}$.

39. Find $f + g$ and fg for the functions f and g given in Exercise 38.

40. Let $f(x) = ax + b$ and $g(x) = cx + d$, where a, b, c, and d are constants. Determine necessary and sufficient conditions on the constants a, b, c, and d so that $f \circ g = g \circ f$.

41. Show that the function $f(x) = ax + b$ from $\mathbf{R}$ to $\mathbf{R}$, where a and b are constants with $a \neq 0$ is invertible, and find the inverse of f.

42. Let f be a function from the set A to the set B. Let S and T be subsets of A. Show that

a) $f(S \cup T) = f(S) \cup f(T)$.

b) $f(S \cap T) \subseteq f(S) \cap f(T)$.

43. a) Give an example to show that the inclusion in part (b) in Exercise 42 may be proper.

b) Show that if f is one-to-one, the inclusion in part (b) in Exercise 42 is an equality.

Let f be a function from the set A to the set B. Let S be a subset of B. We define the **inverse image** of S to be the subset of A whose elements are precisely all preimages of all elements of S. We denote the inverse image of S by $f^{-1}(S)$, so $f^{-1}(S) = \{a \in A \mid f(a) \in S\}$. [*Beware:* The notation f^{-1} is used in two different ways. Do not confuse the notation introduced here with the notation $f^{-1}(y)$ for the value at y of the inverse of the invertible function f. Notice also that $f^{-1}(S)$, the inverse image of the set S, makes sense for all functions f, not just invertible functions.]

44. Let f be the function from $\mathbf{R}$ to $\mathbf{R}$ defined by $f(x) = x^2$. Find

a) $f^{-1}(\{1\})$. b) $f^{-1}(\{x \mid 0 < x < 1\})$.

c) $f^{-1}(\{x \mid x > 4\})$.

45. Let $g(x) = \lfloor x \rfloor$. Find

a) $g^{-1}(\{0\})$. b) $g^{-1}(\{-1, 0, 1\})$.

c) $g^{-1}(\{x \mid 0 < x < 1\})$.

46. Let f be a function from A to B. Let S and T be subsets of B. Show that

a) $f^{-1}(S \cup T) = f^{-1}(S) \cup f^{-1}(T)$.

b) $f^{-1}(S \cap T) = f^{-1}(S) \cap f^{-1}(T)$.

47. Let f be a function from A to B. Let S be a subset of B. Show that $f^{-1}(\overline{S}) = \overline{f^{-1}(S)}$.

48. Show that $\lfloor x + \frac{1}{2} \rfloor$ is the closest integer to the number x, except when x is midway between two integers, when it is the larger of these two integers.

49. Show that $\lceil x - \frac{1}{2} \rceil$ is the closest integer to the number x, except when x is midway between two integers, when it is the smaller of these two integers.

50. Show that if x is a real number, then $\lceil x \rceil - \lfloor x \rfloor = 1$ if x is not an integer and $\lceil x \rceil - \lfloor x \rfloor = 0$ if x is an integer.

51. Show that if x is a real number, then $x - 1 < \lfloor x \rfloor \leq x \leq \lceil x \rceil < x + 1$.

52. Show that if x is a real number and m is an integer, then $\lceil x + m \rceil = \lceil x \rceil + m$.

53. Show that if x is a real number and n is an integer, then

a) $x < n$ if and only if $\lfloor x \rfloor < n$.

b) $n < x$ if and only if $n < \lceil x \rceil$.

54. Show that if x is a real number and n is an integer, then

a) $x \leq n$ if and only if $\lceil x \rceil \leq n$.

b) $n \leq x$ if and only if $n \leq \lfloor x \rfloor$.

55. Prove that if n is an integer, then $\lfloor n/2 \rfloor = n/2$ if n is even and $(n - 1)/2$ if n is odd.

56. Prove that if x is a real number, then $\lfloor -x \rfloor = -\lceil x \rceil$ and $\lceil -x \rceil = -\lfloor x \rfloor$.

57. The function INT is found on some calculators, where $\text{INT}(x) = \lfloor x \rfloor$ when x is a nonnegative real number and $\text{INT}(x) = \lceil x \rceil$ when x is a negative real number. Show that this INT function satisfies the identity $\text{INT}(-x) = -\text{INT}(x)$.

58. Let a and b be real numbers with $a < b$. Use the floor and/or ceiling functions to express the number of integers n that satisfy the inequality $a \leq n \leq b$.

59. Let a and b be real numbers with $a < b$. Use the floor and/or ceiling functions to express the number of integers n that satisfy the inequality $a < n < b$.

60. How many bytes are required to encode n bits of data where n equals

a) 4? b) 10? c) 500? d) 3000?

61. How many bytes are required to encode n bits of data where n equals

a) 7? b) 17? c) 1001? d) 28,800?

62. How many ATM cells (described in Example 30) can be transmitted in 10 seconds over a link operating at the following rates?

a) 128 kilobits per second (1 kilobit = 1000 bits)
b) 300 kilobits per second
c) 1 megabit per second (1 megabit = 1,000,000 bits)

63. Data are transmitted over a particular Ethernet network in blocks of 1500 octets (blocks of 8 bits). How many blocks are required to transmit the following amounts of data over this Ethernet network? (Note that a byte is a synonym for an octet, a kilobyte is 1000 bytes, and a megabyte is 1,000,000 bytes.)

a) 150 kilobytes of data
b) 384 kilobytes of data
c) 1.544 megabytes of data
d) 45.3 megabytes of data

64. Draw the graph of the function $f(n) = 1 - n^2$ from **Z** to **Z**.

65. Draw the graph of the function $f(x) = \lfloor 2x \rfloor$ from **R** to **R**.

66. Draw the graph of the function $f(x) = \lfloor x/2 \rfloor$ from **R** to **R**.

67. Draw the graph of the function $f(x) = \lfloor x \rfloor + \lfloor x/2 \rfloor$ from **R** to **R**.

68. Draw the graph of the function $f(x) = \lceil x \rceil + \lfloor x/2 \rfloor$ from **R** to **R**.

69. Draw graphs of each of these functions.

a) $f(x) = \lfloor x + \frac{1}{2} \rfloor$ **b)** $f(x) = \lfloor 2x + 1 \rfloor$
c) $f(x) = \lceil x/3 \rceil$ **d)** $f(x) = \lceil 1/x \rceil$
e) $f(x) = \lceil x - 2 \rceil + \lfloor x + 2 \rfloor$
f) $f(x) = \lfloor 2x \rfloor \lceil x/2 \rceil$ **g)** $f(x) = \lceil \lfloor x - \frac{1}{2} \rfloor + \frac{1}{2} \rceil$

70. Draw graphs of each of these functions.

a) $f(x) = \lceil 3x - 2 \rceil$ **b)** $f(x) = \lceil 0.2x \rceil$
c) $f(x) = \lfloor -1/x \rfloor$ **d)** $f(x) = \lfloor x^2 \rfloor$
e) $f(x) = \lceil x/2 \rceil \lfloor x/2 \rfloor$ **f)** $f(x) = \lfloor x/2 \rfloor + \lceil x/2 \rceil$
g) $f(x) = \lfloor 2 \lceil x/2 \rceil + \frac{1}{2} \rfloor$

71. Find the inverse function of $f(x) = x^3 + 1$.

72. Suppose that f is an invertible function from Y to Z and g is an invertible function from X to Y. Show that the inverse of the composition $f \circ g$ is given by $(f \circ g)^{-1} = g^{-1} \circ f^{-1}$.

73. Let S be a subset of a universal set U. The **characteristic function** f_S of S is the function from U to the set $\{0, 1\}$ such that $f_S(x) = 1$ if x belongs to S and $f_S(x) = 0$ if x does not belong to S. Let A and B be sets. Show that for all $x \in U$,

a) $f_{A \cap B}(x) = f_A(x) \cdot f_B(x)$
b) $f_{A \cup B}(x) = f_A(x) + f_B(x) - f_A(x) \cdot f_B(x)$
c) $f_{\overline{A}}(x) = 1 - f_A(x)$
d) $f_{A \oplus B}(x) = f_A(x) + f_B(x) - 2f_A(x)f_B(x)$

74. Suppose that f is a function from A to B, where A and B are finite sets with $|A| = |B|$. Show that f is one-to-one if and only if it is onto.

75. Prove or disprove each of these statements about the floor and ceiling functions.

a) $\lceil \lfloor x \rfloor \rceil = \lfloor x \rfloor$ for all real numbers x.
b) $\lfloor 2x \rfloor = 2\lfloor x \rfloor$ whenever x is a real number.
c) $\lceil x \rceil + \lceil y \rceil - \lceil x + y \rceil = 0$ or 1 whenever x and y are real numbers.
d) $\lceil xy \rceil = \lceil x \rceil \lceil y \rceil$ for all real numbers x and y.
e) $\left\lceil \frac{x}{2} \right\rceil = \left\lfloor \frac{x+1}{2} \right\rfloor$ for all real numbers x.

76. Prove or disprove each of these statements about the floor and ceiling functions.

a) $\lfloor \lceil x \rceil \rfloor = \lceil x \rceil$ for all real numbers x.
b) $\lfloor x + y \rfloor = \lfloor x \rfloor + \lfloor y \rfloor$ for all real numbers x and y.
c) $\lceil \lceil x/2 \rceil /2 \rceil = \lceil x/4 \rceil$ for all real numbers x.
d) $\lfloor \sqrt{\lceil x \rceil} \rfloor = \lfloor \sqrt{x} \rfloor$ for all positive real numbers x.
e) $\lfloor x \rfloor + \lfloor y \rfloor + \lfloor x + y \rfloor \leq \lfloor 2x \rfloor + \lfloor 2y \rfloor$ for all real numbers x and y.

77. Prove that if x is a positive real number, then

a) $\lfloor \sqrt{\lfloor x \rfloor} \rfloor = \lfloor \sqrt{x} \rfloor$.
b) $\lceil \sqrt{\lceil x \rceil} \rceil = \lceil \sqrt{x} \rceil$.

78. Let x be a real number. Show that $\lfloor 3x \rfloor = \lfloor x \rfloor + \lfloor x + \frac{1}{3} \rfloor + \lfloor x + \frac{2}{3} \rfloor$.

79. For each of these partial functions, determine its domain, codomain, domain of definition, and the set of values for which it is undefined. Also, determine whether it is a total function.

a) $f: \mathbf{Z} \to \mathbf{R}, f(n) = 1/n$
b) $f: \mathbf{Z} \to \mathbf{Z}, f(n) = \lceil n/2 \rceil$
c) $f: \mathbf{Z} \times \mathbf{Z} \to \mathbf{Q}, f(m, n) = m/n$
d) $f: \mathbf{Z} \times \mathbf{Z} \to \mathbf{Z}, f(m, n) = mn$
e) $f: \mathbf{Z} \times \mathbf{Z} \to \mathbf{Z}, f(m, n) = m - n$ if $m > n$

80. a) Show that a partial function from A to B can be viewed as a function f^* from A to $B \cup \{u\}$, where u is not an element of B and

$$f^*(a) = \begin{cases} f(a) & \text{if } a \text{ belongs to the domain of definition of } f \\ u & \text{if } f \text{ is undefined at } a. \end{cases}$$

b) Using the construction in (a), find the function f^* corresponding to each partial function in Exercise 79.

81. a) Show that if a set S has cardinality m, where m is a positive integer, then there is a one-to-one correspondence between S and the set $\{1, 2, \ldots, m\}$.
b) Show that if S and T are two sets each with m elements, where m is a positive integer, then there is a one-to-one correspondence between S and T.

***82.** Show that a set S is infinite if and only if there is a proper subset A of S such that there is a one-to-one correspondence between A and S.

2.4 Sequences and Summations

2.4.1 Introduction

Sequences are ordered lists of elements used in discrete mathematics in many ways. For example, they can be used to represent solutions to certain counting problems, as we will see in Chapter 8. They are also an important data structure in computer science. We will often need to work with sums of terms of sequences in our study of discrete mathematics. This section reviews the use of summation notation, basic properties of summations, and formulas for the sums of terms of some particular types of sequences.

The terms of a sequence can be specified by providing a formula for each term of the sequence. In this section we describe another way to specify the terms of a sequence using a recurrence relation, which expresses each term as a combination of the previous terms. We will introduce one method, known as iteration, for finding a closed formula for the terms of a sequence specified via a recurrence relation. Identifying a sequence when the first few terms are provided is a useful skill when solving problems in discrete mathematics. We will provide some tips, including a useful tool on the Web, for doing so.

2.4.2 Sequences

A sequence is a discrete structure used to represent an ordered list. For example, 1, 2, 3, 5, 8 is a sequence with five terms and 1, 3, 9, 27, 81, … , 3^n, … is an infinite sequence.

Definition 1 A *sequence* is a function from a subset of the set of integers (usually either the set $\{0, 1, 2, \dots\}$ or the set $\{1, 2, 3, \dots\}$) to a set S. We use the notation a_n to denote the image of the integer n. We call a_n a *term* of the sequence.

We use the notation $\{a_n\}$ to describe the sequence. (Note that a_n represents an individual term of the sequence $\{a_n\}$. Be aware that the notation $\{a_n\}$ for a sequence conflicts with the notation for a set. However, the context in which we use this notation will always make it clear when we are dealing with sets and when we are dealing with sequences. Moreover, although we have used the letter a in the notation for a sequence, other letters or expressions may be used depending on the sequence under consideration. That is, the choice of the letter a is arbitrary.)

We describe sequences by listing the terms of the sequence in order of increasing subscripts.

EXAMPLE 1 Consider the sequence $\{a_n\}$, where

$$a_n = \frac{1}{n}.$$

The list of the terms of this sequence, beginning with a_1, namely,

$$a_1, a_2, a_3, a_4, \dots,$$

starts with

$$1, \frac{1}{2}, \frac{1}{3}, \frac{1}{4}, \dots.$$

◀

Definition 2 A *geometric progression* is a sequence of the form

$$a, ar, ar^2, \dots, ar^n, \dots$$

where the *initial term a* and the *common ratio r* are real numbers.

Remark: A geometric progression is a discrete analogue of the exponential function $f(x) = ar^x$.

EXAMPLE 2 The sequences $\{b_n\}$ with $b_n = (-1)^n$, $\{c_n\}$ with $c_n = 2 \cdot 5^n$, and $\{d_n\}$ with $d_n = 6 \cdot (1/3)^n$ are geometric progressions with initial term and common ratio equal to 1 and -1, 2 and 5, and 6 and $1/3$, respectively, if we start at $n = 0$. The list of terms $b_0, b_1, b_2, b_3, b_4, \dots$ begins with

$$1, -1, 1, -1, 1, \dots;$$

the list of terms $c_0, c_1, c_2, c_3, c_4, \dots$ begins with

$$2, 10, 50, 250, 1250, \dots;$$

and the list of terms $d_0, d_1, d_2, d_3, d_4, \dots$ begins with

$$6, 2, \frac{2}{3}, \frac{2}{9}, \frac{2}{27}, \dots.$$ ◀

Definition 3 An *arithmetic progression* is a sequence of the form

$$a, a + d, a + 2d, \dots, a + nd, \dots$$

where the *initial term a* and the *common difference d* are real numbers.

Remark: An arithmetic progression is a discrete analogue of the linear function $f(x) = dx + a$.

EXAMPLE 3 The sequences $\{s_n\}$ with $s_n = -1 + 4n$ and $\{t_n\}$ with $t_n = 7 - 3n$ are both arithmetic progressions with initial terms and common differences equal to -1 and 4, and 7 and -3, respectively, if we start at $n = 0$. The list of terms $s_0, s_1, s_2, s_3, \dots$ begins with

$$-1, 3, 7, 11, \dots,$$

and the list of terms $t_0, t_1, t_2, t_3, \dots$ begins with

$$7, 4, 1, -2, \dots.$$ ◀

Sequences of the form $a_1, a_2, \dots, a_n$ are often used in computer science. These finite sequences are also called **strings**. This string is also denoted by $a_1a_2 \dots a_n$. (Recall that bit strings, which are finite sequences of bits, were introduced in Section 1.1.) The **length** of a string is the number of terms in this string. The **empty string**, denoted by λ, is the string that has no terms. The empty string has length zero.

EXAMPLE 4 The string *abcd* is a string of length four. ◀

2.4.3 Recurrence Relations

In Examples 1–3 we specified sequences by providing explicit formulas for their terms. There are many other ways to specify a sequence. For example, another way to specify a sequence is to provide one or more initial terms together with a rule for determining subsequent terms from those that precede them.

Definition 4 A *recurrence relation* for the sequence $\{a_n\}$ is an equation that expresses a_n in terms of one or more of the previous terms of the sequence, namely, $a_0, a_1, \ldots, a_{n-1}$, for all integers n with $n \geq n_0$, where n_0 is a nonnegative integer. A sequence is called a *solution* of a recurrence relation if its terms satisfy the recurrence relation. (A recurrence relation is said to *recursively define* a sequence. We will explain this alternative terminology in Chapter 5.)

EXAMPLE 5 Let $\{a_n\}$ be a sequence that satisfies the recurrence relation $a_n = a_{n-1} + 3$ for $n = 2, 3, 4, \ldots$, and suppose that $a_1 = 2$. What are a_2, a_3, and a_4?

Solution: We see from the recurrence relation that $a_2 = a_1 + 3 = 2 + 3 = 5$. It then follows that $a_3 = 5 + 3 = 8$ and $a_4 = 8 + 3 = 11$. ◀

EXAMPLE 6 Let $\{a_n\}$ be a sequence that satisfies the recurrence relation $a_n = a_{n-1} - a_{n-2}$ for $n = 2, 3, 4, \ldots$, and suppose that $a_0 = 3$ and $a_1 = 5$. What are a_2 and a_3?

Solution: We see from the recurrence relation that $a_2 = a_1 - a_0 = 5 - 3 = 2$ and $a_3 = a_2 - a_1 = 2 - 5 = -3$. We can find a_4, a_5, and each successive term in a similar way. ◀

The **initial conditions** for a recursively defined sequence specify the terms that precede the first term where the recurrence relation takes effect. For instance, the initial condition in Example 5 is $a_1 = 2$, and the initial conditions in Example 6 are $a_0 = 3$ and $a_1 = 5$. Using mathematical induction, a proof technique introduced in Chapter 5, it can be shown that a recurrence relation together with its initial conditions determines a unique solution.

Hop along to Chapter 8 to learn how to find a formula for the Fibonacci numbers.

Next, we define a particularly useful sequence defined by a recurrence relation, known as the **Fibonacci sequence**, after the Italian mathematician Fibonacci who was born in the 12th century (see Chapter 5 for his biography). We will study this sequence in depth in Chapters 5 and 8, where we will see why it is important for many applications, including modeling the population growth of rabbits. Fibonacci numbers occur naturally in the structures of plants and animals, such as in the arrangement of sunflower seeds in a seed head and in the shell of the chambered nautilus.

Links ❯

Definition 5 The *Fibonacci sequence*, $f_0, f_1, f_2, \ldots$, is defined by the initial conditions $f_0 = 0, f_1 = 1$, and the recurrence relation

$$f_n = f_{n-1} + f_{n-2}$$

for $n = 2, 3, 4, \ldots$.

EXAMPLE 7 Find the Fibonacci numbers $f_2, f_3, f_4, f_5,$ and f_6.

Solution: The recurrence relation for the Fibonacci sequence tells us that we find successive terms by adding the previous two terms. Because the initial conditions tell us that $f_0 = 0$ and $f_1 = 1$, using the recurrence relation in the definition we find that

$$\begin{aligned} f_2 &= f_1 + f_0 = 1 + 0 = 1, \\ f_3 &= f_2 + f_1 = 1 + 1 = 2, \\ f_4 &= f_3 + f_2 = 2 + 1 = 3, \\ f_5 &= f_4 + f_3 = 3 + 2 = 5, \\ f_6 &= f_5 + f_4 = 5 + 3 = 8. \end{aligned}$$

◀

EXAMPLE 8 Suppose that $\{a_n\}$ is the sequence of integers defined by $a_n = n!$, the value of the factorial function at the integer n, where $n = 1, 2, 3, \ldots$. Because $n! = n((n-1)(n-2)\cdots 2 \cdot 1) = n(n-1)! = na_{n-1}$, we see that the sequence of factorials satisfies the recurrence relation $a_n = na_{n-1}$, together with the initial condition $a_1 = 1$. ◀

We say that we have solved the recurrence relation together with the initial conditions when we find an explicit formula, called a **closed formula**, for the terms of the sequence.

EXAMPLE 9 Determine whether the sequence $\{a_n\}$, where $a_n = 3n$ for every nonnegative integer n, is a solution of the recurrence relation $a_n = 2a_{n-1} - a_{n-2}$ for $n = 2, 3, 4, \ldots$. Answer the same question where $a_n = 2^n$ and where $a_n = 5$.

Solution: Suppose that $a_n = 3n$ for every nonnegative integer n. Then, for $n \geq 2$, we see that $2a_{n-1} - a_{n-2} = 2(3(n-1)) - 3(n-2) = 3n = a_n$. Therefore, $\{a_n\}$, where $a_n = 3n$, is a solution of the recurrence relation.

Suppose that $a_n = 2^n$ for every nonnegative integer n. Note that $a_0 = 1$, $a_1 = 2$, and $a_2 = 4$. Because $2a_1 - a_0 = 2 \cdot 2 - 1 = 3 \neq a_2$, we see that $\{a_n\}$, where $a_n = 2^n$, is not a solution of the recurrence relation.

Suppose that $a_n = 5$ for every nonnegative integer n. Then for $n \geq 2$, we see that $a_n = 2a_{n-1} - a_{n-2} = 2 \cdot 5 - 5 = 5 = a_n$. Therefore, $\{a_n\}$, where $a_n = 5$, is a solution of the recurrence relation. ◀

Many methods have been developed for solving recurrence relations. Here, we will introduce a straightforward method known as iteration via several examples. In Chapter 8 we will study recurrence relations in depth. In that chapter we will show how recurrence relations can be used to solve counting problems and we will introduce several powerful methods that can be used to solve many different recurrence relations.

EXAMPLE 10 Solve the recurrence relation and initial condition in Example 5.

Solution: We can successively apply the recurrence relation in Example 5, starting with the initial condition $a_1 = 2$, and working upward until we reach a_n to deduce a closed formula for the sequence. We see that

$$\begin{aligned} a_2 &= 2 + 3 \\ a_3 &= (2 + 3) + 3 = 2 + 3 \cdot 2 \\ a_4 &= (2 + 2 \cdot 3) + 3 = 2 + 3 \cdot 3 \\ &\vdots \\ a_n &= a_{n-1} + 3 = (2 + 3 \cdot (n-2)) + 3 = 2 + 3(n-1). \end{aligned}$$

We can also successively apply the recurrence relation in Example 5, starting with the term a_n and working downward until we reach the initial condition $a_1 = 2$ to deduce this same formula. The steps are

$$\begin{aligned} a_n &= a_{n-1} + 3 \\ &= (a_{n-2} + 3) + 3 = a_{n-2} + 3 \cdot 2 \\ &= (a_{n-3} + 3) + 3 \cdot 2 = a_{n-3} + 3 \cdot 3 \\ &\vdots \\ &= a_2 + 3(n-2) = (a_1 + 3) + 3(n-2) = 2 + 3(n-1). \end{aligned}$$

At each iteration of the recurrence relation, we obtain the next term in the sequence by adding 3 to the previous term. We obtain the nth term after n iterations of the recurrence relation. Hence, we have added $3(n - 1)$ to the initial term $a_0 = 2$ to obtain a_n. This gives us the closed formula $a_n = 2 + 3(n - 1)$. Note that this sequence is an arithmetic progression. ◀

The technique used in Example 10 is called **iteration**. We have iterated, or repeatedly used, the recurrence relation. The first approach is called **forward substitution**—we found successive terms beginning with the initial condition and ending with a_n. The second approach is called **backward substitution**, because we began with a_n and iterated to express it in terms of falling terms of the sequence until we found it in terms of a_1. Note that when we use iteration, we essentialy guess a formula for the terms of the sequence. To prove that our guess is correct, we need to use mathematical induction, a technique we discuss in Chapter 5.

In Chapter 8 we will show that recurrence relations can be used to model a wide variety of problems. We provide one such example here, showing how to use a recurrence relation to find compound interest.

EXAMPLE 11 **Compound Interest** Suppose that a person deposits \$10,000 in a savings account at a bank yielding 11% per year with interest compounded annually. How much will be in the account after 30 years?

Extra Examples

Solution: To solve this problem, let P_n denote the amount in the account after n years. Because the amount in the account after n years equals the amount in the account after $n - 1$ years plus interest for the nth year, we see that the sequence $\{P_n\}$ satisfies the recurrence relation

$$P_n = P_{n-1} + 0.11P_{n-1} = (1.11)P_{n-1}.$$

The initial condition is $P_0 = 10{,}000$.

We can use an iterative approach to find a formula for P_n. Note that

$$\begin{aligned} P_1 &= (1.11)P_0 \\ P_2 &= (1.11)P_1 = (1.11)^2P_0 \\ P_3 &= (1.11)P_2 = (1.11)^3P_0 \\ &\vdots \\ P_n &= (1.11)P_{n-1} = (1.11)^nP_0. \end{aligned}$$

When we insert the initial condition $P_0 = 10{,}000$, the formula $P_n = (1.11)^n10{,}000$ is obtained.

Inserting $n = 30$ into the formula $P_n = (1.11)^n10{,}000$ shows that after 30 years the account contains

$$P_{30} = (1.11)^{30}10{,}000 = \$228{,}922.97.$$ ◀

2.4.4 Special Integer Sequences

A common problem in discrete mathematics is finding a closed formula, a recurrence relation, or some other type of general rule for constructing the terms of a sequence. Sometimes only a few terms of a sequence solving a problem are known; the goal is to identify the sequence. Even though the initial terms of a sequence do not determine the entire sequence (after all, there are infinitely many different sequences that start with any finite set of initial terms), knowing the first few terms may help you make an educated conjecture about the identity of your sequence. Once you have made this conjecture, you can try to verify that you have the correct sequence.

When trying to deduce a possible formula, recurrence relation, or some other type of rule for the terms of a sequence when given the initial terms, try to find a pattern in these terms. You might also see whether you can determine how a term might have been produced from those preceding it. There are many questions you could ask, but some of the more useful are:

- Are there runs of the same value? That is, does the same value occur many times in a row?
- Are terms obtained from previous terms by adding the same amount or an amount that depends on the position in the sequence?
- Are terms obtained from previous terms by multiplying by a particular amount?
- Are terms obtained by combining previous terms in a certain way?
- Are there cycles among the terms?

EXAMPLE 12 Find formulae for the sequences with the following first five terms: (a) $1, 1/2, 1/4, 1/8, 1/16$ (b) $1, 3, 5, 7, 9$ (c) $1, -1, 1, -1, 1$.

Extra Examples

Solution: (a) We recognize that the denominators are powers of 2. The sequence with $a_n = 1/2^n$, $n = 0, 1, 2, \dots$ is a possible match. This proposed sequence is a geometric progression with $a = 1$ and $r = 1/2$.

(b) We note that each term is obtained by adding 2 to the previous term. The sequence with $a_n = 2n + 1$, $n = 0, 1, 2, \dots$ is a possible match. This proposed sequence is an arithmetic progression with $a = 1$ and $d = 2$.

(c) The terms alternate between 1 and -1. The sequence with $a_n = (-1)^n$, $n = 0, 1, 2 \dots$ is a possible match. This proposed sequence is a geometric progression with $a = 1$ and $r = -1$. ◀

Examples 13–15 illustrate how we can analyze sequences to find how the terms are constructed.

EXAMPLE 13 How can we produce the terms of a sequence if the first 10 terms are 1, 2, 2, 3, 3, 3, 4, 4, 4, 4?

Solution: In this sequence, the integer 1 appears once, the integer 2 appears twice, the integer 3 appears three times, and the integer 4 appears four times. A reasonable rule for generating this sequence is that the integer n appears exactly n times, so the next five terms of the sequence would all be 5, the following six terms would all be 6, and so on. The sequence generated this way is a possible match. ◀

EXAMPLE 14 How can we produce the terms of a sequence if the first 10 terms are 5, 11, 17, 23, 29, 35, 41, 47, 53, 59?

Solution: Note that each of the first 10 terms of this sequence after the first is obtained by adding 6 to the previous term. (We could see this by noticing that the difference between consecutive terms is 6.) Consequently, the nth term could be produced by starting with 5 and adding 6 a total

TABLE 1 Some Useful Sequences.

nth Term	*First 10 Terms*
n^2	1, 4, 9, 16, 25, 36, 49, 64, 81, 100, ...
n^3	1, 8, 27, 64, 125, 216, 343, 512, 729, 1000, ...
n^4	1, 16, 81, 256, 625, 1296, 2401, 4096, 6561, 10000, ...
f_n	1, 1, 2, 3, 5, 8, 13, 21, 34, 55, 89, ...
2^n	2, 4, 8, 16, 32, 64, 128, 256, 512, 1024, ...
3^n	3, 9, 27, 81, 243, 729, 2187, 6561, 19683, 59049, ...
$n!$	1, 2, 6, 24, 120, 720, 5040, 40320, 362880, 3628800, ...

of $n - 1$ times; that is, a reasonable guess is that the nth term is $5 + 6(n - 1) = 6n - 1$. (This is an arithmetic progression with $a = 5$ and $d = 6$.) ◀

EXAMPLE 15 How can we produce the terms of a sequence if the first 10 terms are 1, 3, 4, 7, 11, 18, 29, 47, 76, 123?

Solution: Observe that each successive term of this sequence, starting with the third term, is the sum of the two previous terms. That is, $4 = 3 + 1$, $7 = 4 + 3$, $11 = 7 + 4$, and so on. Consequently, if L_n is the nth term of this sequence, we guess that the sequence is determined by the recurrence relation $L_n = L_{n-1} + L_{n-2}$ with initial conditions $L_1 = 1$ and $L_2 = 3$ (the same recurrence relation as the Fibonacci sequence, but with different initial conditions). This sequence is known as the **Lucas sequence**, after the French mathematician François Édouard Lucas. Lucas studied this sequence and the Fibonacci sequence in the nineteenth century. ◀

Another useful technique for finding a rule for generating the terms of a sequence is to compare the terms of a sequence of interest with the terms of a well-known integer sequence, such as terms of an arithmetic progression, terms of a geometric progression, perfect squares, perfect cubes, and so on. The first 10 terms of some sequences you may want to keep in mind are displayed in Table 1. Note that we have listed these sequences so that the terms of each sequence grow faster than those in the preceding sequence in the list. The rates of growth of these terms will be studied in Section 3.2.

Links

Courtesy of Neil Sloane

NEIL SLOANE (BORN 1939) Neil Sloane studied mathematics and electrical engineering at the University of Melbourne on a scholarship from the Australian state telephone company. He mastered many telephone-related jobs, such as erecting telephone poles, in his summer work. After graduating, he designed minimal-cost telephone networks in Australia. In 1962 he came to the United States and studied electrical engineering at Cornell University. His Ph.D. thesis was on what are now called neural networks. He took a job at Bell Labs in 1969, working in many areas, including network design, coding theory, and sphere packing. He moved to AT&T Labs in 1996 when it was split off from Bell Labs, working there until his retirement in 2012. One of his favorite problems is the **kissing problem** (a name he coined), which asks how many spheres can be arranged in n dimensions so that they all touch a central sphere of the same size. (In two dimensions the answer is 6, because 6 pennies can be placed so that they touch a central penny. In three dimensions, 12 billiard balls can be placed so that they touch a central billiard ball. Two billiard balls that just touch are said to "kiss," giving rise to the terminology "kissing problem" and "kissing number.") Sloane, together with Andrew Odlyzko, showed that in 8 and 24 dimensions, the optimal kissing numbers are, respectively, 240 and 196,560. The kissing number is known in dimensions 1, 2, 3, 4, 8, and 24, but not in any other dimensions. Sloane's books include *Sphere Packings, Lattices and Groups,* 3rd ed., with John Conway; *The Theory of Error-Correcting Codes* with Jessie MacWilliams; *The Encyclopedia of Integer Sequences* with Simon Plouffe (which has grown into the popular OEIS website); and *The Rock-Climbing Guide to New Jersey Crags* with Paul Nick. The last book demonstrates his interest in rock climbing; it includes more than 50 climbing sites in New Jersey.

EXAMPLE 16 Conjecture a simple formula for a_n if the first 10 terms of the sequence $\{a_n\}$ are 1, 7, 25, 79, 241, 727, 2185, 6559, 19681, 59047.

Solution: To attack this problem, we begin by looking at the difference of consecutive terms, but we do not see a pattern. When we form the ratio of consecutive terms to see whether each term is a multiple of the previous term, we find that this ratio, although not a constant, is close to 3. So it is reasonable to suspect that the terms of this sequence are generated by a formula involving 3^n. Comparing these terms with the corresponding terms of the sequence $\{3^n\}$, we notice that the nth term is 2 less than the corresponding power of 3. We see that $a_n = 3^n - 2$ for $1 \leq n \leq 10$ and conjecture that this formula holds for all n. ◀

We will see throughout this text that integer sequences appear in a wide range of contexts in discrete mathematics. Sequences we have encountered or will encounter include the sequence of prime numbers (Chapter 4), the number of ways to order n discrete objects (Chapter 6), the number of moves required to solve the Tower of Hanoi puzzle with n disks (Chapter 8), and the number of rabbits on an island after n months (Chapter 8).

Check out the puzzles at the OEIS site.

Links ›

Integer sequences appear in an amazingly wide range of subject areas besides discrete mathematics, including biology, engineering, chemistry, and physics, as well as in puzzles. An amazing database of over 360,000 different integer sequences (as of 2024) can be found in the *On-Line Encyclopedia of Integer Sequences (OEIS)*. This database was originated by Neil Sloane in 1964 and is now maintained by the OEIS Foundation. The last printed version of this database was published in 1995 ([SIPI95]); the current encyclopedia would occupy more than 900 volumes of the size of the 1995 book with more than 10,000 new submissions a year. You can use a program on the OEIS website to find sequences from the encyclopedia that match initial terms you provide, if there is a match. For instance, when you enter 1, 1, 2, 3, 5, 8, OEIS displays a page that identifies these numbers as successive terms of the Fibonacci sequence, provides the recurrence relation that generates this sequence, lists an extensive set of comments about the many ways the Fibonacci sequence arises including references, and displays information about quite a few other sequences that begin with these same terms.

2.4.5 Summations

Next, we consider the addition of the terms of a sequence. For this we introduce **summation notation**. We begin by describing the notation used to express the sum of the terms

$$a_m, a_{m+1}, \ldots, a_n$$

from the sequence $\{a_n\}$. We use the notation

$$\sum_{j=m}^{n} a_j, \qquad \textstyle\sum_{j=m}^{n} a_j, \qquad \text{or} \qquad \textstyle\sum_{m\leq j\leq n} a_j$$

(read as the sum from $j = m$ to $j = n$ of a_j) to represent

$$a_m + a_{m+1} + \cdots + a_n.$$

Here, the variable j is called the **index of summation**, and the choice of the letter j as the variable is arbitrary; that is, we could have used any other letter, such as i or k. Or, in notation,

$$\sum_{j=m}^{n} a_j = \sum_{i=m}^{n} a_i = \sum_{k=m}^{n} a_k.$$

Here, the index of summation runs through all integers starting with its **lower limit** m and ending with its **upper limit** n. A large uppercase Greek letter sigma, $\sum$, is used to denote summation.

The usual laws for arithmetic apply to summations. For example, when a and b are real numbers, we have $\sum_{j=1}^{n}(ax_j + by_j) = a\sum_{y=1}^{n} x_j + b\sum_{j=1}^{n} y_j$, where $x_1, x_2, \ldots, x_n$ and $y_1, y_2, \ldots, y_n$ are real numbers. (We do not present a formal proof of this identity here. Such a proof can be constructed using mathematical induction, a proof method we introduce in Chapter 5. The proof also uses the commutative and associative laws for addition and the distributive law of multiplication over addition.)

We give some examples of summation notation.

EXAMPLE 17 Use summation notation to express the sum of the first 100 terms of the sequence $\{a_j\}$, where $a_j = 1/j$ for $j = 1, 2, 3, \ldots$.

Extra Examples

Solution: The lower limit for the index of summation is 1, and the upper limit is 100. We write this sum as

$$\sum_{j=1}^{100} \frac{1}{j}.$$

◀

EXAMPLE 18 What is the value of $\sum_{j=1}^{5} j^2$?

Solution: We have

$$\begin{aligned}\sum_{j=1}^{5} j^2 &= 1^2 + 2^2 + 3^2 + 4^2 + 5^2\\ &= 1 + 4 + 9 + 16 + 25\\ &= 55.\end{aligned}$$

◀

EXAMPLE 19 What is the value of $\sum_{k=4}^{8}(-1)^k$?

Solution: We have

$$\begin{aligned}\sum_{k=4}^{8}(-1)^k &= (-1)^4 + (-1)^5 + (-1)^6 + (-1)^7 + (-1)^8\\ &= 1 + (-1) + 1 + (-1) + 1\\ &= 1.\end{aligned}$$

◀

Sometimes it is useful to shift the index of summation in a sum. This is often done when two sums need to be added but their indices of summation do not match. When shifting an index of summation, it is important to make the appropriate changes in the corresponding summand. This is illustrated by Example 20.

EXAMPLE 20 Suppose we have the sum

$$\sum_{j=1}^{5} j^2$$

Extra Examples

but want the index of summation to run between 0 and 4 rather than from 1 to 5. To do this, we let $k = j - 1$. Then the new summation index runs from 0 (because $k = 1 - 0 = 0$ when $j = 1$) to 4 (because $k = 5 - 1 = 4$ when $j = 5$), and the term j^2 becomes $(k + 1)^2$. Hence,

$$\sum_{j=1}^{5} j^2 = \sum_{k=0}^{4} (k+1)^2.$$

It is easily checked that both sums are $1 + 4 + 9 + 16 + 25 = 55$. ◀

Sums of terms of geometric progressions commonly arise (such sums are called **geometric series**). Theorem 1 gives us a formula for the sum of terms of a geometric progression.

THEOREM 1 If a and r are real numbers and $r \neq 0$, then

$$\sum_{j=0}^{n} ar^j = \begin{cases} \dfrac{ar^{n+1} - a}{r - 1} & \text{if } r \neq 1 \\ (n+1)a & \text{if } r = 1. \end{cases}$$

Proof: Let

$$S_n = \sum_{j=0}^{n} ar^j.$$

To compute S_n, first multiply both sides of the equality by r and then manipulate the resulting sum as follows:

$$\begin{aligned} rS_n &= r\sum_{j=0}^{n} ar^j && \text{substituting summation formula for } S_n \\ &= \sum_{j=0}^{n} ar^{j+1} && \text{by the distributive property} \\ &= \sum_{k=1}^{n+1} ar^k && \text{shifting the index of summation, with } k = j+1 \\ &= \left(\sum_{k=0}^{n} ar^k\right) + (ar^{n+1} - a) && \text{removing } k = n+1 \text{ term and adding } k = 0 \text{ term} \\ &= S_n + (ar^{n+1} - a) && \text{substituting } S_n \text{ for summation formula} \end{aligned}$$

From these equalities, we see that

$$rS_n = S_n + (ar^{n+1} - a).$$

Solving for S_n shows that if $r \neq 1$, then

$$S_n = \frac{ar^{n+1} - a}{r - 1}.$$

If $r = 1$, then the $S_n = \sum_{j=0}^{n} ar^j = \sum_{j=0}^{n} a = (n+1)a$. ◁

EXAMPLE 21 Double summations arise in many contexts (as in the analysis of nested loops in computer programs). An example of a double summation is

$$\sum_{i=1}^{4}\sum_{j=1}^{3} ij.$$

To evaluate the double sum, first expand the inner summation and then continue by computing the outer summation:

$$\begin{aligned}\sum_{i=1}^{4}\sum_{j=1}^{3} ij &= \sum_{i=1}^{4}(i + 2i + 3i)\\ &= \sum_{i=1}^{4} 6i\\ &= 6 + 12 + 18 + 24 = 60.\end{aligned}$$

◀

We can also use summation notation to add all values of a function, or terms of an indexed set, where the index of summation runs over all values in a set. That is, we write

$$\sum_{s\in S} f(s)$$

to represent the sum of the values $f(s)$, for all members s of S.

EXAMPLE 22 What is the value of $\sum_{s\in\{0,2,4\}} s$?

Solution: Because $\sum_{s\in\{0,2,4\}} s$ represents the sum of the values of s for all the members of the set $\{0, 2, 4\}$, it follows that

$$\sum_{s\in\{0,2,4\}} s = 0 + 2 + 4 = 6.$$

◀

Certain sums arise repeatedly throughout discrete mathematics. Having a collection of formulae for such sums can be useful; Table 2 provides a small table of formulae for commonly occurring sums.

We derived the first formula in this table in Theorem 1. The next three formulae give us the sum of the first n positive integers, the sum of their squares, and the sum of their cubes. These three formulae can be derived in many different ways (for example, see Exercises 37 and 38). Also note that each of these formulae, once known, can be proved using mathematical induction, the subject of Section 5.1. The last two formulae in the table involve infinite series and will be discussed shortly.

Example 23 illustrates how the formulae in Table 2 can be useful.

EXAMPLE 23 Find $\sum_{k=50}^{100} k^2$.

Solution: First note that because $\sum_{k=1}^{100} k^2 = \sum_{k=1}^{49} k^2 + \sum_{k=50}^{100} k^2$, we have

$$\sum_{k=50}^{100} k^2 = \sum_{k=1}^{100} k^2 - \sum_{k=1}^{49} k^2.$$

TABLE 2 Some Useful Summation Formulae.

Sum	*Closed Form*
$\sum_{k=0}^{n} ar^k \ (r \neq 0)$	$\dfrac{ar^{n+1} - a}{r - 1}, r \neq 1$
$\sum_{k=1}^{n} k$	$\dfrac{n(n+1)}{2}$
$\sum_{k=1}^{n} k^2$	$\dfrac{n(n+1)(2n+1)}{6}$
$\sum_{k=1}^{n} k^3$	$\dfrac{n^2(n+1)^2}{4}$
$\sum_{k=0}^{\infty} x^k, \lvert x\rvert < 1$	$\dfrac{1}{1-x}$
$\sum_{k=1}^{\infty} kx^{k-1}, \lvert x\rvert < 1$	$\dfrac{1}{(1-x)^2}$

Using the formula $\sum_{k=1}^{n} k^2 = n(n+1)(2n+1)/6$ from Table 2 (and proved in Exercise 38), we see that

$$\sum_{k=50}^{100} k^2 = \frac{100 \cdot 101 \cdot 201}{6} - \frac{49 \cdot 50 \cdot 99}{6} = 338{,}350 - 40{,}425 = 297{,}925.$$

◀

SOME INFINITE SERIES Although most of the summations in this book are finite sums, infinite series are important in some parts of discrete mathematics. Infinite series are usually studied in a course in calculus and even the definition of these series requires the use of calculus, but sometimes they arise in discrete mathematics, because discrete mathematics deals with infinite collections of discrete elements. In particular, in our future studies in discrete mathematics, we will find the closed forms for the infinite series in Examples 24 and 25 to be quite useful.

EXAMPLE 24 (*Requires calculus*) Let x be a real number with $|x| < 1$. Find $\sum_{n=0}^{\infty} x^n$.

Extra Examples

Solution: By Theorem 1 with $a = 1$ and $r = x$ we see that $\sum_{n=0}^{k} x^n = \dfrac{x^{k+1} - 1}{x - 1}$. Because $|x| < 1$, x^{k+1} approaches 0 as k approaches infinity. It follows that

$$\sum_{n=0}^{\infty} x^n = \lim_{k\to\infty} \frac{x^{k+1} - 1}{x - 1} = \frac{0 - 1}{x - 1} = \frac{1}{1 - x}.$$

◀

We can produce new summation formulae by differentiating or integrating existing formulae.

EXAMPLE 25 (*Requires calculus*) Differentiating both sides of the equation

$$\sum_{k=0}^{\infty} x^k = \frac{1}{1 - x},$$

from Example 24 we find that

$$\sum_{k=1}^{\infty} kx^{k-1} = \frac{1}{(1-x)^2}.$$

(This differentiation is valid for $|x| < 1$ by a theorem about infinite series.) ◀

Exercises

1. Find these terms of the sequence $\{a_n\}$, where $a_n = 2 \cdot (-3)^n + 5^n$.

a) a_0 **b)** a_1 **c)** a_4 **d)** a_5

2. What is the term a_8 of the sequence $\{a_n\}$ if a_n equals

a) 2^{n-1}? **b)** 7?
c) $1 + (-1)^n$? **d)** $-(-2)^n$?

3. What are the terms a_0, a_1, a_2, and a_3 of the sequence $\{a_n\}$, where a_n equals

a) $2^n + 1$? **b)** $(n+1)^{n+1}$?
c) $\lfloor n/2 \rfloor$? **d)** $\lfloor n/2 \rfloor + \lceil n/2 \rceil$?

4. What are the terms a_0, a_1, a_2, and a_3 of the sequence $\{a_n\}$, where a_n equals

a) $(-2)^n$? **b)** 3?
c) $7 + 4^n$? **d)** $2^n + (-2)^n$?

5. List the first 10 terms of each of these sequences.

a) the sequence that begins with 2 and in which each successive term is 3 more than the preceding term
b) the sequence that lists each positive integer three times, in increasing order
c) the sequence that lists the odd positive integers in increasing order, listing each odd integer twice
d) the sequence whose nth term is $n! - 2^n$
e) the sequence that begins with 3, where each succeeding term is twice the preceding term
f) the sequence whose first term is 2, second term is 4, and each succeeding term is the sum of the two preceding terms
g) the sequence whose nth term is the number of bits in the binary expansion of the number n (defined in Section 4.2)
h) the sequence where the nth term is the number of letters in the English word for the index n

6. List the first 10 terms of each of these sequences.

a) the sequence obtained by starting with 10 and obtaining each term by subtracting 3 from the previous term
b) the sequence whose nth term is the sum of the first n positive integers
c) the sequence whose nth term is $3^n - 2^n$
d) the sequence whose nth term is $\lfloor \sqrt{n} \rfloor$
e) the sequence whose first two terms are 1 and 5 and each succeeding term is the sum of the two previous terms
f) the sequence whose nth term is the largest integer whose binary expansion (defined in Section 4.2) has n bits (Write your answer in decimal notation.)
g) the sequence whose terms are constructed sequentially as follows: start with 1, then add 1, then multiply by 1, then add 2, then multiply by 2, and so on
h) the sequence whose nth term is the largest integer k such that $k! \leq n$

7. Find at least three different sequences beginning with the terms 1, 2, 4 whose terms are generated by a simple formula or rule.

8. Find at least three different sequences beginning with the terms 3, 5, 7 whose terms are generated by a simple formula or rule.

9. Find the first five terms of the sequence defined by each of these recurrence relations and initial conditions.

a) $a_n = 6a_{n-1}, a_0 = 2$
b) $a_n = a_{n-1}^2, a_1 = 2$
c) $a_n = a_{n-1} + 3a_{n-2}, a_0 = 1, a_1 = 2$
d) $a_n = na_{n-1} + n^2 a_{n-2}, a_0 = 1, a_1 = 1$
e) $a_n = a_{n-1} + a_{n-3}, a_0 = 1, a_1 = 2, a_2 = 0$

10. Find the first six terms of the sequence defined by each of these recurrence relations and initial conditions.

a) $a_n = -2a_{n-1}, a_0 = -1$
b) $a_n = a_{n-1} - a_{n-2}, a_0 = 2, a_1 = -1$
c) $a_n = 3a_{n-1}^2, a_0 = 1$
d) $a_n = na_{n-1} + a_{n-2}^2, a_0 = -1, a_1 = 0$
e) $a_n = a_{n-1} - a_{n-2} + a_{n-3}, a_0 = 1, a_1 = 1, a_2 = 2$

11. Let $a_n = 2^n + 5 \cdot 3^n$ for $n = 0, 1, 2, \ldots$.

a) Find a_0, a_1, a_2, a_3, and a_4.
b) Show that $a_2 = 5a_1 - 6a_0$, $a_3 = 5a_2 - 6a_1$, and $a_4 = 5a_3 - 6a_2$.
c) Show that $a_n = 5a_{n-1} - 6a_{n-2}$ for all integers n with $n \geq 2$.

12. Show that the sequence $\{a_n\}$ is a solution of the recurrence relation $a_n = -3a_{n-1} + 4a_{n-2}$ if

a) $a_n = 0$. **b)** $a_n = 1$.
c) $a_n = (-4)^n$. **d)** $a_n = 2(-4)^n + 3$.

13. Is the sequence $\{a_n\}$ a solution of the recurrence relation $a_n = 8a_{n-1} - 16a_{n-2}$ if

a) $a_n = 0$? **b)** $a_n = 1$?
c) $a_n = 2^n$? **d)** $a_n = 4^n$?
e) $a_n = n4^n$? **f)** $a_n = 2 \cdot 4^n + 3n4^n$?
g) $a_n = (-4)^n$? **h)** $a_n = n^2 4^n$?

14. For each of these sequences find a recurrence relation satisfied by this sequence. (The answers are not unique because there are infinitely many different recurrence relations satisfied by any sequence.)

a) $a_n = 3$
b) $a_n = 2n$
c) $a_n = 2n + 3$
d) $a_n = 5^n$
e) $a_n = n^2$
f) $a_n = n^2 + n$
g) $a_n = n + (-1)^n$
h) $a_n = n!$

15. Show that the sequence $\{a_n\}$ is a solution of the recurrence relation $a_n = a_{n-1} + 2a_{n-2} + 2n - 9$ if

a) $a_n = -n + 2$.
b) $a_n = 5(-1)^n - n + 2$.
c) $a_n = 3(-1)^n + 2^n - n + 2$.
d) $a_n = 7 \cdot 2^n - n + 2$.

16. Find the solution to each of these recurrence relations with the given initial conditions. Use an iterative approach such as that used in Example 10.

a) $a_n = -a_{n-1}, a_0 = 5$
b) $a_n = a_{n-1} + 3, a_0 = 1$
c) $a_n = a_{n-1} - n, a_0 = 4$
d) $a_n = 2a_{n-1} - 3, a_0 = -1$
e) $a_n = (n + 1)a_{n-1}, a_0 = 2$
f) $a_n = 2na_{n-1}, a_0 = 3$
g) $a_n = -a_{n-1} + n - 1, a_0 = 7$

17. Find the solution to each of these recurrence relations and initial conditions. Use an iterative approach such as that used in Example 10.

a) $a_n = 3a_{n-1}, a_0 = 2$
b) $a_n = a_{n-1} + 2, a_0 = 3$
c) $a_n = a_{n-1} + n, a_0 = 1$
d) $a_n = a_{n-1} + 2n + 3, a_0 = 4$
e) $a_n = 2a_{n-1} - 1, a_0 = 1$
f) $a_n = 3a_{n-1} + 1, a_0 = 1$
g) $a_n = na_{n-1}, a_0 = 5$
h) $a_n = 2na_{n-1}, a_0 = 1$

18. A person deposits \$1000 in an account that yields 9% interest compounded annually.

a) Set up a recurrence relation for the amount in the account at the end of n years.
b) Find an explicit formula for the amount in the account at the end of n years.
c) How much money will the account contain after 100 years?

19. Suppose that the number of bacteria in a colony triples every hour.

a) Set up a recurrence relation for the number of bacteria after n hours have elapsed.
b) If 100 bacteria are used to begin a new colony, how many bacteria will be in the colony in 10 hours?

20. Assume that the population of the world in 2023 was 7.9 billion and is growing at the rate of 1.05% a year.

Links
a) Set up a recurrence relation for the population of the world n years after 2023.
b) Find an explicit formula for the population of the world n years after 2023.
c) What will the population of the world be in 2050?

21. A factory makes custom sports cars at an increasing rate. In the first month only one car is made, in the second month two cars are made, and so on, with n cars made in the nth month.

a) Set up a recurrence relation for the number of cars produced in the first n months by this factory.
b) How many cars are produced in the first year?
c) Find an explicit formula for the number of cars produced in the first n months by this factory.

22. An employee joined a company in 2023 with a starting salary of \$50,000. Every year this employee receives a raise of \$1000 plus 5% of the salary of the previous year.

a) Set up a recurrence relation for the salary of this employee n years after 2023.
b) What will the salary of this employee be in 2031?
c) Find an explicit formula for the salary of this employee n years after 2023.

23. Find a recurrence relation for the balance $B(k)$ owed at the end of k months on a loan of \$5000 at a rate of 7% if a payment of \$100 is made each month. [*Hint:* Express $B(k)$ in terms of $B(k-1)$; the monthly interest is $(0.07/12)B(k-1)$.]

24. a) Find a recurrence relation for the balance $B(k)$ owed at the end of k months on a loan at a rate of r if a payment P is made on the loan each month. [*Hint:* Express $B(k)$ in terms of $B(k-1)$ and note that the monthly interest rate is $r/12$.]

Links

b) Determine what the monthly payment P should be so that the loan is paid off after T months.

25. For each of these lists of integers, provide a simple formula or rule that generates the terms of an integer sequence that begins with the given list. Assuming that your formula or rule is correct, determine the next three terms of the sequence.

a) 1, 0, 1, 1, 0, 0, 1, 1, 1, 0, 0, 0, 1, ...
b) 1, 2, 2, 3, 4, 4, 5, 6, 6, 7, 8, 8, ...
c) 1, 0, 2, 0, 4, 0, 8, 0, 16, 0, ...
d) 3, 6, 12, 24, 48, 96, 192, ...
e) 15, 8, 1, −6, −13, −20, −27, ...
f) 3, 5, 8, 12, 17, 23, 30, 38, 47, ...
g) 2, 16, 54, 128, 250, 432, 686, ...
h) 2, 3, 7, 25, 121, 721, 5041, 40321, ...

26. For each of these lists of integers, provide a simple formula or rule that generates the terms of an integer sequence that begins with the given list. Assuming that your formula or rule is correct, determine the next three terms of the sequence.

a) 3, 6, 11, 18, 27, 38, 51, 66, 83, 102, ...
b) 7, 11, 15, 19, 23, 27, 31, 35, 39, 43, ...
c) 1, 10, 11, 100, 101, 110, 111, 1000, 1001, 1010, 1011, ...
d) 1, 2, 2, 2, 3, 3, 3, 3, 3, 5, 5, 5, 5, 5, 5, 5, ...
e) 0, 2, 8, 26, 80, 242, 728, 2186, 6560, 19682, ...
f) 1, 3, 15, 105, 945, 10395, 135135, 2027025, 34459425, ...
g) 1, 0, 0, 1, 1, 1, 0, 0, 0, 0, 1, 1, 1, 1, 1, ...
h) 2, 4, 16, 256, 65536, 4294967296, ...

****27.** Show that if a_n denotes the nth positive integer that is not a perfect square, then $a_n = n + \{\sqrt{n}\}$, where $\{x\}$ denotes the integer closest to the real number x.

***28.** Let a_n be the nth term of the sequence 1, 2, 2, 3, 3, 3, 4, 4, 4, 4, 5, 5, 5, 5, 5, 6, 6, 6, 6, 6, 6, ..., constructed by including the integer k exactly k times. Show that $a_n = \lfloor \sqrt{2n} + \frac{1}{2} \rfloor$.

29. What are the values of these sums?

a) $\sum_{k=1}^{5} (k+1)$ **b)** $\sum_{j=0}^{4} (-2)^j$

c) $\sum_{i=1}^{10} 3$ **d)** $\sum_{j=0}^{8} (2^{j+1} - 2^j)$

30. What are the values of these sums, where $S = \{1, 3, 5, 7\}$?

a) $\sum_{j \in S} j$ **b)** $\sum_{j \in S} j^2$

c) $\sum_{j \in S} (1/j)$ **d)** $\sum_{j \in S} 1$

31. What is the value of each of these sums of terms of a geometric progression?

a) $\sum_{j=0}^{8} 3 \cdot 2^j$ **b)** $\sum_{j=1}^{8} 2^j$

c) $\sum_{j=2}^{8} (-3)^j$ **d)** $\sum_{j=0}^{8} 2 \cdot (-3)^j$

32. Find the value of each of these sums.

a) $\sum_{j=0}^{8} (1 + (-1)^j)$ **b)** $\sum_{j=0}^{8} (3^j - 2^j)$

c) $\sum_{j=0}^{8} (2 \cdot 3^j + 3 \cdot 2^j)$ **d)** $\sum_{j=0}^{8} (2^{j+1} - 2^j)$

33. Compute each of these double sums.

a) $\sum_{i=1}^{2} \sum_{j=1}^{3} (i+j)$ **b)** $\sum_{i=0}^{2} \sum_{j=0}^{3} (2i+3j)$

c) $\sum_{i=1}^{3} \sum_{j=0}^{2} i$ **d)** $\sum_{i=0}^{2} \sum_{j=1}^{3} ij$

34. Compute each of these double sums.

a) $\sum_{i=1}^{3} \sum_{j=1}^{2} (i-j)$ **b)** $\sum_{i=0}^{3} \sum_{j=0}^{2} (3i+2j)$

c) $\sum_{i=1}^{3} \sum_{j=0}^{2} j$ **d)** $\sum_{i=0}^{2} \sum_{j=0}^{3} i^2 j^3$

35. Show that $\sum_{j=1}^{n} (a_j - a_{j-1}) = a_n - a_0$, where $a_0, a_1, \ldots, a_n$ is a sequence of real numbers. This type of sum is called **telescoping**.

36. Use the identity $1/(k(k+1)) = 1/k - 1/(k+1)$ and Exercise 35 to compute $\sum_{k=1}^{n} 1/(k(k+1))$.

37. Sum both sides of the identity $k^2 - (k-1)^2 = 2k - 1$ from $k = 1$ to $k = n$ and use Exercise 35 to find

a) a formula for $\sum_{k=1}^{n} (2k-1)$ (the sum of the first n odd natural numbers).

b) a formula for $\sum_{k=1}^{n} k$.

***38.** Use the technique given in Exercise 35, together with the result of Exercise 37b, to derive the formula for $\sum_{k=1}^{n} k^2$ given in Table 2. [*Hint:* Take $a_k = k^3$ in the telescoping sum in Exercise 35.]

39. Find $\sum_{k=100}^{200} k$. (Use Table 2.)

40. Find $\sum_{k=99}^{200} k^3$. (Use Table 2.)

41. Find $\sum_{k=10}^{20} k^2(k-3)$. (Use Table 2.)

42. Find $\sum_{k=10}^{20} (k-1)(2k^2+1)$. (Use Table 2.)

***43.** Find a formula for $\sum_{k=0}^{m} \lfloor \sqrt{k} \rfloor$, when m is a positive integer.

***44.** Find a formula for $\sum_{k=0}^{m} \lfloor \sqrt[3]{k} \rfloor$, when m is a positive integer.

There is also a special notation for products. The product of $a_m, a_{m+1}, \ldots, a_n$ is represented by $\prod_{j=m}^{n} a_j$, read as the product from $j = m$ to $j = n$ of a_j.

45. What are the values of the following products?

a) $\prod_{i=0}^{10} i$ **b)** $\prod_{i=5}^{8} i$

c) $\prod_{i=1}^{100} (-1)^i$ **d)** $\prod_{i=1}^{10} 2$

Recall that the value of the factorial function at a positive integer n, denoted by $n!$, is the product of the positive integers from 1 to n, inclusive. Also, we specify that $0! = 1$.

46. Express $n!$ using product notation.

47. Find $\sum_{j=0}^{4} j!$.

48. Find $\prod_{j=0}^{4} j!$.

2.5 Cardinality of Sets

2.5.1 Introduction

In Definition 4 of Section 2.1 we defined the cardinality of a finite set as the number of elements in the set. We use the cardinalities of finite sets to tell us when they have the same size, or when one is bigger than the other. In this section we extend this notion to infinite sets. That is, we will define what it means for two infinite sets to have the same cardinality, providing us with a way to measure the relative sizes of infinite sets.

We will be particularly interested in countably infinite sets, which are sets with the same cardinality as the set of positive integers. We will establish the surprising result that the set of rational numbers is countably infinite. We will also provide an example of an uncountable set when we show that the set of real numbers is not countable.

The concepts developed in this section have important applications to computer science. A function is called uncomputable if no computer program can be written to find all its values, even with unlimited time and memory. We will use the concepts in this section to explain why uncomputable functions exist.

We now define what it means for two sets to have the same size, or cardinality. In Section 2.1, we discussed the cardinality of finite sets and we defined the size, or cardinality, of such sets. In Exercise 81 of Section 2.3 we showed that there is a one-to-one correspondence between any two finite sets with the same number of elements. We use this observation to extend the concept of cardinality to all sets, both finite and infinite.

Definition 1 The sets A and B have the same *cardinality* if and only if there is a one-to-one correspondence from A to B. When A and B have the same cardinality, we write $|A| = |B|$.

For infinite sets the definition of cardinality provides a relative measure of the sizes of two sets, rather than a measure of the size of one particular set. We can also define what it means for one set to have a smaller cardinality than another set.

Definition 2 If there is a one-to-one function from A to B, the cardinality of A is less than or the same as the cardinality of B and we write $|A| \leq |B|$. Moreover, when $|A| \leq |B|$ and A and B have different cardinality, we say that the cardinality of A is less than the cardinality of B and we write $|A| < |B|$.

Remark: In Definitions 1 and 2 we introduced the notations $|A| = |B|$ and $|A| < |B|$ to denote that A and B have the same cardinality and that the cardinality of A is less than the cardinality of B. However, these definitions do not give any separate meaning to $|A|$ and $|B|$ when A and B are arbitrary infinite sets.

2.5.2 Countable Sets

We will now split infinite sets into two groups, those with the same cardinality as the set of natural numbers and those with a different cardinality.

Definition 3 A set that is either finite or has the same cardinality as the set of positive integers is called *countable.* A set that is not countable is called *uncountable.* When an infinite set S is countable, we denote the cardinality of S by $\aleph_0$ (where $\aleph$ is aleph, the first letter of the Hebrew alphabet). We write $|S| = \aleph_0$ and say that S has cardinality "aleph null."

We illustrate how to show a set is countable in the next example.

EXAMPLE 1 Show that the set of odd positive integers is a countable set.

Solution: To show that the set of odd positive integers is countable, we will exhibit a one-to-one correspondence between this set and the set of positive integers. Consider the function

$$f(n) = 2n - 1$$

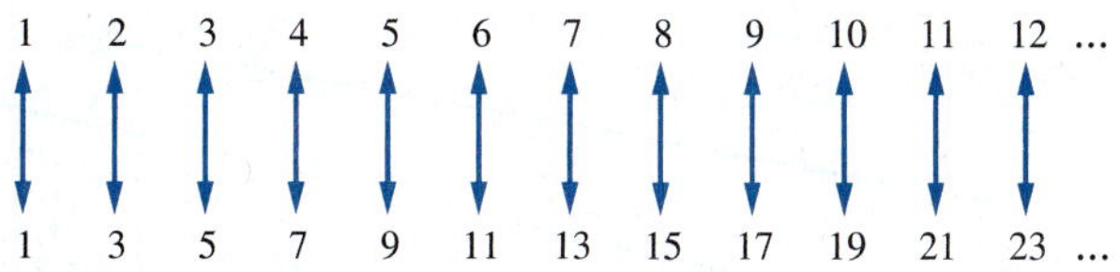

FIGURE 1 **A one-to-one correspondence between $\mathbf{Z}^+$ and the set of odd positive integers.**

from $\mathbf{Z}^+$ to the set of odd positive integers. We show that f is a one-to-one correspondence by showing that it is both one-to-one and onto. To see that it is one-to-one, suppose that $f(n) = f(m)$. Then $2n - 1 = 2m - 1$, so $n = m$. To see that it is onto, suppose that t is an odd positive integer. Then t is 1 less than an even integer $2k$, where k is a natural number. Hence $t = 2k - 1 = f(k)$. We display this one-to-one correspondence in Figure 1. ◀

An infinite set is countable if and only if it is possible to list the elements of the set in a sequence (indexed by the positive integers). The reason for this is that a one-to-one correspondence f from the set of positive integers to a set S can be expressed in terms of a sequence $a_1, a_2, \ldots, a_n, \ldots$, where $a_1 = f(1), a_2 = f(2), \ldots, a_n = f(n), \ldots$.

You can always get a room at Hilbert's Grand Hotel!

Links

HILBERT'S GRAND HOTEL We now describe a paradox that shows that something impossible with finite sets may be possible with infinite sets. The famous mathematician David Hilbert invented the notion of the **Grand Hotel**, which has a countably infinite number of rooms, each occupied by a guest. When a new guest arrives at a hotel with a finite number of rooms, and all rooms are occupied, this guest cannot be accommodated without evicting a current guest. However, we can always accommodate a new guest at the Grand Hotel, even when all rooms are already occupied, as we show in Example 2. Exercises 5 and 8 ask you to show that we can accommodate a finite number of new guests and a countable number of new guests, respectively, at the fully occupied Grand Hotel.

EXAMPLE 2 How can we accommodate a new guest arriving at the fully occupied Grand Hotel without removing any of the current guests?

Solution: Because the rooms of the Grand Hotel are countable, we can list them as Room 1, Room 2, Room 3, and so on. When a new guest arrives, we move the guest in Room 1 to Room 2, the guest in Room 2 to Room 3, and in general, the guest in Room n to Room $n + 1$, for all positive integers n. This frees up Room 1, which we assign to the new guest, and all the current guests still have rooms. We illustrate this situation in Figure 2. ◀

When there are finitely many rooms in a hotel, the notion that all rooms are occupied is equivalent to the notion that no new guests can be accommodated. However, Hilbert's paradox

Links

EMILIO SEGRE VISUAL ARCHIVES/AMERICAN INSTITUTE OF PHYSICS/Science Source

DAVID HILBERT (1862–1943) Hilbert, born in Königsberg, the city famous in mathematics for its seven bridges, was the son of a judge. During his tenure at Göttingen University, from 1892 to 1930, he made many fundamental contributions to a wide range of mathematical subjects. He almost always worked on one area of mathematics at a time, making important contributions, then moving to a new mathematical subject. Some areas in which Hilbert worked are the calculus of variations, geometry, algebra, number theory, logic, and mathematical physics. Besides his many outstanding original contributions, Hilbert is remembered for his important and influential list of 23 unsolved problems. He described these problems at the 1900 International Congress of Mathematicians, as a challenge to mathematicians at the birth of the twentieth century. Since that time, they have spurred a tremendous amount and variety of research. Although many of these problems have now been solved, several remain open, including the Riemann hypothesis, which is part of Problem 8 on Hilbert's list. Hilbert was also the author of several important textbooks in number theory and geometry.

FIGURE 2 A new guest arrives at Hilbert's Grand Hotel.

of the Grand Hotel can be explained by noting that this equivalence no longer holds when there are infinitely many rooms.

EXAMPLES OF COUNTABLE AND UNCOUNTABLE SETS We will now show that certain sets of numbers are countable. We begin with the set of all integers. Note that we can show that the set of all integers is countable by listing its members.

EXAMPLE 3 Show that the set of all integers is countable.

Solution: We can list all integers in a sequence by starting with 0 and alternating between positive and negative integers: $0, 1, -1, 2, -2, \ldots$. Alternatively, we could find a one-to-one correspondence between the set of positive integers and the set of all integers. We leave it to the reader to show that the function $f(n) = n/2$ when n is even and $f(n) = -(n-1)/2$ when n is odd is such a function. Consequently, the set of all integers is countable. ◀

It is not surprising that the set of odd integers and the set of all integers are both countable sets (as shown in Examples 1 and 3). Many people are amazed to learn that the set of rational numbers is countable, as Example 4 demonstrates.

EXAMPLE 4 Show that the set of positive rational numbers is countable.

Solution: It may seem surprising that the set of positive rational numbers is countable, but we will show how we can list the positive rational numbers as a sequence $r_1, r_2, \ldots, r_n, \ldots$. First, note that every positive rational number is the quotient p/q of two positive integers.

We can arrange the positive rational numbers by listing those with denominator $q = 1$ in the first row, those with denominator $q = 2$ in the second row, and so on, as displayed in Figure 3.

The key to listing the rational numbers in a sequence is to first list the positive rational numbers p/q with $p + q = 2$, followed by those with $p + q = 3$, followed by those with $p + q = 4$, and so on, following the path shown in Figure 3. Whenever we encounter a number p/q that is already listed, we do not list it again. For example, when we come to $2/2 = 1$ we do not list it because we have already listed $1/1 = 1$. The initial terms in the list of positive rational numbers we have constructed are $1, 1/2, 2, 3, 1/3, 1/4, 2/3, 3/2, 4, 5$, and so on. These numbers are shown circled; the uncircled numbers in the list are those we leave out because they are

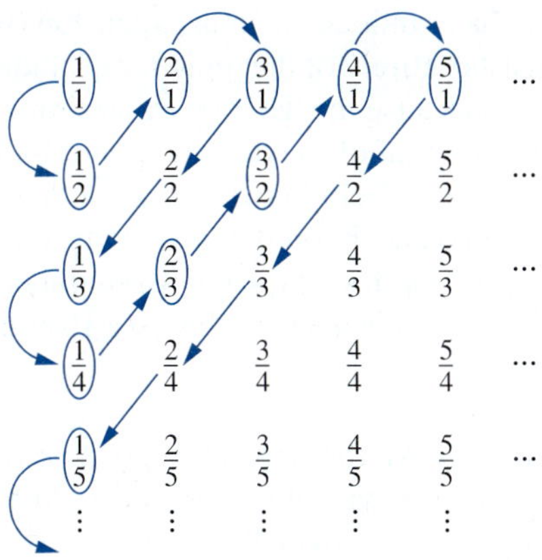

FIGURE 3 The positive rational numbers are countable.

already listed. Because all positive rational numbers are listed once, as the reader can verify, we have shown that the set of positive rational numbers is countable. ◀

2.5.3 An Uncountable Set

Not all infinite sets have the same size!

We have seen that the set of positive rational numbers is a countable set. Do we have a promising candidate for an uncountable set? The first place we might look is the set of real numbers. In Example 5 we use an important proof method, introduced in 1879 by Georg Cantor and known as the **Cantor diagonalization argument**, to prove that the set of real numbers is not countable. This proof method is used extensively in mathematical logic and in the theory of computation.

Links

EXAMPLE 5 Show that the set of real numbers is an uncountable set.

Extra Examples

Solution: To show that the set of real numbers is uncountable, we suppose that the set of real numbers is countable and arrive at a contradiction. Then, the subset of all real numbers that fall between 0 and 1 would also be countable (because any subset of a countable set is also countable; see Exercise 16). Under this assumption, the real numbers between 0 and 1 can be listed in some order, say, $r_1, r_2, r_3, \ldots$. Let the decimal representation of these real numbers be

$$\begin{aligned} r_1 &= 0.d_{11}d_{12}d_{13}d_{14} \ldots \\ r_2 &= 0.d_{21}d_{22}d_{23}d_{24} \ldots \\ r_3 &= 0.d_{31}d_{32}d_{33}d_{34} \ldots \\ r_4 &= 0.d_{41}d_{42}d_{43}d_{44} \ldots \\ &\vdots \end{aligned}$$

where $d_{ij} \in \{0, 1, 2, 3, 4, 5, 6, 7, 8, 9\}$. (For example, if $r_1 = 0.23794102\ldots$, we have $d_{11} = 2$, $d_{12} = 3$, $d_{13} = 7$, and so on.) Then, form a new real number with decimal expansion $r = 0.d_1d_2d_3d_4\ldots$, where the decimal digits are determined by the following rule:

$$d_i = \begin{cases} 4 \text{ if } d_{ii} \neq 4 \\ 5 \text{ if } d_{ii} = 4. \end{cases}$$

(As an example, suppose that $r_1 = 0.23794102\ldots$, $r_2 = 0.44590138\ldots$, $r_3 = 0.09118764\ldots$, $r_4 = 0.80553900\ldots$, and so on. Then we have $r = 0.d_1d_2d_3d_4\ldots = 0.4544\ldots$, where $d_1 = 4$ because $d_{11} \neq 4$, $d_2 = 5$ because $d_{22} = 4$, $d_3 = 4$ because $d_{33} \neq 4$, $d_4 = 4$ because $d_{44} \neq 4$, and so on.)

A number with a decimal expansion that terminates has a second decimal expansion ending with an infinite sequence of 9s because $1 = 0.999\ldots$.

Every real number has a unique decimal expansion (when the possibility that the expansion has a tail end that consists entirely of the digit 9 is excluded). Therefore, the real number r is not equal to any of $r_1, r_2, \ldots$ because the decimal expansion of r differs from the decimal expansion of r_i in the ith place to the right of the decimal point, for each i.

Because there is a real number r between 0 and 1 that is not in the list, the assumption that all the real numbers between 0 and 1 could be listed must be false. Therefore, all the real numbers between 0 and 1 cannot be listed, so the set of real numbers between 0 and 1 is uncountable. Any set with an uncountable subset is uncountable (see Exercise 15). Hence, the set of real numbers is uncountable. ◀

Remark: Another way to prove that the set of positive rational numbers is countable that produces a list of these numbers with no repetition is introduced in Section 11.2. This method generates what is known as the *Calkin-Wilf tree*.

2.5.4 Algebraic and Transcendental Numbers

We have shown that the set of rational numbers is countable and that the set of real numbers is uncountable. Next, we will give an example of a countable set which contains the set of rational numbers and we will give an example of another uncountable set. We begin with a definition.

Definition 4 An **algebraic number** α is a real or complex number that is a root of a polynomial with integer coefficients. That is, α is algebraic if there are integers $a_0, a_1, a_2, \ldots, a_n$ such that $a_0\alpha^n + a_1\alpha^{n-1} + \cdots + a_n = 0$. A **transcendental number** is a real or complex number that is not algebraic.

EXAMPLE 6 Note that every rational number p/q where p and q are integers is algebraic because $\frac{p}{q}$ is a root of $qx - p = 0$.

We see that $\sqrt{2}$ is algebraic, because it is a root of $x^2 - 2$. Similarly, $\frac{1+\sqrt{5}}{2}$ is algebraic, because it is a root of $x^2 - x - 1$. ◀

EXAMPLE 7 Examples of transcendental numbers are π and e. Proofs that these numbers are transcendental can be found in [HaWr08] and on line at https://math.colorado.edu/~rohi1040/expository/eistranscendental.pdf. ◀

We leave the proof that there are countably many algebraic numbers to the exercises (see Exercise 43). Once we know that there are countably many algebraic numbers, it follows that there are an uncountable number of transcendental numbers. If there were a countable number of transcendental numbers, then there would also be a countable number of real numbers. This follows by Theorem 1 which tells us that the union of two countable sets is countable.

2.5.5 Results About Cardinality

We will now discuss some results about the cardinality of sets. First, we will prove that the union of two countable sets is also countable.

THEOREM 1 If A and B are countable sets, then $A \cup B$ is also countable.

This proof uses WLOG and cases.

Proof: Suppose that A and B are both countable sets. Without loss of generality, we can assume that A and B are disjoint. (If they are not, we can replace B by $B - A$, because $A \cap (B - A) = \emptyset$ and $A \cup (B - A) = A \cup B$.) Furthermore, without loss of generality, if one of the two sets is countably infinite and the other finite, we can assume that B is the one that is finite.

There are three cases to consider: (i) A and B are both finite, (ii) A is infinite and B is finite, and (iii) A and B are both countably infinite.

Case (i): Note that when A and B are finite, $A \cup B$ is also finite, and therefore, countable.

Case (ii): Because A is countably infinite, its elements can be listed in an infinite sequence $a_1, a_2, a_3, \ldots, a_n, \ldots$ and because B is finite, its terms can be listed as $b_1, b_2, \ldots, b_m$ for some positive integer m. We can list the elements of $A \cup B$ as $b_1, b_2, \ldots, b_m, a_1, a_2, a_3, \ldots, a_n, \ldots$. This means that $A \cup B$ is countably infinite.

Case (iii): Because both A and B are countably infinite, we can list their elements as $a_1, a_2, a_3, \ldots, a_n, \ldots$ and $b_1, b_2, b_3, \ldots, b_n, \ldots$, respectively. By alternating terms of these two sequences we can list the elements of $A \cup B$ in the infinite sequence $a_1, b_1, a_2, b_2, a_3, b_3, \ldots, a_n, b_n, \ldots$. This means $A \cup B$ must be countably infinite.

We have completed the proof, as we have shown that $A \cup B$ is countable in all three cases. ◁

Because of its importance, we now state a key theorem in the study of cardinality.

THEOREM 2 **SCHRÖDER-BERNSTEIN THEOREM** If A and B are sets with $|A| \leq |B|$ and $|B| \leq |A|$, then $|A| = |B|$. In other words, if there are one-to-one functions f from A to B and g from B to A, then there is a one-to-one correspondence between A and B.

Because Theorem 2 seems to be quite straightforward, we might expect that it has an easy proof. However, this is not the case because when you have an injection from a set A to a set B that may not be onto, and another injection from B to A that also may not be onto, there is no obvious way to construct a bijection from A to B. Moreover, even though it has been proved in a variety of ways without using advanced mathematics, all known proofs are rather subtle and have twists and turns that are tricky to explain. One of these proofs is developed in Exercise 41 and students are invited to complete the details. Motivated readers can also find proofs in [AiZiHo09] and [Ve06]. This result is called the Schröder-Bernstein theorem after Ernst Schröder who published a flawed proof of it in 1898 and Felix Bernstein, a student of Georg Cantor, who presented a proof in 1897. However, a proof of this theorem was found in notes of Richard Dedekind dated 1887. Dedekind was a German mathematician who made important contributions to the foundations of mathematics, abstract algebra, and number theory.

We illustrate the use of Theorem 2 with an example.

EXAMPLE 8 Show that $|(0, 1)| = |(0, 1]|$.

Solution: It is not at all obvious how to find a one-to-one correspondence between $(0, 1)$ and $(0, 1]$ to show that $|(0, 1)| = |(0, 1]|$. Fortunately, we can use the Schröder-Bernstein theorem instead. Finding a one-to-one function from $(0, 1)$ to $(0, 1]$ is simple. Because $(0, 1) \subset (0, 1]$, $f(x) = x$ is a one-to-one function from $(0, 1)$ to $(0, 1]$. Finding a one-to-one function from $(0, 1]$ to $(0, 1)$ is also not difficult. The function $g(x) = x/2$ is clearly one-to-one and maps $(0, 1]$ to $(0, 1/2] \subset (0, 1)$. As we have found one-to-one functions from $(0, 1)$ to $(0, 1]$ and from $(0, 1]$ to $(0, 1)$, the Schröder-Bernstein theorem tells us that $|(0, 1)| = |(0, 1]|$. ◀

UNCOMPUTABLE FUNCTIONS We will now describe an important application of the concepts of this section to computer science. In particular, we will show that there are functions whose values cannot be computed by any computer program.

Definition 5 We say that a function is **computable** if there is a computer program in some programming language that finds the values of this function. If a function is not computable we say it is **uncomputable**.

To show that there are uncomputable functions, we need to establish two results. First, we need to show that the set of all computer programs in any particular programming language is countable. This can be proved by noting that a computer program in a particular language can be thought of as a string of characters from a finite alphabet (see Exercise 37). Next, we show that there are uncountably many different functions from a particular countably infinite set to itself. In particular, Exercise 38 shows that the set of functions from the set of positive integers to itself is uncountable. This is a consequence of the uncountability of the real numbers between 0 and 1 (see Example 5). Putting these two results together (Exercise 39) shows that there are uncomputable functions.

THE CONTINUUM HYPOTHESIS We conclude this section with a brief discussion of a famous open question about cardinality. It can be shown that the power set of $\mathbf{Z}^+$ and the set of real numbers $\mathbf{R}$ have the same cardinality (see Exercise 36). In other words, we know that $|\mathcal{P}(\mathbf{Z}^+)| = |\mathbf{R}| = \mathfrak{c}$, where $\mathfrak{c}$ denotes the cardinality of the set of real numbers.

An important theorem of Cantor (Exercise 40) states that the cardinality of a set is always less than the cardinality of its power set. Hence, $|\mathbf{Z}^+| < |\mathcal{P}(\mathbf{Z}^+)|$. We can rewrite this as $\aleph_0 < 2^{\aleph_0}$, using the notation $2^{|S|}$ to denote the cardinality of the power set of the set S. Also, note that the relationship $|\mathcal{P}(\mathbf{Z}^+)| = |\mathbf{R}|$ can be expressed as $2^{\aleph_0} = \mathfrak{c}$.

This leads us to the **continuum hypothesis**, which asserts that there is no cardinal number X between $\aleph_0$ and $\mathfrak{c}$. In other words, the continuum hypothesis states that there is no set A such that $\aleph_0$, the cardinality of the set of positive integers, is less than $|A|$ and $|A|$ is less than $\mathfrak{c}$, the cardinality of the set of real numbers. It can be shown that the smallest infinite cardinal numbers form an infinite sequence $\aleph_0 < \aleph_1 < \aleph_2 < \dots$. If we assume that the continuum hypothesis is true, it would follow that $\mathfrak{c} = \aleph_1$, so that $2^{\aleph_0} = \aleph_1$.

$\mathfrak{c}$ is the lowercase Fraktur c.

The continuum hypothesis was stated by Cantor in 1877. He labored unsuccessfully to prove it, becoming extremely dismayed that he could not. By 1900, settling the continuum hypothesis was considered to be among the most important unsolved problems in mathematics. It was the first problem posed by David Hilbert in his 1900 list of open problems in mathematics.

The continuum hypothesis is still an open question and remains an area for active research. However, it has been shown that it can be neither proved nor disproved under the standard set theory axioms in modern mathematics, the Zermelo-Fraenkel axioms. The Zermelo-Fraenkel axioms were formulated to avoid the paradoxes of naive set theory, such as Russell's paradox, but there is much controversy whether they should be replaced by some other set of axioms for set theory.

Exercises

1. Determine whether each of these sets is finite, countably infinite, or uncountable. For those that are countably infinite, exhibit a one-to-one correspondence between the set of positive integers and that set.

a) the negative integers

b) the even integers

c) the integers less than 100

d) the real numbers between 0 and $\frac{1}{2}$

e) the positive integers less than 1,000,000,000

f) the integers that are multiples of 7

2. Determine whether each of these sets is finite, countably infinite, or uncountable. For those that are countably infinite, exhibit a one-to-one correspondence between the set of positive integers and that set.

a) the integers greater than 10
b) the odd negative integers
c) the integers with absolute value less than 1,000,000
d) the real numbers between 0 and 2
e) the set $A \times \mathbf{Z}^+$ where $A = \{2, 3\}$
f) the integers that are multiples of 10

3. Determine whether each of these sets is countable or uncountable. For those that are countably infinite, exhibit a one-to-one correspondence between the set of positive integers and that set.

a) all bit strings not containing the bit 0
b) all positive rational numbers that cannot be written with denominators less than 4
c) the real numbers not containing 0 in their decimal representation
d) the real numbers containing only a finite number of 1s in their decimal representation

4. Determine whether each of these sets is countable or uncountable. For those that are countably infinite, exhibit a one-to-one correspondence between the set of positive integers and that set.

a) integers not divisible by 3
b) integers divisible by 5 but not by 7
c) the real numbers with decimal representations consisting of all 1s
d) the real numbers with decimal representations of all 1s or 9s

5. Show that a finite group of guests arriving at Hilbert's fully occupied Grand Hotel can be given rooms without evicting any current guest.

6. Suppose that Hilbert's Grand Hotel is fully occupied, but the hotel closes all the even numbered rooms for maintenance. Show that all guests can remain in the hotel.

7. Suppose that Hilbert's Grand Hotel is fully occupied on the day the hotel expands to a second building which also contains a countably infinite number of rooms. Show that the current guests can be spread out to fill every room of the two buildings of the hotel.

8. Show that a countably infinite number of guests arriving at Hilbert's fully occupied Grand Hotel can be given rooms without evicting any current guest.

***9.** Suppose that a countably infinite number of buses, each containing a countably infinite number of guests, arrive at Hilbert's fully occupied Grand Hotel. Show that all the arriving guests can be accommodated without evicting any current guest.

10. Give an example of two uncountable sets A and B such that $A - B$ is

a) finite.
b) countably infinite.
c) uncountable.

11. Give an example of two uncountable sets A and B such that $A \cap B$ is

a) finite.
b) countably infinite.
c) uncountable.

12. Show that if A and B are sets and $A \subset B$ then $|A| \leq |B|$.

13. Explain why the set A is countable if and only if $|A| \leq |\mathbf{Z}^+|$.

14. Show that if A and B are sets with the same cardinality, then $|A| \leq |B|$ and $|B| \leq |A|$.

15. Show that if A and B are sets, A is uncountable, and $A \subseteq B$, then B is uncountable.

16. Show that a subset of a countable set is also countable.

17. If A is an uncountable set and B is a countable set, must $A - B$ be uncountable?

18. Show that if A and B are sets $|A| = |B|$, then $|\mathcal{P}(A)| = |\mathcal{P}(B)|$.

19. Show that if A, B, C, and D are sets with $|A| = |B|$ and $|C| = |D|$, then $|A \times C| = |B \times D|$.

20. Show that if $|A| = |B|$ and $|B| = |C|$, then $|A| = |C|$.

21. Show that if A, B, and C are sets such that $|A| \leq |B|$ and $|B| \leq |C|$, then $|A| \leq |C|$.

22. Suppose that A is a countable set. Show that the set B is also countable if there is an onto function f from A to B.

23. Show that if A is an infinite set, then it contains a countably infinite subset.

24. Show that there is no infinite set A such that $|A| < |\mathbf{Z}^+| = \aleph_0$.

25. Prove that if it is possible to label each element of an infinite set S with a finite string of keyboard characters, from a finite list characters, where no two elements of S have the same label, then S is a countably infinite set.

26. Use Exercise 25 to provide a proof different from that in the text that the set of rational numbers is countable. [*Hint:* Show that you can express a rational number as a string of digits with a slash and possibly a minus sign.]

***27.** Show that the union of a countable number of countable sets is countable.

28. Show that the set $\mathbf{Z}^+ \times \mathbf{Z}^+$ is countable.

***29.** Show that the set of all finite bit strings is countable.

***30.** Show that the set of real numbers that are solutions of quadratic equations $ax^2 + bx + c = 0$, where a, b, and c are integers, is countable.

***31.** Show that $\mathbf{Z}^+ \times \mathbf{Z}^+$ is countable by showing that the polynomial function $f : \mathbf{Z}^+ \times \mathbf{Z}^+ \to \mathbf{Z}^+$ with $f(m, n) = (m + n - 2)(m + n - 1)/2 + m$ is one-to-one and onto.

***32.** Show that when you substitute $(3n + 1)^2$ for each occurrence of n and $(3m + 1)^2$ for each occurrence of m in the right-hand side of the formula for the function $f(m, n)$ in Exercise 31, you obtain a one-to-one polynomial function $\mathbf{Z} \times \mathbf{Z} \to \mathbf{Z}$. It is an open question whether there is a one-to-one polynomial function $\mathbf{Q} \times \mathbf{Q} \to \mathbf{Q}$.

33. Use the Schröder-Bernstein theorem to show that (0, 1) and [0, 1] have the same cardinality.

34. Show that (0, 1) and $\mathbf{R}$ have the same cardinality by

a) showing that $f(x) = \frac{2x-1}{2x(1-x)}$ is a bijection from (0, 1) to $\mathbf{R}$.
b) using the Schröder-Bernstein theorem.

35. Show that there is no one-to-one correspondence from the set of positive integers to the power set of the set of positive integers. [*Hint:* Assume that there is such a one-to-one correspondence. Represent a subset of the set of positive integers as an infinite bit string with ith bit 1 if i belongs to the subset and 0 otherwise. Suppose that you can list these infinite strings in a sequence indexed by the positive integers. Construct a new bit string with its ith bit equal to the complement of the ith bit of the ith string in the list. Show that this new bit string cannot appear in the list.]

***36.** Show that there is a one-to-one correspondence from the set of subsets of the positive integers to the set real numbers between 0 and 1. Use this result and Exercises 34 and 35 to conclude that $\aleph_0 < |\mathcal{P}(\mathbf{Z}^+)| = |\mathbf{R}|$. [*Hint:* Look at the first part of the hint for Exercise 35.]

***37.** Show that the set of all computer programs in a particular programming language is countable. [*Hint:* A computer program written in a programming language can be thought of as a string of symbols from a finite alphabet.]

***38.** Show that the set of functions from the positive integers to the set $\{0, 1, 2, 3, 4, 5, 6, 7, 8, 9\}$ is uncountable. [*Hint:* First set up a one-to-one correspondence between the set of real numbers between 0 and 1 and a subset of these functions. Do this by associating to the real number $0.d_1d_2 \ldots d_n \ldots$ the function f with $f(n) = d_n$.]

***39.** We say that a function is **computable** if there is a computer program that finds the values of this function. Use Exercises 37 and 38 to show that there are functions that are not computable.

***40.** Show that if S is a set, then there does not exist an onto function f from S to $\mathcal{P}(S)$, the power set of S. Conclude that $|S| < |\mathcal{P}(S)|$. This result is known as **Cantor's theorem**. [*Hint:* Suppose such a function f existed. Let $T = \{s \in S \mid s \notin f(s)\}$ and show that no element s can exist for which $f(s) = T$.]

***41.** In this exercise, we prove the Schröder-Bernstein theorem. Suppose that A and B are sets where $|A| \leq |B|$ and $|B| \leq |A|$. This means that there are injections $f : A \to B$ and $g : B \to A$. To prove the theorem, we must show that there is a bijection $h : A \to B$, implying that $|A| = |B|$.

To build h, we construct the *chain* of an element $a \in A$. This chain contains the elements $a, f(a), g(f(a)), f(g(f(a))), g(f(g(f(a)))), \ldots$. It also may contain more elements that precede a, extending the chain backwards. So, if there is a $b \in B$ with $g(b) = a$, then b will be the term of the chain just before a. Because g may not be a surjection, there may not be any such b, so that a is the first element of the chain. If such a b exists, because g is an injection, it is the unique element of B mapped by g to a; we denote it by $g^{-1}(a)$. (Note that this defines g^{-1} as a partial function from B to A.) We extend the chain backwards as long as possible in the same way, adding $f^{-1}(g^{-1}(a)), g^{-1}(f^{-1}(g^{-1}(a))), \ldots$. To construct the proof, complete these five parts.

a) Show that every element in A or in B belongs to exactly one chain.

b) Show that there are four types of chains: chains that form a loop, that is, carrying them forward from every element in the chain will eventually return to this element (type 1), chains that go backwards without stopping (type 2), chains that go backwards and end in the set A (type 3), and chains that go backwards and end in the set B (type 4).

c) We now define a function $h : A \to B$. We set $h(a) = f(a)$ when a belongs to a chain of type 1, 2, or 3. Show that we can define $h(a)$ when a is in a chain of type 4, by taking $h(a) = g^{-1}(a)$. In parts (d) and (e), we show that this function is a bijection from A to B, proving the theorem.

d) Show that h is one-to-one. (You can consider chains of types 1, 2, and 3 together, but chains of type 4 separately.)

e) Show that h is onto. (You need to consider chains of types 1, 2, and 3 together, but chains of type 4 separately.)

42. Show that the union of a countable number of finite sets is countable.

43. Show that the set of algebraic numbers is countable using the fact that a polynomial of degree n with real coefficients has most n distinct roots. (*Hint*: Show that for every positive integer N that there are finitely many polynomials with integer coefficients $a_0x^n + a_1x^{n-1} + \cdots + a_n$ with $N = n + |a_0| + |a_1| + \cdots + |a_n|$ and hence only finitely many numbers that are solutions of one of these equations. Then use Exercise 42.)

2.6 Matrices

2.6.1 Introduction

Matrices are used throughout discrete mathematics to express relationships between elements in sets. In subsequent chapters we will use matrices in a wide variety of models. For instance, matrices will be used in models of communications networks and transportation systems. Many algorithms will be developed that use these matrix models. This section reviews matrix arithmetic that will be used in these algorithms.

Definition 1 A *matrix* is a rectangular array of numbers. A matrix with m rows and n columns is called an $m \times n$ matrix. The plural of matrix is *matrices*. A matrix with the same number of rows as columns is called *square*. Two matrices are *equal* if they have the same number of rows and the same number of columns and the corresponding entries in every position are equal.

EXAMPLE 1 The matrix $\begin{bmatrix} 1 & 1 \\ 0 & 2 \\ 1 & 3 \end{bmatrix}$ is a 3×2 matrix. ◀

We now introduce some terminology about matrices. Boldface uppercase letters will be used to represent matrices.

Definition 2 Let m and n be positive integers and let

$$\mathbf{A} = \begin{bmatrix} a_{11} & a_{12} & \cdots & a_{1n} \\ a_{21} & a_{22} & \cdots & a_{2n} \\ \cdot & \cdot & & \cdot \\ \cdot & \cdot & & \cdot \\ \cdot & \cdot & & \cdot \\ a_{m1} & a_{m2} & \cdots & a_{mn} \end{bmatrix}.$$

The ith *row* of $\mathbf{A}$ is the $1 \times n$ matrix $[a_{i1}, a_{i2}, \ldots, a_{in}]$. The jth *column* of $\mathbf{A}$ is the $m \times 1$ matrix

$$\begin{bmatrix} a_{1j} \\ a_{2j} \\ \cdot \\ \cdot \\ \cdot \\ a_{mj} \end{bmatrix}.$$

The (i, j)th *element* or *entry* of $\mathbf{A}$ is the element a_{ij}, that is, the number in the ith row and jth column of $\mathbf{A}$. A convenient shorthand notation for expressing the matrix $\mathbf{A}$ is to write $\mathbf{A} = [a_{ij}]$, which indicates that $\mathbf{A}$ is the matrix with its (i, j)th element equal to a_{ij}.

2.6.2 Matrix Arithmetic

The basic operations of matrix arithmetic will now be discussed, beginning with a definition of matrix addition.

Definition 3 Let $\mathbf{A} = [a_{ij}]$ and $\mathbf{B} = [b_{ij}]$ be $m \times n$ matrices. The *sum* of $\mathbf{A}$ and $\mathbf{B}$, denoted by $\mathbf{A} + \mathbf{B}$, is the $m \times n$ matrix that has $a_{ij} + b_{ij}$ as its (i, j)th element. In other words, $\mathbf{A} + \mathbf{B} = [a_{ij} + b_{ij}]$.

The sum of two matrices of the same size is obtained by adding elements in the corresponding positions. Matrices of different sizes cannot be added, because such matrices will not both have entries in some of their positions.

EXAMPLE 2 We have $\begin{bmatrix} 1 & 0 & -1 \\ 2 & 2 & -3 \\ 3 & 4 & 0 \end{bmatrix} + \begin{bmatrix} 3 & 4 & -1 \\ 1 & -3 & 0 \\ -1 & 1 & 2 \end{bmatrix} = \begin{bmatrix} 4 & 4 & -2 \\ 3 & -1 & -3 \\ 2 & 5 & 2 \end{bmatrix}$. ◀

We now discuss matrix products. A product of two matrices is defined only when the number of columns in the first matrix equals the number of rows of the second matrix.

Definition 4 Let **A** be an $m \times k$ matrix and **B** be a $k \times n$ matrix. The *product* of **A** and **B**, denoted by **AB**, is the $m \times n$ matrix with its (i, j)th entry equal to the sum of the products of the corresponding elements from the ith row of **A** and the jth column of **B**. In other words, if $\mathbf{AB} = [c_{ij}]$, then

$$c_{ij} = a_{i1}b_{1j} + a_{i2}b_{2j} + \cdots + a_{ik}b_{kj}.$$

In Figure 1 the colored row of **A** and the colored column of **B** are used to compute the element c_{ij} of **AB**. The product of two matrices is not defined when the number of columns in the first matrix and the number of rows in the second matrix are not the same.

We now give some examples of matrix products.

EXAMPLE 3 Let

$$\mathbf{A} = \begin{bmatrix} 1 & 0 & 4 \\ 2 & 1 & 1 \\ 3 & 1 & 0 \\ 0 & 2 & 2 \end{bmatrix} \quad \text{and} \quad \mathbf{B} = \begin{bmatrix} 2 & 4 \\ 1 & 1 \\ 3 & 0 \end{bmatrix}.$$

Find **AB** if it is defined.

Extra Examples

Solution: Because **A** is a 4×3 matrix and **B** is a 3×2 matrix, the product **AB** is defined and is a 4×2 matrix. To find the elements of **AB**, the corresponding elements of the rows of **A** and the columns of **B** are first multiplied and then these products are added. For instance, the element in the (3, 1)th position of **AB** is the sum of the products of the corresponding elements of the third row of **A** and the first column of **B**; namely, $3 \cdot 2 + 1 \cdot 1 + 0 \cdot 3 = 7$. When all the elements of **AB** are computed, we see that

$$\mathbf{AB} = \begin{bmatrix} 14 & 4 \\ 8 & 9 \\ 7 & 13 \\ 8 & 2 \end{bmatrix}.$$

◀

Although matrix multiplication is associative, as can easily be proved using the associativity of addition and multiplication of real numbers, matrix multiplication is *not* commutative. That

$$\begin{bmatrix} a_{11} & a_{12} & \cdots & a_{1k} \\ a_{21} & a_{22} & \cdots & a_{2k} \\ \vdots & \vdots & & \vdots \\ a_{i1} & a_{i2} & \cdots & a_{ik} \\ \vdots & \vdots & & \vdots \\ a_{m1} & a_{m2} & \cdots & a_{mk} \end{bmatrix} \begin{bmatrix} b_{11} & b_{12} & \cdots & b_{1j} & \cdots & b_{1n} \\ b_{21} & b_{22} & \cdots & b_{2j} & \cdots & b_{2n} \\ \vdots & \vdots & & \vdots & & \vdots \\ b_{k1} & b_{k2} & \cdots & b_{kj} & \cdots & b_{kn} \end{bmatrix} = \begin{bmatrix} c_{11} & c_{12} & \cdots & c_{1n} \\ c_{21} & c_{22} & \cdots & c_{2n} \\ \vdots & \vdots & c_{ij} & \vdots \\ c_{m1} & c_{m2} & \cdots & c_{mn} \end{bmatrix}$$

FIGURE 1 The product of A = $[a_{ij}]$ and B = $[b_{ij}]$.

is, if **A** and **B** are two matrices, it is not necessarily true that **AB** and **BA** are the same. In fact, it may be that only one of these two products is defined. For instance, if **A** is 2×3 and **B** is 3×4, then **AB** is defined and is 2×4; however, **BA** is not defined, because it is impossible to multiply a 3×4 matrix and a 2×3 matrix.

In general, suppose that **A** is an $m \times n$ matrix and **B** is an $r \times s$ matrix. Then **AB** is defined only when $n = r$ and **BA** is defined only when $s = m$. Moreover, even when **AB** and **BA** are both defined, they will not be the same size unless $m = n = r = s$. Hence, if both **AB** and **BA** are defined and are the same size, then both **A** and **B** must be square and of the same size. Furthermore, even with **A** and **B** both $n \times n$ matrices, **AB** and **BA** are not necessarily equal, as Example 4 demonstrates.

EXAMPLE 4 Let

$$\mathbf{A} = \begin{bmatrix} 1 & 1 \\ 2 & 1 \end{bmatrix} \quad \text{and} \quad \mathbf{B} = \begin{bmatrix} 2 & 1 \\ 1 & 1 \end{bmatrix}.$$

Does $\mathbf{AB} = \mathbf{BA}$?

Solution: We find that

$$\mathbf{AB} = \begin{bmatrix} 3 & 2 \\ 5 & 3 \end{bmatrix} \quad \text{and} \quad \mathbf{BA} = \begin{bmatrix} 4 & 3 \\ 3 & 2 \end{bmatrix}.$$

Hence, $\mathbf{AB} \neq \mathbf{BA}$. ◀

2.6.3 Transposes and Powers of Matrices

We now introduce an important matrix with entries that are zeros and ones.

Definition 5 The *identity matrix of order n* is the $n \times n$ matrix $\mathbf{I}_n = [\delta_{ij}]$, (the *Kronecker delta*) where $\delta_{ij} = 1$ if $i = j$ and $\delta_{ij} = 0$ if $i \neq j$. Hence,

$$\mathbf{I}_n = \begin{bmatrix} 1 & 0 & \dots & 0 \\ 0 & 1 & \dots & 0 \\ \cdot & \cdot & & \cdot \\ \cdot & \cdot & & \cdot \\ \cdot & \cdot & & \cdot \\ 0 & 0 & \dots & 1 \end{bmatrix}.$$

Multiplying a matrix by an appropriately sized identity matrix does not change this matrix. In other words, when **A** is an $m \times n$ matrix, we have

$$\mathbf{AI}_n = \mathbf{I}_m\mathbf{A} = \mathbf{A}.$$

Powers of square matrices can be defined because matrix multiplication is associative. When **A** is an $n \times n$ matrix, we have

$$\mathbf{A}^0 = \mathbf{I}_n, \qquad \mathbf{A}^r = \underbrace{\mathbf{AAA} \cdots \mathbf{A}}_{r \text{ times}}.$$

The operation of interchanging the rows and columns of a square matrix arises in many contexts.

Definition 6 Let $\mathbf{A} = [a_{ij}]$ be an $m \times n$ matrix. The *transpose* of $\mathbf{A}$, denoted by $\mathbf{A}^t$, is the $n \times m$ matrix obtained by interchanging the rows and columns of $\mathbf{A}$. In other words, if $\mathbf{A}^t = [b_{ij}]$, then $b_{ij} = a_{ji}$ for $i = 1, 2, \ldots, n$ and $j = 1, 2, \ldots, m$.

EXAMPLE 5 The transpose of the matrix $\begin{bmatrix} 1 & 2 & 3 \\ 4 & 5 & 6 \end{bmatrix}$ is the matrix $\begin{bmatrix} 1 & 4 \\ 2 & 5 \\ 3 & 6 \end{bmatrix}$. ◀

Matrices that do not change when their rows and columns are interchanged are often important.

Definition 7 A square matrix $\mathbf{A}$ is called *symmetric* if $\mathbf{A} = \mathbf{A}^t$. Thus, $\mathbf{A} = [a_{ij}]$ is symmetric if $a_{ij} = a_{ji}$ for all i and j with $1 \le i \le n$ and $1 \le j \le n$.

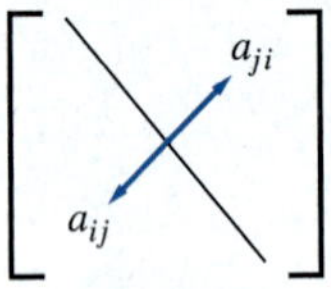

FIGURE 2 A symmetric matrix.

Note that a matrix is symmetric if and only if it is square and it is symmetric with respect to its main diagonal (which consists of entries that are in the ith row and ith column for some i). This symmetry is displayed in Figure 2.

EXAMPLE 6 The matrix $\begin{bmatrix} 1 & 1 & 0 \\ 1 & 0 & 1 \\ 0 & 1 & 0 \end{bmatrix}$ is symmetric. ◀

2.6.4 Zero–One Matrices

A matrix all of whose entries are either 0 or 1 is called a **zero–one matrix**. Zero–one matrices are often used to represent discrete structures, as we will see in Chapters 9 and 10. Algorithms using these structures are based on Boolean arithmetic with zero–one matrices. This arithmetic is based on the Boolean operations $\wedge$ and $\vee$, which operate on pairs of bits, defined by

$$b_1 \wedge b_2 = \begin{cases} 1 & \text{if } b_1 = b_2 = 1 \\ 0 & \text{otherwise,} \end{cases}$$

$$b_1 \vee b_2 = \begin{cases} 1 & \text{if } b_1 = 1 \text{ or } b_2 = 1 \\ 0 & \text{otherwise.} \end{cases}$$

Definition 8 Let $\mathbf{A} = [a_{ij}]$ and $\mathbf{B} = [b_{ij}]$ be $m \times n$ zero–one matrices. Then the *join* of $\mathbf{A}$ and $\mathbf{B}$ is the zero–one matrix with (i, j)th entry $a_{ij} \vee b_{ij}$. The join of $\mathbf{A}$ and $\mathbf{B}$ is denoted by $\mathbf{A} \vee \mathbf{B}$. The *meet* of $\mathbf{A}$ and $\mathbf{B}$ is the zero–one matrix with (i, j)th entry $a_{ij} \wedge b_{ij}$. The meet of $\mathbf{A}$ and $\mathbf{B}$ is denoted by $\mathbf{A} \wedge \mathbf{B}$.

EXAMPLE 7 Find the join and meet of the zero–one matrices

$$\mathbf{A} = \begin{bmatrix} 1 & 0 & 1 \\ 0 & 1 & 0 \end{bmatrix}, \qquad \mathbf{B} = \begin{bmatrix} 0 & 1 & 0 \\ 1 & 1 & 0 \end{bmatrix}.$$

Solution: We find that the join of $\mathbf{A}$ and $\mathbf{B}$ is

$$\mathbf{A} \vee \mathbf{B} = \begin{bmatrix} 1 \vee 0 & 0 \vee 1 & 1 \vee 0 \\ 0 \vee 1 & 1 \vee 1 & 0 \vee 0 \end{bmatrix} = \begin{bmatrix} 1 & 1 & 1 \\ 1 & 1 & 0 \end{bmatrix}.$$

The meet of $\mathbf{A}$ and $\mathbf{B}$ is

$$\mathbf{A} \wedge \mathbf{B} = \begin{bmatrix} 1 \wedge 0 & 0 \wedge 1 & 1 \wedge 0 \\ 0 \wedge 1 & 1 \wedge 1 & 0 \wedge 0 \end{bmatrix} = \begin{bmatrix} 0 & 0 & 0 \\ 0 & 1 & 0 \end{bmatrix}.$$

◀

We now define the **Boolean product** of two matrices.

Definition 9 Let $\mathbf{A} = [a_{ij}]$ be an $m \times k$ zero–one matrix and $\mathbf{B} = [b_{ij}]$ be a $k \times n$ zero–one matrix. Then the *Boolean product* of $\mathbf{A}$ and $\mathbf{B}$, denoted by $\mathbf{A} \odot \mathbf{B}$, is the $m \times n$ matrix with (i, j)th entry c_{ij} where

$$c_{ij} = (a_{i1} \wedge b_{1j}) \vee (a_{i2} \wedge b_{2j}) \vee \cdots \vee (a_{ik} \wedge b_{kj}).$$

Note that the Boolean product of $\mathbf{A}$ and $\mathbf{B}$ is obtained in an analogous way to the ordinary product of these matrices, but with addition replaced with the operation $\vee$ and with multiplication replaced with the operation $\wedge$. We give an example of the Boolean products of matrices.

EXAMPLE 8 Find the Boolean product of $\mathbf{A}$ and $\mathbf{B}$, where

$$\mathbf{A} = \begin{bmatrix} 1 & 0 \\ 0 & 1 \\ 1 & 0 \end{bmatrix}, \qquad \mathbf{B} = \begin{bmatrix} 1 & 1 & 0 \\ 0 & 1 & 1 \end{bmatrix}.$$

Solution: The Boolean product $\mathbf{A} \odot \mathbf{B}$ is given by

$$\mathbf{A} \odot \mathbf{B} = \begin{bmatrix} (1 \wedge 1) \vee (0 \wedge 0) & (1 \wedge 1) \vee (0 \wedge 1) & (1 \wedge 0) \vee (0 \wedge 1) \\ (0 \wedge 1) \vee (1 \wedge 0) & (0 \wedge 1) \vee (1 \wedge 1) & (0 \wedge 0) \vee (1 \wedge 1) \\ (1 \wedge 1) \vee (0 \wedge 0) & (1 \wedge 1) \vee (0 \wedge 1) & (1 \wedge 0) \vee (0 \wedge 1) \end{bmatrix}$$

$$= \begin{bmatrix} 1 \vee 0 & 1 \vee 0 & 0 \vee 0 \\ 0 \vee 0 & 0 \vee 1 & 0 \vee 1 \\ 1 \vee 0 & 1 \vee 0 & 0 \vee 0 \end{bmatrix}$$

$$= \begin{bmatrix} 1 & 1 & 0 \\ 0 & 1 & 1 \\ 1 & 1 & 0 \end{bmatrix}.$$

◀

We can also define the Boolean powers of a square zero–one matrix. These powers will be used in our subsequent studies of paths in graphs, which are used to model such things as communications paths in computer networks.

Definition 10 Let **A** be a square zero–one matrix and let r be a positive integer. The rth *Boolean power* of **A** is the Boolean product of r factors of **A**. The rth Boolean product of **A** is denoted by $\mathbf{A}^{[r]}$. Hence,

$$\mathbf{A}^{[r]} = \underbrace{\mathbf{A} \odot \mathbf{A} \odot \mathbf{A} \odot \cdots \odot \mathbf{A}}_{r \text{ times}}.$$

(This is well defined because the Boolean product of matrices is associative.) We also define $\mathbf{A}^{[0]}$ to be $\mathbf{I}_n$.

EXAMPLE 9 Let $\mathbf{A} = \begin{bmatrix} 0 & 0 & 1 \\ 1 & 0 & 0 \\ 1 & 1 & 0 \end{bmatrix}$. Find $\mathbf{A}^{[n]}$ for all positive integers n.

Solution: We find that

$$\mathbf{A}^{[2]} = \mathbf{A} \odot \mathbf{A} = \begin{bmatrix} 1 & 1 & 0 \\ 0 & 0 & 1 \\ 1 & 0 & 1 \end{bmatrix}.$$

We also find that

$$\mathbf{A}^{[3]} = \mathbf{A}^{[2]} \odot \mathbf{A} = \begin{bmatrix} 1 & 0 & 1 \\ 1 & 1 & 0 \\ 1 & 1 & 1 \end{bmatrix}, \qquad \mathbf{A}^{[4]} = \mathbf{A}^{[3]} \odot \mathbf{A} = \begin{bmatrix} 1 & 1 & 1 \\ 1 & 0 & 1 \\ 1 & 1 & 1 \end{bmatrix}.$$

Additional computation shows that

$$\mathbf{A}^{[5]} = \begin{bmatrix} 1 & 1 & 1 \\ 1 & 1 & 1 \\ 1 & 1 & 1 \end{bmatrix}.$$

The reader can now see that $\mathbf{A}^{[n]} = \mathbf{A}^{[5]}$ for all positive integers n with $n \geq 5$. ◀

Exercises

1. Let $\mathbf{A} = \begin{bmatrix} 1 & 1 & 1 & 3 \\ 2 & 0 & 4 & 6 \\ 1 & 1 & 3 & 7 \end{bmatrix}$.

a) What size is **A**?

b) What is the third column of **A**?

c) What is the second row of **A**?

d) What is the element of **A** in the (3, 2)th position?

e) What is $\mathbf{A}^t$?

2. Find **A** + **B**, where

a) $\mathbf{A} = \begin{bmatrix} 1 & 0 & 4 \\ -1 & 2 & 2 \\ 0 & -2 & -3 \end{bmatrix}$, $\mathbf{B} = \begin{bmatrix} -1 & 3 & 5 \\ 2 & 2 & -3 \\ 2 & -3 & 0 \end{bmatrix}$.

b) $\mathbf{A} = \begin{bmatrix} -1 & 0 & 5 & 6 \\ -4 & -3 & 5 & -2 \end{bmatrix}$, $\mathbf{B} = \begin{bmatrix} -3 & 9 & -3 & 4 \\ 0 & -2 & -1 & 2 \end{bmatrix}$.

3. Find **AB** if

a) $\mathbf{A} = \begin{bmatrix} 2 & 1 \\ 3 & 2 \end{bmatrix}$, $\mathbf{B} = \begin{bmatrix} 0 & 4 \\ 1 & 3 \end{bmatrix}$.

b) $\mathbf{A} = \begin{bmatrix} 1 & -1 \\ 0 & 1 \\ 2 & 3 \end{bmatrix}$, $\mathbf{B} = \begin{bmatrix} 3 & -2 & -1 \\ 1 & 0 & 2 \end{bmatrix}$.

c) $\mathbf{A} = \begin{bmatrix} 4 & -3 \\ 3 & -1 \\ 0 & -2 \\ -1 & 5 \end{bmatrix}$, $\mathbf{B} = \begin{bmatrix} -1 & 3 & 2 & -2 \\ 0 & -1 & 4 & -3 \end{bmatrix}$.

4. Find the product **AB**, where

a) $\mathbf{A} = \begin{bmatrix} 1 & 0 & 1 \\ 0 & -1 & -1 \\ -1 & 1 & 0 \end{bmatrix}$, $\mathbf{B} = \begin{bmatrix} 0 & 1 & -1 \\ 1 & -1 & 0 \\ -1 & 0 & 1 \end{bmatrix}$.

b) $\mathbf{A} = \begin{bmatrix} 1 & -3 & 0 \\ 1 & 2 & 2 \\ 2 & 1 & -1 \end{bmatrix}$, $\mathbf{B} = \begin{bmatrix} 1 & -1 & 2 & 3 \\ -1 & 0 & 3 & -1 \\ -3 & -2 & 0 & 2 \end{bmatrix}$.

c) $\mathbf{A} = \begin{bmatrix} 0 & -1 \\ 7 & 2 \\ -4 & -3 \end{bmatrix}$, $\mathbf{B} = \begin{bmatrix} 4 & -1 & 2 & 3 & 0 \\ -2 & 0 & 3 & 4 & 1 \end{bmatrix}$.

5. Find a matrix **A** such that

$$\begin{bmatrix} 2 & 3 \\ 1 & 4 \end{bmatrix} \mathbf{A} = \begin{bmatrix} 3 & 0 \\ 1 & 2 \end{bmatrix}.$$

[*Hint:* Finding **A** requires that you solve systems of linear equations.]

6. Find a matrix **A** such that

$$\begin{bmatrix} 1 & 3 & 2 \\ 2 & 1 & 1 \\ 4 & 0 & 3 \end{bmatrix} \mathbf{A} = \begin{bmatrix} 7 & 1 & 3 \\ 1 & 0 & 3 \\ -1 & -3 & 7 \end{bmatrix}.$$

7. Let **A** be an $m \times n$ matrix and let **0** be the $m \times n$ matrix that has all entries equal to zero. Show that $\mathbf{A} = \mathbf{0} + \mathbf{A} = \mathbf{A} + \mathbf{0}$.

8. Show that matrix addition is commutative; that is, show that if **A** and **B** are both $m \times n$ matrices, then $\mathbf{A} + \mathbf{B} = \mathbf{B} + \mathbf{A}$.

9. Show that matrix addition is associative; that is, show that if **A**, **B**, and **C** are all $m \times n$ matrices, then $\mathbf{A} + (\mathbf{B} + \mathbf{C}) = (\mathbf{A} + \mathbf{B}) + \mathbf{C}$.

10. Let **A** be a 3×4 matrix, **B** be a 4×5 matrix, and **C** be a 4×4 matrix. Determine which of the following products are defined and find the size of those that are defined.

a) **AB** b) **BA** c) **AC**
d) **CA** e) **BC** f) **CB**

11. What do we know about the sizes of the matrices **A** and **B** if both of the products **AB** and **BA** are defined?

12. In this exercise we show that matrix multiplication is distributive over matrix addition.

a) Suppose that **A** and **B** are $m \times k$ matrices and that **C** is a $k \times n$ matrix. Show that $(\mathbf{A} + \mathbf{B})\mathbf{C} = \mathbf{AC} + \mathbf{BC}$.

b) Suppose that **C** is an $m \times k$ matrix and that **A** and **B** are $k \times n$ matrices. Show that $\mathbf{C}(\mathbf{A} + \mathbf{B}) = \mathbf{CA} + \mathbf{CB}$.

13. In this exercise we show that matrix multiplication is associative. Suppose that **A** is an $m \times p$ matrix, **B** is a $p \times k$ matrix, and **C** is a $k \times n$ matrix. Show that $\mathbf{A}(\mathbf{BC}) = (\mathbf{AB})\mathbf{C}$.

14. The $n \times n$ matrix $\mathbf{A} = [a_{ij}]$ is called a **diagonal matrix** if $a_{ij} = 0$ when $i \neq j$. Show that the product of two $n \times n$ diagonal matrices is again a diagonal matrix. Give a simple rule for determining this product.

15. Let

$$\mathbf{A} = \begin{bmatrix} 1 & 1 \\ 0 & 1 \end{bmatrix}.$$

Find a formula for $\mathbf{A}^n$, whenever n is a positive integer.

16. Show that $(\mathbf{A}^t)^t = \mathbf{A}$.

17. Let **A** and **B** be two $n \times n$ matrices. Show that

a) $(\mathbf{A} + \mathbf{B})^t = \mathbf{A}^t + \mathbf{B}^t$.

b) $(\mathbf{AB})^t = \mathbf{B}^t\mathbf{A}^t$.

If **A** and **B** are $n \times n$ matrices with $\mathbf{AB} = \mathbf{BA} = \mathbf{I}_n$, then **B** is called the **inverse** of **A** (this terminology is appropriate because such a matrix **B** is unique) and **A** is said to be **invertible**. The notation $\mathbf{B} = \mathbf{A}^{-1}$ denotes that **B** is the inverse of **A**.

18. Show that

$$\begin{bmatrix} 2 & 3 & -1 \\ 1 & 2 & 1 \\ -1 & -1 & 3 \end{bmatrix}$$

is the inverse of

$$\begin{bmatrix} 7 & -8 & 5 \\ -4 & 5 & -3 \\ 1 & -1 & 1 \end{bmatrix}.$$

19. Let **A** be the 2×2 matrix

$$\mathbf{A} = \begin{bmatrix} a & b \\ c & d \end{bmatrix}.$$

Show that if $ad - bc \neq 0$, then

$$\mathbf{A}^{-1} = \begin{bmatrix} \dfrac{d}{ad - bc} & \dfrac{-b}{ad - bc} \\ \dfrac{-c}{ad - bc} & \dfrac{a}{ad - bc} \end{bmatrix}.$$

20. Let

$$\mathbf{A} = \begin{bmatrix} -1 & 2 \\ 1 & 3 \end{bmatrix}.$$

a) Find $\mathbf{A}^{-1}$. [*Hint:* Use Exercise 19.]

b) Find $\mathbf{A}^3$.

c) Find $(\mathbf{A}^{-1})^3$.

d) Use your answers to (b) and (c) to show that $(\mathbf{A}^{-1})^3$ is the inverse of $\mathbf{A}^3$.

21. Let **A** be an invertible matrix. Show that $(\mathbf{A}^n)^{-1} = (\mathbf{A}^{-1})^n$ whenever n is a positive integer.

22. Let **A** be a matrix. Show that the matrix $\mathbf{AA}^t$ is symmetric. [*Hint:* Show that this matrix equals its transpose with the help of Exercise 17b.]

23. Suppose that $\mathbf{A}$ is an $n \times n$ matrix where n is a positive integer. Show that $\mathbf{A} + \mathbf{A}^t$ is symmetric.

24. a) Show that the system of simultaneous linear equations

$$\begin{aligned} a_{11}x_1 + a_{12}x_2 + \cdots + a_{1n}x_n &= b_1 \\ a_{21}x_1 + a_{22}x_2 + \cdots + a_{2n}x_n &= b_2 \\ &\vdots \\ a_{n1}x_1 + a_{n2}x_2 + \cdots + a_{nn}x_n &= b_n. \end{aligned}$$

in the variables $x_1, x_2, \ldots, x_n$ can be expressed as $\mathbf{AX} = \mathbf{B}$, where $\mathbf{A} = [a_{ij}]$, $\mathbf{X}$ is an $n \times 1$ matrix with x_i the entry in its ith row, and $\mathbf{B}$ is an $n \times 1$ matrix with b_i the entry in its ith row.

b) Show that if the matrix $\mathbf{A} = [a_{ij}]$ is invertible (as defined in the preamble to Exercise 18), then the solution of the system in part (a) can be found using the equation $\mathbf{X} = \mathbf{A}^{-1}\mathbf{B}$.

25. Use Exercises 18 and 24 to solve the system

$$\begin{aligned} 7x_1 - 8x_2 + 5x_3 &= 5 \\ -4x_1 + 5x_2 - 3x_3 &= -3 \\ x_1 - x_2 + x_3 &= 0 \end{aligned}$$

26. Let

$$\mathbf{A} = \begin{bmatrix} 1 & 1 \\ 0 & 1 \end{bmatrix} \quad \text{and} \quad \mathbf{B} = \begin{bmatrix} 0 & 1 \\ 1 & 0 \end{bmatrix}.$$

Find

a) $\mathbf{A} \vee \mathbf{B}$. **b)** $\mathbf{A} \wedge \mathbf{B}$. **c)** $\mathbf{A} \odot \mathbf{B}$.

27. Let

$$\mathbf{A} = \begin{bmatrix} 1 & 0 & 1 \\ 1 & 1 & 0 \\ 0 & 0 & 1 \end{bmatrix} \quad \text{and} \quad \mathbf{B} = \begin{bmatrix} 0 & 1 & 1 \\ 1 & 0 & 1 \\ 1 & 0 & 1 \end{bmatrix}.$$

Find

a) $\mathbf{A} \vee \mathbf{B}$. **b)** $\mathbf{A} \wedge \mathbf{B}$. **c)** $\mathbf{A} \odot \mathbf{B}$.

28. Find the Boolean product of $\mathbf{A}$ and $\mathbf{B}$, where

$$\mathbf{A} = \begin{bmatrix} 1 & 0 & 0 & 1 \\ 0 & 1 & 0 & 1 \\ 1 & 1 & 1 & 1 \end{bmatrix} \quad \text{and} \quad \mathbf{B} = \begin{bmatrix} 1 & 0 \\ 0 & 1 \\ 1 & 1 \\ 1 & 0 \end{bmatrix}.$$

29. Let

$$\mathbf{A} = \begin{bmatrix} 1 & 0 & 0 \\ 1 & 0 & 1 \\ 0 & 1 & 0 \end{bmatrix}.$$

Find

a) $\mathbf{A}^{[2]}$. **b)** $\mathbf{A}^{[3]}$.
c) $\mathbf{A} \vee \mathbf{A}^{[2]} \vee \mathbf{A}^{[3]}$.

30. Let $\mathbf{A}$ be a zero–one matrix. Show that

a) $\mathbf{A} \vee \mathbf{A} = \mathbf{A}$. **b)** $\mathbf{A} \wedge \mathbf{A} = \mathbf{A}$.

31. In this exercise we show that the meet and join operations are commutative. Let $\mathbf{A}$ and $\mathbf{B}$ be $m \times n$ zero–one matrices. Show that

a) $\mathbf{A} \vee \mathbf{B} = \mathbf{B} \vee \mathbf{A}$. **b)** $\mathbf{B} \wedge \mathbf{A} = \mathbf{A} \wedge \mathbf{B}$.

32. In this exercise we show that the meet and join operations are associative. Let $\mathbf{A}$, $\mathbf{B}$, and $\mathbf{C}$ be $m \times n$ zero–one matrices. Show that

a) $(\mathbf{A} \vee \mathbf{B}) \vee \mathbf{C} = \mathbf{A} \vee (\mathbf{B} \vee \mathbf{C})$.
b) $(\mathbf{A} \wedge \mathbf{B}) \wedge \mathbf{C} = \mathbf{A} \wedge (\mathbf{B} \wedge \mathbf{C})$.

33. We will establish distributive laws of the meet over the join operation in this exercise. Let $\mathbf{A}$, $\mathbf{B}$, and $\mathbf{C}$ be $m \times n$ zero–one matrices. Show that

a) $\mathbf{A} \vee (\mathbf{B} \wedge \mathbf{C}) = (\mathbf{A} \vee \mathbf{B}) \wedge (\mathbf{A} \vee \mathbf{C})$.
b) $\mathbf{A} \wedge (\mathbf{B} \vee \mathbf{C}) = (\mathbf{A} \wedge \mathbf{B}) \vee (\mathbf{A} \wedge \mathbf{C})$.

34. Let $\mathbf{A}$ be an $n \times n$ zero–one matrix. Let $\mathbf{I}$ be the $n \times n$ identity matrix. Show that $\mathbf{A} \odot \mathbf{I} = \mathbf{I} \odot \mathbf{A} = \mathbf{A}$.

35. In this exercise we will show that the Boolean product of zero–one matrices is associative. Assume that $\mathbf{A}$ is an $m \times p$ zero–one matrix, $\mathbf{B}$ is a $p \times k$ zero–one matrix, and $\mathbf{C}$ is a $k \times n$ zero–one matrix. Show that $\mathbf{A} \odot (\mathbf{B} \odot \mathbf{C}) = (\mathbf{A} \odot \mathbf{B}) \odot \mathbf{C}$.

Key Terms and Results

TERMS

algebraic number: a real or complex number that is a root of a polynomial with integer coefficients

transcendental number: a real or complex number that is not algebraic

set: an unordered collection of distinct objects

axiom: a basic assumption of a theory

paradox: a logical inconsistency

element, member of a set: an object in a set

roster method: a method that describes a set by listing its elements

set builder notation: the notation that describes a set by stating a property an element must have to be a member

multiset: an unordered collection of objects where objects can occur multiple times

∅ (empty set, null set): the set with no members

universal set: the set containing all objects under consideration

Venn diagram: a graphical representation of a set or sets

$S = T$ (set equality): S and T have the same elements

$S \subseteq T$ (S is a subset of T): every element of S is also an element of T

$S \subset T$ (S is a proper subset of T): S is a subset of T and $S \neq T$

finite set: a set with n elements, where n is a nonnegative integer

infinite set: a set that is not finite

$|S|$ (the cardinality of S): the number of elements in S

$P(S)$ (the power set of S): the set of all subsets of S

$A \cup B$ (the union of A and B): the set containing those elements that are in at least one of A and B

$A \cap B$ (the intersection of A and B): the set containing those elements that are in both A and B.

$A - B$ (the difference of A and B): the set containing those elements that are in A but not in B

$\overline{A}$ (the complement of A): the set of elements in the universal set that are not in A

$A \oplus B$ (the symmetric difference of A and B): the set containing those elements in exactly one of A and B

membership table: a table displaying the membership of elements in sets

function from A to B: an assignment of exactly one element of B to each element of A

domain of f: the set A, where f is a function from A to B

codomain of f: the set B, where f is a function from A to B

b is the image of a under f: $b = f(a)$

a is a preimage of b under f: $f(a) = b$

range of f: the set of images of f

onto function, surjection: a function from A to B such that every element of B is the image of some element in A

one-to-one function, injection: a function such that the images of elements in its domain are distinct

one-to-one correspondence, bijection: a function that is both one-to-one and onto

inverse of f: the function that reverses the correspondence given by f (when f is a bijection)

$f \circ g$ (composition of f and g): the function that assigns $f(g(x))$ to x

$\lfloor x \rfloor$ (floor function): the largest integer not exceeding x

$\lceil x \rceil$ (ceiling function): the smallest integer greater than or equal to x

partial function: an assignment to each element in a subset of the domain a unique element in the codomain

sequence: a function with domain that is a subset of the set of integers

geometric progression: a sequence of the form $a, ar, ar^2, \ldots,$ where a and r are real numbers

arithmetic progression: a sequence of the form $a, a + d, a + 2d, \ldots,$ where a and d are real numbers

string: a finite sequence

empty string: a string of length zero

recurrence relation: an equation that expresses the nth term a_n of a sequence in terms of one or more of the previous terms of the sequence for all integers n greater than a particular integer

$\sum_{i=1}^{n} a_i$: the sum $a_1 + a_2 + \cdots + a_n$

$\prod_{i=1}^{n} a_i$: the product $a_1 a_2 \cdots a_n$

cardinality: two sets A and B have the same cardinality if there is a one-to-one correspondence from A to B

countable set: a set that either is finite or can be placed in one-to-one correspondence with the set of positive integers

uncountable set: a set that is not countable

$\aleph_0$ (aleph null): the cardinality of a countable set

$\mathfrak{c}$: the cardinality of the set of real numbers

Cantor diagonalization argument: a proof technique used to show that the set of real numbers is uncountable

computable function: a function for which there is a computer program in some programming language that finds its values

uncomputable function: a function for which no computer program in a programming language exists that finds its values

continuum hypothesis: the statement that no set A exists such that $\aleph_0 < |A| < \mathfrak{c}$

matrix: a rectangular array of numbers

matrix addition: see page 188

matrix multiplication: see page 189

$\mathbf{I}_n$ (identity matrix of order n): the $n \times n$ matrix that has entries equal to 1 on its diagonal and 0s elsewhere

$\mathbf{A}^t$ (transpose of $\mathbf{A}$): the matrix obtained from $\mathbf{A}$ by interchanging the rows and columns

symmetric matrix: a matrix is symmetric if it equals its transpose

zero–one matrix: a matrix with each entry equal to either 0 or 1

$\mathbf{A} \vee \mathbf{B}$ (the join of $\mathbf{A}$ and $\mathbf{B}$): see page 191

$\mathbf{A} \wedge \mathbf{B}$ (the meet of $\mathbf{A}$ and $\mathbf{B}$): see page 191

$\mathbf{A} \odot \mathbf{B}$ (the Boolean product of $\mathbf{A}$ and $\mathbf{B}$): see page 192

RESULTS

The set identities given in Table 1 in Section 2.2

The summation formulae in Table 2 in Section 2.4

The set of rational numbers is countable.

The set of real numbers is uncountable.

Review Questions

1. Explain what it means for one set to be a subset of another set. How do you prove that one set is a subset of another set?

2. What is the empty set? Show that the empty set is a subset of every set.

3. a) Define $|S|$, the cardinality of the set S.
   b) Give a formula for $|A \cup B|$, where A and B are sets.

4. a) Define the power set of a set S.
   b) When is the empty set in the power set of a set S?

c) How many elements does the power set of a set S with n elements have?

5. a) Define the union, intersection, difference, and symmetric difference of two sets.
   b) What are the union, intersection, difference, and symmetric difference of the set of positive integers and the set of odd integers?

6. a) Explain what it means for two sets to be equal.
   b) Describe as many of the ways as you can to show that two sets are equal.
   c) Show in at least two different ways that the sets $A - (B \cap C)$ and $(A - B) \cup (A - C)$ are equal.

7. Explain the relationship between logical equivalences and set identities.

8. a) Define the domain, codomain, and range of a function.
   b) Let $f(n)$ be the function from the set of integers to the set of integers such that $f(n) = n^2 + 1$. What are the domain, codomain, and range of this function?

9. a) Define what it means for a function from the set of positive integers to the set of positive integers to be one-to-one.
   b) Define what it means for a function from the set of positive integers to the set of positive integers to be onto.
   c) Give an example of a function from the set of positive integers to the set of positive integers that is both one-to-one and onto.
   d) Give an example of a function from the set of positive integers to the set of positive integers that is one-to-one but not onto.
   e) Give an example of a function from the set of positive integers to the set of positive integers that is not one-to-one but is onto.
   f) Give an example of a function from the set of positive integers to the set of positive integers that is neither one-to-one nor onto.

10. a) Define the inverse of a function.
    b) When does a function have an inverse?
    c) Does the function $f(n) = 10 - n$ from the set of integers to the set of integers have an inverse? If so, what is it?

11. a) Define the floor and ceiling functions from the set of real numbers to the set of integers.
    b) For which real numbers x is it true that $\lfloor x \rfloor = \lceil x \rceil$?

12. Conjecture a formula for the terms of the sequence that begins 8, 14, 32, 86, 248 and find the next three terms of your sequence.

13. Suppose that $a_n = a_{n-1} - 5$ for $n = 1, 2, \ldots$. Find a formula for a_n.

14. What is the sum of the terms of the geometric progression $a + ar + \cdots + ar^n$ when $r \neq 1$?

15. Show that the set of odd integers is countable.

16. Give an example of an uncountable set.

17. Define the product of two matrices **A** and **B**. When is this product defined?

18. Show that matrix multiplication is not commutative.

Supplementary Exercises

1. Let A be the set of English words that contain the letter x, and let B be the set of English words that contain the letter q. Express each of these sets as a combination of A and B.
   a) The set of English words that do not contain the letter x.
   b) The set of English words that contain both an x and a q.
   c) The set of English words that contain an x but not a q.
   d) The set of English words that do not contain either an x or a q.
   e) The set of English words that contain an x or a q, but not both.

2. Show that if A is a subset of B, then the power set of A is a subset of the power set of B.

3. Suppose that A and B are sets such that the power set of A is a subset of the power set of B. Does it follow that A is a subset of B?

4. Let **E** denote the set of even integers and **O** denote the set of odd integers. As usual, let **Z** denote the set of all integers. Determine each of these sets.

   a) $\mathbf{E} \cup \mathbf{O}$ b) $\mathbf{E} \cap \mathbf{O}$ c) $\mathbf{Z} - \mathbf{E}$ d) $\mathbf{Z} - \mathbf{O}$

5. Show that if A and B are sets, then $A - (A - B) = A \cap B$.

6. Let A and B be sets. Show that $A \subseteq B$ if and only if $A \cap B = A$.

7. Let A, B, and C be sets. Show that $(A - B) - C$ is not necessarily equal to $A - (B - C)$.

8. Suppose that A, B, and C are sets. Prove or disprove that $(A - B) - C = (A - C) - B$.

9. Suppose that A, B, C, and D are sets. Prove or disprove that $(A - B) - (C - D) = (A - C) - (B - D)$.

10. Show that if A and B are finite sets, then $|A \cap B| \leq |A \cup B|$. Determine when this relationship is an equality.

11. Let A and B be sets in a finite universal set U. List the following in order of increasing size.
    a) $|A|, |A \cup B|, |A \cap B|, |U|, |\emptyset|$
    b) $|A - B|, |A \oplus B|, |A| + |B|, |A \cup B|, |\emptyset|$

12. Let A and B be subsets of the finite universal set U. Show that $|\overline{A} \cap \overline{B}| = |U| - |A| - |B| + |A \cap B|$.

13. Let f and g be functions from $\{1, 2, 3, 4\}$ to $\{a, b, c, d\}$ and from $\{a, b, c, d\}$ to $\{1, 2, 3, 4\}$, respectively, with $f(1) = d$, $f(2) = c$, $f(3) = a$, and $f(4) = b$, and $g(a) = 2$, $g(b) = 1$, $g(c) = 3$, and $g(d) = 2$.

a) Is f one-to-one? Is g one-to-one?

b) Is f onto? Is g onto?

c) Does either f or g have an inverse? If so, find this inverse.

14. Suppose that f is a function from A to B where A and B are finite sets. Explain why $|f(S)| \leq |S|$ for all subsets S of A.

15. Suppose that f is a function from A to B where A and B are finite sets. Explain why $|f(S)| = |S|$ for all subsets S of A if and only if f is one-to-one.

Suppose that f is a function from A to B. We define the function S_f from $\mathcal{P}(A)$ to $\mathcal{P}(B)$ by the rule $S_f(X) = f(X)$ for each subset X of A. Similarly, we define the function $S_{f^{-1}}$ from $\mathcal{P}(B)$ to $\mathcal{P}(A)$ by the rule $S_{f^{-1}}(Y) = f^{-1}(Y)$ for each subset Y of B. Here, we are using Definition 4, and the definition of the inverse image of a set found in the preamble to Exercise 44, both in Section 2.3.

***16.** Suppose that f is a function from the set A to the set B. Prove that

a) if f is one-to-one, then S_f is a one-to-one function from $\mathcal{P}(A)$ to $\mathcal{P}(B)$.

b) if f is onto function, then S_f is an onto function from $\mathcal{P}(A)$ to $\mathcal{P}(B)$.

c) if f is onto function, then $S_{f^{-1}}$ is a one-to-one function from $\mathcal{P}(B)$ to $\mathcal{P}(A)$.

d) if f is one-to-one, then $S_{f^{-1}}$ is an onto function from $\mathcal{P}(B)$ to $\mathcal{P}(A)$.

e) if f is a one-to-one correspondence, then S_f is a one-to-one correspondence from $\mathcal{P}(A)$ to $\mathcal{P}(B)$ and $S_{f^{-1}}$ is a one-to-one correspondence from $\mathcal{P}(B)$ to $\mathcal{P}(A)$. [*Hint:* Use parts (a)–(d).]

17. Prove that if f and g are functions from A to B and $S_f = S_g$ (using the definition in the preamble to Exercise 16), then $f(x) = g(x)$ for all $x \in A$.

18. Show that if n is an integer, then $n = \lceil n/2 \rceil + \lfloor n/2 \rfloor$.

19. For which real numbers x and y is it true that $\lfloor x + y \rfloor = \lfloor x \rfloor + \lfloor y \rfloor$?

20. For which real numbers x and y is it true that $\lceil x + y \rceil = \lceil x \rceil + \lceil y \rceil$?

21. For which real numbers x and y is it true that $\lceil x + y \rceil = \lceil x \rceil + \lfloor y \rfloor$?

22. Prove that $\lfloor n/2 \rfloor \lceil n/2 \rceil = \lfloor n^2/4 \rfloor$ for all integers n.

23. Prove that if m is an integer, then $\lfloor x \rfloor + \lfloor m - x \rfloor = m - 1$, unless x is an integer, in which case, it equals m.

24. Prove that if x is a real number, then $\lfloor \lfloor x/2 \rfloor /2 \rfloor = \lfloor x/4 \rfloor$.

25. Prove that if n is an odd integer, then $\lceil n^2/4 \rceil = (n^2 + 3)/4$.

26. Prove that if m and n are positive integers and x is a real number, then

$$\left\lfloor \frac{\lfloor x \rfloor + n}{m} \right\rfloor = \left\lfloor \frac{x + n}{m} \right\rfloor.$$

***27.** Prove that if m is a positive integer and x is a real number, then

$$\lfloor mx \rfloor = \lfloor x \rfloor + \left\lfloor x + \frac{1}{m} \right\rfloor + \left\lfloor x + \frac{2}{m} \right\rfloor + \cdots + \left\lfloor x + \frac{m-1}{m} \right\rfloor.$$

***28.** We define the **Ulam numbers** by setting $u_1 = 1$ and $u_2 = 2$. Furthermore, after determining whether the integers less than n are Ulam numbers, we set n equal to the next Ulam number if it can be written uniquely as the sum of two different Ulam numbers. Note that $u_3 = 3$, $u_4 = 4$, $u_5 = 6$, and $u_6 = 8$.

a) Find the first 20 Ulam numbers.

b) Prove that there are infinitely many Ulam numbers.

29. Determine the value of $\prod_{k=1}^{100} \frac{k+1}{k}$. (The notation used here for products is defined in the preamble to Exercise 45 in Section 2.4.)

***30.** Determine a rule for generating the terms of the sequence that begins 1, 3, 4, 8, 15, 27, 50, 92, …, and find the next four terms of the sequence.

***31.** Determine a rule for generating the terms of the sequence that begins 2, 3, 3, 5, 10, 13, 39, 43, 172, 177, 885, 891, …, and find the next four terms of the sequence.

32. Show that the set of irrational numbers is an uncountable set.

33. Show that the set S is a countable set if there is a function f from S to the positive integers such that $f^{-1}(j)$ is countable whenever j is a positive integer.

34. Show that the set of all finite subsets of the set of positive integers is a countable set.

****35.** Show that $|\mathbf{R} \times \mathbf{R}| = |\mathbf{R}|$. [*Hint:* Use the Schröder-Bernstein theorem to show that $|(0, 1) \times (0, 1)| = |(0, 1)|$. To construct an injection from $(0, 1) \times (0, 1)$ to $(0, 1)$, suppose that $(x, y) \in (0, 1) \times (0, 1)$. Map (x, y) to the number with decimal expansion formed by alternating between the digits in the decimal expansions of x and y, which do not end with an infinite string of 9s.]

****36.** Show that **C**, the set of complex numbers, has the same cardinality as **R**, the set of real numbers.

37. Find $\mathbf{A}^n$ if $\mathbf{A}$ is

$$\begin{bmatrix} 0 & 1 \\ -1 & 0 \end{bmatrix}.$$

38. Show that if $\mathbf{A} = c\mathbf{I}$, where c is a real number and $\mathbf{I}$ is the $n \times n$ identity matrix, then $\mathbf{AB} = \mathbf{BA}$ whenever $\mathbf{B}$ is an $n \times n$ matrix.

39. Show that if $\mathbf{A}$ is a 2×2 matrix such that $\mathbf{AB} = \mathbf{BA}$ whenever $\mathbf{B}$ is a 2×2 matrix, then $\mathbf{A} = c\mathbf{I}$, where c is a real number and $\mathbf{I}$ is the 2×2 identity matrix.

40. Show that if $\mathbf{A}$ and $\mathbf{B}$ are invertible matrices and $\mathbf{AB}$ exists, then $(\mathbf{AB})^{-1} = \mathbf{B}^{-1}\mathbf{A}^{-1}$.

41. Let $\mathbf{A}$ be an $n \times n$ matrix and let $\mathbf{0}$ be the $n \times n$ matrix all of whose entries are zero. Show that the following are true.

a) $\mathbf{A} \odot \mathbf{0} = \mathbf{0} \odot \mathbf{A} = \mathbf{0}$
b) $\mathbf{A} \vee \mathbf{0} = \mathbf{0} \vee \mathbf{A} = \mathbf{A}$
c) $\mathbf{A} \wedge \mathbf{0} = \mathbf{0} \wedge \mathbf{A} = \mathbf{0}$

Computer Projects

Write programs with the specified input and output.

1. Given subsets A and B of a set with n elements, use bit strings to find $\overline{A}$, $A \cup B$, $A \cap B$, $A - B$, and $A \oplus B$.

2. Given multisets A and B from the same universal set, find $A \cup B$, $A \cap B$, $A - B$, and $A + B$ (see Section 2.2).

3. Given fuzzy sets A and B, find $\overline{A}$, $A \cup B$, and $A \cap B$ (see preamble to Exercise 73 of Section 2.2).

4. Given a function f from $\{1, 2, \ldots, n\}$ to the set of integers, determine whether f is one-to-one.

5. Given a function f from $\{1, 2, \ldots, n\}$ to itself, determine whether f is onto.

6. Given a bijection f from the set $\{1, 2, \ldots, n\}$ to itself, find f^{-1}.

7. Given an $m \times k$ matrix $\mathbf{A}$ and a $k \times n$ matrix $\mathbf{B}$, find $\mathbf{AB}$.

8. Given a square matrix $\mathbf{A}$ and a positive integer n, find $\mathbf{A}^n$.

9. Given a square matrix, determine whether it is symmetric.

10. Given two $m \times n$ Boolean matrices, find their meet and join.

11. Given an $m \times k$ Boolean matrix $\mathbf{A}$ and a $k \times n$ Boolean matrix $\mathbf{B}$, find the Boolean product of $\mathbf{A}$ and $\mathbf{B}$.

12. Given a square Boolean matrix $\mathbf{A}$ and a positive integer n, find $\mathbf{A}^{[n]}$.

Computations and Explorations

Use a computational program or programs you have written to do these exercises.

1. Given two finite sets, list all elements in the Cartesian product of these two sets.

2. Given a finite set, list all elements of its power set.

3. Calculate the number of one-to-one functions from a set S to a set T, where S and T are finite sets of various sizes. Can you determine a formula for the number of such functions? (We will find such a formula in Chapter 6.)

4. Calculate the number of onto functions from a set S to a set T, where S and T are finite sets of various sizes. Can you determine a formula for the number of such functions? (We will find such a formula in Chapter 8.)

5. Given an positive integer n, generate the first n Fibonacci numbers. Looking at the terms, formulate conjectures about them. (For instance look at their size, divisibility by different factors, or possible identities relating different terms.)

***6.** Develop a collection of different rules for generating the terms of a sequence and a program for randomly selecting one of these rules and the particular sequence generated using these rules. Make this part of an interactive program that prompts for the next term of the sequence and determines whether the response is the intended next term.

Writing Projects

Respond to these with essays using outside sources.

1. Discuss how an axiomatic set theory can be developed to avoid Russell's paradox. (See Exercise 54 of Section 2.1.)

2. Research where the concept of a function first arose, and describe how this concept was first used.

3. Explain the different ways in which the *Encyclopedia of Integer Sequences* has been found useful. Also, describe a few of the more unusual sequences in this encyclopedia and how they arise.

4. Define the EKG (or electrocardiogram) sequence and describe some of its properties and open questions about it.

5. Look up the definition of a transcendental number. Explain how to show that such numbers exist and how such numbers can be constructed. Which famous numbers can be shown to be transcendental and for which famous numbers is it still unknown whether they are transcendental?

6. Expand the discussion of the continuum hypothesis in the text.

CHAPTER

3 Algorithms

3.1 Algorithms

3.2 The Growth of Functions

3.3 Complexity of Algorithms

Many problems can be solved by considering them as special cases of general problems. For instance, consider the problem of locating the largest integer in the sequence 101, 12, 144, 212, 98. This is a specific case of the problem of locating the largest integer in a sequence of integers. To solve this general problem we must give an algorithm, which specifies a sequence of steps used to solve this general problem. We will study algorithms for solving many different types of problems in this book. For example, in this chapter we will introduce algorithms for two of the most important problems in computer science, searching for an element in a list and sorting a list so its elements are in some prescribed order, such as increasing, decreasing, or alphabetic. Later in the book we will develop algorithms that find the greatest common divisor of two integers, that generate all the orderings of a finite set, that find the shortest path between nodes in a network, and for solving many other problems.

We will also introduce the notion of an algorithmic paradigm, which provides a general method for designing algorithms. In particular we will discuss brute-force algorithms, which find solutions using a straightforward approach without introducing any cleverness. We will also discuss greedy algorithms, a class of algorithms used to solve optimization problems. Proofs are important in the study of algorithms. In this chapter we illustrate this by proving that a particular greedy algorithm always finds an optimal solution.

One important consideration concerning an algorithm is its computational complexity, which measures the processing time and computer memory required by the algorithm to solve problems of a particular size. To measure the complexity of algorithms we use big-O and big-Theta notation, which we develop in this chapter. We will illustrate the analysis of the complexity of algorithms in this chapter, focusing on the time an algorithm takes to solve a problem. Furthermore, we will discuss what the time complexity of an algorithm means in practical and theoretical terms.

3.1 Algorithms

3.1.1 Introduction

There are many general classes of problems that arise in discrete mathematics. For instance: given a sequence of integers, find the largest one; given a set, list all its subsets; given a set of integers, put them in increasing order; given a network, find the shortest path between two vertices. When presented with such a problem, the first thing to do is to construct a model that translates the problem into a mathematical context. Discrete structures used in such models include sets, sequences, and functions—structures discussed in Chapter 2—as well as such other structures as permutations, relations, graphs, trees, networks, and finite state machines—concepts that will be discussed in later chapters.

Setting up the appropriate mathematical model is only part of the solution. To complete the solution, a method is needed that will solve the general problem using the model. Ideally, what is required is a procedure that follows a sequence of steps that leads to the desired answer. Such a sequence of steps is called an **algorithm**.

Definition 1 An *algorithm* is a finite sequence of precise instructions for performing a computation or for solving a problem.

The term *algorithm* is a corruption of the name *al-Khowarizmi*, a mathematician of the ninth century, whose book on Hindu numerals is the basis of modern decimal notation. Originally, the word *algorism* was used for the rules for performing arithmetic using decimal notation. *Algorism* evolved into the word *algorithm* by the eighteenth century. With the growing interest in computing machines, the concept of an algorithm was given a more general meaning, to include all definite procedures for solving problems, not just the procedures for performing arithmetic. (We will discuss algorithms for performing arithmetic with integers in Chapter 4.)

Remark: Note that the term "algorithm" is sometimes used to describe specific types of procedures. In particular, *social media algorithms* are the rules, signals, and data that define how a social media platform works. These algorithms specify how a platform filters, ranks, selects, and recommends content to its users. For example, these algorithms may track upvotes and downvotes of particular content in order to present content to users most likely to increase engagement with the platform. To learn more about social media algorithms for various social media, such as Instagram, LinkedIn, Facebook, TikTok, X (previously known as Twitter), YouTube, and Pinterest, consult the web. To find good links, use the search term "social media algorithms."

In this book, we will discuss algorithms that solve a wide variety of problems. In this section, we will use the problem of finding the largest integer in a finite sequence of integers to illustrate the concept of an algorithm and the properties algorithms have. Also, we will describe algorithms for locating a particular element in a finite set. In subsequent sections, procedures for finding the greatest common divisor of two integers, for finding the shortest path between two points in a network, for multiplying matrices, and so on, will be discussed.

EXAMPLE 1 Describe an algorithm for finding the maximum (largest) value in a finite sequence of integers.

Extra Examples

Even though the problem of finding the maximum element in a sequence is relatively trivial, it provides a good illustration of the concept of an algorithm. Also, there are many instances where the largest integer in a finite sequence of integers is required. For instance, a university may need to find the highest score on a competitive exam taken by thousands of students. Or a sports organization may want to identify the member with the highest rating each month. We want to develop an algorithm that can be used whenever the problem of finding the largest element in a finite sequence of integers arises.

We can specify a procedure for solving this problem in several ways. One method is simply to use the English language to describe the sequence of steps used. We now provide such a solution.

Solution of Example 1: We perform the following steps.

1. Set the temporary maximum equal to the first integer in the sequence. (The temporary maximum will be the largest integer examined at any stage of the procedure.)
2. Compare the next integer in the sequence to the temporary maximum, and if it is larger than the temporary maximum, set the temporary maximum equal to this integer.
3. Repeat the previous step if there are more integers in the sequence.
4. Stop when there are no integers left in the sequence. The temporary maximum at this point is the largest integer in the sequence. ◀

Links

dbtravel/Alamy Stock Photo

ABU JA'FAR MOHAMMED IBN MUSA AL-KHOWARIZMI (C. 780–C. 850) al-Khowarizmi, an astronomer and mathematician, was a member of the House of Wisdom, an academy of scientists in Baghdad. The name al-Khowarizmi means "from the town of Kowarizm," which was then part of Persia, but is now called *Khiva* and is part of Uzbekistan. al-Khowarizmi wrote books on mathematics, astronomy, and geography. Western Europeans first learned about algebra from his works. The word *algebra* comes from al-jabr, part of the title of his book *Kitab al-jabr w'al muquabala*. This book was translated into Latin and was a widely used textbook. His book on the use of Hindu numerals describes procedures for arithmetic operations using these numerals. European authors used a Latin corruption of his name, which later evolved to the word *algorithm*, to describe the subject of arithmetic with Hindu numerals.

An algorithm can also be described using a computer language. However, when that is done, only those instructions permitted in the language can be used. This often leads to a description of the algorithm that is complicated and difficult to understand. Furthermore, because many programming languages are in common use, it would be undesirable to choose one particular language. So, instead of using a particular computer language to specify algorithms, a form of **pseudocode**, described in Appendix 3, will be used in this book. (We will also describe algorithms using the English language.) Pseudocode provides an intermediate step between an English language description of an algorithm and an implementation of this algorithm in a programming language. The steps of the algorithm are specified using instructions resembling those used in programming languages. However, in pseudocode, the instructions used can include any well-defined operations or statements. A computer program can be produced in any computer language using the pseudocode description as a starting point.

The pseudocode used in this book is designed to be easily understood. It can serve as an intermediate step in the construction of programs implementing algorithms in one of a variety of different programming languages. Although this pseudocode does not follow the syntax of Java, C, C++, or any other programming language, students familiar with a modern programming language will find it easy to follow. A key difference between this pseudocode and code in a programming language is that we can use any well-defined instruction even if it would take many lines of code to implement this instruction. The details of the pseudocode used in the text are given in Appendix 3. The reader should refer to this appendix whenever the need arises.

A pseudocode description of the algorithm for finding the maximum element in a finite sequence follows.

ALGORITHM 1 Finding the Maximum Element in a Finite Sequence.

procedure $max(a_1, a_2, \ldots, a_n$: integers)
$max := a_1$
for $i := 2$ **to** n
  if $max < a_i$ **then** $max := a_i$
return $max\{max$ is the largest element$\}$

This algorithm first assigns the initial term of the sequence, a_1, to the variable *max*. The "for" loop is used to successively examine terms of the sequence. If a term is greater than the current value of *max*, it is assigned to be the new value of *max*. The algorithm terminates after all terms have been examined. The value of *max* on termination is the maximum element in the sequence.

To gain insight into how an algorithm works it is useful to construct a **trace** that shows its steps when given specific input. For instance, a trace of Algorithm 1 with input 8, 4, 11, 3, 10 begins with the algorithm setting *max* to 8, the value of the initial term. It then compares 4, the second term, with 8, the current value of *max*. Because $4 \leq 8$, *max* is unchanged. Next, the algorithm compares the third term, 11, with 8, the current value of *max*. Because $8 < 11$, *max* is set equal to 11. The algorithm then compares 3, the fourth term, and 11, the current value of *max*. Because $3 \leq 11$, *max* is unchanged. Finally, the algorithm compares 10, the fifth term, and 11, the current value of *max*. As $10 \leq 11$, *max* remains unchanged. Because there are five terms, we have $n = 5$. So after examining 10, the last term, the algorithm terminates, with $max = 11$. When it terminates, the algorithms reports that 11 is the largest term in the sequence.

PROPERTIES OF ALGORITHMS There are several properties that algorithms generally share. They are useful to keep in mind when algorithms are described. These properties are:

- *Input.* An algorithm has input values from a specified set.
- *Output.* From each set of input values an algorithm produces output values from a specified set. The output values are the solution to the problem.

- *Definiteness.* The steps of an algorithm must be defined precisely.
- *Correctness.* An algorithm should produce the correct output values for each set of input values.
- *Finiteness.* An algorithm should produce the desired output after a finite (but perhaps large) number of steps for any input in the set.
- *Effectiveness.* It must be possible to perform each step of an algorithm exactly and in a finite amount of time.
- *Generality.* The procedure should be applicable for all problems of the desired form, not just for a particular set of input values.

EXAMPLE 2 Show that Algorithm 1 for finding the maximum element in a finite sequence of integers has all the properties listed.

Solution: The input to Algorithm 1 is a sequence of integers. The output is the largest integer in the sequence. Each step of the algorithm is precisely defined, because only assignments, a finite loop, and conditional statements occur. To show that the algorithm is correct, we must show that when the algorithm terminates, the value of the variable *max* equals the maximum of the terms of the sequence. To see this, note that the initial value of *max* is the first term of the sequence; as successive terms of the sequence are examined, *max* is updated to the value of a term if the term exceeds the maximum of the terms previously examined. This (informal) argument shows that when all the terms have been examined, *max* equals the value of the largest term. (A rigorous proof of this requires the use of mathematical induction, a proof technique developed in Section 5.1.) The algorithm uses a finite number of steps, because it terminates after all the integers in the sequence have been examined. The algorithm can be carried out in a finite amount of time because each step is either a comparison or an assignment, there are a finite number of these steps, and each of these two operations takes a finite amount of time. Finally, Algorithm 1 is general, because it can be used to find the maximum of any finite sequence of integers. ◀

3.1.2 Searching Algorithms

The problem of locating an element in an ordered list occurs in many contexts. For instance, a program that checks the spelling of words searches for them in a dictionary, which is just an ordered list of words. Problems of this kind are called **searching problems**. We will discuss several algorithms for searching in this section. We will study the number of steps used by each of these algorithms in Section 3.3.

The general searching problem can be described as follows: Locate an element x in a list of distinct elements $a_1, a_2, \ldots, a_n$, or determine that it is not in the list. The solution to this search problem is the location of the term in the list that equals x (that is, i is the solution if $x = a_i$) and is 0 if x is not in the list.

THE LINEAR SEARCH The first algorithm that we will present is called the **linear search**, or **sequential search**, algorithm. The linear search algorithm begins by comparing x and a_1. When $x = a_1$, the solution is the location of a_1, namely, 1. When $x \neq a_1$, compare x with a_2. If $x = a_2$, the solution is the location of a_2, namely, 2. When $x \neq a_2$, compare x with a_3. Continue this process, comparing x successively with each term of the list until a match is found, where the solution is the location of that term, unless no match occurs. If the entire list has been searched without locating x, the solution is 0. The pseudocode for the linear search algorithm is displayed as Algorithm 2.

Links

ALGORITHM 2 The Linear Search Algorithm.

procedure *linear search*(x: integer, $a_1, a_2, \ldots, a_n$: distinct integers)
$i := 1$
while ($i \leq n$ and $x \neq a_i$)
  $i := i + 1$
if $i \leq n$ **then** *location* $:= i$
else *location* $:= 0$
return *location*{*location* is the subscript of the term that equals x, or is 0 if x is not found}

THE BINARY SEARCH We will now consider another searching algorithm. This algorithm can be used when the list has terms occurring in order of increasing size (for instance: if the terms are numbers, they are listed from smallest to largest; if they are words, they are listed in lexicographic, or alphabetic, order). This second searching algorithm is called the **binary search algorithm**. It proceeds by comparing the element to be located to the middle term of the list. The list is then split into two smaller sublists of the same size, or where one of these smaller lists has one fewer term than the other. The search continues by restricting the search to the appropriate sublist based on the comparison of the element to be located and the middle term. In Section 3.3, it will be shown that the binary search algorithm is much more efficient than the linear search algorithm. Example 3 demonstrates how a binary search works.

Links

EXAMPLE 3 To search for 19 in the list

1 2 3 5 6 7 8 10 12 13 15 16 18 19 20 22,

first split this list, which has 16 terms, into two smaller lists with eight terms each, namely,

1 2 3 5 6 7 8 10   12 13 15 16 18 19 20 22.

Then, compare 19 and the largest term in the first list. Because $10 < 19$, the search for 19 can be restricted to the list containing the 9th through the 16th terms of the original list. Next, split this list, which has eight terms, into the two smaller lists of four terms each, namely,

12 13 15 16   18 19 20 22.

Because $16 < 19$ (comparing 19 with the largest term of the first list) the search is restricted to the second of these lists, which contains the 13th through the 16th terms of the original list. The list 18 19 20 22 is split into two lists, namely,

18 19   20 22.

Because 19 is not greater than the largest term of the first of these two lists, which is also 19, the search is restricted to the list: 18 19, which contains the 13th and 14th terms of the original list. Next, this list of two terms is split into two lists of one term each: 18 and 19. Because $18 < 19$, the search is restricted to the second list: the list containing the 14th term of the list, which is 19. Now that the search has been narrowed down to one term, a comparison is made, and 19 is located as the 14th term in the original list. ◀

We now specify the steps of the binary search algorithm. To search for the integer x in the list $a_1, a_2, \ldots, a_n$, where $a_1 < a_2 < \cdots < a_n$, begin by comparing x with the middle term a_m of the list, where $m = \lfloor (n+1)/2 \rfloor$. (Recall that $\lfloor x \rfloor$ is the greatest integer not exceeding x.) If

$x > a_m$, the search for x is restricted to the second half of the list, which is $a_{m+1}, a_{m+2}, \ldots, a_n$. If x is not greater than a_m, the search for x is restricted to the first half of the list, which is $a_1, a_2, \ldots, a_m$.

The search has now been restricted to a list with no more than $\lceil n/2 \rceil$ elements. (Recall that $\lceil x \rceil$ is the smallest integer greater than or equal to x.) Using the same procedure, compare x to the middle term of the restricted list. Then restrict the search to the first or second half of the list. Repeat this process until a list with one term is obtained. Then determine whether this term is x. Pseudocode for the binary search algorithm is displayed as Algorithm 3.

ALGORITHM 3 The Binary Search Algorithm.

```
procedure binary search(x: integer, a_1, a_2, ..., a_n: increasing integers)
i := 1{i is left endpoint of search interval}
j := n {j is right endpoint of search interval}
while i < j
    m := ⌊(i + j)/2⌋
    if x > a_m then i := m + 1
    else j := m
if x = a_i then location := i
else location := 0
return location{location is the subscript i of the term a_i equal to x, or 0 if x is not found}
```

Algorithm 3 proceeds by successively narrowing down the part of the sequence being searched. At any given stage only the terms from a_i to a_j are under consideration. In other words, i and j are the smallest and largest subscripts of the remaining terms, respectively. Algorithm 3 continues narrowing the part of the sequence being searched until only one term of the sequence remains. When this is done, a comparison is made to see whether this term equals x.

3.1.3 Sorting

Demo

Ordering the elements of a list is a problem that occurs in many contexts. For example, to produce a telephone directory it is necessary to alphabetize the names of subscribers. Similarly, producing a directory of songs available for downloading requires that their titles be put in alphabetic order. Putting addresses in order in an email mailing list can determine whether there are duplicated addresses. Creating a useful dictionary requires that words be put in alphabetical order. Similarly, generating a parts list requires that we order them according to increasing part number.

Suppose that we have a list of elements of a set. Furthermore, suppose that we have a way to order elements of the set. (The notion of ordering elements of sets will be discussed in detail in Section 9.6.) **Sorting** is putting these elements into a list in which the elements are in increasing order. For instance, sorting the list 7, 2, 1, 4, 5, 9 produces the list 1, 2, 4, 5, 7, 9. Sorting the list d, h, c, a, f (using alphabetical order) produces the list a, c, d, f, h.

An amazingly large percentage of computing resources is devoted to sorting one thing or another. Hence, much effort has been devoted to the development of sorting algorithms. A surprisingly large number of sorting algorithms have been devised using distinct strategies, with new ones introduced regularly. In the third volume of his fundamental work *The Art of Computer Programming* [Kn98], Donald Knuth devotes close to 400 pages to sorting, covering around 15 different sorting algorithms in depth! More than 100 sorting algorithms have been devised, and it is surprising how often new sorting algorithms are developed. Among the newest sorting algorithms that have caught on is a widely useful algorithm called Timsort, which was invented in 2002, and the library sort, also known as the gapped insertion sort, invented in 2006.

Sorting is thought to hold the record as the problem solved by the most fundamentally different algorithms!

There are many reasons why sorting algorithms interest computer scientists and mathematicians. Among these reasons are that some algorithms are easier to implement, some algorithms are more efficient (either in general, or when given input with certain characteristics, such as lists slightly out of order), some algorithms take advantage of particular computer architectures, and some algorithms are particularly clever. In this section we will introduce two sorting algorithms, the bubble sort and the insertion sort. Two other sorting algorithms, the selection sort and the binary insertion sort, are introduced in the exercises, and the shaker sort is introduced in the Supplementary Exercises. In Section 5.4, we will discuss the merge sort and introduce the quick sort in the exercises in that section; the tournament sort is introduced in the exercise set in Section 11.2. We cover sorting algorithms both because sorting is an important problem and because these algorithms can serve as examples for many important concepts.

Links

THE BUBBLE SORT The **bubble sort** is one of the simplest sorting algorithms, but not one of the most efficient. It puts a list into increasing order by successively comparing adjacent elements, interchanging them if they are in the wrong order. To carry out the bubble sort, we perform the basic operation, that is, interchanging a larger element with a smaller one following it, starting at the beginning of the list, for a full pass. We iterate this procedure until the sort is complete. Pseudocode for the bubble sort is given as Algorithm 4. We can imagine the elements in the list placed in a column. In the bubble sort, the smaller elements ''bubble'' to the top as they are interchanged with larger elements. The larger elements ''sink'' to the bottom. This is illustrated in Example 4.

EXAMPLE 4 Use the bubble sort to put 3, 2, 4, 1, 5 into increasing order.

Solution: The steps of this algorithm are illustrated in Figure 1. Begin by comparing the first two elements, 3 and 2. Because $3 > 2$, interchange 3 and 2, producing the list 2, 3, 4, 1, 5. Because $3 < 4$, continue by comparing 4 and 1. Because $4 > 1$, interchange 1 and 4, producing the list 2, 3, 1, 4, 5. Because $4 < 5$, the first pass is complete. The first pass guarantees that the largest element, 5, is in the correct position.

The second pass begins by comparing 2 and 3. Because these are in the correct order, 3 and 1 are compared. Because $3 > 1$, these numbers are interchanged, producing 2, 1, 3, 4, 5. Because $3 < 4$, these numbers are in the correct order. It is not necessary to do any more comparisons for this pass because 5 is already in the correct position. The second pass guarantees that the two largest elements, 4 and 5, are in their correct positions.

The third pass begins by comparing 2 and 1. These are interchanged because $2 > 1$, producing 1, 2, 3, 4, 5. Because $2 < 3$, these two elements are in the correct order. It is not necessary to do any more comparisons for this pass because 4 and 5 are already in the correct positions. The third pass guarantees that the three largest elements, 3, 4, and 5, are in their correct positions.

The fourth pass consists of one comparison, namely, the comparison of 1 and 2. Because $1 < 2$, these elements are in the correct order. This completes the bubble sort. ◀

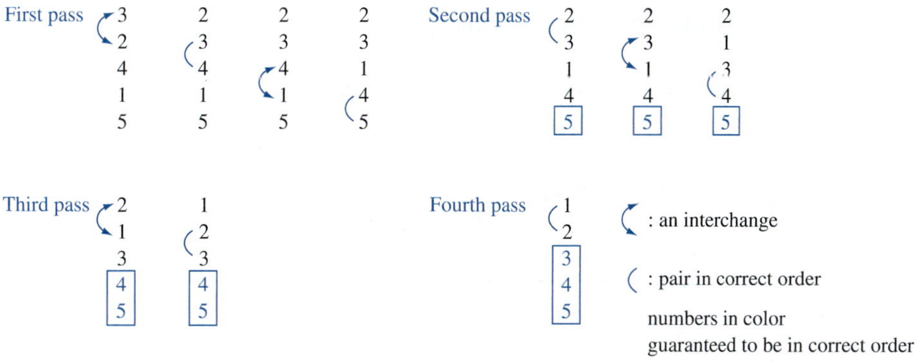

FIGURE 1 The steps of a bubble sort.

ALGORITHM 4 The Bubble Sort.

procedure *bubblesort*($a_1, \ldots, a_n$: real numbers with $n \geq 2$)
for $i := 1$ **to** $n - 1$
  for $j := 1$ **to** $n - i$
    if $a_j > a_{j+1}$ **then** interchange a_j and a_{j+1}
$\{a_1, \ldots, a_n$ is in increasing order$\}$

THE INSERTION SORT The **insertion sort** is a simple sorting algorithm, but it is usually not the most efficient. To sort a list with n elements, the insertion sort begins with the second element. The insertion sort compares this second element with the first element and inserts it before the first element if it does not exceed the first element and after the first element if it exceeds the first element. At this point, the first two elements are in the correct order. The third element is then compared with the first element, and if it is larger than the first element, it is compared with the second element; it is inserted into the correct position among the first three elements.

Links

In general, in the jth step of the insertion sort, the jth element of the list is inserted into the correct position in the list of the previously sorted $j - 1$ elements. To insert the jth element in the list, a linear search technique is used (see Exercise 45); the jth element is successively compared with the already sorted $j - 1$ elements at the start of the list until the first element that is not less than this element is found or until it has been compared with all $j - 1$ elements; the jth element is inserted in the correct position so that the first j elements are sorted. The algorithm continues until the last element is placed in the correct position relative to the already sorted list of the first $n - 1$ elements. The insertion sort is described in pseudocode in Algorithm 5.

EXAMPLE 5 Use the insertion sort to put the elements of the list 3, 2, 4, 1, 5 in increasing order.

Solution: The insertion sort first compares 2 and 3. Because $3 > 2$, it places 2 in the first position, producing the list 2, 3, 4, 1, 5 (the sorted part of the list is shown in color). At this point, 2 and 3 are in the correct order. Next, it inserts the third element, 4, into the already sorted part of the list by making the comparisons $4 > 2$ and $4 > 3$. Because $4 > 3$, 4 remains in the third position. At this point, the list is 2, 3, 4, 1, 5 and we know that the ordering of the first three elements is correct. Next, we find the correct place for the fourth element, 1, among the already sorted elements, 2, 3, 4. Because $1 < 2$, we obtain the list 1, 2, 3, 4, 5. Finally, we insert 5 into the correct position by successively comparing it to 1, 2, 3, and 4. Because $5 > 4$, it stays at the end of the list, producing the correct order for the entire list. ◀

ALGORITHM 5 The Insertion Sort.

procedure *insertion sort*($a_1, a_2, \ldots, a_n$: real numbers with $n \geq 2$)
for $j := 2$ **to** n
  $i := 1$
  while $a_j > a_i$
    $i := i + 1$
  $m := a_j$
  for $k := 0$ **to** $j - i - 1$
    $a_{j-k} := a_{j-k-1}$
  $a_i := m$
$\{a_1, \ldots, a_n$ is in increasing order$\}$

3.1.4 String Matching

Although searching and sorting are the most commonly encountered problems in computer science, many other problems arise frequently. One of these problems asks where a particular string of characters P, called the **pattern**, occurs, if it does, within another string T, called the **text**. For instance, we can ask whether the pattern 101 can be found within the string 11001011. By inspection we can see that the pattern 101 occurs within the text 11001011 at a shift of four characters, because 101 is the string formed by the fifth, sixth, and seventh characters of the text. On the other hand, the pattern 111 does not occur within the text 110110001101.

Finding where a pattern occurs in a text string is called **string matching**. String matching plays an essential role in a wide variety of applications, including text editing, spam filters, systems that look for attacks in a computer network, search engines, plagiarism detection, bioinformatics, and many other important applications. For example, in text editing, the string matching problem arises whenever we need to find all occurrences of a string so that we can replace this string with a different string. Search engines look for matching of search keywords with words on web pages. Many problems in bioinformatics arise in the study of DNA molecules, which are made up of four bases: thymine (T), adenine (A), cytosine (C), and guanine (G). The process of DNA sequencing is the determination of the order of the four bases in DNA. This leads to string matching problems involving strings made up from the four letters T, A, C, and G. For instance, we can ask whether the pattern CAG occurs in the text CATCACAGAGA. The answer is yes, because it occurs with a shift of five characters. Solving questions about the genome requires the use of efficient algorithms for string matching, especially because a string representing a human genome is about 3×10^9 characters long.

We will now describe a brute force algorithm, Algorithm 6, for string matching, called the **naive string matcher**. The input to this algorithm is the pattern we wish to match, $P = p_1p_2 \dots p_m$, and the text, $T = t_1t_2 \dots t_n$. When this pattern begins at position $s + 1$ in the text T, we say that P occurs with **shift** s in T, that is, when $t_{s+1} = p_1, t_{s+2} = p_2, \dots, t_{s+m} = p_m$. To find all valid shifts, the naive string matcher runs through all possible shifts s from $s = 0$ to $s = n - m$, checking whether s is a valid shift. In Figure 2, we display the operation of Algorithm 6 when it is used to search for the pattern $P = eye$ in the text $T = eceyeye$.

ALGORITHM 6 Naive String Matcher.

```
procedure string match (n, m: positive integers, m ≤ n, t1, t2, ..., tn, p1, p2, ..., pm: characters)
    for s := 0 to n − m
        j := 1
        while (j ≤ m and t(s+j) = pj)
            j := j + 1
        if j > m then print "s is a valid shift"
```

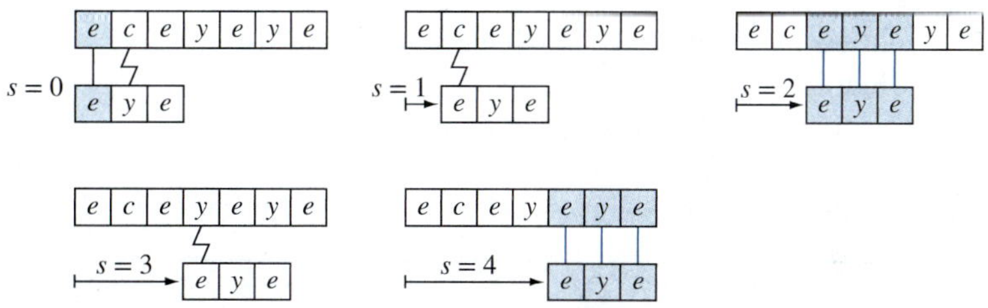

FIGURE 2 The steps of the naive string matcher with $P = eye$ in $T = eceyeye$. Matches are identified with a solid line and mismatches with a jagged line. The algorithm finds two valid shifts, $s = 2$ and $s = 4$.

Many other string matching algorithms have been developed besides the naive string matcher. These algorithms use a surprisingly wide variety of approaches to make them more efficient than the naive string matcher. To learn more about these algorithms, consult [CoLeRiSt09], as well as books on algorithms in bioinformatics.

3.1.5 Greedy Algorithms

"Greed is good ... Greed is right, greed works. Greed clarifies ..." – spoken by the character Gordon Gecko in the film *Wall Street.*

Many algorithms we will study in this book are designed to solve **optimization problems**. The goal of such problems is to find a solution to the given problem that either minimizes or maximizes the value of some parameter. Optimization problems studied later in this text include finding a route between two cities with least total mileage, determining a way to encode messages using the fewest bits possible, and finding a set of fiber links between network nodes using the least amount of fiber.

Links

You have to prove that a greedy algorithm always finds an optimal solution.

Surprisingly, one of the simplest approaches often leads to a solution of an optimization problem. This approach selects the best choice at each step, instead of considering all sequences of steps that may lead to an optimal solution. Algorithms that make what seems to be the "best" choice at each step are called **greedy algorithms**. Once we know that a greedy algorithm finds a feasible solution, we need to determine whether it has found an optimal solution. (Note that we call the algorithm "greedy" whether or not it finds an optimal solution.) To do this, we either prove that the solution is optimal or we show that there is a counterexample where the algorithm yields a nonoptimal solution. To make these concepts more concrete, we will consider the **cashier's algorithm** that makes change using coins. (This algorithm is called the cashier's algorithm because cashiers often used this algorithm for making change in the days before cash registers became electronic.)

EXAMPLE 6 Consider the problem of making n cents change with quarters, dimes, nickels, and pennies, and using the least total number of coins. We can devise a greedy algorithm for making change for n cents by making a locally optimal choice at each step; that is, at each step we choose the coin of the largest denomination possible to add to the pile of change without exceeding n cents. For example, to make change for 67 cents, we first select a quarter (leaving 42 cents). We next select a second quarter (leaving 17 cents), followed by a dime (leaving 7 cents), followed by a nickel (leaving 2 cents), followed by a penny (leaving 1 cent), followed by a penny. ◀

Demo

We display the cashier's algorithm for n cents, using any set of denominations of coins, as Algorithm 7.

ALGORITHM 7 Cashier's Algorithm.

procedure *change*($c_1, c_2, \ldots, c_r$: values of denominations of coins, where
  $c_1 > c_2 > \cdots > c_r$; n: a positive integer)
for $i := 1$ **to** r
  $d_i := 0$ {d_i counts the coins of denomination c_i used}
  while $n \geq c_i$
    $d_i := d_i + 1$ {add a coin of denomination c_i}
    $n := n - c_i$
{d_i is the number of coins of denomination c_i in the change for $i = 1, 2, \ldots, r$}

We have described the cashier's algorithm, a greedy algorithm for making change, using any finite set of coins with denominations $c_1, c_2, \ldots, c_r$. In the particular case where the four denominations are quarters, dimes, nickels, and pennies, we have $c_1 = 25$, $c_2 = 10$, $c_3 = 5$, and $c_4 = 1$. For this case, we will show that this algorithm leads to an optimal solution in the sense that it uses the fewest coins possible. Before we embark on our proof, we show that there are sets of coins for which the cashier's algorithm (Algorithm 7) does not necessarily produce change using the fewest coins possible. For example, if we have only quarters, dimes, and pennies (and no nickels) to use, the cashier's algorithm would make change for 30 cents using six coins—a quarter and five pennies—whereas we could have used three coins, namely, three dimes.

LEMMA 1 If n is a positive integer, then n cents in change using quarters, dimes, nickels, and pennies using the fewest coins possible has at most two dimes, at most one nickel, at most four pennies, and cannot have two dimes and a nickel. The amount of change in dimes, nickels, and pennies cannot exceed 24 cents.

Proof: We use a proof by contradiction. We will show that if we had more than the specified number of coins of each type, we could replace them using fewer coins that have the same value. We note that if we had three dimes we could replace them with a quarter and a nickel, if we had two nickels we could replace them with a dime, if we had five pennies we could replace them with a nickel, and if we had two dimes and a nickel we could replace them with a quarter. Because we can have at most two dimes, one nickel, and four pennies, but we cannot have two dimes and a nickel, it follows that 24 cents is the most money we can have in dimes, nickels, and pennies when we make change using the fewest number of coins for n cents. ◁

THEOREM 1 The cashier's algorithm (Algorithm 7) always makes changes using the fewest coins possible when change is made from quarters, dimes, nickels, and pennies.

Proof: We will use a proof by contradiction. Suppose that there is a positive integer n such that there is a way to make change for n cents using quarters, dimes, nickels, and pennies that uses fewer coins than the greedy algorithm finds. We first note that q', the number of quarters used in this optimal way to make change for n cents, must be the same as q, the number of quarters used by the greedy algorithm. To show this, first note that the greedy algorithm uses the most quarters possible, so $q' \leq q$. However, it is also the case that q' cannot be less than q. If it were, we would need to make up at least 25 cents from dimes, nickels, and pennies in this optimal way to make change. But this is impossible by Lemma 1.

Because there must be the same number of quarters in the two ways to make change, the value of the dimes, nickels, and pennies in these two ways must be the same, and these coins are worth no more than 24 cents. There must be the same number of dimes, because the greedy algorithm used the most dimes possible and by Lemma 1, when change is made using the fewest coins possible, at most one nickel and at most four pennies are used, so that the most dimes possible are also used in the optimal way to make change. Similarly, we have the same number of nickels and, finally, the same number of pennies. ◁

A greedy algorithm makes the best choice at each step according to a specified criterion. The next example shows that it can be difficult to determine which of many possible criteria to choose.

EXAMPLE 7 Suppose we have a group of proposed talks with preset start and end times. Devise a greedy algorithm to schedule as many of these talks as possible in a lecture hall, under the assumptions that once a talk starts, it continues until it ends, no two talks can proceed at the same time, and a talk can begin at the same time another one ends. Assume that talk j begins at time s_j (where s stands for *start*) and ends at time e_j (where e stands for *end*).

Solution: To use a greedy algorithm to schedule the most talks, that is, an optimal schedule, we need to decide how to choose which talk to add at each step. There are many criteria we could use to select a talk at each step, where we chose from the talks that do not overlap talks already selected. For example, we could add talks in order of earliest start time, we could add talks in order of shortest time, we could add talks in order of earliest finish time, or we could use some other criterion.

We now consider these possible criteria. Suppose we add the talk that starts earliest among the talks compatible with those already selected. We can construct a counterexample to see that the resulting algorithm does not always produce an optimal schedule. For instance, suppose that we have three talks: Talk 1 starts at 8 A.M. and ends at 12 noon, Talk 2 starts at 9 A.M. and ends at 10 A.M., and Talk 3 starts at 11 A.M. and ends at 12 noon. We first select the Talk 1 because it starts earliest. But once we have selected Talk 1 we cannot select either Talk 2 or Talk 3 because both overlap Talk 1. Hence, this greedy algorithm selects only one talk. This is not optimal because we could schedule Talk 2 and Talk 3, which do not overlap.

Now suppose we add the talk that is shortest among the talks that do not overlap any of those already selected. Again we can construct a counterexample to show that this greedy algorithm does not always produce an optimal schedule. So, suppose that we have three talks: Talk 1 starts at 8 A.M. and ends at 9:15 A.M., Talk 2 starts at 9 A.M. and ends at 10 A.M., and Talk 3 starts at 9:45 A.M. and ends at 11 A.M. We select Talk 2 because it is shortest, requiring one hour. Once we select Talk 2, we cannot select either Talk 1 or Talk 3 because neither is compatible with Talk 2. Hence, this greedy algorithm selects only one talk. However, it is possible to select two talks, Talk 1 and Talk 3, which are compatible.

However, it can be shown that we schedule the most talks possible if in each step we select the talk with the earliest ending time among the talks compatible with those already selected. We will prove this in Chapter 5 using the method of mathematical induction. The first step we will make is to sort the talks according to increasing finish time. After this sorting, we relabel the talks so that $e_1 \le e_2 \le \cdots \le e_n$. The resulting greedy algorithm is given as Algorithm 8. ◀

ALGORITHM 8 Greedy Algorithm for Scheduling Talks.

procedure *schedule*($s_1, s_2, \ldots, s_n$: start times of talks,
    $e_1, e_2, \ldots, e_n$: ending times of talks)
sort talks by finish time and reorder so that $e_1 \le e_2 \le \cdots \le e_n$
$S := \emptyset$
for $j := 1$ **to** n
    if talk j is compatible with S **then**
        $S := S \cup \{\text{talk } j\}$
return S {S is the set of talks scheduled}

3.1.6 The Halting Problem

Links

We will now describe a proof of one of the most famous theorems in computer science. We will show that there is a problem that cannot be solved using any procedure. That is, we will show there are unsolvable problems. The problem we will study is the **halting problem**. It asks whether there is a procedure that does this: It takes as input a computer program and input to the program and determines whether the program will eventually stop when run with this input. It would be convenient to have such a procedure, if it existed. Certainly being able to test whether a program entered into an infinite loop would be helpful when writing and debugging programs. However, in 1936 Alan Turing showed that no such procedure exists (see his biography in Section 13.4).

Before we present a proof that the halting problem is unsolvable, first note that we cannot simply run a program and observe what it does to determine whether it terminates when run

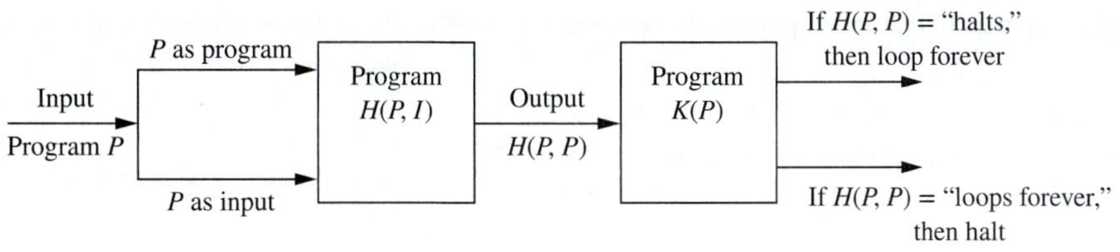

FIGURE 3 Showing that the halting problem is unsolvable.

with the given input. If the program halts, we have our answer, but if it is still running after any fixed length of time has elapsed, we do not know whether it will never halt or we just did not wait long enough for it to terminate. After all, it is not hard to design a program that will stop only after more than a billion years has elapsed.

We will describe Turing's proof that the halting problem is unsolvable; it is a proof by contradiction. (The reader should note that our proof is not completely rigorous, because we have not explicitly defined what a procedure is. To remedy this, the concept of a Turing machine is needed. This concept is introduced in Section 13.5.)

Proof: Assume there is a solution to the halting problem, a procedure called $H(P, I)$. The procedure $H(P, I)$ takes two inputs, one a program P and the other I, an input to the program P. $H(P,I)$ generates the string "halt" as output if H determines that P stops when given I as input. Otherwise, $H(P, I)$ generates the string "loops forever" as output. We will now derive a contradiction.

When a procedure is coded, it is expressed as a string of characters; this string can be interpreted as a sequence of bits. This means that a program itself can be used as data. Therefore, a program can be thought of as input to another program, or even itself. Hence, H can take a program P as both of its inputs, which are a program and input to this program. H should be able to determine whether P will halt when it is given a copy of itself as input.

To show that no procedure H exists that solves the halting problem, we construct a simple procedure $K(P)$, which works as follows, making use of the output $H(P, P)$. If the output of $H(P, P)$ is "loops forever," which means that P loops forever when given a copy of itself as input, then $K(P)$ halts. If the output of $H(P, P)$ is "halt," which means that P halts when given a copy of itself as input, then $K(P)$ loops forever. That is, $K(P)$ does the opposite of what the output of $H(P, P)$ specifies. (See Figure 3.)

Now suppose we provide K as input to K. We note that if the output of $H(K, K)$ is "loops forever," then by the definition of K, we see that $K(K)$ halts. This means that by the definition of H, the output of $H(K, K)$ is "halt," which is a contradiction. Otherwise, if the output of $H(K, K)$ is "halts," then by the definition of K, we see that $K(K)$ loops forever, which means that by the definition of H, the output of $H(K, K)$ is "loops forever." This is also a contradiction. Thus, H cannot always give the correct answers. Consequently, there is no procedure that solves the halting problem. ◁

Exercises

1. List all the steps used by Algorithm 1 to find the maximum of the list 1, 8, 12, 9, 11, 2, 14, 5, 10, 4.

2. Determine which properties of an algorithm described in the text (after Algorithm 1) the following procedures have and which they lack.

a) **procedure** *double*(n: positive integer)
  while $n > 0$
    $n := 2n$

b) **procedure** *divide*(n: positive integer)
  while $n \geq 0$
    $m := 1/n$
    $n := n - 1$

c) **procedure** *sum*(n: positive integer)
  $sum := 0$
  while $i < 10$
    $sum := sum + i$

d) **procedure** *choose*(a, b: integers)
  $x :=$ either a or b

3. Devise an algorithm that finds the sum of all the integers in a list.

4. Describe an algorithm that takes as input a list of n integers and produces as output the largest difference obtained by subtracting an integer in the list from the one following it.

5. Describe an algorithm that takes as input a list of n integers in nondecreasing order and produces the list of all values that occur more than once. (Recall that a list of integers is **nondecreasing** if each integer in the list is at least as large as the previous integer in the list.)

6. Describe an algorithm that takes as input a list of n integers and finds the number of negative integers in the list.

7. Describe an algorithm that takes as input a list of n integers and finds the location of the last even integer in the list or returns 0 if there are no even integers in the list.

8. Describe an algorithm that takes as input a list of n distinct integers and finds the location of the largest even integer in the list or returns 0 if there are no even integers in the list.

9. A **palindrome** is a string that reads the same forward and backward. Describe an algorithm for determining whether a string of n characters is a palindrome.

10. Devise an algorithm to compute x^n, where x is a real number and n is an integer. [*Hint:* First give a procedure for computing x^n when n is nonnegative by successive multiplication by x, starting with 1. Then extend this procedure, and use the fact that $x^{-n} = 1/x^n$ to compute x^n when n is negative.]

11. Describe an algorithm that interchanges the values of the variables x and y, using only assignments. What is the minimum number of assignment statements needed to do this?

12. Describe an algorithm that uses only assignment statements that replaces the triple (x, y, z) with (y, z, x). What is the minimum number of assignment statements needed?

13. List all the steps used to search for 9 in the sequence 1, 3, 4, 5, 6, 8, 9, 11 using

a) a linear search. **b)** a binary search.

14. List all the steps used to search for 7 in the sequence given in Exercise 13 for both a linear search and a binary search.

15. Describe an algorithm that inserts an integer x in the appropriate position into the list $a_1, a_2, \ldots, a_n$ of integers that are in increasing order.

16. Describe an algorithm for finding the smallest integer in a finite sequence of natural numbers.

17. Describe an algorithm that locates the first occurrence of the largest element in a finite list of integers, where the integers in the list are not necessarily distinct.

18. Describe an algorithm that locates the last occurrence of the smallest element in a finite list of integers, where the integers in the list are not necessarily distinct.

19. Describe an algorithm that produces the maximum, median, mean, and minimum of a set of three integers. (The **median** of a set of integers is the middle element in the list when these integers are listed in order of increasing size. The **mean** of a set of integers is the sum of the integers divided by the number of integers in the set.)

20. Describe an algorithm for finding both the largest and the smallest integers in a finite sequence of integers.

21. Describe an algorithm that puts the first three terms of a sequence of integers of arbitrary length in increasing order.

22. Describe an algorithm to find the longest word in an English sentence (where a sentence is a sequence of symbols, either a letter or a blank, which can then be broken into alternating words and blanks).

23. Describe an algorithm that determines whether a function from a finite set of integers to another finite set of integers is onto.

24. Describe an algorithm that determines whether a function from a finite set to another finite set is one-to-one.

25. Describe an algorithm that will count the number of 1s in a bit string by examining each bit of the string to determine whether it is a 1 bit.

26. Change Algorithm 3 so that the binary search procedure compares x to a_m at each stage of the algorithm, with the algorithm terminating if $x = a_m$. What advantage does this version of the algorithm have?

27. The **ternary search algorithm** locates an element in a list of increasing integers by successively splitting the list into three sublists of equal (or as close to equal as possible) size, and restricting the search to the appropriate piece. Specify the steps of this algorithm.

28. Specify the steps of an algorithm that locates an element in a list of increasing integers by successively splitting the list into four sublists of equal (or as close to equal as possible) size, and restricting the search to the appropriate piece.

In a list of elements the same element may appear several times. A **mode** of such a list is an element that occurs at least as often as each of the other elements; a list has more than one mode when more than one element appears the maximum number of times.

29. Devise an algorithm that finds a mode in a list of nondecreasing integers. (Recall that a list of integers is nondecreasing if each term is at least as large as the preceding term.)

30. Devise an algorithm that finds all modes. (Recall that a list of integers is nondecreasing if each term of the list is at least as large as the preceding term.)

31. Two strings are **anagrams** if each can be formed from the other string by rearranging its characters. Devise an algorithm to determine whether two strings are anagrams

a) by first finding the frequency of each character that appears in the strings.

b) by first sorting the characters in both strings.

32. Given n real numbers $x_1, x_2, \ldots, x_n$, find the two that are closest together by

a) a brute force algorithm that finds the distance between every pair of these numbers.

b) sorting the numbers and computing the least number of distances needed to solve the problem.

33. Devise an algorithm that finds the first term of a sequence of integers that equals some previous term in the sequence.

34. Devise an algorithm that finds all terms of a finite sequence of integers that are greater than the sum of all previous terms of the sequence.

35. Devise an algorithm that finds the first term of a sequence of positive integers that is less than the immediately preceding term of the sequence.

36. Use the bubble sort to sort 6, 2, 3, 1, 5, 4, showing the lists obtained at each step.

37. Use the bubble sort to sort 3, 1, 5, 7, 4, showing the lists obtained at each step.

38. Use the bubble sort to sort d, f, k, m, a, b, showing the lists obtained at each step.

***39.** Adapt the bubble sort algorithm so that it stops when no interchanges are required. Express this more efficient version of the algorithm in pseudocode.

40. Use the insertion sort to sort the list in Exercise 36, showing the lists obtained at each step.

41. Use the insertion sort to sort the list in Exercise 37, showing the lists obtained at each step.

42. Use the insertion sort to sort the list in Exercise 38, showing the lists obtained at each step.

The **selection sort** begins by finding the least element in the list. This element is moved to the front. Then the least element among the remaining elements is found and put into the second position. This procedure is repeated until the entire list has been sorted.

Links

43. Sort these lists using the selection sort.

a) 3, 5, 4, 1, 2 **b)** 5, 4, 3, 2, 1

c) 1, 2, 3, 4, 5

44. Write the selection sort algorithm in pseudocode.

45. Describe an algorithm based on the linear search for determining the correct position in which to insert a new element in an already sorted list.

46. Describe an algorithm based on the binary search for determining the correct position in which to insert a new element in an already sorted list.

47. How many comparisons does the insertion sort use to sort the list $1, 2, \ldots, n$?

48. How many comparisons does the insertion sort use to sort the list $n, n-1, \ldots, 2, 1$?

The **binary insertion sort** is a variation of the insertion sort that uses a binary search technique (see Exercise 46) rather than a linear search technique to insert the ith element in the correct place among the previously sorted elements.

49. Show all the steps used by the binary insertion sort to sort the list 3, 2, 4, 5, 1, 6.

50. Compare the number of comparisons used by the insertion sort and the binary insertion sort to sort the list 7, 4, 3, 8, 1, 5, 4, 2.

***51.** Express the binary insertion sort in pseudocode.

52. a) Devise a variation of the insertion sort that uses a linear search technique that inserts the jth element in the correct place by first comparing it with the $(j-1)$st element, then the $(j-2)$th element if necessary, and so on.

b) Use your algorithm to sort 3, 2, 4, 5, 1, 6.

c) Answer Exercise 47 using this algorithm.

d) Answer Exercise 48 using this algorithm.

53. When a list of elements is in close to the correct order, would it be better to use an insertion sort or its variation described in Exercise 52?

54. List all the steps the naive string matcher uses to find all occurrences of the pattern FE in the text COVFEFE.

55. List all the steps the naive string matcher uses to find all occurrences of the pattern ACG in the text TACAGACG.

56. Use the cashier's algorithm to make change using quarters, dimes, nickels, and pennies for

a) 87 cents. **b)** 49 cents.

c) 99 cents. **d)** 33 cents.

57. Use the cashier's algorithm to make change using quarters, dimes, nickels, and pennies for

a) 51 cents. **b)** 69 cents.

c) 76 cents. **d)** 60 cents.

58. Use the cashier's algorithm to make change using quarters, dimes, and pennies (but no nickels) for each of the amounts given in Exercise 56. For which of these amounts does the greedy algorithm use the fewest coins of these denominations possible?

59. Use the cashier's algorithm to make change using quarters, dimes, and pennies (but no nickels) for each of the amounts given in Exercise 57. For which of these amounts does the greedy algorithm use the fewest coins of these denominations possible?

60. Show that if there were a coin worth 12 cents, the cashier's algorithm using quarters, 12-cent coins, dimes, nickels, and pennies would not always produce change using the fewest coins possible.

61. Use Algorithm 8 to schedule the largest number of talks in a lecture hall from a proposed set of talks, if the starting and ending times of the talks are 9:00 A.M. and 9:45 A.M.; 9:30 A.M. and 10:00 A.M.; 9:50 A.M. and 10:15 A.M.; 10:00 A.M. and 10:30 A.M.; 10:10 A.M. and 10:25 A.M.; 10:30 A.M. and 10:55 A.M.; 10:15 A.M. and 10:45 A.M.; 10:30 A.M. and 11:00 A.M.; 10:45 A.M. and 11:30 A.M.; 10:55 A.M. and 11:25 A.M.; 11:00 A.M. and 11:15 A.M.

62. Show that a greedy algorithm that schedules talks in a lecture hall, as described in Example 7, by selecting at each step the talk that overlaps the fewest other talks, does not always produce an optimal schedule.

*63. a) Devise a greedy algorithm that determines the fewest lecture halls needed to accommodate n talks given the starting and ending time for each talk.

b) Prove that your algorithm is optimal.

Suppose we have s men $m_1, m_2, \ldots, m_s$ and s women $w_1, w_2, \ldots, w_s$. We wish to match each person with a member of the opposite gender. Furthermore, suppose that each person ranks, in order of preference, with no ties, the people of the opposite gender. We say that a matching of people of opposite genders to form couples is **stable** if we cannot find a man m and a woman w who are not assigned to each other such that m prefers w over his assigned partner and w prefers m to her assigned partner.

Links

64. Suppose we have three men m_1, m_2, and m_3 and three women w_1, w_2, and w_3. Furthermore, suppose that the preference rankings of the men for the three women, from highest to lowest, are m_1: w_3, w_1, w_2; m_2: w_1, w_2, w_3; m_3: w_2, w_3, w_1; and the preference rankings of the women for the three men, from highest to lowest, are w_1: m_1, m_2, m_3; w_2: m_2, m_1, m_3; w_3: m_3, m_2, m_1. For each of the six possible matchings of men and women to form three couples, determine whether this matching is stable.

The **deferred acceptance algorithm**, also known as the **Gale-Shapley algorithm**, can be used to construct a stable matching of men and women. In this algorithm, members of one gender are the **suitors** and members of the other gender the **suitees**. The algorithm uses a sequence of rounds; in each round every suitor whose proposal was rejected in the previous round proposes to his or her highest ranking suitee who has not already rejected a proposal from this suitor. A suitee rejects all proposals except that from the suitor that this suitee ranks highest among all the suitors who have proposed to this suitee in this round or previous rounds. The proposal of this highest ranking suitor remains pending and is rejected in a later round if a more appealing suitor proposes in that round. The series of rounds ends when every suitor has exactly one pending proposal. All pending proposals are then accepted.

65. Write the deferred acceptance algorithm in pseudocode.

66. Show that the deferred acceptance algorithm terminates.

*67. Show that the deferred acceptance always terminates with a stable assignment.

An element of a sequence is called a **majority element** if it occurs repeatedly for more than half the terms of the sequence. The **Boyer-Moore majority vote algorithm** (named after Robert Boyer and J. Strother Moore) finds the majority element of a sequence, if it exists. The algorithm maintains a counter that is initially set to 0 and a temporary candidate for a majority element initially with no assigned value. The algorithm processes the elements in order. When it processes the first element, this element becomes the majority candidate and the counter is set equal to 1. Then, as it processes the remaining elements in order, if the counter is 0, this element becomes the majority candidate and the counter is set equal to 1, while if the counter is nonzero, the counter is incremented (that is, 1 is added to it) or decremented (that is, 1 is subtracted from it), depending on whether this element equals the current candidate. After all the terms are processed, the majority candidate is the majority element, if it exists.

68. a) Explain why a sequence has at most one majority element.

b) Show all the steps of the Boyer-Moore majority vote algorithm when given the sequence 2, 1, 3, 3, 2, 3.

c) Express the Boyer-Moore majority vote algorithm in pseudocode.

d) Explain how you can determine whether the majority candidate element produced by the Boyer-Moore algorithm is actually a majority element.

*69. a) Prove that the Boyer-Moore majority vote algorithm outputs the majority element of a sequence, if it exists.

b) Prove or disprove that the majority candidate of the Boyer-Moore majority vote algorithm will be a mode of the sequence (that is, its most common element) even when no majority element exists.

70. Show that the problem of determining whether a program with a given input ever prints the digit 1 is unsolvable.

71. Show that the following problem is solvable. Given two programs with their inputs and the knowledge that exactly one of them halts, determine which halts.

72. Show that the problem of deciding whether a specific program with a specific input halts is solvable.

3.2 The Growth of Functions

3.2.1 Introduction

In Section 3.1 we discussed the concept of an algorithm. We introduced algorithms that solve a variety of problems, including searching for an element in a list and sorting a list. In Section 3.3 we will study the number of operations used by these algorithms. In particular, we will estimate the number of comparisons used by the linear and binary search algorithms to find an element in a sequence of n elements. We will also estimate the number of comparisons used by the bubble sort and by the insertion sort to sort a list of n elements. The time required to solve a problem depends on more than only the number of operations it uses. The time also depends on the hardware and software used to run the program that implements the algorithm. However, when we change the hardware and software used to implement an algorithm, we can closely approximate the time required to solve a problem of size n by multiplying the previous time required by a constant. For example, on a supercomputer we might be able to solve a problem of size n a million times faster than we can on a PC. However, this factor of one million will not depend on n (except perhaps in some minor ways). One of the advantages of using **big-O notation**, which we introduce in this section, is that we can estimate the growth of a function without worrying about constant multipliers or smaller order terms. This means that, using big-O notation, we do not have to worry about the hardware and software used to implement an algorithm. Furthermore, using big-O notation, we can assume that the different operations used in an algorithm take the same time, which simplifies the analysis considerably.

Big-O notation is used extensively to estimate the number of operations an algorithm uses as its input grows. With the help of this notation, we can determine whether it is practical to use a particular algorithm to solve a problem as the size of the input increases. Furthermore, using big-O notation, we can compare two algorithms to determine which is more efficient as the size of the input grows. For instance, if we have two algorithms for solving a problem, one using $100n^2 + 17n + 4$ operations and the other using n^3 operations, big-O notation can help us see that the first algorithm uses far fewer operations when n is large, even though it uses more operations for small values of n, such as $n = 10$.

This section introduces big-O notation and the related big-Omega and big-Theta notations. We will explain how big-O, big-Omega, and big-Theta estimates are constructed and establish estimates for some important functions that are used in the analysis of algorithms.

3.2.2 Big-O Notation

The growth of functions is often described using a special notation. Definition 1 describes this notation.

Definition 1 Let f and g be functions from the set of integers or the set of real numbers to the set of real numbers. We say that $f(x)$ is $O(g(x))$ if there are positive constants C and k such that

$$|f(x)| \le C|g(x)|$$

whenever $x > k$. [This is read as "$f(x)$ is big-oh of $g(x)$."]

Remark: Intuitively, the definition that $f(x)$ is $O(g(x))$ says that $f(x)$ grows slower than some fixed multiple of $g(x)$ as x grows without bound.

Assessment

The constants C and k in the definition of big-O notation are called **witnesses** to the relationship $f(x)$ is $O(g(x))$. To establish that $f(x)$ is $O(g(x))$ we need only one pair of witnesses to this relationship. That is, to show that $f(x)$ is $O(g(x))$, we need find only *one* pair of constants C and k, the witnesses, such that $|f(x)| \le C|g(x)|$ whenever $x > k$.

Links

Note that when there is one pair of witnesses to the relationship $f(x)$ is $O(g(x))$, there are *infinitely many* pairs of witnesses. To see this, note that if C and k are one pair of witnesses, then any pair C' and k', where $C < C'$ and $k < k'$, is also a pair of witnesses, because $|f(x)| \le C|g(x)| \le C'|g(x)|$ whenever $x > k' > k$.

THE HISTORY OF BIG-O NOTATION Big-O notation has been used in mathematics for more than a century. In computer science it is widely used in the analysis of algorithms, as will be seen in Section 3.3. The German mathematician Paul Bachmann first introduced big-O notation in 1892 in an important book on number theory. The big-O symbol is sometimes called a **Landau symbol** after the German mathematician Edmund Landau, who used this notation throughout his work. The use of big-O notation in computer science was popularized by Donald Knuth, who also introduced the big-Ω and big-Θ notations defined later in this section.

WORKING WITH THE DEFINITION OF BIG-*O* NOTATION A useful approach for finding a pair of witnesses is to first select a value of k for which the size of $|f(x)|$ can be readily estimated when $x > k$ and to see whether we can use this estimate to find a value of C for which $|f(x)| \le C|g(x)|$ for $x > k$. This approach is illustrated in Example 1.

EXAMPLE 1 Show that $f(x) = x^2 + 2x + 1$ is $O(x^2)$.

Extra Examples

Solution: We observe that we can readily estimate the size of $f(x)$ when $x > 1$ because $x < x^2$ and $1 < x^2$ when $x > 1$. It follows that

$$0 \le x^2 + 2x + 1 \le x^2 + 2x^2 + x^2 = 4x^2$$

whenever $x > 1$, as shown in Figure 1. Consequently, we can take $C = 4$ and $k = 1$ as witnesses to show that $f(x)$ is $O(x^2)$. That is, $f(x) = x^2 + 2x + 1 < 4x^2$ whenever $x > 1$. (Note that it is not necessary to use absolute values here because all functions in these equalities are positive when x is positive.)

Alternatively, we can estimate the size of $f(x)$ when $x > 2$. When $x > 2$, we have $2x \le x^2$ and $1 \le x^2$. Consequently, if $x > 2$, we have

$$0 \le x^2 + 2x + 1 \le x^2 + x^2 + x^2 = 3x^2.$$

It follows that $C = 3$ and $k = 2$ are also witnesses to the relation $f(x)$ is $O(x^2)$.

Observe that in the relationship "$f(x)$ is $O(x^2)$," x^2 can be replaced by any function that has larger values than x^2 for all $x \ge k$ for some positive real number k. For example, $f(x)$ is $O(x^3)$, $f(x)$ is $O(x^2 + x + 7)$, and so on.

It is also true that x^2 is $O(x^2 + 2x + 1)$, because $x^2 < x^2 + 2x + 1$ whenever $x > 1$. This means that $C = 1$ and $k = 1$ are witnesses to the relationship x^2 is $O(x^2 + 2x + 1)$. ◀

Note that in Example 1 we have two functions, $f(x) = x^2 + 2x + 1$ and $g(x) = x^2$, such that $f(x)$ is $O(g(x))$ and $g(x)$ is $O(f(x))$—the latter fact following from the inequality $x^2 \le x^2 + 2x + 1$, which holds for all nonnegative real numbers x. We say that two functions

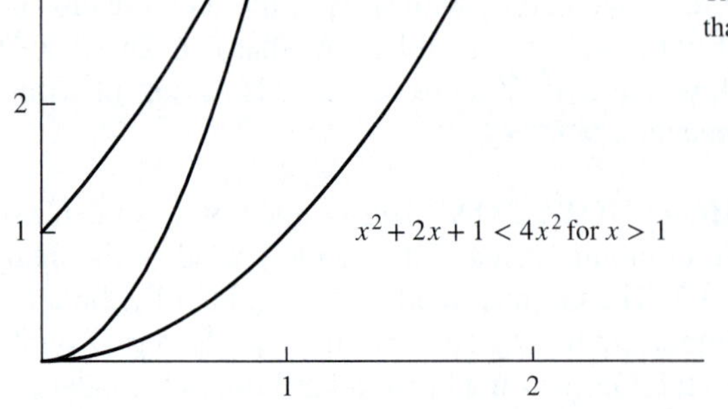

The part of the graph of $f(x) = x^2 + 2x + 1$ that satisfies $f(x) < 4x^2$ is shown in color.

FIGURE 1 **The function $x^2 + 2x + 1$ is $O(x^2)$.**

$f(x)$ and $g(x)$ that satisfy both of these big-O relationships are of the **same order**. We will return to this notion later in this section.

Remark: The fact that $f(x)$ is $O(g(x))$ is sometimes written $f(x) = O(g(x))$. However, the equals sign in this notation does *not* represent a genuine equality. Rather, this notation tells us that an inequality holds relating the values of the functions f and g for sufficiently large numbers in the domains of these functions. However, it is acceptable to write $f(x) \in O(g(x))$ because $O(g(x))$ represents the set of functions that are $O(g(x))$.

When $f(x)$ is $O(g(x))$, and $h(x)$ is a function that has larger absolute values than $g(x)$ does for sufficiently large values of x, it follows that $f(x)$ is $O(h(x))$. In other words, the function $g(x)$ in the relationship $f(x)$ is $O(g(x))$ can be replaced by a function with larger absolute values. To see this, note that if

$$|f(x)| \leq C|g(x)| \qquad \text{if } x > k,$$

and if $|h(x)| > |g(x)|$ for all $x > k$, then

$$|f(x)| \leq C|h(x)| \qquad \text{if } x > k.$$

Hence, $f(x)$ is $O(h(x))$.

When big-O notation is used, the function g in the relationship $f(x)$ is $O(g(x))$ is often chosen to have the smallest growth rate of the functions belonging to a set of reference functions, such as functions of the form x^n, where n is a positive real number. (Important reference functions are discussed later in this section.)

In subsequent discussions, we will almost always deal with functions that take on only positive values. All references to absolute values can be dropped when working with big-O estimates for such functions. Figure 2 illustrates the relationship $f(x)$ is $O(g(x))$.

Links

Source: Bachmann, Paul. Die Arithmetik Der Quadratischen Formen. *Verlag und Druck von BG Teubner. Leipzig. Berlin, 1923.*

PAUL GUSTAV HEINRICH BACHMANN (1837–1920) Paul Bachmann, the son of a Lutheran pastor, shared his father's pious lifestyle and love of music. His mathematical talent was discovered by one of his teachers, even though he had difficulties with some of his early mathematical studies. After recuperating from tuberculosis in Switzerland, Bachmann studied mathematics, first at the University of Berlin and later at Göttingen, where he attended lectures presented by the famous number theorist Dirichlet. He received his doctorate under the German number theorist Kummer in 1862; his thesis was on group theory. Bachmann was a professor at Breslau and later at Münster. After he retired from his professorship, he continued his mathematical writing, played the piano, and served as a music critic for newspapers. Bachmann's mathematical writings include a five-volume survey of results and methods in number theory, a two-volume work on elementary number theory, a book on irrational numbers, and a book on the famous conjecture known as Fermat's Last Theorem. He introduced big-O notation in his 1892 book *Analytische Zahlentheorie.*

Links

Source: Smith Collection, Rare Book & Manuscript Library, Columbia University in the City of New York.

EDMUND LANDAU (1877–1938) Edmund Landau, the son of a Berlin gynecologist, attended high school and university in Berlin. He received his doctorate in 1899, under the direction of Frobenius. Landau first taught at the University of Berlin and then moved to Göttingen, where he was a full professor until the Nazis forced him to stop teaching. Landau's main contributions to mathematics were in the field of analytic number theory. In particular, he established several important results concerning the distribution of primes. He authored a three-volume exposition on number theory as well as other books on number theory and mathematical analysis.

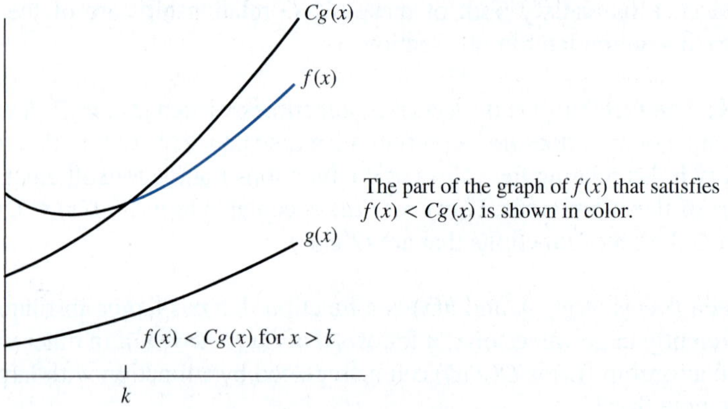

FIGURE 2 The function $f(x)$ is $O(g(x))$.

Example 2 illustrates how big-O notation is used to estimate the growth of functions.

EXAMPLE 2 Show that $7x^2$ is $O(x^3)$.

Solution: Note that when $x > 7$, we have $7x^2 < x^3$. (We can obtain this inequality by multiplying both sides of $x > 7$ by x^2.) Consequently, we can take $C = 1$ and $k = 7$ as witnesses to establish the relationship $7x^2$ is $O(x^3)$. Alternatively, when $x > 1$, we have $7x^2 < 7x^3$, so that $C = 7$ and $k = 1$ are also witnesses to the relationship $7x^2$ is $O(x^3)$. ◄

Links

Courtesy of Stanford University News Service

DONALD E. KNUTH (BORN 1938) Knuth grew up in Milwaukee, where his father taught bookkeeping at a Lutheran high school and owned a small printing business. He was an excellent student, earning academic achievement awards. He applied his intelligence in unconventional ways, winning a contest when he was in the eighth grade by finding over 4500 words that could be formed from the letters in "Ziegler's Giant Bar." This won a television set for his school and a candy bar for everyone in his class.

Knuth had a difficult time choosing physics over music as his major at the Case Institute of Technology. He then switched from physics to mathematics, and in 1960 he received his bachelor of science degree, simultaneously receiving a master of science degree by a special award of the faculty who considered his work outstanding. At Case, he managed the basketball team and applied his talents by constructing a formula for the value of each player. This novel approach was covered by *Newsweek* and by Walter Cronkite on the CBS television network. Knuth began graduate work at the California Institute of Technology in 1960 and received his Ph.D. there in 1963. During this time he worked as a consultant, writing compilers for different computers.

Knuth joined the staff of the California Institute of Technology in 1963, where he remained until 1968, when he took a job as a full professor at Stanford University. He retired as Professor Emeritus in 1992 to concentrate on writing. He is especially interested in updating and completing new volumes of his series *The Art of Computer Programming,* a work that has had a profound influence on the development of computer science, which he began writing as a graduate student in 1962, focusing on compilers. In common jargon, "Knuth," referring to *The Art of Computer Programming*, has come to mean the reference that answers all questions about such topics as data structures and algorithms.

Knuth is the founder of the modern study of computational complexity. He has made fundamental contributions to the subject of compilers. His dissatisfaction with mathematics typography sparked him to invent the now widely used TeX and Metafont systems. TeX has become a standard language for computer typography. Two of the many awards Knuth has received are the 1974 Turing Award and the 1979 National Medal of Technology, awarded to him by President Carter.

Knuth has written for a wide range of professional journals in computer science and in mathematics. However, his first publication, in 1957, when he was a college freshman, was a parody of the metric system called "The Potrzebie Systems of Weights and Measures," which appeared in *MAD Magazine* and has been in reprint several times. He is a church organist, as his father was. He is also a composer of music for the organ. Knuth believes that writing computer programs can be an aesthetic experience, much like writing poetry or composing music.

Knuth pays \$2.56 for the first person to find each error in his books and \$0.32 for significant suggestions. If you send him a letter with an error (you will need to use regular mail, because he has given up reading email), he will eventually inform you whether you were the first person to tell him about this error. Be prepared for a long wait, because he receives an overwhelming amount of mail. (The author received a letter years after sending an error report to Knuth, noting that this report arrived several months after the first report of this error.)

Remark: In Example 2 we did not choose the smallest possible power of x the reference function in the big-O estimate. Note that $7x^2$ is also big-O of x^2 and x^2 grows much slower than x^3. In fact, x^2 would be the smallest possible power of x suitable as the reference function in the big-O estimate.

Example 3 illustrates how to show that a big-O relationship does not hold.

EXAMPLE 3 Show that n^2 is not $O(n)$.

Solution: To show that n^2 is not $O(n)$, we must show that no pair of witnesses C and k exist such that $n^2 \leq Cn$ whenever $n > k$. We will use a proof by contradiction to show this.

Suppose that there are constants C and k for which $n^2 \leq Cn$ whenever $n > k$. Observe that when $n > 0$ we can divide both sides of the inequality $n^2 \leq Cn$ by n to obtain the equivalent inequality $n \leq C$. However, no matter what C and k are, the inequality $n \leq C$ cannot hold for all n with $n > k$. In particular, once we set a value of k, we see that when n is larger than the maximum of k and C, it is not true that $n \leq C$ even though $n > k$. This contradiction shows that n^2 is not $O(n)$. ◀

EXAMPLE 4 Example 2 shows that $7x^2$ is $O(x^3)$. Is it also true that x^3 is $O(7x^2)$?

Solution: To determine whether x^3 is $O(7x^2)$, we need to determine whether witnesses C and k exist, so that $x^3 \leq C(7x^2)$ whenever $x > k$. We will show that no such witnesses exist using a proof by contradiction.

If C and k are witnesses, the inequality $x^3 \leq C(7x^2)$ holds for all $x > k$. Observe that the inequality $x^3 \leq C(7x^2)$ is equivalent to the inequality $x \leq 7C$, which follows by dividing both sides by the positive quantity x^2. However, no matter what C is, it is not the case that $x \leq 7C$ for all $x > k$ no matter what k is, because x can be made arbitrarily large. It follows that no witnesses C and k exist for this proposed big-O relationship. Hence, x^3 is *not* $O(7x^2)$. ◀

3.2.3 Big-*O* Estimates for Some Important Functions

Polynomials can often be used to estimate the growth of functions. Instead of analyzing the growth of polynomials each time they occur, we would like a result that can always be used to estimate the growth of a polynomial. Theorem 1 does this. It shows that the leading term of a polynomial dominates its growth by asserting that a polynomial of degree n or less is $O(x^n)$.

THEOREM 1 Let $f(x) = a_n x^n + a_{n-1} x^{n-1} + \cdots + a_1 x + a_0$, where $a_0, a_1, \ldots, a_{n-1}, a_n$ are real numbers. Then $f(x)$ is $O(x^n)$.

Proof: Using the triangle inequality (see Exercise 9 in Section 1.8), if $x > 1$ we have

$$\begin{aligned}|f(x)| &= |a_n x^n + a_{n-1} x^{n-1} + \cdots + a_1 x + a_0| \\ &\leq |a_n| x^n + |a_{n-1}| x^{n-1} + \cdots + |a_1| x + |a_0| \\ &= x^n \left(|a_n| + |a_{n-1}|/x + \cdots + |a_1|/x^{n-1} + |a_0|/x^n\right) \\ &\leq x^n \left(|a_n| + |a_{n-1}| + \cdots + |a_1| + |a_0|\right).\end{aligned}$$

This shows that

$$|f(x)| \le Cx^n,$$

where $C = |a_n| + |a_{n-1}| + \cdots + |a_0|$ whenever $x > 1$. Hence, the witnesses $C = |a_n| + |a_{n-1}| + \cdots + |a_0|$ and $k = 1$ show that $f(x)$ is $O(x^n)$. ◁

We now give some examples involving functions that have the set of positive integers as their domains.

EXAMPLE 5 How can big-O notation be used to estimate the sum of the first n positive integers?

Solution: Because each of the integers in the sum of the first n positive integers does not exceed n, it follows that

$$1 + 2 + \cdots + n \le n + n + \cdots + n = n^2.$$

From this inequality it follows that $1 + 2 + 3 + \cdots + n$ is $O(n^2)$, taking $C = 1$ and $k = 1$ as witnesses. (In this example the domains of the functions in the big-O relationship are the set of positive integers.) ◀

In Example 6, big-O estimates will be developed for the factorial function and its logarithm. These estimates will be important in the analysis of the number of steps used in sorting procedures.

EXAMPLE 6 Give big-O estimates for the factorial function and the logarithm of the factorial function, where the factorial function $f(n) = n!$ is defined by

$$n! = 1 \cdot 2 \cdot 3 \cdot \cdots \cdot n$$

whenever n is a positive integer, and $0! = 1$. For example,

$$1! = 1, \quad 2! = 1 \cdot 2 = 2, \quad 3! = 1 \cdot 2 \cdot 3 = 6, \quad 4! = 1 \cdot 2 \cdot 3 \cdot 4 = 24.$$

Note that the function $n!$ grows rapidly. For instance,

$$20! = 2{,}432{,}902{,}008{,}176{,}640{,}000.$$

Solution: A big-O estimate for $n!$ can be obtained by noting that each term in the product does not exceed n. Hence,

$$\begin{aligned} n! &= 1 \cdot 2 \cdot 3 \cdot \cdots \cdot n \\ &\le n \cdot n \cdot n \cdot \cdots \cdot n \\ &= n^n. \end{aligned}$$

This inequality shows that $n!$ is $O(n^n)$, taking $C = 1$ and $k = 1$ as witnesses. Taking logarithms of both sides of the inequality established for $n!$, we obtain

$$\log n! \le \log n^n = n \log n.$$

This implies that $\log n!$ is $O(n \log n)$, again taking $C = 1$ and $k = 1$ as witnesses. ◀

EXAMPLE 7 In Section 5.1, we will show that $n < 2^n$ whenever n is a positive integer. Show that this inequality implies that n is $O(2^n)$, and use this inequality to show that $\log n$ is $O(n)$.

Solution: Using the inequality $n < 2^n$, we quickly can conclude that n is $O(2^n)$ by taking $k = C = 1$ as witnesses. Note that because the logarithm function is increasing, taking logarithms (base 2) of both sides of this inequality shows that

$$\log n < n.$$

It follows that

$$\log n \text{ is } O(n).$$

(Again we take $C = k = 1$ as witnesses.)

If we have logarithms to a base b, where b is different from 2, we still have $\log_b n$ is $O(n)$ because

$$\log_b n = \frac{\log n}{\log b} < \frac{n}{\log b}$$

whenever n is a positive integer. We take $C = 1/\log b$ and $k = 1$ as witnesses. (We have used Theorem 3 in Appendix 2 to see that $\log_b n = \log n / \log b$.) ◀

As mentioned before, big-O notation is used to estimate the number of operations needed to solve a problem using a specified procedure or algorithm. The functions used in these estimates often include the following:

$$1,\ \log n,\ n,\ n\log n,\ n^2,\ 2^n,\ n!$$

Using calculus it can be shown that each function in the list is smaller than the succeeding function, in the sense that the ratio of a function and the succeeding function tends to zero as n grows without bound. Figure 3 displays the graphs of these functions, using a scale for the values of the functions that doubles for each successive marking on the graph. That is, the vertical scale in this graph is logarithmic.

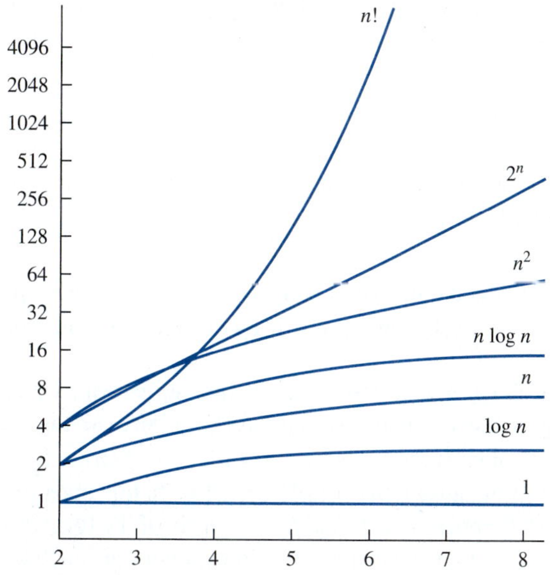

FIGURE 3 **A display of the growth of functions commonly used in big-O estimates.**

USEFUL BIG-*O* ESTIMATES INVOLVING LOGARITHMS, POWERS, AND EXPONENTIAL FUNCTIONS We now give some useful facts that help us determine whether big-O relationships hold between pairs of functions when each of the functions is a power of a logarithm, a power, or an exponential function of the form b^n where $b > 1$. Their proofs are left as Exercises 57–62 for readers skilled with calculus.

Theorem 1 shows that if $f(n)$ is a polynomial of degree d or less, then $f(n)$ is $O(n^d)$. Applying this theorem, we see that if $d > c > 1$, then n^c is $O(n^d)$. We leave it to the reader to show that the reverse of this relationship does not hold. Putting these facts together, we see that if $d > c > 1$, then

$$n^c \text{ is } O(n^d), \text{ but } n^d \text{ is not } O(n^c).$$

In Example 7, we showed that $\log_b n$ is $O(n)$ whenever $b > 1$. More generally, whenever $b > 1$ and c and d are positive, we have

$$(\log_b n)^c \text{ is } O(n^d), \text{ but } n^d \text{ is not } O((\log_b n)^c).$$

This tells us that every positive power of the logarithm of n to the base b, where $b > 1$, is big-O of every positive power of n, but the reverse relationship never holds.

In Example 7, we also showed that n is $O(2^n)$. More generally, whenever d is positive and $b > 1$, we have

$$n^d \text{ is } O(b^n), \text{ but } b^n \text{ is not } O(n^d).$$

This tells us that every power of n is big-O of every exponential function of n with a base that is greater than one, but the reverse relationship never holds. Furthermore, when $c \geq b > 1$ we have

$$b^n \text{ is } O(c^n), \text{ but } c^n \text{ is not } O(b^n).$$

This tells us that if we have two exponential functions with different bases greater than one, one of these functions is big-O of the other if and only if its base is smaller or equal.

Finally, we note that if $c > 1$, we have

$$c^n \text{ is } O(n!), \text{ but } n! \text{ is not } O(c^n).$$

We can use the big-O estimates discussed here to help us order the growth of different functions, as Example 8 illustrates.

EXAMPLE 8 Arrange the functions $f_1(n) = 8\sqrt{n}$, $f_2(n) = (\log n)^2$, $f_3(n) = 2n \log n$, $f_4(n) = n!$, $f_5(n) = (1.1)^n$, and $f_6(n) = n^2$ in a list so that each function is big-O of the next function.

Solution: From the big-O estimates described in this subsection, we see that $f_2(n) = (\log n)^2$ is the slowest growing of these functions. (This follows because $\log n$ grows slower than any positive power of n.) The next three functions, in order, are $f_1(n) = 8\sqrt{n}$, $f_3(n) = 2n \log n$, and $f_6(n) = n^2$. (We know this because $f_1(n) = 8n^{1/2}$, $f_3(n) = 2n \log n$ is a function that grows faster than n but slower than n^c for every $c > 1$, and $f_6(n) = n^2$ is of the form n^c where $c = 2$.) The next function in the list is $f_5(n) = (1.1)^n$, because it is an exponential function with base 1.1. Finally, $f_4(n) = n!$ is the fastest growing function on the list, because $f(n) = n!$ grows faster than any exponential function of n. ◀

3.2.4 The Growth of Combinations of Functions

Many algorithms are made up of two or more separate subprocedures. The number of steps used by a computer to solve a problem with input of a specified size using such an algorithm is the sum of the number of steps used by these subprocedures. To give a big-O estimate for the number of steps needed, it is necessary to find big-O estimates for the number of steps used by each subprocedure and then combine these estimates.

Big-O estimates of combinations of functions can be provided if care is taken when different big-O estimates are combined. In particular, it is often necessary to estimate the growth of the sum and the product of two functions. What can be said if big-O estimates for each of two functions are known? To see what sort of estimates hold for the sum and the product of two functions, suppose that $f_1(x)$ is $O(g_1(x))$ and $f_2(x)$ is $O(g_2(x))$.

From the definition of big-O notation, there are constants C_1, C_2, k_1, and k_2 such that

$$|f_1(x)| \leq C_1|g_1(x)|$$

when $x > k_1$, and

$$|f_2(x)| \leq C_2|g_2(x)|$$

when $x > k_2$. To estimate the sum of $f_1(x)$ and $f_2(x)$, note that

$$\begin{aligned} |(f_1 + f_2)(x)| &= |f_1(x) + f_2(x)| \\ &\leq |f_1(x)| + |f_2(x)| \quad \text{using the triangle inequality } |a + b| \leq |a| + |b|. \end{aligned}$$

When x is greater than both k_1 and k_2, it follows from the inequalities for $|f_1(x)|$ and $|f_2(x)|$ that

$$\begin{aligned} |f_1(x)| + |f_2(x)| &\leq C_1|g_1(x)| + C_2|g_2(x)| \\ &\leq C_1|g(x)| + C_2|g(x)| \\ &= (C_1 + C_2)|g(x)| \\ &= C|g(x)|, \end{aligned}$$

where $C = C_1 + C_2$ and $g(x) = \max(|g_1(x)|, |g_2(x)|)$. [Here $\max(a, b)$ denotes the maximum, or larger, of a and b.]

This inequality shows that $|(f_1 + f_2)(x)| \leq C|g(x)|$ whenever $x > k$, where $k = \max(k_1, k_2)$. We state this useful result as Theorem 2.

THEOREM 2 Suppose that $f_1(x)$ is $O(g_1(x))$ and that $f_2(x)$ is $O(g_2(x))$. Then $(f_1 + f_2)(x)$ is $O(g(x))$, where $g(x) = \max(|g_1(x)|, |g_2(x)|)$ for all x.

We often have big-O estimates for f_1 and f_2 in terms of the same function g. In this situation, Theorem 2 can be used to show that $(f_1 + f_2)(x)$ is also $O(g(x))$, because $\max(g(x), g(x)) = g(x)$. This result is stated in Corollary 1.

COROLLARY 1 Suppose that $f_1(x)$ and $f_2(x)$ are both $O(g(x))$. Then $(f_1 + f_2)(x)$ is $O(g(x))$.

In a similar way big-O estimates can be derived for the product of the functions f_1 and f_2. When x is greater than $\max(k_1, k_2)$ it follows that

$$\begin{aligned}
|(f_1f_2)(x)| &= |f_1(x)||f_2(x)| \\
&\le C_1|g_1(x)|C_2|g_2(x)| \\
&\le C_1C_2|(g_1g_2)(x)| \\
&\le C|(g_1g_2)(x)|,
\end{aligned}$$

where $C = C_1C_2$. From this inequality, it follows that $f_1(x)f_2(x)$ is $O((g_1g_2)(x))$, because there are constants C and k, namely, $C = C_1C_2$ and $k = \max(k_1, k_2)$, such that $|(f_1f_2)(x)| \le C|g_1(x)g_2(x)|$ whenever $x > k$. This result is stated in Theorem 3.

THEOREM 3 Suppose that $f_1(x)$ is $O(g_1(x))$ and $f_2(x)$ is $O(g_2(x))$. Then $(f_1f_2)(x)$ is $O(g_1(x)g_2(x))$.

The goal in using big-O notation to estimate functions is to choose a function $g(x)$ as simple as possible, that grows relatively slowly so that $f(x)$ is $O(g(x))$. Examples 9 and 10 illustrate how to use Theorems 2 and 3 to do this. The type of analysis given in these examples is often used in the analysis of the time used to solve problems using computer programs.

EXAMPLE 9 Give a big-O estimate for $f(n) = 3n\log(n!) + (n^2 + 3)\log n$, where n is a positive integer.

Solution: First, the product $3n\log(n!)$ will be estimated. From Example 6 we know that $\log(n!)$ is $O(n\log n)$. Using this estimate and the fact that $3n$ is $O(n)$, Theorem 3 gives the estimate that $3n\log(n!)$ is $O(n^2\log n)$.

Next, the product $(n^2 + 3)\log n$ will be estimated. Because $(n^2 + 3) < 2n^2$ when $n > 2$, it follows that $n^2 + 3$ is $O(n^2)$. Thus, from Theorem 3 it follows that $(n^2 + 3)\log n$ is $O(n^2\log n)$. Using Theorem 2 to combine the two big-O estimates for the products shows that $f(n) = 3n\log(n!) + (n^2 + 3)\log n$ is $O(n^2\log n)$. ◀

EXAMPLE 10 Give a big-O estimate for $f(x) = (x + 1)\log(x^2 + 1) + 3x^2$.

Solution: First, a big-O estimate for $(x + 1)\log(x^2 + 1)$ will be found. Note that $(x + 1)$ is $O(x)$. Furthermore, $x^2 + 1 \le 2x^2$ when $x > 1$. Hence,

$$\log(x^2 + 1) \le \log(2x^2) = \log 2 + \log x^2 = \log 2 + 2\log x \le 3\log x,$$

if $x > 2$. This shows that $\log(x^2 + 1)$ is $O(\log x)$.

From Theorem 3 it follows that $(x + 1)\log(x^2 + 1)$ is $O(x\log x)$. Because $3x^2$ is $O(x^2)$, Theorem 2 tells us that $f(x)$ is $O(\max(x\log x, x^2))$. Because $x\log x \le x^2$, for $x > 1$, it follows that $f(x)$ is $O(x^2)$. ◀

3.2.5 Big-Omega and Big-Theta Notation

Big-O notation is used extensively to describe the growth of functions, but it has limitations. In particular, when $f(x)$ is $O(g(x))$, we have an upper bound, in terms of $g(x)$, for the size of $f(x)$ for large values of x. However, big-O notation does not provide a lower bound for the size of $f(x)$ for large x. For this, we use **big-Omega (big-Ω) notation**. When we want to give both an upper and a lower bound on the size of a function $f(x)$, relative to a reference function $g(x)$, we use **big-Theta (big-Θ) notation**. Both big-Omega and big-Theta notation were introduced

Ω and Θ are the Greek uppercase letters omega and theta, respectively.

by Donald Knuth in the 1970s. His motivation for introducing these notations was the common misuse of big-O notation when both an upper and a lower bound on the size of a function are needed. We now define big-Omega notation and illustrate its use. After doing so, we will do the same for big-Theta notation.

Definition 2 Let f and g be functions from the set of integers or the set of real numbers to the set of real numbers. We say that $f(x)$ is $\Omega(g(x))$ if there are positive constants C and k such that

$$|f(x)| \geq C|g(x)|$$

whenever $x > k$. [This is read as "$f(x)$ is big-Omega of $g(x)$."]

There is a strong connection between big-O and big-Omega notation. In particular, $f(x)$ is $\Omega(g(x))$ if and only if $g(x)$ is $O(f(x))$. We leave the verification of this fact as a straightforward exercise for the reader.

EXAMPLE 11 The function $f(x) = 8x^3 + 5x^2 + 7$ is $\Omega(g(x))$, where $g(x)$ is the function $g(x) = x^3$. This is easy to see because $f(x) = 8x^3 + 5x^2 + 7 \geq 8x^3$ for all positive real numbers x. This is equivalent to saying that $g(x) = x^3$ is $O(8x^3 + 5x^2 + 7)$, which can be established directly by turning the inequality around. ◀

Often, it is important to know the order of growth of a function in terms of some relatively simple reference function such as x^n when n is a positive integer or c^x, where $c > 1$. Knowing the order of growth requires that we have both an upper bound and a lower bound for the size of the function. That is, given a function $f(x)$, we want a reference function $g(x)$ such that $f(x)$ is $O(g(x))$ and $f(x)$ is $\Omega(g(x))$. Big-Theta notation, defined as follows, is used to express both of these relationships, providing both an upper and a lower bound on the size of a function.

Definition 3 Let f and g be functions from the set of integers or the set of real numbers to the set of real numbers. We say that $f(x)$ is $\Theta(g(x))$ if $f(x)$ is $O(g(x))$ and $f(x)$ is $\Omega(g(x))$. When $f(x)$ is $\Theta(g(x))$, we say that f is big-Theta of $g(x)$, that $f(x)$ is of *order* $g(x)$, and that $f(x)$ and $g(x)$ are of the *same order*.

When $f(x)$ is $\Theta(g(x))$, it is also the case that $g(x)$ is $\Theta(f(x))$. Also note that $f(x)$ is $\Theta(g(x))$ if and only if $f(x)$ is $O(g(x))$ and $g(x)$ is $O(f(x))$ (see Exercise 31). Furthermore, note that $f(x)$ is $\Theta(g(x))$ if and only if there are positive real numbers C_1 and C_2 and a positive real number k such that

$$C_1|g(x)| \leq |f(x)| \leq C_2|g(x)|$$

whenever $x > k$. The existence of the constants C_1, C_2, and k tells us that $f(x)$ is $\Omega(g(x))$ and that $f(x)$ is $O(g(x))$, respectively.

Usually, when big-Theta notation is used, the function $g(x)$ in $\Theta(g(x))$ is a relatively simple reference function, such as x^n, c^x, $\log x$, and so on, while $f(x)$ can be relatively complicated.

EXAMPLE 12 We showed (in Example 5) that the sum of the first n positive integers is $O(n^2)$. Determine whether this sum is of order n^2 without using the summation formula for this sum.

Extra Examples ❯

Solution: Let $f(n) = 1 + 2 + 3 + \cdots + n$. Because we already know that $f(n)$ is $O(n^2)$, to show that $f(n)$ is of order n^2 we need to find a positive constant C such that $f(n) > Cn^2$ for sufficiently large integers n. To obtain a lower bound for this sum, we can ignore the first half of the terms. Summing only the terms greater than $\lceil n/2 \rceil$, we find that

$$\begin{aligned} 1+2+\cdots+n &\geq \lceil n/2 \rceil + (\lceil n/2 \rceil + 1) + \cdots + n \\ &\geq \lceil n/2 \rceil + \lceil n/2 \rceil + \cdots + \lceil n/2 \rceil \\ &= (n - \lceil n/2 \rceil + 1) \lceil n/2 \rceil \\ &\geq (n/2)(n/2) \\ &= n^2/4. \end{aligned}$$

This shows that $f(n)$ is $\Omega(n^2)$. We conclude that $f(n)$ is of order n^2, or in symbols, $f(n)$ is $\Theta(n^2)$. ◀

Remark: Note that we can also show that $f(n) = \sum_{i=1}^{n} i$ is $\Theta(n^2)$ using the closed formula $\sum_{i=1}^{n} i = n(n+1)/2$ from Table 2 in Section 2.4 and derived in Exercise 37(b) of that section.

EXAMPLE 13 Show that $3x^2 + 8x \log x$ is $\Theta(x^2)$.

Extra Examples

Solution: Because $0 \leq 8x \log x \leq 8x^2$, it follows that $3x^2 + 8x \log x \leq 11x^2$ for $x > 1$. Consequently, $3x^2 + 8x \log x$ is $O(x^2)$. Clearly, x^2 is $O(3x^2 + 8x \log x)$. Consequently, $3x^2 + 8x \log x$ is $\Theta(x^2)$. ◀

One useful fact is that the leading term of a polynomial determines its order. For example, if $f(x) = 3x^5 + x^4 + 17x^3 + 2$, then $f(x)$ is of order x^5. This is stated in Theorem 4, whose proof is left as Exercise 50.

THEOREM 4 Let $f(x) = a_n x^n + a_{n-1} x^{n-1} + \cdots + a_1 x + a_0$, where $a_0, a_1, \ldots, a_n$ are real numbers with $a_n \neq 0$. Then $f(x)$ is of order x^n.

EXAMPLE 14 The polynomials $3x^8 + 10x^7 + 221x^2 + 1444$, $x^{19} - 18x^4 - 10{,}112$, and $-x^{99} + 40{,}001x^{98} + 100{,}003x$ are of orders x^8, x^{19}, and x^{99}, respectively. ◀

Unfortunately, as Knuth observed, big-O notation is often used by careless writers and speakers as if it had the same meaning as big-Theta notation. Keep this in mind when you see big-O notation used. The recent trend has been to use big-Theta notation whenever both upper and lower bounds on the size of a function are needed.

Exercises

In Exercises 1–14, to establish a big-O relationship, find witnesses C and k such that $|f(x)| \leq C|g(x)|$ whenever $x > k$.

1. Determine whether each of these functions is $O(x)$.

a) $f(x) = 10$
b) $f(x) = 3x + 7$
c) $f(x) = x^2 + x + 1$
d) $f(x) = 5 \log x$
e) $f(x) = \lfloor x \rfloor$
f) $f(x) = \lceil x/2 \rceil$

2. Determine whether each of these functions is $O(x^2)$.

a) $f(x) = 17x + 11$
b) $f(x) = x^2 + 1000$
c) $f(x) = x \log x$
d) $f(x) = x^4/2$
e) $f(x) = 2^x$
f) $f(x) = \lfloor x \rfloor \cdot \lceil x \rceil$

3. Use the definition of "$f(x)$ is $O(g(x))$" to show that $x^4 + 9x^3 + 4x + 7$ is $O(x^4)$.

4. Use the definition of "$f(x)$ is $O(g(x))$" to show that $2^x + 17$ is $O(3^x)$.
5. Show that $(x^2 + 1)/(x + 1)$ is $O(x)$.
6. Show that $(x^3 + 2x)/(2x + 1)$ is $O(x^2)$.
7. Find the least integer n such that $f(x)$ is $O(x^n)$ for each of these functions.
   a) $f(x) = 2x^3 + x^2 \log x$
   b) $f(x) = 3x^3 + (\log x)^4$
   c) $f(x) = (x^4 + x^2 + 1)/(x^3 + 1)$
   d) $f(x) = (x^4 + 5 \log x)/(x^4 + 1)$
8. Find the least integer n such that $f(x)$ is $O(x^n)$ for each of these functions.
   a) $f(x) = 2x^2 + x^3 \log x$
   b) $f(x) = 3x^5 + (\log x)^4$
   c) $f(x) = (x^4 + x^2 + 1)/(x^4 + 1)$
   d) $f(x) = (x^3 + 5 \log x)/(x^4 + 1)$
9. Show that $x^2 + 4x + 17$ is $O(x^3)$ but that x^3 is not $O(x^2 + 4x + 17)$.
10. Show that x^3 is $O(x^4)$ but that x^4 is not $O(x^3)$.
11. Show that $3x^4 + 1$ is $O(x^4/2)$ and $x^4/2$ is $O(3x^4 + 1)$.
12. Show that $x \log x$ is $O(x^2)$ but that x^2 is not $O(x \log x)$.
13. Show that 2^n is $O(3^n)$ but that 3^n is not $O(2^n)$. (Note that this is a special case of Exercise 60.)
14. Determine whether x^3 is $O(g(x))$ for each of these functions $g(x)$.
   a) $g(x) = x^2$
   b) $g(x) = x^3$
   c) $g(x) = x^2 + x^3$
   d) $g(x) = x^2 + x^4$
   e) $g(x) = 3^x$
   f) $g(x) = x^3/2$
15. Explain what it means for a function to be $O(1)$.
16. Show that if $f(x)$ is $O(x)$, then $f(x)$ is $O(x^2)$.
17. Suppose that $f(x)$, $g(x)$, and $h(x)$ are functions such that $f(x)$ is $O(g(x))$ and $g(x)$ is $O(h(x))$. Show that $f(x)$ is $O(h(x))$.
18. Let k be a positive integer. Show that $1^k + 2^k + \cdots + n^k$ is $O(n^{k+1})$.
19. Determine whether each of the functions 2^{n+1} and 2^{2n} is $O(2^n)$.
20. Determine whether each of the functions $\log(n + 1)$ and $\log(n^2 + 1)$ is $O(\log n)$.
21. Arrange the functions $\sqrt{n}$, $1000 \log n$, $n \log n$, $2n!$, 2^n, 3^n, and $n^2/1{,}000{,}000$ in a list so that each function is big-O of the next function.
22. Arrange the functions $(1.5)^n$, n^{100}, $(\log n)^3$, $\sqrt{n} \log n$, 10^n, $(n!)^2$, and $n^{99} + n^{98}$ in a list so that each function is big-O of the next function.
23. Suppose that you have two different algorithms for solving a problem. To solve a problem of size n, the first algorithm uses exactly $n(\log n)$ operations and the second algorithm uses exactly $n^{3/2}$ operations. As n grows, which algorithm uses fewer operations?
24. Suppose that you have two different algorithms for solving a problem. To solve a problem of size n, the first algorithm uses exactly $n^2 2^n$ operations and the second algorithm uses exactly $n!$ operations. As n grows, which algorithm uses fewer operations?
25. Give as good a big-O estimate as possible for each of these functions.
   a) $(n^2 + 8)(n + 1)$
   b) $(n \log n + n^2)(n^3 + 2)$
   c) $(n! + 2^n)(n^3 + \log(n^2 + 1))$
26. Give a big-O estimate for each of these functions. For the function g in your estimate $f(x)$ is $O(g(x))$, use a simple function g of smallest order.
   a) $(n^3 + n^2 \log n)(\log n + 1) + (17 \log n + 19)(n^3 + 2)$
   b) $(2^n + n^2)(n^3 + 3^n)$
   c) $(n^n + n2^n + 5^n)(n! + 5^n)$
27. Give a big-O estimate for each of these functions. For the function g in your estimate that $f(x)$ is $O(g(x))$, use a simple function g of the smallest order.
   a) $n \log(n^2 + 1) + n^2 \log n$
   b) $(n \log n + 1)^2 + (\log n + 1)(n^2 + 1)$
   c) $n^{2^n} + n^{n^2}$
28. For each function in Exercise 1, determine whether that function is $\Omega(x)$ and whether it is $\Theta(x)$.
29. For each function in Exercise 2, determine whether that function is $\Omega(x^2)$ and whether it is $\Theta(x^2)$.
30. Show that each of these pairs of functions are of the same order.
   a) $3x + 7$, x
   b) $2x^2 + x - 7$, x^2
   c) $\lfloor x + 1/2 \rfloor$, x
   d) $\log(x^2 + 1)$, $\log_2 x$
   e) $\log_{10} x$, $\log_2 x$
31. Show that $f(x)$ is $\Theta(g(x))$ if and only if $f(x)$ is $O(g(x))$ and $g(x)$ is $O(f(x))$.
32. Show that if $f(x)$ and $g(x)$ are functions from the set of real numbers to the set of real numbers, then $f(x)$ is $O(g(x))$ if and only if $g(x)$ is $\Omega(f(x))$.
33. Show that if $f(x)$ and $g(x)$ are functions from the set of real numbers to the set of real numbers, then $f(x)$ is $\Theta(g(x))$ if and only if there are positive constants k, C_1, and C_2 such that $C_1|g(x)| \le |f(x)| \le C_2|g(x)|$ whenever $x > k$.
34. a) Show that $3x^2 + x + 1$ is $\Theta(3x^2)$ by directly finding the constants k, C_1, and C_2 in Exercise 33.
   b) Express the relationship in part (a) using a picture showing the functions $3x^2 + x + 1$, $C_1 \cdot 3x^2$, and $C_2 \cdot 3x^2$, and the constant k on the x-axis, where C_1, C_2, and k are the constants you found in part (a) to show that $3x^2 + x + 1$ is $\Theta(3x^2)$.
35. Express the relationship $f(x)$ is $\Theta(g(x))$ using a picture. Show the graphs of the functions $f(x)$, $C_1|g(x)|$, and $C_2|g(x)|$, as well as the constant k on the x-axis.
36. Explain what it means for a function to be $\Omega(1)$.
37. Explain what it means for a function to be $\Theta(1)$.
38. Give a big-O estimate of the product of the first n odd positive integers.
39. Show that if f and g are real-valued functions such that $f(x)$ is $O(g(x))$, then for every positive integer n, $f^n(x)$ is $O(g^n(x))$. [Note that $f^n(x) = f(x)^n$.]
40. Show that for all real numbers a and b with $a > 1$ and $b > 1$, if $f(x)$ is $O(\log_b x)$, then $f(x)$ is $O(\log_a x)$.

41. Suppose that $f(x)$ is $O(g(x))$, where f and g are increasing and unbounded functions. Show that $\log |f(x)|$ is $O(\log |g(x)|)$.

42. Suppose that $f(x)$ is $O(g(x))$. Does it follow that $2^{f(x)}$ is $O(2^{g(x)})$?

43. Let $f_1(x)$ and $f_2(x)$ be functions from the set of real numbers to the set of positive real numbers. Show that if $f_1(x)$ and $f_2(x)$ are both $\Theta(g(x))$, where $g(x)$ is a function from the set of real numbers to the set of positive real numbers, then $f_1(x) + f_2(x)$ is $\Theta(g(x))$. Is this still true if $f_1(x)$ and $f_2(x)$ can take negative values?

44. Suppose that $f(x)$, $g(x)$, and $h(x)$ are functions such that $f(x)$ is $\Theta(g(x))$ and $g(x)$ is $\Theta(h(x))$. Show that $f(x)$ is $\Theta(h(x))$.

45. If $f_1(x)$ and $f_2(x)$ are functions from the set of positive integers to the set of positive real numbers and $f_1(x)$ and $f_2(x)$ are both $\Theta(g(x))$, is $(f_1 - f_2)(x)$ also $\Theta(g(x))$? Either prove that it is or give a counterexample.

46. Show that if $f_1(x)$ and $f_2(x)$ are functions from the set of positive integers to the set of real numbers and $f_1(x)$ is $\Theta(g_1(x))$ and $f_2(x)$ is $\Theta(g_2(x))$, then $(f_1f_2)(x)$ is $\Theta((g_1g_2)(x))$.

47. Find functions f and g from the set of positive integers to the set of real numbers such that $f(n)$ is not $O(g(n))$ and $g(n)$ is not $O(f(n))$.

48. Express the relationship $f(x)$ is $\Omega(g(x))$ using a picture. Show the graphs of the functions $f(x)$ and $Cg(x)$, as well as the constant k on the real axis.

49. Show that if $f_1(x)$ is $\Theta(g_1(x))$, $f_2(x)$ is $\Theta(g_2(x))$, and $f_2(x) \neq 0$ and $g_2(x) \neq 0$ for all real numbers $x > 0$, then $(f_1/f_2)(x)$ is $\Theta((g_1/g_2)(x))$.

50. Show that if $f(x) = a_nx^n + a_{n-1}x^{n-1} + \cdots + a_1x + a_0$, where $a_0, a_1, \ldots, a_{n-1}$, and a_n are real numbers and $a_n \neq 0$, then $f(x)$ is $\Theta(x^n)$.

Big-O, big-Theta, and big-Omega notation can be extended to functions in more than one variable. For example, the statement $f(x, y)$ is $O(g(x, y))$ means that there exist constants C, k_1, and k_2 such that $|f(x, y)| \leq C|g(x, y)|$ whenever $x > k_1$ and $y > k_2$.

51. Define the statement $f(x, y)$ is $\Theta(g(x, y))$.

52. Define the statement $f(x, y)$ is $\Omega(g(x, y))$.

53. Show that $(x^2 + xy + x \log y)^3$ is $O(x^6y^3)$.

54. Show that $x^5y^3 + x^4y^4 + x^3y^5$ is $\Omega(x^3y^3)$.

55. Show that $\lfloor xy \rfloor$ is $O(xy)$.

56. Show that $\lceil xy \rceil$ is $\Omega(xy)$.

57. (*Requires calculus*) Show that if $c > d > 0$, then n^d is $O(n^c)$, but n^c is not $O(n^d)$.

58. (*Requires calculus*) Show that if $b > 1$ and c and d are positive, then $(\log_b n)^c$ is $O(n^d)$, but n^d is not $O((\log_b n)^c)$.

59. (*Requires calculus*) Show that if d is positive and $b > 1$, then n^d is $O(b^n)$, but b^n is not $O(n^d)$.

60. (*Requires calculus*) Show that if $c > b > 1$, then b^n is $O(c^n)$, but c^n is not $O(b^n)$.

61. (*Requires calculus*) Show that if $c > 1$, then c^n is $O(n!)$, but $n!$ is not $O(c^n)$.

62. (*Requires calculus*) Prove or disprove that $(2n)!$ is $O(n!)$.

The following problems deal with another type of asymptotic notation, called **little-*o*** notation. Because little-o notation is based on the concept of limits, a knowledge of calculus is needed for these problems. We say that $f(x)$ is $o(g(x))$ [read $f(x)$ is "little-oh" of $g(x)$], when

$$\lim_{x\to\infty} \frac{f(x)}{g(x)} = 0.$$

63. (*Requires calculus*) Show that

a) x^2 is $o(x^3)$.
b) $x \log x$ is $o(x^2)$.
c) x^2 is $o(2^x)$.
d) $x^2 + x + 1$ is not $o(x^2)$.

64. (*Requires calculus*)

a) Show that if $f(x)$ and $g(x)$ are functions such that $f(x)$ is $o(g(x))$ and c is a constant, then $cf(x)$ is $o(g(x))$, where $(cf)(x) = cf(x)$.

b) Show that if $f_1(x)$, $f_2(x)$, and $g(x)$ are functions such that $f_1(x)$ is $o(g(x))$ and $f_2(x)$ is $o(g(x))$, then $(f_1 + f_2)(x)$ is $o(g(x))$, where $(f_1 + f_2)(x) = f_1(x) + f_2(x)$.

65. (*Requires calculus*) Represent pictorially that $x \log x$ is $o(x^2)$ by graphing $x \log x$, x^2, and $x \log x/x^2$. Explain how this picture shows that $x \log x$ is $o(x^2)$.

66. (*Requires calculus*) Express the relationship $f(x)$ is $o(g(x))$ using a picture. Show the graphs of $f(x)$, $g(x)$, and $f(x)/g(x)$.

***67.** (*Requires calculus*) Suppose that $f(x)$ is $o(g(x))$. Does it follow that $2^{f(x)}$ is $o(2^{g(x)})$?

***68.** (*Requires calculus*) Suppose that $f(x)$ is $o(g(x))$. Does it follow that $\log |f(x)|$ is $o(\log |g(x)|)$?

69. (*Requires calculus*) The two parts of this exercise describe the relationship between little-o and big-O notation.

a) Show that if $f(x)$ and $g(x)$ are functions such that $f(x)$ is $o(g(x))$, then $f(x)$ is $O(g(x))$.

b) Show that if $f(x)$ and $g(x)$ are functions such that $f(x)$ is $O(g(x))$, then it does not necessarily follow that $f(x)$ is $o(g(x))$.

70. (*Requires calculus*) Show that if $f(x)$ is a polynomial of degree n and $g(x)$ is a polynomial of degree m where $m > n$, then $f(x)$ is $o(g(x))$.

71. (*Requires calculus*) Show that if $f_1(x)$ is $O(g(x))$ and $f_2(x)$ is $o(g(x))$, then $f_1(x) + f_2(x)$ is $O(g(x))$.

72. (*Requires calculus*) Let H_n be the nth **harmonic number**

$$H_n = 1 + \frac{1}{2} + \frac{1}{3} + \cdots + \frac{1}{n}.$$

Show that H_n is $O(\log n)$. [*Hint:* First establish the inequality

$$\sum_{j=2}^{n} \frac{1}{j} < \int_1^n \frac{1}{x}dx$$

by showing that the sum of the areas of the rectangles of height $1/j$ with base from $j-1$ to j, for $j = 2, 3, \ldots, n$, is less than the area under the curve $y = 1/x$ from 2 to n.]

***73.** Show that $n \log n$ is $O(\log n!)$.

74. Determine whether $\log n!$ is $\Theta(n \log n)$. Justify your answer.

***75.** Show that $\log n!$ is greater than $(n \log n)/4$ for $n > 4$. [*Hint:* Begin with the inequality $n! > n(n-1)(n-2) \cdots \lceil n/2 \rceil$.]

Let $f(x)$ and $g(x)$ be functions from the set of real numbers to the set of real numbers. We say that the functions f and g are **asymptotic** and write $f(x) \sim g(x)$ if $\lim_{x\to\infty} f(x)/g(x) = 1$.

76. (*Requires calculus*) For each of these pairs of functions, determine whether f and g are asymptotic.

a) $f(x) = x^2 + 3x + 7$, $g(x) = x^2 + 10$
b) $f(x) = x^2 \log x$, $g(x) = x^3$
c) $f(x) = x^4 + \log(3x^8 + 7)$, $g(x) = (x^2 + 17x + 3)^2$
d) $f(x) = (x^3 + x^2 + x + 1)^4$, $g(x) = (x^4 + x^3 + x^2 + x + 1)^3$.

77. (*Requires calculus*) For each of these pairs of functions, determine whether f and g are asymptotic.

a) $f(x) = \log(x^2 + 1)$, $g(x) = \log x$
b) $f(x) = 2^{x+3}$, $g(x) = 2^{x+7}$
c) $f(x) = 2^{2^x}$, $g(x) = 2^{x^2}$
d) $f(x) = 2^{x^2+x+1}$, $g(x) = 2^{x^2+2x}$

3.3 Complexity of Algorithms

3.3.1 Introduction

When does an algorithm provide a satisfactory solution to a problem? First, it must always produce the correct answer. How this can be demonstrated will be discussed in Chapter 5. Second, it should be efficient. The efficiency of algorithms will be discussed in this section.

How can the efficiency of an algorithm be analyzed? One measure of efficiency is the time used by a computer to solve a problem using the algorithm, when input values are of a specified size. A second measure is the amount of computer memory required to implement the algorithm when input values are of a specified size.

Questions such as these involve the **computational complexity** of the algorithm. An analysis of the time required to solve a problem of a particular size involves the **time complexity** of the algorithm. An analysis of the computer memory required involves the **space complexity** of the algorithm. Considerations of the time and space complexity of an algorithm are essential when algorithms are implemented. It is important to know whether an algorithm will produce an answer in a microsecond, a minute, or a billion years. Likewise, the required memory must be available to solve a problem, so that space complexity must be taken into account.

Considerations of space complexity are tied in with the particular data structures used to implement the algorithm. Because data structures are not dealt with in detail in this book, space complexity will not be considered. We will restrict our attention to time complexity.

3.3.2 Time Complexity

The time complexity of an algorithm can be expressed in terms of the number of operations used by the algorithm when the input has a particular size. The operations used to measure time complexity can be the comparison of integers, the addition of integers, the multiplication of integers, the division of integers, or any other basic operation.

Time complexity is described in terms of the number of operations required instead of actual computer time because of the difference in time needed for different computers to perform basic operations. Moreover, it is quite complicated to break all operations down to the basic bit operations that a computer uses. Furthermore, the fastest computers in existence can perform basic bit operations (for instance, adding, multiplying, comparing, or exchanging two bits) in 10^{-11} second (10 picoseconds), but personal computers may require 10^{-8} second (10 nanoseconds), which is 1000 times as long, to do the same operations.

We illustrate how to analyze the time complexity of an algorithm by considering Algorithm 1 of Section 3.1, which finds the maximum of a finite set of integers.

EXAMPLE 1 Describe the time complexity of Algorithm 1 of Section 3.1 for finding the maximum element in a finite set of integers.

Extra Examples

Solution: The number of comparisons will be used as the measure of the time complexity of the algorithm, because comparisons are the basic operations used.

To find the maximum element of a set with n elements, listed in an arbitrary order, the temporary maximum is first set equal to the initial term in the list. Then, after a comparison $i \leq n$ has been done to determine that the end of the list has not yet been reached, the temporary maximum and second term are compared, updating the temporary maximum to the value of the second term if it is larger. This procedure is continued, using two additional comparisons for each term of the list—one $i \leq n$, to determine that the end of the list has not been reached and another $\max < a_i$, to determine whether to update the temporary maximum. Because two comparisons are used for each of the second through the nth elements and one more comparison is used to exit the loop when $i = n + 1$, exactly $2(n - 1) + 1 = 2n - 1$ comparisons are used whenever this algorithm is applied. Hence, the algorithm for finding the maximum of a set of n elements has time complexity $\Theta(n)$, measured in terms of the number of comparisons used. Note that for this algorithm the number of comparisons is independent of particular input of n numbers. ◀

Next, we will analyze the time complexity of searching algorithms.

EXAMPLE 2 Describe the time complexity of the linear search algorithm (specified as Algorithm 2 in Section 3.1).

Solution: The number of comparisons used by Algorithm 2 in Section 3.1 will be taken as the measure of the time complexity. At each step of the loop in the algorithm, two comparisons are performed—one $i \leq n$, to see whether the end of the list has been reached and one $x \leq a_i$, to compare the element x with a term of the list. Finally, one more comparison $i \leq n$ is made outside the loop. Consequently, if $x = a_i$, $2i + 1$ comparisons are used. The most comparisons, $2n + 2$, are required when the element is not in the list. In this case, $2n$ comparisons are used to determine that x is not a_i, for $i = 1, 2, \ldots, n$, an additional comparison is used to exit the loop, and one comparison is made outside the loop. So when x is not in the list, a total of $2n + 2$ comparisons are used. Hence, a linear search requires $\Theta(n)$ comparisons in the worst case, because $2n + 2$ is $\Theta(n)$. ◀

WORST-CASE COMPLEXITY The type of complexity analysis done in Example 2 is a **worst-case** analysis. By the worst-case performance of an algorithm, we mean the largest number of operations needed to solve the given problem using this algorithm on input of specified size. Worst-case analysis tells us how many operations an algorithm requires to guarantee that it will produce a solution.

EXAMPLE 3 Describe the time complexity of the binary search algorithm (specified as Algorithm 3 in Section 3.1) in terms of the number of comparisons used (and ignoring the time required to compute $m = \lfloor (i + j)/2 \rfloor$ in each iteration of the loop in the algorithm).

Solution: For simplicity, assume there are $n = 2^k$ elements in the list $a_1, a_2, \ldots, a_n$, where k is a nonnegative integer. Note that $k = \log n$. (If n, the number of elements in the list, is not a power of 2, the list can be considered part of a larger list with 2^{k+1} elements, where $2^k < n < 2^{k+1}$. Here 2^{k+1} is the smallest power of 2 larger than n.)

At each stage of the algorithm, i and j, the locations of the first term and the last term of the restricted list at that stage, are compared to see whether the restricted list has more than one term. If $i < j$, a comparison is done to determine whether x is greater than the middle term of the restricted list.

At the first stage the search is restricted to a list with 2^{k-1} terms. So far, two comparisons have been used. This procedure is continued, using two comparisons at each stage to restrict the search to a list with half as many terms. In other words, two comparisons are used at the first stage of the algorithm when the list has 2^k elements, two more when the search has been reduced to a list with 2^{k-1} elements, two more when the search has been reduced to a list with 2^{k-2} elements, and so on, until two comparisons are used when the search has been reduced to a list with $2^1 = 2$ elements. Finally, when one term is left in the list, one comparison tells us that there are no additional terms left, and one more comparison is used to determine if this term is x.

Hence, at most $2k + 2 = 2\log n + 2$ comparisons are required to perform a binary search when the list being searched has 2^k elements. (If n is not a power of 2, the original list is expanded to a list with 2^{k+1} terms, where $k = \lfloor \log n \rfloor$, and the search requires at most $2\lceil \log n \rceil + 2$ comparisons.) It follows that in the worst case, binary search requires $O(\log n)$ comparisons.

Note that in the worst case, $2\log n + 2$ comparisons are used by the binary search. Hence, the binary search uses $\Theta(\log n)$ comparisons in the worst case, because $2\log n + 2 = \Theta(\log n)$. From this analysis it follows that in the worst case, the binary search algorithm is more efficient than the linear search algorithm, because we know by Example 2 that the linear search algorithm has $\Theta(n)$ worst-case time complexity. ◀

AVERAGE-CASE COMPLEXITY Another important type of complexity analysis, besides worst-case analysis, is called **average-case** analysis. The average number of operations used to solve the problem over all possible inputs of a given size is found in this type of analysis. Average-case time complexity analysis is usually much more complicated than worst-case analysis. However, the average-case analysis for the linear search algorithm can be done without difficulty, as shown in Example 4.

EXAMPLE 4 Describe the average-case performance of the linear search algorithm in terms of the average number of comparisons used, assuming that the integer x is in the list and it is equally likely that x is in any position.

Solution: By hypothesis, the integer x is one of the integers $a_1, a_2, \ldots, a_n$ in the list. If x is the first term a_1 of the list, three comparisons are needed, one $i \le n$ to determine whether the end of the list has been reached, one $x \ne a_i$ to compare x and the first term, and one $i \le n$ outside the loop. If x is the second term a_2 of the list, two more comparisons are needed, so that a total of five comparisons are used. In general, if x is the ith term of the list a_i, two comparisons will be used at each of the i steps of the loop, and one outside the loop, so that a total of $2i + 1$ comparisons are needed. Hence, the average number of comparisons used equals

$$\frac{3 + 5 + 7 + \cdots + (2n + 1)}{n} = \frac{2(1 + 2 + 3 + \cdots + n) + n}{n}.$$

Using the formula from line 2 of Table 2 in Section 2.4 (and see Exercise 37(b) of Section 2.4),

$$1 + 2 + 3 + \cdots + n = \frac{n(n + 1)}{2}.$$

Hence, the average number of comparisons used by the linear search algorithm (when x is known to be in the list) is

$$\frac{2[n(n + 1)/2]}{n} + 1 = n + 2,$$

which is $\Theta(n)$. ◀

Remark: In the analysis in Example 4 we assumed that x is in the list being searched. It is also possible to do an average-case analysis of this algorithm when x may not be in the list (see Exercise 23).

Remark: Although we have counted the comparisons needed to determine whether we have reached the end of a loop, these comparisons are often not counted. From this point on we will ignore such comparisons.

WORST-CASE COMPLEXITY OF TWO SORTING ALGORITHMS We analyze the worst-case complexity of the bubble sort and the insertion sort in Examples 5 and 6.

EXAMPLE 5 What is the worst-case complexity of the bubble sort in terms of the number of comparisons made?

Solution: The bubble sort described before Example 4 in Section 3.1 sorts a list by performing a sequence of passes through the list. During each pass the bubble sort successively compares adjacent elements, interchanging them if necessary. When the ith pass begins, the $i - 1$ largest elements are guaranteed to be in the correct positions. During this pass, $n - i$ comparisons are used. Consequently, the total number of comparisons used by the bubble sort to order a list of n elements is

$$(n-1) + (n-2) + \cdots + 2 + 1 = \frac{(n-1)n}{2}$$

using a summation formula from line 2 in Table 2 in Section 2.4 (and Exercise 37(b) in Section 2.4). Note that the bubble sort always uses this many comparisons, because it continues even if the list becomes completely sorted at some intermediate step. Consequently, the bubble sort uses $(n-1)n/2$ comparisons, so it has $\Theta(n^2)$ worst-case complexity in terms of the number of comparisons used. ◀

EXAMPLE 6 What is the worst-case complexity of the insertion sort in terms of the number of comparisons made?

Solution: The insertion sort (described in Section 3.1) inserts the jth element into the correct position among the first $j - 1$ elements that have already been put into the correct order. It does this by using a linear search technique, successively comparing the jth element with successive terms until a term that is greater than or equal to it is found or it compares a_j with itself and stops because a_j is not less than itself. Consequently, in the worst case, j comparisons are required to insert the jth element into the correct position. Therefore, the total number of comparisons used by the insertion sort to sort a list of n elements is

$$2 + 3 + \cdots + n = \frac{n(n+1)}{2} - 1,$$

using the summation formula for the sum of consecutive integers in line 2 of Table 2 of Section 2.4 (and see Exercise 37(b) of Section 2.4), and noting that the first term, 1, is missing in this sum. Note that the insertion sort may use considerably fewer comparisons if the smaller elements started out at the end of the list. We conclude that the insertion sort has worst-case complexity $\Theta(n^2)$. ◀

In Examples 5 and 6 we showed that both the bubble sort and the insertion sort have worst-case time complexity $\Theta(n^2)$. However, the most efficient sorting algorithms can sort n items in $O(n \log n)$ time, as we will show in Sections 8.3 and 11.2 using techniques we develop in those sections. From this point on, we will assume that sorting n items can be done in $O(n \log n)$ time.

Links

You can run animations found on many different websites that simultaneously run different sorting algorithms on the same lists. Doing so will help you gain insights into the efficiency of different sorting algorithms. Among the sorting algorithms that you can find are the bubble sort, the insertion sort, the shell sort, the merge sort, and the quick sort. Some of these animations allow you to test the relative performance of these sorting algorithms on lists of randomly selected items, lists that are nearly sorted, and lists that are in reversed order.

3.3.3 Complexity of Matrix Multiplication

The definition of the product of two matrices can be expressed as an algorithm for computing the product of two matrices. Suppose that $\mathbf{C} = [c_{ij}]$ is the $m \times n$ matrix that is the product of the $m \times k$ matrix $\mathbf{A} = [a_{ij}]$ and the $k \times n$ matrix $\mathbf{B} = [b_{ij}]$. The algorithm based on the definition of the matrix product is expressed in pseudocode in Algorithm 1.

ALGORITHM 1 Matrix Multiplication.

procedure *matrix multiplication*(**A**, **B**: matrices)
for $i := 1$ **to** m
    for $j := 1$ **to** n
        $c_{ij} := 0$
        for $q := 1$ **to** k
            $c_{ij} := c_{ij} + a_{iq}b_{qj}$
return C {$\mathbf{C} = [c_{ij}]$ is the product of **A** and **B**}

We can determine the complexity of this algorithm in terms of the number of additions and multiplications used.

EXAMPLE 7 How many additions of integers and multiplications of integers are used by Algorithm 1 to multiply two $n \times n$ matrices with integer entries?

Solution: There are n^2 entries in the product of **A** and **B**. To find each entry requires a total of n multiplications and n additions. Hence, a total of n^3 multiplications and n^3 additions are used. ◀

Surprisingly, there are more efficient algorithms for matrix multiplication than that given in Algorithm 1. As Example 7 shows, multiplying two $n \times n$ matrices directly from the definition requires $O(n^3)$ multiplications and additions. Using other algorithms, two $n \times n$ matrices can be multiplied using $O(n^{\log 7})$ multiplications and additions. As of January 2024, the best known bound on the asymptotic complexity of the multiplication of two $n \times n$ matrices is $O(n^{2.371552})$. (Details of such algorithms can be found in [CoLeRiSt09].)

We can also analyze the complexity of the algorithm we described in Chapter 2 for computing the Boolean product of two matrices, which we display as Algorithm 2.

ALGORITHM 2 The Boolean Product of Zero–One Matrices.

procedure *Boolean product of Zero–One Matrices* (**A**, **B**: zero–one matrices)
for $i := 1$ **to** m
  for $j := 1$ **to** n
    $c_{ij} := 0$
    for $q := 1$ **to** k
      $c_{ij} := c_{ij} \vee (a_{iq} \wedge b_{qj})$
return C {$\mathbf{C} = [c_{ij}]$ is the Boolean product of **A** and **B**}

The number of bit operations used to find the Boolean product of two $n \times n$ matrices can be easily determined.

EXAMPLE 8 How many bit operations are used to find $\mathbf{A} \odot \mathbf{B}$, where **A** and **B** are $n \times n$ zero–one matrices?

Solution: There are n^2 entries in $\mathbf{A} \odot \mathbf{B}$. Using Algorithm 2, a total of n *OR*s and n *AND*s are used to find an entry of $\mathbf{A} \odot \mathbf{B}$. Hence, $2n$ bit operations are used to find each entry. Therefore, $2n^3$ bit operations are required to compute $\mathbf{A} \odot \mathbf{B}$ using Algorithm 2. ◀

MATRIX-CHAIN MULTIPLICATION There is another important problem involving the complexity of the multiplication of matrices. How should the **matrix-chain** $\mathbf{A}_1\mathbf{A}_2 \cdots \mathbf{A}_n$ be computed using the fewest multiplications of integers, where $\mathbf{A}_1, \mathbf{A}_2, \ldots, \mathbf{A}_n$ are $m_1 \times m_2, m_2 \times m_3, \ldots, m_n \times m_{n+1}$ matrices, respectively, and each has integers as entries? (Because matrix multiplication is associative, as shown in Exercise 13 in Section 2.6, the order of the multiplication used does not change the product.) Note that $m_1m_2m_3$ multiplications of integers are performed to multiply an $m_1 \times m_2$ matrix and an $m_2 \times m_3$ matrix using Algorithm 1. Example 9 illustrates this problem.

Links

EXAMPLE 9 In which order should the matrices $\mathbf{A}_1$, $\mathbf{A}_2$, and $\mathbf{A}_3$—where $\mathbf{A}_1$ is 30×20, $\mathbf{A}_2$ is 20×40, and $\mathbf{A}_3$ is 40×10, all with integer entries—be multiplied to use the least number of multiplications of integers?

Solution: There are two possible ways to compute $\mathbf{A}_1\mathbf{A}_2\mathbf{A}_3$. These are $\mathbf{A}_1(\mathbf{A}_2\mathbf{A}_3)$ and $(\mathbf{A}_1\mathbf{A}_2)\mathbf{A}_3$.

If $\mathbf{A}_2$ and $\mathbf{A}_3$ are first multiplied, a total of $20 \cdot 40 \cdot 10 = 8000$ multiplications of integers are used to obtain the 20×10 matrix $\mathbf{A}_2\mathbf{A}_3$. Then, to multiply $\mathbf{A}_1$ and $\mathbf{A}_2\mathbf{A}_3$ requires $30 \cdot 20 \cdot 10 = 6000$ multiplications. Hence, a total of

$$8000 + 6000 = 14{,}000$$

multiplications are used. On the other hand, if $\mathbf{A}_1$ and $\mathbf{A}_2$ are first multiplied, then $30 \cdot 20 \cdot 40 = 24{,}000$ multiplications are used to obtain the 30×40 matrix $\mathbf{A}_1\mathbf{A}_2$. Then, to multiply $\mathbf{A}_1\mathbf{A}_2$ and $\mathbf{A}_3$ requires $30 \cdot 40 \cdot 10 = 12{,}000$ multiplications. Hence, a total of

$$24{,}000 + 12{,}000 = 36{,}000$$

multiplications are used.

Clearly, the first method is more efficient. ◀

We will return to this problem in Exercise 57 in Section 8.1. Algorithms for determining the most efficient way to carry out matrix-chain multiplication are discussed in [CoLeRiSt09].

3.3.4 Algorithmic Paradigms

In Section 3.1 we introduced the basic notion of an algorithm. We provided examples of many different algorithms, including searching and sorting algorithms. We also introduced the concept of a greedy algorithm, giving examples of several problems that can be solved by greedy algorithms. Greedy algorithms provide an example of an **algorithmic paradigm**, that is, a general approach based on a particular concept that can be used to construct algorithms for solving a variety of problems.

In this book we will construct algorithms for solving many different problems based on a variety of algorithmic paradigms, including the most widely used algorithmic paradigms. These paradigms can serve as the basis for constructing efficient algorithms for solving a wide range of problems.

Some of the algorithms we have already studied are based on an algorithmic paradigm known as brute force, which we will describe in this section. Algorithmic paradigms, studied later in this book, include divide-and-conquer algorithms studied in Chapter 8, dynamic programming, also studied in Chapter 8, backtracking, studied in Chapter 11, and probabilistic algorithms, studied in Chapter 7. There are many important algorithmic paradigms besides those described in this book. Consult books on algorithm design such as [KlTa06] to learn more about them.

BRUTE-FORCE ALGORITHMS Brute force is an important, and basic, algorithmic paradigm. In a **brute-force algorithm**, a problem is solved in the most straightforward manner based on the statement of the problem and the definitions of terms. Brute-force algorithms are designed to solve problems without regard to the computing resources required. For example, in some brute-force algorithms the solution to a problem is found by examining every possible solution, looking for the best possible. In general, brute-force algorithms are naive approaches for solving problems that do not take advantage of any special structure of the problem or clever ideas.

Note that Algorithm 1 in Section 3.1 for finding the maximum number in a sequence is a brute-force algorithm because it examines each of the n numbers in a sequence to find the maximum term. The algorithm for finding the sum of n numbers by adding one additional number at a time is also a brute-force algorithm, as is the algorithm for matrix multiplication based on its definition (Algorithm 1). The bubble, insertion, and selection sorts (described in Section 3.1 in Algorithms 4 and 5 and in the preamble of Exercise 43, respectively) are also considered to be brute-force algorithms; all three of these sorting algorithms are straightforward approaches much less efficient than other sorting algorithms such as the merge sort and the quick sort discussed in Chapters 5 and 8.

Although brute-force algorithms are often inefficient, they are often quite useful. A brute-force algorithm may be able to solve practical instances of problems, particularly when the input is not too large, even if it is impractical to use this algorithm for larger inputs. Furthermore, when designing new algorithms to solve a problem, the goal is often to find a new algorithm that is more efficient than a brute-force algorithm. One such problem of this type is described in Example 10.

EXAMPLE 10 Construct a brute-force algorithm for finding the closest pair of points in a set of n points in the plane and provide a worst-case big-O estimate for the number of bit operations used by the algorithm.

Solution: Suppose that we are given as input the points $(x_1, y_1), (x_2, y_2), \ldots, (x_n, y_n)$. Recall that the distance between (x_i, y_i) and (x_j, y_j) is $\sqrt{(x_j - x_i)^2 + (y_j - y_i)^2}$. A brute-force algorithm can find the closest pair of these points by computing the distances between all pairs of the n points and determining the smallest distance. (We can make one small simplification to make the computation easier; we can compute the square of the distance between pairs of points to find the closest pair, rather than the distance between these points. We can do this because the square of the distance between a pair of points is smallest when the distance between these points is smallest.)

ALGORITHM 3 Brute-Force Algorithm for Closest Pair of Points.

procedure *closest-pair*($(x_1, y_1), (x_2, y_2), \ldots, (x_n, y_n)$: pairs of real numbers)
$min = \infty$
for $i := 2$ **to** n
  for $j := 1$ **to** $i - 1$
    if $(x_j - x_i)^2 + (y_j - y_i)^2 < min$ **then**
      $min := (x_j - x_i)^2 + (y_j - y_i)^2$
      closest pair $:= ((x_i, y_i), (x_j, y_j))$
return *closest pair*

To estimate the number of operations used by the algorithm, first note that there are $n(n - 1)/2$ pairs of points $((x_i, y_i), (x_j, y_j))$ that we loop through (as the reader should verify). For each such pair we compute $(x_j - x_i)^2 + (y_j - y_i)^2$, compare it with the current value of *min*, and if it is smaller than *min*, replace the current value of *min* by this new value. It follows that this algorithm uses $\Theta(n^2)$ operations, in terms of arithmetic operations and comparisons.

In Chapter 8 we will devise an algorithm that determines the closest pair of points when given n points in the plane as input that has $O(n \log n)$ worst-case complexity. The original discovery of such an algorithm, much more efficient than the brute-force approach, was considered quite surprising. ◀

3.3.5 Understanding the Complexity of Algorithms

Table 1 displays some common terminology used to describe the time complexity of algorithms. For example, an algorithm that finds the largest of the first 100 terms of a list of n elements by applying Algorithm 1 of Section 3.1 to the sequence of the first 100 terms, where n is an integer with $n \geq 100$, has **constant complexity** because it uses 99 comparisons no matter what n is (as the reader can verify). The linear search algorithm has **linear** (worst-case or average-case) **complexity** and the binary search algorithm has **logarithmic** (worst-case) **complexity**. Many important algorithms have $n \log n$, or **linearithmic** (worst-case) **complexity**, such as the merge sort, which we will introduce in Chapter 5. (The word *linearithmic* is a combination of the words *linear* and log*arithmic*.)

TABLE 1 Commonly Used Terminology for the Complexity of Algorithms.

Complexity	*Terminology*
$\Theta(1)$	Constant complexity
$\Theta(\log n)$	Logarithmic complexity
$\Theta(n)$	Linear complexity
$\Theta(n \log n)$	Linearithmic complexity
$\Theta(n^b)$	Polynomial complexity
$\Theta(b^n)$, where $b > 1$	Exponential complexity
$\Theta(n!)$	Factorial complexity

An algorithm has **polynomial complexity** if it has complexity $\Theta(n^b)$, where b is an integer with $b \geq 1$. For example, the bubble sort algorithm is a polynomial-time algorithm because it uses $\Theta(n^2)$ comparisons in the worst case. An algorithm has **exponential complexity** if it has time complexity $\Theta(b^n)$, where $b > 1$. The algorithm that determines whether a compound proposition in n variables is satisfiable by checking all possible assignments of truth variables is an algorithm with exponential complexity, because it uses $\Theta(2^n)$ operations. Finally, an algorithm has **factorial complexity** if it has $\Theta(n!)$ time complexity. The algorithm that finds all orders that a traveling salesperson could use to visit n cities has factorial complexity; we will discuss this algorithm in Chapter 10.

TRACTABILITY A problem that is solvable using an algorithm with polynomial (or better) worst-case complexity is called **tractable**, because the expectation is that the algorithm will produce the solution to the problem for reasonably sized input in a relatively short time. However, if the polynomial in the big-Θ estimate has high degree (such as degree 100) or if the coefficients are extremely large, the algorithm may take an extremely long time to solve the problem. Consequently, that a problem can be solved using an algorithm with polynomial worst-case time complexity is no guarantee that the problem can be solved in a reasonable amount of time for even relatively small input values. Fortunately, in practice, the degree and coefficients of polynomials in such estimates are often small.

The situation is much worse for problems that cannot be solved using an algorithm with worst-case polynomial time complexity. Such problems are called **intractable**. Usually, but not always, an extremely large amount of time is required to solve the problem for the worst cases of even small input values. In practice, however, there are situations where an algorithm with a certain worst-case time complexity may be able to solve a problem much more quickly for most cases than for its worst case. When we are willing to allow that some, perhaps small, number of cases may not be solved in a reasonable amount of time, the average-case time complexity is a better measure of how long an algorithm takes to solve a problem. Many problems important in industry are thought to be intractable but can be practically solved for essentially all sets of input that arise in daily life. Another way that intractable problems are handled when they arise in practical applications is that instead of looking for exact solutions of a problem, approximate solutions are sought. It may be the case that fast algorithms exist for finding such approximate solutions, perhaps even with a guarantee that they do not differ by very much from an exact solution.

Some problems even exist for which it can be shown that no algorithm exists for solving them. Such problems are called **unsolvable** (as opposed to **solvable** problems that can be solved using an algorithm). The first proof that there are unsolvable problems was provided by the great English mathematician and computer scientist Alan Turing when he showed that the halting problem is unsolvable. Recall that we proved that the halting problem is unsolvable in Section 3.1. (A biography of Alan Turing and a description of some of his other work can be found in Chapter 13.)

P VERSUS NP The study of the complexity of algorithms goes far beyond what we can describe here. Note, however, that many solvable problems are believed to have the property that no algorithm with polynomial worst-case time complexity solves them, but that a solution, if known, can be checked in polynomial time. Problems for which a solution can be checked in polynomial time are said to belong to the **class NP** (tractable problems are said to belong to **class P**). The abbreviation NP stands for *nondeterministic polynomial* time. The satisfiability problem, discussed in Section 1.3, is an example of an NP problem—we can quickly verify that an assignment of truth values to the variables of a compound proposition makes it true, but no polynomial time algorithm has been discovered for finding such an assignment of truth values. (For example, an exhaustive search of all possible truth values requires $\Omega(2^n)$ bit operations where n is the number of variables in the compound proposition.)

Links

There is also an important class of problems, called **NP-complete problems**, with the property that if any of these problems can be solved by a polynomial worst-case time algorithm, then all problems in the class NP can be solved by polynomial worst-case time algorithms. The satisfiability problem is also an example of an NP-complete problem. It is an NP problem and if a polynomial time algorithm for solving it were known, there would be polynomial time algorithms for all problems known to be in this class of problems (and there are many important problems in this class). This last statement follows from the fact that every problem in NP can be reduced in polynomial time to the satisfiability problem. Although more than 3000 NP-complete problems are now known, the satisfiability problem was the first problem shown to be NP-complete. The theorem that asserts this is known as the **Cook-Levin theorem** after Stephen Cook and Leonid Levin, who independently proved it in the early 1970s.

The **P versus NP problem** asks whether NP, the class of problems for which it is possible to check solutions in polynomial time, equals P, the class of tractable problems. If $P \neq NP$, there would be some problems that cannot be solved in polynomial time, but whose solutions could be verified in polynomial time. The concept of NP-completeness is helpful in research aimed at solving the P versus NP problem, because NP-complete problems are the problems in NP considered most likely not to be in P, as every problem in NP can be reduced to an NP-complete problem in polynomial time. A large majority of theoretical computer scientists believe that $P \neq NP$, which would mean that no NP-complete problem can be solved in polynomial time. One reason for this belief is that despite extensive research, no one has succeeded in showing that $P = NP$. In particular, no one has been able to find an algorithm with worst-case polynomial time complexity that solves any NP-complete problem. The P versus NP problem is one of the most famous unsolved problems in the mathematical sciences (which include theoretical computer science). It is one of the seven famous Millennium Prize Problems, of which six remain unsolved. A prize of $1,000,000 is offered by the Clay Mathematics Institute for its solution.

Links

For more information about the complexity of algorithms, consult the references, including [CoLeRiSt09], for this section listed at the end of this book. (Also, for a more formal discussion of computational complexity in terms of Turing machines, see Section 13.5.)

PRACTICAL CONSIDERATIONS Note that a big-Θ estimate of the time complexity of an algorithm expresses how the time required to solve the problem increases as the input grows in size. In practice, the best estimate (that is, with the smallest reference function) that can be shown is used. However, big-Θ estimates of time complexity cannot be directly translated into the actual amount of computer time used. One reason is that a big-Θ estimate $f(n)$ is $\Theta(g(n))$, where $f(n)$ is the time complexity of an algorithm and $g(n)$ is a reference function, means that $C_1 g(n) \leq f(n) \leq C_2 g(n)$ when $n > k$, where C_1, C_2, and k are constants. So without knowing the constants C_1, C_2, and k in the inequality, this estimate cannot be used to determine a lower bound and an upper bound on the number of operations used in the worst case. As remarked before, the time required for an operation depends on the type of operation and the

Links

Courtesy of Dr. Stephen Cook

STEPHEN COOK (BORN 1939) Stephen Cook was born in Buffalo, where his father worked as an industrial chemist and taught university courses. His mother taught English courses in a community college. While in high school Cook developed an interest in electronics through his work with a famous local inventor noted for inventing the first implantable cardiac pacemaker.

Cook was a mathematics major at the University of Michigan, graduating in 1961. He did graduate work at Harvard, receiving a master's degree in 1962 and a Ph.D. in 1966. Cook was appointed an assistant professor in the Mathematics Department at the University of California, Berkeley, in 1966. He was not granted tenure there, possibly because the members of the Mathematics Department did not find his work on what is now considered to be one of the most important areas of theoretical computer science of sufficient interest. In 1970, he joined the University of Toronto as an assistant professor, holding a joint appointment in the Computer Science Department and the Mathematics Department. He has remained at the University of Toronto, where he was appointed a University Professor in 1985.

Cook is considered to be one of the founders of computational complexity theory. His 1971 paper "The Complexity of Theorem Proving Procedures" formalized the notions of NP-completeness and polynomial-time reduction, showed that NP-complete problems exist by showing that the satisfiability problem is such a problem, and introduced the notorious P versus NP problem.

Cook has received many awards, including the 1982 Turing Award. He is married and has two sons. Among his interests are playing the violin and racing sailboats.

TABLE 2 The Computer Time Used by Algorithms.

Problem Size	*Bit Operations Used*					
n	$\log n$	n	$n \log n$	n^2	2^n	$n!$
10	3×10^{-11} s	10^{-10} s	3×10^{-10} s	10^{-9} s	10^{-8} s	3×10^{-7} s
10^2	7×10^{-11} s	10^{-9} s	7×10^{-9} s	10^{-7} s	4×10^{11} yr	*
10^3	1.0×10^{-10} s	10^{-8} s	1×10^{-7} s	10^{-5} s	*	*
10^4	1.3×10^{-10} s	10^{-7} s	1×10^{-6} s	10^{-3} s	*	*
10^5	1.7×10^{-10} s	10^{-6} s	2×10^{-5} s	0.1 s	*	*
10^6	2×10^{-10} s	10^{-5} s	2×10^{-4} s	0.17 min	*	*

computer being used. Often, instead of a big-Θ estimate on the worst-case time complexity of an algorithm, we have only a big-O estimate. Note that a big-O estimate on the time complexity of an algorithm provides an upper, but not a lower, bound on the worst-case time required for the algorithm as a function of the input size. Nevertheless, for simplicity, we will often use big-O estimates when describing the time complexity of algorithms, with the understanding that big-Θ estimates would provide more information.

Table 2 displays the time needed to solve problems of various sizes with an algorithm using the indicated number n of bit operations, assuming that each bit operation takes 10^{-11} seconds, a reasonable estimate of the time required for a bit operation using the fastest computers as of 2018. Times of more than 10^{100} years are indicated with an asterisk. In the future, these times will decrease as faster computers are developed. We can use the times shown in Table 2 to see whether it is reasonable to expect a solution to a problem of a specified size using an algorithm with known worst-case time complexity when we run this algorithm on a modern computer. Note that we cannot determine the exact time a computer uses to solve a problem with input of a particular size because of a myriad of issues involving computer hardware and the particular software implementation of the algorithm.

It is important to have a reasonable estimate for how long it will take a computer to solve a problem. For instance, if an algorithm requires approximately 10 hours, it may be worthwhile to spend the computer time (and money) required to solve this problem. But, if an algorithm requires approximately 10 billion years to solve a problem, it would be unreasonable to use resources to implement this algorithm. One of the most interesting phenomena of modern technology is the tremendous increase in the speed and memory space of computers. Another important factor that decreases the time needed to solve problems on computers is **parallel processing**, which is the technique of performing sequences of operations simultaneously.

Efficient algorithms, including most algorithms with polynomial time complexity, benefit most from significant technology improvements. However, these technology improvements offer little help in overcoming the complexity of algorithms of exponential or factorial time complexity. Because of the increased speed of computation, increases in computer memory, and the use of algorithms that take advantage of parallel processing, many problems that were considered impossible to solve five years ago are now routinely solved, and certainly five years from now this statement will still be true. This is even true when the algorithms used are intractable.

Exercises

1. Give a big-O estimate for the number of operations (where an operation is an addition or a multiplication) used in this segment of an algorithm.

```
t := 0
for i := 1 to 3
  for j := 1 to 4
    t := t + ij
```

2. Give a big-O estimate for the number additions used in this segment of an algorithm.

```
t := 0
for i := 1 to n
  for j := 1 to n
    t := t + i + j
```

3. Give a big-O estimate for the number of operations, where an operation is a comparison or a multiplication, used in this segment of an algorithm (ignoring comparisons used to test the conditions in the **for** loops, where $a_1, a_2, ..., a_n$ are positive real numbers).

$m := 0$
for $i := 1$ **to** n
  for $j := i + 1$ **to** n
    $m := \max(a_i a_j, m)$

4. Give a big-O estimate for the number of operations, where an operation is an addition or a multiplication, used in this segment of an algorithm (ignoring comparisons used to test the conditions in the **while** loop).

$i := 1$
$t := 0$
while $i \leq n$
  $t := t + i$
  $i := 2i$

5. How many comparisons are used by the algorithm given in Exercise 16 of Section 3.1 to find the smallest natural number in a sequence of n natural numbers?

6. **a)** Use pseudocode to describe the algorithm that puts the first four terms of a list of real numbers of arbitrary length in increasing order using the insertion sort.

b) Show that this algorithm has time complexity $O(1)$ in terms of the number of comparisons used.

7. Suppose that an element is known to be among the first four elements in a list of 32 elements. Would a linear search or a binary search locate this element more rapidly?

8. Given a real number x and a positive integer k, determine the number of multiplications used to find x^{2^k} starting with x and successively squaring (to find x^2, x^4, and so on). Is this a more efficient way to find x^{2^k} than by multiplying x by itself the appropriate number of times?

9. Give a big-O estimate for the number of comparisons used by the algorithm that determines the number of 1s in a bit string by examining each bit of the string to determine whether it is a 1 bit (see Exercise 25 of Section 3.1).

∗10. **a)** Show that this algorithm determines the number of 1 bits in the bit string S:

procedure *bit count*(S: bit string)
$count := 0$
while $S \neq 0$
  $count := count + 1$
  $S := S \wedge (S - 1)$
return $count$ {$count$ is the number of 1s in S}

Here $S - 1$ is the bit string obtained by changing the rightmost 1 bit of S to a 0 and all the 0 bits to the right of this to 1s. [Recall that $S \wedge (S - 1)$ is the bitwise *AND* of S and $S - 1$.]

b) How many bitwise *AND* operations are needed to find the number of 1 bits in a string S using the algorithm in part (a)?

11. **a)** Suppose we have n subsets $S_1, S_2, \ldots, S_n$ of the set $\{1, 2, \ldots, n\}$. Express a brute-force algorithm that determines whether there is a disjoint pair of these subsets. [*Hint:* The algorithm should loop through the subsets; for each subset S_i, it should then loop through all other subsets; and for each of these other subsets S_j, it should loop through all elements k in S_i to determine whether k also belongs to S_j.]

b) Give a big-O estimate for the number of times the algorithm needs to determine whether an integer is in one of the subsets.

12. Consider the following algorithm, which takes as input a sequence of n integers $a_1, a_2, \ldots, a_n$ and produces as output a matrix $\mathbf{M} = \{m_{ij}\}$ where m_{ij} is the minimum term in the sequence of integers $a_i, a_{i+1}, \ldots, a_j$ for $j \geq i$ and $m_{ij} = 0$ otherwise.

initialize $\mathbf{M}$ so that $m_{ij} = a_i$ if $j \geq i$ and $m_{ij} = 0$ otherwise
for $i := 1$ **to** n
  for $j := i + 1$ **to** n
    for $k := i + 1$ **to** j
      $m_{ij} := \min(m_{ij}, a_k)$
return $\mathbf{M} = \{m_{ij}\}$ {m_{ij} is the minimum term of $a_i, a_{i+1}, \ldots, a_j$}

a) Show that this algorithm uses $O(n^3)$ comparisons to compute the matrix $\mathbf{M}$.

b) Show that this algorithm uses $\Omega(n^3)$ comparisons to compute the matrix $\mathbf{M}$. Using this fact and part (a), conclude that the algorithms uses $\Theta(n^3)$ comparisons. [*Hint:* Only consider the cases where $i \leq n/4$ and $j \geq 3n/4$ in the two outer loops in the algorithm.]

13. The conventional algorithm for evaluating a polynomial $a_n x^n + a_{n-1}x^{n-1} + \cdots + a_1 x + a_0$ at $x = c$ can be expressed in pseudocode by

procedure *polynomial*($c, a_0, a_1, \ldots, a_n$: real numbers)
$power := 1$
$y := a_0$
for $i := 1$ **to** n
  $power := power * c$
  $y := y + a_i * power$
return y $\{y = a_n c^n + a_{n-1}c^{n-1} + \cdots + a_1 c + a_0\}$

where the final value of y is the value of the polynomial at $x = c$.

a) Evaluate $3x^2 + x + 1$ at $x = 2$ by working through each step of the algorithm showing the values assigned at each assignment step.

b) Exactly how many multiplications and additions are used to evaluate a polynomial of degree n at $x = c$? (Do not count additions used to increment the loop variable.)

14. There is a more efficient algorithm (in terms of the number of multiplications and additions used) for evaluating polynomials than the conventional algorithm described in the previous exercise. It is called **Horner's method**. This pseudocode shows how to use this method to find the value of $a_n x^n + a_{n-1}x^{n-1} + \cdots + a_1 x + a_0$ at $x = c$.

procedure *Horner*($c, a_0, a_1, a_2, \ldots, a_n$: real numbers)
  $y := a_n$
  for $i := 1$ **to** n
    $y := y * c + a_{n-i}$
  return y $\{y = a_n c^n + a_{n-1}c^{n-1} + \cdots + a_1 c + a_0\}$

a) Evaluate $3x^2 + x + 1$ at $x = 2$ by working through each step of the algorithm showing the values assigned at each assignment step.

b) Exactly how many multiplications and additions are used by this algorithm to evaluate a polynomial of degree n at $x = c$? (Do not count additions used to increment the loop variable.)

15. What is the largest n for which one can solve within one second a problem using an algorithm that requires $f(n)$ bit operations, where each bit operation is carried out in 10^{-9} seconds, with these functions $f(n)$?

a) $\log n$ b) n c) $n \log n$
d) n^2 e) 2^n f) $n!$

16. What is the largest n for which one can solve within a day using an algorithm that requires $f(n)$ bit operations, where each bit operation is carried out in 10^{-11} seconds, with these functions $f(n)$?

a) $\log n$ b) $1000n$ c) n^2
d) $1000n^2$ e) n^3 f) 2^n
g) 2^{2n} h) 2^{2^n}

17. What is the largest n for which one can solve within a minute using an algorithm that requires $f(n)$ bit operations, where each bit operation is carried out in 10^{-12} seconds, with these functions $f(n)$?

a) $\log \log n$ b) $\log n$ c) $(\log n)^2$
d) $1{,}000{,}000n$ e) n^2 f) 2^n
g) 2^{n^2}

18. How much time does an algorithm take to solve a problem of size n if this algorithm uses $2n^2 + 2^n$ operations, each requiring 10^{-9} seconds, with these values of n?

a) 10 b) 20 c) 50 d) 100

19. How much time does an algorithm using 2^{50} operations need if each operation takes these amounts of time?

a) 10^{-6} s b) 10^{-9} s c) 10^{-12} s

20. What is the effect in the time required to solve a problem when you double the size of the input from n to $2n$, assuming that the number of milliseconds the algorithm uses to solve the problem with input size n is each of these functions? [Express your answer in the simplest form possible, either as a ratio or a difference. Your answer may be a function of n or a constant.]

a) $\log \log n$ b) $\log n$ c) $100n$
d) $n \log n$ e) n^2 f) n^3
g) 2^n

21. What is the effect in the time required to solve a problem when you increase the size of the input from n to $n + 1$, assuming that the number of milliseconds the algorithm uses to solve the problem with input size n is each of these functions? [Express your answer in the simplest form possible, either as a ratio or a difference. Your answer may be a function of n or a constant.]

a) $\log n$ b) $100n$ c) n^2
d) n^3 e) 2^n f) 2^{n^2}
g) $n!$

22. Determine the least number of comparisons, or best-case performance,

a) required to find the maximum of a sequence of n integers, using Algorithm 1 of Section 3.1.

b) used to locate an element in a list of n terms with a linear search.

c) used to locate an element in a list of n terms using a binary search.

23. Analyze the average-case performance of the linear search algorithm, if exactly half the time the element x is not in the list, and if x is in the list, it is equally likely to be in any position.

24. An algorithm is called **optimal** for the solution of a problem with respect to a specified operation if there is no algorithm for solving this problem using fewer operations.

a) Show that Algorithm 1 in Section 3.1 is an optimal algorithm with respect to the number of comparisons of integers. [*Note:* Comparisons used for bookkeeping in the loop are not of concern here.]

b) Is the linear search algorithm optimal with respect to the number of comparisons of integers (not including comparisons used for bookkeeping in the loop)?

25. Describe the worst-case time complexity, measured in terms of comparisons, of the ternary search algorithm described in Exercise 27 of Section 3.1.

26. Describe the worst-case time complexity, measured in terms of comparisons, of the search algorithm described in Exercise 28 of Section 3.1.

27. Analyze the worst-case time complexity of the algorithm you devised in Exercise 29 of Section 3.1 for locating a mode in a list of nondecreasing integers.

28. Analyze the worst-case time complexity of the algorithm you devised in Exercise 30 of Section 3.1 for locating all modes in a list of nondecreasing integers.

29. Analyze the worst-case time complexity of the algorithm you devised in Exercise 33 of Section 3.1 for finding the first term of a sequence of integers equal to some previous term.

30. Analyze the worst-case time complexity of the algorithm you devised in Exercise 34 of Section 3.1 for finding all terms of a sequence that are greater than the sum of all previous terms.

31. Analyze the worst-case time complexity of the algorithm you devised in Exercise 35 of Section 3.1 for finding the first term of a sequence less than the immediately preceding term.

32. Determine the worst-case complexity in terms of comparisons of the algorithm from Exercise 5 in Section 3.1 for determining all values that occur more than once in a sorted list of integers.

33. Determine the worst-case complexity in terms of comparisons of the algorithm from Exercise 9 in Section 3.1 for determining whether a string of n characters is a palindrome.

34. How many comparisons does the selection sort (see preamble to Exercise 43 in Section 3.1) use to sort n items? Use your answer to give a big-O estimate of the complexity of the selection sort in terms of number of comparisons for the selection sort.

35. Determine a big-O estimate for the worst-case complexity in terms of number of comparisons used and the number of terms swapped by the binary insertion sort described in the preamble to Exercise 49 in Section 3.1.

36. Determine the number of character comparisons used by the naive string matcher to look for a pattern of m characters in a text with n characters if the first character of the pattern does not occur in the text.

37. Determine a big-O estimate of the number of character comparisons used by the naive string matcher to find all occurrences of a pattern of m characters in a text with n characters, in terms of the parameters m and n.

38. Determine big-O estimates for the algorithms for deciding whether two strings are anagrams from parts (a) and (b) of Exercise 31 of Section 3.1.

39. Determine big-O estimates for the algorithms for finding the closest of n real numbers from parts (a) and (b) of Exercise 32 of Section 3.1.

40. Show that the greedy algorithm for making change for n cents using quarters, dimes, nickels, and pennies has $O(n)$ complexity measured in terms of comparisons needed.

Exercises 41 and 42 deal with the problem of scheduling the most talks possible given the start and end times of n talks.

41. Find the complexity of a brute-force algorithm for scheduling the talks by examining all possible subsets of the talks. [*Hint:* Use the fact that a set with n elements has 2^n subsets.]

42. Find the complexity of the greedy algorithm for scheduling the most talks by adding at each step the talk with the earliest end time compatible with those already scheduled (Algorithm 8 in Section 3.1). Assume that the talks are not already sorted by earliest end time and assume that the worst-case time complexity of sorting is $O(n \log n)$.

43. Describe how the number of comparisons used in the worst case changes when these algorithms are used to search for an element of a list when the size of the list doubles from n to $2n$, where n is a positive integer.

a) linear search **b)** binary search

44. Describe how the number of comparisons used in the worst case changes when the size of the list to be sorted doubles from n to $2n$, where n is a positive integer when these sorting algorithms are used.

a) bubble sort **b)** insertion sort

c) selection sort (described in the preamble to Exercise 43 in Section 3.1)

d) binary insertion sort (described in the preamble to Exercise 49 in Section 3.1)

An $n \times n$ matrix is called **upper triangular** if $a_{ij} = 0$ whenever $i > j$.

45. From the definition of the matrix product, describe an algorithm in English for computing the product of two upper triangular matrices that ignores those products in the computation that are automatically equal to zero.

46. Give a pseudocode description of the algorithm in Exercise 45 for multiplying two upper triangular matrices.

47. How many multiplications of entries are used by the algorithm found in Exercise 45 for multiplying two $n \times n$ upper triangular matrices?

In Exercises 48–49 assume that the number of multiplications of entries used to multiply a $p \times q$ matrix and a $q \times r$ matrix is pqr.

48. What is the best order to form the product **ABC** if **A**, **B**, and **C** are matrices with dimensions 3×9, 9×4, and 4×2, respectively?

49. What is the best order to form the product **ABCD** if **A**, **B**, **C**, and **D** are matrices with dimensions 30×10, 10×40, 40×50, and 50×30, respectively?

Key Terms and Results

TERMS

algorithm: a finite sequence of precise instructions for performing a computation or solving a problem

searching problem: the problem of locating an element in a list

linear search algorithm: a procedure for searching a list element by element

binary search algorithm: a procedure for searching an ordered list by successively splitting the list in half

sorting: the reordering of the elements of a list into prescribed order

string matching: given a string, determining all the occurrences where this string occurs within a longer string

$f(x)$ is $O(g(x))$: the fact that $|f(x)| \leq C|g(x)|$ for all $x > k$ for some constants C and k

witness to the relationship $f(x)$ is $O(g(x))$: a pair C and k such that $|f(x)| \leq C|g(x)|$ whenever $x > k$

$f(x)$ is $\Omega(g(x))$: the fact that $|f(x)| \geq C|g(x)|$ for all $x > k$ for some positive constants C and k

$f(x)$ is $\Theta(g(x))$: the fact that $f(x)$ is both $O(g(x))$ and $\Omega(g(x))$

time complexity: the amount of time required for an algorithm to solve a problem

space complexity: the amount of space in computer memory required for an algorithm to solve a problem

worst-case time complexity: the greatest amount of time required for an algorithm to solve a problem of a given size

average-case time complexity: the average amount of time required for an algorithm to solve a problem of a given size

algorithmic paradigm: a general approach for constructing algorithms based on a particular concept

brute force: the algorithmic paradigm based on constructing algorithms for solving problems in a naive manner from the statement of the problem and definitions

greedy algorithm: an algorithm that makes the best choice at each step according to some specified condition

tractable problem: a problem for which there is a worst-case polynomial-time algorithm that solves it

intractable problem: a problem for which no worst-case polynomial-time algorithm exists for solving it

solvable problem: a problem that can be solved by an algorithm

unsolvable problem: a problem that cannot be solved by an algorithm

RESULTS

linear and binary search algorithms: (given in Section 3.1)

bubble sort: a sorting that uses passes where successive items are interchanged if they are in the wrong order

insertion sort: a sorting that at the jth step inserts the jth element into the correct position in the list, when the first $j-1$ elements of the list are already sorted

The linear search has $O(n)$ worst case time complexity.
The binary search has $O(\log n)$ worst case time complexity.
The bubble and insertion sorts have $O(n^2)$ worst case time complexity.
$\log n!$ is $O(n \log n)$.

If $f_1(x)$ is $O(g_1(x))$ and $f_2(x)$ is $O(g_2(x))$, then $(f_1+f_2)(x)$ is $O(\max(g_1(x), g_2(x)))$ and $(f_1f_2)(x)$ is $O((g_1g_2(x))$.

If $a_0, a_1, \ldots, a_n$ are real numbers with $a_n \neq 0$, then $a_nx^n + a_{n-1}x^{n-1} + \cdots + a_1x + a_0$ is $\Theta(x^n)$, and hence $O(n)$ and $\Omega(n)$.

Review Questions

1. a) Define the term *algorithm.*

b) What are the different ways to describe algorithms?

c) What is the difference between an algorithm for solving a problem and a computer program that solves this problem?

2. a) Describe, using English, an algorithm for finding the largest integer in a list of n integers.

b) Express this algorithm in pseudocode.

c) How many comparisons does the algorithm use?

3. a) State the definition of the fact that $f(n)$ is $O(g(n))$, where $f(n)$ and $g(n)$ are functions from the set of positive integers to the set of real numbers.

b) Use the definition of the fact that $f(n)$ is $O(g(n))$ directly to prove or disprove that $n^2 + 18n + 107$ is $O(n^3)$.

c) Use the definition of the fact that $f(n)$ is $O(g(n))$ directly to prove or disprove that n^3 is $O(n^2 + 18n + 107)$.

4. List these functions so that each function is big-O of the next function in the list: $(\log n)^3$, $n^3/1{,}000{,}000$, $\sqrt{n}$, $100n + 101$, 3^n, $n!$, 2^nn^2.

5. a) How can you produce a big-O estimate for a function that is the sum of different terms where each term is the product of several functions?

b) Give a big-O estimate for the function $f(n) = (n! + 1)(2^n + 1) + (n^{n-2} + 8n^{n-3})(n^3 + 2^n)$. For the function g in your estimate $f(x)$ is $O(g(x))$, use a simple function of smallest possible order.

6. a) Define what the worst-case time complexity, average-case time complexity, and best-case time complexity (in terms of comparisons) mean for an algorithm that finds the smallest integer in a list of n integers.

b) What are the worst-case, average-case, and best-case time complexities, in terms of comparisons, of the algorithm that finds the smallest integer in a list of n integers by comparing each of the integers with the smallest integer found so far?

7. a) Describe the linear search and binary search algorithm for finding an integer in a list of integers in increasing order.

b) Compare the worst-case time complexities of these two algorithms.

c) Is one of these algorithms always faster than the other (measured in terms of comparisons)?

8. a) Describe the bubble sort algorithm.

b) Use the bubble sort algorithm to sort the list 5, 2, 4, 1, 3.

c) Give a big-O estimate for the number of comparisons used by the bubble sort.

9. a) Describe the insertion sort algorithm.

b) Use the insertion sort algorithm to sort the list 2, 5, 1, 4, 3.

c) Give a big-O estimate for the number of comparisons used by the insertion sort.

10. a) Explain the concept of a greedy algorithm.

b) Provide an example of a greedy algorithm that produces an optimal solution and explain why it produces an optimal solution.

c) Provide an example of a greedy algorithm that does not always produce an optimal solution and explain why it fails to do so.

11. Define what it means for a problem to be tractable and what it means for a problem to be solvable.

Supplementary Exercises

1. a) Describe an algorithm for locating the last occurrence of the largest number in a list of integers.
   b) Estimate the number of comparisons used.
2. a) Describe an algorithm for finding the first and second largest elements in a list of integers.
   b) Estimate the number of comparisons used.
3. a) Give an algorithm to determine whether a bit string contains a pair of consecutive zeros.
   b) How many comparisons does the algorithm use?
4. a) Suppose that a list contains integers that are in order of largest to smallest and an integer can appear repeatedly in this list. Devise an algorithm that locates all occurrences of an integer x in the list.
   b) Estimate the number of comparisons used.
5. a) Adapt Algorithm 1 in Section 3.1 to find the maximum and the minimum of a sequence of n elements by employing a temporary maximum and a temporary minimum that is updated as each successive element is examined.
   b) Describe the algorithm from part (a) in pseudocode.
   c) How many comparisons of elements in the sequence are carried out by this algorithm? (Do not count comparisons used to determine whether the end of the sequence has been reached.)
6. a) Describe in detail (and in English) the steps of an algorithm that finds the maximum and minimum of a sequence of n elements by examining pairs of successive elements, keeping track of a temporary maximum and a temporary minimum. If n is odd, both the temporary maximum and temporary minimum should initially equal the first term, and if n is even, the temporary minimum and temporary maximum should be found by comparing the initial two elements. The temporary maximum and temporary minimum should be updated by comparing them with the maximum and minimum of the pair of elements being examined.
   b) Express the algorithm described in part (a) in pseudocode.
   c) How many comparisons of elements of the sequence are carried out by this algorithm? (Do not count comparisons used to determine whether the end of the sequence has been reached.) How does this compare to the number of comparisons used by the algorithm in Exercise 5?

*7. Show that the worst-case complexity in terms of comparisons of an algorithm that finds the maximum and minimum of n elements is at least $\lceil 3n/2 \rceil - 2$.

8. Devise an efficient algorithm for finding the second largest element in a sequence of n elements and determine the worst-case complexity of your algorithm.

9. Devise an algorithm that finds all equal pairs of sums of two terms of a sequence of n numbers, and determine the worst-case complexity of your algorithm.

10. Devise an algorithm that finds the closest pair of integers in a sequence of n integers, and determine the worst-case complexity of your algorithm. [*Hint:* Sort the sequence. Use the fact that sorting can be done with worst-case time complexity $O(n \log n)$.]

The **shaker sort** (or **bidirectional bubble sort**) successively compares pairs of adjacent elements, exchanging them if they are out of order, and alternately passing through the list from the beginning to the end and then from the end to the beginning until no exchanges are needed.

11. Show the steps used by the shaker sort to sort the list 3, 5, 1, 4, 6, 2.

12. Express the shaker sort in pseudocode.

13. Show that the shaker sort has $O(n^2)$ complexity measured in terms of the number of comparisons it uses.

14. Explain why the shaker sort is efficient for sorting lists that are already in close to the correct order.

15. Show that $(n \log n + n^2)^3$ is $O(n^6)$.

16. Show that $8x^3 + 12x + 100 \log x$ is $O(x^3)$.

17. Give a big-O estimate for $(x^2 + x(\log x)^3) \cdot (2^x + x^3)$.

18. Find a big-O estimate for $\sum_{j=1}^{n} j(j+1)$.

*19. Show that $n!$ is not $O(2^n)$.

*20. Show that n^n is not $O(n!)$.

21. Find all pairs of functions of the same order in this list of functions: $n^2 + (\log n)^2$, $n^2 + n$, $n^2 + \log 2^n + 1$, $(n+1)^3 - (n-1)^3$, and $(n + \log n)^2$.

22. Find all pairs of functions of the same order in this list of functions $n^2 + 2^n$, $n^2 + 2^{100}$, $n^2 + 2^{2n}$, $n^2 + n!$, $n^2 + 3^n$, and $(n^2 + 1)^2$.

23. Find an integer n with $n > 2$ for which $n^{2^{100}} < 2^n$.

24. Find an integer n with $n > 2$ for which $(\log n)^{2^{100}} < \sqrt{n}$.

*25. Arrange the functions n^n, $(\log n)^2$, $n^{1.0001}$, $(1.0001)^n$, $2^{\sqrt{\log_2 n}}$, and $n(\log n)^{1001}$ in a list so that each function is big-O of the next function. [*Hint:* To determine the relative size of some of these functions, take logarithms.]

*26. Arrange the function 2^{100n}, 2^{n^2}, $2^{n!}$, 2^{2^n}, $n^{\log n}$, $n \log n \log \log n$, $n^{3/2}$, $n(\log n)^{3/2}$, and $n^{4/3}(\log n)^2$ in a list so that each function is big-O of the next function. [*Hint:* To determine the relative size of some of these functions, take logarithms.]

*27. Give an example of two increasing functions $f(n)$ and $g(n)$ from the set of positive integers to the set of positive integers such that neither $f(n)$ is $O(g(n))$ nor $g(n)$ is $O(f(n))$.

28. Show that if the denominations of coins are $c^0, c^1, \ldots, c^k$, where k is a positive integer and c is a positive integer, $c > 1$, the greedy algorithm always produces change using the fewest coins possible.

29. **a)** Use pseudocode to specify a brute-force algorithm that determines when given as input a sequence of n positive integers whether there are two distinct terms of the sequence that have as sum a third term. The algorithm should loop through all triples of terms of the sequence, checking whether the sum of the first two terms equals the third.

b) Give a big-O estimate for the complexity of the brute-force algorithm from part (a).

30. **a)** Devise a more efficient algorithm for solving the problem described in Exercise 29 that first sorts the input sequence and then checks for each pair of terms whether their difference is in the sequence.

b) Give a big-O estimate for the complexity of this algorithm. Is it more efficient than the brute-force algorithm from Exercise 29?

Suppose we have s men and s women each with their preference lists for the members of the opposite gender, as described in the preamble to Exercise 64 in Section 3.1. We say that a woman w is a **valid partner** for a man m if there is some stable matching in which they are paired. Similarly, a man m is a **valid partner** for a woman w if there is some stable matching in which they are paired. A matching in which each man is assigned his valid partner ranking highest on his preference list is called **male optimal**, and a matching in which each woman is assigned her valid partner ranking lowest on her preference list is called **female pessimal**.

31. Find all valid partners for each man and each woman if there are three men m_1, m_2, and m_3 and three women w_1, w_2, w_3 with these preference rankings of the men for the women, from highest to lowest: m_1: w_3, w_1, w_2; m_2: w_3, w_2, w_1; m_3: w_2, w_3, w_1; and with these preference rankings of the women for the men, from highest to lowest: w_1: m_3, m_2, m_1; w_2: m_1, m_3, m_2; w_3: m_3, m_2, m_1.

***32.** Show that the deferred acceptance algorithm given in the preamble to Exercise 65 of Section 3.1, always produces a male optimal and female pessimal matching.

33. Define what it means for a matching to be female optimal and for a matching to be male pessimal.

***34.** Show that when woman do the proposing in the deferred acceptance algorithm, the matching produced is female optimal and male pessimal.

In Exercises 35 and 36 we consider variations on the problem of finding stable matchings of men and women described in the preamble to Exercise 65 in Section 3.1.

***35.** In this exercise we consider matching problems where there may be different numbers of men and women, so that it is impossible to match everyone with a member of the opposite gender.

a) Extend the definition of a stable matching from that given in the preamble to Exercise 64 in Section 3.1 to cover the case where there are unequal numbers of men and women. Avoid all cases where a man and a woman would prefer each other to their current situation, including those involving unmatched people. (Assume that an unmatched person prefers a match with a member of the opposite gender to remaining unmatched.)

b) Adapt the deferred acceptance algorithm to find stable matchings, using the definition of stable matchings from part (a), when there are different numbers of men and women.

c) Prove that all matchings produced by the algorithm from part (b) are stable, according to the definition from part (a).

***36.** In this exercise we consider matching problems where some man-woman pairs are not allowed.

a) Extend the definition of a stable matching to cover the situation where there are the same number of men and women, but certain pairs of men and women are forbidden. Avoid all cases where a man and a woman would prefer each other to their current situation, including those involving unmatched people.

b) Adapt the deferred acceptance algorithm to find stable matchings when there are the same number of men and women, but certain man-woman pairs are forbidden. Be sure to consider people who are unmatched at the end of the algorithm. (Assume that an unmatched person prefers a match with a member of the opposite gender who is not a forbidden partner to remaining unmatched.)

c) Prove that all matchings produced by the algorithm from (b) are stable, according to the definition in part (a).

Exercises 37–40 deal with the problem of scheduling n jobs on a single processor. To complete job j, the processor must run job j for time t_j without interruption. Each job has a deadline d_j. If we start job j at time s_j, it will be completed at time $e_j = s_j + t_j$. The **lateness** of the job measures how long it finishes after its deadline, that is, the lateness of job j is $\max(0, e_j - d_j)$. We wish to devise a greedy algorithm that minimizes the maximum lateness of a job among the n jobs.

37. Suppose we have five jobs with specified required times and deadlines: $t_1 = 25$, $d_1 = 50$; $t_2 = 15$, $d_2 = 60$; $t_3 = 20$, $d_3 = 60$; $t_4 = 5$, $d_4 = 55$; $t_5 = 10$, $d_5 = 75$. Find the maximum lateness of any job when the jobs are scheduled in this order (and they start at time 0): Job 3, Job 1, Job 4, Job 2, Job 5. Answer the same question for the schedule Job 5, Job 4, Job 3, Job 1, Job 2.

38. The **slackness** of a job requiring time t and with deadline d is $d - t$, the difference between its deadline and the time it requires. Find an example that shows that scheduling jobs by increasing slackness does not always yield a schedule with the smallest possible maximum lateness.

39. Find an example that shows that scheduling jobs in order of increasing time required does not always yield a schedule with the smallest possible maximum lateness.

*40. Prove that scheduling jobs in order of increasing deadlines always produces a schedule that minimizes the maximum lateness of a job. [*Hint:* First show that for a schedule to be optimal, jobs must be scheduled with no idle time between them and so that no job is scheduled before another with an earlier deadline.]

41. Suppose that we have a knapsack with total capacity of W kg. We also have n items where item j has mass w_j. The **knapsack problem** asks for a subset of these n items with the largest possible total mass not exceeding W.

a) Devise a brute-force algorithm for solving the knapsack problem.

b) Solve the knapsack problem when the capacity of the knapsack is 18 kg and there are five items: a 5-kg sleeping bag, an 8-kg tent, a 7-kg food pack, a 4-kg container of water, and an 11-kg portable stove.

In Exercises 42–46 we will study the problem of load balancing. The input to the problem is a collection of p processors and n jobs, t_j is the time required to run job j, jobs run without interruption on a single machine until finished, and a processor can run only one job at a time. The **load** L_k of processor k is the sum over all jobs assigned to processor k of the times required to run these jobs. The **makespan** is the maximum load over all the p processors. The load balancing problem asks for an assignment of jobs to processors to minimize the makespan.

42. Suppose we have three processors and five jobs requiring times $t_1 = 3$, $t_2 = 5$, $t_3 = 4$, $t_4 = 7$, and $t_5 = 8$. Solve the load balancing problem for this input by finding the assignment of the five jobs to the three processors that minimizes the makespan.

43. Suppose that L^* is the minimum makespan when p processors are given n jobs, where t_j is the time required to run job j.

a) Show that $L^* \geq \max_{j=1,2,\ldots,n} t_j$.

b) Show that $L^* \geq \frac{1}{p} \sum_{j=1}^{n} t_j$.

44. Write out in pseudocode the greedy algorithm that goes through the jobs in order and assigns each job to the processor with the smallest load at that point in the algorithm.

45. Run the algorithm from Exercise 44 on the input given in Exercise 42.

An **approximation algorithm** for an optimization problem produces a solution guaranteed to be close to an optimal solution. More precisely, suppose that the optimization problem asks for an input S that minimizes $F(X)$ where F is some function of the input X. If an algorithm always finds an input T with $F(T) \leq cF(S)$, where c is a fixed positive real number, the algorithm is called a ***c*-approximation algorithm** for the problem.

*46. Prove that the algorithm from Exercise 44 is a 2-approximation algorithm for the load balancing problem. [*Hint:* Use both parts of Exercise 43.]

Computer Projects

Write programs with these inputs and outputs.

1. Given a list of n integers, find the largest integer in the list.

2. Given a list of n integers, find the first and last occurrences of the largest integer in the list.

3. Given a list of n distinct integers, determine the position of an integer in the list using a linear search.

4. Given an ordered list of n distinct integers, determine the position of an integer in the list using a binary search.

5. Given a list of n integers, sort them using a bubble sort.

6. Given a list of n integers, sort them using an insertion sort.

7. Given two strings of characters use the naive string matching algorithm to determine whether the shorter string occurs in the longer string.

8. Given an integer n, use the cashier's algorithm to find the change for n cents using quarters, dimes, nickels, and pennies.

9. Given the starting and ending times of n talks, use the appropriate greedy algorithm to schedule the most talks possible in a single lecture hall.

10. Given an ordered list of n integers and an integer x in the list, find the number of comparisons used to determine the position of x in the list using a linear search and using a binary search.

11. Given a list of integers, determine the number of comparisons used by the bubble sort and by the insertion sort to sort this list.

Computations and Explorations

Use a computational program or programs you have written to do these exercises.

1. We know that n^b is $O(d^n)$ when b and d are positive numbers with $d \geq 2$. Give values of the constants C and k such that $n^b \leq Cd^n$ whenever $x > k$ for each of these sets of values: $b = 10$, $d = 2$; $b = 20$, $d = 3$; $b = 1000$, $d = 7$.

2. Compute the change for different values of n with coins of different denominations using the cashier's algorithm and determine whether the smallest number of coins was used. Can you find conditions so that the cashier's algorithm is guaranteed to use the fewest coins possible?

3. Using a generator of random orderings of the integers $1, 2, \ldots, n$, find the number of comparisons used by the bubble sort, insertion sort, binary insertion sort, and selection sort to sort these integers.

4. Collect experimental evidence comparing the number of comparisons used by the sorting algorithms in Question 3 when used to sort sequences that only have a small fraction of terms out of order.

*5. Write a program that animates the progress of all the sorting algorithms in Question 3 when given the numbers from 1 to 100 in random order.

Writing Projects

Respond to these with essays using outside sources.

1. Examine the history of the word *algorithm* and describe the use of this word in early writings.
2. Look up Bachmann's original introduction of big-O notation. Explain how he and others have used this notation.
3. Explain how sorting algorithms can be classified into a taxonomy based on the underlying principle on which they are based.
4. Describe the radix sort algorithm.
5. Describe some of the different algorithms for string matching.
6. Describe some of the different applications of string matching in bioinformatics.
7. Describe the historic trends in how quickly processors can perform operations and use these trends to estimate how quickly processors will be able to perform operations in the next 20 years.
8. Develop a detailed list of algorithmic paradigms and provide examples using each of these paradigms.
9. Explain what the Turing Award is and describe the criteria used to select winners. List six past winners of the award and why they received the award.
10. Describe what is meant by a parallel algorithm. Explain how the pseudocode used in this book can be extended to handle parallel algorithms.
11. Explain how the complexity of parallel algorithms can be measured. Give some examples to illustrate this concept, showing how a parallel algorithm can work more quickly than one that does not operate in parallel.
12. Describe six different NP-complete problems.
13. Demonstrate how one of the many different NP-complete problems can be reduced to the satisfiability problem.
14. Discuss Google's AlphaDev AI project that finds new sorting algorithms.

CHAPTER

4 Number Theory and Cryptography

4.1 Divisibility and Modular Arithmetic

4.2 Integer Representations and Algorithms

4.3 Primes and Greatest Common Divisors

4.4 Solving Congruences

4.5 Applications of Congruences

4.6 Cryptography

The part of mathematics devoted to the study of the set of integers and their properties is known as number theory. In this chapter we will develop some of the important concepts of number theory including many of those used in computer science. As we develop number theory, we will use the proof methods developed in Chapter 1 to prove many theorems.

We will first introduce the notion of divisibility of integers, which we use to introduce modular, or clock, arithmetic. Modular arithmetic operates with the remainders of integers when they are divided by a fixed positive integer, called the modulus. We will prove many important results about modular arithmetic which we will use extensively in this chapter.

Integers can be represented with any positive integer b greater than 1 as a base. In this chapter we discuss base b representations of integers and give an algorithm for finding them. In particular, we will discuss binary, octal, and hexadecimal (base 2, 8, and 16) representations. We will describe algorithms for carrying out arithmetic using these representations and study their complexity. These algorithms were the first procedures called algorithms.

We will discuss prime numbers, the integers greater than 1 that have only 1 and themselves as positive divisors. We will prove that there are infinitely many primes; the proof we give is considered to be one of the most beautiful proofs in mathematics. We will discuss the distribution of primes and many famous open questions concerning primes. We will introduce the concept of greatest common divisors and study the Euclidean algorithm for computing them. This algorithm was first described thousands of years ago. We will introduce the fundamental theorem of arithmetic, a key result which tells us that every positive integer has a unique factorization into primes.

We will explain how to solve linear congruences, as well as systems of linear congruences, which we solve using the famous Chinese remainder theorem. We will introduce the notion of pseudoprimes, which are composite integers masquerading as primes, and show how this notion can help us rapidly generate prime numbers.

This chapter introduces several important applications of number theory. In particular, we will use number theory to generate pseudorandom numbers, to assign memory locations to computer files, and to find check digits used to detect errors in various kinds of identification numbers. We also introduce the subject of cryptography. Number theory plays an essential role both in classical cryptography, first used thousands of years ago, and modern cryptography, which plays an essential role in electronic communication. We will show how the ideas we develop can be used in cryptographical protocols, introducing protocols for sharing keys and for sending signed messages. Number theory, once considered the purest of subjects, has become an essential tool in providing computer and Internet security.

Finally, it should be noted that this chapter is designed to introduce some key aspects of number theory. As with all the topics covered in this book, there is a great deal more to learn. Interested students can consult [Ro23], the author's number theory text, to explore this fascinating subject more fully.

4.1 Divisibility and Modular Arithmetic

4.1.1 Introduction

The ideas that we will develop in this section are based on the notion of divisibility. Division of an integer by a positive integer produces a quotient and a remainder. Working with these remainders leads to modular arithmetic, which plays an important role in mathematics and which is used throughout computer science. We will discuss some important applications of modular arithmetic later in this chapter, including generating pseudorandom numbers, assigning computer memory locations to files, constructing check digits, and encrypting messages.

4.1.2 Division

When one integer is divided by a second nonzero integer, the quotient may or may not be an integer. For example, $12/3 = 4$ is an integer, whereas $11/4 = 2.75$ is not. This leads to Definition 1.

Definition 1 If a and b are integers with $a \neq 0$, we say that a *divides* b if there is an integer c such that $b = ac$ (or equivalently, if $\frac{b}{a}$ is an integer). When a divides b we say that a is a *factor* or *divisor* of b, and that b is a *multiple* of a. The notation $a \mid b$ denotes that a divides b. We write $a \nmid b$ when a does not divide b.

Remark: We can express $a \mid b$ using quantifiers as $\exists c(ac = b)$, where the universe of discourse is the set of integers.

In Figure 1 a number line indicates which integers are divisible by the positive integer d.

EXAMPLE 1 Determine whether $3 \mid 7$ and whether $3 \mid 12$.

Solution: We see that $3 \nmid 7$, because $7/3$ is not an integer. On the other hand, $3 \mid 12$ because $12/3 = 4$. ◀

EXAMPLE 2 Let n and d be positive integers. How many positive integers not exceeding n are divisible by d?

Extra Examples

Solution: The positive integers divisible by d are all the integers of the form dk, where k is a positive integer. Hence, the number of positive integers divisible by d that do not exceed n equals the number of integers k with $0 < dk \leq n$, or with $0 < k \leq n/d$. Therefore, there are $\lfloor n/d \rfloor$ positive integers not exceeding n that are divisible by d. ◀

Some of the basic properties of divisibility of integers are given in Theorem 1.

THEOREM 1 Let a, b, and c be integers, where $a \neq 0$. Then

(*i*) if $a \mid b$ and $a \mid c$, then $a \mid (b + c)$;
(*ii*) if $a \mid b$, then $a \mid bc$ for all integers c;
(*iii*) if $a \mid b$ and $b \mid c$, then $a \mid c$.

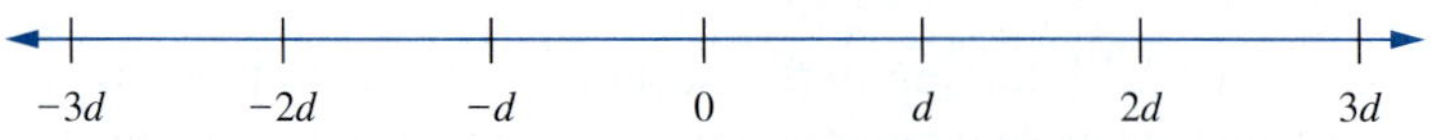

FIGURE 1 Integers divisible by the positive integer d.

Proof: We will give a direct proof of (*i*). Suppose that $a \mid b$ and $a \mid c$. Then, from the definition of divisibility, it follows that there are integers s and t with $b = as$ and $c = at$. Hence,

$$b + c = as + at = a(s + t).$$

Therefore, a divides $b + c$. This establishes part (*i*) of the theorem. The proofs of parts (*ii*) and (*iii*) are left as Exercises 3 and 4. ◁

Theorem 1 has this useful consequence.

COROLLARY 1 If a, b, and c are integers, where $a \neq 0$, such that $a \mid b$ and $a \mid c$, then $a \mid mb + nc$ whenever m and n are integers.

Proof: We will give a direct proof. By part (*ii*) of Theorem 1 we see that $a \mid mb$ and $a \mid nc$ whenever m and n are integers. By part (*i*) of Theorem 1 it follows that $a \mid mb + nc$. ◁

4.1.3 The Division Algorithm

When an integer is divided by a positive integer, there is a quotient and a remainder, as the division algorithm shows.

THEOREM 2 **THE DIVISION ALGORITHM** Let a be an integer and d a positive integer. Then there are unique integers q and r, with $0 \leq r < d$, such that $a = dq + r$.

We defer the proof of the division algorithm to Section 5.2. (See Example 5 and Exercise 41 in that section.)

Remark: Theorem 2 is not really an algorithm. (Why not?) Nevertheless, we use its traditional name.

Definition 2 In the equality given in the division algorithm, d is called the *divisor*, a is called the *dividend*, q is called the *quotient*, and r is called the *remainder*. This notation is used to express the quotient and remainder:

$$q = a \textbf{ div } d, \qquad r = a \textbf{ mod } d.$$

Remark: Note that both a **div** d and a **mod** d for a fixed d are functions on the set of integers. Furthermore, when a is an integer and d is a positive integer, we have $a \textbf{ div } d = \lfloor a/d \rfloor$ and $a \textbf{ mod } d = a - d\lfloor a/d \rfloor$. (See Exercise 24.)

Examples 3 and 4 illustrate the division algorithm.

EXAMPLE 3 What are the quotient and remainder when 101 is divided by 11?

Solution: We have

$$101 = 11 \cdot 9 + 2.$$

Hence, the quotient when 101 is divided by 11 is $9 = 101 \textbf{ div } 11$, and the remainder is $2 = 101 \textbf{ mod } 11$. ◀

EXAMPLE 4 What are the quotient and remainder when -11 is divided by 3?

Extra Examples

Solution: We have

$$-11 = 3(-4) + 1.$$

Hence, the quotient when -11 is divided by 3 is $-4 = -11$ **div** 3, and the remainder is $1 = -11$ **mod** 3.

Note that the remainder cannot be negative. Consequently, the remainder is *not* -2, even though

$$-11 = 3(-3) - 2,$$

because $r = -2$ does not satisfy $0 \leq r < 3$. ◀

Note that the integer a is divisible by the integer d if and only if the remainder is zero when a is divided by d.

Remark: A programming language may have one, or possibly two, operators for modular arithmetic, denoted by mod (in BASIC, Maple, Mathematica, EXCEL, and SQL), % (in C, C++, Java, and Python), rem (in Ada and Lisp), or something else. Be careful when using them, because for $a < 0$, some of these operators return $a - m\lceil a/m \rceil$ instead of a **mod** $m = a - m\lfloor a/m \rfloor$ (as shown in Exercise 24). Also, unlike a **mod** m, some of these operators are defined when $m < 0$, and even when $m = 0$.

4.1.4 Modular Arithmetic

In some situations we care only about the remainder of an integer when it is divided by some specified positive integer. For instance, when we ask what time it will be (on a 24-hour clock) 50 hours from now, we care only about the remainder when 50 plus the current hour is divided by 24. Because we are often interested only in remainders, we have special notations for them. We have already introduced the notation a **mod** m to represent the remainder when an integer a is divided by the positive integer m. We now introduce a different, but related, notation that indicates that two integers have the same remainder when they are divided by the positive integer m.

Definition 3 If a and b are integers and m is a positive integer, then a is *congruent to b modulo m* if m divides $a - b$. We use the notation $a \equiv b \pmod{m}$ to indicate that a is congruent to b modulo m. We say that $a \equiv b \pmod{m}$ is a **congruence** and that m is its **modulus** (plural **moduli**). If a and b are not congruent modulo m, we write $a \not\equiv b \pmod{m}$.

Although both notations $a \equiv b \pmod{m}$ and a **mod** $m = b$ include "mod," they represent fundamentally different concepts. The first represents a relation on the set of integers, whereas the second represents a function. However, the relation $a \equiv b \pmod{m}$ and the **mod** m function are closely related, as described in Theorem 3.

THEOREM 3 Let a and b be integers, and let m be a positive integer. Then $a \equiv b \pmod{m}$ if and only if a **mod** $m = b$ **mod** m.

The proof of Theorem 3 is left as Exercises 21 and 22. Recall that a **mod** m and b **mod** m are the remainders when a and b are divided by m, respectively. Consequently, Theorem 3 also says that $a \equiv b \pmod{m}$ if and only if a and b have the same remainder when divided by m.

EXAMPLE 5 Determine whether 17 is congruent to 5 modulo 6 and whether 24 and 14 are congruent modulo 6.

Solution: Because 6 divides $17 - 5 = 12$, we see that $17 \equiv 5 \pmod{6}$. However, because $24 - 14 = 10$ is not divisible by 6, we see that $24 \not\equiv 14 \pmod{6}$. ◀

The great German mathematician Karl Friedrich Gauss developed the concept of congruences at the end of the eighteenth century. The notion of congruences has played an important role in the development of number theory.

Modular arithmetic has many important and useful applications. In Section 4.5 we will show how modular arithmetic can be used to produce hashing functions and pseudorandom numbers. We will also show how it can be used for check digits, such as for universal product codes and international standard book numbers. In Section 4.6 we will use modular arithmetic to construct secret codes.

Theorem 4 provides a useful way to work with congruences.

THEOREM 4 Let m be a positive integer. The integers a and b are congruent modulo m if and only if there is an integer k such that $a = b + km$.

Proof: If $a \equiv b \pmod{m}$, by the definition of congruence (Definition 3), we know that $m \mid (a - b)$. This means that there is an integer k such that $a - b = km$, so that $a = b + km$. Conversely, if there is an integer k such that $a = b + km$, then $km = a - b$. Hence, m divides $a - b$, so that $a \equiv b \pmod{m}$. ◁

The set of all integers congruent to an integer a modulo m is called the **congruence class** of a modulo m. In Chapter 9 we will show that there are m pairwise disjoint congruence classes modulo m and that the union of these congruence classes is the set of integers.

Theorem 5 shows that additions and multiplications preserve congruences.

THEOREM 5 Let m be a positive integer. If $a \equiv b \pmod{m}$ and $c \equiv d \pmod{m}$, then

$$a + c \equiv b + d \pmod{m} \quad \text{and} \quad ac \equiv bd \pmod{m}.$$

Links ›

Hulton Archive/Getty Images

KARL FRIEDRICH GAUSS (1777–1855) Karl Friedrich Gauss, the son of a bricklayer, was a child prodigy. He demonstrated his potential at the age of 10, when he quickly solved a problem assigned by a teacher to keep the class busy. The teacher asked the students to find the sum of the first 100 positive integers. Gauss realized that this sum could be found by forming 50 pairs, each with the sum 101: $1 + 100, 2 + 99, \ldots, 50 + 51$. This brilliance attracted the sponsorship of patrons, including Duke Ferdinand of Brunswick, who made it possible for Gauss to attend Caroline College and the University of Göttingen. While a student, he invented the method of least squares, which is used to estimate the most likely value of a variable from experimental results. In 1796 Gauss made a fundamental discovery in geometry, advancing a subject that had not advanced since ancient times. He showed that a 17-sided regular polygon could be drawn using just a ruler and compass.

In 1799 Gauss presented the first rigorous proof of the fundamental theorem of algebra, which states that a polynomial of degree n has exactly n roots in the complex numbers (counting multiplicities). Gauss achieved worldwide fame when he successfully calculated the orbit of the first asteroid discovered, Ceres, using scanty data.

Gauss was called the Prince of Mathematics by his contemporary mathematicians. Although Gauss is noted for his many discoveries in geometry, algebra, analysis, astronomy, and physics, he had a special interest in number theory, which can be seen from his statement "Mathematics is the queen of the sciences, and the theory of numbers is the queen of mathematics." Gauss laid the foundations for modern number theory with the publication of his book *Disquisitiones Arithmeticae* in 1801.

Proof: We use a direct proof. Because $a \equiv b \pmod{m}$ and $c \equiv d \pmod{m}$, by Theorem 4 there are integers s and t with $b = a + sm$ and $d = c + tm$. Hence,

$$b + d = (a + sm) + (c + tm) = (a + c) + m(s + t)$$

and

$$bd = (a + sm)(c + tm) = ac + m(at + cs + stm).$$

Hence,

$$a + c \equiv b + d \pmod{m} \qquad \text{and} \qquad ac \equiv bd \pmod{m}.$$ ◁

EXAMPLE 6 Because $7 \equiv 2 \pmod{5}$ and $11 \equiv 1 \pmod{5}$, it follows from Theorem 5 that

$$18 = 7 + 11 \equiv 2 + 1 = 3 \pmod{5}$$

and that

$$77 = 7 \cdot 11 \equiv 2 \cdot 1 = 2 \pmod{5}.$$ ◀

You cannot always divide both sides of a congruence by the same number!

We must be careful working with congruences. Some properties we may expect to be true are not valid. For example, if $ac \equiv bc \pmod{m}$, the congruence $a \equiv b \pmod{m}$ may be false. Similarly, if $a \equiv b \pmod{m}$ and $c \equiv d \pmod{m}$, the congruence $a^c \equiv b^d \pmod{m}$ may be false. (See Exercise 43.)

Corollary 2 shows how to find the values of the **mod** m function at the sum and product of two integers using the values of this function at each of these integers. We will use this result in Section 5.4.

COROLLARY 2 Let m be a positive integer and let a and b be integers. Then

$$(a + b) \textbf{ mod } m = ((a \textbf{ mod } m) + (b \textbf{ mod } m)) \textbf{ mod } m$$

and

$$ab \textbf{ mod } m = ((a \textbf{ mod } m)(b \textbf{ mod } m)) \textbf{ mod } m.$$

Proof: By the definitions of **mod** m and of congruence modulo m, we know that $a \equiv (a \textbf{ mod } m) \pmod{m}$ and $b \equiv (b \textbf{ mod } m) \pmod{m}$. Hence, Theorem 5 tells us that

$$a + b \equiv (a \textbf{ mod } m) + (b \textbf{ mod } m) \pmod{m}$$

and

$$ab \equiv (a \textbf{ mod } m)(b \textbf{ mod } m) \pmod{m}.$$

The equalities in this corollary follow from these last two congruences by Theorem 3. ◁

In Section 4.6 we will carry out a variety of computations using the **mod** function when we study cryptography. Example 7 illustrates a type of computation involving the **mod** function that we will encounter.

EXAMPLE 7 Find the value of $(19^3 \mathbf{\ mod\ } 31)^4 \mathbf{\ mod\ } 23$.

Solution: To compute $(19^3 \mathbf{\ mod\ } 31)^4 \mathbf{\ mod\ } 23$, we will first evaluate $19^3 \mathbf{\ mod\ } 31$. Because $19^3 = 6859$ and $6859 = 221 \cdot 31 + 8$, we have $19^3 \mathbf{\ mod\ } 31 = 6859 \mathbf{\ mod\ } 31 = 8$. So, $(19^3 \mathbf{\ mod\ } 31)^4 \mathbf{\ mod\ } 23 = 8^4 \mathbf{\ mod\ } 23$.

Next, note that $8^4 = 4096$. Because $4096 = 178 \cdot 23 + 2$, we have $4096 \mathbf{\ mod\ } 23 = 2$. Hence, $(19^3 \mathbf{\ mod\ } 31)^4 \mathbf{\ mod\ } 23 = 2$. ◂

4.1.5 Arithmetic Modulo *m*

We can define arithmetic operations on $\mathbf{Z}_m$, the set of nonnegative integers less than m, that is, the set $\{0, 1, \ldots, m-1\}$. In particular, we define addition of these integers, denoted by $+_m$ by

$$a +_m b = (a + b) \mathbf{\ mod\ } m,$$

where the addition on the right-hand side of this equation is the ordinary addition of integers, and we define multiplication of these integers, denoted by $\cdot_m$ by

$$a \cdot_m b = (a \cdot b) \mathbf{\ mod\ } m,$$

where the multiplication on the right-hand side of this equation is the ordinary multiplication of integers. The operations $+_m$ and $\cdot_m$ are called addition and multiplication modulo m and when we use these operations, we are said to be doing **arithmetic modulo** m.

EXAMPLE 8 Use the definition of addition and multiplication in $\mathbf{Z}_m$ to find $7 +_{11} 9$ and $7 \cdot_{11} 9$.

Solution: Using the definition of addition modulo 11, we find that

$$7 +_{11} 9 = (7 + 9) \mathbf{\ mod\ } 11 = 16 \mathbf{\ mod\ } 11 = 5,$$

and

$$7 \cdot_{11} 9 = (7 \cdot 9) \mathbf{\ mod\ } 11 = 63 \mathbf{\ mod\ } 11 = 8.$$

Hence, $7 +_{11} 9 = 5$ and $7 \cdot_{11} 9 = 8$. ◂

The operations $+_m$ and $\cdot_m$ satisfy many of the same properties of ordinary addition and multiplication of integers. In particular, they satisfy these properties:

Closure If a and b belong to $\mathbf{Z}_m$, then $a +_m b$ and $a \cdot_m b$ belong to $\mathbf{Z}_m$.

Associativity If a, b, and c belong to $\mathbf{Z}_m$, then $(a +_m b) +_m c = a +_m (b +_m c)$ and $(a \cdot_m b) \cdot_m c = a \cdot_m (b \cdot_m c)$.

Commutativity If a and b belong to $\mathbf{Z}_m$, then $a +_m b = b +_m a$ and $a \cdot_m b = b \cdot_m a$.

Identity elements The elements 0 and 1 are identity elements for addition and multiplication modulo m, respectively. That is, if a belongs to $\mathbf{Z}_m$, then $a +_m 0 = 0 +_m a = a$ and $a \cdot_m 1 = 1 \cdot_m a = a$.

Additive inverses If $a \neq 0$ belongs to $\mathbf{Z}_m$, then $m - a$ is an additive inverse of a modulo m and 0 is its own additive inverse. That is, $a +_m (m - a) = 0$ and $0 +_m 0 = 0$.

Distributivity If a, b, and c belong to $\mathbf{Z}_m$, then $a \cdot_m (b +_m c) = (a \cdot_m b) +_m (a \cdot_m c)$ and $(a +_m b) \cdot_m c = (a \cdot_m c) +_m (b \cdot_m c)$.

These properties follow from the properties we have developed for congruences and remainders modulo m, together with the properties of integers; we leave their proofs as Exercises 48–50. Note that we have listed the property that every element of $\mathbf{Z}_m$ has an additive inverse, but no analogous property for multiplicative inverses has been included. This is because multiplicative inverses do not always exist modulo m. For instance, there is no multiplicative inverse of 2 modulo 6, as the reader can verify. We will return to the question of when an integer has a multiplicative inverse modulo m later in this chapter.

Remark: Because $\mathbf{Z}_m$ with the operations of addition and multiplication modulo m satisfies the properties listed, $\mathbf{Z}_m$ with modular addition is said to be a **commutative group** and $\mathbf{Z}_m$ with both of these operations is said to be a **commutative ring**. Note that the set of integers with ordinary addition and multiplication also forms a commutative ring. Groups and rings are studied in courses that cover abstract algebra.

Remark: In Exercise 36, and in later sections, we will use the notations $+$ and $\cdot$ for $+_m$ and $\cdot_m$ without the subscript m on the symbol for the operator whenever we work with $\mathbf{Z}_m$.

Exercises

1. Does 17 divide each of these numbers?
a) 68 **b)** 84 **c)** 357 **d)** 1001

2. Prove that if a is an integer other than 0, then
a) 1 divides a. **b)** a divides 0.

3. Prove that part (*ii*) of Theorem 1 is true.

4. Prove that part (*iii*) of Theorem 1 is true.

5. Show that if $a \mid b$ and $b \mid a$, where a and b are integers, then $a = b$ or $a = -b$.

6. Show that if a, b, c, and d are integers, where $a \neq 0$, such that $a \mid c$ and $b \mid d$, then $ab \mid cd$.

7. Show that if a, b, and c are integers, where $a \neq 0$ and $c \neq 0$, such that $ac \mid bc$, then $a \mid b$.

8. Prove or disprove that if $a \mid bc$, where a, b, and c are positive integers and $a \neq 0$, then $a \mid b$ or $a \mid c$.

9. Prove that if a and b are integers and a divides b, then a is odd or b is even.

10. Prove that if a and b are nonzero integers, a divides b, and $a + b$ is odd, then a is odd.

11. Prove that if a is an integer that is not divisible by 3, then $(a + 1)(a + 2)$ is divisible by 3.

12. Prove that if a is a positive integer, then 4 does not divide $a^2 + 2$.

13. What are the quotient and remainder when
a) 19 is divided by 7? **b)** −111 is divided by 11?
c) 789 is divided by 23? **d)** 1001 is divided by 13?
e) 0 is divided by 19? **f)** 3 is divided by 5?
g) −1 is divided by 3? **h)** 4 is divided by 1?

14. What are the quotient and remainder when
a) 44 is divided by 8?
b) 777 is divided by 21?
c) −123 is divided by 19?
d) −1 is divided by 23?
e) −2002 is divided by 87?
f) 0 is divided by 17?
g) 1,234,567 is divided by 1001?
h) −100 is divided by 101?

15. What time does a 12-hour clock read
a) 80 hours after it reads 11:00?
b) 40 hours before it reads 12:00?
c) 100 hours after it reads 6:00?

16. What time does a 24-hour clock read
a) 100 hours after it reads 2:00?
b) 45 hours before it reads 12:00?
c) 168 hours after it reads 19:00?

17. Suppose that a and b are integers, $a \equiv 4 \pmod{13}$, and $b \equiv 9 \pmod{13}$. Find the integer c with $0 \leq c \leq 12$ such that
a) $c \equiv 9a \pmod{13}$.
b) $c \equiv 11b \pmod{13}$.
c) $c \equiv a + b \pmod{13}$.
d) $c \equiv 2a + 3b \pmod{13}$.
e) $c \equiv a^2 + b^2 \pmod{13}$.
f) $c \equiv a^3 - b^3 \pmod{13}$.

18. Suppose that a and b are integers, $a \equiv 11 \pmod{19}$, and $b \equiv 3 \pmod{19}$. Find the integer c with $0 \leq c \leq 18$ such that
a) $c \equiv 13a \pmod{19}$.
b) $c \equiv 8b \pmod{19}$.
c) $c \equiv a - b \pmod{19}$.
d) $c \equiv 7a + 3b \pmod{19}$.
e) $c \equiv 2a^2 + 3b^2 \pmod{19}$.
f) $c \equiv a^3 + 4b^3 \pmod{19}$.

19. Show that if a and d are positive integers, then $(-a)$ **div** $d = -a$ **div** d if and only if d divides a.

20. Prove or disprove that if a, b, and d are integers with $d > 0$, then $(a + b)$ **div** $d = a$ **div** $d + b$ **div** d.

21. Let m be a positive integer. Show that $a \equiv b \pmod{m}$ if a **mod** $m = b$ **mod** m.

22. Let m be a positive integer. Show that a **mod** $m = b$ **mod** m if $a \equiv b \pmod{m}$.

23. Show that if n and k are positive integers, then $\lceil n/k \rceil = \lfloor (n-1)/k \rfloor + 1$.

24. Show that if a is an integer and d is an integer greater than 1, then the quotient and remainder obtained when a is divided by d are $\lfloor a/d \rfloor$ and $a - d\lfloor a/d \rfloor$, respectively.

25. Find a formula for the integer with smallest absolute value that is congruent to an integer a modulo m, where m is a positive integer.

26. Evaluate these quantities.

a) -17 **mod** 2 **b)** 144 **mod** 7
c) -101 **mod** 13 **d)** 199 **mod** 19

27. Evaluate these quantities.

a) 13 **mod** 3 **b)** -97 **mod** 11
c) 155 **mod** 19 **d)** -221 **mod** 23

28. Find a **div** m and a **mod** m when

a) $a = -111$, $m = 99$.
b) $a = -9999$, $m = 101$.
c) $a = 10299$, $m = 999$.
d) $a = 123456$, $m = 1001$.

29. Find a **div** m and a **mod** m when

a) $a = 228$, $m = 119$.
b) $a = 9009$, $m = 223$.
c) $a = -10101$, $m = 333$.
d) $a = -765432$, $m = 38271$.

30. Find the integer a such that

a) $a \equiv 43 \pmod{23}$ and $-22 \le a \le 0$.
b) $a \equiv 17 \pmod{29}$ and $-14 \le a \le 14$.
c) $a \equiv -11 \pmod{21}$ and $90 \le a \le 110$.

31. Find the integer a such that

a) $a \equiv -15 \pmod{27}$ and $-26 \le a \le 0$.
b) $a \equiv 24 \pmod{31}$ and $-15 \le a \le 15$.
c) $a \equiv 99 \pmod{41}$ and $100 \le a \le 140$.

32. List five integers that are congruent to 4 modulo 12.

33. List all integers between -100 and 100 that are congruent to -1 modulo 25.

34. Decide whether each of these integers is congruent to 3 modulo 7.

a) 37 **b)** 66
c) -17 **d)** -67

35. Decide whether each of these integers is congruent to 5 modulo 17.

a) 80 **b)** 103
c) -29 **d)** -122

36. Find each of these values.

a) (177 **mod** 31 + 270 **mod** 31) **mod** 31
b) (177 **mod** 31 · 270 **mod** 31) **mod** 31

37. Find each of these values.

a) (-133 **mod** 23 + 261 **mod** 23) **mod** 23
b) (457 **mod** 23 · 182 **mod** 23) **mod** 23

38. Find each of these values.

a) (19^2 **mod** 41) **mod** 9
b) (32^3 **mod** 13)^2 **mod** 11
c) (7^3 **mod** 23)^2 **mod** 31
d) (21^2 **mod** 15)^3 **mod** 22

39. Find each of these values.

a) (99^2 **mod** 32)^3 **mod** 15
b) (3^4 **mod** 17)^2 **mod** 11
c) (19^3 **mod** 23)^2 **mod** 31
d) (89^3 **mod** 79)^4 **mod** 26

40. Show that if $a \equiv b \pmod{m}$ and $c \equiv d \pmod{m}$, where a, b, c, d, and m are integers with $m \ge 2$, then $a - c \equiv b - d \pmod{m}$.

41. Show that if $n \mid m$, where n and m are integers greater than 1, and if $a \equiv b \pmod{m}$, where a and b are integers, then $a \equiv b \pmod{n}$.

42. Show that if a, b, c, and m are integers such that $m \ge 2$, $c > 0$, and $a \equiv b \pmod{m}$, then $ac \equiv bc \pmod{mc}$.

43. Find counterexamples to each of these statements about congruences.

a) If $ac \equiv bc \pmod{m}$, where a, b, c, and m are integers with $m \ge 2$, then $a \equiv b \pmod{m}$.
b) If $a \equiv b \pmod{m}$ and $c \equiv d \pmod{m}$, where a, b, c, d, and m are integers with c and d positive and $m \ge 2$, then $a^c \equiv b^d \pmod{m}$.

44. Show that if n is an integer then $n^2 \equiv 0$ or $1 \pmod{4}$.

45. Use Exercise 44 to show that if m is a positive integer of the form $4k + 3$ for some nonnegative integer k, then m is not the sum of the squares of two integers.

46. Prove that if n is an odd positive integer, then $n^2 \equiv 1 \pmod{8}$.

47. Show that if a, b, k, and m are integers such that $k \ge 1$, $m \ge 2$, and $a \equiv b \pmod{m}$, then $a^k \equiv b^k \pmod{m}$.

48. Show that $\mathbf{Z}_m$ with addition modulo m, where $m \ge 2$ is an integer, satisfies the closure, associative, and commutative properties, 0 is an additive identity, and for every nonzero $a \in \mathbf{Z}_m$, $m - a$ is an inverse of a modulo m.

49. Show that $\mathbf{Z}_m$ with multiplication modulo m, where $m \ge 2$ is an integer, satisfies the closure, associative, and commutativity properties, and 1 is a multiplicative identity.

50. Show that the distributive property of multiplication over addition holds for $\mathbf{Z}_m$, where $m \ge 2$ is an integer.

51. Write out the addition and multiplication tables for $\mathbf{Z}_5$ (where by addition and multiplication we mean $+_5$ and $\cdot_5$).

52. Write out the addition and multiplication tables for $\mathbf{Z}_6$ (where by addition and multiplication we mean $+_6$ and $\cdot_6$).

53. Determine whether each of the functions $f(a) = a$ **div** d and $g(a) = a$ **mod** d, where d is a fixed positive integer, from the set of integers to the set of integers, is one-to-one, and determine whether each of these functions is onto.

4.2 Integer Representations and Algorithms

4.2.1 Introduction

Integers can be expressed using any integer greater than one as a base, as we will show in this section. Although we commonly use decimal (base 10) representations, binary (base 2), octal (base 8), and hexadecimal (base 16) representations are often used, especially in computer science. Given a base b and an integer n, we will show how to construct the base b representation of this integer. We will also explain how to quickly convert between binary and octal and between binary and hexadecimal notations.

As mentioned in Section 3.1, the term *algorithm* originally referred to procedures for performing arithmetic operations using the decimal representations of integers. These algorithms, adapted for use with binary representations, are the basis for computer arithmetic. They provide good illustrations of the concept of an algorithm and the complexity of algorithms. For these reasons, they will be discussed in this section.

We will also introduce an algorithm for finding a **div** d and a **mod** d where a and d are integers with $d > 1$. Finally, we will describe an efficient algorithm for modular exponentiation, which is a particularly important algorithm for cryptography, as we will see in Section 4.6.

4.2.2 Representations of Integers

In everyday life we use decimal notation to express integers. In decimal notation, an integer n is written as a sum of the form $a_k 10^k + a_{k-1} 10^{k-1} + \cdots + a_1 10 + a_0$, where a_j is an integer with $0 \leq a_j \leq 9$ for $j = 0, 1, \ldots, k$. For example, 965 is used to denote $9 \cdot 10^2 + 6 \cdot 10 + 5$. However, it is often convenient to use bases other than 10. In particular, computers usually use binary notation (with 2 as the base) when carrying out arithmetic, and octal (base 8) or hexadecimal (base 16) notation when expressing characters, such as letters or digits. In fact, we can use any integer greater than 1 as the base when expressing integers. This is stated in Theorem 1.

THEOREM 1 Let b be an integer greater than 1. Then if n is a positive integer, it can be expressed uniquely in the form

$$n = a_k b^k + a_{k-1} b^{k-1} + \cdots + a_1 b + a_0,$$

where k is a nonnegative integer, $a_0, a_1, \ldots, a_k$ are nonnegative integers less than b, and $a_k \neq 0$.

A proof of this theorem can be constructed using mathematical induction, a proof method that is discussed in Section 5.1. It can also be found in [Ro23]. The representation of n given in Theorem 1 is called the **base *b* expansion of *n***. The base b expansion of n is denoted by $(a_k a_{k-1} \ldots a_1 a_0)_b$. For instance, $(245)_8$ represents $2 \cdot 8^2 + 4 \cdot 8 + 5 = 165$. Typically, the subscript 10 is omitted for base 10 expansions of integers because base 10, or **decimal expansions**, are commonly used to represent integers.

BINARY EXPANSIONS Choosing 2 as the base gives **binary expansions** of integers. In binary notation each digit is either a 0 or a 1. In other words, the binary expansion of an integer is just a bit string. Binary expansions (and related expansions that are variants of binary expansions) are used by computers to represent and do arithmetic with integers.

EXAMPLE 1 What is the decimal expansion of the integer that has $(1\ 0101\ 1111)_2$ as its binary expansion?

Solution: We have

$$(1\ 0101\ 1111)_2 = 1 \cdot 2^8 + 0 \cdot 2^7 + 1 \cdot 2^6 + 0 \cdot 2^5 + 1 \cdot 2^4 + 1 \cdot 2^3 + 1 \cdot 2^2 + 1 \cdot 2^1 + 1 \cdot 2^0 = 351.$$ ◀

OCTAL AND HEXADECIMAL EXPANSIONS Among the most important bases in computer science are base 2, base 8, and base 16. Base 8 expansions are called **octal** expansions and base 16 expansions are **hexadecimal** expansions.

EXAMPLE 2 What is the decimal expansion of the number with octal expansion $(7016)_8$?

Solution: Using the definition of a base b expansion with $b = 8$ tells us that

$$(7016)_8 = 7 \cdot 8^3 + 0 \cdot 8^2 + 1 \cdot 8 + 6 = 3598.$$ ◀

Sixteen different digits are required for hexadecimal expansions. Usually, the hexadecimal digits used are 0, 1, 2, 3, 4, 5, 6, 7, 8, 9, A, B, C, D, E, and F, where the letters A through F represent the digits corresponding to the numbers 10 through 15 (in decimal notation).

EXAMPLE 3 What is the decimal expansion of the number with hexadecimal expansion $(2AE0B)_{16}$?

Solution: Using the definition of a base b expansion with $b = 16$ tells us that

$$(2AE0B)_{16} = 2 \cdot 16^4 + 10 \cdot 16^3 + 14 \cdot 16^2 + 0 \cdot 16 + 11 = 175627.$$ ◀

Each hexadecimal digit can be represented using four bits. For instance, we see that $(1110\ \ 0101)_2 = (E5)_{16}$ because $(1110)_2 = (E)_{16}$ and $(0101)_2 = (5)_{16}$. **Bytes**, which are bit strings of length eight, can be represented by two hexadecimal digits.

BASE CONVERSION We will now describe an algorithm for constructing the base b expansion of an integer n. First, divide n by b to obtain a quotient and remainder, that is,

$$n = bq_0 + a_0, \qquad 0 \le a_0 < b.$$

The remainder, a_0, is the rightmost digit in the base b expansion of n. Next, divide q_0 by b to obtain

$$q_0 = bq_1 + a_1, \qquad 0 \le a_1 < b.$$

We see that a_1 is the second digit from the right in the base b expansion of n. Continue this process, successively dividing the quotients by b, obtaining additional base b digits as the remainders. This process terminates when we obtain a quotient equal to zero. It produces the base b digits of n from the right to the left.

EXAMPLE 4 Find the octal expansion of $(12345)_{10}$.

Extra Examples ❯

Solution: First, divide 12345 by 8 to obtain

$$12345 = 8 \cdot 1543 + 1.$$

Successively dividing quotients by 8 gives

$$\begin{aligned} 1543 &= 8 \cdot 192 + 7, \\ 192 &= 8 \cdot 24 + 0, \\ 24 &= 8 \cdot 3 + 0, \\ 3 &= 8 \cdot 0 + 3. \end{aligned}$$

The successive remainders that we have found, 1, 7, 0, 0, and 3, are the digits from the right to the left of 12345 in base 8. Hence,

$$(12345)_{10} = (30071)_8.$$ ◀

EXAMPLE 5 Find the hexadecimal expansion of $(177130)_{10}$.

Solution: First divide 177130 by 16 to obtain

$$177130 = 16 \cdot 11070 + 10.$$

Successively dividing quotients by 16 gives

$$\begin{aligned} 11070 &= 16 \cdot 691 + 14, \\ 691 &= 16 \cdot 43 + 3, \\ 43 &= 16 \cdot 2 + 11, \\ 2 &= 16 \cdot 0 + 2. \end{aligned}$$

The successive remainders that we have found, 10, 14, 3, 11, 2, give us the digits from the right to the left of 177130 in the hexadecimal (base 16) expansion of $(177130)_{10}$. It follows that

$$(177130)_{10} = (2B3EA)_{16}.$$

(Recall that the integers 10, 11, and 14 correspond to the hexadecimal digits A, B, and E, respectively.) ◀

EXAMPLE 6 Find the binary expansion of $(241)_{10}$.

Solution: First divide 241 by 2 to obtain

$$241 = 2 \cdot 120 + 1.$$

Successively dividing quotients by 2 gives

$$\begin{aligned} 120 &= 2 \cdot 60 + 0, \\ 60 &= 2 \cdot 30 + 0, \\ 30 &= 2 \cdot 15 + 0, \\ 15 &= 2 \cdot 7 + 1, \\ 7 &= 2 \cdot 3 + 1, \\ 3 &= 2 \cdot 1 + 1, \\ 1 &= 2 \cdot 0 + 1. \end{aligned}$$

The successive remainders that we have found, 1, 0, 0, 0, 1, 1, 1, 1, are the digits from the right to the left in the binary (base 2) expansion of $(241)_{10}$. Hence,

$$(241)_{10} = (1111\ 0001)_2.$$ ◀

The pseudocode given in Algorithm 1 finds the base b expansion $(a_{k-1} \dots a_1 a_0)_b$ of the integer n.

TABLE 1 Hexadecimal, Octal, and Binary Representation of the Integers 0 through 15.

Decimal	0	1	2	3	4	5	6	7	8	9	10	11	12	13	14	15
Hexadecimal	0	1	2	3	4	5	6	7	8	9	A	B	C	D	E	F
Octal	0	1	2	3	4	5	6	7	10	11	12	13	14	15	16	17
Binary	0	1	10	11	100	101	110	111	1000	1001	1010	1011	1100	1101	1110	1111

ALGORITHM 1 Constructing Base b Expansions.

procedure *base b expansion*(n, b: positive integers with $b > 1$)
$q := n$
$k := 0$
while $q \neq 0$
  $a_k := q$ **mod** b
  $q := q$ **div** b
  $k := k + 1$
return $(a_{k-1}, \ldots, a_1, a_0)$ $\{(a_{k-1} \ldots a_1 a_0)_b$ is the base b expansion of $n\}$

In Algorithm 1, q represents the quotient obtained by successive divisions by b, starting with $q = n$. The digits in the base b expansion are the remainders of these divisions and are given by q **mod** b. The algorithm terminates when a quotient $q = 0$ is reached.

Remark: Note that Algorithm 1 can be thought of as a greedy algorithm, because the base b digits are taken as large as possible in each step.

CONVERSION BETWEEN BINARY, OCTAL, AND HEXADECIMAL EXPANSIONS

Conversion between binary and octal and between binary and hexadecimal expansions is extremely easy because each octal digit corresponds to a block of three binary digits and each hexadecimal digit corresponds to a block of four binary digits, with these correspondences shown in Table 1 without initial 0s shown. (We leave it as Exercises 13–16 to show that this is the case.) This conversion is illustrated in Example 7.

EXAMPLE 7 Find the octal and hexadecimal expansions of $(11\ 1110\ 1011\ 1100)_2$ and the binary expansions of $(765)_8$ and $(\text{A8D})_{16}$.

Solution: To convert $(11\ 1110\ 1011\ 1100)_2$ into octal notation we group the binary digits into blocks of three, adding initial zeros at the start of the leftmost block if necessary. These blocks, from left to right, are 011, 111, 010, 111, and 100, corresponding to 3, 7, 2, 7, and 4, respectively. Consequently, $(11\ 1110\ 1011\ 1100)_2 = (37274)_8$. To convert $(11\ 1110\ 1011\ 1100)_2$ into hexadecimal notation we group the binary digits into blocks of four, adding initial zeros at the start of the leftmost block if necessary. These blocks, from left to right, are 0011, 1110, 1011, and 1100, corresponding to the hexadecimal digits 3, E, B, and C, respectively. Consequently, $(11\ 1110\ 1011\ 1100)_2 = (\text{3EBC})_{16}$.

To convert $(765)_8$ into binary notation, we replace each octal digit by a block of three binary digits. These blocks are 111, 110, and 101. Hence, $(765)_8 = (1\ 1111\ 0101)_2$. To convert $(\text{A8D})_{16}$ into binary notation, we replace each hexadecimal digit by a block of four binary digits. These blocks are 1010, 1000, and 1101. Hence, $(\text{A8D})_{16} = (1010\ 1000\ 1101)_2$. ◀

4.2.3 Algorithms for Integer Operations

The algorithms for performing operations with integers using their binary expansions are extremely important in computer arithmetic. We will describe algorithms for the addition and the multiplication of two integers expressed in binary notation. We will also analyze the computational complexity of these algorithms, in terms of the actual number of bit operations used. Throughout this discussion, suppose that the binary expansions of a and b are

$$a = (a_{n-1}a_{n-2} \dots a_1a_0)_2, b = (b_{n-1}b_{n-2} \dots b_1b_0)_2,$$

so that a and b each have n bits (putting bits equal to 0 at the beginning of one of these expansions if necessary).

We will measure the complexity of algorithms for integer arithmetic in terms of the number of bits in these numbers.

ADDITION ALGORITHM Consider the problem of adding two integers in binary notation. A procedure to perform addition can be based on the usual method for adding numbers with pencil and paper. This method proceeds by adding pairs of binary digits together with carries, when they occur, to compute the sum of two integers. This procedure will now be specified in detail.

To add a and b, first add their rightmost bits. This gives

$$a_0 + b_0 = c_0 \cdot 2 + s_0,$$

where s_0 is the rightmost bit in the binary expansion of $a + b$ and c_0 is the **carry**, which is either 0 or 1. Then add the next pair of bits and the carry,

$$a_1 + b_1 + c_0 = c_1 \cdot 2 + s_1,$$

where s_1 is the next bit (from the right) in the binary expansion of $a + b$, and c_1 is the carry. Continue this process, adding the corresponding bits in the two binary expansions and the carry, to determine the next bit from the right in the binary expansion of $a + b$. At the last stage, add a_{n-1}, b_{n-1}, and c_{n-2} to obtain $c_{n-1} \cdot 2 + s_{n-1}$. The leading bit of the sum is $s_n = c_{n-1}$. This procedure produces the binary expansion of the sum, namely, $a + b = (s_ns_{n-1}s_{n-2} \dots s_1s_0)_2$.

EXAMPLE 8 Add $a = (1110)_2$ and $b = (1011)_2$.

Solution: Following the procedure specified in the algorithm, first note that

$$a_0 + b_0 = 0 + 1 = 0 \cdot 2 + 1,$$

so that $c_0 = 0$ and $s_0 = 1$. Then, because

$$a_1 + b_1 + c_0 = 1 + 1 + 0 = 1 \cdot 2 + 0,$$

it follows that $c_1 = 1$ and $s_1 = 0$. Continuing,

$$a_2 + b_2 + c_1 = 1 + 0 + 1 = 1 \cdot 2 + 0,$$

so that $c_2 = 1$ and $s_2 = 0$. Finally, because

$$a_3 + b_3 + c_2 = 1 + 1 + 1 = 1 \cdot 2 + 1,$$

follows that $c_3 = 1$ and $s_3 = 1$. This means that $s_4 = c_3 = 1$. Therefore, $s = a + b = (1\ 1001)_2$. This addition is displayed in Figure 1, where carries are shown in color. ◀

$$\begin{array}{r} \mathit{1\ 1\ 1} \\ 1\ 1\ 1\ 0 \\ +\ 1\ 0\ 1\ 1 \\ \hline 1\ 1\ 0\ 0\ 1 \end{array}$$

FIGURE 1 Adding $(1110)_2$ and $(1011)_2$.

The algorithm for addition can be described using pseudocode as follows.

ALGORITHM 2 Addition of Integers.

procedure *add*(a, b: positive integers)
{the binary expansions of a and b are $(a_{n-1}a_{n-2} \ldots a_1a_0)_2$
  and $(b_{n-1}b_{n-2} \ldots b_1b_0)_2$, respectively}
$c := 0$
for $j := 0$ **to** $n - 1$
  $d := \lfloor (a_j + b_j + c)/2 \rfloor$
  $s_j := a_j + b_j + c - 2d$
  $c := d$
$s_n := c$
return $(s_0, s_1, \ldots, s_n)$ {the binary expansion of the sum is $(s_ns_{n-1} \ldots s_0)_2$}

Next, the number of additions of bits used by Algorithm 2 will be analyzed.

EXAMPLE 9 How many additions of bits are required to use Algorithm 2 to add two integers with n bits (or less) in their binary representations?

Solution: Two integers are added by successively adding pairs of bits and, when it occurs, a carry. Adding each pair of bits and the carry requires two additions of bits. Thus, the total number of additions of bits used is less than twice the number of bits in the expansion. Hence, the number of additions of bits used by Algorithm 2 to add two n-bit integers is $O(n)$. ◀

MULTIPLICATION ALGORITHM Next, consider the multiplication of two n-bit integers a and b. The conventional algorithm (used when multiplying with pencil and paper) works as follows. Using the distributive law, we see that

$$\begin{aligned} ab &= a(b_0 2^0 + b_1 2^1 + \cdots + b_{n-1} 2^{n-1}) \\ &= a(b_0 2^0) + a(b_1 2^1) + \cdots + a(b_{n-1} 2^{n-1}). \end{aligned}$$

We can compute ab using this equation. We first note that $ab_j = a$ if $b_j = 1$ and $ab_j = 0$ if $b_j = 0$. Each time we multiply a term by 2, we shift its binary expansion one place to the left and add a zero at the tail end of the expansion. Consequently, we can obtain $(ab_j)2^j$ by **shifting** the binary expansion of ab_j by j places to the left, adding j zero bits at the tail end of this binary expansion. Finally, we obtain ab by adding the n integers $ab_j 2^j$, $j = 0, 1, 2, \ldots, n - 1$.

Algorithm 3 displays this procedure for multiplication.

ALGORITHM 3 Multiplication of Integers.

procedure *multiply*(a, b: positive integers)
{the binary expansions of a and b are $(a_{n-1}a_{n-2} \dots a_1a_0)_2$
  and $(b_{n-1}b_{n-2} \dots b_1b_0)_2$, respectively}
for $j := 0$ **to** $n - 1$
  if $b_j = 1$ **then** $c_j := a$ shifted j places
  else $c_j := 0$
{$c_0, c_1, \dots, c_{n-1}$ are the partial products}
$p := 0$
for $j := 0$ **to** $n - 1$
  $p := add(p, c_j)$
return p {p is the value of ab}

Example 10 illustrates the use of this algorithm.

EXAMPLE 10 Find the product of $a = (110)_2$ and $b = (101)_2$.

Solution: First note that

$$ab_0 \cdot 2^0 = (110)_2 \cdot 1 \cdot 2^0 = (110)_2,$$

$$ab_1 \cdot 2^1 = (110)_2 \cdot 0 \cdot 2^1 = (0000)_2,$$

and

$$ab_2 \cdot 2^2 = (110)_2 \cdot 1 \cdot 2^2 = (11000)_2.$$

To find the product, add $(110)_2$, $(0000)_2$, and $(11000)_2$. Carrying out these additions (using Algorithm 2, including initial zero bits when necessary) shows that $ab = (1\ 1110)_2$. This multiplication is displayed in Figure 2. ◀

$$\begin{array}{r} 1\,1\,0 \\ \times\, 1\,0\,1 \\ \hline 1\,1\,0 \\ 0\,0\,0 \\ 1\,1\,0 \\ \hline 1\,1\,1\,1\,0 \end{array}$$

FIGURE 2 **Multiplying $(110)_2$ and $(101)_2$.**

Next, we determine the number of additions of bits and shifts of bits used by Algorithm 3 to multiply two integers.

EXAMPLE 11 How many additions of bits and shifts of bits are used to multiply a and b using Algorithm 3?

Solution: Algorithm 3 computes the products of a and b by adding the partial products $c_0, c_1, c_2, \dots,$ and c_{n-1}. When $b_j = 1$, we compute the partial product c_j by shifting the binary expansion of a by j bits. When $b_j = 0$, no shifts are required because $c_j = 0$. Hence, to find all n of the integers $ab_j2^j, j = 0, 1, \dots, n - 1$, requires at most

$$0 + 1 + 2 + \cdots + n - 1$$

shifts. Hence, by Example 5 in Section 3.2 the number of shifts required is $O(n^2)$.

To add the integers ab_j from $j = 0$ to $j = n - 1$ requires the addition of an n-bit integer, an $(n + 1)$-bit integer, ..., and a $(2n)$-bit integer. We know from Example 9 that each of these additions requires $O(n)$ additions of bits. Consequently, a total of $O(n^2)$ additions of bits are required for all n additions. ◀

Surprisingly, there are more efficient algorithms than the conventional algorithm for multiplying integers. One such algorithm, which uses $O(n^{1.585})$ bit operations to multiply n-bit numbers, will be described in Section 8.3.

ALGORITHM FOR div AND mod Given integers a and d, $d > 0$, we can find $q = a$ **div** d and $r = a$ **mod** d using Algorithm 4. In this brute-force algorithm, when a is positive we subtract d from a as many times as necessary until what is left is less than d. The number of times we perform this subtraction is the quotient and what is left over after all these subtractions is the remainder. Algorithm 4 also covers the case where a is negative. This algorithm finds the quotient q and remainder r when $|a|$ is divided by d. Then, when $a < 0$ and $r > 0$, it uses these to find the quotient $-(q + 1)$ and remainder $d - r$ when a is divided by d. We leave it to the reader (Exercise 65) to show that, assuming that $a > d$, this algorithm uses $O(q \log a)$ bit operations.

ALGORITHM 4 Computing div and mod.

```
procedure division algorithm(a: integer, d: positive integer)
q := 0
r := |a|
while r ≥ d
    r := r − d
    q := q + 1
if a < 0 and r > 0 then
    r := d − r
    q := −(q + 1)
return (q, r) {q = a div d is the quotient, r = a mod d is the remainder}
```

There are more efficient algorithms than Algorithm 4 for determining the quotient $q = a$ **div** d and the remainder $r = a$ **mod** d when a positive integer a is divided by a positive integer d (see [Kn98] for details). These algorithms require $O(\log a \cdot \log d)$ bit operations. If both of the binary expansions of a and d contain n or fewer bits, then we can replace $\log a \cdot \log d$ by n^2. This means that we need $O(n^2)$ bit operations to find the quotient and remainder when a is divided by d.

4.2.4 Modular Exponentiation

In cryptography it is important to be able to find b^n **mod** m efficiently without using an excessive amount of memory, where b, n, and m are large integers. It is impractical to first compute b^n and then find its remainder when divided by m, because b^n can be a huge number and we will need a huge amount of computer memory to store such numbers. Instead, we can avoid time and memory problems by using an algorithm that employs the binary expansion of the exponent n.

Before we present an algorithm for fast modular exponentiation based on the binary expansion of the exponent, first observe that we can avoid using large amount of memory if we compute b^n **mod** m by successively computing b^k **mod** m for $k = 1, 2, \ldots, n$ using the fact that b^{k+1} **mod** $m = b(b^k$ **mod** $m)$ **mod** m (by Corollary 2 of Theorem 5 of Section 4.1). However, this approach is impractical because it requires $n - 1$ multiplications of integers and n might be huge.

To motivate the fast modular exponentiation algorithm, we illustrate its basic idea. We will explain how to use the binary expansion of n, say $n = (a_{k-1} \ldots a_1 a_0)_2$, to compute b^n. First, note that

$$b^n = b^{a_{k-1} \cdot 2^{k-1} + \cdots + a_1 \cdot 2 + a_0} = b^{a_{k-1} \cdot 2^{k-1}} \cdots b^{a_1 \cdot 2} \cdot b^{a_0}.$$

If you are rusty with the laws for exponents, this is the time to review them! See Theorem 1 in Appendix 2.

This shows that to compute b^n, we need only compute the values of b, b^2, $(b^2)^2 = b^4$, $(b^4)^2 = b^8, \ldots, b^{2^k}$. Once we have these values, we multiply the terms b^{2^j} in this list, where $a_j = 1$. (For efficiency and to reduce space requirements, after multiplying by each term, we reduce the result modulo m.)

This gives us b^n. For example, to compute 3^{11} we first note that $11 = (1011)_2$, so that $3^{11} = 3^8 3^2 3^1$. By successively squaring, we find that $3^2 = 9$, $3^4 = 9^2 = 81$, and $3^8 = (81)^2 = 6561$. Consequently, $3^{11} = 3^8 3^2 3^1 = 6561 \cdot 9 \cdot 3 = 177{,}147$.

Note that $(b^{2^n})^2 = b^{2^{n+1}}$ when n is a nonnegative integer.

The algorithm successively finds b **mod** m, b^2 **mod** m, b^4 **mod** $m, \ldots, b^{2^{k-1}}$ **mod** m and multiplies together those terms b^{2^j} **mod** m where $a_j = 1$, finding the remainder of the product when divided by m after each multiplication. Note that we need only perform $O(\log_2(n))$ multiplications. Pseudocode for this algorithm is shown in Algorithm 5. Note that in Algorithm 5 we can use the most efficient algorithm available to compute values of the **mod** function, not necessarily Algorithm 4.

Be sure to reduce modulo m after each multiplication!

ALGORITHM 5 Fast Modular Exponentiation.

procedure *modular exponentiation*(b: integer, $n = (a_{k-1}a_{k-2} \ldots a_1a_0)_2$,
  m: positive integers)
$x := 1$
power := b **mod** m
for $i := 0$ **to** $k - 1$
  if $a_i = 1$ **then** $x := (x \cdot \textit{power})$ **mod** m
  power := (*power* · *power*) **mod** m
return $x\{x$ equals b^n **mod** $m\}$

We illustrate how Algorithm 5 works in Example 12.

EXAMPLE 12 Use Algorithm 5 to find 3^{644} **mod** 645.

Solution: Algorithm 5 initially sets $x = 1$ and *power* $= 3$ **mod** $645 = 3$. In the computation of 3^{644} **mod** 645, this algorithm determines 3^{2^j} **mod** 645 for $j = 1, 2, \ldots, 9$ by successively squaring and reducing modulo 645. If $a_j = 1$ (where a_j is the bit in the jth position in the binary expansion of 644, which is $(1010000100)_2$), it multiplies the current value of x by 3^{2^j} **mod** 645 and reduces the result modulo 645. Here are the steps used:

$i = 0$: Because $a_0 = 0$, we have $x = 1$ and *power* $= 3^2$ **mod** $645 = 9$ **mod** $645 = 9$;
$i = 1$: Because $a_1 = 0$, we have $x = 1$ and *power* $= 9^2$ **mod** $645 = 81$ **mod** $645 = 81$;
$i = 2$: Because $a_2 = 1$, we have $x = 1 \cdot 81$ **mod** $645 = 81$ and *power* $= 81^2$ **mod** $645 = 6561$ **mod** $645 = 111$;
$i = 3$: Because $a_3 = 0$, we have $x = 81$ and *power* $= 111^2$ **mod** $645 = 12{,}321$ **mod** $645 = 66$;
$i = 4$: Because $a_4 = 0$, we have $x = 81$ and *power* $= 66^2$ **mod** $645 = 4356$ **mod** $645 = 486$;
$i = 5$: Because $a_5 = 0$, we have $x = 81$ and *power* $= 486^2$ **mod** $645 = 236{,}196$ **mod** $645 = 126$;
$i = 6$: Because $a_6 = 0$, we have $x = 81$ and *power* $= 126^2$ **mod** $645 = 15{,}876$ **mod** $645 = 396$;
$i = 7$: Because $a_7 = 1$, we find that $x = (81 \cdot 396)$ **mod** $645 = 471$ and *power* $= 396^2$ **mod** $645 = 156{,}816$ **mod** $645 = 81$;
$i = 8$: Because $a_8 = 0$, we have $x = 471$ and *power* $= 81^2$ **mod** $645 = 6561$ **mod** $645 = 111$;
$i = 9$: Because $a_9 = 1$, we find that $x = (471 \cdot 111)$ **mod** $645 = 36$.

This shows that following the steps of Algorithm 5 produces the result 3^{644} **mod** $645 = 36$. ◀

Algorithm 5 is quite efficient; it uses $O((\log m)^2 \log n)$ bit operations to find b^n **mod** m (see Exercise 64).

Exercises

1. Convert the decimal expansion of each of these integers to a binary expansion.

a) 231 **b)** 4532 **c)** 97644

2. Convert the decimal expansion of each of these integers to a binary expansion.

a) 321 **b)** 1023 **c)** 100632

3. Convert the binary expansion of each of these integers to a decimal expansion.

a) $(1\ 1111)_2$ **b)** $(10\ 0000\ 0001)_2$
c) $(1\ 0101\ 0101)_2$ **d)** $(110\ 1001\ 0001\ 0000)_2$

4. Convert the binary expansion of each of these integers to a decimal expansion.

a) $(1\ 1011)_2$ **b)** $(10\ 1011\ 0101)_2$
c) $(11\ 1011\ 1110)_2$ **d)** $(111\ 1100\ 0001\ 1111)_2$

5. Convert the octal expansion of each of these integers to a binary expansion.

a) $(572)_8$ **b)** $(1604)_8$
c) $(423)_8$ **d)** $(2417)_8$

6. Convert the binary expansion of each of these integers to an octal expansion.

a) $(1111\ 0111)_2$
b) $(1010\ 1010\ 1010)_2$
c) $(111\ 0111\ 0111\ 0111)_2$
d) $(101\ 0101\ 0101\ 0101)_2$

7. Convert the hexadecimal expansion of each of these integers to a binary expansion.

a) $(80\text{E})_{16}$ **b)** $(135\text{AB})_{16}$
c) $(\text{ABBA})_{16}$ **d)** $(\text{DEFACED})_{16}$

8. Convert $(\text{BADFACED})_{16}$ from its hexadecimal expansion to its binary expansion.

9. Convert $(\text{ABCDEF})_{16}$ from its hexadecimal expansion to its binary expansion.

10. Convert each of the integers in Exercise 6 from a binary expansion to a hexadecimal expansion.

11. Convert $(1011\ 0111\ 1011)_2$ from its binary expansion to its hexadecimal expansion.

12. Convert $(1\ 1000\ 0110\ 0011)_2$ from its binary expansion to its hexadecimal expansion.

13. Show that the hexadecimal expansion of a positive integer can be obtained from its binary expansion by grouping together blocks of four binary digits, adding initial zeros if necessary, and translating each block of four binary digits into a single hexadecimal digit.

14. Show that the binary expansion of a positive integer can be obtained from its hexadecimal expansion by translating each hexadecimal digit into a block of four binary digits.

15. Show that the octal expansion of a positive integer can be obtained from its binary expansion by grouping together blocks of three binary digits, adding initial zeros if necessary, and translating each block of three binary digits into a single octal digit.

16. Show that the binary expansion of a positive integer can be obtained from its octal expansion by translating each octal digit into a block of three binary digits.

17. Convert $(7345321)_8$ to its binary expansion and $(10\ 1011\ 1011)_2$ to its octal expansion.

18. Give a procedure for converting from the hexadecimal expansion of an integer to its octal expansion using binary notation as an intermediate step.

19. Give a procedure for converting from the octal expansion of an integer to its hexadecimal expansion using binary notation as an intermediate step.

20. Explain how to convert from binary to base 64 expansions and from base 64 expansions to binary expansions and from octal to base 64 expansions and from base 64 expansions to octal expansions.

21. Find the sum and the product of each of these pairs of numbers. Express your answers as a binary expansion.

a) $(100\ 0111)_2$, $(111\ 0111)_2$
b) $(1110\ 1111)_2$, $(1011\ 1101)_2$
c) $(10\ 1010\ 1010)_2$, $(1\ 1111\ 0000)_2$
d) $(10\ 0000\ 0001)_2$, $(11\ 1111\ 1111)_2$

22. Find the sum and product of each of these pairs of numbers. Express your answers as a base 3 expansion.

a) $(112)_3$, $(210)_3$
b) $(2112)_3$, $(12021)_3$
c) $(20001)_3$, $(1111)_3$
d) $(120021)_3$, $(2002)_3$

23. Find the sum and product of each of these pairs of numbers. Express your answers as an octal expansion.

a) $(763)_8$, $(147)_8$
b) $(6001)_8$, $(272)_8$
c) $(1111)_8$, $(777)_8$
d) $(54321)_8$, $(3456)_8$

24. Find the sum and product of each of these pairs of numbers. Express your answers as a hexadecimal expansion.

a) $(1\text{AE})_{16}$, $(\text{BBC})_{16}$
b) $(20\text{CBA})_{16}$, $(\text{A}01)_{16}$
c) $(\text{ABCDE})_{16}$, $(1111)_{16}$
d) $(\text{E}0000\text{E})_{16}$, $(\text{BAAA})_{16}$

25. Use Algorithm 5 to find 7^{644} **mod** 645.

26. Use Algorithm 5 to find 11^{644} **mod** 645.

27. Use Algorithm 5 to find 3^{2003} **mod** 99.

28. Use Algorithm 5 to find 123^{1001} **mod** 101.

29. Show that every positive integer can be represented uniquely as the sum of distinct powers of 2. [*Hint:* Consider binary expansions of integers.]

30. It can be shown that every integer can be uniquely represented in the form

$$e_k 3^k + e_{k-1} 3^{k-1} + \cdots + e_1 3 + e_0,$$

where $e_j = -1, 0,$ or 1 for $j = 0, 1, 2, \ldots, k$. Expansions of this type are called **balanced ternary expansions**. Find the balanced ternary expansions of

a) 5. **b)** 13. **c)** 37. **d)** 79.

31. Show that a positive integer is divisible by 3 if and only if the sum of its decimal digits is divisible by 3.

32. Show that a positive integer is divisible by 11 if and only if the difference of the sum of its decimal digits in even-numbered positions and the sum of its decimal digits in odd-numbered positions is divisible by 11.

33. Show that a positive integer is divisible by 3 if and only if the difference of the sum of its binary digits in even-numbered positions and the sum of its binary digits in odd-numbered positions is divisible by 3.

34. Determine how we can use the decimal expansion of an integer n to determine whether n is divisible by

a) 2 **b)** 5 **c)** 10

35. Determine how we can use the decimal expansion of an integer n to determine whether n is divisible by

a) 4 **b)** 25 **c)** 20

36. Suppose that n and b are positive integers with $b \geq 2$ and the base b expansion of n is $n = (a_m a_{m-1} \dots a_1 a_0)_b$. Find the base b expansion of

a) bn. **b)** $b^2 n$;
c) $\lfloor n/b \rfloor$, **d)** $\lfloor n/b^2 \rfloor$.

37. Prove that if n and b are positive integers with $b \geq 2$ the base b representation of n has $\lfloor \log_b n \rfloor + 1$ digits.

38. Find the decimal expansion of the number with the n-digit base seven expansion $(111 \dots 111)_7$ (with n 1's). [*Hint*: Use the formula for the sum of the terms of a geometric progression.]

39. Find the decimal expansion of the number with the $3n$ bit binary expansion $(101101 \dots .101101)_2$ (so that the binary expansion is made of n copies of 101). [*Hint*: Use the formula for the sum of the terms of a geometric progression.]

One's complement representations of integers are used to simplify computer arithmetic. To represent positive and negative integers with absolute value less than 2^{n-1}, a total of n bits is used. The leftmost bit is used to represent the sign. A 0 bit in this position is used for positive integers, and a 1 bit in this position is used for negative integers. For positive integers, the remaining bits are identical to the binary expansion of the integer. For negative integers, the remaining bits are obtained by first finding the binary expansion of the absolute value of the integer, and then taking the complement of each of these bits, where the complement of a 1 is a 0 and the complement of a 0 is a 1.

40. Find the one's complement representations, using bit strings of length six, of the following integers.

a) 22 **b)** 31 **c)** -7 **d)** -19

41. What integer does each of the following one's complement representations of length five represent?

a) 11001 **b)** 01101
c) 10001 **d)** 11111

42. If m is a positive integer less than 2^{n-1}, how is the one's complement representation of $-m$ obtained from the one's complement of m, when bit strings of length n are used?

43. How is the one's complement representation of the sum of two integers obtained from the one's complement representations of these integers?

44. How is the one's complement representation of the difference of two integers obtained from the one's complement representations of these integers?

45. Show that the integer m with one's complement representation $(a_{n-1} a_{n-2} \dots a_1 a_0)$ can be found using the equation $m = -a_{n-1}(2^{n-1} - 1) + a_{n-2} 2^{n-2} + \dots + a_1 \cdot 2 + a_0$.

Two's complement representations of integers are also used to simplify computer arithmetic and are used more commonly than one's complement representations. To represent an integer x with $-2^{n-1} \leq x \leq 2^{n-1} - 1$ for a specified positive integer n, a total of n bits is used. The leftmost bit is used to represent the sign. A 0 bit in this position is used for positive integers, and a 1 bit in this position is used for negative integers, just as in one's complement expansions. For a positive integer, the remaining bits are identical to the binary expansion of the integer. For a negative integer, the remaining bits are the bits of the binary expansion of $2^{n-1} - |x|$. Two's complement expansions of integers are often used by computers because addition and subtraction of integers can be performed easily using these expansions, where these integers can be either positive or negative.

46. Answer Exercise 40, but this time find the two's complement expansion using bit strings of length six.

47. Answer Exercise 41 if each expansion is a two's complement expansion of length five.

48. Answer Exercise 42 for two's complement expansions.

49. Answer Exercise 43 for two's complement expansions.

50. Answer Exercise 44 for two's complement expansions.

51. Show that the integer m with two's complement representation $(a_{n-1} a_{n-2} \dots a_1 a_0)$ can be found using the equation $m = -a_{n-1} \cdot 2^{n-1} + a_{n-2} 2^{n-2} + \dots + a_1 \cdot 2 + a_0$.

52. Give a simple algorithm for forming the two's complement representation of an integer from its one's complement representation.

53. Sometimes integers are encoded by using four-digit binary expansions to represent each decimal digit. This produces the **binary coded decimal** form of the integer. For instance, 791 is encoded in this way by 011110010001. How many bits are required to represent a number with n decimal digits using this type of encoding?

A **Cantor expansion** is a sum of the form

$$a_n n! + a_{n-1}(n-1)! + \dots + a_2 2! + a_1 1!,$$

where a_i is an integer with $0 \leq a_i \leq i$ for $i = 1, 2, \dots, n$.

54. Find the Cantor expansions of

a) 2. **b)** 7.
c) 19. **d)** 87.
e) 1000. **f)** 1,000,000.

*55. Describe an algorithm that finds the Cantor expansion of an integer.

*56. Describe an algorithm to add two integers from their Cantor expansions.

57. Add $(10111)_2$ and $(11010)_2$ by working through each step of the algorithm for addition given in the text.

58. Multiply $(1110)_2$ and $(1010)_2$ by working through each step of the algorithm for multiplication given in the text.

59. Describe an algorithm for finding the difference of two binary expansions.

60. Estimate the number of bit operations used to subtract two binary expansions.

61. Devise an algorithm that, given the binary expansions of the integers a and b, determines whether $a > b$, $a = b$, or $a < b$.

62. How many bit operations does the comparison algorithm from Exercise 61 use when the larger of a and b has n bits in its binary expansion?

63. Estimate the complexity of Algorithm 1 for finding the base b expansion of an integer n in terms of the number of divisions used.

*64. Show that Algorithm 5 uses $O((\log m)^2 \log n)$ bit operations to find b^n **mod** m.

65. Show that Algorithm 4 uses $O(q \log a)$ bit operations, assuming that $a > d$.

4.3 Primes and Greatest Common Divisors

4.3.1 Introduction

In Section 4.1 we studied the concept of divisibility of integers. One important concept based on divisibility is that of a prime number. A prime is an integer greater than 1 that is divisible by no positive integers other than 1 and itself. The study of prime numbers goes back to ancient times. Thousands of years ago it was known that there are infinitely many primes; the proof of this fact, found in the works of Euclid, is famous for its elegance and beauty.

We will discuss the distribution of primes among the integers. We will describe some of the results about primes found by mathematicians in the last 400 years. In particular, we will introduce an important theorem, the fundamental theorem of arithmetic. This theorem, which asserts that every positive integer can be written uniquely as the product of primes in nondecreasing order, has many interesting consequences. We will also discuss some of the many old conjectures about primes that remain unsettled today.

Primes have become essential in modern cryptographic systems, and we will develop some of their properties important in cryptography. For example, finding large primes is essential in modern cryptography. The length of time required to factor large integers into their prime factors is the basis for the strength of some important modern cryptographic systems.

In this section we will also study the greatest common divisor of two integers, as well as the least common multiple of two integers. We will develop an important algorithm for computing greatest common divisors, called the Euclidean algorithm.

4.3.2 Primes

Every integer greater than 1 is divisible by at least two integers, because a positive integer is divisible by 1 and by itself. Positive integers that have exactly two different positive integer factors are called **primes**.

Definition 1 An integer p greater than 1 is called *prime* if the only positive factors of p are 1 and p. A positive integer that is greater than 1 and is not prime is called *composite*.

Remark: The integer 1 is not prime, because it has only one positive factor. Note also that an integer n is composite if and only if there exists an integer a such that $a \mid n$ and $1 < a < n$.

EXAMPLE 1 The integer 7 is prime because its only positive factors are 1 and 7, whereas the integer 9 is composite because it is divisible by 3. ◀

The primes are the building blocks of positive integers, as the fundamental theorem of arithmetic shows. The proof will be given in Section 5.2.

THEOREM 1 **THE FUNDAMENTAL THEOREM OF ARITHMETIC** Every integer greater than 1 can be written uniquely as a prime or as the product of two or more primes, where the prime factors are written in order of nondecreasing size.

Example 2 gives some prime factorizations of integers.

EXAMPLE 2 The prime factorizations of 100, 641, 999, and 1024 are given by

Extra Examples

$$100 = 2 \cdot 2 \cdot 5 \cdot 5 = 2^2 5^2,$$
$$641 = 641,$$
$$999 = 3 \cdot 3 \cdot 3 \cdot 37 = 3^3 \cdot 37,$$
$$1024 = 2 \cdot 2 \cdot 2 \cdot 2 \cdot 2 \cdot 2 \cdot 2 \cdot 2 \cdot 2 \cdot 2 = 2^{10}.$$
◀

4.3.3 Trial Division

It is often important to show that a given integer is prime. For instance, in cryptography, large primes are used in some methods for making messages secret. One procedure for showing that an integer is prime is based on the following observation.

THEOREM 2 If n is a composite integer, then n has a prime divisor less than or equal to $\sqrt{n}$.

Proof: If n is composite, by the definition of a composite integer, we know that it has a factor a with $1 < a < n$. Hence, by the definition of a factor of a positive integer, we have $n = ab$, where b is a positive integer greater than 1. We will show that $a \leq \sqrt{n}$ or $b \leq \sqrt{n}$. If $a > \sqrt{n}$ and $b > \sqrt{n}$, then $ab > \sqrt{n} \cdot \sqrt{n} = n$, which is a contradiction. Consequently, $a \leq \sqrt{n}$ or $b \leq \sqrt{n}$. Because both a and b are divisors of n, we see that n has a positive divisor not exceeding $\sqrt{n}$. This divisor is either prime or, by the fundamental theorem of arithmetic, has a prime divisor less than itself. In either case, n has a prime divisor less than or equal to $\sqrt{n}$. ◁

From Theorem 2, it follows that an integer is prime if it is not divisible by any prime less than or equal to its square root. This leads to the brute-force algorithm known as **trial division**. To use trial division we divide n by all primes not exceeding $\sqrt{n}$ and conclude that n is prime if it is not divisible by any of these primes. In Example 3 we use trial division to show that 101 is prime.

EXAMPLE 3 Show that 101 is prime.

Solution: The only primes not exceeding $\sqrt{101}$ are 2, 3, 5, and 7. Because 101 is not divisible by 2, 3, 5, or 7 (the quotient of 101 and each of these integers is not an integer), it follows that 101 is prime. ◀

Because every integer has a prime factorization, it would be useful to have a procedure for finding this prime factorization. Consider the problem of finding the prime factorization of n. Begin by dividing n by successive primes, starting with the smallest prime, 2. If n has a prime factor, then by Theorem 2 a prime factor p not exceeding $\sqrt{n}$ will be found. So, if no prime factor not exceeding $\sqrt{n}$ is found, then n is prime. Otherwise, if a prime factor p is found, continue by factoring n/p. Note that n/p has no prime factors less than p. Again, if n/p has no prime factor greater than or equal to p and not exceeding its square root, then it is prime. Otherwise, if it has a prime factor q, continue by factoring $n/(pq)$. This procedure is continued until the factorization has been reduced to a prime. This procedure is illustrated in Example 4.

EXAMPLE 4 Find the prime factorization of 7007.

Solution: To find the prime factorization of 7007, first perform divisions of 7007 by successive primes, beginning with 2. None of the primes 2, 3, and 5 divides 7007. However, 7 divides 7007, with $7007/7 = 1001$. Next, divide 1001 by successive primes, beginning with 7. It is immediately seen that 7 also divides 1001, because $1001/7 = 143$. Continue by dividing 143 by successive primes, beginning with 7. Although 7 does not divide 143, 11 does divide 143, and $143/11 = 13$. Because 13 is prime, the procedure is completed. It follows that $7007 = 7 \cdot 1001 = 7 \cdot 7 \cdot 143 = 7 \cdot 7 \cdot 11 \cdot 13$. Consequently, the prime factorization of 7007 is $7 \cdot 7 \cdot 11 \cdot 13 = 7^2 \cdot 11 \cdot 13$. ◀

Links

Prime numbers were studied in ancient times for philosophical reasons. Today, there are highly practical reasons for their study. In particular, large primes play a crucial role in cryptography, as we will see in Section 4.6.

4.3.4 The Sieve of Eratosthenes

Note that composite integers not exceeding 100 must have a prime factor not exceeding 10. Because the only primes less than 10 are 2, 3, 5, and 7, the primes not exceeding 100 are these four primes and those positive integers greater than 1 and not exceeding 100 that are divisible by none of 2, 3, 5, or 7.

Links

The **sieve of Eratosthenes** is used to find all primes not exceeding a specified positive integer. For instance, the following procedure is used to find the primes not exceeding 100. We begin with the list of all integers between 1 and 100. To begin the sieving process, the integers that are divisible by 2, other than 2, are deleted. Because 3 is the first integer greater than 2 that is left, all those integers divisible by 3, other than 3, are deleted. Because 5 is the next integer left after 3, those integers divisible by 5, other than 5, are deleted. The next integer left is 7, so those integers divisible by 7, other than 7, are deleted. Because all composite integers not exceeding 100 are divisible by 2, 3, 5, or 7, all remaining integers except 1 are prime. In Table 1,

Links

Source: The Picture Art Collection/Alamy StocK Photo

ERATOSTHENES (276 B.C.E.–194 B.C.E.) It is known that Eratosthenes was born in Cyrene, a Greek colony west of Egypt, and spent time studying at Plato's Academy in Athens. We also know that King Ptolemy II invited Eratosthenes to Alexandria to tutor his son and that later Eratosthenes became chief librarian at the famous library at Alexandria, a central repository of ancient wisdom. Eratosthenes was an extremely versatile scholar, writing on mathematics, geography, astronomy, history, philosophy, and literary criticism. Besides his work in mathematics, he is most noted for his chronology of ancient history and for his famous measurement of the size of the earth.

TABLE 1 The Sieve of Eratosthenes.

Integers divisible by 2 other than 2 receive an underline.									
1	2	3	4	5	6	7	8	9	10
11	12	13	14	15	16	17	18	19	20
21	22	23	24	25	26	27	28	29	30
31	32	33	34	35	36	37	38	39	40
41	42	43	44	45	46	47	48	49	50
51	52	53	54	55	56	57	58	59	60
61	62	63	64	65	66	67	68	69	70
71	72	73	74	75	76	77	78	79	80
81	82	83	84	85	86	87	88	89	90
91	92	93	94	95	96	97	98	99	100

Integers divisible by 3 other than 3 receive an underline.									
1	2	3	4	5	6	7	8	9	10
11	12	13	14	15	16	17	18	19	20
21	22	23	24	25	26	27	28	29	30
31	32	33	34	35	36	37	38	39	40
41	42	43	44	45	46	47	48	49	50
51	52	53	54	55	56	57	58	59	60
61	62	63	64	65	66	67	68	69	70
71	72	73	74	75	76	77	78	79	80
81	82	83	84	85	86	87	88	89	90
91	92	93	94	95	96	97	98	99	100

Integers divisible by 5 other than 5 receive an underline.									
1	2	3	4	5	6	7	8	9	10
11	12	13	14	15	16	17	18	19	20
21	22	23	24	25	26	27	28	29	30
31	32	33	34	35	36	37	38	39	40
41	42	43	44	45	46	47	48	49	50
51	52	53	54	55	56	57	58	59	60
61	62	63	64	65	66	67	68	69	70
71	72	73	74	75	76	77	78	79	80
81	82	83	84	85	86	87	88	89	90
91	92	93	94	95	96	97	98	99	100

Integers divisible by 7 other than 7 receive an underline; integers in color are prime.									
1	2	3	4	5	6	7	8	9	10
11	12	13	14	15	16	17	18	19	20
21	22	23	24	25	26	27	28	29	30
31	32	33	34	35	36	37	38	39	40
41	42	43	44	45	46	47	48	49	50
51	52	53	54	55	56	57	58	59	60
61	62	63	64	65	66	67	68	69	70
71	72	73	74	75	76	77	78	79	80
81	82	83	84	85	86	87	88	89	90
91	92	93	94	95	96	97	98	99	100

the panels display those integers deleted at each stage, where each integer divisible by 2, other than 2, is underlined in the first panel, each integer divisible by 3, other than 3, is underlined in the second panel, each integer divisible by 5, other than 5, is underlined in the third panel, and each integer divisible by 7, other than 7, is underlined in the fourth panel. The integers not underlined are the primes not exceeding 100. We conclude that the primes less than 100 are 2, 3, 5, 7, 11, 13, 17, 19, 23, 29, 31, 37, 41, 43, 47, 53, 59, 61, 67, 71, 73, 79, 83, 89, and 97.

THE INFINITUDE OF PRIMES It has long been known that there are infinitely many primes. This means that whenever $p_1, p_2, \ldots, p_n$ are the n smallest primes, we know there is a larger prime not listed. We will prove this fact using a proof given by Euclid in his famous mathematics text, *The Elements*. This simple, yet elegant, proof is considered by many mathematicians to be among the most beautiful proofs in mathematics. It is the first proof presented in the book *Proofs from THE BOOK* ([AiZi18]), where THE BOOK refers to the imagined collection of perfect proofs that the legendary mathematician Paul Erdős claimed is maintained by God. A famous quotation by Paul Erdős is: "God has the Big Book, the beautiful proofs of mathematical theorems are listed here." He also said, "You don't have to believe in God, but you should believe in The Book." By the way, there are a vast number of different proofs that there are an infinitude of primes, and new ones are published surprisingly frequently.

THEOREM 3 There are infinitely many primes.

Proof: We will prove this theorem using a proof by contradiction. We assume that there are only finitely many primes, $p_1, p_2, \ldots, p_n$. Let

$$Q = p_1 p_2 \cdots p_n + 1.$$

By the fundamental theorem of arithmetic, Q is prime or else it can be written as the product of two or more primes. However, none of the primes p_j divides Q, for if $p_j \mid Q$, then p_j divides $Q - p_1 p_2 \cdots p_n = 1$. Hence, there is a prime not in the list $p_1, p_2, \ldots, p_n$. This prime is either Q, if it is prime, or a prime factor of Q. This is a contradiction because we assumed that we have listed all the primes. Consequently, there are infinitely many primes. ◁

Remark: Note that in this proof we do *not* state that Q is prime! Furthermore, in this proof, we have given a nonconstructive existence proof that given any n primes, there is a prime not in this list. For this proof to be constructive, we would have had to explicitly give a prime not in our original list of n primes.

Because there are infinitely many primes, given any positive integer there are primes greater than this integer. There is an ongoing quest to discover larger and larger prime numbers; for almost all the last 300 years, the largest prime known has been an integer of the special form $2^p - 1$, where p is also prime. (Note that $2^n - 1$ cannot be prime when n is not prime; see Exercise 19.) Such primes are called **Mersenne primes**, after the French monk Marin Mersenne, who studied them in the seventeenth century. The reason that the largest known prime has usually been a Mersenne prime is that there is an extremely efficient test, known as the Lucas–Lehmer test, for determining whether $2^p - 1$ is prime. Furthermore, it is not currently possible to test numbers not of this or certain other special forms anywhere near as quickly to determine whether they are prime.

EXAMPLE 5 The numbers $2^2 - 1 = 3$, $2^3 - 1 = 7$, $2^5 - 1 = 31$ and $2^7 - 1 = 127$ are Mersenne primes, while $2^{11} - 1 = 2047$ is not a Mersenne prime because $2047 = 23 \cdot 89$. ◀

Progress in finding Mersenne primes has been steady since computers were invented. As of early 2024, 51 Mersenne primes were known, with 20 found since 1990. The largest Mersenne

Links

Apic/Hulton Archive/Getty Images

MARIN MERSENNE (1588–1648) Mersenne was born in Maine, France, into a family of laborers and attended the College of Mans and the Jesuit College at La Flèche. He continued his education at the Sorbonne, studying theology from 1609 to 1611. He joined the religious order of the Minims in 1611, a group whose name comes from the word *minimi* (the members of this group were extremely humble; they considered themselves the least of all religious orders). Besides prayer, the members of this group devoted their energy to scholarship and study. In 1612 he became a priest at the Place Royale in Paris; between 1614 and 1618 he taught philosophy at the Minim Convent at Nevers. He returned to Paris in 1619, where his cell in the Minims de l'Annociade became a place for meetings of French scientists, philosophers, and mathematicians, including Fermat and Pascal. Mersenne corresponded extensively with scholars throughout Europe, serving as a clearinghouse for mathematical and scientific knowledge, a function later served by mathematical journals (and today also by the Internet). Mersenne wrote books covering mechanics, physics, mathematics, music, and acoustics. He studied prime numbers and tried unsuccessfully to construct a formula representing all primes. In 1644 Mersenne claimed that $2^p - 1$ is prime for $p = 2, 3, 5, 7, 13, 17, 19, 31, 67, 127, 257$ but is composite for all other primes less than 257. It took over 300 years to determine that Mersenne's claim was wrong five times. Specifically, $2^p - 1$ is not prime for $p = 67$ and $p = 257$ but is prime for $p = 61$, $p = 87$, and $p = 107$. It is also noteworthy that Mersenne defended two of the most famous men of his time, Descartes and Galileo, from religious critics. He also helped expose alchemists and astrologers as frauds.

Links prime known (again as of early 2024) is $2^{82,589,933} - 1$, a number with 49,724,095 decimal digits, which was shown to be prime in December 2018. A communal effort, the Great Internet Mersenne Prime Search (GIMPS), is devoted to the search for new Mersenne primes. You can join this search, and if you are lucky, find a new Mersenne prime and possibly even win a cash prize. By the way, even the search for Mersenne primes has practical implications. A commonly used quality control test for supercomputers is to replicate the Lucas–Lehmer test that establishes the primality of a large Mersenne prime. Also, in January 2016, it was reported that a bug in the Intel Skylake processor was found when GIMPS software was run. (See [Ro23] for more information about the quest for finding Mersenne primes.)

THE DISTRIBUTION OF PRIMES Theorem 3 tells us that there are infinitely many primes. However, how many primes are less than a positive number x? This question interested mathematicians for many years; in the late eighteenth century, mathematicians produced large tables of prime numbers to gather evidence concerning the distribution of primes. Using this evidence, the great mathematicians of the day, including Gauss and Legendre, conjectured, but did not prove, Theorem 4.

THEOREM 4 **THE PRIME NUMBER THEOREM** The ratio of $\pi(x)$, the number of primes not exceeding x, and $x/\ln x$ approaches 1 as x grows without bound. (Here $\ln x$ is the natural logarithm of x.)

Links The prime number theorem was first proved in 1896 by the French mathematician Jacques Hadamard and the Belgian mathematician Charles-Jean-Gustave-Nicholas de la Vallée-Poussin using the theory of complex variables. Although proofs not using complex variables have been found, all known proofs of the prime number theorem are quite complicated. Many refinements of the prime number theorem have been proved, with many addressing the error made by estimating $\pi(x)$ with $x/\ln x$, and by estimating $\pi(x)$ with other functions. Many unsolved questions remain in this area of study.

Table 2 displays $\pi(x)$, $x/\ln x$, and their ratio, for $x = 10^n$ where $3 \leq n \leq 10$. A tremendous amount of effort has been devoted to computing $\pi(x)$ for progressively larger values of x. As of early 2024, the number of primes less than or equal to 10^n has been determined for all positive integers n with $n \leq 29$. In particular, it is known that

$$\pi(10^{29}) = 1{,}520{,}698{,}109{,}714{,}272{,}166{,}094{,}258{,}063,$$

to the nearest integer

$$\pi(10^{29}) - (10^{29}/\ln 10^{29}) = 23{,}130{,}930{,}737{,}541{,}725{,}917{,}951{,}446,$$

TABLE 2 Approximating $\pi(x)$ by $x/\ln x$.

x	$\pi(x)$	$x/\ln x$	$\pi(x)/(x/\ln x)$
10^3	168	144.8	1.161
10^4	1229	1085.7	1.132
10^5	9592	8685.9	1.104
10^6	78,498	72,382.4	1.084
10^7	664,579	620,420.7	1.071
10^8	5,761,455	5,428,681.0	1.061
10^9	50,847,534	48,254,942.4	1.054
10^{10}	455,052,512	434,294,481.9	1.048

and up to six decimal places

$$\pi(10^{29})/(10^{29}/\ln(10^{29})) = 1.015446.$$

You can find a great deal of computational data relating to $\pi(x)$ and functions that estimate $\pi(x)$ using the web.

Links

We can use the prime number theorem to estimate the probability that a randomly chosen number is prime. (See Chapter 7 to learn the basics of probability theory.) The prime number theorem tells us that the number of primes not exceeding x can be approximated by $x/\ln x$. Consequently, the odds that a randomly selected positive integer less than n is prime are approximately $(n/\ln n)/n = 1/\ln n$. Sometimes we need to find a prime with a particular number of digits. We would like an estimate of how many integers with a particular number of digits we need to select before we encounter a prime. Using the prime number theorem and calculus, it can be shown that the probability that an integer n is prime is also approximately $1/\ln n$. For example, the odds that an integer near 10^{1000} is prime are approximately $1/\ln 10^{1000}$, which is approximately $1/2300$. (Note that if we choose only odd numbers, we double our chances of finding a prime.)

Using trial division with Theorem 2 gives procedures for factoring and for primality testing. However, these procedures are not efficient algorithms; many much more practical and efficient algorithms for these tasks have been developed. Factoring and primality testing have become important in the applications of number theory to cryptography. This has led to a great interest in developing efficient algorithms for both tasks. Clever procedures have been devised in the last 30 years for efficiently generating large primes. Moreover, in 2002, an important theoretical discovery was made by Manindra Agrawal, Neeraj Kayal, and Nitin Saxena. They showed there is a polynomial-time algorithm in the number of bits in the binary expansion of an integer for determining whether a positive integer is prime. Algorithms based on their work use $O((\log n)^6)$ bit operations to determine whether a positive integer n is prime.

However, even though powerful new factorization methods have been developed in the same time frame, factoring large numbers remains extraordinarily more time-consuming than primality testing. No polynomial-time algorithm for factoring integers is known. Nevertheless, the challenge of factoring large numbers interests many people. There is a communal effort on the Internet to factor large numbers, especially those of the special form $k^n \pm 1$, where k is a small positive integer and n is a large positive integer (such numbers are called *Cunningham numbers*). At any given time, there is a list of the "Ten Most Wanted" large numbers of this type awaiting factorization.

Links

Paul Fearn/Alamy Stock Photo

CHARLES-JEAN-GUSTAVE-NICHOLAS DE LA VALLÉE-POUSSIN (1866–1962) De la Vallée-Poussin, the son of a professor of geology, was born in Louvain, Belgium. He attended the Jesuit College at Mons, first studying philosophy, but then turning to engineering. After graduating, he devoted himself to mathematics instead of engineering. His most important contribution to mathematics was his proof of the prime number theorem. He also established results about the distribution of primes in arithmetic progressions and refined the prime number theorem to include error estimates. De la Vallée-Poussin made important contributions to differential equations, analysis, and approximation theory. He also wrote a textbook, *Cours d'analyse*, which had significant impact on mathematical thought in the first half of the twentieth century.

ullstein bild Dtl/Getty Images

JACQUES HADAMARD (1865–1963) Hadamard, whose father was a Latin teacher and mother a distinguished piano teacher, was born in Versailles, France. After graduating from college, he taught at a secondary school in Paris. After receiving his Ph.D. in 1892, he was a lecturer at the Faculté des Sciences of Bordeaux. Later, he served on the faculties of the Sorbonne, the Collège de France, the École Polytéchnique, and the École Centrale des Arts et Manufacturers. Hadamard made significant contributions to complex analysis, functional analysis, and mathematical physics. He was recognized as an innovative teacher, writing many articles about elementary mathematics that were used in French schools and a widely used elementary geometry book.

PRIMES AND ARITHMETIC PROGRESSIONS Every odd integer is in one of the two arithmetic progressions $4k + 1$ or $4k + 3$, $k = 1, 2, \ldots$. Because we know that there are infinitely many primes, we can ask whether there are infinitely many primes in both of these arithmetic progressions. The primes 5, 13, 17, 29, 37, 41, … are in the arithmetic progression $4k + 1$; the primes 3, 7, 11, 19, 23, 31, 43, … are in the arithmetic progression $4k + 3$. Looking at the evidence hints that there may be infinitely many primes in both progressions. What about other arithmetic progressions $ak + b$, $k = 1, 2, \ldots$, where no integer greater than one divides both a and b? Do they contain infinitely many primes? The answer was provided by the German mathematician G. Lejeune Dirichlet, who proved that every such arithmetic progression contains infinitely many primes. His proof, and all proofs found later, are beyond the scope of this book. However, it is possible to prove special cases of Dirichlet's theorem using the ideas developed in this book. For example, Exercises 54 and 55 ask for proofs that there are infinitely many primes in the arithmetic progressions $3k + 2$ and $4k + 3$, where k is a positive integer. (The hint for each of these exercises supplies the basic idea needed for the proof.)

We have explained that every arithmetic progression $ak + b$, $k = 1, 2, \ldots$, where a and b have no common factor greater than one, contains infinitely many primes. But are there long arithmetic progressions made up of just primes? For example, some exploration shows that 5, 11, 17, 23, 29 is an arithmetic progression of five primes and 199, 409, 619, 829, 1039, 1249, 1459, 1669, 1879, 2089 is an arithmetic progression of ten primes. In the 1930s, the legendary and prolific mathematician Paul Erdős conjectured that for every positive integer n greater than two, there is an arithmetic progression of length n made up entirely of primes. In 2006, Ben Green and Terence Tao were able to prove this conjecture. Their proof, considered to be a mathematical tour de force, is a nonconstructive proof that combines powerful ideas from several advanced areas of mathematics.

4.3.5 Conjectures and Open Problems About Primes

Number theory is noted as a subject for which it is easy to formulate conjectures, some of which are difficult to prove and others that remained open problems for many years. We will describe some conjectures in number theory and discuss their status in Examples 6–9.

EXAMPLE 6 *Extra Examples* It would be useful to have a function $f(n)$ such that $f(n)$ is prime for all positive integers n. If we had such a function, we could find large primes for use in cryptography and other applications. Looking for such a function, we might check out different polynomial functions, as some mathematicians did several hundred years ago. After a lot of computation we may encounter

Links

Courtesy of Reed Hutchinson/UCLA

TERENCE TAO (BORN 1975) Tao was born in Australia. His father is a pediatrician and his mother taught mathematics at a Hong Kong secondary school. Tao was a child prodigy, teaching himself arithmetic at the age of two. At 10, he became the youngest contestant at the International Mathematical Olympiad (IMO); he won an IMO gold medal at 13. Tao received his bachelor's and master's degrees when he was 17, and began graduate studies at Princeton, receiving his Ph.D. in three years. In 1996 he became a faculty member at UCLA, where he continues to work.

Tao is extremely versatile; he enjoys working on problems in diverse areas, including harmonic analysis, partial differential equations, number theory, and combinatorics. You can follow his work by reading his blog, where he discusses progress on various problems. His most famous result is the Green-Tao theorem, which says that there are arbitrarily long arithmetic progressions of primes. Tao has made important contributions to the applications of mathematics, such as developing a method for reconstructing digital images using the least possible amount of information.

Tao has an amazing reputation among mathematicians; he has become a Mr. Fix-It for researchers in mathematics. The well-known mathematician Charles Fefferman, himself a child prodigy, has said that "if you're stuck on a problem, then one way out is to interest Terence Tao." Tao maintains a popular blog that describes his research work and many mathematical problems in great detail. In 2006 Tao was awarded a Fields Medal, the most prestigious award for mathematicians under the age of 40. He was also awarded a MacArthur Fellowship in 2006, and in 2008, he received the Allan T. Waterman award, which came with a $500,000 cash prize to support research work of scientists early in their careers. Tao's wife Laura is an engineer at the Jet Propulsion Laboratory.

the polynomial $f(n) = n^2 - n + 41$. This polynomial has the interesting property that $f(n)$ is prime for all positive integers n not exceeding 40. [We have $f(1) = 41$, $f(2) = 43$, $f(3) = 47$, $f(4) = 53$, and so on.] This can lead us to the conjecture that $f(n)$ is prime for all positive integers n. Can we settle this conjecture?

Solution: Perhaps not surprisingly, this conjecture turns out to be false; we do not have to look far to find a positive integer n for which $f(n)$ is composite, because $f(41) = 41^2 - 41 + 41 = 41^2$. Because $f(n) = n^2 - n + 41$ is prime for all positive integers n with $1 \leq n \leq 40$, we might be tempted to find a different polynomial with the property that $f(n)$ is prime for *all* positive integers n. However, there is no such polynomial. It can be shown that for every polynomial $f(n)$ with integer coefficients, there is a positive integer y such that $f(y)$ is composite. (See Exercise 23 in the Supplementary Exercises.) ◀

Many famous problems about primes still await ultimate resolution by clever people. We describe a few of the most accessible and better known of these open problems in Examples 7–9. Number theory is noted for its wealth of easy-to-understand conjectures that resist attack by all but the most sophisticated techniques, or simply resist all attacks. We present these conjectures to show that many questions that seem relatively simple remain unsettled even in the twenty-first century.

EXAMPLE 7 **Goldbach's Conjecture** In 1742, Christian Goldbach, in a letter to Leonhard Euler, conjectured that every odd integer n, $n > 5$, is the sum of three primes. Euler replied that this conjecture is equivalent to the conjecture that every even integer n, $n > 2$, is the sum of two primes (see Exercise 21 in the Supplementary Exercises). The conjecture that every even integer n, $n > 2$, is the sum of two primes is now called **Goldbach's conjecture**. We can check this conjecture for small even numbers. For example, $4 = 2 + 2$, $6 = 3 + 3$, $8 = 5 + 3$, $10 = 7 + 3$, $12 = 7 + 5$, and so on. Goldbach's conjecture was verified by hand calculations for numbers up to the millions prior to the advent of computers. With computers it can be checked for extremely large numbers. The conjecture has been checked for all positive even integers up to $4 \cdot 10^{18}$.

Links

Although no proof of Goldbach's conjecture has been found, most mathematicians believe it is true. Several theorems have been proved, using complicated methods from analytic number theory far beyond the scope of this book, establishing results weaker than Goldbach's conjecture. Among these are the result that every even integer greater than 2 is the sum of at most six primes (proved in 1995 by O. Ramaré) and that every sufficiently large positive integer is the sum of a prime and a number that is either prime or the product of two primes (proved in 1966 by J. R. Chen). Perhaps Goldbach's conjecture will be settled in the not too distant future. ◀

EXAMPLE 8 There are many conjectures asserting that there are infinitely many primes of certain special forms. A conjecture of this sort is the conjecture that there are infinitely many primes of the form $n^2 + 1$, where n is a positive integer. For example, $5 = 2^2 + 1$, $17 = 4^2 + 1$, $37 = 6^2 + 1$, and so on. The best result currently known is that there are infinitely many positive integers n such that $n^2 + 1$ is prime or the product of at most two primes (proved by Henryk Iwaniec in 1973 using advanced techniques from analytic number theory, far beyond the scope of this book). ◀

Links

Links

CHRISTIAN GOLDBACH (1690–1764) Christian Goldbach was born in Königsberg, Prussia, the city noted for its famous bridge problem (which will be studied in Section 10.5). He became professor of mathematics at the Academy in St. Petersburg in 1725. In 1728 Goldbach went to Moscow to tutor the son of the Tsar. He entered the world of politics when, in 1742, he became a staff member in the Russian Ministry of Foreign Affairs. Goldbach is best known for his correspondence with eminent mathematicians, including Euler and Bernoulli, for his intriguing conjectures in number theory, and for several contributions to analysis. No authentic image has yet been found of Goldbach. Unfortunately, there are some images found on the Internet that are purported to be of Goldbach, but instead are of a Portuguese naval officer. Perhaps there is a misplaced painting of him hiding in an archive.

EXAMPLE 9 **The Twin Prime Conjecture** **Twin primes** are pairs of primes that differ by 2, such as 3 and 5, 5 and 7, 11 and 13, 17 and 19, and 4967 and 4969. The twin prime conjecture asserts that there are infinitely many twin primes. The strongest result proved concerning twin primes is that there are infinitely many pairs p and $p + 2$, where p is prime and $p + 2$ is prime or the product of two primes (proved by J. R. Chen in 1966).

Links

The world's record for twin primes, as of early 2024, consists of the numbers $2{,}996{,}863{,}034{,}895 \cdot 2^{1{,}290{,}000} \pm 1$, which have 388,342 decimal digits.

Let $P(n)$ be the statement that there are infinitely many pairs of primes that differ by exactly n. The twin prime conjecture is the statement that $P(2)$ is true. Mathematicians working on the twin prime conjecture formulated a weaker conjecture, known as the *bounded gap conjecture*, which asserts that there is an integer N for which $P(N)$ is true. The mathematical community was surprised when Yitang Zhang, a 50-year-old professor at the University of New Hampshire, who had not published a paper since 2001, proved the bounded gap conjecture in 2013. In particular, he showed that there is an integer $N < 70{,}000{,}000$ such that $P(N)$ is true. A team of mathematicians, including Terrance Tao, lowered the Zhang's bound by showing that there is an integer $N \leq 246$ for which $P(N)$ is true. Furthermore, they showed that if a certain conjecture was true, it could be shown that $N \leq 6$ and that this is the best possible estimate that could be proved using the methods introduced by Zhang. ◀

4.3.6 Greatest Common Divisors and Least Common Multiples

The largest integer that divides both of two integers is called the **greatest common divisor** of these integers.

Definition 2 Let a and b be integers, not both zero. The largest integer d such that $d \mid a$ and $d \mid b$ is called the *greatest common divisor* of a and b. The greatest common divisor of a and b is denoted by $\gcd(a, b)$.

The greatest common divisor of two integers, not both zero, exists because the set of common divisors of these integers is nonempty and finite. One way to find the greatest common divisor of two integers is to find all the positive common divisors of both integers and then take the largest divisor. This is done in Examples 10 and 11. Later, a more efficient method of finding greatest common divisors will be given.

EXAMPLE 10 What is the greatest common divisor of 24 and 36?

Solution: The positive common divisors of 24 and 36 are 1, 2, 3, 4, 6, and 12. Hence, $\gcd(24, 36) = 12$. ◀

Links

Source: Yitang Zhang

YITANG ZHANG (BORN 1955) Yitang Zhang was born in Shanghai, China, in 1955. When he was ten years old, he learned about famous conjectures, including Fermat's last theorem and the Goldbach conjecture. During the Cultural Revolution, he spent ten years working in the fields instead of attending school. However, once this period was over, he was able to attend Peking University, receiving his bachelor's and master's degree in 1982 and 1984, respectively. He moved to the United States, attending Purdue University and completing the work for his Ph.D. in 1991.

After receiving his Ph.D., Zhang could not find an academic position because of the poor job market and disagreements with his thesis advisor. Instead he did accounting work and delivered food for a Queens, New York restaurant; he later worked in Kentucky at Subway restaurants owned by a friend. He even lived in his car while looking for work, but was finally able to obtain an academic job as a lecturer at the University of New Hampshire. He held this position from 1999 until early 2014. From 2009 to 2013, he worked on the bounded gap conjecture seven days a week, about ten hours a day, until he made his key discovery. His success led the University of New Hampshire to promote him to full professorship. In 2015, however, he accepted the offer of a full professorship at the University of California, Santa Barbara. Zhang was awarded a MacArthur Fellowship, also known as a Genius Award, in 2014.

EXAMPLE 11 What is the greatest common divisor of 17 and 22?

Solution: The integers 17 and 22 have no positive common divisors other than 1, so that $\gcd(17, 22) = 1$. ◀

Because it is often important to specify that two integers have no common positive divisor other than 1, we have Definition 3.

Definition 3 The integers a and b are *relatively prime* if their greatest common divisor is 1.

EXAMPLE 12 By Example 11 it follows that the integers 17 and 22 are relatively prime, because $\gcd(17, 22) = 1$. ◀

Because we often need to specify that no two integers in a set of integers have a common positive divisor greater than 1, we make Definition 4.

Definition 4 The integers $a_1, a_2, \ldots, a_n$ are *pairwise relatively prime* if $\gcd(a_i, a_j) = 1$ whenever $1 \le i < j \le n$.

EXAMPLE 13 Determine whether the integers 10, 17, and 21 are pairwise relatively prime and whether the integers 10, 19, and 24 are pairwise relatively prime.

Solution: Because $\gcd(10, 17) = 1$, $\gcd(10, 21) = 1$, and $\gcd(17, 21) = 1$, we conclude that 10, 17, and 21 are pairwise relatively prime.

Because $\gcd(10, 24) = 2 > 1$, we see that 10, 19, and 24 are not pairwise relatively prime. ◀

Another way to find the greatest common divisor of two positive integers is to use the prime factorizations of these integers. Suppose that the prime factorizations of the positive integers a and b are

$$a = p_1^{a_1} p_2^{a_2} \cdots p_n^{a_n}, \quad b = p_1^{b_1} p_2^{b_2} \cdots p_n^{b_n},$$

where each exponent is a nonnegative integer, and where all primes occurring in the prime factorization of either a or b are included in both factorizations, with zero exponents if necessary. Then $\gcd(a, b)$ is given by

$$\gcd(a, b) = p_1^{\min(a_1, b_1)} p_2^{\min(a_2, b_2)} \cdots p_n^{\min(a_n, b_n)},$$

where $\min(x, y)$ represents the minimum of the two numbers x and y. To show that this formula for $\gcd(a, b)$ is valid, we must show that the integer on the right-hand side divides both a and b, and that no larger integer also does. This integer does divide both a and b, because the power of each prime in the factorization does not exceed the power of this prime in either the factorization of a or that of b. Further, no larger integer can divide both a and b, because the exponents of the primes in this factorization cannot be increased, and no other primes can be included.

EXAMPLE 14 Because the prime factorizations of 120 and 500 are $120 = 2^3 \cdot 3 \cdot 5$ and $500 = 2^2 \cdot 5^3$, the greatest common divisor is

$$\gcd(120, 500) = 2^{\min(3,2)}3^{\min(1,0)}5^{\min(1,3)} = 2^2 3^0 5^1 = 20.$$ ◀

Prime factorizations can also be used to find the **least common multiple** of two integers.

Definition 5 The *least common multiple* of the positive integers a and b is the smallest positive integer that is divisible by both a and b. The least common multiple of a and b is denoted by $\text{lcm}(a, b)$.

The least common multiple exists because the set of integers divisible by both a and b is nonempty (because ab belongs to this set, for instance), and every nonempty set of positive integers has a least element (by the well-ordering property, which will be discussed in Section 5.2). Suppose that the prime factorizations of a and b are as before. Then the least common multiple of a and b is given by

$$\text{lcm}(a, b) = p_1^{\max(a_1, b_1)} p_2^{\max(a_2, b_2)} \cdots p_n^{\max(a_n, b_n)},$$

where $\max(x, y)$ denotes the maximum of the two numbers x and y. This formula is valid because a common multiple of a and b has at least $\max(a_i, b_i)$ factors of p_i in its prime factorization, and the least common multiple has no other prime factors besides those in a and b.

EXAMPLE 15 What is the least common multiple of $2^3 3^5 7^2$ and $2^4 3^3$?

Solution: We have

$$\text{lcm}(2^3 3^5 7^2, 2^4 3^3) = 2^{\max(3,4)}3^{\max(5,3)}7^{\max(2,0)} = 2^4 3^5 7^2.$$ ◀

Theorem 5 gives the relationship between the greatest common divisor and least common multiple of two integers. It can be proved using the formulae we have derived for these quantities. The proof of this theorem is left as Exercise 31.

THEOREM 5 Let a and b be positive integers. Then

$$ab = \gcd(a, b) \cdot \text{lcm}(a, b).$$

4.3.7 The Euclidean Algorithm

Links ❯

Computing the greatest common divisor of two integers directly from the prime factorizations of these integers is inefficient. The reason is that it is time-consuming to find prime factorizations. We will give a more efficient method of finding the greatest common divisor, called the **Euclidean algorithm**. This algorithm has been known since ancient times. It is named after the ancient Greek mathematician Euclid, who included a description of this algorithm in his book *The Elements*.

Before describing the Euclidean algorithm, we will show how it is used to find gcd(91, 287). First, divide 287, the larger of the two integers, by 91, the smaller, to obtain

$$287 = 91 \cdot 3 + 14.$$

Any divisor of 91 and 287 must also be a divisor of $287 - 91 \cdot 3 = 14$. Also, any divisor of 91 and 14 must also be a divisor of $287 = 91 \cdot 3 + 14$. Hence, the greatest common divisor of 91 and 287 is the same as the greatest common divisor of 91 and 14. This means that the problem of finding gcd(91, 287) has been reduced to the problem of finding gcd(91, 14).

Next, divide 91 by 14 to obtain

$$91 = 14 \cdot 6 + 7.$$

Because any common divisor of 91 and 14 also divides $91 - 14 \cdot 6 = 7$ and any common divisor of 14 and 7 divides 91, it follows that gcd(91, 14) = gcd(14, 7).

Continue by dividing 14 by 7, to obtain

$$14 = 7 \cdot 2.$$

Because 7 divides 14, it follows that gcd(14, 7) = 7. Furthermore, because gcd(287, 91) = gcd(91, 14) = gcd(14, 7) = 7, the original problem has been solved.

We now describe how the Euclidean algorithm works in generality. We will use successive divisions to reduce the problem of finding the greatest common divisor of two positive integers to the same problem with smaller integers, until one of the integers is zero.

The Euclidean algorithm is based on the following result about greatest common divisors and the division algorithm.

LEMMA 1 Let $a = bq + r$, where a, b, q, and r are integers. Then $\gcd(a, b) = \gcd(b, r)$.

Proof: If we can show that the common divisors of a and b are the same as the common divisors of b and r, we will have shown that $\gcd(a, b) = \gcd(b, r)$, because both pairs must have the same *greatest* common divisor.

So suppose that d divides both a and b. Then it follows that d also divides $a - bq = r$ (from Theorem 1 of Section 4.1). Hence, any common divisor of a and b is also a common divisor of b and r.

Likewise, suppose that d divides both b and r. Then d also divides $bq + r = a$. Hence, any common divisor of b and r is also a common divisor of a and b.

Consequently, $\gcd(a, b) = \gcd(b, r)$. ◁

Links

bilwissedition Ltd. & Co. KG/Alamy

EUCLID (325 B.C.E.–265 B.C.E.) Euclid was the author of the most successful mathematics book ever written, *The Elements*, which appeared in over 1000 different editions from ancient to modern times. Little is known about Euclid's life, other than that he taught at the famous academy at Alexandria in Egypt. Apparently, Euclid did not stress applications. When a student asked what he would get by learning geometry, Euclid explained that knowledge was worth acquiring for its own sake and told his servant to give the student a coin "because he must make a profit from what he learns."

Suppose that a and b are positive integers with $a \geq b$. Let $r_0 = a$ and $r_1 = b$. When we successively apply the division algorithm, we obtain

$$\begin{aligned}
r_0 &= r_1 q_1 + r_2 & 0 \leq r_2 < r_1,\\
r_1 &= r_2 q_2 + r_3 & 0 \leq r_3 < r_2,\\
&\ \ \vdots\\
r_{n-2} &= r_{n-1} q_{n-1} + r_n & 0 \leq r_n < r_{n-1},\\
r_{n-1} &= r_n q_n.
\end{aligned}$$

Eventually a remainder of zero occurs in this sequence of successive divisions, because the sequence of remainders $a = r_0 > r_1 > r_2 > \cdots \geq 0$ cannot contain more than a terms. Furthermore, it follows from Lemma 1 that

$$\begin{aligned}
\gcd(a, b) = \gcd(r_0, r_1) = \gcd(r_1, r_2) = \cdots &= \gcd(r_{n-2}, r_{n-1})\\
&= \gcd(r_{n-1}, r_n) = \gcd(r_n, 0) = r_n.
\end{aligned}$$

Hence, the greatest common divisor is the last nonzero remainder in the sequence of divisions.

EXAMPLE 16 Find the greatest common divisor of 414 and 662 using the Euclidean algorithm.

Solution: Successive uses of the division algorithm give:

$$\begin{aligned}
662 &= 414 \cdot 1 + 248\\
414 &= 248 \cdot 1 + 166\\
248 &= 166 \cdot 1 + 82\\
166 &= 82 \cdot 2 + 2\\
82 &= 2 \cdot 41.
\end{aligned}$$

Hence, gcd(414, 662) = 2, because 2 is the last nonzero remainder.

We can summarize these steps in tabular form.

j	r_j	r_{j+1}	q_{j+1}	r_{j+2}
0	662	414	1	248
1	414	248	1	166
2	248	166	1	82
3	166	82	2	2
4	82	2	41	0

◀

The Euclidean algorithm is expressed in pseudocode in Algorithm 1.

ALGORITHM 1 The Euclidean Algorithm.

```
procedure gcd(a, b: positive integers)
x := a
y := b
while y ≠ 0
    r := x mod y
    x := y
    y := r
return x{gcd(a, b) is x}
```

In Algorithm 1, the initial values of x and y are a and b, respectively. At each stage of the procedure, x is replaced by y, and y is replaced by x **mod** y, which is the remainder when x is divided by y. This process is repeated as long as $y \neq 0$. The algorithm terminates when $y = 0$, and the value of x at that point, the last nonzero remainder in the procedure, is the greatest common divisor of a and b.

We will study the time complexity of the Euclidean algorithm in Section 5.3, where we will show that the number of divisions required to find the greatest common divisor of a and b, where $a \geq b$, is $O(\log b)$.

4.3.8 gcds as Linear Combinations

An important result we will use throughout the remainder of this section is that the greatest common divisor of two integers a and b can be expressed in the form

$$sa + tb,$$

where s and t are integers. In other words, $\gcd(a, b)$ can be expressed as a **linear combination** with integer coefficients of a and b. For example, $\gcd(6, 14) = 2$, and $2 = (-2) \cdot 6 + 1 \cdot 14$. We state this fact as Theorem 6.

THEOREM 6 **BÉZOUT'S THEOREM** If a and b are positive integers, then there exist integers s and t such that $\gcd(a, b) = sa + tb$.

Definition 6 If a and b are positive integers, then integers s and t such that $\gcd(a, b) = sa + tb$ are called *Bézout coefficients* of a and b (after Étienne Bézout, a French mathematician of the eighteenth century). Also, the equation $\gcd(a, b) = sa + tb$ is called *Bézout's identity*.

We will not give a formal proof of Theorem 6 here (see Exercise 50 in Section 5.2 and [Ro23] for proofs). We will present two different methods that can be used to find a linear combination of two integers equal to their greatest common divisor. (In this section, we will assume that a linear combination has integer coefficients.)

The first method proceeds by working backward through the divisions of the Euclidean algorithm, so this method requires a forward pass and a backward pass through the steps of the Euclidean algorithm. We will illustrate how this method works with an example. The main

Links

Chronicle/Alamy Stock Photo

ÉTIENNE BÉZOUT (1730–1783) Bézout was born in Nemours, France, where his father was a magistrate. Reading the writings of the great mathematician Leonhard Euler enticed him to become a mathematician. In 1758 he was appointed to a position at the Académie des Sciences in Paris; in 1763 he was appointed examiner of the Gardes de la Marine, where he was assigned the task of writing mathematics textbooks. This assignment led to a four-volume textbook completed in 1767. Bézout is well known for his six-volume comprehensive textbook on mathematics. His textbooks were extremely popular and were studied by many generations of students hoping to enter the École Polytechnique, the highly regarded engineering and science school. His books were translated into English and used in North America, including at Harvard.

His most important original work was published in 1779 in the book *Théorie générale des équations algébriques*, where he introduced important methods for solving simultaneous polynomial equations in many unknowns. The most well-known result in this book is now called *Bézout's theorem*, which in its general form tells us that the number of common points on two plane algebraic curves equals the product of the degrees of these curves. Bézout is also credited with inventing the determinant (which was called the Bézoutian by the noted English mathematician James Joseph Sylvester). He was considered to be a kind person with a warm heart, although he had a reserved and somber personality. He was happily married and a father.

advantage of the second method, known as the **extended Euclidean algorithm**, is that it uses one pass through the steps of the Euclidean algorithm to find Bézout coefficients of a and b, unlike the first method, which uses two passes. To run this extended Euclidean algorithm we set $s_0 = 1$, $s_1 = 0$, $t_0 = 0$, and $t_1 = 1$ and let

Links

$$s_j = s_{j-2} - q_{j-1}s_{j-1} \text{ and } t_j = t_{j-2} - q_{j-1}t_{j-1}$$

for $j = 2, 3, \ldots, n$, where the q_j are the quotients in the divisions used when the Euclidean algorithm finds gcd(a, b), as shown in the text. We can prove by strong induction (see Exercise 50 in Section 5.2, or see [Ro23]) that gcd(a, b) $= s_n a + t_n b$.

EXAMPLE 17 Express gcd(252, 198) = 18 as a linear combination of 252 and 198 by working backwards through the steps of the Euclidean algorithm.

Solution: To show that gcd(252, 198) = 18, the Euclidean algorithm uses these divisions:

$$252 = 198 \cdot 1 + 54$$
$$198 = 54 \cdot 3 + 36$$
$$54 = 36 \cdot 1 + 18$$
$$36 = 18 \cdot 2 + 0.$$

We summarize these steps in tabular form:

j	r_j	r_{j+1}	q_{j+1}	r_{j+2}
0	252	198	1	54
1	198	54	3	36
2	54	36	1	18
3	36	18	2	0

Using the next-to-last division (the third division), we can express gcd(252, 198) = 18 as a linear combination of 54 and 36. We find that

$$18 = 54 - 1 \cdot 36.$$

The second division tells us that

$$36 = 198 - 3 \cdot 54.$$

Substituting this expression for 36 into the previous equation, we can express 18 as a linear combination of 54 and 198. We have

$$18 = 54 - 1 \cdot 36 = 54 - 1 \cdot (198 - 3 \cdot 54) = 4 \cdot 54 - 1 \cdot 198.$$

The first division tells us that

$$54 = 252 - 1 \cdot 198.$$

Substituting this expression for 54 into the previous equation, we can express 18 as a linear combination of 252 and 198. We conclude that

$$18 = 4 \cdot (252 - 1 \cdot 198) - 1 \cdot 198 = 4 \cdot 252 - 5 \cdot 198,$$

completing the solution. ◀

The next example shows how to solve the same problem posed in the previous example using the extended Euclidean algorithm.

EXAMPLE 18 Express gcd(252, 198) = 18 as a linear combination of 252 and 198 using the extended Euclidean algorithm.

Solution: Example 17 displays the steps the Euclidean algorithm uses to find gcd(252, 198) = 18. The quotients are $q_1 = 1$, $q_2 = 3$, $q_3 = 1$, and $q_4 = 2$. The desired Bézout coefficients are the values of s_4 and t_4 generated by the extended Euclidean algorithm, where $s_0 = 1$, $s_1 = 0$, $t_0 = 0$, and $t_1 = 1$, and

$$s_j = s_{j-2} - q_{j-1}s_{j-1} \quad \text{and} \quad t_j = t_{j-2} - q_{j-1}t_{j-1}$$

for $j = 2, 3, 4$. We find that

$$\begin{aligned} s_2 &= s_0 - s_1q_1 = 1 - 0 \cdot 1 = 1, &\text{and}\quad t_2 &= t_0 - t_1q_1 = 0 - 1 \cdot 1 = -1,\\ s_3 &= s_1 - s_2q_2 = 0 - 1 \cdot 3 = -3, &\text{and}\quad t_3 &= t_1 - t_2q_2 = 1 - (-1)3 = 4,\\ s_4 &= s_2 - s_3q_3 = 1 - (-3) \cdot 1 = 4, &\text{and}\quad t_4 &= t_2 - t_3q_3 = -1 - 4 \cdot 1 = -5. \end{aligned}$$

Because $s_4 = 4$ and $t_4 = -5$, we see that $18 = \gcd(252, 198) = 4 \cdot 252 - 5 \cdot 198$.

We summarize the steps of the extended Euclidean algorithm in a table:

j	r_j	r_{j+1}	q_{j+1}	r_{j+2}	s_j	t_j
0	252	198	1	54	1	0
1	198	54	3	36	0	1
2	54	36	1	18	1	−1
3	36	18	2	0	−3	4
4					4	−5

◀

We will use Theorem 6 to develop several useful results. One of our goals will be to prove the part of the fundamental theorem of arithmetic asserting that a positive integer has at most one prime factorization. We will show that if a positive integer has a factorization into primes, where the primes are written in nondecreasing order, then this factorization is unique.

First, we need to develop some results about divisibility.

LEMMA 2 If a, b, and c are positive integers such that $\gcd(a, b) = 1$ and $a \mid bc$, then $a \mid c$.

Proof: Because $\gcd(a, b) = 1$, by Bézout's theorem there are integers s and t such that

$$sa + tb = 1.$$

Multiplying both sides of this equation by c, we obtain

$$sac + tbc = c.$$

We can now use Theorem 1 of Section 4.1 to show that $a \mid c$. By part (*ii*) of that theorem, $a \mid tbc$. Because $a \mid sac$ and $a \mid tbc$, by part (*i*) of that theorem, we conclude that a divides $sac + tbc$. Because $sac + tbc = c$, we conclude that $a \mid c$, completing the proof. ◁

We will use the following generalization of Lemma 2 in the proof of uniqueness of prime factorizations. (The proof of Lemma 3 is left as Exercise 64 in Section 5.1, because it can be most easily carried out using the method of mathematical induction, covered in that section.)

LEMMA 3 If p is a prime and $p \mid a_1a_2 \cdots a_n$, where each a_i is an integer, then $p \mid a_i$ for some i.

We can now show that a factorization of an integer into primes is unique. That is, we will show that every integer can be written as the product of primes in nondecreasing order in at most one way. This is part of the fundamental theorem of arithmetic. We will prove the other part, that every integer has a factorization into primes, in Section 5.2.

Proof (of the uniqueness of the prime factorization of a positive integer): We will use a proof by contradiction. Suppose that the positive integer n can be written as the product of primes in two different ways, say, $n = p_1p_2 \cdots p_s$ and $n = q_1q_2 \cdots q_t$, where each p_i and q_j is prime such that $p_1 \leq p_2 \leq \cdots \leq p_s$ and $q_1 \leq q_2 \leq \cdots \leq q_t$.

When we remove all common primes from the two factorizations, we have

$$p_{i_1}p_{i_2} \cdots p_{i_u} = q_{j_1}q_{j_2} \cdots q_{j_v},$$

where no prime occurs on both sides of this equation and u and v are positive integers. By Lemma 3 it follows that p_{i_1} divides q_{j_k} for some k. Because no prime divides another prime, this is impossible. Consequently, there can be at most one factorization of n into primes in nondecreasing order. ◁

Lemma 2 can also be used to prove a result about dividing both sides of a congruence by the same integer. We have shown (Theorem 5 in Section 4.1) that we can multiply both sides of a congruence by the same integer. However, dividing both sides of a congruence by an integer does not always produce a valid congruence, as Example 19 shows.

EXAMPLE 19 The congruence $14 \equiv 8 \pmod 6$ holds, but both sides of this congruence cannot be divided by 2 to produce a valid congruence because $14/2 = 7$ and $8/2 = 4$, but $7 \not\equiv 4 \pmod 6$. ◀

Although we cannot divide both sides of a congruence by any integer to produce a valid congruence, we can if this integer is relatively prime to the modulus. Theorem 7 establishes this important fact. We use Lemma 2 in the proof.

THEOREM 7 Let m be a positive integer and let a, b, and c be integers. If $ac \equiv bc \pmod m$ and $\gcd(c, m) = 1$, then $a \equiv b \pmod m$.

Proof: Because $ac \equiv bc \pmod m$, $m \mid ac - bc = c(a - b)$. By Lemma 2, because $\gcd(c, m) = 1$, it follows that $m \mid a - b$. We conclude that $a \equiv b \pmod m$. ◁

Exercises

1. Determine whether each of these integers is prime.
 a) 21 **b)** 29
 c) 71 **d)** 97
 e) 111 **f)** 143

2. Determine whether each of these integers is prime.
 a) 19 **b)** 27
 c) 93 **d)** 101
 e) 107 **f)** 113

3. Find the prime factorization of each of these integers.
 a) 88 **b)** 126 **c)** 729
 d) 1001 **e)** 1111 **f)** 909,090

4. Find the prime factorization of each of these integers.
 a) 39 **b)** 81 **c)** 101
 d) 143 **e)** 289 **f)** 899

5. Find the prime factorization of 10!.

***6.** How many zeros are there at the end of 100!?

7. Express in pseudocode the trial division algorithm for determining whether an integer is prime.

8. Express in pseudocode the algorithm described in the text for finding the prime factorization of an integer.

9. Show that $a^m + 1$ is composite if a and m are integers greater than 1 and m is odd. [*Hint:* Show that $x + 1$ is a factor of the polynomial $x^m + 1$ if m is odd.]

10. Show that if $2^m + 1$ is an odd prime, then $m = 2^n$ for some nonnegative integer n. [*Hint:* First show that the polynomial identity $x^m + 1 = (x^k + 1)(x^{k(t-1)} - x^{k(t-2)} + \cdots - x^k + 1)$ holds, where $m = kt$ and t is odd.]

***11.** Show that $\log_2 3$ is an irrational number. Recall that an irrational number is a real number x that cannot be written as the ratio of two integers.

12. Prove that for every positive integer n, there are n consecutive composite integers. [*Hint:* Consider the n consecutive integers starting with $(n + 1)! + 2$.]

***13.** Prove or disprove that there are three consecutive odd positive integers that are primes, that is, odd primes of the form p, $p + 2$, and $p + 4$.

14. Which positive integers less than 12 are relatively prime to 12?

15. Which positive integers less than 30 are relatively prime to 30?

16. Determine whether the integers in each of these sets are pairwise relatively prime.

a) 21, 34, 55 **b)** 14, 17, 85
c) 25, 41, 49, 64 **d)** 17, 18, 19, 23

17. Determine whether the integers in each of these sets are pairwise relatively prime.

a) 11, 15, 19 **b)** 14, 15, 21
c) 12, 17, 31, 37 **d)** 7, 8, 9, 11

18. We call a positive integer **perfect** if it equals the sum of its positive divisors other than itself.

a) Show that 6 and 28 are perfect.

b) Show that $2^{p-1}(2^p - 1)$ is a perfect number when $2^p - 1$ is prime.

19. Show that if $2^n - 1$ is prime, then n is prime. [*Hint:* Use the identity $2^{ab} - 1 = (2^a - 1) \cdot (2^{a(b-1)} + 2^{a(b-2)} + \cdots + 2^a + 1)$.]

20. Determine whether each of these integers is prime, verifying some of Mersenne's claims.

a) $2^7 - 1$ **b)** $2^9 - 1$
c) $2^{11} - 1$ **d)** $2^{13} - 1$

The value of the **Euler ϕ-function** at the positive integer n is defined to be the number of positive integers less than or equal to n that are relatively prime to n. For instance, $\phi(6) = 2$ because of the positive integers less or equal to 6, only 1 and 5 are relatively prime to 6. [*Note:* ϕ is the Greek letter phi.]

21. Find these values of the Euler ϕ-function.

a) $\phi(4)$ **b)** $\phi(10)$ **c)** $\phi(13)$

22. Show that n is prime if and only if $\phi(n) = n - 1$.

23. What is the value of $\phi(p^k)$ when p is prime and k is a positive integer?

24. What are the greatest common divisors of these pairs of integers?

a) $2^2 \cdot 3^3 \cdot 5^5, 2^5 \cdot 3^3 \cdot 5^2$
b) $2 \cdot 3 \cdot 5 \cdot 7 \cdot 11 \cdot 13, 2^{11} \cdot 3^9 \cdot 11 \cdot 17^{14}$
c) $17, 17^{17}$ **d)** $2^2 \cdot 7, 5^3 \cdot 13$
e) 0, 5 **f)** $2 \cdot 3 \cdot 5 \cdot 7, 2 \cdot 3 \cdot 5 \cdot 7$

25. What are the greatest common divisors of these pairs of integers?

a) $3^7 \cdot 5^3 \cdot 7^3, 2^{11} \cdot 3^5 \cdot 5^9$
b) $11 \cdot 13 \cdot 17, 2^9 \cdot 3^7 \cdot 5^5 \cdot 7^3$
c) $23^{31}, 23^{17}$
d) $41 \cdot 43 \cdot 53, 41 \cdot 43 \cdot 53$
e) $3^{13} \cdot 5^{17}, 2^{12} \cdot 7^{21}$
f) 1111, 0

26. What is the least common multiple of each pair in Exercise 24?

27. What is the least common multiple of each pair in Exercise 25?

28. Find gcd(1000, 625) and lcm(1000, 625) and verify that $\gcd(1000, 625) \cdot \text{lcm}(1000, 625) = 1000 \cdot 625$.

29. Find gcd(92928, 123552) and lcm(92928, 123552), and verify that $\gcd(92928, 123552) \cdot \text{lcm}(92928, 123552) = 92928 \cdot 123552$. [*Hint:* First find the prime factorizations of 92928 and 123552.]

30. If the product of two integers is $2^7 3^8 5^2 7^{11}$ and their greatest common divisor is $2^3 3^4 5$, what is their least common multiple?

31. Show that if a and b are positive integers, then $ab = \gcd(a, b) \cdot \text{lcm}(a, b)$. [*Hint:* Use the prime factorizations of a and b and the formulae for $\gcd(a, b)$ and $\text{lcm}(a, b)$ in terms of these factorizations.]

32. Use the Euclidean algorithm to find

a) gcd(1, 5). **b)** gcd(100, 101).
c) gcd(123, 277). **d)** gcd(1529, 14039).
e) gcd(1529, 14038). **f)** gcd(11111, 111111).

33. Use the Euclidean algorithm to find

a) gcd(12, 18). **b)** gcd(111, 201).
c) gcd(1001, 1331). **d)** gcd(12345, 54321).
e) gcd(1000, 5040). **f)** gcd(9888, 6060).

34. How many divisions are required to find gcd(21, 34) using the Euclidean algorithm?

35. How many divisions are required to find gcd(34, 55) using the Euclidean algorithm?

***36.** Show that if a and b are both positive integers, then $(2^a - 1) \textbf{ mod } (2^b - 1) = 2^{a \textbf{ mod } b} - 1$.

***37.** Use Exercise 36 to show that if a and b are positive integers, then $\gcd(2^a - 1, 2^b - 1) = 2^{\gcd(a, b)} - 1$. [*Hint:* Show that the remainders obtained when the Euclidean algorithm is used to compute $\gcd(2^a - 1, 2^b - 1)$ are of the form $2^r - 1$, where r is a remainder arising when the Euclidean algorithm is used to find $\gcd(a, b)$.]

38. Use Exercise 37 to show that the integers $2^{35} - 1$, $2^{34} - 1$, $2^{33} - 1$, $2^{31} - 1$, $2^{29} - 1$, and $2^{23} - 1$ are pairwise relatively prime.

39. Using the method followed in Example 17, express the greatest common divisor of each of these pairs of integers as a linear combination of these integers.

a) 10, 11 **b)** 21, 44 **c)** 36, 48
d) 34, 55 **e)** 117, 213 **f)** 0, 223
g) 123, 2347 **h)** 3454, 4666 **i)** 9999, 11111

40. Using the method followed in Example 17, express the greatest common divisor of each of these pairs of integers as a linear combination of these integers.

a) 9, 11 **b)** 33, 44 **c)** 35, 78
d) 21, 55 **e)** 101, 203 **f)** 124, 323
g) 2002, 2339 **h)** 3457, 4669 **i)** 10001, 13422

41. Use the extended Euclidean algorithm to express gcd(26, 91) as a linear combination of 26 and 91.

42. Use the extended Euclidean algorithm to express gcd(252, 356) as a linear combination of 252 and 356.

43. Use the extended Euclidean algorithm to express gcd(144, 89) as a linear combination of 144 and 89.

44. Use the extended Euclidean algorithm to express gcd(1001, 100001) as a linear combination of 1001 and 100001.

45. Describe the extended Euclidean algorithm using pseudocode.

46. Find the smallest positive integer with exactly n different positive factors when n is

a) 3. **b)** 4. **c)** 5.
d) 6. **e)** 10.

47. Can you find a formula or rule for the nth term of a sequence related to the prime numbers or prime factorizations so that the initial terms of the sequence have these values?

a) 0, 1, 1, 0, 1, 0, 1, 0, 0, 0, 1, 0, 1, …
b) 1, 2, 3, 2, 5, 2, 7, 2, 3, 2, 11, 2, 13, 2, …
c) 1, 2, 2, 3, 2, 4, 2, 4, 3, 4, 2, 6, 2, 4, …
d) 1, 1, 1, 0, 1, 1, 1, 0, 0, 1, 1, 0, 1, 1, …
e) 1, 2, 3, 3, 5, 5, 7, 7, 7, 7, 11, 11, 13, 13, …
f) 1, 2, 6, 30, 210, 2310, 30030, 510510, 9699690, 223092870, …

48. Can you find a formula or rule for the nth term of a sequence related to the prime numbers or prime factorizations so that the initial terms of the sequence have these values?

a) 2, 2, 3, 5, 5, 7, 7, 11, 11, 11, 11, 13, 13, …
b) 0, 1, 2, 2, 3, 3, 4, 4, 4, 4, 5, 5, 6, 6, …
c) 1, 0, 0, 1, 0, 1, 0, 1, 1, 1, 0, 1, 0, 1, …
d) 1, −1, −1, 0, −1, 1, −1, 0, 0, 1, −1, 0, −1, 1, 1, …
e) 1, 1, 1, 1, 1, 0, 1, 1, 1, 0, 1, 0, 1, 0, 0, …
f) 4, 9, 25, 49, 121, 169, 289, 361, 529, 841, 961, 1369, …

49. Prove that the product of any three consecutive integers is divisible by 6.

50. Show that if a, b, and m are integers such that $m \geq 2$ and $a \equiv b \pmod{m}$, then $\gcd(a, m) = \gcd(b, m)$.

***51.** Prove or disprove that $n^2 - 79n + 1601$ is prime whenever n is a positive integer.

52. Prove or disprove that $p_1 p_2 \cdots p_n + 1$ is prime for every positive integer n, where $p_1, p_2, \ldots, p_n$ are the n smallest prime numbers.

53. Show that there is a composite integer in every arithmetic progression $ak + b$, $k = 1, 2, \ldots$, where a and b are positive integers.

54. Adapt the proof in the text that there are infinitely many primes to prove that there are infinitely many primes of the form $3k + 2$, where k is a nonnegative integer. [*Hint:* Suppose that there are only finitely many such primes $q_1, q_2, \ldots, q_n$, and consider the number $3q_1 q_2 \cdots q_n - 1$.]

55. Adapt the proof in the text that there are infinitely many primes to prove that there are infinitely many primes of the form $4k + 3$, where k is a nonnegative integer. [*Hint:* Suppose that there are only finitely many such primes $q_1, q_2, \ldots, q_n$, and consider the number $4q_1 q_2 \cdots q_n - 1$.]

***56.** Prove that the set of positive rational numbers is countable by setting up a function that assigns to a rational number p/q with $\gcd(p, q) = 1$ the base 11 number formed by the decimal representation of p followed by the base 11 digit A, which corresponds to the decimal number 10, followed by the decimal representation of q.

***57.** Prove that the set of positive rational numbers is countable by showing that the function K is a one-to-one correspondence between the set of positive rational numbers and the set of positive integers if $K(m/n) = p_1^{2a_1} p_2^{2a_2} \cdot \cdots \cdot p_s^{2a_s} q_1^{2b_1-1} q_2^{2b_2-1} \cdot \cdots \cdot q_t^{2b_t-1}$, where $\gcd(m, n) = 1$ and the prime-power factorizations of m and n are $m = p_1^{a_1} p_2^{a_2} \cdot \cdots \cdot p_s^{a_s}$ and $n = q_1^{b_1} q_2^{b_2} \cdots q_t^{b_t}$.

4.4 Solving Congruences

4.4.1 Introduction

Solving linear congruences, which have the form $ax \equiv b \pmod{m}$, is an essential task in the study of number theory and its applications, just as solving linear equations plays an important role in calculus and linear algebra. To solve linear congruences, we employ inverses modulo m. We explain how to work backwards through the steps of the Euclidean algorithm to find inverses modulo m. Once we have found an inverse of a modulo m, we solve the congruence $ax \equiv b \pmod{m}$ by multiplying both sides of the congruence by this inverse.

Simultaneous systems of linear congruence have been studied since ancient times. For example, the Chinese mathematician Sun-Tsu studied them in the first century. We will show how to solve systems of linear congruences modulo pairwise relatively prime moduli. The result we will prove is called the Chinese remainder theorem, and our proof will give a method to find all solutions of such systems of congruences. We will also show how to use the Chinese remainder theorem as a basis for performing arithmetic with large integers.

We will introduce a useful result of Fermat, known as Fermat's little theorem, which states that if p is prime and p does not divide a, then $a^{p-1} \equiv 1 \pmod{p}$. We will examine the converse of this statement, which will lead us to the concept of a pseudoprime. A pseudoprime m to the base a is a composite integer m that masquerades as a prime by satisfying the congruence $a^{m-1} \equiv 1 \pmod{m}$. We will also give an example of a Carmichael number, which is a composite integer that is a pseudoprime to all bases a relatively prime to it.

We also introduce the notion of discrete logarithms, which are analogous to ordinary logarithms. To define discrete logarithms we must first define primitive roots. A primitive root of a prime p is an integer r such that every integer not divisible by p is congruent to a power of r modulo p. If r is a primitive root of p and $r^e \equiv a \pmod{p}$, then e is the discrete logarithm of a modulo p to the base r. Finding discrete logarithms turns out to be an extremely difficult problem in general. The difficulty of this problem is the basis for the security of many cryptographic systems.

4.4.2 Linear Congruences

A congruence of the form

$$ax \equiv b \pmod{m},$$

where m is a positive integer, a and b are integers, and x is a variable, is called a **linear congruence**. Such congruences arise throughout number theory and its applications.

How can we solve the linear congruence $ax \equiv b \pmod{m}$, that is, how can we find all integers x that satisfy this congruence? One method that we will describe uses an integer $\overline{a}$ such that $\overline{a}a \equiv 1 \pmod{m}$, if such an integer exists. Such an integer $\overline{a}$ is said to be an **inverse** of a modulo m. Theorem 1 guarantees that an inverse of a modulo m exists whenever a and m are relatively prime.

THEOREM 1 If a and m are relatively prime integers and $m > 1$, then an inverse of a modulo m exists. Furthermore, this inverse is unique modulo m. (That is, there is a unique positive integer $\overline{a}$ less than m that is an inverse of a modulo m and every other inverse of a modulo m is congruent to $\overline{a}$ modulo m.)

Proof: By Theorem 6 of Section 4.3, because $\gcd(a, m) = 1$, there are integers s and t such that

$$sa + tm = 1.$$

This implies that

$$sa + tm \equiv 1 \pmod{m}.$$

Because $tm \equiv 0 \pmod{m}$, it follows that

$$sa \equiv 1 \pmod{m}.$$

Consequently, s is an inverse of a modulo m. That this inverse is unique modulo m is left as Exercise 7. ◁

Using inspection to find an inverse of a modulo m is easy when m is small. To find this inverse, we look for a multiple of a that exceeds a multiple of m by 1. For example, to find an inverse of 3 modulo 7, we can find $j \cdot 3$ for $j = 1, 2, \ldots, 6$, stopping when we find a multiple of 3 that is one more than a multiple of 7. We can speed this approach up if we note that $2 \cdot 3 \equiv -1 \pmod{7}$. This means that $(-2) \cdot 3 \equiv 1 \pmod{7}$. Hence, $5 \cdot 3 \equiv 1 \pmod{7}$, so 5 is an inverse of 3 modulo 7.

We can design a more efficient algorithm than brute force to find an inverse of a modulo m when $\gcd(a, m) = 1$ using the steps of the Euclidean algorithm. By reversing these steps as in Example 17 of Section 4.3, we can find a linear combination $sa + tm = 1$, where s and t are integers. Reducing both sides of this equation modulo m tells us that s is an inverse of a modulo m. We illustrate this procedure in Example 1.

EXAMPLE 1 Find an inverse of 3 modulo 7 by first finding Bézout coefficients of 3 and 7. (Note that we have already shown that 5 is an inverse of 3 modulo 7 by inspection.)

Solution: Because $\gcd(3, 7) = 1$, Theorem 1 tells us that an inverse of 3 modulo 7 exists. The Euclidean algorithm ends quickly when used to find the greatest common divisor of 3 and 7:

$$7 = 2 \cdot 3 + 1.$$

From this equation we see that

$$-2 \cdot 3 + 1 \cdot 7 = 1.$$

This shows that -2 and 1 are Bézout coefficients of 3 and 7. We see that -2 is an inverse of 3 modulo 7. Note that every integer congruent to -2 modulo 7 is also an inverse of 3, such as 5, -9, 12, and so on. ◀

EXAMPLE 2 Find an inverse of 101 modulo 4620.

Solution: For completeness, we present all steps used to compute an inverse of 101 modulo 4620. (Only the last step goes beyond methods developed in Section 4.3 and illustrated in Example 17 in that section.) First, we use the Euclidean algorithm to show that $\gcd(101, 4620) = 1$. Then we will reverse the steps to find Bézout coefficients a and b such that $101a + 4620b = 1$. It will then follow that a is an inverse of 101 modulo 4620. The steps used by the Euclidean algorithm to find $\gcd(101, 4620)$ are

$$\begin{aligned}
4620 &= 45 \cdot 101 + 75 \\
101 &= 1 \cdot 75 + 26 \\
75 &= 2 \cdot 26 + 23 \\
26 &= 1 \cdot 23 + 3 \\
23 &= 7 \cdot 3 + 2 \\
3 &= 1 \cdot 2 + 1 \\
2 &= 2 \cdot 1.
\end{aligned}$$

Because the last nonzero remainder is 1, we know that $\gcd(101, 4620) = 1$. We can now find the Bézout coefficients for 101 and 4620 by working backwards through these steps, expressing gcd(101, 4620) = 1 in terms of each successive pair of remainders. In each step we eliminate the remainder by expressing it as a linear combination of the divisor and the dividend. We obtain

$$\begin{aligned}
1 &= 3 - 1 \cdot 2 \\
&= 3 - 1 \cdot (23 - 7 \cdot 3) = -1 \cdot 23 + 8 \cdot 3 \\
&= -1 \cdot 23 + 8 \cdot (26 - 1 \cdot 23) = 8 \cdot 26 - 9 \cdot 23 \\
&= 8 \cdot 26 - 9 \cdot (75 - 2 \cdot 26) = -9 \cdot 75 + 26 \cdot 26 \\
&= -9 \cdot 75 + 26 \cdot (101 - 1 \cdot 75) = 26 \cdot 101 - 35 \cdot 75 \\
&= 26 \cdot 101 - 35 \cdot (4620 - 45 \cdot 101) = -35 \cdot 4620 + 1601 \cdot 101.
\end{aligned}$$

That $-35 \cdot 4620 + 1601 \cdot 101 = 1$ tells us that -35 and 1601 are Bézout coefficients of 4620 and 101, and 1601 is an inverse of 101 modulo 4620. ◀

Once we have an inverse $\overline{a}$ of a modulo m, we can solve the congruence $ax \equiv b \pmod{m}$ by multiplying both sides of the linear congruence by $\overline{a}$, as Example 3 illustrates.

EXAMPLE 3 What are the solutions of the linear congruence $3x \equiv 4 \pmod{7}$?

Solution: By Example 1 we know that -2 is an inverse of 3 modulo 7. Multiplying both sides of the congruence by -2 shows that

$$-2 \cdot 3x \equiv -2 \cdot 4 \pmod{7}.$$

Because $-6 \equiv 1 \pmod{7}$ and $-8 \equiv 6 \pmod{7}$, it follows that if x is a solution, then $x \equiv -8 \equiv 6 \pmod{7}$.

We need to determine whether every x with $x \equiv 6 \pmod{7}$ is a solution. Assume that $x \equiv 6 \pmod{7}$. Then, by Theorem 5 of Section 4.1, it follows that

$$3x \equiv 3 \cdot 6 = 18 \equiv 4 \pmod{7},$$

which shows that all such x satisfy the congruence. We conclude that the solutions to the congruence are the integers x such that $x \equiv 6 \pmod{7}$, namely, 6, 13, 20, … and -1, -8, -15, … . ◀

4.4.3 The Chinese Remainder Theorem

Systems of linear congruences arise in many contexts. For example, as we will see later, they are the basis for a method that can be used to perform arithmetic with large integers. Such systems can even be found as word puzzles in the writings of ancient Chinese and Hindu mathematicians, such as that given in Example 4.

Links

EXAMPLE 4 In the first century, the Chinese mathematician Sun-Tsu asked:

> There are certain things whose number is unknown. When divided by 3, the remainder is 2; when divided by 5, the remainder is 3; and when divided by 7, the remainder is 2. What will be the number of things?

This puzzle can be translated into the following question: What are the solutions of the system of congruences

$$x \equiv 2 \pmod{3},$$
$$x \equiv 3 \pmod{5},$$
$$x \equiv 2 \pmod{7}?$$

We will solve this system, and with it Sun-Tsu's puzzle, later in this section. ◀

The *Chinese remainder theorem*, named after the Chinese heritage of problems involving systems of linear congruences, states that when the moduli of a system of linear congruences are pairwise relatively prime, there is a unique solution of the system modulo the product of the moduli.

THEOREM 2 **THE CHINESE REMAINDER THEOREM** Let $m_1, m_2, \ldots, m_n$ be pairwise relatively prime positive integers greater than one and $a_1, a_2, \ldots, a_n$ arbitrary integers. Then the system

$$x \equiv a_1 \pmod{m_1},$$
$$x \equiv a_2 \pmod{m_2},$$
$$\cdot$$
$$\cdot$$
$$\cdot$$
$$x \equiv a_n \pmod{m_n}$$

has a unique solution modulo $m = m_1 m_2 \cdots m_n$. (That is, there is a solution x with $0 \le x < m$, and all other solutions are congruent modulo m to this solution.)

Proof: To establish this theorem, we need to show that a solution exists and that it is unique modulo m. We will show that a solution exists by describing a way to construct this solution; showing that the solution is unique modulo m is Exercise 30.

To construct a simultaneous solution, first let

$$M_k = m/m_k$$

for $k = 1, 2, \ldots, n$. That is, M_k is the product of the moduli except for m_k. Because m_i and m_k have no common factors greater than 1 when $i \neq k$, it follows that $\gcd(m_k, M_k) = 1$. Consequently, by Theorem 1, we know that there is an integer y_k, an inverse of M_k modulo m_k, such that

$$M_k y_k \equiv 1 \pmod{m_k}.$$

To construct a simultaneous solution, form the sum

$$x = a_1 M_1 y_1 + a_2 M_2 y_2 + \cdots + a_n M_n y_n.$$

We will now show that x is a simultaneous solution. First, note that because $M_j \equiv 0 \pmod{m_k}$ whenever $j \neq k$, all terms except the kth term in this sum are congruent to 0 modulo m_k. Because $M_k y_k \equiv 1 \pmod{m_k}$ we see that

$$x \equiv a_k M_k y_k \equiv a_k \pmod{m_k},$$

for $k = 1, 2, \ldots, n$. We have shown that x is a simultaneous solution to the n congruences. ◁

Example 5 illustrates how to use the construction given in our proof of the Chinese remainder theorem to solve a system of congruences. We will solve the system given in Example 4, which arises in Sun-Tsu's puzzle.

EXAMPLE 5 To solve the system of congruences in Example 4, first let $m = 3 \cdot 5 \cdot 7 = 105$, $M_1 = m/3 = 35$, $M_2 = m/5 = 21$, and $M_3 = m/7 = 15$. We see that 2 is an inverse of $M_1 = 35$ modulo 3, because $35 \cdot 2 \equiv 2 \cdot 2 \equiv 1 \pmod{3}$; 1 is an inverse of $M_2 = 21$ modulo 5, because $21 \equiv 1 \pmod{5}$; and 1 is an inverse of $M_3 = 15 \pmod{7}$, because $15 \equiv 1 \pmod{7}$. The solutions to this system are those x such that

$$\begin{aligned} x \equiv a_1M_1y_1 + a_2M_2y_2 + a_3M_3y_3 &= 2 \cdot 35 \cdot 2 + 3 \cdot 21 \cdot 1 + 2 \cdot 15 \cdot 1 \\ &= 233 \equiv 23 \pmod{105}. \end{aligned}$$

It follows that 23 is the smallest positive integer that is a simultaneous solution. We conclude that 23 is the smallest positive integer that leaves a remainder of 2 when divided by 3, a remainder of 3 when divided by 5, and a remainder of 2 when divided by 7. ◀

Although the construction in Theorem 2 provides a general method for solving systems of linear congruences with pairwise relatively prime moduli, it can be easier to solve a system using a different method. Example 6 illustrates the use of a method known as **back substitution**.

EXAMPLE 6 Use the method of back substitution to find all integers x such that $x \equiv 1 \pmod{5}$, $x \equiv 2 \pmod{6}$, and $x \equiv 3 \pmod{7}$.

Solution: By Theorem 4 in Section 4.1, the first congruence can be rewritten as an equality, $x = 5t + 1$, where t is an integer. Substituting this expression for x into the second congruence tells us that

$$5t + 1 \equiv 2 \pmod{6},$$

which can be solved to show that $t \equiv 5 \pmod{6}$ (as the reader should verify). Using Theorem 4 in Section 4.1 again, we see that $t = 6u + 5$, where u is an integer. Substituting this expression for t back into the equation $x = 5t + 1$ tells us that $x = 5(6u + 5) + 1 = 30u + 26$. We insert this into the third equation to obtain

$$30u + 26 \equiv 3 \pmod{7}.$$

Solving this congruence tells us that $u \equiv 6 \pmod{7}$ (as the reader should verify). Hence, Theorem 4 in Section 4.1 tells us that $u = 7v + 6$, where v is an integer. Substituting this expression for u into the equation $x = 30u + 26$ tells us that $x = 30(7v + 6) + 26 = 210u + 206$. Translating this back into a congruence, we find the solution to the simultaneous congruences,

$$x \equiv 206 \pmod{210}.$$

◀

4.4.4 Computer Arithmetic with Large Integers

Suppose that $m_1, m_2, \dots, m_n$ are pairwise relatively prime moduli and let m be their product. By the Chinese remainder theorem, we can show (see Exercise 28) that an integer a with $0 \le a < m$ can be uniquely represented by the n-tuple consisting of its remainders upon division by m_i, $i = 1, 2, \dots, n$. That is, we can uniquely represent a by

$$(a \textbf{ mod } m_1, a \textbf{ mod } m_2, \dots, a \textbf{ mod } m_n).$$

EXAMPLE 7 What are the pairs used to represent the nonnegative integers less than 12 when they are represented by the ordered pair where the first component is the remainder of the integer upon division by 3 and the second component is the remainder of the integer upon division by 4?

Solution: We have the following representations, obtained by finding the remainder of each integer when it is divided by 3 and by 4:

$$\begin{array}{lll} 0 = (0, 0) & 4 = (1, 0) & 8 = (2, 0) \\ 1 = (1, 1) & 5 = (2, 1) & 9 = (0, 1) \\ 2 = (2, 2) & 6 = (0, 2) & 10 = (1, 2) \\ 3 = (0, 3) & 7 = (1, 3) & 11 = (2, 3). \end{array}$$

◀

To perform arithmetic with large integers, we select moduli $m_1, m_2, \ldots, m_n$, where each m_i is an integer greater than 2, $\gcd(m_i, m_j) = 1$ whenever $i \neq j$, and $m = m_1 m_2 \cdots m_n$ is greater than the results of the arithmetic operations we want to carry out.

Once we have selected our moduli, we carry out arithmetic operations with large integers by performing componentwise operations on the n-tuples representing these integers using their remainders upon division by m_i, $i = 1, 2, \ldots, n$. Once we have computed the value of each component in the result, we recover its value by solving a system of n congruences modulo m_i, $i = 1, 2, \ldots, n$. This method of performing arithmetic with large integers has several valuable features. First, it can be used to perform arithmetic with integers larger than can ordinarily be carried out on a computer. Second, computations with respect to the different moduli can be done in parallel, speeding up the arithmetic.

EXAMPLE 8 Suppose that performing arithmetic with integers less than 100 on a certain processor is much quicker than doing arithmetic with larger integers. We can restrict almost all our computations to integers less than 100 if we represent integers using their remainders modulo pairwise relatively prime integers less than 100. For example, we can use the moduli of 99, 98, 97, and 95. (These integers are pairwise relatively prime, because no two have a common factor greater than 1.)

By the Chinese remainder theorem, every nonnegative integer less than $99 \cdot 98 \cdot 97 \cdot 95 = 89{,}403{,}930$ can be represented uniquely by its remainders when divided by these four moduli. For example, we represent 123,684 as (33, 8, 9, 89), because $123{,}684 \textbf{ mod } 99 = 33$; $123{,}684 \textbf{ mod } 98 = 8$; $123{,}684 \textbf{ mod } 97 = 9$; and $123{,}684 \textbf{ mod } 95 = 89$. Similarly, we represent 413,456 as (32, 92, 42, 16).

To find the sum of 123,684 and 413,456, we work with these 4-tuples instead of these two integers directly. We add the 4-tuples componentwise and reduce each component with respect to the appropriate modulus. This yields

$$\begin{aligned} &(33, 8, 9, 89) + (32, 92, 42, 16) \\ &\qquad = (65 \textbf{ mod } 99, 100 \textbf{ mod } 98, 51 \textbf{ mod } 97, 105 \textbf{ mod } 95) \\ &\qquad = (65, 2, 51, 10). \end{aligned}$$

To find the sum, that is, the integer represented by (65, 2, 51, 10), we need to solve the system of congruences

$$\begin{aligned} x &\equiv 65 \pmod{99}, \\ x &\equiv 2 \pmod{98}, \\ x &\equiv 51 \pmod{97}, \\ x &\equiv 10 \pmod{95}. \end{aligned}$$

It can be shown (see Exercise 53) that 537,140 is the unique nonnegative solution of this system less than 89,403,930. Consequently, 537,140 is the sum. Note that it is only when we have to recover the integer represented by (65, 2, 51, 10) that we have to do arithmetic with integers larger than 100. ◀

Particularly good choices for moduli for arithmetic with large integers are sets of integers of the form $2^k - 1$, where k is a positive integer, because it is easy to do binary arithmetic modulo such integers, and because it is easy to find sets of such integers that are pairwise relatively prime. [The second reason is a consequence of the fact that $\gcd(2^a - 1, 2^b - 1) = 2^{\gcd(a, b)} - 1$, as Exercise 37 in Section 4.3 shows.] Suppose, for instance, that we can do arithmetic with integers less than 2^{35} easily on our computer, but that working with larger integers requires special procedures. We can use pairwise relatively prime moduli less than 2^{35} to perform arithmetic with integers as large as their product. For example, as Exercise 38 in Section 4.3 shows, the integers $2^{35} - 1$, $2^{34} - 1$, $2^{33} - 1$, $2^{31} - 1$, $2^{29} - 1$, and $2^{23} - 1$ are pairwise relatively prime. Because the product of these six moduli exceeds 2^{184}, we can perform arithmetic with integers as large as 2^{184} (as long as the results do not exceed this number) by doing arithmetic modulo each of these six moduli, none of which exceeds 2^{35}.

4.4.5 Fermat's Little Theorem

The French mathematician Pierre de Fermat, one of the leading mathematicians of the first half of the 17th century, made many important discoveries in number theory. One of the most useful of these states that p divides $a^{p-1} - 1$ whenever p is prime and a is an integer not divisible by p. Fermat announced this result in a letter to one of his correspondents. However, he did not include a proof in the letter, stating that he feared the proof would be too long. Although Fermat never published a proof of this fact, there is little doubt that he knew how to prove it, unlike the result known as Fermat's last theorem. The first published proof is credited to Leonhard Euler. We now state this theorem in terms of congruences.

THEOREM 3 **FERMAT'S LITTLE THEOREM** If p is prime and a is an integer not divisible by p, then

$$a^{p-1} \equiv 1 \pmod{p}.$$

Furthermore, for every integer a we have

$$a^p \equiv a \pmod{p}.$$

Remark: Fermat's little theorem tells us that if $a \in \mathbf{Z}_p$, then $a^{p-1} = 1$ in $\mathbf{Z}_p$.

The proof of Theorem 3 is outlined in Exercise 19.

Fermat's little theorem is extremely useful in computing the remainders modulo p of large powers of integers, as Example 9 illustrates.

EXAMPLE 9 Find 7^{222} **mod** 11.

Solution: We can use Fermat's little theorem to evaluate 7^{222} **mod** 11 rather than using the fast modular exponentiation algorithm. By Fermat's little theorem we know that $7^{10} \equiv 1 \pmod{11}$, so $(7^{10})^k \equiv 1 \pmod{11}$ for every positive integer k. To take advantage of this last congruence, we divide the exponent 222 by 10, finding that $222 = 22 \cdot 10 + 2$. We now see that

$$7^{222} = 7^{22 \cdot 10 + 2} = (7^{10})^{22} 7^2 \equiv (1)^{22} \cdot 49 \equiv 5 \pmod{11}.$$

It follows that 7^{222} **mod** $11 = 5$. ◀

Example 9 illustrated how we can use Fermat's little theorem to compute a^n **mod** p, where p is prime and $p \nmid a$. First, we use the division algorithm to find the quotient q and remainder r when n is divided by $p - 1$, so that $n = q(p-1) + r$, where $0 \leq r < p - 1$. It follows that $a^n = a^{q(p-1)+r} = (a^{p-1})^q a^r \equiv 1^q a^r \equiv a^r \pmod{p}$. Hence, to find a^n **mod** p, we only need to compute a^r **mod** p. We will take advantage of this simplification many times in our study of number theory.

4.4.6 Pseudoprimes

In Section 4.2 we showed that an integer n is prime when it is not divisible by any prime p with $p \leq \sqrt{n}$. Unfortunately, using this criterion to show that a given integer is prime is inefficient. It requires that we find all primes not exceeding $\sqrt{n}$ and that we carry out trial division by each such prime to see whether it divides n.

Are there more efficient ways to determine whether an integer is prime? According to some sources, ancient Chinese mathematicians believed that n was an odd prime if and only if

$$2^{n-1} \equiv 1 \pmod{n}.$$

If this were true, it would provide an efficient primality test. Why did they believe this congruence could be used to determine whether an integer $n > 2$ is prime? First, they observed that the congruence holds whenever n is an odd prime. For example, 5 is prime and

$$2^{5-1} = 2^4 = 16 \equiv 1 \pmod{5}.$$

By Fermat's little theorem, we know that this observation was correct, that is, $2^{n-1} \equiv 1 \pmod{n}$ whenever n is an odd prime. Second, they never found a composite integer n for which the congruence holds. However, the ancient Chinese were only partially correct. They were correct in thinking that the congruence holds whenever n is prime, but they were incorrect in concluding that n is necessarily prime if the congruence holds.

Unfortunately, there are composite integers n such that $2^{n-1} \equiv 1 \pmod{n}$. Such integers are called **pseudoprimes** to the base 2.

EXAMPLE 10 The integer 341 is a pseudoprime to the base 2 because it is composite ($341 = 11 \cdot 31$) and as Exercise 37 shows

$$2^{340} \equiv 1 \pmod{341}.$$

◀

We can use an integer other than 2 as the base when we study pseudoprimes.

Links

PHOTOS.com/Getty Images

PIERRE DE FERMAT (1601–1665) Pierre de Fermat, one of the most important mathematicians of the seventeenth century, was a lawyer by profession. He is the most famous amateur mathematician in history. Fermat published little of his mathematical discoveries. It is through his correspondence with other mathematicians that we know of his work. Fermat was one of the inventors of analytic geometry and developed some of the fundamental ideas of calculus. Fermat, along with Pascal, gave probability theory a mathematical basis. Fermat formulated what was the most famous unsolved problem in mathematics. He asserted that the equation $x^n + y^n = z^n$ has no nontrivial positive integer solutions when n is an integer greater than 2. For more than 300 years, no proof (or counterexample) was found. In his copy of the works of the ancient Greek mathematician Diophantus, Fermat wrote that he had a proof but that it would not fit in the margin. Because the first proof, found by Andrew Wiles in 1994, relies on sophisticated, modern mathematics, most people think that Fermat thought he had a proof, but that the proof was incorrect. However, he may have been tempting others to look for a proof, not being able to find one himself.

Definition 1 Let b be a positive integer. If n is a composite positive integer, and $b^{n-1} \equiv 1 \pmod{n}$, then n is called a *pseudoprime to the base b.*

Given a positive integer n, determining whether $2^{n-1} \equiv 1 \pmod{n}$ is a useful test that provides some evidence concerning whether n is prime. In particular, if n satisfies this congruence, then it is either prime or a pseudoprime to the base 2; if n does not satisfy this congruence, it is composite. We can perform similar tests using bases b other than 2 and obtain more evidence as to whether n is prime. If n passes all such tests, it is either prime or a pseudoprime to all the bases b we have chosen. Furthermore, among the positive integers not exceeding x, where x is a positive real number, compared to primes there are relatively few pseudoprimes to the base b, where b is a positive integer. For example, among the positive integers less than 10^{10} there are 455,052,512 primes, but only 14,884 pseudoprimes to the base 2. Unfortunately, we cannot distinguish between primes and pseudoprimes just by choosing sufficiently many bases, because there are composite integers n that pass all tests with bases b such that $\gcd(b, n) = 1$. This leads to Definition 2.

Definition 2 A composite integer n that satisfies the congruence $b^{n-1} \equiv 1 \pmod{n}$ for all positive integers b with $\gcd(b, n) = 1$ is called a *Carmichael number*. (These numbers are named after Robert Carmichael, who studied them in the early twentieth century.)

EXAMPLE 11 The integer 561 is a Carmichael number. To see this, first note that 561 is composite because $561 = 3 \cdot 11 \cdot 17$. Next, note that if $\gcd(b, 561) = 1$, then $\gcd(b, 3) = \gcd(b, 11) = \gcd(b, 17) = 1$.

Using Fermat's little theorem we find that

$$b^2 \equiv 1 \pmod{3},\ b^{10} \equiv 1 \pmod{11},\ \text{and } b^{16} \equiv 1 \pmod{17}.$$

It follows that

$$b^{560} = (b^2)^{280} \equiv 1 \pmod{3},$$
$$b^{560} = (b^{10})^{56} \equiv 1 \pmod{11},$$
$$b^{560} = (b^{16})^{35} \equiv 1 \pmod{17}.$$

By Exercise 29, it follows that $b^{560} \equiv 1 \pmod{561}$ for all positive integers b with $\gcd(b, 561) = 1$. Hence, 561 is a Carmichael number. ◀

Although there are infinitely many Carmichael numbers, more delicate tests, described in the exercise set, can be devised that can be used as the basis for efficient probabilistic primality tests. Such tests can be used to quickly show that it is almost certainly the case that a given

Links

The Mathematical Association of America

ROBERT DANIEL CARMICHAEL (1879–1967) Robert Daniel Carmichael was born in Alabama. He received his undergraduate degree from Lineville College in 1898 and his Ph.D. in 1911 from Princeton. Carmichael held positions at Indiana University from 1911 until 1915 and at the University of Illinois from 1915 until 1947. Carmichael was an active researcher in a wide variety of areas, including number theory, real analysis, differential equations, mathematical physics, and group theory. His Ph.D. thesis, written under the direction of G. D. Birkhoff, is considered the first significant American contribution to the subject of differential equations.

integer is prime. More precisely, if an integer is not prime, then the probability that it passes a series of tests is close to 0. We will describe such a test in Chapter 7 and discuss the notions from probability theory that this test relies on. These probabilistic primality tests can be used, and are used, to find large primes extremely rapidly on computers.

4.4.7 Primitive Roots and Discrete Logarithms

In the set of positive real numbers, if $b > 1$, and $x = b^y$, we say that y is the logarithm of x to the base b. Here, we will show that we can also define the concept of logarithms modulo p of positive integers, where p is a prime. Before we do so, we need a definition.

Definition 3 A *primitive root* modulo a prime p is an integer r in $\mathbf{Z}_p$ such that every nonzero element of $\mathbf{Z}_p$ is a power of r.

EXAMPLE 12 Determine whether 2 and 3 are primitive roots modulo 11.

Solution: When we compute the powers of 2 in $\mathbf{Z}_{11}$, we obtain $2^1 = 2$, $2^2 = 4$, $2^3 = 8$, $2^4 = 5$, $2^5 = 10$, $2^6 = 9$, $2^7 = 7$, $2^8 = 3$, $2^9 = 6$, $2^{10} = 1$. Because every nonzero element of $\mathbf{Z}_{11}$ is a power of 2, 2 is a primitive root of 11.

When we compute the powers of 3 modulo 11, we obtain $3^1 = 3$, $3^2 = 9$, $3^3 = 5$, $3^4 = 4$, $3^5 = 1$. This pattern repeats when we compute higher powers of 3. Because not all nonzero elements of $\mathbf{Z}_{11}$ are powers of 3, we conclude that 3 is not a primitive root of 11. ◀

An important fact in number theory is that there is a primitive root modulo p for every prime p. We refer the reader to [Ro23] for a proof of this fact. Suppose that p is prime and r is a primitive root modulo p. If a is an integer between 1 and $p - 1$, that is, a nonzero element of $\mathbf{Z}_p$, we know that there is an unique exponent e such that $r^e = a$ in $\mathbf{Z}_p$, that is, $r^e \textbf{ mod } p = a$.

Definition 4 Suppose that p is a prime, r is a primitive root modulo p, and a is an integer between 1 and $p - 1$ inclusive. If $r^e \textbf{ mod } p = a$ and $0 \le e \le p - 1$, we say that e is the *discrete logarithm* of a modulo p to the base r and we write $\log_r a = e$ (where the prime p is understood).

EXAMPLE 13 Find the discrete logarithms of 3 and 5 modulo 11 to the base 2.

Solution: When we computed the powers of 2 modulo 11 in Example 12, we found that $2^8 = 3$ and $2^4 = 5$ in $\mathbf{Z}_{11}$. Hence, the discrete logarithms of 3 and 5 modulo 11 to the base 2 are 8 and 4, respectively. (These are the powers of 2 that equal 3 and 5, respectively, in $\mathbf{Z}_{11}$.) We write $\log_2 3 = 8$ and $\log_2 5 = 4$ (where the modulus 11 is understood and not explicitly noted in the notation). ◀

The discrete logarithm problem is hard!

The **discrete logarithm problem** takes as input a prime p, a primitive root r modulo p, and a positive integer $a \in \mathbf{Z}_p$; its output is the discrete logarithm of a modulo p to the base r. Although this problem might seem not to be that difficult, it turns out that no polynomial time algorithm is known for solving it. The difficulty of this problem plays an important role in cryptography, as we will see in Section 4.6.

Exercises

1. Show that 15 is an inverse of 7 modulo 26.
2. Show that 937 is an inverse of 13 modulo 2436.
3. By inspection (as discussed prior to Example 1), find an inverse of 4 modulo 9.
4. By inspection (as discussed prior to Example 1), find an inverse of 2 modulo 17.
5. Find an inverse of a modulo m for each of these pairs of relatively prime integers using the method followed in Example 2.
   a) $a = 4, m = 9$
   b) $a = 19, m = 141$
   c) $a = 55, m = 89$
   d) $a = 89, m = 232$
6. Find an inverse of a modulo m for each of these pairs of relatively prime integers using the method followed in Example 2.
   a) $a = 2, m = 17$
   b) $a = 34, m = 89$
   c) $a = 144, m = 233$
   d) $a = 200, m = 1001$

*7. Show that if a and m are relatively prime positive integers, then the inverse of a modulo m is unique modulo m. [*Hint:* Assume that there are two solutions b and c of the congruence $ax \equiv 1 \pmod{m}$. Use Theorem 7 of Section 4.3 to show that $b \equiv c \pmod{m}$.]

8. Show that an inverse of a modulo m, where a is an integer and $m > 2$ is a positive integer, does not exist if $\gcd(a, m) > 1$.
9. Solve the congruence $4x \equiv 5 \pmod{9}$ using the inverse of 4 modulo 9 found in part (a) of Exercise 5.
10. Solve the congruence $2x \equiv 7 \pmod{17}$ using the inverse of 2 modulo 17 found in part (a) of Exercise 6.
11. Solve each of these congruences using the modular inverses found in parts (b), (c), and (d) of Exercise 5.
    a) $19x \equiv 4 \pmod{141}$
    b) $55x \equiv 34 \pmod{89}$
    c) $89x \equiv 2 \pmod{232}$
12. Solve each of these congruences using the modular inverses found in parts (b), (c), and (d) of Exercise 6.
    a) $34x \equiv 77 \pmod{89}$
    b) $144x \equiv 4 \pmod{233}$
    c) $200x \equiv 13 \pmod{1001}$
13. Find the solutions of the congruence $15x^2 + 19x \equiv 5 \pmod{11}$. [*Hint:* Show the congruence is equivalent to the congruence $15x^2 + 19x + 6 \equiv 0 \pmod{11}$. Factor the left-hand side of the congruence; show that a solution of the quadratic congruence is a solution of one of the two different linear congruences.]
14. Find the solutions of the congruence $12x^2 + 25x \equiv 10 \pmod{11}$. [*Hint:* Show the congruence is equivalence to the congruence $12x^2 + 25x + 12 \equiv 0 \pmod{11}$. Factor the left-hand side of the congruence; show that a solution of the quadratic congruence is a solution of one of two different linear congruences.]

*15. Show that if m is an integer greater than 1 and $ac \equiv bc \pmod{m}$, then $a \equiv b \pmod{m/\gcd(c, m)}$.

16. **a)** Show that the positive integers less than 11, except 1 and 10, can be split into pairs of integers such that each pair consists of integers that are inverses of each other modulo 11.
    b) Use part (a) to show that $10! \equiv -1 \pmod{11}$.
17. Show that if p is prime, the only solutions of $x^2 \equiv 1 \pmod{p}$ are integers x such that $x \equiv 1 \pmod{p}$ or $x \equiv -1 \pmod{p}$.

*18. **a)** Generalize the result in part (a) of Exercise 16; that is, show that if p is a prime, the positive integers less than p, except 1 and $p - 1$, can be split into $(p - 3)/2$ pairs of integers such that each pair consists of integers that are inverses of each other. [*Hint:* Use the result of Exercise 17.]
    b) From part (a) conclude that $(p - 1)! \equiv -1 \pmod{p}$ whenever p is prime. This result is known as **Wilson's theorem**.
    c) What can we conclude if n is a positive integer such that $(n - 1)! \not\equiv -1 \pmod{n}$?

*19. This exercise outlines a proof of Fermat's little theorem.
    a) Suppose that a is not divisible by the prime p. Show that no two of the integers $1 \cdot a, 2 \cdot a, \ldots, (p - 1)a$ are congruent modulo p.
    b) Conclude from part (a) that the product of $1, 2, \ldots, p - 1$ is congruent modulo p to the product of $a, 2a, \ldots, (p - 1)a$. Use this to show that

$$(p - 1)! \equiv a^{p-1}(p - 1)! \pmod{p}.$$

    c) Use Theorem 7 of Section 4.3 to show from part (b) that $a^{p-1} \equiv 1 \pmod{p}$ if $p \not| a$. [*Hint:* Use Lemma 3 of Section 4.3 to show that p does not divide $(p - 1)!$ and then use Theorem 7 of Section 4.3. Alternatively, use Wilson's theorem from Exercise 18(b).]
    d) Use part (c) to show that $a^p \equiv a \pmod{p}$ for all integers a.

20. Use the construction in the proof of the Chinese remainder theorem to find all solutions to the system of congruences $x \equiv 2 \pmod{3}$, $x \equiv 1 \pmod{4}$, and $x \equiv 3 \pmod{5}$.
21. Use the construction in the proof of the Chinese remainder theorem to find all solutions to the system of congruences $x \equiv 1 \pmod{2}$, $x \equiv 2 \pmod{3}$, $x \equiv 3 \pmod{5}$, and $x \equiv 4 \pmod{11}$.
22. Solve the system of congruence $x \equiv 3 \pmod{6}$ and $x \equiv 4 \pmod{7}$ using the method of back substitution.
23. Solve the system of congruences in Exercise 20 using the method of back substitution.
24. Solve the system of congruences in Exercise 21 using the method of back substitution.

25. Write out in pseudocode an algorithm for solving a simultaneous system of linear congruences based on the construction in the proof of the Chinese remainder theorem.

***26.** Find all solutions, if any, to the system of congruences $x \equiv 5 \pmod 6$, $x \equiv 3 \pmod{10}$, and $x \equiv 8 \pmod{15}$.

***27.** Find all solutions, if any, to the system of congruences $x \equiv 7 \pmod 9$, $x \equiv 4 \pmod{12}$, and $x \equiv 16 \pmod{21}$.

28. Use the Chinese remainder theorem to show that an integer a, with $0 \le a < m = m_1 m_2 \cdots m_n$, where the positive integers $m_1, m_2, \ldots, m_n$ are pairwise relatively prime, can be represented uniquely by the n-tuple (a **mod** m_1, a **mod** $m_2, \ldots, a$ **mod** m_n).

***29.** Let $m_1, m_2, \ldots, m_n$ be pairwise relatively prime integers greater than or equal to 2. Show that if $a \equiv b \pmod{m_i}$ for $i = 1, 2, \ldots, n$, then $a \equiv b \pmod m$, where $m = m_1 m_2 \cdots m_n$. (This result will be used in Exercise 30 to prove the Chinese remainder theorem. Consequently, do not use the Chinese remainder theorem to prove it.)

***30.** Complete the proof of the Chinese remainder theorem by showing that the simultaneous solution of a system of linear congruences modulo pairwise relatively prime moduli is unique modulo the product of these moduli. [*Hint:* Assume that x and y are two simultaneous solutions. Show that $m_i \mid x - y$ for all i. Using Exercise 29, conclude that $m = m_1 m_2 \cdots m_n \mid x - y$.]

31. Which integers leave a remainder of 1 when divided by 2 and also leave a remainder of 1 when divided by 3?

32. Which integers are divisible by 5 but leave a remainder of 1 when divided by 3?

33. Use Fermat's little theorem to find 7^{121} **mod** 13.

34. Use Fermat's little theorem to find 23^{1002} **mod** 41.

35. Use Fermat's little theorem to show that if p is prime and $p \nmid a$, then a^{p-2} is an inverse of a modulo p.

36. Use Exercise 35 to find an inverse of 5 modulo 41.

37. **a)** Show that $2^{340} \equiv 1 \pmod{11}$ by Fermat's little theorem and noting that $2^{340} = (2^{10})^{34}$.
b) Show that $2^{340} \equiv 1 \pmod{31}$ using the fact that $2^{340} = (2^5)^{68} = 32^{68}$.
c) Conclude from parts (a) and (b) that $2^{340} \equiv 1 \pmod{341}$.

38. **a)** Use Fermat's little theorem to compute 3^{302} **mod** 5, 3^{302} **mod** 7, and 3^{302} **mod** 11.
b) Use your results from part (a) and the Chinese remainder theorem to find 3^{302} **mod** 385. (Note that $385 = 5 \cdot 7 \cdot 11$.)

39. **a)** Use Fermat's little theorem to compute 5^{2003} **mod** 7, 5^{2003} **mod** 11, and 5^{2003} **mod** 13.
b) Use your results from part (a) and the Chinese remainder theorem to find 5^{2003} **mod** 1001. (Note that $1001 = 7 \cdot 11 \cdot 13$.)

40. Show with the help of Fermat's little theorem that if n is a positive integer, then 42 divides $n^7 - n$.

41. Show that if p is an odd prime, then every divisor of the Mersenne number $2^p - 1$ is of the form $2kp + 1$, where k is a nonnegative integer. [*Hint:* Use Fermat's little theorem and Exercise 37 of Section 4.3.]

42. Use Exercise 41 to determine whether $M_{13} = 2^{13} - 1 = 8191$ and $M_{23} = 2^{23} - 1 = 8{,}388{,}607$ are prime.

43. Use Exercise 41 to determine whether $M_{11} = 2^{11} - 1 = 2047$ and $M_{17} = 2^{17} - 1 = 131{,}071$ are prime.

☞ Let n be a positive integer and let $n - 1 = 2^s t$, where s is a nonnegative integer and t is an odd positive integer. We say that n passes **Miller's test for the base b** if either $b^t \equiv 1 \pmod n$ or $b^{2^j t} \equiv -1 \pmod n$ for some j with $0 \le j \le s - 1$. It can be shown (see [Ro23]) that a composite integer n passes Miller's test for fewer than $n/4$ bases b with $1 < b < n$. A composite positive integer n that passes Miller's test to the base b is called a **strong pseudoprime to the base b**.

***44.** Show that if n is prime and b is a positive integer with $n \nmid b$, then n passes Miller's test to the base b.

45. Show that 2047 is a strong pseudoprime to the base 2 by showing that it passes Miller's test to the base 2, but is composite.

46. Show that 1729 is a Carmichael number.

47. Show that 2821 is a Carmichael number.

***48.** Show that if $n = p_1 p_2 \cdots p_k$, where $p_1, p_2, \ldots, p_k$ are distinct primes that satisfy $p_j - 1 \mid n - 1$ for $j = 1, 2, \ldots, k$, then n is a Carmichael number.

49. **a)** Use Exercise 48 to show that every integer of the form $(6m + 1)(12m + 1)(18m + 1)$, where m is a positive integer and $6m + 1$, $12m + 1$, and $18m + 1$ are all primes, is a Carmichael number.
b) Use part (a) to show that 172,947,529 is a Carmichael number.

50. Find the nonnegative integer a less than 28 represented by each of these pairs, where each pair represents (a **mod** 4, a **mod** 7).

a) (0, 0) **b)** (1, 0) **c)** (1, 1)
d) (2, 1) **e)** (2, 2) **f)** (0, 3)
g) (2, 0) **h)** (3, 5) **i)** (3, 6)

51. Express each nonnegative integer a less than 15 as a pair (a **mod** 3, a **mod** 5).

52. Explain how to use the pairs found in Exercise 51 to add 4 and 7.

53. Solve the system of congruences that arises in Example 8.

54. Show that 2 is a primitive root of 19.

55. Find the discrete logarithms of 5 and 6 to the base 2 modulo 19.

56. Let p be an odd prime and r a primitive root of p. Show that if a and b are positive integers in $\mathbf{Z}_p$, then $\log_r(ab) \equiv \log_r a + \log_r b \pmod{p - 1}$.

57. Write out a table of discrete logarithms modulo 17 with respect to the primitive root 3.

If m is a positive integer, the integer a is a **quadratic residue** of m if $\gcd(a, m) = 1$ and the congruence $x^2 \equiv a \pmod m$ has a solution. In other words, a quadratic residue of m is an integer relatively prime to m that is a perfect square modulo m. If a is not a quadratic residue of m and $\gcd(a, m) = 1$, we say that it is a **quadratic nonresidue** of m. For example, 2 is a quadratic residue of 7 because $\gcd(2, 7) = 1$ and $3^2 \equiv 2 \pmod 7$ and 3 is a quadratic nonresidue of 7 because $\gcd(3, 7) = 1$ and $x^2 \equiv 3 \pmod 7$ has no solution.

58. Which integers are quadratic residues of 11?

59. Show that if p is an odd prime and a is an integer not divisible by p, then the congruence $x^2 \equiv a \pmod{p}$ has either no solutions or exactly two incongruent solutions modulo p.

60. Show that if p is an odd prime, then there are exactly $(p-1)/2$ quadratic residues of p among the integers $1, 2, \ldots, p-1$.

If p is an odd prime and a is an integer not divisible by p, the **Legendre symbol** $\left(\frac{a}{p}\right)$ is defined to be 1 if a is a quadratic residue of p and -1 otherwise.

61. Show that if p is an odd prime and a and b are integers with $a \equiv b \pmod{p}$, then

$$\left(\frac{a}{p}\right) = \left(\frac{b}{p}\right).$$

62. Prove **Euler's criterion**, which states that if p is an odd prime and a is a positive integer not divisible by p, then

$$\left(\frac{a}{p}\right) \equiv a^{(p-1)/2} \pmod{p}.$$

[*Hint:* If a is a quadratic residue modulo p, apply Fermat's little theorem; otherwise, apply Wilson's theorem, given in Exercise 18(b).]

63. Use Exercise 62 to show that if p is an odd prime and a and b are integers not divisible by p, then

$$\left(\frac{ab}{p}\right) = \left(\frac{a}{p}\right)\left(\frac{b}{p}\right).$$

64. Show that if p is an odd prime, then -1 is a quadratic residue of p if $p \equiv 1 \pmod{4}$, and -1 is not a quadratic residue of p if $p \equiv 3 \pmod{4}$. [*Hint:* Use Exercise 62.]

65. Find all solutions of the congruence $x^2 \equiv 29 \pmod{35}$. [*Hint:* Find the solutions of this congruence modulo 5 and modulo 7, and then use the Chinese remainder theorem.]

66. Find all solutions of the congruence $x^2 \equiv 16 \pmod{105}$. [*Hint:* Find the solutions of this congruence modulo 3, modulo 5, and modulo 7, and then use the Chinese remainder theorem.]

67. Describe a brute-force algorithm for solving the discrete logarithm problem and find the worst-case and average-case time complexity of this algorithm.

4.5 Applications of Congruences

Congruences have many applications to discrete mathematics, computer science, and many other disciplines. We will introduce three applications in this section: the use of congruences to assign memory locations to computer files, the generation of pseudorandom numbers, and check digits.

Suppose that a customer identification number is ten digits long. To retrieve customer files quickly, we do not want to assign a memory location to a customer record using the ten-digit identification number. Instead, we want to use a smaller integer associated to the identification number. This can be done using what is known as a hashing function. In this section we will show how we can use modular arithmetic to do hashing.

Constructing sequences of random numbers is important for randomized algorithms, for simulations, and for many other purposes. Constructing a sequence of truly random numbers is extremely difficult, or perhaps impossible, because any method for generating what are supposed to be random numbers may generate numbers with hidden patterns. As a consequence, methods have been developed for finding sequences of numbers that have many desirable properties of random numbers, and which can be used for various purposes in place of random numbers. In this section we will show how to use congruences to generate sequences of pseudorandom numbers. The advantage is that the pseudorandom numbers so generated are constructed quickly; the disadvantage is that they have too much predictability to be used for many tasks.

Congruences also can be used to produce check digits for identification numbers of various kinds, such as code numbers used to identify retail products, numbers used to identify books, airline ticket numbers, and so on. We will explain how to construct check digits using congruences for a variety of types of identification numbers. We will show that these check digits can be used to detect certain kinds of common errors made when identification numbers are printed.

4.5.1 Hashing Functions

Links

The central computer at an insurance company maintains records for each of its customers. How can memory locations be assigned so that customer records can be retrieved quickly? The solution to this problem is to use a suitably chosen **hashing function**. Records are identified using a **key**, which uniquely identifies each customer's records. For instance, customer records are often identified using the Social Security number of the customer as the key. A hashing function h assigns memory location $h(k)$ to the record that has k as its key.

In practice, many different hashing functions are used. One of the most common is the function

$$h(k) = k \textbf{ mod } m$$

where m is the number of available memory locations.

Hashing functions should be easily evaluated so that files can be quickly located. The hashing function $h(k) = k \textbf{ mod } m$ meets this requirement; to find $h(k)$, we need only compute the remainder when k is divided by m. Furthermore, the hashing function should be onto, so that all memory locations are possible. The function $h(k) = k \textbf{ mod } m$ also satisfies this property.

EXAMPLE 1 Find the memory locations assigned by the hashing function $h(k) = k \textbf{ mod } 111$ to the records of customers with Social Security numbers 064212848 and 037149212.

Solution: The record of the customer with Social Security number 064212848 is assigned to memory location 14, because

$$h(064212848) = 064212848 \textbf{ mod } 111 = 14.$$

Similarly, because

$$h(037149212) = 037149212 \textbf{ mod } 111 = 65,$$

the record of the customer with Social Security number 037149212 is assigned to memory location 65. ◀

Because a hashing function is not one-to-one (because there are more possible keys than memory locations), more than one file may be assigned to a memory location. When this happens, we say that a **collision** occurs. One way to resolve a collision is to assign the first free location following the occupied memory location assigned by the hashing function.

EXAMPLE 2 After making the assignments of records to memory locations in Example 1, assign a memory location to the record of the customer with Social Security number 107405723.

Solution: First note that the hashing function $h(k) = k \textbf{ mod } 111$ maps the Social Security number 107405723 to location 14, because

$$h(107405723) = 107405723 \textbf{ mod } 111 = 14.$$

However, this location is already occupied (by the file of the customer with Social Security number 064212848). But, because memory location 15, the first location following memory location 14, is free, we assign the record of the customer with Social Security number 107405723 to this location. ◀

In Example 1 we used a **linear probing function**, namely, $h(k, i) = h(k) + i \textbf{ mod } m$, to look for the first free memory location, where i runs from 0 to $m - 1$. There are many other ways to resolve collisions that are discussed in the references on hashing functions given at the end of the book.

4.5.2 Pseudorandom Numbers

Randomly chosen numbers are often needed for computer simulations. Different methods have been devised for generating numbers that have properties of randomly chosen numbers. Because numbers generated by systematic methods are not truly random, they are called **pseudorandom numbers**.

Links

The most commonly used procedure for generating pseudorandom numbers is the **linear congruential method**. We choose four integers: the **modulus** m, **multiplier** a, **increment** c, and **seed** x_0, with $2 \leq a < m$, $0 \leq c < m$, and $0 \leq x_0 < m$. We generate a sequence of pseudorandom numbers $\{x_n\}$, with $0 \leq x_n < m$ for all n, by successively using the recursively defined function

$$x_{n+1} = (ax_n + c) \textbf{ mod } m.$$

(This is an example of a recursive definition, discussed in Section 5.3. In that section we will show that such sequences are well defined.)

Many computer experiments require the generation of pseudorandom numbers between 0 and 1. To generate such numbers, we divide numbers generated with a linear congruential generator by the modulus: that is, we use the numbers x_n/m.

EXAMPLE 3 Find the sequence of pseudorandom numbers generated by the linear congruential method with modulus $m = 9$, multiplier $a = 7$, increment $c = 4$, and seed $x_0 = 3$.

Solution: We compute the terms of this sequence by successively using the recursively defined function $x_{n+1} = (7x_n + 4) \textbf{ mod } 9$, beginning by inserting the seed $x_0 = 3$ to find x_1. We find that

$$\begin{aligned}
x_1 &= 7x_0 + 4 \textbf{ mod } 9 = 7 \cdot 3 + 4 \textbf{ mod } 9 = 25 \textbf{ mod } 9 = 7,\\
x_2 &= 7x_1 + 4 \textbf{ mod } 9 = 7 \cdot 7 + 4 \textbf{ mod } 9 = 53 \textbf{ mod } 9 = 8,\\
x_3 &= 7x_2 + 4 \textbf{ mod } 9 = 7 \cdot 8 + 4 \textbf{ mod } 9 = 60 \textbf{ mod } 9 = 6,\\
x_4 &= 7x_3 + 4 \textbf{ mod } 9 = 7 \cdot 6 + 4 \textbf{ mod } 9 = 46 \textbf{ mod } 9 = 1,\\
x_5 &= 7x_4 + 4 \textbf{ mod } 9 = 7 \cdot 1 + 4 \textbf{ mod } 9 = 11 \textbf{ mod } 9 = 2,\\
x_6 &= 7x_5 + 4 \textbf{ mod } 9 = 7 \cdot 2 + 4 \textbf{ mod } 9 = 18 \textbf{ mod } 9 = 0,\\
x_7 &= 7x_6 + 4 \textbf{ mod } 9 = 7 \cdot 0 + 4 \textbf{ mod } 9 = 4 \textbf{ mod } 9 = 4,\\
x_8 &= 7x_7 + 4 \textbf{ mod } 9 = 7 \cdot 4 + 4 \textbf{ mod } 9 = 32 \textbf{ mod } 9 = 5,\\
x_9 &= 7x_8 + 4 \textbf{ mod } 9 = 7 \cdot 5 + 4 \textbf{ mod } 9 = 39 \textbf{ mod } 9 = 3.
\end{aligned}$$

Because $x_9 = x_0$ and because each term depends only on the previous term, we see that the sequence

$$3, 7, 8, 6, 1, 2, 0, 4, 5, 3, 7, 8, 6, 1, 2, 0, 4, 5, 3, \ldots$$

is generated. This sequence contains nine different numbers before repeating. ◀

Most computers do use linear congruential generators to generate pseudorandom numbers. Often, a linear congruential generator with increment $c = 0$ is used. Such a generator is called a **pure multiplicative generator**. For example, the pure multiplicative generator with modulus $2^{31} - 1$ and multiplier $7^5 = 16{,}807$ is widely used. With these values, it can be shown that $2^{31} - 2$ numbers are generated before repetition begins.

Pseudorandom numbers generated by linear congruential generators have long been used for many tasks. Unfortunately, it has been shown that sequences of pseudorandom numbers generated in this way do not share some important statistical properties that true random numbers have. Because of this, it is not advisable to use them for some tasks, such as large simulations. For such sensitive tasks, other methods are used to produce sequences of pseudorandom numbers, either using some sort of algorithm or sampling numbers arising from a random physical phenomenon. For more details on pseudorandom number, see [Kn97] and [Re10].

4.5.3 Check Digits

Congruences are used to check for errors in digit strings. A common technique for detecting errors in such strings is to add an extra digit at the end of the string. This final digit, or check digit, is calculated using a particular function. Then, to determine whether a digit string is correct, a check is made to see whether this final digit has the correct value. We begin with an application of this idea for checking the correctness of bit strings.

EXAMPLE 4 **Parity Check Bits** Digital information is represented by bit string, split into blocks of a specified size. Before each block is stored or transmitted, an extra bit, called a **parity check bit**, can be appended to each block. The parity check bit x_{n+1} for the bit string $x_1 x_2 \dots x_n$ is defined by

$$x_{n+1} = x_1 + x_2 + \cdots + x_n \ \mathbf{mod}\ 2.$$

It follows that x_{n+1} is 0 if there are an even number of 1 bits in the block of n bits and it is 1 if there are an odd number of 1 bits in the block of n bits. When we examine a string that includes a parity check bit, we know that there is an error in it if the parity check bit is wrong. However, when the parity check bit is correct, there still may be an error. A parity check can detect an odd number of errors in the previous bits, but not an even number of errors. (See Exercise 14.)

Suppose we receive in a transmission the bit strings 01100101 and 11010110, each ending with a parity check bit. Should we accept these bit strings as correct?

Solution: Before accepting these strings as correct, we examine their parity check bits. The parity check bit of the first string is 1. Because $0 + 1 + 1 + 0 + 0 + 1 + 0 \equiv 1 \pmod 2$, the parity check bit is correct. The parity check bit of the second string is 0. We find that $1 + 1 + 0 + 1 + 0 + 1 + 1 \equiv 1 \pmod 2$, so the parity check is incorrect. We conclude that the first string may have been transmitted correctly and we know for certain that the second string was transmitted incorrectly. We accept the first string as correct (even though it still may contain an even number of errors), but we reject the second string. ◀

Check bits computed using congruences are used extensively to verify the correctness of various kinds of identification numbers. Examples 5 and 6 show how check bits are computed for codes that identify products (Universal Product Codes) and books (International Standard Book Numbers). The preambles to Exercises 18, 28, and 32 introduce the use of congruences to find and use check digits in money order numbers, airline ticket numbers, and identification numbers for periodicals, respectively. Note that congruences are also used to compute check digits for bank account numbers, drivers license numbers, credit card numbers, and many other types of identification numbers.

EXAMPLE 5 **UPCs** Retail products are identified by their **Universal Product Codes (UPCs)**. The most common form of a UPC has 12 decimal digits: the first digit identifies the product category, the next five digits identify the manufacturer, the following five identify the particular product, and the last digit is a check digit. The check digit is determined by the congruence

$$3x_1 + x_2 + 3x_3 + x_4 + 3x_5 + x_6 + 3x_7 + x_8 + 3x_9 + x_{10} + 3x_{11} + x_{12} \equiv 0 \pmod{10}.$$

Answer these questions:

(a) Suppose that the first 11 digits of a UPC are 79357343104. What is the check digit?
(b) Is 041331021641 a valid UPC?

Solution: (a) We insert the digits of 79357343104 into the congruence for UPC check digits. This gives $3 \cdot 7 + 9 + 3 \cdot 3 + 5 + 3 \cdot 7 + 3 + 3 \cdot 4 + 3 + 3 \cdot 1 + 0 + 3 \cdot 4 + x_{12} \equiv 0 \pmod{10}$. Simplifying, we have $21 + 9 + 9 + 5 + 21 + 3 + 12 + 3 + 3 + 0 + 12 + x_{12} \equiv 0 \pmod{10}$. Hence, $98 + x_{12} \equiv 0 \pmod{10}$. It follows that $x_{12} \equiv 2 \pmod{10}$, so the check digit is 2.

(b) To check whether 041331021641 is valid, we insert the digits into the congruence these digits must satisfy. This gives $3 \cdot 0 + 4 + 3 \cdot 1 + 3 + 3 \cdot 3 + 1 + 3 \cdot 0 + 2 + 3 \cdot 1 + 6 + 3 \cdot 4 + 1 \equiv 0 + 4 + 3 + 3 + 9 + 1 + 0 + 2 + 3 + 6 + 12 + 1 \equiv 4 \not\equiv 0 \pmod{10}$. Hence, 041331021641 is not a valid UPC. ◀

EXAMPLE 6 **ISBNs** All books are identified by an **International Standard Book Number (ISBN-10)**, a 10-digit code $x_1x_2 \ldots x_{10}$, assigned by the publisher. (In 2007, a 13-digit code known as ISBN-13 was introduced to identify a larger number of published works; see the preamble to Exercise 42 in the Supplementary Exercises.) An ISBN-10 consists of blocks identifying the language, the publisher, the number assigned to the book by its publishing company, and finally, a check digit that is either a digit or the letter X (used to represent 10). This check digit is selected so that

Remember that the check digit of an ISBN-10 can be an X!

$$x_{10} \equiv \sum_{i=1}^{9} ix_i \pmod{11},$$

or equivalently, so that

$$\sum_{i=1}^{10} ix_i \equiv 0 \pmod{11}.$$

Answer these questions about ISBN-10s:

(a) The first nine digits of the ISBN-10 of the sixth edition of this book are 007288008. What is the check digit?

(b) Is 084930149X a valid ISBN-10?

Solution: (a) The check digit is determined by the congruence $\sum_{i=1}^{10} ix_i \equiv 0 \pmod{11}$. Inserting the digits 007288008 gives $x_{10} \equiv 1 \cdot 0 + 2 \cdot 0 + 3 \cdot 7 + 4 \cdot 2 + 5 \cdot 8 + 6 \cdot 8 + 7 \cdot 0 + 8 \cdot 0 + 9 \cdot 8 \pmod{11}$. This means that $x_{10} \equiv 0 + 0 + 21 + 8 + 40 + 48 + 0 + 0 + 72 \pmod{11}$, so $x_{10} \equiv 189 \equiv 2 \pmod{11}$. Hence, $x_{10} = 2$.

(b) To see whether 084930149X is a valid ISBN-10, we see if $\sum_{i=1}^{10} ix_i \equiv 0 \pmod{11}$. We see that $1 \cdot 0 + 2 \cdot 8 + 3 \cdot 4 + 4 \cdot 9 + 5 \cdot 3 + 6 \cdot 0 + 7 \cdot 1 + 8 \cdot 4 + 9 \cdot 9 + 10 \cdot 10 = 0 + 16 + 12 + 36 + 15 + 0 + 7 + 32 + 81 + 100 = 299 \equiv 2 \not\equiv 0 \pmod{11}$. Hence, 084930149X is not a valid ISBN-10. ◀

Publishers sometimes do not calculate ISBNs correctly for their books, as was done for an earlier edition of this text.

Several kinds of errors often arise in identification numbers. A **single error**, an error in one digit of an identification number, is perhaps the most common type of error. Another common kind of error is a **transposition error**, which occurs when two digits are accidentally interchanged. For each type of identification number, including a check digit, we would like to be able to detect these common types of errors, as well as other types of errors. We will investigate whether the check digit for ISBNs can detect single errors and transposition errors. Whether check digits for UPCs can detect these kinds of errors is left as Exercises 26 and 27.

Suppose that $x_1x_2 \dots x_{10}$ is a valid ISBN (so that $\sum_{i=1}^{10} x_i \equiv 0 \pmod{11}$). We will show that we can detect a single error and a transposition of two digits (where we include the possibility that one of the two digits is the check digit X, representing 10). Suppose that this ISBN has been printed with a single error as $y_1y_2 \dots y_{10}$. If there is a single error, then, for some integer j, $y_i = x_i$ for $i \neq j$ and $y_j = x_j + a$, where $-10 \leq a \leq 10$ and $a \neq 0$. Note that $a = y_j - x_j$ is the error in the jth place. It then follows that

$$\sum_{i=1}^{10} iy_i = \Big(\sum_{i=1}^{10} ix_i\Big) + ja \equiv ja \not\equiv 0 \pmod{11}.$$

These last two congruences hold because $\sum_{i=1}^{10} x_i \equiv 0 \pmod{11}$ and $11 \nmid ja$, because $11 \nmid j$ and $11 \nmid a$. We conclude that $y_1y_2 \dots y_{10}$ is not a valid ISBN. So, we have detected the single error.

Now suppose that two unequal digits have been transposed. It follows that there are distinct integers j and k such that $y_j = x_k$ and $y_k = x_j$, and $y_i = x_i$ for $i \neq j$ and $i \neq k$. Hence,

$$\sum_{i=1}^{10} iy_i = \Big(\sum_{i=1}^{10} ix_i\Big) + (jx_k - jx_j) + (kx_j - kx_k) \equiv (j-k)(x_k - x_j) \not\equiv 0 \pmod{11},$$

because $\sum_{i=1}^{10} x_i \equiv 0 \pmod{11}$ and $11 \nmid (j-k)$ and $11 \nmid (x_k - x_j)$. We see that $y_1y_2 \dots y_{10}$ is not a valid ISBN. Thus, we can detect the interchange of two unequal digits.

Exercises

1. Which memory locations are assigned by the hashing function $h(k) = k$ **mod** 97 to the records of insurance company customers with these Social Security numbers?

a) 034567981 **b)** 183211232
c) 220195744 **d)** 987255335

2. Which memory locations are assigned by the hashing function $h(k) = k$ **mod** 101 to the records of insurance company customers with these Social Security numbers?

a) 104578690 **b)** 432222187
c) 372201919 **d)** 501338753

3. A parking lot has 31 visitor spaces, numbered from 0 to 30. Visitors are assigned parking spaces using the hashing function $h(k) = k$ **mod** 31, where k is the number formed from the first three digits on a visitor's license plate.

a) Which spaces are assigned by the hashing function to cars that have these first three digits on their license plates: 317, 918, 007, 100, 111, 310?

b) Describe a procedure visitors should follow to find a free parking space, when the space they are assigned is occupied.

Another way to resolve collisions in hashing is to use *double hashing*. We use an initial hashing function $h(k) = k \,\textbf{mod}\, p$, where p is prime. We also use a second hashing function $g(k) = (k + 1) \,\textbf{mod}\, (p - 2)$. When a collision occurs, we use a *probing sequence* $h(k, i) = (h(k) + i \cdot g(k)) \,\textbf{mod}\, p$.

4. Use the double hashing procedure we have described with $p = 4969$ to assign memory locations to files for employees with social security numbers $k_1 = 132489971$, $k_2 = 509496993$, $k_3 = 546332190$, $k_4 = 034367980$, $k_5 = 047900151$, $k_6 = 329938157$, $k_7 = 212228844$, $k_8 = 325510778$, $k_9 = 353354519$, $k_{10} = 053708912$.

5. What sequence of pseudorandom numbers is generated using the linear congruential generator $x_{n+1} = (3x_n + 2) \,\textbf{mod}\, 13$ with seed $x_0 = 1$?

6. What sequence of pseudorandom numbers is generated using the linear congruential generator $x_{n+1} = (4x_n + 1) \,\textbf{mod}\, 7$ with seed $x_0 = 3$?

7. What sequence of pseudorandom numbers is generated using the pure multiplicative generator $x_{n+1} = 3x_n \,\textbf{mod}\, 11$ with seed $x_0 = 2$?

8. Write an algorithm in pseudocode for generating a sequence of pseudorandom numbers using a linear congruential generator.

The **middle-square method** for generating pseudorandom numbers begins with an n-digit integer. This number is squared, initial zeros are appended to ensure that the result has $2n$ digits, and its middle n digits are used to form the next number in the sequence. This process is repeated to generate additional terms.

9. Find the first eight terms of the sequence of four-digit pseudorandom numbers generated by the middle square method starting with 2357.

10. Explain why both 3792 and 2916 would be bad choices for the initial term of a sequence of four-digit pseudorandom numbers generated by the middle square method.

The **power generator** is a method for generating pseudorandom numbers. To use the power generator, parameters p and d are specified, where p is a prime, d is a positive integer such that $p \not| d$, and a seed x_0 is specified. The pseudorandom numbers $x_1, x_2, \ldots$ are generated using the recursive definition $x_{n+1} = x_n^d \ \mathbf{mod}\ p$.

11. Find the sequence of pseudorandom numbers generated by the power generator with $p = 7$, $d = 3$, and seed $x_0 = 2$.

12. Find the sequence of pseudorandom numbers generated by the power generator with $p = 11$, $d = 2$, and seed $x_0 = 3$.

13. Suppose you received these bit strings over a communications link, where the last bit is a parity check bit. In which string are you sure there is an error?

a) 00000111111 **b)** 10101010101
c) 11111100000 **d)** 10111101111

14. Prove that a parity check bit can detect an error in a string if and only if the string contains an odd number of errors.

15. The first nine digits of the ISBN-10 of the European version of the fifth edition of this book are 0-07-119881. What is the check digit for that book?

16. The ISBN-10 of the sixth edition of *Elementary Number Theory and Its Applications* is 0-321-500Q1-8, where Q is a digit. Find the value of Q.

17. Determine whether the check digit of the ISBN-10 for this textbook (the 2025 Release of *Discrete Mathematics and Its Applications*) was computed correctly by the publisher.

The United States Postal Service (USPS) sells money orders identified by an 11-digit number $x_1x_2 \ldots x_{11}$. The first ten digits identify the money order; x_{11} is a check digit that satisfies $x_{11} = x_1 + x_2 + \cdots + x_{10} \ \mathbf{mod}\ 9$.

18. Find the check digit for the USPS money orders that have identification number that start with these ten digits.

a) 7555618873 **b)** 6966133421
c) 8018927435 **d)** 3289744134

19. Determine whether each of these numbers is a valid USPS money order identification number.

a) 74051489623 **b)** 88382013445
c) 56152240784 **d)** 66606631178

20. One digit in each of these identification numbers of a postal money order is smudged. Can you recover the smudged digit, indicated by a Q, in each of these numbers?

a) Q1223139784 **b)** 6702120Q988
c) 27Q41007734 **d)** 213279032Q1

21. One digit in each of these identification numbers of a postal money order is smudged. Can you recover the smudged digit, indicated by a Q, in each of these numbers?

a) 493212Q0688 **b)** 850Q9103858
c) 2Q941007734 **d)** 66687Q03201

22. Determine which single digit errors are detected by the USPS money order code.

23. Determine which transposition errors are detected by the USPS money order code.

24. Determine the check digit for the UPCs that have these initial 11 digits.

a) 73232184434 **b)** 63623991346
c) 04587320720 **d)** 93764323341

25. Determine whether each of the strings of 12 digits is a valid UPC code.

a) 036000291452 **b)** 012345678903
c) 782421843014 **d)** 726412175425

26. Does the check digit of a UPC code detect all single errors? Prove your answer or find a counterexample.

27. Determine which transposition errors the check digit of a UPC code finds.

Some airline tickets have a 15-digit identification number $a_1a_2 \ldots a_{15}$, where a_{15} is a check digit that equals $a_1a_2 \ldots a_{14}$ **mod** 7.

28. Find the check digit a_{15} that follows each of these initial 14 digits of an airline ticket identification number.

a) 10237424413392 **b)** 00032781811234
c) 00611232134231 **d)** 00193222543435

29. Determine whether each of these 15-digit numbers is a valid airline ticket identification number.

a) 101333341789013 **b)** 007862342770445
c) 113273438882531 **d)** 000122347322871

30. Which errors in a single digit of a 15-digit airline ticket identification number can be detected?

***31.** Can the accidental transposition of two consecutive digits in an airline ticket identification number be detected using the check digit?

Periodicals are identified using an **International Standard Serial Number (ISSN)**. An ISSN consists of two blocks of four digits. The last digit in the second block is a check digit. This check digit is determined by the congruence $d_8 \equiv 3d_1 + 4d_2 + 5d_3 + 6d_4 + 7d_5 + 8d_6 + 9d_7 \pmod{11}$. When $d_8 \equiv 10 \pmod{11}$, we use the letter X to represent d_8 in the code.

32. For each of these initial seven digits of an ISSN, determine the check digit (which may be the letter X).

a) 1570-868 **b)** 1553-734
c) 1089-708 **d)** 1383-811

33. Are each of these eight-digit codes possible ISSNs? That is, do they end with a correct check digit?

a) 1059-1027 **b)** 0002-9890
c) 1530-8669 **d)** 1007-120X

34. Does the check digit of an ISSN detect every single error in an ISSN? Justify your answer with either a proof or a counterexample.

35. Does the check digit of an ISSN detect every error where two consecutive digits are accidentally interchanged? Justify your answer with either a proof or a counterexample.

4.6 Cryptography

4.6.1 Introduction

Number theory plays a key role in cryptography, the subject of transforming information so that it cannot be easily recovered without special knowledge. Number theory is the basis of many classical ciphers, first used thousands of years ago, and used extensively until the 20th century. These ciphers encrypt messages by changing each letter to a different letter, or each block of letters to a different block of letters. We will discuss some classical ciphers, including shift ciphers, which replace each letter by the letter a fixed number of positions later in the alphabet, wrapping around to the beginning of the alphabet when necessary. The classical ciphers we will discuss are examples of private key ciphers where knowing how to encrypt allows someone to also decrypt messages. With a private key cipher, two parties who wish to communicate in secret must share a secret key. The classical ciphers we will discuss are also vulnerable to cryptanalysis, which seeks to recover encrypted information without access to the secret information used to encrypt the message. We will show how to cryptanalyze messages sent using shift ciphers.

Number theory is also important in public key cryptography, a type of cryptography invented in the 1970s. In public key cryptography, knowing how to encrypt does not also tell someone how to decrypt. The most widely used public key system, called the RSA cryptosystem, encrypts messages using modular exponentiation, where the modulus is the product of two large primes. Knowing how to encrypt requires that someone know the modulus and an exponent. (It does not require that the two prime factors of the modulus be known.) As far as it is known, knowing how to decrypt requires someone to know how to invert the encryption function, which can only be done in a practical amount of time when someone knows these two large prime factors. In this chapter we will explain how the RSA cryptosystem works, including how to encrypt and decrypt messages.

The subject of cryptography also includes the subject of cryptographic protocols, which are exchanges of messages carried out by two or more parties to achieve a specific security goal. We will discuss two important protocols in this chapter. One allows two people to share a common secret key. The other can be used to send signed messages so that a recipient can be sure that they were sent by the purported sender. Finally, we will introduce the notion of a homomorphic cryptosystem, now playing an important role in cloud computing. Data can become vulnerable if they must be decrypted for use as input to programs. Homomorphic encryption eliminates this vulnerability by allowing programs to be run on encrypted data. The output of these programs is then the encryption of the desired output.

4.6.2 Classical Cryptography

One of the earliest known uses of cryptography was by Julius Caesar. He made messages secret by shifting each letter three letters forward in the alphabet (sending the last three letters of the alphabet to the first three). For instance, using this scheme the letter B is sent to E and the letter X is sent to A. This is an example of **encryption**, that is, the process of making a message secret.

To express Caesar's encryption process mathematically, first replace each letter by an element of $\mathbf{Z}_{26}$, that is, an integer from 0 to 25 equal to one less than its position in the alphabet. For example, replace A by 0, K by 10, and Z by 25. Caesar's encryption method can be represented by the function f that assigns to the nonnegative integer p, $p \leq 25$, the integer $f(p)$ in the set $\{0, 1, 2, \ldots, 25\}$ with

$$f(p) = (p + 3) \textbf{ mod } 26.$$

In the encrypted version of the message, the letter represented by p is replaced with the letter represented by $(p + 3)$ **mod** 26.

EXAMPLE 1 What is the secret message produced from the message "MEET YOU IN THE PARK" using the Caesar cipher?

Solution: First replace the letters in the message with numbers. This produces

12 4 4 19   24 14 20   8 13   19 7 4   15 0 17 10.

Now replace each of these numbers p by $f(p) = (p + 3)$ **mod** 26. This gives

15 7 7 22   1 17 23   11 16   22 10 7   18 3 20 13.

Translating this back to letters produces the encrypted message "PHHW BRX LQ WKH SDUN." ◀

To recover the original message from a secret message encrypted by the Caesar cipher, the function f^{-1}, the inverse of f, is used. Note that the function f^{-1} sends an integer p from $\mathbf{Z}_{26}$, to $f^{-1}(p) = (p - 3)$ **mod** 26. In other words, to find the original message, each letter is shifted back three letters in the alphabet, with the first three letters sent to the last three letters of the alphabet. The process of determining the original message from the encrypted message is called **decryption**.

There are various ways to generalize the Caesar cipher. For example, instead of shifting the numerical equivalent of each letter by 3, we can shift the numerical equivalent of each letter by k, so that

$$f(p) = (p + k) \textbf{ mod } 26.$$

Such a cipher is called a *shift cipher*. Note that decryption can be carried out using

$$f^{-1}(p) = (p - k) \textbf{ mod } 26.$$

Here the integer k is called a **key**. We illustrate the use of a shift cipher in Examples 2 and 3.

EXAMPLE 2 Encrypt the plaintext message "STOP GLOBAL WARMING" using the shift cipher with shift $k = 11$.

Solution: To encrypt the message "STOP GLOBAL WARMING" we first translate each letter to the corresponding element of $\mathbf{Z}_{26}$. This produces the string

18 19 14 15   6 11 14 1 0 11   22 0 17 12 8 13 6.

We now apply the shift $f(p) = (p + 11)$ **mod** 26 to each number in this string. We obtain

3 4 25 0   17 22 25 12 11 22   7 11 2 23 19 24 17.

Translating this last string back to letters, we obtain the ciphertext "DEZA RWZMLW HLCXTYR." ◀

EXAMPLE 3 Decrypt the ciphertext message "LEWLYPLUJL PZ H NYLHA ALHJOLY" that was encrypted with the shift cipher with shift $k = 7$.

Solution: To decrypt the ciphertext "LEWLYPLUJL PZ H NYLHA ALHJOLY" we first translate the letters back to elements of $\mathbf{Z}_{26}$. We obtain

11 4 22 11 24 15 11 20 9 11   15 25   7   13 24 11 7 0   0 11 7 9 14 11 24.

Next, we shift each of these numbers by $-k = -7$ modulo 26 to obtain

4 23 15 4 17 8 4 13 2 4   8 18   0   6 17 4 0 19   19 4 0 2 7 4 17.

Finally, we translate these numbers back to letters to obtain the plaintext. We obtain "EXPERIENCE IS A GREAT TEACHER." ◀

We can generalize shift ciphers further to slightly enhance security by using a function of the form

$$f(p) = (ap + b) \textbf{ mod } 26,$$

where a and b are integers, chosen so that f is a bijection. (The function $f(p) = (ap + b)$ **mod** 26 is a bijection if and only if $\gcd(a, 26) = 1$.) Such a mapping is called an *affine transformation*, and the resulting cipher is called an *affine cipher*.

EXAMPLE 4 What letter replaces the letter K when the function $f(p) = (7p + 3)$ **mod** 26 is used for encryption?

Solution: First, note that 10 represents K. Then, using the encryption function specified, it follows that $f(10) = (7 \cdot 10 + 3)$ **mod** $26 = 21$. Because 21 represents V, K is replaced by V in the encrypted message. ◀

We will now show how to decrypt messages encrypted using an affine cipher. Suppose that $c = (ap + b)$ **mod** 26 with $\gcd(a, 26) = 1$. To decrypt we need to show how to express p in terms of c. To do this, we apply the encrypting congruence $c \equiv ap + b \pmod{26}$, and solve it for p. To do this, we first subtract b from both sides, to obtain $c - b \equiv ap \pmod{26}$. Because $\gcd(a, 26) = 1$, we know that there is an inverse $\overline{a}$ of a modulo 26. Multiplying both sides of the last equation by $\overline{a}$ gives us $\overline{a}(c - b) \equiv \overline{a}ap \pmod{26}$. Because $\overline{a}a \equiv 1 \pmod{26}$, this tells us that $p \equiv \overline{a}(c - b) \pmod{26}$. This determines p because p belongs to $\mathbf{Z}_{26}$.

CRYPTANALYSIS The process of recovering plaintext from ciphertext without knowledge of both the encryption method and the key is known as **cryptanalysis** or **breaking codes**. In general, cryptanalysis is a difficult process, especially when the encryption method is unknown.

We will not discuss cryptanalysis in general, but we will explain how to break messages that were encrypted using a shift cipher.

If we know that a ciphertext message was produced by enciphering a message using a shift cipher, we can try to recover the message by shifting all characters of the ciphertext by each of the 26 possible shifts (including a shift of zero characters). One of these is guaranteed to be the plaintext. However, we can use a more intelligent approach, which we can build upon to cryptanalyze ciphertext resulting from other ciphers. The main tool for cryptanalyzing ciphertext encrypted using a shift cipher is the count of the frequency of letters in the ciphertext. The nine most common letters in English text and their approximate relative frequencies are E 13%, T 9%, A 8%, O 8%, I 7%, N 7%, S 7%, H 6%, and R 6%. To cryptanalyze ciphertext that we know was produced using a shift cipher, we first find the relative frequencies of letters in the ciphertext. We list the most common letters in the ciphertext in frequency order; we hypothesize that the most common letter in the ciphertext is produced by encrypting E. Then, we determine the value of the shift under this hypothesis, say k. If the message produced by shifting the ciphertext by $-k$ makes sense, we presume that our hypothesis is correct and that we have the correct value of k. If it does not make sense, we next consider the hypothesis that the most common letter in the ciphertext is produced by encrypting T, the second most common letter in English; we find k under this hypothesis, shift the letters of the message by $-k$, and see whether the resulting message makes sense. If it does not, we continue the process working our way through the letters from most common to least common.

Mathematicians make the best code breakers. Their work in World War II changed the course of the war.

EXAMPLE 5 Suppose that we intercepted the ciphertext message ZNK KGXRE HOXJ MKZY ZNK CUXS that we know was produced by a shift cipher. What was the original plaintext message?

Solution: Because we know that the intercepted ciphertext message was encrypted using a shift cipher, we begin by calculating the frequency of letters in the ciphertext. We find that the most common letter in the ciphertext is K. So, we hypothesize that the shift cipher sent the plaintext letter E to the ciphertext letter K. If this hypothesis is correct, we know that $10 = 4 + k \textbf{ mod } 26$, so $k = 6$. Next, we shift the letters of the message by -6, obtaining THE EARLY BIRD GETS THE WORM. Because this message makes sense, we assume that the hypothesis that $k = 6$ is correct. ◀

Links

BLOCK CIPHERS Shift ciphers and affine ciphers proceed by replacing each letter of the alphabet by another letter in the alphabet. Because of this, these ciphers are called **character** or **monoalphabetic ciphers**. Encryption methods of this kind are vulnerable to attacks based on the analysis of letter frequency in the ciphertext, as we just illustrated. We can make it harder to successfully attack ciphertext by replacing blocks of letters with other blocks of letters instead of replacing individual characters with individual characters; such ciphers are called **block ciphers**.

We will now introduce a simple type of block cipher, called the **transposition cipher**. As a key we use a permutation σ of the set $\{1, 2, \dots, m\}$ for some positive integer m, that is, a one-to-one function from $\{1, 2, \dots, m\}$ to itself. To encrypt a message we first split its letters into blocks of size m. (If the number of letters in the message is not divisible by m we add some random letters at the end to fill out the final block.) We encrypt the block $p_1p_2 \dots p_m$ as $c_1c_2 \dots c_m = p_{\sigma(1)}p_{\sigma(2)} \dots, p_{\sigma(m)}$. To decrypt a ciphertext block $c_1c_2 \dots c_m$, we transpose its letters using the permutation σ^{-1}, the inverse of σ. Example 6 illustrates encryption and decryption for a transposition cipher.

EXAMPLE 6 Using the transposition cipher based on the permutation σ of the set $\{1, 2, 3, 4\}$ with $\sigma(1) = 3$, $\sigma(2) = 1$, $\sigma(3) = 4$, and $\sigma(4) = 2$,

(a) Encrypt the plaintext message PIRATE ATTACK.

(b) Decrypt the ciphertext message SWUE TRAE OEHS, which was encrypted using this cipher.

Solution: (a) We first split the letters of the plaintext into blocks of four letters. We obtain PIRA TEAT TACK. To encrypt each block, we send the first letter to the third position, the second letter to the first position, the third letter to the fourth position, and the fourth letter to the second position. We obtain IAPR ETTA AKTC.

(b) We note that σ^{-1}, the inverse of σ, sends 1 to 2, sends 2 to 4, sends 3 to 1, and sends 4 to 3. Applying $\sigma^{-1}(m)$ to each block gives us the plaintext: USEW ATER HOSE. (Grouping together these letters to form common words, we surmise that the plaintext is USE WATER HOSE.) ◀

CRYPTOSYSTEMS We have defined two families of ciphers: shift ciphers and affine ciphers. We now introduce the notion of a cryptosystem, which provides a general structure for defining new families of ciphers.

Definition 1 A *cryptosystem* is a five-tuple $(\mathcal{P}, \mathcal{C}, \mathcal{K}, \mathcal{E}, \mathcal{D})$, where $\mathcal{P}$ is the set of plaintext strings, $\mathcal{C}$ is the set of ciphertext strings, $\mathcal{K}$ is the *keyspace* (the set of all possible keys), $\mathcal{E}$ is the set of encryption functions, and $\mathcal{D}$ is the set of decryption functions. We denote by E_k the encryption function in $\mathcal{E}$ corresponding to the key k and D_k the decryption function in $\mathcal{D}$ that decrypts ciphertext that was encrypted using E_k, that is, $D_k(E_k(p)) = p$, for all plaintext strings p.

We now illustrate the use of the definition of a cryptosystem.

EXAMPLE 7 Describe the family of shift ciphers as a cryptosytem.

Solution: To encrypt a string of English letters with a shift cipher, we first translate each letter to an integer between 0 and 25, that is, to an element of $\mathbf{Z}_{26}$. We then shift each of these integers by a fixed integer modulo 26, and finally, we translate the integers back to letters. To apply the definition of a cryptosystem to shift ciphers, we assume that our messages are already integers, that is, elements of $\mathbf{Z}_{26}$. That is, we assume that the translation between letters and integers is outside of the cryptosystem. Consequently, both the set of plaintext strings $\mathcal{P}$ and the set of ciphertext strings $\mathcal{C}$ are the set of strings of elements of $\mathbf{Z}_{26}$. The set of keys $\mathcal{K}$ is the set of possible shifts, so $\mathcal{K} = \mathbf{Z}_{26}$. The set $\mathcal{E}$ consists of functions of the form $E_k(p) = (p + k)\ \mathbf{mod}\ 26$, and the set $\mathcal{D}$ of decryption functions is the same as the set of encrypting functions where $D_k(p) = (p - k)\ \mathbf{mod}\ 26$. ◀

The concept of a cryptosystem is useful in the discussion of additional families of ciphers and is used extensively in cryptography.

4.6.3 Public Key Cryptography

All classical ciphers, including shift ciphers and affine ciphers, are examples of **private key cryptosystems**. In a private key cryptosystem, once you know an encryption key, you can quickly find the decryption key. So, knowing how to encrypt messages using a particular key allows you to decrypt messages that were encrypted using this key. For example, when a shift cipher is used with encryption key k, the plaintext integer p is sent to

$$c = (p + k)\ \mathbf{mod}\ 26.$$

Decryption is carried out by shifting by $-k$; that is,

$$p = (c - k)\ \mathbf{mod}\ 26.$$

So knowing how to encrypt with a shift cipher also tells you how to decrypt.

When a private key cryptosystem is used, two parties who wish to communicate in secret must share a secret key. Because anyone who knows this key can both encrypt and decrypt messages, two people who want to communicate securely need to securely exchange this key. (We will introduce a method for doing this later in this section.) The shift cipher and affine cipher cryptosystems are private key cryptosystems. They are quite simple and are extremely vulnerable to cryptanalysis. However, the same is not true of many modern private key cryptosystems. In particular, the current US government standard for private key cryptography, the Advanced Encryption Standard (AES), is extremely complex and is considered to be highly resistant to cryptanalysis. (See [St06] for details on AES and other modern private key cryptosystems.) AES is widely used in government and commercial communications. However, it still shares the property that for secure communications keys be shared. Furthermore, for extra security, a new key is used for each communication session between two parties, which requires a method for generating keys and securely sharing them.

To avoid the need for keys to be shared by every pair of parties that wish to communicate securely, in the 1970s cryptologists introduced the concept of **public key cryptosystems**. When such cryptosystems are used, knowing how to send an encrypted message does not help decrypt messages. In such a system, everyone can have a publicly known encryption key. Only the decryption keys are kept secret, and only the intended recipient of a message can decrypt it, because, as far as it is currently known, knowledge of the encryption key does not let someone recover the plaintext message without an extraordinary amount of work (such as billions of years of computer time).

The first public key cryptosystems were invented in the mid-1970s. Many additional public key cryptosystems have been developed in the ensuing decades. We will introduce the most commonly used public key cryptosystem, known as the RSA system, in this book. Besides RSA, there are several other commonly used public key cryptosystems that are now used for many applications. These other public key cryptosystems will play an important role in the future when advances in computing may make the RSA cryptosystem obsolete, as often happens in cryptography. We will explain why this may be so shortly.

Although public key cryptography has the advantage that two parties who wish to communicate securely do not need to exchange keys, it has the disadvantage that encryption and decryption can be extremely time-consuming. For many applications, this makes public key cryptography impractical. In such situations, private key cryptography is used instead. However, public key cryptography may still be used in the key exchange process.

4.6.4 The RSA Cryptosystem

M.I.T. is also known as the 'Tute.

In 1976, three researchers at the Massachusetts Institute of Technology—Ronald Rivest, Adi Shamir, and Leonard Adleman—introduced to the world a public key cryptosystem, known as the **RSA system**, from the initials of its inventors. As often happens with cryptographic discoveries, the RSA system had been discovered several years earlier in secret government research in the United Kingdom. Clifford Cocks, working in secrecy at the United Kingdom's Government Communications Headquarters (GCHQ), had discovered this cryptosystem in 1973. However, his invention was unknown to the outside world until the late 1990s, when he was allowed to share classified GCHQ documents from the early 1970s. (An excellent account of this earlier discovery, as well as the work of Rivest, Shamir, and Adleman, can be found in [Si99].)

Unfortunately, no one calls this the Cocks cryptosystem.

In the RSA cryptosystem, each individual has an encryption key (n, e), where $n = pq$, the modulus is the product of two large primes p and q, say with 300 digits each, and an exponent e that is relatively prime to $(p - 1)(q - 1)$. To produce a usable key, two large primes must be found. This can be done quickly on a computer using probabilistic primality tests, referred to earlier in this section. However, the product of these primes $n = pq$, with approximately 600 digits, cannot, as far as is currently known, be factored in a reasonable length of time. As we will see, this is an important reason why decryption cannot, as far as is currently known, be done quickly without a separate decryption key.

Remark: With the steady increase of the speed of computers, the recommended size of the primes p and q used to produce a RSA public key has increased. But the larger n is, the slower RSA encryption and decryption become. When considering this tradeoff, the number of years a message needs to remain secret is important. A more important consideration is that the development of quantum computing threatens the security of the RSA cryptosystem, because factorization algorithms have been developed for quantum computers that could then be used to quickly factor large primes. So, once quantum computing becomes practical, perhaps in the next 20 to 30 years, other public key cryptosystems that cannot be broken using quantum computing will need to be used.

4.6.5 RSA Encryption

To encrypt messages using a particular key (n, e), we first translate a plaintext message M into sequences of integers. To do this, we first translate each plaintext letter into a two-digit number, using the same translation we employed for shift ciphers, with one key difference. That is, we include an initial zero for the letters A through J, so that A is translated into 00, B into 01, … , and J into 09. Then, we concatenate these two-digit numbers into strings of digits. Next, we divide this string into equally sized blocks of $2N$ digits, where $2N$ is the largest even number such that the number 2525 … 25 with $2N$ digits does not exceed n. (When necessary, we pad the plaintext message with dummy Xs to make the last block the same size as all other blocks.)

After these steps, we have translated the plaintext message M into a sequence of integers $m_1, m_2, \ldots, m_k$ for some integer k. Encryption proceeds by transforming each block m_i to a ciphertext block c_i. This is done using the function

$$c = m^e \textbf{ mod } n.$$

(To perform the encryption, we use an algorithm for fast modular exponentiation, such as Algorithm 5 in Section 4.2.) We leave the encrypted message as blocks of numbers and send these to the intended recipient. Because the RSA cryptosystem encrypts blocks of characters into blocks of characters, it is a block cipher.

Example 8 illustrates how RSA encryption is performed. For practical reasons we use small primes p and q in this example, rather than primes with 300 or more digits. Although the cipher described in this example is not secure, it does illustrate the techniques used in the RSA cipher.

EXAMPLE 8 Encrypt the message STOP using the RSA cryptosystem with key (2537, 13). Note that $2537 = 43 \cdot 59$, $p = 43$ and $q = 59$ are primes, and

$$\gcd(e, (p-1)(q-1)) = \gcd(13, 42 \cdot 58) = 1.$$

Links

Attribution: U.S. Department of Defense (USDOD)

CLIFFORD COCKS (BORN 1950) Clifford Cocks, born in Cheshire, England, was a talented mathematics student. He attended the Manchester Grammar School. In 1968 he won a silver medal at the International Mathematical Olympiad. Cocks attended King's College, Cambridge, studying mathematics. He also spent a short time at Oxford University working in number theory. In 1973 he decided not to complete his graduate work, instead taking a mathematical job at the Government Communications Headquarters (GCHQ) of British intelligence. Two months after joining GCHQ, Cocks learned about public key cryptography from an internal GCHQ report written by James Ellis. Cocks used his number theory knowledge to invent what is now called the RSA cryptosystem. He quickly realized that a public key cryptosystem could be based on the difficulty of reversing the process of multiplying two large primes. In 1997 he was allowed to reveal declassified GCHQ internal documents describing his discovery. Cocks is also known for his invention of a secure identity-based encryption scheme, which uses information about a user's identity as a public key. In 2001, Cocks became the Chief Mathematician at GCHQ. He has also set up the Heilbronn Institute for Mathematical Research, a partnership between GCHQ and the University of Bristol.

Solution: To encrypt, we first translate the letters in STOP into their numerical equivalents. We then group these numbers into blocks of four digits (because 2525 < 2537 < 252525), to obtain

1819 1415.

We encrypt each block using the mapping

$$c = m^{13} \textbf{ mod } 2537.$$

Computations using fast modular multiplication show that 1819^{13} **mod** 2537 = 2081 and 1415^{13} **mod** 2537 = 2182. The encrypted message is 2081 2182. ◀

4.6.6 RSA Decryption

The plaintext message can be quickly recovered from a ciphertext message when the decryption key d, an inverse of e modulo $(p-1)(q-1)$, is known. [Such an inverse exists because $\gcd(e, (p-1)(q-1)) = 1$.] To see this, note that if $de \equiv 1 \pmod{(p-1)(q-1)}$, there is an integer k such that $de = 1 + k(p-1)(q-1)$. It follows that

$$c^d \equiv (m^e)^d = m^{de} = m^{1+k(p-1)(q-1)} \pmod{n}.$$

By Fermat's little theorem [assuming that $\gcd(m, p) = \gcd(m, q) = 1$, which holds except in rare cases, which we cover in Exercise 28], it follows that $m^{p-1} \equiv 1 \pmod{p}$ and $m^{q-1} \equiv 1 \pmod{q}$. Consequently,

$$c^d \equiv m \cdot (m^{p-1})^{k(q-1)} \equiv m \cdot 1 = m \pmod{p}$$

Links ❯

Courtesy of Ronald L. Rivest

RONALD RIVEST (BORN 1948) Ronald Rivest received a B.A. from Yale in 1969 and his Ph.D. in computer science from Stanford in 1974. Rivest is a computer science professor at M.I.T. and was a cofounder of RSA Data Security, which held the patent on the RSA cryptosystem that he invented together with Adi Shamir and Leonard Adleman. Areas that Rivest has worked in besides cryptography include machine learning, VLSI design, and computer algorithms. He is a coauthor of a popular text on algorithms ([CoLeRiSt09]).

The Asahi Shimbun/Getty Images

ADI SHAMIR (BORN 1952) Adi Shamir was born in Tel Aviv, Israel. His undergraduate degree is from Tel Aviv University (1972) and his Ph.D. is from the Weizmann Institute of Science (1977). Shamir was a research assistant at the University of Warwick and an assistant professor at M.I.T. He is currently a professor in the Applied Mathematics Department at the Weizmann Institute and leads a group studying computer security. Shamir's contributions to cryptography, besides the RSA cryptosystem, include cracking knapsack cryptosystems, cryptanalysis of the Data Encryption Standard (DES), and the design of many cryptographic protocols.

Courtesy of Leonard Adleman

LEONARD ADLEMAN (BORN 1945) Leonard Adleman was born in San Francisco, California. He received a B.S. in mathematics (1968) and his Ph.D. in computer science (1976) from the University of California, Berkeley. Adleman was a member of the mathematics faculty at M.I.T. from 1976 until 1980, where he was a coinventor of the RSA cryptosystem, and in 1980 he took a position in the computer science department at the University of Southern California (USC). He was appointed to a chaired position at USC in 1985. Adleman has worked on computer security, computational complexity, immunology, and molecular biology. He invented the term "computer virus." Adleman's recent work on DNA computing has sparked great interest. He was a technical adviser for the movie *Sneakers*, in which computer security played an important role.

and

$$c^d \equiv m \cdot (m^{q-1})^{k(p-1)} \equiv m \cdot 1 = m \pmod{q}.$$

Because $\gcd(p, q) = 1$, it follows by the Chinese remainder theorem that

$$c^d \equiv m \pmod{pq}.$$

Example 9 illustrates how to decrypt messages sent using the RSA cryptosystem.

EXAMPLE 9 We receive the encrypted message 0981 0461. What is the decrypted message if it was encrypted using the RSA cipher from Example 8?

Solution: The message was encrypted using the RSA cryptosystem with $n = 43 \cdot 59$ and exponent 13. As Exercise 2 in Section 4.4 shows, $d = 937$ is an inverse of 13 modulo $42 \cdot 58 = 2436$. We use 937 as our decryption exponent. Consequently, to decrypt a block c, we compute

$$m = c^{937} \textbf{ mod } 2537.$$

To decrypt the message, we use the fast modular exponentiation algorithm to compute 0981^{937} **mod** $2537 = 0704$ and 0461^{937} **mod** $2537 = 1115$. Consequently, the numerical version of the original message is 0704 1115. Translating this back to English letters, we see that the message is HELP. ◀

4.6.7 RSA as a Public Key System

Links

Why is the RSA cryptosystem suitable for public key cryptography? First, it is possible to rapidly construct a public key by finding two large primes p and q, each with more than 300 digits, and to find an integer e relatively prime to $(p-1)(q-1)$. When we know the factorization of the modulus n, that is, when we know p and q, we can quickly find an inverse d of e modulo $(p-1)(q-1)$. [This is done by using the Euclidean algorithm to find Bézout coefficients s and t for d and $(p-1)(q-1)$, which shows that the inverse of d modulo $(p-1)(q-1)$ is s **mod** $(p-1)(q-1)$.] Knowing d lets us decrypt messages sent using our key. However, no method is known to decrypt messages that is not based on finding a factorization of n, or that does not also lead to the factorization of n.

Factorization is believed to be a difficult problem, as opposed to finding large primes p and q, which can be done quickly. The most efficient factorization methods known (as of 2024) require billions of years to factor 600-digit integers. Consequently, when p and q are 300-digit primes, it is believed that messages encrypted using $n = pq$ as the modulus cannot be decrypted in a reasonable time unless the primes p and q are known.

Although no polynomial-time algorithm is known for factoring large integers, active research is under way to find new ways to efficiently factor integers. Integers that were thought, as recently as several years ago, to be far too large to be factored in a reasonable amount of time can now be factored routinely. Integers with more than 250 digits have been factored using team efforts. When new factorization techniques are found, it will be necessary to use larger primes to ensure the secrecy of messages. Unfortunately, messages that were considered secure earlier can be saved and subsequently decrypted by unintended recipients when it becomes feasible to factor the $n = pq$ in the key used for RSA encryption. (Note that the RSA system will be insecure once quantum computing is available.)

The RSA method is now widely used. However, the most commonly used cryptosystems are private key cryptosystems. The use of public key cryptography, via the RSA system, is growing. Nevertheless, there are applications that use both private key and public key systems. For example, a public key cryptosystem, such as RSA, can be used to distribute private keys to pairs of individuals when they wish to communicate. These people then use a private key system for the encryption and decryption of messages.

4.6.8 Cryptographic Protocols

So far we have shown how cryptography can be used to make messages secure. However, there are many other important applications of cryptography. Among these applications are **cryptographic protocols**, which are exchanges of messages carried out by two or more parties to achieve a particular security goal. In particular, we will show how cryptography can be used to allow two people to exchange a secret key over an insecure communication channel. We will also show how cryptography can be used to send signed secret messages so that the recipient can be sure that the message came from the purported sender. We refer the reader to [St05] for thorough discussions of a variety of cryptographic protocols.

KEY EXCHANGE We now discuss a protocol that two parties can use to exchange a secret key over an insecure communications channel without having shared any information in the past. Generating a key that two parties can share is important for many applications of cryptography. For example, for two people to send secure messages to each other using a private key cryptosystem they need to share a common key. The protocol we will describe is known as the **Diffie-Hellman key agreement protocol**, after Whitfield Diffie and Martin Hellman, who described it in 1976. However, this protocol was invented in 1974 by Malcolm Williamson in secret work at the British GCHQ. It was not until 1997 that his discovery was made public.

Suppose that Alice and Bob want to share a common key. The protocol follows these steps, where the computations are done in $\mathbf{Z}_p$.

(1) Alice and Bob agree to use a prime p and a primitive root a of p.

(2) Alice chooses a secret integer k_1 and sends $a^{k_1} \mathbf{mod}\, p$ to Bob.

(3) Bob chooses a secret integer k_2 and sends $a^{k_2} \mathbf{mod}\, p$ to Alice.

(4) Alice computes $(a^{k_2})^{k_1} \mathbf{mod}\, p$.

(5) Bob computes $(a^{k_1})^{k_2} \mathbf{mod}\, p$.

At the end of this protocol, Alice and Bob have computed their shared key, namely,

$$(a^{k_2})^{k_1} \mathbf{mod}\, p = (a^{k_1})^{k_2} \mathbf{mod}\, p.$$

To analyze the security of this protocol, note that the messages sent in steps (1), (2), and (3) are not assumed to be sent securely. We can even assume that these communications were in the clear and that their contents are public information. So, p, a, $a^{k_1} \mathbf{mod}\, p$, and $a^{k_2} \mathbf{mod}\, p$ are assumed to be public information. The protocol ensures that k_1, k_2, and the common key $(a^{k_2})^{k_1} \mathbf{mod}\, p = (a^{k_1})^{k_2} \mathbf{mod}\, p$ are kept secret. To find the secret information from this public information requires that an adversary solves instances of the discrete logarithm problem, because the adversary would need to find k_1 and k_2 from $a^{k_1} \mathbf{mod}\, p$ and $a^{k_2} \mathbf{mod}\, p$, respectively. Furthermore, no other method is known for finding the shared key using just the public information. We have remarked that this is thought to be computationally infeasible when p and a are sufficiently large. With the computing power available now, this system is considered unbreakable when p has more than 300 decimal digits and k_1 and k_2 have more than 100 decimal digits each.

DIGITAL SIGNATURES Not only can cryptography be used to secure the confidentiality of a message, but it also can be used so that the recipient of the message knows that it came from the person they think it came from. We first show how a message can be sent so that a recipient of the message will be sure that the message came from the purported sender of the message. In particular, we can show how this can be accomplished using the RSA cryptosystem to apply a **digital signature** to a message.

Suppose that Alice's RSA public key is (n, e) and her private key is d. Alice encrypts a plaintext message x using the encryption function $E_{(n,e)}(x) = x^e \mathbf{\ mod\ } n$. She decrypts a ciphertext message y using the decryption function $D_{(n,e)}(y) = y^d \mathbf{\ mod\ } n$. Alice wants to send the message M so that everyone who receives the message knows that it came from her. Just as in RSA encryption, she translates the letters into their numerical equivalents and splits the resulting string into blocks $m_1, m_2, \ldots, m_k$ such that each block is the same size, which is as large as possible so that $0 \le m_i \le n$ for $i = 1, 2, ..., k$. She then applies her *decryption function* $D_{(n,e)}$ to each block, obtaining $D_{n,e}(m_i)$, $i = 1, 2, \ldots, k$. She sends the result to all intended recipients of the message.

When a recipient receives her message, they apply Alice's encryption function $E_{(n,e)}$ to each block, which everyone has available because Alice's key (n, e) is public information. The result is the original plaintext block because $E_{(n,e)}(D_{(n,e)}(x)) = x$. So, Alice can send her message to as many people as she wants and by signing it in this way, every recipient can be sure it came from Alice. Example 10 illustrates this protocol.

EXAMPLE 10 Suppose Alice's public RSA cryptosystem key is the same as in Example 8. That is, $n = 43 \cdot 59 = 2537$ and $e = 13$. Her decryption key is $d = 937$, as described in Example 9. She wants to send the message "MEET AT NOON" to her friends so that they are sure it came from her. What should she send?

Solution: Alice first translates the message into blocks of digits, obtaining 1204 0419 0019 1314 1413 (as the reader should verify). She then applies her decryption transformation $D_{(2537,13)}(x) = x^{937} \mathbf{\ mod\ } 2537$ to each block. Using fast modular exponentiation (with the help of a computational aid), she finds that $1204^{937} \mathbf{\ mod\ } 2537 = 817$, $419^{937} \mathbf{\ mod\ } 2537 = 555$, $19^{937} \mathbf{\ mod\ } 2537 = 1310$, $1314^{937} \mathbf{\ mod\ } 2537 = 2173$, and $1413^{937} \mathbf{\ mod\ } 2537 = 1026$.

So, the message she sends, split into blocks, is 0817 0555 1310 2173 1026. When one of her friends gets this message, they apply her encryption transformation $E_{(2537,13)}$ to each block. When they do this, they obtain the blocks of digits of the original message, which they translate back to English letters. ◀

We have shown that signed messages can be sent using the RSA cryptosystem. We can extend this by sending signed secret messages. To do this, the sender applies RSA encryption using the publicly known encryption key of an intended recipient to each block that was encrypted using the sender's decryption transformation. The recipient then first applies his private decryption transformation and then the sender's public encryption transformation. (Exercise 32 asks for this protocol to be carried out.)

4.6.9 Homomorphic Encryption

A cryptosystem, such as RSA, can be used to encrypt files to keep them secret. Today, many users store encrypted files in the **cloud**, where they reside on remote computers. It is often necessary to run programs using data from files stored on the cloud. These data are vulnerable to users with access to the remote computer where our data are stored if we run programs on the cloud without downloading these data. These data are also vulnerable to eavesdroppers when we download files, run programs on our computer, and upload the output to the cloud. Could we avoid these vulnerabilities by just running programs on encrypted data? Although this seems farfetched at first blush, in 1979, soon after RSA was introduced, the question was proposed whether there was a cryptosystem that allowed arbitrary computations to be run on encrypted data that would produce the encryption of the unencrypted output produced by this unencrypted input. For such a cryptosystem, it would not be necessary to decrypt input data because the program could be run on a remote system without disclosing either the input or the output. So, the search began for a **fully homomorphic cryptosystem** that allows arbitrary computations to be run remotely on encrypted data.

Before we discuss progress toward fully homomorphic encryption, we will show that the RSA cryptosytem is not fully homomorphic, although it allows some computations to be done on encrypted data.

EXAMPLE 11 **RSA is Partially Homomorphic** Let (n, e) be a public key for the RSA cryptosystem and suppose that M_1 and M_2 are plaintext messages, so that $0 \leq M_1 < n$ and $0 \leq M_2 < n$. Then

$$\begin{aligned} E_{(n,e)}(M_1)E_{(n,e)}(M_2) \mathbf{\ mod\ } m &= (M_1^e \mathbf{\ mod\ } m \cdot M_2^e \mathbf{\ mod\ } m) \mathbf{\ mod\ } m \\ &= (M_1M_2)^e \mathbf{\ mod\ } m = E_{(n,e)}(M_1M_2). \end{aligned}$$

From this equation we see that $E_{(n,e)}(M_1) \cdot_n E_{(n,e)}(M_2) = E_{(n,e)}(M_1M_2)$ in $\mathbf{Z}_n$. Because of this, we say that RSA is **multiplicatively homomorphic**. When we have encrypted using RSA, we can carry out multiplications without first decrypting because the encryption of the product of two plaintexts equals the product of their encryptions.

However, it is not true that $E_{(n,e)}(M_1) +_n E_{(n,e)}(M_2) = E_{(n,e)}(M_1 + M_2)$ for all M_1 and M_2 in $\mathbf{Z}_n$. (For instance, this is easy to see when $M_2 = 1$.) That is, when we have encrypted with RSA, we cannot add the encryptions of two numbers to obtain the encryption of their sum. Moreover, there is no method known for determining $E_{(n,e)}(M_1 + M_2)$ from $E_{(n,e)}(M_1)$ and $E_{(n,e)}(M_2)$ without decrypting these two ciphertexts. We say that RSA is not **additively homomorphic**. Because it is multiplicatively homomorphic, but not additively homomorphic, RSA is **partially homomorphic**. ◀

In 2009, Craig Gentry described the first fully homomorphic cryptosystem, based on what is known as lattice-based cryptography. Although a number of fully homomorphic cryptosystems have been developed, running them on standard processing hardware is not currently possible because of massive processing overhead. Perhaps the ongoing project called the Data Protection in Virtual Environments program will succeed in speeding up fully homomorphic encryption to usable levels.

Links

Source: John D. and Catherine T. MacArthur Foundation - used with permission.

CRAIG B. GENTRY (BORN 1972) Craig B. Gentry received his B.S. from Duke University in 1993 and his J.D. from Harvard University Law School in 1998. For two years he worked as an intellectual property lawyer. From 2000 to 2005 he was a senior research engineer at NTT DoCoMo USA Labs. He decided to return to school, and in 2009 he received a Ph.D. in computer science from Stanford University. Gentry joined the Cryptography Research Group of the IBM Watson Research Center in 2009 where he currently works.

Gentry's invention of a fully homomorphic scheme, which resolved an open problem proposed in 1978, earned him the 2010 ACM Grace Murray Hopper Award. In 2013, Gentry with others constructed the first cryptographic multilinear maps, using them to construct the first cryptographic program obfuscation schemes, which many had thought might not exist. Gentry's work on fully homomorphic cryptography and on cryptographic multilinear maps is based on lattice-based cryptography, which cannot be broken by quantum computing, unlike the RSA cryptosystem. Gentry, along with his colleagues, advanced the area of verifiable computation, which allows a computer to offload the computation of functions to other computers while maintaining verifiable results. In 2014, Gentry was awarded a MacArthur Fellowship (also known as a Genuis Grant).

Exercises

1. Encrypt the message DO NOT PASS GO by translating the letters into numbers, applying the given encryption function, and then translating the numbers back into letters.
   a) $f(p) = (p + 3)$ **mod** 26 (the Caesar cipher)
   b) $f(p) = (p + 13)$ **mod** 26
   c) $f(p) = (3p + 7)$ **mod** 26
2. Encrypt the message STOP POLLUTION by translating the letters into numbers, applying the given encryption function, and then translating the numbers back into letters.
   a) $f(p) = (p + 4)$ **mod** 26
   b) $f(p) = (p + 21)$ **mod** 26
   c) $f(p) = (17p + 22)$ **mod** 26
3. Encrypt the message WATCH YOUR STEP by translating the letters into numbers, applying the given encryption function, and then translating the numbers back into letters.
   a) $f(p) = (p + 14)$ **mod** 26
   b) $f(p) = (14p + 21)$ **mod** 26
   c) $f(p) = (-7p + 1)$ **mod** 26
4. Decrypt these messages that were encrypted using the Caesar cipher.
   a) EOXH MHDQV
   b) WHVW WRGDB
   c) HDW GLP VXP
5. Decrypt these messages encrypted using the shift cipher $f(p) = (p + 10)$ **mod** 26.
   a) CEBBOXNOB XYG
   b) LO WI PBSOXN
   c) DSWO PYB PEX
6. Suppose that when a long string of text is encrypted using a shift cipher $f(p) = (p + k)$ **mod** 26, the most common letter in the ciphertext is X. What is the most likely value for k, assuming that the distribution of letters in the text is typical of English text?
7. Suppose that when a string of English text is encrypted using a shift cipher $f(p) = (p + k)$ **mod** 26, the resulting ciphertext is DY CVOOZ ZOBMRKXMO DY NBOKW. What was the original plaintext string?
8. Suppose that the ciphertext DVE CFMV KF NFEUVI, REU KYRK ZJ KYV JVVU FW JTZVETV was produced by encrypting a plaintext message using a shift cipher. What is the original plaintext?
9. Suppose that the ciphertext ERC WYJJMGMIRXPC EHZERGIH XIGLRSPSKC MW MRHMWXM-RKYMWLEFPI JVSQ QEKMG was produced by encrypting a plaintext message using a shift cipher. What is the original plaintext?
10. Determine whether there is a key for which the enciphering function for the shift cipher is the same as the deciphering function.
11. What is the decryption function for an affine cipher if the encryption function is $c = (15p + 13)$ **mod** 26?
*12. Find all pairs of integers keys (a, b) for affine ciphers for which the encryption function $c = (ap + b)$ **mod** 26 is the same as the corresponding decryption function.
13. Suppose that the most common letter and the second most common letter in a long ciphertext produced by encrypting a plaintext using an affine cipher $f(p) = (ap + b)$ **mod** 26 are Z and J, respectively. What are the most likely values of a and b?
14. Encrypt the message GRIZZLY BEARS using blocks of five letters and the transposition cipher based on the permutation of $\{1, 2, 3, 4, 5\}$ with $\sigma(1) = 3$, $\sigma(2) = 5$, $\sigma(3) = 1$, $\sigma(4) = 2$, and $\sigma(5) = 4$. For this exercise, use the letter X as many times as necessary to fill out the final block of fewer then five letters.
15. Decrypt the message EABW EFRO ATMR ASIN, which is the ciphertext produced by encrypting a plaintext message using the transposition cipher with blocks of four letters and the permutation σ of $\{1, 2, 3, 4\}$ defined by $\sigma(1) = 3$, $\sigma(2) = 1$, $\sigma(3) = 4$, and $\sigma(4) = 2$.
*16. Suppose that you know that a ciphertext was produced by encrypting a plaintext message with a transposition cipher. How might you go about breaking it?
17. Suppose you have intercepted a ciphertext message and when you determine the frequencies of letters in this message, you find the frequencies are similar to the frequency of letters in English text. Which type of cipher do you suspect was used?

The **Vigenère cipher** is a block cipher, with a key that is a string of letters with numerical equivalents $k_1k_2 \ldots k_m$, where $k_i \in \mathbf{Z}_{26}$ for $i = 1, 2, \ldots, m$. Suppose that the numerical equivalents of the letters of a plaintext block are $p_1p_2 \ldots p_m$. The corresponding numerical ciphertext block is $(p_1 + k_1)$ **mod** 26 $(p_2 + k_2)$ **mod** 26 ... $(p_m + k_m)$ **mod** 26. Finally, we translate back to letters. For example, suppose that the key string is RED, with numerical equivalents 17 4 3. Then, the plaintext ORANGE, with numerical equivalents 14 17 00 13 06 04, is encrypted by first splitting it into two blocks 14 17 00 and 13 06 04. Then, in each block we shift the first letter by 17, the second by 4, and the third by 3. We obtain 5 21 03 and 04 10 07. The cipherext is FVDEKH.

18. Use the Vigenère cipher with key BLUE to encrypt the message SNOWFALL.
19. The ciphertext OIKYWVHBX was produced by encrypting a plaintext message using the Vigenère cipher with key HOT. What is the plaintext message?
20. Express the Vigenère cipher as a cryptosystem.

To break a Vigenère cipher by recovering a plaintext message from the ciphertext message without having the key, the first step is to figure out the length of the key string. The second step is to figure out each character of the key string by determining the corresponding shift. Exercises 21 and 22 deal with these two aspects.

21. Suppose that when a long string of text is encrypted using a Vigenère cipher, the same string is found in the ciphertext starting at several different positions. Explain how this information can be used to help determine the length of the key.

22. Once the length of the key string of a Vigènere cipher is known, explain how to determine each of its characters. Assume that the plaintext is long enough so that the frequency of its letters is reasonably close to the frequency of letters in typical English text.

***23.** Show that we can easily factor n when we know that n is the product of two primes, p and q, and we know the value of $(p - 1)(q - 1)$.

In Exercises 24–27 first express your answers without computing modular exponentiations. Then use a computational aid to complete these computations.

24. Encrypt the message ATTACK using the RSA system with $n = 43 \cdot 59$ and $e = 13$, translating each letter into integers and grouping together pairs of integers, as done in Example 8.

25. Encrypt the message UPLOAD using the RSA system with $n = 53 \cdot 61$ and $e = 17$, translating each letter into integers and grouping together pairs of integers, as done in Example 8.

26. What is the original message encrypted using the RSA system with $n = 53 \cdot 61$ and $e = 17$ if the encrypted message is 3185 2038 2460 2550? (To decrypt, first find the decryption exponent d, which is the inverse of $e = 17$ modulo $52 \cdot 60$.)

27. What is the original message encrypted using the RSA system with $n = 43 \cdot 59$ and $e = 13$ if the encrypted message is 0667 1947 0671? (To decrypt, first find the decryption exponent d which is the inverse of $e = 13$ modulo $42 \cdot 58$.)

***28.** Suppose that (n, e) is an RSA encryption key, with $n = pq$, where p and q are large primes and $\gcd(e, (p - 1)(q - 1)) = 1$. Furthermore, suppose that d is an inverse of e modulo $(p - 1)(q - 1)$. Suppose that $C \equiv M^e \pmod{pq}$. In the text we showed that RSA decryption, that is, the congruence $C^d \equiv M \pmod{pq}$ holds when $\gcd(M, pq) = 1$. Show that this decryption congruence also holds when $\gcd(M, pq) > 1$. [*Hint:* Use congruences modulo p and modulo q and apply the Chinese remainder theorem.]

29. Describe the steps that Alice and Bob follow when they use the Diffie-Hellman key exchange protocol to generate a shared key. Assume that they use the prime $p = 23$ and take $a = 5$, which is a primitive root of 23, and that Alice selects $k_1 = 8$ and Bob selects $k_2 = 5$. (You may want to use some computational aid.)

30. Describe the steps that Alice and Bob follow when they use the Diffie-Hellman key exchange protocol to generate a shared key. Assume that they use the prime $p = 101$ and take $a = 2$, which is a primitive root of 101, and that Alice selects $k_1 = 7$ and Bob selects $k_2 = 9$. (You may want to use some computational aid.)

In Exercises 31–32 suppose that Alice and Bob have these public keys and corresponding private keys: $(n_{\text{Alice}}, e_{\text{Alice}}) = (2867, 7) = (61 \cdot 47, 7)$, $d_{\text{Alice}} = 1183$ and $(n_{\text{Bob}}, e_{\text{Bob}}) = (3127, 21) = (59 \cdot 53, 21)$, $d_{\text{Bob}} = 1149$. First express your answers without carrying out the calculations. Then, using a computational aid, if available, perform the calculation to get the numerical answers.

31. Alice wants to send to all her friends, including Bob, the message "SELL EVERYTHING" so that he knows that she sent it. What should she send to her friends, assuming she signs the message using the RSA cryptosystem.

32. Alice wants to send to Bob the message "BUY NOW" so that he knows that she sent it and so that only Bob can read it. What should she send to Bob, assuming she signs the message and then encrypts it using Bob's public key?

33. We describe a basic key exchange protocol using private key cryptography upon which more sophisticated protocols for key exchange are based. Encryption within the protocol is done using a private key cryptosystem (such as AES) that is considered secure. The protocol involves three parties, Alice and Bob, who wish to exchange a key, and a trusted third party Cathy. Assume that Alice has a secret key k_{Alice} that only she and Cathy know, and Bob has a secret key k_{Bob} which only he and Cathy know. The protocol has three steps:

(i) Alice sends the trusted third party Cathy the message "request a shared key with Bob" encrypted using Alice's key k_{Alice}.

(ii) Cathy sends back to Alice a key $k_{\text{Alice,Bob}}$, which she generates, encrypted using the key k_{Alice}, followed by this same key $k_{\text{Alice,Bob}}$, encrypted using Bob's key, k_{Bob}.

(iii) Alice sends to Bob the key $k_{\text{Alice,Bob}}$ encrypted using k_{Bob}, known only to Bob and to Cathy.

Explain why this protocol allows Alice and Bob to share the secret key $k_{\text{Alice,Bob}}$, known only to them and to Cathy.

The **Paillier cryptosystem** is a public key cryptosystem described in 1999 by P. Paillier, used in some electronic voting systems. A public key (n, g) and a corresponding private key (p, q) are created by randomly selecting primes p and q so that $\gcd(pq, \lambda) = 1$, where $\lambda = (p - 1)(q - 1)$, and a nonzero element g of $\mathbf{Z}_{n^2}$ with $\gcd((g^{\lambda \,\mathbf{mod}\, n^2} - 1)/n, n) = 1$. To encrypt a message $m \in \mathbf{Z}_n$ a random nonzero element r of $\mathbf{Z}_n$ is first selected, and then used to find $c = g^m r^n \,\mathbf{mod}\, n^2$.

34. **a)** Show that we can take $p = 149$, $q = 179$, and $g = 5$ to generate a public key for the Paillier cryptosystem by checking that all the conditions required for these parameters hold, and find the public and private keys that are generated.

b) Find the ciphertext corresponding to the plaintext $m = 67$ where we choose $r = 81$.

35. Show that the Paillier cryptosystem is additively homomorphic.

Key Terms and Results

TERMS

$a \mid b$ (a divides b): there is an integer c such that $b = ac$

a and b are congruent modulo m: m divides $a - b$

modular arithmetic: arithmetic done modulo an integer $m \geq 2$

prime: an integer greater than 1 with exactly two positive integer divisors

composite: an integer greater than 1 that is not prime

Mersenne prime: a prime of the form $2^p - 1$, where p is prime

gcd(a, b) (greatest common divisor of a and b): the largest integer that divides both a and b

relatively prime integers: integers a and b such that $\gcd(a, b) = 1$

pairwise relatively prime integers: a set of integers with the property that every pair of these integers is relatively prime

lcm(a, b) (least common multiple of a and b): the smallest positive integer that is divisible by both a and b

a mod b: the remainder when the integer a is divided by the positive integer b

$a \equiv b \pmod{m}$ (a is congruent to b modulo m): $a - b$ is divisible by m

$n = (a_k a_{k-1} \ldots a_1 a_0)_b$: the base b representation of n

binary expansion: the base 2 representation of an integer

octal expansion: the base 8 representation of an integer

hexadecimal expansion: the base 16 representation of an integer

linear combination of a and b with integer coefficients: an expression of the form $sa + tb$, where s and t are integers

Bézout coefficients of a and b: integers s and t such that the **Bézout identity** $sa + tb = \gcd(a, b)$ holds

inverse of a modulo m: an integer $\overline{a}$ such that $\overline{a}a \equiv 1 \pmod{m}$

linear congruence: a congruence of the form $ax \equiv b \pmod{m}$, where x is an integer variable

pseudoprime to the base b: a composite integer n such that $b^{n-1} \equiv 1 \pmod{n}$

Carmichael number: a composite integer n such that n is a pseudoprime to the base b for all positive integers b with $\gcd(b, n) = 1$

primitive root of a prime p: an integer r in $\mathbf{Z}_p$ such that every integer not divisible by p is congruent modulo p to a power of r

discrete logarithm of a to the base r modulo p: the integer e with $0 \leq e \leq p - 1$ such that $r^e \equiv a \pmod{p}$

encryption: the process of making a message secret

decryption: the process of returning a secret message to its original form

encryption key: a value that determines which of a family of encryption functions is to be used

shift cipher: a cipher that encrypts the plaintext letter p as $(p + k)$ **mod** m for an integer k

affine cipher: a cipher that encrypts the plaintext letter p as $(ap + b)$ **mod** m for integers a and b with $\gcd(a, 26) = 1$

character cipher: a cipher that encrypts characters one by one

block cipher: a cipher that encrypts blocks of characters of a fixed size

cryptanalysis: the process of recovering the plaintext from ciphertext without knowledge of the encryption method, or with knowledge of the encryption method, but not the key

cryptosystem: a five-tuple $(\mathcal{P}, \mathcal{C}, \mathcal{K}, \mathcal{E}, \mathcal{D})$ where $\mathcal{P}$ is the set of plaintext messages, $\mathcal{C}$ is the set of ciphertext messages, $\mathcal{K}$ is the set of keys, $\mathcal{E}$ is the set of encryption functions, and $\mathcal{D}$ is the set of decryption functions

private key encryption: encryption where both encryption keys and decryption keys must be kept secret

public key encryption: encryption where encryption keys are public knowledge, but decryption keys are kept secret

RSA cryptosystem: the cryptosystem where $\mathcal{P}$ and $\mathcal{C}$ are both $\mathbf{Z}_{26}$, $\mathcal{K}$ is the set of pairs $k = (n, e)$ where $n = pq$ where p and q are large primes and e is a positive integer, $E_k(p) = p^e$ **mod** n, and $D_k(c) = c^d$ **mod** n where d is the inverse of e modulo $(p - 1)(q - 1)$

key exchange protocol: a protocol used for two parties to generate a shared key

digital signature: a method that a recipient can use to determine that the purported sender of a message actually sent the message

fully homomorphic cryptosystem: a cryptosystem that allows arbitrary computations to be run on encrypted data so that the ouptut is the encryption of the unencrypted output of the unencrypted input

RESULTS

division algorithm: Let a and d be integers with d positive. Then there are unique integers q and r with $0 \leq r < d$ such that $a = dq + r$.

Let b be an integer greater than 1. Then if n is a positive integer, it can be expressed uniquely in the form $n = a_k b^k + a_{k-1} b^{k-1} + \cdots + a_1 b + a_0$.

The algorithm for finding the base b expansion of an integer (see Algorithm 1 in Section 4.2)

The conventional algorithms for addition and multiplication of integers (given in Section 4.2)

The fast modular exponentiation algorithm (see Algorithm 5 in Section 4.2)

Euclidean algorithm: for finding greatest common divisors by successively using the division algorithm (see Algorithm 1 in Section 4.3)

Bézout's theorem: If a and b are positive integers, then $\gcd(a, b)$ is a linear combination of a and b.

sieve of Eratosthenes: a procedure for finding all primes not exceeding a specified number n, described in Section 4.3

fundamental theorem of arithmetic: Every positive integer can be written uniquely as the product of primes, where the prime factors are written in order of increasing size.

If a and b are positive integers, then $ab = \gcd(a, b) \cdot \text{lcm}(a, b)$.

If m is a positive integer and $\gcd(a, m) = 1$, then a has a unique inverse modulo m.

Chinese remainder theorem: A system of linear congruences modulo pairwise relatively prime integers has a unique solution modulo the product of these moduli.

Fermat's little theorem: If p is prime and $p \nmid a$, then $a^{p-1} \equiv 1 \pmod{p}$.

Review Questions

1. Find 210 **div** 17 and 210 **mod** 17.
2. a) Define what it means for a and b to be congruent modulo 7.
   b) Which pairs of the integers $-11, -8, -7, -1, 0, 3$, and 17 are congruent modulo 7?
   c) Show that if a and b are congruent modulo 7, then $10a + 13$ and $-4b + 20$ are also congruent modulo 7.
3. Show that if $a \equiv b \pmod{m}$ and $c \equiv d \pmod{m}$, then $a + c \equiv b + d \pmod{m}$.
4. Describe a procedure for converting decimal (base 10) expansions of integers into hexadecimal expansions.
5. Convert $(1101\ 1001\ 0101\ 1011)_2$ to octal and hexadecimal representations.
6. Convert $(7206)_8$ and $(\text{A0EB})_{16}$ to a binary representation.
7. State the fundamental theorem of arithmetic.
8. a) Describe a procedure for finding the prime factorization of an integer.
   b) Use this procedure to find the prime factorization of 80,707.
9. a) Define the greatest common divisor of two integers.
   b) Describe at least three different ways to find the greatest common divisor of two integers. When does each method work best?
   c) Find the greatest common divisor of 1,234,567 and 7,654,321.
   d) Find the greatest common divisor of $2^3 3^5 5^7 7^9 11$ and $2^9 3^7 5^5 7^3 13$.
10. a) How can you find a linear combination (with integer coefficients) of two integers that equals their greatest common divisor?
    b) Express gcd(84, 119) as a linear combination of 84 and 119.
11. a) What does it mean for $\bar{a}$ to be an inverse of a modulo m?
    b) How can you find an inverse of a modulo m when m is a positive integer and $\gcd(a, m) = 1$?
    c) Find an inverse of 7 modulo 19.
12. a) How can an inverse of a modulo m be used to solve the congruence $ax \equiv b \pmod{m}$ when $\gcd(a, m) = 1$?
    b) Solve the linear congruence $7x \equiv 13 \pmod{19}$.
13. a) State the Chinese remainder theorem.
    b) Find the solutions to the system $x \equiv 1 \pmod{4}$, $x \equiv 2 \pmod{5}$, and $x \equiv 3 \pmod{7}$.
14. Suppose that $2^{n-1} \equiv 1 \pmod{n}$. Is n necessarily prime?
15. Use Fermat's little theorem to evaluate 9^{200} **mod** 19.
16. Explain how the check digit is found for a 10-digit ISBN.
17. Encrypt the message APPLES AND ORANGES using a shift cipher with key $k = 13$.
18. a) What is the difference between a public key and a private key cryptosystem?
    b) Explain why using shift ciphers is a private key system.
    c) Explain why the RSA cryptosystem is a public key system.
19. Explain how encryption and decryption are done in the RSA cryptosystem.
20. Describe how two parties can share a secret key using the Diffie-Hellman key exchange protocol.

Supplementary Exercises

1. The odometer on a car goes to up 100,000 miles. The present owner of a car bought it when the odometer read 43,179 miles. The owner now wants to sell it; when you examine the car for possible purchase, you notice that the odometer reads 89,697 miles. What can you conclude about how many miles the owner drove the car, assuming that the odometer always worked correctly?
2. a) Explain why n **div** 7 equals the number of complete weeks in n days.
   b) Explain why n **div** 24 equals the number of complete days in n hours.
3. Find four numbers congruent to 5 modulo 17.
4. Show that if a and d are positive integers, then there are integers q and r such that $a = dq + r$, where $-d/2 < r \leq d/2$.

*5. Show that if $ac \equiv bc \pmod{m}$, where a, b, c, and m are integers with $m > 2$, and $d = \gcd(m, c)$, then $a \equiv b \pmod{m/d}$.

6. Show that the sum of the squares of two odd integers cannot be the square of an integer.
7. Show that if $n^2 + 1$ is a perfect square, where n is an integer, then n is even.

8. Prove that there are no solutions in integers x and y to the equation $x^2 - 5y^2 = 2$. [*Hint:* Consider this equation modulo 5.]

9. Develop a test for divisibility of a positive integer n by 8 based on the binary expansion of n.

10. Develop a test for divisibility of a positive integer n by 3 based on the binary expansion of n.

11. Devise an algorithm for guessing a number between 1 and $2^n - 1$ by successively guessing each bit in its binary expansion.

12. Determine the complexity, in terms of the number of guesses, needed to determine a number between 1 and $2^n - 1$ by successively guessing the bits in its binary expansion.

13. Show that an integer is divisible by 9 if and only if the sum of its decimal digits is divisible by 9.

**14. Show that if a and b are positive irrational numbers such that $1/a + 1/b = 1$, then every positive integer can be uniquely expressed as either $\lfloor ka \rfloor$ or $\lfloor kb \rfloor$ for some positive integer k.

15. Prove there are infinitely many primes by showing that $Q_n = n! + 1$ must have a prime factor greater than n whenever n is a positive integer.

16. Find a positive integer n for which $Q_n = n! + 1$ is not prime.

17. Use Dirichlet's theorem, which states there are infinitely many primes in every arithmetic progression $ak + b$ where $\gcd(a, b) = 1$, to show that there are infinitely many primes that have a decimal expansion ending with a 1.

18. Prove that if n is a positive integer such that the sum of the divisors of n is $n + 1$, then n is prime.

*19. Show that every integer greater than 11 is the sum of two composite integers.

20. Find the five smallest consecutive composite integers.

21. Show that Goldbach's conjecture, which states that every even integer greater than 2 is the sum of two primes, is equivalent to the statement that every integer greater than 5 is the sum of three primes.

22. Find an arithmetic progression of length six beginning with 7 that contains only primes.

*23. Prove that if $f(x)$ is a nonconstant polynomial with integer coefficients, then there is an integer y such that $f(y)$ is composite. [*Hint:* Assume that $f(x_0) = p$ is prime. Show that p divides $f(x_0 + kp)$ for all integers k. Obtain a contradiction of the fact that a polynomial of degree n, where $n > 1$, takes on each value at most n times.]

*24. How many zeros are at the end of the binary expansion of $(100_{10})!$?

25. Use the Euclidean algorithm to find the greatest common divisor of 10,223 and 33,341.

26. How many divisions are required to find $\gcd(144, 233)$ using the Euclidean algorithm?

27. Find $\gcd(2n + 1, 3n + 2)$, where n is a positive integer. [*Hint:* Use the Euclidean algorithm.]

28. **a)** Show that if a and b are positive integers with $a \geq b$, then $\gcd(a, b) = a$ if $a = b$, $\gcd(a, b) = 2\gcd(a/2, b/2)$ if a and b are even, $\gcd(a, b) = \gcd(a/2, b)$ if a is even and b is odd, and $\gcd(a, b) = \gcd(a - b, b)$ if both a and b are odd.
b) Explain how to use part (a) to construct an algorithm for computing the greatest common divisor of two positive integers that uses only comparisons, subtractions, and shifts of binary expansions, without using any divisions.
c) Find $\gcd(1202, 4848)$ using this algorithm.

29. Adapt the proof that there are infinitely many primes (Theorem 3 in Section 4.3) to show that there are infinitely many primes in the arithmetic progression $6k + 5$, $k = 1, 2, \ldots$.

30. Explain why you cannot directly adapt the proof that there are infinitely many primes (Theorem 3 in Section 4.3) to show that there are infinitely many primes in the arithmetic progression $3k + 1$, $k = 1, 2, \ldots$.

31. Explain why you cannot directly adapt the proof that there are infinitely many primes (Theorem 3 in Section 4.3) to show that there are infinitely many primes in the arithmetic progression $4k + 1$, $k = 1, 2, \ldots$.

32. Show that if the smallest prime factor p of the positive integer n is larger than $\sqrt[3]{n}$, then n/p is prime or equal to 1.

A set of integers is called **mutually relatively prime** if the greatest common divisor of these integers is 1.

33. Determine whether the integers in each of these sets are mutually relatively prime.
a) 8, 10, 12
b) 12, 15, 25
c) 15, 21, 28
d) 21, 24, 28, 32

34. Find a set of four mutually relatively prime integers such that no two of them are relatively prime.

*35. For which positive integers n is $n^4 + 4^n$ prime?

36. Show that the system of congruences $x \equiv 2 \pmod{6}$ and $x \equiv 3 \pmod{9}$ has no solutions.

37. Find all solutions of the system of congruences $x \equiv 4 \pmod{6}$ and $x \equiv 13 \pmod{15}$.

*38. **a)** Show that the system of congruences $x \equiv a_1 \pmod{m_1}$ and $x \equiv a_2 \pmod{m_2}$, where a_1, a_2, m_1, and m_2 are integers with $m_1 > 0$ and $m_2 > 0$, has a solution if and only if $\gcd(m_1, m_2) \mid a_1 - a_2$.
b) Show that if the system in part (a) has a solution, then it is unique modulo $\text{lcm}(m_1, m_2)$.

39. Prove that 30 divides $n^9 - n$ for every nonnegative integer n.

40. Prove that $n^{12} - 1$ is divisible by 35 for every integer n for which $\gcd(n, 35) = 1$.

41. Show that if p and q are distinct prime numbers, then $p^{q-1} + q^{p-1} \equiv 1 \pmod{pq}$.

The check digit a_{13} for an ISBN-13 with initial digits $a_1a_2 \ldots a_{12}$ is determined by the congruence $(a_1 + a_3 + \cdots + a_{13}) + 3(a_2 + a_4 + \cdots + a_{12}) \equiv 0 \pmod{10}$.

42. Determine whether each of these 13-digit numbers is a valid ISBN-13.
   a) 978-0-073-20679-1
   b) 978-0-45424-521-1
   c) 978-3-16-148410-0
   d) 978-0-201-10179-9
43. Show that the check digit of an ISBN-13 can always detect a single error.
44. Show that there are transpositions of two digits that are not detected by an ISBN-13.

A **routing transit number (RTN)** is a bank code used in the United States which appears on the bottom of checks. The most common form of an RTN has nine digits, where the last digit is a check digit. If $d_1 d_2 \ldots d_9$ is a valid RTN, the congruence $3(d_1 + d_4 + d_7) + 7(d_2 + d_5 + d_8) + (d_3 + d_6 + d_9) \equiv 0 \pmod{10}$ must hold.

45. Show that if $d_1 d_2 \ldots d_9$ is a valid RTN, then $d_9 = 7(d_1 + d_4 + d_7) + 3(d_2 + d_5 + d_8) + 9(d_3 + d_6) \,\mathbf{mod}\, 10$. Furthermore, use this formula to find the check digit that follows the eight digits 11100002 in a valid RTN.
46. Show that the check digit of an RTN can detect all single errors and determine which transposition errors an RTN check digit can catch and which ones it cannot catch.
47. The encrypted version of a message is LJMKG MGMXF QEXMW. If it was encrypted using the affine cipher $f(p) = (7p + 10) \,\mathbf{mod}\, 26$, what was the original message?

Autokey ciphers are ciphers where the nth letter of the plaintext is shifted by the numerical equivalent of the nth letter of a keystream. The keystream begins with a seed letter; its subsequent letters are constructed using either the plaintext or the ciphertext. When the plaintext is used, each character of the keystream, after the first, is the previous letter of the plaintext. When the ciphertext is used, each subsequent character of the keystream, after the first, is the previous letter of the ciphertext computed so far. In both cases, plaintext letters are encrypted by shifting each character by the numerical equivalent of the corresponding keystream letter.

48. Use the autokey cipher to encrypt the message NOW IS THE TIME TO DECIDE (ignoring spaces) using
   a) the keystream with seed X followed by letters of the plaintext.
   b) the keystream with seed X followed by letters of the ciphertext.
49. Use the autokey cipher to encrypt the message THE DREAM OF REASON (ignoring spaces) using
   a) the keystream with seed X followed by letters of the plaintext.
   b) the keystream with seed X followed by letters of the ciphertext.

Computer Projects

Write programs with these inputs and outputs.

1. Given integers n and b, each greater than 1, find the base b expansion of this integer.
2. Given the positive integers a, b, and m with $m > 1$, find $a^b \,\mathbf{mod}\, m$.
3. Given a positive integer, find the Cantor expansion of this integer (see the preamble to Exercise 54 of Section 4.2).
4. Given a positive integer, determine whether it is prime using trial division.
5. Given a positive integer, find the prime factorization of this integer.
6. Given two positive integers, find their greatest common divisor using the Euclidean algorithm.
7. Given two positive integers, find their least common multiple.
8. Given positive integers a and b, find Bézout coefficients s and t of a and b.
9. Given relatively prime positive integers a and b, find an inverse of a modulo b.
10. Given n linear congruences modulo pairwise relatively prime moduli, find the simultaneous solution of these congruences modulo the product of these moduli.
11. Given a positive integer N, a modulus m, a multiplier a, an increment c, and a seed x_0, where $0 \le a < m$, $0 \le c < m$, and $0 \le x_0 < m$, generate the sequence of N pseudorandom numbers using the linear congruential generator $x_{n+1} = (ax_n + c) \,\mathbf{mod}\, m$.
12. Given a set of identification numbers, use a hash function to assign them to memory locations where there are k memory locations.
13. Compute the check digit when given the first nine digits of an ISBN-10.
14. Given a message and a positive integer k less than 26, encrypt this message using the shift cipher with key k; and given a message encrypted using a shift cipher with key k, decrypt this message.
15. Given a message and positive integers a and b less than 26 with gcd$(a, 26)$, encrypt this message using an affine cipher with key (a, b); and given a message encrypted using the affine cipher with key (a, b), decrypt this message, by first finding the decryption key and then applying the appropriate decryption transformation.
16. Find the original plaintext message from the ciphertext message produced by encrypting the plaintext message

using a shift cipher. Do this using a frequency count of letters in the ciphertext.

*17. Construct a valid RSA encryption key by finding two primes p and q with 200 digits each and an integer $e > 1$ relatively prime to $(p - 1)(q - 1)$.

18. Given a message and an integer $n = pq$ where p and q are odd primes and an integer $e > 1$ relatively prime to $(p - 1)(q - 1)$, encrypt the message using the RSA cryptosystem with key (n, e).

19. Given a valid RSA key (n, e), and the primes p and q with $n = pq$, find the associated decryption key d.

20. Given a message encrypted using the RSA cryptosystem with key (n, e) and the associated decryption key d, decrypt this message.

21. Generate a shared key using the Diffie-Hellman key exchange protocol.

22. Given the RSA public and private keys of two parties, send a signed secret message from one of the parties to the other.

Computations and Explorations

Use a computational program or programs you have written to do these exercises.

1. Determine whether $2^p - 1$ is prime for each of the primes not exceeding 100.
2. Test a range of large Mersenne numbers $2^p - 1$ to determine whether they are prime. (You may want to use software from the GIMPS project.)
3. Determine whether $Q_n = p_1 p_2 \cdots p_n + 1$ is prime where $p_1, p_2, \ldots, p_n$ are the n smallest primes, for as many positive integers n as possible.
4. Look for polynomials in one variables whose values at long runs of consecutive integers are all primes.
5. Find as many primes of the form $n^2 + 1$ where n is a positive integer as you can. It is not known whether there are infinitely many such primes.
6. Find 10 different primes each with 100 digits.
7. How many primes are there less than 1,000,000, less than 10,000,000, and less than 100,000,000? Can you propose an estimate for the number of primes less than x where x is a positive integer?
8. Find a prime factor of each of 10 different 20-digit odd integers, selected at random. Keep track of how long it takes to find a factor of each of these integers. Do the same thing for 10 different 30-digit odd integers, 10 different 40-digit odd integers, and so on, continuing as long as possible.
9. Find all pseudoprimes to the base 2 that do not exceed 10,000.

Writing Projects

Respond to these with essays using outside sources.

1. Describe the Lucas–Lehmer test for determining whether a Mersenne number is prime. Discuss the progress of the GIMPS project in finding Mersenne primes using this test.
2. Explain how probabilistic primality tests are used in practice to produce extremely large numbers that are almost certainly prime. Do such tests have any potential drawbacks?
3. The question of whether there are infinitely many Carmichael numbers was solved recently after being open for more than 75 years. Describe the ingredients that went into the proof that there are infinitely many such numbers.
4. Summarize the current status of factoring algorithms in terms of their complexity and the size of numbers that can currently be factored. When do you think that it will be feasible to factor 200-digit numbers?
5. Describe the algorithms that are actually used by modern computers to add, subtract, multiply, and divide positive integers.
6. Describe the history of the Chinese remainder theorem. Describe some of the relevant problems posed in Chinese and Hindu writings and how the Chinese remainder theorem applies to them.
7. When are the numbers of a sequence truly random numbers, and not pseudorandom? What shortcomings have been observed in simulations and experiments in which pseudorandom numbers have been used? What are the properties that pseudorandom numbers can have that random numbers should not have?
8. Explain how a check digit is found for an International Bank Account Number (IBAN) and discuss the types of errors that can be found using this check digit.
9. Describe the Luhn algorithm for finding the check digit of a credit card number and discuss the types of errors that can be found using this check digit.
10. Show how a congruence can be used to tell the day of the week for any given date.

11. Describe how public key cryptography is being applied. Are the ways it is applied secure given the status of factoring algorithms? Will information kept secure using public key cryptography become insecure in the future?

12. Describe how public key cryptography can be used to produce signed secret messages so that the recipient is relatively sure the message was sent by the person expected to have sent it.

13. Describe the Rabin public key cryptosystem, explaining how to encrypt and how to decrypt messages and why it is suitable for use as a public key cryptosystem.

*14. Explain why it would be unsuitable to use p, where p is a large prime, as the modulus for encryption in the RSA cryptosystem. That is, explain how someone could, without excessive computation, find a private key from the corresponding public key if the modulus were a large prime, rather than the product of two large primes.

15. Explain what is meant by a cryptographic hash function. What are the important properties such a function must have?

*16. Explain the steps that Gentry used to construct a fully homomorphic cryptosystem.

CHAPTER

5 Induction and Recursion

5.1 Mathematical Induction

5.2 Strong Induction and Well-Ordering

5.3 Recursive Definitions and Structural Induction

5.4 Recursive Algorithms

5.5 Program Correctness

Many mathematical statements assert that a property is true for all positive integers. Examples of such statements are that for every positive integer n: $n! \leq n^n$, $n^3 - n$ is divisible by 3; a set with n elements has 2^n subsets; and the sum of the first n positive integers is $n(n + 1)/2$. A major goal of this chapter, and the book, is to provide a thorough understanding of mathematical induction, which is used to prove results of this kind.

Proofs using mathematical induction have two parts. First, they show that the statement holds for the positive integer 1. Second, they show that if the statement holds for a positive integer then it must also hold for the next larger integer. Mathematical induction is based on the rule of inference that tells us that if $P(1)$ and $\forall k(P(k) \to P(k + 1))$ are true for the domain of positive integers, then $\forall nP(n)$ is true. Mathematical induction can be used to prove a tremendous variety of results. Understanding how to read and construct proofs by mathematical induction is a key goal of learning discrete mathematics.

In Chapter 2 we explicitly defined sets and functions. That is, we described sets by listing their elements or by giving some property that characterizes these elements. We gave formulae for the values of functions. There is another important way to define such objects, based on mathematical induction. To define functions, some initial terms are specified, and a rule is given for finding subsequent values from values already known. (We briefly touched on this sort of definition in Chapter 2 when we showed how sequences can be defined using recurrence relations.) Sets can be defined by listing some of their elements and giving rules for constructing elements from those already known to be in the set. Such definitions, called *recursive definitions,* are used throughout discrete mathematics and computer science. Once we have defined a set recursively, we can use a proof method called structural induction to prove results about this set.

When a procedure is specified for solving a problem, this procedure must *always* solve the problem correctly. Just testing to see that the correct result is obtained for a set of input values does not show that the procedure always works correctly. The correctness of a procedure can be guaranteed only by proving that it always yields the correct result. The final section of this chapter contains an introduction to the techniques of program verification. This is a formal technique to verify that procedures are correct. Program verification serves as the basis for attempts under way to prove in a mechanical fashion that programs are correct.

5.1 Mathematical Induction

5.1.1 Introduction

Suppose that we have an infinite ladder, as shown in Figure 1, and we want to know whether we can reach every step on this ladder. We know two things:

1. We can reach the first rung of the ladder.
2. If we can reach a particular rung of the ladder, then we can reach the next rung.

Can we conclude that we can reach every rung? By (1), we know that we can reach the first rung of the ladder. Moreover, because we can reach the first rung, by (2), we can also reach the second rung; it is the next rung after the first rung. Applying (2) again, because we can reach the second rung, we can also reach the third rung. Continuing in this way, we can show that we can reach the fourth rung, the fifth rung, and so on. For example, after 100 uses of (2), we know that we can reach the 101st rung. But can we conclude that we are able to reach every rung of this infinite ladder? The answer is yes, something we can verify using an important proof technique

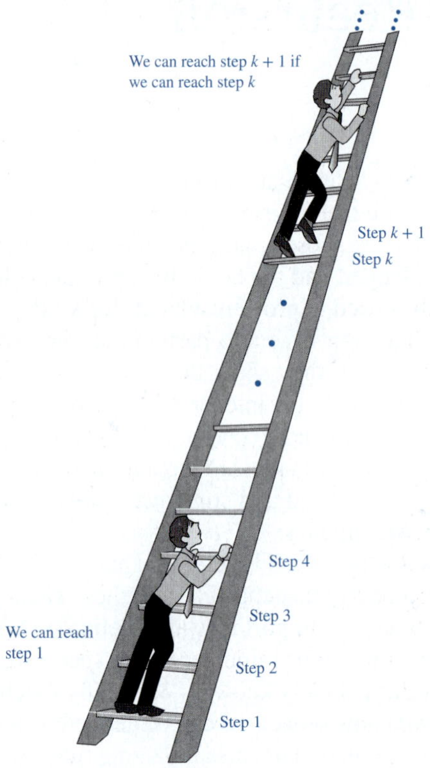

FIGURE 1 Climbing an infinite ladder.

called **mathematical induction**. That is, we can show that the statement that we can reach the nth rung of the ladder is true for all positive integers n.

Remark: In Figure 1, consider whether there is any rung of the ladder that we cannot reach. No matter how high a rung we choose, we can get to the rung below it, because we can get to the one below that rung, and so on, down to the first rung (which we know we can reach). If there are none we cannot reach, then we must be able reach them all! Furthermore, note that we cannot reach the top of the ladder, as it has no top, but we can reach each rung.

Mathematical induction is an extremely important proof technique that can be used to prove assertions of this type. As we will see in this section and in subsequent sections of this chapter and later chapters, mathematical induction is used extensively to prove results about a large variety of discrete objects. For example, it is used to prove results about the complexity of algorithms, the correctness of certain types of computer programs, theorems about graphs and trees, as well as a wide range of identities and inequalities.

In this section, we will describe how mathematical induction can be used and why it is a valid proof technique. It is extremely important to note that mathematical induction can be used only to prove results obtained in some other way. It is *not* a tool for discovering formulae or theorems.

5.1.2 Mathematical Induction

Assessment

In general, mathematical induction* can be used to prove statements that assert that $P(n)$ is true for all positive integers n, where $P(n)$ is a propositional function. A proof by mathematical

*Unfortunately, using the terminology "mathematical induction" clashes with the terminology used to describe different types of reasoning. In logic, **deductive reasoning** uses rules of inference to draw conclusions from premises, whereas **inductive reasoning** makes conclusions only supported, but not ensured, by evidence. Mathematical proofs, including arguments that use mathematical induction, are deductive, not inductive.

induction has two parts, a **basis step**, where we show that $P(1)$ is true, and an **inductive step**, where we show that for all positive integers k, if $P(k)$ is true, then $P(k+1)$ is true.

PRINCIPLE OF MATHEMATICAL INDUCTION: To prove that $P(n)$ is true for all positive integers n, where $P(n)$ is a propositional function, we complete two steps:

BASIS STEP: We verify that $P(1)$ is true.

INDUCTIVE STEP: We show that the conditional statement $P(k) \rightarrow P(k+1)$ is true for all positive integers k.

To complete the inductive step of a proof using the principle of mathematical induction, we assume that $P(k)$ is true for an arbitrary positive integer k and show that under this assumption, $P(k+1)$ must also be true. The assumption that $P(k)$ is true is called the **inductive hypothesis**. Once we complete both steps in a proof by mathematical induction, we have shown that $P(n)$ is true for all positive integers n, that is, we have shown that $\forall n P(n)$ is true where the quantification is over the set of positive integers. In the inductive step, we show that $\forall k(P(k) \rightarrow P(k+1))$ is true, where again, the domain is the set of positive integers.

Expressed as a rule of inference, this proof technique can be stated as

$$(P(1) \wedge \forall k(P(k) \rightarrow P(k+1))) \rightarrow \forall n P(n),$$

when the domain is the set of positive integers. Because mathematical induction is such an important technique, it is worthwhile to explain in detail the steps of a proof using this technique. The first thing we do to prove that $P(n)$ is true for all positive integers n is to show that $P(1)$ is true. This amounts to showing that the particular statement obtained when n is replaced by 1 in $P(n)$ is true. Then we must show that $P(k) \rightarrow P(k+1)$ is true for every positive integer k. To prove that this conditional statement is true for every positive integer k, we need to show that $P(k+1)$ cannot be false when $P(k)$ is true. This can be accomplished by assuming that $P(k)$ is true and showing that *under this hypothesis* $P(k+1)$ must also be true.

Remark: In a proof by mathematical induction it is *not* assumed that $P(k)$ is true for all positive integers! It is only shown that *if it is assumed* that $P(k)$ is true, then $P(k+1)$ is also true. Thus, a proof by mathematical induction is not a case of begging the question, or circular reasoning.

After completing the basis and inductive steps of a proof that $P(n)$ is true for all positive integers n, we know that $P(1)$ is true. That is what is shown in the basis step. We can conclude that $P(2)$ is true, because we know that $P(1)$ is true and from the inductive step we know that $P(1) \rightarrow P(2)$. Furthermore, we know that $P(3)$ is true because $P(2)$ is true and we know that $P(2) \rightarrow P(3)$ from the inductive step. Continuing along these lines using a finite number of implications, we can show that $P(n)$ is true for any particular positive integer n.

Links

HISTORICAL NOTE The first known use of mathematical induction is in the work of the sixteenth-century mathematician Francesco Maurolico (1494–1575). Maurolico wrote extensively on the works of classical mathematics and made many contributions to geometry and optics. In his book *Arithmeticorum Libri Duo*, Maurolico presented a variety of properties of the integers together with proofs of these properties. To prove some of these properties, he devised the method of mathematical induction. His first use of mathematical induction in this book was to prove that the sum of the first n odd positive integers equals n^2. Augustus De Morgan is credited with the first presentation in 1838 of formal proofs using mathematical induction, as well as introducing the terminology "mathematical induction." Maurolico's proofs were informal and he never used the word "induction." See [Gu10] to learn more about the history of the method of mathematical induction.

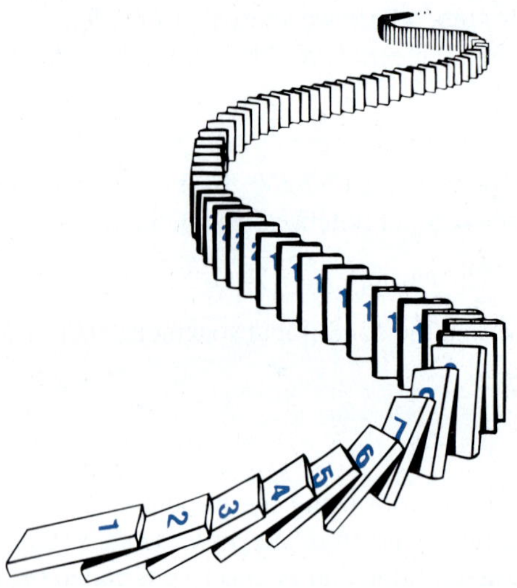

FIGURE 2 Illustrating how mathematical induction works using dominoes.

WAYS TO REMEMBER HOW MATHEMATICAL INDUCTION WORKS Thinking of the infinite ladder and the rules for reaching steps can help you remember how mathematical induction works. Note that statements (1) and (2) for the infinite ladder are exactly the basis step and inductive step, respectively, of the proof that $P(n)$ is true for all positive integers n, where $P(n)$ is the statement that we can reach the nth rung of the ladder. Consequently, we can invoke mathematical induction to conclude that we can reach every rung.

Another way to illustrate the principle of mathematical induction is to consider an infinite row of dominoes, labeled $1, 2, 3, \ldots, n, \ldots$, where each domino is standing up. Let $P(n)$ be the proposition that domino n is knocked over. If the first domino is knocked over—that is, if $P(1)$ is true—and if, whenever the kth domino is knocked over, it also knocks the $(k + 1)$st domino over—that is, if $P(k) \rightarrow P(k + 1)$ is true for all positive integers k—then all the dominoes are knocked over. This is illustrated in Figure 2.

5.1.3 Why Mathematical Induction is Valid

Why is mathematical induction a valid proof technique? The reason comes from the well-ordering property, listed in Appendix 1 as an axiom for the set of positive integers, which states that every nonempty subset of the set of positive integers has a least element. So, suppose we know that $P(1)$ is true and that the proposition $P(k) \rightarrow P(k + 1)$ is true for all positive integers k. To show that $P(n)$ must be true for all positive integers n, assume that there is at least one positive integer n for which $P(n)$ is false. Then the set S of positive integers n for which $P(n)$ is false is nonempty. Thus, by the well-ordering property, S has a least element, which will be denoted by m. We know that m cannot be 1, because $P(1)$ is true. Because m is positive and greater than 1, $m - 1$ is a positive integer. Furthermore, because $m - 1$ is less than m, it is not in S, so $P(m - 1)$ must be true. Because the conditional statement $P(m - 1) \rightarrow P(m)$ is also true, it must be the case that $P(m)$ is true. This contradicts the choice of m. Hence, $P(n)$ must be true for every positive integer n.

Remark: In this book we take the well-ordering property for the positive integers as an axiom. We proved that mathematical induction is a valid proof technique. Instead, we could have taken the principle of mathematical induction as an axiom and proved that the positive integers are well ordered. That is, the well-ordering property for positive integers and the principle of mathematical induction are equivalent. (In Section 5.2 we will present examples of proofs that use the well-ordering property directly. Also, Exercise 45 in that section asks for a proof that the well-ordering property for positive integers is a consequence of the principle of mathematical induction.

Remark: Mathematical induction can be used to prove results about any countable set of integers that begin at an arbitrary integer b. To understand this, imagine the infinite ladder beginning at step b with subsequent steps labelled by the increasing integers greater than b in a countably infinite sequence.

5.1.4 Choosing the Correct Basis Step

Mathematical induction can be used to prove theorems other than those of the form "$P(n)$ is true for all positive integers n." Often, we will need to show that $P(n)$ is true for $n = b, b+1, b+2, \ldots$, where b is an integer other than 1. We can use mathematical induction to accomplish this, as long as we change the basis step by replacing $P(1)$ with $P(b)$. In other words, to use mathematical induction to show that $P(n)$ is true for $n = b, b+1, b+2, \ldots$, where b is an integer other than 1, we show that $P(b)$ is true in the basis step. In the inductive step, we show that the conditional statement $P(k) \to P(k+1)$ is true for $k = b, b+1, b+2, \ldots$. Note that b can be negative, zero, or positive. Following the domino analogy we used earlier, imagine that we begin by knocking down the bth domino (the basis step), and as each domino falls, it knocks down the next domino (the inductive step). We leave it to the reader to show that this form of induction is valid (see Exercise 85).

We will illustrate this notion in Example 3, which states that a summation formula is valid for all nonnegative integers. In this example, we need to prove that $P(n)$ is true for $n = 0, 1, 2, \ldots$. So, the basis step in Example 3 will show that $P(0)$ is true.

5.1.5 Guidelines for Proofs by Mathematical Induction

Examples 1–14 will illustrate how to use mathematical induction to prove a diverse collection of theorems. Each of these examples includes all the elements needed in a proof by mathematical induction. We will also present an example of an invalid proof by mathematical induction. Before we give these proofs, we will provide some useful guidelines for constructing correct proofs by mathematical induction.

Template for Proofs by Mathematical Induction

1. Express the statement that is to be proved in the form "for all $n \geq b$, $P(n)$" for a fixed integer b. For statements of the form "$P(n)$ for all positive integers n," let $b = 1$, and for statements of the form "$P(n)$ for all nonnegative integers n," let $b = 0$. For some statements of the form $P(n)$, such as inequalities, you may need to determine the appropriate value of b by checking the truth values of $P(n)$ for small values of n, as is done in Example 6.
2. Write out the words "Basis Step." Then show that $P(b)$ is true, taking care that the correct value of b is used. This completes the first part of the proof.
3. Write out the words "Inductive Step" and state, and clearly identify, the inductive hypothesis, in the form "Assume that $P(k)$ is true for an arbitrary fixed integer $k > b$."
4. State what needs to be proved under the assumption that the inductive hypothesis is true. That is, write out what $P(k+1)$ says.
5. Prove the statement $P(k+1)$ making use of the assumption $P(k)$. (Generally, this is the most difficult part of a mathematical induction proof. Decide on the most promising proof strategy and look ahead to see how to use the induction hypothesis to build your proof of the inductive step. Also, be sure that your proof is valid for all integers k with $k \geq b$, taking care that the proof works for small values of k, including $k = b$.)
6. Clearly identify the conclusion of the inductive step, such as by saying "This completes the inductive step."
7. After completing the basis step and the inductive step, state the conclusion, namely, "By mathematical induction, $P(n)$ is true for all integers n with $n \geq b$."

Readers will find it worthwhile to see how the steps described in the template are completed in each of the 14 examples. It will also be useful to follow these guidelines in the solutions of the exercises that ask for proofs by mathematical induction. The guidelines that we presented can be adapted for each of the variants of mathematical induction that we introduce in the exercises and later in this chapter.

5.1.6 The Good and the Bad of Mathematical Induction

You can prove a theorem by mathematical induction even if you do not have the slightest idea why it is true!

An important point needs to be made about mathematical induction before we commence a study of its use. The good thing about mathematical induction is that it can be used to prove a conjecture once it is has been made (and is true). The bad thing about it is that it cannot be used to find new theorems. Mathematicians sometimes find proofs by mathematical induction unsatisfying because they do not provide insights as to why theorems are true. Many theorems can be proved in many ways, including by mathematical induction. Proofs of these theorems by methods other than mathematical induction are often preferred because of the insights they bring. (See Example 8 and the subsequent remark for an example of this.)

5.1.7 Examples of Proofs by Mathematical Induction

Many theorems assert that $P(n)$ is true for all positive integers n, where $P(n)$ is a propositional function. Mathematical induction is a technique for proving theorems of this kind. In other words, mathematical induction can be used to prove statements of the form $\forall n\, P(n)$, where the domain is the set of positive integers. Mathematical induction can be used to prove an extremely wide variety of theorems, each of which is a statement of this form. (Remember, many mathematical assertions include an implicit universal quantifier. The statement "if n is a positive integer, then $n^3 - n$ is divisible by 3" is an example of this. Making the implicit universal quantifier explicit yields the statement "for every positive integer n, $n^3 - n$ is divisible by 3.")

Links

We will use a variety of examples to illustrate how theorems are proved using mathematical induction. The theorems we will prove include summation formulae, inequalities, identities for combinations of sets, divisibility results, theorems about algorithms, and some other creative results. In this section, and in later sections, we will employ mathematical induction to prove many other types of results, including the correctness of computer programs and algorithms. Mathematical induction can be used to prove a wide variety of theorems both similar to, and also quite different from, the examples here. (For proofs by mathematical induction of many more interesting and diverse results, see the *Handbook of Mathematical Induction* by David Gunderson [Gu11].)

There are many opportunities for errors in induction proofs. We will describe some incorrect proofs by mathematical induction at the end of this section and in the exercises. To avoid making errors in proofs by mathematical induction, try to follow the guidelines for such proofs provided previously in Subsection 5.1.5.

Look for the $\overset{\text{IH}}{=}$ symbol to see where the inductive hypothesis is used.

SEEING WHERE THE INDUCTIVE HYPOTHESIS IS USED To help the reader understand each of the mathematical induction proofs in this section, we will note where the inductive hypothesis is used. We indicate this use in three different ways: by explicit mention in the text, by inserting the acronym IH (for inductive hypothesis) over an equals sign or a sign for an inequality, or by specifying the inductive hypothesis as the reason for a step in a multi-line display.

PROVING SUMMATION FORMULAE We begin by using mathematical induction to prove several summation formulae. As we will see, mathematical induction is particularly well suited for proving that such formulae are valid. However, summation formulae can be proven in other ways. This is not surprising because there are often different ways to prove a theorem. The major disadvantage of using mathematical induction to prove a summation formula is that you cannot use it to derive this formula. That is, you must already have the formula before you attempt to prove it by mathematical induction.

Examples 1–4 illustrate how to use mathematical induction to prove summation formulae. The first summation formula we will prove by mathematical induction, in Example 1, is a closed formula for the sum of the smallest n positive integers.

EXAMPLE 1 Show that if n is a positive integer, then

$$1 + 2 + \cdots + n = \frac{n(n+1)}{2}.$$

Extra Examples

Solution: Let $P(n)$ be the proposition that the sum of the first n positive integers, $1 + 2 + \cdots n$, is $n(n+1)/2$. We must do two things to prove that $P(n)$ is true for $n = 1, 2, 3, \ldots$. Namely, we must show that $P(1)$ is true and that the conditional statement $P(k)$ implies $P(k+1)$ is true for $k = 1, 2, 3, \ldots$.

BASIS STEP: $P(1)$ is true, because $1 = \frac{1(1+1)}{2}$. (The left-hand side of this equation is 1 because 1 is the sum of the first positive integer. The right-hand side is found by substituting 1 for n in $n(n+1)/2$.)

INDUCTIVE STEP: For the inductive hypothesis we assume that $P(k)$ holds for an arbitrary positive integer k. That is, we assume that

If you are rusty simplifying algebraic expressions, this is the time to do some reviewing!

$$1 + 2 + \cdots + k = \frac{k(k+1)}{2}.$$

Under this assumption, it must be shown that $P(k+1)$ is true, namely, that

$$1 + 2 + \cdots + k + (k+1) = \frac{(k+1)[(k+1)+1]}{2} = \frac{(k+1)(k+2)}{2}$$

is also true.

We now look ahead to see how we might be able to prove that $P(k+1)$ holds under the assumption that $P(k)$ is true. We observe that the summation in the left-hand side of $P(k+1)$ is $k+1$ more than the summation in the left-hand side of $P(k)$. Our strategy will be to add $k+1$ to both sides of the equation in $P(k)$ and simplify the result algebraically to complete the inductive step.

We now return to the proof of the inductive step. When we add $k+1$ to both sides of the equation in $P(k)$, we obtain

$$\begin{aligned}1 + 2 + \cdots + k + (k+1) &\overset{\text{IH}}{=} \frac{k(k+1)}{2} + (k+1)\\ &= \frac{k(k+1) + 2(k+1)}{2}\\ &= \frac{(k+1)(k+2)}{2}.\end{aligned}$$

This last equation shows that $P(k+1)$ is true under the assumption that $P(k)$ is true. This completes the inductive step.

We have completed the basis step and the inductive step, so by mathematical induction we know that $P(n)$ is true for all positive integers n. That is, we have proven that $1 + 2 + \cdots + n = n(n+1)/2$ for all positive integers n. ◀

As we noted, mathematical induction is not a tool for finding theorems about all positive integers. Rather, it is a proof method for proving such results once they are conjectured. In Example 2, using mathematical induction to prove a summation formula, we will both formulate and then prove a conjecture.

EXAMPLE 2 Conjecture a formula for the sum of the first n positive odd integers. Then prove your conjecture using mathematical induction.

Solution: The sums of the first n positive odd integers for $n = 1, 2, 3, 4, 5$ are

$$1 = 1, \quad 1 + 3 = 4, \quad 1 + 3 + 5 = 9,$$
$$1 + 3 + 5 + 7 = 16, \quad 1 + 3 + 5 + 7 + 9 = 25.$$

From these values it is reasonable to conjecture that the sum of the first n positive odd integers is n^2, that is, $1 + 3 + 5 + \cdots + (2n - 1) = n^2$. We need a method to *prove* that this *conjecture* is correct, if in fact it is.

Let $P(n)$ denote the proposition that the sum of the first n odd positive integers is n^2. Our conjecture is that $P(n)$ is true for all positive integers n. To use mathematical induction to prove this conjecture, we must first complete the basis step; that is, we must show that $P(1)$ is true. Then we must carry out the inductive step; that is, we must show that $P(k + 1)$ is true when $P(k)$ is assumed to be true. We now attempt to complete these two steps.

BASIS STEP: $P(1)$ states that the sum of the first one odd positive integer is 1^2. This is true because the sum of the first odd positive integer is 1. The basis step is complete.

INDUCTIVE STEP: To complete the inductive step we must show that the proposition $P(k) \to P(k + 1)$ is true for every positive integer k. To do this, we first assume the inductive hypothesis. The inductive hypothesis is the statement that $P(k)$ is true for an arbitrary positive integer k, that is,

$$1 + 3 + 5 + \cdots + (2k - 1) = k^2.$$

(Note that the kth odd positive integer is $(2k - 1)$, because this integer is obtained by adding 2 a total of $k - 1$ times to 1.)

To show that $\forall k(P(k) \to P(k + 1))$ is true, we must show that if $P(k)$ is true (the inductive hypothesis), then $P(k + 1)$ is true. Note that $P(k + 1)$ is the statement that

$$1 + 3 + 5 + \cdots + (2k - 1) + (2k + 1) = (k + 1)^2.$$

Before we complete the inductive step, we will take a time out to figure out a strategy. At this stage of a mathematical induction proof it is time to look for a way to use the inductive hypothesis to show that $P(k + 1)$ is true. Here we note that $1 + 3 + 5 + \cdots + (2k - 1) + (2k + 1)$ is the sum of its first k terms $1 + 3 + 5 + \cdots + (2k - 1)$ and its last term $2k - 1$. So, we can use our inductive hypothesis to replace $1 + 3 + 5 + \cdots + (2k - 1)$ by k^2.

We now return to our proof. We find that

$$\begin{aligned}
1 + 3 + 5 + \cdots + (2k - 1) + (2k + 1) &= [1 + 3 + \cdots + (2k - 1)] + (2k + 1) \\
&\overset{\text{IH}}{=} k^2 + (2k + 1) \\
&= k^2 + 2k + 1 \\
&= (k + 1)^2.
\end{aligned}$$

This shows that $P(k + 1)$ follows from $P(k)$. Note that we used the inductive hypothesis $P(k)$ in the second equality to replace the sum of the first k odd positive integers by k^2.

We have now completed both the basis step and the inductive step. That is, we have shown that $P(1)$ is true and the conditional statement $P(k) \to P(k+1)$ is true for all positive integers k. Consequently, by the principle of mathematical induction we can conclude that $P(n)$ is true for all positive integers n. That is, we know that $1 + 3 + 5 + \cdots + (2n - 1) = n^2$ for all positive integers n. ◀

EXAMPLE 3 Use mathematical induction to show that

$$1 + 2 + 2^2 + \cdots + 2^n = 2^{n+1} - 1$$

for all nonnegative integers n.

Solution: Let $P(n)$ be the proposition that $1 + 2 + 2^2 + \cdots + 2^n = 2^{n+1} - 1$ for the integer n.

BASIS STEP: $P(0)$ is true because $2^0 = 1 = 2^1 - 1$. This completes the basis step.

INDUCTIVE STEP: For the inductive hypothesis, we assume that $P(k)$ is true for an arbitrary nonnegative integer k. That is, we assume that

$$1 + 2 + 2^2 + \cdots + 2^k = 2^{k+1} - 1.$$

To carry out the inductive step using this assumption, we must show that when we assume that $P(k)$ is true, then $P(k+1)$ is also true. That is, we must show that

$$1 + 2 + 2^2 + \cdots + 2^k + 2^{k+1} = 2^{(k+1)+1} - 1 = 2^{k+2} - 1$$

assuming the inductive hypothesis $P(k)$. Under the assumption of $P(k)$, we see that

$$\begin{aligned} 1 + 2 + 2^2 + \cdots + 2^k + 2^{k+1} &= (1 + 2 + 2^2 + \cdots + 2^k) + 2^{k+1} \\ &\overset{IH}{=} (2^{k+1} - 1) + 2^{k+1} \\ &= 2 \cdot 2^{k+1} - 1 \\ &= 2^{k+2} - 1. \end{aligned}$$

Note that we used the inductive hypothesis in the second equation in this string of equalities to replace $1 + 2 + 2^2 + \cdots + 2^k$ by $2^{k+1} - 1$. We have completed the inductive step.

Because we have completed the basis step and the inductive step, by mathematical induction we know that $P(n)$ is true for all nonnegative integers n. That is, $1 + 2 + \cdots + 2^n = 2^{n+1} - 1$ for all nonnegative integers n. ◀

The formula given in Example 3 is a special case of a general result for the sum of terms of a geometric progression (Theorem 1 in Section 2.4). We will use mathematical induction to provide an alternative proof of this formula.

EXAMPLE 4 **Sums of Geometric Progressions** Use mathematical induction to prove this formula for the sum of a finite number of terms of a geometric progression with initial term a and common ratio r:

$$\sum_{j=0}^{n} ar^j = a + ar + ar^2 + \cdots + ar^n = \frac{ar^{n+1} - a}{r - 1} \qquad \text{when } r \neq 1,$$

where n is a nonnegative integer.

Solution: To prove this formula using mathematical induction, let $P(n)$ be the statement that the sum of the first $n + 1$ terms of a geometric progression in this formula is correct.

BASIS STEP: $P(0)$ is true, because

$$\frac{ar^{0+1} - a}{r - 1} = \frac{ar - a}{r - 1} = \frac{a(r - 1)}{r - 1} = a.$$

INDUCTIVE STEP: The inductive hypothesis is the statement that $P(k)$ is true, where k is an arbitrary nonnegative integer. That is, $P(k)$ is the statement that

$$a + ar + ar^2 + \cdots + ar^k = \frac{ar^{k+1} - a}{r - 1}.$$

To complete the inductive step we must show that if $P(k)$ is true, then $P(k + 1)$ is also true. To show that this is the case, we first add ar^{k+1} to both sides of the equality asserted by $P(k)$. We find that

$$a + ar + ar^2 + \cdots + ar^k + ar^{k+1} \overset{\text{IH}}{=} \frac{ar^{k+1} - a}{r - 1} + ar^{k+1}.$$

Rewriting the right-hand side of this equation shows that

$$\begin{aligned} \frac{ar^{k+1} - a}{r - 1} + ar^{k+1} &= \frac{ar^{k+1} - a}{r - 1} + \frac{ar^{k+2} - ar^{k+1}}{r - 1} \\ &= \frac{ar^{k+2} - a}{r - 1}. \end{aligned}$$

Combining these last two equations gives

$$a + ar + ar^2 + \cdots + ar^k + ar^{k+1} = \frac{ar^{k+2} - a}{r - 1}.$$

This shows that if the inductive hypothesis $P(k)$ is true, then $P(k + 1)$ must also be true. This completes the inductive argument.

We have completed the basis step and the inductive step, so by mathematical induction $P(n)$ is true for all nonnegative integers n. This shows that the formula for the sum of the terms of a geometric series is correct. ◀

As previously mentioned, the formula in Example 3 is the case of the formula in Example 4 with $a = 1$ and $r = 2$. The reader should verify that putting these values for a and r into the general formula gives the same formula as in Example 3.

PROVING INEQUALITIES Mathematical induction can be used to prove a variety of inequalities that hold for all positive integers greater than a particular positive integer, as Examples 5–7 illustrate.

EXAMPLE 5 Use mathematical induction to prove the inequality

$$n < 2^n$$

for all positive integers n.

Solution: Let $P(n)$ be the proposition that $n < 2^n$.

Extra Examples

BASIS STEP: $P(1)$ is true, because $1 < 2^1 = 2$. This completes the basis step.

INDUCTIVE STEP: We first assume the inductive hypothesis that $P(k)$ is true for an arbitrary positive integer k. That is, the inductive hypothesis $P(k)$ is the statement that $k < 2^k$. To complete the inductive step, we need to show that if $P(k)$ is true, then $P(k + 1)$, which is the statement that $k + 1 < 2^{k+1}$ is true. That is, we need to show that if $k < 2^k$, then $k + 1 < 2^{k+1}$. To show that this conditional statement is true for the positive integer k, we first add 1 to both sides of $k < 2^k$, and then note that $1 < 2^k$. This tells us that

$$k + 1 \overset{\text{IH}}{<} 2^k + 1 < 2^k + 2^k = 2 \cdot 2^k = 2^{k+1}.$$

This shows that $P(k + 1)$ is true, namely, that $k + 1 < 2^{k+1}$, based on the assumption that $P(k)$ is true. The induction step is complete.

Therefore, because we have completed both the basis step and the inductive step, by the principle of mathematical induction we have shown that $n < 2^n$ is true for all positive integers n. ◀

EXAMPLE 6 Use mathematical induction to prove that $2^n < n!$ for every integer n with $n \geq 4$. (Note that this inequality is false for $n = 1, 2,$ and 3.)

Solution: Let $P(n)$ be the proposition that $2^n < n!$.

BASIS STEP: To prove the inequality for $n \geq 4$ requires that the basis step be $P(4)$. Note that $P(4)$ is true, because $2^4 = 16 < 24 = 4!$.

INDUCTIVE STEP: For the inductive step, we assume that $P(k)$ is true for an arbitrary integer k with $k \geq 4$. That is, we assume that $2^k < k!$ for the positive integer k with $k \geq 4$. We must show that under this hypothesis, $P(k + 1)$ is also true. That is, we must show that if $2^k < k!$ for an arbitrary positive integer k where $k \geq 4$, then $2^{k+1} < (k + 1)!$. We have

$$\begin{aligned}
2^{k+1} &= 2 \cdot 2^k && \text{by definition of exponent} \\
&\overset{\text{IH}}{<} 2 \cdot k! && \text{by the inductive hypothesis} \\
&< (k + 1)k! && \text{because } 2 < k + 1 \\
&= (k + 1)! && \text{by definition of factorial function}
\end{aligned}$$

This shows that $P(k + 1)$ is true when $P(k)$ is true. This completes the inductive step of the proof.

We have completed the basis step and the inductive step. Hence, by mathematical induction $P(n)$ is true for all integers n with $n \geq 4$. That is, we have proved that $2^n < n!$ is true for all integers n with $n \geq 4$. ◀

An important inequality for the sum of the reciprocals of a set of positive integers will be proved in Example 7.

EXAMPLE 7 **An Inequality for Harmonic Numbers** The **harmonic numbers** $H_j, j = 1, 2, 3, \ldots,$ are defined by

$$H_j = 1 + \frac{1}{2} + \frac{1}{3} + \cdots + \frac{1}{j}.$$

For instance,

$$H_4 = 1 + \frac{1}{2} + \frac{1}{3} + \frac{1}{4} = \frac{25}{12}.$$

Use mathematical induction to show that

$$H_{2^n} \geq 1 + \frac{n}{2},$$

whenever n is a nonnegative integer.

Solution: To carry out the proof, let $P(n)$ be the proposition that $H_{2^n} \geq 1 + \frac{n}{2}$.

BASIS STEP: $P(0)$ is true, because $H_{2^0} = H_1 = 1 \geq 1 + \frac{0}{2}$.

INDUCTIVE STEP: The inductive hypothesis is the statement that $P(k)$ is true, that is, $H_{2^k} \geq 1 + \frac{k}{2}$, where k is an arbitrary nonnegative integer. We must show that if $P(k)$ is true, then $P(k+1)$, which states that $H_{2^{k+1}} \geq 1 + \frac{k+1}{2}$, is also true. So, assuming the inductive hypothesis, it follows that

$$\begin{aligned}
H_{2^{k+1}} &= 1 + \frac{1}{2} + \frac{1}{3} + \cdots + \frac{1}{2^k} + \frac{1}{2^k+1} + \cdots + \frac{1}{2^{k+1}} && \text{by the definition of harmonic number} \\
&= H_{2^k} + \frac{1}{2^k+1} + \cdots + \frac{1}{2^{k+1}} && \text{by the definition of } 2^k\text{th harmonic number} \\
&\overset{\text{IH}}{\geq} \left(1 + \frac{k}{2}\right) + \frac{1}{2^k+1} + \cdots + \frac{1}{2^{k+1}} && \text{by the inductive hypothesis} \\
&\geq \left(1 + \frac{k}{2}\right) + 2^k \cdot \frac{1}{2^{k+1}} && \text{because there are } 2^k \text{ terms each} \geq 1/2^{k+1} \\
&= \left(1 + \frac{k}{2}\right) + \frac{1}{2} && \text{canceling a common factor of } 2^k \text{ in second term} \\
&= 1 + \frac{k+1}{2}.
\end{aligned}$$

This establishes the inductive step of the proof.

We have completed the basis step and the inductive step. Thus, by mathematical induction $P(n)$ is true for all nonnegative integers n. That is, the inequality $H_{2^n} \geq 1 + \frac{n}{2}$ for the harmonic numbers holds for all nonnegative integers n. ◀

Remark: The inequality established here shows that the **harmonic series**

$$1 + \frac{1}{2} + \frac{1}{3} + \cdots + \frac{1}{n} + \cdots$$

is a divergent infinite series. This is an important example in the study of infinite series.

PROVING DIVISIBILITY RESULTS Mathematical induction can be used to prove divisibility results about integers. Although such results are often easier to prove using basic results in number theory, it is instructive to see how to prove such results using mathematical induction, as Examples 8 and 9 illustrate.

EXAMPLE 8 Use mathematical induction to prove that $n^3 - n$ is divisible by 3 whenever n is a positive integer. (Note that this is the statement with $p = 3$ of Fermat's little theorem, which is Theorem 3 of Section 4.4.)

Extra Examples

Solution: To construct the proof, let $P(n)$ denote the proposition: "$n^3 - n$ is divisible by 3."

BASIS STEP: The statement $P(1)$ is true because $1^3 - 1 = 0$ is divisible by 3. This completes the basis step.

INDUCTIVE STEP: For the inductive hypothesis we assume that $P(k)$ is true; that is, we assume that $k^3 - k$ is divisible by 3 for an arbitrary positive integer k. To complete the inductive step, we must show that when we assume the inductive hypothesis, it follows that $P(k + 1)$, the statement that $(k + 1)^3 - (k + 1)$ is divisible by 3, is also true. That is, we must show that $(k + 1)^3 - (k + 1)$ is divisible by 3. Note that

$$\begin{aligned}(k+1)^3 - (k+1) &= (k^3 + 3k^2 + 3k + 1) - (k + 1)\\ &= (k^3 - k) + 3(k^2 + k).\end{aligned}$$

Using the inductive hypothesis, we conclude that the first term $k^3 - k$ is divisible by 3. The second term is divisible by 3 because it is 3 times an integer. So, by part (i) of Theorem 1 in Section 4.1, we know that $(k + 1)^3 - (k + 1)$ is also divisible by 3. This completes the inductive step.

Because we have completed both the basis step and the inductive step, by the principle of mathematical induction we know that $n^3 - n$ is divisible by 3 whenever n is a positive integer. ◀

Remark: We have included Example 8 as an illustration how a divisibility result can be proved by mathematical induction. However, there are simpler proofs. For example, we can prove that $n^3 - n$ is divisible by 3 for all positive integers n using the factorization $n^3 - n = n(n^2 - 1) = n(n - 1)(n + 1) = (n - 1)n(n + 1)$. Hence, $n^3 - 1$ is divisible by 3 because it is the product of three consecutive integers, one of which is divisible by 3.

Example 9 presents a more challenging proof by mathematical induction of a divisibility result.

EXAMPLE 9 Use mathematical induction to prove that $7^{n+2} + 8^{2n+1}$ is divisible by 57 for every nonnegative integer n.

Extra Examples

Solution: To construct the proof, let $P(n)$ denote the proposition: "$7^{n+2} + 8^{2n+1}$ is divisible by 57."

BASIS STEP: To complete the basis step, we must show that $P(0)$ is true, because we want to prove that $P(n)$ is true for every nonnegative integer n. We see that $P(0)$ is true because $7^{0+2} + 8^{2\cdot 0+1} = 7^2 + 8^1 = 57$ is divisible by 57. This completes the basis step.

INDUCTIVE STEP: For the inductive hypothesis we assume that $P(k)$ is true for an arbitrary nonnegative integer k; that is, we assume that $7^{k+2} + 8^{2k+1}$ is divisible by 57. To complete the inductive step, we must show that when we assume that the inductive hypothesis $P(k)$ is true, then $P(k + 1)$, the statement that $7^{(k+1)+2} + 8^{2(k+1)+1}$ is divisible by 57, is also true.

The difficult part of the proof is to see how to use the inductive hypothesis. To take advantage of the inductive hypothesis, we use these steps:

$$\begin{aligned}7^{(k+1)+2} + 8^{2(k+1)+1} &= 7^{k+3} + 8^{2k+3}\\ &= 7 \cdot 7^{k+2} + 8^2 \cdot 8^{2k+1}\\ &= 7 \cdot 7^{k+2} + 64 \cdot 8^{2k+1}\\ &= 7(7^{k+2} + 8^{2k+1}) + 57 \cdot 8^{2k+1}.\end{aligned}$$

We can now use the inductive hypothesis, which states that $7^{k+2} + 8^{2k+1}$ is divisible by 57. We will use parts (*i*) and (*ii*) of Theorem 1 in Section 4.1. By part (*ii*) of this theorem, and the inductive hypothesis, we conclude that the first term in this last sum, $7(7^{k+2} + 8^{2k+1})$, is divisible by 57. By part (*ii*) of this theorem, the second term in this sum, $57 \cdot 8^{2k+1}$, is divisible by 57. Hence, by part (*i*) of this theorem, we conclude that $7(7^{k+2} + 8^{2k+1}) + 57 \cdot 8^{2k+1} = 7^{k+3} + 8^{2k+3}$ is divisible by 57. This completes the inductive step.

Because we have completed both the basis step and the inductive step, by the principle of mathematical induction we know that $7^{n+2} + 8^{2n+1}$ is divisible by 57 for every nonnegative integer n. ◀

PROVING RESULTS ABOUT SETS Mathematical induction can be used to prove many results about sets. In particular, in Example 10 we prove a formula for the number of subsets of a finite set and in Example 11 we establish a set identity.

EXAMPLE 10 **The Number of Subsets of a Finite Set** Use mathematical induction to show that if S is a finite set with n elements, where n is a nonnegative integer, then S has 2^n subsets. (We will prove this result directly in several ways in Chapter 6.)

Solution: Let $P(n)$ be the proposition that a set with n elements has 2^n subsets.

BASIS STEP: $P(0)$ is true, because a set with zero elements, the empty set, has exactly $2^0 = 1$ subset, namely, itself.

INDUCTIVE STEP: For the inductive hypothesis we assume that $P(k)$ is true for an arbitrary nonnegative integer k, that is, we assume that every set with k elements has 2^k subsets. It must be shown that under this assumption, $P(k+1)$, which is the statement that every set with $k+1$ elements has 2^{k+1} subsets, must also be true. To show this, let T be a set with $k+1$ elements. Then, it is possible to write $T = S \cup \{a\}$, where a is one of the elements of T and $S = T - \{a\}$ (and hence $|S| = k$). The subsets of T can be obtained in the following way. For each subset X of S there are exactly two subsets of T, namely, X and $X \cup \{a\}$. (This is illustrated in Figure 3.) These constitute all the subsets of T and are all distinct. We now use the inductive hypothesis to conclude that S has 2^k subsets, because it has k elements. We also know that there are two subsets of T for each subset of S. Therefore, there are $2 \cdot 2^k = 2^{k+1}$ subsets of T. This finishes the inductive argument.

Because we have completed the basis step and the inductive step, by mathematical induction it follows that $P(n)$ is true for all nonnegative integers n. That is, we have proved that a set with n elements has 2^n subsets whenever n is a nonnegative integer. ◀

EXAMPLE 11 Use mathematical induction to prove the following generalization of one of De Morgan's laws:

$$\overline{\bigcap_{j=1}^{n} A_j} = \bigcup_{j=1}^{n} \overline{A_j}$$

whenever $A_1, A_2, \ldots, A_n$ are subsets of a universal set U and $n \geq 2$.

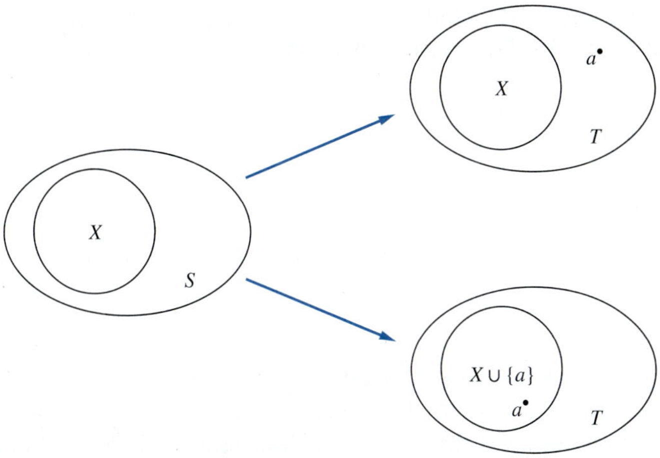

FIGURE 3 **Generating subsets of a set with $k + 1$ elements. Here $T = S \cup \{a\}$.**

Solution: Let $P(n)$ be the identity for n sets.

BASIS STEP: The statement $P(2)$ asserts that $\overline{A_1 \cap A_2} = \overline{A_1} \cup \overline{A_2}$. This is one of De Morgan's laws; it was proved in Example 11 of Section 2.2.

INDUCTIVE STEP: The inductive hypothesis is the statement that $P(k)$ is true, where k is an arbitrary integer with $k \geq 2$; that is, it is the statement that

$$\overline{\bigcap_{j=1}^{k} A_j} = \bigcup_{j=1}^{k} \overline{A_j}$$

whenever $A_1, A_2, \ldots, A_k$ are subsets of the universal set U. To carry out the inductive step, we need to show that this assumption implies that $P(k + 1)$ is true. That is, we need to show that if this equality holds for every collection of k subsets of U, then it must also hold for every collection of $k + 1$ subsets of U. Suppose that $A_1, A_2, \ldots, A_k, A_{k+1}$ are subsets of U. When the inductive hypothesis is assumed to hold, it follows that

$$\begin{aligned}
\overline{\bigcap_{j=1}^{k+1} A_j} &= \overline{\left(\bigcap_{j=1}^{k} A_j\right) \cap A_{k+1}} && \text{by the definition of intersection} \\
&= \overline{\left(\bigcap_{j=1}^{k} A_j\right)} \cup \overline{A_{k+1}} && \text{by De Morgan's law (where the two sets are } \textstyle\bigcap_{j=1}^{k} A_j \text{ and } A_{k+1}) \\
&\overset{\text{IH}}{=} \left(\bigcup_{j=1}^{k} \overline{A_j}\right) \cup \overline{A_{k+1}} && \text{by the inductive hypothesis} \\
&= \bigcup_{j=1}^{k+1} \overline{A_j} && \text{by the definition of union}
\end{aligned}$$

This completes the inductive step.

Because we have completed both the basis step and the inductive step, by mathematical induction we know that $P(n)$ is true whenever n is a positive integer, $n \geq 2$. That is, we know that

$$\overline{\bigcap_{j=1}^{n} A_j} = \bigcup_{j=1}^{n} \overline{A_j}$$

whenever $A_1, A_2, \ldots, A_n$ are subsets of a universal set U and $n \geq 2$. ◀

PROVING RESULTS ABOUT ALGORITHMS Next, we provide an example (somewhat more difficult than previous examples) that illustrates one of many ways mathematical induction is used in the study of algorithms. We will show how mathematical induction can be used to prove that a greedy algorithm we introduced in Section 3.1 always yields an optimal solution.

EXAMPLE 12 Recall the algorithm for scheduling talks discussed in Example 7 of Section 3.1. The input to this algorithm is a group of m proposed talks with preset starting and ending times. The goal is to schedule as many of these lectures as possible in the main lecture hall so that no two talks overlap. Suppose that talk t_j begins at time s_j and ends at time e_j. (No two lectures can proceed in the main lecture hall at the same time, but a lecture in this hall can begin at the same time another one ends.)

Without loss of generality, we assume that the talks are listed in order of nondecreasing ending time, so that $e_1 \leq e_2 \leq \cdots \leq e_m$. The greedy algorithm proceeds by selecting at each stage a talk with the earliest ending time among all those talks that begin no sooner than when the last talk scheduled in the main lecture hall has ended. Note that a talk with the earliest end time is always selected first by the algorithm. We will show that this greedy algorithm is optimal in the sense that it always schedules the most talks possible in the main lecture hall. To prove the optimality of this algorithm we use mathematical induction on the variable n, the number of talks scheduled by the algorithm. We let $P(n)$ be the proposition that if the greedy algorithm schedules n talks in the main lecture hall, then it is not possible to schedule more than n talks in this hall.

BASIS STEP: Suppose that the greedy algorithm managed to schedule just one talk, t_1, in the main lecture hall. This means that no other talk can start at or after e_1, the end time of t_1. Otherwise, the first such talk we come to as we go through the talks in order of nondecreasing end times could be added. Hence, at time e_1 each of the remaining talks needs to use the main lecture hall because they all start before e_1 and end after e_1. It follows that no two talks can be scheduled because both need to use the main lecture hall at time e_1. This shows that $P(1)$ is true and completes the basis step.

INDUCTIVE STEP: The inductive hypothesis is that $P(k)$ is true, where k is an arbitrary positive integer, that is, that the greedy algorithm always schedules the most possible talks when it selects k talks, where k is a positive integer, given any set of talks, no matter how many. We must show that $P(k+1)$ follows from the assumption that $P(k)$ is true, that is, we must show that under the assumption of $P(k)$, the greedy algorithm always schedules the most possible talks when it selects $k+1$ talks.

Now suppose that the greedy algorithm has selected $k+1$ talks. Our first step in completing the inductive step is to show there is a schedule including the most talks possible that contains talk t_1, a talk with the earliest end time. This is easy to see because a schedule that begins with the talk t_i in the list, where $i > 1$, can be changed so that talk t_1 replaces talk t_i. To see this, note that because $e_1 \leq e_i$, all talks that were scheduled to follow talk t_i can still be scheduled.

Once we included talk t_1, scheduling the talks so that as many as possible are scheduled is reduced to scheduling as many talks as possible that begin at or after time e_1. So, if we have scheduled as many talks as possible, the schedule of talks other than talk t_1 is an optimal schedule of the original talks that begin once talk t_1 has ended. Because the greedy algorithm schedules k talks when it creates this schedule, we can apply the inductive hypothesis to conclude that it has scheduled the most possible talks. It follows that the greedy algorithm has scheduled the most possible talks, $k + 1$, when it produced a schedule with $k + 1$ talks, so $P(k + 1)$ is true. This completes the inductive step.

We have completed the basis step and the inductive step. So, by mathematical induction we know that $P(n)$ is true for all positive integers n. This completes the proof of optimality. That is, we have proved that when the greedy algorithm schedules n talks, when n is a positive integer, then it is not possible to schedule more than n talks. ◀

CREATIVE USES OF MATHEMATICAL INDUCTION Mathematical induction can often be used in unexpected ways. We will illustrate two particularly clever uses of mathematical induction here, the first relating to survivors in a pie fight and the second relating to tilings with regular triominoes of checkerboards with one square missing.

EXAMPLE 13 **Odd Pie Fights** An odd number of people stand in a yard at mutually distinct distances. At the same time each person throws a pie at their nearest neighbor, hitting this person. Use mathematical induction to show that there is at least one survivor, that is, at least one person who is not hit by a pie. (This problem was introduced by Carmony [Ca79]. Note that this result is false when there are an even number of people; see Exercise 77.)

Extra Examples ›

Solution: Let $P(n)$ be the statement that there is a survivor whenever $2n + 1$ people stand in a yard at distinct mutual distances and each person throws a pie at their nearest neighbor. To prove this result, we will show that $P(n)$ is true for all positive integers n. This follows because as n runs through all positive integers, $2n + 1$ runs through all odd integers greater than or equal to 3. Note that one person cannot engage in a pie fight because there is no one else to throw the pie at.

BASIS STEP: When $n = 1$, there are $2n + 1 = 3$ people in the pie fight. Of the three people, suppose that the closest pair are A and B, and C is the third person. Because distances between pairs of people are different, the distance between A and C and the distance between B and C are both different from, and greater than, the distance between A and B. It follows that A and B throw pies at each other, while C throws a pie at either A or B, whichever is closer. Hence, C is not hit by a pie. This shows that at least one of the three people is not hit by a pie, completing the basis step.

INDUCTIVE STEP: For the inductive step, assume that $P(k)$ is true for an arbitrary odd integer k with $k \geq 3$. That is, assume that there is at least one survivor whenever $2k + 1$ people stand in a yard at distinct mutual distances and each throws a pie at their nearest neighbor. We must show that if the inductive hypothesis $P(k)$ is true, then $P(k + 1)$, the statement that there is at least one survivor whenever $2(k + 1) + 1 = 2k + 3$ people stand in a yard at distinct mutual distances and each throws a pie at their nearest neighbor, is also true.

So suppose that we have $2(k + 1) + 1 = 2k + 3$ people in a yard with distinct distances between pairs of people. Let A and B be the closest pair of people in this group of $2k + 3$ people. When each person throws a pie at the nearest person, A and B throw pies at each other. We have two cases to consider, *(i)* when someone else throws a pie at either A or B and *(ii)* when no one else throws a pie at either A or B.

Case (i): Because A and B throw pies at each other and someone else throws a pie at either A and B, at least three pies are thrown at A and B, and at most $(2k + 3) - 3 = 2k$ pies are thrown at the remaining $2k + 1$ people. This guarantees that at least one person is a survivor, for if each of these $2k + 1$ people was hit by at least one pie, a total of at least $2k + 1$ pies would have to be thrown at them. (The reasoning used in this last step is an example of the pigeonhole principle discussed further in Section 6.2.)

Case (ii): No one else throws a pie at either A and B. Besides A and B, there are $2k + 1$ people. Because the distances between pairs of these people are all different, we can use the inductive hypothesis to conclude that there is at least one survivor S when these $2k + 1$ people each throws a pie at their nearest neighbor. Furthermore, S is also not hit by either the pie thrown by A or the pie thrown by B because A and B throw their pies at each other, so S is a survivor because S is not hit by any of the pies thrown by these $2k + 3$ people.

We have completed both the basis step and the inductive step, using a proof by cases. So by mathematical induction it follows that $P(n)$ is true for all positive integers n. We conclude that whenever an odd number of people located in a yard at distinct mutual distances each throws a pie at their nearest neighbor, there is at least one survivor. ◀

Links

In Section 1.8 we discussed the tiling of checkerboards by polyominoes. Example 14 illustrates how mathematical induction can be used to prove a result about covering checkerboards with right triominoes, pieces shaped like the letter "L."

EXAMPLE 14 Let n be a positive integer. Show that every $2^n \times 2^n$ checkerboard with one square removed can be tiled using right triominoes, where these pieces cover three squares at a time, as shown in Figure 4.

FIGURE 4 A right triomino.

Solution: Let $P(n)$ be the proposition that every $2^n \times 2^n$ checkerboard with one square removed can be tiled using right triominoes. We can use mathematical induction to prove that $P(n)$ is true for all positive integers n.

BASIS STEP: $P(1)$ is true, because each of the four 2×2 checkerboards with one square removed can be tiled using one right triomino, as shown in Figure 5.

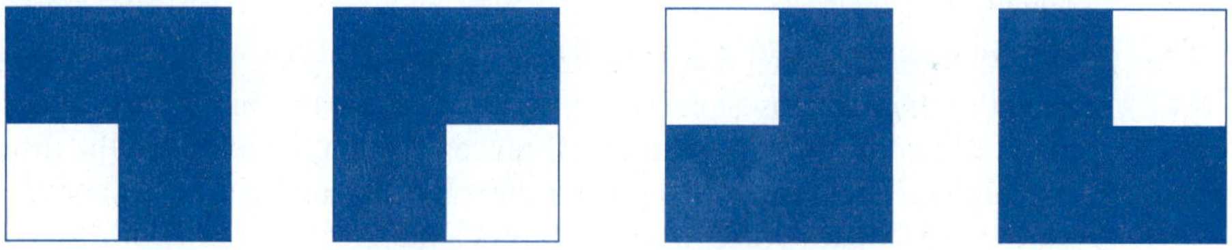

FIGURE 5 Tiling 2×2 checkerboards with one square removed.

INDUCTIVE STEP: The inductive hypothesis is the assumption that $P(k)$ is true for the positive integer k; that is, it is the assumption that every $2^k \times 2^k$ checkerboard with one square removed can be tiled using right triominoes. It must be shown that under the assumption of the inductive hypothesis, $P(k + 1)$ must also be true; that is, any $2^{k+1} \times 2^{k+1}$ checkerboard with one square removed can be tiled using right triominoes.

To see this, consider a $2^{k+1} \times 2^{k+1}$ checkerboard with one square removed. Split this checkerboard into four checkerboards of size $2^k \times 2^k$, by dividing it in half in both directions. This is illustrated in Figure 6. No square has been removed from three of these four checkerboards. The fourth $2^k \times 2^k$ checkerboard has one square removed, so we now use the inductive hypothesis to conclude that it can be covered by right triominoes. Now temporarily remove the square from each of the other three $2^k \times 2^k$ checkerboards that has the center of the original, larger checkerboard as one of its corners, as shown in Figure 7. By the inductive hypothesis, each of these three $2^k \times 2^k$ checkerboards with a square removed can be tiled by right triominoes. Furthermore, the three squares that were temporarily removed can be covered by one right triomino. Hence, the entire $2^{k+1} \times 2^{k+1}$ checkerboard can be tiled with right triominoes.

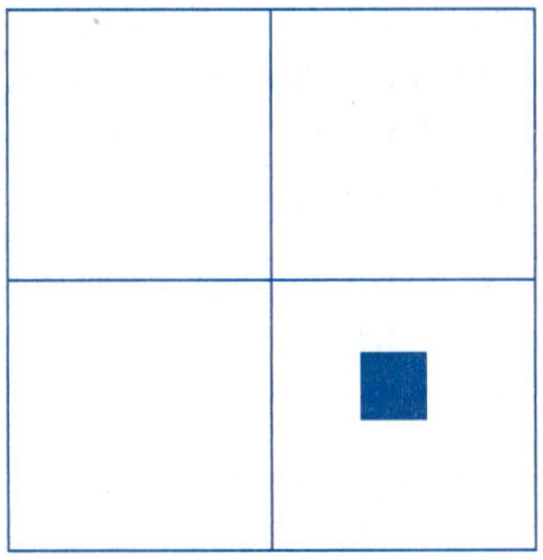

FIGURE 6 **Dividing a $2^{k+1} \times 2^{k+1}$ checkerboard into four $2^k \times 2^k$ checkerboards.**

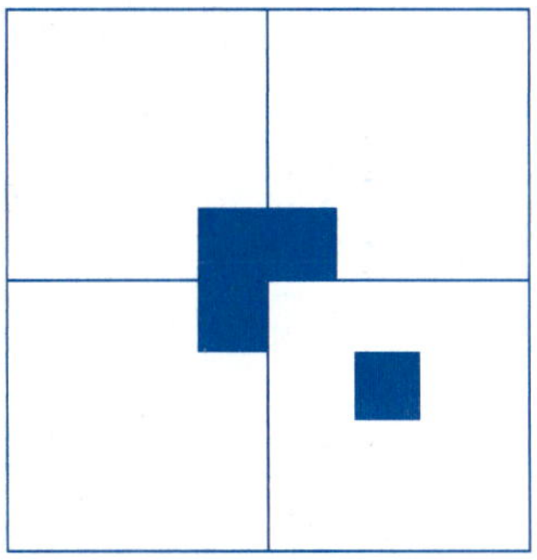

FIGURE 7 **Tiling the $2^{k+1} \times 2^{k+1}$ checkerboard with one square removed.**

We have completed the basis step and the inductive step. Therefore, by mathematical induction $P(n)$ is true for all positive integers n. This shows that we can tile every $2^n \times 2^n$ checkerboard, where n is a positive integer, with one square removed, using right triominoes. ◀

5.1.8 Mistaken Proofs By Mathematical Induction

Consult *Common Errors in Discrete Mathematics* on this book's website for more basic mistakes.

As with every proof method, there are many opportunities for making errors when using mathematical induction. Many well-known mistaken, and often entertaining, proofs by mathematical induction of clearly false statements have been devised, as exemplified by Example 15 and Exercises 49–51. Often, it is not easy to find where the error in reasoning occurs in such mistaken proofs.

To uncover errors in proofs by mathematical induction, remember that in every such proof, both the basis step and the inductive step must be done correctly. Not completing the basis step in a supposed proof by mathematical induction can lead to mistaken proofs of clearly ridiculous statements such as "$n = n + 1$ whenever n is a positive integer." (We leave it to the reader to show that it is easy to construct a correct inductive step in an attempted proof of this statement.) Locating the error in a faulty proof by mathematical induction, as Example 15 illustrates, can be quite tricky, especially when the error is hidden in the basis step.

EXAMPLE 15 Find the error in this "proof" of the clearly false claim that every set of lines in the plane, no two of which are parallel, meet in a common point.

"Proof:" Let $P(n)$ be the statement that every set of n lines in the plane, no two of which are parallel, meet in a common point. We will attempt to prove that $P(n)$ is true for all positive integers $n \geq 2$.

BASIS STEP: The statement $P(2)$ is true because any two lines in the plane that are not parallel meet in a common point (by the definition of parallel lines).

INDUCTIVE STEP: The inductive hypothesis is the statement that $P(k)$ is true for the positive integer k, that is, it is the assumption that every set of k lines in the plane, no two of which are parallel, meet in a common point. To complete the inductive step, we must show that if $P(k)$ is true, then $P(k + 1)$ must also be true. That is, we must show that if every set of k lines in the plane, no two of which are parallel, meet in a common point, then every set of $k + 1$ lines in the plane, no two of which are parallel, meet in a common point. So, consider a set of $k + 1$ distinct lines in the plane. By the inductive hypothesis, the first k of these lines meet in a common point

p_1. Moreover, by the inductive hypothesis, the last k of these lines meet in a common point p_2. We will show that p_1 and p_2 must be the same point. If p_1 and p_2 were different points, all lines containing both of them must be the same line because two points determine a line. This contradicts our assumption that all these lines are distinct. Thus, p_1 and p_2 are the same point. We conclude that the point $p_1 = p_2$ lies on all $k + 1$ lines. We have shown that $P(k + 1)$ is true assuming that $P(k)$ is true. That is, we have shown that if we assume that every k, $k \geq 2$, distinct lines meet in a common point, then every $k + 1$ distinct lines meet in a common point. This completes the inductive step.

We have completed the basis step and the inductive step, and supposedly we have a correct proof by mathematical induction.

Solution: Examining this supposed proof by mathematical induction it appears that everything is in order. However, there is an error, as there must be. The error is rather subtle. Carefully looking at the inductive step shows that this step requires that $k \geq 3$. We cannot show that $P(2)$ implies $P(3)$. When $k = 2$, our goal is to show that every three distinct lines meet in a common point. The first two lines must meet in a common point p_1 and the last two lines must meet in a common point p_2. But in this case, p_1 and p_2 do not have to be the same, because only the second line is common to both sets of lines. Here is where the inductive step fails. ◀

Exercises

1. There are infinitely many stations on a train route. Suppose that the train stops at the first station and suppose that if the train stops at a station, then it stops at the next station. Show that the train stops at all stations.

2. Suppose that you know that a golfer plays the first hole of a golf course with an infinite number of holes and that if this golfer plays one hole, then the golfer goes on to play the next hole. Prove that this golfer plays every hole on the course.

Use mathematical induction in Exercises 3–17 to prove summation formulae. Be sure to identify where you use the inductive hypothesis.

3. Let $P(n)$ be the statement that $1^2 + 2^2 + \cdots + n^2 = n(n + 1)(2n + 1)/6$ for the positive integer n.

a) What is the statement $P(1)$?

b) Show that $P(1)$ is true, completing the basis step of a proof that $P(n)$ is true for all positive integers n.

c) What is the inductive hypothesis of a proof that $P(n)$ is true for all positive integers n?

d) What do you need to prove in the inductive step of a proof that $P(n)$ is true for all positive integers n?

e) Complete the inductive step of a proof that $P(n)$ is true for all positive integers n, identifying where you use the inductive hypothesis.

f) Explain why these steps show that this formula is true whenever n is a positive integer.

4. Let $P(n)$ be the statement that $1^3 + 2^3 + \cdots + n^3 = (n(n + 1)/2)^2$ for the positive integer n.

a) What is the statement $P(1)$?

b) Show that $P(1)$ is true, completing the basis step of the proof of $P(n)$ for all positive integers n.

c) What is the inductive hypothesis of a proof that $P(n)$ is true for all positive integers n?

d) What do you need to prove in the inductive step of a proof that $P(n)$ is true for all positive integers n?

e) Complete the inductive step of a proof that $P(n)$ is true for all positive integers n, identifying where you use the inductive hypothesis.

f) Explain why these steps show that this formula is true whenever n is a positive integer.

5. Prove that $1^2 + 3^2 + 5^2 + \cdots + (2n + 1)^2 = (n + 1)(2n + 1)(2n + 3)/3$ whenever n is a nonnegative integer.

6. Prove that $1 \cdot 1! + 2 \cdot 2! + \cdots + n \cdot n! = (n + 1)! - 1$ whenever n is a positive integer.

7. Prove that $3 + 3 \cdot 5 + 3 \cdot 5^2 + \cdots + 3 \cdot 5^n = 3(5^{n+1} - 1)/4$ whenever n is a nonnegative integer.

8. Prove that $2 - 2 \cdot 7 + 2 \cdot 7^2 - \cdots + 2(-7)^n = (1 - (-7)^{n+1})/4$ whenever n is a nonnegative integer.

9. a) Find a formula for the sum of the first n even positive integers.

b) Prove the formula that you conjectured in part (a).

10. a) Find a formula for

$$\frac{1}{1 \cdot 2} + \frac{1}{2 \cdot 3} + \cdots + \frac{1}{n(n + 1)}$$

by examining the values of this expression for small values of n.

b) Prove the formula you conjectured in part (a).

11. a) Find a formula for

$$\frac{1}{2} + \frac{1}{4} + \frac{1}{8} + \cdots + \frac{1}{2^n}$$

by examining the values of this expression for small values of n.

b) Prove the formula you conjectured in part (a).

12. Prove that

$$\sum_{j=0}^{n}\left(-\frac{1}{2}\right)^j = \frac{2^{n+1}+(-1)^n}{3\cdot 2^n}$$

whenever n is a nonnegative integer.

13. Prove that $1^2 - 2^2 + 3^2 - \cdots + (-1)^{n-1}n^2 = (-1)^{n-1} n(n+1)/2$ whenever n is a positive integer.

14. Prove that for every positive integer n, $\sum_{k=1}^{n} k2^k = (n-1)2^{n+1} + 2$.

15. Prove that for every positive integer n,

$$1\cdot 2 + 2\cdot 3 + \cdots + n(n+1) = n(n+1)(n+2)/3.$$

16. Prove that for every positive integer n,

$$1\cdot 2\cdot 3 + 2\cdot 3\cdot 4 + \cdots + n(n+1)(n+2) = n(n+1)(n+2)(n+3)/4.$$

17. Prove that $\sum_{j=1}^{n} j^4 = n(n+1)(2n+1)(3n^2+3n-1)/30$ whenever n is a positive integer.

Use mathematical induction to prove the inequalities in Exercises 18–29.

18. Let $P(n)$ be the statement that $n! < n^n$, where n is an integer greater than 1.

a) What is the statement $P(2)$?

b) Show that $P(2)$ is true, completing the basis step of a proof by mathematical induction that $P(n)$ is true for all integers n greater than 1.

c) What is the inductive hypothesis of a proof by mathematical induction that $P(n)$ is true for all integers n greater than 1?

d) What do you need to prove in the inductive step of a proof by mathematical induction that $P(n)$ is true for all integers n greater than 1?

e) Complete the inductive step of a proof by mathematical induction that $P(n)$ is true for all integers n greater than 1.

f) Explain why these steps show that this inequality is true whenever n is an integer greater than 1.

19. Let $P(n)$ be the statement that

$$1 + \frac{1}{4} + \frac{1}{9} + \cdots + \frac{1}{n^2} < 2 - \frac{1}{n},$$

where n is an integer greater than 1.

a) What is the statement $P(2)$?

b) Show that $P(2)$ is true, completing the basis step of a proof by mathematical induction that $P(n)$ is true for all integers n greater than 1.

c) What is the inductive hypothesis of a proof by mathematical induction that $P(n)$ is true for all integers n greater than 1?

d) What do you need to prove in the inductive step of a proof by mathematical induction that $P(n)$ is true for all integers n greater than 1?

e) Complete the inductive step of a proof by mathematical induction that $P(n)$ is true for all integers n greater than 1.

f) Explain why these steps show that this inequality is true whenever n is an integer greater than 1.

20. Prove that $3^n < n!$ if n is an integer greater than 6.

21. Prove that $2^n > n^2$ if n is an integer greater than 4.

22. For which nonnegative integers n is $n^2 \le n!$? Prove your answer.

23. For which nonnegative integers n is $2n + 3 \le 2^n$? Prove your answer.

24. Prove that $1/(2n) \le [1\cdot 3\cdot 5\cdot \cdots \cdot (2n-1)]/(2\cdot 4\cdot \cdots \cdot 2n)$ whenever n is a positive integer.

***25.** Prove that if $h > -1$, then $1 + nh \le (1+h)^n$ for all nonnegative integers n. This is called **Bernoulli's inequality**.

***26.** Suppose that a and b are real numbers with $0 < b < a$. Prove that if n is a positive integer, then $a^n - b^n \le na^{n-1}(a-b)$.

***27.** Prove that for every positive integer n,

$$1 + \frac{1}{\sqrt{2}} + \frac{1}{\sqrt{3}} + \cdots + \frac{1}{\sqrt{n}} > 2(\sqrt{n+1} - 1).$$

28. Prove that $n^2 - 7n + 12$ is nonnegative whenever n is an integer with $n \ge 3$.

In Exercises 29 and 30, H_n denotes the nth harmonic number.

***29.** Prove that $H_{2^n} \le 1 + n$ whenever n is a nonnegative integer.

***30.** Prove that

$$H_1 + H_2 + \cdots + H_n = (n+1)H_n - n.$$

Use mathematical induction in Exercises 31–37 to prove divisibility facts.

31. Prove that 2 divides $n^2 + n$ whenever n is a positive integer.

32. Prove that 3 divides $n^3 + 2n$ whenever n is a positive integer.

33. Prove that 5 divides $n^5 - n$ whenever n is a nonnegative integer.

34. Prove that 6 divides $n^3 - n$ whenever n is a nonnegative integer.

***35.** Prove that $n^2 - 1$ is divisible by 8 whenever n is an odd positive integer.

***36.** Prove that 21 divides $4^{n+1} + 5^{2n-1}$ whenever n is a positive integer.

***37.** Prove that if n is a positive integer, then 133 divides $11^{n+1} + 12^{2n-1}$.

Use mathematical induction in Exercises 38–46 to prove results about sets.

38. Prove that if $A_1, A_2, \ldots, A_n$ and $B_1, B_2, \ldots, B_n$ are sets such that $A_j \subseteq B_j$ for $j = 1, 2, \ldots, n$, then

$$\bigcup_{j=1}^{n} A_j \subseteq \bigcup_{j=1}^{n} B_j.$$

39. Prove that if $A_1, A_2, \ldots, A_n$ and $B_1, B_2, \ldots, B_n$ are sets such that $A_j \subseteq B_j$ for $j = 1, 2, \ldots, n$, then

$$\bigcap_{j=1}^{n} A_j \subseteq \bigcap_{j=1}^{n} B_j.$$

40. Prove that if $A_1, A_2, \dots, A_n$ and B are sets, then

$$(A_1 \cap A_2 \cap \cdots \cap A_n) \cup B = (A_1 \cup B) \cap (A_2 \cup B) \cap \cdots \cap (A_n \cup B).$$

41. Prove that if $A_1, A_2, \dots, A_n$ and B are sets, then

$$(A_1 \cup A_2 \cup \cdots \cup A_n) \cap B = (A_1 \cap B) \cup (A_2 \cap B) \cup \cdots \cup (A_n \cap B).$$

42. Prove that if $A_1, A_2, \dots, A_n$ and B are sets, then

$$(A_1 - B) \cap (A_2 - B) \cap \cdots \cap (A_n - B) = (A_1 \cap A_2 \cap \cdots \cap A_n) - B.$$

43. Prove that if $A_1, A_2, \dots, A_n$ are subsets of a universal set U, then

$$\overline{\bigcup_{k=1}^{n} A_k} = \bigcap_{k=1}^{n} \overline{A_k}.$$

44. Prove that if $A_1, A_2, \dots, A_n$ and B are sets, then

$$(A_1 - B) \cup (A_2 - B) \cup \cdots \cup (A_n - B) = (A_1 \cup A_2 \cup \cdots \cup A_n) - B.$$

45. Prove that a set with n elements has $n(n-1)/2$ subsets containing exactly two elements whenever n is an integer greater than or equal to 2.

***46.** Prove that a set with n elements has $n(n-1)(n-2)/6$ subsets containing exactly three elements whenever n is an integer greater than or equal to 3.

In Exercises 47 and 48 we consider the problem of placing towers along a straight road, so that every building on the road receives cellular service. Assume that a building receives cellular service if it is within one mile of a tower.

47. Devise a greedy algorithm that uses the minimum number of towers possible to provide cell service to d buildings located at positions $x_1, x_2, \dots, x_d$ from the start of the road. [*Hint:* At each step, go as far as possible along the road before adding a tower so as not to leave any buildings without coverage.]

***48.** Use mathematical induction to prove that the algorithm you devised in Exercise 47 produces an optimal solution, that is, that it uses the fewest towers possible to provide cellular service to all buildings.

Exercises 49–51 present incorrect proofs using mathematical induction. You will need to identify an error in reasoning in each exercise.

49. What is wrong with this "proof" that all horses are the same color?

Let $P(n)$ be the proposition that all the horses in a set of n horses are the same color.

Basis Step: Clearly, $P(1)$ is true.

Inductive Step: Assume that $P(k)$ is true, so that all the horses in any set of k horses are the same color. Consider any $k+1$ horses; number these as horses $1, 2, 3, \dots, k, k+1$. Now the first k of these horses all must have the same color, and the last k of these must also have the same color. Because the set of the first k horses and the set of the last k horses overlap, all $k+1$ must be the same color. This shows that $P(k+1)$ is true and finishes the proof by induction.

50. What is wrong with this "proof"?

"Theorem" For every positive integer n, $\sum_{i=1}^{n} i = (n+\frac{1}{2})^2/2$.

Basis Step: The formula is true for $n = 1$.

Inductive Step: Suppose that $\sum_{i=1}^{n} i = (n+\frac{1}{2})^2/2$. Then $\sum_{i=1}^{n+1} i = (\sum_{i=1}^{n} i) + (n+1)$. By the inductive hypothesis, we have $\sum_{i=1}^{n+1} i = (n+\frac{1}{2})^2/2 + n + 1 = (n^2+n+\frac{1}{4})/2 + n + 1 = (n^2+3n+\frac{9}{4})/2 = (n+\frac{3}{2})^2/2 = [(n+1)+\frac{1}{2}]^2/2$, completing the inductive step.

51. What is wrong with this "proof"?

"Theorem" For every positive integer n, if x and y are positive integers with $\max(x, y) = n$, then $x = y$.

Basis Step: Suppose that $n = 1$. If $\max(x, y) = 1$ and x and y are positive integers, we have $x = 1$ and $y = 1$.

Inductive Step: Let k be a positive integer. Assume that whenever $\max(x, y) = k$ and x and y are positive integers, then $x = y$. Now let $\max(x, y) = k+1$, where x and y are positive integers. Then $\max(x-1, y-1) = k$, so by the inductive hypothesis, $x-1 = y-1$. It follows that $x = y$, completing the inductive step.

52. Suppose that m and n are positive integers with $m > n$ and f is a function from $\{1, 2, \dots, m\}$ to $\{1, 2, \dots, n\}$. Use mathematical induction on the variable n to show that f is not one-to-one.

***53.** Use mathematical induction to show that n people can divide a cake (where each person gets one or more separate pieces of the cake) so that the cake is divided fairly, that is, in the sense that each person thinks they got at least $(1/n)$th of the cake. [*Hint:* For the inductive step, take a fair division of the cake among the first k people, have each person divide their share into what this person thinks are $k+1$ equal portions, and then have the $(k+1)$st person select a portion from each of the k people. When showing this produces a fair division for $k+1$ people, suppose that person $k+1$ thinks that person i got p_i of the cake, where $\sum_{i=1}^{k} p_i = 1$.]

54. Use mathematical induction to show that given a set of $n+1$ positive integers, none exceeding $2n$, there is at least one integer in this set that divides another integer in the set.

***55.** A knight on a chessboard can move one space horizontally (in either direction) and two spaces vertically (in either direction) or two spaces horizontally (in either direction) and one space vertically (in either direction). Suppose that we have an infinite chessboard, made up of all squares (m, n), where m and n are nonnegative integers that denote the row number and the column number of the square, respectively. Use mathematical induction to show that a knight starting at $(0, 0)$ can visit every square using a finite sequence of moves. [*Hint:* Use induction on the variable $s = m + n$.]

56. Suppose that

$$\mathbf{A} = \begin{bmatrix} a & 0 \\ 0 & b \end{bmatrix},$$

where a and b are real numbers. Show that

$$\mathbf{A}^n = \begin{bmatrix} a^n & 0 \\ 0 & b^n \end{bmatrix}$$

for every positive integer n.

57. (*Requires calculus*) Use mathematical induction to prove that the derivative of $f(x) = x^n$ equals nx^{n-1} whenever n is a positive integer. (For the inductive step, use the product rule for derivatives.)

58. Suppose that **A** and **B** are square matrices with the property $\mathbf{AB} = \mathbf{BA}$. Show that $\mathbf{AB}^n = \mathbf{B}^n\mathbf{A}$ for every positive integer n.

59. Suppose that m is a positive integer. Use mathematical induction to prove that if a and b are integers with $a \equiv b \pmod{m}$, then $a^k \equiv b^k \pmod{m}$ whenever k is a nonnegative integer.

60. Use mathematical induction to show that $\neg(p_1 \vee p_2 \vee \cdots \vee p_n)$ is equivalent to $\neg p_1 \wedge \neg p_2 \wedge \cdots \wedge \neg p_n$ whenever $p_1, p_2, \ldots, p_n$ are propositions.

***61.** Show that

$$[(p_1 \to p_2) \wedge (p_2 \to p_3) \wedge \cdots \wedge (p_{n-1} \to p_n)] \to [(p_1 \wedge p_2 \wedge \cdots \wedge p_{n-1}) \to p_n]$$

is a tautology whenever $p_1, p_2, \ldots, p_n$ are propositions, where $n \geq 2$.

***62.** Show that n lines separate the plane into $(n^2 + n + 2)/2$ regions if no two of these lines are parallel and no three pass through a common point.

****63.** Let $a_1, a_2, \ldots, a_n$ be positive real numbers. The **arithmetic mean** of these numbers is defined by

$$A = (a_1 + a_2 + \cdots + a_n)/n,$$

and the **geometric mean** of these numbers is defined by

$$G = (a_1 a_2 \cdots a_n)^{1/n}.$$

Use mathematical induction to prove that $A \geq G$.

64. Use mathematical induction to prove Lemma 3 of Section 4.3, which states that if p is a prime and $p \mid a_1 a_2 \cdots a_n$, where a_i is an integer for $i = 1, 2, 3, \ldots, n$, then $p \mid a_i$ for some integer i.

65. Show that if n is a positive integer, then

$$\sum_{\{a_1,\ldots,a_k\} \subseteq \{1,2,\ldots,n\}} \frac{1}{a_1 a_2 \cdots a_k} = n.$$

(Here the sum is over all nonempty subsets of the set of the n smallest positive integers.)

***66.** Use the well-ordering property to show that the following form of mathematical induction is a valid method to prove that $P(n)$ is true for all positive integers n.

Basis Step: $P(1)$ and $P(2)$ are true.

Inductive Step: For each positive integer k, if $P(k)$ and $P(k + 1)$ are both true, then $P(k + 2)$ is true.

67. Show that if $A_1, A_2, \ldots, A_n$ are sets where $n \geq 2$, and for all pairs of integers i and j with $1 \leq i < j \leq n$, either A_i is a subset of A_j or A_j is a subset of A_i, then there is an integer i, $1 \leq i \leq n$, such that A_i is a subset of A_j for all integers j with $1 \leq j \leq n$.

***68.** A guest at a party is a **celebrity** if this person is known by every other guest, but knows none of them. There is at most one celebrity at a party, for if there were two, they would know each other. A particular party may have no celebrity. Your assignment is to find the celebrity, if one exists, at a party, by asking only one type of question—asking a guest whether they know a second guest. Everyone must answer your questions truthfully. That is, if Alice and Bob are two people at the party, you can ask Alice whether she knows Bob; she must answer correctly. Use mathematical induction to show that if there are n people at the party, then you can find the celebrity, if there is one, with $3(n - 1)$ questions. [*Hint:* First ask a question to eliminate one person as a celebrity. Then use the inductive hypothesis to identify a potential celebrity. Finally, ask two more questions to determine whether that person is actually a celebrity.]

Suppose there are n people in a group, each aware of a scandal no one else in the group knows about. These people communicate by telephone; when two people in the group talk, they share information about all scandals each knows about. For example, on the first call, two people share information, so by the end of the call, each of these people knows about two scandals. The **gossip problem** asks for $G(n)$, the minimum number of telephone calls that are needed for all n people to learn about all the scandals. Exercises 69–71 deal with the gossip problem.

69. Find $G(1)$, $G(2)$, $G(3)$, and $G(4)$.

70. Use mathematical induction to prove that $G(n) \leq 2n - 4$ for $n \geq 4$. [*Hint:* In the inductive step, have a new person call a particular person at the start and at the end.]

****71.** Prove that $G(n) = 2n - 4$ for $n \geq 4$.

***72.** Show that it is possible to arrange the numbers $1, 2, \ldots, n$ in a row so that the average of any two of these numbers never appears between them. [*Hint:* Show that it suffices to prove this fact when n is a power of 2. Then use mathematical induction to prove the result when n is a power of 2.]

***73.** Show that if $I_1, I_2, \ldots, I_n$ is a collection of open intervals on the real number line, $n \geq 2$, and every pair of these intervals has a nonempty intersection, that is, $I_i \cap I_j \neq \emptyset$ whenever $1 \leq i \leq n$ and $1 \leq j \leq n$, then the intersection of all these sets is nonempty, that is, $I_1 \cap I_2 \cap \cdots \cap I_n \neq \emptyset$. (Recall that an **open interval** is the set of real numbers x with $a < x < b$, where a and b are real numbers with $a < b$.)

Sometimes we cannot use mathematical induction to prove a result we believe to be true, but we can use mathematical induction to prove a stronger result. Because the inductive hypothesis of the stronger result provides more to work with, this process is called **inductive loading**. We use inductive loading in Exercise 74–76.

74. Show that we cannot use mathematical induction to prove that $\sum_{j=1}^{n} 1/j^2 < 2$ for all positive integers n, but that this inequality is a consequence of the inequality proved by mathematical induction in Exercise 19.

75. Suppose that we want to prove that

$$\sum_{j=1}^{n} j/(j+1)! < 1$$

for all positive integers n.

a) Show that if we try to prove this inequality using mathematical induction, the basis step works, but the inductive step fails.

b) Show that mathematical induction can be used to prove the stronger inequality

$$\sum_{j=1}^{n} j/(j+1)! \leq 1 - 1/(n+1)!$$

for all positive integers n, implying that the weaker inequality is also true.

76. Suppose that we want to prove that

$$\frac{1}{2} \cdot \frac{3}{4} \cdots \frac{2n-1}{2n} < \frac{1}{\sqrt{3n}}$$

for all positive integers n.

a) Show that if we try to prove this inequality using mathematical induction, the basis step works, but the inductive step fails.

b) Show that mathematical induction can be used to prove the stronger inequality

$$\frac{1}{2} \cdot \frac{3}{4} \cdots \frac{2n-1}{2n} < \frac{1}{\sqrt{3n+1}}$$

for all integers n greater than 1, which, together with a verification for the case where $n = 1$, establishes the weaker inequality we originally tried to prove using mathematical induction.

77. Let n be an even integer. Show that it is possible for n people to stand in a yard at mutually distinct distances so that when each person throws a pie at their nearest neighbor, everyone is hit by a pie.

78. Construct a tiling using right triominoes of the 4×4 checkerboard with the square in the upper left corner removed.

79. Construct a tiling using right triominoes of the 8×8 checkerboard with the square in the upper left corner removed.

80. Prove or disprove that all checkerboards of these shapes can be completely covered using right triominoes whenever n is a positive integer.

a) 3×2^n **b)** 6×2^n

c) $3^n \times 3^n$ **d)** $6^n \times 6^n$

***81.** Show that a three-dimensional $2^n \times 2^n \times 2^n$ checkerboard with one $1 \times 1 \times 1$ cube missing can be completely covered by $2 \times 2 \times 2$ cubes with one $1 \times 1 \times 1$ cube removed.

***82.** Show that an $n \times n$ checkerboard with one square removed can be completely covered using right triominoes if $n > 5$, n is odd, and $3 \nmid n$.

83. Show that a 5×5 checkerboard with a corner square removed can be tiled using right triominoes.

***84.** Find a 5×5 checkerboard with a square removed that cannot be tiled using right triominoes. Prove that such a tiling does not exist for this board.

85. Use the principle of mathematical induction to show that $P(n)$ is true for $n = b, b+1, b+2, \ldots$, where b is an integer, if $P(b)$ is true and the conditional statement $P(k) \rightarrow P(k+1)$ is true for all integers k with $k \geq b$.

5.2 Strong Induction and Well-Ordering

5.2.1 Introduction

In Section 5.1 we introduced mathematical induction and we showed how to use it to prove a variety of theorems. In this section we will introduce another form of mathematical induction, called **strong induction**, which can often be used when we cannot easily prove a result using mathematical induction. The basis step of a proof by strong induction is the same as a proof of the same result using mathematical induction. That is, in a strong induction proof that $P(n)$ is true for all positive integers n, the basis step shows that $P(1)$ is true. However, the inductive steps in these two proof methods are different. In a proof by mathematical induction, the inductive step shows that if the inductive hypothesis $P(k)$ is true, then $P(k+1)$ is also true. In a proof by strong induction, the inductive step shows that if $P(j)$ is true for all positive integers j not exceeding k, then $P(k+1)$ is true. That is, for the inductive hypothesis we assume that $P(j)$ is true for $j = 1, 2, \ldots, k$.

The validity of both mathematical induction and strong induction follow from the well-ordering property in Appendix 1. In fact, mathematical induction, strong induction, and well-ordering are all equivalent principles (as shown in Exercises 45, 46, and 47). That is, the validity of each can be proved from either of the other two. This means that a proof using one of these two principles can be rewritten as a proof using either of the other two principles. Just as it is sometimes the case that it is much easier to see how to prove a result using strong induction rather than mathematical induction, it is sometimes easier to use well-ordering than one of the two forms of mathematical induction. In this section we will give some examples of how the well-ordering property can be used to prove theorems.

5.2.2 Strong Induction

Before we illustrate how to use strong induction, we state this principle again.

STRONG INDUCTION: To prove that $P(n)$ is true for all positive integers n, where $P(n)$ is a propositional function, we complete two steps:

BASIS STEP: We verify that the proposition $P(1)$ is true.

INDUCTIVE STEP: We show that the conditional statement $[P(1) \wedge P(2) \wedge \cdots \wedge P(k)] \rightarrow P(k+1)$ is true for all positive integers k.

Note that when we use strong induction to prove that $P(n)$ is true for all positive integers n, our inductive hypothesis is the assumption that $P(j)$ is true for $j = 1, 2, \ldots, k$ for an arbitrary positive integer k. That is, the inductive hypothesis includes all k statements $P(1), P(2), \ldots, P(k)$. Because we can use all k statements $P(1), P(2), \ldots, P(k)$ to prove $P(k+1)$, rather than just the statement $P(k)$ as in a proof by mathematical induction, strong induction is a more flexible proof technique. Because of this, some mathematicians prefer to always use strong induction instead of mathematical induction, even when a proof by mathematical induction is easy to find.

You may be surprised that mathematical induction and strong induction are equivalent. That is, each can be shown to be a valid proof technique assuming that the other is valid. In particular, any proof using mathematical induction can also be considered to be a proof by strong induction because the inductive hypothesis of a proof by mathematical induction is part of the inductive hypothesis in a proof by strong induction. That is, if we can complete the inductive step of a proof using mathematical induction by showing that $P(k+1)$ follows from $P(k)$ for every positive integer k, then it also follows that $P(k+1)$ follows from all the statements $P(1), P(2), \ldots, P(k)$, because we are assuming that not only $P(k)$ is true, but also more, namely, that the $k-1$ statements $P(1), P(2), \ldots, P(k-1)$ are true. However, it is much more awkward to convert a proof by strong induction into a proof using the principle of mathematical induction. (See Exercise 46.)

Strong induction is sometimes called the **second principle of mathematical induction** or **complete induction**. When the terminology "complete induction" is used, the principle of mathematical induction is called **incomplete induction**, a technical term that is a somewhat unfortunate choice because there is nothing incomplete about the principle of mathematical induction; after all, it is a valid proof technique.

STRONG INDUCTION AND THE INFINITE LADDER To better understand strong induction, consider the infinite ladder in Section 5.1. Strong induction tells us that we can reach all rungs if

1. we can reach the first rung, and
2. for every positive integer k, if we can reach all the first k rungs, then we can reach the $(k+1)$st rung.

That is, if $P(n)$ is the statement that we can reach the nth rung of the ladder, by strong induction we know that $P(n)$ is true for all positive integers n, because (1) tells us $P(1)$ is true, completing the basis step and (2) tells us that $P(1) \wedge P(2) \wedge \cdots \wedge P(k)$ implies $P(k+1)$, completing the inductive step.

Example 1 illustrates how strong induction can help us prove a result that cannot easily be proved using the principle of mathematical induction.

EXAMPLE 1 Suppose we can reach the first and second rungs of an infinite ladder, and we know that if we can reach a rung, then we can reach two rungs higher. Can we prove that we can reach every rung using the principle of mathematical induction? Can we prove that we can reach every rung using strong induction?

Solution: We first try to prove this result using the principle of mathematical induction.

BASIS STEP: The basis step of such a proof holds; here it simply verifies that we can reach the first rung.

ATTEMPTED INDUCTIVE STEP: The inductive hypothesis is the statement that we can reach the kth rung of the ladder. To complete the inductive step, we need to show that if we assume the inductive hypothesis for the positive integer k, namely, if we assume that we can reach the kth rung of the ladder, then we can show that we can reach the $(k+1)$st rung of the ladder. However, there is no obvious way to complete this inductive step because we do not know from the given information that we can reach the $(k+1)$st rung from the kth rung. After all, we only know that if we can reach a rung we can reach the rung two higher.

Now consider a proof using strong induction.

BASIS STEP: The basis step is the same as before; it simply verifies that we can reach the first rung.

INDUCTIVE STEP: The inductive hypothesis states that we can reach each of the first k rungs. To complete the inductive step, we need to show that if we assume that the inductive hypothesis is true, that is, if we can reach each of the first k rungs, then we can reach the $(k+1)$st rung. We already know that we can reach the second rung. We can complete the inductive step by noting that as long as $k \geq 2$, we can reach the $(k+1)$st rung from the $(k-1)$st rung because we know we can climb two rungs from a rung we can already reach, and because $k-1 \leq k$, by the inductive hypothesis we can reach the $(k-1)$st rung. This completes the inductive step and finishes the proof by strong induction.

We have proved that if we can reach the first two rungs of an infinite ladder and for every positive integer k if we can reach all the first k rungs then we can reach the $(k+1)$st rung, then we can reach all rungs of the ladder. ◀

5.2.3 Examples of Proofs Using Strong Induction

Now that we have both mathematical induction and strong induction, how do we decide which method to apply in a particular situation? Although there is no cut-and-dried answer, we can supply some useful pointers. In practice, you should use mathematical induction when it is straightforward to prove that $P(k) \rightarrow P(k+1)$ is true for all positive integers k. This is the case for all the proofs in the examples in Section 5.1. In general, you should restrict your use of the principle of mathematical induction to such scenarios. Unless you can clearly see that the inductive step of a proof by mathematical induction goes through, you should attempt a proof by strong induction. That is, use strong induction and not mathematical induction when you see how to prove that $P(k+1)$ is true from the assumption that $P(j)$ is true for all positive integers j not exceeding k, but you cannot see how to prove that $P(k+1)$ follows from just $P(k)$. Keep this in mind as you examine the proofs in this section. For each of these proofs, consider why strong induction works better than mathematical induction.

Remark: Proofs by strong induction can begin with a basis step that starts at any integer b, just as proofs by mathematical induction can. We will see that sometimes it is necessary to include more than one integer in the basis case. These considerations are illustrated in forthcoming examples.

We will illustrate how strong induction is employed in Examples 2–4. In these examples, we will prove a diverse collection of results. Pay particular attention to the inductive step in each of these examples, where we show that a result $P(k+1)$ follows under the assumption that $P(j)$ holds for all positive integers j not exceeding k, where $P(n)$ is a propositional function.

Before we present these examples, note that we can slightly modify strong induction to handle a wider variety of situations. In particular, we can adapt strong induction to handle cases where the inductive step is valid only for integers greater than a particular integer. Let b be a fixed integer and j a fixed positive integer. The form of strong induction we need tells us that $P(n)$ is true for all integers n with $n \geq b$ if we can complete these two steps:

BASIS STEP: We verify that the propositions $P(b), P(b+1), \ldots, P(b+j)$ are true.

INDUCTIVE STEP: We show that $[P(b) \wedge P(b+1) \wedge \cdots \wedge P(k)] \to P(k+1)$ is true for every integer $k \geq b+j$.

That this alternative form is equivalent to strong induction is left as Exercise 30.

We begin with one of the most prominent uses of strong induction, the part of the fundamental theorem of arithmetic that tells us that every positive integer can be written as the product of primes.

EXAMPLE 2 Show that if n is an integer greater than 1, then n can be written as the product of primes.

Extra Examples

Solution: Let $P(n)$ be the proposition that n can be written as the product of primes.

BASIS STEP: $P(2)$ is true, because 2 can be written as the product of one prime, itself. (Note that $P(2)$ is the first case we need to establish.)

INDUCTIVE STEP: The inductive hypothesis is the assumption that $P(j)$ is true for all integers j with $2 \leq j \leq k$, that is, the assumption that j can be written as the product of primes whenever j is a positive integer at least 2 and not exceeding k. To complete the inductive step, it must be shown that $P(k+1)$ is true under this assumption, that is, that $k+1$ is the product of primes.

There are two cases to consider, namely, when $k+1$ is prime and when $k+1$ is composite. If $k+1$ is prime, we immediately see that $P(k+1)$ is true. Otherwise, $k+1$ is composite and can be written as the product of two positive integers a and b with $2 \leq a \leq b < k+1$. Because both a and b are integers at least 2 and not exceeding k, we can use the inductive hypothesis to write both a and b as the product of primes. Thus, if $k+1$ is composite, it can be written as the product of primes, namely, those primes in the factorization of a and those in the factorization of b. ◀

Remark: Because 1 can be thought of as an *empty* product of primes, that is, the product of no primes, we could have started the proof in Example 2 with $P(1)$ as the basis step. We chose not to do so because many people find this confusing.

Example 2 completes the proof of the fundamental theorem of arithmetic, which asserts that every nonnegative integer can be written uniquely as the product of primes in nondecreasing order. We showed in Section 4.3 that an integer has at most one such factorization into primes. Example 2 shows there is at least one such factorization.

Next, we show how strong induction can be used to prove that a player has a winning strategy in a game.

EXAMPLE 3 Consider a game in which two players take turns removing any positive number of matches they want from one of two piles of matches. The player who removes the last match wins the game. Show that if the two piles contain the same number of matches initially, the second player can always guarantee a win.

Solution: Let n be the number of matches in each pile. We will use strong induction to prove $P(n)$, the statement that the second player can win when there are initially n matches in each pile.

BASIS STEP: When $n = 1$, the first player has only one choice, removing one match from one of the piles, leaving a single pile with a single match, which the second player can remove to win the game.

INDUCTIVE STEP: The inductive hypothesis is the statement that $P(j)$ is true for all j with $1 \leq j \leq k$, that is, the assumption that the second player can always win whenever there are j matches, where $1 \leq j \leq k$ in each of the two piles at the start of the game. We need to show that $P(k + 1)$ is true, that is, that the second player can win when there are initially $k + 1$ matches in each pile, under the assumption that $P(j)$ is true for $j = 1, 2, \ldots, k$. So suppose that there are $k + 1$ matches in each of the two piles at the start of the game and suppose that the first player removes r matches $(1 \leq r \leq k)$ from one of the piles, leaving $k + 1 - r$ matches in this pile. By removing the same number of matches from the other pile, the second player creates the situation where there are two piles each with $k + 1 - r$ matches. Because $1 \leq k + 1 - r \leq k$, we can now use the inductive hypothesis to conclude that the second player can always win. We complete the proof by noting that if the first player removes all $k + 1$ matches from one of the piles, the second player can win by removing all the remaining matches. ◀

Using the principle of mathematical induction, instead of strong induction, to prove the results in Examples 2 and 3 is difficult. However, as Example 4 shows, some results can be readily proved using either the principle of mathematical induction or strong induction.

EXAMPLE 4 Prove that every amount of postage of 12 cents or more can be formed using just 4-cent and 5-cent stamps.

Solution: We will prove this result using the principle of mathematical induction. Then we will present a proof using strong induction. Let $P(n)$ be the statement that postage of n cents can be formed using 4-cent and 5-cent stamps.

We begin by using the principle of mathematical induction.

BASIS STEP: Postage of 12 cents can be formed using three 4-cent stamps.

INDUCTIVE STEP: The inductive hypothesis is the statement that $P(k)$ is true. That is, under this hypothesis, postage of k cents can be formed using 4-cent and 5-cent stamps. To complete the inductive step, we need to show that when we assume $P(k)$ is true, then $P(k + 1)$ is also true where $k \geq 12$. That is, we need to show that if we can form postage of k cents, then we can form postage of $k + 1$ cents. So, assume the inductive hypothesis is true; that is, assume that we can form postage of k cents using 4-cent and 5-cent stamps. We consider two cases, when at least one 4-cent stamp has been used and when no 4-cent stamps have been used. First, suppose that at least one 4-cent stamp was used to form postage of k cents. Then we can replace this stamp with a 5-cent stamp to form postage of $k + 1$ cents. But if no 4-cent stamps were used, we can form postage of k cents using only 5-cent stamps. Moreover, because $k \geq 12$, we needed at least three 5-cent stamps to form postage of k cents. So, we can replace three 5-cent stamps with four 4-cent stamps to form postage of $k + 1$ cents. This completes the inductive step.

Because we have completed the basis step and the inductive step, we know that $P(n)$ is true for all $n \geq 12$. That is, we can form postage of n cents, where $n \geq 12$ using just 4-cent and 5-cent stamps. This completes the proof by mathematical induction.

Next, we will use strong induction to prove the same result. In this proof, in the basis step we show that $P(12)$, $P(13)$, $P(14)$, and $P(15)$ are true, that is, that postage of 12, 13, 14, or 15 cents can be formed using just 4-cent and 5-cent stamps. In the inductive step we show how to get postage of $k + 1$ cents for $k \geq 15$ from postage of $k - 3$ cents.

BASIS STEP: We can form postage of 12, 13, 14, and 15 cents using three 4-cent stamps, two 4-cent stamps and one 5-cent stamp, one 4-cent stamp and two 5-cent stamps, and three 5-cent stamps, respectively. This shows that $P(12)$, $P(13)$, $P(14)$, and $P(15)$ are true. This completes the basis step.

INDUCTIVE STEP: The inductive hypothesis is the statement that $P(j)$ is true for $12 \leq j \leq k$, where k is an integer with $k \geq 15$. To complete the inductive step, we assume that we can form postage of j cents, where $12 \leq j \leq k$. We need to show that under the assumption that $P(k+1)$ is true, we can also form postage of $k+1$ cents. Using the inductive hypothesis, we can assume that $P(k-3)$ is true because $k-3 \geq 12$, that is, we can form postage of $k-3$ cents using just 4-cent and 5-cent stamps. To form postage of $k+1$ cents, we need only add another 4-cent stamp to the stamps we used to form postage of $k-3$ cents. That is, we have shown that if the inductive hypothesis is true, then $P(k+1)$ is also true. This completes the inductive step.

Because we have completed the basis step and the inductive step of a strong induction proof, we know by strong induction that $P(n)$ is true for all integers n with $n \geq 12$. That is, we know that every postage of n cents, where n is at least 12, can be formed using 4-cent and 5-cent stamps. This finishes the proof by strong induction. ◀

5.2.4 Using Strong Induction in Computational Geometry

Our next example of strong induction will come from **computational geometry**, the part of discrete mathematics that studies computational problems involving geometric objects. Computational geometry is used extensively in computer graphics, computer games, robotics, scientific calculations, and a vast array of other areas. Before we can present this result, we introduce some terminology, possibly familiar from earlier studies in geometry.

A **polygon** is a closed geometric figure consisting of a sequence of line segments $s_1, s_2, \ldots, s_n$, called **sides**. Each pair of consecutive sides, s_i and s_{i+1}, $i = 1, 2, \ldots, n-1$, as well as the last side s_n and the first side s_1, of the polygon meet at a common endpoint, called a **vertex**. A polygon is called **simple** if no two nonconsecutive sides intersect. Every simple polygon divides the plane into two regions: its **interior**, consisting of the points inside the curve, and its **exterior**, consisting of the points outside the curve. This last fact is surprisingly complicated to prove. It is a special case of the deceptively simple Jordan curve theorem, an important result with a rich history, which tells us that every simple curve divides the plane into two regions; see [Or00], for example.

A polygon is called **convex** if every line segment connecting two points in the interior of the polygon lies entirely inside the polygon. (A polygon that is not convex is said to be **nonconvex**.) Figure 1 displays some polygons; polygons (a) and (b) are convex, but polygons (c) and (d) are not. A **diagonal** of a simple polygon is a line segment connecting two nonconsecutive vertices of the polygon, and a diagonal is called an **interior diagonal** if it lies entirely inside the polygon, except for its endpoints. For example, in polygon (d), the line segment connecting a and f is an interior diagonal, but the line segment connecting a and d is a diagonal that is not an interior diagonal.

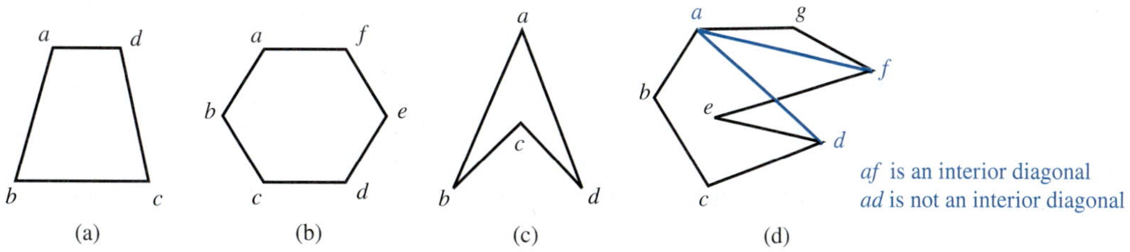

FIGURE 1 Convex and nonconvex polygons.

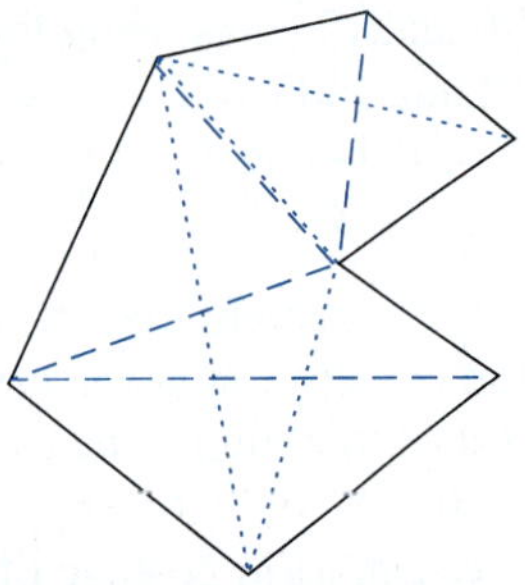

Two different triangulations of a simple polygon with seven sides into five triangles, shown with dotted lines and with dashed lines, respectively

FIGURE 2 Triangulations of a polygon.

One of the most basic operations of computational geometry involves dividing a simple polygon into triangles by adding nonintersecting diagonals. This process is called **triangulation**. Note that a simple polygon can have many different triangulations, as shown in Figure 2. Perhaps the most basic fact in computational geometry is that it is possible to triangulate every simple polygon, as we state in Theorem 1.

THEOREM 1 A simple polygon with n sides, where n is an integer with $n \geq 3$, can be triangulated into $n - 2$ triangles.

It seems obvious that we should be able to triangulate a simple polygon by successively adding interior diagonals. Consequently, a proof by strong induction seems promising. However, such a proof requires this crucial lemma.

LEMMA 1 Every simple polygon with at least four sides has an interior diagonal.

Although Lemma 1 seems particularly simple, it is surprisingly tricky to prove. In fact, as recently as 30 years ago, a variety of incorrect proofs thought to be correct were commonly seen in books and articles. We defer the proof of Lemma 1 until after we prove Theorem 1. It is not uncommon to prove a theorem pending the later proof of an important lemma.

Proof (of Theorem 1): We will prove this result using strong induction. Let $T(n)$ be the statement that every simple polygon with n sides can be triangulated into $n - 2$ triangles.

BASIS STEP: $T(3)$ is true because a simple polygon with three sides is a triangle. We do not need to add any diagonals to triangulate a triangle; it is already triangulated into one triangle, itself. Consequently, every simple polygon with $n = 3$ has can be triangulated into $n - 2 = 3 - 2 = 1$ triangle.

INDUCTIVE STEP: For the inductive hypothesis, we assume that $T(j)$ is true for all integers j with $3 \leq j \leq k$. That is, we assume that we can triangulate a simple polygon with j sides into $j - 2$ triangles whenever $3 \leq j \leq k$. To complete the inductive step, we must show that when we assume the inductive hypothesis, $P(k + 1)$ is true, that is, that every simple polygon with $k + 1$ sides can be triangulated into $(k + 1) - 2 = k - 1$ triangles.

So, suppose that we have a simple polygon P with $k + 1$ sides. Because $k + 1 \geq 4$, Lemma 1 tells us that P has an interior diagonal ab. Now, ab splits P into two simple polygons Q, with s sides, and R, with t sides. The sides of Q and R are the sides of P, together with the side ab, which is a side of both Q and R. Note that $3 \leq s \leq k$ and $3 \leq t \leq k$ because both Q and R have at least one fewer side than P does (after all, each of these is formed from P by deleting at least two sides and replacing these sides by the diagonal ab). Furthermore, the number of sides of P is two less than the sum of the numbers of sides of Q and the number of sides

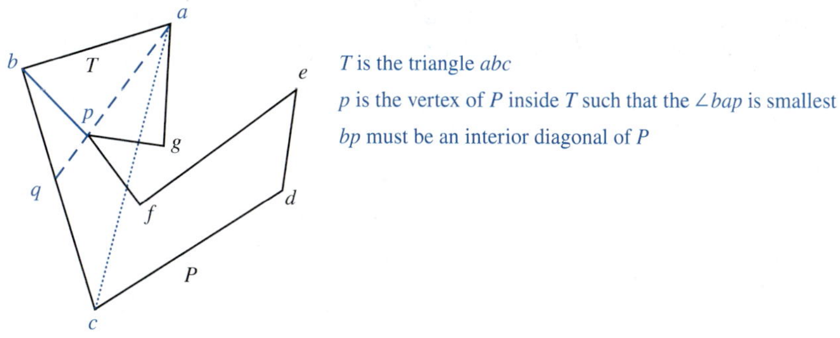

FIGURE 3 **Constructing an interior diagonal of a simple polygon.**

of R, because each side of P is a side of either Q or of R, but not both, and the diagonal ab is a side of both Q and R, but not P. That is, $k + 1 = s + t - 2$.

We now use the inductive hypothesis. Because both $3 \leq s \leq k$ and $3 \leq t \leq k$, by the inductive hypothesis we can triangulate Q and R into $s - 2$ and $t - 2$ triangles, respectively. Next, note that these triangulations together produce a triangulation of P. (Each diagonal added to triangulate one of these smaller polygons is also a diagonal of P.) Consequently, we can triangulate P into a total of $(s - 2) + (t - 2) = s + t - 4 = (k + 1) - 2$ triangles. This completes the proof by strong induction. That is, we have shown that every simple polygon with n sides, where $n \geq 3$, can be triangulated into $n - 2$ triangles. ◁

We now return to our proof of Lemma 1. We present a proof published by Chung-Wu Ho [Ho75]. Note that although this proof may be omitted without loss of continuity, it does provide a correct proof of a result proved incorrectly by many mathematicians.

Proof: Suppose that P is a simple polygon drawn in the plane. Furthermore, suppose that b is the point of P with the least y-coordinate among the points of P with the smallest x-coordinate. Then b must be a vertex of P, for if it is not a vertex, there would have to be a vertex of P with a smaller x-coordinate. Two other vertices each share an edge with b, say a and c. It follows that the angle in the interior of P formed by ab and bc must be less than 180 degrees (otherwise, there would be points of P with smaller x-coordinates than b).

Now let T be the triangle $\triangle abc$. If there are no vertices of P on or inside T, we can connect a and c to obtain an interior diagonal. On the other hand, if there are vertices of P inside T, we will find a vertex p of P on or inside T such that bp is an interior diagonal. (This is the tricky part. Ho noted that in many published proofs of this lemma a vertex p was found such that bp was not necessarily an interior diagonal of P. See Exercise 23.) The key is to select a vertex p such that the angle $\angle bap$ is smallest. To see this, note that the ray starting at a and passing through p hits the line segment bc at a point, say q. It then follows that the triangle $\triangle baq$ cannot contain any vertices of P in its interior. Hence, we can connect b and p to produce an interior diagonal of P. Locating this vertex p is illustrated in Figure 3. ◁

5.2.5 Proofs Using the Well-Ordering Property

The validity of both the principle of mathematical induction and strong induction follows from a fundamental axiom of the set of integers, the **well-ordering property** (see Appendix 1). The well-ordering property states that every nonempty set of nonnegative integers has a least element. We will show how the well-ordering property can be used directly in proofs. Furthermore, it can be shown (see Exercises 45, 46, and 47) that the well-ordering property, the principle of mathematical induction, and strong induction are all equivalent. That is, the validity of each of these three proof techniques implies the validity of the other two techniques. In Section 5.1 we showed that the principle of mathematical induction follows from the well-ordering property. The other parts of this equivalence are left as Exercises 45, 46, and 47.

THE WELL-ORDERING PROPERTY Every nonempty set of nonnegative integers has a least element.

The well-ordering property can be used directly in proofs, as Example 5 illustrates.

EXAMPLE 5 Use the well-ordering property to prove the division algorithm. Recall that the division algorithm states that if a is an integer and d is a positive integer, then there are unique integers q and r with $0 \leq r < d$ and $a = dq + r$.

Extra Examples

Solution: Let S be the set of nonnegative integers of the form $a - dq$, where q is an integer. This set is nonempty because $-dq$ can be made as large as desired (taking q to be a negative integer with large absolute value). By the well-ordering property, S has a least element $r = a - dq_0$.

The integer r is nonnegative. It is also the case that $r < d$. If it were not, then there would be a smaller nonnegative element in S, namely, $a - d(q_0 + 1)$. To see this, suppose that $r \geq d$. Because $a = dq_0 + r$, it follows that $a - d(q_0 + 1) = (a - dq_0) - d = r - d \geq 0$. Consequently, there are integers q and r with $0 \leq r < d$. The proof that q and r are unique is left as Exercise 41. ◀

EXAMPLE 6 In a round-robin tournament every player plays every other player exactly once and each match has a winner and a loser. We say that the players $p_1, p_2, \ldots, p_m$ form a *cycle* if p_1 beats p_2, p_2 beats $p_3, \ldots, p_{m-1}$ beats p_m, and p_m beats p_1. Use the well-ordering property to show that if there is a cycle of length m ($m \geq 3$) among the players in a round-robin tournament, there must be a cycle of three of these players.

Solution: We assume that there is no cycle of three players. Because there is at least one cycle in the round-robin tournament, the set of all positive integers n for which there is a cycle of length n is nonempty. By the well-ordering property, this set of positive integers has a least element k, which by assumption must be greater than three. Consequently, there exists a cycle of players $p_1, p_2, p_3, \ldots, p_k$ and no shorter cycle exists.

Because there is no cycle of three players, we know that $k > 3$. Consider the first three elements of this cycle, p_1, p_2, and p_3. There are two possible outcomes of the match between p_1 and p_3. If p_3 beats p_1, it follows that p_1, p_2, p_3 is a cycle of length three, contradicting our assumption that there is no cycle of three players. Consequently, it must be the case that p_1 beats p_3. This means that we can omit p_2 from the cycle $p_1, p_2, p_3, \ldots, p_k$ to obtain the cycle $p_1, p_3, p_4, \ldots, p_k$ of length $k - 1$, contradicting the assumption that the smallest cycle has length k. We conclude that there must be a cycle of length three. ◀

Exercises

1. Use strong induction to show that if you can run one mile or two miles, and if you can always run two more miles once you have run a specified number of miles, then you can run any number of miles.
2. Use strong induction to show that all dominoes fall in an infinite arrangement of dominoes if you know that the first three dominoes fall, and that when a domino falls, the domino three farther down in the arrangement also falls.
3. Use strong induction to show that in any nonempty finite set of natural numbers there is an integer that is greater than or equal to all the integers in the set.
4. Use strong induction to show that in a country with n cities, if every pair of cities is connected by a one-way road then there is a route that passes through all n cities.
5. Let $P(n)$ be the statement that a postage of n cents can be formed using just 3-cent stamps and 5-cent stamps. The parts of this exercise outline a strong induction proof that $P(n)$ is true for all integers $n \geq 8$.
   a) Show that the statements $P(8)$, $P(9)$, and $P(10)$ are true, completing the basis step of a proof by strong induction that $P(n)$ is true for all integers $n \geq 8$.
   b) What is the inductive hypothesis of a proof by strong induction that $P(n)$ is true for all integers $n \geq 8$?
   c) What do you need to prove in the inductive step of a proof by strong induction that $P(n)$ is true for all integers $n \geq 8$?
   d) Complete the inductive step for $k \geq 10$.
   e) Explain why these steps show that $P(n)$ is true whenever $n \geq 8$.

6. Let $P(n)$ be the statement that a postage of n cents can be formed using just 4-cent stamps and 7-cent stamps. The parts of this exercise outline a strong induction proof that $P(n)$ is true for all integers $n \geq 18$.
   a) Show that the statements $P(18)$, $P(19)$, $P(20)$, and $P(21)$ are true, completing the basis step of a proof by strong induction that $P(n)$ is true for all integers $n \geq 18$.
   b) What is the inductive hypothesis of a proof by strong induction that $P(n)$ is true for all integers $n \geq 18$?
   c) What do you need to prove in the inductive step of a proof that $P(n)$ is true for all integers $n \geq 18$?
   d) Complete the inductive step for $k \geq 21$.
   e) Explain why these steps show that $P(n)$ is true for all integers $n \geq 18$.

7. a) Determine which amounts of postage can be formed using just 4-cent and 11-cent stamps.
   b) Prove your answer to (a) using the principle of mathematical induction. Be sure to state explicitly your inductive hypothesis in the inductive step.
   c) Prove your answer to (a) using strong induction. How does the inductive hypothesis in this proof differ from that in the inductive hypothesis for a proof using mathematical induction?

8. a) Determine which amounts of postage can be formed using just 3-cent and 10-cent stamps.
   b) Prove your answer to (a) using the principle of mathematical induction. Be sure to state explicitly your inductive hypothesis in the inductive step.
   c) Prove your answer to (a) using strong induction. How does the inductive hypothesis in this proof differ from that in the inductive hypothesis for a proof using mathematical induction?

9. Which amounts of money can be formed using just two-dollar bills and five-dollar bills? Prove your answer using strong induction.

10. Suppose that a store offers gift certificates in denominations of 25 dollars and 40 dollars. Determine the possible total amounts you can form using these gift certificates. Prove your answer using strong induction.

*11. Use strong induction to prove that $\sqrt{2}$ is irrational. [*Hint:* Let $P(n)$ be the statement that $\sqrt{2} \neq n/b$ for any positive integer b.]

12. Assume that a chocolate bar consists of n squares arranged in a rectangular pattern. The entire bar, or any smaller rectangular piece of the bar, can be broken along a vertical or a horizontal line separating the squares. Assuming that only one piece can be broken at a time, determine how many breaks you must successively make to break the bar into n separate squares. Use strong induction to prove your answer.

13. Consider this variation of the game of Nim. The game begins with n matches. Two players take turns removing matches, one, two, or three at a time. The player removing the last match loses. Use strong induction to show that if each player plays the best strategy possible, the first player wins if $n = 4j$, $4j + 2$, or $4j + 3$ for some nonnegative integer j and the second player wins in the remaining case when $n = 4j + 1$ for some nonnegative integer j.

14. Use strong induction to show that every positive integer can be written as a sum of distinct powers of two, that is, as a sum of a subset of the integers $2^0 = 1$, $2^1 = 2$, $2^2 = 4$, and so on. [*Hint:* For the inductive step, separately consider the case where $k + 1$ is even and where it is odd. When it is even, note that $(k + 1)/2$ is an integer.]

*15. A jigsaw puzzle is put together by successively joining pieces that fit together into blocks. A move is made each time a piece is added to a block, or when two blocks are joined. Use strong induction to prove that no matter how the moves are carried out, exactly $n - 1$ moves are required to assemble a puzzle with n pieces.

16. Suppose you begin with a pile of n stones and split this pile into n piles of one stone each by successively splitting a pile of stones into two smaller piles. Each time you split a pile you multiply the number of stones in each of the two smaller piles you form, so that if these piles have r and s stones in them, respectively, you compute rs. Show that no matter how you split the piles, the sum of the products computed at each step equals $n(n - 1)/2$.

17. Prove that the first player has a winning strategy for the game of Chomp, introduced in Example 12 in Section 1.8, if the initial board is square. [*Hint:* Use strong induction to show that this strategy works. For the first move, the first player chomps all cookies except those in the left and top edges. On subsequent moves, after the second player has chomped cookies on either the top or left edge, the first player chomps cookies in the same relative positions in the left or top edge, respectively.]

*18. Prove that the first player has a winning strategy for the game of Chomp, introduced in Example 12 in Section 1.8, if the initial board is two squares wide, that is, a $2 \times n$ board. [*Hint:* Use strong induction. The first move of the first player should be to chomp the cookie in the bottom row at the far right.]

19. Use strong induction to show that if a simple polygon with at least four sides is triangulated, then at least two of the triangles in the triangulation have two sides that border the exterior of the polygon.

*20. Use strong induction to show that when a simple polygon P with consecutive vertices $v_1, v_2, \ldots, v_n$ is triangulated into $n - 2$ triangles, the $n - 2$ triangles can be numbered $1, 2, \ldots, n - 2$ so that v_i is a vertex of triangle i for $i = 1, 2, \ldots, n - 2$.

*21. **Pick's theorem** says that the area of a simple polygon P in the plane with vertices that are all lattice points (that is, points with integer coordinates) equals $I(P) + B(P)/2 - 1$, where $I(P)$ and $B(P)$ are the number of lattice points in the interior of P and on the boundary of P, respectively. Use strong induction on the number of vertices of P to prove Pick's theorem. [*Hint:* For the basis step, first prove the theorem for rectangles, then for right triangles, and finally for all triangles by noting that the area of a triangle is the area of a larger rectangle containing it with the areas of at most three triangles subtracted. For the inductive step, take advantage of Lemma 1.]

****22.** Suppose that P is a simple polygon with vertices $v_1, v_2, \ldots, v_n$ listed so that consecutive vertices are connected by an edge, and v_1 and v_n are connected by an edge. A vertex v_i is called an **ear** if the line segment connecting the two vertices adjacent to v_i is an interior diagonal of the simple polygon. Two ears v_i and v_j are called **nonoverlapping** if the interiors of the triangles with vertices v_i and its two adjacent vertices and v_j and its two adjacent vertices do not intersect. Prove that every simple polygon with at least four vertices has at least two nonoverlapping ears.

23. In the proof of Lemma 1 we mentioned that many incorrect methods for finding a vertex p such that the line segment bp is an interior diagonal of P have been published. This exercise presents some of the incorrect ways p has been chosen in these proofs. Show, by considering one of the polygons drawn here, that for each of these choices of p, the line segment bp is not necessarily an interior diagonal of P.

a) p is the vertex of P such that the angle $\angle abp$ is smallest.

b) p is the vertex of P with the least x-coordinate (other than b).

c) p is the vertex of P that is closest to b.

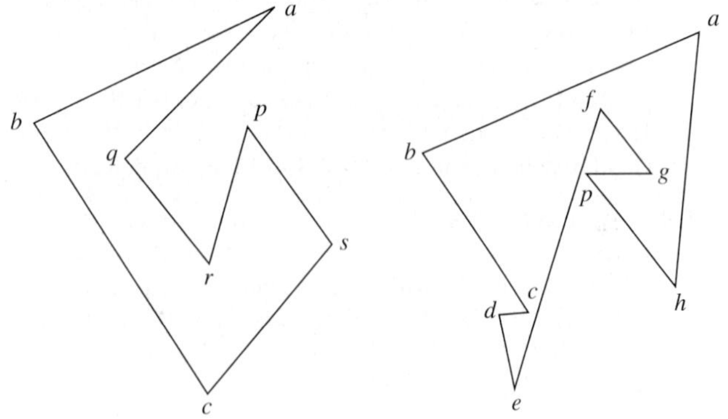

Exercises 24 and 25 present examples that show inductive loading can be used to prove results in computational geometry.

***24.** Let $P(n)$ be the statement that when nonintersecting diagonals are drawn inside a convex polygon with n sides, at least two vertices of the polygon are not endpoints of any of these diagonals.

a) Show that when we attempt to prove $P(n)$ for all integers n with $n \geq 3$ using strong induction, the inductive step does not go through.

b) Show that we can prove that $P(n)$ is true for all integers n with $n \geq 3$ by proving by strong induction the stronger assertion $Q(n)$, for $n \geq 4$, where $Q(n)$ states that whenever nonintersecting diagonals are drawn inside a convex polygon with n sides, at least two *nonadjacent* vertices are not endpoints of any of these diagonals.

25. Let $E(n)$ be the statement that in a triangulation of a simple polygon with n sides, at least one of the triangles in the triangulation has two sides bordering the exterior of the polygon.

a) Explain where a proof using strong induction that $E(n)$ is true for all integers $n \geq 4$ runs into difficulties.

b) Show that we can prove that $E(n)$ is true for all integers $n \geq 4$ by proving by strong induction the stronger statement $T(n)$ for all integers $n \geq 4$, which states that in every triangulation of a simple polygon, at least two of the triangles in the triangulation have two sides bordering the exterior of the polygon.

***26.** A stable assignment, defined in the preamble to Exercise 64 in Section 3.1, is called **optimal for suitors** if no stable assignment exists in which a suitor is paired with a suitee whom this suitor prefers to the person to whom this suitor is paired in this stable assignment. Use strong induction to show that the deferred acceptance algorithm produces a stable assignment that is optimal for suitors.

27. Suppose that $P(n)$ is a propositional function. Determine for which positive integers n the statement $P(n)$ must be true, and justify your answer, if

a) $P(1)$ is true; for all positive integers n, if $P(n)$ is true, then $P(n+2)$ is true.

b) $P(1)$ and $P(2)$ are true; for all positive integers n, if $P(n)$ and $P(n+1)$ are true, then $P(n+2)$ is true.

c) $P(1)$ is true; for all positive integers n, if $P(n)$ is true, then $P(2n)$ is true.

d) $P(1)$ is true; for all positive integers n, if $P(n)$ is true, then $P(n+1)$ is true.

28. Suppose that $P(n)$ is a propositional function. Determine for which nonnegative integers n the statement $P(n)$ must be true if

a) $P(0)$ is true; for all nonnegative integers n, if $P(n)$ is true, then $P(n+2)$ is true.

b) $P(0)$ is true; for all nonnegative integers n, if $P(n)$ is true, then $P(n+3)$ is true.

c) $P(0)$ and $P(1)$ are true; for all nonnegative integers n, if $P(n)$ and $P(n+1)$ are true, then $P(n+2)$ is true.

d) $P(0)$ is true; for all nonnegative integers n, if $P(n)$ is true, then $P(n+2)$ and $P(n+3)$ are true.

29. Show that if the statement $P(n)$ is true for infinitely many positive integers n and $P(n+1) \rightarrow P(n)$ is true for all positive integers n, then $P(n)$ is true for all positive integers n.

30. Let b be a fixed integer and j a fixed positive integer. Show that if $P(b), P(b+1), \ldots, P(b+j)$ are true and $[P(b) \wedge P(b+1) \wedge \cdots \wedge P(k)] \rightarrow P(k+1)$ is true for every integer $k \geq b+j$, then $P(n)$ is true for all integers n with $n \geq b$.

31. What is wrong with this "proof" by strong induction?

"Theorem" For every nonnegative integer n, $5n = 0$.

Basis Step: $5 \cdot 0 = 0$.

Inductive Step: Suppose that $5j = 0$ for all nonnegative integers j with $0 \leq j \leq k$. Write $k + 1 = i + j$, where i and j are natural numbers less than $k+1$. By the inductive hypothesis, $5(k+1) = 5(i+j) = 5i + 5j = 0 + 0 = 0$.

***32.** Find the flaw with the following "proof" that $a^n = 1$ for all nonnegative integers n, whenever a is a nonzero real number.

Basis Step: $a^0 = 1$ is true by the definition of a^0.

Inductive Step: Assume that $a^j = 1$ for all nonnegative integers j with $j \leq k$. Then note that

$$a^{k+1} = \frac{a^k \cdot a^k}{a^{k-1}} = \frac{1 \cdot 1}{1} = 1.$$

33. What is the error in this "proof" by strong induction that every nonnegative integer equals 0? "*Proof*". When n is a nonnegative integer let $S(n)$ be the statement that $n = 0$.

Basis Step: Because $S(0)$ is true, the basis step holds.

Inductive Step: Suppose that k is a nonnegative integer and that $S(j)$ is true for all nonnegative integers not exceeding k. To show that $S(k+1)$ is true, use the assumption that $S(k)$ is true, so that $k = 0$. Since we are also assuming that $S(1)$ is true, $1 = 0$. It follows that $k + 1 = 0 + 0 = 0$, proving that $S(k+1)$ is true. This completes the inductive step, so by strong induction, every nonnegative integer equals 0.

34. What is the error in this "proof" by strong induction that every integral postage of 5 cents or more can be formed using only 3-cent and 5-cent stamps?

Basis Step: Postage of 5 cents can be formed using a single 5-cent stamp.

Inductive Step: Assume postage of 5, 6, …, k cents can be formed using only 3 and 5-cent stamps. Postage of $k + 1$ cents can be formed by combining postage of $k - 2$ cents, which is possible by the induction hypothesis, and a 3-cent stamp.

***35.** Show that strong induction is a valid method of proof by showing that it follows from the well-ordering property.

36. Find the flaw with the following "proof" that every postage of three cents or more can be formed using just 3-cent and 4-cent stamps.

Basis Step: We can form postage of three cents with a single 3-cent stamp and we can form postage of four cents using a single 4-cent stamp.

Inductive Step: Assume that we can form postage of j cents for all nonnegative integers j with $j \leq k$ using just 3-cent and 4-cent stamps. We can then form postage of $k + 1$ cents by replacing one 3-cent stamp with a 4-cent stamp or by replacing two 4-cent stamps by three 3-cent stamps.

37. Show that we can prove that $P(n, k)$ is true for all pairs of positive integers n and k if we show

a) $P(1, 1)$ is true and $P(n, k) \to [P(n+1, k) \wedge P(n, k+1)]$ is true for all positive integers n and k.

b) $P(1, k)$ is true for all positive integers k, and $P(n, k) \to P(n+1, k)$ is true for all positive integers n and k.

c) $P(n, 1)$ is true for all positive integers n, and $P(n, k) \to P(n, k+1)$ is true for all positive integers n and k.

38. Prove that $\sum_{j=1}^{n} j(j+1)(j+2)\cdots(j+k-1) = n(n+1)(n+2)\cdots(n+k)/(k+1)$ for all positive integers k and n. [*Hint*: Use a technique from Exercise 37.]

***39.** Show that if $a_1, a_2, \ldots, a_n$ are n distinct real numbers, exactly $n - 1$ multiplications are used to compute the product of these n numbers no matter how parentheses are inserted into their product. [*Hint:* Use strong induction and consider the last multiplication.]

***40.** The well-ordering property can be used to show that there is a unique greatest common divisor of two positive integers. Let a and b be positive integers, and let S be the set of positive integers of the form $as + bt$, where s and t are integers.

a) Show that S is nonempty.

b) Use the well-ordering property to show that S has a smallest element c.

c) Show that if d is a common divisor of a and b, then d is a divisor of c.

d) Show that $c \mid a$ and $c \mid b$. [*Hint:* First, assume that $c \nmid a$. Then $a = qc + r$, where $0 < r < c$. Show that $r \in S$, contradicting the choice of c.]

e) Conclude from (c) and (d) that the greatest common divisor of a and b exists. Finish the proof by showing that this greatest common divisor is unique.

41. Let a be an integer and d be a positive integer. Show that the integers q and r with $a = dq + r$ and $0 \leq r < d$, which were shown to exist in Example 5, are unique.

42. Use mathematical induction to show that a rectangular checkerboard with an even number of cells and two squares missing, one white and one black, can be covered by dominoes.

****43.** Can you use the well-ordering property to prove the statement: "Every positive integer can be described using no more than fifteen English words"? Assume the words come from a particular dictionary of English. [*Hint:* Suppose that there are positive integers that cannot be described using no more than fifteen English words. By well ordering, *the smallest positive integer that cannot be described using no more than fifteen English words* would then exist.]

44. Use the well-ordering property to show that if x and y are real numbers with $x < y$, then there is a rational number r with $x < r < y$. [*Hint:* Use the Archimedean property, given in Appendix 1, to find a positive integer A with $A > 1/(y - x)$. Then show that there is a rational number r with denominator A between x and y by looking at the numbers $\lfloor x \rfloor + j/A$, where j is a positive integer.]

*45. Show that the well-ordering property can be proved when the principle of mathematical induction is taken as an axiom.

*46. Show that the principle of mathematical induction and strong induction are equivalent; that is, each can be shown to be valid from the other.

*47. Show that we can prove the well-ordering property when we take strong induction as an axiom instead of taking the well-ordering property as an axiom.

Another variant of mathematical induction, known as **downward induction** or **backward induction**, can be used to prove a proposition $S(n)$ is true for all positive integers by showing that (i) $S(n)$ is true for infinitely many positive integers n and (ii) for every positive integer m with $m \geq 2$, $S(m) \rightarrow S(m-1)$.

48. Prove that downward induction is a valid proof method by showing that its validity follows from the principle of mathematical induction.

49. Suppose that there are infinitely many dominoes arranged in a straight line. If you knock over the first domino, the 2^nth domino also falls for all positive integers n. The dominoes are arranged so that if one of them falls, the one preceding it must have fallen. Use the principle of downward induction to show that if you knock down the first domino, all of the dominoes fall.

50. If a and b are positive integers, show that $\gcd(a, b) = s_n a + t_n b$ where s_n and t_n are the nth terms of the sequence defined by $s_0 = 1, s_1 = 0, s_j = s_{j-2} - q_{j-1}s_{j-1}$ and by $t_0 = 0, t_1 = 1, t_j = t_{j-2} - q_{j-1}t_{j-1}$ where q_j for $j = 2, 3, \ldots, n$ are the quotients in the divisions of the Euclidean algorithm when used to find $\gcd(a, b)$.

5.3 Recursive Definitions and Structural Induction

5.3.1 Introduction

Sometimes it is difficult to define an object explicitly. However, it may be easy to define this object in terms of itself. This process is called **recursion**. For instance, the picture shown in Figure 1 is produced recursively. First, an original picture is given. Then a process of successively superimposing centered smaller pictures on top of the previous pictures is carried out.

We can use recursion to define sequences, functions, and sets. In Section 2.4, and in most beginning mathematics courses, the terms of a sequence are specified using an explicit formula. For instance, the sequence of powers of 2 is given by $a_n = 2^n$ for $n = 0, 1, 2, \ldots$. Recall from

FIGURE 1 A recursively defined picture.

Section 2.4 that we can also define a sequence recursively by specifying how terms of the sequence are found from previous terms. The sequence of powers of 2 can also be defined by giving the first term of the sequence, namely, $a_0 = 1$, and a rule for finding a term of the sequence from the previous one, namely, $a_{n+1} = 2a_n$ for $n = 0, 1, 2, \ldots$. When we define a sequence recursively by specifying how terms of the sequence are found from previous terms, we can use induction to prove results about the sequence.

When we define a set recursively, we specify some initial elements in a basis step and provide a rule for constructing new elements from those we already have in the recursive step. To prove results about recursively defined sets we use a method called *structural induction*.

5.3.2 Recursively Defined Functions

We use two steps to define a function with the set of nonnegative integers as its domain:

BASIS STEP: Specify the value of the function at zero.

Assessment

RECURSIVE STEP: Give a rule for finding its value at an integer from its values at smaller integers.

Such a definition is called a **recursive** or **inductive definition**. Note that a function $f(n)$ from the set of nonnegative integers to the set of a real numbers is the same as a sequence $a_0, a_1, \ldots$, where a_i is a real number for every nonnegative integer i. So, defining a real-valued sequence $a_0, a_1, \ldots$ using a recurrence relation, as was done in Section 2.4, is the same as defining a function from the set of nonnegative integers to the set of real numbers.

EXAMPLE 1 Suppose that f is defined recursively by

$$f(0) = 3,$$

Extra Examples

$$f(n+1) = 2f(n) + 3.$$

Find $f(1), f(2), f(3)$, and $f(4)$.

Solution: From the recursive definition it follows that

$$f(1) = 2f(0) + 3 = 2 \cdot 3 + 3 = 9,$$
$$f(2) = 2f(1) + 3 = 2 \cdot 9 + 3 = 21,$$
$$f(3) = 2f(2) + 3 = 2 \cdot 21 + 3 = 45,$$
$$f(4) = 2f(3) + 3 = 2 \cdot 45 + 3 = 93.$$

◀

Recursively defined functions are **well defined**. That is, for every positive integer, the value of the function at this integer is determined in an unambiguous way. This means that given any positive integer, we can use the two parts of the definition to find the value of the function at that integer, and that we obtain the same value no matter how we apply the two parts of the definition. This is a consequence of the principle of mathematical induction. (See Exercise 58.) Additional examples of recursive definitions are given in Examples 2 and 3.

EXAMPLE 2 Give a recursive definition of a^n, where a is a nonzero real number and n is a nonnegative integer.

Solution: The recursive definition contains two parts. First a^0 is specified, namely, $a^0 = 1$. Then the rule for finding a^{n+1} from a^n, namely, $a^{n+1} = a \cdot a^n$, for $n = 0, 1, 2, 3, \ldots$, is given. These two equations uniquely define a^n for all nonnegative integers n. ◀

EXAMPLE 3 Give a recursive definition of

$$\sum_{k=0}^{n} a_k.$$

Solution: The first part of the recursive definition is

$$\sum_{k=0}^{0} a_k = a_0.$$

The second part is

$$\sum_{k=0}^{n+1} a_k = \left(\sum_{k=0}^{n} a_k\right) + a_{n+1}.$$

◀

In some recursive definitions of functions, the values of the function at the first k positive integers are specified, and a rule is given for determining the value of the function at larger integers from its values at some or all of the preceding k integers. That recursive definitions defined in this way produce well-defined functions follows from strong induction (see Exercise 59).

Recall from Section 2.4 that the Fibonacci numbers, $f_0, f_1, f_2, \dots$, are defined by the equations $f_0 = 0, f_1 = 1$, and

$$f_n = f_{n-1} + f_{n-2}$$

Links

for $n = 2, 3, 4, \dots$. [We can think of the Fibonacci number f_n either as the nth term of the sequence of Fibonacci numbers $f_0, f_1, \dots$ or as the value at the integer n of a function $f(n)$.]

We can use the recursive definition of the Fibonacci numbers to prove many properties of these numbers. We give one such property in Example 4.

EXAMPLE 4 Show that whenever $n \geq 3, f_n > \alpha^{n-2}$, where $\alpha = (1 + \sqrt{5})/2$.

Extra Examples

Solution: We can use strong induction to prove this inequality. Let $P(n)$ be the statement $f_n > \alpha^{n-2}$. We want to show that $P(n)$ is true whenever n is an integer greater than or equal to 3.

BASIS STEP: First, note that

$$\alpha < 2 = f_3, \qquad \alpha^2 = (3 + \sqrt{5})/2 < 3 = f_4,$$

so $P(3)$ and $P(4)$ are true.

INDUCTIVE STEP: Assume that $P(j)$ is true, namely, that $f_j > \alpha^{j-2}$, for all integers j with $3 \leq j \leq k$, where $k \geq 4$. We must show that $P(k+1)$ is true, that is, that $f_{k+1} > \alpha^{k-1}$. Because α is a solution of $x^2 - x - 1 = 0$ (as the quadratic formula shows), it follows that $\alpha^2 = \alpha + 1$. Therefore,

$$\alpha^{k-1} = \alpha^2 \cdot \alpha^{k-3} = (\alpha + 1)\alpha^{k-3} = \alpha \cdot \alpha^{k-3} + 1 \cdot \alpha^{k-3} = \alpha^{k-2} + \alpha^{k-3}.$$

By the inductive hypothesis, because $k \geq 4$, we have

$$f_{k-1} > \alpha^{k-3}, \qquad f_k > \alpha^{k-2}.$$

Therefore, it follows that

$$f_{k+1} = f_k + f_{k-1} > \alpha^{k-2} + \alpha^{k-3} = \alpha^{k-1}.$$

Hence, $P(k+1)$ is true. This completes the proof. ◀

Remark: The inductive step of the proof by strong induction in Example 4 shows that whenever $k \geq 4$, $P(k+1)$ follows from the assumption that $P(j)$ is true for $3 \leq j \leq k$. Hence, the inductive step does *not* show that $P(3) \rightarrow P(4)$. Therefore, we had to show that $P(4)$ is true separately.

We can now show that the Euclidean algorithm, introduced in Section 4.3, uses $O(\log b)$ divisions to find the greatest common divisor of the positive integers a and b, where $a \geq b$.

THEOREM 1 **LAMÉ'S THEOREM** Let a and b be positive integers with $a \geq b$. Then the number of divisions used by the Euclidean algorithm to find $\gcd(a, b)$ is less than or equal to five times the number of decimal digits in b.

Proof: Recall that when the Euclidean algorithm is applied to find $\gcd(a, b)$ with $a \geq b$, this sequence of equations (where $a = r_0$ and $b = r_1$) is obtained.

$$\begin{aligned} r_0 &= r_1 q_1 + r_2 && 0 \leq r_2 < r_1, \\ r_1 &= r_2 q_2 + r_3 && 0 \leq r_3 < r_2, \\ &\vdots \\ r_{n-2} &= r_{n-1} q_{n-1} + r_n && 0 \leq r_n < r_{n-1}, \\ r_{n-1} &= r_n q_n. \end{aligned}$$

Here n divisions have been used to find $r_n = \gcd(a, b)$. Note that the quotients $q_1, q_2, \ldots, q_{n-1}$ are all at least 1. Moreover, $q_n \geq 2$, because $r_n < r_{n-1}$. This implies that

$$\begin{aligned} r_n &\geq 1 = f_2, \\ r_{n-1} &\geq 2r_n \geq 2f_2 = f_3, \\ r_{n-2} &\geq r_{n-1} + r_n \geq f_3 + f_2 = f_4, \\ &\vdots \\ r_2 &\geq r_3 + r_4 \geq f_{n-1} + f_{n-2} = f_n, \\ b = r_1 &\geq r_2 + r_3 \geq f_n + f_{n-1} = f_{n+1}. \end{aligned}$$

It follows that if n divisions are used by the Euclidean algorithm to find $\gcd(a, b)$ with $a \geq b$, then $b \geq f_{n+1}$. By Example 4 we know that $f_{n+1} > \alpha^{n-1}$ for $n > 2$, where $\alpha = (1 + \sqrt{5})/2$. Therefore, it follows that $b > \alpha^{n-1}$. Furthermore, because $\log_{10} \alpha \approx 0.208 > 1/5$, we see that

$$\log_{10} b > (n-1) \log_{10} \alpha > (n-1)/5.$$

Hence, $n - 1 < 5 \cdot \log_{10} b$. Now suppose that b has k decimal digits. Then $b < 10^k$ and $\log_{10} b < k$. It follows that $n - 1 < 5k$, and because k is an integer, it follows that $n \leq 5k$. This finishes the proof. ◁

Because the number of decimal digits in b, which equals $\lfloor \log_{10} b \rfloor + 1$, is less than or equal to $\log_{10} b + 1$, Theorem 1 tells us that the number of divisions required to find $\gcd(a, b)$ with

Links

Mondadori Portfolio/ Hulton Fine Art Collection/Getty Images

FIBONACCI (1170–1250) Fibonacci (short for *filius Bonacci,* or "son of Bonacci") was also known as Leonardo of Pisa. He was born in the Italian commercial center of Pisa. Fibonacci was a merchant who traveled extensively throughout the Mideast, where he came into contact with Arabian mathematics. In his book *Liber Abaci,* Fibonacci introduced the European world to Arabic notation for numerals and algorithms for arithmetic. It was in this book that his well known rabbit problem (described in Section 8.1) appeared. Fibonacci also wrote books on geometry and trigonometry and on Diophantine equations, which involve finding integer solutions to equations.

$a > b$ is less than or equal to $5(\log_{10} b + 1)$. Because $5(\log_{10} b + 1)$ is $O(\log b)$, we see that $O(\log b)$ divisions are used by the Euclidean algorithm to find $\gcd(a, b)$ whenever $a > b$.

5.3.3 Recursively Defined Sets and Structures

Assessment

We have explored how functions can be defined recursively. We now turn our attention to how sets can be defined recursively. Just as in the recursive definition of functions, recursive definitions of sets have two parts, a **basis step** and a **recursive step**. In the basis step, an initial collection of elements is specified. In the recursive step, rules for forming new elements in the set from those already known to be in the set are provided. Recursive definitions may also include an **exclusion rule**, which specifies that a recursively defined set contains nothing other than those elements specified in the basis step or generated by applications of the recursive step. In our discussions, we will always tacitly assume that the exclusion rule holds and no element belongs to a recursively defined set unless it is in the initial collection specified in the basis step or can be generated using the recursive step one or more times. Later we will see how we can use a technique known as structural induction to prove results about recursively defined sets.

Examples 5, 6, 8, and 9 illustrate the recursive definition of sets. In each example, we show those elements generated by the first few applications of the recursive step.

EXAMPLE 5 Consider the subset S of the set of integers recursively defined by

BASIS STEP: $3 \in S$.

RECURSIVE STEP: If $x \in S$ and $y \in S$, then $x + y \in S$.

Extra Examples

The new elements found to be in S are 3 by the basis step, $3 + 3 = 6$ at the first application of the recursive step, $3 + 6 = 6 + 3 = 9$ and $6 + 6 = 12$ at the second application of the recursive step, and so on. We will show in Example 10 that S is the set of all positive multiples of 3. ◀

Recursive definitions play an important role in the study of strings. (See Chapter 13 for an introduction to the theory of formal languages, for example.) Recall from Section 2.4 that a string over an alphabet Σ is a finite sequence of symbols from Σ. We can define Σ^*, the set of strings over Σ, recursively, as Definition 1 shows.

Definition 1 The set Σ^* of *strings* over the alphabet Σ is defined recursively by

BASIS STEP: $\lambda \in \Sigma^*$ (where λ is the empty string containing no symbols).

RECURSIVE STEP: If $w \in \Sigma^*$ and $x \in \Sigma$, then $wx \in \Sigma^*$.

Links

Paul Fearn/Alamy Stock Photo

GABRIEL LAMÉ (1795–1870) Gabriel Lamé entered the École Polytechnique in 1813, graduating in 1817. He continued his education at the École des Mines, graduating in 1820.

In 1820 Lamé went to Russia, where he was appointed director of the Schools of Highways and Transportation in St. Petersburg. Not only did he teach, but he also planned roads and bridges while in Russia. He returned to Paris in 1832, where he helped found an engineering firm. However, he soon left the firm, accepting the chair of physics at the École Polytechnique, which he held until 1844. While holding this position, he was active outside academia as an engineering consultant, serving as chief engineer of mines and participating in the building of railways.

Lamé contributed original work to number theory, applied mathematics, and thermodynamics. His best-known work involves the introduction of curvilinear coordinates. His work on number theory includes proving Fermat's last theorem for $n = 7$, as well as providing the upper bound for the number of divisions used by the Euclidean algorithm given in this text.

In the opinion of Gauss, one of the most important mathematicians of all time, Lamé was the foremost French mathematician of his time. However, French mathematicians considered him too practical, whereas French scientists considered him too theoretical.

The basis step of the recursive definition of strings says that the empty string belongs to Σ^*. The recursive step states that new strings are produced by adding a symbol from Σ to the end of strings in Σ^*. At each application of the recursive step, strings containing one additional symbol are generated.

EXAMPLE 6 If $\Sigma = \{0, 1\}$, the strings found to be in Σ^*, the set of all bit strings, are λ, specified to be in Σ^* in the basis step, 0 and 1 formed during the first application of the recursive step, 00, 01, 10, and 11 formed during the second application of the recursive step, and so on. ◀

Recursive definitions can be used to define operations or functions on the elements of recursively defined sets. This is illustrated in Definition 2 of the concatenation of two strings and Example 7 concerning the length of a string.

Definition 2 Two strings can be combined via the operation of *concatenation*. Let Σ be a set of symbols and Σ^* the set of strings formed from symbols in Σ. We can define the concatenation of two strings, denoted by $\cdot$, recursively as follows.

BASIS STEP: If $w \in \Sigma^*$, then $w \cdot \lambda = w$, where λ is the empty string.

RECURSIVE STEP: If $w_1 \in \Sigma^*$ and $w_2 \in \Sigma^*$ and $x \in \Sigma$, then $w_1 \cdot (w_2 x) = (w_1 \cdot w_2)x$.

The concatenation of the strings w_1 and w_2 is often written as $w_1 w_2$ rather than $w_1 \cdot w_2$. By repeated application of the recursive definition, it follows that the concatenation of two strings w_1 and w_2 consists of the symbols in w_1 followed by the symbols in w_2. For instance, the concatenation of $w_1 = abra$ and $w_2 = cadabra$ is $w_1 w_2 = abracadabra$.

EXAMPLE 7 **Length of a String** Give a recursive definition of $l(w)$, the length of the string w.

Solution: The length of a string can be recursively defined by

$$l(\lambda) = 0;$$
$$l(wx) = l(w) + 1 \text{ if } w \in \Sigma^* \text{ and } x \in \Sigma.$$

◀

Another important use of recursive definitions is to define **well-formed formulae** of various types. This is illustrated in Examples 8 and 9.

EXAMPLE 8 **Well-Formed Formulae in Propositional Logic** We can define the set of well-formed formulae in propositional logic involving **T**, **F**, propositional variables, and operators from the set $\{\neg, \wedge, \vee, \rightarrow, \leftrightarrow\}$.

BASIS STEP: **T**, **F**, and s, where s is a propositional variable, are well-formed formulae.

RECURSIVE STEP: If E and F are well-formed formulae, then $(\neg E)$, $(E \wedge F)$, $(E \vee F)$, $(E \rightarrow F)$, and $(E \leftrightarrow F)$ are well-formed formulae.

For example, by the basis step we know that **T**, **F**, p, and q are well-formed formulae, where p and q are propositional variables. From an initial application of the recursive step, we know that $(p \vee q)$, $(p \rightarrow \mathbf{F})$, $(\mathbf{F} \rightarrow q)$, and $(q \wedge \mathbf{F})$ are well-formed formulae. A second application of the recursive step shows that $((p \vee q) \rightarrow (q \wedge \mathbf{F}))$, $(q \vee (p \vee q))$, and $((p \rightarrow \mathbf{F}) \rightarrow \mathbf{T})$ are well-formed formulae. We leave it to the reader to show that $p\neg \wedge q$, $pq\wedge$, and $\neg \wedge pq$ are *not* well-formed formulae, by showing that none can be obtained using the basis step and one or more applications of the recursive step. ◀

EXAMPLE 9 **Well-Formed Formulae of Operators and Operands** We can define the set of well-formed formulae consisting of variables, numerals, and operators from the set $\{+, -, *, /, \uparrow\}$ (where $*$ denotes multiplication and $\uparrow$ denotes exponentiation) recursively.

BASIS STEP: x is a well-formed formula if x is a numeral or a variable.

RECURSIVE STEP: If F and G are well-formed formulae, then $(F + G)$, $(F - G)$, $(F * G)$, (F/G), and $(F \uparrow G)$ are well-formed formulae.

For example, by the basis step we see that x, y, 0, and 3 are well-formed formulae (as is any variable or numeral). Well-formed formulae generated by applying the recursive step once include $(x + 3)$, $(3 + y)$, $(x - y)$, $(3 - 0)$, $(x * 3)$, $(3 * y)$, $(3/0)$, (x/y), $(3 \uparrow x)$, and $(0 \uparrow 3)$. Applying the recursive step twice shows that formulae such as $((x + 3) + 3)$ and $(x - (3 * y))$ are well-formed formulae. [Note that $(3/0)$ is a well-formed formula because we are concerned only with syntax matters here.] We leave it to the reader to show that each of the formulae $x3+$, $y * +x$, and $* \; x/y$ is *not* a well-formed formula by showing that none of them can be obtained from the basis step and one or more applications of the recursive step. ◀

We will study trees extensively in Chapter 11. A tree is a special type of a graph; a graph is made up of vertices and edges connecting some pairs of vertices. We will study graphs in Chapter 10. We will briefly introduce them here to illustrate how they can be defined recursively.

Definition 3 The set of *rooted trees*, where a rooted tree consists of a set of vertices containing a distinguished vertex called the *root*, and edges connecting these vertices, can be defined recursively by these steps:

BASIS STEP: A single vertex r is a rooted tree.

RECURSIVE STEP: Suppose that $T_1, T_2, \dots, T_n$ are disjoint rooted trees with roots $r_1, r_2, \dots, r_n$, respectively. Then the graph formed by starting with a root r, which is not in any of the rooted trees $T_1, T_2, \dots, T_n$, and adding an edge from r to each of the vertices $r_1, r_2, \dots, r_n$, is also a rooted tree.

In Figure 2 we illustrate some of the rooted trees formed starting with the basis step and applying the recursive step one time and two times. Note that infinitely many rooted trees are formed at each application of the recursive definition.

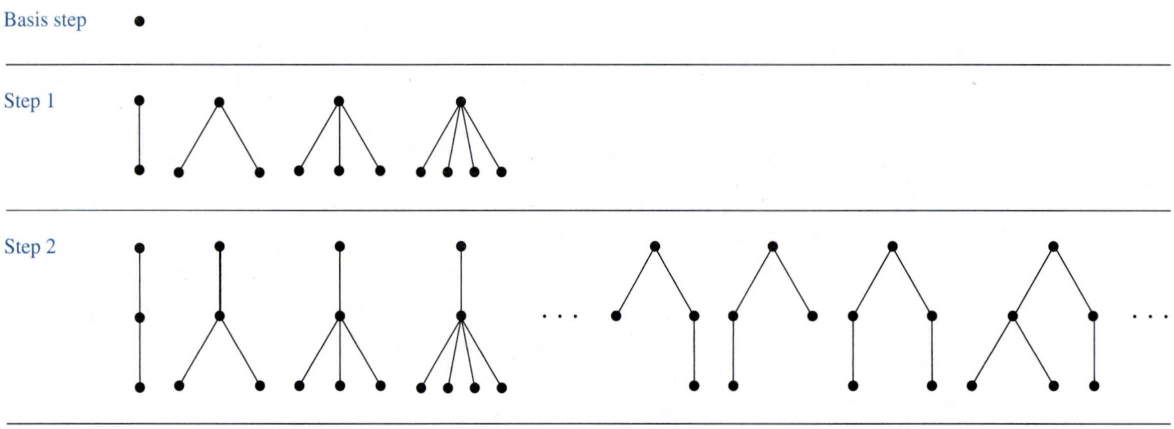

FIGURE 2 **Building up rooted trees.**

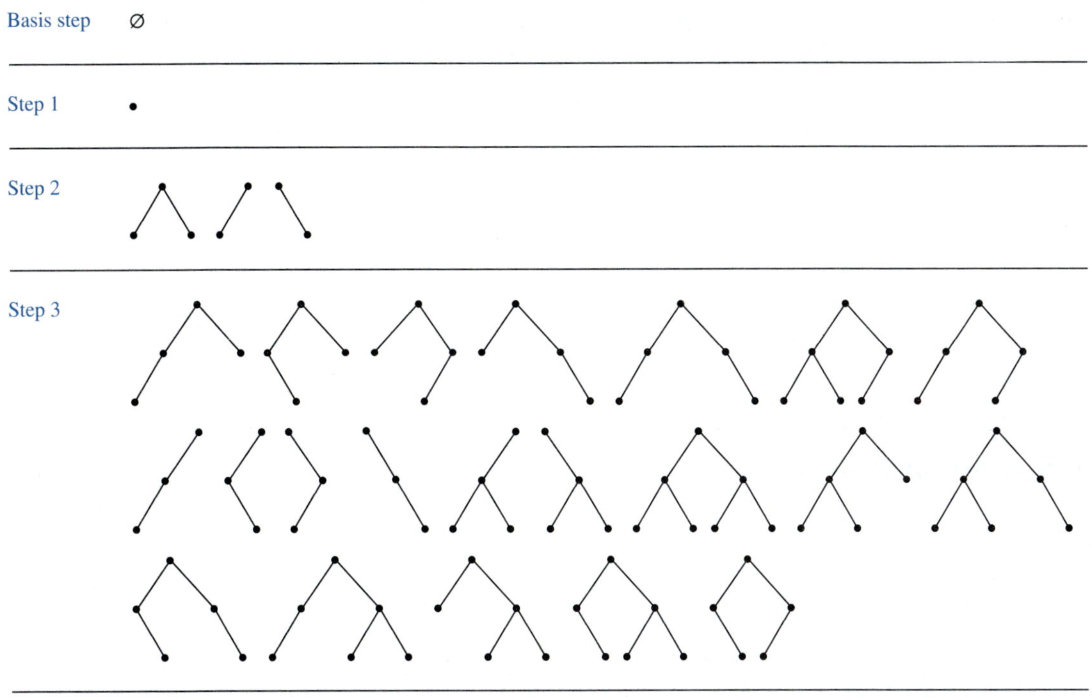

FIGURE 3 Building up extended binary trees.

Binary trees are a special type of rooted trees. We will provide recursive definitions of two types of binary trees—full binary trees and extended binary trees. In the recursive step of the definition of each type of binary tree, two binary trees are combined to form a new tree with one of these trees designated the left subtree and the other the right subtree. In extended binary trees, the left subtree or the right subtree can be empty, but in full binary trees this is not possible. Binary trees are one of the most important types of structures in computer science. In Chapter 11 we will see how they can be used in searching and sorting algorithms, in algorithms for compressing data, and in many other applications. We first define extended binary trees.

Definition 4 The set of *extended binary trees* can be defined recursively by these steps:

BASIS STEP: The empty set is an extended binary tree.

RECURSIVE STEP: If T_1 and T_2 are disjoint extended binary trees, there is an extended binary tree, denoted by $T_1 \cdot T_2$, consisting of a root r together with edges connecting the root to each of the roots of the left subtree T_1 and the right subtree T_2 when these trees are nonempty.

Figure 3 shows how extended binary trees are built up by applying the recursive step from one to three times.

We now show how to define the set of full binary trees. Note that the difference between this recursive definition and that of extended binary trees lies entirely in the basis step.

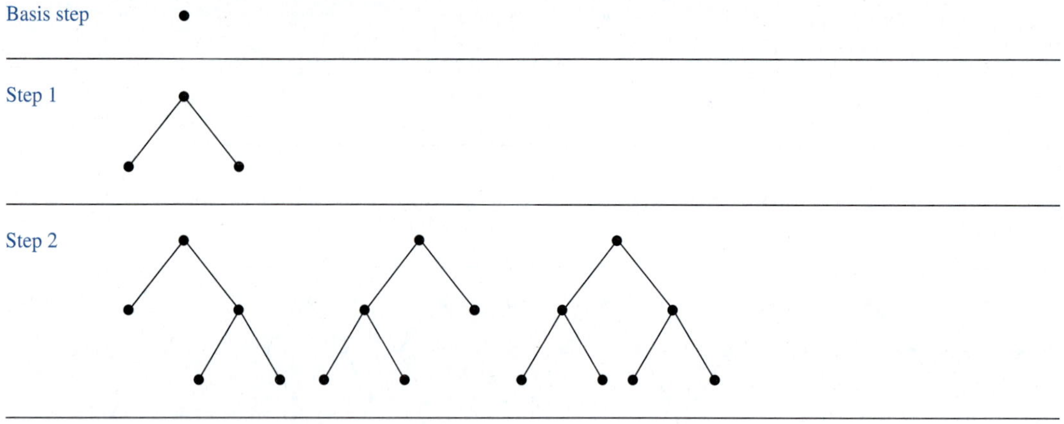

FIGURE 4 Building up full binary trees.

Definition 5

The set of full binary trees can be defined recursively by these steps:

BASIS STEP: There is a full binary tree consisting only of a single vertex r.

RECURSIVE STEP: If T_1 and T_2 are disjoint full binary trees, there is a full binary tree, denoted by $T_1 \cdot T_2$, consisting of a root r together with edges connecting the root to each of the roots of the left subtree T_1 and the right subtree T_2.

Figure 4 shows how full binary trees are built up by applying the recursive step one and two times.

5.3.4 Structural Induction

To prove results about recursively defined sets, we generally use some form of mathematical induction. Example 10 illustrates the connection between recursively defined sets and mathematical induction.

EXAMPLE 10 Show that the set S defined in Example 5 by specifying that $3 \in S$ and that if $x \in S$ and $y \in S$, then $x + y \in S$, is the set of all positive integers that are multiples of 3.

Solution: Let A be the set of all positive integers divisible by 3. To prove that $A = S$, we must show that A is a subset of S and that S is a subset of A. To prove that A is a subset of S, we must show that every positive integer divisible by 3 is in S. We will use mathematical induction to prove this.

Let $P(n)$ be the statement that $3n$ belongs to S. The basis step holds because by the first part of the recursive definition of S, $3 \cdot 1 = 3$ is in S. To establish the inductive step, assume that $P(k)$ is true, namely, that $3k$ is in S. Because $3k$ is in S and because 3 is in S, it follows from the second part of the recursive definition of S that $3k + 3 = 3(k + 1)$ is also in S.

To prove that S is a subset of A, we use the recursive definition of S. First, the basis step of the definition specifies that 3 is in S. Because $3 = 3 \cdot 1$, all elements specified to be in S in this step are divisible by 3 and are therefore in A. To finish the proof, we must show that all integers in S generated using the second part of the recursive definition are in A. This consists of showing that $x + y$ is in A whenever x and y are elements of S also assumed to be in A. Now if x and y are both in A, it follows that $3 \mid x$ and $3 \mid y$. By part (*i*) of Theorem 1 of Section 4.1, it follows that $3 \mid (x + y)$ completing the proof. ◀

In Example 10 we used mathematical induction over the set of positive integers and a recursive definition to prove a result about a recursively defined set. However, instead of using mathematical induction directly to prove results about recursively defined sets, we can use a more convenient form of induction known as **structural induction**. A proof by structural induction consists of two parts. These parts are

BASIS STEP: Show that the result holds for all elements specified in the basis step of the recursive definition to be in the set.

RECURSIVE STEP: Show that if the statement is true for each of the elements used to construct new elements in the recursive step of the definition, the result holds for these new elements.

The validity of structural induction follows from the principle of mathematical induction for the nonnegative integers. To see this, let $P(n)$ state that the claim is true for all elements of the set that are generated by n or fewer applications of the rules in the recursive step of a recursive definition. We will have established that the principle of mathematical induction implies the principle of structural induction if we can show that $P(n)$ is true whenever n is a positive integer. In the basis step of a proof by structural induction we show that $P(0)$ is true. That is, we show that the result is true of all elements specified to be in the set in the basis step of the definition. A consequence of the recursive step is that if we assume $P(k)$ is true, it follows that $P(k + 1)$ is true. When we have completed a proof using structural induction, we have shown that $P(0)$ is true and that $P(k)$ implies $P(k + 1)$. By mathematical induction it follows that $P(n)$ is true for all nonnegative integers n. This also shows that the result is true for all elements generated by the recursive definition, and shows that structural induction is a valid proof technique.

EXAMPLES OF PROOFS USING STRUCTURAL INDUCTION Structural induction can be used to prove that all members of a set constructed recursively have a particular property. We will illustrate this idea by using structural induction to prove results about well-formed formulae, strings, and binary trees. For each proof, we have to carry out the appropriate basis step and the appropriate recursive step. For example, to use structural induction to prove a result about the set of well-formed formulae defined in Example 8, where we specify that **T**, **F**, and every propositional variable s are well-formed formulae and where we specify that if E and F are well-formed formulae, then $(\neg E)$, $(E \wedge F)$, $(E \vee F)$, $(E \to F)$, and $(E \leftrightarrow F)$ are well-formed formulae, we need to complete this basis step and this recursive step.

BASIS STEP: Show that the result is true for **T**, **F**, and s whenever s is a propositional variable.

RECURSIVE STEP: Show that if the result is true for the compound propositions p and q, it is also true for $(\neg p)$, $(p \vee q)$, $(p \wedge q)$, $(p \to q)$, and $(p \leftrightarrow q)$.

Example 11 illustrates how we can prove results about well-formed formulae using structural induction.

EXAMPLE 11 Show that every well-formed formula for compound propositions, as defined in Example 8, contains an equal number of left and right parentheses.

Solution:

BASIS STEP: Each of the formula **T**, **F**, and s contains no parentheses, so clearly they contain an equal number of left and right parentheses.

RECURSIVE STEP: Assume p and q are well-formed formulae, each containing an equal number of left and right parentheses. That is, if l_p and l_q are the number of left parentheses in p and q, respectively, and r_p and r_q are the number of right parentheses in p and q, respectively, then $l_p = r_p$ and $l_q = r_q$. To complete the inductive step, we need to show that each of $(\neg p)$, $(p \vee q)$, $(p \wedge q)$, $(p \to q)$, and $(p \leftrightarrow q)$ also contains an equal number of left and right parentheses. The number of left parentheses in the first of these compound propositions equals

$l_p + 1$ and in each of the other compound propositions equals $l_p + l_q + 1$. Similarly, the number of right parentheses in the first of these compound propositions equals $r_p + 1$ and in each of the other compound propositions equals $r_p + r_q + 1$. Because $l_p = r_p$ and $l_q = r_q$, it follows that each of these compound expressions contains the same number of left and right parentheses. This completes the proof by structural induction. ◀

Suppose that $P(w)$ is a propositional function over the set of strings $w \in \Sigma^*$. To use structural induction to prove that $P(w)$ holds for all strings $w \in \Sigma^*$, we need to complete both a basis step and a recursive step. These steps are:

BASIS STEP: Show that $P(\lambda)$ is true.

RECURSIVE STEP: Assume that $P(w)$ is true, where $w \in \Sigma^*$. Show that if $x \in \Sigma$, then $P(wx)$ must also be true.

Example 12 illustrates how structural induction can be used in proofs about strings.

EXAMPLE 12 Use structural induction to prove that $l(xy) = l(x) + l(y)$, where x and y belong to Σ^*, the set of strings over the alphabet Σ.

Solution: We will base our proof on the recursive definition of the set Σ^* given in Definition 1 and the definition of the length of a string in Example 7, which specifies that $l(\lambda) = 0$ and $l(wx) = l(w) + 1$ when $w \in \Sigma^*$ and $x \in \Sigma$. Let $P(y)$ be the statement that $l(xy) = l(x) + l(y)$ whenever x belongs to Σ^*.

BASIS STEP: To complete the basis step, we must show that $P(\lambda)$ is true. That is, we must show that $l(x\lambda) = l(x) + l(\lambda)$ for all $x \in \Sigma^*$. Because $l(x\lambda) = l(x) = l(x) + 0 = l(x) + l(\lambda)$ for every string x, it follows that $P(\lambda)$ is true.

RECURSIVE STEP: To complete the inductive step, we assume that $P(y)$ is true and show that this implies that $P(ya)$ is true whenever $a \in \Sigma$. What we need to show is that $l(xya) = l(x) + l(ya)$ for every $a \in \Sigma$. To show this, note that by the recursive definition of $l(w)$ (given in Example 7), we have $l(xya) = l(xy) + 1$ and $l(ya) = l(y) + 1$. And, by the inductive hypothesis, $l(xy) = l(x) + l(y)$. We conclude that $l(xya) = l(x) + l(y) + 1 = l(x) + l(ya)$. ◀

We can prove results about trees or special classes of trees using structural induction. For example, to prove a result about full binary trees using structural induction we need to complete this basis step and this recursive step.

BASIS STEP: Show that the result is true for the tree consisting of a single vertex.

RECURSIVE STEP: Show that if the result is true for the full binary trees T_1 and T_2, then it is true for tree $T_1 \cdot T_2$ consisting of a root r, which has T_1 as its left subtree and T_2 as its right subtree.

Before we provide an example showing how structural induction can be used to prove a result about full binary trees, we need some definitions. We will recursively define the height $h(T)$ and the number of vertices $n(T)$ of a full binary tree T. We begin by defining the height of a full binary tree.

Definition 6 We define the height $h(T)$ of a full binary tree T recursively.

BASIS STEP: The height of the full binary tree T consisting of only a root r is $h(T) = 0$.

RECURSIVE STEP: If T_1 and T_2 are full binary trees, then the full binary tree $T = T_1 \cdot T_2$ has height $h(T) = 1 + \max(h(T_1), h(T_2))$.

If we let $n(T)$ denote the number of vertices in a full binary tree, we observe that $n(T)$ satisfies the following recursive formula:

BASIS STEP: The number of vertices $n(T)$ of the full binary tree T consisting of only a root r is $n(T) = 1$.

RECURSIVE STEP: If T_1 and T_2 are full binary trees, then the number of vertices of the full binary tree $T = T_1 \cdot T_2$ is $n(T) = 1 + n(T_1) + n(T_2)$.

We now show how structural induction can be used to prove a result about full binary trees.

THEOREM 2 If T is a full binary tree T, then $n(T) \leq 2^{h(T)+1} - 1$.

Proof: We prove this inequality using structural induction.

BASIS STEP: For the full binary tree consisting of just the root r the result is true because $n(T) = 1$ and $h(T) = 0$, so that $n(T) = 1 \leq 2^{0+1} - 1 = 1$.

RECURSIVE STEP: For the inductive hypothesis we assume that $n(T_1) \leq 2^{h(T_1)+1} - 1$ and $n(T_2) \leq 2^{h(T_2)+1} - 1$ whenever T_1 and T_2 are full binary trees. By the recursive formulae for $n(T)$ and $h(T)$ we have $n(T) = 1 + n(T_1) + n(T_2)$ and $h(T) = 1 + \max(h(T_1), h(T_2))$.

We find that

$$\begin{aligned}
n(T) &= 1 + n(T_1) + n(T_2) && \text{by the recursive formula for } n(T)\\
&\overset{\text{IH}}{\leq} 1 + (2^{h(T_1)+1} - 1) + (2^{h(T_2)+1} - 1) && \text{by the inductive hypothesis}\\
&\leq 2 \cdot \max(2^{h(T_1)+1}, 2^{h(T_2)+1}) - 1 && \text{because the sum of two terms is at most 2 times the larger}\\
&= 2 \cdot 2^{\max(h(T_1), h(T_2))+1} - 1 && \text{because } \max(2^x, 2^y) = 2^{\max(x,y)}\\
&= 2 \cdot 2^{h(T)} - 1 && \text{by the recursive definition of } h(T)\\
&= 2^{h(T)+1} - 1.
\end{aligned}$$

This completes the recursive step. ◁

5.3.5 Generalized Induction

We can extend mathematical induction to prove results about other sets that have the well-ordering property besides the set of integers. Although we will discuss this concept in detail in Section 9.6, we provide an example here to illustrate the usefulness of such an approach.

As an example, note that we can define an ordering on $\mathbf{N} \times \mathbf{N}$, the ordered pairs of nonnegative integers, by specifying that (x_1, y_1) is less than or equal to (x_2, y_2) if either $x_1 < x_2$, or $x_1 = x_2$ and $y_1 < y_2$; this is called the **lexicographic ordering**. The set $\mathbf{N} \times \mathbf{N}$ with this ordering has the property that every subset of $\mathbf{N} \times \mathbf{N}$ has a least element (see Exercise 53 in Section 9.6). This implies that we can recursively define the terms $a_{m,n}$, with $m \in \mathbf{N}$ and $n \in \mathbf{N}$, and prove results about them using a variant of mathematical induction, as illustrated in Example 13.

EXAMPLE 13 Suppose that $a_{m,n}$ is defined recursively for $(m, n) \in \mathbf{N} \times \mathbf{N}$ by $a_{0,0} = 0$ and

$$a_{m,n} = \begin{cases} a_{m-1,n} + 1 & \text{if } n = 0 \text{ and } m > 0 \\ a_{m,n-1} + n & \text{if } n > 0. \end{cases}$$

Show that $a_{m,n} = m + n(n + 1)/2$ for all $(m, n) \in \mathbf{N} \times \mathbf{N}$, that is, for all pairs of nonnegative integers.

Solution: We can prove that $a_{m,n} = m + n(n + 1)/2$ using a generalized version of mathematical induction. The basis step requires that we show that this formula is valid when $(m, n) = (0, 0)$. The induction step requires that we show that if the formula holds for all pairs smaller than (m, n) in the lexicographic ordering of $\mathbf{N} \times \mathbf{N}$, then it also holds for (m, n).

BASIS STEP: Let $(m, n) = (0, 0)$. Then by the basis case of the recursive definition of $a_{m,n}$ we have $a_{0,0} = 0$. Furthermore, when $m = n = 0$, $m + n(n + 1)/2 = 0 + (0 \cdot 1)/2 = 0$. This completes the basis step.

INDUCTIVE STEP: Suppose that $a_{m',n'} = m' + n'(n' + 1)/2$ whenever (m', n') is less than (m, n) in the lexicographic ordering of $\mathbf{N} \times \mathbf{N}$. By the recursive definition, if $n = 0$, then $a_{m,n} = a_{m-1,n} + 1$. Because $(m - 1, n)$ is smaller than (m, n), the inductive hypothesis tells us that $a_{m-1,n} = m - 1 + n(n + 1)/2$, so that $a_{m,n} = m - 1 + n(n + 1)/2 + 1 = m + n(n + 1)/2$, giving us the desired equality. Now suppose that $n > 0$, so $a_{m,n} = a_{m,n-1} + n$. Because $(m, n - 1)$ is smaller than (m, n), the inductive hypothesis tells us that $a_{m,n-1} = m + (n - 1)n/2$, so $a_{m,n} = m + (n - 1)n/2 + n = m + (n^2 - n + 2n)/2 = m + n(n + 1)/2$. This finishes the inductive step. ◀

As mentioned, we will justify this proof technique in Section 9.6.

Exercises

1. Find $f(1), f(2), f(3)$, and $f(4)$ if $f(n)$ is defined recursively by $f(0) = 1$ and for $n = 0, 1, 2, \ldots$

a) $f(n + 1) = f(n) + 2$.
b) $f(n + 1) = 3f(n)$.
c) $f(n + 1) = 2^{f(n)}$.
d) $f(n + 1) = f(n)^2 + f(n) + 1$.

2. Find $f(1), f(2), f(3), f(4)$, and $f(5)$ if $f(n)$ is defined recursively by $f(0) = 3$ and for $n = 0, 1, 2, \ldots$

a) $f(n + 1) = -2f(n)$.
b) $f(n + 1) = 3f(n) + 7$.
c) $f(n + 1) = f(n)^2 - 2f(n) - 2$.
d) $f(n + 1) = 3^{f(n)/3}$.

3. Find $f(2), f(3), f(4)$, and $f(5)$ if f is defined recursively by $f(0) = -1, f(1) = 2$, and for $n = 1, 2, \ldots$

a) $f(n + 1) = f(n) + 3f(n - 1)$.
b) $f(n + 1) = f(n)^2 f(n - 1)$.
c) $f(n + 1) = 3f(n)^2 - 4f(n - 1)^2$.
d) $f(n + 1) = f(n - 1)/f(n)$.

4. Find $f(2), f(3), f(4)$, and $f(5)$ if f is defined recursively by $f(0) = f(1) = 1$ and for $n = 1, 2, \ldots$

a) $f(n + 1) = f(n) - f(n - 1)$.
b) $f(n + 1) = f(n)f(n - 1)$.
c) $f(n + 1) = f(n)^2 + f(n - 1)^3$.
d) $f(n + 1) = f(n)/f(n - 1)$.

5. Determine whether each of these proposed definitions is a valid recursive definition of a function f from the set of nonnegative integers to the set of integers. If f is well defined, find a formula for $f(n)$ when n is a nonnegative integer and prove that your formula is valid.

a) $f(0) = 0, f(n) = 2f(n - 2)$ for $n \geq 1$
b) $f(0) = 1, f(n) = f(n - 1) - 1$ for $n \geq 1$
c) $f(0) = 2, f(1) = 3, f(n) = f(n - 1) - 1$ for $n \geq 2$
d) $f(0) = 1, f(1) = 2, f(n) = 2f(n - 2)$ for $n \geq 2$
e) $f(0) = 1, f(n) = 3f(n - 1)$ if n is odd and $n \geq 1$ and $f(n) = 9f(n - 2)$ if n is even and $n \geq 2$

6. Determine whether each of these proposed definitions is a valid recursive definition of a function f from the set of nonnegative integers to the set of integers. If f is well defined, find a formula for $f(n)$ when n is a nonnegative integer and prove that your formula is valid.

a) $f(0) = 1, f(n) = -f(n - 1)$ for $n \geq 1$
b) $f(0) = 1, f(1) = 0, f(2) = 2, f(n) = 2f(n - 3)$ for $n \geq 3$
c) $f(0) = 0, f(1) = 1, f(n) = 2f(n + 1)$ for $n \geq 2$
d) $f(0) = 0, f(1) = 1, f(n) = 2f(n - 1)$ for $n \geq 1$
e) $f(0) = 2, f(n) = f(n - 1)$ if n is odd and $n \geq 1$ and $f(n) = 2f(n - 2)$ if $n \geq 2$

7. Give a recursive definition of the sequence $\{a_n\}$, $n = 1, 2, 3, \ldots$ if

a) $a_n = 6n$. **b)** $a_n = 2n + 1$.
c) $a_n = 10^n$. **d)** $a_n = 5$.

8. Give a recursive definition of the sequence $\{a_n\}$, $n = 1, 2, 3, \ldots$ if

a) $a_n = 4n - 2$. **b)** $a_n = 1 + (-1)^n$.
c) $a_n = n(n + 1)$. **d)** $a_n = n^2$.

9. Let F be the function such that $F(n)$ is the sum of the first n positive integers. Give a recursive definition of $F(n)$.

10. Give a recursive definition of $S_m(n)$, the sum of the integer m and the nonnegative integer n.

11. Give a recursive definition of $P_m(n)$, the product of the integer m and the nonnegative integer n.

In Exercises 12–19, f_n is the nth Fibonacci number.

12. Prove that $f_1^2 + f_2^2 + \cdots + f_n^2 = f_n f_{n+1}$ when n is a positive integer.

13. Prove that $f_1 + f_3 + \cdots + f_{2n-1} = f_{2n}$ when n is a positive integer.

***14.** Show that $f_{n+1}f_{n-1} - f_n^2 = (-1)^n$ when n is a positive integer.

***15.** Show that $f_0 f_1 + f_1 f_2 + \cdots + f_{2n-1}f_{2n} = f_{2n}^2$ when n is a positive integer.

***16.** Show that $f_0 - f_1 + f_2 - \cdots - f_{2n-1} + f_{2n} = f_{2n-1} - 1$ when n is a positive integer.

17. Determine the number of divisions used by the Euclidean algorithm to find the greatest common divisor of the Fibonacci numbers f_n and f_{n+1}, where n is a nonnegative integer. Verify your answer using mathematical induction.

18. Let

$$\mathbf{A} = \begin{bmatrix} 1 & 1 \\ 1 & 0 \end{bmatrix}.$$

Show that

$$\mathbf{A}^n = \begin{bmatrix} f_{n+1} & f_n \\ f_n & f_{n-1} \end{bmatrix}$$

when n is a positive integer.

19. By taking determinants of both sides of the equation in Exercise 18, prove the identity given in Exercise 14. (Recall that the determinant of the matrix $\begin{vmatrix} a & b \\ c & d \end{vmatrix}$ is $ad - bc$.)

***20.** Give a recursive definition of the functions max and min so that $\max(a_1, a_2, \ldots, a_n)$ and $\min(a_1, a_2, \ldots, a_n)$ are the maximum and minimum of the n numbers $a_1, a_2, \ldots, a_n$, respectively.

***21.** Let $a_1, a_2, \ldots, a_n$, and $b_1, b_2, \ldots, b_n$ be real numbers. Use the recursive definitions that you gave in Exercise 20 to prove these.

a) $\max(-a_1, -a_2, \ldots, -a_n) = -\min(a_1, a_2, \ldots, a_n)$

b) $\max(a_1 + b_1, a_2 + b_2, \ldots, a_n + b_n)$
$\leq \max(a_1, a_2, \ldots, a_n) + \max(b_1, b_2, \ldots, b_n)$

c) $\min(a_1 + b_1, a_2 + b_2, \ldots, a_n + b_n)$
$\geq \min(a_1, a_2, \ldots, a_n) + \min(b_1, b_2, \ldots, b_n)$

22. Show that the set S defined by $1 \in S$ and $s + t \in S$ whenever $s \in S$ and $t \in S$ is the set of positive integers.

23. Give a recursive definition of the set of positive integers that are multiples of 5.

24. Give a recursive definition of

a) the set of odd positive integers.
b) the set of positive integer powers of 3.
c) the set of polynomials with integer coefficients.

25. Give a recursive definition of

a) the set of even integers.
b) the set of positive integers congruent to 2 modulo 3.
c) the set of positive integers not divisible by 5.

26. Let S be the set of positive integers defined by

Basis step: $1 \in S$.

Recursive step: If $n \in S$, then $3n + 2 \in S$ and $n^2 \in S$.

a) Show that if $n \in S$, then $n \equiv 1 \pmod 4$.
b) Show that there exists an integer $m \equiv 1 \pmod 4$ that does not belong to S.

27. Let S be the set of positive integers defined by

Basis step: $5 \in S$.

Recursive step: If $n \in S$, then $3n \in S$ and $n^2 \in S$.

a) Show that if $n \in S$, then $n \equiv 5 \pmod{10}$.
b) Show that there exists an integer $m \equiv 5 \pmod{10}$ that does not belong to S.

28. Let S be the subset of the set of ordered pairs of integers defined recursively by

Basis step: $(0, 0) \in S$.

Recursive step: If $(a, b) \in S$, then $(a + 2, b + 3) \in S$ and $(a + 3, b + 2) \in S$.

a) List the elements of S produced by the first five applications of the recursive definition.
b) Use strong induction on the number of applications of the recursive step of the definition to show that $5 \mid a + b$ when $(a, b) \in S$.
c) Use structural induction to show that $5 \mid a + b$ when $(a, b) \in S$.

29. Let S be the subset of the set of ordered pairs of integers defined recursively by

Basis step: $(0, 0) \in S$.

Recursive step: If $(a, b) \in S$, then $(a, b + 1) \in S$, $(a + 1, b + 1) \in S$, and $(a + 2, b + 1) \in S$.

a) List the elements of S produced by the first four applications of the recursive definition.
b) Use strong induction on the number of applications of the recursive step of the definition to show that $a \leq 2b$ whenever $(a, b) \in S$.
c) Use structural induction to show that $a \leq 2b$ whenever $(a, b) \in S$.

30. Give a recursive definition of each of these sets of ordered pairs of positive integers. [*Hint:* Plot the points in the set in the plane and look for lines containing points in the set.]

a) $S = \{(a, b) \mid a \in \mathbf{Z}^+, b \in \mathbf{Z}^+, \text{ and } a + b \text{ is odd}\}$
b) $S = \{(a, b) \mid a \in \mathbf{Z}^+, b \in \mathbf{Z}^+, \text{ and } a \mid b\}$
c) $S = \{(a, b) \mid a \in \mathbf{Z}^+, b \in \mathbf{Z}^+, \text{ and } 3 \mid a + b\}$

31. Give a recursive definition of each of these sets of ordered pairs of positive integers. Use structural induction to prove that the recursive definition you found is correct. [*Hint:* To find a recursive definition, plot the points in the set in the plane and look for patterns.]

a) $S = \{(a, b) \mid a \in \mathbf{Z}^+, b \in \mathbf{Z}^+, \text{ and } a + b \text{ is even}\}$

b) $S = \{(a, b) \mid a \in \mathbf{Z}^+, b \in \mathbf{Z}^+, \text{ and } a \text{ or } b \text{ is odd}\}$

c) $S = \{(a, b) \mid a \in \mathbf{Z}^+, b \in \mathbf{Z}^+, a + b \text{ is odd, and } 3 \mid b\}$

32. Prove that in a bit string, the string 01 occurs at most one more time than the string 10.

33. Define well-formed formulae of sets, variables representing sets, and operators from $\{\overline{\ }, \cup, \cap, -\}$.

34. **a)** Give a recursive definition of the function *ones*(*s*), which counts the number of ones in a bit string *s*.

b) Use structural induction to prove that $ones(st) = ones(s) + ones(t)$.

35. **a)** Give a recursive definition of the function $m(s)$, which equals the smallest digit in a nonempty string of decimal digits.

b) Use structural induction to prove that $m(st) = \min(m(s), m(t))$.

The **reversal** of a string is the string consisting of the symbols of the string in reverse order. The reversal of the string w is denoted by w^R.

36. Find the reversal of the following bit strings.

a) 0101 **b)** 1 1011 **c)** 1000 1001 0111

37. Give a recursive definition of the reversal of a string. [*Hint:* First define the reversal of the empty string. Then write a string w of length $n + 1$ as xy, where x is a string of length n, and express the reversal of w in terms of x^R and y.]

***38.** Use structural induction to prove that $(w_1w_2)^R = w_2^R w_1^R$.

39. Give a recursive definition of w^i, where w is a string and i is a nonnegative integer. (Here w^i represents the concatenation of i copies of the string w.)

***40.** Give a recursive definition of the set of bit strings that are palindromes.

41. When does a string belong to the set A of bit strings defined recursively by

$$\lambda \in A$$
$$0x1 \in A \text{ if } x \in A,$$

where λ is the empty string?

***42.** Recursively define the set of bit strings that have more zeros than ones.

43. Use Exercise 39 and mathematical induction to show that $l(w^i) = i \cdot l(w)$, where w is a string and i is a nonnegative integer.

***44.** Show that $(w^R)^i = (w^i)^R$ whenever w is a string and i is a nonnegative integer; that is, show that the ith power of the reversal of a string is the reversal of the ith power of the string.

45. Use structural induction to show that $n(T) \geq 2h(T) + 1$, where T is a full binary tree, $n(T)$ equals the number of vertices of T, and $h(T)$ is the height of T.

The set of leaves and the set of internal vertices of a full binary tree can be defined recursively.

Basis step: The root r is a leaf of the full binary tree with exactly one vertex r. This tree has no internal vertices.

Recursive step: The set of leaves of the tree $T = T_1 \cdot T_2$ is the union of the sets of leaves of T_1 and of T_2. The internal vertices of T are the root r of T and the union of the set of internal vertices of T_1 and the set of internal vertices of T_2.

46. Use structural induction to show that $l(T)$, the number of leaves of a full binary tree T, is 1 more than $i(T)$, the number of internal vertices of T.

47. Use generalized induction as was done in Example 13 to show that if $a_{m,n}$ is defined recursively by $a_{0,0} = 0$ and

$$a_{m,n} = \begin{cases} a_{m-1,n} + 1 & \text{if } n = 0 \text{ and } m > 0 \\ a_{m,n-1} + 1 & \text{if } n > 0, \end{cases}$$

then $a_{m,n} = m + n$ for all $(m, n) \in \mathbf{N} \times \mathbf{N}$.

48. Use generalized induction as was done in Example 13 to show that if $a_{m,n}$ is defined recursively by $a_{1,1} = 5$ and

$$a_{m,n} = \begin{cases} a_{m-1,n} + 2 & \text{if } n = 1 \text{ and } m > 1 \\ a_{m,n-1} + 2 & \text{if } n > 1, \end{cases}$$

then $a_{m,n} = 2(m + n) + 1$ for all $(m, n) \in \mathbf{Z}^+ \times \mathbf{Z}^+$.

***49.** A **partition** of a positive integer n is a way to write n as a sum of positive integers where the order of terms in the sum does not matter. For instance, $7 = 3 + 2 + 1 + 1$ is a partition of 7. Let P_m equal the number of different partitions of m, and let $P_{m,n}$ be the number of different ways to express m as the sum of positive integers not exceeding n.

a) Show that $P_{m,m} = P_m$.

b) Show that the following recursive definition for $P_{m,n}$ is correct:

$$P_{m,n} = \begin{cases} 1 & \text{if } m = 1 \\ 1 & \text{if } n = 1 \\ P_{m,m} & \text{if } m < n \\ 1 + P_{m,m-1} & \text{if } m = n > 1 \\ P_{m,n-1} + P_{m-n,n} & \text{if } m > n > 1. \end{cases}$$

c) Find the number of partitions of 5 and of 6 using this recursive definition.

Links

Consider the following inductive definition of a version of **Ackermann's function**. This function was named after Wilhelm Ackermann, a German mathematician who was a student of the great mathematician David Hilbert. Ackermann's function plays an important role in the theory of recursive functions and in the study of the complexity of certain algorithms involving set unions. (There are several different variants of this function. All are called Ackermann's function and have similar properties even though their values do not always agree.)

$$A(m, n) = \begin{cases} 2n & \text{if } m = 0 \\ 0 & \text{if } m \geq 1 \text{ and } n = 0 \\ 2 & \text{if } m \geq 1 \text{ and } n = 1 \\ A(m-1, A(m, n-1)) & \text{if } m \geq 1 \text{ and } n \geq 2 \end{cases}$$

Exercises 50–57 involve this version of Ackermann's function.

50. Find these values of Ackermann's function.

a) $A(1, 0)$ **b)** $A(0, 1)$
c) $A(1, 1)$ **d)** $A(2, 2)$

51. Show that $A(m, 2) = 4$ whenever $m \geq 1$.

52. Show that $A(1, n) = 2^n$ whenever $n \geq 1$.

53. Find these values of Ackermann's function.

a) $A(2, 3)$ ***b)** $A(3, 3)$

***54.** Find $A(3, 4)$.

****55.** Prove that $A(m, n + 1) > A(m, n)$ whenever m and n are nonnegative integers.

***56.** Prove that $A(m + 1, n) \geq A(m, n)$ whenever m and n are nonnegative integers.

57. Prove that $A(i, j) \geq j$ whenever i and j are nonnegative integers.

58. Use mathematical induction to prove that a function F defined by specifying $F(0)$ and a rule for obtaining $F(n + 1)$ from $F(n)$ is well defined.

59. Use strong induction to prove that a function F defined by specifying $F(0)$ and a rule for obtaining $F(n + 1)$ from the values $F(k)$ for $k = 0, 1, 2, \ldots, n$ is well defined.

60. Show that each of these proposed recursive definitions of a function on the set of positive integers does not produce a well-defined function.

a) $F(n) = 1 + F(\lfloor n/2 \rfloor)$ for $n \geq 1$ and $F(1) = 1$.
b) $F(n) = 1 + F(n - 3)$ for $n \geq 2$, $F(1) = 2$, and $F(2) = 3$.
c) $F(n) = 1 + F(n/2)$ for $n \geq 2$, $F(1) = 1$, and $F(2) = 2$.
d) $F(n) = 1 + F(n/2)$ if n is even and $n \geq 2$, $F(n) = 1 - F(n - 1)$ if n is odd, and $F(1) = 1$.
e) $F(n) = 1 + F(n/2)$ if n is even and $n \geq 2$, $F(n) = F(3n - 1)$ if n is odd and $n \geq 3$, and $F(1) = 1$.

61. Show that each of these proposed recursive definitions of a function on the set of positive integers does not produce a well-defined function.

a) $F(n) = 1 + F(\lfloor (n + 1)/2 \rfloor)$ for $n \geq 1$ and $F(1) = 1$.
b) $F(n) = 1 + F(n - 2)$ for $n \geq 2$ and $F(1) = 0$.
c) $F(n) = 1 + F(n/3)$ for $n \geq 3$, $F(1) = 1$, $F(2) = 2$, and $F(3) = 3$.
d) $F(n) = 1 + F(n/2)$ if n is even and $n \geq 2$, $F(n) = 1 + F(n - 2)$ if n is odd, and $F(1) = 1$.
e) $F(n) = 1 + F(F(n - 1))$ if $n \geq 2$ and $F(1) = 2$.

Exercises 62–64 deal with iterations of the logarithm function. Let $\log n$ denote the logarithm of n to the base 2, as usual. The function $\log^{(k)} n$ is defined recursively by

$$\log^{(k)} n = \begin{cases} n & \text{if } k = 0 \\ \log(\log^{(k-1)} n) & \text{if } \log^{(k-1)} n \text{ is defined and positive} \\ \text{undefined} & \text{otherwise.} \end{cases}$$

The **iterated logarithm** is the function $\log^* n$ whose value at n is the smallest nonnegative integer k such that $\log^{(k)} n \leq 1$.

62. Find these values.

a) $\log^{(2)} 16$ **b)** $\log^{(3)} 256$
c) $\log^{(3)} 2^{65536}$ **d)** $\log^{(4)} 2^{2^{65536}}$

63. Find the value of $\log^* n$ for these values of n.

a) 2 **b)** 4 **c)** 8 **d)** 16
e) 256 **f)** 65536 **g)** 2^{2048}

64. Find the largest integer n such that $\log^* n = 5$. Determine the number of decimal digits in this number.

Exercises 65–67 deal with values of iterated functions. Suppose that $f(n)$ is a function from the set of real numbers, or positive real numbers, or some other set of real numbers, to the set of real numbers such that $f(n)$ is monotonically increasing [that is, $f(n) < f(m)$ when $n < m$) and $f(n) < n$ for all n in the domain of f.] The function $f^{(k)}(n)$ is defined recursively by

$$f^{(k)}(n) = \begin{cases} n & \text{if } k = 0 \\ f(f^{(k-1)}(n)) & \text{if } k > 0. \end{cases}$$

Furthermore, let c be a positive real number. The **iterated function** f_c^* is the number of iterations of f required to reduce its argument to c or less, so $f_c^*(n)$ is the smallest nonnegative integer k such that $f^k(n) \leq c$.

65. Let $f(n) = n - a$, where a is a positive integer. Find a formula for $f^{(k)}(n)$. What is the value of $f_0^*(n)$ when n is a positive integer?

66. Let $f(n) = n/2$. Find a formula for $f^{(k)}(n)$. What is the value of $f_1^*(n)$ when n is a positive integer?

67. Let $f(n) = \sqrt{n}$. Find a formula for $f^{(k)}(n)$. What is the value of $f_2^*(n)$ when n is a positive integer?

5.4 Recursive Algorithms

5.4.1 Introduction

Here's a famous humorous quote: "To understand recursion, you must first understand recursion."

Sometimes we can reduce the solution to a problem with a particular set of input values to the solution of the same problem with smaller input values. For instance, the problem of finding the greatest common divisor of two positive integers a and b, where $b > a$, can be reduced to finding the greatest common divisor of a pair of smaller integers, namely, b **mod** a and a, because $\gcd(b \textbf{ mod } a, a) = \gcd(a, b)$.

When such a reduction can be done, the solution to the original problem can be found with a sequence of reductions, until the problem has been reduced to some initial case for which the solution is known. For instance, for finding the greatest common divisor, the reduction continues until the smaller of the two numbers is zero, because $\gcd(a, 0) = a$ when $a > 0$.

We will see that algorithms that successively reduce a problem to the same problem with smaller input are used to solve a wide variety of problems.

Definition 1 An algorithm is called *recursive* if it solves a problem by reducing it to an instance of the same problem with smaller input.

Links

We will describe a variety of different recursive algorithms in this section.

EXAMPLE 1 Give a recursive algorithm for computing $n!$, where n is a nonnegative integer.

Extra Examples

Solution: We can build a recursive algorithm that finds $n!$, where n is a nonnegative integer, based on the recursive definition of $n!$, which specifies that $n! = n \cdot (n-1)!$ when n is a positive integer, and that $0! = 1$. To find $n!$ for a particular integer n, we use the recursive step n times, each time replacing a value of the factorial function with the value of the factorial function at the next smaller integer. At this last step, we insert the value of $0!$. The recursive algorithm we obtain is displayed as Algorithm 1.

To help understand how this algorithm works, we trace the steps used by the algorithm to compute $4!$. First, we use the recursive step to write $4! = 4 \cdot 3!$. We then use the recursive step repeatedly to write $3! = 3 \cdot 2!$, $2! = 2 \cdot 1!$, and $1! = 1 \cdot 0!$. Inserting the value of $0! = 1$, and working back through the steps, we see that $1! = 1 \cdot 1 = 1$, $2! = 2 \cdot 1! = 2$, $3! = 3 \cdot 2! = 3 \cdot 2 = 6$, and $4! = 4 \cdot 3! = 4 \cdot 6 = 24$. ◀

ALGORITHM 1 A Recursive Algorithm for Computing $n!$.

procedure *factorial*(n: nonnegative integer)
if $n = 0$ **then return** 1
else return $n \cdot factorial(n-1)$
{output is $n!$}

Example 2 shows how a recursive algorithm can be constructed to evaluate a function from its recursive definition.

EXAMPLE 2 Give a recursive algorithm for computing a^n, where a is a nonzero real number and n is a nonnegative integer.

Solution: We can base a recursive algorithm on the recursive definition of a^n. This definition states that $a^{n+1} = a \cdot a^n$ for $n > 0$ and the initial condition $a^0 = 1$. To find a^n, successively use the recursive step to reduce the exponent until it becomes zero. We give this procedure in Algorithm 2. ◀

ALGORITHM 2 A Recursive Algorithm for Computing a^n.

procedure *power*(a: nonzero real number, n: nonnegative integer)
if $n = 0$ **then return** 1
else return $a \cdot power(a, n-1)$
{output is a^n}

Next we give a recursive algorithm for finding greatest common divisors.

EXAMPLE 3 Give a recursive algorithm for computing the greatest common divisor of two nonnegative integers a and b with $a < b$.

Solution: We can base a recursive algorithm on the reduction $\gcd(a, b) = \gcd(b \textbf{ mod } a, a)$ and the condition $\gcd(0, b) = b$ when $b > 0$. This produces the procedure in Algorithm 3, which is a recursive version of the Euclidean algorithm.

We illustrate the workings of Algorithm 3 with a trace when the input is $a = 5$, $b = 8$. With this input, the algorithm uses the "else" clause to find that $\gcd(5, 8) = \gcd(8 \textbf{ mod } 5, 5) = \gcd(3, 5)$. It uses this clause again to find that $\gcd(3, 5) = \gcd(5 \textbf{ mod } 3, 3) = \gcd(2, 3)$, then to get $\gcd(2, 3) = \gcd(3 \textbf{ mod } 2, 2) = \gcd(1, 2)$, then to get $\gcd(1, 2) = \gcd(2 \textbf{ mod } 1, 1) = \gcd(0, 1)$. Finally, to find $\gcd(0, 1)$ it uses the first step with $a = 0$ to find that $\gcd(0, 1) = 1$. Consequently, the algorithm finds that $\gcd(5, 8) = 1$. ◀

ALGORITHM 3 A Recursive Algorithm for Computing gcd(*a*, *b*).

procedure *gcd*(*a*, *b*: nonnegative integers with $a < b$)
if $a = 0$ **then return** b
else return *gcd*(b **mod** a, a)
{output is gcd(a, b)}

EXAMPLE 4 Devise a recursive algorithm for computing $b^n \textbf{ mod } m$, where b, n, and m are integers with $m \geq 2$, $n \geq 0$, and $1 \leq b < m$.

Solution: We can base a recursive algorithm on the fact that

$$b^n \textbf{ mod } m = (b \cdot (b^{n-1} \textbf{ mod } m)) \textbf{ mod } m,$$

which follows by Corollary 2 in Section 4.1, and the initial condition $b^0 \textbf{ mod } m = 1$. We leave this as Exercise 12 for the reader.

However, we can devise a much more efficient recursive algorithm based on the observation that

$$b^n \textbf{ mod } m = (b^{n/2} \textbf{ mod } m)^2 \textbf{ mod } m$$

when n is even and

$$b^n \textbf{ mod } m = \left((b^{\lfloor n/2 \rfloor} \textbf{ mod } m)^2 \textbf{ mod } m \cdot b \textbf{ mod } m\right) \textbf{ mod } m$$

when n is odd, which we describe in pseudocode as Algorithm 4.

We trace the execution of Algorithm 4 with input $b = 2$, $n = 5$, and $m = 3$ to illustrate how it works. First, because $n = 5$ is odd we use the "else" clause to see that $mpower(2, 5, 3) = (mpower(2, 2, 3)^2 \textbf{ mod } 3 \cdot 2 \textbf{ mod } 3) \textbf{ mod } 3$. We next use the "else if" clause to see that $mpower(2, 2, 3) = mpower(2, 1, 3)^2 \textbf{ mod } 3$. Using the "else" clause again, we see that $mpower(2, 1, 3) = (mpower(2, 0, 3)^2 \textbf{ mod } 3 \cdot 2 \textbf{ mod } 3) \textbf{ mod } 3$. Finally, using the "if" clause, we see that $mpower(2, 0, 3) = 1$. Working backwards, it follows that $mpower(2, 1, 3) = (1^2 \textbf{ mod } 3 \cdot 2 \textbf{ mod } 3) \textbf{ mod } 3 = 2$, so $mpower(2, 2, 3) = 2^2 \textbf{ mod } 3 = 1$, and finally $mpower(2, 5, 3) = (1^2 \textbf{ mod } 3 \cdot 2 \textbf{ mod } 3) \textbf{ mod } 3 = 2$. ◀

ALGORITHM 4 Recursive Modular Exponentiation.

procedure *mpower*(b, n, m: integers with $b > 0$ and $m \geq 2$, $n \geq 0$)
if $n = 0$ **then**
    return 1
else if n is even **then**
    return *mpower*$(b, n/2, m)^2$ **mod** m
else
    return (*mpower*$(b, \lfloor n/2 \rfloor, m)^2$ **mod** $m \cdot b$ **mod** m) **mod** m
{output is b^n **mod** m}

We will now give recursive versions of searching algorithms that were introduced in Section 3.1.

EXAMPLE 5 Express the linear search algorithm as a recursive procedure.

Solution: To *search* for the first occurrence of x in the sequence $a_1, a_2, \ldots, a_n$, at the ith step of the algorithm, x and a_i are compared. If x equals a_i, then the algorithm returns i, the location of x in the sequence. Otherwise, the search for the first occurrence of x is reduced to a search in a sequence with one fewer element, namely, the sequence $a_{i+1}, \ldots, a_n$. The algorithm returns 0 when x is never found in the sequence after all terms have been examined. We can now give a recursive procedure, which is displayed as pseudocode in Algorithm 5.

Let *search* (i, j, x) be the procedure that searches for the first occurrence of x in the sequence $a_i, a_{i+1}, \ldots, a_j$. The input to the procedure consists of the triple $(1, n, x)$. The algorithm terminates at a step if the first term of the remaining sequence is x or if there is only one term of the sequence and this is not x. If x is not the first term and there are additional terms, the same procedure is carried out but with a search sequence of one fewer term, obtained by deleting the first term of the search sequence. If the algorithm terminates without x having been found, the algorithm returns the value 0. ◀

ALGORITHM 5 A Recursive Linear Search Algorithm.

procedure *search*(i, j, x: integers, $1 \leq i \leq j \leq n$)
if $a_i = x$ **then**
    return i
else if $i = j$ **then**
    return 0
else
    return *search*$(i + 1, j, x)$
{output is the location of x in $a_1, a_2, \ldots, a_n$ if it appears; otherwise it is 0}

EXAMPLE 6 Construct a recursive version of a binary search algorithm.

Solution: Suppose we want to locate x in the sequence $a_1, a_2, \ldots, a_n$ of integers in increasing order. To perform a binary search, we begin by comparing x with the middle term, $a_{\lfloor (n+1)/2 \rfloor}$. Our algorithm will terminate if x equals this term and return the location of this term in the sequence. Otherwise, we reduce the search to a smaller search sequence, namely, the first half of the sequence if x is smaller than the middle term of the original sequence, and the second half otherwise. We have reduced the solution of the search problem to the solution of the same problem with a sequence at most half as long. If we have never encountered the search term x, our algorithm returns the value 0. We express this recursive version of a binary search algorithm as Algorithm 6. ◀

ALGORITHM 6 A Recursive Binary Search Algorithm.

procedure *binary search*(i, j, x: integers, $1 \leq i \leq j \leq n$)
$m := \lfloor (i + j)/2 \rfloor$
if $x = a_m$ **then**
  return m
else if ($x < a_m$ and $i < m$) **then**
  return *binary search*($i, m - 1, x$)
else if ($x > a_m$ and $j > m$) **then**
  return *binary search*($m + 1, j, x$)
else return 0
{output is location of x in $a_1, a_2, \ldots, a_n$ if it appears; otherwise it is 0}

5.4.2 Proving Recursive Algorithms Correct

Mathematical induction, and its variant strong induction, can be used to prove that a recursive algorithm is correct, that is, that it produces the desired output for all possible input values. Examples 7 and 8 illustrate how mathematical induction or strong induction can be used to prove that recursive algorithms are correct. First, we will show that Algorithm 2 is correct.

EXAMPLE 7 Prove that Algorithm 2, which computes powers of real numbers, is correct.

Solution: We use mathematical induction on the exponent n.

BASIS STEP: If $n = 0$, the first step of the algorithm tells us that *power* $(a, 0) = 1$. This is correct because $a^0 = 1$ for every nonzero real number a. This completes the basis step.

INDUCTIVE STEP: The inductive hypothesis is the statement that *power* $(a, k) = a^k$ for all $a \neq 0$ for an arbitrary nonnegative integer k. That is, the inductive hypothesis is the statement that the algorithm correctly computes a^k. To complete the inductive step, we show that if the inductive hypothesis is true, then the algorithm correctly computes a^{k+1}. Because $k + 1$ is a positive integer, when the algorithm computes a^{k+1}, the algorithm sets *power* $(a, k + 1) = a \cdot$ *power* (a, k). By the inductive hypothesis, we have *power* $(a, k) = a^k$, so *power* $(a, k + 1) = a \cdot$ *power* $(a, k) = a \cdot a^k = a^{k+1}$. This completes the inductive step.

We have completed the basis step and the inductive step, so we can conclude that Algorithm 2 always computes a^n correctly when $a \neq 0$ and n is a nonnegative integer. ◀

Generally, we need to use strong induction to prove that recursive algorithms are correct, rather than just mathematical induction. Example 8 illustrates this; it shows how strong induction can be used to prove that Algorithm 4 is correct.

EXAMPLE 8 Prove that Algorithm 4, which computes modular powers, is correct.

Solution: We use strong induction on the exponent n.

BASIS STEP: Let b be an integer and m an integer with $m \geq 2$. When $n = 0$, the algorithm sets *mpower*(b, n, m) equal to 1. This is correct because b^0 **mod** $m = 1$. The basis step is complete.

INDUCTIVE STEP: For the inductive hypothesis we assume that *mpower*$(b, j, m) = b^j$ **mod** m for all integers $0 \leq j < k$ whenever b is a positive integer and m is an integer with $m \geq 2$. To complete the inductive step, we show that if the inductive hypothesis is correct, then *mpower*$(b, k, m) = b^k$ **mod** m. Because the recursive algorithm handles odd and even values of k differently, we split the inductive step into two cases.

When k is even, we have

$$mpower(b, k, m) = (mpower(b, k/2, m))^2 \textbf{ mod } m = (b^{k/2} \textbf{ mod } m)^2 \textbf{ mod } m = b^k \textbf{ mod } m,$$

where we have used the inductive hypothesis to replace *mpower*$(b, k/2, m)$ by $b^{k/2}$ **mod** m.

When k is odd, we have

$$\begin{aligned} mpower(b, k, m) &= ((mpower(b, \lfloor k/2 \rfloor, m))^2 \textbf{ mod } m \cdot b \textbf{ mod } m) \textbf{ mod } m \\ &= ((b^{\lfloor k/2 \rfloor} \textbf{ mod } m)^2 \textbf{ mod } m \cdot b \textbf{ mod } m) \textbf{ mod } m \\ &= b^{2\lfloor k/2 \rfloor + 1} \textbf{ mod } m = b^k \textbf{ mod } m, \end{aligned}$$

using Corollary 2 in Section 4.1, because $2\lfloor k/2 \rfloor + 1 = 2(k-1)/2 + 1 = k$ when k is odd. Here we have used the inductive hypothesis to replace *mpower*$(b, \lfloor k/2 \rfloor, m)$ by $b^{\lfloor k/2 \rfloor}$ **mod** m. This completes the inductive step.

We have completed the basis step and the inductive step, so by strong induction we know that Algorithm 4 is correct. ◀

5.4.3 Recursion and Iteration

A recursive definition expresses the value of a function at a positive integer in terms of the values of the function at smaller integers. This means that we can devise a recursive algorithm to evaluate a recursively defined function at a positive integer. Instead of successively reducing the computation to the evaluation of the function at smaller integers, we can start with the value of the function at one or more integers, the base cases, and successively apply the recursive definition to find the values of the function at successive larger integers. Such a procedure is called **iterative**. Often an iterative approach for the evaluation of a recursively defined sequence requires much less computation than a procedure using recursion (unless special-purpose recursive machines are used). This is illustrated by the iterative and recursive procedures for finding the nth Fibonacci number. The recursive procedure is given first.

ALGORITHM 7 A Recursive Algorithm for Fibonacci Numbers.

procedure *fibonacci*(n: nonnegative integer)
if $n = 0$ **then return** 0
else if $n = 1$ **then return** 1
else return *fibonacci*$(n - 1)$ + *fibonacci*$(n - 2)$
{output is *fibonacci*(n)}

When we use a recursive procedure to find f_n, we first express f_n as $f_{n-1} + f_{n-2}$. Then we replace both of these Fibonacci numbers by the sum of two previous Fibonacci numbers, and so on. When f_1 or f_0 arises, it is replaced by its value.

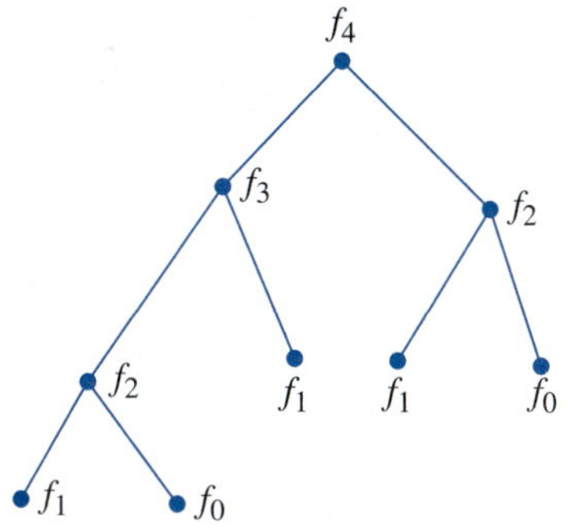

FIGURE 1 Evaluating f_4 recursively.

Note that at each stage of the recursion, until f_1 or f_0 is obtained, the number of Fibonacci numbers to be evaluated has doubled. For instance, when we find f_4 using this recursive algorithm, we must carry out all the computations illustrated in the tree diagram in Figure 1. This tree consists of a root labeled with f_4, and branches from the root to vertices labeled with the two Fibonacci numbers f_3 and f_2 that occur in the reduction of the computation of f_4. Each subsequent reduction produces two branches in the tree. This branching ends when f_0 and f_1 are reached. The reader can verify that this algorithm requires $f_{n+1} - 1$ additions to find f_n.

Now consider the amount of computation required to find f_n using the iterative approach in Algorithm 8.

ALGORITHM 8 An Iterative Algorithm for Computing Fibonacci Numbers.

```
procedure iterative fibonacci(n: nonnegative integer)
if n = 0 then return 0
else
    x := 0
    y := 1
    for i := 1 to n − 1
        z := x + y
        x := y
        y := z
    return y
{output is the nth Fibonacci number}
```

This procedure initializes x as $f_0 = 0$ and y as $f_1 = 1$. When the loop is traversed, the sum of x and y is assigned to the auxiliary variable z. Then x is assigned the value of y and y is assigned the value of the auxiliary variable z. Therefore, after going through the loop the first time, it follows that x equals f_1 and y equals $f_0 + f_1 = f_2$. Furthermore, after going through the loop $n - 1$ times, x equals f_{n-1} and y equals f_n (the reader should verify this statement). Only $n - 1$ additions have been used to find f_n with this iterative approach when $n > 1$. Consequently, this algorithm requires far less computation than does the recursive algorithm.

We have shown that a recursive algorithm may require far more computation than an iterative one when a recursively defined function is evaluated. It is sometimes preferable to use a recursive procedure even if it is less efficient than the iterative procedure. In particular, this is true when the recursive approach is easily implemented and the iterative approach is not. (Also, machines designed to handle recursion may be available that eliminate the advantage of using iteration.)

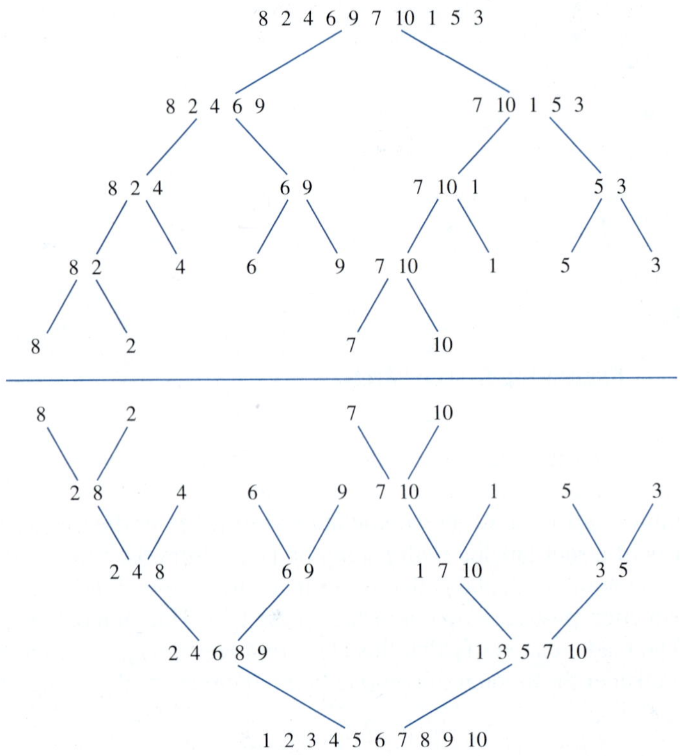

FIGURE 2 The merge sort of 8, 2, 4, 6, 9, 7, 10, 1, 5, 3.

5.4.4 The Merge Sort

Links We now describe a recursive sorting algorithm called the **merge sort** algorithm. We will demonstrate how the merge sort algorithm works with an example before describing it in generality.

EXAMPLE 9 Use the merge sort to put the terms of the list 8, 2, 4, 6, 9, 7, 10, 1, 5, 3 in increasing order.

Solution: A merge sort begins by splitting the list into individual elements by successively splitting lists in two. The progression of sublists for this example is represented with the balanced binary tree of height 4 shown in the upper half of Figure 2.

Sorting is done by successively merging pairs of lists. At the first stage, pairs of individual elements are merged into lists of length two in increasing order. Then successive merges of pairs of lists are performed until the entire list is put into increasing order. The succession of merged lists in increasing order is represented by the balanced binary tree of height 4 shown in the lower half of Figure 2 (note that this tree is displayed "upside down"). ◀

In general, a merge sort proceeds by splitting lists into two sublists of equal length (or where one sublist has one more element than the other) until each sublist contains one element. This succession of sublists can be represented by a binary tree. The procedure continues by successively merging pairs of lists, where both lists are in increasing order, into a larger list with elements in increasing order, until the original list is put into increasing order. The succession of merged lists can be represented by a binary tree.

Demo

We can also describe the merge sort recursively. To do a merge sort, we split a list into two sublists of equal, or approximately equal, size, sorting each sublist using the merge sort algorithm, and then merging the two lists. The recursive version of the merge sort is given in Algorithm 9. This algorithm uses the subroutine *merge*, which is described in Algorithm 10.

ALGORITHM 9 A Recursive Merge Sort.

procedure *mergesort*($L = a_1, \dots, a_n$)
if $n > 1$ **then**
  $m := \lfloor n/2 \rfloor$
  $L_1 := a_1, a_2, \dots, a_m$
  $L_2 := a_{m+1}, a_{m+2}, \dots, a_n$
  $L := merge(mergesort(L_1), mergesort(L_2))$
$\{L$ is now sorted into elements in nondecreasing order$\}$

An efficient algorithm for merging two ordered lists into a larger ordered list is needed to implement the merge sort. We will now describe such a procedure.

EXAMPLE 10 Merge the two lists 2, 3, 5, 6 and 1, 4.

Solution: Table 1 illustrates the steps we use. First, compare the smallest elements in the two lists, 2 and 1, respectively. Because 1 is the smaller, put it at the beginning of the merged list and remove it from the second list. At this stage, the first list is 2, 3, 5, 6, the second is 4, and the combined list is 1.

Next, compare 2 and 4, the smallest elements of the two lists. Because 2 is the smaller, add it to the combined list and remove it from the first list. At this stage the first list is 3, 5, 6, the second is 4, and the combined list is 1, 2.

Continue by comparing 3 and 4, the smallest elements of their respective lists. Because 3 is the smaller of these two elements, add it to the combined list and remove it from the first list. At this stage the first list is 5, 6, and the second is 4. The combined list is 1, 2, 3.

Then compare 5 and 4, the smallest elements in the two lists. Because 4 is the smaller of these two elements, add it to the combined list and remove it from the second list. At this stage the first list is 5, 6, the second list is empty, and the combined list is 1, 2, 3, 4.

Finally, because the second list is empty, all elements of the first list can be appended to the end of the combined list in the order they occur in the first list. This produces the ordered list 1, 2, 3, 4, 5, 6. ◀

We will now consider the general problem of merging two ordered lists L_1 and L_2 into an ordered list L. We will describe an algorithm for solving this problem. Start with an empty list L. Compare the smallest elements of the two lists. Put the smaller of these two elements

TABLE 1 Merging the Two Sorted Lists 2, 3, 5, 6 and 1, 4.

First List	*Second List*	*Merged List*	*Comparison*
2 3 5 6	1 4		$1 < 2$
2 3 5 6	4	1	$2 < 4$
3 5 6	4	1 2	$3 < 4$
5 6	4	1 2 3	$4 < 5$
5 6		1 2 3 4	
		1 2 3 4 5 6	

at the right end of L, and remove it from the list it was in. Next, if one of L_1 and L_2 is empty, append the other (nonempty) list to L, which completes the merging. If neither L_1 nor L_2 is empty, repeat this process. Algorithm 10 gives a pseudocode description of this procedure.

We will need estimates for the number of comparisons used to merge two ordered lists in the analysis of the merge sort. We can easily obtain such an estimate for Algorithm 10. Each time a comparison of an element from L_1 and an element from L_2 is made, an additional element is added to the merged list L. However, when either L_1 or L_2 is empty, no more comparisons are needed. Hence, Algorithm 10 is least efficient when $m + n - 2$ comparisons are carried out, where m and n are the number of elements in L_1 and L_2, respectively, leaving one element in each of L_1 and L_2. The next comparison will be the last one needed, because it will make one of these lists empty. Hence, Algorithm 10 uses no more than $m + n - 1$ comparisons. Lemma 1 summarizes this estimate.

ALGORITHM 10 Merging Two Lists.

procedure *merge*(L_1, L_2 : sorted lists)
L := empty list
while L_1 and L_2 are both nonempty
  remove smaller of first elements of L_1 and L_2 from its list; put it at the right end of L
  if this removal makes one list empty **then** remove all elements from the other list and
    append them to L
return L\{L is the merged list with elements in increasing order\}

LEMMA 1 Two sorted lists with m elements and n elements can be merged into a sorted list using no more than $m + n - 1$ comparisons.

Sometimes two sorted lists of length m and n can be merged using far fewer than $m + n - 1$ comparisons. For instance, when $m = 1$, a binary search procedure can be applied to put the one element in the first list into the second list. This requires only $\lceil \log n \rceil$ comparisons, which is much smaller than $m + n - 1 = n$, for $m = 1$. On the other hand, for some values of m and n, Lemma 1 gives the best possible bound. That is, there are lists with m and n elements that cannot be merged using fewer than $m + n - 1$ comparisons. (See Exercise 47.)

We can now analyze the complexity of the merge sort. Instead of studying the general problem, we will assume that n, the number of elements in the list, is a power of 2, say 2^m. This will make the analysis less complicated, but when this is not the case, various modifications can be applied that will yield the same estimate.

At the first stage of the splitting procedure, the list is split into two sublists, of 2^{m-1} elements each, at level 1 of the tree generated by the splitting. This process continues, splitting the two sublists with 2^{m-1} elements into four sublists of 2^{m-2} elements each at level 2, and so on. In general, there are 2^{k-1} lists at level $k - 1$, each with 2^{m-k+1} elements. These lists at level $k - 1$ are split into 2^k lists at level k, each with 2^{m-k} elements. At the end of this process, we have 2^m lists each with one element at level m.

We start merging by combining pairs of the 2^m lists of one element into 2^{m-1} lists, at level $m - 1$, each with two elements. To do this, 2^{m-1} pairs of lists with one element each are merged. The merger of each pair requires exactly one comparison.

The procedure continues, so that at level k ($k = m, m - 1, m - 2, \ldots, 3, 2, 1$), 2^k lists each with 2^{m-k} elements are merged into 2^{k-1} lists, each with 2^{m-k+1} elements, at level $k - 1$. To do this a total of 2^{k-1} mergers of two lists, each with 2^{m-k} elements, are needed. But, by Lemma 1, each of these mergers can be carried out using at most $2^{m-k} + 2^{m-k} - 1 = 2^{m-k+1} - 1$ comparisons. Hence, going from level k to $k - 1$ can be accomplished using at most $2^{k-1}(2^{m-k+1} - 1)$ comparisons.

Summing all these estimates shows that the number of comparisons required for the merge sort is at most

$$\sum_{k=1}^{m} 2^{k-1}(2^{m-k+1} - 1) = \sum_{k=1}^{m} 2^m - \sum_{k=1}^{m} 2^{k-1} = m2^m - (2^m - 1) = n \log n - n + 1,$$

because $m = \log n$ and $n = 2^m$. (We evaluated $\sum_{k=1}^{m} 2^m$ by noting that it is the sum of m identical terms, each equal to 2^m. We evaluated $\sum_{k=1}^{m} 2^{k-1}$ using the formula for the sum of the terms of a geometric progression from Theorem 1 of Section 2.4.)

Theorem 1 summarizes what we have discovered about the worst-case complexity of the merge sort algorithm.

THEOREM 1 The number of comparisons needed to merge sort a list with n elements is $O(n \log n)$.

In Chapter 11 we will show that the fastest comparison-based sorting algorithms have $O(n \log n)$ time complexity. (A comparison-based sorting algorithm has the comparison of two elements as its basic operation.) Theorem 1 tells us that the merge sort achieves this best possible big-O estimate for the complexity of a sorting algorithm. We describe another efficient algorithm, the quicksort, in the preamble to Exercise 50.

Exercises

1. Trace Algorithm 1 when it is given $n = 5$ as input. That is, show all steps used by Algorithm 1 to find 5!, as is done in Example 1 to find 4!.

2. Trace Algorithm 1 when it is given $n = 6$ as input. That is, show all steps used by Algorithm 1 to find 6!, as is done in Example 1 to find 4!.

3. Trace Algorithm 3 when it finds gcd(8, 13). That is, show all the steps used by Algorithm 3 to find gcd(8, 13).

4. Trace Algorithm 3 when it finds gcd(12, 17). That is, show all the steps used by Algorithm 3 to find gcd(12, 17).

5. Trace Algorithm 4 when it is given $m = 5$, $n = 11$, and $b = 3$ as input. That is, show all the steps Algorithm 4 uses to find 3^{11} **mod** 5.

6. Trace Algorithm 4 when it is given $m = 7$, $n = 10$, and $b = 2$ as input. That is, show all the steps Algorithm 4 uses to find 2^{10} **mod** 7.

7. Give a recursive algorithm for computing nx whenever n is a positive integer and x is an integer, using just addition.

8. Give a recursive algorithm for finding the sum of the first n positive integers.

9. Give a recursive algorithm for finding the sum of the first n odd positive integers.

10. Give a recursive algorithm for finding the maximum of a finite set of integers, making use of the fact that the maximum of n integers is the larger of the last integer in the list and the maximum of the first $n - 1$ integers in the list.

11. Give a recursive algorithm for finding the minimum of a finite set of integers, making use of the fact that the minimum of n integers is the smaller of the last integer in the list and the minimum of the first $n - 1$ integers in the list.

12. Devise a recursive algorithm for finding x^n **mod** m whenever n, x, and m are positive integers based on the fact that x^n **mod** $m = (x^{n-1}$ **mod** $m \cdot x$ **mod** $m)$ **mod** m.

13. Give a recursive algorithm for finding $n!$ **mod** m whenever n and m are positive integers.

14. Give a recursive algorithm for finding a **mode** of a list of integers. (A **mode** is an element in the list that occurs at least as often as every other element.)

15. Devise a recursive algorithm for computing the greatest common divisor of two nonnegative integers a and b with $a < b$ using the fact that $\gcd(a, b) = \gcd(a, b - a)$.

16. Prove that the recursive algorithm for finding the sum of the first n positive integers you found in Exercise 8 is correct.

17. Describe a recursive algorithm for multiplying two nonnegative integers x and y based on the fact that $xy = 2(x \cdot (y/2))$ when y is even and $xy = 2(x \cdot \lfloor y/2 \rfloor) + x$ when y is odd, together with the initial condition $xy = 0$ when $y = 0$.

18. Prove that Algorithm 1 for computing $n!$ when n is a nonnegative integer is correct.

19. Prove that Algorithm 3 for computing $\gcd(a, b)$ when a and b are positive integers with $a < b$ is correct.

20. Prove that the algorithm you devised in Exercise 17 is correct.

21. Prove that the recursive algorithm that you found in Exercise 7 is correct.

22. Prove that the recursive algorithm that you found in Exercise 10 is correct.

23. Devise a recursive algorithm for computing n^2 where n is a nonnegative integer, using the fact that $(n + 1)^2 = n^2 + 2n + 1$. Then prove that this algorithm is correct.

24. Devise a recursive algorithm to find a^{2^n}, where a is a real number and n is a positive integer. [*Hint:* Use the equality $a^{2^{n+1}} = (a^{2^n})^2$.]

25. How does the number of multiplications used by the algorithm in Exercise 24 compare to the number of multiplications used by Algorithm 2 to evaluate a^{2^n}?

***26.** Use the algorithm in Exercise 24 to devise an algorithm for evaluating a^n when n is a nonnegative integer. [*Hint:* Use the binary expansion of n.]

***27.** How does the number of multiplications used by the algorithm in Exercise 26 compare to the number of multiplications used by Algorithm 2 to evaluate a^n?

28. How many additions are used by the recursive and iterative algorithms given in Algorithms 7 and 8, respectively, to find the Fibonacci number f_7?

29. Devise a recursive algorithm to find the nth term of the sequence defined by $a_0 = 1, a_1 = 2$, and $a_n = a_{n-1} \cdot a_{n-2}$, for $n = 2, 3, 4, \ldots$.

30. Devise an iterative algorithm to find the nth term of the sequence defined in Exercise 29.

31. Is the recursive or the iterative algorithm for finding the sequence in Exercise 29 more efficient?

32. Devise a recursive algorithm to find the nth term of the sequence defined by $a_0 = 1$, $a_1 = 2$, $a_2 = 3$, and $a_n = a_{n-1} + a_{n-2} + a_{n-3}$, for $n = 3, 4, 5, \ldots$.

33. Devise an iterative algorithm to find the nth term of the sequence defined in Exercise 32.

34. Is the recursive or the iterative algorithm for finding the sequence in Exercise 32 more efficient?

35. Give iterative and recursive algorithms for finding the nth term of the sequence defined by $a_0 = 1$, $a_1 = 3$, $a_2 = 5$, and $a_n = a_{n-1} \cdot a_{n-2}^2 \cdot a_{n-3}^3$. Which is more efficient?

36. Give a recursive algorithm to find the number of partitions of a positive integer based on the recursive definition given in Exercise 49 in Section 5.3.

37. Give a recursive algorithm for finding the reversal of a bit string. (See the definition of the reversal of a bit string in the preamble of Exercise 36 in Section 5.3.)

38. Give a recursive algorithm for finding the string w^i, the concatenation of i copies of w, when w is a bit string.

39. Prove that the recursive algorithm for finding the reversal of a bit string that you gave in Exercise 37 is correct.

40. Prove that the recursive algorithm for finding the concatenation of i copies of a bit string that you gave in Exercise 38 is correct.

***41.** Give a recursive algorithm for tiling a $2^n \times 2^n$ checkerboard with one square missing using right triominoes.

42. Give a recursive algorithm for triangulating a simple polygon with n sides, using Lemma 1 in Section 5.2.

43. Give a recursive algorithm for computing values of the Ackermann function. (See the definition of the Ackermann function in the preamble to Exercise 50 in Section 5.3.)

44. Use a merge sort to sort 4, 3, 2, 5, 1, 8, 7, 6 into increasing order. Show all the steps used by the algorithm.

45. Use a merge sort to sort *b, d, a, f, g, h, z, p, o, k* into alphabetic order. Show all the steps used by the algorithm.

46. How many comparisons are required to merge these pairs of lists using Algorithm 10?

a) 1, 3, 5, 7, 9; 2, 4, 6, 8, 10
b) 1, 2, 3, 4, 5; 6, 7, 8, 9, 10
c) 1, 5, 6, 7, 8; 2, 3, 4, 9, 10

47. Show that for all positive integers m and n there are sorted lists with m elements and n elements, respectively, such that Algorithm 10 uses $m + n - 1$ comparisons to merge them into one sorted list.

***48.** What is the least number of comparisons needed to merge any two lists in increasing order into one list in increasing order when the number of elements in the two lists are

a) 1, 4? **b)** 2, 4? **c)** 3, 4? **d)** 4, 4?

***49.** Prove that the merge sort algorithm is correct.

Links

The **quicksort** is an efficient algorithm. To sort $a_1, a_2, \ldots, a_n$, this algorithm begins by taking the first element a_1 and forming two sublists, the first containing those elements that are less than a_1, in the order they arise, and the second containing those elements greater than a_1, in the order they arise. Then a_1 is put at the end of the first sublist. This procedure is repeated recursively for each sublist, until all sublists contain one item. The ordered list of n items is obtained by combining the sublists of one item in the order they occur.

50. Sort 3, 5, 7, 8, 1, 9, 2, 4, 6 using the quicksort.

51. Let $a_1, a_2, \ldots, a_n$ be a list of n distinct real numbers. How many comparisons are needed to form two sublists from this list, the first containing elements less than a_1 and the second containing elements greater than a_1?

52. Describe the quicksort algorithm using pseudocode.

53. What is the largest number of comparisons needed to order a list of four elements using the quicksort algorithm?

54. What is the least number of comparisons needed to order a list of four elements using the quicksort algorithm?

55. Determine the worst-case complexity of the quicksort algorithm in terms of the number of comparisons used.

5.5 Program Correctness

5.5.1 Introduction

Suppose that we have designed an algorithm to solve a problem and have written a program to implement it. How can we be sure that the program always produces the correct answer? After all the bugs have been removed so that the syntax is correct, we can test the program with sample input. It is not correct if an incorrect result is produced for any sample input. But even if the program gives the correct answer for all sample input, it may not always produce the correct answer (unless all possible input has been tested). We need a proof to show that the program *always* gives the correct output.

Program verification, the proof of correctness of programs, uses the rules of inference and proof techniques described in this chapter, including mathematical induction. Because an incorrect program can lead to disastrous results, a large amount of methodology has been constructed for verifying programs. Efforts have been devoted to automating program verification so that it can be carried out using a computer. However, only limited progress has been made toward this goal. Indeed, some mathematicians and theoretical computer scientists argue that it will never be realistic to mechanize the proof of correctness of complex programs.

Some of the concepts and methods used to prove that programs are correct will be introduced in this section. Many different methods have been devised for proving that programs are correct. We will discuss a widely used method for program verification introduced by Tony Hoare in this section; several other methods are also commonly used. Furthermore, we will not develop a complete methodology for program verification in this book. This section is meant to be a brief introduction to the area of program verification, which ties together the rules of logic, proof techniques, and the concept of an algorithm.

5.5.2 Program Verification

A correct program should produce the correct output for every possible input. This requirement consists of two elements. The first element is that the correct answer is obtained if the program terminates. This requirement is called **(partial) correctness**. The second element is that the program always terminates. A program that can be shown to always terminate is said to be **totally correct**. However, whether a program terminates depends both on the program and on details of its implementation; for example a program that terminates with the correct output on one computer may not have enough memory to terminate successfully on another machine.

To specify what it means for a program to produce the correct output, two propositions are used. The first is the **initial assertion**, which gives the properties that the input values must have. The second is the **final assertion**, which gives the properties that the output of the program should have, if the program did what was intended. The appropriate initial and final assertions must be provided when a program is checked.

Definition 1 A program, or program segment, S is said to be (partially) *correct with respect to* the initial assertion p and the final assertion q if whenever p is true for the input values of S and S terminates, then q is true for the output values of S. The notation $p\{S\}q$ indicates that the program, or program segment, S is partially correct with respect to the initial assertion p and the final assertion q.

Note: The notation $p\{S\}q$ is known as a *Hoare triple*. Tony Hoare introduced the concept of partial correctness.

Links

Note that the notion of partial correctness has nothing to do with whether a program terminates; it focuses only on whether the program does what it is expected to do if it terminates.

A simple example illustrates the concepts of initial and final assertions.

EXAMPLE 1 Show that the program segment

$y := 2$
$z := x + y$

Extra Examples

is correct with respect to the initial assertion p: $x = 1$ and the final assertion q: $z = 3$.

Solution: Suppose that p is true, so that $x = 1$ as the program begins. Then y is assigned the value 2, and z is assigned the sum of the values of x and y, which is 3. Hence, S is correct with respect to the initial assertion p and the final assertion q. Thus, $p\{S\}q$ is true. ◀

5.5.3 Rules of Inference

A useful rule of inference proves that a program is correct by splitting the program into a sequence of subprograms and then showing that each subprogram is correct.

Suppose that the program S is split into subprograms S_1 and S_2. Write $S = S_1; S_2$ to indicate that S is made up of S_1 followed by S_2. Suppose that the correctness of S_1 with respect to the initial assertion p and final assertion q, and the correctness of S_2 with respect to the initial assertion q and the final assertion r, have been established. It follows that if p is true and S_1 is executed and terminates, then q is true; and if q is true, and S_2 executes and terminates, then r is true. Thus, if p is true and $S = S_1; S_2$ is executed and terminates, then r is true. This rule of inference, called the **composition rule**, can be stated as

$$\begin{array}{l} p\{S_1\}q \\ q\{S_2\}r \\ \hline \therefore p\{S_1; S_2\}r. \end{array}$$

This rule of inference will be used later in this section.

Next, some rules of inference for program segments involving conditional statements and loops will be given. Because programs can be split into segments for proofs of correctness, this will let us verify many different programs.

5.5.4 Conditional Statements

First, rules of inference for conditional statements will be given. Suppose that a program segment has the form

```
if condition then
    S
```

where S is a block of statements. Then S is executed if *condition* is true, and it is not executed when *condition* is false. To verify that this segment is correct with respect to the initial assertion p and final assertion q, two things must be done. First, it must be shown that when p is true and *condition* is also true, then q is true after S terminates. Second, it must be shown that when p is true and *condition* is false, then q is true (because in this case S does not execute).

This leads to the following rule of inference:

$$\begin{array}{l} (p \wedge \mathit{condition})\{S\}q \\ (p \wedge \neg \mathit{condition}) \rightarrow q \\ \hline \therefore p\{\textbf{if } \mathit{condition} \textbf{ then } S\}q. \end{array}$$

Example 2 illustrates how this rule of inference is used.

EXAMPLE 2 Verify that the program segment

```
if x > y then
    y := x
```

is correct with respect to the initial assertion **T** and the final assertion $y \geq x$.

Solution: When the initial assertion is true and $x > y$, the assignment $y := x$ is carried out. Hence, the final assertion, which asserts that $y \geq x$, is true in this case. Moreover, when the initial assertion is true and $x > y$ is false, so that $x \leq y$, the final assertion is again true. Hence, using the rule of inference for program segments of this type, this program is correct with respect to the given initial and final assertions. ◀

Similarly, suppose that a program has a statement of the form

```
if condition then
    S1
else
    S2
```

If *condition* is true, then S_1 executes; if *condition* is false, then S_2 executes. To verify that this program segment is correct with respect to the initial assertion p and the final assertion q, two things must be done. First, it must be shown that when p is true and *condition* is true, then q is true after S_1 terminates. Second, it must be shown that when p is true and *condition* is false, then q is true after S_2 terminates. This leads to the following rule of inference:

$$\begin{array}{l} (p \wedge condition)\{S_1\}q \\ (p \wedge \neg condition)\{S_2\}q \\ \hline \therefore p\{\textbf{if } condition \textbf{ then } S_1 \textbf{ else } S_2\}q. \end{array}$$

Links

Tony Hoare

C. ANTHONY R. HOARE (BORN 1934) Tony Hoare was born in Colombo, Ceylon (now known as Sri Lanka), where his father was a civil servant of the British Empire and his mother's father owned a plantation. He spent his early childhood in Ceylon, moving to England in 1945. Hoare studied philosophy, together with the classics, at the University of Oxford, where he became interested in computing as a result of his fascination with the power of mathematical logic and the certainty of mathematical truth. He received his bachelor's degree from Oxford in 1956.

Hoare learned Russian during his service in the Royal Navy, and later studied the computer translation of natural languages at Moscow State University. He returned to England in 1960, taking a job at a small computer manufacturer, where he wrote a compiler for the programming language Algol. In 1968, he became Professor of Computing Science at the Queen's University, Belfast; in 1977, he moved to the University of Oxford as Professor of Computing; he is now Professor Emeritus. He is a Fellow of the Royal Society and also holds a position at Microsoft Research in Cambridge.

Hoare has made many contributions to the theory of programming languages and to programming methodology. He was first to define a programming language based on how programs could be proved correct with respect to their specifications. Hoare also invented quicksort, one of the most commonly used sorting algorithms (see the preamble to Exercise 50 in Section 5.4). He received the ACM Turing Award in 1980 and in 2000 he was knighted for services to education and computer science. Hoare is a noted writer in the technical and social aspects of computer science.

Example 3 illustrates how this rule of inference is used.

EXAMPLE 3 Verify that the program segment

```
if x < 0 then
    abs := −x
else
    abs := x
```

is correct with respect to the initial assertion **T** and the final assertion $abs = |x|$.

Solution: Two things must be demonstrated. First, it must be shown that if the initial assertion is true and $x < 0$, then $abs = |x|$. This is correct, because when $x < 0$ the assignment statement $abs := -x$ sets $abs = -x$, which is $|x|$ by definition when $x < 0$. Second, it must be shown that if the initial assertion is true and $x < 0$ is false, so that $x \geq 0$, then $abs = |x|$. This is also correct, because in this case the program uses the assignment statement $abs := x$, and x is $|x|$ by definition when $x \geq 0$, so $abs := x$. Hence, using the rule of inference for program segments of this type, this segment is correct with respect to the given initial and final assertions. ◀

5.5.5 Loop Invariants

Links

Next, proofs of correctness of **while** loops will be described. To develop a rule of inference for program segments of the type

```
while condition
    S
```

note that S is repeatedly executed until *condition* becomes false. An assertion that remains true each time S is executed must be chosen. Such an assertion is called a **loop invariant**. In other words, p is a loop invariant if $(p \wedge condition)\{S\}p$ is true.

Suppose that p is a loop invariant. It follows that if p is true before the program segment is executed, p and $\neg condition$ are true after termination, if it occurs. This rule of inference is

$$\frac{(p \wedge condition)\{S\}p}{\therefore p\{\textbf{while } condition \;\; S\}(\neg\, condition \;\wedge p).}$$

The use of a loop invariant is illustrated in Example 4.

EXAMPLE 4 A loop invariant is needed to verify that the program segment

Extra Examples

```
i := 1
factorial := 1
while i < n
    i := i + 1
    factorial := factorial · i
```

terminates with $factorial = n!$ when n is a positive integer.

Let p be the assertion "$factorial = i!$ and $i \leq n$." We first prove that p is a loop invariant. Suppose that, at the beginning of one execution of the **while** loop, p is true and the condition of the **while** loop holds; in other words, assume that $factorial = i!$ and that $i < n$. The new values i_{new} and $factorial_{\text{new}}$ of i and $factorial$ are $i_{\text{new}} = i + 1$ and $factorial_{\text{new}} = factorial \cdot (i + 1) = (i + 1)! = i_{\text{new}}!$. Because $i < n$, we also have $i_{\text{new}} = i + 1 \leq n$. Thus, p is true at the end of the execution of the loop. This shows that p is a loop invariant.

Now we consider the program segment. Just before entering the loop, $i = 1 \leq n$ and $factorial = 1 = 1! = i!$ both hold, so p is true. Because p is a loop invariant, the rule of inference just introduced implied that if the **while** loop terminates, it terminates with p true and with $i < n$ false. In this case, at the end, $factorial = i!$ and $i \leq n$ are true, but $i < n$ is false; in other words, $i = n$ and $factorial = i! = n!$, as desired.

Finally, we need to check that the **while** loop actually terminates. At the beginning of the program i is assigned the value 1, so after $n - 1$ traversals of the loop, the new value of i will be n, and the loop terminates at that point. ◀

A final example will be given to show how the various rules of inference can be used to verify the correctness of a longer program.

EXAMPLE 5 We will outline how to verify the correctness of the program S for computing the product of two integers.

```
procedure multiply(m, n: integers)

S1 { if n < 0 then a := −n
   { else a := n

S2 { k := 0
   { x := 0

S3 { while k < a
   {     x := x + m
   {     k := k + 1

S4 { if n < 0 then product := −x
   { else product := x

return product
{product equals mn}
```

The goal is to prove that after S is executed, $product$ has the value mn. The proof of correctness can be carried out by splitting S into four segments, with $S = S_1; S_2; S_3; S_4$, as shown in the listing of S. The rule of composition can be used to build the correctness proof. Here is how the argument proceeds. The details will be left as an exercise for the reader.

Let p be the initial assertion "m and n are integers." Then, it can be shown that $p\{S_1\}q$ is true, when q is the proposition $p \wedge (a = |n|)$. Next, let r be the proposition $q \wedge (k = 0) \wedge (x = 0)$. It is easily verified that $q\{S_2\}r$ is true. It can be shown that "$x = mk$ and $k \leq a$" is an invariant for the loop in S_3. Furthermore, it is easy to see that the loop terminates after a iterations, with $k = a$, so $x = ma$ at this point. Because r implies that $x = m \cdot 0$ and $0 \leq a$, the loop invariant is true before the loop is entered. Because the loop terminates with $k = a$, it follows that $r\{S_3\}s$ is true, where s is the proposition "$x = ma$ and $a = |n|$." Finally, it can be shown that S_4 is correct with respect to the initial assertion s and final assertion t, where t is the proposition "$product = mn$."

Putting all this together, because $p\{S_1\}q$, $q\{S_2\}r$, $r\{S_3\}s$, and $s\{S_4\}t$ are all true, it follows from the rule of composition that $p\{S\}t$ is true. Furthermore, because all four segments terminate, S does terminate. This verifies the correctness of the program. ◀

Exercises

1. Show that the program segment

while T
  $x := x + 1$

is partially correct but not totally correct with respect to the initial assertion $x = 1$ and final assertion $x = 0$.

2. What should the initial assertion, the final assertion, and the loop invariant be to show that the program finds the largest perfect square not exceeding a positive integer x?

$s := 0$
while $(s + 1)(s + 1) \leq x$
  $s := s + 1$

3. Prove that the program segment

$y := 1$
$z := x + y$

is partially correct with respect to the initial assertion $x = 0$ and the final assertion $z = 1$.

4. Verify that the program segment

if $x < 0$ **then** $x := 0$

is partially correct with respect to the initial assertion **T** and the final assertion $x \geq 0$.

5. Verify that the program segment

$x := 2$
$z := x + y$
if $y > 0$ **then**
  $z := z + 1$
else
  $z := 0$

is partially correct with respect to the initial assertion $y = 3$ and the final assertion $z = 6$.

6. Verify that the program segment

if $x < y$ **then**
  $min := x$
else
  $min := y$

is correct with respect to the initial assertion **T** and the final assertion $(x \leq y \wedge min = x) \vee (x > y \wedge min = y)$.

***7.** Devise a rule of inference for verification of partial correctness of statements of the form

if *condition* 1 **then**
  S_1
else if *condition* 2 **then**
  S_2
  ⋮
else
  S_n

where $S_1, S_2, \ldots, S_n$ are blocks.

8. Use the rule of inference developed in Exercise 7 to verify that the program

if $x < 0$ **then**
  $y := -2|x|/x$
else if $x > 0$ **then**
  $y := 2|x|/x$
else if $x = 0$ **then**
  $y := 2$

is partially correct with respect to the initial assertion **T** and the final assertion $y = 2$.

9. Use a loop invariant to prove that the following program segment for computing the nth power, where n is a positive integer, of a real number x is correct.

$power := 1$
$i := 1$
while $i \leq n$
  $power := power * x$
  $i := i + 1$

***10.** Prove that the iterative program for finding f_n given in Section 5.4 is correct.

11. Provide all the details in the proof of correctness given in Example 5.

12. Suppose that both the conditional statement $p_0 \rightarrow p_1$ and the program assertion $p_1\{S\}q$ are true. Show that $p_0\{S\}q$ also must be true.

13. Suppose that both the program assertion $p\{S\}q_0$ and the conditional statement $q_0 \rightarrow q_1$ are true. Show that $p\{S\}q_1$ also must be true.

14. This program computes quotients and remainders.

$r := a$
$q := 0$
while $r \geq d$
  $r := r - d$
  $q := q + 1$

Verify that it is partially correct with respect to the initial assertion "a and d are positive integers" and the final assertion "q and r are integers such that $a = dq + r$ and $0 \leq r < d$."

15. Use a loop invariant to verify that the Euclidean algorithm (Algorithm 1 in Section 4.3) is partially correct with respect to the initial assertion "a and b are positive integers" and the final assertion "$x = \gcd(a, b)$."

Key Terms and Results

TERMS

sequence: a function with domain that is a subset of the set of integers

geometric progression: a sequence of the form $a, ar, ar^2, \ldots,$ where a and r are real numbers

arithmetic progression: a sequence of the form $a, a+d, a+2d, \ldots,$ where a and d are real numbers

the principle of mathematical induction: The statement $\forall n\, P(n)$ is true if $P(1)$ is true and $\forall k[P(k) \rightarrow P(k+1)]$ is true.

basis step: the proof of $P(1)$ in a proof by mathematical induction of $\forall nP(n)$

inductive step: the proof of $P(k) \rightarrow P(k+1)$ for all positive integers k in a proof by mathematical induction of $\forall nP(n)$

strong induction: The statement $\forall nP(n)$ is true if $P(1)$ is true and $\forall k[(P(1) \wedge \cdots \wedge P(k)) \rightarrow P(k+1)]$ is true.

well-ordering property: Every nonempty set of nonnegative integers has a least element.

downward (or backward) induction: The statement $S(n)$ is true for all positive integers n if $S(n)$ is true for infinitely many positive integers n and for every positive integer $m > 2$, $S(m) \rightarrow S(m-1)$.

recursive definition of a function: a definition of a function that specifies an initial set of values and a rule for obtaining values of this function at integers from its values at smaller integers

recursive definition of a set: a definition of a set that specifies an initial set of elements in the set and a rule for obtaining other elements from those in the set

structural induction: a technique for proving results about recursively defined sets

recursive algorithm: an algorithm that proceeds by reducing a problem to the same problem with smaller input

merge sort: a sorting algorithm that sorts a list by splitting it in two, sorting each of the two resulting lists, and merging the results into a sorted list

iteration: a procedure based on the repeated use of operations in a loop

program correctness: verification that a procedure always produces the correct result

loop invariant: a property that remains true during every traversal of a loop

initial assertion: the statement specifying the properties of the input values of a program

final assertion: the statement specifying the properties the output values should have if the program worked correctly

Review Questions

1. a) Can you use the principle of mathematical induction to find a formula for the sum of the first n terms of a sequence?

b) Can you use the principle of mathematical induction to determine whether a given formula for the sum of the first n terms of a sequence is correct?

c) Find a formula for the sum of the first n even positive integers, and prove it using mathematical induction.

2. a) For which positive integers n is $11n + 17 \leq 2^n$?

b) Prove the conjecture you made in part (a) using mathematical induction.

3. a) Which amounts of postage can be formed using only 5-cent and 9-cent stamps?

b) Prove the conjecture you made using mathematical induction.

c) Prove the conjecture you made using strong induction.

d) Find a proof of your conjecture different from the ones you gave in (b) and (c).

4. Give two different examples of proofs that use strong induction.

5. a) State the well-ordering property for the set of positive integers.

b) Use this property to show that every positive integer greater than one can be written as the product of primes.

6. a) Explain why a function f from the set of positive integers to the set of real numbers is well defined if it is defined recursively by specifying $f(1)$ and a rule for finding $f(n)$ from $f(n-1)$.

b) Provide a recursive definition of the function $f(n) = (n+1)!$.

7. a) Give a recursive definition of the Fibonacci numbers.

b) Show that $f_n > \alpha^{n-2}$ whenever $n \geq 3$, where f_n is the nth term of the Fibonacci sequence and $\alpha = (1 + \sqrt{5})/2$.

8. a) Explain why a sequence a_n is well defined if it is defined recursively by specifying a_1 and a_2 and a rule for finding a_n from $a_1, a_2, \ldots, a_{n-1}$ for $n = 3, 4, 5, \ldots$.

b) Find the value of a_n if $a_1 = 1, a_2 = 2$, and $a_n = a_{n-1} + a_{n-2} + \cdots + a_1$, for $n = 3, 4, 5, \ldots$.

9. Give two examples of how well-formed formulae are defined recursively for different sets of elements and operators.

10. a) Give a recursive definition of the length of a string.
b) Use the recursive definition from part (a) and structural induction to prove that $l(xy) = l(x) + l(y)$.

11. a) What is a recursive algorithm?
b) Describe a recursive algorithm for computing the sum of n numbers in a sequence.

12. Describe a recursive algorithm for computing the greatest common divisor of two positive integers.

13. a) Describe the merge sort algorithm.
b) Use the merge sort algorithm to put the list 4, 10, 1, 5, 3, 8, 7, 2, 6, 9 in increasing order.
c) Give a big-O estimate for the number of comparisons used by the merge sort.

14. a) Does testing a computer program to see whether it produces the correct output for certain input values verify that the program always produces the correct output?
b) Does showing that a computer program is partially correct with respect to an initial assertion and a final assertion verify that the program always produces the correct output? If not, what else is needed?

15. What techniques can you use to show that a long computer program is partially correct with respect to an initial assertion and a final assertion?

16. What is a loop invariant? How is a loop invariant used?

Supplementary Exercises

1. Use mathematical induction to show that $\frac{2}{3} + \frac{2}{9} + \frac{2}{27} + \cdots + \frac{2}{3^n} = 1 - \frac{1}{3^n}$ whenever n is a positive integer.

2. Use mathematical induction to show that $1^3 + 3^3 + 5^3 + \cdots + (2n + 1)^3 = (n + 1)^2(2n^2 + 4n + 1)$ whenever n is a positive integer.

3. Use mathematical induction to show that $1 \cdot 2^0 + 2 \cdot 2^1 + 3 \cdot 2^2 + \cdots + n \cdot 2^{n-1} = (n - 1) \cdot 2^n + 1$ whenever n is a positive integer.

4. Use mathematical induction to show that

$$\frac{1}{1 \cdot 3} + \frac{1}{3 \cdot 5} + \cdots + \frac{1}{(2n - 1)(2n + 1)} = \frac{n}{2n + 1}$$

whenever n is a positive integer.

5. Show that

$$\frac{1}{1 \cdot 4} + \frac{1}{4 \cdot 7} + \cdots + \frac{1}{(3n - 2)(3n + 1)} = \frac{n}{3n + 1}$$

whenever n is a positive integer.

6. Use mathematical induction to show that $2^n > n^2 + n$ whenever n is an integer greater than 4.

7. Use mathematical induction to show that $2^n > n^3$ whenever n is an integer greater than 9.

8. Find an integer N such that $2^n > n^4$ whenever n is an integer greater than N. Prove that your result is correct using mathematical induction.

9. Use mathematical induction to prove that $a - b$ is a factor of $a^n - b^n$ whenever n is a positive integer.

10. Use mathematical induction to prove that 9 divides $n^3 + (n + 1)^3 + (n + 2)^3$ whenever n is a nonnegative integer.

11. Use mathematical induction to prove that 43 divides $6^{n+1} + 7^{2n-1}$ for every positive integer n.

12. Use mathematical induction to prove that 64 divides $3^{2n+2} + 56n + 55$ for every positive integer n.

13. Use mathematical induction to prove this formula for the sum of the terms of an arithmetic progression.

$$a + (a + d) + \cdots + (a + nd) = (n + 1)(2a + nd)/2$$

14. Suppose that $a_j \equiv b_j \pmod{m}$ for $j = 1, 2, \ldots, n$. Use mathematical induction to prove that

a) $\sum_{j=1}^{n} a_j \equiv \sum_{j=1}^{n} b_j \pmod{m}$.

b) $\prod_{j=1}^{n} a_j \equiv \prod_{j=1}^{n} b_j \pmod{m}$.

15. Show that if n is a positive integer, then

$$\sum_{k=1}^{n} \frac{k + 4}{k(k + 1)(k + 2)} = \frac{n(3n + 7)}{2(n + 1)(n + 2)}.$$

16. For which positive integers n is $n + 6 < (n^2 - 8n)/16$? Prove your answer using mathematical induction.

17. (*Requires calculus*) Suppose that $f(x) = e^x$ and $g(x) = xe^x$. Use mathematical induction together with the product rule and the fact that $f'(x) = e^x$ to prove that $g^{(n)}(x) = (x + n)e^x$ whenever n is a positive integer.

18. (*Requires calculus*) Suppose that $f(x) = e^x$ and $g(x) = e^{cx}$, where c is a constant. Use mathematical induction together with the chain rule and the fact that $f'(x) = e^x$ to prove that $g^{(n)} = c^n e^{cx}$ whenever n is a positive integer.

***19.** Formulate a conjecture about which Fibonacci numbers are even, and use a form of mathematical induction to prove your conjecture.

***20.** Determine which Fibonacci numbers are divisible by 3. Use a form of mathematical induction to prove your conjecture.

***21.** Prove that $f_k f_n + f_{k+1} f_{n+1} = f_{n+k+1}$ for all nonnegative integers n and k, where f_i denotes the ith Fibonacci number.

Recall from Example 15 of Section 2.4 that the sequence of **Lucas numbers** is defined by $l_0 = 2$, $l_1 = 1$, and $l_n = l_{n-1} + l_{n-2}$ for $n = 2, 3, 4, \ldots$.

22. Show that $f_n + f_{n+2} = l_{n+1}$ whenever n is a positive integer, where f_i and l_i are the ith Fibonacci number and ith Lucas number, respectively.

23. Show that $l_0^2 + l_1^2 + \cdots + l_n^2 = l_n l_{n+1} + 2$ whenever n is a nonnegative integer and l_i is the ith Lucas number.

***24.** Use mathematical induction to show that the product of any n consecutive positive integers is divisible by $n!$. [*Hint:* Use the identity $m(m + 1) \cdots (m + n - 1)/n! = (m - 1)m(m + 1) \cdots (m + n - 2)/n! + m(m + 1) \cdots (m + n - 2)/(n - 1)!$.]

25. Use mathematical induction to show that $(\cos x + i \sin x)^n = \cos nx + i \sin nx$ whenever n is a positive integer. (Here i is the square root of -1.) [*Hint:* Use the identities $\cos(a+b) = \cos a \cos b - \sin a \sin b$ and $\sin(a+b) = \sin a \cos b + \cos a \sin b$.]

***26.** Use mathematical induction to show that $\sum_{j=1}^{n} \cos jx = \cos[(n+1)x/2] \sin(nx/2)/\sin(x/2)$ whenever n is a positive integer and $\sin(x/2) \neq 0$.

27. Use mathematical induction to prove that $\sum_{j=1}^{n} j^2 2^j = n^2 2^{n+1} - n 2^{n+2} + 3 \cdot 2^{n+1} - 6$ for every positive integer n.

28. (*Requires calculus*) Suppose that the sequence $x_1, x_2, \ldots, x_n, \ldots$ is recursively defined by $x_1 = 0$ and $x_{n+1} = \sqrt{x_n + 6}$.

a) Use mathematical induction to show that $x_1 < x_2 < \cdots < x_n < \cdots$, that is, the sequence $\{x_n\}$ is monotonically increasing.

b) Use mathematical induction to prove that $x_n < 3$ for $n = 1, 2, \ldots$.

c) Show that $\lim_{n\to\infty} x_n = 3$.

29. Show if n is a positive integer with $n \geq 2$, then

$$\sum_{j=2}^{n} \frac{1}{j^2 - 1} = \frac{(n-1)(3n+2)}{4n(n+1)}.$$

30. Use mathematical induction to prove Theorem 1 in Section 4.2, that is, show if b is an integer, where $b > 1$, and n is a positive integer, then n can be expressed uniquely in the form $n = a_k b^k + a_{k-1} b^{k-1} + \cdots + a_1 b + a_0$.

***31.** A **lattice point** in the plane is a point (x, y) where both x and y are integers. Use mathematical induction to show that at least $n + 1$ straight lines are needed to ensure that every lattice point (x, y) with $x \geq 0$, $y \geq 0$, and $x + y \leq n$ lies on one of these lines.

32. (*Requires calculus*) Use mathematical induction and the product rule to show that if n is a positive integer and $f_1(x), f_2(x), \ldots, f_n(x)$, are all differentiable functions, then

$$\frac{(f_1(x) f_2(x) \cdots f_n(x))'}{f_1(x) f_2(x) \cdots f_n(x)} = \frac{f_1'(x)}{f_1(x)} + \frac{f_2'(x)}{f_2(x)} + \cdots + \frac{f_n'(x)}{f_n(x)}.$$

33. (*Requires material in Section 2.6*) Suppose that $\mathbf{B} = \mathbf{M}\mathbf{A}\mathbf{M}^{-1}$, where $\mathbf{A}$ and $\mathbf{B}$ are $n \times n$ matrices and $\mathbf{M}$ is invertible. Show that $\mathbf{B}^k = \mathbf{M}\mathbf{A}^k\mathbf{M}^{-1}$ for all positive integers k. (Consult both the text of Section 2.6 and the preamble to Exercise 18 of Section 2.6.)

34. Use mathematical induction to show that if you draw lines in the plane you only need two colors to color the regions formed so that no two regions that have an edge in common have a common color.

35. Show that $n!$ can be represented as the sum of n of its distinct positive divisors whenever $n \geq 3$. [*Hint:* Use inductive loading. First try to prove this result using mathematical induction. By examining where your proof fails, find a stronger statement that you can easily prove using mathematical induction.]

***36.** Use mathematical induction to prove that if $x_1, x_2, \ldots, x_n$ are positive real numbers with $n \geq 2$, then

$$\left(x_1 + \frac{1}{x_1}\right)\left(x_2 + \frac{1}{x_2}\right)\cdots\left(x_n + \frac{1}{x_n}\right) \geq$$
$$\left(x_1 + \frac{1}{x_2}\right)\left(x_2 + \frac{1}{x_3}\right)\cdots\left(x_{n-1} + \frac{1}{x_n}\right)\left(x_n + \frac{1}{x_1}\right)$$

37. Use mathematical induction to prove that if n people stand in a line, where n is a positive integer, and if the first person in the line is a woman and the last person in line is a man, then somewhere in the line there is a woman directly in front of a man.

***38.** Suppose that for every pair of cities in a country there is a direct one-way road connecting them in one direction or the other. Use mathematical induction to show that there is a city that can be reached from every other city either directly or via exactly one other city.

39. Use mathematical induction to show that when n circles divide the plane into regions, these regions can be colored with two different colors such that no regions with a common boundary are colored the same.

***40.** Suppose that among a group of cars on a circular track there is enough fuel for one car to complete a lap. Use mathematical induction to show that there is a car in the group that can complete a lap by obtaining gas from other cars as it travels around the track.

41. Show that if n is a positive integer, then

$$\sum_{j=1}^{n} (2j-1)\left(\sum_{k=j}^{n} 1/k\right) = n(n+1)/2.$$

42. Use mathematical induction to show that if a, b, and c are the lengths of the sides of a right triangle, where c is the length of the hypotenuse, then $a^n + b^n < c^n$ for all integers n with $n \geq 3$.

***43.** Use mathematical induction to show that if n is a positive integers, the sequence $2 \,\mathbf{mod}\, n$, $2^2 \,\mathbf{mod}\, n$, $2^{2^2} \,\mathbf{mod}\, n$, $2^{2^{2^2}} \,\mathbf{mod}\, n$, … is eventually constant (that is, all terms after a finite number of terms are all the same).

44. A **unit** or **Egyptian fraction** is a fraction of the form $1/n$, where n is a positive integer. In this exercise, we will use strong induction to show that a greedy algorithm can be used to express every rational number p/q with $0 < p/q < 1$ as the sum of distinct unit fractions. At each step of the algorithm, we find the smallest positive integer n such that $1/n$ can be added to the sum without exceeding p/q. For example, to express $5/7$ we first start the sum with $1/2$. Because $5/7 - 1/2 = 3/14$ we add $1/5$ to the sum because 5 is the smallest positive integer k such that $1/k < 3/14$. Because $3/14 - 1/5 = 1/70$, the algorithm terminates, showing that $5/7 = 1/2 + 1/5 + 1/70$. Let $T(p)$ be the statement that this algorithm terminates for all rational numbers p/q with $0 < p/q < 1$. We will prove that the algorithm always terminates by showing that $T(p)$ holds for all positive integers p.

a) Show that the basis step $T(1)$ holds.

b) Suppose that $T(k)$ holds for positive integers k with $k < p$. That is, assume that the algorithm terminates for all rational numbers k/r, where $1 \le k < p$. Show that if we start with p/q and the fraction $1/n$ is selected in the first step of the algorithm, then $p/q = p'/q' + 1/n$, where $p' = np - q$ and $q' = nq$. After considering the case where $p/q = 1/n$, use the inductive hypothesis to show that the greedy algorithm terminates when it begins with p'/q' and complete the inductive step.

The **McCarthy 91 function** (defined by John McCarthy, one of the founders of artificial intelligence) is defined using the rule

$$M(n) = \begin{cases} n - 10 & \text{if } n > 100 \\ M(M(n+11)) & \text{if } n \le 100 \end{cases}$$

for all positive integers n.

45. By successively using the defining rule for $M(n)$, find
a) $M(102)$. b) $M(101)$. c) $M(99)$.
d) $M(97)$. e) $M(87)$. f) $M(76)$.

****46.** Show that the function $M(n)$ is a well-defined function from the set of positive integers to the set of positive integers. [*Hint:* Prove that $M(n) = 91$ for all positive integers n with $n \le 101$.]

47. Is this proof that

$$\frac{1}{1\cdot 2} + \frac{1}{2\cdot 3} + \cdots + \frac{1}{(n-1)n} = \frac{3}{2} - \frac{1}{n},$$

whenever n is a positive integer, correct? Justify your answer.

Basis step: The result is true when $n = 1$ because

$$\frac{1}{1\cdot 2} = \frac{3}{2} - \frac{1}{1}.$$

Inductive step: Assume that the result is true for n. Then

$$\frac{1}{1\cdot 2} + \frac{1}{2\cdot 3} + \cdots + \frac{1}{(n-1)n} + \frac{1}{n(n+1)} = \frac{3}{2} - \frac{1}{n} + \left(\frac{1}{n} - \frac{1}{n+1}\right) = \frac{3}{2} - \frac{1}{n+1}.$$

Hence, the result is true for $n+1$ if it is true for n. This completes the proof.

48. Suppose that $A_1, A_2, \ldots, A_n$ are a collection of sets. Suppose that $R_2 = A_1 \oplus A_2$ and $R_k = R_{k-1} \oplus A_k$ for $k = 3, 4, \ldots, n$. Use mathematical induction to prove that $x \in R_n$ if and only if x belongs to an odd number of the sets $A_1, A_2, \ldots, A_n$. (Recall that $S \oplus T$ is the symmetric difference of the sets S and T defined in the preamble to Exercise 38 of Section 2.2.)

***49.** Show that n circles divide the plane into $n^2 - n + 2$ regions if every two circles intersect in exactly two points and no three circles contain a common point.

***50.** Show that n planes divide three-dimensional space into $(n^3 + 5n + 6)/6$ regions if any three of these planes have exactly one point in common and no four contain a common point.

***51.** Use the well-ordering property to show that $\sqrt{2}$ is irrational. [*Hint:* Assume that $\sqrt{2}$ is rational. Show that the set of positive integers of the form $b\sqrt{2}$ has a least element a. Then show that $a\sqrt{2} - a$ is a smaller positive integer of this form.]

52. A set is **well ordered** if every nonempty subset of this set has a least element. Determine whether each of the following sets is well ordered.

Links

AP Images

JOHN McCARTHY (1927–2011) John McCarthy was born in Boston. He grew up in Boston and in Los Angeles. He studied mathematics as both an undergraduate and a graduate student, receiving his B.S. in 1948 from the California Institute of Technology and his Ph.D. in 1951 from Princeton. After graduating from Princeton, McCarthy held positions at Princeton, Stanford, Dartmouth, and M.I.T. He held a position at Stanford from 1962 until 1994, and was an emeritus professor there. At Stanford, he was the director of the Artificial Intelligence Laboratory, held a named chair in the School of Engineering, and was a senior fellow at the Hoover Institution.

McCarthy was a pioneer in the study of artificial intelligence, a term he coined in 1955. He worked on problems related to the reasoning and information needs required for intelligent computer behavior. McCarthy was among the first computer scientists to design time-sharing computer systems. He developed LISP, a programming language for computing using symbolic expressions. He played an important role in using logic to verify the correctness of computer programs. McCarthy has also worked on the social implications of computer technology. In his later years he worked on the problem of how people and computers make conjectures through assumptions that complications are absent from situations. McCarthy was an advocate of the sustainability of human progress and was optimistic about the future of humanity. In his later years he also wrote science fiction stories. Some of his last work explored the possibility that the world is a computer program written by some higher force.

Among the awards McCarthy won are the Turing Award from the Association for Computing Machinery, the Research Excellence Award of the International Conference on Artificial Intelligence, the Kyoto Prize, and the National Medal of Science.

a) the set of integers
b) the set of integers greater than -100
c) the set of positive rationals
d) the set of positive rationals with denominator less than 100

53. a) Show that if $a_1, a_2, \ldots, a_n$ are positive integers, then $\gcd(a_1, a_2, \ldots, a_{n-1}, a_n) = \gcd(a_1, a_2, \ldots, a_{n-2}, \gcd(a_{n-1}, a_n))$.
b) Use part (a), together with the Euclidean algorithm, to develop a recursive algorithm for computing the greatest common divisor of a set of n positive integers.

***54.** Describe a recursive algorithm for writing the greatest common divisor of n positive integers as a linear combination of these integers.

55. Find an explicit formula for $f(n)$ if $f(1) = 1$ and $f(n) = f(n-1) + 2n - 1$ for $n \geq 2$. Prove your result using mathematical induction.

****56.** Give a recursive definition of the set of bit strings that contain twice as many 0s as 1s.

57. Let S be the set of bit strings defined recursively by $\lambda \in S$ and $0x \in S$, $x1 \in S$ if $x \in S$, where λ is the empty string.
a) Find all strings in S of length not exceeding five.
b) Give an explicit description of the elements of S.

58. Let S be the set of strings defined recursively by $abc \in S$, $bac \in S$, and $acb \in S$, where a, b, and c are fixed letters; and for all $x \in S$, $abcx \in S$; $abxc \in S$, $axbc \in S$, and $xabc \in S$, where x is a variable representing a string of letters.
a) Find all elements of S of length eight or less.
b) Show that every element of S has a length divisible by three.

The set B of all **balanced strings of parentheses** is defined recursively by $\lambda \in B$, where λ is the empty string; $(x) \in B$, $xy \in B$ if $x, y \in B$.

59. Show that (( )( )) is a balanced string of parentheses and (( ))) is not a balanced string of parentheses.

60. Find all balanced strings of parentheses with exactly six symbols.

61. Find all balanced strings of parentheses with four or fewer symbols.

62. Use induction to show that if x is a balanced string of parentheses, then the number of left parentheses equals the number of right parentheses in x.

Define the function N on the set of strings of parentheses by

$$N(\lambda) = 0, N(() = 1, N()) = -1,$$
$$N(uv) = N(u) + N(v),$$

where λ is the empty string, and u and v are strings. It can be shown that N is well defined.

63. Find
a) $N(())$.
b) $N()))())(()$.
c) $N((()(())$.
d) $N()((()))(()))$.

****64.** Show that a string w of parentheses is balanced if and only if $N(w) = 0$ and $N(u) \geq 0$ whenever u is a prefix of w, that is, $w = uv$.

***65.** Give a recursive algorithm for finding all balanced strings of parentheses containing n or fewer symbols.

66. Give a recursive algorithm for finding $\gcd(a, b)$, where a and b are nonnegative integers not both zero, based on these facts: $\gcd(a, b) = \gcd(b, a)$ if $a > b$, $\gcd(0, b) = b$, $\gcd(a, b) = 2\gcd(a/2, b/2)$ if a and b are even, $\gcd(a, b) = \gcd(a/2, b)$ if a is even and b is odd, and $\gcd(a, b) = \gcd(a, b - a)$.

67. Verify the program segment

if $x > y$ **then**
  $x := y$

with respect to the initial assertion **T** and the final assertion $x \leq y$.

***68.** Develop a rule of inference for verifying recursive programs and use it to verify the recursive algorithm for computing factorials given as Algorithm 1 in Section 5.4.

69. Devise a recursive algorithm that counts the number of times the integer 0 occurs in a list of integers.

Exercises 70–77 deal with some unusual sequences, informally called **self-generating sequences**, produced by simple recurrence relations or rules. In particular, Exercises 70–73 deal with the sequence $\{a(n)\}$ defined by $a(n) = n - a(a(n-1))$ for $n \geq 1$ and $a(0) = 0$. (This sequence, as well as those in Exercises 74 and 75, are defined in Douglas Hofstader's fascinating book *Gödel, Escher, Bach* [Ho99].)

70. Find the first 10 terms of the sequence $\{a(n)\}$ defined in the preamble to this exercise.

***71.** Prove that this sequence is well defined. That is, show that $a(n)$ is uniquely defined for all nonnegative integers n.

****72.** Prove that $a(n) = \lfloor (n+1)\mu \rfloor$ where $\mu = (-1 + \sqrt{5})/2$. [*Hint:* First show for all $n > 0$ that $(\mu n - \lfloor \mu n \rfloor) + (\mu^2 n - \lfloor \mu^2 n \rfloor) = 1$. Then show for all real numbers α with $0 \leq \alpha < 1$ and $\alpha \neq 1 - \mu$ that $\lfloor (1+\mu)(1-\alpha) \rfloor + \lfloor \alpha + \mu \rfloor = 1$, considering the cases $0 \leq \alpha < 1 - \mu$ and $1 - \mu < \alpha < 1$ separately.]

***73.** Use the formula from Exercise 72 to show that $a(n) = a(n-1)$ if $\mu n - \lfloor \mu n \rfloor < 1 - \mu$ and $a(n) = a(n-1) + 1$ otherwise.

74. Find the first 10 terms of each of the following self-generating sequences:
a) $a(n) = n - a(a(a(n-1)))$ for $n \geq 1$, $a(0) = 0$
b) $a(n) = n - a(a(a(a(n-1))))$ for $n \geq 1$, $a(0) = 0$
c) $a(n) = a(n - a(n-1)) + a(n - a(n-2))$ for $n \geq 3$, $a(1) = 1$ and $a(2) = 1$

75. Find the first 10 terms of both the sequences $m(n)$ and $f(n)$ defined by the following pair of interwoven recurrence relations: $m(n) = n - f(m(n-1))$, $f(n) = n - m(f(n-1))$ for $n \geq 1$, $f(0) = 1$ and $m(0) = 0$.

Golomb's self-generating sequence is the unique nondecreasing sequence of positive integers $a_1, a_2, a_3, \ldots$ that has the property that it contains exactly a_k occurrences of k for each positive integer k.

76. Find the first 20 terms of Golomb's self-generating sequence.

***77.** Show that if $f(n)$ is the largest integer m such that $a_m = n$, where a_m is the mth term of Golomb's self-generating sequence, then $f(n) = \sum_{k=1}^{n} a_k$ and $f(f(n)) = \sum_{k=1}^{n} ka_k$.

Computer Projects

Write programs with these input and output.

****1.** Given a $2^n \times 2^n$ checkerboard with one square missing, construct a tiling of this checkerboard using right triominoes.

****2.** Generate all well-formed formulae for expressions involving the variables x, y, and z and the operators $\{+, *, /, -\}$ with n or fewer symbols.

****3.** Generate all well-formed formulae for propositions with n or fewer symbols where each symbol is **T**, **F**, one of the propositional variables p and q, or an operator from $\{\neg, \vee, \wedge, \rightarrow, \leftrightarrow\}$.

4. Given a string, find its reversal.

5. Given a real number a and a nonnegative integer n, find a^n using recursion.

6. Given a real number a and a nonnegative integer n, find a^{2^n} using recursion.

***7.** Given a real number a and a nonnegative integer n, find a^n using the binary expansion of n and a recursive algorithm for computing a^{2^k}.

8. Given two integers not both zero, find their greatest common divisor using recursion.

9. Given a list of integers and an element x, locate x in this list using a recursive implementation of a linear search.

10. Given a list of integers and an element x, locate x in this list using a recursive implementation of a binary search.

11. Given a nonnegative integer n, find the nth Fibonacci number using iteration.

12. Given a nonnegative integer n, find the nth Fibonacci number using recursion.

13. Given a positive integer, find the number of partitions of this integer. (See Exercise 49 of Section 5.3.)

14. Given positive integers m and n, find $A(m, n)$, the value of Ackermann's function at the pair (m, n). (See the preamble to Exercise 50 of Section 5.3.)

15. Given a list of integers, sort these integers using the merge sort.

Computations and Explorations

Use a computational program or programs you have written to do these exercises.

1. What are the largest values of n for which $n!$ has fewer than 100 decimal digits and fewer than 1000 decimal digits?

2. Determine which Fibonacci numbers are divisible by 5, which are divisible by 7, and which are divisible by 11. Prove that your conjectures are correct.

3. Construct tilings using right triominoes of various 16×16, 32×32, and 64×64 checkerboards with one square missing.

4. Explore which $m \times n$ checkerboards can be completely covered by right triominoes. Can you make a conjecture that answers this question?

****5.** Implement an algorithm for determining whether a point is in the interior or exterior of a simple polygon.

****6.** Implement an algorithm for triangulating a simple polygon.

7. Which values of Ackermann's function are small enough that you are able to compute them?

8. Compare either the number of operations or the time needed to compute Fibonacci numbers recursively versus that needed to compute them iteratively.

Writing Projects

Respond to these with essays using outside sources.

1. Describe the origins of mathematical induction. Who were the first people to use it and to which problems did they apply it?

2. Explain how to prove the Jordan curve theorem for simple polygons and describe an algorithm for determining whether a point is in the interior or exterior of a simple polygon.

3. Describe how the triangulation of simple polygons is used in some key algorithms in computational geometry.
4. Describe a variety of different applications of the Fibonacci numbers to the biological and the physical sciences.
5. Discuss the uses of Ackermann's function both in the theory of recursive definitions and in the analysis of the complexity of algorithms for set unions.
6. Give the recursive definition of *Knuth's up-arrow notation* and illustrate its use with a variety of examples, including how it can be used to express values of the Ackermann function (defined in the preamble of Exercise 50 in Section 5.3).
7. Discuss some of the various methodologies used to establish the correctness of programs and compare them to Hoare's methods described in Section 5.5.
8. Explain how the ideas and concepts of program correctness can be extended to prove that operating systems are secure.

CHAPTER

6 Counting

6.1 The Basics of Counting

6.2 The Pigeonhole Principle

6.3 Permutations and Combinations

6.4 Binomial Coefficients and Identities

6.5 Generalized Permutations and Combinations

6.6 Generating Permutations and Combinations

Combinatorics, the study of arrangements of objects, is an important part of discrete mathematics. This subject was studied as long ago as the seventeenth century, when combinatorial questions arose in the study of gambling games. Enumeration, the counting of objects with certain properties, is an important part of combinatorics. We must count objects to solve many different types of problems. For instance, counting is used to determine the complexity of algorithms. To see whether a particular password scheme is secure we need to determine the number of possible passwords to make sure this number is large enough to make exhaustive search prohibitive. Counting is also required to determine whether there are enough telephone numbers or Internet Protocol addresses to meet demand. Recently, it has played a key role in mathematical biology, especially in sequencing DNA. Furthermore, counting techniques are used extensively when probabilities of events are computed. Enumeration of the members of particular families of chemicals is important in organic chemistry.

The basic rules of counting, which we will study in Section 6.1, can solve a tremendous variety of problems. For instance, we can use these rules to enumerate the different telephone numbers possible in the United States, the allowable passwords on a computer system, and the different orders in which the runners in a race can finish. Another important combinatorial tool is the pigeonhole principle, which we will study in Section 6.2. This states that when objects are placed in boxes and there are more objects than boxes, then there is a box containing at least two objects. For instance, we can use this principle to show that among a set of 15 or more students, at least 3 were born on the same day of the week.

We can phrase many counting problems in terms of ordered or unordered arrangements of the objects of a set with or without repetitions. These arrangements, called permutations and combinations, are used in many counting problems. For instance, suppose the 100 top finishers on a competitive exam taken by 2000 students are invited to a banquet. We can count the possible sets of 100 students that will be invited, as well as the ways in which the top 10 prizes can be awarded.

Another problem in combinatorics involves generating all the arrangements of a specified kind. This is often important in computer simulations. We will devise algorithms to generate arrangements of various types.

6.1 The Basics of Counting

6.1.1 Introduction

Suppose that a password on a computer system consists of six, seven, or eight characters. Each of these characters must be a digit or a letter of the alphabet. Each password must contain at least one digit. How many such passwords are there? The techniques needed to answer this question and a wide variety of other counting problems will be introduced in this section.

Counting problems arise throughout mathematics and computer science. For example, we must count the successful outcomes of experiments and all the possible outcomes of these experiments to determine probabilities of discrete events. We need to count the number of operations used by an algorithm to study its time complexity.

We will introduce the basic techniques of counting in this section. These methods serve as the foundation for almost all counting techniques.

6.1.2 Basic Counting Principles

Assessment

We first present two basic counting principles, the **product rule** and the **sum rule**. Then we will show how they can be used to solve many different counting problems.

The product rule applies when a procedure is made up of separate tasks.

THE PRODUCT RULE Suppose that a procedure can be broken down into a sequence of two tasks. If there are n_1 ways to do the first task and for each of these ways of doing the first task, there are n_2 ways to do the second task, then there are $n_1 n_2$ ways to do the procedure.

Extra Examples

Examples 1–10 show how the product rule is used.

EXAMPLE 1 A new company with just two employees, Sanchez and Patel, rents a floor of a building with 12 offices. How many ways are there to assign different offices to these two employees?

Solution: The procedure of assigning offices to these two employees consists of assigning an office to Sanchez, which can be done in 12 ways, then assigning an office to Patel different from the office assigned to Sanchez, which can be done in 11 ways. By the product rule, there are $12 \cdot 11 = 132$ ways to assign offices to these two employees. ◀

EXAMPLE 2 The chairs of an auditorium are to be labeled with an uppercase English letter followed by a positive integer not exceeding 100. What is the largest number of chairs that can be labeled differently?

Solution: The procedure of labeling a chair consists of two tasks, namely, assigning to the seat one of the 26 uppercase English letters, and then assigning to it one of the 100 possible integers. The product rule shows that there are $26 \cdot 100 = 2600$ different ways that a chair can be labeled. Therefore, the largest number of chairs that can be labeled differently is 2600. ◀

EXAMPLE 3 There are 32 computers in a data center in the cloud. Each of these computers has 24 ports. How many different computer ports are there in this data center?

Solution: The procedure of choosing a port consists of two tasks, first picking a computer and then picking a port on this computer. Because there are 32 ways to choose the computer and 24 ways to choose the port no matter which computer has been selected, the product rule shows that there are $32 \cdot 24 = 768$ ports. ◀

An extended version of the product rule is often useful. Suppose that a procedure is carried out by performing the tasks $T_1, T_2, \dots, T_m$ in sequence. If each task T_i, $i = 1, 2, \dots, m$, can be done in n_i ways, regardless of how the previous tasks were done, then there are $n_1 \cdot n_2 \cdot \cdots \cdot n_m$ ways to carry out the procedure. This version of the product rule can be proved by mathematical induction from the product rule for two tasks (see Exercise 78).

EXAMPLE 4 How many different bit strings of length seven are there?

Solution: Each of the seven bits can be chosen in two ways, because each bit is either 0 or 1. Therefore, the product rule shows there are a total of $2^7 = 128$ different bit strings of length seven. ◀

EXAMPLE 5 How many different license plates can be made if each plate contains a sequence of three uppercase English letters followed by three digits (and no sequences of letters are prohibited, even if they are obscene)?

$\underbrace{\_\ \_\ \_}_{\text{26 choices for each letter}}\ \underbrace{\_\ \_\ \_}_{\text{10 choices for each digit}}$

Solution: There are 26 choices for each of the three uppercase English letters and 10 choices for each of the three digits. Hence, by the product rule there are a total of $26 \cdot 26 \cdot 26 \cdot 10 \cdot 10 \cdot 10 = 17{,}576{,}000$ possible license plates. ◀

EXAMPLE 6 **Counting Functions** How many functions are there from a set with m elements to a set with n elements?

Solution: A function corresponds to a choice of one of the n elements in the codomain for each of the m elements in the domain. Hence, by the product rule there are $n \cdot n \cdot \cdots \cdot n = n^m$ functions from a set with m elements to one with n elements. For example, there are $5^3 = 125$ different functions from a set with three elements to a set with five elements. ◀

EXAMPLE 7 **Counting One-to-One Functions** How many one-to-one functions are there from a set with m elements to one with n elements?

Solution: First note that when $m > n$ there are no one-to-one functions from a set with m elements to a set with n elements.

Counting the number of onto functions is harder. We'll do this in Chapter 8.

Now let $m \leq n$. Suppose the elements in the domain are $a_1, a_2, \ldots, a_m$. There are n ways to choose the value of the function at a_1. Because the function is one-to-one, the value of the function at a_2 can be picked in $n - 1$ ways (because the value used for a_1 cannot be used again). In general, the value of the function at a_k can be chosen in $n - k + 1$ ways. By the product rule, there are $n(n - 1)(n - 2) \cdots (n - m + 1)$ one-to-one functions from a set with m elements to one with n elements.

For example, there are $5 \cdot 4 \cdot 3 = 60$ one-to-one functions from a set with three elements to a set with five elements. ◀

EXAMPLE 8 **The Telephone Numbering Plan** The *North American numbering plan (NANP)* specifies the format of telephone numbers in the U.S., Canada, and many other parts of North America. A telephone number in this plan consists of 10 digits, which are split into a three-digit area code, a three-digit office code, and a four-digit station code. Because of signaling considerations, there are certain restrictions on some of these digits. To specify the allowable format, let X denote a digit that can take any of the values 0 through 9, let N denote a digit that can take any of the values 2 through 9, and let Y denote a digit that must be a 0 or a 1. Two numbering plans, which will be called the old plan, and the new plan, will be discussed. (The old plan, in use in the 1960s, has been replaced by the new plan, but the recent rapid growth in demand for new numbers for mobile phones and devices will eventually make even this new plan obsolete. In this example, the letters used to represent digits follow the conventions of the *North American Numbering Plan.*) As will be shown, the new plan allows the use of more numbers.

Links ❯

Current projections are that by 2051, it will be necessary to add one or more digits to North American telephone numbers.

In the old plan, the formats of the area code, office code, and station code are *NYX*, *NNX*, and *XXXX*, respectively, so that telephone numbers had the form *NYX-NNX-XXXX*. In the new plan, the formats of these codes are *NXX*, *NXX*, and *XXXX*, respectively, so that telephone numbers have the form *NXX-NXX-XXXX*. How many different North American telephone numbers are possible under the old plan and under the new plan?

Solution: By the product rule, there are $8 \cdot 2 \cdot 10 = 160$ area codes with format *NYX* and $8 \cdot 10 \cdot 10 = 800$ area codes with format *NXX*. Similarly, by the product rule, there are

$8 \cdot 8 \cdot 10 = 640$ office codes with format *NNX*. The product rule also shows that there are $10 \cdot 10 \cdot 10 \cdot 10 = 10{,}000$ station codes with format *XXXX*.

Note that we have ignored restrictions that rule out N11 station codes for most area codes.

Consequently, applying the product rule again, it follows that under the old plan there are

$$160 \cdot 640 \cdot 10{,}000 = 1{,}024{,}000{,}000$$

different numbers available in North America. Under the new plan, there are

$$800 \cdot 800 \cdot 10{,}000 = 6{,}400{,}000{,}000$$

different numbers available. ◀

EXAMPLE 9 What is the value of k after the following code, where $n_1, n_2, \dots, n_m$ are positive integers, has been executed?

```
k := 0
for i_1 := 1 to n_1
  for i_2 := 1 to n_2
      .
      .
      .
    for i_m := 1 to n_m
        k := k + 1
```

Solution: The initial value of k is zero. Each time the nested loop is traversed, 1 is added to k. Let T_i be the task of traversing the ith loop. Then the number of times the loop is traversed is the number of ways to do the tasks $T_1, T_2, \dots, T_m$. The number of ways to carry out the task $T_j, j = 1, 2, \dots, m$, is n_j, because the jth loop is traversed once for each integer i_j with $1 \le i_j \le n_j$. By the product rule, it follows that the nested loop is traversed $n_1 n_2 \cdots n_m$ times. Hence, the final value of k is $n_1 n_2 \cdots n_m$. ◀

EXAMPLE 10 **Counting Subsets of a Finite Set** Use the product rule to show that the number of different subsets of a finite set S is $2^{|S|}$.

Solution: Let S be a finite set. List the elements of S in arbitrary order. Recall from Section 2.2 that there is a one-to-one correspondence between subsets of S and bit strings of length $|S|$. Namely, a subset of S is associated with the bit string with a 1 in the ith position if the ith element in the list is in the subset, and a 0 in this position otherwise. By the product rule, there are $2^{|S|}$ bit strings of length $|S|$. Hence, $|\mathcal{P}(S)| = 2^{|S|}$. (Recall that we used mathematical induction to prove this fact in Example 10 of Section 5.1.) ◀

The product rule is often phrased in terms of sets in this way: If $A_1, A_2, \dots, A_m$ are finite sets, then the number of elements in the Cartesian product of these sets is the product of the number of elements in each set. To relate this to the product rule, note that the task of choosing an element in the Cartesian product $A_1 \times A_2 \times \cdots \times A_m$ is done by choosing an element in A_1, an element in $A_2, \dots,$ and an element in A_m. By the product rule it follows that

$$|A_1 \times A_2 \times \cdots \times A_m| = |A_1| \cdot |A_2| \cdot \cdots \cdot |A_m|.$$

EXAMPLE 11 **DNA and Genomes** The hereditary information of a living organism is encoded using deoxyribonucleic acid (DNA), or in certain viruses, ribonucleic acid (RNA). DNA and RNA are extremely complex molecules, with different molecules interacting in a vast variety of ways to enable living process. For our purposes, we give only the briefest description of how DNA and RNA encode genetic information.

DNA molecules consist of two strands consisting of blocks known as nucleotides. Each nucleotide contains subcomponents called **bases**, each of which is adenine (A), cytosine (C), guanine (G), or thymine (T). The two strands of DNA are held together by hydrogen bonds connecting different bases, with A bonding only with T, and C bonding only with G. Unlike DNA, RNA is single stranded, with uracil (U) replacing thymine as a base. So, in DNA the possible base pairs are A-T and C-G, while in RNA they are A-U, and C-G. The DNA of a living creature consists of multiple pieces of DNA forming separate chromosomes. A **gene** is a segment of a DNA molecule that encodes a particular protein. The entirety of genetic information of an organism is called its **genome**.

Sequences of bases in DNA and RNA encode long chains of proteins called amino acids. There are 22 essential amino acids for human beings. We can quickly see that a sequence of at least three bases are needed to encode these 22 different amino acid. First note, that because there are four possibilities for each base in DNA, A, C, G, and T, by the product rule there are $4^2 = 16 < 22$ different sequences of two bases. However, there are $4^3 = 64$ different sequences of three bases, which provide enough different sequences to encode the 22 different amino acids (even after taking into account that several different sequences of three bases encode the same amino acid).

The DNA of simple living creatures such as algae and bacteria have between 10^5 and 10^7 links, where each link is one of the four possible bases. More complex organisms, such as insects, birds, and mammals, have between 10^8 and 10^{10} links in their DNA. So, by the product rule, there are at least 4^{10^5} different sequences of bases in the DNA of simple organisms and at least 4^{10^8} different sequences of bases in the DNA of more complex organisms. These are both incredibly huge numbers, which helps explain why there is such tremendous variability among living organisms. In the past several decades techniques have been developed for determining the genome of different organisms. The first step is to locate each gene in the DNA of an organism. The next task, called **gene sequencing**, is the determination of the sequence of links on each gene. (The specific sequence of links on these genes depends on the particular individual representative of a species whose DNA is analyzed.) For example, the human genome includes approximately 23,000 genes, each with 1000 or more links. Gene sequencing techniques take advantage of many recently developed algorithms and are based on numerous new ideas in combinatorics. Many mathematicians and computer scientists work on problems involving genomes, taking part in the fast moving fields of bioinformatics and computational biology. ◀

It is now possible to determine your own genetic code at a reasonable price.

We now introduce the sum rule.

THE SUM RULE If a task can be done either in one of n_1 ways or in one of n_2 ways, where none of the set of n_1 ways is the same as any of the set of n_2 ways, then there are $n_1 + n_2$ ways to do the task.

Example 12 illustrates how the sum rule is used.

EXAMPLE 12 Suppose that either a member of the mathematics faculty or a student who is a mathematics major is chosen as a representative to a university committee. How many different choices are there for this representative if there are 37 members of the mathematics faculty and 83 mathematics majors and no one is both a faculty member and a student?

Solution: There are 37 ways to choose a member of the mathematics faculty and there are 83 ways to choose a student who is a mathematics major. Choosing a member of the mathematics faculty is never the same as choosing a student who is a mathematics major because no one is both a faculty member and a student. By the sum rule it follows that there are $37 + 83 = 120$ possible ways to pick this representative. ◀

We can extend the sum rule to more than two tasks. Suppose that a task can be done in one of n_1 ways, in one of n_2 ways, … , or in one of n_m ways, where none of the set of n_i ways of doing the task is the same as any of the set of n_j ways, for all pairs i and j with $1 \leq i < j \leq m$. Then the number of ways to do the task is $n_1 + n_2 + \cdots + n_m$. This extended version of the sum rule is often useful in counting problems, as Examples 13 and 14 show. This version of the sum rule can be proved using mathematical induction from the sum rule for two sets. (See Exercise 77.)

EXAMPLE 13 A student can choose a computer project from one of three lists. The three lists contain 23, 15, and 19 possible projects, respectively. No project is on more than one list. How many possible projects are there to choose from?

Solution: The student can choose a project by selecting a project from the first list, the second list, or the third list. Because no project is on more than one list, by the sum rule there are $23 + 15 + 19 = 57$ ways to choose a project. ◀

EXAMPLE 14 What is the value of k after the following code, where $n_1, n_2, \ldots, n_m$ are positive integers, has been executed?

```
k := 0
for i_1 := 1 to n_1
    k := k + 1
for i_2 := 1 to n_2
    k := k + 1
        .
        .
        .
for i_m := 1 to n_m
    k := k + 1
```

Solution: The initial value of k is zero. This block of code is made up of m different loops. Each time a loop is traversed, 1 is added to k. To determine the value of k after this code has been executed, we need to determine how many times we traverse a loop. Note that there are n_i ways to traverse the ith loop. Because we only traverse one loop at a time, the sum rule shows that the final value of k, which is the number of ways to traverse one of the m loops is $n_1 + n_2 + \cdots + n_m$. ◀

The sum rule can be phrased in terms of sets as: If $A_1, A_2, \ldots, A_m$ are pairwise disjoint finite sets, then the number of elements in the union of these sets is the sum of the numbers of elements in the sets. To relate this to our statement of the sum rule, note there are $|A_i|$ ways to choose an element from A_i for $i = 1, 2, \ldots, m$. Because the sets are pairwise disjoint, when we select an element from one of the sets A_i, we do not also select an element from a different set A_j. Consequently, by the sum rule, because we cannot select an element from two of these sets at the same time, the number of ways to choose an element from one of the sets, which is the number of elements in the union, is

$$|A_1 \cup A_2 \cup \cdots \cup A_m| = |A_1| + |A_2| + \cdots + |A_m| \text{ when } A_i \cap A_j = \text{ for all } i, j.$$

This equality applies only when the sets in question are pairwise disjoint. The situation is much more complicated when these sets have elements in common. That situation will be briefly discussed later in this section and discussed in more depth in Chapter 8.

6.1.3 More Complex Counting Problems

Many counting problems cannot be solved using just the sum rule or just the product rule. However, many complicated counting problems can be solved using both of these rules in combination. We begin by counting the number of variable names in the programming language BASIC. (In the exercises, we consider the number of variable names in Java.) Then we will count the number of valid passwords subject to a particular set of restrictions.

EXAMPLE 15 In a version of the computer language BASIC, the name of a variable is a string of one or two alphanumeric characters, where uppercase and lowercase letters are not distinguished. (An *alphanumeric* character is either one of the 26 English letters or one of the 10 digits.) Moreover, a variable name must begin with a letter and must be different from the five strings of two characters that are reserved for programming use. How many different variable names are there in this version of BASIC?

Extra Examples

Solution: Let V equal the number of different variable names in this version of BASIC. Let V_1 be the number of these that are one character long and V_2 be the number of these that are two characters long. Then by the sum rule, $V = V_1 + V_2$. Note that $V_1 = 26$, because a one-character variable name must be a letter. Furthermore, by the product rule there are $26 \cdot 36$ strings of length two that begin with a letter and end with an alphanumeric character. However, five of these are excluded, so $V_2 = 26 \cdot 36 - 5 = 931$. Hence, there are $V = V_1 + V_2 = 26 + 931 = 957$ different names for variables in this version of BASIC. ◀

EXAMPLE 16 Each user on a computer system has a password, which is six to eight characters long, where each character is an uppercase letter or a digit. Each password must contain at least one digit. How many possible passwords are there?

Solution: Let P be the total number of possible passwords, and let P_6, P_7, and P_8 denote the number of possible passwords of length 6, 7, and 8, respectively. By the sum rule, $P = P_6 + P_7 + P_8$. We will now find P_6, P_7, and P_8. Finding P_6 directly is difficult. To find P_6 it is easier to find the number of strings of uppercase letters and digits that are six characters long, including those with no digits, and subtract from this the number of strings with no digits. By the product rule, the number of strings of six characters is 36^6, and the number of strings with no digits is 26^6. Hence,

$$P_6 = 36^6 - 26^6 = 2{,}176{,}782{,}336 - 308{,}915{,}776 = 1{,}867{,}866{,}560.$$

Similarly, we have

$$P_7 = 36^7 - 26^7 = 78{,}364{,}164{,}096 - 8{,}031{,}810{,}176 = 70{,}332{,}353{,}920$$

and

$$\begin{aligned} P_8 &= 36^8 - 26^8 = 2{,}821{,}109{,}907{,}456 - 208{,}827{,}064{,}576 \\ &= 2{,}612{,}282{,}842{,}880. \end{aligned}$$

Consequently,

$$P = P_6 + P_7 + P_8 = 2{,}684{,}483{,}063{,}360.$$

◀

Bit Number	0	1	2	3	4	8	16	24	31
Class A	0	netid				hostid			
Class B	1	0	netid				hostid		
Class C	1	1	0	netid				hostid	
Class D	1	1	1	0	Multicast Address				
Class E	1	1	1	1	0	Address			

FIGURE 1 **Internet addresses (IPv4).**

Remark: The password scheme in Example 16 provides almost no protection against a hacker using exhaustive search to break into an account. Because of the speed and power of modern computers, a hacker would have little trouble trying every possible password to break into an account where the password belonged to that space of possible passwords. To be secure, a password space should contain an extremely large number of passwords. To guarantee this, the rules for passwords should require that all passwords contain a large number of characters, including digits, lowercase letters, uppercase letters, and special characters. (See Exercise 57.) (It is common today to find requirements that passwords should have at least 12 characters.) Furthermore, rules should be enforced that prohibit users from choosing passwords that make it easier for hackers by forbidding passwords that contain words, names, addresses, or other easy to guess strings. (For an extensive list of rules see [Bi19].)

EXAMPLE 17

Links

Counting Internet Addresses In the Internet, which is made up of interconnected physical networks of computers, each computer (or more precisely, each network connection of a computer) is assigned an *Internet address*. In Version 4 of the Internet Protocol (IPv4), still in use today, an address is a string of 32 bits. It begins with a *network number* (*netid*). The netid is followed by a *host number* (*hostid*), which identifies a computer as a member of a particular network.

Three forms of addresses are used, with different numbers of bits used for netids and hostids. **Class A addresses**, used for the largest networks, consist of 0, followed by a 7-bit netid and a 24-bit hostid. **Class B addresses**, used for medium-sized networks, consist of 10, followed by a 14-bit netid and a 16-bit hostid. **Class C addresses**, used for the smallest networks, consist of 110, followed by a 21-bit netid and an 8-bit hostid. There are several restrictions on addresses because of special uses: 1111111 is not available as the netid of a Class A network, and the hostids consisting of all 0s and all 1s are not available for use in any network. A computer on the Internet has either a Class A, a Class B, or a Class C address. (Besides Class A, B, and C addresses, there are also Class D addresses, reserved for use in multicasting when multiple computers are addressed at a single time, consisting of 1110 followed by 28 bits, and Class E addresses, reserved for future use, consisting of 11110 followed by 27 bits. Neither Class D nor Class E addresses are assigned as the IPv4 address of a computer on the Internet.) Figure 1 illustrates IPv4 addressing. (Limitations on the number of Class A and Class B netids have made IPv4 addressing inadequate; IPv6, a new version of IP, uses 128-bit addresses to solve this problem.)

The lack of available IPv4 address has become a crisis!

How many different IPv4 addresses are available for computers on the Internet?

Solution: Let x be the number of available addresses for computers on the Internet, and let x_A, x_B, and x_C denote the number of Class A, Class B, and Class C addresses available, respectively. By the sum rule, $x = x_A + x_B + x_C$.

To find x_A, note that there are $2^7 - 1 = 127$ Class A netids, recalling that the netid 1111111 is unavailable. For each netid, there are $2^{24} - 2 = 16{,}777{,}214$ hostids, recalling that the hostids consisting of all 0s and all 1s are unavailable. Consequently, $x_A = 127 \cdot 16{,}777{,}214 = 2{,}130{,}706{,}178$.

To find x_B and x_C, note that there are $2^{14} = 16{,}384$ Class B netids and $2^{21} = 2{,}097{,}152$ Class C netids. For each Class B netid, there are $2^{16} - 2 = 65{,}534$ hostids, and for each

Class C netid, there are $2^8 - 2 = 254$ hostids, recalling that in each network the hostids consisting of all 0s and all 1s are unavailable. Consequently, $x_B = 1{,}073{,}709{,}056$ and $x_C = 532{,}676{,}608$.

We conclude that the total number of IPv4 addresses available is $x = x_A + x_B + x_C = 2{,}130{,}706{,}178 + 1{,}073{,}709{,}056 + 532{,}676{,}608 = 3{,}737{,}091{,}842$. ◀

6.1.4 The Subtraction Rule (Inclusion–Exclusion for Two Sets)

Suppose that a task can be done in one of two ways, but some of the ways to do it are common to both ways. In this situation, we cannot use the sum rule to count the number of ways to do the task. If we add the number of ways to do the tasks in these two ways, we get an overcount of the total number of ways to do it, because the ways to do the task that are common to the two ways are counted twice.

Overcounting is perhaps the most common enumeration error.

To correctly count the number of ways to do the two tasks, we must subtract the number of ways that are counted twice. This leads us to an important counting rule.

THE SUBTRACTION RULE If a task can be done in either n_1 ways or n_2 ways, then the number of ways to do the task is $n_1 + n_2$ minus the number of ways to do the task that are common to the two different ways.

The subtraction rule is also known as the **principle of inclusion–exclusion**, especially when it is used to count the number of elements in the union of two sets. Suppose that A_1 and A_2 are sets. Then, there are $|A_1|$ ways to select an element from A_1 and $|A_2|$ ways to select an element from A_2. The number of ways to select an element from A_1 or from A_2, that is, the number of ways to select an element from their union, is the sum of the number of ways to select an element from A_1 and the number of ways to select an element from A_2, minus the number of ways to select an element that is in both A_1 and A_2. Because there are $|A_1 \cup A_2|$ ways to select an element in either A_1 or in A_2, and $|A_1 \cap A_2|$ ways to select an element common to both sets, we have

$$|A_1 \cup A_2| = |A_1| + |A_2| - |A_1 \cap A_2|.$$

This is the formula given in Section 2.2 for the number of elements in the union of two sets.

Example 18 illustrates how we can solve counting problems using the subtraction principle.

EXAMPLE 18 How many bit strings of length eight either start with a 1 bit or end with the two bits 00?

Extra Examples

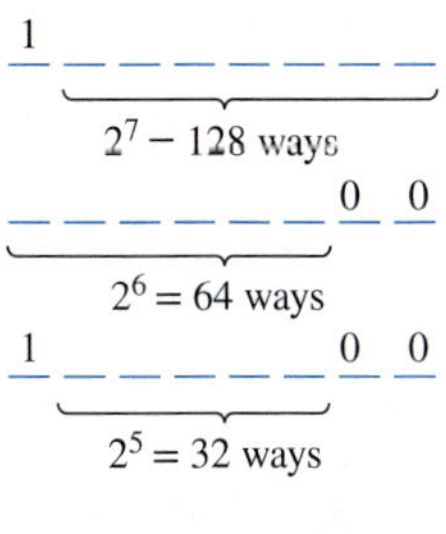

FIGURE 2 8-Bit strings starting with 1 or ending with 00.

Solution: Figure 2 illustrates the three counting problems we need to solve before we can apply the principle of inclusion–exclusion. We can construct a bit string of length eight that either starts with a 1 bit or ends with the two bits 00, by constructing a bit string of length eight beginning with a 1 bit or by constructing a bit string of length eight that ends with the two bits 00. We can construct a bit string of length eight that begins with a 1 in $2^7 = 128$ ways. This follows by the product rule, because the first bit can be chosen in only one way and each of the other seven bits can be chosen in two ways. Similarly, we can construct a bit string of length eight ending with the two bits 00, in $2^6 = 64$ ways. This follows by the product rule, because each of the first six bits can be chosen in two ways and the last two bits can be chosen in only one way.

Some of the ways to construct a bit string of length eight starting with a 1 are the same as the ways to construct a bit string of length eight that ends with the two bits 00. There are $2^5 = 32$ ways to construct such a string. This follows by the product rule, because the first bit can be chosen in only one way, each of the second through the sixth bits can be chosen in two ways, and the last two bits can be chosen in one way. Consequently, the number of bit strings of length eight that begin with a 1 or end with a 00, which equals the number of ways to construct a bit string of length eight that begins with a 1 or that ends with 00, equals $128 + 64 - 32 = 160$. ◀

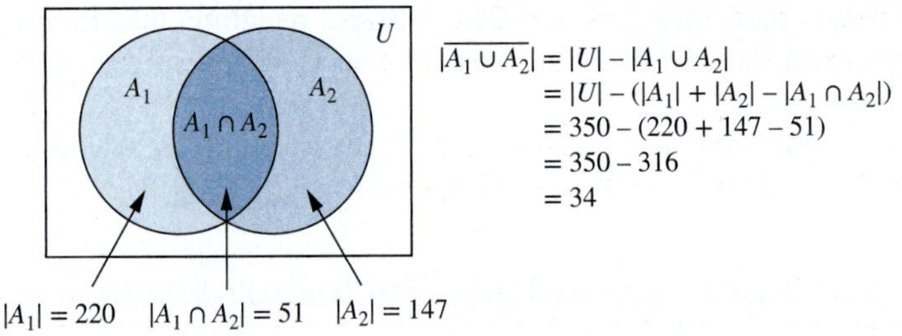

FIGURE 3 Applicants who majored in neither computer science nor business.

We present an example that illustrates how the formulation of the principle of inclusion–exclusion can be used to solve counting problems.

EXAMPLE 19 A computer company receives 350 applications from college graduates for a job planning a line of new web servers. Suppose that 220 of these applicants majored in computer science, 147 majored in business, and 51 majored both in computer science and in business. How many of these applicants majored neither in computer science nor in business?

Solution: To find the number of these applicants who majored neither in computer science nor in business, we can subtract the number of students who majored either in computer science or in business (or both) from the total number of applicants. Let A_1 be the set of students who majored in computer science and A_2 the set of students who majored in business. Then $A_1 \cup A_2$ is the set of students who majored in computer science or business (or both), and $A_1 \cap A_2$ is the set of students who majored both in computer science and in business. By the subtraction rule the number of students who majored either in computer science or in business (or both) equals

$$|A_1 \cup A_2| = |A_1| + |A_2| - |A_1 \cap A_2| = 220 + 147 - 51 = 316.$$

We conclude that $350 - 316 = 34$ of the applicants majored neither in computer science nor in business. A Venn diagram for this example is shown in Figure 3. ◀

The subtraction rule, or the principle of inclusion–exclusion, can be generalized to find the number of ways to do one of n different tasks or, equivalently, to find the number of elements in the union of n sets, whenever n is a positive integer. We will study the inclusion–exclusion principle and some of its many applications in Chapter 8.

6.1.5 The Division Rule

We have introduced the product, sum, and subtraction rules for counting. You may wonder whether there is also a division rule for counting. In fact, there is such a rule, which can be useful when solving certain types of enumeration problems.

THE DIVISION RULE There are n/d ways to do a task if it can be done using a procedure that can be carried out in n ways, and for every way w, exactly d of the n ways correspond to way w.

We can restate the division rule in terms of sets: "If the finite set A is the union of n pairwise disjoint subsets each with d elements, then $n = |A|/d$."

We can also formulate the division rule in terms of functions: "If f is a function from A to B where A and B are finite sets, and that for every value $y \in B$ there are exactly d values $x \in A$ such that $f(x) = y$ (in which case, we say that f is d-to-one), then $|B| = |A|/d$."

Remark: The division rule comes in handy when it appears that a task can be done in n different ways, but it turns out that for each way of doing the task, there are d equivalent ways of doing it. Under these circumstances, we can conclude that there are n/d inequivalent ways of doing the task.

We illustrate the use of the division rule for counting with two examples.

EXAMPLE 20 Suppose that an automated system has been developed that counts the legs of cows in a pasture. Suppose that this system has determined that in a farmer's pasture there are exactly 572 legs. How many cows are there is this pasture, assuming that each cow has four legs and that there are no other animals present?

Solution: Let n be the number of cow legs counted in a pasture. Because each cow has four legs, by the division rule we know that the pasture contains $n/4$ cows. Consequently, the pasture with 572 cow legs has $572/4 = 143$ cows in it. ◀

EXAMPLE 21 How many different ways are there to seat four people around a circular table, where two seatings are considered the same when each person has the same left neighbor and the same right neighbor?

Solution: We arbitrarily select a seat at the table and label it seat 1. We number the rest of the seats in numerical order, proceeding clockwise around the table. Note that are four ways to select the person for seat 1, three ways to select the person for seat 2, two ways to select the person for seat 3, and one way to select the person for seat 4. Thus, there are $4! = 24$ ways to order the given four people for these seats. However, each of the four choices for seat 1 leads to the same arrangement, as we distinguish two arrangements only when one of the people has a different immediate left or immediate right neighbor. Because there are four ways to choose the person for seat 1, by the division rule there are $24/4 = 6$ different seating arrangements of four people around the circular table. ◀

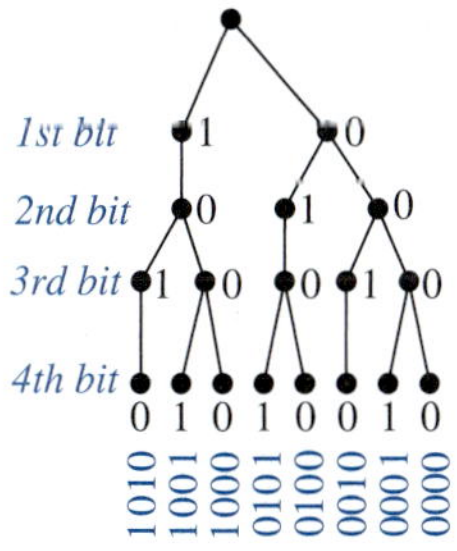

FIGURE 4 Bit strings of length four without consecutive 1s.

6.1.6 Tree Diagrams

Counting problems can be solved using **tree diagrams**. A tree consists of a root, a number of branches leaving the root, and possible additional branches leaving the endpoints of other branches. (We will study trees in detail in Chapter 11.) To use trees in counting, we use a branch to represent each possible choice. We represent the possible outcomes by the leaves, which are the endpoints of branches not having other branches starting at them.

Note that when a tree diagram is used to solve a counting problem, the number of choices of which branch to follow to reach a leaf can vary as in Example 22.

EXAMPLE 22 How many bit strings of length four do not have two consecutive 1s?

Solution: The tree diagram in Figure 4 displays all bit strings of length four without two consecutive 1s. We see that there are eight bit strings of length four without two consecutive 1s. ◀

EXAMPLE 23 A playoff between two teams consists of at most five games. The first team that wins three games wins the playoff. In how many different ways can the playoff occur?

Solution: The tree diagram in Figure 5 displays all the ways the playoff can proceed, with the winner of each game shown. We see that there are 20 different ways for the playoff to occur. ◀

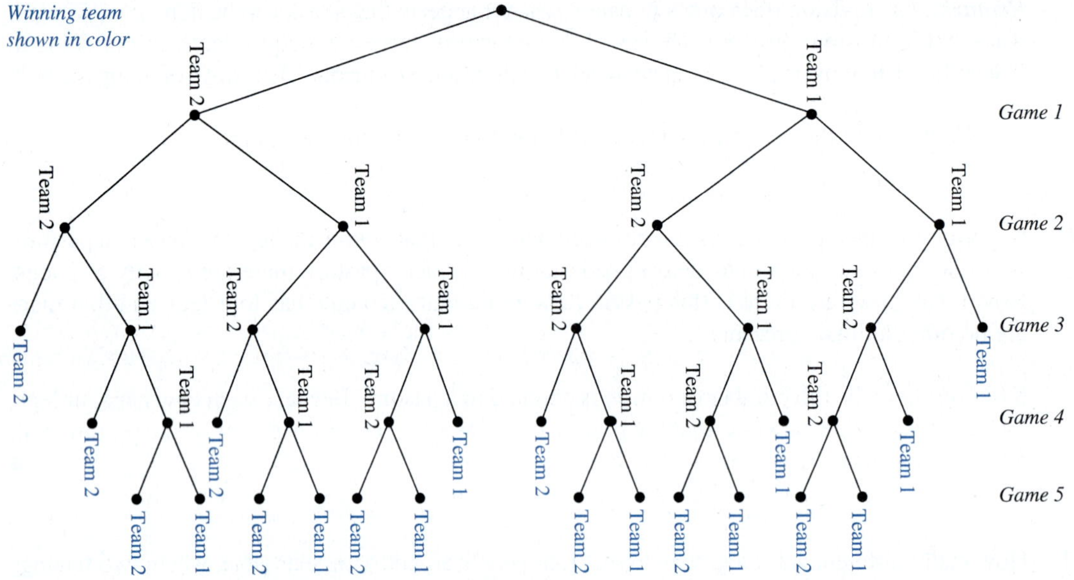

FIGURE 5 Best three games out of five playoffs.

EXAMPLE 24 Suppose that "I Love New Jersey" T-shirts come in five different sizes: S, M, L, XL, and XXL. Further suppose that each size comes in four colors, white, red, green, and black, except for XL, which comes only in red, green, and black, and XXL, which comes only in green and black. How many different shirts does a souvenir shop have to stock to have at least one of each available size and color of the T-shirt?

Solution: The tree diagram in Figure 6 displays all possible size and color pairs. It follows that the souvenir shop owner needs to stock 17 different T-shirts. ◀

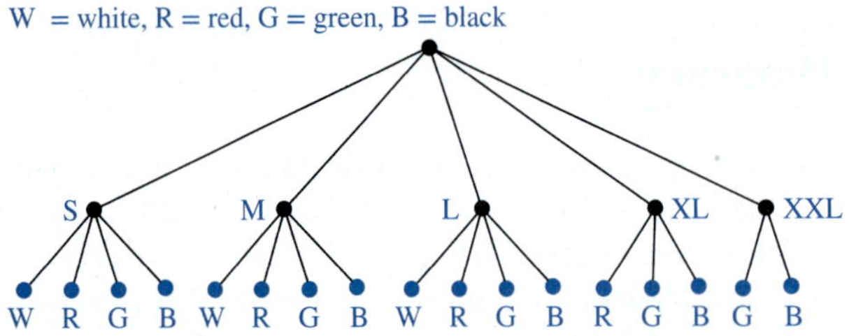

FIGURE 6 Counting varieties of T-shirts.

Exercises

1. There are 18 mathematics majors and 325 computer science majors at a college.
   a) In how many ways can two representatives be picked so that one is a mathematics major and the other is a computer science major?
   b) In how many ways can one representative be picked who is either a mathematics major or a computer science major?
2. An office building contains 27 floors and has 37 offices on each floor. How many offices are in the building?
3. A multiple-choice test contains 10 questions. There are four possible answers for each question.
   a) In how many ways can a student answer the questions on the test if the student answers every question?
   b) In how many ways can a student answer the questions on the test if the student can leave answers blank?
4. A particular brand of shirt comes in 12 colors, has an adult version and a child version, and comes in three sizes for each age group. How many different types of this shirt are made?

5. Six different airlines fly from New York to Denver and seven fly from Denver to San Francisco. How many different pairs of airlines can you choose on which to book a trip from New York to San Francisco via Denver, when you pick an airline for the flight to Denver and an airline for the continuation flight to San Francisco?

6. There are four major auto routes from Boston to Detroit and six from Detroit to Los Angeles. How many major auto routes are there from Boston to Los Angeles via Detroit?

7. How many different three-letter initials can people have?

8. How many different three-letter initials with none of the letters repeated can people have?

9. How many different three-letter initials are there that begin with an A?

10. How many bit strings are there of length eight?

11. How many bit strings of length ten both begin and end with a 1?

12. How many bit strings are there of length six or less, not counting the empty string?

13. How many bit strings with length not exceeding n, where n is a positive integer, consist entirely of 1s, not counting the empty string?

14. How many bit strings of length n, where n is a positive integer, start and end with 1s?

15. How many strings are there of lowercase letters of length four or less, not counting the empty string?

16. How many strings are there of four lowercase letters that have the letter x in them?

17. How many strings of five ASCII characters contain the character @ ("at" sign) at least once? [*Note:* There are 128 different ASCII characters.]

18. How many 5-element DNA sequences

a) end with A?
b) start with T and end with G?
c) contain only A and T?
d) do not contain C?

19. How many 6-element RNA sequences

a) do not contain U?
b) end with GU?
c) start with C?
d) contain only A or U?

20. How many positive integers between 5 and 31

a) are divisible by 3? Which integers are these?
b) are divisible by 4? Which integers are these?
c) are divisible by 3 and by 4? Which integers are these?

21. How many positive integers between 50 and 100

a) are divisible by 7? Which integers are these?
b) are divisible by 11? Which integers are these?
c) are divisible by both 7 and 11? Which integers are these?

22. How many positive integers less than 1000

a) are divisible by 7?
b) are divisible by 7 but not by 11?
c) are divisible by both 7 and 11?
d) are divisible by either 7 or 11?
e) are divisible by exactly one of 7 and 11?
f) are divisible by neither 7 nor 11?
g) have distinct digits?
h) have distinct digits and are even?

23. How many positive integers between 100 and 999 inclusive

a) are divisible by 7?
b) are odd?
c) have the same three decimal digits?
d) are not divisible by 4?
e) are divisible by 3 or 4?
f) are not divisible by either 3 or 4?
g) are divisible by 3 but not by 4?
h) are divisible by 3 and 4?

24. How many positive integers between 1000 and 9999 inclusive

a) are divisible by 9?
b) are even?
c) have distinct digits?
d) are not divisible by 3?
e) are divisible by 5 or 7?
f) are not divisible by either 5 or 7?
g) are divisible by 5 but not by 7?
h) are divisible by 5 and 7?

25. How many strings of three decimal digits

a) do not contain the same digit three times?
b) begin with an odd digit?
c) have exactly two digits that are 4s?

26. How many strings of four decimal digits

a) do not contain the same digit twice?
b) end with an even digit?
c) have exactly three digits that are 9s?

27. A committee is formed consisting of one representative from each of the 50 states in the United States, where the representative from a state is either the governor or one of the two senators from that state. How many ways are there to form this committee?

28. How many license plates can be made using either three digits followed by three uppercase English letters or three uppercase English letters followed by three digits?

29. How many license plates can be made using either two uppercase English letters followed by four digits or two digits followed by four uppercase English letters?

30. How many license plates can be made using either three uppercase English letters followed by three digits or four uppercase English letters followed by two digits?

31. How many license plates can be made using either two or three uppercase English letters followed by either two or three digits?

32. How many strings of eight uppercase English letters are there

a) if letters can be repeated?
b) if no letter can be repeated?
c) that start with X, if letters can be repeated?
d) that start with X, if no letter can be repeated?
e) that start and end with X, if letters can be repeated?
f) that start with the letters BO (in that order), if letters can be repeated?

g) that start and end with the letters BO (in that order), if letters can be repeated?
h) that start or end with the letters BO (in that order), if letters can be repeated?

33. How many strings of eight English letters are there
a) that contain no vowels, if letters can be repeated?
b) that contain no vowels, if letters cannot be repeated?
c) that start with a vowel, if letters can be repeated?
d) that start with a vowel, if letters cannot be repeated?
e) that contain at least one vowel, if letters can be repeated?
f) that contain exactly one vowel, if letters can be repeated?
g) that start with X and contain at least one vowel, if letters can be repeated?
h) that start and end with X and contain at least one vowel, if letters can be repeated?

34. How many different functions are there from a set with 10 elements to sets with the following numbers of elements?
a) 2 b) 3 c) 4 d) 5

35. How many one-to-one functions are there from a set with five elements to sets with the following number of elements?
a) 4 b) 5 c) 6 d) 7

36. How many functions are there from the set $\{1, 2, \ldots, n\}$, where n is a positive integer, to the set $\{0, 1\}$?

37. How many functions are there from the set $\{1, 2, \ldots, n\}$, where n is a positive integer, to the set $\{0, 1\}$
a) that are one-to-one?
b) that assign 0 to both 1 and n?
c) that assign 1 to exactly one of the positive integers less than n?

38. How many partial functions (see Section 2.3) are there from a set with five elements to sets with each of these number of elements?
a) 1 b) 2 c) 5 d) 9

39. How many partial functions (see Definition 13 of Section 2.3) are there from a set with m elements to a set with n elements, where m and n are positive integers?

40. How many subsets of a set with 100 elements have more than one element?

41. A **palindrome** is a string whose reversal is identical to the string. How many bit strings of length n are palindromes?

42. How many 4-element DNA sequences
a) do not contain the base T?
b) contain the sequence ACG?
c) contain all four bases A, T, C, and G?
d) contain exactly three of the four bases A, T, C, and G?

43. How many 4-element RNA sequences
a) contain the base U?
b) do not contain the sequence CUG?
c) do not contain all four bases A, U, C, and G?
d) contain exactly two of the four bases A, U, C, and G?

44. On each of the 22 work days in a particular month, every employee of a start-up venture was sent a company communication. If a total of 4642 total company communications were sent, how many employees does the company have, assuming that no staffing changes were made that month?

45. At a large university, 434 freshmen, 883 sophomores, and 43 juniors are enrolled in an introductory algorithms course. How many sections of this course need to be scheduled to accommodate all these students if each section contains 34 students?

46. How many ways are there to seat four of a group of ten people around a circular table where two seatings are considered the same when everyone has the same immediate left and immediate right neighbor?

47. How many ways are there to seat six people around a circular table where two seatings are considered the same when everyone has the same two neighbors without regard to whether they are right or left neighbors?

48. In how many ways can a photographer at a wedding arrange 6 people in a row from a group of 10 people, where the bride and the groom are among these 10 people, if
a) the bride must be in the picture?
b) both the bride and groom must be in the picture?
c) exactly one of the bride and the groom is in the picture?

49. In how many ways can a photographer at a wedding arrange six people in a row, including the bride and groom, if
a) the bride must be next to the groom?
b) the bride is not next to the groom?
c) the bride is positioned somewhere to the left of the groom?

50. How many bit strings of length seven either begin with two 0s or end with three 1s?

51. How many bit strings of length 10 either begin with three 0s or end with two 0s?

***52.** How many bit strings of length 10 contain either five consecutive 0s or five consecutive 1s?

****53.** How many bit strings of length eight contain either three consecutive 0s or four consecutive 1s?

54. Every student in a discrete mathematics class is either a computer science or a mathematics major or is a joint major in these two subjects. How many students are in the class if there are 38 computer science majors (including joint majors), 23 mathematics majors (including joint majors), and 7 joint majors?

55. How many positive integers not exceeding 100 are divisible either by 4 or by 6?

56. How many different initials can someone have if a person has at least two, but no more than five, different initials? Assume that each initial is one of the 26 uppercase letters of the English language.

57. Suppose that a password for a computer system must have at least 8, but no more than 12, characters, where each character in the password is a lowercase English letter, an uppercase English letter, a digit, or one of the six special characters $*$, $>$, $<$, !, +, and =.

a) How many different passwords are available for this computer system?
b) How many of these passwords contain at least one occurrence of at least one of the six special characters?
c) Using your answer to part (a), determine how long it takes a hacker to try every possible password, assuming that it takes one nanosecond for a hacker to check each possible password.

58. The name of a variable in the C programming language is a string that can contain uppercase letters, lowercase letters, digits, or underscores. Further, the first character in the string must be a letter, either uppercase or lowercase, or an underscore. If the name of a variable is determined by its first eight characters, how many different variables can be named in C? (Note that the name of a variable may contain fewer than eight characters.)

59. The name of a variable in the Java programming language is a string of between 1 and 65,535 characters, inclusive, where each character can be an uppercase or a lowercase letter, a dollar sign, an underscore, or a digit, except the first character must not be a digit. Also, the name of a variable cannot be identical to any of the 53 keywords or literals used in Java. Determine the number of different variable names in Java.

60. The name of a variable in the Python programming language can only contain alphanumeric characters and underscores (that is, the letters A–Z where case matters, the digits 0–9, and the underscore _). These names must start with an uppercase or lowercase letter or an underscore, and they cannot be one of the 33 Python keywords. Variables must be no longer than 64 characters. Determine the number of different allowable Python variable names subject to these requirements.

61. The name of a variable in the Fortran programming language can only contain letters of the alphabet, digits, and underscores. Variable names cannot be longer than 31 characters and their first character must be a letter. Furthermore, the names of variable are case-insensitive, so lowercase and uppercase letters (of the same letter of the alphabet) are considered to the same. Determine the number of different allowable Fortran variable names when these rules are followed.

62. The International Telecommunications Union (ITU) specifies that a telephone number must consist of a country code with between 1 and 3 digits, except that the code 0 is not available for use as a country code, followed by a number with at most 15 digits. How many available possible telephone numbers are there that satisfy these restrictions?

63. Suppose that at some future time every telephone in the world is assigned a number that contains a country code 1 to 3 digits long, that is, of the form *X*, *XX*, or *XXX*, followed by a 10-digit telephone number of the form *NXX*-*NXX*-*XXXX* (as described in Example 8). How many different telephone numbers would be available worldwide under this numbering plan?

64. A key in the Vigenère cryptosystem is a string of English letters, where the case of the letters does not matter. How many different keys for this cryptosystem are there with three, four, five, or six letters?

65. a) The wired equivalent privacy (WEP) standard for Wi-Fi security was introduced in 1997 and discontinued in 2008. WEP keys are strings of either 10, 26, 32, or 58 hexadecimal characters. How many different WEP keys are there?
b) The WEP protocol was replaced by the Wi-Fi protected access standard (WPA) in 1999 when it became apparent that the WEP standard was inadequate. WPA keys are strings of 64 hexadecimal characters. How many different WPA keys are there? (Note that WPA has been superseded by WPA-2 and WPA-3.)

66. Suppose that p and q are prime numbers and that $n = pq$. Use the principle of inclusion–exclusion to find the number of positive integers not exceeding n that are relatively prime to n.

67. Use the principle of inclusion–exclusion to find the number of positive integers less than 1,000,000 that are not divisible by either 4 or by 6.

68. Use a tree diagram to find the number of bit strings of length four with no three consecutive 0s.

69. How many ways are there to arrange the letters a, b, c, and d such that a is not followed immediately by b?

70. Use a tree diagram to find the number of ways that the World Series can occur, where the first team that wins four games out of seven wins the series.

71. Use a tree diagram to determine the number of subsets of $\{3, 7, 9, 11, 24\}$ with the property that the sum of the elements in the subset is less than 28.

72. a) Suppose that a store sells six varieties of soft drinks: cola, ginger ale, orange, root beer, lemonade, and cream soda. Use a tree diagram to determine the number of different types of bottles the store must stock to have all varieties available in all size bottles if all varieties are available in 12-ounce bottles, all but lemonade are available in 20-ounce bottles, only cola and ginger ale are available in 32-ounce bottles, and all but lemonade and cream soda are available in 64-ounce bottles?
b) Answer the question in part (a) using counting rules.

73. a) Suppose that a popular style of running shoe is available for both adults and children. The children's shoe comes in sizes 6, 7, 8, and 9, and the adult's shoe comes in sizes 8, 9, 10, 11, and 12. The adult's shoe comes in white and black, while the children's shoe comes in white, red, and black. Use a tree diagram to determine the number of different shoes that a store has to stock to have at least one pair of this type of running shoe for all available sizes and colors for both adults and children.
b) Answer the question in part (a) using counting rules.

74. Determine the number of matches played in a single-elimination tournament with n players, where for each game between two players the winner goes on, but the loser is eliminated.

75. Determine the minimum and the maximum number of matches that can be played in a double-elimination tournament with n players, where after each game between two players, the winner goes on and the loser goes on if and only if this is not a second loss.

*76. Use the product rule to show that there are 2^{2^n} different truth tables for propositions in n variables.

77. Use mathematical induction to prove the sum rule for m tasks from the sum rule for two tasks.

78. Use mathematical induction to prove the product rule for m tasks from the product rule for two tasks.

79. How many diagonals does a convex polygon with n sides have? (Recall that a polygon is convex if every line segment connecting two points in the interior or boundary of the polygon lies entirely within this set and that a diagonal of a polygon is a line segment connecting two vertices that are not adjacent.)

80. Data are transmitted over the Internet in **datagrams**, which are structured blocks of bits. Each datagram contains header information organized into a maximum of 14 different fields (specifying many things, including the source and destination addresses) and a data area that contains the actual data that are transmitted. One of the 14 header fields is the **header length field** (denoted by HLEN), which is specified by the protocol to be 4 bits long and that specifies the header length in terms of 32-bit blocks of bits. For example, if HLEN = 0110, the header is made up of six 32-bit blocks. Another of the 14 header fields is the 16-bit-long **total length field** (denoted by TOTAL LENGTH), which specifies the length in bits of the entire datagram, including both the header fields and the data area. The length of the data area is the total length of the datagram minus the length of the header.

a) The largest possible value of TOTAL LENGTH (which is 16 bits long) determines the maximum total length in octets (blocks of 8 bits) of an Internet datagram. What is this value?

b) The largest possible value of HLEN (which is 4 bits long) determines the maximum total header length in 32-bit blocks. What is this value? What is the maximum total header length in octets?

c) The minimum (and most common) header length is 20 octets. What is the maximum total length in octets of the data area of an Internet datagram?

d) How many different strings of octets in the data area can be transmitted if the header length is 20 octets and the total length is as long as possible?

6.2 The Pigeonhole Principle

6.2.1 Introduction

Links

Suppose that a flock of 20 pigeons flies into a set of 19 pigeonholes to roost. Because there are 20 pigeons but only 19 pigeonholes, a least one of these 19 pigeonholes must have at least two pigeons in it. To see why this is true, note that if each pigeonhole had at most one pigeon in it, at most 19 pigeons, one per hole, could be accommodated. This illustrates a general principle called the **pigeonhole principle**, which states that if there are more pigeons than pigeonholes, then there must be at least one pigeonhole with at least two pigeons in it (see Figure 1). This principle is extremely useful; it applies to much more than pigeons and pigeonholes.

THEOREM 1 **THE PIGEONHOLE PRINCIPLE** If k is a positive integer and $k + 1$ or more objects are placed into k boxes, then there is at least one box containing two or more of the objects.

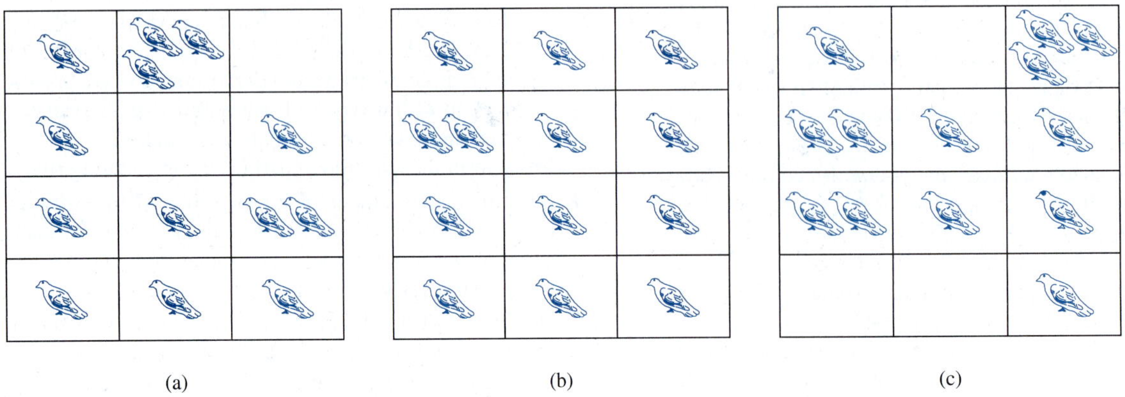

FIGURE 1 **There are more pigeons than pigeonholes.**

Proof: We prove the pigeonhole principle using a proof by contradiction. Suppose that there are $k+1$ objects and that none of the k boxes contains more than one object. Then the total number of objects would be at most k. This is a contradiction, because there are at least $k+1$ objects. ◁

The pigeonhole principle is also called the **Dirichlet drawer principle**, after the nineteenth-century German mathematician G. Lejeune Dirichlet, who often used this principle in his work. (Dirichlet was not the first person to use this principle; a demonstration that there were at least two Parisians with the same number of hairs on their heads dates back to the 17th century—see Exercise 35.) It is an important additional proof technique supplementing those we have developed in earlier chapters. We introduce it in this chapter because of its many important applications to combinatorics.

We will illustrate the usefulness of the pigeonhole principle. We first show that it can be used to prove a useful corollary about functions.

COROLLARY 1 A function f from a set with $k+1$ or more elements to a set with k elements is not one-to-one.

Proof: Suppose that for each element y in the codomain of f we have a box that contains all elements x of the domain of f such that $f(x) = y$. Because the domain contains $k+1$ or more elements and the codomain contains only k elements, the pigeonhole principle tells us that one of these boxes contains two or more elements x of the domain. This means that f cannot be one-to-one. ◁

Examples 1–3 show how the pigeonhole principle is used.

EXAMPLE 1 Among any group of 367 people, there must be at least two with the same birthday, because there are only 366 possible birthdays. ◀

EXAMPLE 2 In any group of 27 English words, there must be at least two that begin with the same letter, because there are 26 letters in the English alphabet. ◀

EXAMPLE 3 How many students must be in a class to guarantee that at least two students receive the same score on the final exam, if the exam is graded on a scale from 0 to 100 points?

Solution: There are 101 possible scores on the final. The pigeonhole principle shows that among any 102 students there must be at least 2 students with the same score. ◀

Links

INTERFOTO/Alamy Stock Photo

G. LEJEUNE DIRICHLET (1805–1859) G. Lejeune Dirichlet was born into a Belgian family living near Cologne, Germany. His father was a postmaster. He became passionate about mathematics at a young age. He was spending all his spare money on mathematics books by the time he entered secondary school in Bonn at the age of 12. At 14 he entered the Jesuit College in Cologne, and at 16 he began his studies at the University of Paris. In 1825 he returned to Germany and was appointed to a position at the University of Breslau. In 1828 he moved to the University of Berlin. In 1855 he was chosen to succeed Gauss at the University of Göttingen. Dirichlet is said to be the first person to master Gauss's *Disquisitiones Arithmeticae*, which appeared 20 years earlier. He is said to have kept a copy at his side even when he traveled. Dirichlet made many important discoveries in number theory, including the theorem that there are infinitely many primes in arithmetical progressions $an + b$ when a and b are relatively prime. He proved the $n = 5$ case of Fermat's last theorem, that there are no nontrivial solutions in integers to $x^5 + y^5 = z^5$. Dirichlet also made many contributions to analysis. Dirichlet was considered to be an excellent teacher who could explain ideas with great clarity. He was married to Rebecka Mendelssohn, one of the sisters of the composer Felix Mendelssohn.

The pigeonhole principle is a useful tool in many proofs, including proofs of surprising results, such as that given in Example 4.

EXAMPLE 4 Show that for every integer n there is a multiple of n that has only 0s and 1s in its decimal expansion.

Extra Examples

Solution: Let n be a positive integer. Consider the $n + 1$ integers 1, 11, 111, …, 11 … 1 (where the last integer in this list is the integer with $n + 1$ 1s in its decimal expansion). Note that there are n possible remainders when an integer is divided by n. Because there are $n + 1$ integers in this list, by the pigeonhole principle there must be two with the same remainder when divided by n. The larger of these integers less the smaller one is a multiple of n, which has a decimal expansion consisting entirely of 0s and 1s. ◀

6.2.2 The Generalized Pigeonhole Principle

The pigeonhole principle states that there must be at least two objects in the same box when there are more objects than boxes. However, even more can be said when the number of objects exceeds a multiple of the number of boxes. For instance, among any set of 21 decimal digits there must be 3 that are the same. This follows because when 21 objects are distributed into 10 boxes, one box must have more than 2 objects.

THEOREM 2 **THE GENERALIZED PIGEONHOLE PRINCIPLE** If N objects are placed into k boxes, then there is at least one box containing at least $\lceil N/k \rceil$ objects.

Proof: We will use a proof by contraposition. Suppose that none of the boxes contains more than $\lceil N/k \rceil - 1$ objects. Then, the total number of objects is at most

$$k\left(\left\lceil \frac{N}{k} \right\rceil - 1\right) < k\left(\left(\frac{N}{k} + 1\right) - 1\right) = N,$$

where the inequality $\lceil N/k \rceil < (N/k) + 1$ has been used. Thus, the total number of objects is less than N. This completes the proof by contraposition. ◁

A common type of problem asks for the minimum number of objects such that at least r of these objects must be in one of k boxes when these objects are distributed among the boxes. When we have N objects, the generalized pigeonhole principle tells us there must be at least r objects in one of the boxes as long as $\lceil N/k \rceil \geq r$. The smallest integer N with $N/k > r - 1$, namely, $N = k(r - 1) + 1$, is the smallest integer satisfying the inequality $\lceil N/k \rceil \geq r$. Could a smaller value of N suffice? The answer is no, because if we had $k(r - 1)$ objects, we could put $r - 1$ of them in each of the k boxes and no box would have at least r objects.

When thinking about problems of this type, it is useful to consider how you can avoid having at least r objects in one of the boxes as you add successive objects. To avoid adding a rth object to any box, you eventually end up with $r - 1$ objects in each box. There is no way to add the next object without putting an rth object in that box.

Examples 5–8 illustrate how the generalized pigeonhole principle is applied.

EXAMPLE 5 Among 100 people there are at least $\lceil 100/12 \rceil = 9$ who were born in the same month. ◀

EXAMPLE 6 What is the minimum number of students required in a discrete mathematics class to be sure that at least six will receive the same grade, if there are five possible grades, A, B, C, D, and F?

Solution: The minimum number of students needed to ensure that at least six students receive the same grade is the smallest integer N such that $\lceil N/5 \rceil = 6$. The smallest such integer is

Extra Examples

$N = 5 \cdot 5 + 1 = 26$. If youhave only 25 students, it is possible for there to be five who have received each grade so that no six students have received the same grade. Thus, 26 is the minimum number of students needed to ensure that at least six students will receive the same grade. ◀

EXAMPLE 7 **a)** How many cards must be selected from a standard deck of 52 cards to guarantee that at least three cards of the same suit are selected?
b) How many must be selected from a standard deck of 52 cards to guarantee that at least three hearts are selected?

A standard deck of 52 cards has 13 kinds of cards, with four cards of each of kind, one in each of the four suits, hearts, diamonds, spades, and clubs.

Solution: **a)** Suppose there are four boxes, one for each suit, and as cards are selected they are placed in the box reserved for cards of that suit. Using the generalized pigeonhole principle, we see that if N cards are selected, there is at least one box containing at least $\lceil N/4 \rceil$ cards. Consequently, we know that at least three cards of one suit are selected if $\lceil N/4 \rceil \geq 3$. The smallest integer N such that $\lceil N/4 \rceil \geq 3$ is $N = 2 \cdot 4 + 1 = 9$, so nine cards suffice. Note that if eight cards are selected, it is possible to have two cards of each suit, so more than eight cards are needed. Consequently, nine cards must be selected to guarantee that at least three cards of one suit are chosen. One good way to think about this is to note that after the eighth card is chosen, there is no way to avoid having a third card of some suit.

b) We do not use the generalized pigeonhole principle to answer this question, because we want to make sure that there are three hearts, not just three cards of one suit. Note that in the worst case, we can select all the clubs, diamonds, and spades, 39 cards in all, before we select a single heart. The next three cards will be all hearts, so we may need to select 42 cards to get three hearts. ◀

EXAMPLE 8 What is the least number of area codes needed to guarantee that the 25 million phones in a state can be assigned distinct 10-digit telephone numbers? (Assume that telephone numbers are of the form *NXX-NXX-XXXX*, where the first three digits form the area code, N represents a digit from 2 to 9 inclusive, and X represents any digit.)

Solution: There are eight million different phone numbers of the form *NXX-XXXX* (as shown in Example 8 of Section 6.1). Hence, by the generalized pigeonhole principle, among 25 million telephones, at least $\lceil 25{,}000{,}000/8{,}000{,}000 \rceil = 4$ of them must have identical phone numbers. Hence, at least four area codes are required to ensure that all 10-digit numbers are different. ◀

Example 9, although not an application of the generalized pigeonhole principle, makes use of similar principles.

EXAMPLE 9 Suppose that a computer science laboratory has 15 workstations and 10 servers. A cable can be used to directly connect a workstation to a server. For each server, only one direct connection to that server can be active at any time. We want to guarantee that at any time any set of 10 or fewer workstations can simultaneously access different servers via direct connections. Although we could do this by connecting every workstation directly to every server (using 150 connections), what is the minimum number of direct connections needed to achieve this goal?

Solution: Suppose that we label the workstations $W_1, W_2, \ldots, W_{15}$ and the servers $S_1, S_2, \ldots, S_{10}$. First, we would like to find a way for there to be far fewer than 150 direct connections between workstations and servers to achieve our goal. One promising approach is to directly connect W_k to S_k for $k = 1, 2, \ldots, 10$ and then to connect each of W_{11}, W_{12}, W_{13}, W_{14}, and W_{15} to all

10 servers. This gives us a total of $10 + 5 \cdot 10 = 60$ direct connections. We need to determine whether with this configuration any set of 10 or fewer workstations can simultaneously access different servers. We note that if workstation W_j is included with $1 \leq j \leq 10$, it can access server S_j, and for each workstation W_k with $k \geq 11$ included, there must be a corresponding workstation W_j with $1 \leq j \leq 10$ not included, so W_k can access server S_j. (This follows because there are at least as many available servers S_j as there are workstations W_j with $1 \leq j \leq 10$ not included.) So, any set of 10 or fewer workstations are able to simultaneously access different servers.

But can we use fewer than 60 direct connections? Suppose there are fewer than 60 direct connections between workstations and servers. Then some server would be connected to at most $\lfloor 59/10 \rfloor = 5$ workstations. (If all servers were connected to at least six workstations, there would be at least $6 \cdot 10 = 60$ direct connections.) This means that the remaining nine servers are not enough for the other 10 or more workstations to simultaneously access different servers. Consequently, at least 60 direct connections are needed. It follows that 60 is the answer. ◀

6.2.3 Some Elegant Applications of the Pigeonhole Principle

In many interesting applications of the pigeonhole principle, the objects to be placed in boxes must be chosen in a clever way. A few such applications will be described here.

EXAMPLE 10 During a month with 30 days, a baseball team plays at least one game a day, but no more than 45 games. Show that there must be a period of some number of consecutive days during which the team must play exactly 14 games.

Solution: Let a_j be the number of games played on or before the jth day of the month. Then $a_1, a_2, \ldots, a_{30}$ is an increasing sequence of distinct positive integers, with $1 \leq a_j \leq 45$. Moreover, $a_1 + 14, a_2 + 14, \ldots, a_{30} + 14$ is also an increasing sequence of distinct positive integers, with $15 \leq a_j + 14 \leq 59$.

The 60 positive integers $a_1, a_2, \ldots, a_{30}, a_1 + 14, a_2 + 14, \ldots, a_{30} + 14$ are all less than or equal to 59. Hence, by the pigeonhole principle two of these integers are equal. Because the integers $a_j, j = 1, 2, \ldots, 30$ are all distinct and the integers $a_j + 14, j = 1, 2, \ldots, 30$ are all distinct, there must be indices i and j with $a_i = a_j + 14$. This means that exactly 14 games were played from day $j + 1$ to day i. ◀

EXAMPLE 11 Show that among any $n + 1$ positive integers not exceeding $2n$ there must be an integer that divides one of the other integers.

Solution: Write each of the $n + 1$ integers $a_1, a_2, \ldots, a_{n+1}$ as a power of 2 times an odd integer. In other words, let $a_j = 2^{k_j} q_j$ for $j = 1, 2, \ldots, n + 1$, where k_j is a nonnegative integer and q_j is odd. The integers $q_1, q_2, \ldots, q_{n+1}$ are all odd positive integers less than $2n$. Because there are only n odd positive integers less than $2n$, it follows from the pigeonhole principle that two of the integers $q_1, q_2, \ldots, q_{n+1}$ must be equal. Therefore, there are distinct integers i and j such that $q_i = q_j$. Let q be the common value of q_i and q_j. Then, $a_i = 2^{k_i} q$ and $a_j = 2^{k_j} q$. It follows that if $k_i < k_j$, then a_i divides a_j; while if $k_i > k_j$, then a_j divides a_i. ◀

A clever application of the pigeonhole principle shows the existence of an increasing or a decreasing subsequence of a certain length in a sequence of distinct integers. We review some definitions before this application is presented. Suppose that $a_1, a_2, \ldots, a_N$ is a sequence of real numbers. A **subsequence** of this sequence is a sequence of the form $a_{i_1}, a_{i_2}, \ldots, a_{i_m}$, where $1 \leq i_1 < i_2 < \cdots < i_m \leq N$. Hence, a subsequence is a sequence obtained from the original sequence by including some of the terms of the original sequence in their original order, and perhaps not including other terms. A sequence is called **strictly increasing** if each term is larger than the one that precedes it, and it is called **strictly decreasing** if each term is smaller than the one that precedes it.

THEOREM 3 Every sequence of $n^2 + 1$ distinct real numbers contains a subsequence of length $n + 1$ that is either strictly increasing or strictly decreasing.

We give an example before presenting the proof of Theorem 3.

EXAMPLE 12 The sequence 8, 11, 9, 1, 4, 6, 12, 10, 5, 7 contains 10 terms. Note that $10 = 3^2 + 1$. There are four strictly increasing subsequences of length four, namely, 1, 4, 6, 12; 1, 4, 6, 7; 1, 4, 6, 10; and 1, 4, 5, 7. There is also a strictly decreasing subsequence of length four, namely, 11, 9, 6, 5. ◀

The proof of the theorem will now be given.

Proof: Let $a_1, a_2, \dots, a_{n^2+1}$ be a sequence of $n^2 + 1$ distinct real numbers. Associate an ordered pair with each term of the sequence, namely, associate (i_k, d_k) to the term a_k, where i_k is the length of the longest increasing subsequence starting at a_k, and d_k is the length of the longest decreasing subsequence starting at a_k.

Suppose that there are no increasing or decreasing subsequences of length $n + 1$. Then i_k and d_k are both positive integers less than or equal to n, for $k = 1, 2, \dots, n^2 + 1$. Hence, by the product rule there are n^2 possible ordered pairs for (i_k, d_k). By the pigeonhole principle, two of these $n^2 + 1$ ordered pairs are equal. In other words, there exist terms a_s and a_t, with $s < t$ such that $i_s = i_t$ and $d_s = d_t$. We will show that this is impossible. Because the terms of the sequence are distinct, either $a_s < a_t$ or $a_s > a_t$. If $a_s < a_t$, then, because $i_s = i_t$, an increasing subsequence of length $i_t + 1$ can be built starting at a_s, by taking a_s followed by an increasing subsequence of length i_t beginning at a_t. This is a contradiction. Similarly, if $a_s > a_t$, the same reasoning shows that d_s must be greater than d_t, which is a contradiction. ◁

Links

The final example shows how the generalized pigeonhole principle can be applied to an important part of combinatorics called **Ramsey theory**, after the English mathematician F. P. Ramsey. In general, Ramsey theory deals with the distribution of subsets of elements of sets.

EXAMPLE 13 Assume that in a group of six people, each pair of individuals consists of two friends or two enemies. Show that there are either three mutual friends or three mutual enemies in the group.

Solution: Let A be one of the six people. Of the five other people in the group, there are either three or more who are friends of A, or three or more who are enemies of A. This follows from

Links

Stephen Frank Burch

FRANK PLUMPTON RAMSEY (1903–1930) Frank Plumpton Ramsey, son of the president of Magdalene College, Cambridge, was educated at Winchester and Trinity Colleges. After graduating in 1923, he was elected a fellow of King's College, Cambridge, where he spent the remainder of his life. Ramsey made important contributions to mathematical logic. What we now call Ramsey theory began with his clever combinatorial arguments, published in the paper "On a Problem of Formal Logic." Ramsey also made contributions to the mathematical theory of economics. He was noted as an excellent lecturer on the foundations of mathematics. According to one of his brothers, he was interested in almost everything, including English literature and politics. Ramsey was married and had two daughters. His death at the age of 26 resulting from chronic liver problems deprived the mathematical community and Cambridge University of a brilliant young scholar.

the generalized pigeonhole principle, because when five objects are divided into two sets, one of the sets has at least $\lceil 5/2 \rceil = 3$ elements. In the former case, suppose that B, C, and D are friends of A. If any two of these three individuals are friends, then these two and A form a group of three mutual friends. Otherwise, B, C, and D form a set of three mutual enemies. The proof in the latter case, when there are three or more enemies of A, proceeds in a similar manner. ◀

The **Ramsey number** $R(m, n)$, where m and n are positive integers greater than or equal to 2, denotes the minimum number of people at a party such that there are either m mutual friends or n mutual enemies, assuming that every pair of people at the party are friends or enemies. Example 13 shows that $R(3, 3) \leq 6$. We conclude that $R(3, 3) = 6$ because in a group of five people where every two people are friends or enemies, there may not be three mutual friends or three mutual enemies (see Exercise 28).

It is possible to prove some useful properties about Ramsey numbers, but for the most part it is difficult to find their exact values. Note that by symmetry it can be shown that $R(m, n) = R(n, m)$ (see Exercise 32). We also have $R(2, n) = n$ for every positive integer $n \geq 2$ (see Exercise 31). The exact values of only nine Ramsey numbers $R(m, n)$ with $3 \leq m \leq n$ are known, including $R(4, 4) = 18$. Only bounds are known for many other Ramsey numbers, including $R(5, 5)$, which is known to satisfy $43 \leq R(5, 5) \leq 48$. The reader interested in learning more about Ramsey numbers should consult [MiRo91] or [GrRoSp90].

Exercises

1. Show that in any set of six classes, each meeting regularly once a week on a particular day of the week, there must be two that meet on the same day, assuming that no classes are held on weekends.

2. Show that if there are 30 students in a class, then at least two have last names that begin with the same letter.

3. A drawer contains a dozen brown socks and a dozen black socks, all unmatched. A person takes socks out at random in the dark.

a) How many socks must be taken out to be sure that there are at least two socks of the same color?

b) How many socks must be taken out to be sure that there are at least two black socks?

4. A bowl contains 10 red balls and 10 blue balls. A woman selects balls at random without looking at them.

a) How many balls must she select to be sure of having at least three balls of the same color?

b) How many balls must she select to be sure of having at least three blue balls?

5. Undergraduate students at a college belong to one of four groups depending on the year in which they are expected to graduate. Each student must choose one of 21 different majors. How many students are needed to assure that there are two students expected to graduate in the same year who have the same major?

6. There are six professors teaching the introductory discrete mathematics class at a university. The same final exam is given by all six professors. If the lowest possible score on the final is 0 and the highest possible score is 100, how many students must there be to guarantee that there are two students with the same professor who earned the same final examination score?

7. Show that among any group of five (not necessarily consecutive) integers, there are two with the same remainder when divided by 4.

8. Let d be a positive integer. Show that among any group of $d + 1$ (not necessarily consecutive) integers there are two with exactly the same remainder when they are divided by d.

9. Let n be a positive integer. Show that in any set of n consecutive integers there is exactly one divisible by n.

10. Show that if f is a function from S to T, where S and T are finite sets with $|S| > |T|$, then there are elements s_1 and s_2 in S such that $f(s_1) = f(s_2)$, or in other words, f is not one-to-one.

11. What is the minimum number of students, each of whom comes from one of the 50 states, who must be enrolled in a university to guarantee that there are at least 100 who come from the same state?

***12.** Let (x_i, y_i), $i = 1, 2, 3, 4, 5$, be a set of five distinct points with integer coordinates in the xy plane. Show that the midpoint of the line joining at least one pair of these points has integer coordinates.

***13.** Let (x_i, y_i, z_i), $i = 1, 2, 3, 4, 5, 6, 7, 8, 9$, be a set of nine distinct points with integer coordinates in xyz space. Show that the midpoint of at least one pair of these points has integer coordinates.

14. How many ordered pairs of integers (a, b) are needed to guarantee that there are two ordered pairs (a_1, b_1) and (a_2, b_2) such that $a_1 \textbf{ mod } 5 = a_2 \textbf{ mod } 5$ and $b_1 \textbf{ mod } 5 = b_2 \textbf{ mod } 5$?

15. a) Show that if five integers are selected from the first eight positive integers, there must be a pair of these integers with a sum equal to 9.

b) Is the conclusion in part (a) true if four integers are selected rather than five?

16. a) Show that if seven integers are selected from the first 10 positive integers, there must be at least two pairs of these integers with the sum 11.

b) Is the conclusion in part (a) true if six integers are selected rather than seven?

17. How many numbers must be selected from the set $\{1, 2, 3, 4, 5, 6\}$ to guarantee that at least one pair of these numbers add up to 7?

18. How many numbers must be selected from the set $\{1, 3, 5, 7, 9, 11, 13, 15\}$ to guarantee that at least one pair of these numbers add up to 16?

19. A company stores products in a warehouse. Storage bins in this warehouse are specified by their aisle, location in the aisle, and shelf. There are 50 aisles, 85 horizontal locations in each aisle, and 5 shelves throughout the warehouse. What is the least number of products the company can have so that at least two products must be stored in the same bin?

20. Suppose that there are nine students in a discrete mathematics class at a small college.

a) Show that the class must have at least five students that are 20 years old or younger or at least five students 21 years old or older.

b) Show that the class must have at least three students 20 years old or younger or at least seven students 21 years old or older.

21. Suppose that every student in a discrete mathematics class of 25 students is a freshman, a sophomore, or a junior.

a) Show that there are at least nine freshmen, at least nine sophomores, or at least nine juniors in the class.

b) Show that there are either at least three freshmen, at least 19 sophomores, or at least five juniors in the class.

22. Find an increasing subsequence of maximal length and a decreasing subsequence of maximal length in the sequence 22, 5, 7, 2, 23, 10, 15, 21, 3, 17.

23. Construct a sequence of 16 positive integers that has no increasing or decreasing subsequence of five terms.

24. Show that if there are 101 people of different heights standing in a line, it is possible to find 11 people in the order they are standing in the line with heights that are either increasing or decreasing.

***25.** Show that whenever 25 vegans and 25 meat eaters are seated around a circular table there is always a person both of whose neighbors are meat eaters.

****26.** Suppose that 21 girls and 21 boys enter a mathematics competition. Furthermore, suppose that each entrant solves at most six questions, and for every boy-girl pair, there is at least one question that they both solved. Show that there is a question that was solved by at least three girls and at least three boys.

***27.** Describe an algorithm in pseudocode for producing the largest increasing or decreasing subsequence of a sequence of distinct integers.

28. Show that in a group of five people (where any two people are either friends or enemies), there are not necessarily three mutual friends or three mutual enemies.

29. Show that in a group of 10 people (where any two people are either friends or enemies), there are either three mutual friends or four mutual enemies, and there are either three mutual enemies or four mutual friends.

30. Use Exercise 29 to show that among any group of 20 people (where any two people are either friends or enemies), there are either four mutual friends or four mutual enemies.

31. Show that if n is an integer with $n \geq 2$, then the Ramsey number $R(2, n)$ equals n. (Recall that Ramsey numbers were discussed after Example 13 in Section 6.2.)

32. Show that if m and n are integers with $m \geq 2$ and $n \geq 2$, then the Ramsey numbers $R(m, n)$ and $R(n, m)$ are equal. (Recall that Ramsey numbers were discussed after Example 13 in Section 6.2.)

33. Show that there are at least six people in California (population: 39 million) with the same three initials who were born on the same day of the year (but not necessarily in the same year). Assume that everyone has three initials.

34. Show that if there are 100,000,000 wage earners in the United States who earn less than 1,000,000 dollars (but at least a penny), then there are two who earned exactly the same amount of money, to the penny, last year.

35. In the 17th century, there were more than 800,000 inhabitants of Paris. At the time, it was believed that no one had more than 200,000 hairs on their head. Assuming these numbers are correct and that everyone has at least one hair on their head (that is, no one is completely bald), use the pigeonhole principle to show, as the French writer Pierre Nicole did, that there had to be two Parisians with the same number of hairs on their heads. Then use the generalized pigeonhole principle to show that there had to be at least five Parisians at that time with the same number of hairs on their heads.

36. Assuming that no one has more than 1,000,000 hairs on their head and that the population of New York City was 8,537,673 in 2016, show there had to be at least nine people in New York City in 2016 with the same number of hairs on their heads.

37. There are 38 different time periods during which classes at a university can be scheduled. If there are 677 different classes, how many different rooms will be needed?

38. A computer network consists of six computers. Each computer is directly connected to at least one of the other computers. Show that there are at least two computers in the network that are directly connected to the same number of other computers.

39. A computer network consists of six computers. Each computer is directly connected to zero or more of the other computers. Show that there are at least two computers in the network that are directly connected to the same number of other computers. [*Hint:* It is impossible to have a computer linked to none of the others and a computer linked to all the others.]

40. Find the least number of cables required to connect eight computers to four printers to guarantee that for every choice of four of the eight computers, these four computers can directly access four different printers. Justify your answer.

41. Find the least number of cables required to connect 100 computers to 20 printers to guarantee that every subset of 20 computers can directly access 20 different printers. (Here, the assumptions about cables and computers are the same as in Example 9.) Justify your answer.

***42.** Prove that at a party where there are at least two people, there are two people who know the same number of other people there.

43. An arm wrestler is the champion for a period of 75 hours. (Here, by an hour, we mean a period starting from an exact hour, such as 1 P.M., until the next hour.) The arm wrestler had at least one match an hour, but no more than 125 total matches. Show that there is a period of consecutive hours during which the arm wrestler had exactly 24 matches.

***44.** Is the statement in Exercise 43 true if 24 is replaced by

a) 2? **b)** 23? **c)** 25? **d)** 30?

45. Show that if f is a function from S to T, where S and T are nonempty finite sets and $m = \lceil |S| / |T| \rceil$, then there are at least m elements of S mapped to the same value of T. That is, show that there are distinct elements $s_1, s_2, \dots, s_m$ of S such that $f(s_1) = f(s_2) = \cdots = f(s_m)$.

46. There are 51 houses on a street. Each house has an address between 1000 and 1099, inclusive. Show that at least two houses have addresses that are consecutive integers.

***47.** Let x be an irrational number. Show that for some positive integer j not exceeding the positive integer n, the absolute value of the difference between jx and the nearest integer to jx is less than $1/n$.

48. Let $n_1, n_2, \dots, n_t$ be positive integers. Show that if $n_1 + n_2 + \cdots + n_t - t + 1$ objects are placed into t boxes, then for some i, $i = 1, 2, \dots, t$, the ith box contains at least n_i objects.

***49.** An alternative proof of Theorem 3 based on the generalized pigeonhole principle is outlined in this exercise. The notation used is the same as that used in the proof in the text.

a) Assume that $i_k \le n$ for $k = 1, 2, \dots, n^2 + 1$. Use the generalized pigeonhole principle to show that there are $n + 1$ terms $a_{k_1}, a_{k_2}, \dots, a_{k_{n+1}}$ with $i_{k_1} = i_{k_2} = \cdots = i_{k_{n+1}}$, where $1 \le k_1 < k_2 < \cdots < k_{n+1}$.

b) Show that $a_{k_j} > a_{k_{j+1}}$ for $j = 1, 2, \dots, n$. [*Hint:* Assume that $a_{k_j} < a_{k_{j+1}}$, and show that this implies that $i_{k_j} > i_{k_{j+1}}$, which is a contradiction.]

c) Use parts (a) and (b) to show that if there is no increasing subsequence of length $n + 1$, then there must be a decreasing subsequence of this length.

6.3 Permutations and Combinations

6.3.1 Introduction

Many counting problems can be solved by finding the number of ways to arrange a specified number of distinct elements of a set of a particular size, where the order of these elements matters. Many other counting problems can be solved by finding the number of ways to select a particular number of elements from a set of a particular size, where the order of the elements selected does not matter. For example, in how many ways can we select three students from a group of five students to stand in line for a picture? How many different committees of three students can be formed from a group of four students? In this section we will develop methods to answer questions such as these.

6.3.2 Permutations

We begin by solving the first question posed in the introduction to this section, as well as related questions.

EXAMPLE 1 In how many ways can we select three students from a group of five students to stand in line for a picture? In how many ways can we arrange all five of these students in a line for a picture?

Extra Examples

Solution: First, note that the order in which we select the students matters. There are five ways to select the first student to stand at the start of the line. Once this student has been selected, there are four ways to select the second student in the line. After the first and second students have been selected, there are three ways to select the third student in the line. By the product rule, there are $5 \cdot 4 \cdot 3 = 60$ ways to select three students from a group of five students to stand in line for a picture.

To arrange all five students in a line for a picture, we select the first student in five ways, the second in four ways, the third in three ways, the fourth in two ways, and the fifth in one way. Consequently, there are $5 \cdot 4 \cdot 3 \cdot 2 \cdot 1 = 120$ ways to arrange all five students in a line for a picture. ◀

Example 1 illustrates how ordered arrangements of distinct objects can be counted. This leads to some terminology.

Links

A **permutation** of a set of distinct objects is an ordered arrangement of these objects. We also are interested in ordered arrangements of some of the elements of a set. An ordered arrangement of r elements of a set is called an ***r*-permutation**.

EXAMPLE 2 Let $S = \{1, 2, 3\}$. The ordered arrangement 3, 1, 2 is a permutation of S. The ordered arrangement 3, 2 is a 2-permutation of S. ◀

The number of r-permutations of a set with n elements is denoted by $P(n, r)$. We can find $P(n, r)$ using the product rule.

EXAMPLE 3 Let $S = \{a, b, c\}$. The 2-permutations of S are the ordered arrangements a, b; a, c; b, a; b, c; c, a; and c, b. Consequently, there are six 2-permutations of this set with three elements. There are always six 2-permutations of a set with three elements. There are three ways to choose the first element of the arrangement. There are two ways to choose the second element of the arrangement, because it must be different from the first element. Hence, by the product rule, we see that $P(3, 2) = 3 \cdot 2 = 6$. ◀

We now use the product rule to find a formula for $P(n, r)$ whenever n and r are positive integers with $1 \le r \le n$.

THEOREM 1 If n is a positive integer and r is an integer with $1 \le r \le n$, then there are

$$P(n, r) = n(n-1)(n-2)\cdots(n-r+1)$$

r-permutations of a set with n distinct elements.

Proof: We will use the product rule to prove that this formula is correct. The first element of the permutation can be chosen in n ways because there are n elements in the set. There are $n - 1$ ways to choose the second element of the permutation, because there are $n - 1$ elements left in the set after using the element picked for the first position. Similarly, there are $n - 2$ ways to choose the third element, and so on, until there are exactly $n - (r - 1) = n - r + 1$ ways to choose the rth element. Consequently, by the product rule, there are

$$n(n-1)(n-2)\cdots(n-r+1)$$

r-permutations of the set. ◁

Note that $P(n, 0) = 1$ whenever n is a nonnegative integer because there is exactly one way to order zero elements. That is, there is exactly one list with no elements in it, namely the empty list.

We now state a useful corollary of Theorem 1.

COROLLARY 1 If n and r are integers with $0 \le r \le n$, then $P(n, r) = \dfrac{n!}{(n-r)!}$.

Proof: When n and r are integers with $1 \le r \le n$, by Theorem 1 we have

$$P(n, r) = n(n-1)(n-2)\cdots(n-r+1) = \frac{n!}{(n-r)!}$$

Because $\dfrac{n!}{(n-0)!} = \dfrac{n!}{n!} = 1$ whenever n is a nonnegative integer, we see that the formula $P(n, r) = \dfrac{n!}{(n-r)!}$ also holds when $r = 0$. ◁

By Theorem 1 we know that if n is a positive integer, then $P(n, n) = n!$. We will illustrate this result with some examples.

EXAMPLE 4 How many ways are there to select a first-prize winner, a second-prize winner, and a third-prize winner from 100 different people who have entered a contest?

Solution: Because it matters which person wins which prize, the number of ways to pick the three prize winners is the number of ordered selections of three elements from a set of 100 elements, that is, the number of 3-permutations of a set of 100 elements. Consequently, the answer is

$$P(100, 3) = 100 \cdot 99 \cdot 98 = 970{,}200.$$

◀

EXAMPLE 5 Suppose that there are eight runners in a race. The winner receives a gold medal, the second-place finisher receives a silver medal, and the third-place finisher receives a bronze medal. How many different ways are there to award these medals, if all possible outcomes of the race can occur and there are no ties?

Solution: The number of different ways to award the medals is the number of 3-permutations of a set with eight elements. Hence, there are $P(8, 3) = 8 \cdot 7 \cdot 6 = 336$ possible ways to award the medals. ◀

EXAMPLE 6 Suppose that a salesperson has to visit eight different cities. The salesperson must begin the trip in a specified city, but the salesperson can visit the other seven cities in any order. How many possible orders can the salesperson use when visiting these cities?

Solution: The number of possible paths between the cities is the number of permutations of seven elements, because the first city is determined, but the remaining seven can be ordered arbitrarily. Consequently, there are $7! = 7 \cdot 6 \cdot 5 \cdot 4 \cdot 3 \cdot 2 \cdot 1 = 5040$ ways for the salesperson to choose the tour. If, for instance, the salesperson wishes to find the path between the cities with minimum distance, and the salesperson computes the total distance for each possible path, the salesperson must consider a total of 5040 paths! ◀

EXAMPLE 7 How many permutations of the letters *ABCDEFGH* contain the string *ABC*?

Solution: Because the letters *ABC* must occur as a block, we can find the answer by finding the number of permutations of six objects, namely, the block *ABC* and the individual letters *D*, *E*, *F*, *G*, and *H*. Because these six objects can occur in any order, there are $6! = 720$ permutations of the letters *ABCDEFGH* in which *ABC* occurs as a block. ◀

6.3.3 Combinations

We now turn our attention to counting unordered selections of objects. We begin by solving a question posed in the introduction to this section of the chapter.

EXAMPLE 8 How many different committees of three students can be formed from a group of four students?

Solution: To answer this question, we need only find the number of subsets with three elements from the set containing the four students. We see that there are four such subsets, one for each of the four students, because choosing three students is the same as choosing one of the four students to leave out of the group. This means that there are four ways to choose the three students for the committee, where the order in which these students are chosen does not matter. ◀

Links

Example 8 illustrates that many counting problems can be solved by finding the number of subsets of a particular size of a set with n elements, where n is a positive integer.

An ***r*-combination** of elements of a set is an unordered selection of r elements from the set. Thus, an r-combination is simply a subset of the set with r elements.

EXAMPLE 9 Let S be the set $\{1, 2, 3, 4\}$. Then $\{1, 3, 4\}$ is a 3-combination from S. (Note that $\{4, 1, 3\}$ is the same 3-combination as $\{1, 3, 4\}$, because the order in which the elements of a set are listed does not matter.) ◀

The number of r-combinations of a set with n distinct elements is denoted by $C(n, r)$. Note that $C(n, r)$ is also denoted by $\binom{n}{r}$ and is called a **binomial coefficient**. We will learn where this terminology comes from in Section 6.4.

EXAMPLE 10 We see that $C(4, 2) = 6$, because the 2-combinations of $\{a, b, c, d\}$ are the six subsets $\{a, b\}$, $\{a, c\}$, $\{a, d\}$, $\{b, c\}$, $\{b, d\}$, and $\{c, d\}$. ◀

We can determine the number of r-combinations of a set with n elements using the formula for the number of r-permutations of a set. To do this, note that the r-permutations of a set can be obtained by first forming r-combinations and then ordering the elements in these combinations. The proof of Theorem 2, which gives the value of $C(n, r)$, is based on this observation.

THEOREM 2 The number of r-combinations of a set with n elements, where n is a nonnegative integer and r is an integer with $0 \leq r \leq n$, equals

$$C(n, r) = \frac{n!}{r!\,(n-r)!}.$$

Proof: The $P(n, r)$ r-permutations of the set can be obtained by forming the $C(n, r)$ r-combinations of the set, and then ordering the elements in each r-combination, which can be done in $P(r, r)$ ways. Consequently, by the product rule,

$$P(n, r) = C(n, r) \cdot P(r, r).$$

This implies that

$$C(n, r) = \frac{P(n, r)}{P(r, r)} = \frac{n!/(n-r)!}{r!/(r-r)!} = \frac{n!}{r!\,(n-r)!}.$$

We can also use the division rule for counting to construct a proof of this theorem. Because the order of elements in a combination does not matter and there are $P(r, r)$ ways to order r elements in an r-combination of n elements, each of the $C(n, r)$ r-combinations of a set with n elements corresponds to exactly $P(r, r)$ r-permutations. Hence, by the division rule, $C(n, r) = \frac{P(n,r)}{P(r,r)}$, which implies as before that $C(n, r) = \frac{n!}{r!\,(n-r)!}$. ◁

The formula in Theorem 2, although explicit, is not helpful when $C(n, r)$ is computed for large values of n and r. The reasons are that it is practical to compute exact values of factorials exactly only for small integer values, and when floating point arithmetic is used, the formula in Theorem 2 may produce a value that is not an integer. When computing $C(n, r)$, first note that when we cancel out $(n - r)!$ from the numerator and denominator of the expression for $C(n, r)$ in Theorem 2, we obtain

$$C(n, r) = \frac{n!}{r!\,(n-r)!} = \frac{n(n-1)\cdots(n-r+1)}{r!}.$$

Consequently, to compute $C(n, r)$ you can cancel out all the terms in the larger factorial in the denominator from the numerator and denominator, then multiply all the terms that do not cancel in the numerator and finally divide by the smaller factorial in the denominator. [When doing this calculation by hand, instead of by machine, it is also worthwhile to factor out common factors in the numerator $n(n - 1) \cdots (n - r + 1)$ and in the denominator $r!$.] Note that many computational programs can be used to find $C(n, r)$. [Such functions may be called *choose*(n, k) or *binom*(n, k).]

Example 11 illustrates how $C(n, k)$ is computed when k is relatively small compared to n and when k is close to n. It also illustrates a key identity enjoyed by the numbers $C(n, k)$.

EXAMPLE 11 How many poker hands of five cards can be dealt from a standard deck of 52 cards? Also, how many ways are there to select 47 cards from a standard deck of 52 cards?

Solution: Because the order in which the five cards are dealt from a deck of 52 cards does not matter, there are

$$C(52, 5) = \frac{52!}{5!47!}$$

different hands of five cards that can be dealt. To compute the value of $C(52, 5)$, first divide the numerator and denominator by 47! to obtain

$$C(52, 5) = \frac{52 \cdot 51 \cdot 50 \cdot 49 \cdot 48}{5 \cdot 4 \cdot 3 \cdot 2 \cdot 1}.$$

This expression can be simplified by first dividing the factor 5 in the denominator into the factor 50 in the numerator to obtain a factor 10 in the numerator, then dividing the factor 4 in the denominator into the factor 48 in the numerator to obtain a factor of 12 in the numerator, then dividing the factor 3 in the denominator into the factor 51 in the numerator to obtain a factor of 17 in the numerator, and finally, dividing the factor 2 in the denominator into the factor 52 in the numerator to obtain a factor of 26 in the numerator. We find that

$$C(52, 5) = 26 \cdot 17 \cdot 10 \cdot 49 \cdot 12 = 2{,}598{,}960.$$

Consequently, there are 2,598,960 different poker hands of five cards that can be dealt from a standard deck of 52 cards.

Note that there are

$$C(52, 47) = \frac{52!}{47!5!}$$

different ways to select 47 cards from a standard deck of 52 cards. We do not need to compute this value because $C(52, 47) = C(52, 5)$. (Only the order of the factors 5! and 47! is different in the denominators in the formulae for these quantities.) It follows that there are also 2,598,960 different ways to select 47 cards from a standard deck of 52 cards. ◀

In Example 11 we observed that $C(52, 5) = C(52, 47)$. This is not surprising because selecting five cards out of 52 is the same as selecting the 47 that we leave out. The identity $C(52, 5) = C(52, 47)$ is a special case of the useful identity for the number of r-combinations of a set given in Corollary 2.

COROLLARY 2 Let n and r be nonnegative integers with $r \leq n$. Then $C(n, r) = C(n, n - r)$.

Proof: From Theorem 2 it follows that

$$C(n, r) = \frac{n!}{r!\,(n - r)!}$$

and

$$C(n, n - r) = \frac{n!}{(n - r)!\,[n - (n - r)]!} = \frac{n!}{(n - r)!\,r!}.$$

Hence, $C(n, r) = C(n, n - r)$. ◁

We can also prove Corollary 2 without relying on algebraic manipulation. Instead, we can use a combinatorial proof. We describe this important type of proof in Definition 1.

Definition 1 A *combinatorial proof* of an identity is a proof that uses counting arguments to prove that both sides of the identity count the same objects but in different ways or a proof that is based on showing that there is a bijection between the sets of objects counted by the two sides of the identity. These two types of proofs are called *double counting proofs* and *bijective proofs*, respectively.

Combinatorial proofs are almost always much shorter and provide more insights than proofs based on algebraic manipulation.

Many identities involving binomial coefficients can be proved using combinatorial proofs. We now show how to prove Corollary 2 using a combinatorial proof. We will provide both a double counting proof and a bijective proof, both based on the same basic idea.

Proof: We will use a bijective proof to show that $C(n, r) = C(n, n - r)$ for all integers n and r with $0 \leq r \leq n$. Suppose that S is a set with n elements. The function that maps a subset A of S to $\overline{A}$ is a bijection between subsets of S with r elements and subsets with $n - r$ elements (as the reader should verify). The identity $C(n, r) = C(n, n - r)$ follows because when there is a bijection between two finite sets, the two sets must have the same number of elements.

Alternatively, we can reformulate this argument as a double counting proof. By definition, the number of subsets of S with r elements equals $C(n, r)$. But each subset A of S is also determined by specifying which elements are not in A, and so are in $\overline{A}$. Because the complement of a subset of S with r elements has $n - r$ elements, there are also $C(n, n - r)$ subsets of S with r elements. It follows that $C(n, r) = C(n, n - r)$. ◁

EXAMPLE 12 How many ways are there to select five players from a 10-member tennis team to make a trip to a match at another school?

Extra Examples

Solution: The answer is given by the number of 5-combinations of a set with 10 elements. By Theorem 2, the number of such combinations is

$$C(10, 5) = \frac{10!}{5!\,5!} = 252.$$ ◀

EXAMPLE 13 A group of 30 people have been trained as astronauts to go on the first mission to Mars. How many ways are there to select a crew of six people to go on this mission (assuming that all crew members have the same job)?

Solution: The number of ways to select a crew of six from the pool of 30 people is the number of 6-combinations of a set with 30 elements, because the order in which these people are chosen does not matter. By Theorem 2, the number of such combinations is

$$C(30, 6) = \frac{30!}{6!\,24!} = \frac{30 \cdot 29 \cdot 28 \cdot 27 \cdot 26 \cdot 25}{6 \cdot 5 \cdot 4 \cdot 3 \cdot 2 \cdot 1} = 593{,}775.$$ ◀

EXAMPLE 14 How many bit strings of length n contain exactly r 1s?

Solution: The positions of r 1s in a bit string of length n form an r-combination of the set $\{1, 2, 3, \ldots, n\}$. Hence, there are $C(n, r)$ bit strings of length n that contain exactly r 1s. ◀

EXAMPLE 15 Suppose that there are 9 faculty members in the mathematics department and 11 in the computer science department. How many ways are there to select a committee to develop a discrete mathematics course at a school if the committee is to consist of three faculty members from the mathematics department and four from the computer science department?

Solution: By the product rule, the answer is the product of the number of 3-combinations of a set with nine elements and the number of 4-combinations of a set with 11 elements. By Theorem 2, the number of ways to select the committee is

$$C(9, 3) \cdot C(11, 4) = \frac{9!}{3!6!} \cdot \frac{11!}{4!7!} = 84 \cdot 330 = 27{,}720.$$ ◀

Exercises

1. List all the permutations of $\{a, b, c\}$.

2. How many different permutations are there of the set $\{a, b, c, d, e, f, g\}$?

3. How many permutations of $\{a, b, c, d, e, f, g\}$ end with a?

4. Let $S = \{1, 2, 3, 4, 5\}$.

a) List all the 3-permutations of S.
b) List all the 3-combinations of S.

5. Find the value of each of these quantities.

a) $P(6, 3)$ b) $P(6, 5)$
c) $P(8, 1)$ d) $P(8, 5)$
e) $P(8, 8)$ f) $P(10, 9)$

6. Find the value of each of these quantities.

a) $C(5, 1)$ b) $C(5, 3)$
c) $C(8, 4)$ d) $C(8, 8)$
e) $C(8, 0)$ f) $C(12, 6)$

7. Find the number of 5-permutations of a set with nine elements.

8. In how many different orders can five runners finish a race if no ties are allowed?

9. How many possibilities are there for the win, place, and show (first, second, and third) positions in a horse race with 12 horses if all orders of finish are possible?

10. There are six different candidates for governor of a state. In how many different orders can the names of the candidates be printed on a ballot?

11. How many bit strings of length 10 contain

a) exactly four 1s?
b) at most four 1s?
c) at least four 1s?
d) an equal number of 0s and 1s?

12. How many bit strings of length 12 contain

a) exactly three 1s?
b) at most three 1s?
c) at least three 1s?
d) an equal number of 0s and 1s?

13. A group contains n students and n faculty. How many ways are there to arrange these people in a row if the students and faculty alternate?

14. In how many ways can a set of two positive integers less than 100 be chosen?

15. In how many ways can a set of five letters be selected from the English alphabet?

16. How many subsets with an odd number of elements does a set with 10 elements have?

17. How many subsets with more than two elements does a set with 100 elements have?

18. A coin is flipped eight times where each flip comes up either heads or tails. How many possible outcomes

a) are there in total?
b) contain exactly three heads?
c) contain at least three heads?
d) contain the same number of heads and tails?

19. A coin is flipped 10 times where each flip comes up either heads or tails. How many possible outcomes

a) are there in total?
b) contain exactly two heads?
c) contain at most three tails?
d) contain the same number of heads and tails?

20. How many bit strings of length 10 have

a) exactly three 0s?
b) more 0s than 1s?
c) at least seven 1s?
d) at least three 1s?

21. How many permutations of the letters *ABCDEFG* contain

a) the string *BCD*?
b) the string *CFGA*?
c) the strings *BA* and *GF*?
d) the strings *ABC* and *DE*?
e) the strings *ABC* and *CDE*?
f) the strings *CBA* and *BED*?

22. How many permutations of the letters *ABCDEFGH* contain

a) the string *ED*?
b) the string *CDE*?
c) the strings *BA* and *FGH*?
d) the strings *AB*, *DE*, and *GH*?
e) the strings *CAB* and *BED*?
f) the strings *BCA* and *ABF*?

23. How many ways are there for eight students and five faculty to stand in a line so that no two faculty stand next to each other? [*Hint:* First position the students and then consider possible positions for the faculty.]

24. How many ways are there for 10 faculty and six students to stand in a line so that no two students stand next to each other? [*Hint:* First position the faculty and then consider possible positions for the students.]

25. How many ways are there for four students and five faculty to stand in a line so that

a) all students stand together?
b) all faculty stand together?

26. How many ways are there for three penguins and six puffins to stand in a line so that

a) all puffins stand together?
b) all penguins stand together?

27. One hundred tickets, numbered 1, 2, 3, …, 100, are sold to 100 different people for a drawing. Four different prizes are awarded, including a grand prize (a trip to Tahiti). How many ways are there to award the prizes if

a) there are no restrictions?
b) the person holding ticket 47 wins the grand prize?
c) the person holding ticket 47 wins one of the prizes?
d) the person holding ticket 47 does not win a prize?
e) the people holding tickets 19 and 47 both win prizes?
f) the people holding tickets 19, 47, and 73 all win prizes?

g) the people holding tickets 19, 47, 73, and 97 all win prizes?

h) none of the people holding tickets 19, 47, 73, and 97 wins a prize?

i) the grand prize winner is a person holding ticket 19, 47, 73, or 97?

j) the people holding tickets 19 and 47 win prizes, but the people holding tickets 73 and 97 do not win prizes?

28. Thirteen people on a softball team show up for a game.

a) How many ways are there to choose 10 players to take the field?

b) How many ways are there to assign the 10 positions by selecting players from the 13 people who show up?

c) Of the 13 people who show up, three are left-handed. How many ways are there to choose 10 players to take the field if at least one of these players must be left-handed?

29. A club has 25 members.

a) How many ways are there to choose four members of the club to serve on an executive committee?

b) How many ways are there to choose a president, vice president, secretary, and treasurer of the club, where no person can hold more than one office?

30. A professor writes 40 discrete mathematics true/false questions. Of the statements in these questions, 17 are true. If the questions can be positioned in any order, how many different answer keys are possible?

***31.** How many 4-permutations of the positive integers not exceeding 100 contain three consecutive integers k, $k+1$, $k+2$, in the correct order

a) where these consecutive integers can perhaps be separated by other integers in the permutation?

b) where they are in consecutive positions in the permutation?

32. Seven women and nine men are on the faculty in the mathematics department at a school.

a) How many ways are there to select a committee of five members of the department if at least one woman must be on the committee?

b) How many ways are there to select a committee of five members of the department if at least one woman and at least one man must be on the committee?

33. The English alphabet contains 21 consonants and five vowels. How many strings of six lowercase letters of the English alphabet contain

a) exactly one vowel?

b) exactly two vowels?

c) at least one vowel?

d) at least two vowels?

34. How many strings of six lowercase letters from the English alphabet contain

a) the letter a?

b) the letters a and b?

c) the letters a and b in consecutive positions with a preceding b, with all the letters distinct?

d) the letters a and b, where a is somewhere to the left of b in the string, with all the letters distinct?

35. Suppose that a department contains 10 left-handed people and 15 right-handed people. How many ways are there to form a committee with six members if it must have the same number of left- and right-handed people?

36. Suppose that a department contains 10 left-handed people and 15 right-handed people. How many ways are there to form a committee with six members if it must have more right-handed than left-handed people?

37. How many bit strings contain exactly eight 0s and 10 1s if every 0 must be immediately followed by a 1?

38. How many bit strings contain exactly five 0s and 14 1s if every 0 must be immediately followed by two 1s?

39. How many bit strings of length 10 contain at least three 1s and at least three 0s?

40. How many ways are there to select 12 countries in the United Nations to serve on a council if 3 are selected from a block of 45, 4 are selected from a block of 57, and the others are selected from the remaining 69 countries?

41. How many license plates consisting of three letters followed by three digits contain no letter or digit twice?

A **circular r-permutation of n** people is a seating of r of these n people around a circular table, where seatings are considered to be the same if they can be obtained from each other by rotating the table.

42. Find the number of circular 3-permutations of 5 people.

43. Find a formula for the number of circular r-permutations of n people.

44. Find a formula for the number of ways to seat r of n people around a circular table, where seatings are considered the same if every person has the same two neighbors without regard to which side these neighbors are sitting on.

45. How many ways are there for a horse race with three horses to finish if ties are possible? [*Note:* Two or three horses may tie.]

***46.** How many ways are there for a horse race with four horses to finish if ties are possible? [*Note:* Any number of the four horses may tie.]

***47.** There are six runners in the 100-yard dash. How many ways are there for three medals to be awarded if ties are possible? (The runner or runners who finish with the fastest time receive gold medals, the runner or runners who finish with exactly one runner ahead receive silver medals, and the runner or runners who finish with exactly two runners ahead receive bronze medals.)

***48.** This procedure is used to break ties in games in the championship round of the World Cup soccer tournament. Each team selects five players in a prescribed order. Each of these players takes a penalty kick, with a player from the first team followed by a player from the second team and so on, following the order of players specified. If the score is still tied at the end of the 10 penalty kicks, this procedure is repeated. If the score is still tied after 20 penalty kicks, a sudden-death shootout occurs, with the first team scoring an unanswered goal victorious.

a) How many different scoring scenarios are possible if the game is settled in the first round of 10 penalty kicks, where the round ends once it is impossible for a team to equal the number of goals scored by the other team?

b) How many different scoring scenarios for the first and second groups of penalty kicks are possible if the game is settled in the second round of 10 penalty kicks?

c) How many scoring scenarios are possible for the full set of penalty kicks if the game is settled with no more than 10 total additional kicks after the two rounds of five kicks for each team?

6.4 Binomial Coefficients and Identities

As we remarked in Section 6.3, the number of r-combinations from a set with n elements is often denoted by $\binom{n}{r}$. This number is also called a **binomial coefficient** because these numbers occur as coefficients in the expansion of powers of binomial expressions such as $(a + b)^n$. We will discuss the **binomial theorem**, which gives a power of a binomial expression as a sum of terms involving binomial coefficients. We will prove this theorem using a combinatorial proof. We will also show how combinatorial proofs can be used to establish some of the many different identities that express relationships among binomial coefficients.

6.4.1 The Binomial Theorem

Links

The binomial theorem gives the coefficients of the expansion of powers of binomial expressions. A **binomial** expression is simply the sum of two terms, such as $x + y$. (The terms can be products of constants and variables, but that does not concern us here.)

Example 1 illustrates how the coefficients in a typical expansion can be found and prepares us for the statement of the binomial theorem.

EXAMPLE 1 The expansion of $(x + y)^3$ can be found using combinatorial reasoning instead of multiplying the three terms out. When $(x + y)^3 = (x + y)(x + y)(x + y)$ is expanded, all products of a term in the first sum, a term in the second sum, and a term in the third sum are added. Terms of the form x^3, x^2y, xy^2, and y^3 arise. To obtain a term of the form x^3, an x must be chosen in each of the sums, and this can be done in only one way. Thus, the x^3 term in the product has a coefficient of 1. To obtain a term of the form x^2y, an x must be chosen in two of the three sums (and consequently a y in the other sum). Hence, the number of such terms is the number of 2-combinations of three objects, namely, $\binom{3}{2}$. Similarly, the number of terms of the form xy^2 is the number of ways to pick one of the three sums to obtain an x (and consequently take a y from each of the other two sums). This can be done in $\binom{3}{1}$ ways. Finally, the only way to obtain a y^3 term is to choose the y for each of the three sums in the product, and this can be done in exactly one way. Consequently, it follows that

$$\begin{aligned}(x + y)^3 &= (x + y)(x + y)(x + y) = (xx + xy + yx + yy)(x + y)\\ &= xxx + xxy + xyx + xyy + yxx + yxy + yyx + yyy\\ &= x^3 + 3x^2y + 3xy^2 + y^3.\end{aligned}$$

◀

We now state the binomial theorem.

THEOREM 1 **THE BINOMIAL THEOREM** Let x and y be variables, and let n be a nonnegative integer. Then

$$(x + y)^n = \sum_{j=0}^{n} \binom{n}{j} x^{n-j}y^j = \binom{n}{0}x^n + \binom{n}{1}x^{n-1}y + \cdots + \binom{n}{n-1}xy^{n-1} + \binom{n}{n}y^n.$$

Proof: We use a combinatorial proof. The terms in the product when it is expanded are of the form $x^{n-j}y^j$ for $j = 0, 1, 2, \ldots, n$. To count the number of terms of the form $x^{n-j}y^j$, note that to obtain such a term it is necessary to choose $n - j$ xs from the n binomial factors (so that the other j terms in the product are ys). Therefore, the coefficient of $x^{n-j}y^j$ is $\binom{n}{n-j}$, which is equal to $\binom{n}{j}$. This proves the theorem. ◁

Some computational uses of the binomial theorem are illustrated in Examples 2–4.

EXAMPLE 2 What is the expansion of $(x + y)^4$?

Extra Examples

Solution: From the binomial theorem it follows that

$$\begin{aligned}(x+y)^4 &= \sum_{j=0}^{4} \binom{4}{j} x^{4-j}y^j \\ &= \binom{4}{0}x^4 + \binom{4}{1}x^3y + \binom{4}{2}x^2y^2 + \binom{4}{3}xy^3 + \binom{4}{4}y^4 \\ &= x^4 + 4x^3y + 6x^2y^2 + 4xy^3 + y^4.\end{aligned}$$

◀

EXAMPLE 3 What is the coefficient of $x^{12}y^{13}$ in the expansion of $(x + y)^{25}$?

Solution: From the binomial theorem it follows that this coefficient is

$$\binom{25}{13} = \frac{25!}{13!\,12!} = 5{,}200{,}300.$$

◀

EXAMPLE 4 What is the coefficient of $x^{12}y^{13}$ in the expansion of $(2x - 3y)^{25}$?

Solution: First, note that this expression equals $(2x + (-3y))^{25}$. By the binomial theorem, we have

$$(2x + (-3y))^{25} = \sum_{j=0}^{25} \binom{25}{j} (2x)^{25-j}(-3y)^j.$$

Consequently, the coefficient of $x^{12}y^{13}$ in the expansion is obtained when $j = 13$, namely,

$$\binom{25}{13} 2^{12}(-3)^{13} = -\frac{25!}{13!\,12!}2^{12}3^{13}.$$

Note that another way to find the solution is to first use the binomial theorem to see that

$$(u + v)^{25} = \sum_{j=0}^{25} \binom{25}{j} u^{25-j}v^j.$$

Setting $u = 2x$ and $v = -3y$ in this equation yields the same result. ◀

We can prove some useful identities using the binomial theorem, as Corollaries 1, 2, and 3 demonstrate.

COROLLARY 1 Let n be a nonnegative integer. Then

$$\sum_{k=0}^{n} \binom{n}{k} = 2^n.$$

Proof: Using the binomial theorem with $x = 1$ and $y = 1$, we see that

$$2^n = (1 + 1)^n = \sum_{k=0}^{n} \binom{n}{k} 1^k 1^{n-k} = \sum_{k=0}^{n} \binom{n}{k}.$$

This is the desired result. ◁

There is also a nice combinatorial proof of Corollary 1, which we now present.

Proof: A set with n elements has a total of 2^n different subsets. Each subset has zero elements, one element, two elements, $\ldots$, or n elements in it. There are $\binom{n}{0}$ subsets with zero elements, $\binom{n}{1}$ subsets with one element, $\binom{n}{2}$ subsets with two elements, $\ldots$, and $\binom{n}{n}$ subsets with n elements. Therefore,

$$\sum_{k=0}^{n} \binom{n}{k}$$

counts the total number of subsets of a set with n elements. By equating the two formulas we have for the number of subsets of a set with n elements, we see that

$$\sum_{k=0}^{n} \binom{n}{k} = 2^n.$$

◁

COROLLARY 2 Let n be a positive integer. Then

$$\sum_{k=0}^{n} (-1)^k \binom{n}{k} = 0.$$

Proof: When we use the binomial theorem with $x = -1$ and $y = 1$, we see that

$$0 = 0^n = ((-1) + 1)^n = \sum_{k=0}^{n} \binom{n}{k} (-1)^k 1^{n-k} = \sum_{k=0}^{n} \binom{n}{k} (-1)^k.$$

This proves the corollary. ◁

Remark: Corollary 2 implies that

$$\binom{n}{0} + \binom{n}{2} + \binom{n}{4} + \cdots = \binom{n}{1} + \binom{n}{3} + \binom{n}{5} + \cdots.$$

COROLLARY 3 Let n be a nonnegative integer. Then

$$\sum_{k=0}^{n} 2^k \binom{n}{k} = 3^n.$$

Proof: We recognize that the left-hand side of this formula is the expansion of $(1+2)^n$ provided by the binomial theorem. Therefore, by the binomial theorem, we see that

$$(1+2)^n = \sum_{k=0}^{n} \binom{n}{k} 1^{n-k} 2^k = \sum_{k=0}^{n} \binom{n}{k} 2^k.$$

Hence

$$\sum_{k=0}^{n} 2^k \binom{n}{k} = 3^n.$$

◁

6.4.2 Pascal's Identity and Triangle

The binomial coefficients satisfy many different identities. We introduce one of the most important of these now.

THEOREM 2 **PASCAL'S IDENTITY** Let n and k be positive integers with $n \geq k$. Then

$$\binom{n+1}{k} = \binom{n}{k-1} + \binom{n}{k}.$$

Proof: We will use a combinatorial proof. Suppose that T is a set containing $n+1$ elements. Let a be an element in T, and let $S = T - \{a\}$. Note that there are $\binom{n+1}{k}$ subsets of T containing k elements. However, a subset of T with k elements either contains a together with $k-1$ elements of S, or contains k elements of S and does not contain a. Because there are $\binom{n}{k-1}$ subsets of $k-1$ elements of S, there are $\binom{n}{k-1}$ subsets of k elements of T that contain a. And there are $\binom{n}{k}$ subsets of k elements of T that do not contain a, because there are $\binom{n}{k}$ subsets of k elements of S. Consequently,

$$\binom{n+1}{k} = \binom{n}{k-1} + \binom{n}{k}.$$

◁

Remark: It is also possible to prove this identity by algebraic manipulation from the formula for $\binom{n}{r}$ (see Exercise 23).

Remark: Pascal's identity, together with the initial conditions $\binom{n}{0} = \binom{n}{n} = 1$ for all integers n, can be used to recursively define binomial coefficients. This recursive definition is useful in the

$$\binom{0}{0}$$
$$\binom{1}{0}\ \binom{1}{1}$$
$$\binom{2}{0}\ \binom{2}{1}\ \binom{2}{2}$$
$$\binom{3}{0}\ \binom{3}{1}\ \binom{3}{2}\ \binom{3}{3}$$
$$\binom{4}{0}\ \binom{4}{1}\ \binom{4}{2}\ \binom{4}{3}\ \binom{4}{4}$$
$$\binom{5}{0}\ \binom{5}{1}\ \binom{5}{2}\ \binom{5}{3}\ \binom{5}{4}\ \binom{5}{5}$$
$$\binom{6}{0}\ \binom{6}{1}\ \binom{6}{2}\ \binom{6}{3}\ \binom{6}{4}\ \binom{6}{5}\ \binom{6}{6}$$
$$\binom{7}{0}\ \binom{7}{1}\ \binom{7}{2}\ \binom{7}{3}\ \binom{7}{4}\ \binom{7}{5}\ \binom{7}{6}\ \binom{7}{7}$$
$$\binom{8}{0}\ \binom{8}{1}\ \binom{8}{2}\ \binom{8}{3}\ \binom{8}{4}\ \binom{8}{5}\ \binom{8}{6}\ \binom{8}{7}\ \binom{8}{8}$$
...

(a)

By Pascal's identity:

$$\binom{6}{4} + \binom{6}{5} = \binom{7}{5}$$

1
1 1
1 2 1
1 3 3 1
1 4 6 4 1
1 5 10 10 5 1
1 6 15 20 15 6 1
1 7 21 35 35 21 7 1
1 8 28 56 70 56 28 8 1
...

(b)

FIGURE 1 Pascal's triangle.

computation of binomial coefficients because only addition, and not multiplication, of integers is needed to use this recursive definition.

Pascal's identity is the basis for a geometric arrangement of the binomial coefficients in a triangle, as shown in Figure 1.

The nth row in the triangle consists of the binomial coefficients

$$\binom{n}{k},\ k = 0, 1, \ldots, n.$$

This triangle is known as **Pascal's triangle**, named after the French mathematician Blaise Pascal. Pascal's identity shows that when two adjacent binomial coefficients in this triangle are added, the binomial coefficient in the next row between these two coefficients is produced.

Pascal's triangle has a long and ancient history, predating Pascal by many centuries. In the East, binomial coefficients and Pascal's identity were known in the second century B.C.E. by the Indian mathematician Pingala. Later, Indian mathematicians included commentaries relating to Pascal's triangle in their books written in the first half of the last millennium. The Persian

Links

Source: National Library of Medicine

BLAISE PASCAL (1623–1662) Blaise Pascal was taught by his father, a tax collector in Rouen, France. He exhibited his talents at an early age, although his father, who had made discoveries in analytic geometry, kept mathematics books away from him to encourage other interests. At 16 Pascal discovered an important result concerning conic sections. At 18 he designed a calculating machine, which he built and sold. Pascal, along with Fermat, laid the foundations for the modern theory of probability. In this work, he made new discoveries concerning what is now called Pascal's triangle. In 1654, Pascal abandoned his mathematical pursuits to devote himself to theology. After this, he returned to mathematics only once. One night, distracted by a severe toothache, he sought comfort by studying the mathematical properties of the cycloid. Miraculously, his pain subsided, which he took as a sign of divine approval of the study of mathematics.

mathematician Al-Karaji and the multitalented Omar Khayyám wrote about Pascal's triangle in the eleventh and twelfth centuries, respectively; in Iran, Pascal's triangle is known as Khayyám's triangle. The triangle was known by the Chinese mathematician Jia Xian in the eleventh century and was written about in the 13th century by Yang Hui; in Chinese Pascal's triangle is often known as Yang Hui's triangle.

In the West, Pascal's triangle appears on the frontispiece of a 1527 book on business calculation written by the German scholar Petrus Apianus. In Italy, Pascal's triangle is called Tartaglia's triangle, after the Italian mathematician Niccolò Fontana Tartaglia who published the first few rows of the triangle in 1556. In his book *Traitè du triangle arithmétique*, published posthumously 1665, Pascal presented results about Pascal's triangle and used them to solve probability theory problems. Later French mathematicians named this triangle after Pascal; in 1730 Abraham de Moivre coined the name "Pascal's Arithmetic Triangle," which later became "Pascal's Triangle."

6.4.3 Other Identities Involving Binomial Coefficients

We conclude this section with combinatorial proofs of two of the many identities enjoyed by the binomial coefficients.

THEOREM 3 **VANDERMONDE'S IDENTITY** Let m, n, and r be nonnegative integers with r not exceeding either m or n. Then

$$\binom{m+n}{r} = \sum_{k=0}^{r} \binom{m}{r-k}\binom{n}{k}.$$

Links

Remark: This identity was discovered by mathematician Alexandre-Théophile Vandermonde in the eighteenth century.

Proof: Suppose that there are m items in one set and n items in a second set. Then the total number of ways to pick r elements from the union of these sets is $\binom{m+n}{r}$.

Another way to pick r elements from the union is to pick k elements from the second set and then $r - k$ elements from the first set, where k is an integer with $0 \leq k \leq r$. Because there are $\binom{n}{k}$ ways to choose k elements from the second set and $\binom{m}{r-k}$ ways to choose $r - k$ elements from the first set, the product rule tells us that this can be done in $\binom{m}{r-k}\binom{n}{k}$ ways. Hence, the total number of ways to pick r elements from the union also equals $\sum_{k=0}^{r} \binom{m}{r-k}\binom{n}{k}$.

We have found two expressions for the number of ways to pick r elements from the union of a set with m items and a set with n items. Equating them gives us Vandermonde's identity. ◁

Corollary 4 follows from Vandermonde's identity.

ALEXANDRE-THÉOPHILE VANDERMONDE (1735–1796) Because Alexandre-Théophile Vandermonde was a sickly child, his physician father directed him to a career in music. However, he later developed an interest in mathematics. His complete mathematical work consists of four papers published in 1771–1772. These papers include fundamental contributions on the roots of equations, on the theory of determinants, and on the knight's tour problem (introduced in the exercises in Section 10.5). Vandermonde's interest in mathematics lasted for only 2 years. Afterward, he published papers on harmony, experiments with cold, and the manufacture of steel. He also became interested in politics, joining the cause of the French revolution and holding several different positions in government.

COROLLARY 4 If n is a nonnegative integer, then

$$\binom{2n}{n} = \sum_{k=0}^{n} \binom{n}{k}^2.$$

Proof: We use Vandermonde's identity with $m = r = n$ to obtain

$$\binom{2n}{n} = \sum_{k=0}^{n} \binom{n}{n-k}\binom{n}{k} = \sum_{k=0}^{n} \binom{n}{k}^2.$$

The last equality was obtained using the identity $\binom{n}{k} = \binom{n}{n-k}$. ◁

We can prove combinatorial identities by counting bit strings with different properties, as the proof of Theorem 4 will demonstrate.

THEOREM 4 Let n and r be nonnegative integers with $r \le n$. Then

$$\binom{n+1}{r+1} = \sum_{j=r}^{n} \binom{j}{r}.$$

Proof: We use a combinatorial proof. By Example 14 in Section 6.3, the left-hand side, $\binom{n+1}{r+1}$, counts the bit strings of length $n + 1$ containing $r + 1$ ones.

We show that the right-hand side counts the same objects by considering the cases corresponding to the possible locations of the final 1 in a string with $r + 1$ ones. This final one must occur at position $r + 1, r + 2, \ldots,$ or $n + 1$. Furthermore, if the last one is the kth bit there must be r ones among the first $k - 1$ positions. Consequently, by Example 14 in Section 6.3, there are $\binom{k-1}{r}$ such bit strings. Summing over k with $r + 1 \le k \le n + 1$, we find that there are

$$\sum_{k=r+1}^{n+1} \binom{k-1}{r} = \sum_{j=r}^{n} \binom{j}{r}$$

bit strings of length n containing exactly $r + 1$ ones. (Note that the last step follows from the change of variables $j = k - 1$.) Because the left-hand side and the right-hand side count the same objects, they are equal. This completes the proof. ◁

Exercises

1. Find the expansion of $(x + y)^4$
   a) using combinatorial reasoning, as in Example 1.
   b) using the binomial theorem.
2. Find the expansion of $(x + y)^5$
   a) using combinatorial reasoning, as in Example 1.
   b) using the binomial theorem.
3. Find the expansion of $(x + y)^6$.
4. Find the coefficient of x^5y^8 in $(x + y)^{13}$.
5. How many terms are there in the expansion of $(x + y)^{100}$ after like terms are collected?
6. What is the coefficient of x^7 in $(1 + x)^{11}$?
7. What is the coefficient of x^9 in $(2 - x)^{19}$?
8. What is the coefficient of x^8y^9 in the expansion of $(3x + 2y)^{17}$?
9. What is the coefficient of $x^{101}y^{99}$ in the expansion of $(2x - 3y)^{200}$?

10. Use the binomial theorem to expand $(3x - y^2)^4$ into a sum of terms of the form $cx^a y^b$, where c is a real number and a and b are nonnegative integers.

11. Use the binomial theorem to expand $(3x^4 - 2y^3)^5$ into a sum of terms of the form $cx^a y^b$, where c is a real number and a and b are nonnegative integers.

12. Use the binomial theorem to find the coefficient of $x^a y^b$ in the expansion of $(5x^2 + 2y^3)^6$, where

a) $a = 6, b = 9$.
b) $a = 2, b = 15$.
c) $a = 3, b = 12$.
d) $a = 12, b = 0$.
e) $a = 8, b = 9$.

13. Use the binomial theorem to find the coefficient of $x^a y^b$ in the expansion of $(2x^3 - 4y^2)^7$, where

a) $a = 9, b = 8$.
b) $a = 8, b = 9$.
c) $a = 0, b = 14$.
d) $a = 12, b = 6$.
e) $a = 18, b = 2$.

***14.** Give a formula for the coefficient of x^k in the expansion of $(x + 1/x)^{100}$, where k is an integer.

***15.** Give a formula for the coefficient of x^k in the expansion of $(x^2 - 1/x)^{100}$, where k is an integer.

16. The row of Pascal's triangle containing the binomial coefficients $\binom{10}{k}$, $0 \le k \le 10$, is:

1 10 45 120 210 252 210 120 45 10 1

Use Pascal's identity to produce the row immediately following this row in Pascal's triangle.

17. What is the row of Pascal's triangle containing the binomial coefficients $\binom{9}{k}$, $0 \le k \le 9$?

18. Show that if n is a positive integer, then $1 = \binom{n}{0} < \binom{n}{1} < \cdots < \binom{n}{\lfloor n/2 \rfloor} = \binom{n}{\lceil n/2 \rceil} > \cdots > \binom{n}{n-1} > \binom{n}{n} = 1$.

19. Show that $\binom{n}{k} \le 2^n$ for all positive integers n and all integers k with $0 \le k \le n$.

20. **a)** Use Exercise 18 and Corollary 1 to show that if n is an integer greater than 1, then $\binom{n}{\lfloor n/2 \rfloor} \ge 2^n/n$.
b) Conclude from part (a) that if n is a positive integer, then $\binom{2n}{n} \ge 4^n/2n$.

21. Show that if n and k are integers with $1 \le k \le n$, then $\binom{n}{k} \le n^k/2^{k-1}$.

22. Suppose that b is an integer with $b \ge 7$. Use the binomial theorem and the appropriate row of Pascal's triangle to find the base-b expansion of $(11)_b^4$ [that is, the fourth power of the number $(11)_b$ in base-b notation].

23. Prove Pascal's identity, using the formula for $\binom{n}{r}$.

24. Suppose that k and n are integers with $1 \le k < n$. Prove the **hexagon identity**

$$\binom{n-1}{k-1}\binom{n}{k+1}\binom{n+1}{k} = \binom{n-1}{k}\binom{n}{k-1}\binom{n+1}{k+1},$$

which relates terms in Pascal's triangle that form a hexagon.

25. Prove that if n and k are integers with $1 \le k \le n$, then $k\binom{n}{k} = n\binom{n-1}{k-1}$,

a) using a combinatorial proof. [*Hint:* Show that the two sides of the identity count the number of ways to select a subset with k elements from a set with n elements and then an element of this subset.]
b) using an algebraic proof based on the formula for $\binom{n}{r}$ given in Theorem 2 in Section 6.3.

26. Prove the identity $\binom{n}{r}\binom{r}{k} = \binom{n}{k}\binom{n-k}{r-k}$, whenever n, r, and k are nonnegative integers with $r \le n$ and $k \le r$,

a) using a combinatorial argument.
b) using an argument based on the formula for the number of r-combinations of a set with n elements.

27. Show that if n and k are positive integers, then

$$\binom{n+1}{k} = (n+1)\binom{n}{k-1}\Big/k.$$

Use this identity to construct an inductive definition of the binomial coefficients.

28. Show that if p is a prime and k is an integer such that $1 \le k \le p - 1$, then p divides $\binom{p}{k}$.

29. Let n be a positive integer. Show that

$$\binom{2n}{n+1} + \binom{2n}{n} = \binom{2n+2}{n+1}\Big/2.$$

***30.** Let n and k be integers with $1 \le k \le n$. Show that

$$\sum_{k=1}^{n} \binom{n}{k}\binom{n}{k-1} = \binom{2n+2}{n+1}\Big/2 - \binom{2n}{n}.$$

***31.** Prove the **hockeystick identity**

$$\sum_{k=0}^{r} \binom{n+k}{k} = \binom{n+r+1}{r}$$

whenever n and r are positive integers,

a) using a combinatorial argument.
b) using Pascal's identity.

32. Show that if n is a positive integer, then $\binom{2n}{2} = 2\binom{n}{2} + n^2$

a) using a combinatorial argument.
b) by algebraic manipulation.

***33.** Give a combinatorial proof that $\sum_{k=1}^{n} k\binom{n}{k} = n2^{n-1}$. [*Hint:* Count in two ways the number of ways to select a committee and to then select a leader of the committee.]

***34.** Give a combinatorial proof that $\sum_{k=1}^{n} k\binom{n}{k}^2 = n\binom{2n-1}{n-1}$. [*Hint:* Count in two ways the number of ways to select a committee, with n members from a group of n mathematics professors and n computer science professors, such that the chairperson of the committee is a mathematics professor.]

35. Show that a nonempty set has the same number of subsets with an odd number of elements as it does subsets with an even number of elements.

***36.** Prove the binomial theorem using mathematical induction.

37. In this exercise we will count the number of paths in the xy plane between the origin $(0, 0)$ and point (m, n), where m and n are nonnegative integers, such that each path is made up of a series of steps, where each step is a move one unit to the right or a move one unit upward. (No moves to the left or downward are allowed.) Two such paths from $(0, 0)$ to $(5, 3)$ are illustrated here.

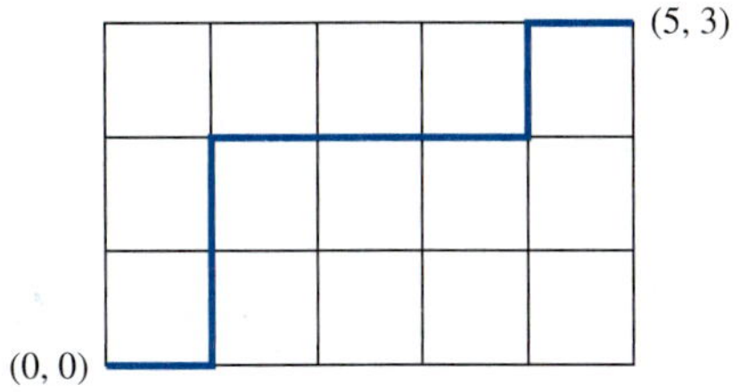

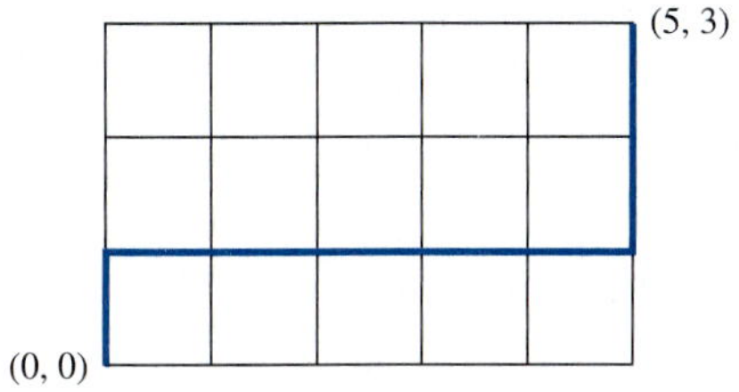

a) Show that each path of the type described can be represented by a bit string consisting of m 0s and n 1s, where a 0 represents a move one unit to the right and a 1 represents a move one unit upward.

b) Conclude from part (a) that there are $\binom{m+n}{n}$ paths of the desired type.

38. Use Exercise 37 to give an alternative proof of Corollary 2 in Section 6.3, which states that $\binom{n}{k} = \binom{n}{n-k}$ whenever k is an integer with $0 \le k \le n$. [*Hint:* Consider the number of paths of the type described in Exercise 37 from $(0, 0)$ to $(n - k, k)$ and from $(0, 0)$ to $(k, n - k)$.]

39. Use Exercise 37 to prove Theorem 4. [*Hint:* Count the number of paths with n steps of the type described in Exercise 37. Every such path must end at one of the points $(n - k, k)$ for $k = 0, 1, 2, \ldots, n$.]

40. Use Exercise 37 to prove Pascal's identity. [*Hint:* Show that a path of the type described in Exercise 37 from $(0, 0)$ to $(n + 1 - k, k)$ passes through either $(n + 1 - k, k - 1)$ or $(n - k, k)$, but not through both.]

41. Use Exercise 37 to prove the hockeystick identity from Exercise 31. [*Hint:* First, note that the number of paths from $(0, 0)$ to $(n + 1, r)$ equals $\binom{n+1+r}{r}$. Second, count the number of paths by summing the number of these paths that start by going k units upward for $k = 0, 1, 2, \ldots, r$.]

42. Give a combinatorial proof that if n is a positive integer then $\sum_{k=0}^{n} k^2\binom{n}{k} = n(n + 1)2^{n-2}$. [*Hint:* Show that both sides count the ways to select a subset of a set of n elements together with two not necessarily distinct elements from this subset. Furthermore, express the right-hand side as $n(n - 1)2^{n-2} + n2^{n-1}$.]

***43.** Determine a formula involving binomial coefficients for the nth term of a sequence if its initial terms are those listed. [*Hint:* Looking at Pascal's triangle will be helpful. Although infinitely many sequences start with a specified set of terms, each of the following lists is the start of a sequence of the type desired.]

a) 1, 3, 6, 10, 15, 21, 28, 36, 45, 55, 66, …

b) 1, 4, 10, 20, 35, 56, 84, 120, 165, 220, …

c) 1, 2, 6, 20, 70, 252, 924, 3432, 12870, 48620, …

d) 1, 1, 2, 3, 6, 10, 20, 35, 70, 126, …

e) 1, 1, 1, 3, 1, 5, 15, 35, 1, 9, …

f) 1, 3, 15, 84, 495, 3003, 18564, 116280, 735471, 4686825, …

6.5 Generalized Permutations and Combinations

6.5.1 Introduction

Links

In many counting problems, elements may be used repeatedly. For instance, a letter or digit may be used more than once on a license plate. When a dozen donuts are selected, each variety can be chosen repeatedly. This contrasts with the counting problems discussed earlier in the chapter where we considered only permutations and combinations in which each item could be used at most once. In this section we will show how to solve counting problems where elements may be used more than once.

Also, some counting problems involve indistinguishable elements. For instance, to count the number of ways the letters of the word *SUCCESS* can be rearranged, the placement of identical letters must be considered. This contrasts with the counting problems discussed earlier where all elements were considered distinguishable. In this section we will describe how to solve counting problems in which some elements are indistinguishable.

Moreover, in this section we will explain how to solve another important class of counting problems, problems involving counting the ways distinguishable elements can be placed in boxes. An example of this type of problem is the number of different ways poker hands can be dealt to four players.

Taken together, the methods described earlier in this chapter and the methods introduced in this section form a useful toolbox for solving a wide range of counting problems. When the additional methods discussed in Chapter 8 are added to this arsenal, you will be able to solve a large percentage of the counting problems that arise in a wide range of areas of study.

6.5.2 Permutations with Repetition

Counting permutations when repetition of elements is allowed can easily be done using the product rule, as Example 1 shows.

EXAMPLE 1 How many strings of length r can be formed from the uppercase letters of the English alphabet?

Solution: By the product rule, because there are 26 uppercase English letters, and because each letter can be used repeatedly, we see that there are 26^r strings of uppercase English letters of length r. ◀

The number of r-permutations of a set with n elements when repetition is allowed is given in Theorem 1.

THEOREM 1 The number of r-permutations of a set of n objects with repetition allowed is n^r.

Proof: There are n ways to select an element of the set for each of the r positions in the r-permutation when repetition is allowed, because for each choice all n objects are available. Hence, by the product rule there are n^r r-permutations when repetition is allowed. ◁

6.5.3 Combinations with Repetition

Consider these examples of combinations with repetition of elements allowed.

EXAMPLE 2 How many ways are there to select four pieces of fruit from a bowl containing apples, oranges, and pears if the order in which the pieces are selected does not matter, only the type of fruit and not the individual piece matters, and there are at least four pieces of each type of fruit in the bowl?

Solution: To solve this problem we list all the ways possible to select the fruit. There are 15 ways:

4 apples	4 oranges	4 pears
3 apples, 1 orange	3 apples, 1 pear	3 oranges, 1 apple
3 oranges, 1 pear	3 pears, 1 apple	3 pears, 1 orange
2 apples, 2 oranges	2 apples, 2 pears	2 oranges, 2 pears
2 apples, 1 orange, 1 pear	2 oranges, 1 apple, 1 pear	2 pears, 1 apple, 1 orange

The solution is the number of 4-combinations with repetition allowed from a three-element set, $\{apple, orange, pear\}$. ◀

To solve more complex counting problems of this type, we need a general method for counting the r-combinations of an n-element set. In Example 3 we will illustrate such a method.

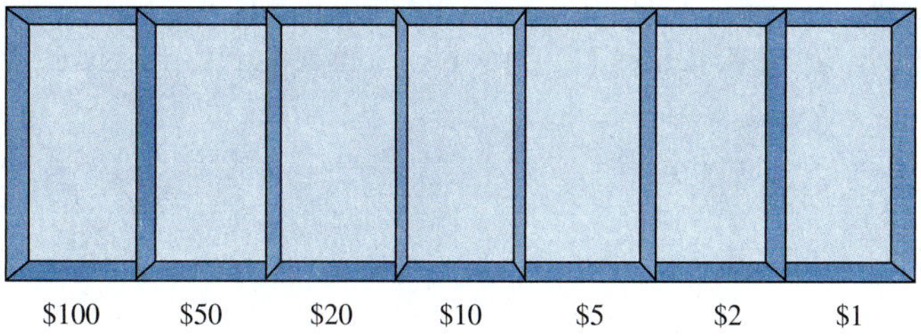

FIGURE 1 **Cash box with seven types of bills.**

EXAMPLE 3 How many ways are there to select five bills from a cash box containing \$1 bills, \$2 bills, \$5 bills, \$10 bills, \$20 bills, \$50 bills, and \$100 bills? Assume that the order in which the bills are chosen does not matter, that the bills of each denomination are indistinguishable, and that there are at least five bills of each type.

Solution: Because the order in which the bills are selected does not matter and seven different types of bills can be selected as many as five times, this problem involves counting 5-combinations with repetition allowed from a set with seven elements. Listing all possibilities would be tedious, because there are a large number of solutions. Instead, we will illustrate the use of a technique for counting combinations with repetition allowed.

Suppose that a cash box has seven compartments, one to hold each type of bill, as illustrated in Figure 1. These compartments are separated by six dividers, as shown in the picture. The choice of five bills corresponds to placing five markers in the compartments holding different types of bills. Figure 2 illustrates this correspondence for three different ways to select five bills, where the six dividers are represented by bars and the five bills by stars.

The number of ways to select five bills corresponds to the number of ways to arrange six bars and five stars in a row with a total of 11 positions. Consequently, the number of ways to select the five bills is the number of ways to select the positions of the five stars from the 11

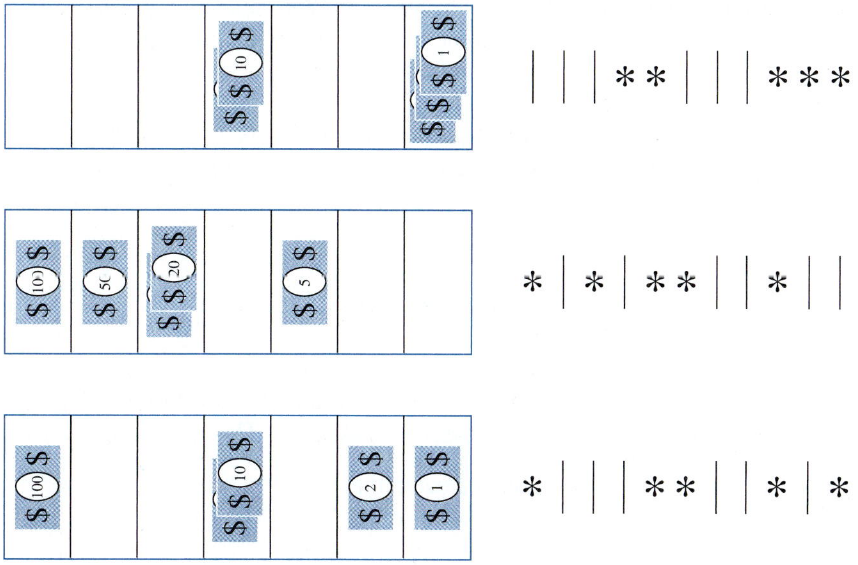

FIGURE 2 **Examples of ways to select five bills.**

positions. This corresponds to the number of unordered selections of 5 objects from a set of 11 objects, which can be done in $C(11, 5)$ ways. Consequently, there are

$$C(11, 5) = \frac{11!}{5!\,6!} = 462$$

ways to choose five bills from the cash box with seven types of bills. ◀

Theorem 2 generalizes this discussion.

THEOREM 2 There are $C(n + r - 1, r) = C(n + r - 1, n - 1)$ r-combinations from a set with n elements when repetition of elements is allowed.

Proof: Each r-combination of a set with n elements when repetition is allowed can be represented by a list of $n - 1$ bars and r stars. The $n - 1$ bars are used to mark off n different cells, with the ith cell containing a star for each time the ith element of the set occurs in the combination. For instance, a 6-combination of a set with four elements is represented with three bars and six stars. Here

|*| |*

represents the combination containing exactly two of the first element, one of the second element, none of the third element, and three of the fourth element of the set.

As we have seen, each different list containing $n - 1$ bars and r stars corresponds to an r-combination of the set with n elements, when repetition is allowed. The number of such lists is $C(n - 1 + r, r)$, because each list corresponds to a choice of the r positions to place the r stars from the $n - 1 + r$ positions that contain r stars and $n - 1$ bars. The number of such lists is also equal to $C(n - 1 + r, n - 1)$, because each list corresponds to a choice of the $n - 1$ positions to place the $n - 1$ bars. ◁

Examples 4–6 show how Theorem 2 is applied.

EXAMPLE 4 Suppose that a cookie shop has four different kinds of cookies. How many different ways can six cookies be chosen? Assume that only the type of cookie, and not the individual cookies or the order in which they are chosen, matters.

Extra Examples

Solution: The number of ways to choose six cookies is the number of 6-combinations of a set with four elements. From Theorem 2 this equals $C(4 + 6 - 1, 6) = C(9, 6)$. Because

$$C(9, 6) = C(9, 3) = \frac{9 \cdot 8 \cdot 7}{1 \cdot 2 \cdot 3} = 84,$$

there are 84 different ways to choose the six cookies. ◀

Theorem 2 can also be used to find the number of solutions of certain linear equations where the variables are integers subject to constraints. This is illustrated by Example 5.

EXAMPLE 5 How many solutions does the equation

$$x_1 + x_2 + x_3 = 11$$

have, where x_1, x_2, and x_3 are nonnegative integers?

Solution: To count the number of solutions, we note that a solution corresponds to a way of selecting 11 items from a set with three elements so that x_1 items of type one, x_2 items of type two, and x_3 items of type three are chosen. Hence, the number of solutions is equal to the number of 11-combinations with repetition allowed from a set with three elements. From Theorem 2 it follows that there are

$$C(3 + 11 - 1, 11) = C(13, 11) = C(13, 2) = \frac{13 \cdot 12}{1 \cdot 2} = 78$$

solutions.

The number of solutions of this equation can also be found when the variables are subject to constraints. For instance, we can find the number of solutions where the variables are integers with $x_1 \geq 1$, $x_2 \geq 2$, and $x_3 \geq 3$. A solution to the equation subject to these constraints corresponds to a selection of 11 items with x_1 items of type one, x_2 items of type two, and x_3 items of type three, where, in addition, there is at least one item of type one, two items of type two, and three items of type three. So, a solution corresponds to a choice of one item of type one, two of type two, and three of type three, together with a choice of five additional items of any type. By Theorem 2 this can be done in

$$C(3 + 5 - 1, 5) = C(7, 5) = C(7, 2) = \frac{7 \cdot 6}{1 \cdot 2} = 21$$

ways. Thus, there are 21 solutions of the equation subject to the given constraints. ◀

Example 6 shows how counting the number of combinations with repetition allowed arises in determining the value of a variable that is incremented each time a certain type of nested loop is traversed.

EXAMPLE 6 What is the value of k after the following pseudocode has been executed?

```
k :– 0
for i_1 := 1 to n
    for i_2 := 1 to i_1
        .
        .
        .
          for i_m := 1 to i_{m-1}
              k := k + 1
```

Solution: Note that the initial value of k is 0 and that 1 is added to k each time the nested loop is traversed with a sequence of integers $i_1, i_2, \ldots, i_m$ such that

$$1 \leq i_m \leq i_{m-1} \leq \cdots \leq i_1 \leq n.$$

The number of such sequences of integers is the number of ways to choose m integers from $\{1, 2, \ldots, n\}$, with repetition allowed. (To see this, note that once such a sequence has been selected, if we order the integers in the sequence in nondecreasing order, this uniquely defines an assignment of $i_m, i_{m-1}, \ldots, i_1$. Conversely, every such assignment corresponds to a unique unordered set.) Hence, from Theorem 2, it follows that $k = C(n + m - 1, m)$ after this code has been executed. ◀

The formulae for the numbers of ordered and unordered selections of r elements, chosen with and without repetition allowed from a set with n elements, are shown in Table 1.

TABLE 1 Combinations and Permutations With and Without Repetition.

Type	*Repetition Allowed?*	*Formula*
r-permutations	No	$\frac{n!}{(n-r)!}$
r-combinations	No	$\frac{n!}{r!\,(n-r)!}$
r-permutations	Yes	n^r
r-combinations	Yes	$\frac{(n+r-1)!}{r!\,(n-1)!}$

6.5.4 Permutations with Indistinguishable Objects

Some elements may be indistinguishable in counting problems. When this is the case, care must be taken to avoid counting things more than once. Consider Example 7.

EXAMPLE 7 How many different strings can be made by reordering the letters of the word *SUCCESS*?

Extra Examples

Solution: Because some of the letters of *SUCCESS* are the same, the answer is *not* given by the number of permutations of seven letters. This word contains three *S*s, two *C*s, one *U*, and one *E*. To determine the number of different strings that can be made by reordering the letters, first note that the three *S*s can be placed among the seven positions in $C(7, 3)$ different ways, leaving four positions free. Then the two *C*s can be placed in $C(4, 2)$ ways, leaving two free positions. The *U* can be placed in $C(2, 1)$ ways, leaving just one position free. Hence *E* can be placed in $C(1, 1)$ way. Consequently, from the product rule, the number of different strings that can be made is

$$\begin{aligned} C(7,3)C(4,2)C(2,1)C(1,1) &= \frac{7!}{3!\,4!}\cdot\frac{4!}{2!\,2!}\cdot\frac{2!}{1!\,1!}\cdot\frac{1!}{1!\,0!} \\ &= \frac{7!}{3!\,2!\,1!\,1!} \\ &= 420. \end{aligned}$$

◀

We can prove Theorem 3 using the same sort of reasoning as in Example 7.

THEOREM 3 The number of different permutations of n objects, where there are n_1 indistinguishable objects of type 1, n_2 indistinguishable objects of type 2, …, and n_k indistinguishable objects of type k, is

$$\frac{n!}{n_1!\,n_2!\cdots n_k!}.$$

Proof: To determine the number of permutations, first note that the n_1 objects of type one can be placed among the n positions in $C(n, n_1)$ ways, leaving $n - n_1$ positions free. Then the objects of type two can be placed in $C(n - n_1, n_2)$ ways, leaving $n - n_1 - n_2$ positions free. Continue placing the objects of type three, …, type $k - 1$, until at the last stage, n_k objects of type k can be placed in $C(n - n_1 - n_2 - \cdots - n_{k-1}, n_k)$ ways. Hence, by the product rule, the total number of different permutations is

$$\begin{aligned} &C(n, n_1)C(n - n_1, n_2) \cdots C(n - n_1 - \cdots - n_{k-1}, n_k) \\ &\quad = \frac{n!}{n_1!\,(n - n_1)!} \frac{(n - n_1)!}{n_2!\,(n - n_1 - n_2)!} \cdots \frac{(n - n_1 - \cdots - n_{k-1})!}{n_k!\,0!} \\ &\quad = \frac{n!}{n_1!\,n_2! \cdots n_k!}. \end{aligned}$$

◁

6.5.5 Distributing Objects into Boxes

Links

Many counting problems can be solved by enumerating the ways objects can be placed into boxes (where the order these objects are placed into the boxes does not matter). The objects can be either *distinguishable*, that is, different from each other, or *indistinguishable*, that is, considered identical. Distinguishable objects are sometimes said to be *labeled*, whereas indistinguishable objects are said to be *unlabeled*. Similarly, boxes can be *distinguishable*, that is, different, or *indistinguishable*, that is, identical. Distinguishable boxes are often said to be *labeled*, while indistinguishable boxes are said to be *unlabeled*. When you solve a counting problem using the model of distributing objects into boxes, you need to determine whether the objects are distinguishable and whether the boxes are distinguishable. Although the context of the counting problem makes these two decisions clear, counting problems are sometimes ambiguous and it may be unclear which model applies. In such a case it is best to state whatever assumptions you are making and explain why the particular model you choose conforms to your assumptions.

Extra Examples

We will see that there are closed formulae for counting the ways to distribute objects, distinguishable or indistinguishable, into distinguishable boxes. We are not so lucky when we count the ways to distribute objects, distinguishable or indistinguishable, into indistinguishable boxes; there are no closed formulae to use in these cases.

Remark: A **closed formula** is an expression that can be evaluated using a finite number of operations and that includes numbers, variables, and values of functions, where the operations and functions belong to a generally accepted set that can depend on the context. In this book, we include the usual arithmetic operations, rational powers, exponential and logarithmic functions, trigonometric functions, and the factorial function. We do not allow infinite series to be included in closed formulae.

DISTINGUISHABLE OBJECTS AND DISTINGUISHABLE BOXES We first consider the case when distinguishable objects are placed into distinguishable boxes. Consider Example 8 in which the objects are cards and the boxes are hands of players.

EXAMPLE 8 How many ways are there to distribute hands of 5 cards to each of four players from the standard deck of 52 cards?

Solution: We will use the product rule to solve this problem. To begin, note that the first player can be dealt 5 cards in $C(52, 5)$ ways. The second player can be dealt 5 cards in $C(47, 5)$ ways, because only 47 cards are left. The third player can be dealt 5 cards in $C(42, 5)$ ways. Finally, the fourth player can be dealt 5 cards in $C(37, 5)$ ways. Hence, the total number of ways to deal four players 5 cards each is

$$\begin{aligned} C(52, 5)C(47, 5)C(42, 5)C(37, 5) &= \frac{52!}{47!\,5!} \cdot \frac{47!}{42!\,5!} \cdot \frac{42!}{37!\,5!} \cdot \frac{37!}{32!\,5!} \\ &= \frac{52!}{5!\,5!\,5!\,5!\,32!}. \end{aligned}$$

◀

Remark: The solution to Example 8 equals the number of permutations of 52 objects, with 5 indistinguishable objects of each of four different types, and 32 objects of a fifth type. This equality can be seen by defining a one-to-one correspondence between permutations of this type and distributions of cards to the players. To define this correspondence, first order the cards from 1 to 52. Then cards dealt to the first player correspond to the cards in the positions assigned to objects of the first type in the permutation. Similarly, cards dealt to the second, third, and fourth players, respectively, correspond to cards in the positions assigned to objects of the second, third, and fourth type, respectively. The cards not dealt to any player correspond to cards in the positions assigned to objects of the fifth type. The reader should verify that this is a one-to-one correspondence.

Example 8 is a typical problem that involves distributing distinguishable objects into distinguishable boxes. The distinguishable objects are the 52 cards, and the five distinguishable boxes are the hands of the four players and the rest of the deck. Counting problems that involve distributing distinguishable objects into boxes can be solved using Theorem 4.

THEOREM 4 The number of ways to distribute n distinguishable objects into k distinguishable boxes so that n_i objects are placed into box i, $i = 1, 2, \ldots, k$, equals

$$\frac{n!}{n_1!\,n_2!\cdots n_k!}.$$

Theorem 4 can be proved using the product rule. We leave the details as Exercise 49. It can also be proved (see Exercise 50) by setting up a one-to-one correspondence between the permutations counted by Theorem 3 and the ways to distribute objects counted by Theorem 4.

INDISTINGUISHABLE OBJECTS AND DISTINGUISHABLE BOXES Counting the number of ways of placing n indistinguishable objects into k distinguishable boxes turns out to be the same as counting the number of n-combinations for a set with k elements when repetitions are allowed. The reason behind this is that there is a one-to-one correspondence between n-combinations from a set with k elements when repetition is allowed and the ways to place n indistinguishable balls into k distinguishable boxes. To set up this correspondence, we put a ball in the ith bin each time the ith element of the set is included in the n-combination.

EXAMPLE 9 How many ways are there to place 10 indistinguishable balls into eight distinguishable bins?

Solution: The number of ways to place 10 indistinguishable balls into eight bins equals the number of 10-combinations from a set with eight elements when repetition is allowed. Consequently, there are

$$C(8 + 10 - 1, 10) = C(17, 10) = \frac{17!}{10!7!} = 19{,}448.$$ ◀

This means that there are $C(n + r - 1, n - 1)$ ways to place r indistinguishable objects into n distinguishable boxes.

DISTINGUISHABLE OBJECTS AND INDISTINGUISHABLE BOXES Counting the ways to place n distinguishable objects into k indistinguishable boxes is more difficult than counting the ways to place objects, distinguishable or indistinguishable objects, into distinguishable boxes. We illustrate this with an example.

EXAMPLE 10 How many ways are there to put four different employees into three indistinguishable offices, when each office can contain any number of employees?

Solution: We will solve this problem by enumerating all the ways these employees can be placed into the offices. We represent the four employees by A, B, C, and D. First, we note that we can distribute employees so that all four are put into one office, three are put into one office and a fourth is put into a second office, two employees are put into one office and two put into a second office, and finally, two are put into one office, and one each put into the other two offices. Each way to distribute these employees to these offices can be represented by a way to partition the elements A, B, C, and D into disjoint subsets.

We can put all four employees into one office in exactly one way, represented by $\{\{A, B, C, D\}\}$. We can put three employees into one office and the fourth employee into a different office in exactly four ways, represented by $\{\{A, B, C\}, \{D\}\}$, $\{\{A, B, D\}, \{C\}\}$, $\{\{A, C, D\}, \{B\}\}$, and $\{\{B, C, D\}, \{A\}\}$. We can put two employees into one office and two into a second office in exactly three ways, represented by $\{\{A, B\}, \{C, D\}\}$, $\{\{A, C\}, \{B, D\}\}$, and $\{\{A, D\}, \{B, C\}\}$. Finally, we can put two employees into one office, and one each into each of the remaining two offices in six ways, represented by $\{\{A, B\}, \{C\}, \{D\}\}$, $\{\{A, C\}, \{B\}, \{D\}\}$, $\{\{A, D\}, \{B\}, \{C\}\}$, $\{\{B, C\}, \{A\}, \{D\}\}$, $\{\{B, D\}\}, \{A\}, \{C\}\}$, and $\{\{C, D\}, \{A\}, \{B\}\}$.

Counting all the possibilities, we find that there are 14 ways to put four different employees into three indistinguishable offices. Another way to look at this problem is to look at the number of offices into which we put employees. Note that there are six ways to put four different employees into three indistinguishable offices so that no office is empty, seven ways to put four different employees into two indistinguishable offices so that no office is empty, and one way to put four employees into one office so that it is not empty. ◀

A **set partition** of a set S is collection P of nonempty pairwise disjoint subsets of S that have S as their union. The elements of P are known as **blocks** of the partition. For example, if $S = \{1, 2, 3, 4, 5, 6, 7, 8, 9, 10\}$, then $P = \{\{1, 3\}, \{2, 4, 5\}, \{6\}, \{7, 8, 9, 10\}\}$ is set partition of S that has four blocks. The ordering of the numbers within blocks and the ordering of the blocks do not matter. With this definition, note that Example 10, which asks for the number of ways to put four employees into three indistinguishable offices, is the same as asking for the number of different set partitions of the set of employees. Given a set partition of the set of employees, a nonempty set of occupants of an office is a block of the set partition. Similarly, the number of ways to place n distinguishable objects into k indistinguishable boxes equals the number of set partitions.

There is no simple closed formula for the number of ways to distribute n distinguishable objects into k indistinguishable boxes. However, there is a formula involving a summation, which we will now describe. Let $S(n, j)$ denote the number of ways to distribute n distinguishable objects into j indistinguishable boxes so that no box is empty. The numbers $S(n, j)$ are called **Stirling numbers of the second kind**. For instance, Example 10 shows that $S(4, 3) = 6$, $S(4, 2) = 7$, and $S(4, 1) = 1$. We see that the number of ways to distribute n distinguishable objects into k indistinguishable boxes (where the number of boxes that are nonempty equals $k, k - 1, \ldots, 2$, or 1) equals $\sum_{j=1}^{k} S(n, j)$.

For instance, following the reasoning in Example 10, the number of ways to distribute four distinguishable objects into three indistinguishable boxes equals $S(4, 1) + S(4, 2) + S(4, 3) = 1 + 7 + 6 = 14$.

Note that $S(n, j)$ is also the number of set partitions of $\{1, 2, 3, \ldots, n\}$ into j blocks, none of which is empty. The total number of set partitions of $\{1, 2, 3, \ldots, n\}$ is denoted by $B(n)$, the nth Bell number, and with $B(0)$ defined to be 1. Even though the Bell numbers were studied in medieval Japan, they are named after Eric Temple Bell who wrote about them in the 1930s. We have $B(n) = \sum_{j=1}^{n} S(n, j)$ for all positive integers n. Using the inclusion–exclusion principle (see Section 8.6) it can be shown that

$$S(n, j) = \frac{1}{j!} \sum_{i=0}^{j-1} (-1)^i \binom{j}{i} (j - i)^n.$$

Consequently, the number of ways to distribute n distinguishable objects into k indistinguishable boxes equals

$$\sum_{j=1}^{k} S(n,j) = \sum_{j=1}^{k} \frac{1}{j!} \sum_{i=0}^{j-1} (-1)^i \binom{j}{i} (j-i)^n.$$

Remark: The reader may be curious about the Stirling numbers of the first kind. A combinatorial definition of the **signless Stirling numbers of the first kind**, the absolute values of the Stirling numbers of the first kind, can be found in the preamble to Exercise 47 in the Supplementary Exercises. For the definition of Stirling numbers of the first kind, for more information about Stirling numbers of the second kind and Bell numbers, and to learn more about Stirling numbers of the first kind and the relationship between Stirling numbers of the first and second kind, see combinatorics textbooks such as [Bó07], [Br99], and [RoTe05], and Chapter 6 in [MiRo91].

INDISTINGUISHABLE OBJECTS AND INDISTINGUISHABLE BOXES Some counting problems can be solved by determining the number of ways to distribute indistinguishable objects into indistinguishable boxes. We illustrate this principle with an example.

EXAMPLE 11 How many ways are there to pack six copies of the same book into four identical boxes, where a box can contain as many as six books?

Solution: We will enumerate all ways to pack the books. For each way to pack the books, we will list the number of books in the box with the largest number of books, followed by the numbers of books in each box containing at least one book, in order of decreasing number of books in a box. The ways we can pack the books are

6	3, 2, 1
5, 1	3, 1, 1, 1
4, 2	2, 2, 2
4, 1, 1	2, 2, 1, 1
3, 3	

For example, 4, 1, 1 indicates that one box contains four books, a second box contains a single book, and a third box contains a single book (and the fourth box is empty). We conclude that there are nine allowable ways to pack the books, because we have listed them all. ◀

Observe that distributing n indistinguishable objects into k indistinguishable boxes is the same as writing n as the sum of at most k positive integers in nonincreasing order. If $a_1 + a_2 + \cdots + a_j = n$, where $a_1, a_2, \ldots, a_j$ are positive integers with $a_1 \geq a_2 \geq \cdots \geq a_j$, we say that $a_1, a_2, \ldots, a_j$ is a **partition** of the positive integer n into j positive integers. We see that if $p_k(n)$ is the number of partitions of n into at most k positive integers, then there are $p_k(n)$ ways to distribute n indistinguishable objects into k indistinguishable boxes. No simple closed formula exists for this number. For more information about partitions of positive integers, see [Ro23].

Exercises

1. In how many different ways can five elements be selected in order from a set with three elements when repetition is allowed?
2. In how many different ways can five elements be selected in order from a set with five elements when repetition is allowed?
3. How many strings of six letters are there?
4. Every day a student randomly chooses a sandwich for lunch from a pile of wrapped sandwiches. If there are six kinds of sandwiches, how many different ways are there for the student to choose sandwiches for the seven days of a week if the order in which the sandwiches are chosen matters?

5. How many ways are there to assign three jobs to five employees if each employee can be given more than one job?

6. How many ways are there to select five unordered elements from a set with three elements when repetition is allowed?

7. How many ways are there to select three unordered elements from a set with five elements when repetition is allowed?

8. How many different ways are there to choose a dozen donuts from the 21 varieties at a donut shop?

9. A bagel shop has onion bagels, poppy seed bagels, egg bagels, salty bagels, pumpernickel bagels, sesame seed bagels, raisin bagels, and plain bagels. How many ways are there to choose

a) six bagels?
b) a dozen bagels?
c) two dozen bagels?
d) a dozen bagels with at least one of each kind?
e) a dozen bagels with at least three egg bagels and no more than two salty bagels?

10. A croissant shop has plain croissants, cherry croissants, chocolate croissants, almond croissants, apple croissants, and broccoli croissants. How many ways are there to choose

a) a dozen croissants?
b) three dozen croissants?
c) two dozen croissants with at least two of each kind?
d) two dozen croissants with no more than two broccoli croissants?
e) two dozen croissants with at least five chocolate croissants and at least three almond croissants?
f) two dozen croissants with at least one plain croissant, at least two cherry croissants, at least three chocolate croissants, at least one almond croissant, at least two apple croissants, and no more than three broccoli croissants?

11. How many ways are there to choose eight coins from a piggy bank containing 100 identical pennies and 80 identical nickels?

12. How many different combinations of pennies, nickels, dimes, quarters, and half dollars can a piggy bank contain if it has 20 coins in it?

13. A book publisher has 3000 copies of a discrete mathematics book. How many ways are there to store these books in their three warehouses if the copies of the book are indistinguishable?

14. How many solutions are there to the equation

$$x_1 + x_2 + x_3 + x_4 = 17,$$

where x_1, x_2, x_3, and x_4 are nonnegative integers?

15. How many solutions are there to the equation

$$x_1 + x_2 + x_3 + x_4 + x_5 = 21,$$

where x_i, $i = 1, 2, 3, 4, 5$, is a nonnegative integer such that

a) $x_1 \geq 1$?
b) $x_i \geq 2$ for $i = 1, 2, 3, 4, 5$?
c) $0 \leq x_1 \leq 10$?
d) $0 \leq x_1 \leq 3$, $1 \leq x_2 < 4$, and $x_3 \geq 15$?

16. How many solutions are there to the equation

$$x_1 + x_2 + x_3 + x_4 + x_5 + x_6 = 29,$$

where x_i, $i = 1, 2, 3, 4, 5, 6$, is a nonnegative integer such that

a) $x_i > 1$ for $i = 1, 2, 3, 4, 5, 6$?
b) $x_1 \geq 1$, $x_2 \geq 2$, $x_3 \geq 3$, $x_4 \geq 4$, $x_5 > 5$, and $x_6 \geq 6$?
c) $x_1 \leq 5$?
d) $x_1 < 8$ and $x_2 > 8$?

17. How many strings of 10 ternary digits (0, 1, or 2) are there that contain exactly two 0s, three 1s, and five 2s?

18. How many strings of 20-decimal digits are there that contain two 0s, four 1s, three 2s, one 3, two 4s, three 5s, two 7s, and three 9s?

19. Suppose that a large family has 14 children, including two sets of identical triplets, three sets of identical twins, and two individual children. How many ways are there to seat these children in a row of chairs if the identical triplets or twins cannot be distinguished from one another?

20. How many solutions are there to the inequality

$$x_1 + x_2 + x_3 \leq 11,$$

where x_1, x_2, and x_3 are nonnegative integers? [*Hint:* Introduce an auxiliary variable x_4 such that $x_1 + x_2 + x_3 + x_4 = 11$.]

21. My Swedish tour guide on a Baltic Sea Cruise has devised a clever way for his clients to recognize him. He owns 13 pairs of shoes of the same style, customized so that each pair has a unique color. How many ways are there for him to choose a left shoe and a right shoe from these 13 pairs

a) without restrictions and which color is on which foot matters?
b) so that the colors of the left and right shoe are different and which color is on which foot matters?
c) so that the colors of the left and right shoe are different but which color is on which foot does not matter?
d) without restrictions, but which color is on which foot does not matter?

***22.** In how many ways can an airplane pilot be scheduled for five days of flying in October if the pilot cannot work on consecutive days?

23. How many ways are there to distribute six indistinguishable balls into nine distinguishable bins?

24. How many ways are there to distribute 12 indistinguishable balls into six distinguishable bins?

25. How many ways are there to distribute 12 distinguishable objects into six distinguishable boxes so that two objects are placed in each box?

26. How many ways are there to distribute 15 distinguishable objects into five distinguishable boxes so that the boxes have one, two, three, four, and five objects in them, respectively.

27. How many positive integers less than 1,000,000 have the sum of their digits equal to 19?

28. How many positive integers less than 1,000,000 have exactly one digit equal to 9 and have a sum of digits equal to 13?

29. There are 10 questions on a discrete mathematics final exam. How many ways are there to assign scores to the problems if the sum of the scores is 100 and each question is worth at least 5 points?

30. Show that there are $C(n + r - q_1 - q_2 - \cdots - q_r - 1, n - q_1 - q_2 - \cdots - q_r)$ different unordered selections of n objects of r different types that include at least q_1 objects of type one, q_2 objects of type two, …, and q_r objects of type r.

31. How many different bit strings can be transmitted if the string must begin with a 1 bit, must include three additional 1 bits (so that a total of four 1 bits is sent), must include a total of 12 0 bits, and must have at least two 0 bits following each 1 bit?

32. How many different strings can be made from the letters in *MISSISSIPPI*, using all the letters?

33. How many different strings can be made from the letters in *ABRACADABRA*, using all the letters?

34. How many different strings can be made from the letters in *AARDVARK*, using all the letters, if all three *A*s must be consecutive?

35. How many different strings can be made from the letters in *ORONO*, using some or all of the letters?

36. How many strings with five or more characters can be formed from the letters in *SEERESS*?

37. How many strings with seven or more characters can be formed from the letters in *EVERGREEN*?

38. How many different bit strings can be formed using six 1s and eight 0s?

39. A student has three mangos, two papayas, and two kiwi fruits. If the student eats one piece of fruit each day, and only the type of fruit matters, in how many different ways can these fruits be consumed?

40. A professor packs a collection of 40 issues of a mathematics journal in four boxes with 10 issues per box. How many ways can the journals be distributed if

a) each box is numbered, so that they are distinguishable?

b) the boxes are identical, so that they cannot be distinguished?

41. How many ways are there to travel in xyz space from the origin (0, 0, 0) to the point (4, 3, 5) by taking steps one unit in the positive x direction, one unit in the positive y direction, or one unit in the positive z direction? (Moving in the negative x, y, or z direction is prohibited, so that no backtracking is allowed.)

42. How many ways are there to travel in $xyzw$ space from the origin (0, 0, 0, 0) to the point (4, 3, 5, 4) by taking steps one unit in the positive x, positive y, positive z, or positive w direction?

43. How many ways are there to deal hands of seven cards to each of five players from a standard deck of 52 cards?

44. In bridge, the 52 cards of a standard deck are dealt to four players. How many different ways are there to deal bridge hands to four players?

45. How many ways are there to deal hands of five cards to each of six players from a deck containing 48 different cards?

46. In how many ways can a dozen books be placed on four distinguishable shelves

a) if the books are indistinguishable copies of the same title?

b) if no two books are the same, and the positions of the books on the shelves matter? [*Hint:* Break this into 12 tasks, placing each book separately. Start with the sequence 1, 2, 3, 4 to represent the shelves. Represent the books by b_i, $i = 1, 2, \ldots, 12$. Place b_1 to the right of one of the terms in 1, 2, 3, 4. Then successively place $b_2, b_3, \ldots,$ and b_{12}.]

47. How many ways can n books be placed on k distinguishable shelves

a) if the books are indistinguishable copies of the same title?

b) if no two books are the same, and the positions of the books on the shelves matter?

48. A shelf holds 12 books in a row. How many ways are there to choose five books so that no two adjacent books are chosen? [*Hint:* Represent the books that are chosen by bars and the books not chosen by stars. Count the number of sequences of five bars and seven stars so that no two bars are adjacent.]

***49.** Use the product rule to prove Theorem 4, by first placing objects in the first box, then placing objects in the second box, and so on.

***50.** Prove Theorem 4 by first setting up a one-to-one correspondence between permutations of n objects with n_i indistinguishable objects of type i, $i = 1, 2, 3, \ldots, k$, and the distributions of n objects in k boxes such that n_i objects are placed in box i, $i = 1, 2, 3, \ldots, k$ and then applying Theorem 3.

***51.** In this exercise we will prove Theorem 2 by setting up a one-to-one correspondence between the set of r-combinations with repetition allowed of $S = \{1, 2, 3, \ldots, n\}$ and the set of r-combinations of the set $T = \{1, 2, 3, \ldots, n + r - 1\}$.

a) Arrange the elements in an r-combination, with repetition allowed, of S into an increasing sequence $x_1 \leq x_2 \leq \cdots \leq x_r$. Show that the sequence formed by adding $k - 1$ to the kth term is strictly increasing. Conclude that this sequence is made up of r distinct elements from T.

b) Show that the procedure described in (a) defines a one-to-one correspondence between the set of r-combinations, with repetition allowed, of S and the r-combinations of T. [*Hint:* Show the correspondence can be reversed by associating to the r-combination $\{x_1, x_2, \ldots, x_r\}$ of T, with $1 \le x_1 < x_2 < \cdots < x_r \le n + r - 1$, the r-combination with repetition allowed from S, formed by subtracting $k - 1$ from the kth element.]

c) Conclude that the number of r-combinations with repetition allowed from a set with n elements is $C(n + r - 1, r)$.

52. How many ways are there to distribute five distinguishable objects into three indistinguishable boxes?

53. How many ways are there to distribute six distinguishable objects into four indistinguishable boxes so that each of the boxes contains at least one object?

54. How many ways are there to put five temporary employees into four identical offices?

55. How many ways are there to put six temporary employees into four identical offices so that there is at least one temporary employee in each of these four offices?

56. How many ways are there to distribute five indistinguishable objects into three indistinguishable boxes?

57. How many ways are there to distribute six indistinguishable objects into four indistinguishable boxes so that each of the boxes contains at least one object?

58. How many ways are there to pack eight identical DVDs into five indistinguishable boxes so that each box contains at least one DVD?

59. How many ways are there to pack nine identical DVDs into three indistinguishable boxes so that each box contains at least two DVDs?

60. How many ways are there to distribute five balls into seven boxes if each box must have at most one ball in it if

a) both the balls and boxes are labeled?

b) the balls are labeled, but the boxes are unlabeled?

c) the balls are unlabeled, but the boxes are labeled?

d) both the balls and boxes are unlabeled?

61. How many ways are there to distribute five balls into three boxes if each box must have at least one ball in it if

a) both the balls and boxes are labeled?

b) the balls are labeled, but the boxes are unlabeled?

c) the balls are unlabeled, but the boxes are labeled?

d) both the balls and boxes are unlabeled?

62. Suppose that a basketball league has 32 teams, split into two conferences of 16 teams each. Each conference is split into three divisions. Suppose that the North Central Division has five teams. Each of the teams in the North Central Division plays four games against each of the other teams in this division, three games against each of the 11 remaining teams in the conference, and two games against each of the 16 teams in the other conference. In how many different orders can the games of one of the teams in the North Central Division be scheduled?

***63.** Suppose that a weapons inspector must inspect each of five different sites twice, visiting one site per day. The inspector is free to select the order in which to visit these sites, but cannot visit site X, the most suspicious site, on two consecutive days. In how many different orders can the inspector visit these sites?

64. How many different terms are there in the expansion of $(x_1 + x_2 + \cdots + x_m)^n$ after all terms with identical sets of exponents are added?

***65.** Prove the **Multinomial Theorem:** If n is a positive integer, then

$$(x_1 + x_2 + \cdots + x_m)^n = \sum_{n_1 + n_2 + \cdots + n_m = n} C(n; n_1, n_2, \ldots, n_m) x_1^{n_1} x_2^{n_2} \cdots x_m^{n_m},$$

where

$$C(n; n_1, n_2, \ldots, n_m) = \frac{n!}{n_1!\, n_2! \cdots n_m!}$$

is a **multinomial coefficient**.

66. Find the expansion of $(x + y + z)^4$

67. Find the coefficient of $x^3y^2z^5$ in $(x + y + z)^{10}$.

68. How many terms are there in the expansion of $(x + y + z)^{100}$?

69. Prove that for all positive integers n and k, $S(n, k) = S(n - 1, k - 1) + kS(n - 1, k)$, where $S(n, j)$ is the Stirling number of the second kind that counts the number of set partitions of $\{1, 2, 3, \ldots, n\}$ into j non-empty subsets.

70. Prove that $B(n) = \sum_{k=0}^{n-1} \binom{n-1}{k} B(n - 1 - k)$ for all positive integers n.

71. Find $B(n)$ for $n = 1, 2, 3$, and 4 by listing all the set partitions of $\{1, 2, 3, \ldots, n\}$.

72. Find $B(5)$ and $B(6)$ using Exercises 70 and 71.

6.6 Generating Permutations and Combinations

6.6.1 Introduction

Methods for counting various types of permutations and combinations were described in the previous sections of this chapter, but sometimes permutations or combinations need to be generated, not just counted. Consider the following three problems. First, suppose that a salesperson must visit six different cities. In which order should these cities be visited to minimize total travel time? One way to determine the best order is to determine the travel time for each of the $6! = 720$ different orders in which the cities can be visited and choose the one with the smallest travel time. Second, suppose we are given a set of six positive integers and wish to find a subset of them that has 100 as their sum, if such a subset exists. One way to find these numbers is to generate all $2^6 = 64$ subsets and check the sum of their elements. Third, suppose a laboratory has 95 employees. A group of 12 of these employees with a particular set of 25 skills is needed for a project. (Each employee can have one or more of these skills.) One way to find such a set of employees is to generate all sets of 12 of these employees and check whether they have the desired skills. These examples show that it is often necessary to generate permutations and combinations to solve problems.

6.6.2 Generating Permutations

Links

Any set with n elements can be placed in one-to-one correspondence with the set $\{1, 2, 3, \ldots, n\}$. We can list the permutations of any set of n elements by generating the permutations of the n smallest positive integers and then replacing these integers with the corresponding elements. Many different algorithms have been developed to generate the $n!$ permutations of this set. We will describe one of these that is based on the **lexicographic** (or **dictionary) ordering** of the set of permutations of $\{1, 2, 3, \ldots, n\}$. In this ordering, the permutation $a_1 a_2 \cdots a_n$ precedes the permutation of $b_1 b_2 \cdots b_n$, if for some k, with $1 \le k \le n$, $a_1 = b_1, a_2 = b_2, \ldots, a_{k-1} = b_{k-1}$, and $a_k < b_k$. In other words, a permutation of the set of the n smallest positive integers precedes (in lexicographic order) a second permutation if the number in this permutation in the first position where the two permutations disagree is smaller than the number in that position in the second permutation.

EXAMPLE 1 The permutation 23415 of the set $\{1, 2, 3, 4, 5\}$ precedes the permutation 23514, because these permutations agree in the first two positions, but the number in the third position in the first permutation, 4, is smaller than the number in the third position in the second permutation, 5. Similarly, the permutation 41532 precedes 52143. ◀

An algorithm for generating the permutations of $\{1, 2, \ldots, n\}$ can be based on a procedure that constructs the next permutation in lexicographic order following a given permutation $a_1 a_2 \cdots a_n$. We will show how this can be done. First, suppose that $a_{n-1} < a_n$. Interchange a_{n-1} and a_n to obtain a larger permutation. No other permutation is both larger than the original permutation and smaller than the permutation obtained by interchanging a_{n-1} and a_n. For instance, the next larger permutation after 234156 is 234165. On the other hand, if $a_{n-1} > a_n$, then a larger permutation cannot be obtained by interchanging these last two terms in the permutation. Look at the last three integers in the permutation. If $a_{n-2} < a_{n-1}$, then the last three integers in the permutation can be rearranged to obtain the next largest permutation. Put the smaller of the two integers a_{n-1} and a_n that is greater than a_{n-2} in position $n - 2$. Then, place the remaining integer and a_{n-2} into the last two positions in increasing order. For instance, the next larger permutation after 234165 is 234516.

On the other hand, if $a_{n-2} > a_{n-1}$ (and $a_{n-1} > a_n$), then a larger permutation cannot be obtained by permuting the last three terms in the permutation. Based on these observations, a general method can be described for producing the next larger permutation in increasing order following a given permutation $a_1 a_2 \cdots a_n$. First, find the integers a_j and a_{j+1} with $a_j < a_{j+1}$ and

$$a_{j+1} > a_{j+2} > \cdots > a_n,$$

that is, the last pair of adjacent integers in the permutation where the first integer in the pair is smaller than the second. Then, the next larger permutation in lexicographic order is obtained by putting in the jth position the least integer among $a_{j+1}, a_{j+2}, \ldots,$ and a_n that is greater than a_j and listing in increasing order the rest of the integers $a_j, a_{j+1}, \ldots, a_n$ in positions $j + 1$ to n. It is easy to see that there is no other permutation larger than the permutation $a_1 a_2 \cdots a_n$ but smaller than the new permutation produced. (The verification of this fact is left as an exercise for the reader.)

EXAMPLE 2 What is the next permutation in lexicographic order after 362541?

Extra Examples

Solution: The last pair of integers a_j and a_{j+1} where $a_j < a_{j+1}$ is $a_3 = 2$ and $a_4 = 5$. The least integer to the right of 2 that is greater than 2 in the permutation is $a_5 = 4$. Hence, 4 is placed in the third position. Then the integers 2, 5, and 1 are placed in order in the last three positions, giving 125 as the last three positions of the permutation. Hence, the next permutation is 364125. ◀

To produce the $n!$ permutations of the integers $1, 2, 3, \ldots, n$, begin with the smallest permutation in lexicographic order, namely, $123 \cdots n$, and successively apply the procedure described for producing the next larger permutation of $n! - 1$ times. This yields all the permutations of the n smallest integers in lexicographic order.

EXAMPLE 3 Generate the permutations of the integers 1, 2, 3 in lexicographic order.

Solution: Begin with 123. The next permutation is obtained by interchanging 3 and 2 to obtain 132. Next, because $3 > 2$ and $1 < 3$, permute the three integers in 132. Put the smaller of 3 and 2 in the first position, and then put 1 and 3 in increasing order in positions 2 and 3 to obtain 213. This is followed by 231, obtained by interchanging 1 and 3, because $1 < 3$. The next larger permutation has 3 in the first position, followed by 1 and 2 in increasing order, namely, 312. Finally, interchange 1 and 2 to obtain the last permutation, 321. We have generated the permutations of 1, 2, 3 in lexicographic order. They are 123, 132, 213, 231, 312, and 321. ◀

Algorithm 1 displays the procedure for finding the next permutation in lexicographic order after a permutation that is not $n\ n-1\ n-2\ \ldots\ 2\ 1$, which is the largest permutation.

ALGORITHM 1 Generating the Next Permutation in Lexicographic Order.

```
procedure next permutation(a1a2 ... an: permutation of
        {1, 2, ..., n} not equal to n n − 1 ... 2 1)
j := n − 1
while aj > aj+1
  j := j − 1
{j is the largest subscript with aj < aj+1}
k := n
while aj > ak
  k := k − 1
{ak is the smallest integer greater than aj to the right of aj}
interchange aj and ak
r := n
s := j + 1
while r > s
  interchange ar and as
  r := r − 1
  s := s + 1
{this puts the tail end of the permutation after the jth position in increasing order}
{a1a2 ... an is now the next permutation}
```

6.6.3 Generating Combinations

Links

How can we generate all the combinations of the elements of a finite set? Because a combination is just a subset, we can use the correspondence between subsets of $\{a_1, a_2, \ldots, a_n\}$ and bit strings of length n.

Recall that the bit string corresponding to a subset has a 1 in position k if a_k is in the subset, and has a 0 in this position if a_k is not in the subset. If all the bit strings of length n can be listed, then by the correspondence between subsets and bit strings, a list of all the subsets is obtained.

Recall that a bit string of length n is also the binary expansion of an integer between 0 and $2^n - 1$. The 2^n bit strings can be listed in order of their increasing size as integers in their binary expansions. To produce all binary expansions of length n, start with the bit string $000 \ldots 00$, with n zeros. Then, successively find the next expansion until the bit string $111 \ldots 11$ is obtained. At each stage the next binary expansion is found by locating the first position from the right that is not a 1, then changing all the 1s to the right of this position to 0s and making this first 0 (from the right) a 1.

EXAMPLE 4 Find the next bit string after 10 0010 0111.

Solution: The first bit from the right that is not a 1 is the fourth bit from the right. Change this bit to a 1 and change all the following bits to 0s. This produces the next larger bit string, 10 0010 1000. ◀

The procedure for producing the next larger bit string after $b_{n-1}b_{n-2}\ldots b_1b_0$ is given as Algorithm 2.

ALGORITHM 2 Generating the Next Larger Bit String.

procedure *next bit string*($b_{n-1}\,b_{n-2}\ldots b_1b_0$: bit string not equal to $11\ldots11$)
$i := 0$
while $b_i = 1$
  $b_i := 0$
  $i := i + 1$
$b_i := 1$
$\{b_{n-1}\,b_{n-2}\ldots b_1b_0$ is now the next bit string$\}$

Next, an algorithm for generating the r-combinations of the set $\{1, 2, 3, \ldots, n\}$ will be given. An r-combination can be represented by a sequence containing the elements in the subset in increasing order. The r-combinations can be listed using lexicographic order on these sequences. In this lexicographic ordering, the first r-combination is $\{1, 2, \ldots, r-1, r\}$ and the last r-combination is $\{n-r+1, n-r+2, \ldots, n-1, n\}$. The next r-combination after $a_1a_2 \cdots a_r$ can be obtained in the following way: First, locate the last element a_i in the sequence such that $a_i \neq n - r + i$. Then, replace a_i with $a_i + 1$ and a_j with $a_i + j - i + 1$, for $j = i + 1, i + 2, \ldots, r$. It is left for the reader to show that this produces the next larger r-combination in lexicographic order. This procedure is illustrated with Example 5.

EXAMPLE 5 Find the next larger 4-combination of the set $\{1, 2, 3, 4, 5, 6\}$ after $\{1, 2, 5, 6\}$.

Solution: The last term among the terms a_i with $a_1 = 1$, $a_2 = 2$, $a_3 = 5$, and $a_4 = 6$ such that $a_i \neq 6 - 4 + i$ is $a_2 = 2$. To obtain the next larger 4-combination, increment a_2 by 1 to obtain

$a_2 = 3$. Then set $a_3 = 3 + 1 = 4$ and $a_4 = 3 + 2 = 5$. Hence the next larger 4-combination is $\{1, 3, 4, 5\}$. ◀

Algorithm 3 displays pseudocode for this procedure.

ALGORITHM 3 Generating the Next r-Combination in Lexicographic Order.

```
procedure next r-combination({a_1, a_2, ..., a_r}: proper subset of
        {1, 2, ..., n} not equal to {n − r + 1, ..., n} with
        a_1 < a_2 < ··· < a_r)
i := r
while a_i = n − r + i
    i := i − 1
a_i := a_i + 1
for j := i + 1 to r
    a_j := a_i + j − i
{{a_1, a_2, ..., a_r} is now the next combination}
```

Exercises

1. Place these permutations of $\{1, 2, 3, 4, 5\}$ in lexicographic order: 43521, 15432, 45321, 23451, 23514, 14532, 21345, 45213, 31452, 31542.

2. Place these permutations of $\{1,2,3,4,5,6\}$ in lexicographic order: 234561, 231456, 165432, 156423, 543216, 541236, 231465, 314562, 432561, 654321, 654312, 435612.

3. The name of a file in a computer directory consists of three uppercase letters followed by a digit, where each letter is either A, B, or C, and each digit is either 1 or 2. List the name of these files in lexicographic order, where we order letters using the usual alphabetic order of letters.

4. Suppose that the name of a file in a computer directory consists of three digits followed by two lowercase letters and each digit is 0, 1, or 2, and each letter is either a or b. List the name of these files in lexicographic order, where we order letters using the usual alphabetic order of letters.

5. Find the next larger permutation in lexicographic order after each of these permutations.

a) 1432 **b)** 54123 **c)** 12453
d) 45231 **e)** 6714235 **f)** 31528764

6. Find the next larger permutation in lexicographic order after each of these permutations.

a) 1342 **b)** 45321 **c)** 13245
d) 612345 **e)** 1623547 **f)** 23587416

7. Use Algorithm 1 to generate the 24 permutations of the first four positive integers in lexicographic order.

8. Use Algorithm 2 to list all the subsets of the set $\{1, 2, 3, 4\}$.

9. Use Algorithm 3 to list all the 3-combinations of $\{1, 2, 3, 4, 5\}$.

10. Show that Algorithm 1 produces the next larger permutation in lexicographic order.

11. Show that Algorithm 3 produces the next larger r-combination in lexicographic order after a given r-combination.

12. Develop an algorithm for generating the r-permutations of a set of n elements.

13. List all 3-permutations of $\{1, 2, 3, 4, 5\}$.

The remaining exercises in this section develop another algorithm for generating the permutations of $\{1, 2, 3, \ldots, n\}$. This algorithm is based on Cantor expansions of integers. Every nonnegative integer less than $n!$ has a unique Cantor expansion

$$a_1 1! + a_2 2! + \cdots + a_{n-1}(n-1)!$$

where a_i is a nonnegative integer not exceeding i, for $i = 1, 2, \ldots, n-1$. The integers $a_1, a_2, \ldots, a_{n-1}$ are called the **Cantor digits** of this integer.

Given a permutation of $\{1, 2, \ldots, n\}$, let $a_{k-1}, k = 2, 3, \ldots, n$, be the number of integers less than k that follow k in the permutation. For instance, in the permutation 43215, a_1 is the number of integers less than 2 that follow 2, so $a_1 = 1$. Similarly, for this example $a_2 = 2$, $a_3 = 3$, and $a_4 = 0$. Consider the function from the set of permutations of $\{1, 2, 3, \ldots, n\}$ to the set of nonnegative integers less than $n!$ that sends a permutation to the integer that has $a_1, a_2, \ldots, a_{n-1}$, defined in this way, as its Cantor digits.

14. Find the Cantor digits $a_1, a_2, \ldots, a_{n-1}$ that correspond to these permutations.

a) 246531 **b)** 12345 **c)** 654321

*15. Show that the correspondence described in the preamble is a bijection between the set of permutations of $\{1, 2, 3, \ldots, n\}$ and the nonnegative integers less than $n!$.

16. Find the permutations of $\{1, 2, 3, 4, 5\}$ that correspond to these integers with respect to the correspondence between Cantor expansions and permutations as described in the preamble to Exercise 14.

a) 3 **b)** 89 **c)** 111

17. Develop an algorithm for producing all permutations of a set of n elements based on the correspondence described in the preamble to Exercise 14.

Key Terms and Results

TERMS

combinatorics: the study of arrangements of objects

enumeration: the counting of arrangements of objects

tree diagram: a diagram made up of a root, branches leaving the root, and other branches leaving some of the endpoints of branches

permutation: an ordered arrangement of the elements of a set

***r*-permutation:** an ordered arrangement of r elements of a set

***P*(*n*,*r*):** the number of r-permutations of a set with n elements

***r*-combination:** an unordered selection of r elements of a set

***C*(*n*,*r*):** the number of r-combinations of a set with n elements

binomial coefficient $\binom{n}{r}$: also the number of r-combinations of a set with n elements

combinatorial proof: a proof that uses counting arguments rather than algebraic manipulation to prove a result

Pascal's triangle: a representation of the binomial coefficients where the ith row of the triangle contains $\binom{i}{j}$ for $j = 0, 1, 2, \ldots, i$

***S*(*n*,*j*):** the Stirling number of the second kind denoting the number of ways to distribute n distinguishable objects into j indistinguishable boxes so that no box is empty

RESULTS

product rule for counting: The number of ways to do a procedure that consists of two tasks is the product of the number of ways to do the first task and the number of ways to do the second task after the first task has been done.

product rule for sets: The number of elements in the Cartesian product of finite sets is the product of the number of elements in each set.

sum rule for counting: The number of ways to do a task in one of two ways is the sum of the number of ways to do these tasks if they cannot be done simultaneously.

sum rule for sets: The number of elements in the union of pairwise disjoint finite sets is the sum of the numbers of elements in these sets.

subtraction rule for counting or **inclusion–exclusion for sets:** If a task can be done in either n_1 ways or n_2 ways, then the number of ways to do the task is $n_1 + n_2$ minus the number of ways to do the task that are common to the two different ways.

subtraction rule or **inclusion–exclusion for sets:** The number of elements in the union of two sets is the sum of the number of elements in these sets minus the number of elements in their intersection.

division rule for counting: There are n/d ways to do a task if it can be done using a procedure that can be carried out in n ways, and for every way w, exactly d of the n ways correspond to way w.

division rule for sets: Suppose that a finite set A is the union of n disjoint subsets each with d elements. Then $n = |A|/d$.

the pigeonhole principle: When more than k objects are placed in k boxes, there must be a box containing more than one object.

the generalized pigeonhole principle: When N objects are placed in k boxes, there must be a box containing at least $\lceil N/k \rceil$ objects.

$$P(n, r) = \frac{n!}{(n-r)!}$$

$$C(n, r) = \binom{n}{r} = \frac{n!}{r!(n-r)!}$$

Pascal's identity: $\binom{n+1}{k} = \binom{n}{k-1} + \binom{n}{k}$

the binomial theorem: $(x+y)^n = \sum_{k=0}^{n} \binom{n}{k} x^{n-k} y^k$

There are n^r r-permutations of a set with n elements when repetition is allowed.

There are $C(n + r - 1, r)$ r-combinations of a set with n elements when repetition is allowed.

There are $n!/(n_1! n_2! \cdots n_k!)$ permutations of n objects of k types where there are n_i indistinguishable objects of type i for $i = 1, 2, 3, \ldots, k$.

the algorithm for generating the permutations of the set $\{1, 2, \ldots, n\}$

Review Questions

1. Explain how the sum and product rules can be used to find the number of bit strings with a length not exceeding 10.

2. Explain how to find the number of bit strings of length not exceeding 10 that have at least one 0 bit.

3. **a)** How can the product rule be used to find the number of functions from a set with m elements to a set with n elements?

b) How many functions are there from a set with five elements to a set with 10 elements?

c) How can the product rule be used to find the number of one-to-one functions from a set with m elements to a set with n elements?

d) How many one-to-one functions are there from a set with five elements to a set with 10 elements?

e) How many onto functions are there from a set with five elements to a set with 10 elements?

4. How can you find the number of possible outcomes of a playoff between two teams where the first team that wins four games wins the playoff?

5. How can you find the number of bit strings of length ten that either begin with 101 or end with 010?

6. a) State the pigeonhole principle.

b) Explain how the pigeonhole principle can be used to show that among any 11 integers, at least two must have the same last digit.

7. a) State the generalized pigeonhole principle.

b) Explain how the generalized pigeonhole principle can be used to show that among any 91 integers, there are at least ten that end with the same digit.

8. a) What is the difference between an r-combination and an r-permutation of a set with n elements?

b) Derive an equation that relates the number of r-combinations and the number of r-permutations of a set with n elements.

c) How many ways are there to select six students from a class of 25 to serve on a committee?

d) How many ways are there to select six students from a class of 25 to hold six different executive positions on a committee?

9. a) What is Pascal's triangle?

b) How can a row of Pascal's triangle be produced from the one above it?

10. What is meant by a combinatorial proof of an identity? How is such a proof different from an algebraic one?

11. Explain how to prove Pascal's identity using a combinatorial argument.

12. a) State the binomial theorem.

b) Explain how to prove the binomial theorem using a combinatorial argument.

c) Find the coefficient of $x^{100}y^{101}$ in the expansion of $(2x + 5y)^{201}$.

13. a) Explain how to find a formula for the number of ways to select r objects from n objects when repetition is allowed and order does not matter.

b) How many ways are there to select a dozen objects from among objects of five different types if objects of the same type are indistinguishable?

c) How many ways are there to select a dozen objects from these five different types if there must be at least three objects of the first type?

d) How many ways are there to select a dozen objects from these five different types if there cannot be more than four objects of the first type?

e) How many ways are there to select a dozen objects from these five different types if there must be at least two objects of the first type, but no more than three objects of the second type?

14. a) Let n and r be positive integers. Explain why the number of solutions of the equation $x_1 + x_2 + \cdots + x_n = r$, where x_i is a nonnegative integer for $i = 1, 2, 3, \ldots, n$, equals the number of r-combinations of a set with n elements.

b) How many solutions in nonnegative integers are there to the equation $x_1 + x_2 + x_3 + x_4 = 17$?

c) How many solutions in positive integers are there to the equation in part (b)?

15. a) Derive a formula for the number of permutations of n objects of k different types, where there are n_1 indistinguishable objects of type one, n_2 indistinguishable objects of type two, …, and n_k indistinguishable objects of type k.

b) How many ways are there to order the letters of the word *INDISCREETNESS*?

16. Describe an algorithm for generating all the permutations of the set of the n smallest positive integers.

17. a) How many ways are there to deal hands of five cards to six players from a standard 52-card deck?

b) How many ways are there to distribute n distinguishable objects into k distinguishable boxes so that n_i objects are placed in box i?

18. Describe an algorithm for generating all the combinations of the set of the n smallest positive integers.

Supplementary Exercises

1. How many ways are there to choose 6 items from 10 distinct items when

a) the items in the choices are ordered and repetition is not allowed?

b) the items in the choices are ordered and repetition is allowed?

c) the items in the choices are unordered and repetition is not allowed?

d) the items in the choices are unordered and repetition is allowed?

2. How many ways are there to choose 10 items from 6 distinct items when

a) the items in the choices are ordered and repetition is not allowed?

b) the items in the choices are ordered and repetition is allowed?

c) the items in the choices are unordered and repetition is not allowed?

d) the items in the choices are unordered and repetition is allowed?

3. A test contains 100 true/false questions. How many different ways can a student answer the questions on the test, if answers may be left blank?

4. How many strings of length 10 either start with 000 or end with 1111?

5. How many bit strings of length 10 over the alphabet $\{a, b, c\}$ have either exactly three as or exactly four bs?

6. The internal telephone numbers in the phone system on a campus consist of five digits, with the first digit not equal to zero. How many different numbers can be assigned in this system?

7. An ice cream parlor has 28 different flavors, 8 different kinds of sauce, and 12 toppings.

a) In how many different ways can a dish of three scoops of ice cream be made where each flavor can be used more than once and the order of the scoops does not matter?

b) How many different kinds of small sundaes are there if a small sundae contains one scoop of ice cream, a sauce, and a topping?

c) How many different kinds of large sundaes are there if a large sundae contains three scoops of ice cream, where each flavor can be used more than once and the order of the scoops does not matter; two kinds of sauce, where each sauce can be used only once and the order of the sauces does not matter; and three toppings, where each topping can be used only once and the order of the toppings does not matter?

8. How many positive integers less than 1000

a) have exactly three decimal digits?

b) have an odd number of decimal digits?

c) have at least one decimal digit equal to 9?

d) have no odd decimal digits?

e) have two consecutive decimal digits equal to 5?

f) are palindromes (that is, read the same forward and backward)?

9. When the numbers from 1 to 1000 are written out in decimal notation, how many of each of these digits are used?

a) 0 **b)** 1 **c)** 2 **d)** 9

10. There are 12 signs of the zodiac. How many people are needed to guarantee that at least six of these people have the same sign?

11. A fortune cookie company makes 213 different fortunes. A student eats at a restaurant that uses fortunes from this company and gives each customer one fortune cookie at the end of a meal. What is the largest possible number of times that the student can eat at the restaurant without getting the same fortune four times?

12. How many people are needed to guarantee that at least two were born on the same day of the week and in the same month (perhaps in different years)?

13. Show that given any set of 10 positive integers not exceeding 50 there exist at least two different five-element subsets of this set that have the same sum.

14. A package of baseball cards contains 20 cards. How many packages must be purchased to ensure that two cards in these packages are identical if there are a total of 550 different cards?

15. a) How many cards must be chosen from a standard deck of 52 cards to guarantee that at least two of the four aces are chosen?

b) How many cards must be chosen from a standard deck of 52 cards to guarantee that at least two of the four aces and at least two of the 13 kinds are chosen?

c) How many cards must be chosen from a standard deck of 52 cards to guarantee that there are at least two cards of the same kind?

d) How many cards must be chosen from a standard deck of 52 cards to guarantee that there are at least two cards of each of two different kinds?

***16.** Show that in any set of $n + 1$ positive integers not exceeding $2n$ there must be two that are relatively prime.

***17.** Show that in a sequence of m integers there exists one or more consecutive terms with a sum divisible by m.

18. Show that if five points are picked in the interior of a square with a side length of 2, then at least two of these points are no farther than $\sqrt{2}$ apart.

19. Show that the decimal expansion of a rational number must repeat itself from some point onward.

20. Once a computer worm infects a personal computer via an infected email message, it sends a copy of itself to 100 email addresses it finds in the electronic message mailbox on this personal computer. What is the maximum number of different computers this one computer can infect in the time it takes for the infected message to be forwarded five times?

21. How many ways are there to choose a dozen donuts from 20 varieties

a) if there are no two donuts of the same variety?

b) if all donuts are of the same variety?

c) if there are no restrictions?

d) if there are at least two varieties among the dozen donuts chosen?

e) if there must be at least six blueberry-filled donuts?

f) if there can be no more than six blueberry-filled donuts?

22. Find n if

a) $P(n, 2) = 110$. **b)** $P(n, n) = 5040$.

c) $P(n, 4) = 12P(n, 2)$.

23. Find n if

a) $C(n, 2) = 45$. **b)** $C(n, 3) = P(n, 2)$.

c) $C(n, 5) = C(n, 2)$.

24. Show that if n and r are nonnegative integers and $n \geq r$, then

$$P(n + 1, r) = P(n, r)(n + 1)/(n + 1 - r).$$

***25.** Suppose that S is a set with n elements. How many ordered pairs (A, B) are there such that A and B are subsets of S with $A \subseteq B$? [*Hint:* Show that each element of S belongs to A, $B - A$, or $S - B$.]

26. Give a combinatorial proof of Corollary 2 of Section 6.4 by setting up a correspondence between the subsets of a set with an even number of elements and the subsets of this set with an odd number of elements. [*Hint:* Take an element a in the set. Set up the correspondence by putting a in the subset if it is not already in it and taking it out if it is in the subset.]

27. Let n and r be integers with $1 \le r < n$. Show that

$$C(n, r-1) = C(n+2, r+1) - 2C(n+1, r+1) + C(n, r+1).$$

28. Prove using mathematical induction that $\sum_{j=2}^{n} C(j, 2) = C(n+1, 3)$ whenever n is an integer greater than 1.

29. Show that if n is an integer then

$$\sum_{k=0}^{n} 3^k \binom{n}{k} = 4^n.$$

30. Show that $\sum_{i=1}^{n-1} \sum_{j=i+1}^{n} 1 = \binom{n}{2}$ if n is an integer with $n \ge 2$.

31. Show that $\sum_{i=1}^{n-2} \sum_{j=i+1}^{n-1} \sum_{k=j+1}^{n} 1 = \binom{n}{3}$ if n is an integer with $n \ge 3$.

32. In this exercise we will derive a formula for the sum of the squares of the n smallest positive integers. We will count the number of triples (i, j, k) where i, j, and k are integers such that $0 \le i < k$, $0 \le j < k$, and $1 \le k \le n$ in two ways.

a) Show that there are k^2 such triples with a fixed k. Deduce that there are $\sum_{k=1}^{n} k^2$ such triples.

b) Show that the number of such triples with $0 \le i < j < k$ and the number of such triples with $0 \le j < i < k$ both equal $C(n+1, 3)$.

c) Show that the number of such triples with $0 \le i = j < k$ equals $C(n+1, 2)$.

d) Combining part (a) with parts (b) and (c), conclude that

$$\sum_{k=1}^{n} k^2 = 2C(n+1, 3) + C(n+1, 2) = n(n+1)(2n+1)/6.$$

***33.** How many bit strings of length n, where $n \ge 4$, contain exactly two occurrences of 01?

34. Let S be a set. We say that a collection of subsets $A_1, A_2, \ldots, A_n$ each containing d elements, where $d \ge 2$, is *2-colorable* if it is possible to assign to each element of S one of two different colors so that in every subset A_i there are elements that have been assigned each color. Let $m(d)$ be the largest integer such that every collection of fewer than $m(d)$ sets each containing d elements is 2-colorable.

a) Show that the collection of all subsets with d elements of a set S with $2d - 1$ elements is not 2-colorable.

b) Show that $m(2) = 3$.

****c)** Show that $m(3) = 7$. [*Hint:* Show that the collection $\{1, 3, 5\}$, $\{1, 2, 6\}$, $\{1, 4, 7\}$, $\{2, 3, 4\}$, $\{2, 5, 7\}$, $\{3, 6, 7\}$, $\{4, 5, 6\}$ is not 2-colorable. Then show that all collections of six sets with three elements each are 2-colorable.]

35. A professor writes 20 multiple-choice questions, each with the possible answer a, b, c, or d, for a discrete mathematics test. If the number of questions with a, b, c, and d as their answer is 8, 3, 4, and 5, respectively, how many different answer keys are possible, if the questions can be placed in any order?

36. How many different arrangements are there of eight people seated at a round table, where two arrangements are considered the same if one can be obtained from the other by a rotation?

37. How many ways are there to assign 24 students to five faculty advisors?

38. How many ways are there to choose a dozen apples from a bushel containing 20 indistinguishable Delicious apples, 20 indistinguishable Macintosh apples, and 20 indistinguishable Granny Smith apples, if at least three of each kind must be chosen?

39. How many solutions are there to the equation $x_1 + x_2 + x_3 = 17$, where x_1, x_2, and x_3 are nonnegative integers with

a) $x_1 > 1$, $x_2 > 2$, and $x_3 > 3$?

b) $x_1 < 6$ and $x_3 > 5$?

c) $x_1 < 4$, $x_2 < 3$, and $x_3 > 5$?

40. a) How many different strings can be made from the word *PEPPERCORN* when all the letters are used?

b) How many of these strings start and end with the letter *P*?

c) In how many of these strings are the three letter *P*s consecutive?

41. How many subsets of a set with ten elements

a) have fewer than five elements?

b) have more than seven elements?

c) have an odd number of elements?

42. A witness to a hit-and-run accident tells the police that the license plate of the car in the accident, which contains three letters followed by three digits, starts with the letters *AS* and contains both the digits 1 and 2. How many different license plates can fit this description?

43. How many ways are there to put n identical objects into m distinct containers so that no container is empty?

44. How many ways are there to seat six adults and eight children in a row of chairs so that no two adults are seated next to each other?

45. How many ways are there to distribute six objects to five boxes if

a) both the objects and boxes are labeled?

b) the objects are labeled, but the boxes are unlabeled?

c) the objects are unlabeled, but the boxes are labeled?

d) both the objects and the boxes are unlabeled?

46. How many ways are there to distribute five objects into six boxes if

a) both the objects and boxes are labeled?

b) the objects are labeled, but the boxes are unlabeled?

c) the objects are unlabeled, but the boxes are labeled?

d) both the objects and the boxes are unlabeled?

The **signless Stirling number of the first kind** $c(n, k)$, where k and n are integers with $1 \leq k \leq n$, equals the number of ways to arrange n people around k circular tables with at least one person seated at each table, where two seatings of m people around a circular table are considered the same if everyone has the same left neighbor and the same right neighbor.

47. Find these signless Stirling numbers of the first kind.

a) $c(3, 2)$ **b)** $c(4, 2)$

c) $c(4, 3)$ **d)** $c(5, 4)$

48. Show that if n is a positive integer, then $\sum_{j=1}^{n} c(n, j) = n!$.

49. Show that if n is a positive integer with $n \geq 3$, then $c(n, n-2) = (3n-1)C(n, 3)/4$.

***50.** Show that if n and k are integers with $1 \leq k < n$, then $c(n+1, k) = c(n, k-1) + nc(n, k)$.

51. Give a combinatorial proof that 2^n divides $n!$ whenever n is an even positive integer. [*Hint:* Use Theorem 3 in Section 6.5 to count the number of permutations of $2n$ objects where there are two indistinguishable objects of n different types.]

52. How many 11-element RNA sequences consist of 4 As, 3 Cs, 2 Us, and 2 Gs, and end with CAA?

Exercises 53 and 54 are based on a discussion in [RoTe09]. A method used in the 1960s for sequencing RNA chains used enzymes to break chains after certain links. Some enzymes break RNA chains after each G link, while others break them after each C or U link. Using these enzymes it is sometimes possible to correctly sequence all the bases in an RNA chain.

***53.** Suppose that when an enzyme that breaks RNA chains after each G link is applied to a 12-link chain, the fragments obtained are G, CCG, AAAG, and UCCG, and when an enzyme that breaks RNA chains after each C or U link is applied, the fragments obtained are C, C, C, C, GGU, and GAAAG. Can you determine the entire 12-link RNA chain from these two sets of fragments? If so, what is this RNA chain?

***54.** Suppose that when an enzyme that breaks RNA chains after each G link is applied to a 12-link chain, the fragments obtained are AC, UG, and ACG and when an enzyme that breaks RNA chains after each C or U link is applied, the fragments obtained are U, GAC, and GAC. Can you determine the entire RNA chain from these two sets of fragments? If so, what is this RNA chain?

55. Devise an algorithm for generating all the r-permutations of a finite set when repetition is allowed.

56. Devise an algorithm for generating all the r-combinations of a finite set when repetition is allowed.

***57.** Show that if m and n are integers with $m \geq 3$ and $n \geq 3$, then $R(m, n) \leq R(m, n-1) + R(m-1, n)$.

***58.** Show that $R(3, 4) \geq 7$ by showing that in a group of six people, where any two people are friends or enemies, there are not necessarily three mutual friends or four mutual enemies.

Computer Projects

Write programs with these input and output.

1. Given a positive integer n and a nonnegative integer not exceeding n, find the number of r-permutations and r-combinations of a set with n elements.

2. Given positive integers n and r, find the number of r-permutations when repetition is allowed and r-combinations when repetition is allowed of a set with n elements.

3. Given a sequence of positive integers, find the longest increasing and the longest decreasing subsequence of the sequence.

***4.** Given an equation $x_1 + x_2 + \cdots + x_n = C$, where C is a constant, and $x_1, x_2, \ldots, x_n$ are nonnegative integers, list all the solutions.

5. Given a positive integer n, list all the permutations of the set $\{1, 2, 3, \ldots, n\}$ in lexicographic order.

6. Given a positive integer n and a nonnegative integer r not exceeding n, list all the r-combinations of the set $\{1, 2, 3, \ldots, n\}$ in lexicographic order.

7. Given a positive integer n and a nonnegative integer r not exceeding n, list all the r-permutations of the set $\{1, 2, 3, \ldots, n\}$ in lexicographic order.

8. Given a positive integer n, list all the combinations of the set $\{1, 2, 3, \ldots, n\}$.

9. Given positive integers n and r, list all the r-permutations, with repetition allowed, of the set $\{1, 2, 3, \ldots, n\}$.

10. Given positive integers n and r, list all the r-combinations, with repetition allowed, of the set $\{1, 2, 3, \ldots, n\}$.

Computations and Explorations

Use a computational program or programs you have written to do these exercises.

1. Find the number of possible outcomes in a two-team playoff when the winner is the first team to win 5 out of 9, 6 out of 11, 7 out of 13, and 8 out of 15.
2. Which binomial coefficients are odd? Can you formulate a conjecture based on numerical evidence?
3. Verify that $C(2n, n)$ is divisible by the square of a prime, when $n \neq 1, 2,$ or 4, for as many positive integers n as you can. [That $C(2n, n)$ is divisible by the square of a prime with $n \neq 1, 2,$ or 4 was proved in 1996 by Andrew Granville and Olivier Ramaré, settling a conjecture made in 1980 by Paul Erdős and Ron Graham.]
4. Find as many odd integers n less than 200 as you can for which $C(n, \lfloor n/2 \rfloor)$ is not divisible by the square of a prime. Formulate a conjecture based on your evidence.

*5. For each integer less than 100 determine whether $C(2n, n)$ is divisible by 3. Can you formulate a conjecture that tells us for which integers n the binomial coefficient $C(2n, n)$ is divisible by 3 based on the digits in the base three expansion of n?

6. Generate all the permutations of a set with eight elements.
7. Generate all the 6-permutations of a set with nine elements.
8. Generate all combinations of a set with eight elements.
9. Generate all 5-combinations with repetition allowed of a set with seven elements.

Writing Projects

Respond to these with essays using outside sources.

1. Describe some of the earliest uses of the pigeonhole principle by Dirichlet and other mathematicians.
2. Discuss ways in which the current telephone numbering plan can be extended to accommodate the rapid demand for more telephone numbers. (See if you can find some of the proposals coming from the telecommunications industry.) For each new numbering plan you discuss, show how to find the number of different telephone numbers it supports.
3. Find the password requirements of at least five different financial applications and determine the number of possible passwords for each of them.
4. One way users can log onto their Android phones is to use a lockscreen pattern. Explain what lockscreen patterns are and determine how many different lockscreen patterns are possible.
5. Discuss the importance of combinatorial reasoning in gene sequencing and related problems involving genomes.
6. Many combinatorial identities are described in this book. Find some sources of such identities and describe important combinatorial identities besides those already introduced in this book. Give some representative proofs, including combinatorial ones, of some of these identities.
7. Describe the different models used to model the distribution of particles in statistical mechanics, including Maxwell–Boltzmann, Bose–Einstein, and Fermi–Dirac statistics. In each case, describe the counting techniques used in the model.
8. Using books about the history of mathematics, describe how set partitions arose in medieval Japanese writings. ([Ma12] is an excellent source for information.)
9. Define the Stirling numbers of the first kind and describe some of their properties and the identities they satisfy.
10. Describe some of the properties and the identities that Stirling numbers of the second kind satisfy, including the connection between Stirling numbers of the first and second kinds.
11. Describe the latest discoveries of values and bounds for Ramsey numbers.
12. Describe additional ways to generate all the permutations of a set with n elements besides those found in Section 6.6. Compare these algorithms and the algorithms described in the text and exercises of Section 6.6 in terms of their computational complexity.
13. Describe at least one way to generate all the partitions of a positive integer n. (See Exercise 49 in Section 5.3.)

CHAPTER

7 Discrete Probability

7.1 An Introduction to Discrete Probability

7.2 Probability Theory

7.3 Bayes' Theorem

7.4 Expected Value and Variance

Combinatorics and probability theory share common origins. The theory of probability was first developed more than 300 years ago, when certain gambling games were analyzed. Although probability theory was originally invented to study gambling, it now plays an essential role in a wide variety of disciplines. For example, probability theory is extensively applied in the study of genetics, where it can be used to help understand the inheritance of traits. Probability theory remains popular both because of its applicability to gambling, an extremely popular human endeavor, and to its use in a wide range of disciplines.

In computer science, probability theory plays an important role in the study of the complexity of algorithms. In particular, ideas and techniques from probability theory are used to determine the average-case complexity of algorithms. Probabilistic algorithms can be used to solve many problems that cannot be easily or practically solved by deterministic algorithms. In a probabilistic algorithm, instead of always following the same steps when given the same input, as a deterministic algorithm does, the algorithm makes one or more random choices, which may lead to different output. In combinatorics, probability theory can even be used to show that objects with certain properties exist. The probabilistic method, a technique in combinatorics introduced by Paul Erdős and Alfréd Rényi, shows that an object with a specified property exists by showing that there is a positive probability that a randomly constructed object has this property. Probability theory can help us answer questions that involve uncertainty, such as determining whether we should reject an incoming mail message as spam based on the words that appear in the message.

7.1 An Introduction to Discrete Probability

7.1.1 Introduction

Probability theory dates back to 1526 when the Italian mathematician, physician, and gambler Girolamo Cardano wrote the first known systematic treatment of the subject in his book *Liber de Ludo Aleae* (*Book on Games of Chance*). (This book was not published until 1663, which may have held back the development of probability theory.) In the seventeenth century, the French mathematician Blaise Pascal determined the odds of winning some popular bets based on the outcome when a pair of dice is repeatedly rolled. In the eighteenth century, the French mathematician Laplace, who also studied gambling, defined the probability of an event as the number of successful outcomes divided by the number of possible outcomes. For instance, the probability that a die comes up an odd number when it is rolled is the number of successful outcomes—namely, the number of ways it can come up odd—divided by the number of possible outcomes—namely, the number of different ways the die can come up. There are a total of six possible outcomes—namely, $1, 2, 3, 4, 5$, and 6—and exactly three of these are successful outcomes—namely, $1, 3$, and 5. Hence, the probability that the die comes up an odd number is $3/6 = 1/2$. (Note that it has been assumed that all possible outcomes are equally likely, or, in other words, that the die is fair.)

In this section we will restrict ourselves to experiments that have finitely many, equally likely, outcomes. This permits us to use Laplace's definition of the probability of an event. We will continue our study of probability in Section 7.2, where we will study experiments with finitely many outcomes that are not necessarily equally likely. In Section 7.2 we will also introduce some key concepts in probability theory, including conditional probability, independence of events, and random variables. In Section 7.4 we will introduce the concepts of the expectation and variance of a random variable.

7.1.2 Finite Probability

An **experiment** is a procedure that yields one of a given set of possible outcomes. The **sample space** of the experiment is the set of possible outcomes. An **event** is a subset of the sample space. Laplace's definition of the probability of an event with finitely many possible outcomes will now be stated.

Definition 1 If S is a finite nonempty sample space of equally likely outcomes, and E is an event, that is, a subset of S, then the *probability* of E is $p(E) = \frac{|E|}{|S|}$.

The probability of an event can never be negative or more than one!

According to Laplace's definition, the probability of an event is between 0 and 1. To see this, note that if E is an event from a finite sample space S, then $0 \leq |E| \leq |S|$, because $E \subseteq S$. Thus, $0 \leq p(E) = |E|/|S| \leq 1$.

Examples 1–7 illustrate how the probability of an event is found.

EXAMPLE 1 An urn contains four blue balls and five red balls. What is the probability that a ball chosen at random from the urn is blue?

Extra Examples

Solution: To calculate the probability, note that there are nine possible outcomes, and four of these possible outcomes produce a blue ball. Hence, the probability that a blue ball is chosen is $4/9$. ◂

EXAMPLE 2 What is the probability that when two dice are rolled, the sum of the numbers on the two dice is 7?

Solution: There are a total of 36 equally likely possible outcomes when two dice are rolled. (The product rule can be used to see this; because each die has six possible outcomes, the total number of outcomes when two dice are rolled is $6^2 = 36$.) There are six successful outcomes, namely, (1, 6), (2, 5), (3, 4), (4, 3), (5, 2), and (6, 1), where the values of the first and second dice

Links

Science History Images/ Alamy Stock Photo

GIROLAMO CARDANO (1501–1576) Cardano, born in Pavia, Italy, was the illegitimate child of Fazio Cardano, a lawyer, mathematician, and friend of Leonardo da Vinci, and Chiara Micheria, a young widow. In spite of illness and poverty, Cardano was able to study at the universities of Pavia and Padua, from where he received his medical degree. Cardano was not accepted into Milan's College of Physicians because of his illegitimate birth, as well as his eccentricity and confrontational style. Nevertheless, his medical skills were highly regarded. One of his main accomplishments as a physician is the first description of typhoid fever.

Cardano published more than 100 books on a diverse range of subjects, including medicine, the natural sciences, mathematics, gambling, physical inventions and experiments, and astrology. He also wrote a fascinating autobiography. In mathematics, Cardano's book *Ars Magna*, published in 1545, established the foundations of abstract algebra. This was the most comprehensive book on abstract algebra for more than a century; it presents many novel ideas of Cardano and of others, including methods for solving cubic and quartic equations from their coefficients. Cardano also made several important contributions to cryptography. Cardano was an advocate of education for the deaf, believing, unlike his contemporaries, that deaf people could learn to read and write before learning to speak, and could use their minds just as well as hearing people.

Cardano was often short of money. However, he kept himself solvent through gambling and winning money by beating others at chess. His book about games of chance, *Liber de Ludo Aleae*, written in 1526 (but published in 1663), offers the first systematic treatment of probability; it also describes effective ways to cheat. Cardano was considered to be a man of dubious moral character; he was often described as a liar, gambler, lecher, and heretic.

are represented by an ordered pair. Hence, the probability that a seven comes up when two fair dice are rolled is $6/36 = 1/6$. ◀

Links

Lotteries are extremely popular throughout the world. We can easily compute the odds of winning different types of lotteries, as illustrated in Examples 3 and 4. (The odds of winning prizes in the popular Mega Millions and Powerball lotteries, as of 2018, are studied in Exercises 38–41. The rules for these games change every few years. The odds for the version of these games in 2012 are studied in the Supplementary Exercises.)

EXAMPLE 3 In a lottery, players win a large prize when they pick four digits that match, in the correct order, four digits selected by a random mechanical process. A smaller prize is won if only three digits are matched. What is the probability that a player wins the large prize? What is the probability that a player wins the small prize?

Solution: There is only one way to choose all four digits correctly. By the product rule, there are $10^4 = 10{,}000$ ways to choose four digits. Hence, the probability that a player wins the large prize is $1/10{,}000 = 0.0001$.

Players win the smaller prize when they correctly choose exactly three of the four digits. Exactly one digit must be wrong to get three digits correct, but not all four correct. By the sum rule, to find the number of ways to choose exactly three digits correctly, we add the number of ways to choose four digits matching the digits picked in all but the ith position, for $i = 1, 2, 3, 4$.

To count the number of successes with the first digit incorrect, note that there are nine possible choices for the first digit (all but the one correct digit), and one choice for each of the other digits, namely, the correct digits for these slots. Hence, there are nine ways to choose four digits where the first digit is incorrect, but the last three are correct. Similarly, there are nine ways to choose four digits where the second digit is incorrect, nine with the third digit incorrect, and nine with the fourth digit incorrect. Hence, there is a total of 36 ways to choose four digits with exactly three of the four digits correct. Thus, the probability that a player wins the smaller prize is $36/10{,}000 = 9/2500 = 0.0036$. ◀

EXAMPLE 4 There are many lotteries now that award enormous prizes to people who correctly choose a set of six numbers out of the first n positive integers, where n is usually between 30 and 60. What is the probability that a person picks the correct six numbers out of 40?

Solution: There is only one winning combination. The total number of ways to choose six numbers out of 40 is

$$C(40, 6) = \frac{40!}{34!\,6!} = 3{,}838{,}380.$$

Links

Georgios Kollidas/ Shutterstock

PIERRE-SIMON LAPLACE (1749–1827) Pierre-Simon Laplace came from humble origins in Normandy. In his childhood he was educated in a school run by the Benedictines. At 16 he entered the University of Caen intending to study theology. However, he soon realized his true interests were in mathematics. After completing his studies, he was named a provisional professor at Caen, and in 1769 he became professor of mathematics at the Paris Military School.

Laplace is best known for his contributions to celestial mechanics, the study of the motions of heavenly bodies. His *Traité de Mécanique Céleste* is considered one of the greatest scientific works of the early nineteenth century. Laplace was one of the founders of probability theory and made many contributions to mathematical statistics. His work in this area is documented in his book *Théorie Analytique des Probabilités*, in which he defined the probability of an event as the ratio of the number of favorable outcomes to the total number of outcomes of an experiment.

Laplace was famous for his political flexibility. He was loyal, in succession, to the French Republic, Napoleon, and King Louis XVIII. This flexibility permitted him to be productive before, during, and after the French Revolution.

Consequently, the probability of picking a winning combination is $1/3{,}838{,}380 \approx 0.00000026$. (Here the symbol $\approx$ means approximately equal to.) ◀

Links

Poker, and other card games, are growing in popularity. To win at these games it helps to know the probability of different hands. We can find the probability of specific hands that arise in card games using the techniques developed so far. A deck of cards contains 52 cards. There are 13 different kinds of cards, with four cards of each kind. (Among the terms commonly used instead of "kind" are "rank," "face value," "denomination," and "value.") These kinds are twos, threes, fours, fives, sixes, sevens, eights, nines, tens, jacks, queens, kings, and aces. There are also four suits: spades, clubs, hearts, and diamonds, each containing 13 cards, with one card of each kind in a suit. In many poker games, a hand consists of five cards.

EXAMPLE 5 Find the probability that a hand of five cards in poker contains four cards of one kind.

Solution: By the product rule, the number of hands of five cards with four cards of one kind is the product of the number of ways to pick one kind, the number of ways to pick the four of this kind out of the four in the deck of this kind, and the number of ways to pick the fifth card. This is

$$C(13, 1)C(4, 4)C(48, 1).$$

By Example 11 in Section 6.3 there are $C(52, 5)$ different hands of five cards. Hence, the probability that a hand contains four cards of one kind is

$$\frac{C(13, 1)C(4, 4)C(48, 1)}{C(52, 5)} = \frac{13 \cdot 1 \cdot 48}{2{,}598{,}960} \approx 0.00024.$$

◀

EXAMPLE 6 What is the probability that a poker hand contains a full house, that is, three of one kind and two of another kind?

Solution: By the product rule, the number of hands containing a full house is the product of the number of ways to pick two kinds in order, the number of ways to pick three out of four for the first kind, and the number of ways to pick two out of four for the second kind. (Note that the order of the two kinds matters, because, for instance, three queens and two aces is different from three aces and two queens.) We see that the number of hands containing a full house is

$$P(13, 2)C(4, 3)C(4, 2) = 13 \cdot 12 \cdot 4 \cdot 6 = 3744.$$

Because there are $C(52, 5) = 2{,}598{,}960$ poker hands, the probability of a full house is

$$\frac{3744}{2{,}598{,}960} \approx 0.0014.$$

◀

EXAMPLE 7 What is the probability that the numbers 11, 4, 17, 39, and 23 are drawn in that order from a bin containing 50 balls labeled with the numbers $1, 2, \ldots, 50$ if (a) the ball selected is not returned to the bin before the next ball is selected and (b) the ball selected is returned to the bin before the next ball is selected?

Solution: (a) By the product rule, there are $50 \cdot 49 \cdot 48 \cdot 47 \cdot 46 = 254{,}251{,}200$ ways to select the balls because each time a ball is drawn there is one fewer ball to choose from. Consequently, the probability that 11, 4, 17, 39, and 23 are drawn in that order is $1/254{,}251{,}200$. This is an example of **sampling without replacement**.

(b) By the product rule, there are $50^5 = 312{,}500{,}000$ ways to select the balls because there are 50 possible balls to choose from each time a ball is drawn. Consequently, the probability that 11, 4, 17, 39, and 23 are drawn in that order is $1/312{,}500{,}000$. This is an example of **sampling with replacement**. ◀

7.1.3 Probabilities of Complements and Unions of Events

We can use counting techniques to find the probability of events derived from other events.

THEOREM 1 Let E be an event in a sample space S. The probability of the event $\overline{E} = S - E$, the complementary event of E, is given by

$$p(\overline{E}) = 1 - p(E).$$

Proof: To find the probability of the event $\overline{E} = S - E$, note that $|\overline{E}| = |S| - |E|$. Hence,

$$p(\overline{E}) = \frac{|S| - |E|}{|S|} = 1 - \frac{|E|}{|S|} = 1 - p(E).$$ ◁

There is an alternative strategy for finding the probability of an event when a direct approach does not work well. Instead of determining the probability of the event, the probability of its complement can be found. This is often easier to do, as Example 8 shows.

EXAMPLE 8 A sequence of 10 bits is randomly generated. What is the probability that at least one of these bits is 0?

Solution: Let E be the event that at least one of the 10 bits is 0. Then $\overline{E}$ is the event that all the bits are 1s. Because the sample space S is the set of all bit strings of length 10, it follows that

$$p(E) = 1 - p(\overline{E}) = 1 - \frac{|\overline{E}|}{|S|} = 1 - \frac{1}{2^{10}}$$
$$= 1 - \frac{1}{1024} = \frac{1023}{1024}.$$

Hence, the probability that the bit string will contain at least one 0 bit is $1023/1024$. It is quite difficult to find this probability directly without using Theorem 1. ◀

We can also find the probability of the union of two events.

THEOREM 2 Let E_1 and E_2 be events in the sample space S. Then

$$p(E_1 \cup E_2) = p(E_1) + p(E_2) - p(E_1 \cap E_2).$$

Proof: Using the formula given in Section 2.2 for the number of elements in the union of two sets, it follows that

$$|E_1 \cup E_2| = |E_1| + |E_2| - |E_1 \cap E_2|.$$

Hence,

$$\begin{aligned} p(E_1 \cup E_2) &= \frac{|E_1 \cup E_2|}{|S|} \\ &= \frac{|E_1| + |E_2| - |E_1 \cap E_2|}{|S|} \\ &= \frac{|E_1|}{|S|} + \frac{|E_2|}{|S|} - \frac{|E_1 \cap E_2|}{|S|} \\ &= p(E_1) + p(E_2) - p(E_1 \cap E_2). \end{aligned}$$

◁

EXAMPLE 9 What is the probability that a positive integer selected at random from the set of positive integers not exceeding 100 is divisible by either 2 or 5?

Extra Examples

Solution: Let E_1 be the event that the integer selected at random is divisible by 2, and let E_2 be the event that it is divisible by 5. Then $E_1 \cup E_2$ is the event that it is divisible by either 2 or 5. Also, $E_1 \cap E_2$ is the event that it is divisible by both 2 and 5, or equivalently, that it is divisible by 10. Because $|E_1| = 50$, $|E_2| = 20$, and $|E_1 \cap E_2| = 10$, it follows that

$$\begin{aligned} p(E_1 \cup E_2) &= p(E_1) + p(E_2) - p(E_1 \cap E_2) \\ &= \frac{50}{100} + \frac{20}{100} - \frac{10}{100} = \frac{3}{5}. \end{aligned}$$

◀

7.1.4 Probabilistic Reasoning

A common problem is determining which of two events is more likely. Analyzing the probabilities of such events can be tricky. Example 10 describes a problem of this type. It discusses a famous problem originating with the television game show *Let's Make a Deal* and named after the host of the show, Monty Hall.

EXAMPLE 10 **The Monty Hall Three-Door Puzzle** Suppose you are a game show contestant. You have a chance to win a large prize. You are asked to select one of three doors to open; the large prize is behind one of the three doors and the other two doors are losers. Once you select a door, the game show host, who knows what is behind each door, does the following. First, whether or not you selected the winning door, he opens one of the other two doors that he knows is a losing door (selecting at random if both are losing doors). Then he asks you whether you would like to switch doors. Which strategy should you use? Should you change doors or keep your original selection, or does it not matter?

Links

Solution: The probability you select the correct door (before the host opens a door and asks you whether you want to change) is $1/3$, because the three doors are equally likely to be the correct door. The probability this is the correct door does not change once the game show host opens one of the other doors, because he will always open a door that the prize is not behind.

The probability that you selected incorrectly is the probability the prize is behind one of the two doors you did not select. Consequently, the probability that you selected incorrectly is $2/3$.

If you selected incorrectly, when the game show host opens a door to show you that the prize is not behind it, the prize is behind the other door. You will always win if your initial choice was incorrect and you change doors. So, by changing doors, the probability you win is $2/3$. In other words, you should always change doors when given the chance to do so by the game show host. This doubles the probability that you will win. (A more rigorous treatment of this puzzle can be found in Exercise 15 of Section 7.3. For much more on this notorious puzzle and its variations, see [Ro09].) ◀

Exercises

1. What is the probability that a card selected at random from a standard deck of 52 cards is an ace?

2. What is the probability that a fair die comes up six when it is rolled?

3. What is the probability that a randomly selected integer chosen from the first 100 positive integers is odd?

4. What is the probability that a randomly selected day of a leap year (with 366 possible days) is in April?

5. What is the probability that the sum of the numbers on two dice is even when they are rolled?

6. What is the probability that a card selected at random from a standard deck of 52 cards is an ace or a heart?

7. What is the probability that when a coin is flipped six times in a row, it lands heads up every time?

8. What is the probability that a five-card poker hand contains the ace of hearts?

9. What is the probability that a five card poker hand does not contain the queen of hearts?

10. What is the probability that a five-card poker hand contains the two of diamonds and the three of spades?

11. What is the probability that a five-card poker hand contains the two of diamonds, the three of spades, the six of hearts, the ten of clubs, and the king of hearts?

12. What is the probability that a five-card poker hand contains exactly one ace?

13. What is the probability that a five-card poker hand contains at least one ace?

14. What is the probability that a five-card poker hand contains cards of five different kinds?

15. What is the probability that a five-card poker hand contains two pairs (that is, two of each of two different kinds and a fifth card of a third kind)?

16. What is the probability that a five-card poker hand contains a flush, that is, five cards of the same suit?

17. What is the probability that a five-card poker hand contains a straight, that is, five cards that have consecutive kinds? (Note that an ace can be considered either the lowest card of an A-2-3-4-5 straight or the highest card of a 10-J-Q-K-A straight.)

18. What is the probability that a five-card poker hand contains a straight flush, that is, five cards of the same suit of consecutive kinds?

***19.** What is the probability that a five-card poker hand contains cards of five different kinds and does not contain a flush or a straight?

20. What is the probability that a five-card poker hand contains a royal flush, that is, the 10, jack, queen, king, and ace of one suit?

21. What is the probability that a fair die never comes up an even number when it is rolled six times?

22. What is the probability that a positive integer not exceeding 100 selected at random is divisible by 3?

23. What is the probability that a positive integer not exceeding 100 selected at random is divisible by 5 or 7?

24. Find the probability of winning a lottery by selecting the correct six integers, where the order in which these integers are selected does not matter, from the positive integers not exceeding

a) 30. **b)** 36. **c)** 42. **d)** 48.

25. Find the probability of winning a lottery by selecting the correct six integers, where the order in which these integers are selected does not matter, from the positive integers not exceeding

a) 50. **b)** 52. **c)** 56. **d)** 60.

26. Find the probability of selecting none of the correct six integers in a lottery, where the order in which these integers are selected does not matter, from the positive integers not exceeding

a) 40. **b)** 48. **c)** 56. **d)** 64.

27. Find the probability of selecting exactly one of the correct six integers in a lottery, where the order in which these integers are selected does not matter, from the positive integers not exceeding

a) 40. **b)** 48. **c)** 56. **d)** 64.

28. In a superlottery, a player selects 7 numbers out of the first 80 positive integers. What is the probability that a person wins the grand prize by picking 7 numbers that are among the 11 numbers selected at random by a computer.

29. In a superlottery, players win a fortune if they choose the eight numbers selected by a computer from the positive integers not exceeding 100. What is the probability that a player wins this superlottery?

30. What is the probability that a player of a lottery wins the prize offered for correctly choosing five (but not six) numbers out of six integers chosen at random from the integers between 1 and 40, inclusive?

31. Suppose that 100 people enter a contest and that different winners are selected at random for first, second, and third prizes. What is the probability that a given contestant wins one of these prizes?

32. Suppose that 100 people enter a contest and that different winners are selected at random for first, second, and third prizes. What is the probability that Kumar, Janice, and Pedro each win a prize if each has entered the contest?

33. What is the probability that Abby, Barry, and Sylvia win the first, second, and third prizes, respectively, in a drawing if 200 people enter a contest and

a) no one can win more than one prize.

b) winning more than one prize is allowed.

34. What is the probability that Bo, Colleen, Jeff, and Rohini win the first, second, third, and fourth prizes, respectively, in a drawing if 50 people enter a contest and

a) no one can win more than one prize.

b) winning more than one prize is allowed.

35. In roulette, a wheel with 38 numbers is spun. Of these, 18 are red, and 18 are black. The other two numbers, which are neither black nor red, are 0 and 00. The probability that when the wheel is spun it lands on any particular number is $1/38$.

a) What is the probability that the wheel lands on a red number?

b) What is the probability that the wheel lands on a black number twice in a row?

c) What is the probability that the wheel lands on 0 or 00?

d) What is the probability that in five spins the wheel never lands on either 0 or 00?

e) What is the probability that the wheel lands on one of the first six integers on one spin, but does not land on any of them on the next spin?

36. Which is more likely: rolling a total of 8 when two dice are rolled or rolling a total of 8 when three dice are rolled?

37. Which is more likely: rolling a total of 9 when two dice are rolled or rolling a total of 9 when three dice are rolled?

38. A player in the Mega Millions lottery picks five different integers between 1 and 70, inclusive, and a sixth integer between 1 and 25, inclusive, which may duplicate one of the earlier five integers. The player wins the jackpot if all six numbers match the numbers drawn.

a) What is the probability that a player wins the jackpot?

b) What is the probability that a player wins \$1,000,000, the prize for matching the first five numbers, but not the sixth number, drawn?

c) What is the probability that a player wins \$500, the prize for matching exactly four of the first five numbers, but not the sixth number, drawn?

d) What is the probability that a player wins \$10, the prize for matching exactly three of the first five numbers but not the sixth number drawn, or for matching exactly two of the first five numbers and the sixth number drawn?

Links

39. When a player buys a Mega Millions ticket in many states (see Exercise 38), the player can also buy the Megaplier, which multiplies the size of a prize other than a jackpot by a multiplier ranging from two to five. The Megaplier is drawn using a pool of 15 balls, with five marked 2X, six marked 3X, three marked 4X, and one marked 5X, where each ball has the same likelihood of being drawn. Find the probability that a player who buys a Mega Millions ticket and the Megaplier wins

a) \$5,000,000? (The only way to do this is to match the first five numbers drawn but not the sixth number drawn, with Megaplier 5X.)

b) \$30,000? (The only way to do this is to match exactly four of the first five numbers drawn and the sixth number drawn, with Megaplier 3X.)

c) \$20? (The three ways to do this are to match exactly three of the first five numbers drawn, but not the sixth number drawn, or exactly two of the first five numbers and the sixth number, with Megaplier 2X, or to match exactly one of the first five numbers and the sixth number, with Megaplier 5X.)

d) \$8? (The two ways to do this are to match exactly one of the first five numbers and the sixth number drawn, with a multiplier of 2X, or to match the sixth number but none of the first five numbers, with Megaplier 4X.)

Links

40. A player in the Powerball lottery picks five different integers between 1 and 69, inclusive, and a sixth integer between 1 and 26, which may duplicate one of the earlier five integers. The player wins the jackpot if all six numbers match the numbers drawn.

a) What is the probability that a player wins the jackpot?

b) What is the probability that a player wins \$1,000,000, which is the prize for matching the first five numbers, but not the sixth number, drawn?

c) What is the probability that a player wins \$100 by matching exactly three of the first five and the sixth numbers drawn, or four of the first five numbers, but not the sixth number, drawn?

d) What is the probability that a player wins a prize of \$4, which is the prize when the player matches the sixth number, and either one or none of the first five numbers drawn?

41. A player in the Powerball lottery (see Exercise 40) can purchase the Power Play option. When this option has been purchased, prizes other than the jackpot are multiplied by a multiplier, chosen using a random number generator with weighted values for the different multipliers. When the jackpot is more than \$150,000,000, the weighted values are 24 for 2X, 13 for 3X, 3 for 4X, and 2 for 5X. When the jackpot does not exceed \$150,000,000, the weighted values are 24 for 2X, 13 for 3X, 3 for 4X, 2 for 5X, and 1 for 10X. All non-jackpot prizes are multiplied by the multiplier chosen, except for the \$1,000,000 prize, which is doubled when the Power Play option is in effect regardless of the multiplier chosen. What is the probability that a play who has purchased a Powerball ticket and Power Play wins

a) \$2,000,000, if the jackpot is more than \$150,000,000?
b) \$2,000,000, if the jackpot does not exceed \$150,000,000?
c) \$1000, if the jackpot does not exceed \$150,000,000? (The two ways to do this are for the Power Play multiplier to be 10X, and to match either exactly four of the first five numbers but not the sixth number drawn, or exactly three of the first five numbers and the sixth number drawn.)
d) \$12, if the jackpot is more than \$150,000,000? (The two ways to do this are for the Power Play multiplier to be 3X and to match the sixth number and either one or none of the first five numbers drawn.)

42. Two events E_1 and E_2 are called **independent** if $p(E_1 \cap E_2) = p(E_1)p(E_2)$. For each of the following pairs of events, which are subsets of the set of all possible outcomes when a coin is tossed three times, determine whether or not they are independent.

a) E_1: tails comes up with the coin is tossed the first time; E_2: heads comes up when the coin is tossed the second time.
b) E_1: the first coin comes up tails; E_2: two, and not three, heads come up in a row.
c) E_1: the second coin comes up tails; E_2: two, and not three, heads come up in a row.

(We will study independence of events in more depth in Section 7.2.)

43. Explain what is wrong with the statement that in the Monty Hall Three-Door Puzzle the probability that the prize is behind the first door you select and the probability that the prize is behind the other of the two doors that Monty does not open are both $1/2$, because there are two doors left.

44. Suppose that instead of three doors, there are four doors in the Monty Hall puzzle. What is the probability that you win by not changing once the host, who knows what is behind each door, opens a losing door and gives you the chance to change doors? What is the probability that you win by changing the door you select to one of the two remaining doors among the three that you did not select?

45. This problem was posed by the Chevalier de Méré and was solved by Blaise Pascal and Pierre de Fermat.

a) Find the probability of rolling at least one six when a fair die is rolled four times.
b) Find the probability that a double six comes up at least once when a pair of dice is rolled 24 times. Answer the query the Chevalier de Méré made to Pascal asking whether this probability was greater than $1/2$.
c) Is it more likely that a six comes up at least once when a fair die is rolled four times or that a double six comes up at least once when a pair of dice is rolled 24 times?

7.2 Probability Theory

7.2.1 Introduction

Links

In Section 7.1 we introduced the notion of the probability of an event. (Recall that an event is a subset of the possible outcomes of an experiment.) We defined the probability of an event E as Laplace did, that is,

$$p(E) = \frac{|E|}{|S|},$$

the number of outcomes in E divided by the total number of outcomes. This definition assumes that all outcomes are equally likely. However, many experiments have outcomes that are not equally likely. For instance, a coin may be biased so that it comes up heads twice as often as tails. Similarly, the likelihood that the input of a linear search is a particular element in a list, or is not in the list, depends on how the input is generated. How can we model the likelihood of events in such situations? In this section we will show how to define probabilities of outcomes to study probabilities of experiments where outcomes may not be equally likely.

Suppose that a fair coin is flipped four times, and the first time it comes up heads. Given this information, what is the probability that heads comes up three times? To answer this and similar questions, we will introduce the concept of *conditional probability*. Does knowing that the first flip comes up heads change the probability that heads comes up three times? If not, these two events are called *independent*, a concept studied later in this section.

Many questions address a particular numerical value associated with the outcome of an experiment. For instance, when we flip a coin 100 times, what is the probability that exactly 40 heads appear? How many heads should we expect to appear? In this section we will introduce *random variables*, which are functions that associate numerical values to the outcomes of experiments.

7.2.2 Assigning Probabilities

Let S be the sample space of an experiment with a finite or countable number of outcomes. We assign a probability $p(s)$ to each outcome s. We require that two conditions be met:

$$(i) \quad 0 \leq p(s) \leq 1 \text{ for each } s \in S$$

and

$$(ii) \quad \sum_{s \in S} p(s) = 1.$$

Condition (i) states that the probability of each outcome is a nonnegative real number no greater than 1. Condition (ii) states that the sum of the probabilities of all possible outcomes should be 1; that is, when we do the experiment, it is a certainty that one of these outcomes occurs. (Note that when the sample space is infinite, $\sum_{s \in S} p(s)$ is a convergent infinite series.) This is a generalization of Laplace's definition in which each of n outcomes is assigned a probability of $1/n$. Indeed, conditions (i) and (ii) are met when Laplace's definition of probabilities of equally likely outcomes is used and S is finite. (See Exercise 4.)

Note that when there are n possible outcomes, $x_1, x_2, \ldots, x_n$, the two conditions to be met are

$$(i) \quad 0 \leq p(x_i) \leq 1 \text{ for } i = 1, 2, \ldots, n$$

and

$$(ii) \quad \sum_{i=1}^{n} p(x_i) = 1.$$

The function p from the set of all outcomes of the sample space S is called a **probability distribution**.

To model an experiment, the probability $p(s)$ assigned to an outcome s should equal the limit of the number of times s occurs divided by the number of times the experiment is performed, because this number grows without bound. (We will assume that all experiments discussed have outcomes that are predictable on the average, so that this limit exists. We also assume that the outcomes of successive trials of an experiment do not depend on past results.)

Remark: We will not discuss probabilities of events when the set of outcomes is not finite or countable, such as when the outcome of an experiment can be any real number. In such cases, integral calculus is usually required for the study of the probabilities of events.

Links

HISTORICAL NOTE The Chevalier de Méré was a French nobleman, a famous gambler, and a bon vivant. He was successful at making bets with odds slightly greater than $1/2$ (such as having at least one six come up in four tosses of a fair die). His correspondence with Pascal asking about the probability of having at least one double six come up when a pair of dice is rolled 24 times led to the development of probability theory. According to one account, Pascal wrote to Fermat about the Chevalier saying something like "He's a good guy but, alas, he's no mathematician."

We can model experiments in which outcomes are either equally likely or not equally likely by choosing the appropriate function $p(s)$, as Example 1 illustrates.

EXAMPLE 1 What probabilities should we assign to the outcomes H (heads) and T (tails) when a fair coin is flipped? What probabilities should be assigned to these outcomes when the coin is biased so that heads comes up twice as often as tails?

Solution: For a fair coin, the probability that heads comes up when the coin is flipped equals the probability that tails comes up, so the outcomes are equally likely. Consequently, we assign the probability $1/2$ to each of the two possible outcomes, that is, $p(H) = p(T) = 1/2$.

For the biased coin we have

$$p(H) = 2p(T).$$

Because

$$p(H) + p(T) = 1,$$

it follows that

$$2p(T) + p(T) = 3p(T) = 1.$$

We conclude that $p(T) = 1/3$ and $p(H) = 2/3$. ◀

Definition 1 Suppose that S is a set with n elements. The *uniform distribution* assigns the probability $1/n$ to each element of S.

We now define the probability of an event as the sum of the probabilities of the outcomes in this event.

Definition 2 The *probability* of the event E is the sum of the probabilities of the outcomes in E. That is,

$$p(E) = \sum_{s \in E} p(s).$$

(Note that when E is an infinite set, $\sum_{s \in E} p(s)$ is a convergent infinite series.)

Note that when there are n outcomes in the event E, that is, if $E = \{a_1, a_2, \ldots, a_n\}$, then $p(E) = \sum_{i=1}^{n} p(a_i)$. Note also that the uniform distribution assigns the same probability to an event that Laplace's original definition of probability assigns to this event. The experiment of selecting an element from a sample space with a uniform distribution is called selecting an element of S **at random**.

EXAMPLE 2 Suppose that a die is biased (or loaded) so that 3 appears twice as often as each other number but that the other five outcomes are equally likely. What is the probability that an odd number appears when we roll this die?

Solution: We want to find the probability of the event $E = \{1, 3, 5\}$. By Exercise 2, we have

$$p(1) = p(2) = p(4) = p(5) = p(6) = 1/7; p(3) = 2/7.$$

It follows that

$$p(E) = p(1) + p(3) + p(5) = 1/7 + 2/7 + 1/7 = 4/7.$$

◀

When possible outcomes are equally likely and there are a finite number of possible outcomes, the definition of the probability of an event given in this section (Definition 2) agrees with Laplace's definition (Definition 1 of Section 7.1). To see this, suppose that there are n equally likely outcomes; each possible outcome has probability $1/n$, because the sum of their probabilities is 1. Suppose the event E contains m outcomes. According to Definition 2,

$$p(E) = \sum_{i=1}^{m} \frac{1}{n} = \frac{m}{n}.$$

Because $|E| = m$ and $|S| = n$, it follows that

$$p(E) = \frac{m}{n} = \frac{|E|}{|S|}.$$

This is Laplace's definition of the probability of the event E.

7.2.3 Probabilities of Complements and Unions of Events

The formulae for probabilities of combinations of events in Section 7.1 continue to hold when we use Definition 2 to define the probability of an event. For example, Theorem 1 of Section 7.1 asserts that

$$p(\overline{E}) = 1 - p(E),$$

where $\overline{E}$ is the complementary event of the event E. This equality also holds when Definition 2 is used. To see this, note that because the sum of the probabilities of the n possible outcomes is 1, and each outcome is either in E or in $\overline{E}$, but not in both, we have

$$\sum_{s \in S} p(s) = 1 = p(E) + p(\overline{E}).$$

Hence, $p(\overline{E}) = 1 - p(E)$.

Under Laplace's definition, by Theorem 2 in Section 7.1, we have

$$p(E_1 \cup E_2) = p(E_1) + p(E_2) - p(E_1 \cap E_2)$$

whenever E_1 and E_2 are events in a sample space S. This also holds when we define the probability of an event as we do in this section. To see this, note that $p(E_1 \cup E_2)$ is the sum of the probabilities of the outcomes in $E_1 \cup E_2$. When an outcome x is in one, but not both, of E_1 and E_2, $p(x)$ occurs in exactly one of the sums for $p(E_1)$ and $p(E_2)$. When an outcome x is in both E_1 and E_2, $p(x)$ occurs in the sum for $p(E_1)$, in the sum for $p(E_2)$, and in the sum for $p(E_1 \cap E_2)$, so it occurs $1 + 1 - 1 = 1$ time on the right-hand side. Consequently, the left-hand side and right-hand side are equal.

Also, note that if the events E_1 and E_2 are disjoint, then $p(E_1 \cap E_2) = 0$, which implies that

$$p(E_1 \cup E_2) = p(E_1) + p(E_2) - p(E_1 \cap E_2) = p(E_1) + p(E_2).$$

Theorem 1 generalizes this last formula by providing a formula for the probability of the union of pairwise disjoint events.

THEOREM 1 If $E_1, E_2, \ldots$ is a sequence of pairwise disjoint events in a sample space S, then

$$p\left(\bigcup_i E_i\right) = \sum_i p(E_i).$$

(Note that this theorem applies when the sequence $E_1, E_2, \ldots$ consists of a finite number or a countably infinite number of pairwise disjoint events.)

We leave the proof of Theorem 1 to the reader (see Exercises 36 and 37).

7.2.4 Conditional Probability

Links

Suppose that we flip a coin three times, and all eight possibilities are equally likely. Moreover, suppose we know that the event F, that the first flip comes up tails, occurs. Given this information, what is the probability of the event E, that an odd number of tails appears? Because the first flip comes up tails, there are only four possible outcomes: TTT, TTH, THT, and THH, where H and T represent heads and tails, respectively. An odd number of tails appears only for the outcomes TTT and THH. Because the eight outcomes have equal probability, each of the four possible outcomes, given that F occurs, should also have an equal probability of $1/4$. This suggests that we should assign the probability of $2/4 = 1/2$ to E, given that F occurs. This probability is called the **conditional probability** of E given F.

In general, to find the conditional probability of E given F, we use F as the sample space. For an outcome from E to occur, this outcome must also belong to $E \cap F$. With this motivation, we make Definition 3.

Definition 3 Let E and F be events with $p(F) > 0$. The *conditional probability* of E given F, denoted by $p(E \mid F)$, is defined as

$$p(E \mid F) = \frac{p(E \cap F)}{p(F)}.$$

EXAMPLE 3 A bit string of length four is generated at random so that each of the 16 bit strings of length four is equally likely. What is the probability that it contains at least two consecutive 0s, given that its first bit is a 0? (We assume that 0 bits and 1 bits are equally likely.)

Extra Examples

Solution: Let E be the event that a bit string of length four contains at least two consecutive 0s, and let F be the event that the first bit of a bit string of length four is a 0. The probability that a bit string of length four has at least two consecutive 0s, given that its first bit is a 0, equals

$$p(E \mid F) = \frac{p(E \cap F)}{p(F)}.$$

Because $E \cap F = \{0000, 0001, 0010, 0011, 0100\}$, we see that $p(E \cap F) = 5/16$. Because there are eight bit strings of length four that start with a 0, we have $p(F) = 8/16 = 1/2$. Consequently,

$$p(E \mid F) = \frac{5/16}{1/2} = \frac{5}{8}.$$

◀

EXAMPLE 4 What is the conditional probability that a family with two children has two children assigned male at birth, given they have at least one child assigned male? Assume that each of the possibilities MM, MF, FM, and FF is equally likely, where M represents a child assigned male and F represents a child assigned female at birth. (Note that MF represents a family with an older child assigned male and a younger child assigned female while FM represents a family with an older child assigned female and a younger child assigned male.)

Solution: Let A be the event that a family with two children has two children assigned male, and let B be the event that a family with two children has at least one child assigned male. It follows that $A = \{MM\}$, $B = \{MM, MF, FM\}$, and $A \cap B = \{MM\}$. Because the four possibilities are equally likely, it follows that $p(B) = 3/4$ and $p(A \cap B) = 1/4$. We conclude that

$$p(A \mid B) = \frac{p(A \cap B)}{p(B)} = \frac{1/4}{3/4} = \frac{1}{3}.$$

◀

7.2.5 Independence

Links

Suppose a coin is flipped three times, as described in the introduction to our discussion of conditional probability. Does knowing that the first flip comes up tails (event F) alter the probability that tails comes up an odd number of times (event E)? In other words, is it the case that $p(E \mid F) = p(E)$? This equality is valid for the events E and F, because $p(E \mid F) = 1/2$ and $p(E) = 1/2$. Because this equality holds, we say that E and F are **independent events**. When two events are independent, the occurrence of one of the events gives no information about the probability that the other event occurs.

Because $p(E \mid F) = p(E \cap F)/p(F)$, asking whether $p(E \mid F) = p(E)$ is the same as asking whether $p(E \cap F) = p(E)p(F)$. This leads to Definition 4.

Definition 4 The events E and F are *independent* if and only if $p(E \cap F) = p(E)p(F)$.

EXAMPLE 5 Suppose E is the event that a randomly generated bit string of length four begins with a 1 and F is the event that this bit string contains an even number of 1s. Are E and F independent, if the 16 bit strings of length four are equally likely?

Extra Examples

Solution: There are eight bit strings of length four that begin with a one: 1000, 1001, 1010, 1011, 1100, 1101, 1110, and 1111. There are also eight bit strings of length four that contain an even number of ones: 0000, 0011, 0101, 0110, 1001, 1010, 1100, 1111. Because there are 16 bit strings of length four, it follows that

$$p(E) = p(F) = 8/16 = 1/2.$$

Because $E \cap F = \{1111, 1100, 1010, 1001\}$, we see that

$$p(E \cap F) = 4/16 = 1/4.$$

Because

$$p(E \cap F) = 1/4 = (1/2)(1/2) = p(E)p(F),$$

we conclude that E and F are independent. ◀

Probability has many applications to genetics, as Examples 6 and 7 illustrate.

EXAMPLE 6 Assume, as in Example 4, that each of the four ways a family can have two children is equally likely. Are the events A, that a family with two children has two children assigned male at birth, and B, that a family with two children has at least one child assigned male at birth, independent?

Solution: Because $A = \{MM\}$, we have $p(A) = 1/4$. In Example 4 we showed that $p(B) = 3/4$ and that $p(A \cap B) = 1/4$. But $p(A)p(B) = \frac{1}{4} \cdot \frac{3}{4} = \frac{3}{16}$. Therefore, $p(A \cap B) \neq p(A)p(B)$, so the events A and B are not independent. ◀

EXAMPLE 7 Are the events A, that a family with three children has children of both sexes, and B, that this family has at most one child assigned male at birth, independent? Assume that the eight ways a family can have three children are equally likely.

Solution: By assumption, each of the eight ways a family can have three children, MMM, MMF, MFM, MFF, FMM, FMF, FFM, and FFF, has a probability of $1/8$. Because $A = \{MMF, MFM, MFF, FMM, FMF, FFM\}$, $B = \{MFF, FMF, FFM, FFF\}$, and $A \cap B = \{MFF, FMF, FFM\}$, it follows that $p(A) = 6/8 = 3/4$, $p(B) = 4/8 = 1/2$, and $p(A \cap B) = 3/8$. Because

$$p(A)p(B) = \frac{3}{4} \cdot \frac{1}{2} = \frac{3}{8},$$

it follows that $p(A \cap B) = p(A)p(B)$, so A and B are independent. (This conclusion may seem surprising. Indeed, if we change the number of children, the conclusion may no longer hold. See Exercise 27.) ◀

Probability theory plays an important role in the study of genetics. As Gregor Mendel discovered in the nineteenth century, the physical traits of all animals and plants are determined by genes, which are strings of nucleotides on chromosomes.

Each gene is made up of two alleles, which may be **dominant** or **recessive**. Dominant genes are represented by uppercase letters and recessive genes are represented by lowercase letters. The next example shows how probability can help us answer questions about the frequency of traits in people.

EXAMPLE 8 One of the genes that is responsible for eye color in humans has two alleles, a dominant one represented by B for brown eyes and a recessive one represented by b for blue eyes. The genotypes associated with alleles are BB, Bb, and bb, where someone who has the BB or Bb genotype has brown eyes and someone with the bb genotype has blue eyes. If a parent has genotype BB, then a child will inherit a B allele from that parent with certainty, and likewise a bb parent will pass down a b allele. However, if a parent has genotype Bb, then a child has equally likely chances of inheriting the B allele and the b allele from that parent. ◀

Suppose that the individuals in a particular area have these probabilities for the three genotypes: $p(BB) = 0.3$, $p(Bb) = 0.5$, and $p(bb) = 0.2$. We want to find the probability that someone born into this community has blue eyes. For someone to have blue eyes, they have to receive the b allele from both their mother and their father. Note that if both genetic parents have blue eyes, then both parents are bb and the child certainly has blue eyes. To answer this question we need to determine the probability of E, the event that this person receives a b allele from their mother, and F, the probability that they receive a b allele from their father. Because these events are independent, the product of these probabilities gives us the answer.

We have $p(E) = p(E|BB)p(BB) + p(E|Bb)p(Bb) + p(E|bb)P(bb) = 0 \cdot 0.3 + 0.5 \cdot 0.5 + 0.2 \cdot 1.0 = 0.45$. Similarly, we can show that $p(F) = 0.45$. It follows that $p(C)$, the probability that this person has blue eyes (which requires that their genotype is bb) is $p(C) = p(E)p(F) = (0.45)^2 = 0.2025$.

PAIRWISE AND MUTUAL INDEPENDENCE We can also define the independence of more than two events. However, there are two different types of independence, given in Definition 5.

Definition 5 The events $E_1, E_2, \ldots, E_n$ are *pairwise independent* if and only if $p(E_i \cap E_j) = p(E_i)p(E_j)$ for all pairs of integers i and j with $1 \leq i < j \leq n$. These events are *mutually independent* if $p(E_{i_1} \cap E_{i_2} \cap \cdots \cap E_{i_m}) = p(E_{i_1})p(E_{i_2}) \cdots p(E_{i_m})$ whenever $i_j, j = 1, 2, \ldots, m$, are integers with $1 \leq i_1 < i_2 < \cdots < i_m \leq n$ and $m \geq 2$.

From Definition 5, we see that every set of n mutually independent events is also pairwise independent. However, n pairwise independent events are not necessarily mutually independent, as we see in Exercise 25 in the Supplementary Exercises. Many theorems about n events include the hypothesis that these events are mutually independent, and not just pairwise independent. We will introduce several such theorems later in this chapter.

7.2.6 Bernoulli Trials and the Binomial Distribution

Links

Suppose that an experiment can have only two possible outcomes. For instance, when a bit is generated at random, the possible outcomes are 0 and 1. When a coin is flipped, the possible outcomes are heads and tails. Each performance of an experiment with two possible outcomes is called a **Bernoulli trial**, after James Bernoulli, who made important contributions to probability theory. In general, a possible outcome of a Bernoulli trial is called a **success** or a **failure**. If p is the probability of a success and q is the probability of a failure, it follows that $p + q = 1$.

Many problems can be solved by determining the probability of k successes when an experiment consists of n mutually independent Bernoulli trials. (Bernoulli trials are **mutually independent** if the conditional probability of success on any given trial is p, given any information whatsoever about the outcomes of the other trials.) Consider Example 9.

EXAMPLE 9 A coin is biased so that the probability of heads is $2/3$. What is the probability that exactly four heads come up when the coin is flipped seven times, assuming that the flips are independent?

Links

INTERFOTO/Alamy Stock Photo

JAMES BERNOULLI (1654–1705) James Bernoulli (also known as Jacob I), was born in Basel, Switzerland. He is one of the eight prominent mathematicians in the Bernoulli family (see Section 11.1 for the Bernoulli family tree of mathematicians). Following his father's wish, James studied theology and entered the ministry. But contrary to the desires of his parents, he also studied mathematics and astronomy. He traveled throughout Europe from 1676 to 1682, learning about the latest discoveries in mathematics and the sciences. Upon returning to Basel in 1682, he founded a school for mathematics and the sciences. He was appointed professor of mathematics at the University of Basel in 1687, remaining in this position for the rest of his life.

James Bernoulli is best known for the work *Ars Conjectandi*, published eight years after his death. In this work, he described the known results in probability theory and in enumeration, often providing alternative proofs of known results. This work also includes the application of probability theory to games of chance and his introduction of the theorem known as the **law of large numbers**. This law states that if $\epsilon > 0$, as n becomes arbitrarily large the probability approaches 1 that the fraction of times an event E occurs during n trials is within ϵ of $p(E)$.

Solution: There are $2^7 = 128$ possible outcomes when a coin is flipped seven times. The number of ways four of the seven flips can be heads is $C(7, 4)$. Because the seven flips are independent, the probability of each of these outcomes (four heads and three tails) is $(2/3)^4(1/3)^3$. Consequently, the probability that exactly four heads appear is

$$C(7, 4)(2/3)^4(1/3)^3 = \frac{35 \cdot 16}{3^7} = \frac{560}{2187}.$$ ◀

Following the same reasoning as was used in Example 9, we can find the probability of k successes in n independent Bernoulli trials.

THEOREM 2 The probability of exactly k successes in n independent Bernoulli trials, with probability of success p and probability of failure $q = 1 - p$, is

$$C(n, k)p^k q^{n-k}.$$

Proof: When n Bernoulli trials are carried out, the outcome is an n-tuple $(t_1, t_2, \ldots, t_n)$, where $t_i = S$ (for success) or $t_i = F$ (for failure) for $i = 1, 2, \ldots, n$. Because the n trials are independent, the probability of each outcome of n trials consisting of k successes and $n - k$ failures (in any order) is $p^k q^{n-k}$. Because there are $C(n, k)$ n-tuples of S's and F's that contain exactly k S's, the probability of exactly k successes is

$$C(n, k)p^k q^{n-k}.$$ ◁

We denote by $b(k; n, p)$ the probability of k successes in n independent Bernoulli trials with probability of success p and probability of failure $q = 1 - p$. Considered as a function of k, we call this function the **binomial distribution**. Theorem 2 tells us that $b(k; n, p) = C(n, k)p^k q^{n-k}$.

EXAMPLE 10 Suppose that the probability that a 0 bit is generated is 0.9, that the probability that a 1 bit is generated is 0.1, and that bits are generated independently. What is the probability that exactly eight 0 bits are generated when 10 bits are generated?

Extra Examples

Solution: By Theorem 2, the probability that exactly eight 0 bits are generated is

$$b(8; 10, 0.9) = C(10, 8)(0.9)^8(0.1)^2 = 0.1937102445.$$ ◀

Note that the sum of the probabilities that there are k successes when n independent Bernoulli trials are carried out, for $k = 0, 1, 2, \ldots, n$, equals

$$\sum_{k=0}^{n} C(n, k)p^k q^{n-k} = (p + q)^n = 1,$$

as should be the case. The first equality in this string of equalities is a consequence of the binomial theorem (see Section 6.4). The second equality follows because $q = 1 - p$.

7.2.7 Random Variables

Many problems are concerned with a numerical value associated with the outcome of an experiment. For instance, we may be interested in the total number of one bits in a randomly generated string of 10 bits; or in the number of times tails come up when a coin is flipped 20 times. To study problems of this type we introduce the concept of a random variable.

Definition 6 A *random variable* is a function from the sample space of an experiment to the set of real numbers. That is, a random variable assigns a real number to each possible outcome.

Remark: Note that a random variable is a function. It is not a variable, and it is not random! The name *random variable* (the translation of *variabile casuale*) was introduced by the Italian mathematician F. P. Cantelli in 1916. In the late 1940s, the mathematicians, W. Feller and J. L. Doob flipped a coin to see whether both would use "random variable" or the more fitting term "chance variable." Feller won; unfortunately "random varible" was used in both books and ever since.

EXAMPLE 11 Suppose that a coin is flipped three times. Let $X(t)$ be the random variable that equals the number of heads that appear when t is the outcome. Then $X(t)$ takes on the following values:

$$X(HHH) = 3,$$
$$X(HHT) = X(HTH) = X(THH) = 2,$$
$$X(TTH) = X(THT) = X(HTT) = 1,$$
$$X(TTT) = 0.$$

◀

Definition 7 The *distribution* of a random variable X on a sample space S is the set of pairs $(r, p(X = r))$ for all $r \in X(S)$, where $p(X = r)$ is the probability that X takes the value r. (The set of pairs in this distribution is determined by the probabilities $p(X = r)$ for $r \in X(S)$.)

EXAMPLE 12 Each of the eight possible outcomes when a fair coin is flipped three times has probability $1/8$. So, the distribution of the random variable $X(t)$ in Example 11 is determined by the probabilities $P(X = 3) = 1/8$, $P(X = 2) = 3/8$, $P(X = 1) = 3/8$, and $P(X = 0) = 1/8$. Consequently, the distribution of $X(t)$ in Example 11 is the set of pairs $(3, 1/8)$, $(2, 3/8)$, $(1, 3/8)$, and $(0, 1/8)$. ◀

EXAMPLE 13 Let X be the sum of the numbers that appear when a pair of dice is rolled. What are the values of this random variable for the 36 possible outcomes (i, j), where i and j are the numbers that appear on the first die and the second die, respectively, when these two dice are rolled?

Solution: The random variable X takes on the following values:

$$X((1, 1)) = 2,$$
$$X((1, 2)) = X((2, 1)) = 3,$$
$$X((1, 3)) = X((2, 2)) = X((3, 1)) = 4,$$
$$X((1, 4)) = X((2, 3)) = X((3, 2)) = X((4, 1)) = 5,$$
$$X((1, 5)) = X((2, 4)) = X((3, 3)) = X((4, 2)) = X((5, 1)) = 6,$$
$$X((1, 6)) = X((2, 5)) = X((3, 4)) = X((4, 3)) = X((5, 2)) = X((6, 1)) = 7,$$
$$X((2, 6)) = X((3, 5)) = X((4, 4)) = X((5, 3)) = X((6, 2)) = 8,$$
$$X((3, 6)) = X((4, 5)) = X((5, 4)) = X((6, 3)) = 9,$$
$$X((4, 6)) = X((5, 5)) = X((6, 4)) = 10,$$
$$X((5, 6)) = X((6, 5)) = 11,$$
$$X((6, 6)) = 12.$$

◀

We will continue our study of random variables in Section 7.4, where we will show how they can be used in a variety of applications.

7.2.8 The Birthday Problem

A famous puzzle asks for the smallest number of people needed in a room so that it is more likely than not that at least two of them have the same day of the year as their birthday. Most people find the answer, which we determine in Example 14, to be surprisingly small. After we solve this famous problem, we will show how similar reasoning can be adapted to solve a question about hashing functions.

EXAMPLE 14 **The Birthday Problem** What is the minimum number of people who need to be in a room so that the probability that at least two of them have the same birthday is greater than $1/2$?

Links

Solution: First, we state some assumptions. We assume that the birthdays of the people in the room are independent. Furthermore, we assume that each birthday is equally likely and that there are 366 days in the year. (In reality, more people are born on some days of the year than others, such as days nine months after some holidays including New Year's Eve, and only leap years have 366 days.)

To find the probability that at least two of n people in a room have the same birthday, we first calculate the probability p_n that these people all have different birthdays. Then, the probability that at least two people have the same birthday is $1-p_n$. To compute p_n, we consider the birthdays of the n people in some fixed order. Imagine them entering the room one at a time; we will compute the probability that each successive person entering the room has a birthday different from those of the people already in the room.

The birthday of the first person certainly does not match the birthday of someone already in the room. The probability that the birthday of the second person is different from that of the first person is $365/366$ because the second person has a different birthday when he or she was born on one of the 365 days of the year other than the day the first person was born. (The assumption that it is equally likely for someone to be born on any of the 366 days of the year enters into this and subsequent steps.)

The probability that the third person has a birthday different from both the birthdays of the first and second people given that these two people have different birthdays is $364/366$. In general, the probability that the jth person, with $2 \le j \le 366$, has a birthday different from the birthdays of the $j - 1$ people already in the room given that these $j - 1$ people have different birthdays is

$$\frac{366 - (j - 1)}{366} = \frac{367 - j}{366}.$$

Because we have assumed that the birthdays of the people in the room are independent, we can conclude that the probability that the n people in the room have different birthdays is

$$p_n = \frac{365}{366}\frac{364}{366}\frac{363}{366}\cdots\frac{367 - n}{366}.$$

It follows that the probability that among n people there are at least two people with the same birthday is

$$1 - p_n = 1 - \frac{365}{366}\frac{364}{366}\frac{363}{366}\cdots\frac{367 - n}{366}.$$

To determine the minimum number of people in the room so that the probability that at least two of them have the same birthday is greater than $1/2$, we use the formula we have found for $1 - p_n$ to compute it for increasing values of n until it becomes greater than $1/2$. (There are more sophisticated approaches using calculus that can eliminate this computation, but we will not use them here.) After considerable computation we find that for $n = 22$, $1 - p_n \approx 0.475$, while for $n = 23$, $1-p_n \approx 0.506$. Consequently, the minimum number of people needed so that the probability that at least two people have the same birthday is greater than $1/2$ is 23. ◂

The solution to the birthday problem leads to the solution of the question in Example 15 about hashing functions.

EXAMPLE 15 **Probability of a Collision in Hashing Functions** Recall from Section 4.5 that a hashing function $h(k)$ is a mapping of the keys (of the records that are to be stored in a database) to storage locations. Hashing functions map a large universe of keys (such as the approximately 300 million Social Security numbers in the United States) to a much smaller set of storage locations. A good hashing function yields few **collisions**, which are mappings of two different keys to the same memory location, when relatively few of the records are in play in a given application. What is the probability that no two keys are mapped to the same location by a hashing function, or, in other words, that there are no collisions?

Solution: To calculate this probability, we assume that the probability that a randomly selected key is mapped to a location is $1/m$, where m is the number of available locations, that is, the hashing function distributes keys uniformly. (In practice, hashing functions may not satisfy this assumption. However, for a good hashing function, this assumption should be close to correct.) Furthermore, we assume that the keys of the records selected have an equal probability to be any of the elements of the key universe and that these keys are independently selected.

Suppose that the keys are $k_1, k_2, \ldots, k_n$. When we add the second record, the probability that it is mapped to a location different from the location of the first record, that $h(k_2) \neq h(k_1)$, is $(m-1)/m$ because there are $m - 1$ free locations after the first record has been placed. The probability that the third record is mapped to a free location after the first and second records have been placed without a collision is $(m-2)/m$. In general, the probability that the jth record is mapped to a free location after the first $j - 1$ records have been mapped to locations $h(k_1), h(k_2), \ldots, h(k_{j-1})$ without collisions is $(m - (j-1))/m$ because $j - 1$ of the m locations are taken.

Because the keys are independent, the probability that all n keys are mapped to different locations is

$$p_n = \frac{m-1}{m} \cdot \frac{m-2}{m} \cdot \cdots \cdot \frac{m-n+1}{m}.$$

It follows that the probability that there is at least one collision, that is, at least two keys are mapped to the same location, is

$$1 - p_n = 1 - \frac{m-1}{m} \cdot \frac{m-2}{m} \cdot \cdots \cdot \frac{m-n+1}{m}.$$

Techniques from calculus can be used to find the smallest value of n given a value of m such that the probability of a collision is greater than a particular threshold. It can be shown that the smallest integer n such that the probability of a collision is greater than $1/2$ is approximately $n = 1.177\sqrt{m}$. For example, when $m = 1{,}000{,}000$, the smallest integer n such that the probability of a collision is greater than $1/2$ is 1178. ◂

7.2.9 Monte Carlo Algorithms

The algorithms discussed so far in this book are all deterministic. That is, each algorithm always proceeds in the same way whenever given the same input. However, there are many situations where we wouldlike an algorithm to make a random choice at one or more steps. Such a situation arises when a deterministic algorithm would have to go through a huge number, or even an unknown number, of possible cases. Algorithms that make random choices at one or more steps are called **probabilistic algorithms**. We will discuss a particular class of probabilistic algorithms in this section, namely, **Monte Carlo algorithms**, for decision problems. Monte Carlo algorithms always produce answers to problems, but a small probability remains that these answers may be incorrect. However, the probability that the answer is incorrect decreases rapidly when the algorithm carries out sufficient computation. Decision problems have either "true" or "false" as their answer. The designation "Monte Carlo" is a reference to the gambling casino in Monaco; the use of randomness and the repetitive processes in these algorithms make them similar to some gambling games. This name was introduced by the inventors of Monte Carlo methods, including Stan Ulam, Enrico Fermi, and John von Neumann.

Monte Carlo methods were invented to help develop the first nuclear weapons.

A Monte Carlo algorithm for a decision problem uses a sequence of tests. The probability that the algorithm answers the decision problem correctly increases as more tests are carried out. At each step of the algorithm, possible responses are "true," which means that the answer is "true" and no additional iterations are needed, or "unknown," which means that the answer could be either "true" or "false." After running all the iterations in such an algorithm, the final answer produced is "true" if at least one iteration yields the answer "true," and the answer is "false" if every iteration yields the answer "unknown." If the correct answer is "false," then the algorithm answers "false," because every iteration will yield "unknown." However, if the correct answer is "true," then the algorithm could answer either "true" or "false," because it may be possible that each iteration produced the response "unknown" even though the correct response was "true." We will show that this possibility becomes extremely unlikely as the number of tests increases.

Suppose that p is the probability that the response of a test is "true," given that the answer is "true." It follows that $1-p$ is the probability that the response is "unknown," given that the answer is "true." Because the algorithm answers "false" when all n iterations yield the answer "unknown" and the iterations perform independent tests, the probability of error is $(1-p)^n$. When $p \neq 0$, this probability approaches 0 as the number of tests increases. Consequently, the probability that the algorithm answers "true" when the answer is "true" approaches 1.

EXAMPLE 16 **Quality Control** (This example is adapted from [AhUl95].) Suppose that a manufacturer orders processor chips in batches of size n, where n is a positive integer. The chip maker has tested only some of these batches to make sure that all the chips in the batch are good (replacing any bad chips found during testing with good ones). In previously untested batches, the probability that a particular chip is bad has been observed to be 0.1 when random testing is done. The PC manufacturer wants to decide whether all the chips in a batch are good. To do this, the PC manufacturer can test each chip in a batch to see whether it is good. However, this requires n tests. Assuming that each test can be carried out in constant time, these tests require $O(n)$ seconds. Can the PC manufacturer determine whether a batch of chips has been tested by the chip maker using less time?

Solution: We can use a Monte Carlo algorithm to determine whether a batch of chips has been tested by the chip maker as long as we are willing to accept some probability of error. The algorithm is set up to answer the question: "Has this batch of chips not been tested by the chip maker?" It proceeds by successively selecting chips at random from the batch and testing them one by one. When a bad chip is encountered, the algorithm answers "true" and stops. If a tested chip is good, the algorithm answers "unknown" and goes on to the next chip. After the algorithm has tested a specified number of chips, say k chips, without getting an answer of "true," the algorithm terminates with the answer "false"; that is, the algorithm concludes that the batch is good, that is, that the chip maker has tested all the chips in the batch.

The only way for this algorithm to answer incorrectly is for it to conclude that an untested batch of chips has been tested by the chip maker. The probability that a chip is good, but that it came from an untested batch, is $1 - 0.1 = 0.9$. Because the events of testing different chips from a batch are independent, the probability that all k steps of the algorithm produce the answer "unknown," given that the batch of chips is untested, is 0.9^k.

By taking k large enough, we can make this probability as small as we like. For example, by testing 66 chips, the probability that the algorithm decides a batch has been tested by the chip maker is 0.9^{66}, which is less than 0.001. That is, the probability is less than 1 in 1000 that the algorithm has answered incorrectly. Note that this probability is independent of n, the number of chips in a batch. That is, the Monte Carlo algorithm uses a constant number, or $O(1)$, tests and requires $O(1)$ seconds, no matter how many chips are in a batch. As long as the PC manufacturer can live with an error rate of less than 1 in 1000, the Monte Carlo algorithm will save the PC manufacturer a lot of testing. If a smaller error rate is needed, the PC manufacturer can test more chips in each batch; the reader can verify that 132 tests lower the error rate to less than 1 in 1,000,000. ◀

EXAMPLE 17 **Probabilistic Primality Testing** In Chapter 4 we remarked that a composite integer, that is, an integer greater than one that is not prime, passes Miller's test (see the preamble to Exercise 44 in Section 4.4) for fewer than $n/4$ bases b with $1 < b < n$. This observation is the basis for a Monte Carlo algorithm to determine whether an integer greater than one is prime. Because large primes play an essential role in public-key cryptography (see Section 4.6), being able to generate large primes quickly has become extremely important.

The goal of the algorithm is to decide the question "Is n composite?" Given an integer n greater than one, we select an integer b at random with $1 < b < n$ and determine whether n passes Miller's test to the base b. If n fails the test, the answer is "true" because n must be composite, and the algorithm ends. Otherwise, we perform the test k times, where k is a positive integer. Each time we select a random integer b and determine whether n passes Miller's test to the base b. If the answer is "unknown" at each step, the algorithm answers "false," that is, it says that n is not composite, so that it is prime. The only possibility for the algorithm to return an incorrect answer occurs when n is composite, and the answer "unknown" is the output at each of the k iterations. The probability that a composite integer n passes Miller's test for a randomly selected base b is less than $1/4$. Because the integer b with $1 < b < n$ is selected at random at each iteration and these iterations are independent, the probability that n is composite but the algorithm responds that n is prime is less than $(1/4)^k$. By taking k to be sufficiently large, we can make this probability extremely small. For example, with 10 iterations, the probability that the algorithm decides that n is prime when it really is composite is less than 1 in 1,000,000. With 30 iterations, this probability drops to less than 1 in 10^{18}, an extremely unlikely event.

A number that passes many iterations of a probabilistic primality test is called an *industrial strength prime*, even though it may be composite.

To generate large primes, say with 200 digits, we randomly choose an integer n with 200 digits and run this algorithm, with 30 iterations. If the algorithm decides that n is prime, we can use it as one of the two primes used in an encryption key for the RSA cryptosystem. If n is actually composite and is used as part of the key, the procedures used to decrypt messages will not produce the original encrypted message. The key is then discarded and two new possible primes are used. ◀

7.2.10 The Probabilistic Method

We discussed existence proofs in Chapter 1 and illustrated the difference between constructive existence proofs and nonconstructive existence proofs. The probabilistic method, introduced by Paul Erdős and Alfréd Rényi, is a powerful technique that can be used to create nonconstructive existence proofs. To use the probabilistic method to prove results about a set S, such as the existence of an element in S with a specified property, we assign probabilities to the elements of S. We then use methods from probability theory to prove results about the elements of S. In particular, we can show that an element with a specified property exists by showing that the probability an element $x \in S$ has this property is positive. The probabilistic method is based on the equivalent statement in Theorem 3.

THEOREM 3 **THE PROBABILISTIC METHOD** If the probability that an element chosen at random from a S does not have a particular property is less than 1, there exists an element in S with this property.

An existence proof based on the probabilistic method is nonconstructive because it does not find a particular element with the desired property.

We illustrate the power of the probabilistic method by finding a lower bound for the Ramsey number $R(k, k)$. Recall from Section 6.2 that $R(k, k)$ equals the minimum number of people at a party needed to ensure that there are at least k mutual friends or k mutual enemies (assuming that any two people are friends or enemies).

THEOREM 4 If k is an integer with $k \geq 2$, then $R(k, k) \geq 2^{k/2}$.

Proof: We note that the theorem holds for $k = 2$ and $k = 3$ because $R(2, 2) = 2$ and $R(3, 3) = 6$, as was shown in Section 6.2. Now suppose that $k \geq 4$. We will use the probabilistic method to show that if there are fewer than $2^{k/2}$ people at a party, it is possible that no k of them are mutual friends or mutual enemies. This will show that $R(k, k)$ is at least $2^{k/2}$.

To use the probabilistic method, we assume that it is equally likely for two people to be friends or enemies. (Note that this assumption does not have to be realistic.) Suppose there are n people at the party. It follows that there are $\binom{n}{k}$ different sets of k people at this party, which we list as $S_1, S_2, \ldots, S_{\binom{n}{k}}$. Let E_i be the event that all k people in S_i are either mutual friends or mutual enemies. The probability that there are either k mutual friends or k mutual enemies among the n people equals $p(\bigcup_{i=1}^{\binom{n}{k}} E_i)$.

According to our assumption it is equally likely for two people to be friends or enemies. The probability that two people are friends equals the probability that they are enemies; both probabilities equal $1/2$. Furthermore, there are $\binom{k}{2} - k(k-1)/2$ pairs of people in S_i because there are k people in S_i. Hence, the probability that all k people in S_i are mutual friends and the probability that all k people in S_i are mutual enemies both equal $(1/2)^{k(k-1)/2}$. It follows that $p(E_i) = 2(1/2)^{k(k-1)/2}$.

The probability that there are either k mutual friends or k mutual enemies in the group of n people equals $p(\bigcup_{i=1}^{\binom{n}{k}} E_i)$. Using Boole's inequality (Exercise 15), it follows that

$$p\left(\bigcup_{i=1}^{\binom{n}{k}} E_i\right) \leq \sum_{i=1}^{\binom{n}{k}} p(E_i) = \binom{n}{k} \cdot 2\left(\frac{1}{2}\right)^{k(k-1)/2}.$$

By Exercise 21 in Section 6.4, we have $\binom{n}{k} \leq n^k/2^{k-1}$. Hence,

$$\binom{n}{k} 2\left(\frac{1}{2}\right)^{k(k-1)/2} \leq \frac{n^k}{2^{k-1}} 2\left(\frac{1}{2}\right)^{k(k-1)/2}.$$

Now if $n < 2^{k/2}$, we have

$$\frac{n^k}{2^{k-1}} 2\left(\frac{1}{2}\right)^{k(k-1)/2} < \frac{2^{k(k/2)}}{2^{k-1}} 2\left(\frac{1}{2}\right)^{k(k-1)/2} = 2^{2-(k/2)} \leq 1,$$

where the last step follows because $k \geq 4$.

We can now conclude that $p(\bigcup_{i=1}^{\binom{n}{k}} E_i) < 1$ when $k \geq 4$. Hence, the probability of the complementary event, that there is no setof either k mutual friends or mutual enemies at the party, is greater than 0. It follows that if $n < 2^{k/2}$, there is at least one set such that no subset of k people are mutual friends or mutual enemies. ◁

Exercises

1. What probability should be assigned to the outcome of heads when a biased coin is tossed, if heads is three times as likely to come up as tails? What probability should be assigned to the outcome of tails?

2. Find the probability of each outcome when a loaded die is rolled, if a 3 is twice as likely to appear as each of the other five numbers on the die.

3. Find the probability of each outcome when a biased die is rolled, if rolling a 2 or rolling a 4 is three times as likely as rolling each of the other four numbers on the die and it is equally likely to roll a 2 or a 4.

4. Show that conditions (*i*) and (*ii*) are met under Laplace's definition of probability, when outcomes are equally likely.

5. A pair of dice is loaded. The probability that a 4 appears on the first die is $2/7$, and the probability that a 3 appears on the second die is $2/7$. Other outcomes for each die appear with probability $1/7$. What is the probability of 7 appearing as the sum of the numbers when the two dice are rolled?

6. What is the probability of these events when we randomly select a permutation of $\{1, 2, 3\}$?

a) 1 precedes 3.

b) 3 precedes 1.

c) 3 precedes 1 and 3 precedes 2.

7. What is the probability of these events when we randomly select a permutation of $\{1, 2, 3, 4\}$?

a) 1 precedes 4.

b) 4 precedes 1.

c) 4 precedes 1 and 4 precedes 2.

d) 4 precedes 1, 4 precedes 2, and 4 precedes 3.

e) 4 precedes 3 and 2 precedes 1.

8. What is the probability of these events when we randomly select a permutation of $\{1, 2, \ldots, n\}$ where $n \geq 4$?

a) 1 precedes 2.

b) 2 precedes 1.

c) 1 immediately precedes 2.

d) n precedes 1 and $n-1$ precedes 2.

e) n precedes 1 and n precedes 2.

9. What is the probability of these events when we randomly select a permutation of the 26 lowercase letters of the English alphabet?

a) The permutation consists of the letters in reverse alphabetic order.

b) z is the first letter of the permutation.

c) z precedes a in the permutation.

d) a immediately precedes z in the permutation.

e) a immediately precedes m, which immediately precedes z in the permutation.

f) m, n, and o are in their original places in the permutation.

10. What is the probability of these events when we randomly select a permutation of the 26 lowercase letters of the English alphabet?

a) The first 13 letters of the permutation are in alphabetical order.

b) a is the first letter of the permutation and z is the last letter.

c) a and z are next to each other in the permutation.

d) a and b are not next to each other in the permutation.

e) a and z are separated by at least 23 letters in the permutation.

f) z precedes both a and b in the permutation.

11. Suppose that E and F are events such that $p(E) = 0.7$ and $p(F) = 0.5$. Show that $p(E \cup F) \geq 0.7$ and $p(E \cap F) \geq 0.2$.

12. Suppose that E and F are events such that $p(E) = 0.8$ and $p(F) = 0.6$. Show that $p(E \cup F) \geq 0.8$ and $p(E \cap F) \geq 0.4$.

13. Show that if E and F are events, then $p(E \cap F) \geq p(E) + p(F) - 1$. This is known as **Bonferroni's inequality**.

14. Use mathematical induction to prove the following generalization of Bonferroni's inequality:

$$p(E_1 \cap E_2 \cap \cdots \cap E_n) \geq p(E_1) + p(E_2) + \cdots + p(E_n) - (n-1),$$

where $E_1, E_2, \ldots, E_n$ are n events.

☞ **15.** Show that if $E_1, E_2, \ldots, E_n$ are events from a finite sample space, then

$$p(E_1 \cup E_2 \cup \cdots \cup E_n) \leq p(E_1) + p(E_2) + \cdots + p(E_n).$$

This is known as **Boole's inequality**.

16. Show that if E and F are independent events, then $\overline{E}$ and $\overline{F}$ are also independent events.

17. If E and F are independent events, prove or disprove that $\overline{E}$ and F are necessarily independent events.

In Exercises 18, 20, and 21 assume that the year has 366 days and all birthdays are equally likely. In Exercise 19 assume it is equally likely that a person is born in any given month of the year.

18. a) What is the probability that two people chosen at random were born on the same day of the week?

b) What is the probability that in a group of n people chosen at random, there are at least two born on the same day of the week?

c) How many people chosen at random are needed to make the probability greater than $1/2$ that there are at least two people born on the same day of the week?

19. a) What is the probability that two people chosen at random were born during the same month of the year?

b) What is the probability that in a group of n people chosen at random, there are at least two born in the same month of the year?

c) How many people chosen at random are needed to make the probability greater than $1/2$ that there are at least two people born in the same month of the year?

20. Find the smallest number of people you need to choose at random so that the probability that at least one of them has a birthday today exceeds $1/2$.

21. Find the smallest number of people you need to choose at random so that the probability that at least two of them were both born on April 1 exceeds $1/2$.

***22.** February 29 occurs only in leap years. Years divisible by 4, but not by 100, are always leap years. Years divisible by 100, but not by 400, are not leap years, but years divisible by 400 are leap years.

a) What probability distribution for birthdays should be used to reflect how often February 29 occurs?

b) Using the probability distribution from part (a), what is the probability that in a group of n people at least two have the same birthday?

23. What is the conditional probability that exactly four heads appear when a fair coin is flipped five times, given that the first flip came up heads?

24. What is the conditional probability that exactly four heads appear when a fair coin is flipped five times, given that the first flip came up tails?

25. What is the conditional probability that a randomly generated bit string of length four contains at least two consecutive 0s, given that the first bit is a 1? (Assume the probabilities of a 0 and a 1 are the same.)

26. Let E be the event that a randomly generated bit string of length three contains an odd number of 1s, and let F be the event that the string starts with 1. Are E and F independent?

27. Let E and F be the events that a family of n children has children of both sexes and has at most one child assigned male at birth, respectively. Are E and F independent if

a) $n = 2$? b) $n = 4$? c) $n = 5$?

28. Assume that the probability a child is assigned male at birth is 0.51 and that the sexes of children born into a family are independent. What is the probability that a family of five children has

a) exactly three children assigned male at birth?

b) at least one child assigned male at birth?

c) at least one child assigned female at birth?

d) all children assigned the same sex at birth?

29. A group of six people play the game of "odd person out" to determine who will buy refreshments. Each person flips a fair coin. If there is a person whose outcome is not the same as that of any other member of the group, this person has to buy the refreshments. What is the probability that there is an odd person out after the coins are flipped once?

30. Find the probability that a randomly generated bit string of length 10 does not contain a 0 if bits are independent and if

a) a 0 bit and a 1 bit are equally likely.

b) the probability that a bit is a 1 is 0.6.

c) the probability that the ith bit is a 1 is $1/2^i$ for $i = 1, 2, 3, \dots, 10$.

31. Find the probability that a family with five children does not have a child assigned male at birth, if the sexes of children are independent and if

a) having a child assigned male at birth and a child assigned female at birth are equally likely.

b) the probability of having a child assigned male at birth is 0.51.

c) the probability that the ith child is assigned male at birth is $0.51 - (i/100)$.

32. Find the probability that a randomly generated bit string of length 10 begins with a 1 or ends with a 00 for the same conditions as in parts (a), (b), and (c) of Exercise 30, if bits are generated independently.

33. Find the probability that the first child of a family with five children is a child assigned male at birth or that the last two children of the family are children assigned female at birth, for the same conditions as in parts (a), (b), and (c) of Exercise 31.

34. Find each of the following probabilities when n independent Bernoulli trials are carried out with probability of success p.

a) the probability of no successes

b) the probability of at least one success

c) the probability of at most one success

d) the probability of at least two successes

35. Find each of the following probabilities when n independent Bernoulli trials are carried out with probability of success p.

a) the probability of no failures

b) the probability of at least one failure

c) the probability of at most one failure

d) the probability of at least two failures

36. Use mathematical induction to prove that if $E_1, E_2, \dots, E_n$ is a sequence of n pairwise disjoint events in a sample space S, where n is a positive integer, then $p(\bigcup_{i=1}^{n} E_i) = \sum_{i=1}^{n} p(E_i)$.

***37.** (*Requires calculus*) Show that if $E_1, E_2, \dots$ is an infinite sequence of pairwise disjoint events in a sample space S, then $p(\bigcup_{i=1}^{\infty} E_i) = \sum_{i=1}^{\infty} p(E_i)$. [*Hint:* Use Exercise 36 and take limits.]

38. A pair of dice is rolled in a remote location and when you ask an honest observer whether at least one die came up six, this honest observer answers in the affirmative.

a) What is the probability that the sum of the numbers that came up on the two dice is seven, given the information provided by the honest observer?

b) Suppose that the honest observer tells us that at least one die came up five. What is the probability the sum of the numbers that came up on the dice is seven, given this information?

39. This exercise employs the probabilistic method to prove a result about round-robin tournaments. In a **round-robin tournament with m players, every two players play one game in which one player wins and the other loses.

We want to find conditions on positive integers m and k with $k < m$ such that it is possible for the outcomes of the tournament to have the property that for every set of k players, there is a player who beats every member in this set. So that we can use probabilistic reasoning to draw conclusions about round-robin tournaments, we assume that when two players compete it is equally likely that either player wins the game and we assume that the outcomes of different games are independent. Let E be the event that for every set S with k players, where k is a positive integer less than m, there is a player who has beaten all k players in S.

a) Show that $p(\overline{E}) \leq \sum_{j=1}^{\binom{m}{k}} p(F_j)$, where F_j is the event that there is no player who beats all k players from the jth set in a list of the $\binom{m}{k}$ sets of k players.

b) Show that the probability of F_j is $(1-2^{-k})^{m-k}$.

c) Conclude from parts (a) and (b) that $p(\overline{E}) \leq \binom{m}{k}(1-2^{-k})^{m-k}$ and, therefore, that there must be a tournament with the described property if $\binom{m}{k}(1-2^{-k})^{m-k} < 1$.

d) Use part (c) to find values of m such that there is a tournament with m players such that for every set S of two players, there is a player who has beaten both players in S. Repeat for sets of three players.

*40. Devise a Monte Carlo algorithm that determines whether a permutation of the integers 1 through n has already been sorted (that is, it is in increasing order), or instead, is a random permutation. A step of the algorithm should answer "true" if it determines the list is not sorted and "unknown" otherwise. After k steps, the algorithm decides that the integers are sorted if the answer is "unknown" in each step. Show that as the number of steps increases, the probability that the algorithm produces an incorrect answer is extremely small. [*Hint:* For each step, test whether certain elements are in the correct order. Make sure these tests are independent.]

41. Use pseudocode to write out the probabilistic primality test described in Example 17.

42. Determine the conditional probability that the father of a child has blue eyes, given that his offspring has blue eyes. (Assume that the probabilities of the genotypes BB, Bb, and bb are those given in Example 8.)

An **albino** is a person having a congenital absence of pigment in the skin and hair (which are white) and in the eyes (which are usually pink). The dominant allele is denoted by A and the recessive allele is denoted by a. People with the genotypes AA and Aa are not albinos and have normal pigmentation. However, those with the genotype aa are albinos and those with the genotype Aa are carriers of albinism.

43. Suppose that a couple, neither of which is an albino, have one child who is an albino. Further, suppose that another child of theirs who is not an albino has a child with a person known to be a carrier of albinism.

a) Determine the probability that their first offspring is an albino.

b) Determine the conditional probability that their second offspring is an albino given that their first child is not an albino.

7.3 Bayes' Theorem

7.3.1 Introduction

There are many times when we want to assess the probability that a particular event occurs on the basis of partial evidence. For example, suppose we know the percentage of people who have a particular disease for which there is a very accurate diagnostic test. People who test positive for this disease would like to know the likelihood that they actually have the disease. In this section we introduce a result that can be used to determine this probability, namely, the probability that a person has the disease given that this person tests positive for it. To use this result, we will need to know the percentage of people who do not have the disease but test positive for it and the percentage of people who have the disease but test negative for it.

Similarly, suppose we know the percentage of incoming email messages that are spam. We will see that we can determine the likelihood that an incoming email message is spam using the occurrence of words in the message. To determine this likelihood, we need to know the percentage of incoming messages that are spam, the percentage of spam messages in which each of these words occurs, and the percentage of messages that are not spam in which each of these words occurs.

The result that we can use to answer questions such as these is called Bayes' theorem and dates back to the eighteenth century. In the past two decades, Bayes' theorem has been extensively applied to estimate probabilities based on partial evidence in areas as diverse as medicine, law, machine learning, engineering, and software development.

7.3.2 Bayes' Theorem

We illustrate the idea behind Bayes' theorem with an example that shows that when extra information is available, we can derive a more realistic estimate that a particular event occurs. That is, suppose we know $p(F)$, the probability that an event F occurs, but we have knowledge that an event E occurs. Then the conditional probability that F occurs given that E occurs, $p(F \mid E)$, is a more realistic estimate than $p(F)$ that F occurs. In Example 1 we will see that we can find $p(F \mid E)$ when we know $p(F)$, $p(E \mid F)$, and $p(E \mid \overline{F})$.

EXAMPLE 1 We have two boxes. The first contains two green balls and seven red balls; the second contains four green balls and three red balls. Bob selects a ball by first choosing one of the two boxes at random and then choosing one of the balls in this box at random. If Bob has selected a red ball, what is the probability that it came from the first box?

Extra Examples

Solution: Let E be the event that Bob has chosen a red ball; $\overline{E}$ is the event that Bob has chosen a green ball. Let F be the event that Bob has chosen a ball from the first box; $\overline{F}$ is the event that Bob has chosen a ball from the second box. We want to find $p(F \mid E)$, the probability that the ball Bob selected came from the first box, given that it is red. By the definition of conditional probability, we have $p(F \mid E) = p(F \cap E)/p(E)$. Can we use the information provided to determine both $p(F \cap E)$ and $p(E)$ so that we can find $p(F \mid E)$?

First, note that because the first box contains seven red balls out of a total of nine balls, we know that $p(E \mid F) = 7/9$. Similarly, because the second box contains three red balls out of a total of seven balls, we know that $p(E \mid \overline{F}) = 3/7$. We assumed that Bob selects a box at random, so $p(F) = p(\overline{F}) = 1/2$. Because $p(E \mid F) = p(E \cap F)/p(F)$, it follows that $p(E \cap F) = p(E \mid F)p(F) = \frac{7}{9} \cdot \frac{1}{2} = \frac{7}{18}$ [as we remarked earlier, this is one of the quantities we need to find to determine $p(F \mid E)$]. Similarly, because $p(E \mid \overline{F}) = p(E \cap \overline{F})/p(\overline{F})$, it follows that $p(E \cap \overline{F}) = p(E \mid \overline{F})p(\overline{F}) = \frac{3}{7} \cdot \frac{1}{2} = \frac{3}{14}$.

We can now find $p(E)$. Note that $E = (E \cap F) \cup (E \cap \overline{F})$, where $E \cap F$ and $E \cap \overline{F}$ are disjoint sets. (If x belongs to both $E \cap F$ and $E \cap \overline{F}$, then x belongs to both F and $\overline{F}$, which is impossible.) It follows that

$$p(E) = p(E \cap F) + p(E \cap \overline{F}) = \frac{7}{18} + \frac{3}{14} = \frac{49}{126} + \frac{27}{126} = \frac{76}{126} = \frac{38}{63}.$$

We have now found both $p(F \cap E) = 7/18$ and $p(E) = 38/63$. We conclude that

$$p(F \mid E) = \frac{p(F \cap E)}{p(E)} = \frac{7/18}{38/63} = \frac{49}{76} \approx 0.645.$$

Before we had any extra information, we assumed that the probability that Bob selected the first box was $1/2$. However, with the extra information that the ball selected at random is red, this probability has increased to approximately 0.645. That is, the probability that Bob selected a ball from the first box increased from $1/2$, when no extra information was available, to 0.645 once we knew that the ball selected was red. ◀

Using the same type of reasoning as in Example 1, we can find the conditional probability that an event F occurs, given that an event E has occurred, when we know $p(E \mid F)$, $p(E \mid \overline{F})$, and $p(F)$. The result we can obtain is called **Bayes' theorem**; it is named after Thomas Bayes, an eighteenth-century British mathematician and minister who introduced this result.

THEOREM 1 **BAYES' THEOREM** Suppose that E and F are events from a sample space S such that $p(E) \neq 0$ and $p(F) \neq 0$. Then

$$p(F \mid E) = \frac{p(E \mid F)p(F)}{p(E \mid F)p(F) + p(E \mid \overline{F})p(\overline{F})}.$$

Proof: The definition of conditional probability tells us that $p(F \mid E) = p(E \cap F)/p(E)$ and $p(E \mid F) = p(E \cap F)/p(F)$. Therefore, $p(E \cap F) = p(F \mid E)p(E)$ and $p(E \cap F) = p(E \mid F)p(F)$. Equating these two expressions for $p(E \cap F)$ shows that

$$p(F \mid E)p(E) = p(E \mid F)p(F).$$

Dividing both sides by $p(E)$, we find that

$$p(F \mid E) = \frac{p(E \mid F)p(F)}{p(E)}.$$

Next, we show that $p(E) = p(E \mid F)p(F) + p(E \mid \overline{F})p(\overline{F})$. To see this, first note that $E = E \cap S = E \cap (F \cup \overline{F}) = (E \cap F) \cup (E \cap \overline{F})$. Furthermore, $E \cap F$ and $E \cap \overline{F}$ are disjoint, because if $x \in E \cap F$ and $x \in E \cap \overline{F}$, then $x \in F \cap \overline{F} = \emptyset$. Consequently, $p(E) = p(E \cap F) + p(E \cap \overline{F})$. We have already shown that $p(E \cap F) = p(E \mid F)p(F)$. Moreover, we have $p(E \mid \overline{F}) = p(E \cap \overline{F})/p(\overline{F})$, which shows that $p(E \cap \overline{F}) = p(E \mid \overline{F})p(\overline{F})$. It now follows that

$$p(E) = p(E \cap F) + p(E \cap \overline{F}) = p(E \mid F)p(F) + p(E \mid \overline{F})p(\overline{F}).$$

To complete the proof we insert this expression for $p(E)$ into the equation $p(F \mid E) = p(E \mid F)p(F)/p(E)$. We have proved that

Links

$$p(F \mid E) = \frac{p(E \mid F)p(F)}{p(E \mid F)p(F) + p(E \mid \overline{F})p(\overline{F})}.$$

◁

APPLYING BAYES' THEOREM Bayes' theorem can be used to solve problems that arise in many disciplines. Next, we will discuss an application of Bayes' theorem to medicine. In particular, we will illustrate how Bayes' theorem can be used to assess the probability that someone testing positive for a disease actually has this disease. The results obtained from Bayes' theorem are often somewhat surprising, as Example 2 shows.

EXAMPLE 2 Suppose that one person in 100,000 has a particular rare disease for which there is a fairly accurate diagnostic test. This test is correct 99.0% of the time when given to a person selected at random who has the disease; it is correct 99.5% of the time when given to a person selected at random who does not have the disease. Given this information, can we find

(a) the probability that a person who tests positive for the disease has the disease?

(b) the probability that a person who tests negative for the disease does not have the disease?

Should a person who tests positive be very concerned that he or she has the disease?

Solution: (a) Let F be the event that a person selected at random has the disease, and let E be the event that a person selected at random tests positive for the disease. We want to compute

$p(F \mid E)$. To use Bayes' theorem to compute $p(F \mid E)$ we need to find $p(E \mid F)$, $p(E \mid \overline{F})$, $p(F)$, and $p(\overline{F})$.

We know that one person in 100,000 has this disease, so $p(F) = 1/100{,}000 = 0.00001$ and $p(\overline{F}) = 1 - 0.00001 = 0.99999$. Because a person who has the disease tests positive 99% of the time, we know that $p(E \mid F) = 0.99$; this is the probability of a true positive, that a person with the disease tests positive. It follows that $p(\overline{E} \mid F) = 1 - p(E \mid F) = 1 - 0.99 = 0.01$; this is the probability of a false negative, that a person who has the disease tests negative.

Furthermore, because a person who does not have the disease tests negative 99.5% of the time, we know that $p(\overline{E} \mid \overline{F}) = 0.995$. This is the probability of a true negative, that a person without the disease tests negative. Finally, we see that $p(E \mid \overline{F}) = 1 - p(\overline{E} \mid \overline{F}) = 1 - 0.995 = 0.005$; this is the probability of a false positive, that a person without the disease tests positive.

The probability that a person who tests positive for the disease actually has the disease is $p(F \mid E)$. By Bayes' theorem, we know that

$$p(F \mid E) = \frac{p(E \mid F)p(F)}{p(E \mid F)p(F) + p(E \mid \overline{F})p(\overline{F})}$$

$$= \frac{(0.99)(0.00001)}{(0.99)(0.00001) + (0.005)(0.99999)} \approx 0.002.$$

(b) The probability that someone who tests negative for the disease does not have the disease is $p(\overline{F} \mid \overline{E})$. By Bayes' theorem, we know that

$$p(\overline{F} \mid \overline{E}) = \frac{p(\overline{E} \mid \overline{F})p(\overline{F})}{p(\overline{E} \mid \overline{F})p(\overline{F}) + p(\overline{E} \mid F)p(F)}$$

$$= \frac{(0.995)(0.99999)}{(0.995)(0.99999) + (0.01)(0.00001)} \approx 0.9999999.$$

Consequently, 99.99999% of the people who test negative really do not have the disease.

In part (a) we showed that only 0.2% of people who test positive for the disease actually have the disease. Because the disease is extremely rare, the number of false positives on the diagnostic test is far greater than the number of true positives, making the percentage of people who test positive who actually have the disease extremely small. People who test positive should not be overly concerned that they actually have the disease. ◀

GENERALIZING BAYES' THEOREM Note that in the statement of Bayes' theorem, the events F and $\overline{F}$ are mutually exclusive and cover the entire sample space S (that is, $F \cup \overline{F} = S$). We can extend Bayes' theorem to any collection of mutually exclusive events that cover the entire sample space S, in the following way.

THEOREM 2 **GENERALIZED BAYES' THEOREM** Suppose that E is an event from a sample space S and that $F_1, F_2, \ldots, F_n$ are mutually exclusive events such that $\bigcup_{i=1}^{n} F_i = S$. Assume that $p(E) \neq 0$ and $p(F_i) \neq 0$ for $i = 1, 2, \ldots, n$. Then

$$p(F_j \mid E) = \frac{p(E \mid F_j)p(F_j)}{\sum_{i=1}^{n} p(E \mid F_i)p(F_i)}.$$

We leave the proof of this generalized version of Bayes' theorem as Exercise 17.

7.3.3 Bayesian Spam Filters

Most electronic mailboxes receive a flood of unwanted and unsolicited messages, known as **spam**. Because spam threatens to overwhelm electronic mail systems, a tremendous amount of work has been devoted to filtering it out. Some of the first tools developed for eliminating spam were based on Bayes' theorem, such as **Bayesian spam filters**.

Links

The use of the word *spam* for unsolicited email comes from a Monty Python comedy sketch about a cafe where the food product Spam comes with everything regardless of whether customers want it.

A Bayesian spam filter uses information about previously seen email messages to guess whether an incoming email message is spam. Bayesian spam filters look for occurrences of particular words in messages. For a particular word w, the probability that w appears in a spam email message is estimated by determining the number of times w appears in a message from a large set of messages known to be spam and the number of times it appears in a large set of messages known not to be spam. When we examine email messages to determine whether they might be spam, we look at words that might be indicators of spam, such as "offer," "special," or "opportunity," as well as words that might indicate that a message is not spam, such as "mom," "lunch," or "Jan" (where Jan is one of your friends). Unfortunately, spam filters sometimes fail to identify a spam message as spam; this is called a false negative. And they sometimes identify a message that is not spam as spam; this is called a false positive. When testing for spam, it is important to minimize false positives, because filtering out wanted email is much worse than letting some spam through.

We will develop some basic Bayesian spam filters. First, suppose we have a set B of messages known to be spam and a set G of messages known not to be spam. (For example, users could classify messages as spam when they examine them in their inboxes.) We next identify the words that occur in B and in G. We count the number of messages in the set containing each word to find $n_B(w)$ and $n_G(w)$, the number of messages containing the word w in the sets B and G, respectively. Then, the empirical probability that a spam message contains the word w is $p(w) = n_B(w)/|B|$, and the empirical probability that a message that is not spam contains the word w is $q(w) = n_G(w)/|G|$. We note that $p(w)$ and $q(w)$ estimate the probabilities that an incoming spam message, and an incoming message that is not spam, contain the word w, respectively.

Now suppose we receive a new email message containing the word w. Let S be the event that the message is spam. Let E be the event that the message contains the word w. The events S,

Links

Stephen Stigler

THOMAS BAYES (1702–1761) Thomas Bayes was the son of a minister in a religious sect known as the Nonconformists. This sect was considered heretical in eighteenth-century Great Britain. Because of the secrecy of the Nonconformists, little is known of Thomas Bayes' life. When Thomas was young, his family moved to London. Thomas was likely educated privately; Nonconformist children generally did not attend school. In 1719 Bayes entered the University of Edinburgh, where he studied logic and theology. He was ordained as a Nonconformist minister like his father and began his work as a minister assisting his father. In 1733 he became minister of the Presbyterian Chapel in Tunbridge Wells, southeast of London, where he remained minister until 1752.

Bayes is best known for his essay on probability published in 1764, three years after his death. This essay was sent to the Royal Society by a friend who found it in the papers left behind when Bayes died. In the introduction to this essay, Bayes stated that his goal was to find a method that could measure the probability that an event happens, assuming that we know nothing about it, but that, under the same circumstances, it has happened a certain proportion of times. Bayes' conclusions were accepted by the great French mathematician Laplace but were later challenged by Boole, who questioned them in his book *Laws of Thought*. Since then Bayes' techniques have been subject to controversy.

Bayes also wrote an article that was published posthumously: "An Introduction to the Doctrine of Fluxions, and a Defense of the Mathematicians Against the Objections of the Author of The Analyst," which supported the logical foundations of calculus. Bayes was elected a Fellow of the Royal Society in 1742, with the support of important members of the Society, even though at that time he had no published mathematical works. Bayes' sole known publication during his lifetime was allegedly a mystical book entitled *Divine Benevolence*, discussing the original causation and ultimate purpose of the universe. Although the book is commonly attributed to Bayes, no author's name appeared on the title page, and the entire work is thought to be of dubious provenance. Evidence for Bayes' mathematical talents comes from a notebook that was almost certainly written by Bayes, which contains much mathematical work, including discussions of probability, trigonometry, geometry, solutions of equations, series, and differential calculus. There are also sections on natural philosophy, in which Bayes looks at topics that include electricity, optics, and celestial mechanics. Bayes is also the author of a mathematical publication on asymptotic series, which appeared after his death.

that the message is spam, and $\overline{S}$, that the message is not spam, partition the set of all messages. Hence, by Bayes' theorem, the probability that the message is spam, given that it contains the word w, is

$$p(S \mid E) = \frac{p(E \mid S)p(S)}{p(E \mid S)p(S) + p(E \mid \overline{S})p(\overline{S})}.$$

To apply this formula, we first estimate $p(S)$, the probability that an incoming message is spam, as well as $p(\overline{S})$, the probability that the incoming message is not spam. Without prior knowledge about the likelihood that an incoming message is spam, for simplicity we assume that the message is equally likely to be spam as it is not to be spam. That is, we assume that $p(S) = p(\overline{S}) = 1/2$. Using this assumption, we find that the probability that a message is spam, given that it contains the word w, is

$$p(S \mid E) = \frac{p(E \mid S)}{p(E \mid S) + p(E \mid \overline{S})}.$$

(Note that if we have some empirical data about the ratio of spam messages to messages that are not spam, we can change this assumption to produce a better estimate for $p(S)$ and for $p(\overline{S})$; see Exercise 22.)

Next, we estimate $p(E \mid S)$, the conditional probability that the message contains the word w given that the message is spam, by $p(w)$. Similarly, we estimate $p(E \mid \overline{S})$, the conditional probability that the message contains the word w, given that the message is not spam, by $q(w)$. Inserting these estimates for $p(E \mid S)$ and $p(E \mid \overline{S})$ tells us that $p(S \mid E)$ can be estimated by

$$r(w) = \frac{p(w)}{p(w) + q(w)};$$

that is, $r(w)$ estimates the probability that the message is spam, given that it contains the word w. If $r(w)$ is greater than a threshold that we set, such as 0.9, then we classify the message as spam.

EXAMPLE 3 Suppose that we have found that the word "Rolex" occurs in 250 of 2000 messages known to be spam and in 5 of 1000 messages known not to be spam. Estimate the probability that an incoming message containing the word "Rolex" is spam, assuming that it is equally likely that an incoming message is spam or not spam. If our threshold for rejecting a message as spam is 0.9, will we reject such messages?

Solution: We use the counts that the word "Rolex" appears in spam messages and messages that are not spam to find that $p(\text{Rolex}) = 250/2000 = 0.125$ and $q(\text{Rolex}) = 5/1000 = 0.005$. Because we are assuming that it is equally likely for an incoming message to be spam as it is not to be spam, we can estimate the probability that an incoming message containing the word "Rolex" is spam by

$$r(\text{Rolex}) = \frac{p(\text{Rolex})}{p(\text{Rolex}) + q(\text{Rolex})} = \frac{0.125}{0.125 + 0.005} = \frac{0.125}{0.130} \approx 0.962.$$

Because $r(\text{Rolex})$ is greater than the threshold 0.9, we reject such messages as spam. ◀

Detecting spam based on the presence of a single word can lead to excessive false positives and false negatives. Consequently, spam filters look at the presence of multiple words.

For example, suppose that the message contains the words w_1 and w_2. Let E_1 and E_2 denote the events that the message contains the words w_1 and w_2, respectively. To make our computations simpler, we assume that E_1 and E_2 are independent events and that $E_1 \mid S$ and $E_2 \mid S$ are independent events and that we have no prior knowledge regarding whether or not the message is spam. (The assumptions that E_1 and E_2 are independent and that $E_1 \mid S$ and $E_2 \mid S$ are independent may introduce some error into our computations; we assume that this error is small.) Using Bayes' theorem and our assumptions, we can show (see Exercise 23) that $p(S \mid E_1 \cap E_2)$, the probability that the message is spam given that it contains both w_1 and w_2, is

$$p(S \mid E_1 \cap E_2) = \frac{p(E_1 \mid S)p(E_2 \mid S)}{p(E_1 \mid S)p(E_2 \mid S) + p(E_1 \mid \overline{S})p(E_2 \mid \overline{S})}.$$

We estimate the probability $p(S \mid E_1 \cap E_2)$ by

$$r(w_1, w_2) = \frac{p(w_1)p(w_2)}{p(w_1)p(w_2) + q(w_1)q(w_2)}.$$

That is, $r(w_1, w_2)$ estimates the probability that the message is spam, given that it contains the words w_1 and w_2. When $r(w_1, w_2)$ is greater than a preset threshold, such as 0.9, we determine that the message is likely spam.

EXAMPLE 4 Suppose that we train a Bayesian spam filter on a set of 2000 spam messages and 1000 messages that are not spam. The word "stock" appears in 400 spam messages and 60 messages that are not spam, and the word "undervalued" appears in 200 spam messages and 25 messages that are not spam. Estimate the probability that an incoming message containing both the words "stock" and "undervalued" is spam, assuming that we have no prior knowledge about whether it is spam. Will we reject such messages as spam when we set the threshold at 0.9?

Solution: Using the counts of each of these two words in messages known to be spam or known not to be spam, we obtain the following estimates: $p(\text{stock}) = 400/2000 = 0.2$, $q(\text{stock}) = 60/1000 = 0.06$, $p(\text{undervalued}) = 200/2000 = 0.1$, and $q(\text{undervalued}) = 25/1000 = 0.025$. Using these probabilities, we can estimate the probability that the message is spam by

$$\begin{aligned} r(\text{stock, undervalued}) &= \frac{p(\text{stock})p(\text{undervalued})}{p(\text{stock})p(\text{undervalued}) + q(\text{stock})q(\text{undervalued})} \\ &= \frac{(0.2)(0.1)}{(0.2)(0.1) + (0.06)(0.025)} \approx 0.930. \end{aligned}$$

Because we have set the threshold for rejecting messages at 0.9, such messages will be rejected by the filter. ◀

The more words we use to estimate the probability that an incoming mail message is spam, the better is our chance that we correctly determine whether it is spam. In general, if E_i is the event that the message contains word w_i, assuming that the number of incoming spam messages is approximately the same as the number of incoming messages that are not spam, and that the events $E_i \mid S$ are independent, then by Bayes' theorem the probability that a message containing all the words $w_1, w_2, \ldots, w_k$ is spam is

$$p(S \mid \bigcap_{i=1}^{k} E_i) = \frac{\prod_{i=1}^{k} p(E_i \mid S)}{\prod_{i=1}^{k} p(E_i \mid S) + \prod_{i=1}^{k} p(E_i \mid \overline{S})}.$$

We can estimate this probability by

$$r(w_1, w_2, \ldots, w_k) = \frac{\prod_{i=1}^{k} p(w_i)}{\prod_{i=1}^{k} p(w_i) + \prod_{i=1}^{k} q(w_i)}.$$

For the most effective spam filter, we choose words for which the probability that each of these words appears in spam is either very high or very low. When we compute this value for a particular message, we reject the message as spam if $r(w_1, w_2, \ldots, w_k)$ exceeds a preset threshold, such as 0.9.

Bayesian poisoning, the insertion of extra words to defeat spam filters, can use random or purposefully selected words.

Another way to improve the performance of a Bayesian spam filter is to look at the probabilities that particular pairs of words appear in spam and in messages that are not spam. We then treat appearances of these pairs of words as appearance of a single block, rather than as the appearance of two separate words. For example, the pair of words "enhance performance" most likely indicates spam, while "operatic performance" indicates a message that is not spam. Similarly, we can assess the likelihood that a message is spam by examining the structure of a message to determine where words appear in it. Also, spam filters look at appearances of certain types of strings of characters rather than just words. For example, a message with the valid email address of one of your friends is less likely to be spam (if not sent by a worm) than one containing an email address that came from a country known to originate a lot of spam. There is an ongoing war between people who create spam and those trying to filter their messages out. This leads to the introduction of many new techniques to defeat spam filters, including inserting into spam messages long strings of words that appear in messages that are not spam, as well as including words inside pictures. The techniques we have discussed here are only the first steps in fighting this war on spam.

Exercises

1. Suppose that E and F are events in a sample space and $p(E) = 1/3$, $p(F) = 1/2$, and $p(E \mid F) = 2/5$. Find $p(F \mid E)$.

2. Suppose that E and F are events in a sample space and $p(E) = 2/3$, $p(F) = 3/4$, and $p(F \mid E) = 5/8$. Find $p(E \mid F)$.

3. Suppose that Frida selects a ball by first picking one of two boxes at random and then selecting a ball from this box at random. The first box contains two white balls and three blue balls, and the second box contains four white balls and one blue ball. What is the probability that Frida picked a ball from the first box if a blue ball was selected?

4. Suppose that Ann selects a ball by first picking one of two boxes at random and then selecting a ball from this box. The first box contains three orange balls and four black balls, and the second box contains five orange balls and six black balls. What is the probability that Ann picked a ball from the second box if an orange ball was selected?

5. Suppose that 8% of all bicycle racers use steroids, that a bicyclist who uses steroids tests positive for steroids 96% of the time, and that a bicyclist who does not use steroids tests positive for steroids 9% of the time. What is the probability that a randomly selected bicyclist who tests positive for steroids actually uses steroids?

6. When a test for steroids is given to soccer players, 98% of the players taking steroids test positive and 12% of the players not taking steroids test positive. Suppose that 5% of soccer players take steroids. What is the probability that a soccer player who tests positive takes steroids?

7. Suppose that a test for opium use has a 2% false positive rate and a 5% false negative rate. That is, 2% of people who do not use opium test positive for opium, and 5% of opium users test negative for opium. Furthermore, suppose that 1% of people actually use opium.

a) Find the probability that someone who tests negative for opium use does not use opium.

b) Find the probability that someone who tests positive for opium use actually uses opium.

8. Suppose that one person in 10,000 people has a rare genetic disease. There is an excellent test for the disease; 99.9% of people with the disease test positive and only 0.02% who do not have the disease test positive.

a) What is the probability that someone who tests positive has the genetic disease?

b) What is the probability that someone who tests negative does not have the disease?

9. Suppose that 8% of the patients tested in a clinic are infected with HIV. Furthermore, suppose that when a blood test for HIV is given, 98% of the patients infected with HIV test positive and that 3% of the patients not infected with HIV test positive. What is the probability that

a) a patient testing positive for HIV with this test is infected with it?
b) a patient testing positive for HIV with this test is not infected with it?
c) a patient testing negative for HIV with this test is infected with it?
d) a patient testing negative for HIV with this test is not infected with it?

10. Suppose that 4% of the patients tested in a clinic are infected with avian influenza. Furthermore, suppose that when a blood test for avian influenza is given, 97% of the patients infected with avian influenza test positive and that 2% of the patients not infected with avian influenza test positive. What is the probability that
a) a patient testing positive for avian influenza with this test is infected with it?
b) a patient testing positive for avian influenza with this test is not infected with it?
c) a patient testing negative for avian influenza with this test is infected with it?
d) a patient testing negative for avian influenza with this test is not infected with it?

11. An electronics company is planning to introduce a new camera phone. The company commissions a marketing report for each new product that predicts either the success or the failure of the product. Of new products introduced by the company, 60% have been successes. Furthermore, 70% of their successful products were predicted to be successes, while 40% of failed products were predicted to be successes. Find the probability that this new camera phone will be successful if its success has been predicted.

*12. A space probe near Neptune communicates with Earth using bit strings. Suppose that in its transmissions it sends a 1 one-third of the time and a 0 two-thirds of the time. When a 0 is sent, the probability that it is received correctly is 0.9, and the probability that it is received incorrectly (as a 1) is 0.1. When a 1 is sent, the probability that it is received correctly is 0.8, and the probability that it is received incorrectly (as a 0) is 0.2.
a) Find the probability that a 0 is received.
b) Use Bayes' theorem to find the probability that a 0 was transmitted, given that a 0 was received.

13. Suppose that E, F_1, F_2, and F_3 are events from a sample space S and that F_1, F_2, and F_3 are pairwise disjoint and their union is S. Find $p(F_1 \mid E)$ if $p(E \mid F_1) = 1/8$, $p(E \mid F_2) = 1/4$, $p(E \mid F_3) = 1/6$, $p(F_1) = 1/4$, $p(F_2) = 1/4$, and $p(F_3) = 1/2$.

14. Suppose that E, F_1, F_2, and F_3 are events from a sample space S and that F_1, F_2, and F_3 are pairwise disjoint and their union is S. Find $p(F_2 \mid E)$ if $p(E \mid F_1) = 2/7$, $p(E \mid F_2) = 3/8$, $p(E \mid F_3) = 1/2$, $p(F_1) = 1/6$, $p(F_2) = 1/2$, and $p(F_3) = 1/3$.

15. In this exercise we will use Bayes' theorem to solve the Monty Hall puzzle (Example 10 in Section 7.1). Recall that in this puzzle you are asked to select one of three doors to open. There is a large prize behind one of the three doors and the other two doors are losers. After you select a door, Monty Hall opens one of the two doors you did not select that he knows is a losing door, selecting at random if both are losing doors. Monty asks you whether you would like to switch doors. Suppose that the three doors in the puzzle are labeled 1, 2, and 3. Let W be the random variable whose value is the number of the winning door; assume that $p(W = k) = 1/3$ for $k = 1, 2, 3$. Let M denote the random variable whose value is the number of the door that Monty opens. Suppose you choose door i.
a) What is the probability that you will win the prize if the game ends without Monty asking you whether you want to change doors?
b) Find $p(M = j \mid W = k)$ for $j = 1, 2, 3$ and $k = 1, 2, 3$.
c) Use Bayes' theorem to find $p(W = j \mid M = k)$ where i and j and k are distinct values.
d) Explain why the answer to part (c) tells you whether you should change doors when Monty gives you the chance to do so.

16. Ramesh can get to work in three different ways: by bicycle, by car, or by bus. Because of commuter traffic, there is a 50% chance that Ramesh will be late when driving. When Ramesh takes the bus, which uses a special lane reserved for buses, there is a 20% chance of being late. The probability that Ramesh is late when riding a bicycle is only 5%. Ramesh arrives late one day. The boss wants to estimate the probability that Ramesh drove to work that day.
a) Suppose the boss assumes that there is a 1/3 chance that Ramesh takes each of the three ways of getting to work. What estimate for the probability that Ramesh drove does the boss obtain from Bayes' theorem under this assumption?
b) Suppose the boss knows that Ramesh drives 30% of the time, takes the bus only 10% of the time, and rides a bicycle 60% of the time. What estimate for the probability that Ramesh drove does the boss obtain from Bayes' theorem using this information?

*17. Prove Theorem 2, the extended form of Bayes' theorem. That is, suppose that E is an event from a sample space S and that $F_1, F_2, \ldots, F_n$ are mutually exclusive events such that $\bigcup_{i=1}^{n} F_i = S$. Assume that $p(E) \neq 0$ and $p(F_i) \neq 0$ for $i = 1, 2, \ldots, n$. Show that

$$p(F_j \mid E) = \frac{p(E \mid F_j)p(F_j)}{\sum_{i=1}^{n} p(E \mid F_i)p(F_i)}.$$

[*Hint:* Use the fact that $E = \bigcup_{i=1}^{n}(E \cap F_i)$.]

18. Suppose that a Bayesian spam filter is trained on a set of 500 spam messages and 200 messages that are not spam. The word "exciting" appears in 40 spam messages and in 25 messages that are not spam. Would an incoming message be rejected as spam if it contains the word "exciting" and the threshold for rejecting spam is 0.9?

19. Suppose that a Bayesian spam filter is trained on a set of 1000 spam messages and 400 messages that are not spam. The word "opportunity" appears in 175 spam messages and 20 messages that are not spam. Would an incoming message be rejected as spam if it contains the word "opportunity" and the threshold for rejecting a message is 0.9?

20. Would we reject a message as spam in Example 4

a) using just the fact that the word "undervalued" occurs in the message?

b) using just the fact that the word "stock" occurs in the message?

21. Suppose that a Bayesian spam filter is trained on a set of 10,000 spam messages and 5000 messages that are not spam. The word "enhancement" appears in 1500 spam messages and 20 messages that are not spam, while the word "herbal" appears in 800 spam messages and 200 messages that are not spam. Estimate the probability that a received message containing both the words "enhancement" and "herbal" is spam. Will the message be rejected as spam if the threshold for rejecting spam is 0.9?

22. Suppose that we have prior information concerning whether a random incoming message is spam. In particular, suppose that over a time period, we find that s spam messages arrive and h messages arrive that are not spam.

a) Use this information to estimate $p(S)$, the probability that an incoming message is spam, and $p(\overline{S})$, the probability an incoming message is not spam.

b) Use Bayes' theorem and part (a) to estimate the probability that an incoming message containing the word w is spam, where $p(w)$ is the probability that w occurs in a spam message and $q(w)$ is the probability that w occurs in a message that is not spam.

23. Suppose that E_1 and E_2 are the events that an incoming mail message contains the words w_1 and w_2, respectively. Assuming that E_1 and E_2 are independent events and that $E_1 \mid S$ and $E_2 \mid S$ are independent events, where S is the event that an incoming message is spam, and that we have no prior knowledge regarding whether or not the message is spam, show that

$$p(S \mid E_1 \cap E_2) = \frac{p(E_1 \mid S)p(E_2 \mid S)}{p(E_1 \mid S)p(E_2 \mid S) + p(E_1 \mid \overline{S})p(E_2 \mid \overline{S})}.$$

7.4 Expected Value and Variance

7.4.1 Introduction

The **expected value** of a random variable is the sum over all elements in a sample space of the product of the probability of the element and the value of the random variable at this element. Consequently, the expected value is a weighted average of the values of a random variable. The expected value of a random variable provides a central point for the distribution of values of this random variable. We can solve many problems using the notion of the expected value of a random variable, such as determining who has an advantage in gambling games and computing the average-case complexity of algorithms. Another useful measure of a random variable is its **variance**, which tells us how spread out the values of this random variable are. We can use the variance of a random variable to help us estimate the probability that a random variable takes values far removed from its expected value.

7.4.2 Expected Values

Links

Many questions can be formulated in terms of the value we expect a random variable to take, or more precisely, the average value of a random variable when an experiment is performed a large number of times. Questions of this kind include: How many heads are expected to appear when a coin is flipped 100 times? What is the expected number of comparisons used to find an element in a list using a linear search? To study such questions we introduce the concept of the expected value of a random variable.

Definition 1 The *expected value*, also called the *expectation* or *mean*, of the random variable X on the sample space S is equal to

$$E(X) = \sum_{s \in S} p(s)X(s).$$

The *deviation* of X at $s \in S$ is $X(s) - E(X)$, the difference between the value of X and the mean of X.

Note that when the sample space S has n elements $S = \{x_1, x_2, \ldots, x_n\}$, $E(X) = \sum_{i=1}^{n} p(x_i)X(x_i)$.

Remark: When there are infinitely many elements of the sample space, the expectation is defined only when the infinite series in the definition is absolutely convergent. In particular, the expectation of a random variable on an infinite sample space is finite if it exists.

EXAMPLE 1 **Expected Value of a Die** Let X be the number that comes up when a fair die is rolled. What is the expected value of X?

Solution: The random variable X takes the values 1, 2, 3, 4, 5, or 6, each with probability $1/6$. It follows that

$$E(X) = \frac{1}{6} \cdot 1 + \frac{1}{6} \cdot 2 + \frac{1}{6} \cdot 3 + \frac{1}{6} \cdot 4 + \frac{1}{6} \cdot 5 + \frac{1}{6} \cdot 6 = \frac{21}{6} = \frac{7}{2}.$$ ◀

EXAMPLE 2 A fair coin is flipped three times. Let S be the sample space of the eight possible outcomes, and let X be the random variable that assigns to an outcome the number of heads in this outcome. What is the expected value of X?

Extra Examples

Solution: In Example 11 of Section 7.2 we listed the values of X for the eight possible outcomes when a coin is flipped three times. Because the coin is fair and the flips are independent, the probability of each outcome is $1/8$. Consequently,

$$\begin{aligned} E(X) &= \frac{1}{8}[X(HHH) + X(HHT) + X(HTH) + X(THH) + X(TTH) \\ &\quad + X(THT) + X(HTT) + X(TTT)] \\ &= \frac{1}{8}(3 + 2 + 2 + 2 + 1 + 1 + 1 + 0) = \frac{12}{8} \\ &= \frac{3}{2}. \end{aligned}$$

Consequently, the expected number of heads that come up when a fair coin is flipped three times is $3/2$. ◀

When an experiment has relatively few outcomes, we can compute the expected value of a random variable directly from its definition, as was done in Example 2. However, when an experiment has a large number of outcomes, it may be inconvenient to compute the expected value of a random variable directly from its definition. Instead, we can find the expected value of a random variable by grouping together all outcomes assigned the same value by the random variable, as Theorem 1 shows.

THEOREM 1 If X is a random variable and $p(X = r)$ is the probability that $X = r$, so that $p(X = r) = \sum_{s\in S, X(s)=r} p(s)$, then

$$E(X) = \sum_{r\in X(S)} p(X = r)r.$$

Proof: Suppose that X is a random variable with range $X(S)$, and let $p(X = r)$ be the probability that the random variable X takes the value r. Consequently, $p(X = r)$ is the sum of the probabilities of the outcomes s such that $X(s) = r$. It follows that

$$E(X) = \sum_{r\in X(S)} p(X = r)r.$$ ◁

Example 3 and the proof of Theorem 2 will illustrate the use of this formula. In Example 3 we will find the expected value of the sum of the numbers that appear on two fair dice when they are rolled. In Theorem 2 we will find the expected value of the number of successes when n Bernoulli trials are performed.

EXAMPLE 3 What is the expected value of the sum of the numbers that appear when a pair of fair dice is rolled?

Solution: Let X be the random variable equal to the sum of the numbers that appear when a pair of dice is rolled. In Example 13 of Section 7.2 we listed the value of X for the 36 outcomes of this experiment. The range of X is $\{2, 3, 4, 5, 6, 7, 8, 9, 10, 11, 12\}$. By Example 13 of Section 7.2 we see that

$$\begin{aligned}
&p(X = 2) = p(X = 12) = 1/36,\\
&p(X = 3) = p(X = 11) = 2/36 = 1/18,\\
&p(X = 4) = p(X = 10) = 3/36 = 1/12,\\
&p(X = 5) = p(X = 9) = 4/36 = 1/9,\\
&p(X = 6) = p(X = 8) = 5/36,\\
&p(X = 7) = 6/36 = 1/6.
\end{aligned}$$

Substituting these values in the formula, we have

$$\begin{aligned}
E(X) &= 2\cdot\frac{1}{36} + 3\cdot\frac{1}{18} + 4\cdot\frac{1}{12} + 5\cdot\frac{1}{9} + 6\cdot\frac{5}{36} + 7\cdot\frac{1}{6}\\
&\quad + 8\cdot\frac{5}{36} + 9\cdot\frac{1}{9} + 10\cdot\frac{1}{12} + 11\cdot\frac{1}{18} + 12\cdot\frac{1}{36}\\
&= 7.
\end{aligned}$$ ◀

THEOREM 2 The expected number of successes when n mutually independent Bernoulli trials are performed, where p is the probability of success on each trial, is np.

Proof: Let X be the random variable equal to the number of successes in n trials. By Theorem 2 of Section 7.2 we see that $p(X = k) = C(n, k)p^k q^{n-k}$. Hence, we have

$$\begin{aligned}
E(X) &= \sum_{k=1}^{n} kp(X = k) && \text{by Theorem 1}\\
&= \sum_{k=1}^{n} kC(n, k)p^k q^{n-k} && \text{by Theorem 2 in Section 7.2}\\
&= \sum_{k=1}^{n} nC(n-1, k-1)p^k q^{n-k} && \text{by Exercise 25 in Section 6.4}\\
&= np\sum_{k=1}^{n} C(n-1, k-1)p^{k-1} q^{n-k} && \text{factoring } np \text{ from each term}\\
&= np\sum_{j=0}^{n-1} C(n-1, j)p^{j} q^{n-1-j} && \text{shifting index of summation with } j = k-1\\
&= np(p+q)^{n-1} && \text{by the binomial theorem}\\
&= np && \text{because } p+q = 1.
\end{aligned}$$

This completes the proof because it shows that the expected number of successes in n mutually independent Bernoulli trials is np. ◁

We will also show that the hypothesis that the Bernoulli trials are mutually independent in Theorem 2 is not necessary.

7.4.3 Linearity of Expectations

Theorem 3 tells us that expected values are linear. For example, the expected value of the sum of random variables is the sum of their expected values. We will find this property exceedingly useful.

THEOREM 3 If X_i, $i = 1, 2, \ldots, n$ with n a positive integer, are random variables on S, and if a and b are real numbers, then

(*i*) $E(X_1 + X_2 + \cdots + X_n) = E(X_1) + E(X_2) + \cdots + E(X_n)$
(*ii*) $E(aX + b) = aE(X) + b$.

Proof: Part (*i*) follows for $n = 2$ directly from the definition of expected value, because

$$\begin{aligned}
E(X_1 + X_2) &= \sum_{s\in S} p(s)(X_1(s) + X_2(s))\\
&= \sum_{s\in S} p(s)X_1(s) + \sum_{s\in S} p(s)X_2(s)\\
&= E(X_1) + E(X_2).
\end{aligned}$$

The case for n random variables follows easily by mathematical induction using the case of two random variables. (We leave it to the reader to complete the proof.)

To prove part (*ii*), note that

$$\begin{aligned}
E(aX + b) &= \textstyle\sum_{s\in S} p(s)(aX(s) + b)\\
&= a\textstyle\sum_{s\in S} p(s)X(s) + b\textstyle\sum_{s\in S} p(s)\\
&= aE(X) + b \text{ because } \textstyle\sum_{s\in S} p(s) = 1.
\end{aligned}$$

◁

Examples 4 and 5 illustrate how to use Theorem 3.

EXAMPLE 4 Use Theorem 3 to find the expected value of the sum of the numbers that appear when a pair of fair dice is rolled. (This was done in Example 3 without the benefit of this theorem.)

Solution: Let X_1 and X_2 be the random variables with $X_1((i,j)) = i$ and $X_2((i,j)) = j$, so that X_1 is the number appearing on the first die and X_2 is the number appearing on the second die. It is easy to see that $E(X_1) = E(X_2) = 7/2$ because both equal $(1+2+3+4+5+6)/6 = 21/6 = 7/2$. The sum of the two numbers that appear when the two dice are rolled is the sum $X_1 + X_2$. By Theorem 3, the expected value of the sum is $E(X_1 + X_2) = E(X_1) + E(X_2) = 7/2 + 7/2 = 7$. ◀

EXAMPLE 5 In the proof of Theorem 2 we found the expected value of the number of successes when n independent Bernoulli trials are performed, where p is the probability of success on each trial by direct computation. Show how Theorem 3 can be used to derive this result where the Bernoulli trials are not necessarily independent.

Solution: Let X_i be the random variable with $X_i((t_1, t_2, \ldots, t_n)) = 1$ if t_i is a success and $X_i((t_1, t_2, \ldots, t_n)) = 0$ if t_i is a failure. The expected value of X_i is $E(X_i) = 1 \cdot p + 0 \cdot (1-p) = p$ for $i = 1, 2, \ldots, n$. Let $X = X_1 + X_2 + \cdots + X_n$, so that X counts the number of successes when these n Bernoulli trials are performed. Theorem 3, applied to the sum of n random variables, shows that $E(X) = E(X_1) + E(X_2) + \cdots + E(X_n) = np$. ◀

We can take advantage of the linearity of expectations to find the solutions of many seemingly difficult problems. The key step is to express a random variable whose expectation we wish to find as the sum of random variables whose expectations are easy to find. Examples 6 and 7 illustrate this technique.

EXAMPLE 6 **Expected Value in the Hatcheck Problem** A new employee checks the hats of n people at a restaurant, forgetting to put claim check numbers on the hats. When customers return for their hats, the checker gives them back hats chosen at random from the remaining hats. What is the expected number of hats that are returned correctly?

Solution: Let X be the random variable that equals the number of people who receive the correct hat from the checker. Let X_i be the random variable with $X_i = 1$ if the ith person receives the correct hat and $X_i = 0$ otherwise. It follows that

$$X = X_1 + X_2 + \cdots + X_n.$$

Because it is equally likely that the checker returns any of the hats to this person, it follows that the probability that the ith person receives the correct hat is $1/n$. Consequently, by Theorem 1, for all i we have

$$E(X_i) = 1 \cdot p(X_i = 1) + 0 \cdot p(X_i = 0) = 1 \cdot 1/n + 0 = 1/n.$$

By the linearity of expectations (Theorem 3), it follows that

$$E(X) = E(X_1) + E(X_2) + \cdots + E(X_n) = n \cdot 1/n = 1.$$

Consequently, the average number of people who receive the correct hat is exactly 1. Note that this answer is independent of the number of people who have checked their hats!

(We will find an explicit formula for the probability that no one receives the correct hat in Example 4 of Section 8.6.) ◀

EXAMPLE 7 **Expected Number of Inversions in a Permutation** The ordered pair (i, j) is called an **inversion** in a permutation of the first n positive integers if $i < j$ but j precedes i in the permutation. For instance, there are six inversions in the permutation 3, 5, 1, 4, 2; these inversions are

$$(1, 3), (1, 5), (2, 3), (2, 4), (2, 5), (4, 5).$$

Let $I_{i,j}$ be the random variable on the set of all permutations of the first n positive integers with $I_{i,j} = 1$ if (i, j) is an inversion of the permutation and $I_{i,j} = 0$ otherwise. It follows that if X is the random variable equal to the number of inversions in the permutation, then

$$X = \sum_{1 \leq i < j \leq n} I_{i,j}.$$

Note that it is equally likely for i to precede j in a randomly chosen permutation as it is for j to precede i. (To see this, note that there are an equal number of permutations with each of these properties.) Consequently, for all pairs i and j we have

$$E(I_{i,j}) = 1 \cdot p(I_{i,j} = 1) + 0 \cdot p(I_{i,j} = 0) = 1 \cdot 1/2 + 0 = 1/2.$$

Because there are $\binom{n}{2}$ pairs i and j with $1 \leq i < j \leq n$ and by the linearity of expectations (Theorem 3), we have

$$E(X) = \sum_{1 \leq i < j \leq n} E(I_{i,j}) = \binom{n}{2} \cdot \frac{1}{2} = \frac{n(n-1)}{4}.$$

It follows that there are an average of $n(n - 1)/4$ inversions in a permutation of the first n positive integers. ◀

7.4.4 Average-Case Computational Complexity

Links Computing the average-case computational complexity of an algorithm can be interpreted as computing the expected value of a random variable. Let the sample space of an experiment be the set of possible inputs $a_j, j = 1, 2, \ldots, n$, and let X be the random variable that assigns to a_j the number of operations used by the algorithm when given a_j as input. Based on our knowledge of the input, we assign a probability $p(a_j)$ to each possible input value a_j. Then, the average-case complexity of the algorithm is

$$E(X) = \sum_{j=1}^{n} p(a_j)X(a_j).$$

This is the expected value of X.

Finding the average-case computational complexity of an algorithm is usually much more difficult than finding its worst-case computational complexity, and often involves the use of sophisticated methods. However, there are some algorithms for which the analysis required to find the average-case computational complexity is not difficult. For instance, in Example 8 we will illustrate how to find the average-case computational complexity of the linear search algorithm under different assumptions concerning the probability that the element for which we search is an element of the list.

EXAMPLE 8 **Average-Case Complexity of the Linear Search Algorithm** We are given a real number x and a list of n distinct real numbers. The linear search algorithm, described in Section 3.1, locates x by successively comparing it to each element in the list, terminating when x is located or when all the elements have been examined and it has been determined that x is not in the list. What is the average-case computational complexity of the linear search algorithm if the probability that x is in the list is p and it is equally likely that x is any of the n elements in the list? (There are $n + 1$ possible types of input: one type for each of the n numbers in the list and a last type for numbers not in the list, which we treat as a single input.)

Solution: In Example 4 of Section 3.3 we showed that $2i + 1$ comparisons are used if x equals the ith element of the list and, in Example 2 of Section 3.3, we showed that $2n + 2$ comparisons are used if x is not in the list. The probability that x equals a_i, the ith element in the list, is p/n, and the probability that x is not in the list is $q = 1 - p$. It follows that the average-case computational complexity of the linear search algorithm is

$$\begin{aligned} E &= \frac{3p}{n} + \frac{5p}{n} + \cdots + \frac{(2n+1)p}{n} + (2n+2)q \\ &= \frac{p}{n}(3 + 5 + \cdots + (2n+1)) + (2n+2)q \\ &= \frac{p}{n}((n+1)^2 - 1) + (2n+2)q \\ &= p(n+2) + (2n+2)q. \end{aligned}$$

(The third equality follows from Example 2 of Section 5.1.) For instance, when x is guaranteed to be in the list, we have $p = 1$ (so the probability that $x = a_i$ is $1/n$ for each i) and $q = 0$. Then $E = n + 2$, as we showed in Example 4 in Section 3.3.

When p, the probability that x is in the list, is $1/2$, it follows that $q = 1 - p = 1/2$, so $E = (n + 2)/2 + n + 1 = (3n + 4)/2$. Similarly, if the probability that x is in the list is $3/4$, we have $p = 3/4$ and $q = 1/4$, so $E = 3(n + 2)/4 + (n + 1)/2 = (5n + 8)/4$.

Finally, when x is guaranteed not to be in the list, we have $p = 0$ and $q = 1$. It follows that $E = 2n + 2$, which is not surprising because we have to search the entire list. ◀

Example 9 illustrates how the linearity of expectations can help us find the average-case complexity of a sorting algorithm, the insertion sort.

EXAMPLE 9 **Average-Case Complexity of the Insertion Sort** What is the average number of comparisons used by the insertion sort to sort n distinct elements?

Solution: We first suppose that X is the random variable equal to the number of comparisons used by the insertion sort (described in Section 3.1) to sort a list $a_1, a_2, \ldots, a_n$ of n distinct elements. Then $E(X)$ is the average number of comparisons used. (Recall that at step i for $i = 2, \ldots, n$, the insertion sort inserts the ith element in the original list into the correct position in the sorted list of the first $i - 1$ elements of the original list.)

We let X_i be the random variable equal to the number of comparisons used to insert a_i into the proper position after the first $i - 1$ elements $a_1, a_2, \ldots, a_{i-1}$ have been sorted. Because

$$X = X_2 + X_3 + \cdots + X_n,$$

we can use the linearity of expectations to conclude that

$$E(X) = E(X_2 + X_3 + \cdots + X_n) = E(X_2) + E(X_3) + \cdots + E(X_n).$$

To find $E(X_i)$ for $i = 2, 3, \ldots, n$, let $p_j(k)$ denote the probability that the largest of the first j elements in the list occurs at the kth position, that is, that $\max(a_1, a_2, \ldots, a_j) = a_k$, where $1 \leq k \leq j$. Because the elements of the list are randomly distributed, it is equally likely for the largest element among the first j elements to occur at any position. Consequently, $p_j(k) = 1/j$. If $X_i(k)$ equals the number of comparisons used by the insertion sort if a_i is inserted into the kth position in the list once $a_1, a_2, \ldots, a_{i-1}$ have been sorted, it follows that $X_i(k) = k$. Because it is possible that a_i is inserted in any of the first i positions, we find that

$$E(X_i) = \sum_{k=1}^{i} p_i(k) \cdot X_i(k) = \sum_{k=1}^{i} \frac{1}{i} \cdot k = \frac{1}{i} \cdot \sum_{k=1}^{i} k = \frac{1}{i} \cdot \frac{i(i+1)}{2} = \frac{i+1}{2}.$$

It follows that

$$\begin{aligned} E(X) &= \sum_{i=2}^{n} E(X_i) = \sum_{i=2}^{n} \frac{i+1}{2} = \frac{1}{2} \sum_{j=3}^{n+1} j \\ &= \frac{1}{2} \frac{(n+1)(n+2)}{2} - \frac{1}{2}(1+2) = \frac{n^2 + 3n - 4}{4}. \end{aligned}$$

To obtain the third of these equalities we shifted the index of summation, setting $j = i + 1$. To obtain the fourth equality, we used the formula $\sum_{k=1}^{m} k = m(m+1)/2$ (from Table 2 in Section 2.4) with $m = n + 1$, subtracting off the missing terms with $j = 1$ and $j = 2$. We conclude that the average number of comparisons used by the insertion sort to sort n elements equals $(n^2 + 3n - 4)/4$, which is $\Theta(n^2)$. ◀

7.4.5 The Geometric Distribution

We now turn our attention to a random variable with infinitely many possible outcomes.

EXAMPLE 10 Suppose that the probability that a coin comes up tails is p. This coin is flipped repeatedly until it comes up tails. What is the expected number of flips until this coin comes up tails?

Links

Solution: We first note that the sample space consists of all sequences that begin with any number of heads, denoted by H, followed by a tail, denoted by T. Therefore, the sample space is the set $\{T, HT, HHT, HHHT, HHHHT, \ldots\}$. Note that this is an infinite sample space. We can determine the probability of an element of the sample space by noting that the coin flips are independent and that the probability of a head is $1 - p$. Therefore, $p(T) = p$, $p(HT) = (1 - p)p$, $p(HHT) = (1 - p)^2 p$, and in general the probability that the coin is flipped n times before a tail comes up, that is, that $n - 1$ heads come up followed by a tail, is $(1 - p)^{n-1} p$. (Exercise 14 asks for a verification that the sum of the probabilities of the points in the sample space is 1.)

Now let X be the random variable equal to the number of flips in an element in the sample space. That is, $X(T) = 1$, $X(HT) = 2$, $X(HHT) = 3$, and so on. Note that $p(X = j) = (1 - p)^{j-1} p$. The expected number of flips until the coin comes up tails equals $E(X)$.

Using Theorem 1, we find that

$$E(X) = \sum_{j=1}^{\infty} j \cdot p(X = j) = \sum_{j=1}^{\infty} j(1 - p)^{j-1} p = p \sum_{j=1}^{\infty} j(1 - p)^{j-1} = p \cdot \frac{1}{p^2} = \frac{1}{p}.$$

[The fourth equality in this chain follows from Table 2 in Section 2.4, which tells us that $\sum_{j=1}^{\infty} j(1 - p)^{j-1} = 1/(1 - (1 - p))^2 = 1/p^2$.] It follows that the expected number of times the coin is flipped until tails comes up is $1/p$. Note that when the coin is fair we have $p = 1/2$, so the expected number of flips until it comes up tails is $1/(1/2) = 2$. ◀

The random variable X that equals the number of flips expected before a coin comes up tails is an example of a random variable with a **geometric distribution**.

Definition 2 A random variable X has a *geometric distribution with parameter p* if $p(X = k) = (1 - p)^{k-1}p$ for $k = 1, 2, 3, \ldots$, where p is a real number with $0 \le p \le 1$.

Geometric distributions arise in many applications because they are used to study the time required before a particular event happens, such as the time required before we find an object with a certain property, the number of attempts before an experiment succeeds, the number of times a product can be used before it fails, and so on.

When we computed the expected value of the number of flips required before a coin comes up tails, we proved Theorem 4.

THEOREM 4 If the random variable X has the geometric distribution with parameter p, then $E(X) = 1/p$.

7.4.6 Independent Random Variables

We have already discussed independent events. We will now define what it means for two random variables to be independent.

Definition 3 The random variables X and Y on a sample space S are *independent* if

$$p(X = r_1 \text{ and } Y = r_2) = p(X = r_1) \cdot p(Y = r_2),$$

or in words, if the probability that $X = r_1$ and $Y = r_2$ equals the product of the probabilities that $X = r_1$ and $Y = r_2$, for all real numbers r_1 and r_2.

EXAMPLE 11 Are the random variables X_1 and X_2 from Example 4 independent?

Extra Examples ❯

Solution: Let $S = \{1, 2, 3, 4, 5, 6\}$, and let $i \in S$ and $j \in S$. Because there are 36 possible outcomes when the pair of dice is rolled and each is equally likely, we have

$$p(X_1 = i \text{ and } X_2 = j) = 1/36.$$

Furthermore, $p(X_1 = i) = 1/6$ and $p(X_2 = j) = 1/6$, because the probability that i appears on the first die and the probability that j appears on the second die are both $1/6$. It follows that

$$p(X_1 = i \text{ and } X_2 = j) = \frac{1}{36} \quad \text{and} \quad p(X_1 = i)p(X_2 = j) = \frac{1}{6} \cdot \frac{1}{6} = \frac{1}{36},$$

so X_1 and X_2 are independent. ◀

EXAMPLE 12 Show that the random variables X_1 and $X = X_1 + X_2$, where X_1 and X_2 are as defined in Example 4, are not independent.

Solution: Note that $p(X_1 = 1 \text{ and } X = 12) = 0$, because $X_1 = 1$ means the number appearing on the first die is 1, which implies that the sum of the numbers appearing on the two dice cannot equal 12. On the other hand, $p(X_1 = 1) = 1/6$ and $p(X = 12) = 1/36$. Hence $p(X_1 = 1 \text{ and } X = 12) \neq p(X_1 = 1) \cdot p(X = 12)$. This counterexample shows that X_1 and X are not independent. ◀

The expected value of the product of two independent random variables is the product of their expected values, as Theorem 5 shows.

THEOREM 5 If X and Y are independent random variables on a sample space S, then $E(XY) = E(X)E(Y)$.

Proof: To prove this formula, we use the key observation that the event $XY = r$ is the disjoint union of the events $X = r_1$ and $Y = r_2$ over all $r_1 \in X(S)$ and $r_2 \in Y(S)$ with $r = r_1 r_2$. We have

$$
\begin{aligned}
E(XY) &= \sum_{r \in XY(S)} r \cdot p(XY = r) && \text{by Theorem 1}\\
&= \sum_{r_1 \in X(S), r_2 \in Y(S)} r_1 r_2 \cdot p(X = r_1 \text{ and } Y = r_2) && \text{expressing } XY = r \text{ as a disjoint union}\\
&= \sum_{r_1 \in X(S)} \sum_{r_2 \in Y(S)} r_1 r_2 \cdot p(X = r_1 \text{ and } Y = r_2) && \text{using a double sum to order the terms}\\
&= \sum_{r_1 \in X(S)} \sum_{r_2 \in Y(S)} r_1 r_2 \cdot p(X = r_1) \cdot p(Y = r_2) && \text{by the independence of } X \text{ and } Y\\
&= \sum_{r_1 \in X(S)} \Big(r_1 \cdot p(X = r_1) \cdot \sum_{r_2 \in Y(S)} r_2 \cdot p(Y = r_2)\Big) && \text{by factoring out } r_1 \cdot p(X = r_1)\\
&= \sum_{r_1 \in X(S)} r_1 \cdot p(X = r_1) \cdot E(Y) && \text{by the definition of } E(Y)\\
&= E(Y)\Big(\sum_{r_1 \in X(S)} r_1 \cdot p(X = r_1)\Big) && \text{by factoring out } E(Y)\\
&= E(Y)E(X) && \text{by the definition of } E(X).
\end{aligned}
$$

We complete the proof by noting that $E(Y)E(X) = E(X)E(Y)$, which is a consequence of the commutative law for multiplication. ◁

Note that when X and Y are random variables that are not independent, we cannot conclude that $E(XY) = E(X)E(Y)$, as Example 13 shows.

EXAMPLE 13 Let X and Y be random variables that count the number of heads and the number of tails when a coin is flipped twice. Because $p(X = 2) = 1/4$, $p(X = 1) = 1/2$, and $p(X = 0) = 1/4$, by Theorem 1 we have

$$E(X) = 2 \cdot \frac{1}{4} + 1 \cdot \frac{1}{2} + 0 \cdot \frac{1}{4} = 1.$$

A similar computation shows that $E(Y) = 1$. We note that $XY = 0$ when either two heads and no tails or two tails and no heads come up and that $XY = 1$ when one head and one tail come up. Hence,

$$E(XY) = 1 \cdot \frac{1}{2} + 0 \cdot \frac{1}{2} = \frac{1}{2}.$$

It follows that

$$E(XY) \neq E(X)E(Y).$$

This does not contradict Theorem 5 because X and Y are not independent, as the reader should verify (see Exercise 16). ◀

7.4.7 Variance

Links ❯

The expected value of a random variable tells us its average value, but nothing about how widely its values are distributed. For example, if X and Y are the random variables on the set $S = \{1, 2, 3, 4, 5, 6\}$, with $X(s) = 0$ for all $s \in S$ and $Y(s) = -1$ if $s \in \{1, 2, 3\}$ and $Y(s) = 1$ if $s \in \{4, 5, 6\}$, then the expected values of X and Y are both zero. However, the random variable X never varies from 0, while the random variable Y always differs from 0 by 1. The variance of a random variable helps us characterize how widely a random variable is distributed. In particular, it provides a measure of how widely X is distributed about its expected value.

Definition 4 Let X be a random variable on a sample space S. The *variance* of X, denoted by $V(X)$, is

$$V(X) = \sum_{s \in S} (X(s) - E(X))^2 p(s).$$

That is, $V(X)$ is the weighted average of the square of the deviation of X. The *standard deviation* of X, denoted $\sigma(X)$, is defined to be $\sqrt{V(X)}$.

Theorem 6 provides a useful simple expression for the variance of a random variable.

THEOREM 6 If X is a random variable on a sample space S, then $V(X) = E(X^2) - E(X)^2$.

Proof: Note that

$$\begin{aligned} V(X) &= \sum_{s \in S} (X(s) - E(X))^2 p(s) \\ &= \sum_{s \in S} X(s)^2 p(s) - 2E(X) \sum_{s \in S} X(s)p(s) + E(X)^2 \sum_{s \in S} p(s) \\ &= E(X^2) - 2E(X)E(X) + E(X)^2 \\ &= E(X^2) - E(X)^2. \end{aligned}$$

We have used the fact that $\sum_{s \in S} p(s) = 1$ in the next-to-last step. ◁

We can use Theorems 3 and 6 to derive an alternative formula for $V(X)$ that provides some insight into the meaning of the variance of a random variable.

COROLLARY 1 If X is a random variable on a sample space S and $E(X) = \mu$, then $V(X) = E((X - \mu)^2)$.

μ is the Greek letter mu.

Proof: If X is a random variable with $E(X) = \mu$, then

$$\begin{aligned}
E((X-\mu)^2) &= E(X^2 - 2\mu X + \mu^2) && \text{expanding } (X-\mu)^2 \\
&= E(X^2) - E(2\mu X) + E(\mu^2) && \text{by part } (i) \text{ of Theorem 3} \\
&= E(X^2) - 2\mu E(X) + E(\mu^2) && \text{by part } (ii) \text{ of Theorem 3, noting that } \mu \text{ is a constant} \\
&= E(X^2) - 2\mu E(X) + \mu^2 && \text{as } E(\mu^2) = \mu^2\text{, because } \mu^2 \text{ is a constant} \\
&= E(X^2) - 2\mu^2 + \mu^2 && \text{because } E(X) = \mu \\
&= E(X^2) - \mu^2 && \text{simplifying} \\
&= V(X) && \text{by Theorem 6 and noting that } E(X) = \mu.
\end{aligned}$$

This completes the proof. ◁

Corollary 1 tells us that the variance of a random variable X is the expected value of the square of the difference between X and its own expected value. This is commonly expressed as saying that the variance of X is the mean of the square of its deviation. We also say that the standard deviation of X is the square root of the mean of the square of its deviation (often read as the "root mean square" of the deviation).

We now compute the variance of some random variables.

EXAMPLE 14 What is the variance of the random variable X with $X(t) = 1$ if a Bernoulli trial is a success and $X(t) = 0$ if it is a failure, where p is the probability of success and q is the probability of failure?

Extra Examples

Solution: Because X takes only the values 0 and 1, it follows that $X^2(t) = X(t)$. Hence,

$$V(X) = E(X^2) - E(X)^2 = p - p^2 = p(1-p) = pq.$$ ◀

EXAMPLE 15 **Variance of the Value of a Die** What is the variance of the random variable X, where X is the number that comes up when a fair die is rolled?

Solution: We have $V(X) = E(X^2) - E(X)^2$. By Example 1 we know that $E(X) = 7/2$. To find $E(X^2)$ note that X^2 takes the values i^2, $i = 1, 2, \ldots, 6$, each with probability $1/6$. It follows that

$$E(X^2) = \frac{1}{6}(1^2 + 2^2 + 3^2 + 4^2 + 5^2 + 6^2) = \frac{91}{6}.$$

We conclude that

$$V(X) = \frac{91}{6} - \left(\frac{7}{2}\right)^2 = \frac{35}{12}.$$ ◀

EXAMPLE 16 What is the variance of the random variable $X((i, j)) = 2i$, where i is the number appearing on the first die and j is the number appearing on the second die, when two fair dice are rolled?

Solution: We will use Theorem 6 to find the variance of X. To do so, we need to find the expected values of X and X^2. Note that because $p(X = k)$ is $1/6$ for $k = 2, 4, 6, 8, 10, 12$ and is 0 otherwise,

$$E(X) = (2 + 4 + 6 + 8 + 10 + 12)/6 = 7,$$

and

$$E(X^2) = (2^2 + 4^2 + 6^2 + 8^2 + 10^2 + 12^2)/6 = 182/3.$$

It follows from Theorem 6 that

$$V(X) = E(X^2) - E(X)^2 = 182/3 - 49 = 35/3.$$

◀

Another useful property is that the variance of the sum of two or more independent random variables is the sum of their variances. The formula that expresses this property is known as **Bienaymé's formula**, after Irenée-Jules Bienaymé, the French mathematician who discovered it in 1853. Bienaymé's formula is useful for computing the variance of the result of n independent Bernoulli trials, for instance.

THEOREM 7 **BIENAYMÉ'S FORMULA** If X and Y are two independent random variables on a sample space S, then $V(X + Y) = V(X) + V(Y)$. Furthermore, if $X_i, i = 1, 2, \dots, n$, with n a positive integer, are pairwise independent random variables on S, then $V(X_1 + X_2 + \cdots + X_n) = V(X_1) + V(X_2) + \cdots + V(X_n)$.

Proof: From Theorem 6, we have

$$V(X + Y) = E((X + Y)^2) - E(X + Y)^2.$$

Links

Paul Fearn/Historic Images/ Alamy Stock Photo

IRENÉE-JULES BIENAYMÉ (1796–1878) Bienaymé, born in Paris, moved with his family to Bruges in 1803 when his father became a government administrator. Bienaymé attended the Lycée impérial in Bruges, and when his family returned to Paris in 1811, the Lycée Louis-le-Grand. As a teenager, he helped defend Paris during the 1814 Napoleonic Wars; in 1815, he became a student at the École Polytechnique. In 1816 he joined the Ministry of Finances to help support his family. In 1819, he left the civil service, taking a job lecturing mathematics at the Académie militaire de Saint-Cyr. Unhappy with conditions there, he soon returned to the Ministry of Finances. He attained the position of inspector general, remaining until 1848, when he was forced to retire when a new regime installed their own people. He was able to return as inspector general in 1850, but he retired a second time in 1852. In 1851 he briefly was professor at the Sorbonne and also served as an expert statistician for Napoleon III. Bienaymé was one of the founders of the Société Mathématique de France, and in 1875 was its president.

Bienaymé was noted for his ingenuity, but his papers frustrated readers by omitting important proofs. He published sparsely, often in obscure journals. However, he made important contributions to probability and statistics, and to their applications to the social sciences and to finance. Among his important contributions are the Bienaymé-Chebyshev inequality, which provides a simple proof of the law of large numbers, a generalization of Laplace's least square method, and Bienaymé's formula for the variance of a sum of random variables. He studied the extinction of aristocratic families, declining despite general population growth. Bienaymé was a skilled linguist; he translated the works of Chebyshev, a close friend, from Russian to French. It has been suggested that his relative obscurity results from his modesty, his lack of interest in asserting the priority of his discoveries, and the fact that his work was often ahead of its time. He and his brother married two sisters who were daughters of a family friend. Bienaymé and his wife had two sons and three daughters.

It follows that

$$V(X+Y) = E(X^2 + 2XY + Y^2) - (E(X) + E(Y))^2$$
$$= E(X^2) + 2E(XY) + E(Y^2) - E(X)^2 - 2E(X)E(Y) - E(Y)^2.$$

Because X and Y are independent, by Theorem 5 we have $E(XY) = E(X)E(Y)$. It follows that

$$V(X+Y) = (E(X^2) - E(X)^2) + (E(Y^2) - E(Y)^2)$$
$$= V(X) + V(Y).$$

We leave the proof of the case for n pairwise independent random variables to the reader (Exercise 34). Such a proof can be constructed by generalizing the proof we have given for the case for two random variables. Note that it is not possible to use mathematical induction in a straightforward way to prove the general case (see Exercise 33). ◁

EXAMPLE 17 Find the variance and standard deviation of the random variable X whose value when two fair dice are rolled is $X((i, j)) = i + j$, where i is the number appearing on the first die and j is the number appearing on the second die.

Solution: Let X_1 and X_2 be the random variables defined by $X_1((i, j)) = i$ and $X_2((i, j)) = j$ for a roll of the dice. Then $X = X_1 + X_2$, and X_1 and X_2 are independent, as Example 11 showed. From Theorem 7 it follows that $V(X) = V(X_1) + V(X_2)$. A simple computation as in Example 16, together with Exercise 29 in the Supplementary Exercises, tells us that $V(X_1) = V(X_2) = 35/12$. Hence, $V(X) = 35/12 + 35/12 = 35/6$ and $\sigma(X) = \sqrt{35/6}$. ◀

We will now find the variance of the random variable that counts the number of successes when n independent Bernoulli trials are carried out.

EXAMPLE 18 What is the variance of the number of successes when n independent Bernoulli trials are performed, where, on each trial, p is the probability of success and q is the probability of failure?

Solution: Let X_i be the random variable with $X_i((t_1, t_2, \ldots, t_n)) = 1$ if trial t_i is a success and $X_i((t_1, t_2, \ldots, t_n)) = 0$ if trial t_i is a failure. Let $X = X_1 + X_2 + \cdots + X_n$. Then X counts the number of successes in the n trials. From Theorem 7 it follows that $V(X) = V(X_1) + V(X_2) + \cdots + V(X_n)$. Using Example 14 we have $V(X_i) = pq$ for $i = 1, 2, \ldots, n$. It follows that $V(X) = npq$. ◀

Links

The History Collection/ Alamy Stock Photo

PAFNUTY LVOVICH CHEBYSHEV (1821–1894) Chebyshev was born into the gentry in Okatovo, Russia. His father was a retired army officer who had fought against Napoleon. In 1832 the family, with its nine children, moved to Moscow, where Pafnuty completed his high school education at home. He entered the Department of Physics and Mathematics at Moscow University. As a student, he developed a new method for approximating the roots of equations. He graduated from Moscow University in 1841 with a degree in mathematics, and he continued his studies, passing his master's exam in 1843 and completing his master's thesis in 1846.

Chebyshev was appointed in 1847 to a position as an assistant at the University of St. Petersburg. He wrote and defended a thesis in 1847. He became a professor at St. Petersburg in 1860, a position he held until 1882. His book on the theory of congruences written in 1849 was influential in the development of number theory. His work on the distribution of prime numbers was seminal. He proved Bertrand's conjecture that for every integer $n > 3$, there is a prime between n and $2n - 2$. Chebyshev helped develop ideas that were later used to prove the prime number theorem. Chebyshev's work on the approximation of functions using polynomials is used extensively when computers are used to find values of functions. Chebyshev was also interested in mechanics. He studied the conversion of rotary motion into rectilinear motion by mechanical coupling. The Chebyshev parallel motion is three linked bars approximating rectilinear motion.

7.4.8 Chebyshev's Inequality

How likely is it that a random variable takes a value far from its expected value? Theorem 8, called Chebyshev's inequality, helps answer this question by providing an upper bound on the probability that the value of a random variable differs from the expected value of the random variable by more than a specified amount.

THEOREM 8 **CHEBYSHEV'S INEQUALITY** Let X be a random variable on a sample space S with probability function p. If r is a positive real number, then

$$p(|X(s) - E(X)| \geq r) \leq V(X)/r^2.$$

Proof: Let A be the event

$$A = \{s \in S \mid |X(s) - E(X)| \geq r\}.$$

What we want to prove is that $p(A) \leq V(X)/r^2$. Note that

$$\begin{aligned} V(X) &= \sum_{s \in S} (X(s) - E(X))^2 p(s) \\ &= \sum_{s \in A} (X(s) - E(X))^2 p(s) + \sum_{s \notin A} (X(s) - E(X))^2 p(s). \end{aligned}$$

The second sum in this expression is nonnegative, because each of its summands is nonnegative. Also, because for each element s in A, $(X(s) - E(X))^2 \geq r^2$, the first sum in this expression is at least $\sum_{s \in A} r^2 p(s)$. Hence, $V(X) \geq \sum_{s \in A} r^2 p(s) = r^2 p(A)$. It follows that $V(X)/r^2 \geq p(A)$, so $p(A) \leq V(X)/r^2$, completing the proof. ◁

EXAMPLE 19 **Deviation from the Mean when Counting Tails** Suppose that X is the random variable that counts the number of tails when a fair coin is tossed n times. Note that X is the number of successes when n independent Bernoulli trials, each with probability of success 1/2, are performed. It follows that $E(X) = n/2$ (by Theorem 2) and $V(X) = n/4$ (by Example 18). Applying Chebyshev's inequality with $r = \sqrt{n}$ shows that

$$p(|X(s) - n/2| \geq \sqrt{n}) \leq (n/4)/(\sqrt{n})^2 = 1/4.$$

Consequently, the probability is no more than 1/4 that the number of tails that come up when a fair coin is tossed n times deviates from the mean by more than $\sqrt{n}$. ◀

Chebyshev's inequality, although applicable to any random variable, often fails to provide a practical estimate for the probability that the value of a random variable exceeds its mean by a large amount. This is illustrated by Example 20.

EXAMPLE 20 Let X be the random variable whose value is the number appearing when a fair die is rolled. We have $E(X) = 7/2$ (see Example 1) and $V(X) = 35/12$ (see Example 15). Because the only possible values of X are 1, 2, 3, 4, 5, and 6, X cannot take a value more than 5/2 from its mean, $E(X) = 7/2$. Hence, $p(|X - 7/2| \geq r) = 0$ if $r > 5/2$. By Chebyshev's inequality we know that $p(|X - 7/2| \geq r) \leq (35/12)/r^2$.

For example, when $r = 3$, Chebyshev's inequality tells us that $p(|X - 7/2| \geq 3) \leq (35/12)/9 = 35/108 \approx 0.324$, which is a poor estimate, because $p(|X - 7/2| \geq 3) = 0$. ◀

Exercises

1. What is the expected number of heads that come up when a fair coin is flipped five times?
2. What is the expected number of heads that come up when a fair coin is flipped 10 times?
3. What is the expected number of times a 6 appears when a fair die is rolled 10 times?
4. A coin is biased so that the probability a head comes up when it is flipped is 0.6. What is the expected number of heads that come up when it is flipped 10 times?
5. What is the expected sum of the numbers that appear on two dice, each biased so that a 3 comes up twice as often as each other number?
6. What is the expected value when a \$1 lottery ticket is bought in which the purchaser wins exactly \$10 million if the ticket contains the six winning numbers chosen from the set $\{1, 2, 3, \ldots, 50\}$ and the purchaser wins nothing otherwise?
7. The final exam of a discrete mathematics course consists of 50 true/false questions, each worth two points, and 25 multiple-choice questions, each worth four points. The probability that Linda answers a true/false question correctly is 0.9, and the probability that Linda answers a multiple-choice question correctly is 0.8. What is Linda's expected score on the final?
8. What is the expected sum of the numbers that appear when three fair dice are rolled?
9. Suppose that the probability that x is in a list of n distinct integers is $2/3$ and that it is equally likely that x equals any element in the list. Find the average number of comparisons used by the linear search algorithm to find x or to determine that it is not in the list.
10. Suppose that we flip a fair coin until either it comes up tails twice or we have flipped it six times. What is the expected number of times we flip the coin?
11. Suppose that we roll a fair die until a 6 comes up or we have rolled it 10 times. What is the expected number of times we roll the die?
12. Suppose that we roll a fair die until a 6 comes up.
    a) What is the probability that we roll the die n times?
    b) What is the expected number of times we roll the die?
13. Suppose that we roll a pair of fair dice until the sum of the numbers on the dice is seven. What is the expected number of times we roll the dice?
14. Show that the sum of the probabilities of a random variable with geometric distribution with parameter p, where $0 < p \leq 1$, equals 1.
15. Show that if the random variable X has the geometric distribution with parameter p, and j is a positive integer, then $p(X \geq j) = (1 - p)^{j-1}$.
16. Let X and Y be the random variables that count the number of heads and the number of tails that come up when two fair coins are flipped. Show that X and Y are not independent.
17. Estimate the expected number of integers with 1000 digits that need to be selected at random to find a prime, if the probability a number with 1000 digits is prime is approximately $1/2302$.
18. Suppose that X and Y are random variables and that X and Y are nonnegative for all points in a sample space S. Let Z be the random variable defined by $Z(s) = \max(X(s), Y(s))$ for all elements $s \in S$. Show that $E(Z) \leq E(X) + E(Y)$.
19. Let X be the number appearing on the first die when two fair dice are rolled and let Y be the sum of the numbers appearing on the two dice. Show that $E(X)E(Y) \neq E(XY)$.

*20. Show that if $X_1, X_2, \ldots, X_n$ are mutually independent random variables, then $E(\prod_{i=1}^{n} X_i) = \prod_{i=1}^{n} E(X_i)$.

The **conditional expectation** of the random variable X given the event A from the sample space S is $E(X|A) = \sum_{r \in X(S)} r \cdot P(X = r|A)$.

21. What is the expected value of the sum of the numbers appearing on two fair dice when they are rolled given that the sum of these numbers is at least nine. That is, what is $E(X|A)$ where X is the sum of the numbers appearing on the two dice and A is the event that $X \geq 9$?

The **law of total expectation** states that if the sample space S is the disjoint union of the events $S_1, S_2, \ldots, S_n$ and X is a random variable, then $E(X) = \sum_{j=1}^{n} E(X|S_j)P(S_j)$.

22. Prove the law of total expectations.
23. Use the law of total expectation to find the average weight of a breeding elephant seal, given that 12% of the breeding elephant seals are male and the rest are female, and the expected weights of a breeding elephant seal is 4200 pounds for a male and 1100 pounds for a female.
24. Let A be an event. Then I_A, the **indicator random variable** of A, equals 1 if A occurs and equals 0 otherwise. Show that the expectation of the indicator random variable of A equals the probability of A, that is, $E(I_A) = p(A)$.

25. A **run** is a maximal sequence of successes in a sequence of Bernoulli trials. For example, in the sequence $S, S, S, F, S, S, F, F, S$, where S represents success and F represents failure, there are three runs consisting of three successes, two successes, and one success, respectively. Let R denote the random variable on the set of sequences of n independent Bernoulli trials that counts the number of runs in this sequence. Find $E(R)$. [*Hint:* Show that $R = \sum_{j=1}^{n} I_j$, where $I_j = 1$ if a run begins at the jth Bernoulli trial and $I_j = 0$ otherwise. Find $E(I_1)$ and then find $E(I_j)$, where $1 < j \leq n$.]

26. Let $X(s)$ be a random variable, where $X(s)$ is a nonnegative integer for all $s \in S$, and let A_k be the event that $X(s) \geq k$. Show that $E(X) = \sum_{k=1}^{\infty} p(A_k)$.

27. What is the variance of the number of heads that come up when a fair coin is flipped 10 times?

28. What is the variance of the number of times a 6 appears when a fair die is rolled 10 times?

29. Let X_n be the random variable that equals the number of tails minus the number of heads when n fair coins are flipped.

a) What is the expected value of X_n?

b) What is the variance of X_n?

30. Show that if X and Y are independent random variables, then $V(XY) = E(X)^2V(Y) + E(Y)^2V(X) + V(X)V(Y)$.

31. Let $A(X) = E(|X - E(X)|)$, the expected value of the absolute value of the deviation of X, where X is a random variable. Prove or disprove that $A(X + Y) = A(X) + A(Y)$ for all random variables X and Y.

32. Provide an example that shows that the variance of the sum of two random variables is not necessarily equal to the sum of their variances when the random variables are not independent.

33. Suppose that X_1 and X_2 are independent Bernoulli trials each with probability $1/2$, and let $X_3 = (X_1 + X_2) \textbf{ mod } 2$.

a) Show that X_1, X_2, and X_3 are pairwise independent, but X_3 and $X_1 + X_2$ are not independent.

b) Show that $V(X_1 + X_2 + X_3) = V(X_1) + V(X_2) + V(X_3)$.

c) Explain why a proof by mathematical induction of Theorem 7 does not work by considering the random variables X_1, X_2, and X_3.

***34.** Prove the general case of Theorem 7. That is, show that if $X_1, X_2, \ldots, X_n$ are pairwise independent random variables on a sample space S, where n is a positive integer, then $V(X_1 + X_2 + \cdots + X_n) = V(X_1) + V(X_2) + \cdots + V(X_n)$. [*Hint:* Generalize the proof given in Theorem 7 for two random variables. Note that a proof using mathematical induction does not work; see Exercise 33.]

35. Use Chebyshev's inequality to find an upper bound on the probability that the number of tails that come up when a fair coin is tossed n times deviates from the mean by more than $5\sqrt{n}$.

36. Use Chebyshev's inequality to find an upper bound on the probability that the number of tails that come up when a biased coin with probability of heads equal to 0.6 is tossed n times deviates from the mean by more than $\sqrt{n}$.

37. Let X be a random variable on a sample space S such that $X(s) \geq 0$ for all $s \in S$. Show that $p(X(s) \geq a) \leq E(X)/a$ for every positive real number a. This inequality is called **Markov's inequality**.

38. Suppose that the number of cans of soda pop filled in a day at a bottling plant is a random variable with an expected value of 10,000 and a variance of 1000.

a) Use Markov's inequality (Exercise 37) to obtain an upper bound on the probability that the plant will fill more than 11,000 cans on a particular day.

b) Use Chebyshev's inequality to obtain a lower bound on the probability that the plant will fill between 9000 and 11,000 cans on a particular day.

39. Suppose that the number of aluminum cans recycled in a day at a recycling center is a random variable with an expected value of 50,000 and a variance of 10,000.

a) Use Markov's inequality (Exercise 37) to find an upper bound on the probability that the center will recycle more than 55,000 cans on a particular day.

b) Use Chebyshev's inequality to provide a lower bound on the probability that the center will recycle 40,000 to 60,000 cans on a certain day.

***40.** Suppose the probability that x is the ith element in a list of n distinct integers is $i/[n(n + 1)]$. Find the average number of comparisons used by the linear search algorithm to find x or to determine that it is not in the list.

***41.** In this exercise we derive an estimate of the average-case complexity of the variant of the bubble sort algorithm that terminates once a pass has been made with no interchanges. Let X be the random variable on the set of permutations of a set of n distinct integers $\{a_1, a_2, \ldots, a_n\}$ with $a_1 < a_2 < \cdots < a_n$ such that $X(P)$ equals the number of comparisons used by the bubble sort to put these integers into increasing order.

a) Show that, under the assumption that the input is equally likely to be any of the $n!$ permutations of these integers, the average number of comparisons used by the bubble sort equals $E(X)$.

b) Use Example 5 in Section 3.3 to show that $E(X) \leq n(n - 1)/2$.

c) Show that the sort makes at least one comparison for every inversion of two integers in the input.

d) Let $I(P)$ be the random variable that equals the number of inversions in the permutation P. Show that $E(X) \geq E(I)$.

e) Let $I_{j,k}$ be the random variable with $I_{j,k}(P) = 1$ if a_k precedes a_j in P and $I_{j,k} = 0$ otherwise. Show that $I(P) = \sum_k \sum_{j<k} I_{j,k}(P)$.

f) Show that $E(I) = \sum_k \sum_{j<k} E(I_{j,k})$.

g) Show that $E(I_{j,k}) = 1/2$. [*Hint:* Show that $E(I_{j,k})$ = probability that a_k precedes a_j in a permutation P. Then show it is equally likely for a_k to precede a_j as it is for a_j to precede a_k in a permutation.]

h) Use parts (f) and (g) to show that $E(I) = n(n - 1)/4$.

i) Conclude from parts (b), (d), and (h) that the average number of comparisons used to sort n integers is $\Theta(n^2)$.

***42.** In this exercise we find the average-case complexity of the quicksort algorithm, described in the preamble to Exercise 50 in Section 5.4, assuming a uniform distribution on the set of permutations.

a) Let X be the number of comparisons used by the quicksort algorithm to sort a list of n distinct integers. Show that the average number of comparisons used by the quicksort algorithm is $E(X)$ (where the sample space is the set of all $n!$ permutations of n integers).

b) Let $I_{j,k}$ denote the random variable that equals 1 if the jth smallest element and the kth smallest element of the initial list are ever compared as the quicksort algorithm sorts the list and equals 0 otherwise. Show that $X = \sum_{k=2}^{n} \sum_{j=1}^{k-1} I_{j,k}$.

c) Show that $E(X) = \sum_{k=2}^{n} \sum_{j=1}^{k-1} p$(the jth smallest element and the kth smallest element are compared).

d) Show that p(the jth smallest element and the kth smallest element are compared), where $k > j$, equals $2/(k-j+1)$.

e) Use parts (c) and (d) to show that $E(X) = 2(n+1)(\sum_{i=2}^{n} 1/i) - 2(n-1)$.

f) Conclude from part (e) and the fact that $\sum_{j=1}^{n} 1/j \approx \ln n + \gamma$, where $\gamma = 0.57721\ldots$ is Euler's constant, that the average number of comparisons used by the quicksort algorithm is $\Theta(n \log n)$.

***43.** What is the variance of the number of **fixed elements**, that is, elements left in the same position, of a randomly selected permutation of n elements? [*Hint:* Let X denote the number of fixed points of a random permutation. Write $X = X_1 + X_2 + \cdots + X_n$, where $X_i = 1$ if the permutation fixes the ith element and $X_i = 0$ otherwise.]

The **covariance** of two random variables X and Y on a sample space S, denoted by $\text{Cov}(X, Y)$, is defined to be the expected value of the random variable $(X - E(X))(Y - E(Y))$. That is, $\text{Cov}(X, Y) = E((X - E(X))(Y - E(Y)))$.

44. Show that $\text{Cov}(X, Y) = E(XY) - E(X)E(Y)$, and use this result to conclude that $\text{Cov}(X, Y) = 0$ if X and Y are independent random variables.

45. Show that $V(X + Y) = V(X) + V(Y) + 2\,\text{Cov}(X, Y)$.

46. Find $\text{Cov}(X, Y)$ if X and Y are the random variables with $X((i, j)) = 2i$ and $Y((i, j)) = i + j$, where i and j are the numbers that appear on the first and second of two dice when they are rolled.

47. When m balls are distributed into n bins uniformly at random, what is the probability that the first bin remains empty?

48. What is the expected number of balls that fall into the first bin when m balls are distributed into n bins uniformly at random?

49. What is the expected number of bins that remain empty when m balls are distributed into n bins uniformly at random?

Key Terms and Results

TERMS

sample space: the set of possible outcomes of an experiment

event: a subset of the sample space of an experiment

probability of an event (Laplace's definition): the number of successful outcomes of this event divided by the number of possible outcomes

probability distribution: a function p from the set of all outcomes of a sample space S for which $0 \le p(x_i) \le 1$ for $i = 1, 2, \ldots, n$ and $\sum_{i=1}^{n} p(x_i) = 1$, where $x_1, \ldots, x_n$ are the possible outcomes

probability of an event E: the sum of the probabilities of the outcomes in E

$p(E|F)$ (conditional probability of E given F): the ratio $p(E \cap F)/p(F)$

independent events: events E and F such that $p(E \cap F) = p(E)p(F)$

pairwise independent events: events $E_1, E_2, \ldots, E_n$ such that $p(E_i \cap E_j) = p(E_i)p(E_j)$ for all pairs of integers i and j with $1 \le j < k \le n$

mutually independent events: events $E_1, E_2, \ldots, E_n$ such that $p(E_{i_1} \cap E_{i_2} \cap \cdots \cap E_{i_m}) = p(E_{i_1})p(E_{i_2}) \cdots p(E_{i_m})$ whenever $i_j, j = 1, 2, \ldots, m$, are integers with $1 \le i_1 < i_2 < \cdots < i_m \le n$ and $m \ge 2$

random variable: a function that assigns a real number to each possible outcome of an experiment

distribution of a random variable X: the set of pairs $(r, p(X = r))$ for $r \in X(S)$

uniform distribution: the assignment of equal probabilities to the elements of a finite set

expected value of a random variable: the weighted average of a random variable, with values of the random variable weighted by the probability of outcomes, that is, $E(X) = \sum_{s \in S} p(s)X(s)$

geometric distribution: the distribution of a random variable X such that $p(X = k) = (1 - p)^{k-1}p$ for $k = 1, 2, \ldots$ for some real number p with $0 \le p \le 1$

independent random variables: random variables X and Y such that $p(X = r_1 \text{ and } Y = r_2) = p(X = r_1)p(Y = r_2)$ for all real numbers r_1 and r_2

variance of a random variable X: the weighted average of the square of the difference between the value of X and its expected value $E(X)$, with weights given by the probability of outcomes, that is, $V(X) = \sum_{s\in S}(X(s) - E(X))^2 p(s)$

standard deviation of a random variable X: the square root of the variance of X, that is, $\sigma(X) = \sqrt{V(X)}$

Bernoulli trial: an experiment with two possible outcomes

probabilistic (or Monte Carlo) algorithm: an algorithm in which random choices are made at one or more steps

probabilistic method: a technique for proving the existence of objects in a set with certain properties that proceeds by assigning probabilities to objects and showing that the probability that an object has these properties is positive

RESULTS

The probability of exactly k successes when n independent Bernoulli trials are carried out equals $C(n, k)p^k q^{n-k}$, where p is the probability of success and $q = 1 - p$ is the probability of failure.

Bayes' theorem: If E and F are events from a sample space S such that $p(E) \neq 0$ and $p(F) \neq 0$, then

$$p(F \mid E) = \frac{p(E \mid F)p(F)}{p(E \mid F)p(F) + p(E \mid \overline{F})p(\overline{F})}.$$

$E(X) = \sum_{r\in X(S)} p(X = r)r.$

linearity of expectations: $E(X_1 + X_2 + \cdots + X_n) = E(X_1) + E(X_2) + \cdots + E(X_n)$ if $X_1, X_2, \ldots, X_n$ are random variables.

If X and Y are independent random variables, then $E(XY) = E(X)E(Y)$.

Bienaymé's formula: If $X_1, X_2, \ldots, X_n$ are independent random variables, then $V(X_1 + X_2 + \cdots + X_n) = V(X_1) + V(X_2) + \cdots + V(X_n)$.

Chebyshev's inequality: $p(|X(s) - E(X)| \geq r) \leq V(X)/r^2$, where X is a random variable with probability function p and r is a positive real number.

Review Questions

1. a) Define the probability of an event when all outcomes are equally likely.

b) What is the probability that you select the six winning numbers in a lottery if the six different winning numbers are selected from the first 50 positive integers?

2. a) What conditions should be met by the probabilities assigned to the outcomes from a finite sample space?

b) What probabilities should be assigned to the outcome of heads and the outcome of tails if heads comes up three times as often as tails?

3. a) Define the conditional probability of an event E given an event F.

b) Suppose E is the event that when a die is rolled it comes up an even number, and F is the event that when a die is rolled it comes up 1, 2, or 3. What is the probability of F given E?

4. a) When are two events E and F independent?

b) Suppose E is the event that an even number appears when a fair die is rolled, and F is the event that a 5 or 6 comes up. Are E and F independent?

5. a) What is a random variable?

b) What are the possible values assigned by the random variable X that assigns to a roll of two dice the larger number that appears on the two dice?

6. a) Define the expected value of a random variable X.

b) What is the expected value of the random variable X that assigns to a roll of two dice the larger number that appears on the two dice?

7. a) Explain how the average-case computational complexity of an algorithm, with finitely many possible input values, can be interpreted as an expected value.

b) What is the average-case computational complexity of the linear search algorithm, if the probability that the element for which we search is in the list is $1/3$, and it is equally likely that this element is any of the n elements in the list?

8. a) What is meant by a Bernoulli trial?

b) What is the probability of k successes in n independent Bernoulli trials?

c) What is the expected value of the number of successes in n independent Bernoulli trials?

9. a) What does the linearity of expectations of random variables mean?

b) How can the linearity of expectations help us find the expected number of people who receive the correct hat when a hatcheck person returns hats at random?

10. a) How can probability be used to solve a decision problem, if a small probability of error is acceptable?

b) How can we quickly determine whether a positive integer is prime, if we are willing to accept a small probability of making an error?

11. State Bayes' theorem and use it to find $p(F \mid E)$ if $p(E \mid F) = 1/3$, $p(E \mid \overline{F}) = 1/4$, and $p(F) = 2/3$, where E and F are events from a sample space S.

12. a) What does it mean to say that a random variable has a geometric distribution with parameter p?

b) What is the mean of a geometric distribution with parameter p?

13. a) What is the variance of a random variable?

b) What is the variance of a Bernoulli trial with probability p of success?

14. a) What is the variance of the sum of n independent random variables?

b) What is the variance of the number of successes when n independent Bernoulli trials, each with probability p of success, are carried out?

15. What does Chebyshev's inequality tell us about the probability that a random variable deviates from its mean by more than a specified amount?

Supplementary Exercises

1. What is the probability that six consecutive integers will be chosen as the winning numbers in a lottery where each number chosen is an integer between 1 and 40 (inclusive)?

2. In 2012, a player in the Mega Millions lottery picks five different integers between 1 and 56, inclusive, and a sixth integer between 1 and 46, which may duplicate one of the earlier five integers. The player wins the jackpot if the first five numbers picked match the first five numbers drawn and the sixth number matches the sixth number drawn.
   a) What is the probability that a player wins the jackpot?
   b) What is the probability that a player wins $250,000, which is the prize for matching the first five numbers, but not the sixth number, drawn?
   c) What is the probability that a player wins $150 by matching exactly three of the first five numbers and the sixth number or by matching four of the first five numbers but not the sixth number?
   d) What is the probability that a player wins a prize, if a prize is given when the player matches at least three of the first five numbers or the last number.

3. In 2012, a player in the Powerball lottery picks five different integers between 1 and 59, inclusive, and a sixth integer between 1 and 39, which may duplicate one of the earlier five integers. The player wins the jackpot if the first five numbers picked match the first five number drawn and the sixth number matches the sixth number drawn.
   a) What is the probability that a player wins the jackpot?
   b) What is the probability that a player wins $200,000, which is the prize for matching the first five numbers, but not the sixth number, drawn?
   c) What is the probability that a player wins $100 by matching exactly three of the first five and the sixth numbers or four of the first five numbers but not the sixth number?
   d) What is the probability that a player wins a prize, if a prize is given when the player matches at least three of the first five numbers or the last number.

4. What is the probability that a hand of 13 cards contains no pairs?

5. What is the probability that a 13-card bridge hand contains
   a) all 13 hearts?
   b) 13 cards of the same suit?
   c) seven spades and six clubs?
   d) seven cards of one suit and six cards of a second suit?
   e) four diamonds, six hearts, two spades, and one club?
   f) four cards of one suit, six cards of a second suit, two cards of a third suit, and one card of the fourth suit?

6. What is the probability that a seven-card poker hand contains
   a) four cards of one kind and three cards of a second kind?
   b) three cards of one kind and pairs of each of two different kinds?
   c) pairs of each of three different kinds and a single card of a fourth kind?
   d) pairs of each of two different kinds and three cards of a third, fourth, and fifth kind?
   e) cards of seven different kinds?
   f) a seven-card flush?
   g) a seven-card straight?
   h) a seven-card straight flush?

An **octahedral die** has eight faces that are numbered 1 through 8.

7. a) What is the expected value of the number that comes up when a fair octahedral die is rolled?
   b) What is the variance of the number that comes up when a fair octahedral die is rolled?

A **dodecahedral die** has 12 faces that are numbered 1 through 12.

8. a) What is the expected value of the number that comes up when a fair dodecahedral die is rolled?
   b) What is the variance of the number that comes up when a fair dodecahedral die is rolled?

9. Suppose that a pair of fair octahedral dice is rolled.
   a) What is the expected value of the sum of the numbers that come up?
   b) What is the variance of the sum of the numbers that come up?

10. Suppose that a pair of fair dodecahedral dice is rolled.
    a) What is the expected value of the sum of the numbers that come up?
    b) What is the variance of the sum of the numbers that come up?

11. Suppose that a fair standard (cubic) die and a fair octahedral die are rolled together.
    a) What is the expected value of the sum of the numbers that come up?
    b) What is the variance of the sum of the numbers that come up?

12. Suppose that a fair octahedral die and a fair dodecahedral die are rolled together.
    a) What is the expected value of the sum of the numbers that come up?
    b) What is the variance of the sum of the numbers that come up?

13. Suppose n people, $n \geq 3$, play "odd person out" to decide who will buy the next round of refreshments. The n people each flip a fair coin simultaneously. If all the coins but one come up the same, the person whose coin comes up different buys the refreshments. Otherwise, the people flip the coins again and continue until just one coin comes up different from all the others.

a) What is the probability that the odd person out is decided in just one coin flip?

b) What is the probability that the odd person out is decided with the kth flip?

c) What is the expected number of flips needed to decide odd person out with n people?

14. Suppose that p and q are primes and $n = pq$. What is the probability that a randomly chosen positive integer less than n is not divisible by either p or q?

***15.** Suppose that m and n are positive integers. What is the probability that a randomly chosen positive integer less than mn is not divisible by either m or n?

16. Suppose that $E_1, E_2, \ldots, E_n$ are n events with $p(E_i) > 0$ for $i = 1, 2, \ldots, n$. Show that

$$\begin{aligned} p(E_1 \cap E_2 \cap \cdots \cap E_n) \\ = p(E_1)p(E_2 \mid E_1)p(E_3 \mid E_1 \cap E_2) \\ \cdots p(E_n \mid E_1 \cap E_2 \cap \cdots \cap E_{n-1}). \end{aligned}$$

17. There are three cards in a box. Both sides of one card are black, both sides of one card are red, and the third card has one black side and one red side. We pick a card at random and observe only one side.

a) If the side is black, what is the probability that the other side is also black?

b) What is the probability that the opposite side is the same color as the one we observed?

18. What is the probability that when a fair coin is flipped n times an equal number of heads and tails appear?

19. What is the probability that a randomly selected bit string of length 10 is a palindrome?

20. What is the probability that a randomly selected bit string of length 11 is a palindrome?

21. Consider the following game. A person flips a coin repeatedly until a head comes up. This person receives a payment of 2^n dollars if the first head comes up at the nth flip.

a) Let X be a random variable equal to the amount of money the person wins. Show that the expected value of X does not exist (that is, it is infinite). Show that a rational gambler, that is, someone willing to pay to play the game as long as the price to play is not more than the expected payoff, should be willing to wager any amount of money to play this game. (This is known as the **St. Petersburg paradox**. Why do you suppose it is called a paradox?)

b) Suppose that the person receives 2^n dollars if the first head comes up on the nth flip where $n < 8$ and $2^8 = 256$ dollars if the first head comes up on or after the eighth flip. What is the expected value of the amount of money the person wins? How much money should a person be willing to pay to play this game?

22. Suppose that n balls are tossed into b bins so that each ball is equally likely to fall into any of the bins and that the tosses are independent.

a) Find the probability that a particular ball lands in a specified bin.

b) What is the expected number of balls that land in a particular bin?

c) What is the expected number of balls tossed until a particular bin contains a ball?

***d)** What is the expected number of balls tossed until all bins contain a ball? [*Hint:* Let X_i denote the number of tosses required to have a ball land in an ith bin once $i - 1$ bins contain a ball. Find $E(X_i)$ and use the linearity of expectations.]

23. Suppose that A and B are events with probabilities $p(A) = 3/4$ and $p(B) = 1/3$.

a) What is the largest $p(A \cap B)$ can be? What is the smallest it can be? Give examples to show that both extremes for $p(A \cap B)$ are possible.

b) What is the largest $p(A \cup B)$ can be? What is the smallest it can be? Give examples to show that both extremes for $p(A \cup B)$ are possible.

24. Suppose that A and B are events with probabilities $p(A) = 2/3$ and $p(B) = 1/2$.

a) What is the largest $p(A \cap B)$ can be? What is the smallest it can be? Give examples to show that both extremes for $p(A \cap B)$ are possible.

b) What is the largest $p(A \cup B)$ can be? What is the smallest it can be? Give examples to show that both extremes for $p(A \cup B)$ are possible.

25. Recall from Definition 5 in Section 7.2 that the events $E_1, E_2, \ldots, E_n$ are **mutually independent** if $p(E_{i_1} \cap E_{i_2} \cap \cdots \cap E_{i_m}) = p(E_{i_1})p(E_{i_2}) \cdots p(E_{i_m})$ whenever $i_j, j = 1, 2, \ldots, m$, are integers with $1 \leq i_1 < i_2 < \cdots < i_m \leq n$ and $m \geq 2$.

a) Write out the conditions required for three events E_1, E_2, and E_3 to be mutually independent.

b) Let E_1, E_2, and E_3 be the events that the first flip comes up heads, that the second flip comes up tails, and that the third flip comes up tails, respectively, when a fair coin is flipped three times. Are E_1, E_2, and E_3 mutually independent?

c) Let E_1, E_2, and E_3 be the events that the first flip comes up heads, that the third flip comes up heads, and that an even number of heads come up, respectively, when a fair coin is flipped three times. Are E_1, E_2, and E_3 pairwise independent? Are they mutually independent?

d) Let E_1, E_2, and E_3 be the events that the first flip comes up heads, that the third flip comes up heads, and that exactly one of the first flip and third flip come up heads, respectively, when a fair coin is flipped three times. Are E_1, E_2, and E_3 pairwise independent? Are they mutually independent?

e) How many conditions must be checked to show that n events are mutually independent?

26. Suppose that A and B are events from a sample space S such that $p(A) \neq 0$ and $p(B) \neq 0$. Show that if $p(B \mid A) < p(B)$, then $p(A \mid B) < p(A)$.

In Exercise 27 we consider the **two children problem**, introduced in 1959 by Martin Garnder is his Mathematical Games column in *Scientific American*. A version of the puzzle asks: "We meet Mr. Smith as he is walking down the street with a young child whom he introduces as his son. He also tells us that he has two children. What is the probability that his other child is a son?" We will show that this puzzle is ambiguous, leading to a paradox, by showing that there are two reasonable answers to this problem and we will describe how to make the puzzle unambiguous.

***27. a)** Solve this puzzle in two different ways. First, answer the problem by considering the probability of the gender of the second child. Then, determine the probability differently, by considering the four different possibilities for a family of two children.

b) Show that the answer to the puzzle becomes unambiguous if we also know that Mr. Smith chose his walking companion at random from his two children.

c) Another variation of this puzzle asks "When we meet Mr. Smith, he tells us that he has two children and at least one is a son. What is the probability that his other child is a son?" Solve this variation of the puzzle, explaining why it is unambiguous.

28. In 2010, the puzzle designer Gary Foshee posed this problem: "Mr. Smith has two children, one of whom is a son born on a Tuesday. What is the probability that Mr. Smith has two sons?" Show that there are two different answers to this puzzle, depending on whether Mr. Smith specifically mentioned his son because he was born on a Tuesday or whether he randomly chose a child and reported its gender and birth day of the week. [*Hint:* For the first possibility, enumerate all the equally likely possibilities for the gender and birth day of the week of the other child. To do, this consider first the cases where the older child is a boy born on a Tuesday and then the case where the older child is not a boy born on a Tuesday.]

29. Let X be a random variable on a sample space S. Show that $V(aX + b) = a^2V(X)$ whenever a and b are real numbers.

30. Use Chebyshev's inequality to show that the probability that more than 10 people get the correct hat back when a hatcheck person returns hats at random does not exceed $1/100$ no matter how many people check their hats. [*Hint:* Use Example 6 and Exercise 43 in Section 7.4.]

31. Suppose that at least one of the events E_j, $j = 1, 2, \ldots, m$, is guaranteed to occur and no more than two can occur. Show that if $p(E_j) = q$ for $j = 1, 2, \ldots, m$ and $p(E_j \cap E_k) = r$ for $1 \leq j < k \leq m$, then $q \geq 1/m$ and $r \leq 2/m$.

32. Show that if m is a positive integer, then the probability that the mth success occurs on the $(m + n)$th trial when independent Bernoulli trials, each with probability p of success, are run, is $\binom{n+m-1}{n}q^n p^m$.

33. There are n different types of collectible cards you can get as prizes when you buy a particular product. Suppose that every time you buy this product it is equally likely that you get any type of these cards. Let X be the random variable equal to the number of products that need to be purchased to obtain at least one of each type of card and let X_j be the random variable equal to the number of additional products that must be purchased after j different cards have been collected until a new card is obtained for $j = 0, 1, \ldots, n - 1$.

a) Show that $X = \sum_{j=0}^{n-1} X_j$.

b) Show that after j distinct types of cards have been obtained, the card obtained with the next purchase will be a card of a new type with probability $(n - j)/n$.

c) Show that X_j has a geometric distribution with parameter $(n - j)/n$.

d) Use parts (a) and (c) to show that $E(X) = n \sum_{j=1}^{n} 1/j$.

e) Use the approximation $\sum_{j=1}^{n} 1/j \approx \ln n + \gamma$, where $\gamma = 0.57721\ldots$ is Euler's constant, to find the expected number of products that you need to buy to get one card of each type if there are 50 different types of cards.

34. The **maximum satisfiability problem** asks for an assignment of truth values to the variables in a compound proposition in conjunctive normal form (which expresses a compound proposition as the conjunction of clauses where each clause is the disjunction of two or more variables or their negations) that makes as many of these clauses true as possible. For example, three but not four of the clauses in

$$(p \vee q) \wedge (p \vee \neg q) \wedge (\neg p \vee r) \wedge (\neg p \vee \neg r)$$

can be made true by an assignment of truth values to p, q, and r. We will show that probabilistic methods can provide a lower bound for the number of clauses that can be made true by an assignment of truth values to the variables.

a) Suppose that there are n variables in a compound proposition in conjunctive normal form. If we pick a truth value for each variable randomly by flipping a coin and assigning true to the variable if the coin comes up heads and false if it comes up tails, what is the probability of each possible assignment of truth values to the n variables?

b) Assuming that each clause is the disjunction of exactly two distinct variables or their negations, what is the probability that a given clause is true, given the random assignment of truth values from part (a)?

c) Suppose that there are D clauses in the compound proposition. What is the expected number of these clauses that are true, given the random assignment of truth values of the variables?

d) Use part (c) to show that for every compound proposition in conjunctive normal form there is an assignment of truth values to the variables that makes at least 3/4 of the clauses true.

35. What is the probability that each player has a hand containing an ace when the 52 cards of a standard deck are dealt to four players?

***36.** The following method can be used to generate a random permutation of a sequence of n terms. First, interchange the nth term and the $r(n)$th term where $r(n)$ is a randomly selected integer with $1 \leq r(n) \leq n$. Next, interchange the $(n-1)$ st term of the resulting sequence with its $r(n-1)$st term where $r(n-1)$ is a randomly selected integer with $1 \leq r(n-1) \leq n-1$. Continue this process until $j = n$, where at the jth step you interchange the $(n-j+1)$st term of the resulting sequence with its $r(n-j+1)$st term, where $r(n-j+1)$ is a randomly selected integer with $1 \leq r(n-j+1) \leq n-j+1$. Show that when this method is followed, each of the $n!$ different permutations of the terms of the sequence is equally likely to be generated. [*Hint:* Use mathematical induction, assuming that the probability that each of the permutations of $n-1$ terms produced by this procedure for a sequence of $n-1$ terms is $1/(n-1)!$.]

Computer Projects

Write programs with these input and output.

1. Given a real number p with $0 \leq p \leq 1$, generate random numbers taken from a Bernoulli distribution with probability p.

2. Given a positive integer n, generate a random permutation of the set $\{1, 2, 3, \ldots, n\}$. (See Exercise 36 in the Supplementary Exercises.)

3. Given positive integers m and n, generate m random permutations of the first n positive integers. Find the number of inversions in each permutation and determine the average number of these inversions.

4. Given a positive integer n, simulate n repeated flips of a biased coin with probability p of heads and determine the number of heads that come up. Display the cumulative results.

5. Given positive integers n and m, generate m random permutations of the first n positive integers. Sort each permutation using the insertion sort, counting the number of comparisons used. Determine the average number of comparisons used over all m permutations.

6. Given positive integers n and m, generate m random permutations of the first n positive integers. Sort each permutation using the version of the bubble sort that terminates when a pass has been made with no interchanges, counting the number of comparisons used. Determine the average number of comparisons used over all m permutations.

7. Given a positive integer m, simulate the collection of cards that come with the purchase of products to find the number of products that must be purchased to obtain a full set of m different collector cards. (See Supplementary Exercise 33.)

8. Given positive integers m and n, simulate the placement of n keys, where a record with key k is placed at location $h(k) = k \textbf{ mod } m$ and determine whether there is at least one collision.

9. Given a positive integer n, find the probability of selecting the six integers from the set $\{1, 2, \ldots, n\}$ that were mechanically selected in a lottery.

10. Simulate repeated trials of the Monty Hall Three-Door problem (Example 10 in Section 7.1) to calculate the probability of winning with each strategy.

11. Given a list of words and the empirical probabilities they occur in spam emails and in emails that are not spam, determine the probability that a new email message is spam.

Computations and Explorations

Use a computational program or programs you have written to do these exercises.

1. Find the probabilities of each type of hand in five-card poker and rank the types of hands by their probability.

2. Find some conditions such that the expected value of buying a \$1 lottery ticket in the New Jersey Pick-6 lottery has an expected value of more than \$1. To win you have to select the six numbers drawn, where order does not matter, from the positive integers 1 to 49, inclusive. The winnings are split evenly among holders of winning tickets. Be sure to consider the total size of the pot going into the drawing and the number of people buying tickets.

3. Estimate the probability that two integers selected at random are relatively prime by testing a large number of randomly selected pairs of integers. Look up the theorem that gives this probability and compare your results with the correct probability.
4. Determine the number of people needed to ensure that the probability at least two of them have the same day of the year as their birthday is at least 70%, at least 80%, at least 90%, at least 95%, at least 98%, and at least 99%.
5. Generate a list of 100 randomly selected permutations of the set of the first 100 positive integers. (See Exercise 36 in the Supplementary Exercises.)
6. Given a collection of email messages, each determined to be spam or not to be spam, develop a Bayesian filter based on the appearance of particular words in these messages.
7. Simulate the odd-person-out procedure (described in Exercise 13 of the Supplementary Exercises) for n people with $3 \leq n \leq 10$. Run a large number of trials for each value of n and use the results to estimate the expected number of flips needed to find the odd person out. Does your result agree with that found in Exercise 29 in Section 7.2? Vary the problem by supposing that exactly one person has a biased coin with probability of heads $p \neq 0.5$.
8. Given a positive integer n, simulate a hatcheck person randomly giving hats back to people. Determine the number of people who get the correct hat back.

Writing Projects

Respond to these with essays using outside sources.

1. Describe the origins of probability theory and the first uses of this theory, including those by Cardano, Pascal, and Laplace.
2. Describe the different bets you can make when you play roulette. Find the probability of each of these bets in the American version where the wheel contains the numbers 0 and 00. Which is the best bet and which is the worst for you?
3. Discuss the probability of winning when you play the game of blackjack versus a casino. Is there a winning strategy for the person playing against the house?
4. Investigate the game of craps and discuss the probability that the shooter wins and how close to a fair game it is.
5. Discuss issues involved in developing successful spam filters and the current situation in the war between spammers and people trying to filter spam out.
6. Discuss the history and solution of what is known as the Newton–Pepys problem, which asks which is most likely: rolling at least one six when six dice are rolled, rolling at least two sixes when 12 dice are rolled, or rolling at least three sixes when 18 dice are rolled.
7. Explain how Erdős and Rényi first used the probabilistic method and describe some other applications of this method.
8. Discuss the different types of probabilistic algorithms and describe some examples of each type.

CHAPTER

8 Advanced Counting Techniques

8.1 Applications of Recurrence Relations

8.2 Solving Linear Recurrence Relations

8.3 Divide-and-Conquer Algorithms and Recurrence Relations

8.4 Generating Functions

8.5 Inclusion–Exclusion

8.6 Applications of Inclusion–Exclusion

Many counting problems cannot be solved easily using the methods discussed in Chapter 6. One such problem is: How many bit strings of length n do not contain two consecutive zeros? To solve this problem, let a_n be the number of such strings of length n. An argument can be given that shows that the sequence $\{a_n\}$ satisfies the recurrence relation $a_{n+1} = a_n + a_{n-1}$ and the initial conditions $a_1 = 2$ and $a_2 = 3$. This recurrence relation and the initial conditions determine the sequence $\{a_n\}$. Moreover, an explicit formula can be found for a_n from the equation relating the terms of the sequence. As we will see, a similar technique can be used to solve many different types of counting problems.

We will discuss two ways that recurrence relations play important roles in the study of algorithms. First, we will introduce an important algorithmic paradigm known as dynamic programming. Algorithms that follow this paradigm break down a problem into overlapping subproblems. The solution to the problem is then found from the solutions to the subproblems through the use of a recurrence relation. Second, we will study another important algorithmic paradigm, divide-and-conquer. Algorithms that follow this paradigm can be used to solve a problem by recursively breaking it into a fixed number of nonoverlapping subproblems until these problems can be solved directly. The complexity of such algorithms can be analyzed using a special type of recurrence relation. In this chapter we will discuss a variety of divide-and-conquer algorithms and analyze their complexity using recurrence relations.

We will also see that many counting problems can be solved using formal power series, called generating functions, where the coefficients of powers of x represent terms of the sequence we are interested in. Besides solving counting problems, we will also be able to use generating functions to solve recurrence relations and to prove combinatorial identities.

Many other kinds of counting problems cannot be solved using the techniques discussed in Chapter 6, such as: How many ways are there to assign seven jobs to three employees so that each employee is assigned at least one job? How many primes are there less than 1000? Both of these problems can be solved by counting the number of elements in the union of sets. We will develop a technique, called the principle of inclusion–exclusion, that counts the number of elements in a union of sets, and we will show how this principle can be used to solve counting problems.

The techniques studied in this chapter, together with the basic techniques of Chapter 6, can be used to solve many counting problems.

8.1 Applications of Recurrence Relations

8.1.1 Introduction

Recall from Chapter 2 that a recursive definition of a sequence specifies one or more initial terms and a rule for determining subsequent terms from those that precede them. Also, recall that a rule of the latter sort (whether or not it is part of a recursive definition) is called a **recurrence relation** and that a sequence is called a *solution* of a recurrence relation if its terms satisfy the recurrence relation.

In this section we will show that such relations can be used to study and to solve counting problems. For example, suppose that the number of bacteria in a colony doubles every hour. If a colony begins with five bacteria, how many will be present in n hours? To solve this problem, let a_n be the number of bacteria at the end of n hours. Because the number of bacteria doubles every hour, the relationship $a_n = 2a_{n-1}$ holds whenever n is a positive integer. This recurrence relation, together with the initial condition $a_0 = 5$, uniquely determines a_n for all nonnegative integers n. We can find a formula for a_n using the iterative approach followed in Chapter 2, namely that $a_n = 5 \cdot 2^n$ for all nonnegative integers n.

Some of the counting problems that cannot be solved using the techniques discussed in Chapter 6 can be solved by finding recurrence relations involving the terms of a sequence, as was done in the problem involving bacteria. In this section we will study a variety of counting problems that can be modeled using recurrence relations. In Chapter 2 we developed methods for solving certain recurrence relation. In Section 8.2 we will study methods for finding explicit formulae for the terms of sequences that satisfy certain types of recurrence relations.

We conclude this section by introducing the algorithmic paradigm of dynamic programming. After explaining how this paradigm works, we will illustrate its use with an example.

8.1.2 Modeling With Recurrence Relations

Assessment

We can use recurrence relations to model a wide variety of problems, such as finding compound interest (see Example 11 in Section 2.4), counting rabbits on an island, determining the number of moves in the Tower of Hanoi puzzle, and counting bit strings with certain properties.

Extra Examples

Example 1 shows how the population of rabbits on an island can be modeled using a recurrence relation.

EXAMPLE 1 **Rabbits and the Fibonacci Numbers** Consider this problem, which was originally posed by Leonardo Pisano, also known as Fibonacci, in the thirteenth century in his book *Liber abaci*. A young pair of rabbits (one of each sex) is placed on an island. A pair of rabbits does not breed until they are 2 months old. After they are 2 months old, each pair of rabbits produces another pair each month, as shown in Figure 1. Find a recurrence relation for the number of pairs of rabbits on the island after n months, assuming that no rabbits ever die.

Links

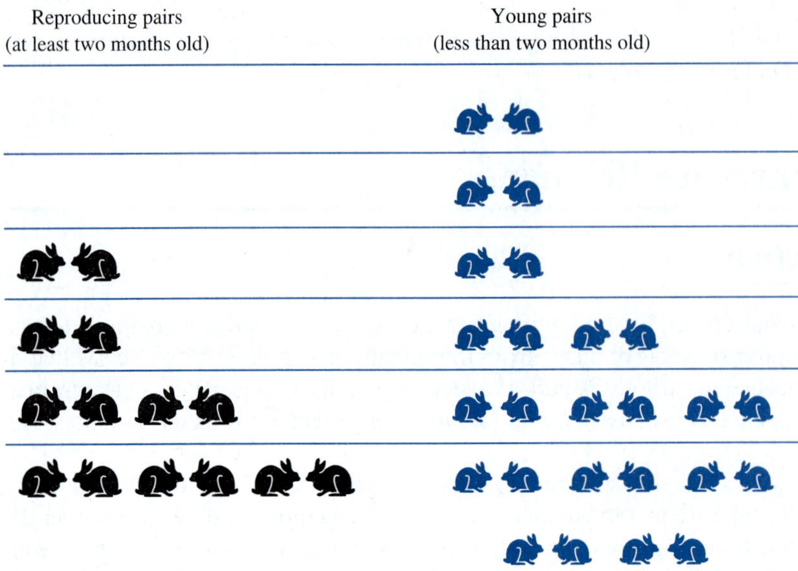

Month	Reproducing pairs	Young pairs	Total pairs
1	0	1	1
2	0	1	1
3	1	1	2
4	1	2	3
5	2	3	5
6	3	5	8

FIGURE 1 **Rabbits on an island.**

Solution: Denote by f_n the number of pairs of rabbits after n months. We will show that f_n, $n = 1, 2, 3, \ldots,$ are the terms of the Fibonacci sequence.

The rabbit population can be modeled using a recurrence relation. At the end of the first month, the number of pairs of rabbits on the island is $f_1 = 1$. Because this pair does not breed during the second month, $f_2 = 1$ also. To find the number of pairs after n months, add the number on the island the previous month, f_{n-1}, and the number of newborn pairs, which equals f_{n-2}, because each newborn pair comes from a pair at least 2 months old.

The Fibonacci numbers appear in many other places in nature, including the number of petals on flowers and the number of spirals on seedheads.

Consequently, the sequence $\{f_n\}$ satisfies the recurrence relation

$$f_n = f_{n-1} + f_{n-2}$$

for $n \geq 3$ together with the initial conditions $f_1 = 1$ and $f_2 = 1$. Because this recurrence relation and the initial conditions uniquely determine this sequence, the number of pairs of rabbits on the island after n months is given by the nth Fibonacci number. ◀

Demo ❯

Example 2 involves a famous puzzle.

EXAMPLE 2 **The Tower of Hanoi Puzzle** A popular puzzle of the late nineteenth century invented by the French mathematician Édouard Lucas, called the Tower of Hanoi, consists of three pegs mounted on a board together with disks of different sizes. Initially these disks are placed on the first peg in order of size, with the largest on the bottom (as shown in Figure 2). The rules of the puzzle allow disks to be moved one at a time from one peg to another as long as a disk is never placed on top of a smaller disk. The goal of the puzzle is to have all the disks on the second peg in order of size, with the largest on the bottom.

Links ❯

Let H_n denote the number of moves needed to solve the Tower of Hanoi puzzle with n disks. Set up a recurrence relation for the sequence $\{H_n\}$.

Solution: Begin with n disks on peg 1. We can transfer the top $n - 1$ disks, following the rules of the puzzle, to peg 3 using H_{n-1} moves (see Figure 3 for an illustration of the pegs and disks at this point). We keep the largest disk fixed during these moves. Then, we use one move to transfer the largest disk to the second peg. Finally, we transfer the $n - 1$ disks on peg 3 to peg 2 using H_{n-1} moves, placing them on top of the largest disk, which always stays fixed on the bottom of peg 2. This shows that we can solve the Tower of Hanoi puzzle for n disks using $2H_{n-1} + 1$ moves.

Schemes for efficiently backing up computer files on multiple tapes or other media are based on the moves used to solve the Tower of Hanoi puzzle.

We now show that we cannot solve the puzzle for n disks using fewer that $2H_{n-1} + 1$ moves. Note that when we move the largest disk, we must have already moved the $n - 1$ smaller disks onto a peg other than peg 1. Doing so requires at least H_{n-1} moves. Another move is needed to

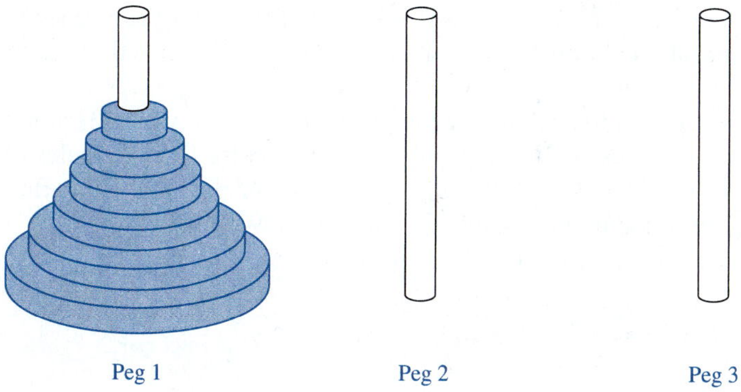

FIGURE 2 **The initial position in the Tower of Hanoi.**

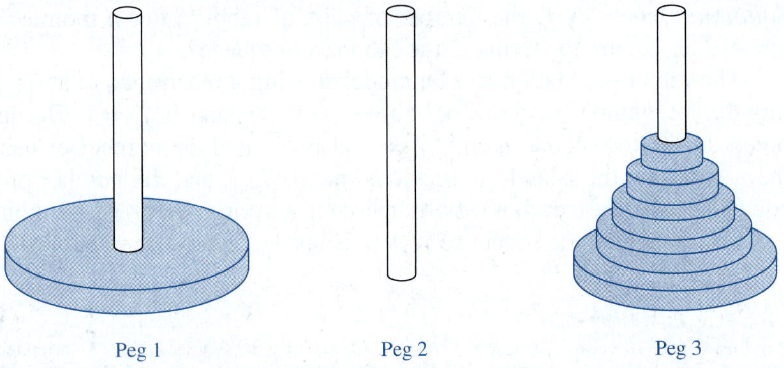

FIGURE 3 An intermediate position in the Tower of Hanoi.

transfer the largest disk. Finally, at least H_{n-1} more moves are needed to put the $n-1$ smallest disks back on top of the largest disk. Adding the number of moves required gives us the desired lower bound.

We conclude that

$$H_n = 2H_{n-1} + 1.$$

The initial condition is $H_1 = 1$, because one disk can be transferred from peg 1 to peg 2, according to the rules of the puzzle, in one move.

We can use an iterative approach to solve this recurrence relation. Note that

$$\begin{aligned}
H_n &= 2H_{n-1} + 1 \\
&= 2(2H_{n-2} + 1) + 1 = 2^2 H_{n-2} + 2 + 1 \\
&= 2^2(2H_{n-3} + 1) + 2 + 1 = 2^3 H_{n-3} + 2^2 + 2 + 1 \\
&\vdots \\
&= 2^{n-1} H_1 + 2^{n-2} + 2^{n-3} + \cdots + 2 + 1 \\
&= 2^{n-1} + 2^{n-2} + \cdots + 2 + 1 \\
&= 2^n - 1.
\end{aligned}$$

We have used the recurrence relation repeatedly to express H_n in terms of previous terms of the sequence. In the next to last equality, the initial condition $H_1 = 1$ has been used. The last equality is based on the formula for the sum of the terms of a geometric series, which can be found in Theorem 1 in Section 2.4.

The iterative approach has produced the solution to the recurrence relation $H_n = 2H_{n-1} + 1$ with the initial condition $H_1 = 1$. This formula can be proved using mathematical induction. This is left for the reader as Exercise 1.

A myth created to accompany the puzzle tells of a tower in Hanoi where monks are transferring 64 gold disks from one peg to another, according to the rules of the puzzle. The myth says that the world will end when they finish the puzzle. How long after the monks started will the world end if the monks take one second to move a disk?

From the explicit formula, the monks require

$$2^{64} - 1 = 18{,}446{,}744{,}073{,}709{,}551{,}615$$

moves to transfer the disks. Making one move per second, it will take them more than 500 billion years to complete the transfer, so the world should survive a while longer than it already has. ◀

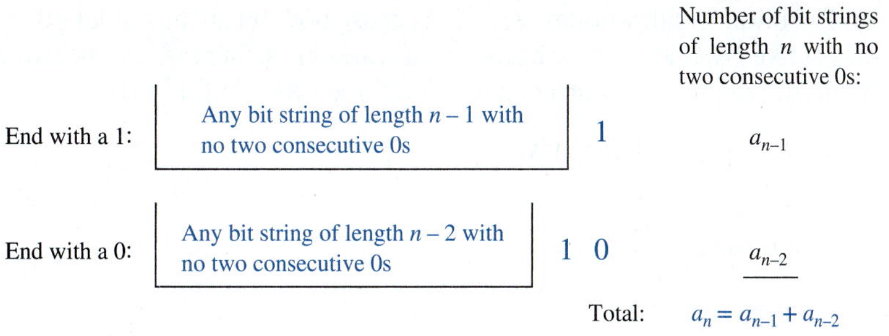

FIGURE 4 Counting bit strings of length *n* with no two consecutive 0s.

Links

Remark: Many people have studied variations of the original Tower of Hanoi puzzle discussed in Example 2. Some variations use more pegs, some allow disks to be of the same size, and some restrict the types of allowable disk moves. One of the oldest and most interesting variations is the **Reve's puzzle**,* proposed in 1907 by Henry Dudeney in his book *The Canterbury Puzzles*. The Reve's puzzle involves pilgrims challenged by the Reve to move a stack of cheese wheels of varying sizes from the first of four stools to another stool without ever placing a cheese wheel on one of smaller diameter. The Reve's puzzle, expressed in terms of pegs and disks, follows the same rules as the Tower of Hanoi puzzle, except that four pegs are used. Similarly, we can generalize the Tower of Hanoi puzzle where there are p pegs, where p is an integer greater than three. You may find it surprising that no one has been able to establish the minimum number of moves required to solve the generalization of this puzzle for p pegs. (Note that there have been some published claims that this problem has been solved, but these are not accepted by experts.) However, in 2014 Thierry Bousch showed that the minimum number of moves required when there are four pegs equals the number of moves used by an algorithm invented by Frame and Stewart in 1939. (See Exercises 38–45 and [St94] and [Bo14] for more information.)

Example 3 illustrates how recurrence relations can be used to count bit strings of a specified length that have a certain property.

EXAMPLE 3 Find a recurrence relation and give initial conditions for the number of bit strings of length n that do not have two consecutive 0s. How many such bit strings are there of length five?

Solution: Let a_n denote the number of bit strings of length n that do not have two consecutive 0s. We assume that $n \geq 3$, so that the bit string has at least three bits. Strings of this sort of length n can be divided into those that end in 1 and those that end in 0. The bit strings of length n ending with 1 that do not have two consecutive 0s are precisely the bit strings of length $n - 1$ with no two consecutive 0s with a 1 added at the end. Consequently, there are a_{n-1} such bit strings.

Bit strings of length n ending with a 0 that do not have two consecutive 0s must have 1 as their $(n - 1)$st bit; otherwise they would end with a pair of 0s. Hence, the bit strings of length n ending with a 0 that have no two consecutive 0s are precisely the bit strings of length $n - 2$ with no two consecutive 0s with 10 added at the end. Consequently, there are a_{n-2} such bit strings.

We conclude, as illustrated in Figure 4, that

$$a_n = a_{n-1} + a_{n-2}$$

for $n \geq 3$.

***Reve*, more commonly spelled *reeve*, is an archaic word for *governor*.

The initial conditions are $a_1 = 2$, because both bit strings of length one, 0 and 1 do not have consecutive 0s, and $a_2 = 3$, because the valid bit strings of length two are 01, 10, and 11. To obtain a_5, we use the recurrence relation three times to find that

$$a_3 = a_2 + a_1 = 3 + 2 = 5,$$
$$a_4 = a_3 + a_2 = 5 + 3 = 8,$$
$$a_5 = a_4 + a_3 = 8 + 5 = 13.$$

◀

Remark: Note that $\{a_n\}$ satisfies the same recurrence relation as the Fibonacci sequence. Because $a_1 = f_3$ and $a_2 = f_4$ it follows that $a_n = f_{n+2}$.

Example 4 shows how a recurrence relation can be used to model the number of codewords that are allowable using certain validity checks.

EXAMPLE 4 **Codeword Enumeration** A computer system considers a string of decimal digits a valid codeword if it contains an even number of 0 digits. For instance, 1230407869 is valid, whereas 120987045608 is not valid. Let a_n be the number of valid n-digit codewords. Find a recurrence relation for a_n.

Solution: Note that $a_1 = 9$ because there are 10 one-digit strings, and only one, namely, the string 0, is not valid. A recurrence relation can be derived for this sequence by considering how a valid n-digit string can be obtained from strings of $n - 1$ digits. There are two ways to form a valid string with n digits from a string with one fewer digit.

First, a valid string of n digits can be obtained by appending a valid string of $n - 1$ digits with a digit other than 0. This appending can be done in nine ways. Hence, a valid string with n digits can be formed in this manner in $9a_{n-1}$ ways.

Second, a valid string of n digits can be obtained by appending a 0 to a string of length $n - 1$ that is not valid. (This produces a string with an even number of 0 digits because the invalid string of length $n - 1$ has an odd number of 0 digits.) The number of ways that this can be done equals the number of invalid $(n - 1)$-digit strings. Because there are 10^{n-1} strings of length $n - 1$, and a_{n-1} are valid, there are $10^{n-1} - a_{n-1}$ valid n-digit strings obtained by appending an invalid string of length $n - 1$ with a 0.

Because all valid strings of length n are produced in one of these two ways, it follows that there are

$$\begin{aligned} a_n &= 9a_{n-1} + (10^{n-1} - a_{n-1}) \\ &= 8a_{n-1} + 10^{n-1} \end{aligned}$$

valid strings of length n. ◀

Example 5 establishes a recurrence relation that appears in many different contexts.

EXAMPLE 5 Find a recurrence relation for C_n, the number of ways to parenthesize the product of $n + 1$ numbers, $x_0 \cdot x_1 \cdot x_2 \cdot \cdots \cdot x_n$, to specify the order of multiplication. For example, $C_3 = 5$ because there are five ways to parenthesize $x_0 \cdot x_1 \cdot x_2 \cdot x_3$ to determine the order of multiplication:

$$((x_0 \cdot x_1) \cdot x_2) \cdot x_3 \qquad (x_0 \cdot (x_1 \cdot x_2)) \cdot x_3 \qquad (x_0 \cdot x_1) \cdot (x_2 \cdot x_3)$$
$$x_0 \cdot ((x_1 \cdot x_2) \cdot x_3) \qquad x_0 \cdot (x_1 \cdot (x_2 \cdot x_3)).$$

Solution: To develop a recurrence relation for C_n, we note that however we insert parentheses in the product $x_0 \cdot x_1 \cdot x_2 \cdot \cdots \cdot x_n$, one "$\cdot$" operator remains outside all parentheses, namely, the operator for the final multiplication to be performed. [For example, in $(x_0 \cdot (x_1 \cdot x_2)) \cdot x_3$, it is the final "$\cdot$", while in $(x_0 \cdot x_1) \cdot (x_2 \cdot x_3)$ it is the second "$\cdot$".] This final operator appears between two of the $n+1$ numbers, say, x_k and x_{k+1}. There are $C_k C_{n-k-1}$ ways to insert parentheses to determine the order of the $n+1$ numbers to be multiplied when the final operator appears between x_k and x_{k+1}, because there are C_k ways to insert parentheses in the product $x_0 \cdot x_1 \cdot \cdots \cdot x_k$ to determine the order in which these $k+1$ numbers are to be multiplied and C_{n-k-1} ways to insert parentheses in the product $x_{k+1} \cdot x_{k+2} \cdot \cdots \cdot x_n$ to determine the order in which these $n-k$ numbers are to be multiplied. Because this final operator can appear between any two of the $n+1$ numbers, it follows that

$$\begin{aligned} C_n &= C_0 C_{n-1} + C_1 C_{n-2} + \cdots + C_{n-2} C_1 + C_{n-1} C_0 \\ &= \sum_{k=0}^{n-1} C_k C_{n-k-1}. \end{aligned}$$

Note that the initial conditions are $C_0 = 1$ and $C_1 = 1$. ◀

The recurrence relation in Example 5 can be solved using the method of generating functions, which will be discussed in Section 8.4. It can be shown that $C_n = C(2n, n)/(n+1)$ (see Exercise 43 in Section 8.4) and that $C_n \sim \frac{4^n}{n^{3/2}\sqrt{\pi}}$ (see [GrKnPa94]). The sequence $\{C_n\}$ is the sequence of **Catalan numbers**, named after Eugène Charles Catalan. This sequence appears as the solution of many different counting problems besides the one considered here (see the chapter on Catalan numbers in [MiRo91] or [RoTe03] for details).

Links

8.1.3 Algorithms and Recurrence Relations

Recurrence relations play an important role in many aspects of the study of algorithms and their complexity. In Section 8.3, we will show how recurrence relations can be used to analyze the complexity of divide-and-conquer algorithms, such as the merge sort algorithm introduced in Section 5.4. As we will see in Section 8.3, divide-and-conquer algorithms recursively divide a problem into a fixed number of nonoverlapping subproblems until they become simple enough to solve directly. We conclude this section by introducing another algorithmic paradigm known as **dynamic programming**, which can be used to solve many optimization problems efficiently.

An algorithm follows the dynamic programming paradigm when it recursively breaks down a problem into simpler overlapping subproblems, and computes the solution using the solutions of the subproblems. Generally, recurrence relations are used to find the overall solution from the solutions of the subproblems. Dynamic programming has been used to solve important problems in such diverse areas as economics, computer vision, speech recognition, artificial intelligence, computer graphics, and bioinformatics. In this section we will illustrate the use of dynamic programming by constructing an algorithm for solving a scheduling problem. Before doing so, we will relate the amusing origin of the name *dynamic programming*, which was introduced by the mathematician Richard Bellman in the 1950s. Bellman was working at the RAND Corporation on projects for the U.S. military, and at that time, the U.S. Secretary of Defense was hostile to mathematical research. Bellman decided that to ensure funding, he needed a name not containing the word mathematics for his method for solving scheduling and planning problems. He decided to use the adjective *dynamic* because, as he said "it's impossible to use the word dynamic in a pejorative sense" and he thought that dynamic programming was "something not even a Congressman could object to."

Links

AN EXAMPLE OF DYNAMIC PROGRAMMING The problem we use to illustrate dynamic programming is related to the problem studied in Example 7 in Section 3.1. In that problem our goal was to schedule as many talks as possible in a single lecture hall. These talks have preset start and end times; once a talk starts, it continues until it ends; no two talks can proceed at the same time; and a talk can begin at the same time another one ends. We developed a greedy algorithm that always produces an optimal schedule, as we proved in Example 12 in Section 5.1. Now suppose that our goal is not to schedule the most talks possible, but rather to have the largest possible combined attendance of the scheduled talks.

We formalize this problem by supposing that we have n talks, where talk j begins at time t_j, ends at time e_j, and will be attended by w_j students. We want a schedule that maximizes the total number of student attendees. That is, we wish to schedule a subset of talks to maximize the sum of w_j over all scheduled talks. (Note that when a student attends more than one talk, this student is counted according to the number of talks attended.) We denote by $T(j)$ the maximum number of total attendees for an optimal schedule from the first j talks, so $T(n)$ is the maximal number of total attendees for an optimal schedule for all n talks.

We first sort the talks in order of increasing end time. After doing this, we renumber the talks so that $e_1 \leq e_2 \leq \cdots \leq e_n$. We say that two talks are **compatible** if they can be part of the same schedule, that is, if the times they are scheduled do not overlap (other than the possibility one ends and the other starts at the same time). We define $p(j)$ to be largest integer i, $i < j$, for which $e_i \leq s_j$, if such an integer exists, and $p(j) = 0$ otherwise. That is, talk $p(j)$ is the talk ending latest among talks compatible with talk j that end before talk j starts, if such a talk exists, and $p(j) = 0$ if there are no such talks.

EXAMPLE 6 Consider seven talks with these start times and end times, as illustrated in Figure 5.

Talk 1: start 8 A.M., end 10 A.M.
Talk 2: start 9 A.M., end 11 A.M.
Talk 3: start 10:30 A.M., end 12 noon
Talk 4: start 9:30 A.M., end 1 P.M.
Talk 5: start 8:30 A.M., end 2 P.M.
Talk 6: start 11 A.M., end 2 P.M.
Talk 7: start 1 P.M., end 2 P.M.

Find $p(j)$ for $j = 1, 2, \ldots, 7$.

Solution: We have $p(1) = 0$ and $p(2) = 0$, because no talks end before either of the first two talks begin. We have $p(3) = 1$ because talk 3 and talk 1 are compatible, but talk 3 and talk 2 are not compatible; $p(4) = 0$ because talk 4 is not compatible with any of talks 1, 2, and 3; $p(5) = 0$

Links

Paul Fearn/The History Collection/Alamy Stock Photo

EUGÈNE CHARLES CATALAN (1814–1894) Eugène Catalan was born in Bruges, then part of France. His father became a successful architect in Paris while Eugène was a boy. Catalan attended a Parisian school for design hoping to follow in his father's footsteps. At 15, he won the job of teaching geometry to his design school classmates. After graduating, Catalan attended a school for the fine arts, but because of his mathematical aptitude his instructors recommended that he enter the École Polytechnique. He became a student there, but after his first year, he was expelled because of his politics. However, he was readmitted, and in 1835, he graduated and won a position at the Collège de Châlons sur Marne.

In 1838, Catalan returned to Paris where he founded a preparatory school with two other mathematicians, Sturm and Liouville. After teaching there for a short time, he was appointed to a position at the École Polytechnique. He received his doctorate from the École Polytechnique in 1841, but his political activity in favor of the French Republic hurt his career prospects. In 1846 Catalan held a position at the Collège de Charlemagne; he was appointed to the Lycée Saint Louis in 1849. However, when Catalan would not take a required oath of allegiance to the new Emperor Louis-Napoleon Bonaparte, he lost his job. For 13 years he held no permanent position. Finally, in 1865 he was appointed to a chair of mathematics at the University of Liège, Belgium, a position he held until his 1884 retirement.

Catalan made many contributions to number theory and to the related subject of continued fractions. He defined what are now known as the Catalan numbers when he solved the problem of dissecting a polygon into triangles using non-intersecting diagonals. Catalan is also well known for formulating what was known as the *Catalan conjecture*. This asserted that 8 and 9 are the only consecutive powers of integers, a conjecture not solved until 2003. Catalan wrote many textbooks, including several that became quite popular and appeared in as many as 12 editions. Perhaps this textbook will have a 12th edition someday!

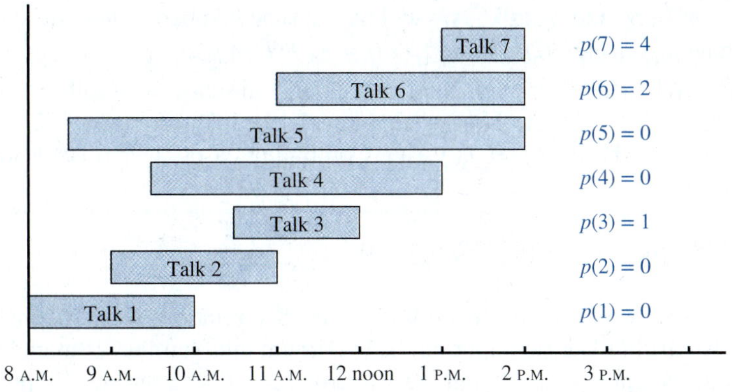

FIGURE 5 A schedule of lectures with the values of $p(n)$ shown.

because talk 5 is not compatible with any of talks 1, 2, 3, and 4; and $p(6) = 2$ because talk 6 and talk 2 are compatible, but talk 6 is not compatible with any of talks 3, 4, and 5. Finally, $p(7) = 4$, because talk 7 and talk 4 are compatible, but talk 7 is not compatible with either of talks 5 or 6. ◀

To develop a dynamic programming algorithm for this problem, we first develop a key recurrence relation. To do this, first note that if $j \leq n$, there are two possibilities for an optimal schedule of the first j talks (recall that we are assuming that the n talks are ordered by increasing end time): *(i)* talk j belongs to the optimal schedule or *(ii)* it does not.

Case (i): We know that talks $p(j) + 1, \ldots, j - 1$ do not belong to this schedule, for none of these other talks are compatible with talk j. Furthermore, the other talks in this optimal schedule must comprise an optimal schedule for talks $1, 2, \ldots, p(j)$. For if there were a better schedule for talks

Links

Alfred Eisenstaedt/The LIFE Picture Collection/ Shutterstock

RICHARD BELLMAN (1920–1984) Richard Bellman, born in Brooklyn, where his father was a grocer, spent many hours in the museums and libraries of New York as a child. After graduating high school, he studied mathematics at Brooklyn College and graduated in 1941. He began postgraduate work at Johns Hopkins University, but because of the war, left to teach electronics at the University of Wisconsin. He was able to continue his mathematics studies at Wisconsin, and in 1943 he received his masters degree there. Later, Bellman entered Princeton University, teaching in a special U.S. Army program. In late 1944, he was drafted into the army. He was assigned to the Manhattan Project at Los Alamos where he worked in theoretical physics. After the war, he returned to Princeton and received his Ph.D. in 1946.

After briefly teaching at Princeton, he moved to Stanford University, where he attained tenure. At Stanford he pursued his fascination with number theory. However, Bellman decided to focus on mathematical questions arising from real-world problems. In 1952, he joined the RAND Corporation, working on multistage decision processes, operations research problems, and applications to the social sciences and medicine. He worked on many military projects while at RAND. In 1965 he left RAND to become professor of mathematics, electrical and biomedical engineering and medicine at the University of Southern California.

In the 1950s Bellman pioneered the use of dynamic programming, a technique invented earlier, in a wide range of settings. He is also known for his work on stochastic control processes, in which he introduced what is now called the Bellman equation. He coined the term *curse of dimensionality* to describe problems caused by the exponential increase in volume associated with adding extra dimensions to a space. He wrote an amazing number of books and research papers with many coauthors, including many on industrial production and economic systems. His work led to the application of computing techniques in a wide variety of areas ranging from the design of guidance systems for space vehicles, to network optimization, and even to pest control.

Tragically, in 1973 Bellman was diagnosed with a brain tumor. Although it was removed successfully, complications left him severely disabled. Fortunately, he managed to continue his research and writing during his remaining ten years of life. Bellman received many prizes and awards, including the first Norbert Wiener Prize in Applied Mathematics and the IEEE Gold Medal of Honor. He was elected to the National Academy of Sciences. He was held in high regard for his achievements, courage, and admirable qualities. Bellman was the father of two children.

1, 2, …, $p(j)$, by adding talk j, we will have a schedule better than the overall optimal schedule. Consequently, in case *(i)*, we have $T(j) = w_j + T(p(j))$.

Case (ii): When talk j does not belong to an optimal schedule, it follows that an optimal schedule from talks 1, 2, …, j is the same as an optimal schedule from talks 1, 2, …, $j-1$. Consequently, in case *(ii)*, we have $T(j) = T(j-1)$. Combining cases *(i)* and *(ii)* leads us to the recurrence relation

$$T(j) = \max(w_j + T(p(j)), T(j-1)).$$

Now that we have developed this recurrence relation, we can construct an efficient algorithm, Algorithm 1, for computing the maximum total number of attendees. We ensure that the algorithm is efficient by storing the value of each $T(j)$ after we compute it. This allows us to compute $T(j)$ only once. If we did not do this, the algorithm would have exponential worst-case complexity. The process of storing the values as each is computed is known as **memoization** and is an important technique for making recursive algorithms efficient.

ALGORITHM 1 Dynamic Programming Algorithm for Scheduling Talks.

```
procedure Maximum Attendees (s_1, s_2, ..., s_n: start times of talks;
  e_1, e_2, ..., e_n: end times of talks; w_1, w_2, ..., w_n: number of attendees to talks)
  sort talks by end time and relabel so that e_1 ≤ e_2 ≤ ··· ≤ e_n
for j := 1 to n
    if no talk i with i < j is compatible with talk j
        p(j) = 0
    else p(j) := max{i | i < j and talk i is compatible with talk j}
    T(0) := 0
for j := 1 to n
    T(j) := max(w_j + T(p(j)), T(j − 1))
return T(n){T(n) is the maximum number of attendees}
```

In Algorithm 1 we determine the maximum number of attendees that can be achieved by a schedule of talks, but we do not find a schedule that achieves this maximum. To find talks we need to schedule, we use the fact that talk j belongs to an optimal solution for the first j talks if and only if $w_j + T(p(j)) \geq T(j-1)$. We leave it as Exercise 53 to construct an algorithm based on this observation that determines which talks should be scheduled to achieve the maximum total number of attendees.

Algorithm 1 is a good example of dynamic programming as the maximum total attendance is found using the optimal solutions of the overlapping subproblems, each of which determines the maximum total attendance of the first j talks for some j with $1 \leq j \leq n-1$. See Exercises 56 and 57 and Supplementary Exercises 14 and 17 for other examples of dynamic programming.

Exercises

1. Use mathematical induction to verify the formula derived in Example 2 for the number of moves required to complete the Tower of Hanoi puzzle.

2. a) Find a recurrence relation for the number of permutations of a set with n elements.

b) Use this recurrence relation to find the number of permutations of a set with n elements using iteration.

3. A vending machine dispensing books of stamps accepts only one-dollar coins, \$1 bills, and \$5 bills.

a) Find a recurrence relation for the number of ways to deposit n dollars in the vending machine, where the order in which the coins and bills are deposited matters.

b) What are the initial conditions?

c) How many ways are there to deposit \$10 for a book of stamps?

4. A country uses as currency coins with values of 1 peso, 2 pesos, 5 pesos, and 10 pesos and bills with values of 5 pesos, 10 pesos, 20 pesos, 50 pesos, and 100 pesos. Find a recurrence relation for the number of ways to pay a bill of n pesos if the order in which the coins and bills are paid matters.

5. How many ways are there to pay a bill of 17 pesos using the currency described in Exercise 4, where the order in which coins and bills are paid matters?

*6. a) Find a recurrence relation for the number of strictly increasing sequences of positive integers that have 1 as their first term and n as their last term, where n is a positive integer. That is, sequences $a_1, a_2, \ldots, a_k$, where $a_1 = 1$, $a_k = n$, and $a_j < a_{j+1}$ for $j = 1, 2, \ldots, k-1$.
b) What are the initial conditions?
c) How many sequences of the type described in (a) are there when n is an integer with $n \geq 2$?

7. a) Find a recurrence relation for the number of bit strings of length n that contain a pair of consecutive 0s.
b) What are the initial conditions?
c) How many bit strings of length seven contain two consecutive 0s?

8. a) Find a recurrence relation for the number of bit strings of length n that contain three consecutive 0s.
b) What are the initial conditions?
c) How many bit strings of length seven contain three consecutive 0s?

9. a) Find a recurrence relation for the number of bit strings of length n that do not contain three consecutive 0s.
b) What are the initial conditions?
c) How many bit strings of length seven do not contain three consecutive 0s?

*10. a) Find a recurrence relation for the number of bit strings of length n that contain the string 01.
b) What are the initial conditions?
c) How many bit strings of length seven contain the string 01?

11. a) Find a recurrence relation for the number of ways to climb n stairs if the person climbing the stairs can take one stair or two stairs at a time.
b) What are the initial conditions?
c) In how many ways can this person climb a flight of eight stairs?

12. a) Find a recurrence relation for the number of ways to climb n stairs if the person climbing the stairs can take one, two, or three stairs at a time.
b) What are the initial conditions?
c) In how many ways can this person climb a flight of eight stairs?

A string that contains only 0s, 1s, and 2s is called a **ternary string**.

13. a) Find a recurrence relation for the number of ternary strings of length n that do not contain two consecutive 0s.
b) What are the initial conditions?
c) How many ternary strings of length six do not contain two consecutive 0s?

14. a) Find a recurrence relation for the number of ternary strings of length n that contain two consecutive 0s.
b) What are the initial conditions?
c) How many ternary strings of length six contain two consecutive 0s?

*15. a) Find a recurrence relation for the number of ternary strings of length n that do not contain two consecutive 0s or two consecutive 1s.
b) What are the initial conditions?
c) How many ternary strings of length six do not contain two consecutive 0s or two consecutive 1s?

*16. a) Find a recurrence relation for the number of ternary strings of length n that contain either two consecutive 0s or two consecutive 1s.
b) What are the initial conditions?
c) How many ternary strings of length six contain two consecutive 0s or two consecutive 1s?

*17. a) Find a recurrence relation for the number of ternary strings of length n that do not contain consecutive symbols that are the same.
b) What are the initial conditions?
c) How many ternary strings of length six do not contain consecutive symbols that are the same?

**18. a) Find a recurrence relation for the number of ternary strings of length n that contain two consecutive symbols that are the same.
b) What are the initial conditions?
c) How many ternary strings of length six contain consecutive symbols that are the same?

19. Messages are transmitted over a communications channel using two signals. The transmittal of one signal requires 1 microsecond, and the transmittal of the other signal requires 2 microseconds.
a) Find a recurrence relation for the number of different messages consisting of sequences of these two signals, where each signal in the message is immediately followed by the next signal, that can be sent in n microseconds.
b) What are the initial conditions?
c) How many different messages can be sent in 10 microseconds using these two signals?

20. A bus driver pays all tolls, using only nickels and dimes, by throwing one coin at a time into the mechanical toll collector.
a) Find a recurrence relation for the number of different ways the bus driver can pay a toll of n cents (where the order in which the coins are used matters).
b) In how many different ways can the driver pay a toll of 45 cents?

21. a) Find the recurrence relation satisfied by R_n, where R_n is the number of regions that a plane is divided into by n lines, if no two of the lines are parallel and no three of the lines go through the same point.
b) Find R_n using iteration.

*22. **a)** Find the recurrence relation satisfied by R_n, where R_n is the number of regions into which the surface of a sphere is divided by n great circles (which are the intersections of the sphere and planes passing through the center of the sphere), if no three of the great circles go through the same point.
b) Find R_n using iteration.

*23. **a)** Find the recurrence relation satisfied by S_n, where S_n is the number of regions into which three-dimensional space is divided by n planes if every three of the planes meet in one point, but no four of the planes go through the same point.
b) Find S_n using iteration.

24. Find a recurrence relation for the number of bit sequences of length n with an even number of 0s.

25. How many bit sequences of length seven contain an even number of 0s?

26. **a)** Find a recurrence relation for the number of ways to completely cover a $2 \times n$ checkerboard with 1×2 dominoes. [*Hint:* Consider separately the coverings where the position in the top right corner of the checkerboard is covered by a domino positioned horizontally and where it is covered by a domino positioned vertically.]
b) What are the initial conditions for the recurrence relation in part (a)?
c) How many ways are there to completely cover a 2×17 checkerboard with 1×2 dominoes?

27. **a)** Find a recurrence relation for the number of ways to lay out a walkway with slate tiles if the tiles are red, green, or gray, so that no two red tiles are adjacent and tiles of the same color are considered indistinguishable.
b) What are the initial conditions for the recurrence relation in part (a)?
c) How many ways are there to lay out a path of seven tiles as described in part (a)?

28. Show that the Fibonacci numbers satisfy the recurrence relation $f_n = 5f_{n-4} + 3f_{n-5}$ for $n = 5, 6, 7, \ldots,$ together with the initial conditions $f_0 = 0, f_1 = 1, f_2 = 1, f_3 = 2,$ and $f_4 = 3$. Use this recurrence relation to show that f_{5n} is divisible by 5, for $n = 1, 2, 3, \ldots$.

*29. Let $S(m, n)$ denote the number of onto functions from a set with m elements to a set with n elements. Show that $S(m, n)$ satisfies the recurrence relation

$$S(m, n) = n^m - \sum_{k=1}^{n-1} C(n, k)S(m, k)$$

whenever $m \geq n$ and $n > 1$, with the initial condition $S(m, 1) = 1$.

30. **a)** Write out all the ways the product $x_0 \cdot x_1 \cdot x_2 \cdot x_3 \cdot x_4$ can be parenthesized to determine the order of multiplication.
b) Use the recurrence relation developed in Example 5 to calculate C_4, the number of ways to parenthesize the product of five numbers so as to determine the order of multiplication. Verify that you listed the correct number of ways in part (a).
c) Check your result in part (b) by finding C_4, using the closed formula for C_n mentioned in the solution of Example 5.

31. **a)** Use the recurrence relation developed in Example 5 to determine C_5, the number of ways to parenthesize the product of six numbers so as to determine the order of multiplication.
b) Check your result with the closed formula for C_5 mentioned in the solution of Example 5.

*32. In the Tower of Hanoi puzzle, suppose our goal is to transfer all n disks from peg 1 to peg 3, but we cannot move a disk directly between pegs 1 and 3. Each move of a disk must be a move involving peg 2. As usual, we cannot place a disk on top of a smaller disk.
a) Find a recurrence relation for the number of moves required to solve the puzzle for n disks with this added restriction.
b) Solve this recurrence relation to find a formula for the number of moves required to solve the puzzle for n disks.
c) How many different arrangements are there of the n disks on three pegs so that no disk is on top of a smaller disk?
d) Show that every allowable arrangement of the n disks occurs in the solution of this variation of the puzzle.

Exercises 33–37 deal with a variation of the **Josephus problem** described by Graham, Knuth, and Patashnik in [GrKnPa94]. This problem is based on an account by the historian Flavius Josephus, who was part of a band of 41 Jewish rebels trapped in a cave by the Romans during the Jewish-Roman war of the first century. The rebels preferred suicide to capture; they decided to form a circle and to repeatedly count off around the circle, killing every third rebel left alive. However, Josephus and another rebel did not want to be killed this way; they determined the positions where they should stand to be the last two rebels remaining alive. The variation we consider begins with n people, numbered 1 to n, standing around a circle. In each stage, every second person still left alive is eliminated until only one survives. We denote the number of the survivor by $J(n)$.

Links

33. Determine the value of $J(n)$ for each integer n with $1 \leq n \leq 16$.

34. Use the values you found in Exercise 33 to conjecture a formula for $J(n)$. [*Hint:* Write $n = 2^m + k$, where m is a nonnegative integer and k is a nonnegative integer less than 2^m.]

35. Show that $J(n)$ satisfies the recurrence relation $J(2n) = 2J(n) - 1$ and $J(2n + 1) = 2J(n) + 1$, for $n \geq 1$, and $J(1) = 1$.

36. Use mathematical induction to prove the formula you conjectured in Exercise 34, making use of the recurrence relation from Exercise 35.

37. Determine $J(100)$, $J(1000)$, and $J(10{,}000)$ from your formula for $J(n)$.

Exercises 38–45 involve the Reve's puzzle, the variation of the Tower of Hanoi puzzle with four pegs and n disks. Before presenting these exercises, we describe the Frame–Stewart algorithm for moving the disks from peg 1 to peg 4 so that no disk is ever on top of a smaller one. This algorithm, given the number of disks n as input, depends on a choice of an integer k with $1 \le k \le n$. When there is only one disk, move it from peg 1 to peg 4 and stop. For $n > 1$, the algorithm proceeds recursively, using these three steps. Recursively move the stack of the $n - k$ smallest disks from peg 1 to peg 2, using all four pegs. Next move the stack of the k largest disks from peg 1 to peg 4, using the three-peg algorithm from the Tower of Hanoi puzzle without using the peg holding the $n - k$ smallest disks. Finally, recursively move the smallest $n - k$ disks to peg 4, using all four pegs. Frame and Stewart showed that to produce the fewest moves using their algorithm, k should be chosen to be the smallest integer such that n does not exceed $t_k = k(k + 1)/2$, the kth triangular number, that is, $t_{k-1} < n \le t_k$. The long-standing conjecture, known as **Frame's conjecture**, that this algorithm uses the fewest number of moves required to solve the puzzle, was proved by Thierry Bousch in 2014.

38. Show that the Reve's puzzle with three disks can be solved using five, and no fewer, moves.

39. Show that the Reve's puzzle with four disks can be solved using nine, and no fewer, moves.

40. Describe the moves made by the Frame–Stewart algorithm, with k chosen so that the fewest moves are required, for

a) 5 disks. **b)** 6 disks. **c)** 7 disks. **d)** 8 disks.

***41.** Show that if $R(n)$ is the number of moves used by the Frame–Stewart algorithm to solve the Reve's puzzle with n disks, where k is chosen to be the smallest integer with $n \le k(k + 1)/2$, then $R(n)$ satisfies the recurrence relation $R(n) = 2R(n - k) + 2^k - 1$, with $R(0) = 0$ and $R(1) = 1$.

***42.** Show that if k is as chosen in Exercise 41, then $R(n) - R(n - 1) = 2^{k-1}$.

***43.** Show that if k is as chosen in Exercise 41, then $R(n) = \sum_{i=1}^{k} i2^{i-1} - (t_k - n)2^{k-1}$.

***44.** Use Exercise 43 to give an upper bound on the number of moves required to solve the Reve's puzzle for all integers n with $1 \le n \le 25$.

***45.** Show that $R(n)$ is $O(\sqrt{n}2^{\sqrt{2n}})$.

Let $\{a_n\}$ be a sequence of real numbers. The **backward differences** of this sequence are defined recursively as shown next. The **first difference** ∇a_n is

$$\nabla a_n = a_n - a_{n-1}.$$

The **(*k* + 1)st difference** $\nabla^{k+1} a_n$ is obtained from $\nabla^k a_n$ by

$$\nabla^{k+1} a_n = \nabla^k a_n - \nabla^k a_{n-1}.$$

46. Find ∇a_n for the sequence $\{a_n\}$, where

a) $a_n = 4$. **b)** $a_n = 2n$.

c) $a_n = n^2$. **d)** $a_n = 2^n$.

47. Find $\nabla^2 a_n$ for the sequences in Exercise 46.

48. Show that $a_{n-1} = a_n - \nabla a_n$.

49. Show that $a_{n-2} = a_n - 2\nabla a_n + \nabla^2 a_n$.

***50.** Prove that a_{n-k} can be expressed in terms of a_n, ∇a_n, $\nabla^2 a_n, \ldots, \nabla^k a_n$.

51. Express the recurrence relation $a_n = a_{n-1} + a_{n-2}$ in terms of a_n, ∇a_n, and $\nabla^2 a_n$.

52. Show that any recurrence relation for the sequence $\{a_n\}$ can be written in terms of a_n, ∇a_n, $\nabla^2 a_n, \ldots$. The resulting equation involving the sequences and its differences is called a **difference equation**.

***53.** Construct the algorithm described in the text after Algorithm 1 for determining which talks should be scheduled to maximize the total number of attendees and not just the maximum total number of attendees determined by Algorithm 1.

54. Use Algorithm 1 to determine the maximum number of total attendees in the talks in Example 6 if w_i, the number of attendees of talk i, $i = 1, 2, \ldots, 7$, is

a) 20, 10, 50, 30, 15, 25, 40.

b) 100, 5, 10, 20, 25, 40, 30.

c) 2, 3, 8, 5, 4, 7, 10.

d) 10, 8, 7, 25, 20, 30, 5.

55. For each part of Exercise 54, use your algorithm from Exercise 53 to find the optimal schedule for talks so that the total number of attendees is maximized.

56. In this exercise we will develop a dynamic programming algorithm for finding the maximum sum of consecutive terms of a sequence of real numbers. That is, given a sequence of real numbers $a_1, a_2, \ldots, a_n$, the algorithm computes the maximum sum $\sum_{i=j}^{k} a_i$ where $1 \le j \le k \le n$.

a) Show that if all terms of the sequence are nonnegative, this problem is solved by taking the sum of all terms. Then, give an example where the maximum sum of consecutive terms is not the sum of all terms.

b) Let $M(k)$ be the maximum of the sums of consecutive terms of the sequence ending at a_k. That is, $M(k) = \max_{1\le j\le k} \sum_{i=j}^{k} a_i$. Explain why the recurrence relation $M(k) = \max(M(k - 1) + a_k, a_k)$ holds for $k = 2, \ldots, n$.

c) Use part (b) to develop a dynamic programming algorithm for solving this problem.

d) Show each step your algorithm from part (c) uses to find the maximum sum of consecutive terms of the sequence 2, −3, 4, 1, −2, 3.

e) Show that the worst-case complexity in terms of the number of additions and comparisons of your algorithm from part (c) is linear.

*57. Dynamic programming can be used to develop an algorithm for solving the matrix-chain multiplication problem introduced in Section 3.3. This is the problem of determining how the product $\mathbf{A}_1\mathbf{A}_2 \cdots \mathbf{A}_n$ can be computed using the fewest integer multiplications, where $\mathbf{A}_1, \mathbf{A}_2, \ldots, \mathbf{A}_n$ are $m_1 \times m_2, m_2 \times m_3, \ldots, m_n \times m_{n+1}$ matrices, respectively, and each matrix has integer entries. Recall that by the associative law, the product does not depend on the order in which the matrices are multiplied.

a) Show that the brute-force method of determining the minimum number of integer multiplications needed to solve a matrix-chain multiplication problem has exponential worst-case complexity. [*Hint:* Do this by first showing that the order of multiplication of matrices is specified by parenthesizing the product. Then, use Example 5 and the result of part (c) of Exercise 43 in Section 8.4.]

b) Denote by $\mathbf{A}_{ij}$ the product $\mathbf{A}_i\mathbf{A}_{i+1} \ldots, \mathbf{A}_j$, and $M(i, j)$ the minimum number of integer multiplications required to find $\mathbf{A}_{ij}$. Show that if the least number of integer multiplications are used to compute $\mathbf{A}_{ij}$, where $i < j$, by splitting the product into the product of $\mathbf{A}_i$ through $\mathbf{A}_k$ and the product of $\mathbf{A}_{k+1}$ through $\mathbf{A}_j$, then the first k terms must be parenthesized so that $\mathbf{A}_{ik}$ is computed in the optimal way using $M(i, k)$ integer multiplications, and $\mathbf{A}_{k+1,j}$ must be parenthesized so that $\mathbf{A}_{k+1,j}$ is computed in the optimal way using $M(k + 1, j)$ integer multiplications.

c) Explain why part (b) leads to the recurrence relation $M(i, j) = \min_{i \le k < j}(M(i, k) + M(k + 1, j) + m_i m_{k+1} m_{j+1})$ if $1 \le i \le j < j \le n$.

d) Use the recurrence relation in part (c) to construct an efficient algorithm for determining the order the n matrices should be multiplied to use the minimum number of integer multiplications. Store the partial results $M(i, j)$ as you find them so that your algorithm will not have exponential complexity.

e) Show that your algorithm from part (d) has $O(n^3)$ worst-case complexity in terms of multiplications of integers.

8.2 Solving Linear Recurrence Relations

8.2.1 Introduction

Links

A wide variety of recurrence relations occur in models. Some of these recurrence relations can be solved using iteration or some other ad hoc technique. However, one important class of recurrence relations can be explicitly solved in a systematic way. These are recurrence relations that express the terms of a sequence as linear combinations of previous terms.

Definition 1 A *linear homogeneous recurrence relation of degree k with constant coefficients* is a recurrence relation of the form

$$a_n = c_1 a_{n-1} + c_2 a_{n-2} + \cdots + c_k a_{n-k},$$

where $c_1, c_2, \ldots, c_k$ are real numbers, and $c_k \neq 0$.

The recurrence relation in the definition is **linear** because the right-hand side is a sum of previous terms of the sequence each multiplied by a function of n. The recurrence relation is **homogeneous** because no terms occur that are not multiples of the a_js. The coefficients of the terms of the sequence are all **constants**, rather than functions that depend on n. The **degree** is k because a_n is expressed in terms of the previous k terms of the sequence.

A consequence of the second principle of mathematical induction is that a sequence satisfying the recurrence relation in the definition is uniquely determined by this recurrence relation and the k initial conditions

$$a_0 = C_0, a_1 = C_1, \ldots, a_{k-1} = C_{k-1}.$$

EXAMPLE 1 The recurrence relation $P_n = (1.11)P_{n-1}$ is a linear homogeneous recurrence relation of degree one. The recurrence relation $f_n = f_{n-1} + f_{n-2}$ is a linear homogeneous recurrence relation of degree two. The recurrence relation $a_n = a_{n-5}$ is a linear homogeneous recurrence relation of degree five. ◀

To help clarify the definition of linear homogeneous recurrence relations with constant coefficients, we will now provide examples of recurrence relations each lacking one of the defining properties.

EXAMPLE 2 The recurrence relation $a_n = a_{n-1} + a_{n-2}^2$ is not linear. The recurrence relation $H_n = 2H_{n-1} + 1$ is not homogeneous. The recurrence relation $B_n = nB_{n-1}$ does not have constant coefficients. ◀

Linear homogeneous recurrence relations with constant coefficients are studied for two reasons. First, they often occur in modeling of problems. Second, they can be systematically solved.

8.2.2 Solving Linear Homogeneous Recurrence Relations with Constant Coefficients

Recurrence relations may be difficult to solve, but fortunately this is not the case for linear homogenous recurrence relations with constant coefficients. We can use two key ideas to find all their solutions. First, these recurrence relations have solutions of the form $a_n = r^n$, where r is a constant. To see this, observe that $a_n = r^n$ is a solution of the recurrence relation $a_n = c_1a_{n-1} + c_2a_{n-2} + \cdots + c_ka_{n-k}$ if and only if

$$r^n = c_1r^{n-1} + c_2r^{n-2} + \cdots + c_kr^{n-k}.$$

When both sides of this equation are divided by r^{n-k} (when $r \neq 0$) and the right-hand side is subtracted from the left, we obtain the equation

$$r^k - c_1r^{k-1} - c_2r^{k-2} - \cdots - c_{k-1}r - c_k = 0.$$

Consequently, the sequence $\{a_n\}$ with $a_n = r^n$ where $r \neq 0$ is a solution if and only if r is a solution of this last equation. We call this the **characteristic equation** of the recurrence relation. The solutions of this equation are called the **characteristic roots** of the recurrence relation. As we will see, these characteristic roots can be used to give an explicit formula for all the solutions of the recurrence relation.

The other key observation is that a linear combination of two solutions of a linear homogeneous recurrence relation is also a solution. To see this, suppose that s_n and t_n are both solutions of $a_n = c_1a_{n-1} + c_2a_{n-2} + \cdots + c_ka_{n-k}$. Then

$$s_n = c_1s_{n-1} + c_2s_{n-2} + \cdots + c_ks_{n-k}$$

and

$$t_n = c_1t_{n-1} + c_2t_{n-2} + \cdots + c_kt_{n-k}.$$

Now suppose that b_1 and b_2 are real numbers. Then

$$\begin{aligned} b_1s_n + b_2t_n &= b_1(c_1s_{n-1} + c_2s_{n-2} + \cdots + c_ks_{n-k}) + b_2(c_1t_{n-1} + c_2t_{n-2} + \cdots + c_kt_{n-k}) \\ &= c_1(b_1s_{n-1} + b_2t_{n-1}) + c_2(b_1s_{n-2} + b_2t_{n-2}) + \cdots + c_k(b_1s_{n-k} + b_kt_{n-k}). \end{aligned}$$

This means that $b_1s_n + b_2t_n$ is also a solution of the same linear homogeneous recurrence relation.

Using these key observations, we will show how to solve linear homogeneous recurrence relations with constant coefficients.

THE DEGREE TWO CASE We now turn our attention to linear homogeneous recurrence relations of degree two. First, consider the case when there are two distinct characteristic roots.

THEOREM 1 Let c_1 and c_2 be real numbers. Suppose that $r^2 - c_1 r - c_2 = 0$ has two distinct roots r_1 and r_2. Then the sequence $\{a_n\}$ is a solution of the recurrence relation $a_n = c_1 a_{n-1} + c_2 a_{n-2}$ if and only if $a_n = \alpha_1 r_1^n + \alpha_2 r_2^n$ for $n = 0, 1, 2, \ldots$, where α_1 and α_2 are constants.

Proof: We must do two things to prove the theorem. First, it must be shown that if r_1 and r_2 are the roots of the characteristic equation, and α_1 and α_2 are constants, then the sequence $\{a_n\}$ with $a_n = \alpha_1 r_1^n + \alpha_2 r_2^n$ is a solution of the recurrence relation. Second, it must be shown that if the sequence $\{a_n\}$ is a solution, then $a_n = \alpha_1 r_1^n + \alpha_2 r_2^n$ for some constants α_1 and α_2.

We now show that if $a_n = \alpha_1 r_1^n + \alpha_2 r_2^n$, then the sequence $\{a_n\}$ is a solution of the recurrence relation. Because r_1 and r_2 are roots of $r^2 - c_1 r - c_2 = 0$, it follows that $r_1^2 = c_1 r_1 + c_2$ and $r_2^2 = c_1 r_2 + c_2$.

From these equations, we see that

$$\begin{aligned} c_1 a_{n-1} + c_2 a_{n-2} &= c_1(\alpha_1 r_1^{n-1} + \alpha_2 r_2^{n-1}) + c_2(\alpha_1 r_1^{n-2} + \alpha_2 r_2^{n-2}) \\ &= \alpha_1 r_1^{n-2}(c_1 r_1 + c_2) + \alpha_2 r_2^{n-2}(c_1 r_2 + c_2) \\ &= \alpha_1 r_1^{n-2} r_1^2 + \alpha_2 r_2^{n-2} r_2^2 \\ &= \alpha_1 r_1^n + \alpha_2 r_2^n \\ &= a_n. \end{aligned}$$

This shows that the sequence $\{a_n\}$ with $a_n = \alpha_1 r_1^n + \alpha_2 r_2^n$ is a solution of the recurrence relation.

To show that every solution $\{a_n\}$ of the recurrence relation $a_n = c_1 a_{n-1} + c_2 a_{n-2}$ has $a_n = \alpha_1 r_1^n + \alpha_2 r_2^n$ for $n = 0, 1, 2, \ldots$, for some constants α_1 and α_2, suppose that $\{a_n\}$ is a solution of the recurrence relation, and the initial conditions $a_0 = C_0$ and $a_1 = C_1$ hold. It will be shown that there are constants α_1 and α_2 such that the sequence $\{a_n\}$ with $a_n = \alpha_1 r_1^n + \alpha_2 r_2^n$ satisfies these same initial conditions. This requires that

$$\begin{aligned} a_0 &= C_0 = \alpha_1 + \alpha_2, \\ a_1 &= C_1 = \alpha_1 r_1 + \alpha_2 r_2. \end{aligned}$$

We can solve these two equations for α_1 and α_2. From the first equation it follows that $\alpha_2 = C_0 - \alpha_1$. Inserting this expression into the second equation gives

$$C_1 = \alpha_1 r_1 + (C_0 - \alpha_1) r_2.$$

Hence,

$$C_1 = \alpha_1 (r_1 - r_2) + C_0 r_2.$$

This shows that

$$\alpha_1 = \frac{C_1 - C_0 r_2}{r_1 - r_2}$$

and

$$\alpha_2 = C_0 - \alpha_1 = C_0 - \frac{C_1 - C_0 r_2}{r_1 - r_2} = \frac{C_0 r_1 - C_1}{r_1 - r_2},$$

where these expressions for α_1 and α_2 depend on the fact that $r_1 \neq r_2$. (When $r_1 = r_2$, this theorem is not true.) Hence, with these values for α_1 and α_2, the sequence $\{a_n\}$ with $\alpha_1 r_1^n + \alpha_2 r_2^n$ satisfies the two initial conditions.

We know that $\{a_n\}$ and $\{\alpha_1 r_1^n + \alpha_2 r_2^n\}$ are both solutions of the recurrence relation $a_n = c_1 a_{n-1} + c_2 a_{n-2}$ and both satisfy the initial conditions when $n = 0$ and $n = 1$. Because there is a unique solution of a linear homogeneous recurrence relation of degree two with two initial conditions, it follows that the two solutions are the same, that is, $a_n = \alpha_1 r_1^n + \alpha_2 r_2^n$ for all nonnegative integers n. We have completed the proof by showing that a solution of the linear homogeneous recurrence relation with constant coefficients of degree two must be of the form $a_n = \alpha_1 r_1^n + \alpha_2 r_2^n$, where α_1 and α_2 are constants. ◁

The characteristic roots of a linear homogeneous recurrence relation with constant coefficients may be complex numbers. Theorem 1 (and also subsequent theorems in this section) still applies in this case. Recurrence relations with complex characteristic roots will not be discussed in the text. Readers familiar with complex numbers may wish to solve Exercises 38 and 39.

Examples 3 and 4 show how to use Theorem 1 to solve recurrence relations.

EXAMPLE 3 What is the solution of the recurrence relation

Extra Examples

$$a_n = a_{n-1} + 2a_{n-2}$$

with $a_0 = 2$ and $a_1 = 7$?

Solution: Theorem 1 can be used to solve this problem. The characteristic equation of the recurrence relation is $r^2 - r - 2 = 0$. Its roots are $r = 2$ and $r = -1$. Hence, the sequence $\{a_n\}$ is a solution to the recurrence relation if and only if

$$a_n = \alpha_1 2^n + \alpha_2 (-1)^n,$$

for some constants α_1 and α_2. From the initial conditions, it follows that

$$a_0 = 2 = \alpha_1 + \alpha_2,$$
$$a_1 = 7 = \alpha_1 \cdot 2 + \alpha_2 \cdot (-1).$$

Solving these two equations shows that $\alpha_1 = 3$ and $\alpha_2 = -1$. Hence, the solution to the recurrence relation and initial conditions is the sequence $\{a_n\}$ with

$$a_n = 3 \cdot 2^n - (-1)^n.$$ ◀

EXAMPLE 4 Find an explicit formula for the Fibonacci numbers.

Solution: Recall that the sequence of Fibonacci numbers satisfies the recurrence relation $f_n = f_{n-1} + f_{n-2}$ and also satisfies the initial conditions $f_0 = 0$ and $f_1 = 1$. The roots of the characteristic equation $r^2 - r - 1 = 0$ are $r_1 = (1 + \sqrt{5})/2$ and $r_2 = (1 - \sqrt{5})/2$. Therefore, from Theorem 1 it follows that the Fibonacci numbers are given by

$$f_n = \alpha_1 \left(\frac{1+\sqrt{5}}{2}\right)^n + \alpha_2 \left(\frac{1-\sqrt{5}}{2}\right)^n,$$

for some constants α_1 and α_2. The initial conditions $f_0 = 0$ and $f_1 = 1$ can be used to find these constants. We have

$$f_0 = \alpha_1 + \alpha_2 = 0,$$
$$f_1 = \alpha_1 \left(\frac{1+\sqrt{5}}{2}\right) + \alpha_2 \left(\frac{1-\sqrt{5}}{2}\right) = 1.$$

The solution to these simultaneous equations for α_1 and α_2 is

$$\alpha_1 = 1/\sqrt{5}, \qquad \alpha_2 = -1/\sqrt{5}.$$

Consequently, the Fibonacci numbers are given by

$$f_n = \frac{1}{\sqrt{5}}\left(\frac{1+\sqrt{5}}{2}\right)^n - \frac{1}{\sqrt{5}}\left(\frac{1-\sqrt{5}}{2}\right)^n.$$

◀

Theorem 1 does not apply when there is one characteristic root of multiplicity two. If this happens, then $a_n = nr_0^n$ is another solution of the recurrence relation when r_0 is a root of multiplicity two of the characteristic equation. Theorem 2 shows how to handle this case.

THEOREM 2 Let c_1 and c_2 be real numbers with $c_2 \neq 0$. Suppose that $r^2 - c_1 r - c_2 = 0$ has only one root r_0. A sequence $\{a_n\}$ is a solution of the recurrence relation $a_n = c_1 a_{n-1} + c_2 a_{n-2}$ if and only if $a_n = \alpha_1 r_0^n + \alpha_2 n r_0^n$, for $n = 0, 1, 2, \ldots$, where α_1 and α_2 are constants.

The proof of Theorem 2 is left as Exercise 10. Example 5 illustrates the use of this theorem.

EXAMPLE 5 What is the solution of the recurrence relation

$$a_n = 6a_{n-1} - 9a_{n-2}$$

with initial conditions $a_0 = 1$ and $a_1 = 6$?

Solution: The only root of $r^2 - 6r + 9 = 0$ is $r = 3$. Hence, the solution to this recurrence relation is

$$a_n = \alpha_1 3^n + \alpha_2 n 3^n$$

for some constants α_1 and α_2. Using the initial conditions, it follows that

$$a_0 = 1 = \alpha_1,$$
$$a_1 = 6 = \alpha_1 \cdot 3 + \alpha_2 \cdot 3.$$

Solving these two equations shows that $\alpha_1 = 1$ and $\alpha_2 = 1$. Consequently, the solution to this recurrence relation and the initial conditions is

$$a_n = 3^n + n3^n.$$

◀

THE GENERAL CASE We will now state the general result about the solution of linear homogeneous recurrence relations with constant coefficients, where the degree may be greater than two, under the assumption that the characteristic equation has distinct roots. The proof of this result will be left as Exercise 16.

THEOREM 3 Let $c_1, c_2, \ldots, c_k$ be real numbers. Suppose that the characteristic equation

$$r^k - c_1 r^{k-1} - \cdots - c_k = 0$$

has k distinct roots $r_1, r_2, \ldots, r_k$. Then a sequence $\{a_n\}$ is a solution of the recurrence relation

$$a_n = c_1 a_{n-1} + c_2 a_{n-2} + \cdots + c_k a_{n-k}$$

if and only if

$$a_n = \alpha_1 r_1^n + \alpha_2 r_2^n + \cdots + \alpha_k r_k^n$$

for $n = 0, 1, 2, \ldots$, where $\alpha_1, \alpha_2, \ldots, \alpha_k$ are constants.

We illustrate the use of the theorem with Example 6.

EXAMPLE 6 Find the solution to the recurrence relation

$$a_n = 6a_{n-1} - 11a_{n-2} + 6a_{n-3}$$

with the initial conditions $a_0 = 2$, $a_1 = 5$, and $a_2 = 15$.

Solution: The characteristic polynomial of this recurrence relation is

$$r^3 - 6r^2 + 11r - 6.$$

The characteristic roots are $r = 1$, $r = 2$, and $r = 3$, because $r^3 - 6r^2 + 11r - 6 = (r-1)(r-2)(r-3)$. Hence, the solutions to this recurrence relation are of the form

$$a_n = \alpha_1 \cdot 1^n + \alpha_2 \cdot 2^n + \alpha_3 \cdot 3^n.$$

To find the constants α_1, α_2, and α_3, use the initial conditions. This gives

$$\begin{aligned} a_0 &= 2 = \alpha_1 + \alpha_2 + \alpha_3, \\ a_1 &= 5 = \alpha_1 + \alpha_2 \cdot 2 + \alpha_3 \cdot 3, \\ a_2 &= 15 = \alpha_1 + \alpha_2 \cdot 4 + \alpha_3 \cdot 9. \end{aligned}$$

When these three simultaneous equations are solved for α_1, α_2, and α_3, we find that $\alpha_1 = 1$, $\alpha_2 = -1$, and $\alpha_3 = 2$. Hence, the unique solution to this recurrence relation and the given initial conditions is the sequence $\{a_n\}$ with

$$a_n = 1 - 2^n + 2 \cdot 3^n.$$ ◀

We now state the most general result about linear homogeneous recurrence relations with constant coefficients, allowing the characteristic equation to have multiple roots. The key point is that for each root r of the characteristic equation, the general solution has a summand of the form $P(n)r^n$, where $P(n)$ is a polynomial of degree $m-1$, with m the multiplicity of this root. We leave the proof of this result as Exercise 51.

THEOREM 4 Let $c_1, c_2, \ldots, c_k$ be real numbers. Suppose that the characteristic equation

$$r^k - c_1 r^{k-1} - \cdots - c_k = 0$$

has t distinct roots $r_1, r_2, \ldots, r_t$ with multiplicities $m_1, m_2, \ldots, m_t$, respectively, so that $m_i \geq 1$ for $i = 1, 2, \ldots, t$ and $m_1 + m_2 + \cdots + m_t = k$. Then a sequence $\{a_n\}$ is a solution of the recurrence relation

$$a_n = c_1 a_{n-1} + c_2 a_{n-2} + \cdots + c_k a_{n-k}$$

if and only if

$$\begin{aligned} a_n = {} & (\alpha_{1,0} + \alpha_{1,1} n + \cdots + \alpha_{1,m_1-1} n^{m_1-1}) r_1^n \\ & + (\alpha_{2,0} + \alpha_{2,1} n + \cdots + \alpha_{2,m_2-1} n^{m_2-1}) r_2^n \\ & + \cdots + (\alpha_{t,0} + \alpha_{t,1} n + \cdots + \alpha_{t,m_t-1} n^{m_t-1}) r_t^n \end{aligned}$$

for $n = 0, 1, 2, \ldots$, where $\alpha_{i,j}$ are constants for $1 \leq i \leq t$ and $0 \leq j \leq m_i - 1$.

Example 7 illustrates how Theorem 4 is used to find the general form of a solution of a linear homogeneous recurrence relation when the characteristic equation has several repeated roots.

EXAMPLE 7 Suppose that the roots of the characteristic equation of a linear homogeneous recurrence relation are 2, 2, 2, 5, 5, and 9 (that is, there are three roots, the root 2 with multiplicity three, the root 5 with multiplicity two, and the root 9 with multiplicity one). What is the form of the general solution?

Solution: By Theorem 4, the general form of the solution is

$$(\alpha_{1,0} + \alpha_{1,1} n + \alpha_{1,2} n^2) 2^n + (\alpha_{2,0} + \alpha_{2,1} n) 5^n + \alpha_{3,0} 9^n.$$

◀

We now illustrate the use of Theorem 4 to solve a linear homogeneous recurrence relation with constant coefficients when the characteristic equation has a root of multiplicity three.

EXAMPLE 8 Find the solution to the recurrence relation

$$a_n = -3a_{n-1} - 3a_{n-2} - a_{n-3}$$

with initial conditions $a_0 = 1$, $a_1 = -2$, and $a_2 = -1$.

Solution: The characteristic equation of this recurrence relation is

$$r^3 + 3r^2 + 3r + 1 = 0.$$

Because $r^3 + 3r^2 + 3r + 1 = (r+1)^3$, there is a single root $r = -1$ of multiplicity three of the characteristic equation. By Theorem 4 the solutions of this recurrence relation are of the form

$$a_n = \alpha_{1,0}(-1)^n + \alpha_{1,1} n(-1)^n + \alpha_{1,2} n^2 (-1)^n.$$

To find the constants $\alpha_{1,0}$, $\alpha_{1,1}$, and $\alpha_{1,2}$, use the initial conditions. This gives

$$a_0 = 1 = \alpha_{1,0},$$
$$a_1 = -2 = -\alpha_{1,0} - \alpha_{1,1} - \alpha_{1,2},$$
$$a_2 = -1 = \alpha_{1,0} + 2\alpha_{1,1} + 4\alpha_{1,2}.$$

The simultaneous solution of these three equations is $\alpha_{1,0} = 1$, $\alpha_{1,1} = 3$, and $\alpha_{1,2} = -2$. Hence, the unique solution to this recurrence relation and the given initial conditions is the sequence $\{a_n\}$ with

$$a_n = (1 + 3n - 2n^2)(-1)^n.$$ ◀

8.2.3 Linear Nonhomogeneous Recurrence Relations with Constant Coefficients

We have seen how to solve linear homogeneous recurrence relations with constant coefficients. Is there a relatively simple technique for solving a linear, but not homogeneous, recurrence relation with constant coefficients, such as $a_n = 3a_{n-1} + 2n$? We will see that the answer is yes for certain families of such recurrence relations.

The recurrence relation $a_n = 3a_{n-1} + 2n$ is an example of a **linear nonhomogeneous recurrence relation with constant coefficients**, that is, a recurrence relation of the form

$$a_n = c_1a_{n-1} + c_2a_{n-2} + \cdots + c_ka_{n-k} + F(n),$$

where $c_1, c_2, \ldots, c_k$ are real numbers and $F(n)$ is a function not identically zero depending only on n. The recurrence relation

$$a_n = c_1a_{n-1} + c_2a_{n-2} + \cdots + c_ka_{n-k}$$

is called the **associated homogeneous recurrence relation**. It plays an important role in the solution of the nonhomogeneous recurrence relation.

EXAMPLE 9 Each of the recurrence relations $a_n = a_{n-1} + 2^n$, $a_n = a_{n-1} + a_{n-2} + n^2 + n + 1$, $a_n = 3a_{n-1} + n3^n$, and $a_n = a_{n-1} + a_{n-2} + a_{n-3} + n!$ is a linear nonhomogeneous recurrence relation with constant coefficients. The associated linear homogeneous recurrence relations are $a_n = a_{n-1}$, $a_n = a_{n-1} + a_{n-2}$, $a_n = 3a_{n-1}$, and $a_n = a_{n-1} + a_{n-2} + a_{n-3}$, respectively. ◀

The key fact about linear nonhomogeneous recurrence relations with constant coefficients is that every solution is the sum of a particular solution and a solution of the associated linear homogeneous recurrence relation, as Theorem 5 shows.

THEOREM 5 If $\{a_n^{(p)}\}$ is a particular solution of the nonhomogeneous linear recurrence relation with constant coefficients

$$a_n = c_1a_{n-1} + c_2a_{n-2} + \cdots + c_ka_{n-k} + F(n),$$

then every solution is of the form $\{a_n^{(p)} + a_n^{(h)}\}$, where $\{a_n^{(h)}\}$ is a solution of the associated homogeneous recurrence relation

$$a_n = c_1a_{n-1} + c_2a_{n-2} + \cdots + c_ka_{n-k}.$$

Proof: Because $\{a_n^{(p)}\}$ is a particular solution of the nonhomogeneous recurrence relation, we know that

$$a_n^{(p)} = c_1 a_{n-1}^{(p)} + c_2 a_{n-2}^{(p)} + \cdots + c_k a_{n-k}^{(p)} + F(n).$$

Now suppose that $\{b_n\}$ is a second solution of the nonhomogeneous recurrence relation, so that

$$b_n = c_1 b_{n-1} + c_2 b_{n-2} + \cdots + c_k b_{n-k} + F(n).$$

Subtracting the first of these two equations from the second shows that

$$b_n - a_n^{(p)} = c_1(b_{n-1} - a_{n-1}^{(p)}) + c_2(b_{n-2} - a_{n-2}^{(p)}) + \cdots + c_k(b_{n-k} - a_{n-k}^{(p)}).$$

It follows that $\{b_n - a_n^p\}$ is a solution of the associated homogeneous linear recurrence, say, $\{a_n^{(h)}\}$. Consequently, $b_n = a_n^{(p)} + a_n^{(h)}$ for all n. ◁

By Theorem 5, we see that the key to solving nonhomogeneous recurrence relations with constant coefficients is finding a particular solution. Then every solution is a sum of this solution and a solution of the associated homogeneous recurrence relation. Although there is no general method for finding such a solution that works for every function $F(n)$, there are techniques that work for certain types of functions $F(n)$, such as polynomials and powers of constants. This is illustrated in Examples 10 and 11.

EXAMPLE 10 Find all solutions of the recurrence relation $a_n = 3a_{n-1} + 2n$. What is the solution with $a_1 = 3$?

Solution: To solve this linear nonhomogeneous recurrence relation with constant coefficients, we need to solve its associated linear homogeneous equation and to find a particular solution for the given nonhomogeneous equation. The associated linear homogeneous equation is $a_n = 3a_{n-1}$. Its solutions are $a_n^{(h)} = \alpha 3^n$, where α is a constant.

We now find a particular solution. Because $F(n) = 2n$ is a polynomial in n of degree one, a reasonable trial solution is a linear function in n, say, $p_n = cn + d$, where c and d are constants. To determine whether there are any solutions of this form, suppose that $p_n = cn + d$ is such a solution. Then the equation $a_n = 3a_{n-1} + 2n$ becomes $cn + d = 3(c(n-1) + d) + 2n$. Simplifying and combining like terms gives $(2 + 2c)n + (2d - 3c) = 0$. It follows that $cn + d$ is a solution if and only if $2 + 2c = 0$ and $2d - 3c = 0$. This shows that $cn + d$ is a solution if and only if $c = -1$ and $d = -3/2$. Consequently, $a_n^{(p)} = -n - 3/2$ is a particular solution.

By Theorem 5 all solutions are of the form

$$a_n = a_n^{(p)} + a_n^{(h)} = -n - \frac{3}{2} + \alpha \cdot 3^n,$$

where α is a constant.

To find the solution with $a_1 = 3$, let $n = 1$ in the formula we obtained for the general solution. We find that $3 = -1 - 3/2 + 3\alpha$, which implies that $\alpha = 11/6$. The solution we seek is $a_n = -n - 3/2 + (11/6)3^n$. ◀

EXAMPLE 11 Find all solutions of the recurrence relation

Extra Examples ❯

$$a_n = 5a_{n-1} - 6a_{n-2} + 7^n.$$

Solution: This is a linear nonhomogeneous recurrence relation. The solutions of its associated homogeneous recurrence relation

$$a_n = 5a_{n-1} - 6a_{n-2}$$

are $a_n^{(h)} = \alpha_1 \cdot 3^n + \alpha_2 \cdot 2^n$, where α_1 and α_2 are constants. Because $F(n) = 7^n$, a reasonable trial solution is $a_n^{(p)} = C \cdot 7^n$, where C is a constant. Substituting the terms of this sequence into the recurrence relation implies that $C \cdot 7^n = 5C \cdot 7^{n-1} - 6C \cdot 7^{n-2} + 7^n$. Factoring out 7^{n-2}, this equation becomes $49C = 35C - 6C + 49$, which implies that $20C = 49$, or that $C = 49/20$. Hence, $a_n^{(p)} = (49/20)7^n$ is a particular solution. By Theorem 5, all solutions are of the form

$$a_n = \alpha_1 \cdot 3^n + \alpha_2 \cdot 2^n + (49/20)7^n.$$

◀

In Examples 10 and 11, we made an educated guess that there are solutions of a particular form. In both cases we were able to find particular solutions. This was not an accident. Whenever $F(n)$ is the product of a polynomial in n and the nth power of a constant, we know exactly what form a particular solution has, as stated in Theorem 6. We leave the proof of Theorem 6 as Exercise 52.

THEOREM 6 Suppose that $\{a_n\}$ satisfies the linear nonhomogeneous recurrence relation

$$a_n = c_1 a_{n-1} + c_2 a_{n-2} + \cdots + c_k a_{n-k} + F(n),$$

where $c_1, c_2, \ldots, c_k$ are real numbers, and

$$F(n) = (b_t n^t + b_{t-1} n^{t-1} + \cdots + b_1 n + b_0) s^n,$$

where $b_0, b_1, \ldots, b_t$ and s are real numbers. When s is not a root of the characteristic equation of the associated linear homogeneous recurrence relation, there is a particular solution of the form

$$(p_t n^t + p_{t-1} n^{t-1} + \cdots + p_1 n + p_0) s^n.$$

When s is a root of this characteristic equation and its multiplicity is m, there is a particular solution of the form

$$n^m (p_t n^t + p_{t-1} n^{t-1} + \cdots + p_1 n + p_0) s^n.$$

Note that in the case when s is a root of multiplicity m of the characteristic equation of the associated linear homogeneous recurrence relation, the factor n^m ensures that the proposed particular solution will not already be a solution of the associated linear homogeneous recurrence relation. We next provide Example 12 to illustrate the form of a particular solution provided by Theorem 6.

EXAMPLE 12 What form does a particular solution of the linear nonhomogeneous recurrence relation $a_n = 6a_{n-1} - 9a_{n-2} + F(n)$ have when $F(n) = 3^n$, $F(n) = n3^n$, $F(n) = n^2 2^n$, and $F(n) = (n^2 + 1)3^n$?

Solution: The associated linear homogeneous recurrence relation is $a_n = 6a_{n-1} - 9a_{n-2}$. Its characteristic equation, $r^2 - 6r + 9 = (r - 3)^2 = 0$, has a single root, 3, of multiplicity two. To apply Theorem 6, with $F(n)$ of the form $P(n)s^n$, where $P(n)$ is a polynomial and s is a constant, we need to ask whether s is a root of this characteristic equation.

Because $s = 3$ is a root with multiplicity $m = 2$ but $s = 2$ is not a root, Theorem 6 tells us that a particular solution has the form $p_0 n^2 3^n$ if $F(n) = 3^n$, the form $n^2(p_1 n + p_0)3^n$ if

$F(n) = n3^n$, the form $(p_2n^2 + p_1n + p_0)2^n$ if $F(n) = n^2 2^n$, and the form $n^2(p_2n^2 + p_1n + p_0)3^n$ if $F(n) = (n^2 + 1)3^n$. ◀

Care must be taken when $s = 1$ when solving recurrence relations of the type covered by Theorem 6. In particular, to apply this theorem with $F(n) = b_t n_t + b_{t-1} n_{t-1} + \cdots + b_1 n + b_0$, the parameter s takes the value $s = 1$ (even though the term 1^n does not explicitly appear). By the theorem, the form of the solution then depends on whether 1 is a root of the characteristic equation of the associated linear homogeneous recurrence relation. This is illustrated in Example 13, which shows how Theorem 6 can be used to find a formula for the sum of the first n positive integers.

EXAMPLE 13 Let a_n be the sum of the first n positive integers, so that

$$a_n = \sum_{k=1}^{n} k.$$

Note that a_n satisfies the linear nonhomogeneous recurrence relation

$$a_n = a_{n-1} + n.$$

(To obtain a_n, the sum of the first n positive integers, from a_{n-1}, the sum of the first $n - 1$ positive integers, we add n.) Note that the initial condition is $a_1 = 1$.

The associated linear homogeneous recurrence relation for a_n is

$$a_n = a_{n-1}.$$

The solutions of this homogeneous recurrence relation are given by $a_n^{(h)} = c(1)^n = c$, where c is a constant. To find all solutions of $a_n = a_{n-1} + n$, we need find only a single particular solution. By Theorem 6, because $F(n) = n = n \cdot (1)^n$ and $s = 1$ is a root of degree one of the characteristic equation of the associated linear homogeneous recurrence relation, there is a particular solution of the form $n(p_1n + p_0) = p_1n^2 + p_0n$.

Inserting this into the recurrence relation gives $p_1n^2 + p_0n = p_1(n-1)^2 + p_0(n-1) + n$. Simplifying, we see that $n(2p_1 - 1) + (p_0 - p_1) = 0$, which means that $2p_1 - 1 = 0$ and $p_0 - p_1 = 0$, so $p_0 = p_1 = 1/2$. Hence,

$$a_n^{(p)} = \frac{n^2}{2} + \frac{n}{2} = \frac{n(n+1)}{2}$$

is a particular solution. Hence, all solutions of the original recurrence relation $a_n = a_{n-1} + n$ are given by $a_n = a_n^{(h)} + a_n^{(p)} = c + n(n+1)/2$. Because $a_1 = 1$, we have $1 = a_1 = c + 1 \cdot 2/2 = c + 1$, so $c = 0$. It follows that $a_n = n(n+1)/2$. (This is the same formula given in Table 2 in Section 2.4 and derived previously.) ◀

Exercises

1. Determine which of these are linear homogeneous recurrence relations with constant coefficients. Also, find the degree of those that are.

a) $a_n = 3a_{n-1} + 4a_{n-2} + 5a_{n-3}$
b) $a_n = 2na_{n-1} + a_{n-2}$
c) $a_n = a_{n-1} + a_{n-4}$
d) $a_n = a_{n-1} + 2$
e) $a_n = a_{n-1}^2 + a_{n-2}$
f) $a_n = a_{n-2}$
g) $a_n = a_{n-1} + n$

2. Determine which of these are linear homogeneous recurrence relations with constant coefficients. Also, find the degree of those that are.

a) $a_n = 3a_{n-2}$
b) $a_n = 3$
c) $a_n = a_{n-1}^2$
d) $a_n = a_{n-1} + 2a_{n-3}$
e) $a_n = a_{n-1}/n$
f) $a_n = a_{n-1} + a_{n-2} + n + 3$
g) $a_n = 4a_{n-2} + 5a_{n-4} + 9a_{n-7}$

3. Solve these recurrence relations together with the initial conditions given.

a) $a_n = 2a_{n-1}$ for $n \geq 1$, $a_0 = 3$
b) $a_n = a_{n-1}$ for $n \geq 1$, $a_0 = 2$
c) $a_n = 5a_{n-1} - 6a_{n-2}$ for $n \geq 2$, $a_0 = 1$, $a_1 = 0$
d) $a_n = 4a_{n-1} - 4a_{n-2}$ for $n \geq 2$, $a_0 = 6$, $a_1 = 8$
e) $a_n = -4a_{n-1} - 4a_{n-2}$ for $n \geq 2$, $a_0 = 0$, $a_1 = 1$
f) $a_n = 4a_{n-2}$ for $n \geq 2$, $a_0 = 0$, $a_1 = 4$
g) $a_n = a_{n-2}/4$ for $n \geq 2$, $a_0 = 1$, $a_1 = 0$

4. Solve these recurrence relations together with the initial conditions given.

a) $a_n = a_{n-1} + 6a_{n-2}$ for $n \geq 2$, $a_0 = 3$, $a_1 = 6$
b) $a_n = 7a_{n-1} - 10a_{n-2}$ for $n \geq 2$, $a_0 = 2$, $a_1 = 1$
c) $a_n = 6a_{n-1} - 8a_{n-2}$ for $n \geq 2$, $a_0 = 4$, $a_1 = 10$
d) $a_n = 2a_{n-1} - a_{n-2}$ for $n \geq 2$, $a_0 = 4$, $a_1 = 1$
e) $a_n = a_{n-2}$ for $n \geq 2$, $a_0 = 5$, $a_1 = -1$
f) $a_n = -6a_{n-1} - 9a_{n-2}$ for $n \geq 2$, $a_0 = 3$, $a_1 = -3$
g) $a_{n+2} = -4a_{n+1} + 5a_n$ for $n \geq 0$, $a_0 = 2$, $a_1 = 8$

5. How many different messages can be transmitted in n microseconds using the two signals described in Exercise 19 in Section 8.1?

6. How many different messages can be transmitted in n microseconds using three different signals if one signal requires 1 microsecond for transmittal, the other two signals require 2 microseconds each for transmittal, and a signal in a message is followed immediately by the next signal?

7. In how many ways can a $2 \times n$ rectangular checkerboard be tiled using 1×2 and 2×2 pieces?

8. A model for the number of lobsters caught per year is based on the assumption that the number of lobsters caught in a year is the average of the number caught in the two previous years.

a) Find a recurrence relation for $\{L_n\}$, where L_n is the number of lobsters caught in year n, under the assumption for this model.
b) Find L_n if 100,000 lobsters were caught in year 1 and 300,000 were caught in year 2.

9. A deposit of \$100,000 is made to an investment fund at the beginning of a year. On the last day of each year two dividends are awarded. The first dividend is 20% of the amount in the account during that year. The second dividend is 45% of the amount in the account in the previous year.

a) Find a recurrence relation for $\{P_n\}$, where P_n is the amount in the account at the end of n years if no money is ever withdrawn.
b) How much is in the account after n years if no money has been withdrawn?

***10.** Prove Theorem 2.

11. The **Lucas numbers** satisfy the recurrence relation

Links

$$L_n = L_{n-1} + L_{n-2},$$

and the initial conditions $L_0 = 2$ and $L_1 = 1$.

a) Show that $L_n = f_{n-1} + f_{n+1}$ for $n = 2, 3, \ldots$, where f_n is the nth Fibonacci number.
b) Find an explicit formula for the Lucas numbers.

12. Find the solution to $a_n = 2a_{n-1} + a_{n-2} - 2a_{n-3}$ for $n = 3, 4, 5, \ldots$, with $a_0 = 3$, $a_1 = 6$, and $a_2 = 0$.

13. Find the solution to $a_n = 7a_{n-2} + 6a_{n-3}$ with $a_0 = 9$, $a_1 = 10$, and $a_2 = 32$.

14. Find the solution to $a_n = 5a_{n-2} - 4a_{n-4}$ with $a_0 = 3$, $a_1 = 2$, $a_2 = 6$, and $a_3 = 8$.

15. Find the solution to $a_n = 2a_{n-1} + 5a_{n-2} - 6a_{n-3}$ with $a_0 = 7$, $a_1 = -4$, and $a_2 = 8$.

***16.** Prove Theorem 3.

17. Prove this identity relating the Fibonacci numbers and the binomial coefficients:

$$f_{n+1} = C(n, 0) + C(n-1, 1) + \cdots + C(n-k, k),$$

where n is a positive integer and $k = \lfloor n/2 \rfloor$. [*Hint:* Let $a_n = C(n, 0) + C(n-1, 1) + \cdots + C(n-k, k)$. Show that the sequence $\{a_n\}$ satisfies the same recurrence relation and initial conditions satisfied by the sequence of Fibonacci numbers.]

18. Solve the recurrence relation $a_n = 6a_{n-1} - 12a_{n-2} + 8a_{n-3}$ with $a_0 = -5$, $a_1 = 4$, and $a_2 = 88$.

19. Solve the recurrence relation $a_n = -3a_{n-1} - 3a_{n-2} - a_{n-3}$ with $a_0 = 5$, $a_1 = -9$, and $a_2 = 15$.

20. Find the general form of the solutions of the recurrence relation $a_n = 8a_{n-2} - 16a_{n-4}$.

21. What is the general form of the solutions of a linear homogeneous recurrence relation if its characteristic equation has roots 1, 1, 1, 1, −2, −2, −2, 3, 3, −4?

22. What is the general form of the solutions of a linear homogeneous recurrence relation if its characteristic equation has the roots −1, −1, −1, 2, 2, 5, 5, 7?

23. Consider the nonhomogeneous linear recurrence relation $a_n = 3a_{n-1} + 2^n$.

a) Show that $a_n = -2^{n+1}$ is a solution of this recurrence relation.
b) Use Theorem 5 to find all solutions of this recurrence relation.
c) Find the solution with $a_0 = 1$.

24. Consider the nonhomogeneous linear recurrence relation $a_n = 2a_{n-1} + 2^n$.

a) Show that $a_n = n2^n$ is a solution of this recurrence relation.
b) Use Theorem 5 to find all solutions of this recurrence relation.
c) Find the solution with $a_0 = 2$.

25. a) Determine values of the constants A and B such that $a_n = An + B$ is a solution of recurrence relation $a_n = 2a_{n-1} + n + 5$.
b) Use Theorem 5 to find all solutions of this recurrence relation.
c) Find the solution of this recurrence relation with $a_0 = 4$.

26. What is the general form of the particular solution guaranteed to exist by Theorem 6 of the linear nonhomogeneous recurrence relation $a_n = 6a_{n-1} - 12a_{n-2} + 8a_{n-3} + F(n)$ if

a) $F(n) = n^2$?
b) $F(n) = 2^n$?
c) $F(n) = n2^n$?
d) $F(n) = (-2)^n$?
e) $F(n) = n^2 2^n$?
f) $F(n) = n^3(-2)^n$?
g) $F(n) = 3$?

27. What is the general form of the particular solution guaranteed to exist by Theorem 6 of the linear nonhomogeneous recurrence relation $a_n = 8a_{n-2} - 16a_{n-4} + F(n)$ if

a) $F(n) = n^3$?
b) $F(n) = (-2)^n$?
c) $F(n) = n2^n$?
d) $F(n) = n^2 4^n$?
e) $F(n) = (n^2 - 2)(-2)^n$?
f) $F(n) = n^4 2^n$?
g) $F(n) = 2$?

28. a) Find all solutions of the recurrence relation $a_n = 2a_{n-1} + 2n^2$.
b) Find the solution of the recurrence relation in part (a) with initial condition $a_1 = 4$.

29. a) Find all solutions of the recurrence relation $a_n = 2a_{n-1} + 3^n$.
b) Find the solution of the recurrence relation in part (a) with initial condition $a_1 = 5$.

30. a) Find all solutions of the recurrence relation $a_n = -5a_{n-1} - 6a_{n-2} + 42 \cdot 4^n$.
b) Find the solution of this recurrence relation with $a_1 = 56$ and $a_2 = 278$.

31. Find all solutions of the recurrence relation $a_n = 5a_{n-1} - 6a_{n-2} + 2^n + 3n$. [*Hint:* Look for a particular solution of the form $qn2^n + p_1 n + p_2$, where q, p_1, and p_2 are constants.]

32. Find the solution of the recurrence relation $a_n = 2a_{n-1} + 3 \cdot 2^n$.

33. Find all solutions of the recurrence relation $a_n = 4a_{n-1} - 4a_{n-2} + (n+1)2^n$.

34. Find all solutions of the recurrence relation $a_n = 7a_{n-1} - 16a_{n-2} + 12a_{n-3} + n4^n$ with $a_0 = -2$, $a_1 = 0$, and $a_2 = 5$.

35. Find the solution of the recurrence relation $a_n = 4a_{n-1} - 3a_{n-2} + 2^n + n + 3$ with $a_0 = 1$ and $a_1 = 4$.

36. Let a_n be the sum of the first n perfect squares, that is, $a_n = \sum_{k=1}^{n} k^2$. Show that the sequence $\{a_n\}$ satisfies the linear nonhomogeneous recurrence relation $a_n = a_{n-1} + n^2$ and the initial condition $a_1 = 1$. Use Theorem 6 to determine a formula for a_n by solving this recurrence relation.

37. Let a_n be the sum of the first n triangular numbers, that is, $a_n = \sum_{k=1}^{n} t_k$, where $t_k = k(k+1)/2$. Show that $\{a_n\}$ satisfies the linear nonhomogeneous recurrence relation $a_n = a_{n-1} + n(n+1)/2$ and the initial condition $a_1 = 1$. Use Theorem 6 to determine a formula for a_n by solving this recurrence relation.

38. a) Find the characteristic roots of the linear homogeneous recurrence relation $a_n = 2a_{n-1} - 2a_{n-2}$. [*Note:* These are complex numbers.]
b) Find the solution of the recurrence relation in part (a) with $a_0 = 1$ and $a_1 = 2$.

***39.** a) Find the characteristic roots of the linear homogeneous recurrence relation $a_n = a_{n-4}$. [*Note:* These include complex numbers.]
b) Find the solution of the recurrence relation in part (a) with $a_0 = 1$, $a_1 = 0$, $a_2 = -1$, and $a_3 = 1$.

***40.** Solve the simultaneous recurrence relations

$$a_n = 3a_{n-1} + 2b_{n-1}$$
$$b_n = a_{n-1} + 2b_{n-1}$$

with $a_0 = 1$ and $b_0 = 2$.

***41.** a) Use the formula found in Example 4 for f_n, the nth Fibonacci number, to show that f_n is the integer closest to

$$\frac{1}{\sqrt{5}}\left(\frac{1+\sqrt{5}}{2}\right)^n.$$

b) Determine for which n f_n is greater than

$$\frac{1}{\sqrt{5}}\left(\frac{1+\sqrt{5}}{2}\right)^n$$

and for which n f_n is less than

$$\frac{1}{\sqrt{5}}\left(\frac{1+\sqrt{5}}{2}\right)^n.$$

42. Show that if $a_n = a_{n-1} + a_{n-2}$, $a_0 = s$ and $a_1 = t$, where s and t are constants, then $a_n = sf_{n-1} + tf_n$ for all positive integers n.

43. Express the solution of the linear nonhomogenous recurrence relation $a_n = a_{n-1} + a_{n-2} + 1$ for $n \geq 2$ where $a_0 = 0$ and $a_1 = 1$ in terms of the Fibonacci numbers. [*Hint:* Let $b_n = a_n + 1$ and apply Exercise 42 to the sequence b_n.]

***44.** (*Linear algebra required*) Let $\mathbf{A}_n$ be the $n \times n$ matrix with 2s on its main diagonal, 1s in all positions next to a diagonal element, and 0s everywhere else. Find a recurrence relation for d_n, the determinant of $\mathbf{A}_n$. Solve this recurrence relation to find a formula for d_n.

45. Suppose that each pair of a genetically engineered species of rabbits left on an island produces two new pairs of rabbits at the age of 1 month and six new pairs of rabbits at the age of 2 months and every month afterward. None of the rabbits ever die or leave the island.
a) Find a recurrence relation for the number of pairs of rabbits on the island n months after one newborn pair is left on the island.
b) By solving the recurrence relation in (a) determine the number of pairs of rabbits on the island n months after one pair is left on the island.

46. Suppose that there are two goats on an island initially. The number of goats on the island doubles every year by natural reproduction, and some goats are either added or removed each year.

a) Construct a recurrence relation for the number of goats on the island at the start of the nth year, assuming that during each year an extra 100 goats are put on the island.

b) Solve the recurrence relation from part (a) to find the number of goats on the island at the start of the nth year.

c) Construct a recurrence relation for the number of goats on the island at the start of the nth year, assuming that n goats are removed during the nth year for each $n \geq 3$.

d) Solve the recurrence relation in part (c) for the number of goats on the island at the start of the nth year.

47. A new employee at an exciting new software company starts with a salary of \$50,000 and is promised that at the end of each year her salary will be double her salary of the previous year, with an extra increment of \$10,000 for each year she has been with the company.

a) Construct a recurrence relation for her salary for her nth year of employment.

b) Solve this recurrence relation to find her salary for her nth year of employment.

Some linear recurrence relations that do not have constant coefficients can be systematically solved. This is the case for recurrence relations of the form $f(n)a_n = g(n)a_{n-1} + h(n)$. Exercises 48–50 illustrate this.

***48. a)** Show that the recurrence relation

$$f(n)a_n = g(n)a_{n-1} + h(n),$$

for $n \geq 1$, and with $a_0 = C$, can be reduced to a recurrence relation of the form

$$b_n = b_{n-1} + Q(n)h(n),$$

where $b_n = g(n+1)Q(n+1)a_n$, with

$$Q(n) = (f(1)f(2)\cdots f(n-1))/(g(1)g(2)\cdots g(n)).$$

b) Use part (a) to solve the original recurrence relation to obtain

$$a_n = \frac{C + \sum_{i=1}^{n} Q(i)h(i)}{g(n+1)Q(n+1)}.$$

***49.** Use Exercise 48 to solve the recurrence relation $(n+1)a_n = (n+3)a_{n-1} + n$, for $n \geq 1$, with $a_0 = 1$.

50. It can be shown that C_n, the average number of comparisons made by the quicksort algorithm (described in preamble to Exercise 50 in Section 5.4), when sorting n elements in random order, satisfies the recurrence relation

$$C_n = n + 1 + \frac{2}{n}\sum_{k=0}^{n-1} C_k$$

for $n = 1, 2, \ldots,$ with initial condition $C_0 = 0$.

a) Show that $\{C_n\}$ also satisfies the recurrence relation $nC_n = (n+1)C_{n-1} + 2n$ for $n = 1, 2, \ldots.$

b) Use Exercise 48 to solve the recurrence relation in part (a) to find an explicit formula for C_n.

****51.** Prove Theorem 4.

****52.** Prove Theorem 6.

53. Solve the recurrence relation $T(n) = nT^2(n/2)$ with initial condition $T(1) = 6$ when $n = 2^k$ for some integer k. [*Hint:* Let $n = 2^k$ and then make the substitution $a_k = \log T(2^k)$ to obtain a linear nonhomogeneous recurrence relation.]

8.3 Divide-and-Conquer Algorithms and Recurrence Relations

8.3.1 Introduction

Links

Many recursive algorithms take a problem with a given input and divide it into one or more smaller problems. This reduction is successively applied until the solutions of the smaller problems can be found quickly. For instance, we perform a binary search by reducing the search for an element in a list to the search for this element in a list half as long. We successively apply this reduction until one element is left. When we sort a list of integers using the merge sort, we split the list into two halves of equal size and sort each half separately. We then merge the two sorted halves. Another example of this type of recursive algorithm is a procedure for multiplying integers that reduces the problem of the multiplication of two integers to three multiplications of pairs of integers with half as many bits. This reduction is successively applied until integers with one bit are obtained. There procedures follow an important algorithmic paradigm known as **divide-and-conquer**, and are called **divide-and-conquer algorithms**, because they *divide* a problem into one or more instances of the same problem of smaller size and they *conquer* the problem by using the solutions of the smaller problems to find a solution of the original problem, perhaps with some additional work.

"Divide et impera" (translation: *"Divide and conquer"*) —Julius Caesar

In this section we will show how recurrence relations can be used to analyze the computational complexity of divide-and-conquer algorithms. We will use these recurrence relations to estimate the number of operations used by many different divide-and-conquer algorithms, including several that we introduce in this section.

8.3.2 Divide-and-Conquer Recurrence Relations

Suppose that a recursive algorithm divides a problem of size n into a subproblems, where each subproblem is of size n/b (for simplicity, assume that n is a multiple of b; in reality, the smaller problems are often of size equal to the nearest integers either less than or equal to, or greater than or equal to, n/b). Also, suppose that a total of $g(n)$ extra operations are required in the conquer step of the algorithm to combine the solutions of the subproblems into a solution of the original problem. Then, if $f(n)$ represents the number of operations required to solve the problem of size n, it follows that f satisfies the recurrence relation

$$f(n) = af(n/b) + g(n).$$

This is called a **divide-and-conquer recurrence relation**.

We will first set up the divide-and-conquer recurrence relations that can be used to study the complexity of some important algorithms. Then we will show how to use these divide-and-conquer recurrence relations to estimate the complexity of these algorithms.

EXAMPLE 1 **Binary Search** We introduced a binary search algorithm in Section 3.1. This binary search algorithm reduces the search for an element in a search sequence of size n to the binary search for this element in a search sequence of size $n/2$, when n is even. (Hence, the problem of size n has been reduced to *one* problem of size $n/2$.) Two comparisons are needed to implement this reduction (one to determine which half of the list to use and the other to determine whether any terms of the list remain). Hence, if $f(n)$ is the number of comparisons required to search for an element in a search sequence of size n, then

Extra Examples

$$f(n) = f(n/2) + 2$$

when n is even. ◀

EXAMPLE 2 **Finding the Maximum and Minimum of a Sequence** Consider the following algorithm for locating the maximum and minimum elements of a sequence $a_1, a_2, \ldots, a_n$. If $n = 1$, then a_1 is the maximum and the minimum. If $n > 1$, split the sequence into two sequences, either where both have the same number of elements or where one of the sequences has one more element than the other. The problem is reduced to finding the maximum and minimum of each of the two smaller sequences. The solution to the original problem results from the comparison of the separate maxima and minima of the two smaller sequences to obtain the overall maximum and minimum.

Let $f(n)$ be the total number of comparisons needed to find the maximum and minimum elements of the sequence with n elements. We have shown that a problem of size n can be reduced into two problems of size $n/2$, when n is even, using two comparisons, one to compare the maxima of the two sequences and the other to compare the minima of the two sequences. This gives the recurrence relation

$$f(n) = 2f(n/2) + 2$$

when n is even. ◀

EXAMPLE 3 **Merge Sort** The merge sort algorithm (introduced in Section 5.4) splits a list to be sorted with n items, where n is even, into two lists with $n/2$ elements each, and uses fewer than n comparisons to merge the two sorted lists of $n/2$ items each into one sorted list. Consequently, the number of comparisons used by the merge sort to sort a list of n elements is less than $M(n)$, where the function $M(n)$ satisfies the divide-and-conquer recurrence relation

$$M(n) = 2M(n/2) + n.$$ ◀

EXAMPLE 4 **Fast Multiplication of Integers** Surprisingly, there are more efficient algorithms than the conventional algorithm (described in Section 4.2) for multiplying integers. One of these algorithms, which uses a divide-and-conquer technique, will be described here. This fast multiplication algorithm proceeds by splitting each of two $2n$-bit integers into two blocks, each with n bits. Then, the original multiplication is reduced from the multiplication of two $2n$-bit integers to three multiplications of n-bit integers, plus shifts and additions.

Links

Suppose that a and b are integers with binary expansions of length $2n$ (add initial bits of zero in these expansions if necessary to make them the same length). Let

$$a = (a_{2n-1}a_{2n-2} \cdots a_1a_0)_2 \quad \text{and} \quad b = (b_{2n-1}b_{2n-2} \cdots b_1b_0)_2.$$

Let

$$a = 2^nA_1 + A_0, \quad b = 2^nB_1 + B_0,$$

where

$$A_1 = (a_{2n-1} \cdots a_{n+1}a_n)_2, \qquad A_0 = (a_{n-1} \cdots a_1a_0)_2,$$
$$B_1 = (b_{2n-1} \cdots b_{n+1}b_n)_2, \qquad B_0 = (b_{n-1} \cdots b_1b_0)_2.$$

The algorithm for fast multiplication of integers is based on the fact that ab can be rewritten as

$$ab = (2^{2n} + 2^n)A_1B_1 + 2^n(A_1 - A_0)(B_0 - B_1) + (2^n + 1)A_0B_0.$$

The important fact about this identity is that it shows that the multiplication of two $2n$-bit integers can be carried out using three multiplications of n-bit integers, together with additions, subtractions, and shifts. This shows that if $f(n)$ is the total number of bit operations needed to multiply two n-bit integers, then

$$f(2n) = 3f(n) + Cn.$$

The reasoning behind this equation is as follows. The three multiplications of n-bit integers are carried out using $3f(n)$-bit operations. Each of the additions, subtractions, and shifts uses a constant multiple of n-bit operations, and Cn represents the total number of bit operations used by these operations. ◀

EXAMPLE 5 **Fast Matrix Multiplication** In Example 7 of Section 3.3 we showed that multiplying two $n \times n$ matrices using the definition of matrix multiplication required n^3 multiplications and n^3 additions. Consequently, computing the product of two $n \times n$ matrices in this way requires $O(n^3)$ operations (multiplications and additions). Surprisingly, there are more efficient divide-and-conquer algorithms for multiplying two $n \times n$ matrices. Such an algorithm, invented by Volker Strassen in 1969, reduces the multiplication of two $n \times n$ matrices, when n is even, to seven multiplications of two $(n/2) \times (n/2)$ matrices and 15 additions of $(n/2) \times (n/2)$ matrices. (See [CoLeRiSt09] for the details of this algorithm.) Hence, if $f(n)$ is the number of operations (multiplications and additions) used, it follows that

Links

$$f(n) = 7f(n/2) + 15n^2/4$$

when n is even. ◀

As Examples 1–5 show, recurrence relations of the form $f(n) = af(n/b) + g(n)$ arise in many different situations. It is possible to derive estimates of the size of functions that satisfy such recurrence relations. Suppose that f satisfies this recurrence relation whenever n is divisible by b. Let $n = b^k$, where k is a positive integer. Then

$$\begin{aligned} f(n) &= af(n/b) + g(n) \\ &= a^2f(n/b^2) + ag(n/b) + g(n) \\ &= a^3f(n/b^3) + a^2g(n/b^2) + ag(n/b) + g(n) \\ &\vdots \\ &= a^kf(n/b^k) + \sum_{j=0}^{k-1} a^j g(n/b^j). \end{aligned}$$

Because $n/b^k = 1$, it follows that

$$f(n) = a^k f(1) + \sum_{j=0}^{k-1} a^j g(n/b^j).$$

We can use this equation for $f(n)$ to estimate the size of functions that satisfy divide-and-conquer relations.

THEOREM 1 Let f be an increasing function that satisfies the recurrence relation

$$f(n) = af(n/b) + c$$

whenever n is divisible by b, where $a \geq 1$, b is an integer greater than 1, and c is a positive real number. Then

$$f(n) \text{ is } \begin{cases} O(n^{\log_b a}) & \text{if } a > 1, \\ O(\log n) & \text{if } a = 1. \end{cases}$$

Furthermore, when $n = b^k$ and $a \neq 1$, where k is a positive integer,

$$f(n) = C_1 n^{\log_b a} + C_2,$$

where $C_1 = f(1) + c/(a-1)$ and $C_2 = -c/(a-1)$.

Proof: First let $n = b^k$. From the expression for $f(n)$ obtained in the discussion preceding the theorem, with $g(n) = c$, we have

$$f(n) = a^k f(1) + \sum_{j=0}^{k-1} a^j c = a^k f(1) + c \sum_{j=0}^{k-1} a^j.$$

When $a = 1$ we have

$$f(n) = f(1) + ck.$$

Because $n = b^k$, we have $k = \log_b n$. Hence,

$$f(n) = f(1) + c \log_b n.$$

When n is not a power of b, we have $b^k < n < b^{k+1}$, for a positive integer k. Because f is increasing, it follows that $f(n) \le f(b^{k+1}) = f(1) + c(k+1) = (f(1)+c) + ck \le (f(1)+c) + c\log_b n$. Therefore, in both cases, $f(n)$ is $O(\log n)$ when $a = 1$.

Now suppose that $a > 1$. First assume that $n = b^k$, where k is a positive integer. From the formula for the sum of terms of a geometric progression (Theorem 1 in Section 2.4), it follows that

$$\begin{aligned} f(n) &= a^k f(1) + c(a^k - 1)/(a-1) \\ &= a^k[f(1) + c/(a-1)] - c/(a-1) \\ &= C_1 n^{\log_b a} + C_2, \end{aligned}$$

because $a^k = a^{\log_b n} = n^{\log_b a}$ (see Exercise 4 in Appendix 2), where $C_1 = f(1) + c/(a-1)$ and $C_2 = -c/(a-1)$.

Now suppose that n is not a power of b. Then $b^k < n < b^{k+1}$, where k is a nonnegative integer. Because f is increasing,

$$\begin{aligned} f(n) &\le f(b^{k+1}) = C_1 a^{k+1} + C_2 \\ &\le (C_1 a)a^{\log_b n} + C_2 \\ &= (C_1 a)n^{\log_b a} + C_2, \end{aligned}$$

because $k \le \log_b n < k+1$.

Hence, we have $f(n)$ is $O(n^{\log_b a})$. ◁

Examples 6–8 illustrate how Theorem 1 is used.

EXAMPLE 6 Let $f(n) = 5f(n/2) + 3$ and $f(1) = 7$. Find $f(2^k)$, where k is a positive integer. Also, estimate $f(n)$ if f is an increasing function.

Extra Examples

Solution: From the proof of Theorem 1, with $a = 5$, $b = 2$, and $c = 3$, we see that if $n = 2^k$, then

$$\begin{aligned} f(n) &= a^k[f(1) + c/(a-1)] + [-c/(a-1)] \\ &= 5^k[7 + (3/4)] - 3/4 \\ &= 5^k(31/4) - 3/4. \end{aligned}$$

Also, if $f(n)$ is increasing, Theorem 1 shows that $f(n)$ is $O(n^{\log_b a}) = O(n^{\log 5})$. ◀

We can use Theorem 1 to estimate the computational complexity of the binary search algorithm and the algorithm given in Example 2 for locating the minimum and maximum of a sequence.

EXAMPLE 7 Give a big-O estimate for the number of comparisons used by a binary search.

Solution: In Example 1 it was shown that $f(n) = f(n/2) + 2$ when n is even, where f is the number of comparisons required to perform a binary search on a sequence of size n. Hence, from Theorem 1, it follows that $f(n)$ is $O(\log n)$. ◀

EXAMPLE 8 Give a big-O estimate for the number of comparisons used to locate the maximum and minimum elements in a sequence using the algorithm given in Example 2.

Solution: In Example 2 we showed that $f(n) = 2f(n/2) + 2$, when n is even, where f is the number of comparisons needed by this algorithm. Hence, from Theorem 1, it follows that $f(n)$ is $O(n^{\log 2}) = O(n)$. ◀

We now state a more general, and more complicated, theorem, which has Theorem 1 as a special case. This theorem (or more powerful versions, including big-Theta estimates) is sometimes known as the master theorem because it is useful in analyzing the complexity of many important divide-and-conquer algorithms.

THEOREM 2 **MASTER THEOREM** Let f be an increasing function that satisfies the recurrence relation

$$f(n) = af(n/b) + cn^d$$

whenever $n = b^k$, where k is a positive integer, $a \geq 1$, b is an integer greater than 1, and c and d are real numbers with c positive and d nonnegative. Then

$$f(n) \text{ is } \begin{cases} O(n^d) & \text{if } a < b^d, \\ O(n^d \log n) & \text{if } a = b^d, \\ O(n^{\log_b a}) & \text{if } a > b^d. \end{cases}$$

The proof of Theorem 2 is left for the reader as Exercises 29–33.

EXAMPLE 9 **Complexity of Merge Sort** In Example 3 we explained that the number of comparisons used by the merge sort to sort a list of n elements is less than $M(n)$, where $M(n) = 2M(n/2) + n$. By the master theorem (Theorem 2) we find that $M(n)$ is $O(n \log n)$, which agrees with the estimate found in Section 5.4. ◀

EXAMPLE 10 Give a big-O estimate for the number of bit operations needed to multiply two n-bit integers using the fast multiplication algorithm described in Example 4.

Solution: Example 4 shows that $f(n) = 3f(n/2) + Cn$, when n is even, where $f(n)$ is the number of bit operations required to multiply two n-bit integers using the fast multiplication algorithm. Hence, from the master theorem (Theorem 2), it follows that $f(n)$ is $O(n^{\log 3})$. Note that log 3 ~ 1.6. Because the conventional algorithm for multiplication uses $O(n^2)$ bit operations, the fast multiplication algorithm is a substantial improvement over the conventional algorithm in terms of time complexity for sufficiently large integers, including large integers that occur in practical applications. ◀

EXAMPLE 11 Give a big-O estimate for the number of multiplications and additions required to multiply two $n \times n$ matrices using the matrix multiplication algorithm referred to in Example 5.

Solution: Let $f(n)$ denote the number of additions and multiplications used by the algorithm mentioned in Example 5 to multiply two $n \times n$ matrices. We have $f(n) = 7f(n/2) + 15n^2/4$, when n is even. Hence, from the master theorem (Theorem 2), it follows that $f(n)$ is $O(n^{\log 7})$.

Note that $\log 7 \sim 2.8$. Because the conventional algorithm for multiplying two $n \times n$ matrices uses $O(n^3)$ additions and multiplications, it follows that for sufficiently large integers n, including those that occur in many practical applications, this algorithm is substantially more efficient in time complexity than the conventional algorithm. ◀

THE CLOSEST-PAIR PROBLEM We conclude this section by introducing a divide-and-conquer algorithm from computational geometry, the part of discrete mathematics devoted to algorithms that solve geometric problems.

EXAMPLE 12 **The Closest-Pair Problem** Consider the problem of determining the closest pair of points in a set of n points $(x_1, y_1), \ldots, (x_n, y_n)$ in the plane, where the distance between two points (x_i, y_i) and (x_j, y_j) is the usual Euclidean distance $\sqrt{(x_i - x_j)^2 + (y_i - y_j)^2}$. This problem arises in many applications such as determining the closest pair of airplanes in the air space at a particular altitude being managed by an air traffic controller. How can this closest pair of points be found in an efficient way?

Links

Solution: To solve this problem we can first determine the distance between every pair of points and then find the smallest of these distances. However, this approach requires $O(n^2)$ computations of distances and comparisons because there are $C(n, 2) = n(n-1)/2$ pairs of points. Surprisingly, there is an elegant divide-and-conquer algorithm that can solve the closest-pair problem for n points using $O(n \log n)$ computations of distances and comparisons. The algorithm we describe here is due to Michael Shamos (see [PrSh85]).

It took researchers more than 10 years to find an algorithm with $O(n \log n)$ complexity that locates the closest pair of points among n points.

For simplicity, we assume that $n = 2^k$, where k is a positive integer. (We avoid some technical considerations that are needed when n is not a power of 2.) When $n = 2$, we have only one pair of points; the distance between these two points is the minimum distance. At the start of the algorithm we use the merge sort twice, once to sort the points in order of increasing x

Links

Michael I. Shamos, Distinguished Career Professor, Language Technologies Institute, and Software and Societal Systems Department, School of Computer Science, Carnegie Mellon University.

MICHAEL SHAMOS (BORN 1947) was born in New York City where his father, an experimental atomic physicist, was chairman of the NYU Physics Department for many years. He attended Princeton, graduating with a bachelor's in physics in 1968. He then worked as a programmer for IBM and attended Vassar College, earning a master's in physics in 1970. From 1970 to 1972, he worked as a Supervisory Programmer at the National Cancer Institute, simultaneously earning an M.S. in technology of management from American University.

In 1972, Shamos began his computer science studies at Yale. He earned an M.S. in 1973 and a M.Phil. in 1974. He developed fundamental algorithms in computational geometry with Dan Hoey for his thesis. In 1975, he wrote the first book on computational geometry with Franco Preparata, *Computational Geometry: An Introduction*. In 1975, he was hired as an assistant professor in computer science and mathematics by Carnegie Mellon. He was awarded his Ph.D. by Yale in 1978 and joined the Statistics department of CMU that year.

In 1978, he founded the company Unilogic to commercialize a document production system developed by a graduate student at CMU. He earned a law degree at Duquesne University and asked CMU for adjunct faculty status at CMU while he worked at Unilogic. After Unilogic was sold in 1987, Shamos spent three years studying the history of billiards. He published four books on billiards and became a contributing editor of *Billiards Digest* magazine.

In 1990, Shamos joined an intellectual property law firm in Pittsburg, where he spent eight years. He then returned to CMU where he was a Director of the Universal Library, a large-scale book scanning project. He also taught courses on algorithm analysis, intellectual property law, the technology of electronic commerce, and internet law. In 2001, Shamos was given the title Distinguished Career Professor by CMU. He developed several new degree programs. He has been a visiting professor at the University of Hong Kong where he has taught courses on electronic payment systems for many years.

Shamos is also interested in electronic voting and inspected e-voting systems in six US states. He has testified before the US Congress and state legislatures on this topic. He also has taught a course at CMU on e-voting.

He continues to play billiards and still writes about the history of billiards. Shamos has also studied bagpipe performance and has put together a large collection of bagpipe music. He met his wife while they were both at Vassar College. They have two children and four grandchildren.

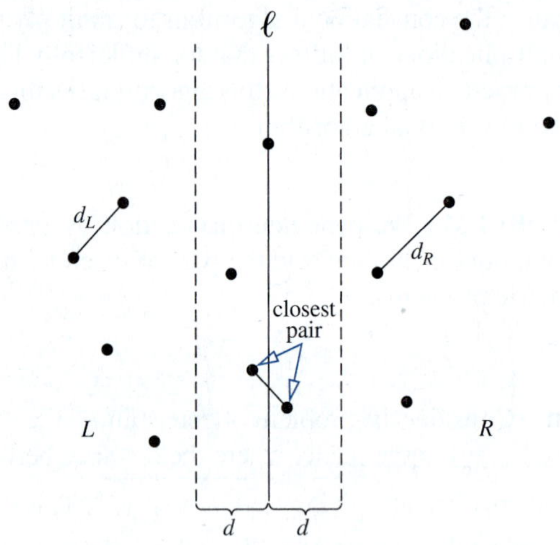

In this illustration the problem of finding the closest pair in a set of 16 points is reduced to two problems of finding the closest pair in a set of eight points *and* the problem of determining whether there are points closer than $d = \min(d_L, d_R)$ within the strip of width $2d$ centered at ℓ.

FIGURE 1 The recursive step of the algorithm for solving the closest-pair problem.

coordinates, and once to sort the points in order of increasing y coordinates. Each of these sorts requires $O(n \log n)$ operations. We will use these sorted lists in each recursive step.

The recursive part of the algorithm divides the problem into two subproblems, each involving half as many points. Using the sorted list of the points by their x coordinates, we construct a vertical line ℓ dividing the n points into two parts, a left part and a right part of equal size, each containing $n/2$ points, as shown in Figure 1. (If any points fall on the dividing line ℓ, we divide them among the two parts if necessary.) At subsequent steps of the recursion we need not sort on x coordinates again, because we can select the corresponding sorted subset of all the points. This selection is a task that can be done with $O(n)$ comparisons.

There are three possibilities concerning the positions of the closest points: (1) they are both in the left region L, (2) they are both in the right region R, or (3) one point is in the left region and the other is in the right region. Apply the algorithm recursively to compute d_L and d_R, where d_L is the minimum distance between points in the left region and d_R is the minimum distance between points in the right region. Let $d = \min(d_L, d_R)$. To successfully divide the problem of finding the closest two points in the original set into the two problems of finding the shortest distances between points in the two regions separately, we have to handle the conquer part of the algorithm, which requires that we consider the case where the closest points lie in different regions, that is, one point is in L and the other in R. Because there is a pair of points at distance d where both points lie in R or both points lie in L, for the closest points to lie in different regions requires that they must be a distance less than d apart.

For a point in the left region and a point in the right region to lie at a distance less than d apart, these points must lie in the vertical strip of width $2d$ that has the line ℓ as its center. (Otherwise, the distance between these points is greater than the difference in their x coordinates, which exceeds d.) To examine the points within this strip, we sort the points so that they are listed in order of increasing y coordinates, using the sorted list of the points by their y coordinates. At each recursive step, we form a subset of the points in the region sorted by their y coordinates from the already sorted set of all points sorted by their y coordinates, which can be done with $O(n)$ comparisons.

Beginning with a point in the strip with the smallest y coordinate, we successively examine each point in the strip, computing the distance between this point and all other points in the strip that have larger y coordinates that could lie at a distance less than d from this point. Note that to examine a point p, we need only consider the distances between p and points in the set that lie within the rectangle of height d and width $2d$ with p on its base and with vertical sides at distance d from ℓ.

We can show that there are at most eight points from the set, including p, in or on this $2d \times d$ rectangle. To see this, note that there can be at most one point in each of the eight $d/2 \times d/2$ squares

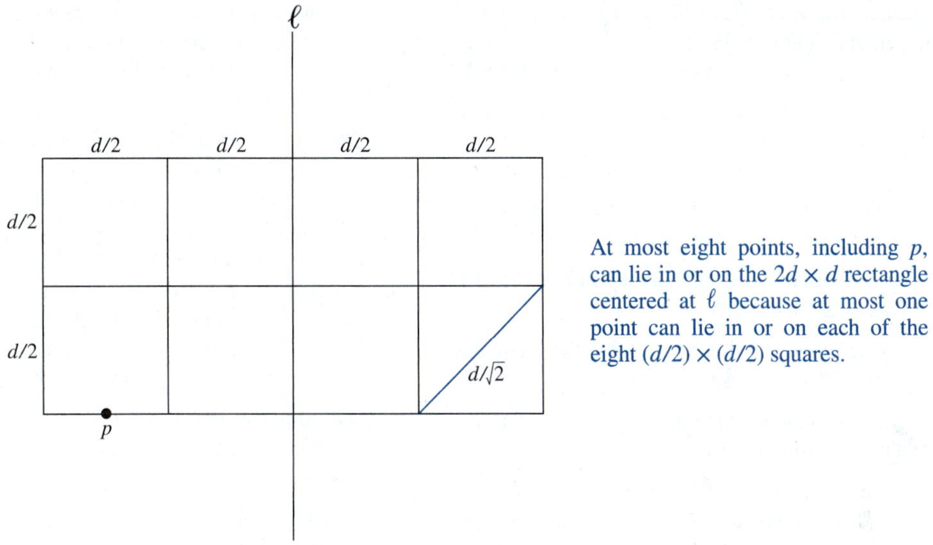

FIGURE 2 Showing that there are at most seven other points to consider for each point in the strip.

shown in Figure 2. This follows because the farthest apart points can be on or within one of these squares is the diagonal length $d/\sqrt{2}$ (which can be found using the Pythagorean theorem), which is less than d, and each of these $d/2 \times d/2$ squares lies entirely within the left region or the right region. This means that at this stage we need only compare at most seven distances, the distances between p and the seven or fewer other points in or on the rectangle, with d.

Because the total number of points in the strip of width $2d$ does not exceed n (the total number of points in the set), at most $7n$ distances need to be compared with d to find the minimum distance between points. That is, there are only $7n$ possible distances that could be less than d. Consequently, once the merge sort has been used to sort the pairs according to their x coordinates and according to their y coordinates, we find that the increasing function $f(n)$ satisfying the recurrence relation

$$f(n) = 2f(n/2) + 7n,$$

where $f(2) = 1$, exceeds the number of comparisons needed to solve the closest-pair problem for n points. By the master theorem (Theorem 2), it follows that $f(n)$ is $O(n \log n)$. The two sorts of points by their x coordinates and by their y coordinates each can be done using $O(n \log n)$ comparisons, by using the merge sort, and the sorted subsets of these coordinates at each of the $O(\log n)$ steps of the algorithm can be done using $O(n)$ comparisons each. Thus, we find that the closest-pair problem can be solved using $O(n \log n)$ comparisons. ◀

Exercises

1. How many comparisons are needed for a binary search in a set of 64 elements?

2. How many comparisons are needed to locate the maximum and minimum elements in a sequence with 128 elements using the algorithm in Example 2?

3. Multiply $(1110)_2$ and $(1010)_2$ using the fast multiplication algorithm.

4. Express the fast multiplication algorithm in pseudocode.

5. Determine a value for the constant C in Example 4 and use it to estimate the number of bit operations needed to multiply two 64-bit integers using the fast multiplication algorithm.

6. How many operations are needed to multiply two 32×32 matrices using the algorithm referred to in Example 5?

7. Suppose that $f(n) = f(n/3) + 1$ when n is a positive integer divisible by 3, and $f(1) = 1$. Find
   a) $f(3)$. b) $f(27)$. c) $f(729)$.

8. Suppose that $f(n) = 2f(n/2) + 3$ when n is an even positive integer, and $f(1) = 5$. Find

a) $f(2)$. **b)** $f(8)$. **c)** $f(64)$. **d)** $f(1024)$.

9. Suppose that $f(n) = f(n/5) + 3n^2$ when n is a positive integer divisible by 5, and $f(1) = 4$. Find

a) $f(5)$. **b)** $f(125)$. **c)** $f(3125)$.

10. Find $f(n)$ when $n = 2^k$, where f satisfies the recurrence relation $f(n) = f(n/2) + 1$ with $f(1) = 1$.

11. Give a big-O estimate for the function f in Exercise 10 if f is an increasing function.

12. Find $f(n)$ when $n = 3^k$, where f satisfies the recurrence relation $f(n) = 2f(n/3) + 4$ with $f(1) = 1$.

13. Give a big-O estimate for the function f in Exercise 12 if f is an increasing function.

14. Suppose that there are $n = 2^k$ teams in an elimination tournament, where there are $n/2$ games in the first round, with the $n/2 = 2^{k-1}$ winners playing in the second round, and so on. Develop a recurrence relation for the number of rounds in the tournament.

15. How many rounds are in the elimination tournament described in Exercise 14 when there are 32 teams?

16. Solve the recurrence relation for the number of rounds in the tournament described in Exercise 14.

17. Suppose that the votes of n people for different candidates (where there can be more than two candidates) for a particular office are the elements of a sequence. A person wins the election if this person receives a majority of the votes.

a) Devise a divide-and-conquer algorithm that determines whether a candidate received a majority and, if so, determine who this candidate is. [*Hint:* Assume that n is even and split the sequence of votes into two sequences, each with $n/2$ elements. Note that a candidate could not have received a majority of votes without receiving a majority of votes in at least one of the two halves.]

b) Use the master theorem to give a big-O estimate for the number of comparisons needed by the algorithm you devised in part (a).

18. Suppose that each person in a group of n people votes for exactly two people from a slate of candidates to fill two positions on a committee. The top two finishers both win positions as long as each receives more than $n/2$ votes.

a) Devise a divide-and-conquer algorithm that determines whether the two candidates who received the most votes each received at least $n/2$ votes and, if so, determine who these two candidates are.

b) Use the master theorem to give a big-O estimate for the number of comparisons needed by the algorithm you devised in part (a).

19. a) Set up a divide-and-conquer recurrence relation for the number of multiplications required to compute x^n, where x is a real number and n is a positive integer, using the recursive algorithm from Exercise 26 in Section 5.4.

b) Use the recurrence relation you found in part (a) to construct a big-O estimate for the number of multiplications used to compute x^n using the recursive algorithm.

20. a) Set up a divide-and-conquer recurrence relation for the number of modular multiplications required to compute a^n **mod** m, where a, m, and n are positive integers, using the recursive algorithms from Example 4 in Section 5.4.

b) Use the recurrence relation you found in part (a) to construct a big-O estimate for the number of modular multiplications used to compute a^n **mod** m using the recursive algorithm.

21. Suppose that the function f satisfies the recurrence relation $f(n) = 2f(\sqrt{n}) + 1$ whenever n is a perfect square greater than 1 and $f(2) = 1$.

a) Find $f(16)$.

b) Give a big-O estimate for $f(n)$. [*Hint:* Make the substitution $m = \log n$.]

22. Suppose that the function f satisfies the recurrence relation $f(n) = 2f(\sqrt{n}) + \log n$ whenever n is a perfect square greater than 1 and $f(2) = 1$.

a) Find $f(16)$.

b) Find a big-O estimate for $f(n)$. [*Hint:* Make the substitution $m = \log n$.]

****23.** This exercise deals with the problem of finding the largest sum of consecutive terms of a sequence of n real numbers. When all terms are positive, the sum of all terms provides the answer, but the situation is more complicated when some terms are negative. For example, the maximum sum of consecutive terms of the sequence $-2, 3, -1, 6, -7, 4$ is $3 + (-1) + 6 = 8$. (This exercise is based on [Be86].) Recall that in Exercise 56 in Section 8.1 we developed a dynamic programming algorithm for solving this problem. Here, we first look at the brute-force algorithm for solving this problem; then we develop a divide-and-conquer algorithm for solving it.

a) Use pseudocode to describe an algorithm that solves this problem by finding the sums of consecutive terms starting with the first term, the sums of consecutive terms starting with the second term, and so on, keeping track of the maximum sum found so far as the algorithm proceeds.

b) Determine the computational complexity of the algorithm in part (a) in terms of the number of sums computed and the number of comparisons made.

c) Devise a divide-and-conquer algorithm to solve this problem. [*Hint:* Assume that there are an even number of terms in the sequence and split the sequence into two halves. Explain how to handle the case when the maximum sum of consecutive terms includes terms in both halves.]

d) Use the algorithm from part (c) to find the maximum sum of consecutive terms of each of the sequences: $-2, 4, -1, 3, 5, -6, 1, 2$; $4, 1, -3, 7, -1, -5, 3, -2$; and $-1, 6, 3, -4, -5, 8, -1, 7$.

e) Find a recurrence relation for the number of sums and comparisons used by the divide-and-conquer algorithm from part (c).

f) Use the master theorem to estimate the computational complexity of the divide-and-conquer algorithm. How does it compare in terms of computational complexity with the algorithm from part (a)?

24. Apply the algorithm described in Example 12 for finding the closest pair of points, using the Euclidean distance between points, to find the closest pair of the points (1, 3), (1, 7), (2, 4), (2, 9), (3, 1), (3, 5), (4, 3), and (4, 7).

25. Apply the algorithm described in Example 12 for finding the closest pair of points, using the Euclidean distance between points, to find the closest pair of the points (1, 2), (1, 6), (2, 4), (2, 8), (3, 1), (3, 6), (3, 10), (4, 3), (5, 1), (5, 5), (5, 9), (6, 7), (7, 1), (7, 4), (7, 9), and (8, 6).

***26.** Use pseudocode to describe the recursive algorithm for solving the closest-pair problem as described in Example 12.

27. Construct a variation of the algorithm described in Example 12 along with justifications of the steps used by the algorithm to find the smallest distance between two points if the distance between two points is defined to be $d((x_i, y_i), (x_j, y_j)) = \max(|x_i - x_j|, |y_i - y_j|)$.

***28.** Suppose someone picks a number x from a set of n numbers. A second person tries to guess the number by successively selecting subsets of the n numbers and asking the first person whether x is in each set. The first person answers either "yes" or "no." When the first person answers each query truthfully, we can find x using $\log n$ queries by successively splitting the sets used in each query in half. Ulam's problem, proposed by Stanislaw Ulam in 1976, asks for the number of queries required to find x, supposing that the first person is allowed to lie exactly once.

Links

a) Show that by asking each question twice, given a number x and a set with n elements, and asking one more question when we find the lie, Ulam's problem can be solved using $2\log n + 1$ queries.

b) Show that by dividing the initial set of n elements into four parts, each with $n/4$ elements, 1/4 of the elements can be eliminated using two queries. [*Hint:* Use two queries, where each of the queries asks whether the element is in the union of two of the subsets with $n/4$ elements and where one of the subsets of $n/4$ elements is used in both queries.]

c) Show from part (b) that if $f(n)$ equals the number of queries used to solve Ulam's problem using the method from part (b) and n is divisible by 4, then $f(n) = f(3n/4) + 2$.

d) Solve the recurrence relation in part (c) for $f(n)$.

e) Is the naive way to solve Ulam's problem by asking each question twice or the divide-and-conquer method based on part (b) more efficient? The most efficient way to solve Ulam's problem has been determined by A. Pelc [Pe87].

In Exercises 29–33, assume that f is an increasing function satisfying the recurrence relation $f(n) = af(n/b) + cn^d$, where $a \geq 1$, b is an integer greater than 1, and c and d are positive real numbers. These exercises supply a proof of Theorem 2.

***29.** Show that if $a = b^d$ and n is a power of b, then $f(n) = f(1)n^d + cn^d \log_b n$.

30. Use Exercise 29 to show that if $a = b^d$, then $f(n)$ is $O(n^d \log n)$.

***31.** Show that if $a \neq b^d$ and n is a power of b, then $f(n) = C_1 n^d + C_2 n^{\log_b a}$, where $C_1 = b^d c/(b^d - a)$ and $C_2 = f(1) + b^d c/(a - b^d)$.

32. Use Exercise 31 to show that if $a < b^d$, then $f(n)$ is $O(n^d)$.

33. Use Exercise 31 to show that if $a > b^d$, then $f(n)$ is $O(n^{\log_b a})$.

34. Find $f(n)$ when $n = 4^k$, where f satisfies the recurrence relation $f(n) = 5f(n/4) + 6n$, with $f(1) = 1$.

35. Give a big-O estimate for the function f in Exercise 34 if f is an increasing function.

36. Find $f(n)$ when $n = 2^k$, where f satisfies the recurrence relation $f(n) = 8f(n/2) + n^2$ with $f(1) = 1$.

37. Give a big-O estimate for the function f in Exercise 36 if f is an increasing function.

8.4 Generating Functions

8.4.1 Introduction

Links

Generating functions are used to represent sequences efficiently by coding the terms of a sequence as coefficients of powers of a variable x in a formal power series. Generating functions can be used to solve many types of counting problems, such as the number of ways to select or distribute objects of different kinds, subject to a variety of constraints, and the number of ways to make change for a dollar using coins of different denominations. Generating functions can be used to solve recurrence relations by translating a recurrence relation for the terms of a sequence into an equation involving a generating function. This equation can then be solved to find a closed formula for the generating function. From this closed formula, the coefficients of the power series for the generating function can be found, solving the original recurrence relation. Generating functions can also be used to prove combinatorial identities by taking advantage of relatively simple relationships between functions that can be translated into identities involving the terms of sequences. Generating functions are a helpful tool for studying many properties of sequences besides those described in this section, such as their use for establishing asymptotic formulae for the terms of a sequence.

We begin with the definition of the generating function for a sequence.

Definition 1 The *generating function for the sequence* $a_0, a_1, \ldots, a_k, \ldots$ of real numbers is the infinite series

$$G(x) = a_0 + a_1 x + \cdots + a_k x^k + \cdots = \sum_{k=0}^{\infty} a_k x^k.$$

Remark: The generating function for $\{a_k\}$ given in Definition 1 is sometimes called the **ordinary generating function** of $\{a_k\}$ to distinguish it from other types of generating functions for this sequence.

EXAMPLE 1 The generating functions for the sequences $\{a_k\}$ with $a_k = 3, a_k = k + 1$, and $a_k = 2^k$ are $\sum_{k=0}^{\infty} 3x^k$, $\sum_{k=0}^{\infty} (k+1)x^k$, and $\sum_{k=0}^{\infty} 2^k x^k$, respectively. ◄

Extra Examples

We can define generating functions for finite sequences of real numbers by extending a finite sequence $a_0, a_1, \ldots, a_n$ into an infinite sequence by setting $a_{n+1} = 0$, $a_{n+2} = 0$, and so on. The generating function $G(x)$ of this infinite sequence $\{a_n\}$ is a polynomial of degree n because no terms of the form $a_j x^j$ with $j > n$ occur, that is,

$$G(x) = a_0 + a_1 x + \cdots + a_n x^n.$$

EXAMPLE 2 What is the generating function for the sequence 1, 1, 1, 1, 1, 1?

Solution: The generating function of 1, 1, 1, 1, 1, 1 is

$$1 + x + x^2 + x^3 + x^4 + x^5.$$

By Theorem 1 of Section 2.4 we have

$$(x^6 - 1)/(x - 1) = 1 + x + x^2 + x^3 + x^4 + x^5$$

when $x \neq 1$. Consequently, $G(x) = (x^6 - 1)/(x - 1)$ is the generating function of the sequence 1, 1, 1, 1, 1, 1. [Because the powers of x are only place holders for the terms of the sequence in a generating function, we do not need to worry that $G(1)$ is undefined.] ◄

EXAMPLE 3 Let m be a positive integer. Let $a_k = C(m, k)$, for $k = 0, 1, 2, \ldots, m$. What is the generating function for the sequence $a_0, a_1, \ldots, a_m$?

Solution: The generating function for this sequence is

$$G(x) = C(m, 0) + C(m, 1)x + C(m, 2)x^2 + \cdots + C(m, m)x^m.$$

The binomial theorem shows that $G(x) = (1 + x)^m$. ◀

8.4.2 Useful Facts About Power Series

When generating functions are used to solve counting problems, they are usually considered to be **formal power series**. As such, they are treated as algebraic objects; questions about their convergence are ignored. However, when formal power series are convergent, valid operations carry over to their use as formal power series. We will take advantage of the power series of particular functions around $x = 0$. These power series are unique and have a positive radius of convergence. Readers familiar with calculus can consult textbooks on this subject for details about power series, including the convergence of the series we use here.

We will now state some widely important facts about infinite series used when working with generating functions. These facts can be found in calculus texts.

EXAMPLE 4 The function $f(x) = 1/(1 - x)$ is the generating function of the sequence 1, 1, 1, 1, …, because

$$1/(1 - x) = 1 + x + x^2 + \cdots$$

for $|x| < 1$. ◀

EXAMPLE 5 The function $f(x) = 1/(1 - ax)$ is the generating function of the sequence $1, a, a^2, a^3, \ldots$, because

$$1/(1 - ax) = 1 + ax + a^2x^2 + \cdots$$

when $|ax| < 1$, or equivalently, for $|x| < 1/|a|$ for $a \neq 0$. ◀

We also will need some results on how to add and how to multiply two generating functions. Proofs of these results can be found in calculus texts.

THEOREM 1 Let $f(x) = \sum_{k=0}^{\infty} a_k x^k$ and $g(x) = \sum_{k=0}^{\infty} b_k x^k$. Then

$$f(x) + g(x) = \sum_{k=0}^{\infty} (a_k + b_k)x^k \quad \text{and} \quad f(x)g(x) = \sum_{k=0}^{\infty} \left(\sum_{j=0}^{k} a_j b_{k-j} \right) x^k.$$

Remark: Theorem 1 is valid only for power series that converge in an interval, as all series considered in this section do. However, the theory of generating functions is not limited to such series. In the case of series that do not converge, the statements in Theorem 1 can be taken as definitions of addition and multiplication of generating functions.

We will illustrate how Theorem 1 can be used with Example 6.

EXAMPLE 6 Let $f(x) = 1/(1 - x)^2$. Use Example 4 to find the coefficients $a_0, a_1, a_2, \ldots$ in the expansion $f(x) = \sum_{k=0}^{\infty} a_k x^k$.

Solution: From Example 4 we see that

$$1/(1-x) = 1 + x + x^2 + x^3 + \cdots.$$

Hence, from Theorem 1, we have

$$1/(1-x)^2 = \sum_{k=0}^{\infty}\left(\sum_{j=0}^{k} 1\right)x^k = \sum_{k=0}^{\infty}(k+1)x^k.$$

◀

Remark: This result also can be derived from Example 4 by differentiation. Taking derivatives is a useful technique for producing new identities from existing identities for generating functions.

To use generating functions to solve many important counting problems, we will need to apply the binomial theorem for exponents that are not positive integers. Before we state an extended version of the binomial theorem, we need to define extended binomial coefficients.

Definition 2 Let u be a real number and k a nonnegative integer. Then the *extended binomial coefficient* $\binom{u}{k}$ is defined by

$$\binom{u}{k} = \begin{cases} u(u-1)\cdots(u-k+1)/k! & \text{if } k > 0, \\ 1 & \text{if } k = 0. \end{cases}$$

EXAMPLE 7 Find the values of the extended binomial coefficients $\binom{-2}{3}$ and $\binom{1/2}{3}$.

Solution: Taking $u = -2$ and $k = 3$ in Definition 2 gives us

$$\binom{-2}{3} = \frac{(-2)(-3)(-4)}{3!} = -4.$$

Similarly, taking $u = 1/2$ and $k = 3$ gives us

$$\begin{aligned}\binom{1/2}{3} &= \frac{(1/2)(1/2-1)(1/2-2)}{3!} \\ &= (1/2)(-1/2)(-3/2)/6 \\ &= 1/16.\end{aligned}$$

◀

Example 8 provides a useful formula for extended binomial coefficients when the top parameter is a negative integer. It will be useful in our subsequent discussions.

EXAMPLE 8 When the top parameter is a negative integer, the extended binomial coefficient can be expressed in terms of an ordinary binomial coefficient. To see that this is the case, note that

$$\binom{-n}{r} = \frac{(-n)(-n-1)\cdots(-n-r+1)}{r!}$$ by definition of extended binomial coefficient

$$= \frac{(-1)^r n(n+1)\cdots(n+r-1)}{r!}$$ factoring out −1 from each term in the numerator

$$= \frac{(-1)^r(n+r-1)(n+r-2)\cdots n}{r!}$$ by the commutative law for multiplication

$$= \frac{(-1)^r(n+r-1)!}{r!(n-1)!}$$ multiplying both the numerator and denominator by $(n-1)!$

$$= (-1)^r\binom{n+r-1}{r}$$ by the definition of binomial coefficients

$$= (-1)^r C(n+r-1, r)$$ using alternative notation for binomial coefficients. ◀

We now state the extended binomial theorem.

THEOREM 2 **THE EXTENDED BINOMIAL THEOREM** Let x be a real number with $|x| < 1$ and let u be a real number. Then

$$(1+x)^u = \sum_{k=0}^{\infty}\binom{u}{k}x^k.$$

Theorem 2 can be proved using the theory of Maclaurin series. We leave its proof to the reader with a familiarity with this part of calculus.

Remark: When u is a positive integer, the extended binomial theorem reduces to the binomial theorem presented in Section 6.4, because in that case $\binom{u}{k} = 0$ if $k > u$.

Example 9 illustrates the use of Theorem 2 when the exponent is a negative integer.

EXAMPLE 9 Find the generating functions for $(1+x)^{-n}$ and $(1-x)^{-n}$, where n is a positive integer, using the extended binomial theorem.

Solution: By the extended binomial theorem, it follows that

$$(1+x)^{-n} = \sum_{k=0}^{\infty}\binom{-n}{k}x^k.$$

Using Example 8, which provides a simple formula for $\binom{-n}{k}$, we obtain

$$(1+x)^{-n} = \sum_{k=0}^{\infty}(-1)^k C(n+k-1, k)x^k.$$

Replacing x by $-x$, we find that

$$(1-x)^{-n} = \sum_{k=0}^{\infty} C(n+k-1, k)x^k.$$

◀

Table 1 presents a useful summary of some generating functions that arise frequently.

TABLE 1 Useful Generating Functions.

$G(x)$	a_k
$(1+x)^n = \sum_{k=0}^{n} C(n,k)x^k = 1 + C(n,1)x + C(n,2)x^2 + \cdots + x^n$	$C(n,k)$
$(1+ax)^n = \sum_{k=0}^{n} C(n,k)a^k x^k = 1 + C(n,1)ax + C(n,2)a^2x^2 + \cdots + a^n x^n$	$C(n,k)a^k$
$(1+x^r)^n = \sum_{k=0}^{n} C(n,k)x^{rk} = 1 + C(n,1)x^r + C(n,2)x^{2r} + \cdots + x^{rn}$	$C(n,k/r)$ if $r \mid k$; 0 otherwise
$\frac{1-x^{n+1}}{1-x} = \sum_{k=0}^{n} x^k = 1 + x + x^2 + \cdots + x^n$	1 if $k \le n$; 0 otherwise
$\frac{1}{1-x} = \sum_{k=0}^{\infty} x^k = 1 + x + x^2 + \cdots$	1
$\frac{1}{1-ax} = \sum_{k=0}^{\infty} a^k x^k = 1 + ax + a^2x^2 + \cdots$	a^k
$\frac{1}{1-x^r} = \sum_{k=0}^{\infty} x^{rk} = 1 + x^r + x^{2r} + \cdots$	1 if $r \mid k$; 0 otherwise
$\frac{1}{(1-x)^2} = \sum_{k=0}^{\infty} (k+1)x^k = 1 + 2x + 3x^2 + \cdots$	$k+1$
$\frac{1}{(1-x)^n} = \sum_{k=0}^{\infty} C(n+k-1,k)x^k = 1 + C(n,1)x + C(n+1,2)x^2 + \cdots$	$C(n+k-1,k) = C(n+k-1,n-1)$
$\frac{1}{(1+x)^n} = \sum_{k=0}^{\infty} C(n+k-1,k)(-1)^k x^k = 1 - C(n,1)x + C(n+1,2)x^2 - \cdots$	$(-1)^k C(n+k-1,k) = (-1)^k C(n+k-1,n-1)$
$\frac{1}{(1-ax)^n} = \sum_{k=0}^{\infty} C(n+k-1,k)a^k x^k = 1 + C(n,1)ax + C(n+1,2)a^2x^2 + \cdots$	$C(n+k-1,k)a^k = C(n+k-1,n-1)a^k$
$e^x = \sum_{k=0}^{\infty} \frac{x^k}{k!} = 1 + x + \frac{x^2}{2!} + \frac{x^3}{3!} + \cdots$	$1/k!$
$\ln(1+x) = \sum_{k=1}^{\infty} \frac{(-1)^{k+1}}{k} x^k = x - \frac{x^2}{2} + \frac{x^3}{3} - \frac{x^4}{4} + \cdots$	$(-1)^{k+1}/k$

Note: The series for the last two generating functions can be found in most calculus books when power series are discussed.

Remark: Note that the second and third formulae in this table can be deduced from the first formula by substituting ax and x^r for x, respectively. Similarly, the sixth and seventh formulae can be deduced from the fifth formula using the same substitutions. The tenth and eleventh can be deduced from the ninth formula by substituting $-x$ and ax for x, respectively. Also, some of the formulae in this table can be derived from other formulae using methods from calculus (such as differentiation and integration). Students are encouraged to know the core formulae in this table (that is, formulae from which the others can be derived, perhaps the first, fourth, fifth, eighth, ninth, twelfth, and thirteenth formulae) and understand how to derive the other formulae from these core formulae.

8.4.3 Counting Problems and Generating Functions

Generating functions can be used to solve a wide variety of counting problems. In particular, they can be used to count the number of combinations of various types. In Chapter 6 we developed techniques to count the r-combinations from a set with n elements when repetition is allowed and additional constraints may exist. Such problems are equivalent to counting the solutions to equations of the form

$$e_1 + e_2 + \cdots + e_n = C,$$

where C is a constant and each e_i is a nonnegative integer that may be subject to a specified constraint. Generating functions can also be used to solve counting problems of this type, as Examples 10–12 show.

EXAMPLE 10 Find the number of solutions of

Extra Examples

$$e_1 + e_2 + e_3 = 17,$$

where e_1, e_2, and e_3 are nonnegative integers with $2 \le e_1 \le 5$, $3 \le e_2 \le 6$, and $4 \le e_3 \le 7$.

Solution: The number of solutions with the indicated constraints is the coefficient of x^{17} in the expansion of

$$(x^2 + x^3 + x^4 + x^5)(x^3 + x^4 + x^5 + x^6)(x^4 + x^5 + x^6 + x^7).$$

This follows because we obtain a term equal to x^{17} in the product by picking a term in the first sum x^{e_1}, a term in the second sum x^{e_2}, and a term in the third sum x^{e_3}, where the exponents e_1, e_2, and e_3 satisfy the equation $e_1 + e_2 + e_3 = 17$ and the given constraints.

It is not hard to see that the coefficient of x^{17} in this product is 3. Hence, there are three solutions. (Note that the calculating of this coefficient involves about as much work as enumerating all the solutions of the equation with the given constraints. However, the method that this illustrates often can be used to solve wide classes of counting problems with special formulae, as we will see. Furthermore, a computer algebra system can be used to do such computations.) ◀

EXAMPLE 11 In how many different ways can eight identical cookies be distributed among three distinct children if each child receives at least two cookies and no more than four cookies?

Solution: Because each child receives at least two but no more than four cookies, for each child there is a factor equal to

$$(x^2 + x^3 + x^4)$$

in the generating function for the sequence $\{c_n\}$, where c_n is the number of ways to distribute n cookies. Because there are three children, this generating function is

$$(x^2 + x^3 + x^4)^3.$$

We need the coefficient of x^8 in this product. The reason is that the x^8 terms in the expansion correspond to the ways that three terms can be selected, with one from each factor, that have exponents adding up to 8. Furthermore, the exponents of the term from the first, second, and third factors are the numbers of cookies the first, second, and third children receive, respectively. Computation shows that this coefficient equals 6. Hence, there are six ways to distribute the cookies so that each child receives at least two, but no more than four, cookies. ◀

EXAMPLE 12 Use generating functions to determine the number of ways to insert tokens worth \$1, \$2, and \$5 into a vending machine to pay for an item that costs r dollars in both the cases when the order in which the tokens are inserted does not matter and when the order does matter. (For example, there are two ways to pay for an item that costs \$3 when the order in which the tokens are inserted does not matter: inserting three \$1 tokens or one \$1 token and a \$2 token. When the order matters, there are three ways: inserting three \$1 tokens, inserting a \$1 token and then a \$2 token, or inserting a \$2 token and then a \$1 token.)

Solution: Consider the case when the order in which the tokens are inserted does not matter. Here, all we care about is the number of each token used to produce a total of r dollars. Because we can use any number of \$1 tokens, any number of \$2 tokens, and any number of \$5 tokens, the answer is the coefficient of x^r in the generating function

$$(1 + x + x^2 + x^3 + \cdots)(1 + x^2 + x^4 + x^6 + \cdots)(1 + x^5 + x^{10} + x^{15} + \cdots).$$

(The first factor in this product represents the \$1 tokens used, the second the \$2 tokens used, and the third the \$5 tokens used.) For example, the number of ways to pay for an item costing \$7 using \$1, \$2, and \$5 tokens is given by the coefficient of x^7 in this expansion, which equals 6.

When the order in which the tokens are inserted matters, the number of ways to insert exactly n tokens to produce a total of r dollars is the coefficient of x^r in

$$(x + x^2 + x^5)^n,$$

because each of the r tokens may be a \$1 token, a \$2 token, or a \$5 token. Because any number of tokens may be inserted, the number of ways to produce r dollars using \$1, \$2, or \$5 tokens, when the order in which the tokens are inserted matters, is the coefficient of x^r in

$$\begin{aligned} 1 + (x + x^2 + x^5) + (x + x^2 + x^5)^2 + \cdots &= \frac{1}{1 - (x + x^2 + x^5)} \\ &= \frac{1}{1 - x - x^2 - x^5}, \end{aligned}$$

where we have added the number of ways to insert 0 tokens, 1 token, 2 tokens, 3 tokens, and so on, and where we have used the identity $1/(1 - x) = 1 + x + x^2 + \cdots$ with x replaced with $x + x^2 + x^5$. For example, the number of ways to pay for an item costing \$7 using \$1, \$2, and \$5 tokens, when the order in which the tokens are used matters, is the coefficient of x^7 in this expansion, which equals 26. [*Hint:* To see that this coefficient equals 26 requires the addition of the coefficients of x^7 in the expansions $(x + x^2 + x^5)^k$ for $2 \leq k \leq 7$. This can be done by hand with considerable computation, or a computer algebra system can be used.] ◀

Example 13 shows the versatility of generating functions when used to solve problems with differing assumptions.

EXAMPLE 13 Use generating functions to find the number of k-combinations of a set with n elements. Assume that the binomial theorem has already been established.

Solution: Each of the n elements in the set contributes the term $(1 + x)$ to the generating function $f(x) = \sum_{k=0}^{n} a_k x^k$. Here $f(x)$ is the generating function for $\{a_k\}$, where a_k represents the number of k-combinations of a set with n elements. Hence,

$$f(x) = (1 + x)^n.$$

But by the binomial theorem, we have

$$f(x) = \sum_{k=0}^{n} \binom{n}{k} x^k,$$

where

$$\binom{n}{k} = \frac{n!}{k!(n-k)!}.$$

Hence, $C(n, k)$, the number of k-combinations of a set with n elements, is

$$\frac{n!}{k!(n-k)!}.$$

◀

Remark: We proved the binomial theorem in Section 6.4 using the formula for the number of r-combinations of a set with n elements. This example shows that the binomial theorem, which can be proved by mathematical induction, can be used to derive the formula for the number of r-combinations of a set with n elements.

EXAMPLE 14 Use generating functions to find the number of r-combinations from a set with n elements when repetition of elements is allowed.

Solution: Let $G(x)$ be the generating function for the sequence $\{a_r\}$, where a_r equals the number of r-combinations of a set with n elements with repetitions allowed. That is, $G(x) = \sum_{r=0}^{\infty} a_r x^r$. Because we can select any number of a particular member of the set with n elements when we form an r-combination with repetition allowed, each of the n elements contributes $(1 + x + x^2 + x^3 + \cdots)$ to a product expansion for $G(x)$. Each element contributes this factor because it may be selected zero times, one time, two times, three times, and so on, when an r-combination is formed (with a total of r elements selected). Because there are n elements in the set and each contributes this same factor to $G(x)$, we have

$$G(x) = (1 + x + x^2 + \cdots)^n.$$

As long as $|x| < 1$, we have $1 + x + x^2 + \cdots = 1/(1 - x)$, so

$$G(x) = 1/(1 - x)^n = (1 - x)^{-n}.$$

Applying the extended binomial theorem (Theorem 2), it follows that

$$(1 - x)^{-n} = (1 + (-x))^{-n} = \sum_{r=0}^{\infty} \binom{-n}{r} (-x)^r.$$

The number of r-combinations of a set with n elements with repetitions allowed, when r is a positive integer, is the coefficient a_r of x^r in this sum. Consequently, using Example 8 we find that a_r equals

$$\begin{aligned}\binom{-n}{r}(-1)^r &= (-1)^r C(n + r - 1, r) \cdot (-1)^r \\ &= C(n + r - 1, r).\end{aligned}$$

◀

Note that the result in Example 14 is the same result we stated as Theorem 2 in Section 6.5.

EXAMPLE 15 Use generating functions to find the number of ways to select r objects of n different kinds if we must select at least one object of each kind.

Solution: Because we need to select at least one object of each kind, each of the n kinds of objects contributes the factor $(x + x^2 + x^3 + \cdots)$ to the generating function $G(x)$ for the sequence $\{a_r\}$, where a_r is the number of ways to select r objects of n different kinds if we need at least one object of each kind. Hence,

$$G(x) = (x + x^2 + x^3 + \cdots)^n = x^n(1 + x + x^2 + \cdots)^n = x^n/(1 - x)^n.$$

Using the extended binomial theorem and Example 8, we have

$$\begin{aligned}G(x) &= x^n/(1 - x)^n \\ &= x^n \cdot (1 - x)^{-n} \\ &= x^n \sum_{r=0}^{\infty} \binom{-n}{r} (-x)^r \\ &= x^n \sum_{r=0}^{\infty} (-1)^r C(n + r - 1, r)(-1)^r x^r \\ &= \sum_{r=0}^{\infty} C(n + r - 1, r) x^{n+r} \\ &= \sum_{t=n}^{\infty} C(t - 1, t - n) x^t \\ &= \sum_{r=n}^{\infty} C(r - 1, r - n) x^r.\end{aligned}$$

We have shifted the summation in the next-to-last equality by setting $t = n + r$ so that $t = n$ when $r = 0$ and $n + r - 1 = t - 1$, and then we replaced t by r as the index of summation in the last equality to return to our original notation. Hence, there are $C(r - 1, r - n)$ ways to select r objects of n different kinds if we must select at least one object of each kind. ◀

8.4.4 Using Generating Functions to Solve Recurrence Relations

We can find the solution to a recurrence relation and its initial conditions by finding an explicit formula for the associated generating function. This is illustrated in Examples 16 and 17.

EXAMPLE 16 Solve the recurrence relation $a_k = 3a_{k-1}$ for $k = 1, 2, 3, \ldots$ and initial condition $a_0 = 2$.

Extra Examples

Solution: Let $G(x)$ be the generating function for the sequence $\{a_k\}$, that is, $G(x) = \sum_{k=0}^{\infty} a_k x^k$. First note that

$$xG(x) = \sum_{k=0}^{\infty} a_k x^{k+1} = \sum_{k=1}^{\infty} a_{k-1} x^k.$$

Using the recurrence relation, we see that

$$\begin{aligned} G(x) - 3xG(x) &= \sum_{k=0}^{\infty} a_k x^k - 3\sum_{k=1}^{\infty} a_{k-1} x^k \\ &= a_0 + \sum_{k=1}^{\infty} (a_k - 3a_{k-1}) x^k \\ &= 2, \end{aligned}$$

because $a_0 = 2$ and $a_k = 3a_{k-1}$. Thus,

$$G(x) - 3xG(x) = (1 - 3x)G(x) = 2.$$

Solving for $G(x)$ shows that $G(x) = 2/(1 - 3x)$. Using the identity $1/(1 - ax) = \sum_{k=0}^{\infty} a^k x^k$, from Table 1, we have

$$G(x) = 2\sum_{k=0}^{\infty} 3^k x^k = \sum_{k=0}^{\infty} 2 \cdot 3^k x^k.$$

Consequently, $a_k = 2 \cdot 3^k$. ◀

EXAMPLE 17 Suppose that a valid codeword is an n-digit number in decimal notation containing an even number of 0s. Let a_n denote the number of valid codewords of length n. In Example 4 of Section 8.1 we showed that the sequence $\{a_n\}$ satisfies the recurrence relation

$$a_n = 8a_{n-1} + 10^{n-1}$$

and the initial condition $a_1 = 9$. Use generating functions to find an explicit formula for a_n.

Solution: To make our work with generating functions simpler, we extend this sequence by setting $a_0 = 1$; when we assign this value to a_0 and use the recurrence relation, we have $a_1 = 8a_0 + 10^0 = 8 + 1 = 9$, which is consistent with our original initial condition. (It also makes sense because there is one code word of length 0—the empty string.)

We multiply both sides of the recurrence relation by x^n to obtain

$$a_n x^n = 8a_{n-1}x^n + 10^{n-1}x^n.$$

Let $G(x) = \sum_{n=0}^{\infty} a_n x^n$ be the generating function of the sequence $a_0, a_1, a_2, \ldots$. We sum both sides of the last equation starting with $n = 1$, to find that

$$\begin{aligned} G(x) - 1 = \sum_{n=1}^{\infty} a_n x^n &= \sum_{n=1}^{\infty} (8a_{n-1}x^n + 10^{n-1}x^n) \\ &= 8\sum_{n=1}^{\infty} a_{n-1}x^n + \sum_{n=1}^{\infty} 10^{n-1}x^n \\ &= 8x\sum_{n=1}^{\infty} a_{n-1}x^{n-1} + x\sum_{n=1}^{\infty} 10^{n-1}x^{n-1} \\ &= 8x\sum_{n=0}^{\infty} a_n x^n + x\sum_{n=0}^{\infty} 10^n x^n \\ &= 8xG(x) + x/(1 - 10x), \end{aligned}$$

where we have used Example 5 to evaluate the second summation. Therefore, we have

$$G(x) - 1 = 8xG(x) + x/(1 - 10x).$$

Solving for $G(x)$ shows that

$$G(x) = \frac{1 - 9x}{(1 - 8x)(1 - 10x)}.$$

Expanding the right-hand side of this equation into partial fractions (as is done in the integration of rational functions studied in calculus) gives

$$G(x) = \frac{1}{2}\left(\frac{1}{1 - 8x} + \frac{1}{1 - 10x}\right).$$

Using Example 5 twice (once with $a = 8$ and once with $a = 10$) gives

$$\begin{aligned} G(x) &= \frac{1}{2}\left(\sum_{n=0}^{\infty} 8^n x^n + \sum_{n=0}^{\infty} 10^n x^n\right) \\ &= \sum_{n=0}^{\infty} \frac{1}{2}(8^n + 10^n)x^n. \end{aligned}$$

Consequently, we have shown that

$$a_n = \frac{1}{2}(8^n + 10^n).$$

◀

8.4.5 Proving Identities via Generating Functions

In Chapter 6 we saw how combinatorial identities could be established using combinatorial proofs. Here we will show that such identities, as well as identities for extended binomial coefficients, can be proved using generating functions. Sometimes the generating function approach is simpler than other approaches, especially when it is simpler to work with the closed formula of a generating function than with the terms of the sequence themselves. We illustrate how generating functions can be used to prove identities with Example 18.

EXAMPLE 18 Use generating functions to show that

$$\sum_{k=0}^{n} C(n, k)^2 = C(2n, n)$$

whenever n is a positive integer.

Solution: First note that by the binomial theorem $C(2n, n)$ is the coefficient of x^n in $(1 + x)^{2n}$. However, we also have

$$\begin{aligned}(1 + x)^{2n} &= [(1 + x)^n]^2 \\ &= [C(n, 0) + C(n, 1)x + C(n, 2)x^2 + \cdots + C(n, n)x^n]^2.\end{aligned}$$

The coefficient of x^n in this expression is

$$C(n, 0)C(n, n) + C(n, 1)C(n, n - 1) + C(n, 2)C(n, n - 2) + \cdots + C(n, n)C(n, 0).$$

This equals $\sum_{k=0}^{n} C(n, k)^2$, because $C(n, n - k) = C(n, k)$. Because both $C(2n, n)$ and $\sum_{k=0}^{n} C(n, k)^2$ represent the coefficient of x^n in $(1 + x)^{2n}$, they must be equal. ◀

Exercises 44 and 45 ask that Pascal's identity and Vandermonde's identity be proved using generating functions.

Exercises

1. Find the generating function for the finite sequence 2, 2, 2, 2, 2, 2.

2. Find the generating function for the finite sequence 1, 4, 16, 64, 256.

In Exercises 3–8, by a **closed formula** we mean an algebraic expression not involving a summation over a range of values or the use of ellipses.

3. Find a closed formula for the generating function for each of these sequences. (For each sequence, use the most obvious choice of a sequence that follows the pattern of the initial terms listed.)

a) 0, 2, 2, 2, 2, 2, 2, 0, 0, 0, 0, 0, …
b) 0, 0, 0, 1, 1, 1, 1, 1, 1, …
c) 0, 1, 0, 0, 1, 0, 0, 1, 0, 0, 1, …
d) 2, 4, 8, 16, 32, 64, 128, 256, …
e) $\binom{7}{0}, \binom{7}{1}, \binom{7}{2}, \ldots, \binom{7}{7}$, 0, 0, 0, 0, 0, …
f) 2, −2, 2, −2, 2, −2, 2, −2, …
g) 1, 1, 0, 1, 1, 1, 1, 1, 1, 1, …
h) 0, 0, 0, 1, 2, 3, 4, …

4. Find a closed formula for the generating function for each of these sequences. (Assume a general form for the terms of the sequence, using the most obvious choice of such a sequence.)

a) −1, −1, −1, −1, −1, −1, −1, 0, 0, 0, 0, 0, 0, …
b) 1, 3, 9, 27, 81, 243, 729, …
c) 0, 0, 3, −3, 3, −3, 3, −3, …
d) 1, 2, 1, 1, 1, 1, 1, 1, 1, …
e) $\binom{7}{0}, 2\binom{7}{1}, 2^2\binom{7}{2}, \ldots, 2^7\binom{7}{7}$, 0, 0, 0, 0, …
f) −3, 3, −3, 3, −3, 3, …
g) 0, 1, −2, 4, −8, 16, −32, 64, …
h) 1, 0, 1, 0, 1, 0, 1, 0, …

5. Find a closed formula for the generating function for the sequence $\{a_n\}$, where

a) $a_n = 5$ for all $n = 0, 1, 2, \ldots$.
b) $a_n = 3^n$ for all $n = 0, 1, 2, \ldots$.
c) $a_n = 2$ for $n = 3, 4, 5, \ldots$ and $a_0 = a_1 = a_2 = 0$.
d) $a_n = 2n + 3$ for all $n = 0, 1, 2, \ldots$.
e) $a_n = \binom{8}{n}$ for all $n = 0, 1, 2, \ldots$.
f) $a_n = \binom{n+4}{n}$ for all $n = 0, 1, 2, \ldots$.

6. Find a closed formula for the generating function for the sequence $\{a_n\}$, where

a) $a_n = -1$ for all $n = 0, 1, 2, \ldots$.
b) $a_n = 2^n$ for $n = 1, 2, 3, 4, \ldots$ and $a_0 = 0$.
c) $a_n = n - 1$ for $n = 0, 1, 2, \ldots$.
d) $a_n = 1/(n+1)!$ for $n = 0, 1, 2, \ldots$.
e) $a_n = \binom{n}{2}$ for $n = 0, 1, 2, \ldots$.
f) $a_n = \binom{10}{n+1}$ for $n = 0, 1, 2, \ldots$.

7. For each of these generating functions, provide a closed formula for the sequence it determines.

a) $(3x - 4)^3$ **b)** $(x^3 + 1)^3$
c) $1/(1 - 5x)$ **d)** $x^3/(1 + 3x)$
e) $x^2 + 3x + 7 + (1/(1 - x^2))$
f) $(x^4/(1 - x^4)) - x^3 - x^2 - x - 1$
g) $x^2/(1 - x)^2$ **h)** $2e^{2x}$

8. For each of these generating functions, provide a closed formula for the sequence it determines.

a) $(x^2 + 1)^3$ **b)** $(3x - 1)^3$
c) $1/(1 - 2x^2)$ **d)** $x^2/(1 - x)^3$
e) $x - 1 + (1/(1 - 3x))$ **f)** $(1 + x^3)/(1 + x)^3$
***g)** $x/(1 + x + x^2)$ **h)** $e^{3x^2} - 1$

9. Find the coefficient of x^{10} in the power series of each of these functions.

a) $(1 + x^5 + x^{10} + x^{15} + \cdots)^3$
b) $(x^3 + x^4 + x^5 + x^6 + x^7 + \cdots)^3$
c) $(x^4 + x^5 + x^6)(x^3 + x^4 + x^5 + x^6 + x^7)(1 + x + x^2 + x^3 + x^4 + \cdots)$
d) $(x^2 + x^4 + x^6 + x^8 + \cdots)(x^3 + x^6 + x^9 + \cdots)(x^4 + x^8 + x^{12} + \cdots)$
e) $(1 + x^2 + x^4 + x^6 + x^8 + \cdots)(1 + x^4 + x^8 + x^{12} + \cdots)(1 + x^6 + x^{12} + x^{18} + \cdots)$

10. Find the coefficient of x^9 in the power series of each of these functions.

a) $(1 + x^3 + x^6 + x^9 + \cdots)^3$
b) $(x^2 + x^3 + x^4 + x^5 + x^6 + \cdots)^3$
c) $(x^3 + x^5 + x^6)(x^3 + x^4)(x + x^2 + x^3 + x^4 + \cdots)$
d) $(x + x^4 + x^7 + x^{10} + \cdots)(x^2 + x^4 + x^6 + x^8 + \cdots)$
e) $(1 + x + x^2)^3$

11. Find the coefficient of x^{10} in the power series of each of these functions.

a) $1/(1 - 2x)$ **b)** $1/(1 + x)^2$
c) $1/(1 - x)^3$ **d)** $1/(1 + 2x)^4$
e) $x^4/(1 - 3x)^3$

12. Find the coefficient of x^{12} in the power series of each of these functions.

a) $1/(1 + 3x)$ **b)** $1/(1 - 2x)^2$
c) $1/(1 + x)^8$ **d)** $1/(1 - 4x)^3$
e) $x^3/(1 + 4x)^2$

13. Use generating functions to determine the number of different ways 10 identical balloons can be given to four children if each child receives at least two balloons.

14. Use generating functions to determine the number of different ways 12 identical action figures can be given to five children so that each child receives at most three action figures.

15. Use generating functions to determine the number of different ways 15 identical stuffed animals can be given to six children so that each child receives at least one but no more than three stuffed animals.

16. Use generating functions to find the number of ways to choose a dozen bagels from three varieties—egg, salty, and plain—if at least two bagels of each kind but no more than three salty bagels are chosen.

17. In how many ways can 25 identical donuts be distributed to four police officers so that each officer gets at least three but no more than seven donuts?

18. Use generating functions to find the number of ways to select 14 balls from a jar containing 100 red balls, 100 blue balls, and 100 green balls so that no fewer than 3 and no more than 10 blue balls are selected. Assume that the order in which the balls are drawn does not matter.

19. What is the generating function for the sequence $\{c_k\}$, where c_k is the number of ways to make change for k dollars using \$1 bills, \$2 bills, \$5 bills, and \$10 bills?

20. What is the generating function for the sequence $\{c_k\}$, where c_k represents the number of ways to make change for k pesos using bills worth 10 pesos, 20 pesos, 50 pesos, and 100 pesos?

21. Give a combinatorial interpretation of the coefficient of x^4 in the expansion $(1 + x + x^2 + x^3 + \cdots)^3$. Use this interpretation to find this number.

22. Give a combinatorial interpretation of the coefficient of x^6 in the expansion $(1 + x + x^2 + x^3 + \cdots)^n$. Use this interpretation to find this number.

23. a) What is the generating function for $\{a_k\}$, where a_k is the number of solutions of $x_1 + x_2 + x_3 = k$ when x_1, x_2, and x_3 are integers with $x_1 \geq 2, 0 \leq x_2 \leq 3$, and $2 \leq x_3 \leq 5$?
b) Use your answer to part (a) to find a_6.

24. a) What is the generating function for $\{a_k\}$, where a_k is the number of solutions of $x_1 + x_2 + x_3 + x_4 = k$ when x_1, x_2, x_3, and x_4 are integers with $x_1 \geq 3, 1 \leq x_2 \leq 5, 0 \leq x_3 \leq 4$, and $x_4 \geq 1$?
b) Use your answer to part (a) to find a_7.

25. Explain how generating functions can be used to find the number of ways in which postage of r cents can be pasted on an envelope using 3-cent, 4-cent, and 20-cent stamps.

a) Assume that the order the stamps are pasted on does not matter.

b) Assume that the order in which the stamps are pasted on matters.

c) Use your answer to part (a) to determine the number of ways 46 cents of postage can be pasted on an envelope using 3-cent, 4-cent, and 20-cent stamps when the order the stamps are pasted on does not matter. (Use of a computer algebra program is advised.)

d) Use your answer to part (b) to determine the number of ways 46 cents of postage can be pasted in a row on an envelope using 3-cent, 4-cent, and 20-cent stamps when the order in which the stamps are pasted on matters. (Use of a computer algebra program is advised.)

26. Explain how generating functions can be used to find the number of ways in which postage of r cents can be pasted on an envelope using 2-cent, 7-cent, 13-cent, and 32-cent stamps.

a) Assume that the order the stamps are pasted on does not matter.

b) Assume that the order the stamps are pasted on matters.

c) Use your answer to part (a) to determine the number of ways 49 cents of postage can be pasted on an envelope using 2-cent, 7-cent, 13-cent, and 32-cent stamps when the order the stamps are pasted on does not matter. (Use of a computer algebra program is advised.)

d) Use your answer to part (b) to determine the number of ways 49 cents of postage can be pasted on an envelope using 2-cent, 7-cent, 13-cent, and 32-cent stamps when the order in which the stamps are pasted on matters. (Use of a computer algebra program is advised.)

27. Customers at a quirky tropical fruit stand can buy at most four mangos, at most two passion fruit, any even number of papayas, three or more coconuts, and carambolas in groups of five.

a) Explain how generating functions can be used to find the number of ways a customer can buy n pieces of these fruits, following the restrictions listed.

b) Use your answer in part (a) to determine the number of ways you can buy a dozen pieces of these fruits.

28. a) Show that $1/(1 - x - x^2 - x^3 - x^4 - x^5 - x^6)$ is the generating function for the number of ways that the sum n can be obtained when a die is rolled repeatedly and the order of the rolls matters.

b) Use part (a) to find the number of ways to roll a total of 8 when a die is rolled repeatedly, and the order of the rolls matters. (Use of a computer algebra package is advised.)

29. Use generating functions (and a computer algebra package, if available) to find the number of ways to make change for \$1 using

a) dimes and quarters.

b) nickels, dimes, and quarters.

c) pennies, dimes, and quarters.

d) pennies, nickels, dimes, and quarters.

30. Use generating functions (and a computer algebra package, if available) to find the number of ways to make change for \$1 using pennies, nickels, dimes, and quarters with

a) no more than 10 pennies.

b) no more than 10 pennies and no more than 10 nickels.

***c)** no more than 10 coins.

31. Use generating functions to find the number of ways to make change for \$100 using

a) \$10, \$20, and \$50 bills.

b) \$5, \$10, \$20, and \$50 bills.

c) \$5, \$10, \$20, and \$50 bills if at least one bill of each denomination is used.

d) \$5, \$10, and \$20 bills if at least one and no more than four of each denomination is used.

32. If $G(x)$ is the generating function for the sequence $\{a_k\}$, what is the generating function for each of these sequences?

a) $2a_0, 2a_1, 2a_2, 2a_3, \ldots$

b) $0, a_0, a_1, a_2, a_3, \ldots$ (assuming that terms follow the pattern of all but the first term)

c) $0, 0, 0, 0, a_2, a_3, \ldots$ (assuming that terms follow the pattern of all but the first four terms)

d) $a_2, a_3, a_4, \ldots$

e) $a_1, 2a_2, 3a_3, 4a_4, \ldots$ [*Hint: Calculus required* here.]

f) $a_0^2,\ 2a_0a_1,\ a_1^2 + 2a_0a_2,\ 2a_0a_3 + 2a_1a_2,\ 2a_0a_4 + 2a_1a_3 + a_2^2, \ldots$

33. If $G(x)$ is the generating function for the sequence $\{a_k\}$, what is the generating function for each of these sequences?

a) $0, 0, 0, a_3, a_4, a_5, \ldots$ (assuming that terms follow the pattern of all but the first three terms)

b) $a_0, 0, a_1, 0, a_2, 0, \ldots$

c) $0, 0, 0, 0, a_0, a_1, a_2, \ldots$ (assuming that terms follow the pattern of all but the first four terms)

d) $a_0, 2a_1, 4a_2, 8a_3, 16a_4, \ldots$

e) $0, a_0, a_1/2, a_2/3, a_3/4, \ldots$ [*Hint: Calculus required* here.]

f) $a_0, a_0 + a_1, a_0 + a_1 + a_2, a_0 + a_1 + a_2 + a_3, \ldots$

34. Use generating functions to solve the recurrence relation $a_k = 7a_{k-1}$ with the initial condition $a_0 = 5$.

35. Use generating functions to solve the recurrence relation $a_k = 3a_{k-1} + 2$ with the initial condition $a_0 = 1$.

36. Use generating functions to solve the recurrence relation $a_k = 3a_{k-1} + 4^{k-1}$ with the initial condition $a_0 = 1$.

37. Use generating functions to solve the recurrence relation $a_k = 5a_{k-1} - 6a_{k-2}$ with initial conditions $a_0 = 6$ and $a_1 = 30$.

38. Use generating functions to solve the recurrence relation $a_k = a_{k-1} + 2a_{k-2} + 2^k$ with initial conditions $a_0 = 4$ and $a_1 = 12$.

39. Use generating functions to solve the recurrence relation $a_k = 4a_{k-1} - 4a_{k-2} + k^2$ with initial conditions $a_0 = 2$ and $a_1 = 5$.

40. Use generating functions to solve the recurrence relation $a_k = 2a_{k-1} + 3a_{k-2} + 4^k + 6$ with initial conditions $a_0 = 20$, $a_1 = 60$.

41. Use generating functions to find an explicit formula for the Fibonacci numbers.

*42. a) Show that if n is a positive integer, then

$$\binom{-1/2}{n} = \binom{2n}{n} \Big/ (-4)^n.$$

b) Use the extended binomial theorem and part (a) to show that the coefficient of x^n in the expansion of $(1-4x)^{-1/2}$ is $\binom{2n}{n}$ for all nonnegative integers n.

*43. (*Calculus required*) Let $\{C_n\}$ be the sequence of Catalan numbers, that is, the solution to the recurrence relation $C_n = \sum_{k=0}^{n-1} C_k C_{n-k-1}$ with $C_0 = C_1 = 1$ (see Example 5 in Section 8.1).

a) Show that if $G(x)$ is the generating function for the sequence of Catalan numbers, then $xG(x)^2 - G(x) + 1 = 0$. Conclude (using the initial conditions) that $G(x) = (1 - \sqrt{1-4x})/(2x)$.

b) Use Exercise 42 to conclude that

$$G(x) = \sum_{n=0}^{\infty} \frac{1}{n+1}\binom{2n}{n} x^n,$$

so that

$$C_n = \frac{1}{n+1}\binom{2n}{n}.$$

c) Show that $C_n \geq 2^{n-1}$ for all positive integers n.

44. Use generating functions to prove Pascal's identity: $C(n, r) = C(n-1, r) + C(n-1, r-1)$ when n and r are positive integers with $r < n$. [*Hint:* Use the identity $(1+x)^n = (1+x)^{n-1} + x(1+x)^{n-1}$.]

45. Use generating functions to prove Vandermonde's identity: $C(m+n, r) = \sum_{k=0}^{r} C(m, r-k)C(n, k)$, whenever m, n, and r are nonnegative integers with r not exceeding either m or n. [*Hint:* Look at the coefficient of x^r in both sides of $(1+x)^{m+n} = (1+x)^m(1+x)^n$.]

46. This exercise shows how to use generating functions to derive a formula for the sum of the first n squares.

a) Show that $(x^2+x)/(1-x)^4$ is the generating function for the sequence $\{a_n\}$, where $a_n = 1^2 + 2^2 + \cdots + n^2$.

b) Use part (a) to find an explicit formula for the sum $1^2 + 2^2 + \cdots + n^2$.

The **exponential generating function** for the sequence $\{a_n\}$ is the series

$$\sum_{n=0}^{\infty} \frac{a_n}{n!} x^n.$$

For example, the exponential generating function for the sequence 1, 1, 1, ... is the function $\sum_{n=0}^{\infty} x^n/n! = e^x$. (You will find this particular series useful in these exercises.) Note that e^x is the (ordinary) generating function for the sequence $1, 1, 1/2!, 1/3!, 1/4!, \ldots$.

47. Find a closed formula for the exponential generating function for the sequence $\{a_n\}$, where

a) $a_n = 2$. b) $a_n = (-1)^n$.
c) $a_n = 3^n$. d) $a_n = n + 1$.
e) $a_n = 1/(n+1)$.

48. Find a closed formula for the exponential generating function for the sequence $\{a_n\}$, where

a) $a_n = (-2)^n$. b) $a_n = -1$.
c) $a_n = n$. d) $a_n = n(n-1)$.
e) $a_n = 1/((n+1)(n+2))$.

49. Find the sequence with each of these functions as its exponential generating function.

a) $f(x) = e^{-x}$ b) $f(x) = 3x^{2x}$
c) $f(x) = e^{3x} - 3e^{2x}$ d) $f(x) = (1-x) + e^{-2x}$
e) $f(x) = e^{-2x} - (1/(1-x))$
f) $f(x) = e^{-3x} - (1+x) + (1/(1-2x))$
g) $f(x) = e^{x^2}$

50. Find the sequence with each of these functions as its exponential generating function.

a) $f(x) = e^{3x}$ b) $f(x) = 2e^{-3x+1}$
c) $f(x) = e^{4x} + e^{-4x}$ d) $f(x) = (1+2x) + e^{3x}$
e) $f(x) = e^x - (1/(1+x))$
f) $f(x) = xe^x$ g) $f(x) = e^{x^3}$

51. A coding system encodes messages using strings of octal (base 8) digits. A codeword is considered valid if and only if it contains an even number of 7s.

a) Find a linear nonhomogeneous recurrence relation for the number of valid codewords of length n. What are the initial conditions?

b) Solve this recurrence relation using Theorem 6 in Section 8.2.

c) Solve this recurrence relation using generating functions.

*52. A coding system encodes messages using strings of base 4 digits (that is, digits from the set {0, 1, 2, 3}). A codeword is valid if and only if it contains an even number of 0s and an even number of 1s. Let a_n equal the number of valid codewords of length n. Furthermore, let b_n, c_n, and d_n equal the number of strings of base 4 digits of length n with an even number of 0s and an odd number of 1s, with an odd number of 0s and an even number of 1s, and with an odd number of 0s and an odd number of 1s, respectively.

a) Show that $d_n = 4^n - a_n - b_n - c_n$. Use this to show that $a_{n+1} = 2a_n + b_n + c_n$, $b_{n+1} = b_n - c_n + 4^n$, and $c_{n+1} = c_n - b_n + 4^n$.

b) What are a_1, b_1, c_1, and d_1?

c) Use parts (a) and (b) to find a_3, b_3, c_3, and d_3.

d) Use the recurrence relations in part (a), together with the initial conditions in part (b), to set up three equations relating the generating functions $A(x)$, $B(x)$, and $C(x)$ for the sequences $\{a_n\}$, $\{b_n\}$, and $\{c_n\}$, respectively.

e) Solve the system of equations from part (d) to get explicit formulae for $A(x)$, $B(x)$, and $C(x)$ and use these to get explicit formulae for a_n, b_n, c_n, and d_n.

Generating functions are useful in studying the number of different types of partitions of an integer n. A **partition** of a positive integer is a way to write this integer as the sum of positive integers where repetition is allowed and the order of the integers in the sum does not matter. For example, the partitions of 5 (with no restrictions) are $1+1+1+1+1$, $1+1+1+2$, $1+1+3$, $1+2+2$, $1+4$, $2+3$, and 5. Exercises 53–58 illustrate some of these uses.

53. Show that the coefficient $p(n)$ of x^n in the formal power series expansion of $1/((1-x)(1-x^2)(1-x^3)\cdots)$ equals the number of partitions of n.

54. Show that the coefficient $p_o(n)$ of x^n in the formal power series expansion of $1/((1-x)(1-x^3)(1-x^5)\cdots)$ equals the number of partitions of n into odd integers, that is, the number of ways to write n as the sum of odd positive integers, where the order does not matter and repetitions are allowed.

55. Show that the coefficient $p_d(n)$ of x^n in the formal power series expansion of $(1+x)(1+x^2)(1+x^3)\cdots$ equals the number of partitions of n into distinct parts, that is, the number of ways to write n as the sum of positive integers, where the order does not matter but no repetitions are allowed.

56. Find $p_o(n)$, the number of partitions of n into odd parts with repetitions allowed, and $p_d(n)$, the number of partitions of n into distinct parts, for $1 \le n \le 8$, by writing each partition of each type for each integer.

57. Show that if n is a positive integer, then the number of partitions of n into distinct parts equals the number of partitions of n into odd parts with repetitions allowed; that is, $p_o(n) = p_d(n)$. [*Hint:* Show that the generating functions for $p_o(n)$ and $p_d(n)$ are equal.]

**58. (*Requires calculus*) Use the generating function of $p(n)$ to show that $p(n) \le e^{C\sqrt{n}}$ for some constant C. [Hardy and Ramanujan showed that $p(n) \sim e^{\pi\sqrt{2/3}\sqrt{n}}/(4\sqrt{3}n)$, which means that the ratio of $p(n)$ and the right-hand side approaches 1 as n approaches infinity.]

Suppose that X is a random variable on a sample space S such that $X(s)$ is a nonnegative integer for all $s \in S$. The **probability generating function** for X is

Links

$$G_X(x) = \sum_{k=0}^{\infty} p(X(s) = k)x^k.$$

59. (*Requires calculus*) Show that if G_X is the probability generating function for a random variable X such that $X(s)$ is a nonnegative integer for all $s \in S$, then

 a) $G_X(1) = 1$. **b)** $E(X) = G_X'(1)$.

 c) $V(X) = G_X''(1) + G_X'(1) - G_X'(1)^2$.

60. Let X be the random variable whose value is n if the first success occurs on the nth trial when independent Bernoulli trials are performed, each with probability of success p.

 a) Find a closed formula for the probability generating function G_X.

 b) Find the expected value and the variance of X using Exercise 59 and the closed formula for the probability generating function found in part (a).

61. Let m be a positive integer. Let X_m be the random variable whose value is n if the mth success occurs on the $(n+m)$th trial when independent Bernoulli trials are performed, each with probability of success p.

 a) Using Exercise 32 in the Supplementary Exercises of Chapter 7, show that the probability generating function G_{X_m} is given by $G_{X_m}(x) = p^m/(1-qx)^m$, where $q = 1-p$.

 b) Find the expected value and the variance of X_m using Exercise 59 and the closed formula for the probability generating function in part (a).

62. Show that if X and Y are independent random variables on a sample space S such that $X(s)$ and $Y(s)$ are nonnegative integers for all $s \in S$, then $G_{X+Y}(x) = G_X(x)G_Y(x)$.

8.5 Inclusion–Exclusion

8.5.1 Introduction

A discrete mathematics class contains 30 computer science majors and 50 sophomores. How many students in the class are either computer science majors or sophomores? This question cannot be answered unless more information is provided. Adding the number of computer science majors in the class and the number of sophomores probably does not give the correct answer, because computer science majors who are sophomores are counted twice. This observation shows that the number of students in the class that are either sophomores or computer science majors is the sum of the number of computer science majors and the number of sophomores in the class minus the number of computer science majors who are sophomores. A technique for solving such counting problems was introduced in Section 6.1. In this section we will generalize the ideas introduced in that section to solve problems that require us to count the number of elements in the union of more than two sets.

8.5.2 The Principle of Inclusion–Exclusion

How many elements are in the union of two finite sets? In Section 2.2 we showed that the number of elements in the union of the two sets A and B is the sum of the numbers of elements in the sets minus the number of elements in their intersection. That is,

$$|A \cup B| = |A| + |B| - |A \cap B|.$$

As we showed in Section 6.1, the formula for the number of elements in the union of two sets is useful in counting problems. Examples 1–3 provide additional illustrations of the usefulness of this formula.

EXAMPLE 1 In a discrete mathematics class every student is a major in computer science or mathematics, or both. The number of students having computer science as a major (possibly along with mathematics) is 25; the number of students having mathematics as a major (possibly along with computer science) is 13; and the number of students majoring in both computer science and mathematics is 8. How many students are in this class?

Solution: Let A be the set of students in the class majoring in computer science and B be the set of students in the class majoring in mathematics. Then $A \cap B$ is the set of students in the class who are joint mathematics and computer science majors. Because every student in the class is majoring in either computer science or mathematics (or both), it follows that the number of students in the class is $|A \cup B|$. Therefore,

$$\begin{aligned} |A \cup B| &= |A| + |B| - |A \cap B| \\ &= 25 + 13 - 8 = 30. \end{aligned}$$

Therefore, there are 30 students in the class. This computation is illustrated in Figure 1. ◀

EXAMPLE 2 How many positive integers not exceeding 1000 are divisible by 7 or 11?

Solution: Let A be the set of positive integers not exceeding 1000 that are divisible by 7, and let B be the set of positive integers not exceeding 1000 that are divisible by 11. Then $A \cup B$ is the set of integers not exceeding 1000 that are divisible by either 7 or 11, and $A \cap B$ is the set of integers not exceeding 1000 that are divisible by both 7 and 11. From Example 2 of Section 4.1, we know that among the positive integers not exceeding 1000 there are $\lfloor 1000/7 \rfloor$ integers divisible by 7 and $\lfloor 1000/11 \rfloor$ divisible by 11. Because 7 and 11 are relatively prime, the integers divisible by both 7 and 11 are those divisible by $7 \cdot 11$. Consequently, there are $\lfloor 1000/(11 \cdot 7) \rfloor$ positive integers not exceeding 1000 that are divisible by both 7 and 11. It follows that there are

$$\begin{aligned} |A \cup B| &= |A| + |B| - |A \cap B| \\ &= \left\lfloor \frac{1000}{7} \right\rfloor + \left\lfloor \frac{1000}{11} \right\rfloor - \left\lfloor \frac{1000}{7 \cdot 11} \right\rfloor \\ &= 142 + 90 - 12 = 220 \end{aligned}$$

positive integers not exceeding 1000 that are divisible by either 7 or 11. This computation is illustrated in Figure 2. ◀

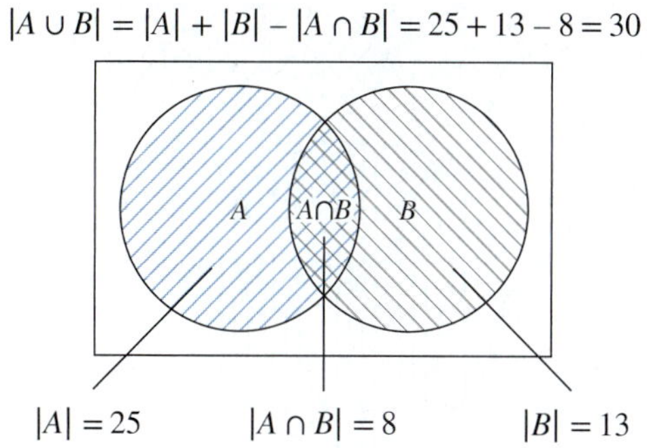

FIGURE 1 The set of students in a discrete mathematics class.

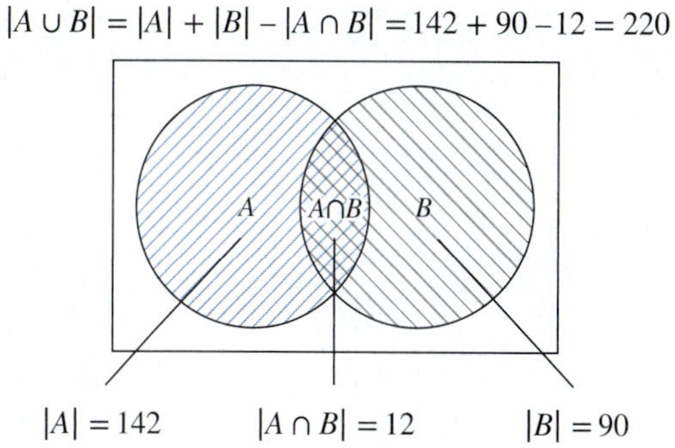

FIGURE 2 The set of positive integers not exceeding 1000 divisible by either 7 or 11.

Example 3 shows how to find the number of elements in a finite universal set that are outside the union of two sets.

EXAMPLE 3 Suppose that there are 1807 freshmen at your school. Of these, 453 are taking a course in computer science, 567 are taking a course in mathematics, and 299 are taking courses in both computer science and mathematics. How many are not taking a course either in computer science or in mathematics?

Solution: To find the number of freshmen who are not taking a course in either mathematics or computer science, subtract the number that are taking a course in either of these subjects from the total number of freshmen. Let A be the set of all freshmen taking a course in computer science, and let B be the set of all freshmen taking a course in mathematics. It follows that $|A| = 453$, $|B| = 567$, and $|A \cap B| = 299$. The number of freshmen taking a course in either computer science or mathematics is

$$|A \cup B| = |A| + |B| - |A \cap B| = 453 + 567 - 299 = 721.$$

Consequently, there are $1807 - 721 = 1086$ freshmen who are not taking a course in computer science or mathematics. ◀

We will now begin our development of a formula for the number of elements in the union of a finite number of sets. The formula we will develop is called the **principle of inclusion–exclusion**. For concreteness, before we consider unions of n sets, where n is any positive integer, we will derive a formula for the number of elements in the union of three sets A, B, and C. To construct this formula, we note that $|A| + |B| + |C|$ counts each element that is in exactly one of the three sets once, elements that are in exactly two of the sets twice, and elements in all three sets three times. This is illustrated in the first panel in Figure 3.

To remove the overcount of elements in more than one of the sets, we subtract the number of elements in the intersections of all pairs of the three sets. We obtain

$$|A| + |B| + |C| - |A \cap B| - |A \cap C| - |B \cap C|.$$

This expression still counts elements that occur in exactly one of the sets once. An element that occurs in exactly two of the sets is also counted exactly once, because this element will occur in one of the three intersections of sets taken two at a time. However, those elements that occur in all three sets will be counted zero times by this expression, because they occur

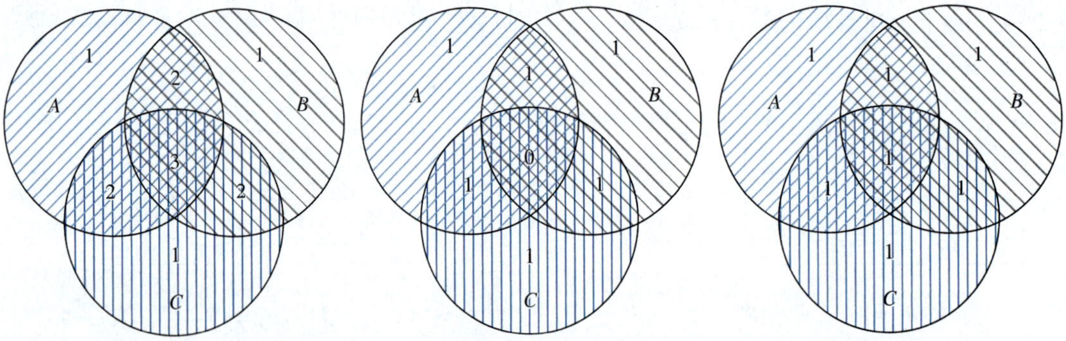

(a) Count of elements by $|A| + |B| + |C|$

(b) Count of elements by $|A| + |B| + |C| - |A \cap B| - |A \cap C| - |B \cap C|$

(c) Count of elements by $|A| + |B| + |C| - |A \cap B| - |A \cap C| - |B \cap C| + |A \cap B \cap C|$

FIGURE 3 Finding a formula for the number of elements in the union of three sets.

in all three intersections of sets taken two at a time. This is illustrated in the second panel in Figure 3.

To remedy this undercount, we add the number of elements in the intersection of all three sets. This final expression counts each element once, whether it is in one, two, or three of the sets. Thus,

$$|A \cup B \cup C| = |A| + |B| + |C| - |A \cap B| - |A \cap C| - |B \cap C| + |A \cap B \cap C|.$$

This formula is illustrated in the third panel of Figure 3.

Example 4 illustrates how this formula can be used.

EXAMPLE 4 A total of 1232 students have taken a course in Spanish, 879 have taken a course in French, and 114 have taken a course in Russian. Further, 103 have taken courses in both Spanish and French, 23 have taken courses in both Spanish and Russian, and 14 have taken courses in both French and Russian. If 2092 students have taken at least one of Spanish, French, and Russian, how many students have taken a course in all three languages?

Solution: Let S be the set of students who have taken a course in Spanish, F the set of students who have taken a course in French, and R the set of students who have taken a course in Russian. Then

$$|S| = 1232, \quad |F| = 879, \quad |R| = 114,$$
$$|S \cap F| = 103, \ |S \cap R| = 23, \ |F \cap R| = 14,$$

and

$$|S \cup F \cup R| = 2092.$$

When we insert these quantities into the equation

$$|S \cup F \cup R| = |S| + |F| + |R| - |S \cap F| - |S \cap R| - |F \cap R| + |S \cap F \cap R|$$

we obtain

$$2092 = 1232 + 879 + 114 - 103 - 23 - 14 + |S \cap F \cap R|.$$

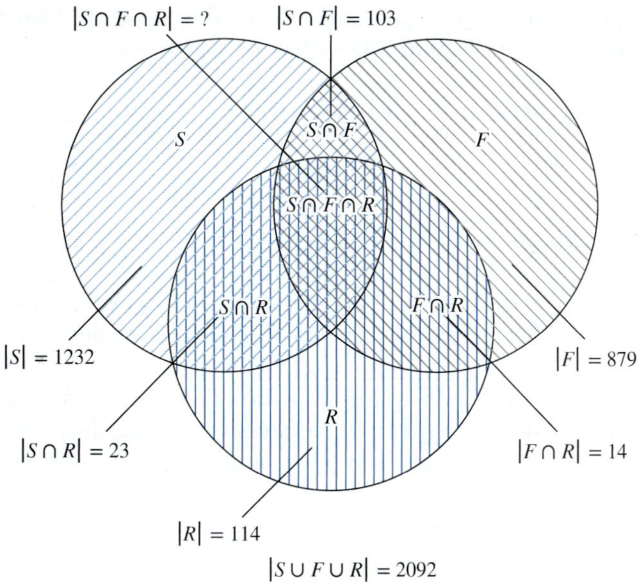

FIGURE 4 The set of students who have taken courses in Spanish, French, and Russian.

We now solve for $|S \cap F \cap R|$. We find that $|S \cap F \cap R| = 7$. Therefore, there are seven students who have taken courses in Spanish, French, and Russian. This is illustrated in Figure 4. ◀

We will now state and prove the **inclusion–exclusion principle** for n sets, where n is a positive integer. This priniciple tells us that we can count the elements in a union of n sets by adding the number of elements in the sets, then subtracting the sum of the number of elements in all intersections of two of these sets, then adding the number of elements in all intersections of three of these sets, and so on, until we reach the number of elements in the intersection of all the sets. It is added when there is an odd number of sets and subtracted when there is an even number of sets.

THEOREM 1 **THE PRINCIPLE OF INCLUSION–EXCLUSION** Let $A_1, A_2, \ldots, A_n$ be finite sets. Then

$$|A_1 \cup A_2 \cup \cdots \cup A_n| = \sum_{1 \le i \le n} |A_i| - \sum_{1 \le i < j \le n} |A_i \cap A_j| + \sum_{1 \le i < j < k \le n} |A_i \cap A_j \cap A_k| - \cdots + (-1)^{n+1} |A_1 \cap A_2 \cap \cdots \cap A_n|.$$

Proof: We will prove the formula by showing that an element in the union is counted exactly once by the right-hand side of the equation. Suppose that a is a member of exactly r of the sets $A_1, A_2, \ldots, A_n$ where $1 \le r \le n$. This element is counted $C(r, 1)$ times by $\Sigma |A_i|$. It is counted $C(r, 2)$ times by $\Sigma |A_i \cap A_j|$. In general, it is counted $C(r, m)$ times by the summation involving m of the sets A_i. Thus, this element is counted exactly

$$C(r, 1) - C(r, 2) + C(r, 3) - \cdots + (-1)^{r+1} C(r, r)$$

times by the expression on the right-hand side of this equation. Our goal is to evaluate this quantity. By Corollary 2 of Section 6.4, we have

$$C(r,0) - C(r,1) + C(r,2) - \cdots + (-1)^r C(r,r) = 0.$$

Hence,

$$1 = C(r,0) = C(r,1) - C(r,2) + \cdots + (-1)^{r+1} C(r,r).$$

Therefore, each element in the union is counted exactly once by the expression on the right-hand side of the equation. This proves the principle of inclusion–exclusion. ◁

The inclusion–exclusion principle gives a formula for the number of elements in the union of n sets for every positive integer n. There are terms in this formula for the number of elements in the intersection of every nonempty subset of the collection of the n sets. Hence, there are $2^n - 1$ terms in this formula.

EXAMPLE 5 Give a formula for the number of elements in the union of four sets.

Extra Examples

Solution: The inclusion–exclusion principle shows that

$$\begin{aligned}|A_1 \cup A_2 \cup A_3 \cup A_4| = {} & |A_1| + |A_2| + |A_3| + |A_4| \\ & - |A_1 \cap A_2| - |A_1 \cap A_3| - |A_1 \cap A_4| - |A_2 \cap A_3| - |A_2 \cap A_4| \\ & - |A_3 \cap A_4| + |A_1 \cap A_2 \cap A_3| + |A_1 \cap A_2 \cap A_4| + |A_1 \cap A_3 \cap A_4| \\ & + |A_2 \cap A_3 \cap A_4| - |A_1 \cap A_2 \cap A_3 \cap A_4|.\end{aligned}$$

Note that this formula contains 15 different terms, one for each nonempty subset of $\{A_1, A_2, A_3, A_4\}$. ◀

Exercises

1. How many elements are in $A_1 \cup A_2$ if there are 12 elements in A_1, 18 elements in A_2, and

a) $A_1 \cap A_2 = \emptyset$? **b)** $|A_1 \cap A_2| = 1$?

c) $|A_1 \cap A_2| = 6$? **d)** $A_1 \subseteq A_2$?

2. There are 345 students at a college who have taken a course in calculus, 212 who have taken a course in discrete mathematics, and 188 who have taken courses in both calculus and discrete mathematics. How many students have taken a course in either calculus or discrete mathematics?

3. A survey of households in the United States reveals that 96% have at least one television set, 98% have telephone service, and 95% have telephone service and at least one television set. What percentage of households in the United States have neither telephone service nor a television set?

4. A marketing report concerning personal computers states that 650,000 owners will buy a printer for their machines next year and 1,250,000 will buy at least one software package. If the report states that 1,450,000 owners will buy either a printer or at least one software package, how many will buy both a printer and at least one software package?

5. Find the number of elements in $A_1 \cup A_2 \cup A_3$ if there are 100 elements in each set and if

a) the sets are pairwise disjoint.

b) there are 50 common elements in each pair of sets and no elements in all three sets.

c) there are 50 common elements in each pair of sets and 25 elements in all three sets.

d) the sets are equal.

6. Find the number of elements in $A_1 \cup A_2 \cup A_3$ if there are 100 elements in A_1, 1000 in A_2, and 10,000 in A_3 if

a) $A_1 \subseteq A_2$ and $A_2 \subseteq A_3$.

b) the sets are pairwise disjoint.

c) there are two elements common to each pair of sets and one element in all three sets.

7. There are 2504 computer science students at a school. Of these, 1876 have taken a course in Java, 999 have taken a course in Linux, and 345 have taken a course in C. Further, 876 have taken courses in both Java and Linux, 231 have taken courses in both Linux and C, and 290 have taken courses in both Java and C. If 189 of these students have taken courses in Linux, Java, and C, how many of these 2504 students have not taken a course in any of these three subjects?

8. In a survey of 270 college students, it is found that 64 like Brussels sprouts, 94 like broccoli, 58 like cauliflower, 26 like both Brussels sprouts and broccoli, 28 like both Brussels sprouts and cauliflower, 22 like both broccoli and cauliflower, and 14 like all three vegetables. How many of the 270 students do not like any of these vegetables?

9. How many students are enrolled in a course either in calculus, discrete mathematics, data structures, or programming languages at a school if there are 507, 292, 312, and 344 students in these courses, respectively; 14 in both calculus and data structures; 213 in both calculus and programming languages; 211 in both discrete mathematics and data structures; 43 in both discrete mathematics and programming languages; and no student may take calculus and discrete mathematics, or data structures and programming languages, concurrently?

10. Find the number of positive integers not exceeding 100 that are not divisible by 5 or by 7.

11. Find the number of positive integers not exceeding 1000 that are not divisible by 3, 17, or 35.

12. Find the number of positive integers not exceeding 10,000 that are not divisible by 3, 4, 7, or 11.

13. Find the number of positive integers not exceeding 100 that are either odd or the square of an integer.

14. Find the number of positive integers not exceeding 1000 that are either the square or the cube of an integer.

15. How many bit strings of length eight do not contain six consecutive 0s?

***16.** How many permutations of the 26 letters of the English alphabet do not contain any of the strings *fish*, *rat* or *bird*?

17. How many permutations of the 10 digits either begin with the 3 digits 987, contain the digits 45 in the fifth and sixth positions, or end with the 3 digits 123?

18. How many elements are in the union of four sets if each of the sets has 100 elements, each pair of the sets shares 50 elements, each three of the sets share 25 elements, and there are 5 elements in all four sets?

19. How many elements are in the union of four sets if the sets have 50, 60, 70, and 80 elements, respectively, each pair of the sets has 5 elements in common, each triple of the sets has 1 common element, and no element is in all four sets?

20. How many terms are there in the formula for the number of elements in the union of 10 sets given by the principle of inclusion–exclusion?

21. Write out the explicit formula given by the principle of inclusion–exclusion for the number of elements in the union of five sets.

22. How many elements are in the union of five sets if the sets contain 10,000 elements each, each pair of sets has 1000 common elements, each triple of sets has 100 common elements, every four of the sets have 10 common elements, and there is 1 element in all five sets?

23. Write out the explicit formula given by the principle of inclusion–exclusion for the number of elements in the union of six sets when it is known that no three of these sets have a common intersection.

***24.** Prove the principle of inclusion–exclusion using mathematical induction.

25. Let E_1, E_2, and E_3 be three events from a sample space S. Find a formula for the probability of $E_1 \cup E_2 \cup E_3$.

26. Find the probability that when a fair coin is flipped five times tails comes up exactly three times, the first and last flips come up tails, or the second and fourth flips come up heads.

27. Find the probability that when four numbers from 1 to 100, inclusive, are picked at random with no repetitions allowed, either all are odd, all are divisible by 3, or all are divisible by 5.

28. Find a formula for the probability of the union of four events in a sample space if no three of them can occur at the same time.

29. Find a formula for the probability of the union of five events in a sample space if no four of them can occur at the same time.

30. Find a formula for the probability of the union of n events in a sample space when no two of these events can occur at the same time.

31. Find a formula for the probability of the union of n events in a sample space.

8.6 Applications of Inclusion–Exclusion

8.6.1 Introduction

Many counting problems can be solved using the principle of inclusion–exclusion. For instance, we can use this principle to find the number of primes less than a positive integer. Many problems can be solved by counting the number of onto functions from one finite set to another. The inclusion–exclusion principle can be used to find the number of such functions. The well-known hatcheck problem can be solved using the principle of inclusion–exclusion. This problem asks for the probability that no person is given the correct hat back by a hatcheck person who gives the hats back randomly.

8.6.2 An Alternative Form of Inclusion–Exclusion

There is an alternative form of the principle of inclusion–exclusion that is useful in counting problems. In particular, this form can be used to solve problems that ask for the number of elements in a set that have none of n properties $P_1, P_2, \dots, P_n$.

Let A_i be the subset containing the elements that have property P_i. The number of elements with all the properties $P_{i_1}, P_{i_2}, \dots, P_{i_k}$ will be denoted by $N(P_{i_1}P_{i_2}\dots P_{i_k})$. Writing these quantities in terms of sets, we have

$$|A_{i_1} \cap A_{i_2} \cap \cdots \cap A_{i_k}| = N(P_{i_1}P_{i_2}\dots P_{i_k}).$$

If the number of elements with none of the properties $P_1, P_2, \dots, P_n$ is denoted by $N(P'_1P'_2\dots P'_n)$ and the number of elements in the set is denoted by N, it follows that

$$N(P'_1P'_2\dots P'_n) = N - |A_1 \cup A_2 \cup \cdots \cup A_n|.$$

From the inclusion–exclusion principle, we see that

$$N(P'_1P'_2\dots P'_n) = N - \sum_{1\le i\le n} N(P_i) + \sum_{1\le i<j\le n} N(P_iP_j) - \sum_{1\le i<j<k\le n} N(P_iP_jP_k) + \cdots + (-1)^n N(P_1P_2\dots P_n).$$

Example 1 shows how the principle of inclusion–exclusion can be used to determine the number of solutions in integers of an equation with constraints.

EXAMPLE 1 How many solutions does

$$x_1 + x_2 + x_3 = 11$$

have, where x_1, x_2, and x_3 are nonnegative integers with $x_1 \le 3$, $x_2 \le 4$, and $x_3 \le 6$?

Solution: To apply the principle of inclusion–exclusion, let a solution have property P_1 if $x_1 > 3$, property P_2 if $x_2 > 4$, and property P_3 if $x_3 > 6$. The number of solutions satisfying the inequalities $x_1 \le 3$, $x_2 \le 4$, and $x_3 \le 6$ is

$$N(P'_1P'_2P'_3) = N - N(P_1) - N(P_2) - N(P_3) + N(P_1P_2) + N(P_1P_3) + N(P_2P_3) - N(P_1P_2P_3).$$

Using the same techniques as in Example 5 of Section 6.5, it follows that

- N = total number of solutions = $C(3 + 11 - 1, 11) = 78$,
- $N(P_1)$ = (number of solutions with $x_1 \ge 4$) = $C(3 + 7 - 1, 7) = C(9, 7) = 36$,
- $N(P_2)$ = (number of solutions with $x_2 \ge 5$) = $C(3 + 6 - 1, 6) = C(8, 6) = 28$,
- $N(P_3)$ = (number of solutions with $x_3 \ge 7$) = $C(3 + 4 - 1, 4) = C(6, 4) = 15$,

- ▶ $N(P_1P_2) =$ (number of solutions with $x_1 \geq 4$ and $x_2 \geq 5$) $= C(3+2-1, 2) = C(4, 2) = 6$,
- ▶ $N(P_1P_3) =$ (number of solutions with $x_1 \geq 4$ and $x_3 \geq 7$) $= C(3+0-1, 0) = 1$,
- ▶ $N(P_2P_3) =$ (number of solutions with $x_2 \geq 5$ and $x_3 \geq 7$) $= 0$,
- ▶ $N(P_1P_2P_3) =$ (number of solutions with $x_1 \geq 4$, $x_2 \geq 5$, and $x_3 \geq 7$) $= 0$.

Inserting these quantities into the formula for $N(P_1'P_2'P_3')$ shows that the number of solutions with $x_1 \leq 3$, $x_2 \leq 4$, and $x_3 \leq 6$ equals

$$N(P_1'P_2'P_3') = 78 - 36 - 28 - 15 + 6 + 1 + 0 - 0 = 6.$$ ◀

8.6.3 The Sieve of Eratosthenes

In Section 4.3 we showed how to use the sieve of Eratosthenes to find all primes less than a specified positive integer n. Using the principle of inclusion–exclusion, we can find the number of primes not exceeding a specified positive integer with the same reasoning as is used in the sieve of Eratosthenes. Recall that a composite integer is divisible by a prime not exceeding its square root. So, to find the number of primes not exceeding 100, first note that composite integers not exceeding 100 must have a prime factor not exceeding 10. Because the only primes not exceeding 10 are 2, 3, 5, and 7, the primes not exceeding 100 are these four primes and those positive integers greater than 1 and not exceeding 100 that are divisible by none of 2, 3, 5, or 7. To apply the principle of inclusion–exclusion, let P_1 be the property that an integer is divisible by 2, let P_2 be the property that an integer is divisible by 3, let P_3 be the property that an integer is divisible by 5, and let P_4 be the property that an integer is divisible by 7. Thus, the number of primes not exceeding 100 is

$$4 + N(P_1'P_2'P_3'P_4').$$

Because there are 99 positive integers greater than 1 and not exceeding 100, the principle of inclusion–exclusion shows that

$$\begin{aligned} N(P_1'P_2'P_3'P_4') = 99 &- N(P_1) - N(P_2) - N(P_3) - N(P_4) \\ &+ N(P_1P_2) + N(P_1P_3) + N(P_1P_4) + N(P_2P_3) + N(P_2P_4) + N(P_3P_4) \\ &- N(P_1P_2P_3) - N(P_1P_2P_4) - N(P_1P_3P_4) - N(P_2P_3P_4) \\ &+ N(P_1P_2P_3P_4). \end{aligned}$$

The number of integers not exceeding 100 (and greater than 1) that are divisible by all the primes in a subset of $\{2, 3, 5, 7\}$ is $\lfloor 100/N \rfloor$, where N is the product of the primes in this subset. (This follows because any two of these primes have no common factor.) Consequently,

$$\begin{aligned} N(P_1'P_2'P_3'P_4') = 99 &- \left\lfloor \frac{100}{2} \right\rfloor - \left\lfloor \frac{100}{3} \right\rfloor - \left\lfloor \frac{100}{5} \right\rfloor - \left\lfloor \frac{100}{7} \right\rfloor \\ &+ \left\lfloor \frac{100}{2 \cdot 3} \right\rfloor + \left\lfloor \frac{100}{2 \cdot 5} \right\rfloor + \left\lfloor \frac{100}{2 \cdot 7} \right\rfloor + \left\lfloor \frac{100}{3 \cdot 5} \right\rfloor + \left\lfloor \frac{100}{3 \cdot 7} \right\rfloor + \left\lfloor \frac{100}{5 \cdot 7} \right\rfloor \\ &- \left\lfloor \frac{100}{2 \cdot 3 \cdot 5} \right\rfloor - \left\lfloor \frac{100}{2 \cdot 3 \cdot 7} \right\rfloor - \left\lfloor \frac{100}{2 \cdot 5 \cdot 7} \right\rfloor - \left\lfloor \frac{100}{3 \cdot 5 \cdot 7} \right\rfloor + \left\lfloor \frac{100}{2 \cdot 3 \cdot 5 \cdot 7} \right\rfloor \\ &= 99 - 50 - 33 - 20 - 14 + 16 + 10 + 7 + 6 + 4 + 2 - 3 - 2 - 1 - 0 + 0 \\ &= 21. \end{aligned}$$

Hence, there are $4 + 21 = 25$ primes not exceeding 100.

8.6.4 The Number of Onto Functions

The principle of inclusion–exclusion can also be used to determine the number of onto functions from a set with m elements to a set with n elements. First consider Example 2.

EXAMPLE 2 How many onto functions are there from a set with six elements to a set with three elements?

Solution: Suppose that the elements in the codomain are b_1, b_2, and b_3. Let P_1, P_2, and P_3 be the properties that b_1, b_2, and b_3 are not in the range of the function, respectively. Note that a function is onto if and only if it has none of the properties P_1, P_2, or P_3. By the inclusion–exclusion principle it follows that the number of onto functions from a set with six elements to a set with three elements is

$$N(P_1'P_2'P_3') = N - [N(P_1) + N(P_2) + N(P_3)] \\ + [N(P_1P_2) + N(P_1P_3) + N(P_2P_3)] - N(P_1P_2P_3),$$

where N is the total number of functions from a set with six elements to one with three elements. We will evaluate each of the terms on the right-hand side of this equation.

From Example 6 of Section 6.1, it follows that $N = 3^6$. Note that $N(P_i)$ is the number of functions that do not have b_i in their range. Hence, there are two choices for the value of the function at each element of the domain. Therefore, $N(P_i) = 2^6$. Furthermore, there are $C(3, 1)$ terms of this kind. Note that $N(P_iP_j)$ is the number of functions that do not have b_i and b_j in their range. Hence, there is only one choice for the value of the function at each element of the domain. Therefore, $N(P_iP_j) = 1^6 = 1$. Furthermore, there are $C(3, 2)$ terms of this kind. Also, note that $N(P_1P_2P_3) = 0$, because this term is the number of functions that have none of b_1, b_2, and b_3 in their range. Clearly, there are no such functions, so the number of onto functions from a set with six elements to one with three elements is

$$3^6 - C(3, 1)2^6 + C(3, 2)1^6 = 729 - 192 + 3 = 540.$$ ◀

The general result that tells us how many onto functions there are from a set with m elements to one with n elements will now be stated. The proof of this result is left as an exercise for the reader.

THEOREM 1 Let m and n be positive integers with $m \geq n$. Then, there are

$$n^m - C(n, 1)(n-1)^m + C(n, 2)(n-2)^m - \cdots + (-1)^{n-1}C(n, n-1) \cdot 1^m$$

onto functions from a set with m elements to a set with n elements.

Counting onto functions is much harder than counting one-to-one functions!

An onto function from a set with m elements to a set with n elements corresponds to a way to distribute the m elements in the domain to n indistinguishable boxes so that no box is empty, and then to associate each of the n elements of the codomain to a box. This means that the number of onto functions from a set with m elements to a set with n elements is the number of ways to distribute m distinguishable objects to n indistinguishable boxes so that no box is empty multiplied by the number of permutations of a set with n elements. Consequently, the number of onto functions from a set with m elements to a set with n elements equals $n!S(m, n)$, where $S(m, n)$ is a *Stirling number of the second kind* defined in Section 6.5. This means that we can use Theorem 1 to deduce the formula given in Section 6.5 for $S(m, n)$. (See Chapter 6 of [MiRo91] for more details about Stirling numbers of the second kind.)

One of the many different applications of Theorem 1 will now be described.

EXAMPLE 3 How many ways are there to assign five different jobs to four different employees if every employee is assigned at least one job?

Solution: Consider the assignment of jobs as a function from the set of five jobs to the set of four employees. An assignment where every employee gets at least one job is the same as an onto function from the set of jobs to the set of employees. Hence, by Theorem 1 it follows that there are

$$4^5 - C(4, 1)3^5 + C(4, 2)2^5 - C(4, 3)1^5 = 1024 - 972 + 192 - 4 = 240$$

ways to assign the jobs so that each employee is assigned at least one job. ◀

8.6.5 Derangements

The principle of inclusion–exclusion will be used to count the permutations of n objects that leave no objects in their original positions. Consider Example 4.

EXAMPLE 4 **The Hatcheck Problem** A new employee checks the hats of n people at a restaurant, forgetting to put claim check numbers on the hats. When customers return for their hats, the checker gives them back hats chosen at random from the remaining hats. What is the probability that no one receives the correct hat? ◀

Remark: The answer is the number of ways the hats can be arranged so that there is no hat in its original position divided by $n!$, the number of permutations of n hats. We will return to this example after we find the number of permutations of n objects that leave no objects in their original position.

Links

A **derangement** is a permutation of objects that leaves no object in its original position. To solve the problem posed in Example 4 we will need to determine the number of derangements of a set of n objects.

EXAMPLE 5 The permutation 21453 is a derangement of 12345 because no number is left in its original position. However, 21543 is not a derangement of 12345, because this permutation leaves 4 fixed. ◀

Let D_n denote the number of derangements of n objects. For instance, $D_3 = 2$, because the derangements of 123 are 231 and 312. We will evaluate D_n, for all positive integers n, using the principle of inclusion–exclusion.

THEOREM 2 The number of derangements of a set with n elements is

$$D_n = n!\left[1 - \frac{1}{1!} + \frac{1}{2!} - \frac{1}{3!} + \cdots + (-1)^n \frac{1}{n!}\right].$$

Proof: Let a permutation have property P_i if it fixes element i. The number of derangements is the number of permutations having none of the properties P_i for $i = 1, 2, \ldots, n$. This means that

$$D_n = N(P'_1 P'_2 \ldots P'_n).$$

Using the principle of inclusion–exclusion, it follows that

$$D_n = N - \sum_i N(P_i) + \sum_{i<j} N(P_i P_j) - \sum_{i<j<k} N(P_i P_j P_k) + \cdots + (-1)^n N(P_1 P_2 \ldots P_n),$$

where N is the number of permutations of n elements. This equation states that the number of permutations that fix no elements equals the total number of permutations, less the number that fix at least one element, plus the number that fix at least two elements, less the number that fix at least three elements, and so on. All the quantities that occur on the right-hand side of this equation will now be found.

First, note that $N = n!$, because N is simply the total number of permutations of n elements. Also, $N(P_i) = (n - 1)!$. This follows from the product rule, because $N(P_i)$ is the number of permutations that fix element i, so the ith position of the permutation is determined, but each of the remaining positions can be filled arbitrarily. Similarly,

$$N(P_i P_j) = (n - 2)!,$$

because this is the number of permutations that fix elements i and j, but where the other $n - 2$ elements can be arranged arbitrarily. In general, note that

$$N(P_{i_1} P_{i_2} \ldots P_{i_m}) = (n - m)!,$$

because this is the number of permutations that fix elements $i_1, i_2, \ldots, i_m$, but where the other $n - m$ elements can be arranged arbitrarily. Because there are $C(n, m)$ ways to choose m elements from n, it follows that

$$\sum_{1 \le i \le n} N(P_i) = C(n, 1)(n - 1)!,$$

$$\sum_{1 \le i < j \le n} N(P_i P_j) = C(n, 2)(n - 2)!,$$

and in general,

$$\sum_{1 \le i_1 < i_2 < \cdots < i_m \le n} N(P_{i_1} P_{i_2} \ldots P_{i_m}) = C(n, m)(n - m)!.$$

Consequently, inserting these quantities into our formula for D_n gives

$$\begin{aligned} D_n &= n! - C(n, 1)(n - 1)! + C(n, 2)(n - 2)! - \cdots + (-1)^n C(n, n)(n - n)! \\ &= n! - \frac{n!}{1!(n-1)!}(n - 1)! + \frac{n!}{2!(n-2)!}(n - 2)! - \cdots + (-1)^n \frac{n!}{n!\,0!} 0!. \end{aligned}$$

Links

HISTORICAL NOTE In *rencontres* (matches), an old French card game, the 52 cards in a deck are laid out in a row. The cards of a second deck are laid out with one card of the second deck on top of each card of the first deck. The score is determined by counting the number of matching cards in the two decks. In 1708 Pierre Raymond de Montmort (1678–1719) posed *le problème de rencontres:* What is the probability that no matches take place in the game of rencontres? The solution to Montmort's problem is the probability that a randomly selected permutation of 52 objects is a derangement, namely, $D_{52}/52!$, which, as we will see, is approximately $1/e$.

TABLE 1 The Probability of a Derangement.

n	2	3	4	5	6	7
$D_n/n!$	0.50000	0.33333	0.37500	0.36667	0.36806	0.36786

Simplifying this expression gives

$$D_n = n!\left[1 - \frac{1}{1!} + \frac{1}{2!} - \cdots + (-1)^n \frac{1}{n!}\right].$$

It is now straightforward to find D_n for a given positive integer n. For instance, using Theorem 2, it follows that

$$D_3 = 3!\left[1 - \frac{1}{1!} + \frac{1}{2!} - \frac{1}{3!}\right] = 6\left(1 - 1 + \frac{1}{2} - \frac{1}{6}\right) = 2,$$

as we have previously remarked.

The solution of the problem in Example 4 can now be given.

Solution: The probability that no one receives the correct hat is $D_n/n!$. By Theorem 2, this probability is

$$\frac{D_n}{n!} = 1 - \frac{1}{1!} + \frac{1}{2!} - \cdots + (-1)^n \frac{1}{n!}.$$

The values of this probability for $2 \leq n \leq 7$ are displayed in Table 1.

By the identity $e^x = \sum_{j=0}^{\infty} x^j/j!$ for all real numbers x (from calculus), we know that

$$e^{-1} = 1 - \frac{1}{1!} + \frac{1}{2!} - \cdots + (-1)^n \frac{1}{n!} + \cdots \approx 0.368.$$

Because this is an alternating series with terms tending to zero, it follows that as n grows without bound, the probability that no one receives the correct hat converges to $e^{-1} \approx 0.368$. In fact, this probability can be shown to be within $1/(n+1)!$ of e^{-1}. ◀

Exercises

1. Suppose that in a bushel of 100 apples there are 20 that have worms in them and 15 that have bruises. Only those apples with neither worms nor bruises can be sold. If there are 10 bruised apples that have worms in them, how many of the 100 apples can be sold?
2. Of 1000 applicants for a mountain-climbing trip in the Himalayas, 450 get altitude sickness, 622 are not in good enough shape, and 30 have allergies. An applicant qualifies if and only if this applicant does not get altitude sickness, is in good shape, and does not have allergies. If there are 111 applicants who get altitude sickness and are not in good enough shape, 14 who get altitude sickness and have allergies, 18 who are not in good enough shape and have allergies, and 9 who get altitude sickness, are not in good enough shape, and have allergies, how many applicants qualify?
3. How many solutions does the equation $x_1 + x_2 + x_3 = 13$ have where x_1, x_2, and x_3 are nonnegative integers less than 6?
4. Find the number of solutions of the equation $x_1 + x_2 + x_3 + x_4 = 17$, where x_i, $i = 1, 2, 3, 4$, are nonnegative integers such that $x_1 \leq 3$, $x_2 \leq 4$, $x_3 \leq 5$, and $x_4 \leq 8$.
5. Find the number of primes less than 200 using the principle of inclusion–exclusion.

6. An integer is called **squarefree** if it is not divisible by the square of a positive integer greater than 1. Find the number of squarefree positive integers less than 100.

7. How many positive integers less than 10,000 are not the second or higher power of an integer?

8. How many onto functions are there from a set with seven elements to one with five elements?

9. How many ways are there to distribute six different toys to three different children such that each child gets at least one toy?

10. In how many ways can eight distinct balls be distributed into three distinct urns if each urn must contain at least one ball?

11. In how many ways can seven different jobs be assigned to four different employees so that each employee is assigned at least one job and the most difficult job is assigned to the best employee?

12. List all the derangements of $\{1, 2, 3, 4\}$.

13. How many derangements are there of a set with seven elements?

14. What is the probability that none of 10 people receives the correct hat if a hatcheck person hands their hats back randomly?

15. A machine that inserts letters into envelopes goes haywire and inserts letters randomly into envelopes. What is the probability that in a group of 100 letters

a) no letter is put into the correct envelope?

b) exactly one letter is put into the correct envelope?

c) exactly 98 letters are put into the correct envelopes?

d) exactly 99 letters are put into the correct envelopes?

e) all letters are put into the correct envelopes?

16. A group of n students is assigned seats for each of two classes in the same classroom. How many ways can these seats be assigned if no student is assigned the same seat for both classes?

***17.** How many ways can the digits 0, 1, 2, 3, 4, 5, 6, 7, 8, 9 be arranged so that no even digit is in its original position?

***18.** Use a combinatorial argument to show that the sequence $\{D_n\}$, where D_n denotes the number of derangements of n objects, satisfies the recurrence relation

$$D_n = (n-1)(D_{n-1} + D_{n-2})$$

for $n \geq 2$. [*Hint:* Note that there are $n - 1$ choices for the first element k of a derangement. Consider separately the derangements that start with k that do and do not have 1 in the kth position.]

***19.** Use Exercise 18 to show that

$$D_n = nD_{n-1} + (-1)^n$$

for $n \geq 1$.

20. Use Exercise 19 to find an explicit formula for D_n.

21. For which positive integers n is D_n, the number of derangements of n objects, even?

22. Suppose that p and q are distinct primes. Use the principle of inclusion–exclusion to find $\phi(pq)$, the number of positive integers not exceeding pq that are relatively prime to pq.

***23.** Use the principle of inclusion–exclusion to derive a formula for $\phi(n)$ when the prime factorization of n is

$$n = p_1^{a_1} p_2^{a_2} \cdots p_m^{a_m}.$$

***24.** Show that if n is a positive integer, then

$$n! = C(n, 0)D_n + C(n, 1)D_{n-1} + \cdots + C(n, n-1)D_1 + C(n, n)D_0,$$

where D_k is the number of derangements of k objects.

25. How many derangements of $\{1, 2, 3, 4, 5, 6\}$ begin with the integers 1, 2, and 3, in some order?

26. How many derangements of $\{1, 2, 3, 4, 5, 6\}$ end with the integers 1, 2, and 3, in some order?

27. Prove Theorem 1.

Key Terms and Results

TERMS

recurrence relation: a formula expressing terms of a sequence, except for some initial terms, as a function of one or more previous terms of the sequence

initial conditions for a recurrence relation: the values of the terms of a sequence satisfying the recurrence relation before this relation takes effect

dynamic programming: an algorithmic paradigm that finds the solution to an optimization problem by recursively breaking down the problem into overlapping subproblems and combining their solutions with the help of a recurrence relation

linear homogeneous recurrence relation with constant coefficients: a recurrence relation that expresses the terms of a sequence, except initial terms, as a linear combination of previous terms

characteristic roots of a linear homogeneous recurrence relation with constant coefficients: the roots of the polynomial associated with a linear homogeneous recurrence relation with constant coefficients

linear nonhomogeneous recurrence relation with constant coefficients: a recurrence relation that expresses the terms of a sequence, except for initial terms, as a linear combination of previous terms plus a function that is not identically zero that depends only on the index

divide-and-conquer algorithm: an algorithm that solves a problem recursively by splitting it into a fixed number of smaller nonoverlapping subproblems of the same type

generating function of a sequence: the formal series that has the nth term of the sequence as the coefficient of x^n

sieve of Eratosthenes: a procedure for finding the primes less than a specified positive integer

derangement: a permutation of objects such that no object is in its original place

RESULTS

the formula for the number of elements in the union of two finite sets:

$$|A \cup B| = |A| + |B| - |A \cap B|$$

the formula for the number of elements in the union of three finite sets:

$$|A \cup B \cup C| = |A| + |B| + |C| - |A \cap B| - |A \cap C| - |B \cap C| + |A \cap B \cap C|$$

the principle of inclusion–exclusion:

$$|A_1 \cup A_2 \cup \cdots \cup A_n| = \sum_{1 \le i \le n} |A_i| - \sum_{1 \le i < j \le n} |A_i \cap A_j| + \sum_{1 \le i < j < k \le n} |A_i \cap A_j \cap A_k| - \cdots + (-1)^{n+1} |A_1 \cap A_2 \cap \cdots \cap A_n|$$

the number of onto functions from a set with m elements to a set with n elements:

$$n^m - C(n, 1)(n-1)^m + C(n, 2)(n-2)^m - \cdots + (-1)^{n-1} C(n, n-1) \cdot 1^m$$

the number of derangements of n objects:

$$D_n = n!\left[1 - \frac{1}{1!} + \frac{1}{2!} - \cdots + (-1)^n \frac{1}{n!}\right]$$

Review Questions

1. a) What is a recurrence relation?

b) Find a recurrence relation for the amount of money that will be in an account after n years if \$1,000,000 is deposited in an account yielding 9% annually.

2. Explain how the Fibonacci numbers are used to solve Fibonacci's problem about rabbits.

3. a) Find a recurrence relation for the number of steps needed to solve the Tower of Hanoi puzzle.

b) Show how this recurrence relation can be solved using iteration.

4. a) Explain how to find a recurrence relation for the number of bit strings of length n not containing two consecutive 1s.

b) Describe another counting problem that has a solution satisfying the same recurrence relation.

5. a) What is dynamic programming and how are recurrence relations used in algorithms that follow this paradigm?

b) Explain how dynamic programming can be used to schedule talks in a lecture hall from a set of possible talks to maximize overall attendance.

6. Define a linear homogeneous recurrence relation of degree k.

7. a) Explain how to solve linear homogeneous recurrence relations of degree 2.

b) Solve the recurrence relation $a_n = 13a_{n-1} - 22a_{n-2}$ for $n \ge 2$ if $a_0 = 3$ and $a_1 = 15$.

c) Solve the recurrence relation $a_n = 14a_{n-1} - 49a_{n-2}$ for $n \ge 2$ if $a_0 = 3$ and $a_1 = 35$.

8. a) Explain how to find $f(b^k)$ where k is a positive integer if $f(n)$ satisfies the divide-and-conquer recurrence relation $f(n) = af(n/b) + g(n)$ whenever b divides the positive integer n.

b) Find $f(256)$ if $f(n) = 3f(n/4) + 5n/4$ and $f(1) = 7$.

9. a) Derive a divide-and-conquer recurrence relation for the number of comparisons used to find a number in a list using a binary search.

b) Give a big-O estimate for the number of comparisons used by a binary search from the divide-and-conquer recurrence relation you gave in (a) using Theorem 1 in Section 8.3.

10. a) Give a formula for the number of elements in the union of three sets.

b) Explain why this formula is valid.

c) Explain how to use the formula from (a) to find the number of integers not exceeding 1000 that are divisible by 6, 10, or 15.

d) Explain how to use the formula from (a) to find the number of solutions in nonnegative integers to the equation $x_1 + x_2 + x_3 + x_4 = 22$ with $x_1 < 8$, $x_2 < 6$, and $x_3 < 5$.

11. a) Give a formula for the number of elements in the union of four sets and explain why it is valid.

b) Suppose the sets A_1, A_2, A_3, and A_4 each contain 25 elements, the intersection of any two of these sets contains 5 elements, the intersection of any three of these sets contains 2 elements, and 1 element is in all four of the sets. How many elements are in the union of the four sets?

12. a) State the principle of inclusion–exclusion.

b) Outline a proof of this principle.

13. Explain how the principle of inclusion–exclusion can be used to count the number of onto functions from a set with m elements to a set with n elements.

14. a) How can you count the number of ways to assign m jobs to n employees so that each employee is assigned at least one job?

b) How many ways are there to assign seven jobs to three employees so that each employee is assigned at least one job?

15. Explain how the inclusion–exclusion principle can be used to count the number of primes not exceeding the positive integer n.

16. a) Define a derangement.

b) Why is counting the number of ways a hatcheck person can return hats to n people, so that no one receives the correct hat, the same as counting the number of derangements of n objects?

c) Explain how to count the number of derangements of n objects.

Supplementary Exercises

1. A group of 10 people begin a chain letter, with each person sending the letter to four other people. Each of these people sends the letter to four additional people.

a) Find a recurrence relation for the number of letters sent at the nth stage of this chain letter, if no person ever receives more than one letter.

b) What are the initial conditions for the recurrence relation in part (a)?

c) How many letters are sent at the nth stage of the chain letter?

2. A nuclear reactor has created 18 grams of a particular radioactive isotope. Every hour 1% of this radioactive isotope decays.

a) Set up a recurrence relation for the amount of this isotope left n hours after its creation.

b) What are the initial conditions for the recurrence relation in part (a)?

c) Solve this recurrence relation.

3. Every hour the U.S. government prints 10,000 more \$1 bills, 4000 more \$5 bills, 3000 more \$10 bills, 2500 more \$20 bills, 1000 more \$50 bills, and the same number of \$100 bills as it did the previous hour. In the initial hour 1000 of each bill were produced.

a) Set up a recurrence relation for the amount of money produced in the nth hour.

b) What are the initial conditions for the recurrence relation in part (a)?

c) Solve the recurrence relation for the amount of money produced in the nth hour.

d) Set up a recurrence relation for the total amount of money produced in the first n hours.

e) Solve the recurrence relation for the total amount of money produced in the first n hours.

4. Suppose that every hour there are two new bacteria in a colony for each bacterium that was present the previous hour, and that all bacteria 2 hours old die. The colony starts with 100 new bacteria.

a) Set up a recurrence relation for the number of bacteria present after n hours.

b) What is the solution of this recurrence relation?

c) When will the colony contain more than 1 million bacteria?

5. Messages are sent over a communications channel using two different signals. One signal requires 2 microseconds for transmittal, and the other signal requires 3 microseconds for transmittal. Each signal of a message is followed immediately by the next signal.

a) Find a recurrence relation for the number of different signals that can be sent in n microseconds.

b) What are the initial conditions of the recurrence relation in part (a)?

c) How many different messages can be sent in 12 microseconds?

6. A small post office has only 4-cent stamps, 6-cent stamps, and 10-cent stamps. Find a recurrence relation for the number of ways to form postage of n cents with these stamps if the order that the stamps are used matters. What are the initial conditions for this recurrence relation?

7. How many ways are there to form these postages using the rules described in Exercise 6?

a) 12 cents **b)** 14 cents
c) 18 cents **d)** 22 cents

8. Find the solutions of the simultaneous system of recurrence relations

$$a_n = a_{n-1} + b_{n-1}$$
$$b_n = a_{n-1} - b_{n-1}$$

with $a_0 = 1$ and $b_0 = 2$.

9. Solve the recurrence relation $a_n = a_{n-1}^2/a_{n-2}$ if $a_0 = 1$ and $a_1 = 2$. [*Hint:* Take logarithms of both sides to obtain a recurrence relation for the sequence $\log a_n$, $n = 0, 1, 2, \ldots$.]

***10.** Solve the recurrence relation $a_n = a_{n-1}^3 a_{n-2}^2$ if $a_0 = 2$ and $a_1 = 2$. (See the hint for Exercise 9.)

11. Find the solution of the recurrence relation $a_n = 3a_{n-1} - 3a_{n-2} + a_{n-3} + 1$ if $a_0 = 2$, $a_1 = 4$, and $a_2 = 8$.

12. Find the solution of the recurrence relation $a_n = 3a_{n-1} - 3a_{n-2} + a_{n-3}$ if $a_0 = 2$, $a_1 = 2$, and $a_2 = 4$.

***13.** Suppose that in Example 1 of Section 8.1 a pair of rabbits leaves the island after reproducing twice. Find a recurrence relation for the number of rabbits on the island in the middle of the nth month.

***14.** In this exercise we construct a dynamic programming algorithm for solving the problem of finding a subset S of items chosen from a set of n items where item i has a weight w_i, which is a positive integer, so that the total weight of the items in S is a maximum but does not exceed a fixed weight limit W. Let $M(j, w)$ denote the maximum total weight of the items in a subset of the first j items such that this total weight does not exceed w. This problem is known as the **knapsack problem**.

a) Show that if $w_j > w$, then $M(j, w) = M(j-1, w)$.

b) Show that if $w_j \leq w$, then $M(j, w) = \max(M(j-1, w), w_j + M(j-1, w-w_j))$.

c) Use (a) and (b) to construct a dynamic programming algorithm for determining the maximum total weight of items so that this total weight does not exceed W. In your algorithm store the values $M(j, w)$ as they are found.

d) Explain how you can use the values $M(j, w)$ computed by the algorithm in part (c) to find a subset of items with maximum total weight not exceeding W.

In Exercises 15–18 we develop a dynamic programming algorithm for finding a longest common subsequence of two sequences $a_1, a_2, \ldots, a_m$ and $b_1, b_2, \ldots, b_n$, an important problem in the comparison of DNA of different organisms.

15. Suppose that $c_1, c_2, \ldots, c_p$ is a longest common subsequence of the sequences $a_1, a_2, \ldots, a_m$ and $b_1, b_2, \ldots, b_n$.

a) Show that if $a_m = b_n$, then $c_p = a_m = b_n$ and $c_1, c_2, \ldots, c_{p-1}$ is a longest common subsequence of $a_1, a_2, \ldots, a_{m-1}$ and $b_1, b_2, \ldots, b_{n-1}$ when $p > 1$.

b) Suppose that $a_m \neq b_n$. Show that if $c_p \neq a_m$, then $c_1, c_2, \ldots, c_p$ is a longest common subsequence of $a_1, a_2, \ldots, a_{m-1}$ and $b_1, b_2, \ldots, b_n$ and also show that if $c_p \neq b_n$, then $c_1, c_2, \ldots, c_p$ is a longest common subsequence of $a_1, a_2, \ldots, a_m$ and $b_1, b_2, \ldots, b_{n-1}$.

16. Let $L(i, j)$ denote the length of a longest common subsequence of $a_1, a_2, \ldots, a_i$ and $b_1, b_2, \ldots, b_j$, where $0 \leq i \leq m$ and $0 \leq j \leq n$. Use parts (a) and (b) of Exercise 15 to show that $L(i, j)$ satisfies the recurrence relation $L(i, j) = L(i-1, j-1) + 1$ if both i and j are nonzero and $a_i = b_i$, and $L(i, j) = \max(L(i, j-1), L(i-1, j))$ if both i and j are nonzero and $a_i \neq b_i$, and the initial condition $L(i, j) = 0$ if $i = 0$ or $j = 0$.

17. Use Exercise 16 to construct a dynamic programming algorithm for computing the length of a longest common subsequence of two sequences $a_1, a_2, \ldots, a_m$ and $b_1, b_2, \ldots, b_n$, storing the values of $L(i, j)$ as they are found.

18. Develop an algorithm for finding a longest common subsequence of two sequences $a_1, a_2, \ldots, a_m$ and $b_1, b_2, \ldots, b_n$ using the values $L(i, j)$ found by the algorithm in Exercise 17.

19. Find the solution to the recurrence relation $f(n) = f(n/2) + n^2$ for $n = 2^k$ where k is a positive integer and $f(1) = 1$.

20. Find the solution to the recurrence relation $f(n) = 3f(n/5) + 2n^4$, when n is divisible by 5, for $n = 5^k$, where k is a positive integer and $f(1) = 1$.

21. Give a big-O estimate for the size of f in Exercise 20 if f is an increasing function.

22. Find a recurrence relation that describes the number of comparisons used by the following algorithm: Find the largest and second largest elements of a sequence of n numbers recursively by splitting the sequence into two subsequences with an equal number of terms, or where there is one more term in one subsequence than in the other, at each stage. Stop when subsequences with two terms are reached.

23. Give a big-O estimate for the number of comparisons used by the algorithm described in Exercise 22.

24. A sequence $a_1, a_2, \ldots, a_n$ is **unimodal** if and only if there is an index m, $1 \leq m \leq n$, such that $a_i < a_{i+1}$ when $1 \leq i < m$ and $a_i > a_{i+1}$ when $m \leq i < n$. That is, the terms of the sequence strictly increase until the mth term and they strictly decrease after it, which implies that a_m is the largest term. In this exercise, a_m will always denote the largest term of the unimodal sequence $a_1, a_2, \ldots, a_n$.

a) Show that a_m is the unique term of the sequence that is greater than both the term immediately preceding it and the term immediately following it.

b) Show that if $a_i < a_{i+1}$ where $1 \leq i < n$, then $i + 1 \leq m \leq n$.

c) Show that if $a_i > a_{i+1}$ where $1 \leq i < n$, then $1 < m < i$.

d) Develop a divide and conquer algorithm for locating the index m. [*Hint:* Suppose that $i < m < j$. Use parts (a), (b), and (c) to determine whether $\lfloor (i+j)/2 \rfloor + 1 \leq m \leq n$, $1 \leq m \leq \lfloor (i+j)/2 \rfloor - 1$, or $m = \lfloor (i+j)/2 \rfloor$.]

25. Show that the algorithm from Exercise 24 has worst-case time complexity $O(\log n)$ in terms of the number of comparisons.

Let $\{a_n\}$ be a sequence of real numbers. The **forward differences** of this sequence are defined recursively as follows: The **first forward difference** is $\Delta a_n = a_{n+1} - a_n$; the **($k$ + 1)st forward difference** $\Delta^{k+1} a_n$ is obtained from $\Delta^k a_n$ by $\Delta^{k+1} a_n = \Delta^k a_{n+1} - \Delta^k a_n$.

26. Find Δa_n, where

a) $a_n = 3$. **b)** $a_n = 4n + 7$. **c)** $a_n = n^2 + n + 1$.

27. Let $a_n = 3n^3 + n + 2$. Find $\Delta^k a_n$, where k equals

a) 2. **b)** 3. **c)** 4.

***28.** Suppose that $a_n = P(n)$, where P is a polynomial of degree d. Prove that $\Delta^{d+1} a_n = 0$ for all nonnegative integers n.

29. Let $\{a_n\}$ and $\{b_n\}$ be sequences of real numbers. Show that

$$\Delta(a_n b_n) = a_{n+1}(\Delta b_n) + b_n(\Delta a_n).$$

30. Show that if $F(x)$ and $G(x)$ are the generating functions for the sequences $\{a_k\}$ and $\{b_k\}$, respectively, and c and d are real numbers, then $(cF + dG)(x)$ is the generating function for $\{ca_k + db_k\}$.

31. (*Requires calculus*) This exercise shows how generating functions can be used to solve the recurrence relation $(n+1)a_{n+1} = a_n + (1/n!)$ for $n \geq 0$ with initial condition $a_0 = 1$.

a) Let $G(x)$ be the generating function for $\{a_n\}$. Show that $G'(x) = G(x) + e^x$ and $G(0) = 1$.

b) Show from part (a) that $(e^{-x}G(x))' = 1$, and conclude that $G(x) = xe^x + e^x$.

c) Use part (b) to find a closed formula for a_n.

32. Suppose that 14 students receive an A on the first exam in a discrete mathematics class, and 18 receive an A on the second exam. If 22 students received an A on either the first exam or the second exam, how many students received an A on both exams?

33. There are 323 farms in Monmouth County that have at least one of horses, cows, and sheep. If 224 have horses, 85 have cows, 57 have sheep, and 18 farms have all three types of animals, how many farms have exactly two of these three types of animals?

34. Queries to a database of student records at a college produced the following data: There are 2175 students at the college, 1675 of these are not freshmen, 1074 students have taken a course in calculus, 444 students have taken a course in discrete mathematics, 607 students are not freshmen and have taken calculus, 350 students have taken calculus and discrete mathematics, 201 students are not freshmen and have taken discrete mathematics, and 143 students are not freshmen and have taken both calculus and discrete mathematics. Can all the responses to the queries be correct?

35. Students in the school of mathematics at a university major in one or more of the following four areas: applied mathematics (AM), pure mathematics (PM), operations research (OR), and computer science (CS). How many students are in this school if (including joint majors) there are 23 students majoring in AM; 17 in PM; 44 in OR; 63 in CS; 5 in AM and PM; 8 in AM and CS; 4 in AM and OR; 6 in PM and CS; 5 in PM and OR; 14 in OR and CS; 2 in PM, OR, and CS; 2 in AM, OR, and CS; 1 in PM, AM, and OR; 1 in PM, AM, and CS; and 1 in all four fields.

36. How many terms are needed when the inclusion–exclusion principle is used to express the number of elements in the union of seven sets if no more than five of these sets have a common element?

37. How many solutions in positive integers are there to the equation $x_1 + x_2 + x_3 = 20$ with $2 < x_1 < 6$, $6 < x_2 < 10$, and $0 < x_3 < 5$?

38. How many positive integers less than 1,000,000 are

a) divisible by 2, 3, or 5?

b) not divisible by 7, 11, or 13?

c) divisible by 3 but not by 7?

39. How many positive integers less than 200 are

a) second or higher powers of integers?

b) either primes or second or higher powers of integers?

c) not divisible by the square of an integer greater than 1?

d) not divisible by the cube of an integer greater than 1?

e) not divisible by three or more primes?

***40.** How many ways are there to assign six different jobs to three different employees if the hardest job is assigned to the most experienced employee and the easiest job is assigned to the least experienced employee?

41. What is the probability that exactly one person is given back the correct hat by a hatcheck person who gives n people their hats back at random?

42. How many bit strings of length six do not contain four consecutive 1s?

43. What is the probability that a bit string of length six chosen at random contains at least four 1s?

Computer Projects

Write programs with these input and output.

1. Given a positive integer n, list all the moves required in the Tower of Hanoi puzzle to move n disks from one peg to another according to the rules of the puzzle.

2. Given a positive integer n and an integer k with $1 \leq k \leq n$, list all the moves used by the Frame–Stewart algorithm (described in the preamble to Exercise 38 of Section 8.1) to move n disks from one peg to another using four pegs according to the rules of the puzzle.

3. Given a positive integer n, list all the bit sequences of length n that do not have a pair of consecutive 0s.

4. Given an integer n greater than 1, write out all ways to parenthesize the product of $n + 1$ variables.

5. Given a set of n talks, their start and end times, and the number of attendees at each talk, use dynamic programming to schedule a subset of these talks in a single lecture hall to maximize total attendance.

6. Given matrices $\mathbf{A_1}, \mathbf{A_2}, \dots, \mathbf{A_n}$, with dimensions $m_1 \times m_2, m_2 \times m_3, \dots, m_n \times m_{n+1}$, respectively, each with integer entries, use dynamic programming, as outlined in Exercise 57 in Section 8.1, to find the minimum number of multiplications of integers needed to compute $\mathbf{A_1 A_2} \cdots \mathbf{A_n}$.

7. Given a recurrence relation $a_n = c_1 a_{n-1} + c_2 a_{n-2}$, where c_1 and c_2 are real numbers, initial conditions $a_0 = C_0$ and $a_1 = C_1$, and a positive integer k, find a_k using iteration.

8. Given a recurrence relation $a_n = c_1 a_{n-1} + c_2 a_{n-2}$ and initial conditions $a_0 = C_0$ and $a_1 = C_1$, determine the unique solution.
9. Given a recurrence relation of the form $f(n) = af(n/b) + c$, where a is a real number, b is a positive integer, and c is a real number, and a positive integer k, find $f(b^k)$ using iteration.
10. Given the number of elements in the intersection of three sets, the number of elements in each pairwise intersection of these sets, and the number of elements in each set, find the number of elements in their union.
11. Given a positive integer n, produce the formula for the number of elements in the union of n sets.
12. Given positive integers m and n, find the number of onto functions from a set with m elements to a set with n elements.
13. Given a positive integer n, list all the derangements of the set $\{1, 2, 3, \ldots, n\}$.

Computations and Explorations

Use a computational program or programs you have written to do these exercises.

1. Find the exact value of f_{100}, f_{500}, and f_{1000}, where f_n is the nth Fibonacci number.
2. Find the smallest Fibonacci number greater than 1,000,000, greater than 1,000,000,000, and greater than 1,000,000,000,000.
3. Find as many prime Fibonacci numbers as you can. It is unknown whether there are infinitely many of these.
4. Write out all the moves required to solve the Tower of Hanoi puzzle with 10 disks.
5. Write out all the moves required to use the Frame–Stewart algorithm to move 20 disks from one peg to another peg using four pegs according to the rules of the Reve's puzzle.
6. Verify the Frame conjecture for solving the Reve's puzzle for n disks for as many integers n as possible by showing that the puzzle cannot be solved using fewer moves than are made by the Frame–Stewart algorithm with the optimal choice of k.
7. Compute the number of operations required to multiply two integers with n bits for various integers n including 16, 64, 256, and 1024 using the fast multiplication described in Example 4 of Section 8.3 and the standard algorithm for multiplying integers (Algorithm 3 in Section 4.2).
8. Compute the number of operations required to multiply two $n \times n$ matrices for various integers n including 4, 16, 64, and 128 using the fast matrix multiplication described in Example 5 of Section 8.3 and the standard algorithm for multiplying matrices (Algorithm 1 in Section 3.3).
9. Find the number of primes not exceeding 10,000 using the method described in Section 8.6 to find the number of primes not exceeding 100.
10. List all the derangements of $\{1, 2, 3, 4, 5, 6, 7, 8\}$.
11. Compute the probability that a permutation of n objects is a derangement for all positive integers not exceeding 20 and determine how quickly these probabilities approach the number $1/e$.

Writing Projects

Respond to these with essays using outside sources.

1. Find the original source where Fibonacci presented his puzzle about modeling rabbit populations. Discuss this problem and other problems posed by Fibonacci and give some information about Fibonacci himself.
2. Explain how the Fibonacci numbers arise in a variety of applications, such as in phyllotaxis, the study of arrangement of leaves in plants, in the study of reflections by mirrors, and so on.
3. Describe different variations of the Tower of Hanoi puzzle, including those with more than three pegs (including the Reve's puzzle discussed in the text and exercises), those where disk moves are restricted, and those where disks may have the same size. Include what is known about the number of moves required to solve each variation.
4. Discuss as many different problems as possible where the Catalan numbers arise.
5. Discuss some of the problems in which Richard Bellman first used dynamic programming.
6. Describe the role dynamic programming algorithms play in bioinformatics including for DNA sequence comparison, gene comparison, and RNA structure prediction.
7. Describe the use of dynamic programming in economics including its use to study optimal consumption and saving.
8. Explain how dynamic programming can be used to solve the egg-dropping puzzle which determines which floors of a multistory building it is safe to drop eggs from without breaking.

9. Describe the solution of Ulam's problem (see Exercise 28 in Section 8.3) involving searching with one lie found by Andrzej Pelc.

10. Discuss variations of Ulam's problem (see Exercise 28 in Section 8.3) involving searching with more than one lie and what is known about this problem.

11. Define the convex hull of a set of points in the plane and describe three different algorithms, including a divide-and-conquer algorithm, for finding the convex hull of a set of points in the plane.

12. Describe how sieve methods are used in number theory. What kind of results have been established using such methods?

13. Look up the rules of the old French card game of *rencontres*. Describe these rules and describe the work of Pierre Raymond de Montmort on *le problème de rencontres*.

14. Describe how exponential generating functions can be used to solve a variety of counting problems.

15. Describe the Polyá theory of counting and the kind of counting problems that can be solved using this theory.

16. The *problème des ménages* (the problem of the households) asks for the number of ways to arrange n couples around a table so that the genders alternate and no husband and wife are seated together. Explain the method used by E. Lucas to solve this problem.

17. Explain how *rook polynomials* can be used to solve counting problems.

CHAPTER

9 Relations

9.1 Relations and Their Properties

9.2 n-ary Relations and Their Applications

9.3 Representing Relations

9.4 Closures of Relations

9.5 Equivalence Relations

9.6 Partial Orderings

Relationships between elements of sets occur in many contexts. Every day we deal with relationships such as those between a business and its telephone number, an employee and his or her salary, a person and a relative, and so on. In mathematics we study relationships such as those between a positive integer and one that it divides, an integer and one that it is congruent to modulo 5, a real number and one that is larger than it, a real number x and the value $f(x)$ where f is a function, and so on. Relationships such as that between a program and a variable it uses, and that between a computer language and a valid statement in this language, often arise in computer science. Relationships between elements of two sets are represented using the structure called a binary relation, which is just a subset of the Cartesian product of the sets. Relations can be used to solve problems such as determining which pairs of cities are linked by airline flights in a network, or finding a viable order for the different phases of a complicated project. We will introduce a number of different properties binary relations may enjoy.

Relationships between elements of more than two sets arise in many contexts. These relationships can be represented by n-ary relations, which are collections of n-tuples. Such relations are the basis of the relational data model, the most common way to store information in computer databases. We will introduce the terminology used to study relational databases, define some important operations on them, and introduce the database query language SQL. We will conclude our brief study of n-ary relations and databases with an important application from data mining. In particular, we will show how databases of transactions, represented by n-ary relations, are used to measure the likelihood that someone buys a particular product from a store when they buy one or more other products.

Two methods for representing relations, using matrices and using directed graphs, consisting of vertices and directed edges, will be introduced and used in later sections of the chapter. We will also study relationships that have certain collections of properties that relations may enjoy. For example, in some computer languages, only the first 31 characters of the name of a variable matter. The relation consisting of ordered pairs of strings in which the first string has the same initial 31 characters as the second string is an example of a special type of relation, known as an equivalence relation. Equivalence relations arise throughout mathematics and computer science. Finally, we will study relations called partial orderings, which generalize the notion of the less than or equal to relation. For example, the set of all pairs of strings of English letters in which the second string is the same as the first string or comes after the first in dictionary order is a partial ordering.

9.1 Relations and Their Properties

9.1.1 Introduction

Links

The most direct way to express a relationship between elements of two sets is to use ordered pairs made up of two related elements. For this reason, sets of ordered pairs are called binary relations. In this section we introduce the basic terminology used to describe binary relations. Later in this chapter we will use relations to solve problems involving communications networks, project scheduling, and identifying elements in sets with common properties.

Definition 1 Let A and B be sets. A *binary relation from A to B* is a subset of $A \times B$.

In other words, a binary relation from A to B is a set R of ordered pairs, where the first element of each ordered pair comes from A and the second element comes from B. We use the notation $a \, R \, b$ to denote that $(a, b) \in R$ and $a \not R b$ to denote that $(a, b) \notin R$. Moreover, when (a, b) belongs to R, a is said to be **related to** b by R.

Binary relations represent relationships between the elements of two sets. We will introduce n-ary relations, which express relationships among elements of more than two sets, later in this chapter. We will omit the word *binary* when there is no danger of confusion.

Examples 1–3 illustrate the notion of a relation.

EXAMPLE 1 Let A be the set of students in your school, and let B be the set of courses. Let R be the relation that consists of those pairs (a, b), where a is a student enrolled in course b. For instance, if Jason Goodfriend and Deborah Sherman are enrolled in CS518, the pairs (Jason Goodfriend, CS518) and (Deborah Sherman, CS518) belong to R. If Jason Goodfriend is also enrolled in CS510, then the pair (Jason Goodfriend, CS510) is also in R. However, if Deborah Sherman is not enrolled in CS510, then the pair (Deborah Sherman, CS510) is not in R.

Note that if a student is not currently enrolled in any courses there will be no pairs in R that have this student as the first element. Similarly, if a course is not currently being offered there will be no pairs in R that have this course as their second element. ◀

EXAMPLE 2 Let A be the set of cities in the U.S.A., and let B be the set of the 50 states in the U.S.A. Define the relation R by specifying that (a, b) belongs to R if a city with name a is in the state b. For instance, (Boulder, Colorado), (Bangor, Maine), (Ann Arbor, Michigan), (Middletown, New Jersey), (Middletown, New York), (Cupertino, California), and (Red Bank, New Jersey) are in R. ◀

EXAMPLE 3 Let $A = \{0, 1, 2\}$ and $B = \{a, b\}$. Then $\{(0, a), (0, b), (1, a), (2, b)\}$ is a relation from A to B. This means, for instance, that $0 \, R \, a$, but that $1 \not R b$. Relations can be represented graphically, as shown in Figure 1, using arrows to represent ordered pairs. Another way to represent this relation is to use a table, which is also done in Figure 1. We will discuss representations of relations in more detail in Section 9.3. ◀

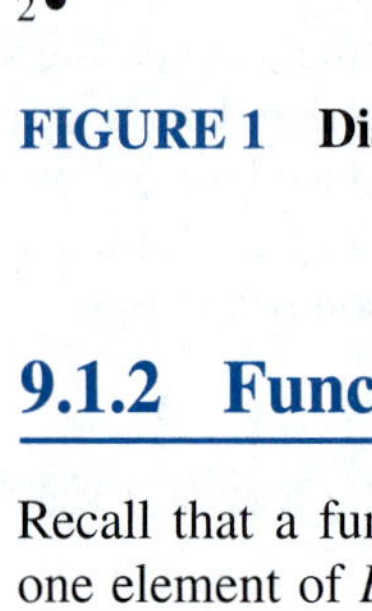

R	a	b
0	×	×
1	×	
2		×

FIGURE 1 Displaying the ordered pairs in the relation R from Example 3.

9.1.2 Functions as Relations

Recall that a function f from a set A to a set B (as defined in Section 2.3) assigns exactly one element of B to each element of A. The graph of f is the set of ordered pairs (a, b) such that $b = f(a)$. Because the graph of f is a subset of $A \times B$, it is a relation from A to B. Moreover, the graph of a function has the property that every element of A is the first element of exactly one ordered pair of the graph.

Conversely, if R is a relation from A to B such that every element in A is the first element of exactly one ordered pair of R, then a function can be defined with R as its graph. This can be done by assigning to an element a of A the unique element $b \in B$ such that $(a, b) \in R$. (Note that the relation R in Example 2 is not the graph of a function because Middletown occurs more than once as the first element of an ordered pair in R.)

A relation can be used to express a one-to-many relationship between the elements of the sets A and B (as in Example 2), where an element of A may be related to more than one element of B. A function represents a relation where exactly one element of B is related to each element of A.

Relations are a generalization of graphs of functions; they can be used to express a much wider class of relationships between sets. (Recall that the graph of the function f from A to B is the set of ordered pairs $(a, f(a))$ for $a \in A$.)

9.1.3 Relations on a Set

Relations from a set A to itself are of special interest.

Definition 2 A *relation on a set A* is a relation from A to A.

In other words, a relation on a set A is a subset of $A \times A$.

EXAMPLE 4 Let A be the set $\{1, 2, 3, 4\}$. Which ordered pairs are in the relation $R = \{(a, b) \mid a \text{ divides } b\}$?

Solution: Because (a, b) is in R if and only if a and b are positive integers not exceeding 4 such that a divides b, we see that

$$R = \{(1, 1), (1, 2), (1, 3), (1, 4), (2, 2), (2, 4), (3, 3), (4, 4)\}.$$

The pairs in this relation are displayed both graphically and in tabular form in Figure 2. ◀

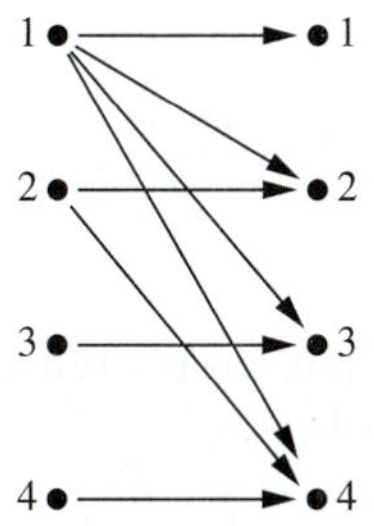

R	1	2	3	4
1	×	×	×	×
2		×		×
3			×	
4				×

FIGURE 2 Displaying the ordered pairs in the relation R from Example 4.

Next, some examples of relations on the set of integers will be given in Example 5.

EXAMPLE 5 Consider these relations on the set of integers:

$$R_1 = \{(a, b) \mid a \leq b\},$$
$$R_2 = \{(a, b) \mid a > b\},$$
$$R_3 = \{(a, b) \mid a = b \text{ or } a = -b\},$$
$$R_4 = \{(a, b) \mid a = b\},$$
$$R_5 = \{(a, b) \mid a = b + 1\},$$
$$R_6 = \{(a, b) \mid a + b \leq 3\}.$$

Which of these relations contain each of the pairs (1, 1), (1, 2), (2, 1), (1, −1), and (2, 2)?

Remark: Unlike the relations in Examples 1–4, these are relations on an infinite set.

Solution: The pair (1, 1) is in R_1, R_3, R_4, and R_6; (1, 2) is in R_1 and R_6; (2, 1) is in R_2, R_5, and R_6; (1, −1) is in R_2, R_3, and R_6; and finally, (2, 2) is in R_1, R_3, and R_4. ◀

It is not hard to determine the number of relations on a finite set, because a relation on a set A is simply a subset of $A \times A$.

EXAMPLE 6 How many relations are there on a set with n elements?

Solution: A relation on a set A is a subset of $A \times A$. Because $A \times A$ has n^2 elements when A has n elements, and a set with m elements has 2^m subsets, there are 2^{n^2} subsets of $A \times A$. Thus, there are 2^{n^2} relations on a set with n elements. For example, there are $2^{3^2} = 2^9 = 512$ relations on the set $\{a, b, c\}$. ◀

9.1.4 Properties of Relations

There are several properties that are used to classify relations on a set. We will introduce the most important of these here. (You may find it instructive to study this material with the contents of Section 9.3. In that section, several methods for representing relations will be introduced that can help you understand each of the properties that we introduce here.)

In some relations an element is always related to itself. For instance, let R be the relation on the set of all people consisting of pairs (x, y) where x and y have the same mother and the same father. Then xRx for every person x.

Definition 3 A relation R on a set A is called *reflexive* if $(a, a) \in R$ for every element $a \in A$.

Remark: Using quantifiers we see that the relation R on the set A is reflexive if $\forall a((a, a) \in R)$, where the universe of discourse is the set of all elements in A.

We see that a relation on A is reflexive if every element of A is related to itself. Examples 7–9 illustrate the concept of a reflexive relation.

EXAMPLE 7 Consider the following relations on $\{1, 2, 3, 4\}$:

$$R_1 = \{(1, 1), (1, 2), (2, 1), (2, 2), (3, 4), (4, 1), (4, 4)\},$$
$$R_2 = \{(1, 1), (1, 2), (2, 1)\},$$
$$R_3 = \{(1, 1), (1, 2), (1, 4), (2, 1), (2, 2), (3, 3), (4, 1), (4, 4)\},$$
$$R_4 = \{(2, 1), (3, 1), (3, 2), (4, 1), (4, 2), (4, 3)\},$$
$$R_5 = \{(1, 1), (1, 2), (1, 3), (1, 4), (2, 2), (2, 3), (2, 4), (3, 3), (3, 4), (4, 4)\},$$
$$R_6 = \{(3, 4)\}.$$

Which of these relations are reflexive?

Solution: The relations R_3 and R_5 are reflexive because they both contain all pairs of the form (a, a), namely, $(1, 1)$, $(2, 2)$, $(3, 3)$, and $(4, 4)$. The other relations are not reflexive because they do not contain all of these ordered pairs. In particular, R_1, R_2, R_4, and R_6 are not reflexive because $(3, 3)$ is not in any of these relations. ◀

EXAMPLE 8 Which of the relations from Example 5 are reflexive?

Solution: The reflexive relations from Example 5 are R_1 (because $a \le a$ for every integer a), R_3, and R_4. For each of the other relations in this example it is easy to find a pair of the form (a, a) that is not in the relation. (This is left as an exercise for the reader.) ◀

EXAMPLE 9 Is the "divides" relation on the set of positive integers reflexive?

Solution: Because $a \mid a$ whenever a is a positive integer, the "divides" relation is reflexive. (Note that if we replace the set of positive integers with the set of all integers the relation is not reflexive because by definition 0 does not divide 0.) ◀

In some relations an element is related to a second element if and only if the second element is also related to the first element. The relation consisting of pairs (x, y), where x and y are students at your school with at least one common class has this property. Other relations have the property that if an element is related to a second element, then this second element is not related to the first. The relation consisting of the pairs (x, y), where x and y are students at your school, where x has a higher grade point average than y has this property.

Definition 4 A relation R on a set A is called *symmetric* if $(b, a) \in R$ whenever $(a, b) \in R$, for all $a, b \in A$. A relation R on a set A such that for all $a, b \in A$, if $(a, b) \in R$ and $(b, a) \in R$, then $a = b$ is called *antisymmetric*.

Remark: Using quantifiers, we see that the relation R on the set A is symmetric if $\forall a \forall b((a, b) \in R \rightarrow (b, a) \in R)$. Similarly, the relation R on the set A is antisymmetric if $\forall a \forall b(((a, b) \in R \wedge (b, a) \in R) \rightarrow (a = b))$.

In other words, a relation is symmetric if and only if a is related to b always implies that b is related to a. For instance, the equality relation is symmetric because $a = b$ if and only if $b = a$. A relation is antisymmetric if and only if there are no pairs of distinct elements a and b with a related to b and b related to a. That is, the only way to have a related to b and b related to a is for a and b to be the same element. For instance, the less than or equal to relation is antisymmetric. To see this, note that $a \leq b$ and $b \leq a$ implies that $a = b$. Additionally, the strictly less than relation is also antisymmetric. The reason is that $a < b$ and $b < a$ is a contradiction, so the definition is true vacuously. The terms *symmetric* and *antisymmetric* are not opposites, because a relation can have both of these properties or may lack both of them (see Exercise 10). A relation cannot be both symmetric and antisymmetric if it contains some pair of the form (a, b) in which $a \neq b$.

Remark: Although relatively few of the 2^{n^2} relations on a set with n elements are symmetric or antisymmetric, as counting arguments can show, many important relations have one of these properties. (See Exercise 49.)

EXAMPLE 10 Which of the relations from Example 7 are symmetric and which are antisymmetric?

Extra Examples

Solution: The relations R_2 and R_3 are symmetric, because in each case (b, a) belongs to the relation whenever (a, b) does. For R_2, the only thing to check is that both $(2, 1)$ and $(1, 2)$ are in the relation. For R_3, it is necessary to check that both $(1, 2)$ and $(2, 1)$ belong to the relation, and $(1, 4)$ and $(4, 1)$ belong to the relation. The reader should verify that none of the other relations is symmetric. This is done by finding a pair (a, b) such that it is in the relation but (b, a) is not.

R_4, R_5, and R_6 are all antisymmetric. For each of these relations there is no pair of elements a and b with $a \neq b$ such that both (a, b) and (b, a) belong to the relation. The reader should verify that none of the other relations is antisymmetric. This is done by finding a pair (a, b) with $a \neq b$ such that (a, b) and (b, a) are both in the relation. ◀

EXAMPLE 11 Which of the relations from Example 5 are symmetric and which are antisymmetric?

Solution: The relations R_3, R_4, and R_6 are symmetric. R_3 is symmetric, for if $a = b$ or $a = -b$, then $b = a$ or $b = -a$. R_4 is symmetric because $a = b$ implies that $b = a$. R_6 is symmetric because $a + b \leq 3$ implies that $b + a \leq 3$. The reader should verify that none of the other relations is symmetric.

The relations R_1, R_2, R_4, and R_5 are antisymmetric. R_1 is antisymmetric because the inequalities $a \leq b$ and $b \leq a$ imply that $a = b$. R_2 is antisymmetric because it is impossible that $a > b$ and $b > a$. R_4 is antisymmetric, because two elements are related with respect to R_4 if and only if they are equal. R_5 is antisymmetric because it is impossible that $a = b + 1$ and $b = a + 1$. The reader should verify that none of the other relations is antisymmetric. ◀

EXAMPLE 12 Is the "divides" relation on the set of positive integers symmetric? Is it antisymmetric?

Solution: This relation is not symmetric because $1 \mid 2$, but $2 \nmid 1$. However, it is antisymmetric. To see this, note that if a and b are positive integers with $a \mid b$ and $b \mid a$, then $a = b$ (the verification of this is left as an exercise for the reader). ◀

Let R be the relation consisting of all pairs (x, y) of students at your school, where x has taken more credits than y. Suppose that x is related to y and y is related to z. This means that x has taken more credits than y and y has taken more credits than z. We can conclude that x has taken more credits than z, so that x is related to z. What we have shown is that R has the transitive property, which is defined as follows.

Definition 5 A relation R on a set A is called *transitive* if whenever $(a, b) \in R$ and $(b, c) \in R$, then $(a, c) \in R$, for all $a, b, c \in A$.

Remark: Using quantifiers we see that the relation R on a set A is transitive if we have $\forall a \forall b \forall c(((a, b) \in R \wedge (b, c) \in R) \rightarrow (a, c) \in R)$.

EXAMPLE 13 Which of the relations in Example 7 are transitive?

Extra Examples

Solution: R_4, R_5, and R_6 are transitive. For each of these relations, we can show that it is transitive by verifying that if (a, b) and (b, c) belong to this relation, then (a, c) also does. For instance, R_4 is transitive, because (3, 2) and (2, 1), (4, 2) and (2, 1), (4, 3) and (3, 1), and (4, 3) and (3, 2) are the only such sets of pairs, and (3, 1), (4, 1), and (4, 2) belong to R_4. The reader should verify that R_5 and R_6 are transitive.

R_1 is not transitive because (3, 4) and (4, 1) belong to R_1, but (3, 1) does not. R_2 is not transitive because (2, 1) and (1, 2) belong to R_2, but (2, 2) does not. R_3 is not transitive because (4, 1) and (1, 2) belong to R_3, but (4, 2) does not. ◀

EXAMPLE 14 Which of the relations in Example 5 are transitive?

Solution: The relations R_1, R_2, R_3, and R_4 are transitive. R_1 is transitive because $a \leq b$ and $b \leq c$ imply that $a \leq c$. R_2 is transitive because $a > b$ and $b > c$ imply that $a > c$. R_3 is transitive because $a = \pm b$ and $b = \pm c$ imply that $a = \pm c$. R_4 is clearly transitive, as the reader should verify. R_5 is not transitive because (2, 1) and (1, 0) belong to R_5, but (2, 0) does not. R_6 is not transitive because (2, 1) and (1, 2) belong to R_6, but (2, 2) does not. ◀

EXAMPLE 15 Is the "divides" relation on the set of positive integers transitive?

Solution: Suppose that a divides b and b divides c. Then there are positive integers k and l such that $b = ak$ and $c = bl$. Hence, $c = a(kl)$, so a divides c. It follows that this relation is transitive. ◀

We can use counting techniques to determine the number of relations with specific properties. Finding the number of relations with a particular property provides information about how common this property is in the set of all relations on a set with n elements.

EXAMPLE 16 How many reflexive relations are there on a set with n elements?

Solution: A relation R on a set A is a subset of $A \times A$. Consequently, a relation is determined by specifying whether each of the n^2 ordered pairs in $A \times A$ is in R. However, if R is reflexive, each of the n ordered pairs (a, a) for $a \in A$ must be in R. Each of the other $n(n - 1)$ ordered pairs of the form (a, b), where $a \neq b$, may or may not be in R. Hence, by the product rule for counting, there are $2^{n(n-1)}$ reflexive relations [this is the number of ways to choose whether each element (a, b), with $a \neq b$, belongs to R]. ◀

Formulas for the number of symmetric relations and the number of antisymmetric relations on a set with n elements can be found using reasoning similar to that in Example 16 (see Exercise 49). However, no general formula is known that counts the transitive relations on a set with n elements. Currently, $T(n)$, the number of transitive relations on a set with n elements, is known only for $0 \leq n \leq 18$. For example, $T(4) = 3{,}994$, $T(5) = 154{,}303$, and $T(6) = 9{,}415{,}189$. (The values of $T(n)$ for $n = 0, 1, 2, \ldots, 18$, are the terms of the sequence A006905 in the OEIS, which is discussed in Section 2.4.)

9.1.5 Combining Relations

Because relations from A to B are subsets of $A \times B$, two relations from A to B can be combined in any way two sets can be combined. Consider Examples 17–19.

EXAMPLE 17 Let $A = \{1, 2, 3\}$ and $B = \{1, 2, 3, 4\}$. The relations $R_1 = \{(1, 1), (2, 2), (3, 3)\}$ and $R_2 = \{(1, 1), (1, 2), (1, 3), (1, 4)\}$ can be combined to obtain

$$\begin{aligned} R_1 \cup R_2 &= \{(1, 1), (1, 2), (1, 3), (1, 4), (2, 2), (3, 3)\}, \\ R_1 \cap R_2 &= \{(1, 1)\}, \\ R_1 - R_2 &= \{(2, 2), (3, 3)\}, \\ R_2 - R_1 &= \{(1, 2), (1, 3), (1, 4)\}, \\ R_1 \oplus R_2 &= (R_1 \cup R_2) - (R_1 \cap R_2) = \{(1, 2), (1, 3), (1, 4), (2, 2), (3, 3)\}. \end{aligned}$$ ◀

EXAMPLE 18 Let A and B be the set of all students and the set of all courses at a school, respectively. Suppose that R_1 consists of all ordered pairs (a, b), where a is a student who has taken course b, and R_2 consists of all ordered pairs (a, b), where a is a student who requires course b to graduate. What are the relations $R_1 \cup R_2$, $R_1 \cap R_2$, $R_1 \oplus R_2$, $R_1 - R_2$, and $R_2 - R_1$?

Solution: The relation $R_1 \cup R_2$ consists of all ordered pairs (a, b), where a is a student who either has taken course b or needs course b to graduate, and $R_1 \cap R_2$ is the set of all ordered pairs (a, b), where a is a student who has taken course b and needs this course to graduate. Also, $R_1 \oplus R_2$ consists of all ordered pairs (a, b), where student a has taken course b but does not need it to graduate or needs course b to graduate but has not taken it. $R_1 - R_2$ is the set of ordered pairs (a, b), where a has taken course b but does not need it to graduate; that is, b is an elective course that a has taken. $R_2 - R_1$ is the set of all ordered pairs (a, b), where b is a course that a needs to graduate but has not taken. ◀

EXAMPLE 19 Let R_1 be the less than relation on the set of real numbers and let R_2 be the greater than relation on the set of real numbers, that is, $R_1 = \{(x, y) \mid x < y\}$ and $R_2 = \{(x, y) \mid x > y\}$. What are $R_1 \cup R_2$, $R_1 \cap R_2$, $R_1 - R_2$, $R_2 - R_1$, and $R_1 \oplus R_2$?

Solution: We note that $(x, y) \in R_1 \cup R_2$ if and only if $(x, y) \in R_1$ or $(x, y) \in R_2$. Hence, $(x, y) \in R_1 \cup R_2$ if and only if $x < y$ or $x > y$. Because the condition $x < y$ or $x > y$ is the same as the condition $x \neq y$, it follows that $R_1 \cup R_2 = \{(x, y) \mid x \neq y\}$. In other words, the union of the less than relation and the greater than relation is the not equals relation.

Next, note that it is impossible for a pair (x, y) to belong to both R_1 and R_2 because it is impossible that $x < y$ and $x > y$. It follows that $R_1 \cap R_2 = \emptyset$. We also see that $R_1 - R_2 = R_1$, $R_2 - R_1 = R_2$, and $R_1 \oplus R_2 = R_1 \cup R_2 - R_1 \cap R_2 = \{(x, y) \mid x \neq y\}$. ◀

There is another way that relations are combined that is analogous to the composition of functions.

Definition 6 Let R be a relation from a set A to a set B and S a relation from B to a set C. The *composite* of R and S is the relation consisting of ordered pairs (a, c), where $a \in A$, $c \in C$, and for which there exists an element $b \in B$ such that $(a, b) \in R$ and $(b, c) \in S$. We denote the composite of R and S by $S \circ R$.

Computing the composite of two relations requires that we find elements that are the second element of ordered pairs in the first relation and the first element of ordered pairs in the second relation, as Examples 20 and 21 illustrate.

EXAMPLE 20 What is the composite of the relations R and S, where R is the relation from $\{1, 2, 3\}$ to $\{1, 2, 3, 4\}$ with $R = \{(1, 1), (1, 4), (2, 3), (3, 1), (3, 4)\}$ and S is the relation from $\{1, 2, 3, 4\}$ to $\{0, 1, 2\}$ with $S = \{(1, 0), (2, 0), (3, 1), (3, 2), (4, 1)\}$?

Solution: $S \circ R$ is constructed using all ordered pairs in R and ordered pairs in S, where the second element of the ordered pair in R agrees with the first element of the ordered pair in S. For example, the ordered pairs $(2, 3)$ in R and $(3, 1)$ in S produce the ordered pair $(2, 1)$ in $S \circ R$. Computing all the ordered pairs in the composite, we find

$$S \circ R = \{(1, 0), (1, 1), (2, 1), (2, 2), (3, 0), (3, 1)\}.$$

Figure 3 illustrates how this composition is found. In the figure, we examine all paths that travel via two directed edges from the leftmost elements to the rightmost elements via an element in the middle.

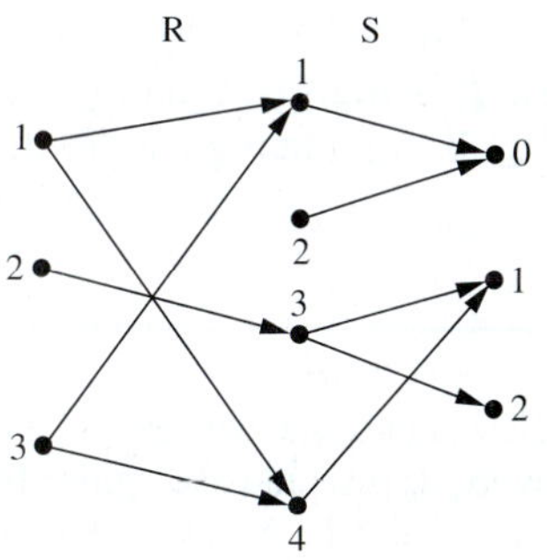

$1 \to 1 \to 0$	$(1, 0)$
$1 \to 4 \to 1$	$(1, 1)$
$2 \to 3 \to 1$	$(2, 1)$
$2 \to 3 \to 2$	$(2, 2)$
$3 \to 1 \to 0$	$(3, 0)$
$3 \to 4 \to 1$	$(3, 1)$

FIGURE 3 **Constructing $S \circ R$.** ◀

EXAMPLE 21 **Composing the Parent Relation with Itself** Let R be the relation on the set of all people such that $(a, b) \in R$ if person a is a parent of person b. Then $(a, c) \in R \circ R$ if and only if there is a person b such that $(a, b) \in R$ and $(b, c) \in R$, that is, if and only if there is a person b such that a is a parent of b and b is a parent of c. In other words, $(a, c) \in R \circ R$ if and only if a is a grandparent of c. ◀

The powers of a relation R can be recursively defined from the definition of a composite of two relations.

Definition 7 Let R be a relation on the set A. The powers R^n, $n = 1, 2, 3, \ldots$, are defined recursively by

$$R^1 = R \qquad \text{and} \qquad R^{n+1} = R^n \circ R.$$

The definition shows that $R^2 = R \circ R$, $R^3 = R^2 \circ R = (R \circ R) \circ R$, and so on.

EXAMPLE 22 Let $R = \{(1, 1), (2, 1), (3, 2), (4, 3)\}$. Find the powers R^n, $n = 2, 3, 4, \ldots$.

Solution: Because $R^2 = R \circ R$, we find that $R^2 = \{(1, 1), (2, 1), (3, 1), (4, 2)\}$. Furthermore, because $R^3 = R^2 \circ R$, $R^3 = \{(1, 1), (2, 1), (3, 1), (4, 1)\}$. Additional computation shows that R^4

is the same as R^3, so $R^4 = \{(1, 1), (2, 1), (3, 1), (4, 1)\}$. It also follows that $R^n = R^3$ for $n = 5, 6, 7, \ldots$. The reader should verify this. ◀

The following theorem shows that the powers of a transitive relation are subsets of this relation. It will be used in Section 9.4.

THEOREM 1 The relation R on a set A is transitive if and only if $R^n \subseteq R$ for $n = 1, 2, 3, \ldots$.

Proof: We first prove the "if" part of the theorem. We suppose that $R^n \subseteq R$ for $n = 1, 2, 3, \ldots$. In particular, $R^2 \subseteq R$. To see that this implies R is transitive, note that if $(a, b) \in R$ and $(b, c) \in R$, then by the definition of composition, $(a, c) \in R^2$. Because $R^2 \subseteq R$, this means that $(a, c) \in R$. Hence, R is transitive.

We will use mathematical induction to prove the only if part of the theorem. Note that this part of the theorem is trivially true for $n = 1$.

Assume that $R^n \subseteq R$, where n is a positive integer. This is the inductive hypothesis. To complete the inductive step we must show that this implies that R^{n+1} is also a subset of R. To show this, assume that $(a, b) \in R^{n+1}$. Then, because $R^{n+1} = R^n \circ R$, there is an element x with $x \in A$ such that $(a, x) \in R$ and $(x, b) \in R^n$. The inductive hypothesis, namely, that $R^n \subseteq R$, implies that $(x, b) \in R$. Furthermore, because R is transitive, and $(a, x) \in R$ and $(x, b) \in R$, it follows that $(a, b) \in R$. This shows that $R^{n+1} \subseteq R$, completing the proof. ◁

Exercises

1. List the ordered pairs in the relation R from $A = \{0, 1, 2, 3, 4\}$ to $B = \{0, 1, 2, 3\}$, where $(a, b) \in R$ if and only if

a) $a = b$. **b)** $a + b = 4$.
c) $a > b$. **d)** $a \mid b$.
e) $\gcd(a, b) = 1$. **f)** $\operatorname{lcm}(a, b) = 2$.

2. a) List all the ordered pairs in the relation $R = \{(a, b) \mid a \text{ divides } b\}$ on the set $\{1, 2, 3, 4, 5, 6\}$.

b) Display this relation graphically, as was done in Example 4.

c) Display this relation in tabular form, as was done in Example 4.

3. For each of these relations on the set $\{1, 2, 3, 4\}$, decide whether it is reflexive, whether it is symmetric, whether it is antisymmetric, and whether it is transitive.

a) $\{(2, 2), (2, 3), (2, 4), (3, 2), (3, 3), (3, 4)\}$
b) $\{(1, 1), (1, 2), (2, 1), (2, 2), (3, 3), (4, 4)\}$
c) $\{(2, 4), (4, 2)\}$
d) $\{(1, 2), (2, 3), (3, 4)\}$
e) $\{(1, 1), (2, 2), (3, 3), (4, 4)\}$
f) $\{(1, 3), (1, 4), (2, 3), (2, 4), (3, 1), (3, 4)\}$

4. Determine whether the relation R on the set of all people is reflexive, symmetric, antisymmetric, and/or transitive, where $(a, b) \in R$ if and only if

a) a is taller than b.
b) a and b were born on the same day.
c) a has the same first name as b.
d) a and b have a common grandparent.

5. Determine whether the relation R on the set of all Web pages is reflexive, symmetric, antisymmetric, and/or transitive, where $(a, b) \in R$ if and only if

a) everyone who has visited Web page a has also visited Web page b.
b) there are no common links found on both Web page a and Web page b.
c) there is at least one common link on Web page a and Web page b.
d) there is a Web page that includes links to both Web page a and Web page b.

6. Determine whether the relation R on the set of all real numbers is reflexive, symmetric, antisymmetric, and/or transitive, where $(x, y) \in R$ if and only if

a) $x + y = 0$. **b)** $x = \pm y$.
c) $x - y$ is a rational number.
d) $x = 2y$. **e)** $xy \ge 0$.
f) $xy = 0$. **g)** $x = 1$.
h) $x = 1$ or $y = 1$.

7. Determine whether the relation R on the set of all integers is reflexive, symmetric, antisymmetric, and/or transitive, where $(x, y) \in R$ if and only if

a) $x \ne y$. **b)** $xy \ge 1$.
c) $x = y + 1$ or $x = y - 1$.
d) $x \equiv y \pmod 7$. **e)** x is a multiple of y.
f) x and y are both negative or both nonnegative.
g) $x = y^2$. **h)** $x \ge y^2$.

8. Show that the relation $R = \emptyset$ on a nonempty set S is symmetric and transitive, but not reflexive.

9. Show that the relation $R = \emptyset$ on the empty set $S = \emptyset$ is reflexive, symmetric, and transitive.

10. Give an example of a relation on a set that is
   a) both symmetric and antisymmetric.
   b) neither symmetric nor antisymmetric.

A relation R on the set A is **irreflexive** if for every $a \in A, (a, a) \notin R$. That is, R is irreflexive if no element in A is related to itself.

11. Which relations in Exercise 3 are irreflexive?

12. Which relations in Exercise 4 are irreflexive?

13. Which relations in Exercise 5 are irreflexive?

14. Which relations in Exercise 6 are irreflexive?

15. Can a relation on a set be neither reflexive nor irreflexive?

16. Use quantifiers to express what it means for a relation to be irreflexive.

17. Give an example of an irreflexive relation on the set of all people.

A relation R is called **asymmetric** if $(a, b) \in R$ implies that $(b, a) \notin R$. Exercises 18–24 explore the notion of an asymmetric relation. Exercise 22 focuses on the difference between asymmetry and antisymmetry.

18. Which relations in Exercise 3 are asymmetric?

19. Which relations in Exercise 4 are asymmetric?

20. Which relations in Exercise 5 are asymmetric?

21. Which relations in Exercise 6 are asymmetric?

22. Must an asymmetric relation also be antisymmetric? Must an antisymmetric relation be asymmetric? Give reasons for your answers.

23. Use quantifiers to express what it means for a relation to be asymmetric.

24. Give an example of an asymmetric relation on the set of all people.

25. How many different relations are there from a set with m elements to a set with n elements?

☞ Let R be a relation from a set A to a set B. The **inverse relation** from B to A, denoted by R^{-1}, is the set of ordered pairs $\{(b, a) \mid (a, b) \in R\}$. The **complementary relation** $\overline{R}$ is the set of ordered pairs $\{(a, b) \mid (a, b) \notin R\}$.

26. Let R be the relation $R = \{(a, b) \mid a < b\}$ on the set of integers. Find
   a) R^{-1}.   **b)** $\overline{R}$.

27. Let R be the relation $R = \{(a, b) \mid a \text{ divides } b\}$ on the set of positive integers. Find
   a) R^{-1}.   **b)** $\overline{R}$.

28. Let R be the relation on the set of all states in the United States consisting of pairs (a, b) where state a borders state b. Find
   a) R^{-1}.   **b)** $\overline{R}$.

29. Suppose that the function f from A to B is a one-to-one correspondence. Let R be the relation that equals the graph of f. That is, $R = \{(a, f(a)) \mid a \in A\}$. What is the inverse relation R^{-1}?

30. Let $R_1 = \{(1, 2), (2, 3), (3, 4)\}$ and $R_2 = \{(1, 1), (1, 2), (2, 1), (2, 2), (2, 3), (3, 1), (3, 2), (3, 3), (3, 4)\}$ be relations from $\{1, 2, 3\}$ to $\{1, 2, 3, 4\}$. Find
   a) $R_1 \cup R_2$.   **b)** $R_1 \cap R_2$.
   c) $R_1 - R_2$.   **d)** $R_2 - R_1$.

31. Let A be the set of students at your school and B the set of books in the school library. Let R_1 and R_2 be the relations consisting of all ordered pairs (a, b), where student a is required to read book b in a course, and where student a has read book b, respectively. Describe the ordered pairs in each of these relations.
   a) $R_1 \cup R_2$ **b)** $R_1 \cap R_2$
   c) $R_1 \oplus R_2$ **d)** $R_1 - R_2$
   e) $R_2 - R_1$

32. Let R be the relation $\{(1, 2), (1, 3), (2, 3), (2, 4), (3, 1)\}$, and let S be the relation $\{(2, 1), (3, 1), (3, 2), (4, 2)\}$. Find $S \circ R$.

33. Let R be the relation on the set of people consisting of pairs (a, b), where a is a parent of b. Let S be the relation on the set of people consisting of pairs (a, b), where a and b are siblings (have the same parents). What are $S \circ R$ and $R \circ S$?

Exercises 34–38 deal with these relations on the set of real numbers:

$R_1 = \{(a, b) \in \mathbf{R}^2 \mid a > b\}$, the greater than relation,
$R_2 = \{(a, b) \in \mathbf{R}^2 \mid a \geq b\}$, the greater than or equal to relation,
$R_3 = \{(a, b) \in \mathbf{R}^2 \mid a < b\}$, the less than relation,
$R_4 = \{(a, b) \in \mathbf{R}^2 \mid a \leq b\}$, the less than or equal to relation,
$R_5 = \{(a, b) \in \mathbf{R}^2 \mid a = b\}$, the equal to relation,
$R_6 = \{(a, b) \in \mathbf{R}^2 \mid a \neq b\}$, the unequal to relation.

34. Find
   a) $R_1 \cup R_3$.   **b)** $R_1 \cup R_5$.
   c) $R_2 \cap R_4$.   **d)** $R_3 \cap R_5$.
   e) $R_1 - R_2$.   **f)** $R_2 - R_1$.
   g) $R_1 \oplus R_3$.   **h)** $R_2 \oplus R_4$.

35. Find
   a) $R_2 \cup R_4$.   **b)** $R_3 \cup R_6$.
   c) $R_3 \cap R_6$.   **d)** $R_4 \cap R_6$.
   e) $R_3 - R_6$.   **f)** $R_6 - R_3$.
   g) $R_2 \oplus R_6$.   **h)** $R_3 \oplus R_5$.

36. Find
   a) $R_1 \circ R_1$.   **b)** $R_1 \circ R_2$.
   c) $R_1 \circ R_3$.   **d)** $R_1 \circ R_4$.
   e) $R_1 \circ R_5$.   **f)** $R_1 \circ R_6$.
   g) $R_2 \circ R_3$.   **h)** $R_3 \circ R_3$.

37. Find
   a) $R_2 \circ R_1$.   **b)** $R_2 \circ R_2$.
   c) $R_3 \circ R_5$.   **d)** $R_4 \circ R_1$.
   e) $R_5 \circ R_3$.   **f)** $R_3 \circ R_6$.
   g) $R_4 \circ R_6$.   **h)** $R_6 \circ R_6$.

38. Find the relations R_i^2 for $i = 1, 2, 3, 4, 5, 6$.

39. Find the relations S_i^2 for $i = 1, 2, 3, 4, 5, 6$ where

$S_1 = \{(a, b) \in \mathbf{Z}^2 \mid a > b\}$, the greater than relation,
$S_2 = \{(a, b) \in \mathbf{Z}^2 \mid a \geq b\}$, the greater than or equal to relation,
$S_3 = \{(a, b) \in \mathbf{Z}^2 \mid a < b\}$, the less than relation,
$S_4 = \{(a, b) \in \mathbf{Z}^2 \mid a \leq b\}$, the less than or equal to relation,
$S_5 = \{(a, b) \in \mathbf{Z}^2 \mid a = b\}$, the equal to relation,
$S_6 = \{(a, b) \in \mathbf{Z}^2 \mid a \neq b\}$, the unequal to relation.

40. Let R be the parent relation on the set of all people (see Example 21). When is an ordered pair in the relation R^3?

41. Let R be the relation on the set of people with doctorates such that $(a, b) \in R$ if and only if a was the thesis advisor of b. When is an ordered pair (a, b) in R^2? When is an ordered pair (a, b) in R^n, when n is a positive integer? (Assume that every person with a doctorate has a thesis advisor.)

42. Let R_1 and R_2 be the "divides" and "is a multiple of" relations on the set of all positive integers, respectively. That is, $R_1 = \{(a, b) \mid a \text{ divides } b\}$ and $R_2 = \{(a, b) \mid a \text{ is a multiple of } b\}$. Find

a) $R_1 \cup R_2$. **b)** $R_1 \cap R_2$.
c) $R_1 - R_2$. **d)** $R_2 - R_1$.
e) $R_1 \oplus R_2$.

43. Let R_1 and R_2 be the "congruent modulo 3" and the "congruent modulo 4" relations, respectively, on the set of integers. That is, $R_1 = \{(a, b) \mid a \equiv b \pmod 3\}$ and $R_2 = \{(a, b) \mid a \equiv b \pmod 4\}$. Find

a) $R_1 \cup R_2$. **b)** $R_1 \cap R_2$.
c) $R_1 - R_2$. **d)** $R_2 - R_1$.
e) $R_1 \oplus R_2$.

44. List the 16 different relations on the set $\{0, 1\}$.

45. How many of the 16 different relations on $\{0, 1\}$ contain the pair $(0, 1)$?

46. Which of the 16 relations on $\{0, 1\}$, which you listed in Exercise 44, are

a) reflexive? **b)** irreflexive?
c) symmetric? **d)** antisymmetric?
e) asymmetric? **f)** transitive?

47. a) How many relations are there on the set $\{a, b, c, d\}$?
b) How many relations are there on the set $\{a, b, c, d\}$ that contain the pair (a, a)?

48. Let S be a set with n elements and let a and b be distinct elements of S. How many relations R are there on S such that

a) $(a, b) \in R$? **b)** $(a, b) \notin R$?
c) no ordered pair in R has a as its first element?
d) at least one ordered pair in R has a as its first element?
e) no ordered pair in R has a as its first element or b as its second element?
f) at least one ordered pair in R either has a as its first element or has b as its second element?

***49.** How many relations are there on a set with n elements that are

a) symmetric? **b)** antisymmetric?
c) asymmetric? **d)** irreflexive?
e) reflexive and symmetric?
f) neither reflexive nor irreflexive?

***50.** How many transitive relations are there on a set with n elements if

a) $n = 1$? **b)** $n = 2$? **c)** $n = 3$?

51. Find the error in the "proof" of the following "theorem."

"*Theorem*": Let R be a relation on a set A that is symmetric and transitive. Then R is reflexive.

"*Proof*": Let $a \in A$. Take an element $b \in A$ such that $(a, b) \in R$. Because R is symmetric, we also have $(b, a) \in R$. Now using the transitive property, we can conclude that $(a, a) \in R$ because $(a, b) \in R$ and $(b, a) \in R$.

52. Suppose that R and S are reflexive relations on a set A. Prove or disprove each of these statements.

a) $R \cup S$ is reflexive.
b) $R \cap S$ is reflexive.
c) $R \oplus S$ is irreflexive.
d) $R - S$ is irreflexive.
e) $S \circ R$ is reflexive.

53. Show that the relation R on a set A is symmetric if and only if $R = R^{-1}$, where R^{-1} is the inverse relation.

54. Show that the relation R on a set A is antisymmetric if and only if $R \cap R^{-1}$ is a subset of the diagonal relation $\Delta = \{(a, a) \mid a \in A\}$.

55. Show that the relation R on a set A is reflexive if and only if the inverse relation R^{-1} is reflexive.

56. Show that the relation R on a set A is reflexive if and only if the complementary relation $\overline{R}$ is irreflexive.

57. Let R be a relation that is reflexive and transitive. Prove that $R^n = R$ for all positive integers n.

58. Let R be the relation on the set $\{1, 2, 3, 4, 5\}$ containing the ordered pairs $(1, 1)$, $(1, 2)$, $(1, 3)$, $(2, 3)$, $(2, 4)$, $(3, 1)$, $(3, 4)$, $(3, 5)$, $(4, 2)$, $(4, 5)$, $(5, 1)$, $(5, 2)$, and $(5, 4)$. Find

a) R^2. **b)** R^3. **c)** R^4. **d)** R^5.

59. Let R be a reflexive relation on a set A. Show that R^n is reflexive for all positive integers n.

***60.** Let R be a symmetric relation. Show that R^n is symmetric for all positive integers n.

61. Suppose that the relation R is irreflexive. Is R^2 necessarily irreflexive? Give a reason for your answer.

62. Derive a big-O estimate for the number of integer comparisons needed to count all transitive relations on a set with n elements using the brute force approach of checking every relation of this set for transitivity.

9.2 n-ary Relations and Their Applications

9.2.1 Introduction

Relationships among elements of more than two sets often arise. For instance, there is a relationship involving the name of a student, the student's major, and the student's grade point average. Similarly, there is a relationship involving the airline, flight number, starting point, destination, departure time, and arrival time of a flight. An example of such a relationship in mathematics involves three integers, where the first integer is larger than the second integer, which is larger than the third. Another example is the betweenness relationship involving points on a line, such that three points are related when the second point is between the first and the third.

We will study relationships among elements from more than two sets in this section. These relationships are called ***n*-ary relations**. These relations are used to represent computer databases. These representations help us answer queries about the information stored in databases, such as: Which flights land at O'Hare Airport between 3 A.M. and 4 A.M.? Which students at your school are sophomores majoring in mathematics or computer science and have greater than a 3.0 average? Which employees of a company have worked for the company less than 5 years and make more than $50,000?

9.2.2 n-ary Relations

We begin with the basic definition on which the theory of relational databases rests.

Definition 1 Let $A_1, A_2, \ldots, A_n$ be sets. An *n-ary relation* on these sets is a subset of $A_1 \times A_2 \times \cdots \times A_n$. The sets $A_1, A_2, \ldots, A_n$ are called the *domains* of the relation, and n is called its *degree*.

EXAMPLE 1 Let R be the relation on $\mathbf{N} \times \mathbf{N} \times \mathbf{N}$ consisting of triples (a, b, c), where a, b, and c are integers with $a < b < c$. Then $(1, 2, 3) \in R$, but $(2, 4, 3) \notin R$. The degree of this relation is 3. Its domains are all equal to the set of natural numbers. ◀

EXAMPLE 2 Let R be the relation on $\mathbf{Z} \times \mathbf{Z} \times \mathbf{Z}$ consisting of all triples of integers (a, b, c) in which a, b, and c form an arithmetic progression. That is, $(a, b, c) \in R$ if and only if there is an integer k such that $b = a + k$ and $c = a + 2k$, or equivalently, such that $b - a = k$ and $c - b = k$. Note that $(1, 3, 5) \in R$ because $3 = 1 + 2$ and $5 = 1 + 2 \cdot 2$, but $(2, 5, 9) \notin R$ because $5 - 2 = 3$ while $9 - 5 = 4$. This relation has degree 3 and its domains are all equal to the set of integers. ◀

EXAMPLE 3 Let R be the relation on $\mathbf{Z} \times \mathbf{Z} \times \mathbf{Z}^+$ consisting of triples (a, b, m), where a, b, and m are integers with $m \geq 1$ and $a \equiv b \pmod{m}$. Then $(8, 2, 3)$, $(-1, 9, 5)$, and $(14, 0, 7)$ all belong to R, but $(7, 2, 3)$, $(-2, -8, 5)$, and $(11, 0, 6)$ do not belong to R because $8 \equiv 2 \pmod{3}$, $-1 \equiv 9 \pmod{5}$, and $14 \equiv 0 \pmod{7}$, but $7 \not\equiv 2 \pmod{3}$, $-2 \not\equiv -8 \pmod{5}$, and $11 \not\equiv 0 \pmod{6}$. This relation has degree 3 and its first two domains are the set of all integers and its third domain is the set of positive integers. ◀

EXAMPLE 4 Let R be the relation consisting of 5-tuples (A, N, S, D, T) representing airplane flights, where A is the airline, N is the flight number, S is the starting point, D is the destination, and T is the departure time. For instance, if Nadir Express Airlines has flight 963 from Newark to Bangor at 15:00, then (Nadir, 963, Newark, Bangor, 15:00) belongs to R. The degree of this relation is 5, and its domains are the set of all airlines, the set of flight numbers, the set of cities, the set of cities (again), and the set of times. ◀

9.2.3 Databases and Relations

Links ❯ The time required to manipulate information in a database depends on how this information is stored. The operations of adding and deleting records, updating records, searching for records, and combining records from overlapping databases are performed millions of times each day in a large database. Because of the importance of these operations, various methods for representing databases have been developed. We will discuss one of these methods, called the **relational data model**, based on the concept of a relation.

A database consists of **records**, which are n-tuples, made up of **fields**. The fields are the entries of the n-tuples. For instance, a database of student records may be made up of fields containing the name, student number, major, and grade point average of the student. The relational data model represents a database of records as an n-ary relation. Thus, student records are represented as 4-tuples of the form (*Student_name, ID_number, Major, GPA*). A sample database of six such records is

(Ackermann, 231455, Computer Science, 3.88)
(Adams, 888323, Physics, 3.45)
(Chou, 102147, Computer Science, 3.49)
(Goodfriend, 453876, Mathematics, 3.45)
(Rao, 678543, Mathematics, 3.90)
(Stevens, 786576, Psychology, 2.99).

Relations used to represent databases are also called **tables**, because these relations are often displayed as tables. Each column of the table corresponds to an *attribute* of the database. For instance, the same database of students is displayed in Table 1. The attributes of this database are Student Name, ID Number, Major, and GPA.

A domain of an n-ary relation is called a **primary key** when the value of the n-tuple from this domain determines the n-tuple. That is, a domain is a primary key when no two n-tuples in the relation have the same value from this domain.

Records are often added to or deleted from databases. Because of this, the property that a domain is a primary key is time-dependent. Consequently, a primary key should be chosen that remains one whenever the database is changed. The current collection of n-tuples in a relation is called the **extension** of the relation. The more permanent part of a database, including the name and attributes of the database, is called its **intension**. When selecting a primary key, the goal should be to select a key that can serve as a primary key for all possible extensions of the database. To do this, it is necessary to examine the intension of the database to understand the set of possible n-tuples that can occur in an extension.

TABLE 1 Students.

Student_name	*ID_number*	*Major*	*GPA*
Ackermann	231455	Computer Science	3.88
Adams	888323	Physics	3.45
Chou	102147	Computer Science	3.49
Goodfriend	453876	Mathematics	3.45
Rao	678543	Mathematics	3.90
Stevens	786576	Psychology	2.99

EXAMPLE 5 *Extra Examples*

Which domains are primary keys for the n-ary relation displayed in Table 1, assuming that no n-tuples will be added in the future?

Solution: Because there is only one 4-tuple in this table for each student name, the domain of student names is a primary key. Similarly, the ID numbers in this table are unique, so the domain of ID numbers is also a primary key. However, the domain of major fields of study is not a primary key, because more than one 4-tuple contains the same major field of study. The domain of grade point averages is also not a primary key, because there are two 4-tuples containing the same GPA. ◀

Combinations of domains can also uniquely identify n-tuples in an n-ary relation. When the values of a set of domains determine an n-tuple in a relation, the Cartesian product of these domains is called a **composite key**.

EXAMPLE 6 Is the Cartesian product of the domain of major fields of study and the domain of GPAs a composite key for the n-ary relation from Table 1, assuming that no n-tuples are ever added?

Solution: Because no two 4-tuples from this table have both the same major and the same GPA, this Cartesian product is a composite key. ◀

Because primary and composite keys are used to identify records uniquely in a database, it is important that keys remain valid when new records are added to the database. Hence, checks should be made to ensure that every new record has values that are different in the appropriate field, or fields, from all other records in this table. For instance, it makes sense to use the student identification number as a key for student records if no two students ever have the same student identification number. A university should not use the name field as a key, because two students may have the same name (such as John Smith).

9.2.4 Operations on n-ary Relations

There are a variety of operations on n-ary relations that can be used to form new n-ary relations. Applied together, these operations can answer queries on databases that ask for all n-tuples that satisfy certain conditions.

The most basic operation on an n-ary relation is determining all n-tuples in the n-ary relation that satisfy certain conditions. For example, we may want to find all the records of all computer science majors in a database of student records. We may want to find all students who have a grade point average above 3.5. We may want to find the records of all computer science majors who have a grade point average above 3.5. To perform such tasks we use the selection operator.

Definition 2 Let R be an n-ary relation and C a condition that elements in R may satisfy. Then the *selection operator* s_C maps the n-ary relation R to the n-ary relation of all n-tuples from R that satisfy the condition C.

EXAMPLE 7 *Extra Examples*

To find the records of computer science majors in the n-ary relation R shown in Table 1, we use the operator s_{C_1}, where C_1 is the condition Major = "Computer Science." The result is the two 4-tuples (Ackermann, 231455, Computer Science, 3.88) and (Chou, 102147, Computer Science, 3.49). Similarly, to find the records of students who have a grade point average above 3.5 in this database, we use the operator s_{C_2}, where C_2 is the condition GPA > 3.5. The result is the two 4-tuples (Ackermann, 231455, Computer Science, 3.88) and (Rao, 678543, Mathematics,

3.90). Finally, to find the records of computer science majors who have a GPA above 3.5, we use the operator s_{C_3}, where C_3 is the condition (Major = "Computer Science" $\wedge$ GPA > 3.5). The result consists of the single 4-tuple (Ackermann, 231455, Computer Science, 3.88). ◀

Projections are used to form new n-ary relations by deleting the same fields in every record of the relation.

Definition 3 The *projection* $P_{i_1 i_2, \ldots, i_m}$ where $i_1 < i_2 < \cdots < i_m$, maps the n-tuple $(a_1, a_2, \ldots, a_n)$ to the m-tuple $(a_{i_1}, a_{i_2}, \ldots, a_{i_m})$, where $m \leq n$.

In other words, the projection $P_{i_1, i_2, \ldots, i_m}$ deletes $n - m$ of the components of an n-tuple, leaving the i_1th, i_2th, ..., and i_mth components.

EXAMPLE 8 What results when the projection $P_{1,3}$ is applied to the 4-tuples (2, 3, 0, 4), (Jane Doe, 234111001, Geography, 3.14), and (a_1, a_2, a_3, a_4)?

Solution: The projection $P_{1,3}$ sends these 4-tuples to (2, 0), (Jane Doe, Geography), and (a_1, a_3), respectively. ◀

Example 9 illustrates how new relations are produced using projections.

EXAMPLE 9 What relation results when the projection $P_{1,4}$ is applied to the relation in Table 1?

Solution: When the projection $P_{1,4}$ is used, the second and third columns of the table are deleted, and pairs representing student names and grade point averages are obtained. Table 2 displays the results of this projection. ◀

Fewer rows may result when a projection is applied to the table for a relation. This happens when some of the n-tuples in the relation have identical values in each of the m components of the projection, and only disagree in components deleted by the projection. For instance, consider the following example.

EXAMPLE 10 What is the table obtained when the projection $P_{1,2}$ is applied to the relation in Table 3?

Solution: Table 4 displays the relation obtained when $P_{1,2}$ is applied to Table 3. Note that there are fewer rows after this projection is applied. ◀

TABLE 2 GPAs.

Student_name	*GPA*
Ackermann	3.88
Adams	3.45
Chou	3.49
Goodfriend	3.45
Rao	3.90
Stevens	2.99

TABLE 3 Enrollments.

Student	*Major*	*Course*
Glauser	Biology	BI 290
Glauser	Biology	MS 475
Glauser	Biology	PY 410
Marcus	Mathematics	MS 511
Marcus	Mathematics	MS 603
Marcus	Mathematics	CS 322
Miller	Computer Science	MS 575
Miller	Computer Science	CS 455

TABLE 4 Majors.

Student	*Major*
Glauser	Biology
Marcus	Mathematics
Miller	Computer Science

TABLE 5 Teaching_assignments.		
Professor	*Department*	*Course_ number*
Cruz	Zoology	335
Cruz	Zoology	412
Farber	Psychology	501
Farber	Psychology	617
Grammer	Physics	544
Grammer	Physics	551
Rosen	Computer Science	518
Rosen	Mathematics	575

TABLE 6 Class_schedule.			
Department	*Course_ number*	*Room*	*Time*
Computer Science	518	N521	2:00 P.M.
Mathematics	575	N502	3:00 P.M.
Mathematics	611	N521	4:00 P.M.
Physics	544	B505	4:00 P.M.
Psychology	501	A100	3:00 P.M.
Psychology	617	A110	11:00 A.M.
Zoology	335	A100	9:00 A.M.
Zoology	412	A100	8:00 A.M.

The **join** operation is used to combine two tables into one when these tables share some identical fields. For instance, a table containing fields for airline, flight number, and gate, and another table containing fields for flight number, gate, and departure time can be combined into a table containing fields for airline, flight number, gate, and departure time.

Definition 4 Let R be a relation of degree m and S a relation of degree n. The *join* $J_p(R, S)$, where $p \leq m$ and $p \leq n$, is a relation of degree $m + n - p$ that consists of all $(m + n - p)$-tuples $(a_1, a_2, \ldots, a_{m-p}, c_1, c_2, \ldots, c_p, b_1, b_2, \ldots, b_{n-p})$, where the m-tuple $(a_1, a_2, \ldots, a_{m-p}, c_1, c_2, \ldots, c_p)$ belongs to R and the n-tuple $(c_1, c_2, \ldots, c_p, b_1, b_2, \ldots, b_{n-p})$ belongs to S.

In other words, the join operator J_p produces a new relation from two relations by combining all m-tuples of the first relation with all n tuples of the second relation, where the last p components of the m-tuples agree with the first p components of the n-tuples.

EXAMPLE 11 What relation results when the join operator J_2 is used to combine the relation displayed in Tables 5 and 6?

Solution: The join J_2 produces the relation shown in Table 7. ◀

There are other operators besides projections and joins that produce new relations from existing relations. A description of these operations can be found in books on database theory.

TABLE 7 Teaching_schedule.				
Professor	*Department*	*Course_number*	*Room*	*Time*
Cruz	Zoology	335	A100	9:00 A.M.
Cruz	Zoology	412	A100	8:00 A.M.
Farber	Psychology	501	A100	3:00 P.M.
Farber	Psychology	617	A110	11:00 A.M.
Grammer	Physics	544	B505	4:00 P.M.
Rosen	Computer Science	518	N521	2:00 P.M.
Rosen	Mathematics	575	N502	3:00 P.M.

TABLE 8 **Flights.**

Airline	*Flight_number*	*Gate*	*Destination*	*Departure_time*
Nadir	122	34	Detroit	08:10
Acme	221	22	Denver	08:17
Acme	122	33	Anchorage	08:22
Acme	323	34	Honolulu	08:30
Nadir	199	13	Detroit	08:47
Acme	222	22	Denver	09:10
Nadir	322	34	Detroit	09:44

9.2.5 SQL

Links

The database query language SQL (short for Structured Query Language) can be used to carry out the operations we have described in this section. Example 12 illustrates how SQL commands are related to operations on n-ary relations.

EXAMPLE 12 We will illustrate how SQL is used to express queries by showing how SQL can be employed to make a query about airline flights using Table 8. The SQL statement

```
SELECT Departure_time
FROM Flights
WHERE Destination='Detroit'
```

is used to find the projection P_5 (on the Departure_time attribute) of the selection of 5-tuples in the Flights database that satisfy the condition: Destination = 'Detroit'. The output would be a list containing the times of flights that have Detroit as their destination, namely, 08:10, 08:47, and 09:44. SQL uses the FROM clause to identify the n-ary relation the query is applied to, the WHERE clause to specify the condition of the selection operation, and the SELECT clause to specify the projection operation that is to be applied. (*Beware:* SQL uses SELECT to represent a projection, rather than a selection operation. This is an unfortunate example of conflicting terminology.) ◀

Example 13 shows how SQL queries can be made involving more than one table.

EXAMPLE 13 The SQL statement

```
SELECT Professor, Time
FROM Teaching_assignments, Class_schedule
WHERE Department='Mathematics'
```

is used to find the projection $P_{1,5}$ of the 5-tuples in the database (shown in Table 7), which is the join J_2 of the Teaching_assignments and Class_schedule databases in Tables 5 and 6, respectively, which satisfy the condition: Department = Mathematics. The output would consist of the single 2-tuple (Rosen, 3:00 P.M.). The SQL FROM clause is used here to find the join of two different databases. ◀

We have only touched on the basic concepts of relational databases in this section. More information can be found in [AhUl95].

9.2.6 Association Rules from Data Mining

Links

We will now introduce concepts from **data mining**, the discipline with the goal of gaining useful information from data. In particular, we will discuss information that can be gleaned from databases of transactions. We will focus on supermarket transactions, but the ideas we present are relevant in a wide range of applications.

By a **transaction** we mean a set of items bought by a customer during a visit to the store, such as {milk, eggs, bread} or {orange juice, bananas, yogurt, cream}. Stores collect large databases of transactions that can be used to help them manage their businesses. We will discuss how these databases can be used to address the question: How likely is it that a customer buys a product given that they also buy a collection of one or more specified products?

We call each product in the store an **item**. A collection of items is known as an **itemset**. A ***k*-itemset** is an itemset that contains exactly k items. The terms **transaction** and **basket** are used synonymously with the word itemset. When a store has n items, $a_1, a_2, \ldots, a_n$, for sale, each transaction can be represented by an n-tuple $b_1, b_2, \ldots, b_n$, where b_i is a binary variable that tells us whether a_i occurs in this transaction. That is, $b_i = 1$ if a_i is in this transaction and $b_i = 0$ otherwise. (Note that we only care whether an item occurs in a transaction and not how many times it occurs.) We can represent a transaction by an $(n+1)$-tuple of the form (*transaction number*, $b_1, b_2, \ldots, b_n$). The collection of all these $(n+1)$-tuples forms a database of transactions.

We now define additional terms used in the study of questions relating to the purchase of particular itemsets.

Definition 5 The *count* of an itemset I, denoted by $\sigma(I)$, in a set of transactions $T = \{t_1, t_2, \ldots, t_k\}$, where k is a positive integer, is the number of transactions that contain this itemset. That is,

$$\sigma(I) = |\{t_i \in T \mid I \subseteq t_i\}|.$$

The *support* of an itemset I is the probability that I is included in a randomly selected transaction from T. That is,

$$\text{support}(I) = \frac{\sigma(I)}{|T|}.$$

The **support threshold** s is specified for a particular application. **Frequent itemset mining** is the process of finding itemsets I with support greater than or equal to s. Such itemsets are said to be **frequent**.

EXAMPLE 14 The morning transactions at a market stand that sells apples, pears, cider, donuts, and mangos are shown in Table 9. In Table 10 we display the corresponding binary database, where each record is an $(n+1)$-tuple consisting of the transaction number followed by binary entries that represent this itemset. Because apples occurs in five of the eight transactions, we see that $\sigma(\{\text{apples}\}) = 5$ and support({apples}) $= 5/8$. Similarly, because the itemset {apples, cider} is a subset of four of the eight transactions, we have $\sigma(\{\text{apples, cider}\}) = 4$ and support({apples, cider}) $= 4/8 = 1/2$.

Extra Examples

If we set the support threshold to be 0.5, an item is frequent if it occurs in at least four of the eight transactions. Consequently, with this support threshold, apples, pears, donuts and cider are the frequent items. The itemset {apples, cider} is a frequent itemset, but the itemset {donuts, pears} is not a frequent itemset. ◀

TABLE 9 A Set of Transactions.

Transaction Number	*Items*
1	{apples, pears, mangos}
2	{pears, cider}
3	{apples, cider, donuts, mangos}
4	{apples, pears, cider, donuts}
5	{apples, cider, donuts}
6	{pears, cider, donuts}
7	{pears, donuts}
8	{apples, pears, cider}

TABLE 10 Binary Database for the Transactions in Table 9.

Transaction Number	*Apples*	*Pears*	*Cider*	*Donuts*	*Mangos*
1	1	1	0	0	1
2	0	1	1	0	0
3	1	0	1	1	1
4	1	1	1	1	0
5	1	0	1	1	0
6	0	1	1	1	0
7	0	1	0	1	0
8	1	1	1	0	0

We can use sets of transactions to help us predict whether a customer will buy a particular item given that they also buy all the items in an itemset (which might just be one item). Before we address a question of this type, we introduce some terminology. If S is a set of items and T is a set of transactions, an **association rule** is an implication of the form $I \to J$, where I and J are disjoint subsets of S. Although this notation borrows the notation for logical implications, its meaning is not entirely analagous. The association rule $I \to J$ is not the statement that whenever I is a subset of a transaction, then J must also be one. Instead, the strength of an association rule is measured in terms of its **support**, the frequency of transactions that contain both I and J, and its **confidence**, the frequency of transactions that contain J when they also contain I.

Definition 6 If I and J are disjoint subsets of a set T of transactions, then

$$\text{support}(I \to J) = \frac{\sigma(I \cup J)}{|T|}$$

and

$$\text{confidence}(I \to J) = \frac{\sigma(I \cup J)}{\sigma(I)}.$$

The support of the association rule $I \to J$, the fraction of transactions that contain both I and J, is a useful measure because a low support value tells us that the basket containing all items in

I and all items in J is seldom purchased, whereas a high value tells us that they are purchased together in a large fraction of transactions. Note that the confidence of the association rule $I \rightarrow J$ is the conditional probability that a transaction will contain all the items in I and in J given that it contains all the items in I. So, the larger the confidence of $I \rightarrow J$, the more likely it is for J to be a subset of a transaction that contains I.

EXAMPLE 15 What are the support and the confidence of the association rule {cider, donuts} → {apples} for the set of transactions in Example 14?

Extra Examples

Solution: The support of this association rule is $\sigma(\{\text{cider, donuts}\} \cup \{\text{apples}\})/|T|$. Because $\sigma(\{\text{cider, donuts}\} \cup \{\text{apples}\}) = \sigma(\{\text{cider, donuts, apples}\}) = 3$ and $|T| = 8$, we see that the support of this rule is $3/8 = 0.375$.

The confidence of this rule is $\sigma(\{\text{cider, donuts}\} \cup \{\text{apples}\})/\sigma(\{\text{cider, donuts}\}) = 3/4 = 0.75$. ◀

An important problem in data mining is to find **strong association rules**, which have support greater than or equal to a minimum support level and confidence greater than or equal to a minimum confidence level. It is important to have efficient algorithms to find strong association rules because the number of available items can be extremely large. For instance, a supermarket may have tens of thousands, or even hundreds of thousands, of items in stock. The brute-force approach of finding association rules with sufficiently large support and confidence by computing the support and confidence of all possible association rules is infeasible because there are an exponential number of such association rules (see Exercise 41). Several widely used algorithms have been developed to solve this problem much more efficiently than brute force. Such algorithms first find frequent itemsets and then turn their attention to finding all the association rules with high confidence from the frequent itemsets that have been found. Consult data mining texts such as [Ag15] for details.

Although we have presented association rules in the context of market baskets, they are useful in a surprisingly wide variety of applications. For instance, association rules can be used to improve medical diagnoses, in which itemsets are collections of test results or symptoms and transactions are the collections of test results and symptoms found on patient records. Association rules, in which itemsets are baskets of key words and transactions are the collections of words on web pages, are used by search engines. Cases of plagiarism can be found using association rules, in which itemsets are collections of sentences and transactions are the contents of documents. Association rules also play helpful roles in various aspects of computer security, including intrusion detection, in which the itemsets are collections of patterns and transactions are the strings transmitted during network attacks. The interested reader will be able to find many more such applications by searching the web.

Exercises

1. List the triples in the relation $\{(a, b, c) \mid a, b, \text{ and } c \text{ are integers with } 0 < a < b < c < 5\}$.

2. Which 4-tuples are in the relation $\{(a, b, c, d) \mid a, b, c, \text{ and } d \text{ are positive integers with } abcd = 6\}$?

3. List the 5-tuples in the relation in Table 8.

4. Assuming that no new n-tuples are added, find all the primary keys for the relations displayed in

a) Table 3. **b)** Table 5.

c) Table 6. **d)** Table 8.

5. Assuming that no new n-tuples are added, find a composite key with two fields containing the Airline field for the database in Table 8.

6. Assuming that no new n-tuples are added, find a composite key with two fields containing the Professor field for the database in Table 7.

7. The 3-tuples in a 3-ary relation represent the following attributes of a student database: student ID number, name, phone number.

a) Is student ID number likely to be a primary key?
b) Is name likely to be a primary key?
c) Is phone number likely to be a primary key?

8. The 4-tuples in a 4-ary relation represent these attributes of published books: title, ISBN, publication date, number of pages.

a) What is a likely primary key for this relation?

b) Under what conditions would (title, publication date) be a composite key?

c) Under what conditions would (title, number of pages) be a composite key?

9. The 5-tuples in a 5-ary relation represent these attributes of all people in the United States: name, Social Security number, street address, city, state.

a) Determine a primary key for this relation.

b) Under what conditions would (name, street address) be a composite key?

c) Under what conditions would (name, street address, city) be a composite key?

10. What do you obtain when you apply the selection operator s_C, where C is the condition Room = A100, to the database in Table 7?

11. What do you obtain when you apply the selection operator s_C, where C is the condition Destination = Detroit, to the database in Table 8?

12. What do you obtain when you apply the selection operator s_C, where C is the condition (Project = 2) $\wedge$ (Quantity $\geq$ 50), to the database in Table 10?

13. What do you obtain when you apply the selection operator s_C, where C is the condition (Airline = Nadir) $\vee$ (Destination = Denver), to the database in Table 8?

14. What do you obtain when you apply the projection $P_{2,3,5}$ to the 5-tuple (a, b, c, d, e)?

15. Which projection mapping is used to delete the first, second, and fourth components of a 6-tuple?

16. Display the table produced by applying the projection $P_{1,2,4}$ to Table 8.

17. Display the table produced by applying the projection $P_{1,4}$ to Table 8.

18. How many components are there in the n-tuples in the table obtained by applying the join operator J_3 to two tables with 5-tuples and 8-tuples, respectively?

19. Construct the table obtained by applying the join operator J_2 to the relations in Tables 11 and 12.

20. Show that if C_1 and C_2 are conditions that elements of the n-ary relation R may satisfy, then $s_{C_1 \wedge C_2}(R) = s_{C_1}(s_{C_2}(R))$.

21. Show that if C_1 and C_2 are conditions that elements of the n-ary relation R may satisfy, then $s_{C_1}(s_{C_2}(R)) = s_{C_2}(s_{C_1}(R))$.

22. Show that if C is a condition that elements of the n-ary relations R and S may satisfy, then $s_C(R \cup S) = s_C(R) \cup s_C(S)$.

23. Show that if C is a condition that elements of the n-ary relations R and S may satisfy, then $s_C(R \cap S) = s_C(R) \cap s_C(S)$.

24. Show that if C is a condition that elements of the n-ary relations R and S may satisfy, then $s_C(R - S) = s_C(R) - s_C(S)$.

25. Show that if R and S are both n-ary relations, then $P_{i_1,i_2,\ldots,i_m}(R \cup S) = P_{i_1,i_2,\ldots,i_m}(R) \cup P_{i_1,i_2,\ldots,i_m}(S)$.

26. Give an example to show that if R and S are both n-ary relations, then $P_{i_1,i_2,\ldots,i_m}(R \cap S)$ may be different from $P_{i_1,i_2,\ldots,i_m}(R) \cap P_{i_1,i_2,\ldots,i_m}(S)$.

27. Give an example to show that if R and S are both n-ary relations, then $P_{i_1,i_2,\ldots,i_m}(R - S)$ may be different from $P_{i_1,i_2,\ldots,i_m}(R) - P_{i_1,i_2,\ldots,i_m}(S)$.

28. a) What are the operations that correspond to the query expressed using this SQL statement?

```
SELECT Supplier
FROM Part_needs
WHERE 1000 ≤ Part_number ≤ 5000
```

b) What is the output of this query given the database in Table 11 as input?

29. a) What are the operations that correspond to the query expressed using this SQL statement?

```
SELECT Supplier, Project
FROM Part_needs, Parts_inventory
WHERE Quantity ≤ 10
```

b) What is the output of this query given the databases in Tables 11 and 12 as input?

30. Determine whether there is a primary key for the relation in Example 2.

31. Determine whether there is a primary key for the relation in Example 3.

32. Show that an n-ary relation with a primary key can be thought of as the graph of a function that maps values of the primary key to $(n - 1)$-tuples formed from values of the other domains.

33. Suppose that the transactions at a convenience store during an evening are {bread, milk, diapers, juice}, {bread, milk, diapers, eggs}, {milk, diapers, beer, eggs}, {bread, beer}, {milk, diapers, eggs, juice}, and {milk, diapers, beer}.

a) Find the count and support of diapers.

b) Find all frequent itemsets if the threshold level is 0.6.

34. Suppose that the key words on eight different web pages are {evolution, primate, Human, Neanderthal, DNA, fossil}, {evolution, Neanderthal, Denisovan, Human, DNA}, {cave, fossil, primate}, {Human, Neanderthal, Denisovan, evolution}, {DNA, genome, evolution, fossil}, {DNA, Human, Neanderthal, Denisovan, genome}, {evolution, primate, cave, fossil}, and {Human, Neanderthal, genome}.

a) Find the count and support of Neanderthal.

b) Find all frequent itemsets if the threshold level is 0.6.

35. Find the support and confidence of the association rule {beer} $\rightarrow$ {diapers} for the set of transactions in Exercise 33. (This association rule has played an important role in the development of the subject.)

36. Find the support and confidence of the association rule {human, DNA} $\rightarrow$ {Neanderthal} for the set of transactions in Exercise 34.

TABLE 11 **Part_needs.**

Supplier	*Part_number*	*Project*
23	1092	1
23	1101	3
23	9048	4
31	4975	3
31	3477	2
32	6984	4
32	9191	2
33	1001	1

TABLE 12 **Parts_inventory.**

Part_number	*Project*	*Quantity*	*Color_code*
1001	1	14	8
1092	1	2	2
1101	3	1	1
3477	2	25	2
4975	3	6	2
6984	4	10	1
9048	4	12	2
9191	2	80	4

37. Suppose that I is an itemset with positive count in a set of transactions. Find the confidence of the association rule $I \to \emptyset$.

38. Suppose that I, J, and K are itemsets. Show that the six association rules $\{I, J\} \to K$, $\{J, K\} \to I$, $\{I, K\} \to J$, $I \to \{J, K\}$, $J \to \{I, K\}$, and $K \to \{I, J\}$ all have the same support.

39. The **lift** of the association rule $I \to J$, where I and J are itemsets with positive support in a set of transactions, equals support($I \cup J$)/(support(I)support(J)).

a) Show that the lift of $I \to J$, when support(I) and support(J) are both positive, equals 1 if and only if the occurrence of I in a transaction and the occurrence of J in a transaction are independent events.

b) Find the lift of the association rule {beer} → {diapers} for the set of transactions in Exercises 33.

c) Find the lift of the association rule {evolution} → {Neanderthals, Denisovans} for the set of transactions in Exercise 34.

40. Show that if an itemset is frequent in a set of transactions, then all its subsets are also frequent itemsets in this set of transactions.

41. Given n unique items, show that there are 3^n possible association rules of the form $I \to J$, where I and J are disjoint subsets of the set of all items. Be sure to allow the association rules where I or J, or both, are empty.

9.3 Representing Relations

9.3.1 Introduction

In this section, and in the remainder of this chapter, all relations we study will be binary relations. Because of this, in this section and in the rest of this chapter, the word relation will always refer to a binary relation. There are many ways to represent a relation between finite sets. As we have seen in Section 9.1, one way is to list its ordered pairs. Another way to represent a relation is to use a table, as we did in Example 3 in Section 9.1. In this section we will discuss two alternative methods for representing relations. One method uses zero–one matrices. The other method uses pictorial representations called directed graphs, which we will discuss later in this section.

Generally, matrices are appropriate for the representation of relations in computer programs. On the other hand, people often find the representation of relations using directed graphs useful for understanding the properties of these relations.

9.3.2 Representing Relations Using Matrices

A relation between finite sets can be represented using a zero–one matrix. Suppose that R is a relation from $A = \{a_1, a_2, \ldots, a_m\}$ to $B = \{b_1, b_2, \ldots, b_n\}$. (Here the elements of the sets A and B have been listed in a particular, but arbitrary, order. Furthermore, when $A = B$ we use the same ordering for A and B.) The relation R can be represented by the matrix $\mathbf{M}_R = [m_{ij}]$, where

$$m_{ij} = \begin{cases} 1 \text{ if } (a_i, b_j) \in R, \\ 0 \text{ if } (a_i, b_j) \notin R. \end{cases}$$

In other words, the zero–one matrix representing R has a 1 as its (i, j) entry when a_i is related to b_j, and a 0 in this position if a_i is not related to b_j. (Such a representation depends on the orderings used for A and B.)

The use of matrices to represent relations is illustrated in Examples 1–6.

EXAMPLE 1 Suppose that $A = \{1, 2, 3\}$ and $B = \{1, 2\}$. Let R be the relation from A to B containing (a, b) if $a \in A$, $b \in B$, and $a > b$. What is the matrix representing R if $a_1 = 1$, $a_2 = 2$, and $a_3 = 3$, and $b_1 = 1$ and $b_2 = 2$?

Solution: Because $R = \{(2, 1), (3, 1), (3, 2)\}$, the matrix for R is

$$\mathbf{M}_R = \begin{bmatrix} 0 & 0 \\ 1 & 0 \\ 1 & 1 \end{bmatrix}.$$

The 1s in $\mathbf{M}_R$ show that the pairs (2, 1), (3, 1), and (3, 2) belong to R. The 0s show that no other pairs belong to R. ◀

EXAMPLE 2 Let $A = \{a_1, a_2, a_3\}$ and $B = \{b_1, b_2, b_3, b_4, b_5\}$. Which ordered pairs are in the relation R represented by the matrix

$$\mathbf{M}_R = \begin{bmatrix} 0 & 1 & 0 & 0 & 0 \\ 1 & 0 & 1 & 1 & 0 \\ 1 & 0 & 1 & 0 & 1 \end{bmatrix}?$$

Solution: Because R consists of those ordered pairs (a_i, b_j) with $m_{ij} = 1$, it follows that

$$R = \{(a_1, b_2), (a_2, b_1), (a_2, b_3), (a_2, b_4), (a_3, b_1), (a_3, b_3), (a_3, b_5)\}.$$ ◀

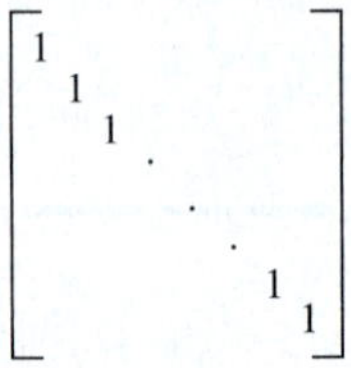

FIGURE 1 The zero–one matrix for a reflexive relation. (Off diagonal elements can be 0 or 1.)

The matrix of a relation on a set, which is a square matrix, can be used to determine whether the relation has certain properties. Recall that a relation R on A is reflexive if $(a, a) \in R$ whenever $a \in A$. Thus, R is reflexive if and only if $(a_i, a_i) \in R$ for $i = 1, 2, \ldots, n$. Hence, R is reflexive if and only if $m_{ii} = 1$, for $i = 1, 2, \ldots, n$. In other words, R is reflexive if all the elements on the main diagonal of $\mathbf{M}_R$ are equal to 1, as shown in Figure 1. Note that the elements off the main diagonal can be either 0 or 1.

The relation R is symmetric if $(a, b) \in R$ implies that $(b, a) \in R$. Consequently, the relation R on the set $A = \{a_1, a_2, \ldots, a_n\}$ is symmetric if and only if $(a_j, a_i) \in R$ whenever $(a_i, a_j) \in R$. In terms of the entries of $\mathbf{M}_R$, R is symmetric if and only if $m_{ji} = 1$ whenever $m_{ij} = 1$. This also means $m_{ji} = 0$ whenever $m_{ij} = 0$. Consequently, R is symmetric if and only if $m_{ij} = m_{ji}$, for all pairs of integers i and j with $i = 1, 2, \ldots, n$ and $j = 1, 2, \ldots, n$. Recalling the definition of the transpose of a matrix from Section 2.6, we see that R is symmetric if and only if

$$\mathbf{M}_R = (\mathbf{M}_R)^t,$$

that is, if $\mathbf{M}_R$ is a symmetric matrix. The form of the matrix for a symmetric relation is illustrated in Figure 2(a).

The relation R is antisymmetric if and only if $(a, b) \in R$ and $(b, a) \in R$ imply that $a = b$. Consequently, the matrix of an antisymmetric relation has the property that if $m_{ij} = 1$ with $i \neq j$, then $m_{ji} = 0$. Or, in other words, either $m_{ij} = 0$ or $m_{ji} = 0$ when $i \neq j$. The form of the matrix for an antisymmetric relation is illustrated in Figure 2(b).

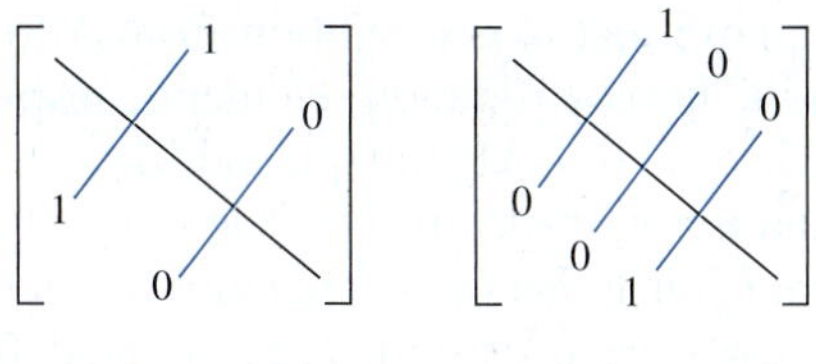

(a) Symmetric (b) Antisymmetric

FIGURE 2 The zero–one matrices for symmetric and antisymmetric relations.

EXAMPLE 3 Suppose that the relation R on a set is represented by the matrix

$$\mathbf{M}_R = \begin{bmatrix} 1 & 1 & 0 \\ 1 & 1 & 1 \\ 0 & 1 & 1 \end{bmatrix}.$$

Is R reflexive, symmetric, and/or antisymmetric?

Solution: Because all the diagonal elements of this matrix are equal to 1, R is reflexive. Moreover, because $\mathbf{M}_R$ is symmetric, it follows that R is symmetric. It is also easy to see that R is not antisymmetric. ◀

The Boolean operations join and meet (discussed in Section 2.6) can be used to find the matrices representing the union and the intersection of two relations. Suppose that R_1 and R_2 are relations on a set A represented by the matrices $\mathbf{M}_{R_1}$ and $\mathbf{M}_{R_2}$, respectively. The matrix representing the union of these relations has a 1 in the positions where either $\mathbf{M}_{R_1}$ or $\mathbf{M}_{R_2}$ has a 1. The matrix representing the intersection of these relations has a 1 in the positions where both $\mathbf{M}_{R_1}$ and $\mathbf{M}_{R_2}$ have a 1. Thus, the matrices representing the union and intersection of these relations are

$$\mathbf{M}_{R_1 \cup R_2} = \mathbf{M}_{R_1} \vee \mathbf{M}_{R_2} \quad \text{and} \quad \mathbf{M}_{R_1 \cap R_2} = \mathbf{M}_{R_1} \wedge \mathbf{M}_{R_2}.$$

EXAMPLE 4 Suppose that the relations R_1 and R_2 on a set A are represented by the matrices

$$\mathbf{M}_{R_1} = \begin{bmatrix} 1 & 0 & 1 \\ 1 & 0 & 0 \\ 0 & 1 & 0 \end{bmatrix} \quad \text{and} \quad \mathbf{M}_{R_2} = \begin{bmatrix} 1 & 0 & 1 \\ 0 & 1 & 1 \\ 1 & 0 & 0 \end{bmatrix}.$$

What are the matrices representing $R_1 \cup R_2$ and $R_1 \cap R_2$?

Solution: The matrices of these relations are

$$\mathbf{M}_{R_1 \cup R_2} = \mathbf{M}_{R_1} \vee \mathbf{M}_{R_2} = \begin{bmatrix} 1 & 0 & 1 \\ 1 & 1 & 1 \\ 1 & 1 & 0 \end{bmatrix},$$

$$\mathbf{M}_{R_1 \cap R_2} = \mathbf{M}_{R_1} \wedge \mathbf{M}_{R_2} = \begin{bmatrix} 1 & 0 & 1 \\ 0 & 0 & 0 \\ 0 & 0 & 0 \end{bmatrix}.$$

◀

We now turn our attention to determining the matrix for the composite of relations. This matrix can be found using the Boolean product of the matrices (discussed in Section 2.6) for these relations. In particular, suppose that R is a relation from A to B and S is a relation from B to C. Suppose that A, B, and C have m, n, and p elements, respectively. Let the zero–one matrices for $S \circ R$, R, and S be $\mathbf{M}_{S \circ R} = [t_{ij}]$, $\mathbf{M}_R = [r_{ij}]$, and $\mathbf{M}_S = [s_{ij}]$, respectively (these matrices have sizes $m \times p$, $m \times n$, and $n \times p$, respectively). The ordered pair (a_i, c_j) belongs to $S \circ R$ if and only if there is an element b_k such that (a_i, b_k) belongs to R and (b_k, c_j) belongs to S. It follows that $t_{ij} = 1$ if and only if $r_{ik} = s_{kj} = 1$ for some k. From the definition of the Boolean product, this means that

$$\mathbf{M}_{S \circ R} = \mathbf{M}_R \odot \mathbf{M}_S.$$

EXAMPLE 5 Find the matrix representing the relations $S \circ R$, where the matrices representing R and S are

$$\mathbf{M}_R = \begin{bmatrix} 1 & 0 & 1 \\ 1 & 1 & 0 \\ 0 & 0 & 0 \end{bmatrix} \quad \text{and} \quad \mathbf{M}_S = \begin{bmatrix} 0 & 1 & 0 \\ 0 & 0 & 1 \\ 1 & 0 & 1 \end{bmatrix}.$$

Solution: The matrix for $S \circ R$ is

$$\mathbf{M}_{S \circ R} = \mathbf{M}_R \odot \mathbf{M}_S = \begin{bmatrix} 1 & 1 & 1 \\ 0 & 1 & 1 \\ 0 & 0 & 0 \end{bmatrix}.$$

◀

The matrix representing the composite of two relations can be used to find the matrix for $\mathbf{M}_{R^n}$. In particular,

$$\mathbf{M}_{R^n} = \mathbf{M}_R^{[n]},$$

from the definition of Boolean powers. Exercise 35 asks for a proof of this formula.

EXAMPLE 6 Find the matrix representing the relation R^2, where the matrix representing R is

$$\mathbf{M}_R = \begin{bmatrix} 0 & 1 & 0 \\ 0 & 1 & 1 \\ 1 & 0 & 0 \end{bmatrix}.$$

Solution: The matrix for R^2 is

$$\mathbf{M}_{R^2} = \mathbf{M}_R^{[2]} = \begin{bmatrix} 0 & 1 & 1 \\ 1 & 1 & 1 \\ 0 & 1 & 0 \end{bmatrix}.$$

◀

9.3.3 Representing Relations Using Digraphs

We have shown that a relation can be represented by listing all of its ordered pairs or by using a zero–one matrix. There is another important way of representing a relation using a pictorial representation. Each element of the set is represented by a point, and each ordered pair is represented using an arc with its direction indicated by an arrow. We use such pictorial representations when we think of relations on a finite set as **directed graphs**, or **digraphs**.

Definition 1 A *directed graph*, or *digraph*, consists of a set V of *vertices* (or *nodes*) together with a set E of ordered pairs of elements of V called *edges* (or *arcs*). The vertex a is called the *initial vertex* of the edge (a, b), and the vertex b is called the *terminal vertex* of this edge.

An edge of the form (a, a) is represented using an arc from the vertex a back to itself. Such an edge is called a **loop**.

EXAMPLE 7 The directed graph with vertices a, b, c, and d, and edges (a, b), (a, d), (b, b), (b, d), (c, a), (c, b), and (d, b) is displayed in Figure 3. ◀

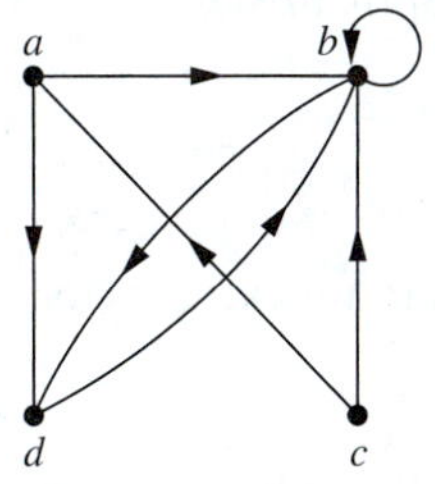

FIGURE 3 **A directed graph.**

The relation R on a set A is represented by the directed graph that has the elements of A as its vertices and the ordered pairs (a, b), where $(a, b) \in R$, as edges. This assignment sets up a one-to-one correspondence between the relations on a set A and the directed graphs with A as their set of vertices. Thus, every statement about relations corresponds to a statement about directed graphs, and vice versa. Directed graphs give a visual display of information about relations. As such, they are often used to study relations and their properties. (Note that relations from a set A to a set B can be represented by a directed graph where there is a vertex for each element of A and a vertex for each element of B, as shown in Section 9.1. However, when $A = B$, such representation provides much less insight than the digraph representations described here.) The use of directed graphs to represent relations on a set is illustrated in Examples 8–10.

EXAMPLE 8 The directed graph of the relation

$$R_1 = \{(1, 1), (1, 3), (2, 1), (2, 3), (2, 4), (3, 1), (3, 2), (4, 1)\}$$

on the set $\{1, 2, 3, 4\}$ is shown in Figure 4. ◀

EXAMPLE 9 What are the ordered pairs in the relation R_2 represented by the directed graph shown in Figure 5?

Solution: The ordered pairs (x, y) in the relation are

$$R_2 = \{(1, 3), (1, 4), (2, 1), (2, 2), (2, 3), (3, 1), (3, 3), (4, 1), (4, 3)\}.$$

Each of these pairs corresponds to an edge of the directed graph, with (2, 2) and (3, 3) corresponding to loops. ◀

We will study directed graphs extensively in Chapter 10.

The directed graph representing a relation can be used to determine whether the relation has various properties. For instance, a relation is reflexive if and only if there is a loop at every vertex of the directed graph, so that every ordered pair of the form (x, x) occurs in the relation. A relation is symmetric if and only if for every edge between distinct vertices in its digraph there is an edge in the opposite direction, so that (y, x) is in the relation whenever (x, y) is in the relation. Similarly, a relation is antisymmetric if and only if there are never two edges in opposite directions between distinct vertices. Finally, a relation is transitive if and only if whenever there is an edge from a vertex x to a vertex y and an edge from a vertex y to a vertex z, there is an edge from x to z (completing a triangle where each side is a directed edge with the correct direction).

Remark: Note that a symmetric relation can be represented by an undirected graph, which is a graph where edges do not have directions. We will study undirected graphs in Chapter 10.

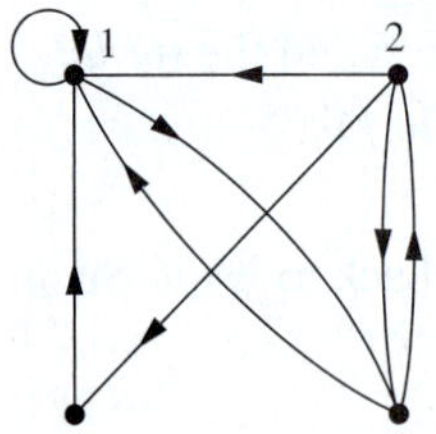

FIGURE 4 **The directed graph of the relation R_1.**

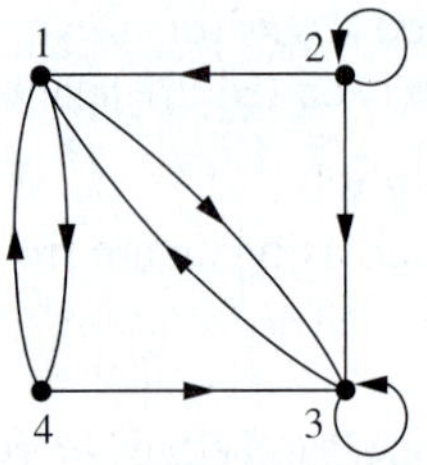

FIGURE 5 **The directed graph of the relation R_2.**

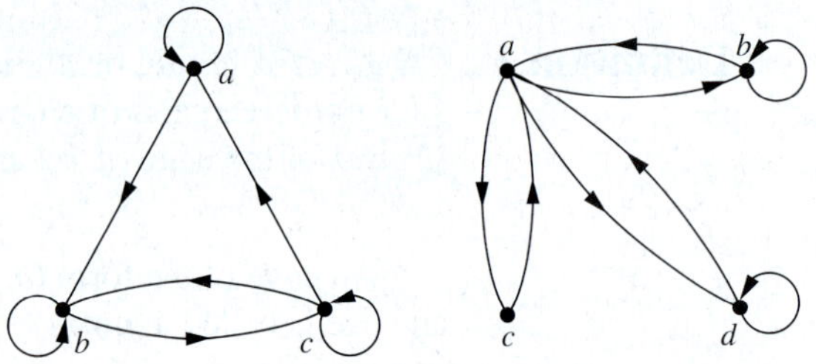

(a) Directed graph of S_1 (b) Directed graph of S_2

FIGURE 6 **The directed graphs of the relations S_1 and S_2.**

EXAMPLE 10 Determine whether the relations for the directed graphs shown in Figure 6 are reflexive, symmetric, antisymmetric, and/or transitive.

Solution: Because there are loops at every vertex of the directed graph of S_1, it is reflexive. The relation S_1 is neither symmetric nor antisymmetric because there is an edge from a to b but not one from b to a, but there are edges in both directions connecting b and c. Finally, S_1 is not transitive because there is an edge from a to b and an edge from b to c, but no edge from a to c.

Because loops are not present at all the vertices of the directed graph of S_2, this relation is not reflexive. It is symmetric and not antisymmetric, because every edge between distinct vertices is accompanied by an edge in the opposite direction. It is also not hard to see from the directed graph that S_2 is not transitive, because (c, a) and (a, b) belong to S_2, but (c, b) does not belong to S_2. ◀

Exercises

1. Represent each of these relations on $\{1, 2, 3\}$ with a matrix (with the elements of this set listed in increasing order).

a) $\{(1, 1), (1, 2), (1, 3)\}$
b) $\{(1, 2), (2, 1), (2, 2), (3, 3)\}$
c) $\{(1, 1), (1, 2), (1, 3), (2, 2), (2, 3), (3, 3)\}$
d) $\{(1, 3), (3, 1)\}$

2. Represent each of these relations on $\{1, 2, 3, 4\}$ with a matrix (with the elements of this set listed in increasing order).

a) $\{(1, 2), (1, 3), (1, 4), (2, 3), (2, 4), (3, 4)\}$
b) $\{(1, 1), (1, 4), (2, 2), (3, 3), (4, 1)\}$
c) $\{(1, 2), (1, 3), (1, 4), (2, 1), (2, 3), (2, 4), (3, 1), (3, 2), (3, 4), (4, 1), (4, 2), (4, 3)\}$
d) $\{(2, 4), (3, 1), (3, 2), (3, 4)\}$

3. List the ordered pairs in the relations on $\{1, 2, 3\}$ corresponding to these matrices (where the rows and columns correspond to the integers listed in increasing order).

a) $\begin{bmatrix} 1 & 0 & 1 \\ 0 & 1 & 0 \\ 1 & 0 & 1 \end{bmatrix}$ **b)** $\begin{bmatrix} 0 & 1 & 0 \\ 0 & 1 & 0 \\ 0 & 1 & 0 \end{bmatrix}$

c) $\begin{bmatrix} 1 & 1 & 1 \\ 1 & 0 & 1 \\ 1 & 1 & 1 \end{bmatrix}$

4. List the ordered pairs in the relations on $\{1, 2, 3, 4\}$ corresponding to these matrices (where the rows and columns correspond to the integers listed in increasing order).

a) $\begin{bmatrix} 1 & 1 & 0 & 1 \\ 1 & 0 & 1 & 0 \\ 0 & 1 & 1 & 1 \\ 1 & 0 & 1 & 1 \end{bmatrix}$ **b)** $\begin{bmatrix} 1 & 1 & 1 & 0 \\ 0 & 1 & 0 & 0 \\ 0 & 0 & 1 & 1 \\ 1 & 0 & 0 & 1 \end{bmatrix}$

c) $\begin{bmatrix} 0 & 1 & 0 & 1 \\ 1 & 0 & 1 & 0 \\ 0 & 1 & 0 & 1 \\ 1 & 0 & 1 & 0 \end{bmatrix}$

5. How can the matrix representing a relation R on a set A be used to determine whether the relation is irreflexive?

6. How can the matrix representing a relation R on a set A be used to determine whether the relation is asymmetric?

7. Determine whether the relations represented by the matrices in Exercise 3 are reflexive, irreflexive, symmetric, antisymmetric, and/or transitive.

8. Determine whether the relations represented by the matrices in Exercise 4 are reflexive, irreflexive, symmetric, antisymmetric, and/or transitive.

9. How many nonzero entries does the matrix representing the relation R on $A = \{1, 2, 3, \ldots, 100\}$ consisting of the first 100 positive integers have if R is

a) $\{(a, b) \mid a > b\}$? **b)** $\{(a, b) \mid a \neq b\}$?
c) $\{(a, b) \mid a = b + 1\}$? **d)** $\{(a, b) \mid a = 1\}$?
e) $\{(a, b) \mid ab = 1\}$?

10. How many nonzero entries does the matrix representing the relation R on $A = \{1, 2, 3, \ldots, 1000\}$ consisting of the first 1000 positive integers have if R is

a) $\{(a, b) \mid a \leq b\}$?
b) $\{(a, b) \mid a = b \pm 1\}$?
c) $\{(a, b) \mid a + b = 1000\}$?
d) $\{(a, b) \mid a + b \leq 1001\}$?
e) $\{(a, b) \mid a \neq 0\}$?

11. How can the matrix for $\overline{R}$, the complement of the relation R, be found from the matrix representing R, when R is a relation on a finite set A?

12. How can the matrix for R^{-1}, the inverse of the relation R, be found from the matrix representing R, when R is a relation on a finite set A?

13. Let R be the relation represented by the matrix

$$\mathbf{M}_R = \begin{bmatrix} 0 & 1 & 1 \\ 1 & 1 & 0 \\ 1 & 0 & 1 \end{bmatrix}.$$

Find the matrix representing

a) R^{-1}. **b)** $\overline{R}$. **c)** R^2.

14. Let R_1 and R_2 be relations on a set A represented by the matrices

$$\mathbf{M}_{R_1} = \begin{bmatrix} 0 & 1 & 0 \\ 1 & 1 & 1 \\ 1 & 0 & 0 \end{bmatrix} \quad \text{and} \quad \mathbf{M}_{R_2} = \begin{bmatrix} 0 & 1 & 0 \\ 0 & 1 & 1 \\ 1 & 1 & 1 \end{bmatrix}.$$

Find the matrices that represent

a) $R_1 \cup R_2$. **b)** $R_1 \cap R_2$. **c)** $R_2 \circ R_1$.
d) $R_1 \circ R_1$. **e)** $R_1 \oplus R_2$.

15. Let R be the relation represented by the matrix

$$\mathbf{M}_R = \begin{bmatrix} 0 & 1 & 0 \\ 0 & 0 & 1 \\ 1 & 1 & 0 \end{bmatrix}.$$

Find the matrices that represent

a) R^2. **b)** R^3. **c)** R^4.

16. Let R be a relation on a set A with n elements. If there are k nonzero entries in $\mathbf{M}_R$, the matrix representing R, how many nonzero entries are there in $\mathbf{M}_{R^{-1}}$, the matrix representing R^{-1}, the inverse of R?

17. Let R be a relation on a set A with n elements. If there are k nonzero entries in $\mathbf{M}_R$, the matrix representing R, how many nonzero entries are there in $\mathbf{M}_{\overline{R}}$, the matrix representing $\overline{R}$, the complement of R?

18. Draw the directed graphs representing each of the relations from Exercise 1.

19. Draw the directed graphs representing each of the relations from Exercise 2.

20. Draw the directed graph representing each of the relations from Exercise 3.

21. Draw the directed graph representing each of the relations from Exercise 4.

22. Draw the directed graph that represents the relation $\{(a, a), (a, b), (b, c), (c, b), (c, d), (d, a), (d, b)\}$.

In Exercises 23–28 list the ordered pairs in the relations represented by the directed graphs.

23.

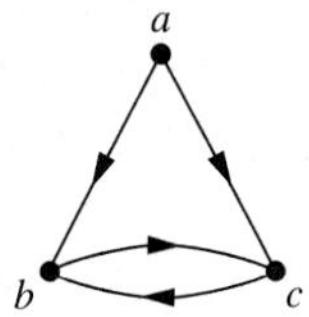

24.

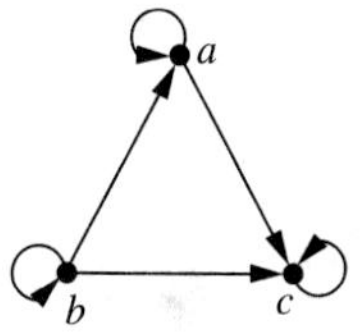

25.

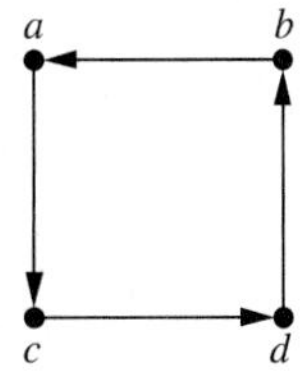

26.

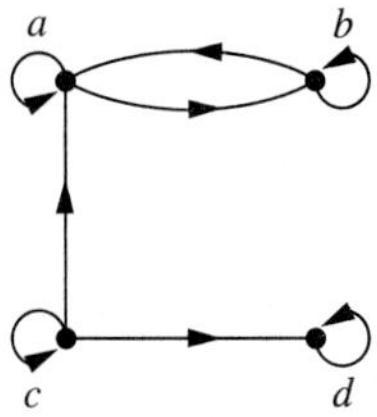

27.

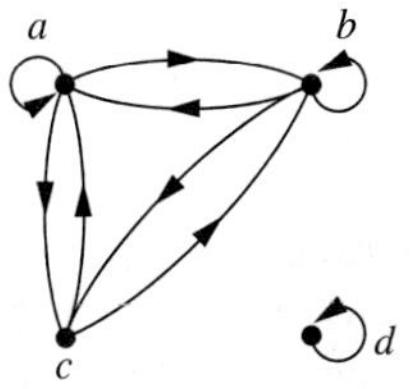

28.

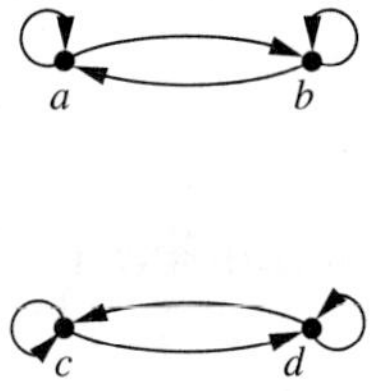

29. How can the directed graph of a relation R on a finite set A be used to determine whether a relation is asymmetric?

30. How can the directed graph of a relation R on a finite set A be used to determine whether a relation is irreflexive?

31. Determine whether the relations represented by the directed graphs shown in Exercises 23–25 are reflexive, irreflexive, symmetric, antisymmetric, and/or transitive.

32. Determine whether the relations represented by the directed graphs shown in Exercises 26–28 are reflexive, irreflexive, symmetric, antisymmetric, asymmetric, and/or transitive.

33. Let R be a relation on a set A. Explain how to use the directed graph representing R to obtain the directed graph representing the inverse relation R^{-1}.

34. Let R be a relation on a set A. Explain how to use the directed graph representing R to obtain the directed graph representing the complementary relation $\overline{R}$.

35. Show that if $\mathbf{M}_R$ is the matrix representing the relation R, then $\mathbf{M}_R^{[n]}$ is the matrix representing the relation R^n.

36. Given the directed graphs representing two relations, how can the directed graph of the union, intersection, symmetric difference, difference, and composition of these relations be found?

9.4 Closures of Relations

9.4.1 Introduction

A computer network has data centers in Boston, Chicago, Denver, Detroit, New York, and San Diego. There are direct, one-way telephone lines from Boston to Chicago, from Boston to Detroit, from Chicago to Detroit, from Detroit to Denver, and from New York to San Diego. Let R be the relation containing (a, b) if there is a telephone line from the data center in a to that in b. How can we determine if there is some (possibly indirect) link composed of one or more telephone lines from one center to another? Because not all links are direct, such as the link from Boston to Denver that goes through Detroit, R cannot be used directly to answer this. In the language of relations, R is not transitive, so it does not contain all the pairs that can be linked. As we will show in this section, we can find all pairs of data centers that have a link by constructing a transitive relation S containing R such that S is a subset of every transitive relation containing R. Here, S is the smallest transitive relation that contains R. This relation is called the **transitive closure** of R.

9.4.2 Different Types of Closures

If R is a relation on a set A, it may or may not have some property **P**, such as reflexivity, symmetry, or transitivity. When R does not enjoy property **P**, we would like to find the smallest relation S on A with property **P** that contains R.

Definition 1 If R is a relation on a set A, then the *closure* of R with respect to **P**, if it exists, is the relation S on A with property **P** that contains R and is a subset of every subset of $A \times A$ containing R with property **P**.

If there is a relation S that is a subset of every relation containing R with property **P**, it must be unique. To see this, suppose that relations S and T both have property **P** and are subsets of every relation with property **P** that contains R. Then, S and T are subsets of each other, and so are equal. Such a relation, if it exists, is the smallest relation with property **P** that contains R because it is a subset of every relation with property **P** that contains R.

We will show how reflexive, symmetric, and transitive closures of relations can be found. In Exercises 15 and 35 we give properties **P** for which the closure of a relation with respect to **P** may not exist.

The relation $R = \{(1, 1), (1, 2), (2, 1), (3, 2)\}$ on the set $A = \{1, 2, 3\}$ is not reflexive. How can we produce a reflexive relation containing R that is as small as possible? This can be done by adding $(2, 2)$ and $(3, 3)$ to R, because these are the only pairs of the form (a, a) that are not in R. This new relation contains R. Furthermore, *any* reflexive relation that contains R must also contain $(2, 2)$ and $(3, 3)$. Because this relation contains R, is reflexive, and is contained within every reflexive relation that contains R, it is called the **reflexive closure** of R.

As this example illustrates, given a relation R on a set A, the reflexive closure of R can be formed by adding to R all pairs of the form (a, a) with $a \in A$, not already in R. The addition of these pairs produces a new relation that is reflexive, contains R, and is contained within any reflexive relation containing R. We see that the reflexive closure of R equals $R \cup \Delta$, where $\Delta = \{(a, a) \mid a \in A\}$ is the **diagonal relation** on A. (The reader should verify this.)

EXAMPLE 1 What is the reflexive closure of the relation $R = \{(a, b) \mid a < b\}$ on the set of integers?

Solution: The reflexive closure of R is

$$R \cup \Delta = \{(a, b) \mid a < b\} \cup \{(a, a) \mid a \in \mathbf{Z}\} = \{(a, b) \mid a \leq b\}.$$

◀

The relation $\{(1, 1), (1, 2), (2, 2), (2, 3), (3, 1), (3, 2)\}$ on $\{1, 2, 3\}$ is not symmetric. How can we produce a symmetric relation that is as small as possible and contains R? To do this, we need only add $(2, 1)$ and $(1, 3)$, because these are the only pairs of the form (b, a) with $(a, b) \in R$ that are not in R. This new relation is symmetric and contains R. Furthermore, *any* symmetric relation that contains R must contain this new relation, because a symmetric relation that contains R must contain $(2, 1)$ and $(1, 3)$. Consequently, this new relation is called the **symmetric closure** of R.

As this example illustrates, the symmetric closure of a relation R can be constructed by adding all ordered pairs of the form (b, a), where (a, b) is in the relation, that are not already present in R. Adding these pairs produces a relation that is symmetric, that contains R, and that is contained in any symmetric relation that contains R. The symmetric closure of a relation can be constructed by taking the union of a relation with its inverse (defined in the preamble of Exercise 26 in Section 9.1); that is, $R \cup R^{-1}$ is the symmetric closure of R, where $R^{-1} = \{(b, a) \mid (a, b) \in R\}$. The reader should verify this statement.

EXAMPLE 2 What is the symmetric closure of the relation $R = \{(a, b) \mid a > b\}$ on the set of positive integers?

Extra Examples

Solution: The symmetric closure of R is the relation

$$R \cup R^{-1} = \{(a, b) \mid a > b\} \cup \{(b, a) \mid a > b\} = \{(a, b) \mid a \neq b\}.$$

This last equality follows because R contains all ordered pairs of positive integers, where the first element is greater than the second element, and R^{-1} contains all ordered pairs of positive integers, where the first element is less than the second. ◀

Suppose that a relation R is not transitive. How can we produce a transitive relation that contains R such that this new relation is contained within any transitive relation that contains R? Can the transitive closure of a relation R be produced by adding all the pairs of the form (a, c), where (a, b) and (b, c) are already in the relation? Consider the relation $R = \{(1, 3), (1, 4), (2, 1), (3, 2)\}$ on the set $\{1, 2, 3, 4\}$. This relation is not transitive because it does not contain all pairs of the form (a, c) where (a, b) and (b, c) are in R. The pairs of this form not in R are $(1, 2)$, $(2, 3)$, $(2, 4)$, and $(3, 1)$. Adding these pairs does *not* produce a transitive relation, because the resulting relation contains $(3, 1)$ and $(1, 4)$ but does not contain $(3, 4)$. This shows that constructing the transitive closure of a relation is more complicated than constructing either the reflexive or symmetric closure. The rest of this section develops algorithms for constructing transitive closures. As will be shown later in this section, the transitive closure of a relation can be found by adding new ordered pairs that must be present and then repeating this process until no new ordered pairs are needed.

9.4.3 Paths in Directed Graphs

We will see that representing relations by directed graphs helps in the construction of transitive closures. We now introduce some terminology that we will use for this purpose.

A path in a directed graph is obtained by traversing along edges (in the same direction as indicated by the arrow on the edge).

Definition 2 A *path* from a to b in the directed graph G is a sequence of edges (x_0, x_1), (x_1, x_2), $(x_2, x_3), \ldots, (x_{n-1}, x_n)$ in G, where n is a nonnegative integer, and $x_0 = a$ and $x_n = b$, that is, a sequence of edges where the terminal vertex of an edge is the same as the initial vertex in the next edge in the path. This path is denoted by $x_0, x_1, x_2, \ldots, x_{n-1}, x_n$ and has *length* n. We view the empty set of edges as a path of length zero from a to a. A path of length $n \geq 1$ that begins and ends at the same vertex is called a *circuit* or *cycle*.

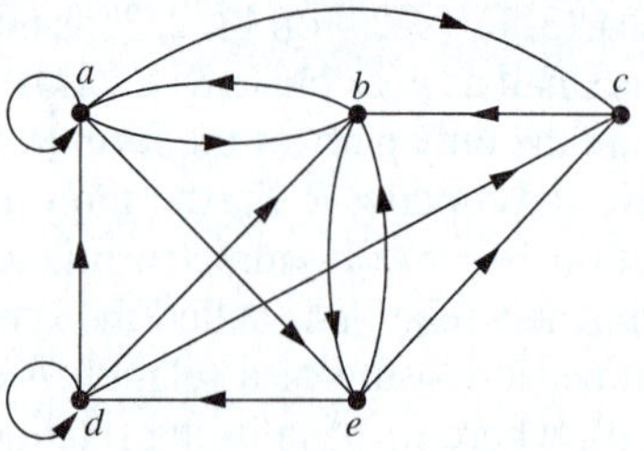

FIGURE 1 **A directed graph.**

A path in a directed graph can pass through a vertex more than once. Moreover, an edge in a directed graph can occur more than once in a path.

EXAMPLE 3 Which of the following are paths in the directed graph shown in Figure 1: a, b, e, d; a, e, c, d, b; b, a, c, b, a, a, b; d, c; c, b, a; e, b, a, b, a, b, e? What are the lengths of those that are paths? Which of the paths in this list are circuits?

Solution: Because each of (a, b), (b, e), and (e, d) is an edge, a, b, e, d is a path of length three. Because (c, d) is not an edge, a, e, c, d, b is not a path. Also, b, a, c, b, a, a, b is a path of length six because (b, a), (a, c), (c, b), (b, a), (a, a), and (a, b) are all edges. We see that d, c is a path of length one, because (d, c) is an edge. Also c, b, a is a path of length two, because (c, b) and (b, a) are edges. All of (e, b), (b, a), (a, b), (b, a), (a, b), and (b, e) are edges, so e, b, a, b, a, b, e is a path of length six.

The two paths b, a, c, b, a, a, b and e, b, a, b, a, b, e are circuits because they begin and end at the same vertex. The paths a, b, e, d; c, b, a; and d, c are not circuits. ◀

The term *path* also applies to relations. Carrying over the definition from directed graphs to relations, there is a **path** from a to b in R if there is a sequence of elements $a, x_1, x_2, \ldots, x_{n-1}, b$ with $(a, x_1) \in R$, $(x_1, x_2) \in R$, $\ldots$, and $(x_{n-1}, b) \in R$. Theorem 1 can be obtained from the definition of a path in a relation.

THEOREM 1 Let R be a relation on a set A. There is a path of length n, where n is a positive integer, from a to b if and only if $(a, b) \in R^n$.

Proof: We will use mathematical induction. By definition, there is a path from a to b of length one if and only if $(a, b) \in R$, so the theorem is true when $n = 1$.

Assume that the theorem is true for the positive integer n. This is the inductive hypothesis. There is a path of length $n + 1$ from a to b if and only if there is an element $c \in A$ such that there is a path of length one from a to c, so $(a, c) \in R$, and a path of length n from c to b. Consequently, by the inductive hypothesis, there is a path of length $n + 1$ from a to b if and only if there is an element c with $(a, c) \in R$ and $(c, b) \in R^n$. But there is such an element if and only if $(a, b) \in R^{n+1}$. Therefore, there is a path of length $n + 1$ from a to b if and only if $(a, b) \in R^{n+1}$. This completes the proof. ◁

9.4.4 Transitive Closures

We now show that finding the transitive closure of a relation is equivalent to determining which pairs of vertices in the associated directed graph are connected by a path. With this in mind, we define a new relation.

Definition 3 Let R be a relation on a set A. The *connectivity relation* R^* consists of the pairs (a, b) such that there is a path of length at least one from a to b in R.

Because R^n consists of the pairs (a, b) such that there is a path of length n from a to b, it follows that R^* is the union of all the sets R^n. In other words,

$$R^* = \bigcup_{n=1}^{\infty} R^n.$$

The connectivity relation is useful in many models.

EXAMPLE 4 Let R be the relation on the set of all people in the world that contains (a, b) if a has met b. What is R^n, where n is a positive integer greater than one? What is R^*?

Solution: The relation R^2 contains (a, b) if there is a person c such that $(a, c) \in R$ and $(c, b) \in R$, that is, if there is a person c such that a has met c and c has met b. Similarly, R^n consists of those pairs (a, b) such that there are people $x_1, x_2, \ldots, x_{n-1}$ such that a has met x_1, x_1 has met $x_2, \ldots,$ and x_{n-1} has met b.

The relation R^* contains (a, b) if there is a sequence of people, starting with a and ending with b, such that each person in the sequence has met the next person in the sequence. (There are many interesting conjectures about R^*. Do you think that this connectivity relation includes the pair with you as the first element and the president of Mongolia as the second element? We will use graphs to model this application in Chapter 10.) ◀

EXAMPLE 5 Let R be the relation on the set of all subway stops in New York City that contains (a, b) if it is possible to travel from stop a to stop b without changing trains. What is R^n when n is a positive integer? What is R^*?

Solution: The relation R^n contains (a, b) if it is possible to travel from stop a to stop b by making at most $n - 1$ changes of trains. The relation R^* consists of the ordered pairs (a, b) where it is possible to travel from stop a to stop b making as many changes of trains as necessary. (The reader should verify these statements.) ◀

EXAMPLE 6 Let R be the relation on the set of all states in the United States that contains (a, b) if state a and state b have a common border. What is R^n, where n is a positive integer? What is R^*?

Solution: The relation R^n consists of the pairs (a, b), where it is possible to go from state a to state b by crossing exactly n state borders. R^* consists of the ordered pairs (a, b), where it is possible to go from state a to state b crossing as many borders as necessary. (The reader should verify these statements.) The only ordered pairs not in R^* are those containing states that are not connected to the continental United States (that is, those pairs containing Alaska or Hawaii). ◀

Theorem 2 shows that the transitive closure of a relation and the associated connectivity relation are the same.

THEOREM 2 The transitive closure of a relation R equals the connectivity relation R^*.

Proof: Note that R^* contains R by definition. To show that R^* is the transitive closure of R we must also show that R^* is transitive and that $R^* \subseteq S$ whenever S is a transitive relation that contains R.

First, we show that R^* is transitive. If $(a, b) \in R^*$ and $(b, c) \in R^*$, then there are paths from a to b and from b to c in R. We obtain a path from a to c by starting with the path from a to b and following it with the path from b to c. Hence, $(a, c) \in R^*$. It follows that R^* is transitive.

Now suppose that S is a transitive relation containing R. Because S is transitive, S^n also is transitive (the reader should verify this) and $S^n \subseteq S$ (by Theorem 1 of Section 9.1). Furthermore, because

$$S^* = \bigcup_{k=1}^{\infty} S^k$$

and $S^k \subseteq S$, it follows that $S^* \subseteq S$. Now note that if $R \subseteq S$, then $R^* \subseteq S^*$, because any path in R is also a path in S. Consequently, $R^* \subseteq S^* \subseteq S$. Thus, any transitive relation that contains R must also contain R^*. Therefore, R^* is the transitive closure of R. ◁

Now that we know that the transitive closure equals the connectivity relation, we turn our attention to the problem of computing this relation. We do not need to examine arbitrarily long paths to determine whether there is a path between two vertices in a finite directed graph. As Lemma 1 shows, it is sufficient to examine paths containing no more than n edges, where n is the number of elements in the set.

LEMMA 1 Let A be a set with n elements, and let R be a relation on A. If there is a path of length at least one in R from a to b, then there is such a path with length not exceeding n. Moreover, when $a \neq b$, if there is a path of length at least one in R from a to b, then there is such a path with length not exceeding $n - 1$.

Proof: Suppose there is a path from a to b in R. Let m be the length of the shortest such path. Suppose that $x_0, x_1, x_2, \ldots, x_{m-1}, x_m$, where $x_0 = a$ and $x_m = b$, is such a path.

Suppose that $a = b$ and that $m > n$, so that $m \geq n + 1$. By the pigeonhole principle, because there are n vertices in A, among the m vertices $x_0, x_1, \ldots, x_{m-1}$, at least two are equal (see Figure 2).

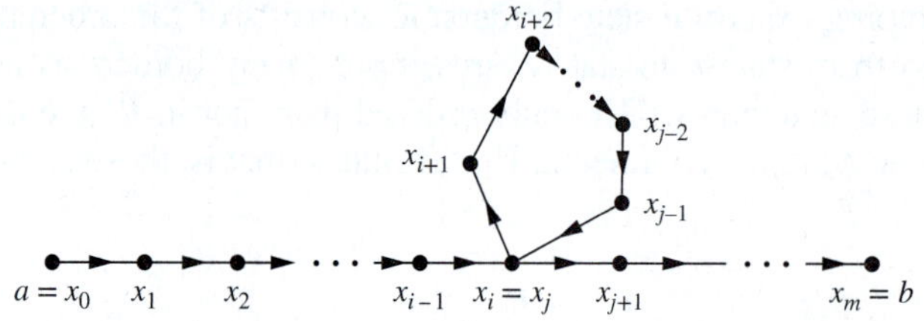

FIGURE 2 Producing a path with length not exceeding *n*.

Suppose that $x_i = x_j$ with $0 \leq i < j \leq m - 1$. Then the path contains a circuit from x_i to itself. This circuit can be deleted from the path from a to b, leaving a path, namely, $x_0, x_1, \ldots, x_i, x_{j+1}, \ldots, x_{m-1}, x_m$, from a to b of shorter length. Hence, the path of shortest length must have length less than or equal to n.

The case where $a \neq b$ is left as an exercise for the reader. ◁

From Lemma 1, we see that the transitive closure of R is the union of R, R^2, R^3, …, and R^n. This follows because there is a path in R^* between two vertices if and only if there is a path between these vertices in R^i, for some positive integer i with $i \leq n$. Because

$$R^* = R \cup R^2 \cup R^3 \cup \cdots \cup R^n$$

and the zero–one matrix representing a union of relations is the join of the zero–one matrices of these relations, the zero–one matrix for the transitive closure is the join of the zero–one matrices of the first n powers of the zero–one matrix of R.

THEOREM 3 Let $\mathbf{M}_R$ be the zero–one matrix of the relation R on a set with n elements. Then the zero–one matrix of the transitive closure R^* is

$$\mathbf{M}_{R^*} = \mathbf{M}_R \vee \mathbf{M}_R^{[2]} \vee \mathbf{M}_R^{[3]} \vee \cdots \vee \mathbf{M}_R^{[n]}.$$

EXAMPLE 7 Find the zero–one matrix of the transitive closure of the relation R where

$$\mathbf{M}_R = \begin{bmatrix} 1 & 0 & 1 \\ 0 & 1 & 0 \\ 1 & 1 & 0 \end{bmatrix}.$$

Solution: By Theorem 3, it follows that the zero–one matrix of R^* is

$$\mathbf{M}_{R^*} = \mathbf{M}_R \vee \mathbf{M}_R^{[2]} \vee \mathbf{M}_R^{[3]}.$$

Because

$$\mathbf{M}_R^{[2]} = \begin{bmatrix} 1 & 1 & 1 \\ 0 & 1 & 0 \\ 1 & 1 & 1 \end{bmatrix} \quad \text{and} \quad \mathbf{M}_R^{[3]} = \begin{bmatrix} 1 & 1 & 1 \\ 0 & 1 & 0 \\ 1 & 1 & 1 \end{bmatrix},$$

it follows that

$$\mathbf{M}_{R^*} = \begin{bmatrix} 1 & 0 & 1 \\ 0 & 1 & 0 \\ 1 & 1 & 0 \end{bmatrix} \vee \begin{bmatrix} 1 & 1 & 1 \\ 0 & 1 & 0 \\ 1 & 1 & 1 \end{bmatrix} \vee \begin{bmatrix} 1 & 1 & 1 \\ 0 & 1 & 0 \\ 1 & 1 & 1 \end{bmatrix} = \begin{bmatrix} 1 & 1 & 1 \\ 0 & 1 & 0 \\ 1 & 1 & 1 \end{bmatrix}.$$

◀

Links ❯ Theorem 3 can be used as a basis for an algorithm for computing the matrix of the relation R^*. To find this matrix, the successive Boolean powers of $\mathbf{M}_R$, up to the nth power, are computed. As each power is calculated, its join with the join of all smaller powers is formed. When this is done with the nth power, the matrix for R^* has been found. This procedure is displayed as Algorithm 1.

ALGORITHM 1 A Procedure for Computing the Transitive Closure.

procedure *transitive closure* ($\mathbf{M}_R$: zero–one $n \times n$ matrix)
$\mathbf{A} := \mathbf{M}_R$
$\mathbf{B} := \mathbf{A}$
for $i := 2$ **to** n
  $\mathbf{A} := \mathbf{A} \odot \mathbf{M}_R$
  $\mathbf{B} := \mathbf{B} \vee \mathbf{A}$
return $\mathbf{B}$\{$\mathbf{B}$ is the zero–one matrix for R^*\}

We can easily find the number of bit operations used by Algorithm 1 to determine the transitive closure of a relation. Computing the Boolean powers $\mathbf{M}_R, \mathbf{M}_R^{[2]}, \ldots, \mathbf{M}_R^{[n]}$ requires that $n-1$ Boolean products of $n \times n$ zero–one matrices be found. Each of these Boolean products can be found using $n^2(2n-1)$ bit operations. Hence, these products can be computed using $n^2(2n-1)(n-1)$ bit operations.

To find $\mathbf{M}_{R^*}$ from the n Boolean powers of $\mathbf{M}_R$, $n-1$ joins of zero–one matrices need to be found. Computing each of these joins uses n^2 bit operations. Hence, $(n-1)n^2$ bit operations are used in this part of the computation. Therefore, when Algorithm 1 is used, the matrix of the transitive closure of a relation on a set with n elements can be found using $n^2(2n-1)(n-1) + (n-1)n^2 = 2n^3(n-1)$, which is $O(n^4)$ bit operations. The remainder of this section describes a more efficient algorithm for finding transitive closures.

9.4.5 Warshall's Algorithm

Links

Warshall's algorithm, named after Stephen Warshall, who described this algorithm in 1960, is an efficient method for computing the transitive closure of a relation. Algorithm 1 can find the transitive closure of a relation on a set with n elements using $2n^3(n-1)$ bit operations. However, the transitive closure can be found by Warshall's algorithm using only $2n^3$ bit operations.

Remark: Warshall's algorithm is sometimes called the Roy–Warshall algorithm, because Bernard Roy described this algorithm in 1959.

Suppose that R is a relation on a set with n elements. Let $v_1, v_2, \ldots, v_n$ be an arbitrary listing of these n elements. The concept of the **interior vertices** of a path is used in Warshall's algorithm. If $a, x_1, x_2, \ldots, x_{m-1}, b$ is a path, its interior vertices are $x_1, x_2, \ldots, x_{m-1}$, that is, all the vertices of the path that occur somewhere other than as the first and last vertices in the path. For instance, the interior vertices of a path a, c, d, f, g, h, b, j in a directed graph are c, d, f, g, h, and b. The interior vertices of a, c, d, a, f, b are c, d, a, and f. (Note that the first vertex in the path is not an interior vertex unless it is visited again by the path, except as the last vertex. Similarly, the last vertex in the path is not an interior vertex unless it was visited previously by the path, except as the first vertex.)

Warshall's algorithm is based on the construction of a sequence of zero–one matrices. These matrices are $\mathbf{W}_0, \mathbf{W}_1, \ldots, \mathbf{W}_n$, where $\mathbf{W}_0 = \mathbf{M}_R$ is the zero–one matrix of this relation, and $\mathbf{W}_k = [w_{ij}^{(k)}]$, where $w_{ij}^{(k)} = 1$ if there is a path from v_i to v_j such that all the interior vertices of this path are in the set $\{v_1, v_2, \ldots, v_k\}$ (the first k vertices in the list) and is 0 otherwise. (The first and last vertices in the path may be outside the set of the first k vertices in the list.) Note that $\mathbf{W}_n = \mathbf{M}_{R^*}$, because the (i, j)th entry of $\mathbf{M}_{R^*}$ is 1 if and only if there is a path from v_i to v_j, with all interior vertices in the set $\{v_1, v_2, \ldots, v_n\}$ (but these are the only vertices in the directed graph). Example 8 illustrates what the matrix $\mathbf{W}_k$ represents.

EXAMPLE 8 Let R be the relation with directed graph shown in Figure 3. Let a, b, c, d be a listing of the elements of the set. Find the matrices $\mathbf{W}_0$, $\mathbf{W}_1$, $\mathbf{W}_2$, $\mathbf{W}_3$, and $\mathbf{W}_4$. The matrix $\mathbf{W}_4$ is the transitive closure of R.

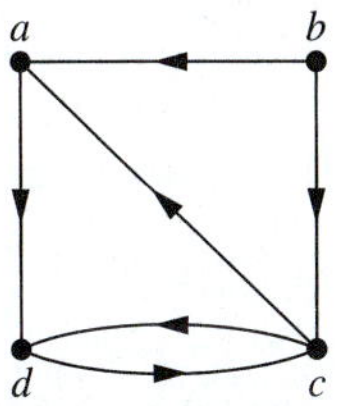

FIGURE 3 The directed graph of the relation *R*.

Solution: Let $v_1 = a$, $v_2 = b$, $v_3 = c$, and $v_4 = d$. $\mathbf{W}_0$ is the matrix of the relation. Hence,

$$\mathbf{W}_0 = \begin{bmatrix} 0 & 0 & 0 & 1 \\ 1 & 0 & 1 & 0 \\ 1 & 0 & 0 & 1 \\ 0 & 0 & 1 & 0 \end{bmatrix}.$$

$\mathbf{W}_1$ has 1 as its (i, j)th entry if there is a path from v_i to v_j that has only $v_1 = a$ as an interior vertex. Note that all paths of length one can still be used because they have no interior vertices. Also, there is now an allowable path from b to d, namely, b, a, d. Hence,

$$\mathbf{W}_1 = \begin{bmatrix} 0 & 0 & 0 & 1 \\ 1 & 0 & 1 & 1 \\ 1 & 0 & 0 & 1 \\ 0 & 0 & 1 & 0 \end{bmatrix}.$$

$\mathbf{W}_2$ has 1 as its (i, j)th entry if there is a path from v_i to v_j that has only $v_1 = a$ and/or $v_2 = b$ as its interior vertices, if any. Because there are no edges that have b as a terminal vertex, no new paths are obtained when we permit b to be an interior vertex. Hence, $\mathbf{W}_2 = \mathbf{W}_1$.

$\mathbf{W}_3$ has 1 as its (i, j)th entry if there is a path from v_i to v_j that has only $v_1 = a$, $v_2 = b$, and/or $v_3 = c$ as its interior vertices, if any. We now have paths from d to a, namely, d, c, a, and from d to d, namely, d, c, d. Hence,

$$\mathbf{W}_3 = \begin{bmatrix} 0 & 0 & 0 & 1 \\ 1 & 0 & 1 & 1 \\ 1 & 0 & 0 & 1 \\ 1 & 0 & 1 & 1 \end{bmatrix}.$$

Finally, $\mathbf{W}_4$ has 1 as its (i, j)th entry if there is a path from v_i to v_j that has $v_1 = a$, $v_2 = b$, $v_3 = c$, and/or $v_4 = d$ as interior vertices, if any. Because these are all the vertices of the graph, this entry is 1 if and only if there is a path from v_i to v_j. Hence,

$$\mathbf{W}_4 = \begin{bmatrix} 1 & 0 & 1 & 1 \\ 1 & 0 & 1 & 1 \\ 1 & 0 & 1 & 1 \\ 1 & 0 & 1 & 1 \end{bmatrix}.$$

This last matrix, $\mathbf{W}_4$, is the matrix of the transitive closure. ◀

Warshall's algorithm computes $\mathbf{M}_{R^*}$ by efficiently computing $\mathbf{W}_1, \mathbf{W}_2, \ldots, \mathbf{W}_n = \mathbf{M}_{R^*}$. This observation shows that we can compute $\mathbf{W}_k$ directly from $\mathbf{W}_{k-1}$: There is a path from v_i to v_j with no vertices other than $v_1, v_2, \ldots, v_k$ as interior vertices if and only if either there is a path from v_i to v_j with its interior vertices among the first $k - 1$ vertices in the list, or there are paths from v_i to v_k and from v_k to v_j that have interior vertices only among the first $k - 1$ vertices in the list. That is, either a path from v_i to v_j already existed before v_k was permitted as an interior vertex, or allowing v_k as an interior vertex produces a path that goes from v_i to v_k and then from v_k to v_j. These two cases are shown in Figure 4.

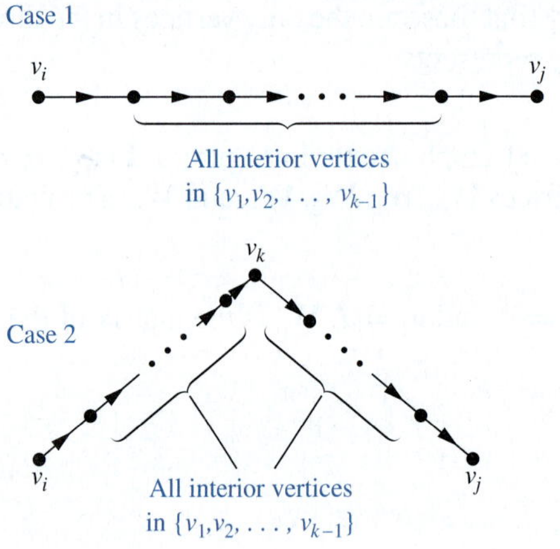

FIGURE 4 Adding v_k to the set of allowable interior vertices.

The first type of path exists if and only if $w_{ij}^{[k-1]} = 1$, and the second type of path exists if and only if both $w_{ik}^{[k-1]}$ and $w_{kj}^{[k-1]}$ are 1. Hence, $w_{ij}^{[k]}$ is 1 if and only if either $w_{ij}^{[k-1]}$ is 1 or both $w_{ik}^{[k-1]}$ and $w_{kj}^{[k-1]}$ are 1. This gives us Lemma 2.

LEMMA 2 Let $\mathbf{W}_k = [w_{ij}^{[k]}]$ be the zero–one matrix that has a 1 in its (i, j)th position if and only if there is a path from v_i to v_j with interior vertices from the set $\{v_1, v_2, \ldots, v_k\}$. Then

$$w_{ij}^{[k]} = w_{ij}^{[k-1]} \vee (w_{ik}^{[k-1]} \wedge w_{kj}^{[k-1]}),$$

whenever i, j, and k are positive integers not exceeding n.

Lemma 2 gives us the means to compute efficiently the matrices $\mathbf{W}_k$, $k = 1, 2, \ldots, n$. We display the pseudocode for Warshall's algorithm, using Lemma 2, as Algorithm 2.

Links

Stephen Warshall

STEPHEN WARSHALL (1935–2006) Stephen Warshall, born in New York City, went to public school in Brooklyn. He attended Harvard University, receiving his degree in mathematics in 1956. He never received an advanced degree, because at that time no programs were available in his areas of interest. However, he took graduate courses at several different universities and contributed to the development of computer science and software engineering.

After graduating from Harvard, Warshall worked at ORO (Operation Research Office), which was set up by Johns Hopkins to do research and development for the U.S. Army. In 1958 he left ORO to take a position at a company called Technical Operations, where he helped build a research and development laboratory for military software projects. In 1961 he left Technical Operations to found Massachusetts Computer Associates. Later, this company became part of Applied Data Research (ADR). After the merger, Warshall sat on the board of directors of ADR and managed a variety of projects and organizations. He retired from ADR in 1982.

During his career Warshall carried out research and development in operating systems, compiler design, language design, and operations research. In the 1971–1972 academic year he presented lectures on software engineering at French universities. There is an interesting anecdote about his proof that the transitive closure algorithm, now known as Warshall's algorithm, is correct. He and a colleague at Technical Operations bet a bottle of rum on who first could determine whether this algorithm always works. Warshall came up with his proof overnight, winning the bet and the rum, which he shared with the loser of the bet. Because Warshall did not like sitting at a desk, he did much of his creative work in unconventional places, such as on a sailboat in the Indian Ocean or in a Greek lemon orchard.

ALGORITHM 2 Warshall Algorithm.

procedure *Warshall* ($\mathbf{M}_R$: $n \times n$ zero–one matrix)
$\mathbf{W} := \mathbf{M}_R$
for $k := 1$ **to** n
  for $i := 1$ **to** n
    for $j := 1$ **to** n
      $w_{ij} := w_{ij} \vee (w_{ik} \wedge w_{kj})$
return $\mathbf{W}\{\mathbf{W} = [w_{ij}]$ is $\mathbf{M}_{R^*}\}$

The computational complexity of Warshall's algorithm can easily be computed in terms of bit operations. To find the entry $w_{ij}^{[k]}$ from the entries $w_{ij}^{[k-1]}$, $w_{ik}^{[k-1]}$, and $w_{kj}^{[k-1]}$ using Lemma 2 requires two bit operations. To find all n^2 entries of $\mathbf{W}_k$ from those of $\mathbf{W}_{k-1}$ requires $2n^2$ bit operations. Because Warshall's algorithm begins with $\mathbf{W}_0 = \mathbf{M}_R$ and computes the sequence of n zero–one matrices $\mathbf{W}_1, \mathbf{W}_2, \ldots, \mathbf{W}_n = \mathbf{M}_{R^*}$, the total number of bit operations used is $n \cdot 2n^2 = 2n^3$.

Exercises

1. Let R be the relation on the set $\{0, 1, 2, 3\}$ containing the ordered pairs $(0, 1)$, $(1, 1)$, $(1, 2)$, $(2, 0)$, $(2, 2)$, and $(3, 0)$. Find the

a) reflexive closure of R. **b)** symmetric closure of R.

2. Let R be the relation $\{(a, b) \mid a \neq b\}$ on the set of integers. What is the reflexive closure of R?

3. Let R be the relation $\{(a, b) \mid a \text{ divides } b\}$ on the set of integers. What is the symmetric closure of R?

4. How can the directed graph representing the reflexive closure of a relation on a finite set be constructed from the directed graph of the relation?

In Exercises 5–7 draw the directed graph of the reflexive closure of the relations with the directed graph shown.

5.

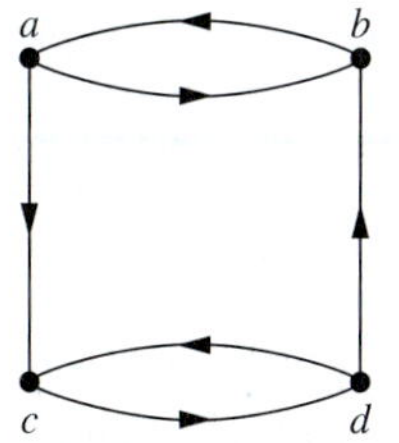

6.

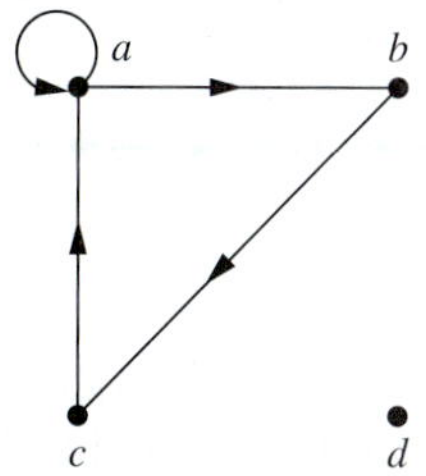

7.

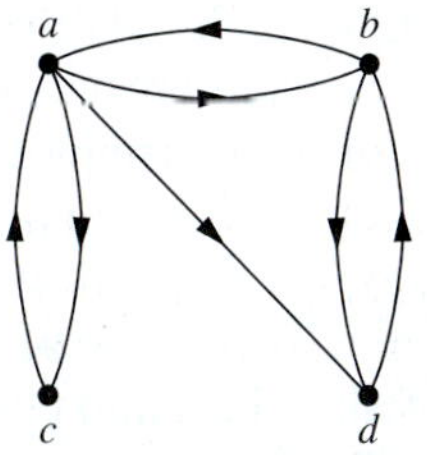

8. How can the directed graph representing the symmetric closure of a relation on a finite set be constructed from the directed graph for this relation?

9. Find the directed graphs of the symmetric closures of the relations with directed graphs shown in Exercises 5–7.

10. Find the smallest relation containing the relation in Example 2 that is both reflexive and symmetric.

11. Find the directed graph of the smallest relation that is both reflexive and symmetric that contains each of the relations with directed graphs shown in Exercises 5–7.

12. Suppose that the relation R on the finite set A is represented by the matrix $\mathbf{M}_R$. Show that the matrix that represents the reflexive closure of R is $\mathbf{M}_R \vee \mathbf{I}_n$.

13. Suppose that the relation R on the finite set A is represented by the matrix $\mathbf{M}_R$. Show that the matrix that represents the symmetric closure of R is $\mathbf{M}_R \vee \mathbf{M}_R^t$.

14. Show that the closure of a relation R with respect to a property **P**, if it exists, is the intersection of all the relations with property **P** that contain R.

15. When is it possible to define the "irreflexive closure" of a relation R, that is, a relation that contains R, is irreflexive, and is contained in every irreflexive relation that contains R?

16. Determine whether these sequences of vertices are paths in this directed graph.

a) a, b, c, e
b) b, e, c, b, e
c) a, a, b, e, d, e
d) b, c, e, d, a, a, b
e) b, c, c, b, e, d, e, d
f) $a, a, b, b, c, c, b, e, d$

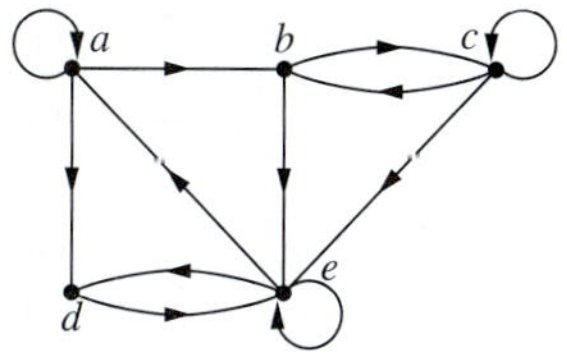

17. Find all circuits of length three in the directed graph in Exercise 16.

18. Determine whether there is a path in the directed graph in Exercise 16 beginning at the first vertex given and ending at the second vertex given.

a) a, b **b)** b, a **c)** b, b
d) a, e **e)** b, d **f)** c, d
g) d, d **h)** e, a **i)** e, c

19. Let R be the relation on the set $\{1, 2, 3, 4, 5\}$ containing the ordered pairs $(1, 3)$, $(2, 4)$, $(3, 1)$, $(3, 5)$, $(4, 3)$, $(5, 1)$, $(5, 2)$, and $(5, 4)$. Find

a) R^2. **b)** R^3. **c)** R^4.
d) R^5. **e)** R^6. **f)** R^*.

20. Let R be the relation that contains the pair (a, b) if a and b are cities such that there is a direct nonstop airline flight from a to b. When is (a, b) in

a) R^2? **b)** R^3? **c)** R^*?

21. Let R be the relation on the set of all students containing the ordered pair (a, b) if a and b are in at least one common class and $a \neq b$. When is (a, b) in

a) R^2? **b)** R^3? **c)** R^*?

22. Suppose that the relation R is reflexive. Show that R^* is reflexive.

23. Suppose that the relation R is symmetric. Show that R^* is symmetric.

24. Suppose that the relation R is irreflexive. Is the relation R^2 necessarily irreflexive?

25. Use Algorithm 1 to find the transitive closures of these relations on $\{1, 2, 3, 4\}$.

a) $\{(1, 2), (2, 1), (2, 3), (3, 4), (4, 1)\}$
b) $\{(2, 1), (2, 3), (3, 1), (3, 4), (4, 1), (4, 3)\}$
c) $\{(1, 2), (1, 3), (1, 4), (2, 3), (2, 4), (3, 4)\}$
d) $\{(1, 1), (1, 4), (2, 1), (2, 3), (3, 1), (3, 2), (3, 4), (4, 2)\}$

26. Use Algorithm 1 to find the transitive closures of these relations on $\{a, b, c, d, e\}$.

a) $\{(a, c), (b, d), (c, a), (d, b), (e, d)\}$
b) $\{(b, c), (b, e), (c, e), (d, a), (e, b), (e, c)\}$
c) $\{(a, b), (a, c), (a, e), (b, a), (b, c), (c, a), (c, b), (d, a), (e, d)\}$
d) $\{(a, e), (b, a), (b, d), (c, d), (d, a), (d, c), (e, a), (e, b), (e, c), (e, e)\}$

27. Use Warshall's algorithm to find the transitive closures of the relations in Exercise 25.

28. Use Warshall's algorithm to find the transitive closures of the relations in Exercise 26.

29. Find the smallest relation containing the relation $\{(1, 2), (1, 4), (3, 3), (4, 1)\}$ that is

a) reflexive and transitive.
b) symmetric and transitive.
c) reflexive, symmetric, and transitive.

30. Finish the proof of the case when $a \neq b$ in Lemma 1.

31. Algorithms have been devised that use $O(n^{2.8})$ bit operations to compute the Boolean product of two $n \times n$ zero–one matrices. Assuming that these algorithms can be used, give big-O estimates for the number of bit operations using Algorithm 1 and using Warshall's algorithm to find the transitive closure of a relation on a set with n elements.

***32.** Devise an algorithm using the concept of interior vertices in a path to find the length of the shortest path between two vertices in a directed graph, if such a path exists.

33. Adapt Algorithm 1 to find the reflexive closure of the transitive closure of a relation on a set with n elements.

34. Adapt Warshall's algorithm to find the reflexive closure of the transitive closure of a relation on a set with n elements.

35. Show that the closure with respect to the property **P** of the relation $R = \{(0, 0), (0, 1), (1, 1), (2, 2)\}$ on the set $\{0, 1, 2\}$ does not exist if **P** is the property

a) "is not reflexive."
b) "has an odd number of elements."

36. Give an example of a relation R on the set $\{a, b, c\}$ such that the symmetric closure of the reflexive closure of the transitive closure of R is not transitive.

9.5 Equivalence Relations

9.5.1 Introduction

In some programming languages the names of variables can contain an unlimited number of characters. However, there is a limit on the number of characters that are checked when a compiler determines whether two variables are equal. For instance, in traditional C, only the first eight characters of a variable name are checked by the compiler. (These characters are uppercase or lowercase letters, digits, or underscores.) Consequently, the compiler considers strings longer than eight characters that agree in their first eight characters the same. Let R be the relation on the set of strings of characters such that sRt, where s and t are two strings, if s and t are at least eight characters long and the first eight characters of s and t agree, or $s = t$. It is easy to see that R is reflexive, symmetric, and transitive. Moreover, R divides the set of all strings into classes, where all strings in a particular class are considered the same by a compiler for traditional C.

The integers a and b are related by the "congruence modulo 4" relation when 4 divides $a - b$. We will show later that this relation is reflexive, symmetric, and transitive. It is not hard to see that a is related to b if and only if a and b have the same remainder when divided by 4. It follows that this relation splits the set of integers into four different classes.

When we care only what remainder an integer leaves when it is divided by 4, we need only know which class it is in, not its particular value.

These two relations, R and congruence modulo 4, are examples of equivalence relations, namely, relations that are reflexive, symmetric, and transitive. In this section we will show that such relations split sets into disjoint classes of equivalent elements. Equivalence relations arise whenever we care only whether an element of a set is in a certain class of elements, instead of caring about its particular identity.

9.5.2 Equivalence Relations

Links

In this section we will study relations with a particular combination of properties that allows them to be used to relate objects that are similar in some way.

Definition 1 A relation on a set A is called an *equivalence relation* if it is reflexive, symmetric, and transitive.

Equivalence relations are important in every branch of mathematics!

Equivalence relations are important throughout mathematics and computer science. One reason for this is that in an equivalence relation, when two elements are related it makes sense to say they are equivalent.

Definition 2 Two elements a and b that are related by an equivalence relation are called *equivalent*. The notation $a \sim b$ is often used to denote that a and b are equivalent elements with respect to a particular equivalence relation.

For the notion of equivalent elements to make sense, every element should be equivalent to itself, as the reflexive property guarantees for an equivalence relation. It makes sense to say that a and b are related (not just that a is related to b) by an equivalence relation, because when a is related to b, by the symmetric property, b is related to a. Furthermore, because an equivalence relation is transitive, if a and b are equivalent and b and c are equivalent, it follows that a and c are equivalent.

Examples 1–5 illustrate the notion of an equivalence relation.

EXAMPLE 1 Let R be the relation on the set of integers such that aRb if and only if $a = b$ or $a = -b$. In Section 9.1 we showed that R is reflexive, symmetric, and transitive. It follows that R is an equivalence relation. ◀

EXAMPLE 2 Let R be the relation on the set of real numbers such that aRb if and only if $a - b$ is an integer. Is R an equivalence relation?

Extra Examples

Solution: Because $a - a = 0$ is an integer for all real numbers a, aRa for all real numbers a. Hence, R is reflexive. Now suppose that aRb. Then $a - b$ is an integer, so $b - a$ is also an integer. Hence, bRa. It follows that R is symmetric. If aRb and bRc, then $a - b$ and $b - c$ are integers. Therefore, $a - c = (a - b) + (b - c)$ is also an integer. Hence, aRc. Thus, R is transitive. Consequently, R is an equivalence relation. ◀

One of the most widely used equivalence relations is congruence modulo m, where m is an integer greater than 1.

EXAMPLE 3 **Congruence Modulo *m*** Let m be an integer with $m > 1$. Show that the relation

$$R = \{(a, b) \mid a \equiv b \pmod{m}\}$$

is an equivalence relation on the set of integers.

Solution: Recall from Section 4.1 that $a \equiv b \pmod{m}$ if and only if m divides $a - b$. Note that $a - a = 0$ is divisible by m, because $0 = 0 \cdot m$. Hence, $a \equiv a \pmod{m}$, so congruence modulo m is reflexive. Now suppose that $a \equiv b \pmod{m}$. Then $a - b$ is divisible by m, so $a - b = km$, where k is an integer. It follows that $b - a = (-k)m$, so $b \equiv a \pmod{m}$. Hence, congruence modulo m is symmetric. Next, suppose that $a \equiv b \pmod{m}$ and $b \equiv c \pmod{m}$. Then m divides both $a - b$ and $b - c$. Therefore, there are integers k and l with $a - b = km$ and $b - c = lm$. Adding these two equations shows that $a - c = (a - b) + (b - c) = km + lm = (k + l)m$. Thus, $a \equiv c \pmod{m}$. Therefore, congruence modulo m is transitive. It follows that congruence modulo m is an equivalence relation. ◀

EXAMPLE 4 Suppose that R is the relation on the set of strings of English letters such that aRb if and only if $l(a) = l(b)$, where $l(x)$ is the length of the string x. Is R an equivalence relation?

Solution: Because $l(a) = l(a)$, it follows that aRa whenever a is a string, so that R is reflexive. Next, suppose that aRb, so that $l(a) = l(b)$. Then bRa, because $l(b) = l(a)$. Hence, R is symmetric. Finally, suppose that aRb and bRc. Then $l(a) = l(b)$ and $l(b) = l(c)$. Hence, $l(a) = l(c)$, so aRc. Consequently, R is transitive. Because R is reflexive, symmetric, and transitive, it is an equivalence relation. ◀

EXAMPLE 5 Let n be a positive integer and S a set of strings. Suppose that R_n is the relation on S such that sR_nt if and only if $s = t$, or both s and t have at least n characters and the first n characters of s and t are the same. That is, a string of fewer than n characters is related only to itself; a string s with at least n characters is related to a string t if and only if t has at least n characters and t begins with the n characters at the start of s. For example, let $n = 3$ and let S be the set of all bit strings. Then sR_3t either when $s = t$ or both s and t are bit strings of length 3 or more that begin with the same three bits. For instance, $01R_301$ and $00111R_300101$, but $01\not{R}_3010$ and $01011\not{R}_301110$.

Show that for every set S of strings and every positive integer n, R_n is an equivalence relation on S.

Solution: The relation R_n is reflexive because $s = s$, so that sR_ns whenever s is a string in S. If sR_nt, then either $s = t$ or s and t are both at least n characters long that begin with the same n characters. This means that tR_ns. We conclude that R_n is symmetric.

Now suppose that sR_nt and tR_nu. Then either $s = t$ or s and t are at least n characters long and s and t begin with the same n characters, and either $t = u$ or t and u are at least n characters long and t and u begin with the same n characters. From this, we can deduce that either $s = u$ or both s and u are n characters long and s and u begin with the same n characters (because in this case we know that s, t, and u are all at least n characters long and both s and u begin with the same n characters as t does). Consequently, R_n is transitive. It follows that R_n is an equivalence relation. ◀

In Examples 6 and 7 we look at two relations that are not equivalence relations.

EXAMPLE 6 Show that the "divides" relation on the set of positive integers is not an equivalence relation.

Solution: By Examples 9 and 15 in Section 9.1, we know that the "divides" relation is reflexive and transitive. However, by Example 12 in Section 9.1, we know that this relation is not symmetric (for instance, $2 \mid 4$ but $4 \nmid 2$). We conclude that the "divides" relation on the set of positive integers is not an equivalence relation. ◀

EXAMPLE 7 Let R be the relation on the set of real numbers such that xRy if and only if x and y are real numbers that differ by less than 1, that is, $|x - y| < 1$. Show that R is not an equivalence relation.

Solution: R is reflexive because $|x - x| = 0 < 1$ whenever $x \in \mathbf{R}$. R is symmetric, for if xRy, where x and y are real numbers, then $|x - y| < 1$, which tells us that $|y - x| = |x - y| < 1$, so that yRx. However, R is not an equivalence relation because it is not transitive. Take $x = 2.8$, $y = 1.9$, and $z = 1.1$, so that $|x - y| = |2.8 - 1.9| = 0.9 < 1$, $|y - z| = |1.9 - 1.1| = 0.8 < 1$, but $|x - z| = |2.8 - 1.1| = 1.7 > 1$. That is, $2.8 R 1.9$, $1.9 R 1.1$, but $2.8 \not{R} 1.1$. ◀

9.5.3 Equivalence Classes

Let A be the set of all students in your school who graduated from high school. Consider the relation R on A that consists of all pairs (x, y), where x and y graduated from the same high school. Given a student x, we can form the set of all students equivalent to x with respect to R. This set consists of all students who graduated from the same high school as x did. This subset of A is called an equivalence class of the relation.

Definition 3 Let R be an equivalence relation on a set A. The set of all elements that are related to an element a of A is called the *equivalence class* of a. The equivalence class of a with respect to R is denoted by $[a]_R$. When only one relation is under consideration, we can delete the subscript R and write $[a]$ for this equivalence class.

In other words, if R is an equivalence relation on a set A, the equivalence class of the element a is

$$[a]_R = \{s \mid (a, s) \in R\}.$$

If $b \in [a]_R$, then b is called a **representative** of this equivalence class. Any element of a class can be used as a representative of this class. That is, there is nothing special about the particular element chosen as the representative of the class.

EXAMPLE 8 What is the equivalence class of an integer for the equivalence relation of Example 1?

Solution: Because an integer is equivalent to itself and its negative in this equivalence relation, it follows that $[a] = \{-a, a\}$. This set contains two distinct integers unless $a = 0$. For instance, $[7] = \{-7, 7\}$, $[-5] = \{-5, 5\}$, and $[0] = \{0\}$. ◀

EXAMPLE 9 What are the equivalence classes of 0, 1, 2, and 3 for congruence modulo 4?

Extra Examples ›

Solution: The equivalence class of 0 contains all integers a such that $a \equiv 0 \pmod 4$. The integers in this class are those divisible by 4. Hence, the equivalence class of 0 for this relation is

$$[0] = \{\dots, -8, -4, 0, 4, 8, \dots\}.$$

The equivalence class of 1 contains all the integers a such that $a \equiv 1 \pmod{4}$. The integers in this class are those that have a remainder of 1 when divided by 4. Hence, the equivalence class of 1 for this relation is

$$[1] = \{\dots, -7, -3, 1, 5, 9, \dots\}.$$

The equivalence class of 2 contains all the integers a such that $a \equiv 2 \pmod{4}$. The integers in this class are those that have a remainder of 2 when divided by 4. Hence, the equivalence class of 2 for this relation is

$$[2] = \{\dots, -6, -2, 2, 6, 10, \dots\}.$$

The equivalence class of 3 contains all the integers a such that $a \equiv 3 \pmod{4}$. The integers in this class are those that have a remainder of 3 when divided by 4. Hence, the equivalence class of 3 for this relation is

$$[3] = \{\dots, -5, -1, 3, 7, 11, \dots\}.$$

Note that every integer is in exactly one of the four equivalence classes and that the integer n is in the class containing n **mod** 4. ◀

In Example 9 the equivalence classes of 0, 1, 2, and 3 with respect to congruence modulo 4 were found. Example 9 can easily be generalized, replacing 4 with any positive integer m. The equivalence classes of the relation congruence modulo m are called the **congruence classes modulo** m. The congruence class of an integer a modulo m is denoted by $[a]_m$, so $[a]_m = \{\dots, a - 2m, a - m, a, a + m, a + 2m, \dots\}$. For instance, from Example 9 we have $[0]_4 = \{\dots, -8, -4, 0, 4, 8, \dots\}$, $[1]_4 = \{\dots, -7, -3, 1, 5, 9, \dots\}$, $[2]_4 = \{\dots, -6, -2, 2, 6, 10, \dots\}$, and $[3]_4 = \{\dots, -5, -1, 3, 7, 11, \dots\}$.

EXAMPLE 10 What is the equivalence class of the string 0111 with respect to the equivalence relation R_3 from Example 5 on the set of all bit strings? (Recall that sR_3t if and only if s and t are bit strings with $s = t$ or s and t are strings of at least three bits that start with the same three bits.)

Solution: The bit strings equivalent to 0111 are the bit strings with at least three bits that begin with 011. These are the bit strings 011, 0110, 0111, 01100, 01101, 01110, 01111, and so on. Consequently,

$$[011]_{R_3} = \{011, 0110, 0111, 01100, 01101, 01110, 01111, \dots\}.$$ ◀

EXAMPLE 11 **Identifiers in the C Programming Language** In the C programming language, an **identifier** is the name of a variable, a function, or another type of entity. Each identifier is a nonempty string of characters where each character is a lowercase or an uppercase English letter, a digit, or an underscore, and the first character is a lowercase or an uppercase English letter or an underscore. Identifiers can be any length. This allows developers to use as many characters as they want to name an entity, such as a variable. However, for compilers for some versions of C, there is a limit on the number of characters checked when two names are compared to see whether they refer to the same thing. For example, Standard C compilers consider two identifiers the same when they agree in their first 31 characters. Consequently, developers must be careful not to use identifiers with the same initial 31 characters for different things. We see that two identifiers are considered the same when they are related by the relation R_{31} in Example 5. Using Example 5, we know that R_{31}, on the set of all identifiers in Standard C, is an equivalence relation.

What are the equivalence classes of each of the identifiers Number_of_tropical_storms, Number_of_named_tropical_storms, and Number_of_named_tropical_storms_in_the_Atlantic_in_2025?

Solution: Note that when an identifier is less than 31 characters long, by the definition of R_{31}, its equivalence class contains only itself. Because the identifier Number_of_tropical_storms is 25 characters long, its equivalence class contains exactly one element, namely, itself.

The identifier Number_of_named_tropical_storms is exactly 31 characters long. An identifier is equivalent to it when it starts with these same 31 characters. Consequently, every identifier at least 31 characters long that starts with Number_of_named_tropical_storms is equivalent to this identifier. It follows that the equivalence class of Number_of_named_tropical_storms is the set of all identifiers that begin with the 31 characters Number_of_named_tropical_storms.

An identifier is equivalent to the Number_of_named_tropical_storms_in_the_Atlantic_in_2025 if and only if it begins with its first 31 characters. Because these characters are Number_of_named_tropical_storms, we see that an identifier is equivalent to Number_of_named_tropical_storms_in_the_Atlantic_in_2025 if and only if it is equivalent to Number_of_named_tropical_storms. It follows that these last two identifiers have the same equivalence class. ◀

9.5.4 Equivalence Classes and Partitions

Let A be the set of students at your school who are majoring in exactly one subject, and let R be the relation on A consisting of pairs (x, y), where x and y are students with the same major. Then R is an equivalence relation, as the reader should verify. We can see that R splits all students in A into a collection of disjoint subsets, where each subset contains students with a specified major. For instance, one subset contains all students majoring (just) in computer science, and a second subset contains all students majoring in history. Furthermore, these subsets are equivalence classes of R. This example illustrates how the equivalence classes of an equivalence relation partition a set into disjoint, nonempty subsets. We will make these notions more precise in the following discussion.

Let R be a relation on the set A. Theorem 1 shows that the equivalence classes of two elements of A are either identical or disjoint.

THEOREM 1 Let R be an equivalence relation on a set A. These statements for elements a and b of A are equivalent:

(*i*) aRb (*ii*) $[a] = [b]$ (*iii*) $[a] \cap [b] \neq \emptyset$

Proof: We first show that (*i*) implies (*ii*). Assume that aRb. We will prove that $[a] = [b]$ by showing $[a] \subseteq [b]$ and $[b] \subseteq [a]$. Suppose $c \in [a]$. Then aRc. Because aRb and R is symmetric, we know that bRa. Furthermore, because R is transitive and bRa and aRc, it follows that bRc. Hence, $c \in [b]$. This shows that $[a] \subseteq [b]$. The proof that $[b] \subseteq [a]$ is similar; it is left as an exercise for the reader.

Second, we will show that (*ii*) implies (*iii*). Assume that $[a] = [b]$. It follows that $[a] \cap [b] \neq \emptyset$ because $[a]$ is nonempty (because $a \in [a]$ because R is reflexive).

Next, we will show that (*iii*) implies (*i*). Suppose that $[a] \cap [b] \neq \emptyset$. Then there is an element c with $c \in [a]$ and $c \in [b]$. In other words, aRc and bRc. By the symmetric property, cRb. Then by transitivity, because aRc and cRb, we have aRb.

Because (*i*) implies (*ii*), (*ii*) implies (*iii*), and (*iii*) implies (*i*), the three statements, (*i*), (*ii*), and (*iii*), are equivalent. ◁

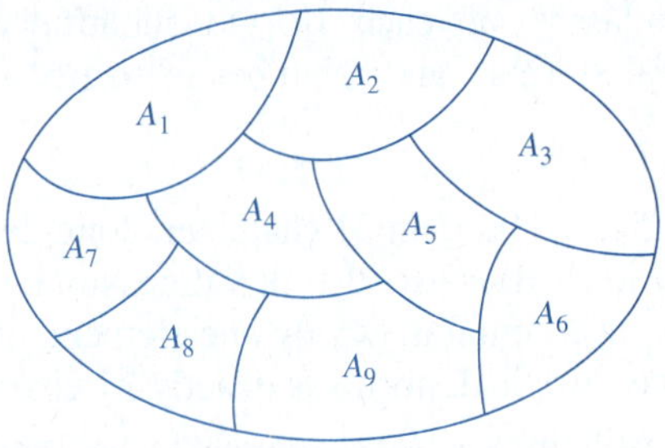

FIGURE 1 A partition of a set.

We are now in a position to show how an equivalence relation *partitions* a set. Let R be an equivalence relation on a set A. The union of the equivalence classes of R is all of A, because an element a of A is in its own equivalence class, namely, $[a]_R$. In other words,

$$\bigcup_{a \in A} [a]_R = A.$$

In addition, from Theorem 1, it follows that these equivalence classes are either equal or disjoint, so

$$[a]_R \cap [b]_R = \emptyset,$$

when $[a]_R \neq [b]_R$.

Recall that an *index set* is a set whose members label, or index, the elements of a set.

These two observations show that the equivalence classes form a partition of A, because they split A into disjoint subsets. More precisely, a **partition** of a set S is a collection of disjoint nonempty subsets of S that have S as their union. In other words, the collection of subsets A_i, $i \in I$ (where I is an index set) forms a partition of S if and only if

$$A_i \neq \emptyset \text{ for } i \in I,$$

$$A_i \cap A_j = \emptyset \text{ when } i \neq j,$$

and

$$\bigcup_{i \in I} A_i = S.$$

(Here the notation $\bigcup_{i \in I} A_i$ represents the union of the sets A_i for all $i \in I$.) Figure 1 illustrates the concept of a partition of a set.

EXAMPLE 12 Suppose that $S = \{1, 2, 3, 4, 5, 6\}$. The collection of sets $A_1 = \{1, 2, 3\}$, $A_2 = \{4, 5\}$, and $A_3 = \{6\}$ forms a partition of S, because these sets are disjoint and their union is S. ◀

We have seen that the equivalence classes of an equivalence relation on a set form a partition of the set. The subsets of S in this partition are the equivalence classes. Conversely, every partition of a set can be used to form an equivalence relation. Two elements are equivalent with respect to this relation if and only if they are in the same subset of S in the partition.

To see this, assume that $\{A_i \mid i \in I\}$ is a partition on S. Let R be the relation on S consisting of the pairs (x, y), where x and y belong to the same subset A_i in the partition. To show that R is an equivalence relation we must show that R is reflexive, symmetric, and transitive.

We see that $(a, a) \in R$ for every $a \in S$, because a is in the same subset of S as itself. Hence, R is reflexive. If $(a, b) \in R$, then b and a are in the same subset of S in the partition, so that

$(b, a) \in R$ as well. Hence, R is symmetric. If $(a, b) \in R$ and $(b, c) \in R$, then a and b are in the same subset X of S in the partition, and b and c are in the same subset Y of S of the partition. Because the subsets of S in the partition are disjoint and b belongs to X and Y, it follows that $X = Y$. Consequently, a and c belong to the same subset of S in the partition, so $(a, c) \in R$. Thus, R is transitive.

It follows that R is an equivalence relation. The equivalence classes of R consist of subsets of S containing related elements, and by the definition of R, these are the subsets of S in the partition. Theorem 2 summarizes the connections we have established between equivalence relations and partitions.

THEOREM 2 Let R be an equivalence relation on a set S. Then the equivalence classes of R form a partition of S. Conversely, given a partition $\{A_i \mid i \in I\}$ of the set S, there is an equivalence relation R that has the sets A_i, $i \in I$, as its equivalence classes.

Example 13 shows how to construct an equivalence relation from a partition.

EXAMPLE 13 List the ordered pairs in the equivalence relation R produced by the partition $A_1 = \{1, 2, 3\}$, $A_2 = \{4, 5\}$, and $A_3 = \{6\}$ of $S = \{1, 2, 3, 4, 5, 6\}$, given in Example 12.

Solution: The subsets of S in the partition are the equivalence classes of R. The pair $(a, b) \in R$ if and only if a and b are in the same subset of the S in the partition. The pairs (1, 1), (1, 2), (1, 3), (2, 1), (2, 2), (2, 3), (3, 1), (3, 2), and (3, 3) belong to R because $A_1 = \{1, 2, 3\}$ is an equivalence class; the pairs (4, 4), (4, 5), (5, 4), and (5, 5) belong to R because $A_2 = \{4, 5\}$ is an equivalence class; and finally the pair (6, 6) belongs to R because $\{6\}$ is an equivalence class. No pair other than those listed belongs to R. ◀

The congruence classes modulo m provide a useful illustration of Theorem 2. There are m different congruence classes modulo m, corresponding to the m different remainders possible when an integer is divided by m. These m congruence classes are denoted by $[0]_m, [1]_m, \dots, [m-1]_m$. They form a partition of the set of integers.

EXAMPLE 14 What are the sets in the partition of the integers arising from congruence modulo 4?

Solution: In Example 9 we found the four congruence classes, $[0]_4$, $[1]_4$, $[2]_4$, and $[3]_4$. They are the sets

$$[0]_4 = \{\dots, -8, -4, 0, 4, 8, \dots\},$$
$$[1]_4 = \{\dots, -7, -3, 1, 5, 9, \dots\},$$
$$[2]_4 = \{\dots, -6, -2, 2, 6, 10, \dots\},$$
$$[3]_4 = \{\dots, -5, -1, 3, 7, 11, \dots\}.$$

These congruence classes are disjoint, and every integer is in exactly one of them. In other words, as Theorem 2 says, these congruence classes form a partition. ◀

We now provide an example of a partition of the set of all strings arising from an equivalence relation on this set.

EXAMPLE 15 Let R_3 be the relation from Example 5. What are the sets in the partition of the set of all bit strings arising from the relation R_3 on the set of all bit strings? (Recall that sR_3t, where s and t are bit strings, if $s = t$ or s and t are bit strings with at least three bits that agree in their first three bits.)

Solution: Note that every bit string of length less than three is equivalent only to itself. Hence $[\lambda]_{R_3} = \{\lambda\}$, $[0]_{R_3} = \{0\}$, $[1]_{R_3} = \{1\}$, $[00]_{R_3} = \{00\}$, $[01]_{R_3} = \{01\}$, $[10]_{R_3} = \{10\}$, and $[11]_{R_3} = \{11\}$. Note that every bit string of length three or more is equivalent to one of the eight bit strings 000, 001, 010, 011, 100, 101, 110, and 111. We have

$$[000]_{R_3} = \{000, 0000, 0001, 00000, 00001, 00010, 00011, \ldots\},$$

$$[001]_{R_3} = \{001, 0010, 0011, 00100, 00101, 00110, 00111, \ldots\},$$

$$[010]_{R_3} = \{010, 0100, 0101, 01000, 01001, 01010, 01011, \ldots\},$$

$$[011]_{R_3} = \{011, 0110, 0111, 01100, 01101, 01110, 01111, \ldots\},$$

$$[100]_{R_3} = \{100, 1000, 1001, 10000, 10001, 10010, 10011, \ldots\},$$

$$[101]_{R_3} = \{101, 1010, 1011, 10100, 10101, 10110, 10111, \ldots\},$$

$$[110]_{R_3} = \{110, 1100, 1101, 11000, 11001, 11010, 11011, \ldots\},$$

$$[111]_{R_3} = \{111, 1110, 1111, 11100, 11101, 11110, 11111, \ldots\}.$$

These 15 equivalence classes are disjoint and every bit string is in exactly one of them. As Theorem 2 tells us, these equivalence classes partition the set of all bit strings. ◀

Exercises

1. Which of these relations on $\{0, 1, 2, 3\}$ are equivalence relations? Determine the properties of an equivalence relation that the others lack.

a) $\{(0, 0), (1, 1), (2, 2), (3, 3)\}$
b) $\{(0, 0), (0, 2), (2, 0), (2, 2), (2, 3), (3, 2), (3, 3)\}$
c) $\{(0, 0), (1, 1), (1, 2), (2, 1), (2, 2), (3, 3)\}$
d) $\{(0, 0), (1, 1), (1, 3), (2, 2), (2, 3), (3, 1), (3, 2), (3, 3)\}$
e) $\{(0, 0), (0, 1), (0, 2), (1, 0), (1, 1), (1, 2), (2, 0), (2, 2), (3, 3)\}$

2. Which of these relations on the set of all people are equivalence relations? Determine the properties of an equivalence relation that the others lack.

a) $\{(a, b) \mid a$ and b are the same age$\}$
b) $\{(a, b) \mid a$ and b have the same parents$\}$
c) $\{(a, b) \mid a$ and b share a common parent$\}$
d) $\{(a, b) \mid a$ and b have met$\}$
e) $\{(a, b) \mid a$ and b speak a common language$\}$

3. Which of these relations on the set of all functions from $\mathbf{Z}$ to $\mathbf{Z}$ are equivalence relations? Determine the properties of an equivalence relation that the others lack.

a) $\{(f, g) \mid f(1) = g(1)\}$
b) $\{(f, g) \mid f(0) = g(0)$ or $f(1) = g(1)\}$
c) $\{(f, g) \mid f(x) - g(x) = 1$ for all $x \in \mathbf{Z}\}$
d) $\{(f, g) \mid$ for some $C \in \mathbf{Z}$, for all $x \in \mathbf{Z}, f(x) - g(x) = C\}$
e) $\{(f, g) \mid f(0) = g(1)$ and $f(1) = g(0)\}$

4. Define three equivalence relations on the set of students in your discrete mathematics class different from the relations discussed in the text. Determine the equivalence classes for each of these equivalence relations.

5. Define three equivalence relations on the set of buildings on a college campus. Determine the equivalence classes for each of these equivalence relations.

6. Define three equivalence relations on the set of classes offered at your school. Determine the equivalence classes for each of these equivalence relations.

7. Show that the relation of logical equivalence on the set of all compound propositions is an equivalence relation. What are the equivalence classes of **F** and of **T**?

8. Let R be the relation on the set of all sets of real numbers such that $S\,R\,T$ if and only if S and T have the same cardinality. Show that R is an equivalence relation. What are the equivalence classes of the sets $\{0, 1, 2\}$ and $\mathbf{Z}$?

9. Suppose that A is a nonempty set, and f is a function that has A as its domain. Let R be the relation on A consisting of all ordered pairs (x, y) such that $f(x) = f(y)$.

a) Show that R is an equivalence relation on A.
b) What are the equivalence classes of R?

10. Suppose that A is a nonempty set and R is an equivalence relation on A. Show that there is a function f with A as its domain such that $(x, y) \in R$ if and only if $f(x) = f(y)$.

11. Show that the relation R consisting of all pairs (x, y) such that x and y are bit strings of length three or more that agree in their first three bits is an equivalence relation on the set of all bit strings of length three or more.

12. Show that the relation R consisting of all pairs (x, y) such that x and y are bit strings of length three or more that agree except perhaps in their first three bits is an equivalence relation on the set of all bit strings of length three or more.

13. Show that the relation R consisting of all pairs (x, y) such that x and y are bit strings that agree in their first and third bits is an equivalence relation on the set of all bit strings of length three or more.

14. Let R be the relation consisting of all pairs (x, y) such that x and y are strings of uppercase and lowercase English letters with the property that for every positive integer n, the nth characters in x and y are the same letter, either uppercase or lowercase. Show that R is an equivalence relation.

15. Let R be the relation on the set of ordered pairs of positive integers such that $((a, b), (c, d)) \in R$ if and only if $a + d = b + c$. Show that R is an equivalence relation.

16. Let R be the relation on the set of ordered pairs of positive integers such that $((a, b), (c, d)) \in R$ if and only if $ad = bc$. Show that R is an equivalence relation.

17. (*Requires calculus*)

a) Show that the relation R on the set of all differentiable functions from **R** to **R** consisting of all pairs (f, g) such that $f'(x) = g'(x)$ for all real numbers x is an equivalence relation.

b) Which functions are in the same equivalence class as the function $f(x) = x^2$?

18. (*Requires calculus*)

a) Let n be a positive integer. Show that the relation R on the set of all polynomials with real-valued coefficients consisting of all pairs (f, g) such that $f^{(n)}(x) = g^{(n)}(x)$ is an equivalence relation. [Here $f^{(n)}(x)$ is the nth derivative of $f(x)$.]

b) Which functions are in the same equivalence class as the function $f(x) = x^4$, where $n = 3$?

19. Let R be the relation on the set of all URLs (or Web addresses) such that $x\,R\,y$ if and only if the Web page at x is the same as the Web page at y. Show that R is an equivalence relation.

20. Let R be the relation on the set of all people who have visited a particular Web page such that $x\,R\,y$ if and only if person x and person y have followed the same set of links starting at this Web page (going from Web page to Web page until they stop using the Web). Show that R is an equivalence relation.

In Exercises 21–23 determine whether the relation with the directed graph shown is an equivalence relation.

21.

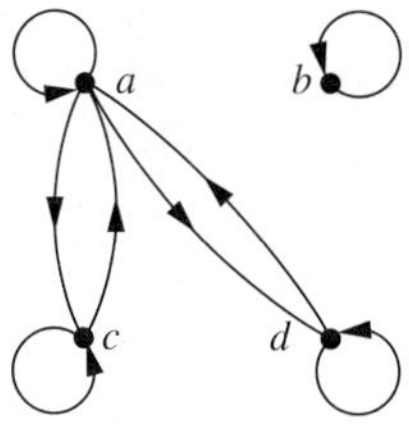

22.

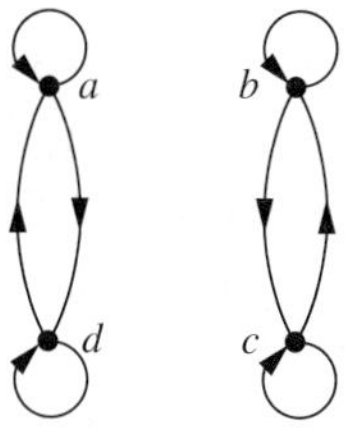

23.

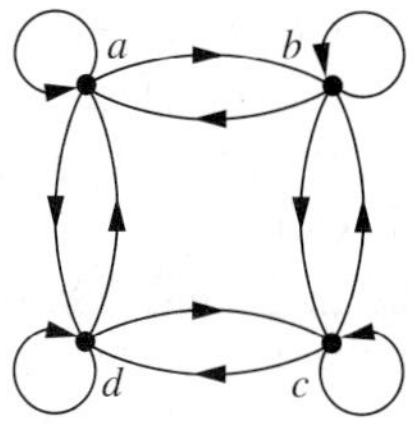

24. Determine whether the relations represented by these zero–one matrices are equivalence relations.

a) $\begin{bmatrix} 1 & 1 & 1 \\ 0 & 1 & 1 \\ 1 & 1 & 1 \end{bmatrix}$ **b)** $\begin{bmatrix} 1 & 0 & 1 & 0 \\ 0 & 1 & 0 & 1 \\ 1 & 0 & 1 & 0 \\ 0 & 1 & 0 & 1 \end{bmatrix}$ **c)** $\begin{bmatrix} 1 & 1 & 1 & 0 \\ 1 & 1 & 1 & 0 \\ 1 & 1 & 1 & 0 \\ 0 & 0 & 0 & 1 \end{bmatrix}$

25. Show that the relation R on the set of all bit strings such that $s\,R\,t$ if and only if s and t contain the same number of 1s is an equivalence relation.

26. What are the equivalence classes of the equivalence relations in Exercise 1?

27. What are the equivalence classes of the equivalence relations in Exercise 2?

28. What are the equivalence classes of the equivalence relations in Exercise 3?

29. What is the equivalence class of the bit string 011 for the equivalence relation in Exercise 25?

30. What are the equivalence classes of these bit strings for the equivalence relation in Exercise 11?

a) 010 **b)** 1011 **c)** 11111 **d)** 01010101

31. What are the equivalence classes of the bit strings in Exercise 30 for the equivalence relation from Exercise 12?

32. What are the equivalence classes of the bit strings in Exercise 30 for the equivalence relation from Exercise 13?

33. What are the equivalence classes of the bit strings in Exercise 30 for the equivalence relation R_4 from Example 5 on the set of all bit strings? (Recall that bit strings s and t are equivalent under R_4 if and only if they are equal or they are both at least four bits long and agree in their first four bits.)

34. What are the equivalence classes of the bit strings in Exercise 30 for the equivalence relation R_5 from Example 5 on the set of all bit strings? (Recall that bit strings s and t are equivalent under R_5 if and only if they are equal or they are both at least five bits long and agree in their first five bits.)

35. What is the congruence class $[n]_5$ (that is, the equivalence class of n with respect to congruence modulo 5) when n is

a) 2? **b)** 3? **c)** 6? **d)** -3?

36. What is the congruence class $[4]_m$ when m is

a) 2? **b)** 3? **c)** 6? **d)** 8?

37. Give a description of each of the congruence classes modulo 6.

38. What is the equivalence class of each of these strings with respect to the equivalence relation in Exercise 14?

a) *No* **b)** *Yes* **c)** *Help*

39. a) What is the equivalence class of (1, 2) with respect to the equivalence relation in Exercise 15?

b) Give an interpretation of the equivalence classes for the equivalence relation R in Exercise 15. [*Hint:* Look at the difference $a - b$ corresponding to (a, b).]

40. a) What is the equivalence class of (1, 2) with respect to the equivalence relation in Exercise 16?

b) Give an interpretation of the equivalence classes for the equivalence relation R in Exercise 16. [*Hint:* Look at the ratio a/b corresponding to (a, b).]

41. Which of these collections of subsets are partitions of $\{1, 2, 3, 4, 5, 6\}$?

a) $\{1, 2\}, \{2, 3, 4\}, \{4, 5, 6\}$

b) $\{1\}, \{2, 3, 6\}, \{4\}, \{5\}$

c) $\{2, 4, 6\}, \{1, 3, 5\}$ **d)** $\{1, 4, 5\}, \{2, 6\}$

42. Which of these collections of subsets are partitions of $\{-3, -2, -1, 0, 1, 2, 3\}$?

a) $\{-3, -1, 1, 3\}, \{-2, 0, 2\}$

b) $\{-3, -2, -1, 0\}, \{0, 1, 2, 3\}$

c) $\{-3, 3\}, \{-2, 2\}, \{-1, 1\}, \{0\}$

d) $\{-3, -2, 2, 3\}, \{-1, 1\}$

43. Which of these collections of subsets are partitions of the set of bit strings of length 8?

a) the set of bit strings that begin with 1, the set of bit strings that begin with 00, and the set of bit strings that begin with 01

b) the set of bit strings that contain the string 00, the set of bit strings that contain the string 01, the set of bit strings that contain the string 10, and the set of bit strings that contain the string 11

c) the set of bit strings that end with 00, the set of bit strings that end with 01, the set of bit strings that end with 10, and the set of bit strings that end with 11

d) the set of bit strings that end with 111, the set of bit strings that end with 011, and the set of bit strings that end with 00

e) the set of bit strings that contain $3k$ ones for some nonnegative integer k, the set of bit strings that contain $3k + 1$ ones for some nonnegative integer k, and the set of bit strings that contain $3k + 2$ ones for some nonnegative integer k.

44. Which of these collections of subsets are partitions of the set of integers?

a) the set of even integers and the set of odd integers

b) the set of positive integers and the set of negative integers

c) the set of integers divisible by 3, the set of integers leaving a remainder of 1 when divided by 3, and the set of integers leaving a remainder of 2 when divided by 3

d) the set of integers less than -100, the set of integers with absolute value not exceeding 100, and the set of integers greater than 100

e) the set of integers not divisible by 3, the set of even integers, and the set of integers that leave a remainder of 3 when divided by 6

45. Which of these are partitions of the set $\mathbf{Z} \times \mathbf{Z}$ of ordered pairs of integers?

a) the set of pairs (x, y), where x or y is odd; the set of pairs (x, y), where x is even; and the set of pairs (x, y), where y is even

b) the set of pairs (x, y), where both x and y are odd; the set of pairs (x, y), where exactly one of x and y is odd; and the set of pairs (x, y), where both x and y are even

c) the set of pairs (x, y), where x is positive; the set of pairs (x, y), where y is positive; and the set of pairs (x, y), where both x and y are negative

d) the set of pairs (x, y), where $3 \mid x$ and $3 \mid y$; the set of pairs (x, y), where $3 \mid x$ and $3 \nmid y$; the set of pairs (x, y), where $3 \nmid x$ and $3 \mid y$; and the set of pairs (x, y), where $3 \nmid x$ and $3 \nmid y$

e) the set of pairs (x, y), where $x > 0$ and $y > 0$; the set of pairs (x, y), where $x > 0$ and $y \leq 0$; the set of pairs (x, y), where $x \leq 0$ and $y > 0$; and the set of pairs (x, y), where $x \leq 0$ and $y \leq 0$

f) the set of pairs (x, y), where $x \neq 0$ and $y \neq 0$; the set of pairs (x, y), where $x = 0$ and $y \neq 0$; and the set of pairs (x, y), where $x \neq 0$ and $y = 0$

46. Which of these are partitions of the set of real numbers?

a) the negative real numbers, $\{0\}$, the positive real numbers

b) the set of irrational numbers, the set of rational numbers

c) the set of intervals $[k, k + 1]$, $k = \ldots, -2, -1, 0, 1, 2, \ldots$

d) the set of intervals $(k, k + 1)$, $k = \ldots, -2, -1, 0, 1, 2, \ldots$

e) the set of intervals $(k, k + 1]$, $k = \ldots, -2, -1, 0, 1, 2, \ldots$

f) the sets $\{x + n \mid n \in \mathbf{Z}\}$ for all $x \in [0, 1)$

47. List the ordered pairs in the equivalence relations produced by these partitions of $\{0, 1, 2, 3, 4, 5\}$.

a) $\{0\}, \{1, 2\}, \{3, 4, 5\}$

b) $\{0, 1\}, \{2, 3\}, \{4, 5\}$

c) $\{0, 1, 2\}, \{3, 4, 5\}$

d) $\{0\}, \{1\}, \{2\}, \{3\}, \{4\}, \{5\}$

48. List the ordered pairs in the equivalence relations produced by these partitions of $\{a, b, c, d, e, f, g\}$.

a) $\{a, b\}, \{c, d\}, \{e, f, g\}$

b) $\{a\}, \{b\}, \{c, d\}, \{e, f\}, \{g\}$

c) $\{a, b, c, d\}, \{e, f, g\}$

d) $\{a, c, e, g\}, \{b, d\}, \{f\}$

A partition P_1 is called a **refinement** of the partition P_2 if every set in P_1 is a subset of one of the sets in P_2.

49. Show that the partition formed from congruence classes modulo 6 is a refinement of the partition formed from congruence classes modulo 3.

50. Show that the partition of the set of people living in the United States consisting of subsets of people living in the same county (or parish) and same state is a refinement of the partition consisting of subsets of people living in the same state.

51. Show that the partition of the set of bit strings of length 16 formed by equivalence classes of bit strings that agree on the last eight bits is a refinement of the partition formed from the equivalence classes of bit strings that agree on the last four bits.

In Exercises 52 and 53, R_n refers to the family of equivalence relations defined in Example 5. Recall that $s\,R_n\,t$, where s and t are two strings if $s = t$ or s and t are strings with at least n characters that agree in their first n characters.

52. Show that the partition of the set of all bit strings formed by equivalence classes of bit strings with respect to the equivalence relation R_4 is a refinement of the partition formed by equivalence classes of bit strings with respect to the equivalence relation R_3.

53. Show that the partition of the set of all identifiers in C formed by the equivalence classes of identifiers with respect to the equivalence relation R_{31} is a refinement of the partition formed by equivalence classes of identifiers with respect to the equivalence relation R_8. (Compilers for "old" C consider identifiers the same when their names agree in their first eight characters, while compilers in standard C consider identifiers the same when their names agree in their first 31 characters.)

54. Suppose that R_1 and R_2 are equivalence relations on a set A. Let P_1 and P_2 be the partitions that correspond to R_1 and R_2, respectively. Show that $R_1 \subseteq R_2$ if and only if P_1 is a refinement of P_2.

55. Find the smallest equivalence relation on the set $\{a, b, c, d, e\}$ containing the relation $\{(a, b), (a, c), (d, e)\}$.

56. Suppose that R_1 and R_2 are equivalence relations on the set S. Determine whether each of these combinations of R_1 and R_2 must be an equivalence relation.

a) $R_1 \cup R_2$ **b)** $R_1 \cap R_2$ **c)** $R_1 \oplus R_2$

57. Consider the equivalence relation from Example 2, namely, $R = \{(x, y) \mid x - y \text{ is an integer}\}$.

a) What is the equivalence class of 1 for this equivalence relation?

b) What is the equivalence class of 1/2 for this equivalence relation?

***58.** Each bead on a bracelet with three beads is either red, white, or blue, as illustrated in the figure shown.

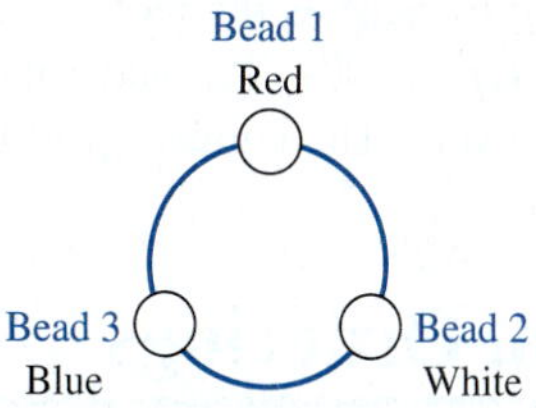

Define the relation R between bracelets as: (B_1, B_2), where B_1 and B_2 are bracelets, belongs to R if and only if B_2 can be obtained from B_1 by rotating it or rotating it and then reflecting it.

a) Show that R is an equivalence relation.

b) What are the equivalence classes of R?

***59.** Let R be the relation on the set of all colorings of the 2×2 checkerboard where each of the four squares is colored either red or blue so that (C_1, C_2), where C_1 and C_2 are 2×2 checkerboards with each of their four squares colored blue or red, belongs to R if and only if C_2 can be obtained from C_1 either by rotating the checkerboard or by rotating it and then reflecting it.

a) Show that R is an equivalence relation.

b) What are the equivalence classes of R?

60. a) Let R be the relation on the set of functions from $\mathbf{Z}^+$ to $\mathbf{Z}^+$ such that (f, g) belongs to R if and only if f is $\Theta(g)$ (see Section 3.2). Show that R is an equivalence relation.

b) Describe the equivalence class containing $f(n) = n^2$ for the equivalence relation of part (a).

61. Determine the number of different equivalence relations on a set with three elements by listing them.

62. Determine the number of different equivalence relations on a set with four elements by listing them.

***63.** Do we necessarily get an equivalence relation when we form the transitive closure of the symmetric closure of the reflexive closure of a relation?

***64.** Do we necessarily get an equivalence relation when we form the symmetric closure of the reflexive closure of the transitive closure of a relation?

65. Suppose we use Theorem 2 to form a partition P from an equivalence relation R. What is the equivalence relation R' that results if we use Theorem 2 again to form an equivalence relation from P?

66. Suppose we use Theorem 2 to form an equivalence relation R from a partition P. What is the partition P' that results if we use Theorem 2 again to form a partition from R?

67. Devise an algorithm to find the smallest equivalence relation containing a given relation.

*68. Let $p(n)$ denote the number of different equivalence relations on a set with n elements (and by Theorem 2 the number of partitions of a set with n elements). Show that $p(n)$ satisfies the recurrence relation $p(n) = \sum_{j=0}^{n-1} C(n-1, j)p(n-j-1)$ and the initial condition $p(0) = 1$. (*Note:* The numbers $p(n)$ are called **Bell numbers** after the American mathematician E. T. Bell.)

69. Use Exercise 68 to find the number of different equivalence relations on a set with n elements, where n is a positive integer not exceeding 10.

9.6 Partial Orderings

9.6.1 Introduction

Links

We often use relations to order some or all of the elements of sets. For instance, we order words using the relation containing pairs of words (x, y), where x comes before y in the dictionary. We schedule projects using the relation consisting of pairs (x, y), where x and y are tasks in a project such that x must be completed before y begins. We order the set of integers using the relation containing the pairs (x, y), where x is less than y. When we add all of the pairs of the form (x, x) to these relations, we obtain a relation that is reflexive, antisymmetric, and transitive. These are properties that characterize relations used to order the elements of sets.

Definition 1 A relation R on a set S is called a *partial ordering* or *partial order* if it is reflexive, antisymmetric, and transitive. A set S together with a partial ordering R is called a *partially ordered set*, or *poset*, and is denoted by (S, R). Members of S are called *elements* of the poset.

We give examples of posets in Examples 1–3.

EXAMPLE 1 Show that the greater than or equal to relation ($\geq$) is a partial ordering on the set of integers.

Extra Examples

Solution: Because $a \geq a$ for every integer a, $\geq$ is reflexive. If $a \geq b$ and $b \geq a$, then $a = b$. Hence, $\geq$ is antisymmetric. Finally, $\geq$ is transitive because $a \geq b$ and $b \geq c$ imply that $a \geq c$. It follows that $\geq$ is a partial ordering on the set of integers and $(\mathbf{Z}, \geq)$ is a poset. ◀

EXAMPLE 2 The divisibility relation $\mid$ is a partial ordering on the set of positive integers, because it is reflexive, antisymmetric, and transitive, as was shown in Section 9.1. We see that $(\mathbf{Z}^+, \mid)$ is a poset. (Recall that $\mathbf{Z}^+$ denotes the set of positive integers.) ◀

EXAMPLE 3 Show that the inclusion relation $\subseteq$ is a partial ordering on the power set of a set S.

Solution: Because $A \subseteq A$ whenever A is a subset of S, $\subseteq$ is reflexive. It is antisymmetric because $A \subseteq B$ and $B \subseteq A$ imply that $A = B$. Finally, $\subseteq$ is transitive, because $A \subseteq B$ and $B \subseteq C$ imply that $A \subseteq C$. Hence, $\subseteq$ is a partial ordering on $\mathcal{P}(S)$, and $(\mathcal{P}(S), \subseteq)$ is a poset. ◀

Example 4 illustrates a relation that is not a partial ordering.

EXAMPLE 4 Let R be the relation on the set of people such that xRy if x and y are people and x is older than y. Show that R is not a partial ordering.

Extra Examples

Solution: Note that R is antisymmetric because if a person x is older than a person y, then y is not older than x. That is, if xRy, then $y \not R x$. The relation R is transitive because if person x is older than person y and y is older than person z, then x is older than z. That is, if xRy

and yRz, then xRz. However, R is not reflexive, because no person is older than himself or herself. That is, $x \not R x$ for all people x. It follows that R is not a partial ordering. ◀

In different posets different symbols such as $\leq$, $\subseteq$, and $|$, are used for a partial ordering. However, we need a symbol that we can use when we discuss the ordering relation in an arbitrary poset. Customarily, the notation $a \preccurlyeq b$ is used to denote that $(a, b) \in R$ in an arbitrary poset (S, R). This notation is used because the less than or equal to relation on the set of real numbers is the most familiar example of a partial ordering and the symbol $\preccurlyeq$ is similar to the $\leq$ symbol. (Note that the symbol $\preccurlyeq$ is used to denote the relation in *any* poset, not just the less than or equal to relation.) The notation $a \prec b$ denotes that $a \preccurlyeq b$, but $a \neq b$. Also, we say "a is less than b" or "b is greater than a" if $a \prec b$.

When a and b are elements of the poset $(S, \preccurlyeq)$, it is not necessary that either $a \preccurlyeq b$ or $b \preccurlyeq a$. For instance, in $(\mathcal{P}(\mathbf{Z}), \subseteq)$, $\{1, 2\}$ is not related to $\{1, 3\}$, and vice versa, because neither set is contained within the other. Similarly, in $(\mathbf{Z}^+, |)$, 2 is not related to 3 and 3 is not related to 2, because $2 \nmid 3$ and $3 \nmid 2$. This leads to Definition 2.

Definition 2 The elements a and b of a poset $(S, \preccurlyeq)$ are called *comparable* if either $a \preccurlyeq b$ or $b \preccurlyeq a$. When a and b are elements of S such that neither $a \preccurlyeq b$ nor $b \preccurlyeq a$, a and b are called *incomparable*.

EXAMPLE 5 In the poset $(\mathbf{Z}^+, |)$, are the integers 3 and 9 comparable? Are 5 and 7 comparable?

Solution: The integers 3 and 9 are comparable, because $3 \mid 9$. The integers 5 and 7 are incomparable, because $5 \nmid 7$ and $7 \nmid 5$. ◀

The adjective "partial" is used to describe partial orderings because pairs of elements may be incomparable. When every two elements in the set are comparable, the relation is called a **total ordering**.

Definition 3 If $(S, \preccurlyeq)$ is a poset and every two elements of S are comparable, S is called a *totally ordered* or *linearly ordered set*, and $\preccurlyeq$ is called a *total order* or a *linear order.* A totally ordered set is also called a *chain.*

EXAMPLE 6 The poset $(\mathbf{Z}, \leq)$ is totally ordered, because $a \leq b$ or $b \leq a$ whenever a and b are integers. ◀

EXAMPLE 7 The poset $(\mathbf{Z}^+, |)$ is not totally ordered because it contains elements that are incomparable, such as 5 and 7. ◀

In Chapter 5 we noted that $(\mathbf{Z}^+, \leq)$ is well-ordered, where $\leq$ is the usual less than or equal to relation. We now define well-ordered sets.

Definition 4 $(S, \preccurlyeq)$ is a *well-ordered set* if it is a poset such that $\preccurlyeq$ is a total ordering and every nonempty subset of S has a least element.

EXAMPLE 8 The set of ordered pairs of positive integers, $\mathbf{Z}^+ \times \mathbf{Z}^+$, with $(a_1, a_2) \preccurlyeq (b_1, b_2)$ if $a_1 < b_1$, or if $a_1 = b_1$ and $a_2 \leq b_2$ (the lexicographic ordering), is a well-ordered set. The verification of this is left as Exercise 53. The set $\mathbf{Z}$, with the usual $\leq$ ordering, is not well-ordered because the set of negative integers, which is a subset of $\mathbf{Z}$, has no least element. ◀

At the end of Section 5.3 we showed how to use the principle of well-ordered induction (called generalized induction in Section 5.3) to prove results about a well-ordered set. We now state and prove that this proof technique is valid.

THEOREM 1 **THE PRINCIPLE OF WELL-ORDERED INDUCTION** Suppose that S is a well-ordered set. Then $P(x)$ is true for all $x \in S$, if

INDUCTIVE STEP: For every $y \in S$, if $P(x)$ is true for all $x \in S$ with $x \prec y$, then $P(y)$ is true.

Proof: Suppose it is not the case that $P(x)$ is true for all $x \in S$. Then there is an element $y \in S$ such that $P(y)$ is false. Consequently, the set $A = \{x \in S \mid P(x) \text{ is false}\}$ is nonempty. Because S is well-ordered, A has a least element a. By the choice of a as a least element of A, we know that $P(x)$ is true for all $x \in S$ with $x \prec a$. This implies by the inductive step $P(a)$ is true. This contradiction shows that $P(x)$ must be true for all $x \in S$. ◁

Remark: We do not need a basis step in a proof using the principle of well-ordered induction because if x_0 is the least element of a well-ordered set, the inductive step tells us that $P(x_0)$ is true. This follows because there are no elements $x \in S$ with $x \prec x_0$, so we know (using a vacuous proof) that $P(x)$ is true for all $x \in S$ with $x \prec x_0$.

The principle of well-ordered induction is a versatile technique for proving results about well-ordered sets. Even when it is possible to use mathematical induction for the set of positive integers to prove a theorem, it may be simpler to use the principle of well-ordered induction, as we saw in Subsection 5.3.5, where we proved a result about the well-ordered set $(\mathbf{N} \times \mathbf{N}, \preccurlyeq)$ where $\preccurlyeq$ is lexicographic ordering on $\mathbf{N} \times \mathbf{N}$.

9.6.2 Lexicographic Order

The words in a dictionary are listed in alphabetic, or lexicographic, order, which is based on the ordering of the letters in the alphabet. This is a special case of an ordering of strings on a set constructed from a partial ordering on the set. We will show how this construction works in any poset.

First, we will show how to construct a partial ordering on the Cartesian product of two posets, $(A_1, \preccurlyeq_1)$ and $(A_2, \preccurlyeq_2)$. The **lexicographic ordering** $\preccurlyeq$ on $A_1 \times A_2$ is defined by specifying that one pair is less than a second pair if the first entry of the first pair is less than (in A_1) the first entry of the second pair, or if the first entries are equal, but the second entry of this pair is less than (in A_2) the second entry of the second pair. In other words, (a_1, a_2) is less than (b_1, b_2), that is,

$$(a_1, a_2) \prec (b_1, b_2),$$

either if $a_1 \prec_1 b_1$ or if both $a_1 = b_1$ and $a_2 \prec_2 b_2$.

We obtain a partial ordering $\preccurlyeq$ by adding equality to the ordering $\prec$ on $A_1 \times A_2$. The verification of this is left as an exercise.

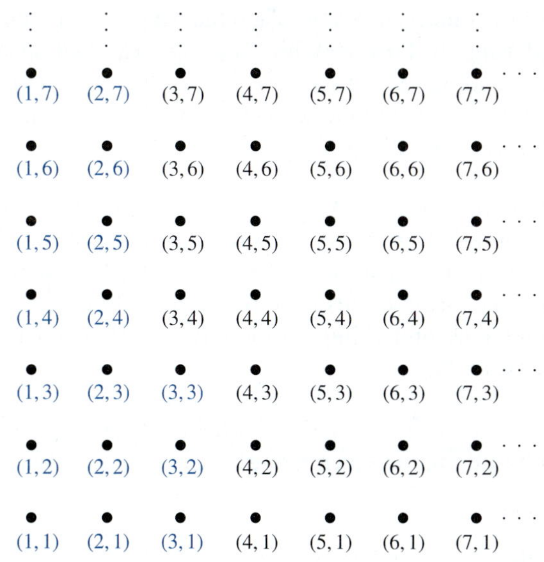

FIGURE 1 The ordered pairs less than (3, 4) in lexicographic order.

EXAMPLE 9 Determine whether $(3, 5) \prec (4, 8)$, whether $(3, 8) \prec (4, 5)$, and whether $(4, 9) \prec (4, 11)$ in the poset $(\mathbf{Z} \times \mathbf{Z}, \preccurlyeq)$, where $\preccurlyeq$ is the lexicographic ordering constructed from the usual $\leq$ relation on $\mathbf{Z}$.

Solution: Because $3 < 4$, it follows that $(3, 5) \prec (4, 8)$ and that $(3, 8) \prec (4, 5)$. We have $(4, 9) \prec (4, 11)$, because the first entries of $(4, 9)$ and $(4, 11)$ are the same but $9 < 11$. ◀

In Figure 1 the ordered pairs in $\mathbf{Z}^+ \times \mathbf{Z}^+$ that are less than $(3, 4)$ are highlighted. A lexicographic ordering can be defined on the Cartesian product of n posets $(A_1, \preccurlyeq_1), (A_2, \preccurlyeq_2), \ldots, (A_n, \preccurlyeq_n)$. Define the partial ordering $\preccurlyeq$ on $A_1 \times A_2 \times \cdots \times A_n$ by

$$(a_1, a_2, \ldots, a_n) \prec (b_1, b_2, \ldots, b_n)$$

if $a_1 \prec_1 b_1$, or if there is an integer $i > 0$ such that $a_1 = b_1, \ldots, a_i = b_i$, and $a_{i+1} \prec_{i+1} b_{i+1}$. In other words, one n-tuple is less than a second n-tuple if the entry of the first n-tuple in the first position where the two n-tuples disagree is less than the entry in that position in the second n-tuple.

EXAMPLE 10 Note that $(1, 2, 3, 5) \prec (1, 2, 4, 3)$, because the entries in the first two positions of these 4-tuples agree, but in the third position the entry in the first 4-tuple, 3, is less than that in the second 4-tuple, 4. (Here the ordering on 4-tuples is the lexicographic ordering that comes from the usual less than or equal to relation on the set of integers.) ◀

We can now define lexicographic ordering of strings. Consider the strings $a_1a_2 \ldots a_m$ and $b_1b_2 \ldots b_n$ on a partially ordered set S. Suppose these strings are not equal. Let t be the minimum of m and n. The definition of lexicographic ordering is that the string $a_1a_2 \ldots a_m$ is less than $b_1b_2 \ldots b_n$ if and only if

$$(a_1, a_2, \ldots, a_t) \prec (b_1, b_2, \ldots, b_t), \text{ or}$$
$$(a_1, a_2, \ldots, a_t) = (b_1, b_2, \ldots, b_t) \text{ and } m < n,$$

where $\prec$ in this inequality represents the lexicographic ordering of S^t. In other words, to determine the ordering of two different strings, the longer string is truncated to the length of the shorter string, namely, to $t = \min(m, n)$ terms. Then the t-tuples made up of the first t terms of each string are compared using the lexicographic ordering on S^t. One string is less than another string if the t-tuple corresponding to the first string is less than the t-tuple of the second string, or if these two t-tuples are the same, but the second string is longer. The verification that this is a partial ordering is left as Exercise 38 for the reader.

EXAMPLE 11 Consider the set of strings of lowercase English letters. Using the ordering of letters in the alphabet, a lexicographic ordering on the set of strings can be constructed. A string is less than a second string if the letter in the first string in the first position where the strings differ comes before the letter in the second string in this position, or if the first string and the second string agree in all positions, but the second string has more letters. This ordering is the same as that used in dictionaries. For example,

$$discreet \prec discrete,$$

because these strings differ first in the seventh position, and $e \prec t$. Also,

$$discreet \prec discreetness,$$

because the first eight letters agree, but the second string is longer. Furthermore,

$$discrete \prec discretion,$$

because

$$discrete \prec discreti.$$

◀

9.6.3 Hasse Diagrams

Many edges in the directed graph for a finite poset do not have to be shown because they must be present. For instance, consider the directed graph for the partial ordering $\{(a, b) \mid a \leq b\}$ on the set $\{1, 2, 3, 4\}$, shown in Figure 2(a). Because this relation is a partial ordering, it is reflexive, and its directed graph has loops at all vertices. Consequently, we do not have to show these loops because they must be present; in Figure 2(b) loops are not shown. Because a partial ordering is transitive, we do not have to show those edges that must be present because of transitivity. For example, in Figure 2(c) the edges $(1, 3)$, $(1, 4)$, and $(2, 4)$ are not shown because they must be present. If we assume that all edges are pointed "upward" (as they are drawn in the figure), we do not have to show the directions of the edges; Figure 2(c) does not show directions.

In general, we can represent a finite poset $(S, \preccurlyeq)$ using this procedure: Start with the directed graph for this relation. Because a partial ordering is reflexive, a loop (a, a) is present at every vertex a. Remove these loops. Next, remove all edges that must be in the partial ordering because

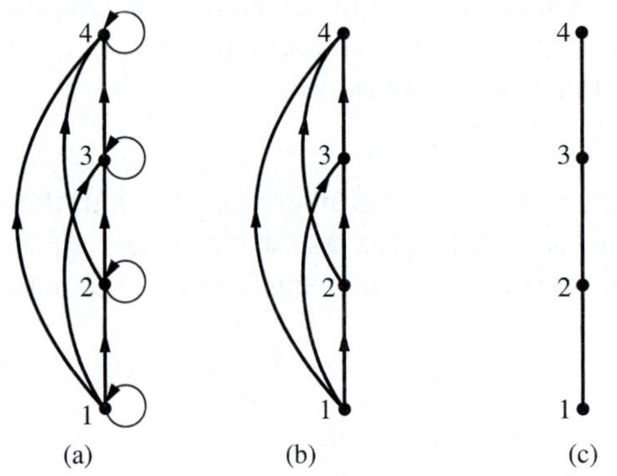

FIGURE 2 Constructing the Hasse diagram for ({1, 2, 3, 4}, ≤).

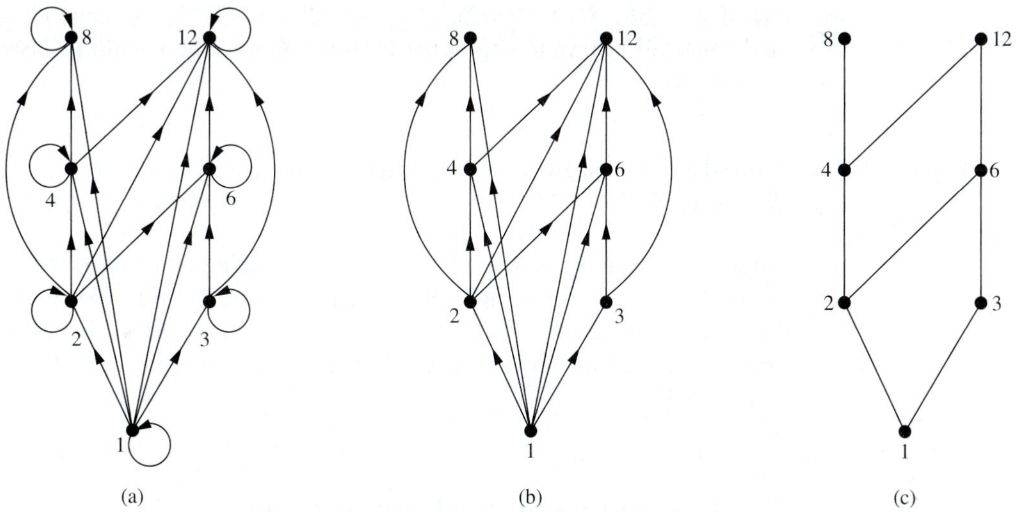

FIGURE 3 **Constructing the Hasse diagram of ({1, 2, 3, 4, 6, 8, 12}, |).**

of the presence of other edges and transitivity. That is, remove all edges (x, y) for which there is an element $z \in S$ such that $x \prec z$ and $z \prec x$. Finally, arrange each edge so that its initial vertex is below its terminal vertex (as it is drawn on paper). Remove all the arrows on the directed edges, because all edges point "upward" toward their terminal vertex.

These steps are well defined, and only a finite number of steps need to be carried out for a finite poset. When all the steps have been taken, the resulting diagram contains sufficient information to find the partial ordering, as we will explain later. The resulting diagram is called the **Hasse diagram** of $(S, \preccurlyeq)$, named after the twentieth-century German mathematician Helmut Hasse who made extensive use of them.

Let $(S, \preccurlyeq)$ be a poset. We say that an element $y \in S$ **covers** an element $x \in S$ if $x \prec y$ and there is no element $z \in S$ such that $x \prec z \prec y$. The set of pairs (x, y) such that y covers x is called the **covering relation** of $(S, \preccurlyeq)$. From the description of the Hasse diagram of a poset, we see that the edges in the Hasse diagram of $(S, \preccurlyeq)$ are upwardly pointing edges corresponding to the pairs in the covering relation of $(S, \preccurlyeq)$. Furthermore, we can recover a poset from its covering relation, because it is the reflexive transitive closure of its covering relation. (Exercise 31 asks for a proof of this fact.) This tells us that we can construct a partial ordering from its Hasse diagram.

EXAMPLE 12 Draw the Hasse diagram representing the partial ordering $\{(a, b) \mid a \text{ divides } b\}$ on $\{1, 2, 3, 4, 6, 8, 12\}$.

Solution: Begin with the digraph for this partial order, as shown in Figure 3(a). Remove all loops, as shown in Figure 3(b). Then delete all the edges implied by the transitive property. These

Links

ullstein bild Dtl./ullstein bild/Getty Images

HELMUT HASSE (1898–1979) Helmut Hasse was born in Kassel, Germany. He served in the German navy after high school. He began his university studies at Göttingen University in 1918, moving in 1920 to Marburg University to study under the number theorist Kurt Hensel. During this time, Hasse made fundamental contributions to algebraic number theory. He became Hensel's successor at Marburg, later becoming director of the renowned mathematical institute at Göttingen in 1934, and took a position at Hamburg University in 1950. Hasse served for 50 years as an editor of *Crelle's Journal*, an illustrious German mathematics periodical, taking over the job of chief editor in 1936 when the Nazis forced Hensel to resign. During World War II Hasse worked on applied mathematics research for the German navy. He was noted for the clarity and personal style of his lectures and was devoted both to number theory and to his students. (Hasse has been controversial for connections with the Nazi party. Investigations have shown he was a strong German nationalist but not an ardent Nazi.)

are (1, 4), (1, 6), (1, 8), (1, 12), (2, 8), (2, 12), and (3, 12). Arrange all edges to point upward, and delete all arrows to obtain the Hasse diagram. The resulting Hasse diagram is shown in Figure 3(c). ◀

EXAMPLE 13 Draw the Hasse diagram for the partial ordering $\{(A, B) \mid A \subseteq B\}$ on the power set $\mathcal{P}(S)$, where $S = \{a, b, c\}$.

Solution: The Hasse diagram for this partial ordering is obtained from the associated digraph by deleting all the loops and all the edges that occur from transitivity, namely, $(\emptyset, \{a, b\})$, $(\emptyset, \{a, c\})$, $(\emptyset, \{b, c\})$, $(\emptyset, \{a, b, c\})$, $(\{a\}, \{a, b, c\})$, $(\{b\}, \{a, b, c\})$, and $(\{c\}, \{a, b, c\})$. Finally, all edges point upward, and arrows are deleted. The resulting Hasse diagram is illustrated in Figure 4. ◀

9.6.4 Maximal and Minimal Elements

Elements of posets that have certain extremal properties are important for many applications. An element of a poset is called maximal if it is not less than any element of the poset. That is, a is **maximal** in the poset $(S, \preccurlyeq)$ if there is no $b \in S$ such that $a \prec b$. Similarly, an element of a poset is called minimal if it is not greater than any element of the poset. That is, a is **minimal** if there is no element $b \in S$ such that $b \prec a$. Maximal and minimal elements are easy to spot using a Hasse diagram. They are the "top" and "bottom" elements in the diagram.

EXAMPLE 14 Which elements of the poset $(\{2, 4, 5, 10, 12, 20, 25\}, \mid)$ are maximal, and which are minimal?

Solution: The Hasse diagram in Figure 5 for this poset shows that the maximal elements are 12, 20, and 25, and the minimal elements are 2 and 5. As this example shows, a poset can have more than one maximal element and more than one minimal element. ◀

Sometimes there is an element in a poset that is greater than every other element. Such an element is called the greatest element. That is, a is the **greatest element** of the poset $(S, \preccurlyeq)$ if $b \preccurlyeq a$ for all $b \in S$. The greatest element is unique when it exists [see Exercise 40(a)]. Likewise, an element is called the least element if it is less than all the other elements in the poset. That is, a is the **least element** of $(S, \preccurlyeq)$ if $a \preccurlyeq b$ for all $b \in S$. The least element is unique when it exists [see Exercise 40(b)].

EXAMPLE 15 Determine whether the posets represented by each of the Hasse diagrams in Figure 6 have a greatest element and a least element.

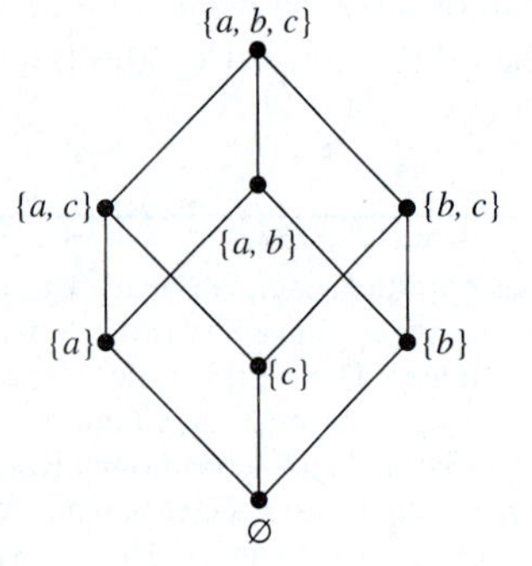

FIGURE 4 **The Hasse diagram of $(P(\{a, b, c\}), \subseteq)$.**

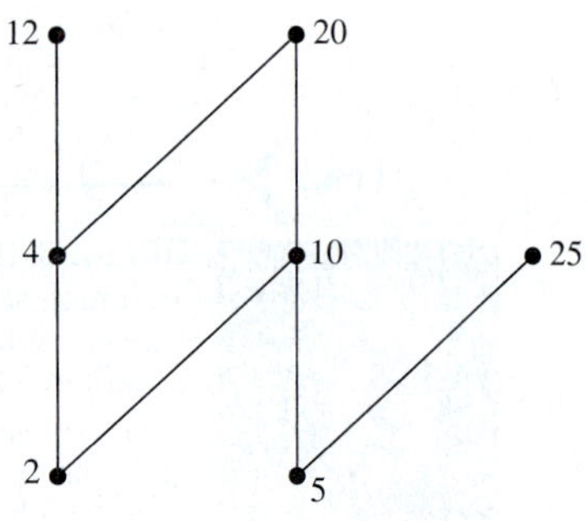

FIGURE 5 **The Hasse diagram of a poset.**

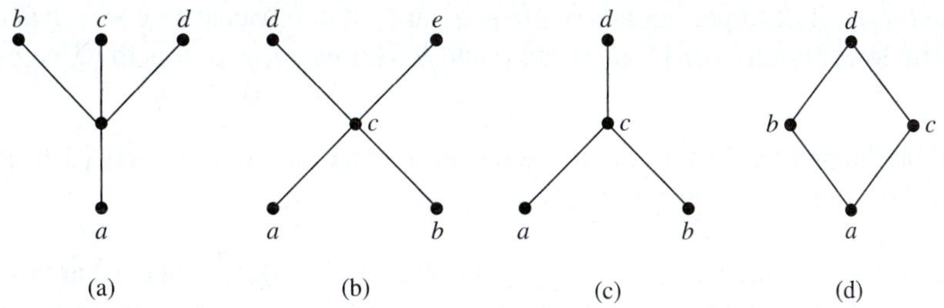

FIGURE 6 Hasse diagrams of four posets.

Solution: The least element of the poset with Hasse diagram (a) is a. This poset has no greatest element. The poset with Hasse diagram (b) has neither a least nor a greatest element. The poset with Hasse diagram (c) has no least element. Its greatest element is d. The poset with Hasse diagram (d) has least element a and greatest element d. ◀

EXAMPLE 16 Let S be a set. Determine whether there is a greatest element and a least element in the poset $(\mathcal{P}(S), \subseteq)$.

Solution: The least element is the empty set, because $\emptyset \subseteq T$ for any subset T of S. The set S is the greatest element in this poset, because $T \subseteq S$ whenever T is a subset of S. ◀

EXAMPLE 17 Is there a greatest element and a least element in the poset $(\mathbf{Z}^+, |)$?

Solution: The integer 1 is the least element because $1|n$ whenever n is a positive integer. Because there is no integer that is divisible by all positive integers, there is no greatest element. ◀

Sometimes it is possible to find an element that is greater than or equal to all the elements in a subset A of a poset $(S, \preccurlyeq)$. If u is an element of S such that $a \preccurlyeq u$ for all elements $a \in A$, then u is called an **upper bound** of A. Likewise, there may be an element less than or equal to all the elements in A. If l is an element of S such that $l \preccurlyeq a$ for all elements $a \in A$, then l is called a **lower bound** of A.

EXAMPLE 18 Find the lower and upper bounds of the subsets $\{a, b, c\}$, $\{j, h\}$, and $\{a, c, d, f\}$ in the poset with the Hasse diagram shown in Figure 7.

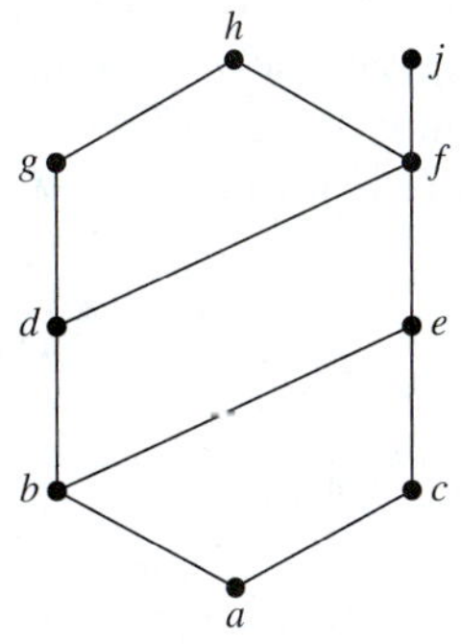

FIGURE 7 The Hasse diagram of a poset.

Solution: The upper bounds of $\{a, b, c\}$ are e, f, j, and h, and its only lower bound is a. There are no upper bounds of $\{j, h\}$, and its lower bounds are a, b, c, d, e, and f. The upper bounds of $\{a, c, d, f\}$ are f, h, and j, and its lower bound is a. ◀

The element x is called the **least upper bound** of the subset A if x is an upper bound that is less than every other upper bound of A. Because there is only one such element, if it exists, it makes sense to call this element *the* least upper bound [see Exercise 42(a)]. That is, x is the least upper bound of A if $a \preccurlyeq x$ whenever $a \in A$, and $x \preccurlyeq z$ whenever z is an upper bound of A. Similarly, the element y is called the **greatest lower bound** of A if y is a lower bound of A and $z \preccurlyeq y$ whenever z is a lower bound of A. The greatest lower bound of A is unique if it exists [see Exercise 42(b)]. The greatest lower bound and least upper bound of a subset A are denoted by glb(A) and lub(A), respectively.

EXAMPLE 19 Find the greatest lower bound and the least upper bound of $\{b, d, g\}$, if they exist, in the poset shown in Figure 7.

Solution: The upper bounds of $\{b, d, g\}$ are g and h. Because $g \prec h$, g is the least upper bound. The lower bounds of $\{b, d, g\}$ are a and b. Because $a \prec b$, b is the greatest lower bound. ◀

EXAMPLE 20 Find the greatest lower bound and the least upper bound of the sets $\{3, 9, 12\}$ and $\{1, 2, 4, 5, 10\}$, if they exist, in the poset $(\mathbf{Z}^+, |)$.

Extra Examples

Solution: An integer is a lower bound of $\{3, 9, 12\}$ if 3, 9, and 12 are divisible by this integer. The only such integers are 1 and 3. Because $1 \mid 3$, 3 is the greatest lower bound of $\{3, 9, 12\}$. The only lower bound for the set $\{1, 2, 4, 5, 10\}$ with respect to | is the element 1. Hence, 1 is the greatest lower bound for $\{1, 2, 4, 5, 10\}$.

An integer is an upper bound for $\{3, 9, 12\}$ if and only if it is divisible by 3, 9, and 12. The integers with this property are those divisible by the least common multiple of 3, 9, and 12, which is 36. Hence, 36 is the least upper bound of $\{3, 9, 12\}$. A positive integer is an upper bound for the set $\{1, 2, 4, 5, 10\}$ if and only if it is divisible by 1, 2, 4, 5, and 10. The integers with this property are those integers divisible by the least common multiple of these integers, which is 20. Hence, 20 is the least upper bound of $\{1, 2, 4, 5, 10\}$. ◀

9.6.5 Lattices

A partially ordered set in which every pair of elements has both a least upper bound and a greatest lower bound is called a **lattice**. Lattices have many special properties. Furthermore, lattices are used in many different applications such as models of information flow and play an important role in Boolean algebra.

EXAMPLE 21 Determine whether the posets represented by each of the Hasse diagrams in Figure 8 are lattices.

Solution: The posets represented by the Hasse diagrams in (a) and (c) are both lattices because in each poset every pair of elements has both a least upper bound and a greatest lower bound, as the reader should verify. On the other hand, the poset with the Hasse diagram shown in (b) is not a lattice, because the elements b and c have no least upper bound. To see this, note that each of the elements d, e, and f is an upper bound, but none of these three elements precedes the other two with respect to the ordering of this poset. ◀

EXAMPLE 22 Is the poset $(\mathbf{Z}^+, |)$ a lattice?

Solution: Let a and b be two positive integers. The least upper bound and greatest lower bound of these two integers are the least common multiple and the greatest common divisor of these integers, respectively, as the reader should verify. It follows that this poset is a lattice. ◀

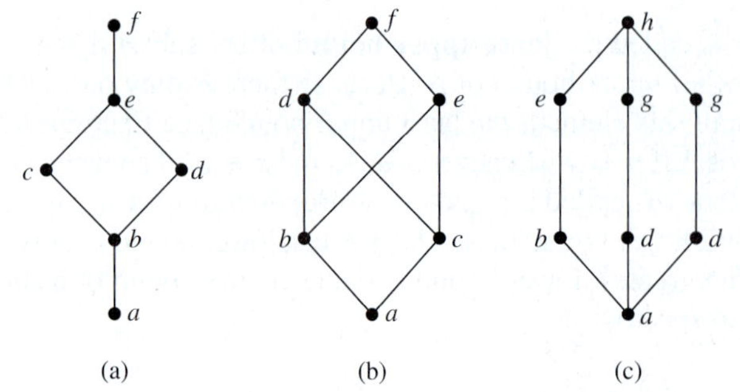

FIGURE 8 Hasse diagrams of three posets.

EXAMPLE 23 Determine whether the posets $(\{1, 2, 3, 4, 5\}, |)$ and $(\{1, 2, 4, 8, 16\}, |)$ are lattices.

Solution: Because 2 and 3 have no upper bounds in $(\{1, 2, 3, 4, 5\}, |)$, they certainly do not have a least upper bound. Hence, the first poset is not a lattice.

Every two elements of the second poset have both a least upper bound and a greatest lower bound. The least upper bound of two elements in this poset is the larger of the elements and the greatest lower bound of two elements is the smaller of the elements, as the reader should verify. Hence, this second poset is a lattice. ◀

EXAMPLE 24 Determine whether $(\mathcal{P}(S), \subseteq)$ is a lattice where S is a set.

Solution: Let A and B be two subsets of S. The least upper bound and the greatest lower bound of A and B are $A \cup B$ and $A \cap B$, respectively, as the reader can show. Hence, $(\mathcal{P}(S), \subseteq)$ is a lattice. ◀

EXAMPLE 25 **The Lattice Model of Information Flow** In many settings the flow of information from one person or computer program to another is restricted via security clearances. We can use a lattice model to represent different information flow policies. For example, one common information flow policy is the *multilevel security policy* used in government and military systems. Each piece of information is assigned to a security class, and each security class is represented by a pair (A, C) where A is an *authority level* and C is a *category*. People and computer programs are then allowed access to information from a specific restricted set of security classes.

Links

There are billions of pages of classified U.S. government documents.

The typical authority levels used in the U.S. government are unclassified (0), confidential (1), secret (2), and top secret (3). (Information is said to be classified if it is confidential, secret, or top secret.) Categories used in security classes are the subsets of a set of all *compartments* relevant to a particular area of interest. Each compartment represents a particular subject area. For example, if the set of compartments is {*spies, moles, double agents*}, then there are eight different categories, one for each of the eight subsets of the set of compartments, such as {*spies, moles*}.

We can order security classes by specifying that $(A_1, C_1) \preccurlyeq (A_2, C_2)$ if and only if $A_1 \leq A_2$ and $C_1 \subseteq C_2$. Information is permitted to flow from security class (A_1, C_1) into security class (A_2, C_2) if and only if $(A_1, C_1) \preccurlyeq (A_2, C_2)$. For example, information is permitted to flow from the security class (*secret*, {*spies, moles*}) into the security class (*top secret*, {*spies, moles, double agents*}), whereas information is not allowed to flow from the security class (*top secret*, {*spies, moles*}) into either of the security classes (*secret*, {*spies, moles, double agents*}) or (*top secret*, {*spies*}).

We leave it to the reader (see Exercise 48) to show that the set of all security classes with the ordering defined in this example forms a lattice. ◀

9.6.6 Topological Sorting

Links

Suppose that a project is made up of 20 different tasks. Some tasks can be completed only after others have been finished. How can an order be found for these tasks? To model this problem we set up a partial order on the set of tasks so that $a \prec b$ if and only if a and b are tasks where b cannot be started until a has been completed. To produce a schedule for the project, we need to produce an order for all 20 tasks that is compatible with this partial order. We will show how this can be done.

We begin with a definition. A total ordering $\preccurlyeq$ is said to be **compatible** with the partial ordering R if $a \preccurlyeq b$ whenever aRb. Constructing a compatible total ordering from a partial ordering is called **topological sorting**.* We will need to use Lemma 1.

LEMMA 1 Every finite nonempty poset $(S, \preccurlyeq)$ has at least one minimal element.

Proof: Choose an element a_0 of S. If a_0 is not minimal, then there is an element a_1 with $a_1 \prec a_0$. If a_1 is not minimal, there is an element a_2 with $a_2 \prec a_1$. Continue this process, so that if a_n is not minimal, there is an element a_{n+1} with $a_{n+1} \prec a_n$. Because there are only a finite number of elements in the poset, this process must end with a minimal element a_n. ◁

The topological sorting algorithm we will describe works for any finite nonempty poset. To define a total ordering on the poset $(A, \preccurlyeq)$, first choose a minimal element a_1; such an element exists by Lemma 1. Next, note that $(A - \{a_1\}, \preccurlyeq)$ is also a poset, as the reader should verify. (Here by $\preccurlyeq$ we mean the restriction of the original relation $\preccurlyeq$ on A to $A - \{a_1\}$.) If it is nonempty, choose a minimal element a_2 of this poset. Then remove a_2 as well, and if there are additional elements left, choose a minimal element a_3 in $A - \{a_1, a_2\}$. Continue this process by choosing a_{k+1} to be a minimal element in $A - \{a_1, a_2, \ldots, a_k\}$, as long as elements remain.

Because A is a finite set, this process must terminate. The end product is a sequence of elements $a_1, a_2, \ldots, a_n$. The desired total ordering $\preccurlyeq_t$ is defined by

$$a_1 \prec_t a_2 \prec_t \cdots \prec_t a_n.$$

This total ordering is compatible with the original partial ordering. To see this, note that if $b \prec c$ in the original partial ordering, c is chosen as the minimal element at a phase of the algorithm where b has already been removed, for otherwise c would not be a minimal element. Pseudocode for this topological sorting algorithm is shown in Algorithm 1.

ALGORITHM 1 Topological Sorting.

procedure *topological sort* ($(S, \preccurlyeq)$: finite poset)
$k := 1$
while $S \neq \emptyset$
    $a_k :=$ a minimal element of S {such an element exists by Lemma 1}
    $S := S - \{a_k\}$
    $k := k + 1$
return $a_1, a_2, \ldots, a_n$ {$a_1, a_2, \ldots, a_n$ is a compatible total ordering of S}

EXAMPLE 26 Find a compatible total ordering for the poset $(\{1, 2, 4, 5, 12, 20\}, |)$.

Solution: The first step is to choose a minimal element. This must be 1, because it is the only minimal element. Next, select a minimal element of $(\{2, 4, 5, 12, 20\}, |)$. There are two minimal elements in this poset, namely, 2 and 5. We select 5. The remaining elements are $\{2, 4, 12, 20\}$. The only minimal element at this stage is 2. Next, 4 is chosen because it is the only minimal

*"Topological sorting" is terminology used by computer scientists; mathematicians use the terminology "linearization of a partial ordering" for the same thing. In mathematics, topology is the branch of geometry dealing with properties of geometric figures that hold for all figures that can be transformed into one another by continuous bijections. In computer science, a topology is any arrangement of objects that can be connected with edges.

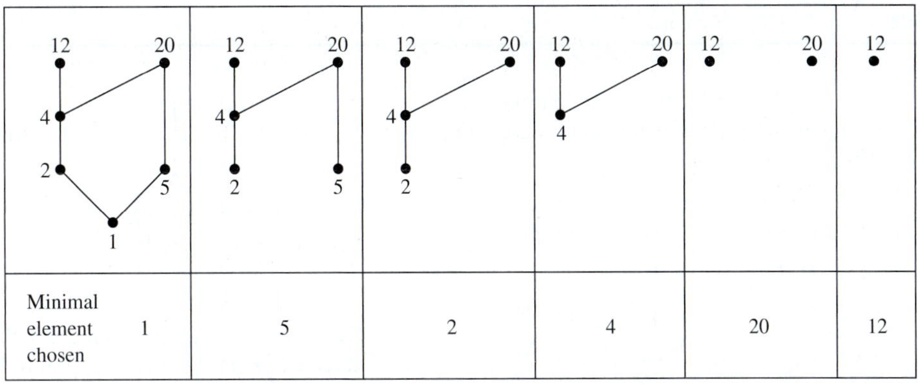

FIGURE 9 **A topological sort of ({1, 2, 4, 5, 12, 20}, |).**

element of ({4, 12, 20}, |). Because both 12 and 20 are minimal elements of ({12, 20}, |), either can be chosen next. We select 20, which leaves 12 as the last element left. This produces the total ordering

$$1 \prec 5 \prec 2 \prec 4 \prec 20 \prec 12.$$

The steps used by this sorting algorithm are displayed in Figure 9. ◀

Topological sorting has an application to the scheduling of projects.

EXAMPLE 27 A development project at a computer company requires the completion of seven tasks. Some of these tasks can be started only after other tasks are finished. A partial ordering on tasks is set up by considering task $X \prec$ task Y if task Y cannot be started until task X has been completed. The Hasse diagram for the seven tasks, with respect to this partial ordering, is shown in Figure 10. Find an order in which these tasks can be carried out to complete the project.

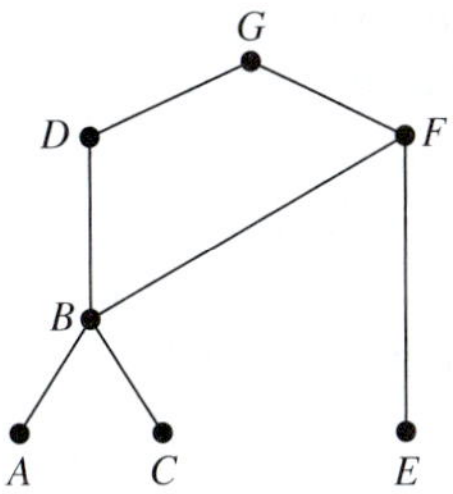

FIGURE 10 **The Hasse diagram for seven tasks.**

Solution: An ordering of the seven tasks can be obtained by performing a topological sort. The steps of a sort are illustrated in Figure 11. The result of this sort, $A \prec C \prec B \prec E \prec F \prec D \prec G$, gives one possible order for the tasks. ◀

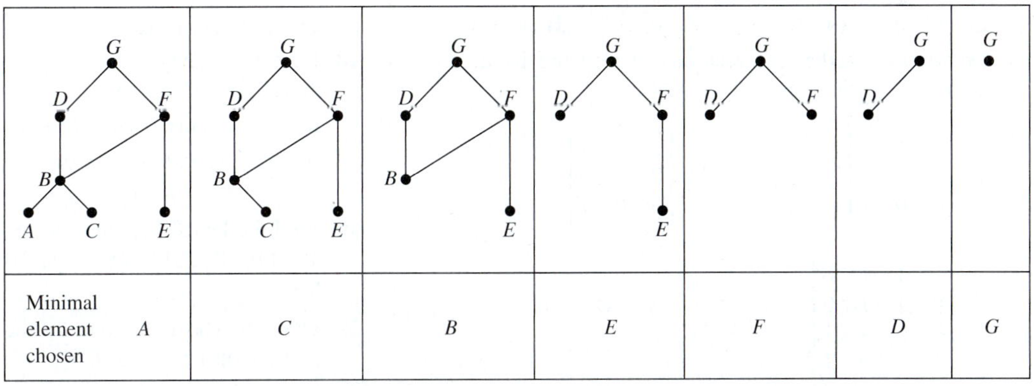

FIGURE 11 **A topological sort of the tasks.**

Exercises

1. Which of these relations on $\{0, 1, 2, 3\}$ are partial orderings? Determine the properties of a partial ordering that the others lack.
   a) $\{(0, 0), (1, 1), (2, 2), (3, 3)\}$
   b) $\{(0, 0), (1, 1), (2, 0), (2, 2), (2, 3), (3, 2), (3, 3)\}$
   c) $\{(0, 0), (1, 1), (1, 2), (2, 2), (3, 3)\}$
   d) $\{(0, 0), (1, 1), (1, 2), (1, 3), (2, 2), (2, 3), (3, 3)\}$
   e) $\{(0, 0), (0, 1), (0, 2), (1, 0), (1, 1), (1, 2), (2, 0), (2, 2), (3, 3)\}$
2. Which of these relations on $\{0, 1, 2, 3\}$ are partial orderings? Determine the properties of a partial ordering that the others lack.
   a) $\{(0, 0), (2, 2), (3, 3)\}$
   b) $\{(0, 0), (1, 1), (2, 0), (2, 2), (2, 3), (3, 3)\}$
   c) $\{(0, 0), (1, 1), (1, 2), (2, 2), (3, 1), (3, 3)\}$
   d) $\{(0, 0), (1, 1), (1, 2), (1, 3), (2, 0), (2, 2), (2, 3), (3, 0), (3, 3)\}$
   e) $\{(0, 0), (0, 1), (0, 2), (0, 3), (1, 0), (1, 1), (1, 2), (1, 3), (2, 0), (2, 2), (3, 3)\}$
3. Is (S, R) a poset if S is the set of all people in the world and $(a, b) \in R$, where a and b are people, if
   a) a is taller than b?
   b) a is not taller than b?
   c) $a = b$ or a is an ancestor of b?
   d) a and b have a common friend?
4. Is (S, R) a poset if S is the set of all people in the world and $(a, b) \in R$, where a and b are people, if
   a) a is no shorter than b?
   b) a weighs more than b?
   c) $a = b$ or a is a descendant of b?
   d) a and b do not have a common friend?
5. Which of these are posets?
   a) $(\mathbf{Z}, =)$ b) $(\mathbf{Z}, \neq)$ c) $(\mathbf{Z}, \geq)$ d) $(\mathbf{Z}, \nmid)$
6. Which of these are posets?
   a) $(\mathbf{R}, =)$ b) $(\mathbf{R}, <)$ c) $(\mathbf{R}, \leq)$ d) $(\mathbf{R}, \neq)$
7. Determine whether the relations represented by these zero–one matrices are partial orders.
   a) $\begin{bmatrix} 1 & 1 & 1 \\ 1 & 1 & 0 \\ 0 & 0 & 1 \end{bmatrix}$ b) $\begin{bmatrix} 1 & 1 & 1 \\ 0 & 1 & 0 \\ 0 & 0 & 1 \end{bmatrix}$
   c) $\begin{bmatrix} 1 & 1 & 1 & 0 \\ 0 & 1 & 1 & 0 \\ 0 & 0 & 1 & 1 \\ 1 & 1 & 0 & 1 \end{bmatrix}$
8. Determine whether the relations represented by these zero–one matrices are partial orders.
   a) $\begin{bmatrix} 1 & 0 & 1 \\ 1 & 1 & 0 \\ 0 & 0 & 1 \end{bmatrix}$ b) $\begin{bmatrix} 1 & 0 & 0 \\ 0 & 1 & 0 \\ 1 & 0 & 1 \end{bmatrix}$
   c) $\begin{bmatrix} 1 & 0 & 1 & 0 \\ 0 & 1 & 1 & 0 \\ 0 & 0 & 1 & 1 \\ 1 & 1 & 0 & 1 \end{bmatrix}$

In Exercises 9–11 determine whether the relation with the directed graph shown is a partial order.

9.
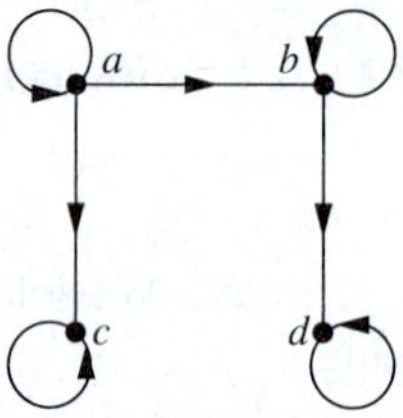

10.
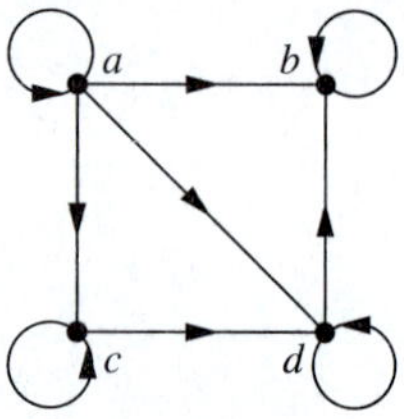

11.

12. Let (S, R) be a poset. Show that (S, R^{-1}) is also a poset, where R^{-1} is the inverse of R. The poset (S, R^{-1}) is called the **dual** of (S, R).
13. Find the duals of these posets.
    a) $(\{0, 1, 2\}, \leq)$ b) $(\mathbf{Z}, \geq)$
    c) $(P(\mathbf{Z}), \supseteq)$ d) $(\mathbf{Z}^+, |)$
14. Which of these pairs of elements are comparable in the poset $(\mathbf{Z}^+, |)$?
    a) 5, 15 b) 6, 9 c) 8, 16 d) 7, 7
15. Find two incomparable elements in these posets.
    a) $(P(\{0, 1, 2\}), \subseteq)$ b) $(\{1, 2, 4, 6, 8\}, |)$
16. Let $S = \{1, 2, 3, 4\}$. With respect to the lexicographic order based on the usual less than elation,
    a) find all pairs in $S \times S$ less than (2, 3).
    b) find all pairs in $S \times S$ greater than (3, 1).
    c) draw the Hasse diagram of the poset $(S \times S, \preccurlyeq)$.
17. Find the lexicographic ordering of these n-tuples:
    a) (1, 1, 2), (1, 2, 1) b) (0, 1, 2, 3), (0, 1, 3, 2)
    c) (1, 0, 1, 0, 1), (0, 1, 1, 1, 0)
18. Find the lexicographic ordering of these strings of lowercase English letters:
    a) *quack, quick, quicksilver, quicksand, quacking*
    b) *open, opener, opera, operand, opened*
    c) *zoo, zero, zoom, zoology, zoological*
19. Find the lexicographic ordering of the bit strings 0, 01, 11, 001, 010, 011, 0001, and 0101 based on the ordering $0 < 1$.
20. Draw the Hasse diagram for the greater than or equal to relation on $\{0, 1, 2, 3, 4, 5\}$.
21. Draw the Hasse diagram for the less than or equal to relation on $\{0, 2, 5, 10, 11, 15\}$.

22. Draw the Hasse diagram for divisibility on the set

a) $\{1, 2, 3, 4, 5, 6\}$. **b)** $\{3, 5, 7, 11, 13, 16, 17\}$.

c) $\{2, 3, 5, 10, 11, 15, 25\}$. **d)** $\{1, 3, 9, 27, 81, 243\}$.

23. Draw the Hasse diagram for divisibility on the set

a) $\{1, 2, 3, 4, 5, 6, 7, 8\}$. **b)** $\{1, 2, 3, 5, 7, 11, 13\}$.

c) $\{1, 2, 3, 6, 12, 24, 36, 48\}$.

d) $\{1, 2, 4, 8, 16, 32, 64\}$.

24. Draw the Hasse diagram for inclusion on the set $P(S)$, where $S = \{a, b, c, d\}$.

In Exercises 25–27 list all ordered pairs in the partial ordering with the accompanying Hasse diagram.

25.

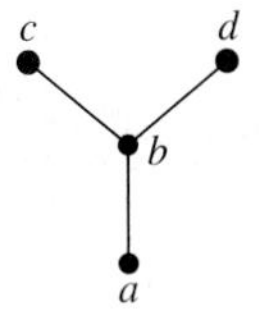

26.

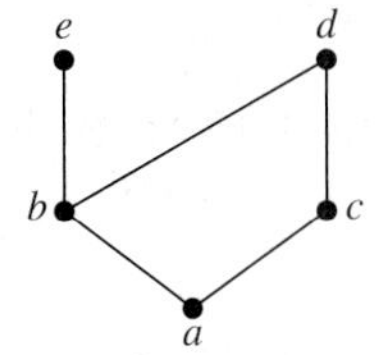

27.

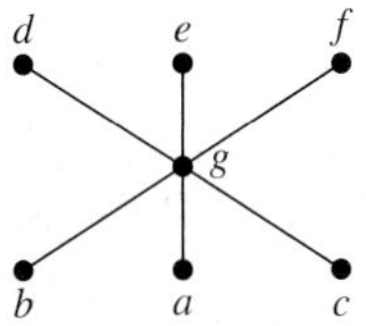

28. What is the covering relation of the partial ordering $\{(a, b) \mid a \text{ divides } b\}$ on $\{1, 2, 3, 4, 6, 12\}$?

29. What is the covering relation of the partial ordering $\{(A, B) \mid A \subseteq B\}$ on the power set of S, where $S = \{a, b, c\}$?

30. What is the covering relation of the partial ordering for the poset of security classes defined in Example 25?

31. Show that a finite poset can be reconstructed from its covering relation. [*Hint:* Show that the poset is the reflexive transitive closure of its covering relation.]

32. Answer these questions for the partial order represented by this Hasse diagram.

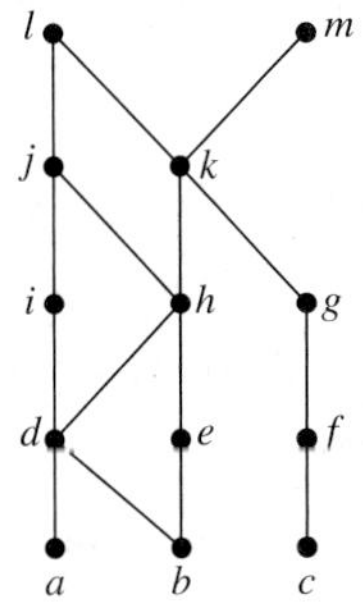

a) Find the maximal elements.

b) Find the minimal elements.

c) Is there a greatest element?

d) Is there a least element?

e) Find all upper bounds of $\{a, b, c\}$.

f) Find the least upper bound of $\{a, b, c\}$, if it exists.

g) Find all lower bounds of $\{f, g, h\}$.

h) Find the greatest lower bound of $\{f, g, h\}$, if it exists.

33. Answer these questions for the poset $(\{3, 5, 9, 15, 24, 45\}, |)$.

a) Find the maximal elements.

b) Find the minimal elements.

c) Is there a greatest element?

d) Is there a least element?

e) Find all upper bounds of $\{3, 5\}$.

f) Find the least upper bound of $\{3, 5\}$, if it exists.

g) Find all lower bounds of $\{15, 45\}$.

h) Find the greatest lower bound of $\{15, 45\}$, if it exists.

34. Answer these questions for the poset $(\{2, 4, 6, 9, 12, 18, 27, 36, 48, 60, 72\}, |)$.

a) Find the maximal elements.

b) Find the minimal elements.

c) Is there a greatest element?

d) Is there a least element?

e) Find all upper bounds of $\{2, 9\}$.

f) Find the least upper bound of $\{2, 9\}$, if it exists.

g) Find all lower bounds of $\{60, 72\}$.

h) Find the greatest lower bound of $\{60, 72\}$, if it exists.

35. Answer these questions for the poset $(\{\{1\}, \{2\}, \{4\}, \{1, 2\}, \{1, 4\}, \{2, 4\}, \{3, 4\}, \{1, 3, 4\}, \{2, 3, 4\}\}, \subseteq)$.

a) Find the maximal elements.

b) Find the minimal elements.

c) Is there a greatest element?

d) Is there a least element?

e) Find all upper bounds of $\{\{2\}, \{4\}\}$.

f) Find the least upper bound of $\{\{2\}, \{4\}\}$, if it exists.

g) Find all lower bounds of $\{\{1, 3, 4\}, \{2, 3, 4\}\}$.

h) Find the greatest lower bound of $\{\{1, 3, 4\}, \{2, 3, 4\}\}$, if it exists.

36. Give a poset that has

a) a minimal element but no maximal element.

b) a maximal element but no minimal element.

c) neither a maximal nor a minimal element.

37. Show that lexicographic order is a partial ordering on the Cartesian product of two posets.

38. Show that lexicographic order is a partial ordering on the set of strings from a poset.

39. Suppose that $(S, \preccurlyeq_1)$ and $(T, \preccurlyeq_2)$ are posets. Show that $(S \times T, \preccurlyeq)$ is a poset where $(s, t) \preccurlyeq (u, v)$ if and only if $s \preccurlyeq_1 u$ and $t \preccurlyeq_2 v$.

40. a) Show that there is exactly one greatest element of a poset, if such an element exists.

b) Show that there is exactly one least element of a poset, if such an element exists.

41. a) Show that there is exactly one maximal element in a poset with a greatest element.

b) Show that there is exactly one minimal element in a poset with a least element.

42. a) Show that the least upper bound of a set in a poset is unique if it exists.

b) Show that the greatest lower bound of a set in a poset is unique if it exists.

43. Determine whether the posets with these Hasse diagrams are lattices.

a)

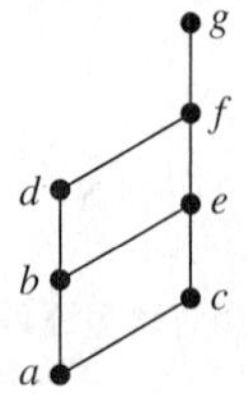

b)

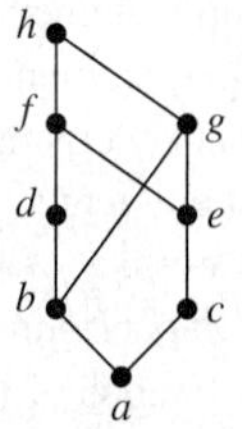

c)

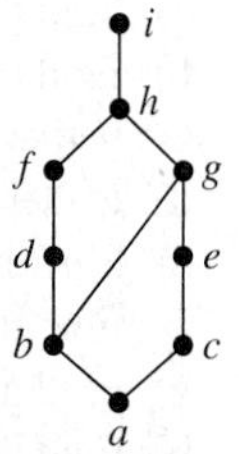

44. Determine whether these posets are lattices.

a) $(\{1, 3, 6, 9, 12\}, |)$ **b)** $(\{1, 5, 25, 125\}, |)$

c) $(\mathbf{Z}, \geq)$

d) $(P(S), \supseteq)$, where $P(S)$ is the power set of a set S

45. Show that every nonempty finite subset of a lattice has a least upper bound and a greatest lower bound.

46. Show that if the poset (S, R) is a lattice then the dual poset (S, R^{-1}) is also a lattice.

47. In a company, the lattice model of information flow is used to control sensitive information with security classes represented by ordered pairs (A, C). Here A is an authority level, which may be nonproprietary (0), proprietary (1), restricted (2), or registered (3). A category C is a subset of the set of all projects $\{$*Cheetah, Impala, Puma*$\}$. (Names of animals are often used as code names for projects in companies.)

a) Is information permitted to flow from (*Proprietary*, $\{$*Cheetah, Puma*$\}$) into (*Restricted*, $\{$*Puma*$\}$)?

b) Is information permitted to flow from (*Restricted*, $\{$*Cheetah*$\}$) into (*Registered*, $\{$*Cheetah, Impala*$\}$)?

c) Into which classes is information from (*Proprietary*, $\{$*Cheetah, Puma*$\}$) permitted to flow?

d) From which classes is information permitted to flow into the security class (*Restricted*, $\{$*Impala, Puma*$\}$)?

48. Show that the set S of security classes (A, C) is a lattice, where A is a positive integer representing an authority class and C is a subset of a finite set of compartments, with $(A_1, C_1) \preccurlyeq (A_2, C_2)$ if and only if $A_1 \leq A_2$ and $C_1 \subseteq C_2$. [*Hint:* First show that $(S, \preccurlyeq)$ is a poset and then show that the least upper bound and greatest lower bound of (A_1, C_1) and (A_2, C_2) are $(\max(A_1, A_2), C_1 \cup C_2)$ and $(\min(A_1, A_2), C_1 \cap C_2)$, respectively.]

***49.** Show that the set of all partitions of a set S with the relation $P_1 \preccurlyeq P_2$ if the partition P_1 is a refinement of the partition P_2 is a lattice. (See the preamble to Exercise 49 of Section 9.5.)

50. Show that every totally ordered set is a lattice.

51. Show that every finite lattice has a least element and a greatest element.

52. Give an example of an infinite lattice with

a) neither a least nor a greatest element.

b) a least but not a greatest element.

c) a greatest but not a least element.

d) both a least and a greatest element.

53. Verify that $(\mathbf{Z}^+ \times \mathbf{Z}^+, \preccurlyeq)$ is a well-ordered set, where $\preccurlyeq$ is lexicographic order, as claimed in Example 8.

54. Determine whether each of these posets is well-ordered.

a) $(S, \leq)$, where $S = \{10, 11, 12, \ldots\}$

b) $(\mathbf{Q} \cap [0, 1], \leq)$ (the set of rational numbers between 0 and 1 inclusive)

c) $(S, \leq)$, where S is the set of positive rational numbers with denominators not exceeding 3

d) $(\mathbf{Z}^-, \geq)$, where $\mathbf{Z}^-$ is the set of negative integers

A poset $(R, \preccurlyeq)$ is **well-founded** if there is no infinite decreasing sequence of elements in the poset, that is, elements $x_1, x_2, \ldots, x_n$ such that $\cdots \prec x_n \prec \cdots \prec x_2 \prec x_1$. A poset $(R, \preccurlyeq)$ is **dense** if for all $x \in S$ and $y \in S$ with $x \prec y$, there is an element $z \in R$ such that $x \prec z \prec y$.

55. Show that the poset $(\mathbf{Z}, \preccurlyeq)$, where $x \prec y$ if and only if $|x| < |y|$ is well-founded but is not a totally ordered set.

56. Show that a dense poset with at least two elements that are comparable is not well-founded.

57. Show that the poset of rational numbers with the usual less than or equal to relation, $(\mathbf{Q}, \leq)$, is a dense poset.

***58.** Show that the set of strings of lowercase English letters with lexicographic order is neither well-founded nor dense.

59. Show that a poset is well-ordered if and only if it is totally ordered and well-founded.

60. Show that a finite nonempty poset has a maximal element.

61. Find a compatible total order for the poset with the Hasse diagram shown in Exercise 32.

62. Find a compatible total order for the divisibility relation on the set $\{1, 2, 3, 6, 8, 12, 24, 36\}$.

63. Find all compatible total orderings for the poset $(\{1, 2, 4, 5, 12, 20\}, |\}$ from Example 26.

64. Find all compatible total orderings for the poset with the Hasse diagram in Exercise 27.

65. Find all possible orders for completing the tasks in the development project in Example 27.

66. Schedule the tasks needed to build a house, by specifying their order, if the Hasse diagram representing these tasks is as shown in the figure.

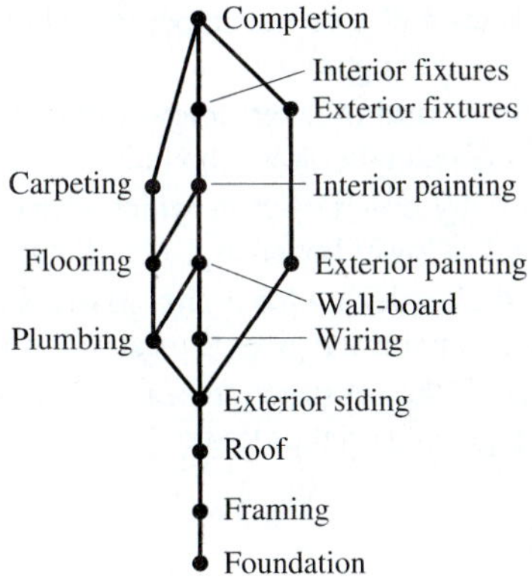

67. Find an ordering of the tasks of a software project if the Hasse diagram for the tasks of the project is as shown.

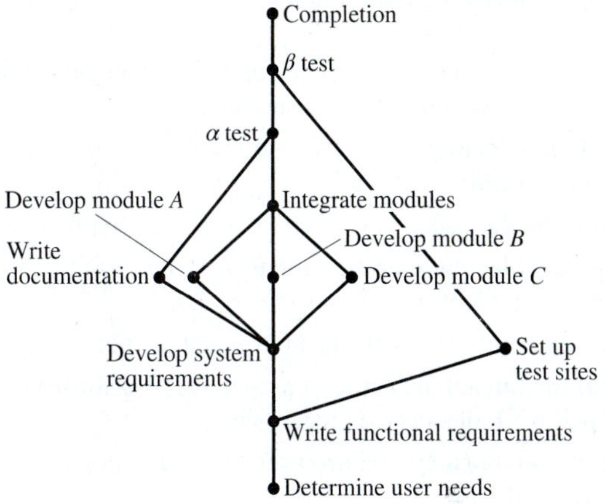

Key Terms and Results

TERMS

binary relation from A to B: a subset of $A \times B$

relation on A: a binary relation from A to itself (that is, a subset of $A \times A$)

$S \circ R$: composite of R and S

R^{-1}: inverse relation of R

R^n: nth power of R

reflexive: a relation R on A is reflexive if $(a, a) \in R$ for all $a \in A$

symmetric: a relation R on A is symmetric if $(b, a) \in R$ whenever $(a, b) \in R$

antisymmetric: a relation R on A is antisymmetric if $a = b$ whenever $(a, b) \in R$ and $(b, a) \in R$

transitive: a relation R on A is transitive if $(a, b) \in R$ and $(b, c) \in R$ implies that $(a, c) \in R$

n-ary relation on $A_1, A_2, \ldots, A_n$: a subset of $A_1 \times A_2 \times \cdots \times A_n$

relational data model: a model for representing databases using n-ary relations

primary key: a domain of an n-ary relation such that an n-tuple is uniquely determined by its value for this domain

composite key: the Cartesian product of domains of an n-ary relation such that an n-tuple is uniquely determined by its values in these domains

selection operator: a function that selects the n-tuples in an n-ary relation that satisfy a specified condition

projection: a function that produces relations of smaller degree from an n-ary relation by deleting fields

join: a function that combines n-ary relations that agree on certain fields

itemset: a collection of items

count of an itemset: the number of transactions that are supersets of the itemset

frequent itemset: an itemset with frequency greater than or equal to the support threshold

support of an itemset: the frequency of transactions that contain the itemset

association rule: an implication of the form $I \rightarrow J$, where I and J are itemsets

support of the association rule $I \rightarrow J$: the fraction of transactions that contain both the itemsets I and J

confidence of an association rule: the conditional probability that J is a subset of a transaction given that I is

directed graph or digraph: a set of elements called vertices and ordered pairs of these elements, called edges

loop: an edge of the form (a, a)

closure of a relation R with respect to a property P: the relation S (if it exists) that contains R, has property **P**, and is contained within any relation that contains R and has property **P**

path in a digraph: a sequence of edges $(a, x_1), (x_1, x_2), \ldots, (x_{n-2}, x_{n-1}), (x_{n-1}, b)$ such that the terminal vertex of each edge is the initial vertex of the succeeding edge in the sequence

circuit (or cycle) in a digraph: a path that begins and ends at the same vertex

R^* (connectivity relation): the relation consisting of those ordered pairs (a, b) such that there is a path from a to b

equivalence relation: a reflexive, symmetric, and transitive relation

equivalent: if R is an equivalence relation, a is equivalent to b if aRb

$[a]_R$ (equivalence class of a with respect to R): the set of all elements of A that are equivalent to a

$[a]_m$ (congruence class modulo m): the set of integers congruent to a modulo m

partition of a set S: a collection of pairwise disjoint nonempty subsets that have S as their union

partial ordering: a relation that is reflexive, antisymmetric, and transitive

poset (S, R): a set S and a partial ordering R on this set

comparable: the elements a and b in the poset $(A, \preccurlyeq)$ are comparable if $a \preccurlyeq b$ or $b \preccurlyeq a$

incomparable: elements in a poset that are not comparable

total (or linear) ordering: a partial ordering for which every pair of elements are comparable

totally (or linearly) ordered set: a poset with a total (or linear) ordering

well-ordered set: a poset $(S, \preccurlyeq)$, where $\preccurlyeq$ is a total order and every nonempty subset of S has a least element

lexicographic order: a partial ordering of Cartesian products or strings

Hasse diagram: a graphical representation of a poset where loops and all edges resulting from the transitive property are not shown, and the direction of the edges is indicated by the position of the vertices

maximal element: an element of a poset that is not less than any other element of the poset

minimal element: an element of a poset that is not greater than any other element of the poset

greatest element: an element of a poset greater than all other elements in this set

least element: an element of a poset less than all other elements in this set

upper bound of a set: an element in a poset greater than all other elements in the set

lower bound of a set: an element in a poset less than all other elements in the set

least upper bound of a set: an upper bound of the set that is less than all other upper bounds

greatest lower bound of a set: a lower bound of the set that is greater than all other lower bounds

lattice: a partially ordered set in which every two elements have a greatest lower bound and a least upper bound

compatible total ordering for a partial ordering: a total ordering that contains the given partial ordering

topological sort: the construction of a total ordering compatible with a given partial ordering

RESULTS

The reflexive closure of a relation R on the set A equals $R \cup \Delta$, where $\Delta = \{(a, a) \mid a \in A\}$.

The symmetric closure of a relation R on the set A equals $R \cup R^{-1}$, where $R^{-1} = \{(b, a) \mid (a, b) \in R\}$.

The transitive closure of a relation equals the connectivity relation formed from this relation.

Warshall's algorithm for finding the transitive closure of a relation

Let R be an equivalence relation. Then the following three statements are equivalent: (1) $a\ R\ b$; (2) $[a]_R \cap [b]_R \neq \emptyset$; (3) $[a]_R = [b]_R$.

The equivalence classes of an equivalence relation on a set A form a partition of A. Conversely, an equivalence relation can be constructed from any partition so that the equivalence classes are the subsets in the partition.

The principle of well-ordered induction

The topological sorting algorithm

Review Questions

1. a) What is a relation on a set?

b) How many relations are there on a set with n elements?

2. a) What is a reflexive relation?

b) What is a symmetric relation?

c) What is an antisymmetric relation?

d) What is a transitive relation?

3. Give an example of a relation on the set $\{1, 2, 3, 4\}$ that is

a) reflexive, symmetric, and not transitive.

b) not reflexive, symmetric, and transitive.

c) reflexive, antisymmetric, and not transitive.

d) reflexive, symmetric, and transitive.

e) reflexive, antisymmetric, and transitive.

4. a) How many reflexive relations are there on a set with n elements?

b) How many symmetric relations are there on a set with n elements?

c) How many antisymmetric relations are there on a set with n elements?

5. a) Explain how an n-ary relation can be used to represent information about students at a university.

b) How can the 5-ary relation containing names of students, their addresses, telephone numbers, majors, and grade point averages be used to form a 3-ary relation containing the names of students, their majors, and their grade point averages?

c) How can the 4-ary relation containing names of students, their addresses, telephone numbers, and majors and the 4-ary relation containing names of students, their student numbers, majors, and numbers of credit hours be combined into a single n-ary relation?

6. a) Explain how to use a zero–one matrix to represent a relation on a finite set.

b) Explain how to use the zero–one matrix representing a relation to determine whether the relation is reflexive, symmetric, and/or antisymmetric.

7. a) Explain how to use a directed graph to represent a relation on a finite set.
   b) Explain how to use the directed graph representing a relation to determine whether a relation is reflexive, symmetric, and/or antisymmetric.

8. a) Define the reflexive closure and the symmetric closure of a relation.
   b) How can you construct the reflexive closure of a relation?
   c) How can you construct the symmetric closure of a relation?
   d) Find the reflexive closure and the symmetric closure of the relation $\{(1, 2), (2, 3), (2, 4), (3, 1)\}$ on the set $\{1, 2, 3, 4\}$.

9. a) Define the transitive closure of a relation.
   b) Can the transitive closure of a relation be obtained by including all pairs (a, c) such that (a, b) and (b, c) belong to the relation?
   c) Describe two algorithms for finding the transitive closure of a relation.
   d) Find the transitive closure of the relation $\{(1,1), (1,3), (2,1), (2,3), (2,4), (3,2), (3,4), (4,1)\}$.

10. a) Define an equivalence relation.
    b) Which relations on the set $\{a, b, c, d\}$ are equivalence relations and contain (a, b) and (b, d)?

11. a) Show that congruence modulo m is an equivalence relation whenever m is a positive integer.
    b) Show that the relation $\{(a, b) \mid a \equiv \pm b \pmod 7\}$ is an equivalence relation on the set of integers.

12. a) What are the equivalence classes of an equivalence relation?
    b) What are the equivalence classes of the "congruent modulo 5" relation?
    c) What are the equivalence classes of the equivalence relation in Question 11(b)?

13. Explain the relationship between equivalence relations on a set and partitions of this set.

14. a) Define a partial ordering.
    b) Show that the divisibility relation on the set of positive integers is a partial order.

15. Explain how partial orderings on the sets A_1 and A_2 can be used to define a partial ordering on the set $A_1 \times A_2$.

16. a) Explain how to construct the Hasse diagram of a partial order on a finite set.
    b) Draw the Hasse diagram of the divisibility relation on the set $\{2, 3, 5, 9, 12, 15, 18\}$.

17. a) Define a maximal element of a poset and the greatest element of a poset.
    b) Give an example of a poset that has three maximal elements.
    c) Give an example of a poset with a greatest element.

18. a) Define a lattice.
    b) Give an example of a poset with five elements that is a lattice and an example of a poset with five elements that is not a lattice.

19. a) Show that every finite subset of a lattice has a greatest lower bound and a least upper bound.
    b) Show that every lattice with a finite number of elements has a least element and a greatest element.

20. a) Define a well-ordered set.
    b) Describe an algorithm for producing a totally ordered set compatible with a given partially ordered set.
    c) Explain how the algorithm from (b) can be used to order the tasks in a project if tasks are done one at a time and each task can be done only after one or more of the other tasks have been completed.

Supplementary Exercises

1. Let S be the set of all strings of English letters. Determine whether these relations are reflexive, irreflexive, symmetric, antisymmetric, and/or transitive.
   a) $R_1 = \{(a, b) \mid a$ and b have no letters in common$\}$
   b) $R_2 = \{(a, b) \mid a$ and b are not the same length$\}$
   c) $R_3 = \{(a, b) \mid a$ is longer than $b\}$

2. Construct a relation on the set $\{a, b, c, d\}$ that is
   a) reflexive, symmetric, but not transitive.
   b) irreflexive, symmetric, and transitive.
   c) irreflexive, antisymmetric, and not transitive.
   d) reflexive, neither symmetric nor antisymmetric, and transitive.
   e) neither reflexive, irreflexive, symmetric, antisymmetric, nor transitive.

3. Show that the relation R on $\mathbf{Z} \times \mathbf{Z}$ defined by $(a, b) R (c, d)$ if and only if $a + d = b + c$ is an equivalence relation.

4. Show that a subset of an antisymmetric relation is also antisymmetric.

5. Let R be a reflexive relation on a set A. Show that $R \subseteq R^2$.

6. Suppose that R_1 and R_2 are reflexive relations on a set A. Show that $R_1 \oplus R_2$ is irreflexive.

7. Suppose that R_1 and R_2 are reflexive relations on a set A. Is $R_1 \cap R_2$ also reflexive? Is $R_1 \cup R_2$ also reflexive?

8. Suppose that R is a symmetric relation on a set A. Is $\overline{R}$ also symmetric?

9. Let R_1 and R_2 be symmetric relations. Is $R_1 \cap R_2$ also symmetric? Is $R_1 \cup R_2$ also symmetric?

10. A relation R is called **circular** if aRb and bRc imply that cRa. Show that R is reflexive and circular if and only if it is an equivalence relation.

11. Show that a primary key in an n-ary relation is a primary key in any projection of this relation that contains this key as one of its fields.

12. Is the primary key in an n-ary relation also a primary key in a larger relation obtained by taking the join of this relation with a second relation?

13. Show that the reflexive closure of the symmetric closure of a relation is the same as the symmetric closure of its reflexive closure.

14. Let R be the relation on the set of all mathematicians that contains the ordered pair (a, b) if and only if a and b have written a published mathematical paper together.

a) Describe the relation R^2.

b) Describe the relation R^*.

c) The **Erdős number** of a mathematician is 1 if this mathematician wrote a paper with the prolific Hungarian mathematician Paul Erdős, it is 2 if this mathematician did not write a joint paper with Erdős but wrote a joint paper with someone who wrote a joint paper with Erdős, and so on (except that the Erdős number of Erdős himself is 0). Give a definition of the Erdős number in terms of paths in R.

15. a) Give an example to show that the transitive closure of the symmetric closure of a relation is not necessarily the same as the symmetric closure of the transitive closure of this relation.

b) Show, however, that the transitive closure of the symmetric closure of a relation must contain the symmetric closure of the transitive closure of this relation.

16. a) Let S be the set of subroutines of a computer program. Define the relation R by $\mathbf{P}R\mathbf{Q}$ if subroutine $\mathbf{P}$ calls subroutine $\mathbf{Q}$ during its execution. Describe the transitive closure of R.

b) For which subroutines $\mathbf{P}$ does $(\mathbf{P}, \mathbf{P})$ belong to the transitive closure of R?

c) Describe the reflexive closure of the transitive closure of R.

17. Suppose that R and S are relations on a set A with $R \subseteq S$ such that the closures of R and S with respect to a property $\mathbf{P}$ both exist. Show that the closure of R with respect to $\mathbf{P}$ is a subset of the closure of S with respect to $\mathbf{P}$.

18. Show that the symmetric closure of the union of two relations is the union of their symmetric closures.

***19.** Devise an algorithm, based on the concept of interior vertices, that finds the length of the longest path between two vertices in a directed graph, or determines that there are arbitrarily long paths between these vertices.

20. Which of these are equivalence relations on the set of all people?

a) $\{(x, y) \mid x$ and y have the same sign of the zodiac$\}$

b) $\{(x, y) \mid x$ and y were born in the same year$\}$

c) $\{(x, y) \mid x$ and y have been in the same city$\}$

***21.** How many different equivalence relations with exactly three different equivalence classes are there on a set with five elements?

22. Show that $\{(x, y) \mid x - y \in \mathbf{Q}\}$ is an equivalence relation on the set of real numbers, where $\mathbf{Q}$ denotes the set of rational numbers. What are $[1]$, $[\frac{1}{2}]$, and $[\pi]$?

Links

George Csicsery/Zala Films

PAUL ERDŐS (1913–1996) Paul Erdős, born in Budapest, Hungary, was the son of two high school mathematics teachers. He was a child prodigy; at age 3 he could multiply three-digit numbers in his head, and at 4 he discovered negative numbers on his own. Because his mother did not want to expose him to contagious diseases, he was mostly home-schooled. At 17 Erdős entered Eőtvős University, graduating four years later with a Ph.D. in mathematics. After graduating he spent four years at Manchester, England, on a postdoctoral fellowship. In 1938 he went to the United States because of the difficult political situation in Hungary, especially for Jews. He spent much of his time in the United States, except for 1954 to 1962, when he was banned as part of the paranoia of the McCarthy era. He also spent considerable time in Israel.

Erdős made many significant contributions to combinatorics and to number theory. One of the discoveries of which he was most proud is his elementary proof (in the sense that it does not use any complex analysis) of the prime number theorem, which provides an estimate for the number of primes not exceeding a fixed positive integer. He also participated in the modern development of the Ramsey theory.

Erdős traveled extensively throughout the world to work with other mathematicians, visiting conferences, universities, and research laboratories. He had no permanent home. He devoted himself almost entirely to mathematics, traveling from one mathematician to the next, proclaiming "My brain is open." Erdős was the author or coauthor of more than 1500 papers and had more than 500 coauthors. Copies of his articles are kept by Ron Graham, a famous discrete mathematician with whom he collaborated extensively and who took care of many of his worldly needs.

Erdős offered rewards, ranging from \$10 to \$10,000, for the solution of problems that he found particularly interesting, with the size of the reward depending on the difficulty of the problem. He paid out close to \$4000. Erdős had his own special language, using such terms as "epsilon" (child), "boss" (woman), "slave" (man), "captured" (married), "liberated" (divorced), "Supreme Fascist" (God), "Sam" (United States), and "Joe" (Soviet Union). Although he was curious about many things, he concentrated almost all his energy on mathematical research. He had no hobbies and no full-time job. He never married and apparently remained celibate. Erdős was extremely generous, donating much of the money he collected from prizes, awards, and stipends for scholarships and to worthwhile causes. He traveled extremely lightly and did not like having many material possessions.

23. Suppose that $P_1 = \{A_1, A_2, \dots, A_m\}$ and $P_2 = \{B_1, B_2, \dots, B_n\}$ are both partitions of the set S. Show that the collection of nonempty subsets of the form $A_i \cap B_j$ is a partition of S that is a refinement of both P_1 and P_2 (see the preamble to Exercise 49 of Section 9.5).

***24.** Show that the transitive closure of the symmetric closure of the reflexive closure of a relation R is the smallest equivalence relation that contains R.

25. Let $\mathbf{R}(S)$ be the set of all relations on a set S. Define the relation $\preccurlyeq$ on $\mathbf{R}(S)$ by $R_1 \preccurlyeq R_2$ if $R_1 \subseteq R_2$, where R_1 and R_2 are relations on S. Show that $(\mathbf{R}(S), \preccurlyeq)$ is a poset.

26. Let $\mathbf{P}(S)$ be the set of all partitions of the set S. Define the relation $\preccurlyeq$ on $\mathbf{P}(S)$ by $P_1 \preccurlyeq P_2$ if P_1 is a refinement of P_2 (see the preamble to Exercise 49 of Section 9.5). Show that $(\mathbf{P}(S), \preccurlyeq)$ is a poset.

27. Schedule the tasks needed to cook a Chinese meal by specifying their order, if the Hasse diagram representing these tasks is as shown here.

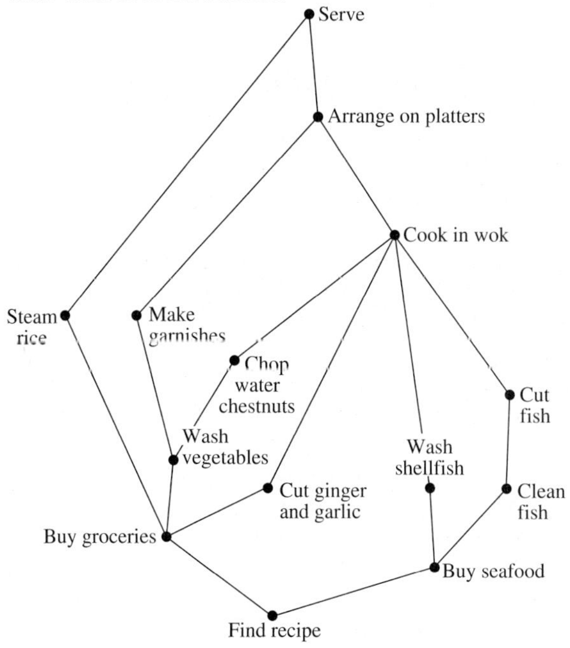

A subset of a poset such that every two elements of this subset are comparable is called a **chain**. A subset of a poset is called an **antichain** if every two elements of this subset are incomparable.

28. Find all chains in the posets with the Hasse diagrams shown in Exercises 25–27 in Section 9.6.

29. Find all antichains in the posets with the Hasse diagrams shown in Exercises 25–27 in Section 9.6.

30. Find an antichain with the greatest number of elements in the poset with the Hasse diagram of Exercise 32 in Section 9.6.

31. Show that every maximal chain in a finite poset $(S, \preccurlyeq)$ contains a minimal element of S. (A maximal chain is a chain that is not a subset of a larger chain.)

****32.** Show that every finite poset can be partitioned into k chains, where k is the largest number of elements in an antichain in this poset.

***33.** Show that in any group of $mn + 1$ people there is either a list of $m + 1$ people where a person in the list (except for the first person listed) is a descendant of the previous person on the list, or there are $n + 1$ people such that none of these people is a descendant of any of the other n people. [*Hint:* Use Exercise 32.]

Suppose that $(S, \preccurlyeq)$ is a well-founded partially ordered set. The *principle of well-founded induction* states that $P(x)$ is true for all $x \in S$ if $\forall x(\forall y(y \prec x \to P(y)) \to P(x))$.

34. Show that no separate basis case is needed for the principle of well-founded induction. That is, $P(u)$ is true for all minimal elements u in S if $\forall x(\forall y(y \prec x \to P(y)) \to P(x))$.

***35.** Show that the principle of well-founded induction is valid.

A relation R on a set A is a **quasi-ordering** on A if R is reflexive and transitive.

36. Let R be the relation on the set of all functions from $\mathbf{Z}^+$ to $\mathbf{Z}^+$ such that (f, g) belongs to R if and only if f is $O(g)$. Show that R is a quasi-ordering.

37. Let R be a quasi-ordering on a set A. Show that $R \cap R^{-1}$ is an equivalence relation.

***38.** Let R be a quasi-ordering and let S be the relation on the set of equivalence classes of $R \cap R^{-1}$ such that (C, D) belongs to S, where C and D are equivalence classes of R, if and only if there are elements c of C and d of D such that (c, d) belongs to R. Show that S is a partial ordering.

Let L be a lattice. Define the **meet** ($\wedge$) and **join** ($\vee$) operations by $x \wedge y = \text{glb}(x, y)$ and $x \vee y = \text{lub}(x, y)$.

39. Show that the following properties hold for all elements x, y, and z of a lattice L.

a) $x \wedge y = y \wedge x$ and $x \vee y = y \vee x$ **(commutative laws)**

b) $(x \wedge y) \wedge z = x \wedge (y \wedge z)$ and $(x \vee y) \vee z = x \vee (y \vee z)$ **(associative laws)**

c) $x \wedge (x \vee y) = x$ and $x \vee (x \wedge y) = x$ **(absorption laws)**

d) $x \wedge x = x$ and $x \vee x = x$ **(idempotent laws)**

40. Show that if x and y are elements of a lattice L, then $x \vee y = y$ if and only if $x \wedge y = x$.

A lattice L is **bounded** if it has both an **upper bound**, denoted by 1, such that $x \preccurlyeq 1$ for all $x \in L$ and a **lower bound**, denoted by 0, such that $0 \preccurlyeq x$ for all $x \in L$.

41. Show that if L is a bounded lattice with upper bound 1 and lower bound 0 then these properties hold for all elements $x \in L$.

a) $x \vee 1 = 1$ **b)** $x \wedge 1 = x$

c) $x \vee 0 = x$ **d)** $x \wedge 0 = 0$

42. Show that every finite lattice is bounded.

A lattice is called **distributive** if $x \vee (y \wedge z) = (x \vee y) \wedge (x \vee z)$ and $x \wedge (y \vee z) = (x \wedge y) \vee (x \wedge z)$ for all x, y, and z in L.

***43.** Give an example of a lattice that is not distributive.

44. Show that the lattice $(P(S), \subseteq)$ where $P(S)$ is the power set of a finite set S is distributive.

45. Is the lattice $(\mathbf{Z}^+, |)$ distributive?

The **complement** of an element a of a bounded lattice L with upper bound 1 and lower bound 0 is an element b such that $a \vee b = 1$ and $a \wedge b = 0$. Such a lattice is **complemented** if every element of the lattice has a complement.

46. Give an example of a finite lattice where at least one element has more than one complement and at least one element has no complement.

47. Show that the lattice $(P(S), \subseteq)$ where $P(S)$ is the power set of a finite set S is complemented.

***48.** Show that if L is a finite distributive lattice, then an element of L has at most one complement.

The game of Chomp, introduced in Example 12 in Section 1.8, can be generalized for play on any finite partially ordered set $(S, \preceq)$ with a least element a. In this game, a move consists of selecting an element x in S and removing x and all elements larger than it from S. The loser is the player who is forced to select the least element a.

49. Show that the game of Chomp with cookies arranged in an $m \times n$ rectangular grid, described in Example 12 in Section 1.8, is the same as the game of Chomp on the poset $(S, |)$, where S is the set of all positive integers that divide $p^{m-1}q^{n-1}$, where p and q are distinct primes.

50. Show that if $(S, \preceq)$ has a greatest element b, then a winning strategy for Chomp on this poset exists. [*Hint:* Generalize the argument in Example 12 in Section 1.8.]

Computer Projects

Write programs with these input and output.

1. Given the matrix representing a relation on a finite set, determine whether the relation is reflexive and/or irreflexive.

2. Given the matrix representing a relation on a finite set, determine whether the relation is symmetric and/or antisymmetric.

3. Given the matrix representing a relation on a finite set, determine whether the relation is transitive.

4. Given a positive integer n, display all the relations on a set with n elements.

***5.** Given a positive integer n, determine the number of transitive relations on a set with n elements.

***6.** Given a positive integer n, determine the number of equivalence relations on a set with n elements.

***7.** Given a positive integer n, display all the equivalence relations on the set of the n smallest positive integers.

8. Given an n-ary relation, find the projection of this relation when specified fields are deleted.

9. Given an m-ary relation and an n-ary relation, and a set of common fields, find the join of these relations with respect to these common fields.

10. Given the matrix representing a relation on a finite set, find the matrix representing the reflexive closure of this relation.

11. Given the matrix representing a relation on a finite set, find the matrix representing the symmetric closure of this relation.

12. Given the matrix representing a relation on a finite set, find the matrix representing the transitive closure of this relation by computing the join of the Boolean powers of the matrix representing the relation.

13. Given the matrix representing a relation on a finite set, find the matrix representing the transitive closure of this relation using Warshall's algorithm.

14. Given the matrix representing a relation on a finite set, find the matrix representing the smallest equivalence relation containing this relation.

15. Given a partial ordering on a finite set, find a total ordering compatible with it using topological sorting.

Computations and Explorations

Use a computational program or programs you have written to do these exercises.

1. Display all the different relations on a set with four elements.

2. Display all the different reflexive and symmetric relations on a set with six elements.

3. Display all the reflexive and transitive relations on a set with five elements.

***4.** Determine how many transitive relations there are on a set with n elements for all positive integers n with $n \leq 7$.

5. Find the transitive closure of a relation of your choice on a set with at least 20 elements. Either use a relation that corresponds to direct links in a particular transportation or communications network or use a randomly generated relation.

6. Compute the number of different equivalence relations on a set with n elements for all positive integers n not exceeding 20.

7. Display all the equivalence relations on a set with seven elements.

*8. Display all the partial orders on a set with five elements.

*9. Display all the lattices on a set with five elements.

Writing Projects

Respond to these with essays using outside sources.

1. Discuss the concept of a fuzzy relation. How are fuzzy relations used?

2. Describe the basic principles of relational databases, going beyond what was covered in Section 9.2. How widely used are relational databases as compared with other types of databases?

3. Explain how the Apriori algorithm is used to find frequent itemsets and strong association rules.

4. Describe some applications of association rules in detail.

5. Look up the original papers by Warshall and by Roy (in French) in which they develop algorithms for finding transitive closures. Discuss their approaches. Why do you suppose that what we call Warshall's algorithm was discovered independently by more than one person?

6. Describe how equivalence classes can be used to define the rational numbers as classes of pairs of integers and how the basic arithmetic operations on rational numbers can be defined following this approach. (See Exercises 16 and 40 in Section 9.5.)

7. Explain how Helmut Hasse used what we now call Hasse diagrams.

8. Describe some of the mechanisms used to enforce information flow policies in computer operating systems.

9. Discuss the use of the Program Evaluation and Review Technique (PERT) to schedule the tasks of a large complicated project. How widely is PERT used?

10. Discuss the use of the Critical Path Method (CPM) to find the shortest time for the completion of a project. How widely is CPM used?

11. Discuss the concept of *duality* in a lattice. Explain how duality can be used to establish new results.

12. Explain what is meant by a *modular lattice*. Describe some of the properties of modular lattices and describe how modular lattices arise in the study of projective geometry.

CHAPTER

10 Graphs

10.1 Graphs and Graph Models

10.2 Graph Terminology and Special Types of Graphs

10.3 Representing Graphs and Graph Isomorphism

10.4 Connectivity

10.5 Euler and Hamilton Paths

10.6 Shortest-Path Problems

10.7 Planar Graphs

10.8 Graph Coloring

Graphs are discrete structures consisting of vertices and edges that connect these vertices. There are different kinds of graphs, depending on whether edges have directions, whether multiple edges can connect the same pair of vertices, and whether loops are allowed. Problems in almost every conceivable discipline can be solved using graph models. We will give examples to illustrate how graphs are used as models in a variety of areas. For instance, we will show how graphs are used to represent the competition of different species in an ecological niche, how graphs are used to represent who influences whom in an organization, and how graphs are used to represent the outcomes of round-robin tournaments. We will describe how graphs can be used to model acquaintanceships between people, collaboration between researchers, telephone calls between telephone numbers, and links between websites. We will show how graphs can be used to model roadmaps and the assignment of jobs to employees of an organization.

Using graph models, we can determine whether it is possible to walk down all the streets in a city without going down a street twice, and we can find the number of colors needed to color the regions of a map. Graphs can be used to determine whether a circuit can be implemented on a planar circuit board. We can distinguish between two chemical compounds with the same molecular formula but different structures using graphs. We can determine whether two computers are connected by a communications link using graph models of computer networks. Graphs with weights assigned to their edges can be used to solve problems such as finding the shortest path between two cities in a transportation network. We can also use graphs to schedule exams and assign channels to television stations. This chapter will introduce the basic concepts of graph theory and present many different graph models. To solve the wide variety of problems that can be studied using graphs, we will introduce many different graph algorithms. We will also study the complexity of these algorithms.

10.1 Graphs and Graph Models

We begin with the definition of a graph.

Definition 1 A *graph* $G = (V, E)$ consists of V, a nonempty set of *vertices* (or *nodes*) and E, a set of *edges*. Each edge has either one or two vertices associated with it, called its *endpoints*. An edge is said to *connect* its endpoints.

Remark: The set of vertices V of a graph G may be infinite. A graph with an infinite vertex set or an infinite number of edges is called an **infinite graph**, and in comparison, a graph with a finite vertex set and a finite edge set is called a **finite graph**. In this book we will usually consider only finite graphs.

Now suppose that a network is made up of data centers and communication links between computers. We can represent the location of each data center by a point and each communications link by a line segment, as shown in Figure 1.

This computer network can be modeled using a graph in which the vertices of the graph represent the data centers and the edges represent communication links. In general, we visualize graphs by using points to represent vertices and line segments, possibly curved, to represent

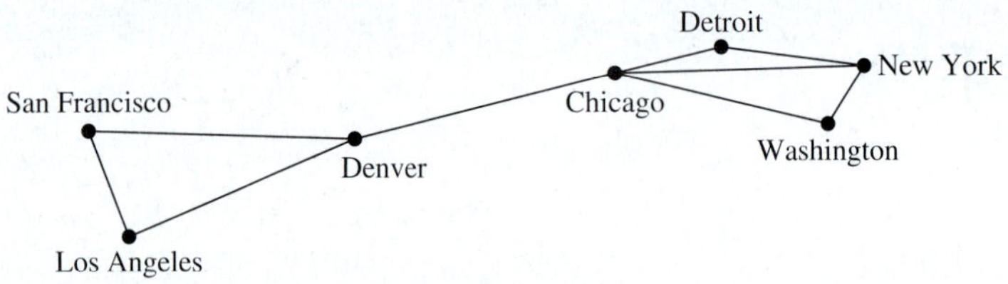

FIGURE 1 A computer network.

edges, where the endpoints of a line segment representing an edge are the points representing the vertices connected by the edge. When we draw a graph, we generally try to draw edges so that they do not cross. However, this is not necessary because any depiction using points to represent vertices and any form of connection between vertices can be used. Indeed, there are some graphs that cannot be drawn in the plane without edges crossing (see Section 10.7). The key point is that the way we draw a graph is arbitrary, as long as the correct connections between vertices are depicted.

Note that each edge of the graph representing this computer network connects two different vertices. That is, no edge connects a vertex to itself. Furthermore, no two different edges connect the same pair of vertices. A graph in which each edge connects two different vertices and where no two edges connect the same pair of vertices is called a **simple graph**. Note that in a simple graph, each edge is associated to an unordered pair of vertices, and no other edge is associated to this same edge. Consequently, when there is an edge of a simple graph associated to $\{u, v\}$, we can also say, without possible confusion, that $\{u, v\}$ is an edge of the graph.

A computer network may contain multiple links between data centers, as shown in Figure 2. To model such networks we need graphs that have more than one edge connecting the same pair of vertices. Graphs that may have **multiple edges** connecting the same vertices are called **multigraphs**. When there are m different edges associated to the same unordered pair of vertices $\{u, v\}$, we also say that $\{u, v\}$ is an edge of multiplicity m. That is, we can think of this set of edges as m different copies of an edge $\{u, v\}$.

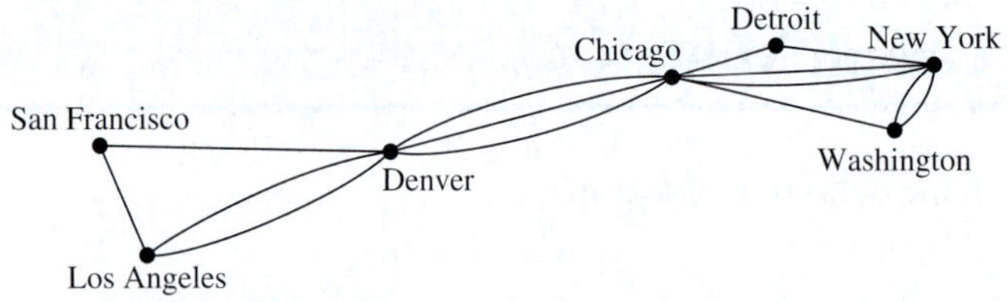

FIGURE 2 A computer network with multiple links between data centers.

Sometimes a communications link connects a data center with itself, perhaps a feedback loop for diagnostic purposes. Such a network is illustrated in Figure 3. To model this network we need to include edges that connect a vertex to itself. Such edges are called **loops**, and sometimes

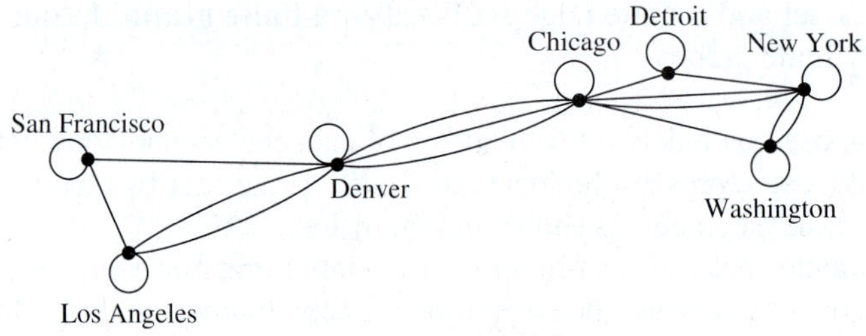

FIGURE 3 A computer network with diagnostic links.

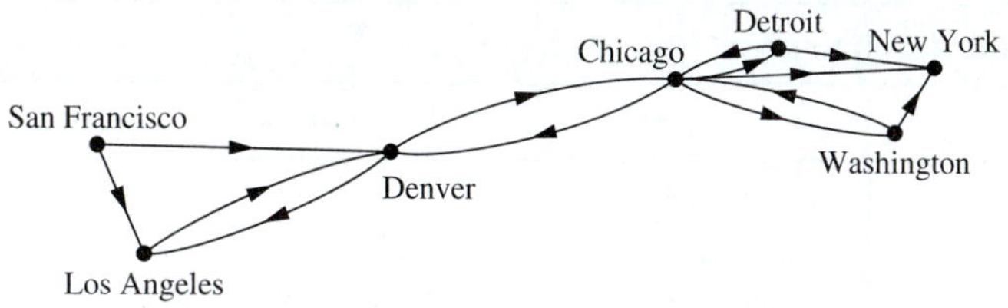

FIGURE 4 A communications network with one-way communications links.

we may even have more than one loop at a vertex. Graphs that may include loops, and possibly multiple edges connecting the same pair of vertices or a vertex to itself, are sometimes called **pseudographs**.

So far the graphs we have introduced are **undirected graphs**. Their edges are also said to be **undirected**. However, to construct a graph model, we may find it necessary to assign directions to the edges of a graph. For example, in a computer network, some links may operate in only one direction (such links are called single duplex lines). This may be the case if there is a large amount of traffic sent to some data centers, with little or no traffic going in the opposite direction. Such a network is shown in Figure 4.

To model such a computer network we use a directed graph. Each edge of a directed graph is associated to an ordered pair. The definition of directed graph we give here is more general than the one we used in Chapter 9, where we used directed graphs to represent relations.

Definition 2 A *directed graph* (or *digraph*) (V, E) consists of a nonempty set of vertices V and a set of *directed edges* (or *arcs*) E. Each directed edge is associated with an ordered pair of vertices. The directed edge associated with the ordered pair (u, v) is said to *start* at u and *end* at v.

When we depict a directed graph with a line drawing, we use an arrow pointing from u to v to indicate the direction of an edge that starts at u and ends at v. A directed graph may contain loops and it may contain multiple directed edges that start and end at the same vertices. A directed graph may also contain directed edges that connect vertices u and v in both directions; that is, when a digraph contains an edge from u to v, it may also contain one or more edges from v to u. Note that we obtain a directed graph when we assign a direction to each edge in an undirected graph. When a directed graph has no loops and has no multiple directed edges, it is called a **simple directed graph**. Because a simple directed graph has at most one edge associated to each ordered pair of vertices (u, v), we call (u, v) an edge if there is an edge associated to it in the graph.

In some computer networks, multiple communication links between two data centers may be present, as illustrated in Figure 5. Directed graphs that may have **multiple directed edges** from a vertex to a second (possibly the same) vertex are used to model such networks. We called such graphs **directed multigraphs**. When there are m directed edges, each associated to an ordered pair of vertices (u, v), we say that (u, v) is an edge of **multiplicity** m.

For some models we may need a graph where some edges are undirected, while others are directed. A graph with both directed and undirected edges is called a **mixed graph**. For example,

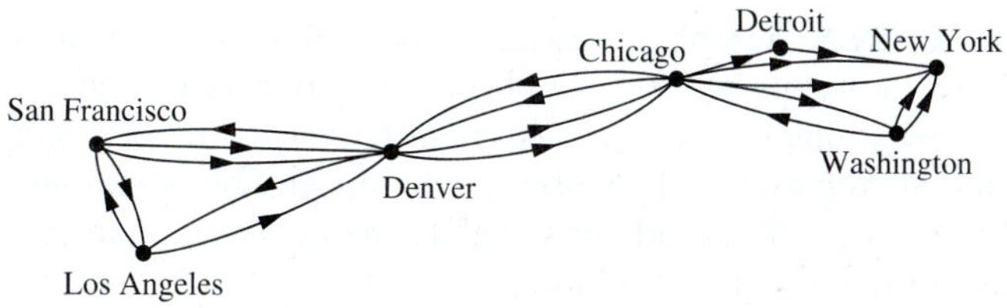

FIGURE 5 A computer network with multiple one-way links.

TABLE 1 Graph Terminology.

Type	*Edges*	*Multiple Edges Allowed?*	*Loops Allowed?*
Simple graph	Undirected	No	No
Multigraph	Undirected	Yes	No
Pseudograph	Undirected	Yes	Yes
Simple directed graph	Directed	No	No
Directed multigraph	Directed	Yes	Yes
Mixed graph	Directed and undirected	Yes	Yes

a mixed graph might be used to model a computer network containing links that operate in both directions and other links that operate only in one direction.

This terminology for the various types of graphs is summarized in Table 1. We will sometimes use the term **graph** as a general term to describe graphs with directed or undirected edges (or both), with or without loops, and with or without multiple edges. At other times, when the context is clear, we will use the term graph to refer only to undirected graphs.

Links

Because of the relatively modern interest in graph theory, and because it has applications to a wide variety of disciplines, many different terminologies of graph theory have been introduced. The reader should determine how such terms are being used whenever they are encountered. The terminology used by mathematicians to describe graphs has been increasingly standardized, but the terminology used to discuss graphs when they are used in other disciplines is still quite varied. Although the terminology used to describe graphs may vary, three key questions can help us understand the structure of a graph:

- Are the edges of the graph undirected or directed (or both)?
- If the graph is undirected, are multiple edges present that connect the same pair of vertices? If the graph is directed, are multiple directed edges present?
- Are loops present?

Answering such questions helps us understand graphs. It is less important to remember the particular terminology used.

10.1.1 Graph Models

Links

Can you find a subject to which graph theory has not been applied?

Graphs are used in a wide variety of models. We began this section by describing how to construct graph models of communications networks linking data centers. We will complete this section by describing some diverse graph models for some interesting applications. We will return to many of these applications later in this chapter and in Chapter 11. We will introduce additional graph models in subsequent sections of this and later chapters. Also, recall that directed graph models for some applications were introduced in Chapter 9. When we build a graph model, we need to make sure that we have correctly answered the three key questions we posed about the structure of a graph.

SOCIAL NETWORKS Graphs are extensively used to model social structures based on different kinds of relationships between people or groups of people. These social structures, and the graphs that represent them, are known as **social networks**. In these graph models, individuals or organizations are represented by vertices; relationships between individuals or organizations are represented by edges. The study of social networks is an extremely active multidisciplinary area, and many different types of relationships between people have been studied using them. We will introduce some of the most commonly studied social networks here. More information about social networks can be found in [Ne10] and [EaKl10].

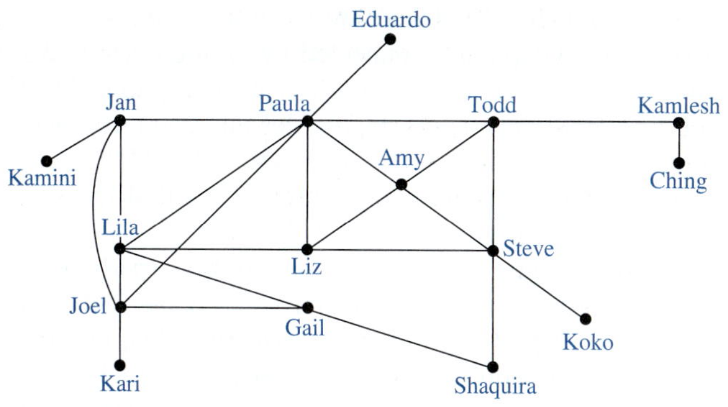

FIGURE 6 **An acquaintanceship graph.**

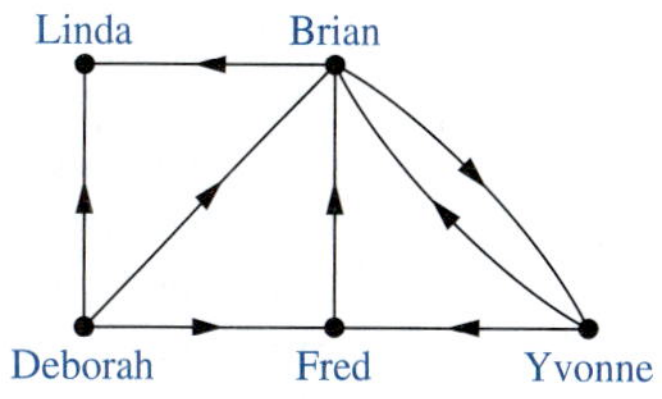

FIGURE 7 **An influence graph.**

EXAMPLE 1 **Acquaintanceship and Friendship Graphs** We can use a simple graph to represent whether two people know each other, that is, whether they are acquainted, or whether they are friends (either in the real world or in the virtual world via a social networking site such as Facebook where friendships are reciprocal). Each person in a particular group of people is represented by a vertex. An undirected edge is used to connect two people when these people know each other, when we are concerned only with acquaintanceship, or whether they are friends. No multiple edges and usually no loops are used. (If we want to include the notion of self-knowledge, we would include loops.) A small acquaintanceship graph is shown in Figure 6. The acquaintanceship graph of all people in the world has more than eight billion vertices and probably more than one trillion edges! We will discuss this graph further in Section 10.4. ◀

Links ❯

EXAMPLE 2 **Influence Graphs** In studies of group behavior it is observed that certain people can influence the thinking of others. A directed graph called an **influence graph** can be used to model this behavior. Each person of the group is represented by a vertex. There is a directed edge from vertex a to vertex b when the person represented by vertex a can influence the person represented by vertex b. This graph does not contain loops and it does not contain multiple directed edges. An example of an influence graph for members of a group is shown in Figure 7. In the group modeled by this influence graph, Deborah cannot be influenced, but Deborah can influence Brian, Fred, and Linda. Also, Yvonne and Brian can influence each other. ◀

Many popular social network applications can be modeled by influence graphs. For example, the social network Instagram is based on the notion of influencers. Instagram can be modeled by a directed graph, where users are represented by vertices and follow relationships are represented by directed edges. If a user B accepts a follow request from user A, A is called a *follower* of B and A receives all the postings of B. Influencers in Instagram are those with a large number of followers.

In X (formerly called Twitter), relationships between users are asymmetric. These relationships can be modeled by the *X follow graph*. In this graph, each user is represented by a node, and edges represent follower relationships. If User A follows User B, there is a directed edge from A to B in this graph. Users with many followers are able to influence large numbers of people. Mutual connections in X can be modeled using the *X undirected mutual graph*. Edges exist only between users who follow each other.

In LinkedIn, if you are connected to someone, you're following them and they are following you. Your followers will receive your content in their homepage feed. However, members don't have to be connected to you to follow you and receive these updates. You can also follow a person if that person has published articles on LinkedIn's platform or changed their privacy settings to enable updates to be publicly followable.

EXAMPLE 3 **Collaboration Graphs** A **collaboration graph** is used to model social networks where two people are related by working together in a particular way. Collaboration graphs are simple graphs, as edges in these graphs are undirected and there are no multiple edges or loops. Vertices in these graphs represent people; two people are connected by an undirected edge when the people have collaborated. There are no loops nor multiple edges in these graphs. The **Hollywood graph** is a collaborator graph that represents actors by vertices and connects two actors with an edge if they have worked together on a movie or television show. The Hollywood graph is a huge graph with more than 2.9 million vertices (as of early 2018). We will discuss some aspects of the Hollywood graph later in Section 10.4.

Links

In an **academic collaboration graph**, vertices represent people (perhaps restricted to members of a certain academic community), and edges link two people if they have jointly published a paper. The collaboration graph for people who have published research papers in mathematics was found in 2004 to have more than 400,000 vertices and 675,000 edges, and these numbers have grown considerably since then. We will have more to say about this graph in Section 10.4. Collaboration graphs have also been used in sports, where two professional athletes are considered to have collaborated if they have ever played on the same team during a regular season of their sport. ◀

COMMUNICATION NETWORKS We can model different communications networks using vertices to represent devices and edges to represent the particular type of communications links of interest. We have already modeled a data network in the first part of this section.

EXAMPLE 4 **Call Graphs** Graphs can be used to model telephone calls made in a network, such as a long-distance telephone network. In particular, a directed multigraph can be used to model calls where each telephone number is represented by a vertex and each telephone call is represented by a directed edge. The edge representing a call starts at the telephone number from which the call was made and ends at the telephone number to which the call was made. We need directed edges because the direction in which the call is made matters. We need multiple directed edges because we want to represent each call made from a particular telephone number to a second number.

Links

A small telephone call graph is displayed in Figure 8(a), representing seven telephone numbers. This graph shows, for instance, that three calls have been made from 732-555-1234 to 732-555-9876 and two in the other direction, but no calls have been made from 732-555-4444 to any of the other six numbers except 732-555-0011. When we care only whether there has been a call connecting two telephone numbers, we use an undirected graph with an edge connecting telephone numbers when there has been a call between these numbers. This version of the call graph is displayed in Figure 8(b).

Call graphs that model actual calling activities can be huge. For example, one call graph studied at AT&T, which models calls during 20 days, has about 290 million vertices and 4 billion edges. We will discuss call graphs further in Section 10.4. ◀

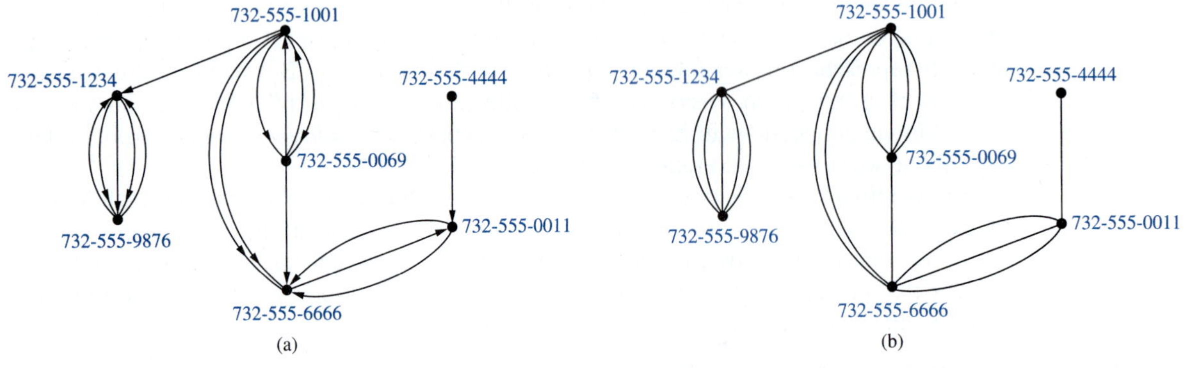

FIGURE 8 **A call graph.**

INFORMATION NETWORKS Graphs can be used to model various networks that link particular types of information. Here, we will describe how to model the web using a graph. We will also describe how to use a graph to model the citations in different types of documents.

EXAMPLE 5 **The Web Graph** The web can be modeled as a directed graph where each web page is represented by a vertex and where an edge starts at the web page *a* and ends at the web page *b* if there is a link on *a* pointing to *b*. Because new web pages are created and others removed somewhere on the web almost every second, the web graph changes on an almost continual basis. Many people are studying the properties of the web graph to better understand the nature of the web. We will return to web graphs in Section 10.4, and in Chapter 11 we will explain how the web graph is used by the web crawlers that search engines use to create indexes of web pages. ◀

Links ❯

EXAMPLE 6 **Citation Graphs** Graphs can be used to represent citations in different types of documents, including academic papers, patents, and legal opinions. In such graphs, each document is represented by a vertex, and there is an edge from one document to a second document if the first document cites the second in its citation list. (In an academic paper, the citation list is the bibliography, or list of references; in a patent it is the list of previous patents that are cited; and in a legal opinion it is the list of previous opinions cited.) A citation graph is a directed graph without loops or multiple edges. ◀

SOFTWARE DESIGN APPLICATIONS Graph models are useful tools in the design of software. We will briefly describe two of these models here.

EXAMPLE 7 **Module Dependency Graphs** One of the most important tasks in designing software is how to structure a program into different parts, or modules. Understanding how the different modules of a program interact is essential not only for program design, but also for testing and maintenance of the resulting software. A **module dependency graph** provides a useful tool for understanding how different modules of a program interact. In a program dependency graph, each module is represented by a vertex. There is a directed edge from a module to a second module if the second module depends on the first. An example of a program dependency graph for a web browser is shown in Figure 9. ◀

EXAMPLE 8 **Precedence Graphs and Concurrent Processing** Computer programs can be executed more rapidly by executing certain statements concurrently. It is important not to execute a statement that requires results of statements not yet executed. The dependence of statements on previous statements can be represented by a directed graph. Each statement is represented by a vertex, and there is an edge from one statement to a second statement if the second statement cannot be executed before the first statement. This resulting graph is called a **precedence graph**. A computer program and its graph are displayed in Figure 10. For instance, the graph shows that statement S_5 cannot be executed before statements S_1, S_2, and S_4 are executed. ◀

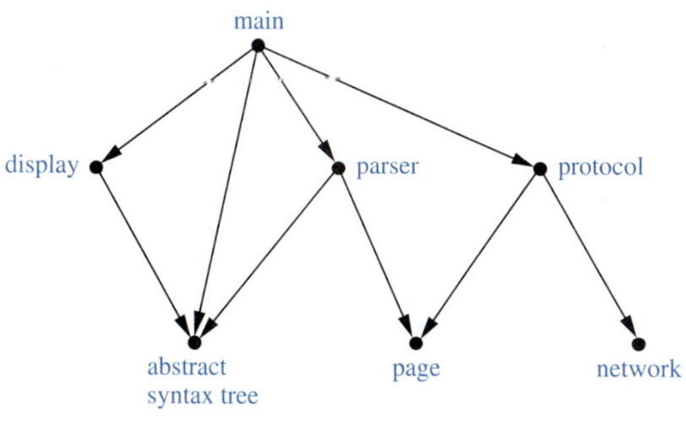

FIGURE 9 **A module dependency graph.**

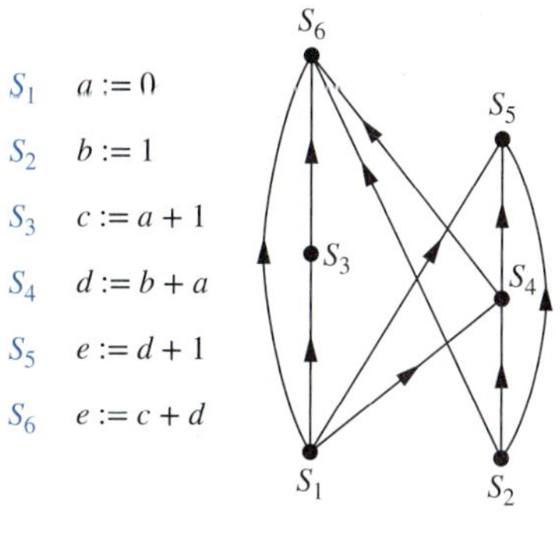

FIGURE 10 **A precedence graph.**

TRANSPORTATION NETWORKS We can use graphs to model many different types of transportation networks, including road, air, and rail networks, as well shipping networks.

EXAMPLE 9 **Airline Routes** We can model airline networks by representing each airport by a vertex. In particular, we can model all the flights by a particular airline each day using a directed edge to represent each flight, going from the vertex representing the departure airport to the vertex representing the destination airport. The resulting graph will generally be a directed multigraph, as there may be multiple flights from one airport to some other airport during the same day. ◀

EXAMPLE 10 **Road Networks** Graphs can be used to model road networks. In such models, vertices represent intersections and edges represent roads. When all roads are two-way and there is at most one road connecting two intersections, we can use a simple undirected graph to model the road network. However, we will often want to model road networks when some roads are one-way and when there may be more than one road between two intersections. To build such models, we use undirected edges to represent two-way roads and we use directed edges to represent one-way roads. Multiple undirected edges represent multiple two-way roads connecting the same two intersections. Multiple directed edges represent multiple one-way roads that start at one intersection and end at a second intersection. Loops represent loop roads. Mixed graphs are needed to model road networks that include both one-way and two-way roads. ◀

BIOLOGICAL NETWORKS Many aspects of the biological sciences can be modeled using graphs.

EXAMPLE 11 **Niche Overlap Graphs in Ecology** Graphs are used in many models involving the interaction of different species of animals. For instance, the competition between species in an ecosystem can be modeled using a **niche overlap graph**. Each species is represented by a vertex. An undirected edge connects two vertices if the two species represented by these vertices compete (that is, some of the food resources they use are the same). A niche overlap graph is a simple graph because no loops or multiple edges are needed in this model. The graph in Figure 11 models the ecosystem of a forest. We see from this graph that squirrels and raccoons compete but that crows and shrews do not. ◀

Links 〉

Extra Examples 〉

EXAMPLE 12 **Protein Interaction Graphs** A protein interaction in a living cell occurs when two or more proteins in that cell bind to perform a biological function. Because protein interactions are crucial for most biological functions, many scientists work on discovering new proteins and understanding interactions between proteins. Protein interactions within a cell can be modeled using a **protein interaction graph** (also called a **protein–protein interaction network**), an undirected graph in which each protein is represented by a vertex, with an edge connecting the vertices representing each pair of proteins that interact. It is a challenging problem to determine genuine protein interactions in a cell, as experiments often produce false positives, which conclude that two proteins interact when they really do not. Protein interaction graphs can be used to deduce important biological information, such as by identifying the most important proteins for various functions and the functionality of newly discovered proteins.

Links 〉

Because there are thousands of different proteins in a typical cell, the protein interaction graph of a cell is extremely large and complex. For example, yeast cells have more than 6,000 proteins, and more than 80,000 interactions between them are known, and human cells have more than 100,000 proteins, with perhaps as many as 1,000,000 interactions between them. Additional vertices and edges are added to a protein interaction graph when new proteins and interactions between proteins are discovered. Because of the complexity of protein interaction graphs, they are often split into smaller graphs called modules that represent groups of proteins that are involved in a particular function of a cell. Figure 12 illustrates a module of the protein interaction graph described in [Bo04], comprising the complex of proteins that degrade RNA in human cells. To learn more about protein interaction graphs, see [Bo04], [Ne10], and [Hu07]. ◀

Semantic Networks Graph models are used extensively in natural language understanding and in information retrieval. **Natural language understanding (NLU)** is the subject of enabling

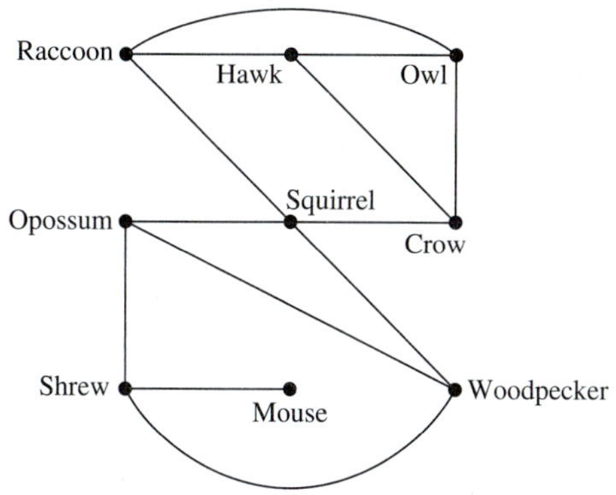

FIGURE 11 **A niche overlap graph.**

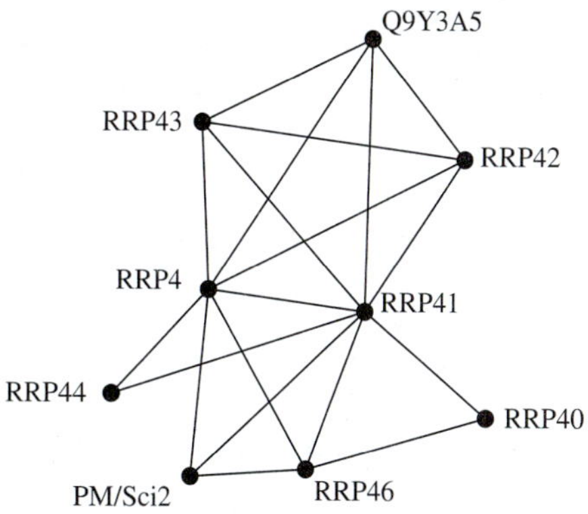

FIGURE 12 **A module of a protein interaction graph.**

machines to disassemble and parse human speech. Its goal is to allow machines to understand and communicate as humans do. **Information retrieval (IR)** is the subject of obtaining information from a collection of sources based on various types of searches. Natural language understanding is the enabling technology when we converse with automated customer service agents. Advances in NLU are evident as communication between humans and machines continually improves. When we do web searches, we take advantage of the many advances in information retrieval made in recent decades.

In graph models for NLU and IR applications, vertices often represent words, phrases, or sentences, and edges represent connections relating to the meaning of these objects.

EXAMPLE 13 In **semantic networks**, vertices are used to represent words, and undirected edges are used to connect vertices when a semantic relation holds between these words. A **semantic relation** is a relationship between two or more words that is based on the meaning of the words. For example, we can build a graph in which vertices represent nouns and two vertices are connected when they have similar meaning. For instance, the names of different countries have similar meaning, as do the names of different vegetables. To determine which nouns have similar meaning, large bodies of text can be examined. Nouns in the text that are separated by a conjunction (such as "or" or "and") or a comma, or that appear in lists, are assumed to have similar meaning. For example, examining books on agriculture, we can determine that nouns representing names of fruits, such as avocado, breadfruit, guava, mango, papaya, and soursop, have similar meaning. Researchers who took this approach using the British National Corpus, a collection of English texts with 100,000,000 words, produced a graph with close to 100,000 vertices, representing nouns, and 500,000 links, connecting vertices representing pairs of words with similar meaning. Figure 13 displays a small graph where the vertices represent nouns and edges connect words with similar meaning. This graph is centered around the word mouse. The graph illustrates that there are two distinct meanings for mouse. It can refer to an animal or it can refer to computer hardware. When a NLU program encounters the word mouse in a sentence, it can see which words with similar meaning would fit the sentence to help determine whether mouse refers to an animal or computer hardware in that sentence. ◀

TOURNAMENTS We now give some examples that show how graphs can also be used to model different kinds of tournaments.

EXAMPLE 14 **Round-Robin Tournaments** A tournament where each team plays every other team exactly once and no ties are allowed is called a **round-robin tournament**. Such tournaments can be modeled using directed graphs where each team is represented by a vertex. Note that (a, b) is an edge if team a beats team b. This graph is a simple directed graph, containing no loops or multiple directed edges (because no two teams play each other more than once). Such a directed graph model is presented in Figure 14. We see that Team 1 is undefeated in this tournament, and Team 3 is winless. ◀

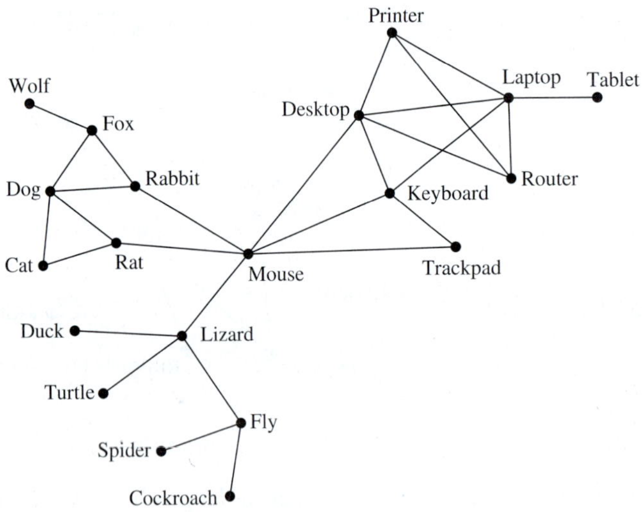

FIGURE 13 **A semantic network of nouns with similar meaning centered of the word mouse.**

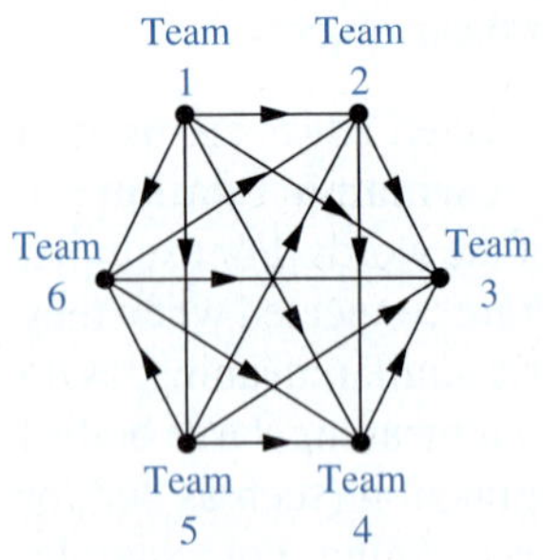

FIGURE 14 **A graph model of a round-robin tournament.**

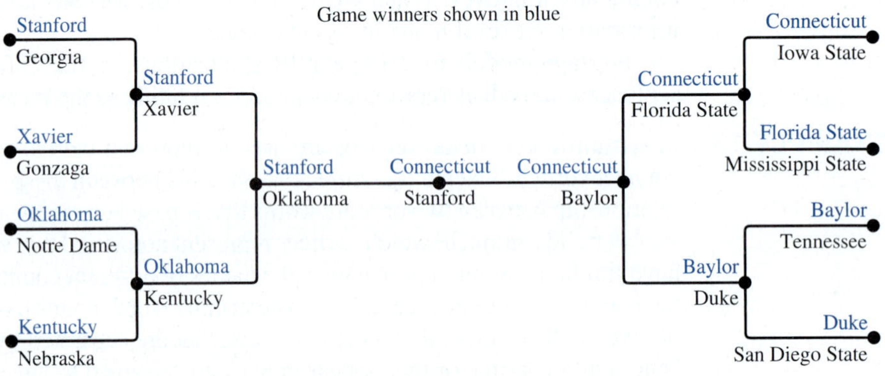

FIGURE 15 **A single-elimination tournament.**

EXAMPLE 15 **Single-Elimination Tournaments** A tournament where each contestant is eliminated after one loss is called a **single-elimination tournament**. Single-elimination tournaments are often used in sports, including tennis championships and the yearly NCAA basketball championship. We can model such a tournament using a vertex to represent each game and a directed edge to connect a game to the next game the winner of this game played in. The graph in Figure 15 represents the games played by the final 16 teams in the 2010 NCAA women's basketball tournament. ◀

Exercises

1. Draw graph models, stating the type of graph (from Table 1) used, to represent airline routes where every day there are four flights from Boston to Newark, two flights from Newark to Boston, three flights from Newark to Miami, two flights from Miami to Newark, one flight from Newark to Detroit, two flights from Detroit to Newark, three flights from Newark to Washington, two flights from Washington to Newark, and one flight from Washington to Miami, with

a) an edge between vertices representing cities that have a flight between them (in either direction).

b) an edge between vertices representing cities for each flight that operates between them (in either direction).

c) an edge between vertices representing cities for each flight that operates between them (in either direction), plus a loop for a special sightseeing trip that takes off and lands in Miami.

d) an edge from a vertex representing a city where a flight starts to the vertex representing the city where it ends.

e) an edge for each flight from a vertex representing a city where the flight begins to the vertex representing the city where the flight ends.

2. What kind of graph (from Table 1) can be used to model a highway system between major cities where

a) there is an edge between the vertices representing cities if there is an interstate highway between them?

b) there is an edge between the vertices representing cities for each interstate highway between them?

c) there is an edge between the vertices representing cities for each interstate highway between them, and there is a loop at the vertex representing a city if there is an interstate highway that circles this city?

For Exercises 3–9, determine whether the graph shown has directed or undirected edges, whether it has multiple edges, and whether it has one or more loops. Use your answers to determine the type of graph in Table 1 this graph is.

3.

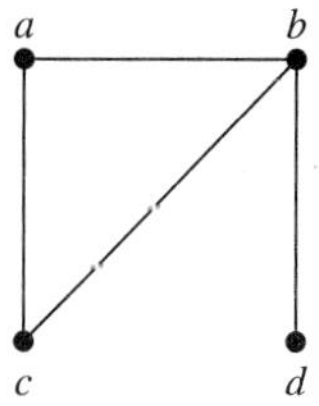

4.

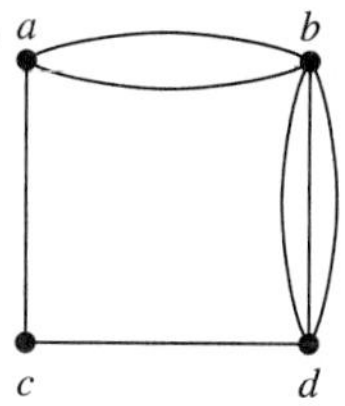

5.

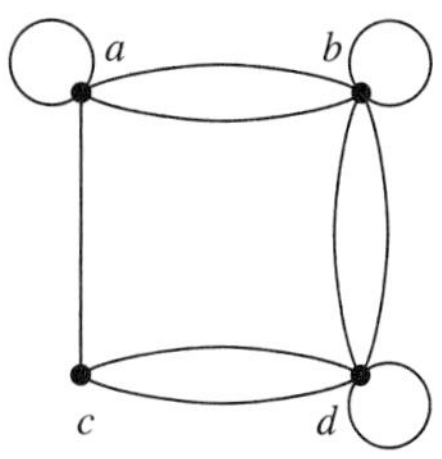

6.

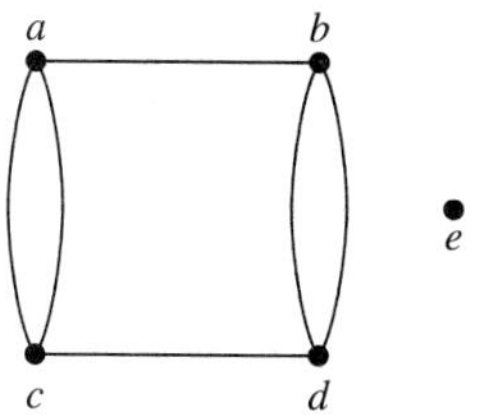

7.

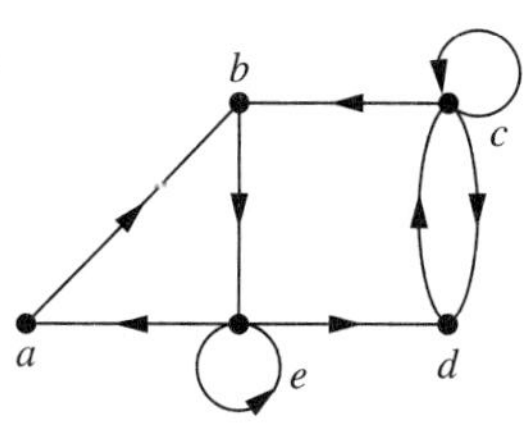

8.

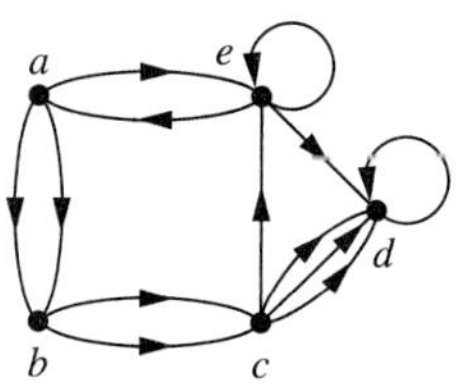

9.

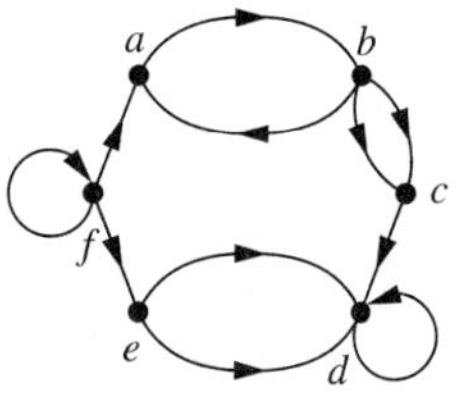

10. For each undirected graph in Exercises 3–9 that is not simple, find a set of edges to remove to make it simple.

11. Let G be a simple graph. Show that the relation R on the set of vertices of G such that uRv if and only if there is an edge associated to $\{u, v\}$ is a symmetric, irreflexive relation on G.

12. Let G be an undirected graph with a loop at every vertex. Show that the relation R on the set of vertices of G such that uRv if and only if there is an edge associated to $\{u, v\}$ is a symmetric, reflexive relation on G.

13. The **intersection graph** of a collection of sets A_1, $A_2, \ldots, A_n$ is the graph that has a vertex for each of these sets and has an edge connecting the vertices representing two sets if these sets have a nonempty intersection. Construct the intersection graph of these collections of sets.

a) $A_1 = \{0, 2, 4, 6, 8\}$, $A_2 = \{0, 1, 2, 3, 4\}$,
$A_3 = \{1, 3, 5, 7, 9\}$, $A_4 = \{5, 6, 7, 8, 9\}$,
$A_5 = \{0, 1, 8, 9\}$

b) $A_1 = \{\ldots, -4, -3, -2, -1, 0\}$,
$A_2 = \{\ldots, -2, -1, 0, 1, 2, \ldots\}$,
$A_3 = \{\ldots, -6, -4, -2, 0, 2, 4, 6, \ldots\}$,
$A_4 = \{\ldots, -5, -3, -1, 1, 3, 5, \ldots\}$,
$A_5 = \{\ldots, -6, -3, 0, 3, 6, \ldots\}$

c) $A_1 = \{x \mid x < 0\}$,
$A_2 = \{x \mid -1 < x < 0\}$,
$A_3 = \{x \mid 0 < x < 1\}$,
$A_4 = \{x \mid -1 < x < 1\}$,
$A_5 = \{x \mid x > -1\}$,
$A_6 = \mathbf{R}$

14. Use the niche overlap graph in Figure 11 to determine the species that compete with hawks.

15. Construct a niche overlap graph for six species of birds, where the hermit thrush competes with the robin and with the blue jay, the robin also competes with the mockingbird, the mockingbird also competes with the blue jay, and the nuthatch competes with the hairy woodpecker.

16. Draw the acquaintanceship graph that represents that Tom and Patricia, Tom and Hope, Tom and Sandy, Tom and Amy, Tom and Marika, Jeff and Patricia, Jeff and Mary, Patricia and Hope, Amy and Hope, and Amy and Marika know each other, but none of the other pairs of people listed know each other.

17. We can use a graph to represent whether two people were alive at the same time. Draw such a graph to represent whether each pair of the mathematicians and computer scientists with biographies in the first five chapters of this book who died before 1900 were contemporaneous. (Assume two people lived at the same time if they were alive during the same year.)

18. Who can influence Fred and whom can Fred influence in the influence graph in Example 2?

19. Construct an influence graph for the board members of a company if the President can influence the Director of Research and Development, the Director of Marketing, and the Director of Operations; the Director of Research and Development can influence the Director of Operations; the Director of Marketing can influence the Director of Operations; and no one can influence, or be influenced by, the Chief Financial Officer.

20. The word apple can refer to a plant, a food, or a computer company. Construct a word graph for these nouns: apple, strawberry, lenovo, cheese, chocolate, ibm, oak, microsoft, hedge, grass, cake, quiche, hp, cider, donut, azalea, pine, dell, fir, raspberry. Connect two vertices by an undirected edge if the nouns they represent have similar meaning.

21. The word rock can refer to a type of music or to something from a mountain. Construct a word graph for these nouns: rock, boulder, jazz, limestone, gravel, folk, bachata, pumice, granite, tango, klezmer, slate, shale, classical, pebbles, sand, rap, marble. Connect two vertices by an undirected edge if the nouns they represent have similar meaning.

22. Which other teams did Team 4 beat and which teams beat Team 4 in the round-robin tournament represented by the graph in Figure 14?

23. In a round-robin tournament the Tigers beat the Blue Jays, the Tigers beat the Cardinals, the Tigers beat the Orioles, the Blue Jays beat the Cardinals, the Blue Jays beat the Orioles, and the Cardinals beat the Orioles. Model this outcome with a directed graph.

24. Construct the call graph for a set of seven telephone numbers 555-0011, 555-1221, 555-1333, 555-8888, 555-2222, 555-0091, and 555-1200 if there were three calls from 555-0011 to 555-8888 and two calls from 555-8888 to 555-0011, two calls from 555-2222 to 555-0091, two calls from 555-1221 to each of the other numbers, and one call from 555-1333 to each of 555-0011, 555-1221, and 555-1200.

25. Explain how the two telephone call graphs for calls made during the month of January and calls made during the month of February can be used to determine the new telephone numbers of people who have changed their telephone numbers.

26. **a)** Explain how graphs can be used to model electronic mail messages in a network. Should the edges be directed or undirected? Should multiple edges be allowed? Should loops be allowed?

b) Describe a graph that models the electronic mail sent in a network in a particular week.

27. How can a graph that models email messages sent in a network be used to find people who have recently started using a new email address?

28. How can a graph that models email messages sent in a network be used to find electronic mail mailing lists used to send the same message to many different email addresses?

29. Describe a graph model that represents whether each person at a party knows the name of each other person at the party. Should the edges be directed or undirected? Should multiple edges be allowed? Should loops be allowed?

30. Describe a graph model that represents a subway system in a large city. Should edges be directed or undirected? Should multiple edges be allowed? Should loops be allowed?

31. For each course at a university, there may be one or more other courses that are its prerequisites. How can a graph be used to model these courses and which courses are prerequisites for which courses? Should edges be directed or undirected? Looking at the graph model, how can we find courses that do not have any prerequisites and how can we find courses that are not the prerequisite for any other courses?

32. Describe a graph model that represents the positive recommendations of movie critics, using vertices to represent both these critics and all movies that are currently being shown.

33. Describe a graph model that represents marriage between people, allowing for polygamy, polyandry, and same gender marriage. When restricted to marriages between one man and one woman, does this graph have any special properties?

34. Which statements must be executed before S_6 is executed in the program in Example 8? (Use the precedence graph in Figure 10.)

35. Construct a precedence graph for the following program:

S_1: $x := 0$
S_2: $x := x + 1$
S_3: $y := 2$
S_4: $z := y$
S_5: $x := x + 2$
S_6: $y := x + z$
S_7: $z := 4$

36. Describe a discrete structure based on a graph that can be used to model airline routes and their flight times. [*Hint:* Add structure to a directed graph.]

37. Describe a discrete structure based on a graph that can be used to model relationships between pairs of individuals in a group, where each individual may either like, dislike, or be neutral about another individual, and the reverse relationship may be different. [*Hint:* Add structure to a directed graph. Treat separately the edges in opposite directions between vertices representing two individuals.]

38. Describe a graph model that can be used to represent all forms of electronic communication between two people in a single graph. What kind of graph is needed?

10.2 Graph Terminology and Special Types of Graphs

10.2.1 Introduction

We introduce some of the basic vocabulary of graph theory in this section. We will use this vocabulary later in this chapter when we solve many different types of problems. One such problem involves determining whether a graph can be drawn in the plane so that no two of its edges cross. Another example is deciding whether there is a one-to-one correspondence between the vertices of two graphs that produces a one-to-one correspondence between the edges of the graphs. We will also introduce several important families of graphs often used as examples and in models. Several important applications will be described where these special types of graphs arise.

Links

10.2.2 Basic Terminology

First, we give some terminology that describes the vertices and edges of undirected graphs.

Definition 1 Two vertices u and v in an undirected graph G are called *adjacent* (or *neighbors*) in G if u and v are endpoints of an edge e of G. Such an edge e is called *incident with* the vertices u and v and e is said to *connect* u and v.

We will also find useful terminology describing the set of vertices adjacent to a particular vertex of a graph.

Definition 2 The set of all neighbors of a vertex v of $G = (V, E)$, denoted by $N(v)$, is called the *neighborhood* of v. If A is a subset of V, we denote by $N(A)$ the set of all vertices in G that are adjacent to at least one vertex in A. So, $N(A) = \bigcup_{v \in A} N(v)$.

To keep track of how many edges are incident to a vertex, we make the following definition.

Definition 3 The *degree of a vertex in an undirected graph* is the number of edges incident with it, except that a loop at a vertex contributes twice to the degree of that vertex. The degree of the vertex v is denoted by $\deg(v)$.

EXAMPLE 1 What are the degrees and what are the neighborhoods of the vertices in the graphs G and H displayed in Figure 1?

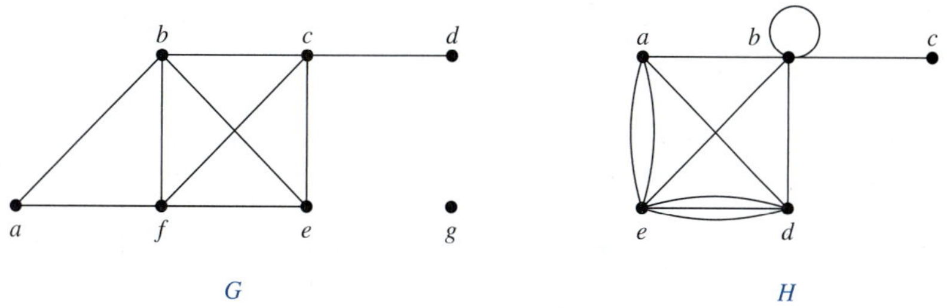

FIGURE 1 **The undirected graphs G and H.**

Solution: In G, $\deg(a) = 2$, $\deg(b) = \deg(c) = \deg(f) = 4$, $\deg(d) = 1$, $\deg(e) = 3$, and $\deg(g) = 0$. The neighborhoods of these vertices are $N(a) = \{b, f\}$, $N(b) = \{a, c, e, f\}$, $N(c) = \{b, d, e, f\}$, $N(d) = \{c\}$, $N(e) = \{b, c, f\}$, $N(f) = \{a, b, c, e\}$, and $N(g) = \emptyset$. In H, $\deg(a) = 4$, $\deg(b) = \deg(e) = 6$, $\deg(c) = 1$, and $\deg(d) = 5$. The neighborhoods of these vertices are $N(a) = \{b, d, e\}$, $N(b) = \{a, b, c, d, e\}$, $N(c) = \{b\}$, $N(d) = \{a, b, e\}$, and $N(e) = \{a, b, d\}$. ◀

A vertex of degree zero is called **isolated**. It follows that an isolated vertex is not adjacent to any vertex. Vertex g in graph G in Example 1 is isolated. A vertex is **pendant** if and only if it has degree one. Consequently, a pendant vertex is adjacent to exactly one other vertex. Vertex d in graph G in Example 1 is pendant.

Examining the degrees of vertices in a graph model can provide useful information about the model, as Example 2 shows.

EXAMPLE 2 What does the degree of a vertex in a niche overlap graph (introduced in Example 11 in Section 10.1) represent? Which vertices in this graph are pendant and which are isolated? Use the niche overlap graph shown in Figure 11 of Section 10.1 to interpret your answers.

Solution: There is an edge between two vertices in a niche overlap graph if and only if the two species represented by these vertices compete. Hence, the degree of a vertex in a niche overlap graph is the number of species in the ecosystem that compete with the species represented by this vertex. A vertex is pendant if the species competes with exactly one other species in the ecosystem. Finally, the vertex representing a species is isolated if this species does not compete with any other species in the ecosystem.

For instance, the degree of the vertex representing the squirrel in the niche overlap graph in Figure 11 in Section 10.1 is four, because the squirrel competes with four other species: the crow, the opossum, the raccoon, and the woodpecker. In this niche overlap graph, the mouse is the only species represented by a pendant vertex, because the mouse competes only with the shrew and all other species compete with at least two other species. There are no isolated vertices in the graph in this niche overlap graph because every species in this ecosystem competes with at least one other species. ◀

What do we get when we add the degrees of all the vertices of a graph $G = (V, E)$? Each edge contributes two to the sum of the degrees of the vertices because an edge is incident with exactly two (possibly equal) vertices. This means that the sum of the degrees of the vertices is twice the number of edges. We have the result in Theorem 1, which is sometimes called the handshaking theorem (and is also often known as the handshaking lemma), because of the analogy between an edge having two endpoints and a handshake involving two hands. (Exercise 6 is based on this analogy.)

THEOREM 1 **THE HANDSHAKING THEOREM** Let $G = (V, E)$ be an undirected graph with m edges. Then

$$2m = \sum_{v \in V} \deg(v).$$

(Note that this applies even if multiple edges and loops are present.)

EXAMPLE 3 How many edges are there in a graph with 10 vertices each of degree six?

Solution: Because the sum of the degrees of the vertices is $6 \cdot 10 = 60$, it follows that $2m = 60$ where m is the number of edges. Therefore, $m = 30$. ◀

Theorem 1 shows that the sum of the degrees of the vertices of an undirected graph is even. This simple fact has many consequences, one of which is given as Theorem 2.

THEOREM 2 An undirected graph has an even number of vertices of odd degree.

Proof: Let V_1 and V_2 be the set of vertices of even degree and the set of vertices of odd degree, respectively, in an undirected graph $G = (V, E)$ with m edges. Then

$$2m = \sum_{v \in V} \deg(v) = \sum_{v \in V_1} \deg(v) + \sum_{v \in V_2} \deg(v).$$

Because $\deg(v)$ is even for $v \in V_1$, the first term in the right-hand side of the last equality is even. Furthermore, the sum of the two terms on the right-hand side of the last equality is even, because this sum is $2m$. Hence, the second term in the sum is also even. Because all the terms in this sum are odd, there must be an even number of such terms. Thus, there are an even number of vertices of odd degree. ◁

Terminology for graphs with directed edges reflects the fact that edges in directed graphs have directions.

Definition 4 When (u, v) is an edge of the graph G with directed edges, u is said to be *adjacent to* v and v is said to be *adjacent from* u. The vertex u is called the *initial vertex* of (u, v), and v is called the *terminal* or *end vertex* of (u, v). The initial vertex and terminal vertex of a loop are the same.

Because the edges in graphs with directed edges are ordered pairs, the definition of the degree of a vertex can be refined to reflect the number of edges with this vertex as the initial vertex and as the terminal vertex.

Definition 5 In a graph with directed edges the *in-degree of a vertex* v, denoted by $\deg^-(v)$, is the number of edges with v as their terminal vertex. The *out-degree of* v, denoted by $\deg^+(v)$, is the number of edges with v as their initial vertex. (Note that a loop at a vertex contributes 1 to both the in-degree and the out-degree of this vertex.)

EXAMPLE 4 Find the in-degree and out-degree of each vertex in the graph G with directed edges shown in Figure 2.

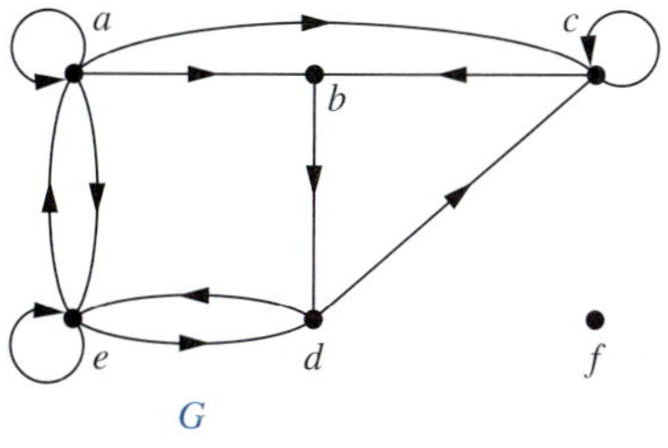

FIGURE 2 **The directed graph *G*.**

Solution: The in-degrees in G are $\deg^-(a) = 2$, $\deg^-(b) = 2$, $\deg^-(c) = 3$, $\deg^-(d) = 2$, $\deg^-(e) = 3$, and $\deg^-(f) = 0$. The out-degrees are $\deg^+(a) = 4$, $\deg^+(b) = 1$, $\deg^+(c) = 2$, $\deg^+(d) = 2$, $\deg^+(e) = 3$, and $\deg^+(f) = 0$. ◀

Because each edge has an initial vertex and a terminal vertex, the sum of the in-degrees and the sum of the out-degrees of all vertices in a graph with directed edges are the same. Both of these sums are the number of edges in the graph. This result is stated as Theorem 3.

THEOREM 3 Let $G = (V, E)$ be a graph with directed edges. Then

$$\sum_{v\in V} \deg^-(v) = \sum_{v\in V} \deg^+(v) = |E|.$$

There are many properties of a graph with directed edges that do not depend on the direction of its edges. Consequently, it is often useful to ignore these directions. The undirected graph that results from ignoring directions of edges is called the **underlying undirected graph**. A graph with directed edges and its underlying undirected graph have the same number of edges.

10.2.3 Some Special Simple Graphs

We will now introduce several classes of simple graphs. These graphs are often used as examples and arise in many applications.

EXAMPLE 5 **Complete Graphs** A **complete graph on n vertices**, denoted by K_n, is a simple graph that contains exactly one edge between each pair of distinct vertices. The graphs K_n, for $n = 1, 2, 3, 4, 5, 6$, are displayed in Figure 3. A simple graph for which there is at least one pair of distinct vertex not connected by an edge is called **noncomplete**. ◀

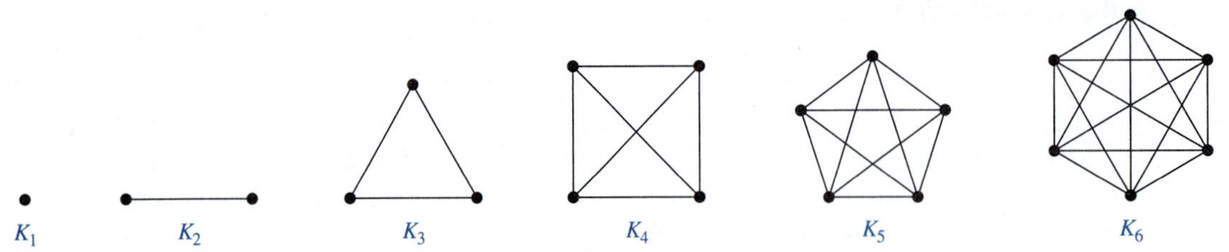

FIGURE 3 **The graphs K_n for $1 \leq n \leq 6$.**

EXAMPLE 6 **Cycles** A **cycle** C_n, $n \geq 3$, consists of n vertices $v_1, v_2, \ldots, v_n$ and edges $\{v_1, v_2\}, \{v_2, v_3\}, \ldots, \{v_{n-1}, v_n\}$, and $\{v_n, v_1\}$. The cycles C_3, C_4, C_5, and C_6 are displayed in Figure 4. ◀

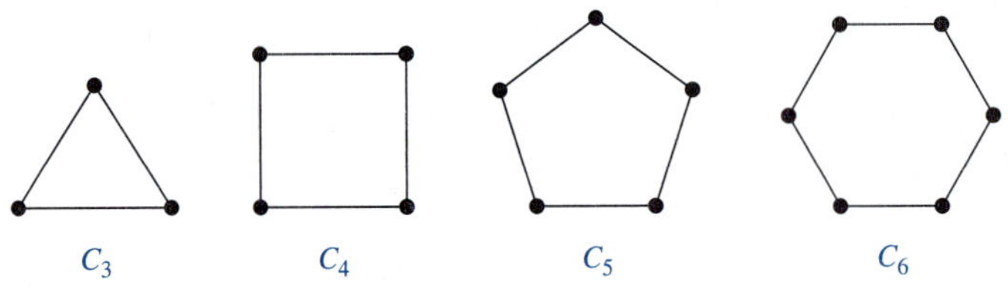

FIGURE 4 **The cycles C_3, C_4, C_5, and C_6.**

EXAMPLE 7 **Wheels** We obtain a **wheel W_n** when we add an additional vertex to a cycle C_n, for $n \geq 3$, and connect this new vertex to each of the n vertices in C_n, by new edges. The wheels W_3, W_4, W_5, and W_6 are displayed in Figure 5. ◀

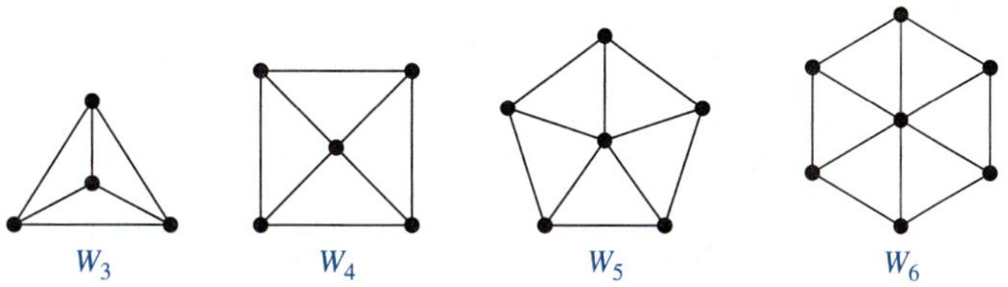

FIGURE 5 **The wheels W_3, W_4, W_5, and W_6.**

EXAMPLE 8 ***n*-Cubes** An ***n*-dimensional hypercube**, or ***n*-cube**, denoted by Q_n, is a graph that has vertices representing the 2^n bit strings of length n. Two vertices are adjacent if and only if the bit strings that they represent differ in exactly one bit position. We display Q_1, Q_2, and Q_3 in Figure 6.

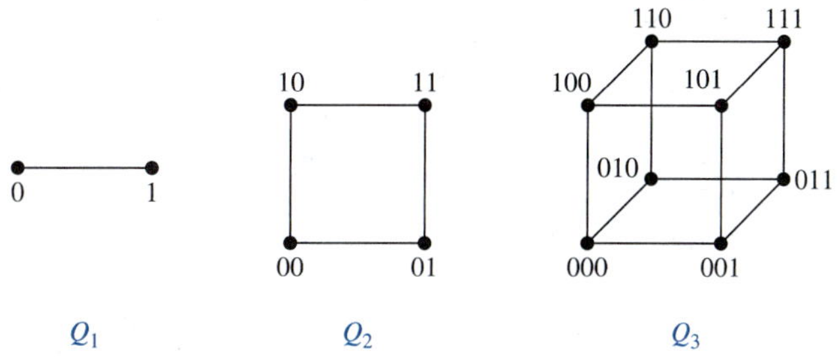

FIGURE 6 **The *n*-cube Q_n, $n = 1, 2, 3$.**

Note that you can construct the $(n + 1)$-cube Q_{n+1} from the n-cube Q_n by making two copies of Q_n, prefacing the labels on the vertices with a 0 in one copy of Q_n and with a 1 in the other copy of Q_n, and adding edges connecting two vertices that have labels differing only in the first bit. In Figure 6, Q_3 is constructed from Q_2 by drawing two copies of Q_2 as the top and bottom faces of Q_3, adding 0 at the beginning of the label of each vertex in the bottom face and 1 at the beginning of the label of each vertex in the top face. (Here, by *face* we mean a face of a cube in three-dimensional space. Think of drawing the graph Q_3 in three-dimensional space with copies of Q_2 as the top and bottom faces of a cube and then drawing the projection of the resulting depiction in the plane.) ◀

10.2.4 Bipartite Graphs

Sometimes a graph has the property that its vertex set can be divided into two disjoint subsets such that each edge connects a vertex in one of these subsets to a vertex in the other subset. For example, consider the graph representing cities served by airlines, where each airline and each city is represented by a vertex, and each edge connects an airline with a city it serves. In this graph, each edge connects a vertex in the subset of vertices representing airlines and a vertex in the subset of vertices representing cities. This leads us to Definition 6.

Links ❯

Definition 6 A simple graph G is called *bipartite* if its vertex set V can be partitioned into two disjoint sets V_1 and V_2 such that every edge in the graph connects a vertex in V_1 and a vertex in V_2 (so that no edge in G connects either two vertices in V_1 or two vertices in V_2). When this condition holds, we call the pair (V_1, V_2) a *bipartition* of the vertex set V of G.

In Example 9 we will show that C_6 is bipartite, and in Example 10 we will show that K_3 is not bipartite.

EXAMPLE 9 C_6 is bipartite, as shown in Figure 7, because its vertex set can be partitioned into the two sets $V_1 = \{v_1, v_3, v_5\}$ and $V_2 = \{v_2, v_4, v_6\}$, and every edge of C_6 connects a vertex in V_1 and a vertex in V_2. ◀

EXAMPLE 10 K_3 is not bipartite. To verify this, note that if we divide the vertex set of K_3 into two disjoint sets, one of the two sets must contain two vertices. If the graph were bipartite, these two vertices could not be connected by an edge, but in K_3 each vertex is connected to every other vertex by an edge. ◀

EXAMPLE 11 Are the graphs G and H displayed in Figure 8 bipartite?

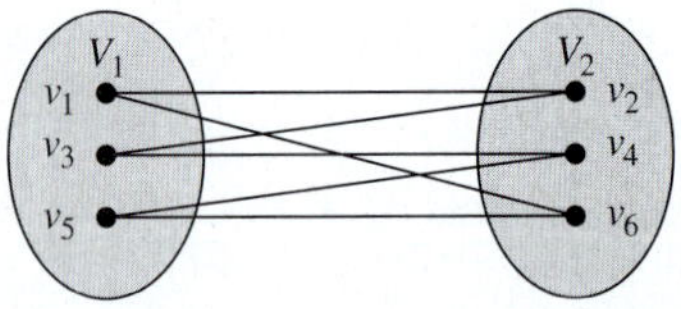

FIGURE 7 **Showing that C_6 is bipartite.**

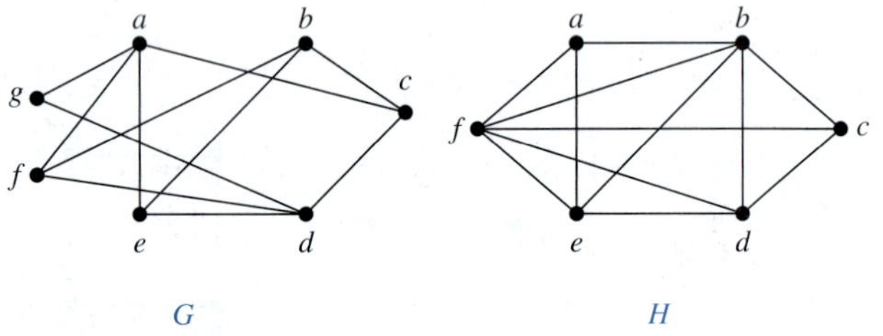

FIGURE 8 **The undirected graphs G and H.**

Solution: Graph G is bipartite because its vertex set is the union of two disjoint sets, $\{a, b, d\}$ and $\{c, e, f, g\}$, and each edge connects a vertex in one of these subsets to a vertex in the other subset. (Note that for G to be bipartite it is not necessary that every vertex in $\{a, b, d\}$ be adjacent to every vertex in $\{c, e, f, g\}$. For instance, b and g are not adjacent.)

Graph H is not bipartite because its vertex set cannot be partitioned into two subsets so that edges do not connect two vertices from the same subset. (The reader should verify this by considering the vertices a, b, and f.) ◀

Theorem 4 provides a useful criterion for determining whether a graph is bipartite.

THEOREM 4 A simple graph is bipartite if and only if it is possible to assign one of two different colors to each vertex of the graph so that no two adjacent vertices are assigned the same color.

Proof: First, suppose that $G = (V, E)$ is a bipartite simple graph. Then $V = V_1 \cup V_2$, where V_1 and V_2 are disjoint sets and every edge in E connects a vertex in V_1 and a vertex in V_2. If we assign one color to each vertex in V_1 and a second color to each vertex in V_2, then no two adjacent vertices are assigned the same color.

Now suppose that it is possible to assign colors to the vertices of the graph using just two colors so that no two adjacent vertices are assigned the same color. Let V_1 be the set of vertices assigned one color and V_2 be the set of vertices assigned the other color. Then, V_1 and V_2 are disjoint and $V = V_1 \cup V_2$. Furthermore, every edge connects a vertex in V_1 and a vertex in V_2 because no two adjacent vertices are either both in V_1 or both in V_2. Consequently, G is bipartite. ◁

We illustrate how Theorem 4 can be used to determine whether a graph is bipartite in Example 12.

EXAMPLE 12 Use Theorem 4 to determine whether the graphs in Example 11 are bipartite.

Solution: We first consider the graph G. We will try to assign one of two colors, say red and blue, to each vertex in G so that no edge in G connects a red vertex and a blue vertex. Without loss of generality we begin by arbitrarily assigning red to a. Then, we must assign blue to c, e, f, and g, because each of these vertices is adjacent to a. To avoid having an edge with two blue endpoints, we must assign red to all the vertices adjacent to either c, e, f, or g. This means that we must assign red to both b and d (and means that a must be assigned red, which it already has been). We have now assigned colors to all vertices, with a, b, and d red and c, e, f, and g blue. Checking all edges, we see that every edge connects a red vertex and a blue vertex. Hence, by Theorem 4 the graph G is bipartite.

Next, we will try to assign either red or blue to each vertex in H so that no edge in H connects a red vertex and a blue vertex. Without loss of generality we arbitrarily assign red to a. Then, we must assign blue to b, e, and f, because each is adjacent to a. But this is not possible because e and f are adjacent, so both cannot be assigned blue. This argument shows that we cannot assign one of two colors to each of the vertices of H so that no adjacent vertices are assigned the same color. It follows by Theorem 4 that H is not bipartite. ◀

Theorem 4 is an example of a result in the part of graph theory known as graph colorings. Graph colorings is an important part of graph theory with important applications. We will study graph colorings further in Section 10.8.

Another useful criterion for determining whether a graph is bipartite is based on the notion of a path, a topic we study in Section 10.4. A graph is bipartite if and only if it is not possible to start at a vertex and return to this vertex by traversing an odd number of distinct edges. We will make this notion more precise when we discuss paths and circuits in graphs in Section 10.4 (see Exercise 63 in that section).

EXAMPLE 13 **Complete Bipartite Graphs** A **complete bipartite graph $K_{m,n}$** is a graph that has its vertex set partitioned into two subsets of m and n vertices, respectively with an edge between two vertices if and only if one vertex is in the first subset and the other vertex is in the second subset. The complete bipartite graphs $K_{2,3}$, $K_{3,3}$, $K_{3,5}$, and $K_{2,6}$ are displayed in Figure 9. ◀

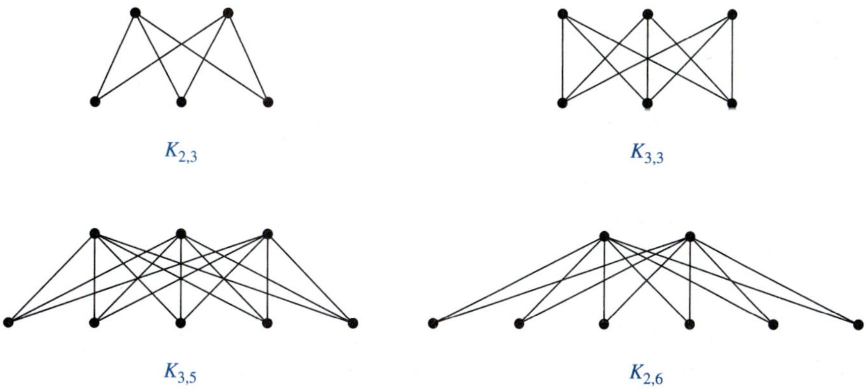

FIGURE 9 **Some complete bipartite graphs.**

10.2.5 Bipartite Graphs and Matchings

Bipartite graphs can be used to model many types of applications that involve matching the elements of one set to elements of another, as Example 14 illustrates.

EXAMPLE 14 **Job Assignments** Suppose that there are m employees in a group and n different jobs that need to be done, where $m \geq n$. Each employee is trained to do one or more of these n jobs. We would like to assign an employee to each job. To help with this task, we can use a graph to model employee capabilities. We represent each employee by a vertex and each job by a vertex. For each employee, we include an edge from that employee to all jobs that the employee has been trained to do. Note that the vertex set of this graph can be partitioned into two disjoint sets, the set of employees and the set of jobs, and each edge connects an employee to a job. Consequently, this graph is bipartite, where the bipartition is (E, J) where E is the set of employees and J is the set of jobs. We now consider two different scenarios.

First, suppose that a group has four employees: Alvarez, Berkowitz, Chen, and Davis; and suppose that four jobs need to be done to complete Project 1: requirements, architecture, implementation, and testing. Suppose that Alvarez has been trained to do requirements and testing; Berkowitz has been trained to do architecture, implementation, and testing; Chen has been trained to do requirements, architecture, and implementation; and Davis has only been trained to do requirements. We model these employee capabilities using the bipartite graph in Figure 10(a).

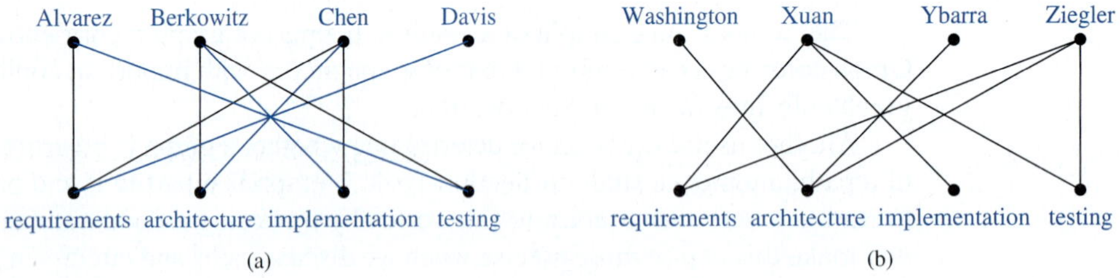

FIGURE 10 **Modeling the jobs for which employees have been trained.**

Second, suppose that a second group also has four employees: Washington, Xuan, Ybarra, and Ziegler; and suppose that the same four jobs need to be done to complete Project 2 as are needed to complete Project 1. Suppose that Washington has been trained to do architecture; Xuan has been trained to do requirements, implementation, and testing; Ybarra has been trained to do architecture; and Ziegler has been trained to do requirements, architecture and testing. We model these employee capabilities using the bipartite graph in Figure 10(b).

To complete Project 1, we must assign an employee to each job so that every job has an employee assigned to it, and so that no employee is assigned more than one job. We can do this by assigning Alvarez to testing, Berkowitz to implementation, Chen to architecture, and Davis to requirements, as shown in Figure 10(a) (where blue lines show this assignment of jobs).

To complete Project 2, we must also assign an employee to each job so that every job has an employee assigned to it and no employee is assigned more than one job. However, this is impossible because there are only two employees, Xuan and Ziegler, who have been trained for at least one of the three jobs of requirements, implementation, and testing. Consequently, there is no way to assign three different employees to these three jobs so that each job is assigned an employee with the appropriate training. ◀

Finding an assignment of jobs to employees can be thought of as finding a matching in the graph model, where a **matching** M in a simple graph $G = (V, E)$ is a subset of the set E of edges of the graph such that no two edges are incident with the same vertex. In other words, a matching is a subset of edges such that if $\{s, t\}$ and $\{u, v\}$ are distinct edges of the matching, then s, t, u, and v are distinct. A vertex that is the endpoint of an edge of a matching M is said to be **matched** in M; otherwise it is said to be **unmatched**. A **maximum matching** is a matching with the largest number of edges. We say that a matching M in a bipartite graph $G = (V, E)$ with bipartition (V_1, V_2) is a **complete matching from** V_1 **to** V_2 if every vertex in V_1 is the endpoint of an edge in the matching, or equivalently, if $|M| = |V_1|$. For example, to assign jobs to employees so that the largest number of jobs are assigned employees, we seek a maximum matching in the graph that models employee capabilities. To assign employees to all jobs we seek a complete matching from the set of jobs to the set of employees. In Example 14, we found a complete matching from the set of jobs to the set of employees for Project 1, and this matching is a maximum matching, and we showed that no complete matching exists from the set of jobs to the employees for Project 2.

We now give an example of how matchings can be used to model marriages.

EXAMPLE 15 **Marriages on an Island** Suppose that there are m men and n women on an island. Each person has a list of members of the opposite gender acceptable as a spouse. We construct a bipartite graph $G = (V_1, V_2)$ where V_1 is the set of men and V_2 is the set of women so that there is an edge between a man and a woman if they find each other acceptable as a spouse. A matching in this graph consists of a set of edges, where each pair of endpoints of an edge is a husband-wife pair. A maximum matching is a largest possible set of married couples, and a complete matching of V_1 is a set of married couples where every man is married, but possibly not all women. ◀

Hall's marriage theorem is an example of a theorem where obvious necessary conditions are sufficient too.

NECESSARY AND SUFFICIENT CONDITIONS FOR COMPLETE MATCHINGS We now turn our attention to the question of determining whether a complete matching from V_1 to V_2 exists when (V_1, V_2) is a bipartition of a bipartite graph $G = (V, E)$. We will introduce a theorem that provides a set of necessary and sufficient conditions for the existence of a complete matching. This theorem was proved by Philip Hall in 1935.

THEOREM 5 **HALL'S MARRIAGE THEOREM** The bipartite graph $G = (V, E)$ with bipartition (V_1, V_2) has a complete matching from V_1 to V_2 if and only if $|N(A)| \geq |A|$ for all subsets A of V_1.

Proof: We first prove the *only if* part of the theorem. To do so, suppose that there is a complete matching M from V_1 to V_2. Then, if $A \subseteq V_1$, for every vertex $v \in A$, there is an edge in M connecting v to a vertex in V_2. Consequently, there are at least as many vertices in V_2 that are neighbors of vertices in V_1 as there are vertices in V_1. It follows that $|N(A)| \geq |A|$.

Links

Courtesy of the Edinburgh Mathematical Society

PHILIP HALL (1904–1982) Philip Hall grew up in London, where his mother was a dressmaker. He won a scholarship for board school reserved for needy children, and later a scholarship to King's College of Cambridge University. He received his bachelor's degree in 1925. In 1926, unsure of his career goals, he took a civil service exam, but decided to continue his studies at Cambridge after failing.

In 1927 Hall was elected to a fellowship at King's College; soon after, he made his first important discovery in group theory. The results he proved are now known as Hall's theorems. In 1933 he was appointed as a Lecturer at Cambridge, where he remained until 1941. During World War II he worked as a cryptographer at Bletchley Park breaking Italian and Japanese codes. At the end of the war, Hall returned to King's College, and was soon promoted. In 1953 he was appointed to the Sadleirian Chair. His work during the 1950s proved to be extremely influential to the rapid development of group theory during the 1960s.

Hall loved poetry and recited it beautifully in Italian and Japanese, as well as English. He was interested in art, music, and botany. He was quite shy and disliked large groups of people. Hall had an incredibly broad and varied knowledge, and was respected for his integrity, intellectual standards, and judgement. He was beloved by his students.

To prove the *if* part of the theorem, the more difficult part, we need to show that if $|N(A)| \geq |A|$ for all $A \subseteq V_1$, then there is a complete matching M from V_1 to V_2. We will use strong induction on $|V_1|$ to prove this.

Basis step: If $|V_1| = 1$, then V_1 contains a single vertex v_0. Because $|N(\{v_0\})| \geq |\{v_0\}| = 1$, there is at least one edge connecting v_0 and a vertex $w_0 \in V_2$. Any such edge forms a complete matching from V_1 to V_2.

Inductive step: We first state the inductive hypothesis.

Inductive hypothesis: Let k be a positive integer. If $G = (V, E)$ is a bipartite graph with bipartition (V_1, V_2), and $|V_1| = j \leq k$, then there is a complete matching M from V_1 to V_2 whenever the condition that $|N(A)| \geq |A|$ for all $A \subseteq V_1$ is met.

Now suppose that $H = (W, F)$ is a bipartite graph with bipartition (W_1, W_2) and $|W_1| = k + 1$. We will prove that the inductive holds using a proof by cases, using two cases. Case *(i)* applies when for all integers j with $1 \leq j \leq k$, the vertices in every set of j elements from W_1 are adjacent to at least $j + 1$ elements of W_2. Case *(ii)* applies when for some j with $1 \leq j \leq k$ there is a subset W_1' of j vertices such that there are exactly j neighbors of these vertices in W_2. Because either Case *(i)* or Case *(ii)* holds, we need only consider these cases to complete the inductive step.

Case (i): Suppose that for all integers j with $1 \leq j \leq k$, the vertices in every subset of j elements from W_1 are adjacent to at least $j + 1$ elements of W_2. Then, we select a vertex $v \in W_1$ and an element $w \in N(\{v\})$, which must exist by our assumption that $|N(\{v\}| \geq |\{v\}| = 1$. We delete v and w and all edges incident to them from H. This produces a bipartite graph H' with bipartition $(W_1 - \{v\}, W_2 - \{w\})$. Because $|W_1 - \{v\}| = k$, the inductive hypothesis tells us there is a complete matching from $W_1 - \{v\}$ to $W_2 - \{w\}$. Adding the edge from v to w to this complete matching produces a complete matching from W_1 to W_2.

Case (ii): Suppose that for some j with $1 \leq j \leq k$, there is a subset W_1' of j vertices such that there are exactly j neighbors of these vertices in W_2. Let W_2' be the set of these neighbors. Then, by the inductive hypothesis there is a complete matching from W_1' to W_2'. Remove these $2j$ vertices from W_1 and W_2 and all incident edges to produce a bipartite graph K with bipartition $(W_1 - W_1', W_2 - W_2')$.

We will show that the graph K satisfies the condition $|N(A)| \geq |A|$ for all subsets A of $W_1 - W_1'$. If not, there would be a subset of t vertices of $W_1 - W_1'$ where $1 \leq t \leq k + 1 - j$ such that the vertices in this subset have fewer than t vertices of $W_2 - W_2'$ as neighbors. Then, the set of $j + t$ vertices of W_1 consisting of these t vertices together with the j vertices we removed from W_1 has fewer than $j + t$ neighbors in W_2, contradicting the hypothesis that $|N(A)| \geq |A|$ for all $A \subseteq W_1$.

Hence, by the inductive hypothesis, the graph K has a complete matching. Combining this complete matching with the complete matching from W_1' to W_2', we obtain a complete matching from W_1 to W_2.

We have shown that in both cases there is a complete matching from W_1 to W_2. This completes the inductive step and completes the proof. ◁

We have used strong induction to prove Hall's marriage theorem. Although our proof is elegant, it does have some drawbacks. In particular, we cannot construct an algorithm based on this proof that finds a complete matching in a bipartite graph. For a constructive proof that can be used as the basis of an algorithm, see [Gi85].

10.2.6 Some Applications of Special Types of Graphs

We conclude this section by introducing some additional graph models that involve the special types of graph we have discussed in this section.

EXAMPLE 16 **Local Area Networks (LANs) and Wireless Local Area Networks (WLANs)** The different devices used by a group of people in an organization, including mobile phones, PCs, network servers, printers, scanners, and so on, can be connected using a *local area network* with physical connections or a *wireless local area network* where connections are made using radio waves. Some of these networks are based on a *star topology*, where all devices are connected to a central control device. A local area network can be represented using a complete bipartite graph $K_{1,n}$, as shown in Figure 11(a). Messages are sent from device to device through the central control device.

Links

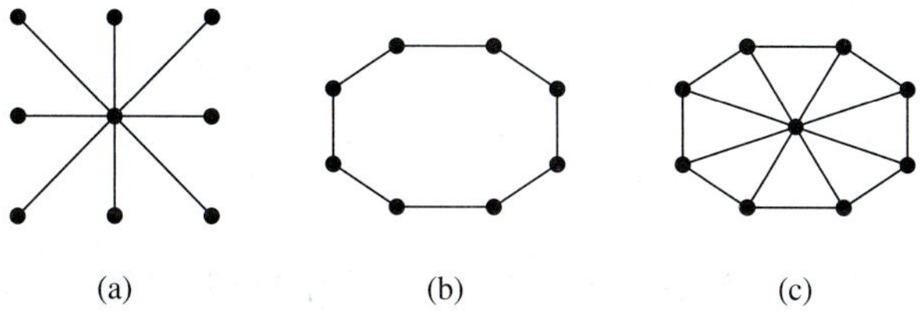

FIGURE 11 **Star, ring, and hybrid topologies for local area networks.**

Other local area networks are based on a *ring topology*, where each device is connected to exactly two others. Local area networks with a ring topology are modeled using n-cycles, C_n, as shown in Figure 11(b). Messages are sent from device to device around the cycle until the intended recipient of a message is reached.

Finally, some local area networks use a hybrid of these two topologies. Messages may be sent around the ring, or through a central device. This redundancy makes the network more reliable. Local area networks with this redundancy can be modeled using wheels W_n, as shown in Figure 11(c). ◀

EXAMPLE 17 **Interconnection Networks for Parallel Computation** For many years, computers executed programs one operation at a time. Consequently, the algorithms written to solve problems were designed to perform one step at a time; such algorithms are called **serial**. (Almost all algorithms described in this book are serial.) However, many computationally intense problems, such as weather simulations, medical imaging, and cryptanalysis, cannot be solved in a reasonable amount of time using serial operations, even on a supercomputer. Furthermore, there is a physical limit to how fast a computer can carry out basic operations, so there will always be problems that cannot be solved in a reasonable length of time using serial operations.

Parallel processing, which uses computers made up of many separate processors, each with its own memory, helps overcome the limitations of computers with a single processor. **Parallel algorithms**, which break a problem into a number of subproblems that can be solved concurrently, can then be devised to rapidly solve problems using a computer with multiple processors. In a parallel algorithm, a single instruction stream controls the execution of the algorithm, sending subproblems to different processors, and directs the input and output of these subproblems to the appropriate processors.

When parallel processing is used, one processor may need output generated by another processor. Consequently, these processors need to be interconnected. We can use the appropriate type of graph to represent the interconnection network of the processors in a computer with multiple processors. In the following discussion, we will describe the most commonly used types of interconnection networks for parallel processors. The type of interconnection network used to implement a particular parallel algorithm depends on the requirements for exchange of data between processors, the desired speed, and, of course, the available hardware.

The simplest, but most expensive, network-interconnecting processors include a two-way link between each pair of processors. This network can be represented by K_n, the complete graph on n vertices, when there are n processors. However, there are serious problems with this type of interconnection network because the required number of connections is so large. In reality, the number of direct connections to a processor is limited, so when there are a large number of

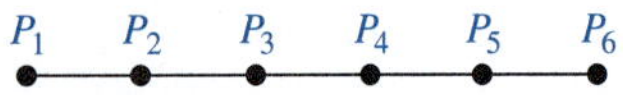

FIGURE 12 **A linear array for six processors.**

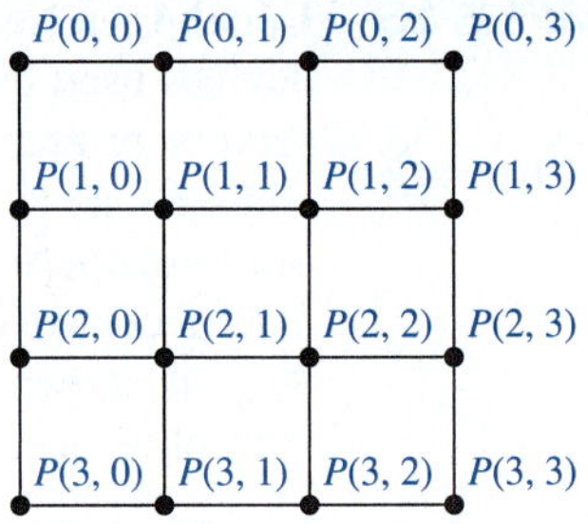

FIGURE 13 **A mesh network for 16 processors.**

processors, a processor cannot be linked directly to all others. For example, when there are 64 processors, $C(64, 2) = 2016$ connections would be required, and each processor would have to be directly connected to 63 others.

On the other hand, perhaps the simplest way to interconnect n processors is to use an arrangement known as a **linear array**. Each processor P_i, other than P_1 and P_n, is connected to its neighbors P_{i-1} and P_{i+1} via a two-way link. P_1 is connected only to P_2, and P_n is connected only to P_{n-1}. The linear array for six processors is shown in Figure 12. The advantage of a linear array is that each processor has at most two direct connections to other processors. The disadvantage is that it is sometimes necessary to use a large number of intermediate links, called **hops**, for processors to share information.

The **mesh network** (or **two-dimensional array**) is a commonly used interconnection network. In such a network, the number of processors is a perfect square, say $n = m^2$. The n processors are labeled $P(i, j)$, $0 \leq i \leq m - 1$, $0 \leq j \leq m - 1$. Two-way links connect processor $P(i, j)$ with its four neighbors, processors $P(i \pm 1, j)$ and $P(i, j \pm 1)$, as long as these are processors in the mesh. (Note that four processors, on the corners of the mesh, have only two adjacent processors, and other processors on the boundaries have only three neighbors. Sometimes a variant of a mesh network in which every processor has exactly four connections is used; see Exercise 78.) The mesh network limits the number of links for each processor. Communication between some pairs of processors requires $O(\sqrt{n}) = O(m)$ intermediate links. (See Exercise 79.) The graph representing the mesh network for 16 processors is shown in Figure 13.

One important type of interconnection network is the hypercube. For such a network, the number of processors is a power of 2, $n = 2^m$. The n processors are labeled $P_0, P_1, \ldots, P_{n-1}$. Each processor has two-way connections to m other processors. Processor P_i is linked to the processors with indices whose binary representations differ from the binary representation of i in exactly one bit. The hypercube network balances the number of direct connections for each processor and the number of intermediate connections required so that processors can communicate. Many computers have been built using a hypercube network, and many parallel algorithms have been devised that use a hypercube network. The graph Q_m, the m-cube, represents the hypercube network with $n = 2^m$ processors. Figure 14 displays the hypercube network for eight processors. (Figure 14 displays a different way to draw Q_3 than was shown in Figure 6.) ◀

10.2.7 New Graphs from Old

Sometimes we need only part of a graph to solve a problem. For instance, we may care only about the part of a large computer network that involves the computer centers in New York, Denver, Detroit, and Atlanta. Then we can ignore the other computer centers and all telephone lines not linking two of these specific four computer centers. In the graph model for the large

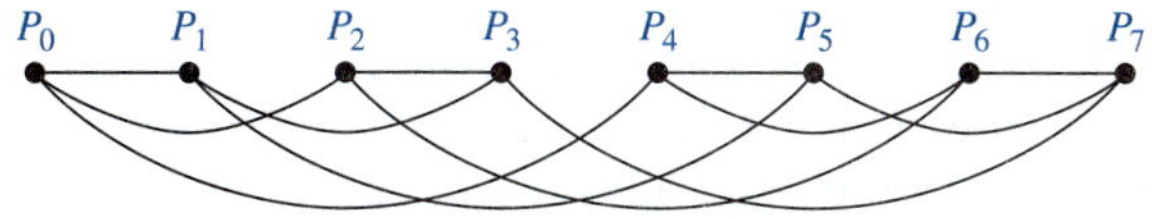

FIGURE 14 A hypercube network for eight processors.

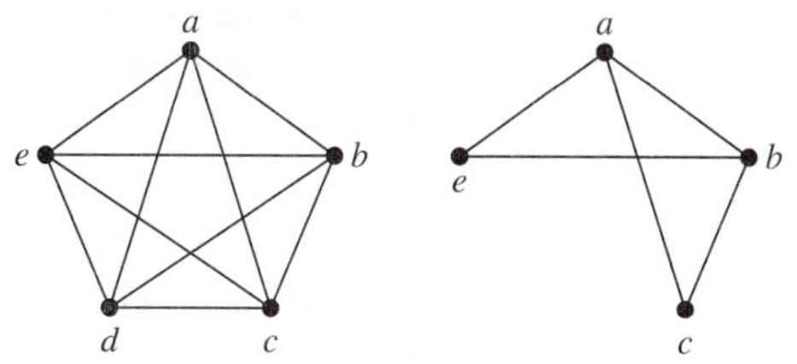

FIGURE 15 A subgraph of K_5.

network, we can remove the vertices corresponding to the computer centers other than the four of interest, and we can remove all edges incident with a vertex that was removed. When edges and vertices are removed from a graph, without removing endpoints of any remaining edges, a smaller graph is obtained. Such a graph is called a **subgraph** of the original graph.

Definition 7 A *subgraph of a graph* $G = (V, E)$ is a graph $H = (W, F)$, where $W \subseteq V$ and $F \subseteq E$. A subgraph H of G is a *proper subgraph* of G if $H \neq G$.

Given a set of vertices of a graph, we can form a subgraph of this graph with these vertices and the edges of the graph that connect them.

Definition 8 Let $G = (V, E)$ be a simple graph. The **subgraph induced** by a subset W of the vertex set V is the graph (W, F), where the edge set F contains an edge in E if and only if both endpoints of this edge are in W.

EXAMPLE 18 The graph G shown in Figure 15 is a subgraph of K_5. If we add the edge connecting c and e to G, we obtain the subgraph induced by $W = \{a, b, c, e\}$. ◀

REMOVING OR ADDING EDGES OF A GRAPH Given a graph $G = (V, E)$ and an edge $e \in E$, we can produce a subgraph of G by removing the edge e. The resulting subgraph, denoted by $G - e$, has the same vertex set V as G. Its edge set is $E - \{e\}$. Hence,

$$G - e = (V, E - \{e\}).$$

Similarly, if E' is a subset of E, we can produce a subgraph of G by removing the edges in E' from the graph. The resulting subgraph has the same vertex set V as G. Its edge set is $E - E'$.

We can also add an edge e to a graph to produce a new larger graph when this edge connects two vertices already in G. We denote by $G + e$ the new graph produced by adding a new edge e, connecting two previously non-adjacent vertices, to the graph G. Hence,

$$G + e = (V, E \cup \{e\}).$$

The vertex set of $G + e$ is the same as the vertex set of G and the edge set is the union of the edge set of G and the set $\{e\}$. (See Example 19 for examples of removing an edge from a graph and adding an edge to a graph.)

EDGE CONTRACTIONS Sometimes when we remove an edge from a graph, we do not want to retain the endpoints of this edge as separate vertices in the resulting subgraph. In such a case we perform an **edge contraction**, which removes an edge e with endpoints u and v and merges u

and v into a new single vertex w, and for each edge with u or v as an endpoint replaces the edge with one with w as endpoint in place of u or v and with the same second endpoint. Hence, the contraction of the edge e with endpoints u and v in the graph $G = (V, E)$ produces a new graph $G' = (V', E')$ (which is not a subgraph of G), where $V' = V - \{u, v\} \cup \{w\}$ and E' contains the edges in E which do not have either u or v as endpoints and an edge connecting w to every neighbor of either u or v in V. For example, the contraction of the edge connecting the vertices e and c in the graph G_1 in Figure 16 produces a new graph G_1' with vertices a, b, d, and w. As in G_1, there is an edge in G_1' connecting a and b and an edge connecting a and d. There also is an edge in G_1' that connects b and w that replaces the edges connecting b and c and connecting b and e in G_1 and an edge in G_1' that connects d and w replacing the edge connecting d and e in G_1. (Also see Example 19 for an example of the contraction of an edge in a graph.)

REMOVING VERTICES FROM A GRAPH When we remove a vertex v and all edges incident to it from $G = (V, E)$, we produce a subgraph, denoted by $G - v$. Observe that $G - v = (V - \{v\}, E')$, where E' is the set of edges of G not incident to v. Similarly, if V' is a subset of V, then the graph $G - V'$ is the subgraph $(V - V', E')$, where E' is the set of edges of G not incident to a vertex in V'. (See Example 19 for an example of the removal of a vertex from a graph.)

EXAMPLE 19 Figure 16 displays an undirected graph G with four different graphs that are the result of different operations on G. These are:

(a) $G - \{b, c\}$, constructed from G by removing the edge $\{b, c\}$
(b) $G + \{e, d\}$, constructed from G by adding the edge $\{e, d\}$
(c) the contraction of G, constructed from G by replacing the edge $\{b, c\}$ with a new vertex f, and replacing the edges $\{c, d\}$, $\{a, b\}$, $\{b, e\}$, and $\{c, e\}$ with the new edges $\{a, f\}$, $\{f, d\}$, and $\{f, e\}$
(d) $G - c$, constructed from G by removing the vertex c and the edges $\{b, c\}$, $\{c, d\}$ and $\{c, e\}$ ◀

GRAPH UNIONS Two or more graphs can be combined in various ways. The new graph that contains all the vertices and edges of these graphs is called the **union** of the graphs. We will give a more formal definition for the union of two simple graphs.

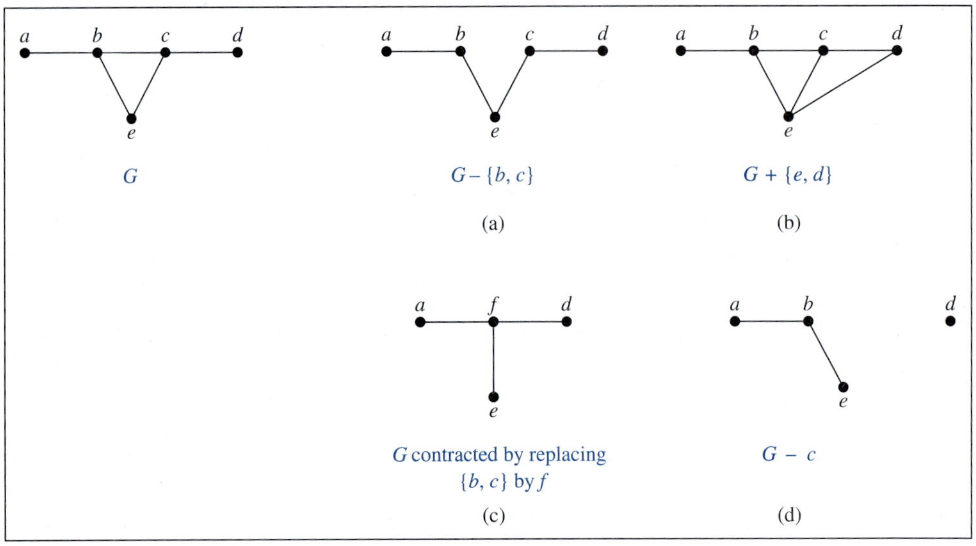

FIGURE 16 The graph G and four graphs resulting from different operations on G.

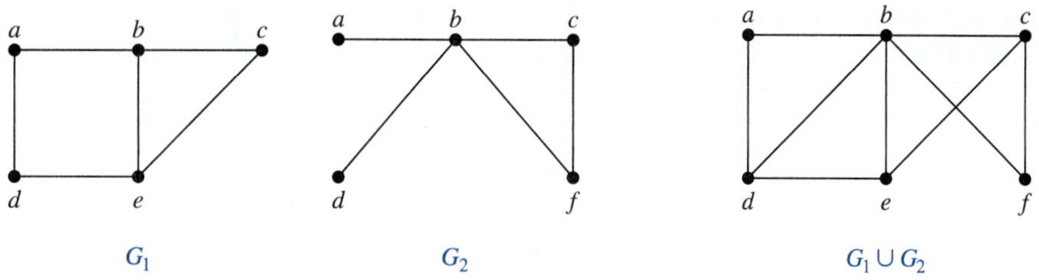

(a) (b)

FIGURE 17 **(a) The simple graphs G_1 and G_2. (b) Their union $G_1 \cup G_2$.**

Definition 9 The *union* of two simple graphs $G_1 = (V_1, E_1)$ and $G_2 = (V_2, E_2)$ is the simple graph with vertex set $V_1 \cup V_2$ and edge set $E_1 \cup E_2$. The union of G_1 and G_2 is denoted by $G_1 \cup G_2$.

EXAMPLE 20 Find the union of the graphs G_1 and G_2 shown in Figure 17(a).

Solution: The vertex set of the union $G_1 \cup G_2$ is the union of the two vertex sets, namely, $\{a, b, c, d, e, f\}$. The edge set of the union is the union of the two edge sets. The union is displayed in Figure 17(b). ◀

Exercises

In Exercises 1–3 find the number of vertices, the number of edges, and the degree of each vertex in the given undirected graph. Identify all isolated and pendant vertices.

1.

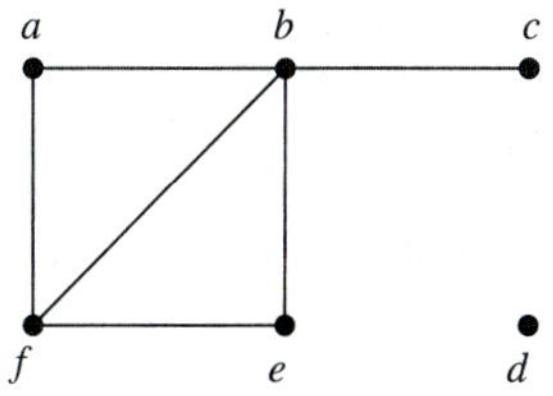

2.

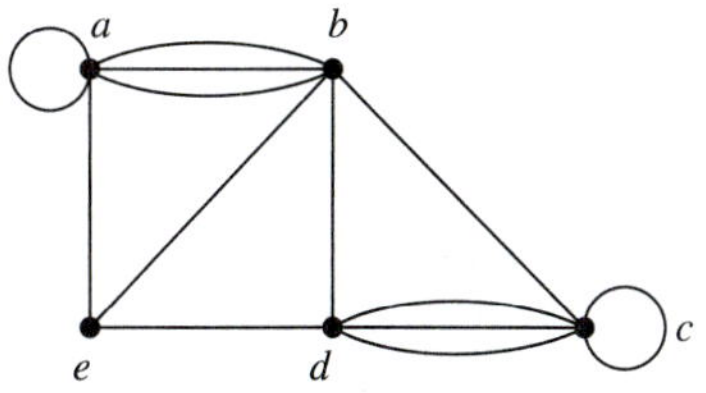

3.

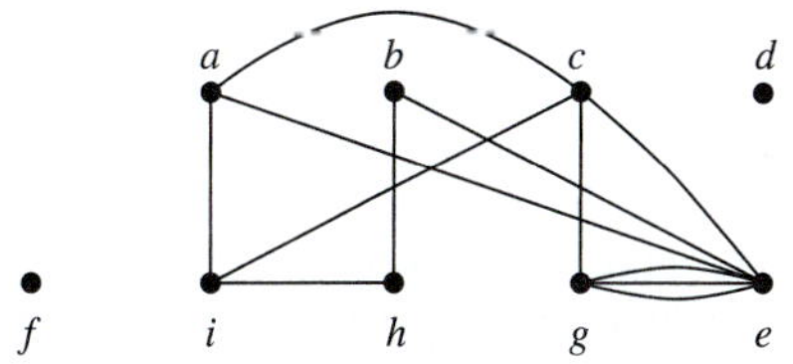

4. Find the sum of the degrees of the vertices of each graph in Exercises 1–3 and verify that it equals twice the number of edges in the graph.

5. Can a simple graph exist with 15 vertices each of degree five?

6. Show that the sum, over the set of people at a party, of the number of people a person has shaken hands with, is even. Assume that no one shakes their own hand.

In Exercises 7–9 determine the number of vertices and edges and find the in-degree and out-degree of each vertex for the given directed multigraph.

7.

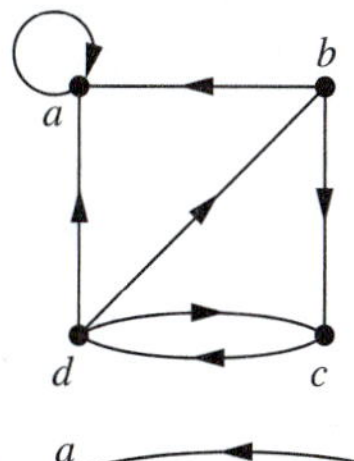

8.

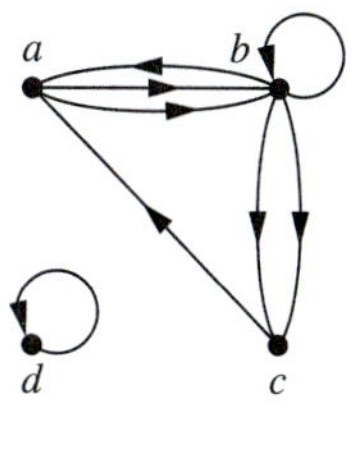

9.

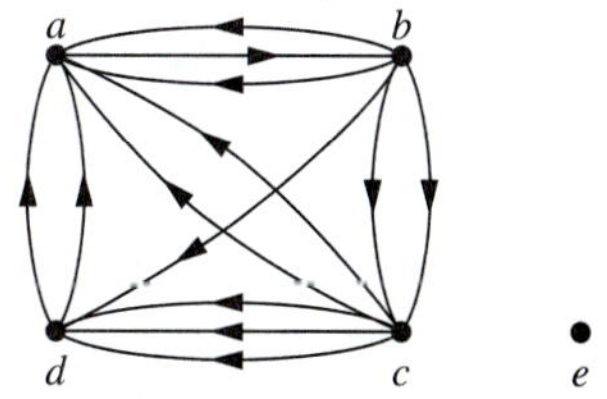

10. For each of the graphs in Exercises 7–9 determine the sum of the in-degrees of the vertices and the sum of the out-degrees of the vertices directly. Show that they are both equal to the number of edges in the graph.

11. Construct the underlying undirected graph for the graph with directed edges in Figure 2.

12. What does the degree of a vertex represent in the acquaintanceship graph, where vertices represent all the people in the world? What does the neighborhood of a vertex in this graph represent? What do isolated and pendant vertices in this graph represent? In one study it was estimated that the average degree of a vertex in this graph is 1000. What does this mean in terms of the model?

13. What does the degree of a vertex represent in an academic collaboration graph? What does the neighborhood of a vertex represent? What do isolated and pendant vertices represent?

14. What does the degree of a vertex in the Hollywood graph represent? What does the neighborhood of a vertex represent? What do the isolated and pendant vertices represent?

15. What do the in-degree and the out-degree of a vertex in a telephone call graph, as described in Example 4 of Section 10.1, represent? What does the degree of a vertex in the undirected version of this graph represent?

16. What do the in-degree and the out-degree of a vertex in the web graph, as described in Example 5 of Section 10.1, represent?

17. What do the in-degree and the out-degree of a vertex in a directed graph modeling a round-robin tournament represent?

18. Show that in a simple graph with at least two vertices there must be two vertices that have the same degree.

19. Use Exercise 18 to show that in a group of people, there must be two people who are friends with the same number of other people in the group.

20. Draw these graphs.

a) K_7 **b)** $K_{1,8}$ **c)** $K_{4,4}$
d) C_7 **e)** W_7 **f)** Q_4

In Exercises 21–25 determine whether the graph is bipartite. You may find it useful to apply Theorem 4 and answer the question by determining whether it is possible to assign either red or blue to each vertex so that no two adjacent vertices are assigned the same color.

21.
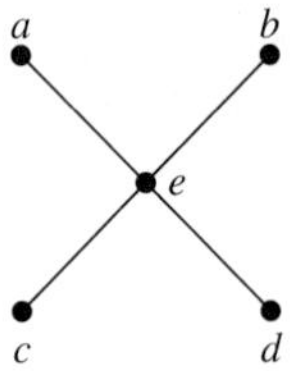

22.
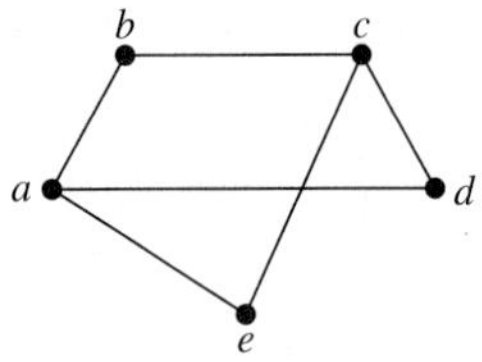

23.
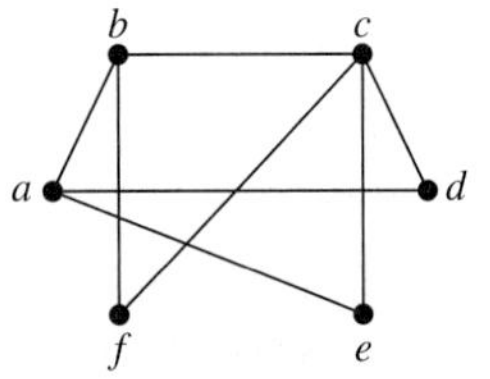

24.
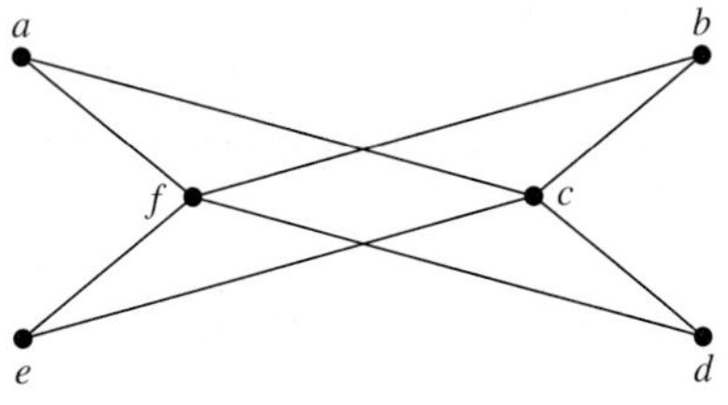

25.
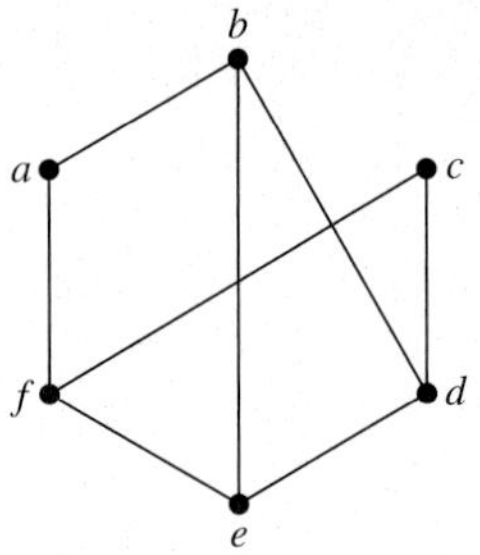

26. For which values of n are these graphs bipartite?

a) K_n **b)** C_n **c)** W_n **d)** Q_n

27. Suppose that there are four employees in the computer support group of the School of Engineering of a large university. Each employee will be assigned to support one of four different areas: hardware, software, networking, and wireless. Suppose that Ping is qualified to support hardware, networking, and wireless; Quiggley is qualified to support software and networking; Ruiz is qualified to support networking and wireless, and Sitea is qualified to support hardware and software.

a) Use a bipartite graph to model the four employees and their qualifications.

b) Use Hall's theorem to determine whether there is an assignment of employees to support areas so that each employee is assigned one area to support.

c) If an assignment of employees to support areas so that each employee is assigned to one support area exists, find one.

28. Suppose that a new company has five employees: Zamora, Agraharam, Smith, Chou, and Macintyre. Each employee will assume one of six responsiblities: planning, publicity, sales, marketing, development, and industry relations. Each employee is capable of doing one or more of these jobs: Zamora could do planning, sales, marketing, or industry relations; Agraharam could do planning or development; Smith could do publicity, sales, or industry relations; Chou could do planning, sales, or industry relations; and Macintyre could do planning, publicity, sales, or industry relations.

a) Model the capabilities of these employees using a bipartite graph.

b) Find an assignment of responsibilites such that each employee is assigned one responsibility.

c) Is the matching of responsibilities you found in part (b) a complete matching? Is it a maximum matching?

29. Suppose that there are five young women and five young men on an island. Each man is willing to marry some of the women on the island and each woman is willing to marry any man who is willing to marry her. Suppose that Sandeep is willing to marry Tina and Vandana; Barry is willing to marry Tina, Xia, and Uma; Teja is willing to marry Tina and Zelda; Anil is willing to marry Vandana and Zelda; and Emilio is willing to marry Tina and Zelda. Use Hall's theorem to show there is no matching of the young men and young women on the island such that each young man is matched with a young woman he is willing to marry.

30. Suppose that there are five young women and six young men on an island. Each woman is willing to marry some of the men on the island and each man is willing to marry any woman who is willing to marry him. Suppose that Anna is willing to marry Jason, Larry, and Matt; Barbara is willing to marry Kevin and Larry; Carol is willing to marry Jason, Nick, and Oscar; Diane is willing to marry Jason, Larry, Nick, and Oscar; and Elizabeth is willing to marry Jason and Matt.

a) Model the possible marriages on the island using a bipartite graph.
b) Find a matching of the young women and the young men on the island such that each young woman is matched with a young man whom she is willing to marry.
c) Is the matching you found in part (b) a complete matching? Is it a maximum matching?

Each of Exercises 31–33 can be solved using Hall's theorem.

***31.** Suppose there is an integer k such that every man on a desert island is willing to marry exactly k of the women on the island and every woman on the island is willing to marry exactly k of the men. Also, suppose that a man is willing to marry a woman if and only if she is willing to marry him. Show that it is possible to match the men and women on the island so that everyone is matched with someone that they are willing to marry.

***32.** Suppose that $2n$ tennis players compete in a round-robin tournament. Every player has exactly one match with every other player during $2n - 1$ consecutive days. Every match has a winner and a loser. Show that it is possible to select a winning player each day without selecting the same player twice.

†33. Suppose that m people are selected as prize winners in a lottery, where each winner can select two prizes from a collection of different prizes. Show if there are $2m$ prizes that every winner wants, then every winner is able to select two prizes that they want.

***34.** In this exercise we prove a theorem of Øystein Ore. Suppose that $G = (V, E)$ is a bipartite graph with bipartition (V_1, V_2) and that $A \subseteq V_1$. Show that the maximum number of vertices of V_1 that are the endpoints of a matching of G equals $|V_1| - \max_{A \subseteq V_1} \text{def}(A)$, where $\text{def}(A) = |A| - |N(A)|$. (Here, def($A$) is called the **deficiency** of A.) [*Hint:* Form a larger graph by adding $\max_{A \subseteq V_1} \text{def}(A)$ new vertices to V_2 and connect all of them to the vertices of V_1.]

35. For the graph G in Exercise 1 find

a) the subgraph induced by the vertices a, b, c, and f.
b) the new graph G_1 obtained from G by contracting the edge connecting b and f, replacing the vertices b and f with a new single vertex x.

36. Let n be a positive integer. Show that a subgraph induced by a nonempty subset of the vertex set of K_n is a complete graph.

37. How many vertices and how many edges do these graphs have?

a) K_n **b)** C_n **c)** W_n
d) $K_{m,n}$ **e)** Q_n

The **density of an undirected graph** is the ratio of the number of edges in the graph to the number of unordered pairs of vertices, and the **density of a directed graph** is the ratio of the number of directed edges in the graph to the number of ordered pairs of distinct vertices.

38. Give a formula for the density of an undirected graph in terms of the number of its edges and the number of its vertices.

39. Give a formula for the density of a directed graph in terms of the number of its edges and the number of its vertices.

40. Find the density of each of these graphs.

a) K_{10} **b)** $K_{5,6}$ **c)** C_9
d) W_8 **e)** Q_6

41. Find formulas for the density of the graphs in each of these families of graphs.

a) K_n **b)** $K_{m,n}$ **c)** C_n
d) W_n **e)** Q_n

The **degree sequence** of a graph is the sequence of the degrees of the vertices of the graph in nonincreasing order. For example, the degree sequence of the graph G in Example 1 is 4, 4, 4, 3, 2, 1, 0.

42. Find the degree sequences for each of the graphs in Exercises 21–25.

43. Find the degree sequence of each of the following graphs.

a) K_4 **b)** C_4 **c)** W_4
d) $K_{2,3}$ **e)** Q_3

44. What is the degree sequence of the bipartite graph $K_{m,n}$ where m and n are positive integers? Explain your answer.

45. What is the degree sequence of K_n, where n is a positive integer? Explain your answer.

46. How many edges does a graph have if its degree sequence is 4, 3, 3, 2, 2? Draw such a graph.

47. How many edges does a graph have if its degree sequence is 5, 2, 2, 2, 2, 1? Draw such a graph.

A sequence $d_1, d_2, \ldots, d_n$ is called **graphic** if it is the degree sequence of a simple graph.

48. Determine whether each of these sequences is graphic. For those that are, draw a graph having the given degree sequence.

a) 5, 4, 3, 2, 1, 0 **b)** 6, 5, 4, 3, 2, 1 **c)** 2, 2, 2, 2, 2, 2
d) 3, 3, 3, 2, 2, 2 **e)** 3, 3, 2, 2, 2, 2 **f)** 1, 1, 1, 1, 1, 1
g) 5, 3, 3, 3, 3, 3 **h)** 5, 5, 4, 3, 2, 1

49. Determine whether each of these sequences is graphic. For those that are, draw a graph having the given degree sequence.

a) 3, 3, 3, 3, 2 **b)** 5, 4, 3, 2, 1 **c)** 4, 4, 3, 2, 1
d) 4, 4, 3, 3, 3 **e)** 3, 2, 2, 1, 0 **f)** 1, 1, 1, 1, 1

***50.** Suppose that $d_1, d_2, \ldots, d_n$ is a graphic sequence. Show that there is a simple graph with vertices $v_1, v_2, \ldots, v_n$ such that $\deg(v_i) = d_i$ for $i = 1, 2, \ldots, n$ and v_1 is adjacent to $v_2, \ldots, v_{d_1+1}$.

***51.** Show that a sequence $d_1, d_2, \ldots, d_n$ of nonnegative integers in nonincreasing order is a graphic sequence if and only if the sequence obtained by reordering the terms of the sequence $d_2 - 1, \ldots, d_{d_1+1} - 1, d_{d_1+2}, \ldots, d_n$ so that the terms are in nonincreasing order is a graphic sequence.

***52.** Use Exercise 51 to construct a recursive algorithm for determining whether a nonincreasing sequence of positive integers is graphic.

53. Show that every nonincreasing sequence of nonnegative integers with an even sum of its terms is the degree sequence of a pseudograph, that is, an undirected graph where loops are allowed. [*Hint:* Construct such a graph by first adding as many loops as possible at each vertex. Then add additional edges connecting vertices of odd degree. Explain why this construction works.]

54. How many subgraphs with at least one vertex does K_2 have?

55. How many subgraphs with at least one vertex does K_3 have?

56. How many subgraphs with at least one vertex does W_3 have?

57. Draw all subgraphs of this graph.

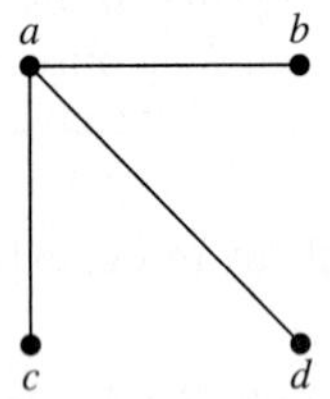

58. Let G be a graph with v vertices and e edges. Let M be the maximum degree of the vertices of G, and let m be the minimum degree of the vertices of G. Show that

a) $2e/v \geq m$. **b)** $2e/v \leq M$.

A simple graph is called **regular** if every vertex of this graph has the same degree. A regular graph is called n-**regular** if every vertex in this graph has degree n.

59. For which values of n are these graphs regular?

a) K_n **b)** C_n **c)** W_n **d)** Q_n

60. For which values of m and n is $K_{m,n}$ regular?

61. How many vertices does a regular graph of degree four with 10 edges have?

In Exercises 62–64 find the union of the given pair of simple graphs. (Assume edges with the same endpoints are the same.)

62.

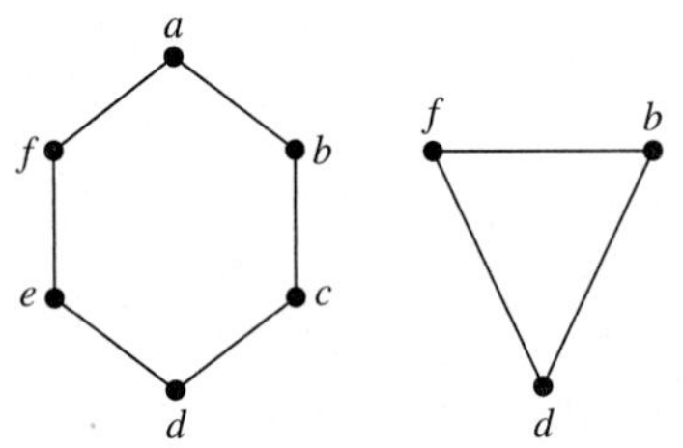

63.

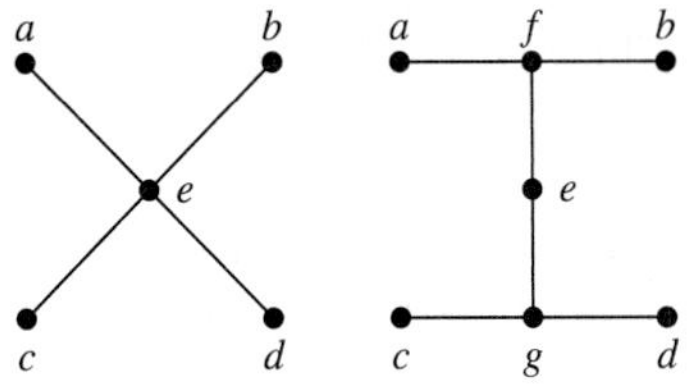

64.

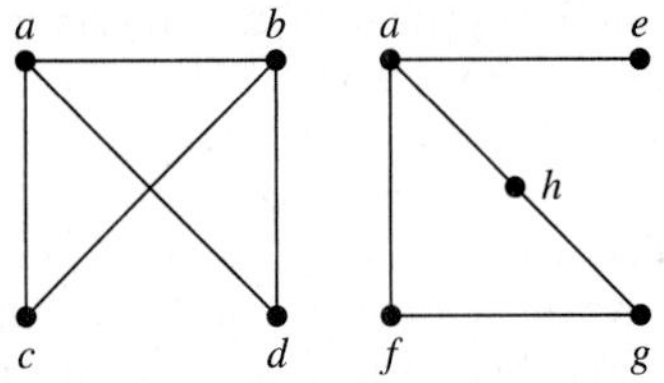

65. The **complementary graph** $\overline{G}$ of a simple graph G has the same vertices as G. Two vertices are adjacent in $\overline{G}$ if and only if they are not adjacent in G. Describe each of these graphs.

a) $\overline{K_n}$ **b)** $\overline{K_{m,n}}$ **c)** $\overline{C_n}$ **d)** $\overline{Q_n}$

66. If G is a simple graph with 15 edges and $\overline{G}$ has 13 edges, how many vertices does G have?

67. If the simple graph G has v vertices and e edges, how many edges does $\overline{G}$ have?

68. If the degree sequence of the simple graph G is 4, 3, 3, 2, 2, what is the degree sequence of $\overline{G}$?

69. If the degree sequence of the simple graph G is $d_1, d_2, \ldots, d_n$, what is the degree sequence of $\overline{G}$?

***70.** Show that if G is a bipartite simple graph with v vertices and e edges, then $e \leq v^2/4$.

71. Show that if G is a simple graph with n vertices, then the union of G and $\overline{G}$ is K_n.

***72.** Describe an algorithm to decide whether a graph is bipartite based on the fact that a graph is bipartite if and only if it is possible to color its vertices two different colors so that no two vertices of the same color are adjacent.

The **converse** of a directed graph $G = (V, E)$, denoted by G^{conv}, is the directed graph (V, F), where the set F of edges of G^{conv} is obtained by reversing the direction of each edge in E.

73. Draw the converse of each of the graphs in Exercises 7–9 in Section 10.1.

74. Show that $(G^{conv})^{conv} = G$ whenever G is a directed graph.

75. Show that the graph G is its own converse if and only if the relation associated with G (see Section 9.3) is symmetric.

76. Show that if a bipartite graph $G = (V, E)$ is n-regular for some positive integer n (see the preamble to Exercise 59) and (V_1, V_2) is a bipartition of V, then $|V_1| = |V_2|$. That is, show that the two sets in a bipartition of the vertex set of an n-regular graph must contain the same number of vertices.

77. Draw the mesh network for interconnecting nine parallel processors.

78. In a variant of a mesh network for interconnecting $n = m^2$ processors, processor $P(i, j)$ is connected to the four processors $P((i \pm 1) \textbf{ mod } m, j)$ and $P(i, (j \pm 1) \textbf{ mod } m)$, so that connections wrap around the edges of the mesh. Draw this variant of the mesh network for 16 processors.

79. Show that every pair of processors in a mesh network of $n = m^2$ processors can communicate using $O(\sqrt{n}) = O(m)$ hops between directly connected processors.

10.3 Representing Graphs and Graph Isomorphism

10.3.1 Introduction

There are many useful ways to represent graphs. As we will see throughout this chapter, in working with a graph it is helpful to be able to choose its most convenient representation. In this section we will show how to represent graphs in several different ways.

Sometimes, two graphs have exactly the same form, in the sense that there is a one-to-one correspondence between their vertex sets that preserves edges. In such a case, we say that the two graphs are **isomorphic**. Determining whether two graphs are isomorphic is an important problem of graph theory that we will study in this section.

10.3.2 Representing Graphs

One way to represent a graph without multiple edges is to list all the edges of this graph. Another way to represent a graph with no multiple edges is to use **adjacency lists**, which specify the vertices that are adjacent to each vertex of the graph.

EXAMPLE 1 Use adjacency lists to describe the simple graph given in Figure 1.

Solution: Table 1 lists those vertices adjacent to each of the vertices of the graph. ◀

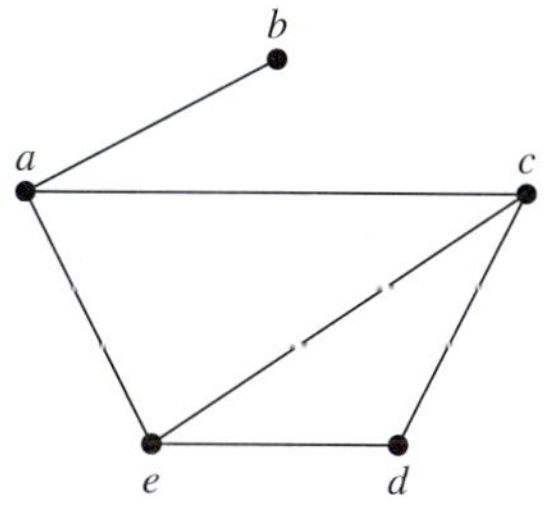

FIGURE 1 **A simple graph.**

TABLE 1 An Adjacency List for a Simple Graph.

Vertex	*Adjacent Vertices*
a	b, c, e
b	a
c	a, d, e
d	c, e
e	a, c, d

EXAMPLE 2 Represent the directed graph shown in Figure 2 by listing all the vertices that are the terminal vertices of edges starting at each vertex of the graph.

Solution: Table 2 represents the directed graph shown in Figure 2. ◀

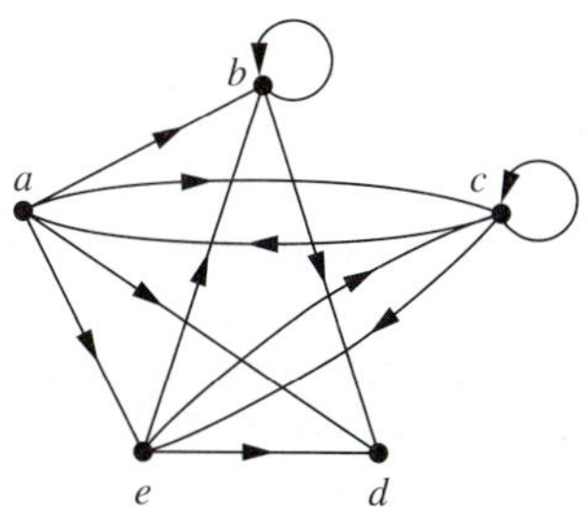

FIGURE 2 **A directed graph.**

TABLE 2 An Adjacency List for a Directed Graph.

Initial Vertex	*Terminal Vertices*
a	b, c, d, e
b	b, d
c	a, c, e
d	
e	b, c, d

10.3.3 Adjacency Matrices

Carrying out graph algorithms using the representation of graphs by lists of edges, or by adjacency lists, can be cumbersome if there are many edges in the graph. To simplify computation, graphs can be represented using matrices. Two types of matrices commonly used to represent graphs will be presented here. One is based on the adjacency of vertices, and the other is based on incidence of vertices and edges.

Links

Suppose that $G = (V, E)$ is a simple graph where $|V| = n$. Suppose that the vertices of G are listed arbitrarily as $v_1, v_2, \ldots, v_n$. The **adjacency matrix A** (or $\mathbf{A}_G$) of G, with respect to this listing of the vertices, is the $n \times n$ zero–one matrix with 1 as its (i, j)th entry when v_i and v_j are adjacent, and 0 as its (i, j)th entry when they are not adjacent. In other words, if its adjacency matrix is $\mathbf{A} = [a_{ij}]$, then

$$a_{ij} = \begin{cases} 1 & \text{if } \{v_i, v_j\} \text{ is an edge of } G, \\ 0 & \text{otherwise.} \end{cases}$$

EXAMPLE 3 Use an adjacency matrix to represent the graph shown in Figure 3.

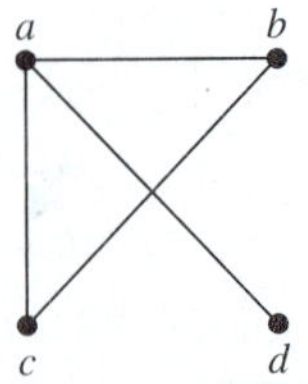

FIGURE 3
A simple graph.

Solution: We order the vertices as a, b, c, d. The matrix representing this graph is

$$\begin{bmatrix} 0 & 1 & 1 & 1 \\ 1 & 0 & 1 & 0 \\ 1 & 1 & 0 & 0 \\ 1 & 0 & 0 & 0 \end{bmatrix}.$$

◀

EXAMPLE 4 Draw a graph with the adjacency matrix

$$\begin{bmatrix} 0 & 1 & 1 & 0 \\ 1 & 0 & 0 & 1 \\ 1 & 0 & 0 & 1 \\ 0 & 1 & 1 & 0 \end{bmatrix}$$

with respect to the ordering of vertices a, b, c, d.

FIGURE 4
A graph with the given adjacency matrix.

Solution: A graph with this adjacency matrix is shown in Figure 4. ◀

Note that an adjacency matrix of a graph is based on the ordering chosen for the vertices. Hence, there may be as many as $n!$ different adjacency matrices for a graph with n vertices, because there are $n!$ different orderings of n vertices.

The adjacency matrix of a simple graph is symmetric, that is, $a_{ij} = a_{ji}$, because both of these entries are 1 when v_i and v_j are adjacent, and both are 0 otherwise. Furthermore, because a simple graph has no loops, each entry a_{ii}, $i = 1, 2, 3, \ldots, n$, is 0.

Adjacency matrices can also be used to represent undirected graphs with loops and with multiple edges. A loop at the vertex v_i is represented by a 1 at the (i, i)th position of the adjacency matrix. When multiple edges connecting the same pair of vertices v_i and v_j, or multiple loops at the same vertex, are present, the adjacency matrix is no longer a zero–one matrix, because the (i, j)th entry of this matrix equals the number of edges that are associated to $\{v_i, v_j\}$. All undirected graphs, including multigraphs and pseudographs, have symmetric adjacency matrices.

EXAMPLE 5 Use an adjacency matrix to represent the pseudograph shown in Figure 5.

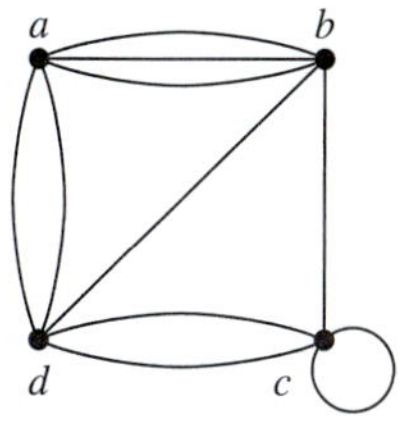

FIGURE 5
A pseudograph.

Solution: The adjacency matrix using the ordering of vertices a, b, c, d is

$$\begin{bmatrix} 0 & 3 & 0 & 2 \\ 3 & 0 & 1 & 1 \\ 0 & 1 & 1 & 2 \\ 2 & 1 & 2 & 0 \end{bmatrix}.$$

◀

We used zero–one matrices in Chapter 9 to represent relations. The matrix for a directed graph $G = (V, E)$ has a 1 in its (i, j)th position if there is an edge from v_i to v_j, where $v_1, v_2, \ldots, v_n$ is an arbitrary listing of the vertices of the directed graph. In other words, if $\mathbf{A} = [a_{ij}]$ is the adjacency matrix for the directed graph with respect to this listing of the vertices, then

$$a_{ij} = \begin{cases} 1 & \text{if } (v_i, v_j) \text{ is an edge of } G, \\ 0 & \text{otherwise.} \end{cases}$$

The adjacency matrix for a directed graph does not have to be symmetric, because there may not be an edge from v_j to v_i when there is an edge from v_i to v_j.

Adjacency matrices can also be used to represent directed multigraphs. Again, such matrices are not zero–one matrices when there are multiple edges in the same direction connecting two vertices. In the adjacency matrix for a directed multigraph, a_{ij} equals the number of edges that are associated to (v_i, v_j).

TRADE-OFFS BETWEEN ADJACENCY LISTS AND ADJACENCY MATRICES When a simple graph contains relatively few edges, that is, when it is **sparse**, it is usually preferable to use adjacency lists rather than an adjacency matrix to represent the graph. For example, if each vertex has degree not exceeding c, where c is a constant much smaller than n, then each adjacency list contains c or fewer vertices. Hence, there are no more than cn items in all these adjacency lists. On the other hand, the adjacency matrix for the graph has n^2 entries. Note, however, that the adjacency matrix of a sparse graph is a **sparse matrix**, that is, a matrix with few nonzero entries, and there are special techniques for representing, and computing with, sparse matrices.

Now suppose that a simple graph is **dense**, that is, suppose that it contains many edges, such as a graph that contains more than half of all possible edges. In this case, using an adjacency matrix to represent the graph is usually preferable over using adjacency lists. To see why, we compare the complexity of determining whether the possible edge $\{v_i, v_j\}$ is present. Using an adjacency matrix, we can determine whether this edge is present by examining the (i, j)th entry in the matrix. This entry is 1 if the graph contains this edge and is 0 otherwise. Consequently, we need make only one comparison, namely, comparing this entry with 0, to determine whether this edge is present. On the other hand, when we use adjacency lists to represent the graph, we need to search the list of vertices adjacent to either v_i or v_j to determine whether this edge is present. This can require $\Theta(|V|)$ comparisons when many edges are present.

10.3.4 Incidence Matrices

Another common way to represent graphs is to use **incidence matrices**. Let $G = (V, E)$ be an undirected graph. Suppose that $v_1, v_2, \ldots, v_n$ are the vertices and $e_1, e_2, \ldots, e_m$ are the edges of G. Then the incidence matrix with respect to this ordering of V and E is the $n \times m$ matrix $\mathbf{M} = [m_{ij}]$, where

$$m_{ij} = \begin{cases} 1 & \text{when edge } e_j \text{ is incident with } v_i, \\ 0 & \text{otherwise.} \end{cases}$$

EXAMPLE 6 Represent the graph shown in Figure 6 with an incidence matrix.

Solution: The incidence matrix is

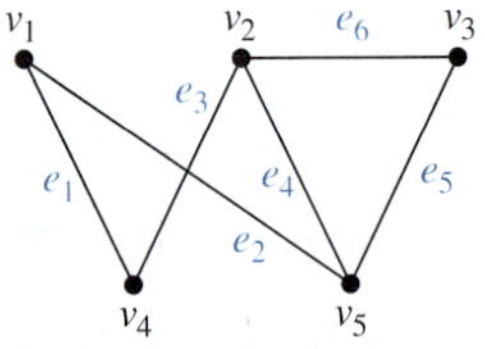

FIGURE 6 An undirected graph.

$$\begin{array}{c} \\ v_1 \\ v_2 \\ v_3 \\ v_4 \\ v_5 \end{array} \begin{array}{c} \begin{matrix} e_1 & e_2 & e_3 & e_4 & e_5 & e_6 \end{matrix} \\ \begin{bmatrix} 1 & 1 & 0 & 0 & 0 & 0 \\ 0 & 0 & 1 & 1 & 0 & 1 \\ 0 & 0 & 0 & 0 & 1 & 1 \\ 1 & 0 & 1 & 0 & 0 & 0 \\ 0 & 1 & 0 & 1 & 1 & 0 \end{bmatrix} \end{array}.$$

◀

Incidence matrices can also be used to represent multiple edges and loops. Multiple edges are represented in the incidence matrix using columns with identical entries, because these edges are incident with the same pair of vertices. Loops are represented using a column with exactly one entry equal to 1, corresponding to the vertex that is incident with this loop.

EXAMPLE 7 Represent the pseudograph shown in Figure 7 using an incidence matrix.

Solution: The incidence matrix for this graph is

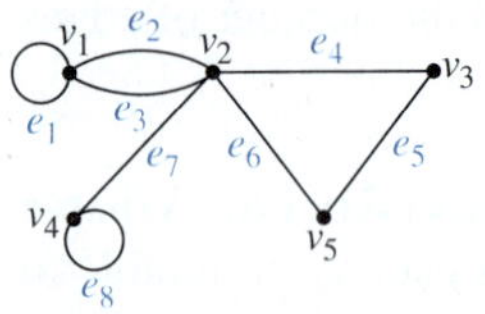

FIGURE 7 A pseudograph.

$$\begin{array}{c} \\ v_1 \\ v_2 \\ v_3 \\ v_4 \\ v_5 \end{array} \begin{array}{c} \begin{matrix} e_1 & e_2 & e_3 & e_4 & e_5 & e_6 & e_7 & e_8 \end{matrix} \\ \begin{bmatrix} 1 & 1 & 1 & 0 & 0 & 0 & 0 & 0 \\ 0 & 1 & 1 & 1 & 0 & 1 & 1 & 0 \\ 0 & 0 & 0 & 1 & 1 & 0 & 0 & 0 \\ 0 & 0 & 0 & 0 & 0 & 0 & 1 & 1 \\ 0 & 0 & 0 & 0 & 1 & 1 & 0 & 0 \end{bmatrix} \end{array}.$$

◀

10.3.5 Isomorphism of Graphs

We often need to know whether it is possible to draw two graphs in the same way. That is, do the graphs have the same structure when we ignore the identities of their vertices? For instance, in chemistry, graphs are used to model chemical compounds (in a way we will describe later). Different compounds can have the same molecular formula but can differ in structure. Such compounds can be represented by graphs that cannot be drawn in the same way. The graphs representing previously known compounds can be used to determine whether a supposedly new compound has been studied before.

There is a useful terminology for graphs with the same structure.

Definition 1 The simple graphs $G_1 = (V_1, E_1)$ and $G_2 = (V_2, E_2)$ are *isomorphic* if there exists a one-to-one and onto function f from V_1 to V_2 with the property that a and b are adjacent in G_1 if and only if $f(a)$ and $f(b)$ are adjacent in G_2, for all a and b in V_1. Such a function f is called an *isomorphism*.* Two simple graphs that are not isomorphic are called *nonisomorphic*.

In other words, when two simple graphs are isomorphic, there is a one-to-one correspondence between vertices of the two graphs that preserves the adjacency relationship. Isomorphism of simple graphs is an equivalence relation. (We leave the verification of this as Exercise 49.)

*The word *isomorphism* comes from the Greek roots *isos* for "equal" and *morphe* for "form."

EXAMPLE 8 Show that the graphs $G = (V, E)$ and $H = (W, F)$, displayed in Figure 8, are isomorphic.

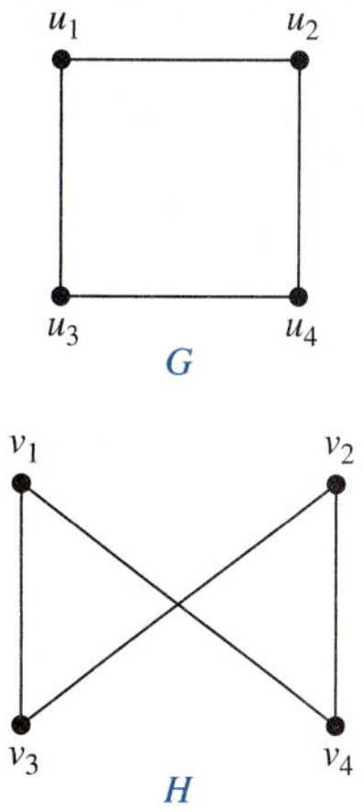

FIGURE 8 **The graphs *G* and *H*.**

Solution: The function f with $f(u_1) = v_1, f(u_2) = v_4, f(u_3) = v_3$, and $f(u_4) = v_2$ is a one-to-one correspondence between V and W. To see that this correspondence preserves adjacency, note that adjacent vertices in G are u_1 and u_2, u_1 and u_3, u_2 and u_4, and u_3 and u_4, and the pairs $f(u_1) = v_1$ and $f(u_2) = v_4$, $f(u_1) = v_1$ and $f(u_3) = v_3$, $f(u_2) = v_4$ and $f(u_4) = v_2$, and $f(u_3) = v_3$ and $f(u_4) = v_2$ are all the adjacent vertices. ◀

10.3.6 Determining whether Two Simple Graphs are Isomorphic

It is often difficult to determine whether two simple graphs are isomorphic. There are $n!$ possible one-to-one correspondences between the vertex sets of two simple graphs with n vertices. Testing each such correspondence to see whether it preserves adjacency and nonadjacency is impractical if n is at all large.

Sometimes it is not hard to show that two graphs are not isomorphic. In particular, we can show that two graphs are not isomorphic if we can find a property only one of the two graphs has, but that is preserved by isomorphism. A property preserved by isomorphism of graphs is called a **graph invariant**. For instance, isomorphic simple graphs must have the same number of vertices, because there is a one-to-one correspondence between the sets of vertices of the graphs.

Links

Isomorphic simple graphs also must have the same number of edges, because the one-to-one correspondence between vertices establishes a one-to-one correspondence between edges. In addition, the degrees of the vertices in isomorphic simple graphs must be the same. That is, a vertex v of degree d in G must correspond to a vertex $f(v)$ of degree d in H, because a vertex w in G is adjacent to v if and only if $f(v)$ and $f(w)$ are adjacent in H.

EXAMPLE 9 Show that the graphs displayed in Figure 9 are not isomorphic.

Extra Examples

Solution: Both G and H have five vertices and six edges. However, H has a vertex of degree one, namely, e, whereas G has no vertices of degree one. It follows that G and H are not isomorphic. ◀

The number of vertices, the number of edges, and the number of vertices of each degree are all invariants under isomorphism. If any of these quantities differ in two simple graphs, these graphs cannot be isomorphic. However, when these invariants are the same, it does not necessarily mean that the two graphs are isomorphic. There are no useful sets of invariants currently known that can be used to determine whether simple graphs are isomorphic.

EXAMPLE 10 Determine whether the graphs shown in Figure 10 are isomorphic.

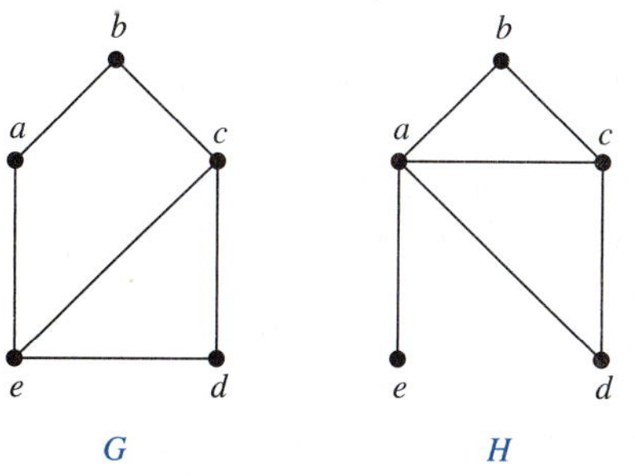

FIGURE 9 **The graphs *G* and *H*.**

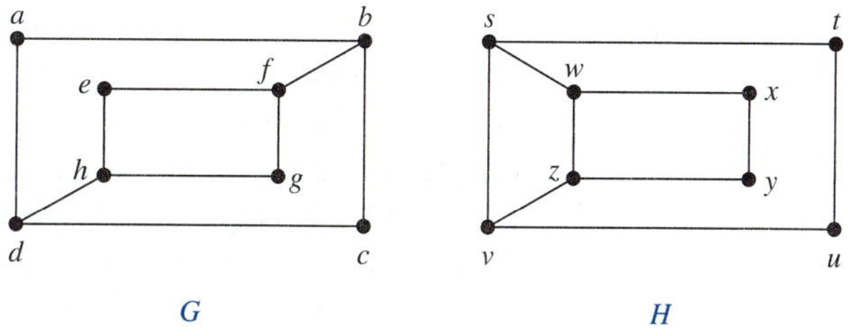

FIGURE 10 **The graphs *G* and *H*.**

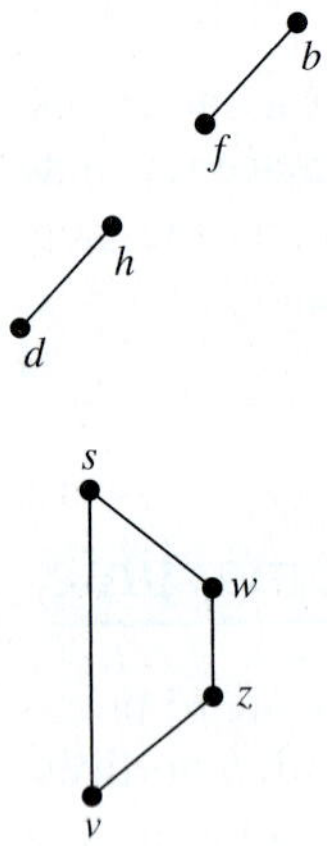

Solution: The graphs G and H both have eight vertices and 10 edges. They also both have four vertices of degree two and four of degree three. Because these invariants all agree, it is still conceivable that these graphs are isomorphic.

However, G and H are not isomorphic. To see this, note that because $\deg(a) = 2$ in G, a must correspond to either t, u, x, or y in H, because these are the vertices of degree two in H. However, each of these four vertices in H is adjacent to another vertex of degree two in H, which is not true for a in G.

Another way to see that G and H are not isomorphic is to note that the subgraphs of G and H made up of vertices of degree three and the edges connecting them must be isomorphic if these two graphs are isomorphic (the reader should verify this). However, these subgraphs, shown in Figure 11, are not isomorphic. ◀

FIGURE 11 The subgraphs of G and H made up of vertices of degree three and the edges connecting them.

To show that a function f from the vertex set of a graph G to the vertex set of a graph H is an isomorphism, we need to show that f preserves the presence and absence of edges. One helpful way to do this is to use adjacency matrices. In particular, to show that f is an isomorphism, we can show that the adjacency matrix of G is the same as the adjacency matrix of H, when rows and columns are labeled to correspond to the images under f of the vertices in G that are the labels of these rows and columns in the adjacency matrix of G. We illustrate how this is done in Example 11.

EXAMPLE 11 Determine whether the graphs G and H displayed in Figure 12 are isomorphic.

Solution: Both G and H have six vertices and seven edges. Both have four vertices of degree two and two vertices of degree three. It is also easy to see that the subgraphs of G and H consisting of all vertices of degree two and the edges connecting them are isomorphic (as the reader should verify). Because G and H agree with respect to these invariants, it is reasonable to try to find an isomorphism f.

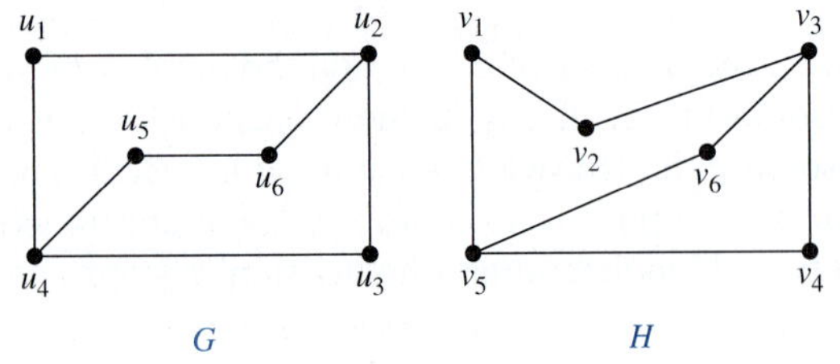

FIGURE 12 Graphs *G* and *H*.

We now will define a function f and then determine whether it is an isomorphism. Because $\deg(u_1) = 2$ and because u_1 is not adjacent to any other vertex of degree two, the image of u_1 must be either v_4 or v_6, the only vertices of degree two in H not adjacent to a vertex of degree two. We arbitrarily set $f(u_1) = v_6$. [If we found that this choice did not lead to isomorphism, we would then try $f(u_1) = v_4$.] Because u_2 is adjacent to u_1, the possible images of u_2 are v_3 and v_5. We arbitrarily set $f(u_2) = v_3$. Continuing in this way, using adjacency of vertices and degrees as a guide, we set $f(u_3) = v_4$, $f(u_4) = v_5$, $f(u_5) = v_1$, and $f(u_6) = v_2$. We now have a one-to-one correspondence between the vertex set of G and the vertex set of H, namely, $f(u_1) = v_6$, $f(u_2) = v_3, f(u_3) = v_4, f(u_4) = v_5, f(u_5) = v_1, f(u_6) = v_2$. To see whether f preserves edges, we examine the adjacency matrix of G,

$$\mathbf{A}_G = \begin{array}{c} \\ u_1 \\ u_2 \\ u_3 \\ u_4 \\ u_5 \\ u_6 \end{array} \begin{array}{c} \begin{array}{cccccc} u_1 & u_2 & u_3 & u_4 & u_5 & u_6 \end{array} \\ \left[\begin{array}{cccccc} 0 & 1 & 0 & 1 & 0 & 0 \\ 1 & 0 & 1 & 0 & 0 & 1 \\ 0 & 1 & 0 & 1 & 0 & 0 \\ 1 & 0 & 1 & 0 & 1 & 0 \\ 0 & 0 & 0 & 1 & 0 & 1 \\ 0 & 1 & 0 & 0 & 1 & 0 \end{array}\right] \end{array},$$

and the adjacency matrix of H with the rows and columns labeled by the images of the corresponding vertices in G,

$$\mathbf{A}_H = \begin{array}{c} \\ v_6 \\ v_3 \\ v_4 \\ v_5 \\ v_1 \\ v_2 \end{array} \begin{array}{c} \begin{array}{cccccc} v_6 & v_3 & v_4 & v_5 & v_1 & v_2 \end{array} \\ \left[\begin{array}{cccccc} 0 & 1 & 0 & 1 & 0 & 0 \\ 1 & 0 & 1 & 0 & 0 & 1 \\ 0 & 1 & 0 & 1 & 0 & 0 \\ 1 & 0 & 1 & 0 & 1 & 0 \\ 0 & 0 & 0 & 1 & 0 & 1 \\ 0 & 1 & 0 & 0 & 1 & 0 \end{array}\right] \end{array}.$$

Because $\mathbf{A}_G = \mathbf{A}_H$, it follows that f preserves edges. We conclude that f is an isomorphism, so G and H are isomorphic. Note that if f turned out not to be an isomorphism, we would *not* have established that G and H are not isomorphic, because another correspondence of the vertices in G and H may be an isomorphism. ◀

Links

ALGORITHMS FOR GRAPH ISOMORPHISM Until recently, the best algorithms known for determining whether two graphs are isomorphic have exponential worst-case time complexity (in the number of vertices of the graphs). However, in early 2017 László Babai announced that he had an algorithm that determines whether two graphs with n vertices are isomorphic in $2^{f(n)}$ time, where $f(n)$ is $O((\log n)^3)$. This big-O estimate means that the algorithm runs in *quasi-polynomial time*, which is between polynomial time and exponential time. This discovery of Babai has not yet been fully peer reviewed, but he was able to correct a serious gap in an algorithm he invented in 2015. Experts believe that his result is now correct. Even though no polynomial-time algorithms have been found for solving this problem, it can be solved by linear average-case time complexity algorithms. There is some hope, but also skepticism, that an algorithm with polynomial worst-case time complexity for determining whether two graphs are isomorphic can be found.

The best practical general purpose software for isomorphism testing, called NAUTY, can be used to determine whether two graphs with as many as 100 vertices are isomorphic in less than a second on a modern PC. NAUTY software can be downloaded over the Internet and experimented with. Practical algorithms for determining whether two graphs are isomorphic exist for graphs that are restricted in various ways, such as when the maximum degree of the vertices is small. The problem of determining whether any two graphs are isomorphic is of special interest because it is one of only a few NP problems (see Exercise 78) not known to be either tractable or NP-complete (see Section 3.3).

APPLICATIONS OF GRAPH ISOMORPHISMS Graph isomorphisms, and functions that are almost graph isomorphisms, arise in applications of graph theory to chemistry and to the design of electronic circuits, and other areas including bioinformatics and computer vision. Chemists use multigraphs, known as molecular graphs, to model chemical compounds. In these graphs, vertices represent atoms and edges represent chemical bonds between these atoms. Two structural isomers, molecules with identical molecular formulas but with atoms bonded differently, have nonisomorphic molecular graphs. When a potentially new chemical compound is synthesized, a database of molecular graphs is checked to see whether the molecular graph of the compound is the same as one already known.

Electronic circuits are modeled using graphs in which vertices represent components and edges represent connections between them. Modern integrated circuits, known as chips, are miniaturized electronic circuits, often with millions of transistors and connections between them. Because of the complexity of modern chips, automation tools are used to design them. Graph isomorphism is the basis for the verification that a particular layout of a circuit produced by an automated tool corresponds to the original schematic of the design. Graph isomorphism can also be used to determine whether a chip from one vendor includes intellectual property from a different vendor. This can be done by looking for large isomorphic subgraphs in the graphs modeling these chips.

Exercises

In Exercises 1–4 use an adjacency list to represent the given graph.

1. 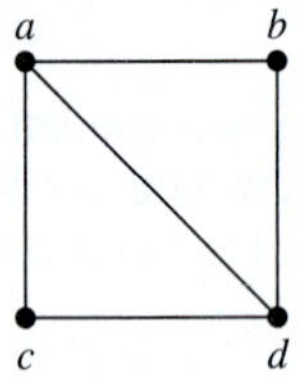

2. 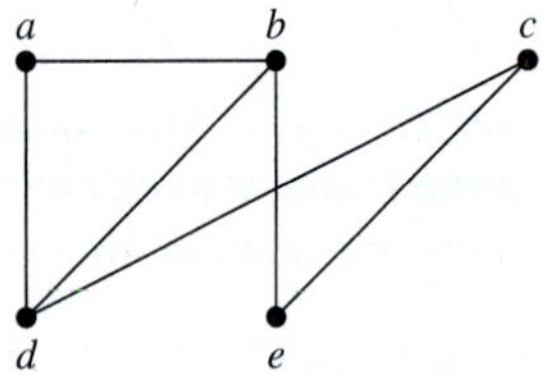

3. 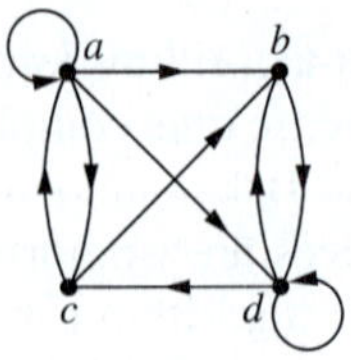

4. 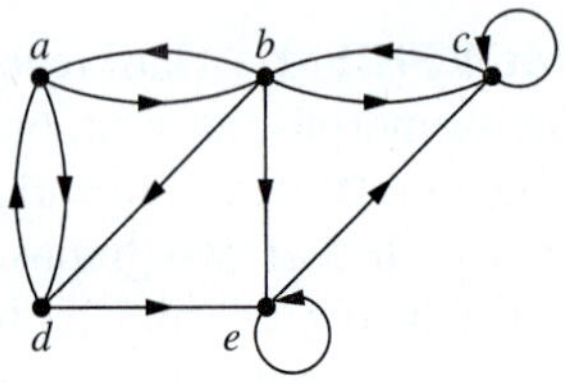

5. Represent the graph in Exercise 1 with an adjacency matrix.

6. Represent the graph in Exercise 2 with an adjacency matrix.

7. Represent the graph in Exercise 3 with an adjacency matrix.

8. Represent the graph in Exercise 4 with an adjacency matrix.

9. Represent each of these graphs with an adjacency matrix.

a) K_4 **b)** $K_{1,4}$ **c)** $K_{2,3}$
d) C_4 **e)** W_4 **f)** Q_3

In Exercises 10–12 draw a graph with the given adjacency matrix.

10. $\begin{bmatrix} 0 & 1 & 0 \\ 1 & 0 & 1 \\ 0 & 1 & 0 \end{bmatrix}$

11. $\begin{bmatrix} 0 & 0 & 1 & 1 \\ 0 & 0 & 1 & 0 \\ 1 & 1 & 0 & 1 \\ 1 & 1 & 1 & 0 \end{bmatrix}$

12. $\begin{bmatrix} 1 & 1 & 1 & 0 \\ 0 & 0 & 1 & 0 \\ 1 & 0 & 1 & 0 \\ 1 & 1 & 1 & 0 \end{bmatrix}$

In Exercises 13–15 represent the given graph using an adjacency matrix.

13. 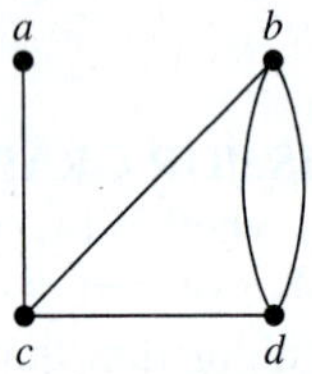

14. 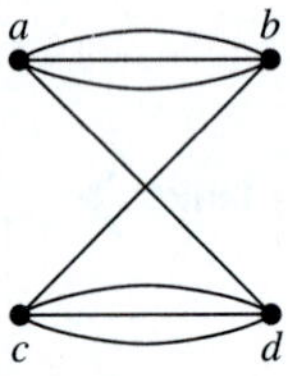

15. 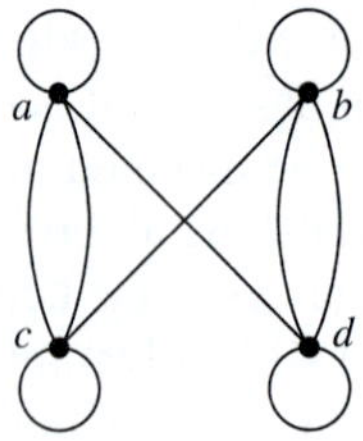

In Exercises 16–18 draw an undirected graph represented by the given adjacency matrix.

16. $\begin{bmatrix} 1 & 3 & 2 \\ 3 & 0 & 4 \\ 2 & 4 & 0 \end{bmatrix}$

17. $\begin{bmatrix} 1 & 2 & 0 & 1 \\ 2 & 0 & 3 & 0 \\ 0 & 3 & 1 & 1 \\ 1 & 0 & 1 & 0 \end{bmatrix}$

18. $\begin{bmatrix} 0 & 1 & 3 & 0 & 4 \\ 1 & 2 & 1 & 3 & 0 \\ 3 & 1 & 1 & 0 & 1 \\ 0 & 3 & 0 & 0 & 2 \\ 4 & 0 & 1 & 2 & 3 \end{bmatrix}$

In Exercises 19–21 find the adjacency matrix of the given directed multigraph with respect to the vertices listed in alphabetic order.

19.

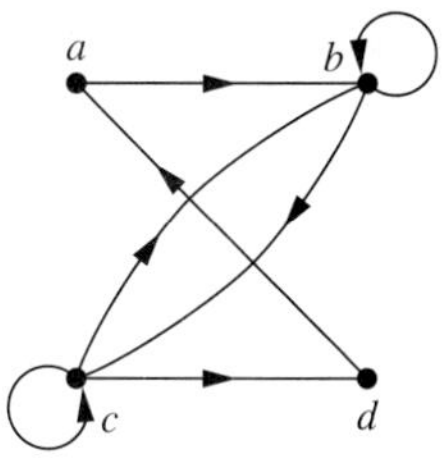

20.

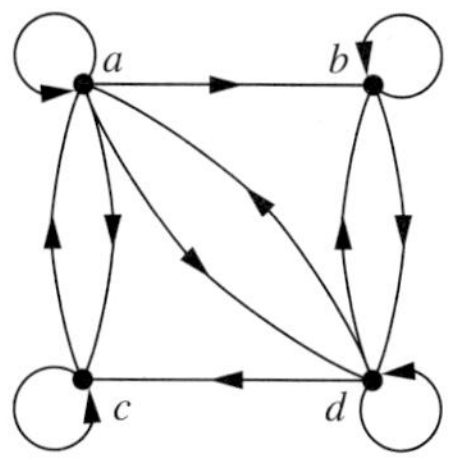

21.

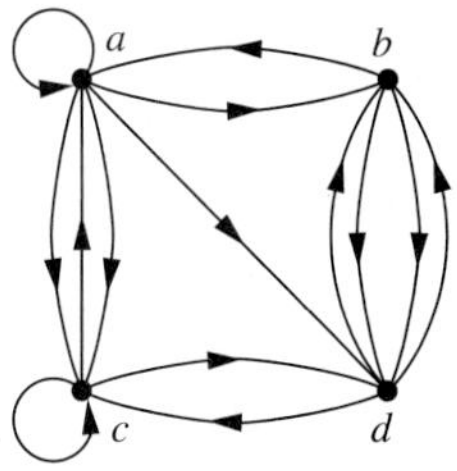

In Exercises 22–24 draw the graph represented by the given adjacency matrix.

22. $\begin{bmatrix} 1 & 0 & 1 \\ 0 & 0 & 1 \\ 1 & 1 & 1 \end{bmatrix}$ **23.** $\begin{bmatrix} 1 & 2 & 1 \\ 2 & 0 & 0 \\ 0 & 2 & 2 \end{bmatrix}$ **24.** $\begin{bmatrix} 0 & 2 & 3 & 0 \\ 1 & 2 & 2 & 1 \\ 2 & 1 & 1 & 0 \\ 1 & 0 & 0 & 2 \end{bmatrix}$

The **density** of an undirected graph G is the number of edges of G divided by the number of possible edges in an undirected graph with $|G|$ vertices. Consequently, the density of $G(V, E)$ is

$$\frac{2|E|}{|V|(|V|-1)}.$$

A family of graphs G_n, $n = 1, 2, \ldots$ is **sparse** if the limit of the density of G_n is zero as n grows without bound, while it is **dense** if this proportion approaches a positive real number. As mentioned in the text, an individual graph is called sparse when it contains relatively few edges and dense if it contains many edges. These terms can be defined precisely depending on the context, but different definitions generally will not agree.

25. Find the density of the graph in

a) Figure 1 of Section 10.1.
b) Figure 6 of Section 10.1.
c) Figure 12 of Section 10.1.

26. Find the density of the graph in

a) Figure 12 of Section 10.2.
b) Figure 13 of Section 10.2.
c) Figure 14 of Section 10.2.

27. (*Requires calculus*) For each of these families of graphs, determine whether the graph family is sparse, dense, or neither. (Refer to the results of Exercise 37 in Section 10.2.)

a) K_n **b)** C_n **c)** $K_{n,n}$
d) Q_n **e)** W_n **f)** $K_{3,n}$

28. Determine whether an undirected graph is sparse, dense, or neither, and explain your answer, if it is used to model

a) the street network in a city (where the vertices are street intersections).

b) whether buildings in a city are within two miles of one another.

c) whether two people in the world are siblings.

d) whether two people work for the same company.

29. Is every zero–one square matrix that is symmetric and has zeros on the diagonal the adjacency matrix of a simple graph?

30. Use an incidence matrix to represent the graphs in Exercises 1 and 2.

31. Use an incidence matrix to represent the graphs in Exercises 13–15.

***32.** What is the sum of the entries in a row of the adjacency matrix for an undirected graph? For a directed graph?

***33.** What is the sum of the entries in a column of the adjacency matrix for an undirected graph? For a directed graph?

34. What is the sum of the entries in a row of the incidence matrix for an undirected graph?

35. What is the sum of the entries in a column of the incidence matrix for an undirected graph?

***36.** Find an adjacency matrix for each of these graphs.

a) K_n **b)** C_n **c)** W_n **d)** $K_{m,n}$ **e)** Q_n

***37.** Find incidence matrices for the graphs in parts (a)–(d) of Exercise 36.

In Exercises 38–48 determine whether the given pair of graphs is isomorphic. Exhibit an isomorphism or provide a rigorous argument that none exists. For additional exercises of this kind, see Exercises 3–5 in the Supplementary Exercises.

38.

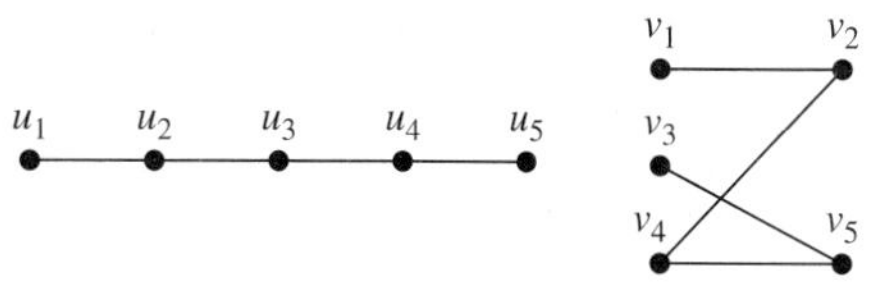

39.

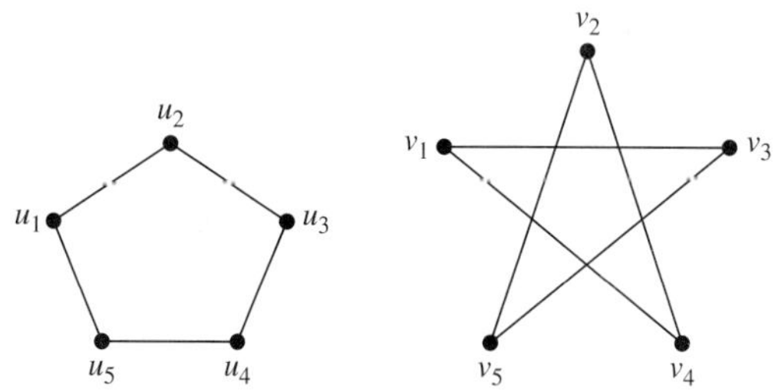

40.

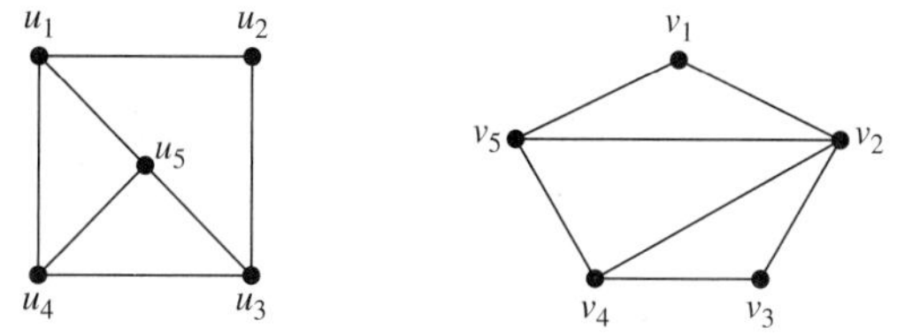

41.

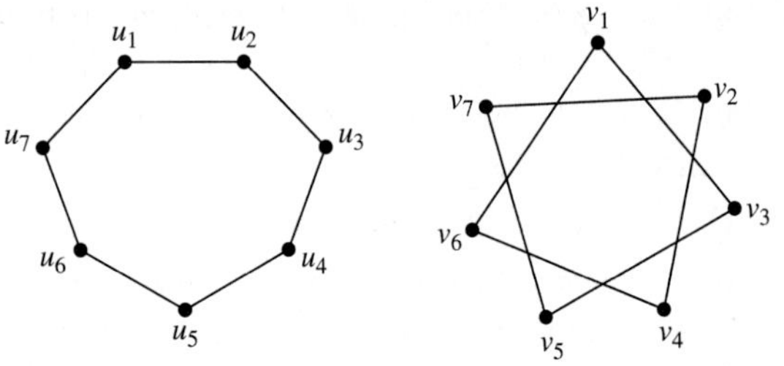

42.

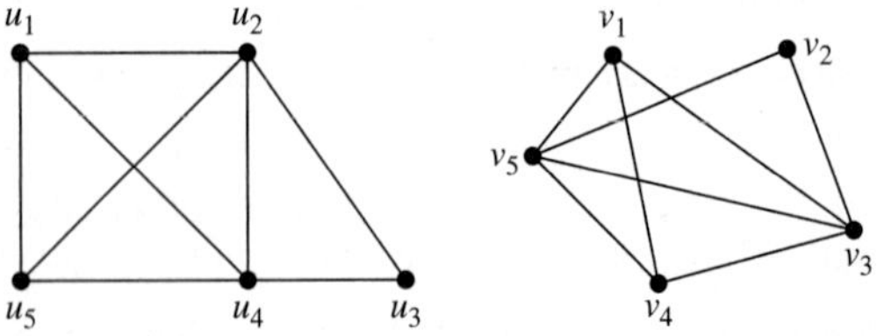

43.

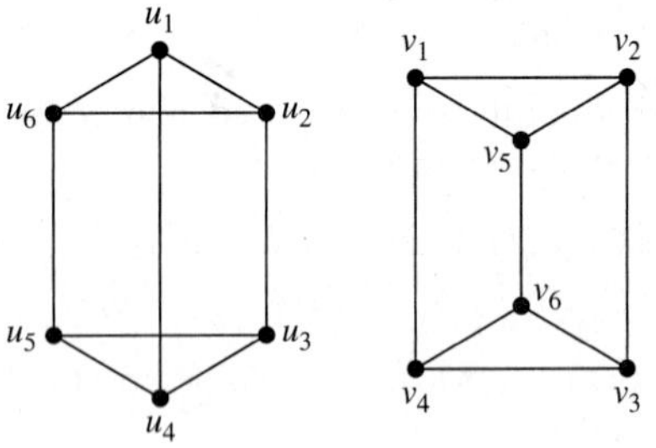

44.

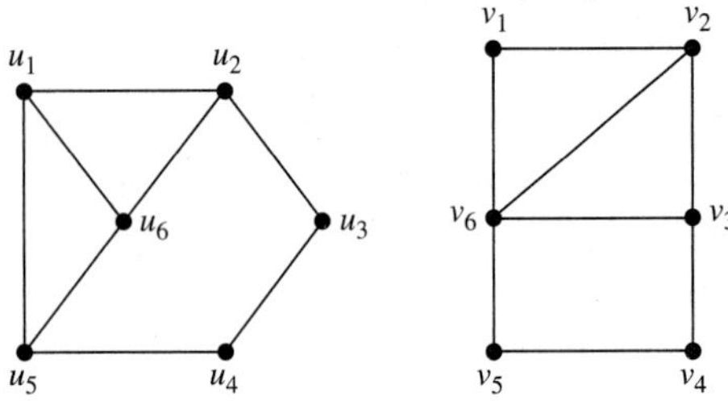

45.

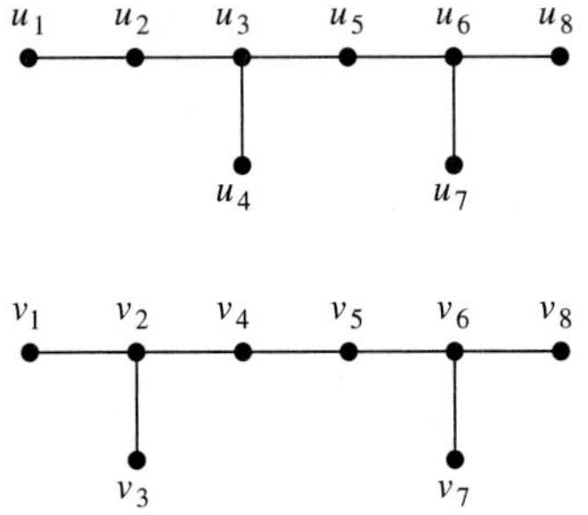

46.

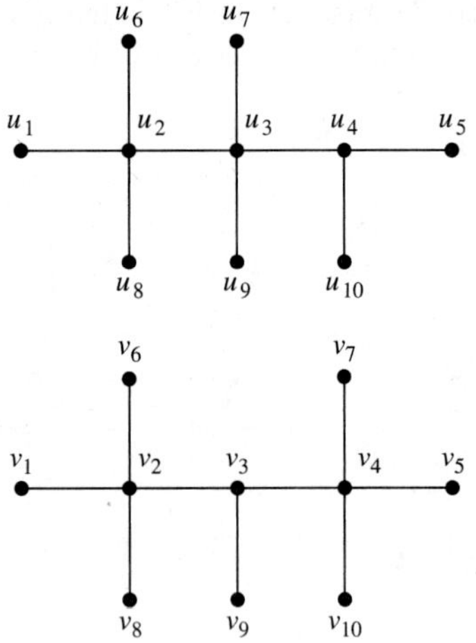

47.

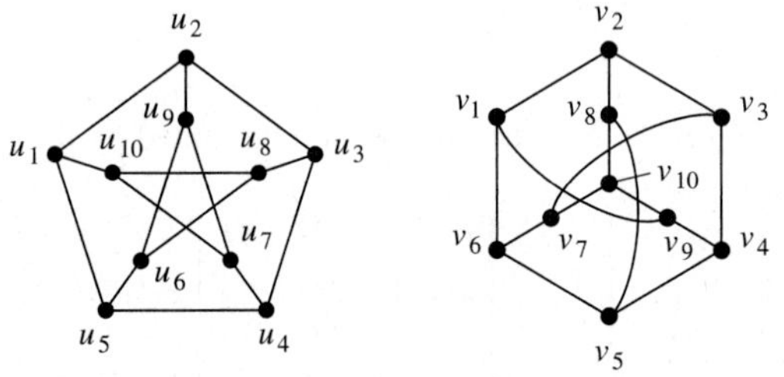

48.

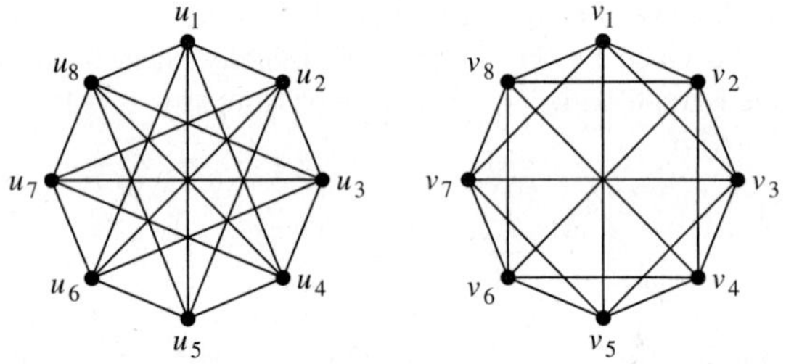

49. Show that isomorphism of simple graphs is an equivalence relation.

50. Suppose that G and H are isomorphic simple graphs. Show that their complementary graphs $\overline{G}$ and $\overline{H}$ are also isomorphic.

51. Describe the row and column of an adjacency matrix of a graph corresponding to an isolated vertex.

52. Describe the row of an incidence matrix of a graph corresponding to an isolated vertex.

53. Show that the vertices of a bipartite graph with two or more vertices can be ordered so that its adjacency matrix has the form

$$\begin{bmatrix} \mathbf{0} & \mathbf{A} \\ \mathbf{B} & \mathbf{0} \end{bmatrix},$$

where the four entries shown are rectangular blocks.

A simple graph G is called **self-complementary** if G and $\overline{G}$ are isomorphic.

54. Show that this graph is self-complementary.

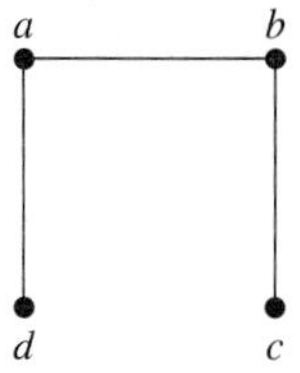

55. Find a self-complementary simple graph with five vertices.

***56.** Show that if G is a self-complementary simple graph with v vertices, then $v \equiv 0$ or $1 \pmod 4$.

57. For which integers n is C_n self-complementary?

58. How many nonisomorphic simple graphs are there with n vertices, when n is

a) 2? **b)** 3? **c)** 4?

59. How many nonisomorphic simple graphs are there with five vertices and three edges?

60. How many nonisomorphic simple graphs are there with six vertices and four edges?

***61.** Find the number of nonisomorphic simple graphs with six vertices in which each vertex has degree three.

62. Find the number of nonisomorphic simple graphs with seven vertices in which each vertex has degree two.

63. Are the simple graphs with the following adjacency matrices isomorphic?

a) $\begin{bmatrix} 0 & 0 & 1 \\ 0 & 0 & 1 \\ 1 & 1 & 0 \end{bmatrix}, \begin{bmatrix} 0 & 1 & 1 \\ 1 & 0 & 0 \\ 1 & 0 & 0 \end{bmatrix}$

b) $\begin{bmatrix} 0 & 1 & 0 & 1 \\ 1 & 0 & 0 & 1 \\ 0 & 0 & 0 & 1 \\ 1 & 1 & 1 & 0 \end{bmatrix}, \begin{bmatrix} 0 & 1 & 1 & 1 \\ 1 & 0 & 0 & 1 \\ 1 & 0 & 0 & 1 \\ 1 & 1 & 1 & 0 \end{bmatrix}$

c) $\begin{bmatrix} 0 & 1 & 1 & 0 \\ 1 & 0 & 0 & 1 \\ 1 & 0 & 0 & 1 \\ 0 & 1 & 1 & 0 \end{bmatrix}, \begin{bmatrix} 0 & 1 & 0 & 1 \\ 1 & 0 & 0 & 0 \\ 0 & 0 & 0 & 1 \\ 1 & 0 & 1 & 0 \end{bmatrix}$

64. Determine whether the graphs without loops with these incidence matrices are isomorphic.

a) $\begin{bmatrix} 1 & 0 & 1 \\ 0 & 1 & 1 \\ 1 & 1 & 0 \end{bmatrix}, \begin{bmatrix} 1 & 1 & 0 \\ 1 & 0 & 1 \\ 0 & 1 & 1 \end{bmatrix}$

b) $\begin{bmatrix} 1 & 1 & 0 & 0 & 0 \\ 1 & 0 & 1 & 0 & 1 \\ 0 & 0 & 0 & 1 & 1 \\ 0 & 1 & 1 & 1 & 0 \end{bmatrix}, \begin{bmatrix} 0 & 1 & 0 & 0 & 1 \\ 0 & 1 & 1 & 1 & 0 \\ 1 & 0 & 0 & 1 & 0 \\ 1 & 0 & 1 & 0 & 1 \end{bmatrix}$

65. Extend the definition of isomorphism of simple graphs to undirected graphs containing loops and multiple edges.

66. Define isomorphism of directed graphs.

In Exercises 67–70 determine whether the given pair of directed graphs are isomorphic. (See Exercise 66.)

67.

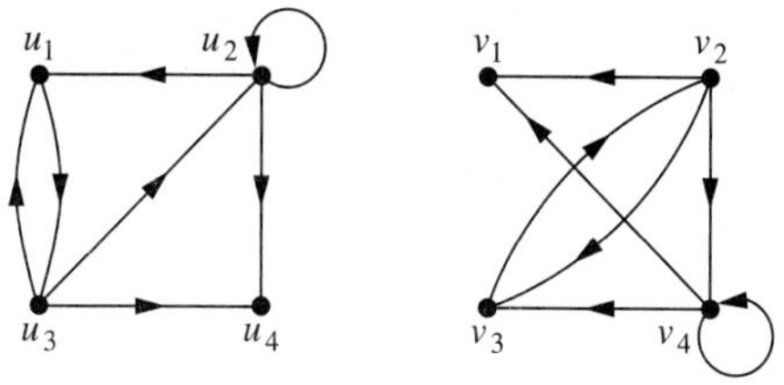

68.

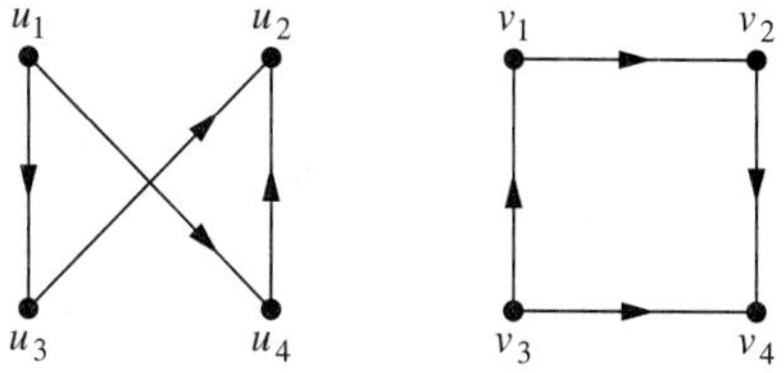

69.

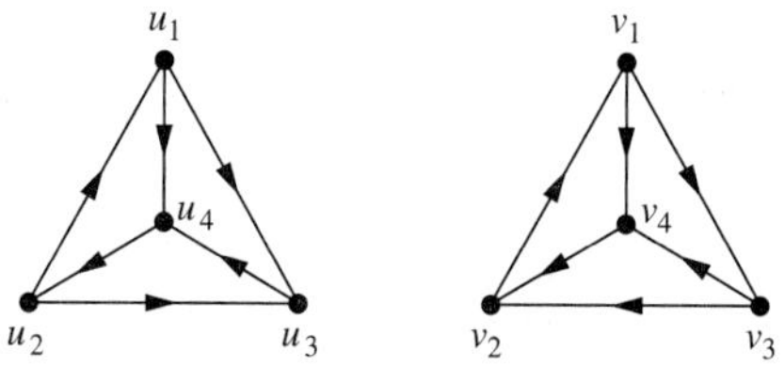

70.

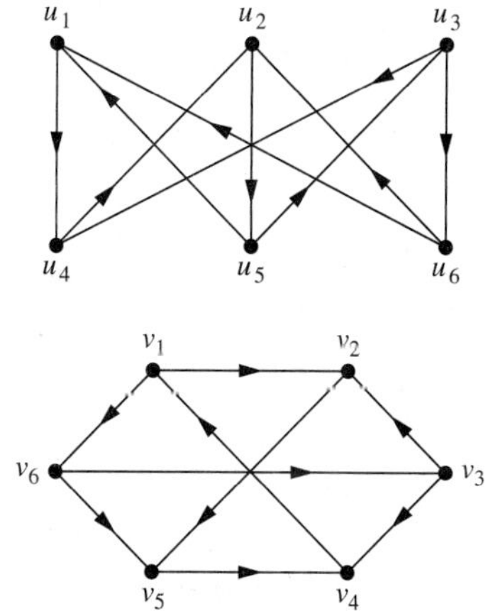

71. Show that if G and H are isomorphic directed graphs, then the converses of G and H (defined in the preamble of Exercise 73 of Section 10.2) are also isomorphic.

72. Show that the property that a graph is bipartite is an isomorphic invariant.

73. Find a pair of nonisomorphic graphs with the same degree sequence (defined in the preamble to Exercise 42 in Section 10.2) such that one graph is bipartite, but the other graph is not bipartite.

***74.** How many nonisomorphic directed simple graphs are there with n vertices, when n is

a) 2? **b)** 3? **c)** 4?

***75.** What is the product of the incidence matrix and its transpose for an undirected graph?

***76.** How much storage is needed to represent a simple graph with n vertices and m edges using

a) adjacency lists?
b) an adjacency matrix?
c) an incidence matrix?

A **devil's pair** for a purported isomorphism test is a pair of nonisomorphic graphs that the test fails to show that they are not isomorphic.

77. Find a devil's pair for the test that checks the degree sequence (defined in the preamble to Exercise 42 in Section 10.2) in two graphs to make sure they agree.

78. Suppose that the function f from V_1 to V_2 is an isomorphism of the graphs $G_1 = (V_1, E_1)$ and $G_2 = (V_2, E_2)$. Show that it is possible to verify this fact in time polynomial in terms of the number of vertices of the graph, in terms of the number of comparisons needed.

10.4 Connectivity

10.4.1 Introduction

Many problems can be modeled with paths formed by traveling along the edges of graphs. For instance, the problem of determining whether a message can be sent between two computers using intermediate links can be studied with a graph model. Problems of efficiently planning routes for mail delivery, garbage pickup, diagnostics in computer networks, and so on can be solved using models that involve paths in graphs.

10.4.2 Paths

Informally, a **path** is a sequence of edges that begins at a vertex of a graph and travels from vertex to vertex along edges of the graph. As the path travels along its edges, it visits the vertices along this path, that is, the endpoints of these edges.

A formal definition of paths and related terminology is given in Definition 1.

Definition 1 Let n be a nonnegative integer and G an undirected graph. A *path* of *length* n from u to v in G is a sequence of n edges $e_1, \dots, e_n$ of G for which there exists a sequence $x_0 = u, x_1, \dots, x_{n-1}, x_n = v$ of vertices such that e_i has, for $i = 1, \dots, n$, the endpoints x_{i-1} and x_i. When the graph is simple, we denote this path by its vertex sequence $x_0, x_1, \dots, x_n$ (because listing these vertices uniquely determines the path). The path is a *circuit* if it begins and ends at the same vertex, that is, if $u = v$, and has length greater than zero. The path or circuit is said to *pass through* the vertices $x_1, x_2, \dots, x_{n-1}$ or *traverse* the edges $e_1, e_2, \dots, e_n$. A path or circuit is *simple* if it does not contain the same edge more than once.

When it is not necessary to distinguish between multiple edges, we will denote a path $e_1, e_2, \dots, e_n$, where e_i is associated with $\{x_{i-1}, x_i\}$ for $i = 1, 2, \dots, n$ by its vertex sequence $x_0, x_1, \dots, x_n$. This notation identifies a path only as far as which vertices it passes through. Consequently, it does not specify a unique path when there is more than one path that passes through this sequence of vertices, which will happen if and only if there are multiple edges between some successive vertices in the list. Note that a path of length zero consists of a single vertex.

Remark: There is considerable variation of terminology concerning the concepts defined in Definition 1. For instance, in some books, the term **walk** is used instead of *path*, where a walk is defined to be an alternating sequence of vertices and edges of a graph, $v_0, e_1, v_1, e_2, \dots, v_{n-1}, e_n, v_n$, where v_{i-1} and v_i are the endpoints of e_i for $i = 1, 2, \dots, n$. When this terminology is used, **closed walk** is used instead of *circuit* to indicate a walk that begins and ends at the same vertex, and **trail** is used to denote a walk that has no repeated edge (replacing the term *simple path*). When this terminology is used, the terminology **path** is often used for a trail with no repeated vertices, conflicting with the terminology in Definition 1. Because of this variation in terminology, you will need to make sure which set of definitions are used in a

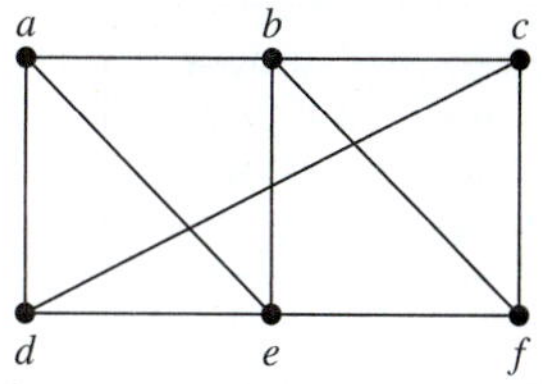

FIGURE 1 A simple graph.

particular book or article when you read about traversing edges of a graph. The text [GrYe06] is a good reference for the alternative terminology described in this remark.

EXAMPLE 1 In the simple graph shown in Figure 1, a, d, c, f, e is a simple path of length 4, because $\{a, d\}$, $\{d, c\}$, $\{c, f\}$, and $\{f, e\}$ are all edges. However, d, e, c, a is not a path, because $\{e, c\}$ is not an edge. Note that b, c, f, e, b is a circuit of length 4 because $\{b, c\}$, $\{c, f\}$, $\{f, e\}$, and $\{e, b\}$ are edges, and this path begins and ends at b. The path a, b, e, d, a, b, which is of length 5, is not simple because it contains the edge $\{a, b\}$ twice. ◀

Paths and circuits in directed graphs were introduced in Chapter 9. We now provide more general definitions.

Definition 2 Let n be a nonnegative integer and G a directed graph. A *path* of length n from u to v in G is a sequence of edges $e_1, e_2, \ldots, e_n$ of G such that e_1 is associated with (x_0, x_1), e_2 is associated with (x_1, x_2), and so on, with e_n associated with (x_{n-1}, x_n), where $x_0 = u$ and $x_n = v$. When there are no multiple edges in the directed graph, this path is denoted by its vertex sequence $x_0, x_1, x_2, \ldots, x_n$. A path of length greater than zero that begins and ends at the same vertex is called a *circuit* or *cycle*. A path or circuit is called *simple* if it does not contain the same edge more than once.

Remark: Terminology other than that given in Definition 2 is often used for the concepts defined there. In particular, the alternative terminology that uses *walk, closed walk, trail*, and *path* (described in the remarks following Definition 1) may be used for directed graphs. See [GrYe05] for details.

Note that the terminal vertex of an edge in a path is the initial vertex of the next edge in the path. When it is not necessary to distinguish between multiple edges, we will denote a path $e_1, e_2, \ldots, e_n$, where e_i is associated with (x_{i-1}, x_i) for $i = 1, 2, \ldots, n$, by its vertex sequence $x_0, x_1, \ldots, x_n$. The notation identifies a path only as far as which the vertices it passes through. There may be more than one path that passes through this sequence of vertices, which will happen if and only if there are multiple edges between two successive vertices in the list.

Paths represent useful information in many graph models, as Examples 2 and 3 demonstrate.

EXAMPLE 2 **Links ›** **Paths in Acquaintanceship Graphs** In an acquaintanceship graph there is a path between two people if there is a chain of people linking these people, where two people adjacent in the chain know one another. For example, in Figure 6 in Section 10.1, there is a chain of six people linking Kamini and Ching. Many social scientists have conjectured that almost every pair of people in the world are linked by a small chain of people, perhaps containing just five or fewer people. This would mean that almost every pair of vertices in the acquaintanceship graph containing all people in the world is linked by a path of length not exceeding four. The play *Six Degrees of Separation* by John Guare is based on this notion. ◀

TABLE 1 The Number of Mathematicians with a Given Erdős Number (as of early 2006).

Erdős Number	*Number of People*
0	1
1	504
2	6,593
3	33,605
4	83,642
5	87,760
6	40,014
7	11,591
8	3,146
9	819
10	244
11	68
12	23
13	5

TABLE 2 The Number of Actors with a Given Bacon Number (as of early 2024).

Bacon Number	*Number of People*
0	1
1	2,195
2	228,904
3	703,856
4	264,564
5	33,936
6	5,311
7	951
8	291
9	98
10	40

EXAMPLE 3 **Paths in Collaboration Graphs** In a collaboration graph, two people a and b are connected by a path when there is a sequence of people starting with a and ending with b such that the endpoints of each edge in the path are people who have collaborated. We will consider two particular collaboration graphs here. First, in the academic collaboration graph of people who have written papers in mathematics, the **Erdős number** of a person m (defined in terms of relations in Supplementary Exercise 14 in Chapter 9) is the length of the shortest path between m and the extremely prolific mathematician Paul Erdős (who died in 1996). That is, the Erdős number of a mathematician is the length of the shortest chain of mathematicians that begins with Paul Erdős and ends with this mathematician, where each adjacent pair of mathematicians have written a joint paper. The number of mathematicians with each Erdős number, last computed in 2006 as part of the Erdős Number Project, is shown in Table 1.

Links

Replace Kevin Bacon by your own favorite actor to invent a new party game.

In the Hollywood graph (see Example 3 in Section 10.1), two actors a and b are linked when there is a chain of actors linking a and b, where every two actors adjacent in the chain have acted in the same movie. In the Hollywood graph, the **Bacon number** of an actor c is defined to be the length of the shortest path connecting c and the well-known actor Kevin Bacon. As new movies are made, including new ones with Kevin Bacon, the Bacon number of actors can change. In Table 2 we show the number of actors with each Bacon number as of early 2024 using data from the Oracle of Bacon website. The origins of the Bacon number of an actor dates back to the early 1990s, when Kevin Bacon remarked that he had worked with everyone in Hollywood or someone who worked with them. This led some people to invent a party game where participants where challenged to find a sequence of movies leading from each actor named to Kevin Bacon. Before the web was born, you needed to be a movie expert to play this game well. Now, you need only consult the Oracle of Bacon. We can find a number similar to a Bacon number using any actor as the center of the acting universe. ◀

10.4.3 Connectedness in Undirected Graphs

When does a computer network have the property that every pair of computers can share information, if messages can be sent through one or more intermediate computers? When a graph is used to represent this computer network, where vertices represent the computers and edges represent the communication links, this question becomes: When is there always a path between two vertices in the graph?

Definition 3 An undirected graph is called *connected* if there is a path between every pair of distinct vertices of the graph. An undirected graph that is not *connected* is called *disconnected*. We say that we *disconnect* a graph when we remove vertices or edges, or both, to produce a disconnected subgraph.

Thus, any two computers in the network can communicate if and only if the graph of this network is connected.

EXAMPLE 4 The graph G_1 in Figure 2 is connected, because for every pair of distinct vertices there is a path between them (the reader should verify this). However, the graph G_2 in Figure 2 is not connected. For instance, there is no path in G_2 between vertices a and d. ◀

We will need the following theorem in Chapter 11.

THEOREM 1 There is a simple path between every pair of distinct vertices of a connected undirected graph.

Proof: Let u and v be two distinct vertices of the connected undirected graph $G = (V, E)$. Because G is connected, there is at least one path between u and v. Let $x_0, x_1, \ldots, x_n$, where $x_0 = u$ and $x_n = v$, be the vertex sequence of a path of least length. This path of least length is simple. To see this, suppose it is not simple. Then $x_i = x_j$ for some i and j with $0 \leq i < j$. This means that there is a path from u to v of shorter length with vertex sequence $x_0, x_1, \ldots, x_{i-1}, x_j, \ldots, x_n$ obtained by deleting the edges corresponding to the vertex sequence $x_i, \ldots, x_{j-1}$. ◁

CONNECTED COMPONENTS A **connected component** of a graph G is a connected subgraph of G that is not a proper subgraph of another connected subgraph of G. That is, a connected component of a graph G is a maximal connected subgraph of G. A graph G that is not connected has two or more connected components that are disjoint and have G as their union.

Links ›

EXAMPLE 5 What are the connected components of the graph H shown in Figure 3?

Solution: The graph H is the union of three disjoint connected subgraphs H_1, H_2, and H_3, shown in Figure 3. These three subgraphs are the connected components of H. ◀

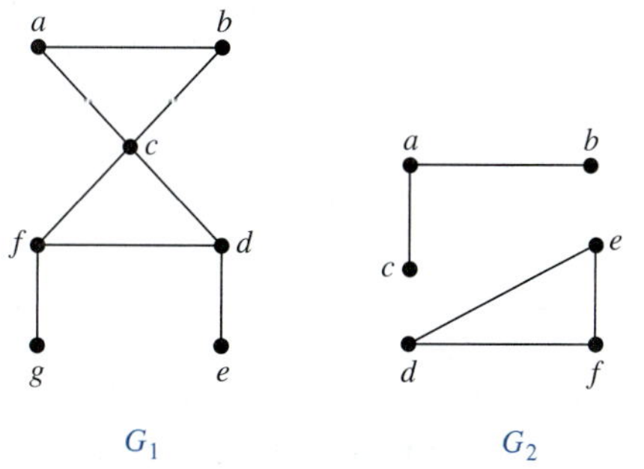

FIGURE 2 **The graphs G_1 and G_2.**

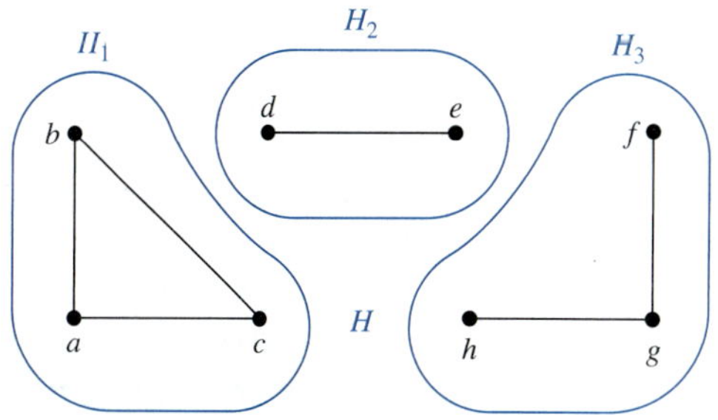

FIGURE 3 **The graph H and its connected components H_1, H_2, and H_3.**

EXAMPLE 6 **Connected Components of Call Graphs** Two vertices x and y are in the same component of a telephone call graph (see Example 4 in Section 10.1) when there is a sequence of telephone calls beginning at x and ending at y. When a call graph for telephone calls made during a particular day in the AT&T network was analyzed, this graph was found to have 53,767,087 vertices, more than 170 million edges, and more than 3.7 million connected components. Most of these components were small; approximately three-fourths consisted of two vertices representing pairs of telephone numbers that called only each other. This graph has one huge connected component with 44,989,297 vertices comprising more than 80% of the total. Furthermore, every vertex in this component can be linked to any other vertex by a chain of no more than 20 calls. ◀

Links

10.4.4 How Connected is a Graph?

Suppose that a graph represents a computer network. Knowing that this graph is connected tells us that any two computers on the network can communicate. However, we would also like to understand how reliable this network is. For instance, will it still be possible for all computers to communicate after a router or a communications link fails? To answer this and similar questions, we now develop some new concepts.

Sometimes the removal from a graph of a vertex and all incident edges produces a subgraph with more connected components. Such vertices are called **cut vertices** (or **articulation points**). The removal of a cut vertex from a connected graph produces a subgraph that is not connected. Analogously, an edge whose removal produces a graph with more connected components than in the original graph is called a **cut edge** or **bridge**. Note that in a graph representing a computer network, a cut vertex and a cut edge represent an essential router and an essential link that cannot fail for all computers to be able to communicate.

EXAMPLE 7 Find the cut vertices and cut edges in the graph G_1 shown in Figure 4.

Solution: The cut vertices of G_1 are b, c, and e. The removal of one of these vertices (and its adjacent edges) disconnects the graph. The cut edges are $\{a, b\}$ and $\{c, e\}$. Removing either one of these edges disconnects G_1. ◀

VERTEX CONNECTIVITY Not all graphs have cut vertices. For example, the complete graph K_n, where $n \geq 3$, has no cut vertices. When you remove a vertex from K_n and all edges incident to it, the resulting subgraph is the complete graph K_{n-1}, a connected graph. Connected graphs without cut vertices are called **nonseparable graphs**, and can be thought of as more connected than those with a cut vertex. We can extend this notion by defining a more granulated measure of graph connectivity based on the minimum number of vertices that can be removed to disconnect a graph.

A subset V' of the vertex set V of $G = (V, E)$ is a **vertex cut**, or **separating set**, if $G - V'$ is disconnected. For instance, in the graph in Figure 1, the set $\{b, c, e\}$ is a vertex cut with three vertices, as the reader should verify. We leave it to the reader (Exercise 51) to show that every connected graph, except a complete graph, has a vertex cut. We define the **vertex connectivity** of a noncomplete graph G, denoted by $\kappa(G)$, as the minimum number of vertices in a vertex cut.

κ is the lowercase Greek letter kappa.

When G is a complete graph, it has no vertex cuts, because removing any subset of its vertices and all incident edges still leaves a complete graph. Consequently, we cannot define $\kappa(G)$ as the minimum number of vertices in a vertex cut when G is complete. Instead, we set $\kappa(K_n) = n - 1$, the number of vertices needed to be removed to produce a graph with a single vertex.

Consequently, for every graph G, $\kappa(G)$ is minimum number of vertices that can be removed from G to either disconnect G or produce a graph with a single vertex. We have

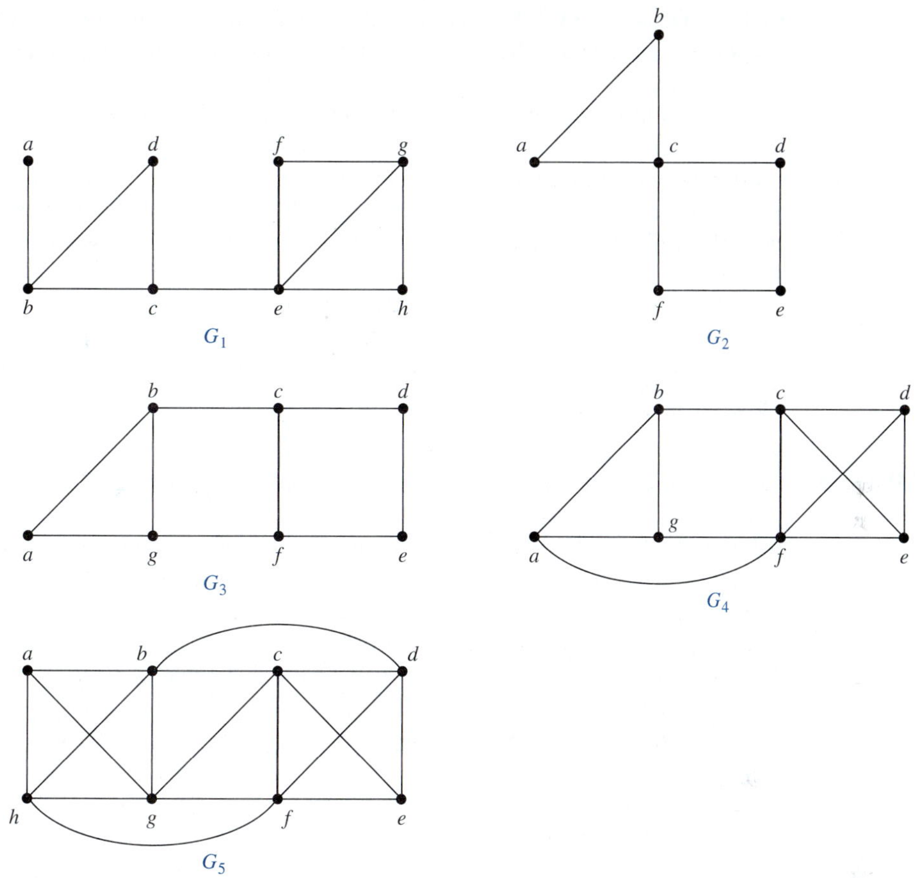

FIGURE 4 **Some connected graphs.**

$0 \leq \kappa(G) \leq n - 1$ if G has n vertices, $\kappa(G) = 0$ if and only if G is disconnected or $G = K_1$, and $\kappa(G) = n - 1$ if and only if G is complete [see Exercise 52(a)].

The larger $\kappa(G)$ is, the more connected we consider G to be. Disconnected graphs and K_1 have $\kappa(G) = 0$, connected graphs with cut vertices and K_2 have $\kappa(G) = 1$, graphs without cut vertices that can be disconnected by removing two vertices and K_3 have $\kappa(G) = 2$, and so on. We say that a graph is ***k*-connected** (or ***k*-vertex-connected**), if $\kappa(G) \geq k$. A graph G is 1-connected if it is connected and not a graph containing a single vertex; a graph is 2-connected, or **biconnected**, if it is nonseparable and has at least three vertices. Note that if G is a k-connected graph, then G is a j-connected graph for all j with $0 \leq j \leq k$.

EXAMPLE 8 Find the vertex connectivity for each of the graphs in Figure 4.

Solution: Each of the five graphs in Figure 4 is connected and has more than vertex, so each of these graphs has positive vertex connectivity. Because G_1 is a connected graph with a cut vertex, as shown in Example 7, we know that $\kappa(G_1) = 1$. Similarly, $\kappa(G_2) = 1$, because c is a cut vertex of G_2.

The reader should verify that G_3 has no cut vertices, but that $\{b, g\}$ is a vertex cut. Hence, $\kappa(G_3) = 2$. Similarly, because G_4 has a vertex cut of size two, $\{c, f\}$, but no cut vertices, it follows that $\kappa(G_4) = 2$. The reader can verify that G_5 has no vertex cut of size two, but $\{b, c, f\}$ is a vertex cut of G_5. Hence, $\kappa(G_5) = 3$. ◀

EDGE CONNECTIVITY We can also measure the connectivity of a connected graph $G = (V, E)$ in terms of the minimum number of edges that we can remove to disconnect it. If a graph has a cut edge, then we need only remove it to disconnect G. If G does not have a cut edge, we look for the smallest set of edges that can be removed to disconnect it. A set of edges E' is called an **edge cut** of G if the subgraph $G - E'$ is disconnected. The **edge connectivity** of a graph G, denoted by $\lambda(G)$, is the minimum number of edges in an edge cut of G. This defines $\lambda(G)$ for all connected graphs with more than one vertex because it is always possible to disconnect such a graph by removing all edges incident to one of its vertices. Note that $\lambda(G) = 0$ if G is not connected. We also specify that $\lambda(G) = 0$ if G is a graph consisting of a single vertex. It follows that if G is a graph with n vertices, then $0 \leq \lambda(G) \leq n - 1$. We leave it to the reader [Exercise 52(b)] to show that $\lambda(G) = n - 1$ where G is a graph with n vertices if and only if $G = K_n$, which is equivalent to the statement that $\lambda(G) \leq n - 2$ when G is not a complete graph.

λ is the lowercase Greek letter lambda.

EXAMPLE 9 Find the edge connectivity of each of the graphs in Figure 4.

Solution: Each of the five graphs in Figure 4 is connected and has more than one vertex, so we know that all of them have positive edge connectivity. As we saw in Example 7, G_1 has a cut edge, so $\lambda(G_1) = 1$.

The graph G_2 has no cut edges, as the reader should verify, but the removal of the two edges $\{a, b\}$ and $\{a, c\}$ disconnects it. Hence, $\lambda(G_2) = 2$. Similarly, $\lambda(G_3) = 2$, because G_3 has no cut edges, but the removal of the two edges $\{b, c\}$ and $\{f, g\}$ disconnects it.

The reader should verify that the removal of no two edges disconnects G_4, but the removal of the three edges $\{b, c\}$, $\{a, f\}$, and $\{f, g\}$ disconnects it. Hence, $\lambda(G_4) = 3$. Finally, the reader should verify that $\lambda(G_5) = 3$, because the removal of any two of its edges does not disconnect it, but the removal of $\{a, b\}$, $\{a, g\}$, and $\{a, h\}$ does. ◀

AN INEQUALITY FOR VERTEX CONNECTIVITY AND EDGE CONNECTIVITY When $G = (V, E)$ is a noncomplete connected graph with at least three vertices, the minimum degree of a vertex of G is an upper bound for both the vertex connectivity of G and the edge connectivity of G. That is, $\kappa(G) \leq \min_{v \in V} \deg(v)$ and $\lambda(G) \leq \min_{v \in V} \deg(v)$. To see this, observe that deleting all the neighbors of a fixed vertex of minimum degree disconnects G, and deleting all the edges that have a fixed vertex of minimum degree as an endpoint disconnects G.

In Exercise 55, we ask the reader to show that $\kappa(G) \leq \lambda(G)$ when G is a connected noncomplete graph. Note also that $\kappa(K_n) = \lambda(K_n) = \min_{v \in V} \deg(v) = n - 1$ when n is a positive integer and that $\kappa(G) = \lambda(G) = 0$ when G is a disconnected graph. Putting these facts together establishes that for all graphs G,

$$\kappa(G) \leq \lambda(G) \leq \min_{v \in V} \deg(v).$$

APPLICATIONS OF VERTEX AND EDGE CONNECTIVITY Graph connectivity plays an important role in many problems involving the reliability of networks. For instance, as we mentioned in our introduction of cut vertices and cut edges, we can model a data network using vertices to represent routers and edges to represent links between them. The vertex connectivity of the resulting graph equals the minimum number of routers that disconnect the network when they are out of service. If fewer routers are down, data transmission between every pair of routers is still possible. The edge connectivity represents the minimum number of links that can be down to disconnect the network. If fewer links are down, it will still be possible for data to be transmitted between every pair of routers.

We can model a highway network, using vertices to represent highway intersections and edges to represent sections of roads running between intersections. The vertex connectivity of the resulting graph represents the minimum number of intersections that can be closed at a particular time that makes it impossible to travel between every two intersections. If fewer intersections are closed, travel between every pair of intersections is still possible. The edge connectivity represents the minimum number of roads that can be closed to disconnect the highway network. If fewer highways are closed, it will still be possible to travel between any two intersections. Clearly, it would be useful for the highway department to take this information into account when planning road repairs.

10.4.5 Connectedness in Directed Graphs

There are two notions of connectedness in directed graphs, depending on whether the directions of the edges are considered.

Definition 4 A directed graph is *strongly connected* if there is a path from a to b and from b to a whenever a and b are vertices in the graph.

For a directed graph to be strongly connected there must be a sequence of directed edges from any vertex in the graph to any other vertex. A directed graph can fail to be strongly connected but still be in "one piece." Definition 5 makes this notion precise.

Definition 5 A directed graph is *weakly connected* if there is a path between every two vertices in the underlying undirected graph.

That is, a directed graph is weakly connected if and only if there is always a path between two vertices when the directions of the edges are disregarded. Clearly, any strongly connected directed graph is also weakly connected.

EXAMPLE 10 Are the directed graphs G and H shown in Figure 5 strongly connected? Are they weakly connected?

Solution: G is strongly connected because there is a path between any two vertices in this directed graph (the reader should verify this). Hence, G is also weakly connected. The graph H is not strongly connected. There is no directed path from a to b in this graph. However, H is weakly connected, because there is a path between any two vertices in the underlying undirected graph of H (the reader should verify this). ◀

STRONG COMPONENTS OF A DIRECTED GRAPH The subgraphs of a directed graph G that are strongly connected but not contained in larger strongly connected subgraphs, that is, the maximal strongly connected subgraphs, are called the **strongly connected components** or **strong components** of G. Note that if a and b are two vertices in a directed graph, their strong components are either the same or disjoint. (We leave the proof of this last fact as Exercise 17.)

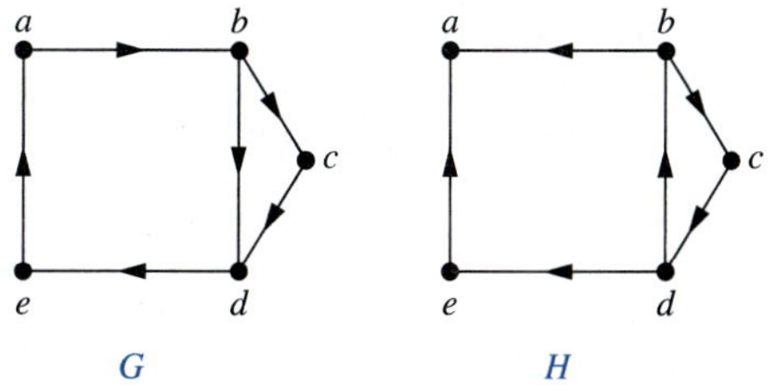

FIGURE 5 **The directed graphs G and H.**

EXAMPLE 11 The graph H in Figure 5 has three strongly connected components, consisting of the vertex a; the vertex e; and the subgraph consisting of the vertices b, c, and d and edges (b, c), (c, d), and (d, b). ◀

EXAMPLE 12 **The Strongly Connected Components of the Web Graph** The web graph introduced in Example 5 of Section 10.1 represents web pages with vertices, and links with directed edges. Determining the size of the web graph is a challenging problem because it is extraordinarily large. A snapshot of the web in 1999 produced a web graph with over 200 million vertices and over 1.5 billion edges. In 2010 the web graph was estimated to have at least 55 billion vertices and one trillion edges. From these limited data (and from what is readily apparent if you have used the web in recent years), the number of web pages has a rapid rate of growth. (See [Br00] and the web resources for details.)

Links

In 2010 it was estimated that 40 terrabytes of computer memory would be needed to represent the adjacency matrix of the web graph, but the storage requirements today are vastly larger.

The underlying undirected graph of the web graph, as it existed in 1999, is not connected, but it has a connected component that includes approximately 90% of the vertices in the graph. The subgraph of the original directed graph corresponding to this connected component of the underlying undirected graph (that is, with the same vertices and all directed edges connecting vertices in this subgraph) has one very large strongly connected component and many small ones. The former is called the **giant strongly connected component (GSCC)** of the directed graph. A web page in this component can be reached following links starting at any other page in this component. The GSCC in the web graph was found to have over 53 million vertices. The remaining vertices in the large connected component of the undirected graph represent three different types of web pages: pages that can be reached from a page in the GSCC, but do not link back to these pages following a series of links; pages that link back to pages in the GSCC following a series of links, but cannot be reached by following links on pages in the GSCC; and pages that cannot reach pages in the GSCC and cannot be reached from pages in the GSCC following a series of links. Each of these three other sets was found to have approximately 44 million vertices. (It is rather surprising that these three sets are close to the same size.) ◀

10.4.6 Paths and Isomorphism

There are several ways that paths and circuits can help determine whether two graphs are isomorphic. For example, the existence of a simple circuit of a particular length is a useful invariant that can be used to show that two graphs are not isomorphic. In addition, paths can be used to construct mappings that may be isomorphisms.

As we mentioned, a useful isomorphic invariant for simple graphs is the existence of a simple circuit of length k, where k is a positive integer greater than 2. (The proof that this is an invariant is left as Exercise 60.) Example 13 illustrates how this invariant can be used to show that two graphs are not isomorphic.

EXAMPLE 13 Determine whether the graphs G and H shown in Figure 6 are isomorphic.

Solution: Both G and H have six vertices and eight edges. Each has four vertices of degree three, and two vertices of degree two. So, the three invariants—number of vertices, number of edges, and degrees of vertices—all agree for the two graphs. However, H has a simple circuit of length three, namely, v_1, v_2, v_6, v_1, whereas G has no simple circuit of length three, as can be determined by inspection (all simple circuits in G have length at least four). Because the existence of a simple circuit of length three is an isomorphic invariant, G and H are not isomorphic. ◀

We have shown how the existence of a type of path, namely, a simple circuit of a particular length, can be used to show that two graphs are not isomorphic. We can also use paths to find mappings that are potential isomorphisms.

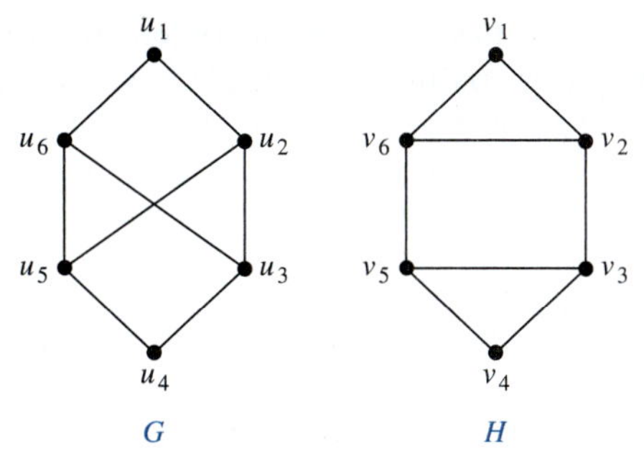

FIGURE 6 The graphs G and H.

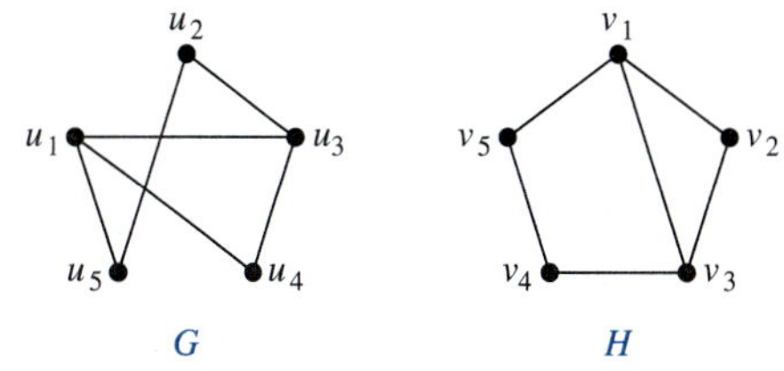

FIGURE 7 The graphs G and H.

EXAMPLE 14 Determine whether the graphs G and H shown in Figure 7 are isomorphic.

Solution: Both G and H have five vertices and six edges, both have two vertices of degree three and three vertices of degree two, and both have a simple circuit of length three, a simple circuit of length four, and a simple circuit of length five. Because all these isomorphic invariants agree, G and H may be isomorphic.

To find a possible isomorphism, we can follow paths that go through all vertices so that the corresponding vertices in the two graphs have the same degree. For example, the paths u_1, u_4, u_3, u_2, u_5 in G and v_3, v_2, v_1, v_5, v_4 in H both go through every vertex in the graph; start at a vertex of degree three; go through vertices of degrees two, three, and two, respectively; and end at a vertex of degree two. By following these paths through the graphs, we define the mapping f with $f(u_1) = v_3, f(u_4) = v_2, f(u_3) = v_1, f(u_2) = v_5$, and $f(u_5) = v_4$. The reader can show that f is an isomorphism, so G and H are isomorphic, either by showing that f preserves edges or by showing that with the appropriate orderings of vertices the adjacency matrices of G and H are the same. ◀

10.4.7 Counting Paths Between Vertices

The number of paths between two vertices in a graph can be determined using its adjacency matrix.

THEOREM 2 Let G be a graph with adjacency matrix $\mathbf{A}$ with respect to the ordering $v_1, v_2, \dots, v_n$ of the vertices of the graph (with directed or undirected edges, with multiple edges and loops allowed). The number of different paths of length r from v_i to v_j, where r is a positive integer, equals the (i, j)th entry of $\mathbf{A}^r$.

Proof: The theorem will be proved using mathematical induction. Let G be a graph with adjacency matrix $\mathbf{A}$ (assuming an ordering $v_1, v_2, \dots, v_n$ of the vertices of G). The number of paths from v_i to v_j of length 1 is the (i, j)th entry of $\mathbf{A}$, because this entry is the number of edges from v_i to v_j.

Assume that the (i, j)th entry of $\mathbf{A}^r$ is the number of different paths of length r from v_i to v_j. This is the inductive hypothesis. Because $\mathbf{A}^{r+1} = \mathbf{A}^r\mathbf{A}$, the (i, j)th entry of $\mathbf{A}^{r+1}$ equals

$$b_{i1}a_{1j} + b_{i2}a_{2j} + \cdots + b_{in}a_{nj},$$

where b_{ik} is the (i, k)th entry of $\mathbf{A}^r$. By the inductive hypothesis, b_{ik} is the number of paths of length r from v_i to v_k.

A path of length $r + 1$ from v_i to v_j is made up of a path of length r from v_i to some intermediate vertex v_k, and an edge from v_k to v_j. By the product rule for counting, the number of such paths is the product of the number of paths of length r from v_i to v_k, namely, b_{ik}, and the number of edges from v_k to v_j, namely, a_{kj}. When these products are added for all possible intermediate vertices v_k, the desired result follows by the sum rule for counting. ◁

EXAMPLE 15 How many paths of length four are there from a to d in the simple graph G in Figure 8?

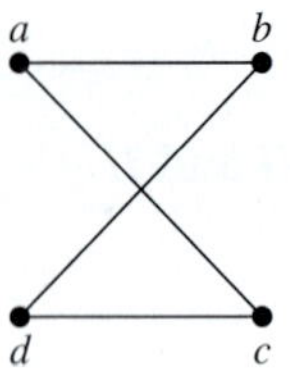

FIGURE 8 The graph G.

Solution: The adjacency matrix of G (ordering the vertices as a, b, c, d) is

$$\mathbf{A} = \begin{bmatrix} 0 & 1 & 1 & 0 \\ 1 & 0 & 0 & 1 \\ 1 & 0 & 0 & 1 \\ 0 & 1 & 1 & 0 \end{bmatrix}.$$

Hence, the number of paths of length four from a to d is the (1, 4)th entry of $\mathbf{A}^4$. Because

$$\mathbf{A}^4 = \begin{bmatrix} 8 & 0 & 0 & 8 \\ 0 & 8 & 8 & 0 \\ 0 & 8 & 8 & 0 \\ 8 & 0 & 0 & 8 \end{bmatrix},$$

Extra Examples

there are exactly eight paths of length four from a to d. By inspection of the graph, we see that a, b, a, b, d; a, b, a, c, d; a, b, d, b, d; a, b, d, c, d; a, c, a, b, d; a, c, a, c, d; a, c, d, b, d; and a, c, d, c, d are the eight paths of length four from a to d. ◀

Theorem 2 can be used to find the length of the shortest path between two vertices of a graph (see Exercise 56), and it can also be used to determine whether a graph is connected (see Exercises 61 and 62).

Exercises

1. Does each of these lists of vertices form a path in the following graph? Which paths are simple? Which are circuits? What are the lengths of those that are paths?

a) a, e, b, c, b **b)** a, e, a, d, b, c, a
c) e, b, a, d, b, e **d)** c, b, d, a, e, c

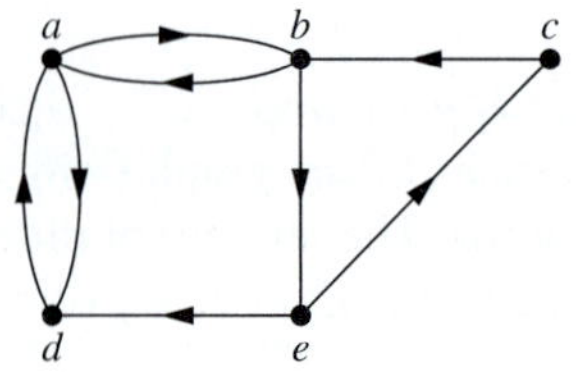

2. Does each of these lists of vertices form a path in the following graph? Which paths are simple? Which are circuits? What are the lengths of those that are paths?

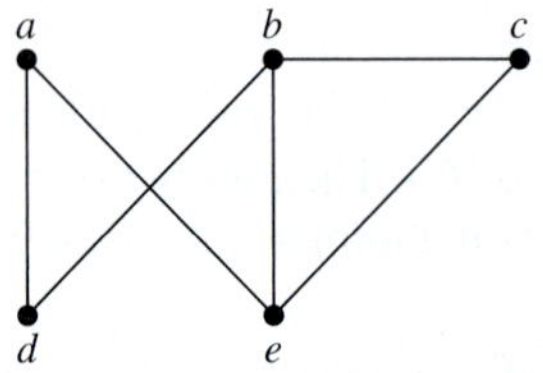

a) a, b, e, c, b **b)** a, d, a, d, a
c) a, d, b, e, a **d)** a, b, e, c, b, d, a

In Exercises 3–5 determine whether the given graph is connected.

3. 

4. 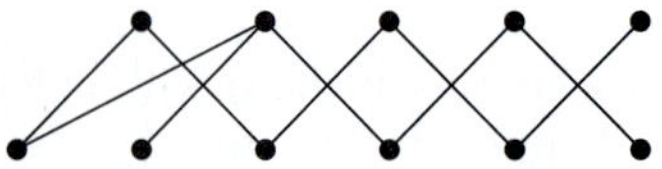

5. 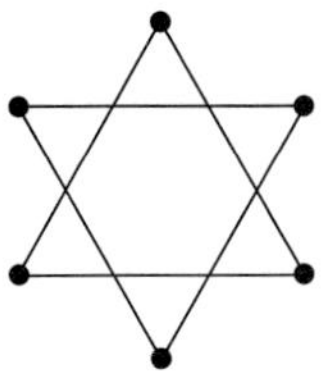

6. How many connected components does each of the graphs in Exercises 3–5 have? For each graph find each of its connected components.

7. What do the connected components of acquaintanceship graphs represent?

8. What do the connected components of a collaboration graph represent?

9. Explain why in the collaboration graph of mathematicians (see Example 3 in Section 10.1) a vertex representing a mathematician is in the same connected component as the vertex representing Paul Erdős if and only if that mathematician has a finite Erdős number.

10. In the Hollywood graph (see Example 3 in Section 10.1), when is the vertex representing an actor in the same connected component as the vertex representing Kevin Bacon?

11. Determine whether each of these graphs is strongly connected and if not, whether it is weakly connected.

a) 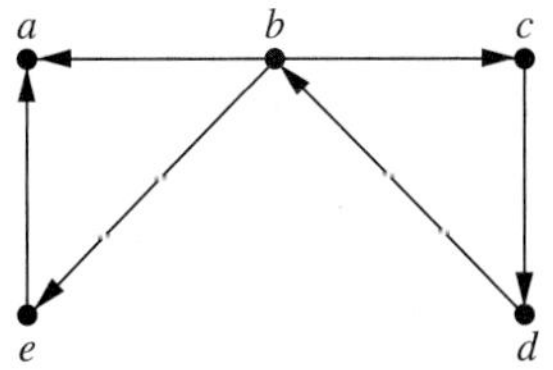

b) 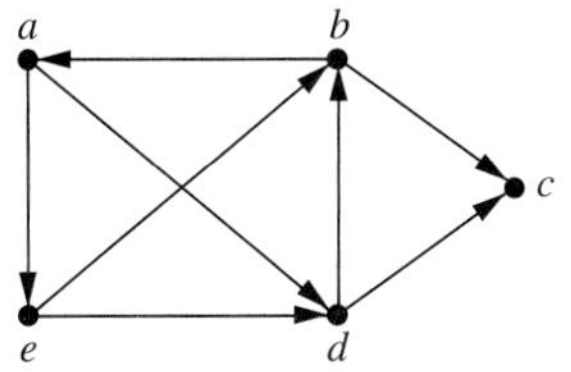

c) 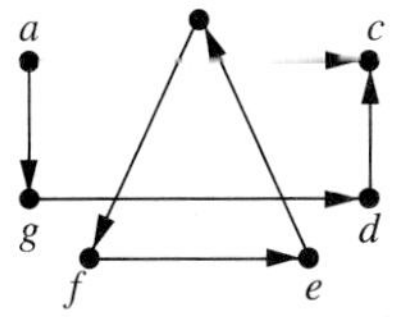

12. Determine whether each of these graphs is strongly connected and if not, whether it is weakly connected.

a) 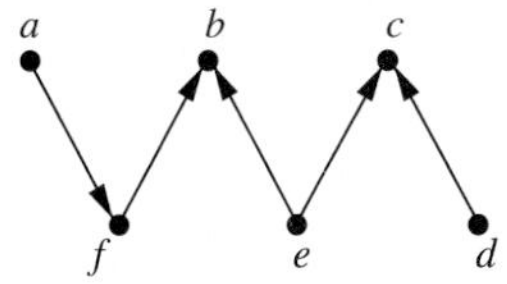

b) 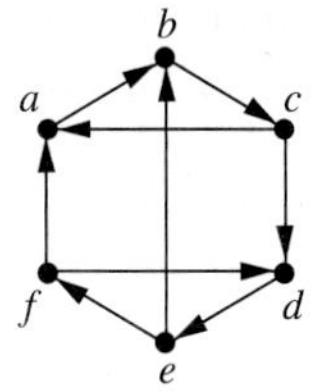

c) 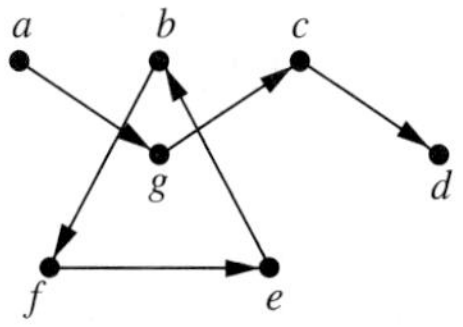

13. What do the strongly connected components of a telephone call graph represent?

14. Find the strongly connected components of each of these graphs.

a) 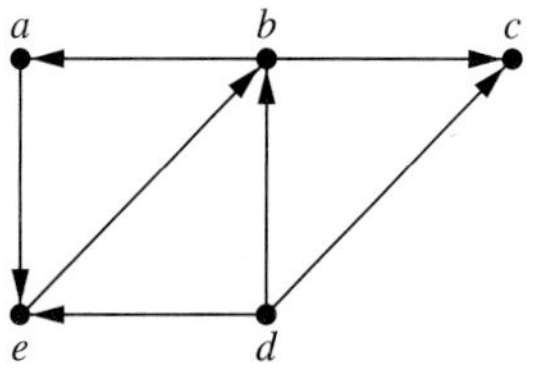

b) 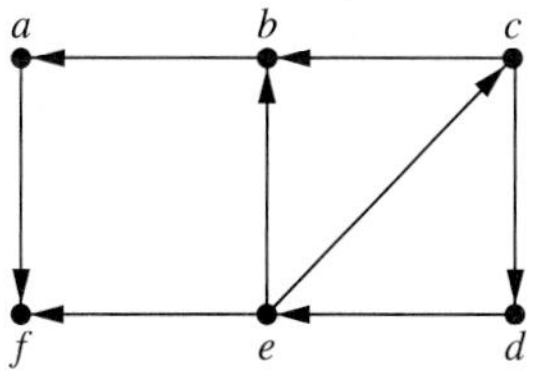

c) 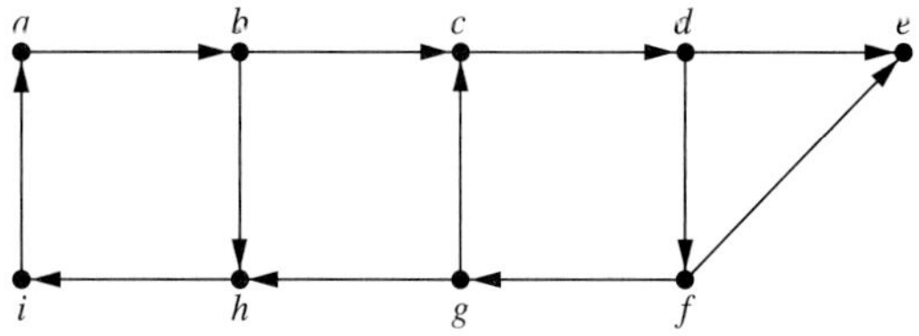

15. Find the strongly connected components of each of these graphs.

a) 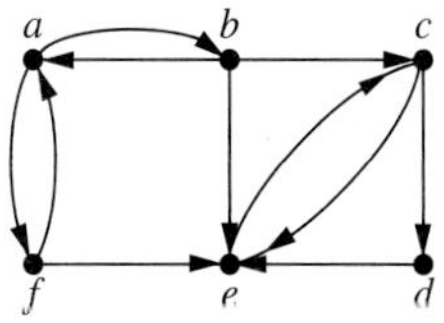

b) 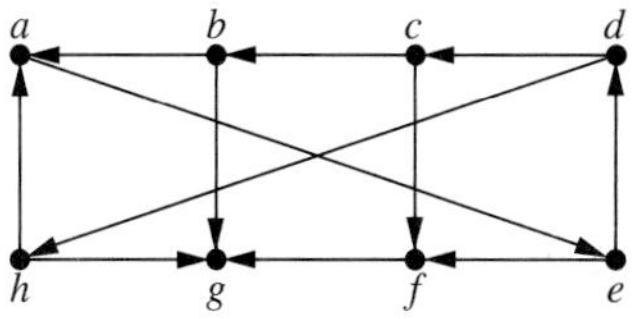

c) 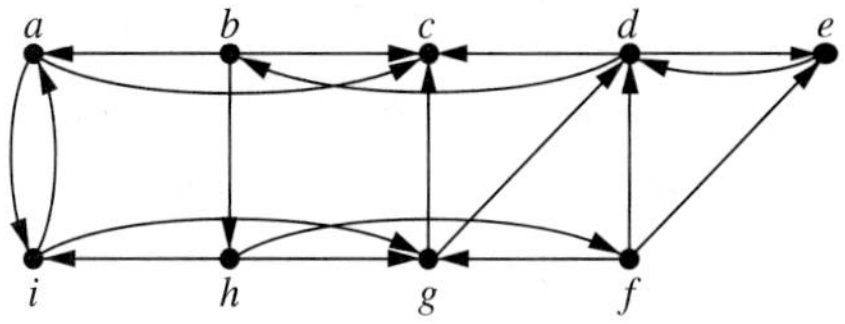

Suppose that $G = (V, E)$ is a directed graph. A vertex $w \in V$ is **reachable** from a vertex $v \in V$ if there is a directed path from v to w. The vertices v and w are **mutually reachable** if there are both a directed path from v to w and a directed path from w to v in G.

16. Show that if $G = (V, E)$ is a directed graph and u, v, and w are vertices in V for which u and v are mutually reachable and v and w are mutually reachable, then u and w are mutually reachable.

17. Show that if $G = (V, E)$ is a directed graph, then the strong components of two vertices u and v of V are either the same or disjoint. [*Hint:* Use Exercise 16.]

18. Show that all vertices visited in a directed path connecting two vertices in the same strongly connected component of a directed graph are also in this strongly connected component.

19. Find the number of paths of length n between two different vertices in K_4 if n is

a) 2. **b)** 3. **c)** 4. **d)** 5.

20. Use paths either to show that these graphs are not isomorphic or to find an isomorphism between these graphs.

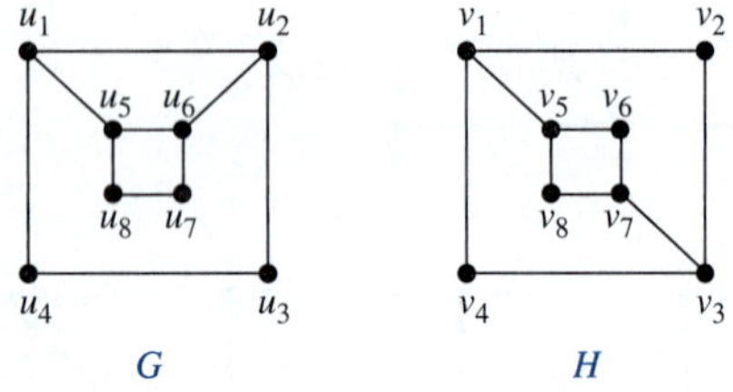

21. Use paths either to show that these graphs are not isomorphic or to find an isomorphism between them.

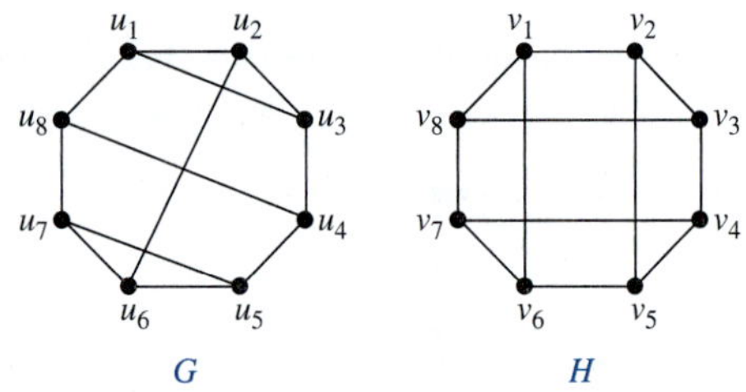

22. Use paths either to show that these graphs are not isomorphic or to find an isomorphism between them.

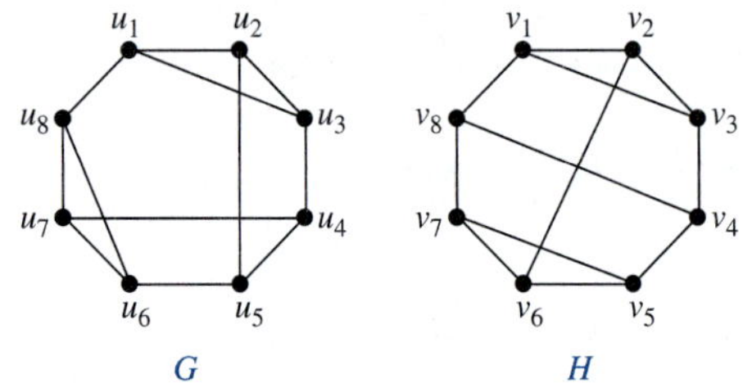

23. Use paths either to show that these graphs are not isomorphic or to find an isomorphism between them.

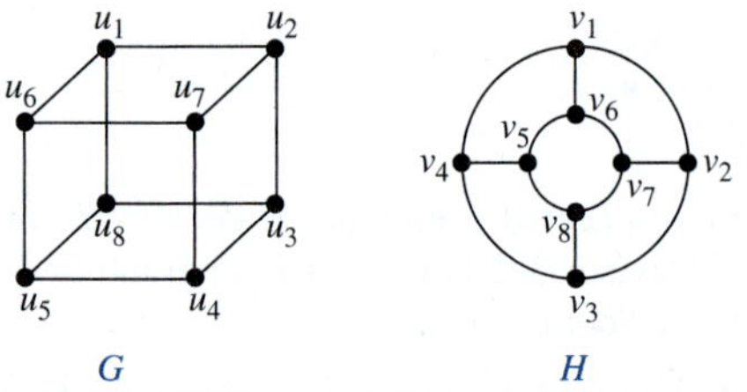

24. Find the number of paths of length n between any two adjacent vertices in $K_{3,3}$ for the values of n in Exercise 19.

25. Find the number of paths of length n between any two nonadjacent vertices in $K_{3,3}$ for the values of n in Exercise 19.

26. Find the number of paths between c and d in the graph in Figure 1 of length

a) 2. **b)** 3. **c)** 4. **d)** 5. **e)** 6. **f)** 7.

27. Find the number of paths from a to e in the directed graph in Exercise 2 of length

a) 2. **b)** 3. **c)** 4. **d)** 5. **e)** 6. **f)** 7.

***28.** Show that every connected graph with n vertices has at least $n - 1$ edges.

29. Let $G = (V, E)$ be a simple graph. Let R be the relation on V consisting of pairs of vertices (u, v) such that there is a path from u to v or such that $u = v$. Show that R is an equivalence relation.

***30.** Show that in every simple graph there is a path from every vertex of odd degree to some other vertex of odd degree.

In Exercises 31–33 find all the cut vertices of the given graph.

31.

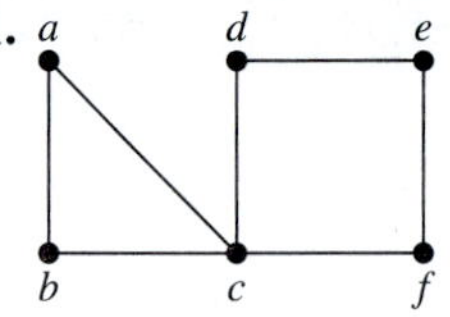

32.

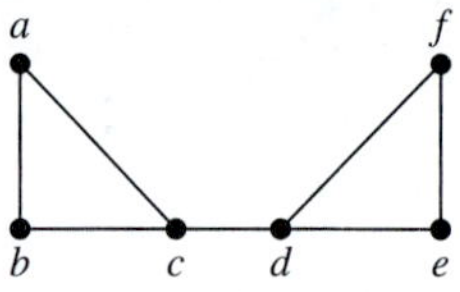

33.

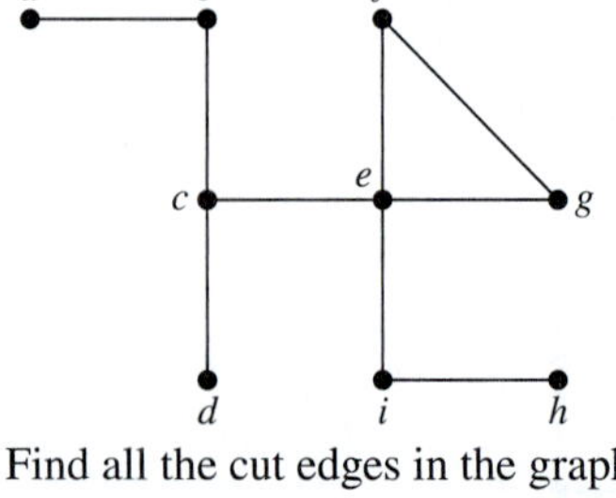

34. Find all the cut edges in the graphs in Exercises 31–33.

***35.** Suppose that v is an endpoint of a cut edge. Prove that v is a cut vertex if and only if this vertex is not pendant.

***36.** Show that a vertex c in the connected simple graph G is a cut vertex if and only if there are vertices u and v, both different from c, such that every path between u and v passes through c.

***37.** Show that a simple graph with at least two vertices has at least two vertices that are not cut vertices.

***38.** Show that an edge in a simple graph is a cut edge if and only if this edge is not part of any simple circuit in the graph.

39. A communications link in a network should be provided with a backup link if its failure makes it impossible for some message to be sent. For each of the communications networks shown here in (a) and (b), determine those links that should be backed up.

a)

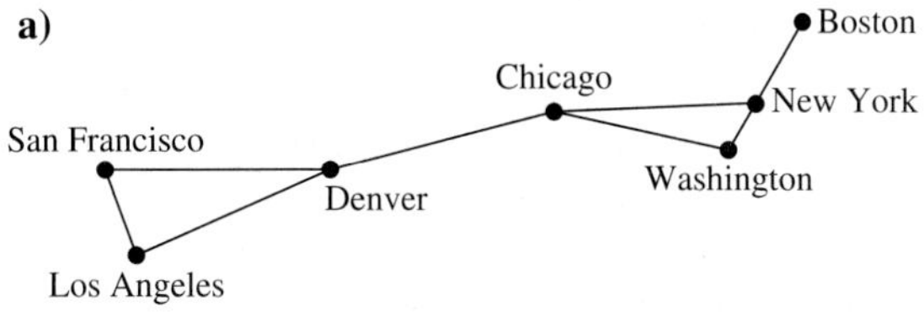

b)

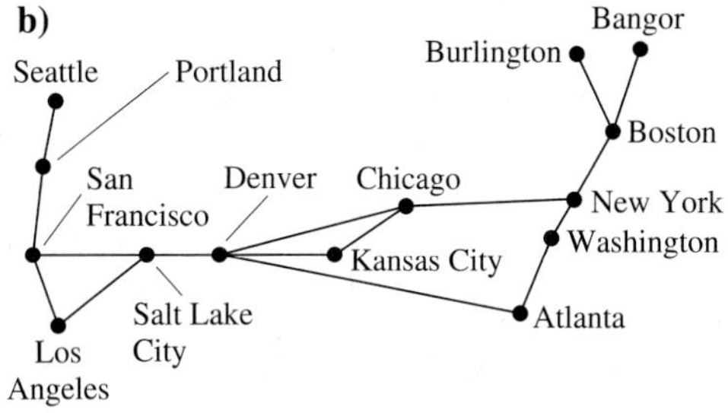

A **vertex basis** in a directed graph G is a minimal set B of vertices of G such that for each vertex v of G not in B there is a path to v from some vertex B.

40. Find a vertex basis for each of the directed graphs in Exercises 7–9 of Section 10.2.

41. What is the significance of a vertex basis in an influence graph (described in Example 2 of Section 10.1)? Find a vertex basis in the influence graph in that example.

42. Show that if a connected simple graph G is the union of the graphs G_1 and G_2, then G_1 and G_2 have at least one common vertex.

***43.** Show that if a simple graph G has k connected components and these components have $n_1, n_2, \ldots, n_k$ vertices, respectively, then the number of edges of G does not exceed

$$\sum_{i=1}^{k} C(n_i, 2).$$

***44.** Use Exercise 43 to show that a simple graph with n vertices and k connected components has at most $(n-k)(n-k+1)/2$ edges. [*Hint:* First show that

$$\sum_{i=1}^{k} n_i^2 \leq n^2 - (k-1)(2n-k),$$

where n_i is the number of vertices in the ith connected component.]

***45.** Show that a simple graph G with n vertices is connected if it has more than $(n-1)(n-2)/2$ edges.

46. Describe the adjacency matrix of a graph with n connected components when the vertices of the graph are listed so that vertices in each connected component are listed successively.

47. How many nonisomorphic connected simple graphs are there with n vertices when n is

a) 2? **b)** 3? **c)** 4? **d)** 5?

48. Show that each of the following graphs has no cut vertices.

a) C_n where $n \geq 3$
b) W_n where $n \geq 3$
c) $K_{m,n}$ where $m \geq 2$ and $n \geq 2$
d) Q_n where $n \geq 2$

49. Show that each of the graphs in Exercise 48 has no cut edges.

50. For each of these graphs, find $\kappa(G)$, $\lambda(G)$, and $\min_{v \in V} \deg(v)$, and determine which of the two inequalities in $\kappa(G) \leq \lambda(G) \leq \min_{v \in V} \deg(v)$ are strict.

a)

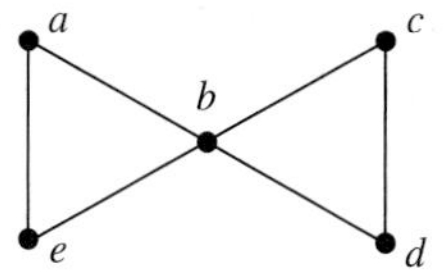

b)

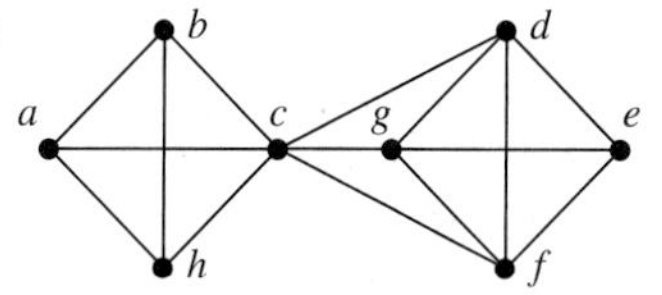

c)

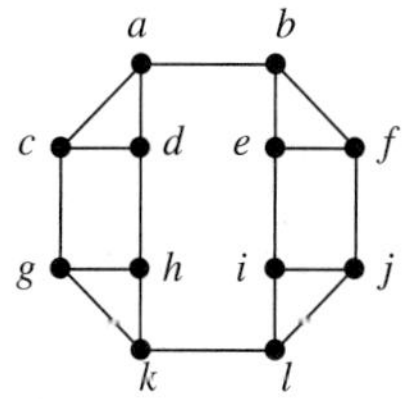

d)

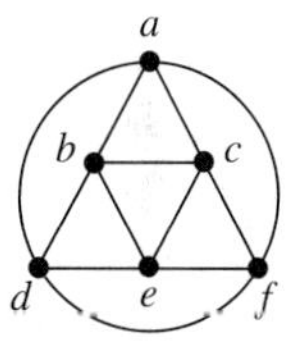

51. Show that if G is a connected graph, then it is possible to remove vertices to disconnect G if and only if G is not a complete graph.

52. Show that if G is a connected graph with n vertices then

a) $\kappa(G) = n - 1$ if and only if $G = K_n$.
b) $\lambda(G) = n - 1$ if and only if $G = K_n$.

53. Find $\kappa(K_{m,n})$ and $\lambda(K_{m,n})$, where m and n are positive integers.

54. Construct a graph G with $\kappa(G) = 1$, $\lambda(G) = 2$, and $\min_{v \in V} \deg(v) = 3$.

***55.** Show that if G is a graph, then $\kappa(G) \leq \lambda(G)$.

56. Explain how Theorem 2 can be used to find the length of the shortest path from a vertex v to a vertex w in a graph.

57. Use Theorem 2 to find the length of the shortest path between a and f in the graph in Figure 1.

58. Use Theorem 2 to find the length of the shortest path from a to c in the directed graph in Exercise 2.

59. Let P_1 and P_2 be two simple paths between the vertices u and v in the simple graph G that do not contain the same set of edges. Show that there is a simple circuit in G.

60. Show that the existence of a simple circuit of length k, where k is an integer greater than 2, is an invariant for graph isomorphism.

61. Explain how Theorem 2 can be used to determine whether a graph is connected.

62. Use Exercise 61 to show that the graph G_1 in Figure 2 is connected whereas the graph G_2 in that figure is not connected.

63. Show that a simple graph G is bipartite if and only if it has no circuits with an odd number of edges.

64. In an old puzzle attributed to Alcuin of York (735–804), a farmer needs to carry a wolf, a goat, and a cabbage across a river. The farmer only has a small boat, which can carry the farmer and only one object (an animal or a vegetable). The farmer can cross the river repeatedly. However, if the farmer is on the other shore, the wolf will eat the goat, and, similarly, the goat will eat the cabbage. We can describe each state by listing what is on each shore. For example, we can use the pair (FG, WC) for the state where the farmer and goat are on the first shore and the wolf and cabbage are on the other shore. [The symbol $\emptyset$ is used when nothing is on a shore, so that $(FWGC, \emptyset)$ is the initial state.]

a) Find all allowable states of the puzzle, where neither the wolf and the goat nor the goat and the cabbage are left on the same shore without the farmer.

b) Construct a graph such that each vertex of this graph represents an allowable state and the vertices representing two allowable states are connected by an edge if it is possible to move from one state to the other using one trip of the boat.

c) Explain why finding a path from the vertex representing $(FWGC, \emptyset)$ to the vertex representing $(\emptyset, FWGC)$ solves the puzzle.

d) Find two different solutions of the puzzle, each using seven crossings.

e) Suppose that the farmer must pay a toll of one dollar whenever he crosses the river with an animal. Which solution of the puzzle should the farmer use to pay the least total toll?

***65.** Use a graph model and a path in your graph, as in Exercise 64, to solve the **jealous husbands problem**. Two married couples, each a husband and a wife, want to cross a river. They can only use a boat that can carry one or two people from one shore to the other shore. Each husband is extremely jealous and is not willing to leave his wife with the other husband, either in the boat or on shore. How can these four people reach the opposite shore?

66. Suppose that you have a three-gallon jug and a five-gallon jug. You may fill either jug with water, you may empty either jug, and you may transfer water from either jug into the other jug. Use a path in a directed graph to show that you can end up with a jug containing exactly one gallon. [*Hint:* Use an ordered pair (a, b) to indicate how much water is in each jug. Represent these ordered pairs by vertices. Add an edge for each allowable operation with the jugs.]

10.5 Euler and Hamilton Paths

10.5.1 Introduction

Can we travel along the edges of a graph starting at a vertex and returning to it by traversing each edge of the graph exactly once? Similarly, can we travel along the edges of a graph starting at a vertex and returning to it while visiting each vertex of the graph exactly once? Although these questions seem to be similar, the first question, which asks whether a graph has an *Euler circuit*, can be easily answered simply by examining the degrees of the vertices of the graph, while the second question, which asks whether a graph has a *Hamilton circuit*, is quite difficult to solve for most graphs. In this section we will study these questions and discuss the difficulty of solving them. Although both questions have many practical applications in many different areas, both arose in old puzzles. We will learn about these old puzzles as well as modern practical applications.

10.5.2 Euler Paths and Circuits

Links

Only five bridges connect Kaliningrad today. Of these, just two remain from Euler's day.

The town of Königsberg, Prussia (now called Kaliningrad and part of the Russian republic), was divided into four sections by the branches of the Pregel River. These four sections included the two regions on the banks of the Pregel, Kneiphof Island, and the region between the two branches of the Pregel. In the eighteenth century seven bridges connected these regions. Figure 1 depicts these regions and bridges.

The townspeople took long walks through town on Sundays. They wondered whether it was possible to start at some location in the town, travel across all the bridges once without crossing any bridge twice, and return to the starting point.

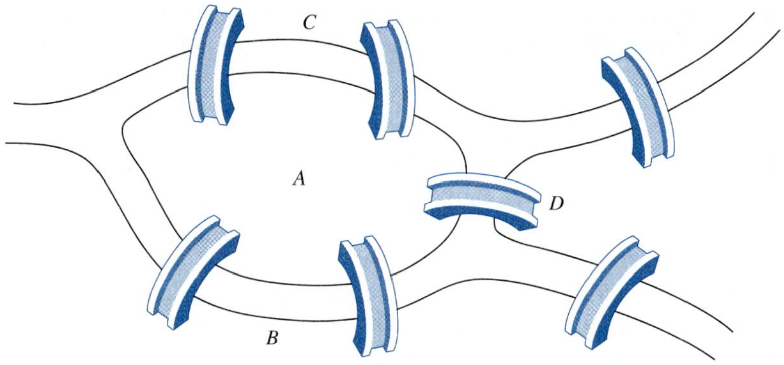

FIGURE 1 **The seven bridges of Königsberg.**

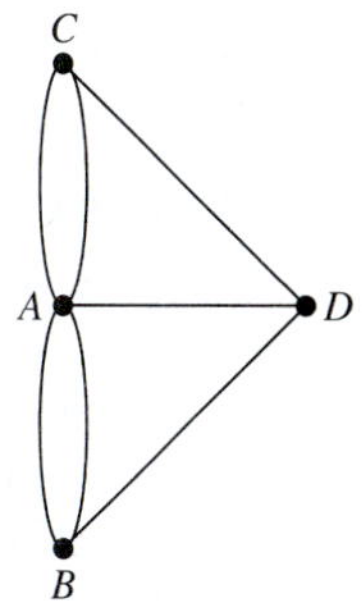

FIGURE 2 **Multigraph model of the town of Königsberg.**

The Swiss mathematician Leonhard Euler solved this problem. His solution, published in 1736, may be the first use of graph theory. (For a translation of Euler's original paper see [BiLlWi99].) Euler studied this problem using the multigraph obtained when the four regions are represented by vertices and the bridges by edges. This multigraph is shown in Figure 2.

The problem of traveling across every bridge without crossing any bridge more than once can be rephrased in terms of this model. The question becomes: Is there a simple circuit in this multigraph that contains every edge?

Definition 1 An *Euler circuit* in a graph G is a simple circuit containing every edge of G. An *Euler path* in G is a simple path containing every edge of G.

Examples 1 and 2 illustrate the concept of Euler circuits and paths.

EXAMPLE 1 Which of the undirected graphs in Figure 3 have an Euler circuit? Of those that do not, which have an Euler path?

Solution: The graph G_1 has an Euler circuit, for example, a, e, c, d, e, b, a. Neither of the graphs G_2 or G_3 has an Euler circuit (the reader should verify this). However, G_3 has an Euler path, namely, a, c, d, e, b, d, a, b. G_2 does not have an Euler path (as the reader should verify). ◀

EXAMPLE 2 Which of the directed graphs in Figure 4 have an Euler circuit? Of those that do not, which have an Euler path?

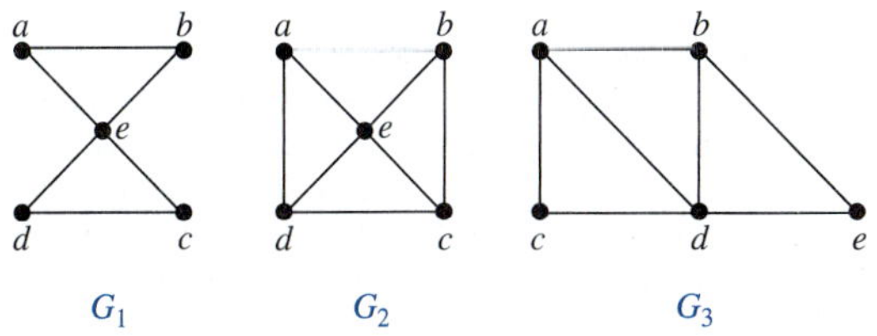

FIGURE 3 **The undirected graphs G_1, G_2, and G_3.**

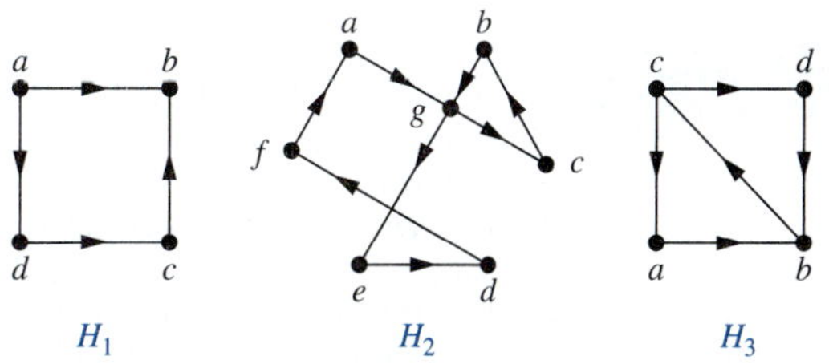

FIGURE 4 **The directed graphs H_1, H_2, and H_3.**

Extra Examples

Solution: The graph H_2 has an Euler circuit, for example, $a, g, c, b, g, e, d, f, a$. Neither H_1 nor H_3 has an Euler circuit (as the reader should verify). H_3 has an Euler path, namely, c, a, b, c, d, b, but H_1 does not (as the reader should verify). ◀

NECESSARY AND SUFFICIENT CONDITIONS FOR EULER CIRCUITS AND PATHS

There are simple criteria for determining whether a multigraph has an Euler circuit or an Euler path. Euler discovered them when he solved the famous Königsberg bridge problem. We will assume that all graphs discussed in this section have a finite number of vertices and edges.

What can we say if a connected multigraph has an Euler circuit? What we can show is that every vertex must have even degree. To do this, first note that an Euler circuit begins with a vertex a and continues with an edge incident with a, say $\{a, b\}$. The edge $\{a, b\}$ contributes one to $\deg(a)$. Each time the circuit passes through a vertex it contributes two to the vertex's degree, because the circuit enters via an edge incident with this vertex and leaves via another such edge. Finally, the circuit terminates where it started, contributing one to $\deg(a)$. Therefore, $\deg(a)$ must be even, because the circuit contributes one when it begins, one when it ends, and two every time it passes through a (if it ever does). A vertex other than a has even degree because the circuit contributes two to its degree each time it passes through the vertex. We conclude that if a connected graph has an Euler circuit, then every vertex must have even degree.

Is this necessary condition for the existence of an Euler circuit also sufficient? That is, must an Euler circuit exist in a connected multigraph if all vertices have even degree? This question can be settled affirmatively with a construction.

Suppose that G is a connected multigraph with at least two vertices and the degree of every vertex of G is even. We will form a simple circuit that begins at an arbitrary vertex a of G, building it edge by edge. Let $x_0 = a$. First, we arbitrarily choose an edge $\{x_0, x_1\}$ incident with a which is possible because G is connected. We continue by building a simple path $\{x_0, x_1\}, \{x_1, x_2\}, \ldots, \{x_{n-1}, x_n\}$, successively adding edges one by one to the path until we cannot add another edge to the path. This happens when we reach a vertex for which we have already included all edges incident with that vertex in the path. For instance, in the graph G in Figure 5 we begin at a and choose in succession the edges $\{a, f\}$, $\{f, c\}$, $\{c, b\}$, and $\{b, a\}$.

The path we have constructed must terminate because the graph has a finite number of edges, so we are guaranteed to eventually reach a vertex for which no edges are available to add to the path. The path begins at a with an edge of the form $\{a, x\}$, and we now show that it must terminate at a with an edge of the form $\{y, a\}$. To see that the path must terminate at a, note that each time the path goes through a vertex with even degree, it uses only one edge to enter

Links

Photos.com/Getty Images

LEONHARD EULER (1707–1783) Leonhard Euler was the son of a Calvinist minister from the vicinity of Basel, Switzerland. At 13 he entered the University of Basel, pursuing a career in theology, as his father wished. At the university Euler was tutored by Johann Bernoulli of the famous Bernoulli family of mathematicians. His interest and skills led him to abandon his theological studies and take up mathematics. Euler obtained his master's degree in philosophy at the age of 16. In 1727 Peter the Great invited him to join the Academy at St. Petersburg. In 1741 he moved to the Berlin Academy, where he stayed until 1766. He then returned to St. Petersburg, where he remained for the rest of his life.

Euler was incredibly prolific, contributing to many areas of mathematics, including number theory, combinatorics, and analysis, as well as its applications to such areas as music and naval architecture. He wrote over 1100 books and papers and left so much unpublished work that it took 47 years after he died for all his work to be published. During his life his papers accumulated so quickly that he kept a large pile of articles awaiting publication. The Berlin Academy published the papers on top of this pile so later results were often published before results they depended on or superseded. Euler had 13 children and was able to continue his work while a child or two bounced on his knees. He was blind for the last 17 years of his life, but because of his fantastic memory this did not diminish his mathematical output. The project of publishing his collected works, undertaken by the Swiss Society of Natural Science, is ongoing and will require more than 75 volumes.

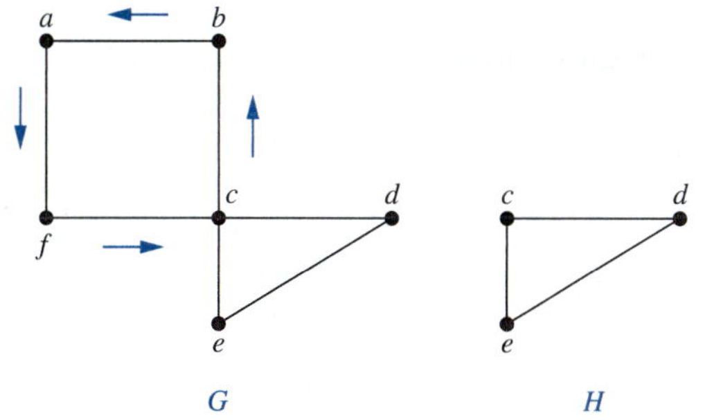

FIGURE 5 **Constructing an Euler circuit in *G*.**

this vertex, so because the degree must be at least two, at least one edge remains for the path to leave the vertex. Furthermore, every time we enter and leave a vertex of even degree, there are an even number of edges incident with this vertex that we have not yet used in our path. Consequently, as we form the path, every time we enter a vertex other than a, we can leave it. This means that the path can end only at a. Next, note that the path we have constructed may use all the edges of the graph, or it may not if we have returned to a for the last time before using all the edges.

An Euler circuit has been constructed if all the edges have been used. Otherwise, consider the subgraph H obtained from G by deleting the edges already used and vertices that are not incident with any remaining edges. When we delete the circuit a, f, c, b, a from the graph in Figure 5, we obtain the subgraph labeled as H.

Because G is connected, H has at least one vertex in common with the circuit that has been deleted. Let w be such a vertex. (In our example, c is the vertex.)

Every vertex in H has even degree (because in G all vertices had even degree, and for each vertex, pairs of edges incident with this vertex have been deleted to form H). Note that H may not be connected. Beginning at w, construct a simple path in H by choosing edges as long as possible, as was done in G. This path must terminate at w. For instance, in Figure 5, c, d, e, c is a path in H. Next, form a circuit in G by splicing the circuit in H with the original circuit in G (this can be done because w is one of the vertices in this circuit). When this is done in the graph in Figure 5, we obtain the circuit a, f, c, d, e, c, b, a.

Continue this process until all edges have been used. (The process must terminate because there are only a finite number of edges in the graph.) This produces an Euler circuit. The construction shows that if the vertices of a connected multigraph all have even degree, then the graph has an Euler circuit.

We summarize these results in Theorem 1.

THEOREM 1 A connected multigraph with at least two vertices has an Euler circuit if and only if each of its vertices has even degree.

We can now solve the Königsberg bridge problem. Because the multigraph representing these bridges, shown in Figure 2, has four vertices of odd degree, it does not have an Euler circuit. There is no way to start at a given point, cross each bridge exactly once, and return to the starting point.

Algorithm 1 gives the constructive procedure for finding Euler circuits given in the discussion preceding Theorem 1. (Because the circuits in the procedure are chosen arbitrarily, there is some ambiguity. We will not bother to remove this ambiguity by specifying the steps of the procedure more precisely.)

ALGORITHM 1 Constructing Euler Circuits.

procedure *Euler*(G: connected multigraph with all vertices of even degree)
circuit := a circuit in G beginning at an arbitrarily chosen vertex with edges successively added to form a path that returns to this vertex
H := G with the edges of this circuit removed
while H has edges
    subcircuit := a circuit in H beginning at a vertex in H that also is an endpoint of an edge of *circuit*
    H := H with edges of *subcircuit* and all isolated vertices removed
    circuit := *circuit* with *subcircuit* inserted at the appropriate vertex
return *circuit* {*circuit* is an Euler circuit}

Algorithm 1 provides an efficient algorithm for finding Euler circuits in a connected multigraph G with all vertices of even degree. We leave it to the reader (Exercise 68) to show that the worst case complexity of this algorithm is $O(m)$, where m is the number of edges of G.

Example 3 shows how Euler paths and circuits can be used to solve a type of puzzle.

EXAMPLE 3 Many puzzles ask you to draw a picture in a continuous motion without lifting a pencil so that no part of the picture is retraced. We can solve such puzzles using Euler circuits and paths. For example, can *Mohammed's scimitars*, shown in Figure 6, be drawn in this way, where the drawing begins and ends at the same point?

Solution: We can solve this problem because the graph G shown in Figure 6 has an Euler circuit. It has such a circuit because all its vertices have even degree. We will use Algorithm 1 to construct an Euler circuit. First, we form the circuit $a, b, d, c, b, e, i, f, e, a$. We obtain the subgraph H by deleting the edges in this circuit and all vertices that become isolated when these edges are removed. Then we form the circuit $d, g, h, j, i, h, k, g, f, d$ in H. After forming this circuit we have used all edges in G. Splicing this new circuit into the first circuit at the appropriate place produces the Euler circuit $a, b, d, g, h, j, i, h, k, g, f, d, c, b, e, i, f, e, a$. This circuit gives a way to draw the scimitars without lifting the pencil or retracing part of the picture. ◀

Another algorithm for constructing Euler circuits, called Fleury's algorithm, is described in the premble to Exercise 50.

We will now show that a connected multigraph has an Euler path (and not an Euler circuit) if and only if it has exactly two vertices of odd degree. First, suppose that a connected multigraph

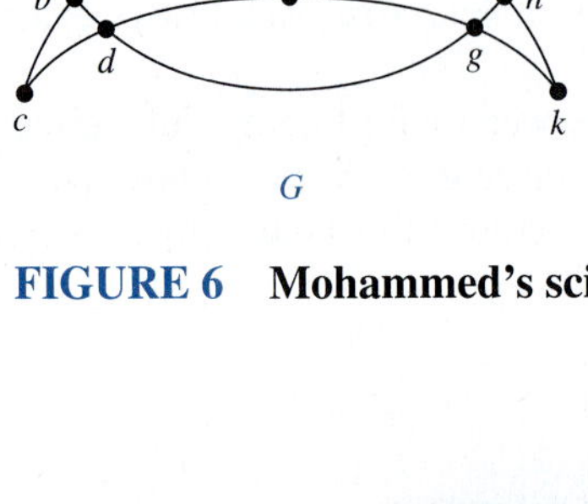

FIGURE 6 Mohammed's scimitars.

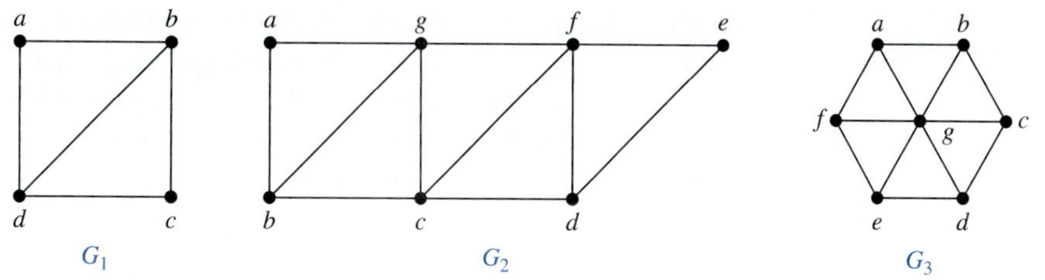

FIGURE 7 Three undirected graphs.

does have an Euler path from a to b, but not an Euler circuit. The first edge of the path contributes one to the degree of a. A contribution of two to the degree of a is made every time the path passes through a. The last edge in the path contributes one to the degree of b. Every time the path goes through b there is a contribution of two to its degree. Consequently, both a and b have odd degree. Every other vertex has even degree, because the path contributes two to the degree of a vertex whenever it passes through it.

Now consider the converse. Suppose that a graph has exactly two vertices of odd degree, say a and b. Consider the larger graph made up of the original graph with the addition of an edge $\{a, b\}$. Every vertex of this larger graph has even degree, so there is an Euler circuit. The removal of the new edge produces an Euler path in the original graph. Theorem 2 summarizes these results.

THEOREM 2 A connected multigraph has an Euler path but not an Euler circuit if and only if it has exactly two vertices of odd degree.

EXAMPLE 4 Which graphs shown in Figure 7 have an Euler path?

Solution: G_1 contains exactly two vertices of odd degree, namely, b and d. Hence, it has an Euler path that must have b and d as its endpoints. One such Euler path is d, a, b, c, d, b. Similarly, G_2 has exactly two vertices of odd degree, namely, b and d. So it has an Euler path that must have b and d as endpoints. One such Euler path is $b, a, g, f, e, d, c, g, b, c, f, d$. G_3 has no Euler path because it has six vertices of odd degree. ◀

Returning to eighteenth-century Königsberg, is it possible to start at some point in the town, travel across all the bridges, and end up at some other point in town? This question can be answered by determining whether there is an Euler path in the multigraph representing the bridges in Königsberg. Because there are four vertices of odd degree in this multigraph, there is no Euler path, so such a trip is impossible.

Necessary and sufficient conditions for Euler paths and circuits in directed graphs are given in Exercises 16 and 17.

Links

APPLICATIONS OF EULER PATHS AND CIRCUITS Euler paths and circuits can be used to solve many practical problems. For example, many applications ask for a path or circuit that traverses each street in a neighborhood, each road in a transportation network, each connection in a utility grid, or each link in a communications network exactly once. Finding an Euler path or circuit in the appropriate graph model can solve such problems. For example, if a postman can find an Euler path in the graph that represents the streets the postman needs to cover, this path produces a route that traverses each street of the route exactly once. If no Euler path exists, some streets will have to be traversed more than once. The problem of finding a circuit in a graph with the fewest edges that traverses every edge at least once is known as the *Chinese postman* *problem* in honor of Guan Meigu, who posed it in 1962. See [MiRo91] for more information on the solution of the Chinese postman problem when no Euler path exists.

Among the other areas where Euler circuits and paths are applied is in the layout of circuits, in network multicasting, and in molecular biology, where Euler paths are used in the sequencing of DNA.

10.5.3 Hamilton Paths and Circuits

Links

We have developed necessary and sufficient conditions for the existence of paths and circuits that contain every edge of a multigraph exactly once. Can we do the same for simple paths and circuits that contain every vertex of the graph exactly once?

Definition 2

A simple path in a graph G that passes through every vertex exactly once is called a *Hamilton path*, and a simple circuit in a graph G that passes through every vertex exactly once is called a *Hamilton circuit*. That is, the simple path $x_0, x_1, \ldots, x_{n-1}, x_n$ in the graph $G = (V, E)$ is a Hamilton path if $V = \{x_0, x_1, \ldots, x_{n-1}, x_n\}$ and $x_i \neq x_j$ for $0 \leq i < j \leq n$, and the simple circuit $x_0, x_1, \ldots, x_{n-1}, x_n, x_0$ (with $n > 0$) is a Hamilton circuit if $x_0, x_1, \ldots, x_{n-1}, x_n$ is a Hamilton path.

This terminology comes from a game, called the *Icosian puzzle*, invented in 1857 by the Irish mathematician Sir William Rowan Hamilton. It consisted of a wooden dodecahedron [a polyhedron with 12 regular pentagons as faces, as shown in Figure 8(a)], with a peg at each vertex of the dodecahedron, and string. The 20 vertices of the dodecahedron were labeled with different cities in the world. The object of the puzzle was to start at a city and travel along the edges of the dodecahedron, visiting each of the other 19 cities exactly once, and end back at the first city. The circuit traveled was marked off using the string and pegs.

Because the author cannot supply each reader with a wooden solid with pegs and string, we will consider the equivalent question: Is there a circuit in the graph shown in Figure 8(b) that passes through each vertex exactly once? This solves the puzzle because this graph is isomorphic to the graph consisting of the vertices and edges of the dodecahedron. A solution of Hamilton's puzzle is shown in Figure 9.

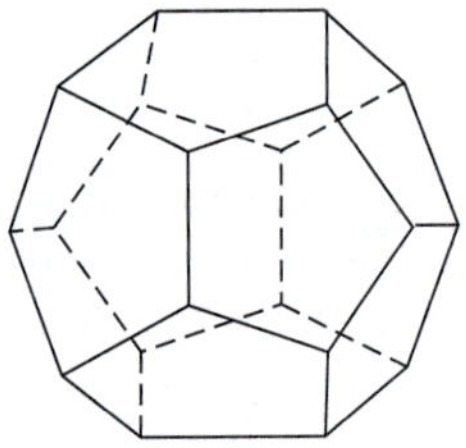

(a)

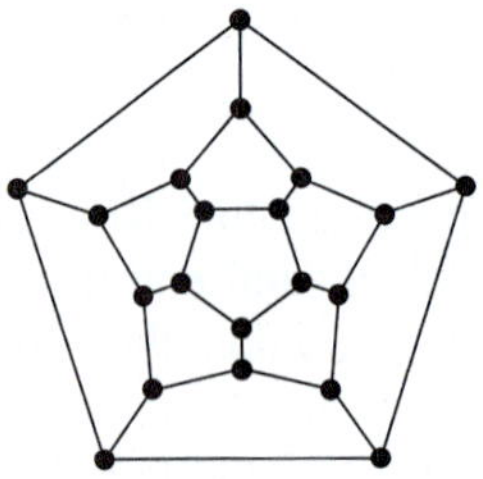

(b)

FIGURE 8 Hamilton's "A Voyage Round the World" puzzle.

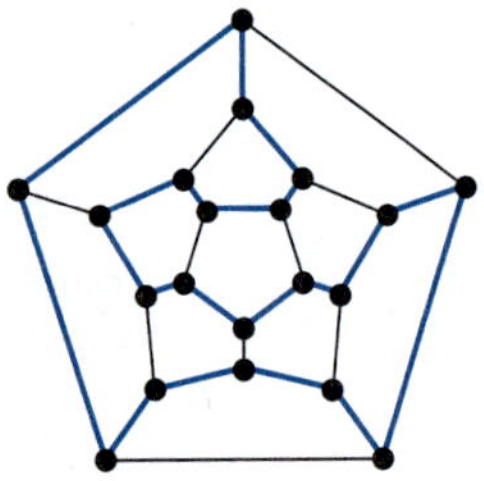

FIGURE 9 A solution to the "A Voyage Round the World" puzzle.

EXAMPLE 5 Which of the simple graphs in Figure 10 have a Hamilton circuit or, if not, a Hamilton path?

Extra Examples

Solution: G_1 has a Hamilton circuit: a, b, c, d, e, a. There is no Hamilton circuit in G_2 (this can be seen by noting that any circuit containing every vertex must contain the edge $\{a, b\}$ twice), but G_2 does have a Hamilton path, namely, a, b, c, d. G_3 has neither a Hamilton circuit nor a Hamilton path, because any path containing all vertices must contain one of the edges $\{a, b\}$, $\{e, f\}$, and $\{c, d\}$ more than once. ◀

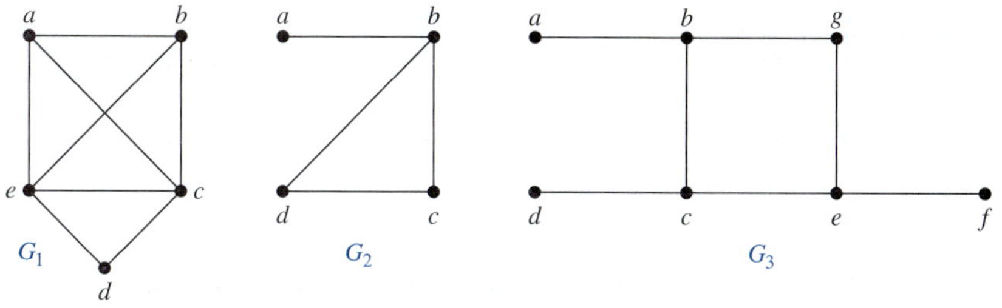

FIGURE 10 **Three simple graphs.**

CONDITIONS FOR THE EXISTENCE OF HAMILTON CIRCUITS Is there a simple way to determine whether a graph has a Hamilton circuit or path? At first, it might seem that there should be an easy way to determine this, because there is a simple way to answer the similar question of whether a graph has an Euler circuit. Surprisingly, there are no known simple necessary and sufficient criteria for the existence of Hamilton circuits. However, many theorems are known that give sufficient conditions for the existence of Hamilton circuits. Also, certain properties can be used to show that a graph has no Hamilton circuit. For instance, a graph with a vertex of degree one cannot have a Hamilton circuit, because in a Hamilton circuit, each vertex is incident with two edges in the circuit. Moreover, if a vertex in the graph has degree two, then both edges that are incident with this vertex must be part of any Hamilton circuit. Also, note that when a Hamilton circuit is being constructed and this circuit has passed through a vertex, then all remaining edges incident with this vertex, other than the two used in the circuit, can be removed from consideration. Furthermore, a Hamilton circuit cannot contain a smaller circuit within it.

EXAMPLE 6 Show that neither graph displayed in Figure 11 has a Hamilton circuit.

Solution: There is no Hamilton circuit in G because G has a vertex of degree one, namely, e. Now consider H. Because the degrees of the vertices a, b, d, and e are all two, every edge incident with these vertices must be part of any Hamilton circuit. It is now easy to see that no Hamilton circuit can exist in H, for any Hamilton circuit would have to contain four edges incident with c, which is impossible. ◀

Links

Hulton-Deutsch/ Corbis/Getty Images

WILLIAM ROWAN HAMILTON (1805–1865) William Rowan Hamilton, the most famous Irish scientist ever to have lived, was born in 1805 in Dublin. His father was a successful lawyer, his mother came from a family noted for their intelligence, and he was a child prodigy. By the age of 3 he was an excellent reader and had mastered advanced arithmetic. Because of his brilliance, he was sent off to live with his uncle James, a noted linguist. By age 8 Hamilton had learned Latin, Greek, and Hebrew; by 10 he had also learned Italian and French and he began his study of eastern languages, including Arabic, Sanskrit, and Persian. During this period he took pride in knowing as many languages as his age. At 17, no longer devoted to learning new languages and having mastered calculus and much mathematical astronomy, he began original work in optics, and he also found an important mistake in Laplace's work on celestial mechanics. Before entering Trinity College, Dublin, at 18, Hamilton had not attended school; rather, he received private tutoring. At Trinity, he was a superior student in both the sciences and the classics. Prior to receiving his degree, because of his brilliance he was appointed the Astronomer Royal of Ireland, beating out several famous astronomers for the post. He held this position until his death, living and working at Dunsink Observatory outside of Dublin. Hamilton made important contributions to optics, abstract algebra, and dynamics. Hamilton invented algebraic objects called quaternions as an example of a noncommutative system. He discovered the appropriate way to multiply quaternions while walking along a canal in Dublin. In his excitement, he carved the formula in the stone of a bridge crossing the canal, a spot marked today by a plaque. Later, Hamilton remained obsessed with quaternions, working to apply them to other areas of mathematics, instead of moving to new areas of research.

In 1857 Hamilton invented "The Icosian Game" based on his work in noncommutative algebra. He sold the idea for 25 pounds to a dealer in games and puzzles. (Because the game never sold well, this turned out to be a bad investment for the dealer.) The "Traveler's Dodecahedron," also called "A Voyage Round the World," the puzzle described in this section, is a variant of that game.

Hamilton married his third love in 1833, but his marriage worked out poorly. He suffered from alcoholism and lived reclusively for the last two decades of his life. He died from gout in 1865, leaving masses of papers containing unpublished research. Mixed in with these papers were a large number of dinner plates, many containing the remains of desiccated, uneaten chops.

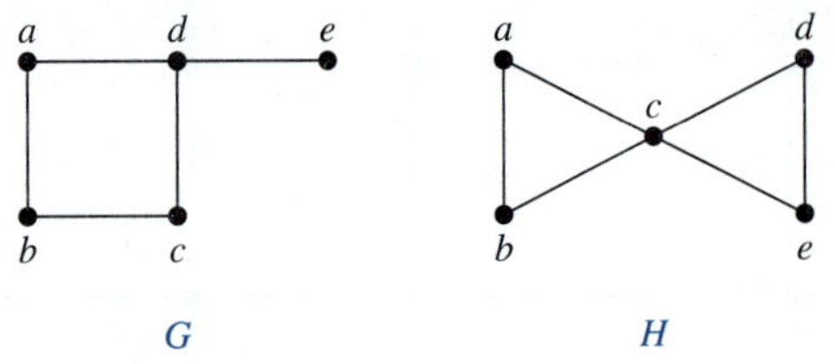

FIGURE 11 **Two graphs that do not have a Hamilton circuit.**

EXAMPLE 7 Show that K_n has a Hamilton circuit whenever $n \geq 3$.

Solution: We can form a Hamilton circuit in K_n beginning at any vertex. Such a circuit can be built by visiting vertices in any order we choose, as long as the path begins and ends at the same vertex and visits each other vertex exactly once. This is possible because there are edges in K_n between any two vertices. ◀

Although no useful necessary and sufficient conditions for the existence of Hamilton circuits are known, quite a few sufficient conditions have been found. Note that the more edges a graph has, the more likely it is to have a Hamilton circuit. Furthermore, adding edges (but not vertices) to a graph with a Hamilton circuit produces a graph with the same Hamilton circuit. So as we add edges to a graph, especially when we make sure to add edges to each vertex, we make it increasingly likely that a Hamilton circuit exists in this graph. Consequently, we would expect there to be sufficient conditions for the existence of Hamilton circuits that depend on the degrees of vertices being sufficiently large. We state two of the most important sufficient conditions here. These conditions were found by Gabriel A. Dirac in 1952 and Øystein Ore in 1960.

THEOREM 3 **DIRAC'S THEOREM** If G is a simple graph with n vertices with $n \geq 3$ such that the degree of every vertex in G is at least $n/2$, then G has a Hamilton circuit.

THEOREM 4 **ORE'S THEOREM** If G is a simple graph with n vertices with $n \geq 3$ such that $\deg(u) + \deg(v) \geq n$ for every pair of nonadjacent vertices u and v in G, then G has a Hamilton circuit.

The proof of Ore's theorem is outlined in Exercise 65. Dirac's theorem can be proved as a corollary to Ore's theorem because the conditions of Dirac's theorem imply those of Ore's theorem.

Both Ore's theorem and Dirac's theorem provide sufficient conditions for a connected simple graph to have a Hamilton circuit. However, these theorems do not provide necessary conditions for the existence of a Hamilton circuit. For example, the graph C_5 has a Hamilton circuit but does not satisfy the hypotheses of either Ore's theorem or Dirac's theorem, as the reader can verify.

Links ❯

The best algorithms known for finding a Hamilton circuit in a graph or determining that no such circuit exists have exponential worst-case time complexity (in the number of vertices of the graph). Finding an algorithm that solves this problem with polynomial worst-case time complexity would be a major accomplishment because it has been shown that this problem is NP-complete (see Section 3.3). Consequently, the existence of such an algorithm would imply that many other seemingly intractable problems could be solved using algorithms with polynomial worst-case time complexity.

10.5.4 Applications of Hamilton Circuits

Hamilton paths and circuits can be used to solve practical problems. For example, many applications ask for a path or circuit that visits each road intersection in a city, each place pipelines intersect in a utility grid, or each node in a communications network exactly once. Finding a Hamilton path or circuit in the appropriate graph model can solve such problems. The famous **traveling salesperson problem** or **TSP** (also known in older literature as the **traveling salesman problem**) asks for the shortest route a traveling salesperson should take to visit a set of cities once each and return to the starting city. This problem reduces to finding a Hamilton circuit in a complete graph such that the total weight of its edges is as small as possible. We will return to this question in Section 10.6.

We now describe a less obvious application of Hamilton circuits to coding.

EXAMPLE 8 **Gray Codes** The position of a rotating pointer can be represented in digital form. One way to do this is to split the circle into 2^n arcs of equal length and to assign a bit string of length n to each arc. Two ways to do this using bit strings of length three are shown in Figure 12.

The digital representation of the position of the pointer can be determined using a set of n contacts. Each contact is used to read one bit in the digital representation of the position. This is illustrated in Figure 13 for the two assignments from Figure 12.

When the pointer is near the boundary of two arcs, a mistake may be made in reading its position. This may result in a major error in the bit string read. For instance, in the coding scheme in Figure 12(a), if a small error is made in determining the position of the pointer, the bit string 100 is read instead of 011. All three bits are incorrect! To minimize the effect of an

Links

Courtesy of Gabriel Dirac

GABRIEL ANDREW DIRAC (1925–1984) Gabriel Dirac was born in Budapest. He moved to England in 1937 when his mother married the illustrious physicist and Nobel Laureate Paul Adrien Maurice Dirac, who adopted him. Gabriel A. Dirac entered Cambridge University in 1942, but his studies were interrupted by wartime service in the aviation industry. He obtained his Ph.D. in mathematics in 1951 from the University of London. He held university positions in England, Canada, Austria, Germany, and Denmark, where he spent his last 14 years. Dirac became interested in graph theory early in his career and help raise its status as an important topic of research. He made important contributions to many aspects of graph theory, including graph coloring and Hamilton circuits. Dirac attracted many students to graph theory and was noted as an excellent lecturer.

Dirac was noted for his penetrating mind and held unconventional views on many topics, including politics and social life. Dirac was a man with many interests and held a great passion for fine art. He had a happy family life with his wife Rosemari and his four children.

Museum of University History (MUV)

ØYSTEIN ORE (1899–1968) Ore was born in Kristiania (the old name for Oslo, Norway). In 1922 he received his bachelor's degree and in 1925 his Ph.D. in mathematics from Kristiania University, after studies in Germany and in Sweden. In 1927 he was recruited to leave his junior position at Kristiania and join Yale University. He was promoted rapidly at Yale, becoming full professor in 1929 and Sterling Professor in 1931, a position he held until 1968.

Ore made many contributions to number theory, ring theory, lattice theory, graph theory, and probability theory. He was a prolific author of papers and books. His interest in the history of mathematics is reflected in his biographies of Abel and Cardano, and in his popular textbook *Number Theory and its History*. He wrote four books on graph theory in the 1960s.

During and after World War II Ore played a major role supporting his native Norway. In 1947 King Haakon VII of Norway gave him the Knight Order of St. Olaf to recognize these efforts. Ore possessed deep knowledge of painting and sculpture and was an ardent collector of ancient maps. He was married and had two children.

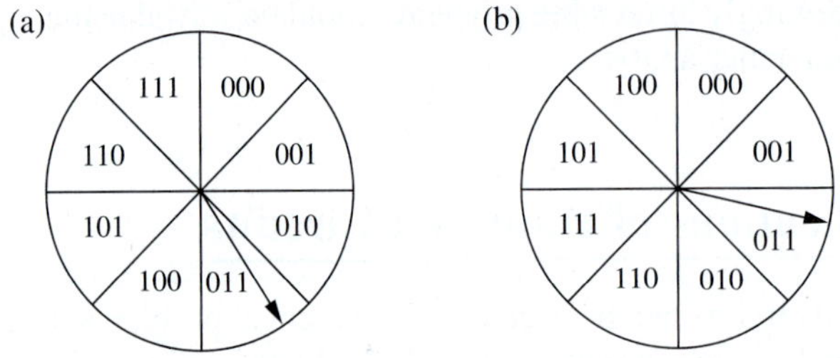

FIGURE 12 Converting the position of a pointer into digital form.

error in determining the position of the pointer, the assignment of the bit strings to the 2^n arcs should be made so that only one bit is different in the bit strings represented by adjacent arcs. This is exactly the situation in the coding scheme in Figure 12(b). An error in determining the position of the pointer gives the bit string 010 instead of 011. Only one bit is wrong.

Links

A **Gray code** is a labeling of the arcs of the circle such that adjacent arcs are labeled with bit strings that differ in exactly one bit. The assignment in Figure 12(b) is a Gray code. We can find a Gray code by listing all bit strings of length n in such a way that each string differs in exactly one position from the preceding bit string, and the last string differs from the first in exactly one position. We can model this problem using the n-cube Q_n. What is needed to solve this problem is a Hamilton circuit in Q_n. Such Hamilton circuits are easily found. For instance, a Hamilton circuit for Q_3 is displayed in Figure 14. The sequence of bit strings differing in exactly one bit produced by this Hamilton circuit is 000, 001, 011, 010, 110, 111, 101, 100.

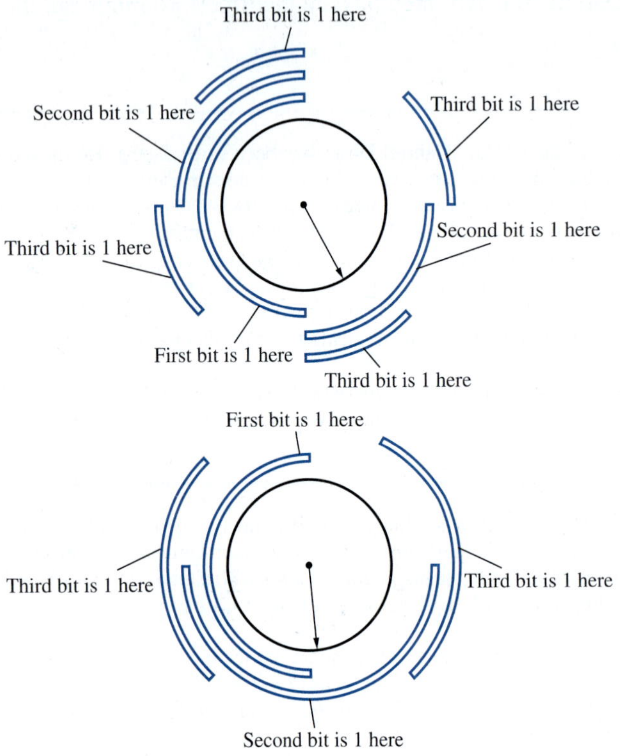

FIGURE 13 The digital representation of the position of the pointer.

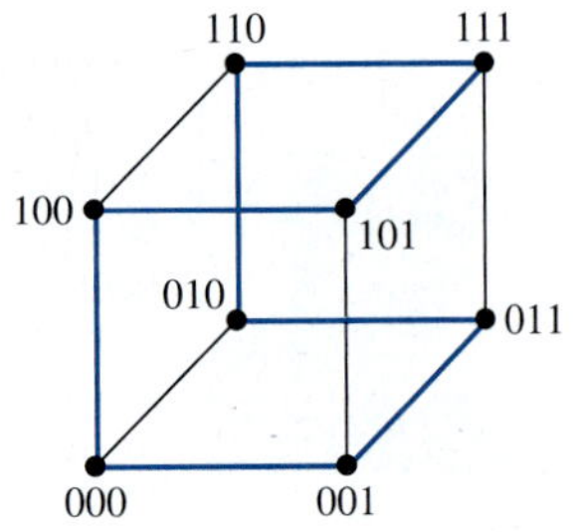

FIGURE 14 A Hamilton circuit for Q_3.

Gray codes are named after Frank Gray, who invented them in the 1940s at AT&T Bell Laboratories to minimize the effect of errors in transmitting digital signals. ◀

Exercises

In Exercises 1–8 determine whether the given graph has an Euler circuit. Construct such a circuit when one exists. If no Euler circuit exists, determine whether the graph has an Euler path and construct such a path if one exists.

1.

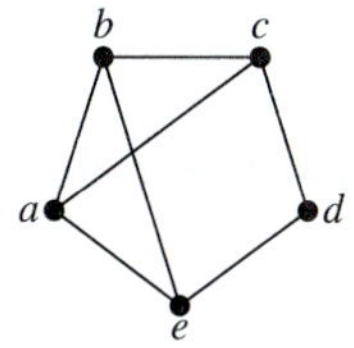

2.

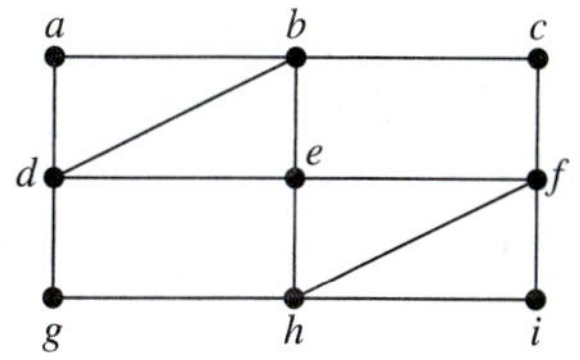

3.

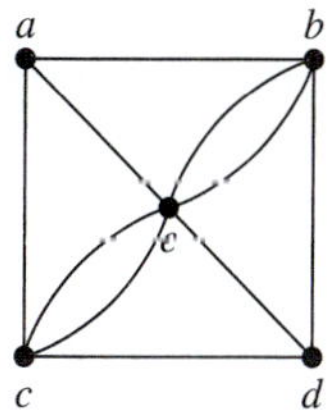

4.

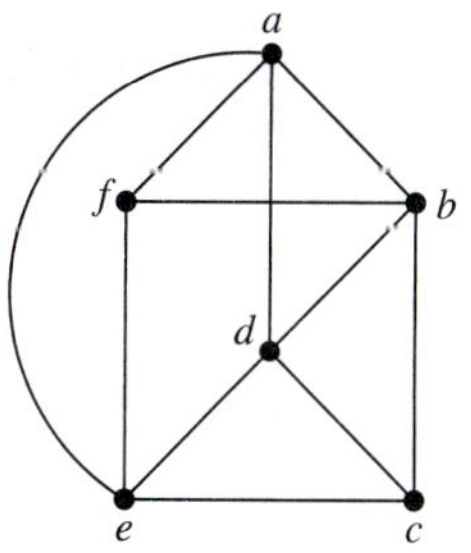

5.

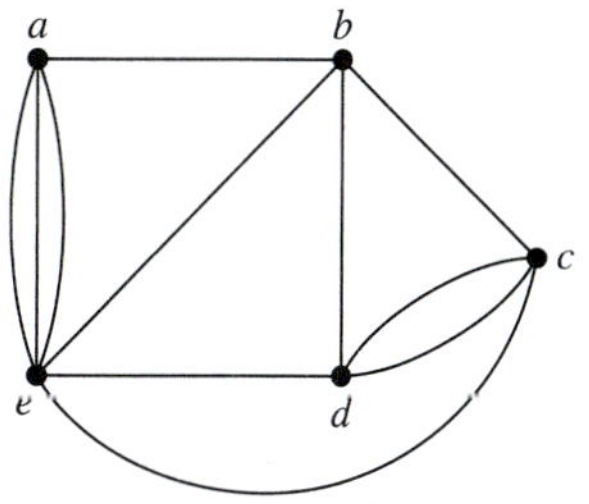

6.

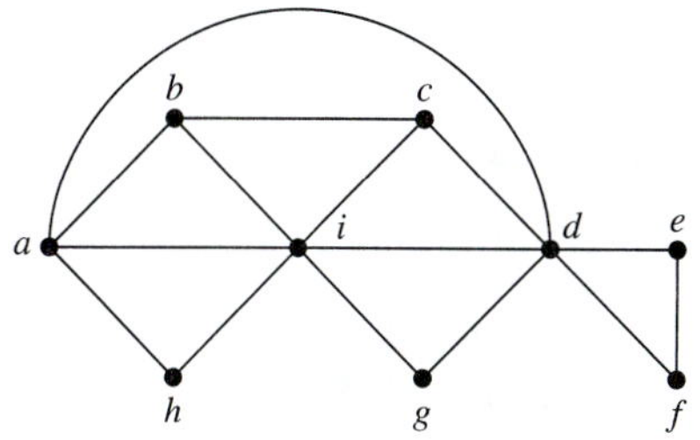

7.

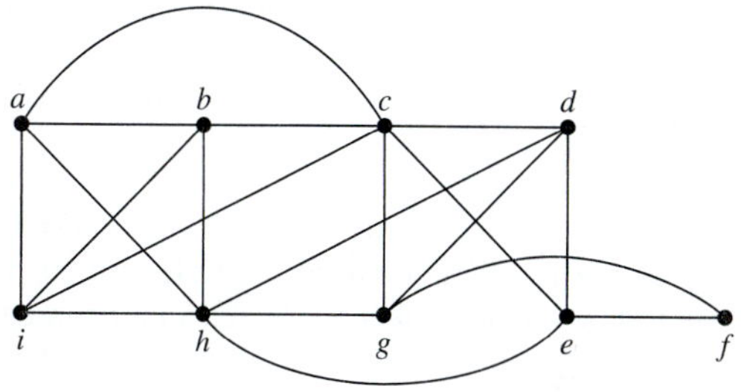

8.

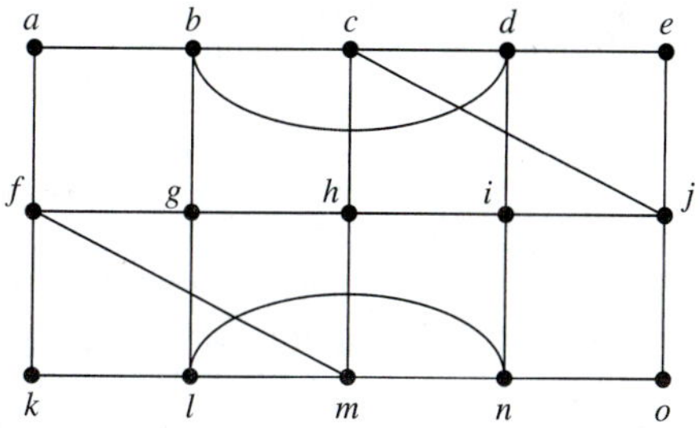

9. Suppose that in addition to the seven bridges of Königsberg (shown in Figure 1) there were two additional bridges, connecting regions B and C and regions B and D, respectively. Could someone cross all nine of these bridges exactly once and return to the starting point?

10. Can someone cross all the bridges shown in this map exactly once and return to the starting point?

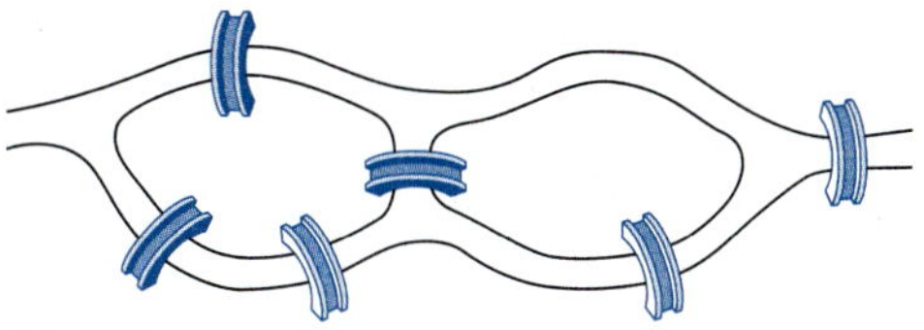

11. When can the centerlines of the streets in a city be painted without traveling a street more than once? (Assume that all the streets are two-way streets.)

12. Devise a procedure, similar to Algorithm 1, for constructing Euler paths in multigraphs.

In Exercises 13–15 determine whether the picture shown can be drawn with a pencil in a continuous motion without lifting the pencil or retracing part of the picture.

13.

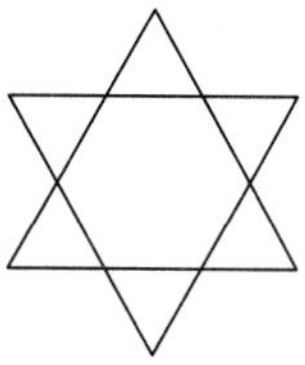

14.

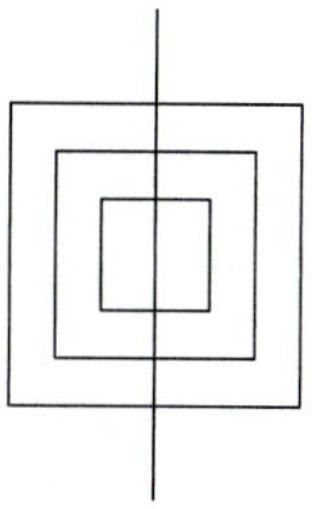

15.

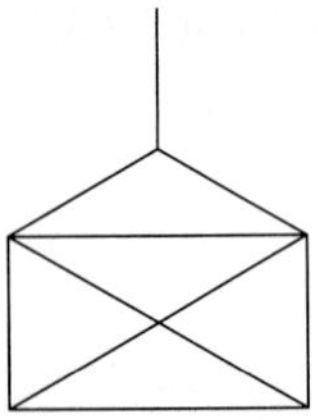

***16.** Show that a directed multigraph having no isolated vertices has an Euler circuit if and only if the graph is weakly connected and the in-degree and out-degree of each vertex are equal.

***17.** Show that a directed multigraph having no isolated vertices has an Euler path but not an Euler circuit if and only if the graph is weakly connected and the in-degree and out-degree of each vertex are equal for all but two vertices, one that has in-degree one larger than its out-degree and the other that has out-degree one larger than its in-degree.

In Exercises 18–23 determine whether the directed graph shown has an Euler circuit. Construct an Euler circuit if one exists. If no Euler circuit exists, determine whether the directed graph has an Euler path. Construct an Euler path if one exists.

18.

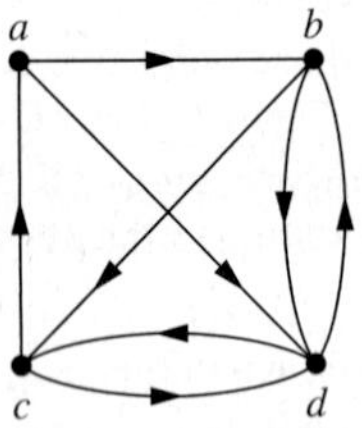

19.

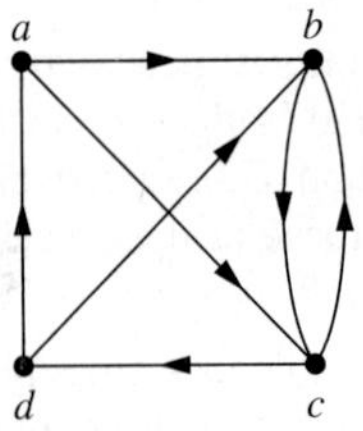

20.

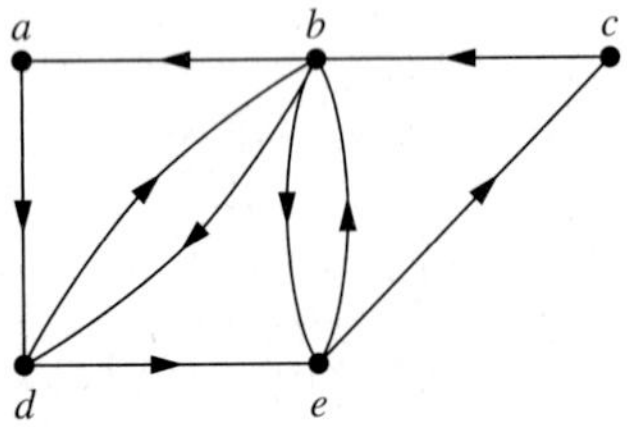

21.

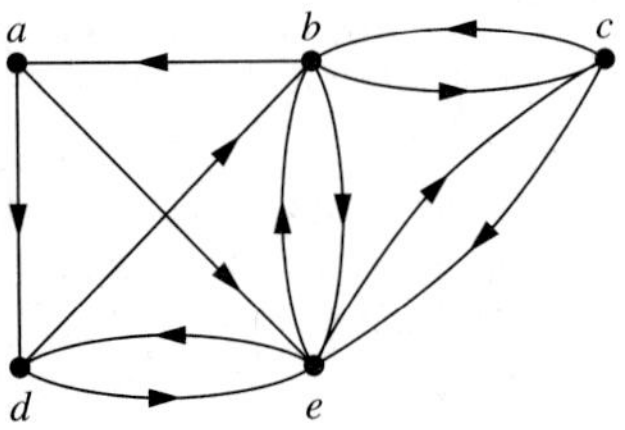

22.

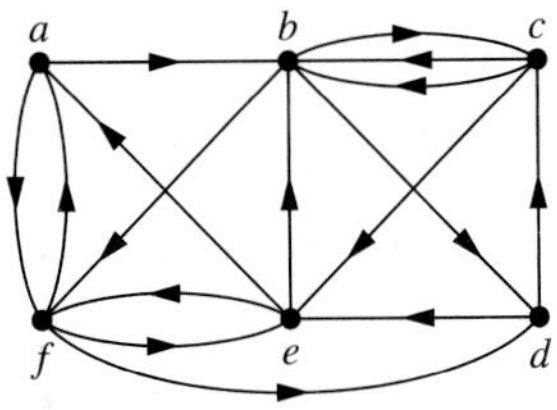

23.

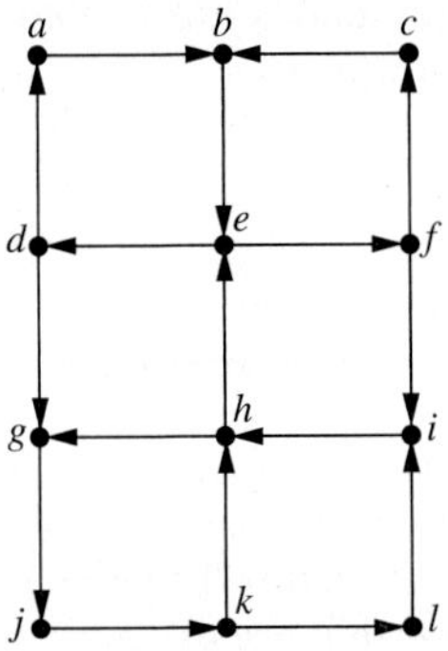

***24.** Devise an algorithm for constructing Euler circuits in directed graphs.

25. Devise an algorithm for constructing Euler paths in directed graphs.

26. For which values of n do these graphs have an Euler circuit?

a) K_n **b)** C_n **c)** W_n **d)** Q_n

27. For which values of n do the graphs in Exercise 26 have an Euler path but no Euler circuit?

28. For which values of m and n does the complete bipartite graph $K_{m,n}$ have an

a) Euler circuit?

b) Euler path?

29. Find the least number of times it is necessary to lift a pencil from the paper when drawing each of the graphs in Exercises 1–7 without retracing any part of the graph.

In Exercises 30–36 determine whether the given graph has a Hamilton circuit. If it does, find such a circuit. If it does not, give an argument to show why no such circuit exists.

30.

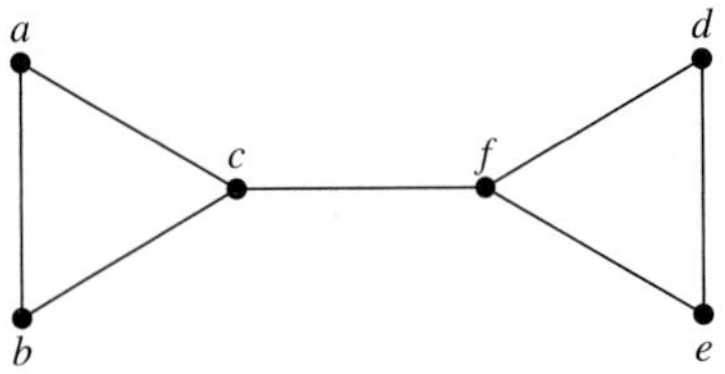

31.

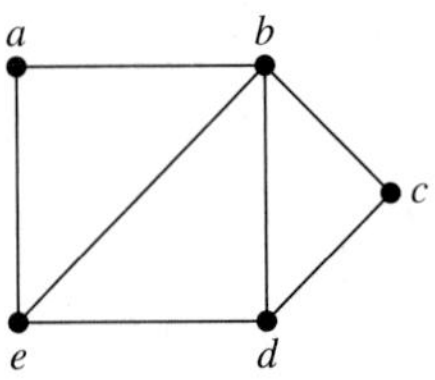

32.

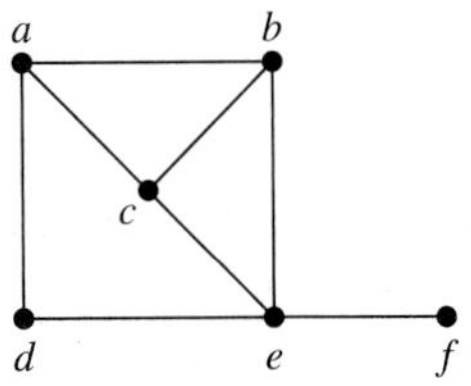

33.

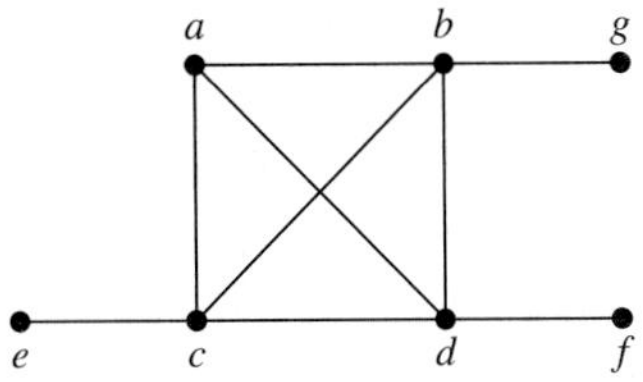

34.

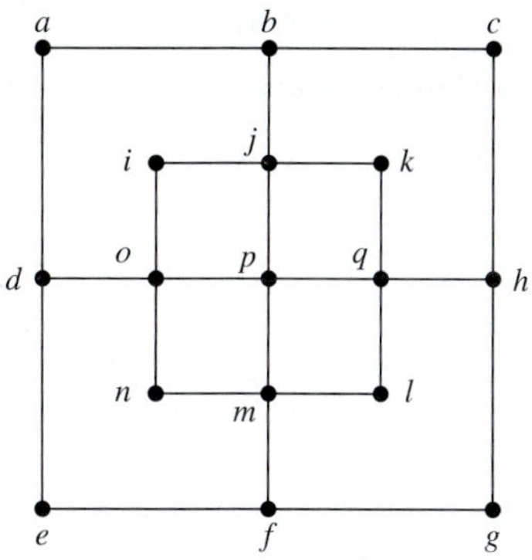

35.

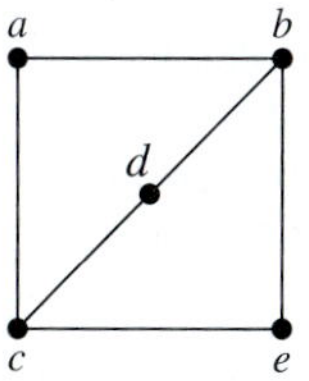

36.

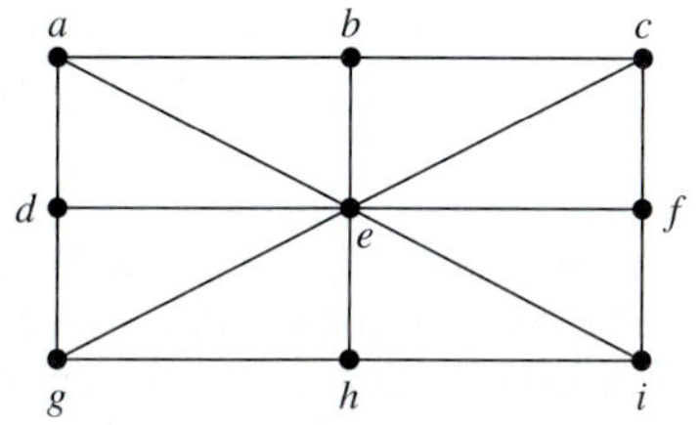

37. Does the graph in Exercise 30 have a Hamilton path? If so, find such a path. If it does not, give an argument to show why no such path exists.

38. Does the graph in Exercise 31 have a Hamilton path? If so, find such a path. If it does not, give an argument to show why no such path exists.

39. Does the graph in Exercise 32 have a Hamilton path? If so, find such a path. If it does not, give an argument to show why no such path exists.

40. Does the graph in Exercise 33 have a Hamilton path? If so, find such a path. If it does not, give an argument to show why no such path exists.

***41.** Does the graph in Exercise 34 have a Hamilton path? If so, find such a path. If it does not, give an argument to show why no such path exists.

42. Does the graph in Exercise 35 have a Hamilton path? If so, find such a path. If it does not, give an argument to show why no such path exists.

43. Does the graph in Exercise 36 have a Hamilton path? If so, find such a path. If it does not, give an argument to show why no such path exists.

44. For which values of n do the graphs in Exercise 26 have a Hamilton circuit?

45. For which values of m and n does the complete bipartite graph $K_{m,n}$ have a Hamilton circuit?

***46.** Show that the **Petersen graph**, shown here, does not have a Hamilton circuit, but that the subgraph obtained by deleting a vertex v, and all edges incident with v, does have a Hamilton circuit.

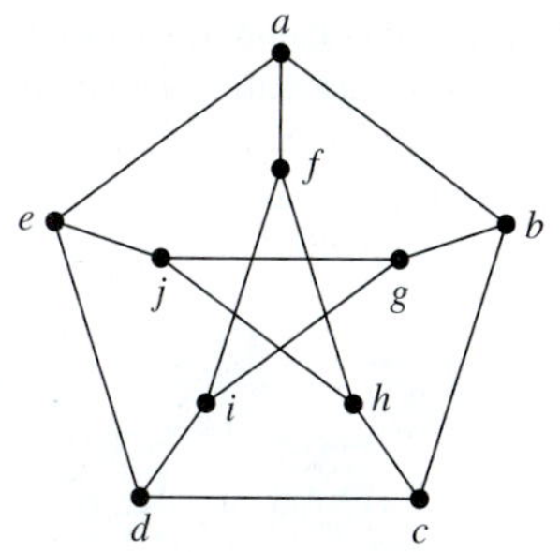

47. For each of these graphs, determine (i) whether Dirac's theorem can be used to show that the graph has a Hamilton circuit, (ii) whether Ore's theorem can be used to show that the graph has a Hamilton circuit, and (iii) whether the graph has a Hamilton circuit.

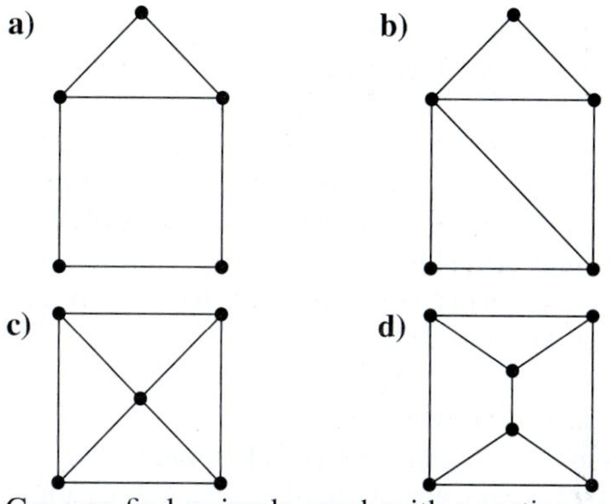

48. Can you find a simple graph with n vertices with $n \geq 3$ that does not have a Hamilton circuit, yet the degree of every vertex in the graph is at least $(n-1)/2$?

***49.** Show that there is a Gray code of order n whenever n is a positive integer, or equivalently, show that the n-cube Q_n, $n > 1$, always has a Hamilton circuit. [*Hint:* Use mathematical induction. Show how to produce a Gray code of order n from one of order $n - 1$.]

Fleury's algorithm, published in 1883, constructs Euler circuits by first choosing an arbitrary vertex of a connected multigraph, and then forming a circuit by choosing edges successively. Once an edge is chosen, it is removed. Edges are chosen successively so that each edge begins where the last edge ends, and so that this edge is not a cut edge unless there is no alternative.

50. Use Fleury's algorithm to find an Euler circuit in the graph G in Figure 5.

***51.** Express Fleury's algorithm in pseudocode.

****52.** Prove that Fleury's algorithm always produces an Euler circuit.

***53.** Give a variant of Fleury's algorithm to produce Euler paths.

54. A diagnostic message can be sent out over a computer network to perform tests over all links and in all devices. What sort of paths should be used to test all links? To test all devices?

55. Show that a bipartite graph with an odd number of vertices does not have a Hamilton circuit.

A **knight** is a chess piece that can move either two spaces horizontally and one space vertically or one space horizontally and two spaces vertically. That is, a knight on square (x, y) can move to any of the eight squares $(x \pm 2, y \pm 1)$, $(x \pm 1, y \pm 2)$, if these squares are on the chessboard, as illustrated here.

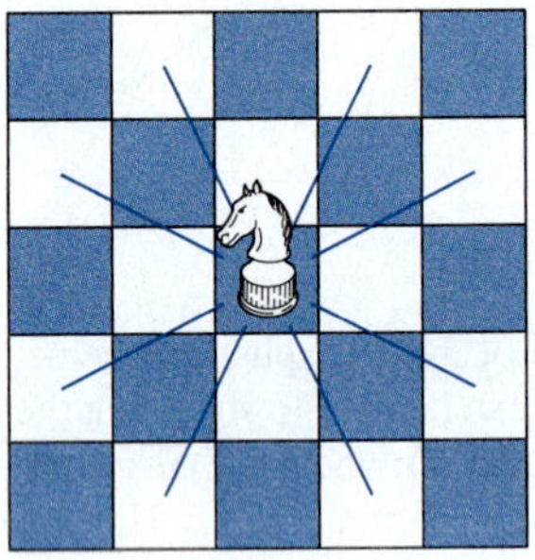

A **knight's tour** is a sequence of legal moves by a knight starting at some square and visiting each square exactly once. A knight's tour is called **reentrant** if there is a legal move that takes the knight from the last square of the tour back to where the tour began. We can model knight's tours using the graph that has a vertex for each square on the board, with an edge connecting two vertices if a knight can legally move between the squares represented by these vertices.

56. Draw the graph that represents the legal moves of a knight on a 3×3 chessboard.

57. Draw the graph that represents the legal moves of a knight on a 3×4 chessboard.

58. a) Show that finding a knight's tour on an $m \times n$ chessboard is equivalent to finding a Hamilton path on the graph representing the legal moves of a knight on that board.

b) Show that finding a reentrant knight's tour on an $m \times n$ chessboard is equivalent to finding a Hamilton circuit on the corresponding graph.

***59.** Show that there is a knight's tour on a 3×4 chessboard.

***60.** Show that there is no knight's tour on a 3×3 chessboard.

***61.** Show that there is no knight's tour on a 4×4 chessboard.

62. Show that the graph representing the legal moves of a knight on an $m \times n$ chessboard, whenever m and n are positive integers, is bipartite.

63. Show that there is no reentrant knight's tour on an $m \times n$ chessboard when m and n are both odd. [*Hint:* Use Exercises 55, 58b, and 62.]

***64.** Show that there is a knight's tour on an 8×8 chessboard. [*Hint:* You can construct a knight's tour using a method invented by H. C. Warnsdorff in 1823: Start in any square, and then always move to a square connected to the fewest number of unused squares. Although this method may not always produce a knight's tour, it often does.]

65. The parts of this exercise outline a proof of Ore's theorem. Suppose that G is a simple graph with n vertices, $n \geq 3$, and $\deg(x) + \deg(y) \geq n$ whenever x and y are nonadjacent vertices in G. Ore's theorem states that under these conditions, G has a Hamilton circuit.

a) Show that if G does not have a Hamilton circuit, then there exists another graph H with the same vertices as G, which can be constructed by adding edges to G, such that the addition of a single edge would produce a Hamilton circuit in H. [*Hint:* Add as many edges as possible at each successive vertex of G without producing a Hamilton circuit.]

b) Show that there is a Hamilton path in H.

c) Let $v_1, v_2, \ldots, v_n$ be a Hamilton path in H. Show that $\deg(v_1) + \deg(v_n) \geq n$ and that there are at most $\deg(v_1)$ vertices not adjacent to v_n (including v_n itself).

Links

The History Collection/ Alamy Stock Photo

JULIUS PETER CHRISTIAN PETERSEN (1839–1910) Julius Petersen was born in the Danish town of Sorø. His father was a dyer. In 1854 his parents were no longer able to pay for his schooling, so he became an apprentice in an uncle's grocery store. When this uncle died, he left Petersen enough money to return to school. After graduating, he began studying engineering at the Polytechnical School in Copenhagen, later deciding to concentrate on mathematics. He published his first textbook, a book on logarithms, in 1858. When his inheritance ran out, he had to teach to make a living. From 1859 until 1871 Petersen taught at a prestigious private high school in Copenhagen. While teaching high school he continued his studies, entering Copenhagen University in 1862. He married Laura Bertelsen in 1862; they had three children, two sons and a daughter.

Petersen obtained a mathematics degree from Copenhagen University in 1866 and finally obtained his doctorate in 1871 from that school. After receiving his doctorate, he taught at a polytechnic and military academy. In 1887 he was appointed to a professorship at the University of Copenhagen. Petersen was well known in Denmark as the author of a large series of textbooks for high schools and universities. One of his books, *Methods and Theories for the Solution of Problems of Geometrical Construction*, was translated into eight languages, with the English language version last reprinted in 1960 and the French version reprinted as recently as 1990, more than a century after the original publication date.

Petersen worked in a wide range of areas, including algebra, analysis, cryptography, geometry, mechanics, mathematical economics, and number theory. His contributions to graph theory, including results on regular graphs, are his best-known work. He was noted for his clarity of exposition, problem-solving skills, originality, sense of humor, vigor, and teaching. One interesting fact about Petersen was that he preferred not to read the writings of other mathematicians. This led him often to rediscover results already proved by others, often with embarrassing consequences. However, he was often angry when other mathematicians did not read his writings!

Petersen's death was front-page news in Copenhagen. A newspaper of the time described him as the Hans Christian Andersen of science—a child of the people who made good in the academic world.

d) Let S be the set of vertices preceding each vertex adjacent to v_1 in the Hamilton path. Show that S contains $\deg(v_1)$ vertices and $v_n \notin S$.

e) Show that S contains a vertex v_k that is adjacent to v_n, implying that there are edges connecting v_1 and v_{k+1} and v_k and v_n.

f) Show that part (e) implies that $v_1, v_2, \ldots, v_{k-1}, v_k, v_n, v_{n-1}, \ldots, v_{k+1}, v_1$ is a Hamilton circuit in G. Conclude from this contradiction that Ore's theorem holds.

66. Show that if u and v are nonadjacent vertices in a graph G with n vertices and $\deg(u) + \deg(v) \geq n$, then G has a Hamilton circuit if and only if $G + \{u, v\}$ has a Hamilton circuit.

***67.** Show that this graph does *not* have a Hamilton circuit.

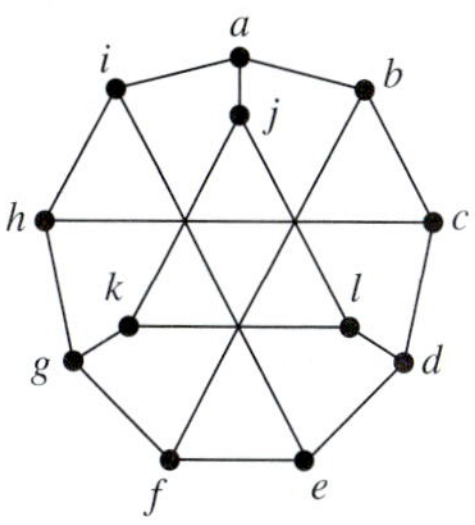

***68.** Show that the worst case computational complexity of Algorithm 1 for finding Euler circuits in a connected graph with all vertices of even degree is $O(m)$, where m is the number of edges of G.

10.6 Shortest-Path Problems

10.6.1 Introduction

Many problems can be modeled using graphs with weights assigned to their edges. As an illustration, consider how an airline system can be modeled. We set up the basic graph model by representing cities by vertices and flights by edges. Problems involving distances can be modeled by assigning distances between cities to the edges. Problems involving flight time can be modeled by assigning flight times to edges. Problems involving fares can be modeled by assigning fares to the edges. Figure 1 displays three different assignments of weights to the edges of a graph representing distances, flight times, and fares, respectively.

Graphs that have a number assigned to each edge are called **weighted graphs**. Weighted graphs are used to model computer networks. Communications costs (such as the monthly cost of leasing a telephone line), the response times of the computers over these lines, or the distance between computers, can all be studied using weighted graphs. Figure 2 displays weighted graphs that represent three ways to assign weights to the edges of a graph of a computer network, corresponding to distance, response time, and cost.

Several types of problems involving weighted graphs arise frequently. Determining a path of least length between two vertices in a network is one such problem. To be more specific, let the **length** of a path in a weighted graph be the sum of the weights of the edges of this path. (The reader should note that this use of the term *length* is different from the use of *length* to denote the number of edges in a path in a graph without weights.) The question is: What is a shortest path, that is, a path of least length, between two given vertices? For instance, in the airline system represented by the weighted graph shown in Figure 1, what is a shortest path in air distance between Boston and Los Angeles? What combinations of flights has the smallest total flight time (that is, total time in the air, not including time between flights) between Boston and Los Angeles? What is the cheapest fare between these two cities? In the computer network shown in Figure 2, what is a least expensive set of network connections needed to connect the computers in San Francisco with those in New York? Which set of network links gives a fastest response time for communications between San Francisco and New York? Which set of network connections has a shortest overall distance?

Another important problem involving weighted graphs asks for a circuit of shortest total length that visits every vertex of a complete graph exactly once. This is the famous *traveling salesperson problem*, which asks for an order in which a salesperson should visit each of the cities on a route exactly once so that the distance travelled is minimized. We will discuss the traveling salesperson problem later in this section.

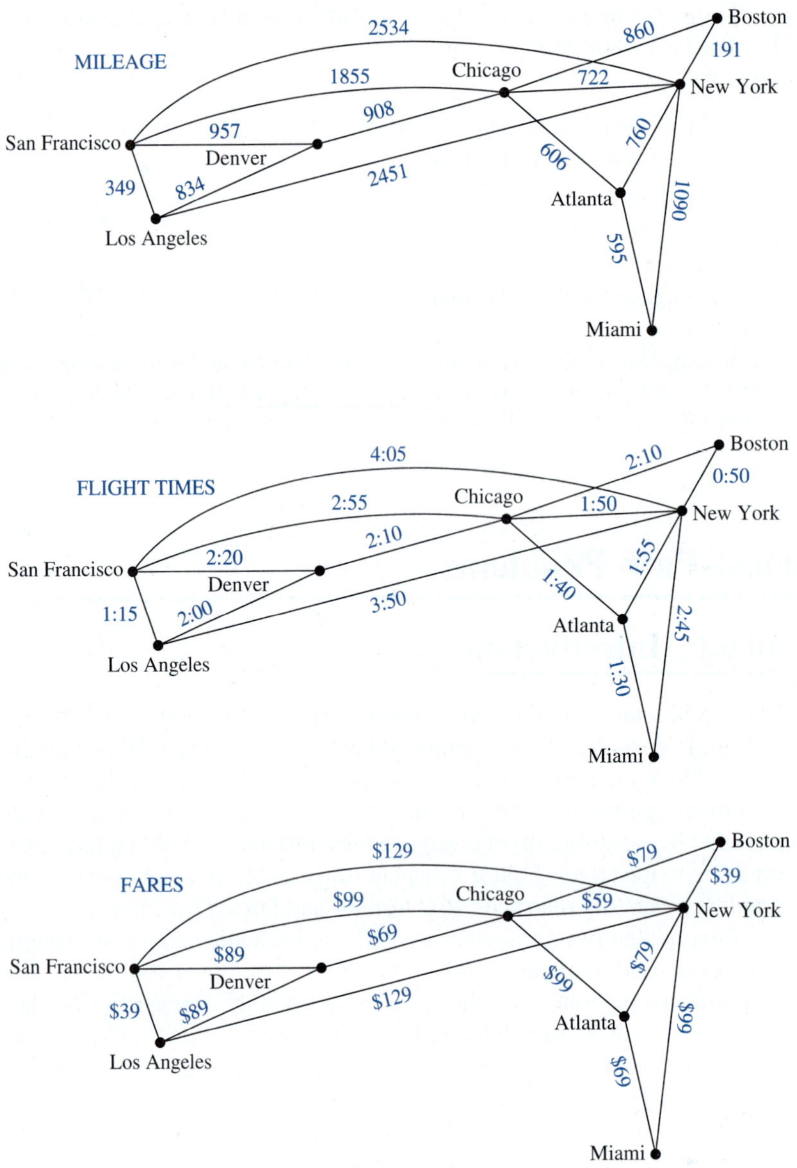

FIGURE 1 Weighted graphs modeling an airline system.

10.6.2 A Shortest-Path Algorithm

Links

There are several different algorithms that find a shortest path between two vertices in a weighted graph. We will present a greedy algorithm discovered by the Dutch mathematician Edsger Dijkstra in 1959. The version we will describe solves this problem in undirected weighted graphs where all the weights are positive. It is easy to adapt it to solve shortest-path problems in directed graphs.

Before giving a formal presentation of the algorithm, we will give an illustrative example.

EXAMPLE 1 What is the length of a shortest path between a and z in the weighted graph shown in Figure 3?

Solution: Although a shortest path is easily found by inspection, we will develop some ideas useful in understanding Dijkstra's algorithm. We will solve this problem by finding the length of a shortest path from a to successive vertices, until z is reached.

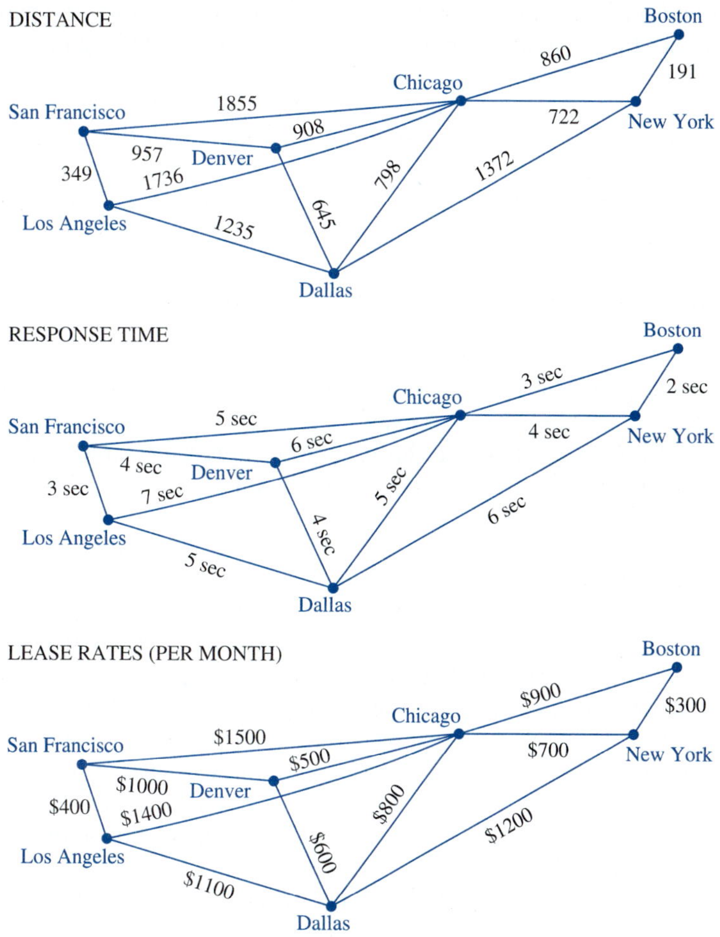

FIGURE 2 **Weighted graphs modeling a computer network.**

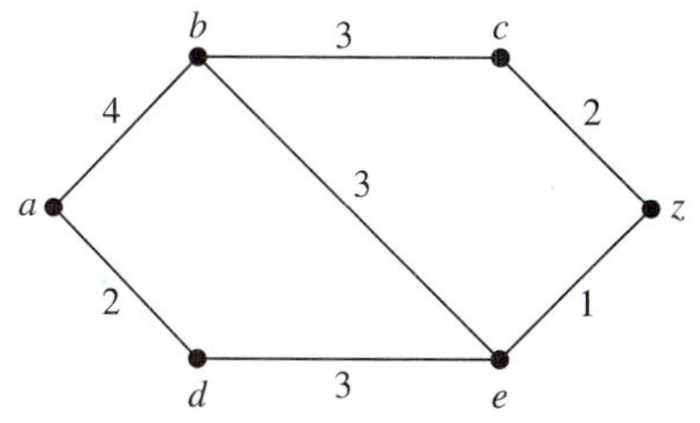

FIGURE 3 **A weighted simple graph.**

A path that starts at a that has no interior vertices other than a must consist of one edge that has a as one of its endpoints. These are the one-edged paths a, b of length 4 and a, d of length 2. It follows that d is the closest vertex to a, and the shortest path from a to d has length 2.

We can find the second closest vertex by examining all paths that begin with the shortest path from a to a vertex in the set $\{a, d\}$, followed by an edge that has one endpoint in $\{a, d\}$ and its other endpoint not in this set. There are two such paths to consider, a, d, e of length 7 and a, b of length 4. Hence, the second closest vertex to a is b, and the shortest path from a to b has length 4.

To find the third closest vertex to a, we need examine only the paths that begin with the shortest path from a to a vertex in the set $\{a, d, b\}$, followed by an edge that has one endpoint in the set $\{a, d, b\}$ and its other endpoint not in this set. There are three such paths, a, b, c of length 7, a, b, e of length 7, and a, d, e of length 5. Because the shortest of these paths is a, d, e, the third closest vertex to a is e, and the length of the shortest path from a to e is 5.

To find the fourth closest vertex to a, we need examine only the paths that begin with the shortest path from a to a vertex in the set $\{a, d, b, e\}$, followed by an edge that has one endpoint in the set $\{a, d, b, e\}$ and its other endpoint not in this set. There are two such paths, a, b, c of length 7 and a, d, e, z of length 6. Because the shorter of these paths is a, d, e, z, the fourth closest vertex to a is z, and the length of the shortest path from a to z is 6. ◀

Example 1 illustrates the general principles used in Dijkstra's algorithm. Note that a shortest path from a to z could have been found by a brute force approach by examining the length of every path from a to z. However, this brute force approach is impractical for humans and even for computers for graphs with a large number of edges.

We will now consider the general problem of finding the length of a shortest path between a and z in an undirected connected simple weighted graph. Dijkstra's algorithm proceeds by finding the length of a shortest path from a to a first vertex, the length of a shortest path from a to a second vertex, and so on, until the length of a shortest path from a to z is found. As a side benefit, this algorithm is easily extended to find the length of the shortest path from a to all other vertices of the graph, and not just to z.

The algorithm relies on a series of iterations. A distinguished set of vertices is constructed by adding one vertex at each iteration. (This is the set in Example 1.) A labeling procedure is carried out at each iteration. In this labeling procedure, a vertex w is labeled with the length of a shortest path from a to w that contains only vertices already in the distinguished set. The vertex added to the distinguished set is one with a minimal label among those vertices not already in the set.

We now give the details of Dijkstra's algorithm. It begins by labeling a with 0 and the other vertices with ∞. We use the notation $L_0(a) = 0$ and $L_0(v) = \infty$ for these labels before any iterations have taken place (the subscript 0 stands for the "0th" iteration). These labels are the lengths of shortest paths from a to the vertices, where the paths contain only the vertex a. (Because no path from a to a vertex different from a exists, ∞ is the length of a shortest path between a and this vertex.)

Dijkstra's algorithm proceeds by forming a distinguished set of vertices. Let S_k denote this set after k iterations of the labeling procedure. We begin with $S_0 = \emptyset$. The set S_k is formed from S_{k-1} by adding a vertex u not in S_{k-1} with the smallest label.

Once u is added to S_k, we update the labels of all vertices not in S_k, so that $L_k(v)$, the label of the vertex v at the kth stage, is the length of a shortest path from a to v that contains vertices only in S_k (that is, vertices that were already in the distinguished set together with u). Note that the way we choose the vertex u to add to S_k at each step is an optimal choice at each step, making

Links

Edsger W. Dijkstra

EDSGER WYBE DIJKSTRA (1930–2002) Edsger Dijkstra, born in the Netherlands, began programming computers in the early 1950s while studying theoretical physics at the University of Leiden. In 1952, realizing that he was more interested in programming than in physics, he quickly completed the requirements for his physics degree and began his career as a programmer, even though programming was not a recognized profession. (In 1957, the authorities in Amsterdam refused to accept "programming" as his profession on his marriage license. However, they did accept "theoretical physicist" when he changed his entry to this.)

Dijkstra was one of the most forceful proponents of programming as a scientific discipline. He has made fundamental contributions to the areas of operating systems, including deadlock avoidance; programming languages, including the notion of structured programming; and algorithms. In 1972 Dijkstra received the Turing Award from the Association for Computing Machinery, one of the most prestigious awards in computer science. Dijkstra became a Burroughs Research Fellow in 1973, and in 1984 he was appointed to a chair in Computer Science at the University of Texas, Austin.

this a greedy algorithm. (We will prove shortly that this greedy algorithm always produces an optimal solution.)

Let v be a vertex not in S_k. To update the label of v, note that $L_k(v)$ is the length of a shortest path from a to v containing only vertices in S_k. The updating can be carried out efficiently when this observation is used: A shortest path from a to v containing only elements of S_k is either a shortest path from a to v that contains only elements of S_{k-1} (that is, the distinguished vertices not including u), or it is a shortest path from a to u at the $(k-1)$st stage with the edge $\{u, v\}$ added. In other words,

$$L_k(a, v) = \min\{L_{k-1}(a, v), L_{k-1}(a, u) + w(u, v)\},$$

where $w(u, v)$ is the length of the edge with u and v as endpoints. This procedure is iterated by successively adding vertices to the distinguished set until z is added. When z is added to the distinguished set, its label is the length of a shortest path from a to z.

Dijkstra's algorithm is given in Algorithm 1. Later we will give a proof that this algorithm is correct. Note that we can find the length of the shortest path from a to all other vertices of the graph if we continue this procedure until all vertices are added to the distinguished set.

Demo

ALGORITHM 1 Dijkstra's Algorithm.

procedure *Dijkstra*(G: weighted connected simple graph, with
  all weights positive)
{G has vertices $a = v_0, v_1, \dots, v_n = z$ and lengths $w(v_i, v_j)$
  where $w(v_i, v_j) = \infty$ if $\{v_i, v_j\}$ is not an edge in G}
for $i := 1$ **to** n
  $L(v_i) := \infty$
$L(a) := 0$
$S := \emptyset$
{the labels are now initialized so that the label of a is 0 and all
  other labels are ∞, and S is the empty set}
while $z \notin S$
  $u :=$ a vertex not in S with $L(u)$ minimal
  $S := S \cup \{u\}$
  for all vertices v not in S
    if $L(u) + w(u, v) < L(v)$ **then** $L(v) := L(u) + w(u, v)$
    {this adds a vertex to S with minimal label and updates the
    labels of vertices not in S}
return $L(z)$ {$L(z)$ = length of a shortest path from a to z}

Example 2 illustrates how Dijkstra's algorithm works. Afterward, we will show that this algorithm always produces the length of a shortest path between two vertices in a weighted graph.

EXAMPLE 2 Use Dijkstra's algorithm to find the length of a shortest path between the vertices a and z in the weighted graph displayed in Figure 4(a).

Solution: The steps used by Dijkstra's algorithm to find a shortest path between a and z are shown in Figure 4. At each iteration of the algorithm the vertices of the set S_k are circled. A shortest path from a to each vertex containing only vertices in S_k is indicated for each iteration. The algorithm terminates when z is circled. We find that a shortest path from a to z is a, c, b, d, e, z, with length 13. ◀

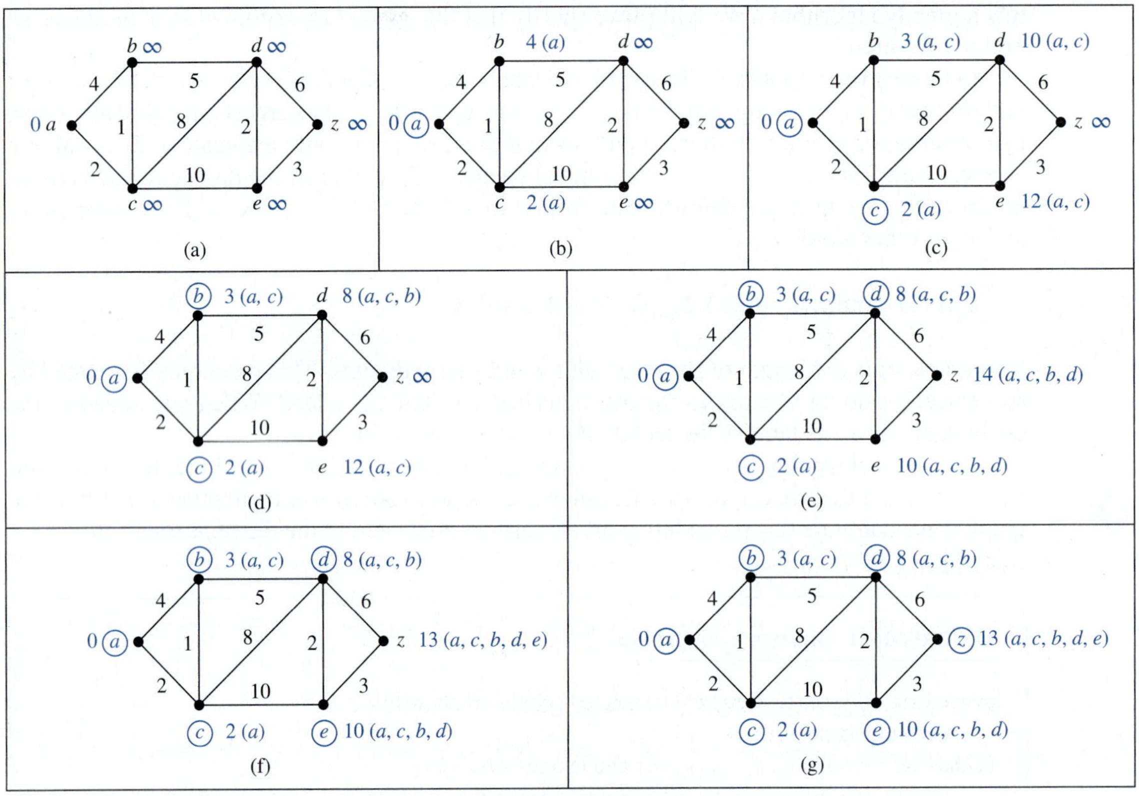

FIGURE 4 Using Dijkstra's algorithm to find a shortest path from a to z.

Remark: In performing Dijkstra's algorithm it is sometimes more convenient to keep track of labels of vertices in each step using a table instead of redrawing the graph for each step.

Next, we use an inductive argument to show that Dijkstra's algorithm produces the length of a shortest path between two vertices a and z in an undirected connected weighted graph. Take as the inductive hypothesis the following assertion: At the kth iteration

(i) the label of every vertex v in S is the length of a shortest path from a to this vertex, and

(ii) the label of every vertex not in S is the length of a shortest path from a to this vertex that contains only (besides the vertex itself) vertices in S.

When $k = 0$, before any iterations are carried out, $S = \emptyset$, so the length of a shortest path from a to a vertex other than a is ∞. Hence, the basis case is true.

Assume that the inductive hypothesis holds for the kth iteration. Let v be the vertex added to S at the $(k + 1)$st iteration, so v is a vertex not in S at the end of the kth iteration with the smallest label (in the case of ties, any vertex with smallest label may be used).

From the inductive hypothesis we see that the vertices in S before the $(k + 1)$st iteration are labeled with the length of a shortest path from a. Also, v must be labeled with the length of a shortest path to it from a. If this were not the case, at the end of the kth iteration there would be a path of length less than $L_k(v)$ containing a vertex not in S [because $L_k(v)$ is the length of a shortest path from a to v containing only vertices in S after the kth iteration]. Let u be the first vertex not in S in such a path. There is a path with length less than $L_k(v)$ from a to u containing only vertices of S. This contradicts the choice of v. Hence, (i) holds at the end of the $(k + 1)$st iteration.

Let u be a vertex not in S after $k + 1$ iterations. A shortest path from a to u containing only elements of S either contains v or it does not. If it does not contain v, then by the inductive hypothesis its length is $L_k(u)$. If it does contain v, then it must be made up of a path from a to v of shortest possible length containing elements of S other than v, followed by the edge from v to u. In this case, its length would be $L_k(v) + w(v, u)$. This shows that (*ii*) is true, because $L_{k+1}(u) = \min\{L_k(u), L_k(v) + w(v, u)\}$.

We now state the thereom that we have proved.

THEOREM 1 Dijkstra's algorithm finds the length of a shortest path between two vertices in a connected simple undirected weighted graph.

We can now estimate the computational complexity of Dijkstra's algorithm (in terms of additions and comparisons). The algorithm uses no more than $n - 1$ iterations where n is the number of vertices in the graph, because one vertex is added to the distinguished set at each iteration. We are done if we can estimate the number of operations used for each iteration. We can identify the vertex not in S_k with the smallest label using no more than $n - 1$ comparisons. Then we use an addition and a comparison to update the label of each vertex not in S_k. It follows that no more than $2(n - 1)$ operations are used at each iteration, because there are no more than $n - 1$ labels to update at each iteration. Because we use no more than $n - 1$ iterations, each using no more than $2(n - 1)$ operations, we have Theorem 2.

THEOREM 2 Dijkstra's algorithm uses $O(n^2)$ operations (additions and comparisons) to find the length of a shortest path between two vertices in a connected simple undirected weighted graph with n vertices.

10.6.3 The Traveling Salesperson Problem

Links

We now discuss an important problem involving weighted graphs. Consider the following problem: A traveling salesperson wants to visit each of n cities exactly once and return to the starting point. For example, suppose that the salesperson wants to visit Detroit, Toledo, Saginaw, Grand Rapids, and Kalamazoo (see Figure 5). In which order should the salesperson visit these cities to travel the minimum total distance? To solve this problem we can assume the salesperson

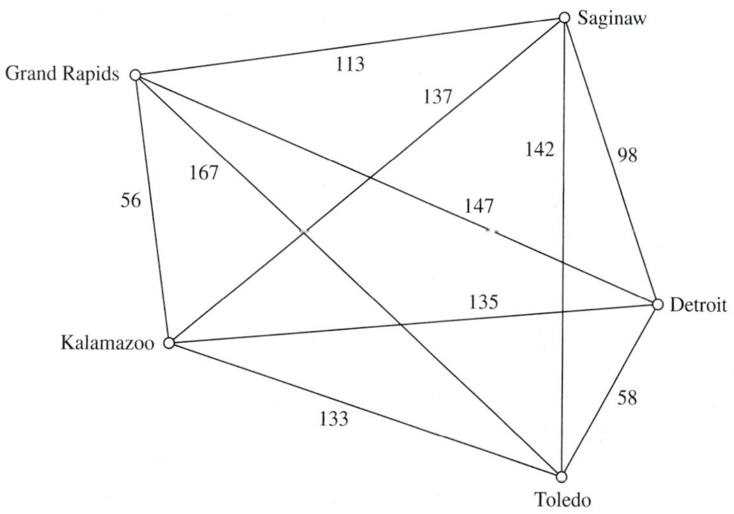

FIGURE 5 The graph showing the distances between five cities.

starts in Detroit (because this must be part of the circuit) and examine all possible ways for the salesperson to visit the other four cities and then return to Detroit (starting elsewhere will produce the same circuits). There are a total of 24 such circuits, but because we travel the same distance when we travel a circuit in reverse order, we need only consider 12 different circuits to find the minimum total distance the salesperson must travel. We list these 12 different circuits and the total distance traveled for each circuit. As can be seen from the list, the minimum total distance of 458 miles is traveled using the circuit Detroit–Toledo–Kalamazoo–Grand Rapids–Saginaw–Detroit (or its reverse).

Route	*Total Distance (miles)*
Detroit–Toledo–Grand Rapids–Saginaw–Kalamazoo–Detroit	610
Detroit–Toledo–Grand Rapids–Kalamazoo–Saginaw–Detroit	516
Detroit–Toledo–Kalamazoo–Saginaw–Grand Rapids–Detroit	588
Detroit–Toledo–Kalamazoo–Grand Rapids–Saginaw–Detroit	458
Detroit–Toledo–Saginaw–Kalamazoo–Grand Rapids–Detroit	540
Detroit–Toledo–Saginaw–Grand Rapids–Kalamazoo–Detroit	504
Detroit–Saginaw–Toledo–Grand Rapids–Kalamazoo–Detroit	598
Detroit–Saginaw–Toledo–Kalamazoo–Grand Rapids–Detroit	576
Detroit–Saginaw–Kalamazoo–Toledo–Grand Rapids–Detroit	682
Detroit–Saginaw–Grand Rapids–Toledo–Kalamazoo–Detroit	646
Detroit–Grand Rapids–Saginaw–Toledo–Kalamazoo–Detroit	670
Detroit–Grand Rapids–Toledo–Saginaw–Kalamazoo–Detroit	728

We just described an instance of the **traveling salesperson problem**. The traveling salesperson problem asks for the circuit of minimum total weight in a weighted, complete, undirected graph that visits each vertex exactly once and returns to its starting point. This is equivalent to asking for a Hamilton circuit with minimum total weight in the complete graph, because each vertex is visited exactly once in the circuit.

An 1832 handbook *Der Handlungsreisende* (*The Traveling Salesperson*) mentions the traveling salesman problem, with sample tours through Germany and Switzerland.

The most straightforward way to solve an instance of the traveling salesperson problem is to examine all possible Hamilton circuits and select one of minimum total length. How many circuits do we have to examine to solve the problem if there are n vertices in the graph? Once a starting point is chosen, there are $(n-1)!$ different Hamilton circuits to examine, because there are $n-1$ choices for the second vertex, $n-2$ choices for the third vertex, and so on. Because a Hamilton circuit can be traveled in reverse order, we need only examine $(n-1)!/2$ circuits to find our answer. Note that $(n-1)!/2$ grows extremely rapidly. Trying to solve a traveling salesperson problem in this way when there are only a few dozen vertices is impractical. For example, with 25 vertices, a total of $24!/2$ (approximately 3.1×10^{23}) different Hamilton circuits would have to be considered. If it took just one nanosecond (10^{-9} second) to examine each Hamilton circuit, a total of approximately ten million years would be required to find a minimum-length Hamilton circuit in this graph by exhaustive search techniques.

Because the traveling salesperson problem has both practical and theoretical importance, a great deal of effort has been devoted to devising efficient algorithms that solve it. However, no algorithm with polynomial worst-case time complexity is known for solving this problem. Furthermore, if a polynomial worst-case time complexity algorithm were discovered for the traveling salesperson problem, many other difficult problems would also be solvable using polynomial worst-case time complexity algorithms (such as determining whether a proposition in n variables is a tautology, discussed in Chapter 1). This follows from the theory of NP-completeness. (For more information about this, consult [GaJo79].)

A practical approach to the traveling salesperson problem when there are many vertices to visit is to use an **approximation algorithm**. These are algorithms that do not necessarily produce the exact solution to the problem but instead are guaranteed to produce a solution that is close to an exact solution. (Also, see the preamble to Exercise 46 in the Supplmentary Exercises of Chapter 3.) That is, they may produce a Hamilton circuit with total weight W' such that $W \leq W' \leq cW$, where W is the total length of an exact solution and c is a constant. For example, there is an algorithm with polynomial worst-case time complexity that works if the weighted graph satisfies the triangle inequality such that $c = 3/2$. For general weighted graphs for every positive real number k no algorithm is known that will always produce a solution at most k times a best solution. If such an algorithm existed, this would show that the class P would be the same as the class NP, perhaps the most famous open question about the complexity of algorithms (see Section 3.3).

In practice, algorithms have been developed that can solve traveling salesperson problems with as many as 1000 vertices within 2% of an exact solution using only a few minutes of computer time. For more information about the traveling salesperson problem, including history, applications, and algorithms, see the chapter on this topic in *Applications of Discrete Mathematics* [MiRo91] also available on the website for this book.

Exercises

1. For each of these problems about a subway system, describe a weighted graph model that can be used to solve the problem.

a) What is the least amount of time required to travel between two stops?

b) What is the minimum distance that can be traveled to reach a stop from another stop?

c) What is the least fare required to travel between two stops if fares between stops are added to give the total fare?

In Exercises 2–4 find the length of a shortest path between a and z in the given weighted graph.

2.

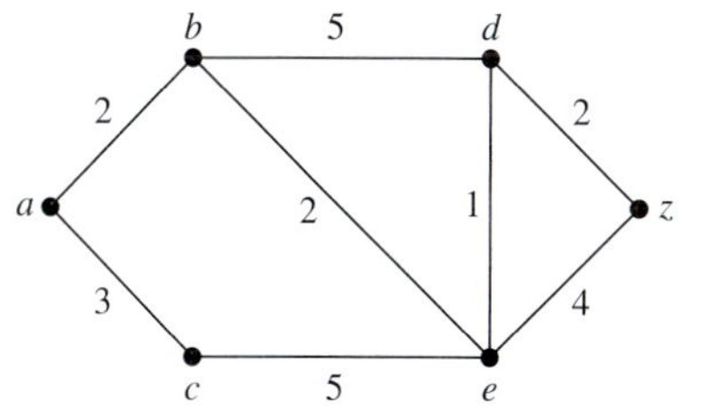

3.

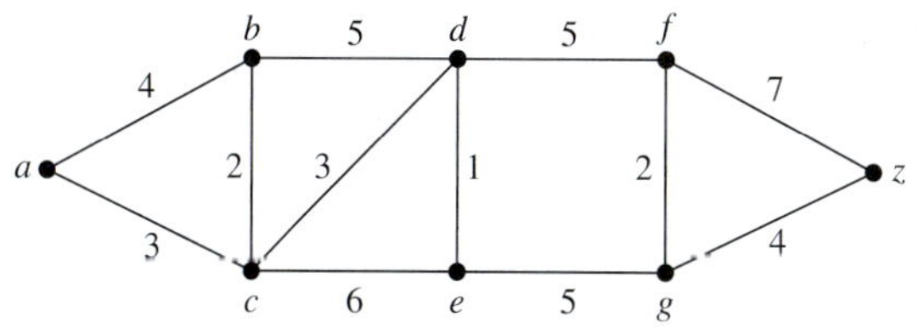

4.

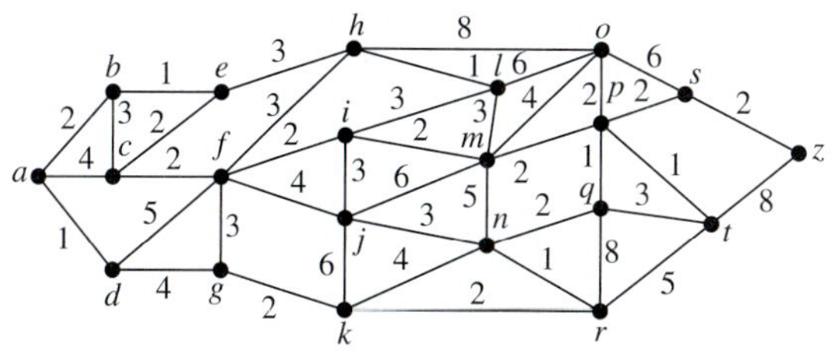

5. Find a shortest path between a and z in each of the weighted graphs in Exercises 2–4.

6. Find the length of a shortest path between these pairs of vertices in the weighted graph in Exercise 3.

a) a and d

b) a and f

c) c and f

d) b and z

7. Find shortest paths in the weighted graph in Exercise 3 between the pairs of vertices in Exercise 6.

8. Find a shortest path (in mileage) between each of the following pairs of cities in the airline system shown in Figure 1.

a) New York and Los Angeles

b) Boston and San Francisco

c) Miami and Denver

d) Miami and Los Angeles

9. Find a combination of flights with the least total air time between the pairs of cities in Exercise 8, using the flight times shown in Figure 1.

10. Find a least expensive combination of flights connecting the pairs of cities in Exercise 8, using the fares shown in Figure 1.

11. Find a shortest route (in distance) between computer centers in each of these pairs of cities in the communications network shown in Figure 2.

a) Boston and Los Angeles

b) New York and San Francisco

c) Dallas and San Francisco

d) Denver and New York

12. Find a route with the shortest response time between the pairs of computer centers in Exercise 11 using the response times given in Figure 2.

13. Find a least expensive route, in monthly lease charges, between the pairs of computer centers in Exercise 11 using the lease charges given in Figure 2.

14. Explain how to find a path with the least number of edges between two vertices in an undirected graph by considering it as a shortest path problem in a weighted graph.

15. Extend Dijkstra's algorithm for finding the length of a shortest path between two vertices in a weighted simple connected graph so that the length of a shortest path between the vertex a and every other vertex of the graph is found.

16. Extend Dijkstra's algorithm for finding the length of a shortest path between two vertices in a weighted simple connected graph so that a shortest path between these vertices is constructed.

17. The weighted graphs in the figures here show some major roads in New Jersey. Part (a) shows the distances between cities on these roads; part (b) shows the tolls.

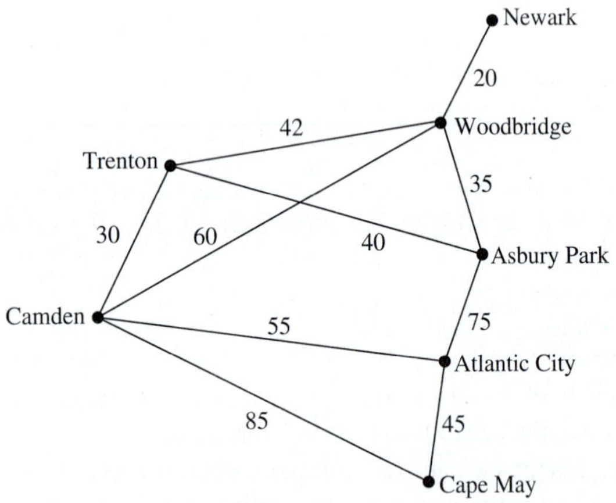

(a)

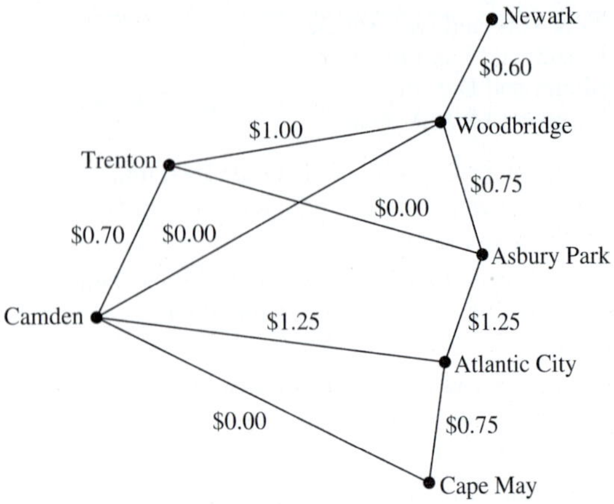

(b)

a) Find a shortest route in distance between Newark and Camden, and between Newark and Cape May, using these roads.

b) Find a least expensive route in terms of total tolls using the roads in the graph between the pairs of cities in part (a) of this exercise.

18. Is a shortest path between two vertices in a weighted graph unique if the weights of edges are distinct?

19. What are some applications where it is necessary to find the length of a longest simple path between two vertices in a weighted graph?

20. What is the length of a longest simple path in the weighted graph in Figure 4 between a and z? Between c and z?

Floyd's algorithm, displayed as Algorithm 2, can be used to find the length of a shortest path between all pairs of vertices in a weighted connected simple graph. However, this algorithm cannot be used to construct shortest paths. (We assign an infinite weight to any pair of vertices not connected by an edge in the graph.)

21. Use Floyd's algorithm to find the distance between all pairs of vertices in the weighted graph in Figure 4(a).

***22.** Prove that Floyd's algorithm determines the shortest distance between all pairs of vertices in a weighted simple graph.

***23.** Give a big-O estimate of the number of operations (comparisons and additions) used by Floyd's algorithm to determine the shortest distance between every pair of vertices in a weighted simple graph with n vertices.

***24.** Show that Dijkstra's algorithm may not work if edges can have negative weights.

ALGORITHM 2 Floyd's Algorithm.

procedure *Floyd*(G: weighted simple graph)
{G has vertices $v_1, v_2, \ldots, v_n$ and weights $w(v_i, v_j)$
  with $w(v_i, v_j) = \infty$ if $\{v_i, v_j\}$ is not an edge}
for $i := 1$ **to** n
  for $j := 1$ **to** n
    $d(v_i, v_j) := w(v_i, v_j)$
for $i := 1$ **to** n
  for $j := 1$ **to** n
    for $k := 1$ **to** n
      if $d(v_j, v_i) + d(v_i, v_k) < d(v_j, v_k)$
        then $d(v_j, v_k) := d(v_j, v_i) + d(v_i, v_k)$
return $[d(v_i, v_j)]$ {$d(v_i, v_j)$ is the length of a shortest path between v_i and v_j for $1 \le i \le n$, $1 \le j \le n$}

25. Solve the traveling salesperson problem for this graph by finding the total weight of all Hamilton circuits and determining a circuit with minimum total weight.

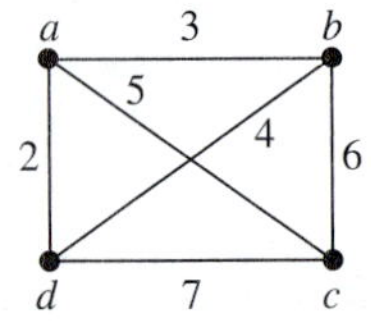

26. Solve the traveling salesperson problem for this graph by finding the total weight of all Hamilton circuits and determining a circuit with minimum total weight.

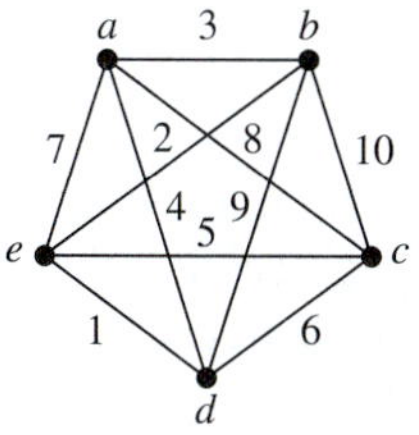

27. Find a route with the least total airfare that visits each of the cities in this graph, where the weight on an edge is the least price available for a flight between the two cities.

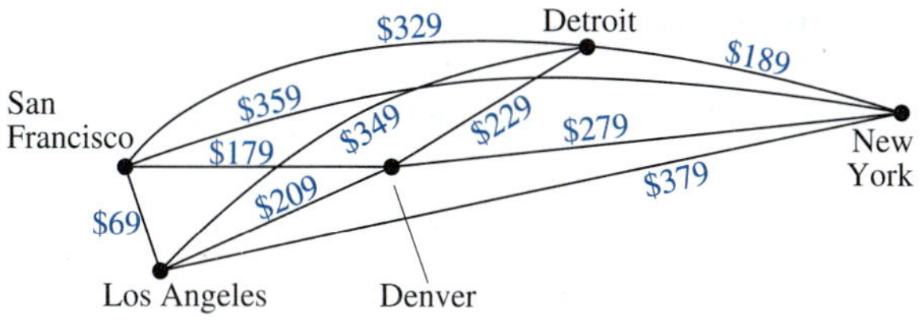

28. Find a route with the least total airfare that visits each of the cities in this graph, where the weight on an edge is the least price available for a flight between the two cities.

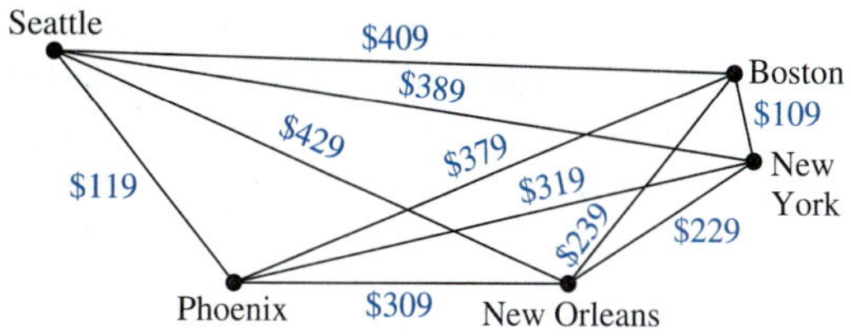

29. Construct a weighted undirected graph such that the total weight of a circuit that visits every vertex at least once is minimized for a circuit that visits some vertices more than once. [*Hint:* There are examples with three vertices.]

30. Show that the problem of finding a circuit of minimum total weight that visits every vertex of a weighted graph at least once can be reduced to the problem of finding a circuit of minimum total weight that visits each vertex of a weighted graph exactly once. Do so by constructing a new weighted graph with the same vertices and edges as the original graph but whose weight of the edge connecting the vertices u and v is equal to the minimum total weight of a path from u to v in the original graph.

***31.** The **longest path problem** in a weighted directed graph with no simple circuits asks for a path in this graph such that the sum of its edge weights is a maximum. Devise an algorithm for solving the longest path problem. [*Hint*: First find a topological ordering of the vertices of the graph.]

10.7 Planar Graphs

10.7.1 Introduction

Links

Consider the problem of joining three houses to each of three separate utilities, as shown in Figure 1. Is it possible to join these houses and utilities so that none of the connections cross? This problem can be modeled using the complete bipartite graph $K_{3,3}$. The original question can be rephrased as: Can $K_{3,3}$ be drawn in the plane so that no two of its edges cross?

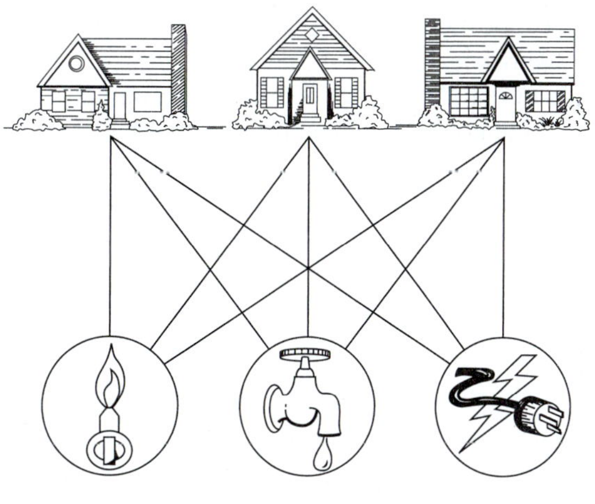

FIGURE 1 **Three houses and three utilities.**

In this section we will study the question of whether a graph can be drawn in the plane without edges crossing. In particular, we will answer the houses-and-utilities problem.

There are always many ways to represent a graph. When is it possible to find at least one way to represent this graph in a plane without any edges crossing?

Definition 1 A graph is called *planar* if it can be drawn in the plane without any edges crossing (where a crossing of edges is the intersection of the lines or arcs representing them at a point other than their common endpoint). Such a drawing is called a *planar representation* of the graph.

A graph may be planar even if it is usually drawn with crossings, because it may be possible to draw it in a different way without crossings.

EXAMPLE 1 Is K_4 (shown in Figure 2 with two edges crossing) planar?

Solution: K_4 is planar because it can be drawn without crossings, as shown in Figure 3. ◀

EXAMPLE 2 Is Q_3, shown in Figure 4, planar?

Solution: Q_3 is planar, because it can be drawn without any edges crossing, as shown in Figure 5. ◀

We can show that a graph is planar by displaying a planar representation. It is harder to show that a graph is nonplanar. We will give an example to show how this can be done in an ad hoc fashion. Later we will develop some general results that can be used to do this.

EXAMPLE 3 Is $K_{3,3}$, shown in Figure 6, planar?

Solution: Any attempt to draw $K_{3,3}$ in the plane with no edges crossing is doomed. We now show why. In any planar representation of $K_{3,3}$, the vertices v_1 and v_2 must be connected to both v_4 and v_5. These four edges form a closed curve that splits the plane into two regions, R_1 and R_2, as shown in Figure 7(a). The vertex v_3 is in either R_1 or R_2. When v_3 is in R_2, the inside of the closed curve, the edges between v_3 and v_4 and between v_3 and v_5 separate R_2 into two subregions, R_{21} and R_{22}, as shown in Figure 7(b).

Next, note that there is no way to place the final vertex v_6 without forcing a crossing. For if v_6 is in R_1, then the edge between v_6 and v_3 cannot be drawn without a crossing. If v_6 is in R_{21}, then the edge between v_2 and v_6 cannot be drawn without a crossing. If v_6 is in R_{22}, then the edge between v_1 and v_6 cannot be drawn without a crossing.

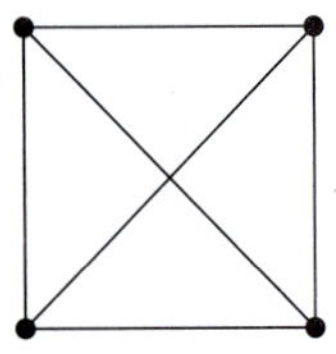

FIGURE 2 The graph K_4.

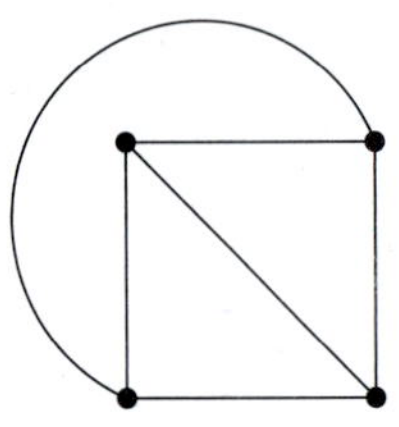

FIGURE 3 K_4 drawn with no crossings.

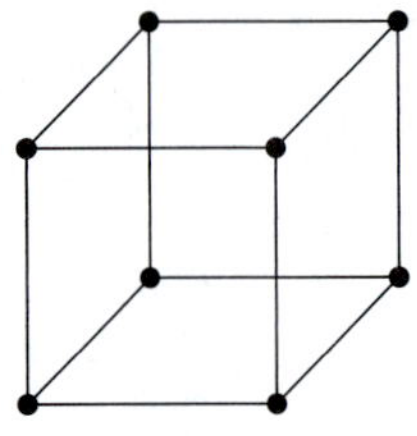

FIGURE 4 The graph Q_3.

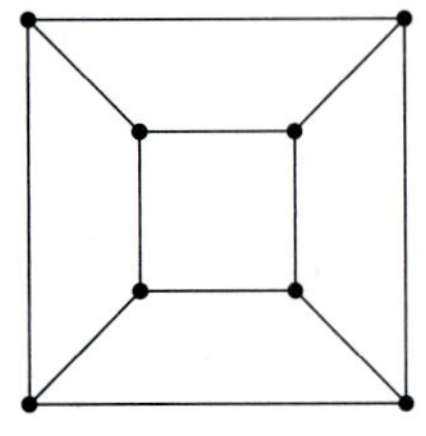

FIGURE 5 A planar representation of Q_3.

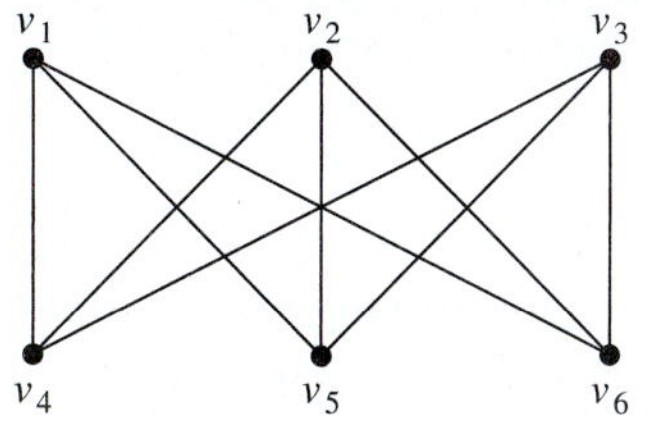

FIGURE 6 **The graph $K_{3,3}$.**

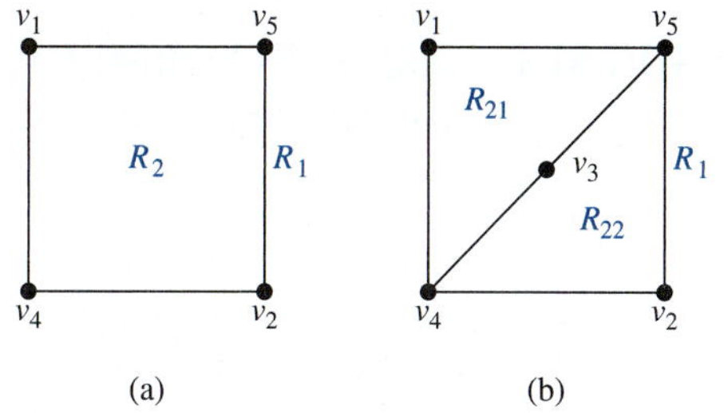

FIGURE 7 **Showing that $K_{3,3}$ is nonplanar.**

A similar argument can be used when v_3 is in R_1. The completion of this argument is left for the reader (see Exercise 10). It follows that $K_{3,3}$ is not planar. ◀

Example 3 solves the utilities-and-houses problem that was described at the beginning of this section. The three houses and three utilities cannot be connected in the plane without a crossing. A similar argument can be used to show that K_5 is nonplanar. (See Exercise 11.)

APPLICATIONS OF PLANAR GRAPHS Planarity of graphs plays an important role in the design of electronic circuits. We can model a circuit with a graph by representing components of the circuit by vertices and connections between them by edges. We can print a circuit on a single board with no connections crossing if the graph representing the circuit is planar. When this graph is not planar, we must turn to more expensive options. For example, we can partition the vertices in the graph representing the circuit into planar subgraphs. We then construct the circuit using multiple layers. (See the preamble to Exercise 30 to learn about the thickness of a graph.) We can construct the circuit using insulated wires whenever connections cross. In this case, drawing the graph with the fewest possible crossings is important. (See the preamble to Exercise 26 to learn about the crossing number of a graph.)

The planarity of graphs is also useful in the design of road networks. Suppose we want to connect a group of cities by roads. We can model a road network connecting these cities using a simple graph with vertices representing the cities and edges representing the highways connecting them. We can built this road network without using underpasses or overpasses if the resulting graph is planar.

10.7.2 Euler's Formula

A planar representation of a graph splits the plane into **regions**, including an unbounded region. For instance, the planar representation of the graph shown in Figure 8 splits the plane into six regions. These are labeled in the figure. Euler showed that all planar representations of a graph split the plane into the same number of regions. He accomplished this by finding a relationship among the number of regions, the number of vertices, and the number of edges of a planar graph.

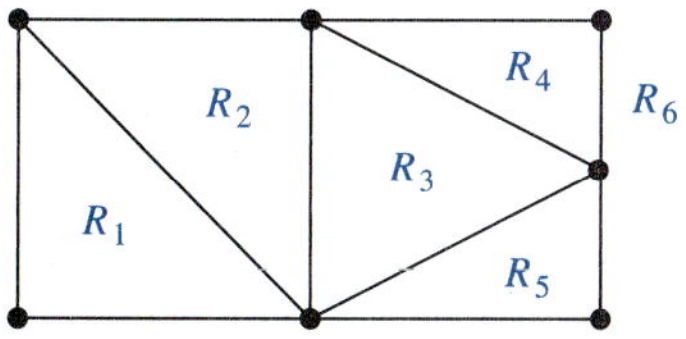

FIGURE 8 **The regions of the planar representation of a graph.**

THEOREM 1 **EULER'S FORMULA** Let G be a connected planar simple graph with e edges and v vertices. Let r be the number of regions in a planar representation of G. Then $r = e - v + 2$.

Proof: First, we specify a planar representation of G. We will prove the theorem by constructing a sequence of subgraphs $G_1, G_2, \ldots, G_e = G$, successively adding an edge at each stage. This is done using the following inductive definition. Arbitrarily pick one edge of G to obtain G_1. Obtain G_n from G_{n-1} by arbitrarily adding an edge that is incident with a vertex already in G_{n-1}, adding the other vertex incident with this edge if it is not already in G_{n-1}. This construction is possible because G is connected. G is obtained after e edges are added. Let r_n, e_n, and v_n represent the number of regions, edges, and vertices of the planar representation of G_n induced by the planar representation of G, respectively.

The proof will now proceed by induction. The relationship $r_1 = e_1 - v_1 + 2$ is true for G_1, because $e_1 = 1$, $v_1 = 2$, and $r_1 = 1$. This is shown in Figure 9.

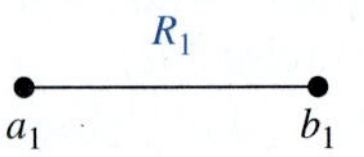

FIGURE 9 The basis case of the proof of Euler's formula.

Now assume that $r_k = e_k - v_k + 2$. Let $\{a_{k+1}, b_{k+1}\}$ be the edge that is added to G_k to obtain G_{k+1}. There are two possibilities to consider. In the first case, both a_{k+1} and b_{k+1} are already in G_k. These two vertices must be on the boundary of a common region R, or else it would be impossible to add the edge $\{a_{k+1}, b_{k+1}\}$ to G_k without two edges crossing (and G_{k+1} is planar). The addition of this new edge splits R into two regions. Consequently, in this case, $r_{k+1} = r_k + 1$, $e_{k+1} = e_k + 1$, and $v_{k+1} = v_k$. Thus, each side of the formula relating the number of regions, edges, and vertices increases by exactly one, so this formula is still true. In other words, $r_{k+1} = e_{k+1} - v_{k+1} + 2$. This case is illustrated in Figure 10(a).

In the second case, one of the two vertices of the new edge is not already in G_k. Suppose that a_{k+1} is in G_k but that b_{k+1} is not. Adding this new edge does not produce any new regions, because b_{k+1} must be in a region that has a_{k+1} on its boundary. Consequently, $r_{k+1} = r_k$. Moreover, $e_{k+1} = e_k + 1$ and $v_{k+1} = v_k + 1$. Each side of the formula relating the number of regions, edges, and vertices remains the same, so the formula is still true. In other words, $r_{k+1} = e_{k+1} - v_{k+1} + 2$. This case is illustrated in Figure 10(b).

We have completed the induction argument. Hence, $r_n = e_n - v_n + 2$ for all n. Because the original graph is the graph G_e, obtained after e edges have been added, the theorem is true. ◁

Euler's formula is illustrated in Example 4.

EXAMPLE 4 Suppose that a connected planar simple graph has 20 vertices, each of degree 3. Into how many regions does a representation of this planar graph split the plane?

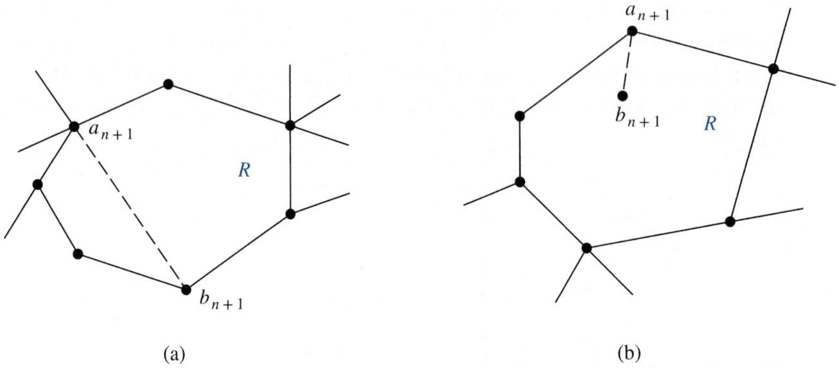

FIGURE 10 Adding an edge to G_n to produce G_{n+1}.

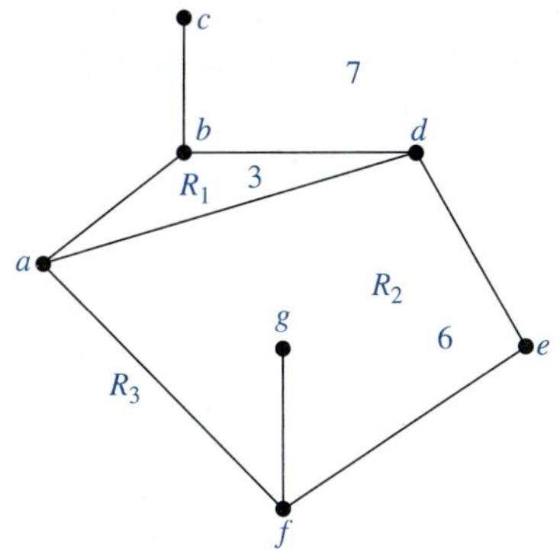

FIGURE 11 The degrees of regions.

Solution: This graph has 20 vertices, each of degree 3, so $v = 20$. Because the sum of the degrees of the vertices, $3v = 3 \cdot 20 = 60$, is equal to twice the number of edges, $2e$, we have $2e = 60$, or $e = 30$. Consequently, from Euler's formula, the number of regions is

$$r = e - v + 2 = 30 - 20 + 2 = 12.$$

◀

Euler's formula can be used to establish some inequalities that must be satisfied by planar graphs. One such inequality is given in Corollary 1.

COROLLARY 1 If G is a connected planar simple graph with e edges and v vertices, where $v \geq 3$, then $e \leq 3v - 6$.

Before we prove Corollary 1 we will use it to prove the following useful result.

COROLLARY 2 If G is a connected planar simple graph, then G has a vertex of degree not exceeding five.

Proof: If G has one or two vertices, the result is true. If G has at least three vertices, by Corollary 1 we know that $e \leq 3v - 6$, so $2e \leq 6v - 12$. If the degree of every vertex were at least six, then because $2e = \sum_{v \in V} \deg(v)$ (by the handshaking theorem), we would have $2e \geq 6v$. But this contradicts the inequality $2e \leq 6v - 12$. It follows that there must be a vertex with degree no greater than five. ◁

The proof of Corollary 1 is based on the concept of the **degree** of a region, which is defined to be the number of edges on the boundary of this region. When an edge occurs twice on the boundary (so that it is traced out twice when the boundary is traced out), it contributes two to the degree. We denote the degree of a region R by $\deg(R)$. The degrees of the regions of the graph shown in Figure 11 are displayed in the figure.

The proof of Corollary 1 can now be given.

Proof: A connected planar simple graph drawn in the plane divides the plane into regions, say r of them. The degree of each region is at least three. (Because the graphs discussed here are simple graphs, no multiple edges that could produce regions of degree two, or loops that could produce regions of degree one, are permitted.) In particular, note that the degree of the unbounded region is at least three because there are at least three vertices in the graph.

Note that the sum of the degrees of the regions is exactly twice the number of edges in the graph, because each edge occurs on the boundary of a region exactly twice (either in two different regions, or twice in the same region). Because each region has degree greater than or equal to three, it follows that

$$2e = \sum_{\text{all regions } R} \deg(R) \geq 3r.$$

Hence,

$$(2/3)e \geq r.$$

Using $r = e - v + 2$ (Euler's formula), we obtain

$$e - v + 2 \leq (2/3)e.$$

It follows that $e/3 \leq v - 2$. This shows that $e \leq 3v - 6$. ◁

This corollary can be used to demonstrate that K_5 is nonplanar.

EXAMPLE 5 Show that K_5 is nonplanar using Corollary 1.

Solution: The graph K_5 has five vertices and 10 edges. However, the inequality $e \leq 3v - 6$ is not satisfied for this graph because $e = 10$ and $3v - 6 = 9$. Therefore, K_5 is not planar. ◀

It was previously shown that $K_{3,3}$ is not planar. Note, however, that this graph has six vertices and nine edges. This means that the inequality $e = 9 \leq 12 = 3 \cdot 6 - 6$ is satisfied. Consequently, the fact that the inequality $e \leq 3v - 6$ is satisfied does *not* imply that a graph is planar. However, the following corollary of Theorem 1 can be used to show that $K_{3,3}$ is nonplanar.

COROLLARY 3 If a connected planar simple graph has e edges and v vertices with $v \geq 3$ and no circuits of length three, then $e \leq 2v - 4$.

The proof of Corollary 3 is similar to that of Corollary 1, except that in this case the fact that there are no circuits of length three implies that the degree of a region must be at least four. The details of this proof are left for the reader (see Exercise 15).

EXAMPLE 6 Use Corollary 3 to show that $K_{3,3}$ is nonplanar.

Solution: Because $K_{3,3}$ has no circuits of length three (this is easy to see because it is bipartite), Corollary 3 can be used. $K_{3,3}$ has six vertices and nine edges. Because $e = 9$ and $2v - 4 = 8$, Corollary 3 shows that $K_{3,3}$ is nonplanar. ◀

10.7.3 Kuratowski's Theorem

We have seen that $K_{3,3}$ and K_5 are not planar. Clearly, a graph is not planar if it contains either of these two graphs as a subgraph. Surprisingly, all nonplanar graphs must contain a subgraph that can be obtained from $K_{3,3}$ or K_5 using certain permitted operations.

If a graph is planar, so will be any graph obtained by removing an edge $\{u, v\}$ and adding a new vertex w together with edges $\{u, w\}$ and $\{w, v\}$. Such an operation is called an **elementary subdivision**. The graphs $G_1 = (V_1, E_1)$ and $G_2 = (V_2, E_2)$ are called **homeomorphic** if they can be obtained from the same graph by a sequence of elementary subdivisions.

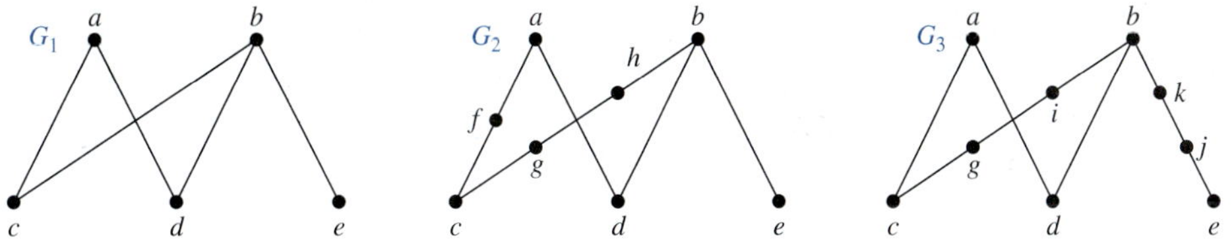

FIGURE 12 Homeomorphic graphs.

EXAMPLE 7 Show that the graphs G_1, G_2, and G_3 displayed in Figure 12 are all homeomorphic.

Solution: These three graphs are homeomorphic because all three can be obtained from G_1 by elementary subdivisions. G_1 can be obtained from itself by an empty sequence of elementary subdivisions. To obtain G_2 from G_1 we can use this sequence of elementary subdivisions: (*i*) remove the edge $\{a, c\}$, add the vertex f, and add the edges $\{a, f\}$ and $\{f, c\}$; (*ii*) remove the edge $\{b, c\}$, add the vertex g, and add the edges $\{b, g\}$ and $\{g, c\}$; and (*iii*) remove the edge $\{b, g\}$, add the vertex h, and add the edges $\{g, h\}$ and $\{b, h\}$. We leave it to the reader to determine the sequence of elementary subdivisions needed to obtain G_3 from G_1. ◀

The Polish mathematician Kazimierz Kuratowski established Theorem 2 in 1930, which characterizes planar graphs using the concept of graph homeomorphism.

THEOREM 2 A graph is nonplanar if and only if it contains a subgraph homeomorphic to $K_{3,3}$ or K_5.

It is clear that a graph containing a subgraph homeomorphic to $K_{3,3}$ or K_5 is nonplanar. However, the proof of the converse, namely that every nonplanar graph contains a subgraph homeomorphic to $K_{3,3}$ or K_5, is complicated and will not be given here. Examples 8 and 9 illustrate how Kuratowski's theorem is used.

EXAMPLE 8 Determine whether the graph G shown in Figure 13 is planar.

Extra Examples

Solution: G has a subgraph H homeomorphic to K_5. H is obtained by deleting h, j, and k and all edges incident with these vertices. H is homeomorphic to K_5 because it can be obtained from K_5 (with vertices a, b, c, g, and i) by a sequence of elementary subdivisions, adding the vertices d, e, and f. (The reader should construct such a sequence of elementary subdivisions.) Hence, G is nonplanar. ◀

Links

Archive PL/Alamy Stock Photo

KAZIMIERZ KURATOWSKI (1896–1980) Kazimierz Kuratowski, the son of a famous Warsaw lawyer, attended secondary school in Warsaw. He studied in Glasgow, Scotland, from 1913 to 1914 but could not return there after the outbreak of World War I. In 1915 he entered Warsaw University, where he was active in the Polish patriotic student movement. He published his first paper in 1919 and received his Ph.D. in 1921. He was an active member of the group known as the Warsaw School of Mathematics, working in the areas of the foundations of set theory and topology. He was appointed associate professor at the Lwów Polytechnical University, where he stayed for seven years, collaborating with the important Polish mathematicians Banach and Ulam. In 1930, while at Lwów, Kuratowski completed his work characterizing planar graphs.

In 1934 he returned to Warsaw University as a full professor. Until the start of World War II, he was active in research and teaching. During the war, because of the persecution of educated Poles, Kuratowski went into hiding under an assumed name and taught at the clandestine Warsaw University. After the war he helped revive Polish mathematics, serving as director of the Polish National Mathematics Institute. He wrote over 180 papers and three widely used textbooks.

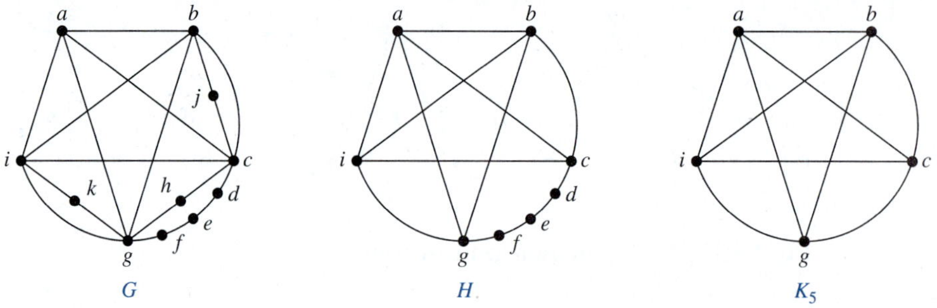

FIGURE 13 **The undirected graph G, a subgraph H homeomorphic to K_5, and K_5.**

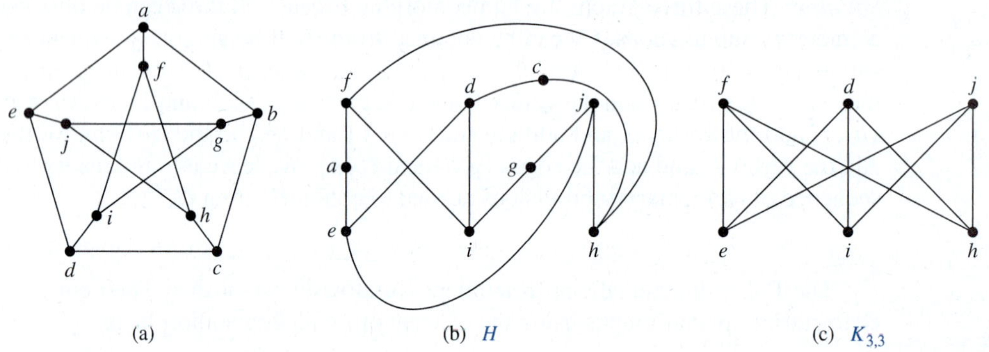

FIGURE 14 **(a) The Petersen graph, (b) a subgraph H homeomorphic to $K_{3,3}$, and (c) $K_{3,3}$.**

EXAMPLE 9 Is the Petersen graph, shown in Figure 14(a), planar? (The Danish mathematician Julius Petersen studied this graph in 1891; it is often used to illustrate various theoretical properties of graphs.)

Solution: The subgraph H of the Petersen graph obtained by deleting b and the three edges that have b as an endpoint, shown in Figure 14(b), is homeomorphic to $K_{3,3}$, with vertex sets $\{f, d, j\}$ and $\{e, i, h\}$, because it can be obtained by a sequence of elementary subdivisions, deleting $\{d, h\}$ and adding $\{c, h\}$ and $\{c, d\}$, deleting $\{e, f\}$ and adding $\{a, e\}$ and $\{a, f\}$, and deleting $\{i, j\}$ and adding $\{g, i\}$ and $\{g, j\}$. Hence, the Petersen graph is not planar. ◀

Exercises

1. Can five houses be connected to two utilities without connections crossing?

In Exercises 2–4 draw the given planar graph without any crossings.

2.

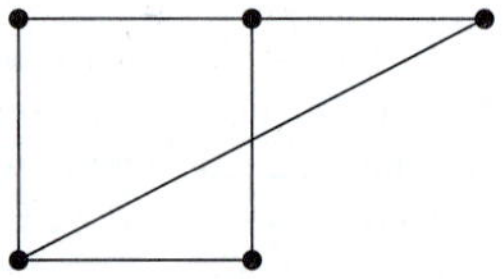

3.

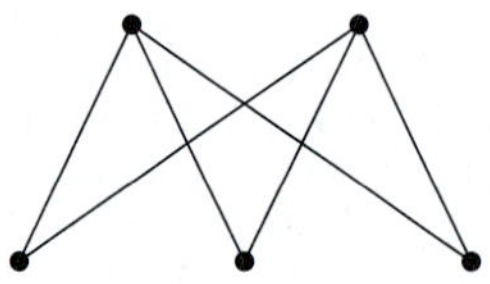

4.

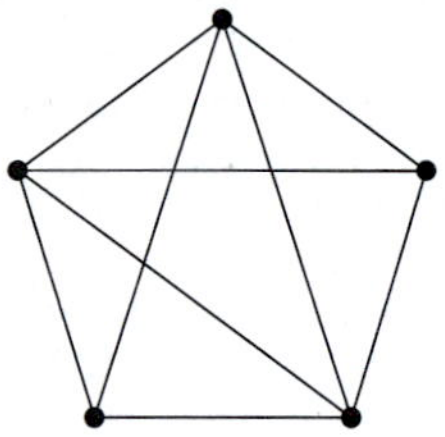

In Exercises 5–9 determine whether the given graph is planar. If so, draw it so that no edges cross.

5.

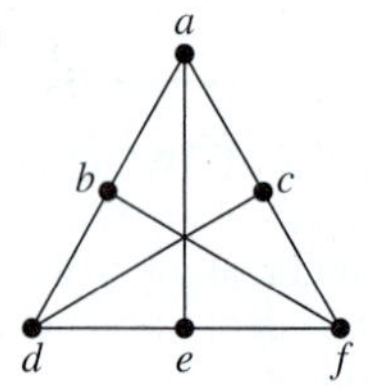

6.

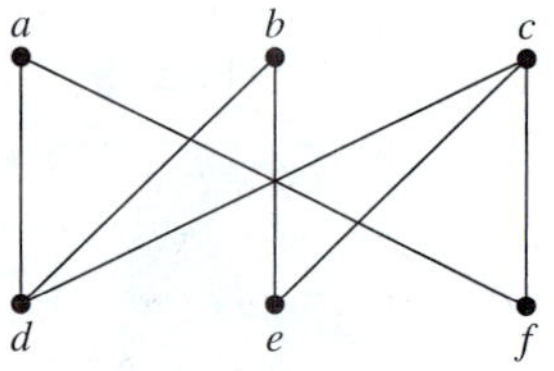

7.

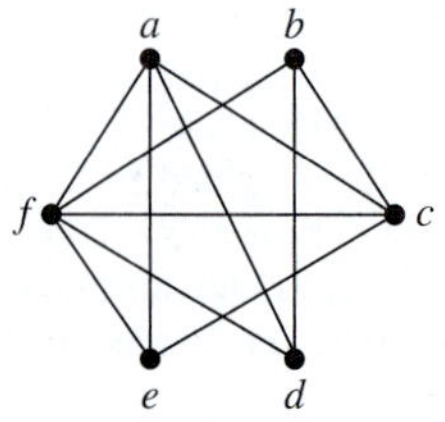

8.

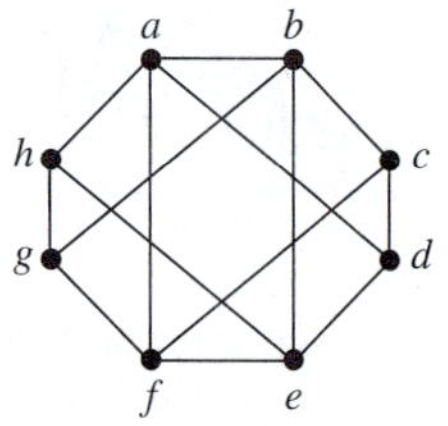

9.

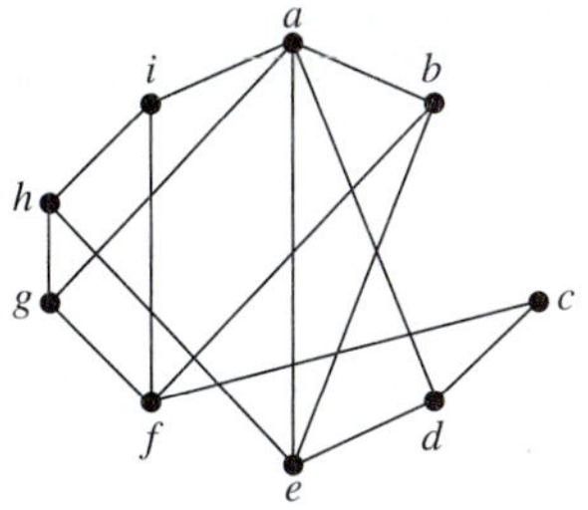

10. Complete the argument in Example 3.

11. Show that K_5 is nonplanar using an argument similar to that given in Example 3.

12. Suppose that a connected planar graph has eight vertices, each of degree three. Into how many regions is the plane divided by a planar representation of this graph?

13. Suppose that a connected planar graph has six vertices, each of degree four. Into how many regions is the plane divided by a planar representation of this graph?

14. Suppose that a connected planar graph has 30 edges. If a planar representation of this graph divides the plane into 20 regions, how many vertices does this graph have?

15. Prove Corollary 3.

16. Suppose that a connected bipartite planar simple graph has e edges and v vertices. Show that $e \leq 2v - 4$ if $v \geq 3$.

***17.** Suppose that a connected planar simple graph with e edges and v vertices contains no simple circuits of length 4 or less. Show that $e \leq (5/3)v - (10/3)$ if $v \geq 4$.

18. Suppose that a planar graph has k connected components, e edges, and v vertices. Also suppose that the plane is divided into r regions by a planar representation of the graph. Find a formula for r in terms of e, v, and k.

19. Which of these nonplanar graphs have the property that the removal of any vertex and all edges incident with that vertex produces a planar graph?

a) K_5 **b)** K_6 **c)** $K_{3,3}$ **d)** $K_{3,4}$

In Exercises 20–22 determine whether the given graph is homeomorphic to $K_{3,3}$.

20.

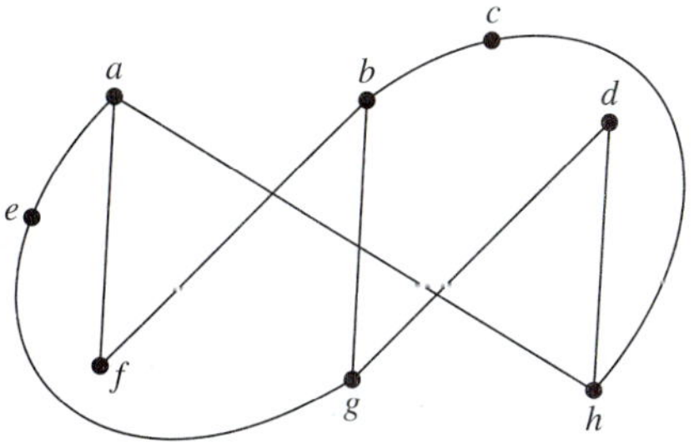

21.

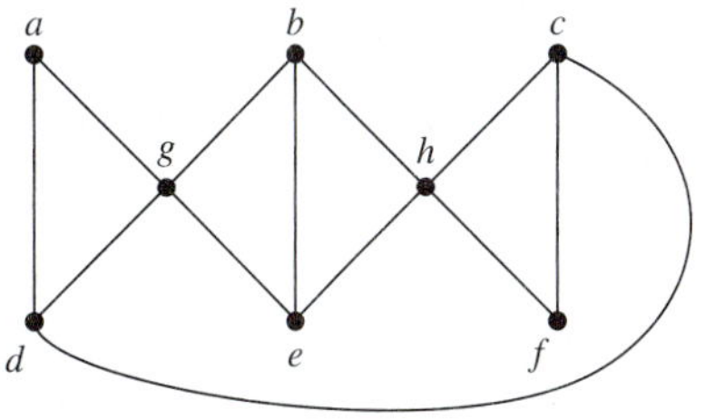

22.

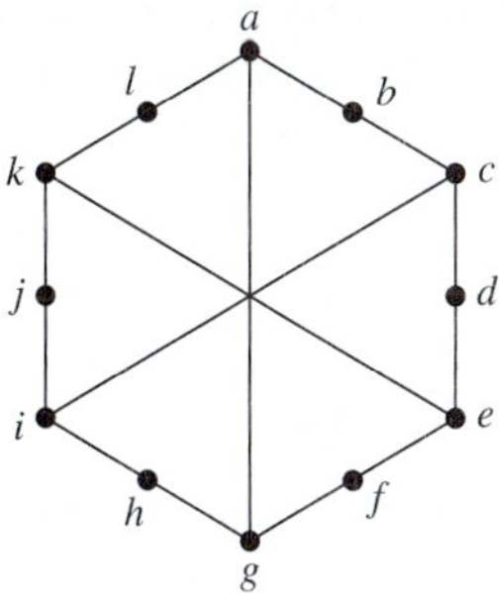

In Exercises 23–25 use Kuratowski's theorem to determine whether the given graph is planar.

23.

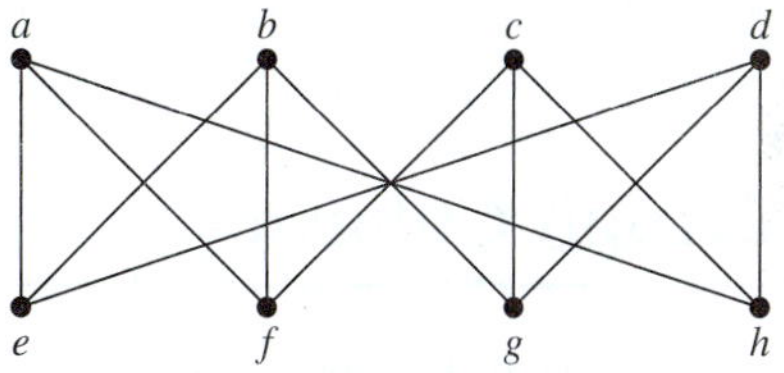

24.

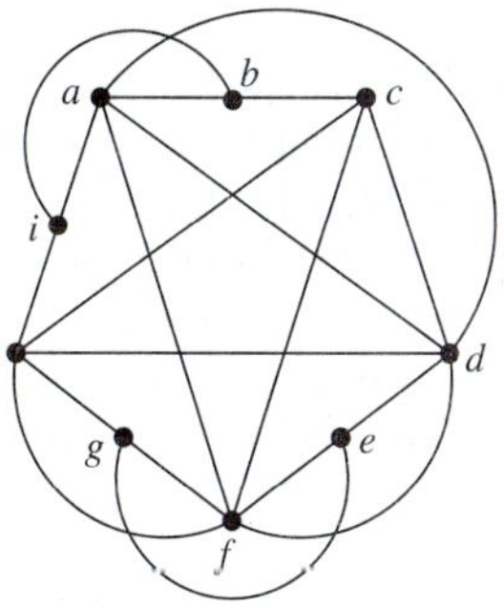

25.

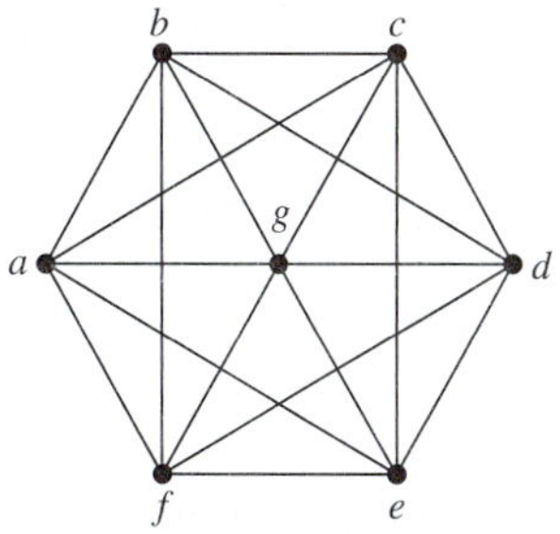

The **crossing number** of a simple graph is the minimum number of crossings that can occur when this graph is drawn in the plane where no three arcs representing edges are permitted to cross at the same point.

26. Show that $K_{3,3}$ has 1 as its crossing number.

****27.** Find the crossing numbers of each of these nonplanar graphs.

a) K_5 **b)** K_6 **c)** K_7

d) $K_{3,4}$ **e)** $K_{4,4}$ **f)** $K_{5,5}$

***28.** Find the crossing number of the Petersen graph.

29. Show that if m and n are even positive integers, the crossing number of $K_{m,n}$ is less than or equal to $mn(m-2)(n-2)/16$. [*Hint:* Place m vertices along the x-axis so that they are equally spaced and symmetric about the origin and place n vertices along the y-axis so that they are equally spaced and symmetric about the origin. Now connect each of the m vertices on the x-axis to each of the vertices on the y-axis and count the crossings.]

The **thickness** of a simple graph G is the smallest number of planar subgraphs of G that have G as their union.

30. Show that $K_{3,3}$ has 2 as its thickness.

***31.** Find the thickness of the graphs in Exercise 27.

32. Show that if G is a connected simple graph with v vertices and e edges, where $v \geq 3$, then the thickness of G is at least $\lceil e/(3v-6) \rceil$.

***33.** Use Exercise 32 to show that the thickness of K_n is at least $\lfloor (n+7)/6 \rfloor$ whenever n is a positive integer.

34. Show that if G is a connected simple graph with v vertices and e edges, where $v \geq 3$, and no circuits of length three, then the thickness of G is at least $\lceil e/(2v-4) \rceil$.

35. Use Exercise 34 to show that the thickness of $K_{m,n}$, where m and n are not both 1, is at least $\lceil mn/(2m+2n-4) \rceil$ whenever m and n are positive integers.

***36.** Draw K_5 on the surface of a torus (a doughnut-shaped solid) so that no edges cross.

***37.** Draw $K_{3,3}$ on the surface of a torus so that no edges cross.

10.8 Graph Coloring

10.8.1 Introduction

Links

Problems related to the coloring of maps of regions, such as maps of parts of the world, have generated many results in graph theory. When a map* is colored, two regions with a common border are customarily assigned different colors. One way to ensure that two adjacent regions never have the same color is to use a different color for each region. However, this is inefficient, and on maps with many regions it would be hard to distinguish similar colors. Instead, a small number of colors should be used whenever possible. Consider the problem of determining the least number of colors that can be used to color a map so that adjacent regions never have the same color. For instance, for the map shown on the left in Figure 1, four colors suffice, but three colors are not enough. (The reader should check this.) In the map on the right in Figure 1, three colors are sufficient (but two are not).

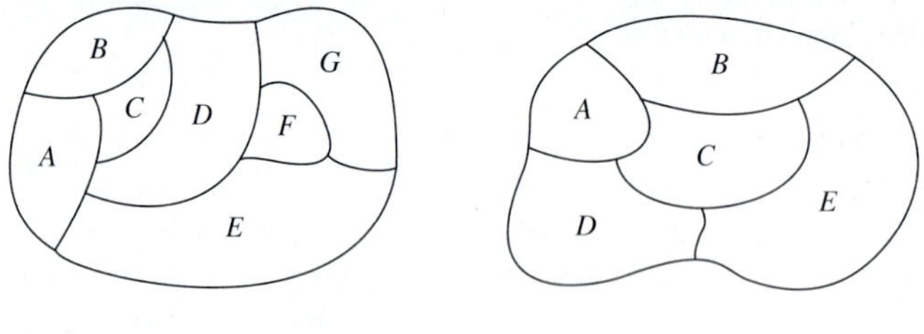

FIGURE 1 **Two maps.**

Each map in the plane can be represented by a graph. To set up this correspondence, each region of the map is represented by a vertex. Edges connect two vertices if the regions represented by these vertices have a common border. Two regions that touch at only one point are not considered adjacent. The resulting graph is called the **dual graph** of the map. By the way in which dual graphs of maps are constructed, it is clear that any map in the plane has a planar dual graph. Figure 2 displays the dual graphs that correspond to the maps shown in Figure 1.

The problem of coloring the regions of a map is equivalent to the problem of coloring the vertices of the dual graph so that no two adjacent vertices in this graph have the same color. We now define a graph coloring.

*We will assume that all regions in a map are connected. This eliminates any problems presented by such geographical entities as Michigan.

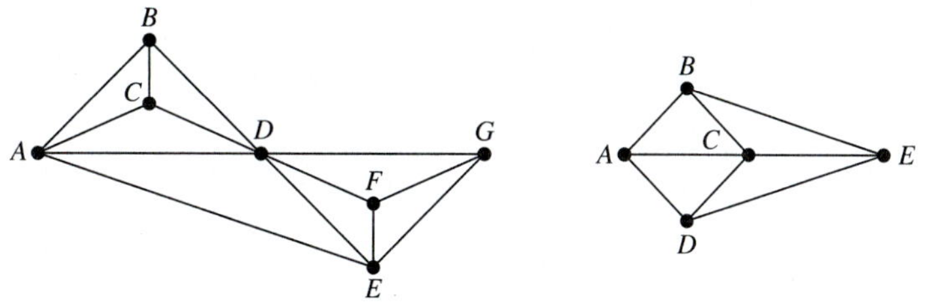

FIGURE 2 **Dual graphs of the maps in Figure 1.**

Definition 1 A *coloring* of a simple graph is the assignment of a color to each vertex of the graph so that no two adjacent vertices are assigned the same color.

A graph can be colored by assigning a different color to each of its vertices. However, for most graphs a coloring can be found that uses fewer colors than the number of vertices in the graph. What is the least number of colors necessary?

Definition 2 The *chromatic number* of a graph is the least number of colors needed for a coloring of this graph. The chromatic number of a graph G is denoted by $\chi(G)$. (Here χ is the Greek letter *chi*.)

Note that asking for the chromatic number of a planar graph is the same as asking for the minimum number of colors required to color a planar map so that no two adjacent regions are assigned the same color. This question has been studied for more than 100 years. The answer is provided by one of the most famous theorems in mathematics.

THEOREM 1 **THE FOUR COLOR THEOREM** The chromatic number of a planar graph is no greater than four.

Links

The four color theorem was originally posed as a conjecture in the 1850s. It was finally proved by the American mathematicians Kenneth Appel and Wolfgang Haken in 1976. Prior to 1976, many incorrect proofs were published, often with hard-to-find errors. In addition, many futile attempts were made to construct counterexamples by drawing maps that require more than four colors. (Proving the five color theorem is not that difficult; see Exercise 41.)

Links

Perhaps the most notorious fallacious proof in all of mathematics is the incorrect proof of the four color theorem published in 1879 by a London barrister and amateur mathematician, Alfred Kempe. Mathematicians accepted his proof as correct until 1890, when Percy Heawood found an error that made Kempe's argument incomplete. However, Kempe's line of reasoning turned out to be the basis of the successful proof given by Appel and Haken. Their proof relies on a careful case-by-case analysis carried out by computer. They showed that if the four color theorem were false, there would have to be a counterexample of one of approximately 2000 different

Courtesy of London Mathematical Society

ALFRED BRAY KEMPE (1849–1922) Kempe was a barrister and a leading authority on ecclesiastical law. However, having studied mathematics at Cambridge University, he retained his interest in it, and later in life he devoted considerable time to mathematical research. Kempe made contributions to kinematics, the branch of mathematics dealing with motion, and to mathematical logic. However, Kempe is best remembered for his fallacious proof of the four color theorem.

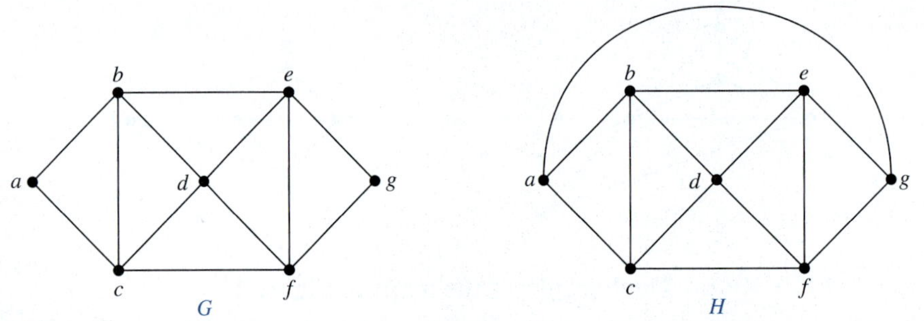

FIGURE 3 **The simple graphs G and H.**

types, and they then showed that none of these types exists. They used over 1000 hours of computer time in their proof. This proof generated a large amount of controversy, because computers played such an important role in it. For example, could there be an error in a computer program that led to incorrect results? Was their argument really a proof if it depended on what could be unreliable computer output? Since their proof appeared, simpler proofs that rely on checking fewer types of possible counterexamples have been found and a proof using an automated proof system has been created. However, no proof not relying on a computer has yet been found.

Note that the four color theorem applies only to planar graphs. Nonplanar graphs can have arbitrarily large chromatic numbers, as will be shown in Example 2.

Two things are required to show that the chromatic number of a graph is k. First, we must show that the graph can be colored with k colors. This can be done by constructing such a coloring. Second, we must show that the graph cannot be colored using fewer than k colors. Examples 1–4 illustrate how chromatic numbers can be found.

EXAMPLE 1 What are the chromatic numbers of the graphs G and H shown in Figure 3?

Extra Examples

Solution: The chromatic number of G is at least three, because the vertices a, b, and c must be assigned different colors. To see if G can be colored with three colors, assign red to a, blue to b, and green to c. Then, d can (and must) be colored red because it is adjacent to b and c. Furthermore, e can (and must) be colored green because it is adjacent only to vertices colored red and blue, and f can (and must) be colored blue because it is adjacent only to vertices colored red and green. Finally, g can (and must) be colored red because it is adjacent only to vertices colored blue and green. This produces a coloring of G using exactly three colors. Figure 4 displays such a coloring.

The graph H is made up of the graph G with an edge connecting a and g. Any attempt to color H using three colors must follow the same reasoning as that used to color G, except at the last stage, when all vertices other than g have been colored. Then, because g is adjacent (in H) to vertices colored red, blue, and green, a fourth color, say brown, needs to be used. Hence, H has a chromatic number equal to 4. A coloring of H is shown in Figure 4. ◀

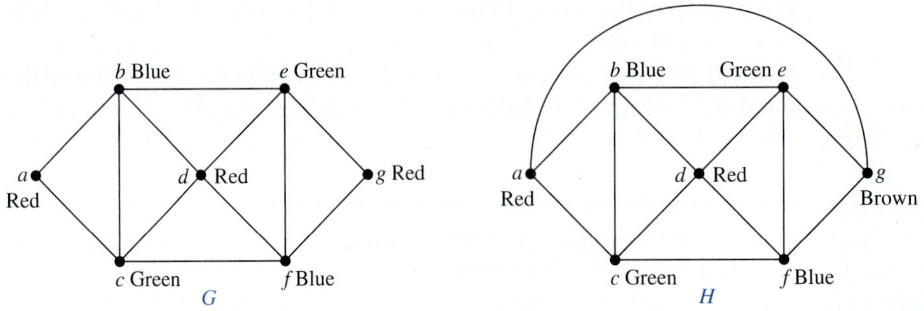

FIGURE 4 **Colorings of the graphs G and H.**

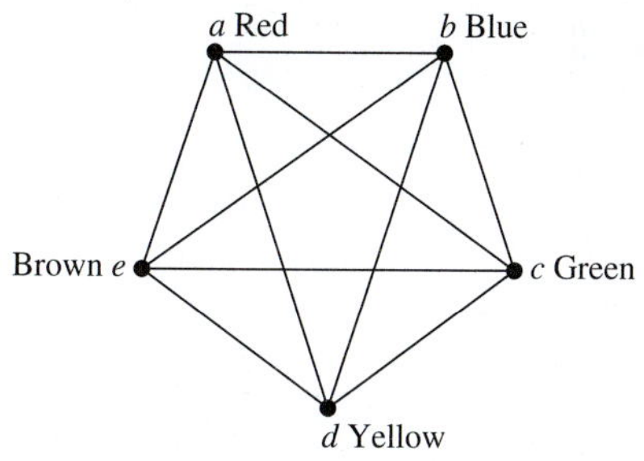

FIGURE 5 **A coloring of K_5.**

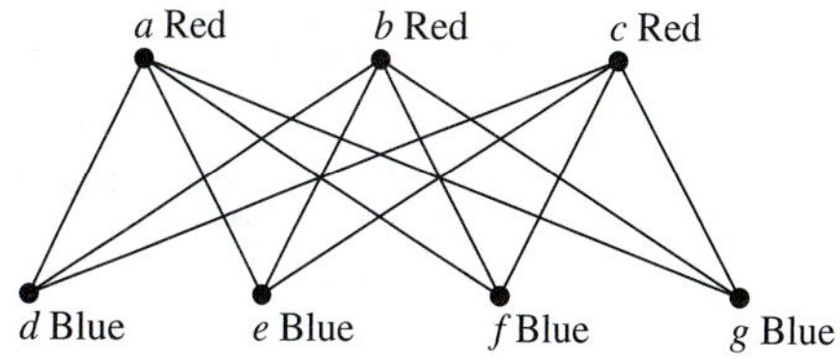

FIGURE 6 **A coloring of $K_{3,4}$.**

EXAMPLE 2 What is the chromatic number of K_n?

Solution: A coloring of K_n can be constructed using n colors by assigning a different color to each vertex. Is there a coloring using fewer colors? The answer is no. No two vertices can be assigned the same color, because every two vertices of this graph are adjacent. Hence, the chromatic number of K_n is n. That is, $\chi(K_n) = n$. (Recall that K_n is not planar when $n \geq 5$, so this result does not contradict the four color theorem.) A coloring of K_5 using five colors is shown in Figure 5. ◀

EXAMPLE 3 What is the chromatic number of the complete bipartite graph $K_{m,n}$, where m and n are positive integers?

Solution: The number of colors needed may seem to depend on m and n. However, as Theorem 4 in Section 10.2 tells us, only two colors are needed, because $K_{m,n}$ is a bipartite graph. Hence, $\chi(K_{m,n}) = 2$. This means that we can color the set of m vertices with one color and the set of n vertices with a second color. Because edges connect only a vertex from the set of m vertices and a vertex from the set of n vertices, no two adjacent vertices have the same color. A coloring of $K_{3,4}$ with two colors is displayed in Figure 6. ◀

EXAMPLE 4 What is the chromatic number of the graph C_n, where $n \geq 3$? (Recall that C_n is the cycle with n vertices.)

Solution: We will first consider some individual cases. To begin, let $n = 6$. Pick a vertex and color it red. Proceed clockwise in the planar depiction of C_6 shown in Figure 7. It is necessary to assign a second color, say blue, to the next vertex reached. Continue in the clockwise direction;

HISTORICAL NOTE In 1852, an ex-student of Augustus De Morgan, Francis Guthrie, noticed that the counties in England could be colored using four colors so that no adjacent counties were assigned the same color. On this evidence, he conjectured that the four color theorem was true. Francis told his brother Frederick, at that time a student of De Morgan, about this problem. Frederick in turn asked his teacher De Morgan about his brother's conjecture. De Morgan was extremely interested in this problem and publicized it throughout the mathematical community. In fact, the first written reference to the conjecture can be found in a letter from De Morgan to Sir William Rowan Hamilton. Although De Morgan thought Hamilton would be interested in this problem, Hamilton apparently was not interested in it, because it had nothing to do with quaternions.

Links ❯ HISTORICAL NOTE Although a simpler proof of the four color theorem was found by Robertson, Sanders, Seymour, and Thomas in 1996, reducing the computational part of the proof to examining 633 configurations, no proof that does not rely on extensive computation has yet been found.

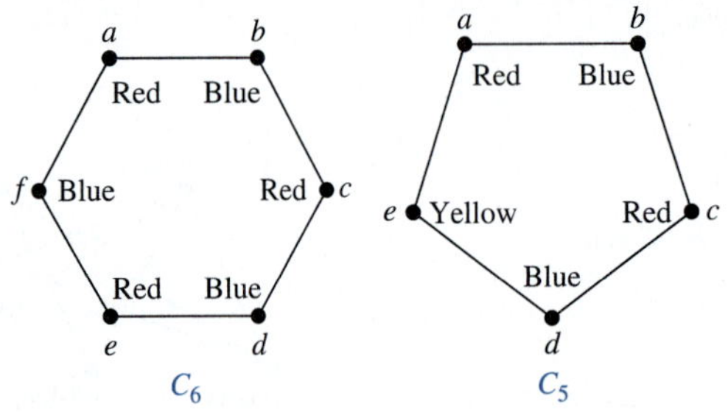

FIGURE 7 **Colorings of C_5 and C_6.**

the third vertex can be colored red, the fourth vertex blue, and the fifth vertex red. Finally, the sixth vertex, which is adjacent to the first, can be colored blue. Hence, the chromatic number of C_6 is 2. Figure 7 displays the coloring constructed here.

Next, let $n = 5$ and consider C_5. Pick a vertex and color it red. Proceeding clockwise, it is necessary to assign a second color, say blue, to the next vertex reached. Continuing in the clockwise direction, the third vertex can be colored red, and the fourth vertex can be colored blue. The fifth vertex cannot be colored either red or blue, because it is adjacent to the fourth vertex and the first vertex. Consequently, a third color is required for this vertex. Thus, the chromatic number of C_5 is 3. A coloring of C_5 using three colors is displayed in Figure 7.

In general, two colors are needed to color C_n when n is even. To construct such a coloring, simply pick a vertex and color it red. Proceed around the graph in a clockwise direction (using a planar representation of the graph) coloring the second vertex blue, the third vertex red, and so on. The nth vertex can be colored blue, because the two vertices adjacent to it, namely the$(n - 1)$st and the first vertices, are both colored red.

When n is odd and $n > 1$, the chromatic number of C_n is 3. To see this, pick an initial vertex. To use only two colors, it is necessary to alternate colors as the graph is traversed in a clockwise direction. However, the nth vertex reached is adjacent to two vertices of different colors, namely, the first and $(n - 1)$st. Hence, a third color must be used.

We have shown that $\chi(C_n) = 2$ if n is an even positive integer with $n \geq 4$ and $\chi(C_n) = 3$ if n is an odd positive integer with $n \geq 3$. ◀

Links ›

The best algorithms known for finding the chromatic number of a graph have exponential worst-case time complexity (in the number of vertices of the graph). Even the problem of finding an approximation to the chromatic number of a graph is difficult. It has been shown that if there were an algorithm with polynomial worst-case time complexity that could approximate the chromatic number of a graph up to a factor of 2 (that is, construct a bound that was no more than double the chromatic number of the graph), then an algorithm with polynomial worst-case time complexity for finding the chromatic number of the graph would also exist.

10.8.2 Applications of Graph Colorings

Graph coloring has a variety of applications to problems involving scheduling and assignments. (Note that because no efficient algorithm is known for graph coloring, this does not lead to efficient algorithms for scheduling and assignments.) Examples of such applications will be given here. The first application deals with the scheduling of final exams.

EXAMPLE 5 **Scheduling Final Exams** How can the final exams at a university be scheduled so that no student has two exams at the same time?

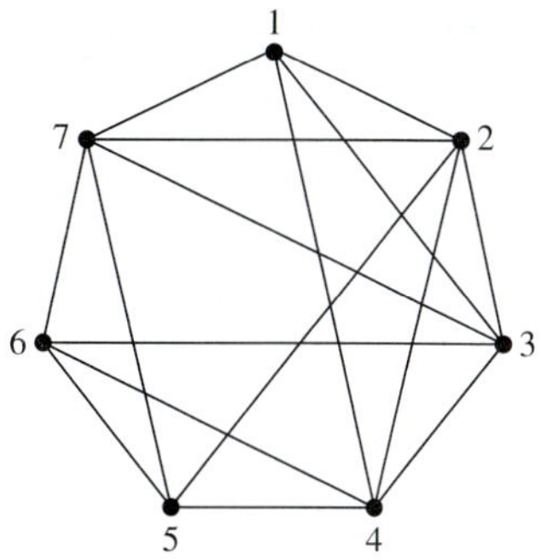

FIGURE 8 The graph representing the scheduling of final exams.

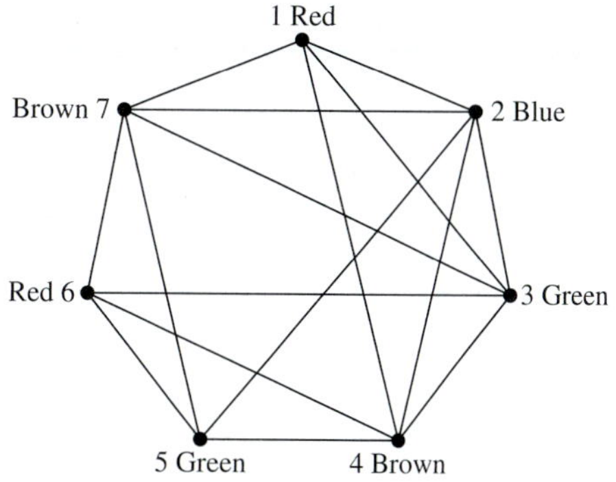

Time Period	Courses
I	1, 6
II	2
III	3, 5
IV	4, 7

FIGURE 9 Using a coloring to schedule final exams.

Solution: This scheduling problem can be solved using a graph model, with vertices representing courses and with an edge between two vertices if there is a common student in the courses they represent. Each time slot for a final exam is represented by a different color. A scheduling of the exams corresponds to a coloring of the associated graph.

For instance, suppose there are seven finals to be scheduled. Suppose the courses are numbered 1 through 7. Suppose that the following pairs of courses have common students: 1 and 2, 1 and 3, 1 and 4, 1 and 7, 2 and 3, 2 and 4, 2 and 5, 2 and 7, 3 and 4, 3 and 6, 3 and 7, 4 and 5, 4 and 6, 5 and 6, 5 and 7, and 6 and 7. In Figure 8 the graph associated with this set of classes is shown. A scheduling consists of a coloring of this graph.

Because the chromatic number of this graph is 4 (the reader should verify this), four time slots are needed. A coloring of the graph using four colors and the associated schedule are shown in Figure 9. ◀

Now consider an application to the assignment of television channels.

EXAMPLE 6 **Frequency Assignments** In broadcast television systems, each television station is broadcast over one of a fixed number of channels. Two stations cannot be assigned the same channel if they are too close. For example, the VHF band in North America has channels 2 through 13 and no two stations within 150 miles can operate on the same channel. How can the assignment of channels be modeled by graph coloring?

Solution: Construct a graph by assigning a vertex to each station. Two vertices are connected by an edge if they are located within 150 miles of each other. An assignment of channels corresponds to a coloring of the graph, where each color represents a different channel. ◀

An application of graph coloring to compilers is considered in Example 7.

EXAMPLE 7 **Index Registers** In efficient compilers the execution of loops is speeded up when frequently used variables are stored temporarily in index registers in the central processing unit, instead of in regular memory. For a given loop, how many index registers are needed? This problem can be addressed using a graph coloring model. To set up the model, let each vertex of a graph represent a variable in the loop. There is an edge between two vertices if the variables they represent must be stored in index registers at the same time during the execution of the loop. Thus, the chromatic number of the graph gives the number of index registers needed, because different registers must be assigned to variables when the vertices representing these variables are adjacent in the graph. ◀

Exercises

In Exercises 1–4 construct the dual graph for the map shown. Then find the number of colors needed to color the map so that no two adjacent regions have the same color.

1.

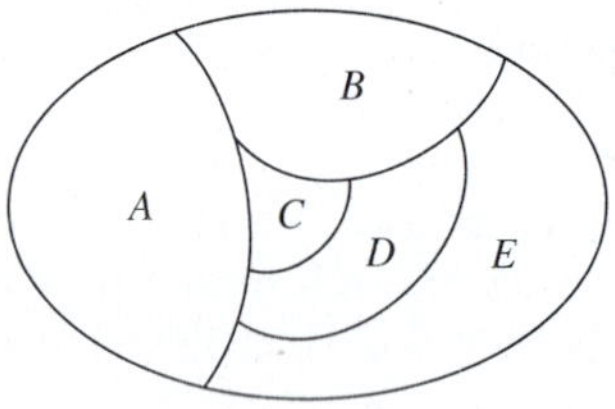

2.

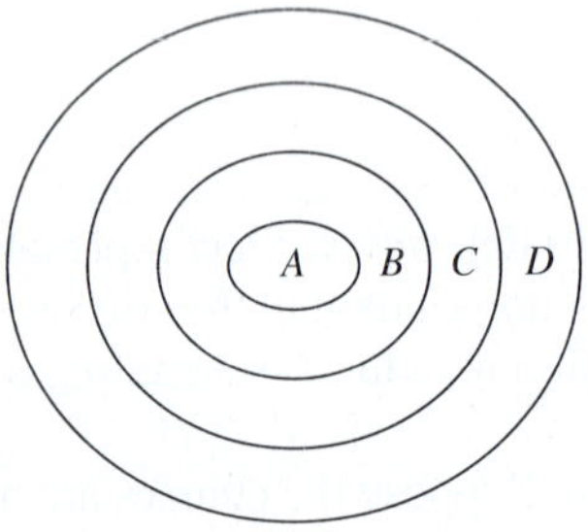

3.

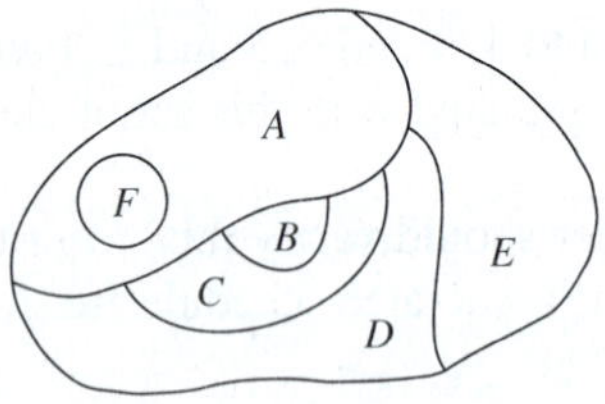

4.

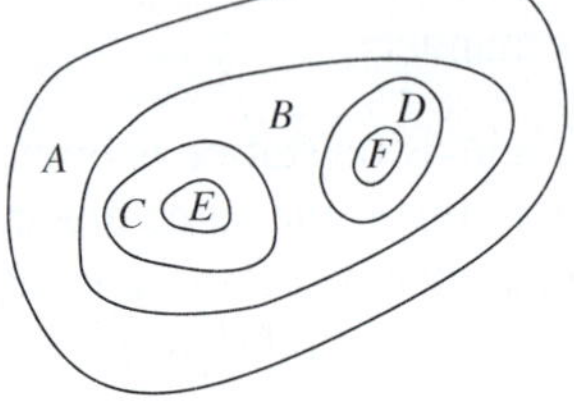

In Exercises 5–11 find the chromatic number of the given graph.

5.

6.

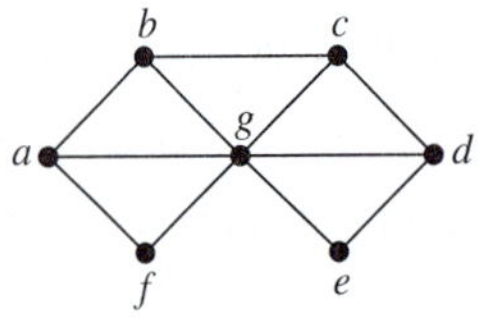

7.

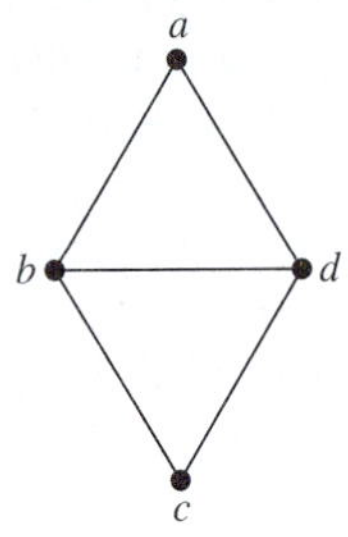

8.

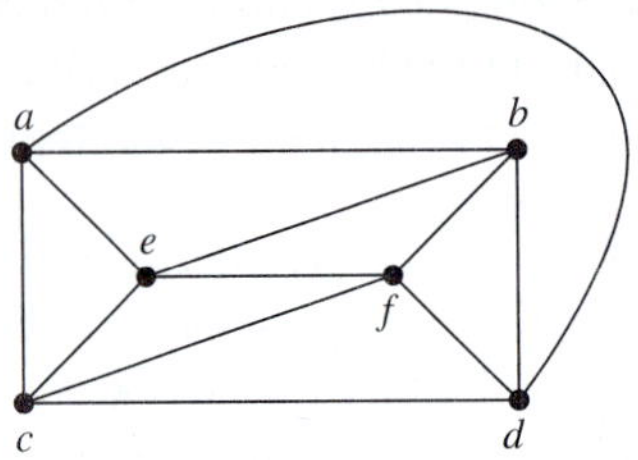

9.

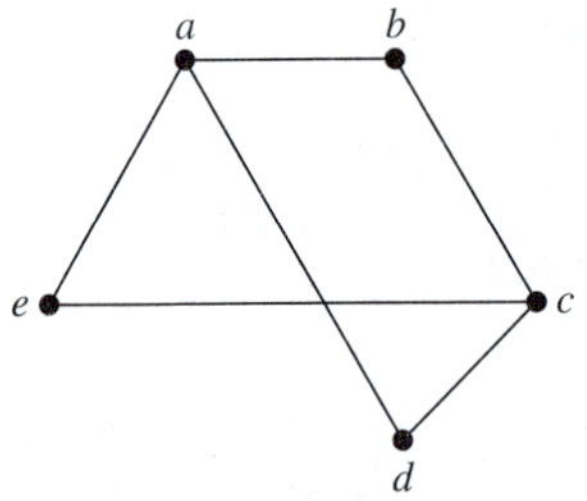

10.

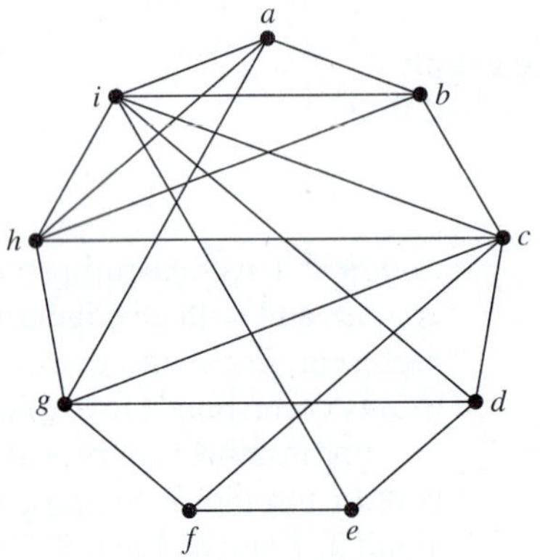

11.

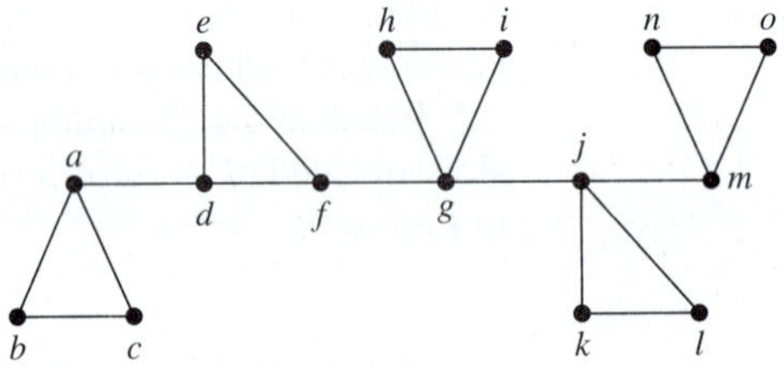

12. For the graphs in Exercises 5–11, decide whether it is possible to decrease the chromatic number by removing a single vertex and all edges incident with it.

13. Which graphs have a chromatic number of 1?

14. What is the least number of colors needed to color a map of the United States? Do not consider adjacent states that meet only at a corner. Suppose that Michigan is one region. Consider the vertices representing Alaska and Hawaii as isolated vertices.

15. What is the chromatic number of W_n?

16. Show that a simple graph that has a circuit with an odd number of vertices in it cannot be colored using two colors.

17. Schedule the final exams for Math 115, Math 116, Math 185, Math 195, CS 101, CS 102, CS 273, and CS 473, using the fewest number of different time slots, if there are no students taking both Math 115 and CS 473, both Math 116 and CS 473, both Math 195 and CS 101, both Math 195 and CS 102, both Math 115 and Math 116, both Math 115 and Math 185, and both Math 185 and Math 195, but there are students in every other pair of courses.

18. How many different channels are needed for six stations located at the distances shown in the table, if two stations cannot use the same channel when they are within 150 miles of each other?

	1	*2*	*3*	*4*	*5*	*6*
1	—	85	175	200	50	100
2	85	—	125	175	100	160
3	175	125	—	100	200	250
4	200	175	100	—	210	220
5	50	100	200	210	—	100
6	100	160	250	220	100	—

19. The mathematics department has six committees, each meeting once a month. How many different meeting times must be used to ensure that no member is scheduled to attend two meetings at the same time if the committees are C_1 = {Arlinghaus, Brand, Zaslavsky}, C_2 = {Brand, Lee, Rosen}, C_3 = {Arlinghaus, Rosen, Zaslavsky}, C_4 = {Lee, Rosen, Zaslavsky}, C_5 = {Arlinghaus, Brand}, and C_6 = {Brand, Rosen, Zaslavsky}?

20. A zoo wants to set up natural habitats in which to exhibit its animals. Unfortunately, some animals will eat some of the others when given the opportunity. How can a graph model and a coloring be used to determine the number of different habitats needed and the placement of the animals in these habitats?

An **edge coloring** of a graph is an assignment of colors to edges so that edges incident with a common vertex are assigned different colors. The **edge chromatic number** of a graph is the smallest number of colors that can be used in an edge coloring of the graph. The edge chromatic number of a graph G is denoted by $\chi'(G)$.

21. Find the edge chromatic number of each of the graphs in Exercises 5–11.

22. Suppose that n devices are on a circuit board and that these devices are connected by colored wires. Express the number of colors needed for the wires, in terms of the edge chromatic number of the graph representing this circuit board, under the requirement that the wires leaving a particular device must be different colors. Explain your answer.

23. Find the edge chromatic numbers of

a) C_n, where $n \geq 3$.

b) W_n, where $n \geq 3$.

24. Show that the edge chromatic number of a graph must be at least as large as the maximum degree of a vertex of the graph.

25. Show that if G is a graph with n vertices, then no more than $n/2$ edges can be colored the same in an edge coloring of G.

***26.** Find the edge chromatic number of K_n when n is a positive integer.

27. Seven variables occur in a loop of a computer program. The variables and the steps during which they must be stored are t: steps 1 through 6; u: step 2; v: steps 2 through 4; w: steps 1, 3, and 5; x: steps 1 and 6; y: steps 3 through 6; and z: steps 4 and 5. How many different index registers are needed to store these variables during execution?

28. What can be said about the chromatic number of a graph that has K_n as a subgraph?

This algorithm can be used to color a simple graph: First, list the vertices $v_1, v_2, v_3, \ldots, v_n$ in order of decreasing degree so that $\deg(v_1) \geq \deg(v_2) \geq \cdots \geq \deg(v_n)$. Assign color 1 to v_1 and to the next vertex in the list not adjacent to v_1 (if one exists), and successively to each vertex in the list not adjacent to a vertex already assigned color 1. Then assign color 2 to the first vertex in the list not already colored. Successively assign color 2 to vertices in the list that have not already been colored and are not adjacent to vertices assigned color 2. If uncolored vertices remain, assign color 3 to the first vertex in the list not yet colored, and use color 3 to successively color those vertices not already colored and not adjacent to vertices assigned color 3. Continue this process until all vertices are colored.

29. Construct a coloring of the graph shown using this algorithm.

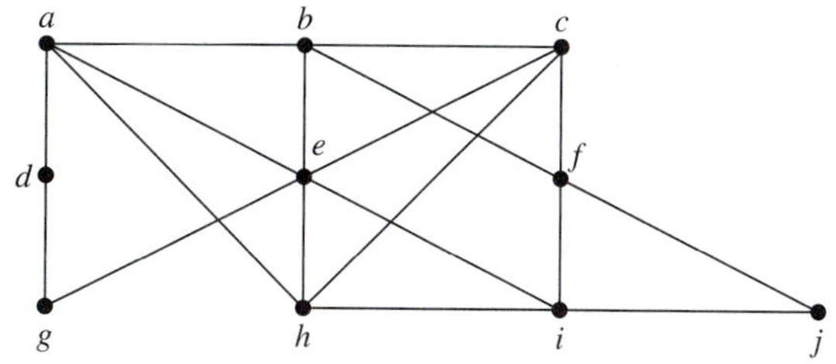

***30.** Use pseudocode to describe this coloring algorithm.

***31.** Show that the coloring produced by this algorithm may use more colors than are necessary to color a graph.

A connected graph G is called **chromatically k-critical** if the chromatic number of G is k, but for every edge of G, the chromatic number of the graph obtained by deleting this edge from G is $k - 1$.

32. Show that C_n is chromatically 3-critical whenever n is an odd positive integer, $n \geq 3$.

33. Show that W_n is chromatically 4-critical whenever n is an odd integer, $n \geq 3$.

34. Show that W_4 is not chromatically 3-critical.

35. Show that if G is a chromatically k-critical graph, then the degree of every vertex of G is at least $k - 1$.

A ***k*-tuple coloring** of a graph G is an assignment of a set of k different colors to each of the vertices of G such that no two adjacent vertices are assigned a common color. We denote by $\chi_k(G)$ the smallest positive integer n such that G has a k-tuple coloring using n colors. For example, $\chi_2(C_4) = 4$. To see this, note that using only four colors we can assign two colors to each vertex of C_4, as illustrated, so that no two adjacent vertices are assigned the same color. Furthermore, no fewer than four colors suffice because the vertices v_1 and v_2 each must be assigned two colors, and a common color cannot be assigned to both v_1 and v_2. (For more information about k-tuple coloring, see [MiRo91].)

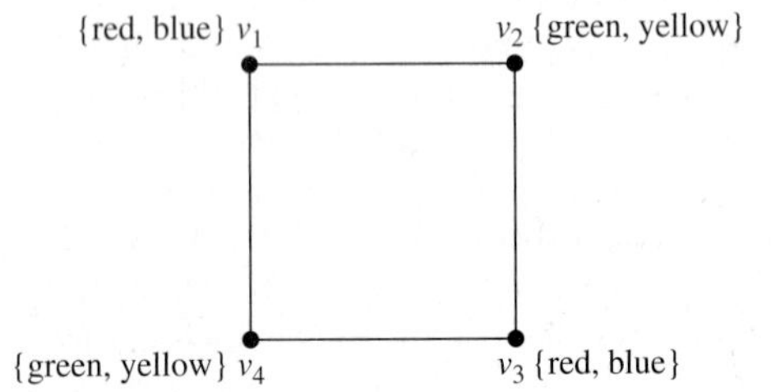

36. Find these values:

a) $\chi_2(K_3)$ **b)** $\chi_2(K_4)$ **c)** $\chi_2(W_4)$
d) $\chi_2(C_5)$ **e)** $\chi_2(K_{3,4})$ **f)** $\chi_3(K_5)$
***g)** $\chi_3(C_5)$ **h)** $\chi_3(K_{4,5})$

***37.** Let G and H be the graphs displayed in Figure 3. Find

a) $\chi_2(G)$. **b)** $\chi_2(H)$.
c) $\chi_3(G)$. **d)** $\chi_3(H)$.

38. What is $\chi_k(G)$ if G is a bipartite graph and k is a positive integer?

39. Frequencies for mobile radio (or cellular) telephones are assigned by zones. Each zone is assigned a set of frequencies to be used by vehicles in that zone. The same frequency cannot be used in different zones when interference can occur between telephones in these zones. Explain how a k-tuple coloring can be used to assign k frequencies to each mobile radio zone in a region.

***40.** Show that every planar graph G can be colored using six or fewer colors. [*Hint:* Use mathematical induction on the number of vertices of the graph. Apply Corollary 2 of Section 10.7 to find a vertex v with $\deg(v) \leq 5$. Consider the subgraph of G obtained by deleting v and all edges incident with it.]

****41.** Show that every planar graph G can be colored using five or fewer colors. [*Hint:* Use the hint provided for Exercise 40.]

The **art gallery problem** asks how many guards are needed to see all parts of an art gallery, where the gallery is the interior and boundary of a polygon with n sides. To state this problem more precisely, we need some terminology. A point x inside or on the boundary of a simple polygon P **covers** or **sees** a point y inside or on P if all points on the line segment xy are in the interior or on the boundary of P. We say that a set of points is a **guarding set** of a simple polygon P if for every point y inside P or on the boundary of P there is a point x in this guarding set that sees y. Denote by $G(P)$ the minimum number of points needed to guard the simple polygon P. The art gallery problem asks for the function $g(n)$, which is the maximum value of $G(P)$ over all simple polygons with n vertices. That is, $g(n)$ is the minimum positive integer for which it is guaranteed that a simple polygon with n vertices can be guarded with $g(n)$ or fewer guards.

42. Show that $g(3) = 1$ and $g(4) = 1$ by showing that all triangles and quadrilaterals can be guarded using one point.

***43.** Show that $g(5) = 1$. That is, show that all pentagons can be guarded using one point. [*Hint:* Show that there are either 0, 1, or 2 vertices with an interior angle greater than 180 degrees and that in each case, one guard suffices.]

***44.** Show that $g(6) = 2$ by first using Exercises 42 and 43 as well as Lemma 1 in Section 5.2 to show that $g(6) \leq 2$ and then find a simple hexagon for which two guards are needed.

***45.** Show that $g(n) \geq \lfloor n/3 \rfloor$. [*Hint:* Consider the polygon with $3k$ vertices that resembles a comb with k prongs, such as the polygon with 15 sides shown here.]

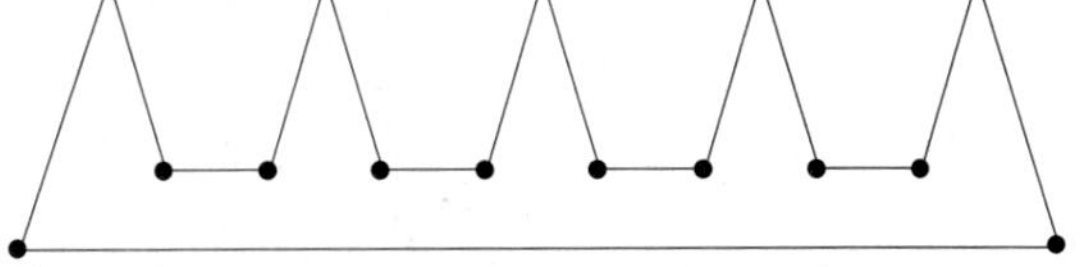

***46.** Solve the art gallery problem by proving the **art gallery theorem**, which states that at most $\lfloor n/3 \rfloor$ guards are needed to guard the interior and boundary of a simple polygon with n vertices. [*Hint:* Use Theorem 1 in Section 5.2 to triangulate the simple polygon into $n - 2$ triangles. Then show that it is possible to color the vertices of the triangulated polygon using three colors so that no two adjacent vertices have the same color. Use induction and Exercise 25 in Section 5.2. Finally, put guards at all vertices that are colored red, where red is the color used least in the coloring of the vertices. Show that placing guards at these points is all that is needed.]

Key Terms and Results

TERMS

undirected edge: an edge associated to a set $\{u, v\}$, where u and v are vertices

directed edge: an edge associated to an ordered pair (u, v), where u and v are vertices

multiple edges: distinct edges connecting the same vertices

multiple directed edges: distinct directed edges associated with the same ordered pair (u, v), where u and v are vertices

loop: an edge connecting a vertex with itself

undirected graph: a set of vertices and a set of undirected edges each of which is associated with a set of one or two of these vertices

simple graph: an undirected graph with no multiple edges or loops

multigraph: an undirected graph that may contain multiple edges but no loops

pseudograph: an undirected graph that may contain multiple edges and loops

directed graph: a set of vertices together with a set of directed edges each of which is associated with an ordered pair of vertices

directed multigraph: a graph with directed edges that may contain multiple directed edges

simple directed graph: a directed graph without loops or multiple directed edges

adjacent: two vertices are adjacent if there is an edge between them

incident: an edge is incident with a vertex if the vertex is an endpoint of that edge

deg v (degree of the vertex v in an undirected graph): the number of edges incident with v with loops counted twice

$\deg^-(v)$ (the in-degree of the vertex v in a graph with directed edges): the number of edges with v as their terminal vertex

$\deg^+(v)$ (the out-degree of the vertex v in a graph with directed edges): the number of edges with v as their initial vertex

underlying undirected graph of a graph with directed edges: the undirected graph obtained by ignoring the directions of the edges

K_n (complete graph on n vertices): the undirected graph with n vertices where each pair of vertices is connected by an edge

bipartite graph: a graph with vertex set that can be partitioned into subsets V_1 and V_2 so that each edge connects a vertex in V_1 and a vertex in V_2. The pair (V_1, V_2) is called a **bipartition** of V.

$K_{m,n}$ (complete bipartite graph): the graph with vertex set partitioned into a subset of m elements and a subset of n elements with two vertices connected by an edge if and only if one is in the first subset and the other is in the second subset

C_n (cycle of size n), $n \geq 3$: the graph with n vertices $v_1, v_2, \ldots, v_n$ and edges $\{v_1, v_2\}, \{v_2, v_3\}, \ldots, \{v_{n-1}, v_n\}, \{v_n, v_1\}$

W_n (wheel of size n), $n \geq 3$: the graph obtained from C_n by adding a vertex and edges from this vertex to the original vertices in C_n

Q_n (n-cube), $n \geq 1$: the graph that has the 2^n bit strings of length n as its vertices and edges connecting every pair of bit strings that differ by exactly one bit

matching in a graph G: a set of edges such that no two edges have a common endpoint

complete matching M from V_1 to V_2: a matching such that every vertex in V_1 is an endpoint of an edge in M

maximum matching: a matching containing the most edges among all matchings in a graph

isolated vertex: a vertex of degree zero

pendant vertex: a vertex of degree one

regular graph: a graph where all vertices have the same degree

subgraph of a graph $G = (V, E)$: a graph (W, F), where W is a subset of V and F is a subset of E

$G_1 \cup G_2$ (union of G_1 and G_2): the graph $(V_1 \cup V_2, E_1 \cup E_2)$, where $G_1 = (V_1, E_1)$ and $G_2 = (V_2, E_2)$

adjacency matrix: a matrix representing a graph using the adjacency of vertices

incidence matrix: a matrix representing a graph using the incidence of edges and vertices

isomorphic simple graphs: the simple graphs $G_1 = (V_1, E_1)$ and $G_2 = (V_2, E_2)$ are isomorphic if there exists a one-to-one correspondence f from V_1 to V_2 such that $\{f(v_1), f(v_2)\} \in E_2$ if and only if $\{v_1, v_2\} \in E_1$ for all v_1 and v_2 in V_1

invariant for graph isomorphism: a property that isomorphic graphs either both have or both do not have

path from u to v in an undirected graph: a sequence of edges $e_1, e_2, \ldots, e_n$, where e_i is associated to $\{x_i, x_{i+1}\}$ for $i = 0, 1, \ldots, n$, where $x_0 = u$ and $x_{n+1} = v$

path from u to v in a graph with directed edges: a sequence of edges $e_1, e_2, \ldots, e_n$, where e_i is associated to (x_i, x_{i+1}) for $i = 0, 1, \ldots, n$, where $x_0 = u$ and $x_{n+1} = v$

simple path: a path that does not contain an edge more than once

circuit: a path of length $n \geq 1$ that begins and ends at the same vertex

connected graph: an undirected graph with the property that there is a path between every pair of vertices

cut vertex of G: a vertex v such that $G - v$ is disconnected

cut edge of G: an edge e such that $G - e$ is disconnected

nonseparable graph: a graph without a cut vertex

vertex cut of G: a subset V' of the set of vertices of G such that $G - V'$ is disconnected

$\kappa(G)$ (the vertex connectivity of G): the size of a smallest vertex cut of G

k-connected graph: a graph that has a vertex connectivity no smaller than k

edge cut of G: a set of edges E' of G such that $G - E'$ is disconnected

$\lambda(G)$ (the edge connectivity of G): the size of a smallest edge cut of G

connected component of a graph G: a maximal connected subgraph of G

strongly connected directed graph: a directed graph with the property that there is a directed path from every vertex to every vertex

strongly connected component of a directed graph G: a maximal strongly connected subgraph of G

Euler path: a path that contains every edge of a graph exactly once

Euler circuit: a circuit that contains every edge of a graph exactly once

Hamilton path: a path in a graph that passes through each vertex exactly once

Hamilton circuit: a circuit in a graph that passes through each vertex exactly once

weighted graph: a graph with numbers assigned to its edges

shortest-path problem: the problem of determining the path in a weighted graph such that the sum of the weights of the edges in this path is a minimum over all paths between specified vertices

traveling salesperson problem: the problem that asks for the circuit of shortest total length that visits every vertex of a weghted graph exactly once

planar graph: a graph that can be drawn in the plane with no crossings

regions of a representation of a planar graph: the regions the plane is divided into by the planar representation of the graph

elementary subdivision: the removal of an edge $\{u, v\}$ of an undirected graph and the addition of a new vertex w together with edges $\{u, w\}$ and $\{w, v\}$

homeomorphic: two undirected graphs are homeomorphic if they can be obtained from the same graph by a sequence of elementary subdivisions

graph coloring: an assignment of colors to the vertices of a graph so that no two adjacent vertices have the same color

chromatic number: the minimum number of colors needed in a coloring of a graph

RESULTS

The handshaking theorem: If $G = (V, E)$ be an undirected graph with m edges, then $2m = \sum_{v \in V} \deg(v)$.

Hall's marriage theorem: The bipartite graph $G = (V, E)$ with bipartition (V_1, V_2) has a complete matching from V_1 to V_2 if and only if $|N(A)| \geq |A|$ for all subsets A of V_1.

There is an Euler circuit in a connected multigraph if and only if every vertex has even degree.

There is an Euler path in a connected multigraph if and only if at most two vertices have odd degree.

Dijkstra's algorithm: a procedure for finding a shortest path between two vertices in a weighted graph (see Section 10.6)

Euler's formula: $r = e - v + 2$ where r, e, and v are the number of regions of a planar representation, the number of edges, and the number of vertices, respectively, of a connected planar graph

Kuratowski's theorem: A graph is nonplanar if and only if it contains a subgraph homeomorphic to $K_{3,3}$ or K_5. (Proof beyond the scope of this book.)

The four color theorem: Every planar graph can be colored using no more than four colors. (Proof far beyond the scope of this book!)

Review Questions

1. a) Define a simple graph, a multigraph, a pseudograph, a directed graph, and a directed multigraph.
   b) Use an example to show how each of the types of graph in part (a) can be used in modeling. For example, explain how to model different aspects of a computer network or airline routes.
2. Give at least four examples of how graphs are used in modeling.
3. What is the relationship between the sum of the degrees of the vertices in an undirected graph and the number of edges in this graph? Explain why this relationship holds.
4. Why must there be an even number of vertices of odd degree in an undirected graph?
5. What is the relationship between the sum of the in-degrees and the sum of the out-degrees of the vertices in a directed graph? Explain why this relationship holds.
6. Describe the following families of graphs.
   a) K_n, the complete graph on n vertices
   b) $K_{m,n}$, the complete bipartite graph on m and n vertices
   c) C_n, the cycle with n vertices
   d) W_n, the wheel of size n
   e) Q_n, the n-cube
7. How many vertices and how many edges are there in each of the graphs in the families in Question 6?
8. a) What is a bipartite graph?
   b) Which of the graphs K_n, C_n, and W_n are bipartite?
   c) How can you determine whether an undirected graph is bipartite?
9. a) Describe three different methods that can be used to represent a graph.

b) Draw a simple graph with at least five vertices and eight edges. Illustrate how it can be represented using the methods you described in part (a).

10. a) What does it mean for two simple graphs to be isomorphic?

b) What is meant by an invariant with respect to isomorphism for simple graphs? Give at least five examples of such invariants.

c) Give an example of two graphs that have the same numbers of vertices, edges, and degrees of vertices, but that are not isomorphic.

d) Is a set of invariants known that can be used to efficiently determine whether two simple graphs are isomorphic?

11. a) What does it mean for a graph to be connected?

b) What are the connected components of a graph?

12. a) Explain how an adjacency matrix can be used to represent a graph.

b) How can adjacency matrices be used to determine whether a function from the vertex set of a graph G to the vertex set of a graph H is an isomorphism?

c) How can the adjacency matrix of a graph be used to determine the number of paths of length r, where r is a positive integer, between two vertices of a graph?

13. a) Define an Euler circuit and an Euler path in an undirected graph.

b) Describe the famous Königsberg bridge problem and explain how to rephrase it in terms of an Euler circuit.

c) How can it be determined whether an undirected graph has an Euler path?

d) How can it be determined whether an undirected graph has an Euler circuit?

14. a) Define a Hamilton circuit in a simple graph.

b) Give some properties of a simple graph that imply that it does not have a Hamilton circuit.

15. Give examples of at least two problems that can be solved by finding the shortest path in a weighted graph.

16. a) Describe Dijkstra's algorithm for finding the shortest path in a weighted graph between two vertices.

b) Draw a weighted graph with at least 10 vertices and 20 edges. Use Dijkstra's algorithm to find the shortest path between two vertices of your choice in the graph.

17. a) What does it mean for a graph to be planar?

b) Give an example of a nonplanar graph.

18. a) What is Euler's formula for connected planar graphs?

b) How can Euler's formula for planar graphs be used to show that a simple graph is nonplanar?

19. State Kuratowski's theorem on the planarity of graphs and explain how it characterizes which graphs are planar.

20. a) Define the chromatic number of a graph.

b) What is the chromatic number of the graph K_n when n is a positive integer?

c) What is the chromatic number of the graph C_n when n is an integer greater than 2?

d) What is the chromatic number of the graph $K_{m,n}$ when m and n are positive integers?

21. State the four color theorem. Are there graphs that cannot be colored with four colors?

22. Explain how graph coloring can be used in modeling. Use at least two different examples.

Supplementary Exercises

1. How many edges does a 50-regular graph with 100 vertices have?

2. How many nonisomorphic subgraphs does K_3 have?

In Exercises 3–5 determine whether two given graphs are isomorphic.

3.

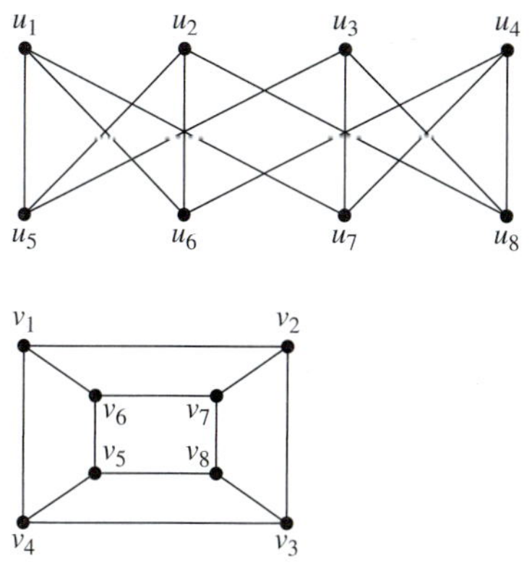

4.

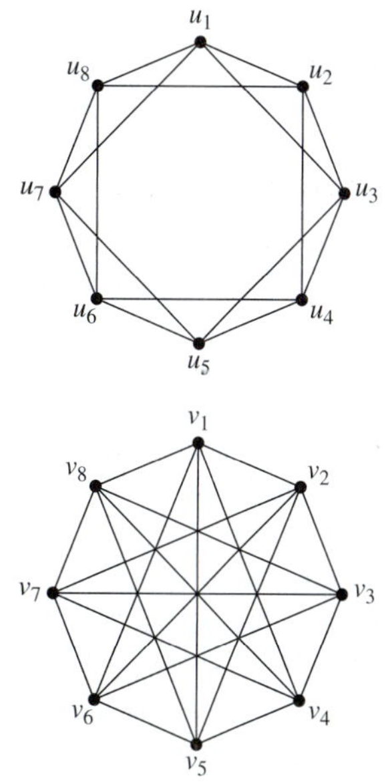

***5.**

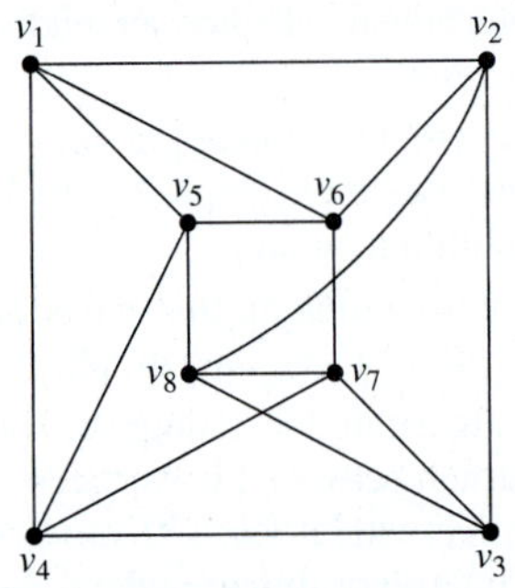

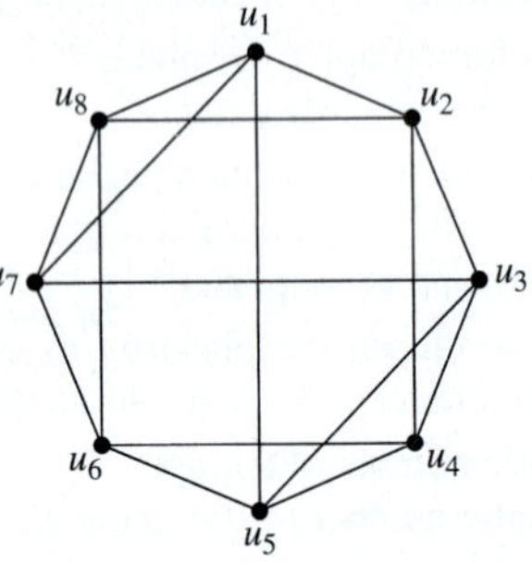

The **complete m-partite graph** $K_{n_1,n_2,\ldots,n_m}$ has vertices partitioned into m subsets of $n_1, n_2, \ldots, n_m$ elements each, and vertices are adjacent if and only if they are in different subsets in the partition.

6. Draw these graphs.

a) $K_{1,2,3}$ **b)** $K_{2,2,2}$ **c)** $K_{1,2,2,3}$

***7.** How many vertices and how many edges does the complete m-partite graph $K_{n_1,n_2,\ldots,n_m}$ have?

8. Prove or disprove that there are always two vertices of the same degree in a finite multigraph having at least two vertices.

9. Let $G = (V, E)$ be an undirected graph and let $A \subseteq V$ and $B \subseteq V$. Show that

a) $N(A \cup B) = N(A) \cup N(B)$.

b) $N(A \cap B) \subseteq N(A) \cap N(B)$, and give an example where $N(A \cap B) \neq N(A) \cap N(B)$.

10. Let $G = (V, E)$ be an undirected graph. Show that

a) $|N(v)| \leq \deg(v)$ for all $v \in V$.

b) $|N(v)| = \deg v$ for all $v \in V$ if and only if G is a simple graph.

Suppose that $S_1, S_2, \ldots, S_n$ is a collection of subsets of a set S where n is a positive integer. A **system of distinct representatives (SDR)** for this family is an ordered n-tuple $(a_1, a_2, \ldots, a_n)$ with the property that $a_i \in S_i$ for $i = 1, 2, \ldots, n$ and $a_i \neq a_j$ for all $i \neq j$.

11. Find a SDR for the sets $S_1 = \{a, c, m, e\}$, $S_2 = \{m, a, c, e\}$, $S_3 = \{a, p, e, x\}$, $S_4 = \{x, e, n, a\}$, $S_5 = \{n, a, m, e\}$, and $S_6 = \{e, x, a, m\}$.

12. Use Hall's marriage theorem to show that a collection of finite subsets $S_1, S_2, \ldots, S_n$ of a set S has a SDR $(a_1, a_2, \ldots, a_n)$ if and only if $|\bigcup_{i \in I} S_i| \geq |I|$ for all subsets I of $\{1, 2, \ldots, n\}$.

13. a) Use Exercise 12 to show that the collection of sets $S_1 = \{a, b, c\}$, $S_2 = \{b, c, d\}$, $S_3 = \{a, b, d\}$, $S_4 = \{b, c, d\}$ has a SDR without finding one explicitly.

b) Find a SDR for the family of four sets in part (a).

14. Use Exercise 12 to show that collection of sets $S_1 = \{a, b, c\}$, $S_2 = \{a, c\}$, $S_3 = \{c, d, e\}$, $S_4 = \{b, c\}$, $S_5 = \{d, e, f\}$, $S_6 = \{a, c, e\}$, and $S_7 = \{a, b\}$ does not have a SDR.

The **clustering coefficient** $C(G)$ of a simple graph G is the probability that if u and v are neighbors and v and w are neighbors, then u and w are neighbors, where u, v, and w are distinct vertices of G.

15. We say that three vertices u, v, and w of a simple graph G form a triangle if there are edges connecting all three pairs of these vertices. Find a formula for $C(G)$ in terms of the number of triangles in G and the number of paths of length two in the graph. [*Hint:* Count each triangle in the graph once for each order of three vertices that form it.]

16. Find the clustering coefficient of each of the graphs in Exercise 20 of Section 10.2.

17. Explain what the clustering coefficient measures in each of these graphs.

a) the Hollywood graph

b) the graph of Facebook friends

c) the academic collaboration graph for researchers in graph theory

d) the protein interaction graph for a human cell

e) the graph representing the routers and communications links that make up the worldwide Internet

18. For each of the graphs in Exercise 17, explain whether you would expect its clustering coefficient to be closer to 0.01 or to 0.10 and why you expect this.

A **clique** in a simple undirected graph is a complete subgraph that is not contained in any larger complete subgraph. In Exercises 19–21 find all cliques in the graph shown.

19.

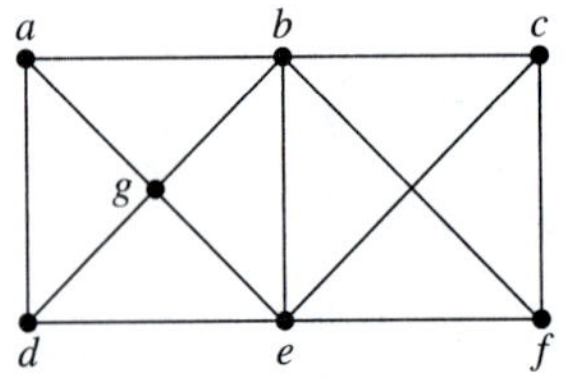

20.

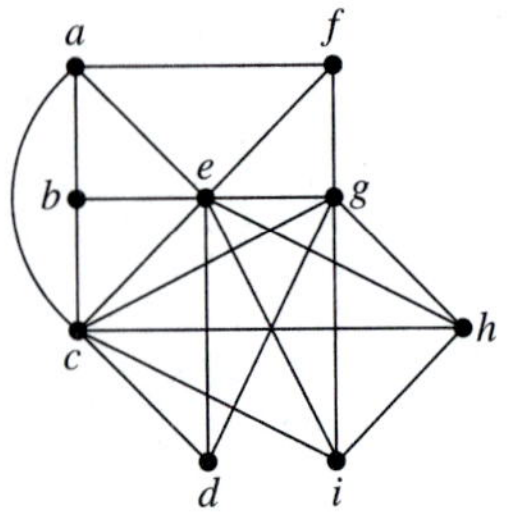

21.

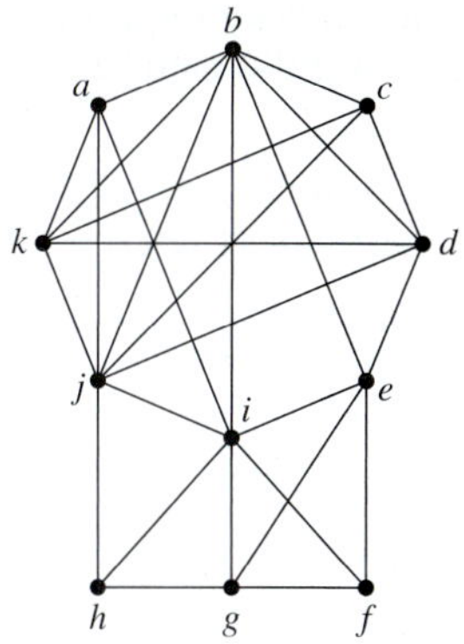

A **dominating set** of vertices in a simple graph is a set of vertices such that every other vertex is adjacent to at least one vertex of this set. A dominating set with the least number of vertices is called a **minimum dominating set**. In Exercises 22–24 find a minimum dominating set for the given graph.

22.

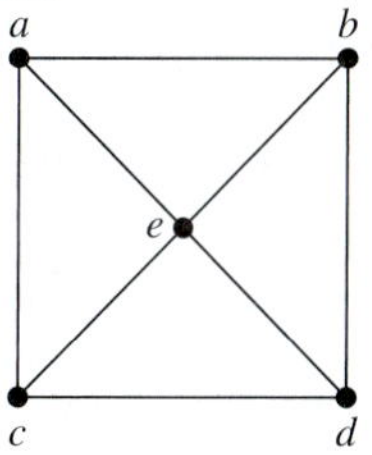

23.

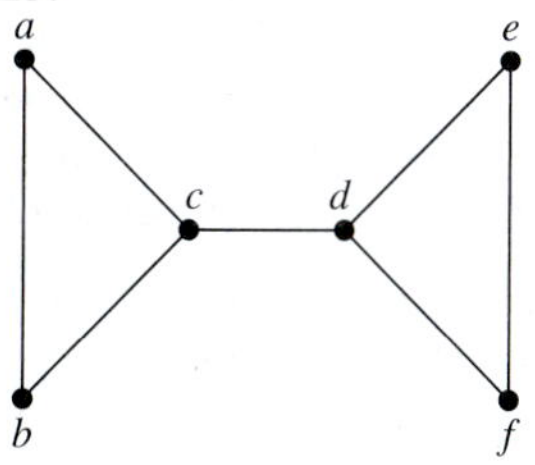

24.

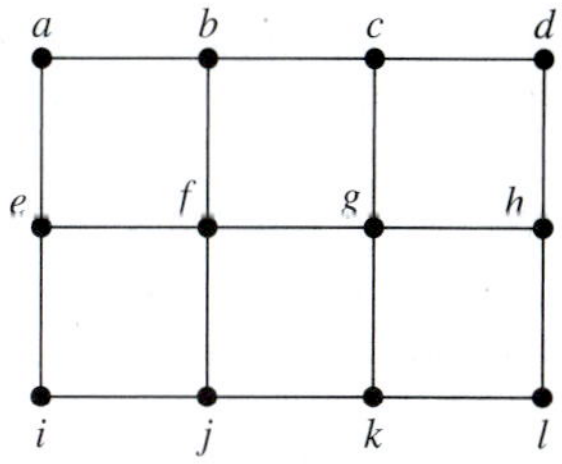

A simple graph can be used to determine the minimum number of queens on a chessboard that control the entire chessboard. An $n \times n$ chessboard has n^2 squares in an $n \times n$ configuration. A queen in a given position controls all squares in the same row, the same column, and on the two diagonals containing this square, as illustrated. The appropriate simple graph has n^2 vertices, one for each square, and two vertices are adjacent if a queen in the square represented by one of the vertices controls the square represented by the other vertex.

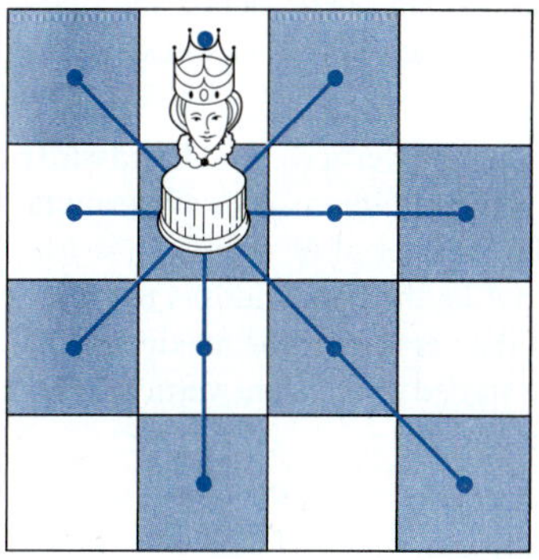

The squares controlled by a queen.

25. Construct the simple graph representing the $n \times n$ chessboard with edges representing the control of squares by queens for

a) $n = 3$. **b)** $n = 4$.

26. Explain how the concept of a minimum dominating set applies to the problem of determining the minimum number of queens controlling an $n \times n$ chessboard.

****27.** Find the minimum number of queens controlling an $n \times n$ chessboard for

a) $n = 3$. **b)** $n = 4$. **c)** $n = 5$.

28. Suppose that G_1 and H_1 are isomorphic and that G_2 and H_2 are isomorphic. Prove or disprove that $G_1 \cup G_2$ and $H_1 \cup H_2$ are isomorphic.

29. Show that each of these properties is an invariant that isomorphic simple graphs either both have or both do not have.

a) connectedness

b) the existence of a Hamilton circuit

c) the existence of an Euler circuit

d) having crossing number C

e) having n isolated vertices

f) being bipartite

30. How can the adjacency matrix of $\overline{G}$ be found from the adjacency matrix of G, where G is a simple graph?

31. How many nonisomorphic connected bipartite simple graphs are there with four vertices?

***32.** How many nonisomorphic simple connected graphs with five vertices are there

a) with no vertex of degree more than two?

b) with chromatic number equal to four?

c) that are nonplanar?

A directed graph is **self-converse** if it is isomorphic to its converse.

33. Determine whether the following graphs are self-converse.

a)

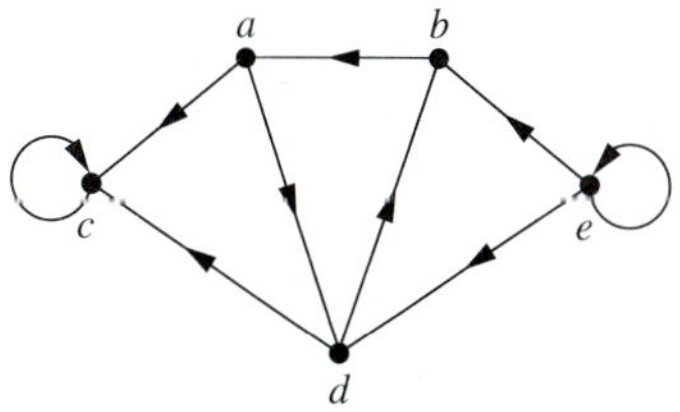

b)

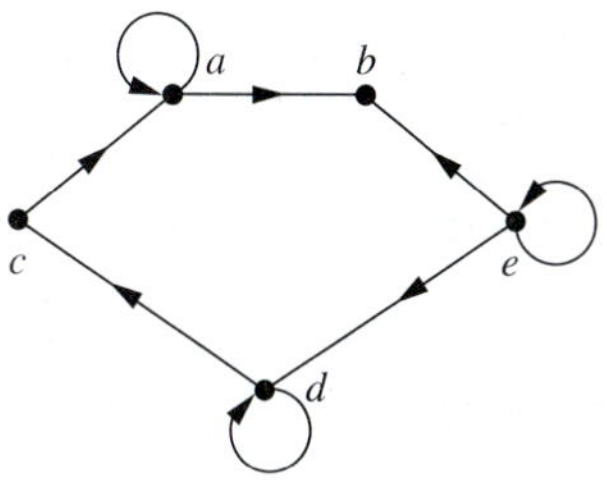

34. Show that if the directed graph G is self-converse and H is a directed graph isomorphic to G, then H is also self-converse.

An **orientation** of an undirected simple graph is an assignment of directions to its edges such that the resulting directed graph is strongly connected. When an orientation of an undirected graph exists, this graph is called **orientable**. In Exercises 35–37 determine whether the given simple graph is orientable.

35.

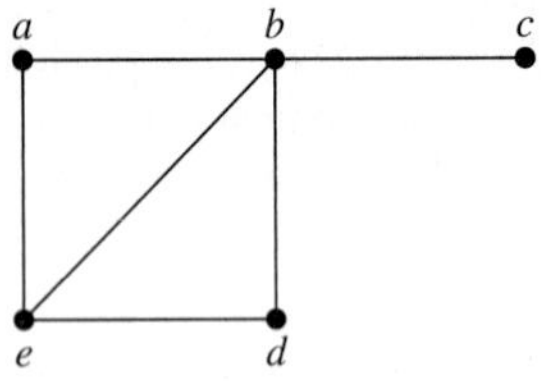

36.

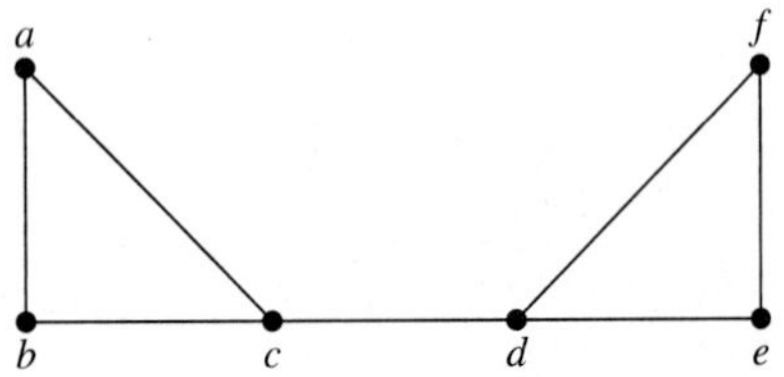

37.

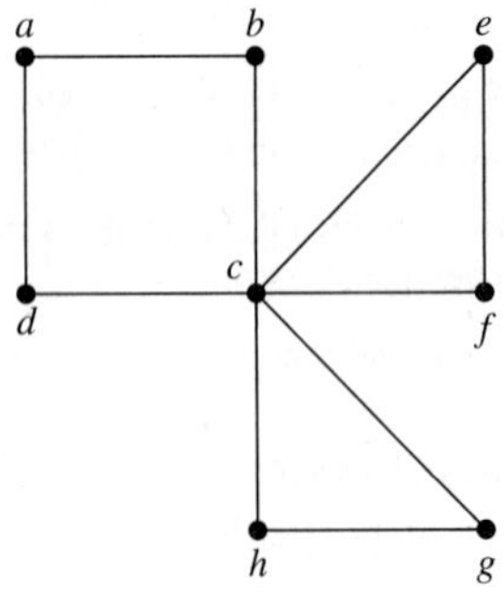

38. Because traffic is growing heavy in the central part of a city, traffic engineers are planning to change all the streets, which are currently two-way, into one-way streets. Explain how to model this problem.

***39.** Show that a graph is not orientable if it has a cut edge.

A **tournament** is a simple directed graph such that if u and v are distinct vertices in the graph, exactly one of (u, v) and (v, u) is an edge of the graph.

40. How many different tournaments are there with n vertices?

41. What is the sum of the in-degree and out-degree of a vertex in a tournament?

***42.** Show that every tournament has a Hamilton path.

43. Given two chickens in a flock, one of them is dominant. This defines the **pecking order** of the flock. How can a tournament be used to model pecking order?

44. Suppose that a connected graph G has n vertices and vertex connectivity $\kappa(G) = k$. Show that G must have at least $\lceil kn/2 \rceil$ edges.

A connected graph $G = (V, E)$ with n vertices and m edges is said to have **optimal connectivity** if $\kappa(G) = \lambda(G) = \min_{v \in V} \deg v = 2m/n$.

45. Show that a connected graph with optimal connectivity must be regular.

46. Show these graphs have optimal connectivity.

a) C_n for $n \geq 3$
b) K_n for $n \geq 3$
c) $K_{r,r}$ for $r \geq 2$

***47.** Find the two nonisomorphic simple graphs with six vertices and nine edges that have optimal connectivity.

48. Suppose that G is a connected multigraph with $2k$ vertices of odd degree. Show that there exist k subgraphs that have G as their union, where each of these subgraphs has an Euler path and where no two of these subgraphs have an edge in common. [*Hint:* Add k edges to the graph connecting pairs of vertices of odd degree and use an Euler circuit in this larger graph.]

In Exercises 49 and 50 we consider a puzzle posed by Petković in [Pe09] (based on a problem in [AvCh80]). Suppose that King Arthur has gathered his $2n$ knights of the Round Table for an important council. Every two knights are either friends or enemies, and each knight has no more than $n - 1$ enemies among the other $2n - 1$ knights. The puzzle asks whether King Arthur can seat his knights around the Round Table so that each knight has two friends for his neighbors.

49. a) Show that the puzzle can be reduced to determining whether there is a Hamilton circuit in the graph in which each knight is represented by a vertex and two knights are connected in the graph if they are friends.

b) Answer the question posed in the puzzle. [*Hint:* Use Dirac's theorem.]

50. Suppose that are eight knights Alynore, Bedivere, Degore, Gareth, Kay, Lancelot, Perceval, and Tristan. Their lists of enemies are A (D, G, P), B (K, P, T), D (A, G, L), G (A, D, T), K (B, L, P), L (D, K, T), P (A, B, K), T (B, G, L), where we have represented each knight by the first letter of his name and shown the list of enemies of that knight following this first letter. Draw the graph representing these eight knights and their friends and find a seating arrangement where each knight sits next to two friends.

***51.** Let G be a simple graph with n vertices. The **bandwidth** of G, denoted by $B(G)$, is the minimum, over all permutations $a_1, a_2, \ldots, a_n$ of the vertices of G, of $\max\{|i - j| \mid a_i$ and a_j are adjacent$\}$. That is, the bandwidth is the minimum over all listings of the vertices of the maximum difference in the indices assigned to adjacent vertices. Find the bandwidths of these graphs.

a) K_5 **b)** $K_{1,3}$ **c)** $K_{2,3}$
d) $K_{3,3}$ **e)** Q_3 **f)** C_5

***52.** The **distance** between two distinct vertices v_1 and v_2 of a connected simple graph is the length (number of edges) of the shortest path between v_1 and v_2. The **radius** of a graph is the minimum over all vertices v of the maximum distance from v to another vertex. The **diameter** of a graph is the maximum distance between two distinct vertices. Find the radius and diameter of

a) K_6. **b)** $K_{4,5}$. **c)** Q_3. **d)** C_6.

***53. a)** Show that if the diameter of the simple graph G is at least four, then the diameter of its complement $\overline{G}$ is no more than two.

b) Show that if the diameter of the simple graph G is at least three, then the diameter of its complement $\overline{G}$ is no more than three.

***54.** Suppose that a multigraph has $2m$ vertices of odd degree. Show that any circuit that contains every edge of the graph must contain at least m edges more than once.

55. Find the second shortest path between the vertices a and z in Figure 3 of Section 10.6.

56. Devise an algorithm for finding the second shortest path between two vertices in a simple connected weighted graph.

57. Find the shortest path between the vertices a and z that passes through the vertex f in the weighted graph in Exercise 3 in Section 10.6.

58. Devise an algorithm for finding the shortest path between two vertices in a simple connected weighted graph that passes through a specified third vertex.

***59.** Show that if G is a simple graph with at least 11 vertices, then either G or $\overline{G}$, the complement of G, is nonplanar.

A set of vertices in a graph is called **independent** if no two vertices in the set are adjacent. The **independence number** of a graph is the maximum number of vertices in an independent set of vertices for the graph.

***60.** What is the independence number of

a) K_n? **b)** C_n? **c)** Q_n? **d)** $K_{m,n}$?

61. Show that the number of vertices in a simple graph is less than or equal to the product of the independence number and the chromatic number of the graph.

62. Show that the chromatic number of a graph is less than or equal to $n - i + 1$, where n is the number of vertices in the graph and i is the independence number of this graph.

63. Suppose that to generate a random simple graph with n vertices we first choose a real number p with $0 \leq p \leq 1$. For each of the $C(n, 2)$ pairs of distinct vertices we generate a random number x between 0 and 1. If $0 \leq x \leq p$, we connect these two vertices with an edge; otherwise these vertices are not connected.

a) What is the probability that a graph with m edges where $0 \leq m \leq C(n, 2)$ is generated?

b) What is the expected number of edges in a randomly generated graph with n vertices if each edge is included with probability p?

c) Show that if $p = 1/2$ then every simple graph with n vertices is equally likely to be generated.

A property retained whenever additional edges are added to a simple graph (without adding vertices) is called **monotone increasing**, and a property that is retained whenever edges are removed from a simple graph (without removing vertices) is called **monotone decreasing**.

64. For each of these properties, determine whether it is monotone increasing and determine whether it is monotone decreasing.

a) The graph G is connected.
b) The graph G is not connected.
c) The graph G has an Euler circuit.
d) The graph G has a Hamilton circuit.
e) The graph G is planar.
f) The graph G has chromatic number four.
g) The graph G has radius three.
h) The graph G has diameter three.

65. Show that the graph property P is monotone increasing if and only if the graph property Q is monotone decreasing where Q is the property of not having property P.

****66.** Suppose that P is a monotone increasing property of simple graphs. Show that the probability a random graph with n vertices has property P is a monotonic nondecreasing function of p, the probability an edge is chosen to be in the graph.

Computer Projects

Write programs with these input and output.

1. Given the vertex pairs associated to the edges of an undirected graph, find the degree of each vertex.

2. Given the ordered pairs of vertices associated to the edges of a directed graph, determine the in-degree and out-degree of each vertex.

3. Given the list of edges of a simple graph, determine whether the graph is bipartite.

4. Given the vertex pairs associated to the edges of a graph, construct an adjacency matrix for the graph. (Produce a version that works when loops, multiple edges, or directed edges are present.)

5. Given an adjacency matrix of a graph, list the edges of this graph and give the number of times each edge appears.

6. Given the vertex pairs associated to the edges of an undirected graph and the number of times each edge appears, construct an incidence matrix for the graph.

7. Given an incidence matrix of an undirected graph, list its edges and give the number of times each edge appears.

8. Given a positive integer n, generate a simple graph with n vertices by producing an adjacency matrix for the graph so that all simple graphs with n vertices are equally likely to be generated.

9. Given a positive integer n, generate a simple directed graph with n vertices by producing an adjacency matrix for the graph so that all simple directed graphs with n vertices are equally likely to be generated.

10. Given the lists of edges of two simple graphs with no more than six vertices, determine whether the graphs are isomorphic.

11. Given an adjacency matrix of a graph and a positive integer n, find the number of paths of length n between two vertices. (Produce a version that works for directed and undirected graphs.)

*12. Given the list of edges of a simple graph, determine whether it is connected and find the number of connected components if it is not connected.

13. Given the vertex pairs associated to the edges of a multigraph, determine whether it has an Euler circuit and, if not, whether it has an Euler path. Construct an Euler path or circuit if it exists.

*14. Given the ordered pairs of vertices associated to the edges of a directed multigraph, construct an Euler path or Euler circuit, if such a path or circuit exists.

**15. Given the list of edges of a simple graph, produce a Hamilton circuit, or determine that the graph does not have such a circuit.

**16. Given the list of edges of a simple graph, produce a Hamilton path, or determine that the graph does not have such a path.

17. Given the list of edges and weights of these edges of a weighted connected simple graph and two vertices in this graph, find the length of a shortest path between them using Dijkstra's algorithm. Also, find a shortest path.

18. Given the list of edges of an undirected graph, find a coloring of this graph using the algorithm given in the preamble to Exercise 29 of Section 10.8.

19. Given a list of students and the courses that they are enrolled in, construct a schedule of final exams.

20. Given the distances between pairs of television stations and the minimum allowable distance between stations, assign frequencies to these stations.

Computations and Explorations

Use a computational program or programs you have written to do these exercises.

1. Display all simple graphs with four vertices.

2. Display a full set of nonisomorphic simple graphs with six vertices.

3. Display a full set of nonisomorphic directed graphs with four vertices.

4. Generate at random 10 different simple graphs each with 20 vertices so that each such graph is equally likely to be generated.

5. Construct a Gray code where the code words are bit strings of length six.

6. Construct knight's tours on chessboards of various sizes.

7. Determine whether each of the graphs you generated in Exercise 4 of this set is planar. If you can, determine the thickness of each of the graphs that are not planar.

8. Determine whether each of the graphs you generated in Exercise 4 of this set is connected. If a graph is not connected, determine the number of connected components of the graph.

9. Generate at random simple graphs with 10 vertices. Stop when you have constructed one with an Euler circuit. Display an Euler circuit in this graph.

10. Generate at random simple graphs with 10 vertices. Stop when you have constructed one with a Hamilton circuit. Display a Hamilton circuit in this graph.

11. Find the chromatic number of each of the graphs you generated in Exercise 4 of this set.

**12. Find the shortest path a traveling salesperson can take to visit each of the capitals of the 50 states in the United States, traveling by air between cities in a straight line.

*13. Estimate the probability that a randomly generated simple graph with n vertices is connected for each positive integer n not exceeding ten by generating a set of random simple graphs and determining whether each is connected.

**14. Work on the problem of determining whether the crossing number of $K_{7,7}$ is 77, 79, or 81. It is known that it equals one of these three values.

Writing Projects

Respond to these with essays using outside sources.

1. Describe the origins and development of graph theory prior to the year 1900.
2. Discuss the applications of graph theory to the study of ecosystems.
3. Discuss the applications of graph theory to sociology and psychology.
4. Discuss what can be learned by investigating the properties of the web graph.
5. Explain what community structure is in a graph representing a network, such as a social network, a computer network, an information network, or a biological network. Define what a community in such a graph is, and explain what communities represent in graphs representing the types of networks listed.
6. Describe some of the algorithms used to detect communities in graphs representing networks of the types listed in Question 5.
7. Describe algorithms for drawing a graph on paper or on a display given the vertices and edges of the graph. What considerations arise in drawing a graph so that it has the best appearance for understanding its properties?
8. Explain how graph theory can help uncover networks of criminals or terrorists by studying relevant social and communication networks.
9. What are some of the capabilities that a software tool for inputting, displaying, and manipulating graphs should have? Which of these capabilities do available tools have?
10. Describe some of the algorithms available for determining whether two graphs are isomorphic and the computational complexity of these algorithms. What is the most efficient such algorithm currently known?
11. What is the subgraph isomorphism problem and what are some of its important applications, including those to chemistry, bioinformatics, electronic circuit design, and computer vision?
12. Explain what the area of graph mining, an important area of data mining, is and describe some of the basic techniques used in graph mining.
13. Describe how Euler paths can be used to help determine DNA sequences.
14. Define *de Bruijn sequences* and discuss how they arise in applications. Explain how de Bruijn sequences can be constructed using Euler circuits.
15. Describe the *Chinese postman problem* and explain how to solve this problem.
16. Describe some of the different conditions that imply that a graph has a Hamilton circuit.
17. Describe some of the strategies and algorithms used to solve the traveling salesperson problem.
18. Describe several different algorithms for determining whether a graph is planar. What is the computational complexity of each of these algorithms?
19. In modeling, very large scale integration (VLSI) graphs are sometimes embedded in a book, with the vertices on the spine and the edges on pages. Define the *book number* of a graph and find the book number of various graphs including K_n for $n = 3, 4, 5$, and 6.
20. Discuss the history of the four color theorem.
21. Describe the role computers played in the proof of the four color theorem. How can we be sure that a proof that relies on a computer is correct?
22. Describe and compare several different algorithms for coloring a graph, in terms of whether they produce a coloring with the least number of colors possible and in terms of their complexity.
23. Explain how graph multicolorings can be used in a variety of different models.
24. Describe some of the applications of edge colorings.
25. Explain how the theory of random graphs can be used in nonconstructive existence proofs of graphs with certain properties.

C H A P T E R

11 Trees

11.1 Introduction to Trees

11.2 Applications of Trees

11.3 Tree Traversal

11.4 Spanning Trees

11.5 Minimum Spanning Trees

A connected graph that contains no simple circuits is called a tree. Trees were used as long ago as 1857, when the English mathematician Arthur Cayley used them to count certain types of chemical compounds. Since that time, trees have been employed to solve problems in a wide variety of disciplines, as the examples in this chapter will show.

Trees are particularly useful in computer science, where they are employed in a wide range of algorithms. For instance, trees are used to construct efficient algorithms for locating items in a list. They can be used in algorithms, such as Huffman coding, that construct efficient codes saving costs in data transmission and storage. Trees can be used to study games such as checkers and chess and can help determine winning strategies for playing these games. Trees can be used to model procedures carried out using a sequence of decisions. Constructing these models can help determine the computational complexity of algorithms based on a sequence of decisions, such as sorting algorithms.

Procedures for building trees containing every vertex of a graph, including depth-first search and breadth-first search, can be used to systematically explore the vertices of a graph. Exploring the vertices of a graph via depth-first search, also known as backtracking, allows for the systematic search for solutions to a wide variety of problems, such as determining how eight queens can be placed on a chessboard so that no queen can attack another.

We can assign weights to the edges of a tree to model many problems. For example, using weighted trees we can develop algorithms to construct networks containing the least expensive set of network connections linking different nodes.

11.1 Introduction to Trees

Links

In Chapter 10 we showed how graphs can be used to model and solve many problems. In this chapter we will focus on a particular type of graph called a **tree**, so named because such graphs resemble trees. For example, *family trees* are graphs that represent genealogical charts. Family trees use vertices to represent the members of a family and edges to represent parent–child relationships. The family tree of the male members of the Bernoulli family of Swiss mathematicians is shown in Figure 1. The undirected graph representing a family tree (restricted to people of just one gender and with no inbreeding) is an example of a tree.

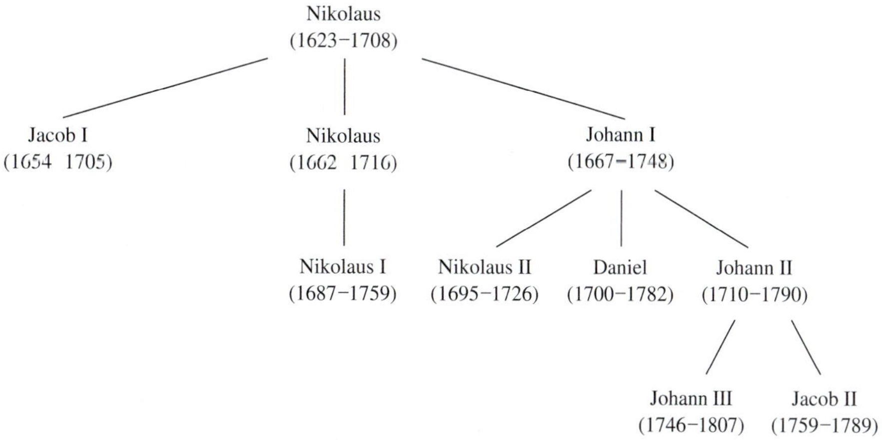

FIGURE 1 **The Bernoulli family of mathematicians.**

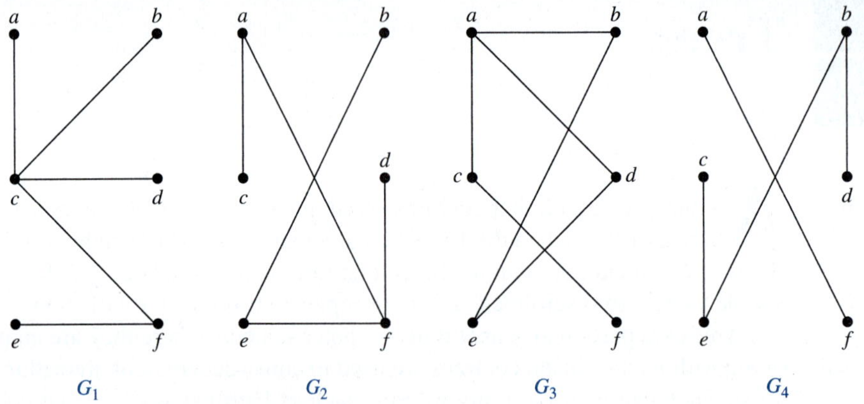

FIGURE 2 **Examples of trees and graphs that are not trees.**

Definition 1 A *tree* is a connected undirected graph with no simple circuits.

Because a tree cannot have a simple circuit, a tree cannot contain multiple edges or loops. Therefore any tree must be a simple graph.

EXAMPLE 1 Which of the graphs shown in Figure 2 are trees?

Solution: G_1 and G_2 are trees, because both are connected graphs with no simple circuits. G_3 is not a tree because e, b, a, d, e is a simple circuit in this graph. Finally, G_4 is not a tree because it is not connected. ◀

Any connected graph that contains no simple circuits is a tree. What about graphs containing no simple circuits that are not necessarily connected? These graphs are called **forests** and have the property that each of their connected components is a tree. Figure 3 displays a forest. Note that the graph G_4 in Figure 2 is also a forest. The graph in Figure 3 is a forest of three trees, while G_4 in Figure 2 is a forest of two trees.

Trees are often defined as undirected graphs with the property that there is a unique simple path between every pair of vertices. Theorem 1 shows that this alternative definition is equivalent to our definition.

THEOREM 1 An undirected graph is a tree if and only if there is a unique simple path between any two of its vertices.

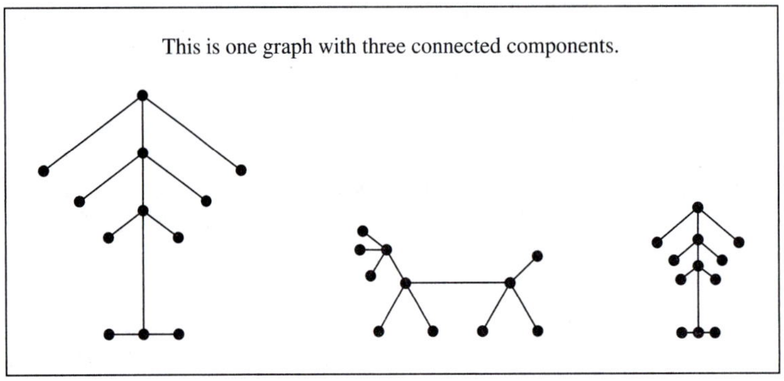

FIGURE 3 **Example of a forest.**

Proof: First assume that T is a tree. Then T is a connected graph with no simple circuits. Let x and y be two vertices of T. Because T is connected, by Theorem 1 of Section 10.4 there is a simple path between x and y. Moreover, this path must be unique, for if there were a second such path, the path formed by combining the first path from x to y followed by the path from y to x obtained by reversing the order of the second path from x to y would form a circuit. This implies, using Exercise 59 of Section 10.4, that there is a simple circuit in T. Hence, there is a unique simple path between any two vertices of a tree.

Now assume that there is a unique simple path between any two vertices of a graph T. Then T is connected, because there is a path between any two of its vertices. Furthermore, T can have no simple circuits. To see that this is true, suppose T had a simple circuit that contained the vertices x and y. Then there would be two simple paths between x and y, because the simple circuit is made up of a simple path from x to y and a second simple path from y to x. Hence, a graph with a unique simple path between any two vertices is a tree. ◁

11.1.1 Rooted Trees

In many applications of trees, a particular vertex of a tree is designated as the **root**. Once we specify a root, we can assign a direction to each edge as follows. Because there is a unique path from the root to each vertex of the graph (by Theorem 1), we direct each edge away from the root. Thus, a tree together with its root produces a directed graph called a **rooted tree**.

Definition 2 A *rooted tree* is a tree in which one vertex has been designated as the root and every edge is directed away from the root.

Rooted trees can also be defined recursively. Refer to Section 5.3 to see how this can be done. We can change an unrooted tree into a rooted tree by choosing any vertex as the root. Note that different choices of the root produce different rooted trees. For instance, Figure 4 displays the rooted trees formed by designating a to be the root and c to be the root, respectively, in the tree T. We usually draw a rooted tree with its root at the top of the graph. The arrows indicating the directions of the edges in a rooted tree can be omitted, because the choice of root determines the directions of the edges.

The terminology for trees has botanical and genealogical origins. Suppose that T is a rooted tree. If v is a vertex in T other than the root, the **parent** of v is the unique vertex u such that there is a directed edge from u to v (the reader should show that such a vertex is unique). When u is the parent of v, v is called a **child** of u. Vertices with the same parent are called **siblings**. The **ancestors** of a vertex other than the root are the vertices in the path from the root to this vertex, excluding the vertex itself and including the root (that is, its parent, its parent's parent, and so on, until the root is reached). The **descendants** of a vertex v are those vertices that have v as an ancestor. A vertex of a rooted tree is called a **leaf** if it has no children. Vertices that have children are called **internal vertices**. The root is an internal vertex unless it is the only vertex in the graph, in which case it is a leaf.

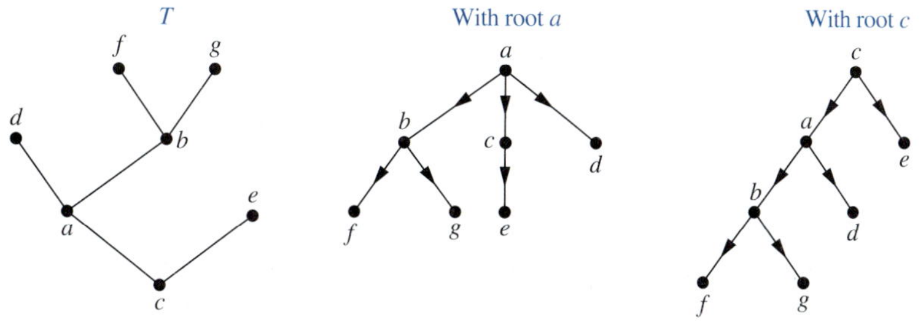

FIGURE 4 **A tree and rooted trees formed by designating two different roots.**

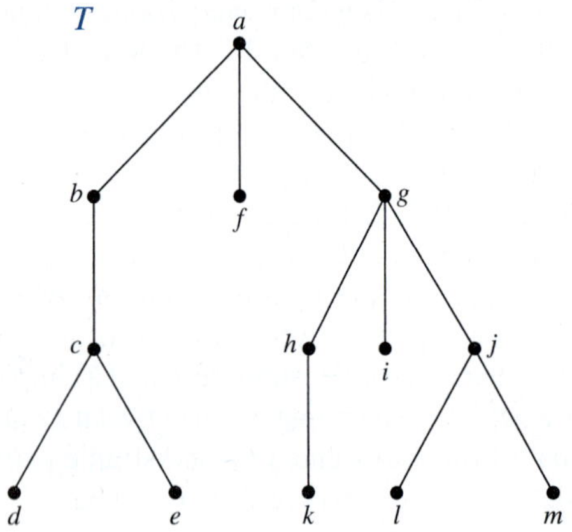

FIGURE 5 **A rooted tree T.**

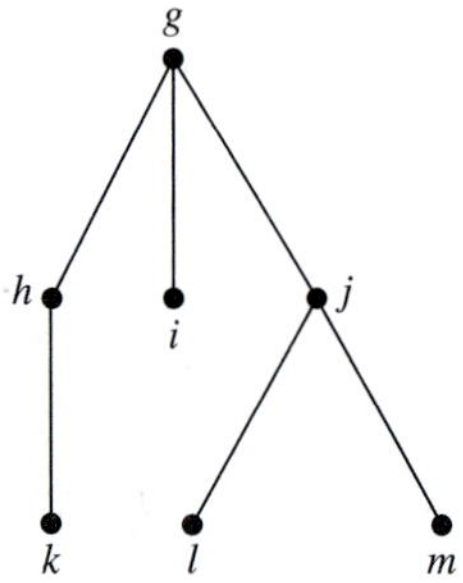

FIGURE 6 **The subtree rooted at g.**

If a is a vertex in a tree, the **subtree** with a as its root is the subgraph of the tree consisting of a and its descendants and all edges incident to these descendants.

EXAMPLE 2 In the rooted tree T (with root a) shown in Figure 5, find the parent of c, the children of g, the siblings of h, all ancestors of e, all descendants of b, all internal vertices, and all leaves. What is the subtree rooted at g?

Extra Examples

Solution: The parent of c is b. The children of g are h, i, and j. The siblings of h are i and j. The ancestors of e are c, b, and a. The descendants of b are c, d, and e. The internal vertices are a, b, c, g, h, and j. The leaves are d, e, f, i, k, l, and m. The subtree rooted at g is shown in Figure 6. ◀

Rooted trees with the property that all of their internal vertices have the same number of children are used in many different applications. Later in this chapter we will use such trees to study problems involving searching, sorting, and coding.

Definition 3 A rooted tree is called an *m-ary tree* if every internal vertex has no more than m children. The tree is called a *full m-ary tree* if every internal vertex has exactly m children. An m-ary tree with $m = 2$ is called a *binary tree*.

Links

EXAMPLE 3 Are the rooted trees in Figure 7 full m-ary trees for some positive integer m?

Solution: T_1 is a full binary tree because each of its internal vertices has two children. T_2 is a full 3-ary tree because each of its internal vertices has three children. In T_3 each internal vertex

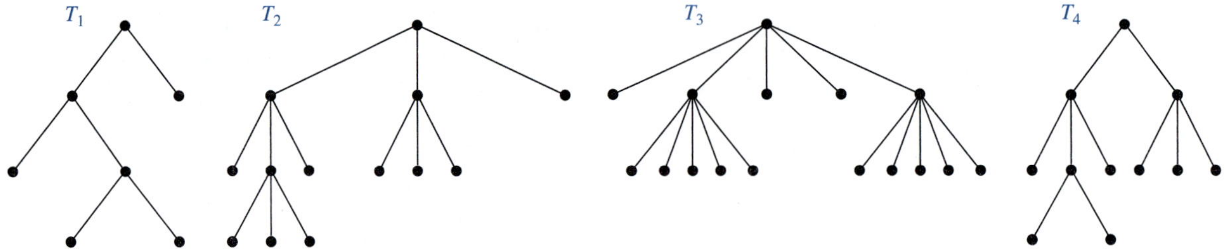

FIGURE 7 **Four rooted trees.**

has five children, so T_3 is a full 5-ary tree. T_4 is not a full m-ary tree for any m because some of its internal vertices have two children and others have three children. ◀

ORDERED ROOTED TREES An **ordered rooted tree** is a rooted tree where the children of each internal vertex are ordered. Ordered rooted trees are drawn so that the children of each internal vertex are shown in order from left to right. Note that a representation of a rooted tree in the conventional way determines an ordering for its edges. We will use such orderings of edges in drawings without explicitly mentioning that we are considering a rooted tree to be ordered.

In an ordered binary tree (usually called just a **binary tree**), if an internal vertex has two children, the first child is called the **left child** and the second child is called the **right child**. The tree rooted at the left child of a vertex is called the **left subtree** of this vertex, and the tree rooted at the right child of a vertex is called the **right subtree** of the vertex. The reader should note that for some applications, every vertex of a binary tree, other than the root, is designated as a right or a left child of its parent. This is done even when some vertices have only one child. We will make such designations whenever it is necessary, but not otherwise.

Ordered rooted trees can be defined recursively. Binary trees, a type of ordered rooted trees, were defined this way in Section 5.3.

EXAMPLE 4 What are the left and right children of d in the binary tree T shown in Figure 8(a) (where the order is that implied by the drawing)? What are the left and right subtrees of c?

Solution: The left child of d is f and the right child is g. We show the left and right subtrees of c in Figures 8(b) and 8(c), respectively. ◀

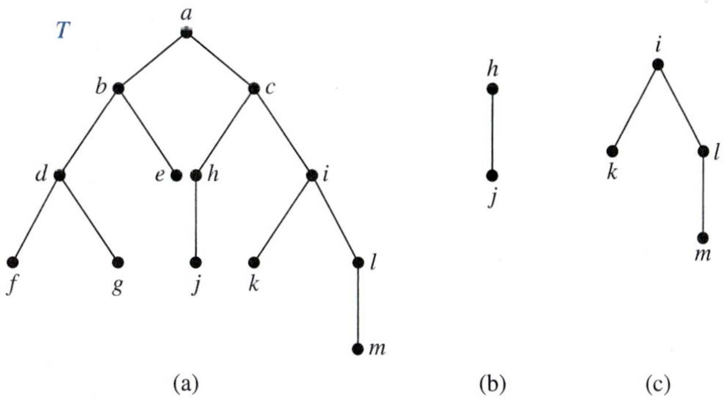

FIGURE 8 **A binary tree T and left and right subtrees of the vertex c.**

Just as in the case of graphs, there is no standard terminology used to describe trees, rooted trees, ordered rooted trees, and binary trees. This nonstandard terminology occurs because trees are used extensively throughout computer science, which is a relatively young field. The reader should carefully check meanings given to terms dealing with trees whenever they occur.

11.1.2 Trees as Models

Trees are used as models in such diverse areas as computer science, chemistry, geology, botany, and psychology. We will describe a variety of such models based on trees.

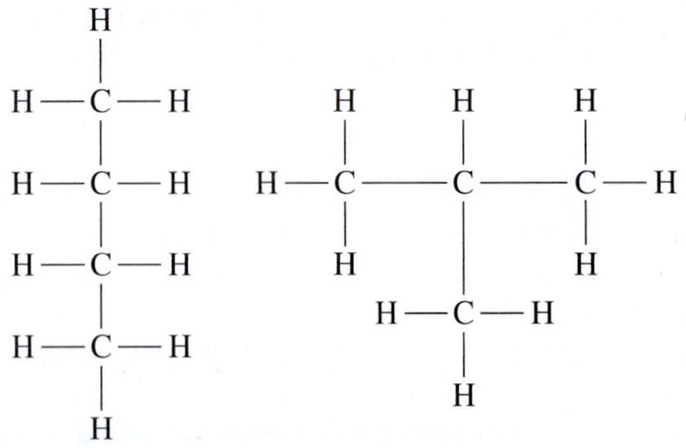

Butane Isobutane

FIGURE 9 The two isomers of butane.

EXAMPLE 5 **Saturated Hydrocarbons and Trees** Graphs can be used to represent molecules, where atoms are represented by vertices and bonds between them by edges. The English mathematician Arthur Cayley discovered trees in 1857 when he was trying to enumerate the isomers of compounds of the form C_nH_{2n+2}, which are called *saturated hydrocarbons*. (Isomers represent compounds with the same chemical formula but different chemical properties.)

In graph models of saturated hydrocarbons, each carbon atom is represented by a vertex of degree 4, and each hydrogen atom is represented by a vertex of degree 1. There are $3n + 2$ vertices in a graph representing a compound of the form C_nH_{2n+2}. The number of edges in such a graph is half the sum of the degrees of the vertices. Hence, there are $(4n + 2n + 2)/2 = 3n + 1$ edges in this graph. Because the graph is connected and the number of edges is one less than the number of vertices, it must be a tree (see Exercise 15).

The nonisomorphic trees with n vertices of degree 4 and $2n + 2$ of degree 1 represent the different isomers of C_nH_{2n+2}. For instance, when $n = 4$, there are exactly two nonisomorphic trees of this type (the reader should verify this). Hence, there are exactly two different isomers of C_4H_{10}. Their structures are displayed in Figure 9. These two isomers are called butane and isobutane (also known as i-butane or methylpropane). ◀

EXAMPLE 6 **Representing Organizations** The structure of a large organization can be modeled using a rooted tree. Each vertex in this tree represents a position in the organization. An edge from one vertex to another indicates that the person represented by the initial vertex is the (direct) boss of the person represented by the terminal vertex. The graph shown in Figure 10 displays

Links

Paul Fearn/The History Collection/Alamy Stock Photo

ARTHUR CAYLEY (1821–1895) Arthur Cayley, the son of a merchant, displayed his mathematical talents at an early age with amazing skill in numerical calculations. Cayley entered Trinity College, Cambridge, when he was 17. While in college he developed a passion for reading novels. Cayley excelled at Cambridge and was elected to a 3-year appointment as Fellow of Trinity and assistant tutor. During this time Cayley began his study of n-dimensional geometry and made a variety of contributions to geometry and to analysis. He also developed an interest in mountaineering, which he enjoyed during vacations in Switzerland. Because no position as a mathematician was available to him, Cayley left Cambridge, entering the legal profession and gaining admittance to the bar in 1849. Although Cayley limited his legal work to be able to continue his mathematics research, he developed a reputation as a legal specialist. During his legal career he was able to write more than 300 mathematical papers. In 1863 Cambridge University established a new post in mathematics and offered it to Cayley. He took this job, even though it paid less money than he made as a lawyer.

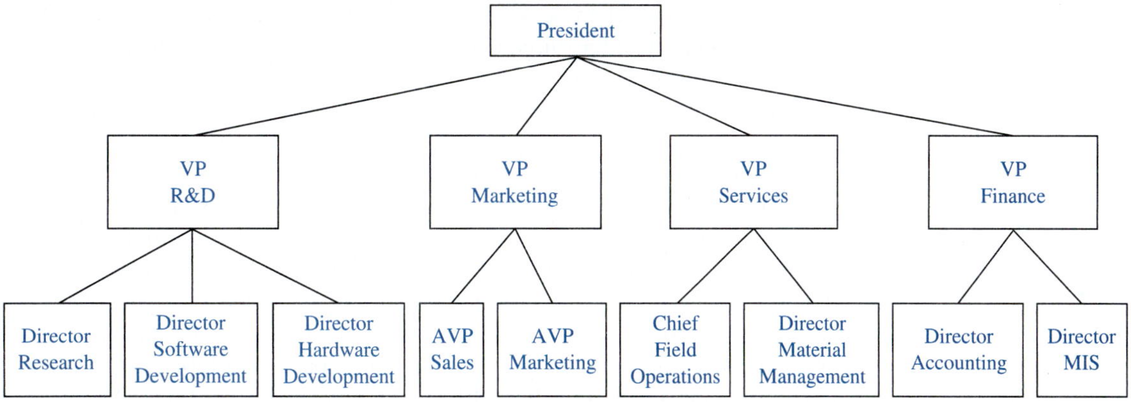

FIGURE 10 **An organizational tree for a computer company.**

such a tree. In the organization represented by this tree, the Director of Hardware Development works directly for the Vice President of R&D. The root of this tree is the vertex representing the President of the organization. ◀

EXAMPLE 7 **Computer File Systems** Files in computer memory can be organized into directories. A directory can contain both files and subdirectories. The root directory contains the entire file system. Thus, a file system may be represented by a rooted tree, where the root represents the root directory, internal vertices represent subdirectories, and leaves represent ordinary files or empty directories. One such file system is shown in Figure 11. In this system, the file khr is in the directory rje. (Note that links to files where the same file may have more than one pathname can lead to circuits in computer file systems.) ◀

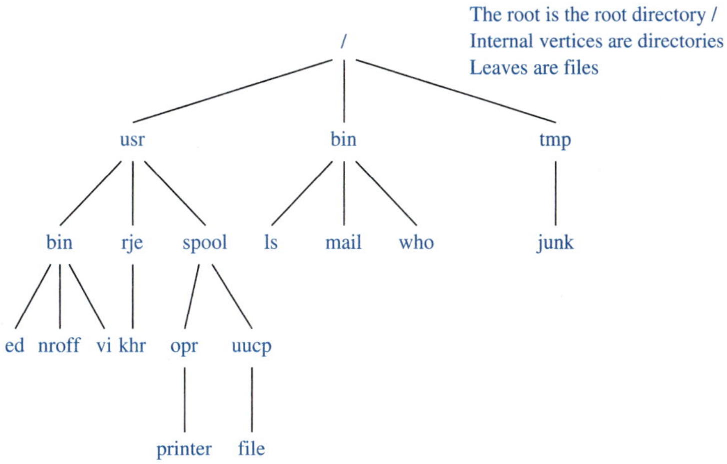

FIGURE 11 **A computer file system.**

EXAMPLE 8 **Tree-Connected Parallel Processors** In Example 17 of Section 10.2 we described several interconnection networks for parallel processing. A **tree-connected network** is another important way to interconnect processors. The graph representing such a network is a complete binary tree, that is, a full binary tree where every root is at the same level. Such a network interconnects

$n = 2^k - 1$ processors, where k is a positive integer. A processor represented by the vertex v that is not a root or a leaf has three two-way connections—one to the processor represented by the parent of v and two to the processors represented by the two children of v. The processor represented by the root has two two-way connections to the processors represented by its two children. A processor represented by a leaf v has a single two-way connection to the parent of v. We display a tree-connected network with seven processors in Figure 12.

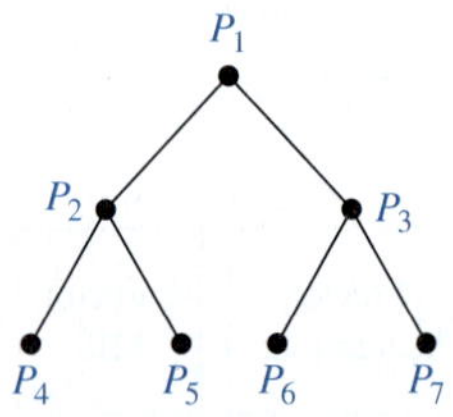

FIGURE 12 A tree-connected network of seven processors.

We now illustrate how a tree-connected network can be used for parallel computation. In particular, we show how the processors in Figure 12 can be used to add eight numbers, using three steps. In the first step, we add x_1 and x_2 using P_4, x_3 and x_4 using P_5, x_5 and x_6 using P_6, and x_7 and x_8 using P_7. In the second step, we add $x_1 + x_2$ and $x_3 + x_4$ using P_2 and $x_5 + x_6$ and $x_7 + x_8$ using P_3. Finally, in the third step, we add $x_1 + x_2 + x_3 + x_4$ and $x_5 + x_6 + x_7 + x_8$ using P_1. The three steps used to add eight numbers compares favorably to the seven steps required to add eight numbers serially, where the steps are the addition of one number to the sum of the previous numbers in the list. ◀

11.1.3 Properties of Trees

We will often need results relating the numbers of edges and vertices of various types in trees.

THEOREM 2 A tree with n vertices has $n - 1$ edges.

Links

Proof: We will use mathematical induction to prove this theorem. Note that for all the trees here we can choose a root and consider the tree rooted.

BASIS STEP: When $n = 1$, a tree with $n = 1$ vertex has no edges. It follows that the theorem is true for $n = 1$.

INDUCTIVE STEP: The inductive hypothesis states that every tree with k vertices has $k - 1$ edges, where k is a positive integer. Suppose that a tree T has $k + 1$ vertices and that v is a leaf of T (which must exist because the tree is finite), and let w be the parent of v. Removing from T the vertex v and the edge connecting w to v produces a tree T' with k vertices, because the resulting graph is still connected and has no simple circuits. By the inductive hypothesis, T' has $k - 1$ edges. It follows that T has k edges because it has one more edge than T', the edge connecting v and w. This completes the inductive step. ◁

Recall that a tree is a connected undirected graph with no simple circuits. So, when G is an undirected graph with n vertices, Theorem 2 tells us that the two conditions (*i*) G is connected and (*ii*) G has no simple circuits, imply (*iii*) G has $n - 1$ edges. Also, when (*i*) and (*iii*) hold, then (*ii*) must also hold, and when (*ii*) and (*iii*) hold, (*i*) must also hold. That is, if G is connected and G has $n - 1$ edges, then G has no simple circuits, so that G is a tree (see Exercise 15(a)), and if G has no simple circuits and G has $n - 1$ edges, then G is connected, and so is a tree (see Exercise 15(b)). Consequently, when two of (*i*), (*ii*), and (*iii*) hold, the third condition must also hold, and G must be a tree.

COUNTING VERTICES IN FULL m-ARY TREES The number of vertices in a full m-ary tree with a specified number of internal vertices is determined, as Theorem 3 shows. As in Theorem 2, we will use n to denote the number of vertices in a tree.

THEOREM 3 A full m-ary tree with i internal vertices contains $n = mi + 1$ vertices.

Proof: Every vertex, except the root, is the child of an internal vertex. Because each of the i internal vertices has m children, there are mi vertices in the tree other than the root. Therefore, the tree contains $n = mi + 1$ vertices. ◁

Suppose that T is a full m-ary tree. Let i be the number of internal vertices and l the number of leaves in this tree. Once one of n, i, and l is known, the other two quantities are determined. Theorem 4 explains how to find the other two quantities from the one that is known.

THEOREM 4 A full m-ary tree with

(*i*) n vertices has $i = (n - 1)/m$ internal vertices and $l = [(m - 1)n + 1]/m$ leaves,
(*ii*) i internal vertices has $n = mi + 1$ vertices and $l = (m - 1)i + 1$ leaves,
(*iii*) l leaves has $n = (ml - 1)/(m - 1)$ vertices and $i = (l - 1)/(m - 1)$ internal vertices.

Proof: Let n represent the number of vertices, i the number of internal vertices, and l the number of leaves. The three parts of the theorem can all be proved using the equality given in Theorem 3, that is, $n = mi + 1$, together with the equality $n = l + i$, which is true because each vertex is either a leaf or an internal vertex. We will prove part (*i*) here. The proofs of parts (*ii*) and (*iii*) are left as exercises for the reader.

Solving for i in $n = mi + 1$ gives $i = (n - 1)/m$. Then inserting this expression for i into the equation $n = l + i$ shows that $l = n - i = n - (n - 1)/m = [(m - 1)n + 1]/m$. ◁

Example 9 illustrates how Theorem 4 can be used.

EXAMPLE 9 Suppose that someone starts a chain letter. Each person who receives the letter is asked to send it on to four other people. Some people do this, but others do not send any letters. How many people have seen the letter, including the first person, if no one receives more than one letter and if the chain letter ends after there have been 100 people who read it but did not send it out? How many people sent out the letter?

Solution: The chain letter can be represented using a full 4-ary tree. The internal vertices correspond to people who sent out the letter, and the leaves correspond to people who did not send it out. Because 100 people did not send out the letter, the number of leaves in this rooted tree is $l = 100$. Hence, part (*iii*) of Theorem 4 shows that the number of people who have seen the letter is $n = (4 \cdot 100 - 1)/(4 - 1) = 133$. Also, the number of internal vertices is $133 - 100 = 33$, so 33 people sent out the letter. ◀

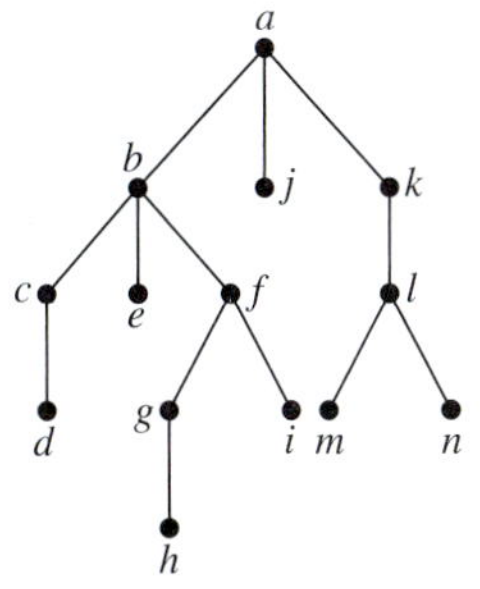

FIGURE 13 A rooted tree.

BALANCED *m*-ARY TREES It is often desirable to use rooted trees that are "balanced" so that the subtrees at each vertex contain paths of approximately the same length. Some definitions will make this concept clear. The **level** of a vertex v in a rooted tree is the length of the unique path from the root to this vertex. The level of the root is defined to be zero. The **height** of a rooted tree is the maximum of the levels of vertices. In other words, the height of a rooted tree is the length of the longest path from the root to any vertex.

EXAMPLE 10 Find the level of each vertex in the rooted tree shown in Figure 13. What is the height of this tree?

Solution: The root a is at level 0. Vertices b, j, and k are at level 1. Vertices c, e, f, and l are at level 2. Vertices d, g, i, m, and n are at level 3. Finally, vertex h is at level 4. Because the largest level of any vertex is 4, this tree has height 4. ◀

A rooted m-ary tree of height h is **balanced** if all leaves are at levels h or $h - 1$.

EXAMPLE 11 Which of the rooted trees shown in Figure 14 are balanced?

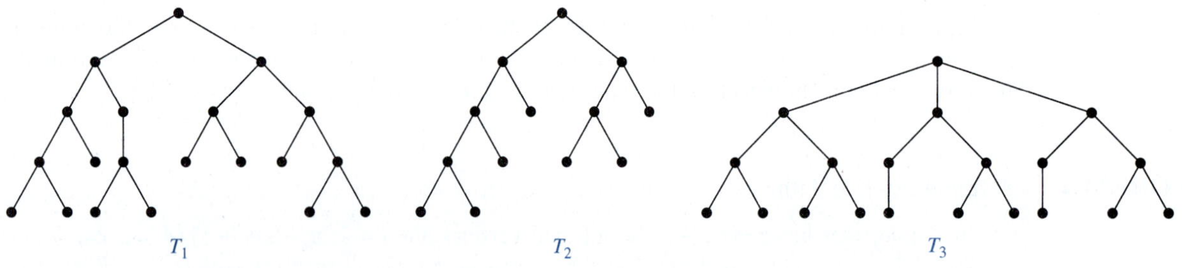

FIGURE 14 Some rooted trees.

Solution: T_1 is balanced, because all its leaves are at levels 3 and 4. However, T_2 is not balanced, because it has leaves at levels 2, 3, and 4. Finally, T_3 is balanced, because all its leaves are at level 3. ◀

A BOUND FOR THE NUMBER OF LEAVES IN AN m-ARY TREE It is often useful to have an upper bound for the number of leaves in an m-ary tree. Theorem 5 provides such a bound in terms of the height of the m-ary tree.

THEOREM 5 There are at most m^h leaves in an m-ary tree of height h.

Proof: The proof uses mathematical induction on the height. First, note that an m-ary tree of height 0 has $m^0 = 1$ leaf, and now consider m-ary trees of height 1. These trees consist of a root with no more than m children, each of which is a leaf. Hence, there are no more than $m^1 = m$ leaves in an m-ary tree of height 1. This is the basis step of the inductive argument.

Now assume that the result is true for all m-ary trees of height less than h; this is the inductive hypothesis. Let T be an m-ary tree of height h. The leaves of T are the leaves of the subtrees of T obtained by deleting the edges from the root to each of the vertices at level 1, as shown in Figure 15.

Each of these subtrees has height less than or equal to $h - 1$. So by the inductive hypothesis, each of these rooted trees has at most m^{h-1} leaves. Because there are at most m such subtrees, each with a maximum of m^{h-1} leaves, there are at most $m \cdot m^{h-1} = m^h$ leaves in the rooted tree. This finishes the inductive argument. ◁

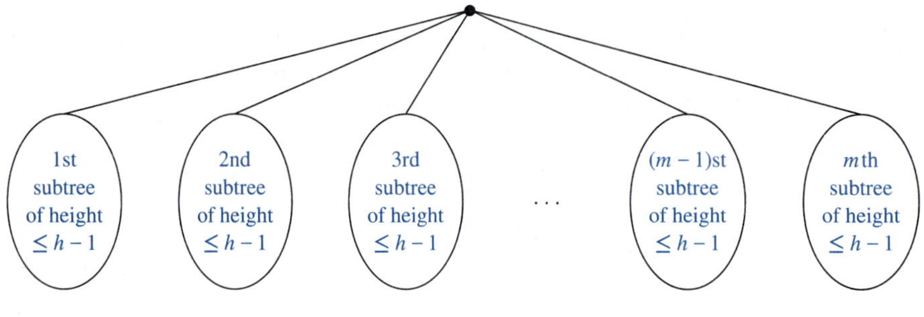

FIGURE 15 The inductive step of the proof.

COROLLARY 1 If an m-ary tree of height h has l leaves, then $h \geq \lceil \log_m l \rceil$. If the m-ary tree is full and balanced, then $h = \lceil \log_m l \rceil$. (We are using the ceiling function here. Recall that $\lceil x \rceil$ is the smallest integer greater than or equal to x.)

Proof: We know that $l \leq m^h$ from Theorem 5. Taking logarithms to the base m shows that $\log_m l \leq h$. Because h is an integer, we have $h \geq \lceil \log_m l \rceil$. Now suppose that the tree is balanced. Then each leaf is at level h or $h - 1$, and because the height is h, there is at least one leaf at level h. It follows that there must be more than m^{h-1} leaves (see Exercise 30). Because $l \leq m^h$, we have $m^{h-1} < l \leq m^h$. Taking logarithms to the base m in this inequality gives $h - 1 < \log_m l \leq h$. Hence, $h = \lceil \log_m l \rceil$. ◁

Exercises

1. Which of these graphs are trees?

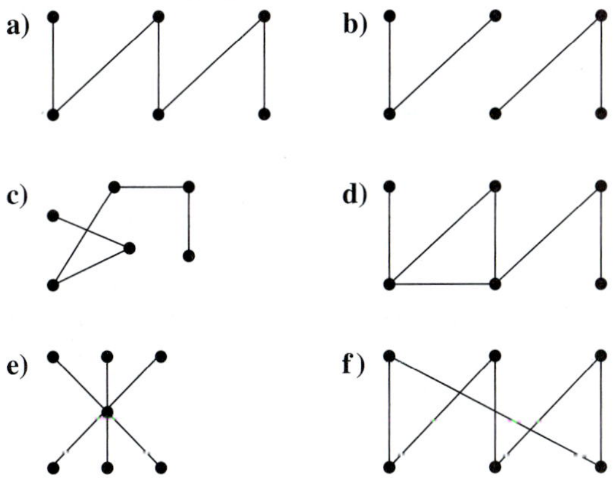

2. Which of these graphs are trees?

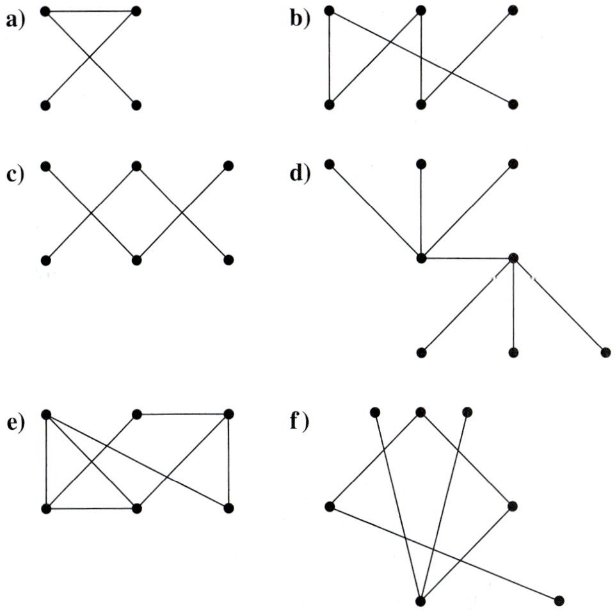

3. Answer these questions about the rooted tree illustrated.

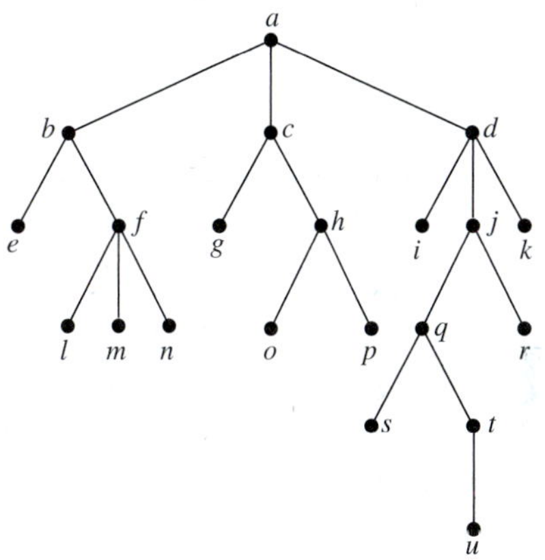

a) Which vertex is the root?

b) Which vertices are internal?

c) Which vertices are leaves?

d) Which vertices are children of j?

e) Which vertex is the parent of h?

f) Which vertices are siblings of o?

g) Which vertices are ancestors of m?

h) Which vertices are descendants of b?

4. Answer the same questions as listed in Exercise 3 for the rooted tree illustrated.

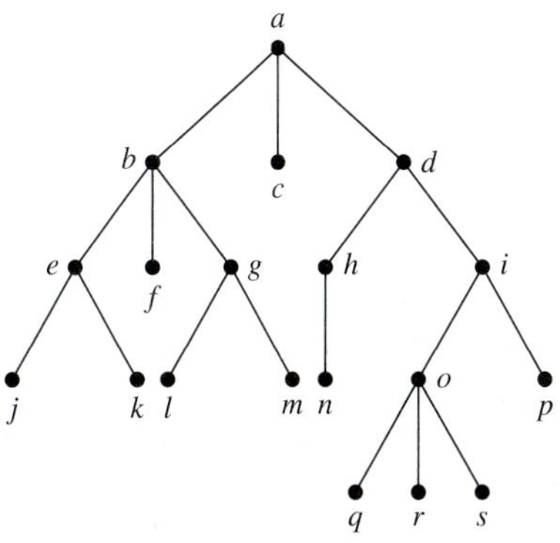

5. Is the rooted tree in Exercise 3 a full m-ary tree for some positive integer m?

6. Is the rooted tree in Exercise 4 a full m-ary tree for some positive integer m?

7. What is the level of each vertex of the rooted tree in Exercise 3?

8. What is the level of each vertex of the rooted tree in Exercise 4?

9. Draw the subtree of the tree in Exercise 3 that is rooted at

a) a. **b)** c. **c)** e.

10. Draw the subtree of the tree in Exercise 4 that is rooted at

a) a. **b)** c. **c)** e.

11. a) How many nonisomorphic unrooted trees are there with three vertices?

b) How many nonisomorphic rooted trees are there with three vertices (using isomorphism for directed graphs)?

***12. a)** How many nonisomorphic unrooted trees are there with four vertices?

b) How many nonisomorphic rooted trees are there with four vertices (using isomorphism for directed graphs)?

***13. a)** How many nonisomorphic unrooted trees are there with five vertices?

b) How many nonisomorphic rooted trees are there with five vertices (using isomorphism for directed graphs)?

***14.** Show that a simple graph is a tree if and only if it is connected but the deletion of any of its edges produces a graph that is not connected.

☞ ***15.** Let G be a simple graph with n vertices. Show that

a) G is a tree if and only if it is connected and has $n - 1$ edges.

b) G is a tree if and only if G has no simple circuits and has $n - 1$ edges. [*Hint:* To show that G is connected if it has no simple circuits and $n - 1$ edges, show that G cannot have more than one connected component.]

16. Which complete bipartite graphs $K_{m,n}$, where m and n are positive integers, are trees?

17. How many edges does a tree with 10,000 vertices have?

18. How many vertices does a full 5-ary tree with 100 internal vertices have?

19. How many edges does a full binary tree with 1000 internal vertices have?

20. How many leaves does a full 3-ary tree with 100 vertices have?

21. Suppose 1000 people enter a chess tournament. Use a rooted tree model of the tournament to determine how many games must be played to determine a champion, if a player is eliminated after one loss and games are played until only one entrant has not lost. (Assume there are no ties.)

22. A chain letter starts when a person sends a letter to five others. Each person who receives the letter either sends it to five other people who have never received it or does not send it to anyone. Suppose that 10,000 people send out the letter before the chain ends and that no one receives more than one letter. How many people receive the letter, and how many do not send it out?

23. A chain letter starts with a person sending a letter out to 10 others. Each person is asked to send the letter out to 10 others, and each letter contains a list of the previous six people in the chain. Unless there are fewer than six names in the list, each person sends one dollar to the first person in this list, removes the name of this person from the list, moves up each of the other five names one position, and inserts his or her name at the end of this list. If no person breaks the chain and no one receives more than one letter, how much money will a person in the chain ultimately receive?

***24.** Either draw a full m-ary tree with 76 leaves and height 3, where m is a positive integer, or show that no such tree exists.

***25.** Either draw a full m-ary tree with 84 leaves and height 3, where m is a positive integer, or show that no such tree exists.

***26.** A full m-ary tree T has 81 leaves and height 4.

a) Give the upper and lower bounds for m.

b) What is m if T is also balanced?

A **complete m-ary tree** is a full m-ary tree in which every leaf is at the same level.

27. Construct a complete binary tree of height 4 and a complete 3-ary tree of height 3.

28. How many vertices and how many leaves does a complete m-ary tree of height h have?

29. Prove

a) part (*ii*) of Theorem 4.

b) part (*iii*) of Theorem 4.

☞ **30.** Show that a full m-ary balanced tree of height h has more than m^{h-1} leaves.

31. How many edges are there in a forest of t trees containing a total of n vertices?

32. Explain how a tree can be used to represent the table of contents of a book organized into chapters, where each chapter is organized into sections, and each section is organized into subsections.

33. How many different isomers do these saturated hydrocarbons have?

a) C_3H_8 **b)** C_5H_{12} **c)** C_6H_{14}

34. What does each of these represent in a tree that represents the structure of an organization?

a) the parent of a vertex

b) a child of a vertex

c) a sibling of a vertex

d) the ancestors of a vertex

e) the descendants of a vertex

f) the level of a vertex

g) the height of the tree

35. Answer the same questions as those given in Exercise 34 for a rooted tree representing a computer file system.

36. a) Draw the complete binary tree with 15 vertices that represents a tree-connected network of 15 processors.

b) Show how 16 numbers can be added using the 15 processors in part (a) using four steps.

37. Let n be a power of 2. Show that n numbers can be added in $\log n$ steps using a tree-connected network of $n - 1$ processors.

***38.** A **labeled tree** is a tree where each vertex is assigned a label. Two labeled trees are considered isomorphic when there is an isomorphism between them that preserves the labels of vertices. How many nonisomorphic trees are there with three vertices labeled with different integers from the set $\{1, 2, 3\}$? How many nonisomorphic trees are there with four vertices labeled with different integers from the set $\{1, 2, 3, 4\}$?

The **eccentricity** of a vertex in an unrooted tree is the length of the longest simple path beginning at this vertex. A vertex is called a **center** if no vertex in the tree has smaller eccentricity than this vertex. In Exercises 39–41 find every vertex that is a center in the given tree.

39.

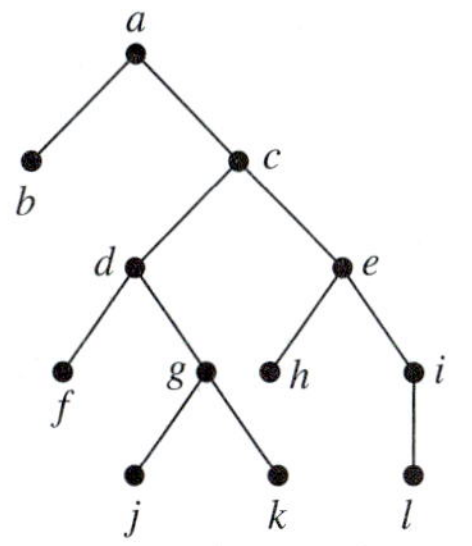

40.

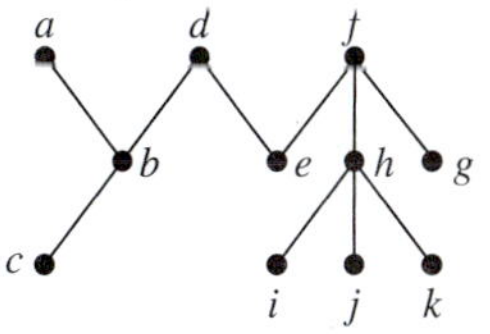

41.

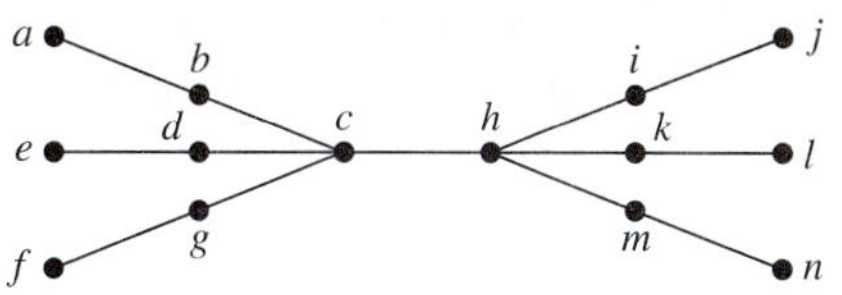

42. Show that a center should be chosen as the root to produce a rooted tree of minimal height from an unrooted tree.

***43.** Show that a tree has either one center or two centers that are adjacent.

44. Show that every tree can be colored using two colors.

The **rooted Fibonacci trees** T_n are defined recursively in the following way. T_1 and T_2 are both the rooted tree consisting of a single vertex, and for $n = 3, 4, \ldots,$ the rooted tree T_n is constructed from a root with T_{n-1} as its left subtree and T_{n-2} as its right subtree.

45. Draw the first seven rooted Fibonacci trees.

***46.** How many vertices, leaves, and internal vertices does the rooted Fibonacci tree T_n have, where n is a positive integer? What is its height?

47. What is wrong with the following "proof" using mathematical induction of the statement that every tree with n vertices has a path of length $n - 1$. *Basis step:* Every tree with one vertex clearly has a path of length 0. *Inductive step:* Assume that a tree with n vertices has a path of length $n - 1$, which has u as its terminal vertex. Add a vertex v and the edge from u to v. The resulting tree has $n + 1$ vertices and has a path of length n. This completes the inductive step.

☞***48.** Show that the average depth of a leaf in a binary tree with n vertices is $\Omega(\log n)$.

11.2 Applications of Trees

11.2.1 Introduction

We will discuss three problems that can be studied using trees. The first problem is: How should items in a list be stored so that an item can be easily located? The second problem is: What series of decisions should be made to find an object with a certain property in a collection of objects of a certain type? The third problem is: How should a set of characters be efficiently coded by bit strings?

11.2.2 Binary Search Trees

Links

Searching for items in a list is one of the most important tasks that arises in computer science. Our primary goal is to implement a searching algorithm that finds items efficiently when the items are totally ordered. This can be accomplished through the use of a **binary search tree**, which is a binary tree in which each child of a vertex is designated as a right or left child, no vertex has more than one right child or left child, and each vertex is labeled with a key, which is one of the items. Furthermore, vertices are assigned keys so that the key of a vertex is both larger than the keys of all vertices in its left subtree and smaller than the keys of all vertices in its right subtree.

This recursive procedure is used to form the binary search tree for a list of items. Start with a tree containing just one vertex, namely, the root. The first item in the list is assigned as the key of the root. To add a new item, first compare it with the keys of vertices already in the tree, starting at the root and moving to the left if the item is less than the key of the respective vertex if this vertex has a left child, or moving to the right if the item is greater than the key of the respective vertex if this vertex has a right child. When the item is less than the respective vertex and this vertex has no left child, then a new vertex with this item as its key is inserted as a new left child. Similarly, when the item is greater than the respective vertex and this vertex has no right child, then a new vertex with this item as its key is inserted as a new right child. We illustrate this procedure with Example 1.

EXAMPLE 1 Form a binary search tree for the words *mathematics, physics, geography, zoology, meteorology, geology, psychology*, and *chemistry* (using alphabetical order).

Solution: Figure 1 displays the steps used to construct this binary search tree. The word *mathematics* is the key of the root. Because *physics* comes after *mathematics* (in alphabetical order), add a right child of the root with key *physics*. Because *geography* comes before *mathematics*, add a left child of the root with key *geography*. Next, add a right child of the vertex with key *physics*, and assign it the key *zoology*, because *zoology* comes after *mathematics* and after *physics*. Similarly, add a left child of the vertex with key *physics* and assign this new vertex the key *meteorology*. Add a right child of the vertex with key *geography* and assign this new vertex the key *geology*. Add a left child of the vertex with key *zoology* and assign it the key *psychology*. Add a left child of the vertex with key *geography* and assign it the key *chemistry*. (The reader should work through all the comparisons needed at each step.) ◀

Once we have a binary search tree, we need a way to locate items in the binary search tree, as well as a way to add new items. Algorithm 1, an insertion algorithm, actually does both of

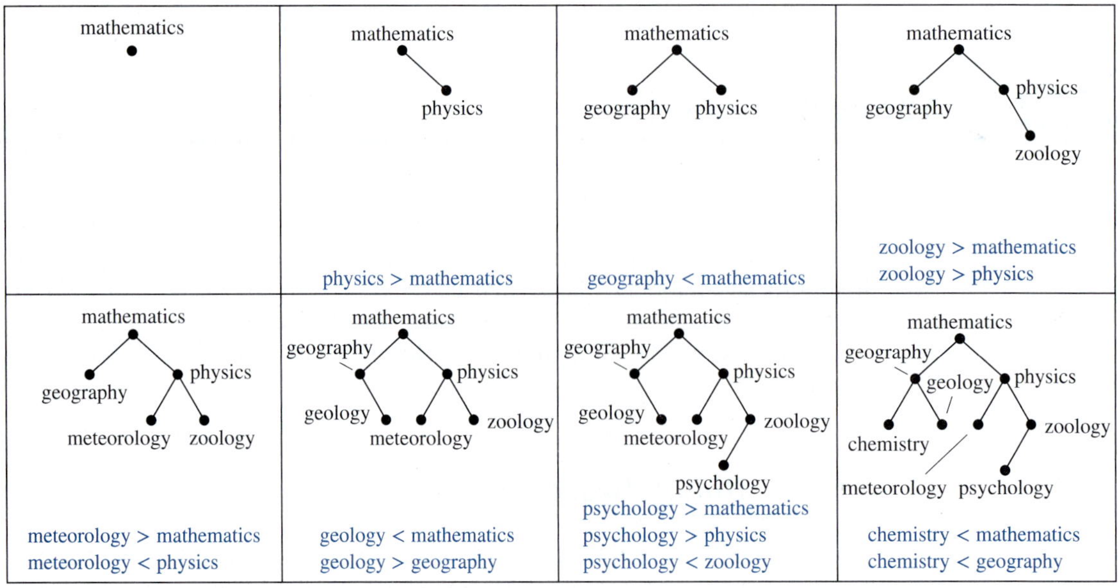

FIGURE 1 Constructing a binary search tree.

these tasks, even though it may appear that it is only designed to add vertices to a binary search tree. That is, Algorithm 1 is a procedure that locates an item x in a binary search tree if it is present, and adds a new vertex with x as its key if x is not present. In the pseudocode, v is the vertex currently under examination and $label(v)$ represents the key of this vertex. The algorithm begins by examining the root. If x equals the key of v, then the algorithm has found the location of x and terminates; if x is less than the key of v, we move to the left child of v and repeat the procedure; and if x is greater than the key of v, we move to the right child of v and repeat the procedure. If at any step we attempt to move to a child that is not present, we know that x is not present in the tree, and we add a new vertex as this child with x as its key.

ALGORITHM 1 Locating an Item in or Adding an Item to a Binary Search Tree.

```
procedure insertion(T: binary search tree, x: item)
v := root of T
{a vertex not present in T has the value null }
while v ≠ null and label(v) ≠ x
  if x < label(v) then
    if left child of v ≠ null then v := left child of v
    else add new vertex as a left child of v and set v := null
  else
    if right child of v ≠ null then v := right child of v
    else add new vertex as a right child of v and set v := null
if root of T = null then add a vertex v to the tree and label it with x
else if v is null or label(v) ≠ x then label new vertex with x and let v be this new vertex
return v {v = location of x}
```

Example 2 illustrates the use of Algorithm 1 to insert a new item into a binary search tree.

EXAMPLE 2 Use Algorithm 1 to insert the word *oceanography* into the binary search tree in Example 1.

Solution: Algorithm 1 begins with v, the vertex under examination, equal to the root of T, so $label(v) = mathematics$. Because $v \neq null$ and $label(v) = mathematics < oceanography$, we next examine the right child of the root. This right child exists, so we set v, the vertex under examination, to be this right child. At this step we have $v \neq null$ and $label(v) = physics > oceanography$, so we examine the left child of v. This left child exists, so we set v, the vertex under examination, to this left child. At this step, we also have $v \neq null$ and $label(v) = meteorology < oceanography$, so we try to examine the right child of v. However, this right child does not exist, so we add a new vertex as the right child of v (which at this point is the vertex with the key *meteorology*) and we set $v := null$. We now exit the **while** loop because $v = null$. Because the root of T is not *null* and $v = null$, we use the **else if** statement at the end of the algorithm to label our new vertex with the key *oceanography*. ◀

We will now determine the computational complexity of this procedure. Suppose we have a binary search tree T for a list of n items. We can form a full binary tree U from T by adding unlabeled vertices whenever necessary so that every vertex with a key has two children. This is illustrated in Figure 2. Once we have done this, we can easily locate or add a new item as a key without adding a vertex.

The most comparisons needed to add a new item is the length of the longest path in U from the root to a leaf. The internal vertices of U are the vertices of T. It follows that U has n internal vertices. We can now use part (*ii*) of Theorem 4 in Section 11.1 to conclude that U has $n + 1$

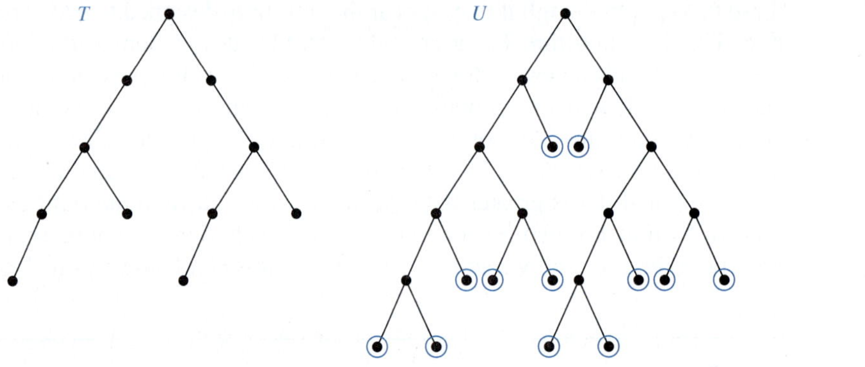

FIGURE 2 Adding unlabeled vertices to make a binary search tree full.

leaves. Using Corollary 1 of Section 11.1, we see that the height of U is greater than or equal to $h = \lceil \log_2(n+1) \rceil$. So, there must exist an item that requires at least $\lceil \log_2(n+1) \rceil$ comparisons to be added to the tree. Note that if U is balanced, its height is $\lceil \log_2(n+1) \rceil$ (by Corollary 1 of Section 11.1). Thus, if a binary search tree is balanced, locating or adding an item requires no more than $\lceil \log_2(n+1) \rceil$ comparisons. A binary search tree can become unbalanced as items are added to it. Because balanced binary search trees give optimal worst-case complexity for binary searching, algorithms have been devised that rebalance binary search trees as items are added. The interested reader can consult references on data structures for the description of such algorithms.

11.2.3 Decision Trees

Links

Rooted trees can be used to model problems in which a series of decisions leads to a solution. For instance, a binary search tree can be used to locate items based on a series of comparisons, where each comparison tells us whether we have located the item, or whether we should go right or left in a subtree. A rooted tree in which each internal vertex corresponds to a decision, with a subtree at these vertices for each possible outcome of the decision, is called a **decision tree**. The possible solutions of the problem correspond to the paths to the leaves of this rooted tree. Example 3 illustrates an application of decision trees.

EXAMPLE 3 Suppose there are seven coins, all with the same weight, and a counterfeit coin that weighs less than the others. How many weighings are necessary using a balance scale to determine which of the eight coins is the counterfeit one? Give an algorithm for finding this counterfeit coin.

Extra Examples

Solution: There are three possibilities for each weighing on a balance scale. The two pans can have equal weight, the first pan can be heavier, or the second pan can be heavier. Consequently, the decision tree for the sequence of weighings is a 3-ary tree. There are at least eight leaves in the decision tree because there are eight possible outcomes (because each of the eight coins can be the counterfeit lighter coin), and each possible outcome must be represented by at least one leaf. The largest number of weighings needed to determine the counterfeit coin is the height of the decision tree. From Corollary 1 of Section 11.1 it follows that the height of the decision tree is at least $\lceil \log_3 8 \rceil = 2$. Hence, at least two weighings are needed.

It is possible to determine the counterfeit coin using two weighings. The decision tree that illustrates how this is done is shown in Figure 3. ◀

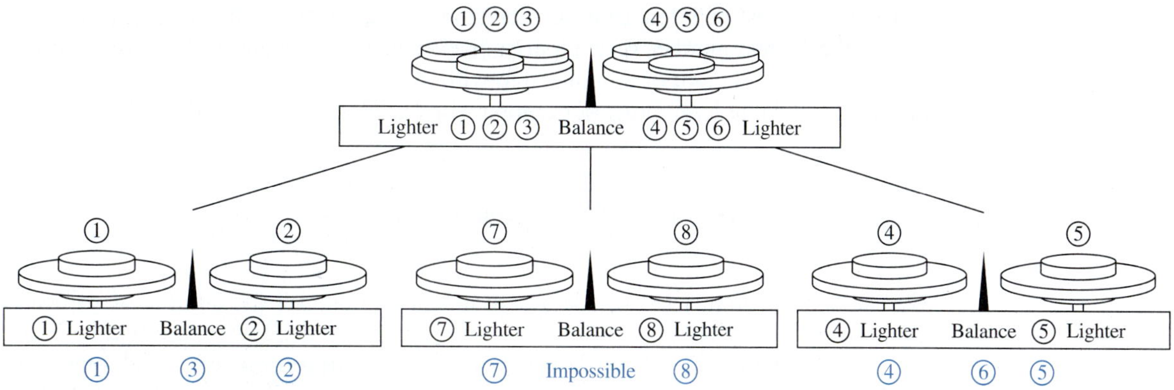

FIGURE 3 A decision tree for locating a counterfeit coin. The counterfeit coin is shown in color below each final weighing.

THE COMPLEXITY OF COMPARISON-BASED SORTING ALGORITHMS Many different sorting algorithms have been developed. To decide whether a particular sorting algorithm is efficient, its complexity is determined. Using decision trees as models, a lower bound for the worst-case complexity of sorting algorithms that are based on binary comparisons can be found.

We can use decision trees to model sorting algorithms and to determine an estimate for the worst-case complexity of these algorithms. Note that given n elements, there are $n!$ possible orderings of these elements, because each of the $n!$ permutations of these elements can be the correct order. The sorting algorithms studied in this book, and most commonly used sorting algorithms, are based on binary comparisons, that is, the comparison of two elements at a time. The result of each such comparison narrows down the set of possible orderings. Thus, a sorting algorithm based on binary comparisons can be represented by a binary decision tree in which each internal vertex represents a comparison of two elements. Each leaf represents one of the $n!$ permutations of n elements.

EXAMPLE 4 We display in Figure 4 a decision tree that orders the elements of the list a, b, c. ◀

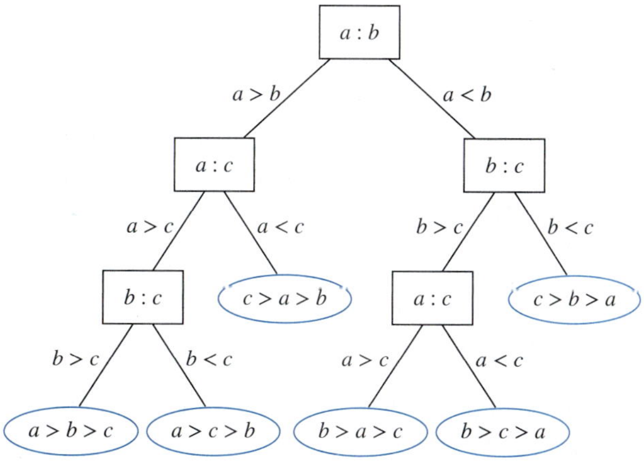

FIGURE 4 A decision tree for sorting three distinct elements.

The complexity of a sort based on binary comparisons is measured in terms of the number of such comparisons used. The largest number of binary comparisons ever needed to sort a list with n elements gives the worst-case performance of the algorithm. The most comparisons used equals the longest path length in the decision tree representing the sorting procedure. In other words, the largest number of comparisons ever needed is equal to the height of the decision tree. Because the height of a binary tree with $n!$ leaves is at least $\lceil \log_2 n! \rceil$ (using Corollary 1 in Section 11.1), at least $\lceil \log_2 n! \rceil$ comparisons are needed, as stated in Theorem 1.

THEOREM 1 A sorting algorithm based on binary comparisons requires at least $\lceil \log_2 n! \rceil$ comparisons.

We can use Theorem 1 to provide a big-Omega estimate for the number of comparisons used by a sorting algorithm based on binary comparison. We need only note that by Exercise 74 in Section 3.2 we know that $\lceil \log n! \rceil$ is $\Theta(n \log n)$, one of the commonly used reference functions for the computational complexity of algorithms. Corollary 1 is a consequence of this estimate.

COROLLARY 1 The number of comparisons used by a sorting algorithm to sort n elements based on binary comparisons is $\Omega(n \log n)$.

A consequence of Corollary 1 is that a sorting algorithm based on binary comparisons that uses $\Theta(n \log n)$ comparisons, in the worst case, to sort n elements is optimal, in the sense that no other such algorithm has better worst-case complexity. Note that by Theorem 1 in Section 5.4 we see that the merge sort algorithm is optimal in this sense.

We can also establish a similar result for the average-case complexity of sorting algorithms. The average number of comparisons used by a sorting algorithm based on binary comparisons is the average depth of a leaf in the decision tree representing the sorting algorithm. By Exercise 48 in Section 11.1 we know that the average depth of a leaf in a binary tree with N vertices is $\Omega(\log N)$. We obtain the following estimate when we let $N = n!$ and note that a function that is $\Omega(\log n!)$ is also $\Omega(n \log n)$ because $\log n!$ is $\Theta(n \log n)$.

THEOREM 2 The average number of comparisons used by a sorting algorithm to sort n elements based on binary comparisons is $\Omega(n \log n)$.

11.2.4 Prefix Codes

Consider the problem of using bit strings to encode the letters of the English alphabet (where no distinction is made between lowercase and uppercase letters). We can represent each letter with a bit string of length five, because there are only 26 letters and there are 32 bit strings of length five. The total number of bits used to encode data is five times the number of characters in the text when each character is encoded with five bits. Is it possible to find a coding scheme of these letters such that, when data are coded, fewer bits are used? We can save memory and reduce transmittal time if this can be done.

Consider using bit strings of different lengths to encode letters. Letters that occur more frequently should be encoded using short bit strings, and longer bit strings should be used to encode rarely occurring letters. When letters are encoded using varying numbers of bits, some method must be used to determine where the bits for each character start and end. For instance, if e were encoded with 0, a with 1, and t with 01, then the bit string 0101 could correspond to *eat, tea, eaea,* or *tt.*

One way to ensure that no bit string corresponds to more than one sequence of letters is to encode letters so that the bit string for a letter never occurs as the first part of the bit string for another letter. Codes with this property are called **prefix codes**. For instance, the encoding of e

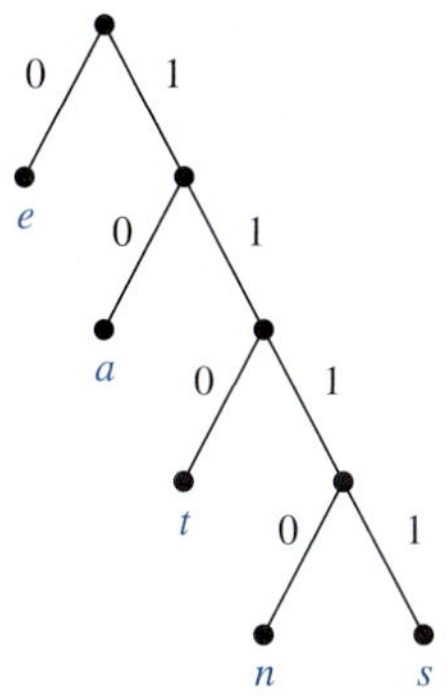

FIGURE 5 A binary tree with a prefix code.

as 0, a as 10, and t as 11 is a prefix code. A word can be recovered from the unique bit string that encodes its letters. For example, the string 10110 is the encoding of *ate*. To see this, note that the initial 1 does not represent a character, but 10 does represent a (and could not be the first part of the bit string of another letter). Then, the next 1 does not represent a character, but 11 does represent t. The final bit, 0, represents e.

A prefix code can be represented using a binary tree, where the characters are the labels of the leaves in the tree. The edges of the tree are labeled so that an edge leading to a left child is assigned a 0 and an edge leading to a right child is assigned a 1. The bit string used to encode a character is the sequence of labels of the edges in the unique path from the root to the leaf that has this character as its label. For instance, the tree in Figure 5 represents the encoding of e by 0, a by 10, t by 110, n by 1110, and s by 1111.

The tree representing a code can be used to decode a bit string. For instance, consider the word encoded by 11111011100 using the code in Figure 5. This bit string can be decoded by starting at the root, using the sequence of bits to form a path that stops when a leaf is reached. Each 0 bit takes the path down the edge leading to the left child of the last vertex in the path, and each 1 bit corresponds to the right child of this vertex. Consequently, the initial 1111 corresponds to the path starting at the root, going right four times, leading to a leaf in the graph that has s as its label, because the string 1111 is the code for s. Continuing with the fifth bit, we reach a leaf next after going right then left, when the vertex labeled with a, which is encoded by 10, is visited. Starting with the seventh bit, we reach a leaf next after going right three times and then left, when the vertex labeled with n, which is encoded by 1110, is visited. Finally, the last bit, 0, leads to the leaf that is labeled with e. Therefore, the original word is *sane*.

We can construct a prefix code from any binary tree where the left edge at each internal vertex is labeled by 0 and the right edge by a 1 and where the leaves are labeled by characters. Characters are encoded with the bit string constructed using the labels of the edges in the unique path from the root to the leaves.

Links

HUFFMAN CODING We now introduce an algorithm that takes as input the frequencies (which are the probabilities of occurrences) of symbols in a string and produces as output a prefix code that encodes the string using the fewest possible bits, among all possible binary prefix codes for these symbols. This algorithm, known as **Huffman coding**, was developed by David Huffman in a term paper he wrote in 1951 while a graduate student at MIT. (Note that this algorithm assumes that we already know how many times each symbol occurs in the string, so we can compute the frequency of each symbol by dividing the number of times this symbol occurs by the length of the string.) Huffman coding is a fundamental algorithm in *data compression*, the subject devoted to reducing the number of bits required to represent information. Huffman coding is extensively used to compress bit strings representing text and it also plays an important role in compressing audio and image files.

Links

Courtesy of California State University

DAVID A. HUFFMAN (1925–1999) David Huffman grew up in Ohio. At the age of 18 he received his B.S. in electrical engineering from The Ohio State University. Afterward he served in the U.S. Navy as a radar maintenance officer on a destroyer that had the mission of clearing mines in Asian waters after World War II. Later, he earned his M.S. from Ohio State and his Ph.D. in electrical engineering from MIT. Huffman joined the MIT faculty in 1953, where he remained until 1967, when he became the founding member of the computer science department at the University of California at Santa Cruz. He played an important role in developing this department and spent the remainder of his career there, retiring in 1994.

Huffman is noted for his contributions to information theory and coding, signal designs for radar and for communications, and design procedures for asynchronous logical circuits. His work on surfaces with zero curvature led him to develop original techniques for folding paper and vinyl into unusual shapes considered works of art by many and publicly displayed in several exhibits. However, Huffman is best known for his development of what is now called Huffman coding, a result of a term paper he wrote during his graduate work at MIT.

Huffman enjoyed exploring the outdoors, hiking, and traveling extensively. He became certified as a scuba diver when he was in his late 60s. He kept poisonous snakes as pets.

Algorithm 2 presents the Huffman coding algorithm. Given symbols and their frequencies, our goal is to construct a rooted binary tree where the symbols are the labels of the leaves. The algorithm begins with a forest of trees each consisting of one vertex, where each vertex has a symbol as its label and where the weight of this vertex equals the frequency of the symbol that is its label. At each step, we combine two trees having the least total weight into a single tree by introducing a new root and placing the tree with larger weight as its left subtree and the tree with smaller weight as its right subtree. Furthermore, we assign the sum of the weights of the two subtrees of this tree as the total weight of the tree. (Although procedures for breaking ties by choosing between trees with equal weights can be specified, we will not specify such procedures here.) The algorithm is finished when it has constructed a tree, that is, when the forest is reduced to a single tree.

Demo

ALGORITHM 2 Huffman Coding.

procedure *Huffman*(C: symbols a_i with frequencies w_i, $i = 1, \dots, n$)
$F :=$ forest of n rooted trees, each consisting of the single vertex a_i and assigned weight w_i
while F is not a tree
  Replace the rooted trees T and T' of least weights from F with $w(T) \geq w(T')$ with a tree having a new root that has T as its left subtree and T' as its right subtree. Label the new edge to T with 0 and the new edge to T' with 1.
  Assign $w(T) + w(T')$ as the weight of the new tree.
{the Huffman coding for the symbol a_i is the concatenation of the labels of the edges in the unique path from the root to the vertex a_i}

Example 5 illustrates how Algorithm 2 is used to encode a set of five symbols.

EXAMPLE 5 Use Huffman coding to encode the following symbols with the frequencies listed: A: 0.08, B: 0.10, C: 0.12, D: 0.15, E: 0.20, F: 0.35. What is the average number of bits used to encode a character?

Extra Examples

Solution: Figure 6 displays the steps used to encode these symbols. The encoding produced encodes A by 111, B by 110, C by 011, D by 010, E by 10, and F by 00. The average number of bits used to encode a symbol using this encoding is

$$3 \cdot 0.08 + 3 \cdot 0.10 + 3 \cdot 0.12 + 3 \cdot 0.15 + 2 \cdot 0.20 + 2 \cdot 0.35 = 2.45.$$ ◀

Note that Huffman coding is a greedy algorithm. Replacing the two subtrees with the smallest weight at each step leads to an optimal code in the sense that no binary prefix code for these symbols can encode these symbols using fewer bits. We leave the proof that Huffman codes are optimal as Exercise 32.

Huffman coding is used in JPEG image coding.

There are many variations of Huffman coding. For example, instead of encoding single symbols, we can encode blocks of symbols of a specified length, such as blocks of two symbols. Doing so may reduce the number of bits required to encode the string (see Exercise 30). We can also use more than two symbols to encode the original symbols in the string (see the preamble to Exercise 28). Furthermore, a variation known as adaptive Huffman coding (see [Sa00]) can be used when the frequency of each symbol in a string is not known in advance, so that encoding is done at the same time the string is being read.

11.2.5 Game Trees

Links

Trees can be used to analyze certain types of games such as tic-tac-toe, nim, checkers, and chess. In each of these games, two players take turns making moves. Each player knows the

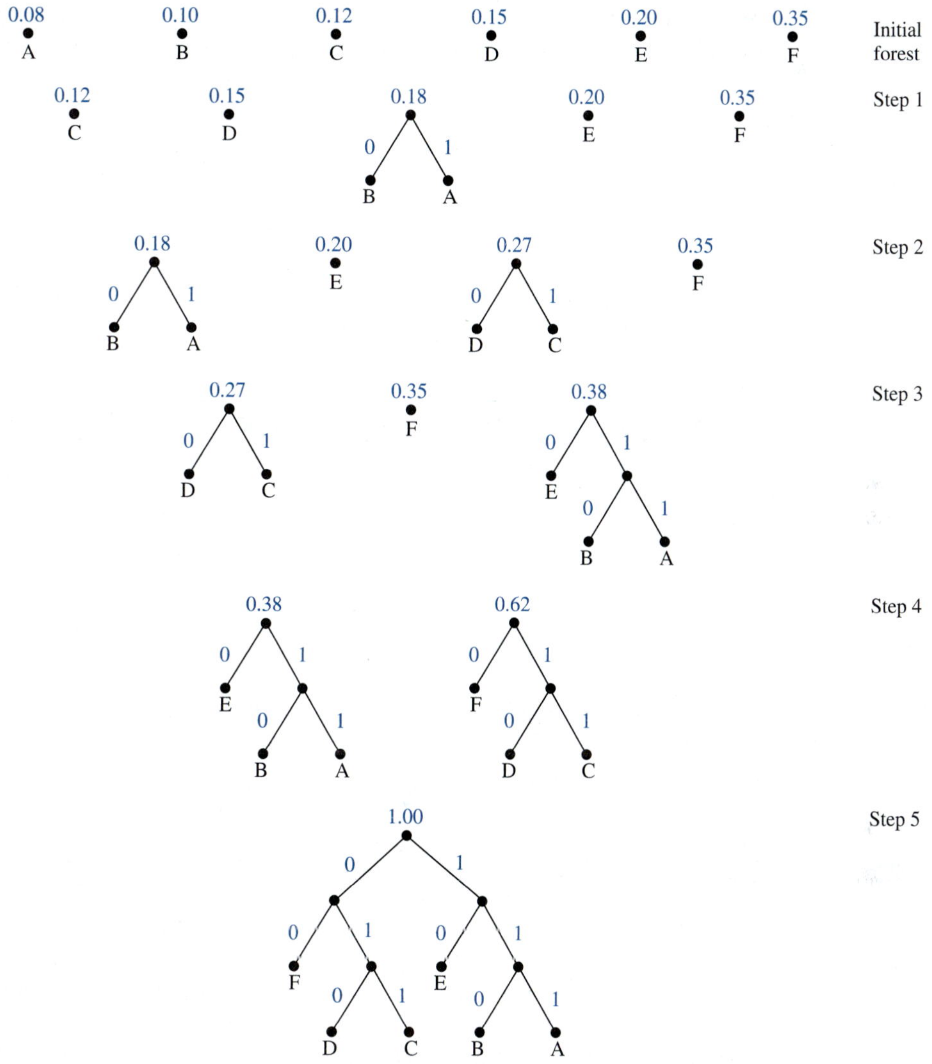

FIGURE 6 **Huffman coding of symbols in Example 5.**

moves made by the other player and no element of chance enters into the game. We model such games using **game trees**; the vertices of these trees represent the positions that a game can be in as it progresses; the edges represent legal moves between these positions. Because game trees are usually large, we simplify game trees by representing all symmetric positions of a game by the same vertex. However, the same position of a game may be represented by different vertices if different sequences of moves lead to this position. The root represents the starting position. The usual convention is to represent vertices at even levels by boxes and vertices at odd levels by circles. When the game is in a position represented by a vertex at an even level, it is the first player's move; when the game is in a position represented by a vertex at an odd level, it is the second player's move. Game trees may be infinite when the games they represent never end, such as games that can enter infinite loops, but for most games there are rules that lead to finite game trees.

The leaves of a game tree represent the final positions of a game. We assign a value to each leaf indicating the payoff to the first player if the game terminates in the position represented by this leaf. For games that are win–lose, we label a terminal vertex represented by a circle with a $+1$ to indicate a win by the first player and we label a terminal vertex represented by a box with a -1 to indicate a win by the second player. For games where draws are allowed, we label a terminal vertex corresponding to a draw position with a 0. Note that for win–lose games, we

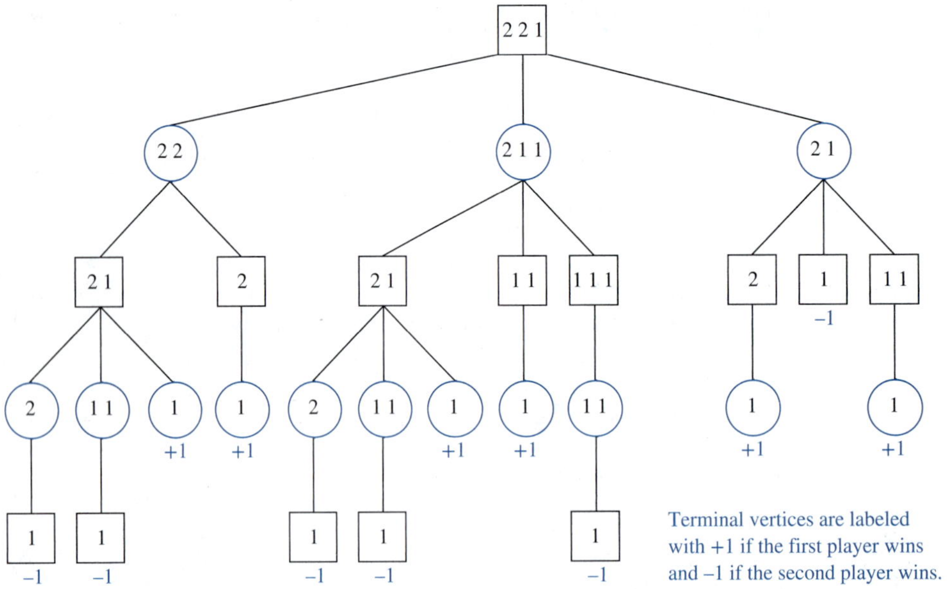

FIGURE 7 The game tree for a game of nim.

have assigned values to terminal vertices so that the larger the value, the better the outcome for the first player.

In Example 6 we display a game tree for a well-known and well-studied game.

EXAMPLE 6 **Nim** In a version of the game of **nim**, at the start of a game there are a number of piles of stones. Two players take turns making moves; a legal move consists of removing one or more stones from one of the piles, without removing all the stones left. A player without a legal move loses. (Another way to look at this is that the player removing the last stone loses because the position with no piles of stones is not allowed.) The game tree shown in Figure 7 represents this version of nim given the starting position where there are three piles of stones containing two, two, and one stone each, respectively. We represent each position with an unordered list of the number of stones in the different piles (the order of the piles does not matter). The initial move by the first player can lead to three possible positions because this player can remove one stone from a pile with two stones (leaving three piles containing one, one, and two stones); two stones from a pile containing two stones (leaving two piles containing two stones and one stone); or one stone from the pile containing one stone (leaving two piles of two stones). When only one pile with one stone is left, no legal moves are possible, so such positions are terminal positions. Because nim is a win–lose game, we label the terminal vertices with +1 when they represent wins for the first player and −1 when they represent wins for the second player. ◀

Although nim is an ancient game, Charles Bouton coined its modern name in 1901 after an archaic English word meaning "to steal."

EXAMPLE 7 **Tic-tac-toe** The game tree for tic-tac-toe is extremely large and cannot be drawn here, although a computer could easily build such a tree. We show a portion of the game tree for tic-tac-toe in Figure 8(a). Note that by considering symmetric positions equivalent, we need only consider three possible initial moves, as shown in Figure 8(a). We also show a subtree of this game tree leading to terminal positions in Figure 8(b), where a player who can win makes a winning move. ◀

We can recursively define the values of all vertices in a game tree in a way that enables us to determine the outcome of this game when both players follow optimal strategies. By a **strategy** we mean a set of rules that tells a player how to select moves to win the game. An optimal strategy for the first player is a strategy that maximizes the payoff to this player and for

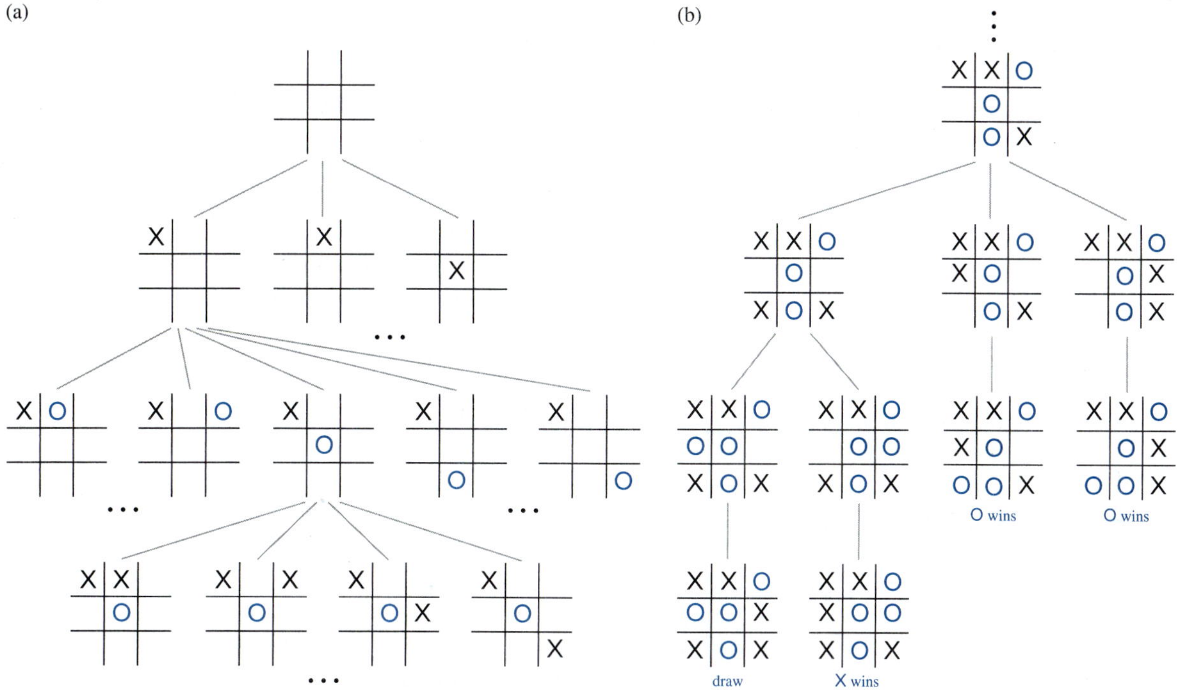

FIGURE 8 **Some of the game tree for tic-tac-toe.**

the second player is a strategy that minimizes this payoff. We now recursively define the value of a vertex.

Definition 1 The *value of a vertex in a game tree* is defined recursively as:

(*i*) the value of a leaf is the payoff to the first player when the game terminates in the position represented by this leaf.

(*ii*) the value of an internal vertex at an even level is the maximum of the values of its children, and the value of an internal vertex at an odd level is the minimum of the values of its children.

The strategy where the first player moves to a position represented by a child with maximum value and the second player moves to a position of a child with minimum value is called the **minmax strategy**. We can determine who will win the game when both players follow the minmax strategy by calculating the value of the root of the tree; this value is called the **value** of the tree. This is a consequence of Theorem 3.

THEOREM 3 The value of a vertex of a game tree tells us the payoff to the first player if both players follow the minmax strategy and play starts from the position represented by this vertex.

Proof: We will use induction to prove this theorem.

BASIS STEP: If the vertex is a leaf, by definition the value assigned to this vertex is the payoff to the first player.

INDUCTIVE STEP: The inductive hypothesis is the assumption that the values of the children of a vertex are the payoffs to the first player, assuming that play starts at each of the positions represented by these vertices. We need to consider two cases, when it is the first player's turn and when it is the second player's turn.

When it is the first player's turn, this player follows the minmax strategy and moves to the position represented by the child with the largest value. By the inductive hypothesis, this value is the payoff to the first player when play starts at the position represented by this child and follows the minmax strategy. By the recursive step in the definition of the value of an internal vertex at an even level (as the maximum value of its children), the value of this vertex is the payoff when play begins at the position represented by this vertex.

When it is the second player's turn, this player follows the minmax strategy and moves to the position represented by the child with the least value. By the inductive hypothesis, this value is the payoff to the first player when play starts at the position represented by this child and both players follow the minmax strategy. By the recursive definition of the value of an internal vertex at an odd level as the minimum value of its children, the value of this vertex is the payoff when play begins at the position represented by this vertex. ◁

Remark: By extending the proof of Theorem 3, it can be shown that the minmax strategy is the optimal strategy for both players.

Example 8 illustrates how the minmax procedure works. It displays the values assigned to the internal vertices in the game tree from Example 6. Note that we can shorten the computation required by noting that for win–lose games, once a child of a square vertex with value $+1$ is found, the value of the square vertex is also $+1$ because $+1$ is the largest possible payoff. Similarly, once a child of a circle vertex with value -1 is found, this is the value of the circle vertex also.

EXAMPLE 8 In Example 6 we constructed the game tree for nim with a starting position where there are three piles containing two, two, and one stones. In Figure 9 we show the values of the vertices of this game tree. The values of the vertices are computed using the values of the leaves and working one level up at a time. In the right margin of this figure we indicate whether we use the maximum or minimum of the values of the children to find the value of an internal vertex at each level. For example, once we have found the values of the three children of the root, which are $1, -1$, and -1, we find the value of the root by computing $\max(1, -1, -1) = 1$. Because the value of the root is 1, it follows that the first player wins when both players follow a minmax strategy. ◀

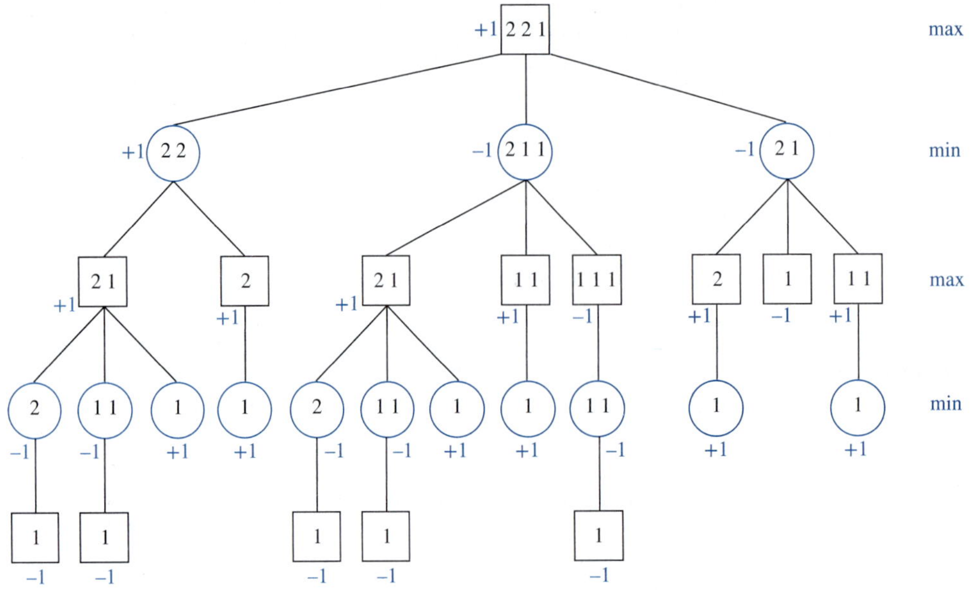

FIGURE 9 **Showing the values of vertices in the game of nim.**

Game trees for some well-known games can be extraordinarily large, because these games have many different possible moves. For example, the game tree for chess has been estimated to have as many as 10^{100} vertices! It may be impossible to use Theorem 3 directly to study a game because of the size of the game tree. Therefore, various approaches have been devised to help determine good strategies and to determine the outcome of such games. One useful technique, called *alpha-beta pruning*, eliminates much computation by pruning portions of the game tree that cannot affect the values of ancestor vertices. (For information about alpha-beta pruning, consult [Gr90].) Another useful approach is to use *evaluation functions*, which estimate the value of internal vertices in the game tree when it is not feasible to compute these values exactly. For example, in the game of tic-tac-toe, as an evaluation function for a position, we may use the number of files (rows, columns, and diagonals) containing no Os (used to indicate moves of the second player) minus the number of files containing no Xs (used to indicate moves of the first player). This evaluation function provides some indication of which player has the advantage in the game. Once the values of an evaluation function are inserted, the value of the game can be computed following the rules used for the minmax strategy. Computer programs created to play chess, such as IBM's Deep Blue, the first chess-playing computer to win a match against a current world champion under regular rules, are based on sophisticated evaluation functions. For more information about how computers play chess see [Le91].

Chess programs on smartphones can now play at the grandmaster level.

Links

The material we have studied comes from **combinatorial game theory**, which deals with games where a player knows all previous moves and selects a move before other players choose theirs. For more information about combinatorial game theory, consult [AlNoWo07], [BeCoGu82a, 82b], or [Be04], and the web links for this subject.

Links

11.2.6 The Tree of Fractions

In Chapter 2 we showed that the set of rational numbers is countable and presented one way to list all positive rational numbers so that each of these numbers is listed once. The construction that we introduced required that we omit rational numbers that we encounter that have already

Links

Courtesy of Peter D. Wilf

HERBERT WILF (1931–2012), born in Philadelphia into a Jewish family, attended public schools in Philadelphia. His father owned a carpet and appliance business. As a child he read mathematics books in bed when his parents thought he was sleeping. His father wanted Wilf to move to Israel, but he decided he wanted to finish his high school education in Philadelphia. Because he was interested in science, he applied to MIT and was accepted. He was awarded his Bachelor of Science degree from MIT in 1952. He attended Columbia University and was awarded his Ph.D. in 1958 for research combining mathematics and physics.

Wilf and his wife Ruth had three children; Susan who became a scholar of Chinese, David who became a lawyer, and Peter who became a paleontologist. To support his family, Wilf worked at the Fairchild Engine Division in 1954 and as head of the Computing Section of Nuclear Development Associates from 1955–1959. In 1959 he assumed the position of assistant professor at the University of Illinois. In 1961 he was appointed to an associate professorship at the University of Pennsylvania where he remained for the rest of his career, holding an endowed chair after 1998.

Wilf was a prolific author. His early research papers included works on Monte Carlo problems, numerical integration, and linear algebra. Wilf is best known for his seminal contributions to discrete mathematics with its applications to computer algorithms. His most important contribution was in "WZ theory" where he worked with Doron Zeilberger on their award-winning paper "Rational Functions Certify Combinatorial Identities" (1990).

His first textbooks included *Mathematical Methods for Digital Computers* (with A Ralston), and *Mathematics for the Physical Sciences*. His later books included *Finite Sections of Some Classical Inequalities*. With Albert Nijenhuis, he published *Combinatorial Algorithms*. His 1984 book *Algorithms and Complexity* had two editions. In 1990 he published *generatingfunctionology*. He also wrote a book with the cryptic title $A = B$ (with Petkovsek and Zeilberger) that describes joint work with Zeilberger.

Wilf was a dedicated teacher who won many teaching awards. He held strong and influential views about the teaching of mathematics. Wilf founded two research journals: *The Journal of Algorithms* (with co-founder Knuth) and *The Electronic Journal of Combinatorics* (with co-founder Calkin). He was the editor-in-chief from 1987 to 1992 of the *American Mathematical Monthly*.

Wilf's most important hobby was flying a single engine airplane. From 1982 until 2002 he and his wife explored much of North America by plane with Wilf flying, without any major mishaps. Wilf also enjoyed playing bridge.

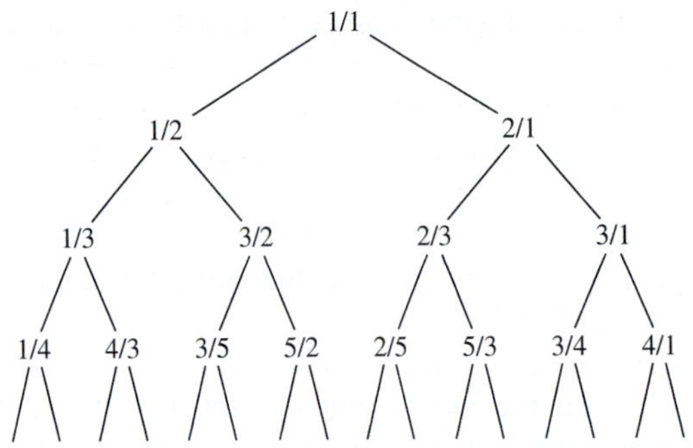

FIGURE 10 **The Tree of Fractions.**

been listed. Here, we will describe a different way of listing all fractions that avoids the problem of eliminating rational numbers that are listed more than once.

The method we now describe is based on the *tree of fractions*. To construct the tree of fractions we begin by placing the fraction 1/1 at the top of the tree. Then to produce the rest of the tree, we use the rule that for each vertex $\frac{a}{b}$ we add two children to this vertex at the next level of the tree, the left child is $a/(a+b)$ and the right child is $(a+b)/b$. The resulting tree is illustrated in Figure 10.

We leave it to the exercises at the end of the section to prove that listing the numbers in order of the level of the tree, beginning with 1/1 at level 1, 1/2 followed by 2/1 at level 2, followed by 1/3, 3/2, 2/3, and 3/1 at level 3, and so on, produces a list of all the positive rational numbers where each number is listed exactly once. (This is the list produced by a breadth-first search of the tree of fractions; breadth-first search is introduced in Subsection 11.4.3.)

Exercises

1. Build a binary search tree for the words *banana, peach, apple, pear, coconut, mango*, and *papaya* using alphabetical order.

2. Build a binary search tree for the words *oenology, phrenology, campanology, ornithology, ichthyology, limnology, alchemy*, and *astrology* using alphabetical order.

3. How many comparisons are needed to locate or to add each of these words in the search tree for Exercise 1, starting fresh each time?

a) *pear* **b)** *banana*
c) *kumquat* **d)** *orange*

4. How many comparisons are needed to locate or to add each of the words in the search tree for Exercise 2, starting fresh each time?

a) *palmistry* **b)** *etymology*
c) *paleontology* **d)** *glaciology*

5. Using alphabetical order, construct a binary search tree for the words in the sentence "*The quick brown fox jumps over the lazy dog.*"

6. How many weighings of a balance scale are needed to find a lighter counterfeit coin among four coins? Describe an algorithm to find the lighter coin using this number of weighings.

7. How many weighings of a balance scale are needed to find a counterfeit coin among four coins if the counterfeit coin may be either heavier or lighter than the others? Describe an algorithm to find the counterfeit coin using this number of weighings.

***8.** How many weighings of a balance scale are needed to find a counterfeit coin among eight coins if the counterfeit coin is either heavier or lighter than the others? Describe an algorithm to find the counterfeit coin using this number of weighings.

***9.** How many weighings of a balance scale are needed to find a counterfeit coin among 12 coins if the counterfeit coin is lighter than the others? Describe an algorithm to find the lighter coin using this number of weighings.

***10.** One of four coins may be counterfeit. If it is counterfeit, it may be lighter or heavier than the others. How many weighings are needed, using a balance scale, to determine whether there is a counterfeit coin, and if there is, whether it is lighter or heavier than the others? Describe an algorithm to find the counterfeit coin and determine whether it is lighter or heavier using this number of weighings.

11. Find the least number of comparisons needed to sort four elements and devise an algorithm that sorts these elements using this number of comparisons.

***12.** Find the least number of comparisons needed to sort five elements and devise an algorithm that sorts these elements using this number of comparisons.

The **tournament sort** is a sorting algorithm that works by building an ordered binary tree. We represent the elements to be sorted by vertices that will become the leaves. We build up the tree one level at a time as we would construct the tree representing the winners of matches in a tournament. Working left to right, we compare pairs of consecutive elements, adding a parent vertex labeled with the larger of the two elements under comparison. We make similar comparisons between labels of vertices at each level until we reach the root of the tree that is labeled with the largest element. The tree constructed by the tournament sort of 22, 8, 14, 17, 3, 9, 27, 11 is illustrated in part (a) of the figure. Once the largest element has been determined, the leaf with this label is relabeled by $-\infty$, which is defined to be less than every element. The labels of all vertices on the path from this vertex up to the root of the tree are recalculated, as shown in part (b) of the figure. This produces the second largest element. This process continues until the entire list has been sorted.

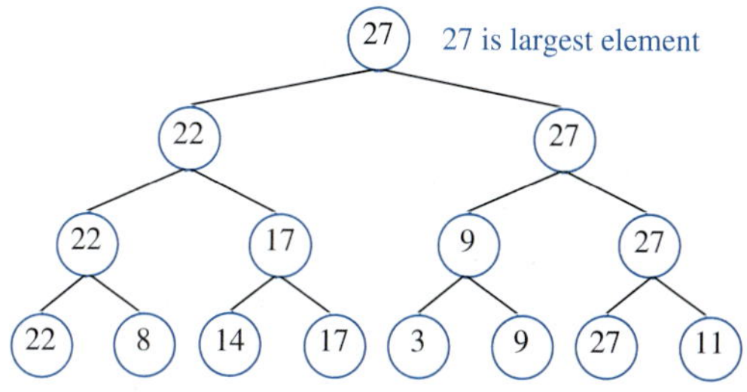

(a)

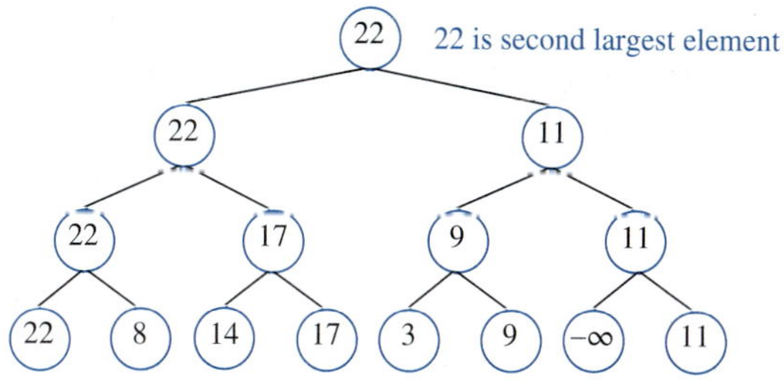

(b)

13. Complete the tournament sort of the list 22, 8, 14, 17, 3, 9, 27, 11. Show the labels of the vertices at each step.

14. Use the tournament sort to sort the list 17, 4, 1, 5, 13, 10, 14, 6.

15. Describe the tournament sort using pseudocode.

16. Assuming that n, the number of elements to be sorted, equals 2^k for some positive integer k, determine the number of comparisons used by the tournament sort to find the largest element of the list using the tournament sort.

17. How many comparisons does the tournament sort use to find the second largest, the third largest, and so on, up to the $(n-1)$st largest (or second smallest) element?

18. Show that the tournament sort requires $\Theta(n \log n)$ comparisons to sort a list of n elements. [*Hint:* By inserting the appropriate number of dummy elements defined to be smaller than all integers, such as $-\infty$, assume that $n = 2^k$ for some positive integer k.]

19. Which of these codes are prefix codes?

a) a: 11, e: 00, t: 10, s: 01

b) a: 0, e: 1, t: 01, s: 001

c) a: 101, e: 11, t: 001, s: 011, n: 010

d) a: 010, e: 11, t: 011, s: 1011, n: 1001, i: 10101

20. Construct the binary tree with prefix codes representing these coding schemes.

a) a: 11, e: 0, t: 101, s: 100

b) a: 1, e: 01, t: 001, s: 0001, n: 00001

c) a: 1010, e: 0, t: 11, s: 1011, n: 1001, i: 10001

21. What are the codes for a, e, i, k, o, p, and u if the coding scheme is represented by this tree?

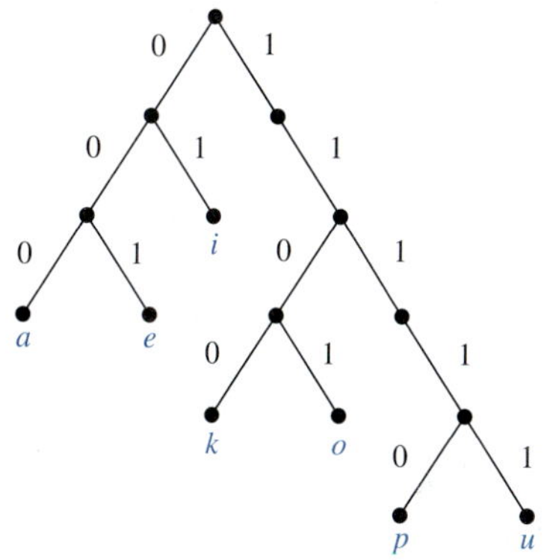

22. Given the coding scheme a: 001, b: 0001, e: 1, r: 0000, s: 0100, t: 011, x: 01010, find the word represented by

a) 01110100011.

b) 0001110000.

c) 0100101010.

d) 01100101010.

23. Use Huffman coding to encode these symbols with given frequencies: a: 0.20, b: 0.10, c: 0.15, d: 0.25, e: 0.30. What is the average number of bits required to encode a character?

24. Use Huffman coding to encode these symbols with given frequencies: A: 0.10, B: 0.25, C: 0.05, D: 0.15, E: 0.30, F: 0.07, G: 0.08. What is the average number of bits required to encode a symbol?

25. Construct two different Huffman codes for these symbols and frequencies: t: 0.2, u: 0.3, v: 0.2, w: 0.3.

26. **a)** Use Huffman coding to encode these symbols with frequencies a: 0.4, b: 0.2, c: 0.2, d: 0.1, e: 0.1 in two different ways by breaking ties in the algorithm differently. First, among the trees of minimum weight select two trees with the largest number of vertices to combine at each stage of the algorithm. Second, among the trees of minimum weight select two trees with the smallest number of vertices at each stage.

b) Compute the average number of bits required to encode a symbol with each code and compute the variances of this number of bits for each code. Which tie-breaking procedure produced the smaller variance in the number of bits required to encode a symbol?

27. Construct a Huffman code for the letters of the English alphabet where the frequencies of letters in typical English text are as shown in this table.

Letter	*Frequency*	*Letter*	*Frequency*
A	0.0817	N	0.0662
B	0.0145	O	0.0781
C	0.0248	P	0.0156
D	0.0431	Q	0.0009
E	0.1232	R	0.0572
F	0.0209	S	0.0628
G	0.0182	T	0.0905
H	0.0668	U	0.0304
I	0.0689	V	0.0102
J	0.0010	W	0.0264
K	0.0080	X	0.0015
L	0.0397	Y	0.0211
M	0.0277	Z	0.0005

Suppose that m is a positive integer with $m \geq 2$. An m-ary Huffman code for a set of N symbols can be constructed analogously to the construction of a binary Huffman code. At the initial step, $((N-1) \textbf{ mod } (m-1)) + 1$ trees consisting of a single vertex with least weights are combined into a rooted tree with these vertices as leaves. At each subsequent step, the m trees of least weight are combined into an m-ary tree.

28. Describe the m-ary Huffman coding algorithm in pseudocode.

29. Using the symbols 0, 1, and 2 use ternary ($m = 3$) Huffman coding to encode these letters with the given frequencies: A: 0.25, E: 0.30, N: 0.10, R: 0.05, T: 0.12, Z: 0.18.

30. Consider the three symbols A, B, and C with frequencies A: 0.80, B: 0.19, C: 0.01.

a) Construct a Huffman code for these three symbols.

b) Form a new set of nine symbols by grouping together blocks of two symbols, AA, AB, AC, BA, BB, BC, CA, CB, and CC. Construct a Huffman code for these nine symbols, assuming that the occurrences of symbols in the original text are independent.

c) Compare the average number of bits required to encode text using the Huffman code for the three symbols in part (a) and the Huffman code for the nine blocks of two symbols constructed in part (b). Which is more efficient?

31. Given $n + 1$ symbols $x_1, x_2, \ldots, x_n, x_{n+1}$ appearing 1, $f_1, f_2, \ldots, f_n$ times in a symbol string, respectively, where f_j is the jth Fibonacci number, what is the maximum number of bits used to encode a symbol when all possible tie-breaking selections are considered at each stage of the Huffman coding algorithm?

***32.** Show that Huffman codes are optimal in the sense that they represent a string of symbols using the fewest bits among all binary prefix codes.

33. Draw a game tree for nim if the starting position consists of two piles with two and three stones, respectively. When drawing the tree represent by the same vertex symmetric positions that result from the same move. Find the value of each vertex of the game tree. Who wins the game if both players follow an optimal strategy?

34. Draw a game tree for nim if the starting position consists of three piles with one, two, and three stones, respectively. When drawing the tree represent by the same vertex symmetric positions that result from the same move. Find the value of each vertex of the game tree. Who wins the game if both players follow an optimal strategy?

35. Suppose that we vary the payoff to the winning player in the game of nim so that the payoff is n dollars when n is the number of legal moves made before a terminal position is reached. Find the payoff to the first player if the initial position consists of

a) two piles with one and three stones, respectively.

b) two piles with two and four stones, respectively.

c) three piles with one, two, and three stones, respectively.

36. Suppose that in a variation of the game of nim we allow a player to either remove one or more stones from a pile or merge the stones from two piles into one pile as long as at least one stone remains. Draw the game tree for this variation of nim if the starting position consists of three piles containing two, two, and one stone, respectively. Find the values of each vertex in the game tree and determine the winner if both players follow an optimal strategy.

37. Draw the subtree of the game tree for tic-tac-toe beginning at each of these positions. Determine the value of each of these subtrees.

a)

O	X	X
X	O	O
		X

b)

X	O	X
O	X	X
		O

c)

X		O
O	O	
X		X

d)

	O	X
	X	O
X	O	

38. Suppose that the first four moves of a tic-tac-toe game are as shown. Does the first player (whose moves are marked by Xs) have a strategy that will always win?

a)

X	O	X
	O	

b)

X	O	X
	O	

c)

	O	X
	X	
O		

d)

X		
X	O	
O		

39. Show that if a game of nim begins with two piles containing the same number of stones, as long as this number is at least two, then the second player wins when both players follow optimal strategies.

40. Show that if a game of nim begins with two piles containing different numbers of stones, the first player wins when both players follow optimal strategies.

41. How many children does the root of the game tree for checkers have? How many grandchildren does it have?

42. How many children does the root of the game tree for nim have and how many grandchildren does it have if the starting position is

a) piles with four and five stones, respectively.
b) piles with two, three, and four stones, respectively.
c) piles with one, two, three, and four stones, respectively.
d) piles with two, two, three, three, and five stones, respectively.

43. Draw the game tree for the game of tic-tac-toe for the levels corresponding to the first two moves. Assign the value of the evaluation function mentioned in the text that assigns to a position the number of files containing no Os minus the number of files containing no Xs as the value of each vertex at this level and compute the value of the tree for vertices as if the evaluation function gave the correct values for these vertices.

44. Use pseudocode to describe an algorithm for determining the value of a game tree when both players follow a minmax strategy.

45. Show that in the tree of fractions the numerator and denominator of each vertex are relatively prime.

46. Show that every positive rational number a/b with $\gcd(a, b) = 1$ occurs at a vertex in the tree of fractions.

47. Show that no positive rational number a/b with $\gcd(a, b) = 1$ occurs at more than one vertex of the tree of fractions.

A hyperbinary expansion of an integer $n \geq 1$ is an expansion of n as a sum of powers of 2, each power being used at most twice. The number of hyperbinary expansions of the positive integer n is denoted by $b(n)$. For example, $b(12) = 5$ because $8 + 4 = 8 + 2 + 2 = 8 + 2 + 1 + 1 = 4 + 4 + 2 + 2 = 4 + 4 + 2 + 1 + 1$. Furthermore, we define $b(0) = 1$.

48. Find $b(n)$ when n equals

a) 5 **b)** 9 **c)** 15 **d)** 20

49. Prove that the sequence $\frac{b(n)}{b(n+1)}$ where $b(n)$ is the number of hyperbinary representations of the integer n for $n = 0, 1, 2, 3, \ldots,$ is the list of the positive rational numbers generated by the tree of fractions.

11.3 Tree Traversal

11.3.1 Introduction

Links

Ordered rooted trees are often used to store information. We need procedures for visiting each vertex of an ordered rooted tree to access data. We will describe several important algorithms for visiting all the vertices of an ordered rooted tree. Ordered rooted trees can also be used to represent various types of expressions, such as arithmetic expressions involving numbers, variables, and operations. The different listings of the vertices of ordered rooted trees used to represent expressions are useful in the evaluation of these expressions.

11.3.2 Universal Address Systems

Procedures for traversing all vertices of an ordered rooted tree rely on the orderings of children. In ordered rooted trees, the children of an internal vertex are shown from left to right in the drawings representing these directed graphs.

We will describe one way we can totally order the vertices of an ordered rooted tree. To produce this ordering, we must first label all the vertices. We do this recursively:

1. Label the root with the integer 0. Then label its k children (at level 1) from left to right with $1, 2, 3, \ldots, k$.
2. For each vertex v at level n with label A, label its k_v children, as they are drawn from left to right, with $A.1, A.2, \ldots, A.k_v$.

Following this procedure, a vertex v at level n, for $n \geq 1$, is labeled $x_1.x_2.\ldots.x_n$, where the unique path from the root to v goes through the x_1st vertex at level 1, the x_2nd vertex at level 2, and so on. This labeling is called the **universal address system** of the ordered rooted tree.

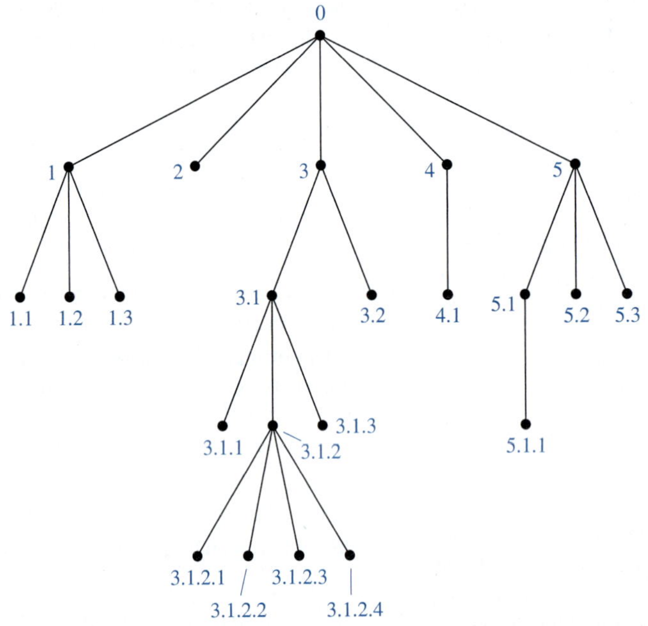

FIGURE 1 **The universal address system of an ordered rooted tree.**

We can totally order the vertices using the lexicographic ordering of their labels in the universal address system. The vertex labeled $x_1.x_2.\ldots.x_n$ is less than the vertex labeled $y_1.y_2.\ldots.y_m$ if there is an i, $0 \le i \le n$, with $x_1 = y_1$, $x_2 = y_2$, …, $x_{i-1} = y_{i-1}$, and $x_i < y_i$; or if $n < m$ and $x_i = y_i$ for $i = 1, 2, \ldots, n$.

EXAMPLE 1 We display the labelings of the universal address system next to the vertices in the ordered rooted tree shown in Figure 1. The lexicographic ordering of the labelings is

Extra Examples

$$0 < 1 < 1.1 < 1.2 < 1.3 < 2 < 3 < 3.1 < 3.1.1 < 3.1.2 < 3.1.2.1 < 3.1.2.2$$
$$< 3.1.2.3 < 3.1.2.4 < 3.1.3 < 3.2 < 4 < 4.1 < 5 < 5.1 < 5.1.1 < 5.2 < 5.3$$ ◀

11.3.3 Traversal Algorithms

Procedures for systematically visiting every vertex of an ordered rooted tree are called **traversal algorithms**. We will describe three of the most commonly used such algorithms, **preorder traversal**, **inorder traversal**, and **postorder traversal**. Each of these algorithms can be defined recursively. We first define preorder traversal.

Definition 1 Let T be an ordered rooted tree with root r. If T consists only of r, then r is the *preorder traversal* of T. Otherwise, suppose that $T_1, T_2, \ldots, T_n$ are the subtrees at r from left to right in T. The *preorder traversal* begins by visiting r. It continues by traversing T_1 in preorder, then T_2 in preorder, and so on, until T_n is traversed in preorder.

The reader should verify that the preorder traversal of an ordered rooted tree gives the same ordering of the vertices as the ordering obtained using a universal address system. Figure 2 indicates how a preorder traversal is carried out.

Example 2 illustrates preorder traversal.

EXAMPLE 2 In which order does a preorder traversal visit the vertices in the ordered rooted tree T shown in Figure 3?

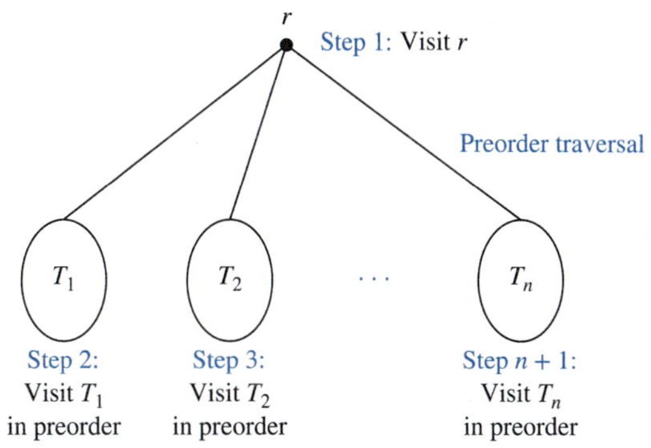

FIGURE 2 Preorder traversal.

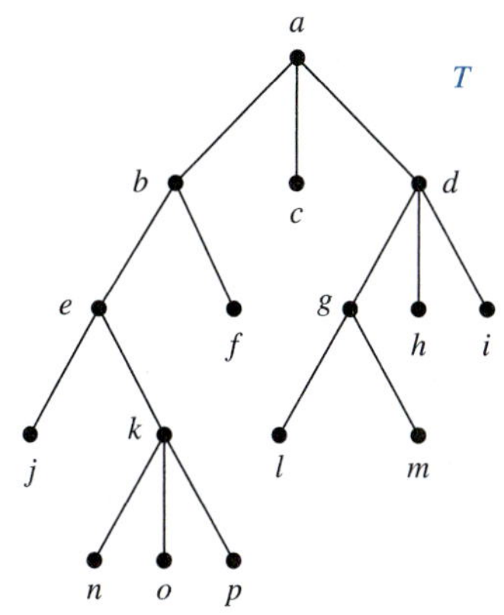

FIGURE 3 The ordered rooted tree T.

Extra Examples

Solution: The steps of the preorder traversal of T are shown in Figure 4. We traverse T in preorder by first listing the root a, followed by the preorder list of the subtree with root b, the preorder list of the subtree with root c (which is just c) and the preorder list of the subtree with root d.

The preorder list of the subtree with root b begins by listing b, then the vertices of the subtree with root e in preorder, and then the subtree with root f in preorder (which is just f). The preorder list of the subtree with root d begins by listing d, followed by the preorder list of the subtree with root g, followed by the subtree with root h (which is just h), followed by the subtree with root i (which is just i).

The preorder list of the subtree with root e begins by listing e, followed by the preorder listing of the subtree with root j (which is just j), followed by the preorder listing of the subtree with root k. The preorder listing of the subtree with root g is g followed by l, followed by m. The preorder listing of the subtree with root k is k, n, o, p. Consequently, the preorder traversal of T is $a, b, e, j, k, n, o, p, f, c, d, g, l, m, h, i$. ◀

We will now define inorder traversal.

Definition 2

Let T be an ordered rooted tree with root r. If T consists only of r, then r is the *inorder traversal* of T. Otherwise, suppose that $T_1, T_2, \ldots, T_n$ are the subtrees at r from left to right. The *inorder traversal* begins by traversing T_1 in inorder, then visiting r. It continues by traversing T_2 in inorder, then T_3 in inorder, …, and finally T_n in inorder.

Figure 5 indicates how inorder traversal is carried out. Example 3 illustrates how inorder traversal is carried out for a particular tree.

EXAMPLE 3 In which order does an inorder traversal visit the vertices of the ordered rooted tree T in Figure 3?

Extra Examples

Solution: The steps of the inorder traversal of the ordered rooted tree T are shown in Figure 6. The inorder traversal begins with an inorder traversal of the subtree with root b, the root a, the inorder listing of the subtree with root c, which is just c, and the inorder listing of the subtree with root d.

The inorder listing of the subtree with root b begins with the inorder listing of the subtree with root e, the root b, and f. The inorder listing of the subtree with root d begins with the inorder listing of the subtree with root g, followed by the root d, followed by h, followed by i.

The inorder listing of the subtree with root e is j, followed by the root e, followed by the inorder listing of the subtree with root k. The inorder listing of the subtree with root g is l, g, m. The inorder listing of the subtree with root k is n, k, o, p. Consequently, the inorder listing of the ordered rooted tree is $j, e, n, k, o, p, b, f, a, c, l, g, m, d, h, i$. ◀

We now define postorder traversal.

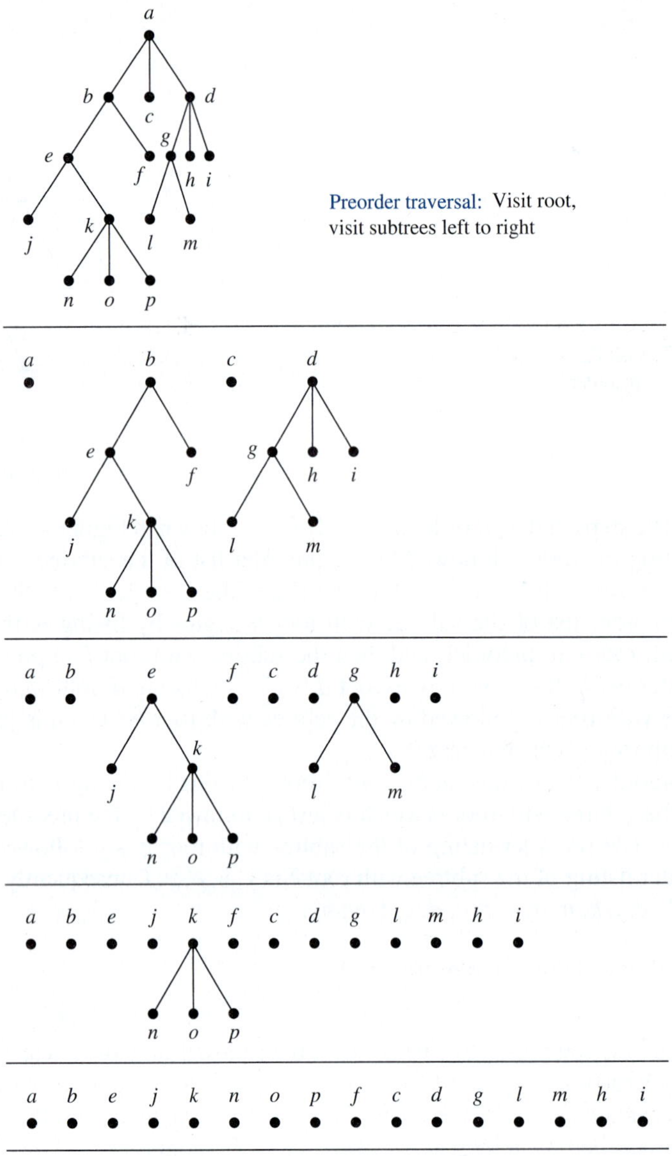

FIGURE 4 **The preorder traversal of T.**

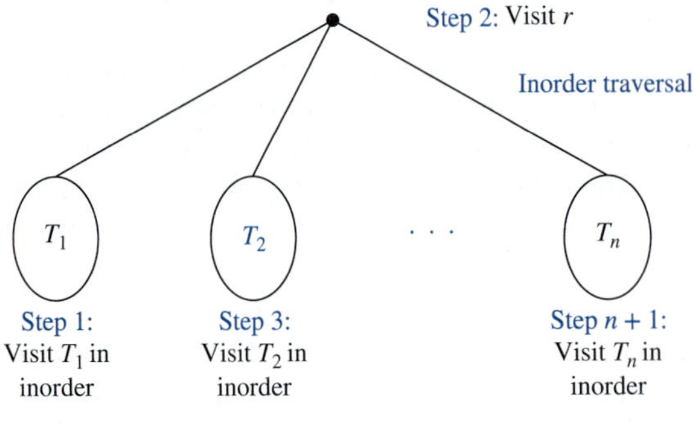

FIGURE 5 **Inorder traversal.**

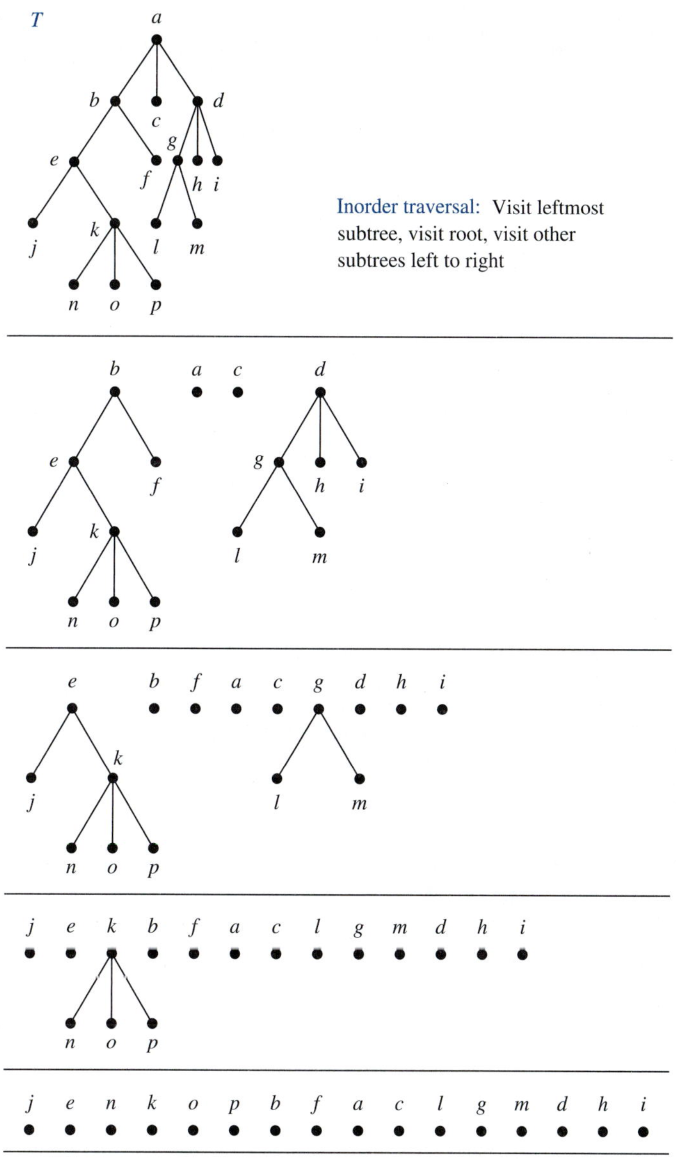

FIGURE 6 **The inorder traversal of *T*.**

Definition 3

Let T be an ordered rooted tree with root r. If T consists only of r, then r is the *postorder traversal* of T. Otherwise, suppose that $T_1, T_2, \ldots, T_n$ are the subtrees at r from left to right. The *postorder traversal* begins by traversing T_1 in postorder, then T_2 in postorder, ..., then T_n in postorder, and ends by visiting r.

Figure 7 illustrates how postorder traversal is done. Example 4 illustrates how postorder traversal works.

EXAMPLE 4 In which order does a postorder traversal visit the vertices of the ordered rooted tree T shown in Figure 3?

Extra Examples

Solution: The steps of the postorder traversal of the ordered rooted tree T are shown in Figure 8. The postorder traversal begins with the postorder traversal of the subtree with root b, the

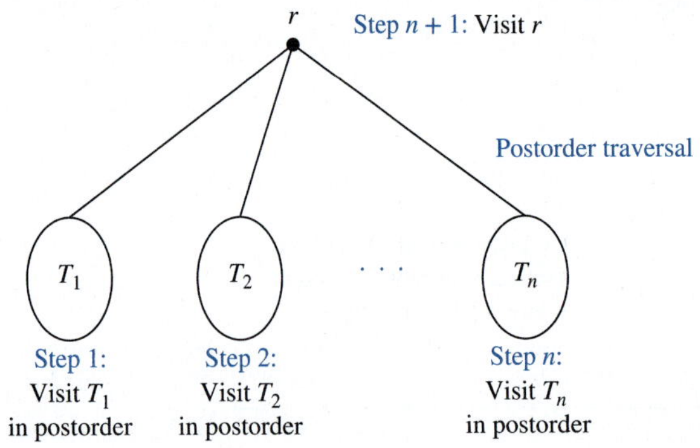

FIGURE 7 Postorder traversal.

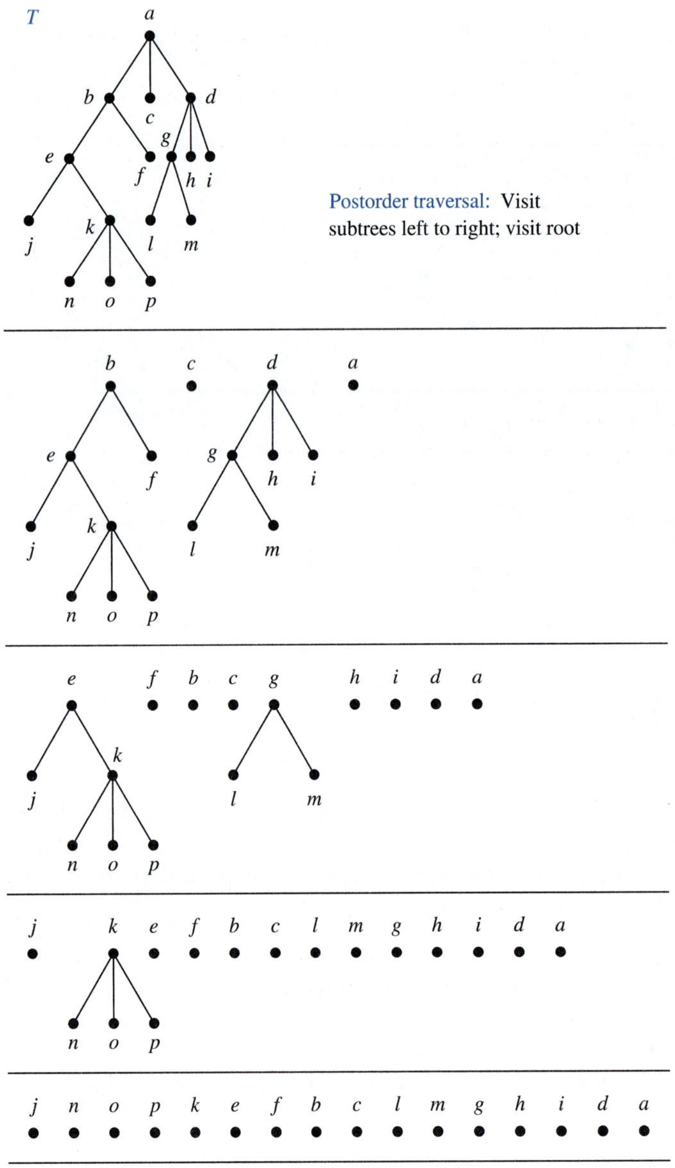

FIGURE 8 The postorder traversal of *T*.

postorder traversal of the subtree with root c, which is just c, the postorder traversal of the subtree with root d, followed by the root a.

The postorder traversal of the subtree with root b begins with the postorder traversal of the subtree with root e, followed by f, followed by the root b. The postorder traversal of the rooted tree with root d begins with the postorder traversal of the subtree with root g, followed by h, followed by i, followed by the root d.

The postorder traversal of the subtree with root e begins with j, followed by the postorder traversal of the subtree with root k, followed by the root e. The postorder traversal of the subtree with root g is l, m, g. The postorder traversal of the subtree with root k is n, o, p, k. Therefore, the postorder traversal of T is j, n, o, p, k, e, f, b, c, l, m, g, h, i, d, a. ◀

There are easy ways to list the vertices of an ordered rooted tree in preorder, inorder, and postorder. To do this, first draw a curve around the ordered rooted tree starting at the root, moving along the edges, as shown in the example in Figure 9. We can list the vertices in preorder by listing each vertex the first time this curve passes it. We can list the vertices in inorder by listing a leaf the first time the curve passes it and listing each internal vertex the second time the curve passes it. We can list the vertices in postorder by listing a vertex the last time it is passed on the way back up to its parent. When this is done in the rooted tree in Figure 9, it follows that the preorder traversal gives a, b, d, h, e, i, j, c, f, g, k; the inorder traversal gives h, d, b, i, e, j, a, f, c, k, g; and the postorder traversal gives h, d, i, j, e, b, f, k, g, c, a.

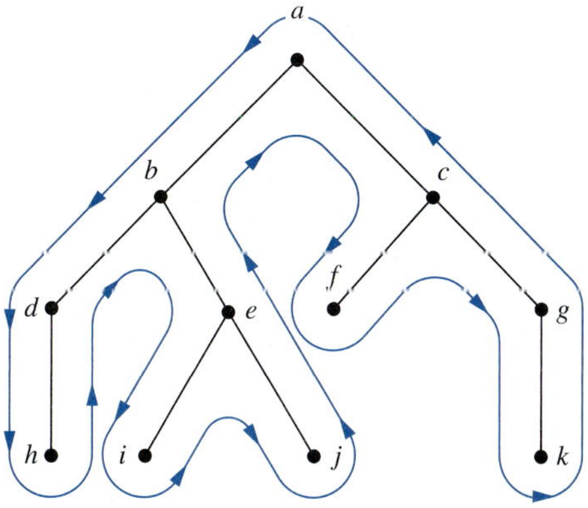

FIGURE 9 A shortcut for traversing an ordered rooted tree in preorder, inorder, and postorder.

Algorithms for traversing ordered rooted trees in preorder, inorder, or postorder are most easily expressed recursively.

ALGORITHM 1 Preorder Traversal.

procedure *preorder*(T: ordered rooted tree)
r := root of T
list r
for each child c of r from left to right
    $T(c)$:= subtree with c as its root
    preorder($T(c)$)

ALGORITHM 2 Inorder Traversal.

procedure *inorder*(T: ordered rooted tree)
$r :=$ root of T
if r is a leaf **then** list r
else
  $l :=$ first child of r from left to right
  $T(l) :=$ subtree with l as its root
  inorder($T(l)$)
  list r
  for each child c of r except for l from left to right
    $T(c) :=$ subtree with c as its root
    inorder($T(c)$)

ALGORITHM 3 Postorder Traversal.

procedure *postorder*(T: ordered rooted tree)
$r :=$ root of T
for each child c of r from left to right
  $T(c) :=$ subtree with c as its root
  postorder($T(c)$)
list r

Both the preorder traversal and the postorder traversal encode the structure of an ordered rooted tree when the number of children of each vertex is specified. That is, an ordered rooted tree is uniquely determined when we specify a list of vertices generated by a preorder traversal or by a postorder traversal of the tree, together with the number of children of each vertex (see Exercises 26 and 27). In particular, both a preorder traversal and a postorder traversal encode the structure of a full ordered m-ary tree. However, when the number of children of vertices is not specified, neither a preorder traversal nor a postorder traversal encodes the structure of an ordered rooted tree (see Exercises 28 and 29).

Uses of Inorder, Preorder, and Postorder Traversals Tree traversals have many applications and they play a key role in the implementation of many algorithms. Binary ordered trees can be used to represent well-formed expressions of objects and operations. Preorder, postorder, and inorder traversals of these trees yield the prefix, postfix, and infix notations of expressions, which can then be used in applications. Before we turn our attention to that use of tree traversals, we will provide useful guidance about the practical use of tree traversals.

When efficiency is not an issue, the vertices of a tree can be visited in any order, so long as each node is visited exactly once. However, for other applications we may need to visit vertices in an order that preserves a specific relationship. Also, when efficiency is important, the most efficient traversal method for that application should be used. A general principle for deciding the traversal to use is to explore the vertices of interest as soon as possible. Preorder traversal is the best choice for applications where internal vertices must be explored before leaves. Preorder traversals are also used to make copies of binary search trees.

An interesting digression is to note that preorder traversal has ancient roots. According to Knuth [Kn98], when a king, duke, or earl dies, his title passes to his first son, then to the descendants of this son; when none of these are alive, it passes to the second son, and then his descendants, and so on. (In more modern times, daughters have also been included in this order.) So, a preorder traversal of the vertices of the relevant family tree produces the order of succession to the throne, once deceased members are removed.

Postorder traversal is the best choice for applications where leaves need to be explored before internal vertices. Postorder explores leaves before internal vertices, so it is the best choice for deleting a tree because the vertices below the root of a subtree can be removed before the root of the subtree. Topological sorting is an example of an algorithm that can be efficiently implemented using postorder traversal. An inorder traversal of a binary search tree, discussed in Section 11.2, visits the vertices in ascending order of their key values. Such a traversal creates a sorted list of the data in a binary tree.

11.3.4 Infix, Prefix, and Postfix Notation

We can represent complicated expressions, such as compound propositions, combinations of sets, and arithmetic expressions using ordered rooted trees. For instance, consider the representation of an arithmetic expression involving the operators + (addition), − (subtraction), $*$ (multiplication), / (division), and $\uparrow$ (exponentiation). We will use parentheses to indicate the order of the operations. An ordered rooted tree can be used to represent such expressions, where the internal vertices represent operations, and the leaves represent the variables or numbers. Each operation operates on its left and right subtrees (in that order).

EXAMPLE 5 What is the ordered rooted tree that represents the expression $((x + y) \uparrow 2) + ((x - 4)/3)$?

Solution: The binary tree for this expression can be built from the bottom up. First, a subtree for the expression $x + y$ is constructed. Then this is incorporated as part of the larger subtree representing $(x + y) \uparrow 2$. Also, a subtree for $x - 4$ is constructed, and then this is incorporated into a subtree representing $(x - 4)/3$. Finally the subtrees representing $(x + y) \uparrow 2$ and $(x - 4)/3$ are combined to form the ordered rooted tree representing $((x + y) \uparrow 2) + ((x - 4)/3)$. These steps are shown in Figure 10. ◀

An inorder traversal of the binary tree representing an expression produces the original expression with the elements and operations in the same order as they originally occurred, except for unary operations, which instead immediately follow their operands. For instance, inorder traversals of the binary trees in Figure 11, which represent the expressions $(x + y)/(x + 3)$, $(x + (y/x)) + 3$, and $x + (y/(x + 3))$, all lead to the infix expression $x + y/x + 3$. To make such expressions unambiguous it is necessary to include parentheses in the inorder traversal whenever we encounter an operation. The fully parenthesized expression obtained in this way is said to be in **infix form**.

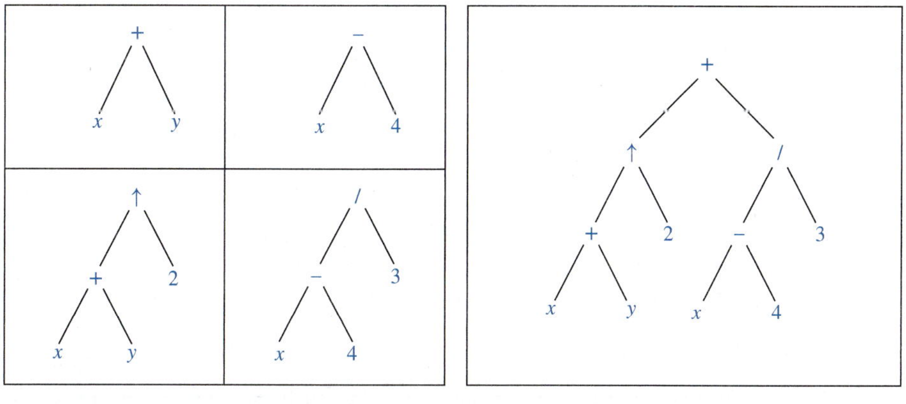

FIGURE 10 **A binary tree representing $((x + y) \uparrow 2) + ((x - 4)/3)$.**

We obtain the **prefix form** of an expression when we traverse its rooted tree in preorder. Expressions written in prefix form are said to be in **Polish notation**, which is named after the Polish logician Jan Łukasiewicz. An expression in prefix notation (where each operation has a specified number of operands), is unambiguous, so no parentheses are needed in such an expression. The verification of this is left as an exercise for the reader.

EXAMPLE 6 What is the prefix form for $((x + y) \uparrow 2) + ((x - 4)/3)$?

Solution: We obtain the prefix form for this expression by traversing the binary tree that represents it in preorder, shown in Figure 10. This produces $+ \uparrow + x\ y\ 2 \,/ - x\ 4\ 3$. ◀

In the prefix form of an expression, a binary operator, such as +, precedes its two operands. Hence, we can evaluate an expression in prefix form by working from right to left. When we encounter an operator, we perform the corresponding operation with the two operands immediately to the right of this operand. Also, whenever an operation is performed, we consider the result a new operand.

EXAMPLE 7 What is the value of the prefix expression $+ - * 2\ 3\ 5 / \uparrow 2\ 3\ 4$?

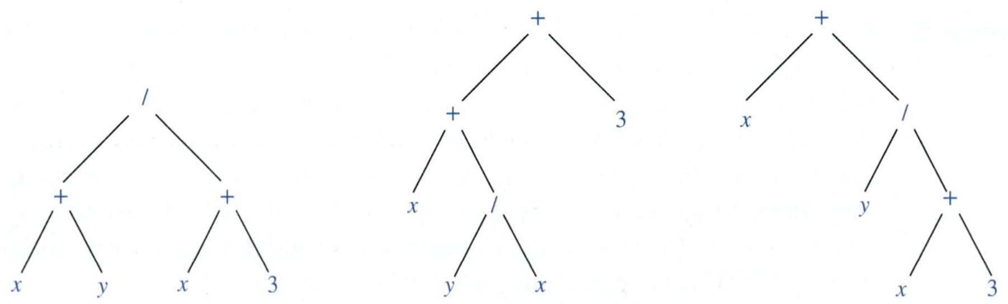

FIGURE 11 **Rooted trees representing $(x + y)/(x + 3)$, $(x + (y/x)) + 3$, and $x + (y/(x + 3))$.**

Solution: The steps used to evaluate this expression by working right to left, and performing operations using the operands on the right, are shown in Figure 12. The value of this expression is 3. ◀

Links

Reverse polish notation was first proposed in 1954 by Burks, Warren, and Wright.

We obtain the **postfix form** of an expression by traversing its binary tree in postorder. Expressions written in postfix form are said to be in **reverse Polish notation**. Expressions in reverse Polish notation are unambiguous, so parentheses are not needed. The verification of this is left to the reader. Reverse polish notation was extensively used in electronic calculators in the 1970s and 1980s.

EXAMPLE 8 What is the postfix form of the expression $((x + y) \uparrow 2) + ((x - 4)/3)$?

Solution: The postfix form of the expression is obtained by carrying out a postorder traversal of the binary tree for this expression, shown in Figure 10. This produces the postfix expression: $x\ y + 2 \uparrow x\ 4 - 3 / +$. ◀

In the postfix form of an expression, a binary operator follows its two operands. So, to evaluate an expression from its postfix form, work from left to right, carrying out operations

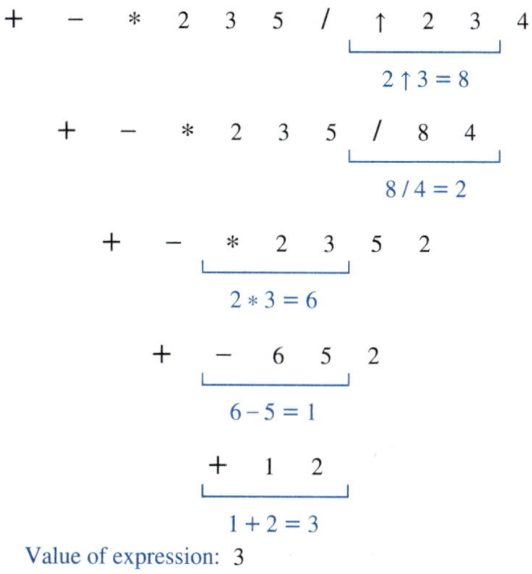

FIGURE 12 **Evaluating a prefix expression.**

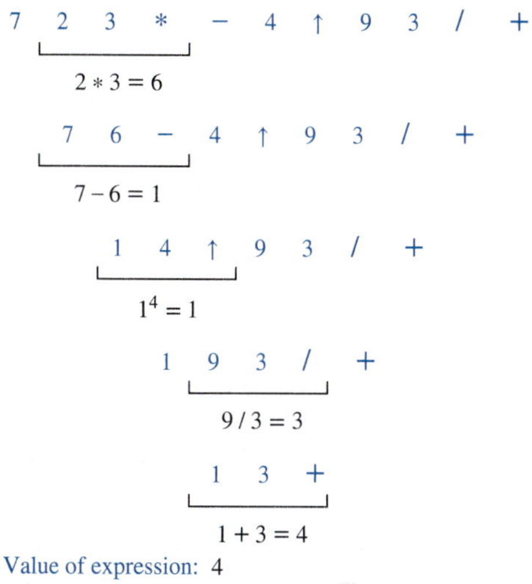

FIGURE 13 **Evaluating a postfix expression.**

whenever an operator follows two operands. After an operation is carried out, the result of this operation becomes a new operand.

EXAMPLE 9 What is the value of the postfix expression $7\ 2\ 3 * -4 \uparrow 9\ 3 / +$?

Solution: The steps used to evaluate this expression by starting at the left and carrying out operations when two operands are followed by an operator are shown in Figure 13. The value of this expression is 4. ◀

Rooted trees can be used to represent other types of expressions, such as those representing compound propositions and combinations of sets. In these examples unary operators, such as the negation of a proposition, occur. To represent such operators and their operands, a vertex representing the operator and a child of this vertex representing the operand are used.

EXAMPLE 10 Find the ordered rooted tree representing the compound proposition $(\neg(p \wedge q)) \leftrightarrow (\neg p \vee \neg q)$. Then use this rooted tree to find the prefix, postfix, and infix forms of this expression.

Links

UtCon Collection/Alamy Stock Photo

JAN ŁUKASIEWICZ (1878–1956) Jan Łukasiewicz was born into a Polish-speaking family in Lvov. At that time Lvov was part of Austria, but it is now in the Ukraine. His father was a captain in the Austrian army. Łukasiewicz became interested in mathematics while in high school. He studied mathematics and philosophy at the University of Lvov at both the undergraduate and graduate levels. After completing his doctoral work he became a lecturer there, and in 1911 he was appointed to a professorship. When the University of Warsaw was reopened as a Polish university in 1915, Łukasiewicz accepted an invitation to join the faculty. In 1919 he served as the Polish Minister of Education. He returned to the position of professor at Warsaw University, where he remained from 1920 to 1939, serving as rector of the university twice.

Łukasiewicz was one of the cofounders of the influential Warsaw School of Logic. He published his innovative text, *Elements of Mathematical Logic*, in 1928. With his influence, mathematical logic was made a required course for mathematics and science undergraduates in Poland. His lectures were considered excellent, even attracting students of the humanities.

Łukasiewicz and his wife experienced great suffering during World War II, which he documented in a posthumously published autobiography. After the war they lived in exile in Belgium. Fortunately, in 1949 he was offered a position at the Royal Irish Academy in Dublin.

Łukasiewicz worked on mathematical logic throughout his career. His work on a three-valued logic was an important contribution to the subject. Nevertheless, he is best known in the mathematical and computer science communities for his introduction of parenthesis-free notation, now called Polish notation.

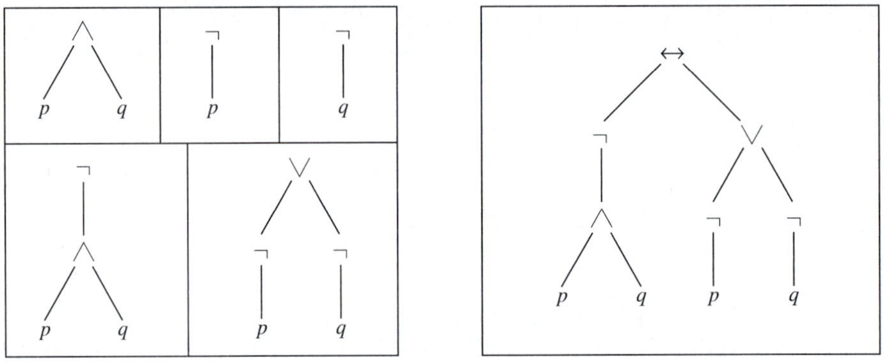

FIGURE 14 Constructing the rooted tree for a compound proposition.

Extra Examples

Solution: The rooted tree for this compound proposition is constructed from the bottom up. First, subtrees for $\neg p$ and $\neg q$ are formed (where $\neg$ is considered a unary operator). Also, a subtree for $p \wedge q$ is formed. Then subtrees for $\neg(p \wedge q)$ and $(\neg p) \vee (\neg q)$ are constructed. Finally, these two subtrees are used to form the final rooted tree. The steps of this procedure are shown in Figure 14.

The prefix, postfix, and infix forms of this expression are found by traversing this rooted tree in preorder, postorder, and inorder (including parentheses), respectively. These traversals give $\leftrightarrow \neg \wedge pq \vee \neg p \neg q$, $pq \wedge \neg p \neg q \neg \vee \leftrightarrow$, and $(\neg(p \wedge q)) \leftrightarrow ((\neg p) \vee (\neg q))$, respectively. ◀

Because prefix and postfix expressions are unambiguous and because they can be evaluated easily without scanning back and forth, they are used extensively in computer science. Such expressions are especially useful in the construction of compilers.

Exercises

In Exercises 1–3 construct the universal address system for the given ordered rooted tree. Then use this to order its vertices using the lexicographic order of their labels.

1.

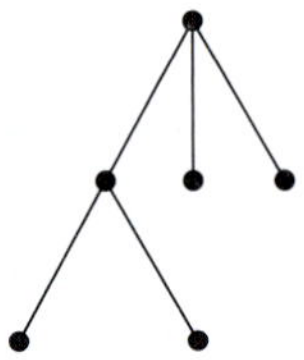

2.

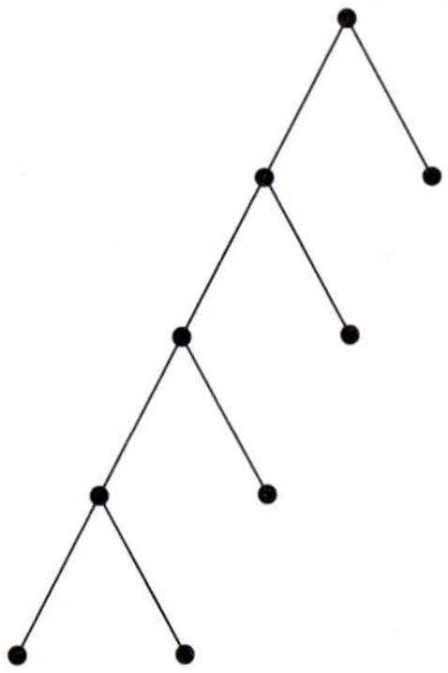

3.

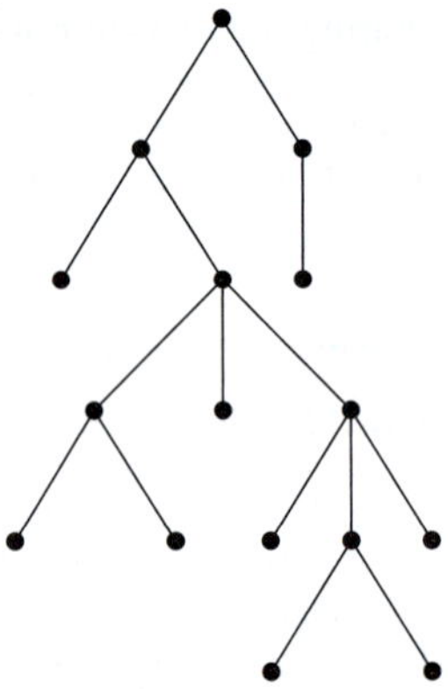

4. Suppose that the address of the vertex v in the ordered rooted tree T is 3.4.5.2.4.

a) At what level is v?
b) What is the address of the parent of v?
c) What is the least number of siblings v can have?
d) What is the smallest possible number of vertices in T if v has this address?
e) Find the other addresses that must occur.

5. Suppose that the vertex with the largest address in an ordered rooted tree T has address 2.3.4.3.1. Is it possible to determine the number of vertices in T?

6. Can the leaves of an ordered rooted tree have the following list of universal addresses? If so, construct such an ordered rooted tree.

a) 1.1.1, 1.1.2, 1.2, 2.1.1.1, 2.1.2, 2.1.3, 2.2, 3.1.1, 3.1.2.1, 3.1.2.2, 3.2

b) 1.1, 1.2.1, 1.2.2, 1.2.3, 2.1, 2.2.1, 2.3.1, 2.3.2, 2.4.2.1, 2.4.2.2, 3.1, 3.2.1, 3.2.2

c) 1.1, 1.2.1, 1.2.2, 1.2.2.1, 1.3, 1.4, 2, 3.1, 3.2, 4.1.1.1

In Exercises 7–9 determine the order in which a preorder traversal visits the vertices of the given ordered rooted tree.

7.

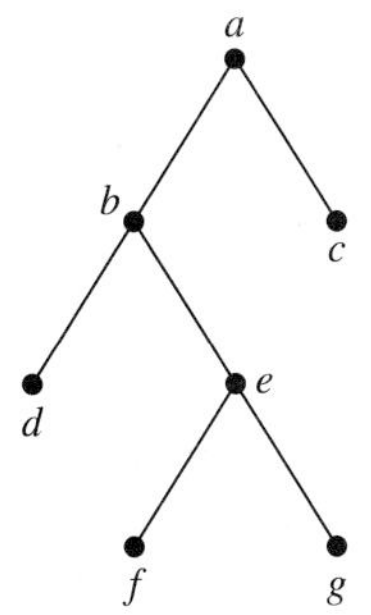

8.

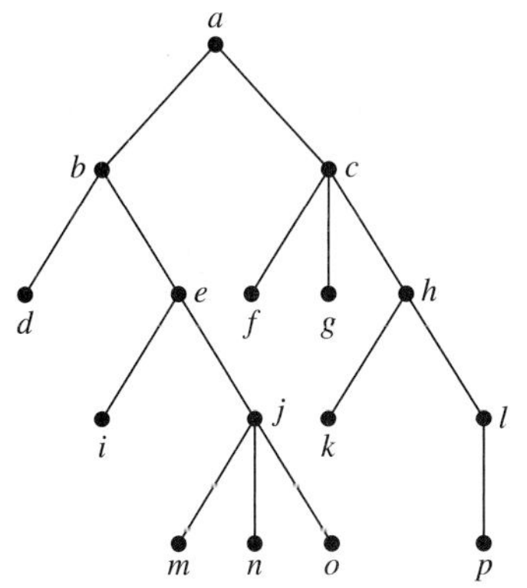

9.

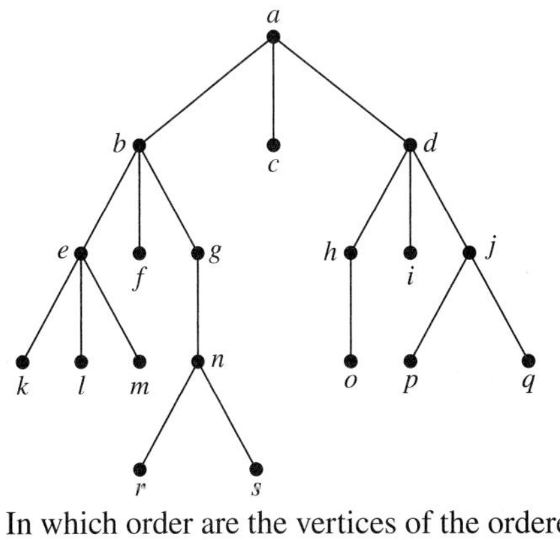

10. In which order are the vertices of the ordered rooted tree in Exercise 7 visited using an inorder traversal?

11. In which order are the vertices of the ordered rooted tree in Exercise 8 visited using an inorder traversal?

12. In which order are the vertices of the ordered rooted tree in Exercise 9 visited using an inorder traversal?

13. In which order are the vertices of the ordered rooted tree in Exercise 7 visited using a postorder traversal?

14. In which order are the vertices of the ordered rooted tree in Exercise 8 visited using a postorder traversal?

15. In which order are the vertices of the ordered rooted tree in Exercise 9 visited using a postorder traversal?

16. a) Represent the expression $((x+2) \uparrow 3) * (y - (3 + x)) - 5$ using a binary tree.

Write this expression in

b) prefix notation.

c) postfix notation.

d) infix notation.

17. a) Represent the expressions $(x + xy) + (x/y)$ and $x + ((xy + x)/y)$ using binary trees.

Write these expressions in

b) prefix notation.

c) postfix notation.

d) infix notation.

18. a) Represent the compound propositions $\neg(p \wedge q) \leftrightarrow (\neg p \vee \neg q)$ and $(\neg p \wedge (q \leftrightarrow \neg p)) \vee \neg q$ using ordered rooted trees.

Write these expressions in

b) prefix notation.

c) postfix notation.

d) infix notation.

19. a) Represent $(A \cap B) - (A \cup (B - A))$ using an ordered rooted tree.

Write this expression in

b) prefix notation.

c) postfix notation.

d) infix notation.

***20.** In how many ways can the string $\neg p \wedge q \leftrightarrow \neg p \vee \neg q$ be fully parenthesized to yield an infix expression?

***21.** In how many ways can the string $A \cap B - A \cap B - A$ be fully parenthesized to yield an infix expression?

22. Draw the ordered rooted tree corresponding to each of these arithmetic expressions written in prefix notation. Then write each expression using infix notation.

a) $+ * + - 5\ 3\ 2\ 1\ 4$

b) $\uparrow + 2\ 3 - 5\ 1$

c) $* / 9\ 3 + * 2\ 4 - 7\ 6$

23. What is the value of each of these prefix expressions?

a) $- * 2 / 8\ 4\ 3$

b) $\uparrow - * 3\ 3 * 4\ 2\ 5$

c) $+ - \uparrow 3\ 2 \uparrow 2\ 3 / 6 - 4\ 2$

d) $* + 3 + 3 \uparrow 3 + 3\ 3\ 3$

24. What is the value of each of these postfix expressions?

a) $5\ 2\ 1 - - 3\ 1\ 4 + + *$

b) $9\ 3 / 5 + 7\ 2 - *$

c) $3\ 2 * 2 \uparrow 5\ 3 - 8\ 4 / * -$

25. Construct the ordered rooted tree whose preorder traversal is *a, b, f, c, g, h, i, d, e, j, k, l*, where *a* has four children, *c* has three children, *j* has two children, *b* and *e* have one child each, and all other vertices are leaves.

***26.** Show that an ordered rooted tree is uniquely determined when a list of vertices generated by a preorder traversal of the tree and the number of children of each vertex are specified.

*27. Show that an ordered rooted tree is uniquely determined when a list of vertices generated by a postorder traversal of the tree and the number of children of each vertex are specified.

28. Show that preorder traversals of the two ordered rooted trees displayed below produce the same list of vertices. Note that this does not contradict the statement in Exercise 26, because the numbers of children of internal vertices in the two ordered rooted trees differ.

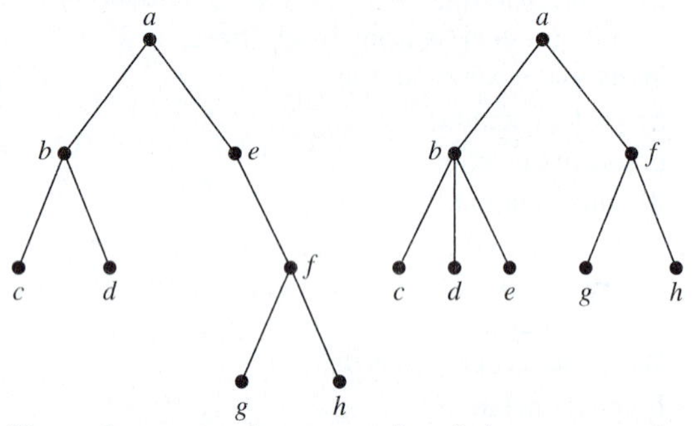

29. Show that postorder traversals of these two ordered rooted trees produce the same list of vertices. Note that this does not contradict the statement in Exercise 27, because the numbers of children of internal vertices in the two ordered rooted trees differ.

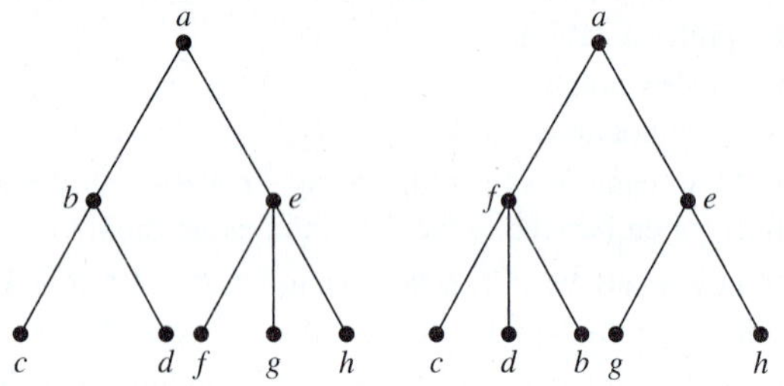

Well-formed formulae in prefix notation over a set of symbols and a set of binary operators are defined recursively by these rules:

(i) if x is a symbol, then x is a well-formed formula in prefix notation;

(ii) if X and Y are well-formed formulae and $*$ is an operator, then $* XY$ is a well-formed formula.

30. Which of these are well-formed formulae over the symbols $\{x, y, z\}$ and the set of binary operators $\{\times, +, \circ\}$?

a) $\times + + x y x$

b) $\circ x y \times x z$

c) $\times \circ x z \times \times x y$

d) $\times + \circ x x \circ x x x$

*31. Show that any well-formed formula in prefix notation over a set of symbols and a set of binary operators contains exactly one more symbol than the number of operators.

32. Give a definition of well-formed formulae in postfix notation over a set of symbols and a set of binary operators.

33. Give six examples of well-formed formulae with three or more operators in postfix notation over the set of symbols $\{x, y, z\}$ and the set of operators $\{+, \times, \circ\}$.

34. Extend the definition of well-formed formulae in prefix notation to sets of symbols and operators where the operators may not be binary.

11.4 Spanning Trees

11.4.1 Introduction

Consider the system of roads in Maine represented by the simple graph shown in Figure 1(a). The only way the roads can be kept open in the winter is by frequently plowing them. The highway department wants to plow the fewest roads so that there will always be cleared roads connecting any two towns. How can this be done?

At least five roads must be plowed to ensure that there is a path between any two towns. Figure 1(b) shows one such set of roads. Note that the subgraph representing these roads is a tree, because it is connected and contains six vertices and five edges.

This problem was solved with a connected subgraph with the minimum number of edges containing all vertices of the original simple graph. Such a graph must be a tree.

Definition 1 Let G be a simple graph. A *spanning tree* of G is a subgraph of G that is a tree containing every vertex of G.

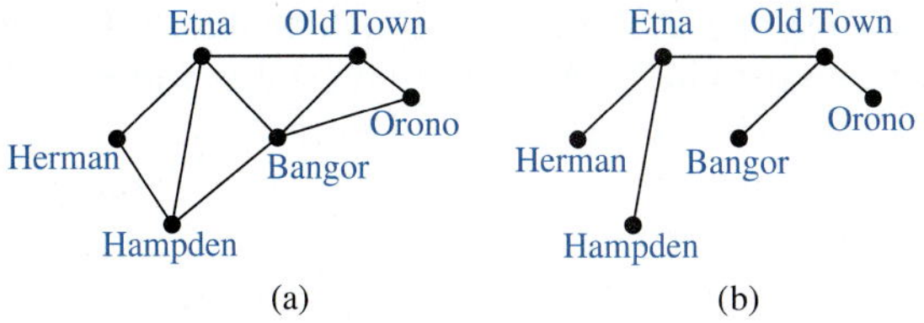

FIGURE 1 (a) A road system and (b) a set of roads to plow.

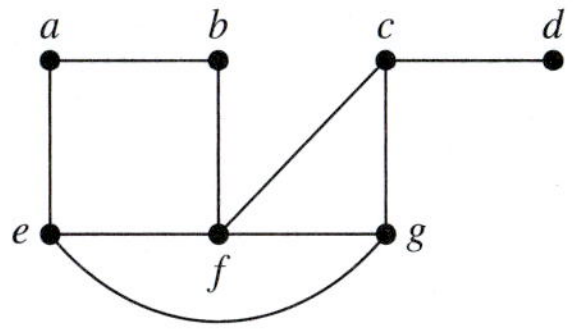

FIGURE 2 The simple graph *G*.

A simple graph with a spanning tree must be connected, because there is a path in the spanning tree between any two vertices. The converse is also true; that is, every connected simple graph has a spanning tree. We will give an example before proving this result.

EXAMPLE 1 Find a spanning tree of the simple graph G shown in Figure 2.

Solution: The graph G is connected, but it is not a tree because it contains simple circuits. Remove the edge $\{a, e\}$. This eliminates one simple circuit, and the resulting subgraph is still connected and still contains every vertex of G. Next remove the edge $\{e, f\}$ to eliminate a second simple circuit. Finally, remove edge $\{c, g\}$ to produce a simple graph with no simple circuits. This subgraph is a spanning tree, because it is a tree that contains every vertex of G. The sequence of edge removals used to produce the spanning tree is illustrated in Figure 3.

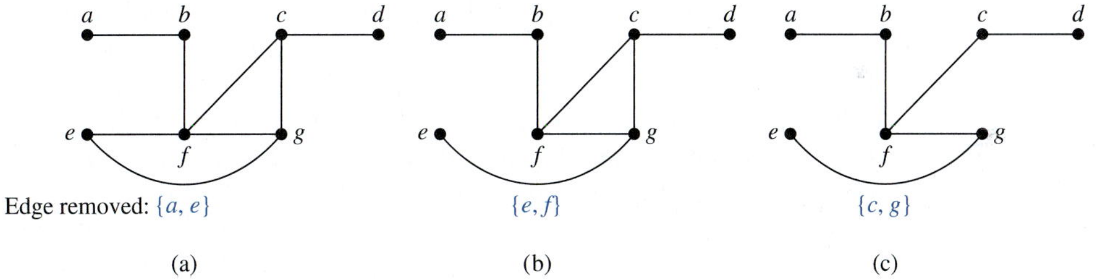

FIGURE 3 Producing a spanning tree for *G* by removing edges that form simple circuits.

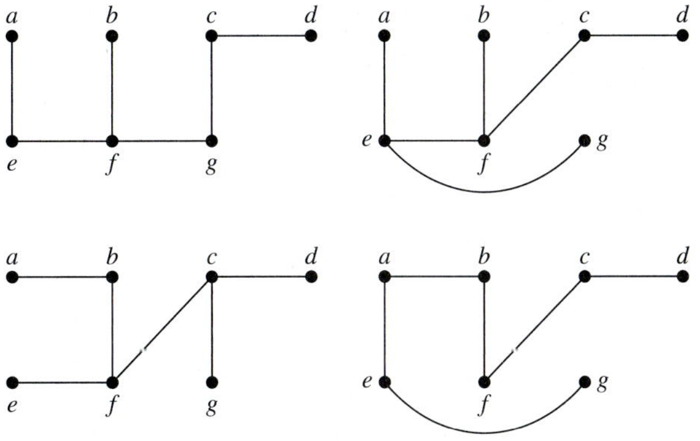

FIGURE 4 Spanning trees of *G*.

The tree shown in Figure 3 is not the only spanning tree of G. For instance, each of the trees shown in Figure 4 is a spanning tree of G. ◀

THEOREM 1 A simple graph is connected if and only if it has a spanning tree.

Proof: First, suppose that a simple graph G has a spanning tree T. Then T contains every vertex of G. Furthermore, there is a path in T between any two of its vertices. Because T is a subgraph of G, there is a path in G between any two of its vertices. Hence, G is connected.

Now suppose that G is connected. If G is not a tree, it must contain a simple circuit. Remove an edge from one of these simple circuits. The resulting subgraph has one fewer edge but still contains all the vertices of G and is connected. This subgraph is still connected because when two vertices are connected by a path containing the removed edge, they are connected by a path not containing this edge. We can construct such a path by inserting into the original path, at the point where the removed edge once was, the simple circuit with this edge removed. If this subgraph is not a tree, it has a simple circuit; so as before, remove an edge that is in a simple circuit. Repeat this process until no simple circuits remain. This is possible because there are only a finite number of edges in the graph. The process terminates when no simple circuits remain. A tree is produced because the graph stays connected as edges are removed. This tree is a spanning tree because it contains every vertex of G. ◁

Spanning trees are important in data networking, as Example 2 shows.

EXAMPLE 2 **IP Multicasting** Spanning trees play an important role in multicasting over Internet Protocol (IP) networks. To send data from a source computer to multiple receiving computers, each of which is a subnetwork, data could be sent separately to each computer. This type of networking, called unicasting, is inefficient, because many copies of the same data are transmitted over the network. To make the transmission of data to multiple receiving computers more efficient, IP multicasting is used. With IP multicasting, a computer sends a single copy of data over the network, and as data reaches intermediate routers, the data are forwarded to one or more other routers so that ultimately all receiving computers in their various subnetworks receive these data. (Routers are computers that are dedicated to forwarding IP datagrams between subnetworks in a network. In multicasting, routers use Class D addresses, each representing a session that receiving computers may join; see Example 17 in Section 6.1.)

Links

For data to reach receiving computers as quickly as possible, there should be no loops (which in graph theory terminology are circuits or cycles) in the path that data take through the network. That is, once data have reached a particular router, data should never return to this router. To avoid loops, the multicast routers use network algorithms to construct a spanning tree in the graph that has the multicast source, the routers, and the subnetworks containing receiving computers as vertices, with edges representing the links between computers and/or routers. The root of this spanning tree is the multicast source. The subnetworks containing receiving computers are leaves of the tree. (Note that subnetworks not containing receiving stations are not included in the graph.) This is illustrated in Figure 5. ◀

11.4.2 Depth-First Search

Links

The proof of Theorem 1 gives an algorithm for finding spanning trees by removing edges from simple circuits. This algorithm is inefficient, because it requires that simple circuits be identified. Instead of constructing spanning trees by removing edges, spanning trees can be built up by successively adding edges. Two algorithms based on this principle will be presented here.

Demo

We can build a spanning tree for a connected simple graph using **depth-first search**. We will form a rooted tree, and the spanning tree will be the underlying undirected graph of this rooted tree. Arbitrarily choose a vertex of the graph as the root. Form a path starting at this vertex by successively adding vertices and edges, where each new edge is incident with the last vertex in the path and a vertex not already in the path. Continue adding vertices and edges to this path as long as possible. If the path goes through all vertices of the graph, the tree consisting of this path is a spanning tree. However, if the path does not go through all vertices, more vertices and edges must be added. Move back to the next to last vertex in the path, and, if possible, form a new path

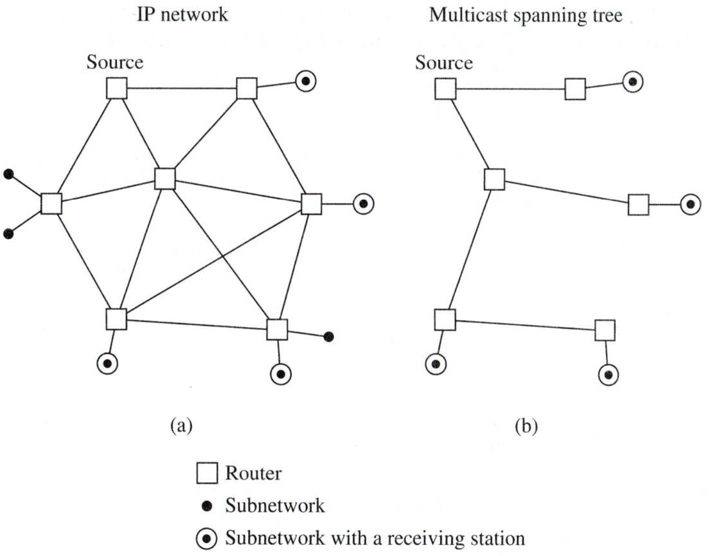

FIGURE 5 A multicast spanning tree.

starting at this vertex passing through vertices that were not already visited. If this cannot be done, move back another vertex in the path, that is, two vertices back in the path, and try again.

Repeat this procedure, beginning at the last vertex visited, moving back up the path one vertex at a time, forming new paths that are as long as possible until no more edges can be added. Because the graph has a finite number of edges and is connected, this process ends with the production of a spanning tree. Each vertex that ends a path at a stage of the algorithm will be a leaf in the rooted tree, and each vertex where a path is constructed starting at this vertex will be an internal vertex.

The reader should note the recursive nature of this procedure. Also, note that if the vertices in the graph are ordered, the choices of edges at each stage of the procedure are all determined when we always choose the first vertex in the ordering that is available. However, we will not always explicitly order the vertices of a graph.

Depth-first search is also called **backtracking**, because the algorithm returns to vertices previously visited to add paths. Example 3 illustrates backtracking.

EXAMPLE 3 Use depth-first search to find a spanning tree for the graph G shown in Figure 6.

Extra Examples

Solution: The steps used by depth-first search to produce a spanning tree of G are shown in Figure 7. We arbitrarily start with the vertex f. A path is built by successively adding edges incident with vertices not already in the path, as long as this is possible. This produces a path f, g, h, k, j (note that other paths could have been built). Next, backtrack to k. There is no path beginning at k containing vertices not already visited. So we backtrack to h. Form the

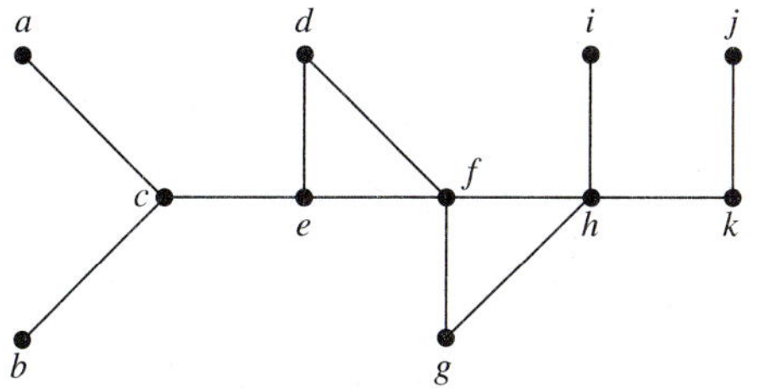

FIGURE 6 The graph *G*.

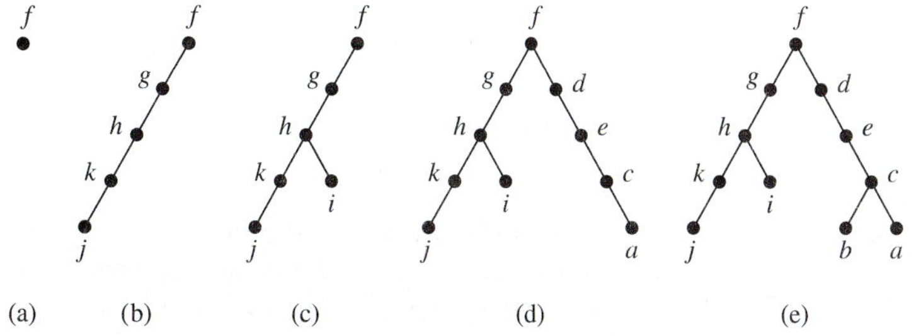

FIGURE 7 Depth-first search of *G*.

path h, i. Then backtrack to h, and then to f. From f build the path f, d, e, c, a. Then backtrack to c and form the path c, b. This produces the spanning tree. ◀

The edges selected by depth-first search of a graph are called **tree edges**. All other edges of the graph must connect a vertex to an ancestor or descendant of this vertex in the tree. These edges are called **back edges**. (Exercise 43 asks for a proof of this fact.)

EXAMPLE 4 In Figure 8 we highlight the tree edges found by depth-first search starting at vertex f by showing them with heavy colored lines. The back edges (e, f) and (f, h) are shown with thinner black lines. ◀

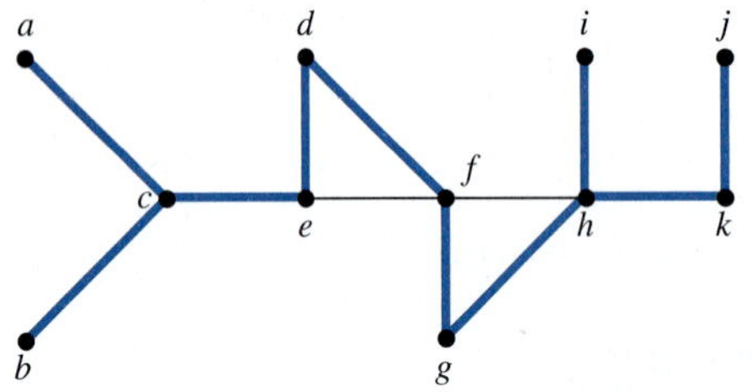

FIGURE 8 The tree edges and back edges of the depth-first search in Example 4.

We have explained how to find a spanning tree of a graph using depth-first search. However, our discussion so far has not brought out the recursive nature of depth-first search. To help make the recursive nature of the algorithm clear, we need a little terminology. We say that we *explore* from a vertex v when we carry out the steps of depth-first search beginning when v is added to the tree and ending when we have backtracked back to v for the last time. The key observation needed to understand the recursive nature of the algorithm is that when we add an edge connecting a vertex v to a vertex w, we finish exploring from w before we return to v to complete exploring from v.

In Algorithm 1 we construct the spanning tree of a graph G with vertices $v_1, \dots, v_n$ by first selecting the vertex v_1 to be the root. We initially set T to be the tree with just this one vertex. At each step we add a new vertex to the tree T together with an edge from a vertex already in T to this new vertex and we explore from this new vertex. Note that at the completion of the algorithm, T contains no simple circuits because no edge is ever added that connects two vertices in the tree. Moreover, T remains connected as it is built. (These last two observations can be easily proved via mathematical induction.) Because G is connected, every vertex in G is visited by the algorithm and is added to the tree (as the reader should verify). It follows that T is a spanning tree of G.

ALGORITHM 1 Depth-First Search.

procedure *DFS*(G: connected graph with vertices $v_1, v_2, \dots, v_n$)
T := tree consisting only of the vertex v_1
visit(v_1)

procedure *visit*(v: vertex of G)
for each vertex w adjacent to v and not yet in T
  add vertex w and edge $\{v, w\}$ to T
  visit(w)

We now analyze the computational complexity of the depth-first search algorithm. The key observation is that for each vertex v, the procedure $visit(v)$ is called when the vertex v is first encountered in the search and it is not called again. Assuming that the adjacency lists for G are available (see Section 10.3), no computations are required to find the vertices adjacent to v. As we follow the steps of the algorithm, we examine each edge at most twice to determine whether to add this edge and one of its endpoints to the tree. Consequently, the procedure *DFS* constructs a spanning tree using $O(e)$, or $O(n^2)$, steps where e and n are the number of edges and vertices in G, respectively. [Note that a step involves examining a vertex to see whether it is already in the spanning tree as it is being built and adding this vertex and the corresponding edge if the vertex is not already in the tree. We have also made use of the inequality $e \leq n(n-1)/2$, which holds for any simple graph.]

Depth-first search can be used as the basis for algorithms that solve many different problems. For example, it can be used to find paths and circuits in a graph, it can be used to determine the connected components of a graph, and it can be used to find the cut vertices of a connected graph. As we will see, depth-first search is the basis of backtracking techniques used to search for solutions of computationally difficult problems. (See [GrYe05], [Ma89], and [CoLeRiSt09] for a discussion of algorithms based on depth-first search.)

11.4.3 Breadth-First Search

Demo

Links

We can also produce a spanning tree of a simple graph by the use of **breadth-first search.** Again, a rooted tree will be constructed, and the underlying undirected graph of this rooted tree forms the spanning tree. Arbitrarily choose a root from the vertices of the graph. Then add all edges incident to this vertex. The new vertices added at this stage become the vertices at level 1 in the spanning tree. Arbitrarily order them. Next, for each vertex at level 1, visited in order, add each edge incident to this vertex to the tree as long as it does not produce a simple circuit. Arbitrarily order the children of each vertex at level 1. This produces the vertices at level 2 in the tree. Follow the same procedure until all the vertices in the tree have been added. The procedure ends because there are only a finite number of edges in the graph. A spanning tree is produced because we have produced a tree containing every vertex of the graph. An example of breadth-first search is given in Example 5.

EXAMPLE 5 Use breadth-first search to find a spanning tree for the graph shown in Figure 9.

Extra Examples

Solution: The steps of the breadth-first search procedure are shown in Figure 10. We choose the vertex e to be the root. Then we add edges incident with all vertices adjacent to e, so edges from e to b, d, f, and i are added. These four vertices are at level 1 in the tree. Next, add the edges from these vertices at level 1 to adjacent vertices not already in the tree. Hence, the edges from b to a and c are added, as are edges from d to h, from f to j and g, and from i to k. The new

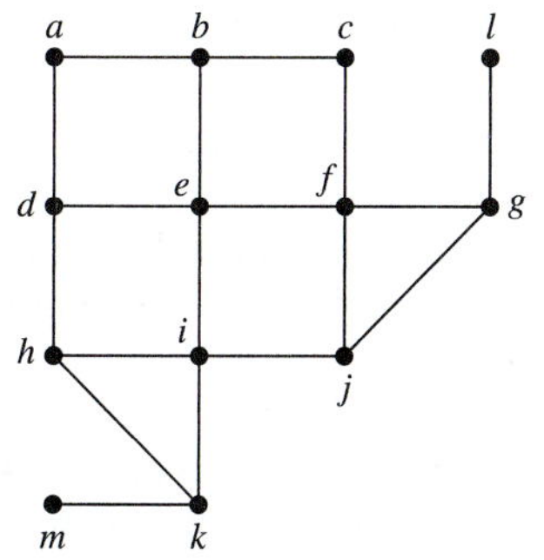

FIGURE 9 **A graph G.**

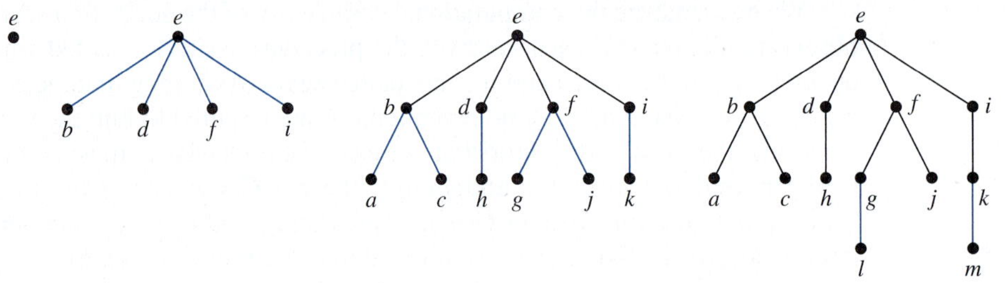

FIGURE 10 **Breadth-first search of *G*.**

vertices a, c, h, j, g, and k are at level 2. Next, add edges from these vertices to adjacent vertices not already in the graph. This adds edges from g to l and from k to m. ◀

We describe breadth-first search in pseudocode as Algorithm 2. In this algorithm, we assume the vertices of the connected graph G are ordered as $v_1, v_2, \ldots, v_n$. In the algorithm we use the term "process" to describe the procedure of adding new vertices, and corresponding edges, to the tree adjacent to the current vertex being processed as long as a simple circuit is not produced.

ALGORITHM 2 Breadth-First Search.

procedure *BFS*(G: connected graph with vertices $v_1, v_2, \ldots, v_n$)
T := tree consisting only of vertex v_1
L := empty list
put v_1 in the list L of unprocessed vertices
while L is not empty
  remove the first vertex, v, from L
  for each neighbor w of v
    if w is not in L and not in T **then**
      add w to the end of the list L
      add w and edge $\{v, w\}$ to T

We now analyze the computational complexity of breadth-first search. For each vertex v in the graph we examine all vertices adjacent to v and we add each vertex not yet visited to the tree T. Assuming we have the adjacency lists for the graph available, no computation is required to determine which vertices are adjacent to a given vertex. As in the analysis of the depth-first search algorithm, we see that we examine each edge at most twice to determine whether we should add this edge and its endpoint not already in the tree. It follows that the breadth-first search algorithm uses $O(e)$ or $O(n^2)$ steps.

Breadth-first search is one of the most useful algorithms in graph theory. In particular, it can serve as the basis for algorithms that solve a wide variety of problems. For example, algorithms that find the connected components of a graph, that determine whether a graph is bipartite, and that find the path with the fewest edges between two vertices in a graph can all be built using breadth-first search.

Breadth-First Search versus Depth-First Search We have introduced two widely used algorithms for constructing spanning trees of graphs, breadth-first search (BFS) and depth-first search (DFS). Both algorithms can be used to produce a spanning tree when given a connected graph. But why might it be better to use one rather than the other?

Although both BFS and DFS can be used to solve the same problems, there are theoretical and practical reasons for choosing between them. To solve a particular type of problem one of these search algorithms can be easier to apply, or can provide more insight, than the other. For some problems, BFS can be easier to use than DFS because BFS partitions the vertices of a graph into levels, which tell us how far away vertices are from the root. Also, edges connect vertices at the same level or at adjoining levels (see Exercise 34). On the other hand, there are many types of problems that are more naturally solved using DFS, such as those discussed in Examples 6–8. Generally, DFS is a better choice when we need to explore a graph by exploring deeper rather than by exploring the graph systematically level by level. The structure obtained when using DFS (see Exercise 51) can also be used when solving problems.

Both BFS and DFS are extensively used in practice. The choice for which to use often depends on implementation details, such as the data structures used. Time and space considerations are paramount, especially when the problems being solved involved huge graphs. Also, keep in mind that we often solve problems using graph searching without completing the task of finding a spanning tree. When we use BFS on a graph that is dense, we spend a lot of time and use a lot of space to explore the graph level by level. In such situations, it can be preferable to use DFS to quickly reach vertices far away from the root. For a sparse graph, however, it may be more efficient to explore a graph level by level.

11.4.4 Backtracking Applications

There are problems that can be solved only by performing an exhaustive search of all possible solutions. One way to search systematically for a solution is to use a decision tree, where each internal vertex represents a decision and each leaf a possible solution. To find a solution via backtracking, first make a sequence of decisions in an attempt to reach a solution as long as this is possible. The sequence of decisions can be represented by a path in the decision tree. Once it is known that no solution can result from any further sequence of decisions, backtrack to the parent of the current vertex and work toward a solution with another series of decisions, if this is possible. The procedure continues until a solution is found, or it is established that no solution exists. Examples 6 to 8 illustrate the usefulness of backtracking.

EXAMPLE 6 **Graph Colorings** How can backtracking be used to decide whether a graph can be colored using n colors?

Solution: We can solve this problem using backtracking in the following way. First pick some vertex a and assign it color 1. Then pick a second vertex b, and if b is not adjacent to a, assign it color 1. Otherwise, assign color 2 to b. Then go on to a third vertex c. Use color 1, if possible, for c. Otherwise use color 2, if this is possible. Only if neither color 1 nor color 2 can be used should color 3 be used. Continue this process as long as it is possible to assign one of the n colors to each additional vertex, always using the first allowable color in the list. If a vertex is reached that cannot be colored by any of the n colors, backtrack to the last assignment made and change the coloring of the last vertex colored, if possible, using the next allowable color in the list. If it is not possible to change this coloring, backtrack farther to previous assignments, one step back at a time, until it is possible to change a coloring of a vertex. Then continue assigning colors of additional vertices as long as possible. If a coloring using n colors exists, backtracking will produce it. (Unfortunately this procedure can be extremely inefficient.)

In particular, consider the problem of coloring the graph shown in Figure 11 with three colors. The tree shown in Figure 11 illustrates how backtracking can be used to construct a 3-coloring. In this procedure, red is used first, then blue, and finally green. This simple example can obviously be done without backtracking, but it is a good illustration of the technique.

In this tree, the initial path from the root, which represents the assignment of red to a, leads to a coloring with a red, b blue, c red, and d green. It is impossible to color e using any of the three colors when a, b, c, and d are colored in this way. So, backtrack to the parent of the

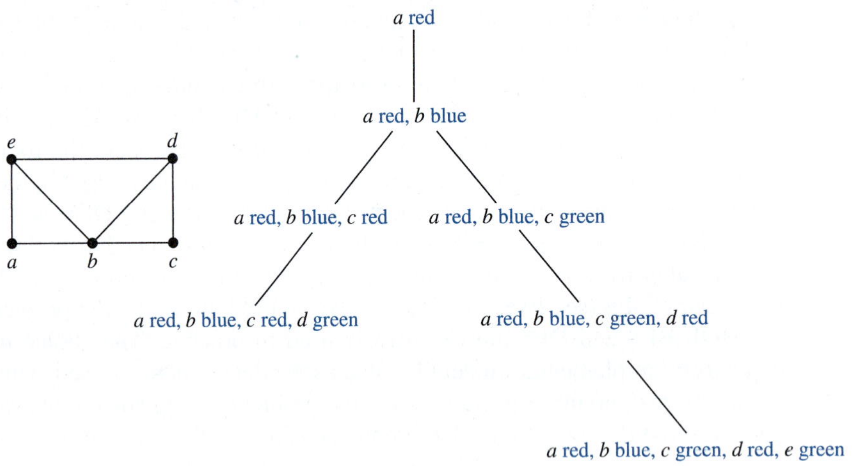

FIGURE 11 **Coloring a graph using backtracking.**

vertex representing this coloring. Because no other color can be used for d, backtrack one more level. Then change the color of c to green. We obtain a coloring of the graph by then assigning red to d and green to e. ◀

EXAMPLE 7 **The n-Queens Problem** The n-queens problem asks how n queens can be placed on an $n \times n$ chessboard so that no two queens can attack one another. How can backtracking be used to solve the n-queens problem?

Links

Solution: To solve this problem we must find n positions on an $n \times n$ chessboard so that no two of these positions are in the same row, same column, or in the same diagonal [a diagonal consists of all positions (i, j) with $i + j = m$ for some m, or $i - j = m$ for some m]. We will use backtracking to solve the n-queens problem. We start with an empty chessboard. At stage $k + 1$ we attempt putting an additional queen on the board in the $(k + 1)$st column, where there are already queens in the first k columns. We examine squares in the $(k + 1)$st column starting with the square in the first row, looking for a position to place this queen so that it is not in the same row or on the same diagonal as a queen already on the board. (We already know it is not in the same column.) If it is impossible to find a position to place the queen in the $(k + 1)$st column, backtrack to the placement of the queen in the kth column, and place this queen in the next allowable row in this column, if such a row exists. If no such row exists, backtrack further.

In particular, Figure 12 displays a backtracking solution to the four-queens problem. In this solution, we place a queen in the first row and column. Then we put a queen in the third row of the second column. However, this makes it impossible to place a queen in the third column. So we backtrack and put a queen in the fourth row of the second column. When we do this, we can place a queen in the second row of the third column. But there is no way to add a queen to the fourth column. This shows that no solution results when a queen is placed in the first row and column. We backtrack to the empty chessboard, and place a queen in the second row of the first column. This leads to a solution as shown in Figure 12. ◀

EXAMPLE 8 **Sums of Subsets** Consider this problem. Given a set of positive integers $x_1, x_2, \ldots, x_n$, find a subset of this set of integers that has M as its sum. How can backtracking be used to solve this problem?

Solution: We start with a sum with no terms. We build up the sum by successively adding terms. An integer in the sequence is included if the sum remains less than M when this integer is added

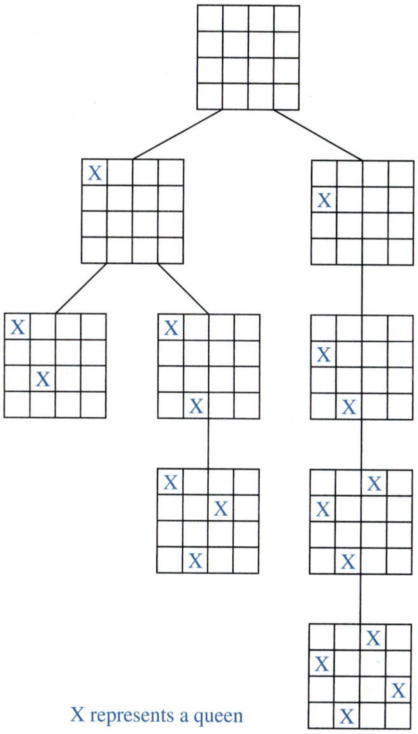

FIGURE 12 **A backtracking solution of the four-queens problem.**

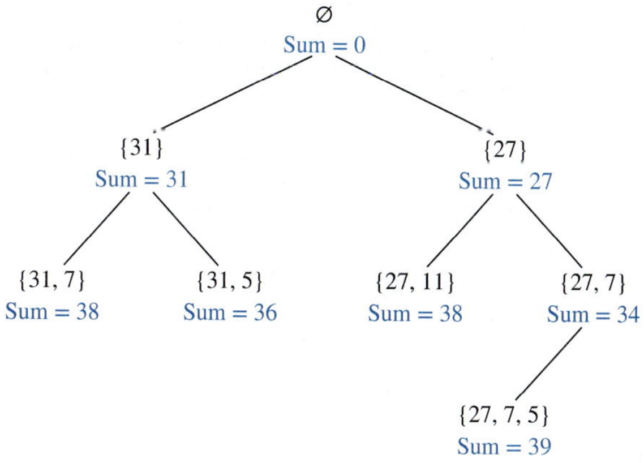

FIGURE 13 **Find a sum equal to 39 using backtracking.**

to the sum. If a sum is reached such that the addition of any term is greater than M, backtrack by dropping the last term of the sum.

Figure 13 displays a backtracking solution to the problem of finding a subset of $\{31, 27, 15, 11, 7, 5\}$ with the sum equal to 39. ◀

11.4.5 Depth-First Search in Directed Graphs

We can easily modify both depth-first search and breadth-first search so that they can run given a directed graph as input. However, the output will not necessarily be a spanning tree, but rather a spanning forest. In both algorithms we can add an edge only when it is directed away from

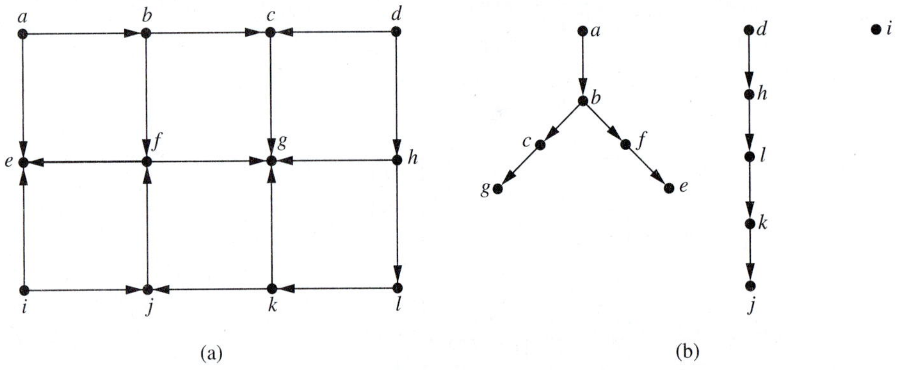

FIGURE 14 Depth-first search of a directed graph.

the vertex that is being visited and to a vertex not yet added. If at a stage of either algorithm we find that no edge exists starting at a vertex already added to one not yet added, the next vertex added by the algorithm becomes the root of a new tree in the spanning forest. This is illustrated in Example 9.

EXAMPLE 9 What is the output of depth-first search given the graph G shown in Figure 14(a) as input?

Solution: We begin the depth-first search at vertex a and add vertices b, c, and g and the corresponding edges where we are blocked. We backtrack to c but we are still blocked, and then backtrack to b, where we add vertices f and e and the corresponding edges. Backtracking takes us all the way back to a. We then start a new tree at d and add vertices h, l, k, and j and the corresponding edges. We backtrack to k, then l, then h, and back to d. Finally, we start a new tree at i, completing the depth-first search. The output is shown in Figure 14(b). ◀

Depth-first search in directed graphs is the basis of many algorithms (see [GrYe05], [Ma89], and [CoLeRiSt09]). It can be used to determine whether a directed graph has a circuit, it can be used to carry out a topological sort of a graph, and it can also be used to find the strongly connected components of a directed graph.

We conclude this section with an application of depth-first search and breadth-first search to search engines on the web.

EXAMPLE 10 **Web Crawlers** To index websites, search engines such as Google and Bing systematically explore the web starting at known sites. These search engines use programs called web crawlers (or spiders or bots) to visit websites and analyze their contents. Web crawlers use both depth-first searching and breadth-first searching to create indices. As described in Example 5 in Section 10.1, web pages and links between them can be modeled by a directed graph called the web graph. Web pages are represented by vertices and links are represented by directed edges. When DFS is used to explore the web, an initial web page is selected, a link is followed to a second web page (if there is such a link), a link on the second web page is followed to a third web page, if there is such a link, and so on, until a page with no new links is found. Backtracking is then used to examine links at the previous level to look for new links, and so on. (Because of practical limitations, web crawlers have limits to the depth they search in depth-first search.) Using BFS an initial web page is selected and a link on this page is followed to a second web page, then a second link on the initial page is followed (if it exists), and so on, until all links of the initial page have been followed. Then links on the pages one level down are followed, page by page, and so on.

Search engines have developed sophisticated strategies for web crawlers that take advantage of BFS and DFS. (Google's web crawler is called Googlebot.) They use BFS, starting on high quality web pages, called **seeds**, and explore all links on these pages. As the web crawler proceeds, it adds the urls of all links on the new pages it visits, adding them to the **crawl frontier**. These urls are ordered according to a particular policy, determining the order in which websites are explored.

When seeds have been selected well, Google's approach has been found to quickly reach high quality pages that have many links pointing to them. However, the quality of pages reached decreases as the web crawl continues. This reaches popular web pages well, but it does not reach many other useful, but less popular, web pages. If seed pages do not yield good results, DFS can be used to find candidates for high quality pages, starting at the seeds or other web pages. Also, DFS can be used to reach parts of the web not reached by BFS when it is restricted to a particular number of levels. ◀

Exercises

1. How many edges must be removed from a connected graph with n vertices and m edges to produce a spanning tree?

In Exercises 2–6 find a spanning tree for the graph shown by removing edges in simple circuits.

2.

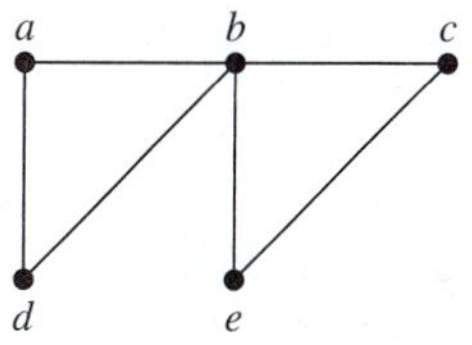

3.

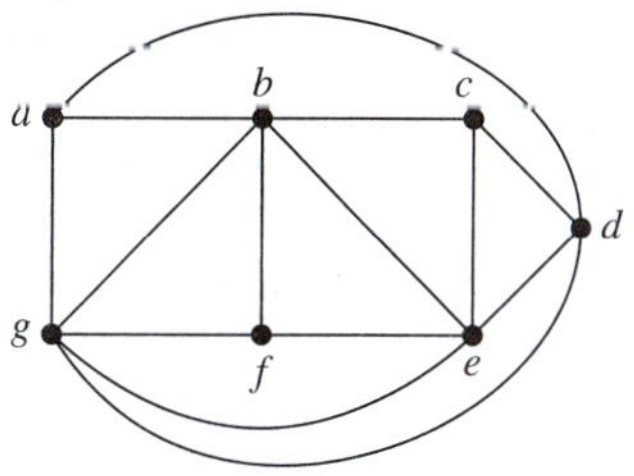

4.

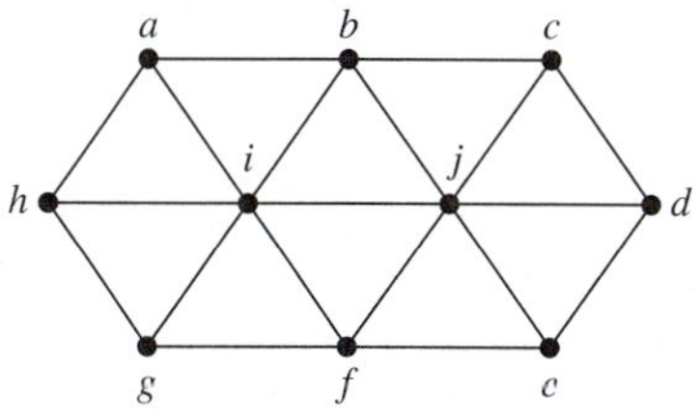

5.

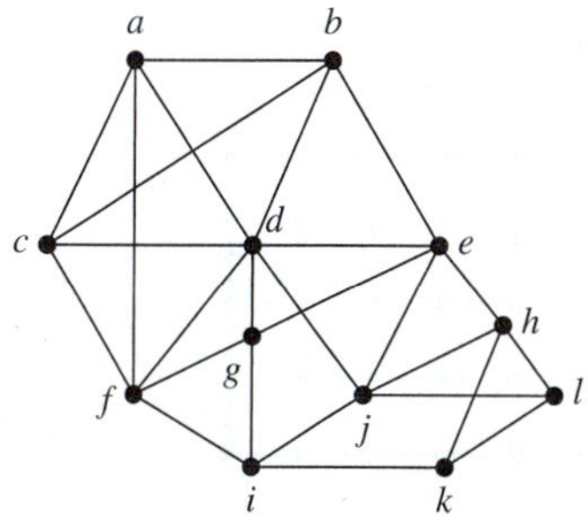

6.

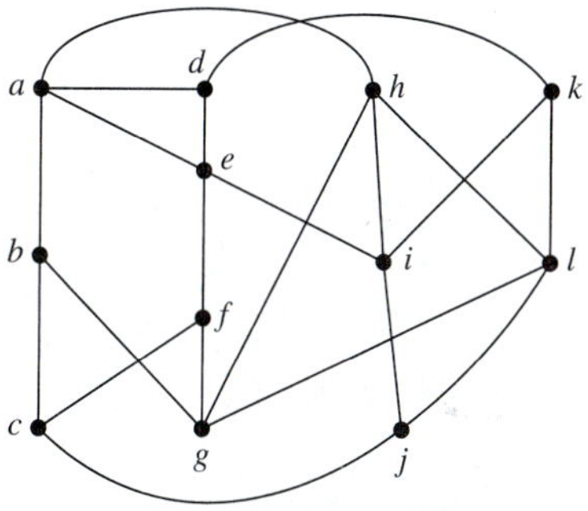

7. Find a spanning tree for each of these graphs.

a) K_5 **b)** $K_{4,4}$ **c)** $K_{1,6}$
d) Q_3 **e)** C_5 **f)** W_5

In Exercises 8–10 draw all the spanning trees of the given simple graphs.

8.

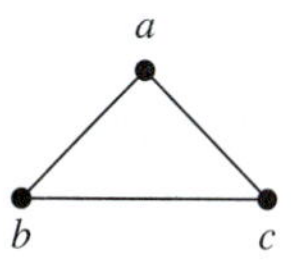

9.

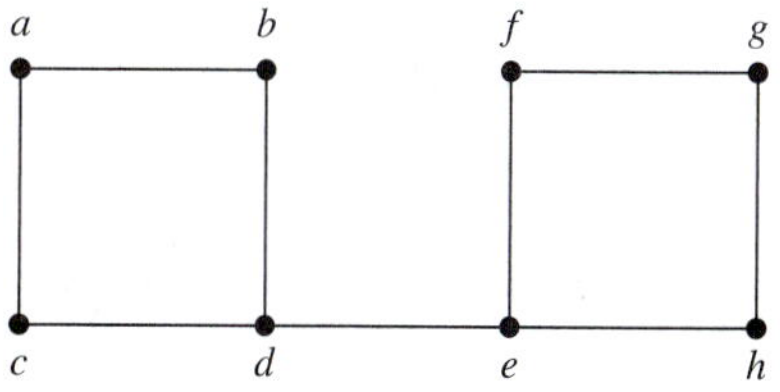

10.

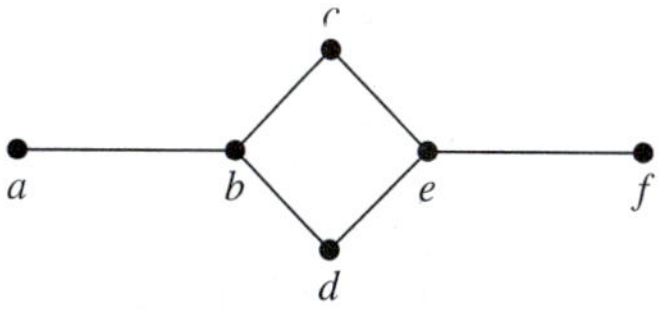

***11.** How many different spanning trees does each of these simple graphs have?

a) K_3 **b)** K_4 **c)** $K_{2,2}$ **d)** C_5

***12.** How many nonisomorphic spanning trees does each of these simple graphs have?

a) K_3 **b)** K_4 **c)** K_5

In Exercises 13–15 use depth-first search to produce a spanning tree for the given simple graph. Choose a as the root of this spanning tree and assume that the vertices are ordered alphabetically.

13.

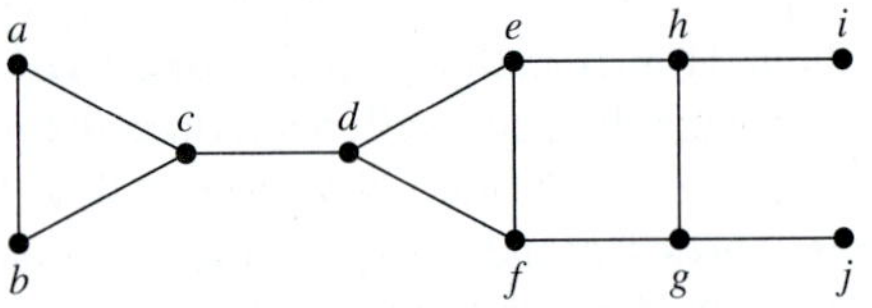

14.

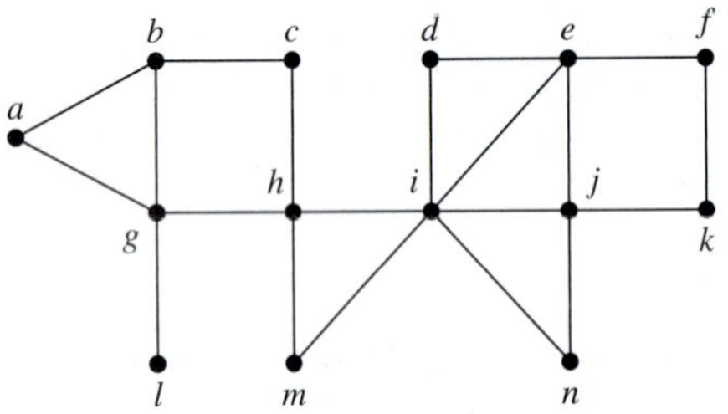

15.

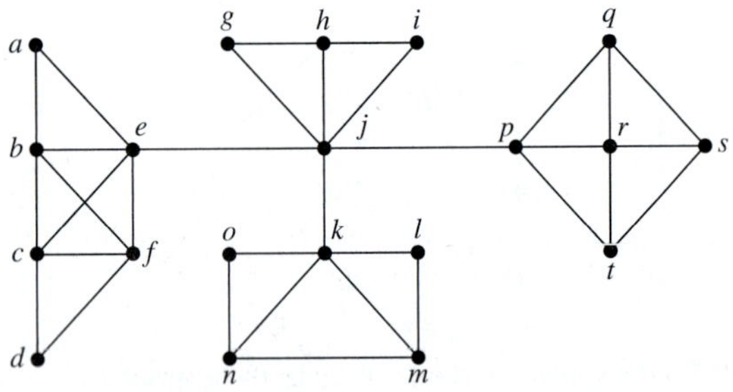

16. Use breadth-first search to produce a spanning tree for each of the simple graphs in Exercises 13–15. Choose a as the root of each spanning tree.

17. Use depth-first search to find a spanning tree of each of these graphs.

a) W_6 (see Example 7 of Section 10.2), starting at the vertex of degree 6

b) K_5

c) $K_{3,4}$, starting at a vertex of degree 3

d) Q_3

18. Use breadth-first search to find a spanning tree of each of the graphs in Exercise 17.

19. Describe the trees produced by breadth-first search and depth-first search of the wheel graph W_n, starting at the vertex of degree n, where n is an integer with $n \geq 3$. (See Example 7 of Section 10.2.) Justify your answers.

20. Describe the trees produced by breadth-first search and depth-first search of the complete graph K_n, where n is a positive integer. Justify your answers.

21. Describe the trees produced by breadth-first search and depth-first search of the complete bipartite graph $K_{m,n}$, starting at a vertex of degree m, where m and n are positive integers. Justify your answers.

22. Describe the tree produced by breadth-first search and depth-first search for the n-cube graph Q_n, where n is a positive integer.

23. Suppose that an airline must reduce its flight schedule to save money. If its original routes are as illustrated here, which flights can be discontinued to retain service between all pairs of cities (where it may be necessary to combine flights to fly from one city to another)?

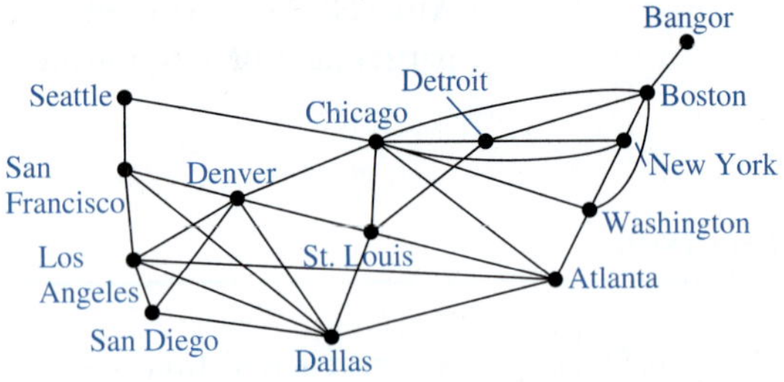

24. Explain how breadth-first search or depth-first search can be used to order the vertices of a connected graph.

***25.** Show that the length of the shortest path between vertices v and u in a connected simple graph equals the level number of u in the breadth-first spanning tree of G with root v.

26. Use backtracking to try to find a coloring of each of the graphs in Exercises 7–9 of Section 10.8 using three colors.

27. Use backtracking to solve the n-queens problem for these values of n.

a) $n = 3$ **b)** $n = 5$ **c)** $n = 6$

28. Use backtracking to find a subset, if it exists, of the set $\{27, 24, 19, 14, 11, 8\}$ with sum

a) 20. **b)** 41. **c)** 60.

29. Explain how backtracking can be used to find a Hamilton path or circuit in a graph.

30. a) Explain how backtracking can be used to find the way out of a maze, given a starting position and the exit position. Consider the maze divided into positions, where at each position the set of available moves includes one to four possibilities (up, down, right, left).

b) Find a path from the starting position marked by X to the exit in this maze.

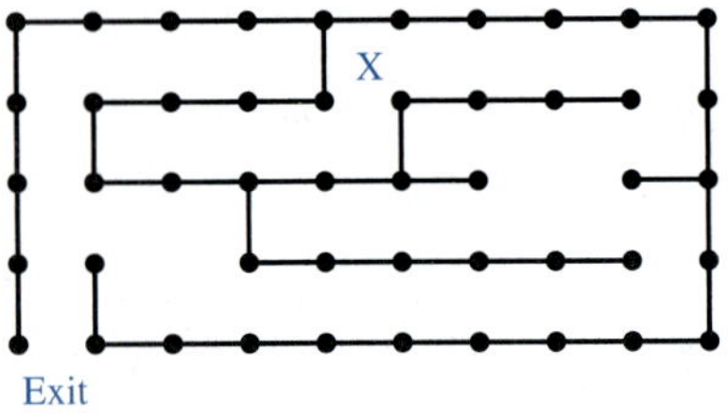

A **spanning forest** of a graph G is a forest that contains every vertex of G such that two vertices are in the same tree of the forest when there is a path in G between these two vertices.

31. Show that every finite simple graph has a spanning forest.

32. How many trees are in the spanning forest of a graph?

33. How many edges must be removed to produce the spanning forest of a graph with n vertices, m edges, and c connected components?

34. Let G be a connected graph. Show that if T is a spanning tree of G constructed using breadth-first search, then an edge of G not in T must connect vertices at the same level or at levels that differ by 1 in this spanning tree.

35. Explain how to use breadth-first search to find the length of a shortest path between two vertices in an undirected graph.

36. Devise an algorithm based on breadth-first search that determines whether a graph has a simple circuit, and if so, finds one.

37. Devise an algorithm based on breadth-first search for finding the connected components of a graph.

38. Explain how breadth-first search and how depth-first search can be used to determine whether a graph is bipartite.

39. Which connected simple graphs have exactly one spanning tree?

40. Devise an algorithm for constructing the spanning forest of a graph based on deleting edges that form simple circuits.

41. Devise an algorithm for constructing the spanning forest of a graph based on depth-first searching.

42. Devise an algorithm for constructing the spanning forest of a graph based on breadth-first searching.

43. Let G be a connected graph. Show that if T is a spanning tree of G constructed using depth-first search, then an edge of G not in T must be a back edge, that is, it must connect a vertex to one of its ancestors or one of its descendants in T.

44. When must an edge of a connected simple graph be in every spanning tree for this graph?

45. For which graphs do depth-first search and breadth-first search produce identical spanning trees no matter which vertex is selected as the root of the tree? Justify your answer.

46. Use Exercise 43 to prove that if G is a connected, simple graph with n vertices and G does not contain a simple path of length k then it contains at most $(k-1)n$ edges.

47. Use mathematical induction to prove that breadth-first search visits vertices in order of their level in the resulting spanning tree.

48. Use pseudocode to describe a variation of depth-first search that assigns the integer n to the nth vertex visited in the search. Show that this numbering corresponds to the numbering of the vertices created by a preorder traversal of the spanning tree.

49. Use pseudocode to describe a variation of breadth-first search that assigns the integer m to the mth vertex visited in the search.

***50.** Suppose that G is a directed graph and T is a spanning tree constructed using breadth-first search. Show that every edge of G has endpoints that are at the same level or one level higher or lower.

51. Show that if G is a directed graph and T is a spanning tree constructed using depth-first search, then every edge not in the spanning tree is a **forward edge** connecting an ancestor to a descendant, a **back edge** connecting a descendant to an ancestor, or a **cross edge** connecting a vertex to a vertex in a previously visited subtree.

***52.** Describe a variation of depth-first search that assigns the smallest available positive integer to a vertex when the algorithm is totally finished with this vertex. Show that in this numbering, each vertex has a larger number than its children and that the children have increasing numbers from left to right.

Let T_1 and T_2 be spanning trees of a graph. The **distance** between T_1 and T_2 is the number of edges in T_1 and T_2 that are not common to T_1 and T_2.

53. Find the distance between each pair of spanning trees shown in Figures 3(c) and 4 of the graph G shown in Figure 2.

***54.** Suppose that T_1, T_2, and T_3 are spanning trees of the simple graph G. Show that the distance between T_1 and T_3 does not exceed the sum of the distance between T_1 and T_2 and the distance between T_2 and T_3.

****55.** Suppose that T_1 and T_2 are spanning trees of a simple graph G. Moreover, suppose that e_1 is an edge in T_1 that is not in T_2. Show that there is an edge e_2 in T_2 that is not in T_1 such that T_1 remains a spanning tree if e_1 is removed from it and e_2 is added to it, and T_2 remains a spanning tree if e_2 is removed from it and e_1 is added to it.

***56.** Show that it is possible to find a sequence of spanning trees leading from any spanning tree to any other by successively removing one edge and adding another.

A **rooted spanning tree** of a directed graph is a rooted tree containing edges of the graph such that every vertex of the graph is an endpoint of one of the edges in the tree.

57. For each of the directed graphs in Exercises 18–23 of Section 10.5 either find a rooted spanning tree of the graph or determine that no such tree exists.

***58.** Show that a connected directed graph in which each vertex has the same in-degree and out-degree has a rooted spanning tree. [*Hint:* Use an Euler circuit.]

***59.** Give an algorithm to build a rooted spanning tree for connected directed graphs in which each vertex has the same in-degree and out-degree.

***60.** Show that if G is a directed graph and T is a spanning tree constructed using depth-first search, then G contains a circuit if and only if G contains a back edge (see Exercise 51) relative to the spanning tree T.

***61.** Use Exercise 60 to construct an algorithm for determining whether a directed graph contains a circuit.

11.5 Minimum Spanning Trees

11.5.1 Introduction

Links

A company plans to build a communications network connecting its five computer centers. Any pair of these centers can be linked with a leased network connection. Which links should be made to ensure that there is a path between any two computer centers so that the total cost of the network is minimized? We can model this problem using the weighted graph shown in Figure 1, where vertices represent computer centers, edges represent possible leased network connections, and the weights on edges are the monthly lease rates of the network connections represented by the edges. We can solve this problem by finding a spanning tree so that the sum of the weights of the edges of the tree is minimized. Such a spanning tree is called a **minimum spanning tree**.

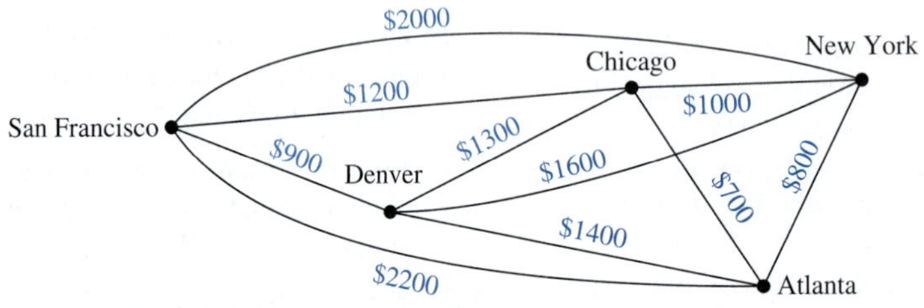

FIGURE 1 A weighted graph showing monthly lease costs for network connections in a computer network.

11.5.2 Algorithms for Minimum Spanning Trees

A wide variety of problems are solved by finding a spanning tree in a weighted graph such that the sum of the weights of the edges in the tree is a minimum.

Definition 1 A *minimum spanning tree* in a connected weighted graph is a spanning tree that has the smallest possible sum of weights of its edges.

Demo

We will present two algorithms for constructing minimum spanning trees. Both proceed by successively adding edges of smallest weight from those edges with a specified property that have not already been used. Both are greedy algorithms. Recall from Section 3.1 that a greedy algorithm is a procedure that makes an optimal choice at each of its steps. Optimizing at each step does not guarantee that the optimal overall solution is produced. However, the two algorithms presented in this section for constructing minimum spanning trees are greedy algorithms that do produce optimal solutions.

Links

The first algorithm that we will discuss was originally discovered by the Czech mathematician Vojtěch Jarník in 1930, who described it in a paper in an obscure Czech journal. The algorithm became well known when it was rediscovered in 1957 by Robert Prim. Because of this, it is known as **Prim's algorithm** (and sometimes as the **Prim-Jarník algorithm**). Begin by choosing any edge with smallest weight, putting it into the spanning tree. Successively add to the tree edges of minimum weight that are incident to a vertex already in the tree, never forming a simple circuit with those edges already in the tree. Stop when $n - 1$ edges have been added.

Later in this section, we will prove that this algorithm produces a minimum spanning tree for any connected weighted graph. Algorithm 1 gives a pseudocode description of Prim's algorithm.

ALGORITHM 1 Prim's Algorithm.

procedure *Prim*(G: weighted connected undirected graph with n vertices)
T := a minimum-weight edge
for $i := 1$ **to** $n - 2$
  e := an edge of minimum weight incident to a vertex in T and not forming a simple circuit in T if added to T
  $T := T$ with e added
return T {T is a minimum spanning tree of G}

Note that the choice of an edge to add at a stage of the algorithm is not determined when there is more than one edge with the same weight that satisfies the appropriate criteria. We need to order the edges to make the choices deterministic. We will not worry about this in the remainder of the section. Also note that there may be more than one minimum spanning tree for a given connected weighted simple graph. (See Exercise 9.) Examples 1 and 2 illustrate how Prim's algorithm is used.

EXAMPLE 1 Use Prim's algorithm to design a minimum-cost communications network connecting all the computers represented by the graph in Figure 1.

Solution: We solve this problem by finding a minimum spanning tree in the graph in Figure 1. Prim's algorithm is carried out by choosing an initial edge of minimum weight and successively adding edges of minimum weight that are incident to a vertex in the tree and that do not form simple circuits. The edges in color in Figure 2 show a minimum spanning tree produced by Prim's algorithm, with the choice made at each step displayed. ◀

EXAMPLE 2 Use Prim's algorithm to find a minimum spanning tree in the graph shown in Figure 3.

Solution: A minimum spanning tree constructed using Prim's algorithm is shown in Figure 4. The successive edges chosen are displayed. ◀

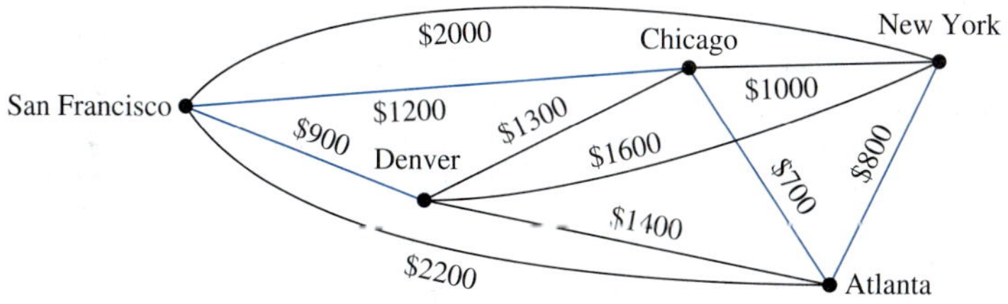

Choice	Edge	Cost
1	{Chicago, Atlanta}	$ 700
2	{Atlanta, New York}	$ 800
3	{Chicago, San Francisco}	$1200
4	{San Francisco, Denver}	$ 900
	Total:	$3600

FIGURE 2 A minimum spanning tree for the weighted graph in Figure 1.

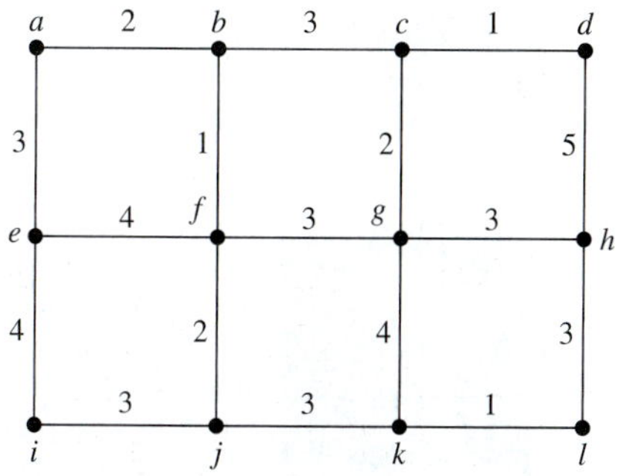

FIGURE 3 A weighted graph.

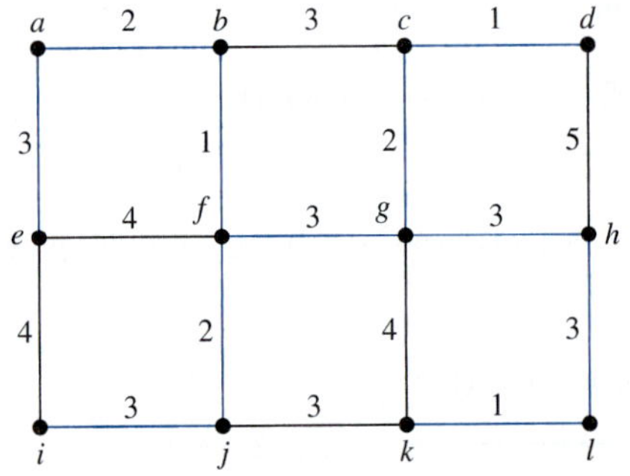

(a)

Choice	Edge	Weight
1	$\{b, f\}$	1
2	$\{a, b\}$	2
3	$\{f, j\}$	2
4	$\{a, e\}$	3
5	$\{i, j\}$	3
6	$\{f, g\}$	3
7	$\{c, g\}$	2
8	$\{c, d\}$	1
9	$\{g, h\}$	3
10	$\{h, l\}$	3
11	$\{k, l\}$	1
	Total:	24

(b)

FIGURE 4 A minimum spanning tree produced using Prim's algorithm.

Links

The second algorithm we will discuss was discovered by Joseph Kruskal in 1956, although the basic ideas it uses were described much earlier. To carry out **Kruskal's algorithm**, choose an edge in the graph with minimum weight.

Successively add edges with minimum weight that do not form a simple circuit with those edges already chosen. Stop after $n - 1$ edges have been selected.

The proof that Kruskal's algorithm produces a minimum spanning tree for every connected weighted graph is left as an exercise. Pseudocode for Kruskal's algorithm is given in Algorithm 2.

ALGORITHM 2 Kruskal's Algorithm.

procedure *Kruskal*(G: weighted connected undirected graph with n vertices)
T := empty graph
for $i := 1$ **to** $n - 1$
  e := any edge in G with smallest weight that does not form a simple circuit
   when added to T
  $T := T$ with e added
return T {T is a minimum spanning tree of G}

Links

Robert Clay Prim

ROBERT CLAY PRIM (BORN 1921) Robert Prim, born in Sweetwater, Texas, received his B.S. in electrical engineering in 1941 and his Ph.D. in mathematics from Princeton University in 1949. He was an engineer at the General Electric Company from 1941 until 1944, an engineer and mathematician at the United States Naval Ordnance Lab from 1944 until 1949, and a research associate at Princeton University from 1948 until 1949. Among the other positions he has held are director of mathematics and mechanics research at Bell Telephone Laboratories from 1958 until 1961 and vice president of research at Sandia Corporation. He is currently retired.

Joseph Kruskal

JOSEPH BERNARD KRUSKAL (1928–2010) Joseph Kruskal was born in New York City, where his father was a fur dealer and his mother promoted the art of origami on early television. Kruskal attended the University of Chicago and received his Ph.D. from Princeton University in 1954. He was an instructor in mathematics at Princeton and at the University of Wisconsin, and later he was an assistant professor at the University of Michigan. In 1959 he became a member of the technical staff at Bell Laboratories, where he worked until his retirement in the late 1990s. Kruskal discovered his algorithm for producing minimum spanning trees when he was a second-year graduate student. He was not sure his $2\frac{1}{2}$-page paper on this subject was worthy of publication, but was convinced by others to submit it. His research interests included statistical linguistics and psychometrics. Besides his work on minimum spanning trees, Kruskal is also known for contributions to multidimensional scaling. It is noteworthy that Joseph Kruskal's two brothers, Martin and William, also were well known mathematicians.

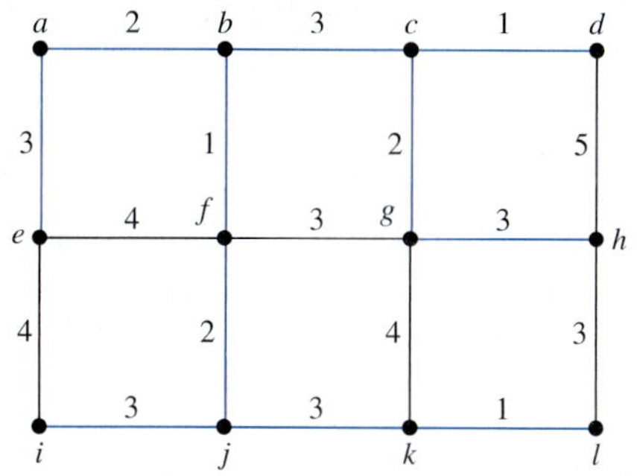

(a)

Choice	Edge	Weight
1	$\{c, d\}$	1
2	$\{k, l\}$	1
3	$\{b, f\}$	1
4	$\{c, g\}$	2
5	$\{a, b\}$	2
6	$\{f, j\}$	2
7	$\{b, c\}$	3
8	$\{j, k\}$	3
9	$\{g, h\}$	3
10	$\{i, j\}$	3
11	$\{a, e\}$	3
	Total:	24

(b)

FIGURE 5 **A minimum spanning tree produced by Kruskal's algorithm.**

The reader should note the difference between Prim's and Kruskal's algorithms. In Prim's algorithm edges of minimum weight that are incident to a vertex already in the tree, and not forming a circuit, are chosen; whereas in Kruskal's algorithm edges of minimum weight that are not necessarily incident to a vertex already in the tree, and that do not form a circuit, are chosen. Note that as in Prim's algorithm, if the edges are not ordered, there may be more than one choice for the edge to add at a stage of this procedure. Consequently, the edges need to be ordered for the procedure to be deterministic. Example 3 illustrates how Kruskal's algorithm is used.

EXAMPLE 3 Use Kruskal's algorithm to find a minimum spanning tree in the weighted graph shown in Figure 3.

Extra Examples

Solution: A minimum spanning tree and the choices of edges at each stage of Kruskal's algorithm are shown in Figure 5. ◀

We will now prove that Prim's algorithm produces a minimum spanning tree of a connected weighted graph.

Proof: Let G be a connected weighted graph. Suppose that the successive edges chosen by Prim's algorithm are $e_1, e_2, \ldots, e_{n-1}$. Let S be the tree with $e_1, e_2, \ldots, e_{n-1}$ as its edges, and let S_k be the tree with $e_1, e_2, \ldots, e_k$ as its edges. Let T be a minimum spanning tree of G containing the edges $e_1, e_2, \ldots, e_k$, where k is the maximum integer with the property that a minimum spanning tree exists containing the first k edges chosen by Prim's algorithm. The theorem follows if we can show that $S = T$.

Suppose that $S \neq T$, so that $k < n - 1$. Consequently, T contains $e_1, e_2, \ldots, e_k$, but not e_{k+1}. Consider the graph made up of T together with e_{k+1}. Because this graph is connected and has n edges, too many edges to be a tree, it must contain a simple circuit. This simple circuit must contain e_{k+1} because there was no simple circuit in T. Furthermore, there must be an

Links

HISTORICAL NOTE Joseph Kruskal and Robert Prim developed their algorithms for constructing minimum spanning trees in the mid-1950s. However, they were not the first people to discover such algorithms. For example, the work of the anthropologist Jan Czekanowski, in 1909, contains many of the ideas required to find minimum spanning trees. In 1926, Otakar Boruvka described methods for constructing minimum spanning trees in work relating to the construction of electric power networks, and as mentioned in the text what is now called Prim's algorithm was discovered by Vojtěch Jarník in 1930.

edge in the simple circuit that does not belong to S_{k+1} because S_{k+1} is a tree. By starting at an endpoint of e_{k+1} that is also an endpoint of one of the edges $e_1, \ldots, e_k$, and following the circuit until it reaches an edge not in S_{k+1}, we can find an edge e not in S_{k+1} that has an endpoint that is also an endpoint of one of the edges $e_1, e_2, \ldots, e_k$.

By deleting e from T and adding e_{k+1}, we obtain a tree T' with $n - 1$ edges (it is a tree because it has no simple circuits). Note that the tree T' contains $e_1, e_2, \ldots, e_k, e_{k+1}$. Furthermore, because e_{k+1} was chosen by Prim's algorithm at the $(k + 1)$st step, and e was also available at that step, the weight of e_{k+1} is less than or equal to the weight of e. From this observation, it follows that T' is also a minimum spanning tree, because the sum of the weights of its edges does not exceed the sum of the weights of the edges of T. This contradicts the choice of k as the maximum integer such that a minimum spanning tree exists containing $e_1, \ldots, e_k$. Hence, $k = n - 1$, and $S = T$. It follows that Prim's algorithm produces a minimum spanning tree. ◁

It can be shown (see [CoLeRiSt09]) that to find a minimum spanning tree of a graph with m edges and n vertices, Kruskal's algorithm can be carried out using $O(m \log m)$ operations and Prim's algorithm can be carried out using $O(m \log n)$ operations. Consequently, it is preferable to use Kruskal's algorithm for graphs that are **sparse**, that is, where m is very small compared to $C(n, 2) = n(n - 1)/2$, the total number of possible edges in an undirected graph with n vertices. Otherwise, there is little difference in the complexity of these two algorithms.

Exercises

1. The roads represented by this graph are all unpaved. The lengths of the roads between pairs of towns are represented by edge weights. Which roads should be paved so that there is a path of paved roads between each pair of towns so that a minimum road length is paved? (*Note:* These towns are in Nevada.)

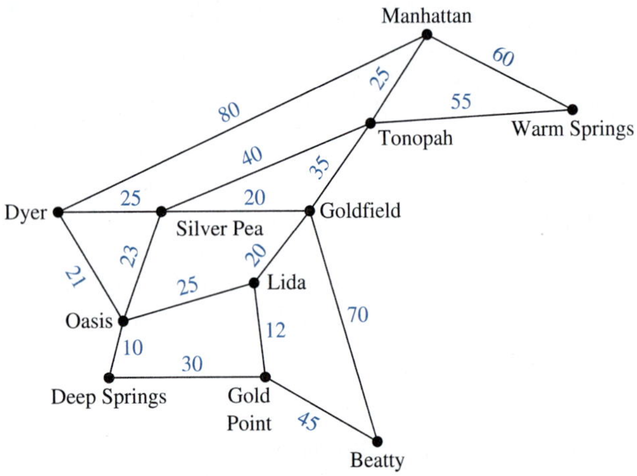

In Exercises 2–4 use Prim's algorithm to find a minimum spanning tree for the given weighted graph.

2.

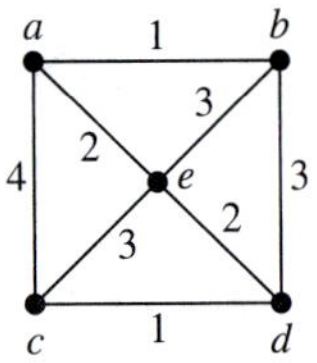

3.

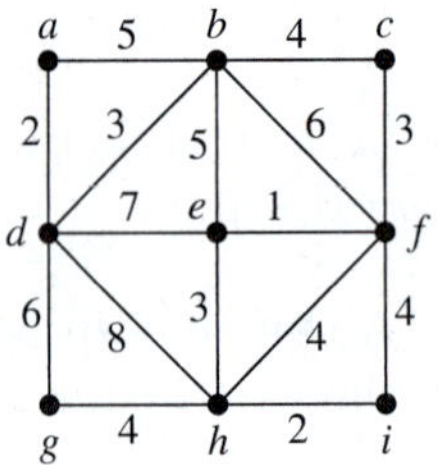

4.

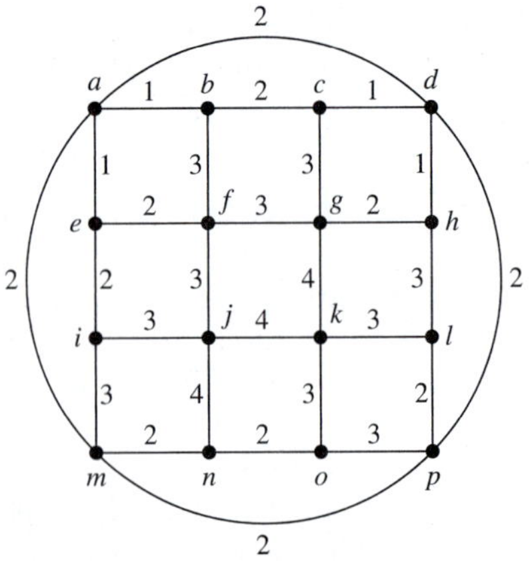

5. Use Kruskal's algorithm to design the communications network described at the beginning of the section.

6. Use Kruskal's algorithm to find a minimum spanning tree for the weighted graph in Exercise 2.

7. Use Kruskal's algorithm to find a minimum spanning tree for the weighted graph in Exercise 3.

8. Use Kruskal's algorithm to find a minimum spanning tree for the weighted graph in Exercise 4.

9. Find a connected weighted simple graph with the fewest edges possible that has more than one minimum spanning tree.

10. A **minimum spanning forest** in a weighted graph is a spanning forest with minimal weight. Explain how Prim's and Kruskal's algorithms can be adapted to construct minimum spanning forests.

A **maximum spanning tree** of a connected weighted undirected graph is a spanning tree with the largest possible weight.

11. Devise an algorithm similar to Prim's algorithm for constructing a maximum spanning tree of a connected weighted graph.

12. Devise an algorithm similar to Kruskal's algorithm for constructing a maximum spanning tree of a connected weighted graph.

13. Find a maximum spanning tree for the weighted graph in Exercise 2.

14. Find a maximum spanning tree for the weighted graph in Exercise 3.

15. Find a maximum spanning tree for the weighted graph in Exercise 4.

16. Find the second least expensive communications network connecting the five computer centers in the problem posed at the beginning of the section.

***17.** Devise an algorithm for finding the second shortest spanning tree in a connected weighted graph.

***18.** Show that an edge with smallest weight in a connected weighted graph must be part of any minimum spanning tree.

19. Show that there is a unique minimum spanning tree in a connected weighted graph if the weights of the edges are all different.

20. Suppose that the computer network connecting the cities in Figure 1 must contain a direct link between New York and Denver. What other links should be included so that there is a link between every two computer centers and the cost is minimized?

21. Find a spanning tree with minimal total weight containing the edges $\{e, i\}$ and $\{g, k\}$ in the weighted graph in Figure 3.

22. Describe an algorithm for finding a spanning tree with minimal weight containing a specified set of edges in a connected weighted undirected simple graph.

23. Express the algorithm devised in Exercise 22 in pseudocode.

Sollin's algorithm produces a minimum spanning tree from a connected weighted simple graph $G = (V, E)$ by successively adding groups of edges. Suppose that the vertices in V are ordered. This produces an ordering of the edges where $\{u_0, v_0\}$ precedes $\{u_1, v_1\}$ if u_0 precedes u_1 or if $u_0 = u_1$ and v_0 precedes v_1. The algorithm begins by simultaneously choosing the edge of least weight incident to each vertex. The first edge in the ordering is taken in the case of ties. This produces a graph with no simple circuits, that is, a forest of trees (Exercise 24 asks for a proof of this fact). Next, simultaneously choose for each tree in the forest the shortest edge between a vertex in this tree and a vertex in a different tree. Again the first edge in the ordering is chosen in the case of ties. (This produces a graph with no simple circuits containing fewer trees than were present before this step; see Exercise 24.) Continue the process of simultaneously adding edges connecting trees until $n - 1$ edges have been chosen. At this stage a minimum spanning tree has been constructed.

***24.** Show that the addition of edges at each stage of Sollin's algorithm produces a forest.

25. Use Sollin's algorithm to produce a minimum spanning tree for the weighted graph shown in

a) Figure 1.
b) Figure 3.

***26.** Express Sollin's algorithm in pseudocode.

****27.** Prove that Sollin's algorithm produces a minimum spanning tree in a connected undirected weighted graph.

***28.** Show that the first step of Sollin's algorithm produces a forest containing at least $\lceil n/2 \rceil$ edges when the input is an undirected graph with n vertices.

***29.** Show that if there are r trees in the forest at some intermediate step of Sollin's algorithm, then at least $\lceil r/2 \rceil$ edges are added by the next iteration of the algorithm.

***30.** Show that when given as input an undirected graph with n vertices, no more than $\lfloor n/2^k \rfloor$ trees remain after the first step of Sollin's algorithm has been carried out and the second step of the algorithm has been carried out $k - 1$ times.

***31.** Show that Sollin's algorithm requires at most $\log n$ iterations to produce a minimum spanning tree from a connected undirected weighted graph with n vertices.

32. Prove that Kruskal's algorithm produces minimum spanning trees.

33. Show that if G is a weighted graph with distinct edge weights, then for every simple circuit of G, the edge of maximum weight in this circuit does not belong to any minimum spanning tree of G.

When Kruskal invented the algorithm that finds minimum spanning trees by adding edges in order of increasing weight as long as they do not form a simple circuit, he also invented another algorithm sometimes called the **reverse-delete algorithm**. This algorithm proceeds by successively deleting edges of maximum weight from a connected graph as long as doing so does not disconnect the graph.

34. Express the reverse-delete algorithm in pseudocode.

35. Prove that the reverse-delete algorithm always produces a minimum spanning tree when given as input a weighted graph with distinct edge weights. [*Hint:* Use Exercise 33.]

Key Terms and Results

TERMS

tree: a connected undirected graph with no simple circuits

forest: an undirected graph with no simple circuits

rooted tree: a directed graph with a specified vertex, called the root, such that there is a unique path to every other vertex from this root

subtree: a subgraph of a tree that is also a tree

parent of v in a rooted tree: the vertex u such that (u, v) is an edge of the rooted tree

child of a vertex v in a rooted tree: any vertex with v as its parent

sibling of a vertex v in a rooted tree: a vertex with the same parent as v

ancestor of a vertex v in a rooted tree: any vertex on the path from the root to v

descendant of a vertex v in a rooted tree: any vertex that has v as an ancestor

internal vertex: a vertex that has children

leaf: a vertex with no children

level of a vertex: the length of the path from the root to this vertex

height of a tree: the largest level of the vertices of a tree

m-ary tree: a tree with the property that every internal vertex has no more than m children

full m-ary tree: a tree with the property that every internal vertex has exactly m children

binary tree: an m-ary tree with $m = 2$ (each child may be designated as a left or a right child of its parent)

ordered rooted tree: a rooted tree in which the children of each internal vertex are linearly ordered

balanced tree: a tree in which every leaf is at level h or $h - 1$, where h is the height of the tree

binary search tree: a binary tree in which the vertices are labeled with items so that a label of a vertex is greater than the labels of all vertices in the left subtree of this vertex and is less than the labels of all vertices in the right subtree of this vertex

decision tree: a rooted tree where each vertex represents a possible outcome of a decision and the leaves represent the possible solutions of a problem

game tree: a rooted tree where vertices represent the possible positions of a game as it progresses and edges represent legal moves between these positions

prefix code: a code that has the property that the code of a character is never a prefix of the code of another character

minmax strategy: the strategy where the first player and second player move to positions represented by a child with maximum and minimum value, respectively

value of a vertex in a game tree: for a leaf, the payoff to the first player when the game terminates in the position represented by this leaf; for an internal vertex, the maximum or minimum of the values of its children, for an internal vertex at an even or odd level, respectively

tree traversal: a listing of the vertices of a tree

preorder traversal: a listing of the vertices of an ordered rooted tree defined recursively—the root is listed, followed by the first subtree, followed by the other in the order they occur from left to right

inorder traversal: a listing of the vertices of an ordered rooted tree defined recursively—the first subtree is listed, followed by the root, followed by the other subtrees in the order they occur from left to right

postorder traversal: a listing of the vertices of an ordered rooted tree defined recursively—the subtrees are listed in the order they occur from left to right, followed by the root

infix notation: the form of an expression (including a full set of parentheses) obtained from an inorder traversal of the binary tree representing this expression

prefix (or Polish) notation: the form of an expression obtained from a preorder traversal of the tree representing this expression

postfix (or reverse Polish) notation: the form of an expression obtained from a postorder traversal of the tree representing this expression

spanning tree: a tree containing all vertices of a graph

minimum spanning tree: a spanning tree with smallest possible sum of weights of its edges

RESULTS

A graph is a tree if and only if there is a unique simple path between every pair of its vertices.

A tree with n vertices has $n - 1$ edges.

A full m-ary tree with i internal vertices has $mi + 1$ vertices.

The relationships among the numbers of vertices, leaves, and internal vertices in a full m-ary tree (see Theorem 4 in Section 11.1)

There are at most m^h leaves in an m-ary tree of height h.

If an m-ary tree has l leaves, its height h is at least $\lceil \log_m l \rceil$. If the tree is also full and balanced, then its height is $\lceil \log_m l \rceil$.

Huffman coding: a procedure for constructing an optimal binary code for a set of symbols, given the frequencies of these symbols

depth-first search, or backtracking: a procedure for constructing a spanning tree by adding edges that form a path until this is not possible, and then moving back up the path until a vertex is found where a new path can be formed

breadth-first search: a procedure for constructing a spanning tree that successively adds all edges incident to the last set of edges added, unless a simple circuit is formed

Prim's algorithm: a procedure for producing a minimum spanning tree in a weighted graph that successively adds edges with minimal weight among all edges incident to a vertex already in the tree so that no edge produces a simple circuit when it is added

Kruskal's algorithm: a procedure for producing a minimum spanning tree in a weighted graph that successively adds edges of least weight that are not already in the tree such that no edge produces a simple circuit when it is added

Review Questions

1. a) Define a tree. b) Define a forest.
2. Can there be two different simple paths between the vertices of a tree?
3. Give at least three examples of how trees are used in modeling.
4. a) Define a rooted tree and the root of such a tree.
   b) Define the parent of a vertex and a child of a vertex in a rooted tree.
   c) What are an internal vertex, a leaf, and a subtree in a rooted tree?
   d) Draw a rooted tree with at least 10 vertices, where the degree of each vertex does not exceed 3. Identify the root, the parent of each vertex, the children of each vertex, the internal vertices, and the leaves.
5. a) How many edges does a tree with n vertices have?
   b) What do you need to know to determine the number of edges in a forest with n vertices?
6. a) Define a full m-ary tree.
   b) How many vertices does a full m-ary tree have if it has i internal vertices? How many leaves does the tree have?
7. a) What is the height of a rooted tree?
   b) What is a balanced tree?
   c) How many leaves can an m-ary tree of height h have?
8. a) What is a binary search tree?
   b) Describe an algorithm for constructing a binary search tree.
   c) Form a binary search tree for the words *vireo, warbler, egret, grosbeak, nuthatch*, and *kingfisher*.
9. a) What is a prefix code?
   b) How can a prefix code be represented by a binary tree?
10. a) Define preorder, inorder, and postorder tree traversal.
    b) Give an example of preorder, postorder, and inorder traversal of a binary tree of your choice with at least 12 vertices.
11. a) Explain how to use preorder, inorder, and postorder traversals to find the prefix, infix, and postfix forms of an arithmetic expression.
    b) Draw the ordered rooted tree that represents $((x - 3) + ((x/4) + (x - y) \uparrow 3))$.
    c) Find the prefix and postfix forms of the expression in part (b).
12. Show that the number of comparisons used by a sorting algorithm to sort a list of n elements is at least $\lceil \log n! \rceil$.
13. a) Describe the Huffman coding algorithm for constructing an optimal code for a set of symbols, given the frequency of these symbols.
    b) Use Huffman coding to find an optimal code for these symbols and frequencies: A: 0.2, B: 0.1, C: 0.3, D: 0.4.
14. Draw the game tree for nim if the starting position consists of two piles with one and four stones, respectively. Who wins the game if both players follow an optimal strategy?
15. a) What is a spanning tree of a simple graph?
    b) Which simple graphs have spanning trees?
    c) Describe at least two different applications that require that a spanning tree of a simple graph be found.
16. a) Describe two different algorithms for finding a spanning tree in a simple graph.
    b) Illustrate how the two algorithms you described in part (a) can be used to find the spanning tree of a simple graph, using a graph of your choice with at least eight vertices and 15 edges.
17. a) Explain how backtracking can be used to determine whether a simple graph can be colored using n colors.
    b) Show, with an example, how backtracking can be used to show that a graph with a chromatic number equal to 4 cannot be colored with three colors, but can be colored with four colors.
18. a) What is a minimum spanning tree of a connected weighted graph?
    b) Describe at least two different applications that require that a minimum spanning tree of a connected weighted graph be found.
19. a) Describe Kruskal's algorithm and Prim's algorithm for finding minimum spanning trees.
    b) Illustrate how Kruskal's algorithm and Prim's algorithm are used to find a minimum spanning tree, using a weighted graph with at least eight vertices and 15 edges.

Supplementary Exercises

*1. Show that a simple graph is a tree if and only if it contains no simple circuits and the addition of an edge connecting two nonadjacent vertices produces a new graph that has exactly one simple circuit (where circuits that contain the same edges are not considered different).

*2. How many nonisomorphic rooted trees are there with six vertices?

3. Show that every tree with at least one edge must have at least two pendant vertices.

4. Show that a tree with n vertices that has $n - 1$ pendant vertices must be isomorphic to $K_{1,n-1}$.

5. What is the sum of the degrees of the vertices of a tree with n vertices?

*6. Suppose that $d_1, d_2, \ldots, d_n$ are n positive integers with sum $2n - 2$. Show that there is a tree that has n vertices such that the degrees of these vertices are $d_1, d_2, \ldots, d_n$.

7. Show that every tree is a planar graph.

8. Show that every tree is bipartite.

9. Show that every forest can be colored using two colors.

A **B-tree of degree *k*** is a rooted tree such that all its leaves are at the same level, its root has at least two and at most k children unless it is a leaf, and every internal vertex other than the root has at least $\lceil k/2 \rceil$, but no more than k, children. Computer files can be accessed efficiently when B-trees are used to represent them.

10. Draw three different B-trees of degree 3 with height 4.

*11. Give an upper bound and a lower bound for the number of leaves in a B-tree of degree k with height h.

*12. Give an upper bound and a lower bound for the height of a B-tree of degree k with n leaves.

The **binomial trees** B_i, $i = 0, 1, 2, \ldots$, are ordered rooted trees defined recursively:

Basis step: The binomial tree B_0 is the tree with a single vertex.

Recursive step: Let k be a nonnegative integer. To construct the binomial tree B_{k+1}, add a copy of B_k to a second copy of B_k by adding an edge that makes the root of the first copy of B_k the leftmost child of the root of the second copy of B_k.

13. Draw B_k for $k = 0, 1, 2, 3, 4$.

14. How many vertices does B_k have? Prove that your answer is correct.

15. Find the height of B_k. Prove that your answer is correct.

16. How many vertices are there in B_k at depth j, where $0 \le j \le k$? Justify your answer.

17. What is the degree of the root of B_k? Prove that your answer is correct.

18. Show that the vertex of largest degree in B_k is the root.

A rooted tree T is called an **S_k-tree** if it satisfies this recursive definition. It is an S_0-tree if it has one vertex. For $k > 0$, T is an S_k-tree if it can be built from two S_{k-1}-trees by making the root of one the root of the S_k-tree and making the root of the other the child of the root of the first S_{k-1}-tree.

19. Draw an S_k-tree for $k = 0, 1, 2, 3, 4$.

20. Show that an S_k-tree has 2^k vertices and a unique vertex at level k. This vertex at level k is called the **handle**.

*21. Suppose that T is an S_k-tree with handle v. Show that T can be obtained from disjoint trees $T_0, T_1, \ldots, T_{k-1}$, with roots $r_0, r_1, \ldots, r_{k-1}$, respectively, where v is not in any of these trees, where T_i is an S_i-tree for $i = 0, 1, \ldots, k-1$, by connecting v to r_0 and r_i to r_{i+1} for $i = 0, 1, \ldots, k-2$.

The listing of the vertices of an ordered rooted tree in **level order** begins with the root, followed by the vertices at level 1 from left to right, followed by the vertices at level 2 from left to right, and so on.

22. List the vertices of the ordered rooted trees in Figures 3 and 9 of Section 11.3 in level order.

23. Devise an algorithm for listing the vertices of an ordered rooted tree in level order.

*24. Devise an algorithm for determining if a set of universal addresses can be the addresses of the leaves of a rooted tree.

25. Devise an algorithm for constructing a rooted tree from the universal addresses of its leaves.

A **cut set** of a graph is a set of edges such that the removal of these edges produces a subgraph with more connected components than in the original graph, but no proper subset of this set of edges has this property.

26. Show that a cut set of a graph must have at least one edge in common with any spanning tree of this graph.

A **cactus** is a connected graph in which no edge is in more than one simple circuit not passing through any vertex other than its initial vertex more than once or its initial vertex other than at its terminal vertex (where two circuits that contain the same edges are not considered different).

27. Which of these graphs are cacti?

a)

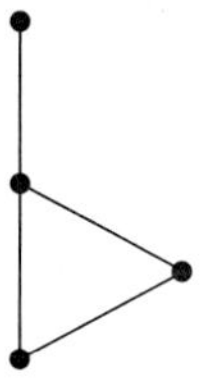

b)

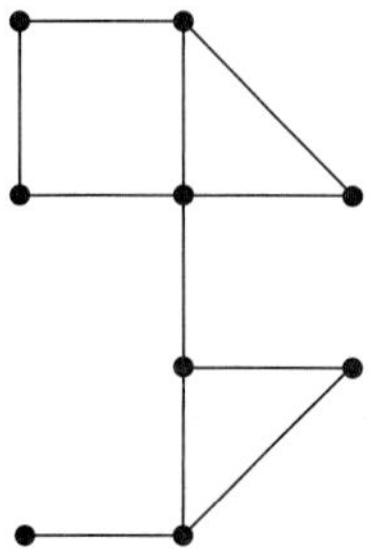

c)

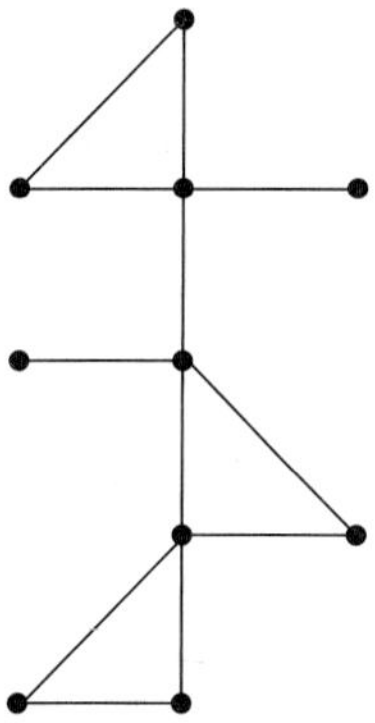

28. Is a tree necessarily a cactus?

29. Show that a cactus is formed if we add a circuit containing new edges beginning and ending at a vertex of a tree.

***30.** Show that if every circuit not passing through any vertex other than its initial vertex more than once in a connected graph contains an odd number of edges, then this graph must be a cactus.

A **degree-constrained spanning tree** of a simple graph G is a spanning tree with the property that the degree of a vertex in this tree cannot exceed some specified bound. Degree-constrained spanning trees are useful in models of transportation systems where the number of roads at an intersection is limited, models of communications networks where the number of links entering a node is limited, and so on.

In Exercises 31–33 find a degree-constrained spanning tree of the given graph where each vertex has degree less than or equal to 3, or show that such a spanning tree does not exist.

31.

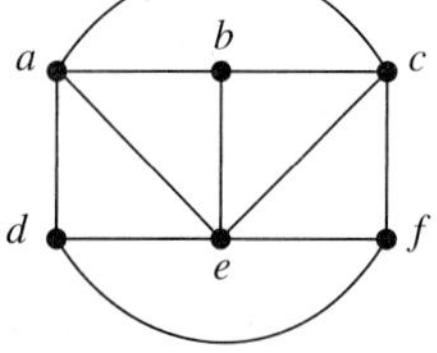

32.

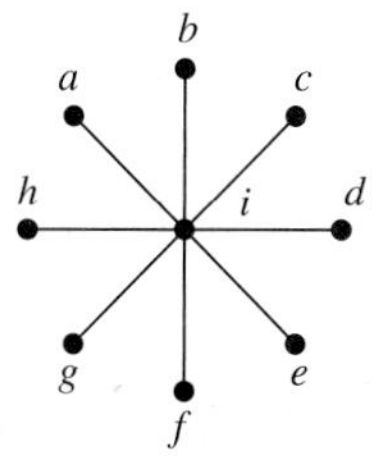

33.

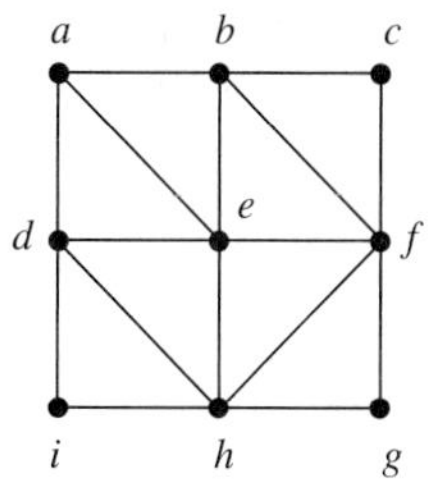

34. Show that a degree-constrained spanning tree of a simple graph in which each vertex has degree not exceeding 2 consists of a single Hamilton path in the graph.

35. A tree with n vertices is called **graceful** if its vertices can be labeled with the integers $1, 2, \ldots, n$ such that the absolute values of the difference of the labels of adjacent vertices are all different. Show that these trees are graceful.

a)

b)

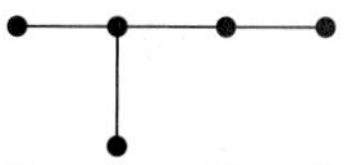

c)

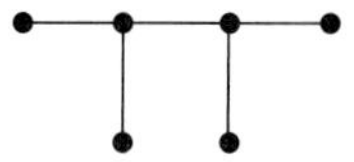

d)

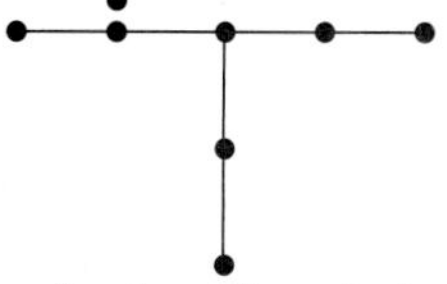

A **caterpillar** is a tree that contains a simple path such that every vertex not contained in this path is adjacent to a vertex in the path.

36. Which of the graphs in Exercise 35 are caterpillars?

37. How many nonisomorphic caterpillars are there with six vertices?

****38.** **a)** Prove or disprove that all trees whose edges form a single path are graceful.

b) Prove or disprove that all caterpillars are graceful.

39. Suppose that in a long bit string the frequency of occurrence of a 0 bit is 0.9 and the frequency of a 1 bit is 0.1 and bits occur independently.

a) Construct a Huffman code for the four blocks of two bits, 00, 01, 10, and 11. What is the average number of bits required to encode a bit string using this code?

b) Construct a Huffman code for the eight blocks of three bits. What is the average number of bits required to encode a bit string using this code?

40. Suppose that G is a directed graph with no circuits. Describe how depth-first search can be used to carry out a topological sort of the vertices of G.

***41.** Suppose that e is an edge in a weighted graph that is incident to a vertex v such that the weight of e does not exceed the weight of any other edge incident to v. Show that there exists a minimum spanning tree containing this edge.

42. Three couples arrive at the bank of a river. Each of the wives is jealous and does not trust her husband when he is with one of the other wives (and perhaps with other people), but not with her. How can six people cross to the other side of the river using a boat that can hold no more than two people so that no husband is alone with a woman other than his wife? Use a graph theory model.

*43. Show that if no two edges in a weighted graph have the same weight, then the edge with least weight incident to a vertex v is included in every minimum spanning tree.

44. Find a minimum spanning tree of each of these graphs where the degree of each vertex in the spanning tree does not exceed 2.

a) 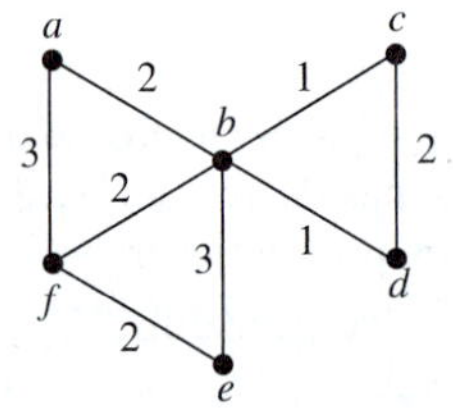

b) 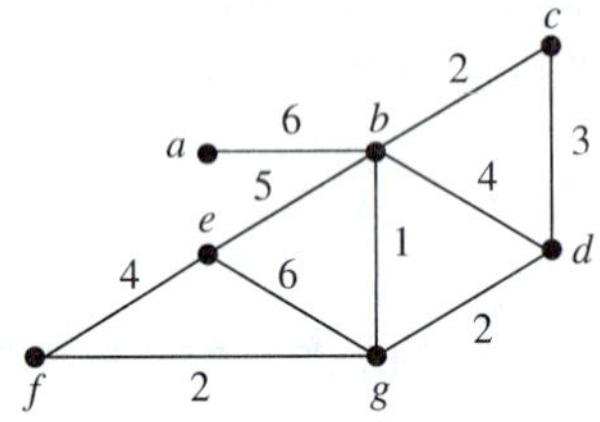

Let $G = (V, E)$ be a directed graph and let r be a vertex in G. An **arborescence** of G rooted at r is a subgraph $T = (V, F)$ of G such that the underlying undirected graph of T is a spanning tree of the underlying undirected graph of G and for every vertex $v \in V$ there is a path from r to v in T (with directions taken into account).

45. Show that a subgraph $T = (V, F)$ of the graph $G = (V, E)$ is an arborescence of G rooted at r if and only if T contains r, T has no simple circuits, and for every vertex $v \in V$ other than r, $\deg^-(v) = 1$ in T.

46. Show that a directed graph $G = (V, E)$ has an arborescence rooted at the vertex r if and only if for every vertex $v \in V$, there is a directed path from r to v.

47. In this exercise we will develop an algorithm to find the strong components of a directed graph $G = (V, E)$. Recall that a vertex $w \in V$ is **reachable** from a vertex $v \in V$ if there is a directed path from v to w.
   a) Explain how to use breadth-first search in the directed graph G to find all the vertices reachable from a vertex $v \in G$.
   b) Explain how to use breadth-first search in G^{conv} to find all the vertices from which a vertex $v \in G$ is reachable. (Recall that G^{conv} is the directed graph obtained from G by reversing the direction of all its edges.)
   c) Explain how to use parts (a) and (b) to construct an algorithm that finds the strong components of a directed graph G, and explain why your algorithm is correct.

Computer Projects

Write programs with these input and output.

1. Given the adjacency matrix of an undirected simple graph, determine whether the graph is a tree.
2. Given the adjacency matrix of a rooted tree and a vertex in the tree, find the parent, children, ancestors, descendants, and level of this vertex.
3. Given the list of edges of a rooted tree and a vertex in the tree, find the parent, children, ancestors, descendants, and level of this vertex.
4. Given a list of items, construct a binary search tree containing these items.
5. Given a binary search tree and an item, locate or add this item to the binary search tree.
6. Given the ordered list of edges of an ordered rooted tree, find the universal addresses of its vertices.
7. Given the ordered list of edges of an ordered rooted tree, list its vertices in preorder, inorder, and postorder.
8. Given an arithmetic expression in prefix form, find its value.
9. Given an arithmetic expression in postfix form, find its value.
10. Given the frequency of symbols, use Huffman coding to find an optimal code for these symbols.
11. Given an initial position in the game of nim, determine an optimal strategy for the first player.
12. Given the adjacency matrix of a connected undirected simple graph, find a spanning tree for this graph using depth-first search.
13. Given the adjacency matrix of a connected undirected simple graph, find a spanning tree for this graph using breadth-first search.
14. Given a set of positive integers and a positive integer N, use backtracking to find a subset of these integers that have N as their sum.

*15. Given the adjacency matrix of an undirected simple graph, use backtracking to color the graph with three colors, if this is possible.

*16. Given a positive integer n, solve the n-queens problem using backtracking.

17. Given the list of edges and their weights of a weighted undirected connected graph, use Prim's algorithm to find a minimum spanning tree of this graph.
18. Given the list of edges and their weights of a weighted undirected connected graph, use Kruskal's algorithm to find a minimum spanning tree of this graph.

Computations and Explorations

Use a computational program or programs you have written to do these exercises.

1. Display all trees with six vertices.
2. Display a full set of nonisomorphic trees with seven vertices.
*3. Construct a Huffman code for the symbols with ASCII codes given the frequency of their occurrence in representative input.
4. Compute the number of different spanning trees of K_n for $n = 1, 2, 3, 4, 5, 6$. Conjecture a formula for the number of such spanning trees whenever n is a positive integer.
5. Compare the number of comparisons needed to sort lists of n elements for $n = 100$, 1000, and 10,000 from the set of positive integers less than 1,000,000, where the elements are randomly selected positive integers, using the selection sort, the insertion sort, the merge sort, and the quick sort.
6. Compute the number of different ways n queens can be arranged on an $n \times n$ chessboard so that no two queens can attack each other for all positive integers n not exceeding 10.
*7. Find a minimum spanning tree of the graph that connects the capital cities of the 50 states in the United States to each other where the weight of each edge is the distance between the cities.
8. Draw the complete game tree for a game of checkers on a 4×4 board.

Writing Projects

Respond to these with essays using outside sources.

1. Explain how Cayley used trees to enumerate the number of certain types of hydrocarbons.
2. Explain how trees are used to represent ancestral relations in the study of evolution.
3. Discuss hierarchical cluster trees and how they are used.
4. Define *AVL-trees* (sometimes also known as *height-balanced trees*). Describe how and why AVL-trees are used in a variety of different algorithms.
5. Define *quad trees* and explain how images can be represented using them. Describe how images can be rotated, scaled, and translated by manipulating the corresponding quad tree.
6. Define a *heap* and explain how trees can be turned into heaps. Why are heaps useful in sorting?
7. Describe dynamic algorithms for data compression based on letter frequencies as they change as characters are successively read, such as adaptive Huffman coding.
8. Explain how *alpha-beta pruning* can be used to simplify the computation of the value of a game tree.
9. Describe the techniques used by chess-playing programs such as Deep Blue.
10. Define the type of graph known as a *mesh of trees*. Explain how this graph is used in applications to very large system integration and parallel computing.
11. Discuss the algorithms used in IP multicasting to avoid loops between routers.
12. Describe an algorithm based on depth-first search for finding the articulation points of a graph.
13. Describe an algorithm based on depth-first search to find the strongly connected components of a directed graph.
14. Describe the search techniques used by Googlebot, and other web crawlers and spiders used by different web search engines.
15. Describe an algorithm for finding the minimum spanning tree of a graph such that the maximum degree of any vertex in the spanning tree does not exceed a fixed constant k.
16. Compare and contrast some of the most important sorting algorithms in terms of their complexity and when they are used.
17. Discuss the history and origins of algorithms for constructing minimum spanning trees.
18. Describe algorithms for producing random trees.

CHAPTER

12 Boolean Algebra

12.1 Boolean Functions

12.2 Representing Boolean Functions

12.3 Logic Gates

12.4 Minimization of Circuits

The circuits in computers and other electronic devices have inputs, each of which is either a 0 or a 1, and produce outputs that are also 0s and 1s. Circuits can be constructed using any basic element that has two different states. Such elements include switches that can be in either the on or the off position and optical devices that can be either lit or unlit. In 1938 Claude Shannon showed how the basic rules of logic, first given by George Boole in 1854 in his *The Laws of Thought*, could be used to design circuits. These rules form the basis for Boolean algebra. In this chapter we develop the basic properties of Boolean algebra. The operation of a circuit is defined by a Boolean function that specifies the value of an output for each set of inputs. The first step in constructing a circuit is to represent its Boolean function by an expression built up using the basic operations of Boolean algebra. We will provide an algorithm for producing such expressions. The expression that we obtain may contain many more operations than are necessary to represent the function. Later in the chapter we will describe methods for finding an expression with the minimum number of sums and products that represents a Boolean function. The procedures that we will develop, Karnaugh maps and the Quine–McCluskey method, are important in the design of efficient circuits.

12.1 Boolean Functions

12.1.1 Introduction

Boolean algebra provides the operations and the rules for working with the set $\{0, 1\}$. Electronic and optical switches can be studied using this set and the rules of Boolean algebra. The three operations in Boolean algebra that we will use most are complementation, the Boolean sum, and the Boolean product. The **complement** of an element, denoted with a bar, is defined by $\overline{0} = 1$ and $\overline{1} = 0$. The Boolean sum, denoted by $+$ or by *OR*, has the following values:

$$1 + 1 = 1, \qquad 1 + 0 = 1, \qquad 0 + 1 = 1, \qquad 0 + 0 = 0.$$

The Boolean product, denoted by $\cdot$ or by *AND*, has the following values:

$$1 \cdot 1 = 1, \qquad 1 \cdot 0 = 0, \qquad 0 \cdot 1 = 0, \qquad 0 \cdot 0 = 0.$$

When there is no danger of confusion, the symbol $\cdot$ can be deleted, just as in writing algebraic products. Unless parentheses are used, the rules of precedence for Boolean operators are: first, all complements are computed, followed by all Boolean products, followed by all Boolean sums. This is illustrated in Example 1.

EXAMPLE 1 Find the value of $1 \cdot 0 + \overline{(0 + 1)}$.

Solution: Using the definitions of complementation, the Boolean sum, and the Boolean product, it follows that

$$\begin{aligned} 1 \cdot 0 + \overline{(0 + 1)} &= 0 + \overline{1} \\ &= 0 + 0 \\ &= 0. \end{aligned}$$

◀

The complement, Boolean sum, and Boolean product correspond to the logical operators, $\neg$, $\vee$, and $\wedge$, respectively, where 0 corresponds to **F** (false) and 1 corresponds to **T** (true). Equalities in Boolean algebra can be directly translated into equivalences of compound propositions. Conversely, equivalences of compound propositions can be translated into equalities in Boolean algebra. We will see later in this section why these translations yield valid logical equivalences and identities in Boolean algebra. Example 2 illustrates the translation from Boolean algebra to propositional logic.

EXAMPLE 2 Translate $1 \cdot 0 + \overline{(0 + 1)} = 0$, the equality found in Example 1, into a logical equivalence.

Solution: We obtain a logical equivalence when we translate each 1 into a **T**, each 0 into an **F**, each Boolean sum into a disjunction, each Boolean product into a conjunction, and each complementation into a negation. We obtain

$$(\mathbf{T} \wedge \mathbf{F}) \vee \neg(\mathbf{T} \vee \mathbf{F}) \equiv \mathbf{F}.$$ ◀

Example 3 illustrates the translation from propositional logic to Boolean algebra.

EXAMPLE 3 Translate the logical equivalence $(\mathbf{T} \wedge \mathbf{T}) \vee \neg\mathbf{F} \equiv \mathbf{T}$ into an identity in Boolean algebra.

Solution: We obtain an identity in Boolean algebra when we translate each **T** into a 1, each **F** into a 0, each disjunction into a Boolean sum, each conjunction into a Boolean product, and each negation into a complementation. We obtain

$$(1 \cdot 1) + \overline{0} = 1.$$ ◀

12.1.2 Boolean Expressions and Boolean Functions

Let $B = \{0, 1\}$. Then $B^n = \{(x_1, x_2, \dots, x_n) \mid x_i \in B \text{ for } 1 \leq i \leq n\}$ is the set of all possible n-tuples of 0s and 1s. The variable x is called a **Boolean variable** if it assumes values only from B, that is, if its only possible values are 0 and 1. A function from B^n to B is called a **Boolean function of degree *n***.

EXAMPLE 4 The function $F(x, y) = x\overline{y}$ from the set of ordered pairs of Boolean variables to the set $\{0, 1\}$ is a Boolean function of degree 2 with $F(1, 1) = 0$, $F(1, 0) = 1$, $F(0, 1) = 0$, and $F(0, 0) = 0$. We display these values of F in Table 1. ◀

Links

Alfred Eisenstaedt/The LIFE Picture Collection/ Shutterstock

CLAUDE ELWOOD SHANNON (1916–2001) Claude Shannon was born in Petoskey, Michigan, and grew up in Gaylord, Michigan. His father was a businessman and a probate judge, and his mother was a language teacher and a high school principal. Shannon attended the University of Michigan, graduating in 1936. He continued his studies at M.I.T., where he took the job of maintaining the differential analyzer, a mechanical computing device consisting of shafts and gears built by his professor, Vannevar Bush. Shannon's master's thesis, written in 1936, studied the logical aspects of the differential analyzer. This master's thesis presents the first application of Boolean algebra to the design of switching circuits; it is perhaps the most famous master's thesis of the twentieth century. He received his Ph.D. from M.I.T. in 1940. Shannon joined Bell Laboratories in 1940, where he worked on transmitting data efficiently. He was one of the first people to use bits to represent information. At Bell Laboratories he worked on determining the amount of traffic that telephone lines can carry. Shannon made many fundamental contributions to information theory. In the early 1950s he was one of the founders of the study of artificial intelligence. He joined the M.I.T. faculty in 1956, where he continued his study of information theory.

Shannon had an unconventional side. He is credited with inventing the rocket-powered Frisbee. He is also famous for riding a unicycle down the hallways of Bell Laboratories while juggling four balls. Shannon retired when he was 50 years old, publishing papers sporadically over the following 10 years. In his later years he concentrated on some pet projects, such as building a motorized pogo stick. One interesting quote from Shannon, published in *Omni Magazine* in 1987, is "I visualize a time when we will be to robots what dogs are to humans. And I am rooting for the machines."

TABLE 1

x	y	$F(x, y)$
1	1	0
1	0	1
0	1	0
0	0	0

Boolean functions can be represented using expressions made up from variables and Boolean operations. The **Boolean expressions** in the variables $x_1, x_2, \dots, x_n$ are defined recursively as

$0, 1, x_1, x_2, \dots, x_n$ are Boolean expressions;

if E_1 and E_2 are Boolean expressions, then $\overline{E}_1$, (E_1E_2), and $(E_1 + E_2)$ are Boolean expressions.

Each Boolean expression represents a Boolean function. The values of this function are obtained by substituting 0 and 1 for the variables in the expression. In Section 12.2 we will show that every Boolean function can be represented by a Boolean expression.

EXAMPLE 5 Find the values of the Boolean function represented by $F(x, y, z) = xy + \overline{z}$.

Solution: The values of this function are displayed in Table 2. ◀

TABLE 2

x	y	z	xy	$\overline{z}$	$F(x, y, z) = xy + \overline{z}$
1	1	1	1	0	1
1	1	0	1	1	1
1	0	1	0	0	0
1	0	0	0	1	1
0	1	1	0	0	0
0	1	0	0	1	1
0	0	1	0	0	0
0	0	0	0	1	1

Note that we can represent a Boolean function graphically by distinguishing the vertices of the n-cube that correspond to the n-tuples of bits where the function has value 1.

EXAMPLE 6 The function $F(x, y, z) = xy + \overline{z}$ from B^3 to B from Example 5 can be represented by distinguishing the vertices that correspond to the five 3-tuples (1, 1, 1), (1, 1, 0), (1, 0, 0), (0, 1, 0), and (0, 0, 0), where $F(x, y, z) = 1$, as shown in Figure 1. These vertices are displayed using solid black circles. ◀

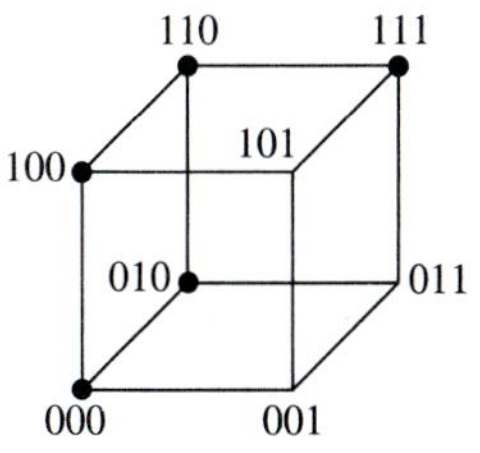

FIGURE 1

Boolean functions F and G of n variables are equal if and only if $F(b_1, b_2, \dots, b_n) = G(b_1, b_2, \dots, b_n)$ whenever $b_1, b_2, \dots, b_n$ belong to B. Two different Boolean expressions that represent the same function are called **equivalent**. For instance, the Boolean expressions xy, $xy + 0$, and $xy \cdot 1$ are equivalent. The **complement** of the Boolean function F is the function $\overline{F}$, where $\overline{F}(x_1, \dots, x_n) = \overline{F(x_1, \dots, x_n)}$. Let F and G be Boolean functions of degree n. The **Boolean sum** $F + G$ and the **Boolean product** FG are defined by

$$(F + G)(x_1, \dots, x_n) = F(x_1, \dots, x_n) + G(x_1, \dots, x_n),$$
$$(FG)(x_1, \dots, x_n) = F(x_1, \dots, x_n)G(x_1, \dots, x_n).$$

A Boolean function of degree two is a function from a set with four elements, namely, pairs of elements from $B = \{0, 1\}$, to B, a set with two elements. Hence, there are 16 different Boolean functions of degree two. In Table 3 we display the values of the 16 different Boolean functions of degree two, labeled $F_1, F_2, \dots, F_{16}$.

TABLE 3 The 16 Boolean Functions of Degree Two.

x	y	F_1	F_2	F_3	F_4	F_5	F_6	F_7	F_8	F_9	F_{10}	F_{11}	F_{12}	F_{13}	F_{14}	F_{15}	F_{16}
1	1	1	1	1	1	1	1	1	1	0	0	0	0	0	0	0	0
1	0	1	1	1	1	0	0	0	0	1	1	1	1	0	0	0	0
0	1	1	1	0	0	1	1	0	0	1	1	0	0	1	1	0	0
0	0	1	0	1	0	1	0	1	0	1	0	1	0	1	0	1	0

EXAMPLE 7 How many different Boolean functions of degree n are there?

Solution: From the product rule for counting, it follows that there are 2^n different n-tuples of 0s and 1s. Because a Boolean function is an assignment of 0 or 1 to each of these 2^n different n-tuples, the product rule shows that there are 2^{2^n} different Boolean functions of degree n. ◀

Table 4 displays the number of different Boolean functions of degrees one through six. The number of such functions grows extremely rapidly.

TABLE 4 The Number of Boolean Functions of Degree *n*.

Degree	*Number*
1	4
2	16
3	256
4	65,536
5	4,294,967,296
6	18,446,744,073,709,551,616

12.1.3 Identities of Boolean Algebra

There are many identities in Boolean algebra. The most important of these are displayed in Table 5. These identities are particularly useful in simplifying the design of circuits. Each of the identities in Table 5 can be proved using a table. We will prove one of the distributive laws in this way in Example 8. The proofs of the remaining properties are left as exercises for the reader.

EXAMPLE 8 Show that the distributive law $x(y + z) = xy + xz$ is valid.

Solution: The verification of this identity is shown in Table 6. The identity holds because the last two columns of the table agree. ◀

The reader should compare the Boolean identities in Table 5 to the logical equivalences in Table 6 of Section 1.3 and the set identities in Table 1 in Section 2.2. All are special cases of the same set of identities in a more abstract structure. Each collection of identities can be obtained by making the appropriate translations. For example, we can transform each of the identities in

TABLE 5 Boolean Identities.

Identity	*Name*
$\overline{\overline{x}} = x$	Law of the double complement
$x + x = x$ $x \cdot x = x$	Idempotent laws
$x + 0 = x$ $x \cdot 1 = x$	Identity laws
$x + 1 = 1$ $x \cdot 0 = 0$	Domination laws
$x + y = y + x$ $xy = yx$	Commutative laws
$x + (y + z) = (x + y) + z$ $x(yz) = (xy)z$	Associative laws
$x + yz = (x + y)(x + z)$ $x(y + z) = xy + xz$	Distributive laws
$\overline{(xy)} = \overline{x} + \overline{y}$ $\overline{(x + y)} = \overline{x}\,\overline{y}$	De Morgan's laws
$x + xy = x$ $x(x + y) = x$	Absorption laws
$x + \overline{x} = 1$	Unit property
$x\overline{x} = 0$	Zero property

Compare these Boolean identities with the logical equivalences in Section 1.3 and the set identities in Section 2.2.

Table 5 into a logical equivalence by changing each Boolean variable into a propositional variable, each 0 into a **F**, each 1 into a **T**, each Boolean sum into a disjunction, each Boolean product into a conjunction, and each complementation into a negation, as we illustrate in Example 9.

EXAMPLE 9 Translate the distributive law $x + yz = (x + y)(x + z)$ in Table 5 into a logical equivalence.

Solution: To translate a Boolean identity into a logical equivalence, we change each Boolean variable into a propositional variable. Here we will change the Boolean variables x, y, and z into the propositional variables p, q, and r. Next, we change each Boolean sum into a disjunction and

TABLE 6 Verifying One of the Distributive Laws.

x	y	z	$y + z$	xy	xz	$x(y + z)$	$xy + xz$
1	1	1	1	1	1	1	1
1	1	0	1	1	0	1	1
1	0	1	1	0	1	1	1
1	0	0	0	0	0	0	0
0	1	1	1	0	0	0	0
0	1	0	1	0	0	0	0
0	0	1	1	0	0	0	0
0	0	0	0	0	0	0	0

each Boolean product into a conjunction. (Note that 0 and 1 do not appear in this identity and complementation also does not appear.) This transforms the Boolean identity into the logical equivalence

$$p \vee (q \wedge r) \equiv (p \vee q) \wedge (p \vee r).$$

This logical equivalence is one of the distributive laws for propositional logic in Table 6 in Section 1.3. ◀

Identities in Boolean algebra can be used to prove further identities. We demonstrate this in Example 10.

EXAMPLE 10 Prove the **absorption law** $x(x + y) = x$ using the other identities of Boolean algebra shown in Table 5. (This is called an absorption law because absorbing $x + y$ into x leaves x unchanged.)

Extra Examples

Solution: We display steps used to derive this identity and the law used in each step:

$$\begin{aligned} x(x+y) &= (x+0)(x+y) && \text{Identity law for the Boolean sum} \\ &= x + 0 \cdot y && \text{Distributive law of the Boolean sum over the Boolean product} \\ &= x + y \cdot 0 && \text{Commutative law for the Boolean product} \\ &= x + 0 && \text{Domination law for the Boolean product} \\ &= x && \text{Identity law for the Boolean sum.} \end{aligned}$$

◀

12.1.4 Duality

The identities in Table 5 come in pairs (except for the law of the double complement and the unit and zero properties). To explain the relationship between the two identities in each pair we use the concept of a dual. The **dual** of a Boolean expression is obtained by interchanging Boolean sums and Boolean products and interchanging 0s and 1s.

EXAMPLE 11 Find the duals of $x(y + 0)$ and $\overline{x} \cdot 1 + (\overline{y} + z)$.

Solution: Interchanging $\cdot$ signs and $+$ signs and interchanging 0s and 1s in these expressions produces their duals. The duals are $x + (y \cdot 1)$ and $(\overline{x} + 0)(\overline{y}z)$, respectively. ◀

The dual of a Boolean function F represented by a Boolean expression is the function represented by the dual of this expression. This dual function, denoted by F^d, does not depend on the particular Boolean expression used to represent F. An identity between functions represented by Boolean expressions remains valid when the duals of both sides of the identity are taken. (See Exercise 30 for the reason why this is true.) This result, called the **duality principle**, is useful for obtaining new identities.

EXAMPLE 12 Construct an identity from the absorption law $x(x + y) = x$ by taking duals.

Solution: Taking the duals of both sides of this identity produces the identity $x + xy = x$, which is also called an absorption law and is shown in Table 5. ◀

12.1.5 The Abstract Definition of a Boolean Algebra

In this section we have focused on Boolean functions and expressions. However, the results we have established can be translated into results about propositions or results about sets. Because of this, it is useful to define Boolean algebras abstractly. Once it is shown that a particular structure is a Boolean algebra, then all results established about Boolean algebras in general apply to this particular structure.

Boolean algebras can be defined in several ways. The most common way is to specify the properties that operations must satisfy, as is done in Definition 1.

Definition 1 A *Boolean algebra* is a set B with two binary operations $\vee$ and $\wedge$, elements 0 and 1, and a unary operation $\bar{}$ such that these properties hold for all x, y, and z in B:

$$\left.\begin{aligned} x \vee 0 &= x \\ x \wedge 1 &= x \end{aligned}\right\} \quad \text{Identity laws}$$

$$\left.\begin{aligned} x \vee \overline{x} &= 1 \\ x \wedge \overline{x} &= 0 \end{aligned}\right\} \quad \text{Complement laws}$$

$$\left.\begin{aligned} (x \vee y) \vee z &= x \vee (y \vee z) \\ (x \wedge y) \wedge z &= x \wedge (y \wedge z) \end{aligned}\right\} \quad \text{Associative laws}$$

$$\left.\begin{aligned} x \vee y &= y \vee x \\ x \wedge y &= y \wedge x \end{aligned}\right\} \quad \text{Commutative laws}$$

$$\left.\begin{aligned} x \vee (y \wedge z) &= (x \vee y) \wedge (x \vee z) \\ x \wedge (y \vee z) &= (x \wedge y) \vee (x \wedge z) \end{aligned}\right\} \quad \text{Distributive laws}$$

Using the laws given in Definition 1, it is possible to prove many other laws that hold for every Boolean algebra, such as idempotent and domination laws. (See Exercises 35–42.)

From our previous discussion, $B = \{0, 1\}$ with the *OR* and *AND* operations and the complement operator, satisfies all these properties. The set of propositions in n variables, with the $\vee$ and $\wedge$ operators, **F** and **T**, and the negation operator, also satisfies all the properties of a Boolean algebra, as can be seen from Table 6 in Section 1.3. Similarly, the set of subsets of a universal set U with the union and intersection operations, the empty set and the universal set, and the set complementation operator, is a Boolean algebra as can be seen by consulting Table 1 in Section 2.2. So, to establish results about each of Boolean expressions, propositions, and sets, we need only prove results about abstract Boolean algebras.

Boolean algebras may also be defined using the notion of a lattice, discussed in Chapter 9. Recall that a lattice L is a partially ordered set in which every pair of elements x, y has a least upper bound, denoted by $\text{lub}(x, y)$ and a greatest lower bound denoted by $\text{glb}(x, y)$. Given two elements x and y of L, we can define two operations $\vee$ and $\wedge$ on pairs of elements of L by $x \vee y = \text{lub}(x, y)$ and $x \wedge y = \text{glb}(x, y)$.

For a lattice L to be a Boolean algebra as specified in Definition 1, it must have two properties. First, it must be **complemented**. For a lattice to be complemented it must have a least element 0 and a greatest element 1 and for every element x of the lattice there must exist an element $\overline{x}$ such that $x \vee \overline{x} = 1$ and $x \wedge \overline{x} = 0$. Second, it must be **distributive**. This means that for every x, y, and z in L, $x \vee (y \wedge z) = (x \vee y) \wedge (x \vee z)$ and $x \wedge (y \vee z) = (x \wedge y) \vee (x \wedge z)$. Showing that a complemented, distributive lattice is a Boolean algebra has been left as Exercise 43. Examples and properties of lattices are also the subject of Supplementary Exercises 39–48 in Chapter 9.

Exercises

1. Find the values of these expressions.

a) $1 \cdot \overline{0}$ **b)** $1 + \overline{1}$ **c)** $\overline{0} \cdot 0$ **d)** $\overline{(1 + 0)}$

2. Find the values, if any, of the Boolean variable x that satisfy these equations.

a) $x \cdot 1 = 0$ **b)** $x + x = 0$
c) $x \cdot 1 = x$ **d)** $x \cdot \overline{x} = 1$

3. a) Show that $(1 \cdot 1) + (\overline{0 \cdot 1} + 0) = 1$.

b) Translate the equation in part (a) into a propositional equivalence by changing each 0 into an **F**, each 1 into a **T**, each Boolean sum into a disjunction, each Boolean product into a conjunction, each complementation into a negation, and the equals sign into a propositional equivalence sign.

4. a) Show that $(\overline{1} \cdot \overline{0}) + (1 \cdot \overline{0}) = 1$.

b) Translate the equation in part (a) into a propositional equivalence by changing each 0 into an **F**, each 1 into a **T**, each Boolean sum into a disjunction, each Boolean product into a conjunction, each complementation into a negation, and the equals sign into a propositional equivalence sign.

5. Use a table to express the values of each of these Boolean functions.

a) $F(x, y, z) = \overline{x}y$
b) $F(x, y, z) = x + yz$
c) $F(x, y, z) = x\overline{y} + \overline{(xyz)}$
d) $F(x, y, z) = x(yz + \overline{y}\,\overline{z})$

6. Use a table to express the values of each of these Boolean functions.

a) $F(x, y, z) = \overline{z}$
b) $F(x, y, z) = \overline{x}y + \overline{y}z$
c) $F(x, y, z) = x\overline{y}z + \overline{(xyz)}$
d) $F(x, y, z) = \overline{y}(xz + \overline{x}\,\overline{z})$

7. Use a 3-cube Q_3 to represent each of the Boolean functions in Exercise 5 by displaying a black circle at each vertex that corresponds to a 3-tuple where this function has the value 1.

8. Use a 3-cube Q_3 to represent each of the Boolean functions in Exercise 6 by displaying a black circle at each vertex that corresponds to a 3-tuple where this function has the value 1.

9. What values of the Boolean variables x and y satisfy $xy = x + y$?

10. How many different Boolean functions are there of degree 7?

11. Prove the absorption law $x + xy = x$ using the other laws in Table 5.

12. Show that $F(x, y, z) = xy + xz + yz$ has the value 1 if and only if at least two of the variables x, y, and z have the value 1.

13. Show that $x\overline{y} + y\overline{z} + \overline{x}z = \overline{x}y + \overline{y}z + x\overline{z}$.

Exercises 14–23 deal with the Boolean algebra $\{0, 1\}$ with addition, multiplication, and complement defined at the beginning of this section. In each case, use a table as in Example 8.

14. Verify the law of the double complement.

15. Verify the idempotent laws.

16. Verify the identity laws.

17. Verify the domination laws.

18. Verify the commutative laws.

19. Verify the associative laws.

20. Verify the first distributive law in Table 5.

21. Verify De Morgan's laws.

22. Verify the unit property.

23. Verify the zero property.

The Boolean operator $\oplus$, called the *XOR* operator, is defined by $1 \oplus 1 = 0$, $1 \oplus 0 = 1$, $0 \oplus 1 = 1$, and $0 \oplus 0 = 0$.

24. Simplify these expressions.

a) $x \oplus 0$ **b)** $x \oplus 1$
c) $x \oplus x$ **d)** $x \oplus \overline{x}$

25. Show that these identities hold.

a) $x \oplus y = (x + y)\overline{(xy)}$
b) $x \oplus y = (x\overline{y}) + (\overline{x}y)$

26. Show that $x \oplus y = y \oplus x$.

27. Prove or disprove these equalities.

a) $x \oplus (y \oplus z) = (x \oplus y) \oplus z$
b) $x + (y \oplus z) = (x + y) \oplus (x + z)$
c) $x \oplus (y + z) = (x \oplus y) + (x \oplus z)$

28. Find the duals of these Boolean expressions.

a) $x + y$ **b)** $\overline{x}\,\overline{y}$
c) $xyz + \overline{x}\,\overline{y}\,\overline{z}$ **d)** $x\overline{z} + x \cdot 0 + \overline{x} \cdot 1$

***29.** Suppose that F is a Boolean function represented by a Boolean expression in the variables $x_1, \dots, x_n$. Show that $F^d(x_1, \dots, x_n) = \overline{F(\overline{x}_1, \dots, \overline{x}_n)}$.

***30.** Show that if F and G are Boolean functions represented by Boolean expressions in n variables and $F = G$, then $F^d = G^d$, where F^d and G^d are the Boolean functions represented by the duals of the Boolean expressions representing F and G, respectively. [*Hint:* Use the result of Exercise 29.]

***31.** How many different Boolean functions $F(x, y, z)$ are there such that $F(\overline{x}, \overline{y}, \overline{z}) = F(x, y, z)$ for all values of the Boolean variables x, y, and z?

***32.** How many different Boolean functions $F(x, y, z)$ are there such that $F(\overline{x}, y, z) = F(x, \overline{y}, z) = F(x, y, \overline{z})$ for all values of the Boolean variables x, y, and z?

33. Show that you obtain De Morgan's laws for propositions (in Table 6 in Section 1.3) when you transform De Morgan's laws for Boolean algebra in Table 5 into logical equivalences.

34. Show that you obtain the absorption laws for propositions (in Table 6 in Section 1.3) when you transform the absorption laws for Boolean algebra in Table 5 into logical equivalences.

In Exercises 35–42, use the laws in Definition 1 to show that the stated properties hold in every Boolean algebra.

35. Show that in a Boolean algebra, the **idempotent laws** $x \vee x = x$ and $x \wedge x = x$ hold for every element x.

36. Show that in a Boolean algebra, every element x has a unique complement $\overline{x}$ such that $x \vee \overline{x} = 1$ and $x \wedge \overline{x} = 0$.

37. Show that in a Boolean algebra, the complement of the element 0 is the element 1 and vice versa.

38. Prove that in a Boolean algebra, the **law of the double complement** holds; that is, $\overline{\overline{x}} = x$ for every element x.

39. Show that **De Morgan's laws** hold in a Boolean algebra. That is, show that for all x and y, $\overline{(x \vee y)} = \overline{x} \wedge \overline{y}$ and $\overline{(x \wedge y)} = \overline{x} \vee \overline{y}$.

40. Show that in a Boolean algebra, the **modular properties** hold. That is, show that $x \wedge (y \vee (x \wedge z)) = (x \wedge y) \vee (x \wedge z)$ and $x \vee (y \wedge (x \vee z)) = (x \vee y) \wedge (x \vee z)$.

41. Show that in a Boolean algebra, if $x \vee y = 0$, then $x = 0$ and $y = 0$, and that if $x \wedge y = 1$, then $x = 1$ and $y = 1$.

42. Show that in a Boolean algebra, the **dual** of an identity, obtained by interchanging the $\vee$ and $\wedge$ operators and interchanging the elements 0 and 1, is also a valid identity.

43. Show that a complemented, distributive lattice is a Boolean algebra.

12.2 Representing Boolean Functions

Two important problems of Boolean algebra will be studied in this section. The first problem is: Given the values of a Boolean function, how can a Boolean expression that represents this function be found? This problem will be solved by showing that any Boolean function can be represented by a Boolean sum of Boolean products of the variables and their complements. The solution of this problem shows that every Boolean function can be represented using the three Boolean operators $\cdot$, $+$, and $^{-}$. The second problem is: Is there a smaller set of operators that can be used to represent all Boolean functions? We will answer this question by showing that all Boolean functions can be represented using only one operator. Both of these problems have practical importance in circuit design.

12.2.1 Sum-of-Products Expansions

We will use examples to illustrate one important way to find a Boolean expression that represents a Boolean function.

EXAMPLE 1 Find Boolean expressions that represent the functions $F(x, y, z)$ and $G(x, y, z)$, which are given in Table 1.

TABLE 1

x	y	z	F	G
1	1	1	0	0
1	1	0	0	1
1	0	1	1	0
1	0	0	0	0
0	1	1	0	0
0	1	0	0	1
0	0	1	0	0
0	0	0	0	0

Solution: An expression that has the value 1 when $x = z = 1$ and $y = 0$, and the value 0 otherwise, is needed to represent F. Such an expression can be formed by taking the Boolean product of x, $\overline{y}$, and z. This product, $x\overline{y}z$, has the value 1 if and only if $x = \overline{y} = z = 1$, which holds if and only if $x = z = 1$ and $y = 0$.

To represent G, we need an expression that equals 1 when $x = y = 1$ and $z = 0$, or $x = z = 0$ and $y = 1$. We can form an expression with these values by taking the Boolean sum of two different Boolean products. The Boolean product $xy\overline{z}$ has the value 1 if and only if $x = y = 1$ and $z = 0$. Similarly, the product $\overline{x}y\overline{z}$ has the value 1 if and only if $x = z = 0$ and $y = 1$. The Boolean sum of these two products, $xy\overline{z} + \overline{x}y\overline{z}$, represents G, because it has the value 1 if and only if $x = y = 1$ and $z = 0$, or $x = z = 0$ and $y = 1$. ◀

Example 1 illustrates a procedure for constructing a Boolean expression representing a function with given values. Each combination of values of the variables for which the function has the value 1 leads to a Boolean product of the variables or their complements.

Definition 1 A *literal* is a Boolean variable or its complement. A *minterm* of the Boolean variables $x_1, x_2, \ldots, x_n$ is a Boolean product $y_1 y_2 \cdots y_n$, where $y_i = x_i$ or $y_i = \overline{x}_i$. Hence, a minterm is a product of n literals, with one literal for each variable.

A minterm has the value 1 for one and only one combination of values of its variables. More precisely, the minterm $y_1 y_2 \ldots y_n$ is 1 if and only if each y_i is 1, and this occurs if and only if $x_i = 1$ when $y_i = x_i$ and $x_i = 0$ when $y_i = \overline{x}_i$.

EXAMPLE 2 Find a minterm that equals 1 if $x_1 = x_3 = 0$ and $x_2 = x_4 = x_5 = 1$, and equals 0 otherwise.

Solution: The minterm $\overline{x}_1 x_2 \overline{x}_3 x_4 x_5$ has the correct set of values. ◀

By taking Boolean sums of distinct minterms we can build up a Boolean expression with a specified set of values. In particular, a Boolean sum of minterms has the value 1 when exactly one of the minterms in the sum has the value 1. It has the value 0 for all other combinations of values of the variables. Consequently, given a Boolean function, a Boolean sum of minterms can be formed that has the value 1 when this Boolean function has the value 1, and has the value 0 when the function has the value 0. The minterms in this Boolean sum correspond to those combinations of values for which the function has the value 1. The sum of minterms that represents the function is called the **sum-of-products expansion** or the **disjunctive normal form** of the Boolean function. (See Exercise 46 in Section 1.3 for the development of disjunctive normal form in propositional calculus.)

Links

EXAMPLE 3 Find the sum-of-products expansion for the function $F(x, y, z) = (x + y)\overline{z}$.

Extra Examples

Solution: We will find the sum-of-products expansion of $F(x, y, z)$ in two ways. First, we will use Boolean identities to expand the product and simplify. We find that

$$
\begin{aligned}
F(x, y, z) &= (x + y)\overline{z} \\
&= x\overline{z} + y\overline{z} && \text{Distributive law} \\
&= x1\overline{z} + 1y\overline{z} && \text{Identity law} \\
&= x(y + \overline{y})\overline{z} + (x + \overline{x})y\overline{z} && \text{Unit property} \\
&= xy\overline{z} + x\overline{y}\,\overline{z} + xy\overline{z} + \overline{x}y\overline{z} && \text{Distributive law} \\
&= xy\overline{z} + x\overline{y}\,\overline{z} + \overline{x}y\,\overline{z}. && \text{Idempotent law}
\end{aligned}
$$

Second, we can construct the sum-of-products expansion by determining the values of F for all possible values of the variables x, y, and z. These values are found in Table 2. The sum-of-products expansion of F is the Boolean sum of three minterms corresponding to the three rows of this table that give the value 1 for the function. This gives

$$F(x, y, z) = xy\overline{z} + x\overline{y}\,\overline{z} + \overline{x}y\overline{z}.$$ ◀

It is also possible to find a Boolean expression that represents a Boolean function by taking a Boolean product of Boolean sums. The resulting expansion is called the **conjunctive normal form** or **product-of-sums expansion** of the function. These expansions can be found from sum-of-products expansions by taking duals. How to find such expansions directly is described in Exercise 10.

TABLE 2

x	y	z	$x+y$	$\overline{z}$	$(x+y)\overline{z}$
1	1	1	1	0	0
1	1	0	1	1	1
1	0	1	1	0	0
1	0	0	1	1	1
0	1	1	1	0	0
0	1	0	1	1	1
0	0	1	0	0	0
0	0	0	0	1	0

12.2.2 Functional Completeness

Every Boolean function can be expressed as a Boolean sum of minterms. Each minterm is the Boolean product of Boolean variables or their complements. This shows that every Boolean function can be represented using the Boolean operators $\cdot$, $+$, and $^{-}$. Because every Boolean function can be represented using these operators we say that the set $\{\cdot, +, ^{-}\}$ is **functionally complete**. Can we find a smaller set of functionally complete operators? We can do so if one of the three operators of this set can be expressed in terms of the other two. This can be done using one of De Morgan's laws. We can eliminate all Boolean sums using the identity

$$x + y = \overline{\overline{x}\,\overline{y}},$$

which is obtained by taking complements of both sides in the second De Morgan law, given in Table 5 in Section 12.1, and then applying the double complementation law. This means that the set $\{\cdot, ^{-}\}$ is functionally complete. Similarly, we could eliminate all Boolean products using the identity

$$xy = \overline{\overline{x} + \overline{y}},$$

which is obtained by taking complements of both sides in the first De Morgan law, given in Table 5 in Section 12.1, and then applying the double complementation law. Consequently $\{+, ^{-}\}$ is functionally complete. Note that the set $\{+, \cdot\}$ is not functionally complete, because it is impossible to express the Boolean function $F(x) = \overline{x}$ using these operators (see Exercise 19).

We have found sets containing two operators that are functionally complete. Can we find a smaller set of functionally complete operators, namely, a set containing just one operator? Such sets exist. Define two operators, the $\mid$ or ***NAND*** operator, defined by $1 \mid 1 = 0$ and $1 \mid 0 = 0 \mid 1 = 0 \mid 0 = 1$; and the $\downarrow$ or ***NOR*** operator, defined by $1 \downarrow 1 = 1 \downarrow 0 = 0 \downarrow 1 = 0$ and $0 \downarrow 0 = 1$. Both of the sets $\{\mid\}$ and $\{\downarrow\}$ are functionally complete. To see that $\{\mid\}$ is functionally complete, because $\{\cdot, ^{-}\}$ is functionally complete, all that we have to do is show that both of the operators $\cdot$ and $^{-}$ can be expressed using just the $\mid$ operator. This can be done as

$$\overline{x} = x \mid x,$$
$$xy = (x \mid y) \mid (x \mid y).$$

The reader should verify these identities (see Exercise 14). We leave the demonstration that $\{\downarrow\}$ is functionally complete for the reader (see Exercises 15 and 16).

Exercises

1. Find a Boolean product of the Boolean variables x, y, and z, or their complements, that has the value 1 if and only if

a) $x = y = 0, z = 1$. **b)** $x = 0, y = 1, z = 0$.
c) $x = 0, y = z = 1$. **d)** $x = y = z = 0$.

2. Find the sum-of-products expansions of these Boolean functions.

a) $F(x, y) = \overline{x} + y$ **b)** $F(x, y) = x\overline{y}$
c) $F(x, y) = 1$ **d)** $F(x, y) = \overline{y}$

3. Find the sum-of-products expansions of these Boolean functions.

a) $F(x, y, z) = x + y + z$
b) $F(x, y, z) = (x + z)y$
c) $F(x, y, z) = x$
d) $F(x, y, z) = x\overline{y}$

4. Find the sum-of-products expansions of the Boolean function $F(x, y, z)$ that equals 1 if and only if

a) $x = 0$. **b)** $xy = 0$.
c) $x + y = 0$. **d)** $xyz = 0$.

5. Find the sum-of-products expansion of the Boolean function $F(w, x, y, z)$ that has the value 1 if and only if an odd number of w, x, y, and z have the value 1.

6. Find the sum-of-products expansion of the Boolean function $F(x_1, x_2, x_3, x_4, x_5)$ that has the value 1 if and only if three or more of the variables x_1, x_2, x_3, x_4, and x_5 have the value 1.

Another way to find a Boolean expression that represents a Boolean function is to form a Boolean product of Boolean sums of literals. Exercises 7–11 are concerned with representations of this kind.

7. Find a Boolean sum containing either x or $\overline{x}$, either y or $\overline{y}$, and either z or $\overline{z}$ that has the value 0 if and only if

a) $x = y = 1, z = 0$. **b)** $x = y = z = 0$.
c) $x = z = 0, y = 1$.

8. Find a Boolean product of Boolean sums of literals that has the value 0 if and only if $x = y = 1$ and $z = 0$, $x = z = 0$ and $y = 1$, or $x = y = z = 0$. [*Hint:* Take the Boolean product of the Boolean sums found in parts (a), (b), and (c) in Exercise 7.]

9. Show that the Boolean sum $y_1 + y_2 + \cdots + y_n$, where $y_i = x_i$ or $y_i = \overline{x}_i$, has the value 0 for exactly one combination of the values of the variables, namely, when $x_i = 0$ if $y_i = x_i$ and $x_i = 1$ if $y_i = \overline{x}_i$. This Boolean sum is called a **maxterm**.

10. Show that a Boolean function can be represented as a Boolean product of maxterms. This representation is called the **product-of-sums expansion** or **conjunctive normal form** of the function. [*Hint:* Include one maxterm in this product for each combination of the variables where the function has the value 0.]

11. Find the product-of-sums expansion of each of the Boolean functions in Exercise 3.

12. Express each of these Boolean functions using the operators $\cdot$ and $^-$.

a) $x + y + z$ **b)** $x + \overline{y}(\overline{x} + z)$
c) $\overline{x + \overline{y}}$ **d)** $\overline{x}(x + \overline{y} + \overline{z})$

13. Express each of the Boolean functions in Exercise 12 using the operators $+$ and $^-$.

14. Show that

a) $\overline{x} = x \mid x$. **b)** $xy = (x \mid y) \mid (x \mid y)$.
c) $x + y = (x \mid x) \mid (y \mid y)$.

15. Show that

a) $\overline{x} = x \downarrow x$.
b) $xy = (x \downarrow x) \downarrow (y \downarrow y)$.
c) $x + y = (x \downarrow y) \downarrow (x \downarrow y)$.

16. Show that $\{\downarrow\}$ is functionally complete using Exercise 15.

17. Express each of the Boolean functions in Exercise 3 using the operator $\mid$.

18. Express each of the Boolean functions in Exercise 3 using the operator $\downarrow$.

19. Show that the set of operators $\{+, \cdot\}$ is not functionally complete.

20. Are these sets of operators functionally complete?
a) $\{+, \oplus\}$ **b)** $\{^-, \oplus\}$ **c)** $\{\cdot, \oplus\}$

12.3 Logic Gates

12.3.1 Introduction

Links

Boolean algebra is used to model the circuitry of electronic devices. Each input and each output of such a device can be thought of as a member of the set $\{0, 1\}$. A computer, or other electronic device, is made up of a number of circuits. Each circuit can be designed using the rules of Boolean algebra that were studied in Sections 12.1 and 12.2. The basic elements of circuits are called **gates**, and were introduced in Section 1.2. Each type of gate implements a Boolean operation. In this section we define several types of gates. Using these gates, we will apply the

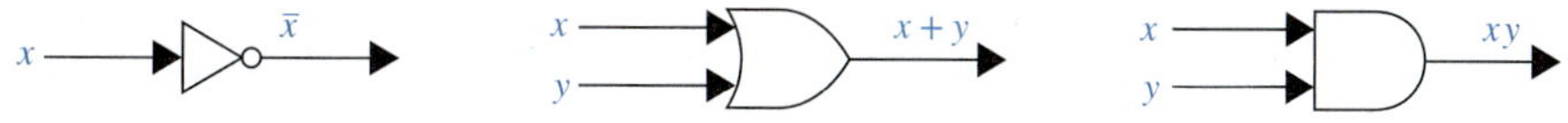

(a) Inverter (b) OR gate (c) AND gate

FIGURE 1 **Basic types of gates.**

rules of Boolean algebra to design circuits that perform a variety of tasks. The circuits that we will study in this chapter give output that depends only on the input, and not on the current state of the circuit. In other words, these circuits have no memory capabilities. Such circuits are called **combinational circuits** or **gating networks**.

We will construct combinational circuits using three types of elements. The first is an **inverter**, which accepts the value of one Boolean variable as input and produces the complement of this value as its output. The symbol used for an inverter is shown in Figure 1(a). The input to the inverter is shown on the left side entering the element, and the output is shown on the right side leaving the element.

The next type of element we will use is the **OR gate**. The inputs to this gate are the values of two or more Boolean variables. The output is the Boolean sum of their values. The symbol used for an OR gate is shown in Figure 1(b). The inputs to the OR gate are shown on the left side entering the element, and the output is shown on the right side leaving the element.

The third type of element we will use is the **AND gate**. The inputs to this gate are the values of two or more Boolean variables. The output is the Boolean product of their values. The symbol used for an AND gate is shown in Figure 1(c). The inputs to the AND gate are shown on the left side entering the element, and the output is shown on the right side leaving the element.

We will permit multiple inputs to AND and OR gates. The inputs to each of these gates are shown on the left side entering the element, and the output is shown on the right side. Examples of AND and OR gates with n inputs are shown in Figure 2.

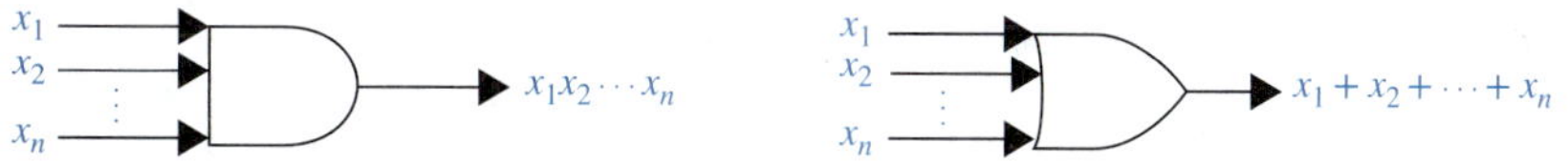

FIGURE 2 **Gates with n inputs.**

12.3.2 Combinations of Gates

Combinational circuits can be constructed using a combination of inverters, OR gates, and AND gates. When combinations of circuits are formed, some gates may share inputs. This is shown in one of two ways in depictions of circuits. One method is to use branchings that indicate all the gates that use a given input. The other method is to indicate this input separately for each gate. Figure 3 illustrates the two ways of showing gates with the same input values. Note also that output from a gate may be used as input by one or more other elements, as shown in Figure 3. Both drawings in Figure 3 depict the circuit that produces the output $xy + \overline{x}y$.

EXAMPLE 1 Construct circuits that produce the following outputs: (a) $(x + y)\overline{x}$, (b) $\overline{\overline{x}\,(y + \overline{z})}$, and (c) $(x + y + z)(\overline{x}\,\overline{y}\,\overline{z})$.

Solution: Circuits that produce these outputs are shown in Figure 4. ◀

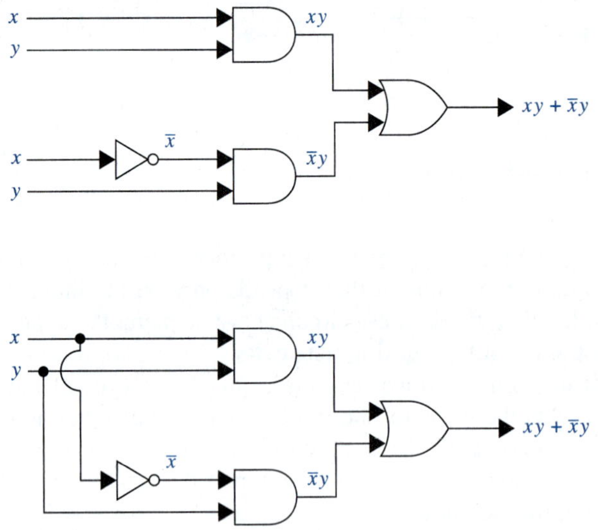

FIGURE 3 Two ways to draw the same circuit.

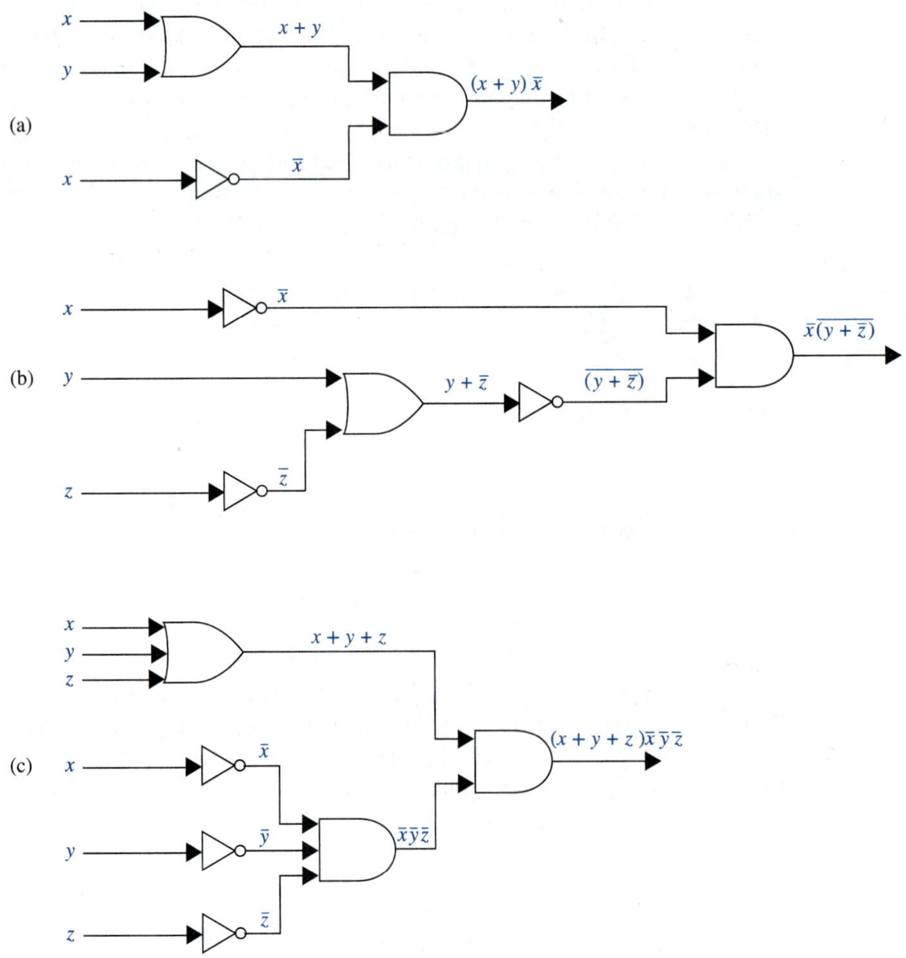

FIGURE 4 Circuits that produce the outputs specified in Example 1.

12.3.3 Examples of Circuits

We will give some examples of circuits that perform some useful functions.

EXAMPLE 2 A committee of three individuals decides issues for an organization. Each individual votes either yes or no for each proposal that arises. A proposal is passed if it receives at least two yes votes. Design a circuit that determines whether a proposal passes.

Extra Examples

Solution: Let $x = 1$ if the first individual votes yes, and $x = 0$ if this individual votes no; let $y = 1$ if the second individual votes yes, and $y = 0$ if this individual votes no; let $z = 1$ if the third individual votes yes, and $z = 0$ if this individual votes no. Then a circuit must be designed that produces the output 1 from the inputs x, y, and z when two or more of x, y, and z are 1. One representation of the Boolean function that has these output values is $xy + xz + yz$ (see Exercise 12 in Section 12.1). The circuit that implements this function is shown in Figure 5. ◀

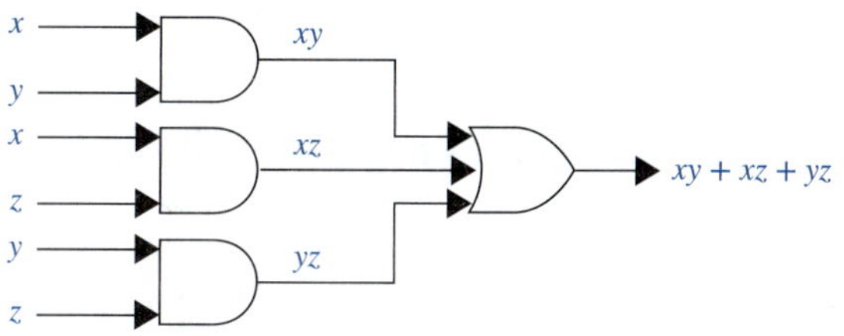

FIGURE 5 **A circuit for majority voting.**

EXAMPLE 3 Sometimes light fixtures are controlled by more than one switch. Circuits need to be designed so that flipping any one of the switches for the fixture turns the light on when it is off and turns the light off when it is on. Design circuits that accomplish this when there are two switches and when there are three switches.

TABLE 1

x	y	$F(x, y)$
1	1	1
1	0	0
0	1	0
0	0	1

Solution: We will begin by designing the circuit that controls the light fixture when two different switches are used. Let $x = 1$ when the first switch is closed and $x = 0$ when it is open, and let $y = 1$ when the second switch is closed and $y = 0$ when it is open. Let $F(x, y) = 1$ when the light is on and $F(x, y) = 0$ when it is off. We can arbitrarily decide that the light will be on when both switches are closed, so that $F(1, 1) = 1$. This determines all the other values of F. When one of the two switches is opened, the light goes off, so $F(1, 0) = F(0, 1) = 0$. When the other switch is also opened, the light goes on, so $F(0, 0) = 1$. Table 1 displays these values. Note that $F(x, y) = xy + \overline{x}\,\overline{y}$. This function is implemented by the circuit shown in Figure 6.

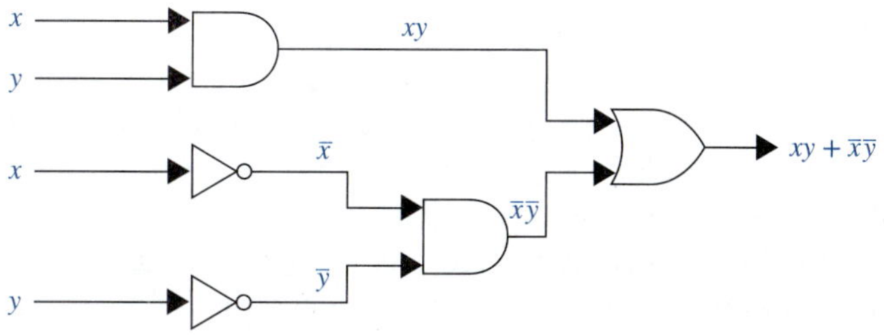

FIGURE 6 **A circuit for a light controlled by two switches.**

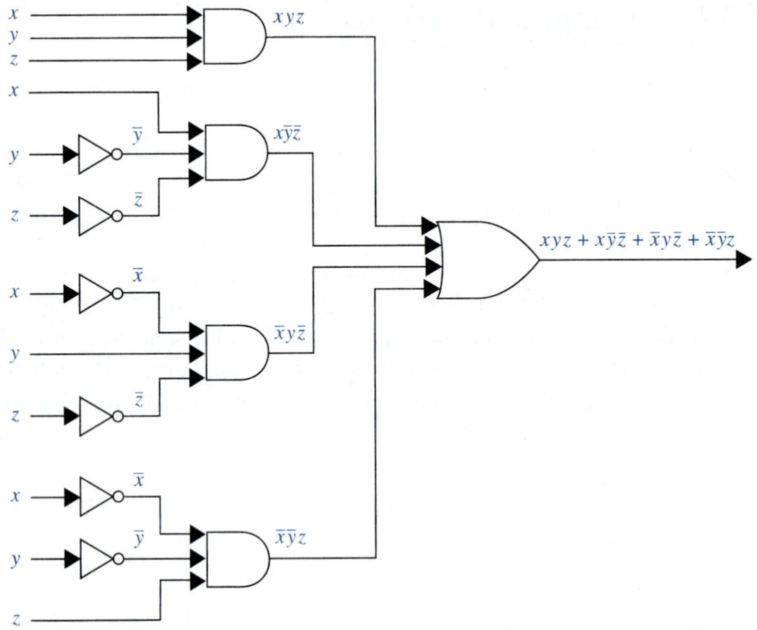

FIGURE 7 **A circuit for a fixture controlled by three switches.**

TABLE 2

x	y	z	$F(x, y, z)$
1	1	1	1
1	1	0	0
1	0	1	0
1	0	0	1
0	1	1	0
0	1	0	1
0	0	1	1
0	0	0	0

We will now design a circuit for three switches. Let x, y, and z be the Boolean variables that indicate whether each of the three switches is closed. We let $x = 1$ when the first switch is closed, and $x = 0$ when it is open; $y = 1$ when the second switch is closed, and $y = 0$ when it is open; and $z = 1$ when the third switch is closed, and $z = 0$ when it is open. Let $F(x, y, z) = 1$ when the light is on and $F(x, y, z) = 0$ when the light is off. We can arbitrarily specify that the light be on when all three switches are closed, so that $F(1, 1, 1) = 1$. This determines all other values of F. When one switch is opened, the light goes off, so $F(1, 1, 0) = F(1, 0, 1) = F(0, 1, 1) = 0$. When a second switch is opened, the light goes on, so $F(1, 0, 0) = F(0, 1, 0) = F(0, 0, 1) = 1$. Finally, when the third switch is opened, the light goes off again, so $F(0, 0, 0) = 0$. Table 2 shows the values of this function.

The function F can be represented by its sum-of-products expansion as $F(x, y, z) = xyz + x\overline{y}\,\overline{z} + \overline{x}y\overline{z} + \overline{x}\,\overline{y}z$. The circuit shown in Figure 7 implements this function. ◀

Links

12.3.4 Adders

TABLE 3 **Input and Output for the Half Adder.**

Input		*Output*	
x	y	s	c
1	1	0	1
1	0	1	0
0	1	1	0
0	0	0	0

We will illustrate how logic circuits can be used to carry out addition of two positive integers from their binary expansions. We will build up the circuitry to do this addition from some component circuits. First, we will build a circuit that can be used to find $x + y$, where x and y are two bits. The input to our circuit will be x and y, because these each have the value 0 or the value 1. The output will consist of two bits, namely, s and c, where s is the sum bit and c is the carry bit. This circuit is called a **multiple output circuit** because it has more than one output. The circuit that we are designing is called the **half adder**, because it adds two bits, without considering a carry from a previous addition. We show the input and output for the half adder in Table 3. From Table 3 we see that $c = xy$ and that $s = x\overline{y} + \overline{x}y = (x + y)\overline{(xy)}$. Hence, the circuit shown in Figure 8 computes the sum bit s and the carry bit c from the bits x and y.

We use the **full adder** to compute the sum bit and the carry bit when two bits and a carry are added. The inputs to the full adder are the bits x and y and the carry c_i. The outputs are the sum bit s and the new carry c_{i+1}. The inputs and outputs for the full adder are shown in Table 4.

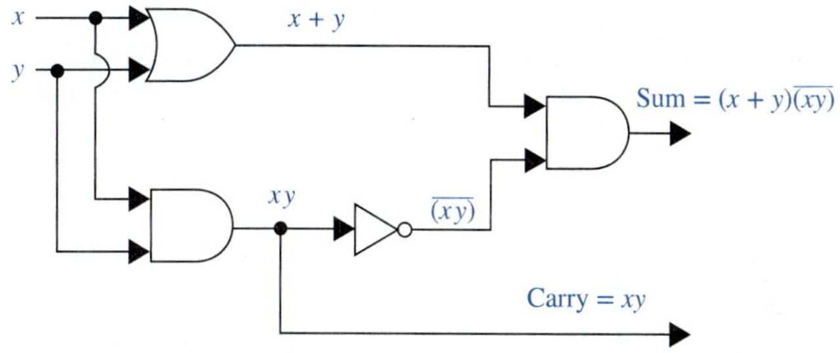

FIGURE 8 **The half adder.**

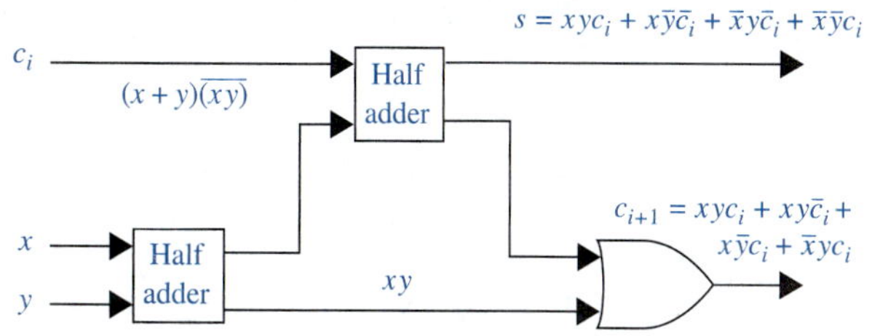

FIGURE 9 **A full adder.**

TABLE 4
Input and Output for the Full Adder.

Input			*Output*	
x	y	c_i	s	c_{i+1}
1	1	1	1	1
1	1	0	0	1
1	0	1	0	1
1	0	0	1	0
0	1	1	0	1
0	1	0	1	0
0	0	1	1	0
0	0	0	0	0

The two outputs of the full adder, the sum bit s and the carry c_{i+1}, are given by the sum-of-products expansions $xyc_i + x\overline{y}\,\overline{c}_i + \overline{x}y\overline{c}_i + \overline{x}\,\overline{y}c_i$ and $xyc_i + xy\overline{c}_i + x\overline{y}c_i + \overline{x}yc_i$, respectively. However, instead of designing the full adder from scratch, we will use half adders to produce the desired output. A full adder circuit using half adders is shown in Figure 9.

Finally, in Figure 10 we show how full and half adders can be used to add the two three-bit integers $(x_2x_1x_0)_2$ and $(y_2y_1y_0)_2$ to produce the sum $(s_3s_2s_1s_0)_2$. Note that s_3, the highest-order bit in the sum, is given by the carry c_2.

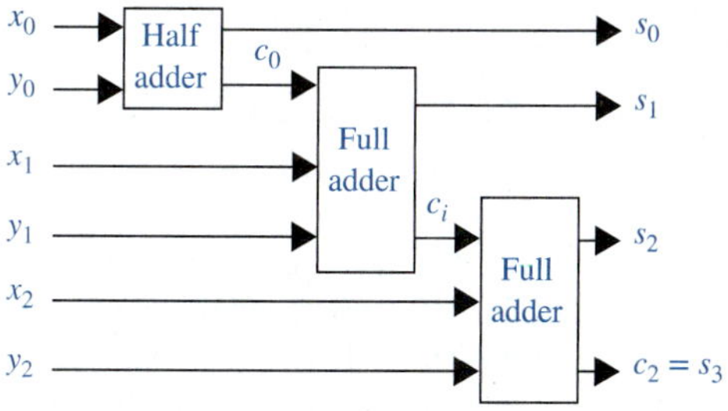

FIGURE 10 **Adding two three-bit integers with full and half adders.**

Exercises

In Exercises 1–5 find the output of the given circuit.

1. 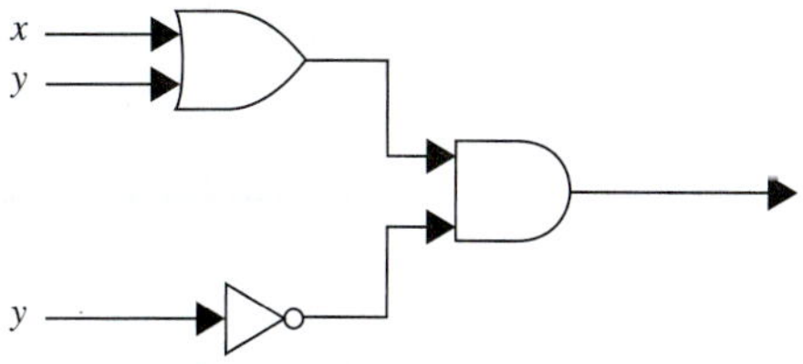

2. 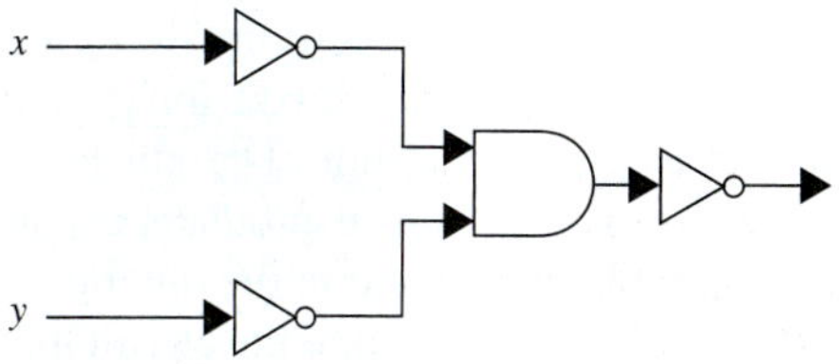

3. 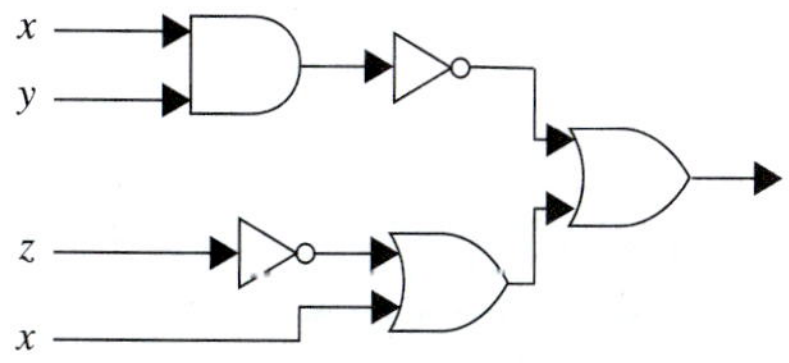

4. 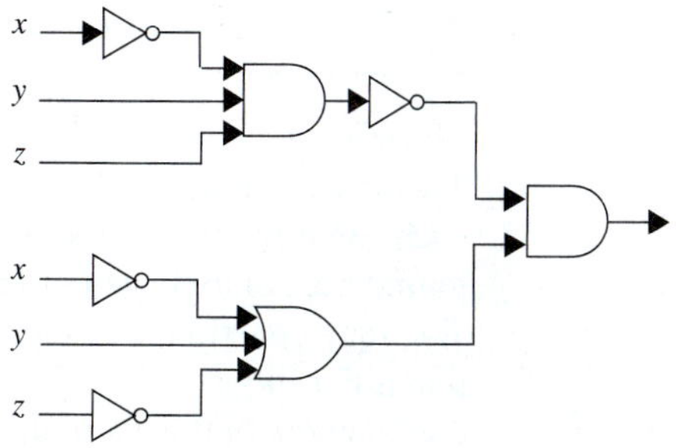

5. 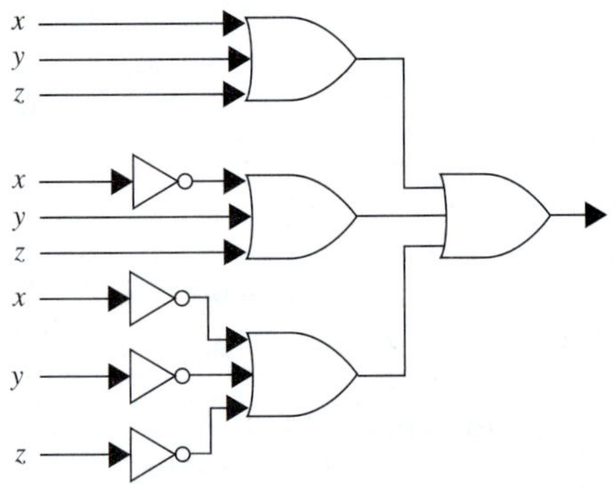

6. Construct circuits from inverters, AND gates, and OR gates to produce these outputs.
   a) $\overline{x} + y$
   b) $\overline{(x + y)}x$
   c) $xyz + \overline{x}\,\overline{y}\,\overline{z}$
   d) $\overline{(\overline{x} + z)(y + \overline{z})}$
7. Design a circuit that implements majority voting for five individuals.
8. Design a circuit for a light fixture controlled by four switches, where flipping one of the switches turns the light on when it is off and turns it off when it is on.
9. Show how the sum of two five-bit integers can be found using full and half adders.
10. Construct a circuit for a half subtractor using AND gates, OR gates, and inverters. A **half subtractor** has two bits as input and produces as output a difference bit and a borrow.
11. Construct a circuit for a full subtractor using AND gates, OR gates, and inverters. A **full subtractor** has two bits and a borrow as input, and produces as output a difference bit and a borrow.
12. Use the circuits from Exercises 10 and 11 to find the difference of two four-bit integers, where the first integer is greater than the second integer.

*13. Construct a circuit that compares the two-bit integers $(x_1x_0)_2$ and $(y_1y_0)_2$, returning an output of 1 when the first of these numbers is larger and an output of 0 otherwise.

*14. Construct a circuit that computes the product of the two-bit integers $(x_1x_0)_2$ and $(y_1y_0)_2$. The circuit should have four output bits for the bits in the product.

Two gates that are often used in circuits are NAND and NOR gates. When NAND or NOR gates are used to represent circuits, no other types of gates are needed. The notation for these gates is as follows:

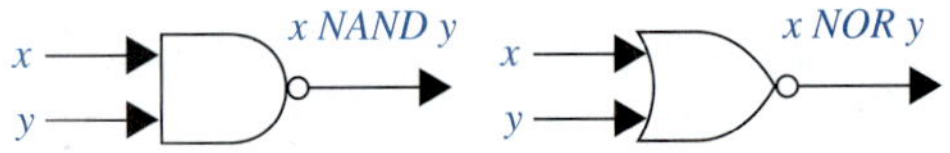

*15. Use NAND gates to construct circuits with these outputs.
   a) $\overline{x}$
   b) $x + y$
   c) xy
   d) $x \oplus y$

*16. Use NOR gates to construct circuits for the outputs given in Exercise 15.

*17. Construct a half adder using NAND gates.

*18. Construct a half adder using NOR gates.

A **multiplexer** is a switching circuit that produces as output one of a set of input bits based on the value of control bits.

19. Construct a multiplexer using AND gates, OR gates, and inverters that has as input the four bits x_0, x_1, x_2, and x_3 and the two control bits c_0 and c_1. Set up the circuit so that x_i is the output, where i is the value of the two-bit integer $(c_1c_0)_2$.

The **depth** of a combinatorial circuit can be defined by specifying that the depth of the initial input is 0 and if a gate has n different inputs at depths $d_1, d_2, \ldots, d_n$, respectively, then its outputs have depth equal to $\max(d_1, d_2, \ldots, d_n) + 1$; this value is also defined to be the depth of the gate. The depth of a combinatorial circuit is the maximum depth of the gates in the circuit.

20. Find the depth of
   a) the circuit constructed in Example 2 for majority voting among three people.
   b) the circuit constructed in Example 3 for a light controlled by two switches.
   c) the half adder shown in Figure 8.
   d) the full adder shown in Figure 9.

12.4 Minimization of Circuits

12.4.1 Introduction

The efficiency of a combinational circuit depends on the number and arrangement of its gates. The process of designing a combinational circuit begins with the table specifying the output for each combination of input values. We can always use the sum-of-products expansion of a circuit to find a set of logic gates that will implement this circuit. However, the sum-of-products expansion may contain many more terms than are necessary. Terms in a sum-of-products expansion that differ in just one variable, so that in one term this variable occurs and in the other term the complement of this variable occurs, can be combined. For instance, consider the circuit that has

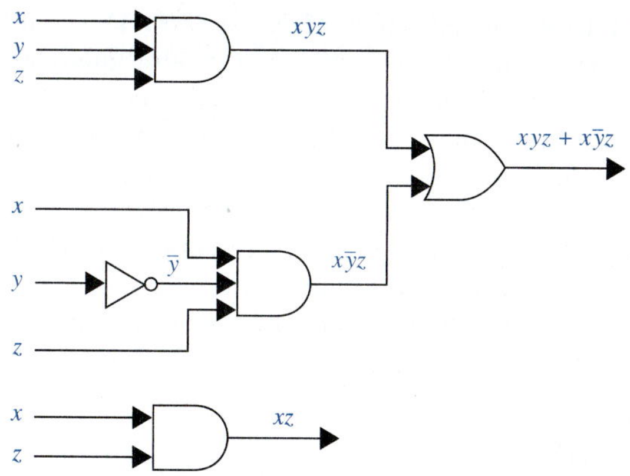

FIGURE 1 **Two circuits with the same output.**

output 1 if and only if $x = y = z = 1$ or $x = z = 1$ and $y = 0$. The sum-of-products expansion of this circuit is $xyz + x\bar{y}z$. The two products in this expansion differ in exactly one variable, namely, y. They can be combined as

$$\begin{aligned} xyz + x\bar{y}z &= (y + \bar{y})(xz) \\ &= 1 \cdot (xz) \\ &= xz. \end{aligned}$$

Hence, xz is a Boolean expression with fewer operators that represents the circuit. We show two different implementations of this circuit in Figure 1. The second circuit uses only one gate, whereas the first circuit uses three gates and an inverter.

This example shows that combining terms in the sum-of-products expansion of a circuit leads to a simpler expression for the circuit. We will describe two procedures that simplify sum-of-products expansions.

The goal of both procedures is to produce Boolean sums of Boolean products that represent a Boolean function with the fewest products of literals such that these products contain the fewest literals possible among all sums of products that represent a Boolean function. Finding such a sum of products is called **minimization of the Boolean function**. Minimizing a Boolean function makes it possible to construct a circuit for this function that uses the fewest gates and fewest inputs to the *AND* gates and *OR* gates in the circuit, among all circuits for the Boolean expression we are minimizing.

Until the early 1960s logic gates were individual components. To reduce costs it was important to use the fewest gates to produce a desired output. However, in the mid-1960s, integrated circuit technology was developed that made it possible to combine gates on a single chip. Even though it is now possible to build increasingly complex integrated circuits on chips at low cost, minimization of Boolean functions remains important.

Reducing the number of gates on a chip can lead to a more reliable circuit and can reduce the cost to produce the chip and reduces the energy consumption of the chip. Also, minimization makes it possible to fit more circuits on the same chip. Furthermore, minimization reduces the number of inputs to gates in a circuit. This reduces the time used by a circuit to compute its output. Moreover, the number of inputs to a gate may be limited because of the particular technology used to build logic gates.

The first procedure we will introduce, known as Karnaugh maps (or K-maps), was designed in the 1950s to help minimize circuits by hand. K-maps are useful in minimizing circuits with up to six variables, although they become rather complex even for five or six variables. The second procedure we will describe, the Quine–McCluskey method, was invented in the 1960s. It automates the process of minimizing combinatorial circuits and can be implemented as a computer program.

COMPLEXITY OF BOOLEAN FUNCTION MINIMIZATION Unfortunately, minimizing Boolean functions with many variables is a computationally intensive problem. It has been shown that this problem is an NP-complete problem (see Section 3.3 and [Ka93]), so the existence of a polynomial-time algorithm for minimizing Boolean circuits is unlikely. The Quine–McCluskey method has exponential complexity. In practice, it can be used only when the number of literals does not exceed ten. Since the 1970s a number of newer algorithms have been developed for minimizing combinatorial circuits (see [Ha93] and [KaBe04]). However, with the best algorithms yet devised, only circuits with no more than 25 variables can be minimized. Also, heuristic (or rule-of-thumb) methods can be used to substantially simplify, but not necessarily minimize, Boolean expressions with a larger number of literals.

12.4.2 Karnaugh Maps

Links

To reduce the number of terms in a Boolean expression representing a circuit, it is necessary to find terms to combine. There is a graphical method, called a **Karnaugh map** or **K-map**, for finding terms to combine for Boolean functions involving a relatively small number of variables. The method we will describe was introduced by Maurice Karnaugh in 1953. His method is based on earlier work by E. W. Veitch. (This method is usually applied only when the function involves six or fewer variables.) K-maps give us a visual method for simplifying sum-of-products expansions; they are not suited for mechanizing this process. We will first illustrate how K-maps are used to simplify expansions of Boolean functions in two variables. We will continue by showing how K-maps can be used to minimize Boolean functions in three variables and then in four variables. Then we will describe the concepts that can be used to extend K-maps to minimize Boolean functions in more than four variables.

	y	$\overline{y}$
x	xy	$x\overline{y}$
$\overline{x}$	$\overline{x}y$	$\overline{x}\,\overline{y}$

FIGURE 2 **K-maps in two variables.**

There are four possible minterms in the sum-of-products expansion of a Boolean function in the two variables x and y. A K-map for a Boolean function in these two variables consists of four cells, where a 1 is placed in the cell representing a minterm if this minterm is present in the expansion. Cells are said to be **adjacent** if the minterms that they represent differ in exactly one literal. For instance, the cell representing $\overline{x}y$ is adjacent to the cells representing xy and $\overline{x}\,\overline{y}$. The four cells and the terms that they represent are shown in Figure 2.

EXAMPLE 1 Find the K-maps for (a) $xy + \overline{x}y$, (b) $x\overline{y} + \overline{x}y$, and (c) $x\overline{y} + \overline{x}y + \overline{x}\,\overline{y}$.

Solution: We include a 1 in a cell when the minterm represented by this cell is present in the sum-of-products expansion. The three K-maps are shown in Figure 3. ◀

We can identify minterms that can be combined from the K-map. Whenever there are 1s in two adjacent cells in the K-map, the minterms represented by these cells can be combined into a product involving just one of the variables. For instance, $x\overline{y}$ and $\overline{x}\,\overline{y}$ are represented by adjacent cells and can be combined into $\overline{y}$, because $x\overline{y} + \overline{x}\,\overline{y} = (x + \overline{x})\overline{y} = \overline{y}$. Moreover, if 1s are in all four cells, the four minterms can be combined into one term, namely, the Boolean expression 1 that involves none of the variables. We circle blocks of cells in the K-map that represent minterms that can be combined and then find the corresponding sum of products. The goal is to identify

Links

Maurice Karnaugh

MAURICE KARNAUGH (1924–2022) Maurice Karnaugh, born in New York City, received his B.S. from the City College of New York and his M.S. and Ph.D. from Yale University. He was a member of the technical staff at Bell Laboratories from 1952 until 1966 and Manager of Research and Development at the Federal Systems Division of AT&T from 1966 to 1970. In 1970 he joined IBM as a member of the research staff. Karnaugh made fundamental contributions to the application of digital techniques in both computing and telecommunications. His interests later in life included knowledge-based systems in computers and heuristic search methods.

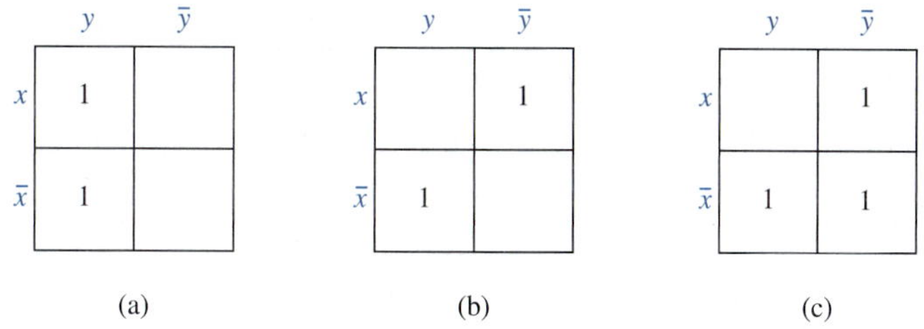

FIGURE 3 K-maps for the sum-of-products expansions in Example 1.

the largest possible blocks, and to cover all the 1s with the fewest blocks using the largest blocks first and always using the largest possible blocks.

EXAMPLE 2 Simplify the sum-of-products expansions given in Example 1.

Solution: The grouping of minterms is shown in Figure 4 using the K-maps for these expansions. Minimal expansions for these sums-of-products are (a) y, (b) $x\bar{y} + \bar{x}y$, and (c) $\bar{x} + \bar{y}$. ◀

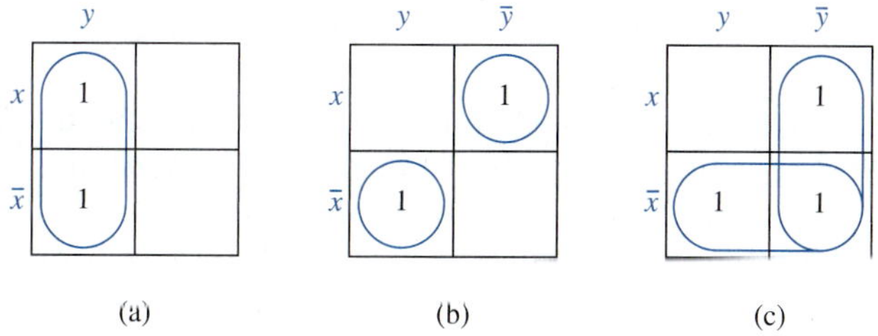

FIGURE 4 Simplifying the sum-of-products expansions from Example 2.

A K-map in three variables is a rectangle divided into eight cells. The cells represent the eight possible minterms in three variables. Two cells are said to be adjacent if the minterms that they represent differ in exactly one literal. One of the ways to form a K-map in three variables is shown in Figure 5(a). This K-map can be thought of as lying on a cylinder, as shown in Figure 5(b). On the cylinder, two cells have a common border if and only if they are adjacent.

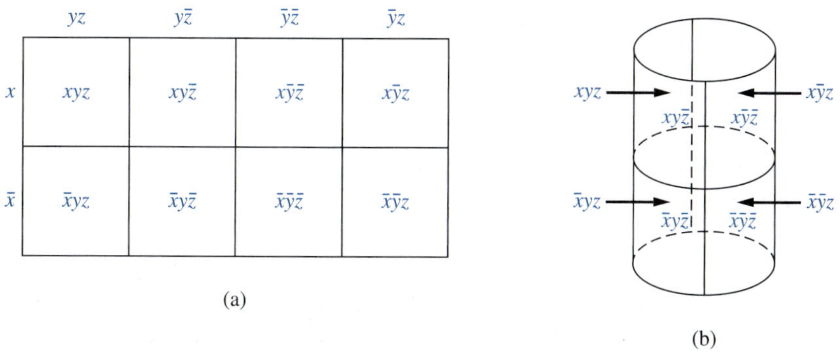

FIGURE 5 K-maps in three variables.

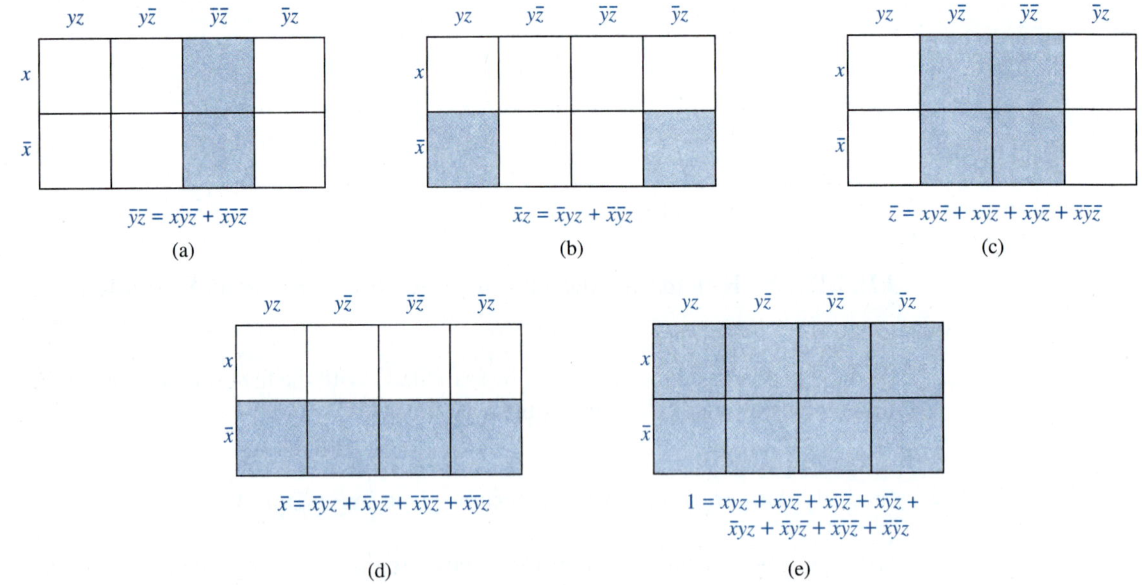

FIGURE 6 Blocks in K-maps in three variables.

To simplify a sum-of-products expansion in three variables, we use the K-map to identify blocks of minterms that can be combined. Blocks of two adjacent cells represent pairs of minterms that can be combined into a product of two literals; 2×2 and 4×1 blocks of cells represent minterms that can be combined into a single literal; and the block of all eight cells represents a product of no literals, namely, the function 1. In Figure 6, 1×2, 2×1, 2×2, 4×1, and 4×2 blocks and the products they represent are shown.

The product of literals corresponding to a block of all 1s in the K-map is called an **implicant** of the function being minimized. It is called a **prime implicant** if this block of 1s is not contained in a larger block of 1s representing the product of fewer literals than in this product.

The goal is to identify the largest possible blocks in the map and cover all the 1s in the map with the least number of blocks, using the largest blocks first. The largest possible blocks are always chosen, but we must always choose a block if it is the only block of 1s covering a 1 in the K-map. Such a block represents an **essential prime implicant**. By covering all the 1s in the map with blocks corresponding to prime implicants we can express the sum of products as a sum of prime implicants. Note that there may be more than one way to cover all the 1s using the least number of blocks.

Example 3 illustrates how K-maps in three variables are used.

EXAMPLE 3 Use K-maps to minimize these sum-of-products expansions.

(a) $xy\bar z + x\bar y\bar z + \bar x yz + \bar x\bar y\bar z$
(b) $x\bar y z + x\bar y\bar z + \bar x yz + \bar x\bar y z + \bar x\bar y\bar z$
(c) $xyz + xy\bar z + x\bar y z + x\bar y\bar z + \bar x yz + \bar x\bar y z + \bar x\bar y\bar z$
(d) $xy\bar z + x\bar y\bar z + \bar x\bar y z + \bar x\bar y\bar z$

Solution: The K-maps for these sum-of-products expansions are shown in Figure 7. The grouping of blocks shows that minimal expansions into Boolean sums of Boolean products are (a) $x\bar z + \bar y\bar z + \bar x yz$, (b) $\bar y + \bar x z$, (c) $x + \bar y + z$, and (d) $x\bar z + \bar x\bar y$. In part (d) note that the prime implicants $x\bar z$ and $\bar x\bar y$ are essential prime implicants, but the prime implicant $\bar y\bar z$ is a

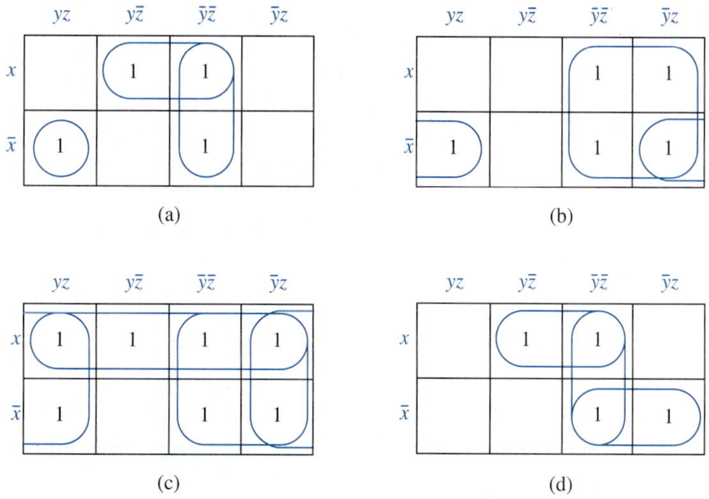

FIGURE 7 Using K-maps in three variables.

prime implicant that is not essential, because the cells it covers are covered by the other two prime implicants. ◀

A K-map in four variables is a square that is divided into 16 cells. The cells represent the 16 possible minterms in four variables. One of the ways to form a K-map in four variables is shown in Figure 8.

Two cells are adjacent if and only if the minterms they represent differ in one literal. Consequently, each cell is adjacent to four other cells. The K-map of a sum-of-products expansion in four variables can be thought of as lying on a torus, so that adjacent cells have a common boundary (see Exercise 28). The simplification of a sum-of-products expansion in four variables is carried out by identifying those blocks of 2, 4, 8, or 16 cells that represent minterms that can be combined. Each cell representing a minterm must either be used to form a product using fewer literals, or be included in the expansion. In Figure 9 some examples of blocks that represent products of three literals, products of two literals, and a single literal are illustrated.

As is the case in K-maps in two and three variables, the goal is to identify the largest blocks of 1s in the map that correspond to the prime implicants and to cover all the 1s using the fewest blocks needed, using the largest blocks first. The largest possible blocks are always used. Example 4 illustrates how K-maps in four variables are used.

	yz	$y\bar{z}$	$\bar{y}\bar{z}$	$\bar{y}z$
wx	$wxyz$	$wxy\bar{z}$	$wx\bar{y}\bar{z}$	$wx\bar{y}z$
$w\bar{x}$	$w\bar{x}yz$	$w\bar{x}y\bar{z}$	$w\bar{x}\bar{y}\bar{z}$	$w\bar{x}\bar{y}z$
$\bar{w}\bar{x}$	$\bar{w}\bar{x}yz$	$\bar{w}\bar{x}y\bar{z}$	$\bar{w}\bar{x}\bar{y}\bar{z}$	$\bar{w}\bar{x}\bar{y}z$
$\bar{w}x$	$\bar{w}xyz$	$\bar{w}xy\bar{z}$	$\bar{w}x\bar{y}\bar{z}$	$\bar{w}x\bar{y}z$

FIGURE 8 K-maps in four variables.

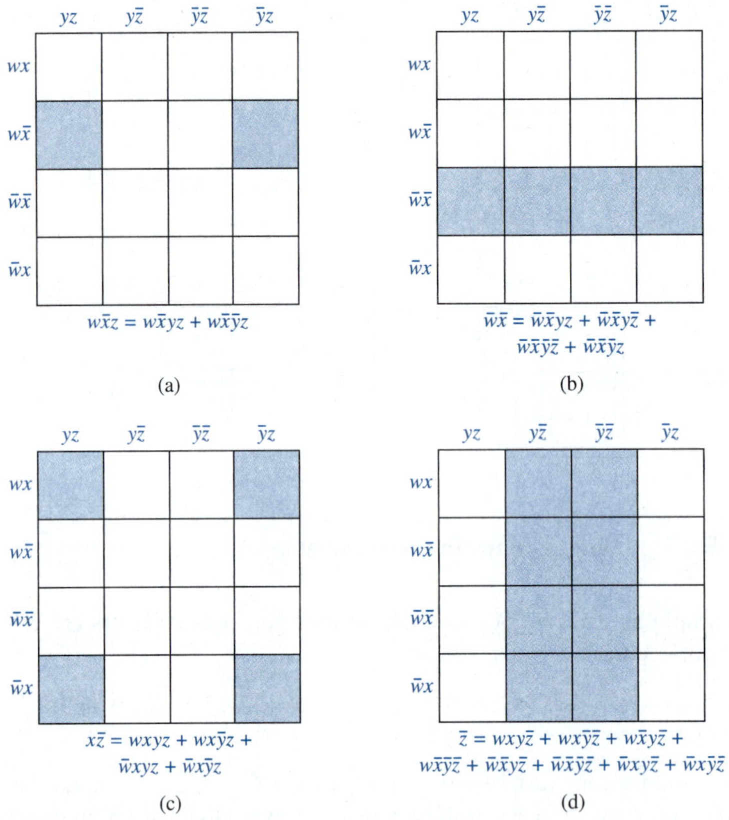

FIGURE 9 **Blocks in K-maps in four variables.**

EXAMPLE 4 Use K-maps to simplify these sum-of-products expansions.

(a) $wxyz + wxy\bar{z} + wx\bar{y}\,\bar{z} + w\bar{x}yz + w\bar{x}\,\bar{y}z + w\bar{x}\,\bar{y}\,\bar{z} + \bar{w}x\bar{y}z + \bar{w}\,\bar{x}yz + \bar{w}\,\bar{x}y\bar{z}$

(b) $wx\bar{y}\,\bar{z} + w\bar{x}yz + w\bar{x}y\bar{z} + w\bar{x}\,\bar{y}\,\bar{z} + \bar{w}x\bar{y}\,\bar{z} + \bar{w}\,\bar{x}y\bar{z} + \bar{w}\,\bar{x}\,\bar{y}\,\bar{z}$

(c) $wxy\bar{z} + wx\bar{y}\,\bar{z} + w\bar{x}yz + w\bar{x}y\bar{z} + w\bar{x}\,\bar{y}\,\bar{z} + \bar{w}xyz + \bar{w}xy\bar{z} + \bar{w}x\bar{y}\,\bar{z} + \bar{w}x\bar{y}z + \bar{w}\,\bar{x}y\bar{z} + \bar{w}\,\bar{x}\,\bar{y}\,\bar{z}$

Solution: The K-maps for these expansions are shown in Figure 10. Using the blocks shown leads to the sum of products (a) $wyz + wx\bar{z} + w\bar{x}\,\bar{y} + \bar{w}\,\bar{x}y + \bar{w}x\bar{y}z$, (b) $\bar{y}\,\bar{z} + w\bar{x}y + \bar{x}\,\bar{z}$, and (c) $\bar{z} + \bar{w}x + w\bar{x}y$. The reader should determine whether there are other choices of blocks in each part that lead to different sums of products representing these Boolean functions. ◀

K-maps can realistically be used to minimize Boolean functions with five or six variables, but beyond that, they are rarely used because they become extremely complicated. However, the concepts used in K-maps play an important role in newer algorithms. Furthermore, mastering these concepts helps you understand these newer algorithms and the computer-aided design (CAD) programs that implement them. As we develop these concepts, we will be able to illustrate them by referring back to our discussion of minimization of Boolean functions in three and in four variables.

The K-maps we used to minimize Boolean functions in two, three, and four variables are built using 2×2, 2×4, and 4×4 rectangles, respectively. Furthermore, corresponding cells in

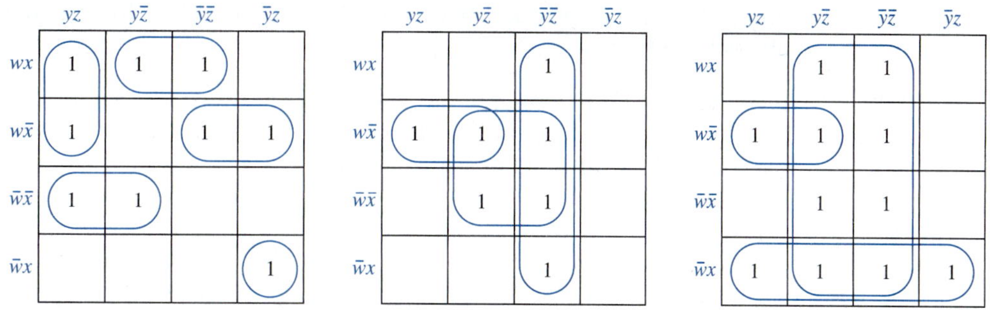

(a) (b) (c)

FIGURE 10 Using K-maps in four variables.

the top row and bottom row and in the leftmost column and rightmost column in each of these cases are considered adjacent because they represent minterms differing in only one literal. We can build K-maps for minimizing Boolean functions in more than four variables in a similar way. We use a rectangle containing $2^{\lfloor n/2 \rfloor}$ rows and $2^{\lceil n/2 \rceil}$ columns. (These K-maps contain 2^n cells because $\lceil n/2 \rceil + \lfloor n/2 \rfloor = n$.) The rows and columns need to be positioned so that the cells representing minterms differing in just one literal are adjacent or are considered adjacent by specifying additional adjacencies of rows and columns. To help (but not entirely) achieve this, the rows and columns of a K-map are arranged using Gray codes (see Section 10.5), where we associate bit strings and products by specifying that a 1 corresponds to the appearance of a variable and a 0 with the appearance of its complement. For example, in a 10-dimensional K-map, the Gray code 01110 used to label a row corresponds to the product $\overline{x}_1 x_2 x_3 x_4 \overline{x}_5$.

EXAMPLE 5 The K-maps we used to minimize Boolean functions with four variables have four rows and four columns. Both the rows and the columns are arranged using the Gray code 11,10,00,01. The rows represent products wx, $w\overline{x}$, $\overline{w}\,\overline{x}$, and $\overline{w}x$, respectively, and the columns correspond to the products yz, $y\overline{z}$, $\overline{y}\,\overline{z}$, and $\overline{y}z$, respectively. Using Gray codes and considering cells adjacent in the first and last rows and in the first and last columns, we ensured that minterms that differ in only one variable are always adjacent. ◀

EXAMPLE 6 To minimize Boolean functions in five variables we use K-maps with $2^3 = 8$ columns and $2^2 = 4$ rows. We label the four rows using the Gray code 11,10,00,01, corresponding to x_1x_2, $x_1\overline{x}_2$, $\overline{x}_1\overline{x}_2$, and $\overline{x}_1x_2$, respectively. We label the eight columns using the Gray code 111,110,100,101,001,000,010,011 corresponding to the terms $x_3x_4x_5$, $x_3x_4\overline{x}_5$, $x_3\overline{x}_4\overline{x}_5$, $x_3\overline{x}_4x_5$, $\overline{x}_3\overline{x}_4x_5$, $\overline{x}_3\overline{x}_4\overline{x}_5$, $\overline{x}_3x_4\overline{x}_5$, and $\overline{x}_3x_4x_5$, respectively. Using Gray codes to label columns and rows ensures that the minterms represented by adjacent cells differ in only one variable. However, to make sure all cells representing products that differ in only one variable are considered adjacent, we consider cells in the top and bottom rows to be adjacent, as well as cells in the first and eighth columns, the first and fourth columns, the second and seventh columns, the third and sixth columns, and the fifth and eighth columns (as the reader should verify). ◀

To use a K-map to minimize a Boolean function in n variables, we first draw a K-map of the appropriate size. We place 1s in all cells corresponding to minterms in the sum-of-products expansion of this function. We then identify all prime implicants of the Boolean function. To do this we look for the blocks consisting of 2^k clustered cells all containing a 1, where $1 \le k \le n$. These blocks correspond to the product of $n - k$ literals. (Exercise 33 asks the reader to verify this.) Furthermore, a block of 2^k cells each containing a 1 not contained in a block of 2^{k+1} cells each containing a 1 represents a prime implicant. The reason that this implicant is a prime implicant is that no product obtained by deleting a literal is also represented by a block of cells all containing 1s.

EXAMPLE 7 A block of eight cells representing a product of two literals in a K-map for minimizing Boolean functions in five variables all containing 1s is a prime implicant if it is not contained in a larger block of 16 cells all containing 1s representing a single literal. ◀

Once all prime implicants have been identified, the goal is to find the smallest possible subset of these prime implicants with the property that the cells representing these prime implicants cover all the cells containing a 1 in the K-map. We begin by selecting the essential prime implicants because each of these is represented by a block that covers a cell containing a 1 that is not covered by any other prime implicant. We add additional prime implicants to ensure that all 1s in the K-map are covered. When the number of variables is large, this last step can become exceedingly complicated.

12.4.3 Don't Care Conditions

Links

In some circuits we care only about the output for some combinations of input values, because other combinations of input values are not possible or never occur. This gives us freedom in producing a simple circuit with the desired output because the output values for all those combinations that never occur can be arbitrarily chosen. The values of the function for these combinations are called ***don't care* conditions**. A *d* is used in a K-map to mark those combinations of values of the variables for which the function can be arbitrarily assigned. In the minimization process we can assign 1s as values to those combinations of the input values that lead to the largest blocks in the K-map. This is illustrated in Example 8.

EXAMPLE 8 One way to code decimal expansions using bits is to use the four bits of the binary expansion of each digit in the decimal expansion. For instance, 873 is encoded as 100001110011. This encoding of a decimal expansion is called a **binary coded decimal expansion**. Because there are 16 blocks of four bits and only 10 decimal digits, there are six combinations of four bits that are not used to encode digits. Suppose that a circuit is to be built that produces an output of 1 if the decimal digit is 5 or greater and an output of 0 if the decimal digit is less than 5. How can this circuit be simply built using OR gates, AND gates, and inverters?

Solution: Let $F(w, x, y, z)$ denote the output of the circuit, where $wxyz$ is a binary expansion of a decimal digit. The values of F are shown in Table 1. The K-map for F, with ds in the

TABLE 1

Digit	w	x	y	z	F
0	0	0	0	0	0
1	0	0	0	1	0
2	0	0	1	0	0
3	0	0	1	1	0
4	0	1	0	0	0
5	0	1	0	1	1
6	0	1	1	0	1
7	0	1	1	1	1
8	1	0	0	0	1
9	1	0	0	1	1

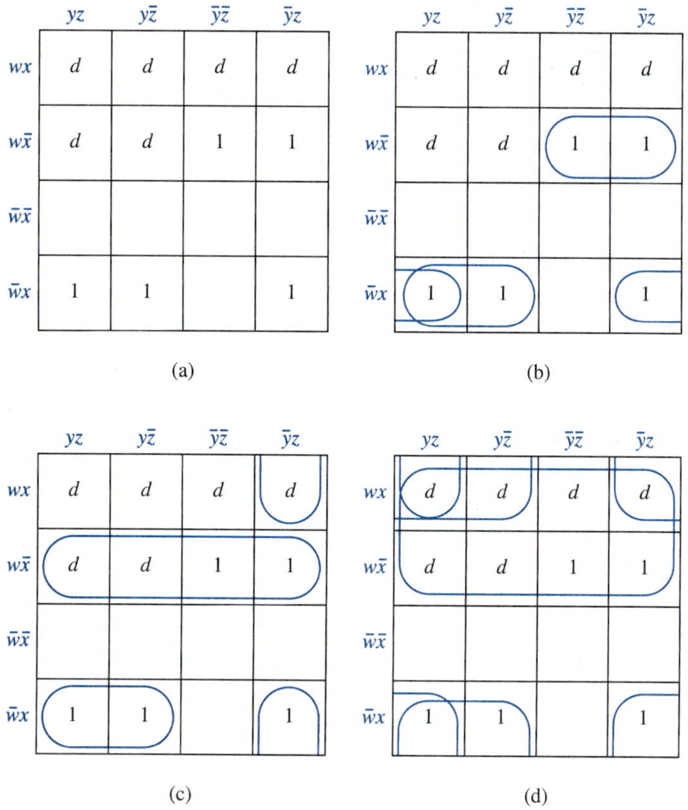

FIGURE 11 **The K-map for *F* showing its *don't care* positions.**

don't care positions, is shown in Figure 11(a). We can either include or exclude squares with *d*s from blocks. This gives us many possible choices for the blocks. For example, excluding all squares with *d*s and forming blocks, as shown in Figure 11(b), produces the expression $w\bar{x}\bar{y} + \bar{w}xy + \bar{w}xz$. Including some of the *d*s and excluding others and forming blocks, as shown in Figure 11(c), produces the expression $w\bar{x} + \bar{w}xy + x\bar{y}z$. Finally, including all the *d*s and using the blocks shown in Figure 11(d) produces the simplest sum-of-products expansion possible, namely, $F(x, y, z) = w + xy + xz$. ◀

12.4.4 The Quine–McCluskey Method

Links

We have seen that K-maps can be used to produce minimal expansions of Boolean functions as Boolean sums of Boolean products. However, K-maps are awkward to use when there are

Links

Stanford University News Service

EDWARD J. McCLUSKEY (1929–2016) Edward McCluskey attended Bowdoin College and M.I.T., where he received his doctorate in electrical engineering in 1956. He joined Bell Telephone Laboratories in 1955, remaining there until 1959. McCluskey was professor of electrical engineering at Princeton University from 1959 until 1966, also serving as director of the Computer Center at Princeton from 1961 to 1966. In 1967 he took a position as professor of computer science and electrical engineering at Stanford University, where he also served as director of the Digital Systems Laboratory from 1969 to 1978. McCluskey worked in a variety of areas in computer science, including fault-tolerant computing, computer architecture, testing, and logic design. He was the director of the Center for Reliable Computing at Stanford University and an ACM Fellow.

TABLE 2

Minterm	*Bit String*	*Number of 1s*
xyz	111	3
$x\overline{y}z$	101	2
$\overline{x}yz$	011	2
$\overline{x}\,\overline{y}z$	001	1
$\overline{x}\,\overline{y}\,\overline{z}$	000	0

more than four variables. Furthermore, the use of K-maps relies on visual inspection to identify terms to group. For these reasons there is a need for a procedure for simplifying sum-of-products expansions that can be mechanized. The Quine–McCluskey method is such a procedure. It can be used for Boolean functions in any number of variables. It was developed in the 1950s by W. V. Quine and E. J. McCluskey. Basically, the Quine–McCluskey method consists of two parts. The first part finds those terms that are candidates for inclusion in a minimal expansion as a Boolean sum of Boolean products. The second part determines which of these terms to actually use. We will use Example 9 to illustrate how, by successively combining implicants into implicants with one fewer literal, this procedure works.

EXAMPLE 9 We will show how the Quine–McCluskey method can be used to find a minimal expansion equivalent to

$$xyz + x\overline{y}z + \overline{x}yz + \overline{x}\,\overline{y}z + \overline{x}\,\overline{y}\,\overline{z}.$$

We will represent the minterms in this expansion by bit strings. The first bit will be 1 if x occurs and 0 if $\overline{x}$ occurs. The second bit will be 1 if y occurs and 0 if $\overline{y}$ occurs. The third bit will be 1 if z occurs and 0 if $\overline{z}$ occurs. We then group these terms according to the number of 1s in the corresponding bit strings. This information is shown in Table 2.

Minterms that can be combined are those that differ in exactly one literal. Hence, two terms that can be combined differ by exactly one in the number of 1s in the bit strings that represent them. When two minterms are combined into a product, this product contains two literals. A product in two literals is represented using a dash to denote the variable that does not occur. For instance, the minterms $x\overline{y}z$ and $\overline{x}\,\overline{y}z$, represented by bit strings 101 and 001, can be combined into $\overline{y}z$, represented by the string –01. All pairs of minterms that can be combined and the product formed from these combinations are shown in Table 3.

Next, all pairs of products of two literals that can be combined are combined into one literal. Two such products can be combined if they contain literals for the same two variables, and literals for only one of the two variables differ. In terms of the strings representing the

TABLE 3

			Step 1			*Step 2*		
	Term	*Bit String*		*Term*	*String*		*Term*	*String*
1	xyz	111	(1,2)	xz	1–1	(1,2,3,4)	z	– –1
2	$x\overline{y}z$	101	(1,3)	yz	–11			
3	$\overline{x}yz$	011	(2,4)	$\overline{y}z$	–01			
4	$\overline{x}\,\overline{y}z$	001	(3,4)	$\overline{x}z$	0–1			
5	$\overline{x}\,\overline{y}\,\overline{z}$	000	(4,5)	$\overline{x}\,\overline{y}$	00–			

TABLE 4

	xyz	$x\overline{y}z$	$\overline{x}yz$	$\overline{x}\,\overline{y}z$	$\overline{x}\,\overline{y}\,\overline{z}$
z	X	X	X	X	
$\overline{x}\,\overline{y}$				X	X

products, these strings must have a dash in the same position and must differ in exactly one of the other two slots. We can combine the products yz and $\overline{y}z$, represented by the strings –11 and –01, into z, represented by the string – –1. We show all the combinations of terms that can be formed in this way in Table 3.

In Table 3 we also indicate which terms have been used to form products with fewer literals; these terms will not be needed in a minimal expansion. The next step is to identify a minimal set of products needed to represent the Boolean function. We begin with all those products that were not used to construct products with fewer literals. Next, we form Table 4, which has a row for each candidate product formed by combining original terms, and a column for each original term; and we put an X in a position if the original term in the sum-of-products expansion was used to form this candidate product. In this case, we say that the candidate product **covers** the original minterm. We need to include at least one product that covers each of the original minterms. Consequently, whenever there is only one X in a column in the table, the product corresponding to the row this X is in must be used. From Table 4 we see that both z and $\overline{x}\,\overline{y}$ are needed. Hence, the final answer is $z + \overline{x}\,\overline{y}$. ◀

As was illustrated in Example 9, the Quine–McCluskey method uses this sequence of steps to simplify a sum-of-products expression.

1. Express each minterm in n variables by a bit string of length n with a 1 in the ith position if x_i occurs and a 0 in this position if $\overline{x}_i$ occurs.

2. Group the bit strings according to the number of 1s in them.

3. Determine all products in $n - 1$ variables that can be formed by taking the Boolean sum of minterms in the expansion. Minterms that can be combined are represented by bit strings that differ in exactly one position. Represent these products in $n - 1$ variables with strings that have a 1 in the ith position if x_i occurs in the product, a 0 in this position if $\overline{x}_i$ occurs, and a dash in this position if there is no literal involving x_i in the product.

Links

Maryclaire M. Quine

WILLARD VAN ORMAN QUINE (1908–2000) Willard Quine, born in Akron, Ohio, attended Oberlin College and later Harvard University, where he received his Ph.D. in philosophy in 1932. He became a Junior Fellow at Harvard in 1933 and was appointed to a position on the faculty there in 1936. He remained at Harvard his entire professional life, except for World War II, when he worked for the U.S. Navy decrypting messages from German submarines. Quine was always interested in algorithms, but not in hardware. He arrived at his discovery of what is now called the Quine–McCluskey method as a device for teaching mathematical logic, rather than as a method for simplifying switching circuits. Quine was one of the most famous philosophers of the twentieth century. He made fundamental contributions to the theory of knowledge, mathematical logic and set theory, and the philosophies of logic and language. His books, including *New Foundations of Mathematical Logic* published in 1937 and *Word and Object* published in 1960, have had a profound impact. Quine retired from Harvard in 1978 but continued to commute from his home in Beacon Hill to his office there. He used the 1927 Remington typewriter on which he prepared his doctoral thesis for his entire life. He even had an operation performed on this machine to add a few special symbols, removing the second period, the second comma, and the question mark. When asked whether he missed the question mark, he replied, "Well, you see, I deal in certainties." There is even a word *quine*, defined in the *New Hacker's Dictionary* as a program that generates a copy of its own source code as its complete output. Producing the shortest possible quine in a given programming language is a popular puzzle for hackers.

4. Determine all products in $n-2$ variables that can be formed by taking the Boolean sum of the products in $n-1$ variables found in the previous step. Products in $n-1$ variables that can be combined are represented by bit strings that have a dash in the same position and differ in exactly one position.
5. Continue combining Boolean products into products in fewer variables as long as possible.
6. Find all the Boolean products that arose that were not used to form a Boolean product in one fewer literal.
7. Find the smallest set of these Boolean products such that the sum of these products represents the Boolean function. This is done by forming a table showing which minterms are covered by which products. Every minterm must be covered by at least one product. The first step in using this table is to find all essential prime implicants. Each essential prime implicant must be included because it is the only prime implicant that covers one of the minterms. Once we have found essential prime implicants, we can simplify the table by eliminating the rows for minterms covered by these prime implicants. Furthermore, we can eliminate any prime implicants that cover a subset of minterms covered by another prime implicant (as the reader should verify). Moreover, we can eliminate from the table the row for a minterm if there is another minterm that is covered by a subset of the prime implicants that cover this minterm. This process of identifying essential prime implicants that must be included, followed by eliminating redundant prime implicants and identifying minterms that can be ignored, is iterated until the table does not change. At this point we use a backtracking procedure to find the optimal solution where we add prime implicants to the cover to find possible solutions, which we compare to the best solution found so far at each step.

A final example will illustrate how this procedure is used to simplify a sum-of-products expansion in four variables.

EXAMPLE 10 Use the Quine–McCluskey method to simplify the sum-of-products expansion $wxy\overline{z} + w\overline{x}yz + w\overline{x}y\overline{z} + \overline{w}x\overline{y}z + \overline{w}x\overline{y}z + \overline{w}\,\overline{x}yz + \overline{w}\,\overline{x}\,\overline{y}z$.

Solution: We first represent the minterms by bit strings and then group these terms together according to the number of 1s in the bit strings. This is shown in Table 5. All the Boolean products that can be formed by taking Boolean sums of these products are shown in Table 6.

The only products that were not used to form products in fewer variables are $\overline{w}z$, $wy\overline{z}$, $w\overline{x}y$, and $\overline{x}yz$. In Table 7 we show the minterms covered by each of these products. To cover these minterms we must include $\overline{w}z$ and $wy\overline{z}$, because these products are the only products that cover $\overline{w}xyz$ and $wxy\overline{z}$, respectively. Once these two products are included, we see that only one of the two products left is needed. Consequently, we can take either $\overline{w}z + wy\overline{z} + w\overline{x}y$ or $\overline{w}z + wy\overline{z} + \overline{x}yz$ as the final answer. ◀

TABLE 5

Term	*Bit String*	*Number of 1s*
$wxy\overline{z}$	1110	3
$w\overline{x}yz$	1011	3
$\overline{w}xyz$	0111	3
$w\overline{x}y\overline{z}$	1010	2
$\overline{w}x\overline{y}z$	0101	2
$\overline{w}\,\overline{x}yz$	0011	2
$\overline{w}\,\overline{x}\,\overline{y}z$	0001	1

TABLE 6

			Step 1			Step 2		
	Term	*Bit String*		*Term*	*String*		*Term*	*String*
1	$wxy\bar{z}$	1110	(1,4)	$wy\bar{z}$	1–10	(3,5,6,7)	$\bar{w}z$	0 – –1
2	$w\bar{x}yz$	1011	(2,4)	$w\bar{x}y$	101–			
3	$\bar{w}xyz$	0111	(2,6)	$\bar{x}yz$	–011			
4	$w\bar{x}y\bar{z}$	1010	(3,5)	$\bar{w}xz$	01–1			
5	$\bar{w}x\bar{y}z$	0101	(3,6)	$\bar{w}yz$	0–11			
6	$\bar{w}\bar{x}yz$	0011	(5,7)	$\bar{w}\bar{y}z$	0–01			
7	$\bar{w}\bar{x}\bar{y}z$	0001	(6,7)	$\bar{w}\bar{x}z$	00–1			

TABLE 7

	$wxy\bar{z}$	$w\bar{x}yz$	$\bar{w}xyz$	$w\bar{x}y\bar{z}$	$\bar{w}x\bar{y}z$	$\bar{w}\bar{x}yz$	$\bar{w}\bar{x}\bar{y}z$
$\bar{w}z$			X		X	X	X
$wy\bar{z}$	X			X			
$w\bar{x}y$		X		X			
$\bar{x}yz$		X				X	

Exercises

1. a) Draw a K-map for a function in two variables and put a 1 in the cell representing $\bar{x}y$.

b) What are the minterms represented by cells adjacent to this cell?

2. Find the sum-of-products expansions represented by each of these K-maps.

a)

	y	$\bar{y}$
x	1	
$\bar{x}$	1	1

b)

	y	$\bar{y}$
x	1	1
$\bar{x}$		

c)

	y	$\bar{y}$
x	1	1
$\bar{x}$	1	1

3. Draw the K-maps of these sum-of-products expansions in two variables.

a) $x\bar{y}$ **b)** $xy + \bar{x}\,\bar{y}$

c) $xy + x\bar{y} + \bar{x}y + \bar{x}\,\bar{y}$

4. Use a K-map to find a minimal expansion as a Boolean sum of Boolean products of each of these functions of the Boolean variables x and y.

a) $\bar{x}y + \bar{x}\,\bar{y}$

b) $xy + x\bar{y}$

c) $xy + x\bar{y} + \bar{x}y + \bar{x}\,\bar{y}$

5. a) Draw a K-map for a function in three variables. Put a 1 in the cell that represents $\bar{x}y\bar{z}$.

b) Which minterms are represented by cells adjacent to this cell?

6. Use K-maps to find simpler circuits with the same output as each of the circuits shown.

a)

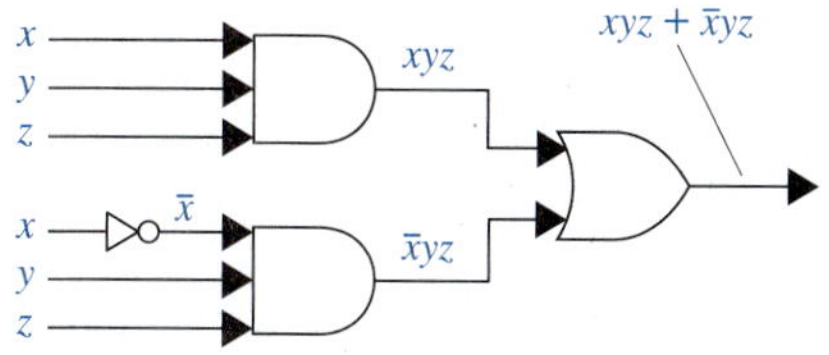

b)

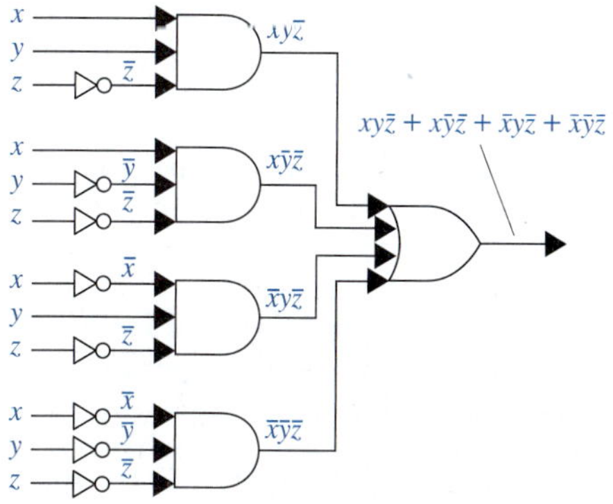

c) 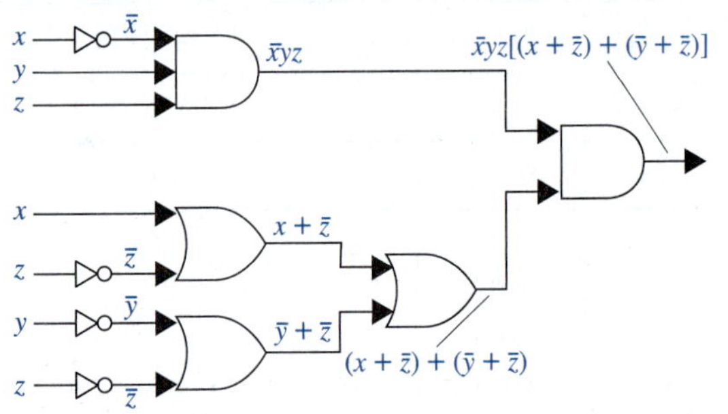

7. Draw the K-maps of these sum-of-products expansions in three variables.

a) $x\overline{y}\,\overline{z}$ **b)** $\overline{x}yz + \overline{x}\,\overline{y}\,\overline{z}$

c) $xyz + xy\overline{z} + \overline{x}y\overline{z} + \overline{x}\,\overline{y}z$

8. Construct a K-map for $F(x, y, z) = xz + yz + xy\overline{z}$. Use this K-map to find the implicants, prime implicants, and essential prime implicants of $F(x, y, z)$.

9. Construct a K-map for $F(x, y, z) = x\overline{z} + xyz + y\overline{z}$. Use this K-map to find the implicants, prime implicants, and essential prime implicants of $F(x, y, z)$.

10. Draw the 3-cube Q_3 and label each vertex with the minterm in the Boolean variables x, y, and z associated with the bit string represented by this vertex. For each literal in these variables indicate the 2-cube Q_2 that is a subgraph of Q_3 and represents this literal.

11. Draw the 4-cube Q_4 and label each vertex with the minterm in the Boolean variables w, x, y, and z associated with the bit string represented by this vertex. For each literal in these variables, indicate which 3-cube Q_3 that is a subgraph of Q_4 represents this literal. Indicate which 2-cube Q_2 that is a subgraph of Q_4 represents the products wz, $\overline{x}y$, and $\overline{y}\,\overline{z}$.

12. Use a K-map to find a minimal expansion as a Boolean sum of Boolean products of each of these functions in the variables x, y, and z.

a) $\overline{x}yz + \overline{x}\,\overline{y}z$

b) $xyz + xy\overline{z} + \overline{x}yz + \overline{x}y\overline{z}$

c) $xy\overline{z} + x\overline{y}z + x\overline{y}\,\overline{z} + \overline{x}yz + \overline{x}\,\overline{y}z$

d) $xyz + x\overline{y}z + x\overline{y}\,\overline{z} + \overline{x}yz + \overline{x}y\overline{z} + \overline{x}\,\overline{y}\,\overline{z}$

13. a) Draw a K-map for a function in four variables. Put a 1 in the cell that represents $\overline{w}xy\overline{z}$.

b) Which minterms are represented by cells adjacent to this cell?

14. Use a K-map to find a minimal expansion as a Boolean sum of Boolean products of each of these functions in the variables w, x, y, and z.

a) $wxyz + wx\overline{y}z + wx\overline{y}\,\overline{z} + w\overline{x}y\overline{z} + w\overline{x}\,\overline{y}z$

b) $wxy\overline{z} + wx\overline{y}z + w\overline{x}yz + \overline{w}x\overline{y}z + \overline{w}\,\overline{x}y\overline{z} + \overline{w}\,\overline{x}\,\overline{y}z$

c) $wxyz + wxy\overline{z} + wx\overline{y}z + w\overline{x}\,\overline{y}z + w\overline{x}\,\overline{y}\,\overline{z} + \overline{w}x\overline{y}z + \overline{w}\,\overline{x}y\overline{z} + \overline{w}\,\overline{x}\,\overline{y}z$

d) $wxyz + wxy\overline{z} + wx\overline{y}z + w\overline{x}yz + w\overline{x}y\overline{z} + \overline{w}xyz + \overline{w}\,\overline{x}yz + \overline{w}\,\overline{x}y\overline{z} + \overline{w}\,\overline{x}\,\overline{y}z$

15. Find the cells in a K-map for Boolean functions with five variables that correspond to each of these products.

a) $x_1x_2x_3x_4$ **b)** $\overline{x}_1x_3x_5$ **c)** x_2x_4

d) $\overline{x}_3\overline{x}_4$ **e)** x_3 **f)** $\overline{x}_5$

16. How many cells in a K-map for Boolean functions with six variables are needed to represent x_1, $\overline{x}_1x_6$, $\overline{x}_1x_2\overline{x}_6$, $x_2x_3x_4x_5$, and $x_1\overline{x}_2x_4\overline{x}_5$, respectively?

17. a) How many cells does a K-map in six variables have?

b) How many cells are adjacent to a given cell in a K-map in six variables?

18. Show that cells in a K-map for Boolean functions in five variables represent minterms that differ in exactly one literal if and only if they are adjacent or are in cells that become adjacent when the top and bottom rows and cells in the first and eighth columns, the first and fourth columns, the second and seventh columns, the third and sixth columns, and the fifth and eighth columns are considered adjacent.

19. Which rows and which columns of a 4×16 map for Boolean functions in six variables using the Gray codes 1111, 1110, 1010, 1011, 1001, 1000, 0000, 0001, 0011, 0010, 0110, 0111, 0101, 0100, 1100, 1101 to label the columns and 11, 10, 00, 01 to label the rows need to be considered adjacent so that cells that represent minterms that differ in exactly one literal are considered adjacent?

***20.** Use K-maps to find a minimal expansion as a Boolean sum of Boolean products of Boolean functions that have as input the binary code for each decimal digit and produce as output a 1 if and only if the digit corresponding to the input is

a) odd. **b)** not divisible by 3.

c) not 4, 5, or 6.

***21.** Suppose that there are five members on a committee, but that Smith and Jones always vote the opposite of Marcus. Design a circuit that implements majority voting of the committee using this relationship between votes.

22. Use the Quine–McCluskey method to simplify the sum-of-products expansions in Example 3.

23. Use the Quine–McCluskey method to simplify the sum-of-products expansions in Exercise 12.

24. Use the Quine–McCluskey method to simplify the sum-of-products expansions in Example 4.

25. Use the Quine–McCluskey method to simplify the sum-of-products expansions in Exercise 14.

***26.** Explain how K-maps can be used to simplify product-of-sums expansions in three variables. [*Hint:* Mark with a 0 all the maxterms in an expansion and combine blocks of maxterms.]

27. Use the method from Exercise 26 to simplify the product-of-sums expansion $(x + y + z)(x + y + \overline{z})(x + \overline{y} + \overline{z})(x + \overline{y} + z)(\overline{x} + y + z)$.

***28.** Draw a K-map for the 16 minterms in four Boolean variables on the surface of a torus.

29. Build a circuit using OR gates, AND gates, and inverters that produces an output of 1 if a decimal digit, encoded using a binary coded decimal expansion, is divisible by 3, and an output of 0 otherwise.

In Exercises 30–32 find a minimal sum-of-products expansion, given the K-map shown with *don't care* conditions indicated with *d*s.

30.

	yz	$y\bar{z}$	$\bar{y}\bar{z}$	$\bar{y}z$
wx	d	1	d	1
$w\bar{x}$		d	d	
$\bar{w}\bar{x}$		d	1	
$\bar{w}x$		1	d	

31.

	yz	$y\bar{z}$	$\bar{y}\bar{z}$	$\bar{y}z$
wx	1			1
$w\bar{x}$		d	1	
$\bar{w}\bar{x}$		1	d	
$\bar{w}x$	d			d

32.

	yz	$y\bar{z}$	$\bar{y}\bar{z}$	$\bar{y}z$
wx		d	d	1
$w\bar{x}$	d	d	1	d
$\bar{w}\bar{x}$				
$\bar{w}x$	1	1	1	d

33. Show that products of k literals correspond to 2^{n-k}-dimensional subcubes of the n-cube Q_n, where the vertices of the cube correspond to the minterms represented by the bit strings labeling the vertices, as described in Example 8 of Section 10.2.

Key Terms and Results

TERMS

Boolean variable: a variable that assumes only the values 0 and 1

$\bar{x}$ **(complement of x):** an expression with the value 1 when x has the value 0 and the value 0 when x has the value 1

$x \cdot y$ **(or xy) (Boolean product** or **conjunction of x and y):** an expression with the value 1 when both x and y have the value 1 and the value 0 otherwise

$x + y$ **(Boolean sum** or **disjunction of x and y):** an expression with the value 1 when either x or y, or both, has the value 1, and 0 otherwise

Boolean expressions: the expressions obtained recursively by specifying that $0, 1, x_1, \ldots, x_n$ are Boolean expressions and $\overline{E}_1, (E_1 + E_2)$, and (E_1E_2) are Boolean expressions if E_1 and E_2 are

dual of a Boolean expression: the expression obtained by interchanging + signs and · signs and interchanging 0s and 1s

Boolean function of degree n: a function from B^n to B where $B = \{0, 1\}$

Boolean algebra: a set B with two binary operations $\vee$ and $\wedge$, elements 0 and 1, and a complementation operator $\bar{}$ that satisfies the identity, complement, associative, commutative, and distributive laws

literal of the Boolean variable x: either x or $\bar{x}$

minterm of $x_1, x_2, \ldots, x_n$: a Boolean product $y_1y_2 \cdots y_n$, where each y_i is either x_i or $\bar{x}_i$

sum-of-products expansion (or **disjunctive normal form**): the representation of a Boolean function as a disjunction of minterms

functionally complete: a set of Boolean operators is called functionally complete if every Boolean function can be represented using these operators

$x \mid y$ **(or x *NAND* y):** the expression that has the value 0 when both x and y have the value 1 and the value 1 otherwise

$x \downarrow y$ **(or x *NOR* y):** the expression that has the value 0 when either x or y or both have the value 1 and the value 0 otherwise

inverter: a device that accepts the value of a Boolean variable as input and produces the complement of the input

OR gate: a device that accepts the values of two or more Boolean variables as input and produces their Boolean sum as output

AND gate: a device that accepts the values of two or more Boolean variables as input and produces their Boolean product as output

half adder: a circuit that adds two bits, producing a sum bit and a carry bit

full adder: a circuit that adds two bits and a carry, producing a sum bit and a carry bit

K-map for n variables: a rectangle divided into 2^n cells where each cell represents a minterm in the variables

minimization of a Boolean function: representing a Boolean function as the sum of the fewest products of literals such that these products contain the fewest literals possible among all sums of products that represent this Boolean function

implicant of a Boolean function: a product of literals with the property that if this product has the value 1, then the value of this Boolean function is 1

prime implicant of a Boolean function: a product of literals that is an implicant of the Boolean function and no product obtained by deleting a literal is also an implicant of this function

essential prime implicant of a Boolean function: a prime implicant of the Boolean function that must be included in a minimization of this function

don't care condition: a combination of input values for a circuit that is not possible or never occurs

RESULTS

The identities for Boolean algebra (see Table 5 in Section 12.1).

An identity between Boolean functions represented by Boolean expressions remains valid when the duals of both sides of the identity are taken.

Every Boolean function can be represented by a sum-of-products expansion.

Each of the sets $\{+, ^{-}\}$ and $\{\cdot, ^{-}\}$ is functionally complete.

Each of the sets $\{\downarrow\}$ and $\{\mid\}$ is functionally complete.

The use of K-maps to minimize Boolean expressions.

The Quine–McCluskey method for minimizing Boolean expressions.

Review Questions

1. Define a Boolean function of degree n.

2. How many Boolean functions of degree two are there?

3. Give a recursive definition of the set of Boolean expressions.

4. **a)** What is the dual of a Boolean expression?
b) What is the duality principle? How can it be used to find new identities involving Boolean expressions?

5. Explain how to construct the sum-of-products expansion of a Boolean function.

6. **a)** What does it mean for a set of operators to be functionally complete?
b) Is the set $\{+, \cdot\}$ functionally complete?
c) Are there sets of a single operator that are functionally complete?

7. Explain how to build a circuit for a light controlled by two switches using OR gates, AND gates, and inverters.

8. Construct a half adder using OR gates, AND gates, and inverters.

9. Is there a single type of logic gate that can be used to build all circuits that can be built using OR gates, AND gates, and inverters?

10. **a)** Explain how K-maps can be used to simplify sum-of-products expansions in three Boolean variables.
b) Use a K-map to simplify the sum-of-products expansion $xyz + x\overline{y}z + x\overline{y}\,\overline{z} + \overline{x}yz + \overline{x}\,\overline{y}\,\overline{z}$.

11. **a)** Explain how K-maps can be used to simplify sum-of-products expansions in four Boolean variables.
b) Use a K-map to simplify the sum-of-products expansion $wxyz + wxy\overline{z} + wx\overline{y}z + wx\overline{y}\,\overline{z} + w\overline{x}yz + w\overline{x}\,\overline{y}z + \overline{w}xyz + \overline{w}\,\overline{x}yz + \overline{w}\,\overline{x}y\overline{z}$.

12. **a)** What is a *don't care* condition?
b) Explain how *don't care* conditions can be used to build a circuit using OR gates, AND gates, and inverters that produces an output of 1 if a decimal digit is 6 or greater, and an output of 0 if this digit is less than 6.

13. **a)** Explain how to use the Quine–McCluskey method to simplify sum-of-products expansions.
b) Use this method to simplify $xy\overline{z} + x\overline{y}\,\overline{z} + \overline{x}y\overline{z} + \overline{x}\,\overline{y}z + \overline{x}\,\overline{y}\,\overline{z}$.

Supplementary Exercises

1. For which values of the Boolean variables x, y, and z does
a) $x + y + z = xyz$?
b) $x(y + z) = x + yz$?
c) $\overline{x}\,\overline{y}\,\overline{z} = x + y + z$?

2. Let x and y belong to $\{0, 1\}$. Does it necessarily follow that $x = y$ if there exists a value z in $\{0, 1\}$ such that
a) $xz = yz$? **b)** $x + z = y + z$?
c) $x \oplus z = y \oplus z$? **d)** $x \downarrow z = y \downarrow z$?
e) $x \mid z = y \mid z$?

A Boolean function F is called **self-dual** if and only if $F(x_1, \ldots, x_n) = \overline{F(\overline{x}_1, \ldots, \overline{x}_n)}$.

3. Which of these functions are self-dual?
a) $F(x, y) = x$ **b)** $F(x, y) = xy + \overline{x}\,\overline{y}$
c) $F(x, y) = x + y$ **d)** $F(x, y) = xy + \overline{x}y$

4. Give an example of a self-dual Boolean function of three variables.

***5.** How many Boolean functions of degree n are self-dual?

We define the relation $\leq$ on the set of Boolean functions of degree n so that $F \leq G$, where F and G are Boolean functions if and only if $G(x_1, x_2, \ldots, x_n) = 1$ whenever $F(x_1, x_2, \ldots, x_n) = 1$.

6. Determine whether $F \leq G$ or $G \leq F$ for the following pairs of functions.
a) $F(x, y) = x$, $G(x, y) = x + y$
b) $F(x, y) = x + y$, $G(x, y) = xy$
c) $F(x, y) = \overline{x}$, $G(x, y) = x + y$

7. Show that if F and G are Boolean functions of degree n, then
a) $F \leq F + G$. **b)** $FG \leq F$.

8. Show that if F, G, and H are Boolean functions of degree n, then $F + G \leq H$ if and only if $F \leq H$ and $G \leq H$.

***9.** Show that the relation $\leq$ is a partial ordering on the set of Boolean functions of degree n.

*10. Draw the Hasse diagram for the poset consisting of the set of the 16 Boolean functions of degree two (shown in Table 3 of Section 12.1) with the partial ordering $\leq$.

*11. For each of these equalities either prove it is an identity or find a set of values of the variables for which it does not hold.

a) $x \mid (y \mid z) = (x \mid y) \mid z$

b) $x \downarrow (y \downarrow z) = (x \downarrow y) \downarrow (x \downarrow z)$

c) $x \downarrow (y \mid z) = (x \downarrow y) \mid (x \downarrow z)$

Define the Boolean operator $\odot$ as follows: $1 \odot 1 = 1$, $1 \odot 0 = 0$, $0 \odot 1 = 0$, and $0 \odot 0 = 1$.

12. Show that $x \odot y = xy + \overline{x}\,\overline{y}$.

13. Show that $x \odot y = \overline{(x \oplus y)}$.

14. Show that each of these identities holds.

a) $x \odot x = 1$ **b)** $x \odot \overline{x} = 0$

c) $x \odot y = y \odot x$

15. Is it always true that $(x \odot y) \odot z = x \odot (y \odot z)$?

*16. Determine whether the set $\{\odot\}$ is functionally complete.

*17. How many of the 16 Boolean functions in two variables x and y can be represented using only the given set of operators, variables x and y, and values 0 and 1?

a) $\{\overline{\ }\}$ **b)** $\{\cdot\}$ **c)** $\{+\}$ **d)** $\{\cdot, +\}$

The notation for an **XOR gate**, which produces the output $x \oplus y$ from x and y, is as follows:

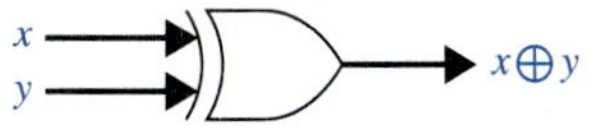

18. Determine the output of each of these circuits.

a)

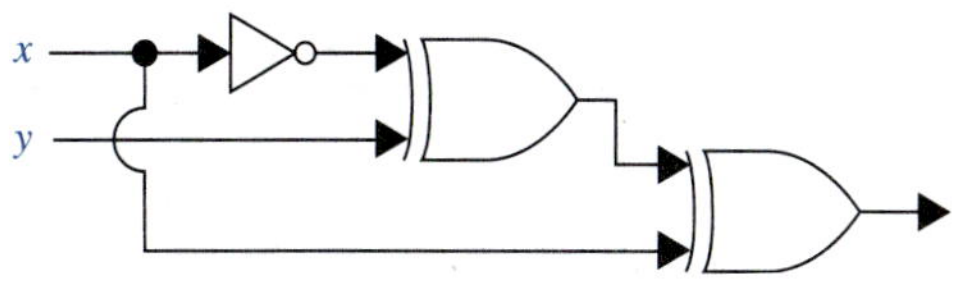

b)

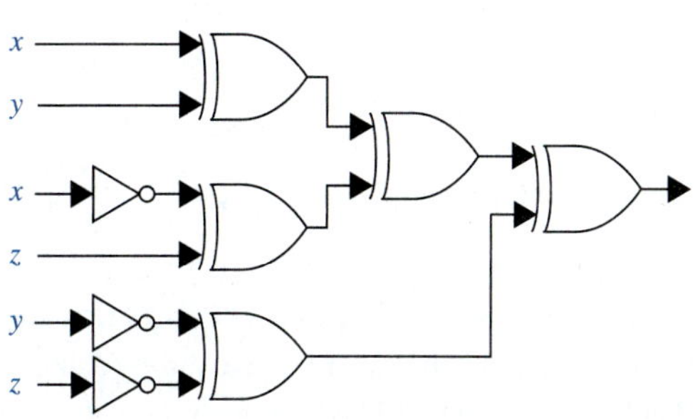

19. Show how a half adder can be constructed using fewer gates than are used in Figure 8 of Section 12.3 when XOR gates can be used in addition to OR gates, AND gates, and inverters.

20. Design a circuit that determines whether three or more of four individuals on a committee vote yes on an issue, where each individual uses a switch for the voting.

A **threshold gate** produces an output y that is either 0 or 1 given a set of input values for the Boolean variables $x_1, x_2, \ldots, x_n$. A threshold gate has a **threshold value** T, which is a real number, and **weights** $w_1, w_2, \ldots, w_n$, each of which is a real number. The output y of the threshold gate is 1 if and only if $w_1x_1 + w_2x_2 + \cdots + w_nx_n \geq T$. The threshold gate with threshold value T and weights $w_1, w_2, \ldots, w_n$ is represented by the following diagram. Threshold gates are useful in modeling in neurophysiology and in artificial intelligence.

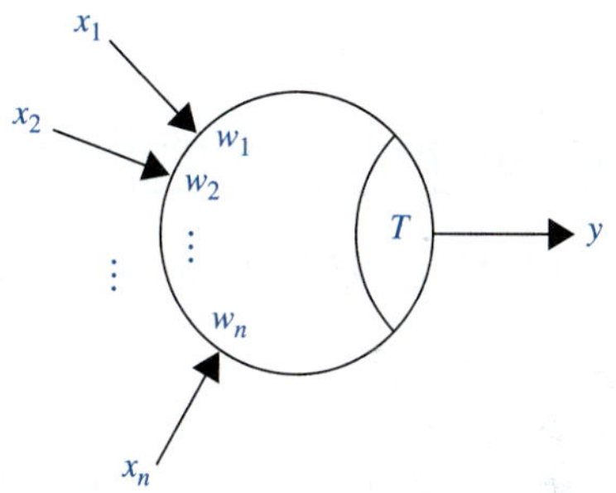

21. A threshold gate represents a Boolean function. Find a Boolean expression for the Boolean function represented by this threshold gate.

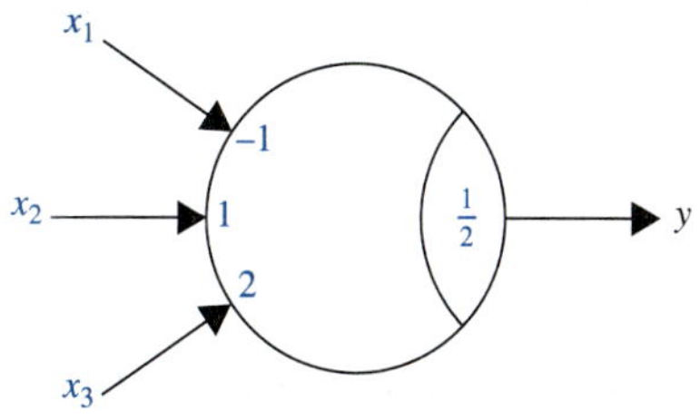

22. A Boolean function that can be represented by a threshold gate is called a **threshold function**. Show that each of these functions is a threshold function.

a) $F(x) = \overline{x}$ **b)** $F(x, y) = x + y$
c) $F(x, y) = xy$ **d)** $F(x, y) = x \mid y$
e) $F(x, y) = x \downarrow y$ **f)** $F(x, y, z) = x + yz$
g) $F(w, x, y, z) = w + xy + z$
h) $F(w, x, y, z) = wxz + x\overline{y}z$

*23. Show that $F(x, y) = x \oplus y$ is not a threshold function.

*24. Show that $F(w, x, y, z) = wx + yz$ is not a threshold function.

Computer Projects

Write programs with these input and output.

1. Given the values of two Boolean variables x and y, find the values of $x + y$, xy, $x \oplus y$, $x \mid y$, and $x \downarrow y$.
2. Construct a table listing the set of values of all 256 Boolean functions of degree three.
3. Given the values of a Boolean function in n variables, where n is a positive integer, construct the sum-of-products expansion of this function.
4. Given the table of values of a Boolean function, express this function using only the operators $\cdot$ and $\overline{}$.
5. Given the table of values of a Boolean function, express this function using only the operators $+$ and $\overline{}$.
*6. Given the table of values of a Boolean function, express this function using only the operator $\mid$.
*7. Given the table of values of a Boolean function, express this function using only the operator $\downarrow$.
8. Given the table of values of a Boolean function of degree three, construct its K-map.
9. Given the table of values of a Boolean function of degree four, construct its K-map.
**10. Given the table of values of a Boolean function, use the Quine–McCluskey method to find a minimal sum-of-products representation of this function.
11. Given a threshold value and a set of weights for a threshold gate and the values of the n Boolean variables in the input, determine the output of this gate.
12. Given a positive integer n, construct a random Boolean expression in n variables in disjunctive normal form.

Computations and Explorations

Use a computational program or programs you have written to do these exercises.

1. Compute the number of Boolean functions of degrees seven, eight, nine, and ten.
2. Construct a table of the Boolean functions of degree three.
3. Construct a table of the Boolean functions of degree four.
4. Express each of the different Boolean expressions in three variables in disjunctive normal form with just the *NAND* operator, using as few *NAND* operators as possible. What is the largest number of *NAND* operators required?
5. Express each of the different Boolean expressions in disjunctive normal form in four variables using just the *NOR* operator, with as few *NOR* operators as possible. What is the largest number of *NOR* operators required?
6. Randomly generate 10 different Boolean expressions in four variables and determine the average number of steps required to minimize them using the Quine–McCluskey method.
7. Randomly generate 10 different Boolean expressions in five variables and determine the average number of steps required to minimize them using the Quine–McCluskey method.

Writing Projects

Respond to these with essays using outside sources.

1. Describe some of the early machines devised to solve problems in logic, such as the Stanhope Demonstrator, Jevons's Logic Machine, and the Marquand Machine.
2. Explain the difference between combinational circuits and sequential circuits. Then explain how *flip-flops* are used to build sequential circuits.
3. Define a *shift register* and discuss how shift registers are used. Show how to build shift registers using flip-flops and logic gates.
4. Show how *multipliers* can be built using logic gates.
5. Find out how logic gates are physically constructed. Discuss whether NAND and NOR gates are used in building circuits.
6. Explain how *dependency notation* can be used to describe complicated switching circuits.
7. Describe how multiplexers are used to build switching circuits.
8. Explain the advantages of using threshold gates to construct switching circuits. Illustrate this by using threshold gates to construct half and full adders.

9. Describe the concept of *hazard-free switching circuits* and give some of the principles used in designing such circuits.
10. Explain how to use K-maps to minimize functions of six variables.
11. Discuss the ideas used by newer methods for minimizing Boolean functions, such as Espresso. Explain how these methods can help solve minimization problems in as many as 25 variables.
12. Describe what is meant by the *functional decomposition* of a Boolean function of n variables and discuss procedures for decomposing Boolean functions into a composition of Boolean functions with fewer variables.

CHAPTER

13 Modeling Computation

13.1 Languages and Grammars

13.2 Finite-State Machines with Output

13.3 Finite-State Machines with No Output

13.4 Language Recognition

13.5 Turing Machines

Computers can perform many tasks. Given a task, two questions arise. The first is: Can it be carried out using a computer? Once we know that this first question has an affirmative answer, we can ask the second question: How can the task be carried out? Models of computation are used to help answer these questions.

We will study three types of structures used in models of computation, namely, grammars, finite-state machines, and Turing machines. Grammars are used to generate the words of a language and to determine whether a word is in a language. Formal languages, which are generated by grammars, provide models both for natural languages, such as English, and for programming languages, such as Pascal, Fortran, Prolog, C, and Java. In particular, grammars are extremely important in the construction and theory of compilers. The grammars that we will discuss were first used by the American linguist Noam Chomsky in the 1950s.

Various types of finite-state machines are used in modeling. All finite-state machines have a set of states, including a starting state, an input alphabet, and a transition function that assigns a next state to every pair of a state and an input. The states of a finite-state machine give it limited memory capabilities. Some finite-state machines produce an output symbol for each transition; these machines can be used to model many kinds of machines, including vending machines, delay machines, binary adders, and language recognizers. We will also study finite-state machines that have no output but do have final states. Such machines are extensively used in language recognition. The strings that are recognized are those that take the starting state to a final state. The concepts of grammars and finite-state machines can be tied together. We will characterize those sets that are recognized by a finite-state machine and show that these are precisely the sets that are generated by a certain type of grammar.

Finally, we will introduce the concept of a Turing machine. We will show how Turing machines can be used to recognize sets. We will also show how Turing machines can be used to compute number-theoretic functions. We will discuss the Church–Turing thesis, which states that every effective computation can be carried out using a Turing machine. We will explain how Turing machines can be used to study the difficulty of solving certain classes of problems. In particular, we will describe how Turing machines are used to classify problems as tractable versus intractable and solvable versus unsolvable.

13.1 Languages and Grammars

13.1.1 Introduction

Words in the English language can be combined in various ways. The grammar of English tells us whether a combination of words is a valid sentence. For instance, *the frog writes neatly* is a valid sentence, because it is formed from a noun phrase, *the frog*, made up of the article *the* and the noun *frog*, followed by a verb phrase, *writes neatly*, made up of the verb *writes* and the adverb *neatly*. We do not care that this is a nonsensical statement, because we are concerned only with the **syntax**, or form, of the sentence, and not its **semantics**, or meaning. We also note that the combination of words *swims quickly mathematics* is not a valid sentence because it does not follow the rules of English grammar.

The syntax of a **natural language**, that is, a spoken language, such as English, French, German, or Spanish, is extremely complicated. In fact, it does not seem possible to specify all the rules of syntax for a natural language. Research in the automatic translation of one language to another has led to the concept of a **formal language**, which, unlike a natural language, is specified by a well-defined set of rules of syntax. Rules of syntax are important not only in linguistics, the study of natural languages, but also in the study of programming languages.

We will describe the sentences of a formal language using a grammar. The use of grammars helps when we consider the two classes of problems that arise most frequently in applications to programming languages: (1) How can we determine whether a combination of words is a valid sentence in a formal language? (2) How can we generate the valid sentences of a formal language? Before giving a technical definition of a grammar, we will describe an example of a grammar that generates a subset of English. This subset of English is defined using a list of rules that describe how a valid sentence can be produced. We specify that

1. a **sentence** is made up of a **noun phrase** followed by a **verb phrase**;
2. a **noun phrase** is made up of an **article** followed by an **adjective** followed by a **noun**, or
3. a **noun phrase** is made up of an **article** followed by a **noun**;
4. a **verb phrase** is made up of a **verb** followed by an **adverb**, or
5. a **verb phrase** is made up of a **verb**;
6. an **article** is *a*, or
7. an **article** is *the*;
8. an **adjective** is *large*, or
9. an **adjective** is *hungry*;
10. a **noun** is *rabbit*, or
11. a **noun** is *mathematician*;
12. a **verb** is *eats*, or
13. a **verb** is *hops*;
14. an **adverb** is *quickly*, or
15. an **adverb** is *wildly*.

From these rules we can form valid sentences using a series of replacements until no more rules can be used. For instance, we can follow the sequence of replacements:

sentence
noun phrase verb phrase
article adjective noun verb phrase
article adjective noun verb adverb
the **adjective noun verb adverb**
the large **noun verb adverb**
the large rabbit **verb adverb**
the large rabbit hops **adverb**
the large rabbit hops quickly

to obtain a valid sentence. It is also easy to see that some other valid sentences are: *a hungry mathematician eats wildly, a large mathematician hops, the rabbit eats quickly*, and so on. Also, we can see that *the quickly eats mathematician* is not a valid sentence.

13.1.2 Phrase-Structure Grammars

Before we give a formal definition of a grammar, we introduce a little terminology.

Definition 1 A *vocabulary* (or *alphabet*) V is a finite, nonempty set of elements called *symbols*. A *word* (or *sentence*) over V is a string of finite length of elements of V. The *empty string* or *null string*, denoted by λ (and sometimes by ϵ), is the string containing no symbols. The set of all words over V is denoted by V^*. A *language over* V is a subset of V^*.

Note that λ, the empty string, is the string containing no symbols. It is different from $\emptyset$, the empty set. It follows that $\{\lambda\}$ is the set containing exactly one string, namely, the empty string.

Languages can be specified in various ways. One way is to list all the words in the language. Another is to give some criteria that a word must satisfy to be in the language. In this section, we describe another important way to specify a language, namely, through the use of a grammar, such as the set of rules we gave in the introduction to this section. A grammar provides a set of symbols of various types and a set of rules for producing words. More precisely, a grammar has a **vocabulary** V, which is a set of symbols used to derive members of the language. Some of the elements of the vocabulary cannot be replaced by other symbols. These are called **terminals**, and the other members of the vocabulary, which can be replaced by other symbols, are called **nonterminals**. The sets of terminals and nonterminals are usually denoted by T and N, respectively. In the example given in the introduction of the section, the set of terminals is {*a, the, rabbit, mathematician, hops, eats, quickly, wildly*}, and the set of nonterminals is {**sentence, noun phrase, verb phrase, adjective, article, noun, verb, adverb**}. There is a special member of the vocabulary called the **start symbol**, denoted by S, which is the element of the vocabulary that we always begin with. In the example in the introduction, the start symbol is **sentence**. The rules that specify when we can replace a string from V^*, the set of all strings of elements in the vocabulary, with another string are called the **productions** of the grammar. We denote by $z_0 \rightarrow z_1$ the production that specifies that z_0 can be replaced by z_1 within a string. The productions in the grammar given in the introduction of this section were listed. The first production, written using this notation, is **sentence → noun phrase verb phrase**. We summarize this terminology in Definition 2.

The notion of a phrase-structure grammar extends the concept of a *rewrite system* devised by Axel Thue in the early 20th century.

Definition 2 A *phrase-structure grammar* $G = (V, T, S, P)$ consists of a vocabulary V, a subset T of V consisting of terminal symbols, a start symbol S from V, and a finite set of productions P. The set $V - T$ is denoted by N. Elements of N are called *nonterminal symbols*. Every production in P must contain at least one nonterminal on its left side.

EXAMPLE 1 Let $G = (V, T, S, P)$, where $V = \{a, b, A, B, S\}$, $T = \{a, b\}$, S is the start symbol, and $P = \{S \rightarrow ABa, A \rightarrow BB, B \rightarrow ab, AB \rightarrow b\}$. G is an example of a phrase-structure grammar. ◀

We will be interested in the words that can be generated by the productions of a phrase structure grammar.

Definition 3 Let $G = (V, T, S, P)$ be a phrase-structure grammar. Let $w_0 = lz_0r$ (that is, the concatenation of l, z_0, and r) and $w_1 = lz_1r$ be strings over V. If $z_0 \rightarrow z_1$ is a production of G, we say that w_1 is *directly derivable* from w_0 and we write $w_0 \Rightarrow w_1$. If $w_0, w_1, \ldots, w_n$ are strings over V such that $w_0 \Rightarrow w_1, w_1 \Rightarrow w_2, \ldots, w_{n-1} \Rightarrow w_n$, then we say that w_n is *derivable from* w_0, and we write $w_0 \stackrel{*}{\Rightarrow} w_n$. The sequence of steps used to obtain w_n from w_0 is called a *derivation*.

EXAMPLE 2 The string $Aaba$ is directly derivable from ABa in the grammar in Example 1 because $B \to ab$ is a production in the grammar. The string $abababa$ is derivable from ABa because $ABa \Rightarrow Aaba \Rightarrow BBaba \Rightarrow Bababa \Rightarrow abababa$, using the productions $B \to ab$, $A \to BB$, $B \to ab$, and $B \to ab$ in succession. ◀

Definition 4 Let $G = (V, T, S, P)$ be a phrase-structure grammar. The *language generated by* G (or the *language of* G), denoted by $L(G)$, is the set of all strings of terminals that are derivable from the starting state S. In other words,

$$L(G) = \{w \in T^* \mid S \stackrel{*}{\Rightarrow} w\}.$$

In Examples 3 and 4 we find the language generated by a phrase-structure grammar.

EXAMPLE 3 Let G be the grammar with vocabulary $V = \{S, A, a, b\}$, set of terminals $T = \{a, b\}$, starting symbol S, and productions $P = \{S \to aA, S \to b, A \to aa\}$. What is $L(G)$, the language of this grammar?

Extra Examples

Solution: From the start state S we can derive aA using the production $S \to aA$. We can also use the production $S \to b$ to derive b. From aA the production $A \to aa$ can be used to derive aaa. No additional words can be derived. Hence, $L(G) = \{b, aaa\}$. ◀

EXAMPLE 4 Let G be the grammar with vocabulary $V = \{S, 0, 1\}$, set of terminals $T = \{0, 1\}$, starting symbol S, and productions $P = \{S \to 11S, S \to 0\}$. What is $L(G)$, the language of this grammar?

Solution: From S we can derive 0 using $S \to 0$, or $11S$ using $S \to 11S$. From $11S$ we can derive either 110 or $1111S$. From $1111S$ we can derive 11110 and $111111S$. At any stage of a derivation we can either add two 1s at the end of the string or terminate the derivation by adding a 0 at the end of the string. We surmise that $L(G) = \{0, 110, 11110, 1111110, \dots\}$, the set of all strings that begin with an even number of 1s and end with a 0. This can be proved using an inductive argument that shows that after n productions have been used, the only strings of terminals generated are those consisting of $n - 1$ concatenations of 11 followed by 0. (This is left as an exercise for the reader.) ◀

The problem of constructing a grammar that generates a given language often arises. Examples 5, 6, and 7 describe problems of this kind.

EXAMPLE 5 Give a phrase-structure grammar that generates the set $\{0^n 1^n \mid n = 0, 1, 2, \dots\}$.

Solution: Two productions can be used to generate all strings consisting of a string of 0s followed by a string of the same number of 1s, including the null string. The first builds up successively longer strings in the language by adding a 0 at the start of the string and a 1 at the end. The second production replaces S with the empty string. The solution is the grammar $G = (V, T, S, P)$, where $V = \{0, 1, S\}$, $T = \{0, 1\}$, S is the starting symbol, and the productions are

$$S \to 0S1$$
$$S \to \lambda.$$

The verification that this grammar generates the correct set is left as an exercise for the reader. ◀

Example 5 involved the set of strings made up of 0s followed by 1s, where the number of 0s and 1s are the same. Example 6 considers the set of strings consisting of 0s followed by 1s, where the number of 0s and 1s may differ.

EXAMPLE 6 Find a phrase-structure grammar to generate the set $\{0^m 1^n \mid m \text{ and } n \text{ are nonnegative integers}\}$.

Solution: We will give two grammars G_1 and G_2 that generate this set. This will illustrate that two different grammars can generate the same language.

The grammar G_1 has alphabet $V = \{S, 0, 1\}$; terminals $T = \{0, 1\}$; and productions $S \rightarrow 0S$, $S \rightarrow S1$, and $S \rightarrow \lambda$. Then, G_1 generates the correct set, because using the first production m times puts m 0s at the beginning of the string, and using the second production n times puts n 1s at the end of the string. Furthermore, every string produced by this grammar is of the form $0^m 1^n$, because the only way to add a symbol to the beginning of a string is to add a 0 by applying the production $S \rightarrow 0S$, and the only way to add a symbol to the end of the string is to add a 1 by applying the production $S \rightarrow S1$.

The grammar G_2 has alphabet $V = \{S, A, 0, 1\}$; terminals $T = \{0, 1\}$; and productions $S \rightarrow 0S$, $S \rightarrow 1A$, $S \rightarrow 1$, $A \rightarrow 1A$, $A \rightarrow 1$, and $S \rightarrow \lambda$. The details that this grammar generates the correct set are left as an exercise for the reader. ◀

Sometimes a set that is easy to describe can be generated only by a complicated grammar. Example 7 illustrates this.

EXAMPLE 7 One grammar that generates the set $\{0^n 1^n 2^n \mid n = 0, 1, 2, 3, \dots\}$ is $G = (V, T, S, P)$ with $V = \{0, 1, 2, S, A, B, C\}$; $T = \{0, 1, 2\}$; starting state S; and productions $S \rightarrow C$, $C \rightarrow 0CAB$, $C \rightarrow 0AB$ $S \rightarrow \lambda$, $BA \rightarrow AB$, $0A \rightarrow 01$, $1A \rightarrow 11$, $1B \rightarrow 12$, and $2B \rightarrow 22$. We leave it as an exercise for the reader (Exercise 12) to show that this statement is correct. The grammar given is the simplest type of grammar that generates this set, in a sense that will be made clear later in this section. ◀

13.1.3 Types of Phrase-Structure Grammars

Links ›

Phrase-structure grammars can be classified according to the types of productions that are allowed. We will describe the classification scheme introduced by Noam Chomsky. In Section 13.4 we will see that the different types of languages defined in this scheme correspond to the classes of languages that can be recognized using different models of computing machines.

A **type 0** grammar has no restrictions on its productions. A **type 1** grammar can have productions of the form $w_1 \rightarrow w_2$, where $w_1 = lAr$ and $w_2 = lwr$, where A is a nonterminal symbol, l and r are strings of zero or more terminal or nonterminal symbols, and w is a nonempty string of terminal or nonterminal symbols. It can also have the production $S \rightarrow \lambda$ as long as S does not appear on the right-hand side of any production. A **type 2** grammar can have productions only of the form $w_1 \rightarrow w_2$, where w_1 is a single nonterminal symbol and w_2 is a string (possibly empty) of terminal or nonterminal symbols. A **type 3** grammar can have productions only of the form $w_1 \rightarrow w_2$ with $w_1 = A$ and either $w_2 = aB$ or $w_2 = a$, where A and B are nonterminal symbols and a is a terminal symbol, or with $w_1 = S$ and $w_2 = \lambda$.

Type 2 grammars are called **context-free grammars** because a nonterminal symbol that is the left side of a production can be replaced in a string whenever it occurs, no matter what else is in the string. A language generated by a type 2 grammar is called a **context-free language**. When there is a production of the form $lw_1r \rightarrow lw_2r$ (but not of the form $w_1 \rightarrow w_2$), the grammar is called type 1 or **context-sensitive** because w_1 can be replaced by w_2 only when it is surrounded by the strings l and r. A language generated by a type 1 grammar is called a **context-sensitive language**. Type 3 grammars are also called **regular grammars**. A language generated by a regular grammar is called **regular**. Section 13.4 deals with the relationship between regular languages and finite-state machines.

Of the four types of grammars we have defined, context-sensitive grammars have the most complicated definition. Sometimes, these grammars are defined in a different way. A production of the form $w_1 \to w_2$ is called **noncontracting** if the length of w_1 is less than or equal to the length of w_2. According to our characterization of context-senstive languages, every production in a type 1 grammar, other than the production $S \to \lambda$, if it is present, is noncontracting. It follows that the lengths of the strings in a derivation in a context-sensitive language are nondecreasing unless the production $S \to \lambda$ is used. This means that the only way for the empty string to belong to the language generated by a context-sensitive grammar is for the production $S \to \lambda$ to be part of the grammar. The other way that context-sensitive grammars are defined is by specifying that all productions are noncontracting. A grammar with this property is called **noncontracting** or **monotonic**. The class of noncontracting grammars is not the same as the class of context-sensitive grammars. However, these two classes are closely related; it can be shown that they define the same set of languages except that noncontracting grammars cannot generate any language containing the empty string λ.

EXAMPLE 8 From Example 6 we know that $\{0^m 1^n \mid m, n = 0, 1, 2, \dots\}$ is a regular language, because it can be generated by a regular grammar, namely, the grammar G_2 in Example 6. ◀

Extra Examples

Context-free and regular grammars play an important role in programming languages. Context-free grammars are used to define the syntax of almost all programming languages. These grammars are strong enough to define a wide range of languages. Furthermore, efficient algorithms can be devised to determine whether and how a string can be generated. Regular grammars are used to search text for certain patterns and in lexical analysis, which is the process of transforming an input stream into a stream of tokens for use by a parser.

EXAMPLE 9 It follows from Example 5 that $\{0^n 1^n \mid n = 0, 1, 2, \dots\}$ is a context-free language, because the productions in this grammar are $S \to 0S1$ and $S \to \lambda$. However, it is not a regular language. This will be shown in Section 13.4. ◀

EXAMPLE 10 The set $\{0^n 1^n 2^n \mid n = 0, 1, 2, \dots\}$ is a context-sensitive language, because it can be generated by a type 1 grammar, as Example 7 shows, but not by any type 2 language. (This is shown in Exercise 28 in the supplementary exercises at the end of the chapter.) ◀

Table 1 summarizes the terminology used to classify phrase-structure grammars.

13.1.4 Derivation Trees

A derivation in the language generated by a context-free grammar can be represented graphically using an ordered rooted tree, called a **derivation tree** or **parse tree**. The root of this tree represents the starting symbol. The internal vertices of the tree represent the nonterminal

TABLE 1 Types of Grammars.

Type	*Restrictions on Productions $w_1 \to w_2$*
0	No restrictions
1	$w_1 = lAr$ and $w_2 = lwr$, where $A \in N$, $l, r, w \in (N \cup T)^*$ and $w \neq \lambda$; or $w_1 = S$ and $w_2 = \lambda$ as long as S is not on the right-hand side of any production
2	$w_1 = A$, where $A \in N$, and $w_2 \in (N \cup T)^*$
3	$w_1 = A$ and $w_2 = aB$ or $w_2 = a$, where $A \in N$, $B \in N$, and $a \in T$; or $w_1 = S$ and $w_2 = \lambda$

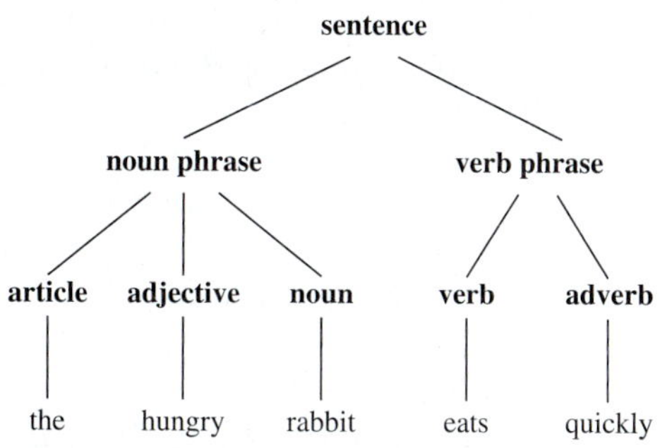

FIGURE 1 A derivation tree.

symbols that arise in the derivation. The leaves of the tree represent the terminal symbols that arise. If the production $A \rightarrow w$ arises in the derivation, the vertex that represents A has children vertices that represent each symbol in w, in order from left to right.

EXAMPLE 11 Construct a derivation tree for the derivation of *the hungry rabbit eats quickly*, given in the introduction of this section.

Solution: The derivation tree is shown in Figure 1. ◀

The problem of determining whether a string is in the language generated by a context-free grammar arises in many applications, such as in the construction of compilers. When given a string, the naive approach for determining whether it is in the language generated by a grammar is to look for a sequence of productions that can be applied, beginning at the start state, that lead to the given string. When following such an approach, it is useful to think a few moves ahead. This approach is known as **top-down parsing**. A second approach, known as **bottom-up parsing**, is to work backward from the given string with the goal of undoing productions one-by-one to reach the start symbol. We illustrate these two approaches to this problem in Example 12.

EXAMPLE 12 Determine whether the word *cbab* belongs to the language generated by the grammar $G = (V, T, S, P)$, where $V = \{a, b, c, A, B, C, S\}$, $T = \{a, b, c\}$, S is the starting symbol, and the productions are

$$
\begin{aligned}
&S \rightarrow AB\\
&A \rightarrow Ca\\
&B \rightarrow Ba\\
&B \rightarrow Cb\\
&B \rightarrow b\\
&C \rightarrow cb\\
&C \rightarrow b.
\end{aligned}
$$

Solution: A top-down approach to this problem is to begin with S and attempt to derive *cbab* using a series of productions. Because there is only one production with S on its left hand side, we must start with $S \Rightarrow AB$. Next we use the only production that has A on its left-hand side,

Links

SASCHA SCHUERMANN/ AFP/Getty Images

AVRAM NOAM CHOMSKY (BORN 1928) Noam Chomsky, born in Philadelphia, is the son of a Hebrew scholar. He received his B.A., M.A., and Ph.D. in linguistics, all from the University of Pennsylvania. He was on the staff of the University of Pennsylvania from 1950 until 1951. In 1955 he joined the faculty at M.I.T., beginning his M.I.T. career teaching engineers French and German. Chomsky is currently the Ferrari P. Ward Professor of foreign languages and linguistics at M.I.T. He is known for his many fundamental contributions to linguistics, including the study of grammars. Chomsky is also widely known for his outspoken political activism.

namely, $A \to Ca$, to obtain $S \Rightarrow AB \Rightarrow CaB$. Because $cbab$ begins with the symbols cb, we use the production $C \to cb$. This gives us $S \Rightarrow AB \Rightarrow CaB \Rightarrow cbaB$. We finish by using the production $B \to b$, to obtain $S \Rightarrow AB \Rightarrow CaB \Rightarrow cbaB \Rightarrow cbab$.

We will solve the same problem using bottom-up parsing. Because $cbab$ is the string to be derived, we can use the production $C \to cb$, so that $Cab \Rightarrow cbab$. Then, we can use the production $A \to Ca$, so that $Ab \Rightarrow Cab \Rightarrow cbab$. Using the production $B \to b$ gives $AB \Rightarrow Ab \Rightarrow Cab \Rightarrow cbab$. Finally, using $S \to AB$ shows that a complete derivation for $cbab$ is $S \Rightarrow AB \Rightarrow Ab \Rightarrow Cab \Rightarrow cbab$. ◀

Example 12 presents the solution of a parsing problem using both top-down and bottom-up parsing. Both approaches were easy to use for that problem. However, parsing problems can be quite challenging. That is, it can be quite challenging to determine whether a string is in the language generated by a context-free grammar. Because parsing is so important, many strategies and algorithms have been devised for top-down and for bottom-up parsing. These algorithms are beyond the scope of this book. The interested reader should consult [AhLaSeUl06] to learn more.

13.1.5 Backus–Naur Form

Links

There is another notation that is sometimes used to specify a type 2 grammar, called the **Backus–Naur form (BNF)**, after John Backus, who invented it, and Peter Naur, who refined

Links

Louis F. Bachrach

JOHN BACKUS (1924–2007) John Backus was born in Philadelphia and grew up in Wilmington, Delaware. He attended the Hill School in Pottstown, Pennsylvania. He needed to attend summer school every year because he disliked studying and was not a serious student. But he enjoyed spending his summers in New Hampshire where he attended summer school and amused himself with summer activities, including sailing. He obliged his father by enrolling at the University of Virginia to study chemistry. But he quickly decided chemistry was not for him, and in 1943 he entered the army, where he received medical training and worked in a neurosurgery ward in an army hospital. Ironically, Backus was soon diagnosed with a bone tumor in his skull and was fitted with a metal plate. His medical work in the army convinced him to try medical school, but he abandoned this after nine months because he disliked the rote memorization required. After dropping out of medical school, he entered a school for radio technicians because he wanted to build his own high fidelity set. A teacher in this school recognized his potential and asked him to help with some mathematical calculations needed for an article in a magazine. Finally, Backus found what he was interested in: mathematics and its applications. He enrolled at Columbia University, from which he received both bachelor's and master's degrees in mathematics. Backus joined IBM as a programmer in 1950. He participated in the design and development of two of IBM's early computers. From 1954 to 1958 he led the IBM group that developed FORTRAN. Backus became a staff member at the IBM Watson Research Center in 1958. He was part of the committees that designed the programming language ALGOL using what is now called the Backus–Naur form for the description of the syntax of this language. (The development of ALGOL led to the development of other programming languages, including Pascal and C.) Later, Backus worked on the mathematics of families of sets and on a functional style of programming. Backus became an IBM Fellow in 1963, and he received the National Medal of Science in 1974 and the prestigious Turing Award from the Association for Computing Machinery in 1977.

©Heidelberg Laureate Forum Foundation

PETER NAUR (1928–2016) Peter Naur was born in Frederiksberg, near Copenhagen. As a boy he became interested in astronomy. Not only did he observe heavenly bodies, but he also computed the orbits of comets and asteroids. Naur attended Copenhagen University, receiving his degree in 1949. He spent 1950 and 1951 in Cambridge, where he used an early computer to calculate the motions of comets and planets. After returning to Denmark, he continued working in astronomy but kept his ties to computing. In 1955 he served as a consultant to the building of the first Danish computer. In 1959 Naur made the switch from astronomy to computing as a full-time activity. His first job as a full-time computer scientist was participating in the development of the programming language ALGOL. From 1960 to 1967 he worked on the development of compilers for ALGOL and COBOL. From 1969 until 1999 he was professor of computer science at Copenhagen University, where he has worked in the area of programming methodology. His research interests included the design, structure, and performance of computer programs. Naur was a pioneer in both the areas of software architecture and software engineering. He rejected the view that computer programming is a branch of mathematics and preferred that computer science be called *datalogy*.

The ancient Indian grammarian Pāṇini specified Sanskrit using 3959 rules; Backus-Naur form is sometimes called Backus-Pāṇini form.

it for use in the specification of the programming language ALGOL. (Surprisingly, a notation quite similar to the Backus–Naur form was used approximately 2500 years ago to describe the grammar of Sanskrit.) The Backus–Naur form is used to specify the syntactic rules of many computer languages, including Java. The productions in a type 2 grammar have a single nonterminal symbol as their left-hand side. Instead of listing all the productions separately, we can combine all those with the same nonterminal symbol on the left-hand side into one statement. Instead of using the symbol → in a production, we use the symbol ::=. We enclose all nonterminal symbols in brackets, ⟨ ⟩, and we list all the right-hand sides of productions in the same statement, separating them by bars. For instance, the productions $A \rightarrow Aa$, $A \rightarrow a$, and $A \rightarrow AB$ can be combined into $\langle A \rangle ::= \langle A \rangle a \mid a \mid \langle A \rangle \langle B \rangle$.

Example 13 illustrates how the Backus–Naur form is used to describe the syntax of programming languages. Our example comes from the original use of Backus–Naur form in the description of ALGOL 60.

EXAMPLE 13 *Extra Examples*

In ALGOL 60 an identifier (which is the name of an entity such as a variable) consists of a string of alphanumeric characters (that is, letters and digits) and must begin with a letter. We can use these rules in Backus–Naur to describe the set of allowable identifiers:

$\langle identifier \rangle ::= \langle letter \rangle \mid \langle identifier \rangle \langle letter \rangle \mid \langle identifier \rangle \langle digit \rangle$
$\langle letter \rangle ::= a \mid b \mid \cdots \mid y \mid z$ the ellipsis indicates that all 26 letters are included
$\langle digit \rangle ::= 0 \mid 1 \mid 2 \mid 3 \mid 4 \mid 5 \mid 6 \mid 7 \mid 8 \mid 9$

For example, we can produce the valid identifier $x99a$ by using the first rule to replace $\langle identifier \rangle$ by $\langle identifier \rangle \langle letter \rangle$, the second rule to obtain $\langle identifier \rangle a$, the first rule twice to obtain $\langle identifier \rangle \langle digit \rangle \langle digit \rangle a$, the third rule twice to obtain $\langle identifier \rangle 99a$, the first rule to obtain $\langle letter \rangle 99a$, and finally the second rule to obtain $x99a$. ◀

EXAMPLE 14 What is the Backus–Naur form of the grammar for the subset of English described in the introduction to this section?

Solution: The Backus–Naur form of this grammar is

$\langle sentence \rangle ::= \langle noun\ phrase \rangle \langle verb\ phrase \rangle$
$\langle noun\ phrase \rangle ::= \langle article \rangle \langle adjective \rangle \langle noun \rangle \mid \langle article \rangle \langle noun \rangle$
$\langle verb\ phrase \rangle ::= \langle verb \rangle \langle adverb \rangle \mid \langle verb \rangle$
$\langle article \rangle ::= a \mid the$
$\langle adjective \rangle ::= large \mid hungry$
$\langle noun \rangle ::= rabbit \mid mathematician$
$\langle verb \rangle ::= eats \mid hops$
$\langle adverb \rangle ::= quickly \mid wildly$

◀

EXAMPLE 15 Give the Backus–Naur form for the production of signed integers in decimal notation. (A **signed integer** is a nonnegative integer preceded by a plus sign or a minus sign.)

Solution: The Backus–Naur form for a grammar that produces signed integers is

$\langle signed\ integer \rangle ::= \langle sign \rangle \langle integer \rangle$
$\langle sign \rangle ::= + \mid -$
$\langle integer \rangle ::= \langle digit \rangle \mid \langle digit \rangle \langle integer \rangle$
$\langle digit \rangle ::= 0 \mid 1 \mid 2 \mid 3 \mid 4 \mid 5 \mid 6 \mid 7 \mid 8 \mid 9$

◀

The Backus–Naur form, with a variety of extensions, is used extensively to specify the syntax of programming languages, such as Java and LISP; database languages, such as SQL; and markup languages, such as XML. Some extensions of the Backus–Naur form that are commonly used in the description of programming languages are introduced in the preamble to Exercise 34.

Exercises

Exercises 1–3 refer to the grammar with start symbol **sentence**, set of terminals $T = \{the, sleepy, happy, tortoise, hare, passes, runs, quickly, slowly\}$, set of nonterminals $N =$ {**noun phrase, transitive verb phrase, intransitive verb phrase, article, adjective, noun, verb, adverb**}, and productions:

sentence → **noun phrase** **transitive verb phrase** **noun phrase**
sentence → **noun phrase** **intransitive verb phrase**
noun phrase → **article** **adjective** **noun**
noun phrase → **article** **noun**
transitive verb phrase → **transitive verb**
intransitive verb phrase → **intransitive verb** **adverb**
intransitive verb phrase → **intransitive verb**
article → *the*
adjective → *sleepy*
adjective → *happy*
noun → *tortoise*
noun → *hare*
transitive verb → *passes*
intransitive verb → *runs*
adverb → *quickly*
adverb → *slowly*

1. Use the set of productions to show that each of these sentences is a valid sentence.

a) *the happy hare runs*

b) *the sleepy tortoise runs quickly*

c) *the tortoise passes the hare*

d) *the sleepy hare passes the happy tortoise*

2. Find five other valid sentences, besides those given in Exercise 1.

3. Show that *the hare runs the sleepy tortoise* is not a valid sentence.

4. Let $G = (V, T, S, P)$ be the phrase-structure grammar with $V = \{0, 1, A, S\}$, $T = \{0, 1\}$, and set of productions P consisting of $S \rightarrow 1S$, $S \rightarrow 00A$, $A \rightarrow 0A$, and $A \rightarrow 0$.

a) Show that 111000 belongs to the language generated by G.

b) Show that 11001 does not belong to the language generated by G.

c) What is the language generated by G?

5. Let $G = (V, T, S, P)$ be the phrase-structure grammar with $V = \{0, 1, A, B, S\}$, $T = \{0, 1\}$, and set of productions P consisting of $S \rightarrow 0A$, $S \rightarrow 1A$, $A \rightarrow 0B$, $B \rightarrow 1A$, $B \rightarrow 1$.

a) Show that 10101 belongs to the language generated by G.

b) Show that 10110 does not belong to the language generated by G.

c) What is the language generated by G?

***6.** Let $V = \{S, A, B, a, b\}$ and $T = \{a, b\}$. Find the language generated by the grammar (V, T, S, P) when the set P of productions consists of

a) $S \rightarrow AB, A \rightarrow ab, B \rightarrow bb.$
b) $S \rightarrow AB, S \rightarrow aA, A \rightarrow a, B \rightarrow ba.$
c) $S \rightarrow AB, S \rightarrow AA, A \rightarrow aB, A \rightarrow ab, B \rightarrow b.$
d) $S \rightarrow AA, S \rightarrow B, A \rightarrow aaA, A \rightarrow aa, B \rightarrow bB, B \rightarrow b.$
e) $S \rightarrow AB, A \rightarrow aAb, B \rightarrow bBa, A \rightarrow \lambda, B \rightarrow \lambda.$

7. Construct a derivation of $0^3 1^3$ using the grammar given in Example 5.

8. Show that the grammar given in Example 5 generates the set $\{0^n 1^n \mid n = 0, 1, 2, \ldots\}$.

9. a) Construct a derivation of $0^2 1^4$ using the grammar G_1 in Example 6.

b) Construct a derivation of $0^2 1^4$ using the grammar G_2 in Example 6.

10. a) Show that the grammar G_1 given in Example 6 generates the set $\{0^m 1^n \mid m, n = 0, 1, 2, \ldots\}$.

b) Show that the grammar G_2 in Example 6 generates the same set.

11. Construct a derivation of $0^2 1^2 2^2$ in the grammar given in Example 7.

***12.** Show that the grammar given in Example 7 generates the set $\{0^n 1^n 2^n \mid n = 0, 1, 2, \ldots\}$.

13. Find a phrase-structure grammar for each of these languages.

a) the set consisting of the bit strings 0, 1, and 11
b) the set of bit strings containing only 1s
c) the set of bit strings that start with 0 and end with 1
d) the set of bit strings that consist of a 0 followed by an even number of 1s

14. Find a phrase-structure grammar for each of these languages.

a) the set consisting of the bit strings 10, 01, and 101
b) the set of bit strings that start with 00 and end with one or more 1s
c) the set of bit strings consisting of an even number of 1s followed by a final 0

d) the set of bit strings that have neither two consecutive 0s nor two consecutive 1s

*15. Find a phrase-structure grammar for each of these languages.

a) the set of all bit strings containing an even number of 0s and no 1s
b) the set of all bit strings made up of a 1 followed by an odd number of 0s
c) the set of all bit strings containing an even number of 0s and an even number of 1s
d) the set of all strings containing 10 or more 0s and no 1s
e) the set of all strings containing more 0s than 1s
f) the set of all strings containing an equal number of 0s and 1s
g) the set of all strings containing an unequal number of 0s and 1s

16. Construct phrase-structure grammars to generate each of these sets.

a) $\{1^n \mid n \geq 0\}$ b) $\{10^n \mid n \geq 0\}$
c) $\{(11)^n \mid n \geq 0\}$

17. Construct phrase-structure grammars to generate each of these sets.

a) $\{0^n \mid n \geq 0\}$ b) $\{1^n0 \mid n \geq 0\}$
c) $\{(000)^n \mid n \geq 0\}$

18. Construct phrase-structure grammars to generate each of these sets.

a) $\{01^{2n} \mid n \geq 0\}$
b) $\{0^n1^{2n} \mid n \geq 0\}$
c) $\{0^n1^m0^n \mid m \geq 0 \text{ and } n \geq 0\}$

19. Let $V = \{S, A, B, a, b\}$ and $T = \{a, b\}$. Determine whether $G = (V, T, S, P)$ is a type 0 grammar but not a type 1 grammar, a type 1 grammar but not a type 2 grammar, or a type 2 grammar but not a type 3 grammar if P, the set of productions, is

a) $S \rightarrow aAB, A \rightarrow Bb, B \rightarrow \lambda$.
b) $S \rightarrow aA, A \rightarrow a, A \rightarrow b$.
c) $S \rightarrow ABa, AB \rightarrow a$.
d) $S \rightarrow ABA, A \rightarrow aB, B \rightarrow ab$.
e) $S \rightarrow bA, A \rightarrow B, B \rightarrow a$.
f) $S \rightarrow aA, aA \rightarrow B, B \rightarrow aA, A \rightarrow b$.
g) $S \rightarrow bA, A \rightarrow b, S \rightarrow \lambda$.
h) $S \rightarrow AB, B \rightarrow aAb, aAb \rightarrow b$.
i) $S \rightarrow aA, A \rightarrow bB, B \rightarrow b, B \rightarrow \lambda$.
j) $S \rightarrow A, A \rightarrow B, B \rightarrow \lambda$.

20. A **palindrome** is a string that reads the same backward as it does forward, that is, a string w, where $w = w^R$, where w^R is the reversal of the string w. Find a context-free grammar that generates the set of all palindromes over the alphabet $\{0, 1\}$.

*21. Let G_1 and G_2 be context-free grammars, generating the languages $L(G_1)$ and $L(G_2)$, respectively. Show that there is a context-free grammar generating each of these sets.

a) $L(G_1) \cup L(G_2)$ b) $L(G_1)L(G_2)$
c) $L(G_1)^*$

22. Find the strings constructed using the derivation trees shown here.

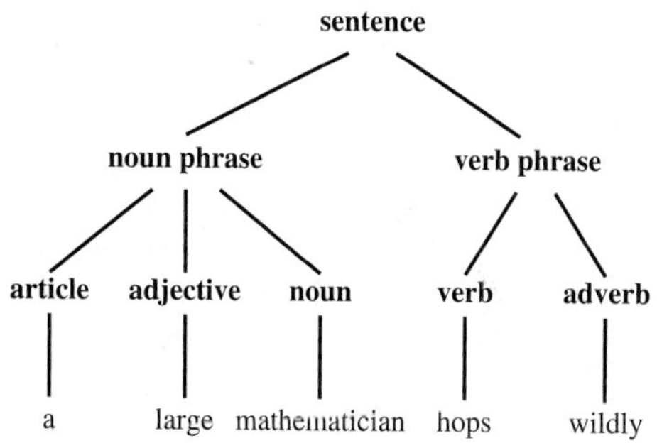

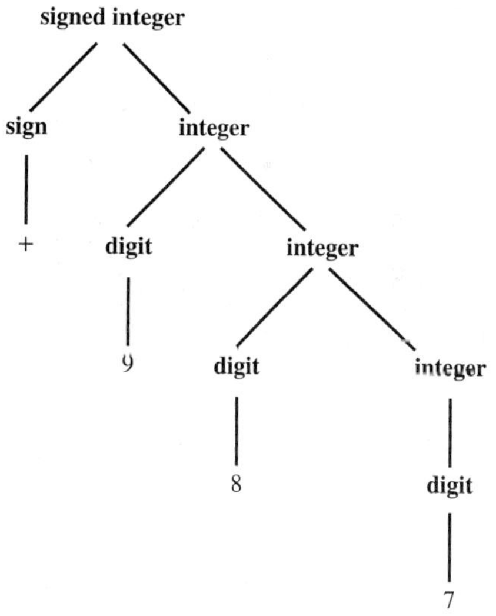

23. Construct derivation trees for the sentences in Exercise 1.

24. Let G be the grammar with $V = \{a, b, c, S\}$; $T = \{a, b, c\}$; starting symbol S; and productions $S \rightarrow abS$, $S \rightarrow bcS$, $S \rightarrow bbS$, $S \rightarrow a$, and $S \rightarrow cb$. Construct derivation trees for

a) $bcbba$. b) $bbbcbba$.
c) $bcabbbbbcb$.

*25. Use top-down parsing to determine whether each of the following strings belongs to the language generated by the grammar in Example 12.

a) $baba$ b) $abab$
c) $cbaba$ d) $bbbcba$

*26. Use bottom-up parsing to determine whether the strings in Exercise 25 belong to the language generated by the grammar in Example 12.

27. Construct a derivation tree for -109 using the grammar given in Example 15.

28. a) Explain what the productions are in a grammar if the Backus–Naur form for productions is as follows:

$$\begin{aligned}\langle expression\rangle &::= (\langle expression\rangle) \mid \\ &\quad \langle expression\rangle + \langle expression\rangle \mid \\ &\quad \langle expression\rangle * \langle expression\rangle \mid \\ &\quad \langle variable\rangle \\ \langle variable\rangle &::= x \mid y\end{aligned}$$

b) Find a derivation tree for $(x * y) + x$ in this grammar.

29. a) Construct a phrase-structure grammar that generates all signed decimal numbers, consisting of a sign, either + or −; a nonnegative integer; and a decimal fraction that is either the empty string or a decimal point followed by a positive integer, where initial zeros in an integer are allowed.

b) Give the Backus–Naur form of this grammar.

c) Construct a derivation tree for -31.4 in this grammar.

30. a) Construct a phrase-structure grammar for the set of all fractions of the form a/b, where a is a signed integer in decimal notation and b is a positive integer.

b) What is the Backus–Naur form for this grammar?

c) Construct a derivation tree for $+311/17$ in this grammar.

31. Give production rules in Backus–Naur form for an identifier if it can consist of

a) one or more lowercase letters.

b) at least three but no more than six lowercase letters.

c) one to six uppercase or lowercase letters beginning with an uppercase letter.

d) a lowercase letter, followed by a digit or an underscore, followed by three or four alphanumeric characters (lower or uppercase letters and digits).

32. Give production rules in Backus–Naur form for the name of a person if this name consists of a first name, which is a string of letters, where only the first letter is uppercase; a middle initial; and a last name, which can be any string of letters.

33. Give production rules in Backus–Naur form that generate all identifiers in the C programming language. In C an identifier starts with a letter or an underscore (_) that is followed by one or more lowercase letters, uppercase letters, underscores, and digits.

Several extensions to Backus–Naur form are commonly used to define phrase-structure grammars. In one such extension, a question mark (?) indicates that the symbol, or group of symbols inside parentheses, to its left can appear zero or once (that is, it is optional), an asterisk (*) indicates that the symbol to its left can appear zero or more times, and a plus (+) indicates that the symbol to its left can appear one or more times. These extensions are part of **extended Backus–Naur form (EBNF)**, and the symbols ?, *, and + are called **metacharacters**. In EBNF the brackets used to denote nonterminals are usually not shown.

34. Describe the set of strings defined by each of these sets of productions in EBNF.

a) $string ::= L{+}D?L{+}$
$L ::= a \mid b \mid c$
$D ::= 0 \mid 1$

b) $string ::= sign\ D{+} \mid D{+}$
$sign ::= + \mid -$
$D ::= 0 \mid 1 \mid 2 \mid 3 \mid 4 \mid 5 \mid 6 \mid 7 \mid 8 \mid 9$

c) $string ::= L{*}(D{+})?L{*}$
$L ::= x \mid y$
$D ::= 0 \mid 1$

35. Give production rules in extended Backus–Naur form that generate all decimal numerals consisting of an optional sign, a nonnegative integer, and a decimal fraction that is either the empty string or a decimal point followed by an optional positive integer optionally preceded by some number of zeros.

36. Give production rules in extended Backus–Naur form that generate a sandwich if a sandwich consists of a lower slice of bread; mustard or mayonnaise; optional lettuce; an optional slice of tomato; one or more slices of either turkey, chicken, or roast beef (in any combination); optionally some number of slices of cheese; and a top slice of bread.

37. Give production rules in extended Backus–Naur form for identifiers in the C programming language (see Exercise 33).

38. Describe how productions for a grammar in extended Backus–Naur form can be translated into a set of productions for the grammar in Backus–Naur form.

This is the Backus–Naur form that describes the syntax of expressions in postfix (or reverse Polish) notation.

$$\begin{aligned}&\langle expression\rangle ::= \langle term\rangle \mid \langle term\rangle\langle term\rangle\langle addOperator\rangle\\ &\langle addOperator\rangle ::= + \mid -\\ &\langle term\rangle ::= \langle factor\rangle \mid \langle factor\rangle\langle factor\rangle\langle mulOperator\rangle\\ &\langle mulOperator\rangle ::= * \mid /\\ &\langle factor\rangle ::= \langle identifier\rangle \mid \langle expression\rangle\\ &\langle identifier\rangle ::= a \mid b \mid \cdots \mid z\end{aligned}$$

39. For each of these strings, determine whether it is generated by the grammar given for postfix notation. If it is, find the steps used to generate the string

a) $abc{*}{+}$
b) $xy{+}{+}$
c) $xy{-}z{*}$
d) $wxyz{-}{*}/$
e) $ade{-}{*}$

40. Use Backus–Naur form to describe the syntax of expressions in infix notation, where the set of operators and identifiers is the same as in the BNF for postfix expressions given in the preamble to Exercise 39, but parentheses must surround expressions being used as factors.

41. For each of these strings, determine whether it is generated by the grammar for infix expressions from Exercise 40. If it is, find the steps used to generate the string.

a) $x + y + z$ **b)** $a/b + c/d$
c) $m * (n + p)$ **d)** $+m - n + p - q$
e) $(m + n) * (p - q)$

42. Let G be a grammar and let R be the relation containing the ordered pair (w_0, w_1) if and only if w_1 is directly derivable from w_0 in G. What is the reflexive transitive closure of R?

13.2 Finite-State Machines with Output

13.2.1 Introduction

Links

Many kinds of machines, including components in computers, can be modeled using a structure called a finite-state machine. Several types of finite-state machines are commonly used in models. All these versions of finite-state machines include a finite set of states, with a designated starting state, an input alphabet, and a transition function that assigns a next state to every state and input pair. Finite-state machines are used extensively in applications in computer science and data networking. For example, finite-state machines are the basis for programs for spell checking, grammar checking, indexing or searching large bodies of text, recognizing speech, transforming text using markup languages such as XML and HTML, and network protocols that specify how computers communicate.

In this section, we will study those finite-state machines that produce output. We will show how finite-state machines can be used to model a vending machine, a machine that delays input, a machine that adds integers, and a machine that determines whether a bit string contains a specified pattern.

Before giving formal definitions, we will show how a vending machine can be modeled. A vending machine accepts nickels (5 cents), dimes (10 cents), and quarters (25 cents). When a total of 30 cents or more has been deposited, the machine immediately returns the amount in excess of 30 cents. When 30 cents has been deposited and any excess refunded, the customer can push an orange button and receive an orange juice or push a red button and receive an apple juice. We can describe how the machine works by specifying its states, how it changes states when input is received, and the output that is produced for every combination of input and current state.

Finite-state machines with output are often called *finite-state transducers.*

The machine can be in any of seven different states s_i, $i = 0, 1, 2, \ldots, 6$, where s_i is the state where the machine has collected $5i$ cents. The machine starts in state s_0, with 0 cents received. The possible inputs are 5 cents, 10 cents, 25 cents, the orange button (O), and the red button (R). The possible outputs are nothing (n), 5 cents, 10 cents, 15 cents, 20 cents, 25 cents, an orange juice, and an apple juice.

We illustrate how this model of the machine works with this example. Suppose that a student puts in a dime followed by a quarter, receives 5 cents back, and then pushes the orange button for an orange juice. The machine starts in state s_0. The first input is 10 cents, which changes the state of the machine to s_2 and gives no output. The second input is 25 cents. This changes the state from s_2 to s_6, and gives 5 cents as output. The next input is the orange button, which changes the state from s_6 back to s_0 (because the machine returns to the start state) and gives an orange juice as its output.

We can display all the state changes and output of this machine in a table. To do this we need to specify for each combination of state and input the next state and the output obtained. Table 1 shows the transitions and outputs for each pair of a state and an input.

Another way to show the actions of a machine is to use a directed graph with labeled edges, where each state is represented by a circle, edges represent the transitions, and edges are labeled

TABLE 1 State Table for a Vending Machine.

	Next State					Output				
	Input					*Input*				
State	5	10	25	*O*	*R*	5	10	25	*O*	*R*
s_0	s_1	s_2	s_5	s_0	s_0	n	n	n	n	n
s_1	s_2	s_3	s_6	s_1	s_1	n	n	n	n	n
s_2	s_3	s_4	s_6	s_2	s_2	n	n	5	n	n
s_3	s_4	s_5	s_6	s_3	s_3	n	n	10	n	n
s_4	s_5	s_6	s_6	s_4	s_4	n	n	15	n	n
s_5	s_6	s_6	s_6	s_5	s_5	n	5	20	n	n
s_6	s_6	s_6	s_6	s_0	s_0	5	10	25	OJ	AJ

with the input and the output for that transition. Figure 1 shows such a directed graph for the vending machine.

13.2.2 Finite-State Machines with Outputs

We will now give the formal definition of a finite-state machine with output.

Definition 1 A *finite-state machine* $M = (S, I, O, f, g, s_0)$ consists of a finite set S of *states*, a finite *input alphabet* I, a finite *output alphabet* O, a *transition function* f that assigns to each state and input pair a new state, an *output function* g that assigns to each state and input pair an output, and an *initial state* s_0.

Let $M = (S, I, O, f, g, s_0)$ be a finite-state machine. We can use a **state table** to represent the values of the transition function f and the output function g for all pairs of states and input. We previously constructed a state table for the vending machine discussed in the introduction to this section.

EXAMPLE 1 The state table shown in Table 2 describes a finite-state machine with $S = \{s_0, s_1, s_2, s_3\}$, $I = \{0, 1\}$, and $O = \{0, 1\}$. The values of the transition function f are displayed in the first two columns, and the values of the output function g are displayed in the last two columns. ◀

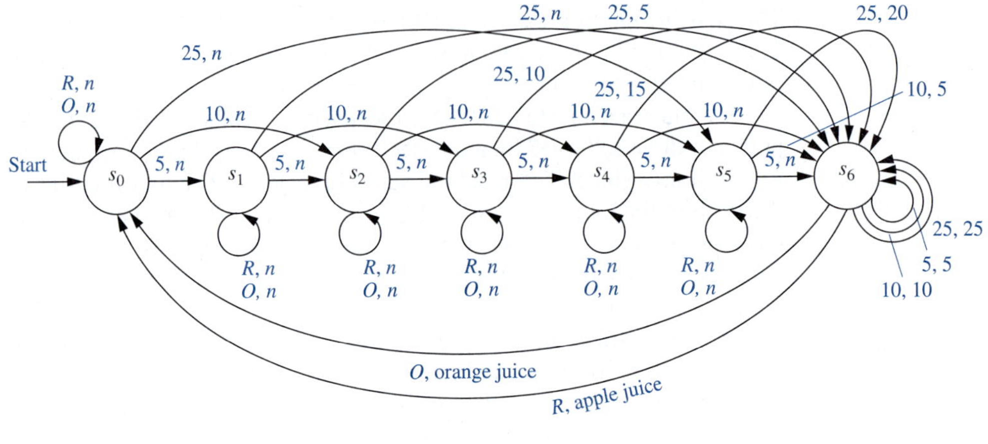

FIGURE 1 A vending machine.

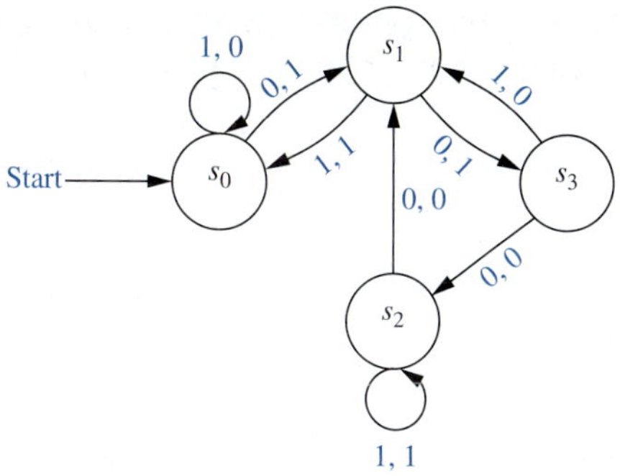

FIGURE 2 **The state diagram for the finite-state machine shown in Table 2.**

TABLE 2

	f		g	
	Input		*Input*	
State	**0**	**1**	**0**	**1**
s_0	s_1	s_0	1	0
s_1	s_3	s_0	1	1
s_2	s_1	s_2	0	1
s_3	s_2	s_1	0	0

Another way to represent a finite-state machine is to use a **state diagram**, which is a directed graph with labeled edges. In this diagram, each state is represented by a circle. Arrows labeled with the input and output pair are shown for each transition.

EXAMPLE 2 Construct the state diagram for the finite-state machine with the state table shown in Table 2.

Solution: The state diagram for this machine is shown in Figure 2. ◀

EXAMPLE 3 Construct the state table for the finite-state machine with the state diagram shown in Figure 3.

Solution: The state table for this machine is shown in Table 3. ◀

An input string takes the starting state through a sequence of states, as determined by the transition function. As we read the input string symbol by symbol (from left to right), each input symbol takes the machine from one state to another. Because each transition produces an output, an input string also produces an output string.

Suppose that the input string is $x = x_1x_2 \ldots x_k$. Then, reading this input takes the machine from state s_0 to state s_1, where $s_1 = f(s_0, x_1)$, then to state s_2, where $s_2 = f(s_1, x_2)$, and so on, with $s_j = f(s_{j-1}, x_j)$ for $j = 1, 2, \ldots, k$, ending at state $s_k = f(s_{k-1}, x_k)$. This sequence of transitions produces an output string $y_1y_2 \ldots y_k$, where $y_1 = g(s_0, x_1)$ is the output corresponding to the transition from s_0 to s_1, $y_2 = g(s_1, x_2)$ is the output corresponding to the transition from s_1 to s_2, and so on. In general, $y_j = g(s_{j-1}, x_j)$ for $j = 1, 2, \ldots, k$. Hence, we can extend the definition of

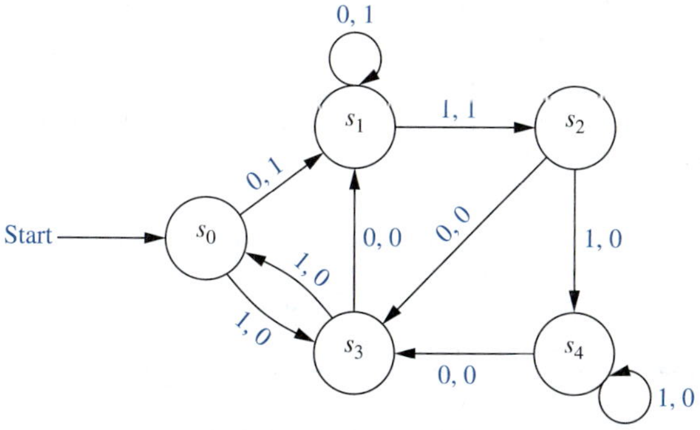

FIGURE 3 **A finite-state machine.**

TABLE 3

	f		g	
	Input		*Input*	
State	**0**	**1**	**0**	**1**
s_0	s_1	s_3	1	0
s_1	s_1	s_2	1	1
s_2	s_3	s_4	0	0
s_3	s_1	s_0	0	0
s_4	s_3	s_4	0	0

the output function g to input strings so that $g(x) = y$, where y is the output corresponding to the input string x. This notation is useful in many applications.

EXAMPLE 4 Find the output string generated by the finite-state machine in Figure 3 if the input string is 101011.

Solution: The output obtained is 001000. The successive states and outputs are shown in Table 4.

We can now look at some examples of useful finite-state machines. Examples 5, 6, and 7 illustrate that the states of a finite-state machine give it limited memory capabilities. The states can be used to remember the properties of the symbols that have been read by the machine. However, because there are only finitely many different states, finite-state machines cannot be used for some important purposes. This will be illustrated in Section 13.4.

EXAMPLE 5 An important element in many electronic devices is a *unit-delay machine*, which produces as output the input string delayed by a specified amount of time. How can a finite-state machine be constructed that delays an input string by one unit of time, that is, produces as output the bit string $0x_1x_2 \dots x_{k-1}$ given the input bit string $x_1x_2 \dots x_k$?

Solution: A delay machine can be constructed that has two possible inputs, namely, 0 and 1. The machine must have a start state s_0. Because the machine has to remember whether the previous input was a 0 or a 1, two other states s_1 and s_2 are needed, where the machine is in state s_1 if the previous input was 1 and in state s_2 if the previous input was 0. An output of 0 is produced for the initial transition from s_0. Each transition from s_1 gives an output of 1, and each transition from s_2 gives an output of 0. The output corresponding to the input of a string $x_1 \dots x_k$ is the string that begins with 0, followed by x_1, followed by $x_2, \dots,$ ending with x_{k-1}. The state diagram for this machine is shown in Figure 4.

EXAMPLE 6 Produce a finite-state machine that adds two positive integers using their binary expansions.

Solution: When $(x_n \dots x_1x_0)_2$ and $(y_n \dots y_1y_0)_2$ are added, the following procedure (as described in Section 4.2) is followed. First, the bits x_0 and y_0 are added, producing a sum bit z_0 and a carry bit c_0. This carry bit is either 0 or 1. Then, the bits x_1 and y_1 are added, together with the carry c_0. This gives a sum bit z_1 and a carry bit c_1. This procedure is continued until the nth stage, where x_n, y_n, and the previous carry c_{n-1} are added to produce the sum bit z_n and the carry bit c_n, which is equal to the sum bit z_{n+1}.

A finite-state machine to carry out this addition can be constructed using just two states. For simplicity we assume that both the initial bits x_n and y_n are 0 (otherwise we have to make special arrangements concerning the sum bit z_{n+1}). The start state s_0 is used to remember that the previous carry is 0 (or for the addition of the rightmost bits). The other state, s_1, is used to remember that the previous carry is 1.

Because the inputs to the machine are pairs of bits, there are four possible inputs. We represent these possibilities by 00 (when both bits are 0), 01 (when the first bit is 0 and the second is 1), 10 (when the first bit is 1 and the second is 0), and 11 (when both bits are 1). The transitions

TABLE 4

Input	1	0	1	0	1	1	—
State	s_0	s_3	s_1	s_2	s_3	s_0	s_3
Output	0	0	1	0	0	0	—

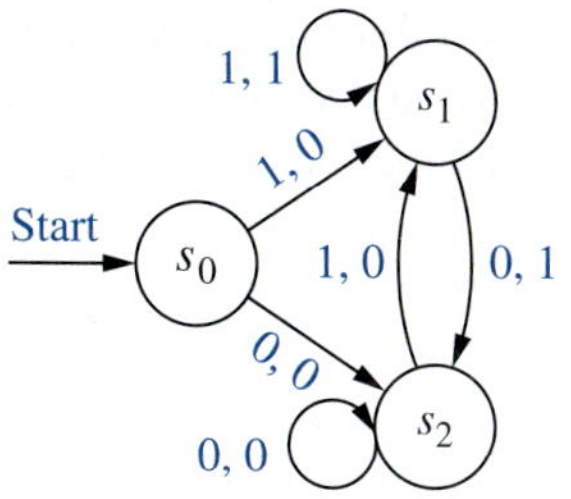

FIGURE 4 A unit-delay machine.

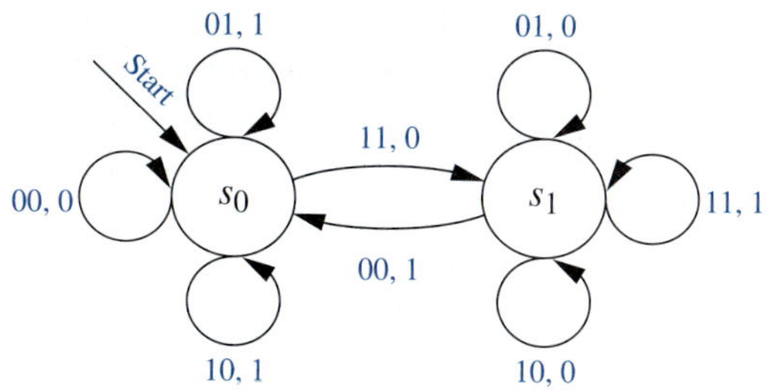

FIGURE 5 A finite-state machine for addition.

and the outputs are constructed from the sum of the two bits represented by the input and the carry represented by the state. For instance, when the machine is in state s_1 and receives 01 as input, the next state is s_1 and the output is 0, because the sum that arises is $0 + 1 + 1 = (10)_2$. The state diagram for this machine is shown in Figure 5. ◀

EXAMPLE 7 In a certain coding scheme, when three consecutive 1s appear in a message, the receiver of the message knows that there has been a transmission error. Construct a finite-state machine that gives a 1 as its current output bit if and only if the last three bits received are all 1s.

Solution: Three states are needed in this machine. The start state s_0 remembers that the previous input value, if it exists, was not a 1. The state s_1 remembers that the previous input was a 1, but the input before the previous input, if it exists, was not a 1. The state s_2 remembers that the previous two inputs were 1s.

An input of 1 takes s_0 to s_1, because now a 1, and not two consecutive 1s, has been read; it takes s_1 to s_2, because now two consecutive 1s have been read; and it takes s_2 to itself, because at least two consecutive 1s have been read. An input of 0 takes every state to s_0, because this breaks up any string of consecutive 1s. The output for the transition from s_2 to itself when a 1 is read is 1, because this combination of input and state shows that three consecutive 1s have been read. All other outputs are 0. The state diagram of this machine is shown in Figure 6. ◀

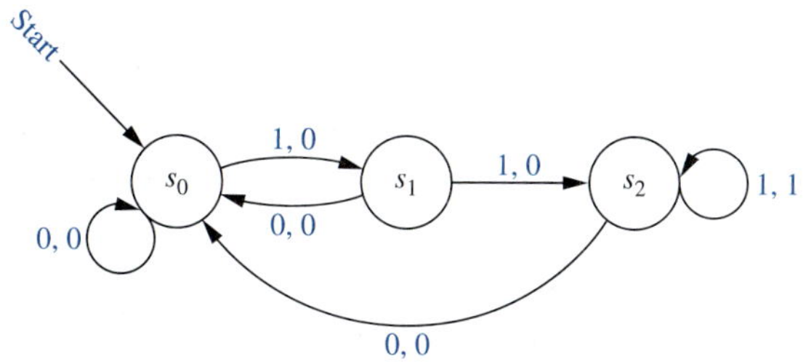

FIGURE 6 A finite-state machine that gives an output of 1 if and only if the input string read so far ends with 111.

The final output bit of the finite-state machine we constructed in Example 7 is 1 if and only if the input string ends with 111. Because of this, we say that this finite-state machine **recognizes** the set of bit strings that end with 111. This leads us to Definition 2.

Definition 2 Let $M = (S, I, O, f, g, s_0)$ be a finite-state machine and $L \subseteq I^*$. We say that M *recognizes* (or *accepts*) L if an input string x belongs to L if and only if the last output bit produced by M when given x as input is a 1.

TYPES OF FINITE-STATE MACHINES Many different kinds of finite-state machines have been developed to model computing machines. In this section we have given a definition of one type of finite-state machine. In the type of machine introduced in this section, outputs correspond to transitions between states. Machines of this type are known as **Mealy machines,** because they were first studied by G. H. Mealy in 1955. There is another important type of finite-state machine with output, where the output is determined only by the state. This type of finite-state machine is known as a **Moore machine**, because E. F. Moore introduced this type of machine in 1956. Moore machines are considered in a sequence of exercises.

In Example 7 we showed how a Mealy machine can be used for language recognition. However, another type of finite-state machine, giving no output, is usually used for this purpose. Finite-state machines with no output, also known as finite-state automata, have a set of final states and recognize a string if and only if it takes the start state to a final state. We will study this type of finite-state machine in Section 13.3.

Exercises

1. Draw the state diagrams for the finite-state machines with these state tables.

a)

State	f Input 0	f Input 1	g Input 0	g Input 1
s_0	s_1	s_0	0	1
s_1	s_0	s_2	0	1
s_2	s_1	s_1	0	0

b)

State	f Input 0	f Input 1	g Input 0	g Input 1
s_0	s_1	s_0	0	0
s_1	s_2	s_0	1	1
s_2	s_0	s_3	0	1
s_3	s_1	s_2	1	0

c)

State	f Input 0	f Input 1	g Input 0	g Input 1
s_0	s_0	s_4	1	1
s_1	s_0	s_3	0	1
s_2	s_0	s_2	0	0
s_3	s_1	s_1	1	1
s_4	s_1	s_0	1	0

2. Give the state tables for the finite-state machines with these state diagrams.

a)

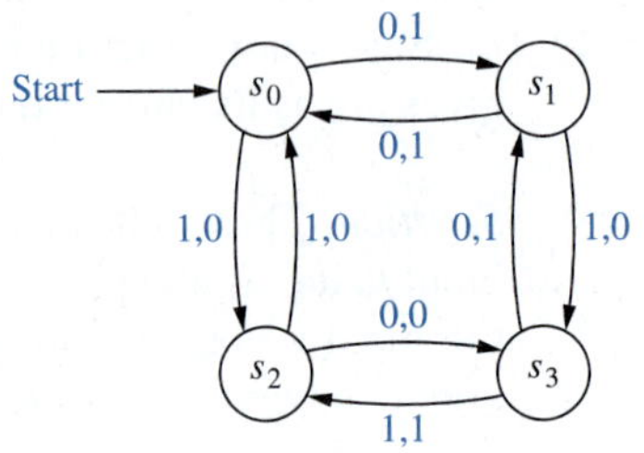

b)

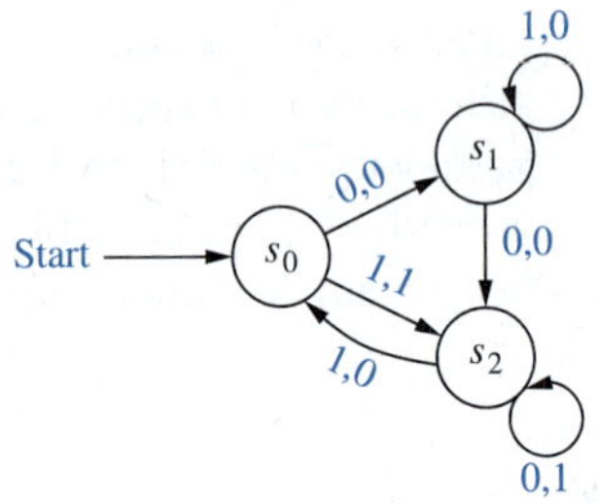

c)

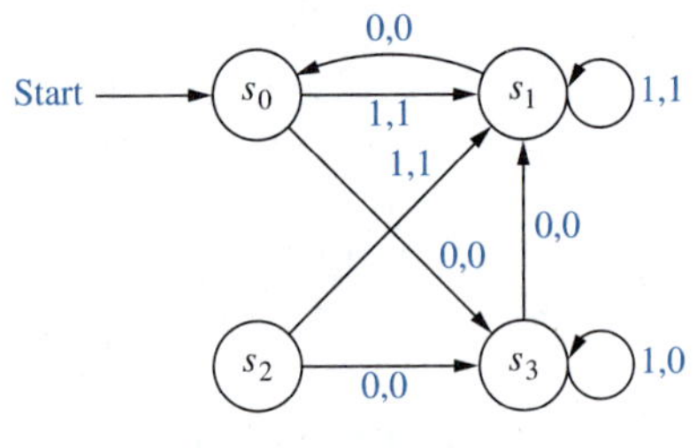

3. Find the output generated from the input string 01110 for the finite-state machine with the state table in

a) Exercise 1(a).

b) Exercise 1(b).

c) Exercise 1(c).

4. Find the output generated from the input string 10001 for the finite-state machine with the state diagram in

a) Exercise 2(a).

b) Exercise 2(b).

c) Exercise 2(c).

5. Find the output for each of these input strings when given as input to the finite-state machine in Example 2.

a) 0111 **b)** 11011011 **c)** 01010101010

6. Find the output for each of these input strings when given as input to the finite-state machine in Example 3.

a) 0000 **b)** 101010 **c)** 11011100010

7. Construct a finite-state machine that models an old-fashioned soda machine that accepts nickels, dimes, and quarters. The soda machine accepts change until 35 cents has been put in. It gives change back for any amount greater than 35 cents. Then the customer can push buttons to receive either a cola, a root beer, or a ginger ale.

8. Construct a finite-state machine that models a newspaper vending machine that has a door that can be opened only after either three dimes (and any number of other coins) or a quarter and a nickel (and any number of other coins) have been inserted. Once the door can be opened, the customer opens it and takes a paper, closing the door. No change is ever returned no matter how much extra money has been inserted. The next customer starts with no credit.

9. Construct a finite-state machine that delays an input string two bits, giving 00 as the first two bits of output.

10. Construct a finite-state machine that changes every other bit, starting with the second bit, of an input string, and leaves the other bits unchanged.

11. Construct a finite-state machine for the log-on procedure for a computer, where the user logs on by entering a user identification number (a string of digits), which is considered to be a single input, and then a password (a string of characters), which is considered to be a single input. If the password is incorrect, the user is asked for the user identification number again.

12. Construct a finite-state machine for a combination lock that contains numbers 1 through 40 and that opens only when the correct combination, 10 right, 8 second left, 37 right, is entered. Each input is a triple consisting of a number, the direction of the turn, and the number of times the lock is turned in that direction.

13. Construct a finite-state machine for a toll machine that opens a gate after 25 cents, in nickels, dimes, or quarters, has been deposited. No change is given for overpayment, and no credit is given to the next driver when more than 25 cents has been deposited.

14. Construct a finite-state machine for entering a security code into an automatic teller machine (ATM) that implements these rules: A user enters a string of four digits, one digit at a time. If the user enters the correct four digits of the password, the ATM displays a welcome screen. When the user enters an incorrect string of four digits, the ATM displays a screen that informs the user that an incorrect password was entered. If a user enters the incorrect password three times, the account is locked.

15. Construct a finite-state machine for a restricted telephone switching system that implements these rules. Only calls to the telephone numbers 0, 911, and the digit 1 followed by 10-digit telephone numbers that begin with 212, 800, 866, 877, and 888 are sent to the network. All other strings of digits are blocked by the system and the user hears an error message.

16. Construct a finite-state machine that gives an output of 1 if the number of input symbols read so far is divisible by 3 and an output of 0 otherwise.

17. Construct a finite-state machine that determines whether the input string has a 1 in the last position and a 0 in the third to the last position read so far.

18. Construct a finite-state machine that determines whether the input string read so far ends in at least five consecutive 1s.

19. Construct a finite-state machine that determines whether the word *computer* has been read as the last eight characters in the input read so far, where the input can be any string of English letters.

A **Moore machine** $M = (S, I, O, f, g, s_0)$ consists of a finite set of states, an input alphabet I, an output alphabet O, a transition function f that assigns a next state to every pair of a state and an input, an output function g that assigns an output to every state, and a starting state s_0. A Moore machine can be represented either by a table listing the transitions for each pair of state and input and the outputs for each state, or by a state diagram that displays the states, the transitions between states, and the output for each state. In the diagram, transitions are indicated with arrows labeled with the input, and the outputs are shown next to the states.

20. Construct the state diagram for the Moore machine with this state table.

	f		
	Input		
State	**0**	**1**	***g***
s_0	s_0	s_2	0
s_1	s_3	s_0	1
s_2	s_2	s_1	1
s_3	s_2	s_0	1

21. Construct the state table for the Moore machine with the state diagram shown here. Each input string to a Moore machine M produces an output string. In particular, the output corresponding to the input string $a_1a_2 \ldots a_k$ is the string $g(s_0)g(s_1) \ldots g(s_k)$, where $s_i = f(s_{i-1}, a_i)$ for $i = 1, 2, \ldots, k$.

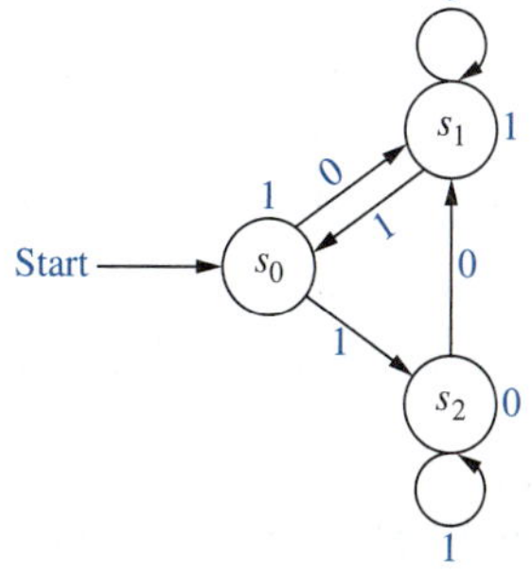

22. Find the output string generated by the Moore machine in Exercise 20 with each of these input strings.

a) 0101 **b)** 111111 **c)** 11101110111

23. Find the output string generated by the Moore machine in Exercise 21 with each of the input strings in Exercise 22.

24. Construct a Moore machine that gives an output of 1 whenever the number of symbols in the input string read so far is divisible by 4 and an output of 0 otherwise.

25. Construct a Moore machine that determines whether an input string contains an even or odd number of 1s. The machine should give 1 as output if an even number of 1s are in the string and 0 as output if an odd number of 1s are in the string.

13.3 Finite-State Machines with No Output

13.3.1 Introduction

Links

One of the most important applications of finite-state machines is in language recognition. This application plays a fundamental role in the design and construction of compilers for programming languages. In Section 13.2 we showed that a finite-state machine with output can be used to recognize a language, by giving an output of 1 when a string from the language has been read and a 0 otherwise. However, there are other types of finite-state machines that are specially designed for recognizing languages. Instead of producing output, these machines have final states. A string is recognized if and only if it takes the starting state to one of these final states.

13.3.2 Set of Strings

Before discussing finite-state machines with no output, we will introduce some important background material on sets of strings. The operations that will be defined here will be used extensively in our discussion of language recognition by finite-state machines.

Definition 1 Suppose that A and B are subsets of V^*, where V is a vocabulary. The *concatenation* of A and B, denoted by AB, is the set of all strings of the form xy, where x is a string in A and y is a string in B.

EXAMPLE 1 Let $A = \{0, 11\}$ and $B = \{1, 10, 110\}$. Find AB and BA.

Solution: The set AB contains every concatenation of a string in A and a string in B. Hence, $AB = \{01, 010, 0110, 111, 1110, 11110\}$. The set BA contains every concatenation of a string in B and a string in A. Hence, $BA = \{10, 111, 100, 1011, 1100, 11011\}$. ◀

Note that it is not necessarily the case that $AB = BA$ when A and B are subsets of V^*, where V is an alphabet, as Example 1 illustrates.

From the definition of the concatenation of two sets of strings, we can define A^n, for $n = 0, 1, 2, \ldots$. This is done recursively by specifying that

$$A^0 = \{\lambda\},$$
$$A^{n+1} = A^n A \quad \text{for } n = 0, 1, 2, \ldots.$$

EXAMPLE 2 Let $A = \{1,00\}$. Find A^n for $n = 0, 1, 2$, and 3.

Solution: We have $A^0 = \{\lambda\}$ and $A^1 = A^0 A = \{\lambda\}A = \{1, 00\}$. To find A^2 we take concatenations of pairs of elements of A. This gives $A^2 = \{11, 100, 001, 0000\}$. To find A^3 we take concatenations of elements in A^2 and A; this gives $A^3 = \{111, 1100, 1001, 10000, 0011, 00100, 00001, 000000\}$. ◀

Definition 2 Suppose that A is a subset of V^*. Then the *Kleene closure* of A, denoted by A^*, is the set consisting of concatenations of arbitrarily many strings from A. That is, $A^* = \bigcup_{k=0}^{\infty} A^k$.

EXAMPLE 3 What are the Kleene closures of the sets $A = \{0\}$, $B = \{0, 1\}$, and $C = \{11\}$?

Solution: The Kleene closure of A is the concatenation of the string 0 with itself an arbitrary finite number of times. Hence, $A^* = \{0^n \mid n = 0, 1, 2, \ldots\}$. The Kleene closure of B is the concatenation of an arbitrary number of strings, where each string is either 0 or 1. This is the set of all strings over the alphabet $V = \{0, 1\}$. That is, $B^* = V^*$. Finally, the Kleene closure of C is the concatenation of the string 11 with itself an arbitrary number of times. Hence, C^* is the set of strings consisting of an even number of 1s. That is, $C^* = \{1^{2n} \mid n = 0, 1, 2, \ldots\}$. ◀

13.3.3 Finite-State Automata

We will now give a definition of a finite-state machine with no output. Such machines are also called **finite-state automata**, and that is the terminology we will use for them here. (*Note:* The singular of *automata* is *automaton*.) These machines differ from the finite-state machines studied in Section 13.2 in that they do not produce output, but they do have a set of final states. As we will see, they recognize strings that take the starting state to a final state.

Definition 3 A *finite-state automaton* $M = (S, I, f, s_0, F)$ consists of a finite set S of *states*, a finite *input alphabet* I, a *transition function* f that assigns a next state to every pair of state and input (so that $f : S \times I \to S$), an *initial* or *start state* s_0, and a subset F of S consisting of *final* (or *accepting*) *states*.

We can represent finite-state automata using either state tables or state diagrams. Final states are indicated in state diagrams by using double circles.

EXAMPLE 4 Construct the state diagram for the finite-state automaton $M = (S, I, f, s_0, F)$, where $S = \{s_0, s_1, s_2, s_3\}$, $I = \{0, 1\}$, $F = \{s_0, s_3\}$, and the transition function f is given in Table 1.

Solution: The state diagram is shown in Figure 1. Note that because both the inputs 0 and 1 take s_2 to s_0, we write 0,1 over the edge from s_2 to s_0. ◀

Links

Courtesy of Bruce Kleene

STEPHEN COLE KLEENE (1909–1994) Stephen Kleene was born in Hartford, Connecticut. His mother, Alice Lena Cole, was a poet, and his father, Gustav Adolph Kleene, was an economics professor. Kleene attended Amherst College and received his Ph.D. from Princeton in 1934, where he studied under the famous logician Alonzo Church. Kleene joined the faculty of the University of Wisconsin in 1935, where he remained except for several leaves, including stays at the Institute for Advanced Study in Princeton. During World War II he was a navigation instructor at the Naval Reserve's Midshipmen's School and later served as the director of the Naval Research Laboratory. Kleene made significant contributions to the theory of recursive functions, investigating questions of computability and decidability, and proved one of the central results of automata theory. He served as the Acting Director of the Mathematics Research Center and as Dean of the College of Letters and Sciences at the University of Wisconsin. Kleene was a student of natural history. He discovered a previously undescribed variety of butterfly that is named after him. He was an avid hiker and climber. Kleene was also noted as a talented teller of anecdotes, using a powerful voice that could be heard several offices away.

TABLE 1

	f	
	Input	
State	**0**	**1**
s_0	s_0	s_1
s_1	s_0	s_2
s_2	s_0	s_0
s_3	s_2	s_1

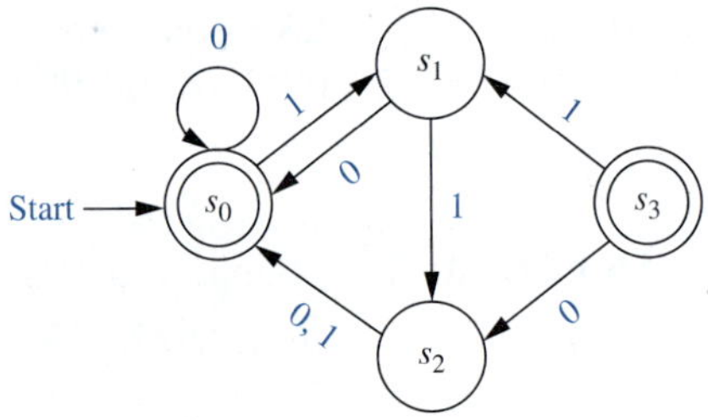

FIGURE 1 The state diagram for a finite-state automaton.

EXTENDING THE TRANSITION FUNCTION The transition function f of a finite-state machine $M = (S, I, f, s_0, F)$ can be extended so that it is defined for all pairs of states and strings; that is, f can be extended to a function $f : S \times I^* \rightarrow S$. Let $x = x_1x_2 \ldots x_k$ be a string in I^*. Then $f(s_1, x)$ is the state obtained by using each successive symbol of x, from left to right, as input, starting with state s_1. From s_1 we go on to state $s_2 = f(s_1, x_1)$, then to state $s_3 = f(s_2, x_2)$, and so on, with $f(s_1, x) = f(s_k, x_k)$. Formally, we can define this extended transition function f recursively for the deterministic finite-state machine $M = (S, I, f, s_0, F)$ by

(*i*) $f(s, \lambda) = s$ for every state $s \in S$; and

(*ii*) $f(s, xa) = f(f(s, x), a)$ for all $s \in S$, $x \in I^*$, and $a \in I$.

We can use structural induction and this recursive definition to prove properties of this extended transition function. For example, in Exercise 15 we ask you to prove that

$$f(s, xy) = f(f(s, x), y)$$

for every state $s \in S$ and strings $x \in I^*$ and $y \in I^*$.

13.3.4 Language Recognition by Finite-State Machines

Next, we define some terms that are used when studying the recognition by finite-state automata of certain sets of strings.

Definition 4 A string x is said to be *recognized* or *accepted* by the machine $M = (S, I, f, s_0, F)$ if it takes the initial state s_0 to a final state, that is, $f(s_0, x)$ is a state in F. The *language recognized* or *accepted* by the machine M, denoted by $L(M)$, is the set of all strings that are recognized by M. Two finite-state automata are called *equivalent* if they recognize the same language.

In Example 5 we will find the languages recognized by several finite-state automata.

EXAMPLE 5 Determine the languages recognized by the finite-state automata M_1, M_2, and M_3 in Figure 2.

Solution: The only final state of M_1 is s_0. The strings that take s_0 to itself are those consisting of zero or more consecutive 1s. Hence, $L(M_1) = \{1^n \mid n = 0, 1, 2, \ldots\}$.

The only final state of M_2 is s_2. The only strings that take s_0 to s_2 are 1 and 01. Hence, $L(M_2) = \{1, 01\}$.

The final states of M_3 are s_0 and s_3. The only strings that take s_0 to itself are λ, 0, 00, 000, ..., that is, any string of zero or more consecutive 0s. The only strings that take s_0 to s_3 are a string of zero or more consecutive 0s, followed by 10, followed by any string. Hence, $L(M_3) = \{0^n, 0^n10x \mid n = 0, 1, 2, \ldots, \text{ and } x \text{ is any string}\}$. ◀

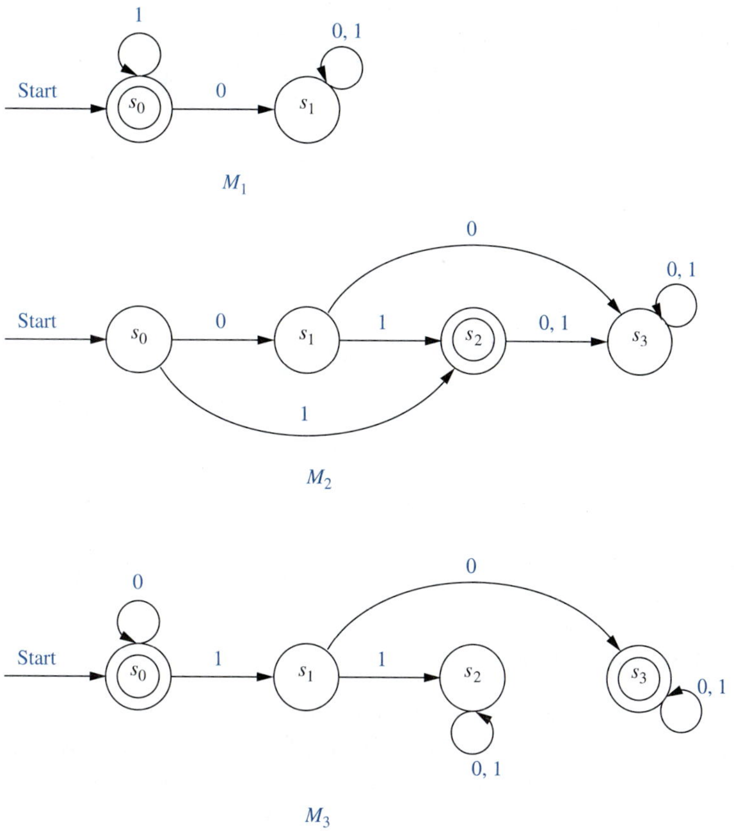

FIGURE 2 **Some finite-state automata.**

DESIGNING FINITE-STATE AUTOMATA We can often construct a finite-state automaton that recognizes a given set of strings by carefully adding states and transitions and determining which of these states should be final states. When appropriate we include states that can keep track of some of the properties of the input string, providing the finite-state automaton with limited memory. Examples 6 and 7 illustrate some of the techniques that can be used to construct finite-state automata that recognize particular types of sets of strings.

EXAMPLE 6 Construct deterministic finite-state automata that recognize each of these languages.

Extra Examples

(a) the set of bit strings that begin with two 0s
(b) the set of bit strings that contain two consecutive 0s
(c) the set of bit strings that do not contain two consecutive 0s
(d) the set of bit strings that end with two 0s
(e) the set of bit strings that contain at least two 0s

Solution: (a) Our goal is to construct a deterministic finite-state automaton that recognizes the set of bit strings that begin with two 0s. Besides the start state s_0, we include a nonfinal state s_1; we move to s_1 from s_0 if the first bit is a 0. Next, we add a final state s_2, which we move to from s_1 if the second bit is a 0. When we have reached s_2 we know that the first two input bits are both 0s, so we stay in the state s_2 no matter what the succeeding bits (if any) are. We move to a nonfinal state s_3 from s_0 if the first bit is a 1 and from s_1 if the second bit is a 1. The reader should verify that the finite-state automaton in Figure 3(a) recognizes the set of bit strings that begin with two 0s.

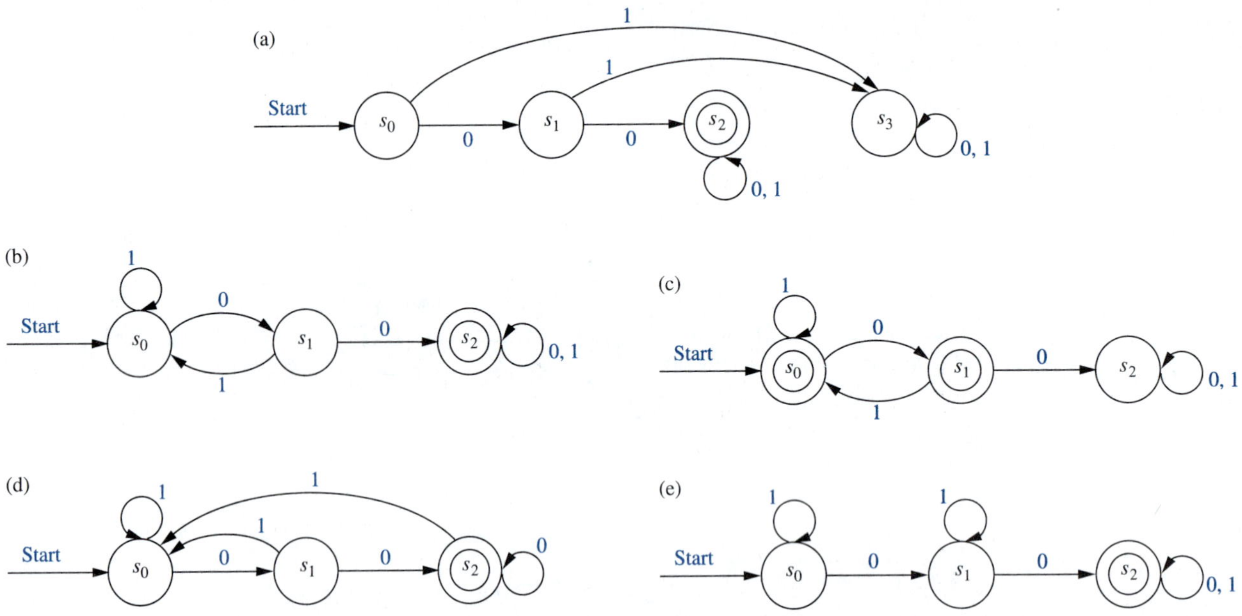

FIGURE 3 Deterministic finite-state automata recognizing the languages in Example 6.

(b) Our goal is to construct a deterministic finite-state automaton that recognizes the set of bit strings that contain two consecutive 0s. Besides the start state s_0, we include a nonfinal state s_1, which tells us that the last input bit seen is a 0, but either the bit before it was a 1, or this bit was the initial bit of the string. We include a final state s_2 that we move to from s_1 when the next input bit after a 0 is also a 0. If a 1 follows a 0 in the string (before we encounter two consecutive 0s), we return to s_0 and begin looking for consecutive 0s all over again. The reader should verify that the finite-state automaton in Figure 3(b) recognizes the set of bit strings that contain two consecutive 0s.

(c) Our goal is to construct a deterministic finite-state automaton that recognizes the set of bit strings that do not contain two consecutive 0s. Besides the start state s_0, which should be a final state, we include a final state s_1, which we move to from s_0 when 0 is the first input bit. When an input bit is a 1, we return to, or stay in, state s_0. We add a state s_2, which we move to from s_1 when the input bit is a 0. Reaching s_2 tells us that we have seen two consecutive 0s as input bits. We stay in state s_2 once we have reached it; this state is not final. The reader should verify that the finite-state automaton in Figure 3(c) recognizes the set of bit strings that do not contain two consecutive 0s. [The astute reader will notice the relationship between the finite-state automaton constructed here and the one constructed in part (b). See Exercise 39.]

(d) Our goal is to construct a deterministic finite-state automaton that recognizes the set of bit strings that end with two 0s. Besides the start state s_0, we include a nonfinal state s_1, which we move to if the first bit is 0. We include a final state s_2, which we move to from s_1 if the next input bit after a 0 is also a 0. If an input of 0 follows a previous 0, we stay in state s_2 because the last two input bits are still 0s. Once we are in state s_2, an input bit of 1 sends us back to s_0, and we begin looking for consecutive 0s all over again. We also return to s_0 if the next input is a 1 when we are in state s_1. The reader should verify that the finite-state automaton in Figure 3(d) recognizes the set of bit strings that end with two 0s.

(e) Our goal is to construct a deterministic finite-state automaton that recognizes the set of bit strings that contain two 0s. Besides the start state, we include a state s_1, which is not final; we stay in s_0 until an input bit is a 0 and we move to s_1 when we encounter the first 0 bit in the input. We add a final state s_2, which we move to from s_1 once we encounter a second 0 bit. Whenever we encounter a 1 as input, we stay in the current state. Once we have reached s_2, we remain there. Here, s_1 and s_2 are used to tell us that we have already seen one or two 0s in the input string so far, respectively. The reader should verify that the finite-state automaton in Figure 3(e) recognizes the set of bit strings that contain two 0s. ◀

EXAMPLE 7 Construct a deterministic finite-state automaton that recognizes the set of bit strings that contain an odd number of 1s and that end with at least two consecutive 0s.

Solution: We can build a deterministic finite-state automaton that recognizes the specified set by including states that keep track of both the parity of the number of 1 bits and whether we have seen no, one, or at least two 0s at the end of the input string.

The start state s_0 can be used to tell us that the input read so far contains an even number of 1s and ends with no 0s (that is, is empty or ends with a 1). Besides the start state, we include five more states. We move to states s_1, s_2, s_3, s_4, and s_5, respectively, when the input string read so far contains an even number of 1s and ends with one 0; when it contains an even number of 1s and ends with at least two 0s; when it contains an odd number of 1s and ends with no 0s; when it contains an odd number of 1s and ends with one 0; and when it contains an odd number of 1s and ends with two 0s. The state s_5 is a final state.

The reader should verify that the finite-state automaton in Figure 4 recognizes the set of bit strings that contain an odd number of 1s and end with at least two consecutive 0s. ◀

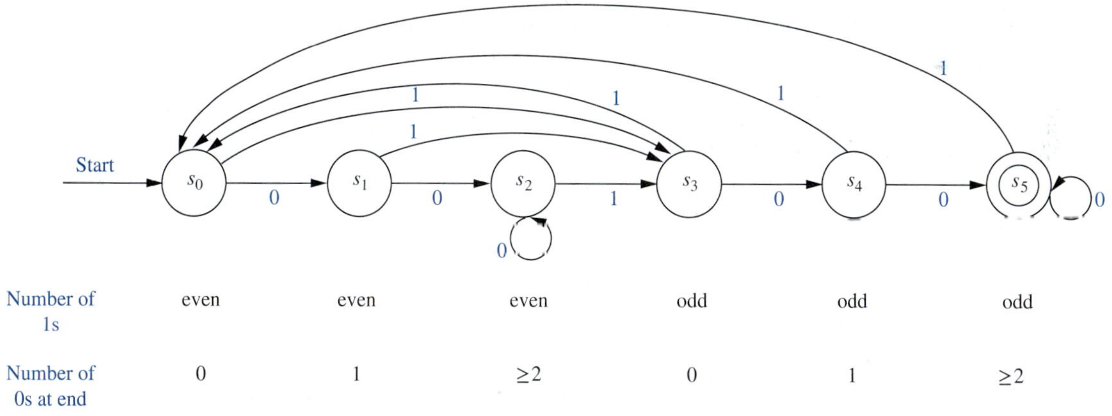

FIGURE 4 A deterministic finite-state automaton recognizing the set of bit strings containing an odd number of 1s and ending with at least two 0s.

EQUIVALENT FINITE-STATE AUTOMATA In Definition 4 we specified that two finite-state automata are equivalent if they recognize the same language. Example 8 provides an example of two equivalent deterministic finite-state machines.

EXAMPLE 8 Show that the two finite-state automata M_0 and M_1 shown in Figure 5 are equivalent.

Solution: For a string x to be recognized by M_0, x must take us from s_0 to the final state s_1 or the final state s_4. The only string that takes us from s_0 to s_1 is the string 1. The strings that take us from s_0 to s_4 are those strings that begin with a 0, which takes us from s_0 to s_2, followed by zero or more additional 0s, which keep the machine in state s_2, followed by a 1, which takes us from state s_2 to the final state s_4. All other strings take us from s_0 to a state that is not final. (We leave it to the reader to fill in the details.) We conclude that $L(M_0)$ is the set of strings of zero or more 0 bits followed by a final 1.

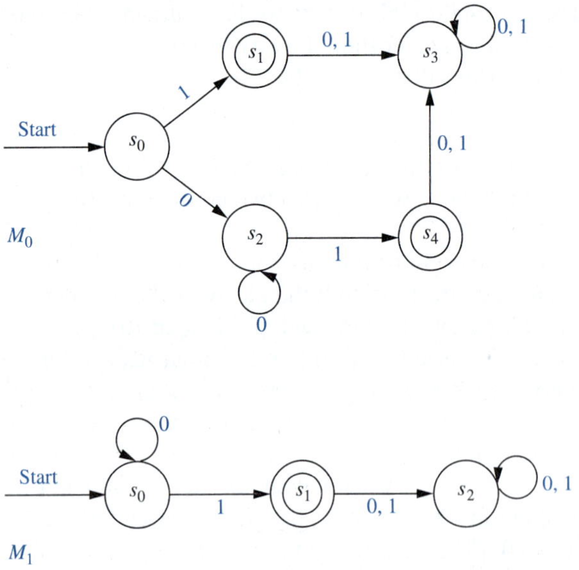

FIGURE 5 M_0 **and** M_1 **are equivalent finite-state automata.**

For a string x to be recognized by M_1, x must take us from s_0 to the final state s_1. So, for x to be recognized, it must begin with some number of 0s, which leave us in state s_0, followed by a 1, which takes us to the final state s_1. A string of all zeros is not recognized because it leaves us in state s_0, which is not final. All strings that contain a 0 after 1 are not recognized because they take us to state s_2, which is not final. It follows that $L(M_1)$ is the same as $L(M_0)$. We conclude that M_0 and M_1 are equivalent.

Note that the finite-state machine M_1 only has three states. No finite state machine with fewer than three states can be used to recognize the set of all strings of zero or more 0 bits followed by a 1 (see Exercise 37). ◀

As Example 8 shows, a finite-state automaton may have more states than one equivalent to it. In fact, algorithms used to construct finite-state automata to recognize certain languages may have many more states than necessary. Using unnecessarily large finite-state machines to recognize languages can make both hardware and software applications inefficient and costly. This problem arises when finite-state automata are used in compilers, which translate computer programs to a language a computer can understand (object code).

Exercises 58–61 develop a procedure that constructs a finite-state automaton with the fewest states possible among all finite-state automata equivalent to a given finite-state automaton. This procedure is known as **machine minimization**. The minimization procedure described in these exercises reduces the number of states by replacing states with equivalence classes of states with respect to an equivalence relation in which two states are equivalent if every input string either sends both states to a final state or sends both to a state that is not final. Before the minimization procedure begins, all states that cannot be reached from the start state using any input string are first removed; removing these does not change the language recognized.

13.3.5 Nondeterministic Finite-State Automata

The finite-state automata discussed so far are **deterministic**, because for each pair of state and input value there is a unique next state given by the transition function. There is another important type of finite-state automaton in which there may be several possible next states for each pair of input value and state. Such machines are called **nondeterministic**. Nondeterministic finite-state automata are important in determining which languages can be recognized by a finite-state automaton.

Definition 5

Links

A *nondeterministic finite-state automaton* $M = (S, I, f, s_0, F)$ consists of a set S of states, an input alphabet I, a transition function f that assigns a set of states to each pair of state and input (so that $f : S \times I \rightarrow P(S)$), a starting state s_0, and a subset F of S consisting of the final states.

We can represent nondeterministic finite-state automata using state tables or state diagrams. When we use a state table, for each pair of state and input value we give a list of possible next states. In the state diagram, we include an edge from each state to all possible next states, labeling edges with the input or inputs that lead to this transition.

EXAMPLE 9 Find the state diagram for the nondeterministic finite-state automaton with the state table shown in Table 2. The final states are s_2 and s_3.

Solution: The state diagram for this automaton is shown in Figure 6. ◀

Links

Bettmann/Getty Images

GRACE BREWSTER MURRAY HOPPER (1906–1992) Grace Hopper, born in New York City, displayed an intense curiosity as a child with how things worked. At the age of seven, she disassembled alarm clocks to discover their mechanisms. She inherited her love of mathematics from her mother, who received special permission to study geometry (but not algebra and trigonometry) at a time when women were actively discouraged from such study. Hopper was inspired by her father, a successful insurance broker, who had lost his legs from circulatory problems. He told his children they could do anything if they put their minds to it. He inspired Hopper to pursue higher education and not conform to the usual roles for females. Her parents made sure that she had an excellent education; she attended private schools for girls in New York. Hopper entered Vassar College in 1924, where she majored in mathematics and physics; she graduated in 1928. She received a master's degree in mathematics from Yale University in 1930. In 1930 she also married an English instructor at the New York School of Commerce; she later divorced and did not have children. Hopper was a mathematics professor at Vassar from 1931 until 1943, earning a Ph.D. from Yale in 1934.

After the attack on Pearl Harbor, Hopper, coming from a family with strong military traditions, decided to leave her academic position and join the Navy WAVES. To enlist, she needed special permission to leave her strategic position as a mathematics professor, as well as a waiver for weighing too little. In December 1943, she was sworn into the Navy Reserve and trained at the Midshipman's School for Women. Hopper was assigned to work at the Naval Ordnance Laboratory at Harvard University. She wrote programs for the world's first large-scale automatically sequenced digital computer, which was used to help aim Navy artillery in varying weather. Hopper has been credited with coining the term "bug" to refer to a hardware glitch, but it was used at Harvard prior to her arrival there. However, it is true that Hopper and her programming team found a moth in one of the relays in the computer hardware that shut the system down. This famous moth was pasted into a lab book. In the 1950s Hopper coined the term "debug" for the process of removing programming errors.

In 1946, when the Navy told her that she was too old for active service, Hopper chose to remain at Harvard as a civilian research fellow. In 1949 she left Harvard to join the Eckert–Mauchly Computer Corporation, where she helped develop UNIVAC, the first commercial computer. Hopper remained with this company when it was taken over by Remington Rand and when Remington Rand merged with the Sperry Corporation. She was a visionary for the potential power of computers; she understood that computers would become widely used if tools that were both programmer-friendly and application-friendly could be developed. In particular, she believed that computer programs could be written in English, rather than using machine instructions. To help achieve this goal, she developed the first compiler. She published the first research paper on compilers in 1952. Hopper is also known as the mother of the computer language COBOL; members of Hopper's staff helped to frame the basic language design for COBOL using their earlier work as a basis.

In 1966, Hopper retired from the Navy Reserve. However, only seven months later, the Navy recalled her from retirement to help standardize high-level naval computer languages. In 1983 she was promoted to the rank of Commodore by special Presidential appointment, and in 1985 she was elevated to the rank of Rear Admiral. Her retirement from the Navy, at the age of 80, was held on the *USS Constitution*.

TABLE 2

	f	
	Input	
State	**0**	**1**
s_0	s_0, s_1	s_3
s_1	s_0	s_1, s_3
s_2		s_0, s_2
s_3	s_0, s_1, s_2	s_1

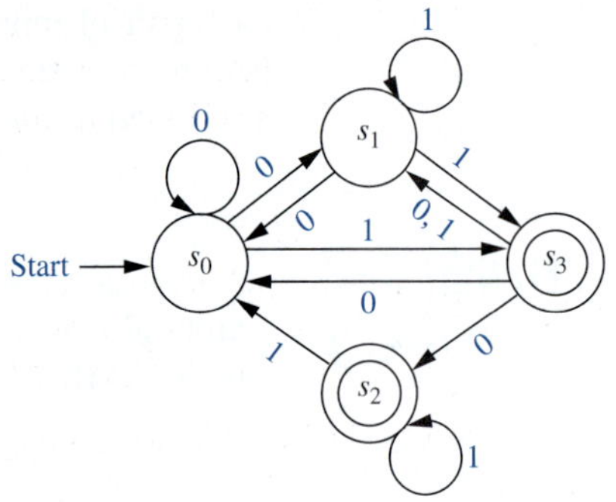

FIGURE 6 The nondeterministic finite-state automaton with state table given in Table 2.

EXAMPLE 10 Find the state table for the nondeterministic finite-state automaton with the state diagram shown in Figure 7.

Solution: The state table is given as Table 3. ◀

What does it mean for a nondeterministic finite-state automaton to recognize a string $x = x_1x_2 \ldots x_k$? The first input symbol x_1 takes the starting state s_0 to a set S_1 of states. The next input symbol x_2 takes each of the states in S_1 to a set of states. Let S_2 be the union of these sets. We continue this process, including, at each stage, all states obtained using a state obtained at the previous stage and the current input symbol. We **recognize**, or **accept**, the string x if there is a final state in the set of all states that can be obtained from s_0 using x. The **language recognized** by a nondeterministic finite-state automaton is the set of all strings recognized by this automaton.

EXAMPLE 11 Find the language recognized by the nondeterministic finite-state automaton shown in Figure 7.

Solution: Because s_0 is a final state, and there is a transition from s_0 to itself when 0 is the input, the machine recognizes all strings consisting of zero or more consecutive 0s. Furthermore, because s_4 is a final state, any string that has s_4 in the set of states that can be reached from s_0 with this input string is recognized. The only such strings are strings consisting of zero or more consecutive 0s followed by 01 or 11. Because s_0 and s_4 are the only final states, the language recognized by the machine is $\{0^n, 0^n01, 0^n11 \mid n \geq 0\}$. ◀

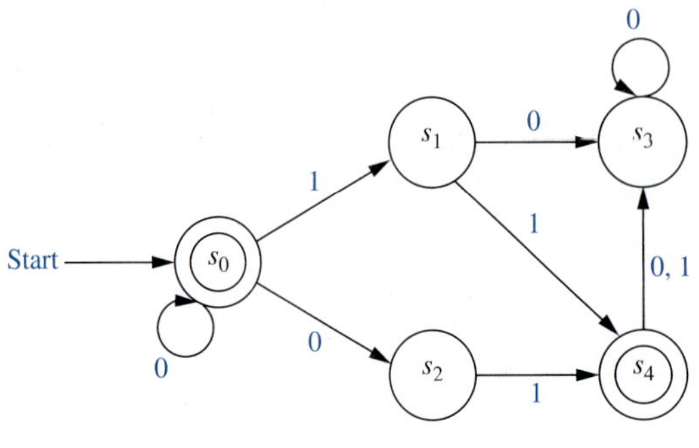

FIGURE 7 A nondeterministic finite-state automaton.

TABLE 3

	f	
	Input	
State	**0**	**1**
s_0	s_0, s_2	s_1
s_1	s_3	s_4
s_2		s_4
s_3	s_3	
s_4	s_3	s_3

One important fact is that a language recognized by a nondeterministic finite-state automaton is also recognized by a deterministic finite-state automaton. We will take advantage of this fact in Section 13.4 when we will determine which languages are recognized by finite-state automata.

THEOREM 1 If the language L is recognized by a nondeterministic finite-state automaton M_0, then L is also recognized by a deterministic finite-state automaton M_1.

Proof: We will describe how to construct the deterministic finite-state automaton M_1 that recognizes L from M_0, the nondeterministic finite-state automaton that recognizes this language. Each state in M_1 will be made up of a set of states in M_0. The start symbol of M_1 is $\{s_0\}$, which is the set containing the start state of M_0. The input set of M_1 is the same as the input set of M_0.

Given a state $\{s_{i_1}, s_{i_2}, \ldots, s_{i_k}\}$ of M_1, the input symbol x takes this state to the union of the sets of next states for the elements of this set, that is, the union of the sets $f(s_{i_1}, x)$, $f(s_{i_2}, x), \ldots, f(s_{i_k}, x)$. The states of M_1 are all the subsets of S, the set of states of M_0, that are obtained in this way starting with s_0. (There are as many as 2^n states in the deterministic machine, where n is the number of states in the nondeterministic machine, because all subsets may occur as states, including the empty set, although usually far fewer states occur.) The final states of M_1 are those sets that contain a final state of M_0.

Suppose that an input string is recognized by M_0. Then one of the states that can be reached from s_0 using this input string is a final state (the reader should provide an inductive proof of this). This means that in M_1, this input string leads from $\{s_0\}$ to a set of states of M_0 that contains a final state. This subset is a final state of M_1, so this string is also recognized by M_1. Also, an input string not recognized by M_0 does not lead to any final states in M_0. (The reader should provide the details that prove this statement.) Consequently, this input string does not lead from $\{s_0\}$ to a final state in M_1. ◁

EXAMPLE 12 Find a deterministic finite-state automaton that recognizes the same language as the nondeterministic finite-state automaton in Example 10.

Solution: The deterministic automaton shown in Figure 8 is constructed from the nondeterministic automaton in Example 10. The states of this deterministic automaton are subsets of the set of all states of the nondeterministic machine. The next state of a subset under an input symbol is the subset containing the next states in the nondeterministic machine of all elements in this subset. For instance, on input of 0, $\{s_0\}$ goes to $\{s_0, s_2\}$, because s_0 has transitions to itself and to s_2 in the nondeterministic machine; the set $\{s_0, s_2\}$ goes to $\{s_1, s_4\}$ on input of 1, because

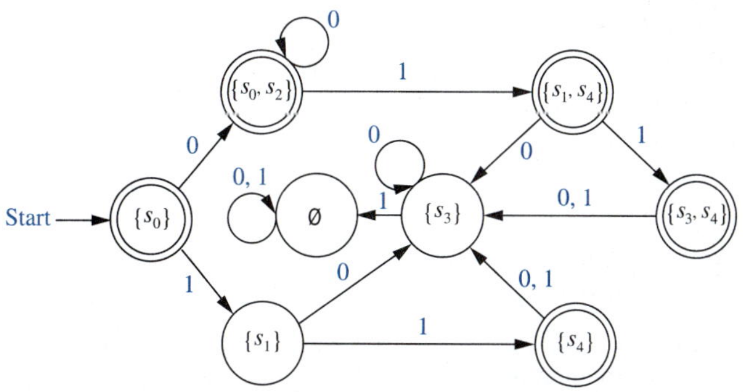

FIGURE 8 A deterministic automaton equivalent to the nondeterministic automaton in Example 10.

s_0 goes just to s_1 and s_2 goes just to s_4 on input of 1 in the nondeterministic machine; and the set $\{s_1, s_4\}$ goes to $\{s_3\}$ on input of 0, because s_1 and s_4 both go to just s_3 on input of 0 in the deterministic machine. All subsets that are obtained in this way are included in the deterministic finite-state machine. Note that the empty set is one of the states of this machine, because it is the subset containing all the next states of $\{s_3\}$ on input of 1. The start state is $\{s_0\}$, and the set of final states are all those that include s_0 or s_4. ◀

Exercises

1. Let $A = \{0, 11\}$ and $B = \{00, 01\}$. Find each of these sets.

a) AB **b)** BA **c)** A^2 **d)** B^3

2. Show that if A is a set of strings, then $A\emptyset = \emptyset A = \emptyset$.

3. Find all pairs of sets of strings A and B for which $AB = \{10, 111, 1010, 1000, 10111, 101000\}$.

4. Show that these equalities hold.

a) $\{\lambda\}^* = \{\lambda\}$

b) $(A^*)^* = A^*$ for every set of strings A

5. Describe the elements of the set A^* for these values of A.

a) $\{10\}$ **b)** $\{111\}$ **c)** $\{0, 01\}$ **d)** $\{1, 101\}$

6. Let V be an alphabet, and let A and B be subsets of V^*. Show that $|AB| \leq |A||B|$.

7. Let V be an alphabet, and let A and B be subsets of V^* with $A \subseteq B$. Show that $A^* \subseteq B^*$.

8. Suppose that A is a subset of V^*, where V is an alphabet. Prove or disprove each of these statements.

a) $A \subseteq A^2$ **b)** if $A = A^2$, then $\lambda \in A$

c) $A\{\lambda\} = A$ **d)** $(A^*)^* = A^*$

e) $A^*A = A^*$ **f)** $|A^n| = |A|^n$

9. Determine whether the string 11101 is in each of these sets.

a) $\{0, 1\}^*$ **b)** $\{1\}^*\{0\}^*\{1\}^*$

c) $\{11\}\{0\}^*\{01\}$ **d)** $\{11\}^*\{01\}^*$

e) $\{111\}^*\{0\}^*\{1\}$ **f)** $\{11, 0\}\{00, 101\}$

10. Determine whether the string 01001 is in each of these sets.

a) $\{0, 1\}^*$ **b)** $\{0\}^*\{10\}\{1\}^*$

c) $\{010\}^*\{0\}^*\{1\}$ **d)** $\{010, 011\}\{00, 01\}$

e) $\{00\}\{0\}^*\{01\}$ **f)** $\{01\}^*\{01\}^*$

11. Determine whether each of these strings is recognized by the deterministic finite-state automaton in Figure 1.

a) 111 **b)** 0011 **c)** 1010111 **d)** 011011011

12. Determine whether each of these strings is recognized by the deterministic finite-state automaton in Figure 1.

a) 010 **b)** 1101 **c)** 1111110 **d)** 010101010

13. Determine whether all the strings in each of these sets are recognized by the deterministic finite-state automaton in Figure 1.

a) $\{0\}^*$ **b)** $\{0\}\{0\}^*$ **c)** $\{1\}\{0\}^*$

d) $\{01\}^*$ **e)** $\{0\}^*\{1\}^*$ **f)** $\{1\}\{0, 1\}^*$

14. Show that if $M = (S, I, f, s_0, F)$ is a deterministic finite-state automaton and $f(s, x) = s$ for the state $s \in S$ and the input string $x \in I^*$, then $f(s, x^n) = s$ for every nonnegative integer n. (Here x^n is the concatenation of n copies of the string x, defined recursively in Exercise 39 in Section 5.3.)

15. Given a deterministic finite-state automaton $M = (S, I, f, s_0, F)$, use structural induction and the recursive definition of the extended transition function f to prove that $f(s, xy) = f(f(s, x), y)$ for all states $s \in S$ and all strings $x \in I^*$ and $y \in I^*$.

In Exercises 16–22 find the language recognized by the given deterministic finite-state automaton.

16.

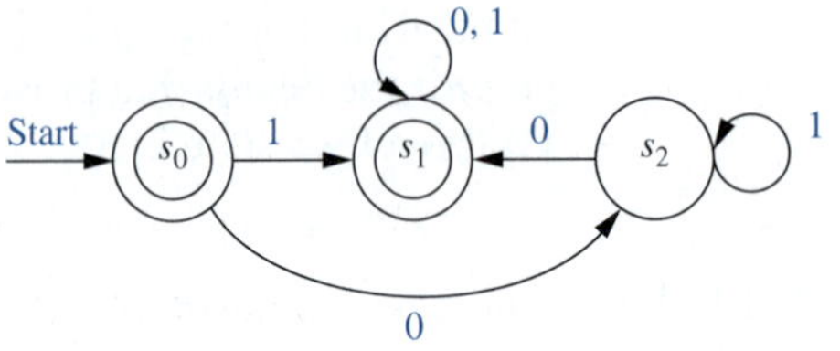

17.

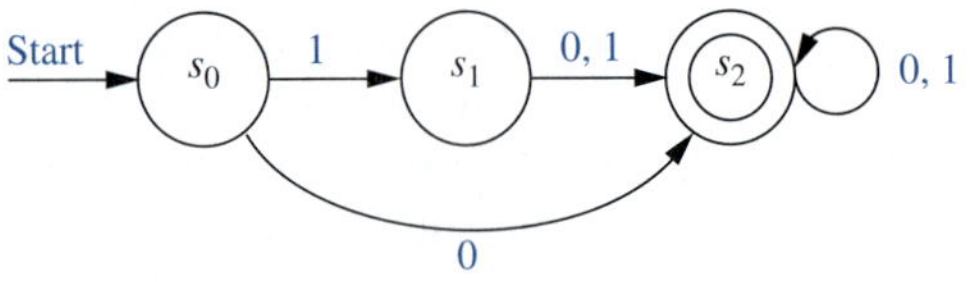

18.

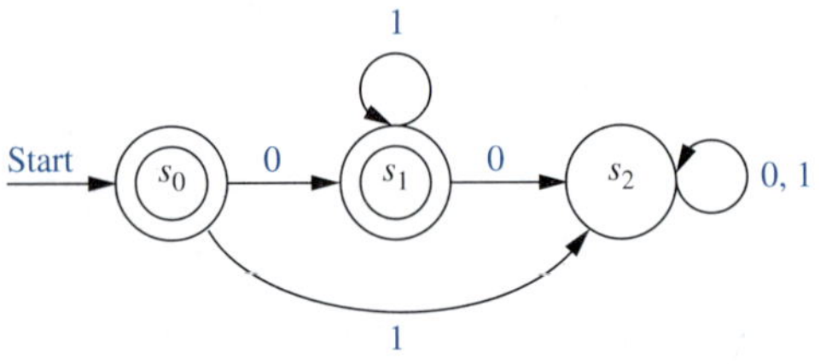

19.

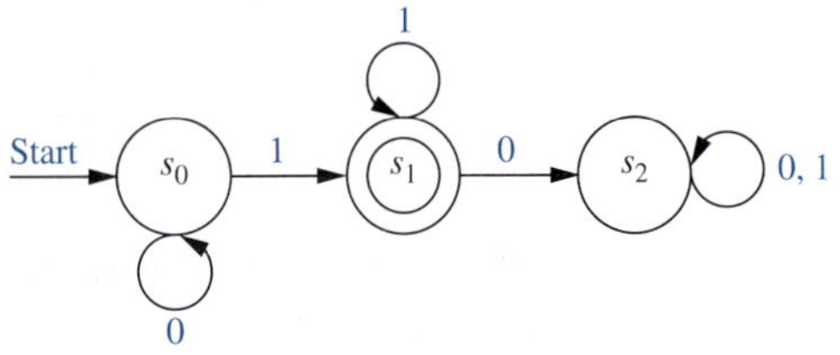

20.

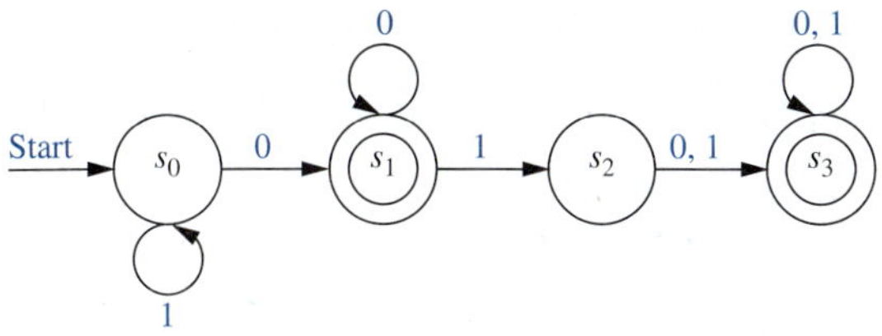

21.

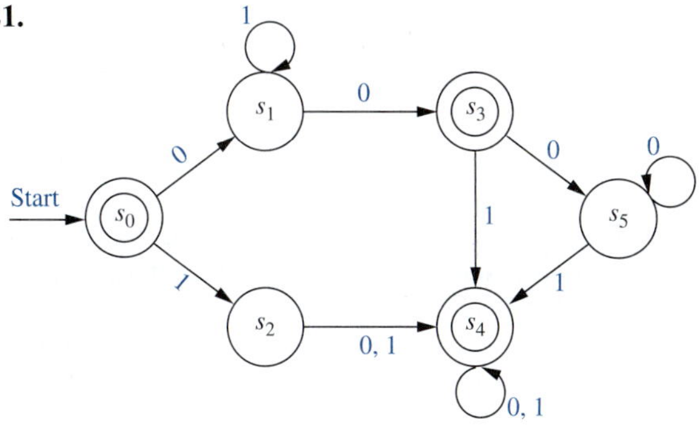

22.

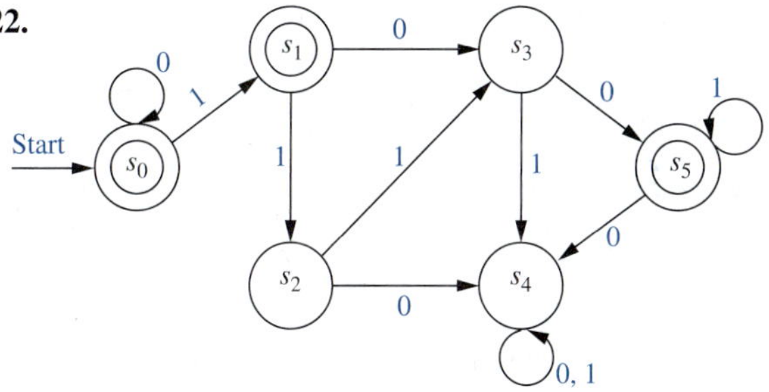

23. Construct a deterministic finite-state automaton that recognizes the set of all bit strings beginning with 01.

24. Construct a deterministic finite-state automaton that recognizes the set of all bit strings that end with 10.

25. Construct a deterministic finite-state automaton that recognizes the set of all bit strings that contain the string 101.

26. Construct a deterministic finite-state automaton that recognizes the set of all bit strings that do not contain three consecutive 0s.

27. Construct a deterministic finite-state automaton that recognizes the set of all bit strings that contain exactly three 0s.

28. Construct a deterministic finite-state automaton that recognizes the set of all bit strings that contain at least three 0s.

29. Construct a deterministic finite-state automaton that recognizes the set of all bit strings that contain three consecutive 1s.

30. Construct a deterministic finite-state automaton that recognizes the set of all bit strings that begin with 0 or with 11.

31. Construct a deterministic finite-state automaton that recognizes the set of all bit strings that begin and end with 11.

32. Construct a deterministic finite-state automaton that recognizes the set of all bit strings that contain an even number of 1s.

33. Construct a deterministic finite-state automaton that recognizes the set of all bit strings that contain an odd number of 0s.

34. Construct a deterministic finite-state automaton that recognizes the set of all bit strings that contain an even number of 0s and an odd number of 1s.

35. Construct a finite-state automaton that recognizes the set of bit strings consisting of a 0 followed by a string with an odd number of 1s.

36. Construct a finite-state automaton with four states that recognizes the set of bit strings containing an even number of 1s and an odd number of 0s.

37. Show that there is no finite-state automaton with two states that recognizes the set of all bit strings that have one or more 1 bits and end with a 0.

38. Show that there is no finite-state automaton with three states that recognizes the set of bit strings containing an even number of 1s and an even number of 0s.

39. Explain how you can change the deterministic finite-state automaton M so that the changed automaton recognizes the set $I^* - L(M)$.

40. Use Exercise 39 and finite-state automata constructed in Example 6 to find deterministic finite-state automata that recognize each of these sets.

a) the set of bit strings that do not begin with two 0s

b) the set of bit strings that do not end with two 0s

c) the set of bit strings that contain at most one 0 (that is, that do not contain at least two 0s)

41. Use the procedure you described in Exercise 39 and the finite-state automata you constructed in Exercise 25 to find a deterministic finite-state automaton that recognizes the set of all bit strings that do not contain the string 101.

42. Use the procedure you described in Exercise 39 and the finite-state automaton you constructed in Exercise 29 to find a deterministic finite-state automaton that recognizes the set of all bit strings that do not contain three consecutive 1s.

In Exercises 43–49 find the language recognized by the given nondeterministic finite-state automaton.

43.

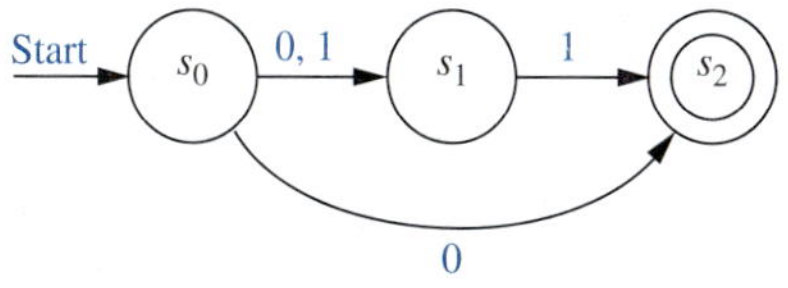

44.

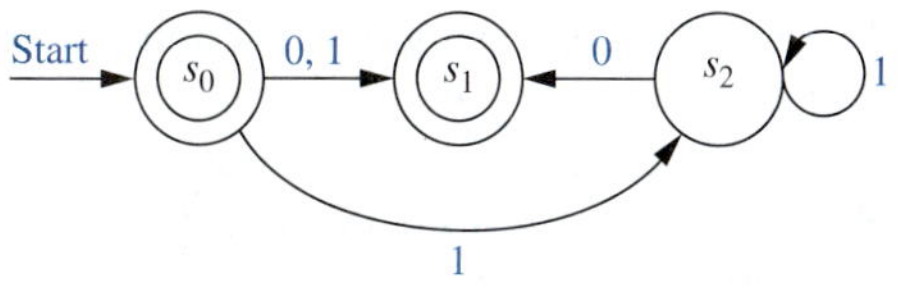

45.

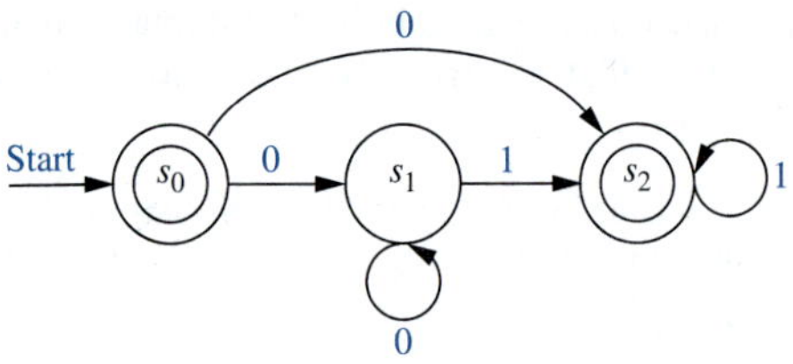

46.

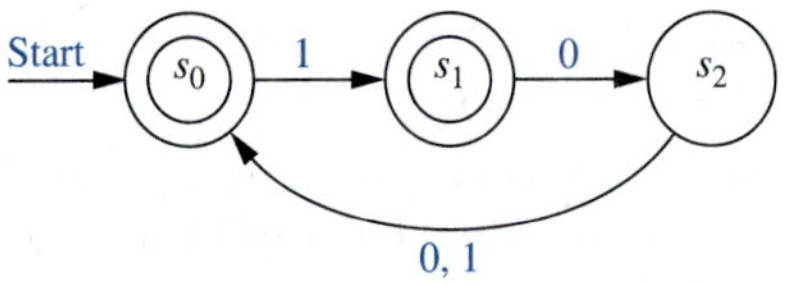

47.

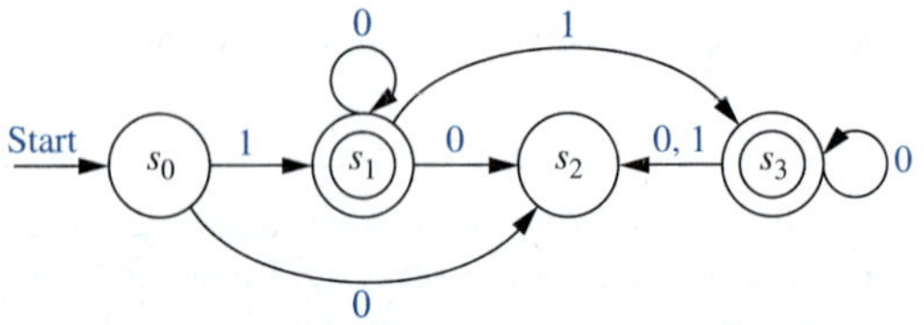

48.

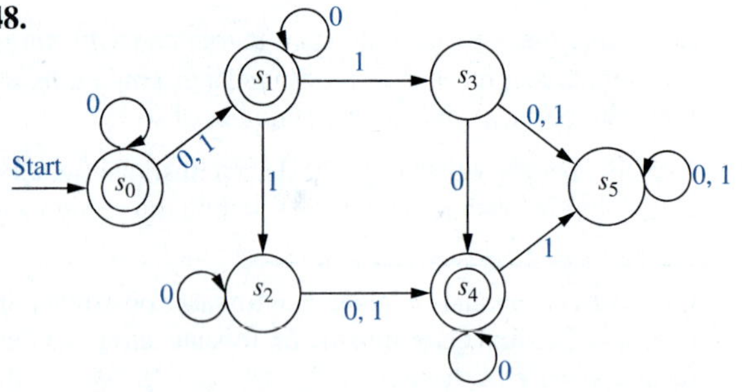

49.

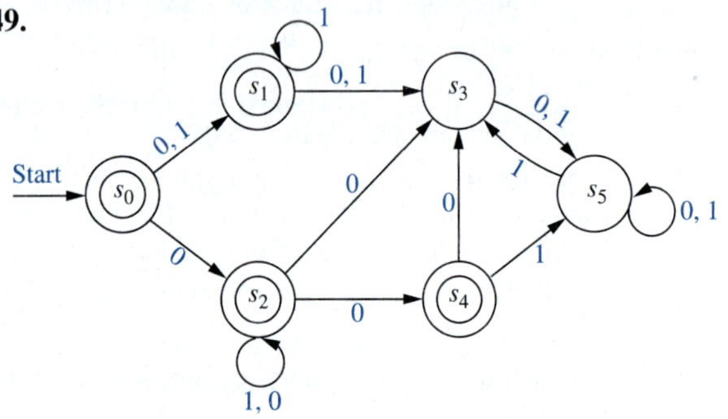

50. Find a deterministic finite-state automaton that recognizes the same language as the nondeterministic finite-state automaton in Exercise 43.

51. Find a deterministic finite-state automaton that recognizes the same language as the nondeterministic finite-state automaton in Exercise 44.

52. Find a deterministic finite-state automaton that recognizes the same language as the nondeterministic finite-state automaton in Exercise 45.

53. Find a deterministic finite-state automaton that recognizes the same language as the nondeterministic finite-state automaton in Exercise 46.

54. Find a deterministic finite-state automaton that recognizes the same language as the nondeterministic finite-state automaton in Exercise 47.

55. Find a deterministic finite-state automaton that recognizes each of these sets.

a) $\{0\}$ **b)** $\{1, 00\}$

c) $\{1^n \mid n = 2, 3, 4, \dots\}$

56. Find a nondeterministic finite-state automaton that recognizes each of the languages in Exercise 55, and has fewer states, if possible, than the deterministic automaton you found in that exercise.

***57.** Show that there is no finite-state automaton that recognizes the set of bit strings containing an equal number of 0s and 1s.

In Exercises 58–62 we introduce a technique for constructing a deterministic finite-state machine equivalent to a given deterministic finite-state machine with the least number of states possible. Suppose that $M = (S, I, f, s_0, F)$ is a finite-state automaton and that k is a nonnegative integer. Let R_k be the relation on the set S of states of M such that sR_kt if and only if for every input string x with $l(x) \leq k$ [where $l(x)$ is the length of x, as usual], $f(s, x)$ and $f(t, x)$ are both final states or both not final states. Furthermore, let R_* be the relation on the set of states of M such that sR_*t if and only if for every input string x, regardless of length, $f(s, x)$ and $f(t, x)$ are both final states or both not final states.

***58.** **a)** Show that for every nonnegative integer k, R_k is an equivalence relation on S. We say that two states s and t are ***k*-equivalent** if sR_kt.

b) Show that R_* is an equivalence relation on S. We say that two states s and t are **$*$-equivalent** if sR_*t.

c) Show that if s and t are two k-equivalent states of M, where k is a positive integer, then s and k are also $(k - 1)$-equivalent.

d) Show that the equivalence classes of R_k are a refinement of the equivalence classes of R_{k-1} if k is a positive integer. (The refinement of a partition of a set is defined in the preamble to Exercise 49 in Section 9.5.)

e) Show that if s and t are k-equivalent for every nonnegative integer k, then they are $*$-equivalent.

f) Show that all states in a given R_*-equivalence class are final states or all are not final states.

g) Show that if s and t are R_*-equivalent, then $f(s, a)$ and $f(t, a)$ are also R_*-equivalent for all $a \in I$.

***59.** Show that there is a nonnegative integer n such that the set of n-equivalence classes of states of M is the same as the set of $(n + 1)$-equivalence classes of states of M. Then show for this integer n, the set of n-equivalence classes of states of M equals the set of $*$-equivalence classes of states of M.

The **quotient automaton** $\overline{M}$ of the deterministic finite-state automaton $M = (S, I, f, s_0, F)$ is the finite-state automaton $(\overline{S}, I, \overline{f}, [s_0]_{R_*}, \overline{F})$, where the set of states $\overline{S}$ is the set of $*$-equivalence classes of S, the transition function $\overline{f}$ is defined by $\overline{f}([s]_{R_*}, a) = [f(s, a)]_{R_*}$ for all states $[s]_{R_*}$ of $\overline{M}$ and input symbols $a \in I$, and $\overline{F}$ is the set consisting of R_*-equivalence classes of final states of M.

*60. a) Show that s and t are 0-equivalent if and only if either both s and t are final states or neither s nor t is a final state. Conclude that each final state of $\overline{M}$, which is an R_*-equivalence class, contains only final states of M.
b) Show that if k is a positive integer, then s and t are k-equivalent if and only if s and t are $(k-1)$-equivalent and for every input symbol $a \in I$, $f(s, a)$ and $f(t, a)$ are $(k-1)$-equivalent. Conclude that the transition function $\overline{f}$ is well-defined.
c) Describe a procedure that can be used to construct the quotient automaton of a finite-automaton M.

**61. a) Show that if M is a finite-state automaton, then the quotient automaton $\overline{M}$ recognizes the same language as M.
b) Show that if M is a finite-state automaton with the property that for every state s of M there is a string $x \in I^*$ such that $f(s_0, x) = s$, then the quotient automaton $\overline{M}$ has the minimum number of states of any finite-state automaton equivalent to M.

62. Answer these questions about the finite-state automaton M shown here.

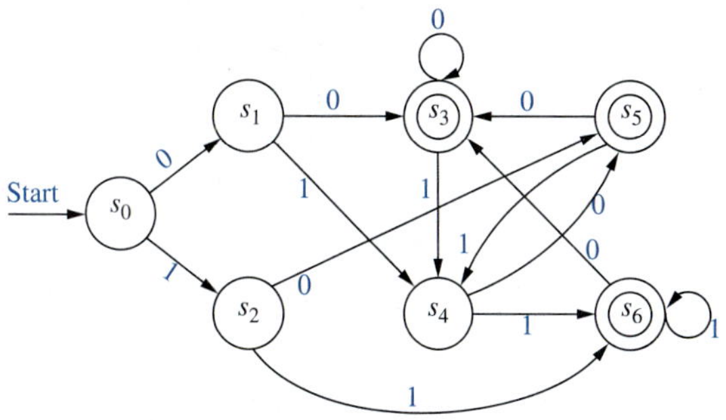

a) Find the k-equivalence classes of M for $k = 0, 1, 2$, and 3. Also, find the $*$-equivalence classes of M.
b) Construct the quotient automaton $\overline{M}$ of M.

13.4 Language Recognition

13.4.1 Introduction

We have seen that finite-state automata can be used as language recognizers. What sets can be recognized by these machines? Although this seems like an extremely difficult problem, there is a simple characterization of the sets that can be recognized by finite state automata. This problem was first solved in 1956 by the American mathematician Stephen Kleene. He showed that there is a finite-state automaton that recognizes a set if and only if this set can be built up from the null set, the empty string, and singleton strings by taking concatenations, unions, and Kleene closures, in arbitrary order. Sets that can be built up in this way are called **regular sets**. Regular grammars were defined in Section 13.1. Because of the terminology used, it is not surprising that there is a connection between regular sets, which are the sets recognized by finite-state automata, and regular grammars. In particular, a set is regular if and only if it is generated by a regular grammar.

Finally, there are sets that cannot be recognized by any finite-state automata. We will give an example of such a set. We will briefly discuss more powerful models of computation, such as pushdown automata and Turing machines, at the end of this section.

To define regular sets we first need to define regular expressions.

Definition 1 The *regular expressions* over a set I are defined recursively by:

the symbol $\emptyset$ is a regular expression;
the symbol λ is a regular expression;
the symbol $\mathbf{x}$ is a regular expression whenever $x \in I$;
the symbols $(\mathbf{AB})$, $(\mathbf{A} \cup \mathbf{B})$, and $\mathbf{A}^*$ are regular expressions whenever $\mathbf{A}$ and $\mathbf{B}$ are regular expressions.

Each regular expression represents a set specified by these rules:

$\emptyset$ represents the empty set, that is, the set with no strings;
λ represents the set $\{\lambda\}$, which is the set containing the empty string;
$\mathbf{x}$ represents the set $\{x\}$ containing the string with one symbol x;
$(\mathbf{AB})$ represents the concatenation of the sets represented by $\mathbf{A}$ and by $\mathbf{B}$;
$(\mathbf{A} \cup \mathbf{B})$ represents the union of the sets represented by $\mathbf{A}$ and by $\mathbf{B}$;
$\mathbf{A}^*$ represents the Kleene closure of the set represented by $\mathbf{A}$.

Sets represented by regular expressions are called regular sets. Henceforth regular expressions will be used to describe regular sets, so when we refer to the regular set $\mathbf{A}$, we will mean the regular set represented by the regular expression $\mathbf{A}$. Note that we will leave out outer parentheses from regular expressions when they are not needed.

Example 1 shows how regular expressions are used to specify regular sets.

EXAMPLE 1 What are the strings in the regular sets specified by the regular expressions $\mathbf{10}^*$, $(\mathbf{10})^*$, $\mathbf{0} \cup \mathbf{01}$, $\mathbf{0}(\mathbf{0} \cup \mathbf{1})^*$, and $(\mathbf{0}^*\mathbf{1})^*$?

Solution: The regular sets represented by these expressions are given in Table 1, as the reader should verify. ◀

Finding a regular expression that specifies a given set can be quite tricky, as Example 2 illustrates.

EXAMPLE 2 Find a regular expression that specifies each of these sets:

(a) the set of bit strings with even length
(b) the set of bit strings ending with a 0 and not containing 11
(c) the set of bit strings containing an odd number of 0s

Solution: (a) To construct a regular expression for the set of bit strings with even length, we use the fact that such a string can be obtained by concatenating zero or more strings each consisting of two bits. The set of strings of two bits is specified by the regular expression $(\mathbf{00} \cup \mathbf{01} \cup \mathbf{10} \cup \mathbf{11})$. Consequently, the set of strings with even length is specified by $(\mathbf{00} \cup \mathbf{01} \cup \mathbf{10} \cup \mathbf{11})^*$.

(b) A bit string ending with a 0 and not containing 11 must be the concatenation of one or more strings where each string is either a 0 or a 10. (To see this, note that such a bit string must consist of 0 bits or 1 bits each followed by a 0; the string cannot end with a single 1 because we know it ends with a 0.) It follows that the regular expression $(\mathbf{0} \cup \mathbf{10})^* (\mathbf{0} \cup \mathbf{10})$ specifies the set of bit strings that do not contain 11 and end with a 0. [Note that the set specified by $(\mathbf{0} \cup \mathbf{10})^*$ includes the empty string, which is not in this set, because the empty string does not end with a 0.]

TABLE 1

Expression	*Strings*
$\mathbf{10}^*$	a 1 followed by any number of 0s (including no zeros)
$(\mathbf{10})^*$	any number of copies of 10 (including the null string)
$\mathbf{0} \cup \mathbf{01}$	the string 0 or the string 01
$\mathbf{0}(\mathbf{0} \cup \mathbf{1})^*$	any string beginning with 0
$(\mathbf{0}^*\mathbf{1})^*$	any string not ending with 0

(c) A bit string containing an odd number of 0s must contain at least one 0, which tells us that it starts with zero or more 1s, followed by a 0, followed by zero or more 1s. That is, each such bit string begins with a string of the form 1^j01^k for nonnegative integers j and k. Because the bit string contains an odd number of 0s, additional bits after this initial block can be split into blocks each starting with a 0 and containing one more 0. Each such block is of the form 01^p01^q, where p and q are nonnegative integers. Consequently, the regular expression **1*01*(01*01*)*** specifies the set of bit strings with an odd number of 0s. ◀

13.4.2 Kleene's Theorem

In 1956 Kleene proved that regular sets are the sets that are recognized by a finite-state automaton. Consequently, this important result is called Kleene's theorem.

THEOREM 1 **KLEENE'S THEOREM** A set is regular if and only if it is recognized by a finite-state automaton.

Links

Kleene's theorem is one of the central results in automata theory. We will prove the *only if* part of this theorem, namely, that every regular set is recognized by a finite-state automaton. The proof of the *if* part, that a set recognized by a finite-state automaton is regular, is left as an exercise for the reader.

Proof: Recall that a regular set is defined in terms of regular expressions, which are defined recursively. We can prove that every regular set is recognized by a finite-state automaton if we can do the following things.

1. Show that $\emptyset$ is recognized by a finite-state automaton.
2. Show that $\{\lambda\}$ is recognized by a finite-state automaton.
3. Show that $\{a\}$ is recognized by a finite-state automaton whenever a is a symbol in I.
4. Show that AB is recognized by a finite-state automaton whenever both A and B are.
5. Show that $A \cup B$ is recognized by a finite-state automaton whenever both A and B are.
6. Show that A^* is recognized by a finite-state automaton whenever A is.

We now consider each of these tasks. First, we show that $\emptyset$ is recognized by a nondeterministic finite-state automaton. To do this, all we need is an automaton with no final states. Such an automaton is shown in Figure 1(a).

Second, we show that $\{\lambda\}$ is recognized by a finite-state automaton. To do this, all we need is an automaton that recognizes λ, the null string, but not any other string. This can be done by making the start state s_0 a final state and having no transitions, so that no other string takes s_0 to a final state. The nondeterministic automaton in Figure 1(b) shows such a machine.

Third, we show that $\{a\}$ is recognized by a nondeterministic finite-state automaton. To do this, we can use a machine with a starting state s_0 and a final state s_1. We have a transition from s_0 to s_1 when the input is a, and no other transitions. The only string recognized by this machine is a. This machine is shown in Figure 1(c).

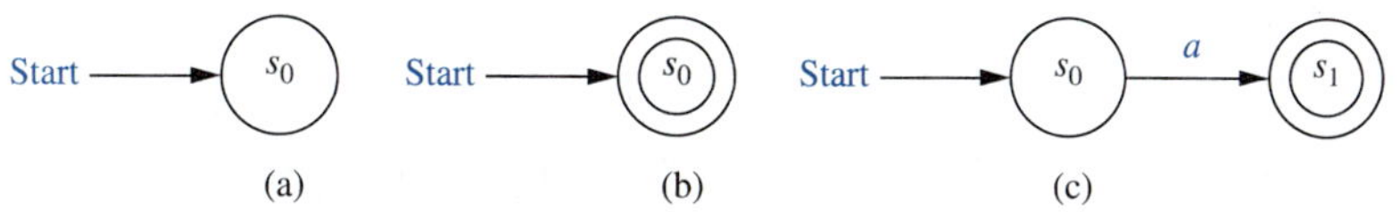

FIGURE 1 **Nondeterministic finite-state automata that recognize some basic sets.**

Next, we show that AB and $A \cup B$ can be recognized by finite-state automata if A and B are languages recognized by finite-state automata. Suppose that A is recognized by $M_A = (S_A, I, f_A, s_A, F_A)$ and B is recognized by $M_B = (S_B, I, f_B, s_B, F_B)$.

We begin by constructing a finite-state machine $M_{AB} = (S_{AB}, I, f_{AB}, s_{AB}, F_{AB})$ that recognizes AB, the concatenation of A and B. We build such a machine by combining the machines for A and B in series, so a string in A takes the combined machine from s_A, the start state of M_A, to s_B, the start state of M_B. A string in B should take the combined machine from s_B to a final state of the combined machine. Consequently, we make the following construction. Let S_{AB} be $S_A \cup S_B$. [Note that we can assume that S_A and S_B are disjoint.] The starting state s_{AB} is the same as s_A. The set of final states, F_{AB}, is the set of final states of M_B with s_{AB} included if and only if $\lambda \in A \cap B$. The transitions in M_{AB} include all transitions in M_A and in M_B, as well as some new transitions. For every transition in M_A that leads to a final state, we form a transition in M_{AB} from the same state to s_B, on the same input. In this way, a string in A takes M_{AB} from s_{AB} to s_B, and then a string in B takes s_B to a final state of M_{AB}. Moreover, for every transition from s_B we form a transition in M_{AB} from s_{AB} to the same state. Figure 2(a) contains an illustration of this construction.

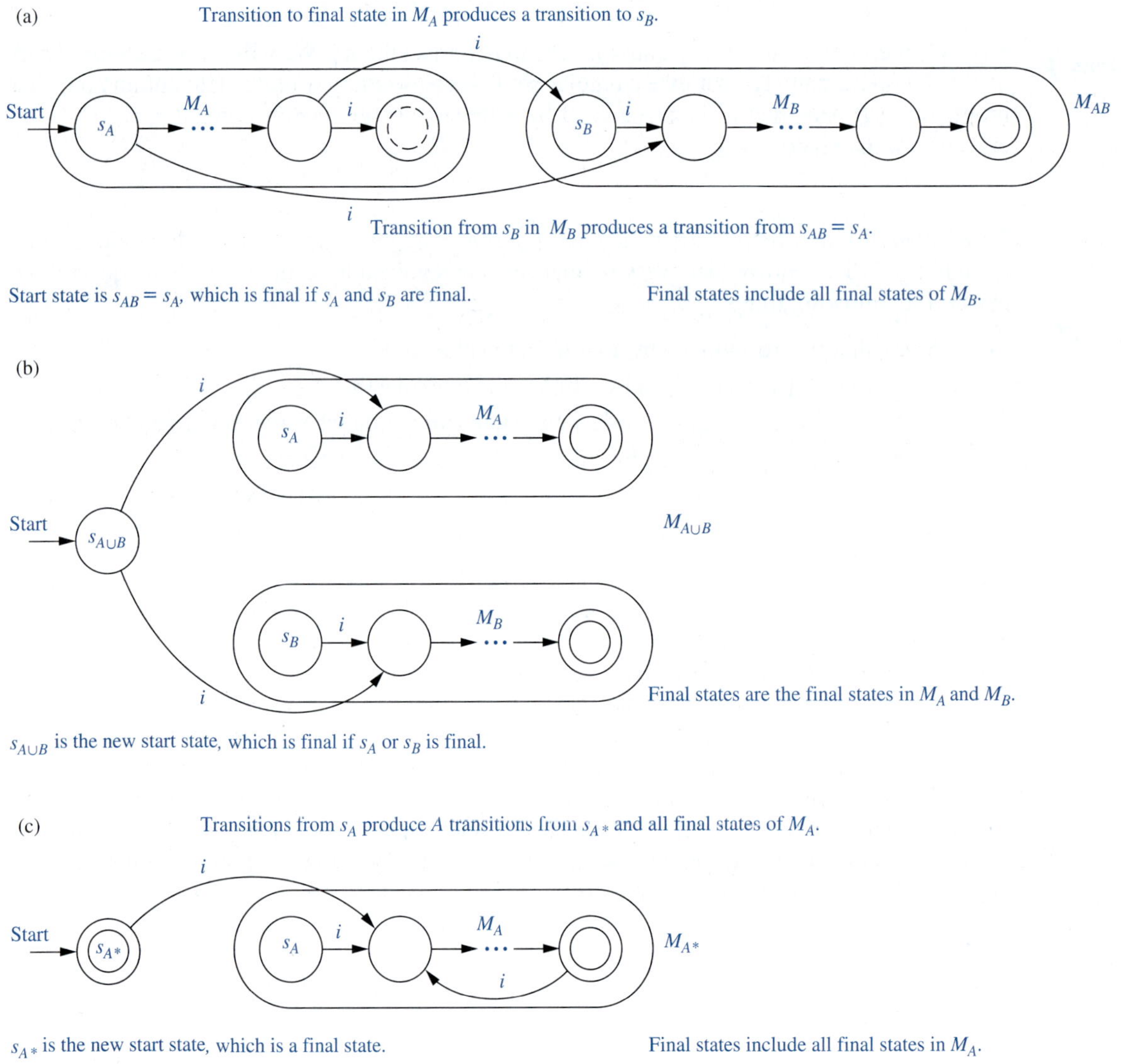

FIGURE 2 Building automata to recognize concatenations, unions, and Kleene closures.

We now construct a machine $M_{A\cup B} = (S_{A\cup B}, I, f_{A\cup B}, s_{A\cup B}, F_{A\cup B})$ that recognizes $A \cup B$. This automaton can be constructed by combining M_A and M_B in parallel, using a new start state that has the transitions that both s_A and s_B have. Let $S_{A\cup B} = S_A \cup S_B \cup \{s_{A\cup B}\}$, where $s_{A\cup B}$ is a new state that is the start state of $M_{A\cup B}$. Let the set of final states $F_{A\cup B}$ be $F_A \cup F_B \cup \{s_{A\cup B}\}$ if $\lambda \in A \cup B$, and $F_A \cup F_B$ otherwise. The transitions in $M_{A\cup B}$ include all those in M_A and in M_B. Also, for each transition from s_A to a state s on input i we include a transition from $s_{A\cup B}$ to s on input i, and for each transition from s_B to a state s on input i we include a transition from $s_{A\cup B}$ to s on input i. In this way, a string in A leads from $s_{A\cup B}$ to a final state in the new machine, and a string in B leads from $s_{A\cup B}$ to a final state in the new machine. Figure 2(b) illustrates the construction of $M_{A\cup B}$.

Finally, we construct $M_{A^*} = (S_{A^*}, I, f_{A^*}, s_{A^*}, F_{A^*})$, a machine that recognizes A^*, the Kleene closure of A. Let S_{A^*} include all states in S_A and one additional state s_{A^*}, which is the starting state for the new machine. The set of final states F_{A^*} includes all states in F_A as well as the start state s_{A^*}, because λ must be recognized. To recognize concatenations of arbitrarily many strings from A, we include all the transitions in M_A, as well as transitions from s_{A^*} that match the transitions from s_A, and transitions from each final state that match the transitions from s_A. With this set of transitions, a string made up of concatenations of strings from A will take s_{A^*} to a final state when the first string in A has been read, returning to a final state when the second string in A has been read, and so on. Figure 2(c) illustrates the construction we used. ◁

A nondeterministic finite-state automaton can be constructed for any regular set using the procedure described in this proof. We illustrate how this is done with Example 3.

EXAMPLE 3 Construct a nondeterministic finite-state automaton that recognizes the regular set $\mathbf{1}^* \cup \mathbf{01}$.

Solution: We begin by building a machine that recognizes $\mathbf{1}^*$. This is done using the machine that recognizes $\mathbf{1}$ and then using the construction for M_{A^*} described in the proof. Next, we build a machine that recognizes $\mathbf{01}$, using machines that recognize $\mathbf{0}$ and $\mathbf{1}$ and the construction in the proof for M_{AB}. Finally, using the construction in the proof for $M_{A\cup B}$, we construct the machine for $\mathbf{1}^* \cup \mathbf{01}$. The finite-state automata used in this construction are shown in Figure 3. The states in the successive machines have been labeled using different subscripts, even when a state is formed from one previously used in another machine. Note that the construction given here does not produce the simplest machine that recognizes $\mathbf{1}^* \cup \mathbf{01}$. A much simpler machine that recognizes this set is shown in Figure 3(b). ◀

13.4.3 Regular Sets and Regular Grammars

In Section 13.1 we introduced phrase-structure grammars and defined different types of grammars. In particular we defined regular, or type 3, grammars, which are grammars of the form $G = (V, T, S, P)$, where each production is of the form $S \to \lambda$, $A \to a$, or $A \to aB$, where a is a terminal symbol, and A and B are nonterminal symbols. As the terminology suggests, there is a close connection between regular grammars and regular sets.

THEOREM 2 A set is generated by a regular grammar if and only if it is a regular set.

Proof: First we show that a set generated by a regular grammar is a regular set. Suppose that $G = (V, T, S, P)$ is a regular grammar generating the set $L(G)$. To show that $L(G)$ is regular we will build a nondeterministic finite-state machine $M = (S, I, f, s_0, F)$ that recognizes $L(G)$. Let S, the set of states, contain a state s_A for each nonterminal symbol A of G and an additional state s_F, which is a final state. The start state s_0 is the state formed from the start symbol S. The transitions of M are formed from the productions of G in the following way. A transition

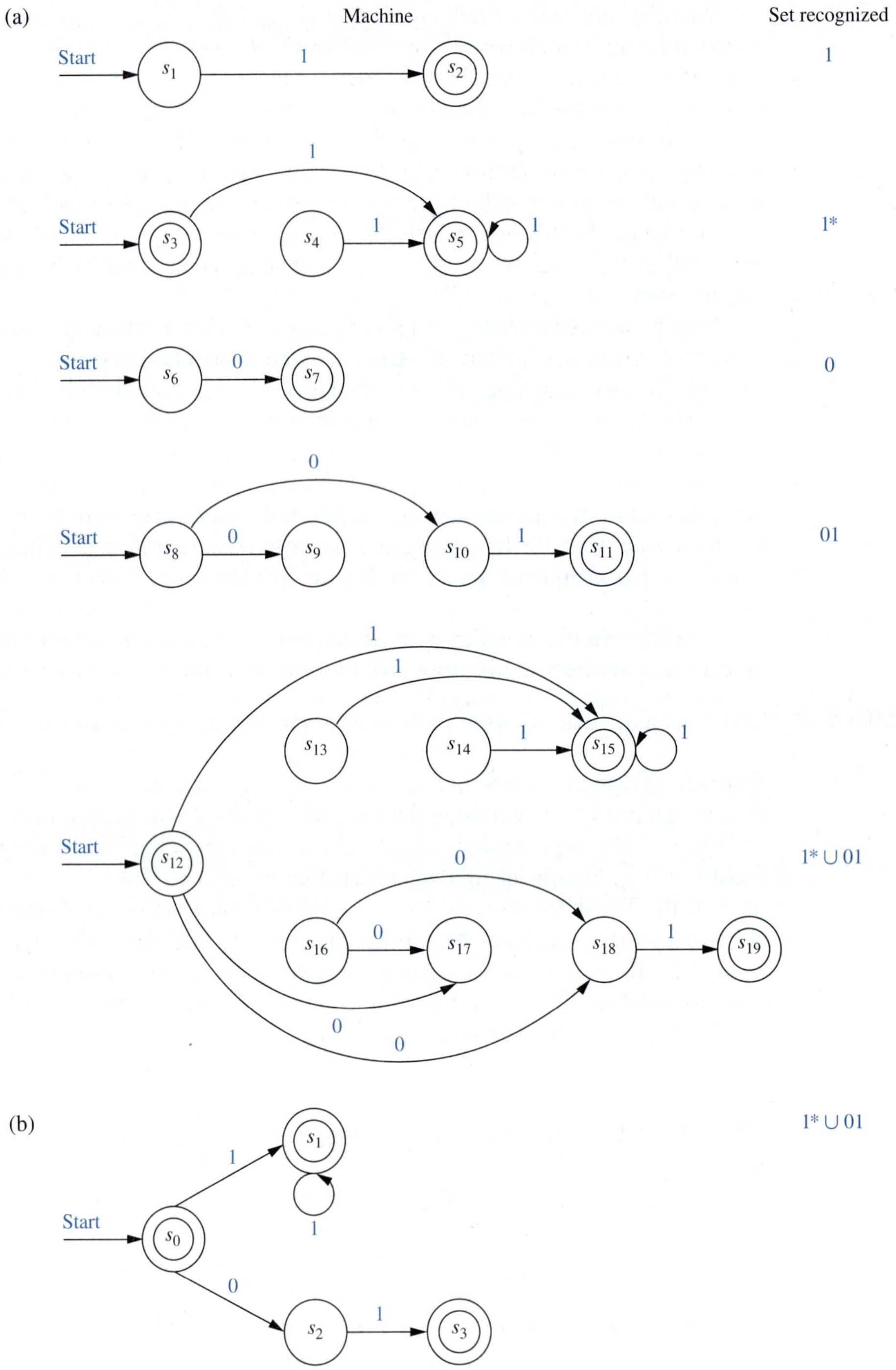

FIGURE 3 **Nondeterministic finite-state automata recognizing $1^* \cup 01$.**

from s_A to s_F on input of a is included if $A \rightarrow a$ is a production, and a transition from s_A to s_B on input of a is included if $A \rightarrow aB$ is a production. The set of final states includes s_F and also includes s_0 if $S \rightarrow \lambda$ is a production in G. It is not hard to show that the language recognized by M equals the language generated by the grammar G, that is, $L(M) = L(G)$. This can be done by determining the words that lead to a final state. The details are left as an exercise for the reader. ◁

Before giving the proof of the converse, we illustrate how a nondeterministic machine is constructed that recognizes the same set as a regular grammar.

EXAMPLE 4 Construct a nondeterministic finite-state automaton that recognizes the language generated by the regular grammar $G = (V, T, S, P)$, where $V = \{0, 1, A, S\}$, $T = \{0, 1\}$, and the productions in P are $S \to 1A$, $S \to 0$, $S \to \lambda$, $A \to 0A$, $A \to 1A$, and $A \to 1$.

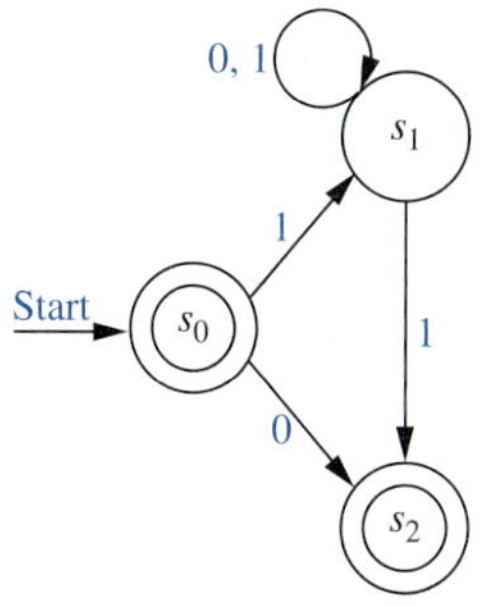

FIGURE 4 A nondeterministic finite-state automaton recognizing $L(G)$.

Solution: The state diagram for a nondeterministic finite-state automaton that recognizes $L(G)$ is shown in Figure 4. This automaton is constructed following the procedure described in the proof. In this automaton, s_0 is the state corresponding to S, s_1 is the state corresponding to A, and s_2 is the final state. ◀

We now complete the proof of Theorem 2.

Proof: We now show that if a set is regular, then there is a regular grammar that generates it. Suppose that M is a finite-state machine that recognizes this set with the property that s_0, the starting state of M, is never the next state for a transition. (We can find such a machine by Exercise 20.) The grammar $G = (V, T, S, P)$ is defined as follows. The set V of symbols of G is formed by assigning a symbol to each state of S and each input symbol in I. The set T of terminal symbols of G is the set I. The start symbol S is the symbol formed from the start state s_0. The set P of productions in G is formed from the transitions in M. In particular, if the state s goes to a final state under input a, then the production $A_s \to a$ is included in P, where A_s is the nonterminal symbol formed from the state s. If the state s goes to the state t on input a, then the production $A_s \to aA_t$ is included in P. The production $S \to \lambda$ is included in P if and only if $\lambda \in L(M)$. Because the productions of G correspond to the transitions of M and the productions leading to terminals correspond to transitions to final states, it is not hard to show that $L(G) = L(M)$. We leave the details as an exercise for the reader. ◁

Example 5 illustrates the construction used to produce a grammar from an automaton that generates the language recognized by this automaton.

EXAMPLE 5 Find a regular grammar that generates the regular set recognized by the finite-state automaton shown in Figure 5.

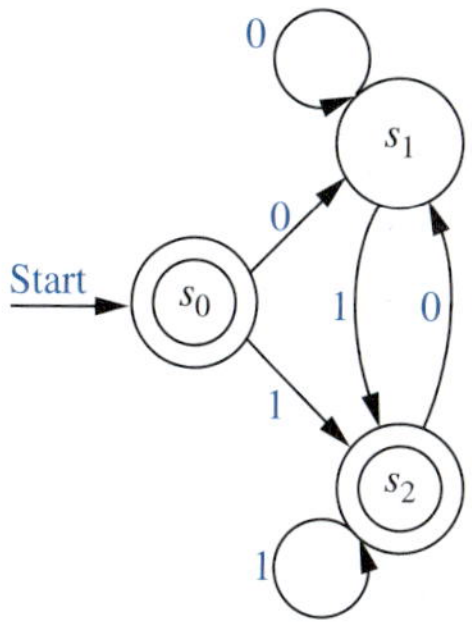

FIGURE 5 A finite-state automaton.

Solution: The grammar $G = (V, T, S, P)$ generates the set recognized by this automaton where $V = \{S, A, B, 0, 1\}$, the symbols S, A, and B correspond to the states s_0, s_1, and s_2, respectively, $T = \{0, 1\}$, S is the start symbol; and the productions are $S \to 0A$, $S \to 1B$, $S \to 1$, $S \to \lambda$, $A \to 0A$, $A \to 1B$, $A \to 1$, $B \to 0A$, $B \to 1B$, and $B \to 1$. ◀

13.4.4 A Set Not Recognized by a Finite-State Automaton

We have seen that a set is recognized by a finite state automaton if and only if it is regular. We will now show that there are sets that are not regular by describing one such set. The technique used to show that this set is not regular illustrates an important method for showing that certain sets are not regular.

EXAMPLE 6 Show that the set $\{0^n 1^n \mid n = 0, 1, 2, \dots\}$, made up of all strings consisting of a block of 0s followed by a block of an equal number of 1s, is not regular.

Solution: Suppose that this set were regular. Then there would be a nondeterministic finite-state automaton $M = (S, I, f, s_0, F)$ recognizing it. Let N be the number of states in this machine,

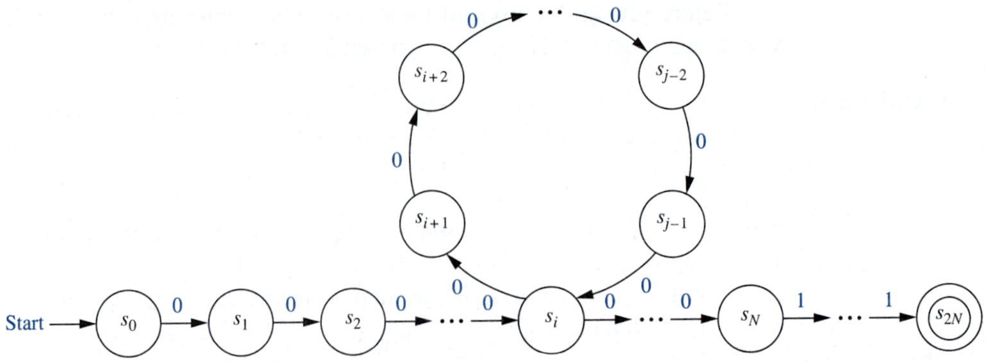

FIGURE 6 The path produced by 0^N1^N.

that is, $N = |S|$. Because M recognizes all strings made up of a number of 0s followed by an equal number of 1s, M must recognize 0^N1^N. Let $s_0, s_1, s_2, \ldots, s_{2N}$ be the sequence of states that is obtained starting at s_0 and using the symbols of 0^N1^N as input, so that $s_1 = f(s_0, 0)$, $s_2 = f(s_1, 0), \ldots, s_N = f(s_{N-1}, 0), s_{N+1} = f(s_N, 1), \ldots, s_{2N} = f(s_{2N-1}, 1)$. Note that s_{2N} is a final state.

Because there are only N states, the pigeonhole principle shows that at least two of the first $N + 1$ of the states, which are $s_0, \ldots, s_N$, must be the same. Say that s_i and s_j are two such identical states, with $0 \le i < j \le N$. This means that $f(s_i, 0^t) = s_j$, where $t = j - i$. It follows that there is a loop leading from s_i back to itself, obtained using the input 0 a total of t times, in the state diagram shown in Figure 6.

Now consider the input string $0^N0^t1^N = 0^{N+t}1^N$. There are t more consecutive 0s at the start of this block than there are consecutive 1s that follow it. Because this string is not of the form 0^n1^n (because it has more 0s than 1s), it is not recognized by M. Consequently, $f(s_0, 0^{N+t}1^N)$ cannot be a final state. However, when we use the string $0^{N+t}1^N$ as input, we end up in the same state as before, namely, s_{2N}. The reason for this is that the extra t 0s in this string take us around the loop from s_i back to itself an extra time, as shown in Figure 6. Then the rest of the string leads us to exactly the same state as before. This contradiction shows that $\{0^n1^n \mid n = 0, 1, 2, \ldots\}$ is not regular. ◀

Links

Album/Archivo ABC/Alamy Stock Photo

ALAN MATHISON TURING (1912–1954) Alan Turing was born in London, although he was conceived in India, where his father was employed in the Indian Civil Service. As a boy, he was fascinated by chemistry, performing a wide variety of experiments, and by machinery. Turing attended Sherborne, an English boarding school. In 1931 he won a scholarship to King's College, Cambridge. After completing his dissertation, which included a rediscovery of the central limit theorem, a famous theorem in statistics, he was elected a fellow of his college. In 1935 Turing became fascinated with the decision problem, a problem posed by the great German mathematician Hilbert, which asked whether there is a general method that can be applied to any assertion to determine whether the assertion is true. Turing enjoyed running (later in life running as a serious amateur in competitions), and one day, while resting after a run, he discovered the key ideas needed to solve the decision problem. In his solution, he invented what is now called a **Turing machine** as the most general model of a computing machine. Using these machines, he found a problem, involving what he called computable numbers, that could not be decided using a general method.

From 1936 to 1938 Turing visited Princeton University to work with Alonzo Church, who had also solved Hilbert's decision problem. In 1939 Turing returned to King's College. However, at the outbreak of World War II, he joined the Foreign Office, performing cryptanalysis of German ciphers. His contribution to the breaking of the code of the Enigma, a mechanical German cipher machine, played an important role in winning the war.

After the war, Turing worked on the development of early computers. He was interested in the ability of machines to think, proposing that if a computer could not be distinguished from a person based on written replies to questions, it should be considered to be "thinking." He was also interested in biology, having written on morphogenesis, the development of form in organisms. In 1954 Turing committed suicide by taking cyanide, without leaving a clear explanation. Legal troubles related to a homosexual relationship and hormonal treatments mandated by the court to lessen his sex drive may have been factors in his decision to end his life.

13.4.5 More Powerful Types of Machines

Links

Finite-state automata are unable to carry out many computations. The main limitation of these machines is their finite amount of memory. This prevents them from recognizing languages that are not regular, such as $\{0^n1^n \mid n = 0, 1, 2, \dots\}$. Because a set is regular if and only if it is the language generated by a regular grammar, Example 6 shows that there is no regular grammar that generates the set $\{0^n1^n \mid n = 0, 1, 2, \dots\}$. However, there is a context-free grammar that generates this set. Such a grammar was given in Example 5 in Section 13.1.

Because of the limitations of finite-state machines, it is necessary to use other, more powerful, models of computation. One such model is the **pushdown automaton**. A pushdown automaton includes everything in a finite-state automaton, as well as a stack, which provides unlimited memory. Symbols can be placed on the top or taken off the top of the stack. A set is recognized in one of two ways by a pushdown automaton. First, a set is recognized if the set consists of all the strings that produce an empty stack when they are used as input. Second, a set is recognized if it consists of all the strings that lead to a final state when used as input. It can be shown that a set is recognized by a pushdown automaton if and only if it is the language generated by a context-free grammar.

However, there are sets that cannot be expressed as the language generated by a context-free grammar. One such set is $\{0^n1^n2^n \mid n = 0, 1, 2, \dots\}$. We will indicate why this set cannot be recognized by a pushdown automaton, but we will not give a proof, because we have not developed the machinery needed. (However, one method of proof is given in Exercise 28 of the supplementary exercises at the end of this chapter.) The stack can be used to show that a string begins with a sequence of 0s followed by an equal number of 1s by placing a symbol on the stack for each 0 (as long as only 0s are read), and removing one of these symbols for each 1 (as long as only 1s following the 0s are read). But once this is done, the stack is empty, and there is no way to determine that there are the same number of 2s in the string as 0s.

There are other machines called **linear bounded automata**, more powerful than pushdown automata, that can recognize sets such as $\{0^n1^n2^n \mid n = 0, 1, 2, \dots\}$. In particular, linear bounded automata can recognize context-sensitive languages. However, these machines cannot recognize all the languages generated by phrase-structure grammars. To avoid the limitations of special types of machines, the model known as a **Turing machine**, named after the British mathematician Alan Turing, is used. A Turing machine is made up of everything included in a finite-state machine together with a tape, which is infinite in both directions. A Turing machine has read and write capabilities on the tape, and it can move back and forth along this tape. Turing machines can recognize all languages generated by phrase-structure grammars. In addition, Turing machines can model all the computations that can be performed on a computing machine. Because of their power, Turing machines are extensively studied in theoretical computer science. We will briefly study them in Section 13.5.

Alan Turing invented Turing machines before modern computers existed!

Exercises

1. Describe in words the strings in each of these regular sets.

a) 1^*0 b) 1^*00^*
c) $111 \cup 001$ d) $(1 \cup 00)^*$
e) $(00^*1)^*$ f) $(0 \cup 1)(0 \cup 1)^*00$

2. Describe in words the strings in each of these regular sets.

a) 001^* b) $(01)^*$
c) $01 \cup 001^*$ d) $0(11 \cup 0)^*$
e) $(101^*)^*$ f) $(0^* \cup 1)11$

3. Determine whether 0101 belongs to each of these regular sets.

a) 01^*0^* b) $0(11)^*(01)^*$
c) $0(10)^*1^*$ d) $0^*10(0 \cup 1)$
e) $(01)^*(11)^*$ f) $0^*(10 \cup 11)^*$
g) $0^*(10)^*11$ h) $01(01 \cup 0)1^*$

4. Determine whether 1011 belongs to each of these regular sets.

a) 10^*1^* b) $0^*(10 \cup 11)^*$
c) $1(01)^*1^*$ d) $1^*01(0 \cup 1)$
e) $(10)^*(11)^*$ f) $1(00)^*(11)^*$
g) $(10)^*1011$ h) $(1 \cup 00)(01 \cup 0)1^*$

5. Express each of these sets using a regular expression.

a) the set consisting of the strings 0, 11, and 010
b) the set of strings of three 0s followed by two or more 0s
c) the set of strings of odd length

d) the set of strings that contain exactly one 1
e) the set of strings ending in 1 and not containing 000

6. Express each of these sets using a regular expression.
a) the set containing all strings with zero, one, or two bits
b) the set of strings of two 0s, followed by zero or more 1s, and ending with a 0
c) the set of strings with every 1 followed by two 0s
d) the set of strings ending in 00 and not containing 11
e) the set of strings containing an even number of 1s

7. Express each of these sets using a regular expression.
a) the set of strings of one or more 0s followed by a 1
b) the set of strings of two or more symbols followed by three or more 0s
c) the set of strings with either no 1 preceding a 0 or no 0 preceding a 1
d) the set of strings containing a string of 1s such that the number of 1s equals 2 modulo 3, followed by an even number of 0s

8. Construct deterministic finite-state automata that recognize each of these sets from I^*, where I is an alphabet.
a) $\emptyset$ **b)** $\{\lambda\}$ **c)** $\{a\}$, where $a \in I$

9. Construct nondeterministic finite-state automata that recognize each of the sets in Exercise 8.

10. Construct nondeterministic finite-state automata that recognize each of these sets.
a) $\{\lambda, 0\}$ **b)** $\{0, 11\}$ **c)** $\{0, 11, 000\}$

***11.** Show that if A is a regular set, then A^R, the set of all reversals of strings in A, is also regular.

12. Using the constructions described in the proof of Kleene's theorem, find nondeterministic finite-state automata that recognize each of these sets.
a) $\mathbf{01^*}$ **b)** $\mathbf{(0 \cup 1)1^*}$ **c)** $\mathbf{00(1^* \cup 10)}$

13. Using the constructions described in the proof of Kleene's theorem, find nondeterministic finite-state automata that recognize each of these sets.
a) $\mathbf{0^*1^*}$ **b)** $\mathbf{(0 \cup 11)^*}$ **c)** $\mathbf{01^* \cup 00^*1}$

14. Construct a nondeterministic finite-state automaton that recognizes the language generated by the regular grammar $G = (V, T, S, P)$, where $V = \{0, 1, S, A, B\}$, $T = \{0, 1\}$, S is the start symbol, and the set of productions is
a) $S \to 0A$, $S \to 1B$, $A \to 0$, $B \to 0$.
b) $S \to 1A$, $S \to 0$, $S \to \lambda$, $A \to 0B$, $B \to 1B$, $B \to 1$.
c) $S \to 1B$, $S \to 0$, $A \to 1A$, $A \to 0B$, $A \to 1$, $A \to 0$, $B \to 1$.

In Exercises 15–17 construct a regular grammar $G = (V, T, S, P)$ that generates the language recognized by the given finite-state machine.

15.
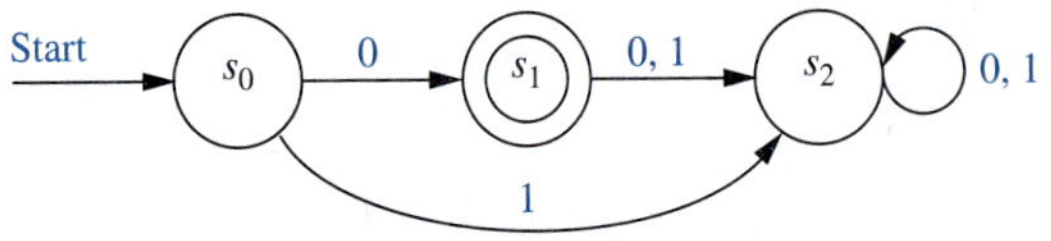

16.
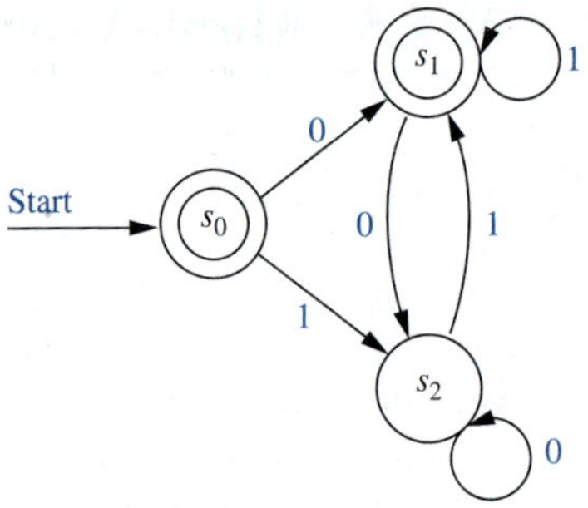

17.
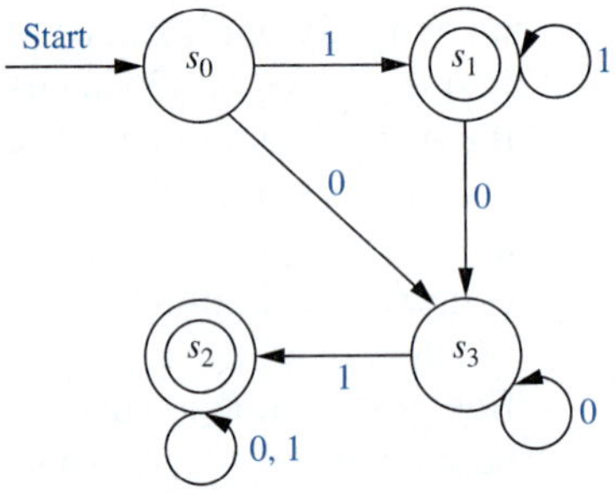

18. Show that the finite-state automaton constructed from a regular grammar in the proof of Theorem 2 recognizes the set generated by this grammar.

19. Show that the regular grammar constructed from a finite-state automaton in the proof of Theorem 2 generates the set recognized by this automaton.

20. Show that every nondeterministic finite-state automaton is equivalent to another such automaton that has the property that its starting state is never revisited.

***21.** Let $M = (S, I, f, s_0, F)$ be a deterministic finite-state automaton. Show that the language recognized by M, $L(M)$, is infinite if and only if there is a word x recognized by M with $l(x) \geq |S|$.

***22.** One important technique used to prove that certain sets are not regular is the **pumping lemma**. The pumping lemma states that if $M = (S, I, f, s_0, F)$ is a deterministic finite-state automaton and if x is a string in $L(M)$, the language recognized by M, with $l(x) \geq |S|$, then there are strings u, v, and w in I^* such that $x = uvw$, $l(uv) \leq |S|$ and $l(v) \geq 1$, and $uv^iw \in L(M)$ for $i = 0, 1, 2, \ldots$. Prove the pumping lemma. [*Hint:* Use the same idea as was used in Example 6.]

Links

***23.** Show that the set $\{0^{2n}1^n \mid n = 0, 1, 2, \ldots\}$ is not regular using the pumping lemma given in Exercise 22.

***24.** Show that the set $\{1^{n^2} \mid n = 0, 1, 2, \ldots\}$ is not regular using the pumping lemma from Exercise 22.

***25.** Show that the set of palindromes over $\{0, 1\}$ is not regular using the pumping lemma given in Exercise 22. [*Hint:* Consider strings of the form 0^N10^N.]

****26.** Show that a set recognized by a finite-state automaton is regular. (This is the *if* part of Kleene's theorem.)

Suppose that L is a subset of I^*, where I is a nonempty set of symbols. If $x \in I^*$, we let $L/x = \{z \in I^* \mid xz \in L\}$. We say that the strings $x \in I^*$ and $y \in I^*$ are **distinguishable with respect to** L if $L/x \neq L/y$. A string z for which $xz \in L$ but $yz \notin L$, or $xz \notin L$, but $yz \in L$ is said to **distinguish** x and y with respect to L. When $L/x = L/y$, we say that x and y are **indistinguishable** with respect to L.

27. Let L be the set of all bit strings that end with 01. Show that 11 and 10 are distinguishable with respect to L and that the strings 1 and 11 are indistinguishable with respect to L.

28. Suppose that $M = (S, I, f, s_0, F)$ is a deterministic finite-state machine. Show that if x and y are two strings in I^* that are distinguishable with respect to $L(M)$, then $f(s_0, x) \neq f(s_0, y)$.

***29.** Suppose that L is a subset of I^* and for some positive integer n there are n strings in I^* such that every two of these strings are distinguishable with respect to L. Prove that every deterministic finite-state automaton recognizing L has at least n states.

***30.** Let L_n be the set of strings with at least n bits in which the nth symbol from the end is a 0. Use Exercise 29 to show that a deterministic finite-state machine recognizing L_n must have at least 2^n states.

***31.** Use Exercise 29 to show that the language consisting of all bit strings that are palindromes (that is, strings that equal their own reversals) is not regular.

13.5 Turing Machines

13.5.1 Introduction

Links

The finite-state automata studied earlier in this chapter cannot be used as general models of computation. They are limited in what they can do. For example, finite-state automata are able to recognize regular sets, but are not able to recognize many easy-to-describe sets, including $\{0^n1^n \mid n \geq 0\}$, which computers recognize using memory. We can use finite-state automata to compute relatively simple functions such as the sum of two numbers, but we cannot use them to compute functions that computers can, such as the product of two numbers. To overcome these deficiencies we can use a more powerful type of machine known as a Turing machine, after Alan Turing, the illustrious mathematician and computer scientist who invented them in the 1930s.

Basically, a Turing machine consists of a control unit, which at any step is in one of finitely many different states, together with a tape divided into cells, which is infinite in both directions. Turing machines have read and write capabilities on the tape as the control unit moves back and forth along this tape, changing states depending on the tape symbol read. Turing machines are more powerful than finite-state machines because they include memory capabilities that finite-state machines lack. We will show how to use Turing machines to recognize sets, including sets that cannot be recognized by finite-state machines. We will also show how to compute functions using Turing machines. Turing machines are the most general models of computation; essentially, they can do whatever a computer can do. Note that Turing machines are much more powerful than real computers, which have finite memory capabilities.

"Machines take me by surprise with great frequency." – Alan Turing

13.5.2 Definition of Turing Machines

We now give the formal definition of a Turing machine. Afterward we will explain how this formal definition can be interpreted in terms of a control head that can read and write symbols on a tape and move either right or left along the tape.

Definition 1 A *Turing machine* $T = (S, I, f, s_0)$ consists of a finite set S of states, an alphabet I containing the blank symbol B, a partial function f from $S \times I$ to $S \times I \times \{R, L\}$, and a starting state s_0.

Recall from Section 2.3 that a partial function is defined only for those elements in its domain of definition. This means that for some (state, symbol) pairs the partial function f may be

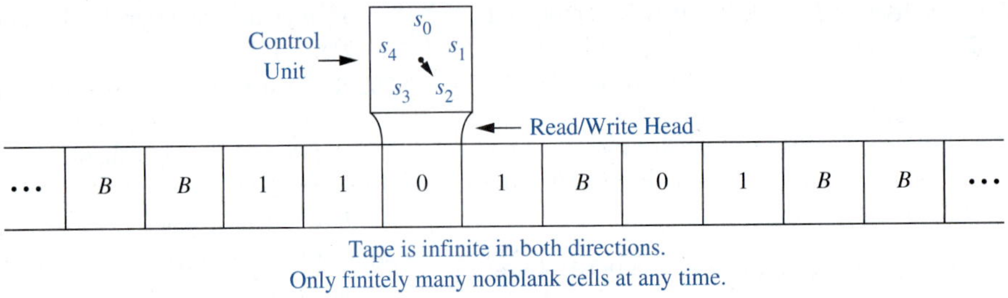

FIGURE 1 A representation of a Turing machine.

undefined, but for a pair for which it is defined, there is a unique (state, symbol, direction) triple associated to this pair. We call the five-tuples corresponding to the partial function in the definition of a Turing machine the **transition rules** of the machine.

To interpret this definition in terms of a machine, consider a control unit and a tape divided into cells, infinite in both directions, having only a finite number of nonblank symbols on it at any given time, as pictured in Figure 1. The action of the Turing machine at each step of its operation depends on the value of the partial function f for the current state and tape symbol.

At each step, the control unit reads the current tape symbol x. If the control unit is in state s and if the partial function f is defined for the pair (s, x) with $f(s, x) = (s', x', d)$, the control unit

1. enters the state s',
2. writes the symbol x' in the current cell, erasing x, and
3. moves right one cell if $d = R$ or moves left one cell if $d = L$.

We write this step as the five-tuple (s, x, s', x', d). If the partial function f is undefined for the pair (s, x), then the Turing machine T will *halt*.

A common way to define a Turing machine is to specify a set of five-tuples of the form (s, x, s', x', d). The set of states and input alphabet is implicitly defined when such a definition is used.

At the beginning of its operation a Turing machine is assumed to be in the initial state s_0 and to be positioned over the leftmost nonblank symbol on the tape. If the tape is all blank, the control head can be positioned over any cell. We will call the positioning of the control head over the leftmost nonblank tape symbol the *initial position* of the machine.

Example 1 illustrates how a Turing machine works.

EXAMPLE 1 What is the final tape when the Turing machine T defined by the seven five-tuples $(s_0, 0, s_0, 0, R)$, $(s_0, 1, s_1, 1, R)$, (s_0, B, s_3, B, R), $(s_1, 0, s_0, 0, R)$, $(s_1, 1, s_2, 0, L)$ (s_1, B, s_3, B, R), and $(s_2, 1, s_3, 0, R)$ is run on the tape shown in Figure 2(a)?

Extra Examples

Solution: We start the operation with T in state s_0 and with T positioned over the leftmost nonblank symbol on the tape. The first step, using the five-tuple $(s_0, 0, s_0, 0, R)$, reads the 0 in the leftmost nonblank cell, stays in state s_0, writes a 0 in this cell, and moves one cell right. The second step, using the five-tuple $(s_0, 1, s_1, 1, R)$, reads the 1 in the current cell, enters state s_1, writes a 1 in this cell, and moves one cell right. The third step, using the five-tuple $(s_1, 0, s_0, 0, R)$, reads the 0 in the current cell, enters state s_0, writes a 0 in this cell, and moves one cell right. The fourth step, using the five-tuple $(s_0, 1, s_1, 1, R)$, reads the 1 in the current cell, enters state s_1, writes a 1 in this cell, and moves right one cell. The fifth step, using the five-tuple $(s_1, 1, s_2, 0, L)$, reads the 1 in the current cell, enters state s_2, writes a 0 in this cell, and moves left one cell. The sixth step, using the five-tuple $(s_2, 1, s_3, 0, R)$, reads the 1 in the current cell, enters the state s_3, writes a 0 in this cell, and moves right one cell. Finally, in the seventh step, the machine halts because there is no five-tuple beginning with the pair $(s_3, 0)$ in the description of the machine. The steps are shown in Figure 2.

Note that T changes the first pair of consecutive 1s on the tape to 0s and then halts. ◀

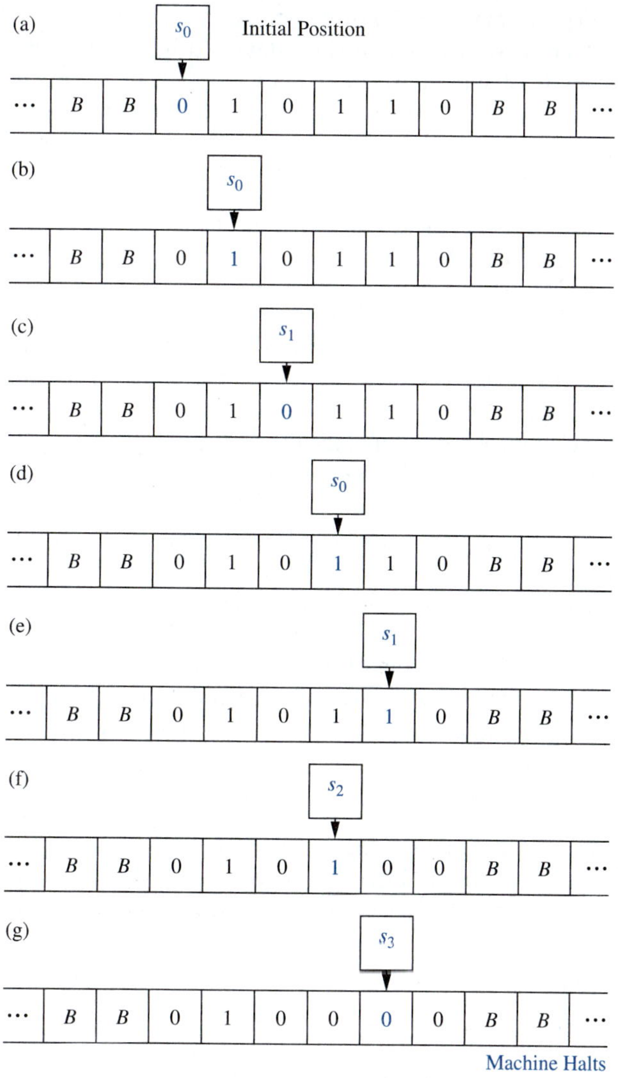

FIGURE 2 **The steps produced by running *T* on the tape in Figure 1.**

13.5.3 Using Turing Machines to Recognize Sets

Turing machines can be used to recognize sets. To do so requires that we define the concept of a final state as follows. A *final state* of a Turing machine T is a state that is not the first state in any five-tuple in the description of T using five-tuples (for example, state s_3 in Example 1).

We can now define what it means for a Turing machine to recognize a string. Given a string, we write consecutive symbols in this string in consecutive cells.

Definition 2 Let V be a subset of an alphabet I. A Turing machine $T = (S, I, f, s_0)$ *recognizes* a string x in V^* if and only if T, starting in the initial position when x is written on the tape, halts in a final state. T is said to recognize a subset A of V^* if x is recognized by T if and only if x belongs to A.

Note that to recognize a subset A of V^* we can use symbols not in V. This means that the input alphabet I may include symbols not in V. These extra symbols are often used as markers (see Example 3).

When does a Turing machine T *not* recognize a string x in V^*? The answer is that x is not recognized if T does not halt or halts in a state that is not final when it operates on a tape containing the symbols of x in consecutive cells, starting in the initial position. (The reader should understand that this is one of many possible ways to define how to recognize sets using Turing machines.)

We illustrate this concept with Example 2.

EXAMPLE 2 Find a Turing machine that recognizes the set of bit strings that have a 1 as their second bit, that is, the regular set $\mathbf{(0 \cup 1)1(0 \cup 1)^*}$.

Solution: We want a Turing machine that, starting at the leftmost nonblank tape cell, moves right, and determines whether the second symbol is a 1. If the second symbol is 1, the machine should move into a final state. If the second symbol is not a 1, the machine should not halt or it should halt in a nonfinal state.

To construct such a machine, we include the five-tuples $(s_0, 0, s_1, 0, R)$ and $(s_0, 1, s_1, 1, R)$ to read in the first symbol and put the Turing machine in state s_1. Next, we include the five-tuples $(s_1, 0, s_2, 0, R)$ and $(s_1, 1, s_3, 1, R)$ to read in the second symbol and either move to state s_2 if this symbol is a 0, or to state s_3 if this symbol is a 1. We do not want to recognize strings that have a 0 as their second bit, so s_2 should not be a final state. We want s_3 to be a final state. So, we can include the five-tuple $(s_2, 0, s_2, 0, R)$. Because we do not want to recognize the empty string or a string with one bit, we also include the five-tuples $(s_0, B, s_2, 0, R)$ and $(s_1, B, s_2, 0, R)$.

The Turing machine T consisting of the seven five-tuples listed here will terminate in the final state s_3 if and only if the bit string has at least two bits and the second bit of the input string is a 1. If the bit string contains fewer than two bits or if the second bit is not a 1, the machine will terminate in the nonfinal state s_2. ◀

Given a regular set, a Turing machine that always moves to the right can be built to recognize this set (as in Example 2). To build the Turing machine, first find a finite-state automaton that recognizes the set and then construct a Turing machine using the transition function of the finite-state machine, always moving to the right.

We will now show how to build a Turing machine that recognizes a nonregular set.

EXAMPLE 3 Find a Turing machine that recognizes the set $\{0^n 1^n \mid n \geq 1\}$.

Solution: To build such a machine, we will use an auxiliary tape symbol M as a marker. We have $V = \{0, 1\}$ and $I = \{0, 1, M\}$. We wish to recognize only a subset of strings in V^*. We will have one final state, s_6. The Turing machine successively replaces a 0 at the leftmost position of the string with an M and a 1 at the rightmost position of the string with an M, sweeping back and forth, terminating in a final state if and only if the string consists of a block of 0s followed by a block of the same number of 1s.

Although this is easy to describe and is easily carried out by a Turing machine, the machine we need to use is somewhat complicated. We use the marker M to keep track of the leftmost and rightmost symbols we have already examined. The five-tuples we use are $(s_0, 0, s_1, M, R)$, $(s_1, 0, s_1, 0, R)$, $(s_1, 1, s_1, 1, R)$, (s_1, M, s_2, M, L), (s_1, B, s_2, B, L), $(s_2, 1, s_3, M, L)$, $(s_3, 1, s_3, 1, L)$, $(s_3, 0, s_4, 0, L)$, (s_3, M, s_5, M, R), $(s_4, 0, s_4, 0, L)$, (s_4, M, s_0, M, R), and (s_5, M, s_6, M, R). For example, the string 000111 would successively become $M00111$, $M0011M$, $MM011M$, $MM01MM$, $MMM1MM$, $MMMMMM$ as the machine operates until it halts. Only the changes are shown, as most steps leave the string unaltered.

We leave it to the reader (Exercise 15) to explain the actions of this Turing machine and to explain why it recognizes the set $\{0^n 1^n \mid n \geq 1\}$. ◀

It can be shown that a set can be recognized by a Turing machine if and only if it can be generated by a type 0 grammar, or in other words, if the set is generated by a phrase-structure grammar. The proof will not be presented here.

13.5.4 Computing Functions with Turing Machines

A Turing machine can be thought of as a computer that finds the values of a partial function. To see this, suppose that the Turing machine T, when given the string x as input, halts with the string y on its tape. We can then define $T(x) = y$. The domain of T is the set of strings for which T halts; $T(x)$ is undefined if T does not halt when given x as input. Thinking of a Turing machine as a machine that computes the values of a function on strings is useful, but how can we use Turing machines to compute functions defined on integers, on pairs of integers, on triples of integers, and so on?

To consider a Turing machine as a computer of functions from the set of k-tuples of nonnegative integers to the set of nonnegative integers (such functions are called **number-theoretic functions**), we need a way to represent k-tuples of integers on a tape. To do so, we use **unary representations** of integers. We represent the nonnegative integer n by a string of $n + 1$ 1s so that, for instance, 0 is represented by the string 1 and 5 is represented by the string 111111. To represent the k-tuple $(n_1, n_2, \dots, n_k)$, we use a string of $n_1 + 1$ 1s, followed by an asterisk, followed by a string of $n_2 + 1$ 1s, followed by an asterisk, and so on, ending with a string of $n_k + 1$ 1s. For example, to represent the four-tuple (2, 0, 1, 3) we use the string$111 * 1 * 11 * 1111$.

We can now consider a Turing machine T as computing a sequence of number-theoretic functions $T, T^2, \dots, T^k, \dots$. The function T^k is defined by the action of T on k-tuples of integers represented by unary representations of integers separated by asterisks.

EXAMPLE 4 Construct a Turing machine for adding two nonnegative integers.

Solution: We need to build a Turing machine T that computes the function $f(n_1, n_2) = n_1 + n_2$. The pair (n_1, n_2) is represented by a string of $n_1 + 1$ 1s followed by an asterisk followed by $n_2 + 1$ 1s. The machine T should take this as input and produce as output a tape with $n_1 + n_2 + 1$ 1s. One way to do this is as follows. The machine starts at the leftmost 1 of the input string, and carries out steps to erase this 1, halting if $n_1 = 0$ so that there are no more 1s before the asterisk, replaces the asterisk with the leftmost remaining 1, and then halts. We can use these five-tuples to do this: $(s_0, 1, s_1, B, R)$, $(s_1, *, s_3, B, R)$, $(s_1, 1, s_2, B, R)$, $(s_2, 1, s_2, 1, R)$, and $(s_2, *, s_3, 1, R)$. ◀

Unfortunately, constructing Turing machines to compute relatively simple functions can be extremely demanding. For example, one Turing machine for multiplying two nonnegative integers found in many books has 31 five-tuples and 11 states. If it is challenging to construct Turing machines to compute even relatively simple functions, what hope do we have of building Turing machines for more complicated functions? One way to simplify this problem is to use a multitape Turing machine that uses more than one tape simultaneously and to build up multitape Turing machines for the composition of functions. It can be shown that for any multitape Turing machine there is a one-tape Turing machine that can do the same thing.

13.5.5 Different Types of Turing Machines

There are many variations on the definition of a Turing machine. We can expand the capabilities of a Turing machine in a wide variety of ways. For example, we can allow a Turing machine to move right, left, or not at all at each step. We can allow a Turing machine to operate on multiple tapes, using $(2 + 3n)$-tuples to describe the Turing machine when n tapes are used. We can allow the tape to be two-dimensional, where at each step we move up, down, right, or left, not just right or left as we do on a one-dimensional tape. We can allow multiple tape heads that read different cells simultaneously. Furthermore, we can allow a Turing machine to be **nondeterministic**, by allowing a (state, tape symbol) pair to possibly appear as the first elements in more than one five-tuple of the Turing machine. So, unlike a deterministic Turing machine, which has at most one prescribed action for a given (state, tape symbol) pair, a nondeterministic Turing machine has a set of prescribed actions for such a pair. Hence, a nondeterministic Turing machine branches into multiple copies as it runs, each of which follows a sequence of possible transitions. We can also reduce the capabilities of a Turing machine in different ways. For example, we can restrict the tape to be infinite in only one direction or we can restrict the tape alphabet to have only two symbols. All these variations of Turing machines have been studied in detail.

The crucial point is that no matter which of these variations we use, or even which combination of variations we use, we never increase or decrease the power of the machine. Anything that one of these variations can do can be done by the Turing machine defined in this section, and vice versa. The reason that these variations are useful is that sometimes they make doing some particular job much easier than if the Turing machine defined in Definition 1 were used. They never extend the capability of the machine. Sometimes it is useful to have a wide variety of Turing machines with which to work. For example, one way to show that for every nondeterministic Turing machine, there is a deterministic Turing machine that recognizes the same language is to use a deterministic Turing machine with three tapes. (For details on variations of Turing machines and demonstrations of their equivalence, see [HoMoUl01].)

Besides introducing the notion of a Turing machine, Turing also showed that it is possible to construct a single Turing machine that can simulate the computations of every Turing machine when given an encoding of this target Turing machine and its input. Such a machine is called a **universal Turing machine**. (See a book on the theory of computation, such as [Si06], for more about universal Turing machines.)

13.5.6 The Church–Turing Thesis

Links

Turing machines are relatively simple. They can have only finitely many states and they can read and write only one symbol at a time on a one-dimensional tape. But it turns out that Turing machines are extremely powerful. We have seen that Turing machines can be built to add numbers and to multiply numbers. Although it may be difficult to actually construct a Turing machine to compute a particular function that can be computed with an algorithm, such a Turing machine can always be found. This was the original goal of Turing when he invented his machines. Furthermore, there is a tremendous amount of evidence for the **Church–Turing thesis**, which states that given any problem that can be solved with an effective algorithm, there is a Turing machine that can solve this problem. The reason this is called a *thesis* rather than a theorem is that the concept of solvability by an effective algorithm is informal and imprecise, as opposed to the notion of solvability by a Turing machine, which is formal and precise. Certainly, though, any problem that can be solved using a computer with a program written in any language, perhaps using an unlimited amount of memory, should be considered effectively solvable. (Note that Turing machines have unlimited memory, unlike computers in the real world, which have only a finite amount of memory.)

Many different formal theories have been developed to capture the notion of effective computability. These include Turing's theory and Church's lambda-calculus, as well as theories proposed by Stephen Kleene and by E. L. Post. These theories seem quite different on the surface. The surprising thing is that they can be shown to be equivalent by demonstrating that they define exactly the same class of functions. With this evidence, it seems that Turing's original ideas, formulated before the invention of modern computers, describe the ultimate capabilities of these machines. The interested reader should consult books on the theory of computation, such as [HoMoUl01] and [Si96], for a discussion of these different theories and their equivalence.

For the remainder of this section we will briefly explore some of the consequences of the Church–Turing thesis and we will describe the importance of Turing machines in the study of the complexity of algorithms. Our goal will be to introduce some important ideas from theoretical computer science to entice the interested student to further study. We will cover a lot of ground quickly without providing explicit details. Our discussion will also tie together some of the concepts discussed in previous parts of the book with the theory of computation.

13.5.7 Computational Complexity, Computability, and Decidability

Throughout this book we have discussed the computational complexity of a wide variety of problems. We described the complexity of these problems in terms of the number of operations used by the most efficient algorithms that solve them. The basic operations used by algorithms differ considerably; we have measured the complexity of different algorithms in terms of bit operations, comparisons of integers, arithmetic operations, and so on. In Section 3.3, we defined various classes of problems in terms of their computational complexity. However, these definitions were not precise, because the types of operations used to measure their complexity vary so drastically. Turing machines provide a way to make the concept of computational complexity precise. If the Church–Turing thesis is true, it would then follow that if a problem can be solved using an effective algorithm, then there is a Turing machine that can solve this problem. When a Turing machine is used to solve a problem, the input to the problem is encoded as a string of symbols that is written on the tape of this Turing machine. How we encode input depends on the domain of this input. For example, as we have seen, we can encode a positive integer using a string of 1s. We can also devise ways to express pairs of integers, negative integers, and so on. Similarly, for graph algorithms, we need a way to encode graphs as strings of symbols. This can be done in many ways and can be based on adjacency lists or adjacency matrices. (We omit the details of how this is done.) However, the way input is encoded does not matter as long as it is relatively efficient, as a Turing machine can always change one encoding into another encoding. We will now use this model to make precise some of the notions concerning computational complexity that were informally introduced in Section 3.3.

The kind of problems that are most easily studied by using Turing machines are those problems that can be answered either by a "yes" or by a "no."

Definition 3 A *decision problem* asks whether statements from a particular class of statements are true. Decision problems are also known as *yes-or-no problems*.

Given a decision problem, we would like to know whether there is an algorithm that can determine whether statements from the class of statements it addresses are true. For example, consider the class of statements each of which asks whether a particular integer n is prime. This is a decision problem because the answer to the question "Is n prime?" is either yes or no. Given this decision problem, we can ask whether there is an algorithm that can decide whether each of the statements in the decision problem is true, that is, given an integer n, deciding whether n is prime. The answer is that there is such an algorithm. In particular, in Section 4.3 we discussed the algorithm that determines whether a positive integer n is prime by checking whether it is divisible by primes not exceeding its square root. (There are many other algorithms for determining whether a positive integer is prime.)

The set of inputs for which the answer to the yes–no problem is "yes" is a subset of the set of possible inputs, that is, it is a subset of the set of strings of the input alphabet. In other words, solving a yes–no problem is the same as recognizing the language consisting of all bit strings that represent input values to the problem leading to the answer "yes." Consequently, solving a yes–no problem is the same as recognizing the language corresponding to the input values for which the answer to the problem is "yes."

DECIDABILITY When there is an effective algorithm that decides whether instances of a decision problem are true, we say that this problem is **solvable** or **decidable**. For instance, the problem of determining whether a positive integer is prime is a solvable problem. However, if no effective algorithm exists for solving a problem, then we say the problem is **unsolvable** or **undecidable**. To show that a decision problem is solvable we need only construct an algorithm that can determine whether statements of the particular class are true. On the other hand, to show that a decision problem is unsolvable we need to prove that no such algorithm exists. (The fact that we tried to find such an algorithm but failed, does not prove the problem is unsolvable.)

By studying only decision problems, it may seem that we are studying only a small set of problems of interest. However, most problems can be recast as decision problems. Recasting the types of problems we have studied in this book as decision problems can be quite complicated, so we will not go into the details of this process here. The interested reader can consult references on the theory of computation, such as [Wo87], which, for example, explains how to recast the traveling salesperson problem (described in Section 10.5) as a decision problem. (To recast the traveling salesperson problem as a decision problem, we first consider the decision problem that asks whether there is a Hamilton circuit of weight not exceeding k, where k is a positive integer. With some additional effort it is possible to use answers to this question for different values of k to find the smallest possible weight of a Hamilton circuit.)

In Section 3.1 we introduced the halting problem and proved that it is an unsolvable problem. That discussion was somewhat informal because the notion of a procedure was not precisely defined. A precise definition of the halting problem can be made in terms of Turing machines.

Definition 4 The *halting problem* is the decision problem that asks whether a Turing machine T eventually halts when given an input string x.

With this definition of the halting problem, we have Theorem 1.

THEOREM 1 The halting problem is an unsolvable decision problem. That is, no Turing machine exists that, when given an encoding of a Turing machine T and its input string x as input, can determine whether T eventually halts when started with x written on its tape.

The proof of Theorem 1 given in Section 3.1 for the informal definition of the halting problem still applies here.

Other examples of unsolvable problems include:

(i) the problem of determining whether two context-free grammars generate the same set of strings;

(ii) the problem of determining whether a given set of tiles can be used with repetition allowed to cover the entire plane without overlap; and

(iii) *Hilbert's Tenth Problem*, which asks whether there are integer solutions to a given polynomial equation with integer coefficients. (This question occurs tenth on the influential list of 23 unsolved problems Hilbert posed in 1900. Hilbert envisioned that the work done to solve these problems would help further the progress of mathematics in the twentieth century. The unsolvability of Hilbert's Tenth Problem was established in 1970 by Yuri Matiyasevich.)

Links

COMPUTABILITY A function that can be computed by a Turing machine is called **computable** and a function that cannot be computed by a Turing machine is called **uncomputable**. It is fairly straightforward, using a countability argument, to show that there are number-theoretic functions that are not computable (see Exercise 39 in Section 2.5). However, it is not so easy to actually produce such a function. The **busy beaver function** defined in the preamble to Exercise 31 is an example of an uncomputable function. One way to show that the busy beaver function is not computable is to show that it grows faster than any computable function. (See Exercise 32.)

Note that every decision problem can be reformulated as the problem of computing a function, namely, the function that has the value 1 when the answer to the problem is "yes" and that has the value 0 when the answer to the problem is "no." A decision problem is solvable if and only if the corresponding function constructed in this way is computable.

THE CLASSES P AND NP In Section 3.3 we informally defined the classes of problems called P and NP. We are now able to define these concepts precisely using the notions of deterministic and nondeterministic Turing machines.

We first elaborate on the difference between a deterministic Turing machine and a nondeterministic Turing machine. The Turing machines we have studied in this section have all been deterministic. In a deterministic Turing machine $T = (S, I, f, s_0)$, transition rules are defined by the partial function f from $S \times I$ to $S \times I \times \{R, L\}$. Consequently, when transition rules of the machine are represented as five-tuples of the form (s, x, s', x', d), where s is the current state, x is the current tape symbol, s' is the next state, x' is the symbol that replaces x on the tape, and d is the direction the machine moves on the tape, no two transition rules begin with the same pair (s, x).

In a nondeterministic Turing machine, allowed steps are defined using a relation consisting of five-tuples rather than using a partial function. The restriction that no two transition rules begin with the same pair (s, x) is eliminated; that is, there may be more than one transition rule beginning with each (state, tape symbol) pair. Consequently, in a nondeterministic Turing machine, there is a choice of transitions for some pairs of the current state and the tape symbol being read. At each step of the operation of a nondeterministic Turing machine, the machine picks one of the different choices of the transition rules that begin with the current state and tape symbol pair. This choice can be considered to be a "guess" of which step to use. Just as for deterministic Turing machines, a nondeterministic Turing machine halts when there is no transition rule in its definition that begins with the current state and tape symbol. Given a nondeterministic Turing machine T, we say that a string x is recognized by T if and only if there exists some sequence of transitions of T that ends in a final state when the machine starts in the initial position with x written on the tape. The nondeterministic Turing machine T recognizes the set A if x is recognized by T if and only if $x \in A$. The nondeterministic Turing machine T is said to solve a decision problem if it recognizes the set consisting of all input values for which the answer to the decision problem is yes.

Definition 5

A decision problem is in P, the *class of polynomial-time problems*, if it can be solved by a deterministic Turing machine in polynomial time in terms of the size of its input. That is, a decision problem is in P if there is a deterministic Turing machine T that solves the decision problem and a polynomial $p(n)$ such that for all integers n, T halts in a final state after no more than $p(n)$ transitions whenever the input to T is a string of length n. A decision problem is in *NP*, the *class of nondeterministic polynomial-time problems*, if it can be solved by a nondeterministic Turing machine in polynomial time in terms of the size of its input. That is, a decision problem is in NP if there is a nondeterministic Turing machine T that solves the problem and a polynomial $p(n)$ such that for all integers n, T halts for every choice of transitions after no more than $p(n)$ transitions whenever the input to T is a string of length n.

Links

©University Archives. Department of Rare Books and Special Collections. Princeton University Library.

ALONZO CHURCH (1903–1995) Alonzo Church was born in Washington, D.C. He studied at Göttingen under Hilbert and later in Amsterdam. He was a member of the faculty at Princeton University from 1927 until 1967, when he moved to UCLA. Church was one of the founding members of the Association for Symbolic Logic. He made many substantial contributions to the theory of computability, including his solution to the decision problem, his invention of the lambda-calculus, and his statement of what is now known as the Church–Turing thesis. Among Church's students were Stephen Kleene and Alan Turing. He published articles past his 90th birthday.

Problems in P are called **tractable**, whereas problems not in P are called **intractable**. For a problem to be in P, a deterministic Turing machine must exist that can decide in polynomial time whether a particular statement of the class addressed by the decision problem is true. For example, determining whether an item is in a list of n elements is a tractable problem. (We will not provide details on how this fact can be shown; the basic ideas used in the analyses of algorithms earlier in the text can be adapted when Turing machines are employed.) For a problem to be in NP, it is necessary only that there be a nondeterministic Turing machine that, when given a true statement from the set of statements addressed by the problem, can verify its truth in polynomial time by making the correct guess at each step from the set of allowable steps corresponding to the current state and tape symbol. The problem of determining whether a given graph has a Hamilton circuit is an NP problem, because a nondeterministic Turing machine can easily verify that a simple circuit in a graph passes through each vertex exactly once. It can do this by making a series of correct guesses corresponding to successively adding edges to form the circuit. Because every deterministic Turing machine can also be considered to be a nondeterministic Turing machine where each (state, tape symbol) pair occurs in exactly one transition rule defining the machine, every problem in P is also in NP. In symbols, $P \subseteq NP$.

One of the most perplexing open questions in theoretical computer science is whether every problem in NP is also in P, that is, whether $P = NP$. As mentioned in Section 3.3, there is an important class of problems, the class of NP-complete problems, such that a problem is in this class if it is in the class NP and if it can be shown that if it is also in the class P, then *every* problem in the class NP must also be in the class P. That is, a problem is NP-complete if the existence of a polynomial-time algorithm for solving it implies the existence of a polynomial-time algorithm for every problem in NP. In this book we have discussed several different NP-complete problems, such as determining whether a simple graph has a Hamilton circuit and determining whether a proposition in n-variables is a tautology.

Exercises

1. Let T be the Turing machine defined by the five-tuples: $(s_0, 0, s_1, 1, R)$, $(s_0, 1, s_1, 0, R)$, $(s_0, B, s_1, 0, R)$, $(s_1, 0, s_2, 1, L)$, $(s_1, 1, s_1, 0, R)$, and $(s_1, B, s_2, 0, L)$. For each of these initial tapes, determine the final tape when T halts, assuming that T begins in initial position.

a)

⋯	B	B	0	0	1	1	B	B	⋯

b)

⋯	B	B	1	0	1	B	B	B	⋯

c)

⋯	B	B	1	1	B	0	1	B	⋯

d)

⋯	B	B	B	B	B	B	B	B	⋯

2. Let T be the Turing machine defined by the five-tuples: $(s_0, 0, s_1, 0, R)$, $(s_0, 1, s_1, 0, L)$, $(s_0, B, s_1, 1, R)$, $(s_1, 0, s_2, 1, R)$, $(s_1, 1, s_1, 1, R)$, $(s_1, B, s_2, 0, R)$, and $(s_2, B, s_3, 0, R)$. For each of these initial tapes, determine the final tape when T halts, assuming that T begins in initial position.

a)

⋯	B	B	0	1	0	1	B	B	⋯

b)

⋯	B	B	1	1	1	B	B	B	⋯

c)

⋯	B	B	0	0	B	0	0	B	⋯

d)

⋯	B	B	B	B	B	B	B	B	⋯

3. What does the Turing machine described by the five-tuples $(s_0, 0, s_0, 0, R)$, $(s_0, 1, s_1, 0, R)$, (s_0, B, s_2, B, R), $(s_1, 0, s_1, 0, R)$, $(s_1, 1, s_0, 1, R)$, and (s_1, B, s_2, B, R) do when given

a) 11 as input?

b) an arbitrary bit string as input?

4. What does the Turing machine described by the five-tuples $(s_0, 0, s_0, 1, R)$, $(s_0, 1, s_0, 1, R)$, (s_0, B, s_1, B, L), $(s_1, 1, s_2, 1, R)$, do when given

a) 101 as input?

b) an arbitrary bit string as input?

5. What does the Turing machine described by the five-tuples $(s_0, 1, s_1, 0, R)$, $(s_1, 1, s_1, 1, R)$, $(s_1, 0, s_2, 0, R)$, $(s_2, 0, s_3, 1, L)$, $(s_2, 1, s_2, 1, R)$, $(s_3, 1, s_3, 1, L)$, $(s_3, 0, s_4, 0, L)$, $(s_4, 1, s_4, 1, L)$, and $(s_4, 0, s_0, 1, R)$ do when given

a) 11 as input?

b) a bit string consisting entirely of 1s as input?

6. Construct a Turing machine with tape symbols 0, 1, and B that, when given a bit string as input, adds a 1 to the end of the bit string and does not change any of the other symbols on the tape.

7. Construct a Turing machine with tape symbols 0, 1, and B that, when given a bit string as input, replaces the first 0 with a 1 and does not change any of the other symbols on the tape.

8. Construct a Turing machine with tape symbols 0, 1, and B that, given a bit string as input, replaces all 0s on the tape with 1s and does not change any of the 1s on the tape.

9. Construct a Turing machine with tape symbols 0, 1, and B that, given a bit string as input, replaces all but the leftmost 1 on the tape with 0s and does not change any of the other symbols on the tape.

10. Construct a Turing machine with tape symbols 0, 1, and B that, given a bit string as input, replaces the first two consecutive 1s on the tape with 0s and does not change any of the other symbols on the tape.

11. Construct a Turing machine that recognizes the set of all bit strings that end with a 0.

12. Construct a Turing machine that recognizes the set of all bit strings that contain at least two 1s.

13. Construct a Turing machine that recognizes the set of all bit strings that contain an even number of 1s.

14. Show at each step the contents of the tape of the Turing machine in Example 3 starting with each of these strings.

a) 0011 **b)** 00011 **c)** 101100 **d)** 000111

15. Explain why the Turing machine in Example 3 recognizes a bit string if and only if this string is of the form 0^n1^n for some positive integer n.

***16.** Construct a Turing machine that recognizes the set $\{0^{2n}1^n \mid n \geq 0\}$.

***17.** Construct a Turing machine that recognizes the set $\{0^n1^n2^n \mid n \geq 0\}$.

18. Construct a Turing machine that computes the function $f(n) = n + 2$ for all nonnegative integers n.

19. Construct a Turing machine that computes the function $f(n) = n - 3$ if $n \geq 3$ and $f(n) = 0$ for $n = 0, 1, 2$ for all nonnegative integers n.

20. Construct a Turing machine that computes the function $f(n) = n \textbf{ mod } 3$ for every nonnegative integer n.

21. Construct a Turing machine that computes the function $f(n) = 3$ if $n \geq 5$ and $f(n) = 0$ if $n = 0, 1, 2, 3$, or 4.

22. Construct a Turing machine that computes the function $f(n) = 2n$ for all nonnegative integers n.

23. Construct a Turing machine that computes the function $f(n) = 3n$ for all nonnegative integers n.

24. Construct a Turing machine that computes the function $f(n_1, n_2) = n_2 + 2$ for all pairs of nonnegative integers n_1 and n_2.

***25.** Construct a Turing machine that computes the function $f(n_1, n_2) = \min(n_1, n_2)$ for all nonnegative integers n_1 and n_2.

26. Construct a Turing machine that computes the function $f(n_1, n_2) = n_1 + n_2 + 1$ for all nonnegative integers n_1 and n_2.

Suppose that T_1 and T_2 are Turing machines with disjoint sets of states S_1 and S_2 and with transition functions f_1 and f_2, respectively. We can define the Turing machine T_1T_2, the **composite** of T_1 and T_2, as follows. The set of states of T_1T_2 is $S_1 \cup S_2$. T_1T_2 begins in the start state of T_1. It first executes the transitions of T_1 using f_1 up to, but not including, the step at which T_1 would halt. Then, for all moves for which T_1 halts, it executes the same transitions of T_1 except that it moves to the start state of T_2. From this point on, the moves of T_1T_2 are the same as the moves of T_2.

27. By finding the composite of the Turing machines you constructed in Exercises 18 and 22, construct a Turing machine that computes the function $f(n) = 2n + 2$.

28. By finding the composite of the Turing machines you constructed in Exercises 18 and 23, construct a Turing machine that computes the function $f(n) = 3(n + 2) = 3n + 6$.

29. Which of the following problems is a decision problem?

a) What is the smallest prime greater than n?
b) Is a graph G bipartite?
c) Given a set of strings, is there a finite-state automaton that recognizes this set of strings?
d) Given a checkerboard and a particular type of polyomino (see Section 1.8), can this checkerboard be tiled using polyominoes of this type?

30. Which of the following problems is a decision problem?

a) Is the sequence $a_1, a_2, \ldots, a_n$ of positive integers in increasing order?
b) Can the vertices of a simple graph G be colored using three colors so that no two adjacent vertices are the same color?
c) What is the vertex of highest degree in a graph G?
d) Given two finite-state machines, do these machines recognize the same language?

Links

Let $B(n)$ be the maximum number of 1s that a Turing machine with n states with the alphabet $\{1, B\}$ may print on a tape that is initially blank. The problem of determining $B(n)$ for particular values of n is known as the **busy beaver problem**. This problem was first studied by Tibor Rado in 1962. Currently it is known that $B(2) = 4$, $B(3) = 6$, and $B(4) = 13$, but $B(n)$ is not known for $n \geq 5$. $B(n)$ grows rapidly; it is known that $B(5) \geq 4098$ and $B(6) \geq 3.5 \times 10^{18267}$.

***31.** Show that $B(2)$ is at least 4 by finding a Turing machine with two states and alphabet $\{1, B\}$ that halts with four consecutive 1s on the tape.

****32.** Show that the function $B(n)$ cannot be computed by any Turing machine. [*Hint:* Assume that there is a Turing machine that computes $B(n)$ in binary. Build a Turing machine T that, starting with a blank tape, writes n down in binary, computes $B(n)$ in binary, and converts $B(n)$ from binary to unary. Show that for sufficiently large n, the number of states of T is less than $B(n)$, leading to a contradiction.]

Key Terms and Results

TERMS

alphabet (or vocabulary): a set that contains elements used to form strings

language: a subset of the set of all strings over an alphabet

phrase-structure grammar (V, T, S, P): a description of a language containing an alphabet V, a set of terminal symbols T, a start symbol S, and a set of productions P

the production $w \to w_1$: w can be replaced by w_1 whenever it occurs in a string in the language

$w_1 \Rightarrow w_2$ (w_2 is directly derivable from w_1): w_2 can be obtained from w_1 using a production to replace a string in w_1 with another string

$w_1 \overset{*}{\Rightarrow} w_2$ (w_2 is derivable from w_1): w_2 can be obtained from w_1 using a sequence of productions to replace strings by other strings

type 0 grammar: any phrase-structure grammar

type 1 grammar: a phrase-structure grammar in which every production is of the form $w_1 \to w_2$, where $w_1 = lAr$ and $w_2 = lwr$, where $A \in N$, $l, r, w \in (N \cup T)*$ and $w \neq \lambda$, or $w_1 = S$ and $w_2 = \lambda$ as long as S is not on the right-hand side of another production

type 2, or context-free, grammar: a phrase-structure grammar in which every production is of the form $A \to w_1$, where A is a nonterminal symbol

type 3, or regular, grammar: a phrase-structure grammar where every production is of the form $A \to aB$, $A \to a$, or $S \to \lambda$, where A and B are nonterminal symbols, S is the start symbol, and a is a terminal symbol

derivation (or parse) tree: an ordered rooted tree where the root represents the starting symbol of a type 2 grammar, internal vertices represent nonterminals, leaves represent terminals, and the children of a vertex are the symbols on the right side of a production, in order from left to right, where the symbol represented by the parent is on the left-hand side

Backus–Naur form: a description of a context-free grammar in which all productions having the same nonterminal as their left-hand side are combined with the different right-hand sides of these productions, each separated by a bar, with nonterminal symbols enclosed in angular brackets and the symbol $\to$ replaced by ::=

finite-state machine (S, I, O, f, g, s_0) (or a Mealy machine): a six-tuple containing a set S of states, an input alphabet I, an output alphabet O, a transition function f that assigns a next state to every pair of a state and an input, an output function g that assigns an output to every pair of a state and an input, and a starting state s_0

AB (concatenation of A and B): the set of all strings formed by concatenating a string in A and a string in B in that order

A^* (Kleene closure of A): the set of all strings made up by concatenating arbitrarily many strings from A

deterministic finite-state automaton (S, I, f, s_0, F): a five-tuple containing a set S of states, an input alphabet I, a transition function f that assigns a next state to every pair of a state and an input, a starting state s_0, and a set of final states F

nondeterministic finite-state automaton (S, I, f, s_0, F): a five-tuple containing a set S of states, an input alphabet I, a transition function f that assigns a set of possible next states to every pair of a state and an input, a starting state s_0, and a set of final states F

language recognized by an automaton: the set of input strings that take the start state to a final state of the automaton

regular expression: an expression defined recursively by specifying that $\emptyset$, λ, and x, for all x in the input alphabet, are regular expressions, and that $(\mathbf{AB})$, $(\mathbf{A} \cup \mathbf{B})$, and $\mathbf{A}^*$ are regular expressions when $\mathbf{A}$ and $\mathbf{B}$ are regular expressions

regular set: a set defined by a regular expression (see page 820)

Turing machine $T = (S, I, f, s_0)$: a four-tuple consisting of a finite set S of states, an alphabet I containing the blank symbol B, a partial function f from $S \times I$ to $S \times I \times \{R, L\}$, and a starting state s_0

nondeterministic Turing machine: a Turing machine that may have more than one transition rule corresponding to each (state, tape symbol) pair

decision problem: a problem that asks whether statements from a particular class of statements are true

solvable problem: a problem with the property that there is an effective algorithm that can solve all instances of the problem

unsolvable problem: a problem with the property that no effective algorithm exists that can solve all instances of the problem

computable function: a function whose values can be computed using a Turing machine

uncomputable function: a function whose values cannot be computed using a Turing machine

P, the class of polynomial-time problems: the class of problems that can be solved by a deterministic Turing machine in polynomial time in terms of the size of the input

NP, the class of nondeterministic polynomial-time problems: the class of problems that can be solved by a nondeterministic Turing machine in polynomial time in terms of the size of the input

NP-complete: a subset of the class of NP problems with the property that if any one of them is in the class P, then all problems in NP are in the class P

RESULTS

For every nondeterministic finite-state automaton there is a deterministic finite-state automaton that recognizes the same set.

Kleene's theorem: A set is regular if and only if there is a finite-state automaton that recognizes it.

A set is regular if and only if it is generated by a regular grammar.

The halting problem is unsolvable.

Review Questions

1. **a)** Define a phrase-structure grammar.

b) What does it mean for a string to be derivable from a string w by a phrase-structure grammar G?

2. **a)** What is the language generated by a phrase-structure grammar G?

b) What is the language generated by the grammar G with vocabulary $\{S, 0, 1\}$, set of terminals $T = \{0, 1\}$, starting symbol S, and productions $S \to 000S$, $S \to 1$?

c) Give a phrase-structure grammar that generates the set $\{01^n \mid n = 0, 1, 2, \dots\}$.

3. **a)** Define a type 1 grammar.

b) Give an example of a grammar that is not a type 1 grammar.

c) Define a type 2 grammar.

d) Give an example of a grammar that is not a type 2 grammar but is a type 1 grammar.

e) Define a type 3 grammar.

f) Give an example of a grammar that is not a type 3 grammar but is a type 2 grammar.

4. **a)** Define a regular grammar.

b) Define a regular language.

c) Show that the set $\{0^m 1^n \mid m, n = 0, 1, 2, \dots\}$ is a regular language.

5. **a)** What is Backus–Naur form?

b) Give an example of the Backus–Naur form of the grammar for a subset of English of your choice.

6. **a)** What is a finite-state machine?

b) Show how a vending machine that accepts only quarters and dispenses a soft drink after 75 cents has been deposited can be modeled using a finite-state machine.

7. Find the set of strings recognized by the deterministic finite-state automaton shown here.

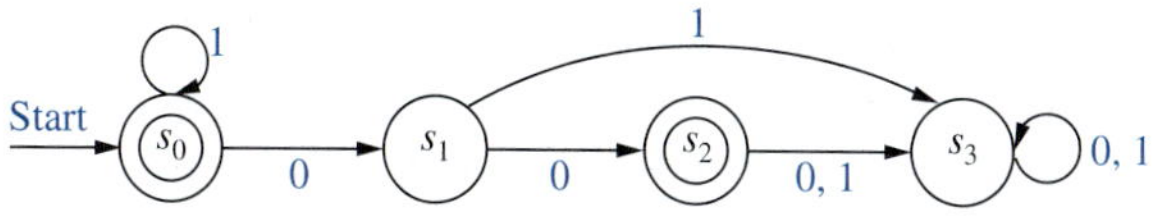

8. Construct a deterministic finite-state automaton that recognizes the set of bit strings that start with 1 and end with 1.

9. **a)** What is the Kleene closure of a set of strings?

b) Find the Kleene closure of the set $\{11, 0\}$.

10. **a)** Define a finite-state automaton.

b) What does it mean for a string to be recognized by a finite-state automaton?

11. **a)** Define a nondeterministic finite-state automaton.

b) Show that given a nondeterministic finite-state automaton, there is a deterministic finite-state automaton that recognizes the same language.

12. **a)** Define the set of regular expressions over a set I.

b) Explain how regular expressions are used to represent regular sets.

13. State Kleene's theorem.

14. Show that a set is generated by a regular grammar if and only if it is a regular set.

15. Give an example of a set not recognized by a finite-state automaton. Show that no finite-state automaton recognizes it.

16. Define a Turing machine.

17. Describe how Turing machines are used to recognize sets.

18. Describe how Turing machines are used to compute number-theoretic functions.

19. What is an unsolvable decision problem? Give an example of such a problem.

Supplementary Exercises

***1.** Find a phrase-structure grammar that generates each of these languages.

a) the set of bit strings of the form $0^{2n}1^{3n}$, where n is a nonnegative integer

b) the set of bit strings with twice as many 0s as 1s

c) the set of bit strings of the form w^2, where w is a bit string

***2.** Find a phrase-structure grammar that generates the set $\{0^{2^n} \mid n \geq 0\}$.

For Exercises 3 and 4, let $G = (V, T, S, P)$ be the context-free grammar with $V = \{(,), S, A, B\}$, $T = \{(,)\}$, starting symbol S, and productions $S \to A$, $A \to AB$, $A \to B$, $B \to (A)$, and $B \to ()$, $S \to \lambda$.

3. Construct the derivation trees of these strings.

a) (()) **b)** ()(()) **c)** ((()()))

***4.** Show that $L(G)$ is the set of all balanced strings of parentheses, defined in the preamble to Supplementary Exercise 59 in Chapter 5.

A context-free grammar is **ambiguous** if there is a word in $L(G)$ with two derivations that produce different derivation trees, considered as ordered, rooted trees.

5. Show that the grammar $G = (V, T, S, P)$ with $V = \{0, S\}$, $T = \{0\}$, starting state S, and productions $S \to 0S$, $S \to S0$, and $S \to 0$ is ambiguous by constructing two different derivation trees for 0^3.

6. Show that the grammar $G = (V, T, S, P)$ with $V = \{0, S\}$, $T = \{0\}$, starting state S, and productions $S \to 0S$ and $S \to 0$ is unambiguous.

7. Suppose that A and B are finite subsets of V^*, where V is an alphabet. Is it necessarily true that $|AB| = |BA|$?

8. Prove or disprove each of these statements for subsets A, B, and C of V^*, where V is an alphabet.

a) $A(B \cup C) = AB \cup AC$
b) $A(B \cap C) = AB \cap AC$
c) $(AB)C = A(BC)$
d) $(A \cup B)^* = A^* \cup B^*$

9. Suppose that A and B are subsets of V^*, where V is an alphabet. Does it follow that $A \subseteq B$ if $A^* \subseteq B^*$?

10. What set of strings with symbols in the set $\{0, 1, 2\}$ is represented by the regular expression $(\mathbf{2}^*)(\mathbf{0} \cup (\mathbf{12}^*))^*$?

The **star height** $h(\mathbf{E})$ of a regular expression over the set I is defined recursively by

$$h(\emptyset) = 0;$$
$$h(\mathbf{x}) = 0 \text{ if } \mathbf{x} \in I;$$
$$h((\mathbf{E}_1 \cup \mathbf{E}_2)) = h((\mathbf{E}_1\mathbf{E}_2)) = \max(h(\mathbf{E}_1), h(\mathbf{E}_2)) \text{ if } \mathbf{E}_1 \text{ and } \mathbf{E}_2 \text{ are regular expressions;}$$
$$h(\mathbf{E}^*) = h(\mathbf{E}) + 1 \text{ if } \mathbf{E} \text{ is a regular expression.}$$

11. Find the star height of each of these regular expressions.

a) $\mathbf{0}^*\mathbf{1}$
b) $\mathbf{0}^*\mathbf{1}^*$
c) $(\mathbf{0}^*\mathbf{01})^*$
d) $((\mathbf{0}^*\mathbf{1})^*)^*$
e) $(\mathbf{010}^*)(\mathbf{1}^*\mathbf{01}^*)^*((\mathbf{01})^*(\mathbf{10})^*)^*$
f) $(((((\mathbf{0}^*)\mathbf{1})^*\mathbf{0})^*)\mathbf{1})^*$

***12.** For each of these regular expressions find a regular expression that represents the same language with minimum star height.

a) $(\mathbf{0}^*\mathbf{1}^*)^*$
b) $(\mathbf{0}(\mathbf{01}^*\mathbf{0})^*)^*$
c) $(\mathbf{0}^* \cup (\mathbf{01})^* \cup \mathbf{1}^*)^*$

13. Construct a finite-state machine with output that produces an output of 1 if the bit string read so far as input contains four or more 1s. Then construct a deterministic finite-state automaton that recognizes this set.

14. Construct a finite-state machine with output that produces an output of 1 if the bit string read so far as input contains four or more consecutive 1s. Then construct a deterministic finite-state automaton that recognizes this set.

15. Construct a finite-state machine with output that produces an output of 1 if the bit string read so far as input ends with four or more consecutive 1s. Then construct a deterministic finite-state automaton that recognizes this set.

16. A state s' in a finite-state machine is said to be **reachable** from state s if there is an input string x such that $f(s, x) = s'$. A state s is called **transient** if there is no nonempty input string x with $f(s, x) = s$. A state s is called a **sink** if $f(s, x) = s$ for all input strings x. Answer these questions about the finite-state machine with the state diagram illustrated here.

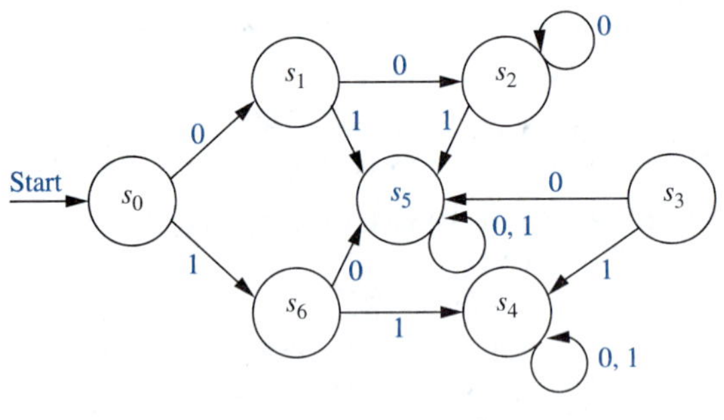

a) Which states are reachable from s_0?
b) Which states are reachable from s_2?
c) Which states are transient?
d) Which states are sinks?

***17.** Suppose that S, I, and O are finite sets such that $|S| = n$, $|I| = k$, and $|O| = m$.

a) How many different finite-state machines (Mealy machines) $M = (S, I, O, f, g, s_0)$ can be constructed, where the starting state s_0 can be arbitrarily chosen?
b) How many different Moore machines $M = (S, I, O, f, g, s_0)$ can be constructed, where the starting state s_0 can be arbitrarily chosen?

***18.** Suppose that S and I are finite sets such that $|S| = n$ and $|I| = k$. How many different finite-state automata $M = (S, I, f, s_0, F)$ are there where the starting state s_0 and the subset F of S consisting of final states can be chosen arbitrarily

a) if the automata are deterministic?
b) if the automata may be nondeterministic? (*Note:* This includes deterministic automata.)

19. Construct a deterministic finite-state automaton that is equivalent to the nondeterministic automaton with the state diagram shown here.

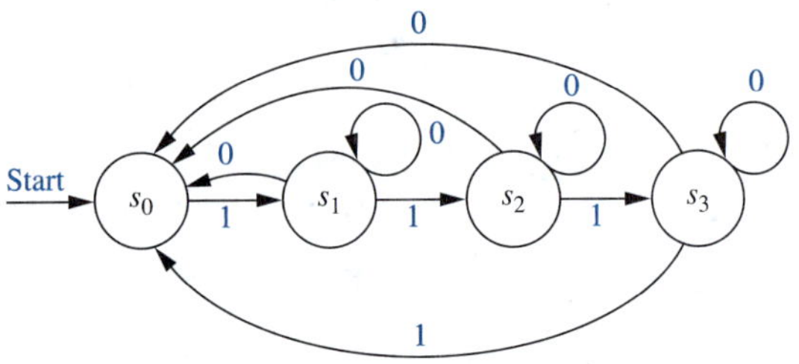

20. What is the language recognized by the automaton in Exercise 19?

21. Construct finite-state automata that recognize these sets.

a) $\mathbf{0}^*(\mathbf{10})^*$
b) $(\mathbf{01} \cup \mathbf{111})^*\mathbf{10}^*(\mathbf{0} \cup \mathbf{1})$
c) $(\mathbf{001} \cup (\mathbf{11})^*)^*$

*22. Find regular expressions that represent the set of all strings of 0s and 1s

a) made up of blocks of even numbers of 1s interspersed with odd numbers of 0s.

b) with at least two consecutive 0s or three consecutive 1s.

c) with no three consecutive 0s or two consecutive 1s.

*23. Show that if A is a regular set, then so is $\overline{A}$.

*24. Show that if A and B are regular sets, then so is $A \cap B$.

*25. Find finite-state automata that recognize these sets of strings of 0s and 1s.

a) the set of all strings that start with no more than three consecutive 0s and contain at least two consecutive 1s

b) the set of all strings with an even number of symbols that do not contain the pattern 101

c) the set of all strings with at least three blocks of two or more 1s and at least two 0s

*26. Show that $\{0^{2^n} \mid n \in \mathbf{N}\}$ is not regular. You may use the pumping lemma given in Exercise 22 of Section 13.4.

*27. Show that $\{1^p \mid p \text{ is prime}\}$ is not regular. You may use the pumping lemma given in Exercise 22 of Section 13.4.

*28. There is a result for context-free languages analogous to the pumping lemma for regular sets. Suppose that $L(G)$ is the language recognized by a context-free language G. This result states that there is a constant N such that if z is a word in $L(G)$ with $l(z) \geq N$, then z can be written as $uvwxy$, where $l(vwx) \leq N$, $l(vx) \geq 1$, and uv^iwx^iy belongs to $L(G)$ for $i = 0, 1, 2, 3, \ldots$. Use this result to show that there is no context-free grammar G with $L(G) = \{0^n1^n2^n \mid n = 0, 1, 2, \ldots\}$.

*29. Construct a Turing machine that computes the function $f(n_1, n_2) = \max(n_1, n_2)$.

*30. Construct a Turing machine that computes the function $f(n_1, n_2) = n_2 - n_1$ if $n_2 \geq n_1$ and $f(n_1, n_2) = 0$ if $n_2 < n_1$.

Computer Projects

Write programs with these input and output.

1. Given the productions in a phrase-structure grammar, determine which type of grammar this is in the Chomsky classification scheme.

2. Given the productions of a phrase-structure grammar, find all strings that are generated using twenty or fewer applications of its production rules.

3. Given the Backus–Naur form of a type 2 grammar, find all strings that are generated using twenty or fewer applications of the rules defining it.

*4. Given the productions of a context-free grammar and a string, produce a derivation tree for this string if it is in the language generated by this grammar.

5. Given the state table of a Moore machine and an input string, produce the output string generated by the machine.

6. Given the state table of a Mealy machine and an input string, produce the output string generated by the machine.

7. Given the state table of a deterministic finite-state automaton and a string, decide whether this string is recognized by the automaton.

8. Given the state table of a nondeterministic finite-state automaton and a string, decide whether this string is recognized by the automaton.

*9. Given the state table of a nondeterministic finite-state automaton, construct the state table of a deterministic finite-state automaton that recognizes the same language.

**10. Given a regular expression, construct a nondeterministic finite-state automaton that recognizes the set that this expression represents.

11. Given a regular grammar, construct a finite-state automaton that recognizes the language generated by this grammar.

12. Given a finite-state automaton, construct a regular grammar that generates the language recognized by this automaton.

*13. Given a Turing machine, find the output string produced by a given input string.

Computations and Explorations

Use a computational program or programs you have written to do these exercises.

1. Solve the busy beaver problem for two states by testing all possible Turing machines with two states and alphabet $\{1, B\}$.

*2. Solve the busy beaver problem for three states by testing all possible Turing machines with three states and alphabet $\{1, B\}$.

**3. Find a busy beaver machine with four states by testing all possible Turing machines with four states and alphabet $\{1, B\}$.

**4. Make as much progress as you can toward finding a busy beaver machine with five states.

**5. Make as much progress as you can toward finding a busy beaver machine with six states.

Writing Projects

Respond to these with essays using outside sources.

1. Describe how the growth of certain types of plants can be modeled using a Lidenmeyer system. Such a system uses a grammar with productions modeling the different ways plants can grow.
2. Describe the Backus–Naur form (and extended Backus–Naur form) rules used to specify the syntax of a programming language, such as Java, LISP, or Ada, or the database language SQL.
3. Explain how finite-state machines are used by spell-checkers.
4. Explain how finite-state machines are used in the study of network protocols.
5. Explain how finite-state machines are used in speech recognition programs.
6. Compare the use of Moore machines versus Mealy machines in the design of hardware systems and computer software.
7. Explain the concept of minimizing finite-state automata. Give an algorithm that carries out this minimization.
8. Give the definition of cellular automata. Explain their applications. Use the Game of Life as an example.
9. Define a pushdown automaton. Explain how pushdown automata are used to recognize sets. Which sets are recognized by pushdown automata? Provide an outline of a proof justifying your answer.
10. Define a linear-bounded automaton. Explain how linear-bounded automata are used to recognize sets. Which sets are recognized by linear-bounded automata? Provide an outline of a proof justifying your answer.
11. Look up Turing's original definition of what we now call a Turing machine. What was his motivation for defining these machines?
12. Describe the concept of the universal Turing machine. Explain how such a machine can be built.
13. Explain the kinds of applications in which nondeterministic Turing machines are used instead of deterministic Turing machines.
14. Show that a Turing machine can simulate any action of a nondeterministic Turing machine.
15. Show that a set is recognized by a Turing machine if and only if it is generated by a phrase-structure grammar.
16. Describe the basic concepts of the lambda-calculus and explain how it is used to study computability of functions.
17. Show that a Turing machine as defined in this chapter can do anything a Turing machine with n tapes can do.
18. Show that a Turing machine with a tape infinite in one direction can do anything a Turing machine with a tape infinite in both directions can do.

APPENDIX

1 Axioms for the Real Numbers and the Positive Integers

In this book we have assumed an explicit set of axioms for the set of real numbers and for the set of positive integers. In this appendix we will list these axioms and we will illustrate how basic facts, also used without proof in the text, can be derived using them.

A1.1 Axioms for Real Numbers

The standard axioms for real numbers include both the **field** (or **algebraic**) **axioms**, used to specify rules for basic arithmetic operations, and the **order axioms**, used to specify properties of the ordering of real numbers.

THE FIELD AXIOMS We begin with the field axioms. As usual, we denote the sum and product of two real numbers x and y by $x + y$ and $x \cdot y$, respectively. (Note that the product of x and y is often denoted by xy without the use of the dot to indicate multiplication. We will not use this abridged notation in this appendix, but will within the text.) Also, by convention, we perform multiplications before additions unless parentheses are used. Although these statements are axioms, they are commonly called *laws* or *rules*. The first two of these axioms tell us that when we add or multiply two real numbers, the result is again a real number; these are the *closure laws*.

- **Closure law for addition** For all real numbers x and y, $x + y$ is a real number.
- **Closure law for multiplication** For all real numbers x and y, $x \cdot y$ is a real number.

The next two axioms tell us that when we add or multiply three real numbers, we get the same result regardless of the order of operations; these are the *associative laws*.

- **Associative law for addition** For all real numbers x, y, and z, $(x + y) + z = x + (y + z)$.
- **Associative law for multiplication** For all real numbers x, y, and z, $(x \cdot y) \cdot z = x \cdot (y \cdot z)$.

Two additional algebraic axioms tell us that the order in which we add or multiply two numbers does not matter; these are the *commutative laws*.

- **Commutative law for addition** For all real numbers x and y, $x + y = y + x$.
- **Commutative law for multiplication** For all real numbers x and y, $x \cdot y = y \cdot x$.

The next two axioms tell us that 0 and 1 are additive and multiplicative identities for the set of real numbers. That is, when we add 0 to a real number or multiply a real number by 1 we do not change this real number. These laws are called *identity laws*.

- **Additive identity law** For every real number x, $x + 0 = 0 + x = x$.
- **Multiplicative identity law** For every real number x, $x \cdot 1 = 1 \cdot x = x$.

Although it seems obvious, we also need the following axiom.

- **Identity elements axiom** The additive identity 0 and the multiplicative identity 1 are distinct, that is $0 \neq 1$.

Two additional axioms tell us that for every real number, there is a real number that can be added to this number to produce 0, and for every nonzero real number, there is a real number by which it can be multiplied to produce 1. These are the *inverse laws*.

- **Inverse law for addition** For every real number x, there exists a real number $-x$ (called the *additive inverse* of x) such that $x + (-x) = (-x) + x = 0$.
- **Inverse law for multiplication** For every nonzero real number x, there exists a real number $1/x$ (called the *multiplicative inverse* of x) such that $x \cdot (1/x) = (1/x) \cdot x = 1$.

The final algebraic axioms for real numbers are the *distributive laws*, which tell us that multiplication distributes over addition; that is, that we obtain the same result when we first add a pair of real numbers and then multiply by a third real number or when we multiply each of these two real numbers by the third real number and then add the two products.

- **Distributive laws** For all real numbers x, y, and z, $x \cdot (y + z) = x \cdot y + x \cdot z$ and $(x + y) \cdot z = x \cdot z + y \cdot z$.

ORDER AXIOMS Next, we will state the *order axioms* for the real numbers, which specify properties of the "greater than" relation, denoted by $>$, on the set of real numbers. We write $x > y$ (and $y < x$) when x is greater than y, and we write $x \geq y$ (and $y \leq x$) when $x > y$ or $x = y$. The first of these axioms tells us that given two real numbers, exactly one of three possibilities occurs: the two numbers are equal, the first is greater than the second, or the second is greater than the first. This rule is called the *trichotomy law*.

- **Trichotomy law** For all real numbers x and y, exactly one of $x = y$, $x > y$, or $y > x$ is true.

Next, we have an axiom, called *the transitivity law*, which tells us that if one number is greater than a second number and this second number is greater than a third, then the first number is greater than the third.

- **Transitivity law** For all real numbers x, y, and z, if $x > y$ and $y > z$, then $x > z$.

We also have two *compatibility laws*, which tell us that when we add a number to both sides in a greater than relationship, the greater than relationship is preserved, and when we multiply both sides of a greater than relationship by a *positive real number* (that is, a real number x with $x > 0$), the greater than relationship is preserved.

- **Additive compatibility law** For all real numbers x, y, and z, if $x > y$, then $x + z > y + z$.
- **Multiplicative compatibility law** For all real numbers x, y, and z, if $x > y$ and $z > 0$, then $x \cdot z > y \cdot z$.

We leave it to the reader (see Exercise 15) to prove that for all real numbers x, y, and z, if $x > y$ and $z < 0$, then $x \cdot z < y \cdot z$. That is, multiplication of an inequality by a negative real number reverses the direction of the inequality.

The final axiom for the set of real numbers is the *completeness property*. Before we state this axiom, we need some definitions. First, given a nonempty set A of real numbers, we say that the real number b is an **upper bound** of A if for every real number a in A, $b \geq a$. A real number s is a **least upper bound** of A if s is an upper bound of A and whenever t is an upper bound of A, then we have $s \leq t$.

- **Completeness property** Every nonempty set of real numbers that is bounded above has a least upper bound.

A1.2 Using Axioms to Prove Basic Facts

The axioms we have listed can be used to prove many properties that are often used without explicit mention. We give several examples of results we can prove using axioms and leave the proof of a variety of other properties as exercises. Although the results we will prove seem quite obvious, proving them using only the axioms we have stated can be challenging.

THEOREM 1 The additive identity element 0 of the real numbers is unique.

Proof: To show that the additive identity element 0 of the real numbers is unique, suppose that $0'$ is also an additive identity for the real numbers. This means that $0' + x = x + 0' = x$ whenever x is a real number. By the additive identity law, it follows that $0 + 0' = 0'$. Because $0'$ is an additive identity, we know that $0 + 0' = 0$. It follows that $0 = 0'$, because both equal $0 + 0'$. This shows that 0 is the unique additive identity for the real numbers. ◁

THEOREM 2 The additive inverse of a real number x is unique.

Proof: Let x be a real number. Suppose that y and z are both additive inverses of x. Then,

$$\begin{aligned} y &= 0 + y && \text{by the additive identity law} \\ &= (z + x) + y && \text{because } z \text{ is an additive inverse of } x \\ &= z + (x + y) && \text{by the associative law for addition} \\ &= z + 0 && \text{because } y \text{ is an additive inverse of } x \\ &= z && \text{by the additive identity law.} \end{aligned}$$

It follows that $y = z$. ◁

Theorems 1 and 2 tell us that the additive identity and additive inverses are unique. Theorems 3 and 4 tell us that the multiplicative identity and multiplicative inverses of nonzero real numbers are also unique. We leave their proofs as exercises.

THEOREM 3 The multiplicative identity element 1 of the real numbers is unique.

THEOREM 4 The multiplicative inverse of a nonzero real number x is unique.

THEOREM 5 For every real number x, $x \cdot 0 = 0$.

Proof: Suppose that x is a real number. By the additive inverse law, there is a real number y that is the additive inverse of $x \cdot 0$, so we have $x \cdot 0 + y = 0$. By the additive identity law, $0 + 0 = 0$. Using the distributive law, we see that $x \cdot 0 = x \cdot (0 + 0) = x \cdot 0 + x \cdot 0$. It follows that

$$0 = x \cdot 0 + y = (x \cdot 0 + x \cdot 0) + y.$$

Next, note that by the associative law for addition and because $x \cdot 0 + y = 0$, it follows that

$$(x \cdot 0 + x \cdot 0) + y = x \cdot 0 + (x \cdot 0 + y) = x \cdot 0 + 0.$$

Finally, by the additive identity law, we know that $x \cdot 0 + 0 = x \cdot 0$. Consequently, $x \cdot 0 = 0$. ◁

THEOREM 6 For all real numbers x and y, if $x \cdot y = 0$, then $x = 0$ or $y = 0$.

Proof: Suppose that x and y are real numbers and $x \cdot y = 0$. If $x \neq 0$, then, by the multiplicative inverse law, x has a multiplicative inverse $1/x$, such that $x \cdot (1/x) = (1/x) \cdot x = 1$. Because $x \cdot y = 0$, we have $(1/x) \cdot (x \cdot y) = (1/x) \cdot 0 = 0$ by Theorem 5. Using the associate law for multiplication, we have $((1/x) \cdot x) \cdot y = 0$. This means that $1 \cdot y = 0$. By the multiplicative identity rule, we see that $1 \cdot y = y$, so $y = 0$. Consequently, either $x = 0$ or $y = 0$. ◁

THEOREM 7 The multiplicative identity element 1 in the set of real numbers is greater than the additive identity element 0.

Proof: By the trichotomy law, either $0 = 1$, $0 > 1$, or $1 > 0$. We know by the identity elements axiom that $0 \neq 1$.

So, assume that $0 > 1$. We will show that this assumption leads to a contradiction. By the additive inverse law, 1 has an additive inverse -1 with $1 + (-1) = 0$. The additive compatibility law tells us that $0 + (-1) > 1 + (-1) = 0$; the additive identity law tells us that $0 + (-1) = -1$. Consequently, $-1 > 0$, and by the multiplicative compatibility law, $(-1) \cdot (-1) > (-1) \cdot 0$. By Theorem 5, the right-hand side of last inequality is 0. By the distributive law, $(-1) \cdot (-1) + (-1) \cdot 1 = (-1) \cdot (-1 + 1) = (-1) \cdot 0 = 0$. Hence, the left-hand side of this last inequality, $(-1) \cdot (-1)$, is the unique additive inverse of -1, so this side of the inequality equals 1. Consequently, this last inequality becomes $1 > 0$, contradicting the trichotomy law because we had assumed that $0 > 1$.

Because we know that $0 \neq 1$ and that it is impossible for $0 > 1$, by the trichotomy law, we conclude that $1 > 0$. ◁

Links

Hulton Deutsch/Corbis Historical/Getty Images

ARCHIMEDES (287 B.C.E.–212 B.C.E.) Archimedes was one of the greatest scientists and mathematicians of ancient times. He was born in Syracuse, a Greek city-state in Sicily. His father, Phidias, was an astronomer. Archimedes was educated in Alexandria, Egypt. After completing his studies, he returned to Syracuse, where he spent the rest of his life. Little is known about his personal life; we do not know whether he was ever married or had children. Archimedes was killed in 212 B.C.E. by a Roman soldier when the Romans overran Syracuse.

Archimedes made many important discoveries in geometry. His method for computing the area under a curve was described two thousand years before his ideas were re-invented as part of integral calculus. Archimedes also developed a method for expressing large integers inexpressible by the usual Greek method. He discovered a method for computing the volume of a sphere, as well as of other solids, and he calculated an approximation of π. Archimedes was also an accomplished engineer and inventor; his machine for pumping water, now called *Archimedes' screw*, is still in use today. Perhaps his best-known discovery is the *principle of buoyancy*, which tells us that an object submerged in liquid becomes lighter by an amount equal to the weight it displaces. Some histories tell us that Archimedes was an early streaker, running naked through the streets of Syracuse shouting "Eureka" (which means "I have found it") when he made this discovery. He is also known for his clever use of machines that held off Roman forces sieging Syracuse for several years during the Second Punic War.

The next theorem tells us that for every real number there is an integer (where by an *integer*, we mean 0, the sum of any number of 1s, and the additive inverses of these sums) greater than this real number. This result is attributed to the Greek mathematician Archimedes. The result can be found in Book V of Euclid's *Elements*.

THEOREM 8 **ARCHIMEDEAN PROPERTY** For every real number x there exists an integer n such that $n > x$.

Proof: Suppose that x is a real number such that $n \leq x$ for every integer n. Then x is an upper bound of the set of integers. By the completeness property it follows that the set of integers has a least upper bound M. Because $M - 1 < M$ and M is a least upper bound of the set of integers, $M - 1$ is not an upper bound of the set of integers. This means that there is an integer n with $n > M - 1$. This implies that $n + 1 > M$, contradicting the fact that M is an upper bound of the set of integers. ◁

A1.3 Axioms for the Set of Positive Integers

The axioms we now list specify the set of positive integers as the subset of the set of integers satisfying four key properties. We assume the truth of these axioms in this textbook.

- **Axiom 1** The number 1 is a positive integer.
- **Axiom 2** If n is a positive integer, then $n + 1$, the *successor* of n, is also a positive integer.
- **Axiom 3** Every positive integer other than 1 is the successor of a positive integer.
- **Axiom 4** **The Well-Ordering Property** Every nonempty subset of the set of positive integers has a least element.

In Sections 5.1 and 5.2 it is shown that the well-ordering principle is equivalent to the principle of mathematical induction.

- **Mathematical induction axiom** If S is a set of positive integers such that $1 \in S$ and for all positive integers n if $n \in S$, then $n + 1 \in S$, then S is the set of positive integers.

Most mathematicians take the real number system as already existing, with the real numbers satisfying the axioms we have listed in this appendix. However, mathematicians in the nineteenth century developed techniques to construct the set of real numbers, starting with more basic sets of numbers. (The process of constructing the real numbers is sometimes studied in advanced undergraduate mathematics classes. A treatment of this can be found in [Mo91], for instance.) The first step in the process is the construction of the set of positive integers using axioms 1–3 and either the well-ordering property or the mathematical induction axiom. Then, the operations of addition and multiplication of positive integers are defined. Once this has been done, the set of integers can be constructed using equivalence classes of pairs of positive integers, where $(a, b) \sim (c, d)$ if and only if $a + d = b + c$; addition and multiplication of integers can be defined using these pairs (see Exercise 21). (Equivalence relations and equivalence classes are discussed in Chapter 9.) Next, the set of rational numbers can be constructed using the equivalence classes of pairs of integers where the second integer in the pair is not zero, where $(a, b) \approx (c, d)$ if and only if $a \cdot d = b \cdot c$; addition and multiplication of rational numbers can be defined in terms of these pairs (see Exercise 22). Using infinite sequences, the set of real numbers can then be constructed from the set of rational numbers. The interested reader will find it worthwhile to read through the many details of the steps of this construction.

Exercises

Use only the axioms and theorems in this appendix in the proofs in your answers to these exercises.

1. Prove Theorem 3, which states that the multiplicative identity element of the real numbers is unique.
2. Prove Theorem 4, which states that for every nonzero real number x, the multiplicative inverse of x is unique.
3. Prove that for all real numbers x and y, $(-x) \cdot y = x \cdot (-y) = -(x \cdot y)$.
4. Prove that for all real numbers x and y, $-(x + y) = (-x) + (-y)$.
5. Prove that for all real numbers x and y, $(-x) \cdot (-y) = x \cdot y$.
6. Prove that for all real numbers x, y, and z, if $x + z = y + z$, then $x = y$.
7. Prove that for every real number x, $-(-x) = x$.

Define the **difference** $x - y$ of real numbers x and y by $x - y = x + (-y)$, where $-y$ is the additive inverse of y, and the **quotient** x/y, where $y \neq 0$, by $x/y = x \cdot (1/y)$, where $1/y$ is the multiplicative inverse of y.

8. Prove that for all real numbers x and y, $x = y$ if and only if $x - y = 0$.
9. Prove that for all real numbers x and y, $-x - y = -(x + y)$.
10. Prove that for all nonzero real numbers x and y, $1/(x/y) = y/x$, where $1/(x/y)$ is the multiplicative inverse of x/y.
11. Prove that for all real numbers w, x, y, and z, if $x \neq 0$ and $z \neq 0$, then $(w/x) + (y/z) = (w \cdot z + x \cdot y)/(x \cdot z)$.
12. Prove that for every positive real number x, $1/x$ is also a positive real number.
13. Prove that for all positive real numbers x and y, $x \cdot y$ is also a positive real number.
14. Prove that for all real numbers x and y, if $x > 0$ and $y < 0$, then $x \cdot y < 0$.
15. Prove that for all real numbers x, y, and z, if $x > y$ and $z < 0$, then $x \cdot z < y \cdot z$.
16. Prove that for every real number x, $x \neq 0$ if and only if $x^2 > 0$.
17. Prove that for all real numbers w, x, y, and z, if $w < x$ and $y < z$, then $w + y < x + z$.
18. Prove that for all positive real numbers x and y, if $x < y$, then $1/x > 1/y$.
19. Prove that for every positive real number x, there exists a positive integer n such that $n \cdot x > 1$.

*20. Prove that between every two distinct real numbers there is a rational number (that is, a number of the form x/y, where x and y are integers with $y \neq 0$).

Exercises 21 and 22 involve the notion of an equivalence relation, discussed in Chapter 9 of the text.

*21. Define a relation $\sim$ on the set of ordered pairs of positive integers by $(w, x) \sim (y, z)$ if and only if $w + z = x + y$. Show that the operations $[(w, x)]_{\sim} + [(y, z)]_{\sim} = [(w + y, x + z)]_{\sim}$ and $[(w, x)]_{\sim} \cdot [(y, z)]_{\sim} = [(w \cdot y + x \cdot z, x \cdot y + w \cdot z)]_{\sim}$ are well-defined, that is, they do not depend on the representative of the equivalence classes chosen for the computation.

*22. Define a relation $\approx$ on ordered pairs of integers with second entry nonzero by $(w, x) \approx (y, z)$ if and only if $w \cdot z = x \cdot y$. Show that the operations $[(w, x)]_{\approx} + [(y, z)]_{\approx} = [(w \cdot z + x \cdot y, x \cdot z)]_{\approx}$ and $[(w, x)]_{\approx} \cdot [(y, z)]_{\approx} = [(w \cdot y, x \cdot z)]_{\approx}$ are well-defined, that is, they do not depend on the representative of the equivalence classes chosen for the computation.

APPENDIX

2 Exponential and Logarithmic Functions

In this appendix we review some of the basic properties of exponential functions and logarithms. These properties are used throughout the text. Students requiring further review of this material should consult precalculus or calculus books, such as those mentioned in the Suggested Readings.

A2.1 Exponential Functions

Let n be a positive integer, and let b be a fixed positive real number. The function $f_b(n) = b^n$ is defined by

$$f_b(n) = b^n = b \cdot b \cdot b \cdot \cdots \cdot b,$$

where there are n factors of b multiplied together on the right-hand side of the equation.

We can define the function $f_b(x) = b^x$ for all real numbers x using techniques from calculus. The function $f_b(x) = b^x$ is called the **exponential function to the base *b***. We will not discuss how to find the values of exponential functions to the base b when x is not an integer.

Two of the important properties satisfied by exponential functions are given in Theorem 1. Proofs of these and other related properties can be found in calculus texts.

THEOREM 1 Let b be a positive real number and x and y real numbers. Then

1. $b^{x+y} = b^x b^y$, and
2. $(b^x)^y = b^{xy}$.

We display the graphs of some exponential functions in Figure 1.

A2.2 Logarithmic Functions

Suppose that b is a real number with $b > 1$. Then the exponential function b^x is strictly increasing (a fact shown in calculus). It is a one-to-one correspondence from the set of real numbers to the set of nonnegative real numbers. Hence, this function has an inverse $\log_b x$, called the **logarithmic function to the base *b***. In other words, if b is a real number greater than 1 and x is a positive real number, then

$$b^{\log_b x} = x.$$

The value of this function at x is called the **logarithm of *x* to the base *b***.

From the definition, it follows that

$$\log_b b^x = x.$$

We give several important properties of logarithms in Theorem 2.

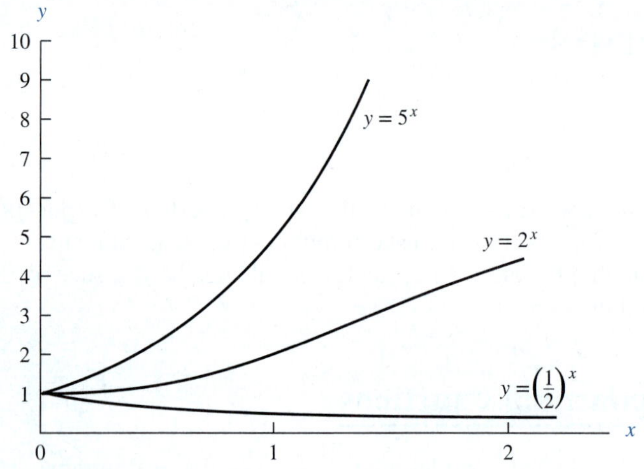

FIGURE 1 **Graphs of the exponential functions to the bases $\frac{1}{2}$, 2, and 5.**

THEOREM 2 Let b be a real number greater than 1. Then

1. $\log_b(xy) = \log_b x + \log_b y$ whenever x and y are positive real numbers, and
2. $\log_b(x^y) = y \log_b x$ whenever x is a positive real number and y is a real number.

Proof: Because $\log_b(xy)$ is the unique real number with $b^{\log_b(xy)} = xy$, to prove part 1 it suffices to show that $b^{\log_b x + \log_b y} = xy$. By part 1 of Theorem 1, we have

$$\begin{aligned} b^{\log_b x + \log_b y} &= b^{\log_b x} b^{\log_b y} \\ &= xy. \end{aligned}$$

To prove part 2, it suffices to show that $b^{y \log_b x} = x^y$. By part 2 of Theorem 1, we have

$$\begin{aligned} b^{y \log_b x} &= (b^{\log_b x})^y \\ &= x^y. \end{aligned}$$ ◁

The following theorem relates logarithms to two different bases.

THEOREM 3 **CHANGE OF BASE FORMULA FOR LOGARITHMS** Let a and b be real numbers greater than 1, and let x be a positive real number. Then

$$\log_a x = \log_b x / \log_b a.$$

Proof: To prove this result, it suffices to show that

$$b^{\log_a x \cdot \log_b a} = x.$$

By part 2 of Theorem 1, we have

$$\begin{aligned} b^{\log_a x \cdot \log_b a} &= (b^{\log_b a})^{\log_a x} \\ &= a^{\log_a x} \\ &= x. \end{aligned}$$

This completes the proof. ◁

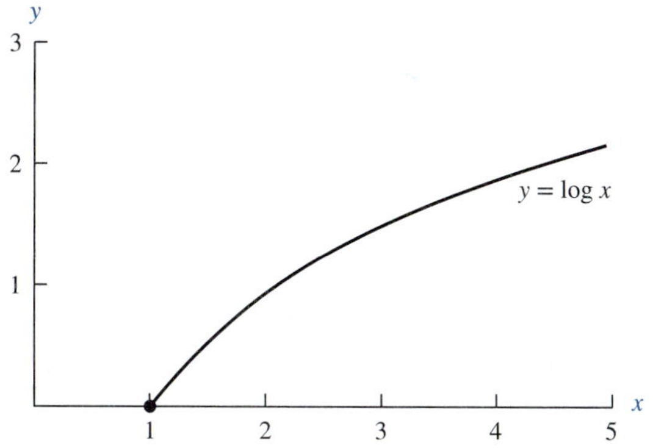

FIGURE 2 **The graph of $f(x) = \log x$.**

Because the base used most often for logarithms in this text is $b = 2$, the notation $\log x$ is used throughout the test to denote $\log_2 x$.

The graph of the function $f(x) = \log x$ is displayed in Figure 2. From Theorem 3, when a base b other than 2 is used, a function that is a constant multiple of the function $\log x$, namely, $(1/\log b)\log x$, is obtained.

Exercises

1. Express each of the following quantities as powers of 2.

a) $2 \cdot 2^2$ **b)** $(2^2)^3$ **c)** $2^{(2^2)}$

2. Find each of the following quantities.

a) $\log_2 1024$ **b)** $\log_2 1/4$ **c)** $\log_4 8$

3. Suppose that $\log_4 x = y$, where x is a positive real number. Find each of the following quantities.

a) $\log_2 x$ **b)** $\log_8 x$ **c)** $\log_{16} x$

4. Let a, b, and c be positive real numbers. Show that $a^{\log_b c} = c^{\log_b a}$.

5. Draw the graph of $f(x) = b^x$ for all real numbers x if b is

a) 3. **b)** $1/3$. **c)** 1.

6. Draw the graph of $f(x) = \log_b x$ for positive real numbers x if b is

a) 4. **b)** 100. **c)** 1000.

APPENDIX

3 Pseudocode

The algorithms in this text are described both in English and in **pseudocode**. Pseudocode is an intermediate step between an English language description of the steps of a procedure and a specification of this procedure using an actual programming language. The advantages of using pseudocode include the simplicity with which it can be written and understood and the ease of producing actual computer code (in a variety of programming languages) from the pseudocode. We will describe the particular types of **statements**, or high-level instructions, of the pseudocode that we will use. Each of these statements in pseudocode can be translated into one or more statements in a particular programming language, which in turn can be translated into one or more (possibly many) low-level instructions for a computer.

Links

This appendix describes the format and syntax of the pseudocode used in the text. This pseudocode is designed so that its basic structure resembles that of commonly used programming languages, such as C++ and Java, which are currently the most commonly taught programming languages. However, the pseudocode we use will be a lot looser than a formal programming language because a lot of English language descriptions of steps will be allowed.

This appendix is not meant for formal study. Rather, it should serve as a reference guide for students when they study the descriptions of algorithms given in the text and when they write pseudocode solutions to exercises.

A3.1 Procedure Statements

The pseudocode for an algorithm begins with a **procedure** statement that gives the name of an algorithm, lists the input variables, and describes what kind of variable each input is. For instance, the statement

procedure *maximum*(L: list of integers)

is the first statement in the pseudocode description of the algorithm, which we have named *maximum*, that finds the maximum of a list L of integers.

A3.2 Assignments and Other Types of Statements

An assignment statement is used to assign values to variables. In an assignment statement the left-hand side is the name of the variable and the right-hand side is an expression that involves constants, variables that have been assigned values, or functions defined by procedures. The right-hand side may contain any of the usual arithmetic operations. However, in the pseudocode in this book it may include any well-defined operation, even if this operation can be carried out only by using a large number of statements in an actual programming language.

The symbol := is used for assignments. Thus, an assignment statement has the form

variable := *expression*

For example, the statement

$max := a$

assigns the value of a to the variable max. A statement such as

$x :=$ largest integer in the list L

can also be used. This sets x equal to the largest integer in the list L. To translate this statement into an actual programming language would require more than one statement. Also, the instruction

interchange a and b

can be used to interchange a and b. We could also express this one statement with several assignment statements (see Exercise 2), but for simplicity, we will often prefer this abbreviated form of pseudocode.

A3.3 Comments

In the pseudocode in this book, statements enclosed in curly braces are not executed. Such statements serve as comments or reminders that help explain how the procedure works. For instance, the statement

$\{x$ is the largest element in $L\}$

can be used to remind the reader that at that point in the procedure the variable x equals the largest element in the list L.

A3.4 Conditional Constructions

The simplest form of the conditional construction that we will use is

if condition **then** statement

or

if condition **then**
    block of statements

Here, the condition is checked, and if it is true, then the statement or block of statements given is carried out. In particular, the pseudocode

```
if condition then
    statement 1
    statement 2
    statement 3
    .
    .
    .
    statement n
```

tells us that the statements in the block are executed sequentially if the condition is true.

For example, in Algorithm 1 in Section 3.1, which finds the maximum of a set of integers, we use a conditional statement to check whether $max < a_i$ for each variable; if it is, we assign the value of a_i to max.

Often, we require the use of a more general type of construction. This is used when we wish to do one thing when the indicated condition is true, but another when it is false. We use the construction

```
if condition then statement 1
else statement 2
```

Note that either one or both of statement 1 and statement 2 can be replaced with a block of statements.

Sometimes, we require the use of an even more general form of a conditional. The general form of the conditional construction that we will use is

```
if condition 1 then statement 1
else if condition 2 then statement 2
else if condition 3 then statement 3
    .
    .
    .
else if condition n then statement n
else statement n + 1
```

When this construction is used, if condition 1 is true, then statement 1 is carried out, and the program exits this construction. In addition, if condition 1 is false, the program checks whether condition 2 is true; if it is, statement 2 is carried out, and so on. Thus, if none of the first $n - 1$ conditions hold, but condition n does, statement n is carried out. Finally, if none of condition 1, condition 2, condition 3, ..., condition n is true, then statement $n + 1$ is executed. Note that any of the $n + 1$ statements can be replaced by a block of statements.

A3.5 Loop Constructions

There are two types of loop construction in the pseudocode in this book. The first is the "for" construction, which has the form

for *variable* := *initial value* **to** *final value*
    statement

or

for *variable* := *initial value* **to** *final value*
    block of statements

where *initial value* and *final value* are integers. Here, at the start of the loop, *variable* is assigned *initial value* if *initial value* is less than or equal to *final value*, and the statements at the end of this construction are carried out with this value of *variable*. Then *variable* is increased by one, and the statement, or the statements in the block, are carried out with this new value of *variable*. This is repeated until *variable* reaches *final value*. After the instructions are carried out with *variable* equal to *final value*, the algorithm proceeds to the next statement. When *initial value* exceeds *final value*, none of the statements in the loop is executed.

We can use the "for" loop construction to find the sum of the positive integers from 1 to n with the following pseudocode.

$sum := 0$
for $i := 1$ **to** n
    $sum := sum + i$

Also, the more general "for" statement, of the form

for all elements with a certain property

is used in this text. This means that the statement or block of statements that follow are carried out successively for the elements with the given property.

The second type of loop construction that we will use is the "while" construction. This has the form

while condition
    statement

or

while condition
    block of statements

When this construction is used, the condition given is checked, and if it is true, the statements that follow are carried out, which may change the values of the variables that are part of the condition. If the condition is still true after these instructions have been carried out, the instructions are carried out again. This is repeated until the condition becomes false. As an example, we can find the sum of the integers from 1 to n using the following block of pseudocode including a "while" construction.

```
sum := 0
while n > 0
    sum := sum + n
    n := n − 1
```

Note that any "for" construction can be turned into a "while" construction (see Exercise 3). However, it is often easier to understand the "for" construction. So, when it makes sense, we will use the "for" construction in preference to the corresponding "while" construction.

A3.6 Loops within Loops

Loops or conditional statements are often used within other loops or conditional statements. In the pseudocode used in this book, we use successive levels of indentation to indicate nested loops, which are loops within loops, and which blocks of commands correspond to which loops.

A3.7 Using Procedures in Other Procedures

We can use a procedure from within another procedure (or within itself in a recursive program) simply by writing the name of this procedure followed by the inputs to this procedure. For instance,

```
max(L)
```

will carry out the procedure *max* with the input list L. After all the steps of this procedure have been carried out, execution carries on with the next statement in the procedure.

A3.8 Return Statements

We use a **return** statement to show where a procedure produces output. A return statement of the form

```
return x
```

produces the current value of x as output. The output x can involve the value of one or more functions, including the same function under evaluation, but at a smaller value. For instance, the statement

```
return f(n − 1)
```

is used to call the algorithm with input of $n - 1$. This means that the algorithm is run again with input equal to $n - 1$.

Exercises

1. What is the difference between the following blocks of two assignment statements?

   $a := b$
   $b := c$

   and

   $b := c$
   $a := b$

2. Give a procedure using assignment statements to interchange the values of the variables x and y. What is the minimum number of assignment statements needed to do this?

3. Show how a loop of the form

   for $i :=$ *initial value* **to** *final value*
   statement

   can be written using the "while" construction.

Suggested Reading

Among the resources available for learning more about the topics covered in this book are printed materials and relevant websites. Printed resources, including those referred to in the text, are described in this section of suggested readings. Readings are listed by chapter and keyed to particular topics of interest. Some general references are also provided. A book you may find particularly useful is the *Handbook of Discrete and Combinatorial Mathematics* by Rosen [Ro18], a comprehensive reference book for the material in this book and many other topics. Additional applications of discrete mathematics can be found in Michaels and Rosen [MiRo91], which is also available online on the companion website for this book. Deeper coverage of many topics in computer science, including those discussed in this book, can be found in Gruska [Gr97]. Biographical information about many of the mathematicians and computer scientists mentioned in this book can be found in Gillispie [Gi70] and on the MacTutor website at *http://www-history.mcs.st-and.ac.uk/*.

To find pertinent websites, consult the links found in the Web Resources Guide on the companion website for this book. Its address is *www.mhhe.com/rosen*.

CHAPTER 1

An entertaining way to study logic is to read Lewis Carroll's book [Ca78]. General references for logic include M. Huth and M. Ryan [HuRy04], Mendelson [Me09], Stoll [St74], and Suppes [Su99]. A comprehensive treatment of logic in discrete mathematics can be found in Gries and Schneider [GrSc93]. System specifications are discussed in Ince [In93]. Smullyan's knights and knaves puzzles were introduced in [Sm78]. He wrote many fascinating books on logic puzzles including [Sm92] and [Sm98]. Prolog is discussed in depth in Nilsson and Maluszynski [NiMa95] and in Clocksin and Mellish [ClMe94]. The basics of proofs are covered in Cupillari [Cu05], Morash [Mo91], Solow [So09], Velleman [Ve06], and Wolf [Wo98]. The science and art of constructing proofs is discussed in a delightful way in three books by Pólya: [Po62], [Po71], and [Po90]. Problems involving tiling checkerboards using dominoes and polyominoes are discussed in Golomb [Go94] and Martin [Ma91].

CHAPTER 2

Lin and Lin [LiLi81] is an easily read text on sets and their applications. Axiomatic developments of set theory can be found in Halmos [Ha60], Monk [Mo69], Pinter [Pi14], and Stoll [St74]. Brualdi [Br09], and Reingold, Nievergelt, and Deo [ReNiDe77] contain introductions to multisets. Fuzzy sets and their application to expert systems and artificial intelligence are treated in Negoita [Ne85] and Zimmerman [Zi91]. Calculus books, such as Apostol [Ap67], Spivak [Sp94], and Thomas and Finney [ThFi96], contain discussions of functions. The best printed source of information about integer sequences is Sloan and Plouffe [SlPl95]. Books on proofs, such as [Ve06], often cover countability in some depth. Stanat and McAllister [StMc77] has a thorough section on countability. Chapter 17 of Aigner and Ziegler [AiZi18] provides an excellent discussion of cardinality and the continuum hypothesis. Discussions of the mathematical foundations needed for computer science can be found in Arbib, Kfoury, and Moll [ArKfMo80]; Bobrow and Arbib [BoAr74]; Beckman [Be80]; and Tremblay and Manohar [TrMa75]. Matrices and their operations are covered in all linear algebra books, such as Curtis [Cu84] and Strang [St09].

CHAPTER 3

The articles by Knuth [Kn77] and Wirth [Wi84] are accessible introductions to the subject of algorithms. Among the best introductions to algorithms are Cormen, Leiserson, Rivest, and Stein [CoLeRiSt09] and Kleinberg and Tardos [KlTa05]. Extensive material on big-O estimates of functions can be found in Knuth [Kn97a]. General references for algorithms and their complexity include Aho, Hopcroft, and Ullman [AhHoUl74]; Baase and Van Gelder [BaGe99]; Cormen, Leiserson, Rivest, and Stein [CoLeRiSt09]; Gonnet [Go84]; Goodman and Hedetniemi [GoHe77]; Harel [Ha87]; Horowitz and Sahni [HoSa82]; Kreher and Stinson [KrSt98]; the famous series of books by Knuth on the art of computer programming [Kn97a], [Kn97b], and [Kn98]; Kronsjö [Kr87]; Levitin [Le06]; Manber [Ma89]; Pohl and Shaw [PoSh81]; Purdom and Brown [PuBr85]; Rawlins [Ra92]; Sedgewick [Se03]; Wilf [Wi02]; and Wirth [Wi76]. Sorting and searching algorithms and their complexity are studied in detail in Knuth [Kn98].

CHAPTER 4

References for number theory include Hardy and Wright [HaWrWiHe08], LeVeque [Le77], Rosen [Ro23], and Stark [St78]. More about the history of number theory can be found in Ore [Or88]. Algorithms for computer arithmetic are discussed in Knuth [Kn97b] and Pohl and Shaw [PoSh81]. More information about algorithms for finding primes and for factorization can be found in

Crandall and Pomerance [CrPo10]. Applications of number theory to cryptography are covered in Denning [De82]; Menezes, van Oorschot, and Vanstone [MeOoVa97]; Rosen [Ro23]; Seberry and Pieprzyk [SePi89]; Sinkov [Si66]; and Stinson [St05]. The RSA public-key system was described by Rivest, Shamir, and Adleman in [RiShAd78]; its discovery by Cocks is described in [Si99], which also provides an appealing account of the history of cryptography.

CHAPTER 5

An accessible introduction to mathematical induction can be found in [Gu10] and in Sominskii [So61]. Books that contain thorough treatments of mathematical induction and recursive definitions include Liu [Li85], Sahni [Sa85], Stanat and McAllister [StMc77], and Tremblay and Manohar [TrMa75]. Computational geometry is covered in [DeOr11] and [Or00]. The Ackermann function, introduced in 1928 by W. Ackermann, arises in the theory of recursive function (see Beckman [Be80] and McNaughton [Mc82], for instance) and in the analysis of the complexity of certain set theoretic algorithms (see Tarjan [Ta83]). Recursion is studied in Roberts [Ro86], Rohl [Ro84], and Wand [Wa80]. Discussions of program correctness and the logical machinery used to prove that programs are correct can be found in Alagic and Arbib [AlAr78], Anderson [An79], Backhouse [Ba86], Sahni [Sa85], and Stanat and McAllister [StMc77].

CHAPTER 6

General references for counting techniques and their applications include Allenby and Slomson [AlSl10]; Anderson [An89]; Berman and Fryer [BeFr72]; Bogart [Bo00]; Bona [Bo07]; Bose and Manvel [BoMa86]; Brualdi [Br09]; Cohen [Co78]; Grimaldi [Gr03]; Gross [Gr07]; Liu [Li68]; Pólya, Tarjan, and Woods [PoTaWo83]; Riordan [Ri58]; Roberts and Tesman [RoTe03]; Tucker [Tu06]; and Williamson [Wi85]. Vilenkin [Vi71] contains a selection of combinatorial problems and their solutions. A selection of more difficult combinatorial problems can be found in Lovász [Lo79]. Information about Internet protocol addresses and datagrams can be found in Comer [Co05]. Applications of the pigeonhole principle can be found in Brualdi [Br09], Liu [Li85], and Roberts and Tesman [RoTe03]. A wide selection of combinatorial identities can be found in Riordan [Ri68] and in Benjamin and Quinn [BeQu03]. Combinatorial algorithms, including algorithms for generating permutations and combinations, are described by Even [Ev73]; Lehmer [Le64]; and Reingold, Nievergelt, and Deo [ReNiDe77].

CHAPTER 7

Useful references for discrete probability theory include Feller [Fe68], Nabin [Na00], and Ross [Ro09a]. Ross [Ro02], which focuses on the application of probability theory to computer science, provides examples of average case complexity analysis and covers the probabilistic method. Aho and Ullman [AhUl95] includes a discussion of various aspects of probability theory important in computer science, including programming applications of probability. The probabilistic method is discussed in a chapter in Aigner and Ziegler [AiZi18], a monograph devoted to clever, insightful, and brilliant proofs, that is, proofs that Paul Erdős described as coming from "*The Book.*" Extensive coverage of the probabilistic method can be found in Alon and Spencer [AlSp00]. Bayes' theorem is covered in [PaPi01]. Additional material on spam filters can be found in [Zd05].

CHAPTER 8

Many different models using recurrence relations can be found in Roberts and Tesman [RoTe03] and Tucker [Tu06]. Exhaustive treatments of linear homogeneous recurrence relations with constant coefficients, and related inhomogeneous recurrence relations, can be found in Brualdi [Br09], Liu [Li68], and Mattson [Ma93]. Divide-and-conquer algorithms and their complexity are covered in Roberts and Tesman [RoTe03] and Stanat and McAllister [StMc77]. Descriptions of fast multiplication of integers and matrices can be found in Aho, Hopcroft, and Ullman [AhHoUl74] and Knuth [Kn97b]. An excellent introduction to generating functions can be found in Pólya, Tarjan, and Woods [PoTaWo83]. Generating functions are studied in detail in Brualdi [Br09]; Cohen [Co78]; Graham, Knuth, and Patashnik [GrKnPa94]; Grimaldi [Gr03]; and Roberts and Tesman [RoTe03]. Additional applications of the principle of inclusion–exclusion can be found in Liu [Li85] and [Li68], Roberts and Tesman [RoTe03], and Ryser [Ry63].

CHAPTER 9

General references for relations, including treatments of equivalence relations and partial orders, include Bobrow and Arbib [BoAr74], Grimaldi [Gr03], Sanhi [Sa85], and Tremblay and Manohar [TrMa75]. Discussions of relational models for databases are given in Date [Da82] and Aho and Ullman [AhUl95]. The original papers by Roy and Warshall for finding transitive closures can be found in [Ro59] and [Wa62], respectively. Directed graphs are studied in Chartrand, Lesniak, and Zhang [ChLeZh15]; Gross and Yellen [GrYe05]; Robinson and Foulds [RoFo80]; Roberts and Tesman [RoTe03]; and Tucker [Tu06]. The application of lattices to information flow is treated in Denning [De82].

CHAPTER 10

General references for graph theory include Agnarsson and Greenlaw [AgGr06]; Aldous, Wilson, and Best [AlWiBe00]; Behzad and Chartrand [BeCh71]; Chartrand, Lesniak, and Zhang [ChLeZh15]; Chartrand and Zhang [ChZh04]; Bondy and Murty [BoMu10]; Chartrand and Oellermann [ChOe93]; Graver and Watkins [GrWa77]; Roberts and Tesman [RoTe03]; Tucker [Tu06]; West [We17]; Wilson [Wi85]; and Wilson and Watkins [WiWa90]. A wide variety of applications of graph theory can be found in Chartrand [Ch77], Deo [De74], Foulds [Fo92], Roberts and Tesman [RoTe03], Roberts [Ro76], Wilson and Beineke [WiBe79], and McHugh [Mc90]. In-depth treatments of the use of graph theory to study social networks, and other types of networks, appears in Easley and Kleinberg [EaKl10] and Newman [Ne10]. Applications involving large graphs, including the Web graph, are discussed in Hayes [Ha00a] and [Ha00b].

A comprehensive description of algorithms in graph theory can be found in Gibbons [Gi85] and in Kocay and Kreher [KoKr04]. Other references for algorithms in graph theory include Buckley and Harary [BuHa90]; Chartrand and Oellermann [ChOe93]; Chachra, Ghare, and Moore [ChGhMo79]; Even [Ev73] and [Ev79]; Hu [Hu82]; and Reingold, Nievergelt, and Deo [ReNiDe77]. A translation of Euler's original paper on the Königsberg bridge problem can be found in Euler [Eu53]. Dijkstra's algorithm is studied in Gibbons [Gi85]; Liu [Li85]; and Reingold, Nievergelt, and Deo [ReNiDe77]. Dijkstra's original paper can be found in [Di59]. A proof of Kuratowski's theorem can be found in Harary [Ha69] and Liu [Li68]. Crossing numbers and thicknesses of graphs are studied in Chartrand, Lesniak, and Zhang [ChLeZh15]. References for graph coloring and the four-color theorem are included in Barnette [Ba83] and Saaty and Kainen [SaKa86]. The original conquest of the four-color theorem is reported in Appel and Haken [ApHa76]. Applications of graph coloring are described by Roberts and Tesman [RoTe03]. The history of graph theory is covered in Biggs, Lloyd, and Wilson [BiLlWi86]. Interconnection networks for parallel processing are discussed in Akl [Ak89] and Siegel and Hsu [SiHs88].

CHAPTER 11

Trees are studied in Deo [De74], Grimaldi [Gr03], Knuth [Kn97a], Roberts and Tesman [RoTe03], and Tucker [Tu06]. The use of trees in computer science is described by Gotlieb and Gotlieb [GoGo78], Horowitz and Sahni [HoSa82], and Knuth [Kn97a] and [Kn98]. Roberts and Tesman [RoTe03] covers applications of trees to many different areas. Combinatorial game theory is studied in [AlNoWo07], [BeCoGu01], and [Be11]. The tree of fractions and its use to show that the set of rational numbers is countable is discussed in [CaWi20]. Prefix codes and Huffman coding are covered in Hamming [Ha80]. Backtracking is an old technique; its use to solve maze puzzles can be found in the 1891 book by Lucas [Lu91]. An extensive discussion of how to solve problems using backtracking can be found in Reingold, Nievergelt, and Deo [ReNiDe77]. Gibbons [Gi85] and Reingold, Nievergelt, and Deo [ReNiDe77] contain discussions of algorithms for constructing spanning trees and minimal spanning trees. The background and history of algorithms for finding minimal spanning trees is covered in Graham and Hell [GrHe85]. Prim and Kruskal described their algorithms for finding minimal spanning trees in [Pr57] and [Kr56], respectively. Sollin's algorithm is an example of an algorithm well suited for parallel processing; although Sollin never published a description of it, his algorithm has been described by Even [Ev73] and Goodman and Hedetniemi [GoHe77].

CHAPTER 12

Boolean algebra is studied in Hohn [Ho66], Kohavi [Ko86], and Tremblay and Manohar [TrMa75]. Applications of Boolean algebra to logic circuits and switching circuits are described by Hayes [Ha93], Hohn [Ho66], Katz and Borriello [KaBo04], and Kohavi [Ko86]. The original papers dealing with the minimization of sum-of-products expansions using maps are Karnaugh [Ka53] and Veitch [Ve52]. The Quine-McCluskey method was introduced in McCluskey [Mc56] and Quine [Qu52] and [Qu55]. Threshold functions are covered in Kohavi [Ko86].

CHAPTER 13

General references for formal grammars, automata theory, and the theory of computation include Davis, Sigal, and Weyuker [DaSiWe94]; Denning, Dennis, and Qualitz [DeDeQu81]; Hopcroft, Motwani, and Ullman [HoMoUl06]; Hopkin and Moss [HoMo76]; Lewis and Papadimitriou [LePa97]; McNaughton [Mc82]; and Sipser [Si06]. Mealy machines and Moore machines were originally introduced in Mealy [Me55] and Moore [Mo56]. The original proof of Kleene's theorem can be found in [Kl56]. Powerful models of computation, including pushdown automata and Turing machines, are discussed in Brookshear [Br89], Hennie [He77], Hopcroft and Ullman [HoUl79], Hopkin and Moss [HoMo76], Martin [Ma03], Sipser [Si06], and Wood [Wo87]. Barwise and Etchemendy [BaEt93] is an excellent introduction to Turing machines. Interesting articles about the history and application of Turing machines and related machines can be found in Herken [He88]. Busy beaver machines were first introduced by Rado in [Ra62], and information about them can be found in Dewdney [De84] and [De93], the article by Brady in Herken [He88], and in Wood [Wo87].

APPENDIXES

A discussion of axioms for the real number and for the integers can be found in Morash [Mo91]. Detailed treatments of exponential and logarithmic functions can be found in calculus books such as Apostol [Ap67], Spivak [Sp94], and Thomas and Finney [ThFi96]. Pohl and Shaw [PoSh81] use a form of pseudocode that has the same features as those described in Appendix 3. Most textbooks on algorithms, such as Cormen, Leiserson, Rivest, and Stein [CoLeRiSt09] and Kleinberg and Tardos [KlTa05], use versions of pseudocode similar to the pseudocode in this text.

REFERENCES

[AgGr06] G. Agnarsson and R. Greenlaw, *Graph Theory: Modeling, Applications, and Algorithms*, Prentice Hall, Englewood Cliffs, NJ, 2006.

[Ag15] C. C. Aggarwal, *Data Mining, The Textbook*, Springer, New York, 2015.

[AhHoUl74] A. V. Aho, J. E. Hopcroft, and J. D. Ullman, *The Design and Analysis of Computer Algorithms*, Addison-Wesley, Reading, MA, 1974.

[AhUl95] Alfred V. Aho and Jeffrey D. Ullman, *Foundations of Computer Science, C Edition*, Computer Science Press, New York, 1995.

[AiZi18] Martin Aigner and Günter M. Ziegler, *Proofs from THE BOOK*, 6th ed., Springer Berlin, Heidelberg, 2018.

[Ak89] S. G. Akl, *The Design and Analysis of Parallel Algorithms*, Prentice Hall, Englewood Cliffs, NJ, 1989.

[AlAr78] S. Alagic and M. A. Arbib, *The Design of Well-Structured and Correct Programs*, Springer-Verlag, New York, 1978.

[AlNoWo07] Michael H. Albert, Richard J. Nowakowski, and David Wolfe, *Lessons in Play: An Introduction to Combinatorial Game Theory*, A. K. Peters, Natick, MA, 2007.

[AlWiBe00] J. M. Aldous, R. J. Wilson, and S. Best, *Graphs and Applications: An Introductory Approach*, Springer, New York, 2000.

[AlSl10] R. B. J. T. Allenby and A. Slomson, *How to Count: An Introduction to Combinatorics*, 2nd ed., Chapman and Hall/CRC, Boca Raton, Florida, 2010.

[AlSp00] Noga Alon and Joel H. Spencer, *The Probabilistic Method*, 2nd ed., Wiley, New York, 2000.

[An89] I. Anderson, *A First Course in Combinatorial Mathematics*, 2nd ed., Oxford University Press, New York, 1989.

[An79] R. B. Anderson, *Proving Programs Correct*, Wiley, New York, 1979.

[Ap67] T. M. Apostol, *Calculus*, Vol. I, 2nd ed., Wiley, New York, 1967.

[ApHa76] K. Appel and W. Haken, "Every Planar Map Is 4-colorable," *Bulletin of the AMS*, 82 (1976), 711–712.

[ArKfMo80] M. A. Arbib, A. J. Kfoury, and R. N. Moll, *A Basis for Theoretical Computer Science*, Springer-Verlag, New York, 1980.

[AvCh90] B. Averbach and O. Chein, *Problem Solving Through Recreational Mathematics*, W. H. Freeman, San Francisco, 1980.

[BaGe99] S. Baase and A. Van Gelder, *Computer Algorithms: Introduction to Design and Analysis*, 3rd ed., Addison-Wesley, Reading, MA, 1999.

[Ba86] R. C. Backhouse, *Program Construction and Verification*, Prentice Hall, Englewood Cliffs, NJ, 1986.

[Ba83] D. Barnette, *Map Coloring, Polyhedra, and the Four-Color Problem*, Mathematical Association of America, Washington, DC, 1983.

[BaEt93] Jon Barwise and John Etchemendy, *Turing's World 3.0 for the Macintosh*, CSLI Publications, Stanford, CA, 1993.

[Be11] József Beck, *Combinatorial Games: Tic-Tac-Toe Theory*, Cambridge, 2011.

[Be80] F. S. Beckman, *Mathematical Foundations of Programming*, Addison-Wesley, Reading, MA, 1980.

[BeCh71] M. Behzad and G. Chartrand, *Introduction to the Theory of Graphs*, Allyn & Bacon, Boston, 1971.

[BeQu03] A. Benjamin and J. J. Quine, *Proofs that Really Count*, Mathematical Association of America, Washington, DC, 2003.

[Be86] J. Bentley, *Programming Pearls*, Addison-Wesley, Reading, MA, 1986.

[BeFr72] G. Berman and K. D. Fryer, *Introduction to Combinatorics*, Academic Press, New York, 1972.

[BeCoGu01] Elwyn R. Berlekamp, John H. Conwayk, and Richard K. Guy, *Winning Ways for Your Mathematical Plays*, Volumes 1–4, 2001–2004.

[Bi19] M. Bishop, *Computer Security: Art and Science*, 2nd ed., Addison-Wesley, Boston, 2019.

[BiLlWi99] N. L. Biggs, E. K. Lloyd, and R. J. Wilson, *Graph Theory 1736–1936*, Oxford University Press, Oxford, England, 1999.

[BoAr74] L. S. Bobrow and M. A. Arbib, *Discrete Mathematics*, Saunders, Philadelphia, 1974.

[Bo00] K. P. Bogart, *Introductory Combinatorics*, 3rd ed. Academic Press, San Diego, 2000.

[Bo07] M. Bona, *Enumerative Combinatorics*, McGraw Hill, New York, 2007.

[BoMu10] J. A. Bondy and U. S. R. Murty, *Graph Theory with Applications*, Springer, New York, 2010.

[Bo04] P. Bork, L. J. Jensen, C. von Mering, A. K. Ramani, I. Lee, and E. M. Marcotte, "Protein interaction networks from yeast to human," *Current Opinion in Structural Biology*, 14 (2004), 292–299.

[BoMa86] R. C. Bose and B. Manvel, *Introduction to Combinatorial Theory*, Wiley, New York, 1986.

[Bo14] T. Bousch, "La quatrieme tour de Hanoi," *Bulletin of the Belgian Mathematical Society Simon Stevin*, 21 (2014) 895–912.

[Bo00] A. Brodera, R. Kumar, F. Maghoula, P. Raghavan, S. Rajagopalan, R. Statac, A. Tomkins, and Janet Wiener, "Graph structure in the Web," *Computer Networks*, 33 (2000), 309–320.

[Br89] J. G. Brookshear, *Theory of Computation*, Benjamin Cummings, Redwood City, CA, 1989.

[Br09] R. A. Brualdi, *Introductory Combinatorics*, 5th ed., Prentice Hall, Englewood Cliffs, NJ, 2009.

[BuHa90] F. Buckley and F. Harary, *Distance in Graphs*, Addison-Wesley, Redwood City, CA, 1990.

[Ca79] L. Carmony, "Odd Pie Fights," *Mathematics Teacher* 72 (January, 1979), 61–64.

[Ca78] L. Carroll, *Symbolic Logic*, Crown, New York, 1978.

[CaWi20] Calkin, Neil J. and Wilf, Herbert S., "Recounting the rationals," *American Mathematical Monthly*, 107 (2000), 360–363.

[ChGhMo79] V. Chachra, P. M. Ghare, and J. M. Moore, *Applications of Graph Theory Algorithms*, North-Holland, New York, 1979.

[Ch77] G. Chartrand, *Graphs as Mathematical Models*, Prindle, Weber & Schmidt, Boston, 1977.

[ChLeZh15] G. Chartrand, L. Lesniak, and P. Zhang, *Graphs and Digraphs*, 6th ed., Chapman and Hall/CRC, Boca Raton, 2015.

[ChOe93] G. Chartrand and O. R. Oellermann, *Applied Algorithmic Graph Theory*, McGraw Hill, New York, 1993.

[ChZh04] G. Chartrand and P. Zhang, *Introduction to Graph Theory*, McGraw Hill, New York, 2004.

[ClMe94] W. F. Clocksin and C. S. Mellish, *Programming in Prolog*, 4th ed., Springer-Verlag, New York, 1994.

[Co78] D. I. A. Cohen, *Basic Techniques of Combinatorial Theory*, Wiley, New York, 1978.

[Co05] D. Comer, *Internetworking with TCP/IP, Principles, Protocols, and Architecture*, Vol. 1, 5th ed., Prentice Hall, Englewood Cliffs, NJ, 2005.

[CoLeRiSt09] T. H. Cormen, C. E. Leierson, R. L. Rivest, and C. Stein, *Introduction to Algorithms*, 3rd ed., MIT Press, Cambridge, MA, 2009.

[CrPo10] Richard Crandall and Carl Pomerance, 2nd ed., *Prime Numbers: A Computational Perspective*, Springer-Verlag, New York, 2010.

[Cu05] Antonella Cupillari, *The Nuts and Bolts of Proofs*, 3rd ed., Academic Press, San Diego, 2005.

[Cu84] C. W. Curtis, *Linear Algebra*, Springer-Verlag, New York, 1984.

[Da82] C. J. Date, *An Introduction to Database Systems*, 3rd ed., Addison-Wesley, Reading, MA, 1982.

[DaSiWe94] M. Davis, R. Sigal, and E. J. Weyuker, *Computability, Complexity, and Languages*, 2nd ed., Academic Press, San Diego, 1994.

[Da10] T. Davis, "The Mathematics of Sudoku," November, 2010, available at *http://geometer.org/mathcircles/sudoku.pdf*

[De82] D. E. R. Denning, *Cryptography and Data Security*, Addison-Wesley, Reading, MA, 1982.

[DeDeQu81] P. J. Denning, J. B. Dennis, and J. E. Qualitz, *Machines, Languages, and Computation*, Prentice Hall, Englewood Cliffs, NJ, 1981.

[De74] N. Deo, *Graph Theory with Applications to Engineering and Computer Science*, Prentice Hall, Englewood Cliffs, NJ, 1974.

[DeOr11] S. L. Devadoss and J. O'Rourke, *Discrete and Computational Geometry*, Princeton University Press, Princeton, NJ, 2011.

[De02] K. Devlin, *The Millennium Problems: The Seven Greatest Unsolved Mathematical Puzzles of Our Time*, Basic Book, New York, 2002.

[De84] A. K. Dewdney, "Computer Recreations," *Scientific American*, 251, no. 2 (August 1984), 19–25; 252, no. 3 (March 1985), 14–23; 251, no. 4 (April 1985), 20–30.

[De93] A. K. Dewdney, *The New Turing Omnibus: Sixty-Six Excursions in Computer Science*, W. H. Freeman, New York, 1993.

[Di59] E. Dijkstra, "Two Problems in Connexion with Graphs," *Numerische Mathematik*, 1 (1959), 269–271.

[EaKl10] D. Easley and J. Kleinberg, *Networks, Crowds, and Markets: Reasoning About a Highly Connected World*, Cambridge University Press, New York, 2010.

[Eu53] L. Euler, "The Koenigsberg Bridges," *Scientific American*, 189, no. 1 (July 1953), 66–70.

[Ev73] S. Even, *Algorithmic Combinatorics*, Macmillan, New York, 1973.

[Ev79] S. Even, *Graph Algorithms*, Computer Science Press, Rockville, MD, 1979.

[Fe68] W. Feller, *An Introduction to Probability Theory and Its Applications*, Vol. 1, 3rd ed., Wiley, New York, 1968.

[Fo92] L. R. Foulds, *Graph Theory Applications*, Springer-Verlag, New York, 1992.

[GaJo79] Michael R. Garey and David S. Johnson, *Computers and Intractability: A Guide to NP-Completeness*, Freeman, New York, 1979.

[Gi85] A. Gibbons, *Algorithmic Graph Theory*, Cambridge University Press, Cambridge, England, 1985.

[Gi70] C. C. Gillispie, ed., *Dictionary of Scientific Biography*, Scribner's, New York, 1970.

[Go94] S. W. Golomb, *Polyominoes*, Princeton University Press, Princeton, NJ, 1994.

[Go84] G. H. Gonnet, *Handbook of Algorithms and Data Structures*, Addison-Wesley, London, 1984.

[GoHe77] S. E. Goodman and S. T. Hedetniemi, *Introduction to the Design and Analysis of Algorithms*, McGraw Hill, New York, 1977.

[GoGo78] C. C. Gotlieb and L. R. Gotlieb, *Data Types and Structures*, Prentice Hall, Englewood Cliffs, NJ, 1978.

[GrHe85] R. L. Graham and P. Hell, "On the History of the Minimum Spanning Tree Problem," *Annals of the History of Computing*, 7 (1985), 43–57.

[GrKnPa94] R. L. Graham, D. E. Knuth, and O. Patashnik, *Concrete Mathematics*, 2nd ed., Addison-Wesley, Reading, MA, 1994.

[GrRoSp90] Ronald L. Graham, Bruce L. Rothschild, and Joel H. Spencer, *Ramsey Theory*, 2nd ed., Wiley, New York, 1990.

[GrWa77] J. E. Graver and M. E. Watkins, *Combinatorics with Emphasis on the Theory of Graphs*, Springer-Verlag, New York, 1977.

[GrSc93] D. Gries and F. B. Schneider, *A Logical Approach to Discrete Math*, Springer-Verlag, New York, 1993.

[Gr03] R. P. Grimaldi, *Discrete and Combinatorial Mathematics*, 5th ed., Addison-Wesley, Reading, MA, 2003.

[Gr07] J. L. Gross, *Combinatorial Methods with Computer Applications*, Chapman and Hall/CRC, Boca Raton, FL, 2007.

[GrYe05] J. L. Gross and J. Yellen, *Graph Theory and Its Applications*, 2nd ed., CRC Press, Boca Raton, FL, 2005.

[GrYe03] J. L. Gross and J. Yellen, *Handbook of Graph Theory*, CRC Press, Boca Raton, FL, 2003.

[Gr90] Jerrold W. Grossman, *Discrete Mathematics: An Introduction to Concepts, Methods, and Applications*, Macmillan, New York, 1990.

[Gr97] J. Gruska, *Foundations of Computing*, International Thomsen Computer Press, London, 1997.

[Gu10] D. A. Gunderson, *Handbook of Mathematical Induction*, Chapman and Hall/CRC, Boca Raton, Florida, 2010.

[Ha60] P. R. Halmos, *Naive Set Theory*, D. Van Nostrand, New York, 1960.

[Ha80] R. W. Hamming, *Coding and Information Theory*, Prentice Hall, Englewood Cliffs, NJ, 1980.

[Ha69] F. Harary, *Graph Theory*, Addison-Wesley, Reading, MA, 1969.

[HaWrWiHe08] G. H. Hardy, E. M. Wright, A. Wiles, and R. Heath-Brown, *An Introduction to the Theory of Numbers*, 6th ed., Oxford University Press, USA, New York, 2008.

[Ha87] D. Harel, *Algorithmics, The Spirit of Computing*, Addison-Wesley, Reading, MA, 1987.

[Ha00a] Brian Hayes, "Graph-Theory in Practice, Part I," *American Scientist*, 88, no. 1 (2000), 9–13.

[Ha00b] Brian Hayes, "Graph-Theory in Practice, Part II," *American Scientist*, 88, no. 2 (2000), 104–109.

[Ha93] John P. Hayes, *Introduction to Digital Logic Design*, Addison-Wesley, Reading, MA, 1993.

[He77] F. Hennie, *Introduction to Computability*, Addison-Wesley, Reading, MA, 1977.

[He88] R. Herken, *The Universal Turing Machine, A Half-Century Survey*, Oxford University Press, New York, 1988.

[Ho76] C. Ho, "Decomposition of a Polygon into Triangles," *Mathematical Gazette*, 60 (1976), 132–134.

[Ho99] D. Hofstadter, *Gödel, Escher, Bach: An Internal Golden Braid*, Basic Books, New York, 1999.

[Ho66] F. E. Hohn, *Applied Boolean Algebra*, 2nd ed., Macmillan, New York, 1966.

[HoMoUl06] J. E. Hopcroft, R. Motwani, and J. D. Ullman, *Introduction to Automata Theory, Languages, and Computation*, 3rd ed., Addison-Wesley, Boston, MA, 2006.

[HoMo76] D. Hopkin and B. Moss, *Automata*, Elsevier, North-Holland, New York, 1976.

[HoSa82] E. Horowitz and S. Sahni, *Fundamentals of Computer Algorithms*, Computer Science Press, Rockville, MD, 1982.

[Hu82] T. C. Hu, *Combinatorial Algorithms*, Addison-Wesley, Reading, MA, 1982.

[Hu07] W. Huber, V. J. Carey, L. Long, S. Falcon, and R. Gentleman, "Graphs in molecular biology," *BMC Bioinformatics*, 8 (Supplement 6) (2007).

[HuRy04] M. Huth and M. Ryan, *Logic in Computer Science*, 2nd ed., Cambridge University Press, Cambridge, England, 2004.

[In93] D. C. Ince, *An Introduction to Discrete Mathematics, Formal System Specification, and Z*, 2nd ed., Oxford, New York, 1993.

[KaMo64] D. Kalish and R. Montague, *Logic: Techniques of Formal Reasoning*, Harcourt, Brace, Jovanovich, New York, 1964.

[Ka53] M. Karnaugh, "The Map Method for Synthesis of Combinatorial Logic Circuits," *Transactions of the AIEE*, part I, 72 (1953), 593–599.

[KaBo04] R. H. Katz and G. Borriello, *Contemporary Logic Design*, Prentice Hall, Englewood Cliffs, NJ, 2004.

[Kl56] S. C. Kleene, "Representation of Events by Nerve Nets," *in Automata Studies*, 3–42, Princeton University Press, Princeton, NJ, 1956.

[KlTa05] J. Kleinberg and E. Tardos, *Algorithm Design*, Addison-Wesley, Boston, 2005.

[Kn77] D. E. Knuth, "Algorithms," *Scientific American*, 236, no. 4 (April 1977), 63–80.

[Kn97a] D. E. Knuth, *The Art of Computer Programming, Vol. I: Fundamental Algorithms*, 3rd ed., Addison-Wesley, Reading, MA, 1997.

[Kn97b] D. E. Knuth, *The Art of Computer Programming, Vol. II: Seminumerical Algorithms*, 3rd ed., Addison-Wesley, Reading, MA, 1997.

[Kn98] D. E. Knuth, *The Art of Computer Programming, Vol. III: Sorting and Searching*, 2nd ed., Addison-Wesley, Reading, MA, 1998.

[KoKr04] W. Kocay and D. L. Kreher, *Graph Algorithms and Optimization*, Chapman and Hall/CRC, Boca Raton, Florida, 2004.

[Ko86] Z. Kohavi, *Switching and Finite Automata Theory*, 2nd ed., McGraw Hill, New York, 1986.

[KrSt98] Donald H. Kreher and Douglas R. Stinson, *Combinatorial Algorithms: Generation, Enumeration, and Search*, CRC Press, Boca Raton, FL, 1998.

[Kr87] L. Kronsjö, *Algorithms: Their Complexity and Efficiency*, 2nd ed., Wiley, New York, 1987.

[Kr56] J. B. Kruskal, "On the Shortest Spanning Subtree of a Graph and the Traveling Salesman Problem," *Proceedings of the AMS*, 1 (1956), 48–50.

[La10] J. C. Lagarias (ed.), *The Ultimate Challenge: The $3x + 1$ Problem*, The American Mathematical Society, Providence, 2010.

[Le06] A. V. Levitin, *Introduction to the Design and Analysis of Algorithms*, 2nd ed., Addison-Wesley, Boston, MA, 2006.

[Le64] D. H. Lehmer, "The Machine Tools of Combinatorics," in E. F. Beckenbach (ed.), *Applied Combinatorial Mathematics*, Wiley, New York, 1964.

[Le77] W. J. LeVeque, *Fundamentals of Number Theory*, Addison-Wesley, Reading, MA, 1977.

[LePa97] H. R. Lewis and C. H. Papadimitriou, *Elements of the Theory of Computation*, 2nd ed., Prentice Hall, Englewood Cliffs, NJ, 1997.

[LiLi81] Y. Lin and S. Y. T. Lin, *Set Theory with Applications*, 2nd ed., Mariner, Tampa, FL, 1981.

[Li68] C. L. Liu, *Introduction to Combinatorial Mathematics*, McGraw Hill, New York, 1968.

[Li85] C. L. Liu, *Elements of Discrete Mathematics*, 2nd ed., McGraw Hill, New York, 1985.

[Lo79] L. Lovász, *Combinatorial Problems and Exercises*, North-Holland, Amsterdam, 1979.

[Lu91] E. Lucas, *Récréations Mathématiques*, Gauthier-Villars, Paris, 1891.

[Ma12] Toufik Mansour, *Combinatorics of Set Partitions*, CRC Press, Boca Raton, 2012.

[Ma89] U. Manber, *Introduction to Algorithms: A Creative Approach*, Addison-Wesley, Reading, MA, 1989.

[Ma91] G. E. Martin, *Polyominoes: A Guide to Puzzles and Problems in Tiling*, Mathematical Association of America, Washington, DC, 1991.

[Ma03] J. C. Martin, *Introduction to Languages and the Theory of Computation*, 3rd ed., McGraw Hill, New York, 2003.

[Ma93] H. F. Mattson, Jr., *Discrete Mathematics with Applications*, Wiley, New York, 1993.

[Mc56] E. J. McCluskey, Jr., "Minimization of Boolean Functions," *Bell System Technical Journal*, 35 (1956), 1417–1444.

[Mc90] J. A. McHugh, *Algorithmic Graph Theory*, Prentice Hall, Englewood Cliffs, NJ, 1990.

[Mc82] R. McNaughton, *Elementary Computability, Formal Languages, and Automata*, Prentice Hall, Englewood Cliffs, NJ, 1982.

[Me55] G. H. Mealy, "A Method for Synthesizing Sequential Circuits," *Bell System Technical Journal*, 34 (1955), 1045–1079.

[Me09] E. Mendelson, *Introduction to Mathematical Logic*, 5th ed., Chapman and Hall/CRC Press, Boca Raton, 2009.

[MeOoVa97] A. J. Menezes, P. C. van Oorschoot, S. A. Vanstone, *Handbook of Applied Cryptography*, CRC Press, Boca Raton, FL, 1997.

[MiRo91] J. G. Michaels and K. H. Rosen, *Applications of Discrete Mathematics*, McGraw Hill, New York, 1991.

[Mo69] J. R. Monk, *Introduction to Set Theory*, McGraw Hill, New York, 1969.

[Mo91] R. P. Morash, *Bridge to Abstract Mathematics*, McGraw Hill, New York, 1991.

[Mo56] E. F. Moore, "Gedanken-Experiments on Sequential Machines," *in Automata Studies*, 129–153, Princeton University Press, Princeton, NJ, 1956.

[Na00] Paul J. Nabin, *Duelling Idiots and Other Probability Puzzlers*, Princeton University Press, Princeton, NJ, 2000.

[Ne85] C. V. Negoita, *Expert Systems and Fuzzy Systems*, Benjamin Cummings, Menlo Park, CA, 1985.

[Ne10] M. Newman, *Networks: An Introduction*, Oxford University Press, New York, 2010.

[NiMa95] Ulf Nilsson and Jan Maluszynski, *Logic, Programming, and Prolog*, 2nd ed., Wiley, Chichester, England, 1995.

[Or00] J. O'Rourke, *Computational Geometry in C*, Cambridge University Press, New York, 2000.

[Or63] O. Ore, *Graphs and Their Uses*, Mathematical Association of America, Washington, DC, 1963.

[Or88] O. Ore, *Number Theory and its History*, Dover, New York, 1988.

[PaPi01] A. Papoulis and S. U. Pillai, *Probability, Random Variables, and Stochastic Processes*, McGraw Hill, New York, 2001.

[Pe87] A. Pelc, "Solution of Ulam's Problem on Searching with a Lie," *Journal of Combinatorial Theory*, Series A, 44 (1987), 129–140.

[Pe09] M. S. Petrović, *Famous Puzzles of Great Mathematicians*, American Mathematical Society, Providence, 2009.

[Pi14] C. Pinter, *A Book of Set Theory*, Dover, New York, 2014.

[PoSh81] I. Pohl and A. Shaw, *The Nature of Computation: An Introduction to Computer Science*, Computer Science Press, Rockville, MD, 1981.

[Po62] George Pólya, *Mathematical Discovery*, Vols. 1 and 2, Wiley, New York, 1962.

[Po71] George Pólya, *How to Solve It*, Princeton University Press, Princeton, NJ, 1971.

[Po90] George Pólya, *Mathematics and Plausible Reasoning*, Princeton University Press, Princeton, NJ, 1990.

[PoTaWo83] G. Pólya, R. E. Tarjan, and D. R. Woods, *Notes on Introductory Combinatorics*, Birkhäuser, Boston, 1983.

[Pr57] R. C. Prim, "Shortest Connection Networks and Some Generalizations," *Bell System Technical Journal*, 36 (1957), 1389–1401.

[PrSh85] F. Preparata and M. Shamos, *Computational Geometry, An Introduction*, Springer-Verlag, New York, 1985.

[PuBr85] P. W. Purdom, Jr. and C. A. Brown, *The Analysis of Algorithms*, Holt, Rinehart & Winston, New York, 1985.

[Qu52] W. V. Quine, "The Problem of Simplifying Truth Functions," *American Mathematical Monthly*, 59 (1952), 521–531.

[Qu55] W. V. Quine, "A Way to Simplify Truth Functions," *American Mathematical Monthly*, 62 (1955), 627–631.

[Ra62] T. Rado, "On Non-Computable Functions," *Bell System Technical Journal* (May 1962), 877–884.

[Ra92] Gregory J. E. Rawlins, *Compared to What? An Introduction to the Analysis of Algorithms*, Computer Science Press, New York, 1992.

[ReNiDe77] E. M. Reingold, J. Nievergelt, and N. Deo, *Combinatorial Algorithms: Theory and Practice*, Prentice Hall, Englewood Cliffs, NJ, 1977.

[Ri58] J. Riordan, *An Introduction to Combinatorial Analysis*, Wiley, New York, 1958.

[Ri68] J. Riordan, *Combinatorial Identities*, Wiley, New York, 1968.

[RiShAd78] R. Rivest, A. Shamir, and L. Adleman, "A Method for Obtaining Digital Signatures and Public-Key Cryptosystems," *Communications of the Association for Computing Machinery*, 31, no. 2 (1978), 120–128.

[Ro86] E. S. Roberts, *Thinking Recursively*, Wiley, New York, 1986.

[Ro76] F. S. Roberts, *Discrete Mathematics Models*, Prentice Hall, Englewood Cliffs, NJ, 1976.

[RoTe03] F. S. Roberts and B. Tesman, *Applied Combinatorics*, 2nd ed., Prentice Hall, Englewood Cliffs, NJ, 2003.

[RoFo80] D. F. Robinson and L. R. Foulds, *Digraphs: Theory and Techniques*, Gordon and Breach, New York, 1980.

[Ro84] J. S. Rohl, *Recursion via Pascal*, Cambridge University Press, Cambridge, England, 1984.

[Ro23] K. H. Rosen, *Elementary Number Theory and Its Applications*, 7th ed., Pearson, Boston, 2023.

[Ro18] K. H. Rosen, *Handbook of Discrete and Combinatorial Mathematics*, 2nd ed., CRC Press, Boca Raton, FL, 2018.

[Ro09] Jason Rosenhouse, *The Monty Hall Problem*, Oxford University Press, New York, 2009.

[Ro09a] Sheldon M. Ross, *A First Course in Probability Theory*, 7th ed., Prentice Hall, Englewood Cliffs, NJ, 2009.

[Ro02] Sheldon M. Ross, *Probability Models for Computer Science*, Harcourt/Academic Press, San Diego, 2002.

[Ro59] B. Roy, "Transitivité et Connexité," *C. R. Acad. Sci. Paris*, 249 (1959), 216.

[Ry63] H. Ryser, *Combinatorial Mathematics*, Mathematical Association of America, Washington, DC, 1963.

[SaKa86] T. L. Saaty and P. C. Kainen, *The Four-Color Problem: Assaults and Conquest*, Dover, New York, 1986.

[Sa85] S. Sahni, *Concepts in Discrete Mathematics*, Camelot, Minneapolis, 1985.

[Sa00] K. Sayood, *Introduction to Data Compression*, 2nd ed., Academic Press, San Diego, 2000.

[SePi89] J. Seberry and J. Pieprzyk, *Cryptography: An Introduction to Computer Security*, Prentice Hall, Englewood Cliffs, NJ, 1989.

[Se03] R. Sedgewick, *Algorithms in Java*, 3rd ed., Addison-Wesley, Reading, MA, 2003.

[SiHs88] H. J. Siegel and W. T. Hsu, "Interconnection Networks," *in Computer Architectures*, V. M. Milutinovic (ed.), North-Holland, New York, 1988, pp. 225–264.

[Si99] S. Singh, *The Code Book*, Doubleday, New York, 1999.

[Si66] A. Sinkov, *Elementary Cryptanalysis*, Mathematical Association of America, Washington, DC, 1966.

[Si06] M. Sipser, *Introduction to the Theory of Computation*, Course Technology, Boston, 2006.

[SlPl95] N. J. A. Sloane and S. Plouffe, *The Encyclopedia of Integer Sequences*, Academic Press, New York, 1995.

[Sm78] Raymond Smullyan, *What Is the Name of This Book?: The Riddle of Dracula and Other Logical Puzzles*, Prentice Hall, Englewood Cliffs, NJ, 1978.

[Sm92] Raymond Smullyan, *Lady or the Tiger? And Other Logic Puzzles Including a Mathematical Novel That Features Godel's Great Discovery*, Times Book, New York, 1992.

[Sm98] Raymond Smullyan, *The Riddle of Scheherazade: And Other Amazing Puzzles, Ancient & Modern*, Harvest Book, Fort Washington, PA, 1998.

[So09] Daniel Solow, *How to Read and Do Proofs: An Introduction to Mathematical Thought Processes*, 5th ed., Wiley, New York, 2009.

[So61] I. S. Sominskii, *Method of Mathematical Induction*, Blaisdell, New York, 1961.

[Sp94] M. Spivak, *Calculus*, 3rd ed., Publish or Perish, Wilmington, DE, 1994.

[StMc77] D. Stanat and D. F. McAllister, *Discrete Mathematics in Computer Science*, Prentice Hall, Englewood Cliffs, NJ, 1977.

[St78] H. M. Stark, *An Introduction to Number Theory*, MIT Press, Cambridge, MA, 1978.

[St05] Douglas R. Stinson, *Cryptography, Theory and Practice*, 3rd ed., Chapman and Hall/CRC, Boca Raton, FL, 2005.

[St94] P. K. Stockmeyer, "Variations on the Four-Post Tower of Hanoi Puzzle," *Congressus Numerantrum*, 102 (1994), 3–12.

[St74] R. R. Stoll, *Sets, Logic, and Axiomatic Theories*, 2nd ed., W. H. Freeman, San Francisco, 1974.

[St09] G. W. Strang, *Linear Algebra and Its Applications*, 4th ed., Wellesley Cambridge Press, Wellesley, MA, 2009.

[Su99] P. Suppes, *Introduction to Logic*, Dover Publications, Inc., Mineola, NY, 1999.

[Ta83] R. E. Tarjan, *Data Structures and Network Algorithms*, Society for Industrial and Applied Mathematics, Philadelphia, 1983.

[ThFi96] G. B. Thomas and R. L. Finney, *Calculus and Analytic Geometry*, 9th ed., Addison-Wesley, Reading, MA, 1996.

[TrMa75] J. P. Tremblay and R. P. Manohar, *Discrete Mathematical Structures with Applications to Computer Science*, McGraw Hill, New York, 1975.

[Tu06] Alan Tucker, *Applied Combinatorics*, 5th ed., Wiley, New York, 2006.

[Ve52] E. W. Veitch, "A Chart Method for Simplifying Truth Functions," *Proceedings of the ACM* (1952), 127–133.

[Ve06] David Velleman, *How to Prove It: A Structured Approach*, Cambridge University Press, New York, 2006.

[Vi71] N. Y. Vilenkin, *Combinatorics*, Academic Press, New York, 1971.

[Wa80] M. Wand, *Induction, Recursion, and Programming*, North-Holland, New York, 1980.

[Wa62] S. Warshall, "A Theorem on Boolean Matrices," *Journal of the ACM*, 9 (1962), 11–12.

[We17] D. B. West, *Introduction to Graph Theory*, 2nd ed., Pearson, Boston, 2017.

[Wi02] Herbert S. Wilf, *Algorithms and Complexity*, 2nd ed., A. K. Peters, Natick, MA, 2002.

[Wi85] S. G. Williamson, *Combinatorics for Computer Science*, Computer Science Press, Rockville, MD, 1985.

[Wi85a] R. J. Wilson, *Introduction to Graph Theory*, 3rd ed., Longman, Essex, England, 1985.

[WiBe79] R. J. Wilson and L. W. Beineke, *Applications of Graph Theory*, Academic Press, London, 1979.

[WiWa90] R. J. Wilson and J. J. Watkins, *Graphs, An Introductory Approach*, Wiley, New York, 1990.

[Wi76] N. Wirth, *Algorithms + Data Structures = Programs*, Prentice Hall, Englewood Cliffs, NJ, 1976.

[Wi84] N. Wirth, "Data Structures and Algorithms," *Scientific American*, 251 (September 1984), 60–69.

[Wo98] Robert S. Wolf, *Proof, Logic, and Conjecture: The Mathematician's Toolbox*, W. H. Freeman, New York, 1998.

[Wo87] D. Wood, *Theory of Computation*, Harper & Row, New York, 1987.

[Zd05] J. Zdziarski, *Ending Spam: Bayesian Content Filtering and the Art of Statistical Language Classification*, No Starch Press, San Francisco, 2005.

[Zi91] H. J. Zimmermann, *Fuzzy Set Theory and Its Applications*, 2nd ed., Kluwer, Boston, 1991.

Answers to Odd-Numbered Exercises

CHAPTER 1

Section 1.1

1. a) Yes, T **b)** Yes, F **c)** Yes, T **d)** Yes, F **e)** No **f)** No **3. a)** Linda is not younger than Sanjay. **b)** Mei does not make more money than Isabella. **c)** Moshe is not taller than Monica. **d)** Abby is not richer than Ricardo. **5. a)** Mei does not have an MP3 player. **b)** There is pollution in New Jersey. **c)** $2 + 1 \neq 3$. **d)** The summer in Maine is not hot or it is not sunny. **7. a)** Steve does not have more than 100 GB free disk space on his laptop. **b)** Zach does not block emails from Jennifer, or he does not block texts from Jennifer. **c)** $7 \cdot 11 \cdot 13 \neq 999$. **d)** Diane did not ride her bike 100 miles on Sunday. **9. a)** F **b)** T **c)** T **d)** T **e)** T **11. a)** Sharks have not been spotted near the shore. **b)** Swimming at the New Jersey shore is allowed, and sharks have been spotted near the shore. **c)** Swimming at the New Jersey shore is not allowed, or sharks have been spotted near the shore. **d)** If swimming at the New Jersey shore is allowed, then sharks have not been spotted near the shore. **e)** If sharks have not been spotted near the shore, then swimming at the New Jersey shore is allowed. **f)** If swimming at the New Jersey shore is not allowed, then sharks have not been spotted near the shore. **g)** Swimming at the New Jersey shore is allowed if and only if sharks have not been spotted near the shore. **h)** Swimming at the New Jersey shore is not allowed, and either swimming at the New Jersey shore is allowed or sharks have not been spotted near the shore. (Note that we were able to incorporate the parentheses by using the word "either" in the second half of the sentence.) **13. a)** $p \wedge q$ **b)** $p \wedge \neg q$ **c)** $\neg p \wedge \neg q$ **d)** $p \vee q$ **e)** $p \rightarrow q$ **f)** $(p \vee q) \wedge (p \rightarrow \neg q)$ **g)** $q \leftrightarrow p$ **15. a)** $\neg p$ **b)** $p \wedge \neg q$ **c)** $p \rightarrow q$ **d)** $\neg p \rightarrow \neg q$ **e)** $p \rightarrow q$ **f)** $q \wedge \neg p$ **g)** $q \rightarrow p$ **17. a)** $r \wedge \neg p$ **b)** $\neg p \wedge q \wedge r$ **c)** $r \rightarrow (q \leftrightarrow \neg p)$ **d)** $\neg q \wedge \neg p \wedge r$ **e)** $(q \rightarrow (\neg r \wedge \neg p)) \wedge \neg((\neg r \wedge \neg p) \rightarrow q)$ **f)** $(p \wedge r) \rightarrow \neg q$ **19. a)** False **b)** True **c)** True **d)** True **21. a)** Exclusive or: You get only one beverage. **b)** Inclusive or: Long passwords can have any combination of symbols. **c)** Inclusive or: A student with both courses is even more qualified. **d)** Either interpretation possible; a traveler might wish to pay with a mixture of the two currencies, or the store may not allow that. **23. a)** Inclusive or: It is allowable to take discrete mathematics if you have had calculus or computer science, or both. Exclusive or: It is allowable to take discrete mathematics if you have had calculus or computer science, but not if you have had both. Most likely the inclusive or is intended. **b)** Inclusive or: You can take the rebate, or you can get a low-interest loan, or you can get both the rebate and a low-interest loan. Exclusive or: You can take the rebate, or you can get a low-interest loan, but you cannot get both the rebate and a low-interest loan. Most likely the exclusive or is intended. **c)** Inclusive or: You can order two items from column A and none from column B, or three items from column B and none from column A, or five items including two from column A and three from column B. Exclusive or: You can order two items from column A or three items from column B, but not both. Almost certainly the exclusive or is intended. **d)** Inclusive or: More than 2 feet of snow or windchill below −100 °F, or both, will close school. Exclusive or: More than 2 feet of snow or windchill below −100 °F, but not both, will close school. Certainly the inclusive or is intended. **25. a)** If the wind blows from the northeast, then it snows. **b)** If it stays warm for a week, then the apple trees will bloom. **c)** If the Pistons win the championship, then they beat the Lakers. **d)** If you get to the top of Long's Peak, then you must have walked 8 miles. **e)** If you are world famous, then you will get tenure as a professor. **f)** If you drive more than 400 miles, then you will need to buy gasoline. **g)** If your guarantee is good, then you must have bought your CD player less than 90 days ago. **h)** If the water is not too cold, then Jan will go swimming. **i)** If people believe in science, then we will have a future. **27. a)** You buy an ice cream cone if and only if it is hot outside. **b)** You win the contest if and only if you hold the only winning ticket. **c)** You get promoted if and only if you have connections. **d)** Your mind will decay if and only if you watch television. **e)** The train runs late if and only if it is a day I take the train. **29. a)** Converse: "I will ski tomorrow only if it snows today." Contrapositive: "If I do not ski tomorrow, then it will not have snowed today." Inverse: "If it does not snow today, then I will not ski tomorrow." **b)** Converse: "If I come to class, then there will be a quiz." Contrapositive: "If I do not come to class, then there will not be a quiz." Inverse: "If there is not going to be a quiz, then I don't come to class." **c)** Converse: "A positive integer is a prime if it has no divisors other than 1 and itself." Contrapositive: "If a positive integer has a divisor other than 1 and itself, then it is not prime." Inverse: "If a positive integer is not prime, then it has a divisor other than 1 and itself." **31. a)** 2 **b)** 16 **c)** 64 **d)** 16

33. a)

p	$\neg p$	$p \wedge \neg p$
T	F	F
F	T	F

b)

p	$\neg p$	$p \vee \neg p$
T	F	T
F	T	T

c)

p	q	$\neg q$	$p \vee \neg q$	$(p \vee \neg q) \rightarrow q$
T	T	F	T	T
T	F	T	T	F
F	T	F	F	T
F	F	T	T	F

d)

p	q	$p \vee q$	$p \wedge q$	$(p \vee q) \to (p \wedge q)$
T	T	T	T	T
T	F	T	F	F
F	T	T	F	F
F	F	F	F	T

e)

p	q	$p \to q$	$\neg q$	$\neg p$	$\neg q \to \neg p$	$(p \to q) \leftrightarrow (\neg q \to \neg p)$
T	T	T	F	F	T	T
T	F	F	T	F	F	T
F	T	T	F	T	T	T
F	F	T	T	T	T	T

f)

p	q	$p \to q$	$q \to p$	$(p \to q) \to (q \to p)$
T	T	T	T	T
T	F	F	T	T
F	T	T	F	F
F	F	T	T	T

35. For parts (a), (b), (c), (d), and (f) we have this table.

p	q	$(p \vee q) \to (p \oplus q)$	$(p \oplus q) \to (p \wedge q)$	$(p \vee q) \oplus (p \wedge q)$	$(p \leftrightarrow q) \oplus (\neg p \leftrightarrow q)$	$(p \oplus q) \to (p \oplus \neg q)$
T	T	F	T	F	T	T
T	F	T	F	T	T	F
F	T	T	F	T	T	F
F	F	T	T	F	T	T

For part (e) we have this table.

p	q	r	$\neg p$	$\neg r$	$p \leftrightarrow q$	$\neg p \leftrightarrow \neg r$	$(p \leftrightarrow q) \oplus (\neg p \leftrightarrow \neg r)$
T	T	T	F	F	T	T	F
T	T	F	F	T	T	F	T
T	F	T	F	F	F	T	T
T	F	F	F	T	F	F	F
F	T	T	T	F	F	F	F
F	T	F	T	T	F	T	T
F	F	T	T	F	T	F	T
F	F	F	T	T	T	T	F

37.

p	q	$p \to \neg q$	$\neg p \leftrightarrow q$	$(p \to q) \vee (\neg p \to q)$	$(p \to q) \wedge (\neg p \to q)$	$(p \leftrightarrow q) \vee (\neg p \leftrightarrow q)$	$(\neg p \leftrightarrow \neg q) \leftrightarrow (p \leftrightarrow q)$
T	T	F	F	T	T	T	T
T	F	T	T	T	F	T	T
F	T	T	T	T	T	T	T
F	F	T	F	T	F	T	T

39.

p	q	r	$p \to (\neg q \vee r)$	$\neg p \to (q \to r)$	$(p \to q) \vee (\neg p \to r)$	$(p \to q) \wedge (\neg p \to r)$	$(p \leftrightarrow q) \vee (\neg q \leftrightarrow r)$	$(\neg p \leftrightarrow \neg q) \leftrightarrow (q \leftrightarrow r)$
T	T	T	T	T	T	T	T	T
T	T	F	F	T	T	T	T	F
T	F	T	T	T	T	F	T	T
T	F	F	T	T	T	F	F	F
F	T	T	T	T	T	T	F	F
F	T	F	T	F	T	F	T	T
F	F	T	T	T	T	T	T	F
F	F	F	T	T	T	F	T	T

41.

p	q	r	s	$p \leftrightarrow q$	$r \leftrightarrow s$	$(p \leftrightarrow q) \leftrightarrow (r \leftrightarrow s)$
T	T	T	T	T	T	T
T	T	T	F	T	F	F
T	T	F	T	T	F	F
T	T	F	F	T	T	T
T	F	T	T	F	T	F
T	F	T	F	F	F	T
T	F	F	T	F	F	T
T	F	F	F	F	T	F
F	T	T	T	F	T	F
F	T	T	F	F	F	T
F	T	F	T	F	F	T
F	T	F	F	F	T	F
F	F	T	T	T	T	T
F	F	T	F	T	F	F
F	F	F	T	T	F	F
F	F	F	F	T	T	T

43. The first clause is true if and only if at least one of p, q, and r is true. The second clause is true if and only if at least one of the three variables is false. Therefore, the entire statement is true if and only if there is at least one T and one F among the truth values of the variables, in other words, that they don't all have the same truth value. **45.** $\left(\bigwedge_{i=1}^{n-1}\bigwedge_{j=i+1}^{n}(\neg p_i \vee \neg p_j)\right) \wedge \left(\bigvee_{i=1}^{n} p_i\right)$ **47. a)** Bitwise *OR* is 111 1111; bitwise *AND* is 000 0000; bitwise *XOR* is 111 1111. **b)** Bitwise *OR* is 1111 1010; bitwise *AND* is 1010 0000; bitwise *XOR* is 0101 1010. **c)** Bitwise *OR* is 10 0111 1001; bitwise *AND* is 00 0100 0000; bitwise *XOR* is 10 0011 1001. **d)** Bitwise *OR* is 11 1111 1111; bitwise *AND* is 00 0000 0000; bitwise *XOR* is 11 1111 1111. **49.** 0.2, 0.6 **51.** 0.8, 0.6 **53. a)** The 99th statement is true and the rest are false. **b)** Statements 1 through 50 are all true and statements 51 through 100 are all false. **c)** This cannot happen; it is a paradox, showing that these cannot be statements.

Section 1.2

1. $e \to a$ **3.** $g \to (r \wedge (\neg m) \wedge (\neg b))$ **5.** $e \to (a \wedge (b \vee p) \wedge r)$ **7. a)** $q \to p$ **b)** $q \wedge \neg p$ **c)** $q \to p$ **d)** $\neg q \to \neg p$ **9.** Not consistent **11.** Consistent **13.** NEW *AND* JERSEY *AND* BEACHES, (JERSEY *AND* BEACHES) *NOT* NEW **15.** "ETHIOPIAN RESTAURANTS" *AND* ("NEW YORK" *OR* "NEW JERSEY") **17. a)** Queen cannot say this. **b)** Queen can say this, but one cannot determine location of treasure. **c)** Queen can say this; treasure is in Trunk 1. **d)** Queen cannot say this. **19.** "If I were to ask you whether the right branch leads to the ruins, would you answer yes?" **21.** If the first professor did not want coffee, then he would know that the answer to the waiter's question was "no." Therefore the waiter and the remaining professors know that the first professor did want coffee. Similarly, the second professor must want coffee. When the third professor said "no," the waiter knows that the third professor does not want coffee. **23.** A is a knight and B is a knave. **25.** A is a knight and B is a knight. **27.** A is a knave and B is a knight. **29.** A is the knight, B is the spy, C is the knave. **31.** A is the knight, B is the spy, C is the knave. **33.** Any of the three can be the knight, any can be the spy, any can be the knave. **35.** No solutions **37.** In order of decreasing salary: Fred, Maggie, Janice **39.** The detective can determine that the butler and cook are lying but cannot determine whether the gardener is telling the truth or whether the property manager is telling the truth. **41.** False. **43.** One honest, 49 corrupt **45. a)** $\neg(p \wedge (q \vee \neg r))$ **b)** $((\neg p) \wedge (\neg q)) \vee (p \wedge r)$

47.

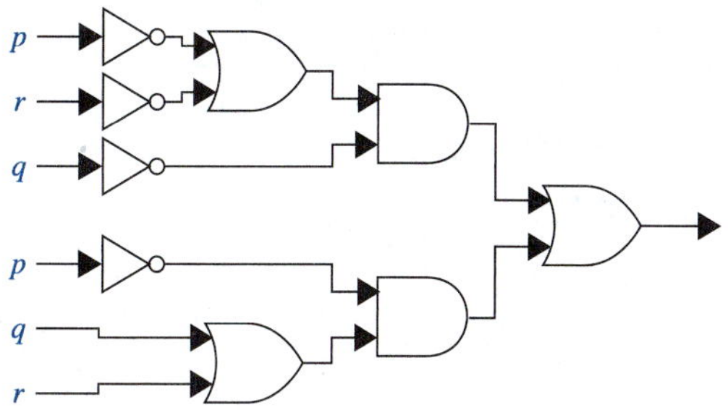

Section 1.3

1. The equivalences follow by showing that the appropriate pairs of columns of this table agree.

p	$p \wedge \mathbf{T}$	$p \vee \mathbf{F}$	$p \wedge \mathbf{F}$	$p \vee \mathbf{T}$	$p \vee p$	$p \wedge p$
T	T	T	F	T	T	T
F	F	F	F	T	F	F

3. a)

p	q	$p \vee q$	$q \vee p$
T	T	T	T
T	F	T	T
F	T	T	T
F	F	F	F

b)

p	q	$p \wedge q$	$q \wedge p$
T	T	T	T
T	F	F	F
F	T	F	F
F	F	F	F

5.

p	q	r	$q \vee r$	$p \wedge (q \vee r)$	$p \wedge q$	$p \wedge r$	$(p \wedge q) \vee (p \wedge r)$
T	T	T	T	T	T	T	T
T	T	F	T	T	T	F	T
T	F	T	T	T	F	T	T
T	F	F	F	F	F	F	F
F	T	T	T	F	F	F	F
F	T	F	T	F	F	F	F
F	F	T	T	F	F	F	F
F	F	F	F	F	F	F	F

7. a) Jan is not rich, or Jan is not happy. **b)** Carlos will not bicycle tomorrow, and Carlos will not run tomorrow. **c)** Mei does not walk to class, and Mei does not take the bus to class. **d)** Ibrahim is not smart, or Ibrahim is not hard working. **9. a)** $\neg p \vee \neg q$ **b)** $(p \wedge \neg q) \vee r$ **c)** $\neg p \vee \neg q$

11. a)

p	q	$p \wedge q$	$(p \wedge q) \to p$
T	T	T	T
T	F	F	T
F	T	F	T
F	F	F	T

b)

p	q	$p \vee q$	$p \to (p \vee q)$
T	T	T	T
T	F	T	T
F	T	T	T
F	F	F	T

c)

p	q	$\neg p$	$p \to q$	$\neg p \to (p \to q)$
T	T	F	T	T
T	F	F	F	T
F	T	T	T	T
F	F	T	T	T

d)

p	q	$p \wedge q$	$p \to q$	$(p \wedge q) \to (p \to q)$
T	T	T	T	T
T	F	F	F	T
F	T	F	T	T
F	F	F	T	T

e)

p	q	$p \to q$	$\neg(p \to q)$	$\neg(p \to q) \to p$
T	T	T	F	T
T	F	F	T	T
F	T	T	F	T
F	F	T	F	T

f)

p	q	$p \to q$	$\neg(p \to q)$	$\neg q$	$\neg(p \to q) \to \neg q$
T	T	T	F	F	T
T	F	F	T	T	T
F	T	T	F	F	T
F	F	T	F	T	T

13. a) If this were not a tautology, then $p \wedge q$ would be true but p would be false. This cannot happen, because the truth of $p \wedge q$ implies the truth of p. **b)** If this were not a tautology, then p would be true but $p \vee q$ would be false. This cannot happen, because the truth of p implies the truth of $p \vee q$. **c)** If this were not a tautology, then p would be false and $p \to q$ would be false. This cannot happen, because $p \to q$ is true when p is false. **d)** If this were not a tautology, then $p \wedge q$ would be true and $p \to q$ would be false. This cannot happen, because $p \to q$ is true when both p and q are true. **e)** If this were not a tautology, then $p \to q$ would be faise and p would be false. This cannot happen, because $p \to q$ is true when p is false. **f)** If this were not a tautology, then $p \to q$ would be false and q would be true. This cannot happen, because $p \to q$ is true when q is true. **15. a)** $(p \wedge q) \to p \equiv \neg(p \wedge q) \vee p \equiv \neg p \vee \neg q \vee p \equiv (p \vee \neg p) \vee \neg q \equiv \mathbf{T} \vee \neg q \equiv \mathbf{T}$ **b)** $p \to (p \vee q) \equiv \neg p \vee (p \vee q) \equiv (\neg p \vee p) \vee q \equiv \mathbf{T} \vee q \equiv \mathbf{T}$ **c)** $\neg p \to (p \to q) \equiv p \vee (p \to q) \equiv p \vee (\neg p \vee q) \equiv (p \vee \neg p) \vee q \equiv \mathbf{T} \vee q \equiv \mathbf{T}$ **d)** $(p \wedge q) \to (p \to q) \equiv \neg(p \wedge q) \vee (\neg p \vee q) \equiv \neg p \vee \neg q \vee \neg p \vee q \equiv (\neg p \vee \neg p) \vee (\neg q \vee q) \equiv \neg p \vee \mathbf{T} \equiv \mathbf{T}$ **e)** $\neg(p \to q) \to p \equiv (p \to q) \vee p \equiv \neg p \vee q \vee p \equiv (\neg p \vee p) \vee q \equiv \mathbf{T} \vee q \equiv \mathbf{T}$ **f)** $\neg(p \to q) \to \neg q \equiv (p \to q) \vee \neg q \equiv \neg p \vee q \vee \neg q \equiv \neg p \vee \mathbf{T} \equiv \mathbf{T}$ **17.** That the fourth column of the truth table shown is identical to the first column proves part (a), and that the sixth column is identical to the first column proves part (b).

p	q	$p \wedge q$	$p \vee (p \wedge q)$	$p \vee q$	$p \wedge (p \vee q)$
T	T	T	T	T	T
T	F	F	T	T	T
F	T	F	F	T	F
F	F	F	F	F	F

19. It is a tautology. **21.** Each of these is true precisely when p and q have opposite truth values. **23.** The proposition $\neg p \leftrightarrow q$ is true when $\neg p$ and q have the same truth values, which means that p and q have different truth values. Similarly, $p \leftrightarrow \neg q$ is true in exactly the same cases. Therefore, these two expressions are logically equivalent. **25.** The proposition $\neg(p \leftrightarrow q)$ is true when $p \leftrightarrow q$ is false, which means that p and q have different truth values. Because this is precisely when $\neg p \leftrightarrow q$ is true, the two expressions are logically equivalent. **27.** For $(p \to r) \wedge (q \to r)$ to be false, one of the two conditional statements must be false, which happens exactly when r is false and at least one of p and q is true. But these are precisely the cases in which $p \vee q$ is true and r is false, which is precisely when $(p \vee q) \to r$ is false. Because the two propositions are false in exactly the same situations, they are logically equivalent. **29.** For $(p \to r) \vee (q \to r)$ to be false, both of the two conditional statements must be false, which happens exactly when r is false and both p and q are true. But this is precisely the case in which $p \wedge q$ is true and r is false, which is precisely when $(p \wedge q) \to r$ is false. Because the two propositions are false in exactly the same situations, they are logically equivalent. **31.** This fact was observed in Section 1 when the biconditional was first defined. Each of these is true precisely when p and q have the same truth values. **33.** The last column is all Ts.

p	q	r	$p \to q$	$q \to r$	$(p \to q) \wedge (q \to r)$	$p \to r$	$(p \to q) \wedge (q \to r) \to (p \to r)$
T	T	T	T	T	T	T	T
T	T	F	T	F	F	F	T
T	F	T	F	T	F	T	T
T	F	F	F	T	F	F	T
F	T	T	T	T	T	T	T
F	T	F	T	F	F	T	T
F	F	T	T	T	T	T	T
F	F	F	T	T	T	T	T

35. These are not logically equivalent because when p, q, and r are all false, $(p \to q) \to r$ is false, but $p \to (q \to r)$ is true. **37.** Many answers are possible. If we let r be true and

p, q, and s be false, then $(p \to q) \to (r \to s)$ will be false, but $(p \to r) \to (q \to s)$ will be true. **39. a)** $p \vee \neg q \vee \neg r$ **b)** $(p \vee q \vee r) \wedge s$ **c)** $(p \wedge \mathbf{T}) \vee (q \wedge \mathbf{F})$ **41.** If we take duals twice, every $\vee$ changes to an $\wedge$ and then back to an $\vee$, every $\wedge$ changes to an $\vee$ and then back to an $\wedge$, every **T** changes to an **F** and then back to a **T**, every **F** changes to a **T** and then back to an **F**. Hence, $(s^*)^* = s$. **43.** Let p and q be equivalent compound propositions involving only the operators $\wedge$, $\vee$, and $\neg$, and **T** and **F**. Note that $\neg p$ and $\neg q$ are also equivalent. Use De Morgan's laws as many times as necessary to push negations in as far as possible within these compound propositions, changing $\vee$s to $\wedge$s, and vice versa, and changing **T**s to **F**s, and vice versa. This shows that $\neg p$ and $\neg q$ are the same as p^* and q^* except that each atomic proposition p_i within them is replaced by its negation. From this we can conclude that p^* and q^* are equivalent because $\neg p$ and $\neg q$ are. **45.** $(p \wedge q \wedge \neg r) \vee (p \wedge \neg q \wedge r) \vee (\neg p \wedge q \wedge r)$ **47.** Given a compound proposition p, form its truth table and then write down a proposition q in disjunctive normal form that is logically equivalent to p. Because q involves only $\neg$, $\wedge$, and $\vee$, this shows that these three operators form a functionally complete set. **49.** By Exercise 47, given a compound proposition p, we can write down a proposition q that is logically equivalent to p and involves only $\neg$, $\wedge$, and $\vee$. By De Morgan's law we can eliminate all the $\wedge$s by replacing each occurrence of $p_1 \wedge p_2 \wedge \cdots \wedge p_n$ with $\neg(\neg p_1 \vee \neg p_2 \vee \cdots \vee \neg p_n)$. **51.** $\neg(p \wedge q)$ is true when either p or q, or both, are false, and is false when both p and q are true. Because this was the definition of $p \mid q$, the two compound propositions are logically equivalent. **53.** $\neg(p \vee q)$ is true when both p and q are false, and is false otherwise. Because this was the definition of $p \downarrow q$, the two are logically equivalent. **55.** $((p \downarrow p) \downarrow q) \downarrow ((p \downarrow p) \downarrow q)$ **57.** This follows immediately from the truth table or definition of $p \mid q$. **59.** 16 **61.** If the database is open, then either the system is in its initial state or the monitor is put in a closed state. **63.** All nine **65. a)** Satisfiable **b)** Not satisfiable **c)** Not satisfiable **67. a)** $\left(\bigwedge_{i=1}^{2} \bigvee_{j=1}^{2} p(i,j)\right) \wedge \left(\bigwedge_{i=1}^{2} \bigwedge_{j=1}^{1} \bigwedge_{k=j+1}^{2} (\neg p(i,j) \vee \neg p(i,k))\right) \wedge \left(\bigwedge_{j=1}^{2} \bigwedge_{i=1}^{1} \bigwedge_{k=i+1}^{2} (\neg p(i,j) \vee \neg p(k,j))\right) \wedge \left(\bigwedge_{i=2}^{2} \bigwedge_{j=1}^{1} \bigwedge_{k=1}^{\min(i-1,2-j)} (\neg p(i,j) \vee \neg p(i-k, k+j))\right) \wedge \left(\bigwedge_{i=1}^{1} \bigwedge_{j=1}^{1} \bigwedge_{k=1}^{\min(2-i,2-j)} (\neg p(i,j) \vee \neg p(i+k, j+k))\right)$; No solutions possible. **b)** $\left(\bigwedge_{i=1}^{3} \bigvee_{j=1}^{3} p(i,j)\right) \wedge \left(\bigwedge_{i=1}^{3} \bigwedge_{j=1}^{2} \bigwedge_{k=j+1}^{3} (\neg p(i,j) \vee \neg p(i,k))\right) \wedge \left(\bigwedge_{j=1}^{3} \bigwedge_{i=1}^{2} \bigwedge_{k=i+1}^{3} (\neg p(i,j) \vee \neg p(k,j))\right) \wedge \left(\bigwedge_{i=2}^{3} \bigwedge_{j=1}^{2} \bigwedge_{k=1}^{\min(i-1,3-j)} (\neg p(i,j) \vee \neg p(i-k, k+j))\right) \wedge \left(\bigwedge_{i=1}^{2} \bigwedge_{j=1}^{2} \bigwedge_{k=1}^{\min(3-i,3-j)} (\neg p(i,j) \vee \neg p(i+k, j+k))\right)$; No solutions possible. **c)** $\left(\bigwedge_{i=1}^{4} \bigvee_{j=1}^{4} p(i,j)\right) \wedge \left(\bigwedge_{i=1}^{4} \bigwedge_{j=1}^{3} \bigwedge_{k=j+1}^{4} (\neg p(i,j) \vee \neg p(i,k))\right) \wedge \left(\bigwedge_{j=1}^{4} \bigwedge_{i=1}^{3} \bigwedge_{k=i+1}^{4} (\neg p(i,j) \vee \neg p(k,j))\right) \wedge \left(\bigwedge_{i=2}^{4} \bigwedge_{j=1}^{3} \bigwedge_{k=1}^{\min(i-1,4-j)} (\neg p(i,j) \vee \neg p(i-k, k+j))\right) \wedge \left(\bigwedge_{i=1}^{3} \bigwedge_{j=1}^{3} \bigwedge_{k=1}^{\min(4-i,4-j)} (\neg p(i,j) \vee \neg p(i+k, j+k))\right)$; (1, 2), (2, 4), (3, 1), (4, 3) or (1, 3), (2, 1), (3, 4), (4, 2) **69.** Use the same propositions as were given in the text for a 9×9 Sudoku puzzle, with the variables indexed from 1 to 4, instead of from 1 to 9, and with a similar change for the propositions for the 2×2 blocks: $\bigwedge_{r=0}^{1} \bigwedge_{s=0}^{1} \bigwedge_{n=1}^{4} \bigvee_{i=1}^{2} \bigvee_{j=1}^{2} p(2r + i, 2s + j, n)$ **71.** $\bigvee_{i=1}^{9} p(i, j, n)$ asserts that column j contains the number n, so $\bigwedge_{n=1}^{9} \bigvee_{i=1}^{9} p(i, j, n)$ asserts that column j contains all 9 numbers; therefore $\bigwedge_{j=1}^{9} \bigwedge_{n=1}^{9} \bigvee_{i=1}^{9} p(i, j, n)$ asserts that every column contains every number.

Section 1.4

1. a) T **b)** T **c)** F **3. a)** T **b)** F **c)** F **d)** F **5. a)** There is a student who spends more than 5 hours every weekday in class. **b)** Every student spends more than 5 hours every weekday in class. **c)** There is a student who does not spend more than 5 hours every weekday in class. **d)** No student spends more than 5 hours every weekday in class. **7. a)** Every comedian is funny. **b)** Every person is a funny comedian. **c)** There exists a person such that if she or he is a comedian, then she or he is funny. **d)** Some comedians are funny. **9. a)** $\exists x(P(x) \wedge Q(x))$ **b)** $\exists x(P(x) \wedge \neg Q(x))$ **c)** $\forall x(P(x) \vee Q(x))$ **d)** $\forall x \neg(P(x) \vee Q(x))$ **11. a)** T **b)** T **c)** F **d)** F **e)** T **f)** F **13. a)** T **b)** T **c)** T **d)** F **15. a)** T **b)** F **c)** T **d)** F **17. a)** $P(0) \vee P(1) \vee P(2) \vee P(3) \vee P(4)$ **b)** $P(0) \wedge P(1) \wedge P(2) \wedge P(3) \wedge P(4)$ **c)** $\neg P(0) \vee \neg P(1) \vee \neg P(2) \vee \neg P(3) \vee \neg P(4)$ **d)** $\neg P(0) \wedge \neg P(1) \wedge \neg P(2) \wedge \neg P(3) \wedge \neg P(4)$ **e)** $\neg(P(0) \vee P(1) \vee P(2) \vee P(3) \vee P(4))$ **f)** $\neg(P(0) \wedge P(1) \wedge P(2) \wedge P(3) \wedge P(4))$ **19. a)** $P(1) \vee P(2) \vee P(3) \vee P(4) \vee P(5)$ **b)** $P(1) \wedge P(2) \wedge P(3) \wedge P(4) \wedge P(5)$ **c)** $\neg(P(1) \vee P(2) \vee P(3) \vee P(4) \vee P(5))$ **d)** $\neg(P(1) \wedge P(2) \wedge P(3) \wedge P(4) \wedge P(5))$ **e)** $(P(1) \wedge P(2) \wedge P(4) \wedge P(5)) \vee (\neg P(1) \vee \neg P(2) \vee \neg P(3) \vee \neg P(4) \vee \neg P(5))$ **21.** Many answers are possible. **a)** All students in your discrete mathematics class; all students in the world **b)** All United States senators; all college football players **c)** George W. Bush and Jeb Bush; all politicians in the United States **d)** Bill Clinton and George W. Bush; all politicians in the United States **23.** Let $C(x)$ be the propositional function "x is in your class." **a)** $\exists x H(x)$ and $\exists x(C(x) \wedge H(x))$, where $H(x)$ is "x can speak Hindi" **b)** $\forall x F(x)$ and $\forall x(C(x) \to F(x))$, where $F(x)$ is "x is friendly" **c)** $\exists x \neg B(x)$ and $\exists x(C(x) \wedge \neg B(x))$, where $B(x)$ is "x was born in California" **d)** $\exists x M(x)$ and $\exists x(C(x) \wedge M(x))$, where $M(x)$ is "x has been in a movie" **e)** $\forall x \neg L(x)$ and $\forall x(C(x) \to \neg L(x))$, where $L(x)$ is "x has taken a course in logic programming" **25.** Let $P(x)$ be "x is perfect"; let $F(x)$ be "x is your friend"; and let the domain be all people. **a)** $\forall x \neg P(x)$ **b)** $\neg \forall x P(x)$ **c)** $\forall x(F(x) \to P(x))$ **d)** $\exists x(F(x) \wedge P(x))$ **e)** $\forall x(F(x) \wedge P(x))$ or $(\forall x F(x)) \wedge (\forall x P(x))$ **f)** $(\neg \forall x F(x)) \vee (\exists x \neg P(x))$ **27.** Let $Y(x)$ be the propositional function that x is in your school or class, as appropriate. **a)** If we let $V(x)$ be "x has lived in Vietnam," then we have $\exists x V(x)$ if the domain is just your schoolmates, or $\exists x(Y(x) \wedge V(x))$ if the domain is all people. If we let $D(x, y)$ mean that person x has lived in country y, then we can rewrite this last one as $\exists x(Y(x) \wedge D(x, \text{Vietnam}))$. **b)** If we let $H(x)$ be "x can speak Hindi," then we have $\exists x \neg H(x)$ if the domain is just your schoolmates, or $\exists x(Y(x) \wedge \neg H(x))$ if the domain is all people. If we let $S(x, y)$ mean that person x can speak language y, then we can rewrite this last one as $\exists x(Y(x) \wedge \neg S(x, \text{Hindi}))$. **c)** If we let $J(x)$, $P(x)$, and $C(x)$ be the propositional functions asserting x's knowledge of Java, Prolog, and C++, respectively, then we have $\exists x(J(x) \wedge P(x) \wedge C(x))$ if the domain is just your schoolmates, or $\exists x(Y(x) \wedge J(x) \wedge P(x) \wedge C(x))$ if the domain is all people. If we let $K(x, y)$ mean that person x knows programming language y, then we can rewrite this last one as $\exists x(Y(x) \wedge K(x, \text{Java}) \wedge K(x, \text{Prolog}) \wedge K(x, \text{C++}))$. **d)** If we let $T(x)$ be "x enjoys Thai food," then we have $\forall x\, T(x)$ if the domain is just your classmates, or $\forall x(Y(x) \rightarrow T(x))$ if the domain is all people. If we let $E(x, y)$ mean that person x enjoys food of type y, then we can rewrite this last one as $\forall x(Y(x) \rightarrow E(x, \text{Thai}))$. **e)** If we let $H(x)$ be "x plays hockey," then we have $\exists x\, \neg H(x)$ if the domain is just your classmates, or $\exists x(Y(x) \wedge \neg H(x))$ if the domain is all people. If we let $P(x, y)$ mean that person x plays game y, then we can rewrite this last one as $\exists x(Y(x) \wedge \neg P(x, \text{hockey}))$. **29.** Let $T(x)$ mean that x is a tautology and $C(x)$ mean that x is a contradiction. **a)** $\exists x\, T(x)$ **b)** $\forall x(C(x) \rightarrow T(\neg x))$ **c)** $\exists x \exists y(\neg T(x) \wedge \neg C(x) \wedge \neg T(y) \wedge \neg C(y) \wedge T(x \vee y))$ **d)** $\forall x \forall y((T(x) \wedge T(y)) \rightarrow T(x \wedge y))$ **31. a)** $Q(0, 0, 0) \wedge Q(0, 1, 0)$ **b)** $Q(0, 1, 1) \vee Q(1, 1, 1) \vee Q(2, 1, 1)$ **c)** $\neg Q(0, 0, 0) \vee \neg Q(0, 0, 1)$ **d)** $\neg Q(0, 0, 1) \vee \neg Q(1, 0, 1) \vee \neg Q(2, 0, 1)$ **33. a)** Let $T(x)$ be the predicate that x can learn new tricks, and let the domain be old dogs. Original is $\exists x\, T(x)$. Negation is $\forall x\, \neg T(x)$: "No old dogs can learn new tricks." **b)** Let $C(x)$ be the predicate that x knows calculus, and let the domain be rabbits. Original is $\neg \exists x\, C(x)$. Negation is $\exists x\, C(x)$: "There is a rabbit that knows calculus." **c)** Let $F(x)$ be the predicate that x can fly, and let the domain be birds. Original is $\forall x\, F(x)$. Negation is $\exists x\, \neg F(x)$: "There is a bird who cannot fly." **d)** Let $T(x)$ be the predicate that x can talk, and let the domain be dogs. Original is $\neg \exists x\, T(x)$. Negation is $\exists x\, T(x)$: "There is a dog that talks." **e)** Let $F(x)$ and $R(x)$ be the predicates that x knows French and knows Russian, respectively, and let the domain be people in this class. Original is $\neg \exists x(F(x) \wedge R(x))$. Negation is $\exists x(F(x) \wedge R(x))$: "There is someone in this class who knows French and Russian." **35. a)** $\exists x(x \leq 1)$ **b)** $\exists x(x > 2)$ **c)** $\forall x(x < 4)$ **d)** $\forall x(x \geq 0)$ **e)** $\exists x((x \geq -1) \wedge (x \leq 2))$ **f)** $\forall x((x \geq 4) \wedge (x \leq 7))$ **37. a)** There is no counterexample. **b)** $x = 0$ **c)** $x = 2$ **39. a)** $\forall x((F(x, 25{,}000) \vee S(x, 25)) \rightarrow E(x))$, where $E(x)$ is "Person x qualifies as an elite flyer in a given year," $F(x, y)$ is "Person x flies more than y miles in a given year," and $S(x, y)$ is "Person x takes more than y flights in a given year" **b)** $\forall x(((M(x) \wedge T(x, 3)) \vee (\neg M(x) \wedge T(x, 3.5))) \rightarrow Q(x))$, where $Q(x)$ is "Person x qualifies for the marathon," $M(x)$ is "Person x is a man," and $T(x, y)$ is "Person x has run the marathon in less than y hours" **c)** $M \rightarrow ((H(60) \vee (H(45) \wedge T)) \wedge \forall y\, G(\text{B}, y))$, where M is the proposition "The student received a masters degree," $H(x)$ is "The student took at least x course hours," T is the proposition "The student wrote a thesis," and $G(x, y)$ is "The person got grade x or higher in course y" **d)** $\exists x((T(x, 21) \wedge G(x, 4.0))$, where $T(x, y)$ is "Person x took more than y credit hours" and $G(x, p)$ is "Person x earned grade point average p" (we assume that we are talking about one given semester) **41. a)** If there is a printer that is both out of service and busy, then some job has been lost. **b)** If every printer is busy, then there is a job in the queue. **c)** If there is a job that is both queued and lost, then some printer is out of service. **d)** If every printer is busy and every job is queued, then some job is lost. **43. a)** $(\exists x\, F(x, 10)) \rightarrow \exists x\, S(x)$, where $F(x, y)$ is "Disk x has more than y kilobytes of free space," and $S(x)$ is "Mail message x can be saved" **b)** $(\exists x\, A(x)) \rightarrow \forall x(Q(x) \rightarrow T(x))$, where $A(x)$ is "Alert x is active," $Q(x)$ is "Message x is queued," and $T(x)$ is "Message x is transmitted" **c)** $\forall x((x \neq \text{main console}) \rightarrow T(x))$, where $T(x)$ is "The diagnostic monitor tracks the status of system x" **d)** $\forall x(\neg L(x) \rightarrow B(x))$, where $L(x)$ is "The host of the conference call put participant x on a special list" and $B(x)$ is "Participant x was billed" **45.** They are not equivalent. Let $P(x)$ be any propositional function that is sometimes true and sometimes false, and let $Q(x)$ be any propositional function that is always false. Then $\forall x(P(x) \rightarrow Q(x))$ is false but $\forall x P(x) \rightarrow \forall x Q(x)$ is true. **47.** Both statements are true precisely when at least one of $P(x)$ and $Q(x)$ is true for at least one value of x in the domain. **49. a)** If A is true, then both sides are logically equivalent to $\forall x P(x)$. If A is false, the left-hand side is clearly false. Furthermore, for every x, $P(x) \wedge A$ is false, so the right-hand side is false. Hence, the two sides are logically equivalent. **b)** If A is true, then both sides are logically equivalent to $\exists x\, P(x)$. If A is false, the left-hand side is clearly false. Furthermore, for every x, $P(x) \wedge A$ is false, so $\exists x(P(x) \wedge A)$ is false. Hence, the two sides are logically equivalent. **51.** We can establish these equivalences by arguing that one side is true if and only if the other side is true. **a)** Suppose that A is true. Then for each x, $P(x) \rightarrow A$ is true; therefore, the left-hand side is always true in this case. By similar reasoning the right-hand side is always true in this case. Therefore, the two propositions are logically equivalent when A is true. On the other hand, suppose that A is false. There are two subcases. If $P(x)$ is false for every x, then $P(x) \rightarrow A$ is vacuously true, so the left-hand side is vacuously true. The same reasoning shows that the right-hand side is also true, because in this subcase $\exists x P(x)$ is false. For the second subcase, suppose that $P(x)$ is true for some x. Then for that x, $P(x) \rightarrow A$ is false, so the left-hand side is false. The right-hand side is also false, because in this subcase $\exists x P(x)$ is true but A is false. Thus, in all cases, the two propositions have the same truth value. **b)** If A is true, then both sides are trivially true, because the conditional statements have true conclusions. If A is false, then there are two subcases. If $P(x)$ is false for some x, then $P(x) \rightarrow A$ is vacuously true for that x, so the left-hand side is true. The same reasoning shows that the right-hand side is true, because in this subcase $\forall x P(x)$ is false. For the second subcase, suppose that $P(x)$ is true for every x. Then for every x, $P(x) \rightarrow A$ is false, so the left-hand side is false (there is no x making the conditional statement true). The right-hand side is also false, because it is a conditional statement with a true hypothesis and a false conclusion. Thus, in all cases, the two propositions have the same truth value. **53.** To show these are not logically equivalent, let $P(x)$ be the statement "x is positive," and let $Q(x)$ be the statement "x is negative" with domain the set of integers. Then $\exists x\ P(x) \land \exists x\ Q(x)$ is true, but $\exists x(P(x) \land Q(x))$ is false. **55. a)** True **b)** False, unless the domain consists of just one element or the hypothesis is false. **c)** True **57. a)** Yes **b)** No **c)** juana, kiko **d)** math273, cs301 **e)** juana, kiko **59.** `sibling(X,Y) :- mother(M,X), mother(M,Y), father(F,X), father(F,Y)` **61. a)** $\forall x(P(x) \to \neg Q(x))$ **b)** $\forall x(Q(x) \to R(x))$ **c)** $\forall x(P(x) \to \neg R(x))$ **d)** The conclusion does not follow. There may be vain professors, because the premises do not rule out the possibility that there are other vain people besides ignorant ones. **63. a)** $\forall x(P(x) \to \neg Q(x))$ **b)** $\forall x(R(x) \to \neg S(x))$ **c)** $\forall x(\neg Q(x) \to S(x))$ **d)** $\forall x(P(x) \to \neg R(x))$ **e)** The conclusion follows. Suppose x is a baby. Then, by the first premise, x is illogical, so by the third premise, x is despised. The second premise says that if x could manage a crocodile, then x would not be despised. Therefore, x cannot manage a crocodile.

Section 1.5

1. a) For every real number x there exists a real number y such that x is less than y. **b)** For every real number x and real number y, if x and y are both nonnegative, then their product is nonnegative. **c)** For every real number x and real number y, there exists a real number z such that $xy = z$. **3. a)** There is some student in your class who has sent a message to some student in your class. **b)** There is some student in your class who has sent a message to every student in your class. **c)** Every student in your class has sent a message to at least one student in your class. **d)** There is a student in your class who has been sent a message by every student in your class. **e)** Every student in your class has been sent a message from at least one student in your class. **f)** Every student in the class has sent a message to every student in the class. **5. a)** Sarah Smith has visited www.att.com. **b)** At least one person has visited www.imdb.org. **c)** Jose Orez has visited at least one website. **d)** There is a website that both Ashok Puri and Cindy Yoon have visited. **e)** There is a person besides David Belcher who has visited all the websites that David Belcher has visited. **f)** There are two different people who have visited exactly the same websites. **7. a)** Abdallah Hussein does not like Japanese cuisine. **b)** Some student at your school likes Korean cuisine, and everyone at your school likes Mexican cuisine. **c)** There is some cuisine that either Monique Arsenault or Jay Johnson likes. **d)** For every pair of distinct students at your school, there is some cuisine that at least one them does not like. **e)** There are two students at your school who like exactly the same set of cuisines. **f)** For every pair of students at your school, there is some cuisine about which they have the same opinion (either they both like it or they both do not like it). **9. a)** $\forall x L(x, \text{Jerry})$ **b)** $\forall x \exists y L(x, y)$ **c)** $\exists y \forall x L(x, y)$ **d)** $\forall x \exists y \neg L(x, y)$ **e)** $\exists x \neg L(\text{Lydia}, x)$ **f)** $\exists x \forall y \neg L(y, x)$ **g)** $\exists x(\forall y L(y, x) \land \forall z((\forall w L(w, z)) \to z = x))$ **h)** $\exists x \exists y(x \neq y \land L(\text{Lynn}, x) \land L(\text{Lynn}, y) \land \forall z(L(\text{Lynn}, z) \to (z = x \lor z = y)))$ **i)** $\forall x L(x, x)$ **j)** $\exists x\ \forall y\ (L(x, y) \leftrightarrow x = y)$ **11. a)** $A(\text{Lois, Professor Michaels})$ **b)** $\forall x(S(x) \to A(x, \text{Professor Gross}))$ **c)** $\forall x(F(x) \to (A(x, \text{Professor Miller}) \lor A(\text{Professor Miller}, x)))$ **d)** $\exists x(S(x) \land \forall y(F(y) \to \neg A(x, y)))$ **e)** $\exists x(F(x) \land \forall y(S(y) \to \neg A(y, x)))$ **f)** $\forall y(F(y) \to \exists x(S(x) \lor A(x, y)))$ **g)** $\exists x(F(x) \land \forall y((F(y) \land (y \neq x)) \to A(x, y)))$ **h)** $\exists x(S(x) \land \forall y(F(y) \to \neg A(y, x)))$ **13. a)** $\neg M(\text{Chou, Koko})$ **b)** $\neg M(\text{Arlene, Sarah}) \land \neg T(\text{Arlene, Sarah})$ **c)** $\neg M(\text{Deborah, Jose})$ **d)** $\forall x M(x, \text{Ken})$ **e)** $\forall x \neg T(x, \text{Nina})$ **f)** $\forall x(T(x, \text{Avi}) \lor M(x, \text{Avi}))$ **g)** $\exists x \forall y(y \neq x \to M(x, y))$ **h)** $\exists x \forall y(y \neq x \to (M(x, y) \lor T(x, y)))$ **i)** $\exists x \exists y(x \neq y \land M(x, y) \land M(y, x))$ **j)** $\exists x M(x, x)$ **k)** $\exists x \forall y(x \neq y \to (\neg M(x, y) \land \neg T(y, x)))$ **l)** $\forall x(\exists y(x \neq y \land (M(y, x) \lor T(y, x))))$ **m)** $\exists x \exists y(x \neq y \land M(x, y) \land T(y, x))$ **n)** $\exists x \exists y(x \neq y \land \forall z((z \neq x \land z \neq y) \to (M(x, z) \lor M(y, z) \lor T(x, z) \lor T(y, z))))$ **15. a)** $\forall x P(x)$, where $P(x)$ is "x needs a course in discrete mathematics" and the domain consists of all computer science students **b)** $\exists x P(x)$, where $P(x)$ is "x owns a personal computer" and the domain consists of all students in this class **c)** $\forall x \exists y P(x, y)$, where $P(x, y)$ is "x has taken y," the domain for x consists of all students in this class, and the domain for y consists of all computer science classes **d)** $\exists x \exists y P(x, y)$, where $P(x, y)$ and domains are the same as in part (c) **e)** $\forall x \forall y P(x, y)$, where $P(x, y)$ is "x has been in y," the domain for x consists of all students in this class, and the domain for y consists of all buildings on campus **f)** $\exists x \exists y \forall z(P(z, y) \to Q(x, z))$, where $P(z, y)$ is "z is in y" and $Q(x, z)$ is "x has been in z"; the domain for x consists of all students in the class, the domain for y consists of all buildings on campus, and the domain of z consists of all rooms. **g)** $\forall x \forall y \exists z(P(z, y) \land Q(x, z))$, with same environment as in part (f) **17. a)** $\forall u \exists m(A(u, m) \land \forall n(n \neq m \to \neg A(u, n)))$, where $A(u, m)$ means that user u has access to mailbox m **b)** $\exists p \forall e(H(e) \land S(p, \text{running})) \to S(\text{kernel, working correctly})$, where $H(e)$ means that error condition e is in effect and $S(x, y)$ means that the status of x is y **c)** $\forall u \forall s(E(s, \text{.edu}) \to A(u, s))$, where $E(s, x)$ means that website s has extension x, and $A(u, s)$ means that user u can access website s **d)** $\exists x \exists y(x \neq y \land \forall z((\forall s M(z, s)) \leftrightarrow (z = x \lor z = y)))$, where $M(a, b)$ means that system a monitors remote server b **19. a)** $\forall x \forall y((x < 0) \land (y < 0) \to (x + y < 0))$ **b)** $\neg \forall x \forall y\ ((x > 0) \land (y > 0) \to (x - y > 0))$ **c)** $\forall x \forall y\ (x^2 + y^2 \geq (x + y)^2)$ **d)** $\forall x \forall y\ (|xy| = |x||y|)$ **21.** $\forall x \exists a \exists b \exists c \exists d\ ((x > 0) \to x = a^2 + b^2 + c^2 + d^2)$, where the domain consists of all integers **23. a)** $\forall x\ \forall y\ ((x < 0) \land (y < 0) \to (xy > 0))$ **b)** $\forall x(x - x = 0)$ **c)** $\forall x \exists a \exists b(a \neq b \land \forall c(c^2 = x \leftrightarrow (c = a \lor c = b)))$ **d)** $\forall x((x < 0) \to \neg \exists y(x = y^2))$ **25. a)** There is a multiplicative identity for the real numbers. **b)** The product of two negative real numbers is always a positive real number. **c)** There exist real numbers x and y such that x^2 exceeds y but x is less than y. **d)** The real numbers are closed under the operation of addition. **27. a)** True **b)** True **c)** True **d)** True **e)** True **f)** False **g)** False **h)** True **i)** False **29. a)** $P(1, 1) \land P(1, 2) \land P(1, 3) \land P(2, 1) \land P(2, 2) \land P(2, 3) \land P(3, 1) \land P(3, 2) \land P(3, 3)$ **b)** $P(1, 1) \lor P(1, 2) \lor P(1, 3) \lor P(2, 1) \lor P(2, 2) \lor P(2, 3) \lor P(3,1) \lor P(3, 2) \lor P(3, 3)$ **c)** $(P(1, 1) \land P(1, 2) \land P(1, 3)) \lor$

$(P(2, 1) \wedge P(2, 2) \wedge P(2, 3)) \vee (P(3, 1) \wedge P(3, 2) \wedge P(3, 3))$ **d)** $(P(1, 1) \vee P(2, 1) \vee P(3, 1)) \wedge (P(1, 2) \vee P(2, 2) \vee P(3, 2)) \wedge (P(1, 3) \vee P(2, 3) \vee P(3, 3))$ **31. a)** $\exists x \forall y \exists z \neg T(x, y, z)$ **b)** $\exists x \forall y \neg P(x, y) \wedge \exists x \forall y \neg Q(x, y)$ **c)** $\exists x \forall y (\neg P(x, y) \vee \forall z \neg R(x, y, z))$ **d)** $\exists x \forall y (P(x, y) \wedge \neg Q(x, y))$ **33. a)** $\exists x \exists y \neg P(x, y)$ **b)** $\exists y \forall x \neg P(x, y)$ **c)** $\exists y \exists x (\neg P(x, y) \wedge \neg Q(x, y))$ **d)** $(\forall x \forall y P(x, y)) \vee (\exists x \exists y \neg Q(x, y))$ **e)** $\exists x (\forall y \exists z \neg P(x, y, z) \vee \forall z \exists y \neg P(x, y, z))$ **35.** Any domain with four or more members makes the statement true; any domain with three or fewer members makes the statement false. **37. a)** There is someone in this class such that for every two different math courses, these are not the two and only two math courses this person has taken. **b)** Every person has either visited Libya or has not visited a country other than Libya. **c)** Someone has climbed every mountain in the Himalayas. **d)** There is someone who has neither been in a movie with Kevin Bacon nor has been in a movie with someone who has been in a movie with Kevin Bacon. **39. a)** $x = 2, y = -2$ **b)** $x = -4$ **c)** $x = 17, y = -1$ **41.** $\forall x \forall y \forall z((x \cdot y) \cdot z = x \cdot (y \cdot z))$ **43.** $\forall m \forall b (m \neq 0 \rightarrow \exists x(mx + b = 0 \wedge \forall w(mw + b = 0 \rightarrow w = x)))$ **45. a)** True **b)** False **c)** True **47.** $\neg(\exists x \forall y P(x, y)) \leftrightarrow \forall x(\neg \forall y P(x, y)) \leftrightarrow \forall x \exists y \neg P(x,y)$ **49. a)** Suppose that $\forall x P(x) \wedge \exists x Q(x)$ is true. Then $P(x)$ is true for all x and there is an element y for which $Q(y)$ is true. Because $P(x) \wedge Q(y)$ is true for all x and there is a y for which $Q(y)$ is true, $\forall x \exists y(P(x) \wedge Q(y))$ is true. Conversely, suppose that the second proposition is true. Let x be an element in the domain. There is a y such that $Q(y)$ is true, so $\exists x Q(x)$ is true. Because $\forall x P(x)$ is also true, it follows that the first proposition is true. **b)** Suppose that $\forall x P(x) \vee \exists x Q(x)$ is true. Then either $P(x)$ is true for all x, or there exists a y for which $Q(y)$ is true. In the former case, $P(x) \vee Q(y)$ is true for all x, so $\forall x \exists y(P(x) \vee Q(y))$ is true. In the latter case, $Q(y)$ is true for a particular y, so $P(x) \vee Q(y)$ is true for all x and consequently $\forall x \exists y(P(x) \vee Q(y))$ is true. Conversely, suppose that the second proposition is true. If $P(x)$ is true for all x, then the first proposition is true. If not, $P(x)$ is false for some x, and for this x there must be a y such that $P(x) \vee Q(y)$ is true. Hence, $Q(y)$ must be true, so $\exists y Q(y)$ is true. It follows that the first proposition must hold. **51.** We will show how an expression can be put into prenex normal form (PNF) if subexpressions in it can be put into PNF. Then, working from the inside out, any expression can be put in PNF. (To formalize the argument, it is necessary to use the method of structural induction that will be discussed in Section 5.3.) By Exercise 49 of Section 1.3, we can assume that the proposition uses only $\vee$ and $\neg$ as logical connectives. Now note that any proposition with no quantifiers is already in PNF. (This is the basis case of the argument.) Now suppose that the proposition is of the form $QxP(x)$, where Q is a quantifier. Because $P(x)$ is a shorter expression than the original proposition, we can put it into PNF. Then Qx followed by this PNF is again in PNF and is equivalent to the original proposition. Next, suppose that the proposition is of the form $\neg P$. If P is already in PNF, we slide the negation sign past all the quantifiers using the equivalences in Table 2 in Section 1.4. Finally, assume that proposition is of the form $P \vee Q$, where each of P and Q is in PNF. If only one of P and Q has quantifiers, then we can use Exercise 48 in Section 1.4 to bring the quantifier in front of both. If both P and Q have quantifiers, we can use Exercise 47 in Section 1.4, Exercise 48, or part (b) of Exercise 49 to rewrite $P \vee Q$ with two quantifiers preceding the disjunction of a proposition of the form $R \vee S$, and then put $R \vee S$ into PNF.

Section 1.6

1. Modus ponens; valid; the conclusion is true, because the hypotheses are true. **3. a)** Addition **b)** Simplification **c)** Modus ponens **d)** Modus tollens **e)** Hypothetical syllogism **5.** Let w be "Randy works hard," let d be "Randy is a dull boy," and let j be "Randy will get the job." The hypotheses are w, $w \rightarrow d$, and $d \rightarrow \neg j$. Using modus ponens and the first two hypotheses, d follows. Using modus ponens and the last hypothesis, $\neg j$, which is the desired conclusion, "Randy will not get the job," follows. **7.** Universal instantiation is used to conclude that "If Socrates is a man, then Socrates is mortal." Modus ponens is then used to conclude that Socrates is mortal. **9. a)** Valid conclusions are "I did not take Tuesday off," "I took Thursday off," and "It rained on Thursday." **b)** "I did not eat spicy foods and it did not thunder" is a valid conclusion. **c)** "I am clever" is a valid conclusion. **d)** "Ralph is not a CS major" is a valid conclusion. **e)** "That you buy lots of stuff is good for the U.S. and is good for you" is a valid conclusion. **f)** "Mice gnaw their food" and "Rabbits are not rodents" are valid conclusions. **11.** Suppose that $p_1, p_2, \ldots, p_n$ are true. We want to establish that $q \rightarrow r$ is true. If q is false, then we are done, vacuously. Otherwise, q is true, so by the validity of the given argument form (that whenever $p_1, p_2, \ldots, p_n, q$ are true, then r must be true), we know that r is true. **13. a)** Let $c(x)$ be "x is in this class," $j(x)$ be "x knows how to write programs in JAVA," and $h(x)$ be "x can get a high-paying job." The premises are c(Doug), j(Doug), $\forall x(j(x) \rightarrow h(x))$. Using universal instantiation and the last premise, j(Doug) $\rightarrow h$(Doug) follows. Applying modus ponens to this conclusion and the second premise, h(Doug) follows. Using conjunction and the first premise, c(Doug) $\wedge h$(Doug) follows. Finally, using existential generalization, the desired conclusion, $\exists x(c(x) \wedge h(x))$ follows. **b)** Let $c(x)$ be "x is in this class," $w(x)$ be "x enjoys whale watching," and $p(x)$ be "x cares about ocean pollution." The premises are $\exists x(c(x) \wedge w(x))$ and $\forall x(w(x) \rightarrow p(x))$. From the first premise, $c(y) \wedge w(y)$ for a particular person y. Using simplification, $w(y)$ follows. Using the second premise and universal instantiation, $w(y) \rightarrow p(y)$ follows. Using modus ponens, $p(y)$ follows, and by conjunction, $c(y) \wedge p(y)$ follows. Finally, by existential generalization, the desired conclusion, $\exists x(c(x) \wedge p(x))$, follows. **c)** Let $c(x)$ be "x is in this class," $p(x)$ be "x owns a PC," and $w(x)$ be "x can use a word-processing program." The premises are c(Zeke), $\forall x(c(x) \rightarrow p(x))$, and

$\forall x(p(x) \to w(x))$. Using the second premise and universal instantiation, $c(\text{Zeke}) \to p(\text{Zeke})$ follows. Using the first premise and modus ponens, $p(\text{Zeke})$ follows. Using the third premise and universal instantiation, $p(\text{Zeke}) \to w(\text{Zeke})$ follows. Finally, using modus ponens, $w(\text{Zeke})$, the desired conclusion, follows. **d)** Let $j(x)$ be "x is in New Jersey," $f(x)$ be "x lives within 50 miles of the ocean," and $s(x)$ be "x has seen the ocean." The premises are $\forall x(j(x) \to f(x))$ and $\exists x(j(x) \wedge \neg s(x))$. The second hypothesis and existential instantiation imply that $j(y) \wedge \neg s(y)$ for a particular person y. By simplification, $j(y)$ for this person y. Using universal instantiation and the first premise, $j(y) \to f(y)$, and by modus ponens, $f(y)$ follows. By simplification, $\neg s(y)$ follows from $j(y) \wedge \neg s(y)$. So $f(y) \wedge \neg s(y)$ follows by conjunction. Finally, the desired conclusion, $\exists x(f(x) \wedge \neg s(x))$, follows by existential generalization. **15. a)** Correct, using universal instantiation and modus ponens **b)** Invalid; fallacy of affirming the conclusion **c)** Invalid; fallacy of denying the hypothesis **d)** Correct, using universal instantiation and modus tollens **17.** We know that *some* x exists that makes $H(x)$ true, but we cannot conclude that Lola is one such x. **19. a)** Fallacy of affirming the conclusion **b)** Valid argument using modus tollens **c)** Fallacy of denying the hypothesis **21.** By the second premise, there is some lion that does not drink coffee. Let Leo be such a creature. By simplification we know that Leo is a lion. By modus ponens we know from the first premise that Leo is fierce. Hence, Leo is fierce and does not drink coffee. By the definition of the existential quantifier, there exist fierce creatures that do not drink coffee, that is, some fierce creatures do not drink coffee. **23.** The error occurs in step (5), because we cannot assume, as is being done here, that the c that makes P true is the same as the c that makes Q true. **25.** We are given the premises $\forall x(P(x) \to Q(x))$ and $\neg Q(a)$. We want to show $\neg P(a)$. Suppose, to the contrary, that $\neg P(a)$ is not true. Then $P(a)$ is true. Therefore, by universal modus ponens, we have $Q(a)$. But this contradicts the given premise $\neg Q(a)$. Therefore, our supposition must have been wrong, and so $\neg P(a)$ is true, as desired.

27.

Step	**Reason**
1. $\forall x(P(x) \wedge R(x))$	Premise
2. $P(a) \wedge R(a)$	Universal instantiation from (1)
3. $P(a)$	Simplification from (2)
4. $\forall x(P(x) \to (Q(x) \wedge S(x)))$	Premise
5. $Q(a) \wedge S(a)$	Universal modus ponens from (3) and (4)
6. $S(a)$	Simplification from (5)
7. $R(a)$	Simplification from (2)
8. $R(a) \wedge S(a)$	Conjunction from (7) and (6)
9. $\forall x(R(x) \wedge S(x))$	Universal generalization from (5)

29.

Step	**Reason**
1. $\exists x \neg P(x)$	Premise
2. $\neg P(c)$	Existential instantiation from (1)
3. $\forall x(P(x) \vee Q(x))$	Premise
4. $P(c) \vee Q(c)$	Universal instantiation from (3)
5. $Q(c)$	Disjunctive syllogism from (4) and (2)
6. $\forall x(\neg Q(x) \vee S(x))$	Premise
7. $\neg Q(c) \vee S(c)$	Universal instantiation from (6)
8. $S(c)$	Disjunctive syllogism from (5) and (7)
9. $\forall x(R(x) \to \neg S(x))$	Premise
10. $R(c) \to \neg S(c)$	Universal instantiation from (9)
11. $\neg R(c)$	Modus tollens from (8) and (10)
12. $\exists x \neg R(x)$	Existential generalization from (11)

31. Let p be "It is raining"; let q be "Yvette has her umbrella"; let r be "Yvette gets wet." Assumptions are $\neg p \vee q$, $\neg q \vee \neg r$, and $p \vee \neg r$. Resolution on the first two gives $\neg p \vee \neg r$. Resolution on this and the third assumption gives $\neg r$, as desired. **33.** Assume that this proposition is satisfiable. Using resolution on the first two clauses enables us to conclude $q \vee q$; in other words, we know that q has to be true. Using resolution on the last two clauses enables us to conclude $\neg q \vee \neg q$; in other words, we know that $\neg q$ has to be true. This is a contradiction. So this proposition is not satisfiable. **35.** Valid

Section 1.7

1. Let $n = 2k + 1$ and $m = 2l + 1$ be odd integers. Then $n + m = 2(k + l + 1)$ is even. **3.** Suppose that n is even. Then $n = 2k$ for some integer k. Therefore, $n^2 = (2k)^2 = 4k^2 = 2(2k^2)$. Because we have written n^2 as 2 times an integer, we conclude that n^2 is even. **5.** Direct proof: Suppose that $m + n$ and $n + p$ are even. Then $m + n = 2s$ for some integer s and $n + p = 2t$ for some integer t. If we add these, we get $m + p + 2n = 2s + 2t$. Subtracting $2n$ from both sides and factoring, we have $m + p = 2s + 2t - 2n = 2(s + t - n)$. Because we have written $m + p$ as 2 times an integer, we conclude that $m + p$ is even. **7.** Because n is odd, we can write $n = 2k + 1$ for some integer k. Then $(k + 1)^2 - k^2 = k^2 + 2k + 1 - k^2 = 2k + 1 = n$. **9.** Suppose that r is rational and i is irrational and $s = r + i$ is rational. Then by Example 8, $s + (-r) = i$ is rational, which is a contradiction. **11.** Because $\sqrt{2} \cdot \sqrt{2} = 2$ is rational and $\sqrt{2}$ is irrational, the product of two irrational numbers is not necessarily irrational. **13.** Proof by contraposition: If $1/x$ were rational, then by definition $1/x = p/q$ for some integers p and q with $q \neq 0$. Because $1/x$ cannot be 0 (if it were, then we'd have the contradiction $1 = x \cdot 0$ by multiplying both sides by x), we know that $p \neq 0$. Now $x = 1/(1/x) = 1/(p/q) = q/p$ by the usual rules of algebra and arithmetic. Hence, x can be written as the quotient of two integers with the denominator nonzero. Thus, by definition, x is rational. **15.** Assume that $\sqrt{x}$ were rational. Then, because the product of two rational numbers is rational, $(\sqrt{x})^2 = x$ is also rational. This contradicts the hypothesis that x is irrational. **17.** Assume that it is not true that $x \geq 1$ or $y \geq 1$. Then $x < 1$ and $y < 1$. Adding these two inequalities, we obtain $x + y < 2$, which is the negation of $x + y \geq 2$. **19. a)** Assume that n is odd, so $n = 2k + 1$ for some integer k. Then $n^3 + 5 = 2(4k^3 + 6k^2 + 3k + 3)$. Because $n^3 + 5$ is two times some integer, it is even. **b)** Suppose that $n^3 + 5$ is odd and n is odd. Because n is odd and the product of two odd numbers is odd, it follows that n^2 is odd and then that n^3 is odd. But then $5 = (n^3 + 5) - n^3$ would have to be even because it is the difference of two odd numbers. Therefore, the supposition that $n^3 + 5$ and n were both odd is wrong. **21.** The proposition is vacuously true because 0 is not a positive integer. Vacuous proof. **23.** $P(1)$ is true because $(a+b)^1 = a+b \geq a^1 + b^1 = a+b$. Direct proof. **25.** If we chose 9 or fewer days on each day of the week, this would account for at most $9 \cdot 7 = 63$ days. But we chose 64 days. This contradiction shows that at least 10 of the days we chose must be on the same day of the week. **27.** Suppose by way of contradiction that a/b is a rational root, where a and b are integers and this fraction is in lowest terms (that is, a and b have no common divisor greater than 1). Plug this proposed root into the equation to obtain $a^3/b^3 + a/b + 1 = 0$. Multiply through by b^3 to obtain $a^3 + ab^2 + b^3 = 0$. If a and b are both odd, then the left-hand side is the sum of three odd numbers and therefore must be odd. If a is odd and b is even, then the left-hand side is odd+even+even, which is again odd. Similarly, if a is even and b is odd, then the left-hand side is even+even+odd, which is again odd. Because the fraction a/b is in simplest terms, it cannot happen that both a and b are even. Thus, in all cases, the left-hand side is odd, and therefore cannot equal 0. This contradiction shows that no such root exists. **29.** First, assume that n is odd, so that $n = 2k + 1$ for some integer k. Then $5n+6 = 5(2k+1)+6 = 10k+11 = 2(5k+5)+1$. Hence, $5n + 6$ is odd. To prove the converse, suppose that n is even, so that $n = 2k$ for some integer k. Then $5n + 6 = 10k + 6 = 2(5k + 3)$, so $5n + 6$ is even. Hence, n is odd if and only if $5n+6$ is odd. **31.** This proposition is true. Suppose that m is neither 1 nor -1. Then mn has a factor m larger than 1. On the other hand, $mn = 1$, and 1 has no such factor. Hence, $m = 1$ or $m = -1$. In the first case $n = 1$, and in the second case $n = -1$, because $n = 1/m$. **33.** We prove that all these are equivalent to x being even. If x is even, then $x = 2k$ for some integer k. Therefore, $3x+2 = 3 \cdot 2k+2 = 6k+2 = 2(3k+1)$, which is even, because it has been written in the form $2t$, where $t = 3k + 1$. Similarly, $x+5 = 2k+5 = 2k+4+1 = 2(k+2)+1$, so $x+5$ is odd; and $x^2 = (2k)^2 = 2(2k^2)$, so x^2 is even. For the converses, we will use a proof by contraposition. So assume that x is not even; thus, x is odd and we can write $x = 2k+1$ for some integer k. Then $3x+2 = 3(2k+1)+2 = 6k+5 = 2(3k+2)+1$, which is odd (i.e., not even), because it has been written in the form $2t+1$, where $t = 3k+2$. Similarly, $x+5 = 2k+1+5 = 2(k+3)$, so $x + 5$ is even (i.e., not odd). That x^2 is odd was already proved in Example 1. **35.** We give proofs by contraposition of $(i) \rightarrow (ii)$, $(ii) \rightarrow (i)$, $(i) \rightarrow (iii)$, and $(iii) \rightarrow (i)$. For the first of these, suppose that $3x + 2$ is rational, namely, equal to p/q for some integers p and q with $q \neq 0$. Then we can write $x = ((p/q) - 2)/3 = (p - 2q)/(3q)$, where $3q \neq 0$. This shows that x is rational. For the second conditional statement, suppose that x is rational, namely, equal to p/q for some integers p and q with $q \neq 0$. Then we can write $3x + 2 = (3p + 2q)/q$, where $q \neq 0$. This shows that $3x + 2$ is rational. For the third conditional statement, suppose that $x/2$ is rational, namely, equal to p/q for some integers p and q with $q \neq 0$. Then we can write $x = 2p/q$, where $q \neq 0$. This shows that x is rational. And for the fourth conditional statement, suppose that x is rational, namely, equal to p/q for some integers p and q with $q \neq 0$. Then we can write $x/2 = p/(2q)$, where $2q \neq 0$. This shows that $x/2$ is rational. **37.** No **39.** Suppose that $p_1 \rightarrow p_4 \rightarrow p_2 \rightarrow p_5 \rightarrow p_3 \rightarrow p_1$. To prove that one of these propositions implies any of the others, just use hypothetical syllogism repeatedly. **41.** We will give a proof by contradiction. Suppose that $a_1, a_2, \ldots, a_n$ are all less than A, where A is the average of these numbers. Then $a_1 + a_2 + \cdots + a_n < nA$. Dividing both sides by n shows that $A = (a_1 + a_2 + \cdots + a_n)/n < A$, which is a contradiction. **43.** We will show that the four statements are equivalent by showing that (i) implies (ii), (ii) implies (iii), (iii) implies (iv), and (iv) implies (i). First, assume that n is even. Then $n = 2k$ for some integer k. Then $n + 1 = 2k + 1$, so $n + 1$ is odd. This shows that (i) implies (ii). Next, suppose that $n + 1$ is odd, so $n + 1 = 2k + 1$ for some integer k. Then $3n + 1 = 2n + (n + 1) = 2(n + k) + 1$, which shows that $3n + 1$ is odd, showing that (ii) implies (iii). Next, suppose that $3n + 1$ is odd, so $3n + 1 = 2k + 1$ for some integer k. Then $3n = (2k + 1) - 1 = 2k$, so $3n$ is even. This shows that (iii) implies (iv). Finally, suppose that n is not even. Then n is odd, so $n = 2k + 1$ for some integer k. Then $3n = 3(2k + 1) = 6k + 3 = 2(3k + 1) + 1$, so $3n$ is odd. This completes a proof by contraposition that (iv) implies (i).

Section 1.8

1. $1^2 + 1 = 2 \geq 2 = 2^1$; $2^2 + 1 = 5 \geq 4 = 2^2$; $3^2 + 1 = 10 \geq 8 = 2^3$; $4^2 + 1 = 17 \geq 16 = 2^4$ **3.** We must show that for all positive integers x it is not true that $x^3 = 100$. *Case (i):* If $x \leq 4$, then $x^3 \leq 64$, so $x^3 \neq 100$. *Case (ii):* If $x \geq 5$, then $x^3 \geq 125$, so $x^3 \neq 100$. **5.** If $x \leq y$, then $\max(x, y) + \min(x, y) = y + x = x + y$. If $x \geq y$, then $\max(x, y) + \min(x, y) = x + y$. Because these are the only two cases, the equality always holds. **7.** Because $|x - y| = |y - x|$, the values of x and y are interchangeable. Therefore, without loss of generality, we can assume that $x \geq y$. Then $(x + y - (x - y))/2 = (x + y - x + y)/2 = 2y/2 = y = \min(x, y)$. Similarly, $(x + y + (x - y))/2 = (x + y + x - y)/2 = 2x/2 = x = \max(x, y)$. **9.** There are four cases. *Case 1:* $x \geq 0$ and $y \geq 0$. Then $|x| + |y| = x + y = |x + y|$. *Case 2:* $x < 0$ and $y < 0$. Then $|x| + |y| = -x + (-y) = -(x + y) = |x + y|$ because $x + y < 0$. *Case 3:* $x \geq 0$ and $y < 0$. Then $|x| + |y| = x + (-y)$. If $x \geq -y$, then $|x + y| = x + y$. But because $y < 0$, $-y > y$, so $|x| + |y| = x + (-y) > x + y = |x + y|$. If $x < -y$, then $|x + y| = -(x + y) = -x + (-y)$. But because $x \geq 0$, $x \geq -x$, so $|x| + |y| = x + (-y) \geq -x + (-y) = |x + y|$. *Case 4:* $x < 0$ and $y \geq 0$. Identical to Case 3 with the roles of x and y reversed. **11.** 10,001, 10,002, …, 10,100 are all nonsquares, because $100^2 = 10{,}000$ and $101^2 = 10{,}201$; constructive. **13.** $8 = 2^3$ and $9 = 3^2$ **15.** Let $x = 2$ and $y = \sqrt{2}$. If $x^y = 2^{\sqrt{2}}$ is irrational, we are done. If not, then let $x = 2^{\sqrt{2}}$ and $y = \sqrt{2}/4$. Then $x^y = (2^{\sqrt{2}})^{\sqrt{2}/4} = 2^{\sqrt{2}\cdot(\sqrt{2})/4} = 2^{1/2} = \sqrt{2}$. **17. a)** This statement asserts the existence of x with a certain property. If we let $y = x$, then we see that $P(x)$ is true. If y is anything other than x, then $P(x)$ is not true. Thus, x is the unique element that makes P true. **b)** The first clause here says that there is an element that makes P true. The second clause says that whenever two elements both make P true, they are in fact the same element. Together these say that P is satisfied by exactly one element. **c)** This statement asserts the existence of an x that makes P true and has the further property that whenever we find an element that makes P true, that element is x. In other words, x is the unique element that makes P true. **19.** The equation $|a-c| = |b-c|$ is equivalent to the disjunction of two equations: $a - c = b - c$ or $a - c = -b + c$. The first of these is equivalent to $a = b$, which contradicts the assumptions made in this problem, so the original equation is equivalent to $a - c = -b + c$. By adding $b + c$ to both sides and dividing by 2, we see that this equation is equivalent to $c = (a + b)/2$. Thus, there is a unique solution. Furthermore, this c is an integer, because the sum of the odd integers a and b is even. **21.** We are being asked to solve $n = (k-2)+(k+3)$ for k. Using the usual, reversible, rules of algebra, we see that this equation is equivalent to $k = (n-1)/2$. In other words, this is the one and only value of k that makes our equation true. Because n is odd, $n - 1$ is even, so k is an integer. **23.** If x is itself an integer, then we can take $n = x$ and $\epsilon = 0$. No other solution is possible in this case, because if the integer n is greater than x, then n is at least $x + 1$, which would make $\epsilon \geq 1$. If x is not an integer, then round it up to the next integer, and call that integer n. Let $\epsilon = n - x$. Clearly $0 \leq \epsilon < 1$; this is the only ϵ that will work with this n, and n cannot be any larger, because ϵ is constrained to be less than 1. **25.** The harmonic mean of distinct positive real numbers x and y is always less than their geometric mean. To prove $2xy/(x + y) < \sqrt{xy}$, multiply both sides by $(x + y)/(2\sqrt{xy})$ to obtain the equivalent inequality $\sqrt{xy} < (x + y)/2$, which is proved in Example 14. **27.** The parity (oddness or evenness) of the sum of the numbers written on the board never changes, because $j + k$ and $|j - k|$ have the same parity (and at each step we reduce the sum by $j + k$ but increase it by $|j - k|$). Therefore the integer at the end of the process must have the same parity as $1 + 2 + \cdots + (2n) = n(2n + 1)$, which is odd because n is odd. **29.** Without loss of generality we can assume that n is nonnegative, because the fourth power of an integer and the fourth power of its negative are the same. We divide an arbitrary positive integer n by 10, obtaining a quotient k and remainder l, whence $n = 10k + l$, and l is an integer between 0 and 9, inclusive. Then we compute n^4 in each of these 10 cases. We get the following values, where X is some integer that is a multiple of 10, whose exact value we do not care about. $(10k + 0)^4 = 10{,}000k^4 = 10{,}000k^4 + 0$, $(10k + 1)^4 = 10{,}000k^4 + X \cdot k^3 + X \cdot k^2 + X \cdot k + 1$, $(10k + 2)^4 = 10{,}000k^4 + X \cdot k^3 + X \cdot k^2 + X \cdot k + 16$, $(10k + 3)^4 = 10{,}000k^4 + X \cdot k^3 + X \cdot k^2 + X \cdot k + 81$, $(10k + 4)^4 = 10{,}000k^4 + X \cdot k^3 + X \cdot k^2 + X \cdot k + 256$, $(10k + 5)^4 = 10{,}000k^4 + X \cdot k^3 + X \cdot k^2 + X \cdot k + 625$, $(10k + 6)^4 = 10{,}000k^4 + X \cdot k^3 + X \cdot k^2 + X \cdot k + 1296$, $(10k + 7)^4 = 10{,}000k^4 + X \cdot k^3 + X \cdot k^2 + X \cdot k + 2401$, $(10k + 8)^4 = 10{,}000k^4 + X \cdot k^3 + X \cdot k^2 + X \cdot k + 4096$, $(10k + 9)^4 = 10{,}000k^4 + X \cdot k^3 + X \cdot k^2 + X \cdot k + 6561$. Because each coefficient indicated by X is a multiple of 10, the corresponding term has no effect on the ones digit of the answer. Therefore, the ones digits are 0, 1, 6, 1, 6, 5, 6, 1, 6, 1, respectively, so it is always a 0, 1, 5, or 6. **31.** Because $n^3 > 100$ for all $n > 4$, we need only note that $n = 1$, $n = 2$, $n = 3$, and $n = 4$ do not satisfy $n^2 + n^3 = 100$. **33.** Because $5^4 = 625$, both x and y must be less than 5. Then $x^4 + y^4 \leq 4^4 + 4^4 = 512 < 625$. **35.** If it is not true that $a \leq \sqrt[3]{n}$, $b \leq \sqrt[3]{n}$, or $c \leq \sqrt[3]{n}$, then $a > \sqrt[3]{n}$, $b > \sqrt[3]{n}$, and $c > \sqrt[3]{n}$. Multiplying these inequalities of positive numbers together we obtain $abc > (\sqrt[3]{n})^3 = n$, which implies the negation of our hypothesis that $n = abc$. **37.** By finding a common denominator, we can assume that the given rational numbers are a/b and c/b, where b is a positive integer and a and c are integers with $a < c$. In particular, $(a + 1)/b \leq c/b$. Thus, $x = (a + \frac{1}{2}\sqrt{2})/b$ is between the two given rational numbers, because $0 < \sqrt{2} < 2$. Furthermore, x is irrational, because if x were rational, then $2(bx - a) = \sqrt{2}$ would be as well, in violation of Example 11 in Section 1.7. **39. a)** Without loss of generality, we can assume that the x sequence is already sorted into nondecreasing order, because we can relabel the indices. There are only a finite number of possible orderings for the y sequence, so if we can show that we can increase the sum (or at least keep it the same) whenever we find y_i and y_j that are out of order (i.e., $i < j$ but $y_i > y_j$) by switching them, then we will have shown that the sum is largest when the y sequence is in nondecreasing order. Indeed, if we perform the swap, then we have added $x_iy_j + x_jy_i$ to the sum and subtracted $x_iy_i + x_jy_j$. The net effect is to have added $x_iy_j + x_jy_i - x_iy_i - x_jy_j = (x_j - x_i)(y_i - y_j)$, which is nonnegative by our ordering assumptions. **b)** Similar to part (a) **41. a)** $6 \to 3 \to 10 \to 5 \to 16 \to 8 \to 4 \to 2 \to 1$ **b)** $7 \to 22 \to 11 \to 34 \to 17 \to 52 \to 26 \to 13 \to 40 \to 20 \to 10 \to 5 \to 16 \to 8 \to 4 \to 2 \to 1$ **c)** $17 \to 52 \to 26 \to 13 \to 40 \to 20 \to 10 \to 5 \to 16 \to 8 \to 4 \to 2 \to 1$ **d)** $21 \to 64 \to 32 \to 16 \to 8 \to 4 \to 2 \to 1$ **43.** Without loss of generality, assume that the upper left and upper right corners of the board are removed. Place three dominoes horizontally to fill the remaining portion of the first row, and fill each of the other seven rows with four horizontal dominoes. **45.** Because there is an even number of squares in all, either there is an even number of squares in each row or there is an even number of squares in each column. In the former case, tile the board in the obvious way by placing the dominoes horizontally, and in the latter case, tile the board in the obvious way by placing the dominoes vertically. **47.** We can rotate the board if necessary to make the removed squares be 1 and 16. Square 2 must be covered by a domino. If that domino is placed to cover squares 2 and 6, then the following domino placements are forced in succession: 5-9, 13-14, and 10-11, at which point there is no way to cover square 15. Otherwise, square 2 must be covered by a domino placed at 2-3. Then the following domino placements are forced: 4-8, 11-12, 6-7, 5-9, and 10-14, and again there is no way to cover square 15. **49.** Remove the two black squares adjacent to a white corner, and remove two white squares other than that corner. Then no domino can cover that white corner.

51. a)

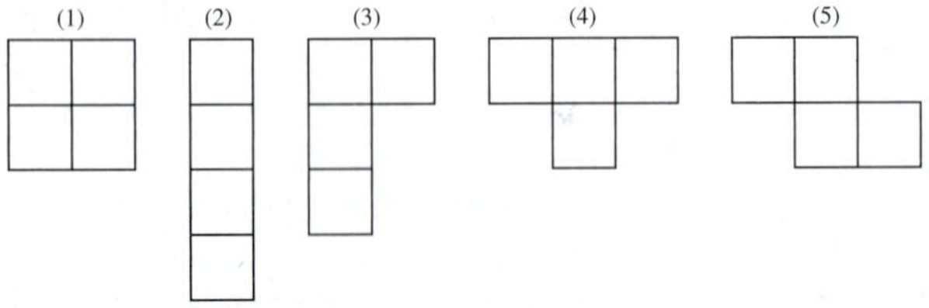

b) The picture shows tilings for the first four patterns.

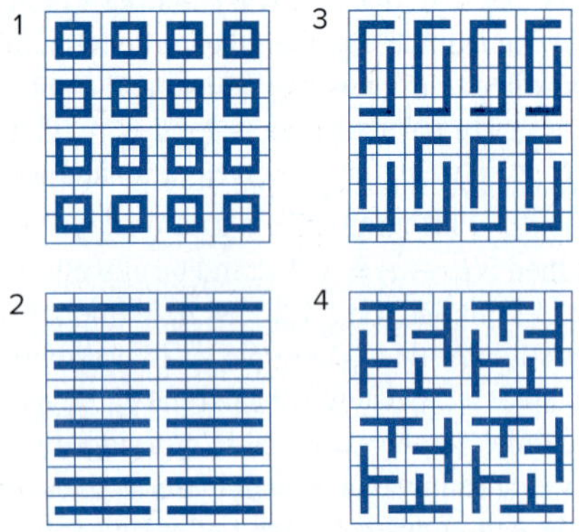

To show that pattern 5 cannot tile the checkerboard, label the squares from 1 to 64, one row at a time from the top, from left to right in each row. Thus, square 1 is the upper left corner, and square 64 is the lower right. Suppose we did have a tiling. By symmetry and without loss of generality, we may suppose that the tile is positioned in the upper left corner, covering squares 1, 2, 10, and 11. This forces a tile to be adjacent to it on the right, covering squares 3, 4, 12, and 13. Continue in this manner and we are forced to have a tile covering squares 6, 7, 15, and 16. This makes it impossible to cover square 8. Thus, no tiling is possible.

Supplementary Exercises

1. a) $q \rightarrow p$ **b)** $q \wedge p$ **c)** $\neg q \vee \neg p$ **d)** $q \leftrightarrow p$ **3. a)** The proposition cannot be false unless $\neg p$ is false, so p is true. If p is true and q is true, then $\neg q \wedge (p \rightarrow q)$ is false, so the conditional statement is true. If p is true and q is false, then $p \rightarrow q$ is false, so $\neg q \wedge (p \rightarrow q)$ is false and the conditional statement is true. **b)** The proposition cannot be false unless q is false. If q is false and p is true, then $(p \vee q) \wedge \neg p$ is false, and the conditional statement is true. If q is false and p is false, then $(p \vee q) \wedge \neg p$ is false, and the conditional statement is true. **5.** $\neg q \rightarrow \neg p$; $p \rightarrow q$; $\neg p \rightarrow \neg q$ **7.** $(p \wedge q \wedge r \wedge \neg s) \vee (p \wedge q \wedge \neg r \wedge s) \vee (p \wedge \neg q \wedge r \wedge s) \vee (\neg p \wedge q \wedge r \wedge s)$ **9.** Translating these statements into symbols, using the obvious letters, we have $\neg t \rightarrow \neg g$, $\neg g \rightarrow \neg q$, $r \rightarrow q$, and $\neg t \wedge r$. Assume the statements are consistent. The fourth statement tells us that $\neg t$ must be true. Therefore, by modus ponens with the first statement, we know that $\neg g$ is true, hence (from the second statement), that $\neg q$ is true. Also, the fourth statement tells us that r must be true, and so again modus ponens (third statement) makes q true. This is a contradiction: $q \wedge \neg q$. Thus, the statements are inconsistent. **11.** Reject-accept-reject-accept, accept-accept-accept-accept, accept-accept-reject-accept, reject-reject-reject-reject, reject-reject-accept-reject, and reject-accept-accept-accept **13.** Aaron is a knave and Crystal is a knight; it cannot be determined what Bohan is. **15.** Brenda **17.** The premises cannot both be true, because they are contradictory. Therefore, it is (vacuously) true that whenever all the premises are true, the conclusion is also true, which by definition makes this a valid argument. Because the premises are not both true, we cannot conclude that the conclusion is true. **19.** Use the same propositions as were given in Section 1.3 for a 9×9 Sudoku puzzle, with the variables indexed from 1 to 16, instead of from 1 to 9, and with a similar change for the propositions for the 4×4 blocks: $\bigwedge_{r=0}^{3} \bigwedge_{s=0}^{3} \bigwedge_{n=1}^{16} \bigvee_{i=1}^{4} \bigvee_{j=1}^{4} p(4r+i, 4s+j, n)$. **21. a)** F **b)** T **c)** F **d)** T **e)** F **f)** T **23.** Many answers are possible. One example is United States senators. **25.** $\forall x \exists y \exists z\, (y \neq z \wedge \forall w(P(w,x) \leftrightarrow (w = y \vee w = z)))$ **27. a)** $\neg \exists x P(x)$ **b)** $\exists x(P(x) \wedge \forall y(P(y) \rightarrow y = x))$ **c)** $\exists x_1 \exists x_2(P(x_1) \wedge P(x_2) \wedge x_1 \neq x_2 \wedge \forall y(P(y) \rightarrow (y = x_1 \vee y = x_2)))$ **d)** $\exists x_1 \exists x_2 \exists x_3(P(x_1) \wedge P(x_2) \wedge P(x_3) \wedge x_1 \neq x_2 \wedge x_1 \neq x_3 \wedge x_2 \neq x_3 \wedge \forall y(P(y) \rightarrow (y = x_1 \vee y = x_2 \vee y = x_3)))$ **29.** Suppose that $\exists x(P(x) \rightarrow Q(x))$ is true. Then either $Q(x_0)$ is true for some x_0, in which case $\forall x P(x) \rightarrow \exists x\, Q(x)$ is true; or $P(x_0)$ is false for some x_0, in which case $\forall x P(x) \rightarrow \exists x Q(x)$ is true. Conversely, suppose that $\exists x(P(x) \rightarrow Q(x))$ is false. That means that $\forall x(P(x) \wedge \neg Q(x))$ is true, which implies $\forall x P(x)$ and $\forall x(\neg Q(x))$. This latter proposition is equivalent to $\neg \exists x Q(x)$. Thus, $\forall x P(x) \rightarrow \exists x Q(x)$ is false. **31.** No **33.** $\forall x\ \forall z\ \exists y\ T(x, y, z)$, where $T(x, y, z)$ is the statement that student x has taken class y in department z, where the domains are the set of students in the class, the set of courses at this university, and the set of departments in the school of mathematical sciences **35.** $\exists! x \exists! y\ T(x, y)$ and $\exists x \forall z((\exists y \forall w(T(z, w) \leftrightarrow w = y)) \leftrightarrow z = x)$, where $T(x, y)$ means that student x has taken class y and the domain is all students in this class **37.** $P(a) \rightarrow Q(a)$ and $Q(a) \rightarrow R(a)$ by universal instantiation; then $\neg Q(a)$ by modus tollens and $\neg P(a)$ by modus tollens **39.** This is not true. Let $x = 2^{1/3}$. Then $x^2 = 4^{1/3}$ is irrational (the proof of this is very similar to the proof in Example 11 in Section 1.7), but $x^3 = 2$ is rational. **41.** We can give a constructive proof by letting $m = 10^{500} + 1$. Then $m^2 = (10^{500} + 1)^2 > (10^{500})^2 = 10^{1000}$. **43.** 23 cannot be written as the sum of eight cubes. **45.** 223 cannot be written as the sum of 36 fifth powers.

CHAPTER 2

Section 2.1

1. a) $\{-1, 1\}$ **b)** $\{1, 2, 3, 4, 5, 6, 7, 8, 9, 10, 11\}$ **c)** $\{0, 1, 4, 9, 16, 25, 36, 49, 64, 81\}$ **d)** $\emptyset$ **3. a)** $\{1, 4, 9, 16, 25, 36, 49, 64, 81, 100, 121\}$ **b)** $\{3, 5, 6, 7, 9, 10, 11, 12, 13, 14, 15, 17, 18, 19\}$ **c)** $\{11, 25, 39, 53, 46\}$ **5. a)** $[0, 5)$, $[0, 5]$ **b)** $(0, 5)$, $(0, 5]$, $[0, 5)$, $[0, 5]$ **c)** $(0, 5)$, $(0, 5]$, $[0, 5)$, $[0, 5]$, $(1, 4]$, $[2, 3]$ **d)** $(0, 5)$, $(0, 5]$, $[0, 5)$, $[0, 5]$, $(1, 4]$, $[2, 3]$ **e)** $(0, 5)$, $(0, 5]$, $[0, 5)$, $[0, 5]$, $(1, 4]$ **f)** $(0, 5]$, $[0, 5]$ **7. a)** $\{(a, b) \mid a \in \mathbf{Z}^+, b \in \mathbf{Z}^+, \text{and } a \geq b\}$ **b)** $\{(a, b) \mid a \in \mathbf{Z}^+, b \in \mathbf{Z}^+, \text{and } a \text{ and } b \text{ have different partity}\}$ **c)** $\{(a, b) \mid a \in \mathbf{Z}^+, b \in \mathbf{Z}^+, \text{ and } b > 2a\}$ **9. a)** The second is a subset of the first, but the first is not a subset of the second. **b)** Neither is a subset of the other. **c)** The first is a subset of the second, but the second is not a subset of the first. **11. a)** Yes **b)** No **c)** No **13. a)** Yes **b)** No **c)** Yes **d)** No **e)** No **f)** No **15. a)** False **b)** False **c)** False **d)** True **e)** False **f)** False **g)** True **17. a)** True **b)** True **c)** False **d)** True **e)** True **f)** False

19.

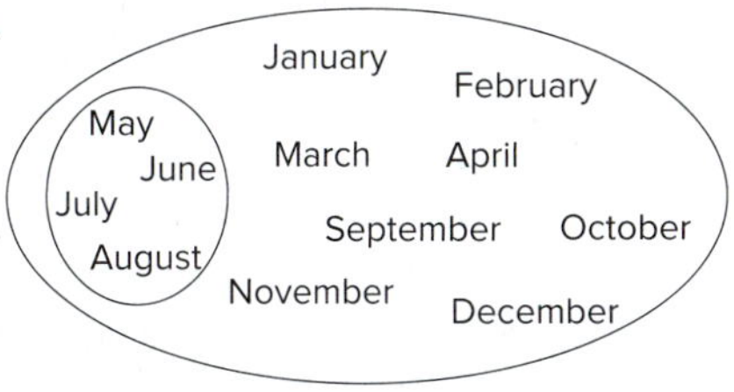

21. The dots in certain regions indicate that those regions are not empty.

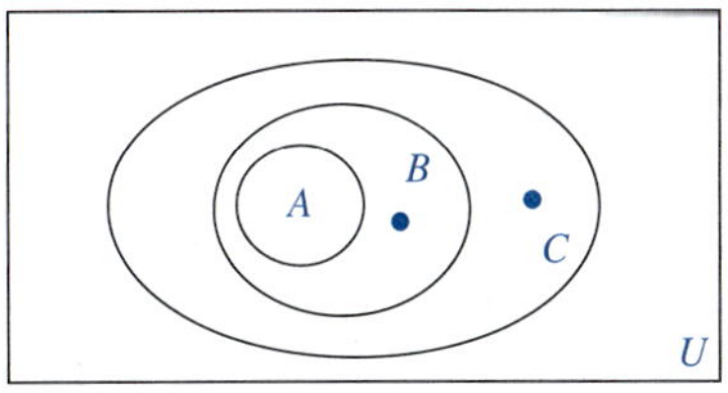

23. Suppose that $x \in A$. Because $A \subseteq B$, this implies that $x \in B$. Because $B \subseteq C$, we see that $x \in C$. Because $x \in A$ implies that $x \in C$, it follows that $A \subseteq C$. **25. a)** 1 **b)** 1 **c)** 2 **d)** 3 **27. a)** $\{\emptyset, \{a\}\}$ **b)** $\{\emptyset, \{a\}, \{b\}, \{a, b\}\}$ **c)** $\{\emptyset, \{\emptyset\}, \{\{\emptyset\}\}, \{\emptyset, \{\emptyset\}\}\}$ **29. a)** 8 **b)** 16 **c)** 2 **31.** For the "if" part, given $A \subseteq B$, we want to show that $\mathcal{P}(A) \subseteq \mathcal{P}(B)$, i.e., if $C \subseteq A$ then $C \subseteq B$. But this follows directly from Exercise 23. For the "only if" part, given that $\mathcal{P}(A) \subseteq \mathcal{P}(B)$, we want to show that $A \subseteq B$. Suppose $a \in A$. Then $\{a\} \subseteq A$, so $\{a\} \in \mathcal{P}(A)$. Since $\mathcal{P}(A) \subseteq \mathcal{P}(B)$, it follows that $\{a\} \in \mathcal{P}(B)$, which means that $\{a\} \subseteq B$. But this implies $a \in B$, as desired. **33. a)** $\{(a, y), (b, y), (c, y), (d, y), (a, z), (b, z), (c, z), (d, z)\}$ **b)** $\{(y, a), (y, b), (y, c), (y, d), (z, a), (z, b), (z, c), (z, d)\}$ **35.** The set of triples (a, b, c), where a is an airline and b and c are cities. A useful subset of this set is the set of triples (a, b, c) for which a flies between b and c. **37.** $\emptyset \times A = \{(x, y) \mid x \in \emptyset \text{ and } y \in A\} = \emptyset = \{(x, y) \mid x \in A \text{ and } y \in \emptyset\} = A \times \emptyset$ **39. a)** $\{(0, 0), (0, 1), (0, 3), (1, 0), (1, 1), (1, 3), (3, 0), (3, 1), (3, 3)\}$ **b)** $\{(1, 1), (1, 2), (1, a), (1, b), (2, 1), (2, 2), (2, a), (2, b), (a, 1), (a, 2), (a, a), (a, b), (b, 1), (b, 2), (b, a), (b, b)\}$ **41.** mn **43.** m^n **45.** The elements of $A \times B \times C$ consist of 3-tuples (a, b, c), where $a \in A$, $b \in B$, and $c \in C$, whereas the elements of $(A \times B) \times C$ look like $((a, b), c)$—ordered pairs, the first coordinate of which is again an ordered pair. **47.** This is not true. The simplest counterexample is to let $A = B = \emptyset$. Then $A \times B = \emptyset$ and $\mathcal{P}(A \times B) = \{\emptyset\}$, whereas $\mathcal{P}(A) = \mathcal{P}(B) = \{\emptyset\}$ and $\mathcal{P}(A) \times \mathcal{P}(B) = \{(\emptyset, \emptyset)\}$. Thus, $\mathcal{P}(A \times B) \neq \mathcal{P}(A) \times \mathcal{P}(B)$. **49. a)** The square of a real number is never -1. True **b)** There exists an integer whose square is 2. False **c)** The square of every integer is positive. False **d)** There is a real number equal to its own square. True **51. a)** $\{-1, 0, 1\}$ **b)** $\mathbf{Z} - \{0, 1\}$ **c)** $\emptyset$ **53.** We must show that $\{\{a\}, \{a, b\}\} = \{\{c\}, \{c, d\}\}$ if and only if $a = c$ and $b = d$. The "if" part is immediate. So assume these two sets are equal. First, consider the case when $a \neq b$. Then $\{\{a\}, \{a, b\}\}$ contains exactly two elements, one of which contains one element. Thus, $\{\{c\}, \{c, d\}\}$ must have the same property, so $c \neq d$ and $\{c\}$ is the element containing exactly one element. Hence, $\{a\} = \{c\}$, which implies that $a = c$. Also, the two-element sets $\{a, b\}$ and $\{c, d\}$ must be equal. Because $a = c$ and $a \neq b$, it follows that $b = d$. Second, suppose that $a = b$. Then $\{\{a\}, \{a, b\}\} = \{\{a\}\}$, a set with one element. Hence, $\{\{c\}, \{c, d\}\}$ has only one element, which can happen only when $c = d$, and the set is $\{\{c\}\}$. It then follows that $a = c$ and $b = d$. **55.** Let $S = \{a_1, a_2, \ldots, a_n\}$. Represent each subset of S with a bit string of length n, where the ith bit is 1 if and only if $a_i \in S$. To generate all subsets of S, list all 2^n bit strings of length n (for instance, in increasing order), and write down the corresponding subsets.

Section 2.2

1. a) The set of students who live within one mile of school and walk to classes **b)** The set of students who live within one mile of school or walk to classes (or do both) **c)** The set of students who live within one mile of school but do not walk to classes **d)** The set of students who walk to classes but live more than one mile away from school **3. a)** $\{0, 1, 2, 3, 4, 5, 6\}$ **b)** $\{3\}$ **c)** $\{1, 2, 4, 5\}$ **d)** $\{0, 6\}$ **5.** $\overline{\overline{A}} = \{x \mid \neg(x \in \overline{A})\} = \{x \mid \neg(\neg x \in A)\} = \{x \mid x \in A\} = A$ **7. a)** $A \cup U = \{x \mid x \in A \vee x \in U\} = \{x \mid x \in A \vee \mathbf{T}\} = \{x \mid \mathbf{T}\} = U$ **b)** $A \cap \emptyset = \{x \mid x \in A \wedge x \in \emptyset\} = \{x \mid x \in A \wedge \mathbf{F}\} = \{x \mid \mathbf{F}\} = \emptyset$ **9. a)** $A \cup \overline{A} = \{x \mid x \in A \vee x \notin A\} = U$ **b)** $A \cap \overline{A} = \{x \mid x \in A \wedge x \notin A\} = \emptyset$ **11. a)** $A \cup B = \{x \mid x \in A \vee x \in B\} = \{x \mid x \in B \vee x \in A\} = B \cup A$ **b)** $A \cap B = \{x \mid x \in A \wedge x \in B\} = \{x \mid x \in B \wedge x \in A\} = B \cap A$ **13.** Suppose $x \in A \cap (A \cup B)$. Then $x \in A$ and $x \in A \cup B$ by the definition of intersection. Because $x \in A$, we have proved that the left-hand side is a subset of the right-hand side. Conversely, let $x \in A$. Then by the definition of union, $x \in A \cup B$ as well. Therefore, $x \in A \cap (A \cup B)$ by the definition of intersection, so the right-hand side is a subset of the left-hand side. **15. a)** $x \in \overline{A \cup B} \equiv x \notin A \cup B \equiv \neg(x \in A \vee x \in B) \equiv \neg(x \in A) \wedge \neg(x \in B) \equiv x \notin A \wedge x \notin B \equiv x \in \overline{A} \wedge x \in \overline{B} \equiv x \in \overline{A} \cap \overline{B}$

b)

A	B	$A \cup B$	$\overline{A \cup B}$	$\overline{A}$	$\overline{B}$	$\overline{A} \cap \overline{B}$
1	1	1	0	0	0	0
1	0	1	0	0	1	0
0	1	1	0	1	0	0
0	0	0	1	1	1	1

17. Suppose $A \subseteq B$. We must show that every element x of U is an element of $\overline{A} \cup B$. Either $x \in \overline{A}$ or $x \in A$, and if $x \in A$ then $x \in B$. Thus, $x \in \overline{A} \cup B$ in all cases. Conversely, suppose that $\overline{A} \cup B = U$, and let $x \in A$. Then $x \notin \overline{A}$, so it must be that $x \in B$. This shows that $A \subseteq B$, and the proof is complete. **19. a)** $x \in \overline{A \cap B \cap C} \equiv x \notin A \cap B \cap C \equiv x \notin A \vee x \notin B \vee x \notin C \equiv x \in \overline{A} \vee x \in \overline{B} \vee x \in \overline{C} \equiv x \in \overline{A} \cup \overline{B} \cup \overline{C}$

b)

A	B	C	$A \cap B \cap C$	$\overline{A \cap B \cap C}$	$\overline{A}$	$\overline{B}$	$\overline{C}$	$\overline{A} \cup \overline{B} \cup \overline{C}$
1	1	1	1	0	0	0	0	0
1	1	0	0	1	0	0	1	1
1	0	1	0	1	0	1	0	1
1	0	0	0	1	0	1	1	1
0	1	1	0	1	1	0	0	1
0	1	0	0	1	1	0	1	1
0	0	1	0	1	1	1	0	1
0	0	0	0	1	1	1	1	1

21. a) Both sides equal $\{x \mid x \in A \wedge x \notin B\}$. **b)** $A = A \cap U = A \cap (B \cup \overline{B}) = (A \cap B) \cup (A \cap \overline{B})$ **23.** $x \in A \cup (B \cup C) \equiv (x \in A) \vee (x \in (B \cup C)) \equiv (x \in A) \vee (x \in B \vee x \in C) \equiv (x \in A \vee x \in B) \vee (x \in C) \equiv x \in (A \cup B) \cup C$ **25.** $x \in A \cup (B \cap C) \equiv (x \in A) \vee (x \in (B \cap C)) \equiv (x \in A) \vee (x \in B \wedge x \in C) \equiv (x \in A \vee x \in B) \wedge (x \in A \vee x \in C) \equiv x \in (A \cup B) \cap (A \cup C)$ **27. a)** $\{4,6\}$ **b)** $\{0,1,2,3,4,5,6,7,8,9,10\}$ **c)** $\{4, 5, 6, 8, 10\}$ **d)** $\{0,2,4,5,6,7,8,9,10\}$ **29. a)** The double-shaded portion is the desired set.

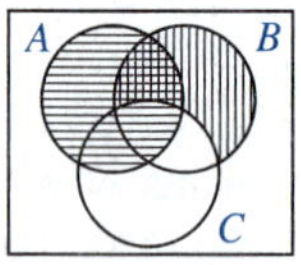

b) The desired set is the entire shaded portion.

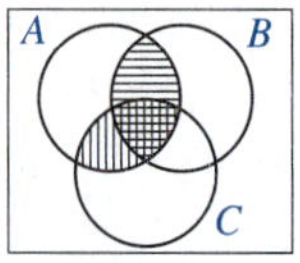

c) The desired set is the entire shaded portion.

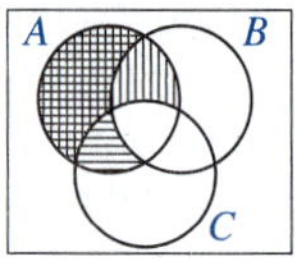

31. a) $B \subseteq A$ **b)** $A \subseteq B$ **c)** $A \cap B = \emptyset$ **d)** Nothing, because this is always true **e)** $A = B$ **33.** $A \subseteq B \equiv \forall x(x \in A \rightarrow x \in B) \equiv \forall x(x \notin B \rightarrow x \notin A) \equiv \forall x(x \in \overline{B} \rightarrow x \in \overline{A}) \equiv \overline{B} \subseteq \overline{A}$ **35.** By De Morgan's law, the left-hand side equals $(\overline{A} \cap \overline{B}) \cap (\overline{B} \cap \overline{C}) \cap (\overline{A} \cap \overline{C})$. By the commutative, associative, and idempotent laws, this simplifies to the right-hand side. **37. a)** Let $(x, y) \in A \times (B - C)$, which means that $x \in A$ and y is an element of B but not C. Thus, $(x, y) \in A \times B$ and $(x, y) \notin A \times C$, so by the definition of set difference, $(x, y) \in (A \times B) - (A \times C)$. Conversely, let $(x, y) \in (A \times B) - (A \times C)$. Then $(x, y) \in A \times B$ and $(x, y) \notin A \times C$. Thus, $x \in A$ and $y \in B$, and because $x \in A$, it must be that $y \notin C$. This implies that $y \in B - C$, so indeed $(x, y) \in A \times (B - C)$. **b)** Note that the complement in the right-hand side must mean with respect to $U \times U$. This is not true. For example, let $U = \{a, b\}$, $A = \{a\}$, $B = \{b\}$, and $C = \emptyset$. Then the left-hand side is $\{(b, a)\}$, whereas the right-hand side is $\{(a, a), (b, a), (b, b)\}$. **39.** The set of students who are computer science majors but not mathematics majors or who are mathematics majors but not computer science majors **41.** An element is in $(A \cup B) - (A \cap B)$ if it is in the union of A and B but not in the intersection of A and B, which means that it is in either A or B but not in both A and B. This is exactly what it means for an element to belong to $A \oplus B$. **43. a)** $A \oplus A = (A - A) \cup (A - A) = \emptyset \cup \emptyset = \emptyset$ **b)** $A \oplus \emptyset = (A - \emptyset) \cup (\emptyset - A) = A \cup \emptyset = A$ **c)** $A \oplus U = (A - U) \cup (U - A) = \emptyset \cup \overline{A} = \overline{A}$ **d)** $A \oplus \overline{A} = (A - \overline{A}) \cup (\overline{A} - A) = A \cup \overline{A} = U$ **45.** $B = \emptyset$ **47.** Yes **49.** Yes **51.** If $A \cup B$ were finite, then it would have n elements for some natural number n. But A already has more than n elements, because it is infinite, and $A \cup B$ has all the elements that A has, so $A \cup B$ has more than n elements. This contradiction shows that $A \cup B$ must be infinite. **53. a)** $\{1, 2, 3, \ldots, n\}$ **b)** $\{1\}$ **55. a)** A_n **b)** $\{0, 1\}$ **57. a)** **Z**, $\{-1, 0, 1\}$ **b)** $\mathbf{Z} - \{0\}$, $\emptyset$ **c)** **R**, $[-1, 1]$ **d)** $[1, \infty)$, $\emptyset$ **59. a)** $\{1, 2, 3, 4, 7, 8, 9, 10\}$ **b)** $\{2, 4, 5, 6, 7\}$ **c)** $\{1, 10\}$ **61.** The bit in the ith position of the bit string of the difference of two sets is 1 if the ith bit of the first string is 1 and the ith bit of the second string is 0, and is 0 otherwise. **63. a)** 11 1110 0000 0000 0000 0000 0000 $\vee$ 01 1100 1000 0000 0100 0101 0000 = 11 1110 1000 0000 0100 0101 0000, representing $\{a, b, c, d, e, g, p, t, v\}$ **b)** 11 1110 0000 0000 0000 0000 0000 $\wedge$ 01 1100 1000 0000 0100 0101 0000 = 01 1100 0000 0000 0000 0000 0000, representing $\{b, c, d\}$ **c)** (11 1110 0000 0000 0000 0000 0000 $\vee$ 00 0110 0110 0001 1000 0110 0110) $\wedge$ (01 1100 1000 0000 0100 0101 0000 $\vee$ 00 1010 0010 0000 1000 0010 0111) = 11 1110 0110 0001 1000 0110 0110 $\wedge$ 01 1110 1010 0000 1100 0111 0111 = 01 1110 0010 0000 1000 0110 0110, representing $\{b, c, d, e, i, o, t, u, x, y\}$ **d)** 11 1110 0000 0000 0000 0000 0000 $\vee$ 01 1100 1000 0000 0100 0101 0000 $\vee$ 00 1010 0010 0000 1000 0010 0111 $\vee$ 00 0110 0110 0001 1000 0110 0110 = 11 1110 1110 0001 1100 0111 0111, representing $\{a,b,c,d,e,g,h,i,n,o,p,t,u,v,x,y,z\}$ **65. a)** $\{1, 2, 3, \{1, 2, 3\}\}$ **b)** $\{\emptyset\}$ **c)** $\{\emptyset, \{\emptyset\}\}$ **d)** $\{\emptyset, \{\emptyset\}, \{\emptyset, \{\emptyset\}\}\}$ **67. a)** $\{3 \cdot a, 3 \cdot b, 1 \cdot c, 4 \cdot d\}$ **b)** $\{2 \cdot a, 2 \cdot b\}$ **c)** $\{1 \cdot a, 1 \cdot c\}$ **d)** $\{1 \cdot b, 4 \cdot d\}$ **e)** $\{5 \cdot a, 5 \cdot b, 1 \cdot c, 4 \cdot d\}$ **69.** These all follow from the definitions of the multiset operations. **a)** False, because the union is actually $\{a, a, a\}$ **b)** False, because the union is actually $\{a, a, a\}$ **c)** True **d)** False, because the correct intersection is as stated in part (c) **e)** True **71. a)** 0, 1 **b)** 1/3, 2/3 **c)** 1, 0 **d)** 1/6, 5/6 **73.** $\overline{F}$ = {0.4 Alice, 0.1 Brian, 0.6 Fred, 0.9 Oscar, 0.5 Rita}, $\overline{R}$ = {0.6 Alice, 0.2 Brian, 0.8 Fred, 0.1 Oscar, 0.3 Rita} **75.** {0.4 Alice, 0.8 Brian, 0.2 Fred, 0.1 Oscar, 0.5 Rita}

Section 2.3

1. a) $f(0)$ is not defined. **b)** $f(x)$ is not defined for $x < 0$. **c)** $f(x)$ is not well defined because there are two distinct values assigned to each x. **3. a)** Not a function **b)** A function **c)** Not a function **5. a)** Domain the set of bit strings; range the set of integers **b)** Domain the set of bit strings; range the set of even nonnegative integers **c)** Domain the set of bit strings; range the set of nonnegative integers not exceeding 7 **d)** Domain the set of positive integers; range the set of squares of positive integers = $\{1, 4, 9, 16, \dots\}$ **7. a)** Domain $\mathbf{Z}^+ \times \mathbf{Z}^+$; range $\mathbf{Z}^+$ **b)** Domain $\mathbf{Z}^+$; range $\{0, 1, 2, 3, 4, 5, 6, 7, 8, 9\}$ **c)** Domain the set of bit strings; range **N** **d)** Domain the set of bit strings; range **N** **9. a)** 1 **b)** 0 **c)** 0 **d)** −1 **e)** 3 **f)** −1 **g)** 2 **h)** 1 **11.** Only the function in part (a) **13.** Only the functions in parts (a) and (d) **15. a)** Onto **b)** Not onto **c)** Onto **d)** Not onto **e)** Onto **17. a)** Depends on whether teachers share offices **b)** One-to-one assuming only one teacher per bus **c)** Most likely not one-to-one, especially if salary is set by a collective bargaining agreement **d)** One-to-one **19.** Answers will vary. **a)** Set of offices at the school; probably not onto **b)** Set of buses going on the trip; onto, assuming every bus gets a teacher chaperone **c)** Set of real numbers; not onto **d)** Set of strings of nine digits with hyphens after third and fifth digits; not onto **21. a)** The function $f(x)$ with $f(x) = 3x + 1$ when $x \geq 0$ and $f(x) = -3x + 2$ when $x < 0$ **b)** $f(x) = |x| + 1$ **c)** The function $f(x)$ with $f(x) = 2x + 1$ when $x \geq 0$ and $f(x) = -2x$ when $x < 0$ **d)** $f(x) = x^2 + 1$ **23. a)** Yes **b)** No **c)** Yes **d)** No **25.** Suppose that f is strictly decreasing. This means that $f(x) > f(y)$ whenever $x < y$. To show that g is strictly increasing, suppose that $x < y$. Then $g(x) = 1/f(x) < 1/f(y) = g(y)$. Conversely, suppose that g is strictly increasing. This means that $g(x) < g(y)$ whenever $x < y$. To show that f is strictly decreasing, suppose that $x < y$. Then $f(x) = 1/g(x) > 1/g(y) = f(y)$. **27. a)** Let f be a given strictly decreasing function from **R** to itself. If $a < b$, then $f(a) > f(b)$; if $a > b$, then $f(a) < f(b)$. Thus, if $a \neq b$, then $f(a) \neq f(b)$. **b)** Answers will vary; for example, $f(x) = 0$ for $x < 0$ and $f(x) = -x$ for $x \geq 0$. **29.** The function is not one-to-one, so it is not invertible. On the restricted domain, the function is the identity function on the nonnegative real numbers, $f(x) = x$, so it is its own inverse. **31. a)** $f(S) = \{0, 1, 3\}$ **b)** $f(S) = \{0, 1, 3, 5, 8\}$ **c)** $f(S) = \{0, 8, 16, 40\}$ **d)** $f(S) = \{1, 12, 33, 65\}$ **33. a)** Let x and y be distinct elements of A. Because g is one-to-one, $g(x)$ and $g(y)$ are distinct elements of B. Because f is one-to-one, $f(g(x)) = (f \circ g)(x)$ and $f(g(y)) = (f \circ g)(y)$ are distinct elements of C. Hence, $f \circ g$ is one-to-one. **b)** Let $y \in C$. Because f is onto, $y = f(b)$ for some $b \in B$. Now because g is onto, $b = g(x)$ for some $x \in A$. Hence, $y = f(b) = f(g(x)) = (f \circ g)(x)$. It follows that $f \circ g$ is onto. **35.** Let $A = \{a\}$, $B = \{b_1, b_2\}$, $C = \{c\}$, $g(a) = b_1$, and $f(b_1) = f(b_2) = c$. **37.** No. For example, suppose that $A = \{a\}$, $B = \{b, c\}$, and $C = \{d\}$. Let $g(a) = b$, $f(b) = d$, and $f(c) = d$. Then f and $f \circ g$ are onto, but g is not. **39.** $(f + g)(x) = x^2 + x + 3$, $(fg)(x) = x^3 + 2x^2 + x + 2$ **41.** f is one-to-one because $f(x_1) = f(x_2) \rightarrow ax_1 + b = ax_2 + b \rightarrow ax_1 = ax_2 \rightarrow x_1 = x_2$. f is onto because $f((y - b)/a) = y$. $f^{-1}(y) = (y - b)/a$. **43. a)** $A = B = \mathbf{R}$, $S = \{x \mid x > 0\}$, $T = \{x \mid x < 0\}$, $f(x) = x^2$ **b)** It suffices to show that $f(S) \cap f(T) \subseteq f(S \cap T)$. Let $y \in B$ be an element of $f(S) \cap f(T)$. Then $y \in f(S)$, so $y = f(x_1)$ for some $x_1 \in S$. Similarly, $y = f(x_2)$ for some $x_2 \in T$. Because f is one-to-one, it follows that $x_1 = x_2$. Therefore, $x_1 \in S \cap T$, so $y \in f(S \cap T)$. **45. a)** $\{x \mid 0 \leq x < 1\}$ **b)** $\{x \mid -1 \leq x < 2\}$ **c)** $\emptyset$ **47.** $f^{-1}(\overline{S}) = \{x \in A \mid f(x) \notin S\} = \overline{\{x \in A \mid f(x) \in S\}} = \overline{f^{-1}(S)}$ **49.** Let $x = \lfloor x \rfloor + \epsilon$, where ϵ is a real number with $0 \leq \epsilon < 1$. If $\epsilon < \frac{1}{2}$, then $\lfloor x \rfloor - 1 < x - \frac{1}{2} < \lfloor x \rfloor$, so $\lceil x - \frac{1}{2} \rceil = \lfloor x \rfloor$ and this is the integer closest to x. If $\epsilon > \frac{1}{2}$, then $\lfloor x \rfloor < x - \frac{1}{2} < \lfloor x \rfloor + 1$, so $\lceil x - \frac{1}{2} \rceil = \lfloor x \rfloor + 1$ and this is the integer closest to x. If $\epsilon = \frac{1}{2}$, then $\lceil x - \frac{1}{2} \rceil = \lfloor x \rfloor$, which is the smaller of the two integers that surround x and are the same distance from x. **51.** Write the real number x as $\lfloor x \rfloor + \epsilon$, where ϵ is a real number with $0 \leq \epsilon < 1$. Because $\epsilon = x - \lfloor x \rfloor$, it follows that $0 \leq -\lfloor x \rfloor < 1$. The first two inequalities, $x - 1 < \lfloor x \rfloor$ and $\lfloor x \rfloor \leq x$, follow directly. For the other two inequalities, write $x = \lceil x \rceil - \epsilon'$, where $0 \leq \epsilon' < 1$. Then $0 \leq \lceil x \rceil - x < 1$, and the desired inequality follows. **53. a)** If $x < n$, because $\lfloor x \rfloor \leq x$, it follows that $\lfloor x \rfloor < n$. Suppose that $x \geq n$. By the definition of the floor function, it follows that $\lfloor x \rfloor \geq n$. This means that if $\lfloor x \rfloor < n$, then $x < n$. **b)** If $n < x$, then because $x \leq \lceil x \rceil$, it follows that $n \leq \lceil x \rceil$. Suppose that $n \geq x$. By the definition of the ceiling function, it follows that $\lceil x \rceil \leq n$. This means that if $n < \lceil x \rceil$, then $n < x$. **55.** If n is even, then $n = 2k$ for some integer k. Thus, $\lfloor n/2 \rfloor = \lfloor k \rfloor = k = n/2$. If n is odd, then $n = 2k + 1$ for some integer k. Thus, $\lfloor n/2 \rfloor = \lfloor k + \frac{1}{2} \rfloor = k = (n - 1)/2$. **57.** Assume that $x \geq 0$. The left-hand side is $\lceil -x \rceil$ and the right-hand side is $-\lfloor x \rfloor$. If x is an integer, then both sides equal $-x$. Otherwise, let $x = n + \epsilon$, where n is a natural number and ϵ is a real number with $0 \leq \epsilon < 1$. Then $\lceil -x \rceil = \lceil -n - \epsilon \rceil = -n$ and $-\lfloor x \rfloor = -\lfloor n + \epsilon \rfloor = -n$ also. When $x < 0$, the equation also holds because it can be obtained by substituting $-x$ for x. **59.** $\lceil b \rceil - \lfloor a \rfloor - 1$ **61. a)** 1 **b)** 3 **c)** 126 **d)** 3600 **63. a)** 100 **b)** 256 **c)** 1030 **d)** 30,200

65.

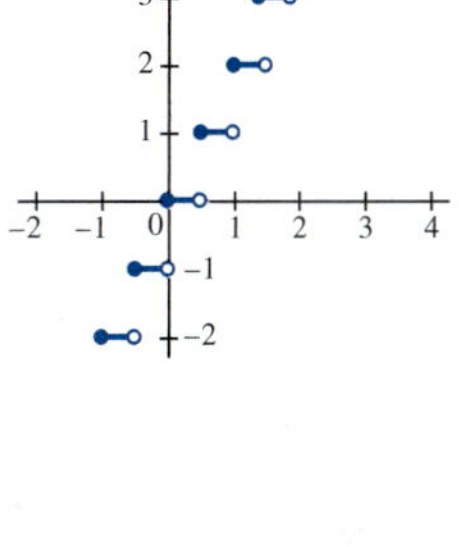

67.

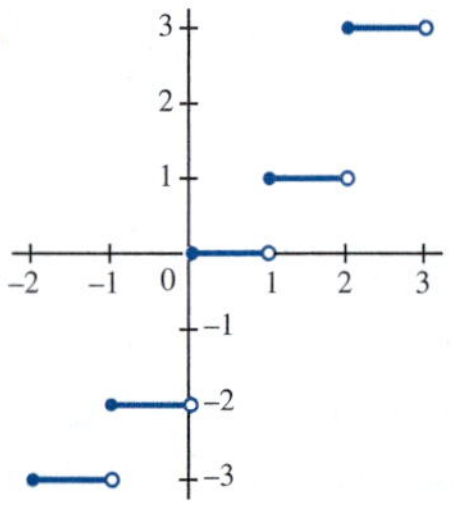

69. a)

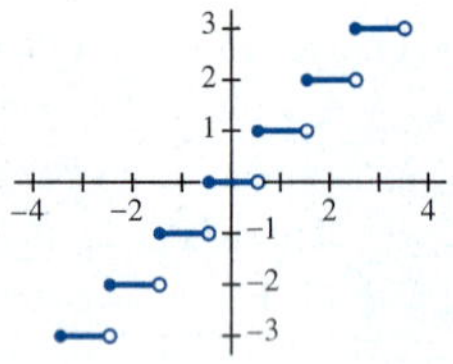

b)

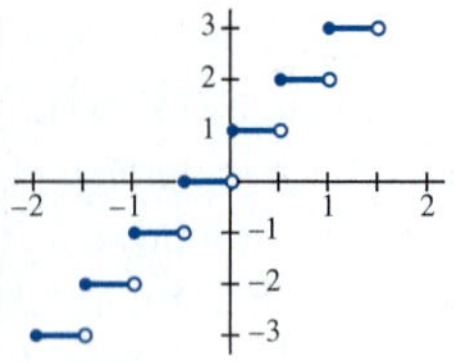

c)

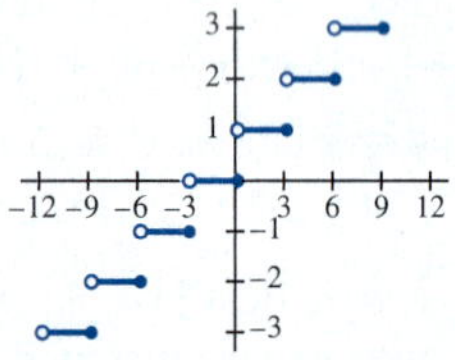

d)

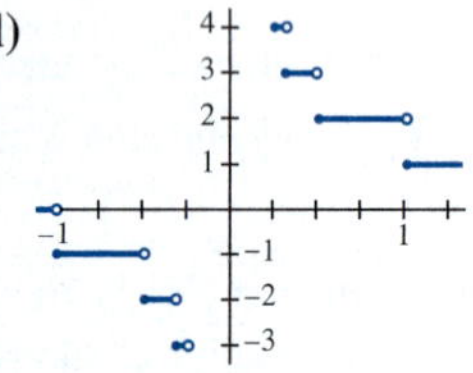

e)

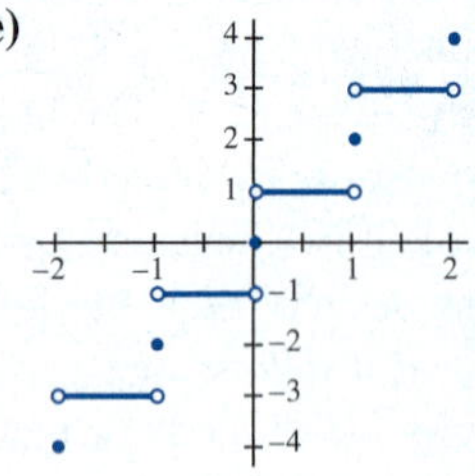

f)

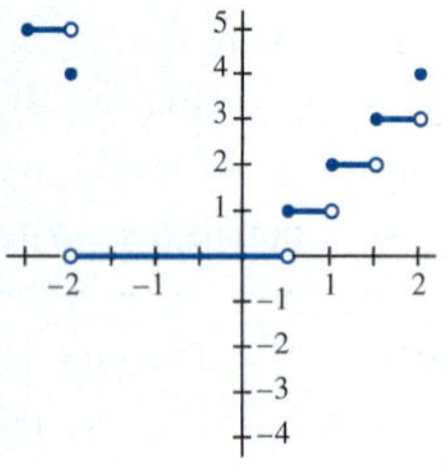

g) See part (a). **71.** $f^{-1}(y) = (y-1)^{1/3}$ **73. a)** $f_{A\cap B}(x) = 1 \leftrightarrow x \in A \cap B \leftrightarrow x \in A$ and $x \in B \leftrightarrow f_A(x) = 1$ and $f_B(x) = 1 \leftrightarrow f_A(x)f_B(x) = 1$ **b)** $f_{A\cup B}(x) = 1 \leftrightarrow x \in A \cup B \leftrightarrow x \in A$ or $x \in B \leftrightarrow f_A(x) = 1$ or $f_B(x) = 1 \leftrightarrow f_A(x)+f_B(x)-f_A(x)f_B(x) = 1$ **c)** $f_{\overline{A}}(x) = 1 \leftrightarrow x \in \overline{A} \leftrightarrow x \notin A \leftrightarrow f_A(x) = 0 \leftrightarrow 1 - f_A(x) = 1$ **d)** $f_{A\oplus B}(x) = 1 \leftrightarrow x \in A \oplus B \leftrightarrow (x \in A$ and $x \notin B)$ or $(x \notin A$ and $x \in B) \leftrightarrow f_A(x) + f_B(x) - 2f_A(x)f_B(x) = 1$ **75. a)** True; because $\lfloor x \rfloor$ is already an integer, $\lceil \lfloor x \rfloor \rceil = \lfloor x \rfloor$. **b)** False; $x = \frac{1}{2}$ is a counterexample. **c)** True; if x or y is an integer, then by property 4b in Table 1, the difference is 0. If neither x nor y is an integer, then $x = n + \epsilon$ and $y = m + \delta$, where n and m are integers and ϵ and δ are positive real numbers less than 1. Then $m + n < x + y < m + n + 2$, so $\lceil x + y \rceil$ is either $m + n + 1$ or $m + n + 2$. Therefore, the given expression is either $(n + 1) + (m + 1) - (m + n + 1) = 1$ or $(n + 1) + (m + 1) - (m + n + 2) = 0$, as desired. **d)** False; $x = \frac{1}{4}$ and $y = 3$ is a counterexample. **e)** False; $x = \frac{1}{2}$ is a counterexample. **77. a)** If x is a positive integer, then the two sides are equal. So suppose that $x = n^2 + m + \epsilon$, where n^2 is the largest perfect square less than x, m is a nonnegative integer, and $0 < \epsilon \le 1$. Then both $\sqrt{x}$ and $\sqrt{\lfloor x \rfloor} = \sqrt{n^2 + m}$ are between n and $n + 1$, so both sides equal n. **b)** If x is a positive integer, then the two sides are equal. So suppose that $x = n^2 - m - \epsilon$, where n^2 is the smallest perfect square greater than x, m is a nonnegative integer, and ϵ is a real number with $0 < \epsilon \le 1$. Then both $\sqrt{x}$ and $\sqrt{\lceil x \rceil} = \sqrt{n^2 - m}$ are between $n - 1$ and n. Therefore, both sides of the equation equal n. **79. a)** Domain is **Z**; codomain is **R**; domain of definition is the set of nonzero integers; the set of values for which f is undefined is $\{0\}$; not a total function. **b)** Domain is **Z**; codomain is **Z**; domain of definition is **Z**; set of values for which f is undefined is $\emptyset$; total function. **c)** Domain is $\mathbf{Z}\times\mathbf{Z}$; codomain is **Q**; domain of definition is $\mathbf{Z} \times (\mathbf{Z} - \{0\})$; set of values for which f is undefined is $\mathbf{Z}\times\{0\}$; not a total function. **d)** Domain is $\mathbf{Z} \times \mathbf{Z}$; codomain is **Z**; domain of definition is $\mathbf{Z}\times\mathbf{Z}$; set of values for which f is undefined is $\emptyset$; total function. **e)** Domain is $\mathbf{Z} \times \mathbf{Z}$; codomain is **Z**; domain of definitions is $\{(m, n) \mid m > n\}$; set of values for which f is undefined is $\{(m, n) \mid m \le n\}$; not a total function. **81. a)** By definition, to say that S has cardinality m is to say that S has exactly m distinct elements. Therefore we can assign the first object to 1, the second to 2, and so on. This provides the one-to-one correspondence. **b)** By part (a), there is a bijection f from S to $\{1, 2, \dots, m\}$ and a bijection g from T to $\{1, 2, \dots, m\}$. Then the composition $g^{-1} \circ f$ is the desired bijection from S to T.

Section 2.4

1. a) 3 **b)** −1 **c)** 787 **d)** 2639 **3. a)** $a_0 = 2, a_1 = 3, a_2 = 5, a_3 = 9$ **b)** $a_0 = 1, a_1 = 4, a_2 = 27, a_3 = 256$ **c)** $a_0 = 0, a_1 = 0, a_2 = 1, a_3 = 1$ **d)** $a_0 = 0, a_1 = 1, a_2 = 2, a_3 = 3$ **5. a)** 2, 5, 8, 11, 14, 17, 20, 23, 26, 29 **b)** 1, 1, 1, 2, 2, 2, 3, 3, 3, 4 **c)** 1, 1, 3, 3, 5, 5, 7, 7, 9, 9 **d)** −1, −2, −2, 8, 88, 656, 4912, 40064, 362368, 3627776 **e)** 3, 6, 12, 24, 48, 96, 192, 384, 768, 1536 **f)** 2, 4, 6, 10, 16, 26, 42, 68, 110, 178 **g)** 1, 2, 2, 3, 3, 3, 3, 4, 4, 4 **h)** 3, 3, 5, 4, 4, 3, 5, 5, 4, 3 **7.** Each term could be twice the previous term; the nth term could be obtained from the previous term by adding $n - 1$; the terms could be the positive integers that are not multiples of 3; there are infinitely many other possibilities. **9. a)** 2, 12, 72, 432, 2592 **b)** 2, 4, 16, 256, 65, 536 **c)** 1, 2, 5, 11, 26 **d)** 1, 1, 6, 27, 204 **e)** 1, 2, 0, 1, 3 **11. a)** 6, 17, 49, 143, 421 **b)** $49 = 5\cdot17-6\cdot6$, $143 = 5 \cdot 49 - 6 \cdot 17$, $421 = 5 \cdot 143 - 6 \cdot 49$ **c)** $5a_{n-1} - 6a_{n-2} = 5(2^{n-1} + 5\cdot3^{n-1}) - 6(2^{n-2} + 5\cdot3^{n-2}) = 2^{n-2}(10-6) + 3^{n-2}(75 - 30) = 2^{n-2} \cdot 4 + 3^{n-2} \cdot 9 \cdot 5 = 2^n + 3^n \cdot 5 = a_n$ **13. a)** Yes **b)** No **c)** No **d)** Yes **e)** Yes **f)** Yes **g)** No **h)** No **15. a)** $a_{n-1} + 2a_{n-2} + 2n - 9 = -(n - 1) + 2 + 2[-(n - 2) + 2] + 2n - 9 = -n + 2 = a_n$ **b)** $a_{n-1} + 2a_{n-2} + 2n - 9 = 5(-1)^{n-1} - (n - 1) + 2 + 2[5(-1)^{n-2} - (n - 2) + 2] + 2n - 9 = 5(-1)^{n-2}(-1 + 2) - n + 2 = a_n$ **c)** $a_{n-1} + 2a_{n-2} + 2n - 9 = 3(-1)^{n-1} + 2^{n-1} - (n - 1) + 2 + 2[3(-1)^{n-2} + 2^{n-2} - (n - 2) + 2] + 2n - 9 = 3(-1)^{n-2}(-1 + 2) + 2^{n-2}(2 + 2) - n + 2 = a_n$ **d)** $a_{n-1} + 2a_{n-2} + 2n - 9 = 7 \cdot 2^{n-1} - (n - 1) + 2 + 2[7 \cdot 2^{n-2} - (n - 2) + 2] + 2n - 9 = 2^{n-2}(7 \cdot 2 + 2 \cdot 7) - n + 2 = a_n$ **17. a)** $a_n = 2 \cdot 3^n$ **b)** $a_n = 2n + 3$ **c)** $a_n = 1 + n(n + 1)/2$ **d)** $a_n = n^2 + 4n + 4$ **e)** $a_n = 1$ **f)** $a_n = (3^{n+1} - 1)/2$ **g)** $a_n = 5n!$ **h)** $a_n = 2^n n!$ **19. a)** $a_n = 3a_{n-1}$ **b)** 5,904,900 **21. a)** $a_n = n + a_{n-1}, a_0 = 0$ **b)** $a_{12} = 78$ **c)** $a_n = n(n + 1)/2$

23. $B(k) = [1 + (0.07/12)]B(k-1) - 100$, with $B(0) = 5000$ **25. a)** One 1 and one 0, followed by two 1s and two 0s, followed by three 1s and three 0s, and so on; 1, 1, 1 **b)** The positive integers are listed in increasing order with each even positive integer listed twice; 9, 10, 10. **c)** The terms in odd-numbered locations are the successive powers of 2; the terms in even-numbered locations are all 0; 32, 0, 64. **d)** $a_n = 3 \cdot 2^{n-1}$; 384, 768, 1536 **e)** $a_n = 15 - 7(n-1) = 22 - 7n$; −34, −41, −48 **f)** $a_n = (n^2 + n + 4)/2$; 57, 68, 80 **g)** $a_n = 2n^3$; 1024, 1458, 2000 **h)** $a_n = n! + 1$; 362881, 3628801, 39916801 **27.** Among the integers $1, 2, \ldots, a_n$, where a_n is the nth positive integer not a perfect square, the nonsquares are $a_1, a_2, \ldots, a_n$ and the squares are $1^2, 2^2, \ldots, k^2$, where k is the integer with $k^2 < n+k < (k+1)^2$. Consequently, $a_n = n + k$, where $k^2 < a_n < (k+1)^2$. To find k, first note that $k^2 < n+k < (k+1)^2$, so $k^2 + 1 \le n + k \le (k+1)^2 - 1$. Hence, $(k - \frac{1}{2})^2 + \frac{3}{4} = k^2 - k + 1 \le n \le k^2 + k = (k + \frac{1}{2})^2 - \frac{1}{4}$. It follows that $k - \frac{1}{2} < \sqrt{n} < k + \frac{1}{2}$, so $k = \{\sqrt{n}\}$ and $a_n = n + k = n + \{\sqrt{n}\}$. **29. a)** 20 **b)** 11 **c)** 30 **d)** 511 **31. a)** 1533 **b)** 510 **c)** 4923 **d)** 9842 **33. a)** 21 **b)** 78 **c)** 18 **d)** 18 **35.** $\sum_{j=1}^{n}(a_j - a_{j-1}) = a_n - a_0$ **37. a)** n^2 **b)** $n(n+1)/2$ **39.** 15150 **41.** 34320 **43.** $\frac{n(n+1)(2n+1)}{3} + \frac{n(n+1)}{2} + (n+1)(m - (n+1)^2 + 1)$, where $n = \lfloor\sqrt{m}\rfloor - 1$ **45. a)** 0 **b)** 1680 **c)** 1 **d)** 1024 **47.** 34

Section 2.5

1. a) Countably infinite, −1, −2, −3, −4, … **b)** Countably infinite, 0, 2, −2, 4, −4, … **c)** Countably infinite, 99, 98, 97, … **d)** Uncountable **e)** Finite **f)** Countably infinite, 0, 7, −7, 14, −14, … **3. a)** Countable: match n with the string of n 1s. **b)** Countable. To find a correspondence, follow the path in Example 4, but omit fractions in the top three rows (as well as continuing to omit fractions not in lowest terms). **c)** Uncountable **d)** Uncountable **5.** Suppose m new guests arrive at the fully occupied hotel. Move the guest in Room n to Room $m + n$ for $n = 1, 2, 3, \ldots$; then the new guests can occupy rooms 1 to m. **7.** For $n = 1, 2, 3, \ldots$, put the guest currently in Room $2n$ into Room n, and the guest currently in Room $2n - 1$ into Room n of the new building. **9.** Move the guest currently in Room i to Room $2i + 1$ for $i = 1, 2, 3, \ldots$. Put the jth guest from the kth bus into Room $2^k(2j + 1)$. **11. a)** $A = [1, 2]$ (closed interval of real numbers from 1 to 2), $B = [3, 4]$ **b)** $A = [1, 2] \cup \mathbf{Z}^+$, $B = [3, 4] \cup \mathbf{Z}^+$ **c)** $A = [1, 3]$, $B = [2, 4]$ **13.** Suppose that A is countable. Then either A has cardinality n for some nonnegative integer n, in which case there is a one-to-one function from A to a subset of $\mathbf{Z}^+$ (the range is the first n positive integers), or there exists a one-to-one correspondence f from A to $\mathbf{Z}^+$; in either case we have satisfied Definition 2. Conversely, suppose that $|A| \le |\mathbf{Z}^+|$. By definition, this means that there is a one-to-one function from A to $\mathbf{Z}^+$, so A has the same cardinality as a subset of $\mathbf{Z}^+$ (namely the range of that function). By Exercise 16 we conclude that A is countable. **15.** Assume that B is countable. Then the elements of B can be listed as $b_1, b_2, b_3, \ldots$. Because A is a subset of B, taking the subsequence of $\{b_n\}$ that contains the terms that are in A gives a listing of the elements of A. Because A is uncountable, this is impossible. **17.** Assume that $A - B$ is countable. Then, because $A = (A - B) \cup (A \cap B)$, the elements of A can be listed in a sequence by alternating elements of $A - B$ and elements of $A \cap B$. This contradicts the uncountability of A. **19.** We are given bijections f from A to B and g from C to D. Then the function from $A \times C$ to $B \times D$ that sends (a, c) to $(f(a), g(c))$ is a bijection. **21.** By the definition of $|A| \le |B|$, there is a one-to-one function $f: A \to B$. Similarly, there is a one-to-one function $g: B \to C$. By Exercise 33 in Section 2.3, the composition $g \circ f: A \to C$ is one-to-one. Therefore, by definition $|A| \le |C|$. **23.** Using the Axiom of Choice from set theory, choose distinct elements $a_1, a_2, a_3, \ldots$ of A one at a time (this is possible because A is infinite). The resulting set $\{a_1, a_2, a_3, \ldots\}$ is the desired infinite subset of A. **25.** The set of finite strings of characters over a finite alphabet is countably infinite, because we can list these strings in alphabetical order by length. Therefore, the infinite set S can be identified with an infinite subset of this countable set, which by Exercise 16 is also countably infinite. **27.** Suppose that $A_1, A_2, A_3, \ldots$ are countable sets. Because A_i is countable, we can list its elements in a sequence as $a_{i1}, a_{i2}, a_{i3}, \ldots$. The elements of the set $\bigcup_{i=1}^{n} A_i$ can be listed by listing all terms a_{ij} with $i + j = 2$, then all terms a_{ij} with $i + j = 3$, then all terms a_{ij} with $i + j = 4$, and so on. **29.** There are a finite number of bit strings of length m, namely, 2^m. The set of all bit strings is the union of the sets of bit strings of length m for $m = 0, 1, 2, \ldots$. Because the union of a countable number of countable sets is countable (see Exercise 27), there are a countable number of bit strings. **31.** It is clear from the formula that the range of values the function takes on for a fixed value of $m + n$, say $m + n = x$, is $(x-2)(x-1)/2 + 1$ through $(x-2)(x-1)/2 + (x-1)$, because m can assume the values $1, 2, 3, \ldots, (x-1)$ under these conditions, and the first term in the formula is a fixed positive integer when $m + n$ is fixed. To show that this function is one-to-one and onto, we merely need to show that the range of values for $x + 1$ picks up precisely where the range of values for x left off, i.e., that $f(x-1, 1) + 1 = f(1, x)$. We have $f(x-1, 1) + 1 = \frac{(x-2)(x-1)}{2} + (x-1) + 1 = \frac{x^2 - x + 2}{2} = \frac{(x-1)x}{2} + 1 = f(1, x)$. **33.** By the Schröder-Bernstein theorem, it suffices to find one-to-one functions $f: (0, 1) \to [0, 1]$ and $g: [0, 1] \to (0, 1)$. Let $f(x) = x$ and $g(x) = (x + 1)/3$. **35.** Each element A of the power set of the set of positive integers (i.e., $A \subseteq \mathbf{Z}^+$) can be represented uniquely by the bit string $a_1a_2a_3\ldots$, where $a_i = 1$ if $i \in A$ and $a_i = 0$ if $i \notin A$. Assume there were a one-to-one correspondence $f: \mathbf{Z}^+ \to \mathcal{P}(\mathbf{Z}^+)$. Form a new bit string $s = s_1s_2s_3\ldots$ by setting s_i to be 1 minus the ith bit of $f(i)$. Then because s differs in the i bit from $f(i)$, s is not in the range of f, a contradiction. **37.** For any finite alphabet there are a finite number of strings of length n, whenever n is a positive integer. It follows by the result of Exercise 27 that there are only a countable number of strings from any given finite alphabet. Because the set of all computer programs in a particular language is a subset of the set of all strings of a finite alphabet, which is a countable set by the result from Exercise 16, it is itself a countable set. **39.** Exercise 37 shows that there are only a countable number of computer programs. Consequently, there are only a countable number of computable functions. Because, as Exercise 38 shows, there are an uncountable number of functions, not all functions are computable. **41. a)** Note that if x is in the chain generated by y, then by the way the chains are generated, y is in the chain generated by x, so these two chains are the same. Thus, if x is in both the chain generated by y_1 and the chain generated by y_2, then the chain generated by y_1 and the chain generated by y_2 are both the same as the chain generated by x and are therefore the same chain. **b)** A moment's reflection will show that by the way the chains are constructed, this is true. **c)** Again, this is clear from the construction. **d)** Because the chains are disjoint and every element of A appears in exactly one chain and every element of B appears in exactly one chain, the function h cannot map two different elements of A to the same element of B. **e)** If an element b of B appears in a chain of types 1, 2, or 3, then it is preceded by an element of A, which maps to it under h. If b appears in a chain of type 4, then it is followed by an element of A, which maps to it under h. **43.** For a fixed positive integer N, let P_N be the set of polynomials with integer coefficients $a_0x^n + a_1x^{n-1} + \cdots + a_n$ such that $a_0 \neq 0$ and $n + |a_0| + |a_1| + \cdots + |a_n| = N$. Note that $n \leq N - 1$. Let S_N be the set $\{-N, -N+1, \ldots, -1, 0, 1, \ldots, N-1, N\}^{N-1}$ and note that S_N is finite (specifically, it contains $(2N+1)^{N-1}$ members by the Product Rule, discussed in Section 6.1). The function $f : P_N \to S_N$ that maps $a_0x^n + a_1x^{n-1} + \ldots + a_n = 0x^{N-1} + 0x^{N-2} + \ldots + 0x^{n+1} + a_0x^n + a_1x^{n-1} + \ldots + a_n$ to $(0, 0, \ldots, 0, a_0, a_1, \ldots, a_n) \in S_N$ is an injection and thus $|P_N| \leq |S_N|$ and so P_N is finite. Now, define R_N to be the set of all (real or complex) numbers that are the root of a polynomial in P_N. Since a polynomial of degree n with real coefficients has at most n distinct roots, the set R_N is the union of a finite number of finite sets and thus is finite. The set of all roots of polynomials with integer coefficients is $\bigcup_{N=1}^{\infty} R_N$, which is the countable union of finite sets and thus is countable by Exercise 42. Hence, the algebraic numbers are countable.

Section 2.6

1. a) 3×4 **b)** $\begin{bmatrix} 1 \\ 4 \\ 3 \end{bmatrix}$ **c)** $\begin{bmatrix} 2 & 0 & 4 & 6 \end{bmatrix}$ **d)** 1 **e)** $\begin{bmatrix} 1 & 2 & 1 \\ 1 & 0 & 1 \\ 1 & 4 & 3 \\ 3 & 6 & 7 \end{bmatrix}$ **3. a)** $\begin{bmatrix} 1 & 11 \\ 2 & 18 \end{bmatrix}$ **b)** $\begin{bmatrix} 2 & -2 & -3 \\ 1 & 0 & 2 \\ 9 & -4 & 4 \end{bmatrix}$ **c)** $\begin{bmatrix} -4 & 15 & -4 & 1 \\ -3 & 10 & 2 & -3 \\ 0 & 2 & -8 & 6 \\ 1 & -8 & 18 & -13 \end{bmatrix}$ **5.** $\begin{bmatrix} 9/5 & -6/5 \\ -1/5 & 4/5 \end{bmatrix}$

7. $\mathbf{0} + \mathbf{A} = [0 + a_{ij}] = [a_{ij} + 0] = \mathbf{0} + \mathbf{A}$ **9.** $\mathbf{A} + (\mathbf{B} + \mathbf{C}) = [a_{ij} + (b_{ij} + c_{ij})] = [(a_{ij} + b_{ij}) + c_{ij}] = (\mathbf{A} + \mathbf{B}) + \mathbf{C}$ **11.** The number of rows of **A** equals the number of columns of **B**, and the number of columns of **A** equals the number of rows of **B**. **13.** $\mathbf{A}(\mathbf{BC}) = \left[\sum_q a_{iq}\left(\sum_r b_{qr}c_{rl}\right)\right] = \left[\sum_q \sum_r a_{iq}b_{qr}c_{rl}\right] = \left[\sum_r \sum_q a_{iq}b_{qr}c_{rl}\right] = \left[\sum_r \left(\sum_q a_{iq}b_{qr}\right)c_{rl}\right] = (\mathbf{AB})\mathbf{C}$ **15.** $\mathbf{A}^n = \begin{bmatrix} 1 & n \\ 0 & 1 \end{bmatrix}$ **17. a)** Let $\mathbf{A} = [a_{ij}]$ and $\mathbf{B} = [b_{ij}]$. Then $\mathbf{A} + \mathbf{B} = [a_{ij} + b_{ij}]$. We have $(\mathbf{A} + \mathbf{B})^t = [a_{ji} + b_{ji}] = [a_{ji}] + [b_{ji}] = \mathbf{A}^t + \mathbf{B}^t$. **b)** Using the same notation as in part (a), we have $\mathbf{B}^t\mathbf{A}^t = \left[\sum_q b_{qi}a_{jq}\right] = \left[\sum_q a_{jq}b_{qi}\right] = (\mathbf{AB})^t$, because the (i, j)th entry is the (j, i)th entry of **AB**. **19.** The result follows because $\begin{bmatrix} a & b \\ c & d \end{bmatrix}\begin{bmatrix} d & -b \\ -c & a \end{bmatrix} = \begin{bmatrix} ad-bc & 0 \\ 0 & ad-bc \end{bmatrix} = (ad - bc)\mathbf{I}_2 = \begin{bmatrix} d & -b \\ -c & a \end{bmatrix}\begin{bmatrix} a & b \\ c & d \end{bmatrix}$. **21.** $\mathbf{A}^n(\mathbf{A}^{-1})^n = \mathbf{A}(\mathbf{A} \cdots (\mathbf{A}(\mathbf{A}\mathbf{A}^{-1})\mathbf{A}^{-1}) \cdots \mathbf{A}^{-1})\mathbf{A}^{-1}$ by the associative law. Because $\mathbf{A}\mathbf{A}^{-1} = \mathbf{I}$, working from the inside shows that $\mathbf{A}^n(\mathbf{A}^{-1})^n = \mathbf{I}$. Similarly $(\mathbf{A}^{-1})^n\mathbf{A}^n = \mathbf{I}$. Therefore, $(\mathbf{A}^n)^{-1} = (\mathbf{A}^{-1})^n$. **23.** The (i, j)th entry of $\mathbf{A} + \mathbf{A}^t$ is $a_{ij} + a_{ji}$, which equals $a_{ji} + a_{ij}$, the (j, i)th entry of $\mathbf{A} + \mathbf{A}^t$, so by definition $\mathbf{A} + \mathbf{A}^t$ is symmetric. **25.** $x_1 = 1, x_2 = -1, x_3 = -2$

27. a) $\begin{bmatrix} 1 & 1 & 1 \\ 1 & 1 & 1 \\ 1 & 0 & 1 \end{bmatrix}$ **b)** $\begin{bmatrix} 0 & 0 & 1 \\ 1 & 0 & 0 \\ 0 & 0 & 1 \end{bmatrix}$ **c)** $\begin{bmatrix} 1 & 1 & 1 \\ 1 & 1 & 1 \\ 1 & 0 & 1 \end{bmatrix}$ **29. a)** $\begin{bmatrix} 1 & 0 & 0 \\ 1 & 1 & 0 \\ 1 & 0 & 1 \end{bmatrix}$ **b)** $\begin{bmatrix} 1 & 0 & 0 \\ 1 & 0 & 1 \\ 1 & 1 & 0 \end{bmatrix}$ **c)** $\begin{bmatrix} 1 & 0 & 0 \\ 1 & 1 & 1 \\ 1 & 1 & 1 \end{bmatrix}$

31. a) $\mathbf{A} \vee \mathbf{B} = [a_{ij} \vee b_{ij}] = [b_{ij} \vee a_{ij}] = \mathbf{B} \vee \mathbf{A}$ **b)** $\mathbf{A} \wedge \mathbf{B} = [a_{ij} \wedge b_{ij}] = [b_{ij} \wedge a_{ij}] = \mathbf{B} \wedge \mathbf{A}$ **33. a)** $\mathbf{A} \vee (\mathbf{B} \wedge \mathbf{C}) = [a_{ij}] \vee [b_{ij} \wedge c_{ij}] = [a_{ij} \vee (b_{ij} \wedge c_{ij})] = [(a_{ij} \vee b_{ij}) \wedge (a_{ij} \vee c_{ij})] = [a_{ij} \vee b_{ij}] \wedge [a_{ij} \vee c_{ij}] = (\mathbf{A} \vee \mathbf{B}) \wedge (\mathbf{A} \vee \mathbf{C})$ **b)** $\mathbf{A} \wedge (\mathbf{B} \vee \mathbf{C}) = [a_{ij}] \wedge [b_{ij} \vee c_{ij}] = [a_{ij} \wedge (b_{ij} \vee c_{ij})] = [(a_{ij} \wedge b_{ij}) \vee (a_{ij} \wedge c_{ij})] = [a_{ij} \wedge b_{ij}] \vee [a_{ij} \wedge c_{ij}] = (\mathbf{A} \wedge \mathbf{B}) \vee (\mathbf{A} \wedge \mathbf{C})$ **35.** $\mathbf{A} \odot (\mathbf{B} \odot \mathbf{C}) = \left[\bigvee_q a_{iq} \wedge \left(\bigvee_r (b_{qr} \wedge c_{rl})\right)\right] = \left[\bigvee_q \bigvee_r (a_{iq} \wedge b_{qr} \wedge c_{rl})\right] = \left[\bigvee_r \bigvee_q (a_{iq} \wedge b_{qr} \wedge c_{rl})\right] = \left[\bigvee_r \left(\bigvee_q (a_{iq} \wedge b_{qr})\right) \wedge c_{rl}\right] = (\mathbf{A} \odot \mathbf{B}) \odot \mathbf{C}$

Supplementary Exercises

1. a) $\overline{A}$ **b)** $A \cap B$ **c)** $A - B$ **d)** $\overline{A} \cap \overline{B}$ **e)** $A \oplus B$ **3.** Yes **5.** $A - (A - B) = A - (A \cap \overline{B}) = A \cap (\overline{A \cap \overline{B}}) = A \cap (\overline{A} \cup B) = (A \cap \overline{A}) \cup (A \cap B) = \emptyset \cup (A \cap B) = A \cap B$ **7.** Let $A = \{1\}$, $B = \emptyset$, $C = \{1\}$. Then $(A - B) - C = \emptyset$, but $A - (B - C) = \{1\}$. **9.** No. For example, let $A = B = \{a, b\}$, $C = \emptyset$, and $D = \{a\}$. Then $(A - B) - (C - D) = \emptyset - \emptyset = \emptyset$, but $(A - C) - (B - D) = \{a, b\} - \{b\} = \{a\}$. **11. a)** $|\emptyset| \leq |A \cap B| \leq |A| \leq |A \cup B| \leq |U|$ **b)** $|\emptyset| \leq |A - B| \leq |A \oplus B| \leq |A \cup B| \leq |A| + |B|$ **13. a)** Yes, no **b)** Yes, no **c)** f has inverse with $f^{-1}(a) = 3$, $f^{-1}(b) = 4$, $f^{-1}(c) = 2$, $f^{-1}(d) = 1$; g has no inverse. **15.** If f is one-to-one, then f provides a bijection between S and $f(S)$, so they have the same cardinality. If f is not one-to-one, then there exist elements x and y in S such that $f(x) = f(y)$. Let $S = \{x, y\}$. Then $|S| = 2$ but $|f(S)| = 1$. **17.** Let $x \in A$. Then $S_f(\{x\}) = \{f(y) \mid y \in \{x\}\} = \{f(x)\}$. By the same reasoning, $S_g(\{x\}) = \{g(x)\}$. Because $S_f = S_g$, we can conclude that $\{f(x)\} = \{g(x)\}$, and so necessarily $f(x) = g(x)$. **19.** The equation is true if and only if the sum of the fractional parts of x and y is less than 1. **21.** The equation is true if and only if either both x and y are integers, or x is not an integer but the sum of the fractional parts of x and y is less than or equal to 1. **23.** If x is an integer, then $\lfloor x\rfloor + \lfloor m - x\rfloor = x + m - x = m$. Otherwise, write x in terms of its integer and fractional parts: $x = n + \epsilon$, where $n = \lfloor x\rfloor$ and $0 < \epsilon < 1$. In this case $\lfloor x\rfloor + \lfloor m - x\rfloor = \lfloor n + \epsilon\rfloor + \lfloor m - n - \epsilon\rfloor = n + m - n - 1 = m - 1$. **25.** Write $n = 2k+1$ for some integer k. Then $n^2 = 4k^2+4k+1$, so $n^2/4 = k^2 + k + \frac{1}{4}$. Therefore, $\lceil n^2/4\rceil = k^2 + k + 1$. But $(n^2 + 3)/4 = (4k^2 + 4k + 1 + 3)/4 = k^2 + k + 1$. **27.** Let $x = n + (r/m) + \epsilon$, where n is an integer, r is a nonnegative integer less than m, and ϵ is a real number with $0 \le \epsilon < 1/m$. The left-hand side is $\lfloor nm + r + m\epsilon\rfloor = nm + r$. On the right-hand side, the terms $\lfloor x\rfloor$ through $\lfloor x + (m + r - 1)/m\rfloor$ are all just n and the terms from $\lfloor x + (m - r)/m\rfloor$ on are all $n + 1$. Therefore, the right-hand side is $(m - r)n + r(n + 1) = nm + r$, as well. **29.** 101 **31.** $a_1 = 1$; $a_{2n+1} = n \cdot a_{2n}$ for all $n > 0$; and $a_{2n} = n + a_{2n-1}$ for all $n > 0$. The next four terms are 5346, 5353, 37, 471, and 37, 479. **33.** If each $f^{-1}(j)$ is countable, then $S = f^{-1}(1) \cup f^{-1}(2) \cup \cdots$ is the countable union of countable sets and is therefore countable by Exercise 27 in Section 2.5. **35.** Because there is a one-to-one correspondence between **R** and the open interval (0, 1) (given by $f(x) = 2\arctan(x)/\pi$), it suffices to shows that $|(0, 1)\times(0, 1)| = |(0, 1)|$. By the Schröder-Bernstein theorem it suffices to find injective functions $f : (0, 1) \to (0, 1)\times(0, 1)$ and $g : (0, 1) \times (0, 1) \to (0, 1)$. Let $f(x) = (x, \frac{1}{2})$. For g we follow the hint. Suppose $(x, y) \in (0, 1) \times (0, 1)$, and represent x and y with their decimal expansions $x = 0.x_1x_2x_3\ldots$ and $y = 0.y_1y_2y_3\ldots$, never choosing the expansion of any number that ends in an infinite string of 9s. Let $g(x, y)$ be the decimal expansion obtained by interweaving these two strings, namely $0.x_1y_1x_2y_2x_3y_3\ldots$. **37.** $\mathbf{A}^{4n} = \begin{bmatrix}1 & 0\\ 0 & 1\end{bmatrix}$, $\mathbf{A}^{4n+1} = \begin{bmatrix}0 & 1\\ -1 & 0\end{bmatrix}$, $\mathbf{A}^{4n+2} = \begin{bmatrix}-1 & 0\\ 0 & -1\end{bmatrix}$, $\mathbf{A}^{4n+3} = \begin{bmatrix}0 & -1\\ 1 & 0\end{bmatrix}$, for $n \ge 0$ **39.** Suppose that $\mathbf{A} = \begin{bmatrix}a & b\\ c & d\end{bmatrix}$. Let $\mathbf{B} = \begin{bmatrix}0 & 1\\ 0 & 0\end{bmatrix}$. Because $\mathbf{AB} = \mathbf{BA}$, it follows that $c = 0$ and $a = d$. Let $\mathbf{B} = \begin{bmatrix}0 & 0\\ 1 & 0\end{bmatrix}$. Because $\mathbf{AB} = \mathbf{BA}$, it follows that $b = 0$. Hence, $\mathbf{A} = \begin{bmatrix}a & 0\\ 0 & a\end{bmatrix} = a\mathbf{I}$. **41. a)** Let $\mathbf{A} \odot \mathbf{0} = [b_{ij}]$. Then $b_{ij} = (a_{i1} \wedge 0) \vee \cdots \vee (a_{ip} \wedge 0) = 0$. Hence, $\mathbf{A} \odot \mathbf{0} = \mathbf{0}$. Similarly, $\mathbf{0} \odot \mathbf{A} = 0$. **b)** $\mathbf{A} \vee \mathbf{0} = [a_{ij} \vee 0] = [a_{ij}] = \mathbf{A}$. Hence, $\mathbf{A} \vee \mathbf{0} = \mathbf{A}$. Similarly, $\mathbf{0} \vee \mathbf{A} = \mathbf{A}$. **c)** $\mathbf{A} \wedge \mathbf{0} = [a_{ij} \wedge 0] = [0] = \mathbf{0}$. Hence, $\mathbf{A} \wedge \mathbf{0} = \mathbf{0}$. Similarly, $\mathbf{0} \wedge \mathbf{A} = \mathbf{0}$.

CHAPTER 3

Section 3.1

1. $max := 1, i := 2, max := 8, i := 3, max := 12, i := 4, i := 5, i := 6, i := 7, max := 14, i := 8, i := 9, i := 10, i := 11$

3.
```
procedure AddUp(a_1, ..., a_n: integers)
  sum := a_1
  for i := 2 to n
     sum := sum + a_i
  return sum
```

5.
```
procedure duplicates(a_1, a_2, ..., a_n: integers in
     nondecreasing order)
  k := 0 {this counts the duplicates}
  j := 2
  while j ≤ n
    if a_j = a_{j-1} then
      k := k + 1
      c_k := a_j
      while j ≤ n and a_j = c_k
        j := j + 1
    j := j + 1
  {c_1, c_2, ..., c_k is the desired list}
```

7.
```
procedure last even location(a_1,a_2, ...,a_n: integers)
  k := 0
  for i := 1 to n
    if a_i is even then k := i
  return k {k = 0 if there are no evens}
```

9.
```
procedure palindrome check(a_1a_2 ... a_n: string)
  answer := true
  for i := 1 to ⌊n/2⌋
    if a_i ≠ a_{n+1-i} then answer := false
  return answer
```

11.
```
procedure interchange(x, y: real numbers)
  z := x
  x := y
  y := z
```
The minimum number of assignments needed is three.

13. Linear search: $i := 1, i := 2, i := 3, i := 4, i := 5, i := 6, i := 7$, $location := 7$; binary search: $i := 1, j := 8, m := 4, i := 5, m := 6, i := 7, m := 7, j := 7$, $location := 7$

15.
```
procedure insert(x, a_1, a_2, ..., a_n: integers)
  {the list is in order: a_1 ≤ a_2 ≤ ··· ≤ a_n}
  a_{n+1} := x + 1
  i := 1
  while x > a_i
    i := i + 1
  for j := 0 to n − i
    a_{n-j+1} := a_{n-j}
  a_i := x
  {x has been inserted into correct position}
```

17.
```
procedure first largest(a_1, ..., a_n: integers)
  max := a_1
  location := 1
```

for $i := 2$ **to** n
if $max < a_i$ **then**
$max := a_i$
$location := i$
return *location*

19. **procedure** *mean-median-max-min*(a, b, c: integers)
$mean := (a + b + c)/3$
{the six different orderings of a, b, c with respect to $\geq$ will be handled separately}
if $a \geq b$ **then**
if $b \geq c$ **then** $median := b$; $max := a$; $min := c$
⋮
(The rest of the algorithm is similar.)

21. **procedure** *first-three*($a_1, a_2, \ldots, a_n$: integers)
if $a_1 > a_2$ **then** interchange a_1 and a_2
if $a_2 > a_3$ **then** interchange a_2 and a_3
if $a_1 > a_2$ **then** interchange a_1 and a_2

23. **procedure** *onto*(f: function from A to B, where $A = \{a_1, \ldots, a_n\}$, $B = \{b_1, \ldots, b_m\}$, $a_1, \ldots, a_n$, $b_1, \ldots, b_m$ are integers)
for $i := 1$ **to** m
$hit(b_i) := 0$
$count := 0$
for $j := 1$ **to** n
if $hit(f(a_j)) = 0$ **then**
$hit(f(a_j)) := 1$
$count := count + 1$
if $count = m$ **then return true else return false**

25. **procedure** *ones*(a: bit string, $a = a_1 a_2 \ldots a_n$)
$count := 0$
for $i := 1$ **to** n
if $a_i := 1$ **then**
$count := count + 1$
return *count*

27. **procedure** *ternary search*(s: integer, $a_1, a_2, \ldots, a_n$: increasing integers)
$i := 1$
$j := n$
while $i < j - 1$
$l := \lfloor (i + j)/3 \rfloor$
$u := \lfloor 2(i + j)/3 \rfloor$
if $x > a_u$ **then** $i := u + 1$
else if $x > a_l$ **then**
$i := l + 1$
$j := u$
else $j := l$
if $x = a_i$ **then** $location := i$
else if $x = a_j$ **then** $location := j$
else $location := 0$
return *location* {0 if not found}

29. **procedure** *find a mode*($a_1, a_2, \ldots, a_n$: nondecreasing integers)
$modecount := 0$
$i := 1$
while $i \leq n$
$value := a_i$
$count := 1$
while $i \leq n$ and $a_i = value$
$count := count + 1$
$i := i + 1$
if $count > modecount$ **then**
$modecount := count$
$mode := value$
return *mode*

31. Assume the input is strings $a_1 a_2 \ldots a_n$ and $b_1 b_2 \ldots b_n$, where each character is a letter, A through Z. Assume also that a function *index* is available, such that *index*(x) is the position of the letter x in the alphabet (*index*('A') = 1, …, *index*('Z') = 26). **a)** Initialize *a-count* and *b-count* to be lists of length 26 with all values equal to 0. For i from 1 to n increment *a-count*(*index*(a_i)) and *b-count*(*index*(b_i)). If *a-count*(i) = *b-count*(i) for all i from 1 to 26, then return "true"; otherwise return "false." **b)** Sort both strings into alphabetical order. Then the two original strings were anagrams if and only if the sorted strings are identical.

33. **procedure** *find duplicate*($a_1, a_2, \ldots, a_n$: integers)
$location := 0$
$i := 2$
while $i \leq n$ and $location = 0$
$j := 1$
while $j < i$ and $location = 0$
if $a_i = a_j$ **then** $location := i$
else $j := j + 1$
$i := i + 1$
return *location*
{*location* is the subscript of the first value that repeats a previous value in the sequence}

35. **procedure** *find decrease*($a_1, a_2, \ldots, a_n$: positive integers)
$location := 0$
$i := 2$
while $i \leq n$ and $location = 0$
if $a_i < a_{i-1}$ **then** $location := i$
else $i := i + 1$
return *location*
{*location* is the subscript of the first value less than the immediately preceding one}

37. At the end of the first pass: 1, 3, 5, 4, 7; at the end of the second pass: 1, 3, 4, 5, 7; at the end of the third pass: 1, 3, 4, 5, 7; at the end of the fourth pass: 1, 3, 4, 5, 7

39. **procedure** *better bubblesort*($a_1, \ldots, a_n$: integers)
$i := 1$; $done :=$ **false**
while $i < n$ and $done =$ **false**
$done :=$ **true**
for $j := 1$ **to** $n - i$
if $a_j > a_{j+1}$ **then**
interchange a_j and a_{j+1}

done : = **false**
$i := i + 1$
$\{a_1, \ldots, a_n$ is in increasing order$\}$

41. At the end of the first, second, and third passes: 1, 3, 5, 7, 4; at the end of the fourth pass: 1, 3, 4, 5, 7 **43. a)** 1, 5, 4, 3, 2; 1, 2, 4, 3, 5; 1, 2, 3, 4, 5; 1, 2, 3, 4, 5 **b)** 1, 4, 3, 2, 5; 1, 2, 3, 4, 5; 1, 2, 3, 4, 5; 1, 2, 3, 4, 5 **c)** 1, 2, 3, 4, 5; 1, 2, 3, 4, 5; 1, 2, 3, 4, 5; 1, 2, 3, 4, 5 **45.** We carry out the linear search algorithm given as Algorithm 2 in this section, except that we replace $x \neq a_i$ by $x < a_i$, and we replace the **else** clause with **else** *location* $:= n + 1$. **47.** $2 + 3 + 4 + \cdots + n = (n^2+n-2)/2$ **49.** Find the location for the 2 in the list 3 (one comparison), and insert it in front of the 3, so the list now reads 2, 3, 4, 5, 1, 6. Find the location for the 4 (compare it to the 2 and then the 3), and insert it, leaving 2, 3, 4, 5, 1, 6. Find the location for the 5 (compare it to the 3 and then the 4), and insert it, leaving 2, 3, 4, 5, 1, 6. Find the location for the 1 (compare it to the 3 and then the 2 and then the 2 again), and insert it, leaving 1, 2, 3, 4, 5, 6. Find the location for the 6 (compare it to the 3 and then the 4 and then the 5), and insert it, giving the final answer 1, 2, 3, 4, 5, 6.

51. procedure *binary insertion sort*($a_1, a_2, \ldots, a_n$: real numbers with $n \geq 2$)
for $j := 2$ **to** n
{binary search for insertion location i}
left := 1
right := $j - 1$
while *left* < *right*
middle := $\lfloor (left + right)/2 \rfloor$
if $a_j > a_{middle}$ **then** *left* := *middle* + 1
else *right* := *middle*
if $a_j < a_{left}$ **then** $i :=$ *left* **else** $i :=$ *left* + 1
{insert a_j in location i by moving a_i through a_{j-1} toward back of list}
$m := a_j$
for $k := 0$ **to** $j - i - 1$
$a_{j-k} := a_{j-k-1}$
$a_i := m$
$\{a_1, a_2, \ldots, a_n$ are sorted$\}$

53. The variation from Exercise 52 **55.** $m = 3, n = 8; s = 0, j = 1$, no match; $s = 1, j = 1, j = 2, j = 3$, no match; $s = 2, j = 1$, no match; $s = 3, j = 1, j = 2$, no match; $s = 4, j = 1$, no match; $s = 5, j = 1, j = 2, j = 3, j = 4$, "5 is a valid shift" **57. a)** Two quarters, one penny **b)** Two quarters, one dime, one nickel, four pennies **c)** Three quarters, one penny **d)** Two quarters, one dime **59.** Cashier's algorithm uses fewest coins in parts (a), (c), and (d). **a)** Two quarters, one penny **b)** Two quarters, one dime, nine pennies **c)** Three quarters, one penny **d)** Two quarters, one dime **61.** The 9:00–9:45 talk, the 9:50–10:15 talk, the 10:15–10:45 talk, the 11:00–11:15 talk **63. a)** Order the talks by starting time. Number the lecture halls 1, 2, 3, and so on. For each talk, assign it to lowest numbered lecture hall that is currently available. **b)** If this algorithm uses n lecture halls, then at the point the nth hall was first assigned, it *had* to be used (otherwise a lower-numbered hall would have been assigned), which means that n talks were going on simultaneously (this talk just assigned and the $n - 1$ talks currently in halls 1 through $n - 1$). **65.** Here we assume that the men are the suitors and the women the suitees.
procedure *stable*($M_1, M_2, \ldots, M_s, W_1, W_2, \ldots, W_s$: preference lists)
for $i := 1$ **to** s
mark man i as rejected
for $i := 1$ **to** s
set man i's rejection list to be empty
for $j := 1$ **to** s
set woman j's proposal list to be empty
while rejected men remain
for $i := 1$ **to** s
if man i is marked rejected **then** add i to the proposal list for the woman j who ranks highest on his preference list but does not appear on his rejection list, and mark i as not rejected
for $j := 1$ **to** s
if woman j's proposal list is nonempty **then** remove from j's proposal list all men i except the man i_0 who ranks highest on her preference list, and for each such man i mark him as rejected and add j to his rejection list
for $j := 1$ **to** s
match j with the one man on j's proposal list
{This matching is stable.}

67. If the assignment is not stable, then there is a man m and a woman w such that m prefers w to the woman w' with whom he is matched, and w prefers m to the man with whom she is matched. But m must have proposed to w before he proposed to w', because he prefers the former. Because m did not end up matched with w, she must have rejected him. Women reject a suitor only when they get a better proposal, and they eventually get matched with a pending suitor, so the woman with whom w is matched must be better in her eyes than m, contradicting our original assumption. Therefore, the marriage is stable. **69. a)** (adapted from Wikipedia) Let m be the majority element. Let n be a number defined at any step of the algorithm to be either the counter, if the majority candidate is m, or the negative of the counter otherwise. Then at each step in which the algorithm encounters a value equal to m, the value of n will increase by 1, and at each step at which it encounters a different value, the value of n may either increase or decrease by one. If m truly is the majority element, there will be more increases than decreases, and n will be positive at the end of the algorithm. But this can be true only when the final stored element is m, the majority element. **b)** A counterexample is ABABC. **71.** Run the two programs on their inputs concurrently and report which one halts.

Section 3.2

1. The choices of C and k are not unique. **a)** $C = 1, k = 10$ **b)** $C = 4, k = 7$ **c)** No **d)** $C = 5, k = 1$ **e)** $C = 1, k = 0$ **f)** $C = 1, k = 2$ **3.** $x^4 + 9x^3 + 4x + 7 \leq 4x^4$ for all $x > 9$;

witnesses $C = 4, k = 9$ **5.** $(x^2+1)/(x+1) = x-1+2/(x+1) < x$ for all $x > 1$; witnesses $C = 1, k = 1$ **7.** The choices of C and k are not unique. **a)** $n = 3, C = 3, k = 1$ **b)** $n = 3, C = 4, k = 1$ **c)** $n = 1, C = 2, k = 1$ **d)** $n = 0, C = 2, k = 1$ **9.** $x^2 + 4x + 17 \le 3x^3$ for all $x > 17$, so $x^2 + 4x + 17$ is $O(x^3)$, with witnesses $C = 3, k = 17$. However, if x^3 were $O(x^2 + 4x + 17)$, then $x^3 \le C(x^2 + 4x + 17) \le 3Cx^2$ for some C, for all sufficiently large x, which implies that $x \le 3C$ for all sufficiently large x, which is impossible. Hence, x^3 is not $O(x^2 + 4x + 17)$. **11.** $3x^4 + 1 \le 4x^4 = 8(x^4/2)$ for all $x > 1$, so $3x^4 + 1$ is $O(x^4/2)$, with witnesses $C = 8$, $k = 1$. Also $x^4/2 \le 3x^4 + 1$ for all $x > 0$, so $x^4/2$ is $O(3x^4 + 1)$, with witnesses $C = 1, k = 0$. **13.** Because $2^n \le 3^n$ for all $n > 0$, it follows that 2^n is $O(3^n)$, with witnesses $C = 1, k = 0$. However, if 3^n were $O(2^n)$, then for some C, $3^n \le C \cdot 2^n$ for all sufficiently large n. This says that $C \ge (3/2)^n$ for all sufficiently large n, which is impossible. Hence, 3^n is not $O(2^n)$. **15.** All functions for which there exist real numbers k and C with $|f(x)| \le C$ for $x > k$. These are the functions $f(x)$ that are bounded for all sufficiently large x. **17.** There are constants C_1, C_2, k_1, and k_2 such that $|f(x)| \le C_1|g(x)|$ for all $x > k_1$ and $|g(x)| \le C_2|h(x)|$ for all $x > k_2$. Hence, for $x > \max(k_1, k_2)$ it follows that $|f(x)| \le C_1|g(x)| \le C_1C_2|h(x)|$. This shows that $f(x)$ is $O(h(x))$. **19.** 2^{n+1} is $O(2^n)$; 2^{2n} is not. **21.** $1000 \log n$; $\sqrt{n}$; $n \log n$; $n^2/1{,}000{,}000$; 2^n; 3^n; $2n!$ **23.** The algorithm that uses $n \log n$ operations **25. a)** $O(n^3)$ **b)** $O(n^5)$ **c)** $O(n^3 \cdot n!)$ **27. a)** $O(n^2 \log n)$ **b)** $O(n^2(\log n)^2)$ **c)** $O(n^{2^n})$ **29. a)** Neither $\Theta(x^2)$ nor $\Omega(x^2)$ **b)** $\Theta(x^2)$ and $\Omega(x^2)$ **c)** Neither $\Theta(x^2)$ nor $\Omega(x^2)$ **d)** $\Omega(x^2)$, but not $\Theta(x^2)$ **e)** $\Omega(x^2)$, but not $\Theta(x^2)$ **f)** $\Omega(x^2)$ and $\Theta(x^2)$ **31.** If $f(x)$ is $\Theta(g(x))$, then there exist constants C_1 and C_2 with $C_1|g(x)| \le |f(x)| \le C_2|g(x)|$. It follows that $|f(x)| \le C_2|g(x)|$ and $|g(x)| \le (1/C_1)|f(x)|$ for $x > k$. Thus, $f(x)$ is $O(g(x))$ and $g(x)$ is $O(f(x))$. Conversely, suppose that $f(x)$ is $O(g(x))$ and $g(x)$ is $O(f(x))$. Then there are constants C_1, C_2, k_1, and k_2 such that $|f(x)| \le C_1|g(x)|$ for $x > k_1$ and $|g(x)| \le C_2|f(x)|$ for $x > k_2$. We can assume that $C_2 > 0$ (we can always make C_2 larger). Then we have $(1/C_2)|g(x)| \le |f(x)| \le C_1|g(x)|$ for $x > \max(k_1, k_2)$. Hence, $f(x)$ is $\Theta(g(x))$. **33.** If $f(x)$ is $\Theta(g(x))$, then $f(x)$ is both $O(g(x))$ and $\Omega(g(x))$. Hence, there are positive constants C_1, k_1, C_2, and k_2 such that $|f(x)| \le C_2|g(x)|$ for all $x > k_2$ and $|f(x)| \ge C_1|g(x)|$ for all $x > k_1$. It follows that $C_1|g(x)| \le |f(x)| \le C_2|g(x)|$ whenever $x > k$, where $k = \max(k_1,k_2)$. Conversely, if there are positive constants C_1, C_2, and k such that $C_1|g(x)| \le |f(x)| \le C_2|g(x)|$ for $x > k$, then taking $k_1 = k_2 = k$ shows that $f(x)$ is both $O(g(x))$ and $\Theta(g(x))$.

35.

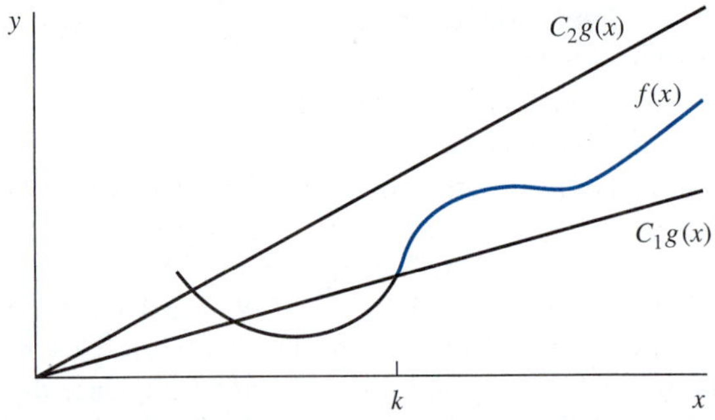

37. If $f(x)$ is $\Theta(1)$, then $|f(x)|$ is bounded between positive constants C_1 and C_2. In other words, $f(x)$ cannot grow larger than a fixed bound or smaller than the negative of this bound and must not get closer to 0 than some fixed bound. **39.** Because $f(x)$ is $O(g(x))$, there are constants C and k such that $|f(x)| \le C|g(x)|$ for $x > k$. Hence, $|f^n(x)| \le C^n|g^n(x)|$ for $x > k$, so $f^n(x)$ is $O(g^n(x))$ by taking the constant to be C^n. **41.** Because $f(x)$ and $g(x)$ are increasing and unbounded, we can assume $f(x) \ge 1$ and $g(x) \ge 1$ for sufficiently large x. There are constants C and k with $f(x) \le Cg(x)$ for $x > k$. This implies that $\log f(x) \le \log C + \log g(x) < 2\log g(x)$ for sufficiently large x. Hence, $\log f(x)$ is $O(\log g(x))$. **43.** By definition there are positive constraints C_1, C_1', C_2, C_2', k_1, k_1', k_2, and k_2' such that $f_1(x) \ge C_1|g(x)|$ for all $x > k_1$, $f_1(x) \le C_1'|g(x)|$ for all $x > k_1'$, $f_2(x) \ge C_2|g(x)|$ for all $x > k_2$, and $f_2(x) \le C_2'|g(x)|$ for all $x > k_2'$. Adding the first and third inequalities shows that $f_1(x) + f_2(x) \ge (C_1 + C_2)|g(x)|$ for all $x > k$ where $k = \max(k_1, k_2)$. Adding the second and fourth inequalities shows that $f_1(x) + f_2(x) \le (C_1' + C_2')|g(x)|$ for all $x > k'$ where $k' = \max(k_1', k_2')$. Hence, $f_1(x) + f_2(x)$ is $\Theta(g(x))$. This is no longer true if f_1 and f_2 can assume negative values. **45.** This is false. Let $f_1 = x^2 + 2x$, $f_2(x) = x^2 + x$, and $g(x) = x^2$. Then $f_1(x)$ and $f_2(x)$ are both $O(g(x))$, but $(f_1 - f_2)(x)$ is not. **47.** Take $f(n)$ to be the function with $f(n) = n$ if n is an odd positive integer and $f(n) = 1$ if n is an even positive integer and $g(n)$ to be the function with $g(n) = 1$ if n is an odd positive integer and $g(n) = n$ if n is an even positive integer. **49.** There are positive constants C_1, C_2, C_1', C_2', k_1, k_1', k_2, and k_2' such that $|f_1(x)| \ge C_1|g_1(x)|$ for all $x > k_1$, $|f_1(x)| \le C_1'|g_1(x)|$ for all $x \ge k_1'$, $|f_2(x)| > C_2|g_2(x)|$ for all $x > k_2$, and $|f_2(x)| \le C_2'|g_2(x)|$ for all $x > k_2'$. Because f_2 and g_2 are never zero, the last two inequalities can be rewritten as $|1/f_2(x)| \le (1/C_2)|1/g_2(x)|$ for all $x > k_2$ and $|1/f_2(x)| \ge (1/C_2')|1/g_2(x)|$ for all $x > k_2'$. Multiplying the first and rewritten fourth inequalities shows that $|f_1(x)/f_2(x)| \ge (C_1/C_2')|g_1(x)/g_2(x)|$ for all $x > \max(k_1, k_2')$, and multiplying the second and rewritten third inequalities gives $|f_1(x)/f_2(x)| \le (C_1'/C_2)|g_1(x)/g_2(x)|$ for all $x > \max(k_1', k_2)$. It follows that f_1/f_2 is big-Theta of g_1/g_2. **51.** There exist positive constants C_1, C_2, k_1, k_2, k_1', k_2' such that $|f(x, y)| \le C_1|g(x, y)|$ for all $x > k_1$ and $y > k_2$ and $|f(x, y)| \ge C_2|g(x, y)|$ for all $x > k_1'$ and $y > k_2'$. **53.** $(x^2 + xy + x\log y)^3 < (3x^2y^3) = 27x^6y^3$ for $x > 1$ and $y > 1$, because $x^2 < x^2y$, $xy < x^2y$, and $x\log y < x^2y$. Hence, $(x^2 + xy + x\log y)^3$ is $O(x^6y^3)$. **55.** For all positive real numbers x and y, $\lfloor xy \rfloor \leq xy$. Hence, $\lfloor xy \rfloor$ is $O(xy)$ from the definition, taking $C = 1$ and $k_1 = k_2 = 0$. **57.** Clearly $n^d < n^c$ for all $n \geq 2$; therefore, n^d is $O(n^c)$. The ratio $n^d/n^c = n^{d-c}$ is unbounded so there is no constant C such that $n^d \leq Cn^c$ for large n. **59.** If f and g are positive-valued functions such that $\lim_{n\to\infty} f(x)/g(x) = C < \infty$, then $f(x) < (C+1)g(x)$ for large enough x, so $f(n)$ is $O(g(n))$. If that limit is ∞, then $f(n)$ is not $O(g(n))$. Here repeated applications of L'Hôpital's rule shows that $\lim_{x\to\infty} x^d/b^x = 0$ and $\lim_{x\to\infty} b^x/x^d = \infty$. **61.** To show that c^n is $O(n!)$, assume WLOG that c is an integer greater than 1. Claim that if $n \geq c^2+c$, then $n! \geq c^n$. Both $n!$ and c^n have n factors. Replacing each of the factors from c^2+1 to c^2+c in $n!$ by c^2 only makes the product smaller. But then each of these factors c^2 can be factored as $c \cdot c$ and one of those factors of c moved to pair with the factors 1 through c in $n!$. At this point every factor of $n!$ is greater than or equal to c, so the product is greater than or equal to c^n. To show that $n!$ is not $O(c^n)$, look at the ratio $n!/c^n$ and write this as $(c!/c^c)\cdot((c+1)/c)\cdot((c+2)/c)\cdot((c+3)/c)\cdots((2c)/c)\cdot((2c+1)/c)\cdots(n/c)$. Treat the product of the first $c+1$ factors as a constant and note that every other factor is greater than 2. Thus, the product is greater than an arbitrarily large power of 2 as $n \to \infty$, which therefore approaches ∞. Therefore, $n!$ cannot be bounded by a constant times c^n. **63. a)** $\lim_{x\to\infty} x^2/x^3 = \lim_{x\to\infty} 1/x = 0$ **b)** $\lim_{x\to\infty} \frac{x\log x}{x^2} = \lim_{x\to\infty} \frac{\log x}{x} = \lim_{x\to\infty} \frac{1}{x \ln 2} = 0$ (using L'Hôpital's rule) **c)** $\lim_{x\to\infty} \frac{x^2}{2^x} = \lim_{x\to\infty} \frac{2x}{2^x \cdot \ln 2} = \lim_{x\to\infty} \frac{2}{2^x \cdot (\ln 2)^2} = 0$ (using L'Hôpital's rule) **d)** $\lim_{x\to\infty} \frac{x^2+x+1}{x^2} = \lim_{x\to\infty}\left(1 + \frac{1}{x} + \frac{1}{x^2}\right) = 1 \neq 0$

65.

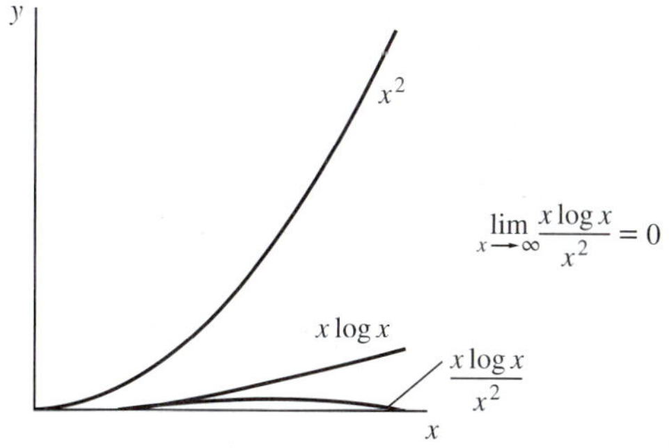

67. No. Take $f(x) = 1/x^2$ and $g(x) = 1/x$. **69. a)** Because $\lim_{x\to\infty} f(x)/g(x) = 0$, $|f(x)|/|g(x)| < 1$ for sufficiently large x. Hence, $|f(x)| < |g(x)|$ for $x > k$ for some constant k. Therefore, $f(x)$ is $O(g(x))$. **b)** Let $f(x) = g(x) = x$. Then $f(x)$ is $O(g(x))$, but $f(x)$ is not $o(g(x))$ because $f(x)/g(x) = 1$. **71.** Because $f_2(x)$ is $o(g(x))$, from Exercise 69(a) it follows that $f_2(x)$ is $O(g(x))$. By Corollary 1, we have $f_1(x) + f_2(x)$ is $O(g(x))$. **73.** We can show that $(n-i)(i+1) \geq n$ for $i = 0, 1, \ldots, n-1$. Hence, $(n!)^2 = (n \cdot 1)((n-1)\cdot 2)\cdot((n-2)\cdot 3)\cdots(2\cdot(n-1))\cdot(1\cdot n) \geq n^n$. Therefore, $2\log n! \geq n\log n$. **75.** Compute that $\log 5! \approx 6.9$ and $(5\log 5)/4 \approx 2.9$, so the inequality holds for $n = 5$. Assume $n \geq 6$. Because $n!$ is the product of all the integers from n down to 1, we have $n! > n(n-1)(n-2)\cdots\lceil n/2\rceil$ (because at least the term 2 is missing). Note that there are more than $n/2$ terms in this product, and each term is at least as big as $n/2$. Therefore, the product is greater than $(n/2)^{(n/2)}$. Taking the log of both sides of the inequality, we have $\log n! > \log\left(\frac{n}{2}\right)^{n/2} = \frac{n}{2}\log\frac{n}{2} = \frac{n}{2}(\log n - 1) > (n\log n)/4$, because $n > 4$ implies $\log n - 1 > (\log n)/2$. **77.** All are not asymptotic.

Section 3.3

1. $O(1)$ **3.** $O(n^2)$ **5.** $2n-1$ **7.** Linear **9.** $O(n)$ **11. a)** **procedure** *disjointpair*($S_1, S_2, \ldots, S_n$: subsets of $\{1, 2, \ldots, n\}$)

```
answer := false
for i := 1 to n
  for j := i + 1 to n
    disjoint := true
    for k := 1 to n
      if k ∈ S_i and k ∈ S_j then disjoint := false
    if disjoint then answer := true
return answer
```

b) $O(n^3)$ **13. a)** *power* := 1, y := 1; i := 1, *power* := 2, y := 3; i := 2, *power* := 4, y := 15 **b)** $2n$ multiplications and n additions **15. a)** $2^{10^9} \approx 10^{3\times 10^8}$ **b)** 10^9 **c)** 3.96×10^7 **d)** 3.16×10^4 **e)** 29 **f)** 12 **17. a)** $2^{2^{60\cdot 10^{12}}}$ **b)** $2^{60\cdot 10^{12}}$ **c)** $\lfloor 2^{\sqrt{60\cdot 10^6}} \rfloor \approx 2 \times 10^{2331768}$ **d)** 60,000,000 **e)** 7,745,966 **f)** 45 **g)** 6 **19. a)** 36 years **b)** 13 days **c)** 19 minutes **21. a)** Less than 1 millisecond more **b)** 100 milliseconds more **c)** $2n+1$ milliseconds more **d)** $3n^2+3n+1$ milliseconds more **e)** Twice as much time **f)** 2^{2n+1} times as many milliseconds **g)** $n+1$ times as many milliseconds **23.** The average number of comparisons is $(3n+4)/2$. **25.** $O(\log n)$ **27.** $O(n)$ **29.** $O(n^2)$ **31.** $O(n)$ **33.** $O(n)$ **35.** $O(\log n)$ comparisons; $O(n^2)$ swaps **37.** $O(m(n-m))$ **39. a)** $O(n^2)$ **b)** $O(n \log n)$ **41.** $O(n^2 2^n)$ **43. a)** doubles **b)** increases by 1 **45.** Use Algorithm 1, where **A** and **B** are now $n \times n$ upper triangular matrices, by replacing m by n in line 1, and having q iterate only from i to j, rather than from 1 to k. **47.** $n(n+1)(n+2)/6$ **49.** **A**((**BC**)**D**)

Supplementary Exercises

1. a) **procedure** *last max*($a_1, \ldots, a_n$: integers)

```
max := a_1
last := 1
i := 2
while i ≤ n
  if a_i ≥ max then
    max := a_i
    last := i
  i := i + 1
return last
```

b) $2n - 1 = O(n)$ comparisons

3. a) procedure *pair zeros*($b_1b_2 \ldots b_n$: bit string, $n \geq 2$)
$x := b_1$
$y := b_2$
$k := 2$
while $k < n$ and ($x \neq 0$ or $y \neq 0$)
$k := k + 1$
$x := y$
$y := b_k$
if $x = 0$ and $y = 0$ **then** print "YES"
else print "NO"

b) $O(n)$

5. a) and **b)**

procedure *smallest and largest*($a_1, a_2, \ldots, a_n$: integers)
$min := a_1$
$max := a_1$
for $i := 2$ **to** n
if $a_i < min$ **then** $min := a_i$
if $a_i > max$ **then** $max := a_i$
{*min* is the smallest integer among the input, and *max* is the largest}

c) $2n - 2$

7. Before any comparisons are done, there is a possibility that each element could be the maximum and a possibility that it could be the minimum. This means that there are $2n$ different possibilities, and $2n - 2$ of them have to be eliminated through comparisons of elements, because we need to find the unique maximum and the unique minimum. We classify comparisons of two elements as "virgin" or "nonvirgin," depending on whether or not both elements being compared have been in any previous comparison. A virgin comparison eliminates the possibility that the larger one is the minimum and that the smaller one is the maximum; thus, each virgin comparison eliminates two possibilities, but it clearly cannot do more. A nonvirgin comparison must be between two elements that are still in the running to be the maximum or two elements that are still in the running to be the minimum, and at least one of these elements must *not* be in the running for the other category. For example, we might be comparing x and y, where all we know is that x has been eliminated as the minimum. If we find that $x > y$ in this case, then only one possibility has been ruled out—we now know that y is not the maximum. Thus, in the worst case, a nonvirgin comparison eliminates only one possibility. (The cases of other nonvirgin comparisons are similar.) Now there are at most $\lfloor n/2 \rfloor$ comparisons of elements that have not been compared before, each removing two possibilities; they remove $2\lfloor n/2 \rfloor$ possibilities altogether. Therefore, we need $2n-2-2\lfloor n/2 \rfloor$ more comparisons that, as we have argued, can remove only one possibility each, in order to find the answers in the worst case, because $2n - 2$ possibilities have to be eliminated. This gives us a total of $2n - 2 - 2\lfloor n/2 \rfloor + \lfloor n/2 \rfloor$ comparisons in all. But $2n - 2 - 2\lfloor n/2 \rfloor + \lfloor n/2 \rfloor = 2n - 2 - \lfloor n/2 \rfloor = 2n - 2 + \lceil -n/2 \rceil = \lceil 2n - n/2 \rceil - 2 = \lceil 3n/2 \rceil - 2$, as desired. **9.** The following algorithm has worst-case complexity $O(n^4)$.

procedure *equal sums*($a_1, a_2, \ldots, a_n$)
for $i := 1$ to n
for $j := i + 1$ **to** n {since we want $i < j$}
for $k := 1$ **to** n
for $l := k + 1$ **to** n {since we want $k < l$}
if $a_i + a_j = a_k + a_l$ and $(i, j) \neq (k, l)$
then output these pairs

11. At end of first pass: 3, 1, 4, 5, 2, 6; at end of second pass: 1, 3, 2, 4, 5, 6; at end of third pass: 1, 2, 3, 4, 5, 6; fourth pass finds nothing to exchange and algorithm terminates **13.** There are possibly as many as n passes through the list, and each pass uses $O(n)$ comparisons. Thus, there are $O(n^2)$ comparisons in all. **15.** Because $\log n < n$, we have $(n \log n + n^2)^3 \leq (n^2 + n^2)^3 \leq (2n^2)^3 = 8n^6$ for all $n > 0$. This proves that $(n \log n + n^2)^3$ is $O(n^6)$, with witnesses $C = 8$ and $k = 0$. **17.** $O(x^2 2^x)$ **19.** Note that $\frac{n!}{2^n} = \frac{n}{2} \cdot \frac{n-1}{2} \cdots \frac{3}{2} \cdot \frac{2}{2} \cdot \frac{1}{2} > \frac{n}{2} \cdot 1 \cdot 1 \cdots 1 \cdot \frac{1}{2} = \frac{n}{4}$. **21.** All of these functions are of the same order. **23.** 2^{107} **25.** $(\log n)^2$, $2^{\sqrt{\log_2 n}}$, $n(\log n)^{1001}$, $n^{1.0001}$, 1.0001^n, n^n **27.** For example, $f(n) = n^{2\lfloor n/2 \rfloor + 1}$ and $g(n) = n^{2\lceil n/2 \rceil}$

29. a) procedure *brute*($a_1, a_2, \ldots, a_n$: integers)
for $i := 1$ **to** $n - 1$
for $j := i + 1$ **to** n
for $k := 1$ **to** n
if $a_i + a_j = a_k$ **then return true**
else return false

b) $O(n^3)$

31. For m_1: w_1 and w_2; for m_2: w_1 and w_3; for m_3: w_2 and w_3; for w_1: m_1 and m_2; for w_2: m_1 and m_3; for w_3: m_2 and m_3 **33.** A matching in which each woman is assigned her valid partner ranking highest on her preference list is female optimal; a matching in which each man is assigned his valid partner ranking lowest on his preference list is male pessimal. **35. a)** Modify the preamble to Exercise 64 in Section 3.1 so that there are s men $m_1, m_2, \ldots, m_s$ and t women $w_1, w_2, \ldots, w_t$. A matching will contain $\min(s, t)$ marriages. The definition of "stable marriage" is the same, with the understanding that each person prefers any mate to being unmatched. **b)** Create $|s - t|$ fictitious people (men or women, whichever is in shorter supply) so that the number of men and the number of women become the same, and put these fictitious people at the bottom of everyone's preference lists. **c)** This follows immediately from Exercise 67 in Section 3.1. **37.** 5; 15 **39.** The first situation in Exercise 37 **41. a)** For each subset S of $\{1, 2, \ldots, n\}$, compute $\sum_{j \in S} w_j$. Keep track of the subset giving the largest such sum that is less than or equal to W, and return that subset as the output of the algorithm. **b)** The food pack and the portable stove **43. a)** The makespan is always at least as large as the load on the processor assigned to do the lengthiest job, which must be at least $\max_{j=1,2,\ldots,n} t_j$. Therefore, the minimum makespan satisfies this inequality. **b)** The total amount of time the processors need to spend working on the jobs (the total load) is $\sum_{j=1}^{n} t_j$. Therefore, the average load per processor is $\frac{1}{p} \sum_{j=1}^{n} t_j$. The maximum load cannot be any smaller than the average, so the minimum makespan is always at least this large. **45.** Processor 1: jobs 1, 4; processor 2: job 2; processor 3: jobs 3, 5

CHAPTER 4

Section 4.1

1. a) Yes **b)** No **c)** Yes **d)** No **3.** Suppose that $a \mid b$. Then there exists an integer k such that $ka = b$. Because $a(ck) = bc$ it follows that $a \mid bc$. **5.** If $a \mid b$ and $b \mid a$, there are integers c and d such that $b = ac$ and $a = bd$. Hence, $a = acd$. Because $a \neq 0$ it follows that $cd = 1$. Thus, either $c = d = 1$ or $c = d = -1$. Hence, either $a = b$ or $a = -b$. **7.** Because $ac \mid bc$ there is an integer k such that $ack = bc$. Hence, $ak = b$, so $a \mid b$. **9.** It is given that $b = m \cdot a$ for some integer m. If a is not odd, then $a = 2k$ for some integer k, whence $b = 2km$. This means by definition that b is even. **11.** If a is an integer that is not divisible by 3, then it must leave a remainder of either 1 or 2 when divided by 3. In the first case $a = 3k + 1$ for some integer k, so $(a+1)(a+2) = (3k+2)(3k+3) = 3(3k+2)(k+1)$ and so is divisible by 3. In the second case $a = 3k+2$ for some integer k, so $(a+1)(a+2) = (3k+3)(3k+4) = 3(k+1)(3k+4)$ and so is divisible by 3. **13. a)** 2, 5 **b)** −11, 10 **c)** 34, 7 **d)** 77, 0 **e)** 0, 0 **f)** 0, 3 **g)** −1, 2 **h)** 4, 0 **15. a)** 7:00 **b)** 8:00 **c)** 10:00 **17. a)** 10 **b)** 8 **c)** 0 **d)** 9 **e)** 6 **f)** 11 **19.** If $d \mid a$, then $a = md$ for some integer m, and it follows that $-a = (-m)d$ as well. Therefore, $-(a$ **div** $d) = -m$ and $(-a)$ **div** $d = -m$, as desired. Conversely, if d does not divide a, then $a = qd + r$ for some r with $0 < r < d$. Here $q = a$ **div** d. It follows that $-a = (-q)d - r = (-q-1)d + (d-r)$. Note that $0 < d - r < d$, so by definition $(-a)$ **div** $d = -q - 1$. But $-(a$ **div** $d) = -q$, so $(-a)$ **div** $d \neq -(a$ **div** $d)$. **21.** If a **mod** $m = b$ **mod** m, then a and b have the same remainder when divided by m. Hence, $a = q_1 m + r$ and $b = q_2 m + r$, where $0 \le r < m$. It follows that $a - b = (q_1 - q_2)m$, so $m \mid (a - b)$. It follows that $a \equiv b \pmod{m}$. **23.** There is some b with $(b-1)k < n \le bk$. Hence, $(b-1)k \le n - 1 < bk$. Divide by k to obtain $b - 1 < n/k \le b$ and $b - 1 \le (n-1)/k < b$. Hence, $\lceil n/k \rceil = b$ and $\lfloor (n-1)/k \rfloor = b - 1$. **25.** x **mod** m if x **mod** $m \le \lceil m/2 \rceil$ and $(x$ **mod** $m) - m$ if x **mod** $m > \lceil m/2 \rceil$ **27. a)** 1 **b)** 2 **c)** 3 **d)** 9 **29. a)** 1, 109 **b)** 40, 89 **c)** −31, 222 **d)** −21, 38259 **31. a)** −15 **b)** −7 **c)** 140 **33.** −1, −26, −51, −76, 24, 49, 74, 99 **35. a)** No **b)** No **c)** Yes **d)** No **37. a)** 13 **a)** 6 **39. a)** 9 **b)** 4 **c)** 25 **d)** 0 **41.** Let $m = tn$. Because $a \equiv b \pmod{m}$ there exists an integer s such that $a = b + sm$. Hence, $a = b + (st)n$, so $a \equiv b \pmod{n}$. **43. a)** Let $m = c = 2$, $a = 0$, and $b = 1$. Then $0 = ac \equiv bc = 2 \pmod{2}$, but $0 = a \not\equiv b = 1 \pmod{2}$. **b)** Let $m = 5$, $a = b = 3$, $c = 1$, and $d = 6$. Then $3 \equiv 3 \pmod{5}$ and $1 \equiv 6 \pmod{5}$, but $3^1 = 3 \not\equiv 4 \equiv 729 = 3^6 \pmod{5}$. **45.** By Exercise 44 the sum of two squares must be either $0 + 0 = 0$, $0 + 1 = 1$, or $1 + 1 = 2$, modulo 4, never 3, and therefore not of the form $4k + 3$. **47.** Because $a \equiv b \pmod{m}$, there exists an integer s such that $a = b + sm$, so $a - b = sm$. Then $a^k - b^k = (a - b)(a^{k-1} + a^{k-2}b + \cdots + ab^{k-2} + b^{k-1})$, $k \ge 2$, is also a multiple of m. It follows that $a^k \equiv b^k \pmod{m}$. **49.** To prove closure, note that $a \cdot_m b = (a \cdot b)$ **mod** m, which by definition is an element of $\mathbf{Z}_m$. Multiplication is associative because $(a \cdot_m b) \cdot_m c$ and $a \cdot_m (b \cdot_m c)$ both equal $(a \cdot b \cdot c)$ **mod** m and multiplication of integers is associative. Similarly, multiplication in $\mathbf{Z}_m$ is commutative because multiplication in $\mathbf{Z}$ is commutative, and 1 is the multiplicative identity for $\mathbf{Z}_m$ because 1 is the multiplicative identity for $\mathbf{Z}$. **51.** $0 +_5 0 = 0$, $0 +_5 1 = 1$, $0 +_5 2 = 2$, $0 +_5 3 = 3$, $0 +_5 4 = 4$; $1 +_5 1 = 2$, $1 +_5 2 = 3$, $1 +_5 3 = 4$, $1 +_5 4 = 0$; $2 +_5 2 = 4$, $2 +_5 3 = 0$, $2 +_5 4 = 1$; $3 +_5 3 = 1$, $3 +_5 4 = 2$; $4 +_4 4 = 3$ and $0 \cdot_5 0 = 0$, $0 \cdot_5 1 = 0$, $0 \cdot_5 2 = 0$, $0 \cdot_5 3 = 0$, $0 \cdot_5 4 = 0$; $1 \cdot_5 1 = 1$, $1 \cdot_5 2 = 2$, $1 \cdot_5 3 = 3$, $1 \cdot_5 4 = 4$; $2 \cdot_5 2 = 4$, $2 \cdot_5 3 = 1$, $2 \cdot_5 4 = 3$; $3 \cdot_5 3 = 4$, $3 \cdot_5 4 = 2$; $4 \cdot_5 4 = 1$ **53.** f is onto but not one-to-one (unless $d = 1$); g is neither.

Section 4.2

1. a) 1110 0111 **b)** 1 0001 1011 0100 **c)** 1 0111 11010110 1100 **3. a)** 31 **b)** 513 **c)** 341 **d)** 26,896 **5. a)** 1 0111 1010 **b)** 11 1000 0100 **c)** 1 0001 0011 **d)** 101 0000 1111 **7. a)** 1000 0000 1110 **b)** 1 0011 0101 1010 1011 **c)** 10101011 1011 1010 **d)** 1101 1110 1111 1010 11001110 1101 **9.** 1010 1011 1100 1101 1110 1111 **11.** $(\text{B7B})_{16}$ **13.** Adding up to three leading 0s if necessary, write the binary expansion as $(\ldots b_{23}b_{22}b_{21}b_{20}b_{13}b_{12}b_{11}b_{10}b_{03}b_{02}b_{01}b_{00})_2$. The value of this numeral is $b_{00} + 2b_{01} + 4b_{02} + 8b_{03} + 2^4 b_{10} + 2^5 b_{11} + 2^6 b_{12} + 2^7 b_{13} + 2^8 b_{20} + 2^9 b_{21} + 2^{10} b_{22} + 2^{11} b_{23} + \cdots$, which we can rewrite as $b_{00} + 2b_{01} + 4b_{02} + 8b_{03} + (b_{10} + 2b_{11} + 4b_{12} + 8b_{13}) \cdot 2^4 + (b_{20} + 2b_{21} + 4b_{22} + 8b_{23}) \cdot 2^8 + \cdots$. Now $(b_{i3}b_{i2}b_{i1}b_{i0})_2$ translates into the hexadecimal digit h_i. So our number is $h_0 + h_1 \cdot 2^4 + h_2 \cdot 2^8 + \cdots = h_0 + h_1 \cdot 16 + h_2 \cdot 16^2 + \cdots$, which is the hexadecimal expansion $(\ldots h_1 h_1 h_0)_{16}$. **15.** Adding up to two leading 0s if necessary, write the binary expansion as $(\ldots b_{22}b_{21}b_{20}b_{12}b_{11}b_{10}b_{02}b_{01}b_{00})_2$. The value of this numeral is $b_{00} + 2b_{01} + 4b_{02} + 2^3 b_{10} + 2^4 b_{11} + 2^5 b_{12} + 2^6 b_{20} + 2^7 b_{21} + 2^8 b_{22} + \cdots$, which we can rewrite as $b_{00} + 2b_{01} + 4b_{02} + (b_{10} + 2b_{11} + 4b_{12}) \cdot 2^3 + (b_{20} + 2b_{21} + 4b_{22}) \cdot 2^6 + \cdots$. Now $(b_{i2}b_{i1}b_{i0})_2$ translates into the octal digit h_i. So our number is $h_0 + h_1 \cdot 2^3 + h_2 \cdot 2^6 + \cdots = h_0 + h_1 \cdot 8 + h_2 \cdot 8^2 + \cdots$, which is the octal expansion $(\ldots h_1 h_1 h_0)_8$. **17.** 1 1101 1100 1010 1101 0001, $1273)_8$ **19.** Convert the given octal numeral to binary, then convert from binary to hexadecimal using Example 7. **21. a)** 1011 1110, 10 0001 0000 0001 **b)** 1 1010 1100, 1011 0000 0111 0011 **c)** 100 1001 1010, 101 0010 1001 0110 0000 **d)** 110 0000 0000, 1000 0000 0001 1111 1111 **23. a)** 1132, 144,305 **b)** 6273, 2,134,272 **c)** 2110, 1,107,667 **d)** 57,777, 237,326,216 **25.** 436 **27.** 27 **29.** The binary expansion of the integer is the unique such sum. **31.** Let $a = (a_{n-1}a_{n-2} \ldots a_1 a_0)_{10}$. Then $a = 10^{n-1}a_{n-1} + 10^{n-2}a_{n-2} + \cdots + 10a_1 + a_0 \equiv a_{n-1} + a_{n-2} + \cdots + a_1 + a_0 \pmod{3}$, because $10^j \equiv 1 \pmod{3}$ for all nonnegative integers j. It follows that $3 \mid a$ if and only if 3 divides the sum of the decimal digits of a. **33.** Let $a = (a_{n-1}a_{n-2} \ldots a_1 a_0)_2$. Then $a = a_0 + 2a_1 + 2^2 a_2 + \cdots + 2^{n-1}a_{n-1} \equiv a_0 - a_1 + a_2 - a_3 + \cdots \pm a_{n-1} \pmod{3}$. It follows that a is divisible by

3 if and only if the sum of the binary digits in the even-numbered positions minus the sum of the binary digits in the odd-numbered positions is divisible by 3. **35. a)** n is divisible by 4 if and only if the two rightmost digits of the decimal expansion, viewed as a two-digit integer (or the single digit if it's a one-digit integer), is divisible by 4. **b)** n is divisible by 25 if and only if the two rightmost digits are 00 (or 0 if it's a one-digit integer), 25, 50, or 75. **c)** n is divisible by 20 if and only if the rightmost digit of the decimal expansion is 0 and the digit in the tens place (if it's not a one-digit integer) is 0, 2, 4, 6, or 8. **37.** The base b representation of n has k digits iff $b^{k-1} \le n < b^k$ iff $k-1 \le \log_b n < k$ iff $\lfloor \log_b n \rfloor + 1 = k$. **39.** $5(8^n - 1)/7$ **41. a)** -6 **b)** 13 **c)** -14 **d)** 0 **43.** The one's complement of the sum is found by adding the one's complements of the two integers except that a carry in the leading bit is used as a carry to the last bit of the sum. **45.** If $m \ge 0$, then the leading bit a_{n-1} of the one's complement expansion of m is 0 and the formula reads $m = \sum_{i=0}^{n-2} a_i 2^i$. This is correct because the right-hand side is the binary expansion of m. When m is negative, the leading bit a_{n-1} of the one's complement expansion of m is 1. The remaining $n-1$ bits can be obtained by subtracting $-m$ from $111\ldots1$ (where there are $n-1$ 1s), because subtracting a bit from 1 is the same as complementing it. Hence, the bit string $a_{n-2}\ldots a_0$ is the binary expansion of $(2^{n-1}-1)-(-m)$. Solving the equation $(2^{n-1}-1)-(-m) = \sum_{i=0}^{n-2} a_i 2^i$ for m gives the desired equation because $a_{n-1} = 1$. **47. a)** -7 **b)** 13 **c)** -15 **d)** -1 **49.** To obtain the two's complement representation of the sum of two integers, add their two's complement representations (as binary integers are added) and ignore any carry out of the leftmost column. However, the answer is invalid if an overflow has occurred. This happens when the leftmost digits in the two's complement representation of the two terms agree and the leftmost digit of the answer differs. **51.** If $m \ge 0$, then the leading bit a_{n-1} is 0 and the formula reads $m = \sum_{i=0}^{n-2} a_i 2^i$. This is correct because the right-hand side is the binary expansion of m. If $m < 0$, its two's complement expansion has 1 as its leading bit and the remaining $n-1$ bits are the binary expansion of $2^{n-1} - (-m)$. This means that $(2^{n-1}) - (-m) = \sum_{i=0}^{n-2} a_i 2^i$. Solving for m gives the desired equation because $a_{n-1} = 1$. **53.** $4n$

55.
```
procedure Cantor(x: positive integer)
n := 1; f := 1
while (n + 1) · f ≤ x
    n := n + 1
    f := f · n
y := x
while n > 0
    a_n := ⌊y/f⌋
    y := y − a_n · f
    f := f/n
    n := n − 1
{x = a_n n! + a_{n−1}(n − 1)! + ··· + a_1 1!}
```

57. First step: $c = 0, d = 0, s_0 = 1$; second step: $c = 0, d = 1, s_1 = 0$; third step: $c = 1, d = 1, s_2 = 0$; fourth step: $c = 1, d = 1, s_3 = 0$; fifth step: $c = 1, d = 1, s_4 = 1$; sixth step: $c = 1, s_5 = 1$

59.
```
procedure subtract(a, b: positive integers, a > b,
        a = (a_{n−1}a_{n−2} ... a_1a_0)_2,
        b = (b_{n−1}b_{n−2} ... b_1b_0)_2)
    B := 0 {B is the borrow}
    for j := 0 to n − 1
        if a_j ≥ b_j + B then
            s_j := a_j − b_j − B
            B := 0
        else
            s_j := a_j + 2 − b_j − B
            B := 1
    {(s_{n−1}s_{n−2} ... s_1s_0)_2 is the difference}
```

61.
```
procedure compare(a, b: positive integers,
        a = (a_n a_{n−1} ... a_1a_0)_2, b = (b_n b_{n−1} ... b_1b_0)_2)
    k := n
    while a_k = b_k and k > 0
        k := k − 1
    if a_k = b_k then print "a equals b"
    if a_k > b_k then print "a is greater than b"
    if a_k < b_k then print "a is less than b"
```

63. $O(\log n)$ **65.** The only time-consuming part of the algorithm is the **while** loop, which is iterated q times. The work done inside is a subtraction of integers no bigger than a, which has $\log a$ bits. The result now follows from Example 9.

Section 4.3

1. 29, 71, 97 prime; 21, 111, 143 not prime **3. a)** $2^3 \cdot 11$ **b)** $2 \cdot 3^2 \cdot 7$ **c)** 3^6 **d)** $7 \cdot 11 \cdot 13$ **e)** $11 \cdot 101$ **f)** $2 \cdot 3^3 \cdot 5 \cdot 7 \cdot 13 \cdot 37$ **5.** $2^8 \cdot 3^4 \cdot 5^2 \cdot 7$

7.
```
procedure primetester(n : integer greater than 1)
    isprime := true
    d := 2
    while isprime and d ≤ √n
        if n mod d = 0 then isprime := false
        else d := d + 1
    return isprime
```

9. Write $n = rs$, where $r > 1$ and $s > 1$. Then $2^n - 1 = 2^{rs} - 1 = (2^r)^s - 1 = (2^r - 1)((2^r)^{s-1} + (2^r)^{s-2} + (2^r)^{s-3} + \cdots + 1)$. The first factor is at least $2^2 - 1 = 3$ and the second factor is at least $2^2 + 1 = 5$. This provides a factoring of $2^n - 1$ into two factors greater than 1, so $2^n - 1$ is composite. **11.** Suppose that $\log_2 3 = a/b$ where $a, b \in \mathbf{Z}^+$ and $b \ne 0$. Then $2^{a/b} = 3$, so $2^a = 3^b$. This violates the fundamental theorem of arithmetic. Hence, $\log_2 3$ is irrational. **13.** 3, 5, and 7 are primes of the desired form. **15.** 1, 7, 11, 13, 17, 19, 23, 29 **17. a)** Yes **b)** No **c)** Yes **d)** Yes **19.** Suppose that n is not prime, so that $n = ab$, where a and b are integers greater than 1. Because $a > 1$, by the identity in the hint, $2^a - 1$ is a factor of $2^n - 1$ that is greater than 1, and the second factor in this identity is also greater than 1. Hence, $2^n - 1$ is not prime.

21. a) 2 b) 4 c) 12 **23.** $\phi(p^k) = p^k - p^{k-1}$ **25.** a) $3^5 \cdot 5^3$ b) 1 c) 23^{17} d) $41 \cdot 43 \cdot 53$ e) 1 f) 1111 **27.** a) $2^{11} \cdot 3^7 \cdot 5^9 \cdot 7^3$ b) $2^9 \cdot 3^7 \cdot 5^5 \cdot 7^3 \cdot 11 \cdot 13 \cdot 17$ c) 23^{31} d) $41 \cdot 43 \cdot 53$ e) $2^{12}3^{13}5^{17}7^{21}$ f) Undefined **29.** gcd (92928, 123552) = 1056; lcm(92928, 123552) = 10,872,576; both products are 11,481,440,256. **31.** Because $\min(x, y) + \max(x, y) = x + y$, the exponent of p_i in the prime factorization of $\gcd(a, b) \cdot \operatorname{lcm}(a, b)$ is the sum of the exponents of p_i in the prime factorizations of a and b. **33.** a) 6 b) 3 c) 11 d) 3 e) 40 f) 12 **35.** 9 **37.** By Exercise 36 it follows that $\gcd(2^b - 1, (2^a - 1) \mathbf{\ mod\ } (2^b - 1)) = \gcd(2^b - 1, 2^{a \mathbf{\ mod\ } b} - 1)$. Because the exponents involved in the calculation are b and a **mod** b, the same as the quantities involved in computing $\gcd(a, b)$, the steps used by the Euclidean algorithm to compute $\gcd(2^a - 1, 2^b - 1)$ run in parallel to those used to compute $\gcd(a, b)$ and show that $\gcd(2^a - 1, 2^b - 1) = 2^{\gcd(a,b)} - 1$. **39.** a) $1 = (-1) \cdot 10 + 1 \cdot 11$ b) $1 = 21 \cdot 21 + (-10) \cdot 44$ c) $12 = (-1) \cdot 36 + 48$ d) $1 = 13 \cdot 55 + (-21) \cdot 34$ e) $3 = 11 \cdot 213 + (-20) \cdot 117$ f) $223 = 1 \cdot 0 + 1 \cdot 223$ g) $1 = 37 \cdot 2347 + (-706) \cdot 123$ h) $2 = 1128 \cdot 3454 + (-835) \cdot 4666$ i) $1 = 2468 \cdot 9999 + (-2221) \cdot 11111$ **41.** $(-3) \cdot 26 + 1 \cdot 91 = 13$ **43.** $34 \cdot 144 + (-55) \cdot 89 = 1$

45. **procedure** *extended Euclidean*(a, b: positive integers)

```
x := a
y := b
oldolds := 1
olds := 0
oldoldt := 0
oldt := 1
while y ≠ 0
  q := x div y
  r := x mod y
  x := y
  y := r
  s := oldolds − q · olds
  t := oldoldt − q · oldt
  oldolds := olds
  oldoldt := oldt
  olds := s
  oldt := t
{gcd(a, b) is x, and (oldolds)a + (oldoldt)b = x}
```

47. a) $a_n = 1$ if n is prime and $a_n = 0$ otherwise. b) a_n is the smallest prime factor of n with $a_1 = 1$. c) a_n is the number of positive divisors of n. d) $a_n = 1$ if n has no divisors that are perfect squares greater than 1 and $a_n = 0$ otherwise. e) a_n is the largest prime less than or equal to n. f) a_n is the product of the first $n - 1$ primes. **49.** Because every second integer is divisible by 2, the product is divisible by 2. Because every third integer is divisible by 3, the product is divisible by 3. Therefore, the product has both 2 and 3 in its prime factorization and is therefore divisible by $3 \cdot 2 = 6$. **51.** $n = 1601$ is a counterexample. **53.** Setting $k = a + b + 1$ will produce the composite number $a(a + b + 1) + b = a^2 + ab + a + b = (a+1)(a+b)$. **55.** Suppose that there are only finitely many primes of the form $4k+3$, namely, $q_1, q_2, \ldots, q_n$, where $q_1 = 3$, $q_2 = 7$, and so on. Let $Q = 4q_1q_2 \cdots q_n - 1$. Note that Q is of the form $4k+3$ (where $k = q_1q_2 \cdots q_n - 1$). If Q is prime, then we have found a prime of the desired form different from all those listed. If Q is not prime, then Q has at least one prime factor not in the list $q_1, q_2, \ldots, q_n$, because the remainder when Q is divided by q_j is $q_j - 1$, and $q_j - 1 \neq 0$. Because all odd primes are either of the form $4k+1$ or of the form $4k+3$, and the product of primes of the form $4k + 1$ is also of this form (because $(4k + 1)(4m + 1) = 4(4km + k + m) + 1$), there must be a factor of Q of the form $4k + 3$ different from the primes we listed. **57.** Given a positive integer x, we show that there is exactly one positive rational number m/n (in lowest terms) such that $K(m/n) = x$. From the prime factorization of x, read off the m and n such that $K(m/n) = x$. The primes that occur to even powers are the primes that occur in the prime factorization of m, with the exponents being half the corresponding exponents in x; and the primes that occur to odd powers are the primes that occur in the prime factorization of n, with the exponents being half of one more than the exponents in x.

Section 4.4

1. $15 \cdot 7 = 105 \equiv 1 \pmod{26}$ **3.** 7 **5.** a) 7 b) 52 c) 34 d) 73 **7.** Suppose that b and c are both inverses of a modulo m. Then $ba \equiv 1 \pmod{m}$ and $ca \equiv 1 \pmod{m}$. Hence, $ba \equiv ca \pmod{m}$. Because $\gcd(a, m) = 1$ it follows by Theorem 7 in Section 4.3 that $b \equiv c \pmod{m}$. **9.** 8 **11.** a) 67 b) 88 c) 146 **13.** 3 and 6 **15.** Let $m' = m/\gcd(c, m)$. Because all the common factors of m and c are divided out of m to obtain m', it follows that m' and c are relatively prime. Because m divides $ac - bc = (a - b)c$, it follows that m' divides $(a - b)c$. By Lemma 2 in Section 4.3, we see that m' divides $a - b$, so $a \equiv b \pmod{m'}$. **17.** Suppose that $x^2 \equiv 1 \pmod{p}$. Then p divides $x^2 - 1 = (x + 1)(x - 1)$. By Lemma 3 in Section 4.3 it follows that $p \mid x + 1$ or $p \mid x - 1$, so $x \equiv -1 \pmod{p}$ or $x \equiv 1 \pmod{p}$. **19.** a) Suppose that $ia \equiv ja \pmod{p}$, where $1 \leq i < j < p$. Then p divides $ja - ia = a(j - i)$. By Lemma 3 in Section 4.3, because a is not divisible by p, p divides $j - i$, which is impossible because $j - i$ is a positive integer less than p. b) By part (a), because no two of $a, 2a, \ldots, (p - 1)a$ are congruent modulo p, each must be congruent to a different number from 1 to $p - 1$. It follows that $a \cdot 2a \cdot 3a \cdot \cdots \cdot (p - 1) \cdot a \equiv 1 \cdot 2 \cdot 3 \cdot \cdots \cdot (p - 1) \pmod{p}$. It follows that $(p-1)! \cdot a^{p-1} \equiv p - 1 \pmod{p}$. c) By Wilson's theorem and part (b), if p does not divide a, it follows that $(-1) \cdot a^{p-1} \equiv -1 \pmod{p}$. Hence, $a^{p-1} \equiv 1 \pmod{p}$. d) If $p \mid a$, then $p \mid a^p$. Hence, $a^p \equiv a \equiv 0 \pmod{p}$. If p does not divide a, then $a^{p-1} \equiv a \pmod{p}$, by part (c). Multiplying both sides of this congruence by a gives $a^p \equiv a \pmod{p}$. **21.** All integers of the form $323 + 330k$, where k is an integer **23.** All integers of the form $53 + 60k$, where k is an integer **25.** **procedure** *chinese*($m_1, m_2, \ldots, m_n$: relatively prime positive integers ; $a_1, a_2, \ldots, a_n$: integers)

```
m := 1
for k := 1 to n
  m := m · m_k
```

```
for k := 1 to n
  M_k := m/m_k
  y_k := M_k^{-1} mod m_k
  x := 0
for k := 1 to n
  x := x + a_k M_k y_k
while x ≥ m
  x := x − m
return x {the smallest solution to the system
     {x ≡ a_k (mod m_k), k = 1, 2, …, n }}
```

27. All integers of the form $16 + 252k$, where k is an integer **29.** Suppose that p is a prime appearing in the prime factorization of $m_1m_2 \cdots m_n$. Because the m_is are relatively prime, p is a factor of exactly one of the m_is, say m_j. Because m_j divides $a-b$, it follows that $a-b$ has the factor p in its prime factorization to a power at least as large as the power to which it appears in the prime factorization of m_j. It follows that $m_1m_2 \cdots m_n$ divides $a-b$, so $a \equiv b \pmod{m_1m_2 \cdots m_n}$. **31.** $x \equiv 1 \pmod 6$ **33.** 7 **35.** $a^{p-2} \cdot a = a \cdot a^{p-2} = a^{p-1} \equiv 1 \pmod p$ **37. a)** By Fermat's little theorem, we have $2^{10} \equiv 1 \pmod{11}$. Hence, $2^{340} = (2^{10})^{34} \equiv 1^{34} = 1 \pmod{11}$. **b)** Because $32 \equiv 1 \pmod{31}$, it follows that $2^{340} = (2^5)^{68} = 32^{68} \equiv 1^{68} = 1 \pmod{31}$. **c)** Because 11 and 31 are relatively prime, and $11 \cdot 31 = 341$, it follows by parts (a) and (b) and Exercise 29 that $2^{340} \equiv 1 \pmod{341}$. **39. a)** 3, 4, 8 **b)** 983 **41.** Suppose that q is an odd prime with $q \mid 2^p - 1$. By Fermat's little theorem, $q \mid 2^{q-1} - 1$. From Exercise 37 in Section 4.3, $\gcd(2^p - 1, 2^{q-1} - 1) = 2^{\gcd(p,q-1)} - 1$. Because q is a common divisor of $2^p - 1$ and $2^{q-1} - 1$, $\gcd(2^p - 1, 2^{q-1} - 1) > 1$. Hence, $\gcd(p, q - 1) = p$, because the only other possibility, namely, $\gcd(p, q - 1) = 1$, gives us $\gcd(2^p - 1, 2^{q-1} - 1) = 1$. Hence, $p \mid q - 1$, and therefore there is a positive integer m such that $q - 1 = mp$. Because q is odd, m must be even, say, $m = 2k$, and so every prime divisor of $2^p - 1$ is of the form $2kp + 1$. Furthermore, the product of numbers of this form is also of this form. Therefore, all divisors of $2^p - 1$ are of this form. **43.** M_{11} is not prime; M_{17} is prime. **45.** First, $2047 = 23 \cdot 89$ is composite. Write $2047 - 1 = 2046 = 2 \cdot 1023$, so $s = 1$ and $t = 1023$ in the definition. Then $2^{1023} = (2^{11})^{93} = 2048^{93} \equiv 1^{93} = 1 \pmod{2047}$, as desired. **47.** We must show that $b^{2820} \equiv 1 \pmod{2821}$ for all b relatively prime to 2821. Note that $2821 = 7 \cdot 13 \cdot 31$, and if $\gcd(b, 2821) = 1$, then $\gcd(b, 7) = \gcd(b, 13) = \gcd(b, 31) = 1$. Using Fermat's little theorem we find that $b^6 \equiv 1 \pmod 7$, $b^{12} \equiv 1 \pmod{13}$, and $b^{30} \equiv 1 \pmod{31}$. It follows that $b^{2820} \equiv (b^6)^{470} \equiv 1 \pmod 7$, $b^{2820} \equiv (b^{12})^{235} \equiv 1 \pmod{13}$, and $b^{2820} \equiv (b^{30})^{94} \equiv 1 \pmod{31}$. By Exercise 29 (or the Chinese remainder theorem) it follows that $b^{2820} \equiv 1 \pmod{2821}$, as desired. **49. a)** If we multiply out this expression, we get $n = 1296m^3 + 396m^2 + 36m + 1$. Clearly $6m \mid n - 1$, $12m \mid n - 1$, and $18m \mid n - 1$. Therefore, the conditions of Exercise 48 are met, and we conclude that n is a Carmichael number. **b)** Letting $m = 51$ gives $n = 172{,}947{,}529$. **51.** $0 = (0, 0)$, $1 = (1, 1)$, $2 = (2, 2)$, $3 = (0, 3)$, $4 = (1, 4)$, $5 = (2, 0)$, $6 = (0, 1)$, $7 = (1, 2)$, $8 = (2, 3)$, $9 = (0, 4)$, $10 = (1, 0)$, $11 = (2, 1)$, $12 = (0, 2)$, $13 = (1, 3)$, $14 = (2, 4)$ **53.** We have $m_1 = 99$, $m_2 = 98$, $m_3 = 97$, and $m_4 = 95$, so $m = 99 \cdot 98 \cdot 97 \cdot 95 = 89{,}403{,}930$. We find that $M_1 = m/m_1 = 903{,}070$, $M_2 = m/m_2 = 912{,}285$, $M_3 = m/m_3 = 921{,}690$, and $M_4 = m/m_4 = 941{,}094$. Using the Euclidean algorithm, we compute that $y_1 = 37$, $y_2 = 33$, $y_3 = 24$, and $y_4 = 4$ are inverses of M_k modulo m_k for $k = 1, 2, 3, 4$, respectively. It follows that the solution is $65 \cdot 903{,}070 \cdot 37 + 2 \cdot 912{,}285 \cdot 33 + 51 \cdot 921{,}690 \cdot 24 + 10 \cdot 941{,}094 \cdot 4 = 3{,}397{,}886{,}480 \equiv 537{,}140 \pmod{89{,}403{,}930}$. **55.** $\log_2 5 = 16$, $\log_2 6 = 14$ **57.** $\log_3 1 = 0$, $\log_3 2 = 14$, $\log_3 3 = 1$, $\log_3 4 = 12$, $\log_3 5 = 5$, $\log_3 6 = 15$, $\log_3 7 = 11$, $\log_3 8 = 10$, $\log_3 9 = 2$, $\log_3 10 = 3$, $\log_3 11 = 7$, $\log_3 12 = 13$, $\log_3 13 = 4$, $\log_3 14 = 9$, $\log_3 15 = 6$, $\log_3 16 = 8$ **59.** Assume that s is a solution of $x^2 \equiv a \pmod p$. Then because $(-s)^2 = s^2$, $-s$ is also a solution. Furthermore, $s \not\equiv -s \pmod p$. Otherwise, $p \mid 2s$, which implies that $p \mid s$, and this implies, using the original assumption, that $p \mid a$, which is a contradiction. Furthermore, if s and t are incongruent solutions modulo p, then because $s^2 \equiv t^2 \pmod p$, $p \mid s^2 - t^2$. This implies that $p \mid (s + t)(s - t)$, and by Lemma 3 in Section 4.3, $p \mid s - t$ or $p \mid s + t$, so $s \equiv t \pmod p$ or $s \equiv -t \pmod p$. Hence, there are at most two solutions. **61.** The value of $\left(\frac{a}{p}\right)$ depends only on whether a is a quadratic residue modulo p, that is, whether $x^2 \equiv a \pmod p$ has a solution. Because this depends only on the equivalence class of a modulo p, it follows that $\left(\frac{a}{p}\right) = \left(\frac{b}{p}\right)$ if $a \equiv b \pmod p$. **63.** By Exercise 62, $\left(\frac{a}{p}\right)\left(\frac{b}{p}\right) = a^{(p-1)/2}b^{(p-1)/2} = (ab)^{(p-1)/2} \equiv \left(\frac{ab}{p}\right) \pmod p$. **65.** $x \equiv 8, 13, 22$, or $27 \pmod{35}$ **67.** Compute r^e **mod** p for $e = 0, 1, 2, \ldots, p - 2$ until we get the answer a. Worst case and average case time complexity are $O(p \log p)$.

Section 4.5

1. 91, 57, 21, 5 **3. a)** 7, 19, 7, 7, 18, 0 **b)** Take the next available space **mod** 31. **5.** 1, 5, 4, 1, 5, 4, 1, 5, 4, … **7.** 2, 6, 7, 10, 8, 2, 6, 7, 10, 8, … **9.** 2357, 5554, 8469, 7239, 4031, 2489, 1951, 8064 **11.** 2, 1, 1, 1, … **13.** Only string (d) **15.** 4 **17.** Correctly, of course **19. a)** Not valid **b)** Valid **c)** Valid **d)** Not valid **21. a)** No **b)** 5 **c)** 7 **d)** 8 **23.** Transposition errors involving the last digit **25. a)** Yes **b)** No **c)** Yes **d)** No **27.** Transposition errors will be detected if and only if the transposed digits are an odd number of positions apart and do not differ by 5. **29. a)** Valid **b)** Not valid **c)** Valid **d)** Valid **31.** Yes, as long as the two digits do not differ by 7 **33. a)** Not valid **b)** Valid **c)** Valid **d)** Not valid **35.** The given congruence is equivalent to $3d_1 + 4d_2 + 5d_3 + 6d_4 + 7d_5 + 8d_6 + 9d_7 + 10d_8 \equiv 0 \pmod{11}$. Transposing adjacent digits x and y (with x on the left) causes the left-hand side to increase by $x - y$. Because $x \not\equiv y \pmod{11}$, the congruence will no longer hold. Therefore, errors of this type are always detected.

Section 4.6

1. a) GR QRW SDVV JR **b)** QB ABG CNFF TB **c)** QX UXM AHJJ ZX **3. a)** KOHQV MCIF GHSD **b)** RVBXP TJPZ NBZX **c)** DBYNE PHRM FYZA **5. a)** SURRENDER NOW **b)** BE MY FRIEND **c)** TIME FOR FUN **7.** TO SLEEP PERCHANCE TO DREAM **9.** ANY SUFFICIENTLY ADVANCED TECHNOLOGY IS INDISTINGUISHABLE FROM MAGIC **11.** $p = 7c + 13 \mathbf{mod}\ 26$ **13.** $a = 18$, $b = 5$ **15.** BEWARE OF MARTIANS **17.** Presumably something like an affine cipher **19.** HURRICANE **21.** The length of the key may well be the greatest common divisor of the distances between the starts of the repeated string (or a factor of the gcd). **23.** Suppose we know both $n = pq$ and $(p-1)(q-1)$. To find p and q, first note that $(p-1)(q-1) = pq - p - q + 1 = n - (p+q) + 1$. From this we can find $s = p + q$. Because $q = s - p$, we have $n = p(s-p)$. Hence, $p^2 - ps + n = 0$. We now can use the quadratic formula to find p. Once we have found p, we can find q because $q = n/p$. **25.** 2545 2757 1211 **27.** SILVER **29.** Alice sends $5^8 \mathbf{mod}\ 23 = 16$ to Bob. Bob sends $5^5 \mathbf{mod}\ 23 = 20$ to Alice. Alice computes $20^8 \mathbf{mod}\ 23 = 6$ and Bob computes $16^5 \mathbf{mod}\ 23 = 6$. The shared key is 6. **31.** 2186 2087 1279 1251 0326 0816 1948 **33.** Alice can decrypt the first part of Cathy's message to learn the key, and Bob can decrypt the second part of Cathy's message, which Alice forwarded to him, to learn the key. No one else besides Cathy can learn the key, because all of these communications use secure private keys. **35.** Working modulo n^2, $E(m_1 + m_2) = g^{m_1+m_2}(r_1r_2)^n = (g^{m_1}r_1^n)(g^{m_2}r_2^n) = E(m_1) \cdot E(m_2)$.

Supplementary Exercises

1. The actual number of miles driven is $46{,}518 + 100{,}000k$ for some natural number k. **3.** 5, 22, −12, −29 **5.** Because $ac \equiv bc \pmod{m}$, there is an integer k such that $ac = bc + km$. Hence, $a - b = km/c$. Because $a - b$ is an integer, $c \mid km$. Letting $d = \gcd(m, c)$, write $c = de$. Because no factor of e divides m/d, it follows that $d \mid m$ and $e \mid k$. Thus, $a - b = (k/e)(m/d)$, where $k/e \in \mathbf{Z}$ and $m/d \in \mathbf{Z}$. Therefore, $a \equiv b \pmod{m/d}$. **7.** Proof of the contrapositive: If n is odd, then $n = 2k + 1$ for some integer k. Therefore, $n^2 + 1 = (2k+1)^2 + 1 = 4k^2 + 4k + 2 \equiv 2 \pmod{4}$. But perfect squares of even numbers are congruent to 0 modulo 4 (because $(2m)^2 = 4m^2$), and perfect squares of odd numbers are congruent to 1 or 3 modulo 4, so $n^2 + 1$ is not a perfect square. **9.** n is divisible by 8 if and only if the binary expansion of n ends with 000. **11.** We assume that someone has chosen a positive integer less than 2^n, which we are to guess. We ask the person to write the number in binary, using leading 0s if necessary to make it n bits long. We then ask, "Is the first bit a 1?", "Is the second bit a 1?", "Is the third bit a 1?", and so on. After we know the answers to these n questions, we will know the number, because we will know its binary expansion. **13.** $(a_na_{n-1} \ldots a_1a_0)_{10} = \sum_{k=0}^{n} 10^k a_k \equiv \sum_{k=0}^{n} a_k \pmod{9}$ because $10^k \equiv 1 \pmod{9}$ for every nonnegative integer k. **15.** Because for all $k \le n$, when Q_n is divided by k the remainder will be 1, it follows that no prime number less than or equal to n is a factor of Q_n. Thus, by the fundamental theorem of arithmetic, Q_n must have a prime factor greater than n. **17.** Take $a = 10$ and $b = 1$ in Dirichlet's theorem. **19.** Every number greater than 11 can be written as either $8+2n$ or $9+2n$ for some $n \ge 2$. **21.** Assume that every even integer greater than 2 is the sum of two primes, and let n be an integer greater than 5. If n is odd, write $n = 3 + (n-3)$ and decompose $n - 3 = p + q$ into the sum of two primes; if n is even, then write $n = 2 + (n-2)$ and decompose $n - 2 = p + q$ into the sum of two primes. For the converse, assume that every integer greater than 5 is the sum of three primes, and let n be an even integer greater than 2. Write $n+2$ as the sum of three primes, one of which is necessarily 2, so $n + 2 = 2 + p + q$, whence $n = p + q$. **23.** Recall that a nonconstant polynomial can take on the same value only a finite number of times. Thus, f can take on the values 0 and ± 1 only finitely many times, so if there is not some y such that $f(y)$ is composite, then there must be some x_0 such that $\pm f(x_0)$ is prime, say p. Look at $f(x_0 + kp)$. When we plug $x_0 + kp$ in for x in the polynomial and multiply it out, every term will contain a factor of p except for the terms that form $f(x_0)$. Therefore, $f(x_0 + kp) = f(x_0) + mp = (m \pm 1)p$ for some integer m. As k varies, this value can be 0, p, or $-p$ only finitely many times; therefore, it must be a composite number for some values of k. **25.** 1 **27.** 1 **29.** If not, then suppose that $q_1, q_2, \ldots, q_n$ are all the primes of the form $6k + 5$. Let $Q = 6q_1q_2 \cdots q_n - 1$. Note that Q is of the form $6k + 5$, where $k = q_1q_2 \cdots q_n - 1$. Let $Q = p_1p_2 \cdots p_t$ be the prime factorization of Q. No p_i is 2, 3, or any q_j, because the remainder when Q is divided by 2 is 1, by 3 is 2, and by q_j is $q_j - 1$. All odd primes other than 3 are of the form $6k + 1$ or $6k + 5$, and the product of primes of the form $6k + 1$ is also of this form. Therefore, at least one of the p_is must be of the form $6k + 5$, a contradiction. **31.** The product of numbers of the form $4k + 1$ is of the form $4k + 1$, but numbers of this form might have numbers not of this form as their only prime factors. For example, $49 = 4 \cdot 12 + 1$, but the prime factorization of 49 is $7 \cdot 7 = (4 \cdot 1 + 3)(4 \cdot 1 + 3)$. **33. a)** Not mutually relatively prime **b)** Mutually relatively prime **c)** Mutually relatively prime **d)** Mutually relatively prime **35.** 1 **37.** $x \equiv 28 \pmod{30}$ **39.** By the Chinese remainder theorem, it suffices to show that $n^9 - n \equiv 0 \pmod{2}$, $n^9 - n \equiv 0 \pmod{3}$, and $n^9 - n \equiv 0 \pmod{5}$. Each in turn follows from applying Fermat's little theorem. **41.** By Fermat's little theorem, $p^{q-1} \equiv 1 \pmod{q}$ and clearly $q^{p-1} \equiv 0 \pmod{q}$. Therefore, $p^{q-1} + q^{p-1} \equiv 1 + 0 = 1 \pmod{q}$. Similarly, $p^{q-1} + q^{p-1} \equiv 1 \pmod{p}$. It follows from the Chinese remainder theorem that $p^{q-1} + q^{p-1} \equiv 1 \pmod{pq}$. **43.** If a_i is changed from x to y, then the change in the left-hand side of the congruence is either $y - x$ or $3(y - x)$, modulo 10, neither of which can be 0 because 1 and 3 are relatively prime to 10. Therefore, the sum can no longer be 0 modulo 10. **45.** Working modulo 10, solve for d_9. The check digit for 11100002 is 5.

47. PLEASE SEND MONEY **49. a)** QAL HUVEM AT WVESGB **b)** QXB EVZZL ZEVZZRFS

CHAPTER 5

Section 5.1

1. Let $P(n)$ be the statement that the train stops at station n. *Basis step:* We are told that $P(1)$ is true. *Inductive step:* We are told that $P(n)$ implies $P(n+1)$ for each $n \geq 1$. Therefore, by the principle of mathematical induction, $P(n)$ is true for all positive integers n. **3. a)** $1^2 = 1 \cdot 2 \cdot 3/6$ **b)** Both sides of $P(1)$ shown in part (a) equal 1. **c)** $1^2 + 2^2 + \cdots + k^2 = k(k+1)(2k+1)/6$ **d)** For each $k \geq 1$ that $P(k)$ implies $P(k+1)$; in other words, that assuming the inductive hypothesis [see part (c)] we can show $1^2 + 2^2 + \cdots + k^2 + (k+1)^2 = (k+1)(k+2)(2k+3)/6$ **e)** $(1^2 + 2^2 + \cdots + k^2) + (k+1)^2 = [k(k+1)(2k+1)/6] + (k+1)^2 = [(k+1)/6][k(2k+1) + 6(k+1)] = [(k+1)/6](2k^2 + 7k + 6) = [(k+1)/6](k+2)(2k+3) = (k+1)(k+2)(2k+3)/6$ **f)** We have completed both the basis step and the inductive step, so by the principle of mathematical induction, the statement is true for every positive integer n. **5.** Let $P(n)$ be "$1^2 + 3^2 + \cdots + (2n+1)^2 = (n+1)(2n+1)(2n+3)/3$." *Basis step:* $P(0)$ is true because $1^2 = 1 = (0+1)(2 \cdot 0+1)(2 \cdot 0+3)/3$. *Inductive step:* Assume that $P(k)$ is true. Then $1^2 + 3^2 + \cdots + (2k+1)^2 + [2(k+1)+1]^2 = (k+1)(2k+1)(2k+3)/3 + (2k+3)^2 = (2k+3)[(k+1)(2k+1)/3 + (2k+3)] = (2k+3)(2k^2+9k+10)/3 = (2k+3)(2k+5)(k+2)/3 = [(k+1)+1][2(k+1)+1][2(k+1)+3]/3$. **7.** Let $P(n)$ be "$\sum_{j=0}^{n} 3 \cdot 5^j = 3(5^{n+1} - 1)/4$." *Basis step:* $P(0)$ is true because $\sum_{j=0}^{0} 3 \cdot 5^j = 3 = 3(5^1 - 1)/4$. *Inductive step:* Assume that $\sum_{j=0}^{k} 3 \cdot 5^j = 3(5^{k+1} - 1)/4$. Then $\sum_{j=0}^{k+1} 3 \cdot 5^j = (\sum_{j=0}^{k} 3 \cdot 5^j) + 3 \cdot 5^{k+1} = 3(5^{k+1} - 1)/4 + 3 \cdot 5^{k+1} = 3(5^{k+1} + 4 \cdot 5^{k+1} - 1)/4 = 3(5^{k+2} - 1)/4$. **9. a)** $2 + 4 + 6 + \cdots + 2n = n(n+1)$ **b)** *Basis step:* $2 = 1 \cdot (1+1)$ is true. *Inductive step:* Assume that $2 + 4 + 6 + \cdots + 2k = k(k+1)$. Then $(2 + 4 + 6 + \cdots + 2k) + 2(k+1) = k(k+1) + 2(k+1) = (k+1)(k+2)$. **11. a)** $\sum_{j=1}^{n} 1/2^j = (2^n - 1)/2^n$ **b)** *Basis step:* $P(1)$ is true because $\frac{1}{2} = (2^1 - 1)/2^1$. *Inductive step:* Assume that $\sum_{j=1}^{k} 1/2^j = (2^k - 1)/2^k$. Then $\sum_{j=1}^{k+1} \frac{1}{2^j} = (\sum_{j=1}^{k} \frac{1}{2^j}) + \frac{1}{2^{k+1}} = \frac{2^k - 1}{2^k} + \frac{1}{2^{k+1}} = \frac{2^{k+1} - 2 + 1}{2^{k+1}} = \frac{2^{k+1} - 1}{2^{k+1}}$. **13.** Let $P(n)$ be "$1^2 - 2^2 + 3^2 - \cdots + (-1)^{n-1} n^2 = (-1)^{n-1} n(n+1)/2$." *Basis step:* $P(1)$ is true because $1^2 = 1 = (-1)^0 1^2$. *Inductive step:* Assume that $P(k)$ is true. Then $1^2 - 2^2 + 3^2 - \cdots + (-1)^{k-1} k^2 + (-1)^k (k+1)^2 = (-1)^{k-1} k(k+1)/2 + (-1)^k (k+1)^2 = (-1)^k (k+1)[-k/2 + (k+1)] = (-1)^k (k+1)[(k/2) + 1] = (-1)^k (k+1)(k+2)/2$. **15.** Let $P(n)$ be "$1 \cdot 2 + 2 \cdot 3 + \cdots + n(n+1) = n(n+1)(n+2)/3$." *Basis step:* $P(1)$ is true because $1 \cdot 2 = 2 = 1(1+1)(1+2)/3$. *Inductive step:* Assume that $P(k)$ is true. Then $1 \cdot 2 + 2 \cdot 3 + \cdots + k(k+1) + (k+1)(k+2) = [k(k+1)(k+2)/3] + (k+1)(k+2) = (k+1)(k+2)[(k/3) + 1] = (k+1)(k+2)(k+3)/3$. **17.** Let $P(n)$ be the statement that $1^4 + 2^4 + 3^4 + \cdots + n^4 = n(n+1)(2n+1)(3n^2+3n-1)/30$. $P(1)$ is true because $1 \cdot 2 \cdot 3 \cdot 5/30 = 1$. Assume that $P(k)$ is true. Then $(1^4 + 2^4 + 3^4 + \cdots + k^4) + (k+1)^4 = k(k+1)(2k+1)(3k^2 + 3k - 1)/30 + (k+1)^4 = [(k+1)/30][k(2k+1)(3k^2+3k-1) + 30(k+1)^3] = [(k+1)/30](6k^4 + 39k^3 + 91k^2 + 89k + 30) = [(k+1)/30](k+2)(2k+3)[3(k+1)^2 + 3(k+1) - 1]$. This demonstrates that $P(k+1)$ is true. **19. a)** $1 + \frac{1}{4} < 2 - \frac{1}{2}$ **b)** This is true because $5/4$ is less than $6/4$. **c)** $1 + \frac{1}{4} + \cdots + \frac{1}{k^2} < 2 - \frac{1}{k}$ **d)** For each $k \geq 2$ that $P(k)$ implies $P(k+1)$; in other words, we want to show that assuming the inductive hypothesis [see part (c)] we can show $1 + \frac{1}{4} + \cdots + \frac{1}{k^2} + \frac{1}{(k+1)^2} < 2 - \frac{1}{k+1}$ **e)** $1 + \frac{1}{4} + \cdots + \frac{1}{k^2} + \frac{1}{(k+1)^2} < 2 - \frac{1}{k} + \frac{1}{(k+1)^2} = 2 - \left[\frac{1}{k} - \frac{1}{(k+1)^2}\right] = 2 - \left[\frac{k^2+2k+1-k}{k(k+1)^2}\right] = 2 - \frac{k^2+k}{k(k+1)^2} - \frac{1}{k(k+1)^2} = 2 - \frac{1}{k+1} - \frac{1}{k(k+1)^2} < 2 - \frac{1}{k+1}$ **f)** We have completed both the basis step and the inductive step, so by the principle of mathematical induction, the statement is true for every integer n greater than 1. **21.** Let $P(n)$ be "$2^n > n^2$." *Basis step:* $P(5)$ is true because $2^5 = 32 > 25 = 5^2$. *Inductive step:* Assume that $P(k)$ is true, that is, $2^k > k^2$. Then $2^{k+1} = 2 \cdot 2^k > k^2 + k^2 > k^2 + 4k \geq k^2 + 2k + 1 = (k+1)^2$ because $k > 4$. **23.** By inspection we find that the inequality $2n + 3 \leq 2^n$ does not hold for $n = 0, 1, 2, 3$. Let $P(n)$ be the proposition that this inequality holds for the positive integer n. $P(4)$, the basis case, is true because $2 \cdot 4 + 3 = 11 \leq 16 = 2^4$. For the inductive step assume that $P(k)$ is true. Then, by the inductive hypothesis, $2(k+1) + 3 = (2k+3) + 2 < 2^k + 2$. But because $k \geq 1$, $2^k + 2 \leq 2^k + 2^k = 2^{k+1}$. This shows that $P(k+1)$ is true. **25.** Let $P(n)$ be "$1 + nh \leq (1+h)^n$, $h > -1$." *Basis step:* $P(0)$ is true because $1 + 0 \cdot h = 1 \leq 1 = (1+h)^0$. *Inductive step:* Assume $1 + kh \leq (1+h)^k$. Then because $(1+h) > 0$, $(1+h)^{k+1} = (1+h)(1+h)^k \geq (1+h)(1+kh) = 1 + (k+1)h + kh^2 \geq 1 + (k+1)h$. **27.** Let $P(n)$ be "$1/\sqrt{1} + 1/\sqrt{2} + 1/\sqrt{3} + \cdots + 1/\sqrt{n} > 2\left(\sqrt{n+1} - 1\right)$." *Basis step:* $P(1)$ is true because $1 > 2\left(\sqrt{2} - 1\right)$. *Inductive step:* Assume that $P(k)$ is true. Then $1 + 1/\sqrt{2} + \cdots + 1/\sqrt{k} + 1/\sqrt{k+1} > 2\left(\sqrt{k+1} - 1\right) + 1/\sqrt{k+1}$. If we show that $2\left(\sqrt{k+1} - 1\right) + 1/\sqrt{k+1} > 2\left(\sqrt{k+2} - 1\right)$, it follows that $P(k+1)$ is true. This inequality is equivalent to $2\left(\sqrt{k+2} - \sqrt{k+1}\right) < 1/\sqrt{k+1}$, which is equivalent to $2\left(\sqrt{k+2} - \sqrt{k+1}\right)\left(\sqrt{k+2} + \sqrt{k+1}\right) < \sqrt{k+1}/\sqrt{k+1} + \sqrt{k+2}/\sqrt{k+1}$. This is equivalent to $2 < 1 + \sqrt{k+2}/\sqrt{k+1}$, which is clearly true. **29.** Let $P(n)$ be "$H_{2^n} \leq 1 + n$." *Basis step:* $P(0)$ is true because $H_{2^0} = H_1 = 1 \leq 1 + 0$. *Inductive step:* Assume that $H_{2^k} \leq 1 + k$. Then $H_{2^{k+1}} = H_{2^k} + \sum_{j=2^k+1}^{2^{k+1}} \frac{1}{j} \leq 1 + k + 2^k\left(\frac{1}{2^{k+1}}\right) < 1 + k + 1 = 1 + (k+1)$. **31.** *Basis step:* $1^2 + 1 = 2$ is divisible by 2. *Inductive step:* Assume the inductive hypothesis, that $k^2 + k$ is divisible by 2. Then $(k+1)^2 + (k+1) = k^2 + 2k + 1 + k + 1 = (k^2 + k) + 2(k+1)$, the sum of a multiple of 2 (by the inductive hypothesis) and a multiple of 2 (by definition), hence, divisible by 2. **33.** Let $P(n)$ be "$n^5 - n$ is divisible by 5." *Basis step:* $P(0)$ is true because $0^5 - 0 = 0$ is divisible by 5. *Inductive step:* Assume that $P(k)$ is true, that is, $k^5 - 5$ is divisible by 5. Then $(k+1)^5 - (k+1) = (k^5 + 5k^4 + 10k^3 + 10k^2 + 5k + 1) - (k+1) = (k^5 - k) + 5(k^4 + 2k^3 + 2k^2 + k)$ is also divisible by 5, because both terms in this sum are divisible by 5. **35.** Let $P(n)$ be the proposition that $(2n-1)^2 - 1$ is divisible by 8. The basis case $P(1)$ is true because $8 \mid 0$. Now assume that $P(k)$ is true. Because $[(2(k+1)-1]^2 - 1 = [(2k-1)^2 - 1] + 8k$, $P(k+1)$ is true because both terms on the right-hand side are divisible by 8. This shows that $P(n)$ is true for all positive integers n, so $m^2 - 1$ is divisible by 8 whenever m is an odd positive integer. **37.** *Basis step:* $11^{1+1} + 12^{2\cdot 1-1} = 121 + 12 = 133$ *Inductive step:* Assume the inductive hypothesis, that $11^{n+1} + 12^{2n-1}$ is divisible by 133. Then $11^{(n+1)+1} + 12^{2(n+1)-1} = 11 \cdot 11^{n+1} + 144 \cdot 12^{2n-1} = 11 \cdot 11^{n+1} + (11 + 133) \cdot 12^{2n-1} = 11(11^{n+1} + 12^{2n-1}) + 133 \cdot 12^{2n-1}$. The expression in parentheses is divisible by 133 by the inductive hypothesis, and obviously the second term is divisible by 133, so the entire quantity is divisible by 133, as desired. **39.** *Basis step:* $A_1 \subseteq B_1$ tautologically implies that $\bigcap_{j=1}^{1} A_j \subseteq \bigcap_{j=1}^{1} B_j$. *Inductive step:* Assume the inductive hypothesis that if $A_j \subseteq B_j$ for $j = 1, 2, \ldots, k$, then $\bigcap_{j=1}^{k} A_j \subseteq \bigcap_{j=1}^{k} B_j$. We want to show that if $A_j \subseteq B_j$ for $j = 1, 2, \ldots, k+1$, then $\bigcap_{j=1}^{k+1} A_j \subseteq \bigcap_{j=1}^{k+1} B_j$. Let x be an arbitrary element of $\bigcap_{j=1}^{k+1} A_j = (\bigcap_{j=1}^{k} A_j) \cap A_{k+1}$. Because $x \in \bigcap_{j=1}^{k} A_j$, we know by the inductive hypothesis that $x \in \bigcap_{j=1}^{k} B_j$; because $x \in A_{k+1}$, we know from the given fact that $A_{k+1} \subseteq B_{k+1}$ that $x \in B_{k+1}$. Therefore, $x \in \left(\bigcap_{j=1}^{k} B_j\right) \cap B_{k+1} = \bigcap_{j=1}^{k+1} B_j$. **41.** Let $P(n)$ be "$(A_1 \cup A_2 \cup \cdots \cup A_n) \cap B = (A_1 \cap B) \cup (A_2 \cap B) \cup \cdots \cup (A_n \cap B)$." *Basis step:* $P(1)$ is trivially true. *Inductive step:* Assume that $P(k)$ is true. Then $(A_1 \cup A_2 \cup \cdots \cup A_k \cup A_{k+1}) \cap B = [(A_1 \cup A_2 \cup \cdots \cup A_k) \cup A_{k+1}] \cap B = [(A_1 \cup A_2 \cup \cdots \cup A_k) \cap B] \cup (A_{k+1} \cap B) = [(A_1 \cap B) \cup (A_2 \cap B) \cup \cdots \cup (A_k \cap B)] \cup (A_{k+1} \cap B) = (A_1 \cap B) \cup (A_2 \cap B) \cup \cdots \cup (A_k \cap B) \cup (A_{k+1} \cap B)$. **43.** Let $P(n)$ be "$\overline{\bigcup_{k=1}^{n} A_k} = \bigcap_{k=1}^{n} \overline{A_k}$." *Basis step:* $P(1)$ is trivially true. *Inductive step:* Assume that $P(k)$ is true. Then $\overline{\bigcup_{j=1}^{k+1} A_j} = \overline{\left(\bigcup_{j=1}^{k} A_j\right) \cup A_{k+1}} = \overline{\left(\bigcup_{j=1}^{k} A_j\right)} \cap \overline{A_{k+1}} = \left(\bigcap_{j=1}^{k} \overline{A_j}\right) \cap \overline{A_{k+1}} = \bigcap_{j=1}^{k+1} \overline{A_j}$. **45.** Let $P(n)$ be the statement that a set with n elements has $n(n-1)/2$ two-element subsets. $P(2)$, the basis case, is true, because a set with two elements has one subset with two elements—namely, itself—and $2(2-1)/2 = 1$. Now assume that $P(k)$ is true. Let S be a set with $k+1$ elements. Choose an element a in S and let $T = S - \{a\}$. A two-element subset of S either contains a or does not. Those subsets not containing a are the subsets of T with two elements; by the inductive hypothesis there are $k(k-1)/2$ of these. There are k subsets of S with two elements that contain a, because such a subset contains a and one of the k elements in T. Hence, there are $k(k-1)/2 + k = (k+1)k/2$ two-element subsets of S. This completes the inductive proof. **47.** Reorder the locations if necessary so that $x_1 \le x_2 \le x_3 \le \cdots \le x_d$. Place the first tower at position $t_1 = x_1 + 1$. Assume tower k has been placed at position t_k. Then place tower $k+1$ at position $t_{k+1} = x + 1$, where x is the smallest x_i greater than $t_k + 1$. **49.** The two sets do not overlap if $n + 1 = 2$. In fact, the conditional statement $P(1) \to P(2)$ is false. **51.** The mistake is in applying the inductive hypothesis to look at $\max(x-1, y-1)$, because even though x and y are positive integers, $x - 1$ and $y - 1$ need not be (one or both could be 0). **53.** For the basis step ($n = 2$) the first person cuts the cake into two portions that they think are each $1/2$ of the cake, and the second person chooses the portion they think is at least $1/2$ of the cake (at least one of the pieces must satisfy that condition). For the inductive step, suppose there are $k + 1$ people. By the inductive hypothesis, we can suppose that the first k people have divided the cake among themselves so that each person is satisfied that they got at least a fraction $1/k$ of the cake. Each of them now cuts their piece into $k+1$ pieces of equal size. The last person gets to choose one piece from each of the first k people's portions. After this is done, each of the first k people is satisfied that they still have $(1/k)(k/(k+1)) = 1/(k+1)$ of the cake. To see that the last person is satisfied, suppose that they thought that the ith person ($1 \le i \le k$) had a portion p_i of the cake, where $\sum_{i=1}^{k} p_i = 1$. By choosing what they think is the largest piece from each person, they are satisfied that they have at least $\sum_{i=1}^{k} p_i/(k+1) = (1/(k+1)) \sum_{i=1}^{k} p_i = 1/(k+1)$ of the cake. **55.** We use the notation (i, j) to mean the square in row i and column j and use induction on $i+j$ to show that every square can be reached by the knight. *Basis step:* There are six base cases, for the cases when $i + j \le 2$. The knight is already at $(0, 0)$ to start, so the empty sequence of moves reaches that square. To reach $(1, 0)$, the knight moves $(0, 0) \to (2, 1) \to (0, 2) \to (1, 0)$. Similarly, to reach $(0, 1)$, the knight moves $(0, 0) \to (1, 2) \to (2, 0) \to (0, 1)$. Note that the knight has reached $(2, 0)$ and $(0, 2)$ in the process. For the last basis step there is $(0, 0) \to (1, 2) \to (2, 0) \to (0, 1) \to (2, 2) \to (0, 3) \to (1, 1)$. *Inductive step:* Assume the inductive hypothesis, that the knight can reach any square (i, j) for which $i + j = k$, where k is an integer greater than 1. We must show how the knight can reach each square (i, j) when $i + j = k + 1$. Because $k + 1 \ge 3$, at least one of i and j is at least 2. If $i \ge 2$, then by the inductive hypothesis, there is a sequence of moves ending at $(i-2, j+1)$, because $i - 2 + j + 1 = i + j - 1 = k$; from there it is just one step to (i, j); similarly, if $j \ge 2$. **57.** *Basis step:* The base cases $n = 0$ and $n = 1$ are true because the derivative of x^0 is 0 and the derivative of $x^1 = x$ is 1. *Inductive step:* Using the product rule, the inductive hypothesis, and the basis step shows that $\frac{d}{dx}x^{k+1} = \frac{d}{dx}(x \cdot x^k) = x \cdot \frac{d}{dx}x^k + x^k \frac{d}{dx}x = x \cdot kx^{k-1} + x^k \cdot 1 = kx^k + x^k = (k+1)x^k$. **59.** *Basis step:* For $k = 0$, $1 \equiv 1 \pmod{m}$. *Inductive step:* Suppose that $a \equiv b \pmod{m}$ and $a^k \equiv b^k \pmod{m}$; we must show that $a^{k+1} \equiv b^{k+1} \pmod{m}$. By Theorem 5 from Section 4.1, $a \cdot a^k \equiv b \cdot b^k \pmod{m}$, which by definition says that $a^{k+1} \equiv b^{k+1} \pmod{m}$. **61.** Let $P(n)$ be "$[(p_1 \to p_2) \wedge (p_2 \to p_3) \wedge \cdots \wedge (p_{n-1} \to p_n)] \to [(p_1 \wedge \cdots \wedge p_{n-1}) \to p_n]$." *Basis step:* $P(2)$ is true because $(p_1 \to p_2) \to (p_1 \to p_2)$ is a tautology. *Inductive step:* Assume $P(k)$ is true. To show $[(p_1 \to p_2) \wedge \cdots \wedge (p_{k-1} \to p_k) \wedge (p_k \to p_{k+1})] \to [(p_1 \wedge \cdots \wedge p_{k-1} \wedge p_k) \to p_{k+1}]$ is a tautology, assume that the hypothesis of this conditional statement is true. Because both the hypothesis and $P(k)$ are true, it follows that $(p_1 \wedge \cdots \wedge p_{k-1}) \to p_k$ is true. Because this is true, and because $p_k \to p_{k+1}$ is true (it is part of the assumption) it follows by hypothetical syllogism that $(p_1 \wedge \cdots \wedge p_{k-1}) \to p_{k+1}$ is true. The weaker statement $(p_1 \wedge \cdots \wedge p_{k-1} \wedge p_k) \to p_{k+1}$ follows from this. **63.** We will first prove the result when n is a power of 2, that is, if $n = 2^k$, $k = 1, 2, \ldots$. Let $P(k)$ be the statement $A \geq G$, where A and G are the arithmetic and geometric means, respectively, of a set of $n = 2^k$ positive real numbers. *Basis step:* $k = 1$ and $n = 2^1 = 2$. Note that $(\sqrt{a_1} - \sqrt{a_2})^2 \geq 0$. Expanding this shows that $a_1 - 2\sqrt{a_1 a_2} + a_2 \geq 0$, that is, $(a_1 + a_2)/2 \geq (a_1 a_2)^{1/2}$. *Inductive step:* Assume that $P(k)$ is true, with $n = 2^k$. We will show that $P(k+1)$ is true. We have $2^{k+1} = 2n$. Now $(a_1 + a_2 + \cdots + a_{2n})/(2n) = [(a_1 + a_2 + \cdots + a_n)/n + (a_{n+1} + a_{n+2} + \cdots + a_{2n})/n]/2$ and similarly $(a_1 a_2 \cdots a_{2n})^{1/(2n)} = [(a_1 \cdots a_n)^{1/n}(a_{n+1} \cdots a_{2n})^{1/n}]^{1/2}$. To simplify the notation, let $A(x, y, \ldots)$ and $G(x, y, \ldots)$ denote the arithmetic mean and geometric mean of $x, y, \ldots$, respectively. Also, if $x \leq x'$, $y \leq y'$, and so on, then $A(x, y, \ldots) \leq A(x', y', \ldots)$ and $G(x, y, \ldots) \leq G(x', y', \ldots)$. Hence, $A(a_1, \ldots, a_{2n}) = A(A(a_1, \ldots, a_n), A(a_{n+1}, \ldots, a_{2n})) \geq A(G(a_1, \ldots, a_n), G(a_{n+1}, \ldots, a_{2n})) \geq G(G(a_1, \ldots, a_n), G(a_{n+1}, \ldots, a_{2n})) = G(a_1, \ldots, a_{2n})$. This finishes the proof for powers of 2. Now if n is not a power of 2, let m be the next higher power of 2, and let $a_{n+1}, \ldots, a_m$ all equal $A(a_1, \ldots, a_n) = \bar{a}$. Then we have $[(a_1 a_2 \cdots a_n)\bar{a}^{m-n}]^{1/m} \leq A(a_1, \ldots, a_m)$, because m is a power of 2. Because $A(a_1, \ldots, a_m) = \bar{a}$, it follows that $(a_1 \cdots a_n)^{1/m}\bar{a}^{1-n/m} \leq \bar{a}^{n/m}$. Raising both sides to the (m/n)th power gives $G(a_1, \ldots, a_n) \leq A(a_1, \ldots, a_n)$. **65.** *Basis step:* For $n = 1$, the left-hand side is just $\frac{1}{1}$, which is 1. For $n = 2$, there are three nonempty subsets $\{1\}$, $\{2\}$, and $\{1, 2\}$, so the left-hand side is $\frac{1}{1} + \frac{1}{2} + \frac{1}{1 \cdot 2} = 2$. *Inductive step:* Assume that the statement is true for k. The set of the first $k + 1$ positive integers has many nonempty subsets, but they fall into three categories: a nonempty subset of the first k positive integers together with $k + 1$, a nonempty subset of the first k positive integers, or just $\{k + 1\}$. By the inductive hypothesis, the sum of the first category is k. For the second category, we can factor out $1/(k + 1)$ from each term of the sum and what remains is just k by the inductive hypothesis, so this part of the sum is $k/(k + 1)$. Finally, the third category simply yields $1/(k + 1)$. Hence, the entire summation is $k + k/(k+1) + 1/(k+1) = k+1$. **67.** *Basis step:* If $A_1 \subseteq A_2$, then A_1 satisfies the condition of being a subset of each set in the collection; otherwise $A_2 \subseteq A_1$, so A_2 satisfies the condition. *Inductive step:* Assume the inductive hypothesis, that the conditional statement is true for k sets, and suppose we are given $k + 1$ sets that satisfy the given conditions. By the inductive hypothesis, there must be a set A_i for some $i \leq k$ such that $A_i \subseteq A_j$ for $1 \leq j \leq k$. If $A_i \subseteq A_{k+1}$, then we are done. Otherwise, we know that $A_{k+1} \subseteq A_i$, and this tells us that A_{k+1} satisfies the condition of being a subset of A_j for $1 \leq j \leq k + 1$. **69.** $G(1) = 0$, $G(2) = 1$, $G(3) = 3$, $G(4) = 4$ **71.** To show that $2n - 4$ calls are sufficient to exchange all the gossip, select persons 1, 2, 3, and 4 to be the central committee. Every person outside the central committee calls one person on the central committee. At this point the central committee members *as a group* know all the scandals. They then exchange information among themselves by making the calls 1-2, 3-4, 1-3, and 2-4 in that order. At this point, *every* central committee member knows all the scandals. Finally, again every person outside the central committee calls one person on the central committee, at which point everyone knows all the scandals. [The total number of calls is $(n-4)+4+(n-4) = 2n-4$.] That this cannot be done with fewer than $2n-4$ calls is much harder to prove; see Sandra M. Hedetniemi, Stephen T. Hedetniemi, and Arthur L. Liestman, "A survey of gossiping and broadcasting in communication networks," *Networks* **18** (1988), no. 4, 319–349, for details. **73.** We prove this by mathematical induction. The basis step ($n = 2$) is true tautologically. For $n = 3$, suppose that the intervals are (a, b), (c, d), and (e, f), where without loss of generality we can assume that $a \leq c \leq e$. Because $(a, b) \cap (e, f) \neq \emptyset$, we must have $e < b$; for a similar reason, $e < d$. It follows that the number halfway between e and the smaller of b and d is common to all three intervals. Now for the inductive step, assume that whenever we have k intervals that have pairwise nonempty intersections then there is a point common to all the intervals, and suppose that we are given intervals $I_1, I_2, \ldots, I_{k+1}$ that have pairwise nonempty intersections. For each i from 1 to k, let $J_i = I_i \cap I_{k+1}$. We claim that the collection $J_1, J_2, \ldots, J_k$ satisfies the inductive hypothesis, that is, that $J_{i_1} \cap J_{i_2} \neq \emptyset$ for each choice of subscripts i_1 and i_2. This follows from the $n = 3$ case proved above, using the sets I_{i_1}, I_{i_2}, and I_{k+1}. We can now invoke the inductive hypothesis to conclude that there is a number common to all of the sets J_i for $i = 1, 2, \ldots, k$, which perforce is in the intersection of all the sets I_i for $i = 1, 2, \ldots, k + 1$. **75. a)** The basis step is $\sum_{j=1}^{1} j/(j + 1)! < 1$, which is true because the left-hand side equals $1/2$. It does not follow from $\sum_{j=1}^{k} j/(j + 1)! < 1$ that $\sum_{j=1}^{k+1} j/(j + 1)! < 1$ because the amount by which $\sum_{j=1}^{k} j/(j + 1)!$ is less than 1 could be less than $(k + 1)/(k + 2)!$, the new term being added in. **b)** Basis step is the true statement that $1/2! \leq 1 - 1/2!$. Assuming inductive hypothesis $P(k)$ that $\sum_{j=1}^{k} j/(j+1)! \leq 1 - 1/(k+1)!$, it follows that $\sum_{j=1}^{k+1} j/(j+1)! = \sum_{j=1}^{k} j/(j+1)! + (k+1)/(k+2)! \leq 1 - 1/(k+1)! + (k+1)/(k+2)! = 1 - (k+2)/(k+2)! + (k+1)/(k+2)! = 1 - 1/(k+2)!$, which is $P(k + 1)$. **77.** Pair up the people. Have the people stand at mutually distinct small distances from their partners but far away from everyone else. Then each person throws a pie at his or her partner, so everyone gets hit.

79.

81. Let $P(n)$ be the statement that every $2^n \times 2^n \times 2^n$ checkerboard with a $1 \times 1 \times 1$ cube removed can be covered by tiles that are $2 \times 2 \times 2$ cubes each with a $1 \times 1 \times 1$ cube removed. The basis step, $P(1)$, holds because one tile coincides with the solid to be tiled. Now assume that $P(k)$ holds. Now consider a $2^{k+1} \times 2^{k+1} \times 2^{k+1}$ cube with a $1 \times 1 \times 1$ cube removed. Split this object into eight pieces using planes parallel to its faces and running through its center. The missing $1 \times 1 \times 1$ piece occurs in one of these eight pieces. Now position one tile with its center at the center of the large object so that the missing $1 \times 1 \times 1$ cube lies in the octant in which the large object is missing a $1 \times 1 \times 1$ cube. This creates eight $2^k \times 2^k \times 2^k$ cubes, each missing a $1 \times 1 \times 1$ cube. By the inductive hypothesis we can fill each of these eight objects with tiles. Putting these tilings together produces the desired tiling.

83.

85. Let $Q(n)$ be $P(n + b - 1)$. The statement that $P(n)$ is true for $n = b, b + 1, b + 2, \ldots$ is the same as the statement that $Q(m)$ is true for all positive integers m. We are given that $P(b)$ is true [i.e., that $Q(1)$ is true], and that $P(k) \to P(k + 1)$ for all $k \geq b$ [i.e., that $Q(m) \to Q(m + 1)$ for all positive integers m]. Therefore, by the principle of mathematical induction, $Q(m)$ is true for all positive integers m.

Section 5.2

1. *Basis step:* We are told we can run one mile, so $P(1)$ is true. *Inductive step:* Assume the inductive hypothesis, that we can run any number of miles from 1 to k. We must show that we can run $k + 1$ miles. If $k = 1$, then we are already told that we can run two miles. If $k > 1$, then the inductive hypothesis tells us that we can run $k - 1$ miles, so we can run $(k - 1) + 2 = k + 1$ miles. **3.** Let $P(n)$ be the statement that for any set A of natural numbers with $|A| = n$, there is a member that is greater than or equal to all elements in A. *Basis step:* We know $P(1)$ is true because the sole member of the set is greater than or equal to itself. *Inductive step:* Assume that $P(j)$ is true for $1 \leq j \leq k$ where $k \geq 1$. Let A be a set of $k + 1$ natural numbers. Pick a nonempty proper subset $S \subseteq A$; then $1 \leq |S| \leq k$ and $1 \leq |A - S| \leq k$. By the inductive hypothesis, there is an $m \in S$ such that $m \geq x$ for all $x \in S$ and an $n \in A - S$ such that $n \geq x$ for all $x \in A - S$. If $m \geq n$, then m is greater than or equal to all members of A; and if $n \geq m$, then n is greater than or equal to all members of A. In either case, there is a member of A greater than or equal to all integers in A, which completes the induction. **5. a)** $P(8)$ is true, because we can form 8 cents of postage with one 3-cent stamp and one 5-cent stamp. $P(9)$ is true, because we can form 9 cents of postage with three 3-cent stamps. $P(10)$ is true, because we can form 10 cents of postage with two 5-cent stamps. **b)** The statement that using just 3-cent and 5-cent stamps we can form j cents postage for all j with $8 \leq j \leq k$, where we assume that $k \geq 10$ **c)** Assuming the inductive hypothesis, we can form $k + 1$ cents postage using just 3-cent and 5-cent stamps **d)** Because $k \geq 10$, we know that $P(k - 2)$ is true, that is, that we can form $k - 2$ cents of postage. Put one more 3-cent stamp on the envelope, and we have formed $k + 1$ cents of postage. **e)** We have completed both the basis step and the inductive step, so by the principle of strong induction, the statement is true for every integer n greater than or equal to 8. **7. a)** 4, 8, 11, 12, 15, 16, 19, 20, 22, 23, 24, 26, 27, 28, and all values greater than or equal to 30 **b)** Let $P(n)$ be the statement that we can form n cents of postage using just 4-cent and 11-cent stamps. We want to prove that $P(n)$ is true for all $n \geq 30$. For the basis step, $30 = 11 + 11 + 4 + 4$. Assume that we can form k cents of postage (the inductive hypothesis); we will show how to form $k + 1$ cents of postage. If the k cents included an 11-cent stamp, then replace it by three 4-cent stamps. Otherwise, k cents was formed from just 4-cent stamps. Because $k \geq 30$, there must be at least eight 4-cent stamps involved. Replace eight 4-cent stamps by three 11-cent stamps, and we have formed $k + 1$ cents in postage. **c)** $P(n)$ is the same as in part (b). To prove that $P(n)$ is true for all $n \geq 30$, we check for the basis step that $30 = 11 + 11 + 4 + 4$, $31 = 11 + 4 + 4 + 4 + 4 + 4$, $32 = 4 + 4 + 4 + 4 + 4 + 4 + 4 + 4$, and $33 = 11 + 11 + 11$. For the inductive step, assume the inductive hypothesis, that $P(j)$ is true for all j with $30 \leq j \leq k$, where k is an arbitrary integer greater than or equal to 33. We want to show that $P(k + 1)$ is true. Because $k - 3 \geq 30$, we know that $P(k - 3)$ is true, that is, that we can form $k - 3$ cents of postage. Put one more 4-cent stamp on the envelope, and we have formed $k + 1$ cents of postage. In this proof, our inductive hypothesis was that $P(j)$ was true for all values of j between 30 and k inclusive, rather than just that $P(30)$ was true. **9.** We can form all amounts except \$1 and \$3. Let $P(n)$ be the statement that we can form n dollars using just 2-dollar and 5-dollar bills. We want to prove that $P(n)$ is true for all $n \geq 5$. (It is clear that \$1 and \$3 cannot be formed and that \$2 and \$4 can be formed.) For the basis step, note that $5 = 5$ and $6 = 2 + 2 + 2$. Assume the inductive hypothesis, that $P(j)$ is true for all j with $5 \leq j \leq k$, where k is an arbitrary integer greater than or equal to 6. We want to show that $P(k + 1)$ is true. Because $k - 1 \geq 5$, we know that $P(k - 1)$ is true, that is, that we can form $k - 1$ dollars. Add another 2-dollar bill, and we have formed $k + 1$ dollars. **11.** Let $P(n)$ be the statement that there is no positive integer b such that $\sqrt{2} = n/b$. *Basis step:* $P(1)$ is true because $\sqrt{2} > 1 \geq 1/b$ for all positive integers b. *Inductive step:* Assume that $P(j)$ is true for all $j \leq k$, where k is an arbitrary positive integer; we prove that $P(k + 1)$ is true by contradiction. Assume that $\sqrt{2} = (k + 1)/b$ for some positive integer b. Then $2b^2 = (k + 1)^2$, so $(k + 1)^2$ is even, and hence, $k + 1$ is even. So write $k + 1 = 2t$ for some positive integer t, whence $2b^2 = 4t^2$ and $b^2 = 2t^2$. By the same reasoning as before, b is even, so $b = 2s$ for some positive integer s. Then $\sqrt{2} = (k+1)/b = (2t)/(2s) = t/s$. But $t \le k$, so this contradicts the inductive hypothesis, and our proof of the inductive step is complete. **13.** *Basis step:* There are four base cases. If $n = 1 = 4 \cdot 0 + 1$, then clearly the second player wins. If there are two, three, or four matches ($n = 4 \cdot 0 + 2$, $n = 4 \cdot 0 + 3$, or $n = 4 \cdot 1$), then the first player can win by removing all but one match. *Inductive step:* Assume the strong inductive hypothesis, that in games with k or fewer matches, the first player can win if $k \equiv 0, 2,$ or $3 \pmod 4$ and the second player can win if $k \equiv 1 \pmod 4$. Suppose we have a game with $k+1$ matches, with $k \ge 4$. If $k+1 \equiv 0 \pmod 4$, then the first player can remove three matches, leaving $k-2$ matches for the other player. Because $k-2 \equiv 1 \pmod 4$, by the inductive hypothesis, this is a game that the second player at that point (who is the first player in our game) can win. Similarly, if $k+1 \equiv 2 \pmod 4$, then the first player can remove one match; and if $k+1 \equiv 3 \pmod 4$, then the first player can remove two matches. Finally, if $k+1 \equiv 1 \pmod 4$, then the first player must leave k, $k-1$, or $k-2$ matches for the other player. Because $k \equiv 0 \pmod 4$, $k-1 \equiv 3 \pmod 4$, and $k-2 \equiv 2 \pmod 4$, by the inductive hypothesis, this is a game that the first player at that point (who is the second player in our game) can win. **15.** Let $P(n)$ be the statement that exactly $n-1$ moves are required to assemble a puzzle with n pieces. Now $P(1)$ is trivially true. Assume that $P(j)$ is true for all $j \le k$, and consider a puzzle with $k+1$ pieces. The final move must be the joining of two blocks, of size j and $k+1-j$ for some integer j with $1 \le j \le k$. By the inductive hypothesis, it required $j-1$ moves to construct the one block, and $k+1-j-1 = k-j$ moves to construct the other. Therefore, $1+(j-1)+(k-j) = k$ moves are required in all, so $P(k+1)$ is true. **17.** Let the Chomp board have n rows and n columns. We claim that the first player can win the game by making the first move to leave just the top row and leftmost column. Let $P(n)$ be the statement that if a player has presented his opponent with a Chomp configuration consisting of just n cookies in the top row and n cookies in the leftmost column, then he can win the game. We will prove $\forall n P(n)$ by strong induction. We know that $P(1)$ is true, because the opponent is forced to take the poisoned cookie at his first turn. Fix $k \ge 1$ and assume that $P(j)$ is true for all $j \le k$. We claim that $P(k+1)$ is true. It is the opponent's turn to move. If she picks the poisoned cookie, then the game is over and she loses. Otherwise, assume she picks the cookie in the top row in column j, or the cookie in the left column in row j, for some j with $2 \le j \le k+1$. The first player now picks the cookie in the left column in row j, or the cookie in the top row in column j, respectively. This leaves the position covered by $P(j-1)$ for his opponent, so by the inductive hypothesis, he can win. **19.** Let $P(n)$ be the statement that if a simple polygon with n sides is triangulated, then at least two of the triangles in the triangulation have two sides that border the exterior of the polygon. We will prove $\forall n \ge 4\, P(n)$. The statement is true for $n = 4$, because there is only one diagonal, leaving two triangles with the desired property. Fix $k \ge 4$ and assume that $P(j)$ is true for all j with $4 \le j \le k$. Consider a polygon with $k+1$ sides, and some triangulation of it. Pick one of the diagonals in this triangulation. First suppose that this diagonal divides the polygon into one triangle and one polygon with k sides. Then the triangle has two sides that border the exterior. Furthermore, the k-gon has, by the inductive hypothesis, two triangles that have two sides that border the exterior of that k-gon, and only one of these triangles can fail to be a triangle that has two sides that border the exterior of the original polygon. The only other case is that this diagonal divides the polygon into two polygons with j sides and $k+3-j$ sides for some j with $4 \le j \le k-1$. By the inductive hypothesis, each of these two polygons has two triangles that have two sides that border their exterior, and in each case only one of these triangles can fail to be a triangle that has two sides that border the exterior of the original polygon. **21.** Let $P(n)$ be the statement that the area of a simple polygon with n sides and vertices all at lattice points is given by $I(P) + B(P)/2 - 1$. We will prove $P(n)$ for all $n \ge 3$. We begin with an additivity lemma: If P is a simple polygon with all vertices at lattice points, divided into polygons P_1 and P_2 by a diagonal, then $I(P) + B(P)/2 - 1 = [I(P_1) + B(P_1)/2 - 1] + [I(P_2) + B(P_2)/2 - 1]$. To prove this, suppose there are k lattice points on the diagonal, not counting its endpoints. Then $I(P) = I(P_1) + I(P_2) + k$ and $B(P) = B(P_1) + B(P_2) - 2k - 2$; and the result follows by simple algebra. What this says in particular is that if Pick's formula gives the correct area for P_1 and P_2, then it must give the correct formula for P, whose area is the sum of the areas for P_1 and P_2; and similarly if Pick's formula gives the correct area for P and one of the P_i's, then it must give the correct formula for the other P_i. Next we prove the theorem for rectangles whose sides are parallel to the coordinate axes. Such a rectangle necessarily has vertices at (a, b), (a, c), (d, b), and (d, c), where a, b, c, and d are integers with $b < c$ and $a < d$. Its area is $(c-b)(d-a)$. Also, $B = 2(c-b+d-a)$ and $I = (c-b-1)(d-a-1) = (c-b)(d-a)-(c-b)-(d-a)+1$. Therefore, $I+B/2-1 = (c-b)(d-a)-(c-b)-(d-a)+1+(c-b+d-a)-1 = (c-b)(d-a)$, which is the desired area. Next consider a right triangle whose legs are parallel to the coordinate axes. This triangle is half a rectangle of the type just considered, for which Pick's formula holds, so by the additivity lemma, it holds for the triangle as well. (The values of B and I are the same for each of the two triangles, so if Picks's formula gave an answer that was either too small or too large, then it would give a correspondingly wrong answer for the rectangle.) For the next step, consider an arbitrary triangle with vertices at lattice points that is not of the type already considered. Embed it in as small a rectangle as possible. There are several possible ways this can happen, but in any case (and adding one more edge in one case), the rectangle will have been partitioned into the given triangle and two or three right triangles with sides parallel to the coordinate axes. Again by the additivity lemma, we are guaranteed that Pick's formula gives the correct area for the given triangle. This completes the proof of $P(3)$, the basis step in our strong induction proof. For the inductive step, given an arbitrary polygon, use Lemma 1 in the text to split it into two polygons. Then by the additivity lemma above and the inductive hypothesis, we know that Pick's formula gives the correct area for this polygon. **23. a)** In the left figure $\angle abp$ is smallest, but $\overline{bp}$ is not an interior diagonal. **b)** In the right figure $\overline{bd}$ is not an interior diagonal. **c)** In the right figure $\overline{bd}$ is not an interior diagonal. **25. a)** When we try to prove the inductive step and find a triangle in each subpolygon with at least two sides bordering the exterior, it may happen in each case that the triangle we are guaranteed in fact borders the diagonal (which is part of the boundary of that polygon). This leaves us with no triangles guaranteed to touch the boundary of the *original* polygon. **b)** We proved the stronger statement $\forall n \geq 4\, T(n)$ in Exercise 19. **27. a)** The inductive step here allows us to conclude that $P(3)$, $P(5)$, ... are all true, but we can conclude nothing about $P(2)$, $P(4)$, ... **b)** $P(n)$ is true for all positive integers n, using strong induction. **c)** The inductive step here enables us to conclude that $P(2)$, $P(4)$, $P(8)$, $P(16)$, ... are all true, but we can conclude nothing about $P(n)$ when n is not a power of 2. **d)** This is mathematical induction; we can conclude that $P(n)$ is true for all positive integers n. **29.** Suppose, for a proof by contradiction, that there is some positive integer n such that $P(n)$ is not true. Let m be the smallest positive integer greater than n for which $P(m)$ *is* true; we know that such an m exists because $P(m)$ is true for infinitely many values of m. But we know that $P(m) \rightarrow P(m-1)$, so $P(m-1)$ is also true. Thus, $m-1$ cannot be greater than n, so $m-1 = n$ and $P(n)$ is in fact true. This contradiction shows that $P(n)$ is true for all n. **31.** The error is in going from the base case $n = 0$ to the next case, $n = 1$; we cannot write 1 as the sum of two smaller natural numbers. **33.** The inductive step must work for all nonnegative values of k, including when $k = 0$, so the assertion that "we are also assuming $S(1)$ is true" is not correct. **35.** Assume that the well-ordering property holds. Suppose that $P(1)$ is true and that the conditional statement $[P(1) \wedge P(2) \wedge \cdots \wedge P(n)] \rightarrow P(n+1)$ is true for every positive integer n. Let S be the set of positive integers n for which $P(n)$ is false. We will show $S = \emptyset$. Assume that $S \neq \emptyset$. Then by the well-ordering property there is a least integer m in S. We know that m cannot be 1 because $P(1)$ is true. Because $n = m$ is the least integer such that $P(n)$ is false, $P(1)$, $P(2)$, ..., $P(m-1)$ are true, and $m - 1 \geq 1$. Because $[P(1) \wedge P(2) \wedge \cdots \wedge P(m-1)] \rightarrow P(m)$ is true, it follows that $P(m)$ must also be true, which is a contradiction. Hence, $S = \emptyset$. **37.** In each case, give a proof by contradiction based on a "smallest counterexample," that is, values of n and k such that $P(n, k)$ is not true and n and k are smallest in some sense. **a)** Choose a counterexample with $n + k$ as small as possible. We cannot have $n = 1$ and $k = 1$, because we are given that $P(1, 1)$ is true. Therefore, either $n > 1$ or $k > 1$. In the former case, by our choice of counterexample, we know that $P(n-1, k)$ is true. But the inductive step then forces $P(n, k)$ to be true, a contradiction. The latter case is similar. So our supposition that there is a counterexample must be wrong, and $P(n, k)$ is true in all cases. **b)** Choose a counterexample with n as small as possible. We cannot have $n = 1$, because we are given that $P(1, k)$ is true for all k. Therefore, $n > 1$. By our choice of counterexample, we know that $P(n-1, k)$ is true. But the inductive step then forces $P(n, k)$ to be true, a contradiction. **c)** Choose a counterexample with k as small as possible. We cannot have $k = 1$, because we are given that $P(n, 1)$ is true for all n. Therefore, $k > 1$. By our choice of counterexample, we know that $P(n, k-1)$ is true. But the inductive step then forces $P(n, k)$ to be true, a contradiction. **39.** Let $P(n)$ be the statement that if $x_1, x_2, \ldots, x_n$ are n distinct real numbers, then $n - 1$ multiplications are used to find the product of these numbers no matter how parentheses are inserted in the product. We will prove that $P(n)$ is true using strong induction. The basis case $P(1)$ is true because $1 - 1 = 0$ multiplications are required to find the product of x_1, a product with only one factor. Suppose that $P(k)$ is true for $1 \leq k \leq n$. The last multiplication used to find the product of the $n + 1$ distinct real numbers $x_1, x_2, \ldots, x_n, x_{n+1}$ is a multiplication of the product of the first k of these numbers for some k and the product of the last $n + 1 - k$ of them. By the inductive hypothesis, $k - 1$ multiplications are used to find the product of k of the numbers, no matter how parentheses were inserted in the product of these numbers, and $n - k$ multiplications are used to find the product of the other $n + 1 - k$ of them, no matter how parentheses were inserted in the product of these numbers. Because one more multiplication is required to find the product of all $n + 1$ numbers, the total number of multiplications used equals $(k-1) + (n-k) + 1 = n$. Hence, $P(n+1)$ is true. **41.** Assume that $a = dq + r = dq' + r'$ with $0 \leq r < d$ and $0 \leq r' < d$. Then $d(q - q') = r' - r$. It follows that d divides $r' - r$. Because $-d < r' - r < d$, we have $r' - r = 0$. Hence, $r' = r$. It follows that $q = q'$. **43.** This is a paradox caused by self-reference. The answer is "no." There are a finite number of English words, so only a finite number of strings of 15 words or fewer; therefore, only a finite number of positive integers can be so described, not all of them. **45.** Suppose that the well-ordering property were false. Let S be a nonempty set of nonnegative integers that has no least element. Let $P(n)$ be the statement "$i \notin S$ for $i = 0, 1, \ldots, n$." $P(0)$ is true because if $0 \in S$ then S has a least element, namely, 0. Now suppose that $P(n)$ is true. Thus, $0 \notin S$, $1 \notin S$, ..., $n \notin S$. Now, $n + 1$ cannot be in S, for if it were, it would be its least element. Thus, $P(n+1)$ is true. So by the principle of mathematical induction, $n \notin S$ for all nonnegative integers n. Thus, $S = \emptyset$, a contradiction. **47.** Strong induction implies the principle of mathematical induction, for if one has shown that $P(k) \rightarrow P(k+1)$ is true, then one has also shown that $[P(1) \wedge \cdots \wedge P(k)] \rightarrow P(k+1)$ is true. By Exercise 45, the principle of mathematical induction implies the well-ordering property. Therefore, by assuming strong induction as an axiom, we can prove the well-ordering property. **49.** Let $S(n)$ be the assertion that if the first domino is knocked over, then the nth domino will fall. For the basis step, we need an infinite collection of positive integers for which $S(n)$ is true. It is given that $S(2^n)$ is true for all positive integers n, so $\{2^n \mid n \in \mathbf{Z}^+\}$ is the needed infinite set. For the induction step, we must show that $S(m) \rightarrow S(m-1)$ for all $m \geq 2$.

This is the assertion that if a domino falls, then the one before it must have fallen, which is given to us as fact. Hence, by downward induction, all the dominoes fall.

Section 5.3

1. a) $f(1) = 3, f(2) = 5, f(3) = 7, f(4) = 9$ **b)** $f(1) = 3, f(2) = 9, f(3) = 27, f(4) = 81$ **c)** $f(1) = 2, f(2) = 4, f(3) = 16, f(4) = 65{,}536$ **d)** $f(1) = 3, f(2) = 13, f(3) = 183, f(4) = 33{,}673$ **3. a)** $f(2) = -1, f(3) = 5, f(4) = 2, f(5) = 17$ **b)** $f(2) = -4, f(3) = 32, f(4) = -4096, f(5) = 536{,}870{,}912$ **c)** $f(2) = 8, f(3) = 176, f(4) = 92{,}672, f(5) = 25{,}764{,}174{,}848$ **d)** $f(2) = -\frac{1}{2}, f(3) = -4, f(4) = \frac{1}{8}, f(5) = -32$ **5. a)** Not valid **b)** $f(n) = 1 - n$. *Basis step:* $f(0) = 1 = 1 - 0$. *Inductive step:* if $f(k) = 1 - k$, then $f(k+1) = f(k) - 1 = 1 - k - 1 = 1 - (k+1)$. **c)** $f(n) = 4 - n$ if $n > 0$, and $f(0) = 2$. *Basis step:* $f(0) = 2$ and $f(1) = 3 = 4 - 1$. *Inductive step* (with $k \geq 1$): $f(k+1) = f(k) - 1 = (4 - k) - 1 = 4 - (k+1)$. **d)** $f(n) = 2^{\lfloor (n+1)/2 \rfloor}$. *Basis step:* $f(0) = 1 = 2^{\lfloor (0+1)/2 \rfloor}$ and $f(1) = 2 = 2^{\lfloor (1+1)/2 \rfloor}$. *Inductive step* (with $k \geq 1$): $f(k+1) = 2f(k-1) = 2 \cdot 2^{\lfloor k/2 \rfloor} = 2^{\lfloor k/2 \rfloor + 1} = 2^{\lfloor ((k+1)+1)/2 \rfloor}$. **e)** $f(n) = 3^n$. *Basis step:* Trivial. *Inductive step:* For odd n, $f(n) = 3f(n-1) = 3 \cdot 3^{n-1} = 3^n$; and for even $n > 1$, $f(n) = 9f(n-2) = 9 \cdot 3^{n-2} = 3^n$. **7.** There are many possible correct answers. We will supply relatively simple ones. **a)** $a_{n+1} = a_n + 6$ for $n \geq 1$ and $a_1 = 6$ **b)** $a_{n+1} = a_n + 2$ for $n \geq 1$ and $a_1 = 3$ **c)** $a_{n+1} = 10a_n$ for $n \geq 1$ and $a_1 = 10$ **d)** $a_{n+1} = a_n$ for $n \geq 1$ and $a_1 = 5$ **9.** $F(0) = 0$, $F(n) = F(n-1) + n$ for $n \geq 1$ **11.** $P_m(0) = 0$, $P_m(n+1) = P_m(n) + m$ **13.** Let $P(n)$ be "$f_1 + f_3 + \cdots + f_{2n-1} = f_{2n}$." *Basis step:* $P(1)$ is true because $f_1 = 1 = f_2$. *Inductive step:* Assume that $P(k)$ is true. Then $f_1 + f_3 + \cdots + f_{2k-1} + f_{2k+1} = f_{2k} + f_{2k+1} = f_{2k+2} + f_{2(k+1)}$. **15.** *Basis step:* $f_0 f_1 + f_1 f_2 = 0 \cdot 1 + 1 \cdot 1 = 1^2 = f_2^2$. *Inductive step:* Assume that $f_0 f_1 + f_1 f_2 + \cdots + f_{2k-1} f_{2k} = f_{2k}^2$. Then $f_0 f_1 + f_1 f_2 + \cdots + f_{2k-1} f_{2k} + f_{2k} f_{2k+1} + f_{2k+1} f_{2k+2} = f_{2k}^2 + f_{2k} f_{2k+1} + f_{2k+1} f_{2k+2} = f_{2k}(f_{2k} + f_{2k+1}) + f_{2k+1} f_{2k+2} = f_{2k} f_{2k+2} + f_{2k+1} f_{2k+2} = (f_{2k} + f_{2k+1}) f_{2k+2} = f_{2k+2}^2$. **17.** The number of divisions used by the Euclidean algorithm to find $\gcd(f_{n+1}, f_n)$ is 0 for $n = 0$, 1 for $n = 1$, and $n - 1$ for $n \geq 2$. To prove this result for $n \geq 2$ we use mathematical induction. For $n = 2$, one division shows that $\gcd(f_3, f_2) = \gcd(2, 1) = \gcd(1, 0) = 1$. Now assume that $k - 1$ divisions are used to find $\gcd(f_{k+1}, f_k)$. To find $\gcd(f_{k+2}, f_{k+1})$, first divide f_{k+2} by f_{k+1} to obtain $f_{k+2} = 1 \cdot f_{k+1} + f_k$. After one division we have $\gcd(f_{k+2}, f_{k+1}) = \gcd(f_{k+1}, f_k)$. By the inductive hypothesis it follows that exactly $k - 1$ more divisions are required. This shows that k divisions are required to find $\gcd(f_{k+2}, f_{k+1})$, finishing the inductive proof. **19.** $|A| = -1$. Hence, $|A^n| = (-1)^n$. It follows that $f_{n+1} f_{n-1} - f_n^2 = (-1)^n$. **21. a)** Proof by induction. *Basis step:* For $n = 1$, $\max(-a_1) = -a_1 = -\min(a_1)$. For $n = 2$, there are two cases. If $a_2 \geq a_1$, then $-a_1 \geq -a_2$, so $\max(-a_1, -a_2) = -a_1 = -\min(a_1, a_2)$. If $a_2 < a_1$, then $-a_1 < -a_2$, so $\max(-a_1, -a_2) = -a_2 = -\min(a_1, a_2)$. *Inductive step:* Assume true for k with $k \geq 2$. Then $\max(-a_1, -a_2, \ldots, -a_k, -a_{k+1}) = \max(\max(-a_1, \ldots, -a_k), -a_{k+1}) = \max(-\min(a_1, \ldots, a_k), -a_{k+1}) = -\min(\min(a_1, \ldots, a_k), a_{k+1}) = -\min(a_1, \ldots, a_{k+1})$. **b)** Proof by mathematical induction. *Basis step:* For $n = 1$, the result is the identity $a_1 + b_1 = a_1 + b_1$. For $n = 2$, first consider the case in which $a_1 + b_1 \geq a_2 + b_2$. Then $\max(a_1 + b_1, a_2 + b_2) = a_1 + b_1$. Also note that $a_1 \leq \max(a_1, a_2)$ and $b_1 \leq \max(b_1, b_2)$, so $a_1 + b_1 \leq \max(a_1, a_2) + \max(b_1, b_2)$. Therefore, $\max(a_1 + b_1, a_2 + b_2) = a_1 + b_1 \leq \max(a_1, a_2) + \max(b_1, b_2)$. The case with $a_1 + b_1 < a_2 + b_2$ is similar. *Inductive step:* Assume that the result is true for k. Then $\max(a_1 + b_1, a_2 + b_2, \ldots, a_k + b_k, a_{k+1} + b_{k+1}) = \max(\max(a_1 + b_1, a_2 + b_2, \ldots, a_k + b_k), a_{k+1} + b_{k+1}) \leq \max(\max(a_1, a_2, \ldots, a_k) + \max(b_1, b_2, \ldots, b_k), a_{k+1} + b_{k+1}) \leq \max(\max(a_1, a_2, \ldots, a_k), a_{k+1}) + \max(\max(b_1, b_2, \ldots, b_k), b_{k+1}) = \max(a_1, a_2, \ldots, a_k, a_{k+1}) + \max(b_1, b_2, \ldots, b_k, b_{k+1})$. **c)** Same as part (b), but replace every occurrence of "max" by "min" and invert each inequality. **23.** $5 \in S$, and $x + y \in S$ if $x, y \in S$. **25. a)** $0 \in S$, and if $x \in S$, then $x + 2 \in S$ and $x - 2 \in S$. **b)** $2 \in S$, and if $x \in S$, then $x + 3 \in S$. **c)** $1 \in S, 2 \in S, 3 \in S, 4 \in S$, and if $x \in S$, then $x + 5 \in S$. **27. a)** *Basis step:* $5 \equiv 5 \pmod{10}$. *Inductive step:* If $n \equiv 5 \pmod{10}$, then $3n \equiv 3 \cdot 5 = 15 \equiv 5 \pmod{10}$ and $n^2 \equiv 5^2 = 25 \equiv 5 \pmod{10}$. **b)** $35 \notin S$ because 35 is not a multiple of 3 nor a perfect square. **29. a)** (0, 1), (1, 1), (2, 1); (0, 2), (1, 2), (2, 2), (3, 2), (4, 2); (0, 3), (1, 3), (2, 3), (3, 3), (4, 3), (5, 3), (6, 3); (0, 4), (1, 4), (2, 4), (3, 4), (4, 4), (5, 4), (6, 4), (7, 4), (8, 4) **b)** Let $P(n)$ be the statement that $a \leq 2b$ whenever $(a, b) \in S$ is obtained by n applications of the recursive step. *Basis step:* $P(0)$ is true, because the only element of S obtained with no applications of the recursive step is (0, 0), and indeed $0 \leq 2 \cdot 0$. *Inductive step:* Assume that $a \leq 2b$ whenever $(a, b) \in S$ is obtained by k or fewer applications of the recursive step, and consider an element obtained with $k + 1$ applications of the recursive step. Because the final application of the recursive step to an element (a, b) must be applied to an element obtained with fewer applications of the recursive step, we know that $a \leq 2b$. Add $0 \leq 2$, $1 \leq 2$, and $2 \leq 2$, respectively, to obtain $a \leq 2(b+1)$, $a + 1 \leq 2(b+1)$, and $a + 2 \leq 2(b+1)$, as desired. **c)** This holds for the basis step, because $0 \leq 0$. If this holds for (a, b), then it also holds for the elements obtained from (a, b) in the recursive step, because adding $0 \leq 2$, $1 \leq 2$, and $2 \leq 2$, respectively, to $a \leq 2b$ yields $a \leq 2(b+1)$, $a + 1 \leq 2(b+1)$, and $a + 2 \leq 2(b+1)$. **31. a)** Define S by $(1, 1) \in S$, and if $(a, b) \in S$, then $(a + 2, b) \in S$, $(a, b + 2) \in S$, and $(a + 1, b + 1) \in S$. All elements put in S satisfy the condition, because (1, 1) has an even sum of coordinates, and if (a, b) has an even sum of coordinates, then so do $(a + 2, b)$, $(a, b + 2)$, and $(a + 1, b + 1)$. Conversely, we show by induction on the sum of the coordinates that if $a + b$ is even, then $(a, b) \in S$. If the sum is 2, then $(a, b) = (1, 1)$, and the basis step put (a, b) into S. Otherwise the sum is at least 4, and at least one of $(a - 2, b)$, $(a, b - 2)$, and $(a - 1, b - 1)$ must have positive integer coordinates whose sum is an even number smaller than $a + b$, and therefore must be in S. Then one application of the recursive step shows that $(a, b) \in S$. **b)** Define S by (1, 1), (1, 2), and (2, 1) are in S, and if $(a, b) \in S$, then $(a + 2, b)$ and

$(a, b+2)$ are in S. To prove that our definition works, we note first that (1, 1), (1, 2), and (2, 1) all have an odd coordinate, and if (a, b) has an odd coordinate, then so do $(a+2, b)$ and $(a, b+2)$. Conversely, we show by induction on the sum of the coordinates that if (a, b) has at least one odd coordinate, then $(a, b) \in S$. If $(a, b) = (1, 1)$ or $(a, b) = (1, 2)$ or $(a, b) = (2, 1)$, then the basis step put (a, b) into S. Otherwise either a or b is at least 3, so at least one of $(a-2, b)$ and $(a, b-2)$ must have positive integer coordinates whose sum is smaller than $a+b$, and therefore must be in S. Then one application of the recursive step shows that $(a, b) \in S$. **c)** $(1, 6) \in S$ and $(2, 3) \in S$, and if $(a, b) \in S$, then $(a+2, b) \in S$ and $(a, b+6) \in S$. To prove that our definition works, we note first that (1, 6) and (2, 3) satisfy the condition, and if (a, b) satisfies the condition, then so do $(a+2, b)$ and $(a, b+6)$. Conversely we show by induction on the sum of the coordinates that if (a, b) satisfies the condition, then $(a, b) \in S$. For sums 5 and 7, the only points are (1, 6), which the basis step put into S, (2, 3), which the basis step put into S, and $(4, 3) = (2+2, 3)$, which is in S by one application of the recursive definition. For a sum greater than 7, either $a \geq 3$, or $a \leq 2$ and $b \geq 9$, in which case either $(a-2, b)$ or $(a, b-6)$ must have positive integer coordinates whose sum is smaller than $a+b$ and satisfy the condition for being in S. Then one application of the recursive step shows that $(a, b) \in S$. **33.** If x is a set or a variable representing a set, then x is a well-formed formula. If x and y are well-formed formulae, then so are $\overline{x}$, $(x \cup y)$, $(x \cap y)$, and $(x - y)$. **35. a)** If $x \in D = \{0, 1, 2, 3, 4, 5, 6, 7, 8, 9\}$, then $m(x) = x$; if $s = tx$, where $t \in D^*$ and $x \in D$, then $m(s) = \min(m(s), x)$. **b)** Let $t = wx$, where $w \in D^*$ and $x \in D$. If $w = \lambda$, then $m(st) = m(sx) = \min(m(s), x) = \min(m(s), m(x))$ by the recursive step and the basis step of the definition of m. Otherwise, $m(st) = m((sw)x) = \min(m(sw), x)$ by the definition of m. Now $m(sw) = \min(m(s), m(w))$ by the inductive hypothesis of the structural induction, so $m(st) = \min(\min(m(s), m(w)), x) = \min(m(s), \min(m(w), x))$ by the meaning of min. But $\min(m(w), x) = m(wx) = m(t)$ by the recursive step of the definition of m. Thus, $m(st) = \min(m(s), m(t))$. **37.** $\lambda^R = \lambda$ and $(ux)^R = xu^R$ for $x \in \Sigma$, $u \in \Sigma^*$. **39.** $w^0 = \lambda$ and $w^{n+1} = ww^n$. **41.** When the string consists of n 0s followed by n 1s for some nonnegative integer n **43.** Let $P(i)$ be "$l(w^i) = i \cdot l(w)$." $P(0)$ is true because $l(w^0) = 0 = 0 \cdot l(w)$. Assume $P(i)$ is true. Then $l(w^{i+1}) = l(ww^i) = l(w) + l(w^i) = l(w) + i \cdot l(w) = (i+1) \cdot l(w)$. **45.** *Basis step:* For the full binary tree consisting of just a root the result is true because $n(T) = 1$ and $h(T) = 0$, and $1 \geq 2 \cdot 0 + 1$. *Inductive step:* Assume that $n(T_1) \geq 2h(T_1) + 1$ and $n(T_2) \geq 2h(T_2) + 1$. By the recursive definitions of $n(T)$ and $h(T)$, we have $n(T) = 1 + n(T_1) + n(T_2)$ and $h(T) = 1 + \max(h(T_1), h(T_2))$. Therefore, $n(T) = 1 + n(T_1) + n(T_2) \geq 1 + 2h(T_1) + 1 + 2h(T_2) + 1 \geq 1 + 2 \cdot \max(h(T_1), h(T_2)) + 2 = 1 + 2(\max(h(T_1), h(T_2)) + 1) = 1 + 2h(T)$. **47.** *Basis step:* $a_{0,0} = 0 = 0 + 0$. *Inductive step:* Assume that $a_{m',n'} = m' + n'$ whenever (m', n') is less than (m, n) in the lexicographic ordering of $\mathbf{N} \times \mathbf{N}$. If $n = 0$ then $a_{m,n} = a_{m-1,n} + 1 = m - 1 + n + 1 = m + n$. If $n > 0$, then $a_{m,n} = a_{m,n-1} + 1 = m + n - 1 + 1 = m + n$.

49. a) $P_{m,m} = P_m$ because a number exceeding m cannot be used in a partition of m. **b)** Because there is only one way to partition 1, namely, $1 = 1$, it follows that $P_{1,n} = 1$. Because there is only one way to partition m into 1s, $P_{m,1} = 1$. When $n > m$ it follows that $P_{m,n} = P_{m,m}$ because a number exceeding m cannot be used. $P_{m,m} = 1 + P_{m,m-1}$ because one extra partition, namely, $m = m$, arises when m is allowed in the partition. $P_{m,n} = P_{m,n-1} + P_{m-n,n}$ if $m > n$ because a partition of m into integers not exceeding n either does not use any ns and hence, is counted in $P_{m,n-1}$ or else uses an n and a partition of $m - n$, and hence, is counted in $P_{m-n,n}$. **c)** $P_5 = 7$, $P_6 = 11$ **51.** Let $P(n)$ be "$A(n, 2) = 4$." *Basis step:* $P(1)$ is true because $A(1, 2) = A(0, A(1, 1)) = A(0, 2) = 2 \cdot 2 = 4$. *Inductive step:* Assume that $P(n)$ is true, that is, $A(n, 2) = 4$. Then $A(n+1, 2) = A(n, A(n+1, 1)) = A(n, 2) = 4$. **53. a)** 16 **b)** 65,536 **55.** Use a double induction argument to prove the stronger statement: $A(m, k) > A(m, l)$ when $k > l$. *Basis step:* When $m = 0$ the statement is true because $k > l$ implies that $A(0, k) = 2k > 2l = A(0, l)$. *Inductive step:* Assume that $A(m, x) > A(m, y)$ for all nonnegative integers x and y with $x > y$. We will show that this implies that $A(m+1, k) > A(m+1, l)$ if $k > l$. *Basis steps:* When $l = 0$ and $k > 0$, $A(m+1, l) = 0$ and either $A(m+1, k) = 2$ or $A(m+1, k) = A(m, A(m+1, k-1))$. If $m = 0$, this is $2A(1, k-1) = 2^k$. If $m > 0$, this is greater than 0 by the inductive hypothesis. In all cases, $A(m+1, k) > 0$, and in fact, $A(m+1, k) \geq 2$. If $l = 1$ and $k > 1$, then $A(m+1, l) = 2$ and $A(m+1, k) = A(m, A(m+1, k-1))$, with $A(m+1, k-1) \geq 2$. Hence, by the inductive hypothesis, $A(m, A(m+1, k-1)) \geq A(m, 2) > A(m, 1) = 2$. *Inductive step:* Assume that $A(m+1, r) > A(m+1, s)$ for all $r > s$, $s = 0, 1, \ldots, l$. Then if $k+1 > l+1$ it follows that $A(m+1, k+1) = A(m, A(m+1, k)) > A(m, A(m+1, k)) = A(m+1, l+1)$. **57.** From Exercise 56 it follows that $A(i, j) \geq A(i-1, j) \geq \cdots \geq A(0, j) = 2j \geq j$. **59.** Let $P(n)$ be "$F(n)$ is well defined." Then $P(0)$ is true because $F(0)$ is specified. Assume that $P(k)$ is true for all $k < n$. Then $F(n)$ is well defined at n because $F(n)$ is given in terms of $F(0), F(1), \ldots, F(n-1)$. So $P(n)$ is true for all integers n. **61. a)** The value of $F(1)$ is ambiguous. **b)** $F(2)$ is not defined because $F(0)$ is not defined. **c)** $F(3)$ is ambiguous and $F(4)$ is not defined because $F(\frac{4}{3})$ makes no sense. **d)** The definition of $F(1)$ is ambiguous because both the second and third clause seem to apply. **e)** $F(2)$ cannot be computed because trying to compute $F(2)$ gives $F(2) = 1 + F(F(1)) = 1 + F(2)$. **63. a)** 1 **b)** 2 **c)** 3 **d)** 3 **e)** 4 **f)** 4 **g)** 5 **65.** $f_0^*(n) = \lceil n/a \rceil$ **67.** $f_2^*(n) = \lceil \log \log n \rceil$ for $n \geq 2$, $f_2^*(1) = 0$

Section 5.4

1. First, we use the recursive step to write $5! = 5 \cdot 4!$. We then use the recursive step repeatedly to write $4! = 4 \cdot 3!$, $3! = 3 \cdot 2!$, $2! = 2 \cdot 1!$, and $1! = 1 \cdot 0!$. Inserting the value of $0! = 1$, and working back through the steps, we see that $1! = 1 \cdot 1 = 1$, $2! = 2 \cdot 1! = 2 \cdot 1 = 2$, $3! = 3 \cdot 2! = 3 \cdot 2 = 6$, $4! = 4 \cdot 3! = 4 \cdot 6 = 24$, and $5! = 5 \cdot 4! = 5 \cdot 24 = 120$. **3.** With this input, the algorithm uses the **else** clause to find that gcd(8, 13) = gcd(13 **mod** 8, 8) = gcd(5, 8). It uses this clause again to find that gcd(5, 8) = gcd(8 **mod** 5, 5) = gcd(3, 5), then to get gcd(3, 5) = gcd(5 **mod** 3, 3) = gcd(2, 3), then gcd(2, 3) = gcd(3 **mod** 2, 2) = gcd(1, 2), and once more to get gcd(1, 2) = gcd(2 **mod** 1, 1) = gcd(0, 1). Finally, to find gcd(0, 1) it uses the first step with $a = 0$ to find that gcd(0, 1) = 1. Consequently, the algorithm finds that gcd(8, 13) = 1. **5.** First, because $n = 11$ is odd, we use the **else** clause to see that $mpower(3, 11, 5) = (mpower(3, 5, 5)^2$ **mod** $5 \cdot 3$ **mod** 5) **mod** 5. We next use the **else** clause again to see that $mpower(3, 5, 5) = (mpower(3, 2, 5)^2$ **mod** $5 \cdot 3$ **mod** 5) **mod** 5. Then we use the **else if** clause to see that $mpower(3, 2, 5) = mpower(3, 1, 5)^2$ **mod** 5. Using the **else** clause again, we have $mpower(3, 1, 5) = (mpower(3, 0, 5)^2$ **mod** $5 \cdot 3$ **mod** 5) **mod** 5. Finally, using the **if** clause, we see that $mpower(3, 0, 5) = 1$. Working backward it follows that $mpower(3, 1, 5) = (1^2$ **mod** $5 \cdot 3$ **mod** 5) **mod** $5 = 3$, $mpower(3, 2, 5) = 3^2$ **mod** $5 = 4$, $mpower(3, 5, 5) = (4^2$ **mod** $5 \cdot 3$ **mod** 5) **mod** $5 = 3$, and finally $mpower(3, 11, 5) = (3^2$ **mod** $5 \cdot 3$ **mod** 5) **mod** $5 = 2$. We conclude that 3^{11} **mod** $5 = 2$.

7. **procedure** *mult*(n: positive integer, x: integer)
  if $n = 1$ **then return** x
  else return $x + mult(n - 1, x)$

9. **procedure** *sum of odds*(n: positive integer)
  if $n = 1$ **then return** 1
  else return *sum of odds* $(n - 1) + 2n - 1$

11. **procedure** *smallest*($a_1, \dots, a_n$: integers)
  if $n = 1$ **then return** a_1
  else return
    $\min(smallest(a_1, \dots, a_{n-1}), a_n)$

13. **procedure** *modfactorial*(n, m: positive integers)
  if $n = 1$ **then return** 1
  else return
    $(n \cdot modfactorial(n - 1, m))$ **mod** m

15. **procedure** *gcd*(a, b: nonnegative integers)
  $\{a < b$ assumed to hold$\}$
  if $a = 0$ **then return** b
  else if $a = b - a$ **then return** a
  else if $a < b - a$ **then return** $gcd(a, b - a)$
  else return $gcd(b - a, a)$

17. **procedure** *multiply*(x, y: nonnegative integers)
  if $y = 0$ **then return** 0
  else if y is even **then**
    return $2 \cdot multiply(x, y/2)$
  else return $2 \cdot multiply(x, (y-1)/2) + x$

19. We use strong induction on a. *Basis step:* If $a = 0$, we know that gcd(0, b) = b for all $b > 0$, and that is precisely what the **if** clause does. *Inductive step:* Fix $k > 0$, assume the inductive hypothesis—that the algorithm works correctly for all values of its first argument less than k—and consider what happens with input (k, b), where $k < b$. Because $k > 0$, the **else** clause is executed, and the answer is whatever the algorithm gives as output for inputs (b **mod** k, k). Because b **mod** $k < k$, the input pair is valid. By our inductive hypothesis, this output is in fact gcd(b **mod** k, k), which equals gcd(k, b) by Lemma 1 in Section 4.3. **21.** If $n = 1$, then $nx = x$, and the algorithm correctly returns x. Assume that the algorithm correctly computes kx. To compute $(k+1)x$ it recursively computes the product of $k + 1 - 1 = k$ and x, and then adds x. By the inductive hypothesis, it computes that product correctly, so the answer returned is $kx + x = (k + 1)x$, which is correct.

23. **procedure** *square*(n: nonnegative integer)
  if $n = 0$ **then return** 0
  else return $square(n - 1) + 2(n - 1) + 1$

Let $P(n)$ be the statement that this algorithm correctly computes n^2. Because $0^2 = 0$, the algorithm works correctly (using the **if** clause) if the input is 0. Assume that the algorithm works correctly for input k. Then for input $k + 1$, it gives as output (because of the **else** clause) its output when the input is k, plus $2(k + 1 - 1) + 1$. By the inductive hypothesis, its output at k is k^2, so its output at $k + 1$ is $k^2 + 2(k + 1 - 1) + 1 = k^2 + 2k + 1 = (k + 1)^2$, as desired. **25.** n multiplications versus 2^n **27.** $O(\log n)$ versus n

29. **procedure** a(n: nonnegative integer)
  if $n = 0$ **then return** 1
  else if $n = 1$ **then return** 2
  else return $a(n - 1) \cdot a(n - 2)$

31. Iterative

33. **procedure** *iterative*(n: nonnegative integer)
  if $n = 0$ **then** $z := 1$
  else if $n = 1$ **then** $z := 2$
  else
    $x := 1$
    $y := 2$
    $z := 3$
    for $i := 1$ **to** $n - 2$
      $w := x + y + z$
      $x := y$
      $y := z$
      $z := w$
  return z $\{z$ is the nth term of the sequence$\}$

35. We first give a recursive procedure and then an iterative procedure.

procedure r(n: nonnegative integer)
if $n < 3$ **then return** $2n + 1$
else return $r(n - 1) \cdot (r(n - 2))^2 \cdot (r(n - 3))^3$

procedure i(n: nonnegative integer)
if $n = 0$ **then** $z := 1$
else if $n = 1$ **then** $z := 3$
else
  $x := 1$
  $y := 3$
  $z := 5$
  for $i := 1$ **to** $n - 2$
    $w := z \cdot y^2 \cdot x^3$
    $x := y$
    $y := z$
    $z := w$

return z {z is the nth term of the sequence}
The iterative version is more efficient.

37. **procedure** *reverse*(w: bit string)
n := length(w)
if $n \leq 1$ **then return** w
else return
substr(w, n, n)*reverse*(*substr*(w, 1, $n - 1$))
{*substr*(w, a, b) is the substring of w consisting of the symbols in the ath through bth positions}

39. The procedure correctly gives the reversal of λ as λ (basis step), and because the reversal of a string consists of its last character followed by the reversal of its first $n - 1$ characters (see Exercise 37 in Section 5.3), the algorithm behaves correctly when $n > 0$ by the inductive hypothesis. **41.** The algorithm implements the idea of Example 14 in Section 5.1. If $n = 1$ (basis step), place the one right triomino so that its armpit corresponds to the hole in the 2×2 board. If $n > 1$, then divide the board into four boards, each of size $2^{n-1} \times 2^{n-1}$, notice which quarter the hole occurs in, position one right triomino at the center of the board with its armpit in the quarter where the missing square is (see Figure 7 in Section 5.1), and invoke the algorithm recursively four times—once on each of the $2^{n-1} \times 2^{n-1}$ boards, each of which has one square missing (either because it was missing to begin with, or because it is covered by the central triomino).

43. **procedure** A(m, n: nonnegative integers)
if $m = 0$ **then return** $2n$
else if $n = 0$ **then return** 0
else if $n = 1$ **then return** 2
else return $A(m - 1, A(m, n - 1))$

45.

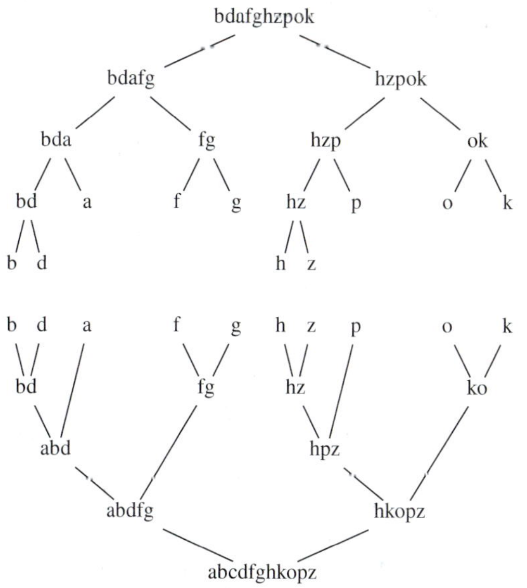

47. Let the two lists be 1, 2, …, $m - 1$, $m + n - 1$ and $m, m + 1, \ldots, m + n - 2, m + n$, respectively. **49.** If $n = 1$, then the algorithm does nothing, which is correct because a list with one element is already sorted. Assume that the algorithm works correctly for $n = 1$ through $n = k$. If $n = k + 1$, then the list is split into two lists, L_1 and L_2. By the inductive hypothesis, *mergesort* correctly sorts each of these sublists; furthermore, *merge* correctly merges two sorted lists into one because with each comparison the smallest element in $L_1 \cup L_2$ not yet put into L is put there. **51.** $O(n)$ **53.** 6 **55.** $O(n^2)$

Section 5.5

1. In addition to being the loop condition, "$x > 0$" is also a loop invariant, so the program segment will not terminate. This means that the segment is not totally correct, since the correct output is not produced, but it is partially correct, since the program does not terminate. **3.** Suppose that $x = 0$. The program segment first assigns the value 1 to y and then assigns the value $x + y = 0 + 1 = 1$ to z. **5.** Suppose that $y = 3$. The program segment assigns the value 2 to x and then assigns the value $x + y = 2 + 3 = 5$ to z. Because $y = 3 > 0$ it then assigns the value $z + 1 = 5 + 1 = 6$ to z.

7.
$(p \wedge condition1)\{S_1\}q$
$(p \wedge \neg condition1 \wedge condition2)\{S_2\}q$
.
.
.
$(p \wedge \neg condition1 \wedge \neg condition2$
$\cdots \wedge \neg condition(n-1))\{S_n\}q$
―――――――――――――――
$\therefore p\{$**if** $condition1$ **then** S_1;
else if $condition2$ **then** S_2; … ; **else** $S_n\}q$

9. We will show that p = "$power = x^{i-1}$ and $i \leq n + 1$" is a loop invariant. Note that p is true initially, because before the loop starts, $i = 1$ and $power = 1 = x^0 = x^{1-1}$. Next, we must show that if p is true and $i \leq n$ after an execution of the loop, then p remains true after one more execution. The loop increments i by 1. Hence, because $i \leq n$ before this pass, $i \leq n + 1$ after this pass. Also the loop assigns $power \cdot x$ to $power$. By the inductive hypothesis we see that $power$ is assigned the value $x^{i-1} \cdot x = x^i$. Hence, p remains true. Furthermore, the loop terminates after n traversals of the loop with $i = n + 1$ because i is assigned the value 1 prior to entering the loop, is incremented by 1 on each pass, and the loop terminates when $i > n$. Consequently, at termination $power = x^n$, as desired. **11.** Suppose that p is "m and n are integers." Then if the condition $n < 0$ is true, $a = -n = |n|$ after S_1 is executed. If the condition $n < 0$ is false, then $a = n = |n|$ after S_1 is executed. Hence, $p\{S_1\}q$ is true where q is $p \wedge (a = |n|)$. Because S_2 assigns the value 0 to both k and x, it is clear that $q\{S_2\}r$ is true where r is $q \wedge (k = 0) \wedge (x = 0)$. Suppose that r is true. Let $P(k)$ be "$x = mk$ and $k \leq a$." We can show that $P(k)$ is a loop invariant for the loop in S_3. $P(0)$ is true because before the loop is entered $x = 0 = m \cdot 0$ and $0 \leq a$. Now assume $P(k)$ is true and $k < a$. Then $P(k + 1)$ is true because x is assigned the value $x + m = mk + m = m(k + 1)$. The loop terminates when $k = a$, and at that point $x = ma$. Hence, $r\{S_3\}s$ is true where s is "$a = |n|$ and $x = ma$." Now assume that s is true. Then if $n < 0$ it follows that $a = -n$, so $x = -mn$. In this case S_4 assigns $-x = mn$ to *product*. If $n > 0$ then $x = ma = mn$, so S_4 assigns mn to *product*. Hence, $s\{S_4\}t$ is true. **13.** Suppose that the initial assertion p is true. Then because $p\{S\}q_0$ is true, q_0 is true after the segment S is executed. Because $q_0 \rightarrow q_1$ is true, it also follows that q_1 is true after S is executed. Hence, $p\{S\}q_1$ is true. **15.** We will use the proposition p, "gcd(a, b) = gcd(x, y) and $y \geq 0$," as the loop invariant. Note that p is true before the loop is entered, because at that point $x = a$, $y = b$, and y is a positive integer, using the initial assertion. Now assume that p is true and $y > 0$; then the loop will be executed again. Inside the loop, x and y are replaced by y and x **mod** y, respectively. By Lemma 1 of Section 4.3, gcd(x, y) = gcd(y, x **mod** y). Therefore, after execution of the loop, the value of gcd(x, y) is the same as it was before. Moreover, because y is the remainder, it is at least 0. Hence, p remains true, so it is a loop invariant. Furthermore, if the loop terminates, then $y = 0$. In this case, we have gcd(x, y) = x, the final assertion. Therefore, the program, which gives x as its output, has correctly computed gcd(a, b). Finally, we can prove the loop must terminate, because each iteration causes the value of y to decrease by at least 1. Therefore, the loop can be iterated at most b times.

Supplementary Exercises

1. Let $P(n)$ be the statement that this equation holds. *Basis step:* $P(1)$ says $2/3 = 1 - (1/3^1)$, which is true. *Inductive step:* Assume that $P(k)$ is true. Then $2/3+2/9+2/27+\cdots+2/3^n+2/3^{n+1} = 1-1/3^n+2/3^{n+1}$ (by the inductive hypothesis), and this equals $1-1/3^{n+1}$, as desired. **3.** Let $P(n)$ be "$1\cdot 1+2\cdot 2+\cdots+n\cdot 2^{n-1} = (n-1)2^n+1$." *Basis step:* $P(1)$ is true because $1\cdot 1 = 1 = (1-1)2^1+1$. *Inductive step:* Assume that $P(k)$ is true. Then $1\cdot 1+2\cdot 2+\cdots+k\cdot 2^{k-1}+(k+1)\cdot 2^k = (k-1)2^k+1+(k+1)2^k = 2k\cdot 2^k+1 = [(k+1)-1]2^{k+1}+1$. **5.** Let $P(n)$ be "$1/(1\cdot 4)+\cdots+1/[(3n-2)(3n+1)] = n/(3n+1)$." *Basis step:* $P(1)$ is true because $1/(1\cdot 4) = 1/4$. *Inductive step:* Assume $P(k)$ is true. Then $1/(1\cdot 4)+\cdots+1/[(3k-2)(3k+1)]+1/[(3k+1)(3k+4)] = k/(3k+1)+1/[(3k+1)(3k+4)] = [k(3k+4)+1]/[(3k+1)(3k+4)] = [(3k+1)(k+1)]/[(3k+1)(3k+4)] = (k+1)/(3k+4)$. **7.** Let $P(n)$ be "$2^n > n^3$." *Basis step:* $P(10)$ is true because $1024 > 1000$. *Inductive step:* Assume $P(k)$ is true. Then $(k+1)^3 = k^3+3k^2+3k+1 \leq k^3+9k^2 \leq k^3+k^3 = 2k^3 < 2\cdot 2^k = 2^{k+1}$. **9.** Let $P(n)$ be "$a-b$ is a factor of $a^n - b^n$." *Basis step:* $P(1)$ is trivially true. Assume $P(k)$ is true. Then $a^{k+1}-b^{k+1} = a^{k+1}-ab^k+ab^k-b^{k+1} = a(a^k-b^k)+b^k(a-b)$. Then because $a-b$ is a factor of a^k-b^k and $a-b$ is a factor of $a-b$, it follows that $a-b$ is a factor of $a^{k+1}-b^{k+1}$. **11.** *Basis step:* When $n = 1$, $6^{n+1}+7^{2n-1} = 36+7 = 43$. *Inductive step:* Assume the inductive hypothesis, that 43 divides $6^{n+1}+7^{2n-1}$; we must show that 43 divides $6^{n+2}+7^{2n+1}$. We have $6^{n+2}+7^{2n+1} = 6\cdot 6^{n+1}+49\cdot 7^{2n-1} = 6\cdot 6^{n+1}+6\cdot 7^{2n-1}+43\cdot 7^{2n-1} = 6(6^{n+1}+7^{2n-1})+43\cdot 7^{2n-1}$. By the inductive hypothesis the first term is divisible by 43, and the second term is divisible by 43; therefore, the sum is divisible by 43. **13.** Let $P(n)$ be "$a+(a+d)+\cdots+(a+nd) = (n+1)(2a+nd)/2$." *Basis step:* $P(1)$ is true because $a+(a+d) = 2a+d = 2(2a+d)/2$. *Inductive step:* Assume that $P(k)$ is true. Then $a+(a+d)+\cdots+(a+kd)+[a+(k+1)d] = (k+1)(2a+kd)/2+a+(k+1)d = \frac{1}{2}(2ak+2a+k^2d+kd+2a+2kd+2d) = \frac{1}{2}(2ak+4a+k^2d+3kd+2d) = \frac{1}{2}(k+2)[2a+(k+1)d]$. **15.** *Basis step:* This is true for $n = 1$ because $5/6 = 10/12$. *Inductive step:* Assume that the equation holds for $n = k$, and consider $n = k+1$. Then $\sum_{i=1}^{k+1} \frac{i+4}{i(i+1)(i+2)} = \sum_{i=1}^{k} \frac{i+4}{i(i+1)(i+2)} + \frac{k+5}{(k+1)(k+2)(k+3)} = \frac{k(3k+7)}{2(k+1)(k+2)} + \frac{k+5}{(k+1)(k+2)(k+3)}$ (by the inductive hypothesis) $= \frac{1}{(k+1)(k+2)}\cdot(\frac{k(3k+7)}{2}+\frac{k+5}{k+3}) = \frac{1}{2(k+1)(k+2)(k+3)}\cdot[k(3k+7)(k+3)+2(k+5)] = \frac{1}{2(k+1)(k+2)(k+3)}\cdot(3k^3+16k^2+23k+10) = \frac{1}{2(k+1)(k+2)(k+3)}\cdot(3k+10)(k+1)^2 = \frac{1}{2(k+2)(k+3)}\cdot(3k+10)(k+1) = \frac{(k+1)(3(k+1)+7)}{2((k+1)+1)((k+1)+2)}$, as desired. **17.** *Basis step:* The statement is true for $n = 1$ because the derivative of $g(x) = xe^x$ is $x\cdot e^x+e^x = (x+1)e^x$ by the product rule. *Inductive step:* Assume that the statement is true for $n = k$, i.e., the kth derivative is given by $g^{(k)} = (x+k)e^x$. Differentiating by the product rule gives the $(k+1)$st derivative: $g^{(k+1)} = (x+k)e^x+e^x = [x+(k+1)]e^x$, as desired. **19.** We will use strong induction to show that f_n is even if $n \equiv 0 \pmod 3$ and is odd otherwise. *Basis step:* This follows because $f_0 = 0$ is even and $f_1 = 1$ is odd. *Inductive step:* Assume that if $j \leq k$, then f_j is even if $j \equiv 0 \pmod 3$ and is odd otherwise. Now suppose $k+1 \equiv 0 \pmod 3$. Then $f_{k+1} = f_k + f_{k-1}$ is even because f_k and f_{k-1} are both odd. If $k+1 \equiv 1 \pmod 3$, then $f_{k+1} = f_k+f_{k-1}$ is odd because f_k is even and f_{k-1} is odd. Finally, if $k+1 \equiv 2 \pmod 3$, then $f_{k+1} = f_k+f_{k-1}$ is odd because f_k is odd and f_{k-1} is even. **21.** Let $P(n)$ be the statement that $f_kf_n+f_{k+1}f_{n+1} = f_{n+k+1}$ for every nonnegative integer k. *Basis step:* This consists of showing that $P(0)$ and $P(1)$ both hold. $P(0)$ is true because $f_kf_0+f_{k+1}f_1 = f_{k+1}\cdot 0+f_{k+1}\cdot 1 = f_1$. Because $f_kf_1+f_{k+1}f_2 = f_k+f_{k+1} = f_{k+2}$, it follows that $P(1)$ is true. *Inductive step:* Now assume that $P(j)$ holds. Then, by the inductive hypothesis and the recursive definition of the Fibonacci numbers, it follows that $f_{k+1}f_{j+1}+f_{k+2}f_{j+2} = f_k(f_{j-1}+f_j)+f_{k+1}(f_j+f_{j+1}) = (f_kf_{j-1}+f_{k+1}f_j)+(f_kf_j+f_{k+1}f_{j+1}) = f_{j-1+k+1}+f_{j+k+1} = f_{j+k+2}$. This shows that $P(j+1)$ is true. **23.** Let $P(n)$ be the statement $l_0^2+l_1^2+\cdots+l_n^2 = l_nl_{n+1}+2$. *Basis step:* $P(0)$ and $P(1)$ both hold because $l_0^2 = 2^2 = 2\cdot 1+2 = l_0l_1+2$ and $l_0^2+l_1^2 = 2^2+1^2 = 1\cdot 3+2 = l_1l_3+2$. *Inductive step:* Assume that $P(k)$ holds. Then by the inductive hypothesis $l_0^2+l_1^2+\cdots+l_k^2+l_{k+1}^2 = l_kl_{k+1}+2+l_{k+1}^2 = l_{k+1}(l_k+l_{k+1})+2 = l_{k+1}l_{k+2}+2$. This shows that $P(k+1)$ holds. **25.** Let $P(n)$ be the statement that the identity holds for the integer n. *Basis step:* $P(1)$ is obviously true. *Inductive step:* Assume that $P(k)$ is true. Then $\cos((k+1)x)+i\sin((k+1)x) = \cos(kx+x)+i\sin(kx+x) = \cos kx\cos x - \sin kx\sin x + i(\sin kx\cos x+\cos kx\sin x) = \cos x(\cos kx+i\sin kx)(\cos x+i\sin x) = (\cos x+i\sin x)^k(\cos x+i\sin x) = (\cos x+i\sin x)^{k+1}$. It follows that $P(k+1)$ is true. **27.** Rewrite the right-hand side as $2^{n+1}(n^2-2n+3)-6$. For $n = 1$ we have $2 = 4\cdot 2-6$. Assume that the equation holds for $n = k$, and consider $n = k+1$. Then $\sum_{j=1}^{k+1} j^2 2^j = \sum_{j=1}^{k} j^2 2^j+(k+1)^2 2^{k+1} = 2^{k+1}(k^2-2k+3)-6+(k^2+2k+1)2^{k+1}$ (by the inductive hypothesis) $= 2^{k+1}(2k^2+4)-6 = 2^{k+2}(k^2+2)-6 = 2^{k+2}[(k+1)^2-2(k+1)+3]-6$. **29.** Let $P(n)$ be the statement that this equation holds. *Basis step:* In $P(2)$ both sides reduce to $1/3$. *Inductive step:* Assume that $P(k)$ is true. Then $\sum_{j=1}^{k+1} 1/(j^2-1) = \left(\sum_{j=1}^{k} 1/(j^2-1)\right) + 1/[(k+1)^2-1] = (k-1)(3k+2)/[4k(k+1)] + 1/[(k+1)^2-1]$ by the inductive hypothesis. This simplifies to $(k-1)(3k+2)/[4k(k+1)]+1/(k^2+2k) = (3k^3+5k^2)/[4k(k+1)(k+2)] = \{[(k+1)-1][3(k+1)+2]\}/[4(k+1)(k+2)]$, which is exactly what $P(k+1)$ asserts. **31.** Let $P(n)$ be the assertion that at least $n+1$ lines are needed to cover the lattice points in the given triangular region. *Basis step:* $P(0)$ is true, because we need at least one line to cover the one point at $(0, 0)$. *Inductive step:* Assume the inductive hypothesis, that at least $k+1$ lines are needed to cover the lattice points with $x \geq 0$, $y \geq 0$, and $x + y \leq k$. Consider the triangle of lattice points defined by $x \geq 0$, $y \geq 0$, and $x + y \leq k + 1$. By way of contradiction, assume that $k+1$ lines could cover this set. Then these lines must cover the $k+2$ points on the line $x + y = k + 1$. But only the line $x + y = k + 1$ itself can cover more than one of these points, because two distinct lines intersect in at most one point. Therefore, none of the $k+1$ lines that are needed (by the inductive hypothesis) to cover the set of lattice points within the triangle but not on this line can cover more than one of the points on this line, and this leaves at least one point uncovered. Therefore, our assumption that $k+1$ lines could cover the larger set is wrong, and our proof is complete. **33.** Let $P(n)$ be $\mathbf{B}^k = \mathbf{M}\mathbf{A}^k\mathbf{M}^{-1}$. *Basis step:* Part of the given conditions. *Inductive step:* Assume the inductive hypothesis. Then $\mathbf{B}^{k+1} = \mathbf{B}\mathbf{B}^k = \mathbf{M}\mathbf{A}\mathbf{M}^{-1}\mathbf{B}^k = \mathbf{M}\mathbf{A}\mathbf{M}^{-1}\mathbf{M}\mathbf{A}^k\mathbf{M}^{-1}$ (by the inductive hypothesis) $= \mathbf{M}\mathbf{A}\mathbf{I}\mathbf{A}^k\mathbf{M}^{-1} = \mathbf{M}\mathbf{A}\mathbf{A}^k\mathbf{M}^{-1} = \mathbf{M}\mathbf{A}^{k+1}\mathbf{M}^{-1}$. **35.** We prove by mathematical induction the following stronger statement: For every $n \geq 3$, we can write $n!$ as the sum of n of its distinct positive divisors, one of which is 1. That is, we can write $n! = a_1 + a_2 + \cdots + a_n$, where each a_i is a divisor of $n!$, the divisors are listed in strictly decreasing order, and $a_n = 1$. *Basis step:* $3! = 3 + 2 + 1$. *Inductive step:* Assume that we can write $k!$ as a sum of the desired form, say $k! = a_1 + a_2 + \cdots + a_k$, where each a_i is a divisor of $n!$, the divisors are listed in strictly decreasing order, and $a_n = 1$. Consider $(k+1)!$. Then we have $(k+1)! = (k+1)k! = (k+1)(a_1 + a_2 + \cdots + a_k) = (k+1)a_1 + (k+1)a_2 + \cdots + (k+1)a_k = (k+1)a_1 + (k+1)a_2 + \cdots + k \cdot a_k + a_k$. Because each a_i was a divisor of $k!$, each $(k+1)a_i$ is a divisor of $(k+1)!$. Furthermore, $k \cdot a_k = k$, which is a divisor of $(k+1)!$, and $a_k = 1$, so the new last summand is again 1. (Notice also that our list of summands is still in strictly decreasing order.) Thus, we have written $(k+1)!$ in the desired form. **37.** When $n = 1$ the statement is vacuously true. Assume that the statement is true for $n = k$, and consider $k+1$ people standing in a line, with a woman first and a man last. If the kth person is a woman, then we have that woman standing in front of the man at the end. If the kth person is a man, then the first k people in line satisfy the conditions of the inductive hypothesis for the first k people in line, so again we can conclude that there is a woman directly in front of a man somewhere in the line. **39.** *Basis step:* When $n = 1$ there is one circle, and we can color the inside blue and the outside red to satisfy the conditions. *Inductive step:* Assume the inductive hypothesis that if there are k circles, then the regions can be 2-colored such that no regions with a common boundary have the same color, and consider a situation with $k+1$ circles. Remove one of the circles, producing a picture with k circles, and invoke the inductive hypothesis to color it in the prescribed manner. Then replace the removed circle and change the color of every region inside this circle. The resulting figure satisfies the condition, because if two regions have a common boundary, then either that boundary involved the new circle, in which case the regions on either side used to be the same region and now the inside portion is different from the outside, or else the boundary did not involve the new circle, in which case the regions are colored differently because they were colored differently before the new circle was restored. **41.** If $n = 1$ then the equation reads $1 \cdot 1 = 1 \cdot 2/2$, which is true. Assume that the equation is true for n and consider it for $n+1$. Then $\sum_{j=1}^{n+1}(2j-1)\left(\sum_{k=j}^{n+1}\frac{1}{k}\right) = \sum_{j=1}^{n}(2j-1)\left(\sum_{k=j}^{n+1}\frac{1}{k}\right) + [2(n+1)-1]\cdot\frac{1}{n+1} = \sum_{j=1}^{n}(2j-1)\left(\frac{1}{n+1} + \sum_{k=j}^{n}\frac{1}{k}\right) + \frac{2n+1}{n+1} = \left(\frac{1}{n+1}\sum_{j=1}^{n}(2j-1)\right) + \left(\sum_{j=1}^{n}(2j-1)\sum_{k=j}^{n}\frac{1}{k}\right) + \frac{2n+1}{n+1} = \left(\frac{1}{n+1}\cdot n^2\right) + \frac{n(n+1)}{2} + \frac{2n+1}{n+1}$ (by the inductive hypothesis) $= \frac{2n^2+n(n+1)^2+(4n+2)}{2(n+1)} = \frac{2(n+1)^2+n(n+1)^2}{2(n+1)} = \frac{(n+1)(n+2)}{2}$.

43. Let $T(n)$ be the statement that the sequence of towers of 2 is eventually constant modulo n. We use strong induction to prove that $T(n)$ is true for all positive integers n. *Basis step:* When $n = 1$ (and $n = 2$), the sequence of towers of 2 modulo n is the sequence of all 0s. *Inductive step:* Suppose that k is an integer with $k \geq 2$. Suppose that $T(j)$ is true for $1 \leq j \leq k-1$. In the proof of the inductive step we denote the rth term of the sequence modulo n by a_r. First suppose k is even. Let $k = 2^s q$ where $s \geq 1$ and $q < k$ is odd. When j is large enough, $a_{j-2} \geq s$, and for such j, $a_j = 2^{2^{a_{j-2}}}$ is a multiple of 2^s. It follows that for sufficiently large j, $a_j \equiv 0 \pmod{2^s}$. Hence, for large enough i, 2^s divides $a_{i+1} - a_i$. By the inductive hypothesis $T(q)$ is true, so the sequence $a_1, a_2, a_3, \ldots$ is eventually constant modulo q. This implies that for large enough i, q divides $a_{i+1} - a_i$. Because $\gcd(q, 2^s) = 1$ and for sufficiently large i both q and 2^s divide $a_{i+1} - a_i$, $k = 2^s q$ divides $a_{i+1} - a_i$ for sufficiently large i. Hence, for sufficiently large i, $a_{i+1} - a_i \equiv 0 \pmod{k}$. This means that the sequence is eventually constant modulo k. Finally, suppose k is odd. Then $\gcd(2, k) = 1$, so by Euler's theorem (found in elementary number theory books, such as [Ro10]), we know that $2^{\phi(k)} \equiv 1 \pmod{k}$. Let $r = \phi(k)$. Because $r < k$, by the inductive hypothesis $T(r)$, the sequence $a_1, a_2, a_3, \ldots$ is eventually constant modulo r, say equal to c. Hence, for large enough i, for some integer t_i, $a_i = t_i r + c$. Hence, $a_{i+1} = 2^{a_i} = 2^{t_i r + c} = (2^r)^{t_i} 2^c \equiv 2^c \pmod{k}$. This shows that $a_1, a_2, \ldots$ is eventually constant modulo k. **45. a)** 92 **b)** 91 **c)** 91 **d)** 91 **e)** 91 **f)** 91 **47.** The basis step is incorrect because $n \neq 1$ for the sum shown. **49.** Let $P(n)$ be "the plane is divided into $n^2 - n + 2$ regions by n circles if every two of these circles have two common points but no three have a common point." *Basis step:* $P(1)$ is true because a circle divides the plane into $2 = 1^2 - 1 + 2$ regions. *Inductive step:*

Assume that $P(k)$ is true, that is, k circles with the specified properties divide the plane into $k^2 - k + 2$ regions. Suppose that a $(k + 1)$st circle is added. This circle intersects each of the other k circles in two points, so these points of intersection form $2k$ new arcs, each of which splits an old region. Hence, there are $2k$ regions split, which shows that there are $2k$ more regions than there were previously. Hence, $k + 1$ circles satisfying the specified properties divide the plane into $k^2 - k + 2 + 2k = (k^2 + 2k + 1) - (k + 1) + 2 = (k + 1)^2 - (k + 1) + 2$ regions. **51.** Suppose $\sqrt{2}$ were rational. Then $\sqrt{2} = a/b$, where a and b are positive integers. It follows that the set $S = \{n\sqrt{2} \mid n \in \mathbf{N}\} \cap \mathbf{N}$ is a nonempty set of positive integers, because $b\sqrt{2} = a$ belongs to S. Let t be the least element of S, which exists by the well-ordering property. Then $t = s\sqrt{2}$ for some integer s. We have $t - s = s\sqrt{2} - s = s(\sqrt{2} - 1)$, so $t - s$ is a positive integer because $\sqrt{2} > 1$. Hence, $t - s$ belongs to S. This is a contradiction because $t - s = s\sqrt{2} - s < s$. Hence, $\sqrt{2}$ is irrational. **53. a)** Let $d = \gcd(a_1, a_2, \ldots, a_n)$. Then d is a divisor of each a_i and so must be a divisor of $\gcd(a_{n-1}, a_n)$. Hence, d is a common divisor of $a_1, a_2, \ldots, a_{n-2}$, and $\gcd(a_{n-1}, a_n)$. To show that it is the greatest common divisor of these numbers, suppose that c is a common divisor of them. Then c is a divisor of a_i for $i = 1, 2, \ldots, n - 2$ and a divisor of $\gcd(a_{n-1}, a_n)$, so it is a divisor of a_{n-1} and a_n. Hence, c is a common divisor of $a_1, a_2, \ldots, a_{n-1}$, and a_n. Hence, it is a divisor of d, the greatest common divisor of $a_1, a_2, \ldots, a_n$. It follows that d is the greatest common divisor, as claimed. **b)** If $n = 2$, apply the Euclidean algorithm. Otherwise, apply the Euclidean algorithm to a_{n-1} and a_n, obtaining $d = \gcd(a_{n-1}, a_n)$, and then apply the algorithm recursively to $a_1, a_2, \ldots, a_{n-2}, d$. **55.** $f(n) = n^2$. Let $P(n)$ be "$f(n) = n^2$." *Basis step:* $P(1)$ is true because $f(1) = 1 = 1^2$, which follows from the definition of f. *Inductive step:* Assume $f(n) = n^2$. Then $f(n + 1) = f((n + 1) - 1) + 2(n + 1) - 1 = f(n) + 2n + 1 = n^2 + 2n + 1 = (n + 1)^2$. **57. a)** λ, 0, 1, 00, 01, 11, 000, 001, 011, 111, 0000, 0001, 0011, 0111, 1111, 00000, 00001, 00011, 00111, 01111, 11111 **b)** $S = \{\alpha\beta \mid \alpha$ is a string of m 0s and β is a string of n 1s, $m \geq 0, n \geq 0\}$ **59.** Apply the first recursive step to λ to get () $\in B$. Apply the second recursive step to this string to get ()() $\in B$. Apply the first recursive step to this string to get (()()) $\in B$. By Exercise 62, (())) is not in B because the number of left parentheses does not equal the number of right parentheses. **61.** λ, (), (()), ()() **63. a)** 0 **b)** −2 **c)** 2 **d)** 0

65.

procedure *generate*(n: nonnegative integer)
if n is odd **then**
  $S := S(n - 1)$ {the S constructed by *generate*$(n - 1)$}
  $T := T(n - 1)$ {the T constructed by *generate*$(n - 1)$}
else if $n = 0$ **then**
  $S := \emptyset$
  $T := \{\lambda\}$
else
  $S' := S(n - 2)$ {the S constructed by *generate*$(n - 2)$}
  $T' := T(n - 2)$ {the T constructed by *generate*$(n - 2)$}
  $T := T' \cup \{(x) | x \in T' \cup S' \wedge \text{length}(x) = n - 2\}$
  $S := S' \cup \{xy | x \in T' \wedge y \in T' \cup S' \wedge \text{length}(xy) = n\}$
{$T \cup S$ is the set of balanced strings of length at most n}

67. If $x \leq y$ initially, then $x := y$ is not executed, so $x \leq y$ is a true final assertion. If $x > y$ initially, then $x := y$ is executed, so $x \leq y$ is again a true final assertion.

69. procedure *zerocount*($a_1, a_2, \ldots, a_n$: list of integers)
 if $n = 1$ **then**
  if $a_1 = 0$ **then return** 1
  else return 0
 else
  if $a_n = 0$ **then return** *zerocount*$(a_1, a_2, \ldots, a_{n-1}) + 1$
  else return *zerocount*$(a_1, a_2, \ldots, a_{n-1})$

71. We will prove that $a(n)$ is a natural number and $a(n) \leq n$. This is true for the base case $n = 0$ because $a(0) = 0$. Now assume that $a(n - 1)$ is a natural number and $a(n - 1) \leq n - 1$. Then $a(a(n - 1))$ is a applied to a natural number less than or equal to $n - 1$. Hence, $a(a(n - 1))$ is also a natural number less than or equal to $n - 1$. Therefore, $n - a(a(n - 1))$ is n minus some natural number less than or equal to $n - 1$, which is a natural number less than or equal to n. **73.** From Exercise 72, $a(n) = \lfloor (n + 1)\mu \rfloor$ and $a(n - 1) = \lfloor n\mu \rfloor$. Because $\mu < 1$, these two values are equal or they differ by 1. First suppose that $\mu n - \lfloor \mu n \rfloor < 1 - \mu$. This is equivalent to $\mu(n+1) < 1 + \lfloor \mu n \rfloor$. If this is true, then $\lfloor \mu(n+1) \rfloor = \lfloor \mu n \rfloor$. On the other hand, if $\mu n - \lfloor \mu n \rfloor \geq 1 - \mu$, then $\mu(n+1) \geq 1 + \lfloor \mu n \rfloor$, so $\lfloor \mu(n+1) \rfloor = \lfloor \mu n \rfloor + 1$, as desired. **75.** $f(0) = 1, m(0) = 0$; $f(1) = 1, m(1) = 0$; $f(2) = 2, m(2) = 1$; $f(3) = 2, m(3) = 2$; $f(4) = 3, m(4) = 2$; $f(5) = 3, m(5) = 3$; $f(6) = 4$, $m(6) = 4$; $f(7) = 5, m(7) = 4$; $f(8) = 5, m(8) = 5$; $f(9) = 6$, $m(9) = 6$ **77.** The last occurrence of n is in the position for which the total number of 1s, 2s, …, ns all together is that position number. But because a_k is the number of occurrences of k, this is just $\sum_{k=1}^{n} a_k$, as desired. Because $f(n)$ is the sum of the first n terms of the sequence, $f(f(n))$ is the sum of the first $f(n)$ terms of the sequence. But because $f(n)$ is the last term whose value is n, this means that the sum is the sum of all terms of the sequence whose value is at most n. Because there are a_k terms of the sequence whose value is k, this sum is $\sum_{k=1}^{n} k \cdot a_k$, as desired.

CHAPTER 6

Section 6.1

1. a) 5850 **b)** 343 **3. a)** 4^{10} **b)** 5^{10} **5.** 42 **7.** 26^3 **9.** 676 **11.** 2^8 **13.** n (not counting the empty string) **15.** 475,254 (not counting the empty string) **17.** 1,321,368,961 **19. a)** 729 **b)** 256 **c)** 1024 **d)** 64 **21. a)** Seven: 56, 63, 70, 77, 84, 91, 98 **b)** Five: 55, 66, 77, 88, 99 **c)** One: 77 **23. a)** 128 **b)** 450 **c)** 9 **d)** 675 **e)** 450 **f)** 450 **g)** 225 **h)** 75 **25. a)** 990 **b)** 500 **c)** 27 **27.** 3^{50} **29.** 52,457,600 **31.** 20,077,200 **33. a)** 37,822,859,361

b) 8,204,716,800 **c)** 40,159,050,880 **d)** 12,113,640,000 **e)** 171,004,205,215 **f)** 72,043,541,640 **g)** 6,230,721,635 **h)** 223,149,655 **35. a)** 0 **b)** 120 **c)** 720 **d)** 2520 **37. a)** 2 if $n = 1$, 2 if $n = 2$, 0 if $n \geq 3$ **b)** 2^{n-2} for $n > 1$; 1 if $n = 1$ **c)** $2(n - 1)$ **39.** $(n + 1)^m$ **41.** If n is even, $2^{n/2}$; if n is odd, $2^{(n+1)/2}$ **43. a)** 175 **b)** 248 **c)** 232 **d)** 84 **45.** 40 **47.** 60 **49. a)** 240 **b)** 480 **c)** 360 **51.** 352 **53.** 147 **55.** 33 **57. a)** 9,920,671,339,261,325,541,376 $\approx 9.9 \times 10^{21}$ **b)** 6,641,514,961,387,068,437,760 $\approx 6.6 \times 10^{21}$ **c)** About 314,000 years **59.** $54(64^{65535} - 1)/63 - 53$ **61.** $26(37^{31} - 1)/36$ **63.** 7,104,000,000,000 **65. a)** $16^{10} + 16^{26} + 16^{32} + 16^{58}$ **b)** 16^{64} **67.** 666,667 **69.** 18 **71.** 17 **73.** 22 **75.** $2n-2, 2n-1$ **77.** Let $P(m)$ be the sum rule for m tasks. For the basis case take $m = 2$. This is just the sum rule for two tasks. Now assume that $P(m)$ is true. Consider $m + 1$ tasks, $T_1, T_2, \ldots, T_m, T_{m+1}$, which can be done in $n_1, n_2, \ldots, n_m, n_{m+1}$ ways, respectively, such that no two of these tasks can be done at the same time. To do one of these tasks, we can either do one of the first m of these or do task T_{m+1}. By the sum rule for two tasks, the number of ways to do this is the sum of the number of ways to do one of the first m tasks, plus n_{m+1}. By the inductive hypothesis, this is $n_1 + n_2 + \cdots + n_m + n_{m+1}$, as desired. **79.** $n(n-3)/2$

Section 6.2

1. Because there are six classes, but only five weekdays, the pigeonhole principle shows that at least two classes must be held on the same day. **3. a)** 3 **b)** 14 **5.** 85 **7.** Because there are four possible remainders when an integer is divided by 4, the pigeonhole principle implies that given five integers, at least two have the same remainder. **9.** Let $a, a + 1, \ldots, a + n - 1$ be the integers in the sequence. The integers $(a + i)$ **mod** n, $i = 0, 1, 2, \ldots, n - 1$, are distinct, because $0 < (a + j) - (a + k) < n$ whenever $0 \leq k < j \leq n - 1$. Because there are n possible values for $(a+i)$ **mod** n and there are n different integers in the set, each of these values is taken on exactly once. It follows that there is exactly one integer in the sequence that is divisible by n. **11.** 4951 **13.** The midpoint of the segment joining the points (a, b, c) and (d, e, f) is $((a+d)/2, (b+e)/2, (c+f)/2)$. It has integer coefficients if and only if a and d have the same parity, b and e have the same parity, and c and f have the same parity. Because there are eight possible triples of parity [such as $(even, odd, even)$], by the pigeonhole principle at least two of the nine points have the same triple of parities. The midpoint of the segment joining two such points has integer coefficients. **15. a)** Group the first eight positive integers into four subsets of two integers each so that the integers of each subset add up to 9: $\{1, 8\}$, $\{2, 7\}$, $\{3, 6\}$, and $\{4, 5\}$. If five integers are selected from the first eight positive integers, by the pigeonhole principle at least two of them come from the same subset. Two such integers have a sum of 9, as desired. **b)** No. Take $\{1, 2, 3, 4\}$, for example. **17.** 4 **19.** 21,251 **21. a)** If there were fewer than 9 freshmen, fewer than 9 sophomores, and fewer than 9 juniors in the class, there would be no more than 8 with each of these three class standings, for a total of at most 24 students, contradicting the fact that there are 25 students in the class. **b)** If there were fewer than 3 freshmen, fewer than 19 sophomores, and fewer than 5 juniors, then there would be at most 2 freshmen, at most 18 sophomores, and at most 4 juniors, for a total of at most 24 students. This contradicts the fact that there are 25 students in the class. **23.** 4, 3, 2, 1, 8, 7, 6, 5, 12, 11, 10, 9, 16, 15, 14, 13 **25.** Number the seats around the table from 1 to 50, and think of seat 50 as being adjacent to seat 1. There are 25 seats with odd numbers and 25 seats with even numbers. If no more than 12 meat eaters occupied the odd-numbered seats, then at least 13 meat eaters would occupy the even-numbered seats, and vice versa. Without loss of generality, assume that at least 13 meat eaters occupy the 25 odd-numbered seats. Then at least two of those meat eaters must be in consecutive odd-numbered seats, and the person sitting between them will have meat eaters as both of their neighbors.

27.
```
procedure long(a_1, ..., a_n: positive integers)
  {first find longest increasing subsequence}
  max := 0; set := 00...00 {n bits}
  for i := 1 to 2^n
    last := 0; count := 0, OK := true
    for j := 1 to n
      if set(j) = 1 then
        if a_j > last then last := a_j
          count := count + 1
      else OK := false
    if count > max then
      max := count
      best := set
    set := set + 1 (binary addition)
  {max is length and best indicates the sequence}
  {repeat for decreasing subsequence with only
    changes being a_j < last instead of a_j > last
    and last := ∞ instead of last := 0}
```

29. By symmetry we need prove only the first statement. Let A be one of the people. Either A has at least four friends, or A has at least six enemies among the other nine people (because $3 + 5 < 9$). Suppose, in the first case, that B, C, D, and E are all A's friends. If any two of these are friends with each other, then we have found three mutual friends. Otherwise $\{B, C, D, E\}$ is a set of four mutual enemies. In the second case, let $\{B, C, D, E, F, G\}$ be a set of enemies of A. By Example 13, among B, C, D, E, F, and G there are either three mutual friends or three mutual enemies, who form, with A, a set of four mutual enemies. **31.** We need to show two things: that if we have a group of n people, then among them we must find either a pair of friends or a subset of n of them all of whom are mutual enemies; and that there exists a group of $n - 1$ people for which this is not possible. For the first statement, if there is any pair of friends, then the condition is satisfied, and if not, then every pair of people are enemies, so the second condition is satisfied. For the second statement, if we have a group of $n - 1$ people all of whom are enemies of each other, then there is neither a pair of friends nor a subset of n of them all of whom are mutual enemies. **33.** There are 6,432,816 possibilities for the three initials and a birthday. So, by the generalized pigeonhole principle, there are at least $\lceil 39{,}000{,}000/6{,}432{,}816 \rceil = 7$ people who share the same initials and birthday. **35.** Because $800{,}001 > 200{,}000$, the pigeonhole principle guarantees that there are at least two Parisians with the same number of hairs on their heads. The generalized pigeonhole principle guarantees that there are at least $\lceil 800{,}001/200{,}000 \rceil = 5$ Parisians with the same number of hairs on their heads. **37.** 18 **39.** Because there are six computers, the number of other computers a computer is connected to is an integer between 0 and 5, inclusive. However, 0 and 5 cannot both occur. To see this, note that if some computer is connected to no others, then no computer is connected to all five others, and if some computer is connected to all five others, then no computer is connected to no others. Hence, by the pigeonhole principle, because there are at most five possibilities for the number of computers a computer is connected to, there are at least two computers in the set of six connected to the same number of others. **41.** Label the computers C_1 through C_{100}, and label the printers P_1 through P_{20}. If we connect C_k to P_k for $k = 1, 2, \ldots, 20$ and connect each of the computers C_{21} through C_{100} to *all* the printers, then we have used a total of $20 + 80 \cdot 20 = 1620$ cables. This is sufficient, because if computers C_1 through C_{20} need printers, then they can use the printers with the same subscripts, and if any computers with higher subscripts need a printer instead of one or more of these, then they can use the printers that are not being used, because they are connected to all the printers. Now we must show that 1619 cables is not enough. Because there are 1619 cables and 20 printers, the average number of computers per printer is $1619/20$, which is less than 81. Therefore, some printer must be connected to fewer than 81 computers. That means it is connected to 80 or fewer computers, so there are 20 computers that are not connected to it. If those 20 computers all needed a printer simultaneously, then they would be out of luck, because they are connected to at most the 19 other printers. **43.** Let a_i be the number of matches completed by hour i. Then $1 \le a_1 < a_2 < \cdots < a_{75} \le 125$. Also $25 \le a_1 + 24 < a_2 + 24 < \cdots < a_{75} + 24 \le 149$. There are 150 numbers $a_1, \ldots, a_{75}, a_1 + 24, \ldots, a_{75} + 24$. By the pigeonhole principle, at least two are equal. Because all the a_is are distinct and all the $(a_i + 24)$s are distinct, it follows that $a_i = a_j + 24$ for some $i > j$. Thus, in the period from the $(j+1)$st to the ith hour, there are exactly 24 matches. **45.** Use the generalized pigeonhole principle, placing the $|S|$ objects $f(s)$ for $s \in S$ in $|T|$ boxes, one for each element of T. **47.** Let d_j be $jx - N(jx)$, where $N(jx)$ is the integer closest to jx for $1 \le j \le n$. Each d_j is an irrational number between $-1/2$ and $1/2$. We will assume that n is even; the case where n is odd is messier. Consider the n intervals $\{x \mid j/n < x < (j+1)/n\}$, $\{x \mid -(j+1)/n < x < -j/n\}$ for $j = 0, 1, \ldots, (n/2) - 1$. If d_j belongs to the interval $\{x \mid 0 < x < 1/n\}$ or to the interval $\{x \mid -1/n < x < 0\}$ for some j, we are done. If not, because there are $n-2$ intervals and n numbers d_j, the pigeonhole principle tells us that there is an interval $\{x \mid (k-1)/n < x < k/n\}$ containing d_r and d_s with $r < s$. The proof can be finished by showing that $(s - r)x$ is within $1/n$ of its nearest integer. **49. a)** Assume that $i_k \le n$ for all k. Then by the generalized pigeonhole principle, at least $\lceil (n^2 + 1)/n \rceil = n + 1$ of the numbers $i_1, i_2, \ldots, i_{n^2+1}$ are equal. **b)** If $a_{k_j} < a_{k_{j+1}}$, then the subsequence consisting of a_{k_j} followed by the increasing subsequence of length $i_{k_{j+1}}$ starting at $a_{k_{j+1}}$ contradicts the fact that $i_{k_j} = i_{k_{j+1}}$. Hence, $a_{k_j} > a_{k_{j+1}}$. **c)** If there is no increasing subsequence of length greater than n, then parts (a) and (b) apply. Therefore, we have $a_{k_{n+1}} > a_{k_n} > \cdots > a_{k_2} > a_{k_1}$, a decreasing sequence of length $n + 1$.

Section 6.3

1. *abc, acb, bac, bca, cab, cba* **3.** 720 **5. a)** 120 **b)** 720 **c)** 8 **d)** 6720 **e)** 40,320 **f)** 3,628,800 **7.** 15,120 **9.** 1320 **11. a)** 210 **b)** 386 **c)** 848 **d)** 252 **13.** $2(n!)^2$ **15.** 65,780 **17.** $2^{100} - 5051$ **19. a)** 1024 **b)** 45 **c)** 176 **d)** 252 **21. a)** 120 **b)** 24 **c)** 120 **d)** 24 **e)** 6 **f)** 0 **23.** 609,638,400 **25. a)** 17,280 **b)** 14,400 **27. a)** 94,109,400 **b)** 941,094 **c)** 3,764,376 **d)** 90,345,024 **e)** 114,072 **f)** 2328 **g)** 24 **h)** 79,727,040 **i)** 3,764,376 **j)** 109,440 **29. a)** 12,650 **b)** 303,600 **31. a)** 37,927 **b)** 18,915 **33. a)** 122,523,030 **b)** 72,930,375 **c)** 223,149,655 **d)** 100,626,625 **35.** 54,600 **37.** 45 **39.** 912 **41.** 11,232,000 **43.** $\frac{n!}{r(n-r)!}$ **45.** 13 **47.** 873

Section 6.4

1. $x^4 + 4x^3y + 6x^2y^2 + 4xy^3 + y^4$ **3.** $x^6 + 6x^5y + 15x^4y^2 + 20x^3y^3 + 15x^2y^4 + 6xy^5 + y^6$ **5.** 101 **7.** $-2^{10}\binom{19}{9} = -94{,}595{,}072$ **9.** $-2^{101}3^{99}\binom{200}{99}$ **11.** $\sum_{j=0}^{5}\binom{5}{j}(3x^4)^{5-j}(-2y^3)^j = 243x^{20} - 810x^{16}y^3 + 1080x^{12}y^6 - 720x^8y^9 + 240x^4y^{12} - 32y^{15}$ **13. a)** 71,680 **b)** 0 **c)** $-16{,}384$ **d)** $-35{,}840$ **e)** $-1{,}792$ **15.** $(-1)^{(200-k)/3}\binom{100}{(200-k)/3}$ if $k \equiv 2 \pmod 3$ and $-100 \le k \le 200$; 0 otherwise **17.** 1 9 36 84 126 126 84 36 9 1 **19.** The sum of *all* the positive numbers $\binom{n}{k}$, as k runs from 0 to n, is 2^n, so each one of them is no bigger than this sum. **21.** $\binom{n}{k} = \frac{n(n-1)(n-2)\cdots(n-k+1)}{k(k-1)(k-2)\cdots 2} \le \frac{n \cdot n \cdot \cdots \cdot n}{2 \cdot 2 \cdot \cdots \cdot 2} = n^k/2^{k-1}$ **23.** $\binom{n}{k-1} + \binom{n}{k} = \frac{n!}{(k-1)!(n-k+1)!} + \frac{n!}{k!(n-k)!} = \frac{n!}{k!(n-k+1)!} \cdot [k + (n - k + 1)] = \frac{(n+1)!}{k!(n+1-k)!} = \binom{n+1}{k}$ **25. a)** We show that each side counts the number of ways to choose from a set with n elements a subset with k elements and a distinguished element of that set. For the left-hand side, first choose the k-set (this can be done in $\binom{n}{k}$ ways) and then choose one of the k elements in this subset to be the distinguished element (this can be done in k ways). For the right-hand side, first choose the distinguished element out of the entire n-set (this can be done in n ways), and then choose the remaining $k - 1$ elements of the subset from the remaining $n - 1$ elements of the set (this can be done in $\binom{n-1}{k-1}$ ways). **b)** $k\binom{n}{k} = k \cdot \frac{n!}{k!(n-k)!} = \frac{n\cdot(n-1)!}{(k-1)!(n-k)!} = n\binom{n-1}{k-1}$ **27.** $\binom{n+1}{k} = \frac{(n+1)!}{k!(n+1-k)!} = \frac{(n+1)}{k}\frac{n!}{(k-1)![n-(k-1)]!} = (n+1)\binom{n}{k-1}/k$. This identity together with $\binom{n}{0} = 1$ gives a recursive definition. **29.** $\binom{2n}{n+1} + \binom{2n}{n} = \binom{2n+1}{n+1} = \frac{1}{2}\left[\binom{2n+1}{n+1} + \binom{2n+1}{n+1}\right] = \frac{1}{2}\left[\binom{2n+1}{n+1} + \binom{2n+1}{n}\right] = \frac{1}{2}\binom{2n+2}{n+1}$ **31. a)** $\binom{n+r+1}{r}$ counts the number of ways to choose a sequence of r 0s and $n+1$ 1s by choosing the positions of the 0s. Alternately, suppose that the $(j+1)$st term is the last term equal to 1, so that $n \le j \le n+r$. Once we have determined where the last 1 is, we decide where the 0s are to be placed in the j spaces before the last 1. There are n 1s and $j-n$ 0s in this range. By the sum rule it follows that there are $\sum_{j=n}^{n+r}\binom{j}{j-n} = \sum_{k=0}^{r}\binom{n+k}{k}$ ways to do this. **b)** Let $P(r)$ be the statement to be proved. The basis step is the equation $\binom{n}{0} = \binom{n+1}{0}$, which is just $1 = 1$. Assume that $P(r)$ is true. Then $\sum_{k=0}^{r+1}\binom{n+k}{k} = \sum_{k=0}^{r}\binom{n+k}{k} + \binom{n+r+1}{r+1} = \binom{n+r+1}{r} + \binom{n+r+1}{r+1} = \binom{n+r+2}{r+1}$, using the inductive hypothesis and Pascal's identity. **33.** We can choose the leader first in n different ways. We can then choose the rest of the committee in 2^{n-1} ways. Hence, there are $n2^{n-1}$ ways to choose the committee and its leader. Meanwhile, the number of ways to select a committee with k people is $\binom{n}{k}$. Once we have chosen a committee with k people, there are k ways to choose its leader. Hence, there are $\sum_{k=1}^{n} k\binom{n}{k}$ ways to choose the committee and its leader. Hence, $\sum_{k=1}^{n} k\binom{n}{k} = n2^{n-1}$. **35.** Let the set have n elements. From Corollary 2 we have $\binom{n}{0} - \binom{n}{1} + \binom{n}{2} - \cdots + (-1)^n\binom{n}{n} = 0$. It follows that $\binom{n}{0} + \binom{n}{2} + \binom{n}{4} + \cdots = \binom{n}{1} + \binom{n}{3} + \binom{n}{5} + \cdots$. The left-hand side gives the number of subsets with an even number of elements, and the right-hand side gives the number of subsets with an odd number of elements. **37. a)** A path of the desired type consists of m moves to the right and n moves up. Each such path can be represented by a bit string of length $m+n$ with m 0s and n 1s, where a 0 represents a move to the right and a 1 a move up. **b)** The number of bit strings of length $m+n$ containing exactly n 1s equals $\binom{m+n}{n} = \binom{m+n}{m}$ because such a string is determined by specifying the positions of the n 1s or by specifying the positions of the m 0s. **39.** By Exercise 37 the number of paths of length n of the type described in that exercise equals 2^n, the number of bit strings of length n. On the other hand, a path of length n of the type described in Exercise 37 must end at a point that has n as the sum of its coordinates, say $(n-k, k)$ for some k between 0 and n, inclusive. By Exercise 37, the number of such paths ending at $(n-k, k)$ equals $\binom{n-k+k}{k} = \binom{n}{k}$. Hence, $\sum_{k=0}^{n}\binom{n}{k} = 2^n$. **41.** By Exercise 37 the number of paths from $(0, 0)$ to $(n+1, r)$ of the type described in that exercise equals $\binom{n+r+1}{r}$. But such a path starts by going j steps vertically for some j with $0 \le j \le r$. The number of these paths beginning with j vertical steps equals the number of paths of the type described in Exercise 37 that go from $(1, j)$ to $(n+1, r)$. This is the same as the number of such paths that go from $(0, 0)$ to $(n, r-j)$, which by Exercise 37 equals $\binom{n+r-j}{r-j}$. Because $\sum_{j=0}^{r}\binom{n+r-j}{r-j} = \sum_{k=0}^{r}\binom{n+k}{k}$, it follows that $\sum_{k=1}^{r}\binom{n+k}{k} = \binom{n+r-1}{r}$. **43. a)** $\binom{n+1}{2}$ **b)** $\binom{n+2}{3}$ **c)** $\binom{2n-2}{n-1}$ **d)** $\binom{n-1}{\lfloor(n-1)/2\rfloor}$ **e)** Largest odd entry in nth row of Pascal's triangle **f)** $\binom{3n-3}{n-1}$

Section 6.5

1. 243 **3.** 26^6 **5.** 125 **7.** 35 **9. a)** 1716 **b)** 50,388 **c)** 2,629,575 **d)** 330 **11.** 9 **13.** 4,504,501 **15. a)** 10,626 **b)** 1,365 **c)** 11,649 **d)** 106 **17.** 2,520 **19.** 302,702,400 **21. a)** 169 **b)** 156 **c)** 78 **d)** 91 **23.** 3003 **25.** 7,484,400 **27.** 30,492 **29.** $C(59, 50)$ **31.** 35 **33.** 83,160 **35.** 63 **37.** 19,635 **39.** 210 **41.** 27,720 **43.** $52!/(7!^5 17!)$ **45.** Approximately 6.5×10^{32} **47. a)** $C(k+n-1, n)$ **b)** $(k+n-1)!/(k-1)!$ **49.** There are $C(n, n_1)$ ways to choose n_1 objects for the first box. Once these objects are chosen, there are $C(n-n_1, n_2)$ ways to choose objects for the second box. Similarly, there are $C(n-n_1-n_2, n_3)$ ways to choose objects for the third box. Continue in this way until there is $C(n-n_1-n_2-\cdots-n_{k-1}, n_k) = C(n_k, n_k) = 1$ way to choose the objects for the last box (because $n_1+n_2+\cdots+n_k = n$). By the product rule, the number of ways to make the entire assignment is $C(n, n_1)C(n-n_1, n_2)C(n-n_1-n_2, n_3)\cdots C(n-n_1-n_2-\cdots-n_{k-1}, n_k)$, which equals $n!/(n_1!n_2!\cdots n_k!)$, as straightforward simplification shows. **51. a)** Because $x_1 \le x_2 \le \cdots \le x_r$, it follows that $x_1 + 0 < x_2 + 1 < \cdots < x_r + r - 1$. The inequalities are strict because $x_j + j - 1 < x_{j+1} + j$ as long as $x_j \le x_{j+1}$. Because $1 \le x_j \le n + r - 1$, this sequence is made up of r distinct elements from T. **b)** Suppose that $1 \le x_1 < x_2 < \cdots < x_r < n + r - 1$. Let $y_k = x_k - (k-1)$. Then it is not hard to see that $y_k \le y_{k+1}$ for $k = 1, 2, \ldots, r-1$ and that $1 \le y_k \le n$ for $k = 1, 2, \ldots r$. It follows that $\{y_1, y_2, \ldots, y_r\}$ is an r-combination with repetitions allowed of S. **c)** From parts (a) and (b) it follows that there is a one-to-one correspondence of r-combinations with repetitions allowed of S and r-combinations of T, a set with $n+r-1$ elements. We conclude that there are $C(n+r-1, r)$ r-combinations with repetitions allowed of S. **53.** 65 **55.** 65 **57.** 2 **59.** 3 **61. a)** 150 **b)** 25 **c)** 6 **d)** 2 **63.** 90,720 **65.** The terms in the expansion are of the form $x_1^{n_1}x_2^{n_2}\cdots x_m^{n_m}$, where $n_1+n_2+\cdots+n_m = n$. Such a term arises from choosing the x_1 in n_1 factors, the x_2 in n_2 factors, $\ldots$, and the x_m in n_m factors. This can be done in $C(n; n_1, n_2, \ldots, n_m)$ ways, because a choice is a permutation of n_1 labels "1," n_2 labels "2," $\ldots$, and n_m labels "m." **67.** 2520 **69.** For given positive integers n and k, $S(n, k)$ is the number of set partitions of $\{1, 2, \ldots, n\}$ into k nonempty subsets. Any such partition either does or does not include the singleton $\{n\}$ as one of the blocks. There are $S(n-1, k-1)$ partitions of $\{1, 2, \ldots, n\}$ that do include $\{n\}$ as a block, since any such partition is a partition of $\{1, 2, \ldots, n-1\}$ into $k-1$ blocks together with $\{n\}$ as an separate block. And there are $kS(n-1, k)$ partitions of $\{1, 2, \ldots, n\}$ into k blocks that do not have $\{n\}$ as a block, since each partition of $\{1, 2, \ldots, n-1\}$ into k blocks gives rise to k different partitions of $\{1, 2, \ldots, n\}$

by choosing one of the existing blocks to insert n into. Thus, $S(n-1,k-1)+kS(n-1,k)$ counts the number of partitions of n into k blocks and hence equals $S(n,k)$. **71.** $n = 1$: $\{\{1\}\}$, $B(1) = 1$; $n = 2$: $\{\{1,2\}\}$, $\{\{1\},\{2\}\}$, $B(2) = 2$; $n = 3$: $\{\{1,2,3\}\}$, $\{\{1,2\},\{3\}\}$, $\{\{1,3\},\{2\}\}$, $\{\{1\},\{2,3\}\}$, $\{\{1\},\{2\},\{3\}\}$, $B(3) = 5$; $n = 4$: $\{\{1,2,3,4\}\}$, $\{\{1,2,3\},\{4\}\}$, $\{\{1,2,4\},\{3\}\}$, $\{\{1,2\},\{3,4\}\}$, $\{\{1,2\},\{3\},\{4\}\}$, $\{\{1,3,4\},\{2\}\}$, $\{\{1,3\},\{2,4\}\}$, $\{\{1,3\},\{2\},\{4\}\}$, $\{\{1,4\},\{2,3\}\}$, $\{\{1\},\{2,3,4\}\}$, $\{\{1\},\{2,3\},\{4\}\}$, $\{\{1,4\},\{2\},\{3\}\}$, $\{\{1\},\{2,4\},\{3\}\}$, $\{\{1\},\{2\},\{3,4\}\}$, $\{\{1\},\{2\},\{3\},\{4\}\}$, $B(4) = 15$

Section 6.6

1. 14532, 15432, 21345, 23451, 23514, 31452, 31542, 43521, 45213, 45321 **3.** AAA1, AAA2, AAB1, AAB2, AAC1, AAC2, ABA1, ABA2, ABB1, ABB2, ABC1, ABC2, ACA1, ACA2, ACB1, ACB2, ACC1, ACC2, BAA1, BAA2, BAB1, BAB2, BAC1, BAC2, BBA1, BBA2, BBB1, BBB2, BBC1, BBC2, BCA1, BCA2, BCB1, BCB2, BCC1, BCC2, CAA1, CAA2, CAB1, CAB2, CAC1, CAC2, CBA1, CBA2, CBB1, CBB2, CBC1, CBC2, CCA1, CCA2, CCB1, CCB2, CCC1, CCC2 **5. a)** 2134 **b)** 54132 **c)** 12534 **d)** 45312 **7.** 1234, 1243, 1324, 1342, 1423, 1432, 2134, 2143, 2314, 2341, 2413, 2431, 3124, 3142, 3214, 3241, 3412, 3421, 4123, 4132, 4213, 4231, 4312, 4321 **9.** $\{1,2,3\}$, $\{1,2,4\}$, $\{1,2,5\}$, $\{1,3,4\}$, $\{1,3,5\}$, $\{1,4,5\}$, $\{2,3,4\}$, $\{2,3,5\}$, $\{2,4,5\}$, $\{3,4,5\}$ **11.** The bit string representing the next larger r-combination must differ from the bit string representing the original one in position i because positions $i+1, \ldots, r$ are occupied by the largest possible numbers. Also $a_i + 1$ is the smallest possible number we can put in position i if we want a combination greater than the original one. Then $a_i+2, \ldots, a_i+r-i+1$ are the smallest allowable numbers for positions $i+1$ to r. Thus, we have produced the next r-combination. **13.** 123, 132, 213, 231, 312, 321, 124, 142, 214, 241, 412, 421, 125, 152, 215, 251, 512, 521, 134, 143, 314, 341, 413, 431, 135, 153, 315, 351, 513, 531, 145, 154, 415, 451, 514, 541, 234, 243, 324, 342, 423, 432, 235, 253, 325, 352, 523, 532, 245, 254, 425, 452, 524, 542, 345, 354, 435, 453, 534, 543 **15.** We will show that it is a bijection by showing that it has an inverse. Given a positive integer less than $n!$, let $a_1, a_2, \ldots, a_{n-1}$ be its Cantor digits. Put n in position $n-a_{n-1}$; then a_{n-1} is the number of integers less than n that follow n in the permutation. Then put $n-1$ in free position $(n-1)-a_{n-2}$, where we have numbered the free positions $1, 2, \ldots, n-1$ (excluding the position that n is already in). Continue until 1 is placed in the only free position left. Because we have constructed an inverse, the correspondence is a bijection.
17.

```
procedure Cantor permutation(n, i: integers with
    n ≥ 1 and 0 ≤ i < n!)
  x := n
  for j := 1 to n
    p_j := 0
  for k := 1 to n − 1
    c := ⌊x/(n − k)!⌋; x := x − c(n − k)!; h := n
    while p_h ≠ 0
      h := h − 1
    for j := 1 to c
      h := h − 1
      while p_h ≠ 0
        h := h − 1
    p_h := n − k + 1
  h := 1
  while p_h ≠ 0
    h := h + 1
  p_h := 1
  {p_1 p_2 ... p_n is the permutation corresponding to i}
```

Supplementary Exercises

1. a) 151,200 **b)** 1,000,000 **c)** 210 **d)** 5005 **3.** 3^{100} **5.** 24,600 **7. a)** 4060 **b)** 2688 **c)** 25,009,600 **9. a)** 192 **b)** 301 **c)** 300 **d)** 300 **11.** 639 **13.** The maximum possible sum is 240, and the minimum possible sum is 15. So the number of possible sums is 226. Because there are 252 subsets with five elements of a set with 10 elements, by the pigeonhole principle it follows that at least two have the same sum. **15. a)** 50 **b)** 50 **c)** 14 **d)** 17 **17.** Let $a_1, a_2, \ldots, a_m$ be the integers, and let $d_i = \sum_{j=1}^{i} a_j$. If $d_i \equiv 0 \pmod{m}$ for some i, we are done. Otherwise d_1 **mod** m, d_2 **mod** $m, \ldots, d_m$ **mod** m are m integers with values in $\{1, 2, \ldots, m-1\}$. By the pigeonhole principle $d_k = d_l$ for some $1 \le k < l \le m$. Then $\sum_{j=k+1}^{l} a_j = d_l - d_k \equiv 0 \pmod{m}$. **19.** The decimal expansion of the rational number a/b can be obtained by division of b into a, where a is written with a decimal point and an arbitrarily long string of 0s following it. The basic step is finding the next digit of the quotient, namely, $\lfloor r/b \rfloor$, where r is the remainder with the next digit of the dividend brought down. The current remainder is obtained from the previous remainder by subtracting b times the previous digit of the quotient. Eventually the dividend has nothing but 0s to bring down. Furthermore, there are only b possible remainders. Thus, at some point, by the pigeonhole principle, we will have the same situation as had previously arisen. From that point onward, the calculation must follow the same pattern. In particular, the quotient will repeat. **21. a)** 125,970 **b)** 20 **c)** 141,120,525 **d)** 141,120,505 **e)** 177,100 **f)** 141,078,021 **23. a)** 10 **b)** 8 **c)** 7 **25.** 3^n **27.** $C(n+2, r+1) = C(n+1, r+1) + C(n+1, r) = 2C(n+1, r+1) - C(n+1, r+1) + C(n+1, r) = 2C(n+1, r+1) - (C(n, r+1) + C(n, r)) + (C(n, r) + C(n, r-1)) = 2C(n+1, r+1) - C(n, r+1) + C(n, r-1)$ **29.** Substitute $x = 1$ and $y = 3$ into the binomial theorem. **31.** Both sides count the number of ways to choose a subset of three distinct numbers $\{i, j, k\}$ with $i < j < k$ from $\{1, 2, \ldots, n\}$. **33.** $C(n+1, 5)$ **35.** 3,491,888,400 **37.** 5^{24} **39. a)** 45 **b)** 57 **c)** 12 **41. a)** 386 **b)** 56 **43.** 0 if $n < m$; $C(n-1, n-m)$ if $n \geq m$ **45. a)** 15,625 **b)** 202 **c)** 210 **d)** 10 **47. a)** 3 **b)** 11 **c)** 6 **d)** 10 **49.** There are two possibilities: three people seated at one table with everyone else sitting alone, which can be done in $2C(n, 3)$ ways (choose the three people and seat them in one of two arrangements), or two groups of two people seated together with everyone else sitting alone, which can be done in $3C(n, 4)$ ways (choose four people and then choose one of the three ways to pair them up). Both $2C(n, 3) + 3C(n, 4)$ and $(3n - 1)C(n, 3)/4$ equal $n^4/8 - 5n^3/12 + 3n^2/8 - n/12$. **51.** The number of permutations of $2n$ objects of n different types, two of each type, is $(2n)!/2^n$. Because this must be an integer, the denominator must divide the numerator. **53.** CCGGUCCGAAAG **55.**

```
procedure next permutation(n: positive integer,
    a_1, a_2, ..., a_r: positive integers not exceeding
    n with a_1a_2 ... a_r ≠ nn ... n)
  i := r
  while a_i = n
    a_i := 1
    i := i − 1
  a_i := a_i + 1
  {a_1a_2 ... a_r is the next permutation in lexicographic
    order}
```

57. We must show that if there are $R(m, n-1) + R(m-1, n)$ people at a party, then there must be at least m mutual friends or n mutual enemies. Consider one person; let's call him Jerry. Then there are $R(m-1, n) + R(m, n-1) - 1$ other people at the party, and by the pigeonhole principle there must be at least $R(m-1, n)$ friends of Jerry or $R(m, n-1)$ enemies of Jerry among these people. First let's suppose there are $R(m-1, n)$ friends of Jerry. By the definition of R, among these people we are guaranteed to find either $m-1$ mutual friends or n mutual enemies. In the former case, these $m-1$ mutual friends together with Jerry are a set of m mutual friends; and in the latter case, we have the desired set of n mutual enemies. The other situation is similar: Suppose there are $R(m, n-1)$ enemies of Jerry; we are guaranteed to find among them either m mutual friends or $n-1$ mutual enemies. In the former case, we have the desired set of m mutual friends, and in the latter case, these $n-1$ mutual enemies together with Jerry are a set of n mutual enemies.

CHAPTER 7

Section 7.1

1. 1/13 **3.** 1/2 **5.** 1/2 **7.** 1/64 **9.** 47/52 **11.** $1/C(52, 5)$ **13.** $1 - [C(48, 5)/C(52, 5)]$ **15.** $C(13, 2)C(4, 2)C(4, 2)C(44, 1)/C(52, 5)$ **17.** $10{,}240/C(52, 5)$ **19.** $1{,}302{,}540/C(52, 5)$ **21.** 1/64 **23.** 8/25 **25. a)** $1/C(50, 6) = 1/15{,}890{,}700$ **b)** $1/C(52, 6) = 1/20{,}358{,}520$ **c)** $1/C(56, 6) = 1/32{,}468{,}436$ **d)** $1/C(60, 6) = 1/50{,}063{,}860$ **27. a)** 139,128/319,865 **b)** 212,667/511,313 **c)** 151,340/386,529 **d)** 163,647/446,276 **29.** $1/C(100, 8)$ **31.** 3/100 **33. a)** 1/7,880,400 **b)** 1/8,000,000 **35. a)** 9/19 **b)** 81/361 **c)** 1/19 **d)** 1,889,568/2,476,099 **e)** 48/361 **37.** Three dice **39. a)** 4/756,438,375 **b)** 13/30,257,535 **c)** 4888/2,750,685 **d)** 90,272/9,823,875 **41. a)** 25/292,201,338 **b)** 25/292,201,338 **c)** 1280/571,120,797 **d)** 7564/661,089 **43.** The door the contestant chooses is chosen at random without knowing where the prize is, but the door chosen by the host is not chosen at random, because he always avoids opening the door with the prize. This makes any argument based on symmetry invalid. **45. a)** 671/1296 **b)** $1 - 35^{24}/36^{24}$; no **c)** The former

Section 7.2

1. $p(T) = 1/4, p(H) = 3/4$ **3.** $p(1) = p(3) = p(5) = p(6) = 1/16; p(2) = p(4) = 3/8$ **5.** 9/49 **7. a)** 1/2 **b)** 1/2 **c)** 1/3 **d)** 1/4 **e)** 1/4 **9. a)** 1/26! **b)** 1/26 **c)** 1/2 **d)** 1/26 **e)** 1/650 **f)** 1/15,600 **11.** Clearly, $p(E \cup F) \geq p(E) = 0.7$. Also, $p(E \cup F) \leq 1$. If we apply Theorem 2 from Section 7.1, we can rewrite this as $p(E) + p(F) - p(E \cap F) \leq 1$, or $0.7 + 0.5 - p(E \cap F) \leq 1$. Solving for $p(E \cap F)$ gives $p(E \cap F) \geq 0.2$. **13.** Because $p(E \cup F) = p(E) + p(F) - p(E \cap F)$ and $p(E \cup F) \leq 1$, it follows that $1 \geq p(E) + p(F) - p(E \cap F)$. From this inequality we conclude that $p(E) + p(F) \leq 1 + p(E \cap F)$. **15.** We will use mathematical induction to prove that the inequality holds for $n \geq 2$. Let $P(n)$ be the statement that $p(\bigcup_{j=1}^{n} E_j) \leq \sum_{j=1}^{n} p(E_j)$. *Basis step:* $P(2)$ is true because $p(E_1 \cup E_2) = p(E_1) + p(E_2) - p(E_1 \cap E_2) \leq p(E_1) + p(E_2)$. *Inductive step:* Assume that $P(k)$ is true. Using the basis case and the inductive hypothesis, it follows that $p(\bigcup_{j=1}^{k+1} E_j) \leq p(\bigcup_{j=1}^{k} E_j) + p(E_{k+1}) \leq \sum_{j=1}^{k+1} p(E_j)$. This shows that $P(k+1)$ is true, completing the proof by mathematical induction. **17.** Because $E \cup \overline{E}$ is the entire sample space S, the event F can be split into two disjoint events: $F = S \cap F = (E \cup \overline{E}) \cap F = (E \cap F) \cup (\overline{E} \cap F)$, using the distributive law. Therefore, $p(F) = p((E \cap F) \cup (\overline{E} \cap F)) = p(E \cap F) + p(\overline{E} \cap F)$, because these two events are disjoint. Subtracting $p(E \cap F)$ from both sides, using the fact that $p(E \cap F) = p(E) \cdot p(F)$ (the hypothesis that E and F are independent), and factoring, we have $p(F)[1 - p(E)] = p(\overline{E} \cap F)$. Because $1 - p(E) = p(\overline{E})$, this says that $p(\overline{E} \cap F) = p(\overline{E}) \cdot p(F)$, as desired. **19. a)** 1/12 **b)** $1 - \frac{11}{12} \cdot \frac{10}{12} \cdot \cdots \cdot \frac{13-n}{12}$ **c)** 5 **21.** 614 **23.** 1/4 **25.** 3/8 **27. a)** Not independent **b)** Not independent **c)** Not independent **29.** 3/16 **31. a)** $1/32 = 0.03125$ **b)** $0.49^5 \approx 0.02825$ **c)** 0.03795012 **33. a)** 5/8 **b)** 0.627649 **c)** 0.6431 **35. a)** p^n **b)** $1 - p^n$ **c)** $p^n + n \cdot p^{n-1} \cdot (1-p)$ **d)** $1 - [p^n + n \cdot p^{n-1} \cdot (1-p)]$ **37.** $p(\bigcup_{i=1}^{\infty} E_i)$ is the sum of $p(s)$ for each outcome s in $\bigcup_{i=1}^{\infty} E_i$. Because the E_is are pairwise disjoint, this is the sum of the probabilities of all the outcomes in any of the E_is, which is what $\sum_{i=1}^{\infty} p(E_i)$ is. (We can rearrange the summands and still get the same answer because

this series converges absolutely.) **39. a)** $\overline{E} = \bigcup_{j=1}^{\binom{m}{k}} F_j$, so the given inequality now follows from Boole's inequality (Exercise 15). **b)** The probability that a particular player not in the jth set beats all k of the players in the jth set is $(1/2)^k = 2^{-k}$. Therefore, the probability that this player does not do so is $1 - 2^{-k}$, so the probability that all $m - k$ of the players not in the jth set are unable to boast of a perfect record against everyone in the jth set is $(1 - 2^{-k})^{m-k}$. That is precisely $p(F_j)$. **c)** The first inequality follows immediately, because all the summands are the same and there are $\binom{m}{k}$ of them. If this probability is less than 1, then it must be possible that $\overline{E}$ fails, i.e., that E happens. So there is a tournament that meets the conditions of the problem as long as the second inequality holds. **d)** $m \geq 21$ for $k = 2$, and $m \geq 91$ for $k = 3$
41. **procedure** *probabilistic prime*(n, k)

```
composite := false
i := 0
while composite = false and i < k
    i := i + 1
    choose b uniformly at random with 1 < b < n
    apply Miller's test to base b
    if n fails the test then composite := true
if composite = true then print ("composite")
else print ("probably prime")
```

43. a) 1/6 **b)** 3/20

Section 7.3

NOTE: In the answers for Section 7.3, all probabilities given in decimal form are rounded to three decimal places. **1.** 3/5 **3.** 3/4 **5.** 0.481 **7. a)** 0.999 **b)** 0.324 **9. a)** 0.740 **b)** 0.260 **c)** 0.002 **d)** 0.998 **11.** 0.724 **13.** 3/17 **15. a)** 1/3 **b)** $p(M = j \mid W = k) = 1$ if i, j, and k are distinct; $p(M = j \mid W = k) = 0$ if $j = k$ or $j = i$; $p(M = j \mid W = k) = 1/2$ if $i = k$ and $j \neq i$ **c)** 2/3 **d)** You should change doors, because you now have a 2/3 chance to win by switching. **17.** The definition of conditional probability tells us that $p(F_j \mid E) = p(E \cap F_j)/p(E)$. For the numerator, again using the definition of conditional probability, we have $p(E \cap F_j) = p(E \mid F_j)p(F_j)$, as desired. For the denominator, we show that $p(E) = \sum_{i=1}^{n} p(E \mid F_i)p(F_i)$. The events $E \cap F_i$ partition the event E; that is, $(E \cap F_{i_1}) \cap (E \cap F_{i_2}) = \emptyset$ when $i_i \neq i_2$ (because the F_i's are mutually exclusive), and $\bigcup_{i=1}^{n}(E \cap F_{i_1}) = E$ (because the $\bigcup_{i=1}^{n} F_i = S$). Therefore, $p(E) = \sum_{i=1}^{n} p(E \cap F_i) = \sum_{i=1}^{n} p(E \mid F_i)p(F_i)$. **19.** No **21.** Yes **23.** By Bayes' theorem, $p(S \mid E_1 \cap E_2) = p(E_1 \cap E_2 \mid S)p(S)/[p(E_1 \cap E_2 \mid S)p(S) + p(E_1 \cap E_2 \mid \overline{S})p(\overline{S})]$. Because we are assuming no prior knowledge about whether a message is or is not spam, we set $p(S) = p(\overline{S}) = 0.5$, and so the equation above simplifies to $p(S \mid E_1 \cap E_2) = p(E_1 \cap E_2 \mid S)/[p(E_1 \cap E_2 \mid S) + p(E_1 \cap E_2 \mid \overline{S})]$. Because of the assumed independence of E_1, E_2, and S, we have $p(E_1 \cap E_2 \mid S) = p(E_1 \mid S) \cdot p(E_2 \mid S)$, and similarly for $\overline{S}$.

Section 7.4

1. 2.5 **3.** 5/3 **5.** 336/49 **7.** 170 **9.** $(4n + 6)/3$ **11.** 50,700,551/10,077,696 ≈ 5.03 **13.** 6 **15.** $p(X \geq j) = \sum_{k=j}^{\infty} p(X = k) = \sum_{k=j}^{\infty}(1-p)^{k-1}p = p(1-p)^{j-1}\sum_{k=0}^{\infty}(1-p)^k = p(1 - p)^{j-1}/(1 - (1 - p)) = (1 - p)^{j-1}$ **17.** 2302 **19.** $(7/2) \cdot 7 \neq 329/12$ **21.** 10 **23.** 1472 pounds **25.** $p + (n-1)p(1-p)$ **27.** 5/2 **29. a)** 0 **b)** n **31.** This is not true. For example, let X be the number of heads in one flip of a fair coin, and let Y be the number of heads in one flip of a second fair coin. Then $A(X) + A(Y) = 1$ but $A(X + Y) = 0.5$. **33. a)** We are told that X_1 and X_2 are independent. To see that X_1 and X_3 are independent, we enumerate the eight possibilities for (X_1, X_2, X_3) and find that (0, 0, 0), (1, 0, 1), (0, 1, 1), (1, 1, 0) each have probability 1/4 and the others have probability 0 (because of the definition of X_3). Thus, $p(X_1 = 0 \wedge X_3 = 0) = 1/4$, $p(X_1 = 0) = 1/2$, and $p(X_3 = 0) = 1/2$, so it is true that $p(X_1 = 0 \wedge X_3 = 0) = p(X_1 = 0)p(X_3 = 0)$. Essentially the same calculation shows that $p(X_1 = 0 \wedge X_3 = 1) = p(X_1 = 0)p(X_3 = 1)$, $p(X_1 = 1 \wedge X_3 = 0) = p(X_1 = 1)p(X_3 = 0)$, and $p(X_1 = 1 \wedge X_3 = 1) = p(X_1 = 1)p(X_3 = 1)$. Therefore, by definition, X_1 and X_3 are independent. The same reasoning shows that X_2 and X_3 are independent. To see that X_3 and $X_1 + X_2$ are not independent, we observe that $p(X_3 = 1 \wedge X_1 + X_2 = 2) = 0$. But $p(X_3 = 1)p(X_1 + X_2 = 2) = (1/2)(1/4) = 1/8$. **b)** We see from the calculation in part (a) that X_1, X_2, and X_3 are all Bernoulli random variables, so the variance of each is $(1/2)(1/2) = 1/4$. Therefore, $V(X_1) + V(X_2) + V(X_3) = 3/4$. We use the calculations in part (a) to see that $E(X_1 + X_2 + X_3) = 3/2$, and then $V(X_1 + X_2 + X_3) = 3/4$. **c)** In order to use the first part of Theorem 7 to show that $V((X_1 + X_2 + \cdots + X_k) + X_{k+1}) = V(X_1 + X_2 + \cdots + X_k) + V(X_{k+1})$ in the inductive step of a proof by mathematical induction, we would have to know that $X_1 + X_2 + \cdots + X_k$ and X_{k+1} are independent, but we see from part (a) that this is not necessarily true. **35.** 1/100 **37.** $E(X)/a = \sum_r (r/a) \cdot p(X = r) \geq \sum_{r \geq a} 1 \cdot p(X = r) = p(X \geq a)$ **39. a)** 10/11 **b)** 0.9999 **41. a)** Each of the $n!$ permutations occurs with probability $1/n!$, so $E(X)$ is the number of comparisons, averaged over all these permutations. **b)** Even if the algorithm continues $n - 1$ rounds, X will be at most $n(n-1)/2$. It follows from the formula for expectation that $E(X) \leq n(n-1)/2$. **c)** The algorithm proceeds by comparing adjacent elements and then swapping them if necessary. Thus, the only way that inverted elements can become uninverted is for them to be compared and swapped. **d)** Because $X(P) \geq I(P)$ for all P, it follows from the definition of expectation that $E(X) \geq E(I)$. **e)** This summation counts 1 for every instance of an inversion. **f)** This follows from Theorem 3. **g)** By Theorem 2 with $n = 1$, the expectation of $I_{j,k}$ is the probability that a_k precedes a_j in the permutation. This is clearly 1/2 by symmetry. **h)** The summation in part (f) consists of $C(n, 2) = n(n-1)/2$ terms, each equal to $1/2$, so the sum is $n(n-1)/4$. **i)** From part (a) and part (b) we know that $E(X)$, the object of interest, is at most $n(n-1)/2$, and from part (d) and part (h) we know that $E(X)$ is at least $n(n-1)/4$, both of which are $\Theta(n^2)$. **43.** 1 **45.** $V(X+Y) = E((X+Y)^2) - E(X+Y)^2 = E(X^2+2XY+Y^2)-[E(X)+E(Y)]^2 = E(X^2)+2E(XY)+E(Y^2)-E(X)^2 - 2E(X)E(Y) - E(Y)^2 = E(X^2) - E(X)^2 + 2[E(XY) - E(X)E(Y)] + E(Y^2) - E(Y)^2 = V(X) + 2\,\mathrm{Cov}(X, Y) + V(Y)$ **47.** $[(n-1)/n]^m$ **49.** $(n-1)^m/n^{m-1}$

Supplementary Exercises

1. 1/109,668 **3. a)** 1/195,249,054 **b)** 1/5,138,133 **c)** 45/357,599 **d)** 18,285/18,821 **5. a)** $1/C(52, 13)$ **b)** $4/C(52, 13)$ **c)** $2{,}944{,}656/C(52, 13)$ **d)** $35{,}335{,}872/C(52, 13)$ **7. a)** 9/2 **b)** 21/4 **9. a)** 9 **b)** 21/2 **11. a)** 8 **b)** 49/6 **13. a)** $n/2^{n-1}$ **b)** $p(1-p)^{k-1}$, where $p = n/2^{n-1}$ **c)** $2^{n-1}/n$ **15.** $\frac{(m-1)(n-1)+\gcd(m,n)-1}{mn-1}$ **17. a)** 2/3 **b)** 2/3 **19.** 1/32 **21. a)** The probability that one wins 2^n dollars is $1/2^n$, because that happens precisely when the player gets $n-1$ tails followed by a head. The expected value of the winnings is therefore the sum of 2^n times $1/2^n$ as n goes from 1 to infinity. Because each of these terms is 1, the sum is infinite. In other words, one should be willing to wager any amount of money and expect to come out ahead in the long run. **b)** \$9, \$9 **23. a)** 1/3 when $S = \{1, 2, 3, 4, 5, 6, 7, 8, 9, 10, 11, 12\}$, $A = \{1, 2, 3, 4, 5, 6, 7, 8, 9\}$, and $B = \{1, 2, 3, 4\}$; 1/12 when $S = \{1, 2, 3, 4, 5, 6, 7, 8, 9, 10, 11, 12\}$, $A = \{4, 5, 6, 7, 8, 9, 10, 11, 12\}$, and $B = \{1, 2, 3, 4\}$ **b)** 1 when $S = \{1, 2, 3, 4, 5, 6, 7, 8, 9, 10, 11, 12\}$, $A = \{4, 5, 6, 7, 8, 9, 10, 11, 12\}$, and $B = \{1, 2, 3, 4\}$; 3/4 when $S = \{1, 2, 3, 4, 5, 6, 7, 8, 9, 10, 11, 12\}$, $A = \{1, 2, 3, 4, 5, 6, 7, 8, 9\}$, and $B = \{1, 2, 3, 4\}$ **25. a)** $p(E_1 \cap E_2) = p(E_1)p(E_2)$, $p(E_1 \cap E_3) = p(E_1)p(E_3)$, $p(E_2 \cap E_3) = p(E_2)p(E_3)$, $p(E_1 \cap E_2 \cap E_3) = p(E_1)p(E_2)p(E_3)$ **b)** Yes **c)** Yes; yes **d)** Yes; no **e)** $2^n - n - 1$ **27. a)** 1/2 under first interpretation; 1/3 under second interpretation **b)** Let M be the event that both of Mr. Smith's children are boys and let B be the event that Mr. Smith chose a boy for today's walk. Then $p(M) = 1/4$, $p(B \mid M) = 1$, and $p(B \mid \overline{M}) = 1/3$. Apply Bayes' theorem to compute $p(M \mid B) = 1/2$. **c)** This variation is equivalent to the second interpretation discussed in part (a), so the answer is unambiguously 1/3. **29.** $V(aX+b) = E((aX+b)^2) - E(aX+b)^2 = E(a^2X^2 + 2abX + b^2) - [aE(X) + b]^2 = E(a^2X^2) + E(2abX) + E(b^2) - [a^2E(X)^2 + 2abE(X) + b^2] = a^2E(X^2) + 2abE(X) + b^2 - a^2E(X)^2 - 2abE(X) - b^2 = a^2[E(X^2) - E(X)^2] = a^2V(X)$ **31.** To count every element in the sample space exactly once, we must include every element in each of the sets and then take away the double counting of the elements in the intersections. Thus, $p(E_1 \cup E_2 \cup \cdots \cup E_m) = p(E_1)+p(E_2)+\cdots+p(E_m)-p(E_1 \cap E_2)-p(E_1 \cap E_3)-\cdots-p(E_1 \cap E_m)-p(E_2 \cap E_3)-p(E_2 \cap E_4)-\cdots-p(E_2 \cap E_m)-\cdots-p(E_{m-1} \cap E_m) = qm-(m(m-1)/2)r$, because $C(m, 2)$ terms are being subtracted. But $p(E_1 \cup E_2 \cup \cdots \cup E_m) = 1$, so we have $qm-[m(m-1)/2]r = 1$. Because $r \ge 0$, this equation tells us that $qm \ge 1$, so $q \ge 1/m$. Because $q \le 1$, this equation also implies that $[m(m-1)/2]r = qm-1 \le m-1$, from which it follows that $r \le 2/m$. **33. a)** We purchase the cards until we have gotten one of each type. That means we have purchased X cards in all. On the other hand, that also means that we purchased X_0 cards until we got the first type we got, and then purchased X_1 more cards until we got the second type we got, and so on. Thus, X is the sum of the X_js. **b)** Once j distinct types have been obtained, there are $n-j$ new types available out of a total of n types available. Because it is equally likely that we get each type, the probability of success on the next purchase (getting a new type) is $(n-j)/n$. **c)** This follows immediately from the definition of geometric distribution, the definition of X_j, and part (b). **d)** From part (c) it follows that $E(X_j) = n/(n-j)$. Thus, by linearity of expectation and part (a), we have $E(X) = E(X_0) + E(X_1) + \cdots + E(X_{n-1}) = \frac{n}{n} + \frac{n}{n-1} + \cdots + \frac{n}{1} = n\left(\frac{1}{n} + \frac{1}{n-1} + \cdots + \frac{1}{1}\right)$. **e)** About 224.46 **35.** $24 \cdot 13^4/(52 \cdot 51 \cdot 50 \cdot 49)$

CHAPTER 8

Section 8.1

1. Let $P(n)$ be "$H_n = 2^n - 1$." *Basis step:* $P(1)$ is true because $H_1 = 1$. *Inductive step:* Assume that $H_n = 2^n - 1$. Then because $H_{n+1} = 2H_n + 1$, it follows that $H_{n+1} = 2(2^n-1)+1 = 2^{n+1} - 1$. **3. a)** $a_n = 2a_{n-1} + a_{n-5}$ for $n \ge 5$ **b)** $a_0 = 1$, $a_1 = 2$, $a_2 = 4$, $a_3 = 8$, $a_4 = 16$ **c)** 1217 **5.** 9494 **7. a)** $a_n = a_{n-1}+a_{n-2}+2^{n-2}$ for $n \ge 2$ **b)** $a_0 = 0$, $a_1 = 0$ **c)** 94 **9. a)** $a_n = a_{n-1}+a_{n-2}+a_{n-3}$ for $n \ge 3$ **b)** $a_0 = 1$, $a_1 = 2$, $a_2 = 4$ **c)** 81 **11. a)** $a_n = a_{n-1} + a_{n-2}$ for $n \ge 2$ **b)** $a_0 = 1$, $a_1 = 1$ **c)** 34 **13. a)** $a_n = 2a_{n-1}+2a_{n-2}$ for $n \ge 2$ **b)** $a_0 = 1$, $a_1 = 3$ **c)** 448 **15. a)** $a_n = 2a_{n-1}+a_{n-2}$ for $n \ge 2$ **b)** $a_0 = 1$, $a_1 = 3$ **c)** 239 **17. a)** $a_n = 2a_{n-1}$ for $n \ge 2$ **b)** $a_1 = 3$ **c)** 96 **19. a)** $a_n = a_{n-1} + a_{n-2}$ for $n \ge 2$ **b)** $a_0 = 1$, $a_1 = 1$ **c)** 89 **21. a)** $R_n = n + R_{n-1}$, $R_0 = 1$ **b)** $R_n = n(n+1)/2+1$ **23. a)** $S_n = S_{n-1}+(n^2-n+2)/2$, $S_0 = 1$ **b)** $S_n = (n^3+5n+6)/6$ **25.** 64 **27. a)** $a_n = 2a_{n-1}+2a_{n-2}$ **b)** $a_0 = 1$, $a_1 = 3$ **c)** 1224 **29.** Clearly, $S(m, 1) = 1$ for $m \ge 1$. If $m \ge n$, then a function that is not onto from the set with m elements to the set with n elements can be specified by picking the size of the range, which is an integer between 1 and $n-1$ inclusive, picking the elements of the range, which can be done in $C(n, k)$ ways, and picking an onto function onto the range, which can be done in $S(m, k)$ ways. Hence, there are $\sum_{k=1}^{n-1} C(n, k)S(m, k)$ functions that are not onto. But there are n^m functions altogether, so $S(m, n) = n^m - \sum_{k=1}^{n-1} C(n, k)S(m, k)$. **31. a)** $C_5 = C_0C_4 + C_1C_3+C_2C_2+C_3C_1+C_4C_0 = 1\cdot14+1\cdot5+2\cdot2+5\cdot1+14\cdot1 = 42$ **b)** $C(10, 5)/6 = 42$ **33.** $J(1) = 1$, $J(2) = 1$, $J(3) = 3$,

$J(4) = 1, J(5) = 3, J(6) = 5, J(7) = 7, J(8) = 1, J(9) = 3, J(10) = 5, J(11) = 7, J(12) = 9, J(13) = 11, J(14) = 13, J(15) = 15, J(16) = 1$ **35.** First, suppose that the number of people is even, say $2n$. After going around the circle once and returning to the first person, because the people at locations with even numbers have been eliminated, there are exactly n people left and the person currently at location i is the person who was originally at location $2i - 1$. Therefore, the survivor [originally in location $J(2n)$] is now in location $J(n)$; this was the person who was at location $2J(n) - 1$. Hence, $J(2n) = 2J(n) - 1$. Similarly, when there are an odd number of people, say $2n + 1$, then after going around the circle once and then eliminating person 1, there are n people left and the person currently at location i is the person who was at location $2i + 1$. Therefore, the survivor will be the player currently occupying location $J(n)$, namely, the person who was originally at location $2J(n) + 1$. Hence, $J(2n + 1) = 2J(n) + 1$. The basis step is $J(1) = 1$. **37.** 73, 977, 3617 **39.** These nine moves solve the puzzle: Move disk 1 from peg 1 to peg 2; move disk 2 from peg 1 to peg 3; move disk 1 from peg 2 to peg 3; move disk 3 from peg 1 to peg 2; move disk 4 from peg 1 to peg 4; move disk 3 from peg 2 to peg 4; move disk 1 from peg 3 to peg 2; move disk 2 from peg 3 to peg 4; move disk 1 from peg 2 to peg 4. To see that at least nine moves are required, first note that at least seven moves are required no matter how many pegs are present: three to unstack the disks, one to move the largest disk 4, and three more moves to restack them. At least two other moves are needed, because to move disk 4 from peg 1 to peg 4 the other three disks must be on pegs 2 and 3, so at least one move is needed to restack them and one move to unstack them. **41.** The base cases are obvious. If $n > 1$, the algorithm consists of three stages. In the first stage, by the inductive hypothesis, $R(n - k)$ moves are used to transfer the smallest $n - k$ disks to peg 2. Then using the usual three-peg Tower of Hanoi algorithm, it takes $2^k - 1$ moves to transfer the rest of the disks (the largest k disks) to peg 4, avoiding peg 2. Then again by the inductive hypothesis, it takes $R(n - k)$ moves to transfer the smallest $n - k$ disks to peg 4; all the pegs are available for this, because the largest disks, now on peg 4, do not interfere. This establishes the recurrence relation. **43.** First note that $R(n) = \sum_{j=1}^{n} [R(j) - R(j-1)]$ [which follows because the sum is telescoping and $R(0) = 0$]. By Exercise 42, this is the sum of $2^{k'-1}$ for this range of values of j. Therefore, the sum is $\sum_{i=1}^{k} i2^{i-1}$, except that if n is not a triangular number, then the last few values when $i = k$ are missing, and that is what the final term in the given expression accounts for. **45.** By Exercise 43, $R(n)$ is no larger than $\sum_{i=1}^{k} i2^{i-1}$. It can be shown that this sum equals $(k + 1)2^k - 2^{k+1} + 1$, so it is no greater than $(k+1)2^k$. Because $n > k(k-1)/2$, the quadratic formula can be used to show that $k < 1 + \sqrt{2n}$ for all $n > 1$. Therefore, $R(n)$ is bounded above by $(1 + \sqrt{2n} + 1)2^{1+\sqrt{2n}} < 8\sqrt{n}2^{\sqrt{2n}}$ for all $n > 2$. Hence, $R(n)$ is $O(\sqrt{n}2^{\sqrt{2n}})$. **47. a)** 0 **b)** 0 **c)** 2 **d)** $2^{n-1} - 2^{n-2}$ **49.** $a_n - 2\nabla a_n + \nabla^2 a_n = a_n - 2(a_n - a_{n-1}) + (\nabla a_n - \nabla a_{n-1}) = -a_n + 2a_{n-1} + [(a_n - a_{n-1}) - (a_{n-1} - a_{n-2})] = -a_n + 2a_{n-1} + (a_n - 2a_{n-1} + a_{n-2}) = a_{n-2}$ **51.** $a_n = a_{n-1} + a_{n-2} = (a_n - \nabla a_n) + (a_n - 2\nabla a_n + \nabla^2 a_n) = 2a_n - 3\nabla a_n + \nabla^2 a_n$, or $a_n = 3\nabla a_n - \nabla^2 a_n$ **53.** Insert $S(0) := \emptyset$ after $T(0) := 0$ (where $S(j)$ will record the optimal set of talks among the first j talks), and replace the statement $T(j) := \max(w_j + T(p(j)), T(j-1))$ with the following code:

```
if w_j + T(p(j)) > T(j − 1) then
    T(j) := w_j + T(p(j))
    S(j) := S(p(j)) ∪ {j}
else
    T(j) := T(j − 1)
    S(j) := S(j − 1)
```

55. a) Talks 1, 3, and 7 **b)** Talks 1 and 6, or talks 1, 3, and 7 **c)** Talks 1, 3, and 7 **d)** Talks 1 and 6 **57. a)** This follows immediately from Example 5 and Exercise 43c in Section 8.4. **b)** The last step in computing $\mathbf{A}_{ij}$ is to multiply $\mathbf{A}_{ik}$ by $\mathbf{A}_{k+1,j}$ for some k between i and $j - 1$ inclusive, which will require $m_i m_{k+1} m_{j+1}$ integer multiplications, independent of the manner in which $\mathbf{A}_{ik}$ and $\mathbf{A}_{k+1,j}$ are computed. Therefore, to minimize the total number of integer multiplications, each of those two factors must be computed in the most efficient manner. **c)** This follows immediately from part (b) and the definition of $M(i, j)$. **d)**

```
procedure matrix order(m_1, ..., m_{n+1}:
            positive integers)
for i := 1 to n
    M(i, i) := 0
for d := 1 to n − 1
    for i := 1 to n − d
        min := 0
        for k := i to i + d
            new := M(i, k) + M(k + 1, i + d) + m_i m_{k+1} m_{i+d+1}
            if new < min then
                min := new
                where(i, i + d) := k
        M(i, i + d) := min
```

e) The algorithm has three nested loops, each of which is indexed over at most n values.

Section 8.2

1. a) Degree 3 **b)** No **c)** Degree 4 **d)** No **e)** No **f)** Degree 2 **g)** No **3. a)** $a_n = 3 \cdot 2^n$ **b)** $a_n = 2$ **c)** $a_n = 3 \cdot 2^n - 2 \cdot 3^n$ **d)** $a_n = 6 \cdot 2^n - 2 \cdot n2^n$ **e)** $a_n = n(-2)^{n-1}$ **f)** $a_n = 2^n - (-2)^n$ **g)** $a_n = (1/2)^{n+1} - (-1/2)^{n+1}$ **5.** $a_n = \frac{1}{\sqrt{5}}\left(\frac{1+\sqrt{5}}{2}\right)^{n+1} - \frac{1}{\sqrt{5}}\left(\frac{1-\sqrt{5}}{2}\right)^{n+1}$ **7.** $[2^{n+1} + (-1)^n]/3$ **9. a)** $P_n = 1.2P_{n-1} + 0.45P_{n-2}$, $P_0 = 100{,}000$, $P_1 = 120{,}000$ **b)** $P_n = (250{,}000/3)(3/2)^n + (50{,}000/3)(-3/10)^n$ **11. a)** *Basis step:* For $n = 1$ we have $1 = 0 + 1$, and for $n = 2$ we have $3 = 1 + 2$. *Inductive step:* Assume true for $k \le n$. Then $L_{n+1} = L_n + L_{n-1} = f_{n-1} + f_{n+1} + f_{n-2} + f_n = (f_{n-1} + f_{n-2}) + (f_{n+1} + f_n) = f_n + f_{n+2}$. **b)** $L_n = \left(\frac{1+\sqrt{5}}{2}\right)^n + \left(\frac{1-\sqrt{5}}{2}\right)^n$ **13.** $a_n = 8(-1)^n - 3(-2)^n + 4 \cdot 3^n$ **15.** $a_n = 5 + 3(-2)^n - 3^n$

17. Let $a_n = C(n, 0) + C(n-1, 1) + \cdots + C(n-k, k)$ where $k = \lfloor n/2 \rfloor$. First, assume that n is even, so that $k = n/2$, and the last term is $C(k, k)$. By Pascal's identity we have $a_n = 1 + C(n-2, 0) + C(n-2, 1) + C(n-3, 1) + C(n-3, 2) + \cdots + C(n-k, k-2) + C(n-k, k-1) + 1 = 1 + C(n-2, 1) + C(n-3, 2) + \cdots + C(n-k, k-1) + C(n-2, 0) + C(n-3, 1) + \cdots + C(n-k, k-2) + 1 = a_{n-1} + a_{n-2}$ because $\lfloor (n-1)/2 \rfloor = k - 1 = \lfloor (n-2)/2 \rfloor$. A similar calculation works when n is odd. Hence, $\{a_n\}$ satisfies the recurrence relation $a_n = a_{n-1} + a_{n-2}$ for all positive integers n, $n \geq 2$. Also, $a_1 = C(1, 0) = 1$ and $a_2 = C(2, 0) + C(1, 1) = 2$, which are f_2 and f_3. It follows that $a_n = f_{n+1}$ for all positive integers n. **19.** $a_n = (n^2 + 3n + 5)(-1)^n$ **21.** $(a_{1,0} + a_{1,1}n + a_{1,2}n^2 + a_{1,3}n^3) + (a_{2,0} + a_{2,1}n + a_{2,2}n^2)(-2)^n + (a_{3,0} + a_{3,1}n)3^n + a_{4,0}(-4)^n$ **23. a)** $3a_{n-1} + 2^n = 3(-2)^n + 2^n = 2^n(-3+1) = -2^{n+1} = a_n$ **b)** $a_n = \alpha 3^n - 2^{n+1}$ **c)** $a_n = 3^{n+1} - 2^{n+1}$ **25. a)** $A = -1, B = -7$ **b)** $a_n = \alpha 2^n - n - 7$ **c)** $a_n = 11 \cdot 2^n - n - 7$ **27. a)** $p_3n^3 + p_2n^2 + p_1n + p_0$ **b)** $n^2p_0(-2)^n$ **c)** $n^2(p_1n + p_0)2^n$ **d)** $(p_2n^2 + p_1n + p_0)4^n$ **e)** $n^2(p_2n^2 + p_1n + p_0)(-2)^n$ **f)** $n^2(p_4n^4 + p_3n^3 + p_2n^2 + p_1n + p_0)2^n$ **g)** p_0 **29. a)** $a_n = \alpha 2^n + 3^{n+1}$ **b)** $a_n = -2 \cdot 2^n + 3^{n+1}$ **31.** $a_n = \alpha 2^n + \beta 3^n - n \cdot 2^{n+1} + 3n/2 + 21/4$ **33.** $a_n = (\alpha + \beta n + n^2 + n^3/6)2^n$ **35.** $a_n = -4 \cdot 2^n - n^2/4 - 5n/2 + 1/8 + (39/8)3^n$ **37.** $a_n = n(n+1)(n+2)/6$ **39. a)** $1, -1, i, -i$ **b)** $a_n = \frac{1}{4} - \frac{1}{4}(-1)^n + \frac{2+i}{4}i^n + \frac{2-i}{4}(-i)^n$ **41. a)** Using the formula for f_n, we see that $\left|f_n - \frac{1}{\sqrt{5}}\left(\frac{1+\sqrt{5}}{2}\right)^n\right| = \left|\frac{1}{\sqrt{5}}\left(\frac{1-\sqrt{5}}{2}\right)^n\right| < 1/\sqrt{5} < 1/2$. This means that f_n is the integer closest to $\frac{1}{\sqrt{5}}\left(\frac{1+\sqrt{5}}{2}\right)^n$. **b)** Less when n is even; greater when n is odd **43.** $a_n = f_{n-1} + 2f_n - 1$ **45. a)** $a_n = 3a_{n-1} + 4a_{n-2}$, $a_0 = 2$, $a_1 = 6$ **b)** $a_n = [4^{n+1} + (-1)^n]/5$ **47. a)** $a_n = 2a_{n+1} + (n-1)10{,}000$ **b)** $a_n = 70{,}000 \cdot 2^{n-1} - 10{,}000n - 10{,}000$ **49.** $a_n = 5n^2/12 + 13n/12 + 1$ **51.** See Chapter 11, Section 5 in [Ma93]. **53.** $6^n \cdot 4^{n-1}/n$

Section 8.3

1. 14 **3.** The first step is $(1110)_2(1010)_2 = (2^4 + 2^2)(11)_2\ (10)_2 + 2^2[(11)_2 - (10)_2][(10)_2 - (10)_2] + (2^2 + 1)(10)_2 \cdot (10)_2$. The product is $(10001100)_2$. **5.** $C = 50{,}665C + 729 = 33{,}979$ **7. a)** 2 **b)** 4 **c)** 7 **9. a)** 79 **b)** 48,829 **c)** 30,517,579 **11.** $O(\log n)$ **13.** $O(n^{\log_3 2})$ **15.** 5 **17. a)** *Basis step:* If the sequence has just one element, then the one person on the list is the winner. *Recursive step:* Divide the list into two parts—the first half and the second half—as equally as possible. Apply the algorithm recursively to each half to come up with at most two names. Then run through the entire list to count the number of occurrences of each of those names to decide which, if either, is the winner. **b)** $O(n \log n)$ **19. a)** $f(n) = f(n/2) + 2$ **b)** $O(\log n)$ **21. a)** 7 **b)** $O(\log n)$
23. a) procedure *largest sum*$(a_1, \ldots, a_n)$

```
    best := 0 {empty subsequence has sum 0}
    for i := 1 to n
        sum := 0
        for j := i + 1 to n
            sum := sum + a_j
            if sum > best then best := sum
    {best is the maximum possible sum of numbers
        in the list}
```

b) $O(n^2)$ **c)** We divide the list into a first half and a second half and apply the algorithm recursively to find the largest sum of consecutive terms for each half. The largest sum of consecutive terms in the entire sequence is either one of these two numbers or the sum of a sequence of consecutive terms that crosses the middle of the list. To find the largest possible sum of a sequence of consecutive terms that crosses the middle of the list, we start at the middle and move forward to find the largest possible sum in the second half of the list, and move backward to find the largest possible sum in the first half of the list; the desired sum is the sum of these two quantities. The final answer is then the largest of this sum and the two answers obtained recursively. The base case is that the largest sum of a sequence of one term is the larger of that number and 0. **d)** 11, 9, 14 **e)** $S(n) = 2S(n/2) + n$, $C(n) = 2C(n/2) + n + 2$, $S(1) = 0$, $C(1) = 1$ **f)** $O(n \log n)$, better than $O(n^2)$ **25.** (1, 6) and (3, 6) at distance 2 **27.** The algorithm is essentially the same as the algorithm given in Example 12. The central strip still has width $2d$ but we need to consider just two boxes of size $d \times d$ rather than eight boxes of size $(d/2) \times (d/2)$. The recurrence relation is the same as the recurrence relation in Example 12, except that the coefficient 7 is replaced by 1. **29.** With $k = \log_b n$, it follows that $f(n) = a^k f(1) + \sum_{j=0}^{k-1} a^j c(n/b^j)^d = a^k f(1) + \sum_{j=0}^{k-1} cn^d = a^k f(1) + kcn^d = a^{\log_b n} f(1) + c(\log_b n)n^d = n^{\log_b a} f(1) + cn^d \log_b n = n^d f(1) + cn^d \log_b n$. **31.** Let $k = \log_b n$ where n is a power of b. *Basis step:* If $n - 1$ and $k = 0$, then $c_1n^d + c_2n^{\log_b a} = c_1 + c_2 = b^dc/(b^d - a) + f(1) + b^dc/(a - b^d) = f(1)$. *Inductive step:* Assume true for k, where $n = b^k$. Then for $n = b^{k+1}$, $f(n) = af(n/b) + cn^d = a\{[b^dc/(b^d - a)](n/b)^d + [f(1) + b^dc/(a - b^d)] \cdot (n/b)^{\log_b a})\} + cn^d = b^dc/(b^d - a)n^d a/b^d + [f(1) + b^dc/(a - b^d)]n^{\log_b a} + cn^d = n^d[ac/(b^d - a) + c(b^d - a)/(b^d - a)] + [f(1) + b^dc/(a - b^dc)]n^{\log_b a} = [b^dc/(b^d - a)]n^d + [f(1) + b^dc/(a - b^d)]n^{\log_b a}$. **33.** If $a > b^d$, then $\log_b a > d$, so the second term dominates, giving $O(n^{\log_b a})$. **35.** $O(n^{\log_4 5})$ **37.** $O(n^3)$

Section 8.4

1. $f(x) = 2(x^6 - 1)/(x - 1)$ **3. a)** $f(x) = 2x(1 - x^6)/(1 - x)$ **b)** $x^3/(1 - x)$ **c)** $x/(1 - x^3)$ **d)** $2/(1 - 2x)$ **e)** $(1 + x)^7$ **f)** $2/(1 + x)$ **g)** $[1/(1 - x)] - x^2$ **h)** $x^3/(1 - x)^2$ **5. a)** $5/(1 - x)$ **b)** $1/(1 - 3x)$ **c)** $2x^3/(1 - x)$ **d)** $(3 - x)/(1 - x)^2$ **e)** $(1 + x)^8$ **7. a)** $a_0 = -64$, $a_1 = 144$, $a_2 = -108$, $a_3 = 27$, and $a_n = 0$ for all $n \geq 4$ **b)** The only nonzero coefficients are $a_0 = 1$, $a_3 = 3$, $a_6 = 3$, $a_9 = 1$. **c)** $a_n = 5^n$ **d)** $a_n = (-3)^{n-3}$ for $n \geq 3$, and $a_0 = a_1 = a_2 = 0$ **e)** $a_0 = 8$, $a_1 = 3$, $a_2 = 2$, $a_n = 0$ for odd n greater than 2 and $a_n = 1$ for even n greater than 2 **f)** $a_n = 1$ if n is a positive multiple 4, $a_n = -1$ if

$n < 4$, and $a_n = 0$ otherwise **g)** $a_n = n - 1$ for $n \geq 2$ and $a_0 = a_1 = 0$ **h)** $a_n = 2^{n+1}/n!$ **9. a)** 6 **b)** 3 **c)** 9 **d)** 0 **e)** 5 **11. a)** 1024 **b)** 11 **c)** 66 **d)** 292,864 **e)** 20,412 **13.** 10 **15.** 50 **17.** 20 **19.** $f(x) = 1/[(1-x)(1-x^2)(1-x^5)(1-x^{10})]$ **21.** 15 **23. a)** $x^4(1+x+x^2+x^3)^2/(1-x)$ **b)** 6 **25. a)** The coefficient of x^r in the power series expansion of $1/[(1-x^3)(1-x^4)(1-x^{20})]$ **b)** $1/(1-x^3-x^4-x^{20})$ **c)** 7 **d)** 3224 **27. a)** The generating function is $(1+x+x^2+x^3+x^4)(1+x+x^2)(1+x^2+x^4+x^6+\cdots)(x^3+x^4+x^5+x^6+\cdots)(1+x^5+x^{10}+x^{15}+\cdots) = x^3(1+x+x^2+x^3+x^4)(1+x+x^2)/[(1-x^2)(1-x)(1-x^5)] = x^3+3x^4+7x^5+12x^6+19x^7+27x^8+37x^9+48x^{10}+61x^{11}+75x^{12}+\cdots$. The coefficient of x^n is the answer. **b)** 75 **29. a)** 3 **b)** 29 **c)** 29 **d)** 242 **31. a)** 10 **b)** 49 **c)** 2 **d)** 4 **33. a)** $G(x) - a_0 - a_1x - a_2x^2$ **b)** $G(x^2)$ **c)** $x^4G(x)$ **d)** $G(2x)$ **e)** $\int_0^x G(t)dt$ **f)** $G(x)/(1-x)$ **35.** $a_k = 2 \cdot 3^k - 1$ **37.** $a_k = 18 \cdot 3^k - 12 \cdot 2^k$ **39.** $a_k = k^2 + 8k + 20 + (6k - 18)2^k$ **41.** Let $G(x) = \sum_{k=0}^{\infty} f_k x^k$. After shifting indices of summation and adding series, we see that $G(x) - xG(x) - x^2G(x) = f_0 + (f_1 - f_0)x + \sum_{k=2}^{\infty}(f_k - f_{k-1} - f_{k-2})x^k = 0 + x + \sum_{k=2}^{\infty} 0x^k$. Hence, $G(x) - xG(x) - x^2G(x) = x$. Solving for $G(x)$ gives $G(x) = x/(1-x-x^2)$. By the method of partial fractions, it can be shown that $x/(1-x-x^2) = (1/\sqrt{5})[1/(1-\alpha x) - 1/(1-\beta x)]$, where $\alpha = (1+\sqrt{5})/2$ and $\beta = (1-\sqrt{5})/2$. Using the fact that $1/(1-\alpha x) = \sum_{k=0}^{\infty} \alpha^k x^k$, it follows that $G(x) = (1/\sqrt{5}) \cdot \sum_{k=0}^{\infty}(\alpha^k - \beta^k)x^k$. Hence, $f_k = (1/\sqrt{5}) \cdot (\alpha^k - \beta^k)$. **43. a)** Let $G(x) = \sum_{n=0}^{\infty} C_n x^n$ be the generating function for $\{C_n\}$. Then $G(x)^2 = \sum_{n=0}^{\infty}(\sum_{k=0}^{n} C_k C_{n-k})x^n = \sum_{n=1}^{\infty}(\sum_{k=0}^{n-1} C_k C_{n-1-k})x^{n-1} = \sum_{n=1}^{\infty} C_n x^{n-1}$. Hence, $xG(x)^2 = \sum_{n=1}^{\infty} C_n x^n$, which implies that $xG(x)^2 - G(x) + 1 = 0$. Applying the quadratic formula shows that $G(x) = \frac{1 \pm \sqrt{1-4x}}{2x}$. We choose the minus sign in this formula because the choice of the plus sign leads to a division by zero. **b)** By Exercise 42, $(1-4x)^{-1/2} = \sum_{n=0}^{\infty}\binom{2n}{n}x^n$. Integrating term by term (which is valid by a theorem from calculus) shows that $\int_0^x (1-4t)^{-1/2}dt = \sum_{n=0}^{\infty}\frac{1}{n+1}\binom{2n}{n}x^{n+1} = x\sum_{n=0}^{\infty}\frac{1}{n+1}\binom{2n}{n}x^n$. Because $\int_0^x (1-4t)^{-1/2}dt = \frac{1-\sqrt{1-4x}}{2} = xG(x)$, equating coefficients shows that $C_n = \frac{1}{n+1}\binom{2n}{n}$. **c)** Verify the basis step for $n = 1, 2, 3, 4, 5$. Assume the inductive hypothesis that $C_j \geq 2^{j-1}$ for $1 \leq j < n$, where $n \geq 6$. Then $C_n = \sum_{k=0}^{n-1} C_k C_{n-k-1} \geq \sum_{k=1}^{n-2} C_k C_{n-k-1} \geq (n-2)2^{k-1}2^{n-k-2} = (n-2)2^{n-1}/4 \geq 2^{n-1}$. **45.** Applying the binomial theorem to the equality $(1+x)^{m+n} = (1+x)^m(1+x)^n$, shows that $\sum_{r=0}^{m+n} C(m+n, r)x^r = \sum_{r=0}^{m} C(m, r)x^r \cdot \sum_{r=0}^{n} C(n, r)x^r = \sum_{r=0}^{m+n}\left[\sum_{k=0}^{r} C(m, r-k)\, C(n, k)\right] x^r$. Comparing coefficients gives the desired identity. **47. a)** $2e^x$ **b)** e^{-x} **c)** e^{3x} **d)** $xe^x + e^x$ **49. a)** $a_n = (-1)^n$ **b)** $a_n = 3 \cdot 2^n$ **c)** $a_n = 3^n - 3 \cdot 2^n$ **d)** $a_n = (-2)^n$ for $n \geq 2$, $a_1 = -3$, $a_0 = 2$ **e)** $a_n = (-2)^n + n!$ **f)** $a_n = (-3)^n + n! \cdot 2^n$ for $n \geq 2$, $a_0 = 1$, $a_1 = -2$ **g)** $a_n = 0$ if n is odd and $a_n = n!/(n/2)!$ if n is even **51. a)** $a_n = 6a_{n-1} + 8^{n-1}$ for $n \geq 1$, $a_0 = 1$ **b)** The general solution of the associated linear homogeneous recurrence relation is $a_n^{(h)} = \alpha 6^n$. A particular solution is $a_n^{(p)} = \frac{1}{2} \cdot 8^n$. Hence, the general solution is $a_n = \alpha 6^n + \frac{1}{2} \cdot 8^n$. Using the initial condition, it follows that $\alpha = \frac{1}{2}$. Hence, $a_n = (6^n + 8^n)/2$. **c)** Let $G(x) = \sum_{k=0}^{\infty} a_k x^k$. Using the recurrence relation for $\{a_k\}$, it can be shown that $G(x) - 6xG(x) = (1-7x)/(1-8x)$. Hence, $G(x) = (1-7x)/[(1-6x)(1-8x)]$. Using partial fractions, it follows that $G(x) = (1/2)/(1-6x) + (1/2)/(1-8x)$. With the help of Table 1, it follows that $a_n = (6^n + 8^n)/2$. **53.** $\frac{1}{1-x} \cdot \frac{1}{1-x^2} \cdot \frac{1}{1-x^3} \cdots$ **55.** $(1+x)(1+x)^2(1+x)^3 \cdots$ **57.** The generating functions obtained in Exercises 54 and 55 are equal because $(1+x)(1+x^2)(1+x^3)\cdots = \frac{1-x^2}{1-x} \cdot \frac{1-x^4}{1-x^2} \cdot \frac{1-x^6}{1-x^3} \cdots = \frac{1}{1-x} \cdot \frac{1}{1-x^3} \cdot \frac{1}{1-x^5} \cdots$. **59. a)** $G_X(1) = \sum_{k=0}^{\infty} p(X=k) \cdot 1^k = \sum_{k=0}^{\infty} P(X=k) = 1$ **b)** $G_X'(1) = \frac{d}{dx}\sum_{k=0}^{\infty} p(X=k) \cdot x^k|_{x=1} = \sum_{k=0}^{\infty} p(X=k) \cdot k \cdot x^{k-1}|_{x=1} = \sum_{k=0}^{\infty} p(X=k) \cdot k = E(X)$ **c)** $G_X''(1) = \frac{d^2}{dx^2}\sum_{k=0}^{\infty} p(X=k) \cdot x^k|_{x=1} = \sum_{k=0}^{\infty} p(X=k) \cdot k(k-1) \cdot x^{k-2}|_{x=1} = \sum_{k=0}^{\infty} p(X=k) \cdot (k^2 - k) = V(X) + E(X)^2 - E(X)$. Combining this with part (b) gives the desired results. **61. a)** $G(x) = p^m/(1-qx)^m$ **b)** $V(x) = mq/p^2$

Section 8.5

1. a) 30 **b)** 29 **c)** 24 **d)** 18 **3.** 1% **5. a)** 300 **b)** 150 **c)** 175 **d)** 100 **7.** 492 **9.** 974 **11.** 610 **13.** 55 **15.** 248 **17.** 50,138 **19.** 234 **21.** $|A_1 \cup A_2 \cup A_3 \cup A_4 \cup A_5| = |A_1| + |A_2| + |A_3| + |A_4| + |A_5| - |A_1 \cap A_2| - |A_1 \cap A_3| - |A_1 \cap A_4| - |A_1 \cap A_5| - |A_2 \cap A_3| - |A_2 \cap A_4| - |A_2 \cap A_5| - |A_3 \cap A_4| - |A_3 \cap A_5| - |A_4 \cap A_5| + |A_1 \cap A_2 \cap A_3| + |A_1 \cap A_2 \cap A_4| + |A_1 \cap A_2 \cap A_5| + |A_1 \cap A_3 \cap A_4| + |A_1 \cap A_3 \cap A_5| + |A_1 \cap A_4 \cap A_5| + |A_2 \cap A_3 \cap A_4| + |A_2 \cap A_3 \cap A_5| + |A_2 \cap A_4 \cap A_5| + |A_3 \cap A_4 \cap A_5| - |A_1 \cap A_2 \cap A_3 \cap A_4| - |A_1 \cap A_2 \cap A_3 \cap A_5| - |A_1 \cap A_2 \cap A_4 \cap A_5| - |A_1 \cap A_3 \cap A_4 \cap A_5| - |A_2 \cap A_3 \cap A_4 \cap A_5| + |A_1 \cap A_2 \cap A_3 \cap A_4 \cap A_5|$ **23.** $|A_1 \cup A_2 \cup A_3 \cup A_4 \cup A_5 \cup A_6| = |A_1| + |A_2| + |A_3| + |A_4| + |A_5| + |A_6| - |A_1 \cap A_2| - |A_1 \cap A_3| - |A_1 \cap A_4| - |A_1 \cap A_5| - |A_1 \cap A_6| - |A_2 \cap A_3| - |A_2 \cap A_4| - |A_2 \cap A_5| - |A_2 \cap A_6| - |A_3 \cap A_4| - |A_3 \cap A_5| - |A_3 \cap A_6| - |A_4 \cap A_5| - |A_4 \cap A_6| - |A_5 \cap A_6|$ **25.** $p(E_1 \cup E_2 \cup E_3) = p(E_1) + p(E_2) + p(E_3) - p(E_1 \cap E_2) - p(E_1 \cap E_3) - p(E_2 \cap E_3) + p(E_1 \cap E_2 \cap E_3)$ **27.** 4972/71,295 **29.** $p(E_1 \cup E_2 \cup E_3 \cup E_4 \cup E_5) = p(E_1) + p(E_2) + p(E_3) + p(E_4) + p(E_5) - p(E_1 \cap E_2) - p(E_1 \cap E_3) - p(E_1 \cap E_4) - p(E_1 \cap E_5) - p(E_2 \cap E_3) - p(E_2 \cap E_4) - p(E_2 \cap E_5) - p(E_3 \cap E_4) - p(E_3 \cap E_5) - p(E_4 \cap E_5) + p(E_1 \cap E_2 \cap E_3) + p(E_1 \cap E_2 \cap E_4) + p(E_1 \cap E_2 \cap E_5) + p(E_1 \cap E_3 \cap E_4) + p(E_1 \cap E_3 \cap E_5) + p(E_1 \cap E_4 \cap E_5) + p(E_2 \cap E_3 \cap E_4) + P(E_2 \cap E_3 \cap E_5) + p(E_2 \cap E_4 \cap E_5) + p(E_3 \cap E_4 \cap E_5)$ **31.** $p\left(\bigcup_{i=1}^{n} E_i\right) = \sum_{1 \leq i \leq n} p(E_i) - \sum_{1 \leq i < j \leq n} p(E_i \cap E_j) + \sum_{1 \leq i < j < k \leq n} p(E_i \cap E_j \cap E_k) - \cdots + (-1)^{n+1} p\left(\bigcap_{i=1}^{n} E_i\right)$

Section 8.6

1. 75 **3.** 6 **5.** 46 **7.** 9875 **9.** 540 **11.** 2100 **13.** 1854 **15. a)** $D_{100}/100!$ **b)** $100D_{99}/100!$ **c)** $C(100,2)/100!$ **d)** 0 **e)** $1/100!$ **17.** 2,170,680 **19.** By Exercise 18 we have $D_n - nD_{n-1} = -[D_{n-1} - (n-1)D_{n-2}]$. Iterating, we have $D_n - nD_{n-1} = -[D_{n-1} - (n-1)D_{n-2}] = -[-(D_{n-2} - (n-2)D_{n-3})] = D_{n-2} - (n-2)D_{n-3} = \cdots = (-1)^n(D_2 - 2D_1) = (-1)^n$ because $D_2 = 1$ and $D_1 = 0$. **21.** When n is odd

23. $\phi(n) = n - \sum_{i=1}^{m} \frac{n}{p_i} + \sum_{1\le i<j\le m} \frac{n}{p_i p_j} - \cdots \pm \frac{n}{p_1 p_2 \cdots p_m} = n\prod_{i=1}^{m}\left(a - \frac{1}{p_i}\right)$ **25.** 4 **27.** There are n^m functions from a set with m elements to a set with n elements, $C(n, 1)(n-1)^m$ functions from a set with m elements to a set with n elements that miss exactly one element, $C(n, 2)(n-2)^m$ functions from a set with m elements to a set with n elements that miss exactly two elements, and so on, with $C(n, n-1)\cdot 1^m$ functions from a set with m elements to a set with n elements that miss exactly $n-1$ elements. Hence, by the principle of inclusion–exclusion, there are $n^m - C(n, 1)(n-1)^m + C(n, 2)(n-2)^m - \cdots + (-1)^{n-1}C(n, n-1)\cdot 1^m$ onto functions.

Supplementary Exercises

1. a) $A_n = 4A_{n-1}$ **b)** $A_1 = 40$ **c)** $A_n = 10\cdot 4^n$ **3. a)** $M_n = M_{n-1} + 160{,}000$ **b)** $M_1 = 186{,}000$ **c)** $M_n = 160{,}000n + 26{,}000$ **d)** $T_n = T_{n-1} + 160{,}000n + 26{,}000$ **e)** $T_n = 80{,}000n^2 + 106{,}000n$ **5. a)** $a_n = a_{n-2} + a_{n-3}$ **b)** $a_1 = 0, a_2 = 1, a_3 = 1$ **c)** $a_{12} = 12$ **7. a)** 2 **b)** 5 **c)** 8 **d)** 16 **9.** $a_n = 2^n$ **11.** $a_n = 2 + 4n/3 + n^2/2 + n^3/6$ **13.** $a_n = a_{n-2} + a_{n-3}$ **15. a)** Under the given conditions, one longest common subsequence ends at the last term in each sequence, so $a_m = b_n = c_p$. Furthermore, a longest common subsequence of what is left of the a-sequence and the b-sequence after those last terms are deleted has to form the beginning of a longest common subsequence of the original sequences. **b)** If $c_p \neq a_m$, then the longest common subsequence's appearance in the a-sequence must terminate before the end; therefore, the c-sequence must be a longest common subsequence of $a_1, a_2, \ldots, a_{m-1}$ and $b_1, b_2, \ldots, b_n$. The other half is similar.

17.
```
procedure howlong(a_1, ..., a_m, b_1, ..., b_n: sequences)
for i := 1 to m
    L(i, 0) := 0
for j := 1 to n
    L(0, j) := 0
for i := 1 to m
    for j := 1 to n
        if a_i = b_j then L(i, j) := L(i − 1, j − 1) + 1
        else L(i, j) := max(L(i, j − 1), L(i − 1, j))
return L(m, n)
```

19. $f(n) = (4n^2 - 1)/3$ **21.** $O(n^4)$ **23.** $O(n)$ **25.** Using just two comparisons, the algorithm is able to narrow the search for m down to the first half or the second half of the original sequence. Since the length of the sequence is cut in half each time, only about $2\log_2 n$ comparisons are needed in all. **27. a)** $18n + 18$ **b)** 18 **c)** 0 **29.** $\Delta(a_n b_n) = a_{n+1}b_{n+1} - a_n b_n = a_{n+1}(b_{n+1} - b_n) + b_n(a_{n+1} - a_n) = a_{n+1}\Delta b_n + b_n\Delta a_n$ **31. a)** Let $G(x) = \sum_{n=0}^{\infty} a_n x^n$. Then $G'(x) = \sum_{n=1}^{\infty} n a_n x^{n-1} = \sum_{n=0}^{\infty}(n+1)a_{n+1}x^n$. Therefore, $G'(x) - G(x) = \sum_{n=0}^{\infty}[(n+1)a_{n+1} - a_n]x^n = \sum_{n=0}^{\infty} x^n/n! = e^x$, as desired. That $G(0) = a_0 = 1$ is given. **b)** We have $[e^{-x}G(x)]' = e^{-x}G'(x) - e^{-x}G(x) = e^{-x}[G'(x) - G(x)] = e^{-x}\cdot e^x = 1$. Hence, $e^{-x}G(x) = x + c$, where c is a constant. Consequently, $G(x) = xe^x + ce^x$. Because $G(0) = 1$, it follows that $c = 1$. **c)** We have $G(x) = \sum_{n=0}^{\infty} x^{n+1}/n! + \sum_{n=0}^{\infty} x^n/n! = \sum_{n=1}^{\infty} x^n/(n-1)! + \sum_{n=0}^{\infty} x^n/n!$. Therefore, $a_n = 1/(n-1)! + 1/n!$ for all $n \ge 1$, and $a_0 = 1$. **33.** 7 **35.** 110 **37.** 0 **39. a)** 19 **b)** 65 **c)** 122 **d)** 167 **e)** 168 **41.** $D_{n-1}/(n-1)!$ **43.** 11/32

CHAPTER 9

Section 9.1

1. a) {(0, 0), (1, 1), (2, 2), (3, 3)} **b)** {(1, 3), (2, 2), (3, 1), (4, 0)} **c)** {(1, 0), (2, 0), (2, 1), (3, 0), (3, 1), (3, 2), (4, 0), (4, 1), (4, 2), (4, 3)} **d)** {(1, 0), (1, 1), (1, 2), (1, 3), (2, 0), (2, 2), (3, 0), (3, 3), (4, 0)} **e)** {(0, 1), (1, 0), (1, 1), (1, 2), (1, 3), (2, 1), (2, 3), (3, 1), (3, 2), (4, 1), (4, 3)} **f)** {(1, 2), (2, 1), (2, 2)} **3. a)** Transitive **b)** Reflexive, symmetric, transitive **c)** Symmetric **d)** Antisymmetric **e)** Reflexive, symmetric, antisymmetric, transitive **f)** None of these properties **5. a)** Reflexive, transitive **b)** Symmetric **c)** Symmetric **d)** Symmetric **7. a)** Symmetric **b)** Symmetric, transitive **c)** Symmetric **d)** Reflexive, symmetric, transitive **e)** Reflexive, transitive **f)** Reflexive, symmetric, transitive **g)** Antisymmetric **h)** Antisymmetric, transitive **9.** Each of the three properties is vacuously satisfied. **11.** (c), (d), (f) **13. a)** Not irreflexive **b)** Not irreflexive **c)** Not irreflexive **d)** Not irreflexive **15.** Yes, for instance {(1, 1)} on {1, 2} **17.** $(a, b) \in R$ if and only if a is taller than b **19.** (a) **21.** None **23.** $\forall a \forall b\,[(a, b) \in R \rightarrow (b, a) \notin R]$ **25.** 2^{mn} **27. a)** $\{(a, b) \mid b$ divides $a\}$ **b)** $\{(a, b) \mid a$ does not divide $b\}$ **29.** The graph of f^{-1} **31. a)** $\{(a, b) \mid a$ is required to read or has read $b\}$ **b)** $\{(a, b) \mid a$ is required to read and has read $b\}$ **c)** $\{(a, b) \mid$ either a is required to read b but has not read it or a has read b but is not required to$\}$ **d)** $\{(a, b) \mid a$ is required to read b but has not read it$\}$ **e)** $\{(a, b) \mid a$ has read b but is not required to$\}$ **33.** $S \circ R = \{(a, b) \mid a$ is a parent of b and b has a sibling$\}$, $R \circ S = \{(a, b) \mid a$ is a sibling of one of b's parents$\}$ **35. a)** $\mathbf{R}^2$ **b)** R_6 **c)** R_3 **d)** R_3 **e)** $\emptyset$ **f)** R_1 **g)** R_4 **h)** R_4 **37. a)** R_1 **b)** R_2 **c)** R_3 **d)** $\mathbf{R}^2$ **e)** R_3 **f)** $\mathbf{R}^2$ **g)** $\mathbf{R}^2$ **h)** $\mathbf{R}^2$ **39.** $S_1^2 = \{(a, b) \in \mathbf{Z}^2 \mid a > b + 1\}$, $S_2^2 = S_2$, $S_3^2 = \{(a, b) \in \mathbf{Z}^2 \mid a < b - 1\}$, $S_4^2 = S_4$, $S_5^2 = S_5$, $S_6^2 = \mathbf{Z}^2$ **41.** b got his or her doctorate under someone who got his or her doctorate under a; there is a sequence of $n + 1$ people, starting with a and ending with b, such that each is the advisor of the next person in the sequence **43. a)** $\{(a, b) \mid a - b \equiv 0, 3, 4, 6, 8,$ or $9 \pmod{12}\}$ **b)** $\{(a, b) \mid a \equiv b \pmod{12}\}$ **c)** $\{(a, b) \mid a - b \equiv 3, 6,$ or $9 \pmod{12}\}$ **d)** $\{(a, b) \mid a - b \equiv 4$ or $8 \pmod{12}\}$ **e)** $\{(a, b) \mid a - b \equiv 3, 4, 6, 8,$ or $9 \pmod{12}\}$ **45.** 8 **47. a)** 65,536 **b)** 32,768 **49. a)** $2^{n(n+1)/2}$ **b)** $2^n 3^{n(n-1)/2}$ **c)** $3^{n(n-1)/2}$ **d)** $2^{n(n-1)}$ **e)** $2^{n(n-1)/2}$ **f)** $2^{n^2} - 2\cdot 2^{n(n-1)}$ **51.** There may be no such b. **53.** If R is symmetric and $(a, b) \in R$, then $(b, a) \in R$, so $(a, b) \in R^{-1}$. Hence, $R \subseteq R^{-1}$. Similarly, $R^{-1} \subseteq R$. So $R = R^{-1}$. Conversely, if $R = R^{-1}$ and $(a, b) \in R$, then $(a, b) \in R^{-1}$, so $(b, a) \in R$. Thus, R is symmetric. **55.** R is reflexive if and only if $(a, a) \in R$ for all $a \in A$ if and only if $(a, a) \in R^{-1}$ [because $(a, a) \in R$ if and only if $(a, a) \in R^{-1}$] if and only if R^{-1} is reflexive. **57.** Use mathematical induction. The result is trivial for $n = 1$. Assume R^n is reflexive and transitive. By Theorem 1, $R^{n+1} \subseteq R$. To see that $R \subseteq R^{n+1} = R^n \circ R$, let $(a, b) \in R$. By the inductive hypothesis, $R^n = R$ and hence, is reflexive. Thus, $(b, b) \in R^n$. Therefore, $(a, b) \in R^{n+1}$. **59.** Use mathematical induction. The result is trivial for $n = 1$. Assume R^n is reflexive. Then $(a, a) \in R^n$ for all $a \in A$ and $(a, a) \in R$. Thus, $(a, a) \in R^n \circ R = R^{n+1}$ for all $a \in A$. **61.** No, for instance, take $R = \{(1, 2), (2, 1)\}$.

Section 9.2

1. $\{(1, 2, 3), (1, 2, 4), (1, 3, 4), (2, 3, 4)\}$ **3.** (Nadir, 122, 34, Detroit, 08:10), (Acme, 221, 22, Denver, 08:17), (Acme, 122, 33, Anchorage, 08:22), (Acme, 323, 34, Honolulu 08:30), (Nadir, 199, 13, Detroit, 08:47), (Acme, 222, 22, Denver, 09:10), (Nadir, 322, 34, Detroit, 09:44) **5.** Airline and flight number, airline and departure time **7. a)** Yes **b)** No **c)** No **9. a)** Social Security number **b)** There are no two people with the same name who happen to have the same street address. **c)** There are no two people with the same name living together. **11.** (Nadir, 122, 34, Detroit, 08:10), (Nadir, 199, 13, Detroit, 08:47), (Nadir, 322, 34, Detroit, 09:44) **13.** (Nadir, 122, 34, Detroit, 08:10), (Nadir, 199, 13, Detroit, 08:47), (Nadir, 322, 34, Detroit, 09:44), (Acme, 221, 22, Denver, 08:17), (Acme, 222, 22, Denver, 09:10) **15.** $P_{3,5,6}$

17.

Airline	*Destination*
Nadir	Detroit
Acme	Denver
Acme	Anchorage
Acme	Honolulu

19.

Supplier	*Part_number*	*Project*	*Quantity*	*Color_code*
23	1092	1	2	2
23	1101	3	1	1
23	9048	4	12	2
31	4975	3	6	2
31	3477	2	25	2
32	6984	4	10	1
32	9191	2	80	4
33	1001	1	14	8

21. Both sides of this equation pick out the subset of R consisting of those n-tuples satisfying both conditions C_1 and C_2. **23.** Both sides of this equation pick out the set of n-tuples that are in R, are in S, and satisfy condition C. **25.** Both sides of this equation pick out the m-tuples consisting of i_1th, i_2th, … , i_mth components of n-tuples in either R or S. **27.** Let $R = \{(a, b)\}$ and $S = \{(a, c)\}$, $n = 2$, $m = 1$, and $i_1 = 1$; $P_1(R - S) = \{(a)\}$, but $P_1(R) - P_1(S) = \emptyset$. **29. a)** J_2 followed by $P_{1,3}$ **b)** (23, 1), (23, 3), (31, 3), (32, 4) **31.** There is no primary key. **33. a)** count 5, support 5/6 **b)** {diapers}, {milk}, and {diapers, milk} **35.** support 1/3, confidence 2/3 **37.** 1 **39. a)** $P(I) = \sigma(I)/|T|$, $P(J) = \sigma(J)/|T|$, $P(I \text{ and } J) = \sigma(I \cup J)/|T|$, so $P(I \text{ and } J) = P(I) \cdot P(J)$ if and only if $\sigma(I \cup J)/|T| = (\sigma(I)/|T|) \cdot (\sigma(J)/|T|)$, which means that the numerator and denominator of the lift are equal. **b)** 4/5 **c)** 16/15 **41.** For each of the n items, there are 3 choices: the item can be in I or in J or in neither. By the product rule, there are 3^n ways to make these decisions.

Section 9.3

1. a) $\begin{bmatrix} 1 & 1 & 1 \\ 0 & 0 & 0 \\ 0 & 0 & 0 \end{bmatrix}$ **b)** $\begin{bmatrix} 0 & 1 & 0 \\ 1 & 1 & 0 \\ 0 & 0 & 1 \end{bmatrix}$ **c)** $\begin{bmatrix} 1 & 1 & 1 \\ 0 & 1 & 1 \\ 0 & 0 & 1 \end{bmatrix}$ **d)** $\begin{bmatrix} 0 & 0 & 1 \\ 0 & 0 & 0 \\ 1 & 0 & 0 \end{bmatrix}$

3. a) (1, 1), (1, 3), (2, 2), (3, 1), (3, 3) **b)** (1, 2), (2, 2), (3, 2) **c)** (1, 1), (1, 2), (1, 3), (2, 1), (2, 3), (3, 1), (3, 2), (3, 3) **5.** The relation is irreflexive if and only if the main diagonal of the matrix contains only 0s. **7. a)** Reflexive, symmetric, transitive **b)** Antisymmetric, transitive **c)** Symmetric **9. a)** 4950 **b)** 9900 **c)** 99 **d)** 100 **e)** 1 **11.** Change each 0 to a 1 and each 1 to a 0.

13. a) $\begin{bmatrix} 0 & 1 & 1 \\ 1 & 1 & 0 \\ 1 & 0 & 1 \end{bmatrix}$ **b)** $\begin{bmatrix} 1 & 0 & 0 \\ 0 & 0 & 1 \\ 0 & 1 & 0 \end{bmatrix}$ **c)** $\begin{bmatrix} 1 & 1 & 1 \\ 1 & 1 & 1 \\ 1 & 1 & 1 \end{bmatrix}$

15. a) $\begin{bmatrix} 0 & 0 & 1 \\ 1 & 1 & 0 \\ 0 & 1 & 1 \end{bmatrix}$ **b)** $\begin{bmatrix} 1 & 1 & 0 \\ 0 & 1 & 1 \\ 1 & 1 & 1 \end{bmatrix}$ **c)** $\begin{bmatrix} 0 & 1 & 1 \\ 1 & 1 & 1 \\ 1 & 1 & 1 \end{bmatrix}$

17. $n^2 - k$

19.

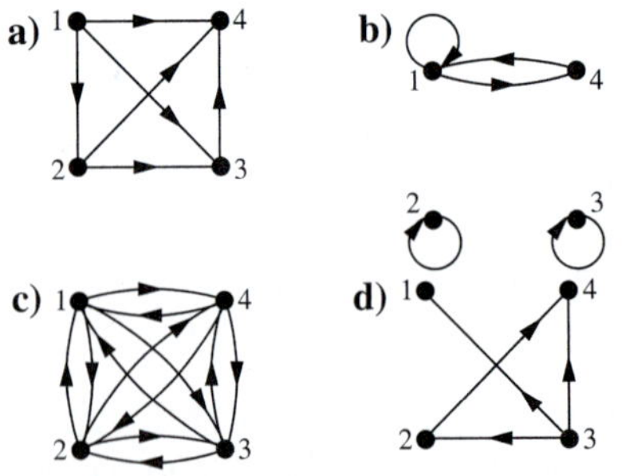

21. For simplicity we have indicated pairs of edges between the same two vertices in opposite directions by using a double arrowhead, rather than drawing two separate lines.

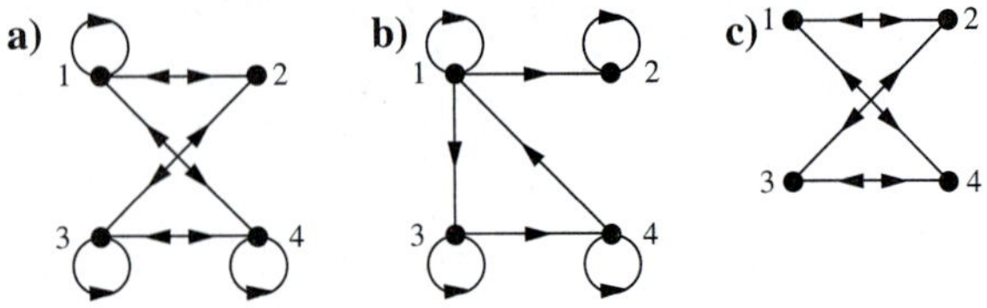

23. $\{(a, b), (a, c), (b, c), (c, b)\}$ **25.** (a, c), (b, a), (c, d), (d, b) **27.** $\{(a, a), (a, b), (a, c), (b, a), (b, b), (b, c), (c, a),$

$(c, b), (d, d)\}$ **29.** The relation is asymmetric if and only if the directed graph has no loops and no closed paths of length 2. **31.** Exercise 23: irreflexive. Exercise 24: reflexive, antisymmetric, transitive. Exercise 25: irreflexive, antisymmetric. **33.** Reverse the direction on every edge in the digraph for R. **35.** Proof by mathematical induction. *Basis step:* Trivial for $n = 1$. *Inductive step:* Assume true for k. Because $R^{k+1} = R^k \circ R$, its matrix is $\mathbf{M}_R \odot \mathbf{M}_{R^k}$. By the inductive hypothesis this is $\mathbf{M}_R \odot \mathbf{M}_R^{[k]} = \mathbf{M}_R^{[k+1]}$.

Section 9.4

1. a) $\{(0, 0), (0, 1), (1, 1), (1, 2), (2, 0), (2, 2), (3, 0), (3, 3)\}$ **b)** $\{(0, 1), (0, 2), (0, 3), (1, 0), (1, 1), (1, 2), (2, 0), (2, 1), (2, 2), (3, 0)\}$ **3.** $\{(a, b) \mid a \text{ divides } b \text{ or } b \text{ divides } a\}$

5. 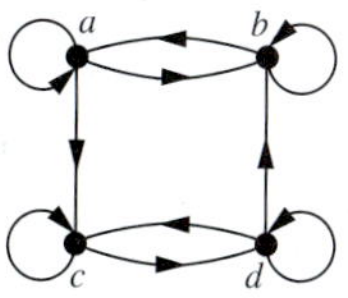

7. 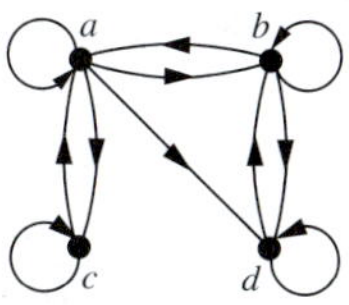

9. a) 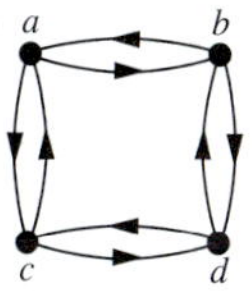

b) 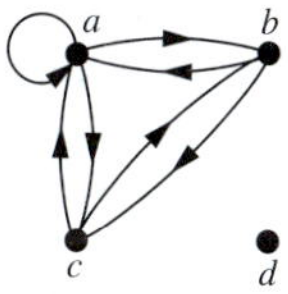

c) 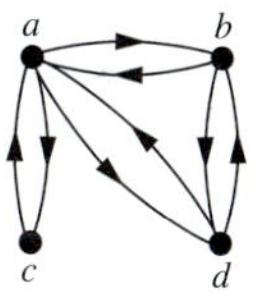

11. a) 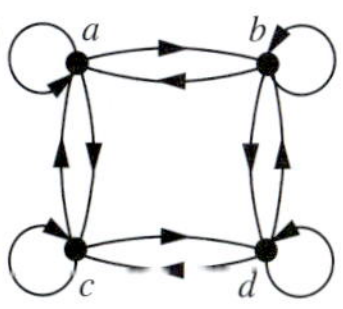

b) 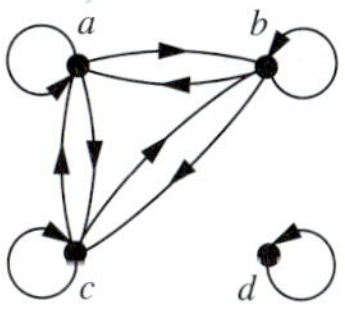

c) 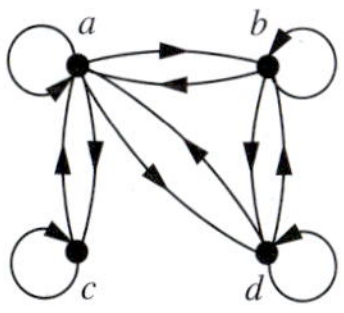

13. The symmetric closure of R is $R \cup R^{-1}$. $\mathbf{M}_{R \cup R^{-1}} = \mathbf{M}_R \vee \mathbf{M}_{R^{-1}} = \mathbf{M}_R \vee \mathbf{M}_R^t$. **15.** Only when R is irreflexive, in which case it is its own closure. **17.** a, a, a, a; a, b, e, a; a, d, e, a; b, c, c, b; b, e, a, b; c, b, c, c; c, c, b, c; c, c, c, c; d, e, a, d; d, e, e, d; e, a, b, e; e, a, d, e; e, d, e, e; e, e, d, e; e, e, e, e **19. a)** $\{(1, 1), (1, 5), (2, 3), (3, 1), (3, 2), (3, 3), (3, 4), (4, 1), (4, 5), (5, 3), (5, 4)\}$ **b)** $\{(1, 1), (1, 2), (1, 3), (1, 4), (2, 1), (2, 5), (3, 1), (3, 3), (3, 4), (3, 5), (4, 1), (4, 2), (4, 3), (4, 4), (5, 1), (5, 3), (5, 5)\}$ **c)** $\{(1, 1), (1, 3), (1, 4), (1, 5), (2, 1), (2, 2), (2, 3), (2, 4), (3, 1), (3, 2), (3, 3), (3, 4), (3, 5), (4, 1), (4, 3), (4, 4), (4, 5), (5, 1), (5, 2), (5, 3), (5, 4), (5, 5)\}$ **d)** $\{(1, 1), (1, 2), (1, 3), (1, 4), (1, 5), (2, 1), (2, 3), (2, 4), (2, 5), (3, 1), (3, 2), (3, 3), (3, 4), (3, 5), (4, 1), (4, 2), (4, 3), (4, 4), (4, 5), (5, 1), (5, 2), (5, 3), (5, 4), (5, 5)\}$ **e)** $\{(1, 1), (1, 2), (1, 3), (1, 4), (1, 5), (2, 1), (2, 2), (2, 3), (2, 4), (2, 5), (3, 1), (3, 2), (3, 3), (3, 4), (3, 5), (4, 1), (4, 2), (4, 3), (4, 4), (4, 5), (5, 1), (5, 2), (5, 3), (5, 4), (5, 5)\}$ **f)** $\{(1, 1), (1, 2), (1, 3), (1, 4), (1, 5), (2, 1), (2, 2), (2, 3), (2, 4), (2, 5), (3, 1), (3, 2), (3, 3), (3, 4), (3, 5), (4, 1), (4, 2), (4, 3), (4, 4), (4, 5), (5, 1), (5, 2), (5, 3), (5, 4), (5, 5)\}$ **21. a)** If there is a student c who shares a class with a and a class with b **b)** If there are two students c and d such that a and c share a class, c and d share a class, and d and b share a class **c)** If there is a sequence $s_0, \ldots, s_n$ of students with $n \geq 1$ such that $s_0 = a$, $s_n = b$, and for each $i = 1, 2, \ldots, n$, s_i and s_{i-1} share a class **23.** The result follows from $(R^*)^{-1} = \left(\bigcup_{n=1}^{\infty} R^n\right)^{-1} = \bigcup_{n=1}^{\infty} (R^n)^{-1} = \bigcup_{n=1}^{\infty} R^n = R^*$.

25. a) $\begin{bmatrix} 1 & 1 & 1 & 1 \\ 1 & 1 & 1 & 1 \\ 1 & 1 & 1 & 1 \\ 1 & 1 & 1 & 1 \end{bmatrix}$ **b)** $\begin{bmatrix} 0 & 0 & 0 & 0 \\ 1 & 0 & 1 & 1 \\ 1 & 0 & 1 & 1 \\ 1 & 0 & 1 & 1 \end{bmatrix}$

c) $\begin{bmatrix} 0 & 1 & 1 & 1 \\ 0 & 0 & 1 & 1 \\ 0 & 0 & 0 & 1 \\ 0 & 0 & 0 & 0 \end{bmatrix}$ **d)** $\begin{bmatrix} 1 & 1 & 1 & 1 \\ 1 & 1 & 1 & 1 \\ 1 & 1 & 1 & 1 \\ 1 & 1 & 1 & 1 \end{bmatrix}$

27. Answers same as for Exercise 25. **29. a)** $\{(1, 1), (1, 2), (1, 4), (2, 2), (3, 3), (4, 1), (4, 2), (4, 4)\}$ **b)** $\{(1, 1), (1, 2), (1, 4), (2, 1), (2, 2), (2, 4), (3, 3), (4, 1), (4, 2), (4, 4)\}$ **c)** $\{(1, 1), (1, 2), (1, 4), (2, 1), (2, 2), (2, 4), (3, 3), (4, 1), (4, 2), (4, 4)\}$ **31.** Algorithm 1: $O(n^{3.8})$; Algorithm 2: $O(n^3)$ **33.** Initialize with $\mathbf{A} := \mathbf{M}_R \vee \mathbf{I}_n$ and loop only for $i := 2$ to $n - 1$. **35. a)** Because R is reflexive, every relation containing it must also be reflexive. **b)** Both $\{(0, 0), (0, 1), (0, 2), (1, 1), (2, 2)\}$ and $\{(0, 0), (0, 1), (1, 0), (1, 1), (2, 2)\}$ contain R and have an odd number of elements, but neither is a subset of the other.

Section 9.5

1. a) Equivalence relation **b)** Not reflexive, not transitive **c)** Equivalence relation **d)** Not transitive **e)** Not symmetric, not transitive **3. a)** Equivalence relation **b)** Not transitive **c)** Not reflexive, not symmetric, not transitive **d)** Equivalence relation **e)** Not reflexive, not transitive **5.** Many answers are possible. (1) Two buildings are equivalent if they were opened during the same year; an equivalence class consists of the set of buildings opened in a given year (as long as there was at least one building opened that year). (2) Two buildings are equivalent if they have the same number of stories; the equivalence classes are the set of 1-story buildings, the set of 2-story buildings, and so on (one class for each n for which there is at least one n-story building). (3) Every building in which you have a class is equivalent to every building in which you have a class (including itself), and every building in which you don't have a class is equivalent to every building in which you don't have a class (including itself); there are two equivalence classes—the set of buildings in which you have a class and the set of buildings in which you don't have a class (assuming these are nonempty). **7.** The statement "p is equivalent to q" means that p and q have the same entries in their truth tables. R is reflexive, because p has the same truth table as p. R is symmetric, for if p and q have the same truth table, then q and

p have the same truth table. If p and q have the same entries in their truth tables and q and r have the same entries in their truth tables, then p and r also do, so R is transitive. The equivalence class of **T** is the set of all tautologies; the equivalence class of **F** is the set of all contradictions. **9. a)** $(x, x) \in R$ because $f(x) = f(x)$. Hence, R is reflexive. $(x, y) \in R$ if and only if $f(x) = f(y)$, which holds if and only if $f(y) = f(x)$ if and only if $(y, x) \in R$. Hence, R is symmetric. If $(x, y) \in R$ and $(y, z) \in R$, then $f(x) = f(y)$ and $f(y) = f(z)$. Hence, $f(x) = f(z)$. Thus, $(x, z) \in R$. It follows that R is transitive. **b)** The sets $f^{-1}(b)$ for b in the range of f **11.** Let x be a bit string of length 3 or more. Because x agrees with itself in the first three bits, $(x, x) \in R$. Hence, R is reflexive. Suppose that $(x, y) \in R$. Then x and y agree in the first three bits. Hence, y and x agree in the first three bits. Thus, $(y, x) \in R$. If (x, y) and (y, z) are in R, then x and y agree in the first three bits, as do y and z. Hence, x and z agree in the first three bits. Hence, $(x, z) \in R$. It follows that R is transitive. **13.** This follows from Exercise 9, where f is the function that takes a bit string of length 3 or more to the ordered pair with its first bit as the first component and the third bit as its second component. **15.** For reflexivity, $((a, b), (a, b)) \in R$ because $a + b = b + a$. For symmetry, if $((a, b), (c, d)) \in R$, then $a + d = b + c$, so $c + b = d + a$, so $((c, d), (a, b)) \in R$. For transitivity, if $((a, b), (c, d)) \in R$ and $((c, d), (e, f)) \in R$, then $a+d = b+c$ and $c+e = d+f$, so $a+d+c+e = b+c+d+f$, so $a+e = b+f$, so $((a, b), (e, f)) \in R$. An easier solution is to note that by algebra, the given condition is the same as the condition that $f((a, b)) = f((c, d))$, where $f((x, y)) = x - y$; therefore, by Exercise 9 this is an equivalence relation. **17. a)** This follows from Exercise 9, where the function f from the set of differentiable functions (from **R** to **R**) to the set of functions (from **R** to **R**) is the differentiation operator. **b)** The set of all functions of the form $g(x) = x^2 + C$ for some constant C **19.** This follows from Exercise 9, where the function f from the set of all URLs to the set of all Web pages is the function that assigns to each URL the Web page for that URL. **21.** No **23.** No **25.** R is reflexive because a bit string s has the same number of 1s as itself. R is symmetric because s and t having the same number of 1s implies that t and s do. R is transitive because s and t having the same number of 1s, and t and u having the same number of 1s implies that s and u have the same number of 1s. **27. a)** The sets of people of the same age **b)** The sets of people with the same two parents **29.** The set of all bit strings with exactly two 1s **31. a)** The set of all bit strings of length 3 **b)** The set of all bit strings of length 4 that end with a 1 **c)** The set of all bit strings of length 5 that end 11 **d)** The set of all bit strings of length 8 that end 10101 **33.** Each of the 15 bit strings of length less than four is in an equivalence class by itself: $[\lambda]_{R_4} = \{\lambda\}$, $[0]_{R_4} = \{0\}$, $[1]_{R_4} = \{1\}$, $[00]_{R_4} = \{00\}$, $[01]_{R_4} = \{01\}$, …, $[111]_{R_4} = \{111\}$. The remaining 16 equivalence classes are determined by the bit strings of length 4: $[0000]_{R_4} = \{0000, 00000, 00001, 000000, 000001, 000010, 000011, 0000000, \ldots\}$, $[0001]_{R_4} = \{0001, 00010, 00011, 000100, 000101, 000110, 000111, 0001000, \ldots\}$, …, $[1111]_{R_4} = \{1111, 11110, 11111, 111100, 111101, 111110, 111111, 1111000, \ldots\}$ **35. a)** $[2]_5 = \{i \mid i \equiv 2 \pmod 5\} = \{\ldots, -8, -3, 2, 7, 12, \ldots\}$ **b)** $[3]_5 = \{i \mid i \equiv 3 \pmod 5\} = \{\ldots, -7, -2, 3, 8, 13, \ldots\}$ **c)** $[6]_5 = \{i \mid i \equiv 6 \pmod 5\} = \{\ldots, -9, -4, 1, 6, 11, \ldots\}$ **d)** $[-3]_5 = \{i \mid i \equiv -3 \pmod 5\} = \{\ldots, -8, -3, 2, 7, 12, \ldots\}$ **37.** $\{6n + k \mid n \in \mathbf{Z}\}$ for $k \in \{0, 1, 2, 3, 4, 5\}$ **39. a)** $[(1, 2)] = \{(a, b) \mid a - b = -1\} = \{(1, 2), (3, 4), (4, 5), (5, 6), \ldots\}$ **b)** Each equivalence class can be interpreted as an integer (negative, positive, or zero); specifically, $[(a, b)]$ can be interpreted as $a - b$. **41. a)** No **b)** Yes **c)** Yes **d)** No **43.** (a), (c), (e) **45.** (b), (d), (e) **47. a)** $\{(0, 0), (1, 1), (1, 2), (2, 1), (2, 2), (3, 3), (3, 4), (3, 5), (4, 3), (4, 4), (4, 5), (5, 3), (5, 4), (5, 5)\}$ **b)** $\{(0, 0), (0, 1), (1, 0), (1, 1), (2, 2), (2, 3), (3, 2), (3, 3), (4, 4), (4, 5), (5, 4), (5, 5)\}$ **c)** $\{(0, 0), (0, 1), (0, 2), (1, 0), (1, 1), (1, 2), (2, 0), (2, 1), (2, 2), (3, 3), (3, 4), (3, 5), (4, 3), (4, 4), (4, 5), (5, 3), (5, 4), (5, 5)\}$ **d)** $\{(0, 0), (1, 1), (2, 2), (3, 3), (4, 4), (5, 5)\}$ **49.** $[0]_6 \subseteq [0]_3$, $[1]_6 \subseteq [1]_3$, $[2]_6 \subseteq [2]_3$, $[3]_6 \subseteq [0]_3$, $[4]_6 \subseteq [1]_3$, $[5]_6 \subseteq [2]_3$ **51.** Let A be a set in the first partition. Pick a particular element x of A. The set of all bit strings of length 16 that agree with x on the last four bits is one of the sets in the second partition, and clearly every string in A is in that set. **53.** We claim that each equivalence class $[x]_{R_{31}}$ is a subset of the equivalence class $[x]_{R_8}$. To show this, choose an arbitrary element $y \in [x]_{R_{31}}$. Then y is equivalent to x under R_{31}, so either $y = x$ or y and x are each at least 31 characters long and agree on their first 31 characters. Because strings that are at least 31 characters long and agree on their first 31 characters perforce are at least 8 characters long and agree on their first 8 characters, we know that either $y = x$ or y and x are each at least 8 characters long and agree on their first 8 characters. This means that y is equivalent to x under R_8, so $y \in [x]_{R_8}$. **55.** $\{(a, a), (a, b), (a, c), (b, a), (b, b), (b, c), (c, a), (c, b), (c, c), (d, d), (d, e), (e, d), (e, e)\}$ **57. a) Z** **b)** $\{n+\frac{1}{2} \mid n \in \mathbf{Z}\}$ **59. a)** R is reflexive because any coloring can be obtained from itself via a 360-degree rotation. To see that R is symmetric and transitive, use the fact that each rotation is the composition of two reflections and conversely the composition of two reflections is a rotation. Hence, (C_1, C_2) belongs to R if and only if C_2 can be obtained from C_1 by a composition of reflections. So if (C_1, C_2) belongs to R, so does (C_2, C_1) because the inverse of the composition of reflections is also a composition of reflections (in the opposite order). Hence, R is symmetric. To see that R is transitive, suppose (C_1, C_2) and (C_2, C_3) belong to R. Taking the composition of the reflections in each case yields a composition of reflections, showing that (C_1, C_3) belongs to R. **b)** We express colorings with sequences of length four, with r and b denoting red and blue, respectively. We list letters denoting the colors of the upper left square, upper right square, lower left square, and lower right square, in that order. The equivalence classes are: $\{rrrr\}$, $\{bbbb\}$, $\{rrrb, rrbr, rbrr, brrr\}$, $\{bbbr, bbrb, brbb, rbbb\}$, $\{rbbr, brrb\}$, $\{rrbb, brbr, bbrr, rbrb\}$. **61.** 5 **63.** Yes **65.** R **67.** First form the reflexive closure of R, then form the symmetric closure of the reflexive closure, and finally form the transitive closure of the symmetric closure of the reflexive closure. **69.** $p(0) = 1$, $p(1) = 1$, $p(2) = 2$, $p(3) = 5$,

$p(4) = 15, p(5) = 52, p(6) = 203, p(7) = 877, p(8) = 4140,$ $p(9) = 21147, p(10) = 115975$

Section 9.6

1. a) Is a partial ordering **b)** Not antisymmetric, not transitive **c)** Is a partial ordering **d)** Is a partial ordering **e)** Not antisymmetric, not transitive **3. a)** No **b)** No **c)** Yes **5. a)** Yes **b)** No **c)** Yes **d)** No **7. a)** No **b)** Yes **c)** No **9.** No **11.** Yes **13. a)** $\{(0, 0), (1, 0), (1, 1), (2, 0), (2, 1), (2, 2)\}$ **b)** $(\mathbf{Z}, \leq)$ **c)** $(P(\mathbf{Z}), \subseteq)$ **d)** $(\mathbf{Z}^+$, "is a multiple of") **15. a)** $\{0\}$ and $\{1\}$, for instance **b)** 4 and 6, for instance **17. a)** $(1, 1, 2) < (1, 2, 1)$ **b)** $(0, 1, 2, 3) < (0, 1, 3, 2)$ **c)** $(0, 1, 1, 1, 0) < (1, 0, 1, 0, 1)$ **19.** $0 < 0001 < 001 < 01 < 010 < 0101 < 011 < 11$

21.

23. a)
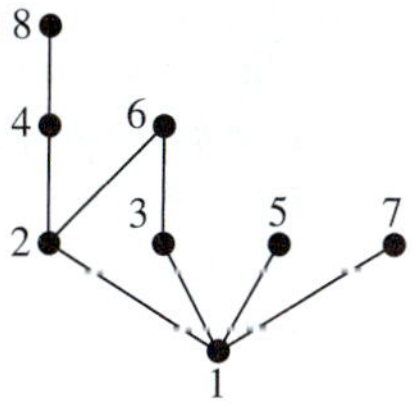

b)
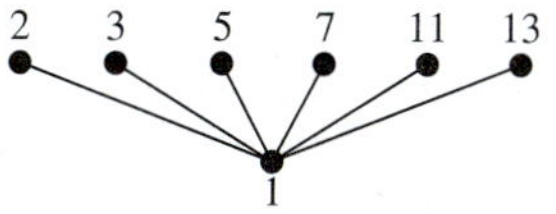

c)
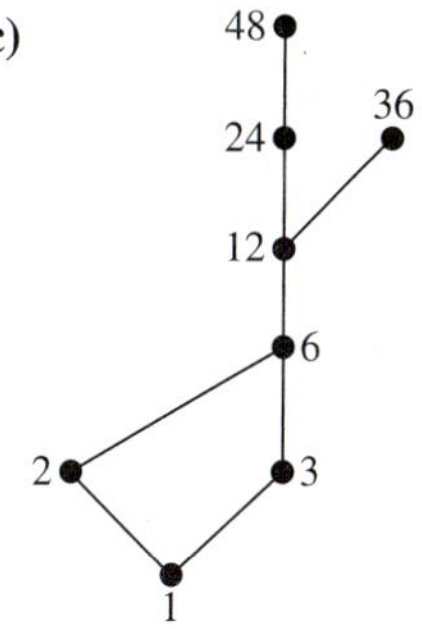

d)

25. $(a, b), (a, c), (a, d), (b, c), (b, d), (a, a), (b, b), (c, c), (d, d)$ **27.** $(a, a), (a, g), (a, d), (a, e), (a, f), (b, b), (b, g), (b, d), (b, e), (b, f), (c, c), (c, g), (c, d), (c, e), (c, f), (g, d), (g, e), (g, f), (g, g), (d, d), (e, e), (f, f)$ **29.** $(\emptyset, \{a\}), (\emptyset, \{b\}), (\emptyset, \{c\}), (\{a\}, \{a, b\}), (\{a\}, \{a, c\}), (\{b\}, \{a, b\}), (\{b\}, \{b, c\}), (\{c\}, \{a, c\}), (\{c\}, \{b, c\}), (\{a, b\}, \{a, b, c\}), (\{a, c\}, \{a, b, c\})(\{b, c\}, \{a, b, c\})$ **31.** Let $(S, \preccurlyeq)$ be a finite poset. We will show that this poset is the reflexive transitive closure of its covering relation. Suppose that (a, b) is in the reflexive transitive closure of the covering relation. Then $a = b$ or $a \prec b$, so $a \preccurlyeq b$, or else there is a sequence $a_1, a_2, \ldots, a_n$ such that $a \prec a_1 \prec a_2 \prec \cdots \prec a_n \prec b$, in which case again $a \preccurlyeq b$ by the transitivity of $\preccurlyeq$. Conversely, suppose that $a \prec b$. If $a = b$ then (a, b) is in the reflexive transitive closure of the covering relation. If $a \prec b$ and there is no z such that $a \prec z \prec b$, then (a, b) is in the covering relation and therefore in its reflexive transitive closure. Otherwise, let $a \prec a_1 \prec a_2 \prec \cdots \prec a_n \prec b$ be a longest possible sequence of this form (which exists because the poset is finite). Then no intermediate elements can be inserted, so each pair $(a, a_1), (a_1, a_2), \ldots, (a_n, b)$ is in the covering relation, so again (a, b) is in its reflexive transitive closure. **33. a)** 24, 45 **b)** 3, 5 **c)** No **d)** No **e)** 15, 45 **f)** 15 **g)** 15, 5, 3 **h)** 15 **35. a)** $\{1, 2\}, \{1, 3, 4\}, \{2, 3, 4\}$ **b)** $\{1\}, \{2\}, \{4\}$ **c)** No **d)** No **e)** $\{2, 4\}, \{2, 3, 4\}$ **f)** $\{2, 4\}$ **g)** $\{3, 4\}, \{4\}$ **h)** $\{3, 4\}$ **37.** Because $(a, b) \preccurlyeq (a, b)$, $\preccurlyeq$ is reflexive. If $(a_1, a_2) \preccurlyeq (b_1, b_2)$ and $(a_1, a_2) \neq (b_1, b_2)$, either $a_1 \prec b_1$, or $a_1 = b_1$ and $a_2 \prec b_2$. In either case, (b_1, b_2) is not less than or equal to (a_1, a_2). Hence, $\preccurlyeq$ is antisymmetric. Suppose that $(a_1, a_2) \prec (b_1, b_2) \prec (c_1, c_2)$. Then if $a_1 \prec b_1$ or $b_1 \prec c_1$, we have $a_1 \prec c_1$, so $(a_1, a_2) \prec (c_1, c_2)$, but if $a_1 = b_1 = c_1$, then $a_2 \prec b_2 \prec c_2$, which implies that $(a_1, a_2) \prec (c_1, c_2)$. Hence, $\preccurlyeq$ is transitive. **39.** Because $(s, t) \preccurlyeq (s, t)$, $\preccurlyeq$ is reflexive. If $(s, t) \preccurlyeq (u, v)$ and $(u, v) \preccurlyeq (s, t)$, then $s \preccurlyeq u \preccurlyeq s$ and $t \preccurlyeq v \preccurlyeq t$; hence, $s = u$ and $t = v$. Hence, $\preccurlyeq$ is antisymmetric. Suppose that $(s, t) \preccurlyeq (u, v) \preccurlyeq (w, x)$. Then $s \preccurlyeq u$, $t \preccurlyeq v$, $u \preccurlyeq w$, and $v \preccurlyeq x$. It follows that $s \preccurlyeq w$ and $t \preccurlyeq x$. Hence, $(s, t) \preccurlyeq (w, x)$. Hence, $\preccurlyeq$ is transitive. **41. a)** Suppose that x is maximal and that y is the largest element. Then $x \preccurlyeq y$. Because x is not less than y, it follows that $x = y$. By Exercise 40(a) y is unique. Hence, x is unique. **b)** Suppose that x is minimal and that y is the smallest element. Then $x \succcurlyeq y$. Because x is not greater than y, it follows that $x = y$. By Exercise 40(b) y is unique. Hence, x is unique. **43. a)** Yes **b)** No **c)** Yes **45.** Use mathematical induction. Let $P(n)$ be "Every subset with n elements from a lattice has a least upper bound and a greatest lower bound." *Basis step:* $P(1)$ is true because the least upper bound and greatest lower bound of $\{x\}$ are both x. *Inductive step:* Assume that $P(k)$ is true. Let S be a set with $k + 1$ elements. Let $x \in S$ and $S' = S - \{x\}$. Because S' has k elements, by the inductive hypothesis, it has a least upper bound y and a greatest lower bound a. Now because we are in a lattice, there are elements $z = \text{lub}(x, y)$ and $b = \text{glb}(x, a)$. We are done if we can show that z is the least upper bound of S and b is the greatest lower bound of S. To show that z is the least upper bound of S, first note that if $w \in S$, then $w = x$ or $w \in S'$. If $w = x$ then $w \preccurlyeq z$ because z is the least upper bound of x and y. If $w \in S'$, then $w \preccurlyeq z$ because $w \preccurlyeq y$, which is true because y is the least upper bound of S', and $y \preccurlyeq z$, which is true because $z = \text{lub}(x, y)$. To see that z is the least upper bound of S, suppose that u is an upper bound of S. Note that such an element u must be an upper bound of x and y, but because $z = \text{lub}(x, y)$, it follows that $z \preccurlyeq u$. We omit the similar argument that b is the greatest lower bound of S. **47. a)** No **b)** Yes **c)** (*Proprietary*, {*Cheetah*, *Puma*}), (*Restricted*, {*Cheetah*, *Puma*}), (*Registered*, {*Cheetah*, *Puma*}), (*Proprietary*, {*Cheetah*, *Puma*,

Impala}), (*Restricted*, {*Cheetah*, *Puma*, *Impala*}), (*Registered*, {*Cheetah*, *Puma*, *Impala*}) **d)** (*Nonproprietary*, {*Impala*, *Puma*}), (*Proprietary*, {*Impala*, *Puma*}), (*Restricted*, {*Impala*, *Puma*}), (*Nonproprietary*, {*Impala*}), (*Proprietary*, {*Impala*}), (*Restricted*, {*Impala*}), (*Nonproprietary*, {*Puma*}), (*Proprietary*, {*Puma*}), (*Restricted*, {*Puma*}), (*Nonproprietary*, ∅), (*Proprietary*, ∅), (*Restricted*, ∅) **49.** Let Π be the set of all partitions of a set S with $P_1 \preccurlyeq P_2$ if P_1 is a refinement of P_2, that is, if every set in P_1 is a subset of a set in P_2. First, we show that $(\Pi, \preccurlyeq)$ is a poset. Because $P \preccurlyeq P$ for every partition P, $\preccurlyeq$ is reflexive. Now suppose that $P_1 \preccurlyeq P_2$ and $P_2 \preccurlyeq P_1$. Let $T \in P_1$. Because $P_1 \preccurlyeq P_2$, there is a set $T' \in P_2$ such that $T \subseteq T'$. Because $P_2 \preccurlyeq P_1$ there is a set $T'' \in P_1$ such that $T' \subseteq T''$. It follows that $T \subseteq T''$. But because P_1 is a partition, $T = T''$, which implies that $T = T'$ because $T \subseteq T' \subseteq T''$. Thus, $T \in P_2$. By reversing the roles of P_1 and P_2 it follows that every set in P_2 is also in P_1. Hence, $P_1 = P_2$ and $\preccurlyeq$ is antisymmetric. Next, suppose that $P_1 \preccurlyeq P_2$ and $P_2 \preccurlyeq P_3$. Let $T \in P_1$. Then there is a set $T' \in P_2$ such that $T \subseteq T'$. Because $P_2 \preccurlyeq P_3$ there is a set $T'' \in P_3$ such that $T' \subseteq T''$. This means that $T \subseteq T''$. Hence, $P_1 \preccurlyeq P_3$. It follows that $\preccurlyeq$ is transitive. The greatest lower bound of the partitions P_1 and P_2 is the partition P whose subsets are the nonempty sets of the form $T_1 \cap T_2$ where $T_1 \in P_1$ and $T_2 \in P_2$. We omit the justification of this statement here. The least upper bound of the partitions P_1 and P_2 is the partition that corresponds to the equivalence relation in which $x \in S$ is related to $y \in S$ if there is a sequence $x = x_0, x_1, x_2, \ldots, x_n = y$ for some nonnegative integer n such that for each i from 1 to n, x_{i-1} and x_i are in the same element of P_1 or of P_2. We omit the details that this is an equivalence relation and the details of the proof that this is the least upper bound of the two partitions. **51.** By Exercise 45 there is a least upper bound and a greatest lower bound for the entire finite lattice. By definition these elements are the greatest and least elements, respectively. **53.** The least element of a subset of $\mathbf{Z}^+ \times \mathbf{Z}^+$ is that pair that has the smallest possible first coordinate, and, if there is more than one such pair, that pair among those that has the smallest second coordinate. **55.** If x is an integer in a decreasing sequence of elements of this poset, then at most $|x|$ elements can follow x in the sequence, namely, integers whose absolute values are $|x| - 1, |x| - 2, \ldots, 1, 0$. Therefore, there can be no infinite decreasing sequence. This is not a totally ordered set, because 5 and −5, for example, are incomparable. **57.** To find which of two rational numbers is larger, write them with a positive common denominator and compare numerators. To show that this set is dense, suppose that $x < y$ are two rational numbers. Then their average, i.e., $(x+y)/2$, is a rational number between them. **59.** Let $(S, \preccurlyeq)$ be a partially ordered set. It is enough to show that every nonempty subset of S contains a least element if and only if there is no infinite decreasing sequence of elements $a_1, a_2, a_3, \ldots$ in S (i.e., where $a_{i+1} \prec a_i$ for all i). An infinite decreasing sequence of elements clearly has no least element. Conversely, let A be any nonempty subset of S that has no least element. Because A is nonempty, choose $a_1 \in A$. Because a_1 is not the least element of A, choose $a_2 \in A$ with $a_2 \prec a_1$. Because a_2 is not the least element of A, choose $a_3 \in A$ with $a_3 \prec a_2$. Continue in this manner, producing an infinite decreasing sequence in S. **61.** $a \prec_t b \prec_t c \prec_t d \prec_t e \prec_t f \prec_t g \prec_t h \prec_t i \prec_t j \prec_t k \prec_t l \prec_t m$ **63.** $1 \prec 5 \prec 2 \prec 4 \prec 12 \prec 20$, $1 \prec 2 \prec 5 \prec 4 \prec 12 \prec 20$, $1 \prec 2 \prec 4 \prec 5 \prec 12 \prec 20$, $1 \prec 2 \prec 4 \prec 12 \prec 5 \prec 20$, $1 \prec 5 \prec 2 \prec 4 \prec 20 \prec 12$, $1 \prec 2 \prec 5 \prec 4 \prec 20 \prec 12$, $1 \prec 2 \prec 4 \prec 5 \prec 20 \prec 12$ **65.** $A \prec C \prec E \prec B \prec D \prec F \prec G$, $A \prec E \prec C \prec B \prec D \prec F \prec G$, $C \prec A \prec E \prec B \prec D \prec F \prec G$, $C \prec E \prec A \prec B \prec D \prec F \prec G$, $E \prec A \prec C \prec B \prec D \prec F \prec G$, $E \prec C \prec A \prec B \prec D \prec F \prec G$, $A \prec C \prec B \prec E \prec D \prec F \prec G$, $C \prec A \prec B \prec E \prec D \prec F \prec G$, $A \prec C \prec B \prec D \prec E \prec F \prec G$, $C \prec A \prec B \prec D \prec E \prec F \prec G$, $A \prec C \prec E \prec B \prec F \prec D \prec G$, $A \prec E \prec C \prec B \prec F \prec D \prec G$, $C \prec A \prec E \prec B \prec F \prec D \prec G$, $C \prec E \prec A \prec B \prec F \prec D \prec G$, $E \prec A \prec C \prec B \prec F \prec D \prec G$, $E \prec C \prec A \prec B \prec F \prec D \prec G$, $A \prec C \prec B \prec E \prec F \prec D \prec G$, $C \prec A \prec B \prec E \prec F \prec D \prec G$ **67.** Determine user needs ≺ Write functional requirements ≺ Set up test sites ≺ Develop system requirements ≺ Write documentation ≺ Develop module A ≺ Develop module B ≺ Develop module C ≺ Integrate modules ≺ α test ≺ β test ≺ Completion

Supplementary Exercises

1. a) Irreflexive (we do not include the empty string), symmetric **b)** Irreflexive, symmetric **c)** Irreflexive, antisymmetric, transitive **3.** $((a, b), (a, b)) \in R$ because $a+b = a+b$. Hence, R is reflexive. If $((a, b), (c, d)) \in R$ then $a + d = b + c$, so that $c + b = d + a$. It follows that $((c, d), (a, b)) \in R$. Hence, R is symmetric. Suppose that $((a, b), (c, d))$ and $((c, d), (e, f))$ belong to R. Then $a+d = b+c$ and $c+f = d+e$. Adding these two equations and subtracting $c + d$ from both sides gives $a + f = b+e$. Hence, $((a, b), (e, f))$ belongs to R. Hence, R is transitive. **5.** Suppose that $(a, b) \in R$. Because $(b, b) \in R$ it follows that $(a, b) \in R^2$. **7.** Yes, yes **9.** Yes, yes **11.** Two records with identical keys in the projection would have identical keys in the original. **13.** $(\Delta \cup R)^{-1} = \Delta^{-1} \cup R^{-1} = \Delta \cup R^{-1}$ **15. a)** $R = \{(a, b), (a, c)\}$. The transitive closure of the symmetric closure of R is $\{(a, a), (a, b), (a, c), (b, a), (b, b), (b, c), (c, a), (c, b), (c, c)\}$ and is different from the symmetric closure of the transitive closure of R, which is $\{(a, b), (a, c), (b, a), (c, a)\}$. **b)** Suppose that (a, b) is in the symmetric closure of the transitive closure of R. We must show that (a, b) is in the transitive closure of the symmetric closure of R. We know that at least one of (a, b) and (b, a) is in the transitive closure of R. Hence, there is either a path from a to b in R or a path from b to a in R (or both). In the former case, there is a path from a to b in the symmetric closure of R. In the latter case, we can form a path from a to b in the symmetric closure of R by reversing the directions of all the edges in a path from b to a, going backward. Hence, (a, b) is in the transitive closure of the symmetric closure of R. **17.** The closure of S with respect to property **P** is a relation with property **P** that contains R because $R \subseteq S$. Hence, the closure of S with respect to property **P** contains the closure of R with respect to property **P**. **19.** Use the basic idea of Warshall's algorithm, except let $w_{ij}^{[k]}$ equal the length of the longest path from v_i to v_j using interior vertices with subscripts not exceeding k, and equal to -1 if there is no such path. To find $w_{ij}^{[k]}$ from the entries of $\mathbf{W}_{k-1}$, determine for each pair (i, j) whether there are paths from v_i to v_k and from v_k to v_j using no vertices labeled greater than k. If either $w_{ik}^{[k-1]}$ or $w_{kj}^{[k-1]}$ is -1, then such a pair of paths does not exist, so set $w_{ij}^{[k]} = w_{ij}^{[k-1]}$. If such a pair of paths exists, then there are two possibilities. If $w_{kk}^{[k-1]} > 0$, there are paths of arbitrary long length from v_i to v_j, so set $w_{ij}^{[k]} = \infty$. If $w_{kk}^{[k-1]} = 0$, set $w_{ij}^{[k-1]} = \max(w_{ij}^{[k-1]}, w_{ik}^{[k-1]} + w_{kj}^{[k-1]})$. (Initially take $\mathbf{W}_0 = \mathbf{M}_R$.) **21.** 25 **23.** Because $A_i \cap B_j$ is a subset of A_i and of B_j, the collection of subsets is a refinement of each of the given partitions. We must show that it is a partition. By construction, each of these sets is nonempty. To see that their union is S, suppose that $s \in S$. Because P_1 and P_2 are partitions of S, there are sets A_i and B_j such that $s \in A_i$ and $s \in B_j$. Therefore, $s \in A_i \cap B_j$. Hence, the union of these sets is S. To see that they are pairwise disjoint, note that unless $i = i'$ and $j = j'$, $(A_i \cap B_j) \cap (A_{i'} \cap B_{j'}) = (A_i \cap A_{i'}) \cap (B_j \cap B_{j'}) = \emptyset$. **25.** The subset relation is a partial ordering on any collection of sets, because it is reflexive, antisymmetric, and transitive. Here the collection of sets is $\mathbf{R}(S)$. **27.** Find recipe $\prec$ Buy seafood $\prec$ Buy groceries $\prec$ Wash shellfish $\prec$ Cut ginger and garlic $\prec$ Clean fish $\prec$ Steam rice $\prec$ Cut fish $\prec$ Wash vegetables $\prec$ Chop water chestnuts $\prec$ Make garnishes $\prec$ Cook in wok $\prec$ Arrange on platter $\prec$ Serve **29. a)** The only antichain with more than one element is $\{c, d\}$. **b)** The only antichains with more than one element are $\{b, c\}$, $\{c, e\}$, and $\{d, e\}$. **c)** The only antichains with more than one element are $\{a, b\}$, $\{a, c\}$, $\{b, c\}$, $\{a, b, c\}$, $\{d, e\}$, $\{d, f\}$, $\{e, f\}$, and $\{d, e, f\}$. **31.** Let $(S, \preccurlyeq)$ be a finite poset, and let A be a maximal chain. Because $(A, \preccurlyeq)$ is also a poset it must have a minimal element m. Suppose that m is not minimal in S. Then there would be an element a of S with $a \prec m$. However, this would make the set $A \cup \{a\}$ a larger chain than A. To show this, we must show that a is comparable with every element of A. Because m is comparable with every element of A and m is minimal, it follows that $m \prec x$ when x is in A and $x \neq m$. Because $a \prec m$ and $m \prec x$, the transitive law shows that $a \prec x$ for every element of A. **33.** Let aRb denote that a is a descendant of b. By Exercise 32, if no set of $n+1$ people none of whom is a descendant of any other (an antichain) exists, then $k \leq n$, so the set can be partitioned into $k \leq n$ chains. By the pigeonhole principle, at least one of these chains contains at least $m+1$ people. **35.** We prove by contradiction that if S has no infinite decreasing sequence and $\forall x\left(\{\forall y[y \prec x \rightarrow P(y)]\} \rightarrow P(x)\right)$, then $P(x)$ is true for all $x \in S$. If it does not hold that $P(x)$ is true for all $x \in S$, let x_1 be an element of S such that $P(x_1)$ is not true. Then by the conditional statement already given, it must be the case that $\forall y[y \prec x_1 \rightarrow P(y)]$ is not true. This means that there is some x_2 with $x_2 \prec x_1$ such that $P(x_2)$ is not true. Again invoking the conditional statement, we get an $x_3 \prec x_2$ such that $P(x_3)$ is not true, and so on forever. This contradicts the well-foundedness of our poset. Therefore, $P(x)$ is true for all $x \in S$. **37.** Suppose that R is a quasi-ordering. Because R is reflexive, if $a \in A$, then $(a, a) \in R$. This implies that $(a, a) \in R^{-1}$. Hence, $a \in R \cap R^{-1}$. It follows that $R \cap R^{-1}$ is reflexive. $R \cap R^{-1}$ is symmetric for any relation R because, for any relation R, if $(a, b) \in R$ then $(b, a) \in R^{-1}$ and vice versa. To show that $R \cap R^{-1}$ is transitive, suppose that $(a, b) \in R \cap R^{-1}$ and $(b, c) \in R \cap R^{-1}$. Because $(a, b) \in R$ and $(b, c) \in R$, $(a, c) \in R$, because R is transitive. Similarly, because $(a, b) \in R^{-1}$ and $(b, c) \in R^{-1}$, $(b, a) \in R$ and $(c, b) \in R$, so $(c, a) \in R$ and $(a, c) \in R^{-1}$. Hence, $(a, c) \in R \cap R^{-1}$. It follows that $R \cap R^{-1}$ is an equivalence relation. **39. a)** Because $\text{glb}(x, y) = \text{glb}(y, x)$ and $\text{lub}(x, y) = \text{lub}(y, x)$, it follows that $x \wedge y = y \wedge x$ and $x \vee y = y \vee x$. **b)** Using the definition, $(x \wedge y) \wedge z$ is a lower bound of x, y, and z that is greater than every other lower bound. Because x, y, and z play interchangeable roles, $x \wedge (y \wedge z)$ is the same element. Similarly, $(x \vee y) \vee z$ is an upper bound of x, y, and z that is less than every other upper bound. Because x, y, and z play interchangeable roles, $x \vee (y \vee z)$ is the same element. **c)** To show that $x \wedge (x \vee y) = x$ it is sufficient to show that x is the greatest lower bound of x, and $x \vee y$. Note that x is a lower bound of x, and because $x \vee y$ is by definition greater than x, x is a lower bound for it as well. Therefore, x is a lower bound. But any lower bound of x has to be less than x, so x is the greatest lower bound. The second statement is the dual of the first; we omit its proof. **d)** x is a lower, and an upper, bound for itself and itself, and the greatest, and least, such bound. **41. a)** Because 1 is the only element greater than or equal to 1, it is the only upper bound for 1 and therefore the only possible value of the least upper bound of x and 1. **b)** Because $x \preccurlyeq 1$, x is a lower bound for both x and 1 and no other lower bound can be greater than x, so $x \wedge 1 = x$. **c)** Because $0 \preccurlyeq x$, x is an upper bound for both x and 0 and no other bound can be less than x, so $x \vee 0 = x$. **d)** Because 0 is the only element less than or equal to 0, it is the only lower bound for 0 and therefore the only possible value of the greatest lower bound of x and 0. **43.** $L = (S, \subseteq)$ where $S = \{\emptyset, \{1\}, \{2\}, \{3\}, \{1, 2\}, \{2, 3\}, \{1, 2, 3\}\}$ **45.** Yes **47.** The complement of a subset $X \subseteq S$ is its complement $S - X$. To prove this, note that $X \vee (S - X) = 1$ and $X \wedge (S - X) = 0$ because $X \cup (S - X) = S$ and $X \cap (S - X) = \emptyset$. **49.** Think of the rectangular grid as representing elements in a matrix. Thus, we number from top to bottom and within that from left to right. The partial order is that $(a, b) \preceq (c, d)$ iff $a \leq c$ and $b \leq d$. Note that $(1, 1)$ is the least element under this relation. The rules for Chomp as explained in Chapter 1 coincide with the rules stated in the preamble here. But now we can identify the point (a, b) with the natural number $p^{a-1}q^{b-1}$ for all a and b with $1 \leq a \leq m$ and $1 \leq b \leq n$. This identifies the points in the rectangular grid with the set S in this exercise, and the partial order $\preceq$ just described is the same as the divides relation, because $p^{a-1}q^{b-1} \mid p^{c-1}q^{d-1}$ if and only if the exponent of p on the left does not exceed the exponent of p on the right, and similarly for q.

CHAPTER 10

Section 10.1

1. a)

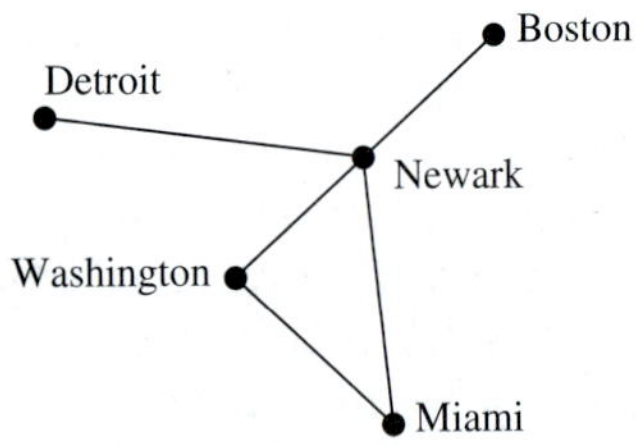

b)

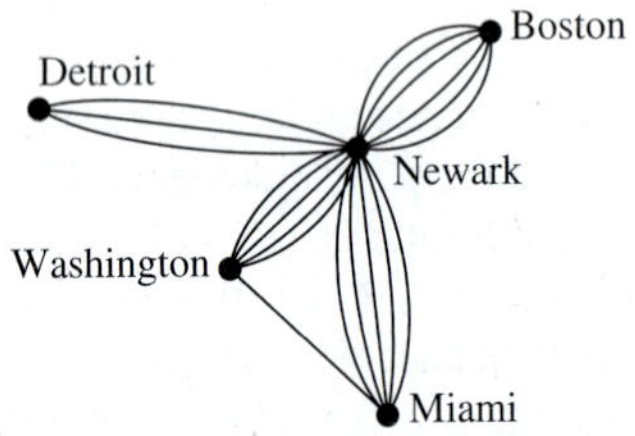

c)

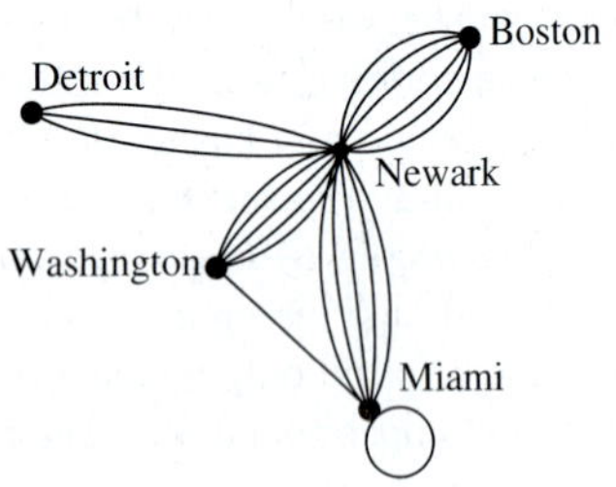

d)

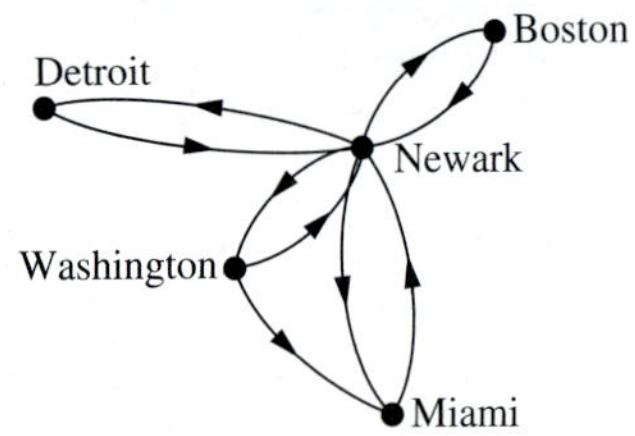

e)

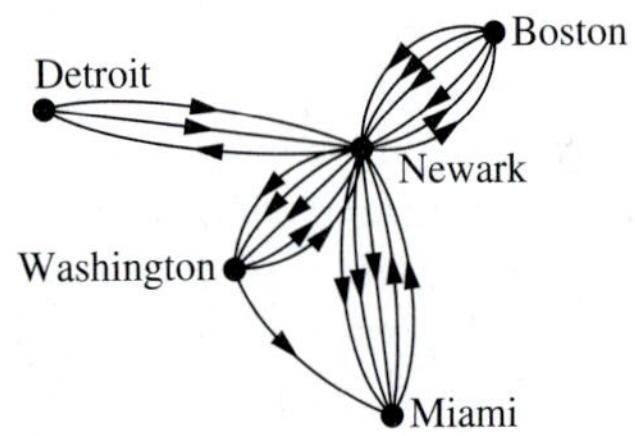

3. Simple graph **5.** Pseudograph **7.** Directed graph **9.** Directed multigraph **11.** If uRv, then there is an edge associated with $\{u, v\}$. But $\{u, v\} = \{v, u\}$, so this edge is associated with $\{v, u\}$ and therefore vRu. Thus, by definition, R is a symmetric relation. A simple graph does not allow loops; therefore, uRu never holds, and so by definition R is irreflexive.

13. a)

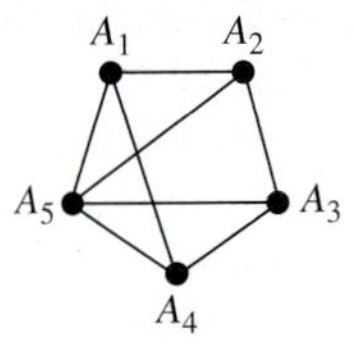

b)

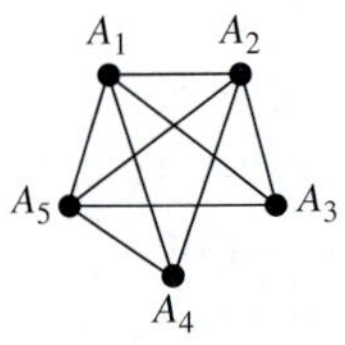

c)

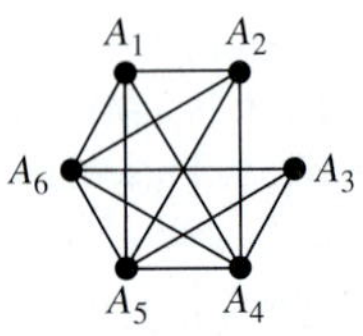

15.

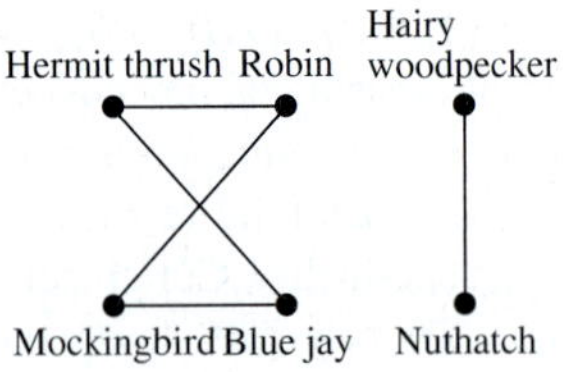

17.

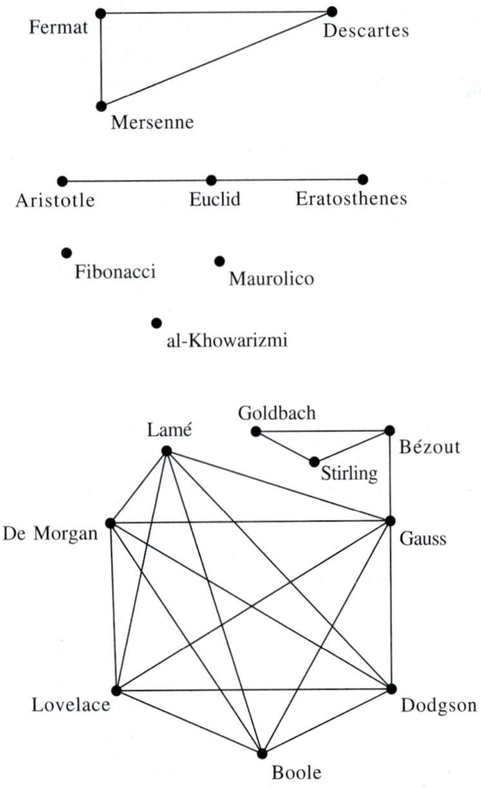

19.

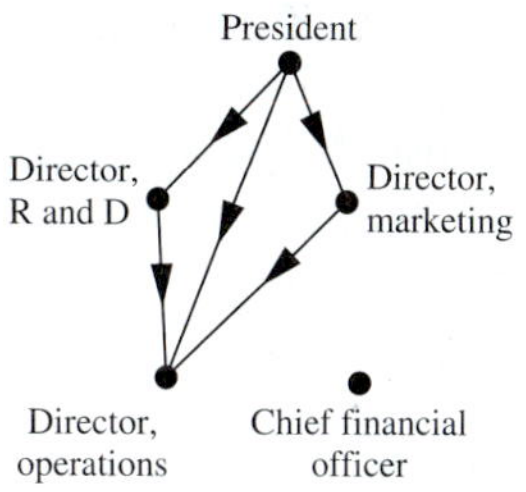

21.

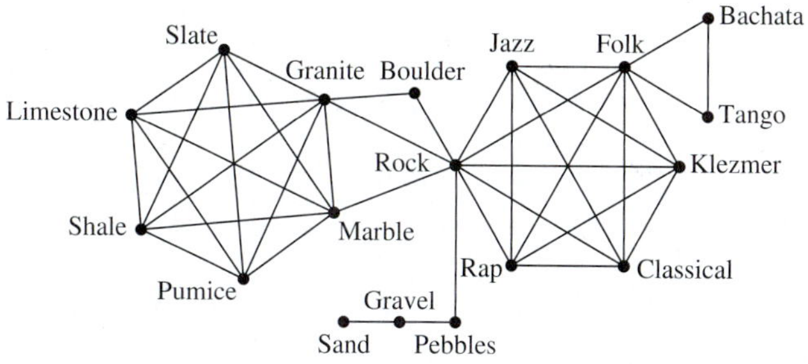

23.

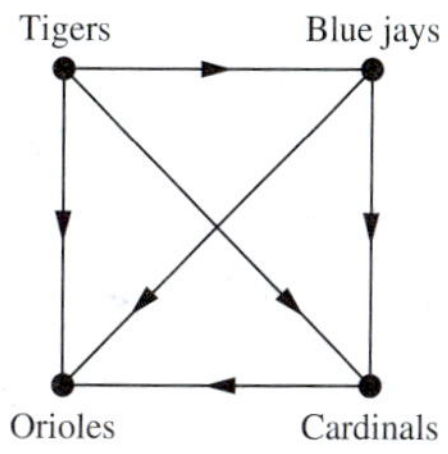

25. We find the telephone numbers in the call graph for February that are not present in the call graph for January and vice versa. For each number we find, we make a list of the numbers they called or were called by using the edges in the call graph. We examine these lists to find new telephone numbers in February that had similar calling patterns to defunct telephone numbers in January. **27.** We use the graph model that has email addresses as vertices and for each message sent, an edge from the email address of the sender to the email address of the recipient. For each email address, we can make a list of other addresses they sent messages to and a list of other addresses from which they received messages. If two email addresses had almost the same pattern, we conclude that these addresses might have belonged to the same person who had recently changed his or her email address. **29.** Let V be the set of people at the party. Let E be the set of ordered pairs (u, v) in $V \times V$ such that u knows v's name. The edges are directed, but multiple edges are not allowed. Literally, there is a loop at each vertex, but for simplicity, the model could omit the loops. **31.** Vertices are the courses; edges are directed; edge uv means that course u is prerequisite for course v; courses without prerequisites are vertices with in-degree 0; courses that are not prerequisite for any other courses are vertices with out-degree 0. **33.** Let the set of vertices be a set of people, and two vertices are joined by an edge if the two people were ever married. Restricted to monogamous marriage between cisgendered heterosexuals, this graph has the property that there are two types of vertices (men and women), and every edge joins vertices of opposite types.

35.

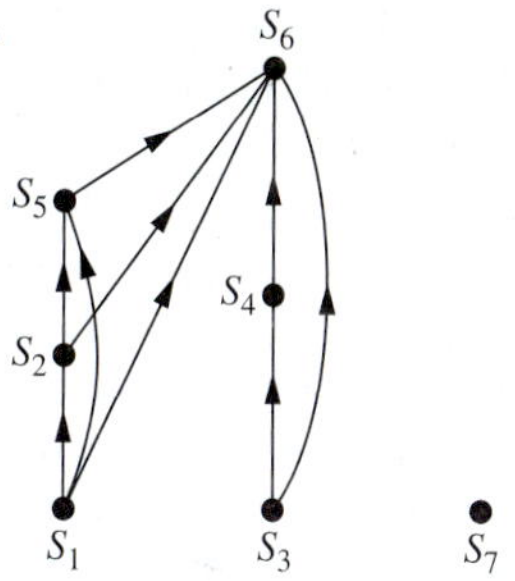

37. Represent people in the group by vertices. Put a directed edge into the graph for every pair of vertices. Label the edge from the vertex representing A to the vertex representing B with a + (plus) if A likes B, a − (minus) if A dislikes B, and a 0 if A is neutral about B.

Section 10.2

1. $v = 6$; $e = 6$; $\deg(a) = 2$, $\deg(b) = 4$, $\deg(c) = 1$, $\deg(d) = 0$, $\deg(e) = 2$, $\deg(f) = 3$; c is pendant; d is isolated. **3.** $v = 9$; $e = 12$; $\deg(a) = 3$, $\deg(b) = 2$, $\deg(c) = 4$, $\deg(d) = 0$, $\deg(e) = 6$, $\deg(f) = 0$; $\deg(g) = 4$; $\deg(h) = 2$; $\deg(i) = 3$; d and f are isolated. **5.** No **7.** $v = 4$; $e = 7$; $\deg^-(a) = 3$, $\deg^-(b) = 1$, $\deg^-(c) = 2$, $\deg^-(d) = 1$, $\deg^+(a) = 1$, $\deg^+(b) = 2$, $\deg^+(c) = 1$, $\deg^+(d) = 3$ **9.** 5 vertices, 13 edges; $\deg^-(a) = 6$, $\deg^+(a) = 1$, $\deg^-(b) = 1$, $\deg^+(b) = 5$, $\deg^-(c) = 2$, $\deg^+(c) = 5$, $\deg^-(d) = 4$, $\deg^+(d) = 2$, $\deg^-(e) = 0$, $\deg^+(e) = 0$

11.

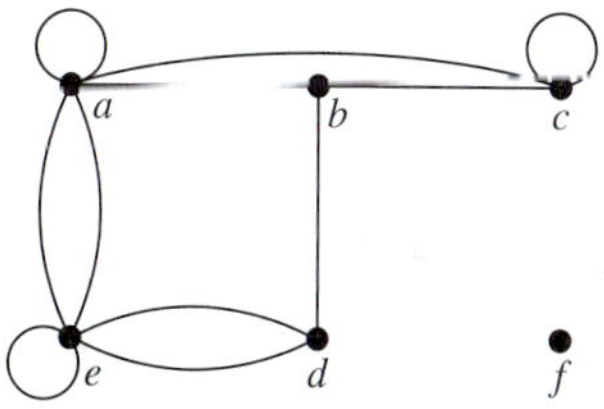

13. The number of coauthors that person has; that person's coauthors; a person who has no coauthors; a person who has only one coauthor **15.** In the directed graph $\deg^-(v)$ = number of calls v received, $\deg^+(v)$ = number of calls v made; in the undirected graph, $\deg(v)$ is the number of calls either made or received by v. **17.** $(\deg^+(v), \deg^-(v))$ is the win–loss record of v. **19.** In the undirected graph model in which the vertices are people in the group and two vertices are adjacent if those two people are friends, the degree of a vertex is the number of friends in the group that person has. By Exercise 18, there are two vertices with the same degree, which means that there are two people in the group with the same number of friends in the group. **21.** Bipartite **23.** Not bipartite **25.** Not bipartite **27. a)** Parts $\{h, s, n, w\}$ and $\{P, Q, R, S\}$, $E = \{\{P, n\}, \{P, w\}, \{Q, s\}, \{Q, n\}, \{R, n\}, \{R, w\}, \{S, h\}, \{S, s\}\}$ **b)** There is. **c)** $\{Pw, Qs, Rn, Sh\}$ among others **29.** Only Barry is willing to marry Uma and Xia. **31.** Model this with an undirected bipartite graph, with an edge between a man and a woman if they are willing to marry each other. By

Hall's theorem, it is enough to show that for every set S of women, the set $N(S)$ of men willing to marry them has cardinality at least $|S|$. Let m be the number of edges between S and $N(S)$. Since every vertex in S has degree k, it follows that $m = k|S|$. Because these edges are incident to $N(S)$, it follows that $m \leq k|N(S)|$. Therefore, $k|S| \leq k|N(S)|$, so $|N(S)| \geq |S|$. **33.** Model this with the bipartite graph where $V_1 = \{(p, i) \mid p \text{ is a winner and } i = 1, 2\}$ is the set in which each prize winner is represented twice, V_2 is the set of prizes, and an edge represents a player wanting a prize. For any subset A of V_1, $N(A) = V_2$, since every winner wants all of the $2m$ prizes. So $|N(A)| = 2m = |V_1| \geq |A|$ and Hall's theorem applies. **35. a)** $(\{a, b, c, f\}, \{\{a, b\}, \{a, f\}, \{b, c\}, \{b, f\}\})$ **b)** $(\{a, x, c, d, e\}, \{\{a, x\}, \{c, x\}, \{e, x\}\})$ **37. a)** n vertices, $n(n-1)/2$ edges **b)** n vertices, n edges **c)** $n+1$ vertices, $2n$ edges **d)** $m+n$ vertices, mn edges **e)** 2^n vertices, $n2^{n-1}$ edges **39.** $|E|/(|V|(|V|-1))$ **41. a)** 1 **b)** $\frac{2nm}{(n+m)(n+m-1)}$ **c)** $\frac{2}{n-1}$ **d)** $\frac{4}{n+1}$ **e)** $\frac{n}{2^n-1}$ **43. a)** 3, 3, 3, 3 **b)** 2, 2, 2, 2 **c)** 4, 3, 3, 3, 3 **d)** 3, 3, 2, 2, 2 **e)** 3, 3, 3, 3, 3, 3, 3, 3 **45.** Each of the n vertices is adjacent to each of the other $n-1$ vertices, so the degree sequence is $n-1, n-1, \ldots, n-1$ (n terms).

47. 7 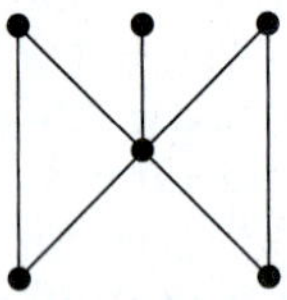

49. a) Yes 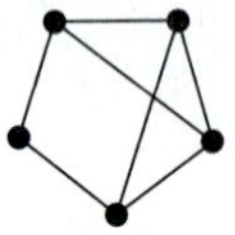

b) No **c)** No **d)** No

e) Yes 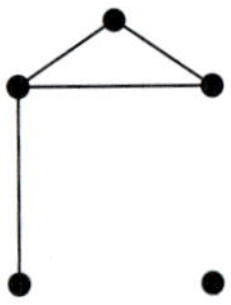

f) No **51.** First, suppose that $d_1, d_2, \ldots, d_n$ is graphic. We must show that the sequence whose terms are $d_2-1, d_3-1, \ldots, d_{d_1+1}-1, d_{d_1+2}, d_{d_1+3}, \ldots, d_n$ is graphic once it is put into nonincreasing order. In Exercise 50 it is proved that if the original sequence is graphic, then in fact there is a graph having this degree sequence in which the vertex of degree d_1 is adjacent to the vertices of degrees $d_2, d_3, \ldots, d_{d_1+1}$. Remove from this graph the vertex of highest degree (d_1). The resulting graph has the desired degree sequence. Conversely, suppose that $d_1, d_2, \ldots, d_n$ is a nonincreasing sequence such that the sequence $d_2-1, d_3-1, \ldots, d_{d_1+1}-1, d_{d_1+2}, d_{d_1+3}, \ldots, d_n$ is graphic once it is put into nonincreasing order. Take a graph with this latter degree sequence, where vertex v_i has degree $d_i - 1$ for $2 \leq i \leq d_1 + 1$ and vertex v_i has degree d_i for $d_1 + 2 \leq i \leq n$. Adjoin one new vertex (call it v_1), and put in an edge from v_1 to each of the vertices $v_2, v_3, \ldots, v_{d_1+1}$. The resulting graph has degree sequence $d_1, d_2, \ldots, d_n$. **53.** Let $d_1, d_2, \ldots, d_n$ be a nonincreasing sequence of nonnegative integers with an even sum. Construct a graph as follows: Take vertices $v_1, v_2, \ldots, v_n$ and put $\lfloor d_i/2 \rfloor$ loops at vertex v_i, for $i = 1, 2, \ldots, n$. For each i, vertex v_i now has degree either d_i or $d_i - 1$. Because the original sum was even, the number of vertices for which $\deg(v_i) = d_i - 1$ is even. Pair them up arbitrarily, and put in an edge joining the vertices in each pair. **55.** 17

57. 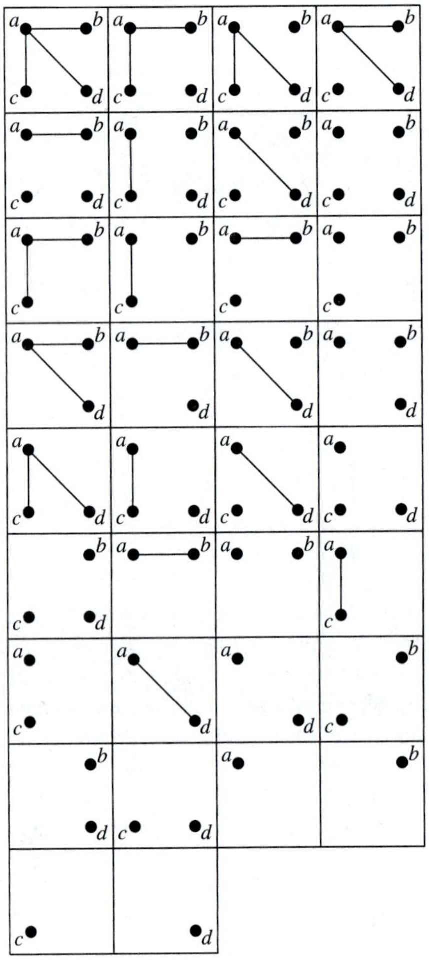

59. a) For all $n \geq 1$ **b)** For all $n \geq 3$ **c)** For $n = 3$ **d)** For all $n \geq 0$ **61.** 5

63. 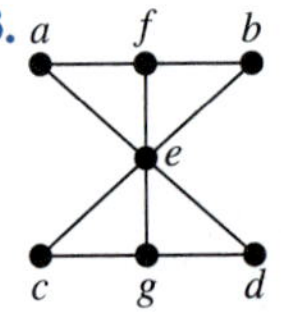

65. a) The graph with n vertices and no edges **b)** The disjoint union of K_m and K_n **c)** The graph with vertices $\{v_1, \ldots, v_n\}$ with an edge between v_i and v_j unless $i \equiv j \pm 1 \pmod{n}$ **d)** The graph whose vertices are represented by bit strings of length n with an edge between two vertices if the associated bit strings differ in more than one bit **67.** $v(v-1)/2 - e$ **69.** $n-1-d_n, n-1-d_{n-1}, \ldots, n-1-d_2, n-1-d_1$ **71.** The union of G and $\overline{G}$ contains an edge between each pair of the n vertices. Hence, this union is K_n.

73. Exercise 7:

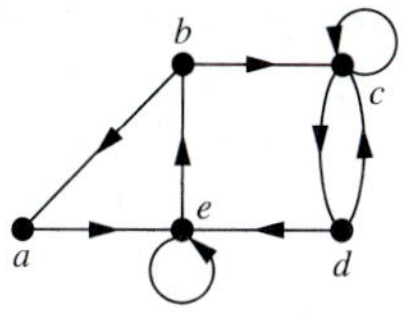

Exercise 8:

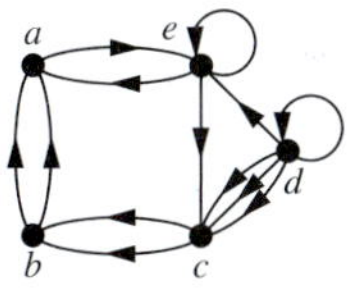

Exercise 9:

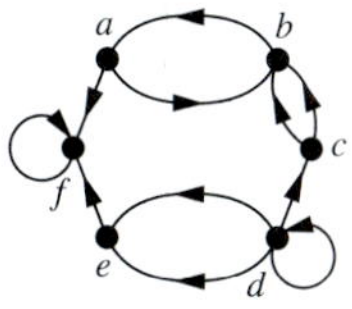

75. A directed graph $G = (V, E)$ is its own converse if and only if it satisfies the condition $(u, v) \in E$ if and only if $(v, u) \in E$. But this is precisely the condition that the associated relation must satisfy to be symmetric.

77.

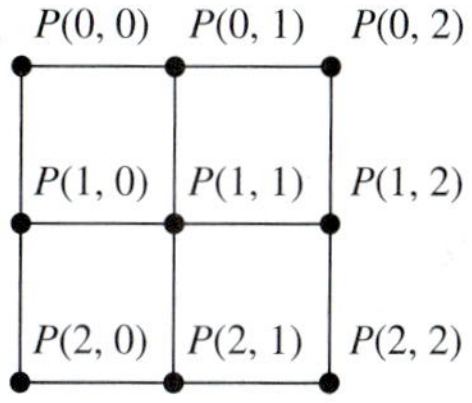

79. We can connect $P(i, j)$ and $P(k, l)$ by using $|i - k|$ hops to connect $P(i, j)$ and $P(k, j)$ and $|j - l|$ hops to connect $P(k, j)$ and $P(k, l)$. Hence, the total number of hops required to connect $P(i, j)$ and $P(k, l)$ does not exceed $|i - k| + |j - l|$. This is less than or equal to $m + m = 2m$, which is $O(m)$.

Section 10.3

1.

Vertex	Adjacent Vertices
a	b, c, d
b	a, d
c	a, d
d	a, b, c

3.

Vertex	Terminal Vertices
a	a, b, c, d
b	d
c	a, b
d	b, c, d

5. $\begin{bmatrix} 0 & 1 & 1 & 1 \\ 1 & 0 & 0 & 1 \\ 1 & 0 & 0 & 1 \\ 1 & 1 & 1 & 0 \end{bmatrix}$

7. $\begin{bmatrix} 1 & 1 & 1 & 1 \\ 0 & 0 & 0 & 1 \\ 1 & 1 & 0 & 0 \\ 0 & 1 & 1 & 1 \end{bmatrix}$ **9. a)** $\begin{bmatrix} 0 & 1 & 1 & 1 \\ 1 & 0 & 1 & 1 \\ 1 & 1 & 0 & 1 \\ 1 & 1 & 1 & 0 \end{bmatrix}$

b) $\begin{bmatrix} 0 & 1 & 1 & 1 & 1 \\ 1 & 0 & 0 & 0 & 0 \\ 1 & 0 & 0 & 0 & 0 \\ 1 & 0 & 0 & 0 & 0 \\ 1 & 0 & 0 & 0 & 0 \end{bmatrix}$ **c)** $\begin{bmatrix} 0 & 0 & 1 & 1 & 1 \\ 0 & 0 & 1 & 1 & 1 \\ 1 & 1 & 0 & 0 & 0 \\ 1 & 1 & 0 & 0 & 0 \\ 1 & 1 & 0 & 0 & 0 \end{bmatrix}$

d) $\begin{bmatrix} 0 & 1 & 0 & 1 \\ 1 & 0 & 1 & 0 \\ 0 & 1 & 0 & 1 \\ 1 & 0 & 1 & 0 \end{bmatrix}$ **e)** $\begin{bmatrix} 0 & 1 & 0 & 1 & 1 \\ 1 & 0 & 1 & 0 & 1 \\ 0 & 1 & 0 & 1 & 1 \\ 1 & 0 & 1 & 0 & 1 \\ 1 & 1 & 1 & 1 & 0 \end{bmatrix}$

f) $\begin{bmatrix} 0 & 1 & 1 & 0 & 1 & 0 & 0 & 0 \\ 1 & 0 & 0 & 1 & 0 & 1 & 0 & 0 \\ 1 & 0 & 0 & 1 & 0 & 0 & 1 & 0 \\ 0 & 1 & 1 & 0 & 0 & 0 & 0 & 1 \\ 1 & 0 & 0 & 0 & 0 & 1 & 1 & 0 \\ 0 & 1 & 0 & 0 & 1 & 0 & 0 & 1 \\ 0 & 0 & 1 & 0 & 1 & 0 & 0 & 1 \\ 0 & 0 & 0 & 1 & 0 & 1 & 1 & 0 \end{bmatrix}$

11.

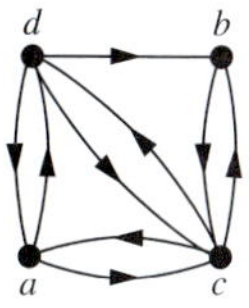

13. $\begin{bmatrix} 0 & 0 & 1 & 0 \\ 0 & 0 & 1 & 2 \\ 1 & 1 & 0 & 1 \\ 0 & 2 & 1 & 0 \end{bmatrix}$

15. $\begin{bmatrix} 1 & 0 & 2 & 1 \\ 0 & 1 & 1 & 2 \\ 2 & 1 & 1 & 0 \\ 1 & 2 & 0 & 1 \end{bmatrix}$

17.

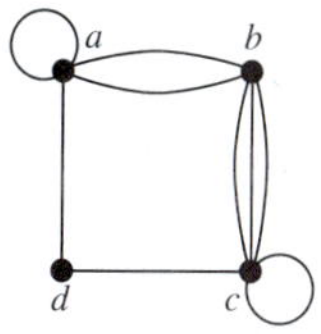

19. $\begin{bmatrix} 0 & 1 & 0 & 0 \\ 0 & 1 & 1 & 0 \\ 0 & 1 & 1 & 1 \\ 1 & 0 & 0 & 0 \end{bmatrix}$ **21.** $\begin{bmatrix} 1 & 1 & 2 & 1 \\ 1 & 0 & 0 & 2 \\ 1 & 0 & 1 & 1 \\ 0 & 2 & 1 & 0 \end{bmatrix}$

23.

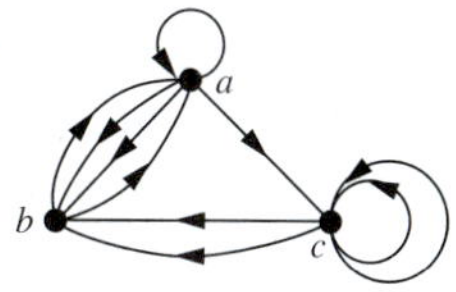

25. a) 3/7 **b)** 5/24 **c)** 19/36 **27. a)** dense **b)** sparse **c)** dense **d)** sparse **e)** sparse **f)** sparse **29.** Yes

31. Exercise 13: $\begin{bmatrix} 1 & 0 & 0 & 0 & 0 \\ 0 & 1 & 1 & 1 & 0 \\ 1 & 1 & 0 & 0 & 1 \\ 0 & 0 & 1 & 1 & 1 \end{bmatrix}$

Exercise 14: $\begin{bmatrix} 1 & 1 & 1 & 1 & 0 & 0 & 0 & 0 \\ 1 & 1 & 1 & 0 & 1 & 0 & 0 & 0 \\ 0 & 0 & 0 & 0 & 1 & 1 & 1 & 1 \\ 0 & 0 & 0 & 1 & 0 & 1 & 1 & 1 \end{bmatrix}$

Exercise 15: $\begin{bmatrix} 1 & 1 & 1 & 1 & 0 & 0 & 0 & 0 & 0 & 0 \\ 0 & 0 & 0 & 0 & 1 & 1 & 1 & 1 & 0 & 0 \\ 0 & 1 & 1 & 0 & 0 & 1 & 0 & 0 & 1 & 0 \\ 0 & 0 & 0 & 1 & 0 & 0 & 1 & 1 & 0 & 1 \end{bmatrix}$

33. $\deg(v)$ – number of loops at v; $\deg^-(v)$ **35.** 2 if e is not a loop, 1 if e is a loop

37. a)
$$\begin{bmatrix} 1 & 1 & \cdots & 1 & 0 & \cdots & 0 \\ 1 & 0 & \cdots & 0 & 1 & \cdots & 0 \\ 0 & 1 & \cdots & 0 & 1 & \cdots & 0 \\ \vdots & \vdots & & \vdots & \vdots & & \vdots \\ 0 & 0 & \cdots & 0 & 0 & \cdots & 1 \\ 0 & 0 & \cdots & 1 & 0 & \cdots & 1 \end{bmatrix}$$

b)
$$\begin{bmatrix} 1 & 0 & \cdots & 0 & 1 \\ 1 & 1 & \cdots & 0 & 0 \\ 0 & 1 & \cdots & 0 & 0 \\ \vdots & \vdots & & \vdots & \vdots \\ 0 & 0 & \cdots & 1 & 0 \\ 0 & 0 & \cdots & 1 & 1 \end{bmatrix}$$

c)
$$\begin{bmatrix} 0 \quad 0 \quad \cdots \quad 0 & 1 & 1 & \cdots & 1 \\ & 1 & 0 & \cdots & 0 \\ \mathbf{B} & 0 & 1 & \cdots & 0 \\ & \vdots & \vdots & & \vdots \\ & 0 & 0 & \cdots & 1 \end{bmatrix}$$

where **B** is the answer to (b)

d)
$$\begin{bmatrix} 1 & 1 & \cdots & 1 & 0 & \cdots & 0 \\ 0 & 0 & \cdots & 0 & 1 & \cdots & 0 \\ \vdots & \vdots & & \vdots & \vdots & & \vdots \\ 0 & 0 & \cdots & 0 & 0 & \cdots & 1 \\ 1 & 0 & \cdots & 0 & 1 & \cdots & 0 \\ 0 & 1 & \cdots & 0 & 0 & \cdots & 1 \\ \vdots & \vdots & & \vdots & \vdots & & \vdots \\ 0 & 0 & \cdots & 1 & 0 & \cdots & 0 \end{bmatrix}$$

39. Isomorphic **41.** Isomorphic **43.** Isomorphic **45.** Not isomorphic **47.** Isomorphic **49.** G is isomorphic to itself by the identity function, so isomorphism is reflexive. Suppose that G is isomorphic to H. Then there exists a one-to-one correspondence f from G to H that preserves adjacency and nonadjacency. It follows that f^{-1} is a one-to-one correspondence from H to G that preserves adjacency and nonadjacency. Hence, isomorphism is symmetric. If G is isomorphic to H and H is isomorphic to K, then there are one-to-one correspondences f and g from G to H and from H to K that preserve adjacency and nonadjacency. It follows that $g \circ f$ is a one-to-one correspondence from G to K that preserves adjacency and nonadjacency. Hence, isomorphism is transitive. **51.** All zeros **53.** Label the vertices in order so that all of the vertices in the first set of the partition of the vertex set come first. Because no edges join vertices in the same set of the partition, the matrix has the desired form. **55.** C_5 **57.** $n = 5$ only **59.** 4 **61.** 2 **63. a)** Yes **b)** No **c)** No **65.** $G = (V_1, E_1)$ is isomorphic to $H = (V_2, E_2)$ if and only if there exist functions f from V_1 to V_2 and g from E_1 to E_2 such that each is a one-to-one correspondence and for every edge e in E_1 the endpoints of $g(e)$ are $f(v)$ and $f(w)$ where v and w are the endpoints of e. **67.** Yes **69.** Yes **71.** If f is an isomorphism from a directed graph G to a directed graph H, then f is also an isomorphism from G^{conv} to H^{conv}. To see this note that (u, v) is an edge of G^{conv} if and only if (v, u) is an edge of G if and only if $(f(v), f(u))$ is an edge of H if and only if $(f(u), f(v))$ is an edge of H^{conv}. **73.** Many answers are possible; for example, C_6 and $C_3 \cup C_3$. **75.** The product is $[a_{ij}]$ where a_{ij} is the number of edges from v_i to v_j when $i \neq j$ and a_{ii} is the number of edges incident to v_i. **77.** The graphs in Exercise 45 are a devil's pair.

Section 10.4

1. a) Path of length 4; not a circuit; not simple **b)** Not a path **c)** Not a path **d)** Simple circuit of length 5 **3.** No **5.** No **7.** Maximal sets of people with the property that for any two of them, we can find a string of acquaintances that takes us from one to the other **9.** If a person has Erdős number n, then there is a path of length n from that person to Erdős in the collaboration graph, so by definition, that means that that person is in the same component as Erdős. If a person is in the same component as Erdős, then there is a path from that person to Erdős, and the length of the shortest such path is that person's Erdős number. **11. a)** Weakly connected **b)** Weakly connected **c)** Not strongly or weakly connected **13.** The maximal sets of phone numbers for which it is possible to find directed paths between every two different numbers in the set **15. a)** $\{a, b, f\}$, $\{c, d, e\}$ **b)** $\{a, b, c, d, e, h\}$, $\{f\}$, $\{g\}$ **c)** $\{a, b, d, e, f, g, h, i\}$, $\{c\}$ **17.** Suppose the strong components of u and v are not disjoint, say with vertex w in both. Suppose x is a vertex in the strong component of u. Then x is also in the strong component of v, because there is a path from x to v (namely the path from x to u followed by the path from u to w followed by the path from w to v) and vice versa. Thus, x is in the strong component of v. This shows that the strong component of u is a subgraph of the strong component of v, and equality follows by symmetry. **19. a)** 2 **b)** 7 **c)** 20 **d)** 61 **21.** Not isomorphic (G has a triangle; H does not) **23.** Isomorphic (the path $u_1, u_2, u_7, u_6, u_5, u_4, u_3, u_8, u_1$ corresponds to the path $v_1, v_2, v_3, v_4, v_5, v_8, v_7, v_6, v_1$) **25. a)** 3 **b)** 0 **c)** 27 **d)** 0 **27. a)** 1 **b)** 0 **c)** 2 **d)** 1 **e)** 5 **f)** 3 **29.** R is reflexive by definition. Assume that $(u, v) \in R$; then there is a path from u to v. Then $(v, u) \in R$ because there is a path from v to u, namely, the path from u to v traversed backward. Assume that $(u, v) \in R$ and $(v, w) \in R$; then there are paths from u to v and from v to w. Putting these two paths together gives a path from u to w. Hence, $(u, w) \in R$. It follows that R is transitive. **31.** c **33.** b, c, e, i **35.** If a vertex is pendant it is clearly not a cut vertex. So an endpoint of a cut edge that is a cut vertex is not pendant. Removal of a cut edge produces a graph with more connected components than in the original graph. If an endpoint of a cut edge is not pendant, the connected component it is in after the removal of the cut edge contains more than just this vertex. Consequently, removal of that vertex and all edges incident to it, including the original cut edge, produces a graph with more connected components than were in the original graph. Hence, an endpoint of a cut edge that is not pendant is a cut vertex. **37.** Assume there exists a connected graph G with at most one vertex that is not a cut vertex. Define the distance between the vertices u and v, denoted by $d(u, v)$, to be the length of the shortest path between u and v in G. Let s and t be vertices in G such that $d(s, t)$ is a maximum. Either s or t (or both) is a cut vertex, so without loss of generality suppose that s is a cut vertex. Let w belong to the connected component that does not contain t of the graph obtained by deleting s and all edges incident to s from G. Because every path from w to t contains s, $d(w, t) > d(s, t)$, which is a contradiction. **39. a)** Denver–Chicago, Boston–New York **b)** Seattle–Portland, Portland–San Francisco, Salt Lake City–Denver, New York–Boston, Boston–Burlington, Boston–Bangor **41.** A minimal set of people who collectively influence everyone (directly or indirectly); {Deborah} **43.** An edge cannot connect two vertices in different connected components. Because there are at most $C(n_i, 2)$ edges in the connected component with n_i vertices, it follows that there are at most $\sum_{i=1}^{k} C(n_i, 2)$ edges in the graph. **45.** Suppose that G is not connected. Then it has a component of k vertices for some k, $1 \le k \le n-1$. The most edges G could have is $C(k, 2) + C(n-k, 2) = [k(k-1) + (n-k)(n-k-1)]/2 = k^2 - nk + (n^2 - n)/2$. This quadratic function of f is minimized at $k = n/2$ and maximized at $k = 1$ or $k = n-1$. Hence, if G is not connected, the number of edges does not exceed the value of this function at 1 and at $n-1$, namely, $(n-1)(n-2)/2$. **47. a)** 1 **b)** 2 **c)** 6 **d)** 21 **49. a)** Removing an edge from a cycle leaves a path, which is still connected. **b)** Removing an edge from the cycle portion of the wheel leaves that portion still connected and the central vertex still connected to it as well. Removing a spoke leaves the cycle intact and the central vertex still connected to it as well. **c)** Any four vertices, two from each part of the bipartition, are connected by a 4-cycle; removing one edge does not disconnect them. **d)** Deleting the edge joining $(b_1, b_2, \dots, b_{i-1}, 0, b_{i+1}, \dots, b_n)$ and $(b_1, b_2, \dots, b_{i-1}, 1, b_{i+1}, \dots, b_n)$ does not disconnect the graph because these two vertices are still joined via the path $(b_1, b_2, \dots, b_{i-1}, 0, b_{i+1}, \dots, 0)$, $(b_1, b_2, \dots, b_{i-1}, 0, b_{i+1}, \dots, 1)$, $(b_1, b_2, \dots, b_{i-1}, 1, b_{i+1}, \dots, 1)$, $(b_1, b_2, \dots, b_{i-1}, 1, b_{i+1}, \dots, 0)$ if $n < 2$ and $b_n = 0$, and similarly in the other three cases. **51.** If G is complete, then removing vertices one by one leaves a complete graph at each step, so we never get a disconnected graph. Conversely, if edge uv is missing from G, then removing all the vertices except u and v creates a disconnected graph. **53.** Both equal $\min(m, n)$. **55.** Let G be a graph with n vertices; then $\kappa(G) \le n-1$. Let C be a smallest edge cut, leaving a nonempty proper subset S of the vertices of G disconnected from the complementary set $S' = V - S$. If xy is an edge of G for every $x \in S$ and $y \in S'$, then the size of C is $|S||S'|$, which is at least $n-1$, so $\kappa(G) \le \lambda(G)$. Otherwise, let $x \in S$ and $y \in S'$ be nonadjacent vertices. Let T consist of all neighbors of x in S' together with all vertices of $S - \{x\}$ with neighbors in S'. Then T is a vertex cut, because it separates x and y. Now look at the edges from x to $T \cap S'$ and one edge from each vertex of $T \cap S$ to S'; this gives us $|T|$ distinct edges that lie in C, so $\lambda(G) = |C| \ge |T| \ge \kappa(G)$. **57.** 2 **59.** Let the simple paths P_1 and P_2 be $u = x_0, x_1, \dots, x_n = v$ and $u = y_0, y_1, \dots, y_m = v$, respectively. The paths thus start out at the same vertex. Since the paths do not contain the same set of edges, they must diverge eventually. If they diverge only after one of them has ended, then the rest of the other path is a simple circuit from v to v. Otherwise we can suppose that $x_0 = y_0, x_1 = y_1, \dots, x_i = y_i$, but $x_{i+1} \ne y_{i+1}$. To form our simple circuit, we follow the path y_i, y_{i+1}, y_{i+2}, and so on, until it once again first encounters a vertex on P_1 (possibly as early as y_{i+1}, no later than y_m). Once we are back on P_1, we follow it along—forwards or backwards, as necessary—to return to x_i. Since $x_i = y_i$, this certainly forms a circuit. It must be a simple circuit, since no edge among the x_ks or the y_ls can be repeated (P_1 and P_2 are simple by hypothesis) and no edge among the x_ks can equal one of the edges y_l that we used, since we abandoned P_2 for P_1 as soon as we hit P_1. **61.** The graph G is connected if and only if every off-diagonal entry of $\mathbf{A} + \mathbf{A}^2 + \mathbf{A}^3 + \cdots + \mathbf{A}^{n-1}$ is positive, where $\mathbf{A}$ is the adjacency matrix of G. **63.** If the graph is bipartite, say with parts A and B, then the vertices in every path must alternately lie in A and B. Therefore, a path that starts in A, say, will end in B after an odd number of steps and in A after an even number of steps. Because a circuit ends at the same vertex where it starts, the length must be even. Conversely, suppose that all circuits have even length; we must show that the graph is bipartite. We can assume that the graph is connected, because if it is not, then we can just work on one component at a time. Let v be a vertex of the graph, and let A be the set of all vertices to which there is a path of odd length starting at v, and let B be the set of all vertices to which there is a path of even length starting at v. Because the component is connected, every vertex lies in A or B. No vertex can lie in both A and B, because if one did, then following the odd-length path from v to that vertex and then back along the even-length path from that vertex to v would produce an odd circuit, contrary to the hypothesis. Thus, the set of vertices has been partitioned into two sets. To show that every edge has endpoints in different parts, suppose that xy is an edge, where $x \in A$. Then the odd-length path from v to x followed by xy produces an even-length path from v to y, so $y \in B$. (Similarly, if $x \in B$.) **65.** $(H_1W_1H_2W_2\langle\text{boat}\rangle, \emptyset) \to (H_2W_2, H_1W_1\langle\text{boat}\rangle) \to (H_1H_2W_2\langle\text{boat}\rangle, W_1) \to (W_2, H_1W_1H_2\langle\text{boat}\rangle) \to (H_2W_2\langle\text{boat}\rangle, H_1W_1) \to (\emptyset, H_1W_1H_2W_2\langle\text{boat}\rangle)$

Section 10.5

1. Neither **3.** No Euler circuit; $a, e, c, e, b, e, d, b, a, c, d$ **5.** $a, b, c, d, c, e, d, b, e, a, e, a$ **7.** $a, i, h, g, d, e, f, g, c, e, h, d, c, a, b, i, c, b, h, a$ **9.** No, A still has odd degree. **11.** When the graph in which vertices represent intersections and edges streets has an Euler path **13.** Yes **15.** No **17.** If there is an Euler path, then as we follow it each vertex except the starting and ending vertices must have equal in-degree and out-degree, because whenever we come to a vertex along an edge, we leave it along another edge. The starting vertex must have out-degree 1 larger than its in-degree, because we use one edge leading out of this vertex and whenever we visit it again we use one edge leading into it and one leaving it. Similarly, the ending vertex must have in-degree 1 greater than its out-degree. Because the Euler path with directions erased produces a path between any two vertices, in the underlying undirected graph, the graph is weakly connected. Conversely, suppose the graph meets the degree conditions stated. If we add one more edge from the vertex of deficient out-degree to the vertex of deficient in-degree, then the graph has every vertex with equal in-degree and out-degree. Because the graph is still weakly connected, by Exercise 16 this new graph has an Euler circuit. Now delete the added edge to obtain the Euler path. **19.** Neither **21.** No Euler circuit; *a, d, e, d, b, a, e, c, e, b, c, b, e* **23.** Neither **25.** Follow the same procedure as Algorithm 1, taking care to follow the directions of edges. **27. a)** $n = 2$ **b)** None **c)** None **d)** $n = 1$ **29.** Exercise 1:1 time; Exercises 2–7: 0 times **31.** *a, b, c, d, e, a* is a Hamilton circuit. **33.** No Hamilton circuit exists, because once a purported circuit has reached *e* it would have nowhere to go. **35.** No Hamilton circuit exists, because every edge in the graph is incident to a vertex of degree 2 and therefore must be in the circuit. **37.** *a, b, c, f, d, e* is a Hamilton path. **39.** *f, e, d, a, b, c* is a Hamilton path. **41.** No Hamilton path exists. There are eight vertices of degree 2, and only two of them can be end vertices of a path. For each of the other six, their two incident edges must be in the path. It is not hard to see that if there is to be a Hamilton path, exactly one of the inside corner vertices must be an end, and that this is impossible. **43.** *a, b, c, f, i, h, g, d, e* is a Hamilton path. **45.** $m = n \geq 2$ **47. a)** (i) No, (ii) No, (iii) Yes **b)** (i) No, (ii) No, (iii) Yes **c)** (i) Yes, (ii) Yes, (iii) Yes **d)** (i) Yes, (ii) Yes, (iii) Yes **49.** The result is trivial for $n = 1$: code is 0, 1. Assume we have a Gray code of order n. Let $c_1, \ldots, c_k$, $k = 2^n$ be such a code. Then $0c_1, \ldots, 0c_k, 1c_k, \ldots, 1c_1$ is a Gray code of order $n + 1$.

51. **procedure** *Fleury*($G = (V, E)$: connected multigraph with the degrees of all vertices even, $V = \{v_1, \ldots, v_n\}$)

```
v := v_1
circuit := v
H := G
while H has edges
    e := first edge with endpoint v in H (with respect to
        listing of V) such that e is not a cut edge of H, if
        one exists, and simply the first edge in H with
        endpoint v otherwise
    w := other endpoint of e
    circuit := circuit with e, w added
    v := w
    H := H − e
return circuit {circuit is an Euler circuit}
```

53. If G has an Euler circuit, then it also has an Euler path. If not, add an edge between the two vertices of odd degree and apply the algorithm to get an Euler circuit. Then delete the new edge. **55.** Suppose $G = (V, E)$ is a bipartite graph with $V = V_1 \cup V_2$, where $V_1 \cap V_2 = \emptyset$ and no edge connects a vertex in V_1 and a vertex in V_2. Suppose that G has a Hamilton circuit. Such a circuit must be of the form $a_1, b_1, a_2, b_2, \ldots, a_k, b_k, a_1$, where $a_i \in V_1$ and $b_i \in V_2$ for $i = 1, 2, \ldots, k$. Because the Hamilton circuit visits each vertex exactly once, except for v_1, where it begins and ends, the number of vertices in the graph equals $2k$, an even number. Hence, a bipartite graph with an odd number of vertices cannot have a Hamilton circuit.

57.

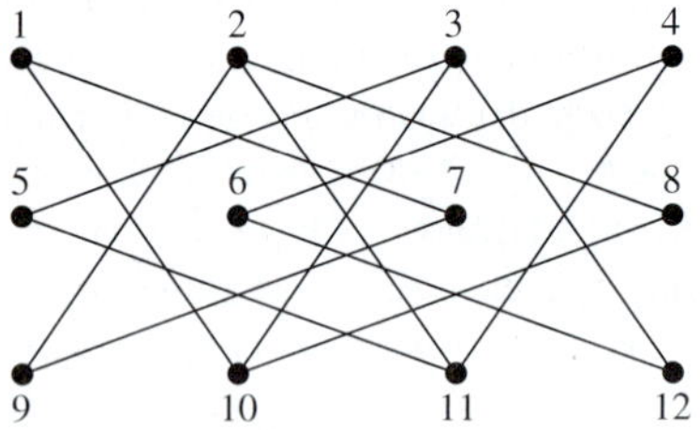

59. We represent the squares of a 3×4 chessboard as follows:

1	2	3	4
5	6	7	8
9	10	11	12

A knight's tour can be made by following the moves 8, 10, 1, 7, 9, 2, 11, 5, 3, 12, 6, 4. **61.** We represent the squares of a 4×4 chessboard as follows:

1	2	3	4
5	6	7	8
9	10	11	12
13	14	15	16

There are only two moves from each of the four corner squares. If we include all the edges 1–10, 1–7, 16–10, and 16–7, a circuit is completed too soon, so at least one of these edges must be missing. Without loss of generality, assume the path starts 1–10, 10–16, 16–7. Now the only moves from square 3 are to squares 5, 10, and 12, and square 10 already has two incident edges. Therefore, 3–5 and 3–12 must be in the Hamilton circuit. Similarly, edges 8–2 and 8–15 must be in the circuit. Now the only moves from square 9 are to squares 2, 7, and 15. If there were edges from square 9 to both squares 2 and 15, a circuit would be completed too soon. Therefore, the edge 9–7 must be in the circuit giving square 7 its full complement of edges. But now square 14 is forced to be joined to squares 5 and 12, completing a circuit too soon (5–14–12–3–5). This contradiction shows that there is no knight's tour on the 4×4 board. **63.** Because there are mn squares on an $m \times n$ board, if both m and n are odd, there are an odd number of squares.

Because by Exercise 62 the corresponding graph is bipartite, by Exercise 55 it has no Hamilton circuit. Hence, there is no reentrant knight's tour. **65. a)** If G does not have a Hamilton circuit, continue as long as possible adding missing edges one at a time in such a way that we do not obtain a graph with a Hamilton circuit. This cannot go on forever, because once we've formed the complete graph by adding all missing edges, there is a Hamilton circuit. Whenever the process stops, we have obtained a (necessarily noncomplete) graph H with the desired property. **b)** Add one more edge to H. This produces a Hamilton circuit, which uses the added edge. The path consisting of this circuit with the added edge omitted is a Hamilton path in H. **c)** Clearly v_1 and v_n are not adjacent in H, because H has no Hamilton circuit. Therefore, they are not adjacent in G. But the hypothesis was that the sum of the degrees of vertices not adjacent in G was at least n. This inequality can be rewritten as $n - \deg(v_n) \leq \deg(v_1)$. But $n - \deg(v_n)$ is just the number of vertices not adjacent to v_n. **d)** Because there is no vertex following v_n in the Hamilton path, v_n is not in S. Each one of the $\deg(v_1)$ vertices adjacent to v_1 gives rise to an element of S, so S contains $\deg(v_1)$ vertices. **e)** By part (c) there are at most $\deg(v_1)-1$ vertices other than v_n not adjacent to v_n, and by part (d) there are $\deg(v_1)$ vertices in S, none of which is v_n. Therefore, at least one vertex of S is adjacent to v_n. By definition, if v_k is this vertex, then H contains edges $v_k v_n$ and $v_1 v_{k+1}$, where $1 < k < n - 1$. **f)** Now $v_1, v_2, \ldots, v_{k-1}, v_k, v_n, v_{n-1}, \ldots, v_{k+1}, v_1$ is a Hamilton circuit in H, contradicting the construction of H. Therefore, our assumption that G did not originally have a Hamilton circuit is wrong, and our proof by contradiction is complete. **67.** Assume that there is a Hamilton circuit H. First assume that edge $aj \notin H$. Then ab, ai, jk, and jl are forced to be in H, which means that $kl \notin H$, and then gk and dl are in H. If ch were not in H, then bc, cd, gh, and hi would all have to be in H, completing a circuit without all the vertices. It follows that $ch \in H$. A circuit would be completed too soon if either both bc and cd were in H, or both gh and hi were in H, so by symmetry WOLOG assume that $cd \in H$ and $hi \in H$ and the other two edges are not. This then forces $bf \in H$ and $fg \in H$, and we have completed a circuit without vertex e. Going back to the beginning, we now know that $aj \in H$. Because of the 120° rotational symmetry of the figure, it then follows that dl and gk are in H as well. It is clearly now impossible to include all three of the vertices i, j, and k in the Hamilton circuit, and our proof by contradiction is complete.

Section 10.6

1. a) Vertices are the stops, edges join adjacent stops, weights are the times required to travel between adjacent stops. **b)** Same as part (a), except weights are distances between adjacent stops. **c)** Same as part (a), except weights are fares between stops. **3.** 16 **5.** Exercise 2: *a, b, e, d, z*; Exercise 3: *a, c, d, e, g, z*; Exercise 4: *a, b, e, h, l, m, p, s, z* **7. a)** *a, c, d* **b)** *a, c, d, f* **c)** *c, d, f* **e)** *b, d, e, g, z* **9. a)** Direct **b)** Via New York **c)** Via Atlanta and Chicago **d)** Via New York **11. a)** Via Chicago **b)** Via Chicago **c)** Via Los Angeles **d)** Via Chicago **13. a)** Via Chicago **b)** Via Chicago **c)** Via Los Angeles **d)** Via Chicago **15.** Do not stop the algorithm when z is added to the set S. **17. a)** Via Woodbridge, via Woodbridge and Camden **b)** Via Woodbridge, via Woodbridge and Camden **19.** For instance, sightseeing tours, street cleaning

21.

	a	*b*	*c*	*d*	*e*	*z*
a	4	3	2	8	10	13
b	3	2	1	5	7	10
c	2	1	2	6	8	11
d	8	5	6	4	2	5
e	10	7	8	2	4	3
z	13	10	11	5	3	6

23. $O(n^3)$ **25.** *a–c–b–d–a* (or the same circuit starting at some other point and/or traversing the vertices in reverse order) **27.** San Francisco–Denver–Detroit–New York–Los Angeles–San Francisco (or the same circuit starting at some other point and/or traversing the vertices in reverse order) **29.** Consider this graph:

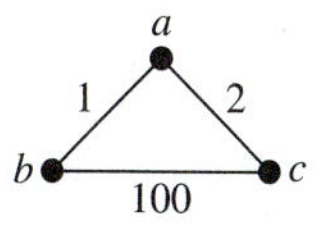

The circuit *a-b-a-c-a* visits each vertex at least once (and the vertex *a* twice) and has total weight 6. Every Hamilton circuit has total weight 103. **31.** Let $v_1, v_2, \ldots, v_n$ be a topological ordering of the vertices of the given directed acyclic graph. Let $w(i, j)$ be the weight of edge $v_i v_j$. Iteratively define $P(i)$ with the intent that it will be the weight of a longest path ending at v_i and $C(i)$ with the intent that it will be the vertex preceding v_i in some longest path: For i from 1 to n, let $P(i)$ be the maximum of $P(j) + w(j, i)$ over all $j < i$ such that $v_j v_i$ is an edge in the directed graph (and if such a j exists let $C(i)$ be a value of j for which this maximum is achieved) and let $P(i) = 0$ if there are no such values of j. At the conclusion of this loop, a longest path can be found by choosing i that maximizes $P(i)$ and following the C links back to the start of the path.

Section 10.7

1. Yes **3.**

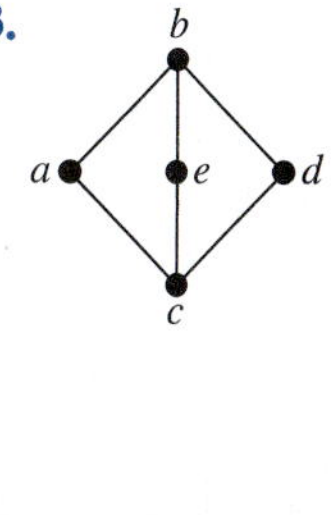

5. No

7. Yes

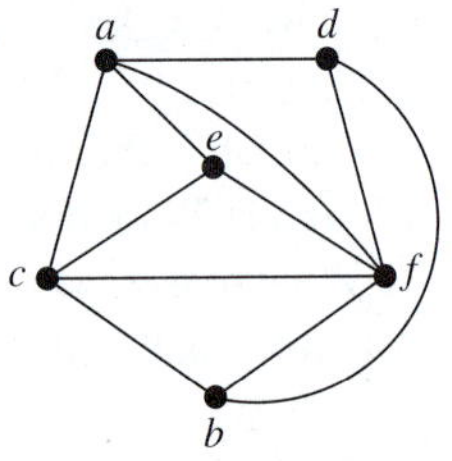

9. No **11.** A triangle is formed by the planar representation of the subgraph of K_5 consisting of the edges connecting v_1, v_2, and v_3. The vertex v_4 must be placed either within the triangle or outside of it. We will consider only the case when v_4 is inside the triangle; the other case is similar. Drawing the three edges from v_1, v_2, and v_3 to v_4 forms four regions. No matter which of these four regions v_5 is in, it is possible to join it to only three, and not all four, of the other vertices. **13.** 8 **15.** Because there are no loops or multiple edges and no simple circuits of length 3, and the degree of the unbounded region is at least 4, each region has degree at least 4. Thus, $2e \geq 4r$, or $r \leq e/2$. But $r = e - v + 2$, so we have $e - v + 2 \leq e/2$, which implies that $e \leq 2v - 4$. **17.** As in the argument in the proof of Corollary 1, we have $2e \geq 5r$ and $r = e - v + 2$. Thus, $e - v + 2 \leq 2e/5$, which implies that $e \leq (5/3)v - (10/3)$. **19.** Only (a) and (c) **21.** Not homeomorphic to $K_{3,3}$ **23.** Planar **25.** Nonplanar **27. a)** 1 **b)** 3 **c)** 9 **d)** 2 **e)** 4 **f)** 16 **29.** Draw $K_{m,n}$ as described in the hint. The number of crossings is four times the number in the first quadrant. The vertices on the x-axis to the right of the origin are (1, 0), (2, 0), …, $(m/2, 0)$ and the vertices on the y-axis above the origin are (0, 1), (0, 2), …, $(0, n/2)$. We obtain all crossings by choosing any two numbers a and b with $1 \leq a < b \leq m/2$ and two numbers r and s with $1 \leq r < s \leq n/2$; we get exactly one crossing in the graph between the edge connecting $(a, 0)$ and $(0, s)$ and the edge connecting $(b, 0)$ and $(0, r)$. Hence, the number of crossings in the first quadrant is $C\left(\frac{m}{2}, 2\right) \cdot C\left(\frac{n}{2}, 2\right) = \frac{(m/2)(m/2-1)}{2} \cdot \frac{(n/2)(n/2-1)}{2}$. Hence, the total number of crossings is $4 \cdot mn(m-2)(n-2)/64 = mn(m-2)(n-2)/16$. **31. a)** 2 **b)** 2 **c)** 2 **d)** 2 **e)** 2 **f)** 2 **33.** The formula is valid for $n \leq 4$. If $n > 4$, by Exercise 32 the thickness of K_n is at least $C(n, 2)/(3n - 6) = (n + 1 + \frac{2}{n-2})/6$ rounded up. Because this quantity is never an integer, it equals $\lfloor (n + 7)/6 \rfloor$. **35.** This follows from Exercise 34 because $K_{m,n}$ has mn edges and $m + n$ vertices and has no triangles because it is bipartite.

37.

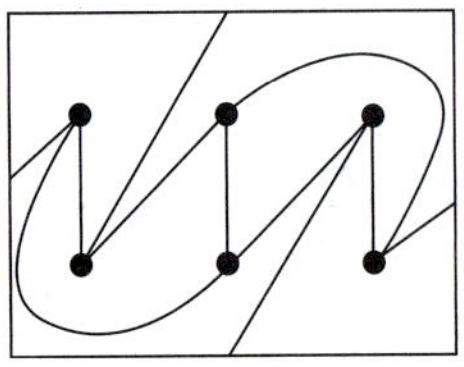

Section 10.8

1. Four colors

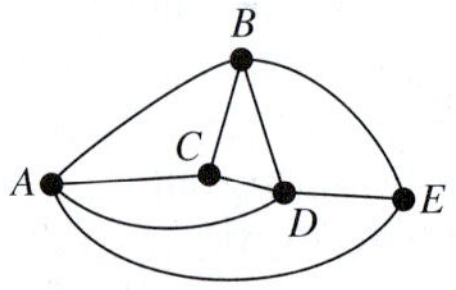

3. Three colors

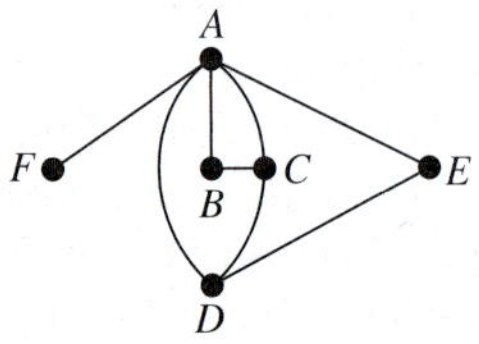

5. 3 **7.** 3 **9.** 2 **11.** 3 **13.** Graphs with no edges **15.** 3 if n is even, 4 if n is odd **17.** Period 1: Math 115, Math 185; period 2: Math 116, CS 473; period 3: Math 195, CS 101; period 4: CS 102; period 5: CS 273 **19.** 5 **21.** Exercise 5: 3 Exercise 6: 6 Exercise 7: 3 Exercise 8: 4 Exercise 9: 3 Exercise 10: 6 Exercise 11: 4 **23. a)** 2 if n is even, 3 if n is odd **b)** n **25.** Two edges that have the same color share no endpoints. Therefore, if more than $n/2$ edges were colored the same, the graph would have more than $2(n/2) = n$ vertices. **27.** 5 **29.** Color 1: e, f, d; color 2: c, a, i, g; color 3: h, b, j **31.** Color C_6 **33.** Four colors are needed to color W_n when n is an odd integer greater than 1, because three colors are needed for the rim (see Example 4), and the center vertex, being adjacent to all the rim vertices, will require a fourth color. To see that the graph obtained from W_n by deleting one edge can be colored with three colors, consider two cases. If we remove a rim edge, then we can color the rim with two colors, by starting at an endpoint of the removed edge and using the colors alternately around the portion of the rim that remains. The third color is then assigned to the center vertex. If we remove a spoke edge, then we can color the rim by assigning color #1 to the rim endpoint of the removed edge and colors #2 and #3 alternately to the remaining vertices on the rim, and then assign color #1 to the center. **35.** Suppose that G is chromatically k-critical but has a vertex v of degree $k - 2$ or less. Remove from G one of the edges incident to v. By definition of "k-critical," the resulting graph can be colored with $k - 1$ colors. Now restore the missing edge and use this coloring for all vertices except v. Because we had a proper coloring of the smaller graph, no two adjacent vertices have the same color. Furthermore, v has at most $k-2$ neighbors, so we can color v with an unused color to obtain a proper $(k-1)$-coloring of G. This contradicts the fact that G has chromatic number k. Therefore, our assumption was wrong, and every vertex of G must have degree at least $k - 1$. **37. a)** 6 **b)** 7 **c)** 9 **d)** 11 **39.** Represent frequencies by colors and zones by vertices. Join two vertices with an edge if the zones these vertices represent interfere with one another. Then a k-tuple coloring is precisely an assignment of frequencies that avoids interference. **41.** We use induction on the number of vertices of the graph. Every graph with five or fewer vertices can be colored with five or fewer colors, because each vertex can get a different color. That takes care of the basis case(s). So we assume that all graphs with k vertices can be 5-colored and consider a graph G with $k + 1$ vertices. By Corollary 2 in Section 10.7, G has a vertex v with degree at most 5. Remove v to form the graph G'. Because G' has only k vertices, we

5-color it by the inductive hypothesis. If the neighbors of v do not use all five colors, then we can 5-color G by assigning to v a color not used by any of its neighbors. The difficulty arises if v has five neighbors, and each has a different color in the 5-coloring of G'. Suppose that the neighbors of v, when considered in clockwise order around v, are a, b, c, m, and p. (This order is determined by the clockwise order of the curves representing the edges incident to v.) Suppose that the colors of the neighbors are azure, blue, chartreuse, magenta, and purple, respectively. Consider the azure-chartreuse subgraph (i.e., the vertices in G colored azure or chartreuse and all the edges between them). If a and c are not in the same component of this graph, then in the component containing a we can interchange these two colors (make the azure vertices chartreuse and vice versa), and G' will still be properly colored. That makes a chartreuse, so we can now color v azure, and G has been properly colored. If a and c are in the same component, then there is a path of vertices alternately colored azure and chartreuse joining a and c. This path together with edges av and vc divides the plane into two regions, with b in one of them and m in the other. If we now interchange blue and magenta on all the vertices in the same region as b, we will still have a proper coloring of G', but now blue is available for v. In this case, too, we have found a proper coloring of G. This completes the inductive step, and the theorem is proved. **43.** We follow the hint. Because the measures of the interior angles of a pentagon total 540°, there cannot be as many as three interior angles of measure more than 180° (reflex angles). If there are no reflex angles, then the pentagon is convex, and a guard placed at any vertex can see all points. If there is one reflex angle, then the pentagon must look essentially like figure (a) below, and a guard at vertex v can see all points. If there are two reflex angles, then they can be adjacent or nonadjacent (figures (b) and (c)); in either case, a guard at vertex v can see all points. [In figure (c), choose the reflex vertex closer to the bottom side.] Thus, for all pentagons, one guard suffices, so $g(5) = 1$.

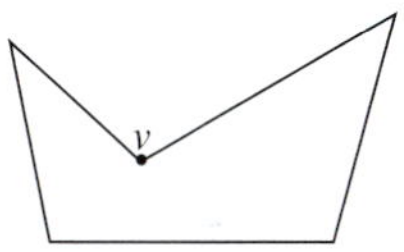

(a)

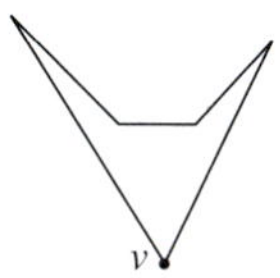

(b)

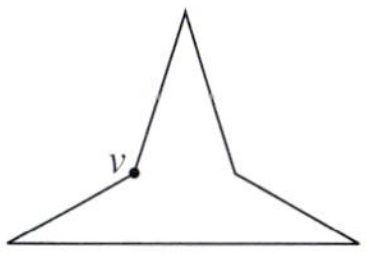

(c)

45. The figure suggested in the hint (generalized to have k prongs for any $k \geq 1$) has $3k$ vertices. The sets of locations from which the tips of different prongs are visible are disjoint. Therefore, a separate guard is needed for each of the k prongs, so at least k guards are needed. This shows that $g(3k) \geq k = \lfloor 3k/3 \rfloor$. If $n = 3k + i$, where $0 \leq i \leq 2$, then $g(n) \geq g(3k) \geq k = \lfloor n/3 \rfloor$.

Supplementary Exercises

1. 2500 **3.** Yes **5.** Yes **7.** $\sum_{i=1}^{m} n_i$ vertices, $\sum_{i<j} n_i n_j$ edges **9. a)** If $x \in N(A \cup B)$, then x is adjacent to some vertex $v \in A \cup B$. WOLOG suppose $v \in A$; then $x \in N(A)$ and therefore also in $N(A) \cup N(B)$. Conversely, if $x \in N(A) \cup N(B)$, then WOLOG suppose $x \in N(A)$. Thus, x is adjacent to some vertex $v \in A \subseteq A \cup B$, so $x \in N(A \cup B)$. **b)** If $x \in N(A \cap B)$, then x is adjacent to some vertex $v \in A \cap B$. Since both $v \in A$ and $v \in B$, it follows that $x \in N(A)$ and $x \in N(B)$, whence $x \in N(A) \cap N(B)$. For the counterexample, let $G = (\{u, v, w\}, \{\{u, v\}, \{v, w\}\})$, $A = \{u\}$, and $B = \{w\}$. **11.** (c, a, p, x, n, m) and many others **13.** (c, d, a, b) and many others **15.** 6 times the number of triangles divided by the number of paths of length 2 **17. a)** The probability that two actors each of whom has appeared in a film with a randomly chosen actor have appeared in a film together **b)** The probability that two of a randomly chosen person's Facebook friends are themselves Facebook friends **c)** The probability that two of a randomly chosen person's coauthors are themselves coauthors **d)** The probability that two proteins that each interact with a randomly chosen protein interact with each other **e)** The probability that two routers each of which has a communications link to a randomly chosen router are themselves linked **19.** Complete subgraphs containing the following sets of vertices: $\{b, c, e, f\}$, $\{a, b, g\}$, $\{a, d, g\}$, $\{d, e, g\}$, $\{b, e, g\}$ **21.** Complete subgraphs containing the following sets of vertices: $\{b, c, d, j, k\}$, $\{a, b, j, k\}$, $\{e, f, g, i\}$, $\{a, b, i\}$, $\{a, i, j\}$, $\{b, d, e\}$, $\{b, e, i\}$, $\{b, i, j\}$, $\{g, h, i\}$, $\{h, i, j\}$ **23.** $\{c, d\}$ is a minimum dominating set **25. a)**

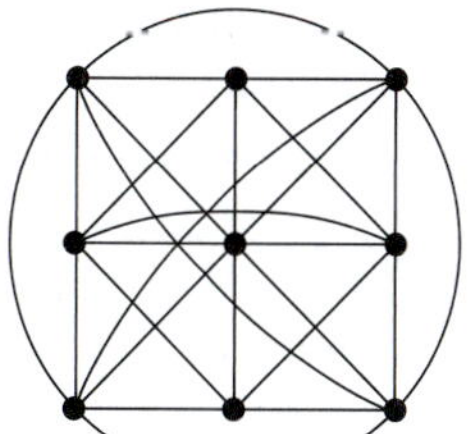

b)

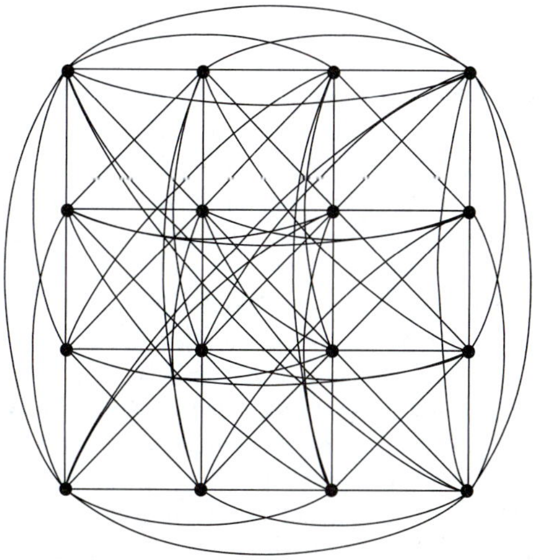

27. a) 1 **b)** 2 **c)** 3 **29. a)** A path from u to v in a graph G induces a path from $f(u)$ to $f(v)$ in an isomorphic graph H. **b)** Suppose f is an isomorphism from G to H. If $v_0, v_1, \ldots, v_n, v_0$ is a Hamilton circuit in G, then $f(v_0), f(v_1), \ldots, f(v_n), f(v_0)$ must be a Hamilton circuit in H because it is still a circuit and $f(v_i) \neq f(v_j)$ for $0 \leq i < j \leq n$. **c)** Suppose f is an isomorphism from G to H. If $v_0, v_1, \ldots, v_n, v_0$ is an Euler circuit in G, then $f(v_0), f(v_1), \ldots, f(v_n), f(v_0)$ must be an Euler circuit in H because it is a circuit that contains each edge exactly once. **d)** Two isomorphic graphs must have the same crossing number because they can be drawn exactly the same way in the plane. **e)** Suppose f is an isomorphism from G to H. Then v is isolated in G if and only if $f(v)$ is isolated in H. Hence, the graphs must have the same number of isolated vertices. **f)** Suppose f is an isomorphism from G to H. If G is bipartite, then the vertex set of G can be partitioned into V_1 and V_2 with no edge connecting vertices within V_1 or vertices within V_2. Then the vertex set of H can be partitioned into $f(V_1)$ and $f(V_2)$ with no edge connecting vertices within $f(V_1)$ or vertices within $f(V_2)$. **31.** 3 **33. a)** Yes **b)** No **35.** No **37.** Yes **39.** If e is a cut edge with endpoints u and v, then if we direct e from u to v, there will be no path in the directed graph from v to u, or else e would not have been a cut edge. Similar reasoning works if we direct e from v to u. **41.** $n - 1$ **43.** Let the vertices represent the chickens. We include the edge (u, v) in the graph if and only if chicken u dominates chicken v. **45.** By the handshaking theorem, the average vertex degree is $2m/n$, which equals the minimum degree; it follows that all the vertex degrees are equal. **47.** $K_{3,3}$ and the skeleton of a triangular prism **49. a)** A Hamilton circuit in the graph exactly corresponds to a seating of the knights at the Round Table such that adjacent knights are friends. **b)** The degree of each vertex in this graph is at least $2n-1-(n-1) = n \geq (2n/2)$, so by Dirac's theorem, this graph has a Hamilton circuit. **c)** a, b, d, f, g, z **51. a)** 4 **b)** 2 **c)** 3 **d)** 4 **e)** 4 **f)** 2 **53. a)** Suppose that $G = (V, E)$. Let $a, b \in V$. We must show that the distance between a and b in $\overline{G}$ is at most 2. If $\{a, b\} \notin E$ this distance is 1, so assume $\{a, b\} \in E$. Because the diameter of G is greater than 3, there are vertices u and v such that the distance in G between u and v is greater than 3. Either u or v, or both, is not in the set $\{a, b\}$. Assume that u is different from both a and b. Either $\{a, u\}$ or $\{b, u\}$ belongs to E; otherwise a, u, b would be a path in $\overline{G}$ of length 2. So, without loss of generality, assume $\{a, u\} \in E$. Thus, v cannot be a or b, and by the same reasoning either $\{a, v\} \in E$ or $\{b, v\} \in E$. In either case, this gives a path of length less than or equal to 3 from u to v in G, a contradiction. **b)** Suppose $G = (V, E)$. Let $a, b \in V$. We must show that the distance between a and b in $\overline{G}$ does not exceed 3. If $\{a, b\} \notin E$, the result follows, so assume that $\{a, b\} \in E$. Because the diameter of G is greater than or equal to 3, there exist vertices u and v such that the distance in G between u and v is greater than or equal to 3. Either u or v, or both, is not in the set $\{a, b\}$. Assume u is different from both a and b. Either $\{a, u\} \in E$ or $\{b, u\} \in E$; otherwise a, u, b is a path of length 2 in $\overline{G}$. So, without loss of generality, assume $\{a, u\} \in E$. Thus, v is different from a and from b. If $\{a, v\} \in E$, then u, a, v is a path of length 2 in G, so $\{a, v\} \notin E$ and thus $\{b, v\} \in E$ (or else there would be a path a, v, b of length 2 in $\overline{G}$). Hence, $\{u, b\} \notin E$; otherwise u, b, v is a path of length 2 in G. Thus, a, v, u, b is a path of length 3 in $\overline{G}$, as desired. **55.** a, b, e, z **57.** a, c, d, f, g, z **59.** If G is planar, then because $e \leq 3v - 6$, G has at most 27 edges. (If G is not connected it has even fewer edges.) Similarly, $\overline{G}$ has at most 27 edges. But the union of G and $\overline{G}$ is K_{11}, which has 55 edges, and $55 > 27 + 27$. **61.** Suppose that G is colored with k colors and has independence number i. Because each color class must be an independent set, each color class has no more than i elements. Thus, there are at most ki vertices. **63. a)** $C(n, m)p^m(1-p)^{n-m}$ **b)** np **c)** To generate a labeled graph G, as we apply the process to pairs of vertices, the random number x chosen must be less than or equal to $1/2$ when G has an edge between that pair of vertices and greater than $1/2$ when G has no edge there. Hence, the probability of making the correct choice is $1/2$ for each edge and $1/2^{C(n,2)}$ overall. Hence, all labeled graphs are equally likely. **65.** Suppose P is monotone increasing. If the property of not having P were not retained whenever edges are removed from a simple graph, there would be a simple graph G not having P and another simple graph G' with the same vertices but with some of the edges of G missing that has P. But P is monotone increasing, so because G' has P, so does G obtained by adding edges to G'. This is a contradiction. The proof of the converse is similar.

CHAPTER 11

Section 11.1

1. (a), (c), (e) **3. a)** a **b)** $a, b, c, d, f, h, j, q, t$ **c)** $e, g, i, k, l, m, n, o, p, r, s, u$ **d)** q, r **e)** c **f)** p **g)** f, b, a **h)** e, f, l, m, n **5.** No **7.** Level 0: a; level 1: b, c, d; level 2: e through k (in alphabetical order); level 3: l through r; level 4: s, t; level 5: u **9. a)** The entire tree **b)** c, g, h, o, p and the four edges cg, ch, ho, hp **c)** e alone **11. a)** 1 **b)** 2 **13. a)** 3 **b)** 9 **15. a)** The "only if" part is Theorem 2 and the definition of a tree. Suppose G is a connected simple graph with n vertices and $n - 1$ edges. If G is not a tree, it contains, by Exercise 14, an edge whose removal produces a graph G', which is still connected. If G' is not a tree, remove an edge to produce a connected graph G''. Repeat this procedure until the result is a tree. This requires at most $n-1$ steps because there are only $n-1$ edges. By Theorem 2, the resulting graph has $n - 1$ edges because it has n vertices. It follows that no edges were deleted, so G was already a tree. **b)** Suppose that G is a tree. By part (a), G has $n - 1$ edges, and by definition, G has no simple circuits. Conversely, suppose that G has no simple circuits and has $n - 1$ edges. Let c equal the number of components of G, each of which is necessarily a tree, say with n_i vertices, where $\sum_{i=1}^{c} n_i = n$. By part (a), the total number of edges in G is $\sum_{i=1}^{c}(n_i - 1) = n - c$. Since we are given that this equals $n - 1$, it follows that $c = 1$, i.e., G is connected and therefore satisfies the definition of a tree. **17.** 9999 **19.** 2000 **21.** 999 **23.** 1,000,000 dollars **25.** No such tree exists by Theorem 4 because it is impossible for $m = 2$ or $m = 84$.

27. Complete binary tree of height 4:

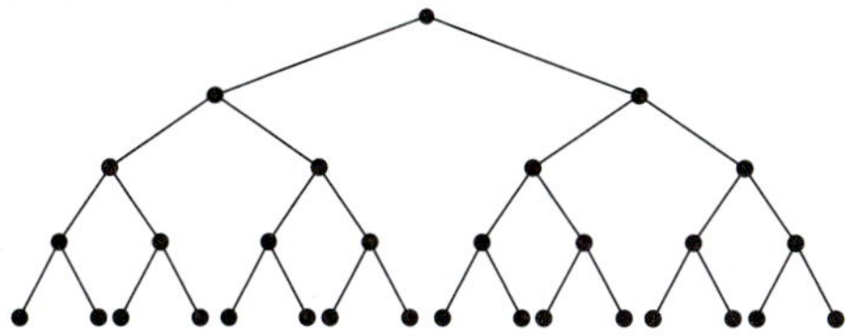

Complete 3-ary tree of height 3:

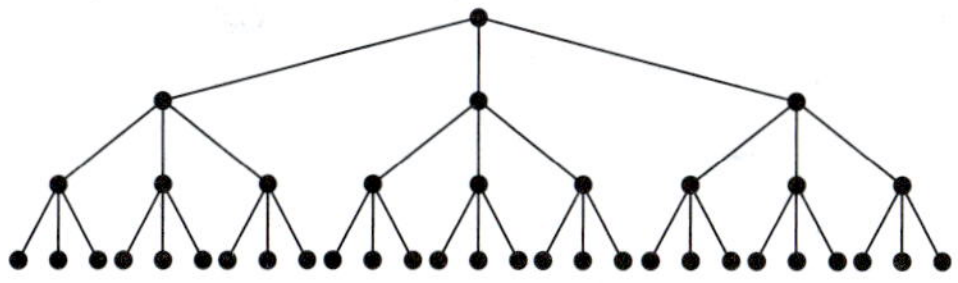

29. a) By Theorem 3 it follows that $n = mi + 1$. Because $i+l = n$, we have $l = n-i$, so $l = (mi+1)-i = (m-1)i+1$. **b)** We have $n = mi+1$ and $i+l = n$. Hence, $i = n-l$. It follows that $n = m(n-l)+1$. Solving for n gives $n = (ml-1)/(m-1)$. From $i = n-l$ we obtain $i = [(ml-1)/(m-1)]-l = (l-1)/(m-1)$. **31.** $n-t$ **33. a)** 1 **b)** 3 **c)** 5 **35. a)** The parent directory **b)** A subdirectory or contained file **c)** A subdirectory or contained file in the same parent directory **d)** All directories in the path name **e)** All subdirectories and files continued in the directory or a subdirectory of this directory, and so on **f)** The length of the path to this directory or file **g)** The depth of the system, i.e., the length of the longest path **37.** Let $n = 2^k$, where k is a positive integer. If $k = 1$, there is nothing to prove because we can add two numbers with $n - 1 = 1$ processor in $\log 2 = 1$ step. Assume we can add $n = 2^k$ numbers in $\log n$ steps using a tree-connected network of $n - 1$ processors. Let $x_1, x_2, \ldots, x_{2n}$ be $2n = 2^{k+1}$ numbers that we wish to add. The tree-connected network of $2n - 1$ processors consists of the tree-connected network of $n - 1$ processors together with two new processors as children of each leaf. In one step we can use the leaves of the larger network to find $x_1 + x_2, x_3 + x_4, \ldots, x_{2n-1} + x_{2n}$, giving us n numbers, which, by the inductive hypothesis, we can add in $\log n$ steps using the rest of the network. Because we have used $\log n + 1$ steps and $\log(2n) = \log 2 + \log n = 1 + \log n$, this completes the proof. **39.** c only **41.** c and h **43.** Suppose a tree T has at least two centers. Let u and v be distinct centers, both with eccentricity e, with u and v not adjacent. Because T is connected, there is a simple path P from u to v. Let c be any other vertex on this path. Because the eccentricity of c is at least e, there is a vertex w such that the unique simple path from c to w has length at least e. Clearly, this path cannot contain both u and v or else there would be a simple circuit. In fact, this path from c to w leaves P and does not return to P once it, possibly, follows part of P toward either u or v. Without loss of generality, assume this path does not follow P toward u. Then the path from u to c to w is simple and of length more than e, a contradiction. Hence, u and v are adjacent. Now because any two centers are adjacent, if there were more than two centers, T would contain K_3, a simple circuit, as a subgraph, which is a contradiction.

45.

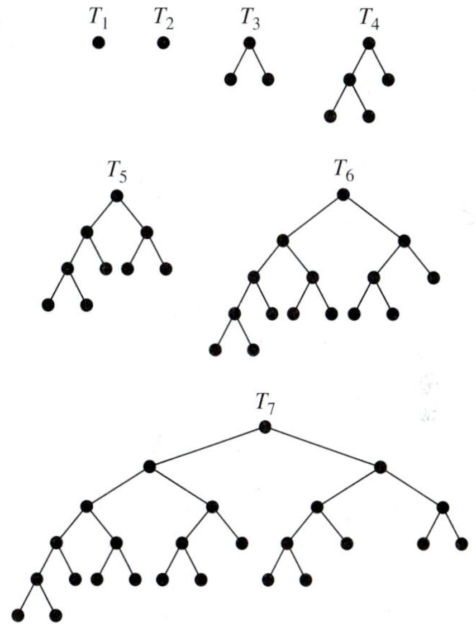

47. The statement is that *every* tree with n vertices has a path of length $n - 1$, and it was shown only that there exists a tree with n vertices having a path of length $n - 1$.

Section 11.2

1.

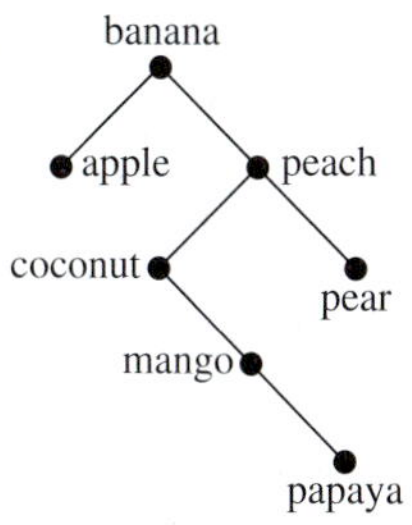

3. a) 3 **b)** 1 **c)** 4 **d)** 5

5.

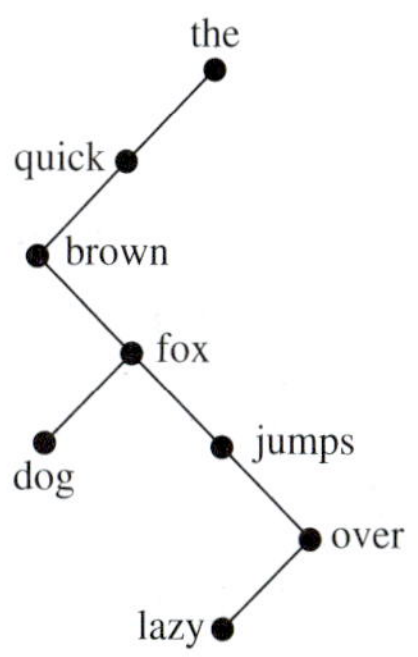

7. At least $\lceil \log_3 4 \rceil = 2$ weighings are needed, because there are only four outcomes (because it is not required to determine whether the coin is lighter or heavier). In fact, two weighings suffice. Begin by weighing coin 1 against coin 2. If they balance, weigh coin 1 against coin 3. If coin 1 and coin 3 are the same weight, coin 4 is the counterfeit coin, and if they are not the same weight, then coin 3 is the counterfeit coin. If coin 1 and coin 2 are not the same weight, again weigh coin 1 against coin 3. If they balance, coin 2 is the counterfeit coin; if they do not balance, coin 1 is the counterfeit coin. **9.** At least $\lceil \log_3 13 \rceil = 3$ weighings are needed. In fact, three weighings suffice. Start by putting coins 1, 2, and 3 on the left-hand side of the balance and coins 4, 5, and 6 on the right-hand side. If equal, apply Example 3 to coins 1, 2, 7, 8, 9, 10, 11, and 12. If unequal, apply Example 3 to 1, 2, 3, 4, 5, 6, 7, and 8. **11.** The least number is five. Call the elements a, b, c, and d. First compare a and b; then compare c and d. Without loss of generality, assume that $a < b$ and $c < d$. Next compare a and c. Whichever is smaller is the smallest element of the set. Again without loss of generality, suppose $a < c$. Finally, compare b with both c and d to completely determine the ordering. **13.** The first two steps are shown in the text. After 22 has been identified as the second largest element, we replace the leaf 22 by $-\infty$ in the tree and recalculate the winner in the path from the leaf where 22 used to be up to the root. Next, we see that 17 is the third largest element, so we repeat the process: replace the leaf 17 by $-\infty$ and recalculate. Next, we see that 14 is the fourth largest element, so we repeat the process: replace the leaf 14 by $-\infty$ and recalculate. Next, we see that 11 is the fifth largest element, so we repeat the process: replace the leaf 11 by $-\infty$ and recalculate. The process continues in this manner. We determine that 9 is the sixth largest element, 8 is the seventh largest element, and 3 is the eighth largest element. The trees produced in all steps, except the second to last, are shown here.

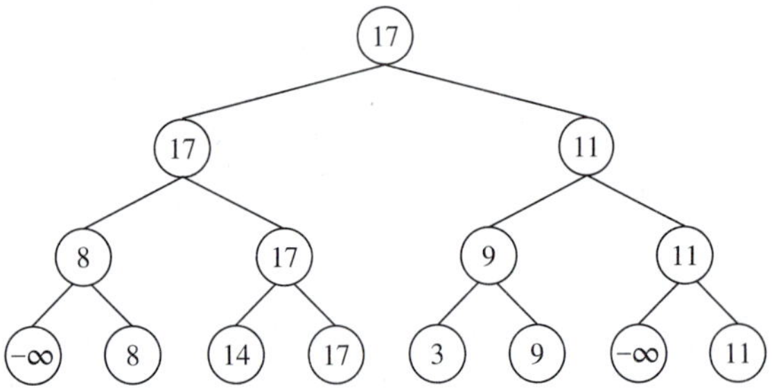

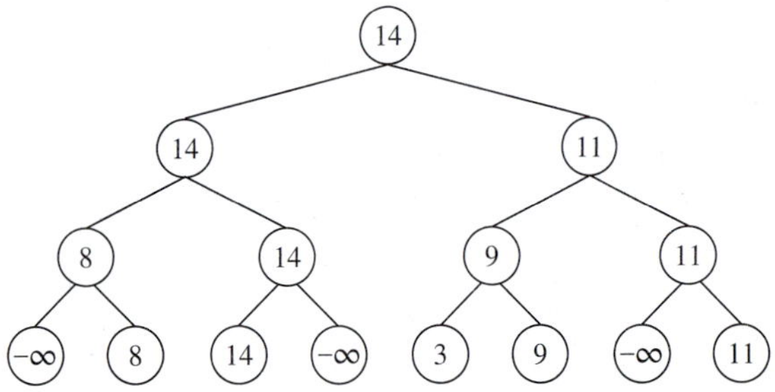

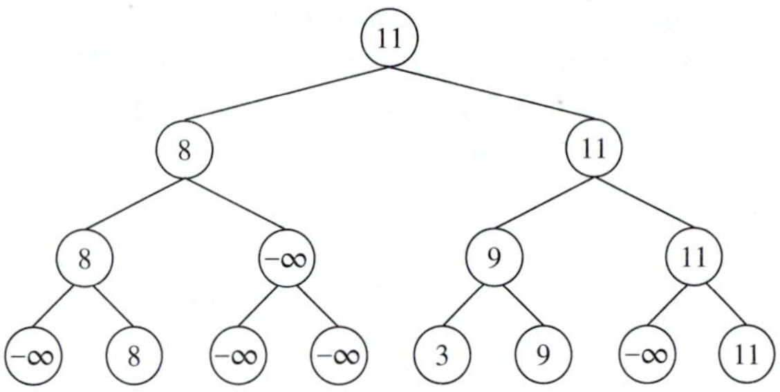

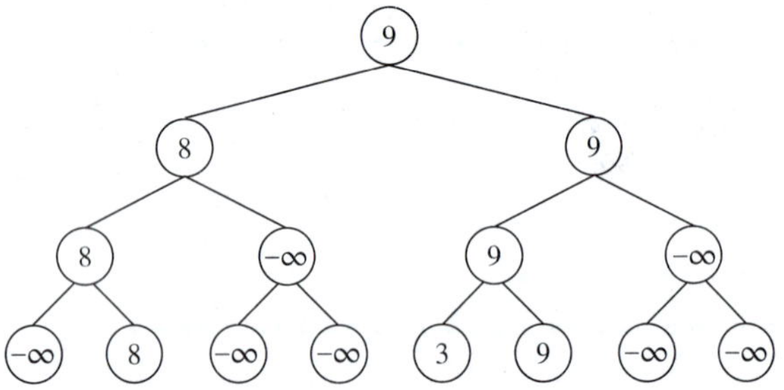

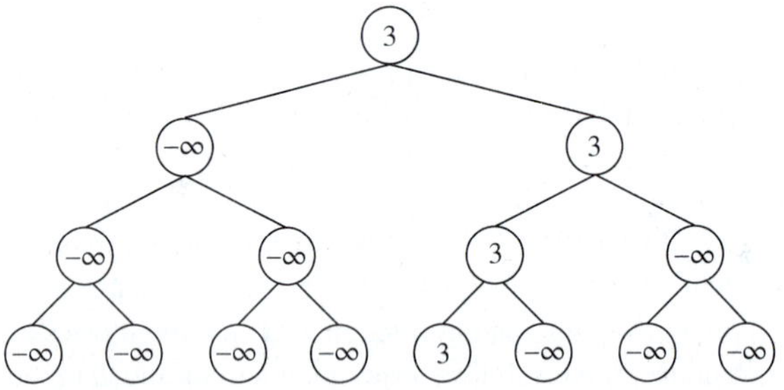

15. The value of a vertex is the list element currently there, and the label is the name (i.e., location) of the leaf responsible for that value.

```
procedure tournament sort(a_1, ..., a_n)
k := ⌈log n⌉
build a binary tree of height k
for i := 1 to n
    set the value of the ith leaf to be a_i and its label to
        be itself
for i := n + 1 to 2^k
    set the value of the ith leaf to be −∞ and its label to
        be itself
for i := k − 1 downto 0
    for each vertex v at level i
        set the value of v to the larger of the values of its
            children and its label to be the label of the child
            with the larger value
for i := 1 to n
    c_i := value at the root
    let v be the label of the root
    set the value of v to be −∞
        while the label at the root is still v
            v := parent(v)
            set the value of v to the larger of the values of its
                children and its label to be the label of the child
                with the larger value
{c_1, ..., c_n is the list in nonincreasing order}
```

17. $k - 1$, where $n = 2^k$ **19. a)** Yes **b)** No **c)** Yes **d)** Yes **21.** a: 000, e: 001, i: 01, k: 1100, o: 1101, p: 11110, u: 11111 **23.** a: 11; b: 101; c: 100; d: 01; e: 00; 2.25 bits (Note: This coding depends on how ties are broken, but the average number of bits is always the same.) **25.** There are four possible answers in all, the one shown here and three more obtained from this one by swapping t and v and/or swapping u and w.

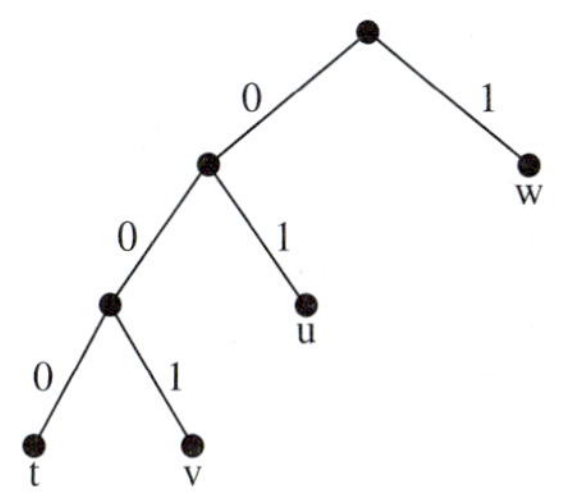

27. A:0001; B:101001; C:11001; D:00000; E:100; F:001100; G:001101; H:0101; I:0100; J:110100101; K:1101000; L:00001; M:10101; N:0110; O:0010; P:101000; Q:1101001000; R:1011; S:0111; T:111; U:00111; V:110101; W:11000; X:11010011; Y:11011; Z:1101001001 **29.** A:2; E:1; N:010; R:011; T:02; Z:00 **31.** n **33.** Because the tree is rather large, we have indicated in some places to "see text." Refer to Figure 9; the subtree rooted at these square or circle vertices is exactly the same as the corresponding subtree in Figure 9. First player wins.

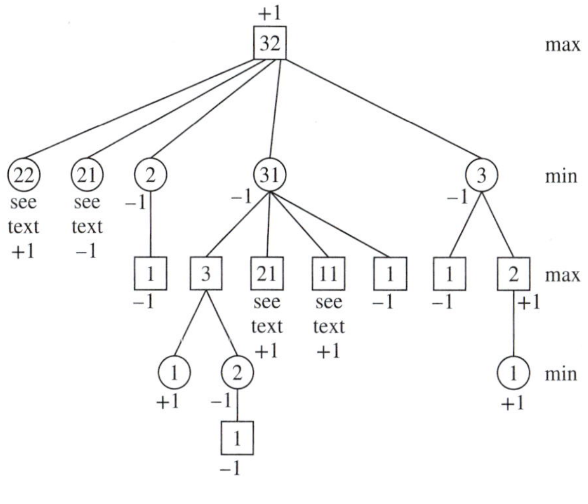

35. a) \$1 **b)** \$3 **c)** −\$3 **37.** See the figures shown next. **a)** 0 **b)** 0 **c)** 1 **d)** This position cannot have occurred in a game; this picture is impossible.

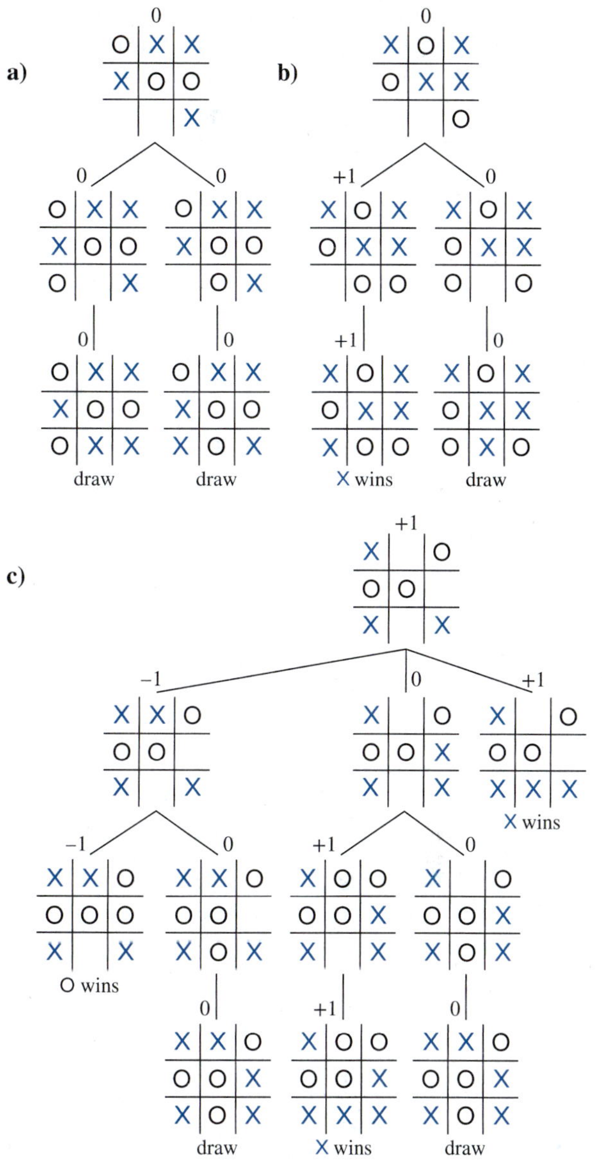

39. Proof by strong induction: *Basis step:* When there are $n = 2$ stones in each pile, if first player takes two stones from a pile, then second player takes one stone from the remaining pile and wins. If first player takes one stone from a pile, then second player takes two stones from the other pile and wins. *Inductive step:* Assume inductive hypothesis that second player can always win if the game starts with two piles of j stones for all $2 \le j \le k$, where $k \ge 2$, and consider a game with two piles containing $k + 1$ stones each. If first player takes all the stones from one of the piles, then second player takes all but one stone from the remaining pile and wins. If first player takes all but one stone from one of the piles, then second player takes all the stones from the other pile and wins. Otherwise first player leaves j stones in one pile, where $2 \le j \le k$, and $k + 1$ stones in the other pile. Second player takes the same number of stones from the larger pile, also leaving j stones there. At this point the game consists of two piles of j stones each. By the inductive hypothesis, the second player in that game, who is also the second player in our actual game, can win, and the proof by strong induction is complete. **41.** 7; 49 **43.** Value of tree is 1. Note: The second and third trees are the subtrees of the two children of the root in the first tree whose subtrees are not shown because of space limitations. They should be thought of as spliced into the first picture.

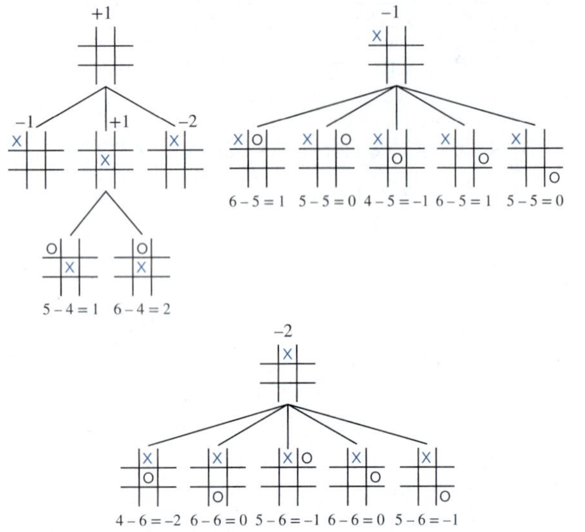

45. The numerator and denominator of the fraction at the root of the tree, 1/1, are relatively prime. Suppose that a/b is a fraction in the tree with $\gcd(a, b) = 1$. Lemma 1 of Section 4.3 implies that $\gcd(a, a + b) = \gcd(a, b)$ and $\gcd(a + b, b) = \gcd(a, b)$, and hence the numerators and denominators of the children of a/b, namely $a/(a + b)$ and $(a + b)/b$, are also relatively prime. Hence, by induction, every fraction that appears in the tree of fractions has relatively prime numerator and denominator. **47.** Suppose, for the sake of contradiction, that some positive rational numbers do appear more than once in the tree of fractions. Let a/b with $\gcd(a, b) = 1$ be the rational number appearing in the tree more than once with the smallest denominator and with the smallest numerator among those with denominator b. We consider three cases: $a = b = 1$, $a > b$, or $a < b$. Since the parent of 1/1 would be either 0/1 or 1/0, the only time 1/1 can appear in the tree is at the root, and $a = b = 1$ is not possible. If $a > b$, then each such vertex would have $(a - b)/b$ as its parent, so that $(a - b)/b$ would also appear in the tree multiple times, contradicting the minimality of a; so $a > b$ is not possible. Finally, if $a < b$, then the parent vertex $a/(b - a)$ would occur in the tree more than once, contradicting the minimality of b. Since all cases lead to contradiction, we conclude that no positive rational occurs in the tree of fractions more the once. **49.** We will first show that for all $n \geq 0$, the children in the tree of the rational that appears at position n appear at positions $2n + 1$ and $2n + 2$. For $n = 0$, this is true since the two children of the root appear at positions 1 and 2 in the list. Assuming that $n \geq 0$ and that the children in the tree of the rational at position n appear in positions $2n + 1$ and $2n + 2$, then the children of the rational at position $n + 1$ are the next two vertices in the tree and thus appear in the list at positions $2n + 2$ and $2n + 3$, completing this induction.

Note that the rational in position 0 is 1/1 and that $b(0) = 1$ by definition and $b(1) = 1$ since there is only 1 hyperbinary expansion of 1. Using strong induction, we will now show that for all $n \geq 0$, the rationals at positions $2n + 1$ and $2n + 2$ are $b(2n+1)/b(2n+2)$ and $b(2n+2)/b(2n+3)$. For $n = 0$, the children of the root are 1/2 and 2/1 and $b(1) = 1$, $b(2) = 2$ since $2 = 2 = 1 + 1$, and $b(3) = 1$ since $3 = 2 + 1$ is the only hyperbinary expansion of 3. So the basis step is true. Assume the claim is true for all $0 \leq k \leq n - 1$ where $n \geq 1$. Since the rationals at positions $2n + 1$ and $2n + 2$ have parent at position n, which equals $2k + 1$ or $2k + 2$ with $k = \frac{n}{2} - 1$, the inductive hypothesis implies that the rational at position n is $b(n)/b(n + 1)$. By definition of the tree, its children are $\frac{b(n)}{b(n)+b(n+1)}$ at position $2n+1$ and $\frac{b(n)+b(n+1)}{b(n+1)}$ at position $2n + 2$. Now observe that $b(2n + 1) = b(n)$, since any hyperbinary expansion of $2n+1$ must include a single 1 and thus can be made into an expansion of n by removing the 1 and dividing all other terms by 2, and conversely an expansion of n becomes an expansion of $2n+1$ by multiplying by 2 and adding 1. And consequently, $b(2n+3) = b(2(n+1)+1) = b(n+1)$. We also observe that $b(2n + 2) = b(n) + b(n + 1)$. A hyperbinary expansion of $2n+2$ may have zero or two 1s in it. If there are zero 1s, then dividing the expansion by 2 yields an expansion of $n+1$; if there are two 1s, then removing them and dividing by 2 produces an expansion of n. Conversely, an expansion of $n+1$ can be multiplied by 2 to produce one of $2n+2$, and an expansion of n, after multiplying by 2 and adding two 1s yields an expansion of $2n+2$. Hence, the rationals at positions $2n+1$ and $2n + 2$ are $\frac{b(n)}{b(n)+b(n+1)} = \frac{b(2n+1)}{b(2n+2)}$ and $\frac{b(n)+b(n+1)}{b(n+1)} = \frac{b(2n+2)}{n(2n+3)}$. This completes the induction, and the theorem follows.

Section 11.3

1.

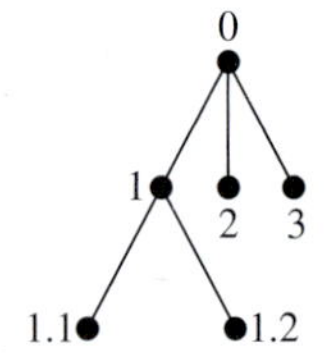

$0 < 1 < 1.1 < 1.2 < 2 < 3$

3.

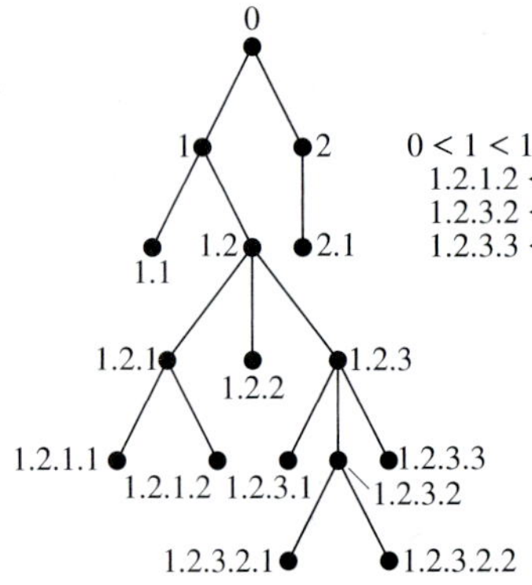

$0 < 1 < 1.1 < 1.2 < 1.2.1 < 1.2.1.1 <$ $1.2.1.2 < 1.2.2 < 1.2.3 < 1.2.3.1 <$ $1.2.3.2 < 1.2.3.2.1 < 1.2.3.2.2 <$ $1.2.3.3 < 2 < 2.1$

5. No **7.** *a, b, d, e, f, g, c* **9.** *a, b, e, k, l, m, f, g, n, r, s, c, d, h, o, i, j, p, q* **11.** *d, b, i, e, m, j, n, o, a, f, c, g, k, h, p, l* **13.** *d, f, g, e, b, c, a* **15.** *k, l, m, e, f, r, s, n, g, b, c, o, h, i, p, q, j, d, a*

17. a)

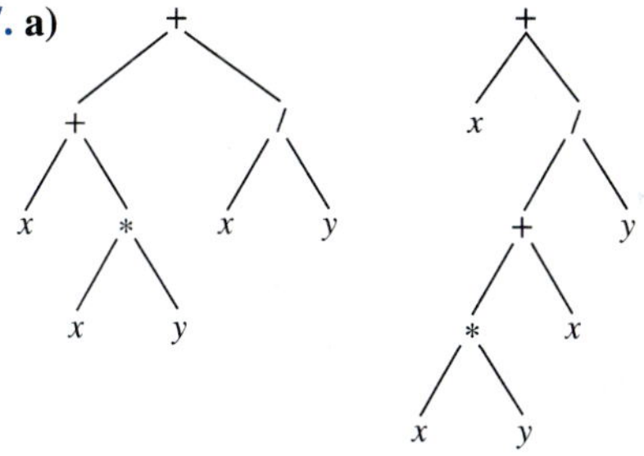

b) $++x*xy/xy$, $+x/+*xyxy$ **c)** $xxy*+xy/+$, $xxy*x+y/+$
d) $((x+(x*y))+(x/y))$, $(x+(((x*y)+x)/y))$

19. a)

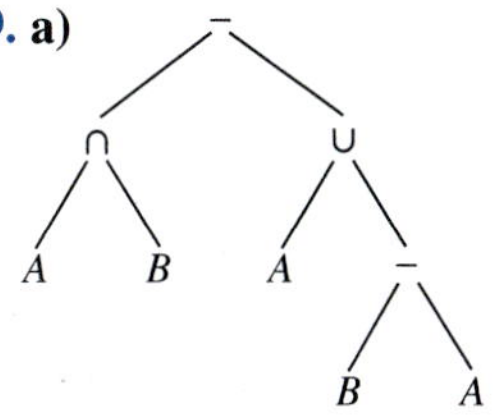

b) $-\cap AB\cup A-BA$ **c)** $AB\cap ABA-\cup-$
d) $((A\cap B)-(A\cup(B-A)))$ **21.** 14 **23. a)** 1 **b)** 1 **c)** 4 **d)** 2205

25.

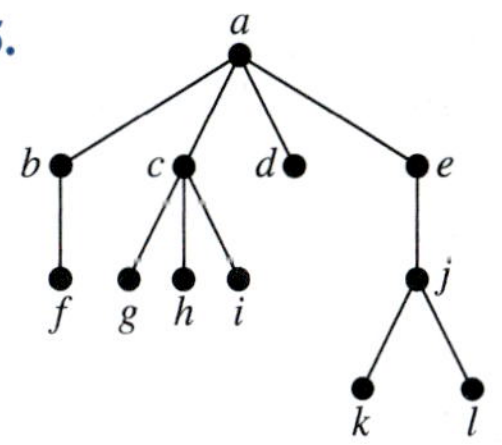

27. Use mathematical induction. The result is trivial for a list with one element. Assume the result is true for a list with n elements. For the inductive step, start at the end. Find the sequence of vertices at the end of the list starting with the last leaf, ending with the root, each vertex being the last child of the one following it. Remove this leaf and apply the inductive hypothesis. **29.** *c, d, b, f, g, h, e, a* in each case **31.** Proof by mathematical induction. Let $S(X)$ and $O(X)$ represent the number of symbols and number of operators in the well-formed formula X, respectively. The statement is true for well-formed formulae of length 1, because they have 1 symbol and 0 operators. Assume the statement is true for all well-formed formulae of length less than n. A well-formed formula of length n must be of the form $*XY$, where $*$ is an operator and X and Y are well-formed formulae of length less than n. Then by the inductive hypothesis $S(*XY) = S(X)+S(Y) = [O(X)+1]+[O(Y)+1] = O(X)+O(Y)+2$. Because $O(*XY) = 1+O(X)+O(Y)$, it follows that $S(*XY) = O(*XY)+1$. **33.** $xy+zx\circ+x\circ$, $xyz++yx++$, $xyxy\circ\circ xy\circ\circ z\circ+$, $xz\times$, $zz+\circ$, $yyyy\circ\circ\circ$, $zx+yz+\circ$, for instance

Section 11.4

1. $m-n+1$

3.

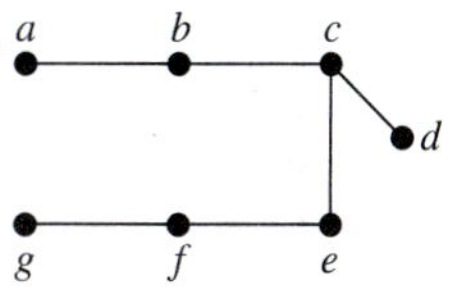

5.

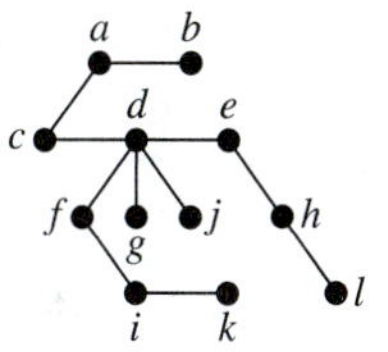

7. a)

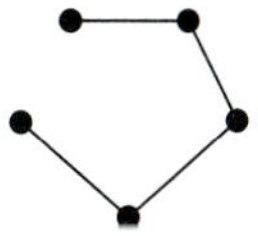

b)

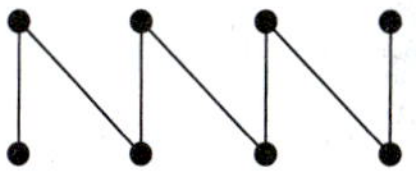

c)

d)

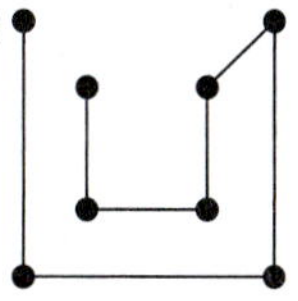

e)

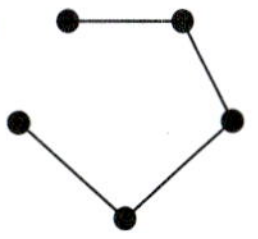

f)

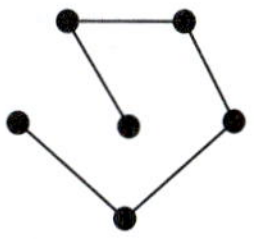

9.

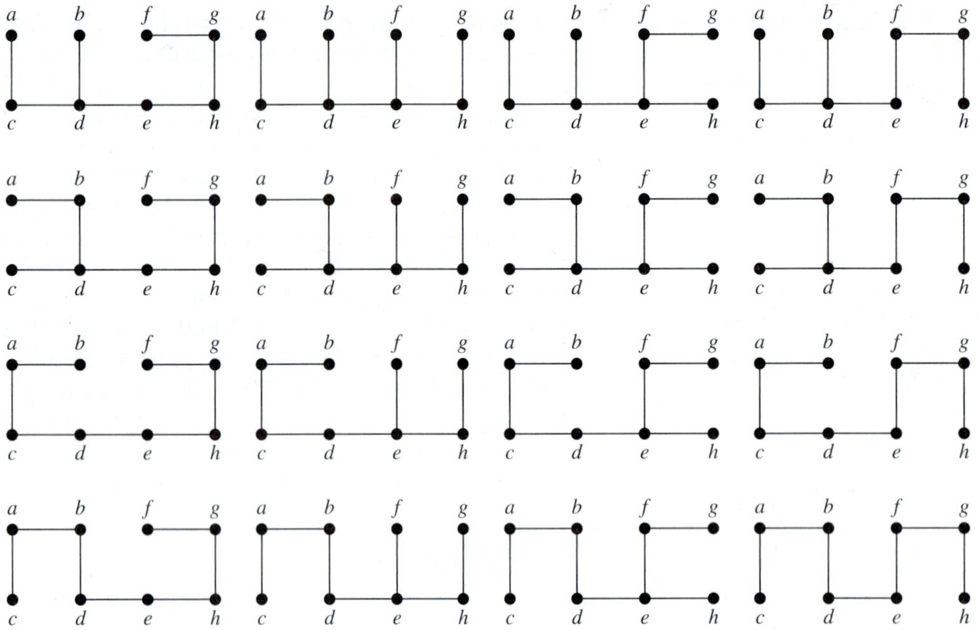

11. a) 3 **b)** 16 **c)** 4 **d)** 5

13.

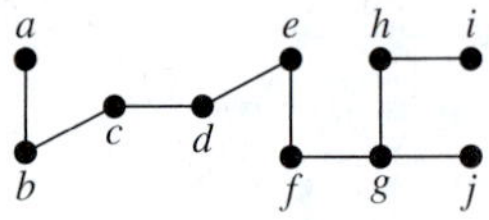

15.

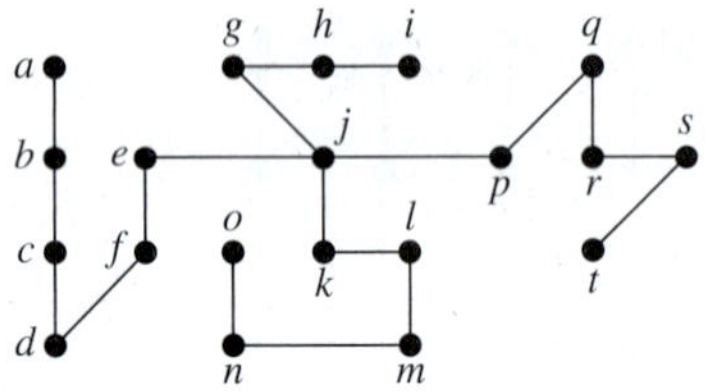

17. a) A path of length 6 **b)** A path of length 5 **c)** A path of length 6 **d)** Depends on order chosen to visit the vertices; may be a path of length 7 **19.** With breadth-first search, the initial vertex is the middle vertex, and the n spokes are added to the tree as this vertex is processed. Thus, the resulting tree is $K_{1,n}$. With depth-first search, we start at the vertex in the middle of the wheel and visit a neighbor—one of the vertices on the rim. From there we move to an adjacent vertex on the rim, and so on all the way around until we have reached every vertex. Thus, the resulting spanning tree is a path of length n. **21.** With breadth-first search, we fan out from a vertex of degree m to all the vertices of degree n as the first step. Next, a vertex of degree n is processed, and the edges from it to all the remaining vertices of degree m are added. The result is a $K_{1,n-1}$ and a $K_{1,m-1}$ with their centers joined by an edge. With depth-first search, we travel back and forth from one partite set to the other until we can go no further. If $m = n$ or $m = n - 1$, then we get a path of length $m + n - 1$. Otherwise, the path ends while some vertices in the larger partite set have not been visited, so we back up one link in the path to a vertex v and then successively visit the remaining vertices in that set from v. The result is a path with extra pendant edges coming out of one end of the path. **23.** A possible set of flights to discontinue are: Boston–New York, Detroit–Boston, Boston–Washington, New York–Washington, New York–Chicago, Atlanta–Washington, Atlanta–Dallas, Atlanta–Los Angeles, Atlanta–St. Louis, St. Louis–Dallas, St. Louis–Detroit, St. Louis–Denver, Dallas–San Diego, Dallas–Los Angeles, Dallas–San Francisco, San Diego–Los Angeles, Los Angeles–San Francisco, San Francisco–Seattle. **25.** Proof by induction on the length of the path: If the path has length 0, then the result is trivial. If the length is 1, then u is adjacent to v, so u is at level 1 in the breadth-first spanning tree. Assume that the result is true for paths of length l. If the length of a path is $l + 1$, let u' be the next-to-last vertex in a shortest path from v to u. By the inductive hypothesis, u' is at level l in the breadth-first spanning tree. If u were at a level not exceeding l, then clearly the length of the shortest path from v to u would also not exceed l. So u has not been added to the breadth-first spanning tree yet after the vertices of level l have been added. Because u is adjacent to u', it will be added at level $l + 1$ (although the edge connecting u' and u is not necessarily added). **27. a)** No solution

b)

c)

29. Start at a vertex and proceed along a path without repeating vertices as long as possible, allowing the return to the start after all vertices have been visited. When it is impossible to continue along a path, backtrack and try another extension of the current path. **31.** Take the union of the spanning trees of the connected components of G. They are disjoint, so the result is a forest. **33.** $m-n+c$ **35.** Assume that we wish to find the length of a shortest path from v_1 to every other vertex of G using Algorithm 2. In line 2 of that algorithm, add $L(v_1) := 0$, and add the following as a third step in the **then** clause at the end: $L(w) := 1 + L(v)$. **37.** Add an instruction to the BFS algorithm to mark each vertex as it is encountered. When BFS terminates we have found (all the vertices of) one component of the graph. Repeat, starting at an unmarked vertex, and continue in this way until all vertices have been marked. **39.** Trees **41.** Use depth-first search on each component. **43.** If an edge uv is not followed while we are processing vertex u during the depth-first search process, then it must be the case that the vertex v had already been visited. There are two cases. If vertex v was visited after we started processing u, then, because we are not finished processing u yet, v must appear in the subtree rooted at u (and hence, must be a descendant of u). On the other hand, if the processing of v had already begun before we started processing u, then why wasn't this edge followed at that time? It must be that we had not finished processing v, in other words, that we are still forming the subtree rooted at v, so u is a descendant of v, and hence, v is an ancestor of u. **45.** Certainly these two procedures produce the identical spanning trees if the graph we are working with is a tree itself, because in this case there is only one spanning tree (the whole graph). This is the only case in which that happens, however. If the original graph has any other edges, then by Exercise 43 they must be back edges and hence, join a vertex to an ancestor or descendant, whereas by Exercise 34, they must connect vertices at the same level or at levels that differ by 1. Clearly these two possibilities are mutually exclusive. Therefore, there can be no edges other than tree edges if the two spanning trees are to be the same. **47.** Because the edges not in the spanning tree are not followed in the process, we can ignore them. Thus, we can assume that the graph was a rooted tree to begin with. The basis step is trivial (there is only one vertex), so we assume the inductive hypothesis that breadth-first search applied to trees with n vertices have their vertices visited in order of their level in the tree and consider a tree T with $n+1$ vertices. The last vertex to be visited during breadth-first search of this tree, say v, is the one that was added last to the list of vertices waiting to be processed. It was added when its parent, say u, was being processed. We must show that v is at the lowest (bottom-most, i.e., numerically greatest) level of the tree. Suppose not; say vertex x, whose parent is vertex w, is at a lower level. Then w is at a lower level than u. Clearly v must be a leaf, because any child of v could not have been seen before v is seen. Consider the tree T' obtained from T by deleting v. By the inductive hypothesis, the vertices in T' must be processed in order of their level in T' (which is the same as their level in T, and the absence of v in T' has no effect on the rest of the algorithm). Therefore, u must have been processed before w, and therefore v would have joined the waiting list before x did, a contradiction. Therefore, v is at the bottom-most level of the tree, and the proof is complete. **49.** We modify the pseudocode given in Algorithm 2 by initializing m to be 0 at the beginning of the algorithm, and adding the statements "$m := m + 1$" and "assign m to vertex v" after the statement that removes vertex v from L. **51.** If a directed edge uv is not followed while we are processing its tail u during the depth-first search process, then it must be the case that its head v had already been visited. There are three cases. If vertex v was visited after we started processing u, then, because we are not finished processing u yet, v must appear in the subtree rooted at u (and hence, must be a descendant of u), so we have a forward edge. Otherwise, the processing of v must have already begun before we started processing u. If it had not yet finished (i.e., we are still forming the subtree rooted at v), then u is a descendant of v, and hence, v is an ancestor of u (we have a back edge). Finally, if the processing of v had already finished, then by definition we have a cross edge. **53.** Let T be the spanning tree constructed in Figure 3 and T_1, T_2, T_3, and T_4 the spanning trees in Figure 4. Denote by $d(T', T'')$ the distance between T' and T''. Then $d(T, T_1) = 6$, $d(T, T_2) = 4$, $d(T, T_3) = 4$, $d(T, T_4) = 2$, $d(T_1, T_2) = 4$, $d(T_1, T_3) = 4$, $d(T_1, T_4) = 6$, $d(T_2, T_3) = 4$, $d(T_2, T_4) = 2$, and $d(T_3, T_4) = 4$. **55.** Suppose $e_1 = \{u, v\}$ is as specified. Then $T_2 \cup \{e_1\}$ contains a simple circuit C containing e_1. The graph $T_1 - \{e_1\}$ has two connected components; the endpoints of e_1 are in different components. Travel C from u in the direction opposite to e_1 until you come to the first vertex in the same component as v. The edge just crossed is e_2. Clearly, $T_2 \cup \{e_1\} - \{e_2\}$ is a tree, because e_2 was on C. Also $T_1 - \{e_1\} \cup \{e_2\}$ is a tree, because e_2 reunited the two components.

57. Exercise 18: Exercise 19: Exercise 20:

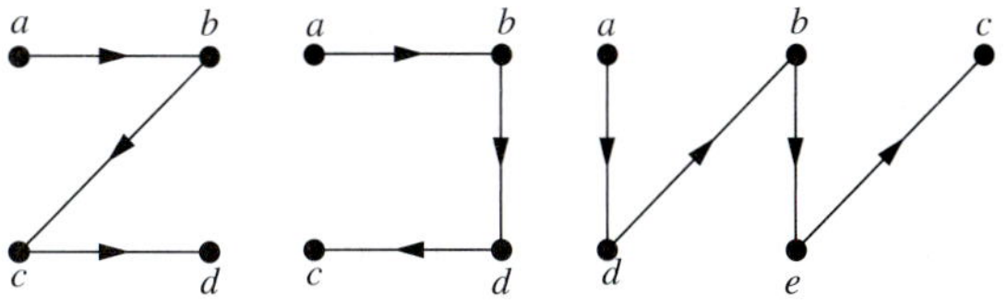

Exercise 21: Exercise 22: Exercise 23:

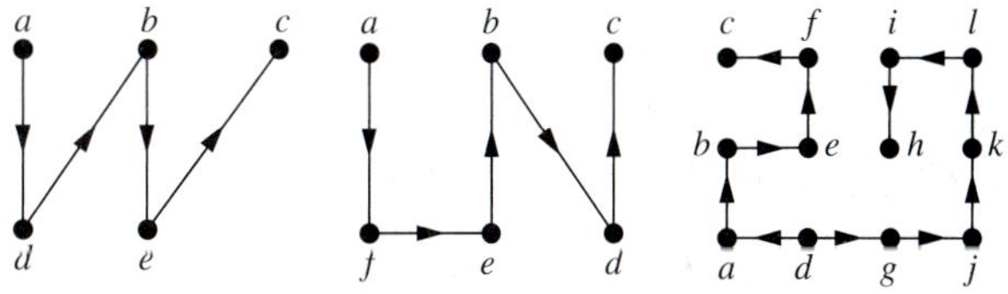

59. First construct an Euler circuit in the directed graph. Then delete from this circuit every edge that goes to a vertex previously visited. **61.** According to Exercise 60, a directed graph contains a circuit if and only if there are any back edges. We can detect back edges as follows. Add a marker on each vertex v to indicate what its status is: not yet seen (the initial situation), seen (i.e., put into T) but not yet finished (i.e., $visit(v)$ has not yet terminated), or finished (i.e., $visit(v)$ has terminated). A few extra lines in Algorithm 1 will accomplish this bookkeeping. Then to determine whether a directed graph has a circuit, we just have to check when looking at edge uv whether the status of v is "seen." If that ever happens, then we know there is a circuit; if not, then there is no circuit.

Section 11.5

1. Deep Springs–Oasis, Oasis–Dyer, Oasis–Silver Peak, Silver Peak–Goldfield, Lida–Gold Point, Gold Point–Beatty, Lida–Goldfield, Goldfield–Tonopah, Tonopah–Manhattan, Tonopah–Warm Springs **3.** $\{e, f\}$, $\{c, f\}$, $\{e, h\}$, $\{h, i\}$, $\{b, c\}$, $\{b, d\}$, $\{a, d\}$, $\{g, h\}$

5.

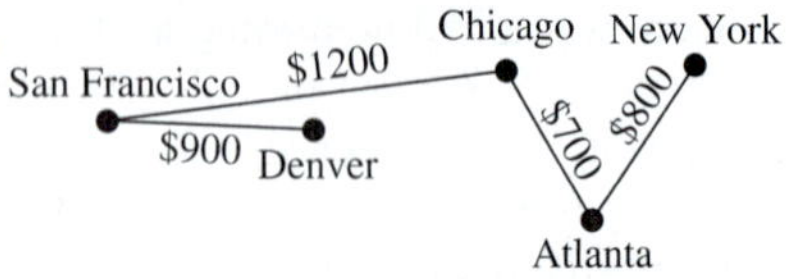

7. $\{e,f\}$, $\{a, d\}$, $\{h, i\}$, $\{b, d\}$, $\{c,f\}$, $\{e, h\}$, $\{b, c\}$, $\{g, h\}$

9.

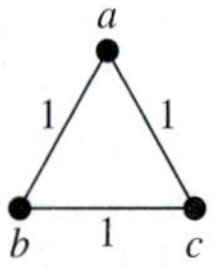

11. Instead of choosing minimum-weight edges at each stage, choose maximum-weight edges at each stage with the same properties.

13.

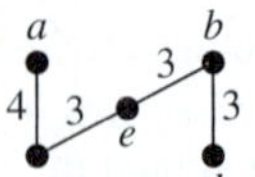

15.

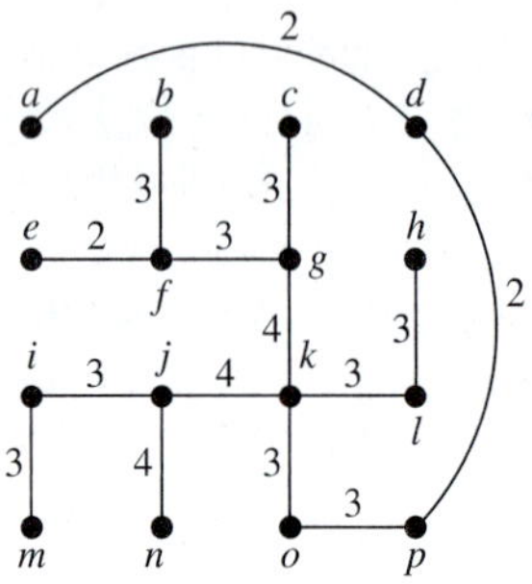

17. First find a minimum spanning tree T of the graph G with n edges. Then for $i = 1$ to $n - 1$, delete only the ith edge of T from G and find a minimum spanning tree of the remaining graph. Pick the one of these $n-1$ trees with the shortest length. **19.** If all edges have different weights, then a contradiction is obtained in the proof that Prim's algorithm works when an edge e_{k+1} is added to T and an edge e is deleted, instead of possibly producing another spanning tree.

21.

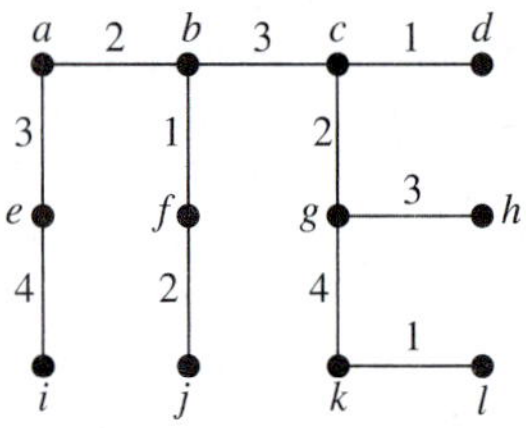

23. Same as Kruskal's algorithm, except start with $T :=$ this set of edges and iterate from $i = 1$ to $i = n - 1 - s$, where s is the number of edges you start with.

25. a)

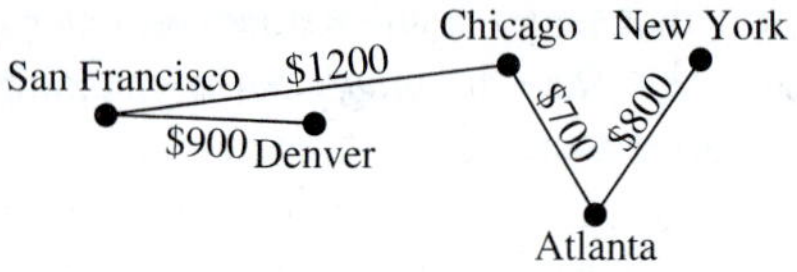

b)

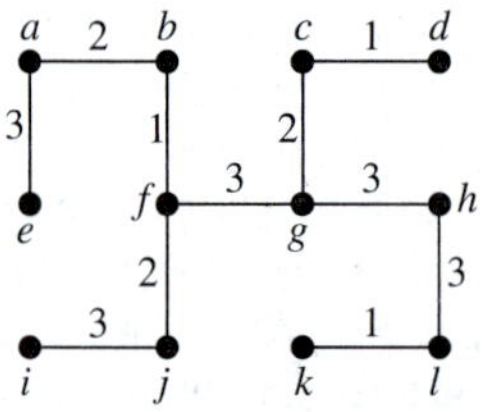

27. By Exercise 24, at each stage of Sollin's algorithm a forest results. Hence, after $n - 1$ edges are chosen, a tree results. It remains to show that this tree is a minimum spanning tree. Let T be a minimum spanning tree with as many edges in common with Sollin's tree S as possible. If $T \neq S$, then there is an edge $e \in S - T$ added at some stage in the algorithm, where prior to that stage all edges in S are also in T. $T \cup \{e\}$ contains a unique simple circuit. Find an edge $e' \in S - T$ and an edge $e'' \in T - S$ on this circuit and "adjacent" when viewing the trees of this stage as "supervertices." Then by the algorithm, $w(e') \leq w(e'')$. So replace T by $T - \{e''\} \cup \{e'\}$ to produce a minimum spanning tree closer to S than T was. **29.** Each of the r trees is joined to at least one other tree by a new edge. Hence, there are at most $r/2$ trees in the result (each new tree contains two or more old trees). To accomplish this, we need to add $r - (r/2) = r/2$ edges. Because the number of edges added is integral, it is at least $\lceil r/2 \rceil$. **31.** If $k \geq \log n$, then $n/2^k \leq 1$, so $\lceil n/2^k \rceil = 1$, so by Exercise 30 the algorithm is finished after at most $\log n$ iterations. **33.** Suppose that a minimum spanning tree T contains edge $e = uv$ that is the maximum weight edge in simple circuit C. Delete e from T. This creates a forest with two components, one containing u and the other containing v. Follow the edges of the path $C-e$, starting at u. At some point this path must jump from the component of $T-e$ containing u to the component of $T-e$ containing v, say using edge f. This edge cannot be in T, because e can be the only edge of T joining the two components (otherwise there would be a simple circuit in T). Because e is the edge of greatest weight in C, the weight of f is smaller. The tree formed by replacing e by f in T therefore has smaller weight, a contradiction. **35.** The reverse-delete algorithm must terminate and produce a spanning tree, because the algorithm never disconnects the graph and upon termination there can be no more simple circuits. The edge deleted at each stage of the algorithm must have been the edge of maximum weight in whatever circuits it was a part of. Therefore, by Exercise 33 it cannot be in any minimum spanning tree. Since only edges that could not have been in any minimum spanning tree have been deleted, the result must be a minimum spanning tree.

Supplementary Exercises

1. Suppose T is a tree. Then clearly T has no simple circuits. If we add an edge e connecting two nonadjacent vertices u and

v, then obviously a simple circuit is formed, because when e is added to T the resulting graph has too many edges to be a tree. The only simple circuit formed is made up of the edge e together with the unique path in T from v to u. Suppose T satisfies the given conditions. All that is needed is to show that T is connected, because there are no simple circuits in the graph. Assume that T is not connected. Then let u and v be in separate connected components. Adding $e = \{u, v\}$ does not satisfy the conditions. **3.** Suppose that a tree T has n vertices of degrees $d_1, d_2, \ldots, d_n$, respectively. Because $2e = \sum_{i=1}^{n} d_i$ and $e = n - 1$, we have $2(n - 1) = \sum_{i=1}^{n} d_i$. Because each $d_i \geq 1$, it follows that $2(n - 1) = n + \sum_{i=1}^{n}(d_i - 1)$, or that $n - 2 = \sum_{i=1}^{n}(d_i - 1)$. Hence, at most $n - 2$ of the terms of this sum can be 1 or more. Hence, at least two of them are 0. It follows that $d_i = 1$ for at least two values of i. **5.** $2n - 2$ **7.** A tree has no circuits, so it cannot have a subgraph homeomorphic to $K_{3,3}$ or K_5. **9.** Color each connected component separately. For each of these connected components, first root the tree, then color all vertices at even levels red and all vertices at odd levels blue. **11.** Upper bound: k^h; lower bound: $2\lceil k/2 \rceil^{h-1}$

13.

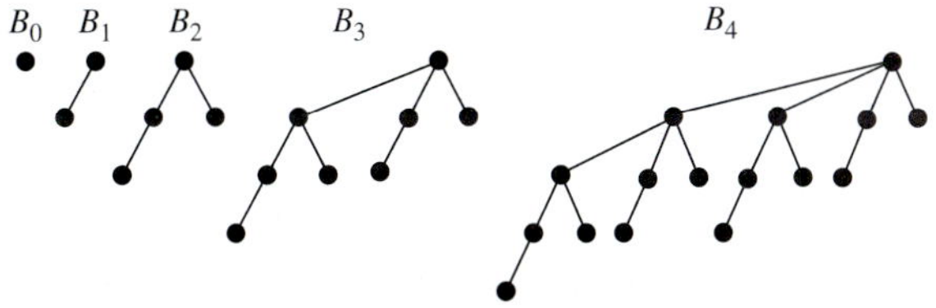

15. Because B_{k+1} is formed from two copies of B_k, one shifted down one level, the height increases by 1 as k increases by 1. Because B_0 had height 0, it follows by induction that B_k has height k. **17.** Because the root of B_{k+1} is the root of B_k with one additional child (namely the root of the other B_k), the degree of the root increases by 1 as k increases by 1. Because B_0 had a root with degree 0, it follows by induction that B_k has a root with degree k.

19.

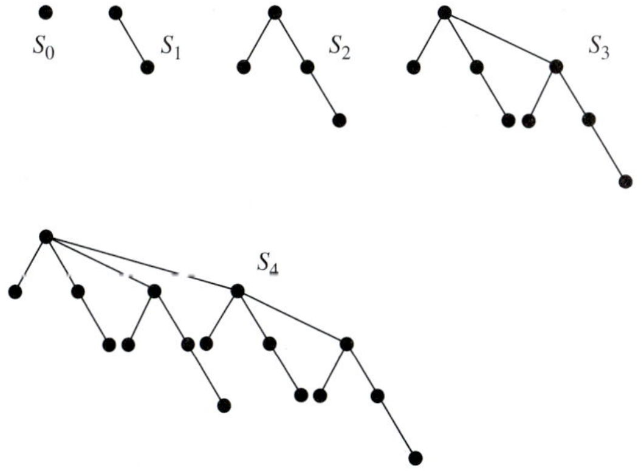

21. Use mathematical induction. The result is trivial for $k = 0$. Suppose it is true for $k-1$. T_{k-1} is the parent tree for T. By induction, the child tree for T can be obtained from $T_0, \ldots, T_{k-2}$ in the manner stated. The final connection of r_{k-2} to r_{k-1} is as stated in the definition of S_k-tree.

23. **procedure** *level*(T: ordered rooted tree with root r)
queue := sequence consisting of just the root r
while *queue* contains at least one term
  v := first vertex in queue
  list v
  remove v from queue and put children of v onto the end of queue

25. Build the tree by inserting a root for the address 0, and then inserting a subtree for each vertex labeled i, for i a positive integer, built up from subtrees for each vertex labeled $i.j$ for j a positive integer, and so on. **27. a)** Yes **b)** No **c)** Yes **29.** The resulting graph has no edge that is in more than one simple circuit of the type described. Hence, it is a cactus.

31.

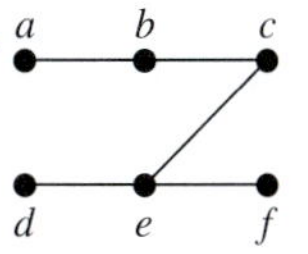

33.

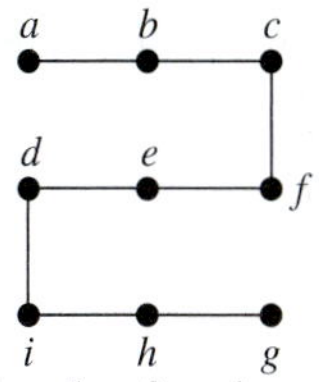

35. a)

b)

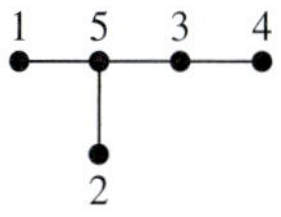

c)

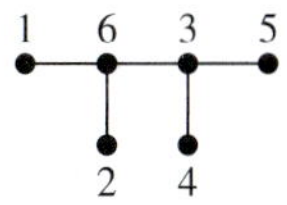

d)

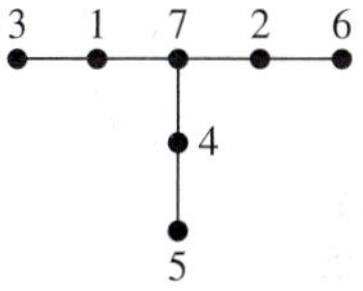

37. 6 **39. a)** 0 for 00, 11 for 01, 100 for 10, 101 for 11 (exact coding depends on how ties were broken, but all versions are equivalent); $0.645n$ for string of length n **b)** 0 for 000, 100 for 001, 101 for 010, 110 for 100, 11100 for 011, 11101 for 101, 11110 for 110, 11111 for 111 (exact coding depends on how ties were broken, but all versions are equivalent); $0.532\overline{6}n$ for string of length n **41.** Let G' be the graph obtained by deleting from G the vertex v and all edges incident to v. A minimum spanning tree of G can be obtained by taking an edge of minimal weight incident to v together with a minimum spanning tree of G'. **43.** Suppose that edge e is the edge of least weight incident to vertex v, and suppose that T is a spanning tree that does not include e. Add e to T, and delete from the simple circuit formed thereby the other edge of the circuit that contains v. The result will be a spanning tree of strictly smaller weight (because the deleted edge has weight greater than the weight of e). This is a contradiction, so T must include e. **45.** Because paths in trees are unique, an arborescence T of a directed graph G is just a subgraph of G that is a tree rooted at r, containing all the vertices of G, with all the edges directed away from the root. Thus, the in-degree of each vertex other than r is 1. For the converse, it is enough to show that for each $v \in V$ there is a unique directed path from r to v. Because the in-degree of each vertex other than r is 1, we can follow the edges of T backwards from v. This path can never return to a previously visited vertex, because that would create a simple circuit. Therefore, the path must eventually stop, and it can stop only at r, whose in-degree is not necessarily 1. Following this path forward gives the path from r to v required by the definition of arborescence. **47. a)** Run the breadth-first search algorithm, starting from v and respecting the directions of the edges, marking each vertex encountered as reachable. **b)** Running breadth-first search on G^{conv}, again starting at v, respecting the directions of the edges, and marking each vertex encountered, will identify all the vertices from which v is reachable. **c)** Choose a vertex v_1 and using parts (a) and (b) find the strong component containing v_1, namely all vertices w such that w is reachable from v_1 and v_1 is reachable from w. Then choose another vertex v_2 not yet in a strong component and find the strong component of v_2. Repeat until all vertices have been included. The correctness of this algorithm follows from the definition of strong component and Exercise 17 in Section 10.4.

CHAPTER 12

Section 12.1

1. a) 1 **b)** 1 **c)** 0 **d)** 0 **3. a)** $(1 \cdot 1) + (\overline{0 \cdot 1} + 0) = 1 + (\bar{0} + 0) = 1 + (1 + 0) = 1 + 1 = 1$ **b)** $(\mathbf{T} \wedge \mathbf{T}) \vee (\neg(\mathbf{F} \wedge \mathbf{T}) \vee \mathbf{F}) \equiv \mathbf{T}$

5. a)

x	y	z	$\bar{x}y$
1	1	1	0
1	1	0	0
1	0	1	0
1	0	0	0
0	1	1	1
0	1	0	1
0	0	1	0
0	0	0	0

b)

x	y	z	$x + yz$
1	1	1	1
1	1	0	1
1	0	1	1
1	0	0	1
0	1	1	1
0	1	0	0
0	0	1	0
0	0	0	0

c)

x	y	z	$x\bar{y} + \overline{xyz}$
1	1	1	0
1	1	0	1
1	0	1	1
1	0	0	1
0	1	1	1
0	1	0	1
0	0	1	1
0	0	0	1

d)

x	y	z	$x(yz + \bar{y}\bar{z})$
1	1	1	1
1	1	0	0
1	0	1	0
1	0	0	1
0	1	1	0
0	1	0	0
0	0	1	0
0	0	0	0

7. a)

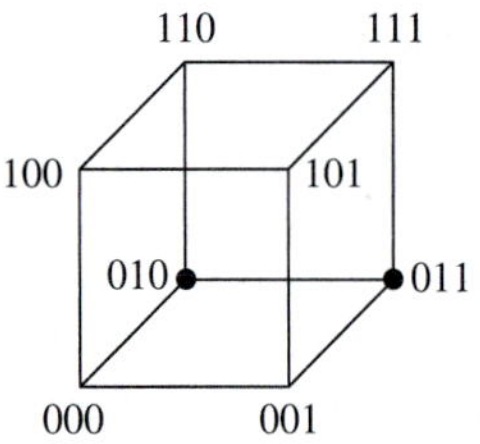

b)

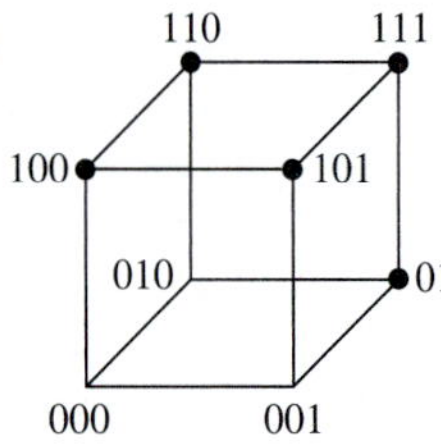

c)

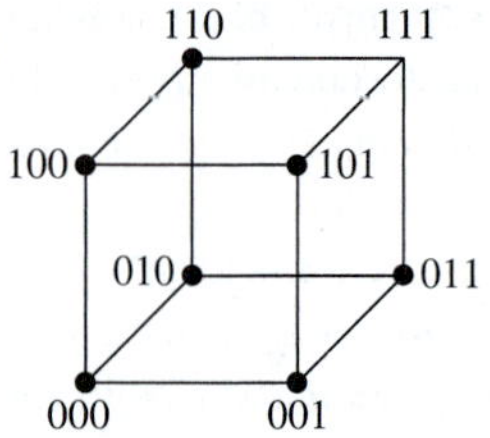

d)

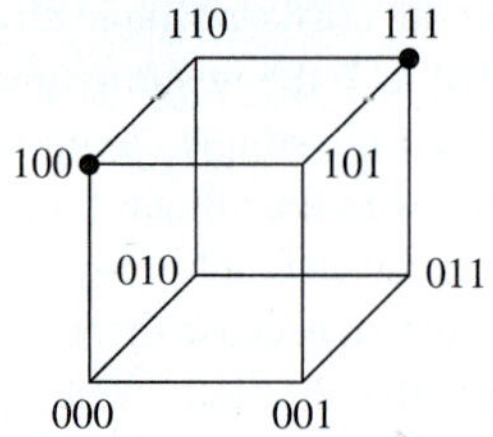

9. (0, 0) and (1, 1) **11.** $x + xy = x \cdot 1 + xy = x(1 + y) = x(y + 1) = x \cdot 1 = x$

13.

x	y	z	$x\bar{y}$	$y\bar{z}$	$\bar{x}z$	$x\bar{y} + y\bar{z} + \bar{x}z$	$\bar{x}y$	$\bar{y}z$	$x\bar{z}$	$\bar{x}y + \bar{y}z + x\bar{z}$
1	1	1	0	0	0	0	0	0	0	0
1	1	0	0	1	0	1	0	0	1	1
1	0	1	1	0	0	1	0	1	0	1
1	0	0	1	0	0	1	0	0	1	1
0	1	1	0	0	1	1	1	0	0	1
0	1	0	0	1	0	1	1	0	0	1
0	0	1	0	0	1	1	0	1	0	1
0	0	0	0	0	0	0	0	0	0	0

15.

x	$x + x$	$x \cdot x$
0	0	0
1	1	1

17.

x	$x + 1$	$x \cdot 0$
0	1	0
1	1	0

19.

x	y	z	$y + z$	$x + (y + z)$	$x + y$	$(x + y) + z$	yz	$x(yz)$	xy	$(xy)z$
1	1	1	1	1	1	1	1	1	1	1
1	1	0	1	1	1	1	0	0	1	0
1	0	1	1	1	1	1	0	0	0	0
1	0	0	0	1	1	1	0	0	0	0
0	1	1	1	1	1	1	1	0	0	0
0	1	0	1	1	1	1	0	0	0	0
0	0	1	1	1	0	1	0	0	0	0
0	0	0	0	0	0	0	0	0	0	0

21.

x	y	xy	$\overline{(xy)}$	$\bar{x}$	$\bar{y}$	$\bar{x} + \bar{y}$	$x + y$	$\overline{(x + y)}$	$\bar{x}\bar{y}$
1	1	1	0	0	0	0	1	0	0
1	0	0	1	0	1	1	1	0	0
0	1	0	1	1	0	1	1	0	0
0	0	0	1	1	1	1	0	1	1

23. $0 \cdot \bar{0} = 0 \cdot 1 = 0$; $1 \cdot \bar{1} = 1 \cdot 0 = 0$

25.

x	y	$x \oplus y$	$x + y$	xy	$\overline{(xy)}$	$(x + y)\overline{(xy)}$	$x\bar{y}$	$\bar{x}y$	$x\bar{y} + \bar{x}y$
1	1	0	1	1	0	0	0	0	0
1	0	1	1	0	1	1	1	0	1
0	1	1	1	0	1	1	0	1	1
0	0	0	0	0	1	0	0	0	0

27. a) True, as a table of values can show **b)** False; take $x = 1$, $y = 1$, $z = 1$, for instance **c)** False; take $x = 1$, $y = 1$, $z = 0$, for instance **29.** By De Morgan's laws, the complement of an expression is like the dual except that the complement of each variable has been taken. **31.** 16 **33.** If we replace each 0 by **F**, 1 by **T**, Boolean sum by $\vee$, Boolean product by $\wedge$, and $\overline{}$ by $\neg$ (and x by p and y by q so that the variables look like they represent propositions, and the equals sign by the logical equivalence symbol), then $\overline{xy} = \overline{x} + \overline{y}$ becomes $\neg(p \wedge q) \equiv \neg p \vee \neg q$ and $\overline{x+y} = \overline{x}\,\overline{y}$ becomes $\neg(p \vee q) \equiv \neg p \wedge \neg q$. **35.** By the domination, distributive, and identity laws, $x \vee x = (x \vee x) \wedge 1 = (x \vee x) \wedge (x \vee \overline{x}) = x \vee (x \wedge \overline{x}) = x \vee 0 = x$. Similarly, $x \wedge x = (x \wedge x) \vee 0 = (x \wedge x) \vee (x \wedge \overline{x}) = x \wedge (x \vee \overline{x}) = x \wedge 1 = x$. **37.** Because $0 \vee 1 = 1$ and $0 \wedge 1 = 0$ by the identity and commutative laws, it follows that $\overline{0} = 1$. Similarly, because $1 \vee 0 = 1$ and $1 \wedge 0 = 1$, it follows that $\overline{1} = 0$. **39.** First, note that $x \wedge 0 = 0$ and $x \vee 1 = 1$ for all x, as can easily be proved. To prove the first identity, it is sufficient to show that $(x \vee y) \vee (\overline{x} \wedge \overline{y}) = 1$ and $(x \vee y) \wedge (\overline{x} \wedge \overline{y}) = 0$. By the associative, commutative, distributive, domination, and identity laws, $(x \vee y) \vee (\overline{x} \wedge \overline{y}) = y \vee [x \vee (\overline{x} \wedge \overline{y})] = y \vee [(x \vee \overline{x}) \wedge (x \vee \overline{y})] = y \vee [1 \wedge (x \vee \overline{y})] = y \vee (x \vee \overline{y}) = (y \vee \overline{y}) \vee x = 1 \vee x = 1$ and $(x \vee y) \wedge (\overline{x} \wedge \overline{y}) = \overline{y} \wedge [\overline{x} \wedge (x \vee y)] = \overline{y} \wedge [(\overline{x} \wedge x) \vee (\overline{x} \wedge y)] = \overline{y} \wedge [0 \vee (\overline{x} \wedge y)] = \overline{y} \wedge (\overline{x} \wedge y) = \overline{x} \wedge (y \wedge \overline{y}) = \overline{x} \wedge 0 = 0$. The second identity is proved in a similar way. **41.** Using the hypotheses, Exercise 35, and the distributive law it follows that $x = x \vee 0 = x \vee (x \vee y) = (x \vee x) \vee y = x \vee y = 0$. Similarly, $y = 0$. To prove the second statement, note that $x = x \wedge 1 = x \wedge (x \wedge y) = (x \wedge x) \wedge y = x \wedge y = 1$. Similarly, $y = 1$. **43.** Use Exercises 39 and 41 in the Supplementary Exercises in Chapter 9 and the definition of a complemented, distributed lattice to establish the five pairs of laws in the definition.

Section 12.2

1. a) $\overline{x}\,\overline{y}z$ **b)** $\overline{x}y\overline{z}$ **c)** $\overline{x}yz$ **d)** $\overline{x}\,\overline{y}\,\overline{z}$ **3. a)** $xyz + xy\overline{z} + x\overline{y}z + x\overline{y}\,\overline{z} + \overline{x}yz + \overline{x}y\overline{z} + \overline{x}\,\overline{y}z$ **b)** $xyz + xy\overline{z} + \overline{x}yz$ **c)** $xyz + xy\overline{z} + x\overline{y}z + x\overline{y}\,\overline{z}$ **d)** $x\overline{y}z + x\overline{y}\,\overline{z}$ **5.** $wxy\overline{z} + wx\overline{y}z + w\overline{x}yz + \overline{w}xyz + \overline{w}x\overline{y}\,\overline{z} + \overline{w}\,\overline{x}\,\overline{y}z + \overline{w}\,\overline{x}y\overline{z} + w\overline{x}\,\overline{y}\,\overline{z}$ **7. a)** $\overline{x}+\overline{y}+z$ **b)** $x+y+z$ **c)** $x+\overline{y}+z$ **9.** $y_1 + y_2 + \cdots + y_n = 0$ if and only if $y_i = 0$ for $i = 1, 2, \ldots, n$. This holds if and only if $x_i = 0$ when $y_i = x_i$ and $x_i = 1$ when $y_i = \overline{x}_i$. **11. a)** $x+y+z$ **b)** $(x+y+z)(x+y+\overline{z})(x+\overline{y}+z)(\overline{x}+y+z)(\overline{x}+y+\overline{z})$ **c)** $(x+y+z)(x+y+\overline{z})(x+\overline{y}+z)(x+\overline{y}+\overline{z})$ **d)** $(x+y+z)(x+y+\overline{z})(x+\overline{y}+z)(x+\overline{y}+\overline{z})(\overline{x}+\overline{y}+z)(\overline{x}+\overline{y}+\overline{z})$ **13. a)** $x+y+z$ **b)** $x + [y + \overline{(\overline{x}+z)}]$ **c)** $\overline{(x+\overline{y})}$ **d)** $[x + \overline{(x+\overline{y}+\overline{z})}]$

15. a)

x	$\overline{x}$	$x \downarrow x$
1	0	0
0	1	1

b)

x	y	xy	$x \downarrow x$	$y \downarrow y$	$(x \downarrow x) \downarrow (y \downarrow y)$
1	1	1	0	0	1
1	0	0	0	1	0
0	1	0	1	0	0
0	0	0	1	1	0

c)

x	y	$x+y$	$(x \downarrow y)$	$(x \downarrow y) \downarrow (x \downarrow y)$
1	1	1	0	1
1	0	1	0	1
0	1	1	0	1
0	0	0	1	0

17. a) $\{[(x \mid x) \mid (y \mid y)] \mid [(x \mid x) \mid (y \mid y)]\} \mid (z \mid z)$ **b)** $\{[(x \mid x) \mid (z \mid z)] \mid y\} \mid \{[(x \mid x) \mid (z \mid z)] \mid y\}$ **c)** x **d)** $[x \mid (y \mid y)] \mid [x \mid (y \mid y)]$ **19.** It is impossible to represent $\overline{x}$ using + and · because there is no way to get the value 0 if the input is 1.

Section 12.3

1. $(x+y)\overline{y}$ **3.** $\overline{(xy)}+(\overline{z}+x)$ **5.** $(x+y+z) + (\overline{x}+y+z)+(\overline{x}+\overline{y}+\overline{z})$

7.

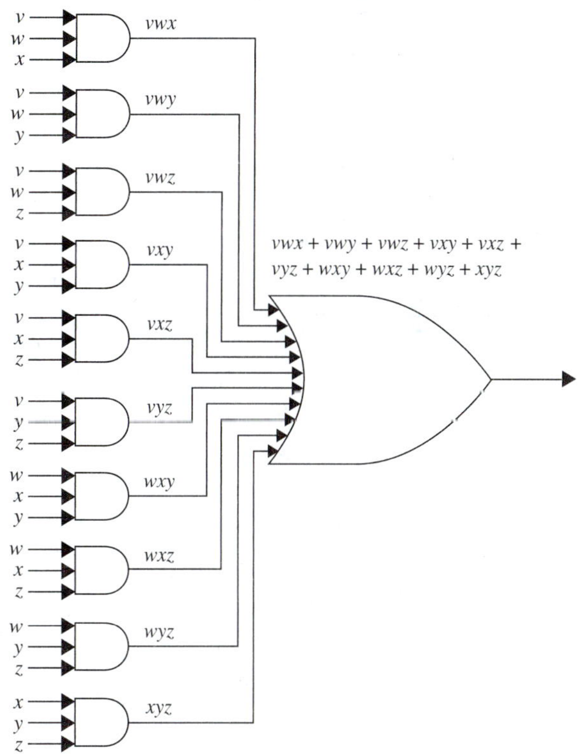

9.

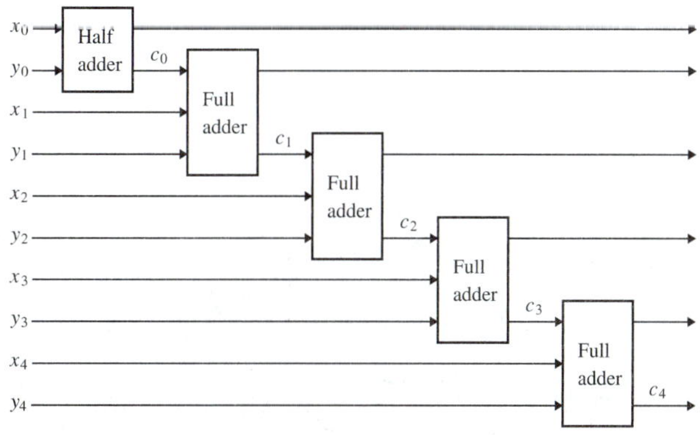

11.

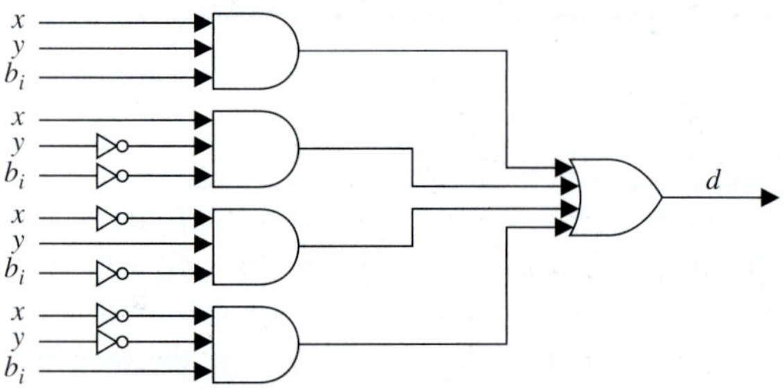

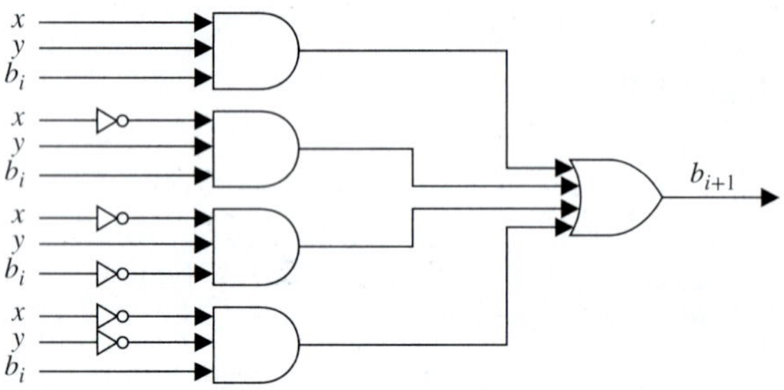

13.

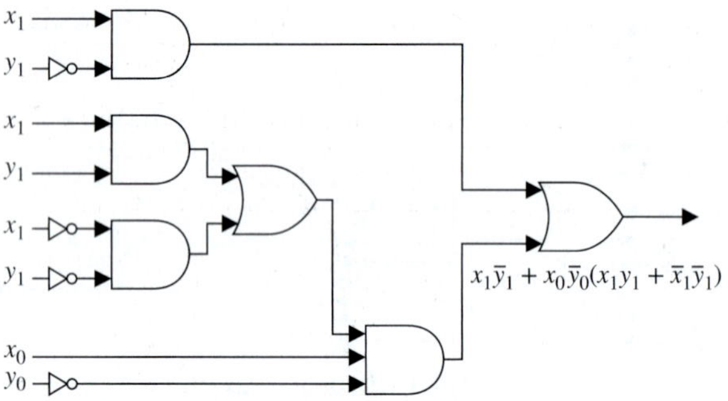

15. a)

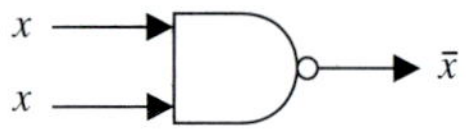

b)

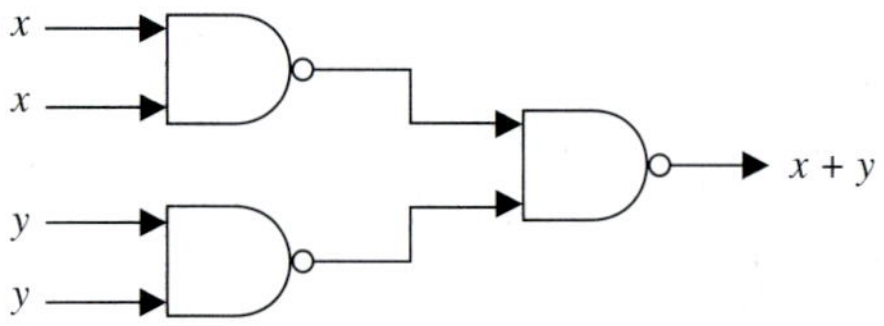

c)

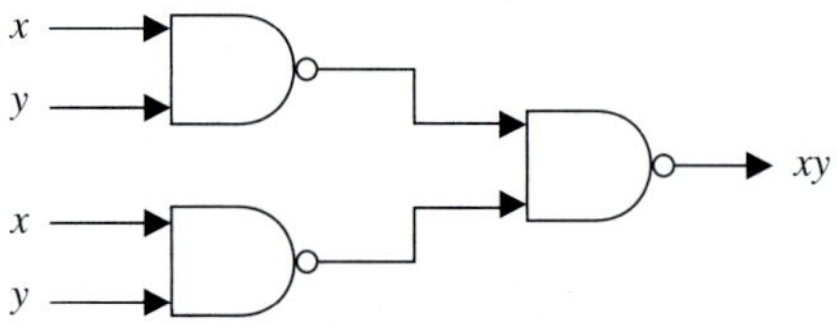

d)

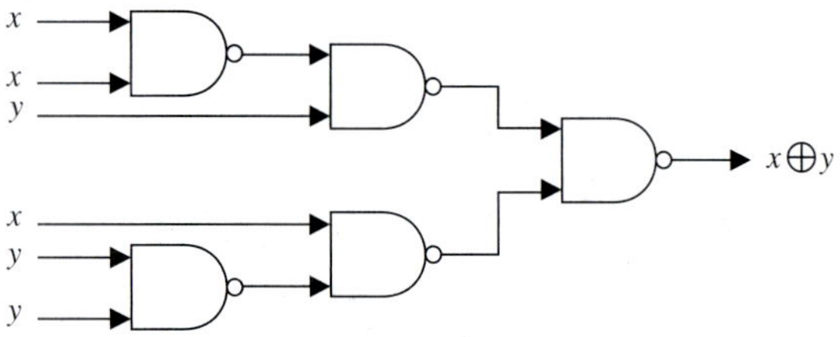

17.

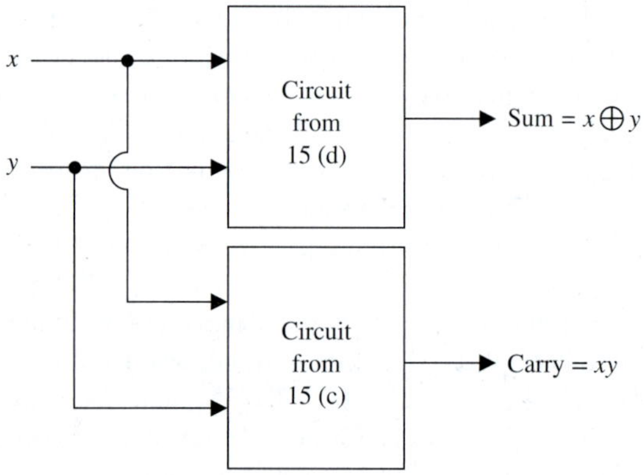

19.

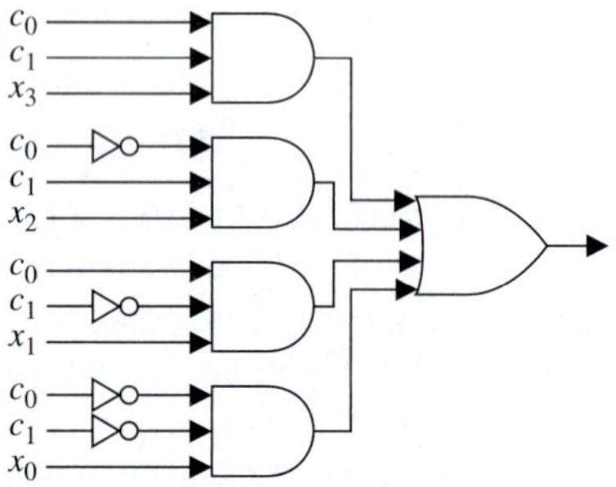

Section 12.4

1. a)

	y	$\bar{y}$
x		
$\bar{x}$	1	

b) xy and $\bar{x}\,\bar{y}$

3. a)

	y	$\bar{y}$
x		1
$\bar{x}$		

b)

	y	$\bar{y}$
x	1	
$\bar{x}$		1

c)

	y	$\bar{y}$
x	1	1
$\bar{x}$	1	1

5. a)

	yz	$y\bar{z}$	$\bar{y}\bar{z}$	$\bar{y}z$
x				
$\bar{x}$		1		

b) $\bar{x}yz$, $\bar{x}\,\bar{y}\,\bar{z}$, $xy\bar{z}$

7. a)

	yz	$y\bar{z}$	$\bar{y}\bar{z}$	$\bar{y}z$
x			1	
$\bar{x}$				

b)

	yz	$y\bar{z}$	$\bar{y}\bar{z}$	$\bar{y}z$
x				
$\bar{x}$	1		1	

c)

	yz	$y\bar{z}$	$\bar{y}\bar{z}$	$\bar{y}z$
x	1	1		
$\bar{x}$		1		1

9. Implicants: xyz, $xy\bar{z}$, $x\bar{y}\bar{z}$, $\bar{x}y\bar{z}$, xy, $x\bar{z}$, $y\bar{z}$; prime implicants: xy, $x\bar{z}$, $y\bar{z}$; essential prime implicants: xy, $x\bar{z}$, $y\bar{z}$

	yz	$y\bar{z}$	$\bar{y}\bar{z}$	$\bar{y}z$
x	1	1	1	
$\bar{x}$		1		

11. The 3-cube on the right corresponds to w; the 3-cube given by the top surface of the whole figure represents x; the 3-cube given by the back surface of the whole figure represents y; the 3-cube given by the right surfaces of both the left and the right 3-cube represents z. In each case, the opposite 3-face represents the complemented literal. The 2-cube that represents wz is the right face of the 3-cube on the right; the 2-cube that represents $\bar{x}y$ is bottom rear; the 2-cube that represents $\bar{y}\,\bar{z}$ is front left.

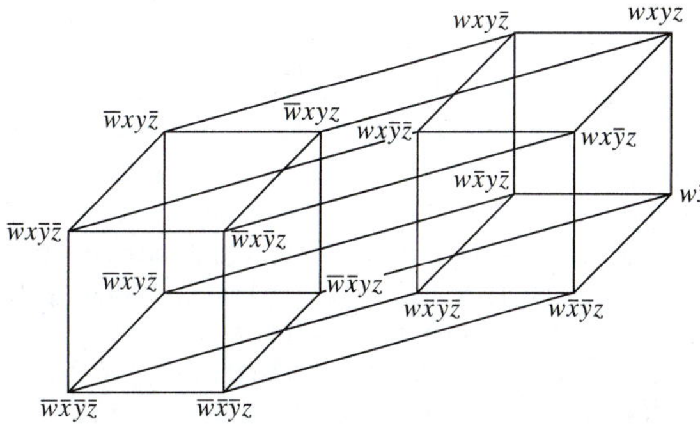

13. a)

	yz	$y\bar{z}$	$\bar{y}\bar{z}$	$\bar{y}z$
wx				
$w\bar{x}$				
$\bar{w}\bar{x}$				
$\bar{w}x$		1		

b) $\bar{w}xyz$, $\bar{w}\,\bar{x}y\bar{z}$, $\bar{w}x\bar{y}\,\bar{z}$, $wxy\,\bar{z}$

15. a)

	$x_3x_4x_5$	$x_3x_4\bar{x}_5$	$x_3\bar{x}_4\bar{x}_5$	$x_3\bar{x}_4x_5$	$\bar{x}_3\bar{x}_4x_5$	$\bar{x}_3\bar{x}_4\bar{x}_5$	$\bar{x}_3x_4\bar{x}_5$	$\bar{x}_3x_4x_5$
x_1x_2	1	1						
$x_1\bar{x}_2$								
$\bar{x}_1\bar{x}_2$								
$\bar{x}_1x_2$								

b)

	$x_3x_4x_5$	$x_3x_4\bar{x}_5$	$x_3\bar{x}_4\bar{x}_5$	$x_3\bar{x}_4x_5$	$\bar{x}_3\bar{x}_4x_5$	$\bar{x}_3\bar{x}_4\bar{x}_5$	$\bar{x}_3x_4\bar{x}_5$	$\bar{x}_3x_4x_5$
x_1x_2								
$x_1\bar{x}_2$								
$\bar{x}_1\bar{x}_2$	1			1				
$\bar{x}_1x_2$	1			1				

c)

	$x_3x_4x_5$	$x_3x_4\bar{x}_5$	$x_3\bar{x}_4\bar{x}_5$	$x_3\bar{x}_4x_5$	$\bar{x}_3\bar{x}_4x_5$	$\bar{x}_3\bar{x}_4\bar{x}_5$	$\bar{x}_3x_4\bar{x}_5$	$\bar{x}_3x_4x_5$
x_1x_2	1	1					1	1
$x_1\bar{x}_2$								
$\bar{x}_1\bar{x}_2$								
$\bar{x}_1x_2$	1	1					1	1

d)

	$x_3x_4x_5$	$x_3x_4\bar{x}_5$	$x_3\bar{x}_4\bar{x}_5$	$x_3\bar{x}_4x_5$	$\bar{x}_3\bar{x}_4x_5$	$\bar{x}_3\bar{x}_4\bar{x}_5$	$\bar{x}_3x_4\bar{x}_5$	$\bar{x}_3x_4x_5$
x_1x_2					1	1		
$x_1\bar{x}_2$					1	1		
$\bar{x}_1\bar{x}_2$					1	1		
$\bar{x}_1x_2$					1	1		

e)

	$x_3x_4x_5$	$x_3x_4\bar{x}_5$	$x_3\bar{x}_4\bar{x}_5$	$x_3\bar{x}_4x_5$	$\bar{x}_3\bar{x}_4x_5$	$\bar{x}_3\bar{x}_4\bar{x}_5$	$\bar{x}_3x_4\bar{x}_5$	$\bar{x}_3x_4x_5$
x_1x_2	1	1	1	1				
$x_1\bar{x}_2$	1	1	1	1				
$\bar{x}_1\bar{x}_2$	1	1	1	1				
$\bar{x}_1x_2$	1	1	1	1				

f)

	$x_3x_4x_5$	$x_3x_4\bar{x}_5$	$x_3\bar{x}_4\bar{x}_5$	$x_3\bar{x}_4x_5$	$\bar{x}_3\bar{x}_4x_5$	$\bar{x}_3\bar{x}_4\bar{x}_5$	$\bar{x}_3x_4\bar{x}_5$	$\bar{x}_3x_4x_5$
x_1x_2		1	1			1	1	
$x_1\bar{x}_2$		1	1			1	1	
$\bar{x}_1\bar{x}_2$		1	1			1	1	
$\bar{x}_1x_2$		1	1			1	1	

17. a) 64 **b)** 6 **19.** Rows 1 and 4 are considered adjacent. The pairs of columns considered adjacent are: columns 1 and 4, 1 and 12, 1 and 16, 2 and 11, 2 and 15, 3 and 6, 3 and 10, 4 and 9, 5 and 8, 5 and 16, 6 and 15, 7 and 10, 7 and 14, 8 and 13, 9 and 12, 11 and 14, 13 and 16.

21.

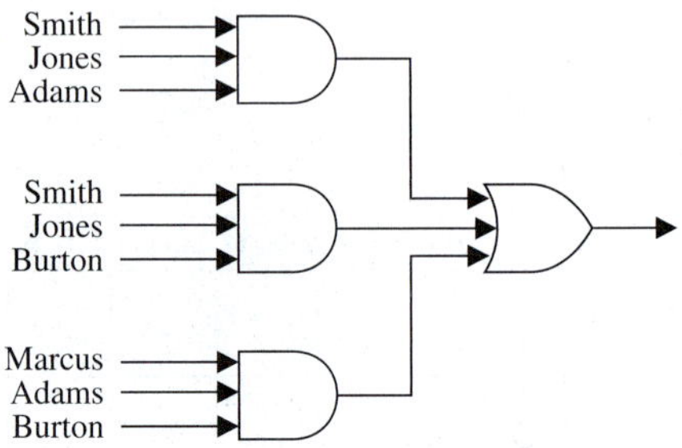

23. a) $\bar{x}z$ **b)** y **c)** $x\bar{z}+\bar{x}z+\bar{y}z$ **d)** $xz+\bar{x}y+\bar{y}\bar{z}$ **25. a)** $wxz + wx\bar{y} + w\bar{y}z + w\bar{x}y\bar{z}$ **b)** $x\bar{y}z + \bar{w}\,\bar{y}z + wxy\bar{z} + w\bar{x}yz + \bar{w}\,\bar{x}y\bar{z}$ **c)** $\bar{y}z + wxy + w\bar{x}\,\bar{y} + \bar{w}\,\bar{x}y\bar{z}$ **d)** $wy + yz + \bar{x}y + wxz + \bar{w}\,\bar{x}z$ **27.** $x(y+z)$

29.

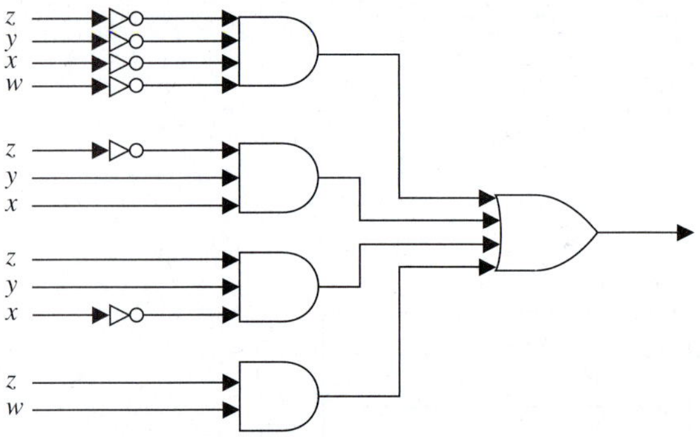

31. $\bar{x}\,\bar{z} + xz$ **33.** We use induction on n. If $n = 1$, then we are looking at a line segment, labeled 0 at one end and 1 at the other end. The only possible value of k is also 1, and if the literal is x_1, then the subcube we have is the 0-dimensional subcube consisting of the endpoint labeled 1, and if the literal is $\bar{x}_1$, then the subcube we have is the 0-dimensional subcube consisting of the endpoint labeled 0. Now assume that the statement is true for n; we must show that it is true for $n + 1$. If the literal x_{n+1} (or its complement) is not part of the product, then by the inductive hypothesis, the product when viewed in the setting of n variables corresponds to an $(n - k)$-dimensional subcube of the n-dimensional cube, and the Cartesian product of that subcube with the line segment $[0, 1]$ gives us a subcube one dimension higher in our given $(n+1)$-dimensional cube, namely having dimension $(n+1)-k$, as desired. On the other hand, if the literal x_{n+1} (or its complement) is part of the product, then the product of the remaining $k - 1$ literals corresponds to a subcube of dimension $n - (k - 1) = (n + 1) - k$ in the n-dimensional cube, and that slice, at either the 1-end or the 0-end in the last variable, is the desired subcube.

Supplementary Exercises

1. a) $x = 0,\ y = 0,\ z = 0$; $x = 1,\ y = 1,\ z = 1$ **b)** $x = 0,\ y = 0,\ z = 0$; $x = 0,\ y = 0,\ z = 1$; $x = 0,\ y = 1,\ z = 0$; $x = 1,\ y = 0,\ z = 1$; $x = 1,\ y = 1,\ z = 0$; $x = 1,\ y = 1,\ z = 1$ **c)** No values **3. a)** Yes **b)** No **c)** No **d)** Yes **5.** $2^{2^{n-1}}$ **7. a)** If $F(x_1, \ldots, x_n) = 1$, then $(F + G)(x_1, \ldots, x_n) = F(x_1, \ldots, x_n) + G(x_1, \ldots, x_n) = 1$ by the dominance law. Hence, $F \le F + G$. **b)** If $(FG)(x_1, \ldots, x_n) = 1$, then $F(x_1, \ldots, x_n) \cdot G(x_1, \ldots, x_n) = 1$. Hence, $F(x_1, \ldots, x_n) = 1$. It follows that $FG \le F$. **9.** Because $F(x_1, \ldots, x_n) = 1$ implies that $F(x_1, \ldots, x_n) = 1$, $\le$ is reflexive. Suppose that $F \le G$ and $G \le F$. Then $F(x_1, \ldots, x_n) = 1$ if and only if $G(x_1, \ldots, x_n) = 1$. This implies that $F = G$. Hence, $\le$ is antisymmetric. Suppose that $F \le G \le H$. Then if $F(x_1, \ldots, x_n) = 1$, it follows that $G(x_1, \ldots, x_n) = 1$, which implies that $H(x_1, \ldots, x_n) = 1$. Hence, $F \le H$, so $\le$ is transitive. **11. a)** $x = 1, y = 0, z = 0$ **b)** $x = 1, y = 0, z = 0$ **c)** $x = 1, y = 0, z = 0$

13.

x	y	$x \odot y$	$x \oplus y$	$\overline{(x \oplus y)}$
1	1	1	0	1
1	0	0	1	0
0	1	0	1	0
0	0	1	0	1

15. Yes, as a truth table shows **17. a)** 6 **b)** 5 **c)** 5 **d)** 6

19.

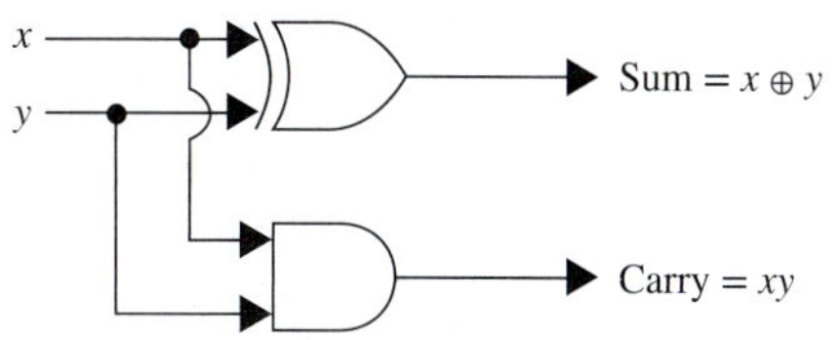

21. $x_3 + x_2\bar{x}_1$ **23.** Suppose it were with weights a and b. Then there would be a real number T such that $xa + yb \geq T$ for (1,0) and (0,1), but with $xa + yb < T$ for (0,0) and (1,1). Hence, $a \geq T$, $b \geq T$, $0 < T$, and $a + b < T$. Thus, a and b are positive, which implies that $a + b > a \geq T$, a contradiction.

CHAPTER 13

Section 13.1

1. a) sentence ⇒ **noun phrase intransitive verb phrase** ⇒ **article adjective noun intransitive verb phrase** ⇒ **article adjective noun intransitive verb** ⇒ … (after 3 steps) … ⇒ *the happy hare runs.*
b) sentence ⇒ **noun phrase intransitive verb phrase** ⇒ **article adjective noun intransitive verb phrase** ⇒ **article adjective noun intransitive verb adverb**… (after 4 steps)… ⇒ *the sleepy tortoise runs quickly*
c) sentence ⇒ **noun phrase transitive verb phrase noun phrase** ⇒ **article noun transitive verb phrase noun phrase** ⇒ **article noun transitive verb noun phrase** ⇒ **article noun transitive verb article noun** ⇒ … (after 4 steps)… ⇒ *the tortoise passes the hare*
d) sentence ⇒ **noun phrase transitive verb phrase noun phrase** ⇒ **article adjective noun transitive verb phrase noun phrase** ⇒ **article adjective noun transitive verb noun phrase** ⇒ **article adjective noun transitive verb article adjective noun** ⇒ … (after 6 steps)… ⇒ *the sleepy hare passes the happy tortoise*
3. The only way to get a noun, such as *tortoise*, at the end is to have a noun phrase at the end, which can be achieved only via the production **sentence** → **noun phrase transitive verb phrase noun phrase**. However, **transitive verb phrase** → **transitive verb** → *passes*, and this sentence does not contain *passes*.
5. a) $S \Rightarrow 1A \Rightarrow 10B \Rightarrow 101A \Rightarrow 1010B \Rightarrow 10101$ **b)** Because of the productions in this grammar, every 1 must be followed by a 0 unless it occurs at the end of the string. **c)** All strings consisting of a 0 or a 1 followed by one or more repetitions of 01
7. $S \Rightarrow 0S1 \Rightarrow 00S11 \Rightarrow 000S111 \Rightarrow 000111$
9. a) $S \Rightarrow 0S \Rightarrow 00S \Rightarrow 00S1 \Rightarrow 00S11 \Rightarrow 00S111 \Rightarrow 00S1111 \Rightarrow 001111$ **b)** $S \Rightarrow 0S \Rightarrow 00S \Rightarrow 001A \Rightarrow 0011A \Rightarrow 00111A \Rightarrow 001111$ **11.** $S \Rightarrow C \Rightarrow 0CAB \Rightarrow 00ABAB \Rightarrow 00AABB \Rightarrow 001ABB \Rightarrow 0011BB \Rightarrow 00112B \Rightarrow 001122$ **13. a)** $S \to 0$, $S \to 1$, $S \to 11$ **b)** $S \to 1S$, $S \to \lambda$ **c)** $S \to 0A1$, $A \to 1A$, $A \to 0A$, $A \to \lambda$ **d)** $S \to 0A$, $A \to 11A$, $A \to \lambda$
15. a) $S \to 00S$, $S \to \lambda$ **b)** $S \to 10A$, $A \to 00A$, $A \to \lambda$ **c)** $S \to AAS$, $S \to BBS$, $AB \to BA$, $BA \to AB$, $S \to \lambda$, $A \to 0$, $B \to 1$ **d)** $S \to 0000000000A$, $A \to 0A$, $A \to \lambda$ **e)** $S \to AS$, $S \to ABS$, $S \to A$, $AB \to BA$, $BA \to AB$, $A \to 0$, $B \to 1$ **f)** $S \to ABS$, $S \to \lambda$, $AB \to BA$, $BA \to AB$, $A \to 0$, $B \to 1$ **g)** $S \to ABS$, $S \to T$, $S \to U$, $T \to AT$, $T \to A$, $U \to BU$, $U \to B$, $AB \to BA$, $BA \to AB$, $A \to 0$, $B \to 1$ **17. a)** $S \to 0S$, $S \to \lambda$ **b)** $S \to A0$, $A \to 1A$, $A \to \lambda$ **c)** $S \to 000S$, $S \to \lambda$ **19. a)** Type 2, not type 3 **b)** Type 3 **c)** Type 0, not type 1 **d)** Type 2, not type 3 **e)** Type 2, not type 3 **f)** Type 0, not type 1 **g)** Type 3 **h)** Type 0, not type 1 **i)** Type 2, not type 3 **j)** Type 2, not type 3 **21.** Let S_1 and S_2 be the start symbols of G_1 and G_2, respectively. Let S be a new start symbol. **a)** Add S and productions $S \to S_1$ and $S \to S_2$. **b)** Add S and production $S \to S_1S_2$. **c)** Add S and production $S \to \lambda$ and $S \to S_1S$.

23. a)

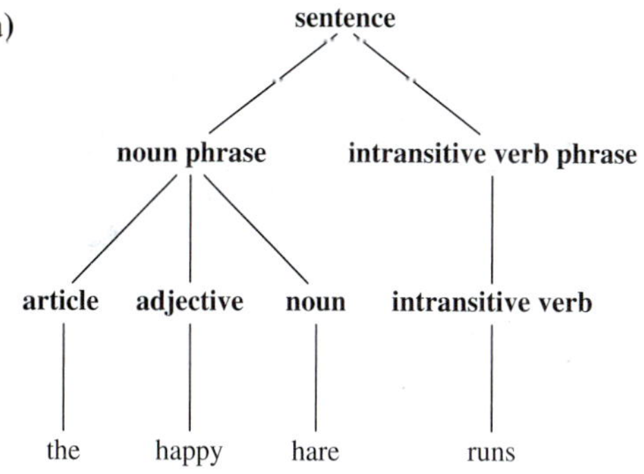

b)

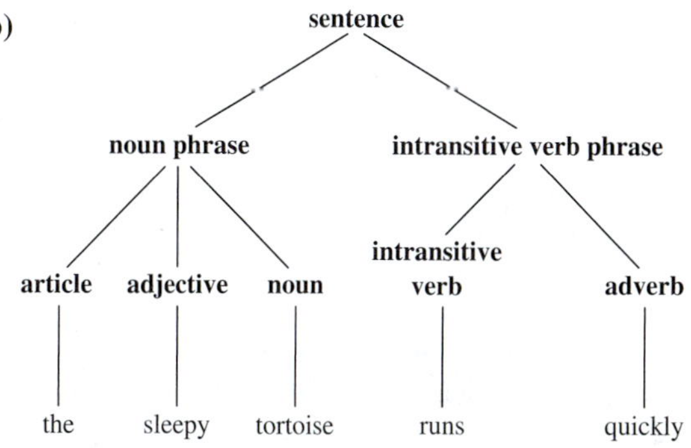

c)

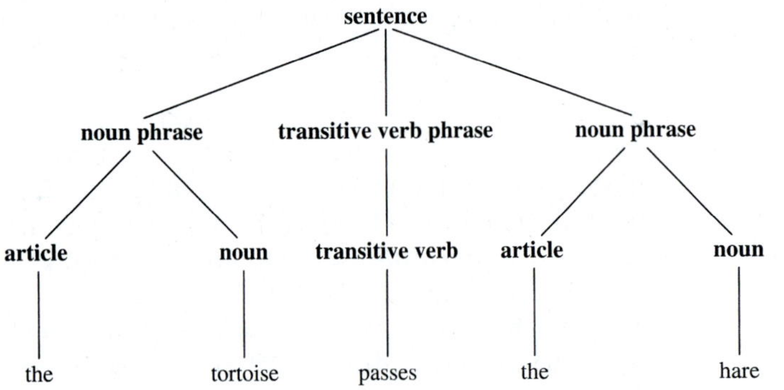

d)

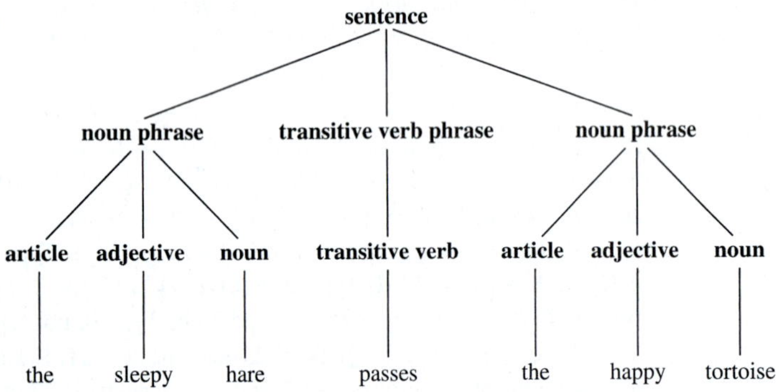

25. a) Yes **b)** No **c)** Yes **d)** No

27.

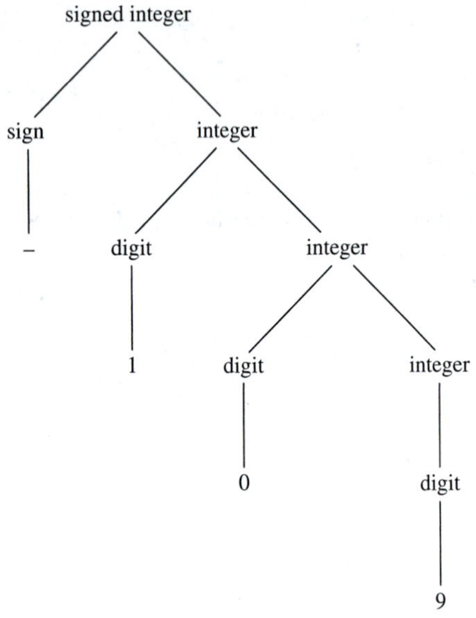

29. a) $S \rightarrow \langle sign\rangle\langle integer\rangle$
$S \rightarrow \langle sign\rangle\langle integer\rangle\,.\,\langle positive\ integer\rangle$
$\langle sign\rangle \rightarrow +$
$\langle sign\rangle \rightarrow -$
$\langle integer\rangle \rightarrow \langle digit\rangle$
$\langle integer\rangle \rightarrow \langle integer\rangle\langle digit\rangle$
$\langle digit\rangle \rightarrow i, i = 1, 2, 3, 4, 5, 6, 7, 8, 9, 0$
$\langle positive\ integer\rangle \rightarrow \langle integer\rangle\langle nonzero\ digit\rangle$
$\langle positive\ integer\rangle \rightarrow \langle nonzero\ digit\rangle\langle integer\rangle$
$\langle positive\ integer\rangle \rightarrow \langle integer\rangle\langle nonzero\ digit\rangle\langle integer\rangle$
$\langle positive\ integer\rangle \rightarrow \langle nonzero\ digit\rangle$
$\langle nonzero\ digit\rangle \rightarrow i, i = 1, 2, 3, 4, 5, 6, 7, 8, 9$

b) $\langle signed\ decimal\ number\rangle ::= \langle sign\rangle\langle integer\rangle \mid \langle sign\rangle\langle integer\rangle\,.\,\langle positive\ integer\rangle$
$\langle sign\rangle ::= + \mid -$
$\langle integer\rangle ::= \langle digit\rangle \mid \langle integer\rangle\langle digit\rangle$
$\langle digit\rangle ::= 0 \mid 1 \mid 2 \mid 3 \mid 4 \mid 5 \mid 6 \mid 7 \mid 8 \mid 9$
$\langle nonzero\ digit\rangle ::= 1 \mid 2 \mid 3 \mid 4 \mid 5 \mid 6 \mid 7 \mid 8 \mid 9$
$\langle positive\ integer\rangle ::= \langle integer\rangle\langle nonzero\ digit\rangle \mid \langle nonzero\ digit\rangle\langle integer\rangle \mid \langle integer\rangle\langle nonzero\ integer\rangle\langle integer\rangle \mid \langle nonzero\ digit\rangle$

c)

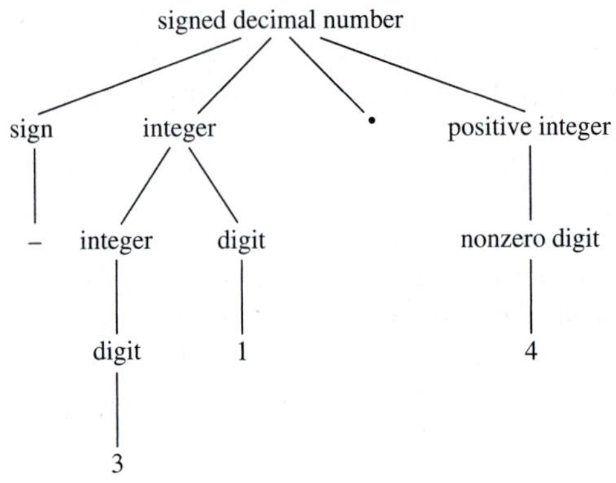

31. a) ⟨*identifier*⟩ ::= ⟨*lcletter*⟩ | ⟨*identifier*⟩⟨*lcletter*⟩
⟨*lcletter*⟩ ::= *a* | *b* | *c* | ⋯ | *z*

b) ⟨*identifier*⟩ ::= ⟨*lcletter*⟩⟨*lcletter*⟩⟨*lcletter*⟩ | ⟨*lcletter*⟩⟨*lcletter*⟩⟨*lcletter*⟩⟨*lcletter*⟩ |
⟨*lcletter*⟩⟨*lcletter*⟩⟨*lcletter*⟩⟨*lcletter*⟩⟨*lcletter*⟩ |
⟨*lcletter*⟩⟨*lcletter*⟩⟨*lcletter*⟩⟨*lcletter*⟩⟨*lcletter*⟩⟨*lcletter*⟩
⟨*lcletter*⟩ ::= *a* | *b* | *c* | ⋯ | *z*

c) ⟨*identifier*⟩ ::= ⟨*ucletter*⟩ | ⟨*ucletter*⟩⟨*letter*⟩ | ⟨*ucletter*⟩⟨*letter*⟩⟨*letter*⟩ |
⟨*ucletter*⟩⟨*letter*⟩⟨*letter*⟩⟨*letter*⟩ | ⟨*ucletter*⟩⟨*letter*⟩⟨*letter*⟩⟨*letter*⟩⟨*letter*⟩ |
⟨*ucletter*⟩⟨*letter*⟩⟨*letter*⟩⟨*letter*⟩⟨*letter*⟩⟨*letter*⟩
⟨*letter*⟩ ::= ⟨*lcletter*⟩ | ⟨*ucletter*⟩
⟨*lcletter*⟩ ::= *a* | *b* | *c* | ⋯ | *z*
⟨*ucletter*⟩ ::= *A* | *B* | *C* | ⋯ | *Z*

d) ⟨*identifier*⟩ ::= ⟨*lcletter*⟩⟨*digitorus*⟩⟨*alphanumeric*⟩⟨*alphanumeric*⟩⟨*alphanumeric*⟩ |
⟨*lcletter*⟩⟨*digitorus*⟩⟨*alphanumeric*⟩⟨*alphanumeric*⟩⟨*alphanumeric*⟩⟨*alphanumeric*⟩
⟨*digitorus*⟩ ::= ⟨*digit*⟩ | _
⟨*alphanumeric*⟩ ::= ⟨*letter*⟩ | ⟨*digit*⟩
⟨*letter*⟩ ::= ⟨*lcletter*⟩ | ⟨*ucletter*⟩
⟨*lcletter*⟩ ::= *a* | *b* | *c* | ⋯ | *z*
⟨*ucletter*⟩ ::= *A* | *B* | *C* | ⋯ | *Z*
⟨*digit*⟩ ::= 0 | 1 | 2 | ⋯ | 9

33. ⟨*identifier*⟩ ::= ⟨*letterorus*⟩ | ⟨*identifier*⟩⟨*symbol*⟩
⟨*letterorus*⟩ ::= ⟨*letter*⟩ | _
⟨*symbol*⟩ ::= ⟨*letterorus*⟩ | ⟨*digit*⟩
⟨*letter*⟩ ::= ⟨*lcletter*⟩ | ⟨*ucletter*⟩
⟨*lcletter*⟩ ::= *a* | *b* | *c* | ⋯ | *z*
⟨*ucletter*⟩ ::= *A* | *B* | *C* | ⋯ | *Z*
⟨*digit*⟩ ::= 0 | 1 | 2 | ⋯ | 9

35. *numeral* ::= *sign*? *nonzerodigit digit*∗ *decimal*? | *sign*? 0 *decimal*?
sign ::= + | −
nonzerodigit ::= 1 | 2 | ⋯ | 9
digit ::= 0 | *nonzerodigit*
decimal ::= *.digit*∗

37. *identifier* ::= *letterorus symbol* ∗
letterorus ::= *letter* | _
symbol ::= *letterorus* | *digit*
letter ::= *lcletter* | *ucletter*
lcletter ::= *a* | *b* | *c* | ⋯ | *z*
ucletter ::= *A* | *B* | *C* | ⋯ | *Z*
digit ::= 0 | 1 | 2 | ⋯ | 9

39. a) ⟨*expression*⟩
⟨*term*⟩⟨*term*⟩⟨*addOperator*⟩
⟨*factor*⟩⟨*factor*⟩⟨*factor*⟩⟨*mulOperator*⟩⟨*addOperator*⟩
⟨*identifier*⟩⟨*identifier*⟩⟨*identifier*⟩⟨*mulOperator*⟩⟨*addOperator*⟩
a b c ∗ +

b) Not generated

c) ⟨*expression*⟩
⟨*term*⟩
⟨*factor*⟩⟨*factor*⟩⟨*mulOperator*⟩
⟨*expression*⟩⟨*factor*⟩⟨*mulOperator*⟩
⟨*term*⟩⟨*term*⟩⟨*addOperator*⟩⟨*factor*⟩⟨*mulOperator*⟩
⟨*factor*⟩⟨*factor*⟩⟨*addOperator*⟩⟨*factor*⟩⟨*mulOperator*⟩
⟨*identifier*⟩⟨*identifier*⟩⟨*addOperator*⟩⟨*identifier*⟩⟨*mulOperator*⟩
x y − *z* ∗

d) ⟨*expression*⟩
⟨*term*⟩
⟨*factor*⟩⟨*factor*⟩⟨*mulOperator*⟩
⟨*factor*⟩⟨*expression*⟩⟨*mulOperator*⟩
⟨*factor*⟩⟨*term*⟩⟨*mulOperator*⟩
⟨*factor*⟩⟨*factor*⟩⟨*factor*⟩⟨*mulOperator*⟩⟨*mulOperator*⟩
⟨*factor*⟩⟨*factor*⟩⟨*expression*⟩⟨*mulOperator*⟩⟨*mulOperator*⟩
⟨*factor*⟩⟨*factor*⟩⟨*term*⟩⟨*term*⟩⟨*addOperator*⟩⟨*mulOperator*⟩⟨*mulOperator*⟩
⟨*factor*⟩⟨*factor*⟩⟨*factor*⟩⟨*factor*⟩⟨*addOperator*⟩⟨*mulOperator*⟩⟨*mulOperator*⟩
⟨*identifier*⟩⟨*identifier*⟩⟨*identifier*⟩⟨*identifier*⟩⟨*addOperator*⟩⟨*mulOperator*⟩⟨*mulOperator*⟩
$w\,x\,y\,z - * /$
e) ⟨*expression*⟩
⟨*term*⟩
⟨*factor*⟩⟨*factor*⟩⟨*mulOperator*⟩
⟨*factor*⟩⟨*expression*⟩⟨*mulOperator*⟩
⟨*factor*⟩⟨*term*⟩⟨*term*⟩⟨*addOperator*⟩⟨*mulOperator*⟩
⟨*factor*⟩⟨*factor*⟩⟨*factor*⟩⟨*addOperator*⟩⟨*mulOperator*⟩
⟨*identifier*⟩⟨*identifier*⟩⟨*identifier*⟩⟨*addOperator*⟩⟨*mulOperator*⟩
$a\,d\,e - *$

41. a) Not generated
b) ⟨*expression*⟩
⟨*term*⟩⟨*addOperator*⟩⟨*term*⟩
⟨*factor*⟩⟨*mulOperator*⟩⟨*factor*⟩⟨*addOperator*⟩⟨*factor*⟩⟨*mulOperator*⟩⟨*factor*⟩
⟨*identifier*⟩⟨*mulOperator*⟩⟨*identifier*⟩⟨*addOperator*⟩⟨*identifier*⟩⟨*mulOperator*⟩⟨*identifier*⟩
$a/b + c/d$
c) ⟨*expression*⟩
⟨*term*⟩
⟨*factor*⟩⟨*mulOperator*⟩⟨*factor*⟩
⟨*factor*⟩⟨*mulOperator*⟩(⟨*expression*⟩)
⟨*factor*⟩⟨*mulOperator*⟩(⟨*term*⟩⟨*addOperator*⟩⟨*term*⟩)
⟨*factor*⟩⟨*mulOperator*⟩(⟨*factor*⟩⟨*addOperator*⟩⟨*factor*⟩)
⟨*identifier*⟩⟨*mulOperator*⟩(⟨*identifier*⟩⟨*addOperator*⟩⟨*identifier*⟩)
$m * (n + p)$
d) Not generated
e) ⟨*expression*⟩
⟨*term*⟩
⟨*factor*⟩⟨*mulOperator*⟩⟨*factor*⟩
(⟨*expression*⟩)⟨*mulOperator*⟩(⟨*expression*⟩)
(⟨*term*⟩⟨*addOperator*⟩⟨*term*⟩)⟨*mulOperator*⟩(⟨*term*⟩⟨*addOperator*⟩⟨*term*⟩)
(⟨*factor*⟩⟨*addOperator*⟩⟨*factor*⟩)⟨*mulOperator*⟩(⟨*factor*⟩⟨*addOperator*⟩⟨*factor*⟩)
(⟨*identifier*⟩⟨*addOperator*⟩⟨*identifier*⟩)⟨*mulOperator*⟩(⟨*identifier*⟩⟨*addOperator*⟩⟨*identifier*⟩)
$(m + n) * (p - q)$

Section 13.2

42. a)

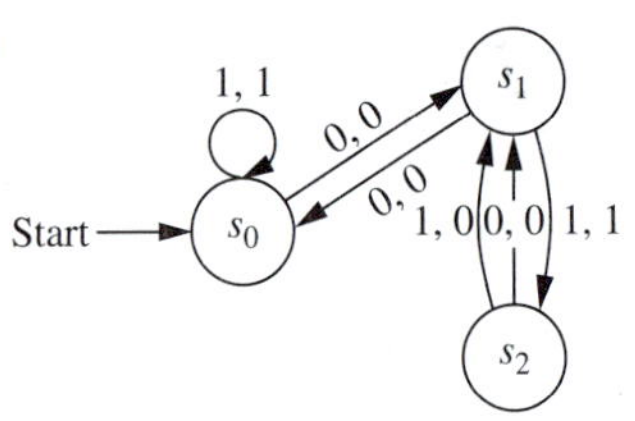

b)

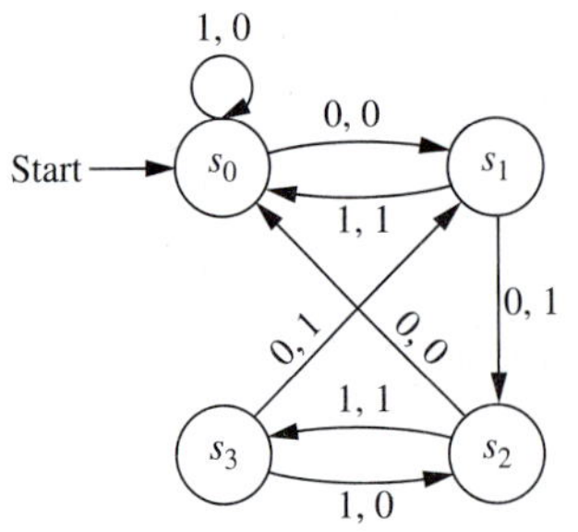

c)

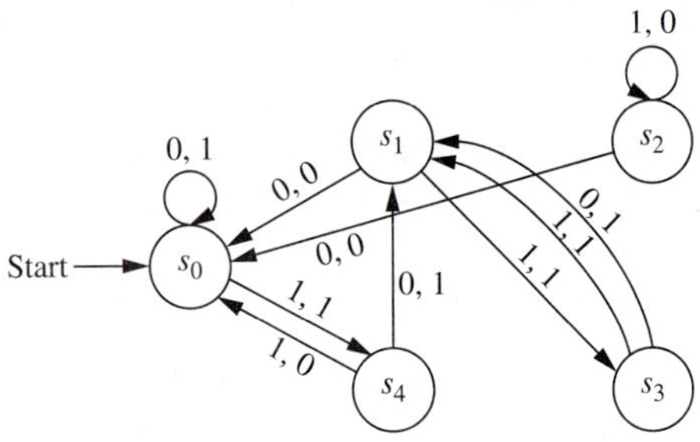

3. **a)** 01010 **b)** 01000 **c)** 11011 **5.** **a)** 1100 **b)** 00110110 **c)** 11111111111

7.

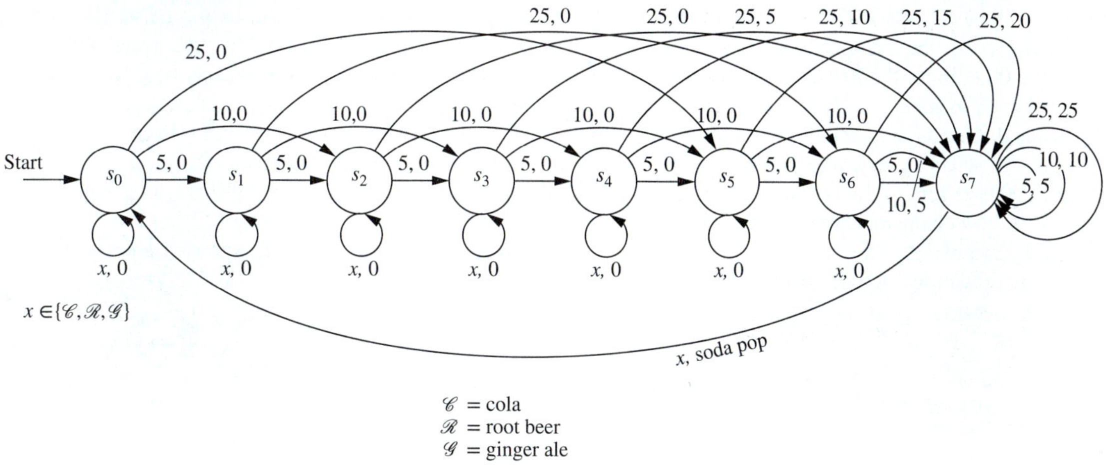

9.

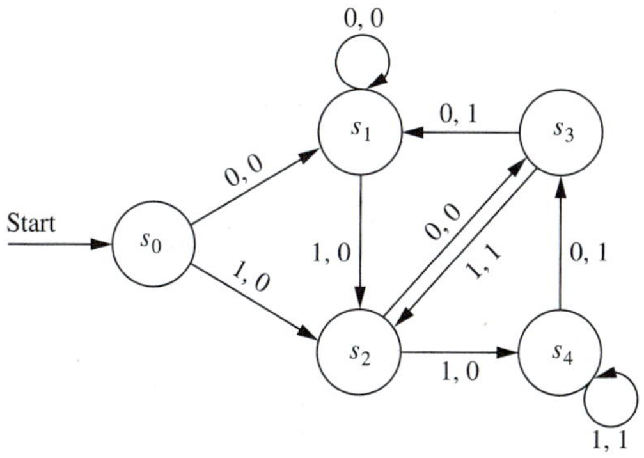

11.

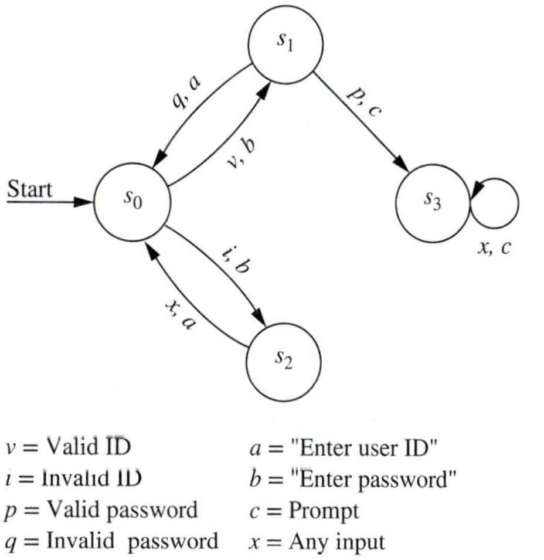

13.

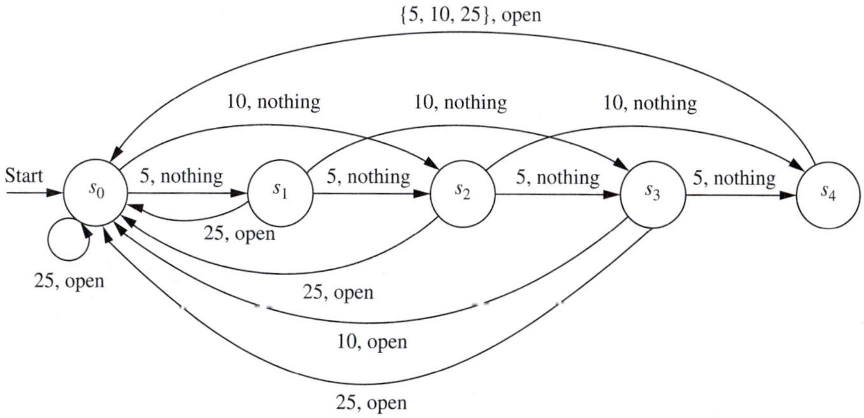

15. Let s_0 be the start state and let s_1 be the state representing a successful call. From s_0, inputs of 2, 3, 4, 5, 6, 7, or 8 send the machine back to s_0 with output of an error message for the user. From s_0 an input of 0 sends the machine to state s_1, with the output being that the 0 is sent to the network. From s_0 an input of 9 sends the machine to state s_2 with no output; from there an input of 1 sends the machine to state s_3 with no output; from there an input of 1 sends the machine to state s_1 with the output being that the 911 is sent to the network. All other inputs while in states s_2 or s_3 send the machine back to s_0 with output of an error message for the user. From s_0 an input of 1 sends the machine to state s_4 with no output; from s_4 an input of 2 sends the machine to state s_5 with no output; and this path continues in a similar manner to the 911 path, looking next for 1, then 2, then any seven digits, at which point the machine goes to state s_1 with the output being that the ten-digit input is sent to the network. Any "incorrect" input while in states s_5 or s_6 (that is, anything except a 1 while in s_5 or a 2 while in s_6) sends the machine back to s_0 with output of an error message for the user. Similarly, from s_4 an input of 8 followed by appropriate successors drives us eventually to s_1, but inappropriate outputs drive us back to s_0 with an error message. Also, inputs while in state s_4 other than 2 or 8 send the machine back to state s_0 with output of an error message for the user.

17.

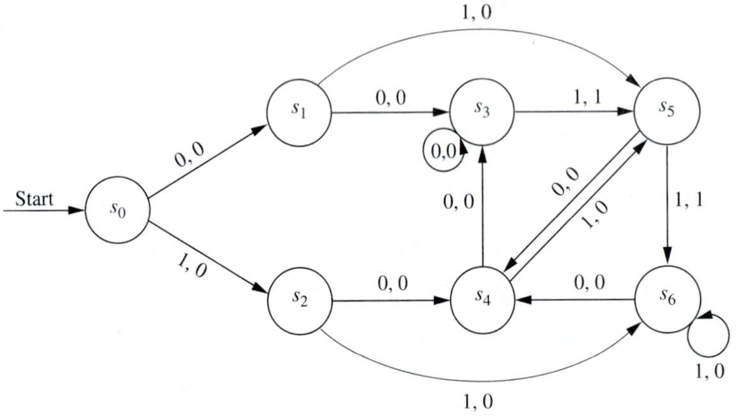

19.

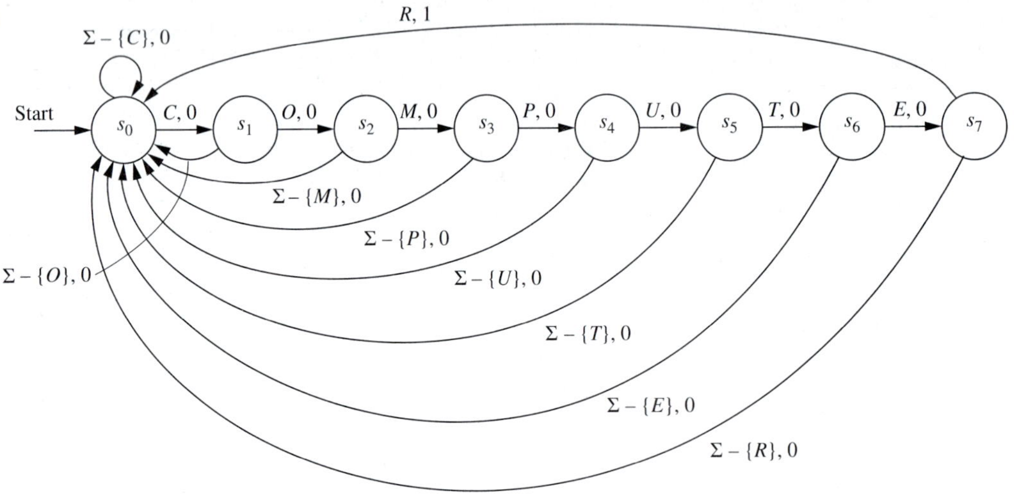

21.

	f		
	Input		
State	**0**	**1**	*g*
s_0	s_1	s_2	1
s_1	s_1	s_0	1
s_2	s_1	s_2	0

23. a) 11111 **b)** 1000000 **c)** 100011001100

25.

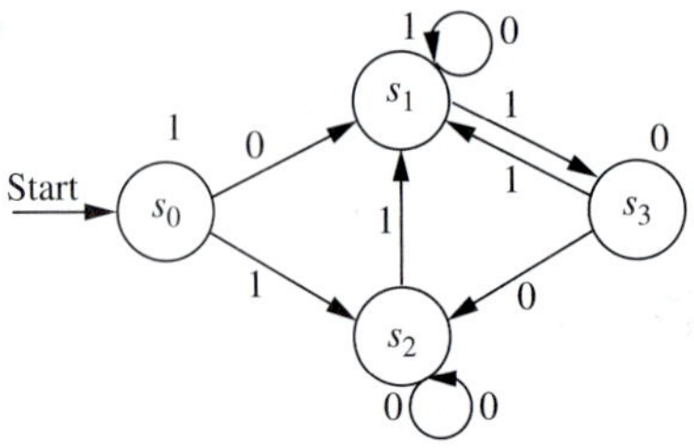

Section 13.3

1. a) {000,001,1100,1101} **b)** {000,0011,010,0111} **c)** {00, 011, 110, 1111} **d)** {000000, 000001, 000100, 000101, 010000, 010001, 010100, 010101} **3.** $A = \{1, 101\}$, $B = \{0, 11, 000\}$; $A = \{10, 111, 1010, 1000, 10111, 101000\}$, $B = \{\lambda\}$; $A = \{\lambda, 10\}$, $B = \{10, 111, 1000\}$ or $A = \{\lambda\}$, $B = \{10, 111, 1010, 1000, 10111, 101000\}$ **5. a)** The set of strings consisting of zero or more consecutive bit pairs 10 **b)** The set of strings consisting of all 1s such that the number of 1s is divisible by 3, including the null string **c)** The set of strings in which every 1 is immediately preceded by a 0 **d)** The set of strings that begin and end with a 1 and have at least two 1s between every pair of 0s **7.** A string is in A^* if and only if it is a concatenation of an arbitrary number of strings in A. Because each string in A is also in B, it follows that a string in A^* is also a concatenation of strings in B. Hence, $A^* \subseteq B^*$. **9. a)** Yes **b)** Yes **c)** No **d)** No **e)** Yes **f)** Yes **11. a)** Yes **b)** No **c)** Yes **d)** No **13. a)** Yes **b)** Yes **c)** No **d)** No **e)** No **f)** No **15.** We use structural induction on the input string y. The basis step considers $y = \lambda$, and for the inductive step we write $y = wa$, where $w \in I^*$ and $a \in I$. For the basis step, we have $xy = x$, so we must show that $f(s, x) = f(f(s, x), \lambda)$. But part (i) of the definition of the extended transition function says that this is true. We then assume the inductive hypothesis that the equation holds for w and prove that $f(s, xwa) = f(f(s, x), wa)$. By part (ii) of the definition, the left hand side of this equation equals $f(f(s, xw), a)$. By the inductive hypothesis, $f(s, xw) = f(f(s, x), w)$, so $f(f(s, xw), a) = f(f(f(s, x), w), a)$. The right-hand side of our desired equality is, by part (ii) of the definition, also equal to $f(f(f(s, x), w), a)$, as desired. **17.** $\{0, 10, 11\}\{0, 1\}^*$ **19.** $\{0^m1^n \mid m \geq 0 \text{ and } n \geq 1\}$ **21.** $\{\lambda\} \cup \{0\}\{1\}^*\{0\} \cup \{10, 11\}\{0, 1\}^* \cup \{0\}\{1\}^*\{01\}\{0, 1\}^* \cup \{0\}\{1\}^*\{00\}\{0\}^*\{1\}\{0, 1\}^*$ **23.** Let s_2 be the only final state, and put transitions from s_2 to itself on either input. Put a transition from the start state s_0 to s_1 on input 0, and a transition from s_1 to s_2 on input 1. Create state s_3, and have the other transitions from s_0 and s_1 (as well as both transitions from s_3) lead to s_3. **25.** Start state s_0, only final state s_3; transitions from s_0 to s_0 on 0, from s_0 to s_1 on 1, from s_1 to s_2 on 0, from s_1 to s_1 on 1, from s_2 to s_0 on 0, from s_2 to s_3 on 1, from s_3 to s_3 on 0, from s_3 to s_3 on 1 **27.** Have five states, with only s_3 final. For $i = 0, 1, 2, 3$, transition from s_i to itself on input 1 and to s_{i+1} on input 0. Both transitions from s_4 are to itself. **29.** Have four states, with only s_3 final. For $i = 0, 1, 2$, transition from s_i to s_{i+1} on input 1 but back to s_0 on input 0. Both transitions from s_3 are to itself.

31.

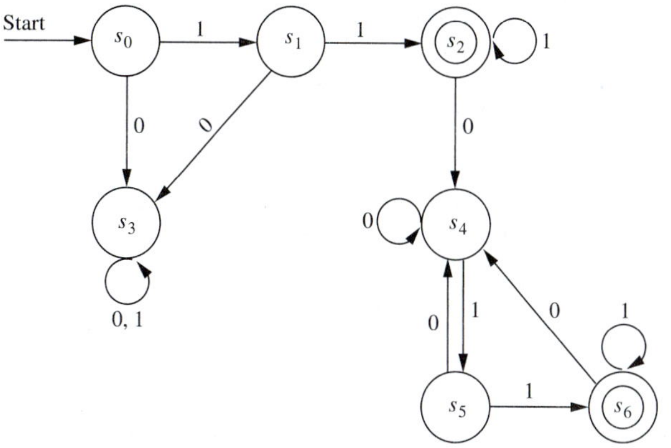

33. Start state s_0, only final state s_1; transitions from s_0 to s_0 on 1, from s_0 to s_1 on 0, from s_1 to s_1 on 1; from s_1 to s_0 on 0

35.

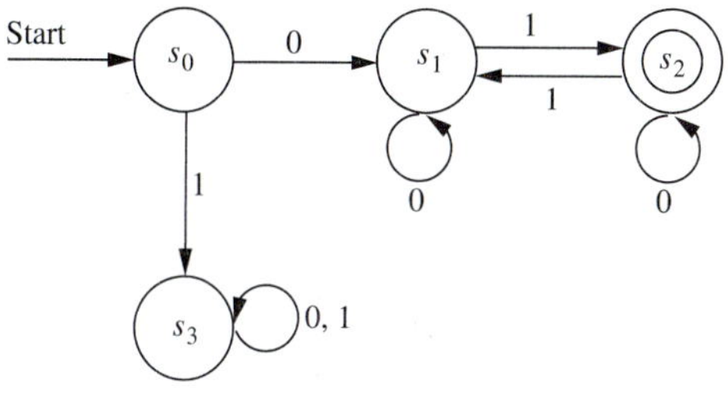

37. Suppose that such a machine exists, with start state s_0 and other state s_1. Because the empty string is not in the language but some strings are accepted, we must have s_1 as the only final state, with at least one transition from s_0 to s_1. Because the string 0 is not in the language, the transition from s_0 on input 0 must be to itself, so the transition from s_0 on input 1 must be to s_1. But this contradicts the fact that 1 is not in the language. **39.** Change each final state to a nonfinal state and vice versa. **41.** Same machine as in Exercise 25, but with s_0, s_1, and s_2 as the final states **43.** {0,01,11} **45.** $\{\lambda, 0\} \cup \{0^m1^n \mid m \geq 1, n \geq 1\}$ **47.** $\{10^n \mid n \geq 0\} \cup \{10^n10^m \mid n, m \geq 0\}$ **49.** The union of the set of all strings that start with a 0 and the set of all strings that have no 0s

51.

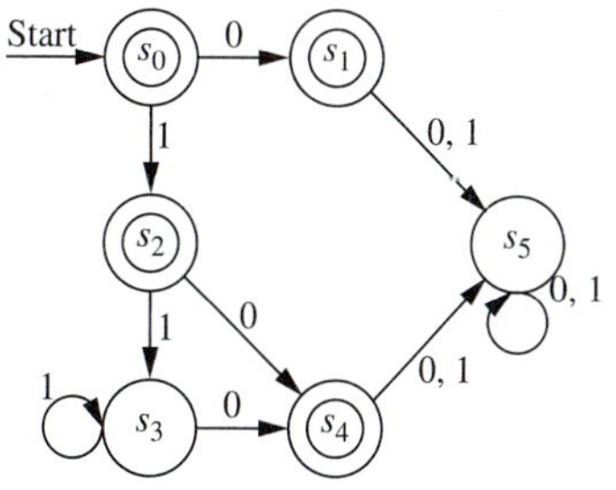

53. Add a nonfinal state s_3 with transitions to s_3 from s_0 on input 0, from s_1 on input 1, and from s_3 on input 0 or 1.

55. a)

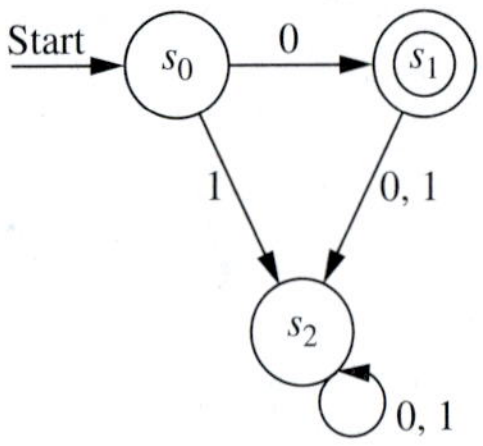

b)

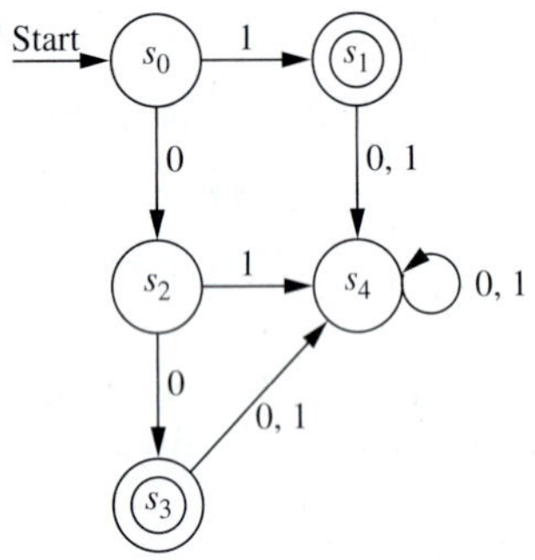

c)

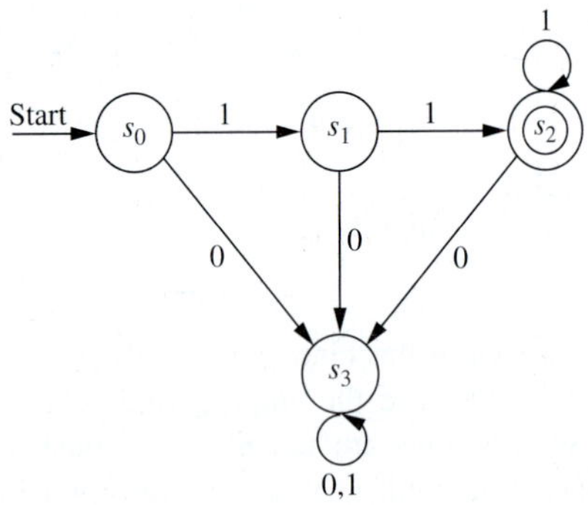

57. Suppose that M is a finite-state automaton that accepts the set of bit strings containing an equal number of 0s and 1s. Suppose M has n states. Consider the string $0^{n+1}1^{n+1}$. By the pigeonhole principle, as M processes this string, it must encounter the same state more than once as it reads the first $n+1$ 0s; so let s be a state it hits at least twice. Then k 0s in the input takes M from state s back to itself for some positive integer k. But then M ends up exactly at the same place after reading $0^{n+1+k}1^{n+1}$ as it will after reading $0^{n+1}1^{n+1}$. Therefore, because M accepts $0^{n+1}1^{n+1}$ it also accepts $0^{n+k+1}1^{n+1}$, which is a contradiction. **59.** We know from Exercise 58d that the equivalence classes of R_k are a refinement of the equivalence classes of R_{k-1} for each positive integer k. The equivalence classes are finite sets, and finite sets cannot be refined indefinitely (the most refined they can be is for each equivalence class to contain just one state). Therefore, this sequence of refinements must remain unchanged from some point onward. It remains to show that as soon as we have $R_n = R_{n+1}$, then $R_n = R_m$ for all $m > n$, from which it follows that $R_n = R_*$, and so the equivalence classes for these two relations will be the same. By induction, it suffices to show that if $R_n = R_{n+1}$, then $R_{n+1} = R_{n+2}$. Suppose that $R_{n+1} \neq R_{n+2}$. This means that there are states s and t that are $(n+1)$-equivalent but not $(n+2)$-equivalent. Thus, there is a string x of length $n+2$ such that, say, $f(s, x)$ is final but $f(t, x)$ is nonfinal. Write $x = aw$, where $a \in I$. Then $f(s, a)$ and $f(t, a)$ are not $(n+1)$-equivalent, because w drives the first to a final state and the second to a nonfinal state. But $f(s, a)$ and $f(t, a)$ are n-equivalent, because s and t are $(n+1)$-equivalent. This contradicts the fact that $R_n = R_{n+1}$. **61. a)** By the way the machine $\overline{M}$ was constructed, a string will drive M from the start state to a final state if and only if that string drives $\overline{M}$ from the start state to a final state. **b)** For a proof of this theorem, see a source such as *Introduction to Automata Theory, Languages, and Computation* (2nd Edition) by John E. Hopcroft, Rajeev Motwani, and Jeffrey D. Ullman (Addison-Wesley, 2000).

Section 13.4

1. a) Any number of 1s followed by a 0 **b)** Any number of 1s followed by one or more 0s **c)** 111 or 001 **d)** A string of any number of 1s or 00s or some of each in a row **e)** λ or a string that ends with a 1 and has one or more 0s before each 1 **f)** A string of length at least 3 that ends with 00 **3. a)** No **b)** No **c)** Yes **d)** Yes **e)** Yes **f)** No **g)** No **h)** Yes **5. a)** $\mathbf{0 \cup 11 \cup 010}$ **b)** $\mathbf{000000^*}$ **c)** $\mathbf{(0 \cup 1)((0 \cup 1)(0 \cup 1))^*}$ **d)** $\mathbf{0^*10^*}$ **e)** $\mathbf{(1 \cup 01 \cup 001)^*}$ **7. a)** $\mathbf{00^*1}$ **b)** $\mathbf{(0 \cup 1)(0 \cup 1)(0 \cup 1)^*0000^*}$ **c)** $\mathbf{0^*1^* \cup 1^*0^*}$ **d)** $\mathbf{11(111)^*(00)^*}$ **9. a)** Have the start state s_0, nonfinal, with no transitions. **b)** Have the start state s_0, final, with no transitions. **c)** Have the nonfinal start state s_0 and a final state s_1 and the transition from s_0 to s_1 on input a. **11.** Use an inductive proof. If the regular expression for A is $\emptyset$, λ, or $\mathbf{x}$, the result is trivial. Otherwise, suppose the regular expression for A is $\mathbf{BC}$. Then $A = BC$ where B is the set generated by $\mathbf{B}$ and C is the set generated by $\mathbf{C}$. By the inductive hypothesis there are regular expressions $\mathbf{B'}$ and $\mathbf{C'}$ that generate B^R and C^R, respectively. Because $A^R = (BC)^R = C^R B^R$, $\mathbf{C'B'}$ is a regular expression for A^R. If the regular expression for A is $\mathbf{B \cup C}$, then the regular expression for A^R is $\mathbf{B' \cup C'}$ because $(B \cup C)^R = (B^R) \cup (C^R)$. Finally, if the regular expression for A is $\mathbf{B^*}$, then it is easy to see that $\mathbf{(B')^*}$ is a regular expression for A^R.

13. a)

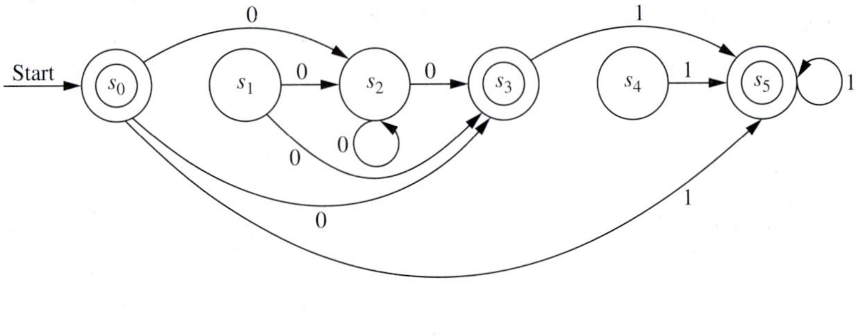

b)

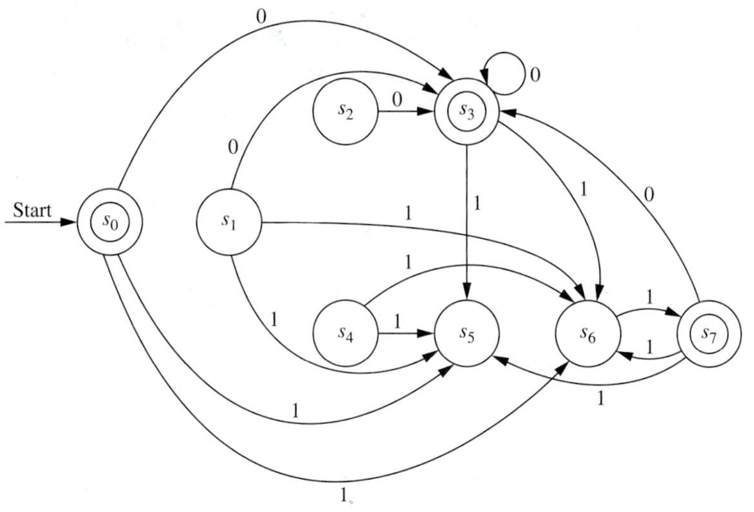

c)

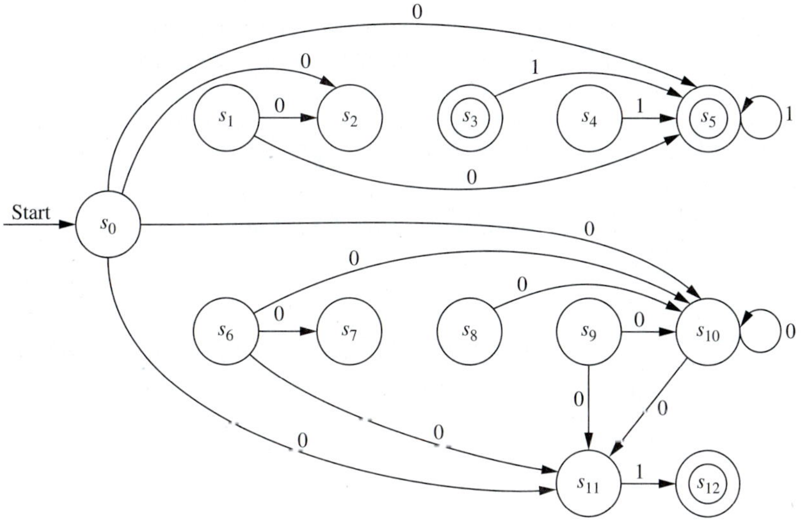

15. $S \rightarrow 0A$, $S \rightarrow 1B$, $S \rightarrow 0$, $A \rightarrow 0B$, $A \rightarrow 1B$, $B \rightarrow 0B$, $B \rightarrow 1B$ **17.** $S \rightarrow 0C$, $S \rightarrow 1A$, $S \rightarrow 1$, $A \rightarrow 1A$, $A \rightarrow 0C$, $A \rightarrow 1$, $B \rightarrow 0B$, $B \rightarrow 1B$, $B \rightarrow 0$, $B \rightarrow 1$, $C \rightarrow 0C$, $C \rightarrow 1B$, $C \rightarrow 1$. **19.** This follows because input that leads to a final state in the automaton corresponds uniquely to a derivation in the grammar. **21.** The "only if" part is clear because I is finite. For the "if" part let the states be $s_{i_0}, s_{i_1}, s_{i_2}, \ldots, s_{i_n}$, where $n = l(x)$. Because $n \geq |S|$, some state is repeated by the pigeonhole principle. Let y be the part of x that causes the loop, so that $x = uyv$ and y sends s_j to s_j, for some j. Then $uy^kv \in L(M)$ for all k. Hence, $L(M)$ is infinite. **23.** Suppose that $L = \{0^{2n}1^n, n = 0, 1, 2 \ldots\}$ were regular. Let S be the set of states of a finite-state machine recognizing this set. Let $z = 0^{2n}1^n$ where $3n \geq |S|$. Then by the pumping lemma, $z = 0^{2n}1^n = uvw$, $l(v) \geq 1$, and $uv^iw \in \{0^{2n}1^n \mid n \geq 0\}$. Obviously v cannot contain both 0 and 1, because v^2 would then contain 10. So v is all 0s or all 1s, and hence, uv^2w contains too many 0s or too many 1s, so it is not in L. This contradiction shows that L is not regular. **25.** Suppose that the set of palindromes over $\{0, 1\}$ were regular. Let S be the set of states of a finite-state machine recognizing this set. Let $z = 0^n10^n$, where $n > |S|$. Apply the pumping lemma to get $uv^iw \in L$ for all nonnegative integers i where $l(v) \geq 1$, and $l(uv) \leq |S|$, and $z = 0^n10^n = uvw$. Then v must be a string of 0s (because $n > |S|$), so uv^2w is not a palindrome. Hence, the set of palindromes is not regular. **27.** Let $z = 1$; then $111 \notin L$ but $101 \in L$, so 11 and 10 are distinguishable. For the second question, the only way for $1z$ to be in L is for z to end with 01, and that is also the only way for $11z$ to be in L, so 1 and 11 are indistinguishable. **29.** This follows immediately from Exercise 28, because the n distinguishable strings must drive the machine from the start state to n different states. **31.** Any two distinct strings of the same length are distinguishable with respect to the language P of all palindromes, because if x and y are distinct strings of length n, then $xx^R \in P$ but $yx^R \notin P$. Because there are 2^n different strings of length n, Exercise 29 tells us that any deterministic finite-state automaton for recognizing palindromes must have at least 2^n states. Because n is arbitrary, this is impossible.

Section 13.5

1. a) The nonblank portion of the tape contains the string 1111 when the machine halts. **b)** The nonblank portion of the tape contains the string 011 when the machine halts. **c)** The nonblank portion of the tape contains the string 00001 when the machine halts. **d)** The nonblank portion of the tape contains the string 00 when the machine halts. **3. a)** The machine halts (and accepts) at the blank following the input, having changed the tape from 11 to 01. **b)** The machine changes every other occurrence of a 1, if any, starting with the first, to a 0, and otherwise leaves the string unchanged; it halts (and accepts) when it comes to the end of the string. **5. a)** Halts with 01 on the tape, and does not accept **b)** The first 1 (if any) is changed to a 0 and the others are left alone. The input is not accepted.
7. $(s_0, 0, s_1, 1, R)$, $(s_0, 1, s_0, 1, R)$ **9.** $(s_0, 0, s_0, 0, R)$, $(s_0, 1, s_1, 1, R)$, $(s_1, s_1, 0, R)$, $(s_1, 1, s_1, 0, R)$ **11.** $(s_0, 0, s_1, 0, R)$, $(s_0, 1, s_0, 0, R)$, $(s_1, 0, s_1, 0, R)$, $(s_1, 1, s_0, 0, R)$, (s_1, B, s_2, B, R) **13.** $(s_0, 0, s_0, 0, R)$, $(s_0, 1, s_1, 1, R)$, $(s_1, 0, s_1, 0, R)$, $(s_1, 1, s_0, 1, R)$, (s_0, B, s_2, B, R) **15.** If the input string is blank or starts with a 1 the machine halts in nonfinal state s_0. Otherwise, the initial 0 is changed to an M and the machine skips past all the intervening 0s and 1s until it either comes to the end of the input string or else comes to an M. At this point, it backs up one square and enters state s_2. Because the acceptable strings must have a 1 at the right for each 0 at the left, there must be a 1 here if the string is acceptable. Therefore, the only transition out of s_2 occurs when this square contains a 1. If it does, the machine replaces it with an M and makes its way back to the left; if it does not, the machine halts in nonfinal state s_2. On its way back, it stays in s_3 as long as it sees 1s, then stays in s_4 as long as it sees 0s. Eventually either it encouters a 1 while in s_4 at which point it halts without accepting or else it reaches the rightmost M that had been written over a 0 at the start of the string. If it is in s_3 when this happens, then there are no more 0s in the string, so it had better be the case that there are no more 1s either; this is accomplished by the transitions (s_3, M, s_5, M, R) and (s_5, M, s_6, M, R), and s_6 is a final state. Otherwise the machine halts in nonfinal state s_5. If it is in s_4 when this M is encountered, things start all over again, except now the string will have had its leftmost remaining 0 and its rightmost remaining 1 replaced by Ms. So the machine moves, staying in state s_4, to the leftmost remaining 0 and goes back into state s_0 to repeat the process.
17. (s_0, B, s_9, B, L), $(s_0, 0, s_1, 0, L)$, (s_1, B, s_2, E, R), (s_2, M, s_2, M, R), $(s_2, 0, s_3, M, R)$, $(s_3, 0, s_3, 0, R)$, (s_3, M, s_3, M, R), $(s_3, 1, s_4, M, R)$, $(s_4, 1, s_4, 1, R)$, (s_4, M, s_4, M, R), $(s_4, 2, s_5, M, R)$, $(s_5, 2, s_5, 2, R)$, (s_5, B, s_6, B, L), (s_6, M, s_8, M, L), $(s_6, 2, s_7, 2, L)$, $(s_7, 0, s_7, 0, L)$, $(s_7, 1, s_7, 1, L)$, $(s_7, 2, s_7, 2, L)$, (s_7, M, s_7, M, L), (s_7, E, s_2, E, R), (s_8, M, s_8, M, L), (s_8, E, s_9, E, L) where M and E are markers, with E marking the left end of the input
19. $(s_0, 1, s_1, B, R)$, $(s_1, 1, s_2, B, R)$, $(s_2, 1, s_3, B, R)$, $(s_3, 1, s_4, 1, R)$, $(s_1, B, s_4, 1, R)$, $(s_2, B, s_4, 1, R)$, $(s_3, B, s_4, 1, R)$
21. $(s_0, 1, s_1, B, R)$, $(s_1, 1, s_2, B, R)$, (s_1, B, s_6, B, R), $(s_2, 1, s_3, B, R)$, (s_2, B, s_6, B, R), $(s_3, 1, s_4, B, R)$, (s_3, B, s_6, B, R), $(s_4, 1, s_5, B, R)$, (s_4, B, s_6, B, R), $(s_6, B, s_{10}, 1, R)$, $(s_5, 1, s_5, B, R)$, $(s_5, B, s_7, 1, R)$, $(s_7, B, s_8, 1, R)$, $(s_8, B, s_9, 1, R)$, $(s_9, B, s_{10}, 1, R)$
23. $(s_0, 1, s_0, 1, R)$, (s_0, B, s_1, B, L), $(s_1, 1, s_2, 0, L)$, $(s_2, 0, s_2, 0, L)$, $(s_2, 1, s_3, 0, R)$, (s_2, B, s_6, B, R), $(s_3, 0, s_3, 0, R)$, $(s_3, 1, s_3, 1, R)$, $(s_3, B, s_4, 1, R)$, $(s_4, B, s_5, 1, L)$, $(s_5, 1, s_5, 1, L)$, $(s_5, 0, s_2, 0, L)$, $(s_6, 0, s_6, 1, R)$, $(s_6, 1, s_7, 1, R)$, (s_6, B, s_7, B, R)
25. $(s_0, 0, s_0, 0, R)$, $(s_0, *, s_5, B, R)$, $(s_3, *, s_3, *, L)$, $(s_3, 0, s_3, 0, L)$, $(s_3, 1, s_3, 1, L)$, (s_3, B, s_0, B, R), $(s_5, 1, s_5, B, R)$, $(s_5, 0, s_5, B, R)$, (s_5, B, s_6, B, L), (s_6, B, s_6, B, L), $(s_6, 0, s_7, 1, L)$, $(s_7, 0, s_7, 1, L)$, $(s_0, 1, s_1, 0, R)$, $(s_1, 1, s_1, 1, R)$, $(s_1, *, s_2, *, R)$, $(s_2, 0, s_2, 0, R)$, $(s_2, 1, s_3, 0, L)$, (s_2, B, s_4, B, L), $(s_4, 0, s_4, 1, L)$, $(s_4, *, s_8, B, L)$, $(s_8, 0, s_8, B, L)$, $(s_8, 1, s_8, B, L)$
27. Suppose that s_m is the only halt state for the Turing machine in Exercise 22, where m is the largest state number, and suppose that we have designed that machine so that when the machine halts the tape head is reading the leftmost 1 of the answer. Renumber each state in the machine for Exercise 18 by adding m to each subscript, and take the union of the two sets of five-tuples. **29. a)** No **b)** Yes **c)** Yes **d)** Yes
31. $(s_0, B, s_1, 1, L)$, $(s_0, 1, s_1, 1, R)$, $(s_1, B, s_0, 1, R)$

Supplementary Exercises

1. a) $S \rightarrow 00S111$, $S \rightarrow \lambda$ **b)** $S \rightarrow AABS$, $AB \rightarrow BA$, $BA \rightarrow AB$, $A \rightarrow 0$, $B \rightarrow 1$, $S \rightarrow \lambda$ **c)** $S \rightarrow ET$, $T \rightarrow 0TA$, $T \rightarrow 1TB$, $T \rightarrow \lambda$, $0A \rightarrow A0$, $1A \rightarrow A1$, $0B \rightarrow B0$, $1B \rightarrow B1$, $EA \rightarrow E0$, $EB \rightarrow E1$, $E \rightarrow \lambda$

3.

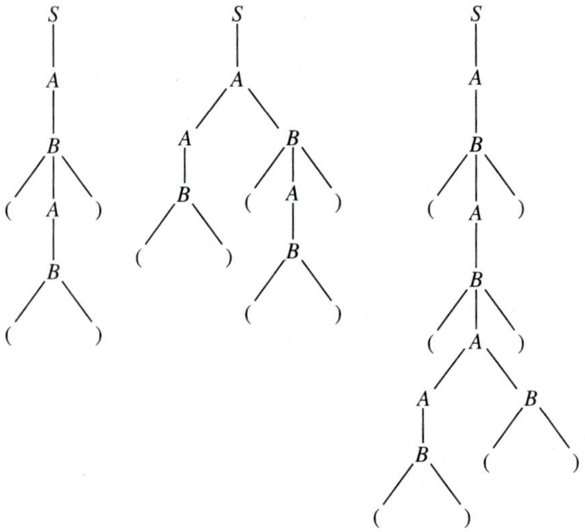

5.

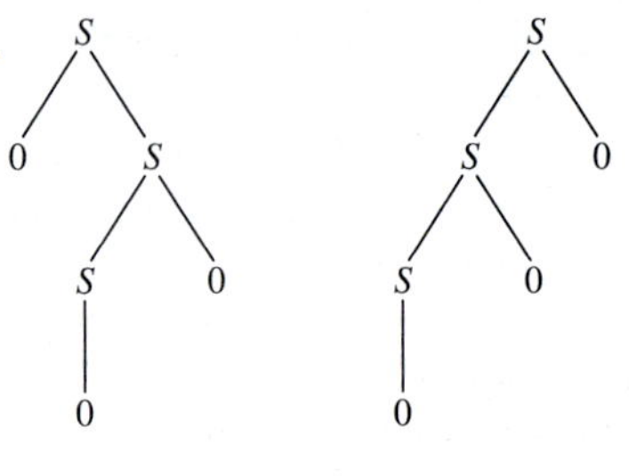

7. No, take $A = \{1, 10\}$ and $B = \{0, 00\}$. **9.** No, take $A = \{00, 000, 00000\}$ and $B = \{00, 000\}$. **11. a)** 1 **b)** 1 **c)** 2 **d)** 3 **e)** 2 **f)** 4

13.

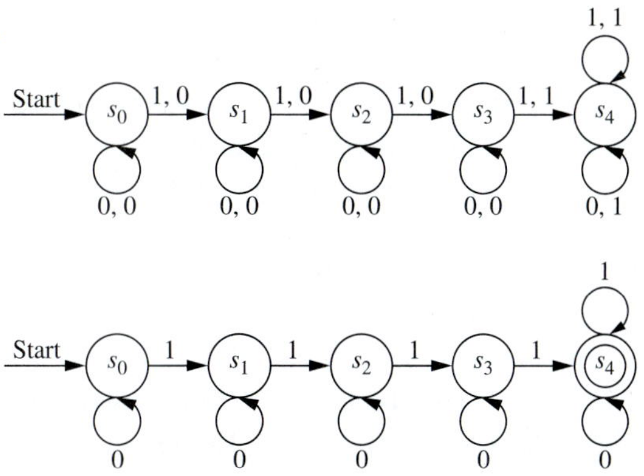

15.

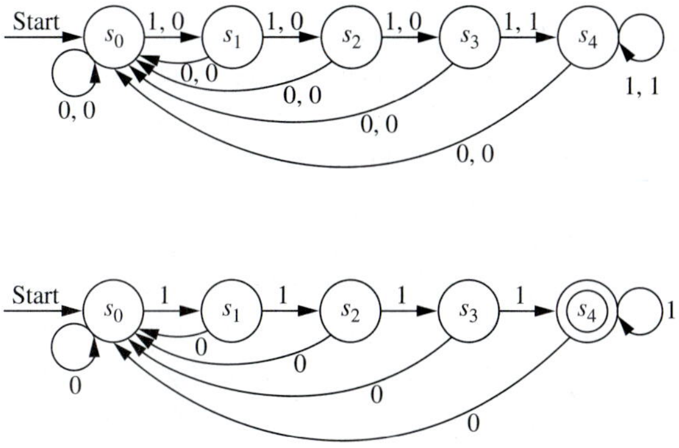

17. a) $n^{nk+1}m^{nk}$ **b)** $n^{nk+1}m^{n}$

19.

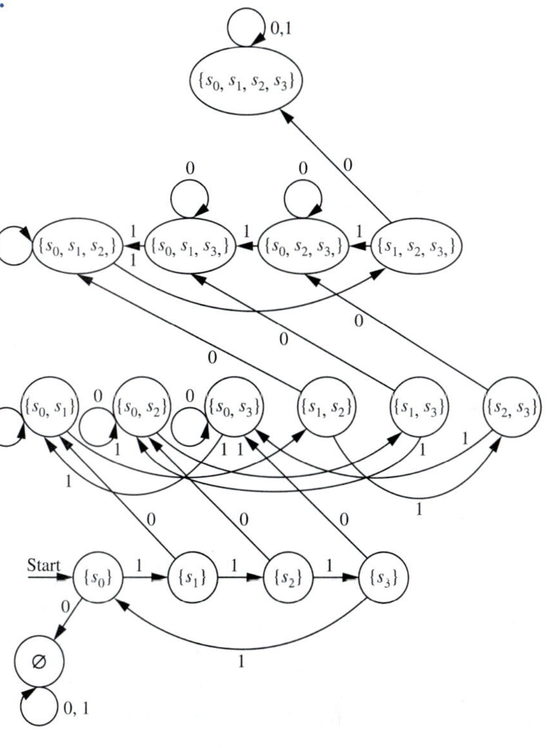

21. a)

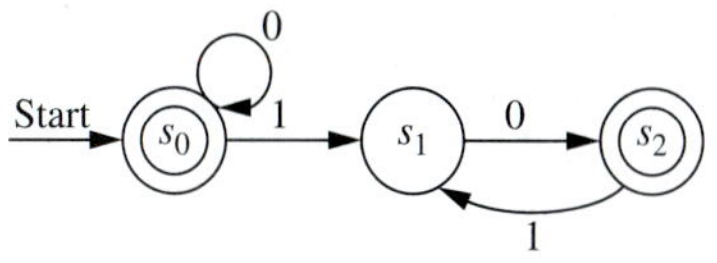

b)

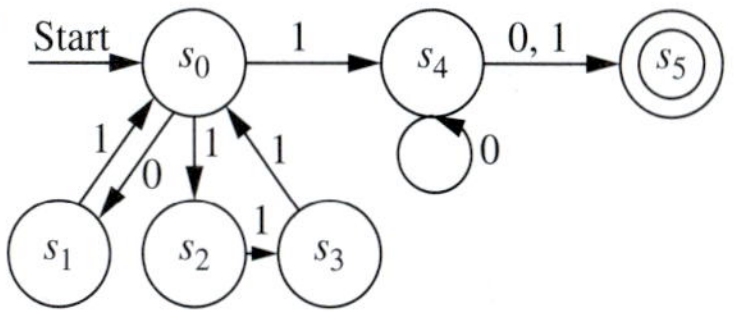

c)

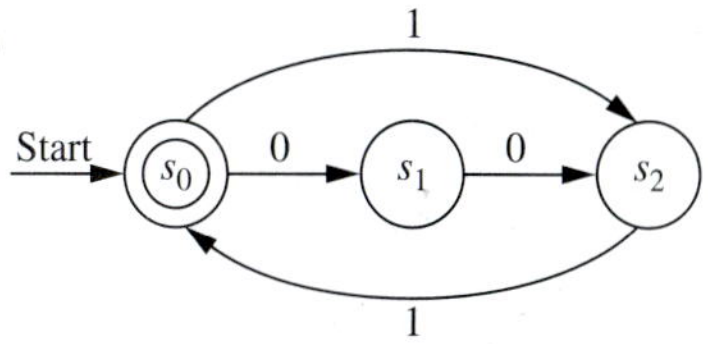

23. Construct the deterministic finite automaton for A with states S and final states F. For $\overline{A}$ use the same automaton but with final states $S - F$.

25. a)

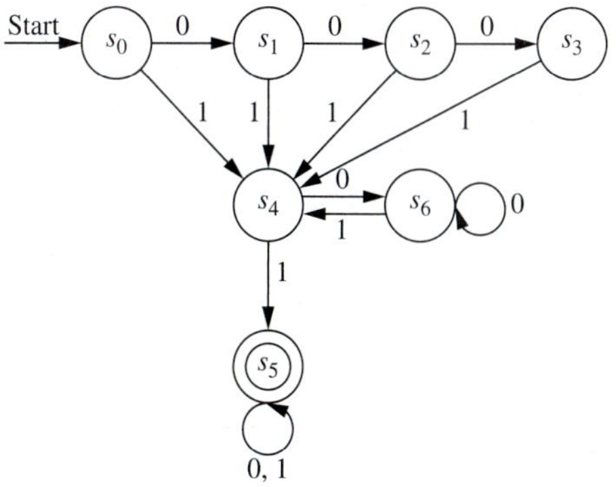

b)

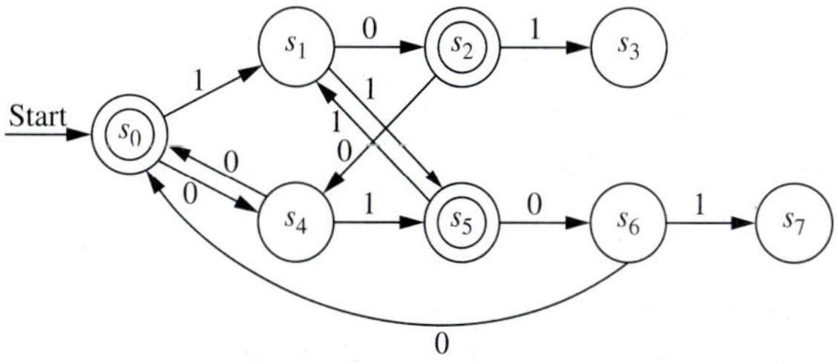

c)

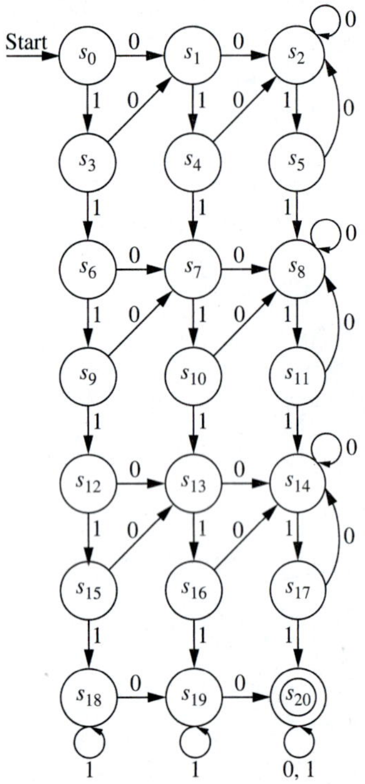

27. Suppose that $L = \{1^p \mid p \text{ is prime}\}$ is regular, and let S be the set of states in a finite-state automaton recognizing L. Let $z = 1^p$ where p is a prime with $p > |S|$ (such a prime exists because there are infinitely many primes). By the pumping lemma it must be possible to write $z = uvw$ with $l(uv) \leq |S|$, $l(v) \geq 1$, and for all nonnegative integers i, $uv^iw \in L$. Because z is a string of all 1s, $u = 1^a$, $v = 1^b$, and $w = 1^c$, where $a + b + c = p$, $a + b \leq n$, and $b \geq 1$. This means that $uv^iw = 1^a1^{bi}1^c = 1^{(a+b+c)+b(i-1)} = 1^{p+b(i-1)}$. Now take $i = p + 1$. Then $uv^iw = 1^{p(1+b)}$. Because $p(1 + b)$ is not prime, $uv^iw \notin L$, which is a contradiction. **29.** $(s_0, *, s_5, B, L)$, $(s_0, 0, s_0, 0, R)$, $(s_0, 1, s_1, 0, R)$, $(s_1, *, s_2, *, R)$, $(s_1, 1, s_1, 1, R)$, $(s_2, 0, s_2, 0, R)$, $(s_2, 1, s_3, 0, L)$, (s_2, B, s_4, B, L), $(s_3, *, s_3, *, L)$, $(s_3, 0, s_3, 0, L)$, $(s_3, 1, s_3, 1, L)$, (s_3, B, s_0, B, R), $(s_4, *, s_8, B, L)$, $(s_4, 0, s_4, B, L)$, $(s_5, 0, s_5, B, L)$, (s_5, B, s_6, B, R), $(s_6, 0, s_7, 1, R)$, (s_6, B, s_6, B, R), $(s_7, 0, s_7, 1, R)$, $(s_7, 1, s_7, 1, R)$, $(s_8, 0, s_8, 1, L)$, $(s_8, 1, s_8, 1, L)$

APPENDIXES

Appendix 1

1. Suppose that $1'$ is also a multiplicative identity for the real numbers. Then, by definition, we have both $1 \cdot 1' = 1$ and $1 \cdot 1' = 1'$, so $1' = 1$. **3.** For the first part, it suffices to show that $[(-x) \cdot y] + (x \cdot y) = 0$, because Theorem 2 guarantees that additive inverses are unique. Thus, $[(-x) \cdot y] + (x \cdot y) = (-x + x) \cdot y$ (by the distributive law) $= 0 \cdot y$ (by the inverse law) $= y \cdot 0$ (by the commutative law) $= 0$ (by Theorem 5). The second part is almost identical. **5.** It suffices to show that $[(-x) \cdot (-y)] + [-(x \cdot y)] = 0$, because Theorem 2 guarantees that additive inverses are unique: $[(-x) \cdot (-y)] + [-(x \cdot y)] = [(-x) \cdot (-y)] + [(-x) \cdot y]$ (by Exercise 3) $= (-x) \cdot [(-y) + y]$ (by the distributive law) $= (-x) \cdot 0$ (by the inverse law) $= 0$ (by Theorem 5). **7.** By definition, $-(-x)$ is the additive inverse of $-x$. But $-x$ is the additive inverse of x, so x is the additive inverse of $-x$. Therefore, $-(-x) = x$ by Theorem 2. **9.** It suffices to show that $(-x - y) + (x + y) = 0$, because Theorem 2 guarantees that additive inverses are unique: $(-x - y) + (x + y) = [(-x) + (-y)] + (x + y)$ (by definition of subtraction) $= [(-y) + (-x)] + (x + y)$ (by the commutative law) $= (-y) + [(-x) + (x + y)]$ (by the associative law) $= (-y) + [(-x + x) + y]$ (by the associative law) $= (-y) + (0 + y)$ (by the inverse law) $= (-y) + y$ (by the identity law) $= 0$ (by the inverse law). **11.** By definition of division and uniqueness of multiplicative inverses (Theorem 4) it suffices to prove that $[(w/x) + (y/z)] \cdot (x \cdot z) = w \cdot z + x \cdot y$. But this follows after several steps, using the distributive law, the associative and commutative laws for multiplication, and the definition that division is the same as multiplication by the inverse. **13.** We must show that if $x > 0$ and $y > 0$, then $x \cdot y > 0$. By the multiplicative compatibility law, the commutative law, and Theorem 5, we have $x \cdot y > 0 \cdot y = 0$. **15.** First note that if $z < 0$, then $-z > 0$ (add $-z$ to both sides of the hypothesis). Now given $x > y$ and $-z > 0$, we have $x \cdot (-z) > y \cdot (-z)$ by the multiplicative compatibility law. But by Exercise 3 this is equivalent to $-(x \cdot z) > -(y \cdot z)$. Then add $x \cdot z$ and $y \cdot z$ to both sides and apply the various laws in the obvious ways to yield $x \cdot z < y \cdot z$. **17.** The additive compatibility law tells us that $w + y < x + y$ and (together with the commutative law) that $x + y < x + z$. By the transitivity law, this gives the desired conclusion. **19.** By Theorem 8, applied to $1/x$ in place of x, there is an integer n (necessarily positive, because $1/x$ is positive) such that $n > 1/x$. By the multiplicative compatibility law, this means that $n \cdot x > 1$. **21.** We must show that if $(w, x) \sim (w', x')$ and $(y, z) \sim (y', z')$, then $(w + y, x + z) \sim (w' + y', x' + z')$ and that $(w \cdot y + x \cdot z, x \cdot y + w \cdot z) \sim (w' \cdot y' + x' \cdot z', x' \cdot y' + w' \cdot z')$. Thus, we are given that $w + x' = x + w'$ and that $y + z' = z + y'$, and we want to show that $w + y + x' + z' = x + z + w' + y'$ and that $w \cdot y + x \cdot z + x' \cdot y' + w' \cdot z' = x \cdot y + w \cdot z + w' \cdot y' + x' \cdot z'$. For the first of the desired conclusions, add the two given equations. For the second, rewrite the given equations as $w - x = w' - x'$ and $y - z = y' - z'$, multiply them, and do the algebra.

Appendix 2

1. a) 2^3 **b)** 2^6 **c)** 2^4 **3. a)** $2y$ **b)** $2y/3$ **c)** $y/2$

5.

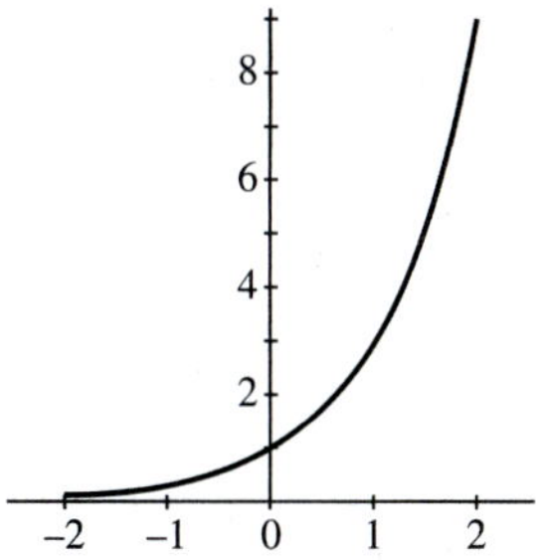

(a)

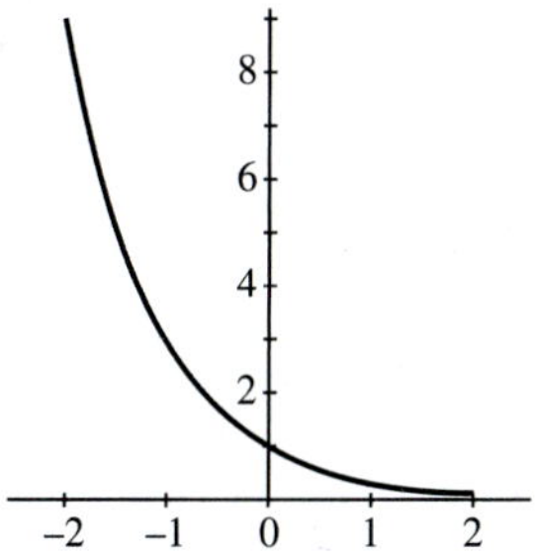

(b)

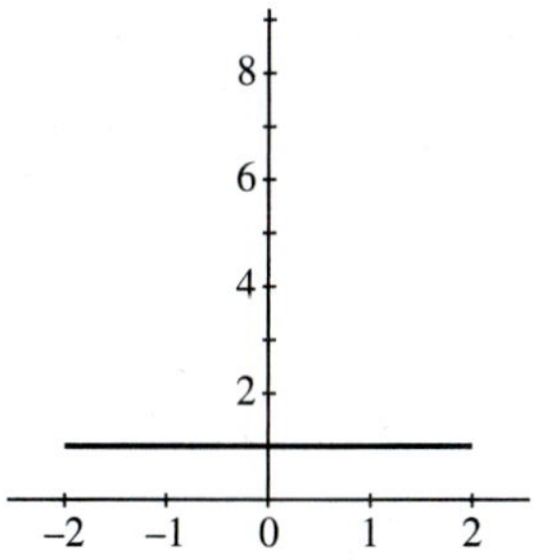

(c)

Appendix 3

1. After the first block is executed, a has been assigned the original value of b and b has been assigned the original value of c, whereas after the second block is executed, b is assigned the original value of c and a the original value of c as well. **3.** The following construction does the same thing.

i := *initial value*
while $i \leq$ *final value*
  statement
  $i := i + 1$

Index of Biographies

Ada, Augusta (Countess of Lovelace), 32
Adleman, Leonard, 321
Al-Khowarizmi, Abu Ja'Far Mohammed Ibn Musa, 206
Archimedes, A–4
Aristotle, 2

Bachmann, Paul Gustav Heinrich, 223
Backus, John, 900
Bayes, Thomas, 504
Bellman, Richard, 541
Bernoulli, James, 490
Bézout, Étienne, 289
Bhaskaracharya, 147
Bienaymé, Irenée-Jules, 521
Boole, George, 5

Cantor, Georg, 126
Cardano, Girolamo, 476
Carmichael, Robert Daniel, 303
Carroll, Lewis (Charles Dodgson), 54
Catalan, Eugéne Charles, 540
Cayley, Arthur, 792
Chebyshev, Pafnuty Lvovich, 522
Chomsky, Avram Noam, 899
Church, Alonzo, 943
Cocks, Clifford, 320
Colmerauer, Alain, 55
Cook, Stephen, 244

De La Vallée-Poussin, Charles-Jean-Gustave-Nicholas, 281
De Morgan, Augustus, 31
Descartes, René, 132
Dijkstra, Edsger Wybe, 752
Dirac, Gabriel Andrew, 743
Dirichlet, G. Lejeune, 427
Dodgson, Charles (Lewis Carroll), 54

Eratosthenes, 277
Erdős, Paul, 674
Euclid, 287
Euler, Leonhard, 736

Fermat, Pierre de, 302
Fibonacci, 373

Gauss, Karl Friedrich, 259
Gentry, Craig B., 325
Goldbach, Christian, 283

Hadamard, Jacques, 281
Hall, Philip, 699
Hamilton, William Rowan, 741
Hardy, Godfrey Harold, 103
Hasse, Helmut, 661
Hilbert, David, 185
Hoare, C. Anthony R., 399
Hopper, Grace Brewster Murray, 919
Huffman, David A., 805

Karnaugh, Maurice, 874
Kempe, Alfred Bray, 769
Kleene, Stephen Cole, 913
Knuth, Donald E., 224
Kruskal, Joseph Bernard, 844
Kuratowski, Kazimierz, 765

Lamé, Gabriel, 374
Landau, Edmund, 223
Laplace, Pierre-Simon, 477
Łukasiewicz, Jan, 825

McCarthy, John, 406
McCluskey, Edward J., 881
Mersenne, Marin, 279
Naur, Peter, 900

Ore, Øystein, 743

Pascal, Blaise, 447
Peirce, Charles Sanders, 42
Petersen, Julius Peter Christian, 748
Prim, Robert Clay, 844

Quine, Willard Van Orman, 883

Ramanujan, Srinivasa, 105
Ramsey, Frank Plumpton, 431
Rivest, Ronald, 321
Russell, Bertrand, 128

Shamir, Adi, 321
Shamos, Michael, 565
Shannon, Claude Elwood, 856
Sheffer, Henry Maurice, 37
Sloane, Neil, 175
Smullyan, Raymond, 21
Stirling, James, 164

Tao, Terence, 282
Tukey, John Wilder, 12
Turing, Alan Mathison, 932

Vandermonde, Alexandre-Théophile, 448
Venn, John, 129

Warshall, Stephen, 642
Wiles, Andrew, 113
Wilf, Herbert, 811

Zhang, Yitang, 284

Index

$3x + 1$ conjecture, 113
25 problems,
  Hilbert's list, 185, 942

Absorption laws
  for Boolean algebra, 859, 860
  for lattices, 675
  for propositions, 29
  for sets, 140
Academic collaboration graph, 684
Academic paper, 685
Ackermann's function, 384
Acquaintanceship graph, 683, 721
Ada, Augusta (Countess of Lovelace), 31, 32
Adapting, existing proofs, 107
Adders, 870–871
  full, 870, 871
  half, 870, 871
Addition
  algorithm, 268–269
  associative laws for, A-1
  Boolean, 857
  closure laws for, A-1
  commutative laws for, A-1
  finite-state machine for, 908–909
  inverse laws for, A-2
  of matrices, 193
  of multisets, 147
  rules of inference, 76
  of subsets, 836–837
Additive compatibility law, A-2
Additive inverses, for integers modulo m, 261
Additively homomorphic, 325
Adjacency lists, 709
Adjacency matrix, 710–711
  counting paths between vertices by, 729–730
Adjacent cells, 874
Adjacent vertices, 691
Adjective, 894, 895
Adleman, Leonard, 319
Advanced counting techniques, 533–597
  recurrence relations, 539–542
Advanced Encryption Standard (AES), 319
Adverb, 894, 895
Affine cipher, 316–318
Affine transformation, 316
Affirming the conclusion, fallacies of, 79
Airline routes, 686
Aleph null, 184
Algebraic axioms, A-1–A-2
Algebraic numbers, 188
Algebraic number theory, 113
ALGOL, 900, 901
Algorithmic paradigm, 241–242
  brute-force, 241–242
Algorithms, 205–245
  addition, 268–269
  approximation, 252, 756–757
  average-case computational complexity of, 514–516
  binary search, 209–210
  binary search tree, 799–802
  Boolean product of zero-one matrices, 240
  Boyer-Moore majority vote, 220
  breadth-first search, 834
  brute-force, 241–242
  bubble sort, 211–212
  c-approximation, 252
  cashier's, 214–215
  complexity of, 235–245
  computer time used by, 245
  correctness of, 208
  deferred acceptance, 220
  definiteness of, 208
  definition of, 205, 264
  depth-first search, 832
  Dijkstra's, 750–755
  for **div** function, 271
  divide-and-conquer, 559
  division, 257–258
  effectiveness of, 208
  Euclidean, 286–289, 373–374
  Euler circuits, 738
  finding maximum element, 207
  finiteness of, 208
  Fleury's, 738, 747
  Floyd's, 758
  Gale-Shapley, 220
  generality of, 204
  for graph isomorphism, 715
  greedy, 214–216, 350–351, 806
  Hamilton circuit, 742–743
  Huffman coding, 805–806
  input, 207
  insertion sort, 212
  for integer operations, 268–271
  Kruskal's, 844–846
  linear, 208–209
  merge sort, 392–395
  for minimum spanning trees, 842–846
  for **mod** function, 271
  modular exponentiation, 272
  Monte Carlo, 494–496, 500
  multiplication, 269–270
  naive string matcher, 213–214
  optimal, 247
  origin of term, 206
  output, 208
  parallel, 701
  preorder traversal, 821
  Prim's, 842–844
  probabilistic, 475, 495
  properties of, 207–208
  pseudocode, 207
  and recurrence relations, 539–542
  recursive. *See* Recursive algorithms
  reverse-delete, 847
  searching, 208–210
  sequential, 208–209
  serial, 701
  shortest-path, 750–755
  social media, 206
  Sollin's, 847
  sorting, 210–212
  string matching, 213–214
  topological sorting, 666
  traversal, 816–823
  Warshall's, 640–643
Algorithms and Complexity (Wilf), 811
Alice in Wonderland, 54
Al-Khowarizmi, Abu Ja'far Mohammed ibn Musa, 206
Alphabet, 895
Alpha-beta pruning, 811
Alphanumeric character, 417
American Association of Artificial Intelligence, 55
Anagrams, 218
Analytic Engine, 32
Analytische Zahlentheorie, 223
Ancestors, of vertex, 789
AND gate, 22, 867
AND operator, 13
Antecedent, of conditional statement, 6
Antichain, 675
Antisymmetric relation, 609–610
Appel, Kenneth, 769
Applications
  of backtracking, 835–837
  of Bayes' theorem, 502–503
  of congruences, 307–312
  of Euler circuits, 739–740
  of Euler paths, 739–740
  of graphs colorings, 772–773
  of pigeonhole principle, 430–432
  of propositional logic, 17–23

of satisfiability, 33–37
of trees, 799–812
Approximation algorithm, 252, 756–757
Arborescence, 852
Archimedean property, A-5
Archimedes, A-4
Archimedes' screw, A-4
Argument, 73
building, rules of inference and, 77–78
Cantor diagonalization, 187
premises of, 54, 73, 74
in propositional logic, 74–75
valid, 73–75
Argument form, 75
valid, 73–75
Aristotle, 2, 3
Arithmetic
fundamental theorem of, 276
mean, 106, 357
modulo m, 261–262
Arithmeticorum Libri Duo, 337
Arithmetic progression, 170
primes and, 282
Array
linear, 702
two-dimensional, 702
Ars Conjectandi (Bernoulli), 490
Art gallery problem, 776
Art gallery theorem, 776
Article, 894, 895
Articulation points, 724
The Art of Computer Programming, 210, 224
Assertion
final, 397
initial, 397
Assignments, frequency, 773
Assignment statements, A-11–A-12
Associated homogeneous recurrence relations, 553
Association of Logic Programming, 55
Association rules
from data mining, 623–625
lift of, 627
strong, 625
Associative laws
for addition, A-1
for Boolean algebra, 859
for disjunction, 30
for lattices, 675
for multiplication, A-1
for propositions, 29
for sets, 140
Associativity, for integers modulo m, 261
Asymmetric relations, 615
Asymptotic functions, 235
Asynchronous transfer mode (ATM), 162
Atomic propositions, 3
Atomic sets, 140
Autokey ciphers, 331
Automation
deterministic, 918
finite-state, 913–922
linear bounded automata, 933
nondeterministic, 919, 927
pushdown, 933
quotient, 924
Average-case complexity, 514–516
time, 237–238
Axioms, 86
field/algebraic, A-1–A-2
mathematical induction, A-5
order, A-1–A-2
for real numbers, A-1–A-2
for set of positive integers, A-5
used to prove basic facts, A-3–A-5
Zermelo-Fraenkel, 190

Bachmann, Paul Gustav Heinrich, 223
Back edges, 832, 841
Back substitution, 299
Backtracking, 831–833
applications of, 835–837
Backus, John, 900
Backus-Naur form (BNF), 900–902
Backward differences, 545
Backward reasoning, 106–107
Backward substitution, 173
Bacon, Kevin, 722
Bacon number, 722
Balanced m-ary trees, 795–796
Balanced strings of parentheses, 407
Balanced ternary expansions, 273
Bandwidth, of graph, 782
Base b
expansions, 264, 267
Miller's test for, 306
pseudoprimes to, 303
strong pseudoprime to, 306
Base conversion, 265–267
Bases, of DNA molecules, 415
Basis step, 337, 339, 374
proof by strong induction, 358
Basket, 623
Bayesian spam filters, 504–507
Bayes' theorem, 500–507
generalized, 503
Bayes, Thomas, 504
Begging the question, 95
Bell, E. T., 656
Bellman, Richard, 539, 541
Bell numbers, 656
Bernoulli family tree, 787
Bernoulli, James, 490
Bernoulli's inequality, 355
Bernoulli trial, 490
Bézout, Étienne, 289
Bézout's coefficients, 289
Bézout's identity, 289
Bézout's theorem, 289–290
Bhaskaracharya, 147
Biconditional statement, 10
logical equivalences involving, 29
Biconnected graphs, 725
Bidirectional bubble sort, 250
Bienaymé, Irenée-Jules, 521
Bienaymé's formula, 521
Big-O estimates, 225–228
for combinations of functions, 229–230
for exponential functions, 228
for functions, 225–228
for logarithms, 228
for polynomials, 225–228
Big-Omega (big-Ω) notation, 230–232
Big-O notation, 221–225
definition of, 222–225, 248
history of, 221
Big-Theta (big-Θ) notation, 230–232
Bi-implications, 10
Bijection, 156
Bijective function, 156
Bijective proofs, 439
BilliardsDigest magazine, 565
Binary coded decimal expansion, 880
Binary coded decimal form, 274
Binary expansions, 264–265
conversion between octal/hexadecimal expansions and, 267
Binary insertion sort, 219
Binary relation, 605
Binary representation, 267
Binary search algorithm, 209–210, 560
recursive, 388–389
worst-case complexity, 236–237
Binary search trees, 799–802
Binary trees, 377–378, 790, 791
extended, 377
full, 377–378
with prefix codes, 804–806
Binomial coefficients, 437, 439, 443–449
extended, 572
identities involving, 448–449
Binomial distribution, 491
Binomial expression, 443
Binomial theorem, 443–446
extended, 573
Binomial trees, 850
Bipartite graphs, 695–697
coloring of, 696
complete, 697
and matchings, 698–700
Birthday problem, 493–494
Bit operations, 12

Bits, 12
parity check, 310
Bit string, 12
encoding, 804
Bitwise operations, 13
Bitwise operators, 13
Block ciphers, 317–318
Block of statements, A-14–A-15
Blocks, Sudoku, 35
Board, 109
Bonferroni's inequality, 498
Boolean algebra, 5, 37, 855–885
abstract definition of, 861
Boolean functions. *See* Boolean functions
identities of, 858–860
Boolean expressions, 856–858
dual of, 860, 863
Boolean functions, 855–861
Boolean expressions and, 856–858
complement of, 857
of degree *n*, 856, 858
dual of, 860
functionally complete set of operators for, 865
implicant of, 876
minimization of, 872–885
representing, 857–858, 863–865
self-dual, 888
threshold, 889
values of, 857
Boolean power, of square zero-one matrix, 198
Boolean product, 855–857
Boolean product of zero-one matrix, 197, 240
Boolean searches, 19–20
Boolean sum, 855–856
Boolean variable, 12, 856, 864
Boole, George, 3, 5, 855
Boole's inequality, 497, 498
Boruvka, Otakar, 845
Bottom-up parsing, 899
Bound
greatest lower, 663
least upper, 663
lower. *See* Lower bound
for number of leaves in *m*-ary tree, 796–797
upper. *See* Upper bound
Bounded lattices, 675
Bound variable, 48
Boxes, distributing objects into, 457–460
distinguishable, 457–460
indistinguishable, 457–460
Boyer-Moore majority vote algorithm, 220
Breadth-first search, 833–835
depth-first search *vs.*, 834–835
Breaking codes, 316
Bridge, in graphs, 724
Brute-force algorithm, 241–242
B-tree of degree *k*, 850
Bubble sort, 211–212
bidirectional, 250
worst-case complexity of, 238–239
Buoyancy, principle of, A-4
Busy beaver function, 942
Busy beaver problem, 945
Butane, 792
Bytes, 265

Cactus, 850
Caesar, Julius, 315
Calculus, predicate, 43
Call graphs, 684
connected components of, 724
Cantelli, F. P., 492
The Canterbury Puzzles (Dudeney), 537
Cantor diagonalization argument, 187
Cantor digits, 467
Cantor expansion, 274
Cantor, Georg, 126, 127
c-approximation algorithm, 252
Cardano, Girolamo, 475, 476
Cardinality, 130, 183–190
of multisets, 147
Cardinal numbers, 130
Carmichael number, 303–304
Carmichael, Robert Daniel, 303
Carroll, Lewis (C. L. Dodgson), 54
Carry, 268
Cartesian products, 131–134
Cases, proof by, 98–100
errors with, 101–102
Cashier's algorithm, 214–215
Catalan conjecture, 540
Catalan, Eugéne Charles, 540
Catalan numbers, 539
Caterpillar, 851
Cayley, Arthur, 787, 792
Ceiling function, 161–163
properties of, 163
Celebrity problem, 357
Cells, adjacent, 874
Center vertex, 799
Chain, 657
Character cipher, 317
Characteristic equation, 547
Characteristic function, 168
Characteristic roots, 547
Chebyshev, Pafnuty Lvovich, 522
Chebyshev's inequality, 523–524
Check digits, 310–312
ISBN, 311–312
ISSN, 314
parity check bits, 310
Universal Product Code, 310–311
Checkerboards
definition of, 109
standard, 109
tilings of, 109–112, 352–353
Chessboard
knight's tour on, 748
n-queens problem on, 836
Child, of vertex, 789
Chinese postman problem, 739–740
Chinese remainder theorem, 297–299
Chomp, 103–104, 367
Chomsky, Noam, 893, 899
Chromatically *k*-critical graphs, 775
Chromatic number, 767–771, 779
Church, Alonzo, 943
Church-Turing thesis, 940–941
Ciphers
affine, 316
autokey, 331
block, 317–318
character, 317
monoalphabetic, 317
shift, 315–317
transposition, 317
Vigenére, 326
Circuits
combinatorial, 22–23, 867
in directed graph, 635
Euler, 734–740
in graphs, 720
Hamilton, 734, 740–745
Circular reasoning, 95
Circular relation, 673
Circular *r*-permutation, 442
Citation graphs, 685
Class
congruence, 259
equivalence, 647
NP, 243, 943–944
P, 243, 943–944
Class A addresses, 418
Class B addresses, 418
Class C addresses, 418
Classical cryptography, 315–318
Clauses, 79
Clique, 780
Closed formula, 172, 457, 581
Closed interval, 125
Closed walk, 720

Closest-pair problem, 565–567
Closure laws
for addition, A-1
for multiplication, A-1
Closures
for integers modulo *m*, 261
Kleene, 913
Closures, of relations, 634–643
reflexive, 634–635
symmetric, 635
transitive, 634, 636–640
Cloud, 324
Clustering coefficient, 780
COBOL, 900
Cocks, Clifford, 319, 320
Codeword enumeration, 538
Codomain
of functions, 151
of partial function, 165
Coefficients, 289
Bézout's, 289
binomial, 437, 439, 443–449
clustering, 780
constant. *See* Constant coefficients
multinomial, 463
Collaboration graphs, 683–684, 722
Collatz problem, 113
Collision
in hashing, 308
probability in hashing functions, 494
Colmerauer Alain, 55
Coloring
of bipartite graphs, 696
edge, 775
of graphs, 768–773, 835–836
k-tuple, 776
of maps of regions, 768
of simple graphs, 769
Combinational circuits, 867–868
depth of, 872
examples of, 869–870
minimization of, 872–885
Combinations, 437–440
of events, 479–480, 486–487
generalized, 451–460
generating, 466–467
with repetition, 452–456
Combinatorial Algorithms (Wilf), 811
Combinatorial circuit, 22–23
Combinatorial game theory, 811
Combinatorial proof, 439
Combinatorics, 411, 475
Comments, in pseudocode, A-12
Common difference, 170
Common ratio, 170
Communication networks, 684
Commutative group, 262
Commutative laws
for addition, A-1
for Boolean algebra, 859
for lattices, 675
for multiplication, A-1
for propositions, 29
for sets, 140
Commutative ring, 262
Commutativity, for integers modulo *m*, 261
Comparable elements, 657
Compatibility laws, A-2
Compatible talks, 540
Compatible total ordering, 666
Complement
of Boolean function, 857
of elements, 676, 855
of fuzzy sets, 150
of sets, 138, 139
Complementary event, 479
Complementary graph, 708
Complementary relations, 615
Complementation law, 140
Complemented lattices, 676, 861
Complement laws, 140
Complete bipartite graphs, 697
Complete graphs, 694
Complete induction, 359
Complete *m*-ary tree, 798
Complete matching, 699
Complete *m*-partite graph, 780
Completeness property, A-2
Complexity, of algorithms, 235–245
average-case, 237–238, 514–516
computational, 235
constant, 242
Dijkstra's, 755
exponential, 242, 243
factorial, 242, 243
linear, 242
linearithmic, 242
logarithmic, 242
merge sort, 564
polynomial, 242, 243
practical considerations, 244–245
space, 235
time, 235–239
worst-case, 236–237
Complex numbers, sets of, 125
Composite, 275, 945
of relations, 612–613
Composite key, 619
Composition rule, 398
Compositions, of functions, 157–160
Compound interest, 173
Compound propositions, 3, 11, 26, 27
equivalences, in Boolean algebra, 856
equivalent, 9
logically equivalent, 27–30
satisfiable, 33
truth tables of, 11
unsatisfiable, 33
Computable functions, 190, 942
Computational complexity, 235, 941–944
of depth-first search algorithms, 833
Dijkstra's algorithm, 755
Computational Geometry: An Introduction, 565
Computational geometry, strong induction in, 363–365
Computation models, 893–944
finite-state machines, 893, 905–910, 912–922
language, 933
pushdown automaton, 933
Turing machine, 893, 932, 933
Computer arithmetic with large integers, 299–301
Computer file systems, 793
Computer networks, 679–680
local area networks, 701
multicasting over, 830
Computer representation, of sets, 145–146
Concatenation, 375, 912
Conclusion, 54, 74, 75
of conditional statement, 6
fallacy of affirming, 79
Concurrent processing, 685
Conditional constructions, A-12–A-13
Conditional-disjunction equivalence, 28
Conditional expectation, of random variables, 524
Conditional probability, 483, 487–488
Conditional statements, 6–10
contrapositive of, 9
converse of, 9
definition of, 6
direct proof of, 87–88
inverse of, 9
logical equivalences involving, 29
for program correctness, 398–400
truth table for, 6
Conditions, don't care, 880–881
Confidence, 624
Congruence class, 259
Congruence classes modulo *m*, 648
Congruence modulo *m*, 646
Congruences, 258
applications of, 307–312
linear, solving, 294–304

Congruent to b modulo m, 258
Conjectures, 86
  $3x + 1$, 113
  Catalan, 539
  Frame's, 545
  Goldbach's, 283
  twin primes, 284
Conjunction, 4
  over, 28
  rules of inference, 76
  truth table for, 4
Conjunctive normal form, 864, 866
Connected components, of graphs, 723–724
Connected graphs, 720–730
  directed, 727–728
  strongly, 727
  undirected, 722–724
  weakly, 727
Connecting vertices, 691
Connectives, 4
Consequence, of conditional statement, 6
Consistent system specifications, 18–19
Constant coefficients
  linear homogeneous recurrent relations with, 547–553
  linear nonhomogeneous recurrent relations with, 553–556
Constant complexity, 242
Constants, 546
Constructions
  conditional, A-12–A-13
  loop, A-14–A-15
Constructive existence proofs, 102
Context-free grammars, 897
Context-free language, 897
Context-sensitive grammars, 897
Context-sensitive language, 897
Contingency, 26
Continuum hypothesis, 190
Contradiction, 26, 27
  proofs by, 91–94
Contraposition, proof by, 88–91
Contrapositive, of conditional statements, 9
Converse
  of conditional statements, 9
  of directed graphs, 708
Conversion
  base, 265–267
  between binary and hexadecimal expansions, 267
  between binary and octal expansions, 267
Convex polygon, 363
Cook-Levin theorem, 244
Cook, Stephen, 244
Corollary, 86
Correctness
  of algorithm, 208
  of recursive algorithms, 389–390
Correspondence, one-to-one, 156–158, 160, 185
Countable sets, 184–187
Counterexamples, 44, 94, 108
Countess of Lovelace (Augusta Ada), 31, 32
Counting, 411–467
  advanced techniques, 533–600
  bit strings, 414
  by distributing objects into boxes, 457–460
  functions, 413
  generating functions and, 575–579
  Internet addresses, 418–419
  one-to-one functions, 413
  passwords, 417–418
  paths between vertices, 729–730
  permutations, 434–437
  pigeonhole principle and, 426–432
  principles, 412–417
  problems, 411, 412, 417–419
  subsets of finite sets, 348–350, 414
Count of an itemset, 623
A Course of Pure Mathematics, 103
Covariance, of random variables, 526
Covering relation, 661
Covers, 661, 776, 883
C programming language, 648–649
Crawl frontier, 839
Cross edge, 841
Crossing number, 767
Cryptanalysis, 316–317
Cryptographic protocols, 323–324
Cryptography, 314–325
  classical, 315–318
  public key, 314, 318–319
Cryptosystems, 318
  fully homomorphic, 325
  Paillier, 327
  private key, 318–319
  public key, 319
  RSA system, 319–320
Curse of dimensionality, 541
Cut edge, 724
Cut set, 850
Cut vertices, 724
Cycle, 721
  in directed graph, 635, 694
Czekanowski, Jan, 845

Databases
  attribute of, 618
  composite key, 619
  fields, 618
  intension, 618
  primary key, 618
  records, 618
  and relations, 618–619
Data compression, 805
Datalogy, 900
Data mining, association rules from, 623–625
Datatype, 125
Decidable problem, 941–942
Decimal expansions, 264
Decision problem, 941
Decision trees, 802–804
Decreasing functions, 155
Decryption, 315
  RSA system, 321–322
Deductive reasoning, 336
de Fermat, Pierre, 302
Deferred acceptance algorithm, 220
Deficiency, 707
Definiteness, of algorithm, 208
Degree
  of linear homogenous recurrence relations, 546
  n-ary relations, 617
  of region, 763
  of vertex in niche overlap graph, 692
  of vertex in undirected graph, 691
Degree-constrained spanning tree, 851
Degree of membership, 150
De La Valleé-Poussin, Charles-Jean-Gustave-Nicholas, 281
Delay machine, 908, 909
de Méré, Chevalier, 484
De Morgan, Augustus, 27, 30, 31, 771
De Morgan's laws, 27
  for Boolean algebra, 859, 863
  for propositions, 29, 30
  proving by mathematical induction, 348–350
  for quantifiers, 50–51
  for sets, 140
Dense
  graph, 711, 717
  partially ordered set, 670
Density, of undirected graphs, 717
Denying the hypothesis, fallacies of, 80
Depth-first search, 830–833
  breadth-first search *vs.*, 834–835
  in directed graphs, 837–839
Depth of combinational circuits, 872
Derangements, 595–597
Derivable from, 895
Derivation, 895
Derivation trees, 898–900
Descartes, René, 132

Descendants, of vertex, 789
Deterministic finite-state automata, 918
Deviation, 510
  standard, 519
Devil's pair, 720
Diagonal matrix, 199
Diagonal, of simple polygon, 363
Diagonal relation, 634
Diagrams
  Hasse, 660–662
  state, 907
  Venn, 127
Diameter, of graphs, 783
Dictionary ordering, 464
Die, 520–521
  dodecahedral, 528
  octahedral, 528
Differences, A-6
  backward, 545
  common, 170
  equation, 545
  first, 545
  first forward, 601
  forward, 601
  $(k + 1)$st, 545
  of multisets, 147
  of sets, 138, 139
  symmetric, 149
Diffie-Hellman key agreement
    protocol, 323
Digital circuits, 22–23
Digital signatures, 324
Digraphs/directed graphs, 682
  circuit/cycle in, 635, 694
  connectedness in, 727–728
  converse of, 708
  dense, 711
  depth-first search in, 837–839
  edges of, 631, 693–694
  Euler circuit of, 735
  loops in, 631
  paths in, 635–636
  representing relations using, 630–632
  self-converse, 781
  simple, 681
  strong components of, 727–728
  vertices, 631
Dijkstra, Edsger Wybe, 750, 752
Dijkstra's algorithm, 750–755
Diophantus, 112, 302
Dirac, Gabriel A., 742, 743
Dirac's theorem, 742
Directed edges, 681, 693–694
Directed graphs, 682
Directed multigraphs, 681, 682
Directly derivable, 895
Direct proof, 87–88
Dirichlet drawer principle, 427
Dirichlet, G. Lejeune, 427
Discrete logarithm, 304
Discrete logarithm problem, 304
Discrete probability, 475–481
  assigning, 484–486
  of collision in hashing functions, 494
  combinations of events, 479–480,
    486–487
  conditional, 483, 487–488
  finite, 476–479
Disjoint sets, 138
Disjunction, 4–5
  associative law for, 30
  definition of, 4
  distributive law over conjunction, 28
  truth table for, 4
Disjunctive normal form, 864
Disquisitiones Arithmeticae, 259
Distance, 841
Distinguishable boxes
  distinguishable objects in, 457–460
  indistinguishable objects in, 458–459
Distribution
  binomial, 491
  geometric, 516–517
  probability, 484
  of random variable, 492
  uniform, 485
Distributive lattice, 675, 861
Distributive laws, 859–860
  for Boolean algebra, 859
  for propositions, 29
  for real numbers, A-2
  for sets, 140
Distributivity, for integers modulo
    m, 262
div function, 257
Divide-and-conquer algorithms, 559
Divide-and-conquer recurrence relations,
    560–567
Dividend, 257
Divisibility relation, 656
Divisibility, results, proving by
    mathematical induction, 346–348
Division
  algorithm, 257–258
  of integers, 255–258
  rule, 420–421
  trial, 276–277
Divisor, 256, 257
  greatest common, 284–286
DNA (deoxyribonucleic acid), 415
Dodecahedral die, 528
Dodgson, Charles Lutwidge (Lewis
    Carroll), 54
Domain, 44, 46
  of discourse, 44
  finite, 47
  of function, 151
  n-ary relations, 617
  of partial function, 165
  restricted, quantifiers with, 48
Dominating set, of vertices, 781
Domination laws
  for Boolean algebra, 859
  for propositions, 29
  for sets, 140
Dominos, 109, 338
Don't care conditions,
    880–881
Doob J. L., 492
Double counting proofs, 439
Double hashing, 312
Double negation law, 29
Dual
  of Boolean expression,
    860, 863
  graph, 768
  of partially ordered set, 668
Duality principle, 860
Dudeney, Henry, 537
Dynamic programming, 539
  example of, 540–542

Ear, 368
Eccentricity, of vertex, 799
Edge chromatic number, 775
Edge coloring, 775
Edge contraction, 703–704
Edge cut, 726
Edges, 679
  adding from graph, 703
  back, 832, 841
  connectivity, 726–727
  cross, 841
  cut, 724
  of digraphs, 631, 693–694
  directed, 681
  endpoints of, 679
  forward, 841
  incident, 691
  matched, 699
  of multigraphs, 680
  multiple, 680
  multiple directed, 681
  of multiplicity, 681
  noncomplete, 694
  removing from graph, 703
  tree, 832
  undirected, 683
  unmatched, 699
Effectiveness, of algorithm, 208
Egyptian fraction, 405
*The Electronic Journal of
    Combinatorics*, 811
Elementary subdivision, 764

Elements
  comparable, 657
  complement of, 676, 855
  equivalent, 645
  fixed, 526
  identity, for integers modulo *m*, 261
  incomparable, 657
  majority, 220
  maximal, in partially ordered set, 662–664
  maximum, finding in finite sequence, 207
  minimal, in partially ordered set, 662–664
  of partially ordered set, 656
  relationships between. *See* Relations
  of sets, 124
The Elements, 286, A-5
Ellipses (. . .), 124
Empty set, 126
Empty string, 170, 895
Encryption, 315
  homomorphic, 314, 324–325
  key, 318, 319, 324
  RSA, 320–321
Endpoints, of edges, 679
Enumeration, 411–467, 538
  codeword, 538
Equality
  of functions, 152
  of sets, 126
Equal matrices, 193
Equation
  characteristic, 547
  difference, 545
Equivalence
  logical. *See* Logical equivalences
  proof of, 93–94
Equivalence classes, 647–652
  definition of, 647
  and partitions, 649–652
  representative of, 647
Equivalence relations, 644–652
*-equivalent, 924
Equivalent
  Boolean expressions, 857
  compound propositions, 9
  elements, 645
Eratosthenes, 277
Erdős number, 674, 722
Erdős, Paul, 674
Errors
  with exhaustive proof, 101–102
  with proof by cases, 101–102
  in proofs, 94–96
  single, 312
  transposition, 312
Essential prime implicant, 876
Euclid, 275, 278, 286, 287, A-5
Euclidean algorithm, 286–289, 373–374
Euler circuit, 734–740
Euler, Leonhard, 735, 736
Euler paths, 734–740
Euler's criterion, 307
Euler's formula, 761–764
Evaluation functions, 811
Even integer, 87
Events, 476
  combinations of, 479–480, 486–487
  complementary, 479
  independent, 483, 488–490
  probability of, 476
Exams, scheduling, 772–773
Exclusion rule, 374
Exclusive or
  definition of, 5
  truth table for, 6
Exhaustive proofs, 98–99
  errors with, 101–102
Existence proofs
  constructive, 102
  definition of, 102
  nonconstructive, 102–103
Existential generalization, 80–81
Existential instantiation, 80, 81
Existential quantification, 43, 45–48, 50, 102
Existential quantifier, 45–46
Expansions
  balanced ternary, 273
  base-*b*, 264, 267
  binary, 264–265
  binary coded decimal, 880
  Cantor, 274
  decimal, 264
  hexadecimal, 265
  octal, 265
  sum-of-products, 863–865
Expected values, 509–512
  in hatcheck problem, 513–514
  of inversions in permutation, 514
  linearity of, 512–514
Experiment, 476
Exponential complexity, 242, 243
Exponential functions, A-7
  big-*O* estimates for, 228
Exponential generating function, 584
Expressions
  Boolean, 856–858
  infix form of, 823
  prefix form of, 824
  regular, 925
Extended Backus-Naur form (EBNF), 904
Extended binary trees, 377
Extended binomial coefficients, 572
Extended binomial theorem, 573
Extended Euclidean algorithm, 290
Extension, of relations, 618
Exterior, of simple polygon, 363

Factor, 256
Factorial complexity, 242, 243
Factorial function, 164
Facts, 86
Failure, Bernoulli trial, 490
Fallacies, 73, 79–80
  of affirming the conclusion, 79
  of denying the hypothesis, 80
Family trees, 787
Fefferman, Charles, 282
Feller, W., 492
Fermat's last theorem, 112–113, 223
Fermat's little theorem, 301–302
Fibonacci, 373
Fibonacci numbers
  formula for, 549–550
  iterative algorithm for, 391
  rabbits and, 534–535
  recursive algorithm for, 390
Fibonacci sequence, 171–172
Field axioms, A-1–A-2
Fields, databases, 618
Final assertion, 397
Final exams, scheduling, 772–773
Final state, finite-state automaton, 913
Final value, A-14
Finite domain, quantifiers, 47
Finite graph, 679
Finiteness, of algorithms, 208
Finite probability, 476–479
Finite Sections of Some Classical Inequalities (Wilf), 811
Finite sets, 130
  Jaccard similarity of, 150
  number of subsets of, 348–350, 414
  union of, 586
Finite-state automata, 913–918
  designing, 915–917
  deterministic, 917–918
  equivalent, 917–918
  nondeterministic, 918–922
  set not recognized by, 931–932
Finite-state machine, 893, 905–910
  for addition, 908–909
  language recognition by, 914–918
  with no output, 912–922
  with outputs, 906–910
  types of, 910
First difference, 545
First forward difference, 601
Fixed elements, 526
Fleury's algorithm, 738, 747

Floor function, 161–163
properties of, 163
Floyd's algorithm, 758
Forests, 798
minimum spanning, 847
spanning, 841
Formal language, 894
Formal power series, 571
Formulas
Bienaymé's, 521
closed, 172
Euler's, 761–764
for Fibonacci numbers, 549–550
Stirling's, 164
summations, 180, 340–344
well-formed, 375–376, 828
Fortran, 900
Forward differences, 601
Forward edge, 841
Forward reasoning, 106–107
Forward substitution, 173
Four color theorem, 769–772
Frame's conjecture, 545
Free variable, 48
Frend, Sophia, 31
Frequency assignments, 773
Frequent, 623
Frequent itemset mining, 623
Friendship graph, 683
Frisbee, rocket-powered, 856
Full adder, 870, 871
Full binary trees, 377–378
Full *m*-ary trees, 790
counting vertices in, 794–795
Full subtractor, 872
Fully homomorphic cryptosystem, 325
Functional completeness, 865
Functionally complete set of operators, for Boolean functions, 865
Functions, 123, 151–165
Ackermann's, 384
asymptotic, 235
big-*O* estimates for, 225–228
bijective, 156
Boolean. *See* Boolean functions
busy beaver, 942
ceiling, 161–163
characteristic, 168
codomain of, 151
compositions of, 157–160
computable, 190
counting, 413
decreasing, 155
definition of, 151
div, 257, 271
domain of, 151
equality of, 152
Euler ϕ-function, 293
evaluation, 811
exponential, A-7
factorial, 164
floor, 161–163
generating. *See* Generating functions
graphs of, 160–161
greatest integer, 161
growth of, 220–232
hashing, 308–309
increasing, 155
integer-valued, 153
inverse, 157–160
invertible, 158
iterated, 385
linear probing, 309
logarithmic, A-7–A-9
McCarthy 91, 406
mod, 257, 271
not invertible, 158
number-theoretic, 939
one-to-one, 154–157, 413
one-to-one correspondence, 156–158, 160, 185
onto, 154–157, 594–595
output, 906
partial, 165
propositional, 41, 42
range of, 151
real-valued, 153
recursive/inductive definitions, 371–374
as relations, 606–607
strictly decreasing, 155
strictly increasing, 155
surjective, 155
total, 165
transition, 906, 913, 914
uncomputable, 190, 942
Fundamental theorem of arithmetic, 276, 361
Fuzzy sets, 146
complement of, 150
intersection of, 150
union of, 150

Gale-Shapley algorithm, 220
Gambling, 475
Game trees, 806–811
Gates, logic, 22–23, 866–871
AND, 22, 867
NOT, 22
OR, 22, 867
threshold, 889
XOR, 889
Gating networks, 867
examples of, 869–870
Gauss, Karl Friedrich, 259
Gene, 415
General case, 551–553
Generality, of algorithm, 208
Generalization
existential, 80–81
universal, 80, 81
Generalized combinations, 451–460
Generalized induction, 381–382
Generalized permutations, 451–460
Generalized pigeonhole principle, 428–430
generatingfunctionology (Wilf), 811
Generating functions, 569–581
and counting problems, 575–579
exponential, 584
ordinary, 570
probability, 585
proving identities via, 581
for recurrence relations, 579–580
Gene sequencing, 415
Genome, 415
Gentry, Craig B., 325
Geometric distribution, 516–517
Geometric mean, 106, 357
Geometric progression, 170
sums of, 343–344
sums of terms, 178
Geometric series, 178
Giant strongly connected component (GSCC), 728
Givens, Sudoku, 35
Goldbach, Christian, 283
Goldbach's conjecture, 283
Golomb's self-generating sequence, 407–408
Gossip problem, 357
Graceful tree, 851
Grammars, 897, 898. *See also* Language
Backus-Naur form, 900–902
context-free, 897
context-sensitive, 897
language generated by, 896
monotonic, 898
noncontracting, 898
phrase-structure, 894–898
productions of, 895
regular, 897, 929–931
type 1, 897, 898
type 2, 897, 898
type 3, 897, 898
use of, 894
Grand Hotel, Hilbert's, 185–186
Graphic sequence, 707
Graphs, 679–776
academic collaboration, 684
acquaintanceship, 683, 721
airline routes, 686
bandwidth of, 782
biconnected, 725

Graphs (*cont.*)
bipartite, 695–700
call, 684
chromatically k-critical, 775
chromatic number of, 769
circuits in, 720
citation, 685
collaboration, 683–684, 722
coloring of, 768–773, 835–836
complementary, 708
complete, 694
complete m-partite, 780
connected component of, 723–724
connectivity. *See* Connected graphs
cut set of, 850
definition of, 679
degree sequence of, 707
with diagnostic lines, 680
diameter of, 783
directed. *See* Digraphs
dual, 768
edges. *See* Edges
Euler circuit in, 734–740
Euler paths in, 734–740
finite, 679
friendship, 683
Hollywood, 684
homeomorphic, 764–766
infinite, 679
influence, 683
intersection, 689
isomorphic, 709, 712–716
k-connected, 725
k-tuple coloring, 776
mixed, 681, 682
models, 682–688
module dependency, 685
monotone decreasing property of, 783
monotone increasing property of, 783
multigraphs. *See* Multigraphs
with multiple lines, 680
with multiple one-way lines, 681
neighbors in, 691
niche overlap, 686, 687
nonisomorphic, 712
nonplanar, 764
nonseparable, 724
paths in, 720–722
Petersen, 747
planar, 759–766
precedence, 685
protein interaction, 686, 687
pseudographs, 681, 682
radius of, 783
regular, 708
representation of, 709–716
simple. *See* Simple graphs
simple directed, 681, 682
spanning forest of, 841
sparse, 711, 717
strongly connected, 727
subgraph of, 703
terminology, 682, 691–694
underlying undirected, 694
undirected. *See* Undirected graphs
union of, 704–705
vertices. *See* Vertex (vertices)
weakly connected, 727
web, 685, 728
weighted. *See* Weighted graphs
Graphs of functions, 160–161
Gray codes, 743–745
Gray, Frank, 745
"Greater than or equal" relation, 656
Greatest common divisor (gcd), 284–286
as linear combinations, 289–292
recursive algorithm for computing, 387
Greatest element, of partially ordered set, 662
Greatest integer function, 161
Greatest lower bound, 663
Great Internet Mersenne Prime Links Search (GIMPS), 280
Greedy algorithms, 214–216, 350–351, 806
Growth of functions, 220–232
big-O notation, 221–225
Guarding set, 776
Guthrie, Francis, 771

Hadamard, Jacques, 281
Haken, Wolfgang, 769
Half adder, 870, 871
Half subtractor, 872
Hall, Philip, 699
Hall's marriage theorem, 699–700
Halting problem, 216–217, 942
Hamilton circuits, 734, 740–745
Hamilton paths, 740–745
Hamilton, William Rowan, 740, 741, 771
Handle vertex, 850
Handshaking theorem, 692–693
Hardy, Godfrey Harold, 103
Hardy-Weinberg law, 103
Harmonic number, 234–235
inequality of, 345–346
Harmonic series, 346
Hashing/hashing function, 308–309
collision in, 308
double, 312
probability of collision in, 494
Hasse diagrams, 660–662
Hasse, Helmut, 661
Hasse's algorithm, 113
Hatcheck problem, 513–514, 595–597
Heawood, Percy, 769
Height, of rooted trees, 795
Hexadecimal expansions, 265
conversion between binary expansion and, 267
Hexadecimal representation, 267
Hexagon identity, 450
Hilbert, David, 185, 190
Hilbert's Grand Hotel, 185–186
Hilbert's Tenth Problem, 942
Ho, Chung-Wu, 365
Hoare, C. Anthony R., 399
Hoare triple, 397
Hockeystick identity, 450
Hoey, Dan, 565
Hollywood graphs, 684
Homeomorphic graphs, 764–766
Homomorphic encryption, 314, 324–325
Hopper, Grace Brewster Murray, 919
Hops, 702
Horner's method, 246–247
Host number (hostid), 418
Houses-and-utilities problem, 760
Huffman coding, 805–806
Huffman, David A., 805
Hydrocarbons, trees modeling, 792
Hypothesis
of conditional statement, 6
continuum, 190
inductive, 337, 340, 352–354, 700
Hypothetical syllogism, 76, 77

Icosian puzzle, 740
Idempotent laws
for Boolean algebra, 859, 863
for lattices, 675
for propositions, 29
for sets, 140
Identification numbers, single error in, 312
Identifiers, in C Programming Language, 648–649
Identity
Bézout's, 289
hexagon, 450
hockeystick, 450
involving binomial coefficients, 448–449
Pascal's, 446–448
proving, generating functions for, 581
Vandermonde's, 448–449
Identity elements
axiom, A-1
for integers modulo m, 261
Identity laws
additive, A-1
for Boolean algebra, 859
multiplicative, A-1
for propositions, 29
for sets, 140
Identity matrix, 195

if then statement, 398–400, A-12–A-13
Image
of element, 151
inverse, 167
of subsets, 153
Implicant
essential prime, 876
prime, 876
Implication, 6
Incidence matrices, 711–712
Incident edge, 691
Inclusion-exclusion principle. *See* Principle of inclusion-exclusion
Inclusive or, 5
Incomparable elements, 657
Incomplete induction, 359
Increasing functions, 155
Increment, for linear congruential method, 309
In-degree of vertex, 693
Independence number, 783
Independent events, 483, 488–490
Independent random variables, 517–519
Independent vertices, 783
Index of summation, 176–177
Index registers, 773
Indicator random variable, 524
Indirect proofs, 88
Indistinguishable boxes
distinguishable objects in, 458–459
indistinguishable objects in, 460
Induction
complete, 359
generalized, 381–382
incomplete, 359
mathematical, 31, 335–358
strong, 358–366
structural, 371, 378–381
Inductive definitions, 371
of functions, 371–374
of sets, 374–378
of strings, 374
of structures, 374–378
Inductive hypothesis, 337, 340, 352–354, 700
Inductive loading, 357
Inductive reasoning, 336
Inductive step, 337
Inequality
Bernoulli's, 355
Bonferroni's, 497, 498
Boole's, 497, 498
Chebyshev's, 523–524
harmonic numbers, 345–346
Markov's, 525
proving by mathematical induction, 344–346
triangle, 150
Infinite graph, 679
Infinite ladder, 336
strong induction and, 359–360
Infinite series, 180–181
Infinite set, 123, 130
Infinitude of primes, 278–280
Infix form, 823
Infix notation, 823–826
Influence graph, 683
Informal proofs, 85
Information flow, lattice model of, 665
Information networks, 685
Information retrieval (IR), 686–687
Initial assertion, 397
Initial conditions, 171
Initial state, 906, 913
Initial term, 170
Initial value, A-14
Initial vertex, 631
Injection, 154
Inorder traversal, 816–818, 822–823
Input, algorithms, 207
Input alphabet, 906, 913
Insertion sort, 212
average-case complexity of, 515–516
binary, 219
worst-case complexity of, 238–239
Instantiation
existential, 80, 81
universal, 80, 81
Integers, 255, 264
binary coded decimal form, 274
composite, 275
division of, 255–258
even, 87
fast multiplication of, 561
large, computer arithmetic with, 299–301
least common multiple, 286
nondecreasing, 218
odd, 87
one's complement representations, 274
pairwise relatively prime, 285
positive, 125, 255, 292, 335
primes, 275–276
relatively prime, 285
representations of, 264–267
sequences, 174–176
sets of, 125
signed, 901
squarefree, 598
two's complement representations, 274
unary representations, 939
Integer sequences, 174–176
Integers modulo *m*, 261–262
additive inverses, 261
associativity, 261
closure, 261
commutativity, 261
distributivity, 262
identity elements, 261
Integer-valued functions, 153
Intension, of database, 618
Interest, compound, 173
Interior diagonal, of simple polygon, 363
Interior, of simple polygon, 363
Interior vertices, 640
Internal vertices, 789
International Standard Book Number (ISBN),
ISBN-10, 311–312
ISBN-13, 330–331
International Standard Serial Number (ISSN), 314
Internet addresses, counting, 418–419
Internet Protocol (IP) multicasting, 830
Internet, search engines on, 838–839
Intersection
of fuzzy sets, 150
graph, 689
of multisets, 147
of sets, 138, 143–145
Intervals, 125
closed, 125
open, 125
Intractable problems, 243, 944
Invariants
for graph isomorphism, 713
loop, 400–402
Inverse, 295
of conditional statements, 9
functions, 157–160
multiplicative, 65
relations, 615
Inverse laws
for addition, A-2
for multiplication, A-2
Inversions, in permutation, expected number of, 514
Inverter, 22, 867
Invertible function, 158
IP multicasting, 830
IR. *See* Information retrieval (IR)
Irrational numbers, 90
Irreflexive relations, 615
Isobutane, 792
Isolated vertex, 692
Isomorphic graphs, 709, 712–716
paths and, 728–729
Items, 623
Itemset, 623
Iterated function, 385
Iterated logarithm, 385
Iteration, 169, 172, 173, 390–392

Iterative, 390
Iterative algorithm, for Fibonacci numbers, 391

Jaccard distance, 150
Jaccard similarity, of finite sets, 150
James, William, 42
Jealous husbands problem, 734
Jigsaw puzzle, 367
Jobs
  lateness of, 251
  slackness of, 251
Join
  in lattice, 675
  n-ary relations, 621
  of zero-one matrices, 197
Josephus problem, 544
The Journal of Algorithms, 811

Kakutani's problem, 113
Karnaugh maps, 873–880
Karnaugh, Maurice, 874
k-connected graphs, 725
Kempe, Alfred Bray, 769
k-equivalent, 924
Key, 308
  composite, 619
  primary, 618
Key exchange, 323
Key exchange protocol, 327
Keyspace, 318
Kissing problem, 175
Kitab al-jabr w'al muquabala, 206
k-itemset, 623
$(k + 1)$st difference, 545
Kleene closure, 913
Kleene, Stephen, 913, 925
Kleene's theorem, 927–929
K-map, 873–880
Knapsack problem, 252, 601
Knight, 748
Knight's tour, 748
Knuth, Donald E., 210, 221, 224, 231
Kronecker delta, 195
Kruskal, Joseph Bernard, 844, 845
Kruskal's algorithm, 844–846
k-tuple coloring, 776
Kuratowski, Kazimierz, 765
Kuratowski's theorem, 764–766
k-vertex-connected, 725

Labeled objects, 457
Labeled tree, 799
Lady Byron (Annabella Milbanke), 32
Lagarias, Jeffrey, 113
Lamé, Gabriel, 374
Landau, Edmund, 221
Landau symbol, 221
Language, 893–902. *See also* Grammars
  accepted, 914, 920
  context-free, 897
  context-sensitive, 897
  formal, 894
  generated by grammar, 896
  natural, 893
  recognition by finite-state Machines, 914–918
  recognized, 914, 920
  regular, 897
  syntax, 893
Language over, 895
Laplace, Pierre Simon, 475–477
Laplace's definition, of probability, 476
Large integers, computer arithmetic with, 299–301
Lateness, of job, 251
Lattice point, 405
Lattices, 664–665
  associative laws for, 675
  bounded, 675
  commutative laws for, 675
  complemented, 676
  distributive, 675
  idempotent laws for, 675
  join in, 675
  lower bound of, 675
  meet in, 675
  upper bound of, 675
Law of detachment, 75
Law of large numbers, 490
Law of the double complement, in Boolean algebra, 859, 863
Law of total expectation, 524
Laws
  absorption. *See* Absorption laws
  associative. *See* Associative laws
  commutative. *See* Commutative laws
  compatibility, A-2
  De Morgan's. *See* De Morgan's laws
  distributive. *See* Distributive laws
  domination. *See* Domination laws
  of double complement, in Boolean algebra, 859, 863
  double negation, 29
  Hardy-Weinberg, 103
  idempotent. *See* Idempotent laws
  identity. *See* Identity laws
  inverse, A-2
  of large numbers, 490
  negation, 29
  of total expectation, 524
  transitivity, A-2
  trichotomy, A-2
  used to prove basic facts, A-3–A-5
The Laws of Thought (Boole), 3, 847, 855
Leaf, 789
Least common multiple, 286
Least element, of partially ordered set, 662
Least upper bound, 663, A-2
Left child, of vertex, 791
Left subtree, 791
Legendre symbol, 307
Lemma, 86
Length of path, 720
  in directed graphs, 635
  in weighted graph, 749
Length of string, 170, 375
Le probléme de rencontres, 596
Level, of vertex, 795
Level order of vertex, 850
Lexicographic ordering, 381, 464, 658–660
Liber Abaci, 373
Lift, of association rules, 627
Limits, definition of, 65
Linear array, 702
Linear bounded automata, 933
Linear combinations, greatest common divisor as, 289–292
Linear complexity, 242
Linear congruences, solving, 294–304
  Chinese remainder theorem, 297–299
Linear congruential method, 309
Linear homogeneous recurrence relations, 547–553
  degree two case, 548–550
  general case, 551–553
Linearithmic complexity, 242
Linearity of expected values, 512–514
Linearly ordered sets, 657
Linear nonhomogeneous recurrence relations, 553–556
Linear probing function, 309
Linear search algorithm, 208–209
  average-case complexity of, 237–238, 514–516
  recursive, 388
Lists, merging, 393–394
Literal, 864
Load balancing problem, 252
Local Area Networks (LANs), 701
Logarithmic complexity, 242
Logarithmic functions, A-7–A-9
Logarithms
  big-*O* estimates for, 228
  discrete, 304
  iterated, 385
Logical equivalences, 27–30, 860
  constructing, 31–32
  quantifiers, 49
Logical expressions
  translating from English into, 17–18, 51–53, 66–67
  translating mathematical statements into, 64–65

Logically equivalent compound propositions, 27
Logical operators, 3
  precedence of, 11
Logic circuits, 22–23
Logic gates. *See* Gates, logic
Logic programming, 55–56
Logic puzzles, 20–22
Longest path problem, 759
Loop constructions, A-14–A-15
Loop invariants, 400–402
Loops/looping
  in digraphs, 631, 680
  within loops, A-15
  nested, 62
  quantification and, 47, 62
Lord Byron, 32
Lotteries, 477
Lower bound
  greatest, 663
  of lattices, 675
  of partially ordered set, 663
Lower limit, of summation, 177
Lucas, Édouard, 535
Lucas, François Édouard, 175
Lucas numbers, 404, 557
Lucas sequence, 175
Lukasiewicz, Jan, 825

Machine minimization, 918
Machines
  finite-state, 893, 905–910, 912–922
  Mealy, 910
  Moore, 910, 911
  Turing, 893, 932, 933, 935–944
  unit-delay, 908
Majority element, 220
Makespan, 252
Mappings, 151. *See also* Functions
Maps, coloring of, 769
Markov's inequality, 525
m-ary trees, 790
  balanced, 795–796
  complete, 798
  full, 790, 794–795
  number of leaves in, bound for, 796–797
Master theorem, 564
Matched edges, 699
Matchings
  bipartite graphs and, 698–700
  complete, 699
  maximum, 699
The Mathematical Analysis of Logic (Boole), 5
Mathematical induction, 31, 335–358
  axiom, A-5
  bad of, 340
  creative uses of, 351–353
  definition of, 335–336
  good of, 340
  proofs by. *See* Proofs, by mathematical induction
  validity of, 358–359
  as valid proof technique, 338–339
Mathematical Methods for Digital Computers (Wilf), 811
Mathematical statements, translated into logical expressions, 64–65
A Mathematician's Apology, 102
Mathematics for the Physical Sciences (Wilf), 811
Matrix-chain multiplication, 240
Matrix/matrices, 192–198
  addition of, 193
  adjacency, 710–711, 729–730
  arithmetic, 193–195
  Boolean product of, 197
  definition of, 193
  diagonal, 199
  equal, 193
  identity, 195
  incidence, 711–712
  multiplication, 239–240, 561–563
  powers of, 195–196
  product of, 194–195
  representing relations using, 627–630
  sparse, 711
  square, 193, 196
  symmetric, 196
  transpose of, 195–196
  upper triangular, 248
  zero-one, 196–198
Maurolico, Francesco, 337
Maximal elements, 662–664
Maximum matching, 699
Maximum satisfiability problem, 530–531
Maximum spanning tree, 847
Maxterm, 866
McCarthy 91 function, 406
McCarthy, John, 406
McCluskey, Edward, 881
Mealy, G. H., 910
Mealy machines, 910
Mean, 218
  arithmetic, 106, 357
  geometric, 106, 357
Median, 218
Meet
  in lattice, 675
  of zero-one matrix, 197
Mellon, Carnegie, 565
Membership tables, 142
Members, of sets, 124
Memoization, 542
Merge sort, 392–395, 560
  complexity of, 564
  recursive, 393
Mersenne, Marin, 279
Mersenne primes, 279, 280
Mesh network, 702
Metacharacters, 904
Methods
  middle-square, 313
  probabilistic, 496–498
  Quine-McCluskey, 873, 874, 881–885
  roster, 124
Methods of proof, 86–87
  by cases, 98–100
  by contradiction, 91–94
  by contraposition, 88–91
  direct, 87–88
  by exhaustion, 98–99
  proofs of equivalence, 93–94
  trivial, 89
  vacuous, 89
Methodus Differentialis, 164
Metrics, 150
Middle-square method, 313
Milbanke, Annabella, 32
Miller's test for base *b*, 306
Minimal elements, 662–664
Minimization of combinational circuits, 872–885
Minimum dominating set, 781
Minimum spanning forest, 847
Minimum spanning trees, 842–846
  algorithms for, 842–846
  definition of, 842
Minmax strategy, 809
Minterm, 864
Mistakes, in proofs, 94–96
Mixed graph, 681, 682
Mode, 218
Modeling
  computation. *See* Computation models
  with graphs, 682–688
  with recurrence relations, 534–539
  with trees, 791–794
mod function, 257, 271
Modular arithmetic, 255–256, 258–261
Modular exponentiation, 271–272
  algorithm, 272
  recursive algorithm for, 388
Modular properties, in Boolean algebra, 863
Module dependency graph, 685
Modulus, 258
  for linear congruential method, 309
Modus ponens, 75, 76
  universal, 82

Modus tollens, 76
universal, 82
Mohammed's scimitars, 738
Molecules, trees modeling, 792
Monoalphabetic ciphers, 317
Monotone decreasing, 783
Monotone increasing, 783
Monotonic grammars, 898
Monte Carlo algorithms, 494–496, 500
Monty Hall Three Door Puzzle, 480–481
Moore, E. F., 910
Moore machine, 910, 911
Muddy children puzzle, 21
Multicasting, Internet Protocol, 830
Multigraphs, 680, 682
directed, 681, 682
Euler path of, 738–739
Multilevel security policy, 665
Multinomial coefficient, 463
Multinomial theorem, 463
Multiple directed edges, 681
Multiple edges, 680
Multiple output circuit, 870
Multiples, 256
Multiplexer, 872
Multiplication
algorithm, 269–270
associative laws for, A-1
closure laws for, A-1
commutative laws for, A-1
integers, 561
inverse laws for, A-2
matrix, 239–240, 561–563
matrix-chain, 240
Multiplicative compatibility law, A-2
Multiplicative inverse, 65
Multiplicatively homomorphic, 325
Multiplicity
edges of, 681
of multisets elements, 147
Multiplier, for linear congruential method, 309
Multisets, 147–148
cardinality of, 147
definition of, 147
difference of, 147
intersection of, 147
multiplicities of elements, 147
sum of, 147
union of, 147
Mutually independent events, 490, 529
Mutually reachable vertices, 732
Mutually relatively prime, 330

Naive set theory, 126–127
Naive string matcher, 213–214
NAND operator, 865
n-ary predicate, 42
n-ary relations, 617–625
degree, 617
domains, 617
operations on, 619–622
Natural language, 893
Natural language understanding (NLU), 686–687
Natural numbers, 125
sets of, 125
Naur, Peter, 900
NAUTY (software for isomorphism testing), 715
n-cube, 695
n-dimensional hypercube, 695
Negation
laws, 29
of nested quantifiers, 67–69
operator, 4
of propositions, 3, 4
of quantified expression, 49–51
truth table for, 4
Neighbors, in graphs, 691
Nested loops, 62
Nested quantifiers, 61–69
negation of, 67–69
statements involving, 61–62
translating into English, 66
translating mathematical statements into statements involving, 64–65
Network number (netid), 418
Networks
communication, 684
computer. *See* Computer network
gating, 867
information, 685
mesh, 702
semantic, 687
social, 682
tree-connected, 793–794
Niche overlap graph, 686, 687
Nijenhuis, Albert, 811
Nim, 808
NLU. *See* Natural language understanding (NLU)
Nodes. *See* Vertex (vertices)
Noncomplete edges, 694
Nonconstructive existence proofs, 102–103
Noncontracting grammars, 898
Nonconvex polygon, 363
Nondecreasing integers, 218
Nondeterministic finite-state automata, 918–922
Nondeterministic Turing machine, 939–940
Nonisomorphic graphs, 712
Nonoverlapping, 368
Nonplanar graphs, 764
Nonseparable graphs, 724
Nonterminals, 895
Nonterminal symbols, 895
NOR operator, 865
Notation
big-*O*, 221–225
big-Omega (big-Ω), 230–232, 248
big-Theta (big-Θ), 230–232, 248
infix, 823–826
Polish, 824
postfix, 823–826
prefix, 823–826
reverse Polish, 824
sequence, 169
set builder, 124
summation, 176
NOT gate, 22
Not invertible function, 158
Noun, 894, 895
Noun phrase, 894, 895
NP (class of nondeterministic polynomial-time problems), 243, 943–944
NP-complete problems, 243–244
n-place predicate, 42
n-queens problem, 33–35, 836
n-tuples, 618
ordered, 131
Null quantification, 60, 67
Null set, 126
Null string, 895
Numbering plan, 413–414
Numbers
algebraic, 188
Bacon, 722
Bell, 656
cardinal, 130
Carmichael, 303–304
Catalan, 539
chromatic, 769
complex, 125
crossing, 767
edge chromatic, 775
Erdős, 674, 722
Fibonacci. *See* Fibonacci numbers
harmonic, 234–235, 345–346
independence, 783
large, law of, 490
Lucas, 404, 557
natural, 125
pseudorandom, 309–310
Ramsey, 432
rational, 125
real, 125
routing transit, 331
transcendental, 188
Ulam, 203

Number-theoretic functions, 939
Number theory, 255, 314

Objects, 126
  distinguishable, 457–460
  indistinguishable, 458–460
Obligato game, 118
Octahedral die, 528
Octal expansions, 265
  conversion between binary expansion and, 267
Octal representation, 267
Odd integer, 87
Odd pie fights, 351–352
Odlyzko, Andrew, 175
One's complement, 274
One-to-one correspondence, 156–158, 160, 185
One-to-one function, 154–157, 413
On-Line Encyclopedia of Integer Sequences (OEIS), 176
Onto functions, 154–157, 594–595
Open interval, 125, 357
Open problems, 112–114
  about primes, 282–284
Operands, well-formed formula of, 376–378
Operators
  bitwise, 13
  logical. *See* Logical operators
  negation, 4
  selection, 619
  well-formed formulae of, 376–378
Opposite parity, 87
Optimal algorithm, 247
Optimal connectivity, 782
Optimal for suitors, 368
Optimization problems, 214
Order axioms, A-1–A-2
Ordered n-tuples, 131
Ordered pairs, 131
Ordered rooted tree, 791
Ordering
  lexicographic, 381, 464, 658–660
  partial, 656–667
  quasi-ordering, 675
  total, 657, 666
Order, of quantifiers, 62–64
Ordinary generating function, 570
Ore, Øystein, 742
Ore's theorem, 742
Organizational tree, 792–793
OR gate, 22, 867
Orientation of undirected graph, 782
OR operator, 13, 19
Out-degree of vertex, 693
Output, algorithms, 207
Output alphabet, 906
Output function, 906

P (class of polynomial-time problems), 243, 943–944
Paillier cryptosystem, 327
Pairwise independent events, 490
Pairwise relatively prime integers, 285
Palindrome, 218, 903
Paradoxes, 126
  Russell's, 136
  St. Petersburg, 529
Parallel algorithms, 701
Parallel processing, 245, 701
  tree-connected, 793–794
Parent, of vertex, 789
Parent relations, 613
Parity
  opposite, 87
  same, 87
Parity check bits, 310
Parse trees, 898–900
Parsing
  bottom-up, 899
  top-down, 899
Partial correctness, 397
Partial functions, 165
  codomain of, 165
  domain of, 165
  domain of definition, 165
  undefined values of, 165
Partially homomorphic, 325
Partially ordered set, 656
  comparable elements of, 657
  dense, 670
  dual of, 668
  elements of, 656
  greatest element of, 662
  Hasse diagram of, 660–662
  least element of, 662
  maximal elements in, 662–664
  minimal elements in, 662–664
  well-founded, 670
Partial orderings, 656–667
Partitions
  of positive integers, 384, 460, 584
  refinement of, 655
  of sets, 649–652
Pascal, Blaise, 447, 475, 484
Pascal's identity, 446–448
Pascal's triangle, 446–448
Passwords, 417–418
Paths, 720–722
  in acquaintanceship graphs, 721
  in collaboration graphs, 722
  counting between vertices, 729–730
  in directed graphs, 635–636
  Euler, 734–740
  Hamilton, 740–745
  and isomorphism, 728–729
  length of. *See* Length of path
Pattern, strings, 213
Payoff, 807
Pecking order, 782
Peirce, Charles Sanders, 42
Pendant vertex, 692
Perfect power, 98
Perfect square, 88
Permutations, 434–437
  generalized, 451–460
  generating, 464–465
  with indistinguishable objects, 456–457
  inversions in, expected number of, 514
  with repetition, 452
Pessimal, for suitors, 251
Petersen graph, 747
Petersen, Julius Peter Christian, 748
Phrase-structure grammars, 894–898
  types of, 897–898
Pick's theorem, 367
Pigeonhole principle, 91, 426–432
  applications of, 430–432
  generalized, 428–430
Planar graphs, 759–766
  applications of, 761
  Euler's formula and, 761–764
Polish notation, 824
Pólya, George, 114
Polygon
  convex, 363
  definition of, 363
  nonconvex, 363
  sides, 363
  simple, 363
  triangulations, 364
  vertex, 363
Polynomial complexity, 242, 243
Polynomials, big-O estimates for, 225–228
Polyominoes, 111
Poset. *See* Partially ordered set
Positive integers, 125, 255, 335
  partition of, 384, 460, 584
  prime factorization of, 292
Positive real numbers, 125
Postconditions, 43
Postfix form, 824
Postfix notation, 823–826
Postorder traversal, 816, 819–820, 822–823
Postulates, 86
Power generator, 313
Power series, 571–575
  formal, 571
Power sets, 131

Powers, of matrices, 195–196
Precedence
  graphs, 685
  of logical operators, 11
  of quantifiers, 48
Preconditions, 43
Predicate calculus, 43
Predicate logic, 40, 41
Predicates, 40–43
  logically equivalent, 49
  n-ary, 42
  n-place, 42
  truth set of, 134
Prefix codes, 804–806
Prefix form, 824
Prefix notation, 823–826
Preimage, of element, 151
Premises
  of argument, 54, 73, 74
  of conditional statement, 6
Preorder traversal, 816–818, 821–823
Preparata, Franco, 565
Primary key, 618
Prime implicant, 876
Prime number theorem, 280–281
Primes, 275–276
  and arithmetic progressions, 282
  conjectures about, 282–284
  distribution of, 280–281
  factorization, 277
  infinitude of, 278–280
  mutually relatively, 330
  open problems about, 282–284
  primitive root of, 304
Primitive root, of primes, 304
Prim-Jarnk algorithm, 842–844
Prim, Robert Clay, 844, 845
Prim's algorithm, 842–844
Principia Mathematica, 37, 128
Principles
  buoyancy, A-4
  counting, 412–417
  Dirichlet drawer, 427
  duality, 860
  inclusion-exclusion, 138, 419, 585–590
    alternative form of, 592–593
    applications of, 591–597
  of mathematical induction, 337
  of well-ordered induction, 658
  pigeonhole, 91, 426–432
Private key cryptosystems, 318–319
Probabilistic algorithms, 475, 495
Probabilistic method, 496–498
Probabilistic primality testing, 496
Probabilistic reasoning, 480–481
Probability
  assigning, 484–486
  of collision in hashing functions, 494
  of combinations of events, 479–480, 486–487
  conditional, 483, 487–488
  discrete. *See* Discrete probability
  distribution, 484
  of events, 476
  finite, 476–479
  Laplace's definition of, 476
Probability generating function, 585
Probability theory, 475, 483–498
Problems
  art gallery, 776
  birthday, 493–494
  busy beaver, 945
  celebrity, 357
  Chinese postman, 739–740
  closest-pair, 565–567
  Collatz, 113
  counting, 411, 412, 417–419
  decidable, 941–942
  decision, 941
  discrete logarithm, 304
  gossip, 357
  halting, 216–217, 942
  hatcheck, 513–514, 595–597
  Hilbert's Tenth, 942
  houses-and-utilities, 760
  intractable, 243, 944
  jealous husbands, 734
  Josephus, 544
  Kakutani's, 113
  kissing, 175
  knapsack, 252, 601
  longest path, 759
  maximum satisfiability, 530–531
  NP-complete, 243–244
  n-queens, 836
  open, 112–114, 282–284
  optimization, 214
  P versus NP, 243–244
  satisfiability, 37
  searching, 208
  shortest-path, 749–757
  solvable, 243, 941–942
  Syracuse, 113
  tractable, 243, 944
  traveling salesperson, 743, 749, 755–757
  two children, 530
  Ulam's, 113
  undecidable, 942
  unsolvable, 243, 942
Procedure statements, A-11
Product
  Boolean, 855–857
  of matrix, 194–195
Productions, of grammars, 895
Product-of-sums expansion, 866
Product rule, 412–415, 417, 419
Program correctness, 397–402
  conditional statements for, 398–400
  loop invariants for, 400–402
  partial, 397
  rules of inference for, 398
  verification, 397–402
Programming, logic, 55–56
Projection, 620–621
Prolog, 55, 56, 79
Prolog facts, 55
Prolog III, 55
Prolog rules, 55
Proofs, 85–96
  adapting existing, 107
  bijective, 439
  by cases, 98–100
  combinatorial, 439
  by contradiction, 91–94
  by contraposition, 88–91
  definition of, 86
  direct, 87–88
  double counting, 439
  of equivalence, 93–94
  exhaustive, 98–99
  existence. *See* Existence proofs
  indirect, 88
  informal, 85
  by mathematical induction. *See* Proofs, by mathematical induction
  methods of. *See* Methods of proof
  mistakes in, 94–96
  trivial, 89
  uniqueness, 104–105
  using strong induction, 360–363
  vacuous, 89
Proofs, by mathematical induction, 339–340
  of divisibility results, 346–348
  examples of, 340–353
  guidelines for, 339–340
  of inequalities, 344–346
  mistaken, 353–354
  of results about algorithms, 350–351
  of results about sets, 348–350
  summation formulae, 340–344
Proof strategy, 90–91, 105–107
  in action, 108–109
Proper subgraph, 703
Proper subset, 129
Properties
  Archimedean, A-5
  relations, 608–611
  of trees, 794–797
  well-ordering, 365–366, A-5
Propositional calculus, 3
Propositional equivalences, 26–37, 140

Propositional function, 41, 42
Propositional logic, 1–13, 3, 40
  applications of, 17–23
  argument in, 74–75
  rules of inference for, 75–77
  valid arguments in, 73–75
  well-formed formulae in, 375
Propositional variables, 2
Propositions, 2–6, 86
  atomic, 3
  compound. *See* Compound propositions
  conjunction of, 4
  definition of, 2, 86
  disjunction of, 4–5
  negation of, 3, 4
  truth value of, 2, 3
Protein interaction graphs, 686, 687
Protocols
  cryptographic, 323–324
  Diffie-Hellman key agreement, 323
Pseudocode, 207, A-11
Pseudographs, 681, 682
Pseudoprimes, 302–304
  to base b, 303
Pseudorandom numbers, 309–310
  generated by linear congruential method, 309–310
  middle-square generator, 313
  power generator for, 313
  pure multiplicative generator, 309–310
Public key cryptography, 314, 318–319
Public key cryptosystems, 319
Pumping lemma, 934
Pure multiplicative generator, 309–310
Pushdown automaton, 933
Puzzles
  Birthday Problem, 493–494
  Icosian, 740
  jigsaw, 367
  logic, 20–22
  Monty Hall Three Door, 480–481
  muddy children, 21
  Reve's, 537
  Sudoku, 35–37
  Tower of Hanoi, 535–537
P versus NP problem, 243–244
Pythagorean triples, 112

Q-Systems, 55
Quadratic nonresidue, 306
Quadratic residue, 306
Quality control, 495–496
Quantification, 43
  existential, 43, 45–48, 50, 102
  and looping, 47, 62
  null, 60, 67
  universal, 43–45, 47–50
Quantified expression, negation of, 49–51
Quantified statements
  restricted domain, 134
  rules of inference for, 80–81
Quantifiers, 43–46
  De Morgan's laws for, 50–51
  existential, 45–46
  finite domain, 47
  logical equivalences, 49
  nested. *See* Nested quantifiers
  order of, 62–64
  precedence of, 48
  with restricted domains, 48
  scope of, 48
  in system specifications, 53
  truth sets and, 134
  universal, 44–45
Quasi-ordering, 675
Quicksort, 396
Quine-McCluskey method, 873, 874, 881–885
Quine, Willard Van Orman, 883
Quotient, 257, A-6
Quotient automaton, 924

Radius, of graphs, 783
Ramanujan, Srinivasa, 102, 103, 105
Ramsey, Frank Plumpton, 431
Ramsey number, 432
Ramsey theory, 431
Random variables, 484, 491–493
  conditional expectation of, 524
  covariance of, 526
  distribution of, 492
  expected values of, 509–512
  independent, 517–519
  indicator, 524
  standard deviation of, 519
  variance of, 509–512
Range, of function, 151
Ratio, common, 170
Rational numbers, 90, 125
r-combinations, 443, 454, 456
Reachable vertex, 732, 852
Real numbers, 125
  axioms for, A-1–A-2
  positive, 125
  upper bound, A-2
Real-valued functions, 153
Reasoning
  backward, 106–107
  circular, 95
  deductive, 336
  forward, 106–107
  inductive, 336
  probabilistic, 480–481
Recognizes, by finite-state machine, 909
Records, databases, 618
Recurrence relations, 171–173
  algorithms and, 539–542
  applications of, 533–542
  associated homogeneous, 553
  divide-and-conquer, 560–567
  linear homogeneous, 547–553
  linear nonhomogeneous, 553–556
  linear, solving, 546–556
  modeling with, 534–539
  simultaneous, 558
  solution of, 171, 533
Recursion, 370, 390–392
Recursive algorithms, 385–395
  binary search, 388–389
  for computing a^n, 386
  for computing greatest common divisor, 387
  for computing $n!$, 386
  correctness of, 389–390
  for Fibonacci numbers, 390
  linear search, 388
  for modular exponentiation, 388
  trace of, 386
Recursive definitions, 335, 371
  of extended binary trees, 377
  of functions, 371–374
  of sets, 374–378
  of strings, 374
  of structures, 374–378
Recursive merge sort, 393
Recursive modular exponentiation, 388
Recursive step, 374
Reentrant knight's tour, 748
Refinement, of partitions, 655
Reflexive closures, 634–635
Reflexive relation, 608, 611
Regions, of planar representation graphs, 761
Regular expressions, 925
Regular grammars, 897
Regular graph, 708
Regular language, 897
Regular sets, 925, 926, 929–931
Relational data model, 618
Relations
  antisymmetric, 609–610
  asymmetric, 615
  binary, 605
  circular, 673
  closures of. *See* Closures, of relations
  combining, 612–614
  complementary, 615
  composite of, 612–613
  covering, 661
  databases and, 618–619
  diagonal, 634
  divisibility, 656
  equivalence, 644–652

Relations (*cont.*)
extension of, 618
functions as, 606–607
"greater than or equal," 656
inverse, 615
irreflexive, 615
n-ary. *See n*-ary relations
parent, 613
properties of, 608–611
recurrence. *See* Recurrence relations
reflexive, 608, 611
representing. *See* Relations representation of
on set, 607–608
of sets, 133
symmetric, 609–610
transitive, 610–611, 614
Relations, representation of, 627–632
using digraphs, 630–632
using matrices, 627–630
Relatively prime integers, 285
Remainder, 257
Rencontres, 596
Repetition
combinations with, 452–456
permutations with, 452
Replacement
sampling with, 479
sampling without, 479
Representative, of equivalence class, 647
Resolution (rules of inference), 76, 79
Resolvent, 79
Restricted domains, quantifiers with, 48, 134
Results, 86
return statement, A-15
Reverse-delete algorithm, 847
Reverse Polish notation, 824
Reve's puzzle, 537
Right child, of vertex, 791
Right triominoes, 111, 353
Ring topology, 701
Rivest, Ronald, 319
RNA (ribonucleic acid), 415
Rocket-powered Frisbee, 856
Root, 376, 789
Rooted Fibonacci trees, 799
Rooted spanning tree, 841
Rooted trees, 376, 789–791
decision trees, 802–804
definition of, 789
height of, 795
ordered, 791
Roster method, 124
Round-robin tournament, 366, 500, 687–688
Roussel, Philippe, 55
Routing transit number (RTN), 331
Roy, Bernard, 640
Roy-Warshall algorithm, 640–643
r-permutation, 435, 437, 438
RSA, 319
cryptosystem, 314, 319–320
decryption, 321–322
digital signatures, 324
encryption, 320–321
as public key system, 322–323
Rule
division, 420–421
product, 412–415, 417, 419
sum, 412, 415–417
Rules of inference, 73–82
addition, 76
and argument building, 77–78
conjunction, 76
disjunctive syllogism, 76
hypothetical syllogism, 76
modus ponens, 75, 76
modus tollens, 76
for program correctness, 398
for propositional logic, 75–77
for quantified statements, 80–81
resolution, 76, 79
simplification, 76
Rules of syntax, 893–894
Run, 525
Russell, Bertrand, 128
Russell's paradox, 136

Same order, big-*O* relationships, 223
Same parity, 87
Sample space, of experiment, 476
Sampling without replacement, 479
Sampling with replacement, 479
Satisfiability, 33
applications of, 33–37
problems, solving, 37
Satisfiable compound propositions, 33
Saturated hydrocarbons, trees modeling, 792
Scheduling problems
final exams, 772–773
greedy algorithm for, 350–351
Schröder-Bernstein theorem, 189
Scope, of quantifiers, 48
Search algorithms, 208–210
binary, 210–211
breadth-first, 833–835
depth-first, 830–833
linear, 208–209
ternary, 218
Search engines, 838–839
Searches
Boolean, 19–20
Web page, 19–20
Searching problems, 208
Second principle of mathematical induction, 359
Seeds, 839
for linear congruential method, 309
Sees, 776
Selection operator, 619
Selection sort, 219
Self-complementary graph, 718
Self-converse directed graph, 781
Self-dual Boolean function, 888
Self-generating sequences, 407
Semantic networks, 687
Semantic relation, 687
Semantics, 893
Sentence, 894, 895
Sentential variables, 2
Separating set, 724
Sequences, 123, 169–176
definition of, 169
Fibonacci, 171–172
finding maximum and minimum of, 560
generating functions for. *See* Generating functions
Golomb's self-generating, 407–408
graphic, 707
integer, 174–176
Lucas, 175
notation, 169
recurrence relations. *See* Recurrence relations
self-generating, 407
strictly decreasing, 431
strictly increasing, 430
unimodal, 601
Sequential search algorithm, 208
Serial algorithms, 701
Series
geometric, 178
harmonic, 346
infinite, 180–181
power, 571–575
Set builder notation, 124
Set identities, 140–143
absorption laws for, 140
associative laws for, 140
commutative laws for, 140
complementation law for, 140
complement laws for, 140
De Morgan's laws for, 140
distributive laws for, 140
domination laws for, 140
idempotent laws for, 140
identity laws for, 140
methods of proving, 143
Sets, 123–134
atomic, 140
cardinality of, 130, 183–190
Cartesian products of, 131–134

complement of, 138, 139
of complex numbers, 125
computer representation of, 145–146
countable, 184–187
definition of, 124
difference of, 138, 139
disjoint, 138
elements, 124
empty, 126
equality of, 126
finite. *See* Finite sets
fuzzy, 150
guarding, 776
identities. *See* Set identities
infinite, 123, 130
of integers, 125
intersection of, 138, 143–145
language of, 123–124
linearly ordered, 657
members of, 124
naive theory, 126–127
of natural numbers, 125
not recognized by finite-state automaton, 931–932
null, 126
operations, 137–148
partially ordered. *See* Partially ordered set
partition of, 649–652
power, 131
recognization, Turing machine and, 937–938
recursive/inductive definitions of, 374–378
regular, 925, 926, 929–931
relation of, 133
relation on, 607–608
roster method, 124
separating, 724
singleton, 126
size of, 130
of strings, 912–913
successor of, 150
totally ordered, 657
uncountable, 184, 186–188
union of, 137–138, 143–145
universal, 127
well-ordered, 406–407, 657
Shaker sort, 250
Shamir, Adi, 319
Shamos, Michael, 565
Shannon, Claude Elwood, 856
Sheffer, Henry Maurice, 37
Sheffer stroke, 37
Shift ciphers, 315–317
Shortest-path algorithm, 750–755
Shortest-path problems, 749–757
Siblings, of vertex, 789
Sides, of polygon, 363
Sieve of Eratosthenes, 277–282, 593
Signatures, digital, 324
Signed integer, 901
Signless Stirling numbers of the first kind, 460, 472
Simple directed graph, 681, 682
Simple graphs, 680, 682, 694–695
coloring of, 769
crossing number of, 767
dense, 711
isomorphic graphs, 712–716
orientation of, 782
paths in, 721
self-complementary, 718
with spanning trees, 828–830
sparse, 711, 717, 846
thickness of, 768
union of, 704–705
vertices of, 704, 709
Simple path, 720
Simple polygon, 363
Simplification (rules of inference), 76
Single-elimination tournament, 688
Singleton set, 126
Size, of sets, 130
S_k-tree, 850
Slackness, of job, 251
Sloane, Neil, 175
Smullyan, Raymond, 20–21
Social media algorithms, 206
Social networks, 682
Sollin's algorithm, 847
Solution, 33
of recurrence relations, 171, 533
Solvable problem, 243, 941–942
Sorting
definition of, 210
topological, 665–667
Sorting algorithms, 210–212
bidirectional bubble sort, 250
bubble sort, 211–212
insertion sort, 212
merge sort, 392–395, 560
quicksort, 396
selection sort, 219
shaker sort, 250
Space complexity, 235
Spam, 504–507
Spanning forest, of graph, 841
Spanning trees, 828–839
breadth-first search and, 833–835
definition of, 828
degree-constrained, 851
depth-first search and, 830–833
in IP multicasting, 830
maximum, 847
minimum, 842–846
rooted, 841
Sparse graphs, 711, 717, 846
Sparse matrix, 711
Specifications system, 18–19
SQL (Structured Query Language), 622
Squarefree integer, 598
Square matrix, 193, 196
Stable matching, 220
Standard checkerboard, 109
Standard deviation, 519
Star topology, 701
Start state, 913
State diagram, 907
Statements, A-11
assignment, A-11–A-12
biconditional, 10, 29
block of, A-14–A-15
conditional. *See* Conditional statements
if then, 398–400, A-12–A-13
involving nested quantifiers, 61–62
procedure, A-11
return, A-15
States, 906, 913
State table, 906
Stirling, James, 164
Stirling numbers of the second kind, 460
Stirling's formula, 164
St. Petersburg paradox, 529
Straight triomino, 111
Strategy
minmax, 809
proofs, 90–91, 105–107
Strictly decreasing functions, 155
Strictly decreasing sequence, 431
Strictly increasing functions, 155
Strictly increasing sequence, 430
Strings, 170
bit, 12
counting, 414
distinguishable, 935
empty, 170, 895
indistinguishable, 935
length of, 170, 375
lexicographic ordering of, 658
matching, 213–214
null, 895
pattern, 213
recognized/accepted, 914
searching, 244
set of, 912–913
ternary, 543
Strong association rules, 625
Strong induction, 358–366
in computational geometry, 363–365
and infinite ladder, 359–360
proofs using, 360–363
validity of, 358–359

Strongly connected components
of directed graph, 727–728
of web graph, 728
Strongly connected graphs, 727
Strong pseudoprime to base *b*, 306
Structural induction, 371, 378–381
examples of proofs using, 379–381
Structured Query Language (SQL), 622
Structures, recursive/inductive definitions of, 374–378
Subgraph, 703
Subgraph induced, 703
Subsequence, 430
Subsets, 127–130
counting number, of finite set, 348–350, 414
image of, 153
inverse image of, 167
proper, 129
sums of, 836–837
Subtraction rule, 419–420
Subtractors
full, 872
half, 872
Subtree, 784
Success, Bernoulli trial, 490
Successor, of sets, 150
Sudoku, 35–37
Suitees, 220
Suitors, 220
Summations, 176–181
formulae, 180, 340–344
index of, 176–177
lower limit, 177
notation, 176
upper limit, 177
Sum-of-products expansions, 863–865
Sum rule, 412, 415–417
Sums
Boolean, 857
of geometric progressions, 343–344
of matrices, 193
of multisets, 147
of subsets, 836–837
Sun-Tsu, 299
Superset, 128
Support, 624
of association rule, 624–625
Support threshold, 623
Surjection, 155
Surjective functions, 155
Symbolic Logic, 129
Symbols, 895
Symmetric closure, 635
Symmetric difference, 149
Symmetric matrix, 196
Symmetric relations, 609–610
Syntax, 893
Syracuse problem, 113
System of distinct representatives (SDR), 780
System specifications, 18–19
quantifiers in, 53

Tables, 618
state, 906
Tao, Terence, 282
Tautology, 26, 27, 37, 75
Telephone calls, 684
Telephone numbering plan, 413–414
Terminals, 895
Terminal vertex, 631
Ternary search algorithm, 218
Ternary string, 543
Test/testing
Miller's, 306
probabilistic primality, 496
Text, 213
Theorems
art gallery, 776
Bayes', 500–507
Bézout's, 289
binomial, 443–446
Chinese remainder, 297–299
Cook-Levin, 244
definition of, 86
Dirac's, 742
extended binomial, 573
Fermat's last, 112–113, 223
Fermat's little, 301–302
four color, 769–772
fundamental theorem of arithmetic, 276, 361
Hall's marriage, 699–700
handshaking, 692–693
Kleene's, 927–929
Kuratowski's, 764–766
master, 564
methods of proof, 86–87
multinomial, 463
Ore's, 742
Pick's, 367
prime number, 280–281
Schröder-Bernstein, 189
stating, 86
Wilson's, 305
Théorie Analytique des Probabilités, 477
Theory
algebraic number, 113
combinatorial game, 811
Naive set, 126–127
number, 255, 314
probability, 475, 483–498, 483–498
Ramsey, 431
Thickness, of simple graphs, 768
Threshold function, 889
Threshold gate, 889
Threshold value, 889
Tic-tac-toe, 808–809
Tilings, of checkerboards, 109–112, 352–353
Time complexity, 235–239
average-case, 237–238
worst-case, 236–237
Top-down parsing, 899
Topological sorting, 665–667
Topology
ring, 701
star, 701
Total expectation, law of, 524
Total function, 165
Totally ordered sets, 657
Total ordering, 657
compatible, 666
Tournament
round-robin, 366, 500, 687–688
single-elimination, 688
sort, 812–813
Tower of Hanoi, 535–537
Trace, algorithm, 207, 386
Tractable problems, 243, 944
Trail, 720
Traité deMécanique Céleste, 477
Transaction, 623
Transcendental numbers, 188
Transformations, 151. *See also* Functions
affine, 316
Transition function, 906, 913, 914
Transition rules, 936
Transitive closures, 634, 636–640
computing procedure, 640
Transitive relations, 610–611, 614
Transitivity law, A-2
Translation
English sentences to logical expressions, 17–18, 51–53, 66–67
mathematical statements to logical expressions, 64–65
nested quantifiers into English, 66
Transpose, of matrix, 195–196
Transposition cipher, 317
Transposition error, 312
Traveling salesperson problem (TSP), 743, 749, 755–757
Traversal algorithms. *See* Tree traversal
Tree-connected network, 793–794
Tree diagrams, 421–422
Tree edges, 832
Trees, 787–846
applications of, 799–812
binary. *See* Binary trees
binary search, 799–802
caterpillar, 851
decision, 802–804

definition of, 787, 788
derivation/parse, 898–900
family, 787
fractions of, 811–812
game, 806–811
graceful, 851
labeled, 799
m-ary, 790
as models, 791–794
properties of, 794–797
rooted. *See* Rooted trees
rooted Fibonacci, 799
spanning. *See* Spanning trees
terminology for, 789
value of, 809
Tree traversal
inorder, 816–818, 822–823
postorder, 816, 819–820, 822–823
preorder, 816–818, 821–823
universal address system, 815–816
Trial division, 276–277
Triangle inequality, 150
Triangulations, of polygon, 364
Trichotomy law, A-2
Triomino
right, 111, 353
straight, 111
Trivial proof, 89
Truth set, of predicate, 134
Truth tables
for biconditional statement, 10
of compound propositions, 11
for conditional statements, 6
for conjunction, 4
for disjunction, 4
for exclusive or, 6
for negation, 4
Truth value, 2, 3
Tukey, John Wilder, 12
Turing, Alan Mathison, 932
Turing machine, 893, 932, 933, 935–944
computing functions with, 939
definition of, 935–937
nondeterministic, 939–940
representation of, 936
and sets recognization, 937–938
transition rules, 936
types of, 939–940
universal, 940
Twin primes conjecture, 284
Two children problem, 530
Two's complement, 274
Type 0 grammar, 897, 898
Type 1 grammar, 897, 898
Type 2 grammar, 897, 898
Type 3 grammar, 897, 898
Type (datatype), 125
Ulam numbers, 203
Ulam's problem, 113
Unary representations, of integers, 939
Uncomputable functions, 190, 942
Uncountable sets, 184, 186–188
Undecidable problem, 942
Undefined values, of partial function, 165
Underlying undirected graph, 694
Undirected edge, 683
Undirected graphs, 681, 682, 691, 693
connectedness in, 722–724
density of, 717
Euler circuit of, 735
orientation of, 782
underlying, 694
Unicasting, 830
Unicycle, 856
Uniform distribution, 485
Unimodal sequence, 601
Union
of finite sets, 586
of fuzzy sets, 150
of graphs, 704–705
of multisets, 147
of sets, 137–138, 143–145
Uniqueness proof, 104–105
Uniqueness quantifier, 46
Unit-delay machine, 908
Unit (Egyptian) fraction, 405
Unit property, in Boolean algebra, 859
Universal address system, 815–816
Universal generalization, 80, 81
Universal instantiation, 80, 81
Universal modus ponens, 82
Universal modus tollens, 82
Universal Product Code (UPC), 310–311
Universal quantification, 43–45, 47–50
Universal quantifier, 44–45
Universal set, 127
degree of membership, 150
Universal transitivity, 84
Universal Turing machine, 940
Universe of discourse, 44
Unlabeled objects, 457
Unmatched edges, 699
Unsatisfiable compound propositions, 33
Unsolvable problem, 243, 942
UPC check digit, 310–311
Upper bound
of lattice, 675
least, 663, A-2
of partially ordered set, 663
of real numbers, A-2
Upper limit, of summation, 177
Upper triangular matrix, 248
U.S. Coast Survey, 42
Vacuous proof, 89
Valid argument, 73–75
Valid argument form, 75
Valid partner, 251
Values
of Boolean function, 857
expected, 509–512
final, A-14
initial, A-14
threshold, 889
of trees, 809
truth, 2, 3
of vertex in game tree, 809–810
Vandermonde, Alexandre-Théophile, 448
Vandermonde's identity, 448–449
Variables, A-14
binding, 48–49
Boolean, 12, 856, 864
free, 48
propositional, 2
random. *See* Random variables
sentential, 2
Variance, 509, 519–522
Veitch, E. W., 874
Vending machine, 905–906
Venn diagrams, 127
complement of sets, 139
difference of sets, 139
intersection of sets, 138
union of sets, 138
Venn, John, 129
Verb, 894, 895
Verb phrase, 894, 895
Vertex basis, 733
Vertex cut, 724
Vertex (vertices), 631
adjacent, 691
ancestors of, 789
bipartition of, 695
child of, 789
in collaboration graph, 683–684
connecting, 691
connectivity, 724–727
counting in full *m*-ary trees, 794–795
counting paths between, 729–730
cut, 724
degree, in undirected graph, 691
descendants of, 789
distance between, 783
dominating set of, 781
eccentricity of, 799
handle, 850
in-degree of, 693
independent, 783
in influence graph, 683
initial, 631
interior, 640
internal, 789

Vertex (*cont.*)
- isolated, 692
- left child, 791
- left subtree, 791
- level of, 795
- level order of, 850
- mutually reachable, 732
- out-degree of, 693
- parent of, 789
- pendant, 692
- of polygon, 363
- reachable, 732, 852
- removing from graph, 704
- right child, 791
- right subtree, 791
- siblings of, 789
- of simple graphs, 704, 709
- terminal, 631
- value in game tree, 809–810

Vigenére cipher, 326
Vocabulary, 895

Walk, 720
Warshall's algorithm, 640–643
Warshall, Stephen, 640, 642
Weakly connected graphs, 727
Web crawlers, 838–839
Web graphs, 685
- strongly connected components of, 728

Web pages, searching, 19–20
Weighted graphs
- definition of, 749
- shortest path between, 749–757

Weights, 889
Well defined, 371
Well-formed formulae, 375–376, 828
Well-founded partially ordered set, 670
Well-ordered induction, principle of, 658
Well-ordered set, 406–407, 657
Well-ordering property, 365–366, A-5
Wheels, 695
Wiles, Andrew, 113
Wilf, Herbert, 811
Wilson's theorem, 305
Wireless Local Area Networks (WLANs), 701
Without loss of generality (WLOG), 101
Witnesses
- big-O notation, 221
- and existence proof, 102

WLOG (without loss of generality), 101
Word, 895
Word and Object, 883
Worst-case complexity, 236–237
- of binary search, 236–237
- of bubble sort, 238–239
- of insertion sort, 238–239

X follow graph, 683
XOR gate, 889
XOR operator, 13
X undirected mutual graph, 683

Zermelo-Fraenkel axioms, 190
Zero-one matrix, 196–198
- Boolean product of, 196–198, 240
- join of, 197
- meet of, 197
- representing relations using, 627–630
- of transitive closure, 639–640
- Warshall's algorithm and, 640–643

Zero property, in Boolean algebra, 859
Zhang, Yitang, 284

LIST OF SYMBOLS

TOPIC	SYMBOL	MEANING	PAGE
LOGIC	$\neg p$	negation of p	3
	$p \wedge q$	conjunction of p and q	4
	$p \vee q$	disjunction of p and q	4
	$p \oplus q$	exclusive or of p and q	5
	$p \rightarrow q$	implication p implies q	6
	$p \leftrightarrow q$	biconditional of p and q	10
	$p \equiv q$	equivalence of p and q	27
	$\mathbf{T}$	tautology	29
	$\mathbf{F}$	contradiction	29
	$P(x_1, \ldots, x_n)$	propositional function	42
	$\forall x P(x)$	universal quantification of $P(x)$	44
	$\exists x P(x)$	existential quantification of $P(x)$	45
	$\exists! x P(x)$	uniqueness quantification of $P(x)$	46
	$\therefore$	therefore	73
	$p\{S\}q$	partial correctness of S	397
SETS	$x \in S$	x is a member of S	124
	$x \notin S$	x is not a member of S	124
	$\{a_1, \ldots, a_n\}$	list of elements of a set	124
	$\{x \mid P(x)\}$	set builder notation	124
	$\mathbf{N}$	set of natural numbers	125
	$\mathbf{Z}$	set of integers	125
	$\mathbf{Z}^+$	set of positive integers	125
	$\mathbf{Q}$	set of rational numbers	125
	$\mathbf{R}$	set of real numbers	125
	$[a, b]$, (a, b)	closed, open intervals	125
	$S = T$	set equality	126
	$\emptyset$	empty (or null) set	126
	$S \subseteq T$	S is a subset of T	128
	$S \subset T$	S is a proper subset of T	128
	$\|S\|$	cardinality of S	130
	$\mathcal{P}(S)$	power set of S	131
	$(a_1, \ldots, a_n)$	n-tuple	131
	(a, b)	ordered pair	131
	$A \times B$	Cartesian product of A and B	132
	$A \cup B$	union of A and B	137
	$A \cap B$	intersection of A and B	137
	$A - B$	difference of A and B	138
	$\overline{A}$	complement of A	139
	$\bigcup_{i=1}^{n} A_i$	union of A_i, $i = 1, 2, \ldots, n$	144
	$\bigcap_{i=1}^{n} A_i$	intersection of A_i, $i = 1, 2, \ldots, n$	144
	$A \oplus B$	symmetric difference of A and B	149
	$\aleph_0$	cardinality of a countable set	184
	$\mathfrak{c}$	cardinality of $\mathbf{R}$	190

TOPIC	*SYMBOL*	*MEANING*	*PAGE*
FUNCTIONS	$f(a)$	value of function f at a	151
	$f{:}A \to B$	function from A to B	151
	$f_1 + f_2$	sum of functions f_1 and f_2	153
	$f_1 f_2$	product of functions f_1 and f_2	153
	$f(S)$	image of set S under function f	153
	$\iota_A(s)$	identity function on A	157
	$f^{-1}(x)$	inverse of f	157
	$f \circ g$	composition of f and g	159
	$\lfloor x \rfloor$	floor function of x	161
	$\lceil x \rceil$	ceiling function of x	161
	a_n	term of $\{a_i\}$ with subscript n	169
	$\sum_{i=1}^{n} a_i$	sum of $a_1, a_2, \ldots, a_n$	176
	$\sum_{\alpha \in S} a_\alpha$	sum of a_α over $\alpha \in S$	179
	$\prod_{i=1}^{n} a_n$	product of $a_1, a_2, \ldots, a_n$	183
	$f(x)$ is $O(g(x))$	$f(x)$ is big-O of $g(x)$	221
	$n!$	n factorial	164
	$f(x)$ is $\Omega(g(x))$	$f(x)$ is big-Omega of $g(x)$	231
	$f(x)$ is $\Theta(g(x))$	$f(x)$ is big-Theta of $g(x)$	231
	$\sim$	asymptotic to	235
	$\min(x, y)$	minimum of x and y	285
	$\max(x, y)$	maximum of x and y	286
	$\approx$	approximately equal to	478
INTEGERS	$a \mid b$	a divides b	256
	$a \nmid b$	a does not divide b	256
	a **div** b	quotient when a is divided by b	257
	a **mod** b	remainder when a is divided by b	257
	$a \equiv b \pmod{m}$	a is congruent to b modulo m	258
	$a \not\equiv b \pmod{m}$	a is not congruent to b modulo m	258
	$\mathbf{Z}_m$	integers modulo m	261
	$(a_k a_{k-1} \ldots a_1 a_0)_b$	base b representation	264
	$\gcd(a, b)$	greatest common divisor of a and b	284
	$\operatorname{lcm}(a, b)$	least common multiple of a and b	286
MATRICES	$[a_{ij}]$	matrix with entries a_{ij}	193
	$\mathbf{A} + \mathbf{B}$	matrix sum of **A** and **B**	193
	$\mathbf{AB}$	matrix product of **A** and **B**	194
	$\mathbf{I}_n$	identity matrix of order n	195
	$\mathbf{A}^t$	transpose of **A**	196
	$\mathbf{A} \vee \mathbf{B}$	join of **A** and **B**	197
	$\mathbf{A} \wedge \mathbf{B}$	meet of **A** and **B**	197
	$\mathbf{A} \odot \mathbf{B}$	Boolean product of **A** and **B**	197
	$\mathbf{A}^{[n]}$	nth Boolean power of **A**	198

(List of Symbols continued at back of book)

LIST OF SYMBOLS

TOPIC	SYMBOL	MEANING	PAGE
COUNTING AND PROBABILITY	$P(n, r)$	number of r-permutations of a set with n elements	435
	$C(n, r)$	number of r-combinations of a set with n elements	437
	$\binom{n}{r}$	binomial coefficient n choose r	437
	$C(n; n_1, n_2, \ldots, n_m)$	multinomial coefficient	463
	$p(E)$	probability of E	476
	$p(E \mid F)$	conditional probability of E given F	487
	$E(X)$	expected value of random variable X	510
	$V(X)$	variance of random variable X	519
	C_n	Catalan number	539
	$N(P_{i_1} \ldots P_{i_n})$	number of elements having all the properties $P_{i_j}, j = 1, \ldots, n$	592
	$N(P'_{i_1} \ldots P'_{i_n})$	number of elements having none of the properties $P_{i_j}, j = 1, \ldots, n$	592
	D_n	number of derangements of n objects	595
RELATIONS	$S \circ R$	composite of relations R and S	612
	R^n	nth power of relation R	613
	R^{-1}	inverse relation	615
	s_C	selection operator for condition C	619
	$P_{i_1, i_2, \ldots, i_m}$	projection	620
	$J_p(R, S)$	join	621
	Δ	diagonal relation	634
	R^*	connectivity relation of R	637
	$a \sim b$	a is equivalent to b	645
	$[a]_R$	equivalence class of a with respect to R	647
	$[a]_m$	congruence class modulo m	648
	(S, R)	poset consisting of set S and partial ordering R	656
	$a \prec b$	a is less than b	657
	$a \succ b$	a is greater than b	657
	$a \preccurlyeq b$	a is less than or equal to b	657
	$a \succcurlyeq b$	a is greater than or equal to b	657
GRAPHS AND TREES	(u, v)	directed edge	631
	$G = (V, E)$	graph with vertex set V and edge set E	679
	$\{u, v\}$	undirected edge	680
	$\deg(v)$	degree of vertex v	691
	$\deg^-(v)$	in-degree of vertex v	693
	$\deg^+(v)$	out-degree of vertex v	693
	K_n	complete graph on n vertices	694
	C_n	cycle of size n	694
	W_n	wheel of size n	695
	Q_n	n-cube	695
	$K_{m,n}$	complete bipartite graph of size m, n	697
	$G - e$	subgraph of G with edge e removed	703
	$G + e$	graph produced by adding edge e to graph G	703

TOPIC	SYMBOL	MEANING	PAGE
GRAPHS AND TREES (cont.)	$G_1 \cup G_2$	union of G_1 and G_2	705
	$a, x_1, \ldots, x_{n-1}, b$	path from a to b	720
	$a, x_1, \ldots, x_{n-1}, a$	circuit	720
	$\kappa(G)$	vertex connectivity of G	724
	$\lambda(G)$	edge connectivity of G	726
	r	number of regions of the plane	762
	$\deg(R)$	degree of region R	763
	$\chi(G)$	chromatic number of G	769
	m	greatest number of children of an internal vertex in a rooted tree	790
	n	number of vertices of a rooted tree	795
	i	number of internal vertices of a rooted tree	795
	l	number of leaves of a rooted tree	795
	h	height of a rooted tree	796
BOOLEAN ALGEBRA	$\overline{x}$	complement of Boolean variable x	855
	$x + y$	Boolean sum of x and y	855
	$x \cdot y$ (or xy)	Boolean product of x and y	855
	B	$\{0, 1\}$	856
	F^d	dual of F	860
	$x \mid y$	x *NAND* y	865
	$x \downarrow y$	x *NOR* y	865
	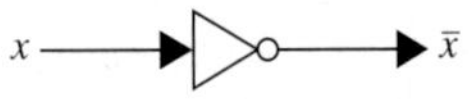	inverter	867
	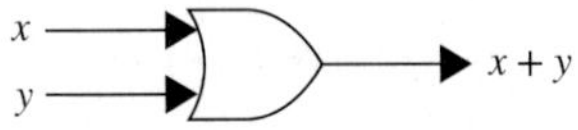	OR gate	867
	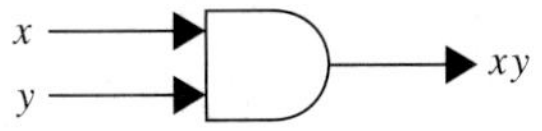	AND gate	867
LANGUAGES AND FINITE-STATE MACHINES	λ	empty string	166, 887
	xy	concatenation of x and y	374
	$l(x)$	length of string x	374
	w^R	reversal of w	384
	(V, T, S, P)	phrase-structure grammar	887
	S	start symbol	887
	$w \to w_1$	production	887
	$w_1 \Rightarrow w_2$	w_2 is directly derivable from w_1.	887
	$w_1 \overset{*}{\Rightarrow} w_2$	w_2 is derivable from w_1.	887
	$\langle A \rangle ::= \langle B \rangle c \mid d$	Backus–Naur form	893
	(S, I, O, f, g, s_0)	finite-state machine with output	898
	s_0	initial or start state	898
	AB	concatenation of sets A and B	904
	A^*	Kleene closure of A	905
	(S, I, f, s_0, F)	finite-state machine automaton with no output	905
	(S, I, f, s_0)	Turing machine	927